MINI Cooper

Service Manual

2002, 2003, 2004, 2005, 2006

Cooper, Cooper S, including convertible

with Diagnostic Trouble Code Manual

[B] BentleyPublishers®
.com

BENTLEY PUBLISHERS® | Automotive Reference™

Bentley Publishers, a division of Robert Bentley, Inc.
1734 Massachusetts Avenue
Cambridge, MA 02138 USA
800-423-4595 / 617-547-4170

Information that makes the difference®

BentleyPublishers®.com

Technical contact information

We welcome your feedback. Please submit corrections and additions to our MINI technical discussion forum at:

http://www.BentleyPublishers.com

Updates and corrections

We will evaluate submissions and post appropriate editorial changes online as text errata or tech discussion. Appropriate errata will be incorporated with the book text in future printings. Check for updates and corrections for this book before beginning work on your vehicle. See the following web address for additional information:

http://www.BentleyPublishers.com/updates/

WARNING—IMPORTANT SAFETY NOTICE

Do not use this manual unless you are familiar with automotive repair procedures and safe workshop practices. This manual illustrates the workshop procedures required for some maintenance and service work. It is not a substitute for full and up-to-date information from the vehicle manufacturer or for proper training as an automotive technician. Note that it is not possible for us to anticipate all of the ways or conditions under which vehicles may be serviced or to provide cautions as to all of the possible hazards that may result.

We have endeavored to ensure the accuracy of the information in this manual. Please note, however, that considering the vast quantity and the complexity of the service information involved, we cannot warrant the accuracy or completeness of the information contained in this manual.

FOR THESE REASONS, NEITHER THE PUBLISHER NOR THE AUTHOR MAKES ANY WARRANTIES, EXPRESS OR IMPLIED, THAT THE INFORMATION IN THIS BOOK IS FREE OF ERRORS OR OMISSIONS, AND WE EXPRESSLY DISCLAIM THE IMPLIED WARRANTIES OF MERCHANTABILITY AND OF FITNESS FOR A PARTICULAR PURPOSE, EVEN IF THE PUBLISHER OR AUTHOR HAVE BEEN ADVISED OF A PARTICULAR PURPOSE, AND EVEN IF A PARTICULAR PURPOSE IS INDICATED IN THE MANUAL. THE PUBLISHER AND AUTHOR ALSO DISCLAIM ALL LIABILITY FOR DIRECT, INDIRECT, INCIDENTAL OR CONSEQUENTIAL DAMAGES THAT RESULT FROM ANY USE OF THE EXAMPLES, INSTRUCTIONS OR OTHER INFORMATION IN THIS BOOK. IN NO EVENT SHALL OUR LIABILITY WHETHER IN TORT, CONTRACT OR OTHERWISE EXCEED THE COST OF THIS MANUAL.

Before attempting any work on your MINI, read **001 General Warnings and Cautions** and any **WARNING** or **CAUTION** that accompanies a procedure in the service manual. Review the **WARNINGS** and **CAUTIONS** each time you prepare to work on your MINI.

Your common sense and good judgment are crucial to safe and successful service work. Read procedures through before starting them. Think about whether the condition of your car, your level of mechanical skill, or your level of reading comprehension might result in or contribute in some way to an occurrence which might cause you injury, damage your car, or result in an unsafe repair. If you have doubts for these or other reasons about your ability to perform safe repair work on your car, have the work done at an authorized MINI dealer or other qualified shop.

Part numbers listed in this manual are for identification purposes only, not for ordering. Always check with your authorized MINI dealer to verify part numbers and availability before beginning service work that may require new parts.

Special tools required to perform certain service operations are identified in the manual and are recommended for use. Use of tools other than those recommended in this manual may be detrimental to the car's safe operation as well as the safety of the person servicing the car.

The vehicle manufacturer will continue to issue service information updates and parts retrofits after the editorial closing of this manual. Some of these updates and retrofits will apply to procedures and specifications in this manual. We regret that we cannot supply updates to purchasers of this manual.

This manual is prepared, published and distributed by Bentley Publishers, 1734 Massachusetts Avenue, Cambridge, Massachusetts 02138 USA. All information contained in this manual is based on the information available to the publisher at the time of editorial closing. MINI has not reviewed and does not vouch for the accuracy or completeness of the technical specifications and work procedures described and given in this manual.

ISBN 978-0-8376-1639-1 **Job code: BM06-04** Editorial closing 06/2006

Library of Congress Cataloging-in-Publication Data for 2007 edition

Mini Cooper service manual : 2002, 2003, 2004, 2005, 2006, Cooper, Cooper S, including convertible.

p. cm.

Includes index.

ISBN: 978-0-8376-1511-0 (alk. paper)

1. Mini-Cooper automobiles--Maintenance and repair--Handbooks, manuals, etc. I. Robert Bentley, Inc.

TL215.M465M55 2007

629.28'722--dc26

2004037877

The paper used in this publication is acid free and meets the requirements of the National Standard for Information Sciences-Permanence of Paper for Printed Library Materials. ∞

Manufactured in the United States of America

202605R001

Foreword

For the MINI owner with automotive mechanical skills, and for independent service professionals, this manual includes maintenance and repair procedures the MINI is likely to need during its service life, including many procedures and specifications that were available in an authorized MINI dealer service department at the time this manual was prepared. The MINI owner with no intention of working on his or her car will find that owning and referring to this manual makes it possible to be better informed and to more knowledgeably discuss repairs with a professional automotive technician.

For those intending to do maintenance and repair on their MINI, it is essential that safety equipment be used and safety precautions observed when working on your vehicle. A minimum safety equipment list includes hand protection, eye protection, and a fire extinguisher. A selection of good quality hand tools is also needed. This includes a torque wrench to ensure that fasteners are tightened in accordance with specifications. In some cases, the text refers to special tools that are recommended or required to accomplish adjustments or repairs. These tools are often identified by their MINI / BMW special tool number and illustrated.

Disclaimer

We have endeavored to ensure the accuracy of the information in this manual. When the vast array of data presented in the manual is taken into account, however, no claim to infallibility can be made. We therefore cannot be responsible for the result of any errors that may have crept into the text. Please also read **WARNING–Important Safety Notice** on the copyright page at the beginning of this book.

Prior to starting a repair procedure, read the procedure, **001 General Warnings and Cautions** and the **Warnings** and **Cautions** that accompany the procedure. Reading a procedure before beginning work helps you determine in advance the need for specific skills, identify hazards, prepare for appropriate capture and handling of hazardous materials, and the need for particular tools and replacement parts such as gaskets.

Bentley Publishers encourages comments from the readers of this manual with regard to errors, and suggestions for improvement of our product. These communications have been and will be carefully considered in the preparation of this and other manuals. If you identify inconsistencies in the manual, you may have found an error. Please contact the publisher and we will endeavor to post applicable corrections on our website. Review corrections (errata) that we have posted before beginning work. Please see the following web address:

`http://www.BentleyPublishers.com/updates/`

MINI continues to issue service information and parts retrofits after the editorial closing of this manual. Some of this updated information may apply to procedures and specifications in this manual. For the latest information, please see the following web address:

`http://www.minitechinfo.com/`

MINI offers extensive warranties, especially on components of the fuel delivery and emission control systems. Therefore, before deciding to repair a MINI that may be covered wholly or in part by any warranties issued by BMW of North America, LLC, consult your authorized MINI dealer. You may find that the dealer can make the repair either free or at minimum cost. Regardless of its age, or whether it is under warranty, your MINI is both an easy car to service and an easy car to get serviced. So if at any time a repair is needed that you feel is too difficult to do yourself, a trained MINI technician is ready to do the job for you.

Bentley Publishers

001 General Warnings and Cautions

PLEASE READ THESE WARNINGS AND CAUTIONS BEFORE PROCEEDING WITH MAINTENANCE AND REPAIR WORK.

WARNINGS—
See also CAUTIONS

- Some repairs may be beyond your capability. If you lack the skills, tools and equipment, or a suitable workplace for any procedure described in this manual, we suggest you leave such repairs to an authorized MINI dealer service department or other qualified shop.
- Do not reuse any fasteners that are worn or deformed. Many fasteners are designed to be used only once and become unreliable and may fail when used a second time. This includes, but is not limited to, nuts, bolts, washers, self-locking nuts or bolts, circlips and cotter pins. Always replace these fasteners with new parts.
- Never work under a lifted car unless it is solidly supported on stands designed for the purpose. Do not support a car on cinder blocks, hollow tiles or other props that may crumble under continuous load. Never work under a car that is supported solely by a jack. Never work under the car while the engine is running.
- If you are going to work under a car on the ground, make sure that the ground is level. Block the wheels to keep the car from rolling. Disconnect the battery negative (–) terminal to prevent others from starting the car while you are under it.
- Never run the engine unless the work area is well ventilated. Carbon monoxide kills.
- Rings, bracelets and other jewelry should be removed so that they cannot cause electrical shorts, get caught in running machinery, or be crushed by heavy parts.
- Tie long hair behind your head. Do not wear a necktie, a scarf, loose clothing, or a necklace when you work near machine tools or running engines. If your hair, clothing, or jewelry were to get caught in the machinery, severe injury could result.
- Do not attempt to work on your car if you do not feel well. You increase the danger of injury to yourself and others if you are tired, upset or have taken medication or any other substance that may keep you from being fully alert.
- Illuminate your work area adequately but safely. Use a portable safety light for working inside or under the car. Make sure the bulb is enclosed by a wire cage. The hot filament of an accidentally broken bulb can ignite spilled fuel, vapors or oil.
- Catch draining fuel, oil, or brake fluid in suitable containers. Do not use food or beverage containers that might mislead someone into drinking from them. Store flammable fluids away from fire hazards. Wipe up spills at once, but do not store the oily rags, which can ignite and burn spontaneously.
- Always observe good workshop practices. Wear goggles when you operate machine tools or work with battery acid. Gloves or other protective clothing should be worn whenever the job requires working with harmful substances.
- Greases, lubricants and other automotive chemicals contain toxic substances, many of which are absorbed directly through the skin. Read the manufacturer's instructions and warnings carefully. Use hand and eye protection. Avoid direct skin contact.
- Disconnect the battery negative (–) terminal whenever you work on the fuel system or the electrical system. Do not smoke or work near heaters or other fire hazards. Keep an approved fire extinguisher handy.
- Friction materials (such as brake pads or shoes or clutch discs) contain asbestos fibers or other friction materials. Do not create dust by grinding, sanding, or by cleaning with compressed air. Avoid breathing dust. Breathing any friction material dust can lead to serious diseases and may result in death.
- Batteries give off explosive hydrogen gas during charging. Keep sparks, lighted matches and open flame away from the top of the battery. If hydrogen gas escaping from the cap vents is ignited, it will ignite gas trapped in the cells and cause the battery to explode.
- Connect and disconnect battery cables, jumper cables or a battery charger only with the ignition switched off. Do not disconnect the battery while the engine is running.
- Do not quick-charge the battery (for boost starting) for longer than one minute. Wait at least one minute before boosting the battery a second time.
- Do not allow battery charging voltage to exceed 16.5 volts. If the battery begins producing gas or boiling violently, reduce the charging rate. Boosting a sulfated battery at a high charging rate can cause an explosion.
- The air conditioning system is filled with chemical refrigerant, which is hazardous. The A/C system should be serviced only by trained technicians using approved refrigerant recovery/recycling equipment, trained in related safety precautions, and familiar with regulations governing the discharging and disposal of automotive chemical refrigerants.
- Do not expose any part of the A/C system to high temperatures such as open flame. Excessive heat will increase system pressure and may cause the system to burst.

Continued on next page

WARNINGS (continued)

- Some aerosol tire inflators are highly flammable. Be extremely cautious when repairing a tire that may have been inflated using an aerosol tire inflator. Keep sparks, open flame or other sources of ignition away from the tire repair area. Inflate and deflate the tire at least four times before breaking the bead from the rim. Completely remove the tire from the rim before attempting any repair.
- Cars covered by this manual are equipped with a supplemental restraint system (SRS), that automatically deploys airbags and pyrotechnic seat belt tensioners in the event of a frontal or side impact. These are explosive devices. Handled improperly or without adequate safeguards, they can be accidently activated and cause serious injury.
- The ignition system produces high voltages that can be fatal. Avoid contact with exposed terminals and use extreme care when working on a car with the engine running or the ignition switched on.
- Place jack stands only at locations specified by the manufacturer. The vehicle lifting jack supplied with the vehicle is intended for tire changes only. A heavy duty floor jack should be used to lift vehicle before installing jack stands. See **010 General**.
- Battery acid (electrolyte) can cause severe burns. Flush contact area with water, seek medical attention.
- Aerosol cleaners and solvents may contain hazardous or deadly vapors and are highly flammable. Use only in a well ventilated area. Do not use on hot surfaces (engines, brakes, etc.).
- Do not remove coolant reservoir or radiator cap with the engine hot. Danger of burns and engine damage.

CAUTIONS—
See also WARNINGS

- If you lack the skills, tools and equipment, or a suitable workshop for any procedure described in this manual, leave such repairs to an authorized MINI dealer or other qualified shop.
- MINI is constantly improving its cars and sometimes these changes, both in parts and specifications, are made applicable to earlier models. Therefore, part numbers listed in this manual are for reference only. Always check with your authorized MINI dealer parts department for the latest information.
- Before starting a job, make certain that you have all the necessary tools and parts on hand. Read all the instructions thoroughly, and do not attempt shortcuts. Use tools appropriate to the work and use only replacement parts meeting MINI specifications.
- Use pneumatic and electric tools only to loosen threaded parts and fasteners. Never use these tools to tighten fasteners, especially on light alloy parts. Always use a torque wrench to tighten fasteners to the tightening torque specification listed.
- Be mindful of the environment and ecology. Before you drain the crankcase, find out the proper way to dispose of the oil. Do not pour oil onto the ground, down a drain, or into a stream, pond or lake. Dispose of waste in accordance with federal, state and local laws.
- The control module for the anti-lock brake system (ABS) cannot withstand temperatures from a paint-drying booth or a heat lamp in excess of 203°F (95°C) and should not be subjected to temperatures in excess of 185°F (85°C) for more than two hours.
- Before doing any electrical welding on cars equipped with ABS, disconnect the battery negative (–) terminal (ground strap) and the ABS control module connector.
- Always make sure ignition is off before disconnecting battery.
- Label battery cables before disconnecting. On some models, battery cables are not color coded.
- Disconnecting the battery may erase fault code(s) stored in control module memory. Using special BMW diagnostic equipment, check for fault codes prior to disconnecting the battery cables. If the Malfunction Indicator Light (MIL) is illuminated, see **OBD On Board Diagnostics** at the back of this manual. This light is identified as the Service Engine Soon light. If any other system faults have been detected (indicated by an illuminated warning light), see an authorized MINI dealer.
- If a normal or rapid charger is used to charge battery, the battery must be disconnected and removed from the vehicle in order to avoid damaging the vehicle.
- Do not quick-charge the battery (for boost starting) for longer than one minute. Wait at least one minute before boosting the battery a second time.
- Connect and disconnect a battery charger only with the battery charger switched off.
- Sealed or "maintenance free" batteries should be slow-charged only, at an amperage rate that is approximately 10% of the battery's ampere-hour (Ah) rating.
- Do not allow battery charging voltage to exceed 16.5 volts. If the battery begins producing gas or boiling violently, reduce the charging rate. Boosting a sulfated battery at a high charging rate can cause an explosion.

002 Vehicle Identification and VIN Decoder

Vehicle Identification Number (VIN), decoding

Some of the information in this manual applies only to cars of a particular model year or range of years. For example, 2005 refers to the 2005 model year but does not necessarily match the calendar year in which the car was manufactured or sold. To be sure of the model year of a particular car, check the Vehicle Identification Number (VIN) on the car.

The VIN is a unique sequence of 17 characters assigned by MINI to identify each individual car. When decoded, the VIN tells the country and year of manufacture; make, model and serial number; assembly plant and even some equipment specifications.

The MINI VIN is located on the top of the dashboard, on the driver's side where the number can be seen through the windshield. The table below explains some of the codes in the VIN for 2002 - 2006 MINI models.

Three platform codes are used by the manufacturer to identify the various MINI models:

- R50: Cooper coupe, MINI One coupe
- R52: Cooper convertible, Cooper S convertible
- R53: Cooper S coupe

Sample VIN: WMWRC33506TK12345

position: 1 2 3 4 5 6 7 8 9 10 11 12-17

VIN position	Description	Decoding information	
1	Manufacturing Country	W	Germany
2	Manufacturer	M	MINI
3	Vehicle Type	W	Passenger vehicle
4	Series	R	MINI
5	Model	A B C D E F H	MINI One MINI One/D MINI Cooper MINI One convertible MINI Cooper S MINI Cooper convertible MINI Cooper S convertible
6	Engine	3 1	Gasoline engine Diesel engine
7	Drive position or destination	1 2 3	Left hand drive Right hand drive Destined for USA
8	Restraint system		
9	Check digit		0 - 9 or X, calculated by NHTSA
10	Model year	2 3 4 5 6	2002 2003 2004 2005 2006
11	Assembly plant	T	Oxford, UK
12-17	Serial number		Sequential production number for specific vehicle

010 Product Familiarization

General

The information included in this section is based on introductory information for MINI vehicles sold in the USA and Canada. The content provided here is intended to serve as a product familiarization guide. Note that the information presented is subject to change and should be used as a general reference only.

> ***WARNING—***
> - *Always check the MINI factory repair information or the publisher's website at www.bentleypublishers.com for information that may supersede any information included in this section.*

Product Overview

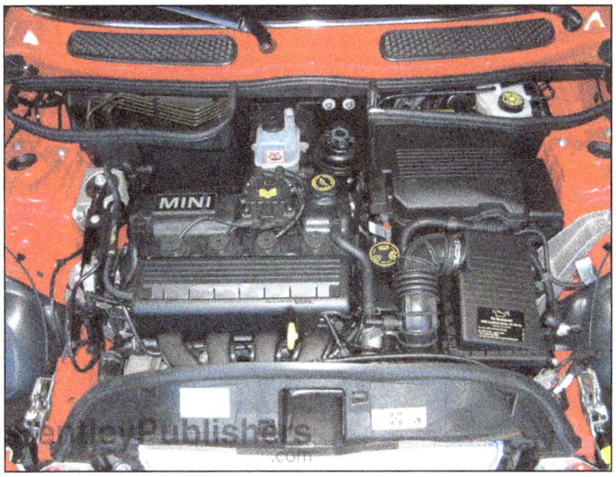

The MINI Cooper was introduced in model year 2002 in coupe form with a 1.6 liter engine and front-wheel drive.

The MINI Cooper's distinctive styling and clean lines are modern, yet the classic original Mini is clearly recognizable in the design. The wheel at each corner design with very little overhang, for example, is more than a styling feature: It gives the car a dynamic advantage. The layout provides excellent road handling, making the MINI both inherently safe and also fun to drive.

From the front, the shape of the hood and the large round headlights give the MINI Cooper its characteristic face. The large, deeply drawn hood is a masterpiece of design and engineering as it sweeps seamlessly down to the wheel arches.

From the rear, the car has the Classic Mini lines, with an extremely short tail, a steeply rising side section and the characteristic taillights. Additionally, the flared wheel arches, wide track and horizontal lines add to the low road-hugging look.

2005 model year changes

2005 saw the debut of the MINI convertible and all models received a modest facelift.

The new look entailed both front and rear light redesigns, including new xenon headlights and LED rear lighting.

Additional visual changes included:

- Rod antenna standard; for satellite, telephone and radio functions (shark fin discontinued)
- Larger interior rear mirror
- Anthracite headliner (dark grey)
- New storage areas under toggle switches and parking brake
- Clock moved from headliner to beside odometer
- Larger rear cup holder
- Extra sun visor on driver side
- Optional chrome package
- Dashboard redesigned from a 5 piece unit to a 3 piece one. There was also an optional gauge package offered.

Technical data (MINI Cooper)

		MINI COOPER	MINI COOPER
Transmission type		5 speed	CVT & Steptronic
No. of doors / seats	-	3 / 4	3 / 4
Vehicle length / width / heights unl.	in.	142.8 / 66.5 / 55.4 / 55.9	142.8 / 66.5 / 55.4 / 55.9
Wheelbase / turning circle	in. / ft.	97.1 / 35	97.1 / 35
Overhang front / rear	in.	24.5 / 21.1	24.5 / 21.1
Track front / rear	in.	57.4 / 57.7	57.4 / 57.7
Width at shoulder height front/rear	in.	50.9 / 44.7	50.9 / 44.7
Elbow room front/rear	in.	53.9 / 44.5	53.9 / 44.5
Head room front/rear	in.	38.8 / 37.6	38.8 / 37.6
Leg Room front/rear	in.	41.3/31.3	41.3/31.3
Cargo Volume – seats up/seats folded	cubic ft.	5.3/23.7	5.3/23.7
Total interior volume EPA	cubic ft.	77.0	77.0
Appr. tank capacity	gal	13.2	13.2
Unladen weight	lbs	2524	2557
Axle load ratio / rear	%	37.1	37.1
Gross vehicle weight / payload	Lbs	3263 / 948	3318 / 948
Axle load limit front / rear	Lbs	1918 / 1543	1918 / 1543
Towing capacity (trailer w/brakes) on(12%) / (8%) slope	Lbs	1433 / 1764	1433 / 1764
Towing capacity (trailer w/brakes) / Roof load limit	Lbs	1102	1102
Engine / No. Of cylinders / valves p. cylinder	-	In-line 4cyl 16v SOC	In-line 4cyl 16v SOC
Fuel management	-	Siemens EMS 2000	Siemens EMS 2000
Displacement	cm^3	1598	1598
Stroke / Bore	in.	3.37/3.03	3.37/3.03
Nominal power / engine speed	hp/rpm	115/ 6000	115/ 6000
Max. Torque / engine speed	ft/lbs	110/4500	110/4500
Compression ratio / fuel type	-	10.6:1 / 91-98 ROZ	10.6:1 / 91-98 ROZ
Transmission type	-	R65 5C39	GACVT16Z
1st, 2nd, 3rd gear ratio	-	3.417/1.947/1.333	variable
4th, 5th, 6th gear ratio	-	1.054/0.846	variable
Reverse gear ratio	-	3.58	3.82
Final drive ratio	-	3.94	4.1
Steering type / Steering ratio	- /	EHPAS / 13,18	EHPAS / 13,18
Brakes front / diameter	- / in.	Vented Disc 10.9x.9	Vented Disc 10.9x.9
Brakes rear / diameter	- / in.	Disc 10.2 x .4	Disc 10.2 x .4
Drag cx / A / cx x A	- / m^2 / m^2	0.36/1.97/0.689	0.36/1.97/0.689

Powertrain

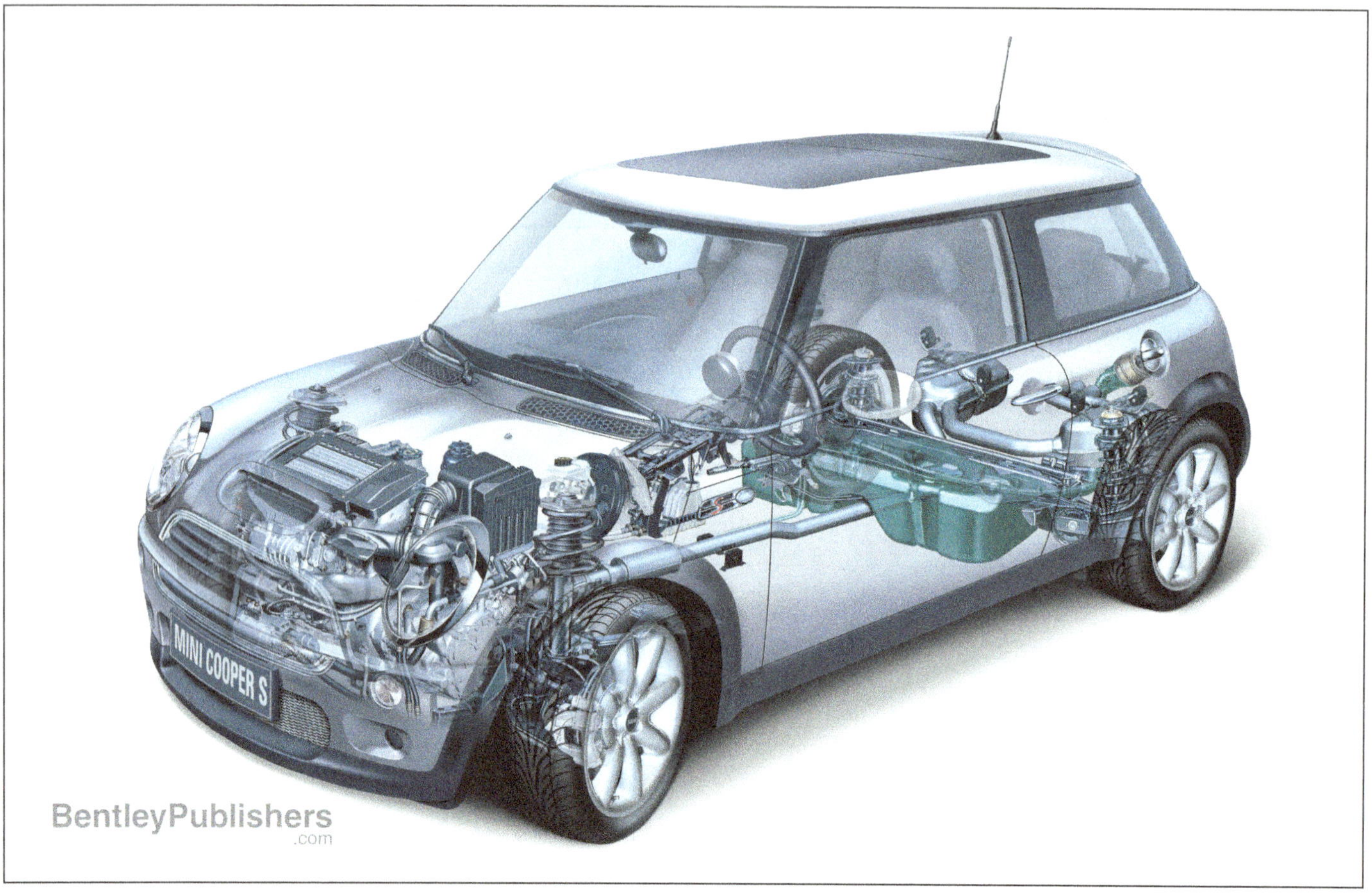

Engines

MINI Cooper is powered by a four-cylinder, 1.6 liter, 16-valve engine producing 115 hp, with a 0 - 60 mph time of 8.5 seconds and a top speed, where allowable, of 126 mph. The MINI Cooper comes with a 13-gallon fuel tank, bigger than all similar sized cars, and offers a driving range of up to 500 miles.

The MINI Cooper engine has an overhead camshaft, four valves per cylinder, and hydraulic valve adjustment elements. The cast iron engine block ensures low levels of noise and vibration. The cylinder head is aluminum.

The engine is managed by a Siemens control unit (EMS 2000). This unit controls the drive-by-wire throttle.

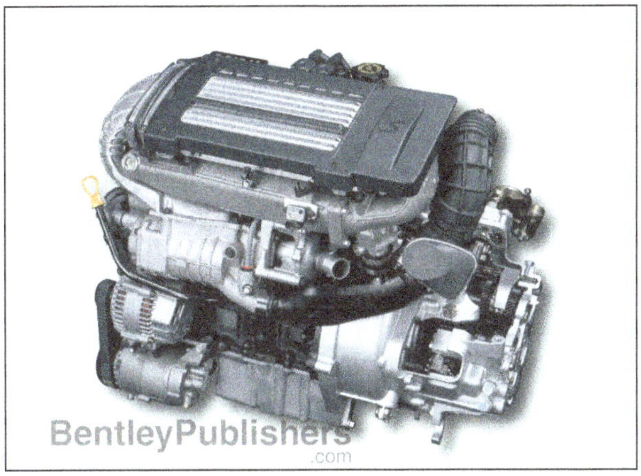

The MINI Cooper S is powered by a four-cylinder, 163 horsepower engine (168 hp from model year 2005), which features a supercharger with an intercooler. The MINI Cooper S offers outstanding performance having a top speed, where allowable, of 136 mph.

The MINI Cooper S reaches 60 mph in 6.9 seconds. The engine's broad torque delivery, peaking at 155 lb-ft @ 4000 rpm (162 lb-ft from model year 2005), provides good power throughout the entire speed range.

A number of components such as the crankshaft, pistons, valves, radiator, and engine management parameters have been specially modified for the S model, owing to the increased performance requirements.

Transmissions

MINI models are available with several manual and automatic transmissions. The transmission is mounted in line with the engine on the driver side of the engine bay. The final drive/differential assembly is integral with the transmission housing and provides drive to the front wheels.

For automatic transmissions, a 5-speed continuously variable transmission (CVT) was available for the MINI Cooper. The CVT transmission is compact and lighter than a conventional automatic transmission. The CVT uses an oil bath multi-disc coupling that is electronically-controlled. The transmission itself uses a fixed-length steel drive belt to connect two double cone-shaped belt pulleys which transmit the drive output from the engine, essentially offering infinitely variable transmission ratios.

The MINI Cooper was available with a 5-speed manual transmission, originally known as the R65 (from model year 2005, a Getrag 252 was used).

The manual transmission fitted to the MINI Coopewr S is a 6-speed unit manufactured by Getrag. This transmission is rated for a maximum torque input of 210 Nm, enough to handle the output of the supercharged engine. This gearbox uses the reference code 285.

Beginning with model year 2005, the Cooper S model was available with an optional six-speed automatic with "Agitronic". Manual-shift mode is also available via steering column-mounted paddles.

The Cooper S automatic is a conventional gear-type transmission with a performance tuned torque converter and fixed gear ratios that are tuned specifically to match the sporty character of the MINI Cooper S. This automatic transmission is built by Aisin, and features the following gear ratios:

Aisin Automatic transmission (MINI Cooper S)	
Gear ratios 1st / 2nd / 3rd	4.044 / 2.371 / 1.556
Gear ratios 4th / 5th / 6th	1.159 / 0.852 / 0.672
Reverse gear ratio	3.193
Final drive ratio	3.683

Body Exterior

The body of the MINI Cooper is exceptionally rigid, offering two to three times the torsional stiffness of other cars of the same size. The car uses variable thickness panel pressings to increase strength where required and reduce weight where possible. To increasing its rigidity, the car has 3,800 spotwelds, many more than cars of a similar size. This helps give the car its exceptional, go-cart like handling and also minimizes body vibration.

An extremely strong and rigid passenger safety cell, including front and rear crumple zones and side impact door beams, also means that occupants are well protected in case of an accident.

From the exterior, the profile is accentuated by a roofline that is angled down slightly at the rear. This characteristic stance is emphasized by the car's distinctive "shoulders" and almost vertical side sections.

All models are three-door. The front doors can be opened extremely wide – up to 80° – making for easy passenger entry and excellent access to the rear seats. When the doors are fully open, the car is wider than it is long.

The doors, with their frameless electric windows are elegant yet unconventional. When the doors open, the windows automatically slide down slightly and when they are closed, they move back up to ensure a perfect seal.

The wide-opening tailgate provides easy access to the rear cargo area and allows someone six feet tall to stand upright under it. The latch has an electric lock and is opened via a button in the handle – another unusual feature in a car of this class.

The use of glass on the car is particularly creative. The upright windshield is reminiscent of the original, but from there backwards the structure is pure 21st century. The A-pillar is concealed behind a black, high-gloss plastic panel and the B- and C-pillars are also hidden behind glass, giving the impression of a continuous window extending all round the car, thus accentuating the roofline.

A heated rear window and windshield wipers are standard and an electrically heated windshield is available as an option. Heated door mirrors and headlight washer jets are also an option.

In the spirit of the Classic Mini, the MINI Cooper roof and electric outside mirrors can be ordered in white or black – contrasting the wide range of body colors. Customers can also choose a body colored roof as a no cost option. In this case, the door mirrors are finished in black. There is also an optional panoramic sunroof.

The MINI Cooper is fitted as standard with 15 in. aluminum wheels in white or silver, 16 in. rims are also available as factory options with 17 in. available from the dealer.

MINI Cooper has door handles, radiator grille, tailgate handle, exhaust and trim rings around the front and taillights all finished in chrome, emphasizing the high-quality appearance of this vehicle.

MINIs are perfect for personalization and a wide range of trim packages and accessories have been made available.

Body dimensions

MINI COOPER AND MINI COOPER S

Where different, MINI Cooper S dimensions in parentheses.

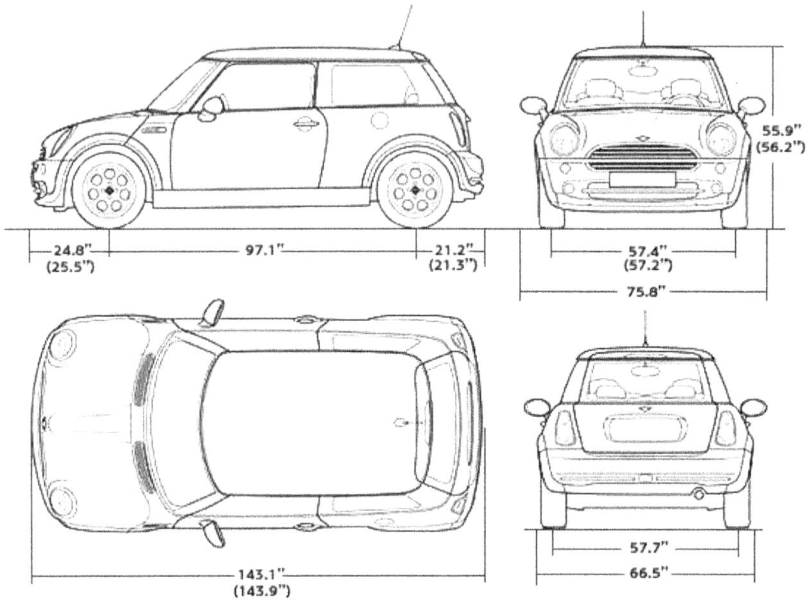

MINI COOPER AND MINI COOPER S CONVERTIBLE

Where different, MINI Cooper S Convertible dimensions in parentheses.

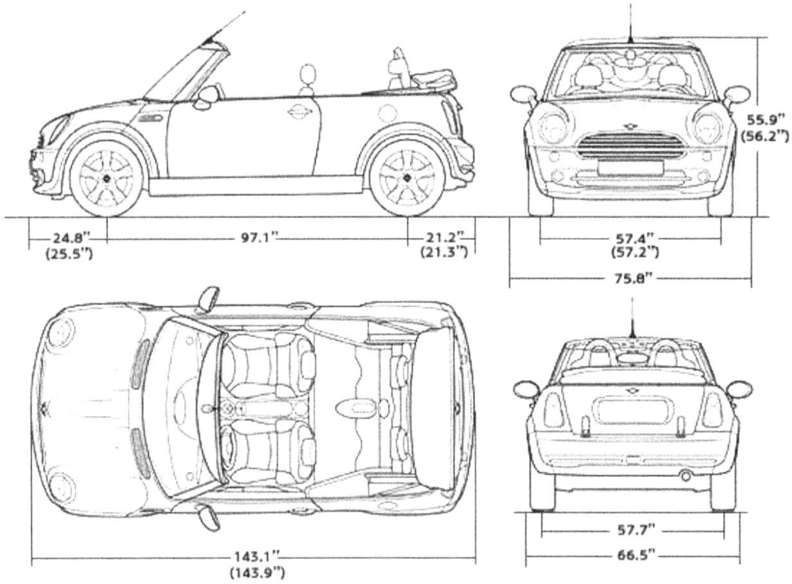

Body Interior

The car's interior has been engineered to provide outstanding ergonomics, comfort and safety.

Modern materials such as aluminum tone facings on the dashboard add to the design. All instruments are framed in silver and have concave anti-reflective glass.

The MINI Cooper tachometer is placed on the steering column in front of the driver to maintain the sporty emphasis. If navigation is ordered, the speedometer is moved to the steering column and located next to the tach, with the navigation screen placed centrally on the dashboard. The dashboard also featured an air-conditioned glove box.

A hi-fi CD sound system was standard equipment. A Harman Kardon speaker system with digital sound was offered as an option. Other options included a multifunction steering wheel, rain sensor windshield wipers and navigation system.

Toggle switches echo the switch gear of the original Mini.

The steering column is height-adjustable and the standard, leather-look steering wheel sports a distinctive two-spoke design (three-spoke on S models). Leather or wood steering wheels with a multi-function system, which allows fingertip control of the stereo and cruise control systems, are available as options.

The dashboard is available in various trims with a choice of silver or wood insert. The appeal of the original Mini was again the influence for the center console, which is framed and supported by vertical struts.

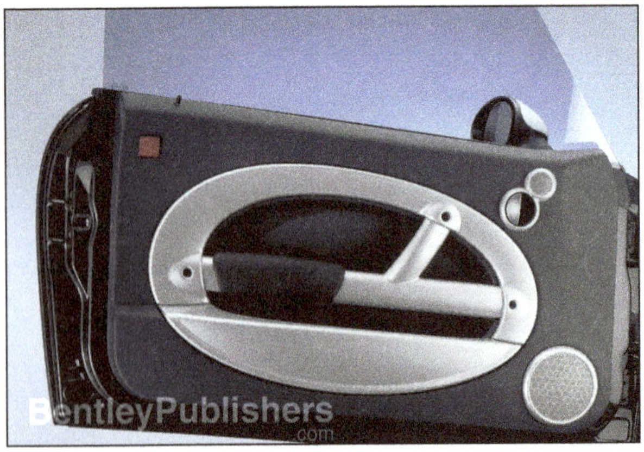

The MINI Cooper door design is a major feature of the interior. The design of the armrest matches the design of the center console. The armrest and storage compartment in the door are stylishly framed in silver and the surface of the armrest is covered with black leather.

Front seats have integrated side airbags. Height adjustment is provided as standard for the driver and optional for the passenger. Sports seats, lumbar support and two-level seat heating are all options. Seats are available in leather, leatherette or cloth.

The rear seat splits 50:50. With both seat-backs folded, the luggage area capacity increases from 5.6 cubic feet to 25 cubic feet. The rear seats have releases conveniently located in the luggage area and all handles are hidden from the outside – a subtle security feature. The seats also fold more easily than in any other small car due to a special kinematic mechanism.

Safety

Passive safety includes six standard airbags, including front and side airbags. Additionally, BMW Group's AHPS-2 (Advanced Head Protection System) is also fitted. This system features side impact head protection airbags. Further enhancing the car's active safety, anti lock brakes (ABS) are standard with dynamic stability control (DSC) and xenon headlights available as options.

Dashboard and controls

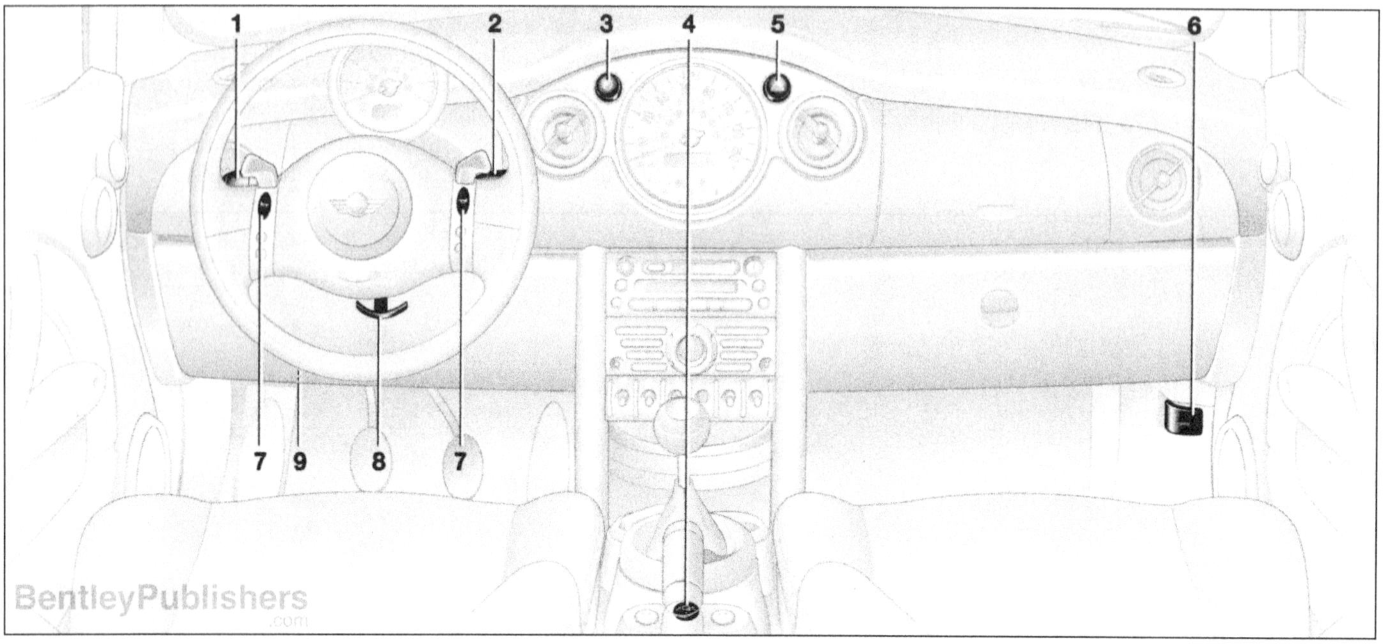

1. Light switch
   Head light flasher
   Turn signal indicators
   Onboard computer
2. Wiper/washer switch
3. Instrument lighting control
4. Outside mirror adjustment switch
5. Hazard warning flasher switch
6. Hood release
7. Horn button
8. Steering wheel adjustment control
9. OBD socket

Display elements

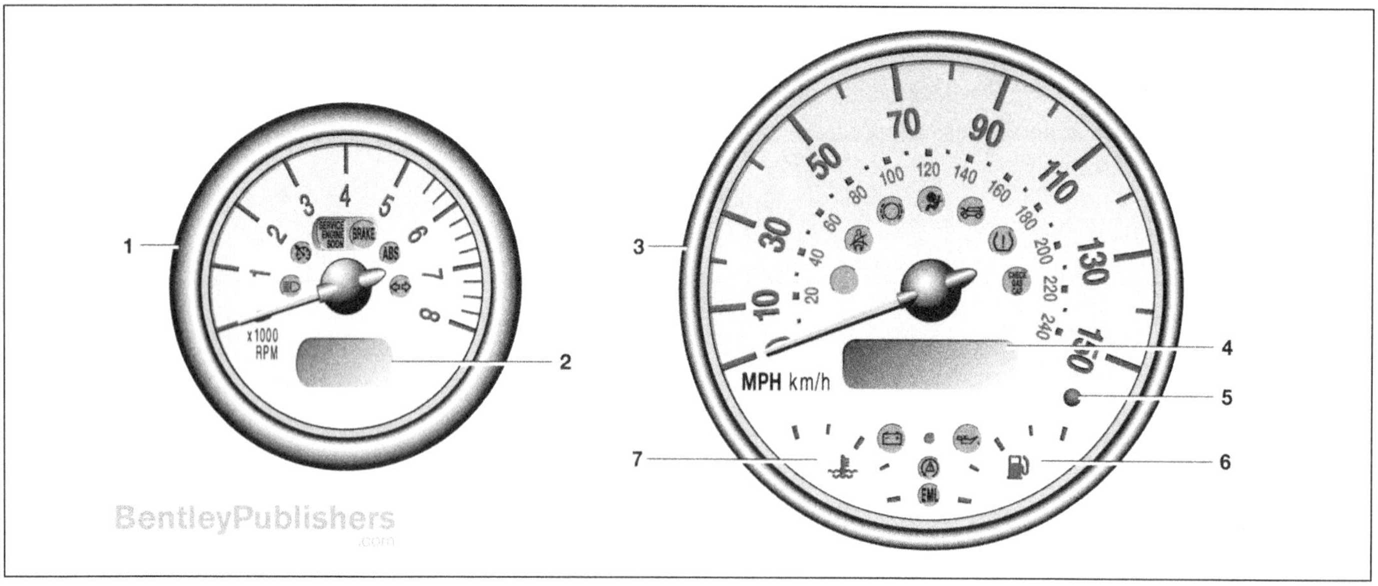

1. Tachometer with indicator and warning lights
2. Display for onboard computer, additional indicator for warning lights (cars with navigation)
3. Speedometer with indicator and warning lights
4. Indicator for odometer, trip odometer, service interval, program display for automatic transmission
5. Trip odometer reset
6. Fuel gauge
7. Engine coolant temperature gauge

Suspension, Steering, Brakes

The suspension systems used on the MINI provide a firm and responsive ride with superb levels of comfort and good acoustic properties. The weight distribution between the front and rear axles is biased to the front (63% front / 37% rear) due the front wheel drive configuration.

Power assisted steering using an electric pump.

The braking system includes large disc brakes all round, with 11 in. ventilated front disc brakes and 10 in. at the rear, with a diagonal dual-circuit arrangement. The system includes four-sensor ABS, electronic braking distribution (EBD) and cornering brake control (CBC) as standard. Automatic stability control (ASC) is standard on the Cooper S, BMW Group's DSC (dynamic stability control) is also available as an option.

The front suspension uses McPherson struts, which offer both weight and space advantages. The design features anti-dive and anti-squat geometry via the anti-roll bar and 1:1 strut movement ratio. The layout of the design minimizes camber loss due to side forces, thus improving handling and steering response.

Advantages of MINI front suspension

- Firm and responsive
- Positive driver feedback through steering
- Exceptional camber control

The multi-link rear suspension is used on all MINI models, making it the only car in its class to use such a sophisticated system. Benefits of this suspension are the exceptional handling characteristics and ride comfort, with good acoustic properties.

Advantages of MINI rear suspension

- Firm and responsive
- Rear passenger comfort
- Noise optimization

MINI Convertible

The MINI Convertible was introduced in 2005 in both the MINI Cooper and the MINI Cooper S.

To enhance the chassis of the MINI Convertible, the floor and side sills are reinforced with larger and sturdier components, primarily through the use of additional lower B-pillar crossbars and floor pan stiffeners. High strength door beams are also incorporated into the vehicle structure to help reduce intrusion into the passenger compartment during a side impact.

High strength aluminum roll-over hoops behind the rear seats along with a reinforced A-pillar windshield surround help to protect occupants in the event of a roll-over. Through these additional reinforcements, the Convertible offers a high level of passenger protection while retaining exceptional driving dynamics.

The MINI Convertible features a fully automatic roof with a unique integrated power sliding sunroof function.

The sliding sunroof is designed with a rigid panel under the first section of the canvas top. When the top button is first pressed, the section slides back along the top rails to create an opening overhead of 16 inches. The sliding sunroof can be operated while the vehicle is in motion up to speeds of 75 mph.

The MINI Convertible roof opens in just over 15 seconds, without the manual release of any latches and can be controlled by the remote locking system. Powered by an electro-hydraulic system, the top's frame mechanism automatically unlatches and retracts. The four side windows lower automatically as the soft top folds neatly to the rear.

The Z-shaped folding mechanism lets the roof fold compactly to store behind the rear seats. There is no need for a tonneau cover because the front section of the roof provides protection when the soft top is down.

The MINI Convertible top comes standard with a heated glass rear window. The top is available in three colors, blue, black and green.

John Cooper Works

The original John Cooper Works (JCW) kit was sold and professionally installed by MINI dealers in the U.S. The original kit took the regular 163 hp MINI Cooper S to over 200 hp. For 2005, a newer higher output kit was developed, boosting power to 207 hp.

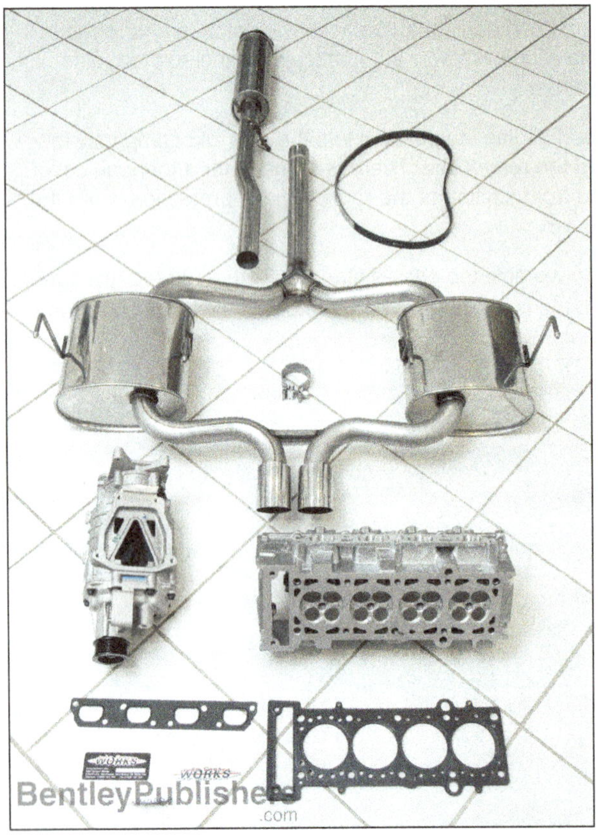

Key elements of the JCW kit were a redesigned air box with an electronically-controlled air induction system, a set of higher flow fuel injectors, a reprogrammed electronic control unit, a stainless steel sport exhaust system, a high performance cylinder head, and an enhanced Eaton supercharger with specially-coated vanes that generate increased flow pressure. The supercharger was fitted with a smaller pulley, allowing the unit to spin at a higher rpm.

The kit originally retailed for US $4,500, plus an estimated 9 hours of labor for installation (US $4,650, plus 10 hours labor for the 2005 higher output kit).

For the 2006 model year, the John Cooper Works kit for the MINI Cooper S and Cooper S Convertible was made available as a factory-installed option. The kit is installed on the production line at the MINI plant in Oxford, England. The package included a limited-slip differential and the John Cooper Works brake kit and retailed for US $6,300.

020 Maintenance

TABLES

GENERAL

This repair group explains the structure of this repair manual and details basic information regarding your vehicle and repair procedures for it. Also included are service and maintenance procedures.

Carry out the maintenance work described in this repair group at the factory specified time or mileage interval shown in **Maintenance tables** in this repair group. Following these intervals helps ensure safe and dependable operation.

The owner's manual, maintenance record and warranty booklet originally supplied with the vehicle contain maintenance schedules that apply to your MINI. Aside from keeping your vehicle in the best possible condition, proper maintenance plays a role in maintaining full protection under MINI's new-car warranty coverage. If in doubt about the terms and conditions of your vehicle's warranty, an authorized MINI dealer should be able to explain them.

MINI is constantly updating their recommended maintenance procedures and requirements. The information contained here may not include updates or revisions made by MINI since the publication of the documents supplied with the vehicle. If there is any doubt about what procedures apply to a specific model or model year, or what intervals to follow, consult an authorized MINI dealer.

How to use this manual

This manual is divided into 11 main sections or partitions:

0 General, maintenance
1 Engine
2 Transmission, driveshaft, differential
3 Suspension, brakes, steering
4 Body
5 Body equipment
6 Electrical system
7 Safety equipment and accessories
ECL Electrical Component Locations
EWD Electrical wiring diagrams
OBD On Board Diagnostics

A master listing of the 11 partitions and the corresponding specific repair groups can be found on the inside front cover.

Thumb tabs are used on the first page of each repair group to help locate the groups quickly. Page numbers throughout the manual are organized according to the repair group system. A comprehensive **Index** is at the end of the manual.

Warnings, Cautions and Notes

Throughout this manual are many passages with the headings **WARNING**, **CAUTION**, or **NOTE**. These headings have different meanings.

> ***WARNING —***
> *Text under this heading warns of unsafe practices that are very likely to cause injury, either by direct threat to the person(s) performing the work or by increased risk of accident or mechanical failure while driving.*

> ***CAUTION —***
> *Text under this heading calls attention to important precautions to be observed during the repair work that will help prevent accidentally damaging the vehicle or its parts.*

NOTE —
A note contains helpful information, tips or pointers which help in doing a better job and completing it more easily.

Read **WARNING, CAUTION** and **NOTE** headings before you begin repair work. See also **001 Warnings and Cautions**.

BASIC SERVICE INFORMATION

Diagnostic trouble codes (DTCs), accessing

With a generic OBD II scan tool connected to the 16-pin OBD II plug, diagnostic trouble codes (DTCs) can be obtained, along with the conditions associated with the illumination of the Malfunction Indicator Light (MIL), also commonly known as the check engine light. A more advanced or MINI-dedicated diagnostic scan tool accesses additional proprietary information.

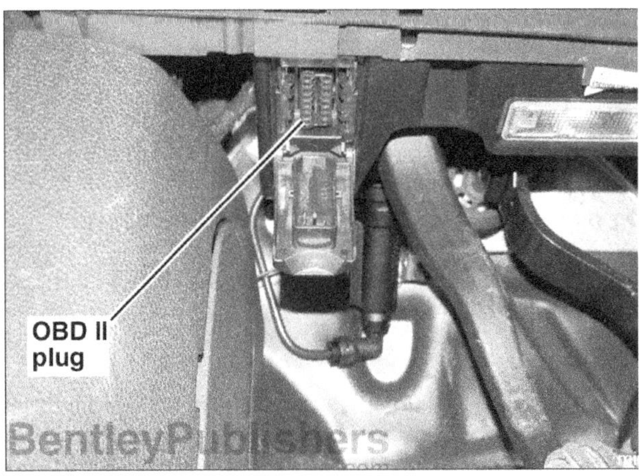

- Connect diagnostic scan tool to OBD II connector located under left side of dashboard.
- Place transmission selector lever in PARK or NUETRAL. Engage parking brake. Make sure ignition is OFF.
- Start engine and let idle.
- Follow scan tool instructions as they appear on screen.

For a list of scan tool suppliers and more OBD II information, see **OBD On Board Diagnostics** section at the back of this manual.

Tightening fasteners

When tightening the bolts or nuts on a component, it is always good practice to tighten the bolts gradually and evenly to avoid misalignment or over-stressing any one portion of the component. For components sealed with gaskets, this method helps to ensure that the gasket will seal properly.

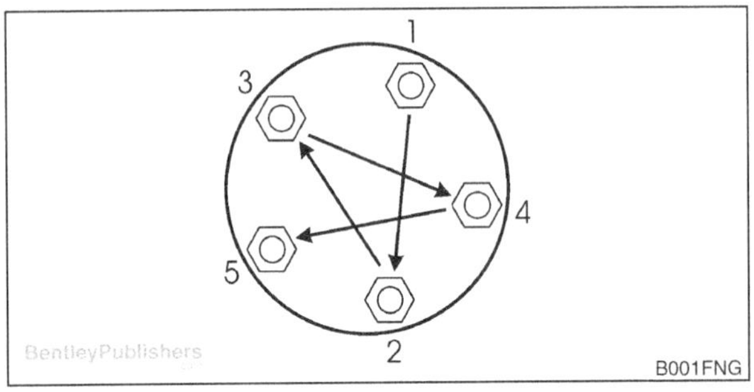

Where there are several fasteners, tighten them in a sequence alternating between opposite sides of the component. Repeat the sequence until all the bolts are evenly tightened to the proper specification.

For some repairs a specific tightening sequence is necessary, or particular order of assembly is required. Such special conditions are noted in the text, and the necessary sequence is described or illustrated. Where no specific torque is listed, the table **General bolt tightening torques in Nm (max. permissible)** can be used as a general guide for tightening fasteners.

NOTE —
- *Metric bolt classes or grades are marked on the bolt head.*
- *Do not confuse wrench size with bolt diameter.*
- *Values in table are for reference only.*

General bolt tightening torques in Nm (max. permissible)						
Bolt class (according to DIN 267)						
Bolt diameter	**5.6**	**5.8**	**6.8**	**8.8**	**10.9**	**12.9**
M5	2.5	3.5	4.5	6	8	10
M6	4.5	6	7.5	10	14	17
M8	11	15	18	24	34	40
M10	23	30	36	47	66	79
M12	39	52	62	82	115	140
M14	62	82	98	130	180	220
M16	94	126	150	200	280	340
M18	130	174	210	280	390	470

Non-reusable fasteners

Many fasteners used on the vehicles covered by this manual must be replaced with new ones once they are removed. These include but are not limited to: bolts, nuts (self-locking, nylock, etc.), roll pins, clips and washers. Genuine MINI parts should be the only replacement parts used for this purpose.

Some bolts are designed to stretch during assembly and are permanently altered, rendering them unreliable once removed. These are known as torque-to-yield fasteners. Always replace fasteners where instructed to do so. Failure to replace these fasteners could cause personal injury or vehicle damage. See an authorized MINI dealer for applications and ordering information.

Parts

Many of the maintenance and repair tasks in this manual call for the installation of new parts, or the use of new gaskets and other materials when reinstalling parts. Most often, the parts that will be needed should be on hand before beginning the job. Read the introductory text and the complete procedure to determine which parts will be needed.

NOTE —
For some bigger jobs, partial disassembly and inspection is required to determine a complete parts list. Read the procedure carefully and, if necessary, make other arrangements to get the necessary parts while your vehicle is disassembled.

Model. When ordering parts it is important that you know the correct model designation for your vehicle. Models covered in this manual are the MINI Cooper and the MINI Cooper S.

Model Year. This is not necessarily the same as date of manufacture or date of sale. A 2003 model may have been manufactured in late 2002, and perhaps not sold until early 2004. It is still a 2003 model. Model years covered by this manual are 2002 to 2006.

Date of manufacture. This information is necessary when ordering replacement parts or determining if any of the warranty recalls are applicable to your vehicle. The label on the driver's door below the door latch specifies the month and year that the vehicle was built.

Genuine MINI parts

Replacement parts from an authorized MINI dealer are designed and manufactured to the same high standards as the original parts. They will be the correct material, manufactured to same specifications, and guaranteed to fit and work as intended. Most genuine MINI parts carry a limited warranty.

MINI is constantly updating and improving their vehicles, often making improvements during a given model year. MINI may recommend a newer, improved part as a replacement, and your authorized dealer parts department will know about it and provide it. The MINI parts organization is best equipped to deal with your parts needs.

Non-returnable parts

Some parts cannot be returned, even for credit. The most common example is electrical parts, which are almost universally considered non-returnable. Buy electrical parts carefully, and be as sure as possible that a replacement is needed, especially for expensive parts such as electronic control units. It may be wise to let an authorized MINI dealer or other qualified shop confirm your diagnosis before replacing an expensive non-returnable part.

Tools

Most maintenance can be accomplished with a small selection of tools. Tools range in quality from inexpensive junk, which may break at first use, to very expensive and well-made tools for the professional. The best tools for most do-it-yourself MINI owners lie somewhere in between.

Many reputable tool manufacturers offer good quality, moderately priced tools with a lifetime guarantee. These are your best buy. They cost a little more, but they are good quality tools that will do what is expected of them. Sears' Craftsman® line is one such source of good quality tools.

Some of the repairs covered in this manual require the use of special tools, such as a custom puller or specialized electrical test equipment. These special tools are called out in the text and can be purchased through an authorized MINI dealer. As an alternative, some special tools mentioned may be purchased from the following tool manufacturers or distributors:

Specialty tool suppliers

Assenmacher Specialty Tools, Inc.
6440 Odell Place
Boulder, CO 80301
800-525-2943
www.asttool.com

Baum Tools Unlimited, Inc.
PO Box 5867
Sarasota, FL 34277
800-848-6657
www.baumtools.com

Mac Tools
4635 Hilton Corporate Drive
Columbus, OH 43232
800-622-8665
www.mactools.com

Metalnerd
509 Crestview Drive, Suite B
Greensburg, PA, 15601 USA
412-601-4270
www.metalnerd.com

Samstag Sales
115 Main St. N., Suite 216
Carthage, TN 37030
615-735-3388
www.samstagsales.com

Shade Tree Software
4186 Culebra Ct.
Boulder, CO 80301
303-449-1664
303-940-2468
www.shadetreesoftware.com

Snap-On Technologies, Inc.
2801 80th St.
Kenosha, WI 53141-1410
262-656-5200
www.snapon.com

Zelenda Automotive, Inc.
66-02 Austin St.
Forest Hills, NY 11374
888-892-8348
www.zelenda.com

Special tools

Some special tools are required for performing engine maintenance tasks. Be sure to have the necessary equipment on hand before starting any job.

00 2 080

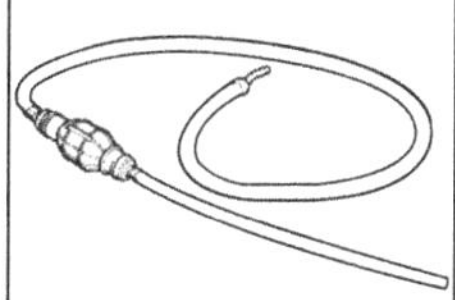

◀ Hand pump for manual transmission lifetime oil (BMW tool no. 00 2 080)

11 8 390

◀ Lever for relieving accessory belt tension (Cooper) (BMW tool no. 11 8 390)

11 8 410

◀ Lever for relieving accessory belt tension (Cooper S) (BMW tool no. 11 8 410)

11 8 280
11 8 470

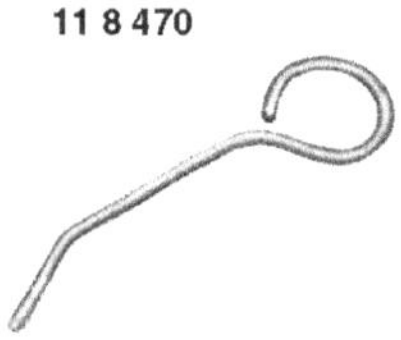

◀ Accessory belt tensioner locking tool (BMW tool no. 11 8 280 for Cooper, BMW tool no. 11 8 470 for Cooper S)

34 1 260

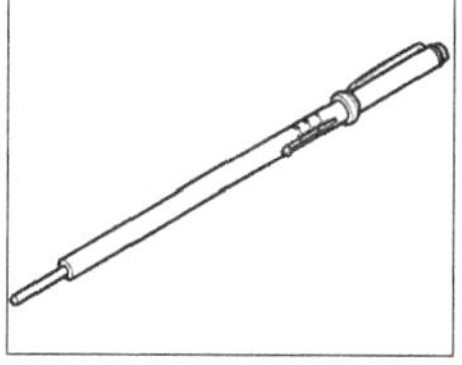

◀ Brake pad wear gauge (BMW tool 34 1 260)

51 2 160

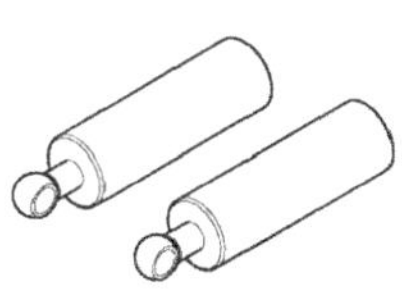

◀ Service position hood props (BMW tool no. 51 2 160)

Identification Plates and Labels

Vehicle Identification Number (VIN)

This is a combination of letters and numbers that identify the particular vehicle. The VIN appears on the state registration document, and on the vehicle itself. One location is on the right front strut tower in the engine compartment (**arrow**), another in the lower left corner of the windshield.

Date of manufacture

This information is necessary when ordering replacement parts or determining if any of the warranty recalls are applicable to your vehicle. The label on the driver's door below the door latch will specify the month and year that the vehicle was built.

Engine and transmission codes

For information on engine codes and engine applications, see **100 Engine–General**.

The transmission type with its identifying code may be important when buying clutch parts, seals, gaskets, and other transmission-related parts. For information on transmission codes and applications, see **200 Transmission–General**.

Raising Vehicle

Lifting vehicle

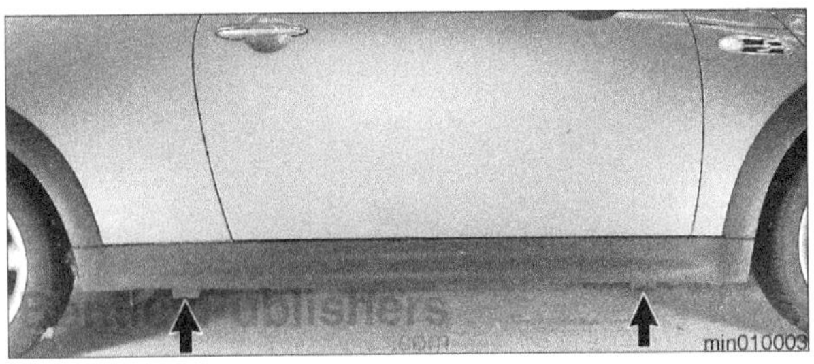

Use proper jacking points (**arrows**) to raise vehicle safely and avoid damage.

> ***WARNING —***
> - *Never work under a lifted vehicle unless it is solidly supported on jack stands that are intended for that purpose.*
> - *When raising the vehicle using a floor jack or hydraulic lift, carefully position the jack pad to prevent damaging the vehicle body.*
> - *Watch the jack closely. Make sure it stays stable and does not shift or tilt. As the vehicle is raised, it may roll slightly and the jack may shift.*

NOTE —
Models equipped with safety ("run-flat") tires, which includes all MINI Cooper S models, may not be equipped with a spare tire, jack or tire changing tools.

Raising vehicle safely using vehicle jack

> ***WARNING —***
> - *When raising the vehicle using a floor jack or hydraulic lift, carefully position the jack pad to prevent damaging the vehicle body.*
> - *Watch the jack closely. Make sure it stays stable and does not shift or tilt. As the vehicle is raised, it may roll slightly and the jack may shift.*

– Park vehicle on flat, level surface.

– If changing a tire, loosen lug bolts before raising vehicle. See **Changing a tire** later in this group.

– Block wheel that is opposite and farthest from jack to prevent vehicle from unexpectedly rolling.

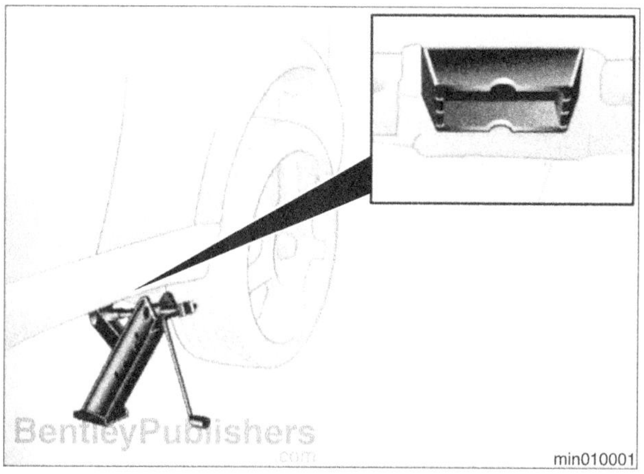

< Place jack into position (**inset**). Make sure jack is resting on flat, solid surface. Use a board or other support to provide a firm surface for jack, if necessary.

– Raise vehicle slowly while constantly checking position of jack and vehicle.

> ***WARNING —***
> - *Do not rely on the transmission or the parking brake to keep the vehicle from rolling. They are not a substitute for positively blocking the opposite wheel.*
> - *Never work under a vehicle that is supported only by a jack. Use jack stands designed to support the vehicle. See* **Working under vehicle safely**.

Working under vehicle safely

> ***WARNING —***
> - *A jack is a temporary lifting device. Do not use a jack alone to support the vehicle while you are under it.*
> - *Do not work under a lifted vehicle unless it is solidly supported on jack stands that are intended for that purpose.*
> - *Do not use wood, concrete blocks, or bricks to support a vehicle. Wood may split. Blocks and bricks, while strong, are not designed for that kind of load, and may break or collapse.*
> - *Use care when removing major (heavy) components from one end of the vehicle. The sudden change in weight and balance can cause vehicle to tip off lift or jack stands.*
> - *Do not support vehicle at engine oil pan, transmission, fuel tank, or on front or rear axle. Serious damage may result.*

– Disconnect negative (–) cable from battery so that vehicle cannot be started. Let others know what you are doing.

> ***CAUTION —***
> *Prior to disconnecting the battery, read the battery disconnection cautions given in* **001 Cautions and Warnings**.

– Raise vehicle slowly. See **Raising vehicle safely using vehicle jack** in this repair group.

Use at least two jack stands to support vehicle. Use jack stands designed for the purpose of supporting a vehicle.

– Place jack stands on firm, solid surface. If necessary, use a flat board or similar solid object to provide a firm footing.

– Lower vehicle slowly until its weight is fully supported by jack stands. Watch to make sure that the jack stands do not tip or lean as the vehicle settles on them.

– Observe jacking precautions again when raising vehicle to remove jack stands.

EMERGENCIES

Flat tire monitor

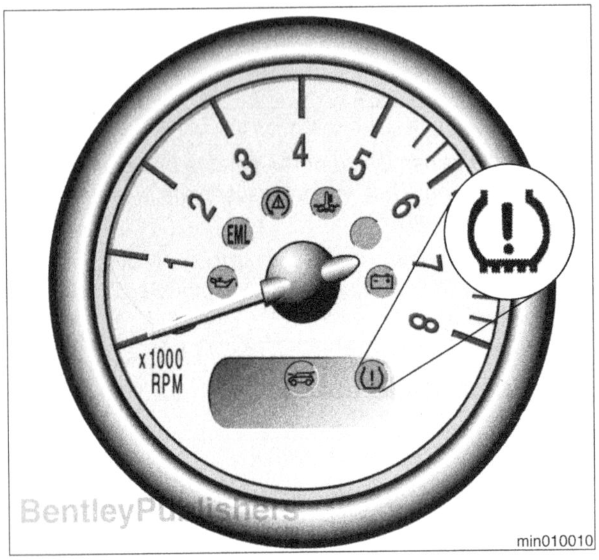

If flat tire monitor (**inset**) in tachometer face flashes yellow:

- Vehicle with regular tires: Reduce speed immediately and stop vehicle quickly and safely. Change tire as described below.
- Vehicle with safety (run-flat) tires: Reduce speed, continue driving at or below 50 mph (80 kph).

– In either case, avoid hard brake applications or quick steering maneuvers until you can investigate and remedy the problem.

– To reset flat tire monitor, initialize the system according to the instructions in your owner's manual.

Changing a tire

NOTE *—*
Models equipped with safety ("run-flat") tires, which includes MINI Cooper S models, may not be equipped with a spare tire, jack or tire changing tools.

– Stop vehicle on as flat a surface as possible, in a place where you can be easily seen by other drivers. Avoid stopping just over the crest of a hill.

> ***WARNING*** *—*
> *If a tire goes flat while driving, pull well off the road. Changing a tire on a busy street or highway is very dangerous. If necessary, drive a short distance on the flat tire to get to a safe place. It is much better to ruin a tire or rim than to risk being injured.*

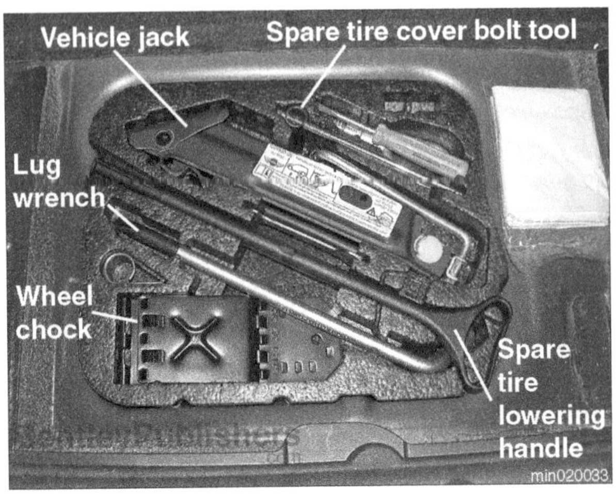

– Turn on emergency flashers, and set out flares or emergency markers well behind vehicle. Use wheel chock in vehicle tool kit to block wheel diagonally opposite to the one being changed. Passengers should get out of vehicle and stand well away from road.

Remove trunk liner and tool kit from rear cargo area.

- Use spare tire cover bolt tool to remove cover for spare tire mounting bolt located under tool kit.
- Thread spare tire lowering handle onto spare tire mounting stud.

< Hold lowering handle and release retaining clips (**arrows**) and lower spare tire to ground with handle.

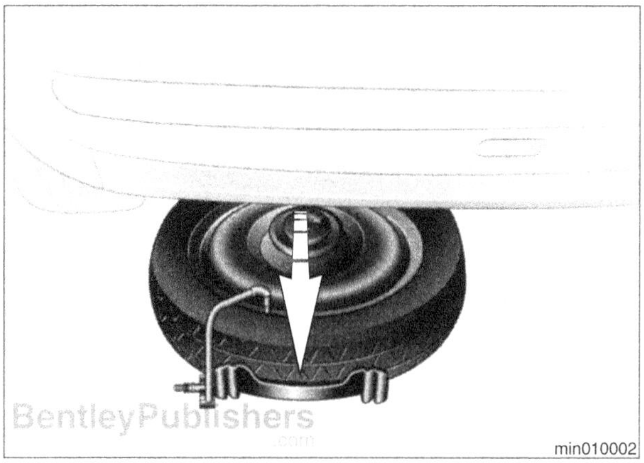

< Remove spare tire from tire storage tray underneath cargo area.

- Loosen wheel bolts while vehicle is on ground, but leave them a little snug.

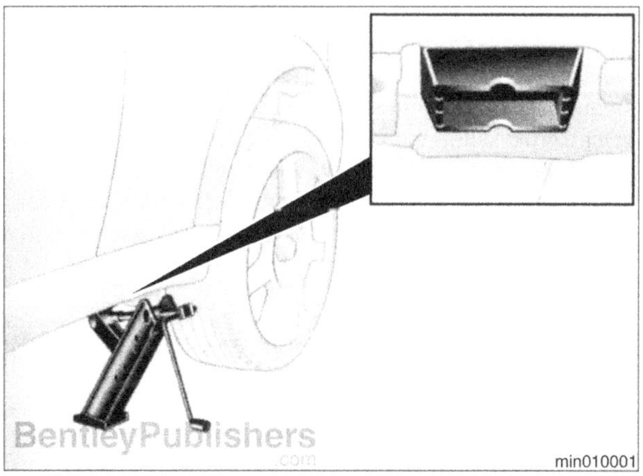

< Place jack in lifting point (**inset**) nearest wheel being changed. Use a board to provide a firm footing for jack if ground is soft. Raise vehicle only far enough so that the wheel is fully off ground and then remove wheel bolts and wheel.

- Remove air pressure valve extension from spare tire, if applicable.
- Install spare tire. Install wheel bolts and tighten them hand tight using lug wrench.

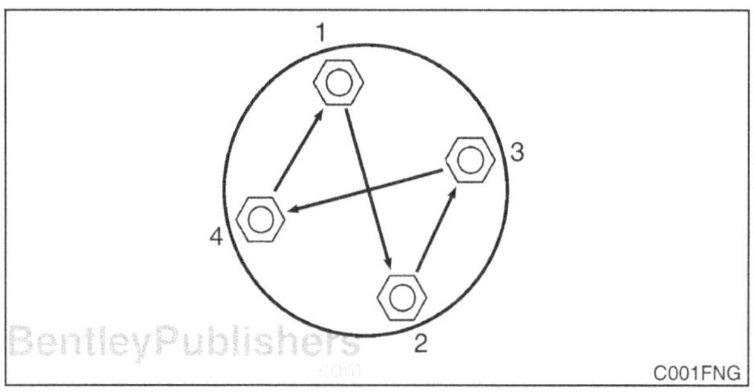

< Lower vehicle. With all wheels on ground, fully tighten nuts in a crisscross pattern. Torque wheel nuts when installing wheel. Check inflation pressure of spare tire.

Tightening torque	
Wheel to wheel hub	120 ± 10 Nm (89 ± 7 ft-lb)

- Replace tools in tool kit and store flat tire in rear cargo area. Plastic bag provided in tool kit can be used to store tire.

Jump starting

vehicles with discharged or dead batteries can be jump-started using the good battery from another vehicle. See owner's manual for proper instruction on jump-starting. When jump-starting the engine, always note the following warnings.

> ***WARNING*** —
> - *Battery acid (electrolyte) can cause severe burns, and will damage the vehicle, clothing, and you. If electrolyte is spilled, wash surface with large quantities of water. If it gets into eyes, flush them with water for several minutes and call a doctor.*
> - *Batteries produce explosive and noxious gasses. Keeps sparks and flames away. Do not smoke near batteries.*
> - *Do not jump-start the engine if you suspect that battery is frozen. Trapped gas may explode. Allow the battery to thaw first.*
> - *Do not quick-charge the battery (for boost starting) for longer than one minute, and do not exceed 16.5 volts at the battery with the boosting cables attached. Wait at least one minute before boosting the battery a second time.*

Indicator and warning lights

Many vehicle systems are self-monitored for faults while driving. Generally, a red warning lamp that comes on during driving should be considered serious. If you cannot immediately determine the seriousness of the warning light, stop the vehicle in a safe place and turn the engine off as soon as possible. Consult the owner's manual for additional information on the warning lamp and the recommended action.

If the Malfunction Indicator Light (MIL) ("Service Engine Soon" warning light) comes on or flashes, it indicates that an emissions-related fault has occurred. Faults such as a bad oxygen sensor or a dead fuel injector can cause the exhaust or evaporative emissions to exceed a specified limit. When these limits are exceeded, the MIL will be turned on. The engine can be safety driven with the light on, although the emission systems should be checked as soon as possible. See **OBD On Board Diagnostics** for more information on the MIL and the on-board diagnostic system.

Towing

vehicles covered by this manual should be towed with a tow truck using wheel lift. Lift front wheels, allowing the rear wheels to trail. Alternatively, use a flat-bed truck for transportation.

> ***CAUTION*** —
> - *Do not tow the vehicle on all four wheels except for very short distances to move it to a safe place*
> - *Do not tow with sling-type equipment. The front spoilers and bumper covers may sustain damage.*

To access threaded towing eye socket, pry open trim on front lower grille or on rear bumper.

NOTE —
A towing eye is provided in the luggage compartment tool kit. The towing eye can be screwed into the front or rear of the vehicle.

< Install towing eye into threaded hole. Screw in until it bottoms firmly.

– Standard transmission: Do not tow with front wheels on ground unless absolutely necessary.

– Automatic transmission: Have front wheels placed on dollies, or use flat bed truck. Use towing eye for winching vehicle on flat bed.

NOTE —
Always tow vehicle with transmission lever in N (neutral).

Spare parts kit

Carrying a basic set of spare parts can prevent a minor breakdown from turning into a major repair annoyance. The list below is a good place to start.

Spare parts kit - basic

- Ribbed accessory belt
- Specified engine oil: one or two quarts
- Specified engine coolant: 1 gallon of premixed 50/50 anti-freeze and water
- Fuse assortment: 7.5A, 10A, 15A, 20A, 30A, 50A
- Radiator hoses: upper and lower

Spare parts kit - additional contents

- Exterior lighting bulbs: headlight, brake light, turn signal, and tail-light
- Wiper blades or blade inserts
- Brake fluid: new, unopened bottle, DOT 4 specification
- Main relay for DME system

Engine Compartment Components

Engine compartment components.
Cooper S shown, Cooper is similar as noted below.

1. Windshield washer reservoir
2. Engine oil dipstick
   - Cooper: dipstick located in front center of engine compartment
3. Coolant expansion tank
4. Engine oil filler
5. Power steering fluid reservoir
6. Auxiliary terminal for jump starting
   - Cooper S: battery located in trunk
7. Air filter housing
   - Cooper: air filter housing is located in front of battery
8. Engine Control Module (ECM) housing
9. Brake fluid reservoir
10. Head lamp washer reservoir
    - Optional equipment

Engine covers, removing and installing

Cooper

Remove front engine cover (fuel rail cover)

- Squeeze locking tabs (**A**) to detach rear of cover.
- Tilt and pull cover forward until plastic clips (**arrows**) detach from fuel rail.

> ***CAUTION—***
> *Plastic clips can break easily.*

– Install in reverse order of removal.

Cooper S

Remove front engine cover (intercooler cover)

- Remove retaining screws (**arrows**) and remove cover.

> ***CAUTION—***
> *Intercooler fins are easily damaged. Be careful working around intercooler. Use intercooler protector (BMW special tool 11 8 480) when working in engine compartment.*

– Install in reverse order of removal.

Air Filter Service

Air filter, replacing (Cooper)

NOTE—
The specified replacement intervals for the air filter (Inspection II service) are based on normal use. If the vehicle is operated primarily in dusty conditions, service air filter more frequently.

– Switch off ignition.

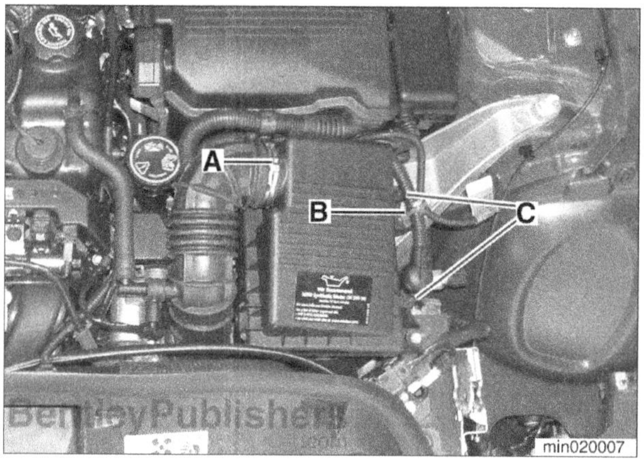

min020007

◄ Loosen or remove air duct clamp (**A**). Detach air duct from air filter housing.

– Detach wire harness (**B**).

– Remove screws (**C**) retaining air filter cover to filter housing.

– Tilt up air filter cover and remove filter element.

– Clean out air filter housing.

– Replace air filter and upper air filter housing. Reattach air duct using new clamp.

Air filter, replacing (Cooper S)

NOTE—
The specified replacement intervals for the air filter (Inspection II service) are based on normal use. If the vehicle is operated primarily in dusty conditions, service air filter more frequently.

– Switch off ignition.

min020008

◄ Pull up (**arrows**) on battery positive connection point (Battery +) to release from right side of air filter housing.

– Unfasten air duct clamp (**A**). Detach air duct from air filter housing.

– Remove screws (**B**) retaining air filter upper housing to filter housing.

– Swing up air filter upper housing and remove filter element.

– Clean out air filter housing.

– Replace air filter and upper air filter housing. Reattach air duct using new clamp.

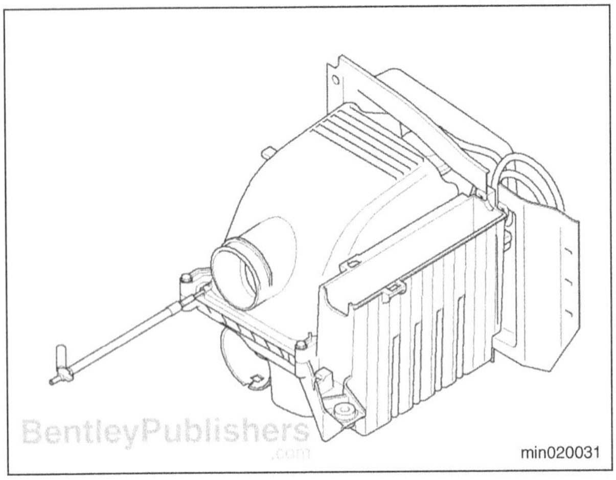

Air filter, assembly (JCW)

JCW (John Cooper Works) air filter assembly replacement instructions are similar to the Cooper S model air filter.

> ***CAUTION—***
> *In order to avoid breaking plastic tabs, raise housing slightly and push back to clear tabs at rear.*

BATTERY SERVICE

Battery, checking, cleaning and replacing

Cooper: Battery is located in left rear of engine compartment in an insulating box with a heat resistant cover.

Cooper S: Battery located in rear cargo compartment, under cargo liner.

Simple maintenance of the battery and its terminal connections will ensure maximum starting performance, especially in winter when colder temperatures reduce battery power.

– Make sure that:
  - Battery cables are tight.
  - Battery terminals, cable clamps and battery case are free of white deposits that indicate corrosion and acid salts. Even a thin layer of dust containing conductive acid salts can cause battery discharge.

– Remove battery corrosion:
  - Disconnect battery cables. Disconnect negative (–) cable first.
  - Remove battery from vehicle.
  - Clean terminal posts and cable clamps with a wire brush.
  - Clean main chassis ground terminal next to battery. Wash away corrosion with a baking soda and water solution to neutralizes battery acid.
  - Reinstall battery and reconnect cable clamps, positive (+) cable first.
  - Lightly coat outside of terminals, hold down screws, and clamps with petroleum jelly, grease, or a commercial battery terminal corrosion inhibitor.

> ***WARNING—***
> *Battery acid is extremely dangerous. Take care to keep it from contacting eyes, skin, or clothing. Wear eye protection. Extinguish all smoking materials and do not work near any open flames.*

> ***CAUTION—***
> *Prior to disconnecting the battery, read the battery disconnection cautions given in* **001 General Cautions and Warnings.**

NOTE—
The original equipment battery in MINI models is maintenance free. The original electrolyte will normally last the entire service life of the battery under moderate climate conditions.

– Maintain battery electrolyte at correct level, just above battery plates and their separators. Correct level is approximately 5 mm (¼ in) above top of battery plates or to top of indicator marks (if applicable). Battery plates and indicator marks can be seen once filler caps are removed. If electrolyte level is low, replenish it by adding distilled water only.

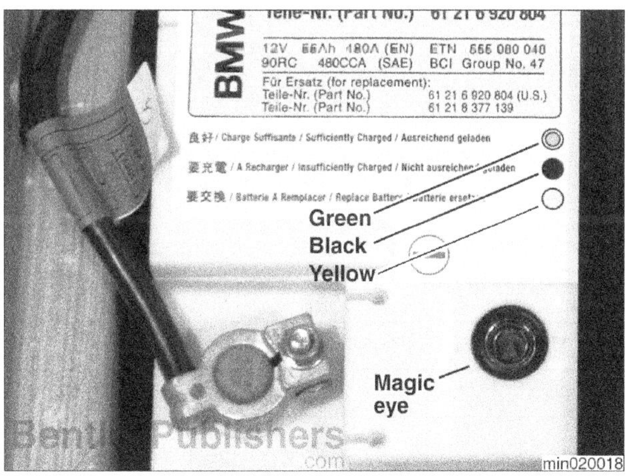

min020018

◄ Original equipment BMW battery is equipped with built-in hydrometer "magic eye". Battery condition is determined by eye color:
  - Green: Adequate charge
  - Black: Inadequate charge, recharge
  - Yellow: Defective battery, replace

Batteries are rated by ampere hours (Ah), the number of hours a specific current drain can be sustained before complete discharge, or by cold cranking amps (CCA), the number of amps available to crank the engine in cold weather conditions.

– Replace battery with one rated equal or higher than the original battery.

Battery rating	
Cooper Cooper S	12 volt, 55 Ah

The battery is held in place by a hold-down strap. A secure battery hold-down is important in order to prevent vibrations and road shock from damaging the battery.

NOTE—
- *Disconnect the negative (–) cable first and connect it last. While changing the battery, clean away any corrosion in or around the battery tray.*
- *More battery and charging system information is in* **121 Battery, Alternator, Starter**.

BRAKE SERVICE

Brake fluid, replacing

Brake fluid absorbs moisture easily, and moisture in the fluid affects brake performance and reliability. This is why MINI strictly recommends replacing the brake fluid every two years.

When replacing or adding brake fluid, use only new fluid from a previously unopened container. Do not use brake fluid that has been bled from the system, even if it is brand new. Use only DOT 4 brake fluid.

NOTE—
See **340 Brakes** *for brake fluid flushing procedures.*

Brake pad / rotor wear, checking

Disc brakes are fitted at all four wheels.

Check brake pad thickness whenever wheels are off or brake work is being done. Also inspect:
- Brake rotor
- Condition of caliper seal
- Condition of caliper slider bolts

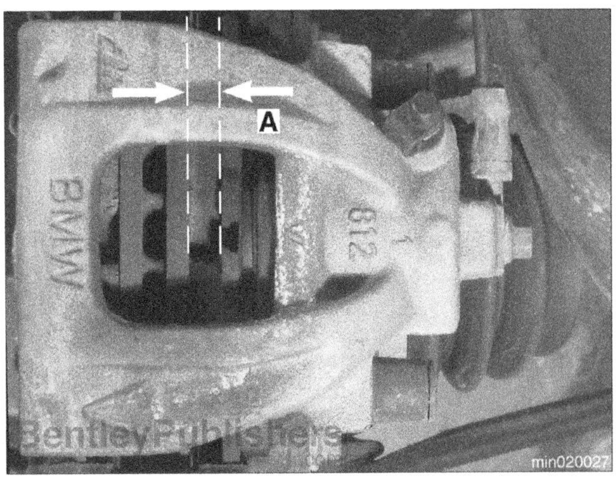

Thickness of inner brake pad can be estimated without removing brake caliper.

Brake pad lining minimum thickness (A)	
Front or rear pad	3.0 mm (0.12 in)

NOTE—
Brake service procedures are given in **340 Brakes**.

Brake system, inspecting

Routine maintenance of the brake system includes maintaining brake fluid in the reservoir, checking brake pads for wear, checking parking brake function, and inspecting the system for fluid leaks or other damage.

- Check that brake hoses are routed to avoid chafing or kinking.
- Inspect unions and brake calipers for signs of fluid leaks.
- Inspect rigid lines for corrosion, dents, or other damage.
- Inspect flexible hoses for cracking.
- Replace faulty hoses or lines as described in **340 Brakes**.

> ***WARNING —***
> *Incorrect installation or over tightened hoses, lines, and unions may cause chafing or leakage. This can lead to partial or complete brake system failure.*

Parking brake, checking

The parking brake system is independent of the main braking system and may require periodic adjustment depending on use.

- Adjust parking brake if lever can be pulled up more than 8 clicks.
- Check that parking brake cables move freely.

Parking brake description and adjustment procedure are in **340 Brakes**.

COOLING SYSTEM, CHECKING

Antifreeze raises the boiling point and lowers the freezing point of engine coolant. The manufacturer recommends coolant that is a 50 / 50 mixture of distilled water and ethylene glycol-based antifreeze.

Routine cooling system maintenance consists of maintaining the coolant level and inspecting hoses. Because the coolant's anticorrosion and antifreeze additives gradually lose their effectiveness, replace coolant every 4 years.

> ***WARNING —***
> *Due to risk of personal injury, be sure the engine is cold before opening any part of the cooling system or removing the coolant reservoir cap.*

> ***CAUTION —***
> *Use only MINI approved phosphate-free antifreeze when filling the cooling system. Use of antifreeze containing phosphates is harmful to the cooling system.*

Maximum and minimum coolant levels are marked on coolant overflow reservoir.

NOTE —
Cooper S engine compartment is illustrated. Cooper engine compartment layout is similar.

– Add coolant to Cooper engine:
  - Open cooling system pressure cap (to left of engine) by turning slowly counterclockwise to allow any pressure to escape safely. Then open cap.
  - If coolant is low, slowly add coolant to correct level. Do not overfill.

– Add coolant to Cooper S engine:
  - Open reservoir cap by turning slowly counterclockwise to allow any pressure to escape safely. Then open cap.
  - If coolant is low, slowly add coolant to correct level. Do not overfill.

Engine coolant		
Model	**Approximate capacity liter (US qt.)**	**Coolant specification**
Cooper	5.3 (5.6)	50% mixture of BMW phosphate / nitrite free antifreeze and water
Cooper S	6 (6.3)	

– Check for tight, dry hose connections. Coolant seepage at clamped connection indicates either that hose clamp is loose, hose is damaged, or connection is dirty or corroded. Dried coolant has a chalky appearance.

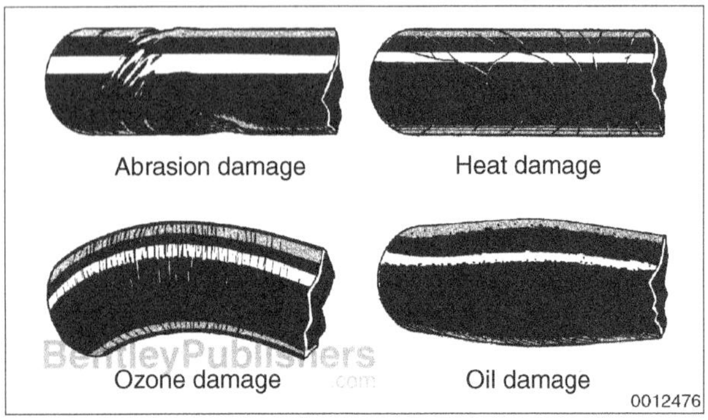

< Check for firm and springy hoses. Replace any hose that is cracked, has become soft and limp, or has been contaminated by oil. The illustration shows examples of damage to coolant hoses. Any of these conditions is cause for replacement. *(Illustration courtesy of Gates Rubber Company, Inc.)*

As a preventive measure, replace cooling system hoses every four years.

Engine Accessory Belt Service

The accessory belt and pulleys transfer power from the engine crankshaft to various accessories. MINI models use one serpentine belt to operate belt-driven components.

Inspect accessory belt with engine off. If belt shows signs of wear, cracking, glazing, or missing sections, replace immediately.

> ***WARNING —***
> *Observe care when replacing the belt. Personal injury could result if a tensioner springs back into position uncontrolled.*

> ***CAUTION —***
> *Mark belt rotation direction if planning to reuse an old belt.*

NOTE —
- *To reduce the chance of belt failure while driving, replace belt every four years.*
- *When a belt is replaced, store the old one in the cargo compartment for emergency use.*

Accessory belt, replacing (Cooper)

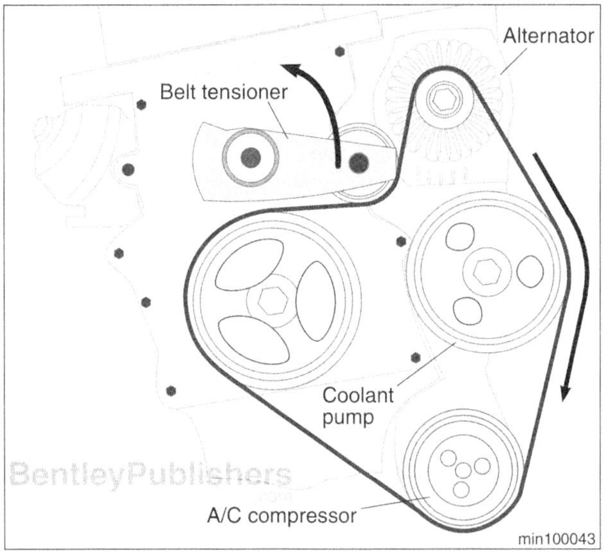

◄ Before removing, note layout of accessory belt.

NOTE—
If belt is to be reused, mark direction of travel and reinstall belt in same direction of rotation.

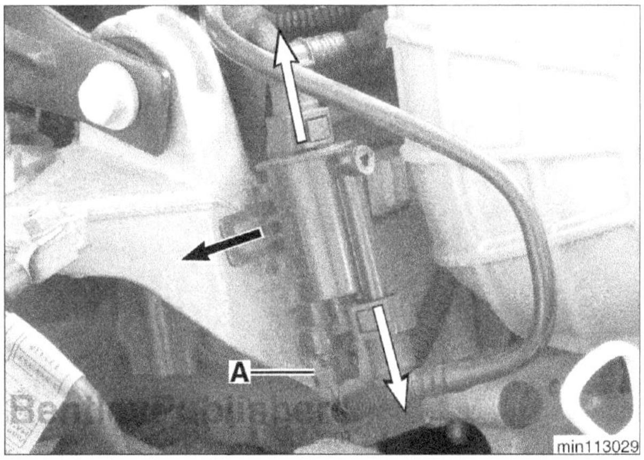

◄ Remove fuel tank vent valve.

- Disconnect vent hoses (**white arrows**) via quick fit connectors.
- Straighten retaining tag and slide vent valve off bracket (**black arrow**).
- Remove electrical harness connector (**A**).

NOTE—
Cooper S engine is illustrated. Cooper engine layout is similar.

– Use special tool 11 8 390 to release accessory belt tension. Use lock pin (special tool 11 8 280) to lock belt tensioner.

> ***CAUTION—***
> *Belt tensioner is under high tension. Check that lock pin is secure.*

– Remove splash shield under engine.

– Remove accessory belt.

> ***CAUTION—***
> - *Check belt for cracks, coolant and oil residue. Replace if necessary.*
> - *Belt must be replaced if contaminated by oil.*

– When reinstalling, make sure belt is correctly seated in pulley grooves.

– Check for fault codes and clear and reset DME control module memory.

Accessory belt, replacing (Cooper S)

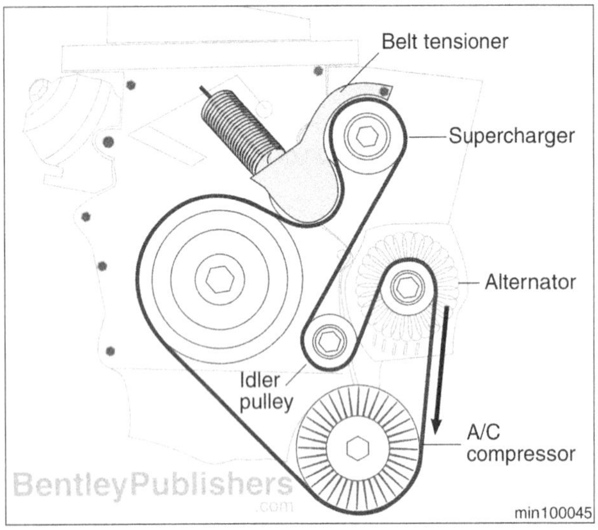

◀ Before removing, note layout of accessory belt.

NOTE —
If belt is to be reused, mark direction of travel and reinstall belt in same direction of rotation.

– Remove dipstick from dipstick tube.

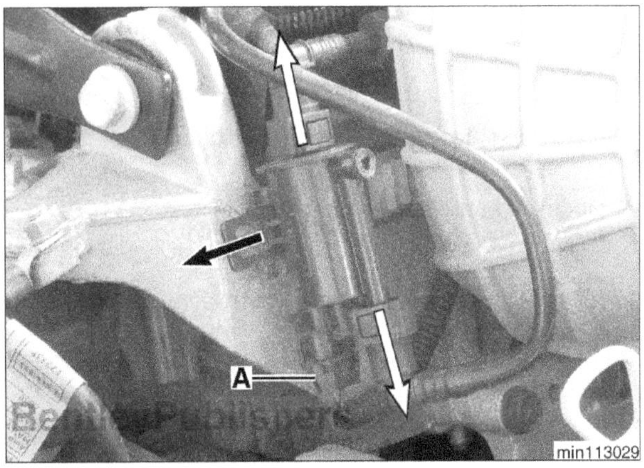

◀ Remove fuel tank vent valve.

- Disconnect vent hoses (**white arrows**) via quick fit connectors.
- Straighten retaining tag and slide vent valve off bracket (**black arrow**).
- Remove electrical harness connector (**A**).

◀ Use special tool 11 8 410 to release accessory belt tension.

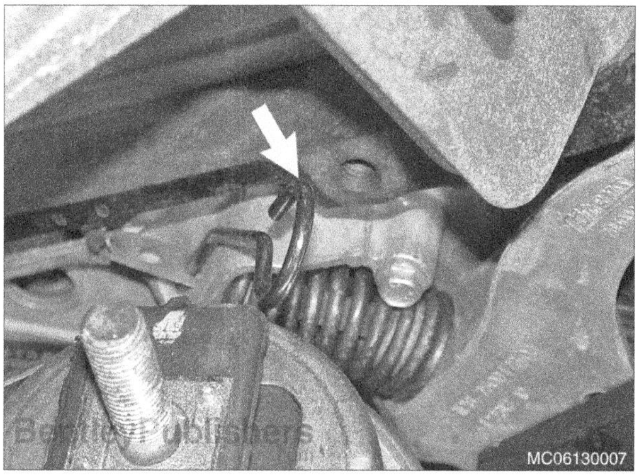

Use lock pin (**arrow**) (special tool 11 8 470, or equivalent) to lock belt tensioner.

> ***CAUTION—***
> *Belt tensioner is under high tension. Check that lock pin is secure.*

– Remove splash shield under engine.

– Remove right front wheel.

– Remove right front wheel housing liner.

– Remove accessory belt.

> ***CAUTION—***
> - *Check belt for cracks, coolant and oil residue. Replace if necessary.*
> - *Belt must be replaced if contaminated by oil.*

– When reinstalling, make sure belt is correctly seated in pulley grooves.

– Check for fault codes and clear and reset DME control module memory.

ENGINE OIL SERVICE

With the introduction of synthetic oil and extended oil change intervals, a new oil filter with improved filter paper design was introduced. The new filter paper resists deterioration caused by high oil temperatures over an extended time.

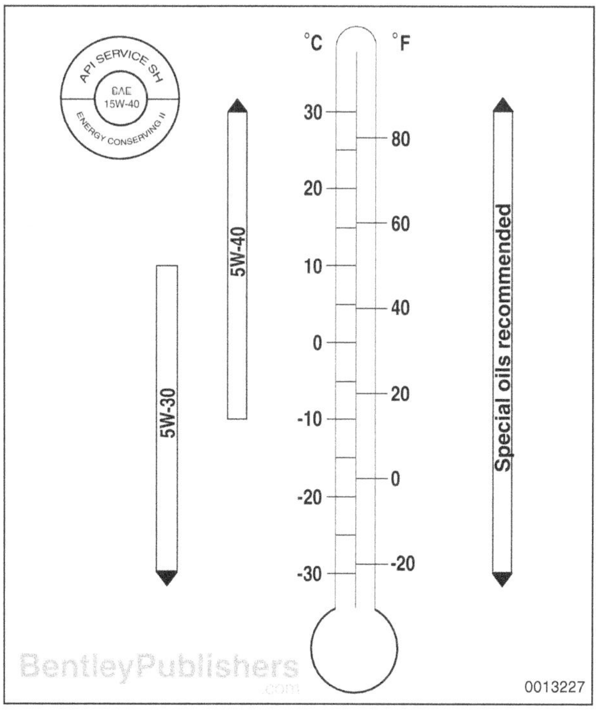

Engine oil viscosity (SAE grade) vs. operating temperature range for the MINI engines covered in this manual.

Use MINI specified oil to top off the engine oil level between oil changes. MINI recommends the following engine oils:

Engine oil with filter change		
Model	**Approximate capacity**	**Oil specification**
Cooper Cooper S	4.7 liters (4.5 US qt.)	BMW Longlife-01 or BMW Longlife-98 ACEA A3

> ***CAUTION—***
> - *In the case of vehicles that accumulate very low mileage per year, MINI recommends that the engine oil should be changed once a year at a minimum.*
> - *Oil change intervals recommended by the manufacturer are based on "normal" vehicle operating conditions. For severe driving conditions, be sure to change the engine oil twice as frequently as recommended by the manufacturer.*

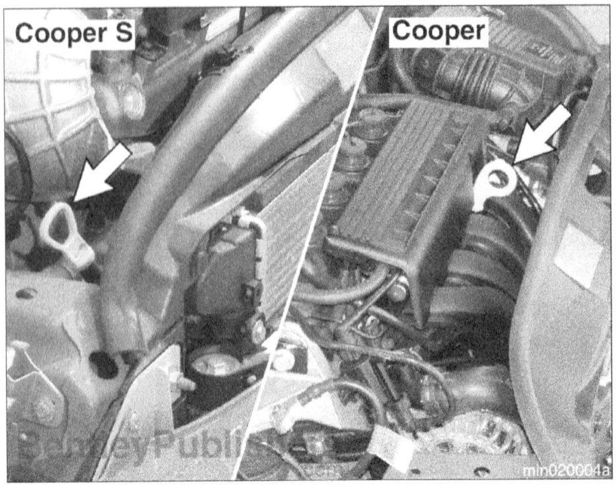

Engine oil, checking level

< Engine oil level is checked with a dipstick (**arrow**) in the engine block.

– Check oil level with vehicle on level surface, after engine has been stopped for at least a few minutes.

– Check level by pulling out dipstick and wiping it clean. Reinsert all the way and withdraw again.

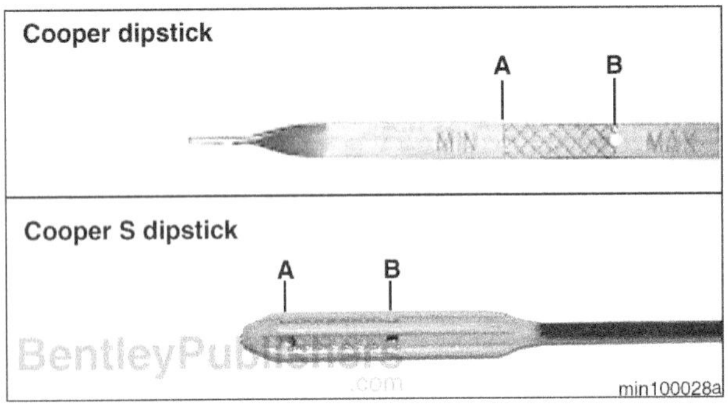

< Oil level is correct if it is between two marks (**A** and **B**) near end of dipstick.

< If oil level is low, add oil by removing oil filler cap (**arrow**) on top of cylinder head cover.

– Add only amount needed to bring oil level to upper mark on dipstick, using an oil of correct viscosity and grade.

Engine oil and filter, changing

A complete oil change requires new oil, a new filter insert kit, and a new drain plug sealing washer. The tools needed are a 13 mm drain plug socket or box wrench, a 36 mm socket or wrench, and a drain pan (6 US qt. capacity).

NOTE —
If using a "fast lube" service facility for oil changes, make sure the technician installs and tightens the engine oil drain plug using only hand tools. Power tools can strip the threads of the plug and oil pan.

– Run engine for a few minutes to warm engine oil. Shut engine off.

– With vehicle on level ground, place drain pan under oil drain plug.

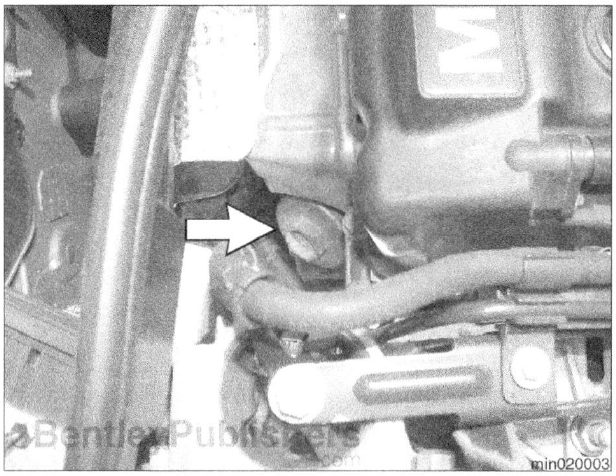

◄ Using a 36 mm socket, loosen oil filter housing cover (**arrow**) slowly.

- Wait for oil to drain back into crankcase.
- Remove oil filter cover
- Remove filter cartridge and discard O-ring seal.

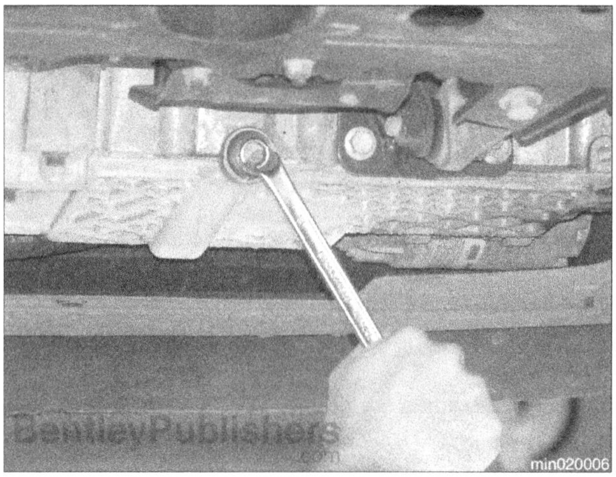

◄ Using a socket or box wrench, loosen drain plug at oil drain pan. Remove plug by hand and let oil drain into pan.

> ***WARNING —***
> *Pull the loose plug away from the hole quickly to avoid being scalded by hot oil. It will run out quickly when the plug is removed. If possible, use gloves to protect your hands.*

– When oil flow has diminished to an occasional drip, reinstall drain plug with a new metal sealing washer and torque plug.

Tightening torque	
Engine oil drain plug to oil pan	25 Nm (18 ft-lb)

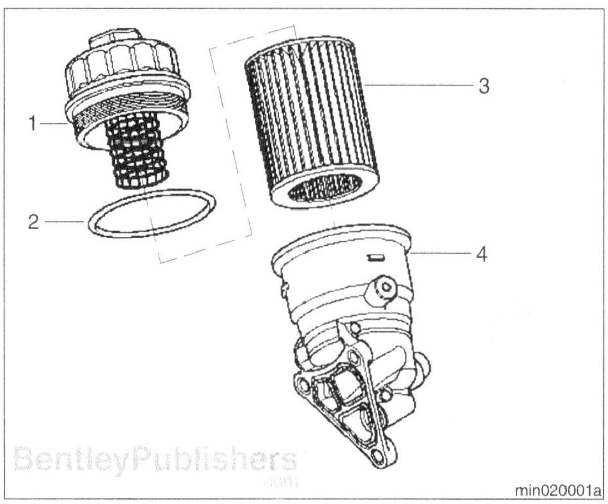

◄ Working at oil filter housing (**4**):

- Lubricate and install new sealing O-ring (**2**).
- Install a new filter cartridge (**3**).
- Install and tighten housing cover (**1**).

Tightening torque	
Cover to oil filter housing	25 Nm (18 ft-lb)

– Refill crankcase with new oil. Approximate oil capacity is listed above. Use dipstick to check for correct oil level.

– Start engine and check that oil pressure warning light immediately goes out.

– Allow engine to run for a few minutes to circulate new oil, then check for leaks at drain plug and oil filter. Stop engine and recheck oil level.

Fuel System Service

Fuel filter

The fuel filter, located in the fuel tank, is integrated with the right side fuel tank sensor assembly. No service or replacement interval is specified for the fuel filter.

Fuel tank and fuel lines, inspecting

Inspect the fuel tank, fuel lines and fuel system for damage or leaks. Check for fuel leaks in the engine compartment or fuel odors in the passenger compartment. Check for any evaporative emissions hoses that may have become disconnected, checking carefully at charcoal canister and evaporative emissions purge system. See **130 Fuel Injection** and **160 Fuel Tank and Fuel Pump** for component locations and additional information.

> ***WARNING —***
> *When checking for fuel leaks, the engine must be cold. A hot exhaust manifold or exhaust system could cause the fuel to ignite or explode causing serious personal injury. Ventilate the work area and clean up spilled fuel immediately.*

Interior Ventilation Microfilter, Replacing

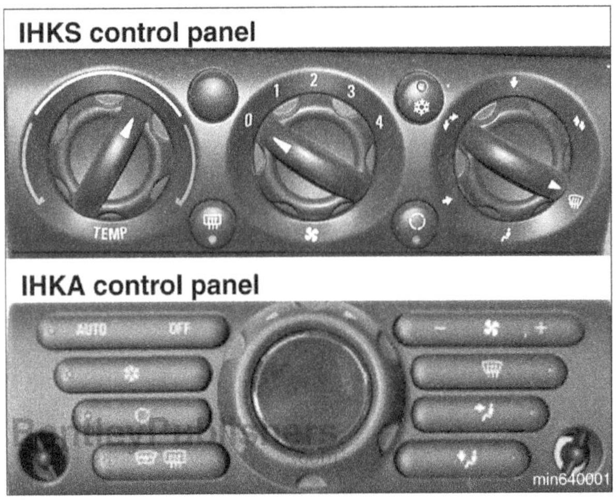

Two heating and A/C systems are available on MINI models:

- **IHKS**: Manually controlled system. Driver input is required to regulate temperature, air direction and air speed.
- **IHKA**: Optional automatic climate control. System controls air outlet temperature, air direction and air speed according to program set by driver. Automatic functions may be overridden by driver.

The ventilation microfilter is in right side of heater and air-conditioning housing, behind glove compartment.

IHKS: Working under glove compartment:

- Remove filter housing cover screws, if necessary.
- Pull down (**arrow**) on cover.
- Pull microfilter down and out of guides and replace.

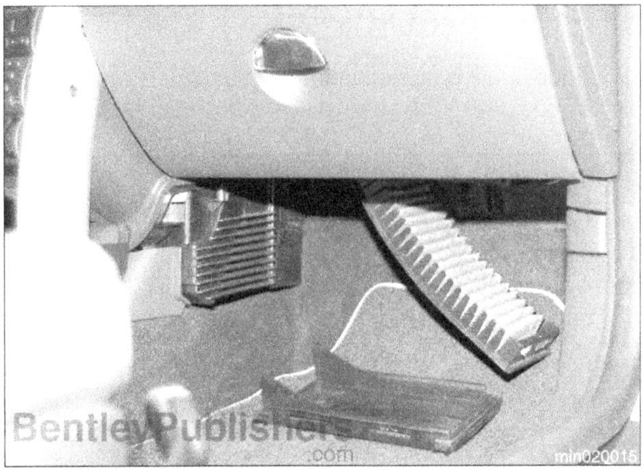

IHKA: Working under glove compartment:

- Unclip plastic cover by pulling straight down.
- Pull right end of microfilter down.
- Slide filter out of housing and replace.

Power Steering Fluid, Checking

> **CAUTION—**
> *The MINI power steering system is filled with Pentosin CHF 11S hydraulic fluid. DO NOT top up with ATF or other types of steering fluid. Mixing hydraulic fluid types will damage the seals in the steering system and cause them to leak.*

The power steering system is permanently filled and does not have a drain. Routinely adding hydraulic fluid is not required unless the system is leaking.

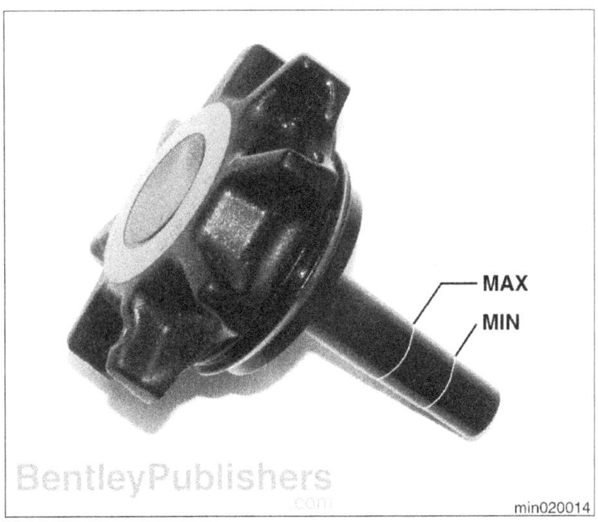

To check power steering fluid level in reservoir:

- Park vehicle on level ground with engine off.
- Unscrew reservoir cap and inspect attached dipstick.
- Level is correct if it is between **MIN** and **MAX** marks on dipstick.
- If level is below **MIN** mark, add fluid to bring level up.
- Hand-tighten cap.

Steering hydraulic fluid specification	
Power steering fluid	Pentosin CHF 11 S

SPARK PLUGS, REPLACING

The recommended spark plug replacement interval is 100,000 miles. The electrode gap in the spark plugs is not adjustable.

Spark plug application	
Cooper Cooper S John Cooper Works (JCW)	NGK BKR 6 EQUP NGK BKR 6 EQUP NGK BKR 7 EQUP

> ***WARNING—***
> *Remove sparks plugs only after the engine has cooled.*

– Turn off ignition.

◀ Working at top of engine, pull up on each spark plug wire connector while gently twisting from side to side.

– Use special spark plug socket (16 mm or 5/8") to remove each spark plug.

NOTE—
Keep spark plugs in the order that they were removed from the cylinders. If diagnosing engine misfire or other malfunctions, knowing which spark plug came out of which cylinder may be helpful.

– Installation is reverse of removal.

- Lightly lubricate new spark plug threads with copper-based anti-seize compound.
- Thread plugs into cylinder head by hand to prevent cross-threading.

Tightening torque	
Spark plug to cylinder head	27 Nm (20 ft-lb)

◀ When reinstalling ignition wires, make sure that they snap securely into place.

TRANSMISSION SERVICE

Transmission service, automatic (CVT)

CAUTION— *The CVT (automatic transmission) is filled with special transmission fluid. DO NOT top up with ATF.*

CVT (automatic transmission) oil change is scheduled for every INSPECTION I service.

CVT (automatic transmission) fluid			
Model, transmission	**Approximate capacity liter (US qt)**	**Refill capacity liter (US qt)**	**Oil specification**
MINI Cooper GACVT16Z	4.5 (4.8)	4.5 (4.8)	ESSO CVT E2L 799

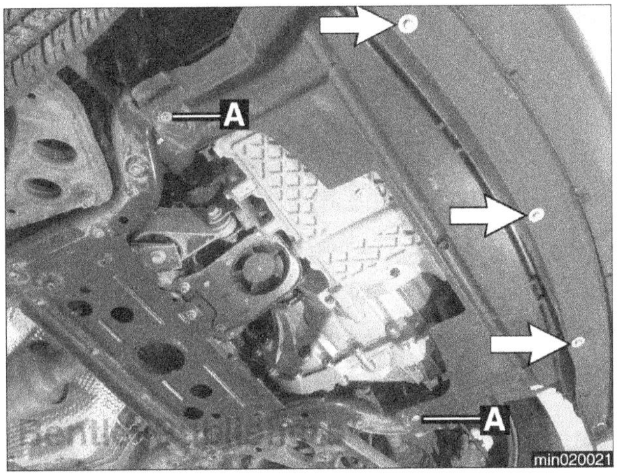

To gain access to transmission fill/level plug, remove splash shield under engine:

- Remove 3 bolts (**arrows**) at bottom of front bumper cover.
- Unlock 2 splash shield mounting screws (**A**).
- Remove splash shield.

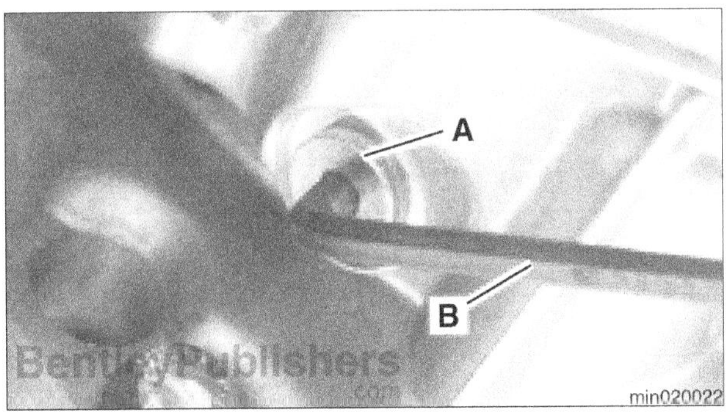

Use 6 mm Allen key (**B**) to remove fill/level plug (**A**). If transmission is full, some fluid will drip out.

Tightening torque	
Fill/level plug to transmission housing (M14)	21 Nm (15 ft-lb)

For additional CVT service information, including changing transmission fluid and ATF filter, see **240 Automatic Transmission**.

Transmission service, automatic (Agitronic)

There is no scheduled automatic transmission fluid change interval. The automatic transmission should be checked for fluid leaks.

For additional automatic transmission service information, including changing transmission fluid, see **240 Automatic Transmission**.

Automatic transmission fluid			
Model, transmission	**Approximate capacity liter (US qt.)**	**When replacing oil sump liter (US qt.)**	**Oil specification**
Cooper S Aisin GA6F21	6.0 (4.8)	4.5 (4.8)	ESSO JWS-3309 ATF

Transmission service, manual

There is no scheduled manual transmission fluid change interval.

The MINI manual transmission is filled with a special fluid.

Manual transmission oil			
Model, transmission	**Approximate capacity liter (US qt.)**	**Refill capacity liter (US qt.)**	**Oil specification**
Cooper GS5S-65 BH/SH (Up to 07/04)	2 (2.1)	2 (2.1)	Long life transmission oil MTF94
Cooper Getrag GS5-52 BG (From 07/04)	1.7 (1.8)	1.5 (1.4)	
Cooper S Getrag GS6-85BG/DG	1.7 (1.8)	1.5 (1.4)	

Manual transmission service consists of inspecting for leaks and checking the fluid. Evidence of transmission leaks is likely to be seen around the driveshaft mounting flange and at the bottom of the bell-housing.

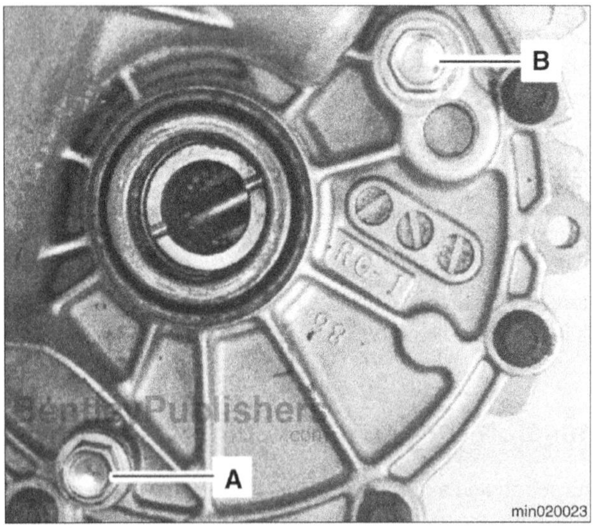

Cooper: Check 5-speed manual transmission fluid level at fill plug (**B**). With vehicle on ground, correct fluid quantity is level with bottom of fill plug.

Tightening torques	
Drain plug to transmission housing	32 - 40 Nm (24 - 30 ft-lb)
Fill plug to transmission housing	20 - 30 Nm (15 - 22 ft-lb)

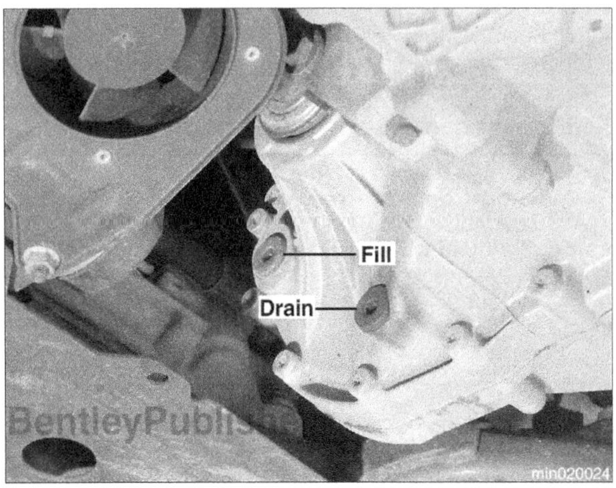

Cooper S: Check 6-speed manual transmission fluid level at **fill** plug. With vehicle on ground, correct fluid quantity is level with bottom of fill plug.

Tightening torque	
Drain or fill plug to transmission housing	43 Nm (32 ft-lb)

For additional manual transmission service information, see **230 Manual Transmission** and **210 Clutch**.

Other Mechanical Maintenance

Catalytic converter and oxygen sensors

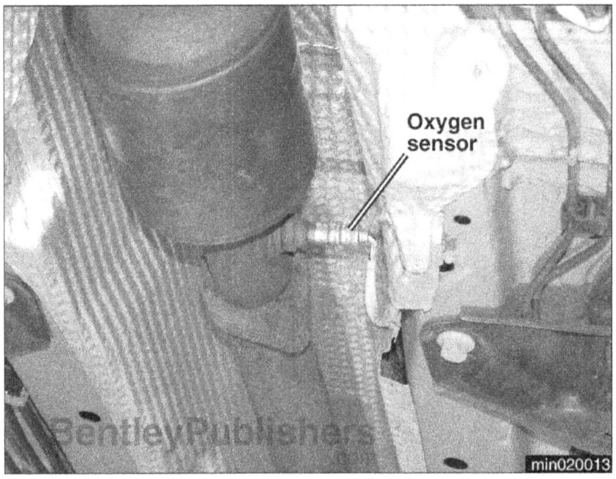

OBD II enhanced emission standards require the engine control module (ECM) to monitor the oxygen content in the exhaust with oxygen sensors placed before and after the catalytic converter. This allows for tighter control of tail pipe emissions and also allows the ECM to diagnose converter problems. If the ECM detects that catalytic converter or oxygen sensor efficiency has degraded past a certain preprogrammed limit, it will turn on the Malfunction Indicator Light (MIL), and store a diagnostic trouble code (DTC) in the ECM. See **OBD On Board Diagnostics** for more information on OBD II systems.

Replacement of oxygen sensors at specified intervals ensures that engine and emission control system continue to operate as designed. Extending replacement interval may void emission control warranty coverage. See **180 Exhaust System** for information on replacing oxygen sensor.

Tightening torque	
Oxygen sensor to exhaust manifold	50 Nm (37 ft-lb)

NOTE—
A special socket for replacing the oxygen sensor is available from most automotive parts stores. The socket has a groove cut down one side to allow the sensor to be installed without damaging the wire harness.

Clutch fluid, checking

In vehicles equipped with manual transmission and clutch, the hydraulic clutch and the brake system share the same reservoir and the same fluid. Clutch fluid level and brake fluid level are checked at the same time.

NOTE—
- *See* **340 Brakes** *for more information.*
- *See* **210 Clutch** *for information on the clutch and the hydraulic clutch operating system.*

Drive axle joint (CV joint) boots, inspecting

Inspect CV joint protective boots closely for cracks and any other damage that allows contaminants to get into CV joints. If a rubber boot fails, water and dirt will quickly damage the joint.

NOTE—
Replacement of the CV joint boots and inspection of the joints are described in **310 Front Suspension, Drive Axles**.

Exhaust system, inspecting

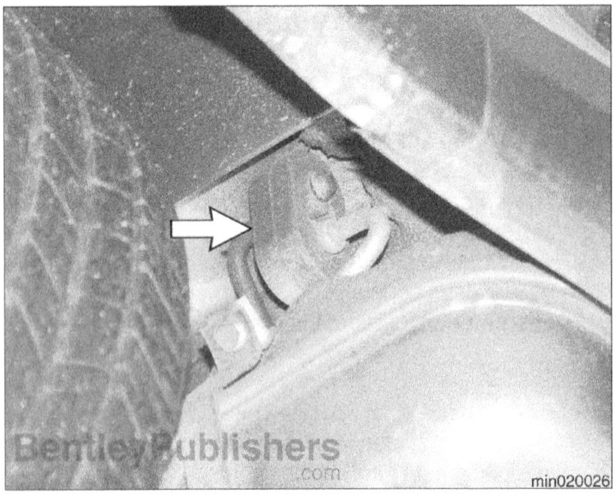

Exhaust system life varies widely according to driving habits and environmental conditions. If short-distance driving predominates, the moisture and condensation in the system will not fully dry out. This will lead to early corrosion damage and more frequent replacement.

Scheduled maintenance of the exhaust system is limited to inspection:

- Check to see that all the hangers (**arrow**) are in place and properly supporting the system and that the system does not strike the body.
- Check for restrictions due to dents or kinks.
- Check for weakness or perforation due to rust.

NOTE—
Alignment of the system and the location of the hangers are described in **180 Exhaust System**.

Idle speed

The idle speed is electronically controlled and adaptive. It is not adjustable. See **130 Fuel Injection** for more information.

Suspension, inspecting

Inspection of the front suspension and steering includes a check of all moving parts for wear and excessive play. Inspect ball joint and tie-rod rubber seals and boots for cracks or tears that could allow the entry of dirt, water, and other contaminants. See **310 Front Suspension, Drive Axles**.

Inspection of the rear suspension includes a check of all moving parts for wear and excessive play. Inspect suspension bushings for wear or deterioration. See **330 Rear Suspension**.

Tires, checking inflation pressure

Correct tire pressures are important to handling and stability, fuel economy, and tire wear. Tire pressures change with temperature. Pressures should be checked often during seasonal temperature changes. Correct inflation pressures can be found on the driver's door pillar and in the owner's manual. Note that tire pressures should be higher when the vehicle is more heavily loaded.

WARNING—
Do not inflate any tire to a pressure higher than the tire's maximum inflation pressure listed on the sidewall. Use care when adding air to warm tires. Warm tire pressures can increase as much as 4 psi (0.3 bar) over their cold pressures.

Tires, rotating

MINI does not recommend tire rotation. Rotating the tires may adversely affect road handling and tire grip.

Wheels, aligning

MINI recommends checking the front and rear alignment once a year and whenever new tires are installed.

NOTE—
See **320 Steering and Wheel Alignment** *for a more detailed discussion of alignment requirements and specifications.*

BODY AND INTERIOR MAINTENANCE

Body and hinges, lubricating

– Lubricate door locks and lock cylinders with an oil that contains graphite.

– Lubricate body and door hinges, hood latch and door check rods with SAE 30 or SAE 40 engine oil.

– Lubricate seat runners with multipurpose grease.

– If door weather-strips are sticking, lubricate them with silicone spray or talcum powder. Do not apply any oil to rubber parts.

– Lubricate hood release cable.

– Use minimum amount of winter lock deicer. Alcohol in deicer will wash grease out of lock assemblies and may cause locks to corrode internally or to become difficult to operate.

Exterior washing

The longer dirt is left on the paint, the greater the risk of damaging the glossy finish, either by scratching or by the chemical effect dirt particles may have on the painted surface.

Do not wash the vehicle in direct sunlight. If the engine hood is warm, allow it to cool. Beads of water not only leave spots when dried rapidly by the sun or heat from the engine, but also can act as small magnifying glasses and burn spots into the finish.

– Wash vehicle with a mixture of lukewarm water and a car wash product.

– Rinse using plenty of clear water.

– Wipe body dry with soft cloth towel or chamois to prevent water-spotting.

Interior care

– Remove dirt spots with lukewarm soapy water or a dry foam cleaner.

– Use spot remover for grease and oil spots. Do not pour liquid directly on carpet or fabric, but dampen a clean cloth and rub carefully, starting at edge of spot and working inward. Do not use gasoline, naphtha, or other flammable substances.

Leather upholstery and trim

- Periodically clean leather upholstery and trim. Use a slightly damp cotton or wool cloth to get rid of dirt in creases and pores that can cause brittleness and premature aging.

- On heavily soiled areas, use mild detergent (such as Woolite®) or other specially formulated leather cleaners.

- Dry trim and upholstery completely using a soft cloth.

Regular use of a good quality leather conditioner will reduce drying and cracking of the leather.

Polishing

- Use paint polish only if finish assumes a dull look after long service.

- Use polish to remove tar spots and tarnish, but afterwards apply coat of wax to protect finish.

Do not use abrasive polish or cleaners on aluminum trim or accessories.

Seat belts

Dirt and other abrasive particles will damage seat belt webbing.

- If it is necessary to clean seat belts, use a mild soap solution. Avoid bleach and other strong cleaning agents, which may weaken belt webbing.

> ***WARNING —***
> *Do not clean the seat belt webbing using dry cleaning or other chemicals. Allow wet belts to dry before allowing them to retract.*

- Inspect condition of belt webbing and function of retractor mechanisms. See **720 Seat Belts** for seat belt troubleshooting.

Special cleaning

- Remove tar spots with a bug and tar remover. Never use gasoline, kerosene, nail polish remover, or other unsuitable solvents.

- Remove insect spots with tar remover or a bit of baking soda dissolved in the wash water. This method can also be used to remove tree sap spots.

Washing chassis

- Periodically wash underside of vehicle, especially in winter, to help prevent accumulation of road salt and rust. The best time to wash the underside is just after the vehicle has been driven in wet conditions.

- Spray chassis with a powerful jet of water. Commercial or self-service car washes may not be best for this, as they may recycle salt-contaminated water.

Waxing

– For a long-lasting, protective and glossy finish, apply a hard wax after vehicle has been washed and dried. Use carnauba or synthetic based products.

Waxing is not needed after every washing. You can tell when waxing is required by looking at the finish when it is wet. If the water coats the paint in smooth sheets instead of forming beads that roll off, a new coat of wax is needed.

Do not wax black trim pieces, rubber or other plastic parts.

Windshield wiper blade maintenance

Common problems with the windshield wipers include streaking or sheeting, water drops after wiping, and blade chatter. Streaking is usually caused when wiper blades are coated with road film or car wash wax.

– Clean blades using soapy water. If cleaning blades does not cure problem, replace them.

– In general, replace wiper blades twice a year, before and after cold season.

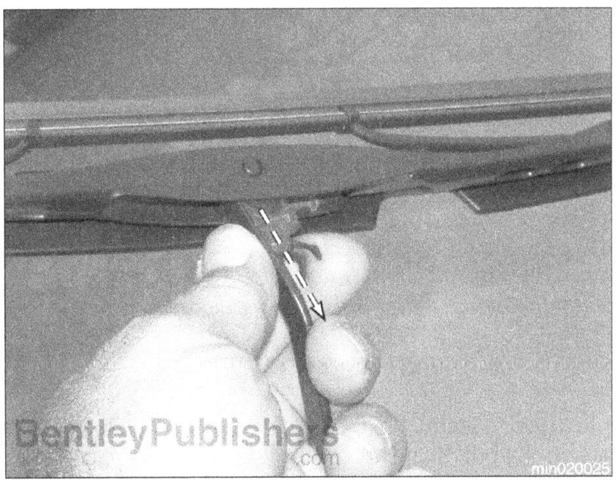

◄ To replace wiper blade, squeeze retaining tab and slide blade out of arm (direction of **arrow**).

– Check tension spring that forces wiper against glass. Replace wiper arm if spring tension is weak.

– Drops that remain behind after wiping are caused by oil, road film, or diesel exhaust coating windshield. Use alcohol or ammonia solution or nonabrasive cleanser to clean windshield.

– Wiper blade chatter may be caused by dirty or worn blades, by a dirty windshield, or by bent or twisted wiper arms.
- Clean blades and windshield as described above.
- Adjust wiper arm so that there is even pressure along blade, and so that blade is perpendicular to windshield at rest.
- If problem persists, blades are excessively aged or worn. Replace them.

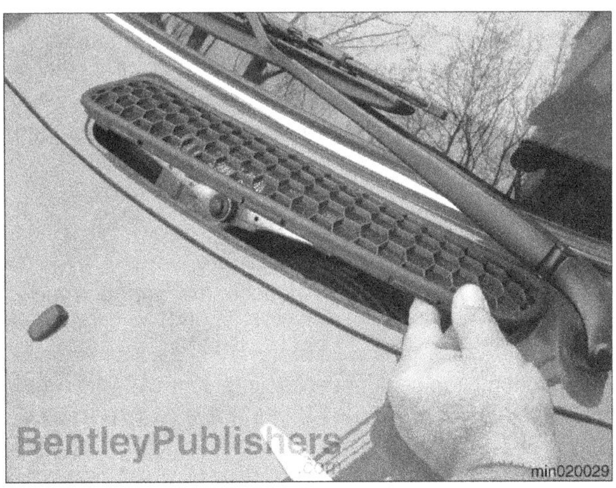

◄ To access wiper linkage:
- Pry off plastic cowl grille (fresh air intake) gently to access wiper assembly.

> ***CAUTION—***
> - *Pry front of grille up, then tilt rearward to remove.*
> - *To avoid marring the paintwork, use a plastic prying tool or wrap screwdriver tip with masking tape.*

– Lubricate wiper linkage with a light oil or spray grease.

– See **611 Wipers and Washers** for more information.

SERVICE INTERVALS

Service interval display (SIA)

The MINI service interval display (SIA) in the center of the speedometer face notifies the driver when oil change and other service is required.

MINI maintenance requirements depend upon vehicle operating conditions. Many factors are monitored by the SIA to determine maintenance intervals.

Inputs to the SIA microprocessor include engine coolant temperature, high or low engine speeds, short or long-distance driving information, and number of vehicle starts.

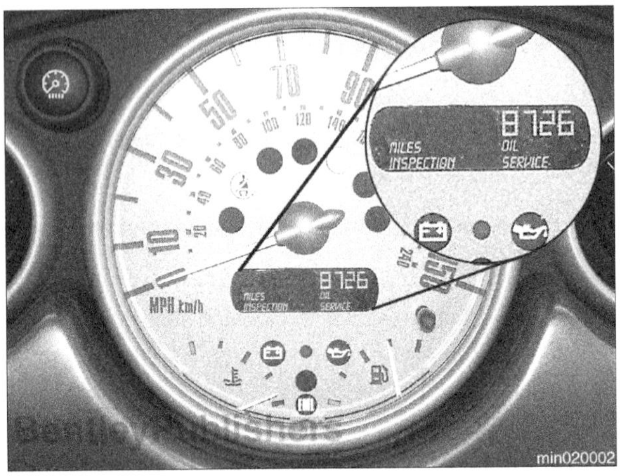

min020002

◀ SIA display appears for a few seconds when the ignition key is in the ON position or after the engine is started.

The message INSPECTION or OIL SERVICE appears together with the distance (in miles or kilometers) remaining before the next scheduled service. The SIA microprocessor bases its calculations on the vehicle driving history since the last service.

A flashing message and a negative number mean that the recommended service interval has already been exceeded by the distance shown on the display.

SIA, resetting

After the specified maintenance has been carried out, reset SIA display memory using BMW scan tool or the procedure given below.

NOTE—
Resetting the SIA is possible only after 20% of the expected fuel usage has been consumed following the previous reset.

- Turn ignition to position 0 (steering locked).
- Press and hold trip reset button, then turn ignition switch to position 1 (steering unlocked). Within 5 seconds display shows current SIA status (OIL SERVICE or INSPECTION).
- Release trip reset button.
- Press and hold trip reset button for 5 seconds to change to reset mode. Display flashes **rst**.
- Press reset button again. Current service requirement is reset and display shows new status for 5 seconds.

Inspection I, Inspection II

The service interval display signals the need for maintenance and inspection. There are two sets of inspection requirements, alternating throughout a vehicle's maintenance history. If the last inspection interval was Inspection I, the next inspection interval (following an oil service) will be Inspection II, the next after that will be Inspection I, and so on.

Inspection I tasks are listed in **Table b. Inspection I service**. Inspection II includes the tasks from Inspection I with additional Inspection II tasks. A complete listing of Inspection II tasks is in **Table c. Inspection II service**.

Maintenance tables

The intervals for most tasks listed in the maintenance tables below are determined by the service interval display.

NOTE—
MINI is constantly upgrading recommended maintenance procedures and requirements. The information contained here is as accurate as possible at the time of publication. If there is any doubt about what procedures apply to a specific model or year, or what intervals should be followed, remember that an authorized MINI dealer has the latest information on factory-recommended maintenance.

Except where noted, the maintenance items listed apply to all models and model years covered by this manual. The number in the "additional repair information" column refers to the repair group in this manual where additional information can be found.

Table a. Oil service

	Tools required	New parts required	Warm engine required	Dealer service recommended	Additional repair information
Engine compartment or vehicle interior maintenance					
Change oil and oil filter.	*	*	*		020
Reset service interval display.					020
Replace interior ventilation microfilter.		*			020
Under vehicle maintenance					
Check overall thickness of front and rear brake pads. If replacement is necessary: Examine brake disc surface. Clean brake pad contact points in calipers. Grease wheel centering hubs. Check thickness of parking brake linings only when replacing rear brake pads. Check operation of parking brake and adjust as necessary.	*				340
Check and adjust tire pressures, including spare.	*				020

Table b. Inspection I service	Tools required	New parts required	Warm engine required	Dealer service recommended	Additional repair information
Under vehicle maintenance					
Change oil and oil filter.	*	*	*		020
Check transmission and differential for external leaks.	*				230 240
Check engine accessory belt					
Change CVT (automatic transmission) fluid.	*	*	*		240
Check CV joint boots for damage or leaks.					310
Visually check fuel tank, fuel lines and connections for leaks.					160
Check condition, position and mounting of exhaust system. Visually check and adjust if necessary.					180
Check struts and shock absorbers for leaks.					310
Check front control arm bushings for damage or wear.					310
Check power steering system for leaks. Check power steering fluid level and adjust if necessary.					320
Check steering rack and tie rods for tightness. Check condition of steering linkage and steering shaft joints.					320
Check overall thickness of front and rear brake pads using BMW special tool. If replacement is necessary: Examine brake disc surface. Clean brake pad contact points in calipers. Grease wheel centering hubs. Check thickness of parking brake linings only when replacing rear brake shoes. Check operation of parking brake and adjust as necessary.	*				340
Check brake system connections and lines for leaks, damage, and incorrect positioning.					340
Check parking brake actuator. Adjust if necessary.	*				340
Check all tire pressures (including spare if equipped) and correct if necessary. Check condition of tires (outer tread surface), tread wear pattern. In case of uneven wear, perform wheel alignment.	*				020
Inspect entire body according to terms of rust perforation limited warranty. (Must be performed at least every two years.)					
Engine compartment or vehicle interior maintenance					
Read out on-board diagnostic (OBD II) fault codes.	*				130 / OBD
Check engine coolant system / heater hose connections for leaks. Check coolant level and antifreeze protection level. Add coolant as necessary.					170
Check windshield washer fluid level and antifreeze protection. Add washer fluid as necessary.					611
Check air-conditioner for operation.					640
Replace ventilation microfilter. (Reduce replacement interval in dusty conditions.)		*			020
Body/electrical					

Table b. Inspection I service

	Tools required	New parts required	Warm engine required	Dealer service recommended	Additional repair information
Check battery state of charge.	*				121
Check operation of: headlights, foglights, parking lights, back-up lights, turn signals, emergency flashers, stop lights, license plate lights, interior lights, glove compartment light, make-up mirror light, flashlight, engine compartment light, trunk light.					630
Check instrument panel, dashboard illumination.					620
Check warning and indicator lights, check control.					
Check operation of horn, headlight dimmer and flasher switch.					630
Check wipers and windshield washer system. Check aim of washer jets and adjust if necessary.					611
Check condition and function of seat belts.					720
Visually examine all airbag units for torn covers, obvious damage or attachment of decals, decorations or accessories.					721
Check central locking system.					515
Check operation and condition of door, engine hood and rear hatch latches.					410 515
Check heater, air-conditioner blower and rear window defogger operation.					640
Check operation of rear view mirrors.					
Check operation of convertible top, as applicable.					541
Reset service interval display.					020
Road test					
Check braking performance, steering, heating and air-conditioner operation.					
Check manual transmission and clutch operation or automatic transmission operation.					

Table c. Inspection II service					
Includes all items listed under **Inspection I** and the items listed in this table.	**Tools required**	**New parts required**	**Warm engine required**	**Dealer service recommended**	**Additional repair information**
Engine compartment maintenance					
Replace air filter element. (Reduce replacement interval in dusty conditions.)		*			**020**
Replace engine accessory belt: Cooper at 100,000 miles Cooper S at 60,000 miles	*	*			**020**
Brake system maintenance					
Replace brake fluid every 2 years (time interval begins from vehicle production date).	*	*		*	**340**
Cooling system service					
Replace coolant every 4 years (time interval begins from vehicle production date).	*	*			**170**
Oxygen sensor service					
Replace oxygen sensor every 160,000 km (100,000 miles).	*	*			**180**
Spark plug service					
Replace spark plugs every 160,000 km (100,000 miles).	*	*			**020**

100 Engine–General

General

This section covers system descriptions and general information on engines and engine management systems. Also covered is basic engine troubleshooting.

For specific repair procedures, refer to the appropriate repair group:

- **110 Engine Removal and Installation**
- **113 Cylinder Head Removal and Installation**
- **116 Cylinder Head and Valvetrain**
- **117 Camshaft Timing Chain**
- **119 Lubrication System**
- **120 Ignition System**
- **130 Fuel Injection**
- **170 Radiator and Cooling System**

MINI Cooper and MINI Cooper S cars are equipped with variants of a 4-cylinder engine, mounted transversely across the front of the car. See **Engine specifications** table.

Engine specifications

Model	Engine code, type	Displacement cc (cu. in.)	Engine management (DME)	Bore / stroke mm (in.)	Compression ratio	Torque lb-ft / rpm	Horsepower (SAE) / rpm
MINI Cooper							
2002 - 2006	W10B16A normally aspirated 4-cylinder	1598 (97.5)	Siemens EMS 2000	77 / 85.8 (3.03 / 3.38)	10.6 : 1	110 / 4500	115 / 6000
MINI Cooper S							
2002 - 2006	W11B16A supercharged 4-cylinder	1598 (97.5)	Siemens EMS 2000	77 / 85.8 (3.03 / 3.38)	8.3 : 1	155 / 4000	163 / 6000
MINI Cooper S with John Cooper Works (JCW) kit							
2003 - 2004	W11B16A supercharged 4-cylinder	1598 (97.5)	Siemens EMS 2000	77 / 85.8 (3.03 / 3.38)	n/a	177 / 4000	200 / 6950
2005 - 2006					n/a	181 / 4500	210 / 6950

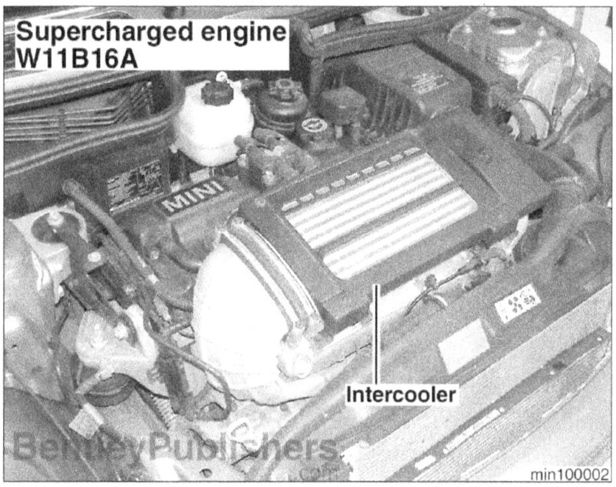

Engine identifying features

The same engine, jointly developed by BMW Group and DaimlerChrysler, is used in both MINI versions. The main visual difference between the two is that the supercharged engine (Cooper S) is equipped with the supercharger on the lower right front of the engine and the intercooler housing on the top and front of the engine.

ENGINE CONSTRUCTION

The engine block and bearing support ladder are constructed from cast iron with an aluminum alloy cylinder head. The oil pan is manufactured from aluminum alloy to reduce weight. Despite the iron block and bearing ladder, the engine is very light at 129.22 kg (285 lb). Main features of the engine include:

- 16 valves
- Single overhead chain driven camshaft
- Hydraulic lifters
- Engine drive belt with automatic tension adjustment
- Supercharger (Cooper S models)

Engine block

The engine block is manufactured in two halves from sand cast nodular iron. The top portion (main cylinder block) includes the cylinder bores and has provisions for five main bearing top shells. The lower portion (bearing support ladder) incorporates the lower main bearing shells and support for the rear main oil seal.

The engine block and ladder are machined as a matched pair and are not serviced as individual components. Three locating dowels are used to ensure perfect alignment between the bearing ladder and the engine block. A number is stamped on the engine block and the same number is stamped on the bearing ladder. This ensures that the matched pair of components remain together during engine assembly. The engine plant serial number is also stamped on the block.

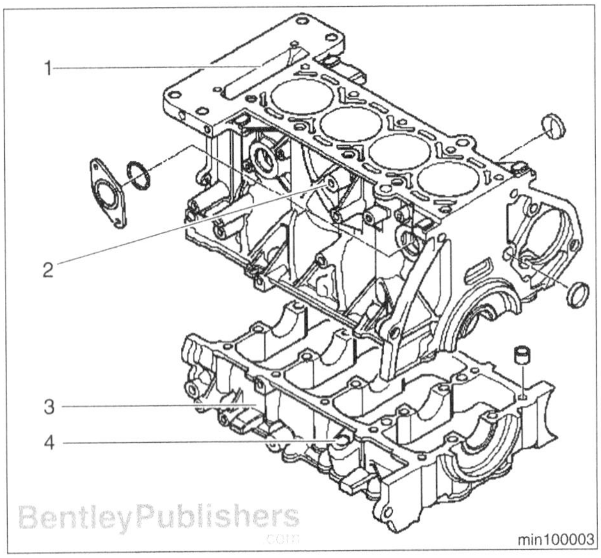

Engine block and ladder, intake side (front of car):

1. Timing chain housing
2. Knock sensor location
3. Bearing support ladder
4. Dipstick tube location

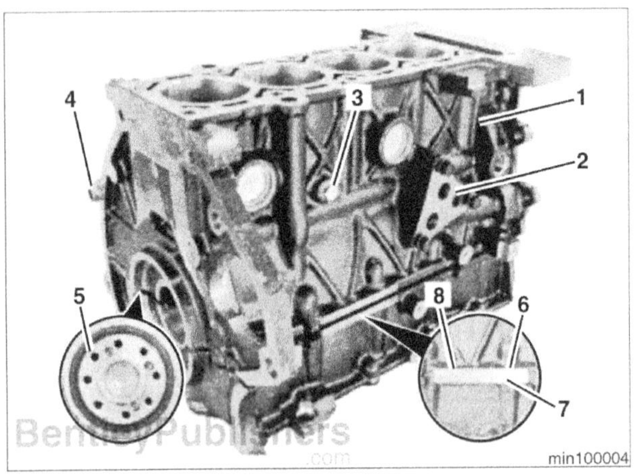

Engine block and ladder, exhaust side (bulkhead side):

1. Cylinder block
2. Oil filter mount
3. Cylinder block coolant drain plug
4. Clutch housing alignment dowels
5. Rear main oil seal
6. Block identification (matched to #7)
7. Ladder identification (matched to #6)
8. Engine serial number
   (Note: This is not the engine number)

Crankshaft assembly

The MINI Cooper crankshaft is machined from nodular cast iron. The MINI Cooper S crankshaft is machined from forged steel. Both crankshafts provide a mounting point for the crankshaft sensor reluctor ring that is retained by three bolts. The drive for the oil pump is provided by machined flats towards the front of the crankshaft. The auxiliary drive belt pulley is a press fit on the crankshaft and retained by a bolt.

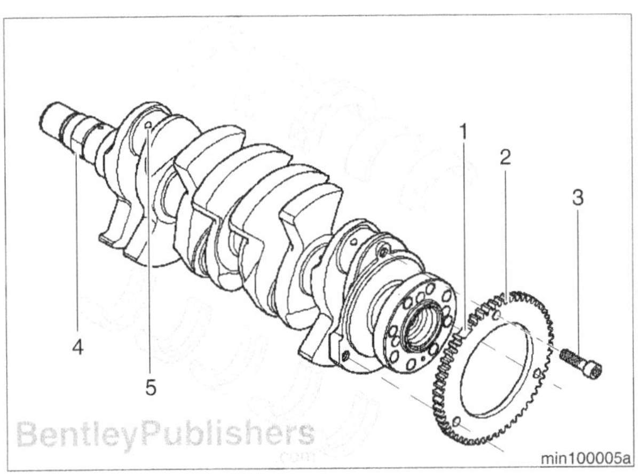

Crankshaft components:

1. Impulse wheel gap
2. Impulse wheel (reluctor)
3. Impulse wheel mounting bolt
4. Oil pump drive flat
5. Oil feed drilling

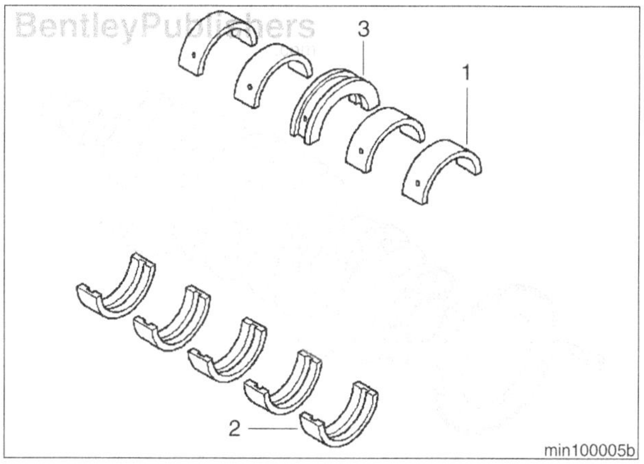

Crankshaft bearings

All MINI Cooper engines use five main bearings. Lubrication is supplied through holes in the upper shell directly from the main oil gallery.

1. Upper shell grooved to transport oil to lower shell
2. Lower shell located in bearing ladder
3. Thrust washer built into center upper main bearing shell to control crankshaft end float

The connecting rod and main bearing shells are made of an aluminum base that is rolled onto a low carbon steel backing.

Connecting rods

MINI Cooper connecting rods are manufactured from non-coplanar powder metal. They are manufactured in one piece and then fractured across the big end journal.

MINI Cooper S connecting rods are manufactured from forged steel to provide additional strength. They are then fractured.

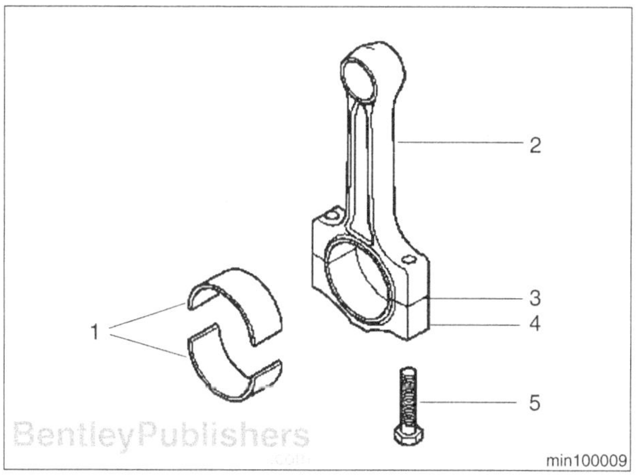

The big end bearings are of a conventional plain shell design, with oil supplied from a hole in the crankshaft.

1. Bearing shells
2. Connecting rod assembly
3. Fracture break
4. Big end cap
5. Connecting rod bolt

Fracture process

In the fracture splitting process, the connecting rod and big end bearing cap are designed to separate close to the theoretical center line with no loss of material. This is achieved by applying a load between the big end bearing cap and the connecting rod via a wedge in a split mandrel. The big end bore in the connecting rod is premachined with a notch introduced at the required joint plane to initiate the fracture. The separation is accurately determined by careful consideration of the geometry of the forging and material selection.

Fracturing of the connecting rod takes place immediately before the bolts are fitted and correctly tightened. This keeps the matching cap and connecting rod together for subsequent finish machining of the bore. The fractured surfaces form a unique multifaceted joint which provides a contact area much greater than that of a normally ground surface. The multifaceted joint also promotes precise mating between the big end cap and the connecting rod. No further machining of the faces is required, and no additional means of locating the big end bearing to connecting rod is necessary.

The main benefits of the fracturing process are:

- Reduction in manufacturing time and cost
- Each rod and bearing cap with a unique fracture, reducing the possibility of mismatched pairs
- Improved rod weight control

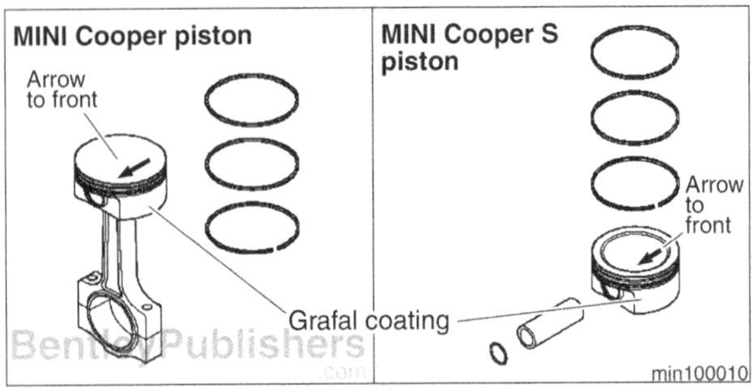

Pistons

The pistons are of aluminum construction with a Grafal coating applied to the skirt to reduce noise, friction, and scuffing.

MINI Cooper pistons have flat tops.

MINI Cooper S has a concave piston top with a volume of 1.66 cc (0.1 cu. in.) to reduce the compression ratio.

Grafal coating

Grafal consists of a fine colloidal graphite which is bonded with resin. It is between 10 and 20 micrometers thick (0.010 - 0.020 mm / 0.0004 - 0.0008 in) and is applied by means of a printing process, followed by curing. Improved adhesion properties are achieved by a thin metallic phosphate layer or other proven methods which are applied prior to coating.

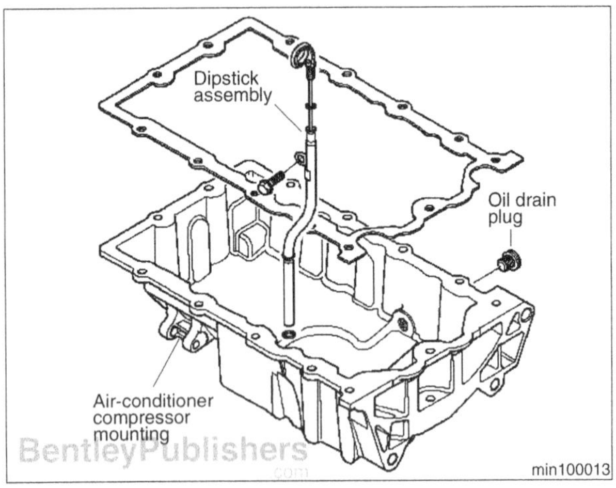

Oil pan

The oil pan is constructed of die cast aluminum. It is secured to the bearing ladder by 13 bolts.

The oil pan provides a mounting position for the air-conditioner compressor on the right side of the engine (viewed from the crankshaft pulley) and for the engine stabilizer on the left side of the engine.

The seal between the oil pan and bearing ladder has a washer fitted to each bolt location to prevent overtightening and distortion of the seal. A lip on the oil pan seal ensures correct location to the bearing ladder.

Oil pan removal and installation is covered in **119 Lubrication System**.

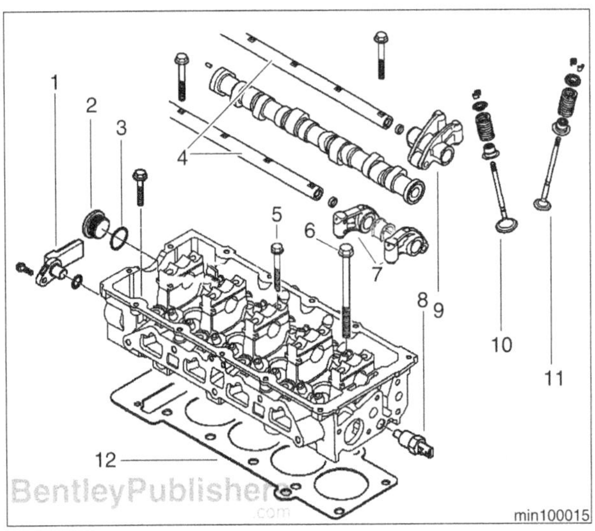

Cylinder head

The crossflow design cylinder head includes a single overhead camshaft, two rocker shafts and four valves per cylinder. The valves are arranged in two inline banks, the intake side facing towards the radiator, the exhaust facing towards the engine compartment rear bulkhead.

1. Camshaft pulse generator
2. Camshaft bore end-plug
3. Sealing O-ring
4. Rocker arm shafts
5. Camshaft bearing cap bolt (M8)
6. Cylinder head bolt (M10x150 mm)
7. Intake rocker arms
8. Coolant temperature sensor
9. Exhaust rocker arm
10. Intake valve assembly
11. Exhaust valve assembly
12. Cylinder head gasket

Cylinder head gasket

The multilayered steel head gasket is constructed from three layers of sheet metal. Four small rivets on the outer edge of the gasket hold the three layers together.

The head gasket contains an oil restrictor that controls the oil flow to the cylinder head. The standard thickness of the gasket is 0.65 mm (0.026 in). A thicker gasket, 0.95 mm (0.037 in) thick, is also available.

NOTE—
The gasket does not have any markings to indicate correct orientation. This is determined by visually lining up the location dowels and oil transfer gallery.

Cylinder head bolts

Discard a head bolt should there be any evidence of thinning at any point along its length. Replace with new.

For further information, see

- **113 Cylinder Head Removal and Installation**
- **116 Cylinder Head and Valvetrain**

Positive crankcase ventilation

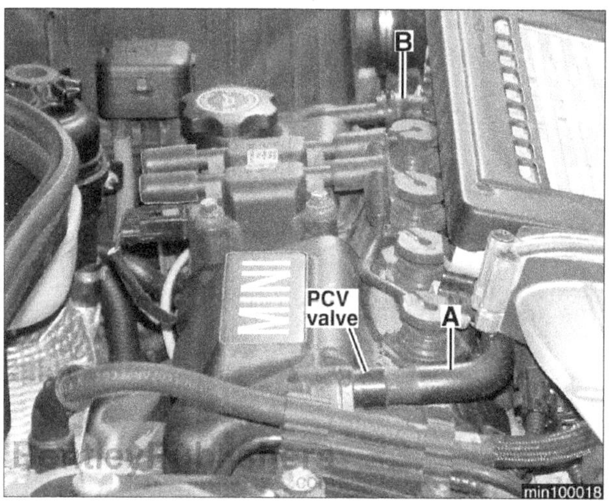

The **PCV valve** in the valve cover has pipe **A** connecting it to the intake manifold. This connection is downstream of the throttle valve (high vacuum area).

Breather pipe **B** connects the valve cover to the intake system rubber bellows between the air cleaner and throttle body. This connection is upstream of the throttle valve. Pipe B has no restrictions and allows air to travel in both directions depending on crankcase pressure.

Under normal driving conditions (negative crankcase pressure), air is drawn into the crankcase via pipe **B** and mixes with blow-by gases in the crankcase. The gases pass back up through the crankcase and reenter the valve cover. Negative pressure (vacuum) in the manifold is sufficient to open the PCV valve and allow the gases to enter the intake manifold downstream of the throttle valve through pipe **A** and be drawn into the combustion chambers.

When engine speed is high (positive crankcase pressure), the volume of blow-by gases may be too great for the PCV valve to handle alone. Vacuum in the intake manifold is also greatly reduced. Under these conditions blow-by gases also flow through pipe **B** and enter the air intake system upstream of the throttle valve, where they are drawn into the combustion chambers.

Valvetrain

Camshaft

The camshaft is machined from nodular iron. Nodular iron combines many advantages including good castability, excellent machinability, wear resistance, and weight savings. The camshaft consists of five bearing journals and three valve lift lobes per cylinder.

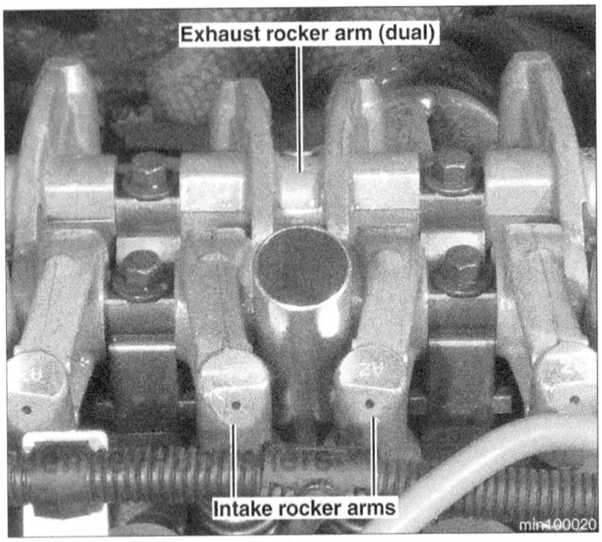

◀ The intake side uses one rocker arm per valve, while on the exhaust side a dual rocker arm operates both valves.

◀ A machined recess (**arrow**) in the cylinder head next to camshaft bearing cap 5 controls the camshaft end float.

– The camshaft for the normally aspirated model is identical to that in the supercharged engine.

Rocker arms and rocker shafts

◀ The rocker arm shafts are hollow to allow an oil supply to the hydraulic lifter mounted in the end of each rocker arm.

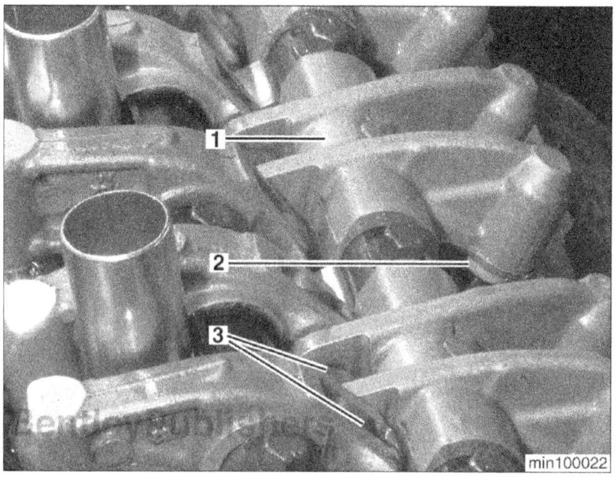

The valves are operated via rockers which pivot on the rocker arm shafts. Each rocker contacts the camshaft via a roller. At the other end of the rocker, a hydraulic valve lifter contacts the valve stem.

1. Rocker
2. Hydraulic lifter
3. Roller

Valves

Powder metal valve guides and seats are installed on both engine versions. Valves, springs and retainers are of conventional design.

Intake valves

The intake valves are made of carbon steel. The carbon content allows the valve to be hardened and tempered to increase strength and also to be locally hardened to improve wear resistance. The MINI Cooper S uses an upgraded material.

Intake valve seat inserts

Powder metal technology is used for valve seat inserts as the sintered part requires little or no machining and any number of material compositions can be developed to satisfy particular engine demands.

Exhaust valves

The exhaust valve specification is an austenitic steel, a particular type of steel with characteristics that are ideal for exhaust valve manufacture. The MINI Cooper S has upgraded exhaust valves.

Exhaust valve seat inserts

Many of the characteristics for the intake valve seat insert carry over to the exhaust valve seat insert. In addition, the exhaust valve seat uses what is known as grade J steel containing molybdenum and tungsten. This provide high heat hardness and increased resistance to indentation and wear.

Powder metal technology

A valve seat is produced from powdered metal by filling a rigid die with a blended powder and applying pressure. The pressure forces the powder particles to interlock, similar to a weld.

After being pressure formed, the parts are heated to 80% of the boiling point of the metal. The heat increases the bonds between the particles and further strengthens the part. To increase thermal conductivity the pores of the powder compact are infiltrated with copper during the sintering process.

See **116 Cylinder Head and Valvetrain** for repair information.

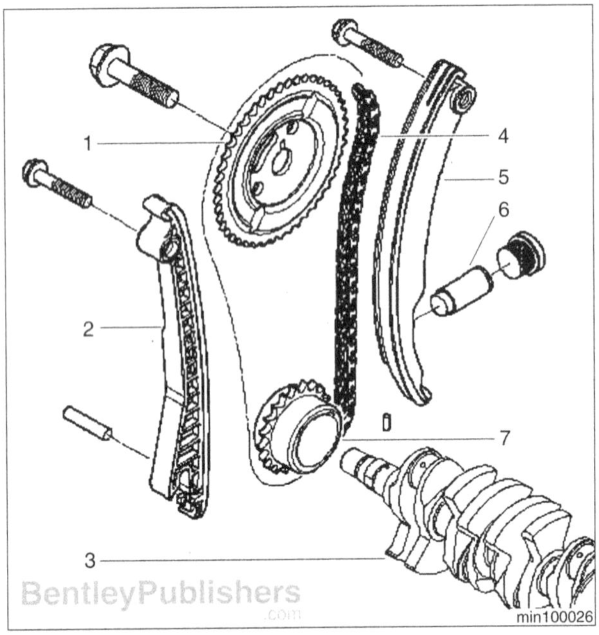

Timing chain

The camshaft sprocket is driven by a roller timing chain.

1. Camshaft sprocket
2. Fixed chain guide
3. Crankshaft
4. Timing chain
5. Semifloating chain guide
6. Chain tensioner
7. Crankshaft sprocket

There is a fixed chain guide on the intake side of the engine. The exhaust side has a semifloating guide that is spring-loaded and contains a self-ratcheting tensioner to retain the adjusted position. Engine oil pressure fine-tunes chain free play using a hydraulic tensioner.

1. Timing marks on crankshaft and camshaft sprockets
2. Copper colored links
3. Crankshaft sprocket locating pin

See **117 Camshaft Timing Chain** for repair information.

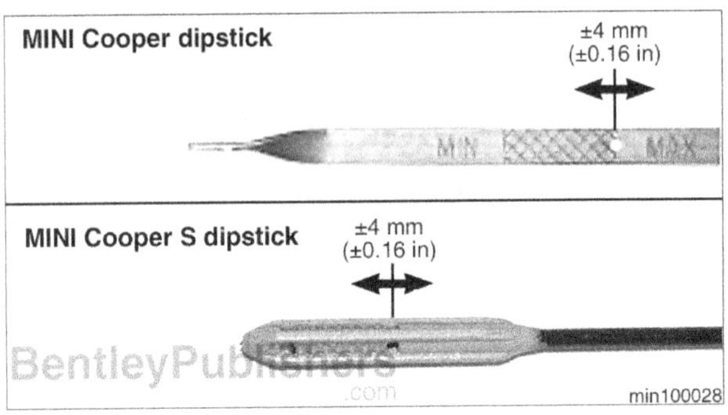

Lubrication system

The engine lubrication system is a full flow filtration pressure feed type.

The oil fill process at the factory allows for a tolerance of ±4 mm (±0.16 in) from the maximum mark on the dipstick. Oil level depends on oil temperature and length of time since the engine was last turned off.

See **119 Lubrication System** for repair information.

John Cooper Works (JCW) Kit

The John Cooper Works performance package was initially introduced to the U.S. market in April, 2003. Produced by the John Cooper Works Company of England, this package is sold through MINI dealers and is covered under MINI's comprehensive new car warranty. It is available as an add-on to MINI Cooper S models and can be retrofitted to earlier Cooper S models.

Engine performance is increased to 200 hp by utilizing a reconfigured cylinder head, uprated supercharger, special engine electronics and a performance exhaust.

Power was further increased to 210 hp in 2005 with the addition of a new air filter housing and special fuel injectors. The new JCW air filter housing has an extra intake flap to allow greater airflow at high revs (above 4500 rpm). This upgrade kit can be installed on engines that already have the earlier JCW kit.

Driveability Troubleshooting

MINI Cooper vehicles are equipped with a sophisticated self-diagnostic engine management system. This system monitors and stores diagnostic fault information.

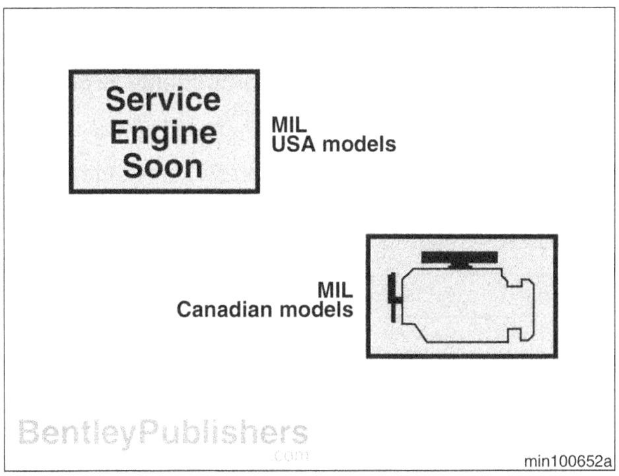

If the malfunction indicator light (MIL) comes on or flashes, it indicates that an emissions-related fault has occurred and that fault information is stored in memory within the ECM. In this situation, the first diagnostic test should be to connect a dedicated BMW scan tool and interrogate the fault memory.

The diagnostic capabilities of these systems have the potential to save hours of diagnostic time and prevent incorrect component replacement. See **OBD On Board Diagnostics**.

Two common causes of driveability problems are incorrect system voltage and bad grounds.

System voltage

Digital motor electronics (DME) requires that the system (battery) voltage be maintained within a narrow range. DC voltage levels beyond or below the operating range, or any A/C voltage in the electrical system can cause havoc. When troubleshooting an illuminated MIL, make sure the battery is fully charged and capable of delivering all its power to the electrical system. An undercharged battery can amplify A/C alternator output fluctuations.

To make a quick check of the battery charge, measure the voltage across the battery terminals with all cables attached and the ignition off. A fully charged battery will measure 12.6 volts or slightly more, compared to 12.15 volts for a battery with a 25% charge.

The DME system operates at low voltage and current levels, making it sensitive to small increases in resistance. The electrical system is routinely subjected to corrosion, vibration and wear, so faults or corrosion in the wiring harness and connectors are not uncommon. Check the battery terminals for corrosion or loose cable connections. See **121 Battery, Starter, Alternator** for additional information.

If a battery cable connection has no visible faults but is still suspect, measure the voltage drop across the connection. A large drop indicates excessive resistance, meaning that the connection is corroded, dirty, or damaged. Clean or repair the connection and retest.

NOTE—
For instructions on conducting a voltage drop test and other general electrical troubleshooting information, see **600 Electrical System–General**.

Visually inspect all wiring, connectors, switches and fuses in the system. Loose or damaged connectors can cause intermittent problems, especially the small terminals in the engine control module (ECM) connectors. Disconnect the wiring harness connectors to check for corrosion, and use electrical cleaning spray to remove contaminants.

Main grounds

Good grounds are critical to proper DME operation. If a ground connection has no visible faults but is still suspect, measure the voltage drop across the connection. A large voltage drop means high resistance. Clean or repair the connection and retest.

The main grounds for the fuel and ignition circuits of the DME system are illustrated below. See **ECL Electrical Component Locations** for additional ground and component locations.

Grounds for engine management system (**arrow**) at left front strut tower.

Fuel pump ground (**arrow**) in left side of luggage compartment behind trim panel.

110 Engine Removal and Installation

GENERAL

Engine removal and installation for the Cooper and Cooper S are covered in this repair group. See **100 Engine–General** for engine details and engine codes.

Cooper models are available with either a 5-speed manual transmission or an automatic Continuously Variable Transmission (CVT). Cooper S models are available with either a 6-speed manual transmission or an automatic (Agitronic) transmission. The engine and transmission are removed as a complete unit.

For additional procedures required during engine removal, refer to the following repair groups:

- **020 Maintenance**
- **121 Battery, Alternator, Starter**
- **170 Radiator and Cooling System**
- **180 Exhaust System**
- **230 Manual Transmission**
- **240 Automatic Transmission**
- **510 Bumpers, External Trim**

Special tools

Some special tools are required for engine removal and installation. Be sure to have the necessary equipment on hand before starting the job.

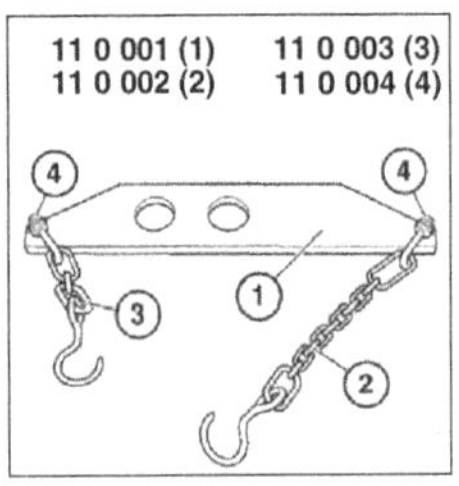

Engine hoisting harness (BMW tool no. 11 0 000)

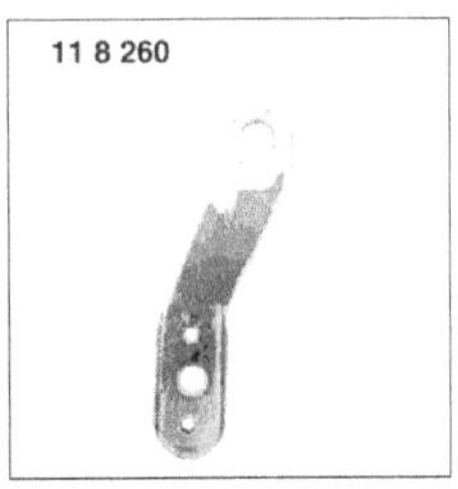

Engine lift attachment bracket (CVT) (BMW tool no. 11 8 260)

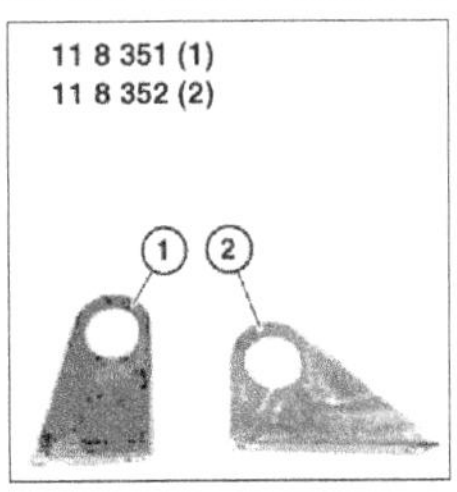

Engine lift attachment brackets (BMW tool no. 11 8 350)

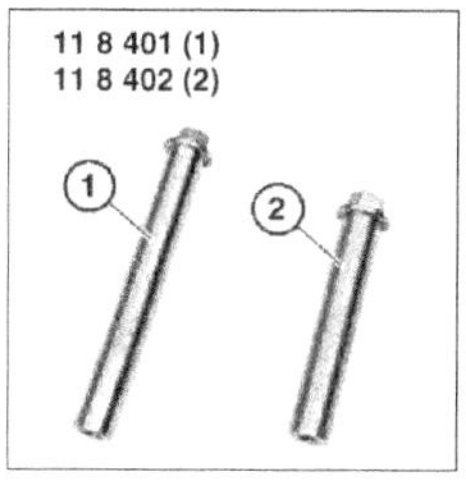

Modular front end extensions (BMW tool no. 11 8 400)

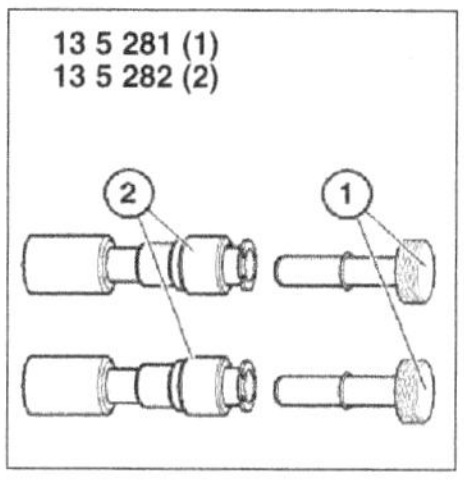

Plugs for fuel lines and fittings (BMW tool no. 13 5 280)

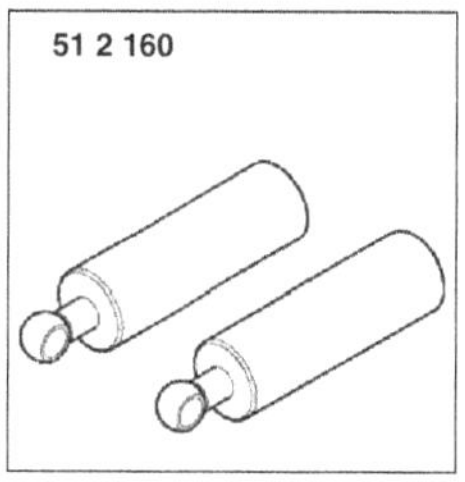

Service position hood prop extensions (BMW tool no. 51 2 160)

> ***CAUTION—***
> - *If the malfunction indicator light (MIL) is illuminated, see* **OBD On Board Diagnostics** *for DME fault code information.*
> - *If other system faults have been detected, as indicated by an illuminated ABS, SRS or ASC/DSC warning light, see the appropriate repair group in this manual or an authorized BMW dealer for more information on fault codes.*

Engine Removal and Installation (Cooper)

> ***WARNING —***
> *Due to risk of personal injury, be sure the engine is cold before beginning the removal procedure.*

Protect painted surfaces before beginning the removal procedure. As an aid to installation, label all components, wires, and hoses before removing them. Do not reuse gaskets,
O-rings or seals during reassembly.

Engine, removing and installing (Cooper)

– Remove battery box cover.

– Disconnect negative (-) cable from battery and remove battery. See **121 Battery, Starter, Alternator** for more information.

> ***CAUTION —***
> *Disconnecting the battery may erase fault code(s) stored in memory. Check for fault codes prior to disconnecting the battery cables.*

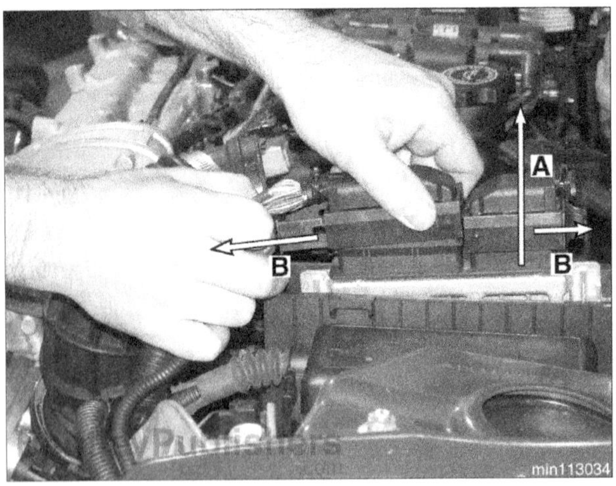

◄ Remove DME control module:

- Remove cover and lift DME control module upwards (**A**).
- Slide connector locking mechanisms (**B**) outward and disconnect control module connectors.

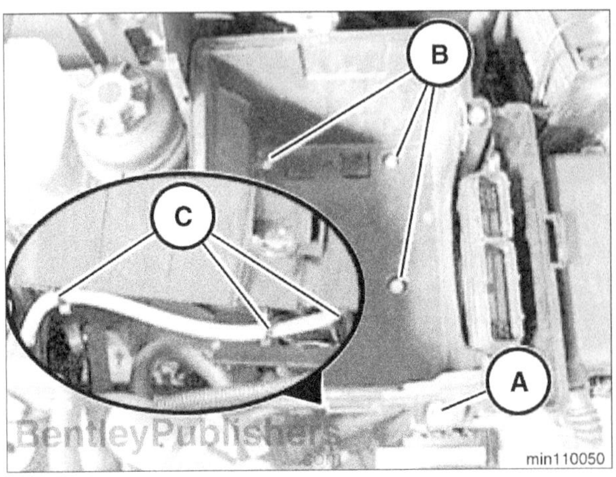

◄ Remove battery box:

- Remove battery box retaining bolt (**A**).
- Release battery box clips (**B**).
- Disconnect harness clips (**C**) from bottom of battery box.

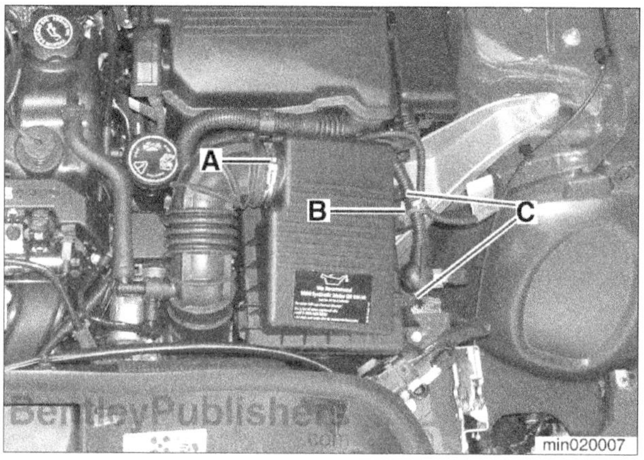

⮜ Remove air filter and air filter housing:

- Release hose clamp (**A**) and remove outlet duct from air filter housing.
- Remove wire tie (**B**) if necessary.
- Remove air filter retaining bolts (**C**).

– Remove air intake duct and remove air filter housing.

– Raise car and support in a safe manner. See **020 Maintenance** for more information.

> ***WARNING—***
> *Make sure the car is stable and well supported at all times. Use a professional automotive lift or jack stands designed for the purpose. A floor jack is not adequate support.*

– Remove exhaust manifold with catalytic converter. See **180 Exhaust System**.

– Models with manual transmission: drain transmission oil. See **230 Manual Transmission**.

– Models with CVT: drain transmission oil. See **240 Automatic Transmission**.

– Remove drive shafts. See **310 Front Suspension, Drive Axles**.

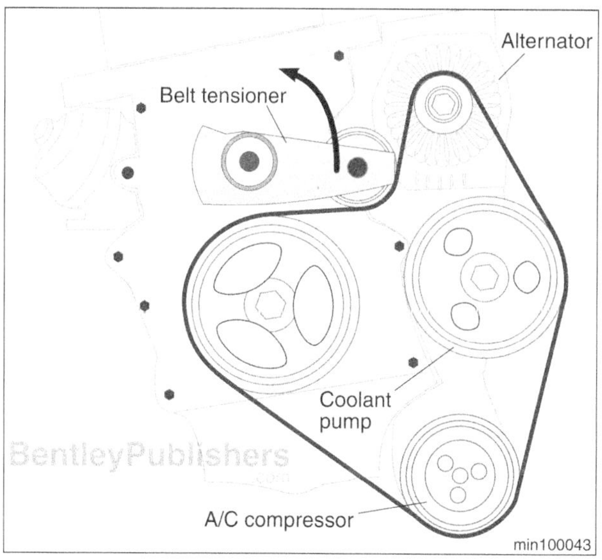

⮜ Make note of accessory belt layout.

NOTE—
If belt is to be reused, mark direction of travel and reinstall belt in same direction of rotation.

– Remove tank venting valve.

– Use special tool 11 8 390 to release accessory belt tension and special tool 11 8 280 to lock belt tensioner.

> ***CAUTION—***
> *Belt tensioner is under high tension. Check that lock pin is secure.*

– Remove accessory belt.

◀ Drain coolant by first releasing pressure cap (**arrow**) on coolant expansion tank.

> ***WARNING —***
> *Cooling system is under high pressure when hot. Coolant may cause burns when hot. Allow engine to cool before opening cooling system.*

– Working from below, remove engine splash shield.

◀ Release clamp (**arrow**) and remove lower radiator hose.

– Allow coolant to drain into suitable catch pan.

NOTE —
Save coolant and reuse, or dispose of properly.

> ***WARNING —***
> *Use extreme caution when draining and disposing of engine coolant. Coolant is poisonous and lethal to humans and pets. Pets are attracted to coolant because of its sweet smell and taste. Seek medical attention immediately if coolant is ingested.*

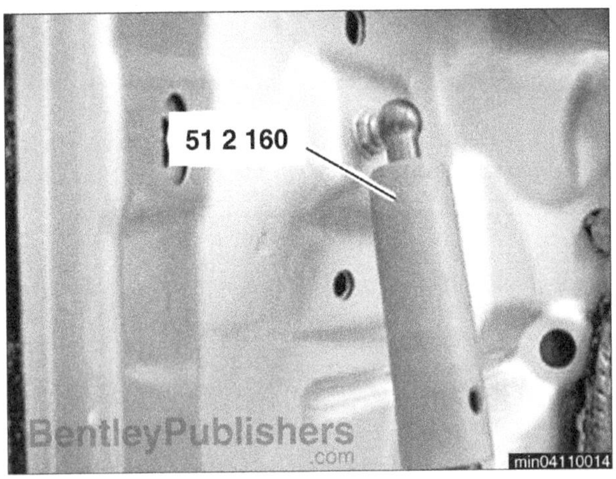

◀ For additional clearance, detach hood struts from hood and insert special tools (BMW 51 2 160) to support hood.

– Remove engine breather valve retaining bolt and remove engine breather valve from inspection cover.

Working underneath car:

- Remove bolt (**A**) holding crush tube to modular front end (MFE).
- Remove bolts (**B**) holding crush tube to front subframe.

NOTE—
MFE removed in photo for purposes of illustration.

Loosen modular front end. See **510 Exterior Trim, Bumpers** for more information.

- Remove right front wheel housing liner.
- Detach left front wheel housing liner from front bumper cover trim.
- Remove front bumper cover trim.
- Remove bumper.
- Remove MFE mounting bolts. Install two 100 mm (4 in) M8 bolts (**arrow**) in left and right bumper support members. Slide modular front end forward, supported on long bolts.
- Remove crush tubes.

> ***CAUTION—***
> *For ease of component alignment when reassembling front bumper assembly, do not loosen or remove bumper alignment bosses.*

NOTE—
- *Cooper S engine is illustrated. Cooper engine layout is similar.*
- *If available, use BMW special tools 11 8 401 and 11 8 402 instead of long bolts to support MFE.*
- *Removal of MFE is a complicated job. It is covered in* ***510 Exterior Trim, Bumpers****.*

Engine Removal and Installation (Cooper)

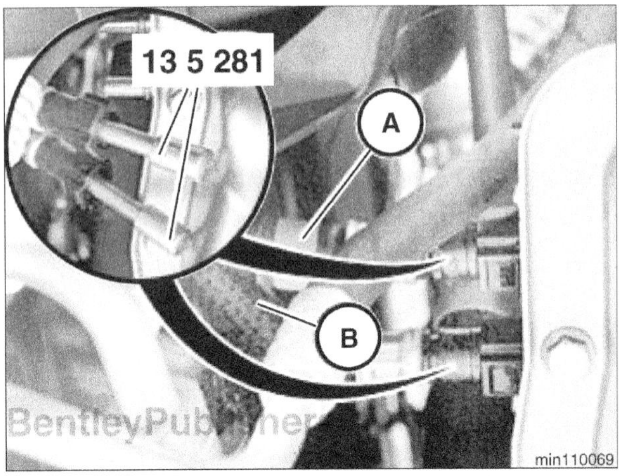

Models with CVT: disconnect oil cooler lines (**A**, **B**) from transmission and fit special tool (BMW 13 5 281) to plug lines.

NOTE —

- *Oil cooler lines are color-coded green and black. Make note of routing and replace in same location.*
- *Be prepared to catch dripping transmission fluid.*

Models with CVT: disconnect oil cooler lines (**arrows**) from oil cooler and fit special tool (BMW 13 5 282) to plug oil cooler fittings.

– Remove oil cooler lines.

NOTE —

Be prepared to catch dripping transmission fluid.

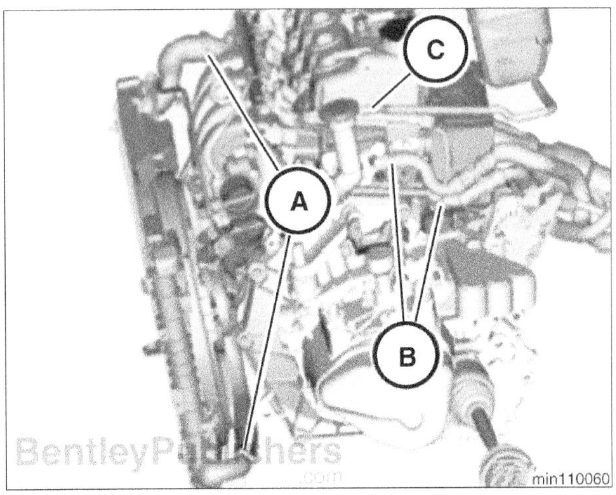

Remove upper and lower radiator hoses (**A**).

– Remove heater hoses (**B**).

– Remove overflow hose (**C**).

◀ Unhook fuel injector electrical harness from fuel rail mounting brackets.

– Disconnect top fuel tank vent line and unclip at fuel rail.

> ***WARNING —***
> *If disconnecting fuel line from fuel rail, fuel will be expelled under pressure. Loosen fuel filler cap to release fuel tank pressure. Do not smoke or work near heaters or other fire hazards. Keep a fire extinguisher handy. Before disconnecting fuel hose, wrap a cloth around hose to absorb any leaking fuel. Plug all open fuel connections.*

> ***CAUTION —***
> *Do not allow fuel to drip on alternator.*

NOTE —
- *Follow the procedure below to avoid disconnecting the fuel hose from the fuel rail.*
- *If it becomes necessary to disconnect the fuel hose from the fuel rail, follow the procedure in* **130 Fuel Injection**. *Be sure to read the warnings and cautions regarding working with fuel.*

◀ Detach fuel rail from top of engine:
- Disconnect fuel injector electrical harness connectors (**A**).
- Working at fuel rail, unclip hoses and ducts. Cut wire ties as needed.
- Disconnect vacuum line at base of fuel pressure regulator (**B**).
- Remove fuel rail mounting bolts (**arrows**).

> ***CAUTION —***
> *Use compressed air to blow away accumulated debris at the base of each fuel injector.*

– Remove fuel rail together with fuel injectors from intake manifold.
- Unclip fuel line from engine vibration damper bracket.
- Carefully fold fuel hose to right and rear of engine. Protect fuel rail and injectors by storing in plastic bag.

> ***CAUTION —***
> *Plug fuel injector bores in intake manifold.*

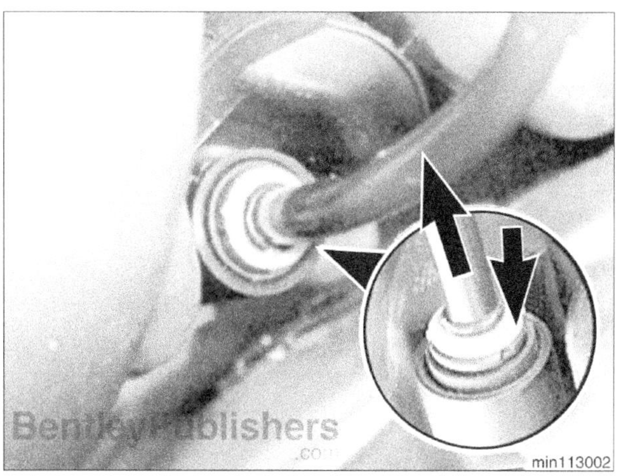

◀ Working at intake manifold, press locking ring down to detach brake booster vacuum line.

Engine Removal and Installation (Cooper)

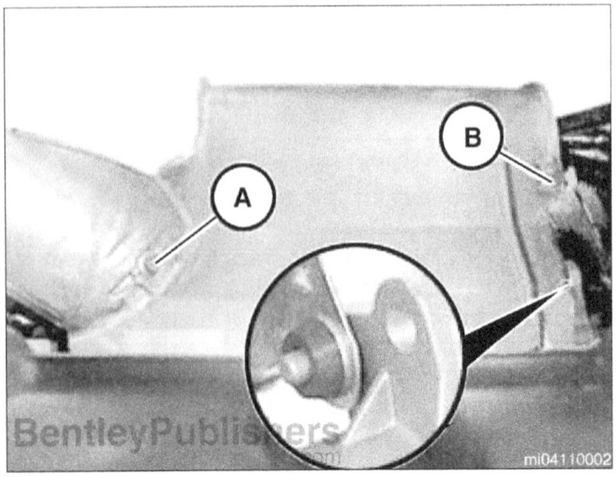

Working underneath car at back of engine, remove starter motor heat shield:

- Remove retaining bolt (**A**).
- Remove oxygen sensor from wire clip (**B**).

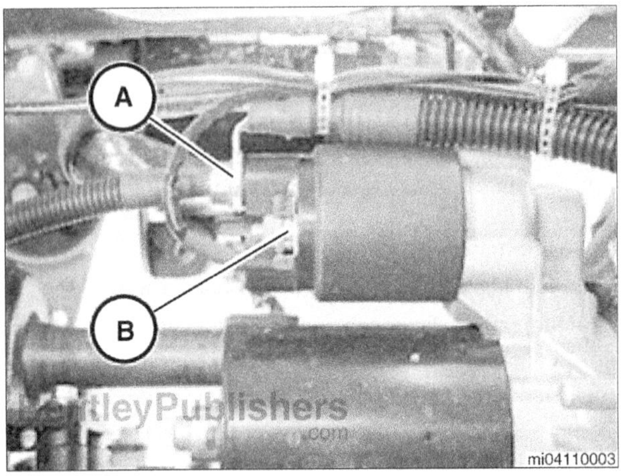

Disconnect wires (**A**, **B**) at starter motor and unclip wire harness from frame. Make note of harness routing.

– Remove protective cover from underneath fuel injector rail.

– Remove tank venting valve lower pipe from lower engine vibration damper.

– Remove ground wire from engine mounting bracket.

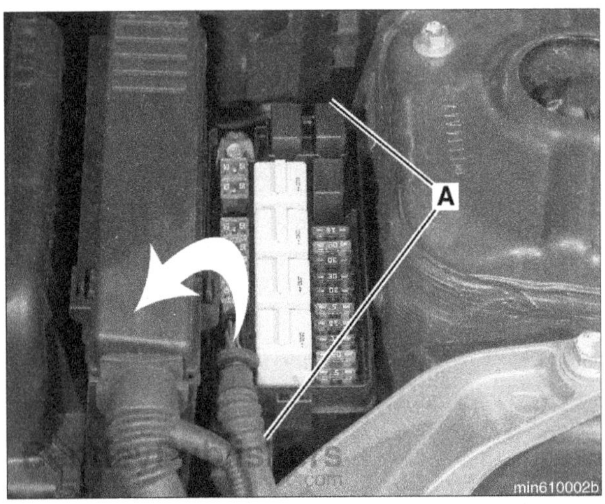

Working in engine compartment on right side of engine:

- Remove cover to fuse and relay panel 3.
- Disconnect harness connector (**arrow**).
- Remove fuse panel retaining bolts (**A**) and move panel aside.

◀ Remove ground wire from left strut tower (**A**).

– Disconnect round engine harness connector (**B**) by twisting and pulling apart.

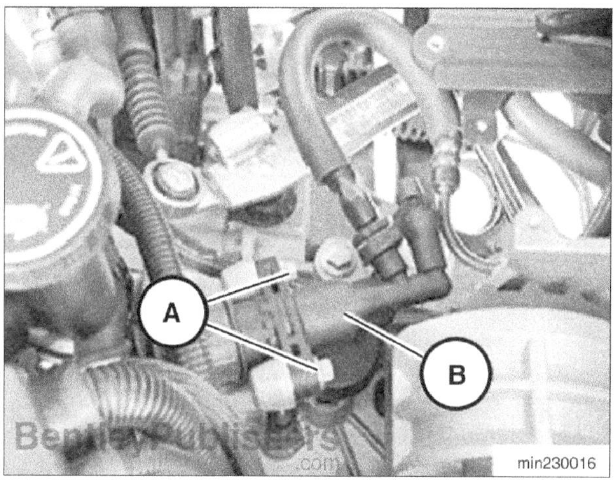

◀ Models with manual transmission: remove clutch slave cylinder retaining bolts (**A**) from top of transmission and remove clutch slave cylinder (**B**).

◀ Models with manual transmission: remove clips from gearshift cable bracket and remove gearshift cables from ball joints by carefully prying off with a pair of screwdrivers.

– Remove gearshift cable bracket bolts and remove gearshift bracket.

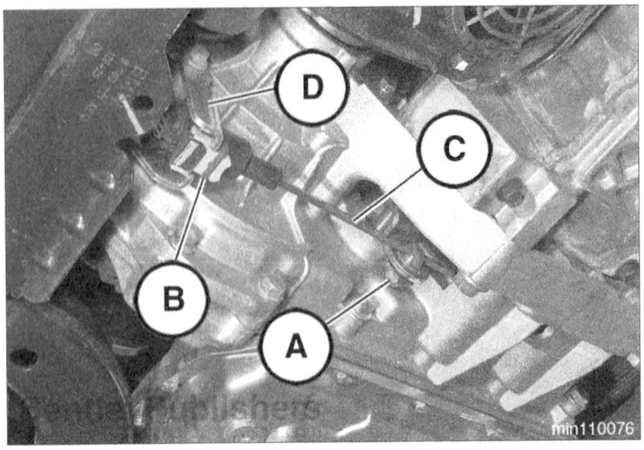

◀ Models with CVT:

- Loosen transmission shift cable retaining nut (**A**).
- Squeeze cable retaining clip (**B**) and remove cable (**D**) from bracket (**C**).

– Tie cable clear of engine compartment with stiff wire.

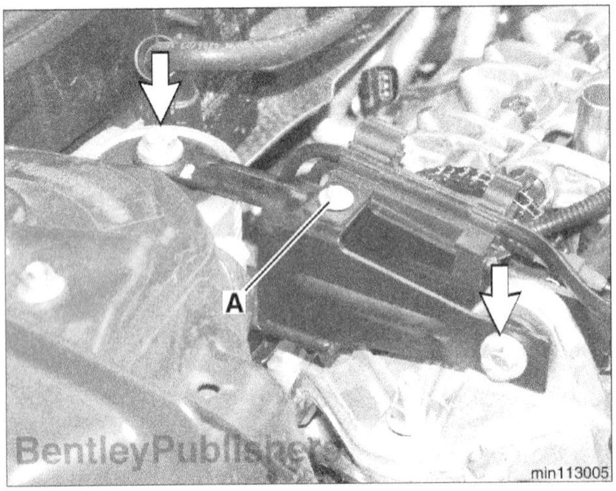

◀ Working on right side of engine, remove bolts from top engine vibration damper bracket (**arrows**), if applicable.

NOTE—
Later model vehicles may not have vibration damper bracket.

– Unclip hoses and remove plastic bracket (**A**).

◀ Working underneath car, remove lower engine vibration damper bracket bolts (**A**) and remove bracket (**B**).

Working underneath car, disconnect electrohydraulic power steering (EHPS) pump electrical connectors (**arrows**) at steering rack at rear of engine oil pan.

Unbolt A/C compressor from engine pan on front of engine:

- Disconnect wiring harness (**A**).
- Remove compressor mounting bolts (**B**).
- Detach A/C high pressure line from engine block in order to allow compressor free movement. Do not detach high pressure line from compressor.

> ***CAUTION—***
> *A/C system does not need to be evacuated for this procedure. However, take care when removing the compressor to avoid personal injury or damage to the system.*

– Pull compressor away from engine oil pan and secure to MFE with stiff wire.

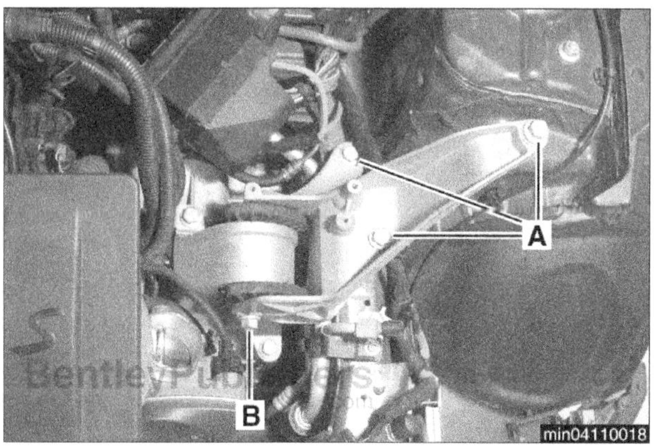

Remove transmission upper mount bracket bolts (**A**).

> ***CAUTION—***
> *Support transmission from below with floor jack before removing bracket mounting bolts.*

– Remove transmission mount to bracket mounting nut and bolt (**B**). Remove bracket.

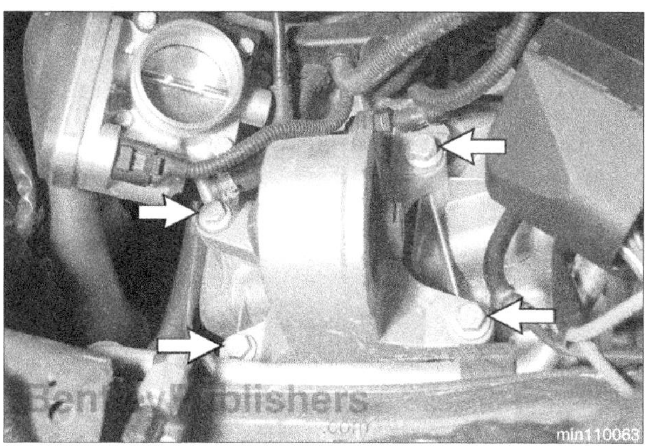

Remove bolts (**arrows**) from transmission mount. Remove mount.

Models with manual transmission: Attach engine hoist bracket (BMW 11 8 352) to transmission.

NOTE—
Models with CVT: Attach engine hoist bracket (BMW 11 8 260) to transmission.

Attach engine hoist bracket (BMW 11 8 351) to front of engine.

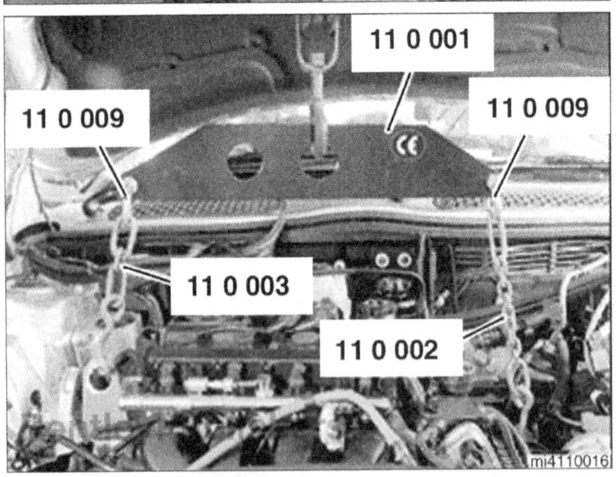

Attach engine hoisting harness (BMW 11 0 001, 11 0 002, 11 0 003, 11 0 009) to engine and transmission brackets and support firmly.

- Raise hoist slightly to remove load from engine and transmission mounts.

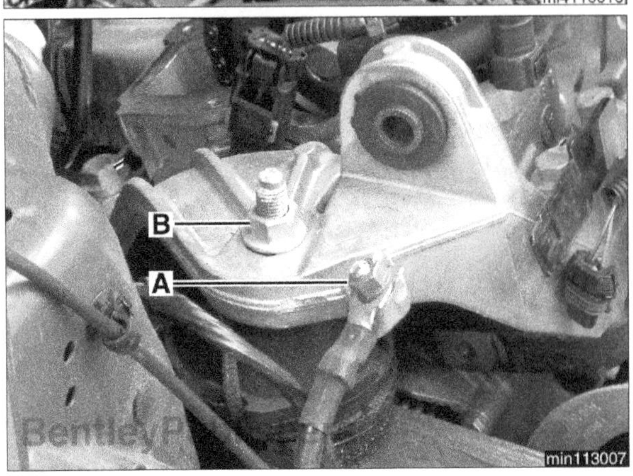

Remove ground wire (**A**) from engine mount bracket and remove nut from hydraulic mount (**B**).

- Carefully remove engine, checking for any wiring, fuel lines or mechanical parts that might become snagged.

– Installation is reverse of removal, noting the following:
  - Replace all gaskets, O-rings and seals.
  - Models with manual transmission: Refill transmission and check all other fluid levels. See **020 Maintenance**.
  - Models with CVT: Refill transmission following procedure in **240 Automatic Transmission**. Check all other fluids.
  - Refill and bleed cooling system. See **170 Radiator and Cooling System**.
  - Check that engine accessory belt properly engages pulley grooves.

> ***CAUTION—***
> *Check accessory belt for cracks, coolant and oil residue. Replace if contaminated by oil.*

  - Install exhaust manifold using new gaskets. Use copper paste on threads. See **180 Exhaust System**.
  - After reattaching battery cables, initialize power windows by raising each window and keeping power window switch in raised position for about 5 seconds.

Tightening torques	
A/C compressor to engine pan mount	25 Nm (18 ft-lb)
Alternator lead to starter motor	14 Nm (10 ft-lb)
Crush tubes to front end	100 Nm (75 ft-lb)
Crush tubes to subframe	100 Nm (75 ft-lb)
CVT shift cable to gearshift lever	12 Nm (9 ft-lb)
Engine mount bracket to cylinder head	100 Nm (74 ft-lb)
Engine mount bracket to engine mount	68 Nm (50 ft-lb)
Engine mount to body	68 Nm (50 ft-lb)
Fusebox retaining bolts to chassis	5 Nm (4 ft-lb)
Gearshift cable bracket to transmission	22 Nm (16 ft-lb)
Ground wire to left strut tower	9 Nm (7 ft-lb)
Heat shield to starter motor	9 Nm (7 ft-lb)
Lower engine vibration damper to holder	100 Nm (74 ft-lb)
Lower engine vibration damper to oil pan	38 Nm (28 ft-lb)
Slave cylinder to bracket	24 Nm (18 ft-lb)
Top engine vibration damper bracket to body	100 Nm (74 ft-lb)
Top engine vibration damper bracket to engine	100 Nm (74 ft-lb)
Transmission mount bolts to gearbox	66 Nm (49 ft-lb)
Transmission mount bracket to transmission	38 Nm (28 ft-lb)
Transmission mount retaining bolt to upper bracket	66 Nm (49 ft-lb)
Wiring to starter solenoid	8 Nm (6 ft-lb)

ENGINE REMOVAL AND INSTALLATION (COOPER S)

> ***WARNING—***
> *Due to risk of personal injury, be sure the engine is cold before beginning the removal procedure.*

Protect painted surfaces before beginning the removal procedure. As an aid to installation, label all components, wires, and hoses before removing them. Do not reuse gaskets, O-rings or seals during reassembly.

Engine, removing and installing (Cooper S)

– Working in luggage compartment, disconnect negative (-) cable from battery. See **121 Battery, Starter, Alternator** for more information.

> ***CAUTION—***
> *Disconnecting the battery may erase fault code(s) stored in memory. Check for fault codes prior to disconnecting the battery cables.*

Detach air filter housing from throttle housing:

- Pull up on battery positive connection point (Batt+) (**arrows**) to release from right side of air filter housing.
- Loosen or remove hose clamp (**A**) and detach outlet duct from air filter housing.
- Remove air filter housing retaining bolt (**B**).

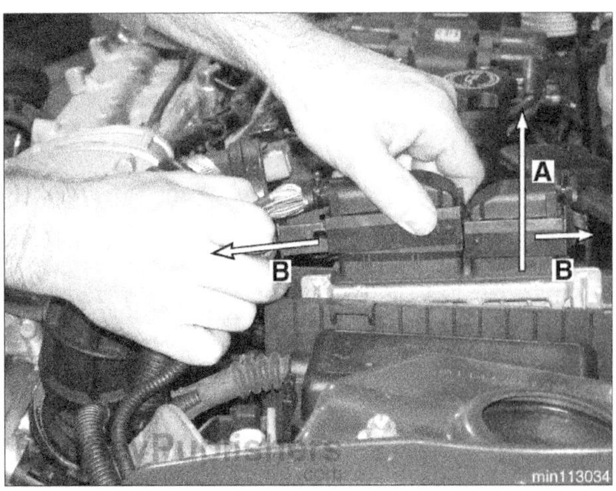

Remove DME control module:

- Remove control module container cover and lift control unit upward (**A**).
- Pull connector locking sliders outward (**B**).
- Disconnect control module connectors.
- Place control module in safe storage location.

NOTE—
There are 2 control module harness connectors.

◀ Unclip air filter inlet duct from modular front end. Wriggle duct while pulling it toward front of car to detach from air filter housing.

– Remove air filter housing from engine compartment.

– Raise car and support in a safe manner. See **020 Maintenance** for more information.

> ***WARNING —***
> *Make sure the car is stable and well supported at all times. Usa a professional automotive lift or jack stands designed for the purpose. A floor jack is not adequate support.*

– Remove exhaust manifold with catalytic converter. See **180 Exhaust System**.

– Drain gearbox oil. See **230 Manual Transmission**.

– Remove drive shafts. See **310 Front Suspension, Drive Axles**.

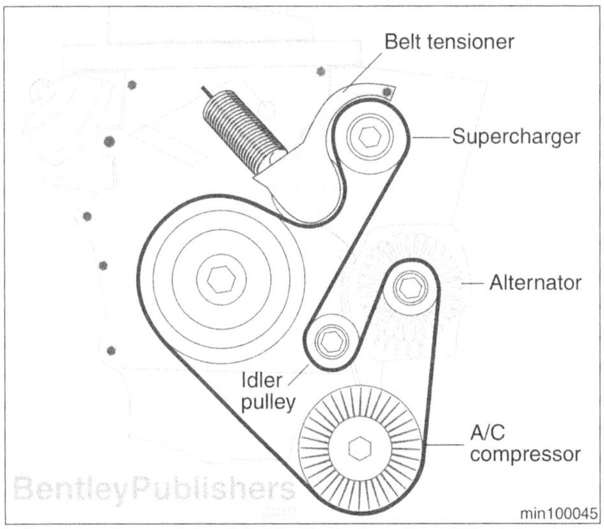

◀ Make note of accessory belt layout.

NOTE —
If belt is to be reused, mark direction of travel and reinstall belt in same direction of rotation.

– Remove tank venting valve.

– Use special tool 11 8 390 to release accessory belt tension and special tool 11 8 280 to lock belt tensioner.

> ***CAUTION —***
> *Belt tensioner is under high tension. Check that lock pin is secure.*

– Remove accessory belt.

– Remove intercooler. See **113 Cylinder Head Removal and Installation**.

◀ Drain coolant by first releasing pressure cap (**arrow**) on coolant expansion tank.

> ***WARNING —***
> *Cooling system under high pressure when hot. Coolant may cause burns when hot. Allow engine to cool before servicing cooling system.*

– Working from below, remove engine splash shield.

Engine Removal and Installation (Cooper S)

Release clamp and remove lower radiator hose (**arrow**).

– Allow coolant to drain into suitable catch pan.

NOTE —
Save coolant and reuse, or dispose of properly.

> ***WARNING —***
> *Use extreme caution when draining and disposing of engine coolant. Coolant is poisonous and lethal to humans and pets. Pets are attracted to coolant because of its sweet smell and taste. Seek medical attention immediately if coolant is ingested.*

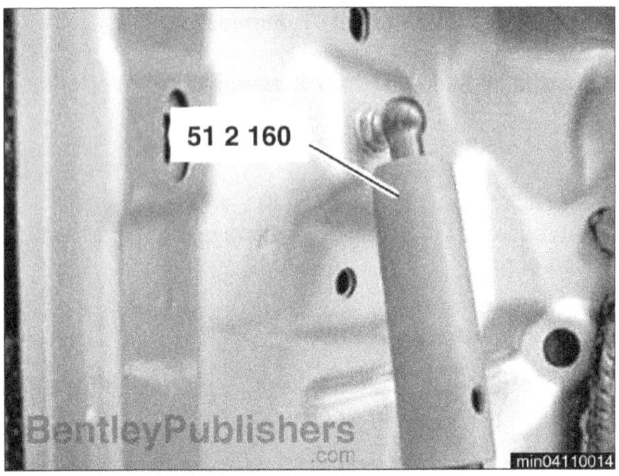

For additional clearance, detach hood struts from hood and insert special tools (BMW 51 2 160) to support hood.

– Remove engine breather valve retaining bolt and remove engine breather valve from inspection cover.

Working underneath car:

- Remove bolt (**A**) holding crush tube to modular front end (MFE).
- Remove bolts (**B**) holding crush tube to front subframe.

NOTE —
MFE removed in photo for purposes of illustration.

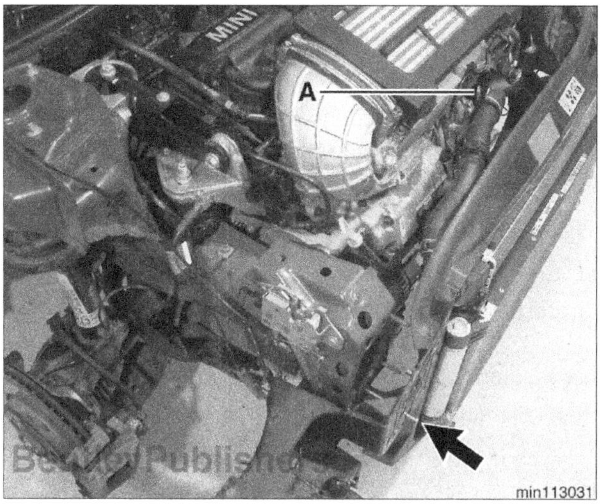

◄ Loosen modular front end. See **510 Exterior Trim, Bumpers** for more information.

- Remove right front wheel housing liner.
- Detach left front wheel housing liner from front bumper cover trim.
- Remove front bumper cover trim.
- Remove bumper.
- Detach radiator upper hose support clamp (**A**) from intake manifold.
- Remove MFE mounting bolts. Install two 100 mm (4 in) M8 bolts (**arrow**) in left and right bumper support members. Slide modular front end forward, supported on long bolts.
- Remove crush tubes.

> ***CAUTION—***
> *For ease of component alignment when reassembling front bumper assembly, do not loosen or remove bumper alignment bosses.*

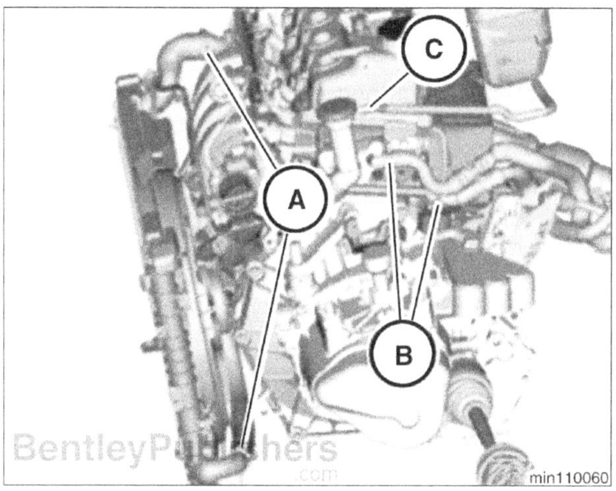

NOTE—
- *If available, use BMW special tools 11 8 401 and 11 8 402 instead of long bolts to support MFE.*
- *Removal of MFE is a complicated job. It is covered in* **510 Bumpers, External Trim**.

◄ Remove upper and lower radiator hoses (**A**).

– Remove heating hoses (**B**).

– Remove overflow hose (**C**).

NOTE—
Cooper coolant hose layout illustrated. Cooper S is similar.

◄ Unhook fuel injector electrical harness from fuel rail mounting brackets.

– Disconnect top fuel tank vent line and unclip at fuel rail.

> ***WARNING—***
> *If disconnecting fuel line from fuel rail, fuel will be expelled under pressure. Loosen fuel filler cap to release fuel tank pressure. Do not smoke or work near heaters or other fire hazards. Keep a fire extinguisher handy. Before disconnecting fuel hose, wrap a cloth around hose to absorb any leaking fuel. Plug all open fuel connections.*

> ***CAUTION—***
> *Do not allow fuel to drip on alternator.*

NOTE—
- *Follow the procedure below to avoid disconnecting the fuel hose from the fuel rail.*
- *If it becomes necessary to disconnect the fuel hose from the fuel rail, follow the procedure in* **130 Fuel Injection**. *Be sure to read the warnings and cautions regarding working with fuel.*

Detach fuel rail from top of engine:
- Disconnect fuel injector electrical harness connectors (**A**).
- Working at fuel rail, unclip hoses and ducts. Cut wire ties as needed.
- Disconnect vacuum line at base of fuel pressure regulator (**B**).
- Remove fuel rail mounting bolts (**arrows**).

CAUTION—
Use compressed air to blow away accumulated debris at the base of each fuel injector.

– Remove fuel rail together with fuel injectors from intake manifold.
- Unclip fuel line from engine vibration damper bracket.
- Carefully fold fuel hose to right and rear of engine. Protect fuel rail and injectors by storing in plastic bag.

CAUTION—
Plug fuel injector bores in intake manifold.

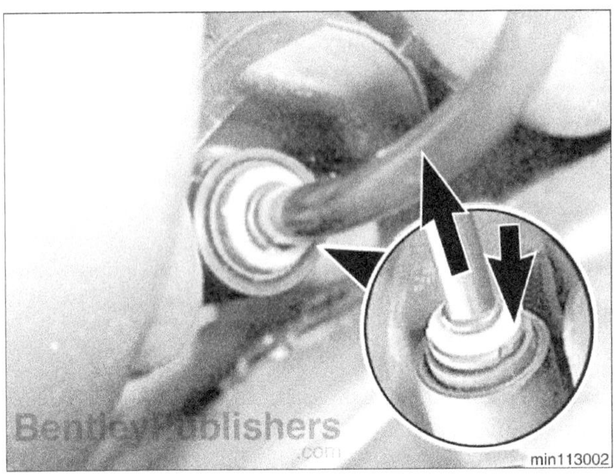

Working at intake manifold, press locking ring down to detach brake booster vacuum line.

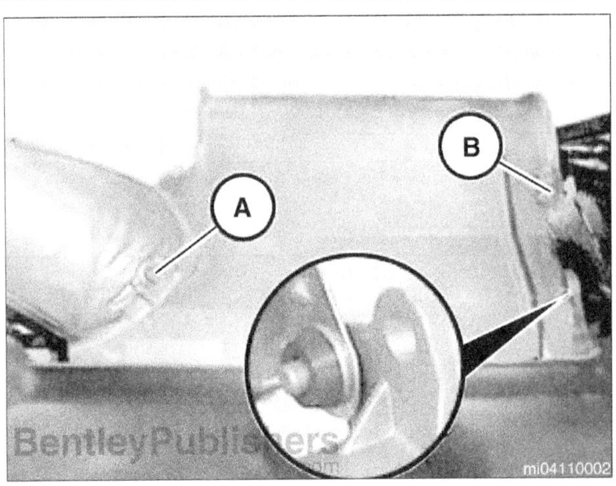

Working underneath car at back of engine, remove starter motor heat shield:
- Remove retaining bolt (**A**).
- Remove oxygen sensor from wire clip (**B**).

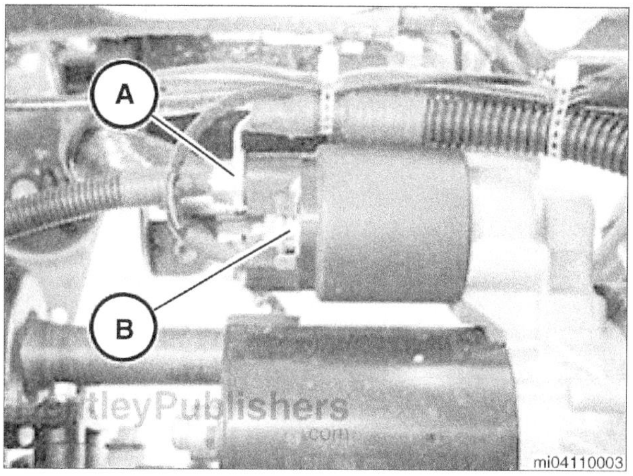

< Disconnect wires (**A**, **B**) at starter motor and unclip wire harness from frame (note harness routing).

– Remove protective cover from underneath fuel injector rail.

– Remove tank venting valve lower pipe from lower engine vibration damper bracket.

– Remove ground wire from engine mount bracket.

– On vehicles with automatic transmission, remove starter.

< Working in engine compartment on right side of engine:

- Remove cover to fuse and relay panel 3.
- Disconnect harness connector (**arrow**).

– Remove fuse panel retaining bolts (**A**) and move panel aside.

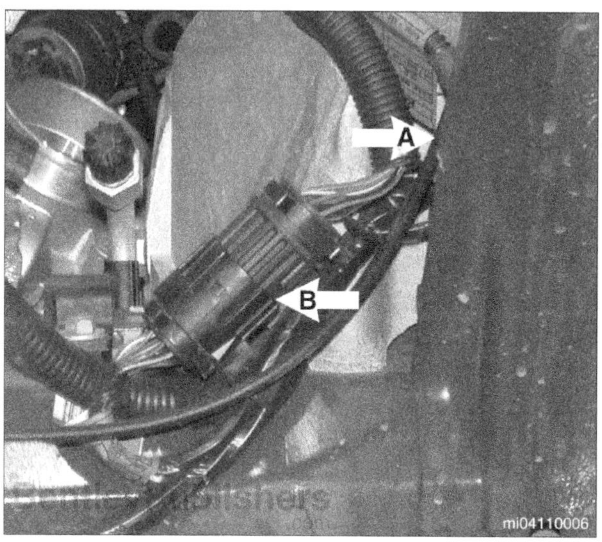

< Remove ground wire from left strut tower (**A**).

– Disconnect round engine harness connector (**B**) by twisting and pulling apart.

Vehicles with manual transmission

- Remove clutch slave cylinder retaining bolts (**arrows**) from front of transmission and remove clutch slave cylinder.

- Remove clips from gearshift cable bracket and remove gearshift cables from ball joints by carefully prying off with a pair of screwdrivers.

– Remove gearshift cable bracket bolts and remove gearshift bracket.

Vehicles with automatic transmission

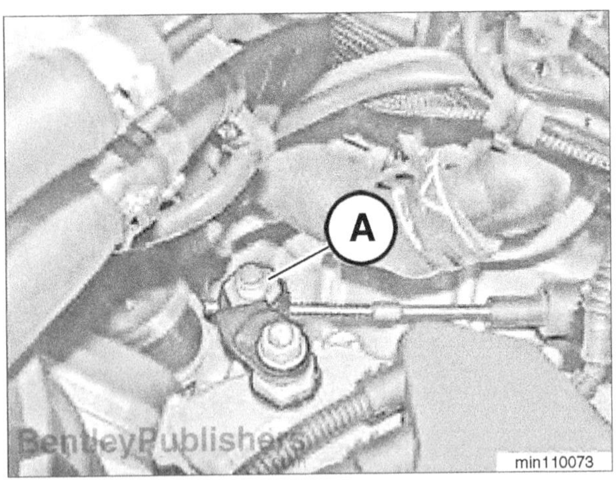

- Loosen lock nut (**A**) for gear selector cable and remove cable from bracket.

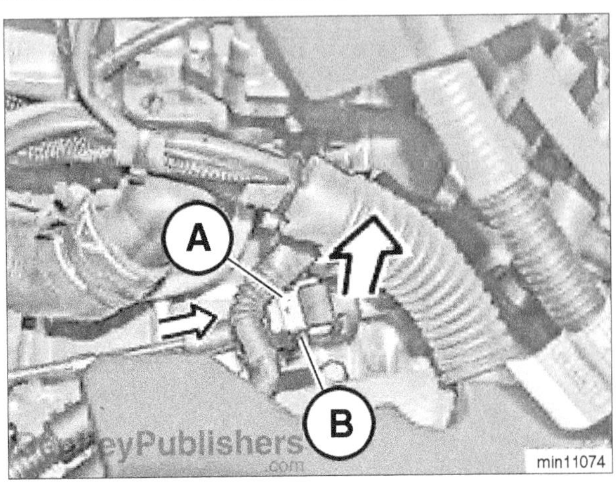

- Press retaining clip (**A**) together and remove +/– gear selector cable from bracket (**B**).

Continued for all vehicles

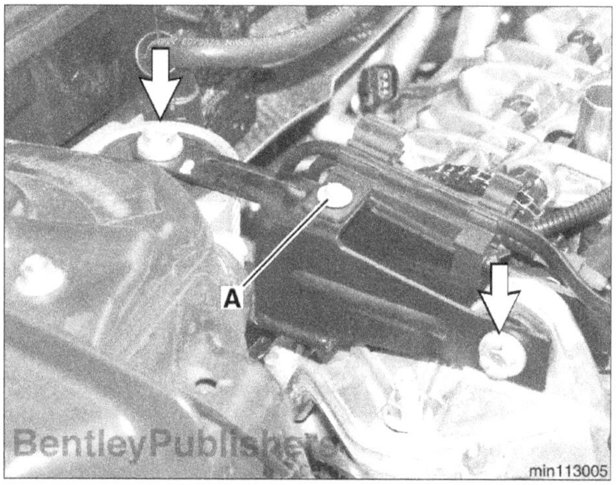

- Working on right side of engine, remove bolts from top engine vibration damper bracket (**arrows**), if applicable.

NOTE —
Later model vehicles may not have vibration damper bracket.

- Unclip hoses and remove plastic bracket (**A**).

- Working underneath car, remove lower engine vibration damper bracket bolts (**A**) and remove bracket (**B**).

- Working underneath car, disconnect electrohydraulic power steering (EHPS) pump electrical connectors (**arrows**) at steering rack at rear of engine oil pan.

Engine Removal and Installation (Cooper S)

◀ Unbolt A/C compressor from engine pan on front of engine:

- Disconnect wiring harness (**A**).
- Remove compressor mounting bolts (**B**).
- Detach A/C high pressure line from engine block in order to allow compressor free movement. Do not detach high pressure line from compressor.

> ***CAUTION—***
> *A/C system does not need to be evacuated for this procedure. However, take care when removing the compressor to avoid personal injury or damage to the system.*

– Pull compressor away from engine oil pan and secure to MFE with stiff wire.

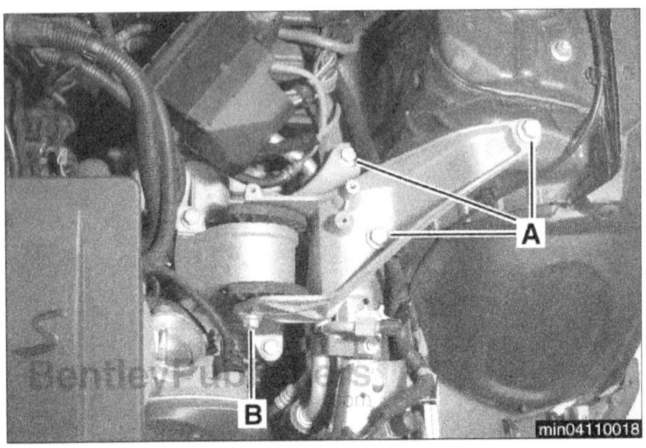

◀ Remove transmission upper mount bracket bolts (**A**).

> ***CAUTION—***
> *Support transmission from below with floor jack before removing bracket mounting bolts.*

– Remove transmission mount to bracket mounting nut and bolt (**B**). Remove bracket.

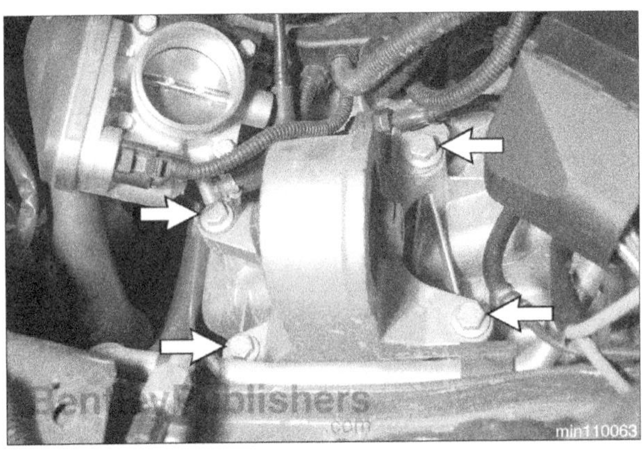

◀ Remove bolts (**arrows**) from transmission mount. Remove mount.

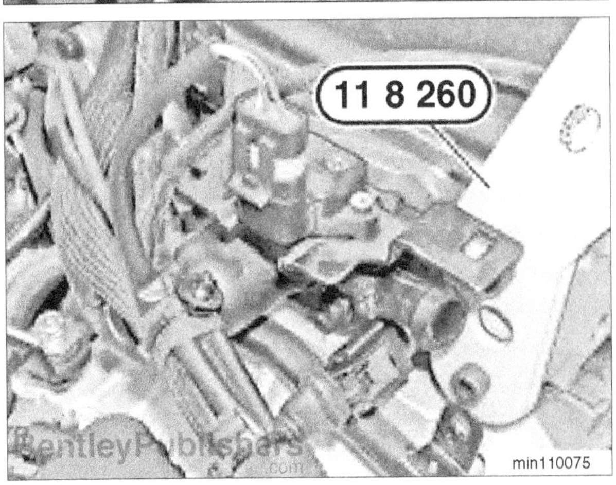

◀ Attach hoist bracket (BMW 11 8 260) to transmission.

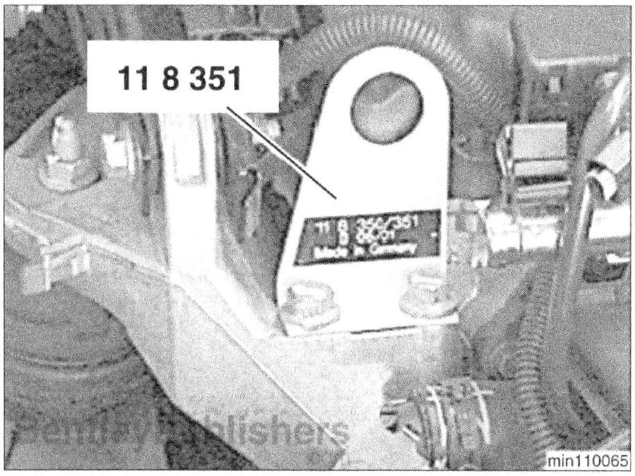

< Attach hoist bracket (BMW 11 8 351) to front of engine.

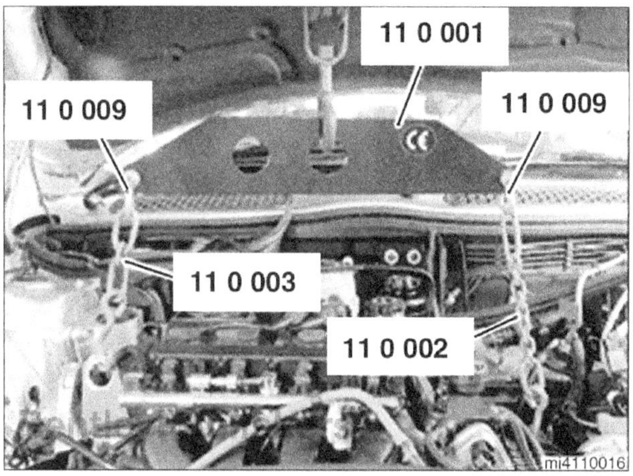

< Attach engine hoisting harness (BMW 11 0 001, 11 0 002, 11 0 003, 11 0 009) to engine and transmission brackets and support firmly.

– Raise hoist slightly to remove load from engine and gearbox mounts.

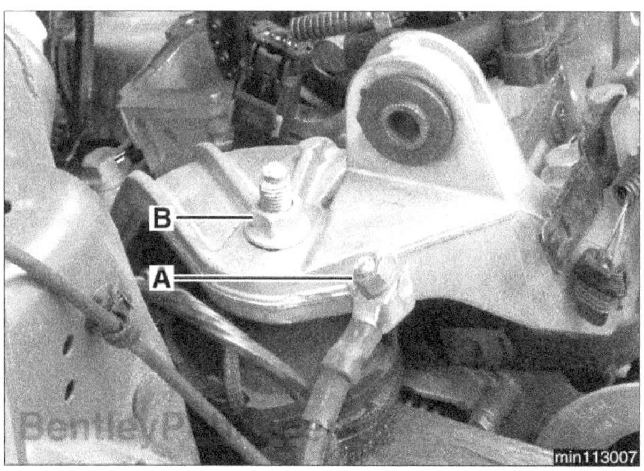

< Remove ground wire (**A**) from engine mount bracket and remove nut from hydraulic mount (**B**).

- Early version shown, later is similar.

– Carefully raise engine out of car, checking for any wiring, fuel lines, or mechanical parts that might become snagged as engine is removed.

– Installation is reverse of removal, noting the following:

- Replace all gaskets, O-rings and seals.
- Refill transmission and check all other fluid levels. See **020 Maintenance**.
- Refill and bleed cooling system. See **170 Radiator and Cooling System**.
- Check that engine accessory belt properly engages pulley grooves.
- Install exhaust manifold using new gaskets. Use copper paste on threads. See **180 Exhaust System**.
- After reattaching battery cables, initialize power windows by raising each window and keeping power window switch in raised position for about 5 seconds.

> ***CAUTION—***
> *Check accessory belt for cracks, coolant and oil residue. Replace if contaminated by oil.*

Tightening torques	
A/C compressor to engine pan mount	25 Nm (18 ft-lb)
Alternator lead to starter motor	14 Nm (10 ft-lb)
Crush tubes to front end	100 Nm (75 ft-lb)
Crush tubes to subframe	100 Nm (75 ft-lb)
Engine mount bracket to engine mount	68 Nm (50 ft-lb)
Engine mount bracket to transmission	38 Nm (28 ft-lb)
Engine mount to body	68 Nm (50 ft-lb)
Fusebox retaining bolts to chassis	5 Nm (4 ft-lb)
Gearshift cable bracket to manual transmission	22 Nm (16 ft-lb)
Gear selector cable to bracket on automatic transmission	12 Nm (9 ft-lb)
Ground wire to left strut tower	9 Nm (7 ft-lb)
Heat shield to starter motor	9 Nm (7 ft-lb)
Intercooler cover bracket to intercooler	9 Nm (7 ft-lb)
Intercooler cover to intercooler	9 Nm (7 ft-lb)
Lower engine vibration damper to holder	100 Nm (74 ft-lb)
Lower engine vibration damper to oil pan	38 Nm (28 ft-lb)
Mount bracket to cylinder head	100 Nm (74 ft-lb)
Slave cylinder to manual transmission bracket	24 Nm (18 ft-lb)
Starter to automatic transmission	82 Nm (60 ft-lb)
Throttle body to intake manifold	9 Nm (7 ft-lb)
Top engine vibration damper bracket to body	100 Nm (74 ft-lb)
Top engine vibration damper bracket to engine	100 Nm (74 ft-lb)
Transmission mount bolts to transmission	66 Nm (49 ft-lb)
Transmission mount retaining bolt to upper bracket	66 Nm (49 ft-lb)
Upper sealing ring bracket to lower bracket	9 Nm (7 ft-lb)
Wiring to starter solenoid	8 Nm (6 ft-lb)

113 Cylinder Head Removal and Installation

GENERAL

This repair group covers MINI cylinder head removal and installation as well as cylinder head / valve diagnosis procedures.

The information given in this repair group assumes that the engine is installed in the engine bay. For cylinder head and valvetrain reconditioning information, see **116 Cylinder Head and Valvetrain**.

NOTE—
If a head gasket problem is suspected, a compression test or leak-down test will usually detect the fault. See **Diagnostic Testing** *in this repair group.*

Special tools

Special service tools are required to properly remove and install the cylinder head. Read the entire procedure through before beginning the job.

00 9 120

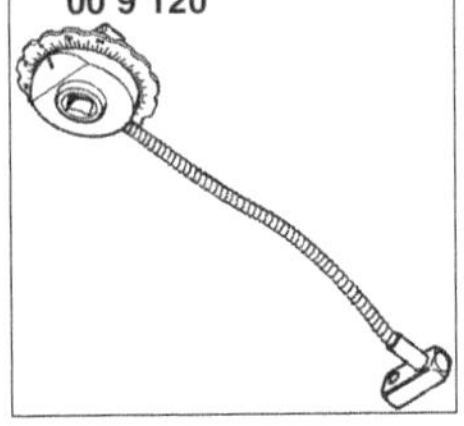

Torque angle measuring gauge
(BMW special tool 00 9 120)

11 0 220

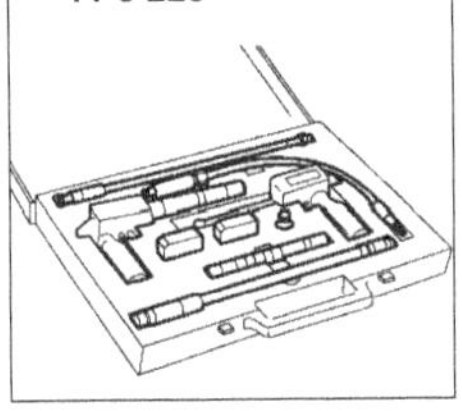

Compression test kit
(BMW special tool 11 0 220)

11 8 250

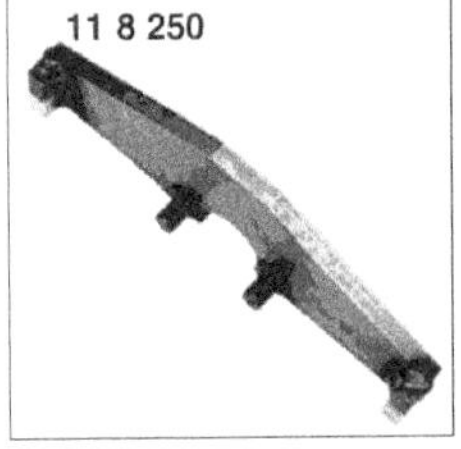

Camshaft locking tool
(BMW special tool 11 8 250)

11 8 370

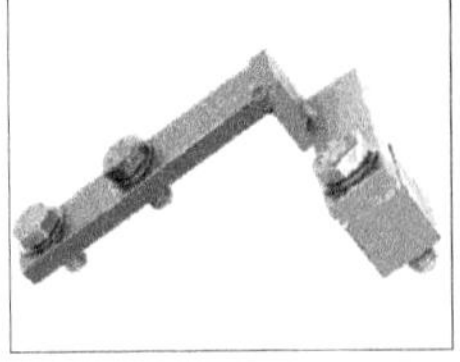

Engine block support tool
(BMW special tool 11 8 370)

Hydraulic engine mount removal and installation tool (BMW special tool 11 8 380)

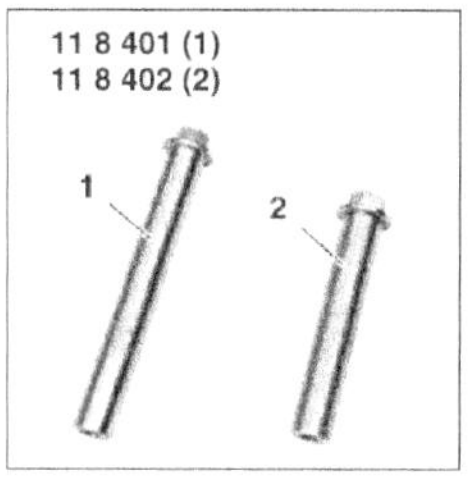

Modular front end supports (BMW special tool 11 8 400)

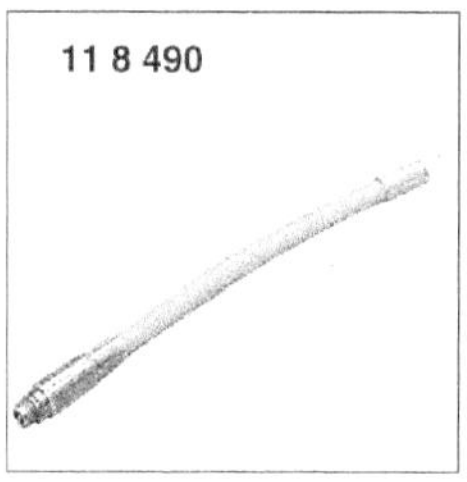

Compression tester adapter (BMW special tool 11 8 490)

DIAGNOSTIC TESTING

Cylinder compression, checking

A compression gauge is needed to make a compression test. For an accurate test, the battery and starter must be capable of cranking the engine at least 300 rpm, and the engine should be at normal operating temperature.

NOTE—
Performing a compression test may cause a fault to set in the ECM and may illuminate the Malfunction Indicator Light (MIL). The light can only be turned out using either BMW special service scan tool or equivalent. Disconnecting the battery will not erase the fault memory or turn out the light. See **OBD On Board Diagnostics**.

– Use BMW service scan tool (DISplus, GT1, MoDiC) or equivalent to read out DME control module fault memory.

– Turn ignition OFF.

Working at top of valve cover, disconnect electrical harness connector to ignition coil pack.

- Press securing clip in direction of **white arrow**.
- Pull off coil pack connector in direction of **black arrow**.

WARNING —
- *The ignition system produces high voltages that can be fatal. Avoid contact with exposed terminals and use extreme caution when working on a car with the ignition switched on or the engine running.*
- *Do not touch or disconnect ignition components while the engine is running or being cranked by the starter.*

Pull up on each spark plug wire connector while gently twisting from side to side to remove. Label wires.

NOTE —
Cooper S engine is illustrated. Cooper engine layout is similar.

– Remove spark plugs from all cylinders.

NOTE —
Check spark plugs for oil deposits that may indicate poor cylinder sealing, then set them aside in order.

Screw compression tester adapter (BMW special tool 11 8 490 or equivalent) into cylinder 1 spark plug thread tight enough to form a good seal. Attach compression gauge (BMW special tool 11 0 224 or equivalent).

– With parking brake set and clutch pressed to floor or transmission in PARK or NEUTRAL, crank engine with starter. Record highest value indicated by gauge.

NOTE —
- *The compression gauge reading should increase with each compression stroke and reach near its maximum reading in about 4 - 6 strokes.*
- *All cylinders should reach maximum compression in the same number of strokes. If a cylinder needs significantly more strokes to reach maximum compression, there is a problem.*

– Release pressure at compression gauge valve, then remove gauge from spark plug hole. Repeat test for each cylinder and compare results with values given in **Compression specifications** table.

Compression specifications	
Compression range Cooper Cooper S	 11.5 - 17.0 bar (167 - 247 psi) 9 - 13.5 bar (131 -196 psi)
Maximum difference between cylinders	0.5 bar (7 psi)

– Compression readings may be interpreted as follows:
  - Low compression in one cylinder indicates a poorly sealed combustion chamber.
  - Relatively even pressures that are below specification normally indicate worn piston rings and/or cylinder walls.
  - Erratic values tend to indicate valve leakage.
  - Dramatic differences between cylinders are often a sign of a failed head gasket, burned valve, or broken piston ring.

– If readings are within specifications, reinstall spark plugs.

NOTE—
Used spark plugs should be reinstalled in the same cylinder from which they were removed.

Tightening torque	
Spark plug to cylinder head	27 Nm (20 ft-lb)

– Remainder of installation is reverse of removal. Be sure to reattach coil pack electrical connector.

– Connect BMW service scan tool or equivalent and delete stored faults.

Wet compression test

To further help analyze the source of poor compression, a wet compression test is the next step.

– Repeat compression test, but this time squirt a teaspoon of oil into each cylinder. The oil will temporarily help seal between piston rings and cylinder wall, practically eliminating leakage past rings for a short time.

– If this test yields a higher reading than "dry" compression test, there is probably leakage between piston rings and cylinder walls, due either to wear or to broken piston rings.

– Little or no change in compression reading indicates other leakage, probably from valves.

Cylinder leak-down test

The most conclusive diagnosis of low compression symptoms requires a cylinder leak-down test. Using a special tester and compressed air, each cylinder, in turn, is pressurized. The rate at which the air leaks out of the cylinder, as well as where the air leaks out, can accurately pinpoint the magnitude and location of the leakage.

Before attempting any repair that requires major engine disassembly, use a leak-down test to confirm low compression.

Cylinder leak-down test equipment can be procured from an automotive parts and equipment supplier.

CYLINDER HEAD REMOVAL AND INSTALLATION (COOPER)

Cylinder head removal and installation is a complicated job and requires special tools.

The job has been divided up into several procedures. Read the procedures before beginning the repair.

> ***WARNING —***
> *Due to risk of personal injury, be sure the engine is cold before beginning any of the procedures.*

> ***CAUTION —***
> *Cover all painted surfaces before beginning the removal procedure. As an aid to installation, label all components, wires, and hoses before removing them. Do not reuse gaskets, O-rings or seals during reassembly.*

Valve cover, removing and installing (Cooper)

– Make sure ignition is OFF.

< Working at top of engine, disconnect **PCV valve** and crankcase breather hose (**arrow**) from valve cover.

Remove fuel rail cover:

- Squeeze locking tabs (**A**) to detach rear of cover.
- Tilt and pull cover forward until plastic clips (**arrows**) detach from fuel rail.

> ***CAUTION—***
> *Plastic clips can break easily.*

Pull up on spark plug wire connector while gently twisting from side to side.

NOTE—
Cooper S engine is illustrated. Cooper engine layout is similar.

Working at top of valve cover, disconnect electrical harness connector to ignition coil pack.

- Press securing clip in direction of **white arrow**.
- Pull off coil pack connector in direction of **black arrow**.

– Remove harness clips from valve cover bolts.

– Remove coil pack mounting bolts. Lift coil pack and spark plug wires off valve cover.

Cylinder Head Removal and Installation (Cooper)

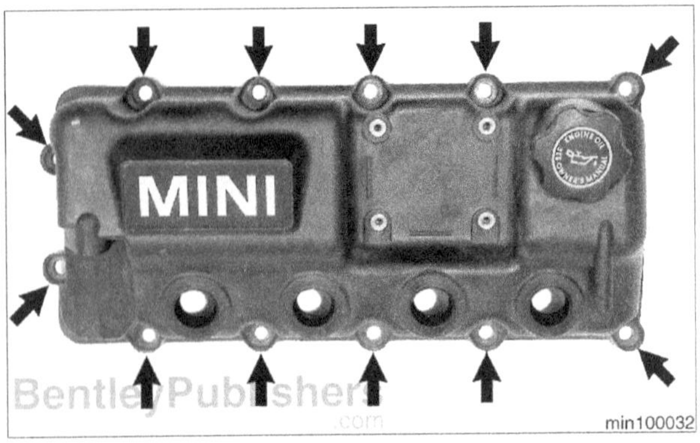

⋖ Remove 12 valve cover bolts (**arrows**).

– Remove valve cover.

– Installation is reverse of removal. Note the following:
- Replace valve cover gasket.
- Replace spark plug hole sealing grommets in valve cover.
- Clean gasket mating faces. Coat lightly with oil.
- Replace valve cover mounting bolts or insulators if damaged.
- Tighten valve cover mounting bolts in sequence, from inside to outside.
- Inspect ignition coil pack insulators. Replace if damaged.

Tightening torques	
Ignition coil pack to valve cover (M6)	12 Nm (9 ft-lb)
Valve cover to cylinder head (M6)	12 Nm (9 ft-lb)

⋖ When reinstalling ignition wires, make sure that they snap securely into place.

Intake manifold, removing and installing (Cooper)

– Working in engine compartment, remove battery box cover. Disconnect negative (-) cable from battery. See **121 Battery, Starter, Alternator** for more information.

> ***CAUTION—***
> *Disconnecting the battery may erase fault code(s) stored in memory. Check for fault codes prior to disconnecting the battery cables.*

⋖ Working at top of engine, loosen or remove air duct clamp (**A**). Detach air duct from air filter housing.
- Detach wire harness (**B**).
- Remove air filter housing mounting screws (**C**).
- Detach crankcase breather hose (**D**) from air duct.

– Remove complete air filter housing.

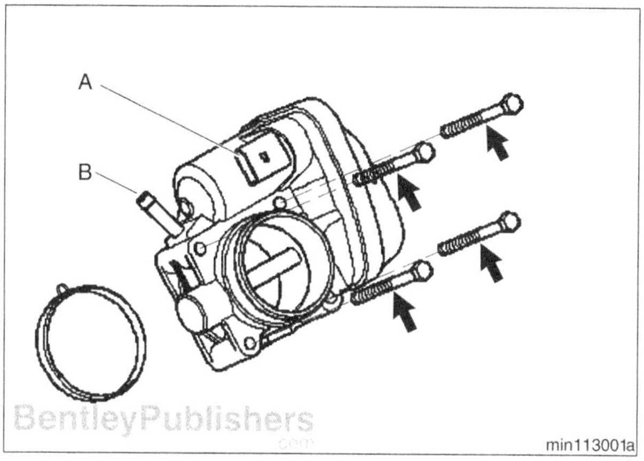

Remove throttle valve assembly:

- Loosen or remove air duct clamp at throttle assembly. Detach air duct.
- Disconnect electrical harness connector (**A**).
- Detach fuel tank vent line (**B**).
- Loosen and remove throttle assembly mounting screws (**arrows**).
- Lift off throttle assembly.

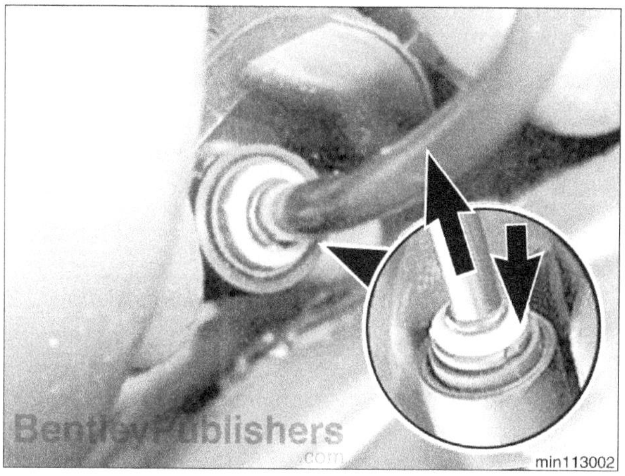

Working at intake manifold, press brake booster vacuum line locking ring downward to detach line from manifold.

– Working at top right of valve cover, remove hose from PCV valve.

Remove fuel rail cover:

- Squeeze locking tabs (**A**) to detach rear of cover.
- Tilt and pull cover forward until plastic clips (**arrows**) detach from fuel rail.

> ***CAUTION—***
> *Plastic clips can break easily.*

– Detach electrical harness connector from intake air temperature / manifold absolute pressure (T-MAP) sensor.

– Detach knock sensor electrical harness connector from fuel rail wiring harness.

– Disconnect top fuel tank vent line and unclip at fuel rail.

> ***WARNING—***
> - *If disconnecting fuel line, fuel will be expelled under pressure. Loosen fuel filler cap to release tank pressure.*
> - *Do not smoke or work near heaters or other fire hazards.*
> - *Keep a fire extinguisher handy.*
> - *Before disconnecting fuel hose, wrap a cloth around hose to absorb any leaking fuel.*
> - *Plug all open fuel connections.*

Cylinder Head Removal and Installation (Cooper)

CAUTION—
Do not allow fuel to drip on alternator.

NOTE—
- *Follow the procedure below to avoid disconnecting the fuel hose from the fuel rail.*
- *If it becomes necessary to disconnect the fuel hose from the fuel rail, follow the procedure in* **130 Fuel Injection**. *Be sure to read the warnings and cautions regarding working with fuel.*

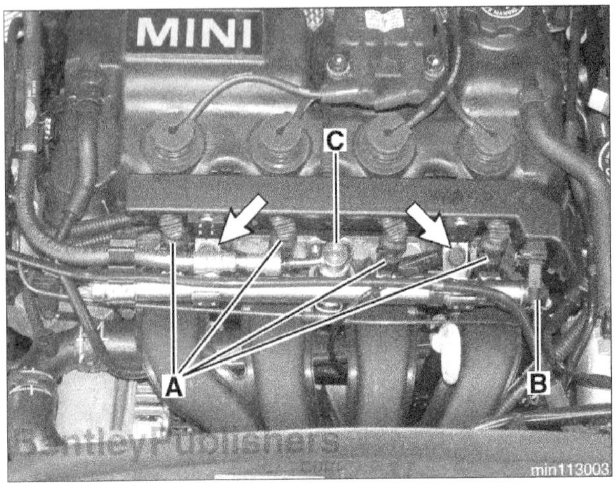

◀ Detach fuel rail from top of engine:
- Disconnect fuel injector electrical harness connectors (**A**).
- Disconnect harness connector (**B**).
- Remove cover from harness loom and swing harness out of way to left side of engine.
- Working at fuel rail, unclip hoses and ducts. Cut wire ties as needed.
- Disconnect vacuum line at base of fuel pressure regulator (**C**).
- Remove fuel rail mounting bolts (**arrows**).

CAUTION—
Use compressed air to blow away accumulated debris at the base of each fuel injector.

– Remove fuel rail together with fuel injectors from intake manifold.
- Unclip fuel line from engine vibration damper bracket.
- Carefully fold fuel hose to right and rear of engine. Protect fuel rail and injectors by storing in plastic bag.

CAUTION—
Plug fuel injector bores in intake manifold.

– Remove engine oil dipstick.

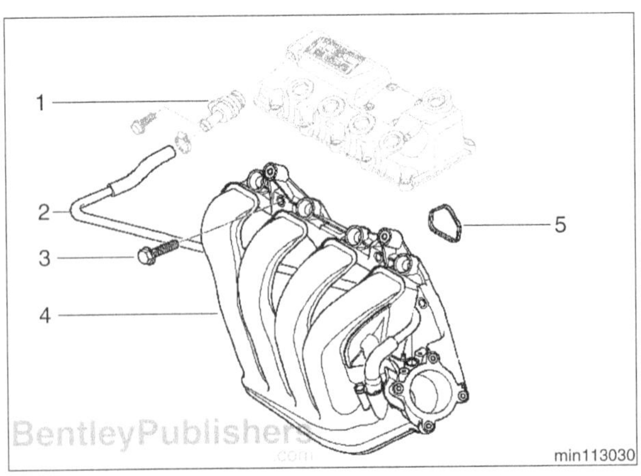

◀ Remove intake manifold mounting bolts in sequence, working from outside to inside. Remove manifold along with PCV valve hose.
1. PCV valve
2. PCV valve hose
3. Bolt, M8 x 35 mm
   -tighten to 26 Nm (19 ft-lb)
4. Intake manifold
5. Intake manifold profile gasket (always replace)

CAUTION—
Stuff clean rags in cylinder head intake ports.

– Unclip coolant line below manifold.

– When reinstalling intake manifold, replace intake manifold profile gaskets.

– Tighten down intake manifold mounting bolts in sequence, working from inside to outside.

Tightening torque	
Intake manifold to cylinder head (M8)	26 Nm (19 ft-lb)

– When reinstalling fuel rail, coat fuel injector sealing O-rings with antiseize agent.

Tightening torque	
Fuel rail to intake manifold (M8)	25 Nm (18 ft-lb)

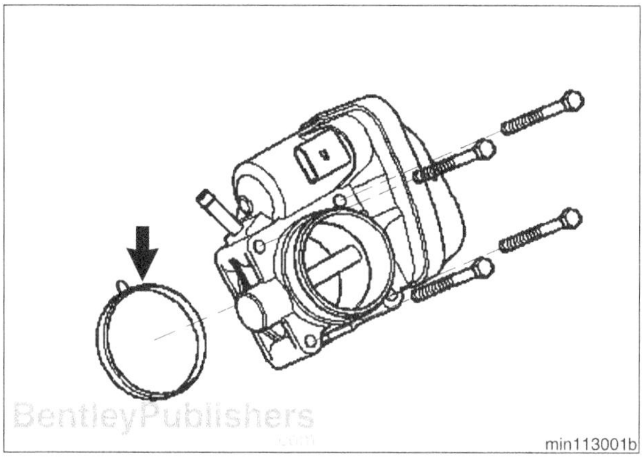

◄ When installing throttle valve assembly, replace sealing ring (**arrow**) between assembly and intake manifold.

Tightening torque	
Throttle valve assembly to intake manifold (M6)	9 Nm (7 ft-lb)

– Reinstall air filter housing.

Tightening torque	
Air filter housing to body (M6)	7 - 10 Nm (5 - 7 ft-lb)

– Reattach air duct using new clamp.

NOTE—
If throttle valve assembly is replaced, be sure to reset DME control unit adaptation values using BMW scan tool.

Cylinder head, removing and installing (Cooper)

– Working in engine compartment, remove battery box cover and detach battery vent. Disconnect battery cables, starting with negative (-) cable. Remove battery See **121 Battery, Starter, Alternator** for more information.

> ***CAUTION—***
> *Disconnecting the battery may erase fault code(s) stored in memory. Check for fault codes prior to disconnecting the battery cables.*

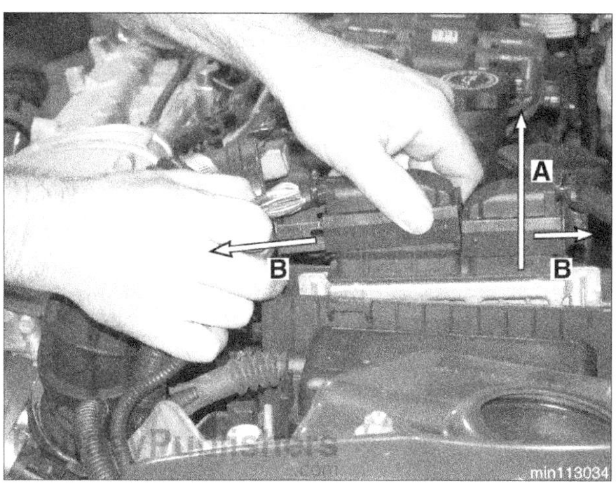

◄ Remove DME control module:

- Remove control module container cover and lift control unit upward (**A**).
- Pull connector locking sliders outward (**B**).
- Disconnect control module connectors.
- Place control module in safe storage location.

NOTE—
There are 2 control module harness connectors.

Cylinder Head Removal and Installation (Cooper)

– Remove battery box:
  - Remove battery box retaining bolts.
  - Release battery box clips.
  - Disconnect harness clips from bottom of battery box.

– Drain coolant. Begin by releasing pressure at coolant filler cap. See **170 Cooling System**.

– Working at rear of cylinder head, loosen or remove heater hose clamps and detach heater hoses.

– Loosen or remove top radiator hose clamp at thermostat housing. Detach hose from housing.

– Remove right front wheel housing liner.

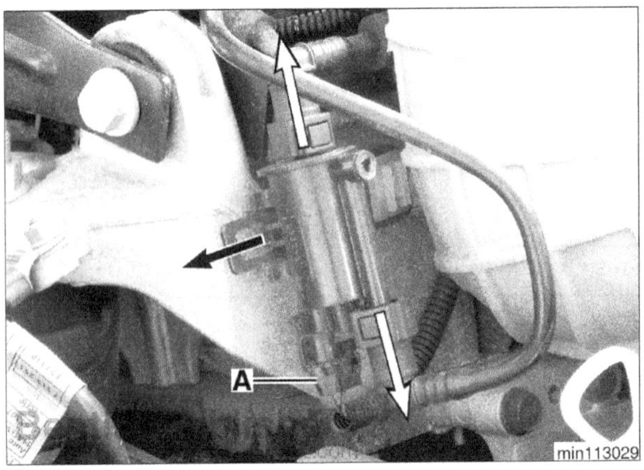

< Remove fuel tank vent valve.
  - Disconnect vent hoses (**white arrows**) via quick fit connectors.
  - Straighten retaining tab and slide vent valve off bracket (**black arrow**).
  - Remove electrical harness connector (**A**).

NOTE—
Cooper S engine is illustrated. Cooper engine layout is similar.

– Remove valve cover. See **Valve cover, removing and installing (Cooper)** in this repair group.

– Working in back of cylinder head, unbolt exhaust manifold from cylinder head:
  - Working underneath car, detach oxygen sensor harness connectors.
  - Detach exhaust system from exhaust manifold. See **180 Exhaust System**.

– Remove spark plugs.

– Remove dipstick.

– Remove intake manifold. See **Intake manifold, removing and installing (Cooper)** in this repair group.
  - Remove air filter housing.
  - Remove throttle assembly.
  - Detach brake booster vacuum line.
  - Remove fuel rail together with fuel injectors from intake manifold.
  - Unclip fuel line from engine vibration damper bracket.
  - Carefully fold fuel hose to right and rear of engine. Protect fuel rail and injectors by storing in plastic bag.

WARNING—
- *If disconnecting fuel line, fuel will be expelled under pressure. Loosen fuel filler cap to release fuel tank pressure.*
- *Do not smoke or work near heaters or other fire hazards.*
- *Keep a fire extinguisher handy.*
- *Before disconnecting fuel hose, wrap a cloth around hose to absorb any leaking fuel.*
- *Plug all open fuel connections.*

CAUTION—
Do not allow fuel to drip on alternator.

NOTE—
- *If the procedure for removing the fuel rail in* **Intake manifold, removing and installing (Cooper)** *is followed, there is no need to disconnect the fuel hose from the fuel rail.*
- *If it becomes necessary to disconnect the fuel hose from the fuel rail, follow the procedure in* **130 Fuel Injection**. *Be sure to read the warnings and cautions regarding working with fuel.*

– Disconnect coolant line from radiator filler neck to coolant expansion tank. This allows engine wiring harness to be routed around thermostat housing.

– Working to left of engine:
  - Detach oxygen sensor harness connector.
  - Remove oxygen sensor harness connector mounting bracket from cylinder head.
  - Remove coolant distributor pipe mounting screw.
  - Detach coolant temperature sensor harness connector.

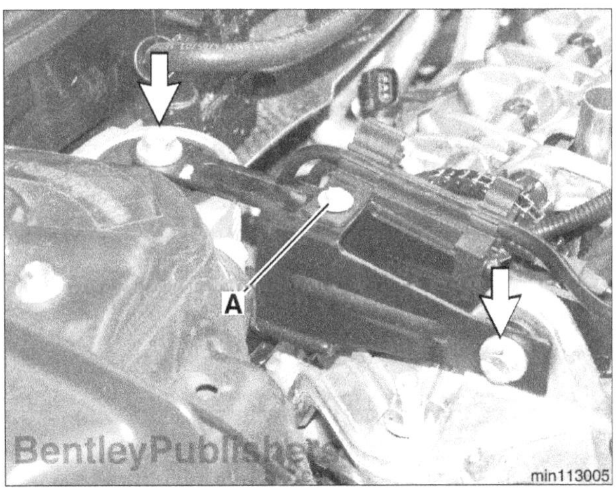

◄ Remove engine vibration damper bracket:
  - Remove fuel line bracket mounting bolt (**A**). Detach any fuel lines still attached. Lift off bracket and set aside.
  - Remove engine vibration damper bracket bolts (**arrows**).
  - Remove bracket.

– Support engine under oil pan.

CAUTION—
To avoid damaging the oil pan, use a rubber pad on top of the engine supporting jack.

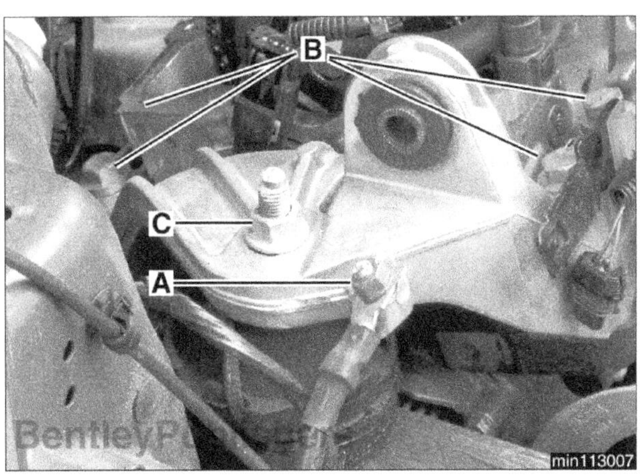

◄ Remove engine mounting bracket:
  - Remove engine ground cable mounting nut (**A**) from bracket. Detach ground cable.
  - Remove 4 engine mounting bolts (**B**) from engine block.
  - Unscrew mounting nut (**C**) from hydraulic front engine mount stud. Lift off engine mount bracket.

– Use special tool 11 8 380 to remove hydraulic engine mount. Counterhold Torx head mounting bolt from below.

Cylinder Head Removal and Installation (Cooper)

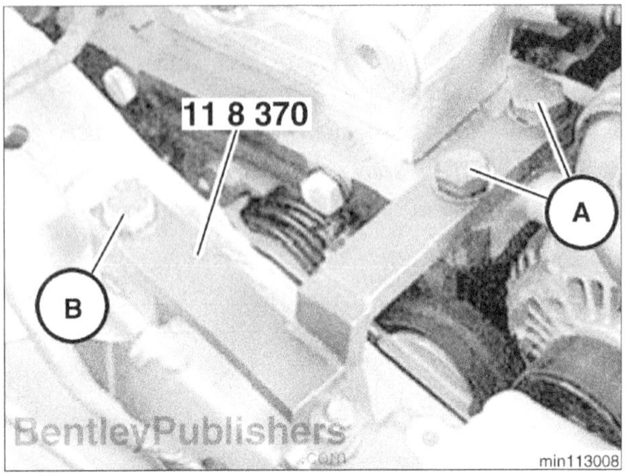

Use BMW special tool 11 8 370 to support engine.

- Attach tool to cylinder head using engine mount bracket bolts (**A**).
- Use bolt (**B**) or long stud to secure tool to frame extension.
- Remove jack from underneath engine oil pan.

Working at right end of cylinder head:

- Disconnect electrical harness connector (**A**) at camshaft position sensor.
- Remove sensor mounting bolt (**B**).
- Remove sensor from cylinder head.
- Remove threaded plugs (**C**) from cylinder head.

Rotate crankshaft until triangular adjustment mark on camshaft sprocket (**dashed line**) is at 12 o'clock.

> ***CAUTION—***
> - *For ease of assembly, mark timing chain at triangular adjustment mark (**B**) with paint.*
> - *Mark crankshaft vibration damper and timing case cover with paint.*

NOTE—
- *Because the engine is tilted slightly toward the back of the car, 12 o'clock on the camshaft sprocket and crankshaft vibration damper is NOT straight up and down.*
- *The brass-colored timing chain links are of no importance to the timing.*

Attach BMW special tool 11 8 250 to camshaft sprocket. Loosen but do not remove camshaft sprocket bolt (**A**).

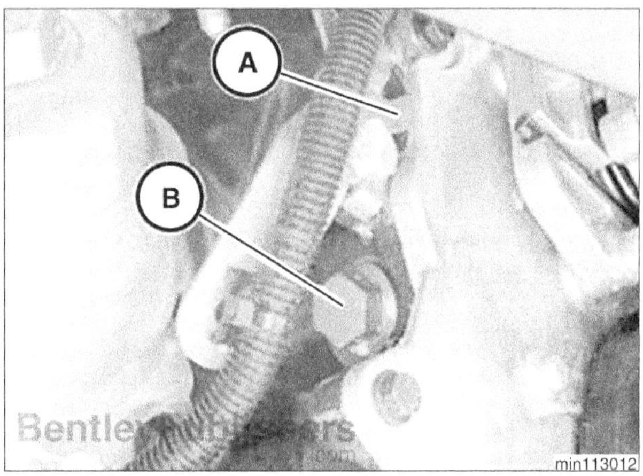

Working in back of engine block, remove wiring harness bracket (**A**).

– Loosen and remove timing chain tensioner plug (**B**). Remove tensioner plunger from inside timing chain housing cover.

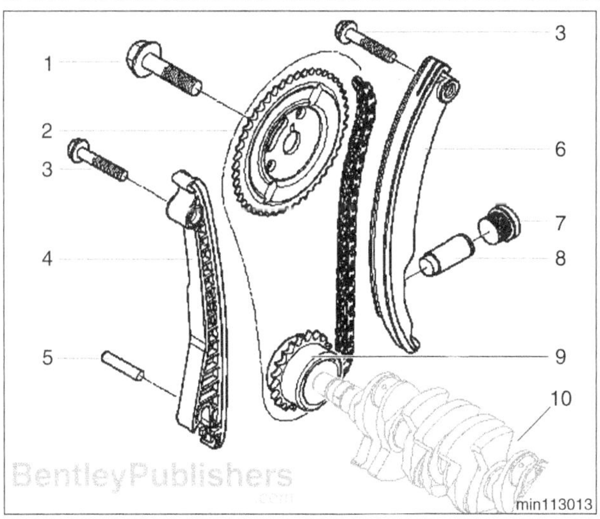

Timing chain components:

1. Bolt, M12
   -tighten to 102 Nm (75 ft-lb)
2. Camshaft sprocket
3. Bolt, M8
   -tighten to 28 Nm (21 ft lb)
4. Timing chain fixed guide
5. Stud
6. Timing chain tensioning guide
7. Timing chain tensioner plug with sealing O-ring
8. Timing chain tensioner plunger
9. Crankshaft sprocket
10. Crankshaft

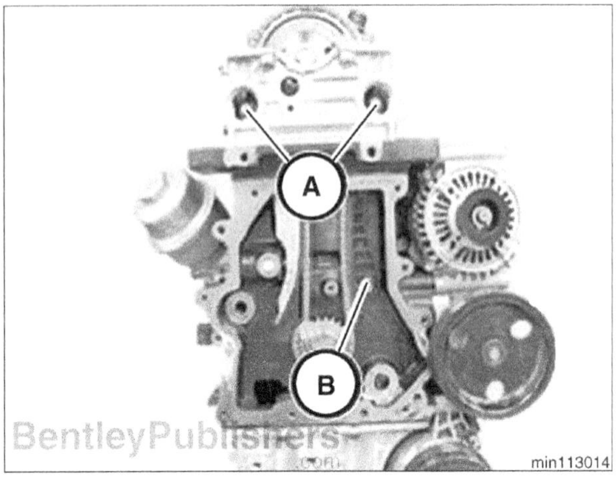

Detach timing chain from camshaft:

- Unscrew camshaft sprocket bolt.
- Remove camshaft sprocket from timing chain and secure timing chain to prevent it from dropping down.
- Remove timing chain fixed and tensioning guide mounting bolts (**A**).
- Pull timing chain guides (**B**) up out of timing chain housing.

> ***CAUTION—***
> *Do not rotate crankshaft. The timing chain cover is designed in such a way that the timing chain can remain on the crankshaft gear without gear teeth being skipped, as long as the crankshaft remains stationary.*

NOTE—
Timing chain housing cover does not need to be removed for this procedure. Illustration shows arrangement of chain guides inside housing.

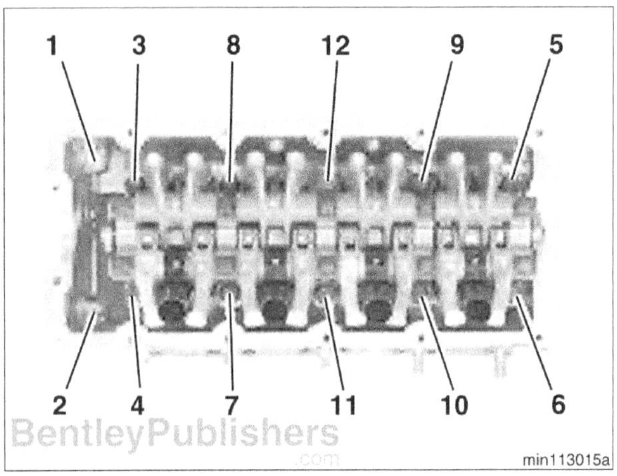

Remove cylinder head to timing chain housing mounting bolts **1** and **2**.

– Loosen and remove cylinder head bolts **3** to **12**, working from outside to inside. Discard bolts.

– Lift cylinder head off engine block.

> ***CAUTION—***
> *When placing cylinder head on work bench, do not rest on sealing surface. There is risk of damage to valves.*

– Clean cylinder head and engine block mating surfaces.

> ***CAUTION—***
> - *Multilayer steel cylinder head gasket (MLS) requires scratch-free sealing surfaces.*
> - *Do not use a metal scraper or wire brush to clean the aluminum cylinder head or pistons. If necessary, use a hard wooden or plastic scraper. Also available are abrasive discs to be used in conjunction with an electric drill. Be sure to use the correct disc for the type of metal being cleaned.*

– Use straight edge to check diagonal and horizontal surface evenness of cylinder head.

Cylinder head distortion specification	
Maximum warpage allowed	0.1 mm (0.004 in)

– Clean engine block oil ducts.

– Check aligning dowel sleeves in engine block for damage. Replace if necessary.

– Thoroughly clean oil and coolant out of cylinder head bolt tapped holes in engine block. If necessary, clean out holes with compressed air.

– Replace cylinder head gasket.

NOTE —
- *The standard thickness of the cylinder head gasket is 0.65 mm (0.026 in). A thicker one of 0.95 mm (0.037 in) is available.*
- *The gasket does not have any markings to indicate correct orientation. This is determined by visually lining up the location dowels and oil transfer gallery.*
- *Larger holes in head gasket slip over location dowels in engine block.*

– Reinstall cylinder head. Insert new cylinder head bolts.

CAUTION —
Do not wash off coating on new bolts.

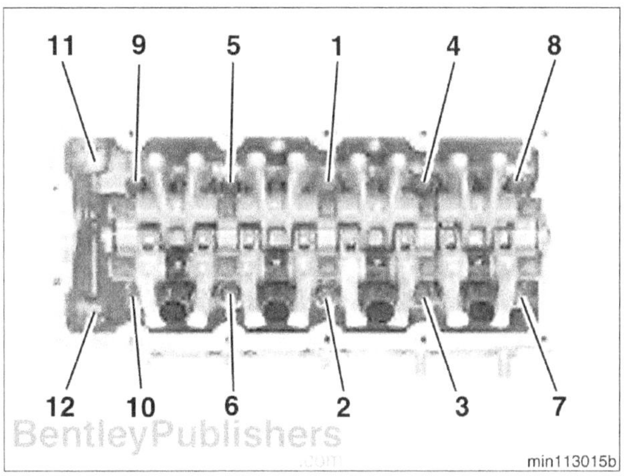

min113015b

◄ Tighten head bolts in sequence **1** to **10**, working from inside to outside.
- Head bolt are tightened in 2 stages.
- Final (stage 2) torque is applied with special tool 00 9 120 or equivalent protractor.

Tightening torque	
Cylinder head to engine block (use new)(M10): Stage 1 Stage 2	 40 Nm (30 ft-lb) rotate 90°

– Install and torque cylinder head to timing chain housing mounting bolts **11** and **12**.

Tightening torque	
Cylinder head to timing chain housing (M8)	28 Nm (21 ft-lb)

– Slide timing chain guides down into chain housing and reinstall mounting screws.

Tightening torque	
Timing chain guide to cylinder head (M8)	28 Nm (21 ft-lb)

min113010

◄ Place timing chain on camshaft sprocket. Use previously made paint mark to align sprocket and timing chain to each other.

– Place sprocket on camshaft and insert mounting bolt.

Cylinder Head Removal and Installation (Cooper)

Attach BMW special tool 11 8 250 to camshaft sprocket. Torque camshaft sprocket bolt.

> ***CAUTION—***
> *Be careful not to damage timing chain during this step.*

Tightening torque	
Camshaft sprocket to camshaft (M12)	102 Nm (75 ft-lb)

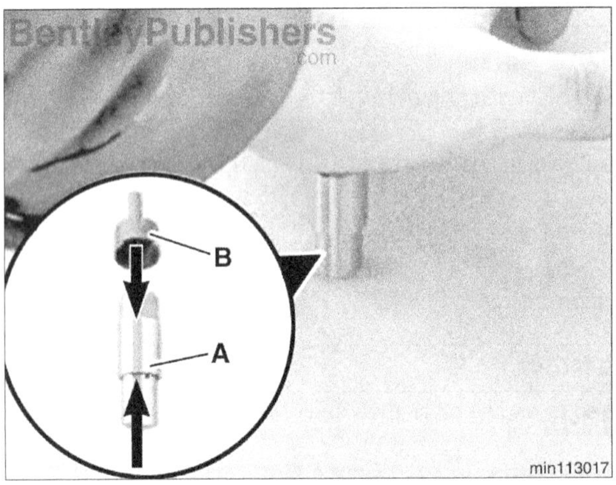

Squeeze timing chain tensioner plunger into retracted position.

- Place plunger (**A**) on level surface.
- Remove plunger cap (**B**).
- Exert continuous hand pressure on plunger until it is completely compressed.
- Replace cap.

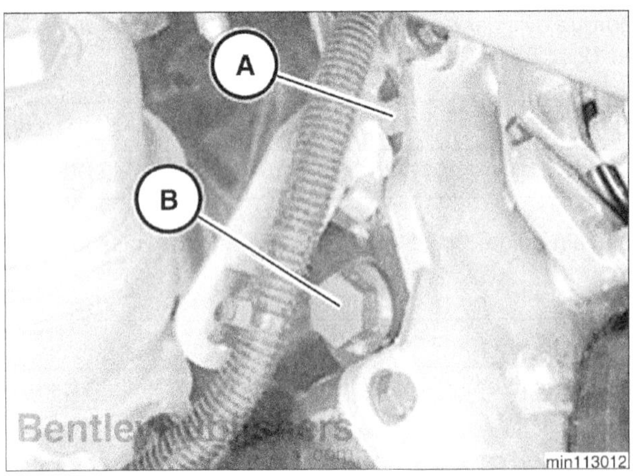

Working in back of engine block, install timing chain tensioner plunger. Install tensioner plug (**B**).

Tightening torque	
Timing chain tensioner to engine block	63 Nm (46 ft-lb)

– Install wiring harness bracket in back of engine block (**A**).

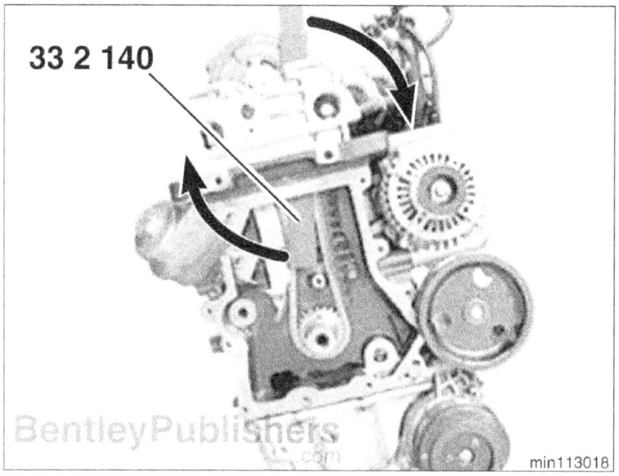

Insert large prying bar (BMW special tool 33 2 140 or equivalent) into timing chain housing and lever clockwise. This releases timing chain tensioner plunger and tensions chain.

> ***CAUTION—***
> - *Do not lever directly at timing chain. Damage may result.*
> - *Make sure that the timing chain is properly positioned within the channel of the timing chain guides.*

– Continue to reassemble engine, paying attention to the following:
  - Use new sealing washers when installing threaded plugs at front of cylinder head.
  - Use new hose clamps, seals and gaskets, as necessary.
  - Fill cooling system and check for leaks.
  - Change engine oil if necessary.
  - Check DME control module adaptations using BMW scan tool. Check for fault codes and clear and reset control module memory.

Tightening torques	
Air filter housing to body (M6)	7 - 10 Nm (5 - 7 ft-lb)
Battery box to body	8 Nm (6 ft-lb)
Battery cable (- or +) to battery (M6)	5 Nm (44 in-lb)
Battery securing strap to body (M6)	6 Nm (53 ft-lb)
Camshaft position sensor to cylinder head (M6)	9 Nm (7 ft-lb)
Engine front mounting bracket to top of engine block (M12)	100 Nm (74 ft-lb)
Engine splash shield to modular front end (M16 x 16 mm)	6 Nm (53 in-lb)
Exhaust manifold to cylinder head (M8)	24 Nm (18 ft-lb)
Hydraulic engine mount to body	100 Nm (74 ft-lb)
Hydraulic engine mount to body (replace Torx bolt): Initial torque Torque angle	 56 Nm (41 ft-lb) 90°
Hydraulic engine mount to engine mounting bracket	68 Nm (50 ft-lb)
Ignition coil pack to valve cover (M6)	12 Nm (9 ft-lb)
Intake manifold to cylinder head (M8)	26 Nm (19 ft-lb)
Sealing plug to cylinder head	18 Nm (13 ft-lb)
Spark plug to cylinder head	27 Nm (20 ft-lb)
Throttle valve assembly to intake manifold (M6)	9 Nm (7 ft-lb)
Upper engine vibration damper bracket to damper and to engine mounting bracket (M12)	100 Nm (74 ft-lb)
Valve cover to cylinder head (M6)	12 Nm (9 ft-lb)

Cylinder Head Removal and Installation (Cooper S)

Cylinder head removal and installation is a complicated job and requires special tools.

The job has been divided up into several procedures. Read the procedures before beginning the repair. If necessary, refer to **Cooper S air intake components** graphic for component locations.

> ***WARNING—***
> *Due to risk of personal injury, be sure the engine is cold before beginning any of the procedures.*

> ***CAUTION—***
> *Cover all painted surfaces before beginning the removal procedure. As an aid to installation, label all components, wires, and hoses before removing them. Do not reuse gaskets, O-rings or seals during reassembly.*

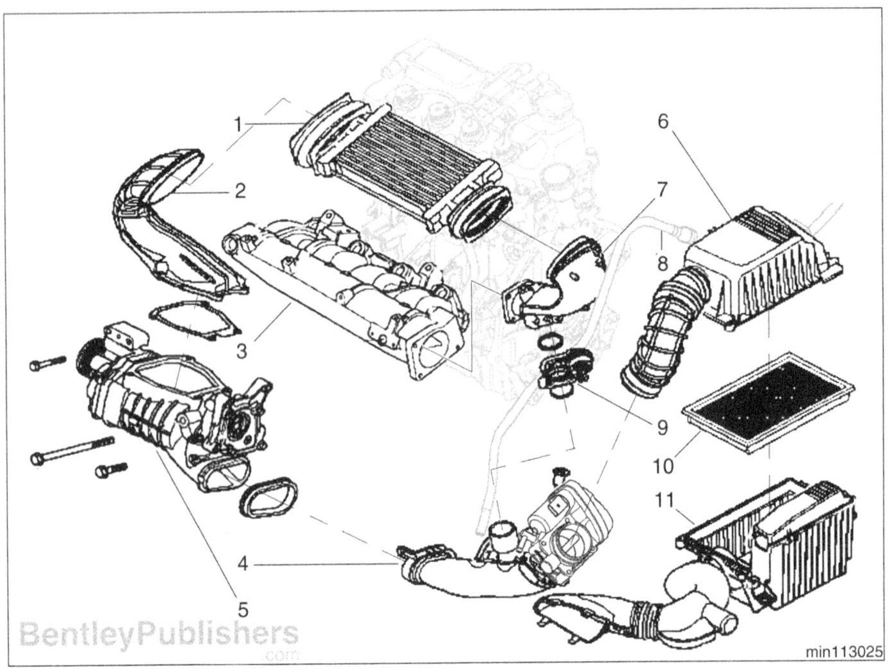

Cooper S air intake components

1. **Intercooler**
2. **Supercharger output air duct**
3. **Intake manifold**
4. **Supercharger intake air duct with throttle housing**
5. **Supercharger**
6. **Air filter cover**
7. **Intercooler output air duct**
8. **Brake booster vacuum hose**
9. **Air intake bypass valve**
10. **Air filter element**
11. **Air filter housing**

Intercooler, removing and installing

– Make sure ignition is OFF.

◄ Remove intercooler cover retaining screws (**arrows**). Remove cover.

> ***CAUTION—***
> *Intercooler fins are easily damaged. Be careful working on or around intercooler. Use intercooler protector (BMW special tool 11 8 480) when working in the engine compartment.*

◄ Remove intercooler sealing bellows clamping bolts (**A**). Remove upper clamps.

– Remove intercooler cover mounting bracket bolts (**B**). Remove brackets.

◄ Detach sealing bellows and tilt intercooler to right to remove.

– Before reinstalling intercooler, check intercooler sealing bellows for splits or cracks. Replace if necessary.

◀ Note positioning of intercooler sealing bellows clamps with bolt heads (**arrows**) pointing up.

> ***CAUTION —***
> *Make sure sealing bellows are clean, dry and correctly located. Do not lubricate.*

– Remainder of intercooler installation is reverse of removal.

Tightening torques	
Intercooler cover to intercooler	9 Nm (7 ft-lb)
Intercooler cover bracket to intercooler	9 Nm (7 ft-lb)
Upper sealing bellows clamp to lower clamp	9 Nm (7 ft-lb)

Valve cover, removing and installing (Cooper S)

– Make sure ignition is OFF.

– Remove intercooler. See **Intercooler, removing and installing** in this repair group.

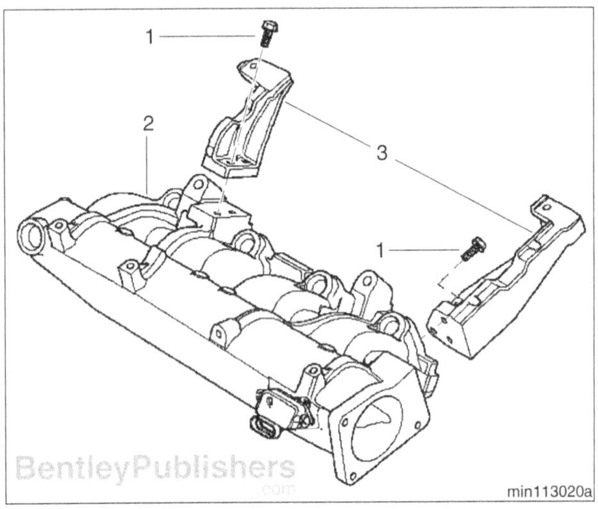

◀ Remove left and right intercooler bracket mounting bolts from intake manifold. Remove brackets.

1. Intercooler bracket mounting bolt, M6 -tighten to 9 Nm (7 ft-lb)
2. Intake manifold
3. Intercooler mounting brackets

◀ Working at top of valve cover, disconnect electrical harness connector to ignition coil pack.

- Press securing clip in direction of **white arrow**.
- Pull off coil pack connector in direction of **black arrow**.

Cylinder Head Removal and Installation (Cooper S)

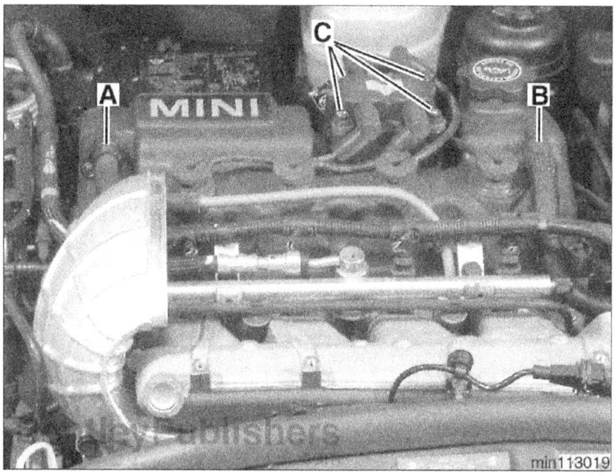

Working at top of engine:

- Disconnect crankcase ventilation hoses (**A**, **B**).
- Remove coil pack mounting bolts (**C**).

Pull up on each spark plug wire connector while gently twisting from side to side.

– Remove harness clips from valve cover studs.

– Remove coil pack and spark plug wires from valve cover.

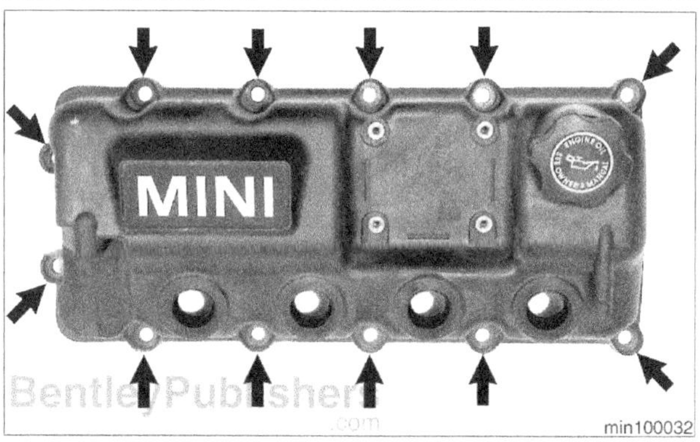

Remove 12 valve cover bolts (**arrows**).

– Remove valve cover.

– Valve cover installation is reverse of removal. Note the following:

- Replace valve cover gasket.
- Replace spark plug hole sealing grommets in valve cover.
- Clean gasket mating faces. Coat lightly with oil.
- Replace valve cover mounting bolts or insulators if damaged.
- Tighten valve cover mounting bolts in sequence, from inside to outside.
- Inspect ignition coil pack insulators. Replace if damaged.

Tightening torques	
Valve cover to cylinder head (M6)	12 Nm (9 ft-lb)
Ignition coil pack to valve cover (M6)	12 Nm (9 ft-lb)

When reinstalling ignition wires, make sure that they snap securely into place.

– Reinstall left and right intercooler brackets on intake manifold.

Tightening torque	
Intercooler bracket to intake manifold (M6)	9 Nm (7 ft-lb)

– Reinstall intercooler.

Intake manifold, removing and installing (Cooper S)

– Working in cargo compartment, lift floor trim cover. Disconnect negative (-) cable from battery. See **121 Battery, Starter, Alternator** for more information.

> ***CAUTION—***
> *Disconnecting the battery may erase fault code(s) stored in memory. Check for fault codes prior to disconnecting the battery cables.*

– Remove intercooler. See **Intercooler, removing and installing** in this repair group.

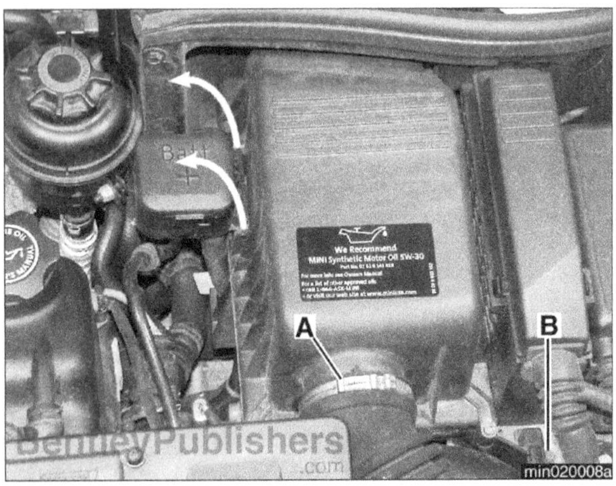

◄ Detach air filter housing from throttle housing:

- Pull up on battery positive connection point (Batt+) (**arrows**) to release from right side of air filter housing.
- Loosen or remove hose clamp (**A**) and detach outlet duct from air filter housing.
- Remove air filter housing retaining bolt (**B**).

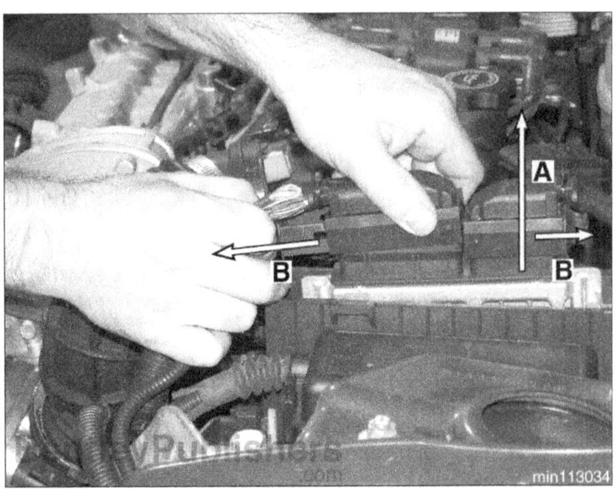

◄ Remove DME control module:

- Remove control module container cover and lift control unit upward (**A**).
- Pull connector locking sliders outward (**B**).
- Disconnect control module connectors.
- Place control module in safe storage location.

NOTE—
There are 2 control module harness connectors.

∢ Unclip air filter inlet duct from modular front end. Wriggle duct while pulling it toward front of car to detach from air filter housing.

– Remove air filter housing from engine compartment.

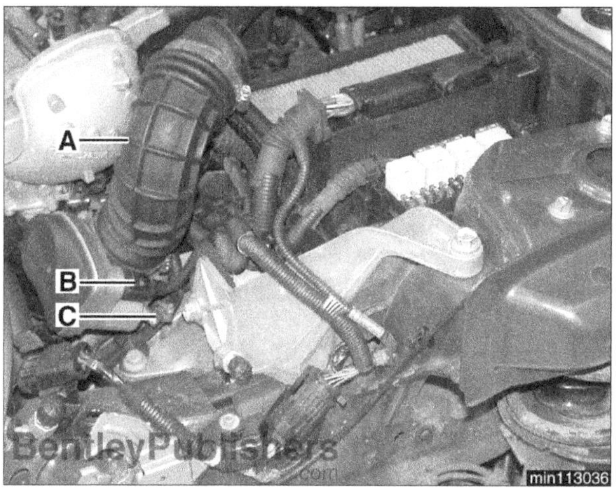

∢ Remove throttle valve assembly:

- Loosen or remove air duct clamp at throttle assembly. Detach air duct (**A**).
- Disconnect electrical harness connector (**B**).
- Detach fuel tank vent line (**B**).
- Loosen and remove throttle assembly mounting screws.
- Lift throttle assembly off supercharger intake duct.

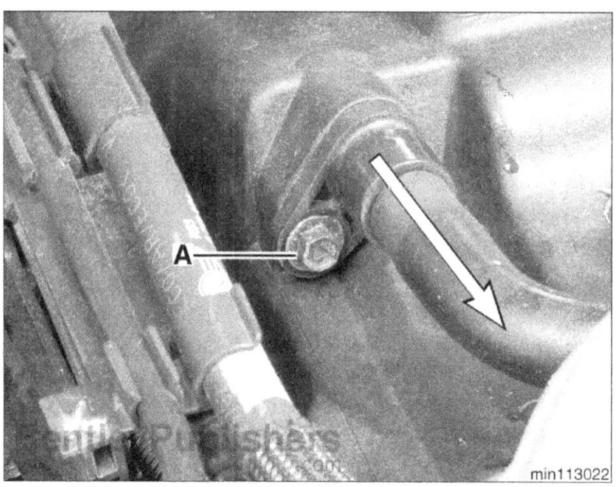

∢ Working at top right of valve cover, remove PCV valve and hose:

- Remove valve retaining screw (**arrow**).
- Pull valve out of valve cover.

– Detach crankcase ventilation hose from left end of valvo cover.

Cylinder Head Removal and Installation (Cooper S)

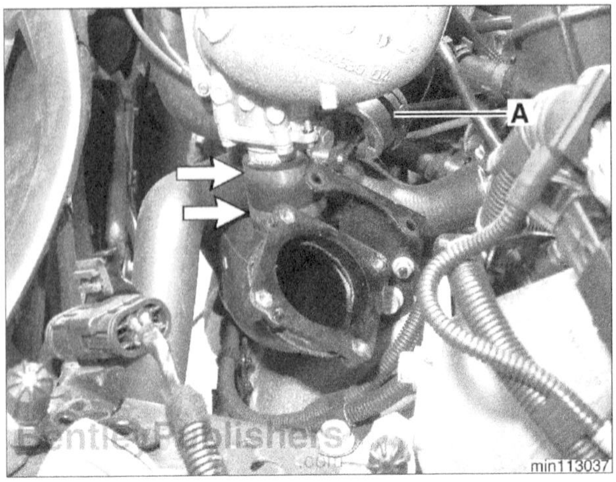

- Working below left end of intake manifold, loosen or remove hose clamps (**arrows**) from air intake bypass hose. Detach hose from air bypass valve (**A**).
- Detach knock sensor electrical harness connector from fuel rail wiring harness.
- Detach electrical harness connectors from intake manifold.

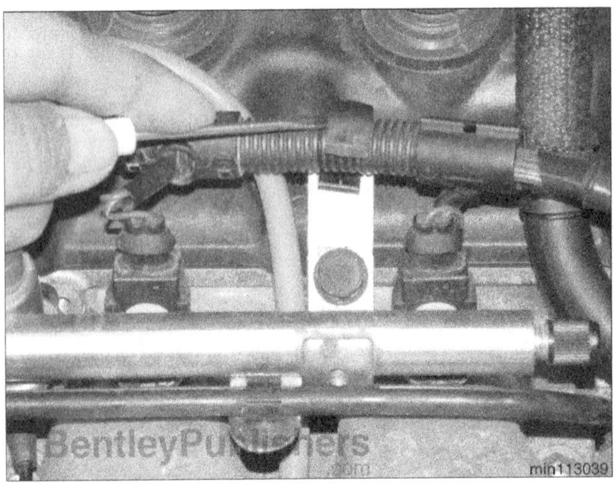

- Unhook fuel injector electrical harness from fuel rail mounting brackets.
- Disconnect top fuel tank vent line and unclip at fuel rail.

> ***WARNING —***
> - *If disconnecting fuel line, fuel will be expelled under pressure. Loosen fuel filler cap to release fuel tank pressure.*
> - *Do not smoke or work near heaters or other fire hazards.*
> - *Keep a fire extinguisher handy.*
> - *Before disconnecting fuel hose, wrap a cloth around hose to absorb any leaking fuel.*
> - *Plug all open fuel connections.*

> ***CAUTION —***
> *Do not allow fuel to drip on alternator.*

NOTE —
- *Follow the procedure below to avoid disconnecting the fuel hose from the fuel rail.*
- *If it becomes necessary to disconnect the fuel hose from the fuel rail, follow the procedure in* **130 Fuel Injection**. *Be sure to read the warnings and cautions regarding working with fuel.*

- Detach fuel rail from top of engine:
  - Disconnect fuel injector electrical harness connectors (**A**).
  - Working at fuel rail, unclip hoses and ducts. Cut wire ties as needed.
  - Disconnect vacuum line at base of fuel pressure regulator (**B**).
  - Remove fuel rail mounting bolts (**arrows**).

> ***CAUTION —***
> *Use compressed air to blow away accumulated debris at the base of each fuel injector.*

– Remove fuel rail together with fuel injectors from intake manifold.
  - Unclip fuel line from engine vibration damper bracket.
  - Carefully fold fuel hose to right and rear of engine. Protect fuel rail and injectors by storing in plastic bag.

> ***CAUTION—***
> *Plug fuel injector bores in intake manifold.*

– Detach radiator upper hose support clamp from intake manifold.

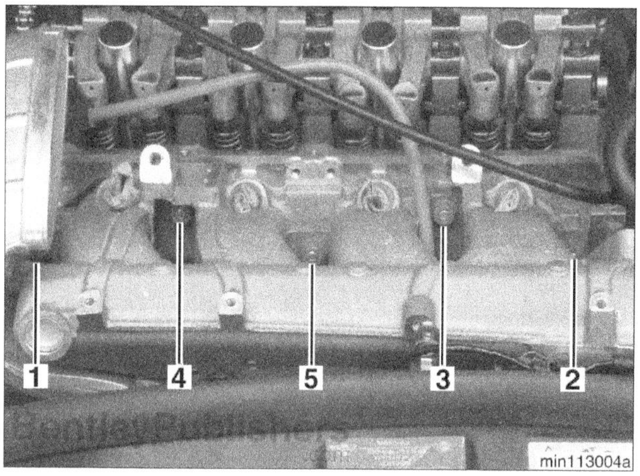

◀ Remove intake manifold mounting nuts in sequence, working from outside to inside (**1** to **5**). Remove manifold.

> ***CAUTION—***
> *Stuff clean rags in cylinder head intake ports.*

– When reinstalling intake manifold, replace intake manifold gasket.

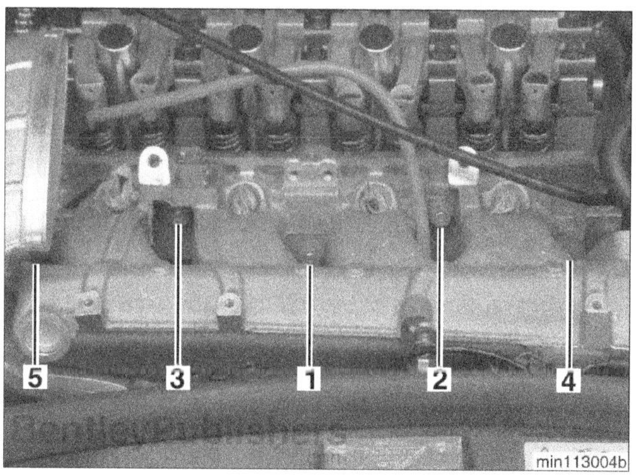

◀ Tighten down intake manifold mounting bolts in sequence, working from inside to outside (**1** - **5**).

Tightening torque	
Intake manifold to cylinder head (M8)	26 Nm (19 ft-lb)

– When reinstalling fuel rail, coat fuel injector sealing O-rings with antiseize agent.

Tightening torque	
Fuel rail to intake manifold (M8)	25 Nm (18 ft-lb)

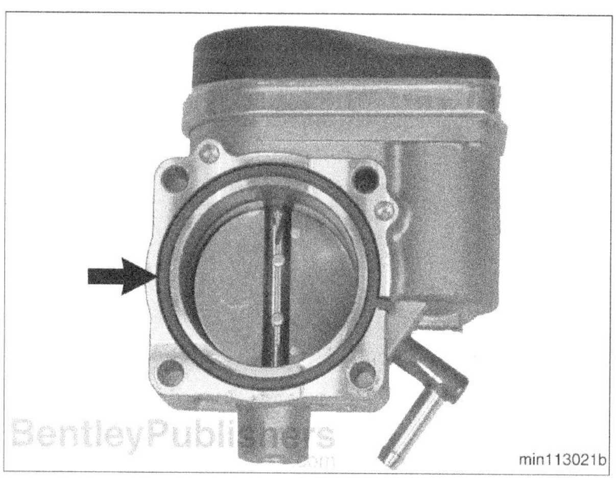

◀ When reinstalling throttle valve assembly, replace sealing O-ring (**arrow**).

◀ Make sure locating pins (**arrows**) on supercharger intake duct align with corresponding bores on throttle valve assembly. Check locating pins for damage. Replace if necessary.

Tightening torque	
Throttle valve assembly to supercharger intake duct (M6)	9 Nm (7 ft-lb)

– Reinstall air filter housing.

Tightening torque	
Air filter housing to body (M6)	7 - 10 Nm (5 - 7 ft-lb)

– Reinstall ECM and attach connectors.

– Reattach air duct using new clamps.

– Reinstall intercooler.

NOTE—
If throttle valve assembly is replaced, be sure to reset DME control unit adaptation values using BMW scan tool.

Cylinder head, removing and installing (Cooper S)

– Working in cargo compartment, lift floor trim cover. Disconnect negative (-) cable from battery. See **121 Battery, Starter, Alternator** for more information.

> ***CAUTION—***
> *Disconnecting the battery may erase fault code(s) stored in memory. Check for fault codes prior to disconnecting the battery cables.*

– Remove intercooler. See **Intercooler, removing and installing** in this repair group.

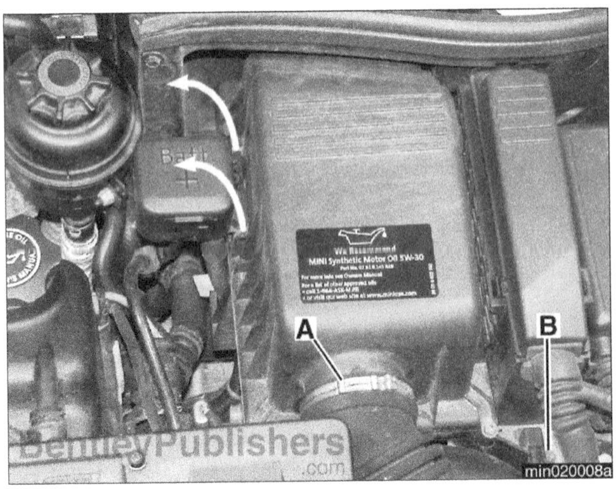

◀ Detach air filter housing from throttle housing:

- Pull up on battery positive connection point (Batt+) (**arrows**) to release from right side of air filter housing.
- Loosen or remove hose clamp (**A**) and detach outlet duct from air filter housing.
- Remove air filter housing retaining bolt (**B**).

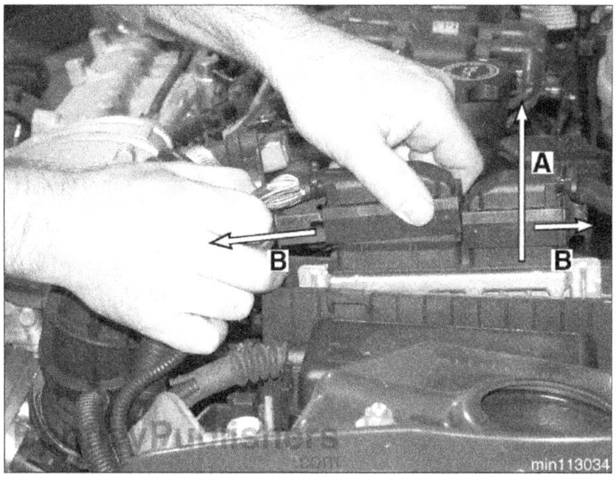

◄ Remove DME control module:

- Remove control module container cover and lift control unit upward (**A**).
- Pull connector locking sliders outward (**B**).
- Disconnect control module connectors.
- Place control module in safe storage location.

NOTE —
There are 2 control module harness connectors.

◄ Unclip air filter inlet duct from modular front end. Wriggle duct while pulling it toward front of car to detach from air filter housing.

– Remove air filter housing from engine compartment.

– Drain coolant. Begin by releasing pressure at coolant reservoir cap. See **170 Cooling System**.

◄ Loosen modular front end (MFE):

- Remove right front wheel housing liner.
- Detach left front wheel housing liner from front bumper cover trim.
- Remove front bumper cover trim.
- Remove bumper.
- Detach radiator upper hose support clamp (**A**) from intake manifold.
- Remove MFE mounting bolts. Install two 100 mm (4 in) M8 bolts (**arrow**) in left and right bumper support members. Slide MFE forward, supported on long bolts.

> ***CAUTION —***
> *For ease of component alignment when reassembling front bumper assembly, do not loosen or remove bumper alignment bosses.*

NOTE —

- *If available, use BMW special tools 11 8 401 and 11 8 402 instead of long bolts to support MFE.*
- *Removal of MFE is a complicated job. It is covered in* **510 Bumpers, External Trim**.

Cylinder Head Removal and Installation (Cooper S)

< Working below left side of engine compartment, loosen or remove lower radiator hose clamp (**arrow**). Detach hose from radiator.

– Allow coolant to drain into suitable catch pan.

NOTE —
Save coolant and reuse, or dispose of properly.

> ***WARNING —***
> *Use extreme caution when draining and disposing of engine coolant. Coolant is poisonous and lethal to humans and pets. Pets are attracted to coolant because of its sweet smell and taste. Seek medical attention immediately if coolant is ingested.*

< Loosen or remove upper radiator hose clamp (**arrow**). Detach hose from radiator.

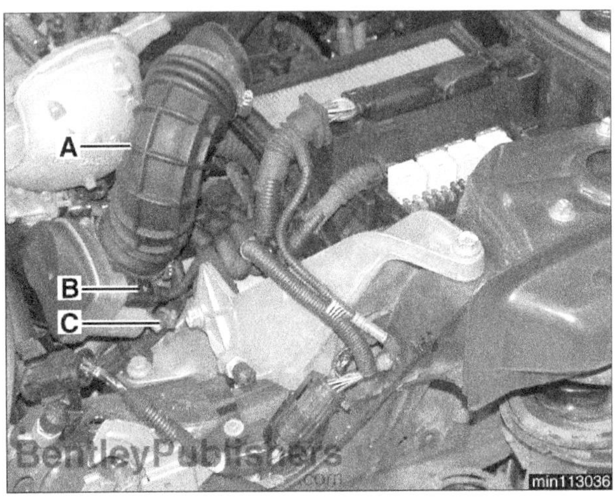

< Remove throttle valve assembly:

- Loosen or remove air duct clamp at throttle assembly. Detach air duct (**A**).
- Disconnect electrical harness connector (**B**).
- Detach fuel tank vent line (**C**).
- Loosen and remove throttle assembly mounting screws.
- Lift throttle assembly off supercharger intake duct.

– Remove valve cover. See **Valve cover, removing and installing (Cooper S)** in this repair group.

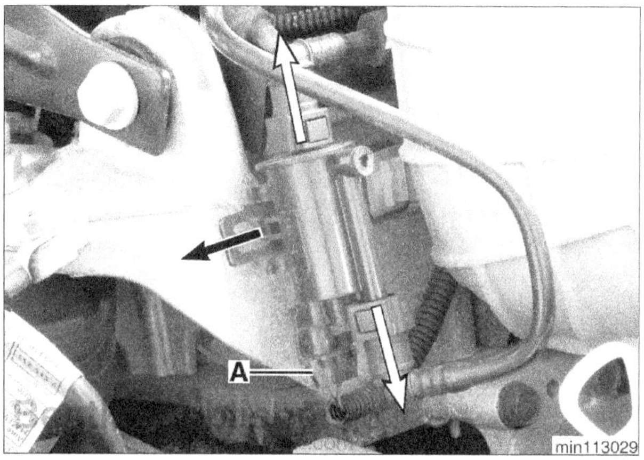

< Remove fuel tank vent valve.

- Disconnect vent hoses (**white arrows**) via quick fit connectors.
- Straighten retaining tab and slide vent valve off bracket (**black arrow**).
- Remove electrical harness connector (**A**).

– Remove intake manifold. See **Intake manifold, removing and installing (Cooper S)** in this repair group.

- Detach brake booster vacuum line.
- Remove fuel rail together with fuel injectors from intake manifold.
- Unclip fuel line from engine vibration damper bracket.
- Carefully fold fuel hose to right and rear of engine. Protect fuel rail and injectors by storing in plastic bag.
- Unbolt and remove intake manifold.

WARNING —
- *If disconnecting fuel line, fuel will be expelled under pressure. Loosen fuel filler cap to release fuel tank pressure.*
- *Do not smoke or work near heaters or other fire hazards.*
- *Keep a fire extinguisher handy.*
- *Before disconnecting fuel hose, wrap a cloth around hose to absorb any leaking fuel.*
- *Plug all open fuel connections.*

CAUTION —
Do not allow fuel to drip on alternator.

NOTE —
- *If the procedure for removing the fuel rail in* **Intake manifold, removing and installing (Cooper S)** *is followed, there is no need to disconnect the fuel hose from the fuel rail.*
- *If it becomes necessary to disconnect the fuel hose from the fuel rail, follow the procedure in* **130 Fuel Injection**. *Be sure to read the warnings and cautions regarding working with fuel.*

– Remove supercharger outlet pipe.

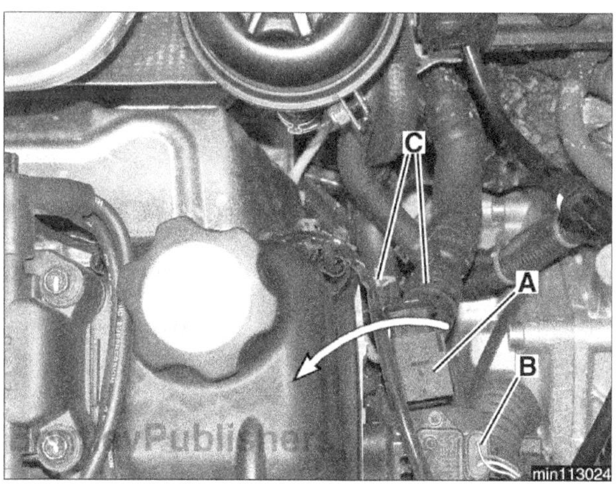

< Working at left end of cylinder head:

- Disconnect oxygen sensor connector (**A**). Twist in direction of **arrow** to detach from support bracket.
- Disconnect MAP sensor connector (**B**).
- Loosen coolant hose clamps (**C**) and detach hoses from coolant manifold.

Cylinder Head Removal and Installation (Cooper S)

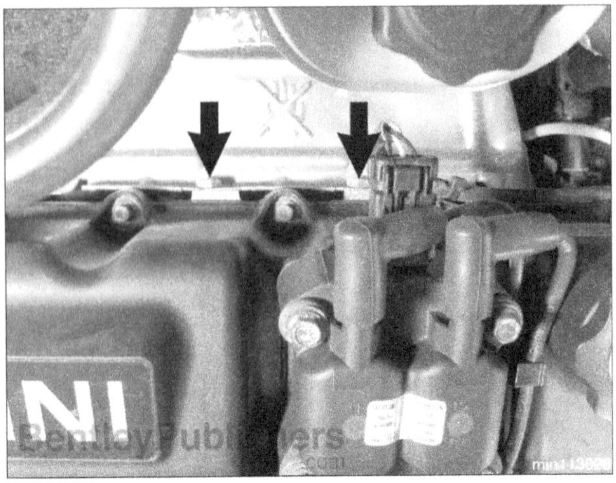

< Working in back of cylinder head, remove exhaust manifold heat shield fasteners (**arrows**).

– Unbolt exhaust manifold from cylinder head:

- Working underneath car, detach oxygen sensor harness connectors.
- Detach exhaust system from exhaust manifold. See **180 Exhaust System**.

– Remove spark plugs.

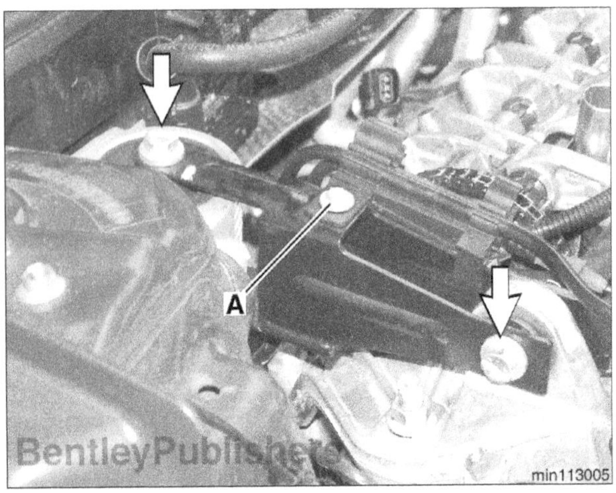

< Remove engine vibration damper bracket:

- Remove fuel line bracket mounting bolt (**A**). Detach any fuel lines still attached. Lift off bracket and set aside.
- Remove engine vibration damper bracket bolts (**arrows**).
- Remove bracket.

– Support engine under oil pan.

> ***CAUTION—***
> *To avoid damaging the oil pan, use a rubber pad on top of the engine supporting jack.*

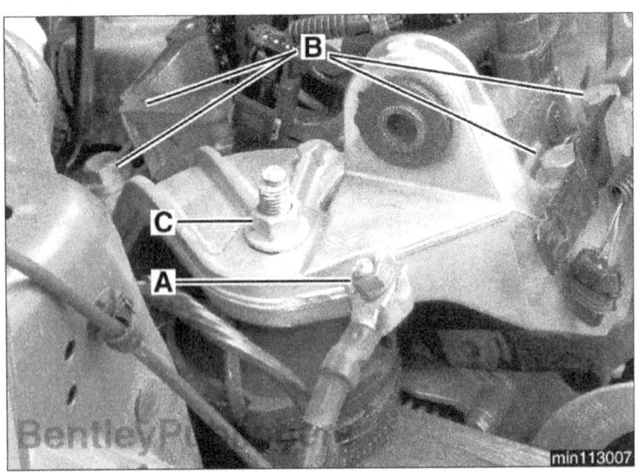

< Remove engine mounting bracket:

- Remove engine ground cable mounting nut (**A**) from bracket. Detach ground cable.
- Remove 4 engine mounting bolts (**B**) from engine block.
- Unscrew mounting nut (**C**) from hydraulic front engine mount stud. Lift off engine mount bracket.

– Use special tool 11 8 380 to remove hydraulic engine mount. Counterhold Torx head mounting bolt from below.

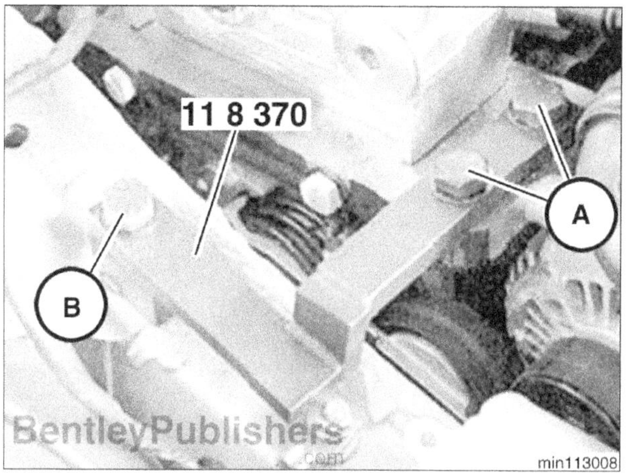

◄ Use BMW special tool 11 8 370 to support engine.

- Attach tool to cylinder head using engine mount bracket bolts (**A**).
- Use bolt (**B**) or long stud to secure tool to frame extension.

– Remove jack from underneath engine oil pan.

◄ Working at right end of cylinder head:

- Disconnect electrical harness connector (**A**) at camshaft position sensor.
- Remove sensor mounting bolt (**B**).
- Remove sensor from cylinder head.
- Remove threaded plugs (**C**) from cylinder head.

◄ Rotate crankshaft until triangular adjustment mark on camshaft sprocket (**dashed line**) is at 12 o'clock.

> ***CAUTION—***
> - *For ease of assembly, mark timing chain at triangular adjustment mark (**B**) with paint.*
> - *Mark crankshaft vibration damper and timing case cover with paint.*

NOTE—
- *Because the engine is tilted slightly toward the back of the car, 12 o'clock on the camshaft sprocket and crankshaft vibration damper is NOT straight up and down.*
- *The brass-colored timing chain links are of no importance to the timing.*

◀ Attach BMW special tool 11 8 250 to camshaft sprocket. Loosen but do not remove camshaft sprocket bolt (**A**).

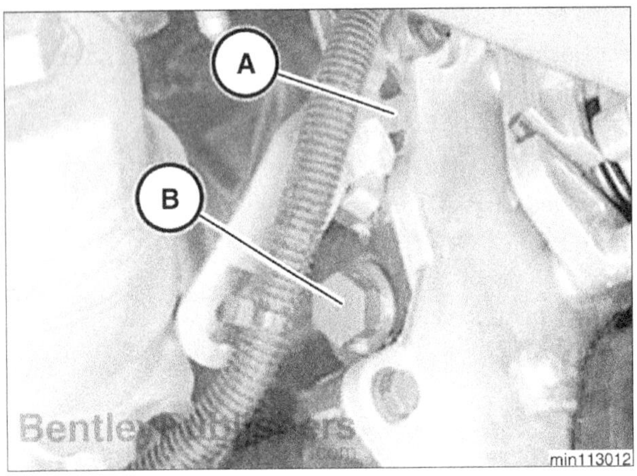

◀ Working in back of engine block, remove wiring harness bracket (**A**).

– Loosen and remove timing chain tensioner plug (**B**). Remove tensioner plunger from inside timing chain housing cover.

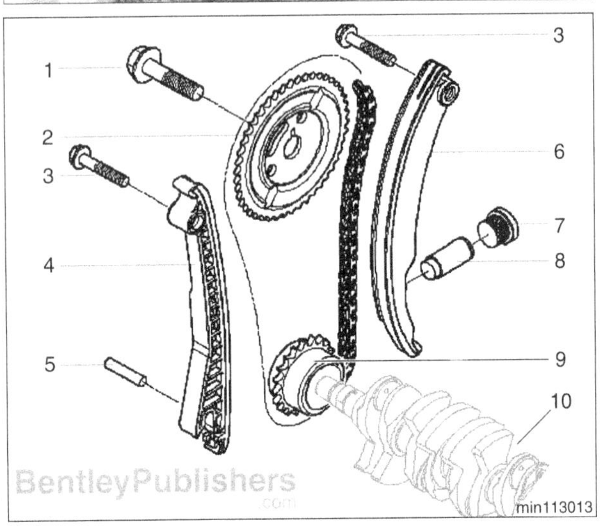

◀ Timing chain components:

1. Bolt, M12
   -tighten to 102 Nm (75 ft-lb)
2. Camshaft sprocket
3. Bolt, M8
   -tighten to 28 Nm (21 ft-lb)
4. Timing chain fixed guide
5. Stud
6. Timing chain tensioning guide
7. Timing chain tensioner plug with sealing O-ring
8. Timing chain tensioner plunger
9. Crankshaft sprocket
10. Crankshaft

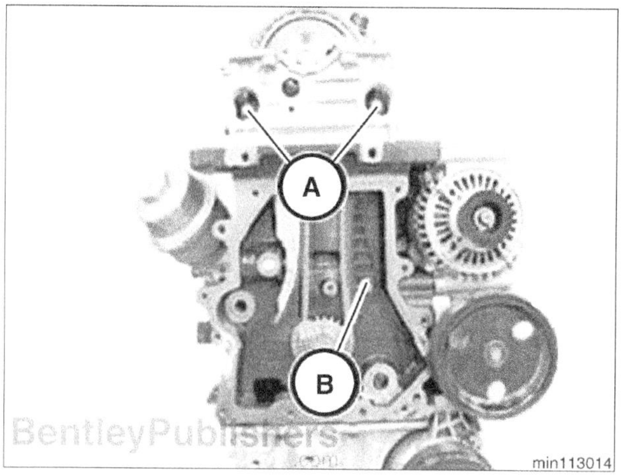

Detach timing chain from camshaft:

- Unscrew camshaft sprocket bolt.
- Remove camshaft sprocket from timing chain and secure timing chain to prevent it from dropping down.
- Remove timing chain fixed and tensioning guide mounting bolts (**A** and **B**).
- Pull timing chain guides up out of timing chain housing.

> ***CAUTION—***
> *Do not rotate crankshaft. The timing chain cover is designed in such a way that the timing chain can remain on the crankshaft gear without gear teeth being skipped, as long as the crankshaft remains stationary.*

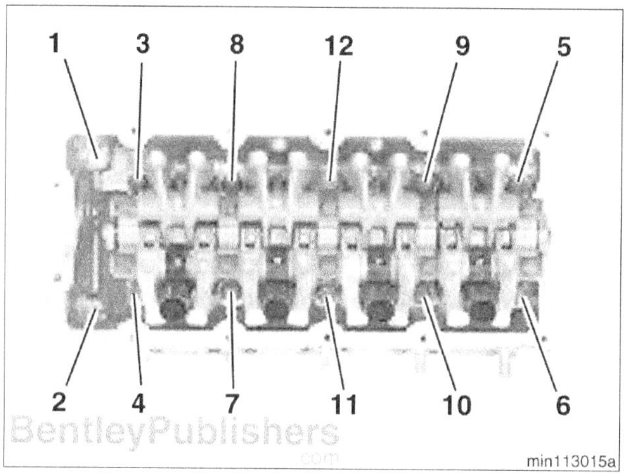

Remove cylinder head to timing chain housing mounting bolts **1** and **2**.

– Loosen and remove cylinder head bolts **3** to **12**, working from outside to inside. Discard bolts.

– Lift cylinder head off engine block.

> ***CAUTION—***
> *When placing cylinder head on work bench, do not rest on sealing surface. There is risk of damage to valves.*

– Clean cylinder head and engine block mating surfaces.

> ***CAUTION—***
> - *Multilayer steel cylinder head gasket (MLS) requires scratch-free sealing surfaces.*
> - *Do not use a metal scraper or wire brush to clean the aluminum cylinder head or pistons. If necessary, use a hard wooden or plastic scraper. Also available are abrasive discs to be used in conjunction with an electric drill. Be sure to use the correct disc for the type of metal being cleaned.*

– Use straight edge to check diagonal and horizontal surface evenness of cylinder head.

Cylinder head distortion specification	
Maximum warpage allowed	0.1 mm (0.004 in)

– Clean engine block oil ducts.

– Check aligning dowel sleeves in engine block for damage. Replace if necessary.

– Thoroughly clean oil and coolant out of cylinder head bolt tapped holes in engine block. If necessary, clean out holes with compressed air.

– Replace cylinder head gasket.

NOTE—

- *The standard thickness of the cylinder head gasket is 0.65 mm (0.026 in). A thicker one of 0.95 mm (0.037in) is available.*
- *The gasket does not have any markings to indicate correct orientation. This is determined by visually lining up the location dowels and oil transfer gallery.*
- *Larger holes in head gasket slip over location dowels in engine block.*

– Reinstall cylinder head. Insert new cylinder head bolts.

> ***CAUTION—***
> *Do not wash off coating on new bolts.*

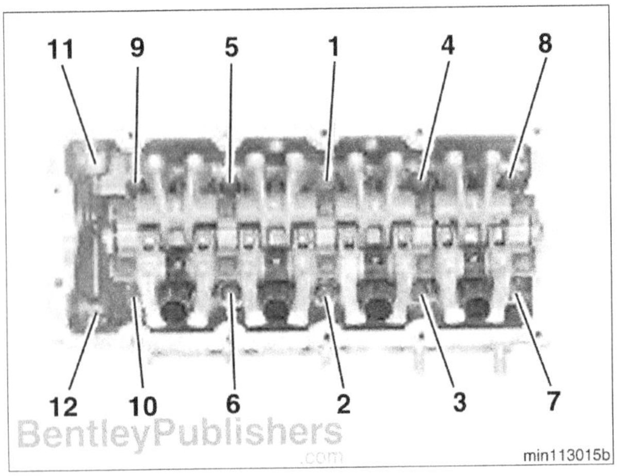

Tighten head bolts in sequence **1** to **10**, working from inside to outside.

- Head bolt are tightened in 2 stages.
- Final (stage 2) torque is applied with special tool 00 9 120 or equivalent protractor.

Tightening torque	
Cylinder head to engine block (use new)(M10): Stage 1 Stage 2	 40 Nm (30 ft-lb) rotate 90°

– Install and torque cylinder head to timing chain housing mounting bolts **11** and **12**.

Tightening torque	
Cylinder head to timing chain housing (M8)	28 Nm (21 ft-lb)

– Slide timing chain guides down into chain housing and reinstall mounting screws.

Tightening torque	
Timing chain guide to cylinder head (M8)	28 Nm (21 ft-lb)

Place timing chain on camshaft sprocket. Use previously made paint mark to align sprocket and timing chain to each other.

– Place sprocket on camshaft and insert mounting bolt.

Attach BMW special tool 11 8 250 to camshaft sprocket. Torque camshaft sprocket bolt (**A**).

> ***CAUTION—***
> *Be careful not to damage timing chain during this step.*

Tightening torque	
Camshaft sprocket to camshaft (M12)	102 Nm (75 ft-lb)

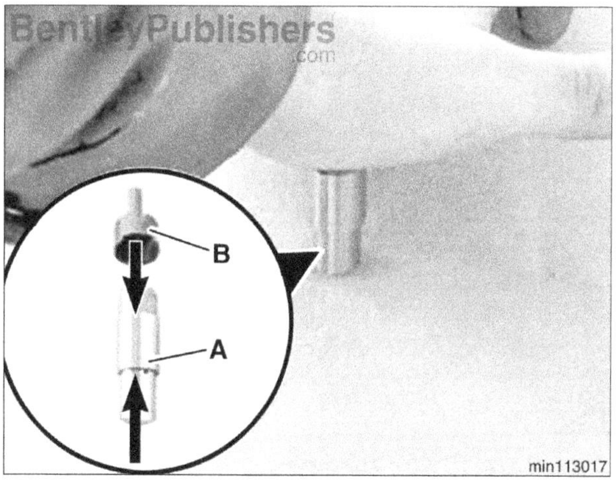

Squeeze timing chain tensioner plunger into retracted position.

- Place plunger (**A**) on level surface.
- Remove plunger cap (**B**).
- Exert continuous hand pressure on plunger until it is completely compressed.
- Replace cap.

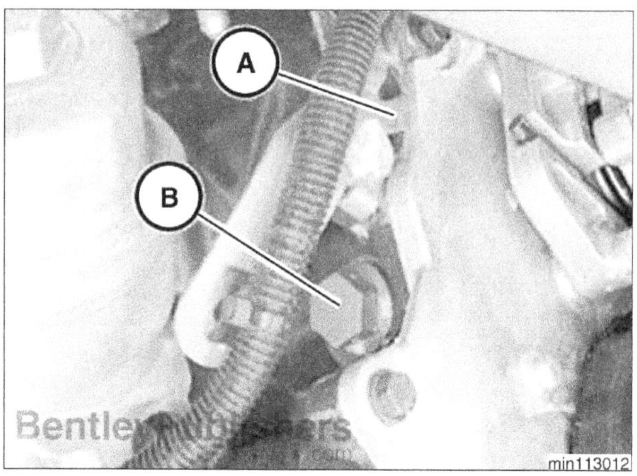

Working in back of engine block, install timing chain tensioner plunger. Install tensioner plug (**B**).

Tightening torque	
Timing chain tensioner to engine block	63 Nm (46 ft-lb)

– Install wiring harness bracket in back of engine block (**A**).

Cylinder Head Removal and Installation (Cooper S)

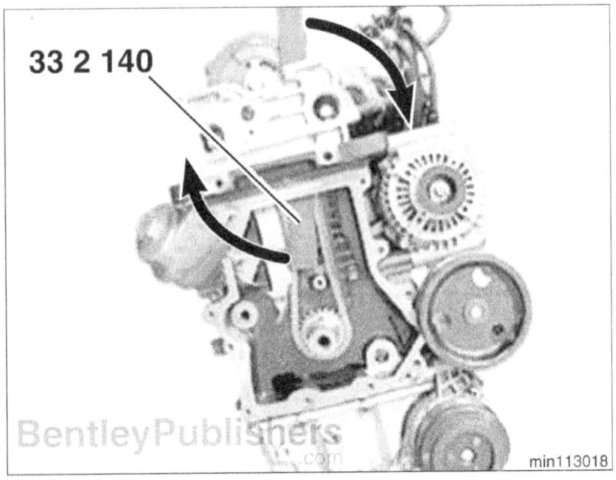

Insert large prying bar (BMW special tool 33 2 140 or equivalent) into timing chain housing and lever clockwise. This releases timing chain tensioner plunger and tensions chain.

> ***CAUTION—***
> - *Do not lever directly at timing chain. Damage may result.*
> - *Make sure that the timing chain is properly positioned within the channel of the timing chain guides.*

– Continue to reassemble engine, paying attention to the following:
  - Use new sealing washers when installing threaded plugs at front of cylinder head.
  - Use new hose clamps, seals and gaskets, as necessary.
  - Fill cooling system and check for leaks.
  - Fill engine with fresh oil.
  - Check DME control module adaptations using diagnostic scan tool. Check for fault codes and clear and reset control module memory.

Tightening torques	
Air filter housing to body (M6)	7 - 10 Nm (5 - 7 ft-lb)
Battery cable (- or +) to battery (M6)	5 Nm (44 in-lb)
Camshaft position sensor to cylinder head (M6)	9 Nm (7 ft-lb)
Crush tube to subframe (M12 x 1.5 x 85 mm)	100 Nm (74 ft-lb)
Engine front mounting bracket to top of engine block (M12)	100 Nm (74 ft-lb)
Engine splash shield to modular front end (M6 x 16 mm)	6 Nm (53 in-lb)
Exhaust manifold to cylinder head (M8)	24 Nm (18 ft-lb)
Exhaust manifold heat shield to cylinder head (M8)	13 Nm (10 ft-lb)
Front bumper to crush tube or to bumper carrier (M8 x 30 mm)	22 Nm (16 ft-lb)
Front bumper to modular front end (M6 x 16 mm)	5 Nm (44 in-lb)
Hydraulic engine mount to body	100 Nm (74 ft-lb)
Hydraulic engine mount to body (replace bolt) Initial torque Torque angle	 56 Nm (41 ft-lb) 90°
Hydraulic engine mount to engine mounting bracket	68 Nm (50 ft-lb)
Ignition coil pack to valve cover (M6)	12 Nm (9 ft-lb)
Intake manifold to cylinder head (M8)	26 Nm (19 ft-lb)
Intercooler outlet pipe to intake manifold (M7)	16 Nm (12 ft-lb)
Modular front end to crush tube (M6)	5 Nm (44 in-lb)
Sealing plug to cylinder head	18 Nm (13 ft-lb)
Spark plug to cylinder head	27 Nm (20 ft-lb)
Supercharger outlet pipe to supercharger (M8)	25 Nm (18 ft--lb)
Throttle valve assembly to intake manifold (M6)	9 Nm (7 ft-lb)
Upper engine vibration damper bracket to damper or to engine mounting bracket (M12)	100 Nm (74 ft-lb)
Valve cover to cylinder head (M6)	12 Nm (9 ft-lb)

116 Cylinder Head and Valvetrain

TABLES

GENERAL

This repair group covers cylinder head and valvetrain service and repair. The information given applies equally to Cooper and Cooper S.

NOTE—
- *For timing chain repair information, see* **117 Camshaft Timing Chain**.
- *If the cylinder head requires significant reconditioning work, a remanufactured cylinder head, available from an authorized MINI dealer, is a good alternative.*

Special tools

Special service tools are required for most cylinder head service described in this repair group. Many of these tools are expensive and only available through an authorized MINI dealer. If the special tools are not available, have the cylinder head disassembled and removed by an authorized MINI dealer.

Be sure to read each procedure thoroughly before starting a job to determine which special tools and equipment will be necessary.

Most of the repairs to a cylinder head require precision machine work to specific tolerances. This type of work should be performed by an authorized MINI repair facility or an ASE certified machinist.

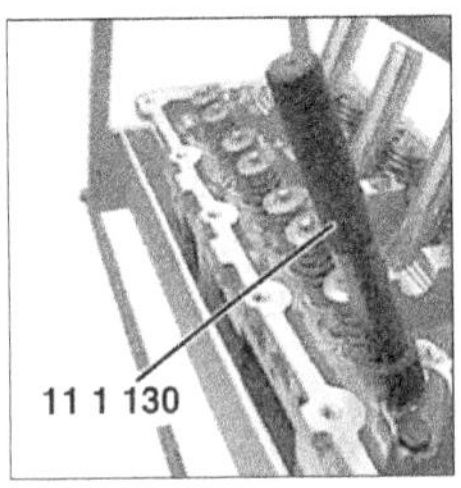

Valve seal installation drift (BMW special tool 11 1 130)

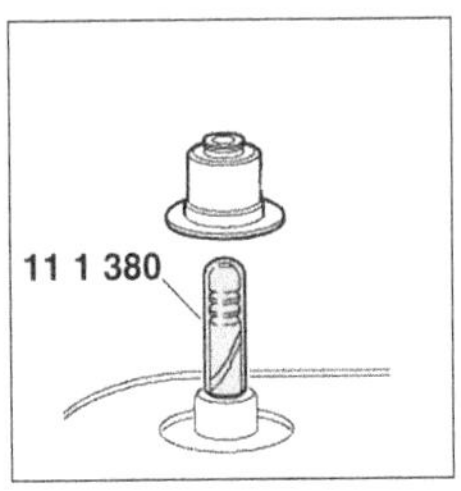

Valve seal installation sleeve (BMW special tool 11 1 380)

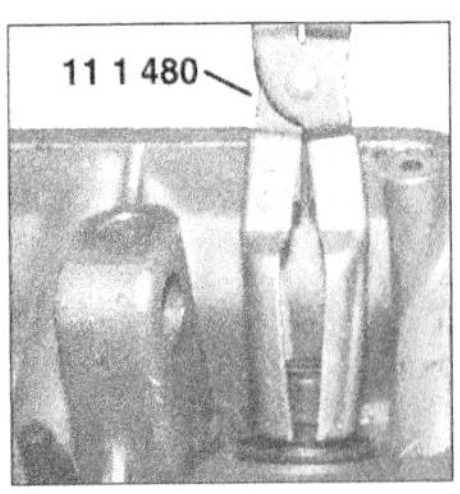

Valve seal removal tool (BMW special tool 11 1 480)

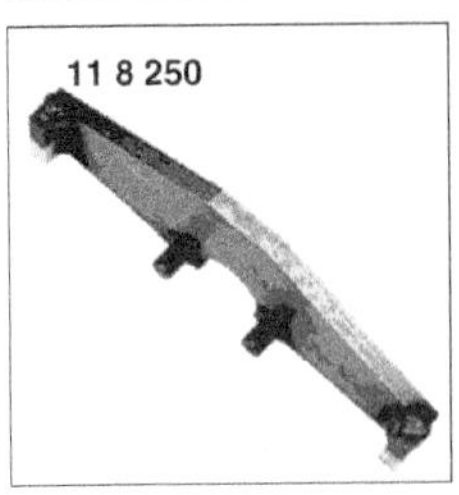

Camshaft locking tool (BMW special tool 11 8 250)

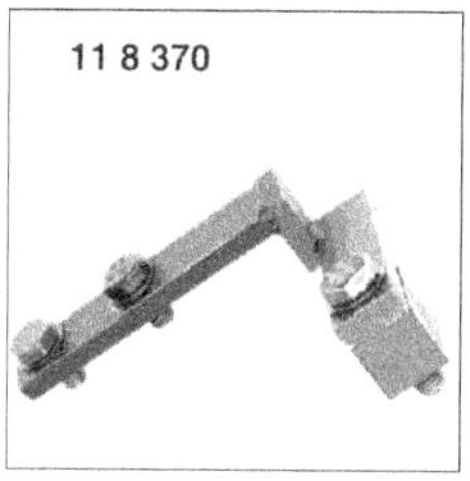

Engine block support tool
(BMW special tool 11 8 370)

Hydraulic engine mount removal and installation tool
(BMW special tool 11 8 380)

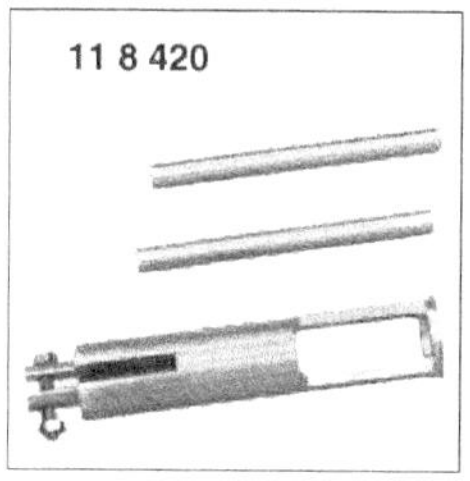

Valve removal and installation tool
(BMW special tool 11 8 420)

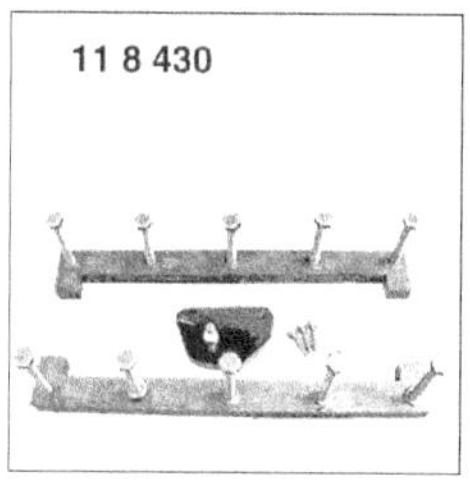

Cylinder head pressure testing jig
(BMW special tool 11 8 430)

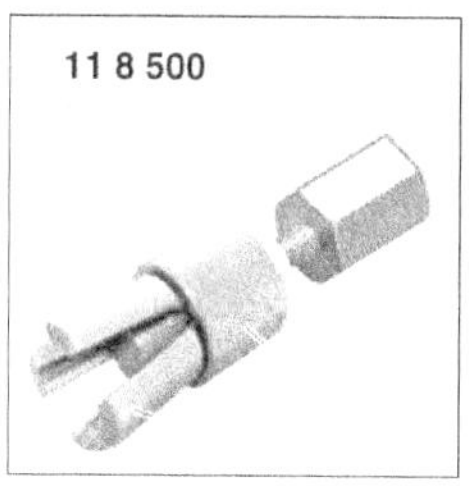

Spark plug sleeve puller
(BMW special tool 11 8 500)

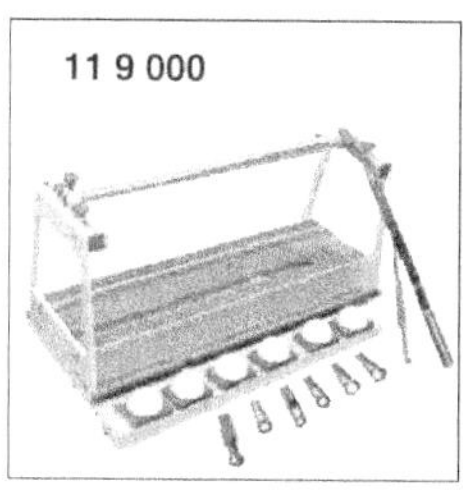

Cylinder head assembly stand
(BMW special tool 11 9 000)

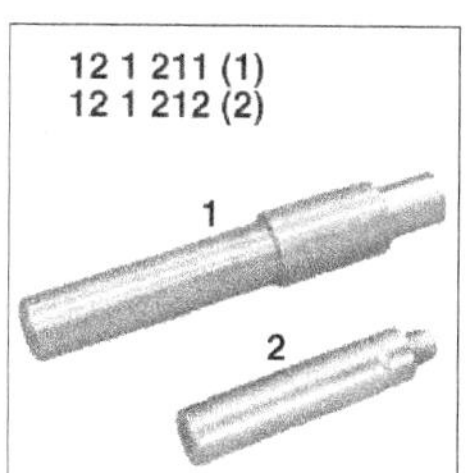

Spark plug sleeve installer
(BMW special tool 12 1 210)

Cylinder Head Disassembly

The procedures described here can be performed without removing the cylinder head from the engine.

Valves are actuated by hydraulic lifters which are integral with valve rocker arms. For information on removing rocker arms and hydraulic lifters, see **Rocker shafts and rocker arms, removing and installing** in this repair group.

> ***WARNING —***
> *Due to risk of personal injury, be sure the engine is cold before beginning any of the procedures.*

> ***CAUTION —***
> *Cover all painted surfaces before beginning the repair procedure. As an aid to installation, label all components, wires, and hoses before removing them. Do not reuse gaskets, O-rings or seals during reassembly.*

Spark plug sleeve, removing and installing

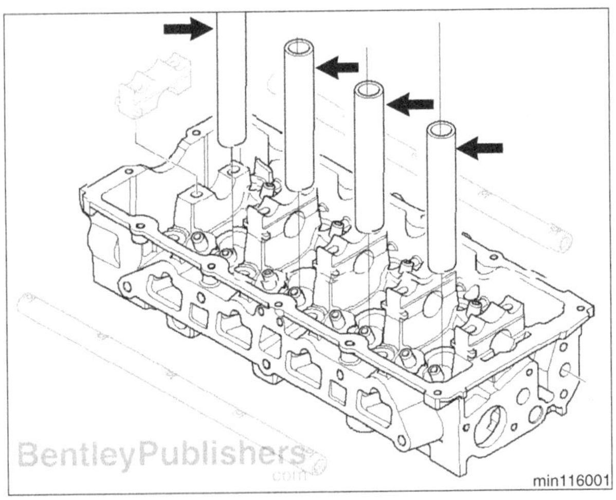

Spark plug sleeves (**arrows**) are pressed into cylinder head.

- Remove valve cover. See **113 Cylinder Head Removal and Installation**.
- Remove spark plug.
- Fit special tool 11 8 500 to spark plug sleeve to be removed and lock in place. Attach slide hammer adapter to special tool and pull out spark plug sleeve.
- If spark plug sleeve is removed, always replace with new.
- Clean area around spark plug bore.
- Apply 3 mm (1/8 in) thick bead of Loctite® 277 sealing compound at distance of about 1 mm (0.04 in) from end of new spark plug sleeve.
- Screw special tool 12 1 212 or equivalent guide into spark plug hole.
- Carefully install spark plug sleeve:
  - Fit with special tool 12 1 211 or equivalent driver.
  - Drive spark plug sleeve into cylinder head bore.

NOTE —
Sleeve is installed correctly when special tool 12 1 211 makes contact with special tool 12 1 212.

- Assemble engine.

Tightening torques	
Spark plug to cylinder head	27 Nm (20 ft-lb)
Ignition coil pack to valve cover (M6)	12 Nm (9 ft-lb)
Valve cover to cylinder head (M6)	12 Nm (9 ft-lb)

Rocker arm shafts and rocker arms, removing and installing

– Disconnect negative (-) cable from battery. See **121 Battery, Starter, Alternator** for more information.

> ***CAUTION—***
> *Disconnecting the battery may erase fault code(s) stored in memory. Check for fault codes prior to disconnecting the battery cables.*

– Remove valve cover. See **113 Cylinder Head Removal and Installation**.

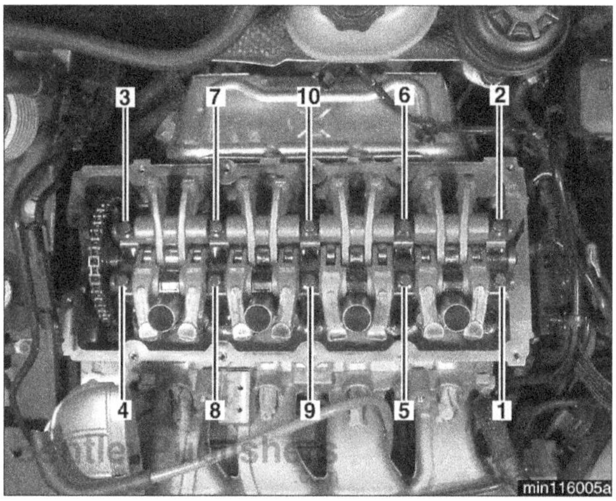

◄ Remove rocker arm shaft / camshaft bearing cap mounting bolts in sequence **1** to **10**, working from outside to inside.

> ***CAUTION—***
> *To avoid damaging the rocker arm shafts, loosen each rocker arm shaft / camshaft bearing cap bolt by one turn in the loosening sequence. Repeat until all bolts are free of tension.*

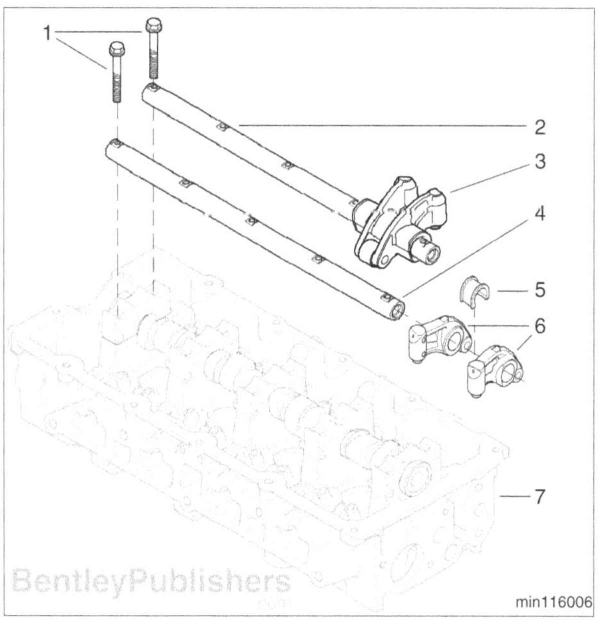

◄ Remove rocker arm shafts.

- Slide rocker arms off and replace if necessary.
- Note rocker arm spacers on intake rocker arm shaft.

NOTE—
Intake and exhaust rocker arm shafts are not interchangeable. Note location of grooves on shafts.

1. Bolt, M8 x 65 mm
   - Tighten to 30 Nm (22 ft-lb)
2. Exhaust rocker arm shaft
3. Exhaust rocker arm (dual)
4. Intake rocker arm shaft
5. Spacer
6. Intake rocker arms
7. Cylinder head and camshaft

– If hydraulic valve lifter is defective, replace together with rocker arm.

– When reassembling rocker arms and rocker arm shafts, lubricate rolling area or rocker arm and shaft with clean engine oil.

– Before tightening rocker arm shaft / camshaft bearing cap bolts make sure bearing caps are correctly seated and rocker arms are in correct installation position.

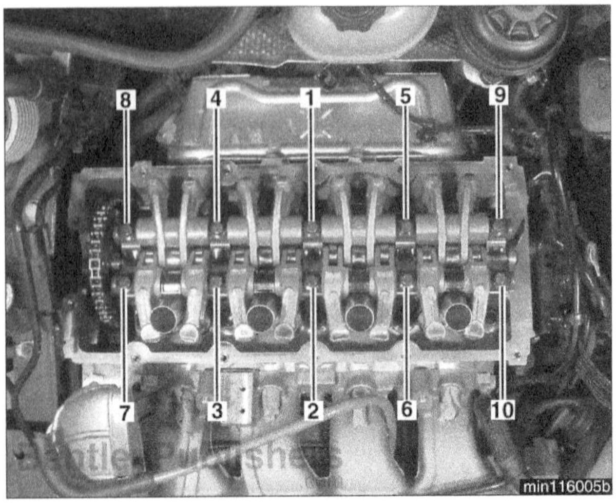

– Initially, tighten all bolts hand-tight.

◄ Torque bolts in sequence **1** to **10**, working from inside to outside.

> ***CAUTION—***
> *To avoid damaging the rocker arm shafts, tighten each rocker arm shaft / camshaft bearing cap bolt by one turn in the tightening sequence. Repeat until all bolts are torqued properly.*

Tightening torques	
Ignition coil pack to valve cover (M6)	12 Nm (9 ft-lb)
Rocker arm shaft / camshaft bearing cap to cylinder head (M8)	30 Nm (22 ft-lb)
Valve cover to cylinder head (M6)	12 Nm (9 ft-lb)

Camshaft, removing and installing

– Disconnect negative (-) cable from battery. See **121 Battery, Starter, Alternator** for more information.

> ***CAUTION—***
> *Disconnecting the battery may erase fault code(s) stored in memory. Check for fault codes prior to disconnecting the battery cables.*

– Remove spark plugs.

– Remove right front wheel housing liner.

– Remove valve cover. See **113 Cylinder Head Removal and Installation**.

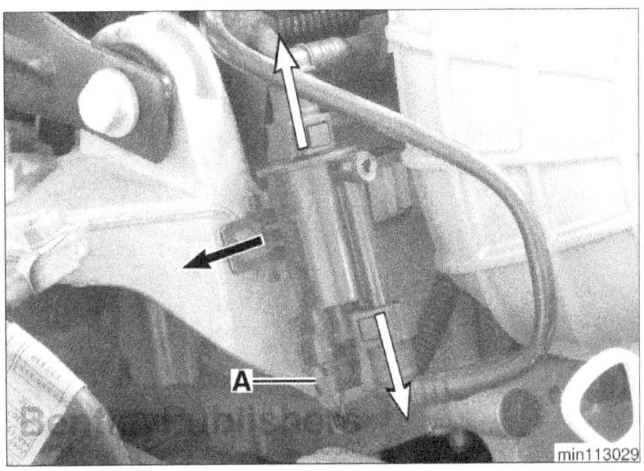

◄ Remove fuel tank vent valve.
- Disconnect vent hoses (**white arrows**) via quick fit connectors.
- Straighten retaining tab and slide vent valve off bracket (**black arrow**).
- Disconnect electrical harness connector (**A**).

NOTE—
Cooper S engine compartment is illustrated. Cooper engine compartment is similar.

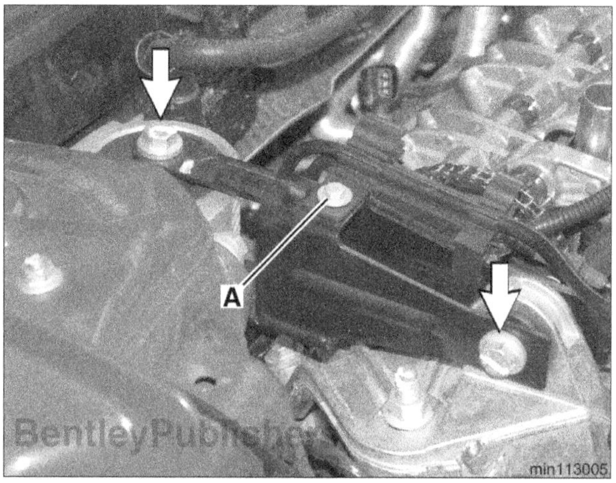

Remove vibration damper bracket:

- Remove fuel line bracket mounting bolt (**A**). Detach any fuel lines still attached. Lift off bracket and set aside.
- Remove engine vibration damper bracket bolts (**arrows**).
- Remove bracket.

– Support engine under oil pan.

> ***CAUTION—***
> *To avoid damaging the oil pan, use a rubber pad on top of the engine supporting jack.*

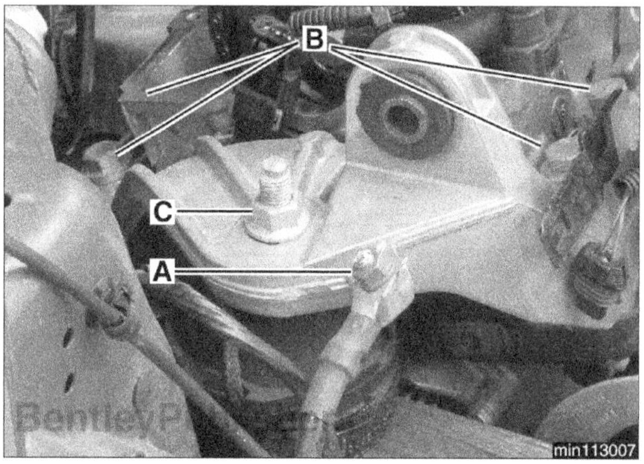

Remove engine mounting bracket:

- Remove engine ground cable mounting nut (**A**) from bracket. Detach ground cable.
- Remove 4 engine mounting bolts (**B**) from engine block.
- Unscrew mounting nut (**C**) from hydraulic front engine mount stud. Lift off engine mount bracket.

– Use special tool 11 8 380 to remove hydraulic engine mount. Counterhold Torx head mounting bolt from below.

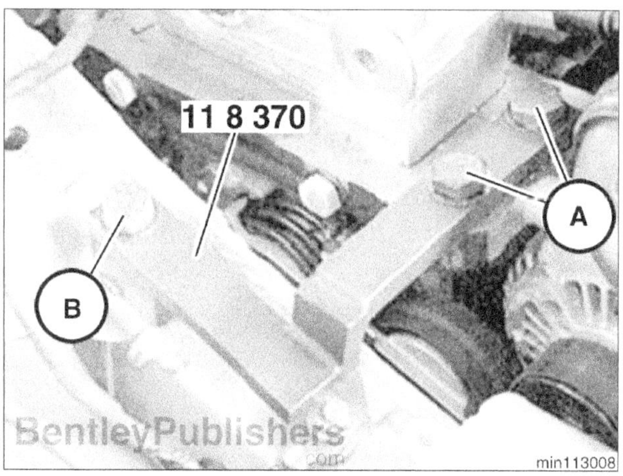

Use BMW special tool 11 8 370 to support engine.

- Attach tool to cylinder head using engine mount bracket bolts (**A**).
- Use bolt (**B**) or long stud to secure tool to frame extension.
- Remove jack from underneath engine oil pan.

Working at right end of cylinder head:

- Disconnect electrical harness connector (**A**) at camshaft position sensor.
- Remove sensor mounting bolt (**B**).
- Remove sensor from cylinder head.
- Remove threaded plugs (**C**) from cylinder head.

Rotate crankshaft until triangular adjustment mark on camshaft sprocket (**dashed line**) is at 12 o'clock.

> ***CAUTION—***
> - *For ease of assembly, mark timing chain at triangular adjustment mark (**B**) with paint.*
> - *Mark crankshaft vibration damper and timing case cover with paint.*

NOTE—
- *Because the engine is tilted slightly toward the back of the car, 12 o'clock on the camshaft sprocket and crankshaft vibration damper is NOT straight up and down.*
- *The brass-colored timing chain links are of no importance to the timing.*

Attach special tool 11 8 250 to camshaft sprocket. Loosen but do not remove camshaft sprocket bolt (**A**).

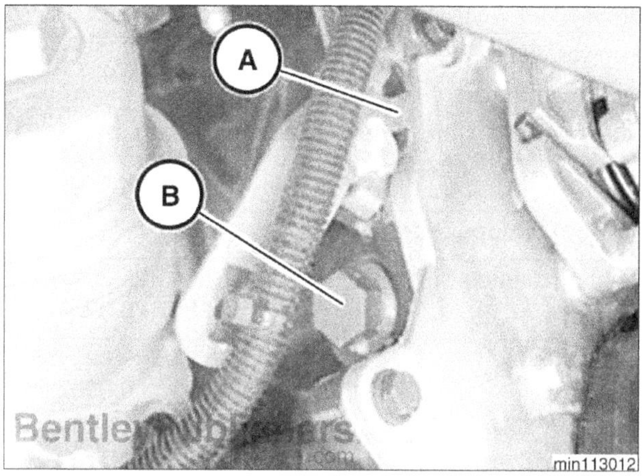

Working in back of engine block, loosen and remove timing chain tensioner plug (**B**). Remove tensioner plunger from inside timing chain housing cover.

– Detach timing chain from camshaft:
  - Unscrew camshaft sprocket bolt.
  - Remove camshaft sprocket from timing chain and secure timing chain to prevent it from dropping down.

> **CAUTION—**
> *Do not rotate crankshaft. The timing chain cover is designed in such a way that the timing chain can remain on the crankshaft gear without gear teeth being skipped, as long as the crankshaft remains stationary.*

Use feeler gauge to measure axial play between end of camshaft and cylinder head.

Camshaft axial play (Cooper engine)	
Camshaft axial play	0.05 - 0.095 mm (0.0020 - 0.00374 in)
Wear limit	0.40 mm (0.0157 in)

– Remove rocker arm shafts. See **Rocker arm shafts and rocker arms, removing and installing** in this repair group.

> **CAUTION—**
> - *Remove rocker arm shaft / camshaft bearing cap mounting bolts in sequence, working from outside to inside.*
> - *To avoid damaging the rocker arm shafts, loosen each rocker arm shaft / camshaft bearing cap bolt by one turn in the loosening sequence. Repeat until all bolts are free of tension.*

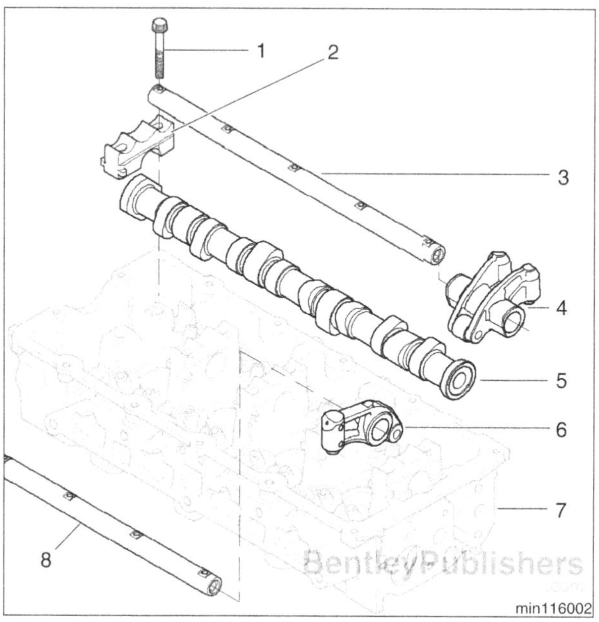

Remove camshaft bearing caps and lift out camshaft.

> **CAUTION—**
> *Note the alignment and numbering of the camshaft bearing caps. Be sure to replace each in original position and orientation.*

1. Bolt, M8 x 65 mm
   - Tighten to 30 Nm (22 ft-lb)
2. Camshaft bearing cap
3. Exhaust rocker arm shaft
4. Exhaust rocker arm (dual)
5. Camshaft
6. Intake rocker arm
7. Cylinder head
8. Intake rocker arm shaft

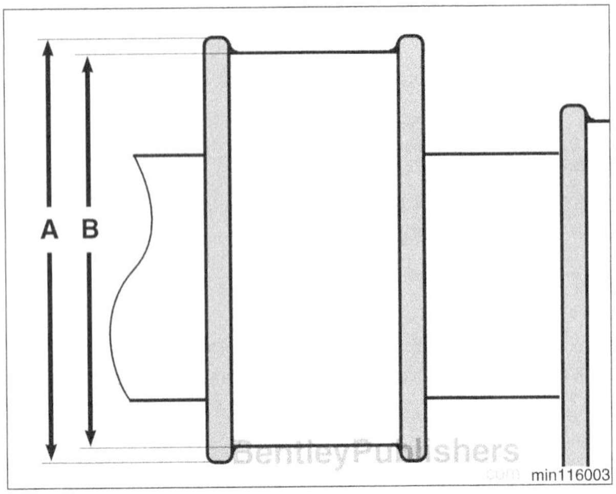

Measure camshaft for wear at widest point of each cam lobe:

- Measure unworn area at edge of cam lobe (**A**).
- Measure worn area at center of lobe (**B**).
- Replace camshaft if difference between measurements **A** and **B** exceeds limit specified in table.

Camshaft runout (Cooper engine)	
Camshaft bearing diameter	25.290 - 25.309 mm (0.9957 - 0.9964 in)
Camshaft radial runout	0.053 - 0.093 mm (0.0021 - 0.0037 in)
Wear limit	0.12 mm (0.0047 in)

> ***CAUTION —***
> *If the camshaft is replaced because of wear or damage, also replace the rocker arms.*

– When reinstalling camshaft, pay attention to the following:

- Lubricate camshaft bearing journals with clean engine oil.
- Install camshaft bearing caps in correct installation sequence and orientation.
- Lubricate rocker arm rolling area with clean engine oil.

– Initially, tighten all rocker arm shaft / camshaft bearing cap bolts hand-tight.

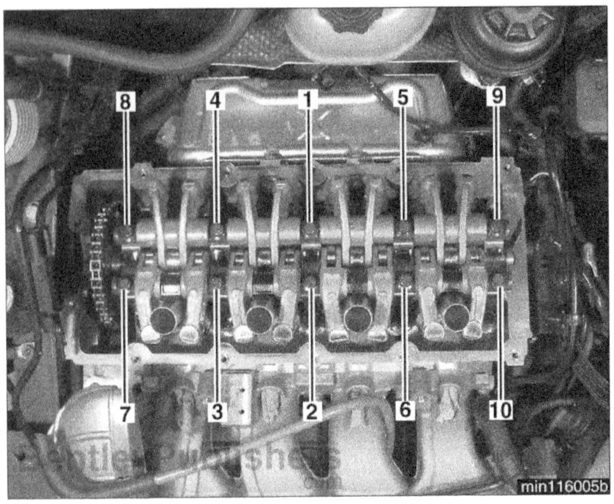

Torque bolts in sequence **1** to **10**, working from inside to outside.

> ***CAUTION —***
> *To avoid damaging the rocker arm shafts, tighten each rocker arm shaft / camshaft bearing cap bolt by one turn in the tightening sequence. Repeat until all bolts are torqued properly.*

Tightening torque	
Rocker arm shaft / camshaft bearing cap to cylinder head (M8)	30 Nm (22 ft-lb)

Place timing chain on camshaft sprocket. Use previously made paint mark to align sprocket and timing chain to each other.

– Place sprocket on camshaft and insert mounting bolt.

Attach special tool 11 8 250 to camshaft sprocket. Torque camshaft sprocket bolt.

> ***CAUTION—***
> *Be careful not to damage timing chain during this step.*

Tightening torque	
Camshaft sprocket to camshaft (M12)	102 Nm (75 ft-lb)

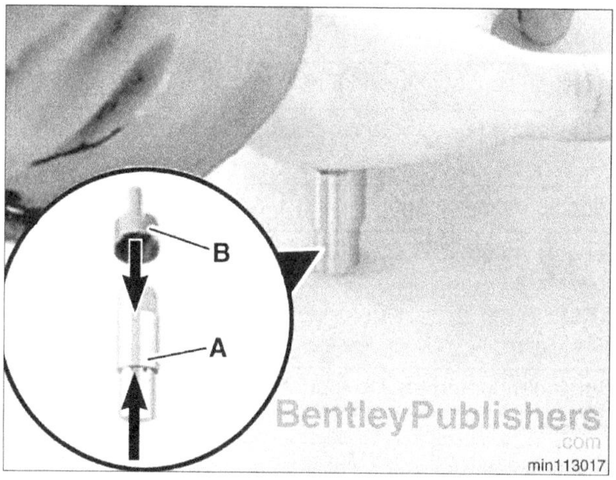

Squeeze timing chain tensioner plunger into retracted position.

- Place plunger (**A**) on level surface.
- Remove plunger cap (**B**).
- Exert continuous hand pressure on plunger until it is completely compressed.
- Replace cap.

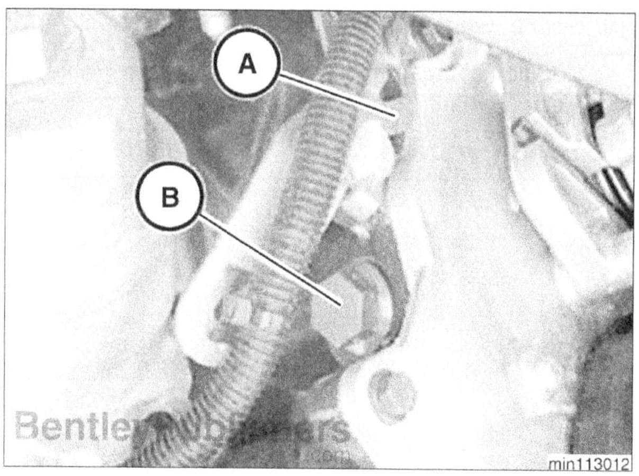

Working in back of engine block, install timing chain tensioner plunger. Install tensioner plug (**B**).

Tightening torque	
Timing chain tensioner to engine block	63 Nm (46 ft-lb)

Cylinder Head Disassembly

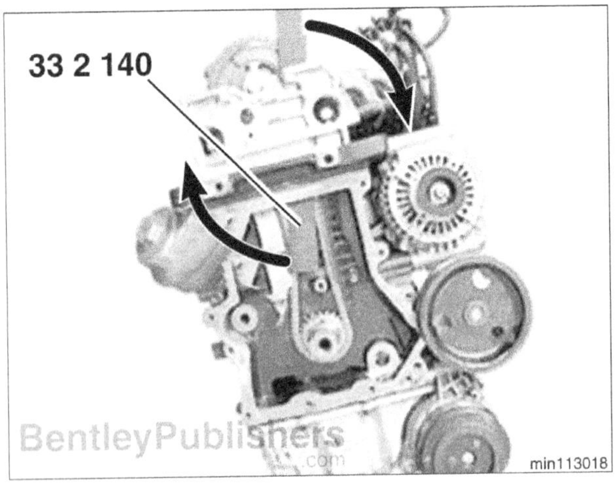

Insert large prying bar (BMW special tool 33 2 140 or equivalent) into timing chain housing and lever clockwise. This releases timing chain tensioner plunger and tensions chain.

> ***CAUTION—***
> - *Do not lever directly at timing chain. Damage may result.*
> - *Make sure that the timing chain is properly positioned within the channel of the timing chain guides.*

– Continue to reassemble engine, paying attention to the following:
- Change engine oil if necessary.
- Check for fault codes and clear and reset DME control module memory.

Tightening torques	
Camshaft position sensor to cylinder head (M6)	9 Nm (7 ft-lb)
Engine front mounting bracket to top of engine block (M12)	100 Nm (74 ft-lb)
Hydraulic engine mount to body	100 Nm (74 ft-lb)
Hydraulic engine mount to body (replace Torx bolt): Initial torque Torque angle	 56 Nm (41 ft-lb) 90°
Hydraulic engine mount to engine mounting bracket	68 Nm (50 ft-lb)
Ignition coil pack to valve cover (M6)	12 Nm (9 ft-lb)
Spark plug to cylinder head	27 Nm (20 ft-lb)
Upper engine vibration damper bracket to damper or to engine mounting bracket (M12)	100 Nm (74 ft-lb)
Valve cover to cylinder head (M6)	12 Nm (9 ft-lb)

Cylinder Head Service

The procedures described here are performed with the cylinder head off the engine and on the work bench.

Cylinder head removal and installation is described in **113 Cylinder Head Removal and Installation**.

Cylinder head, inspecting

- Remove rocker arm shafts and camshaft. See **Cylinder Head Disassembly** in this repair group.
- Check camshaft carrier bearing surfaces for warpage.
- Decarbonize and clean head before inspecting.
- Inspect disassembled and cleaned cylinder head for cracks.
  - Visually inspect cylinder head.
  - If a cracked cylinder head is suspected and no cracks are detected through visual inspection, have head further tested for cracks by authorized MINI dealer or ASE certified machinist.
  - Replace cracked cylinder head.
- Use straight edge to check diagonal and horizontal surface evenness of cylinder head.

Cylinder head distortion limits	
Maximum warpage	0.1 mm (0.004 in)

- Check valve guides and valve seats for wear before machining a warped head.
- Machine warped cylinder head.
  - Do not remove more than 0.3 mm (0.012 in.) of material.
  - If further machining is required, replace cylinder head.

NOTE—
- *Removing more than 0.3 mm (0.012 in.) will reduce the size of the combustion chamber and adversely affect engine performance.*
- *The standard thickness of the cylinder head gasket is 0.65 mm (0.026 in). A thicker one of 0.95 mm (0.037in) is available for machined heads.*

Valves, leak test

- Remove cylinder head. Disassemble rocker arm shafts and camshaft. See **Cylinder Head Disassembly** in this repair group.
- Install a spark plug in each cylinder.
- Place cylinder head on work bench with combustion chamber facing upward.
- Fill each combustion chamber with thin non-flammable liquid, such as parts cleaning fluid. After fifteen minutes, check level of fluid. If fluid level in any cylinder drops, that cylinder is not sealing properly.

Cylinder head, coolant leak test

- Remove and disassemble cylinder head. See **Cylinder Head Disassembly** in this repair group. Clean sealing surfaces.
- Place special tool 11 8 431 on sealing surface of cylinder head. Secure with special tool 11 8 433.
- Close off coolant opening with special tool 11 8 432.
- Connect compressed air supply to special tool 11 8 432. Set test pressure to 4.5 bar (65 psi).
- Immerse cylinder head in a water bath. Check for escaping bubbles.

NOTE—
If necessary, soften the bath water with a cleaning agent.

VALVES

The procedures described here are performed with the cylinder head off the engine and disassembled on the work bench.

Cylinder head removal and installation is described in **113 Cylinder Head Removal and Installation**. Also see **Cylinder Head Disassembly** in this repair group.

Valves, removing and installing

- Mount disassembled cylinder head on cylinder head assembly stand (special tool 11 9 000 or equivalent).
- Press down on valve springs with special tool 11 8 421 and remove valve spring keepers.
- Remove valve springs.
- Use special tool 11 1 480 to remove valve stem seals. Discard seals.
- Remove valves from bottom of cylinder head.

CAUTION—
Label each valve and valve spring assembly as it is removed so it can be installed in its original position.

- Valve dimension specifications are in **Table a**.
- When installing, lightly oil valve stem and valve stem seal.

NOTE—
- *Do not reuse valve stem oil seals. If valves are removed, use new valve stem oil seals.*
- *BMW special tools are available to remove the valve stem oil seals. As an alternative, standard valve seal removal tools are available from most automotive parts stores.*

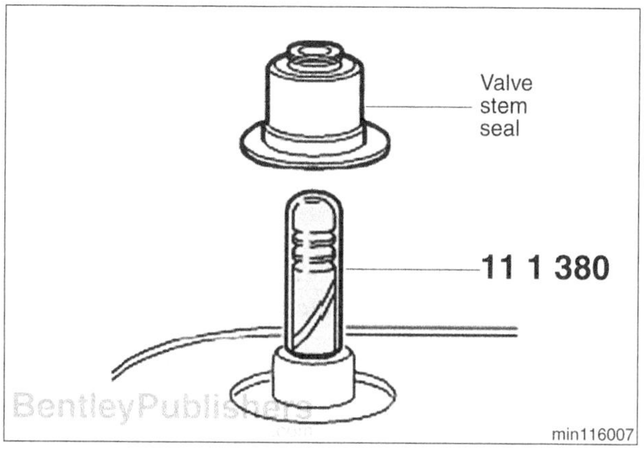

Use installation sleeve (special tool 11 1 380) over end of valve to prevent damage to valve stem seal.

- Press on valve stem seal with special installation drift 11 1 130 as far as it will go, until flush with surface of cylinder head.
- Install valve springs.

Table a. Valve specifications	
Parameter	**Specification**
Valve length intake exhaust	 108.72 - 109.22 mm (4.280 - 4.300 in) 117.57 - 118.07 mm (4.629 - 4.648 in)
Valve stem dia.	5.934 - 5.952 mm (0.2336 - 0.2343 in)

Valve stem oil seals

The purpose of the valve stem oil seal is to prevent excess oil from entering the combustion chamber. The sign of faulty valve stem oil seals is excessive oil consumption and smoke from the exhaust immediately after starting and during deceleration.

NOTE —
- *Do not reuse valve stem oil seals. If valves are removed, use new valve stem oil seals.*
- *BMW special tools are available to remove the valve stem oil seals. As an alternative, standard valve seal removal tools are available from most automotive parts stores.*

- Mount disassembled cylinder head on cylinder head assembly stand (special tool 11 9 000 or equivalent).
- Press down on valve springs with special tool 11 8 421 and remove valve keepers.
- Remove valve springs.
- Use special tool 11 1 480 to remove valve stem seals. Discard seals.
- When installing, lightly oil valve stem and valve stem seal.

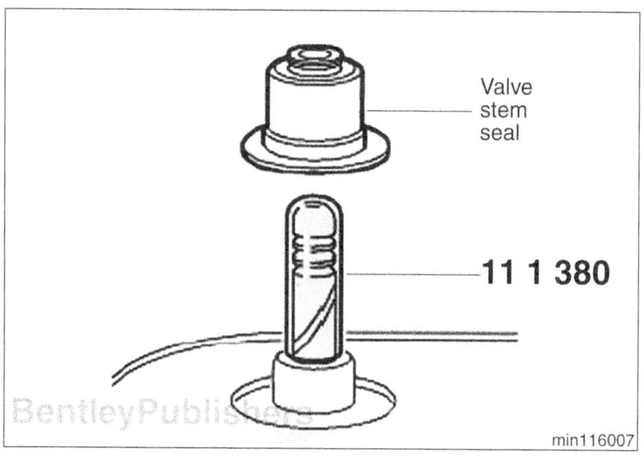

Use installation sleeve (special tool 11 1 380) over end of valve to prevent damage to valve stem seal.

- Press on valve stem seal with special installation drift 11 1 130 as far as it will go, until flush with surface of cylinder head.
- Install valve springs.

Valve guides

Check valve guides for wear using a new valve. Be sure to thoroughly inspect the cylinder head to ensure that it can be reused before reworking the guides.

Valve guide specifications are listed in **Table b**.

Table b. Valve guide specifications

Parameter	Specification
Valve guide inside diameter	5.975 - 6.000 mm (0.2352 - 0.2362 in)
Running clearance between valve and valve guide intake exhaust	 0.048 - 0.066 mm (0.0019 - 0.0026 in) 0.074 - 0.094 mm (0.0029 - 0.0037 in)
Maximum running clearance between valve and valve guide intake exhaust	 0.076 mm (0.0030 in) 0.101 mm (0.0040 in)
Valve guide wear, maximum (valve tilt clearance measured with new valve)	0.25 mm (0.0098 in)

Valve seats

– Resurface valve seats whenever new valves or valve guides are installed. Special cutters are required to resurface seats.

– Always check valves for leaks after reconditioning a valve seat as described above. **Table c** and **Table d** list valve seat dimensions.

Table c. Valve seat dimensions (Cooper)

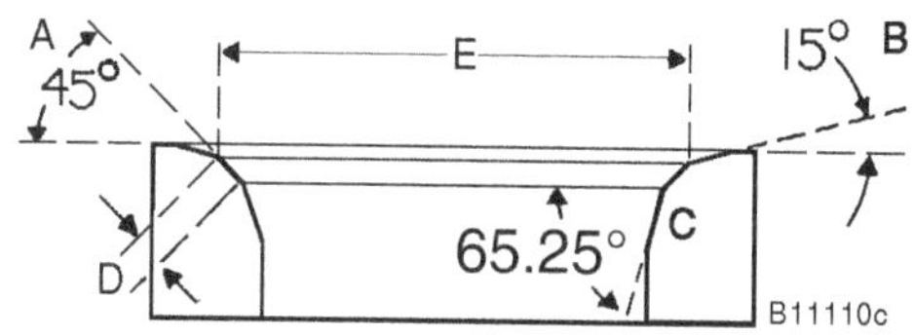

Parameter	Specification
A Valve seat angle	45°
B Correction angle, outside	15°
C Correction angle, inside	65.25°
D Valve seat width intake exhaust	 1.15 - 1.48 mm (0.045 - 0.058 in) 1.47 - 1.80 mm (0.058 - 0.071 in)
E Valve seat outside dia. intake exhaust	 30.000 mm (1.181 in) 23.000 mm (0.906 in)

Table d. Valve seat dimensions (Cooper S)

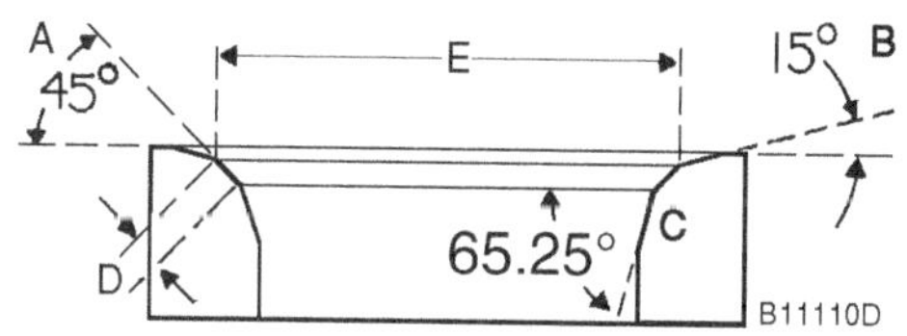

Parameter	Specification
A Valve seat angle	45°
B Correction angle, outside	15°
C Correction angle, inside (not machineable)	65.25°
D Valve seat width intake or exhaust	 0.900 - 1.300 mm (0.035 - 0.051 in)
E Valve seat outside dia. intake exhaust	 30.000 mm (1.181 in) 23.000 mm (0.906 in)

Valve springs

– Check valve springs for fatigue:
  - Line up springs in a row.
  - Place straightedge across top of springs.
  - Replace any spring that is significantly shorter than others.

117 Camshaft Timing Chain

117

GENERAL

This repair group covers timing chain repair information. Timing chain components and operation are identical for Cooper and Cooper S engines.

The timing chain is lubricated by engine oil and does not require maintenance. A worn timing chain or sprockets can lead to noisy operation and erratic valve timing. A faulty tensioner can also cause timing chain noises.

Timing chain replacement is a complicated job. The job has been divided into several procedures. Be sure to read the procedures before beginning the repairs.

NOTE —
Front crankshaft seal replacement is covered in **119 Lubrication System**.

Special tools

Special service tools are needed to remove and disassemble the timing chain. Many of these tools are expensive and only available through an authorized MINI dealer.

Be sure to read each procedure thoroughly before starting a job to determine which special tools will be necessary.

Gear puller
(BMW special tool 11 2 000)

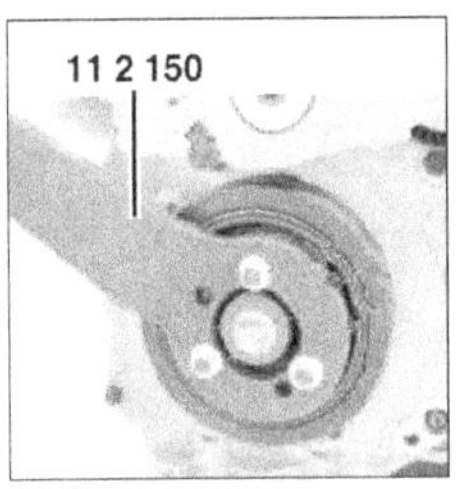

Crankshaft vibration damper holder (Cooper)
(BMW special tool 11 2 150)

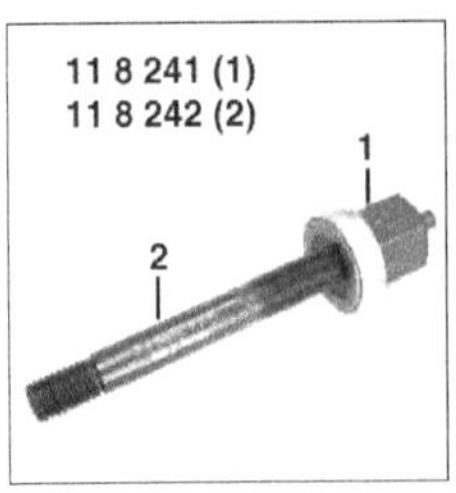

Crankshaft vibration damper installation tools
(BMW special tool 11 8 240)

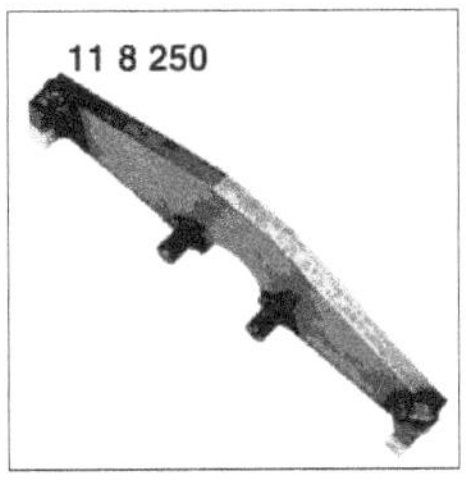

Camshaft locking tool
(BMW special tool 11 8 250)

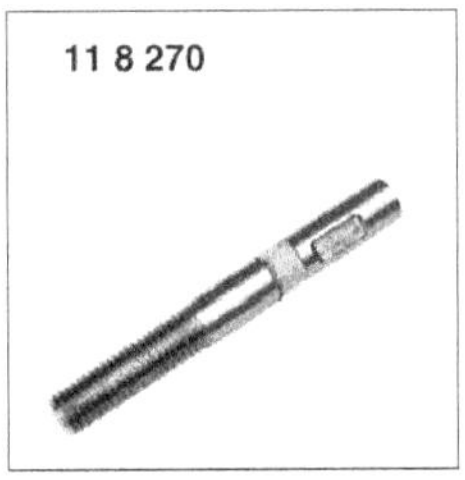

Crankshaft thread protector
(BMW special tool 11 8 270)

Gear puller jig for crankshaft chain sprocket
(BMW special tool 11 8 300)

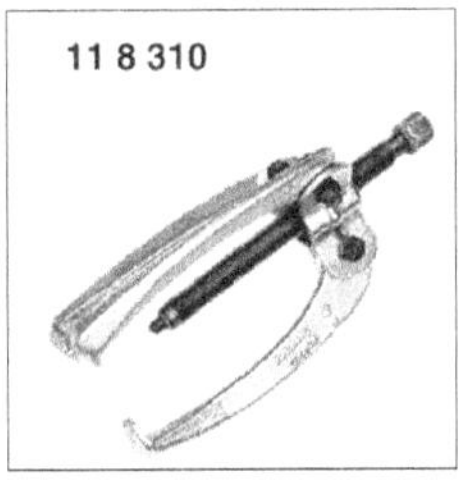

Puller
(BMW special tool 11 8 310)

Crankshaft vibration damper counterholding studs
(BMW special tool 11 8 320)

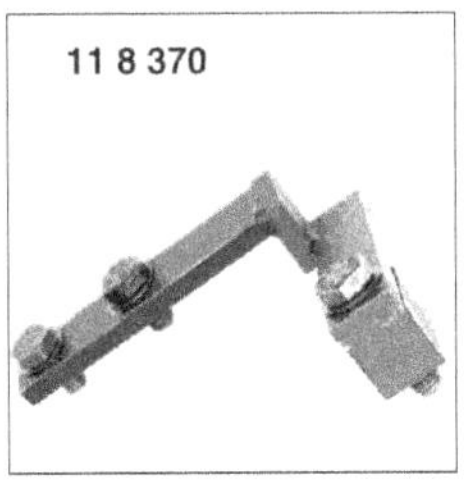

Engine block support tool
(BMW special tool 11 8 370)

Hydraulic engine mount removal and installation tool
(BMW special tool 11 8 380)

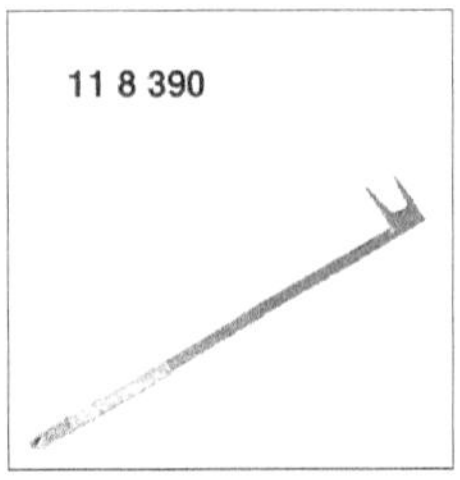

Lever for relieving drive belt tension (Cooper)
(BMW tool no. 11 8 390)

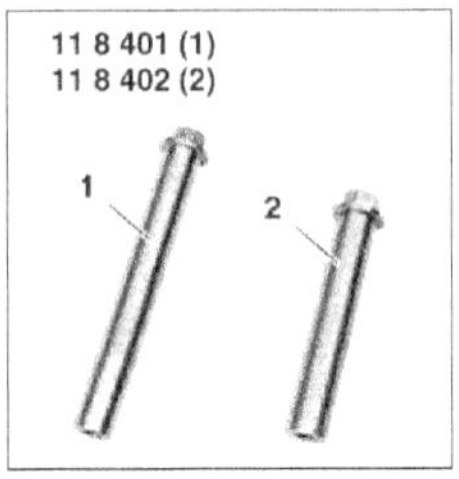

Modular front end extensions
(BMW special tool 11 8 400)

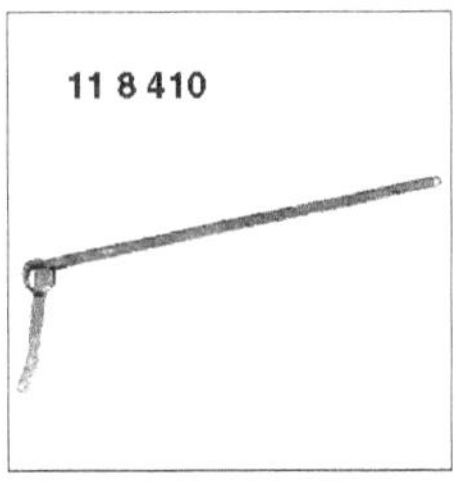

Lever for relieving drive belt tension (Cooper S)
(BMW tool no. 11 8 410)

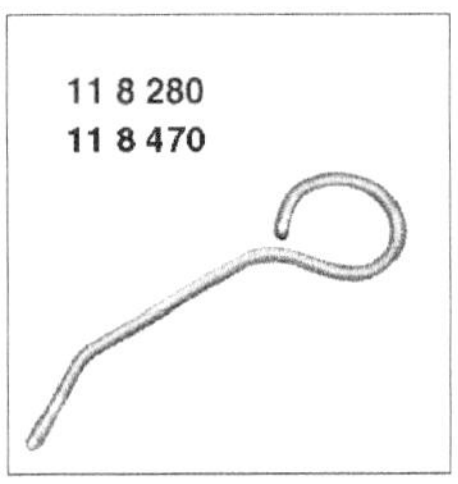

Drive belt tensioner locking tool
(BMW tool no. 11 8 280 for Cooper,
BMW tool no. 11 8 470 for Cooper S)

Crankshaft vibration damper puller (Cooper S)
(BMW tool no. 11 8 450)

Crankshaft vibration damper holder (Cooper S)
(BMW tool no. 11 8 460)

Timing Chain Housing Cover

To access the timing chain, it is necessary to remove the timing chain housing cover.

> ***WARNING —***
> *Due to risk of personal injury, be sure the engine is cold before beginning any of the procedures.*

> ***CAUTION —***
> *Cover all painted surfaces before beginning the removal procedure. As an aid to installation, label all components, wires, and hoses before removing them. Do not reuse gaskets, O-rings or seals during reassembly.*

Crankshaft vibration damper, removing and installing (Cooper)

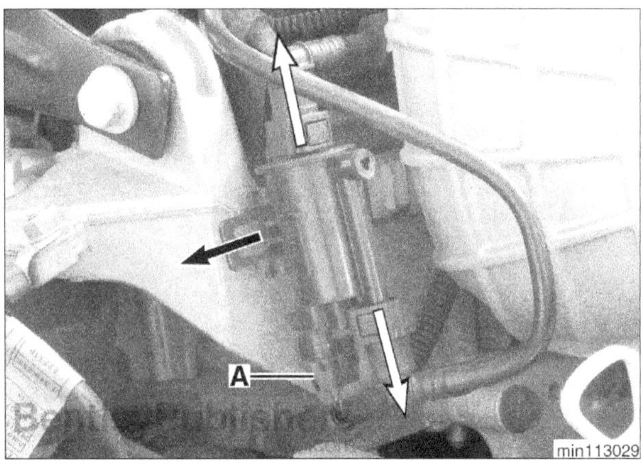

- Remove fuel tank vent valve.
  - Disconnect vent hoses (**white arrows**) via quick fit connectors.
  - Straighten retaining tab and slide vent valve off bracket (**black arrow**).
  - Remove electrical harness connector (**A**).

NOTE —
Cooper S engine is illustrated. Cooper engine layout is similar.

– Remove splash shield under engine.

– Remove right front wheel.

– Remove right front wheel housing liner.

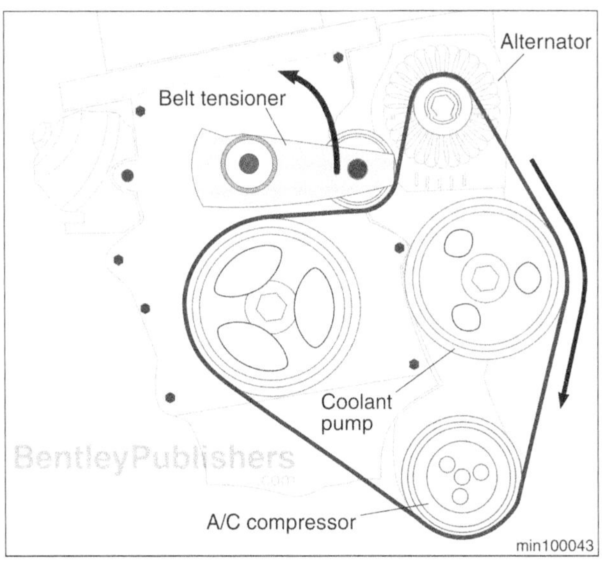

- Note engine drive belt layout. Then remove belt.
  - Use special tool 11 8 390 to release drive belt tension.
  - Use lock pin (special tool 11 8 280) to lock drive belt tensioner.
  - Lift off drive belt.

> ***CAUTION —***
> - *Drive belt tensioner is under high tension. Check that lock pin is secure.*
> - *Take care to avoid damage to paintwork.*

NOTE —
- *Drive belt removal and installation is covered in* **020 Maintenance**.
- *If drive belt is to be reused, mark direction of travel and reinstall drive belt in same direction of rotation.*

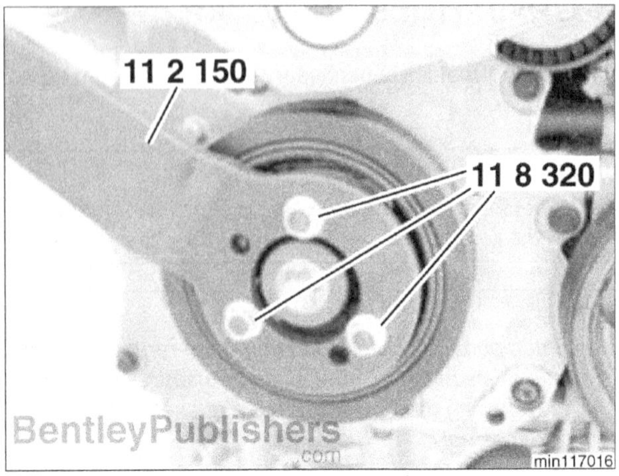

◀ Fit counterholding studs (BMW special tools 11 8 320) to crankshaft vibration damper holder (BMW special tool 11 2 150). Counterhold vibration damper to loosen and remove retaining bolt.

◀ Insert thread protector (BMW special tool 11 8 270) into crankshaft and tighten.

◀ Attach three-claw puller (BMW special tool 11 8 310 or equivalent) and remove vibration damper.

– Remove thread protector (BMW special tool 11 8 270) from crankshaft.

– To replace front crankshaft seal, see **119 Lubrication System**.

Reinstall vibration damper using special tools 11 8 241 and 11 8 242:

- Install threaded stud 11 8 242 in crankshaft threads.
- Place vibration damper over end of crankshaft.
- Use deep 24 mm socket to tighten special tool 11 8 241 to pull vibration damper on crankshaft.

Tightening torque	
Vibration damper to crankshaft (use new M12 bolt)	115 Nm (85 ft-lb)

– Remainder of installation is reverse of removal.

- Replace drive belt if contaminated with oil.
- Check for fault codes and clear and reset DME control module memory.
- Start engine and check for leaks.

Timing chain housing cover, removing and installing (Cooper)

– Remove engine drive belt and crankshaft vibration damper. See **Crankshaft vibration damper, removing and installing (Cooper)** in this repair group.

Remove belt tensioner retaining screws (**A**). Remove tensioner.

> ***WARNING*** —
> *Do not release pretensioned spring.*

Models with CVT or automatic transmission: Disconnect transmission fluid lines (**arrows**) from cooler in front of radiator. Use BMW plugs 113 5 281 / 13 5 282 or equivalent to plug lines and connections.

NOTE —
Be prepared to catch dripping fluid.

Timing Chain Housing Cover

min113031a

Loosen modular front end (MFE):

- Remove right front wheel housing liner.
- Detach left front wheel housing liner from front bumper cover trim.
- Remove front bumper cover trim.
- Remove bumper.
- Remove MFE mounting bolts. Install two 100 mm (4 in) M8 bolts (**arrow**) in left and right bumper support members. Slide MFE forward, supported on long bolts.

> ***CAUTION—***
> *For ease of component alignment when reassembling front bumper assembly, do not loosen or remove bumper alignment bosses.*

NOTE—
- *Cooper S engine is illustrated. Cooper engine layout is similar.*
- *If available, use BMW special tools 11 8 401 and 11 8 402 instead of long bolts to support MFE.*
- *Removal of MFE is a complicated job. It is covered in* **510 Bumpers, External Trim**.

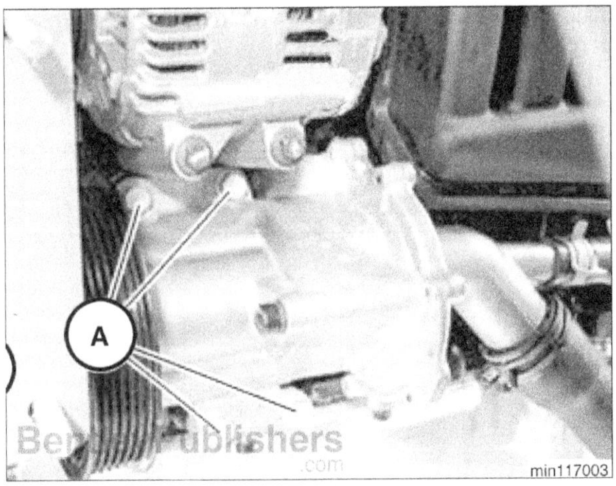

min117003

Remove coolant pump mounting bolts (**A**) and shift coolant pump so that timing chain housing cover is unobstructed.

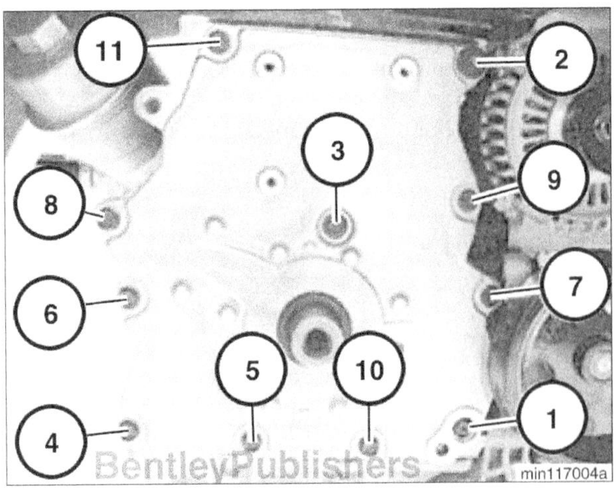

min117004a

Remove timing chain housing cover bolts in specified order (**1** to **11**).

– Remove cover.

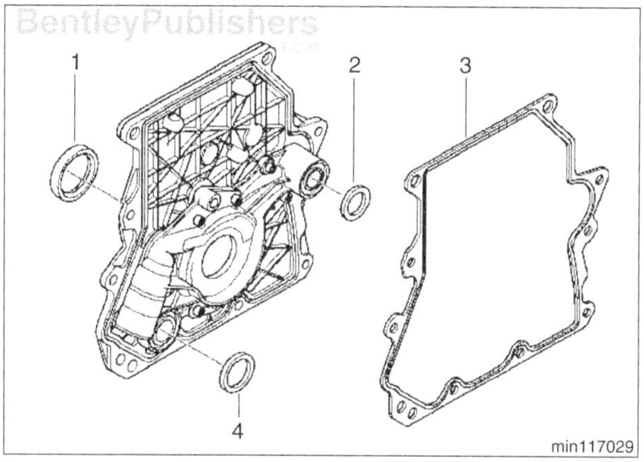

Clean sealing surfaces on timing case cover and crankcase. Install new seals and gaskets:

1. Crankshaft seal
2. Oil pump outlet seal
3. Timing chain housing cover sealing gasket
4. Oil pump inlet seal

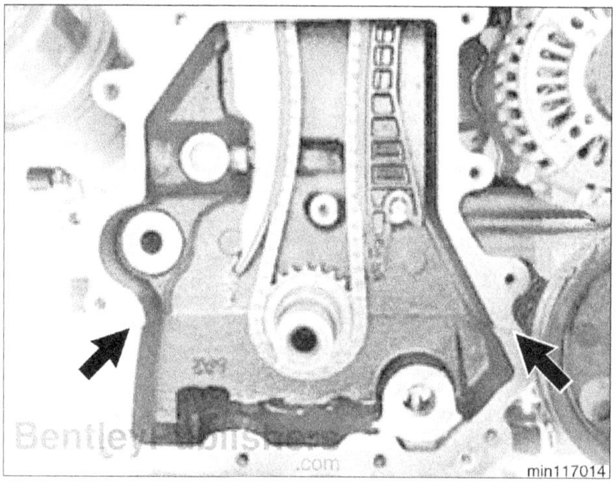

Apply 3 mm (1/8 in) bead of Loctite® RTV 5999 sealer or equivalent to joint (**arrows**) between engine block and bearing ladder.

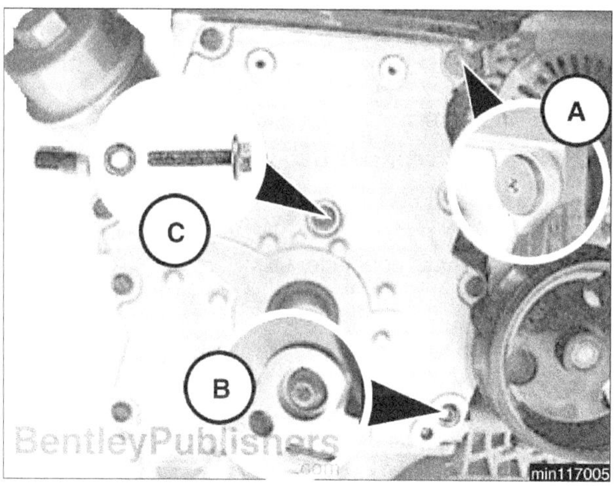

Note location of Torx shoulder bolt (**A**), Torx round head bolt (**B**) and bolt with sealing O-ring (**C**).

- Replace O-ring and lubricate with clean engine oil.

– Before timing housing cover installation, prime oil pump by filling with engine oil.

– While installing timing housing cover, align oil pump rotor flats with flats on crankshaft.

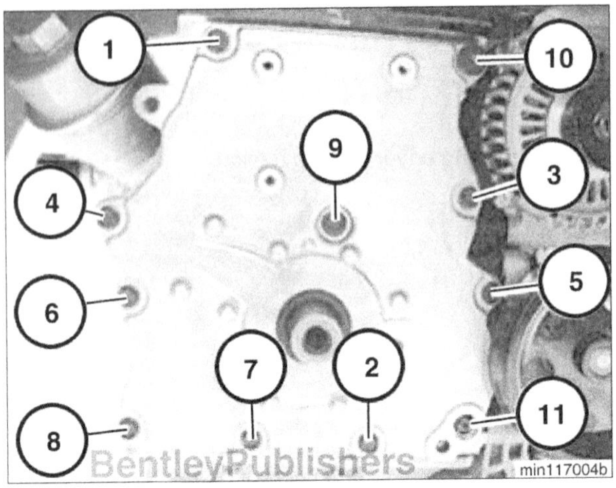

Tighten timing chain cover fasteners in specified order (**1** to **11**).

Tightening torques	
Timing chain housing cover to engine block M6 M8 oval head	 12 Nm (9 ft-lb) 18 Nm (13 ft-lb)

– Remainder of installation is reverse of removal.
  - Replace drive belt if contaminated with oil.
  - Check for fault codes and clear and reset DME control module memory.
  - Start engine and check for leaks.

Tightening torques	
Coolant pump to engine block (M8)	30 Nm (22 ft-lb)
Crush tube to subframe (M12 x 1.5 x 85 mm)	100 Nm (74 ft-lb)
Drive belt tensioner to timing chain housing cover (M6)	9 Nm (7 ft-lb)
Front bumper to crush tube or to bumper carrier (M8 x 30 mm)	22 Nm (16 ft-lb)
Front bumper to modular front end (M6 x 16 mm)	5 Nm (44 in-lb)
Modular front end to crush tube (M6)	5 Nm (44 in-lb)

Crankshaft vibration damper, removing and installing (Cooper S)

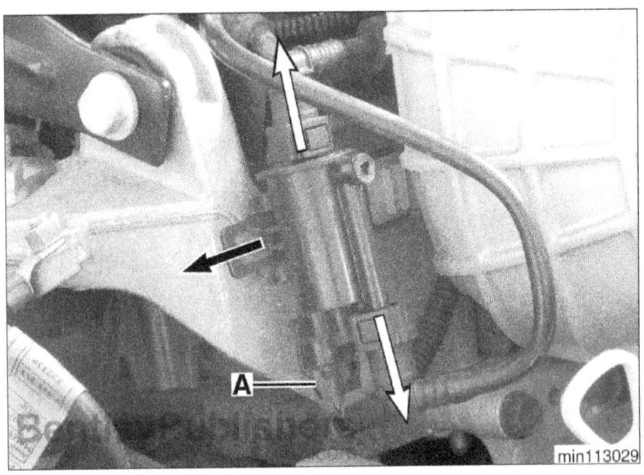

Remove fuel tank vent valve.
  - Disconnect vent hoses (**white arrows**) via quick fit connectors.
  - Straighten retaining tab and slide vent valve off bracket (**black arrow**).
  - Remove electrical harness connector (**A**).

– Remove splash shield under engine.

– Remove right front wheel.

– Remove right front wheel housing liner.

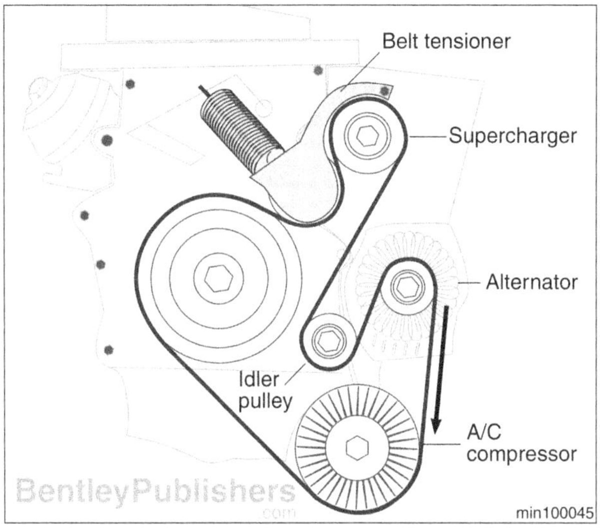

Note engine drive belt layout. Then remove belt.

- Use BMW special tool 11 8 410 to release drive belt tension.
- Use lock pin (BMW special tool 11 8 470) to lock drive belt tensioner.
- Lift off drive belt.

CAUTION—
- *Drive belt tensioner is under high tension. Check that lock pin is secure.*
- *Take care to avoid damage to paintwork.*

NOTE—
- *Drive belt removal and installation is covered in* **020 Maintenance**.
- *If drive belt is to be reused, mark direction of travel and reinstall drive belt in same direction of rotation.*

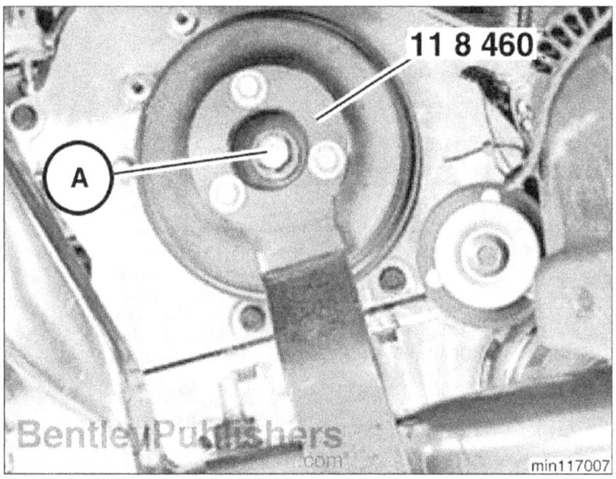

Fit BMW special tool 11 8 460 to crankshaft vibration damper and remove mounting bolt (**A**).

Insert thread protector (BMW special tool 11 8 270) into crankshaft and tighten.

Use puller (BMW special tool 11 8 450 or equivalent) to remove vibration damper.

– Remove thread protector (BMW special tool 11 8 270) from crankshaft and remove puller from vibration damper.

– To replace crankshaft seal, see **119 Lubrication System**.

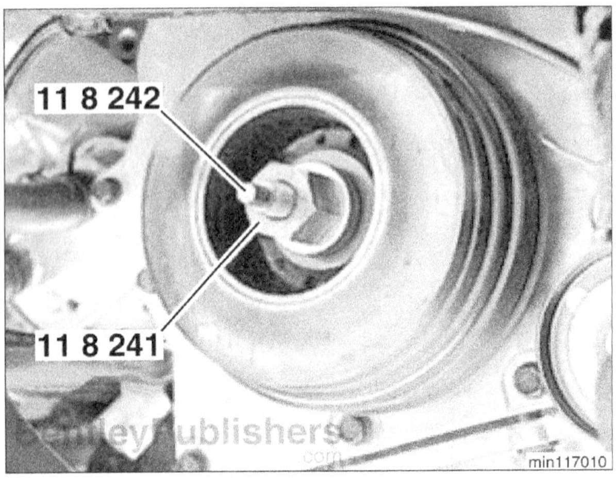

Reinstall vibration damper using BMW special tools 11 8 242 and 11 8 241:

- Install threaded stud (BMW special tool 11 8 242) in crankshaft threads.
- Place vibration damper over end of crankshaft.
- Use deep 24 mm socket to tighten BMW special tool 11 8 24. This presses vibration damper on crankshaft.

Tightening torque	
Vibration damper to crankshaft (use new M12 bolt)	115 Nm (85 ft-lb)

– Remainder of installation is reverse of removal.

- Replace drive belt if contaminated with oil.
- Check for fault codes and clear and reset DME control module memory.
- Start engine and check for leaks.

Timing chain housing cover, removing and installing (Cooper S)

– Remove engine drive belt and crankshaft vibration damper. See **Crankshaft vibration damper, removing and installing (Cooper S)** in this repair group.

Remove belt tensioner mounting bolts (**A** and **B**). Remove tensioner (**C**).

Remove idler pulley (**A**) from timing chain cover.

Models with CVT or automatic transmission: Disconnect transmission fluid lines (**arrows**) from cooler in front of radiator. Use BMW plugs 113 5 281 / 13 5 282 or equivalent to plug lines and connections.

NOTE —
Be prepared to catch dripping fluid.

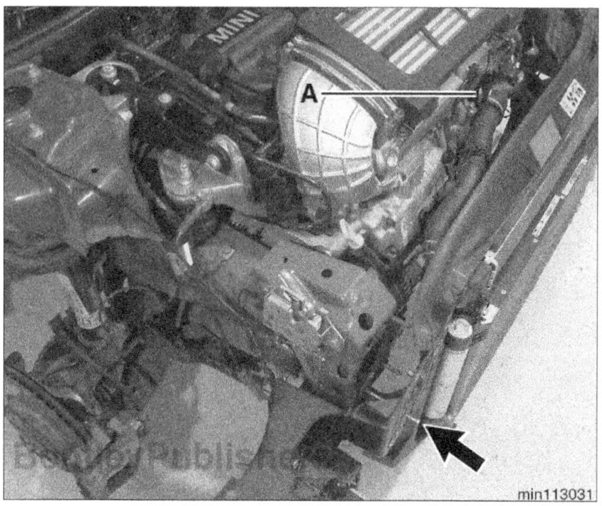

Loosen modular front end (MFE):

- Remove right front wheel housing liner.
- Detach left front wheel housing liner from front bumper cover trim.
- Remove front bumper cover trim.
- Remove bumper.
- Detach radiator upper hose support clamp (**A**) from intake manifold.
- Remove MFE mounting bolts. Install two 100 mm (4 in) M8 bolts (**arrow**) in left and right bumper support members. Slide MFE forward, supported on long bolts.

CAUTION —
For ease of component alignment when reassembling front bumper assembly, do not loosen or remove bumper alignment bosses.

NOTE —
- *If available, use BMW special tools 11 8 401 and 11 8 402 instead of long bolts to support MFE.*
- *Removal of MFE is a complicated job. It is covered in* **510 Bumpers, External Trim**.

Timing Chain Housing Cover

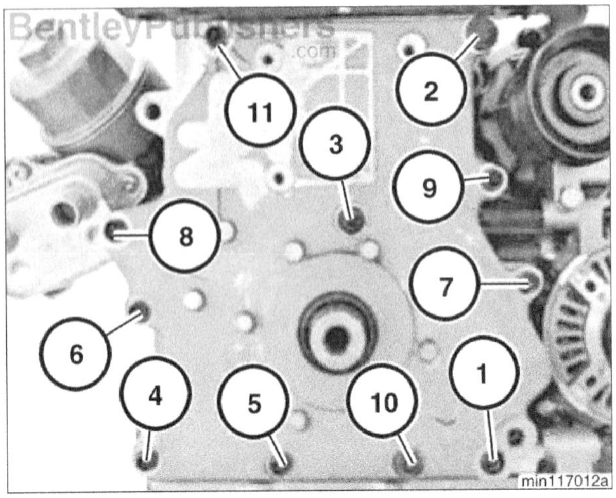

< Remove timing cover fasteners in specified order (**1** to **11**).

– Remove cover.

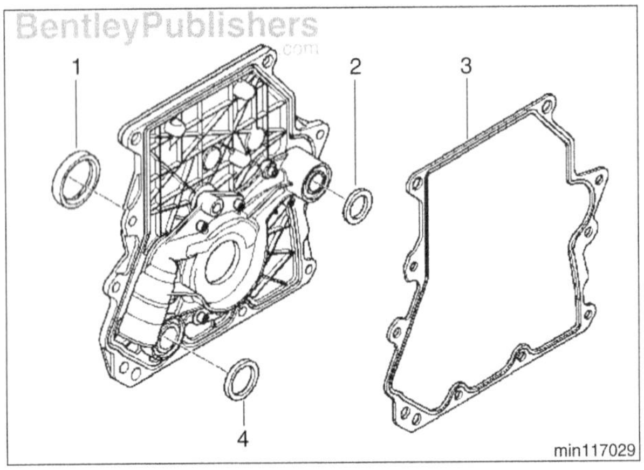

< Clean sealing surfaces on timing case cover and crankcase. Install new seals and gaskets:

1. Crankshaft seal
2. Oil pump outlet seal
3. Timing chain housing cover sealing gasket
4. Oil pump inlet seal

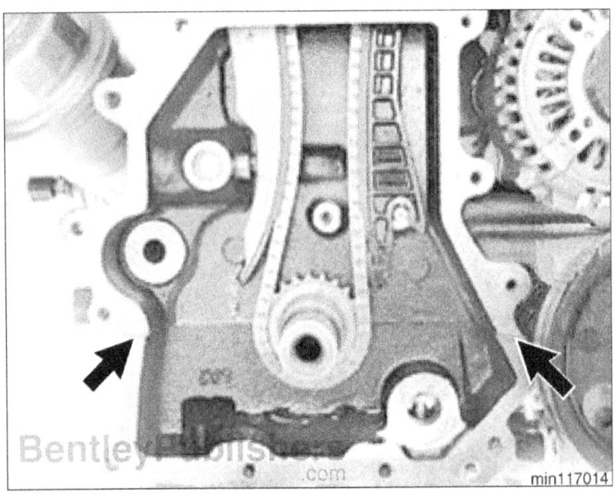

< Apply 3 mm (1⁄8 in) bead of Loctite® RTV 5999 sealer or equivalent to joint (**arrows**) between engine block and bearing ladder.

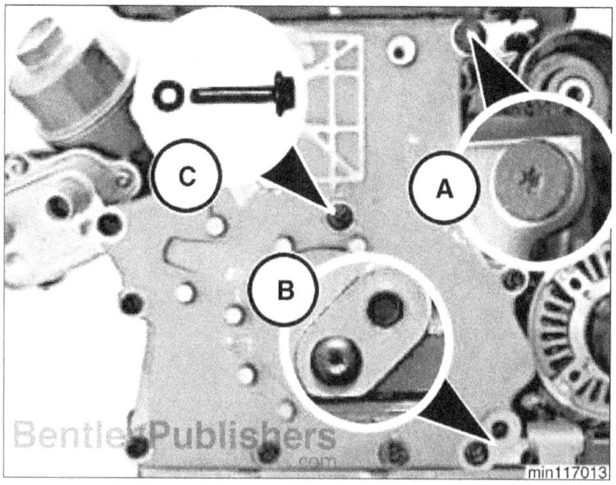

- Note location of Torx shoulder bolt (**A**), Torx round head bolt (**B**) and bolt with sealing O-ring (**C**).
  - Replace sealing O-ring and lubricate with clean engine oil.

– Before timing housing cover installation, prime oil pump by filling with engine oil.

– While installing timing housing cover, align oil pump rotor flats with flats on crankshaft.

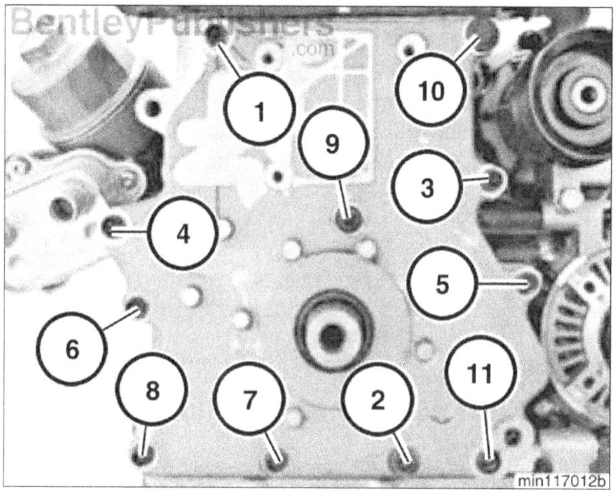

- Tighten timing chain cover fasteners in specified order (**1** to **11**).

Tightening torques	
Timing chain housing cover to engine block M6 M8 oval head	 12 Nm (9 ft-lb) 18 Nm (13 ft-lb)

– Remainder of installation is reverse of removal.
  - Replace drive belt if contaminated with oil.
  - Check for fault codes and clear and reset DME control module memory.
  - Start engine and check for leaks.

Tightening torques	
Crush tube to subframe (M12 x 1.5 x 85 mm)	100 Nm (74 ft-lb)
Drive belt idler pulley to timing chain housing cover (M8)	45 Nm (33 ft-lb)
Drive belt tensioner to timing chain housing cover (M6)	9 Nm (7 ft-lb)
Drive belt tensioner to timing chain housing cover or to supercharger (M10)	45 Nm (33 ft-lb)
Front bumper to crush tube or to bumper carrier (M8 x 30 mm)	22 Nm (16 ft-lb)
Front bumper to modular front end (M6 x 16 mm)	5 Nm (44 in-lb)
Modular front end to crush tube (M6)	5 Nm (44 in-lb)

TIMING CHAIN

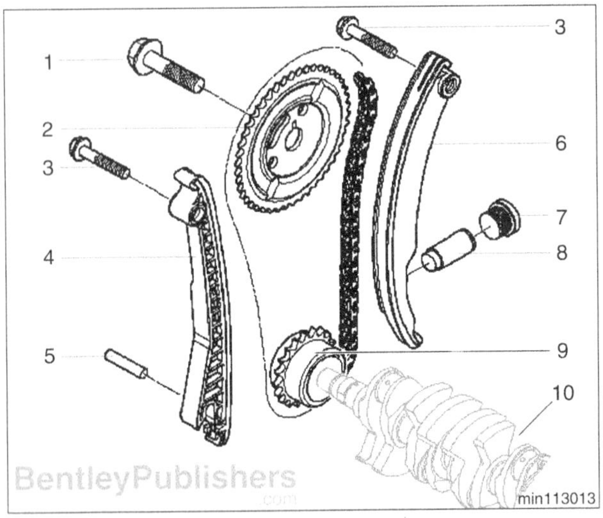

Timing chain components:

1. Bolt, M12
   - Tighten to 102 Nm (75 ft-lb)
2. Camshaft sprocket
3. Bolt, M8
   - Tighten to 28 Nm (21 ft-lb)
4. Timing chain fixed guide
5. Stud
6. Timing chain tensioning guide
7. Timing chain tensioner plug with sealing O-ring
8. Timing chain tensioner plunger
9. Crankshaft sprocket
10. Crankshaft

The timing chain incorporates three copper-coated links that are used to set valve timing. Both the crankshaft and camshaft sprockets incorporate timing marks which are used in conjunction with the copper-coated chain links.

NOTE —
The copper-coated links are used to set initial camshaft -to-crankshaft relative positions. They are not relevant to valve timing later on.

Timing chain, removing and installing

A gear puller and other special tools are necessary for this procedure. See **Special tools** in this repair group.

– Disconnect negative (-) cable from battery. See **121 Battery, Starter, Alternator** for more information.

> ***CAUTION —***
> *Disconnecting the battery may erase fault code(s) stored in memory. Check for fault codes prior to disconnecting the battery cables.*

– Unscrew all spark plugs.

– Remove valve cover. See **113 Cylinder Head Removal and Installation**.

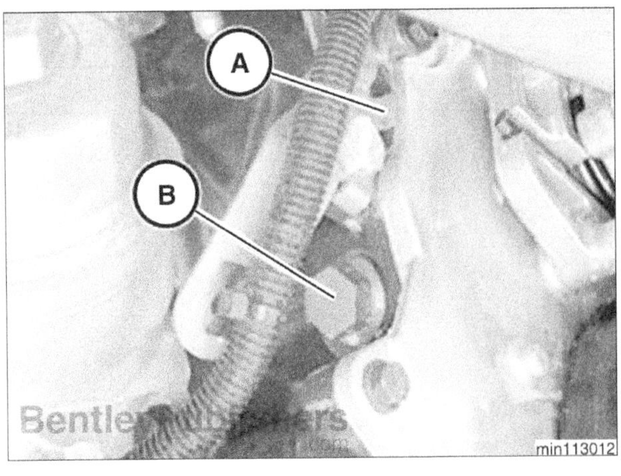

Working in back of engine block, loosen and remove timing chain tensioner plug (**B**). Remove tensioner plunger from inside timing chain housing cover.

– Cooper:

- Remove crankshaft vibration damper. See **Crankshaft vibration damper, removing and installing (Cooper)** in this repair group.
- Remove timing chain housing cover. See **Timing chain housing cover, removing and installing (Cooper)** in this repair group.

– Cooper S:

- Remove crankshaft vibration damper. See **Crankshaft vibration damper, removing and installing (Cooper S)** in this repair group.
- Remove timing chain housing cover. See **Timing chain housing cover, removing and installing (Cooper S)** in this repair group.

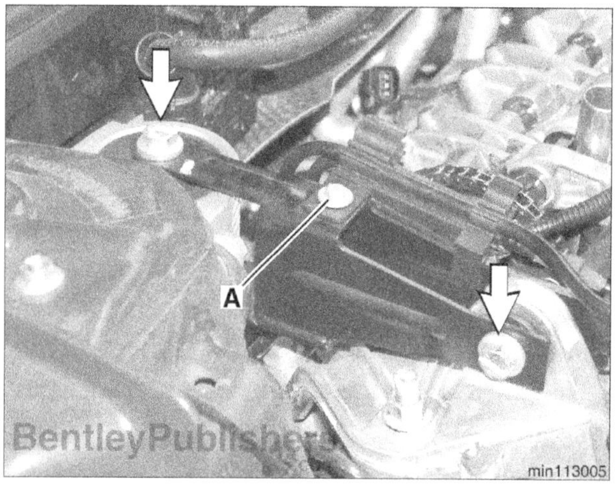

◄ Remove engine vibration damper bracket:

- Remove fuel line bracket mounting bolt (**A**). Detach any fuel lines still attached. Lift off bracket and set aside.
- Remove engine vibration damper bracket bolts (**arrows**).
- Remove bracket.

– Support engine under oil pan.

> **CAUTION—**
> *To avoid damaging the oil pan, use a rubber pad on top of the engine supporting jack.*

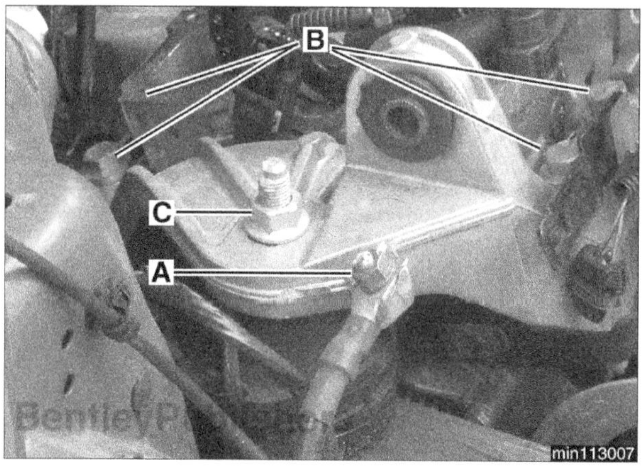

◄ Remove engine mounting bracket:

- Remove engine ground cable mounting nut (**A**) from bracket. Detach ground cable.
- Remove 4 engine mounting bolts (**B**) from engine block.
- Unscrew mounting nut (**C**) from hydraulic front engine mount stud. Lift off engine mount bracket.

– Use BMW special tool 11 8 380 to remove hydraulic engine mount. Counterhold Torx head mounting bolt from below.

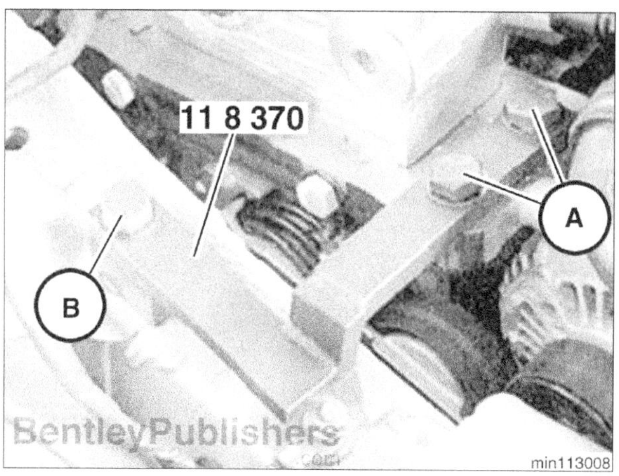

◄ Use BMW special tool 11 8 370 to support engine. Remove jack from underneath engine oil pan.

- Insert bolts **A** in engine block.
- Insert bolt **B** In hydraulic engine mount bolt hole.

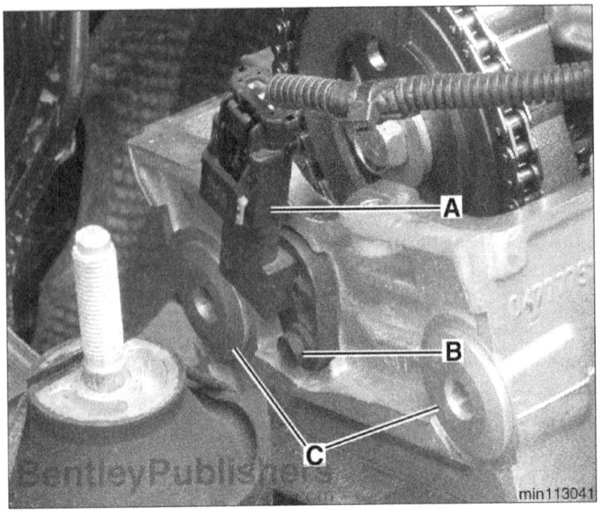

Working at right end of cylinder head:

- Disconnect electrical harness connector (**A**) at camshaft position sensor.
- Remove sensor mounting bolt (**B**).
- Remove sensor from cylinder head.
- Remove threaded plugs (**C**) from cylinder head.

Attach BMW special tool 11 8 250 to camshaft sprocket.

– Loosen and remove camshaft sprocket bolt (**A**).

– Remove camshaft sprocket and chain from camshaft.

– Remove camshaft sprocket from timing chain.

> ***CAUTION—***
> *Prevent timing chain from dropping into timing case.*

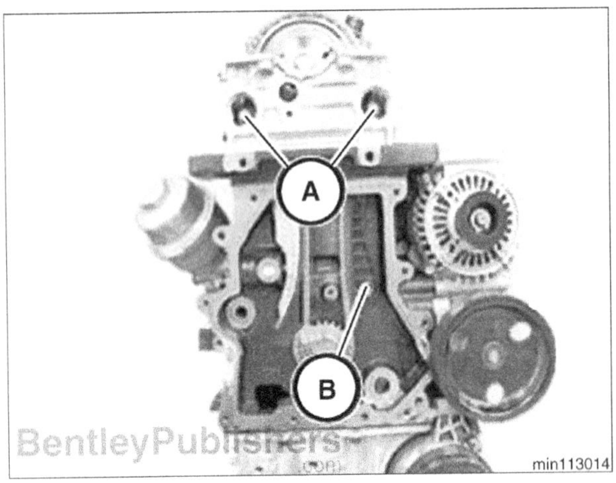

Remove timing chain fixed and tensioning guide mounting bolts (**A** and **B**).

- Pull timing chain guides up out of timing chain housing.

– Remove chain from crankshaft sprocket.

NOTE—
If the timing chain is replaced, also replace the camshaft and crankshaft sprockets.

– Remove crankshaft sprocket using gear pulling jig (BMW special tool 11 8 300) together with gear puller (BMW special tool 11 2 000).

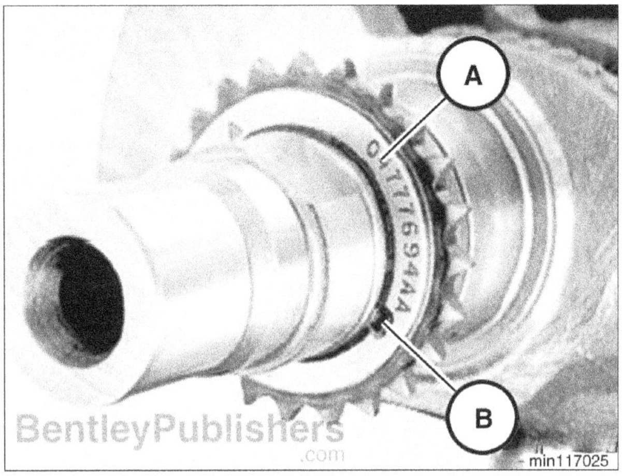

Shrink-fit crankshaft sprocket (**A**) on crankshaft.

- Note correct position of locating pin (**B**).
- Use hotplate or oil bath to heat crankshaft sprocket to a maximum temperature of 150°C (302°F).
- Slide sprocket on crankshaft as far as it will go.

> ***WARNING—***
> *Protect hands with heat-resistant gloves.*

> ***CAUTION—***
> *When heating the sprocket, do not exceed the specified temperature.*

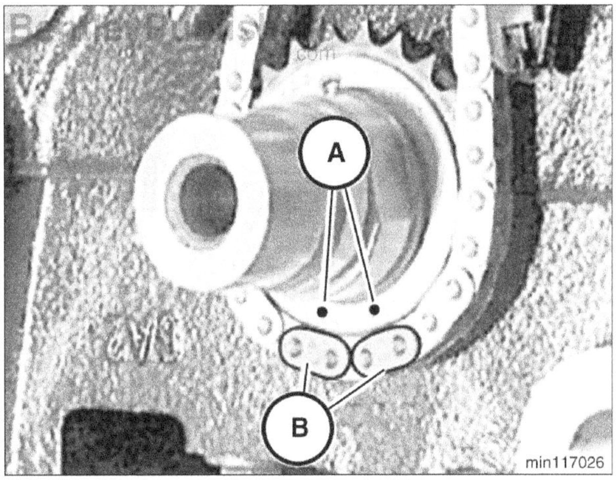

Install timing chain on crankshaft sprocket. Make sure that crankshaft timing marks (**A**) line up with two copper-colored chain links (**B**).

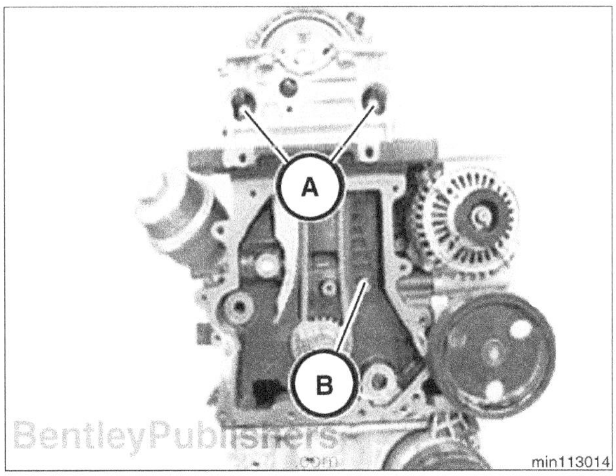

Reinstall timing chain fixed and tensioning guides.

- Install tensioning guide mounting bolts (**arrows**).

Tightening torque	
Timing chain guide to cylinder head (M8)	28 Nm (21 ft-lb)

Timing chain

Place timing chain on camshaft sprocket. Make sure that triangular timing mark on camshaft sprocket is aligned with copper-colored timing chain link (**dashed line**).

– Place sprocket on camshaft and insert mounting bolt.

Attach BMW special tool 11 8 250 to camshaft sprocket. Torque camshaft sprocket bolt.

Tightening torque	
Camshaft sprocket to camshaft (M12)	102 Nm (75 ft-lb)

CAUTION—
Be careful not to damage timing chain during this step.

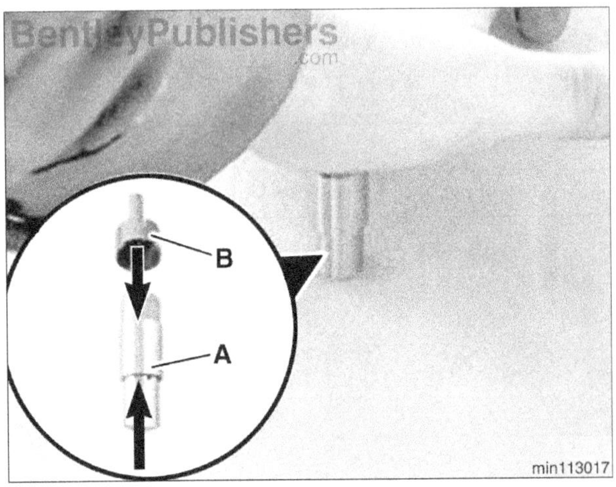

Squeeze timing chain tensioner plunger into retracted position.

- Place plunger (**A**) on level surface.
- Remove plunger cap (**B**).
- Exert continuous hand pressure on plunger until it is completely compressed.
- Replace cap.

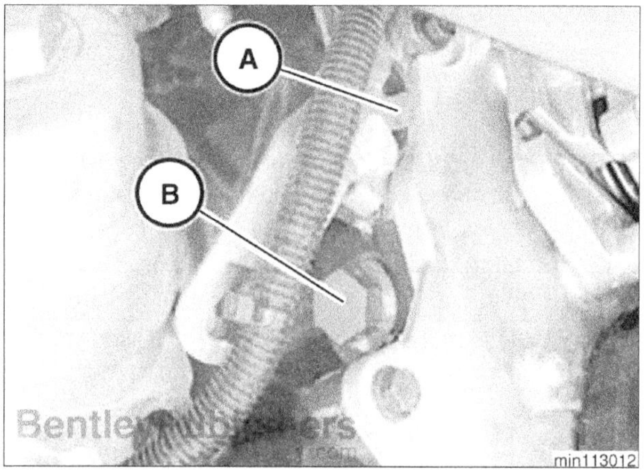

◀ Working in back of engine block, install timing chain tensioner plunger. Install tensioner plug (**B**).

Tightening torque	
Timing chain tensioner to engine block	63 Nm (46 ft-lb)

– If applicable: Install wiring harness bracket in back of engine block (**A**).

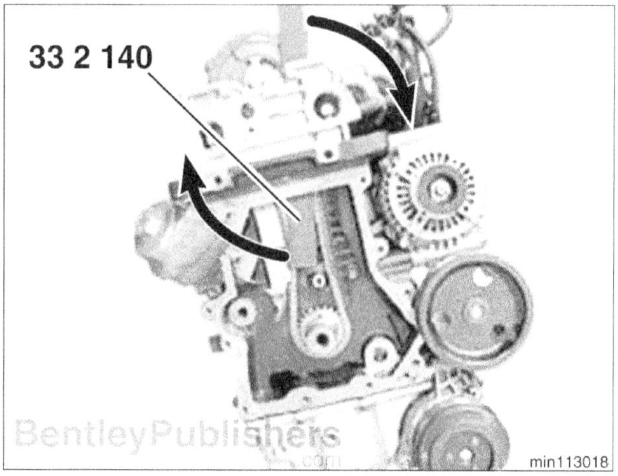

◀ Insert large prying bar (BMW special tool 33 2 140 or equivalent) into timing chain housing and turn clockwise. This releases timing chain tensioner plunger and tensions chain.

> ***CAUTION—***
> - *Do not pry directly on timing chain. Damage may result.*
> - *Make sure that the timing chain is properly positioned within the channels of the timing chain guides.*

– Reinstall timing chain housing cover, crankshaft vibration damper, alternator drive belt tensioner, and drive belt.

Tightening torques	
Drive belt idler pulley to timing chain housing cover (M8)	45 Nm (33 ft-lb)
Drive belt tensioner to timing chain housing cover (M6)	9 Nm (7 ft-lb)
Drive belt tensioner to timing chain housing cover or to supercharger (M10)	45 Nm (33 ft-lb)
Timing chain housing cover to engine block (M6)	12 Nm (9 ft-lb)
Timing chain housing cover to engine block (M8 oval head)	18 Nm (13 ft-lb)
Vibration damper to crankshaft (use new M12 bolt)	115 Nm (85 ft-lb)

– Continue to reassemble engine, paying attention to the following:
  - Use new sealing washers when installing threaded plugs at front of cylinder head.
  - Use new hose clamps, seals and gaskets, as necessary.
  - Check DME control module adaptations using diagnostic scan tool. Check for fault codes and clear and reset control module memory.

Tightening torques	
Battery cable (- or +) to battery (M6)	5 Nm (44 in-lb)
Camshaft position sensor to cylinder head (M6)	9 Nm (7 ft-lb)
Engine front mounting bracket to top of engine block (M12)	100 Nm (74 ft-lb)
Hydraulic engine mount to body	100 Nm (74 ft-lb)
Hydraulic engine mount to body (replace Torx bolt): Initial torque Torque angle	 56 Nm (41 ft-lb) 90°
Hydraulic engine mount to engine mounting bracket	68 Nm (50 ft-lb)
Ignition coil pack to valve cover (M6)	12 Nm (9 ft-lb)
Sealing plug to cylinder head	18 Nm (13 ft-lb)
Spark plug to cylinder head	27 Nm (20 ft-lb)
Upper engine vibration damper bracket to damper and to engine mounting bracket (M12)	100 Nm (74 ft-lb)
Valve cover to cylinder head (M6)	12 Nm (9 ft-lb)

119 Lubrication System

GENERAL

This repair group covers lubrication system troubleshooting as well as oil pan removal and oil pump replacement.

NOTE —
Oil change procedure and oil filter replacement are covered in **020 Maintenance**.

Special tools

Some special tools are required for servicing the lubrication system. Be sure to have the necessary equipment on hand before starting the job.

A standard oil pressure gauge may be used for measuring oil pressure.

Handle for drift
(BMW tool no. 00 5 500)

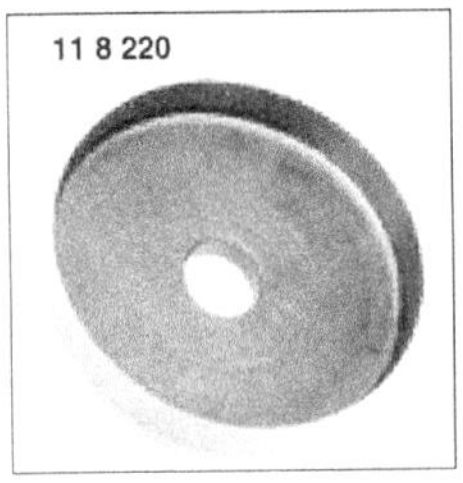

Slip sleeve for rear main seal
(BMW tool no. 11 8 220)

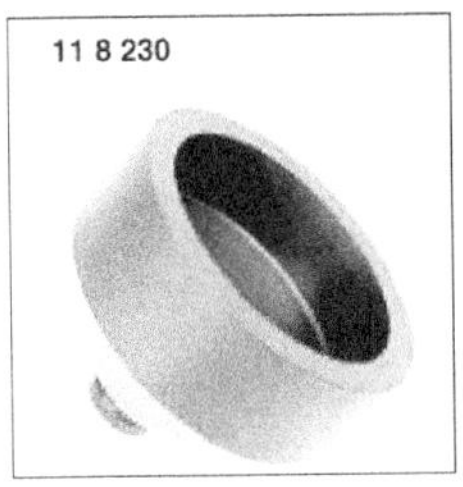

Drift for rear main seal
(BMW tool no. 11 8 230)

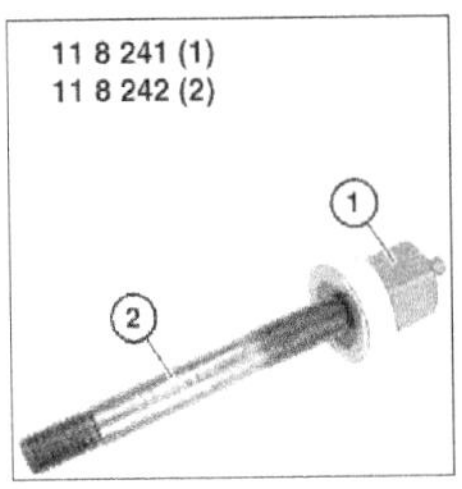

Crankshaft front seal insertion tool
(BMW tool no. 11 8 240)

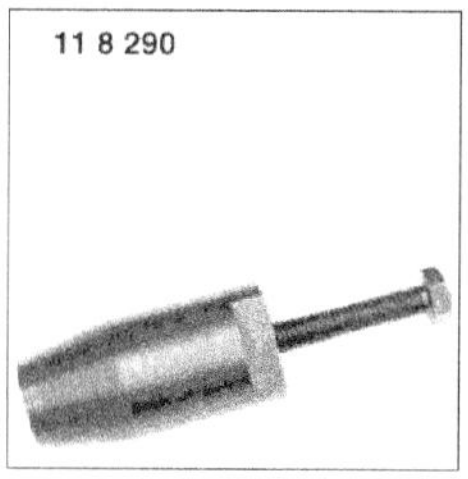

Extraction tool for rear main seal
(BMW tool no. 11 8 290)

Oil pressure tool fitting
(BMW tool no. 11 8 360)

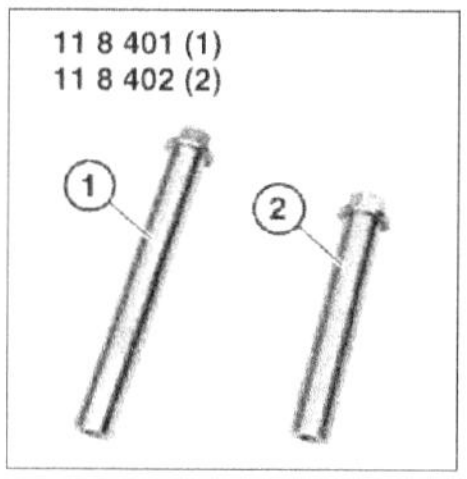

Modular front end extensions
(BMW tool no. 11 8 400)

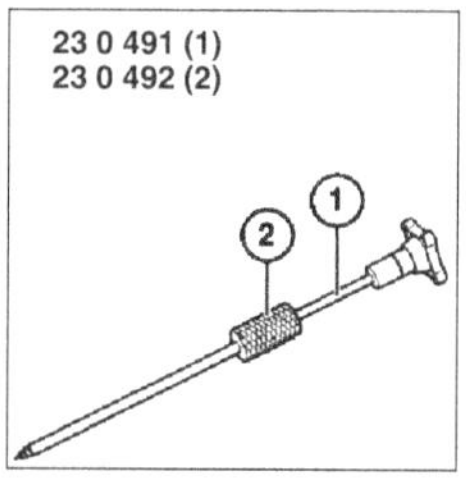

Seal extractor
(BMW tool no. 490)

Engine lubrication

Oil pressure is generated by a gear-type pump located on the inside of the timing cover. The oil pump is driven directly off the front of the crankshaft. Servicing the oil pump requires removal of the timing cover. See **Table a** for oil recommendations and capacity.

Table a. Engine oil with filter change

Model	Approximate capacity	Oil specification
MINI Cooper MINI Cooper S	4.7 liters (4.5 US qt)	BMW Longlife-01 or BMW Longlife-98 ACEA A3

TROUBLESHOOTING

The oil pressure warning system consists of an oil pressure switch mounted on the oil filter housing and an instrument panel warning light. Other safety features include:

- A filter bypass to provide lubrication should the oil filter become clogged.
- An oil pressure relief valve installed in the oil pump housing to prevent excessive system pressure.

> ***CAUTION—***
> *If the red oil pressure light comes on while driving, always assume that oil pressure is low. Stop engine immediately and make arrangements to test oil pressure.*

Oil pressure, checking

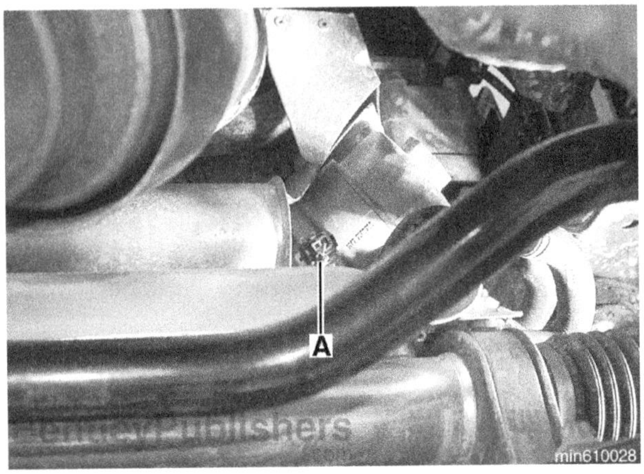

The oil pressure switch (**A**) is located at the base of the oil filter housing.

- Unscrew oil filter cover slowly to allow engine oil to drain back down into oil pan.
- Disconnect harness connector from oil pressure switch.

NOTE—
- *Thoroughly clean around the oil pressure switch before removing it.*
- *Be prepared to catch leaking oil with a shop towel.*

- Install threaded adapter with O-ring (BMW tool no. 11 8 360) (if necessary) and an oil pressure gauge in place of oil pressure switch.
- With gauge installed and filter cover retightened, start engine and allow to reach operating temperature. Check oil pressure. See **Table b**.

NOTE—
For the most accurate test results, the engine oil and filter should be new. The oil should be the correct grade.

Table b. Engine oil pressure	
At idle	0.25 bar (3.6 psi)
At 3000 RPM	1.7 - 5.5 bar (25 - 80 psi)

> ***CAUTION—***
> *If oil pressure is zero at idle, do not run engine at 3000 RPM.*

- Remove pressure gauge and reinstall pressure switch with new sealing washer.

Tightening torque	
Oil pressure switch to oil filter housing	20 Nm (15 ft-lb)

If testing shows low oil pressure, one or more of the following conditions may be indicated:

- Worn or faulty oil pump or faulty pump pressure relief valve
- Worn or damaged engine bearings
- Severe engine wear

Any of these conditions indicate the need for major repairs.

Oil pressure warning system, testing

When the ignition is turned on, the oil pressure warning light comes on. When the engine is started and the oil pressure rises slightly, the oil pressure switch (**A**) contact opens and the warning light goes out. Make sure the oil level is correct before making tests.

– Turn ignition switch on.
  - Warning light on instrument panel must light.

– Remove connector from oil pressure switch.
  - Warning light on instrument panel must go out.

NOTE—
If the light does not go out, the wiring to the switch is most likely grounded somewhere between the switch terminal and the warning light. See **ELE Electrical Wiring Diagrams** *for electrical schematics.*

– If warning light does not light when ignition is on, remove connector from oil pressure switch and use a jumper wire to ground connector terminal to a clean metal surface. The oil pressure warning light should come on.

– If the warning light comes on, check the switch as described in the next step. If the warning light does not come on, the wiring to the instrument cluster or to the light itself is faulty.

– To test the switch, connect an ohmmeter between terminal in switch body and ground. With engine off, there should be continuity. With engine running, Oil pressure should open switch and there should be no continuity.

CAUTION—
Keep in mind that low oil pressure may be preventing the switch from turning the light out. If the light remains on while the engine is running, check the oil pressure as described earlier. Do not drive the car until the problem is corrected. The engine may be severely damaged.

Tightening torque	
Oil pressure switch to oil filter housing	20 Nm (15 ft-lb)

CRANKSHAFT SEAL

Crankshaft front seal, removing and installing

- Remove drive belt.
- Remove crankshaft vibration damper. See **117 Camshaft Timing Chain** for more information.

< Working at the timing cover, fit special tool (BMW 11 8 270) to crankshaft.

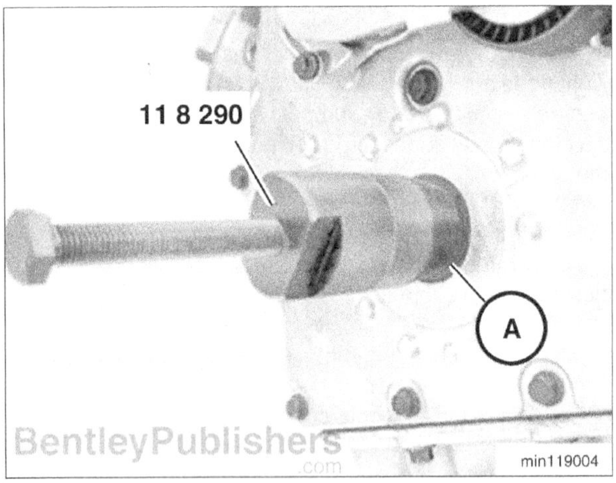

< Fit special tool (BMW 11 8 290) on crankshaft and screw in until it connects solidly with crankshaft seal.

- Remove crankshaft seal (**A**) by tightening bolt.

< To replace crankshaft seal, fit special tool (BMW 11 8 242) to crankshaft and tighten.

- Coat lip of new seal with oil.

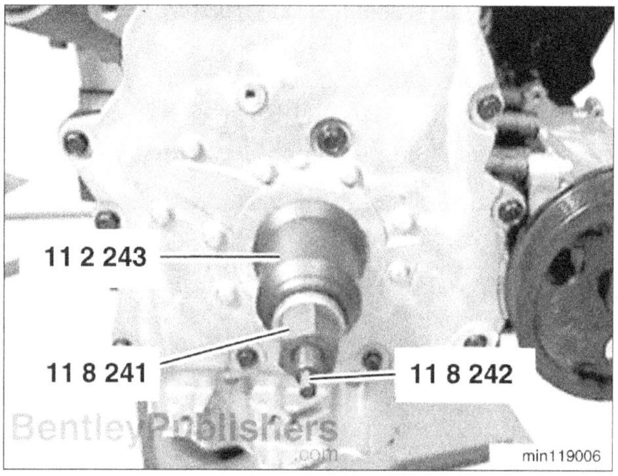

◄ Using special tool (BMW 11 8 243) and (BMW 11 8 241) carefully install crankshaft seal until flush with timing case cover.

– Remainder of installation is reverse of removal.

Crankshaft rear seal, removing and installing

Crankshaft rear main seal replacement requires removal of the transmission and flywheel.

– Remove transmission. See **230 Manual Transmission** or **240 Automatic Transmission**.

– Manual transmission models: Remove clutch pressure plate and disc. See **210 Clutch**.

– Remove flywheel. See **210 Clutch** or **240 Automatic Transmission**.

◄ Remove rear main seal with special tool (BMW 23 0 490).

– Clean seal surfaces on engine block and crankshaft.

◄ Fit special tool (BMW 11 8 220) over crankshaft and slide new seal over special tool.

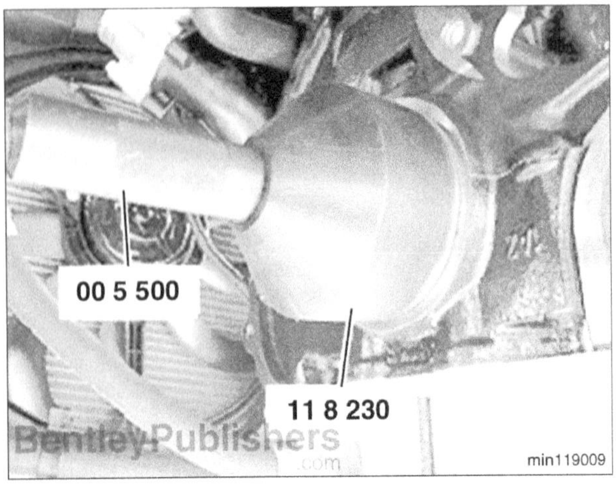

min119009

Use special tool (BMW 00 5 500) or suitable drift and special tool (BMW 11 8 230) to drive rear main seal flush with cylinder block.

> ***CAUTION—***
> *The seal is very delicate and must not be kinked or damaged. Do not touch seal lip with fingers.*

Engine Oil Pan

Oil pan, removing and installing

- Disconnect battery.
- Remove drive belt.
- Remove engine splash guard.
- Drain engine oil.

min113031a

Remove crush tubes and fit special tools (BMW 11 8 401 and 11 8 402) to extend front end way from chassis. See **110 Engine Removal and Installation**.

min119013

Working at front of engine, disconnect wire connector (**1**) from A/C compressor.

- Remove compressor retaining bolts (**2**) and secure compressor (**3**) to modular front end with stiff wire.
- Unclip A/C high pressure pipe from oil pan.

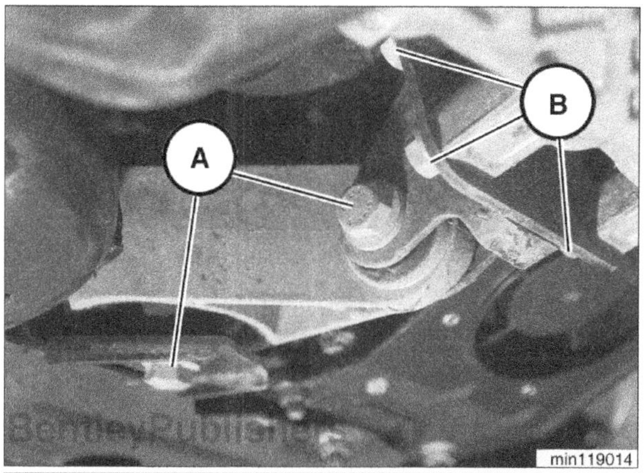

◀ Remove lower engine vibration damper bolts (**A**) and remove damper.

– Remove bracket retaining bolts (**B**) and remove bracket from oil pan.

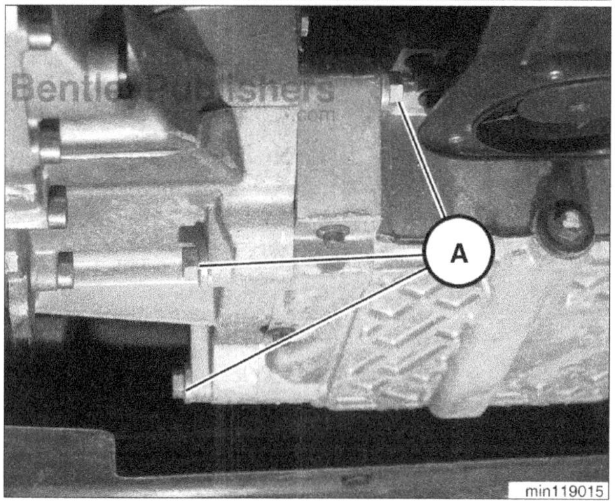

◀ Remove transmission retaining bolts (**A**).

NOTE —
The two shorter bolts are installed in oil pan.

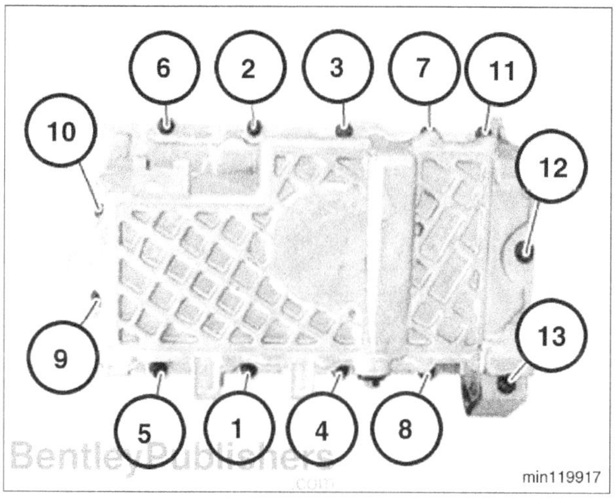

◀ Remove oil pan bolts in sequence (13 - 1) and remove pan.

– Clean surface of oil pan and cylinder block and check for surface damage.

– Installation is reverse of removal noting the following:
  - Replace gasket.
  - Tighten oil pan bolts in reverse sequence to removal (1-13).
  - Replace engine oil and check for leaks.

Tightening torque	
Oil pan to cylinder block	31 Nm (23 ft-lb)

Component Replacement

Oil pressure warning switch, replacing

Oil pressure warning switch is located on side of oil filter housing.

– Remove right front wheel and fender lining to gain access to oil filter housing.

– Using a 36 mm socket, loosen oil filter housing cover slowly to allow oil to drain back into crankcase.

– Tighten oil cover.

Tightening torque	
Cover to oil filter housing	25 Nm (18 ft-lb)

◄ Disconnect electrical connector (**A**) and remove oil pressure switch.

NOTE —
Replace sealing washer anytime oil pressure warning switch is removed.

Tightening torque	
Oil pressure warning switch to oil filter housing	20 Nm (15 ft-lb)

Oil pump, removing and installing

– Remove timing case cover. See **117 Camshaft Timing Chain**.

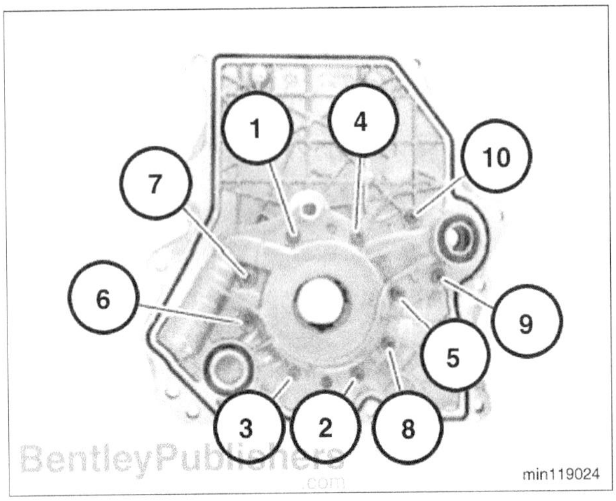

◄ Remove oil pump cover retaining bolts in sequence (10-1).

– Remove oil pump cover.

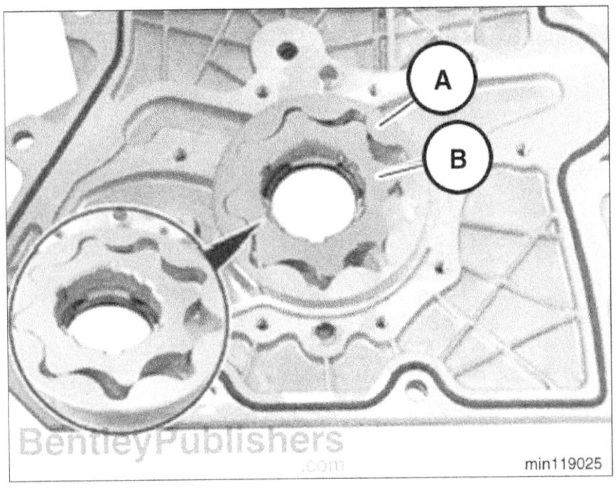

◄ Remove outer gear (**A**) and inner gear (**B**).

NOTE —
Outer and inner oil pump gears form a matched pair and must always be replaced as a set.

– Installation is reverse of removal noting the following:
- Tighten oil pump cover retaining bolts in reverse order (1-10).
- Replace timing cover and check for leaks.

Oil pressure relief valve, removing and installing

– Remove timing case cover. See **117 Camshaft Timing Chain**.

– Remove oil pump. See **Oil pump, removing and installing** in this repair group.

◄ Using a suitable drift, press down on valve spacer to remove load from circlip (**arrow**) and remove circlip.

> ***CAUTION—***
> *Relief valve is under strong spring pressure. Use care when removing.*

– Remove oil pressure relief valve.

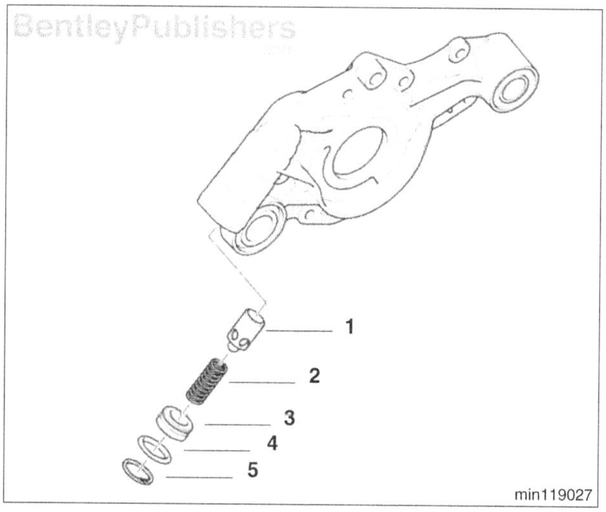

◄ Replace relief valve components in the proper order:

1. Relief valve
2. Spring
3. Spacer
4. O-ring
5. Circlip

– Check oil pump housing and relief valve bore for scoring or damage before replacing.

120 Ignition System

TABLE

GENERAL

This repair group covers component troubleshooting and replacement information for the ignition system.

When diagnosing engine management problems, including on-board diagnostics (OBD II) fault code analysis, also refer to these repair groups:

- **130 Fuel Injection**
- **ELE Electrical Wiring Diagrams**
- **OBD On Board Diagnostics**

Spark plug replacement is covered in **020 Maintenance**.

Special tools

Diagnosis and testing of the ignition system requires special test equipment. If you lack the necessary tools to perform a procedure, repairs are best left to an authorized MINI dealer or other qualified repair facility.

> ***CAUTION—***
> *Only use a digital multimeter for electrical tests.*

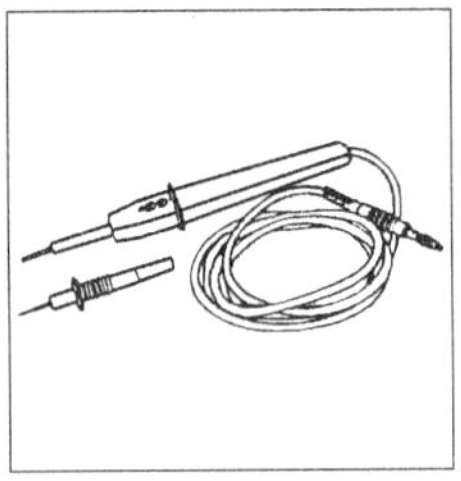

◄ LED test light
(Baum tool 1115)

◄ Automotive digital multimeter

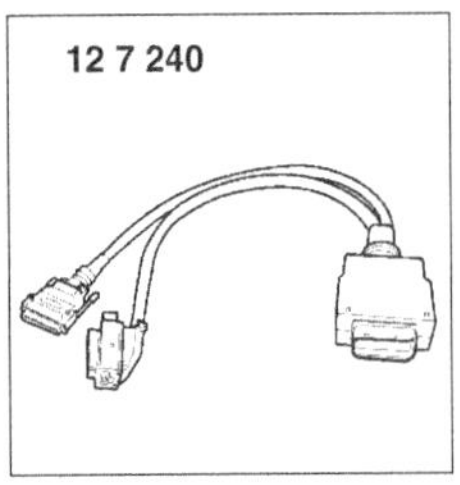

◄ Test lead harness
(BMW special tool 12 7 240)

Engine management

The MINI engine use an advanced engine management system known as EMS2000. This system incorporates on-board diagnostics, fuel injection, ignition and other engine control functions into a single engine control module (ECM).

Second generation on-board diagnostics (OBD II) is incorporated into the engine management system. Use a BMW-specific or compatible scan tool to access diagnostic trouble codes (DTCs) and help pinpoint ignition and other engine management problems.

Additional information about DTCs and engine management system electronic system diagnosis is provided in **OBD On Board Diagnostics.**

Ignition system

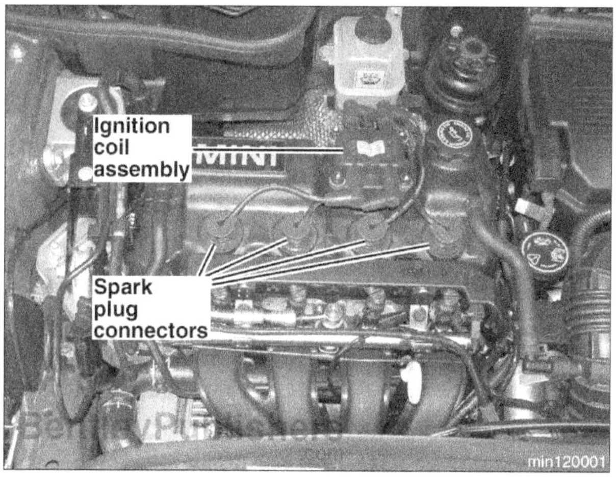

The MINI engine uses a distributorless ignition system. There is no distributor cap or ignition rotor.

A knock sensor (**arrow**) monitors the combustion chambers for engine-damaging knock. If engine knock is detected, the ignition timing is retarded by the ECM.

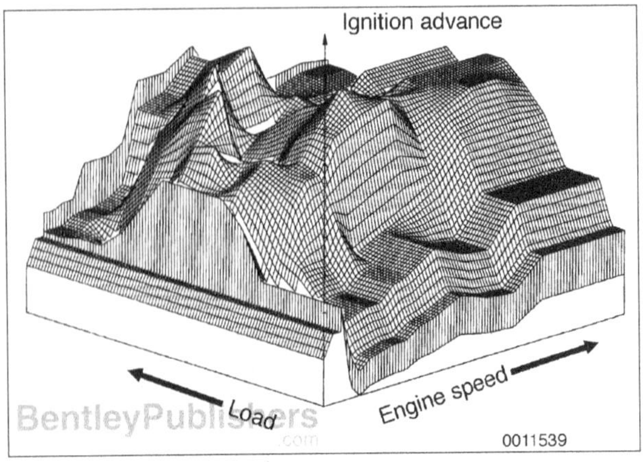

Ignition timing is electronically mapped and not adjustable. The ECM uses engine load, engine speed, coolant temperature, knock detection, and intake air temperature as the basic inputs for timing control. A three dimensional map similar to the one shown is digitally stored in the ECM.

The initial ignition point is determined by the crankshaft speed sensor during cranking. Once the engine is running, timing is continually adjusted based on operating conditions.

Troubleshooting

Troubleshooting ignition system faults should begin with an interrogation of the ECM fault memory. OBD II hardware and software monitors detect ignition system misfire and other related faults. When faults are detected, the ECM stores DTCs along with other pertinent fault information.

> ***WARNING—***
> *Ignition misfires can cause high hydrocarbon exhaust emissions and catalytic converter damage. For this reason, if a severe misfire is detected, the fuel injector will be turned off to the specific cylinder and the MIL will be illuminated. A misfire may also produce an overheated catalytic converter, which can be a fire hazard.*

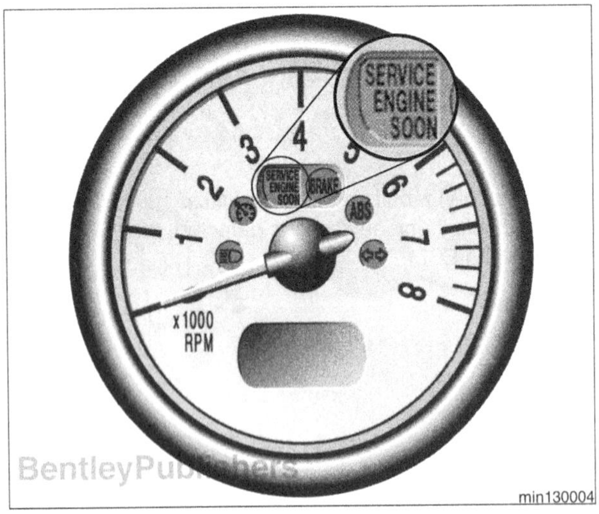

The malfunction indicator light (MIL) illuminates if an emissions related fault is detected. Additional OBD II information, including a DTC lookup table, can be found in **OBD On Board Diagnostics**.

NOTE—
In MINI models, the MIL reads "Service engine soon".

Oscilloscope diagnostic diagrams

One way to diagnose a faulty coil assembly is to use an oscilloscope to analyze spark quality with the engine running.

See **Table a** for a list of common ignition coil voltage faults and related causes.

Table a. Ignition secondary voltage diagnostics

Test point	Secondary voltage low	Secondary voltage high
Spark plug electrode gap	Too small	Too big
Spark plug electrode condition		Worn/burnt
Spark plug electrode temperature	Too high	Too low
Engine compression	Too low	Too high
Spark plug wires		Faulty
Fuel air mixture		Too lean

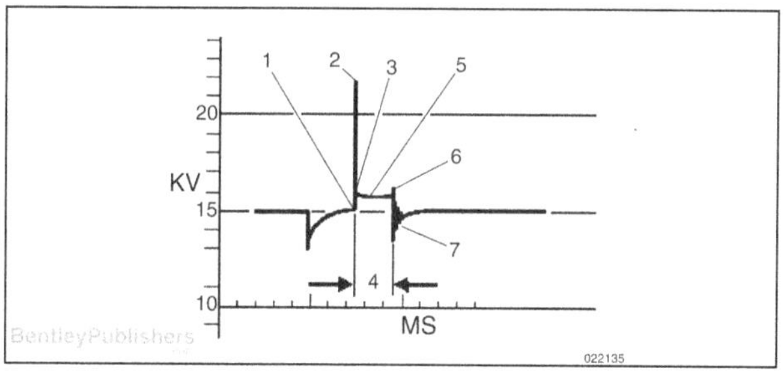

The illustration shows normal 'scope trace of spark at idle.

1. Start of ignition voltage peak
2. Level of ignition voltage
3. Level of combustion voltage
4. Period of combustion
5. Combustion curve characteristics
6. Start of decay process
7. Termination oscillations

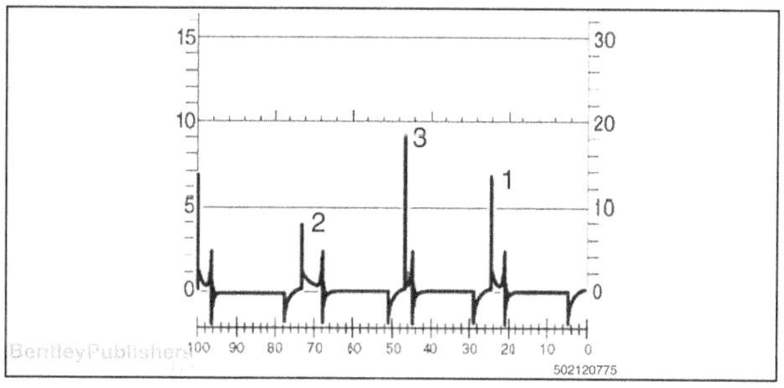

Ignition spark at idle speed.

1. Normal ignition voltage peak indicates spark plug is good.
2. Low voltage peak indicates plug gap too narrow.
3. High voltage peak indicates plug gap too wide.

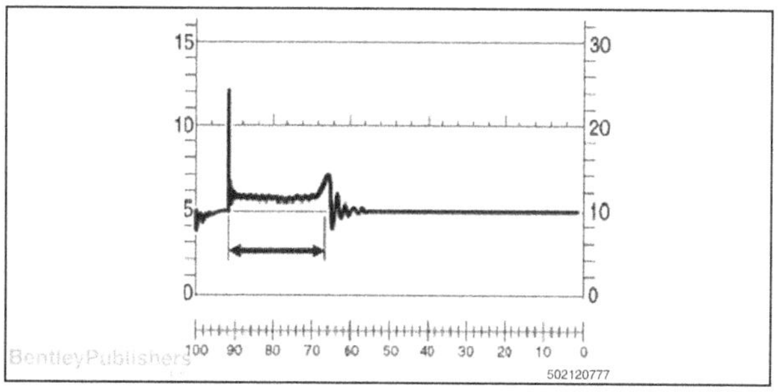

Long combustion period indicates spark plug gap too narrow.

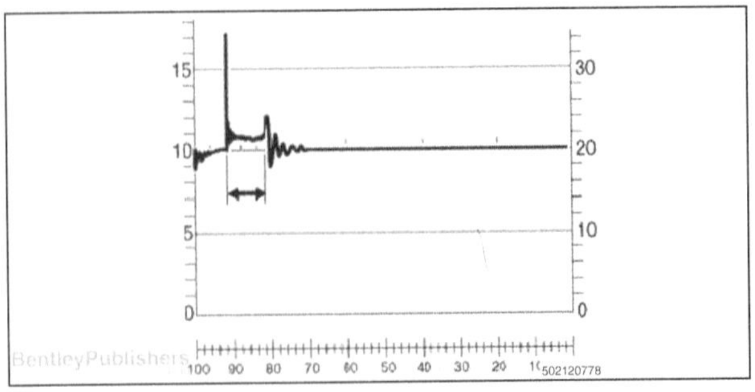

Short combustion period indicates spark plug gap too wide.

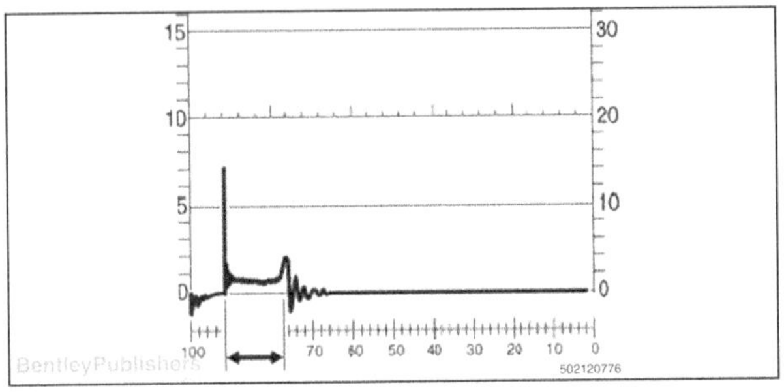

Normal combustion period at idle.

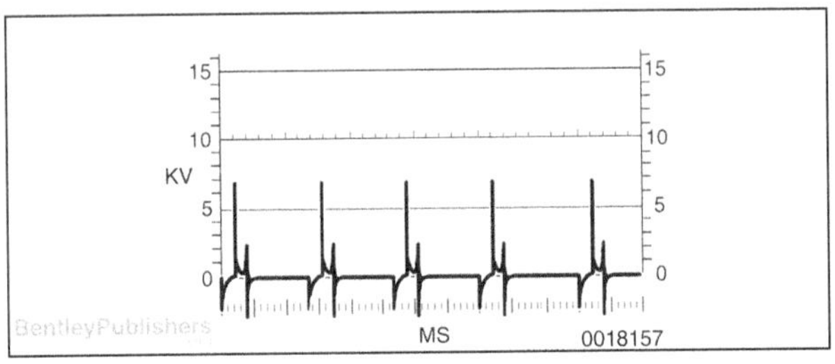

Normal oscilloscope pattern for ignition system at idle.

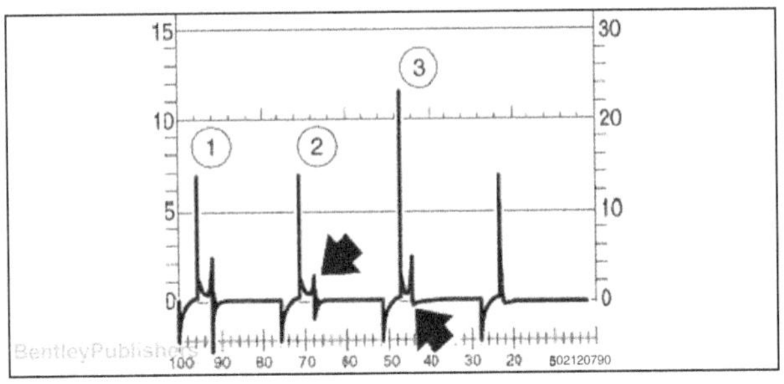

Ignition voltage peaks at idle.

1. Normal ignition peaks.
2. Downward peak (**arrow**) is shortened (ignition coil is defective).
3. Downward peak (**arrow**) is missing completely (ignition coil is defective).

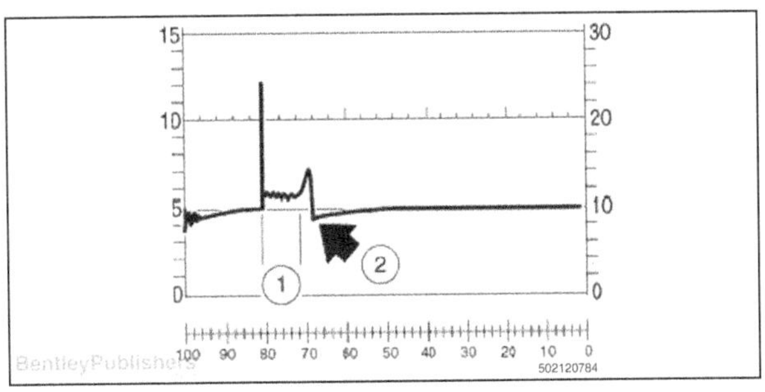

Defective ignition coil.

1. Short spark period.
2. Spark voltage line (**arrow**) is very slight.

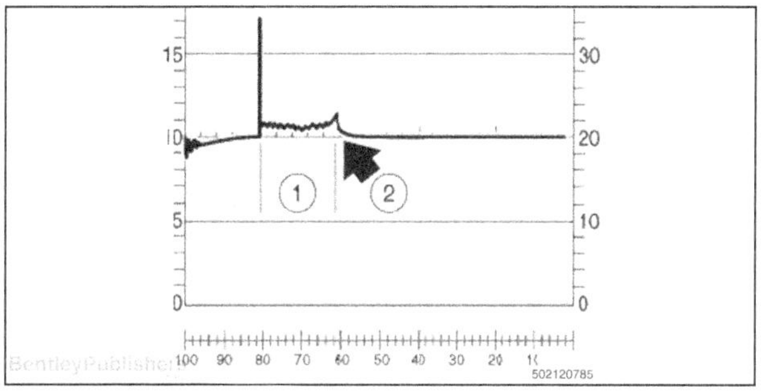

Defective ignition coil.

1. Normal combustion period.
2. Absence of spark voltage line (**arrow**).

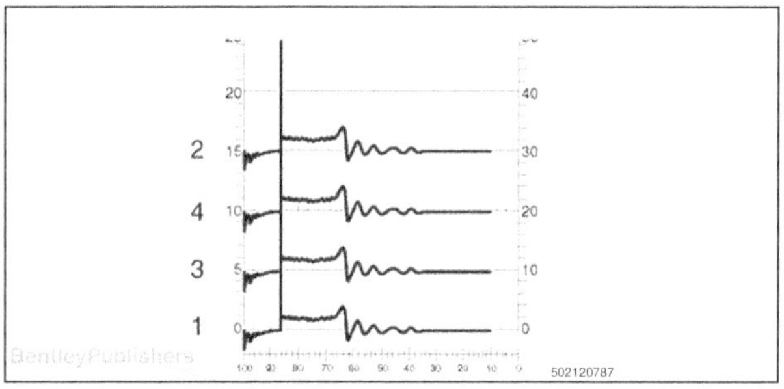

Normal secondary voltage patterns.

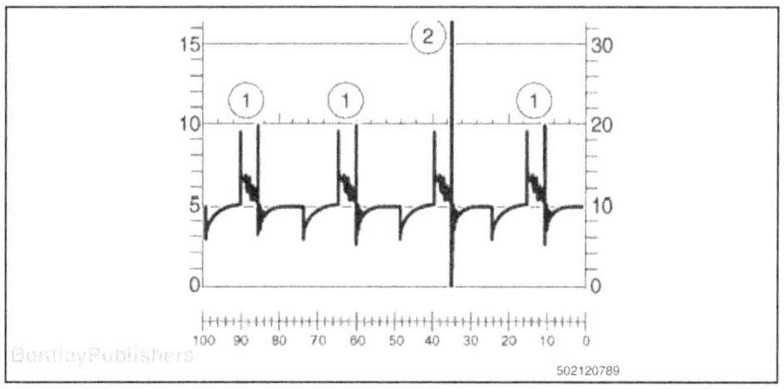

Ignition voltage peaks in response to sudden acceleration load:

1. Normal ignition pattern: Beginning of dying out pattern is not much higher than ignition voltage peak.
2. Faulty ignition system: Beginning of dying out pattern considerably higher than ignition voltage peak. Possible causes:
   - Lean fuel mixture
   - Defective fuel injector
   - Low compression in cylinder

Warnings and cautions

The ignition system produces dangerous high voltage. In addition, sensitive electronic components can be damaged if proper precautions are not adhered to.

> ***WARNING —***
> - *Do not touch or disconnect any cables from the coil assembly while the engine is running or being cranked by the starter.*
> - *The ignition system produces high voltages that can be fatal. Avoid contact with exposed terminals. Use extreme caution when working on a car with the ignition switched on or the engine running.*
> - *Connect and disconnect the DME system wiring and test equipment leads only when the ignition is OFF.*
> - *Before operating the starter without starting the engine (for example when making a compression test) always disable the ignition. See* **Disabling ignition system** *in this repair group.*

> ***CAUTION*** *—*
> - *Connect or disconnect ignition system wires, multiple wire connectors, and ignition test equipment leads only while the ignition is off. Switch multimeter functions or measurement ranges only with the test probes disconnected.*
> - *Do not disconnect the battery while the engine is running.*
> - *Use a high impedance digital multimeter for all voltage and resistance tests. Use an LED test light in place of an incandescent-type test lamp.*
> - *In general, make test connections only as specified by BMW, as described in this manual, or as described by the instrument manufacturer.*

IGNITION SYSTEM SERVICE

Ignition firing order

Cylinder 1 is located at right (drive belt) end of engine.

Firing order	
Cooper, Cooper S	1-3-4-2

Disabling ignition system

The ignition system operates in a lethal voltage range. Disable ignition any time engine service or repair work is being done that requires the ignition to be switched on.

Working at top of valve cover, disconnect electrical harness connector to ignition coil assembly.

- Press securing clip in direction of **white arrow**.
- Pull off coil assembly connector in direction of **black arrow**.

Ignition wire, replacing

- Switch off ignition.
- Pull up on spark plug wire connector while gently twisting from side to side.
- Detach other end of wire from coil terminal.

- When installing, make sure cable snaps securely into place at coil terminal and at spark plug opening in valve cover.

Ignition coil assembly, replacing

- Read out engine control module (ECM) fault memory. Then switch off ignition.

- Working at top of valve cover, disconnect electrical harness connector to ignition coil assembly.
  - Press securing clip in direction of **white arrow**.
  - Pull off coil assembly connector in direction of **black arrow**.
- Detach ignition cables from coil assembly.

Unbolt coil assembly mounting bolts (**arrows**).

- When installing, check coil assembly insulating rubber spacers. Replace mounting bolts and spacers as necessary.

Tightening torque	
Ignition coil assembly to valve cover (M6)	12 Nm (9 ft-lb)

- Clear ECM fault memory.

Crankshaft position sensor, replacing

- Read out engine control module (ECM) fault memory. Then switch off ignition.
- Cooper: Remove intake manifold. See **113 Cylinder Head Removal and Installation**.
- Cooper S: Remove front bumper and loosen modular front end. Pull assembly forward to gain access to front of crankcase.

NOTE —
Removal of modular front end is a complicated job. It is covered in **510 Bumpers, External Trim**.

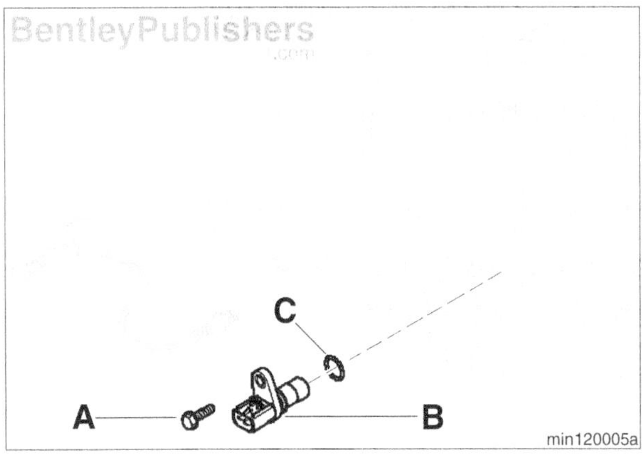

Working at crankshaft position sensor, in front of crankcase at clutch end of engine, unlock and detach electrical harness connector.

- Unscrew mounting bolt (**A**).
- Remove crankshaft sensor (**B**).
- When installing, replace sealing O-ring (**C**).

Tightening torque	
Crankshaft sensor to crankcase (M6)	9 Nm (7 ft-lb)

- After reassembling engine, clear ECM fault memory.

Camshaft position sensor, replacing

– Read out engine control module (ECM) fault memory.

– Switch off ignition.

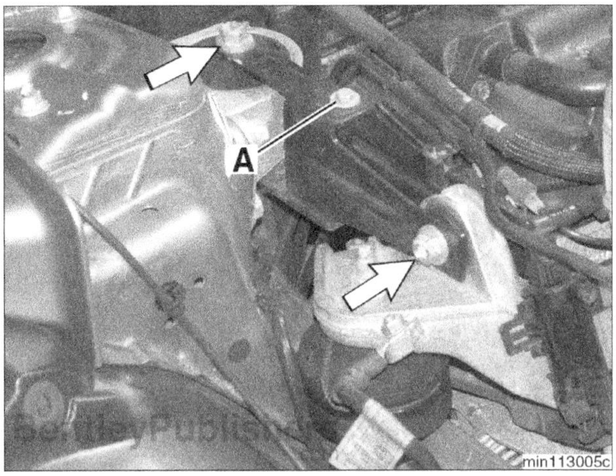

◄ Remove engine vibration damper bracket:

- Remove fuel line bracket mounting bolt (**A**). Detach any fuel lines still attached. Lift off bracket and set aside.
- Remove engine vibration damper bracket bolts (**arrows**).
- Remove bracket.

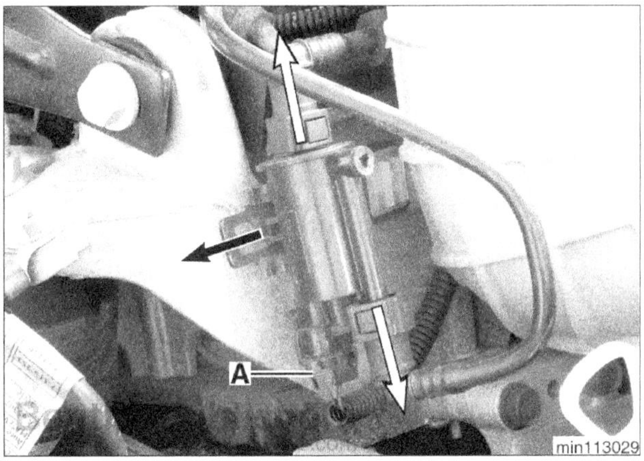

◄ Remove fuel tank vent valve.

- Disconnect vent hoses (**white arrows**) via quick fit connectors.
- Straighten retaining tab and slide vent valve off bracket (**black arrow**).
- Remove electrical harness connector (**A**).

– Support engine under oil pan.

> ***CAUTION—***
> *To avoid damaging the oil pan, use a rubber pad on top of the engine supporting jack.*

– Remove right side engine mounting bracket:

- Remove engine ground cable mounting nut from bracket. Detach ground cable.
- Remove 4 engine mounting bolts from engine block.
- Unscrew mounting nut from hydraulic front engine mount stud. Lift off engine mount bracket.

◄ Working at right end of cylinder head, unlock and detach camshaft sensor electrical harness connector (**arrow**).

– Unscrew sensor mounting bolt and remove sensor.

– When installing, lubricate sealing O-ring with antiseize paste.

Tightening torque	
Camshaft sensor to cylinder head (M6)	9 Nm (7 ft-lb)

– After reassembling engine, clear ECM fault memory.

Knock sensor, replacing

- Read out engine control module (ECM) fault memory. Then switch off ignition.
- Cooper: Remove intake manifold. See **113 Cylinder Head Removal and Installation**.
- Cooper S: Remove supercharger. See **130 Fuel Injection**.
- Working at center of engine block just below cylinder head seam, unlock and detach knock sensor electrical harness connector.

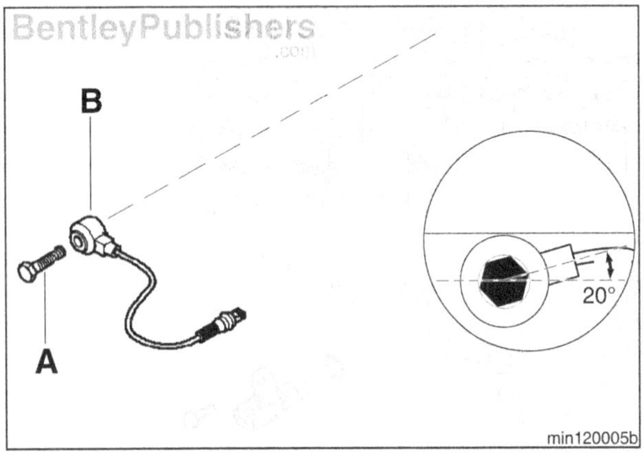

- Note knock sensor orientation before removal. Unscrew sensor mounting bolt (**A**) and remove sensor (**B**).
- When installing:
  - Clean knock sensor and crankcase contact surfaces.
  - Orient knock sensor in relation to engine block with tilt of 20° from horizontal. Refer to insert in illustration.

Tightening torque	
Knock sensor to crankcase (M8)	22 Nm (16 ft-lb)

- After reassembling engine, clear ECM fault memory.

121 Battery, Starter, Alternator

TABLES

General

This section covers battery, starter, alternator and associated components of the electrical system.

Troubleshooting information for these components is found in **Table a**. For additional electrical troubleshooting information, see **600 Electrical System - General**.

Special tools

Automotive digital multimeter

> ***CAUTION —***
> *Only use a digital multimeter when testing automotive electrical components.*

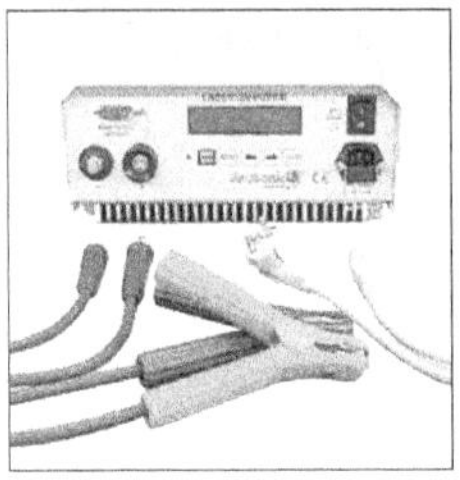

Electronic battery charger
(Deutronic DBL 1000-14)

61 2 410

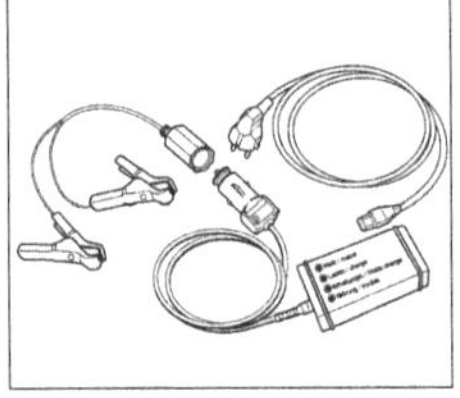

Cigarette lighter trickle charger
(BMW special tool 61 2 410)

Engine electrical system

The alternator and starter are wired directly to the battery. To prevent accidental shorts that might blow a fuse or damage wires and electrical components, always disconnect the negative (-) battery lead before working on the electrical system.

Various versions of alternators, voltage regulators, starters, and batteries have been used in the MINI models covered by this manual. Replace components according to the original equipment specification. When in doubt, consult an authorized MINI parts department.

Battery safety terminal (BST)

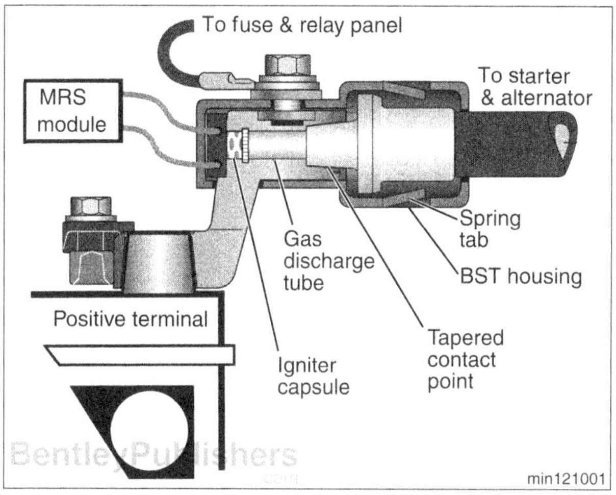

In Cooper S models, due to battery location in the cargo compartment, a battery safety terminal (BST) is installed at the battery positive pole. This feature and a comprehensive electrical fusing system aim to minimize the danger of short circuits in case of a severe accident. To achieve this, the vehicle's electrical system is divided into the starter-alternator circuit and the vehicle system supply circuit:

- The vehicle system supply circuit is protected against short circuits by means of a fusible link.
- The starter and the alternator supply circuit, since it must carry very high current when the starter is engaged and is permanently subjected to voltage, is protected by BST.

The multiple restraint system (MRS) controls the BST. In case of an accident, the MRS control module fires the encapsulated pyrotechnic device in the BST, disconnecting power to the starter and alternator, but maintaining power to the exterior lights and interior of the vehicle.

Troubleshooting

Tests for individual electrical system components are described under different components in this repair group.
Table a gives some general troubleshooting ideas.

Table a. Battery, starter and alternator troubleshooting

Symptom	Probable cause	Corrective action
Engine does not crank.	Fault in immobilizer system (EWS).	Try another ignition key. If problem persists, contact your authorized dealer.
Engine cranks slowly or not at all, solenoid clicks when starter is operated.	Battery cables loose, dirty or corroded.	Clean or replace cables. See **020 Maintenance**.
	Battery discharged.	Charge battery and test. Replace if necessary.
	Battery to body ground cable loose, dirty or corroded.	Inspect ground cable. Clean, tighten or replace if necessary.
	Poor connection at starter motor terminal 30.	Check connections, test for voltage at starter. Test for voltage at neutral safety or clutch interlock switch.
	Starter motor or solenoid faulty.	Test starter.
Battery will not stay charged more than a few days.	Short circuit draining battery.	Test for excessive current drain with everything electrical off.
	Short driving trips and high electrical drain on charging system.	Evaluate driving style. Where possible, reduce electrical consumption when making short trips.
	Engine drive belt loose, worn, damaged.	Inspect or replace belt. See **020 Maintenance**.
	Battery faulty.	Test battery and replace if necessary.
	Battery cables loose, dirty or corroded.	Clean or replace cables. See **020 Maintenance**.
	Alternator or voltage regulator faulty.	Test alternator and voltage regulator.

Warnings and cautions

WARNING —

- *Wear goggles, rubber gloves, and a rubber apron when working around the battery or battery acid (electrolyte).*
- *Battery acid contains sulfuric acid and can cause skin irritation and burning. If acid is spilled on your skin or clothing, flush the area at once with large quantities of water. If electrolyte gets into your eyes, flush them with large quantities of clean water for several minutes and call a physician.*
- *Batteries that are being charged or are fully charged give off explosive hydrogen gas. Keep sparks and open flames away. Do not smoke.*

CAUTION —

- *Only use a digital multimeter when testing automotive electrical components.*
- *If a repair procedure specifies disconnecting the battery, follow the instruction for safety reasons.*
- *Prior to disconnecting the battery, read the battery disconnection cautions in* **001 General Warnings and Cautions**.
- *Before disconnecting battery, switch ignition OFF. If the ignition is not turned off when the battery is disconnected, diagnostic troubles codes (DTCs) may be set in some electronic control modules.*
- *Disconnecting the battery cables may erase DTCs stored in ECM memory.*
- *Disconnecting the battery erases the radio presets. Therefore, note stored stations and restore them after connecting the battery.*
- *Stored settings of the on-board computer and clock will also be lost.*
- *Always disconnect the negative (–) battery cable first and reconnect it last. Cover the battery post with an insulating material whenever the cable is removed.*
- *Do not disconnect battery, alternator or starter wires while the engine is running.*
- *Never reverse the battery cables. Even a momentary wrong connection can damage the alternator and other electrical components.*
- *Do not depend on the color of insulation to tell battery positive and negative cables apart. Label cables before removing.*
- *After disconnecting battery, reinitialize power windows and electric sunroof. See* **512 Door Windows** *and* **540 Sunroof**.

Battery

MINI models are equipped with a 12-volt negative ground electrical system.

Cooper: Battery is located in left rear of engine compartment in an insulating box with a heat resistant cover (**A**).

Cooper S: Battery is located in center of cargo compartment, under a trim cover.

Battery capacity

Required battery capacity is determined by the amount of current consumed by the electrical system.

MINI batteries are rated by ampere-hours (Ah), determined by the average amount of current the battery can deliver over time without dropping below a specified voltage. Always replace the battery with one having the same or higher rating.

Battery capacity	
Cooper, Cooper S	55 Ah

Battery charging

Recharge discharged batteries using a battery charger. Prolonged charging causes electrolyte evaporation to a level that can damage the battery. It is best to use a low-current charger (6 amperes or less) to prevent battery damage caused by overheating.

Cooper S: When charging battery, connect positive charger cable to B+ terminal (**arrow**) in engine compartment. Connect negative cable to suitable exposed metal in engine compartment (such as engine or transmission mounting bracket).

- Before charging battery, test battery as described in **Battery open-circuit voltage test** in this repair group.
  - If voltage is 10 vdc or less, one or more cells may be faulty or battery may already be damaged.
  - In this case, remove battery from vehicle to recharge, because escaping battery gases could damage interior equipment and trim.
  - Remove battery cap covers and top up cells with distilled water. Leave caps off while charging.
  - Attempt to revive faulty cells with low charging current.

WARNING —
- *Hydrogen gas given off by the battery during charging is explosive. Do not smoke. Keep open flames away from the top of the battery, and prevent electrical sparks by turning off the battery charger before connecting or disconnecting it.*
- *When charging battery in the Cooper, make sure battery box cover is off.*
- *When charging battery in the Cooper S, make sure the cargo compartment door is open and battery trim cover in bottom of compartment is lifted up.*

CAUTION —
- *Battery electrolyte (sulfuric acid) can damage the car. If electrolyte is spilled, clean the area with a solution of baking soda and water.*
- *Always allow a frozen battery to thaw before attempting to recharge it.*
- *If a quick charger is used to charge the battery, disconnect battery from vehicle electrical system and remove. This prevents damage to paintwork and upholstery.*

Battery maintenance

Check battery acid level at least once a year. If necessary, top up with distilled water.

Electrical consumers require power supply from the battery even when the vehicle is not being driven. The battery also self-discharges. Battery self-discharging time is dependent on vehicle model and vehicle equipment.

If a long time passes without the battery being recharged, it will eventually result in concealed damage. This will in turn lead to early failure of the battery.

- If battery is connected to vehicle circuits: Recharge every 6 weeks.
- If battery is not connected to vehicle circuits: Recharge every 12 weeks.
- For vehicle in continual but light use, use cigarette lighter trickle charger, BMW special tool 61 2 410, to maintain optimum battery charge. This charger, with electronic circuitry for controlling voltage and current, switches charge voltage off at 13.8 vdc and back on again at 12.6 vdc. Charge current is no greater than 5.5 A.

Battery testing

Battery testing determines the state of battery charge. On conventional or low-maintenance batteries, the most common method of testing the battery is to check the specific gravity of the electrolyte using a hydrometer. Before testing the battery, check that the cables are tight and free of corrosion.

NOTE —
In several battery tests given below, it is assumed that the battery cell caps are removed. This is not possible in some "maintenance-free" batteries.

Hydrometer testing

Before hydrometer testing, load the battery with 15 amperes for one minute. If the battery is installed in the vehicle, this can be done by turning on the headlights without the engine running. The state of the battery charge based on specific gravity values are given in **Table b**.

The hydrometer indicates the specific gravity of the electrolyte. The more dense the concentration of sulfuric acid in the electrolyte, the higher the state of charge.

NOTE —
Electrolyte temperature affects hydrometer reading. Check the electrolyte temperature with a thermometer. Add 0.004 to the hydrometer reading for every 6° C (10° F) that the electrolyte is above 27° C (80° F). Subtract 0.004 from the reading for every 6° C (10° F) that the electrolyte is below 27° C (80° F).

Table b. Specific gravity of battery electrolyte at 27°C (80°F)

Specific gravity	State of charge
1.265	Fully charged
1.225	75% charged
1.190	50% charged
1.155	25% charged
1.120	Fully discharged

– If electrolyte specific gravity is at or above 1.225, but battery lacks power for starting, determine battery service condition with a load voltage test. See **Battery load voltage test** in this repair group.

– If average specific gravity of battery cells is below 1.225, recharge battery.

– After charging several hours, if electrolyte specific gravity remains low or battery lacks starting power, replace battery.

Battery open-circuit voltage test

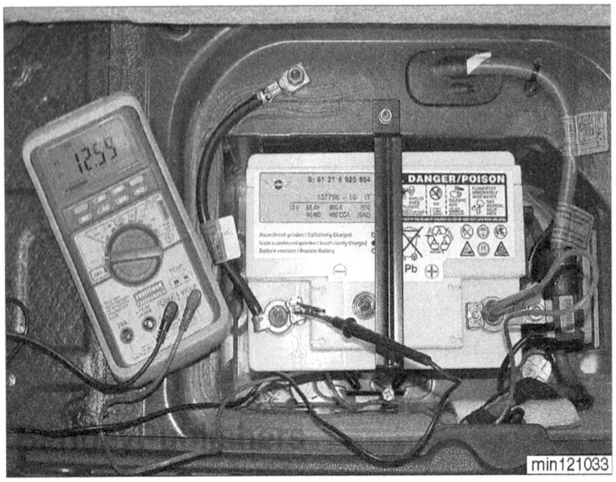
min121033

Before testing, load battery with 15 amperes for one minute with battery load-tester or turn on headlights for about one minute without engine running. Connect digital voltmeter across battery terminals. Open-circuit voltage levels are given in **Table c**.

– If open-circuit voltage is OK but battery still lacks power for starting, perform load voltage test.

– If open-circuit voltage is below 12.4 vdc, recharge battery and retest.

Table c. Open-circuit voltage and battery charge	
Open-circuit voltage	**State of charge**
12.6 vdc or more	Fully charged
12.4 vdc	75% charged
12.2 vdc	50% charged
11.7 vdc or less	Fully discharged

Battery load voltage test

– Disconnect battery cables before making test.

> ***CAUTION—***
> *Prior to disconnecting the battery, read the battery disconnection cautions in* **001 General Warnings and Cautions**.

– Using battery load tester, apply high resistive load (approx. 200 amps) to battery terminals. Measure battery voltage.

NOTE—
The battery should be fully charged for the most accurate results.

– Check results against data in **Table d**. Replace battery if voltage is below minimum.

Table d. Battery load test - minimum voltage	
Ambient temperature	**Voltage***
27°C (80°F)	9.6 vdc
16°C (60°F)	9.5 vdc
4°C (40°F)	9.3 vdc
-7°C (20°F)	8.9 vdc
-18°C (0°F)	8.5 vdc
*Measured after applying a 200 amp load for 15 seconds	

Battery, removing and installing (Cooper)

Cooper battery is in engine compartment, in insulated box.

- Working in engine compartment, pry off rubber seal (**A**).
- Release battery box cover clips (**B**) by pressing in. Remove cover.

- First disconnect negative (–) battery cable, then positive (+) cable.

> ***CAUTION—***
> *Prior to disconnecting the battery, read the battery disconnection cautions in* **001 General Warnings and Cautions**.

- Detach battery ventilation hose (**arrow**).
- Remove battery hold down retaining bolts (**A** and **B**). Remove hold down.
- Lift out battery.
  - It may be helpful to use a battery carrying strap to lift battery out.
- When reinstalling, first connect positive (+) cable, then negative (–) cable. Ensure that both leads are positioned 90° to battery box and routed through recesses in battery box cover.

> ***CAUTION—***
> *Do not apply excessive force when fitting cover to box.*

Tightening torques	
Battery hold down to battery box (M6)	6 Nm (53 in-lb)
Battery terminal clamp (M6)	5 Nm (44 in-lb)

- Reattach ventilation hose.
- Reinitialize power windows and electric sunroof. See **512 Door Windows** and **540 Sunroof**.

Battery box, removing and installing (Cooper)

– Turn ignition off.

– Remove battery box cover and battery. See **Battery, removing and installing (Cooper)** in this repair group.

> ***CAUTION—***
> *Disconnecting the battery may erase fault code(s) stored in memory. Check for fault codes prior to disconnecting the battery cables.*

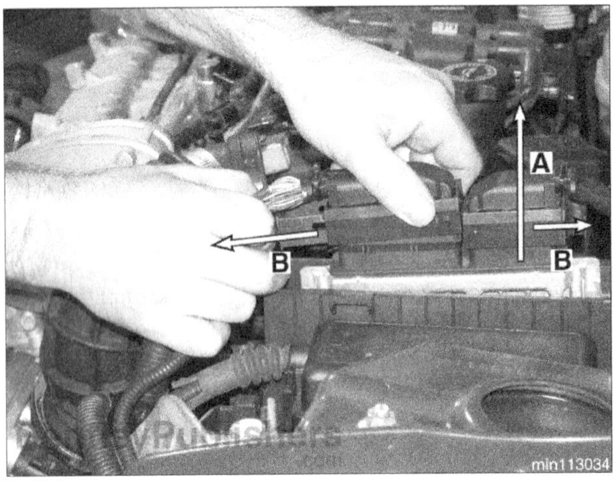

◄ Remove DME control module:

- Remove cover and lift DME control module upwards (**A**).
- Slide connector locking mechanisms (**B**) outward and disconnect control module connectors.

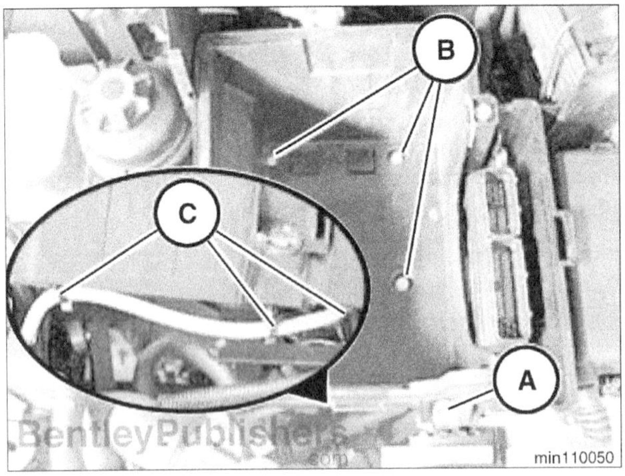

◄ Remove battery box:

- Remove battery box retaining bolt (**A**).
- Release battery box clips (**B**).
- Disconnect harness clips (**C**) from bottom of battery box.

– Installation is reverse of removal.

Tightening torques	
Battery box to support bracket (M6 Torx)	8 Nm (6 ft-lb)
Battery hold down to battery box (M6)	6 Nm (53 in-lb)
Battery terminal clamp (M6)	5 Nm (44 in-lb)

Battery, removing and installing (Cooper S)

Cooper S battery is in center of cargo compartment, under floor trim panel.

- Working in cargo compartment, lift up floor trim.

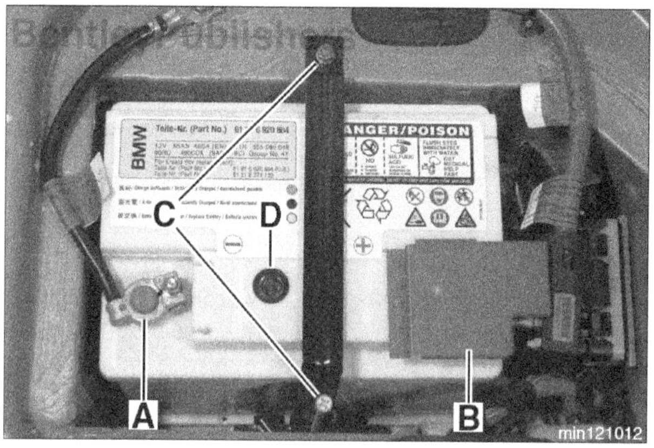

- First disconnect negative (–) battery cable (**A**). Lift positive (+) terminal plastic cover (**B**), then disconnect positive (+) cable.

> ***CAUTION—***
> *Prior to disconnecting the battery, read the battery disconnection cautions in* **001 General Warnings and Cautions**.

- Detach battery ventilation hose.
- Remove battery hold down fasteners (**C**). Remove hold down.
- Lift out battery.
- When reinstalling, first connect positive (+) cable, then negative (–) cable.

Tightening torques	
Battery hold down to battery box (M6)	6 Nm (53 in-lb)
Battery terminal clamp (M6)	5 Nm (44 ft-lb)

- Make sure battery condition indicator (**D**) is visible next to battery hold down.
- Reattach ventilation hose.
- Reinitialize power windows and electric sunroof. See **512 Door Windows** and **540 Sunroof**.

Battery safety terminal (BST) replacement (Cooper S)

If BST has been triggered, investigate and correct cause prior to replacement. Read out and reset multiple restraint system (MRS) control module fault memory.

BST replacement requires replacement of one piece power supply cable to alternator and starter. This cable is routed through grommet in front of cargo compartment just ahead of battery, then underneath vehicle between rear exhaust heat shield and floor pan. It is then routed to left side of floor pan and on to engine compartment components including positive battery connection (B+).

ALTERNATOR (GENERATOR)

The alternator assembly with integrated voltage regulator is bolted to the front of the engine block.

Depending on level of equipment, one of two different alternators have been fitted to MINI models covered by this manual.

Alternator output	
Normal output	105 amp
High output	120 amp

Charging system troubleshooting

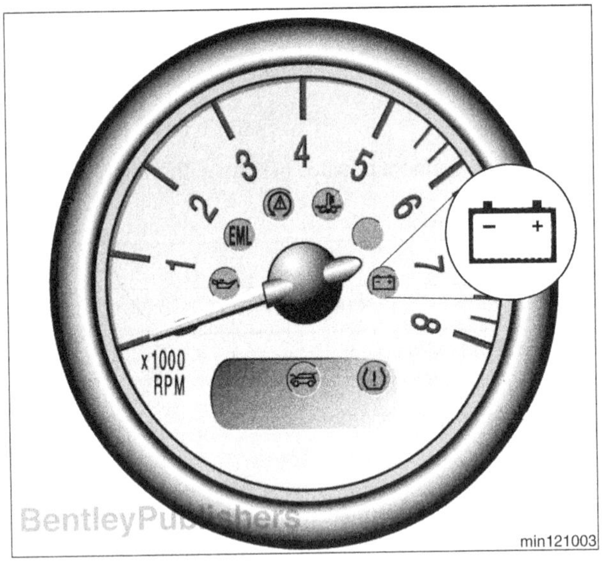

min121003

The alternator is not connected to the battery warning light. Alternator faults are communicated to the engine control module (ECM) which illuminates the battery warning light if a charging system fault is detected.

Charging system diagnostics requires special test equipment. If the test equipment is not available, charging system fault diagnosis can be performed by an authorized MINI dealer or other qualified repair shop.

– Before checking alternator:
  - Make sure battery is fully charged and capable of holding a charge.
  - Check for clean and tight battery cables.
  - Check ground cable running from negative (-) battery terminal to chassis and ground cable running from engine to chassis.
  - Check engine drive belt condition and tension.

See **Table a** for general electrical component troubleshooting.

Charging system quick check

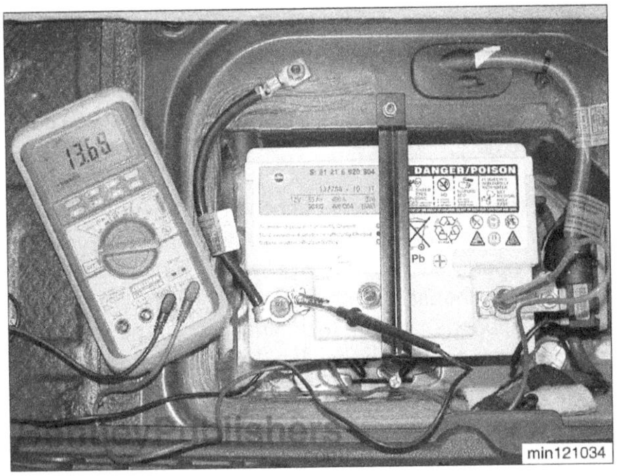

min121034

Use a digital multimeter to measure voltage across the battery terminals with key off and then again with engine running. Battery voltage should be about 12.6 volts with key off and between 13.5 and 14.5 volts with engine running.

NOTE —
The regulated voltage (engine running) should be between 13.2 and 14.5 volts, depending on temperature and operating conditions. If the voltage is higher than 14.8, the voltage regulator is most likely faulty.

Charging system, checking

> ***CAUTION—***
> - *Do not disconnect the battery while the engine is running. Damage to the alternator and/or engine electronic system may result.*
> - *Only use high impedance electronic test equipment when testing charging system components.*

– Read out engine control module (ECM) fault memory.

– Connect BMW diagnostic scan tool or oscilloscope to check alternator function.

NOTE—
An alternator output pattern other than "normal" indicates that the alternator needs rebuilding or replacement.

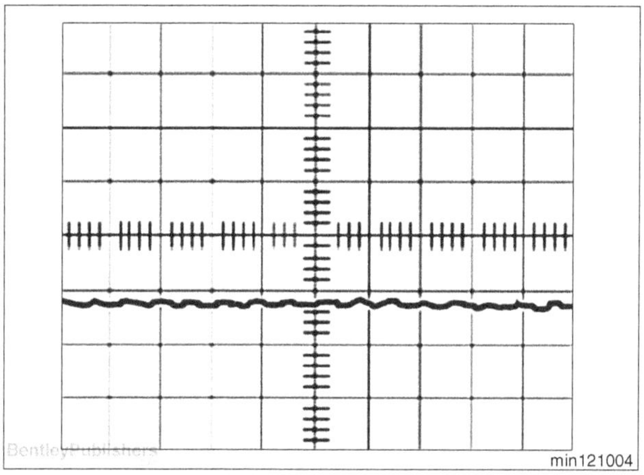

Normal alternator

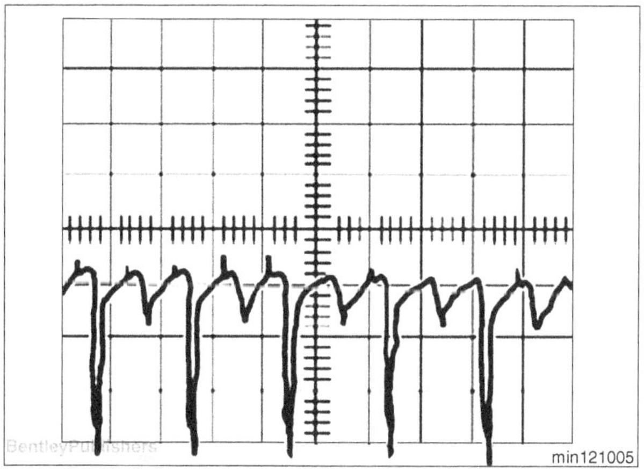

One phase interrupted

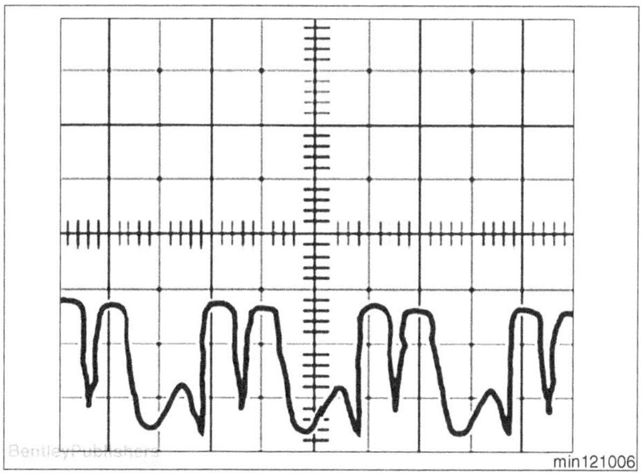

Interturn fault

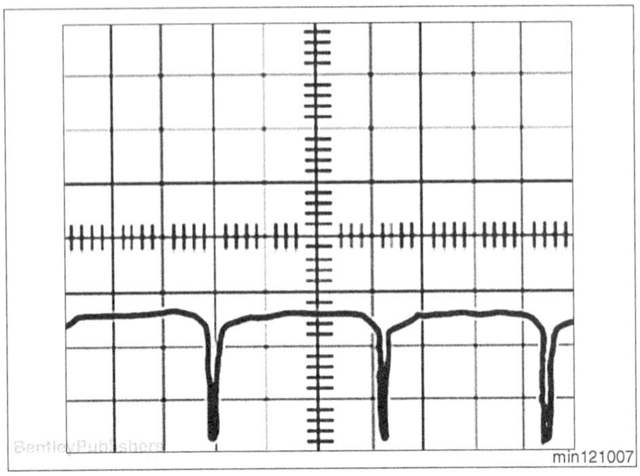

Open circuit in negative diode

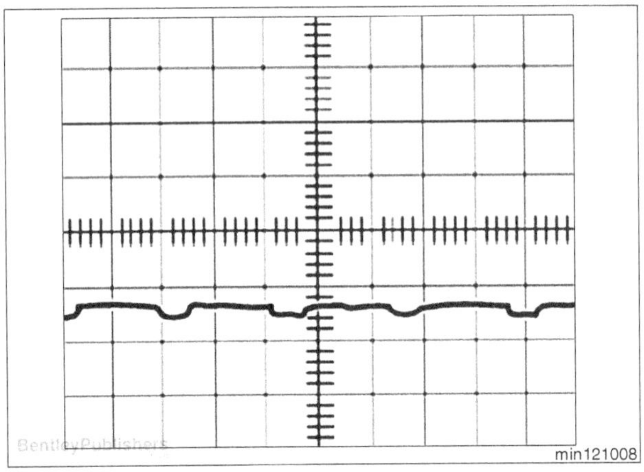

Short circuit in positive diode

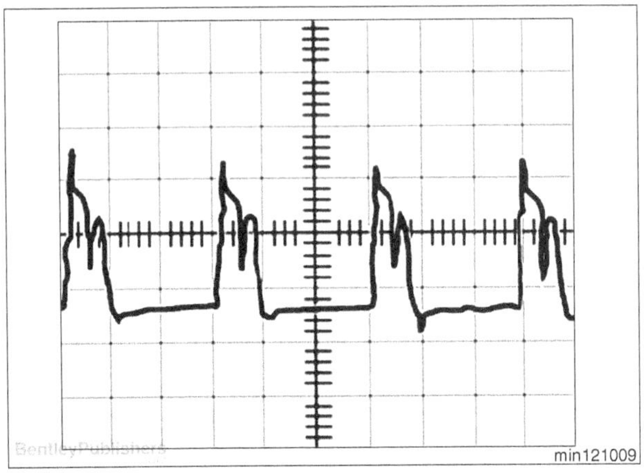

Open circuit in positive diode

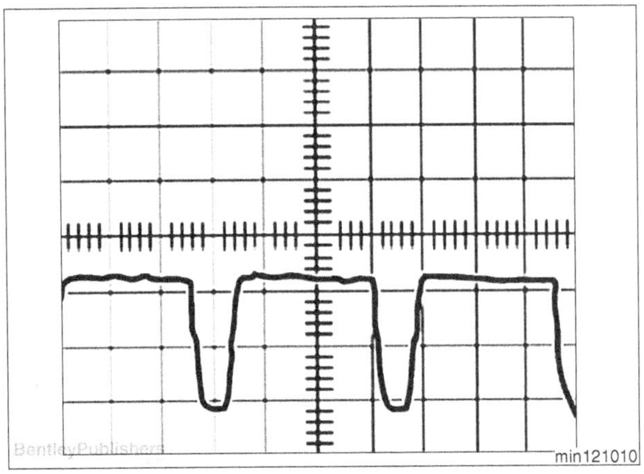

Open circuit in exciter diode

Alternator, removing and installing (Cooper)

– Read out engine control module (ECM) fault memory. Then switch off ignition.

– Disconnect battery.

> ***CAUTION—***
> *Prior to disconnecting the battery, read the battery disconnection cautions in* **001 General Warnings and Cautions**.

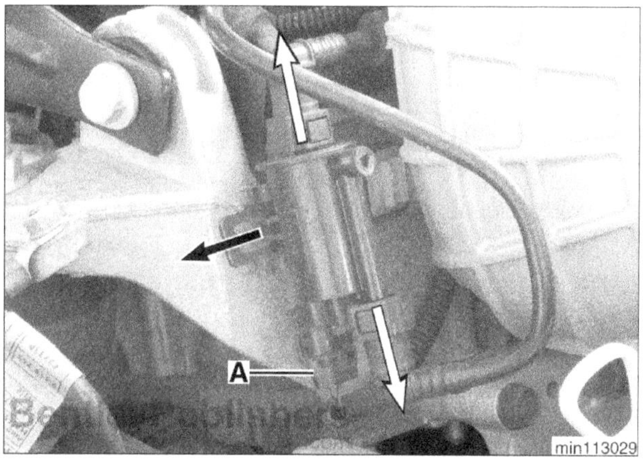

◄ Working at right side engine mount, remove fuel tank vent valve.

- Disconnect vent hoses (**white arrows**) quick fit connectors.
- Straighten retaining tab and slide vent valve off bracket (**black arrow**).
- Remove electrical harness connector (**A**).

NOTE—
Cooper S engine shown. Cooper is similar.

– Remove splash shield under engine.

– Remove right front wheel.

– Remove right front wheel housing liner.

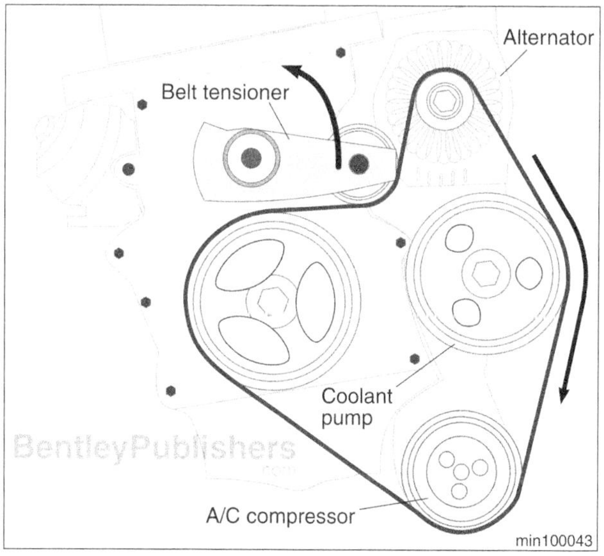

◄ Note engine drive belt layout. Remove belt.

- Use special tool 11 8 390 to release drive belt tension.
- Use lock pin (special tool 11 8 280) to lock drive belt tensioner.
- Lift off drive belt.

> ***CAUTION—***
> - *Drive belt tensioner is under high tension. Check that lock pin is secure.*
> - *Take care to avoid damage to paintwork.*

NOTE—
- *Drive belt removal and installation is covered in* **020 Maintenance**.
- *If drive belt is to be reused, mark direction of travel and reinstall drive belt in same direction of rotation.*

– Remove intake manifold. See **113 Cylinder Head Removal and Installation**.

– Disconnect alternator electrical harness connectors.

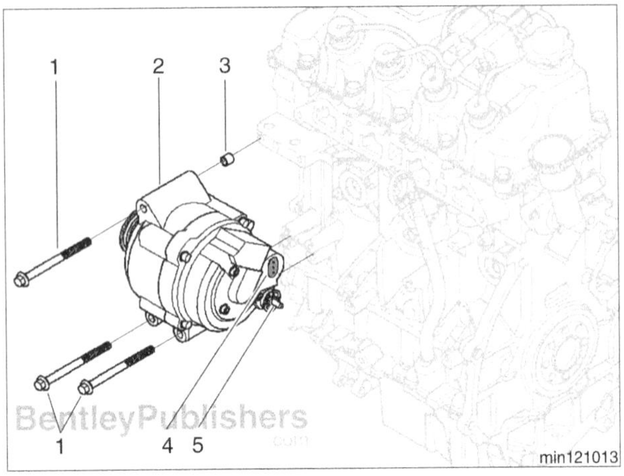

Remove mounting bolts and remove alternator.

1. Bolt M8 x 100 mm
   - Tighten to 25 Nm (18 ft-lb)
2. Alternator
3. Spacer bushing
4. Regulator connector
5. B+ connector

– Installation is reverse of removal.

Tightening torques	
Alternator to engine block (M8)	25 Nm (18 ft-lb)
Alternator harness to alternator B+ connector (M6)	10 Nm (7 ft-lb)

– After reassembling engine, clear ECM fault memory.

Alternator, removing and installing (Cooper S)

– Read out engine control module (ECM) fault memory. Then switch off ignition.

– Disconnect battery.

> ***CAUTION—***
> *Prior to disconnecting the battery, read the battery disconnection cautions in* **001 General Warnings and Cautions**.

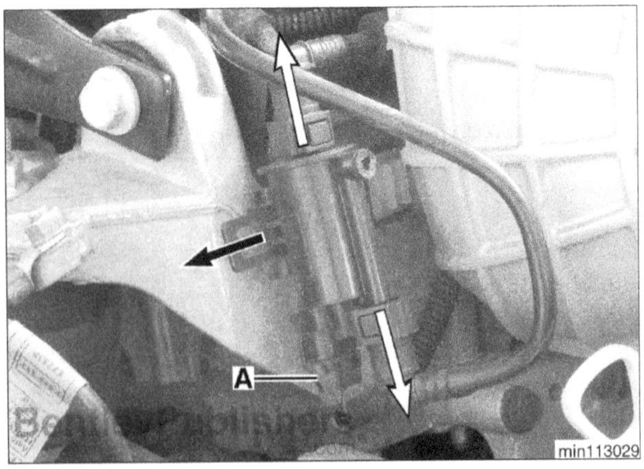

Working at right side engine mount, remove fuel tank vent valve.

- Disconnect vent hoses (**white arrows**) quick fit connectors.
- Straighten retaining tab and slide vent valve off bracket (**black arrow**).
- Remove electrical harness connector (**A**).

– Remove splash shield under engine.

– Remove right front wheel.

– Remove right front wheel housing liner.

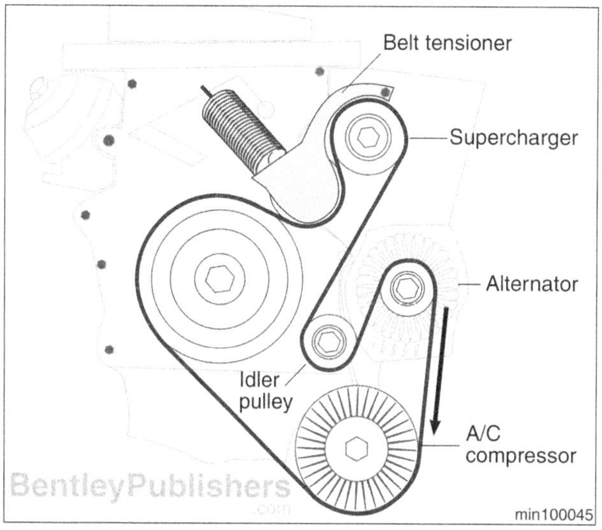

min100045

Note engine drive belt layout. Remove belt.

- Use special tool 11 8 410 to release drive belt tension.
- Use lock pin (special tool 11 8 470) to lock drive belt tensioner.
- Lift off drive belt.

> ***CAUTION—***
> - *Drive belt tensioner is under high tension. Check that lock pin is secure.*
> - *Take care to avoid damage to paintwork.*

NOTE—
- *Drive belt removal and installation is covered in* **020 Maintenance**.
- *If drive belt is to be reused, mark direction of travel and reinstall drive belt in same direction of rotation.*

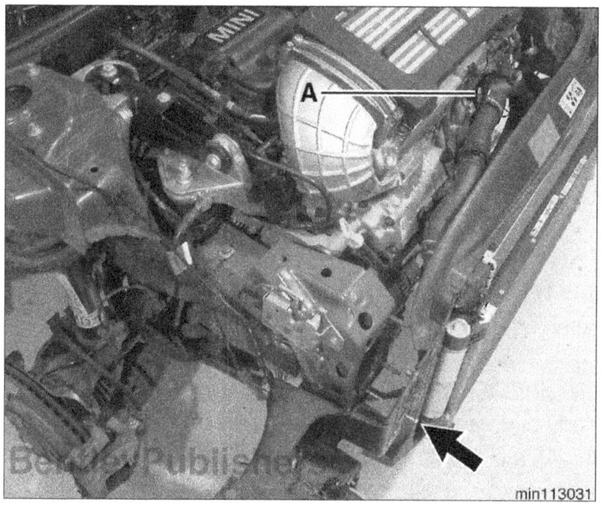

min113031

Loosen modular front end:

- Detach front wheel housing liners from front bumper cover trim.
- Remove front bumper cover trim.
- Working underneath car, loosen and remove crush tubes that support modular front end to front subframe.
- Detach radiator upper hose support clamp (**A**) from intake manifold.
- Remove modular front end mounting bolts. Install two 100 mm (4 in) M8 bolts (**arrow**) in left and right bumper support members. Slide modular front end forward, supported on long bolts.

> ***CAUTION—***
> *For ease of component alignment when reassembling front bumper assembly, do not loosen or remove bumper alignment bosses.*

NOTE—
- *If available, use BMW special tools 11 8 401 and 11 8 402 instead of long bolts to support modular front end.*
- *Removal of modular front end is a complicated job. It is covered in* **510 Exterior Trim, Bumpers**.

– Working underneath vehicle, disconnect alternator electrical harness connectors.

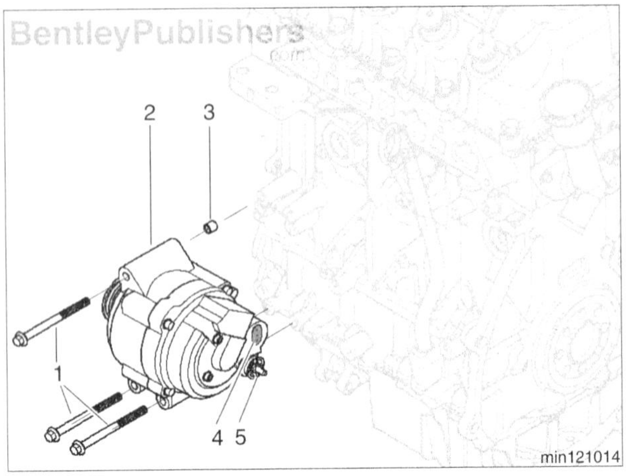

Remove mounting bolts and remove alternator.

1. Bolt M8 x 100 mm
   -tighten to 25 Nm (18 ft-lb)
2. Alternator
3. Spacer bushing
4. Regulator connector
5. B+ connector

– Installation is reverse of removal.

Tightening torques	
Alternator to engine block (M8)	25 Nm (18 ft-lb)
Alternator harness to alternator B+ connector (M6)	10 Nm (7 ft-lb)

– After reassembling engine, clear ECM fault memory.

STARTER

Starter troubleshooting

Large wire at starter terminal 30 is direct battery voltage. Smaller wire at terminal 50 operates starter solenoid via ignition switch.

– If starter turns engine slowly when ignition is in START position:
  - Check battery state of charge.
  - Inspect starter wires, terminals and ground connections for good contact. In particular, make sure ground connections between battery, body and engine are completely clean and tight.
  - If no faults are found, starter may be faulty.

– If starter fails to operate:
  - Check EWS (drive-away protection system). Try another ignition key. If no faults can be found, have the EWS system checked using BMW scan tool equipment.
  - Check clutch pedal operated starter lock-out switch or gear position switch (automatic).

NOTE—
- *A factory-installed drive-away protection system or EWS is used on MINI cars. This system prevents operation of the starter if a specially coded ignition key is not used.*
  See **515 Central Locking and Anti-theft**.
- *On automatic transmission cars, the transmission gear position switch signals EWS to prevent the engine from starting in gear positions other than PARK or NEUTRAL. See* **240 Automatic Transmission**.
- *On manual transmission cars, a starter immobilization switch at the clutch pedal is used to prevent the starter from operating unless the clutch pedal is pushed fully to the floor. See* **612 Switches**.

- Check for battery voltage at terminal 50 of starter motor with key in START position.
  - If voltage is not present, check wiring between ignition switch and starter terminal. Check EWS system and other inputs that disrupt input to starter. See **ELE Electrical Wiring Diagrams**.
  - If voltage is present and no other visible wiring faults can be found, problem is most likely in starter motor.

Starter, removing and installing

- Switch off ignition.
- Disconnect battery.

> ***CAUTION—***
> *Prior to disconnecting the battery, read the battery disconnection cautions in* **001 General Warnings and Cautions**.

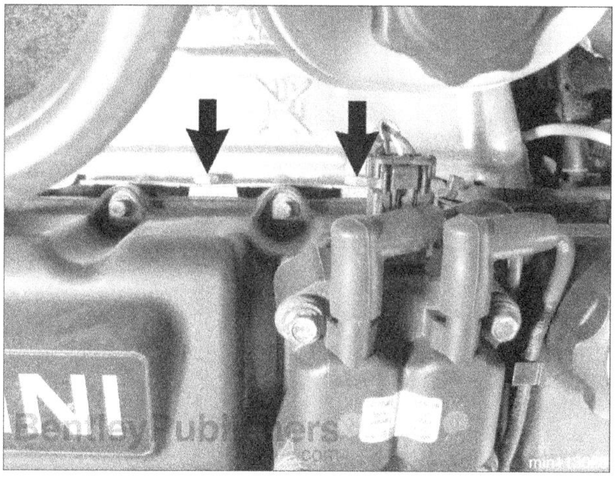

Working between engine and engine compartment rear bulkhead, remove exhaust manifold heat shield mounting bolts (**arrows**) and lift out heat shield.

- Remove exhaust manifold. See **180 Exhaust System**.
- Remove starter motor heat shield.
- Disconnect wires to starter.
- Remove starter mounting bolts and lift out starter.
- Installation is reverse of removal.

Tightening torques	
Exhaust manifold heat shield to cylinder head (M8)	13 Nm (10 ft-lb)
Exhaust manifold to cylinder head (M8)	24 Nm (18 ft-lb)
Starter heat shield to engine block (M6)	9 Nm (7 ft-lb)
Starter to transmission (M12)	85 Nm (63 ft-lb)
Terminal 30 to starter	14 Nm (10 ft-lb)
Terminal 50 to starter	8 Nm (6 ft-lb)

NOTE—
Route starter wires as they were before disassembly.

Starter solenoid, removing and installing

- Switch off ignition.
- Disconnect battery.

> ***CAUTION—***
> *Prior to disconnecting the battery, read the battery disconnection cautions in* **001 General Warnings and Cautions**.

- Remove starter. See **Starter, removing and installing** in this repair group.
- Disconnect field winding strap between starter motor and solenoid switch.

NOTE—
If the field winding strap is damaged or burned, a new or rebuilt starter motor is needed.

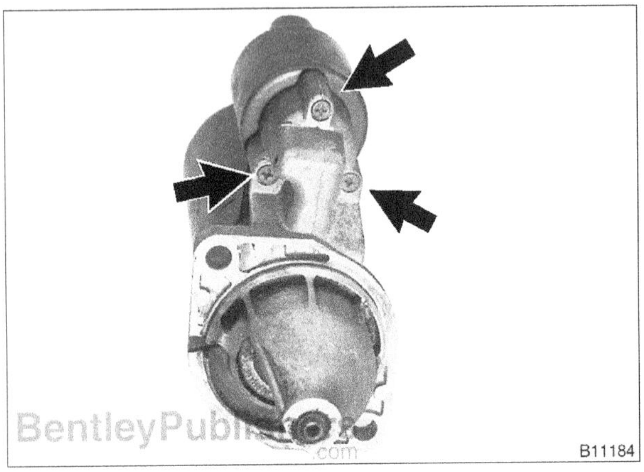

- Remove solenoid switch mounting screws (**arrows**), and separate solenoid from starter.
  - Unhook solenoid plunger from lever in starter drive.
- Installation is reverse of removal. Lubricate solenoid piston with light grease.

> ***CAUTION—***
> *When installing field winding strap to starter, position it so that it does not contact the starter body.*

Tightening torque	
Field strap to starter	13 Nm (10 ft-lb)

130 Fuel Injection

TABLES

GENERAL

This repair group describes the Siemens EMS2000 engine management system used in MINI Cooper and Cooper S models. Fuel injection component replacement is included and pin assignments for the EMS2000 engine control module (ECM) are listed.

The engine management system and software interact with many vehicle systems. Related topics can be found in the following repair groups:

- **120 Ignition System**
- **160 Fuel Tank and Fuel Pump**
- **170 Radiator and Cooling System**
- **180 Exhaust System**
- **640 Heating and Air-conditioning**
- **OBD On Board Diagnostics**

Second generation on-board diagnostics (OBD II) software and hardware is incorporated in the engine management control systems. OBD II monitors components that influence exhaust and evaporative emissions. If a problem is detected, OBD II stores the associated diagnostic trouble code (DTC) and the conditions under which the fault occurred in the engine control module (ECM).

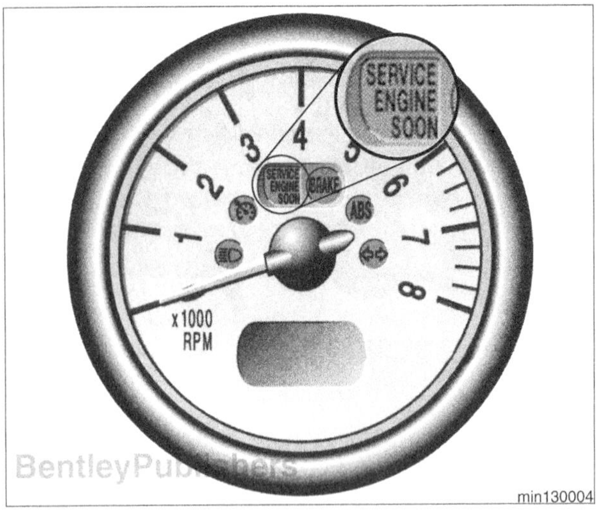

If vehicle emission levels exceed 1.5 times Federally mandated criteria, OBD II illuminates a malfunction indicator light (MIL) in the instrument cluster.

NOTE—
In MINI models, the MIL reads "Service engine soon".

When faults arise, or if the malfunction indicator lamp (MIL) is illuminated, begin troubleshooting by connecting BMW service tester DISplus, MoDiC, GT1 or equivalent scan tool. The capabilities of OBD II software has the potential to save hours of diagnostic time and to help avoid incorrect component replacement and possible damage to system components.

NOTE—
- *In addition to the MINI / BMW factory scan tools, Baum Tools, Snap-on, and Assenmacher Tool are specialized tool companies that may offer MINI-specific scan tools. Check with the tool manufacturer for availability.*
- *Generic OBD II scan tool software programs and handheld units are readily available. Though limited in capabilities compared to the factory tools, they are nonetheless powerful diagnostic devices which read live data streams, freeze-frame data and other valuable diagnostic information.*
- *A simple aftermarket DTC reader is available from Peake Research. This tool is capable of checking for DTCs as well as turning off the illuminated MIL.*

Special tools

Some special tools are needed to diagnose, repair or replace fuel injection system components. If you lack the necessary tools to perform a procedure, repairs are best left to an authorized MINI dealer or other qualified repair facility.

> ***CAUTION—***
> *Only use a digital multimeter for electrical tests.*

Digital multimeter

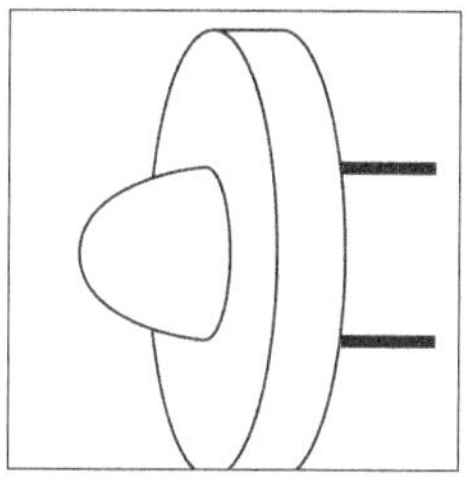

Low current test light ("noid")
(Baum Tools 1115a)

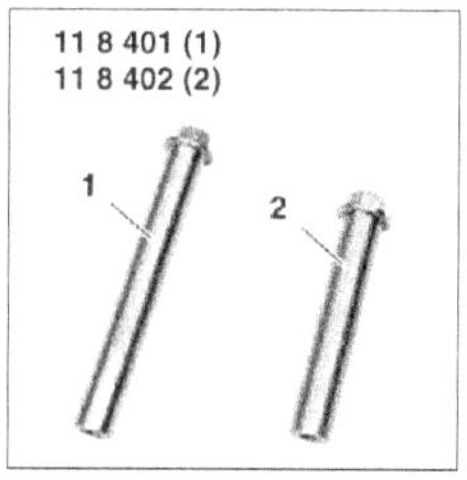

Modular front end supports
(BMW special tool 11 8 400)

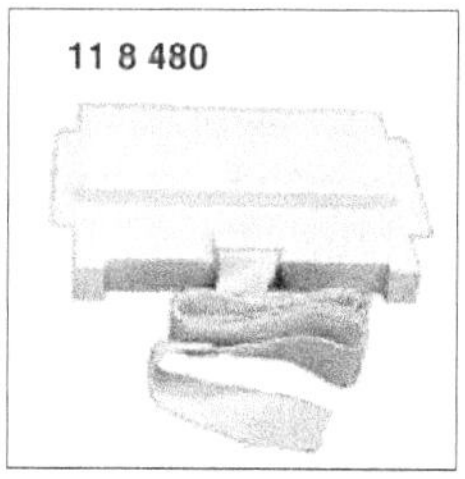

Intercooler protective cover
(BMW special tool 11 8 480)

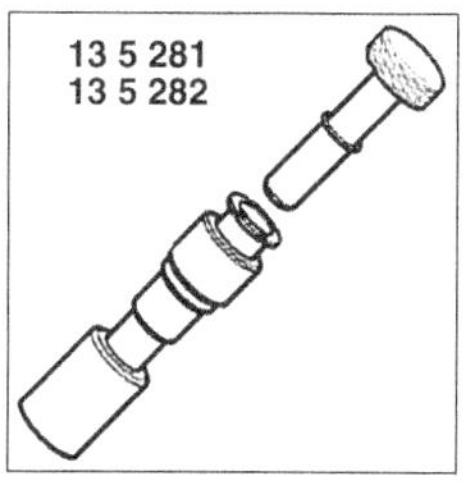

Fuel line plugs
(BMW special tools 13 5 281, 13 5 282)

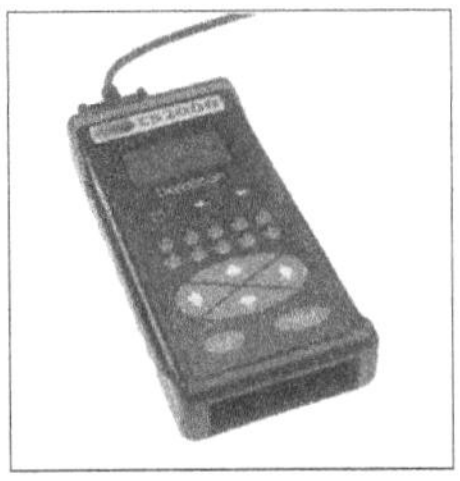

Aftermarket scan tool

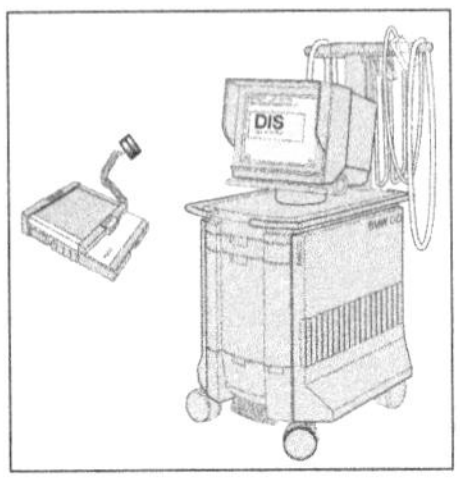

Factory scan tools
(BMW MoDiC and DISplus)

Engine management

The basic function of the EMS2000 engine management system is to produce smooth and efficient engine operation over varied driving conditions and engine loads. It also functions to monitor emissions systems and interfaces with other vehicle systems to enhance the driving experience.

Inputs from sensors, switches and monitoring devices are received by the engine management system (EMS) and processed. Using these inputs, the EMS makes operating decisions and outputs control signals to manage engine, transmission and emission operation.

Most communications among EMS components and with the rest of the vehicle occur over bus connections. See **600 Electrical System–General** for a schematic diagram of MINI bus network.

EMS inputs and outputs are listed in **Table a**.

Table a. EMS2000 inputs / outputs

Inputs	Outputs
Battery	Main relay
Main relay	Electronic throttle control (EDR)
Accelerator pedal sensor (PWG)	EDR warning LED
T-MAP sensor	Inertia switch
MAP sensor (Cooper S)	Fuel pump relay
Coolant temperature sensor	Fuel injectors
Electronic throttle control (EDR)	Ignition coils
Crankshaft sensor	Fuel leak detection system
Knock sensor	Purge valve
Camshaft sensor	Oxygen sensor heaters
Fuel leak detection system	Shift interlock relay
Oxygen sensors	Automatic transmission interface (GIU)
Brake light switch	A/C compressor
Clutch switch	Engine cooling fan
Transmission speed sensor	OBD II plug
Oil temperature sensor	Instrument cluster (IKE)
Alternator (generator)	
Immobilizer (EWS)	
Automatic transmission interface (GIU)	
ABS/DSC	
A/C pressure transducer	
Instrument cluster (IKE)	
Multifunction steering wheel (MFL)	

Engine control module (ECM)

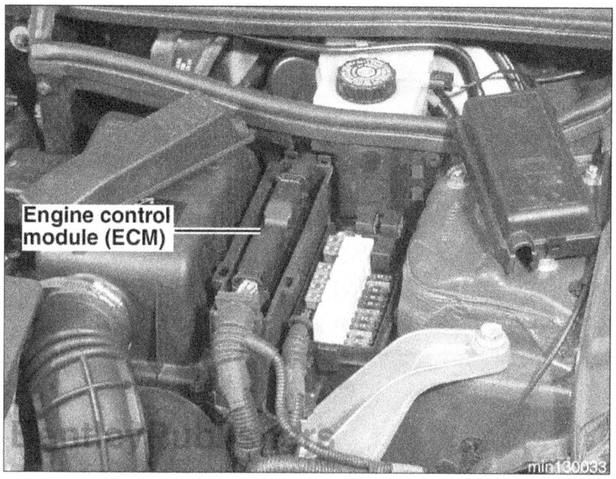

The engine control module (ECM) is the center of the engine management system. It receives and processes all inputs and issues corresponding control commands.

ECM location	
MINI Cooper	Left compartment of battery box next to engine compartment fuse and relay panel
MINI Cooper S	Left of air filter housing

In addition to engine controls, the ECM has direct control over the following systems:

- Drive by wire (DBW) throttle control
- Automatic transmission
- Cruise control
- Air-conditioning clutch relay
- Engine cooling fan relays

The ECM also has an interface with the following systems:

- Air-conditioning (IHKS/IHKA)
- Automatic stability control (ASC)
- Dynamic stability control (DSC)
- Immobilizer (EWS)

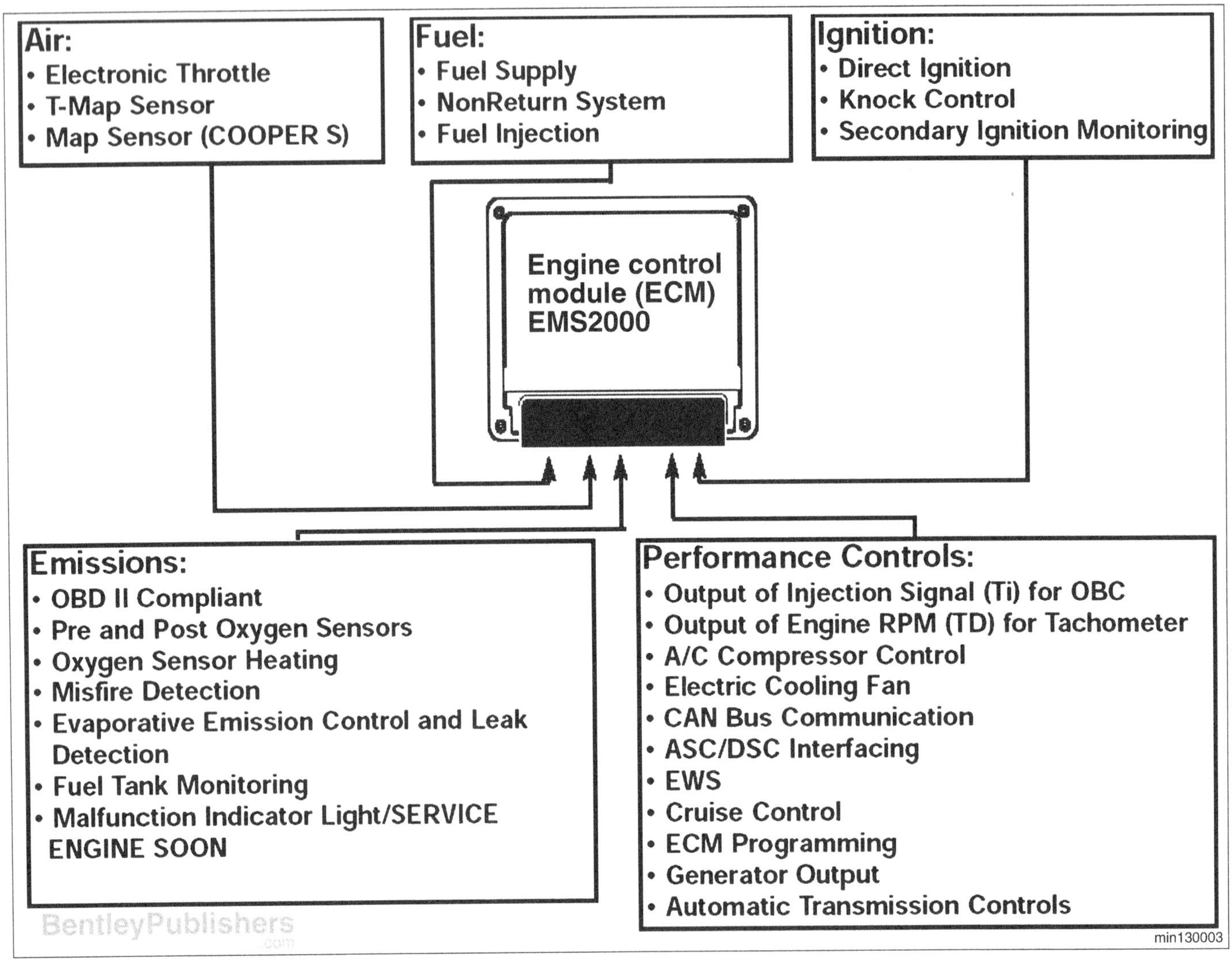

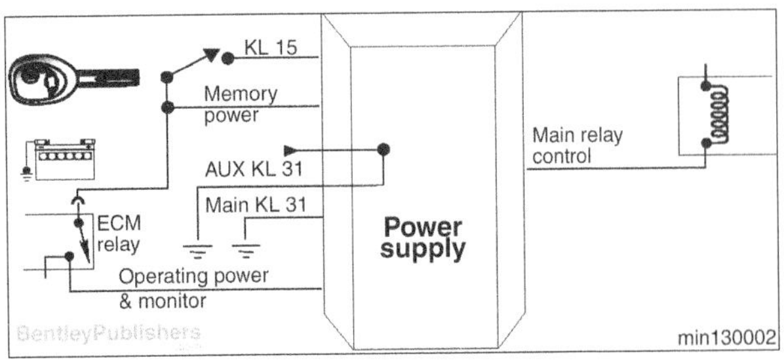

Principles of operation

Power supply

When ignition switch is placed in KL15 (ignition ON) or KL50 (START) positions, fuse 34 is provided with power. Fuse 34 supplies wake up or ON signal to ECM. Upon receipt of ON signal, ECM supplies a ground signal to main relay. This energizes main relay, supplying operating power to engine compartment fuses and engine management components.

At this point the engine is ready to start.

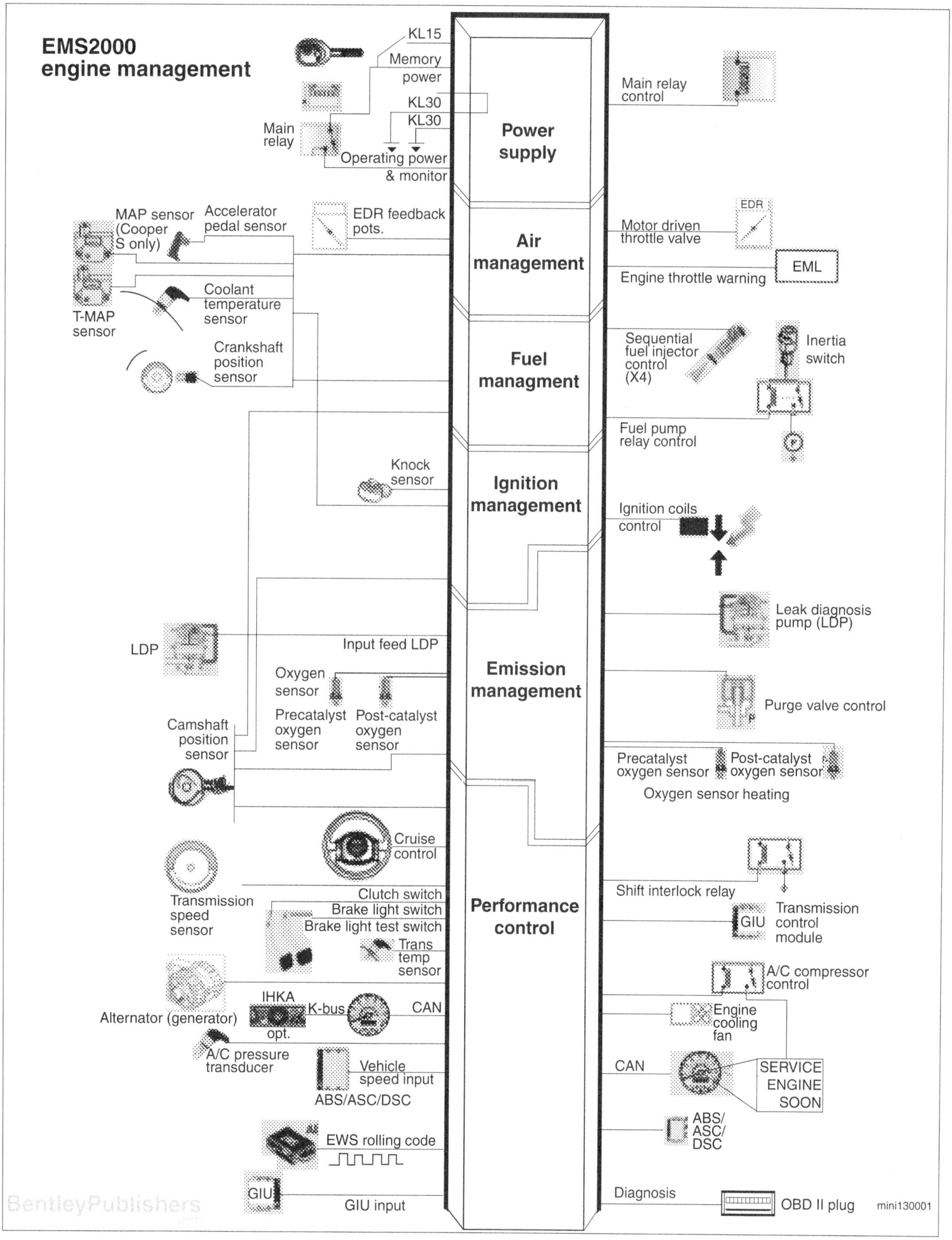
EMS2000
engine management
KL15
Memory
power
KL30
KL30
Main
relay
Operating power
& monitor
Power
supply
Main relay
control
MAP sensor
(Cooper
S only)
Accelerator
pedal sensor
EDR feedback
pots.
Air
management
EDR
Motor driven
throttle valve
EML
Engine throttle warning
T-MAP
sensor
Coolant
temperature
sensor
Crankshaft
position
sensor
Fuel
managment
Sequential
fuel injector
control
(X4)
Inertia
switch
Fuel pump
relay control
Knock
sensor
Ignition
management
Ignition coils
control
Leak diagnosis
pump (LDP)
LDP
Input feed LDP
Emission
management
Oxygen
sensor
Precatalyst
oxygen
sensor
Post-catalyst
oxygen
sensor
Purge valve control
Camshaft
position
sensor
Precatalyst
oxygen sensor
Post-catalyst
oxygen sensor
Oxygen sensor heating
Cruise
control
Transmission
speed
sensor
Clutch switch
Brake light switch
Brake light test switch
Trans
temp
sensor
Performance
control
Shift interlock relay
GIU
Transmission
control
module
A/C compressor
control
Alternator (generator)
IHKA
K-bus
opt.
CAN
Engine
cooling
fan
A/C pressure
transducer
Vehicle
speed input
ABS/ASC/DSC
CAN
SERVICE
ENGINE
SOON
ABS/
ASC/
DSC
EWS rolling code
GIU
GIU input
Diagnosis
OBD II plug
mini130001

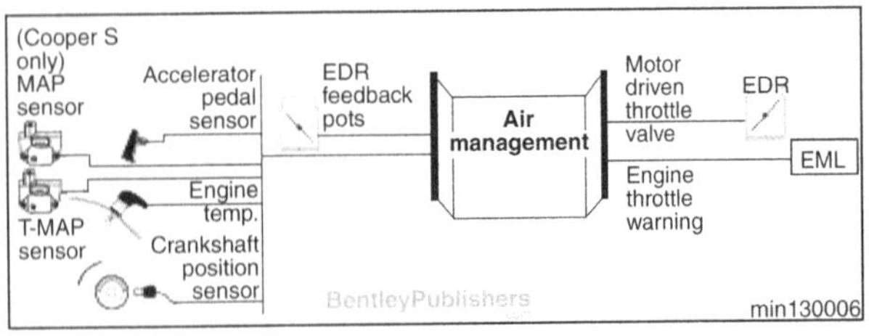

Air management

ECM detects that the engine is cranking via crankshaft sensor signal.

Accelerator pedal sensors (PWG) signal 0.5 vdc, indicating a request for idle. Feedback potentiometers in the electronic throttle control (EDR) are checked to confirm closed throttle valve position. Signals of 0.5 vdc from potentiometer 1 and 4.5 vdc from potentiometer 2 indicate that the throttle valve is in LL or idle position.

With engine cranking, ECM monitors input from T-MAP (and MAP sensor, if Cooper S). T-MAP output drops from 4 vdc (low vacuum) to 1 vdc (high vacuum).

Air volume and density are calculated from T-MAP sensor signal and intake air temperature.

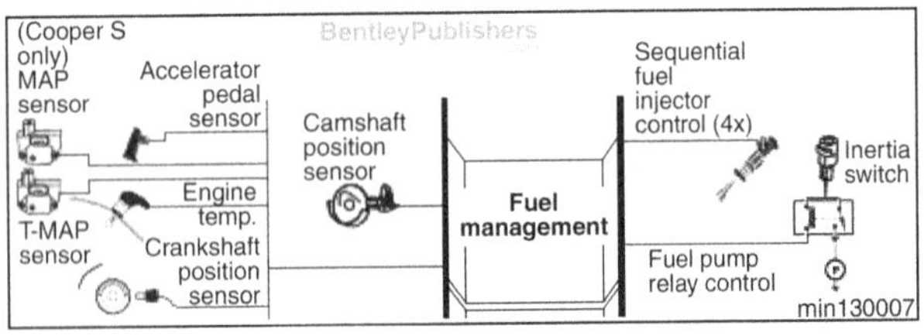

Fuel management

When ECM detects engine revolutions, it provides ground signal to fuel pump relay.

Fuel pump relay is fused. In early models, fuel pump circuit is further protected by inertia switch. If inertia switch is not triggered (triggered = open) power is provided to fuel pump relay.

NOTE—
The inertia switch is eliminated and its functions are assigned to the multiple restraint system (MRS) control module.

Fuel pump is mounted in left side of saddle type fuel tank. When fuel pump receives power, it picks up fuel through fuel filter and passes it to right side tank.

Fuel at 3.5 bar (51 psi) is pumped to engine mounted fuel rail assembly. Fuel rail contains pressure regulator and damper to smooth out fluctuations in fuel pressure during high load situations.

Based on volume and density of air, engine load, engine rpm and temperature, the ECM calculates correct volume of fuel for injection.

The ECM monitors crankshaft and camshaft sensors to time fully sequential fuel injector pulses.

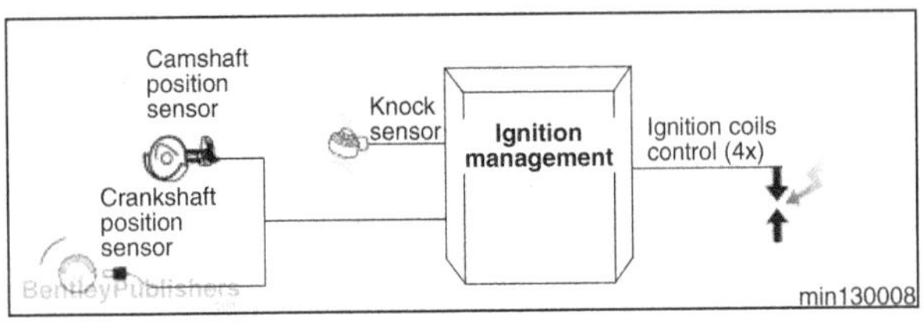

Ignition management

The ECM relies on previously analyzed sensor inputs to time ignition coil firing. As each cylinder approaches TDC, ECM grounds corresponding ignition output stage and fires ignition coil.

The ECM monitors knock sensors for variations in engine sound. If preignition knock is detected, ignition timing is retarded until the knocking stops.

Spark plugs introduce ignition energy into combustion chambers. High voltage arcs across air gap in spark plug, creating spark which ignites air/fuel mixture.

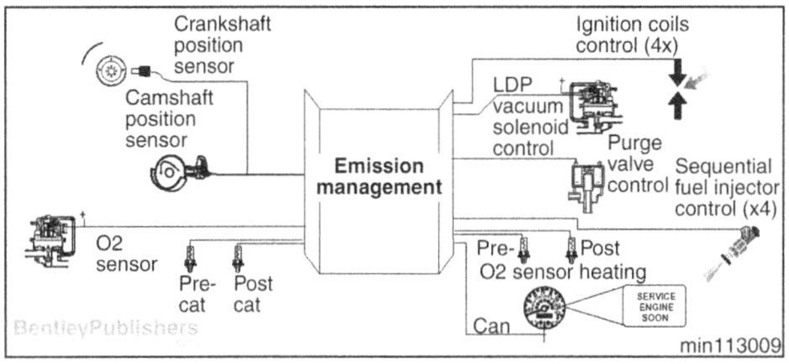

Emission management

Oxygen sensor heating. As soon as engine starts, pulse-width-modulated ground signal from ECM is supplied to oxygen sensor heaters. Duty cycle is increased to approximately 98% until oxygen sensors are fully heated. Afterwards duty cycle is varied to maintain sensor temperature. *Example:* During engine deceleration, duty cycle is increased to compensate for decrease in exhaust temperatures.

Once fully heated, oxygen sensors provide information about oxygen content in exhaust.

Pre-catalyst oxygen sensor. ECM adjusts injector ON-time based on input from pre-catalyst oxygen sensor. High voltage reading from pre-catalyst sensor indicates low oxygen content in exhaust (rich mixture). ECM reduces injector ON-time until voltage reading drops. Injector ON-time is then increased again. This oscillation is referred to as “closed loop”.

Post-catalyst oxygen sensor. ECM monitors catalyst condition using post-catalyst oxygen sensor signal during closed loop operation. Oscillations in oxygen content, detected by precatalyst oxygen sensor, are dampened by oxygen storage capacity of catalyst. Normal post-catalyst sensor has fairly stable signal, indicating oxygen has been consumed.

See **OBD On Board Diagnostics** for additional information on catalyst monitoring.

Evaporative emissions monitoring. ECM also monitors fuel tank evaporative losses via fuel tank leak diagnosis pump (LDP or DMTL).

Adaptation. ECM adapts to driving conditions in order to maintain a stoichiometric air / fuel ratio (14.7 : 1). It is capable of adapting to various environmental conditions encountered while the vehicle is in operation (changes in altitude, humidity, ambient temperature, fuel quality, etc.). These ECM adaptations can only make slight corrections and can not compensate for large changes which may be encountered as a result of incorrect airflow or incorrect fuel supply to engine.

Within allowed adaptation range, ECM modifies injection rate during these states of engine operation:

- During idle and low load mid range speeds (additive adaptation)
- During operation under normal load to higher load at higher engine speeds (multiplicative adaptation)

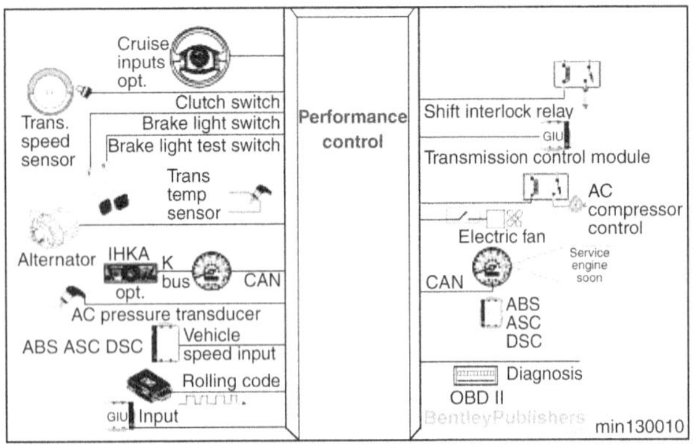

Performance controls

EWS Interface. ECM communicates with immobilizer (EWS) module prior to releasing injection and ignition.

Upon receiving correct rolling code from EWS, ECM allows fuel injectors and ignition coils to operate.

Automatic transmission control. As driver moves shift selector lever from PARK, ECM checks for brake pedal sensor signal before releasing shift interlock relay. Upon receiving a valid brake pedal sensor input, shift interlock relay is released and gear shift lever may be moved into DRIVE position.

ECM uses inputs from crankshaft sensor, transmission temperature sensor and transmission output speed sensor as well as gear shift lever switches to determine gear selection and programming. Instructions for gear control are transmitted from ECM through transmission control module (GIU) to transmission.

A/C compressor control. A request from the IHKS or IHKA is passed via the K-bus to the instrument cluster (IKE) and then over the CAN-bus to ECM. ECM checks signal from A/C pressure transducer and, if within range, activates compressor relay. ECM signals IHKS/IHKA that compressor is ON.

In response to compressor activation, ECM increases idle speed slightly to compensate for increased engine load.

Torque management. To prevent adversely affecting catalyst life or raising combustion chamber and piston temperatures, engine torque is changed by altering only ignition timing or throttle position.

Ignition timing can be altered rapidly and gives an instant torque change because under normal circumstances ECM always ensures that engine runs at peak efficiency.

Throttle position (airflow) change can increase or decease engine torque. If airflow is increased or decreased, ECM automatically maintains correct fuel mixture by balancing fuel input.

Unlike rapid torque change achieved by altering ignition, changes in throttle position take longer to achieve torque variation. Each system can demand either a slow or fast torque variation.

Torque control. EMS2000 can vary engine torque output in response to several conditions.

- Engine running demands. Torque variation is requested internally from within ECM to support the following:
  Idle control
  Catalyst 'light up' and overheat protection
  Limp home control
- Powertrain and chassis demands. Torque variation is requested externally from the following:
  Dynamic Stability Control (DSC)
  Automatic Stability Control + Traction Control (ASC)
  Automatic transmission (ECVT)
  Cruise control
- Driver demand.

Cruise control is integrated into EMS2000 because of drive-by-wire throttle operation:

- ECM controls vehicle speed by activation of electronic throttle valve (EDR).
- Clutch switch disengages cruise control to prevent over-revving during gear changes.
- Brake light switch and brake light test switch are input to ECM to disengage cruise control.

Cruise control functions are activated directly by multifunction (MFL) steering wheel signal to ECM. Individual buttons are digitally encoded in MFL switch and are input to ECM over a serial data wire.

Road speed information for cruise control is supplied from DSC module via CAN-bus to ECM.

Engine cooling fan. ECM determines cooling fan operating speed based on engine coolant temperature sensor and A/C pressure transducer signals.

ECM reprogramming. EMS2000 control module reprogramming is possible. Be sure to confirm viability of software upgrades prior to attempting reprogramming. Failure to do this may result in a no-start condition.

System inputs

Power supply

◄ **Battery power** (KL30) is supplied through engine compartment fuse F01 to provide engine control module (ECM) memory with uninterrupted power.

Battery power (KL30) through engine compartment fuse F02 is provided by main relay to ECM.

Battery positive pole (B+) is main supply of operating voltage to ECM. Battery voltage is monitored by ECM for fluctuations. ECM adjusts output functions to compensate for lower (6 vdc) or high (14 vdc) voltage values.

NOTE —
Based on available battery voltage, the ECM adjusts:
Fuel injection pulse width
Ignition system dwell

◄ **Main relay (K6300)** is located in engine compartment fuse and relay panel.

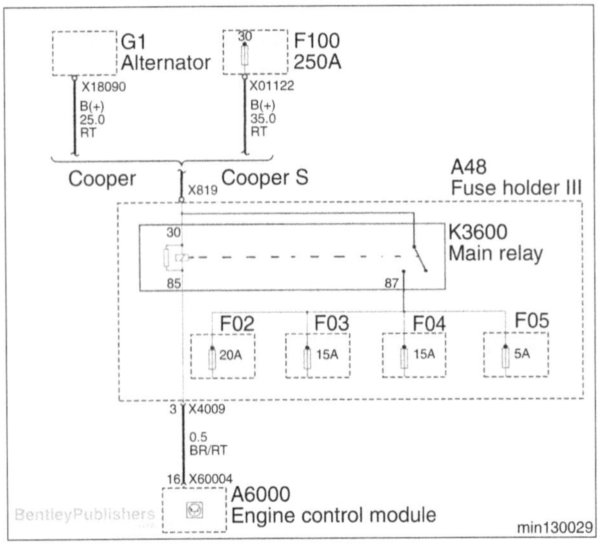

Main relay provides power to ECM and other system components. When ignition switch is in ON (KL15) or START (KL50) positions, fuse 34 supplies ON signal to ECM. Upon receipt of ON signal, ECM supplies ground signal to main relay. Ground signal energizes main relay, supplying operating power to following engine compartment fuses:

- F02: ECM, fuel injectors, crankshaft sensor, ignition coils
- F03: Camshaft sensor, oxygen sensor heaters, engine cooling fan, A/C compressor relay, purge system
- F04: Automatic transmission (CVT) controls
- F05: Engine cooling fan

Main relay supplies power to these components:

- ECM
- Fuel injectors
- Crankshaft sensor
- Ignition coils
- Fuel leakage detection pump (LDP or DMTL)
- Engine cooling fan
- Camshaft sensor
- Transmission control module (GIU)
- Oxygen sensor heaters

The main relay also protects system components in case of reversed battery polarity.

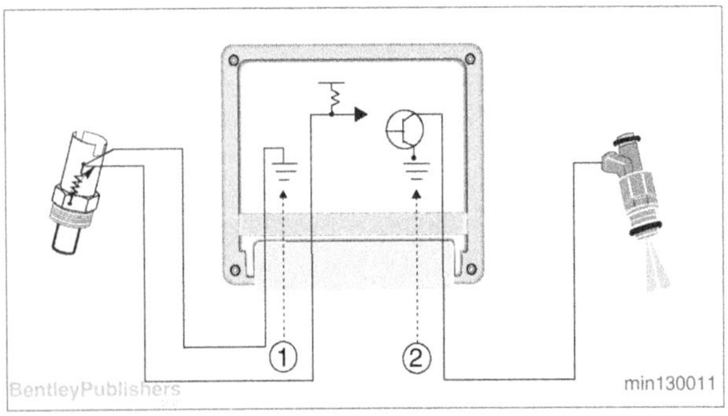

Grounds. Multiple ground points are supplied to complete current path through ECM.

1. Sensors are supplied with constant ground.
2. Components are activated via switched ground.

Accelerator pedal position sensor (PWG)

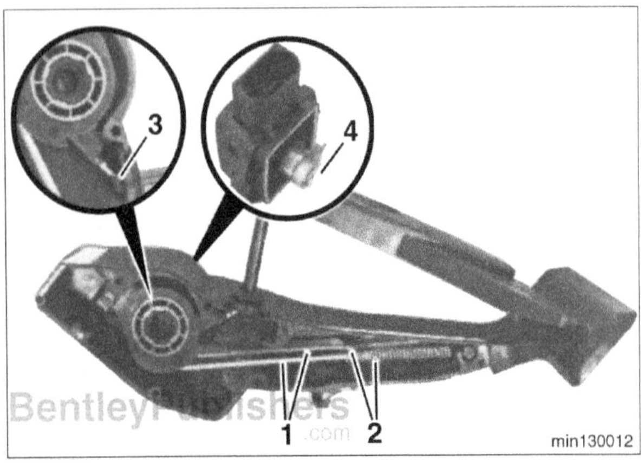

Accelerator pedal position sensor (PWG) is mounted in passenger compartment, clipped to bracket bolted to floor.

1. Twin cables
2. Return springs
3. Ball joint
4. Throttle potentiometer

PWG is monitored by ECM for input or speed requests. PWG uses two Hall sensors with different voltage characteristics and independent power supplies to detect driver input requests.

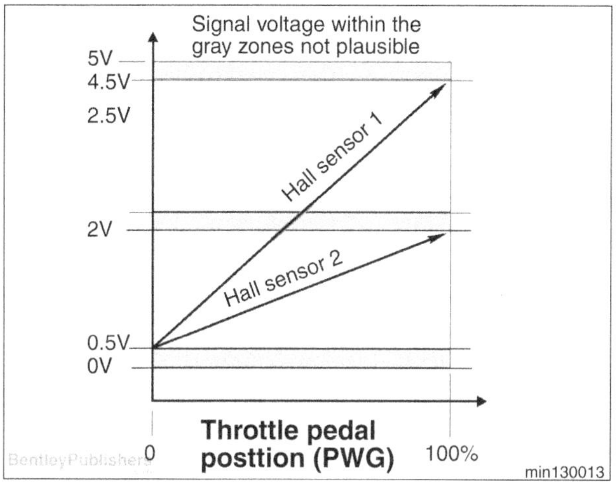

Hall sensors receive power (5 vdc) and ground from ECM and produce linear voltage signals as pedal is pressed from LL position (idle) to VL position (full throttle).

Hall sensor 1 range: 0.5 - 4.5 vdc, driver request

Hall sensor 2 range: 0.5 - 2.0 vdc, plausibility check

PWG is monitored by ECM for pedal angle position and rate of movement. As accelerator pedal is moved, rising voltage signals from Hall sensors request acceleration and determine rate of acceleration. ECM increases volume of fuel injected into engine, advances ignition timing and opens electronic throttle valve (EDR). Full throttle position indicates maximum acceleration to ECM. Full throttle also affects air conditioning compressor activation.

As accelerator pedal is released and returned to rest position by return springs, a decrease in voltage signals ECM to activate fuel shut off if RPM is above idle speed (coasting). Throttle valve closes, then opens just enough to maintain idle speed.

Sensor inputs

ECM uses input from T-MAP sensor (and MAP sensor if Cooper S), ECT sensor and throttle position potentiometers to calculate air volume consumed by engine. This calculated measurement is used by ECM to determine amount of fuel to be injected.

T-MAP sensor (Cooper)

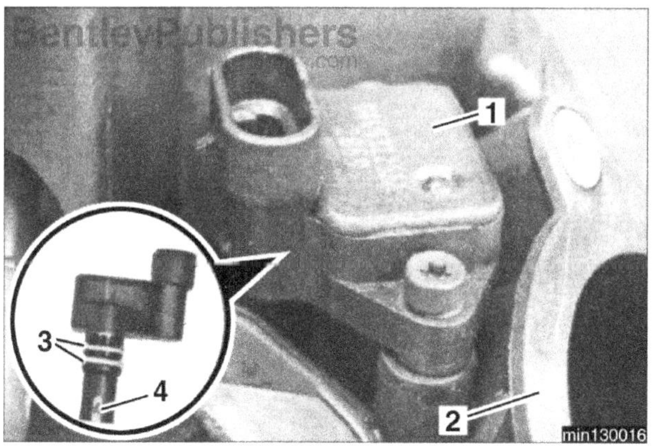

Temperature and manifold absolute pressure (T-MAP) sensor is located in intake airstream next to electronic throttle valve housing (EDR).

1. T-MAP sensor
2. Throttle valve housing (EDR)
3. Sealing O-rings
4. Sensor tip

Manifold pressure sensor portion of T-MAP sensor is a piezoresistive pressure gauge that supplies analog input to ECM. Manifold pressure indicates engine load, used for internal engine control via ECM and for ASC/DSC (traction control). ECM uses manifold pressure signals to determine correct fuel delivery and ignition timing.

- Pressure sensor has maximum pressure range of 120 kpa (17.4 psi).
- At idle, manifold pressure is low (high vacuum). Sensor output voltage is 1 to 2 vdc.
- At higher engine speeds or wide-open throttle (WOT), manifold pressure is higher (low vacuum). Sensor output voltage is about 4 vdc.

Air intake temperature sensor is built into T-MAP sensor. Sensor is a negative temperature coefficient (NTC) type, supplied 5 vdc by ECM.

Air temperature signal in conjunction with intake manifold pressure signal enable ECM to calculate volume of air being consumed by engine:

- At low intake air temperature, sensor resistance is high.
- At high intake air temperature, sensor resistance is low.

T-MAP and MAP sensors (Cooper S)

Pressure differential is measured across supercharger to determine manifold air density. Two sensors are fitted, one on either side of supercharger:

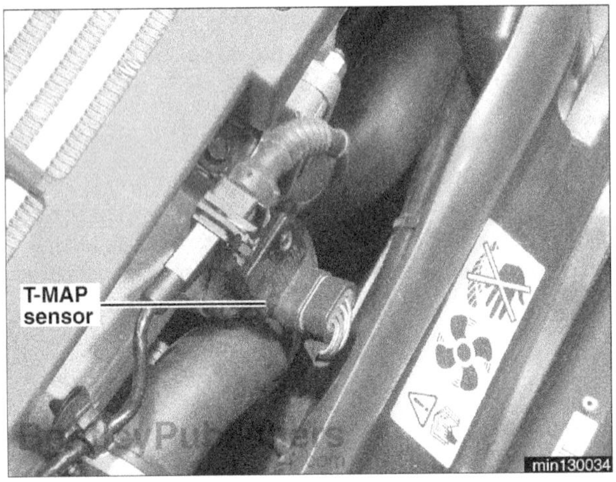

T-MAP sensor is fitted on manifold pressure side and has same hardware specification as Cooper T-MAP sensor, but maximum pressure range is 250 kpa (36.2 psi). This sensor is exposed to higher than atmospheric pressures as produced by supercharger.

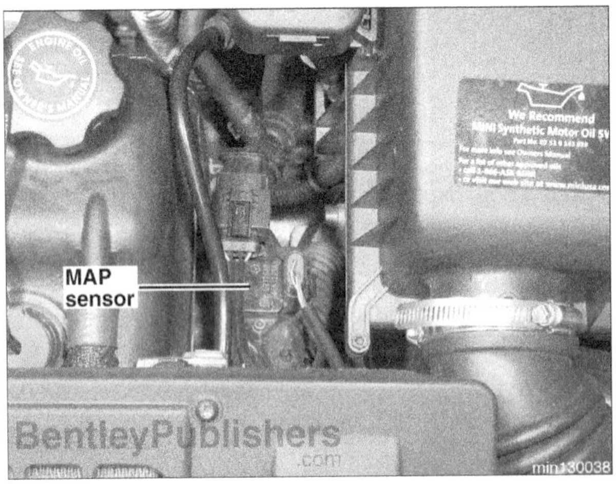

MAP sensor is fitted at left end of cylinder head. It is connected by vacuum line to supercharger supply duct and is not affected by supercharger air pressure. This sensor is same as Cooper T-MAP sensor but without temperature sensor.

Whenever ignition key is ON and ECM is supplied power from main relay without engine running, MAP sensor detects barometric pressure and provides a high voltage signal to ECM.

Once engine starts, MAP sensor measures absolute manifold pressure. This is barometric pressure minus vacuum created by operation of engine pistons. *Example:* If barometer is reading 1.5 bar (22 psi) at sea level and manifold vacuum is 1.0 bar (14.5 psi) at idle, manifold absolute pressure is 0.5 bar (7.3 psi).

ECM compares voltage from MAP sensor (ahead of supercharger) to voltage from T-MAP sensor (subject to increased pressures from supercharger) and calculates air volume drawn into engine.

Voltage reading of 0.6 - 1.5 vdc indicates high vacuum condition (idle or no load).Voltage reading of 4 vdc indicates low vacuum condition (full throttle).

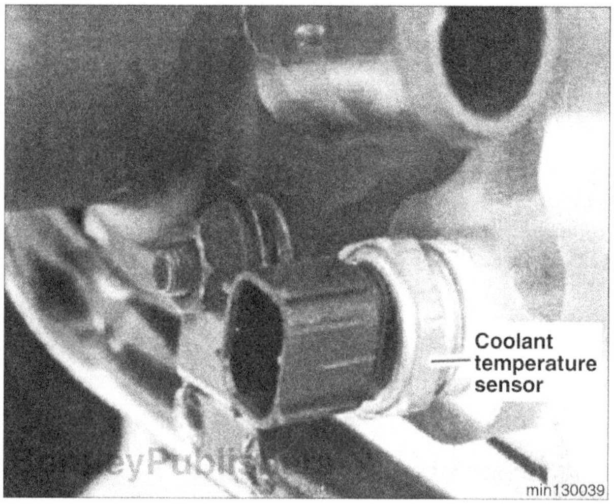

Engine coolant temperature (ECT) sensor

Engine coolant temperature (ECT) sensor is a negative temperature coefficient (NTC) thermistor located in coolant passage in cylinder head next to thermostat housing. Utilized to monitor engine coolant temperature, ECT sensor has two wires connected to ECM. ECT sensor signal is used for following functions:

- Fuel control and ignition timing
- Cooling fan speed
- Coolant temperature gauge in instrument cluster

A 5 vdc reference signal is supplied to ECT sensor by ECM.

- High voltage readings at ECM indicate high sensor resistance or low temperature.
- Low voltage readings indicate low sensor resistance or high temperature.

Throttle position feedback

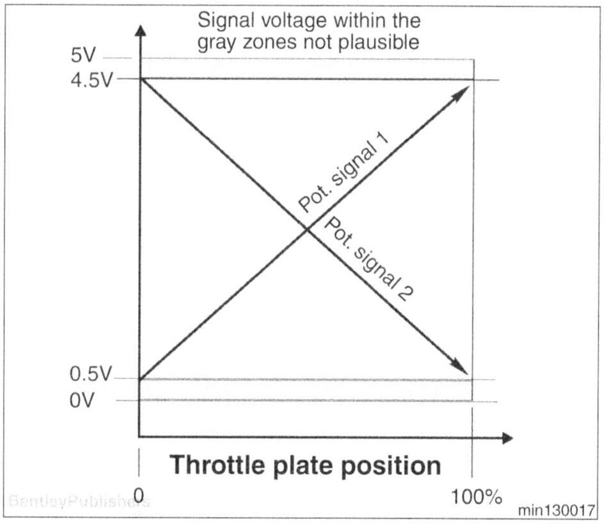

Electronic throttle valve (EDR) position is monitored by two integrated potentiometers, providing feedback voltage signals to ECM for throttle and idle control functions.

- Feedback potentiometer 1: Signal ranges from 0.5 vdc to 4.5 vdc.
- Feedback potentiometer 2: Signal ranges from 4.5 vdc to 0.5 vdc.

Potentiometer signal 1 is primary signal. Potentiometer signal 2 is used as a plausibility cross-check through total range of throttle plate movement. If there is an open or short in signal 1, signal 2 is used as a temporary substitute providing fail-safe operation. In this situation, faults are stored in ECM.

Crankshaft sensor

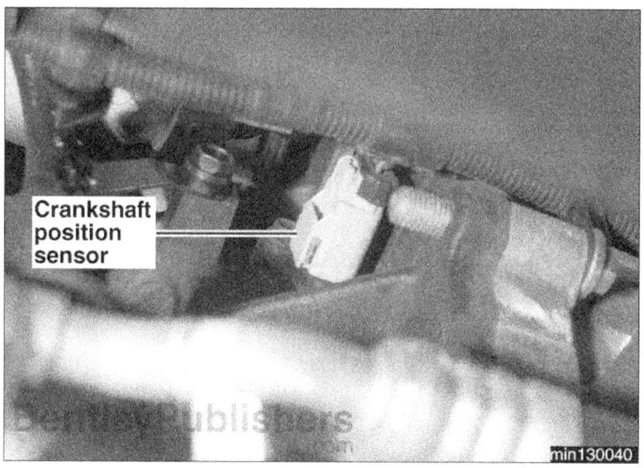

Crankshaft sensor, located at flywheel end and in front of crankcase, detects crankshaft speed and position and transmits this information to ECM. Crankshaft sensor is a Hall effect device providing digital electrical signal that is created as reluctor ring targets pass sensor.

Reluctor ring, bolted to end of crankshaft, is housed inside engine block.

1. Reluctor ring
2. "Missing" teeth

Reluctor ring tooth pattern consists of 58 targets and a space of two missing teeth grouped together (6° tooth intervals). As engine rotates, sensor output enables ECM to determine crankshaft position and speed.

Missing reluctor ring teeth are essential for correct engine operation. They are used by ECM as angular reference point. This information is used in combination with information from camshaft sensor to determine correct timing for spark and fuel delivery.

Crankshaft position sensor signals ECM to start injection as well as providing information about engine operation. This input is used in combination with other inputs to determine engine load which increases / decreases injector pulse duration. Without crankshaft position sensor input, ECM does not activate injectors.

Camshaft sensor

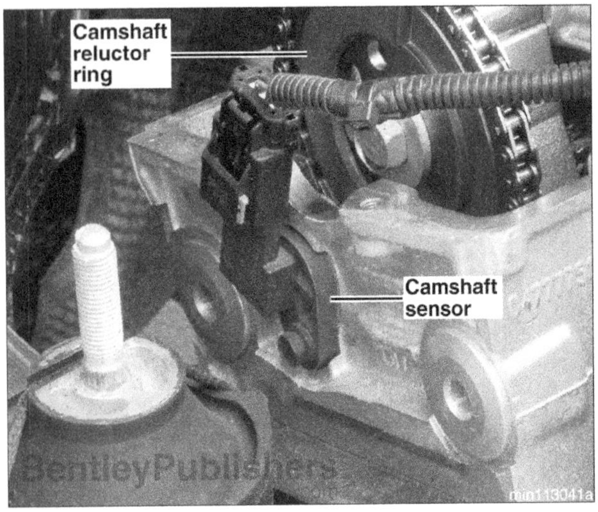

Camshaft sensor is located in front of cylinder head just below valve cover. Sensor reluctor ring is bolted to front of camshaft. Digital signal is provided by sensor in 0 - 5 vdc range. Sensor is Hall effect device which produces one pulse for each camshaft revolution.

Signal from camshaft sensor enables ECM to detect camshaft position in relation to crankshaft position. This allows ECM to synchronize fuel injector pulses and ignition spark.

Reluctor ring is half moon with a single "tooth" that extends over 180° of camshaft rotation.

Camshaft (cylinder ID) sensor affects injection timing (semisequential vs. full sequential). To accomplish this, ECM contains four final stage output transistors that activate injectors individually. Engine operates sufficiently on semisequential injection (two groups of two injector pulses), but more efficiently on full sequential injection (four individual injector pulses). If one fuel injector circuit is defective, engine can still operate on limited power from remaining fuel injector circuits.

Knock sensor

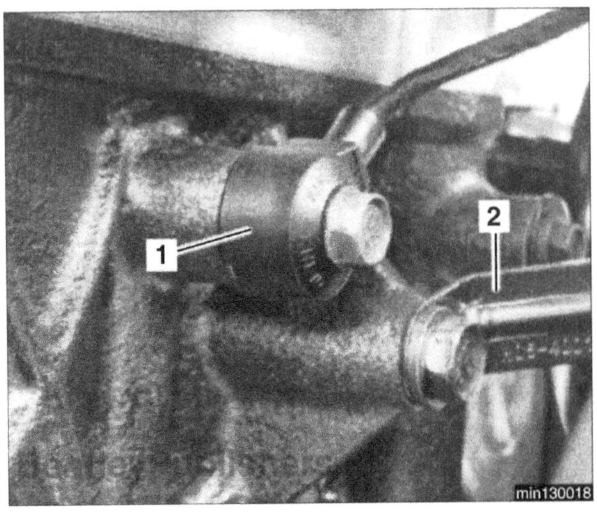

Knock sensor is bolted to front of crankcase just below intake manifold.

1. Knock sensor
2. Intake manifold support bracket

Knock sensor is a piezoelectric accelerometer producing voltage output proportional to mechanical vibration (knock) produced by engine caused by high pressure waves of uncontrolled spontaneous combustion of gasses in cylinder.

ECM control of engine timing based on knock sensor signal prevents detonation (pinging) from damaging engine. ECM response is cylinder selective.

Further information about knock sensor is in **120 Ignition System**.

Fuel leakage diagnosis system (LDP or DMTL)

Clean air legislation in US demands on-board monitoring of fuel system sealing on all internal combustion engine vehicles. Leak diagnosis pump (LDP) system and other controls on fuel evaporative losses are described in **160 Fuel Tank and Fuel Pump**.

Oxygen sensors

Cooper and Cooper S vehicles are equipped with two oxygen sensors each, one positioned upstream and the other downstream of catalyst.

Oxygen sensors are of zirconium dioxide type. They monitor exhaust gas oxygen content and signal ECM. This enables ECM to provide "closed loop" operation and maintain stoichiometric control of air/fuel ratio (14.7 : 1 air / fuel ratio by mass). This allows catalyst to work efficiently and reduce emissions of carbon monoxide (CO), hydrocarbons (HC) and oxides of nitrogen (NO_x) to acceptable levels.

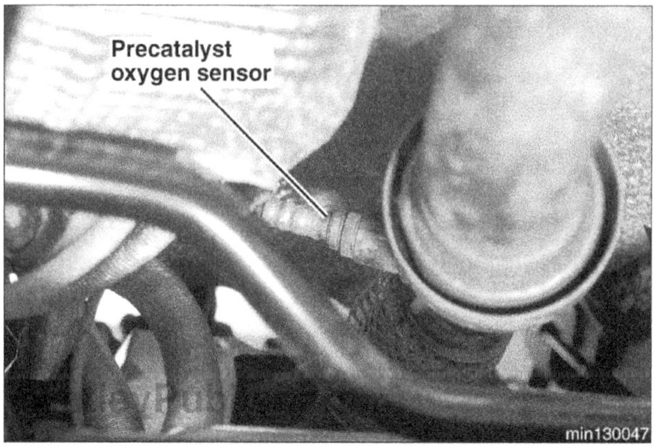

Pre-catalyst oxygen sensor measures residual oxygen content of exhaust gas. Sensor is mounted in hot exhaust stream directly in front of catalytic converter. Sensor produces low voltage (0 - 1000 mv) proportional to exhaust oxygen content. Sensor output allows ECM to monitor air / fuel ratio. Sensor signal constantly changes due to combustion variations and normal exhaust pulsations. ECM monitors length of time sensor is in lean, rich (including time of rise and fall) and rest conditions.

Evaluation period of sensor is over a predefined number of oscillation cycles. This oscillation is efficient when oxygen sensor is hot (250° - 300°C or 480° - 570°F). For this reason, sensor contains heating element to reduce warm up time and retain heat during low engine speeds when exhaust is cooler.

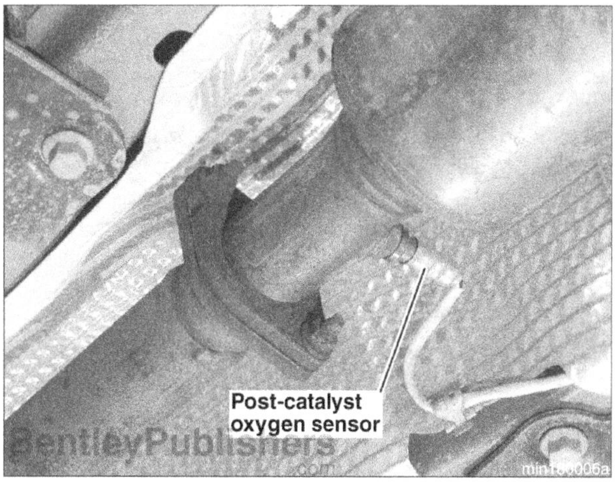

Post-catalyst sensor is used to monitor catalyst performance. Post-catalyst oxygen sensor output is evaluated over course of several pre-catalyst sensor oscillations. During evaluation period, signal of post-catalyst sensor must remain within relatively constant high voltage range (700 - 800 mv) with a very slight fluctuation. If this signal is low in voltage or fluctuates rapidly, fault code is set for catalyst efficiency. MIL illuminates when OBD II limit criteria are triggered.

Brake light switch

Brake light switch is clipped to pedal cluster bracket above brake pedal.

Two separate Hall effect brake switch inputs into ECM allow redundant integrity checking. Main brake switch signal is used to control both cruise control and drive-by-wire systems.

If brake is applied, ECM suspends cruise control, but retains cruise target speed in memory.

Second switch is used in safety plausibility check. If at any time the two brake signals are inconsistent, a fault condition is assumed and any accelerator demand will result in no throttle change. Engine will remain at idle.

For more information on brake light switch, see **612 Switches**.

Clutch switch

Clutch switch module is clipped to clutch master cylinder above clutch pedal.

Clutch switch is Hall effect device used in manual transmission vehicles. Clutch switch signal to ECM is used as follows:

- Prevents starting of engine unless clutch pedal is pressed to floor.
- Suspends cruise control when clutch is operated.

For more information on clutch switch, see **612 Switches**.

Automatic transmission (CVT) sensors

CVT has a dedicated secondary speed sensor located in differential housing and a two wire temperature sensor in valve body, both connected to ECM.

For more information about automatic transmission control, see **240 Automatic Transmission**.

Alternator (generator)

Dedicated output from alternator is provided to ECM to determine electrical load on engine. When electrical loads are switched on, alternator is required to generate more electrical energy, which in turn creates greater load on engine. If alternator signal fails engine may exhibit poor idle speed stability as electrical loads are increased.

For more information on alternator, see **121 Battery, Starter, Alternator**.

Immobilizer (EWS)

ECM communicates with immobilizer system (EWS) to provide theft protection for vehicle.

On vehicles with automatic transmissions, there is a park / neutral inhibit signal that EWS control module receives directly from transmission. EWS allows engine to crank only if automatic transmission is in N or P position.

For more information on EWS system, see **515 Central Locking and Anti-Theft**.

Transmission control module (GIU)

Transmission control module (GIU) functions are as follows:

- Converts inputs from gear selector lever switches (and steering wheel switches if fitted) into CAN instructions for ECM.
- Illuminates LEDs to display transmission mode.
- Converts CAN instruction for ECM into electrical signals to drive transmission ratio control motor and clutch and secondary pressure solenoids.

For more information about GIU, see **240 Automatic Transmission**.

ABS/ASC/DSC

ECM receives road speed signals from ABS/ASC/DSC control module for cruise functions (if equipped) and maximum vehicle speed limiting. Road speed signals arrive at ECM over CAN-bus. Requests for torque modification are also received from ASC/DSC module over CAN-bus.

For more information about ABS/ASC/DSC, see **340 Brakes**.

A/C pressure transducer

Pressure transducer protects A/C system from extremes of system pressure. In conjunction with engine coolant temperature (ECT) sensor, transducer controls engine cooling fan speed. Transducer is fitted in high pressure/temperature line of refrigerant circuit, on AC pipe connecting condenser and thermostatic expansion valve, in left rear corner of engine compartment.

For more information about A/C system controls, see **640 Heating and Air-conditioning**.

Electronic instrument cluster (IKE)

Instrument cluster (IKE) communicates with ECM over CAN-bus. Low fuel level signal is passed to ECM for evaluation of misfires. IKE also transfers requests for the A/C compressor from IHKS/IHKA to the ECM. Any vehicle system not on CAN-bus that communicates with ECM does so through IKE.

For more information about IKE, see **620 Instruments**.

Multifunction steering wheel (MFL)

Vehicles equipped with cruise control have single wire digital communication system between multifunction steering wheel (MFL) and ECM. MFL transmits requests for cruise operation to ECM.

For more information about MFL, see **320 Steering and Wheel Alignment**.

System outputs

Main relay

Upon receiving wake up signal from ignition switch (switch ON or KL15) supplying power to fuse 34, engine control module (ECM) provides switched ground signal via pin 97 to main relay. This energizes main relay, providing operating power to ECM. Main relay output is monitored so that if relay is energized and power is not received by ECM, fault code is set.

Electronic throttle control (EDR)

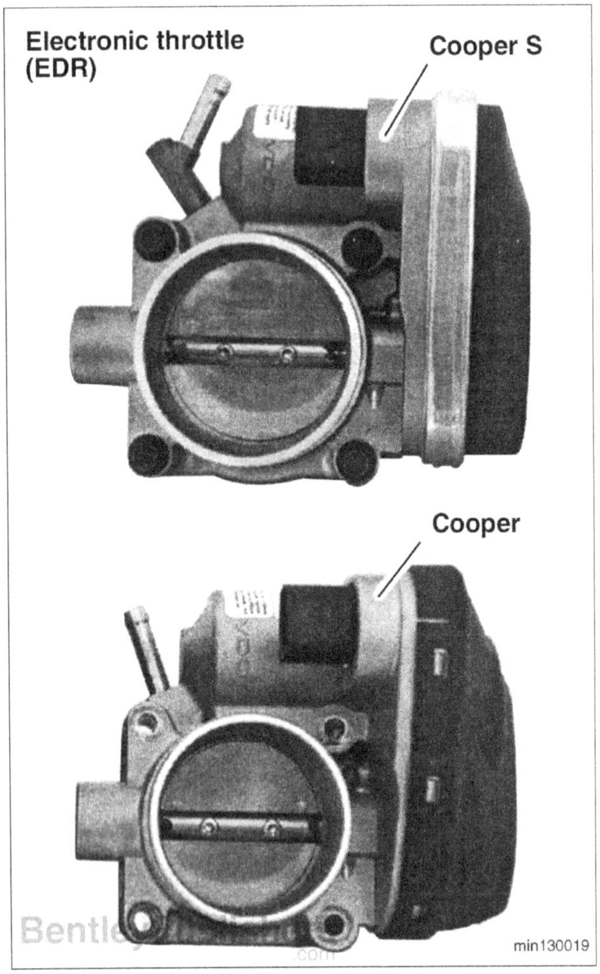

An electronic throttle (EDR) actuator is used to adjust engine load based on throttle position requests received by ECM from accelerator pedal position sensor. A DC motor electrically positions throttle plate from idle to full load. Two potentiometers, integrated in throttle housing, provide feedback on throttle plate position.

- MINI COOPER throttle body is 52 mm internal diameter.
- MINI COOPER S throttle body is 57 mm internal diameter.

EDR is operated by ECM for opening and closing of throttle based on accelerator pedal position, DSC intervention and cruise control function. EDR is a DC motor operating a gear-driven throttle plate. Variable duty-cycle fixed frequency signal is sent to EDR motor by ECM, which, to maintain throttle position, switches EDR signal polarity at rate of 600 Hz. Position and movement of throttle plates is confirmed through dual feedback potentiometers control functions.

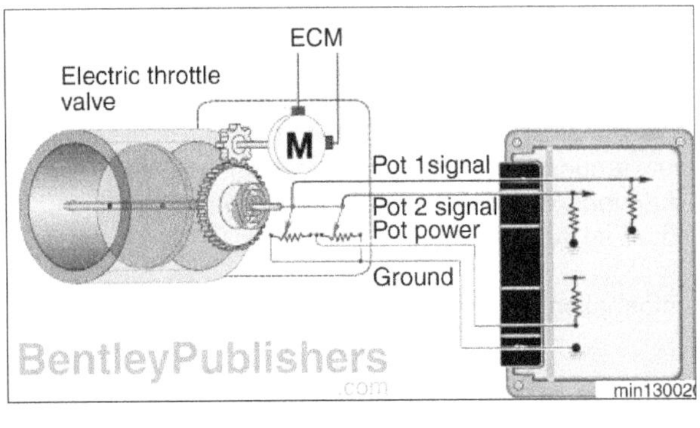

ECM provides power and ground for feedback potentiometers.

Idle speed control

Idle speed is controlled by ECM and EDR with no idle speed motor. For smooth driveability, engine speed remains constant when at idle no matter what varying loads may be. Engine idle does not fluctuate under following conditions:

- Engine cold start (increase in idle speed setting)
- Switching of different electrical loads (e.g. headlights, electrohydraulic steering)
- Air-conditioning compressor engagement

When there is a rapid change in electrical power demand on the alternator, there is corresponding rapid increase in alternator mechanical load upon engine. This has significant effect on idle speed stability. There is delay between electrical demand being made on alternator and mechanical demand on engine during which alternator transmits load change signal to ECM. This allows ECM to control throttle demand at idle to prevent flares and dips in engine speed.

Target idle speed setting for all MINI models is 750 rpm.

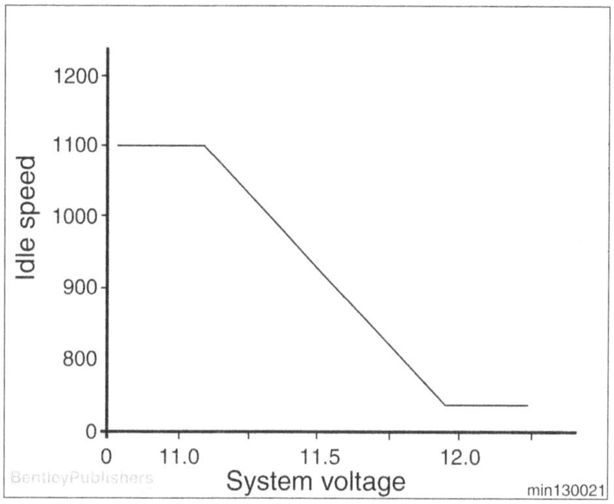

If system voltage decreases below minimum threshold (approx. 11 vdc), idle speed is increased to compensate with increased alternator output. Idle speed remains high for a period before returning to normal set point, when system voltage has recovered.

EML warning LED

If the ECM detects an engine safety related fault, not emissions related, in ECM itself, EML warning light illuminates in instrument cluster (IKE). This fault could be caused by faulty throttle motor, defective transmission or a sticking throttle valve. EML warning LED is amber in color and is activated through CAN-bus message to IKE.

Fuel pump relay and inertia switch

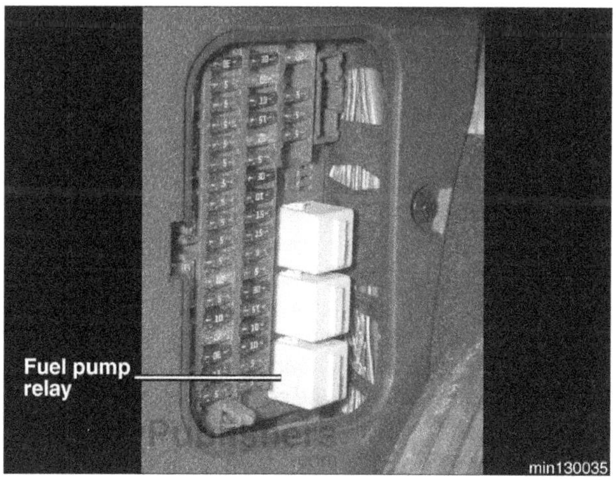

Fuel pump relay is located in passenger compartment fuse and relay panel in left kick panel.

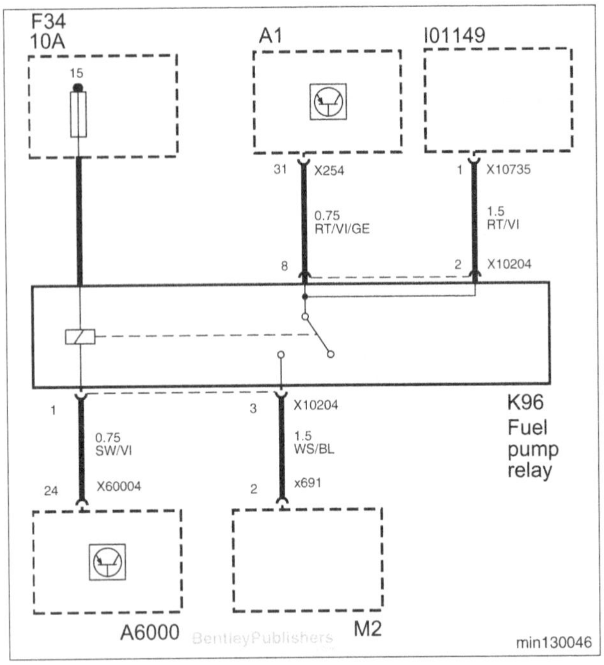

Fuel pump relay is normally open contact relay used to control fuel pump.

Pump is initially energized when ignition is switched ON. It remains energized for set time to pressurize fuel system. Relay remains energized if engine is detected to be running. Otherwise it is switched off after time delay.

Relay is switched OFF immediately if the engine has stopped. It is switched OFF after a time delay if ignition key is turned OFF.

Relay gets its B+ feed from ignition switch through fuse F20 and is energized when ECM provides ground.

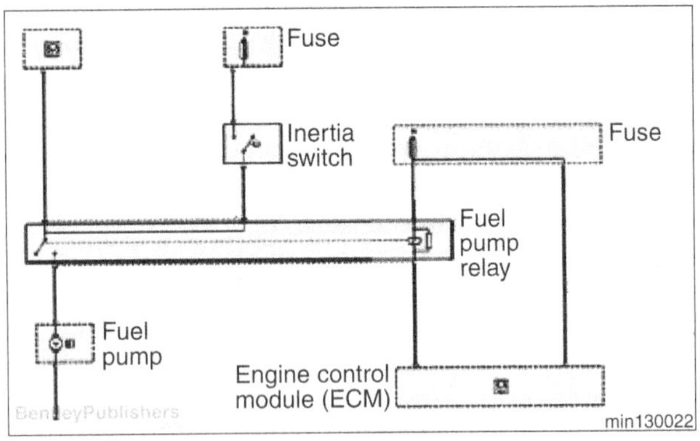

In early MINI models, an inertia switch is installed in the B+ line to relay between fuse and fuel pump relay. If vehicle undergoes an impact greater than 14 G, switch opens, interrupting power to relay.

If triggered, reset inertia switch manually.

Fuel Injectors

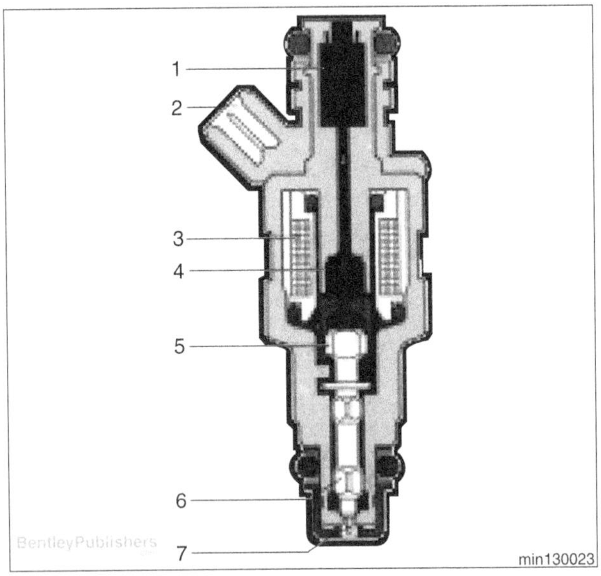

Fuel injectors are electronically controlled solenoid valves that provide precise metered and atomized fuel into engine intake ports.

1. Fuel Strainer
2. Electrical Connector
3. Solenoid Winding
4. Closing Spring
5. Solenoid Armature
6. Needle Valve
7. Pintle

Fuel is supplied from fuel rail to injector body. Fuel is channeled through injector body to needle valve and seat at tip of injector. Without electrical current, needle valve is sprung closed against seat.

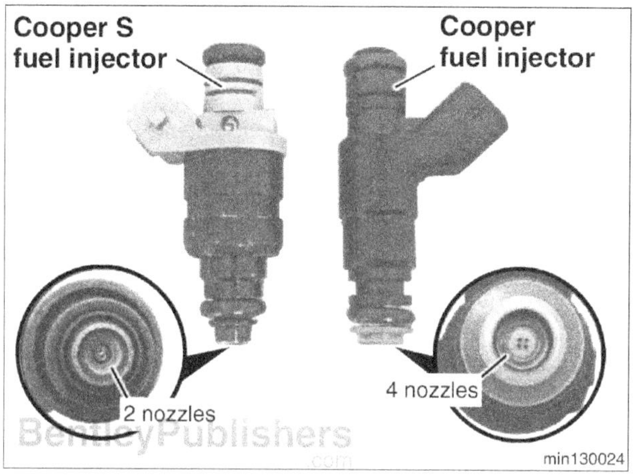

min130024

There are two types of injectors used. Cooper uses an injector with 4 nozzles, while Cooper S uses an injector with a higher flow rate and 2 nozzles.

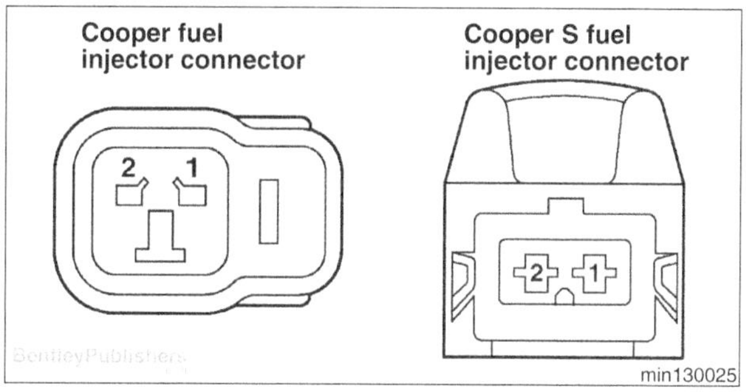

min130025

Injectors for both models are 62 mm in length and operate on fuel pressure of 3.5 Bar (51 psi). Electrical connectors are different for each type of injector, preventing accidental installation of wrong injector.

Fuel injectors receive voltage from main relay. ECM activates fuel injector by providing ground in pulses measured in milliseconds (ms). ECM varies pulse length to regulate air/fuel ratio (mixture).

Injection pulse length is regulated based on the following:

- **Battery voltage**. When cranking engine, battery voltage is low and ECM increases injector pulse duration to compensate for injector lag time. When engine is running and battery voltage is higher, ECM decreases pulse duration due to faster injector reaction time.
- **Cold starting** requires additional fuel to compensate for poor mixture and loss of fuel as it condenses on cold intake ports, valves and cylinder walls. Cold start fuel quantity is based on engine coolant temperature (ECT) sensor input during start up. As engine warms up, coolant temperature input allows ECM to adjust injection duration to compensate.
- **Idle**. With engine at idle, minimum injection is required. Additional fuel is added if ECM detects low engine rpm.
- **Acceleration**. As throttle is opened, ECM recognizes acceleration and increases fuel volume by increasing injection duration (acceleration enrichment).
- **Full throttle** position indicates maximum acceleration and ECM adds more fuel (full load enrichment).
- **Deceleration**. As throttle is closed, ECM decreases injection duration if engine rpm is above idle speed (coasting). This decreases fuel consumption and lowers emissions. When engine rpm approaches idle speed, injection duration is increased (cut-in) to prevent engine stalling. Cut-in rpm is dependent upon engine temperature and rate of deceleration.

Selective injector cut-out

ECM selectively deactivates injectors to control maximum engine rpm (regardless of vehicle speed). When engine speed reaches 6500 rpm, injectors are individually deactivated, as required, to protect engine from over-revving. As engine speed drops below 6500 rpm, normal injector activation is resumed.

> ***CAUTION—***
> *The injector cut-out feature does not protect the engine from driver error, such as forced over-revving from an improper manual transmission downshift.*

Maximum vehicle speed is also limited by ECM selectively deactivating injectors (regardless of engine rpm).

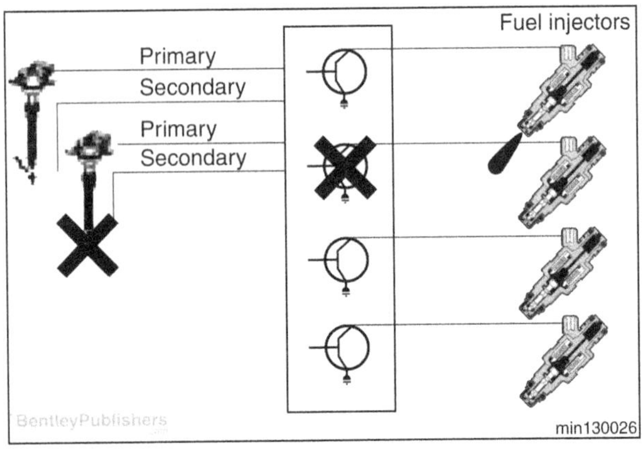

ECM will also protect catalytic converter by deactivating injectors if it detects a misfire because of ignition, injection or combustion fault(s).

Ignition system

ECM regulates ignition by calculating engine load based on a combination of inputs:

- Battery voltage
- Accelerator pedal position
- Calculated air volume and mass
- Engine coolant temperature
- Crankshaft and camshaft position
- Knock sensor

For more information on ignition coil, see **120 Ignition System**.

Evaporative emissions control

Evaporative emission purging is regulated by ECM via fuel tank vent valve (evaporative emission valve). This solenoid that regulates purge flow from active carbon canister into intake manifold. Main relay provides operating voltage, and ECM controls valve by regulating ground circuit.

For more information on evaporative emission controls, see **160 Fuel Tank and Fuel Pump**.

Oxygen sensor heating

Oxygen sensor efficiency is high when hot (250° - 300°C or 480° - 570°F). Oxygen sensor heater reduces warm-up time and maintains sensor temperature during low engine speed when exhaust temperature is cooler. OBD II requires monitoring of oxygen sensor heating.

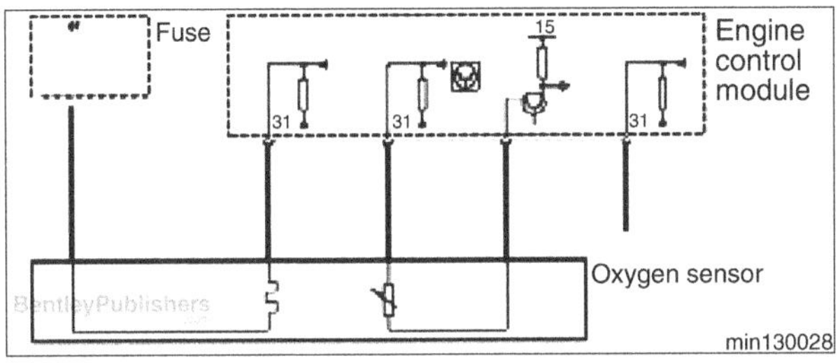

Oxygen sensor heating circuits receive operating voltage from main relay when ignition is switched ON (KL15). Each heater is controlled through a separate final stage transistor by pulse-width modulated ground from engine control module (ECM) during cold start. This allows sensors to be brought up to operating temperature without thermal shock. Heating circuit duty cycle is then varied to maintain sensor heating.

When decelerating (closed throttle), ECM increases heating circuit duty cycle to compensate for decreased exhaust temperature.

Transmission control module (GIU)

ECM receives inputs from main automatic transmission sensors, communicates with transmission control module (GIU) to control transmission, accepts driver inputs, and provides information to driver via instrument cluster.

For more information about GIU, see **240 Automatic Transmission**.

A/C compressor control

ECM receives inputs from air-conditioning or climate control sensors, accepts driver inputs and controls operation of A/C compressor.

For more information about A/C system controls, see **640 Heating and Air-conditioning**.

Engine cooling fan

The engine cooling fan is controlled by ECM through a relay and relay pack at two different speeds: Low and High. In addition to maintaining coolant temperature, cooling fan is used to cool A/C refrigerant and, where applicable, CVT gearbox oil.

For more information about engine cooling fan controls, see **170 Radiator and Cooling System**.

OBD II plug

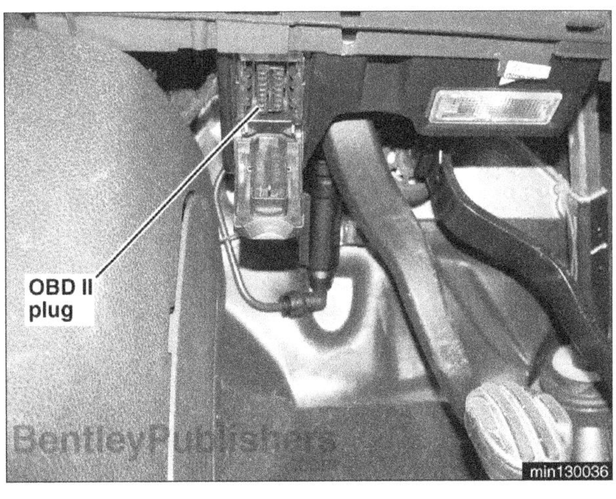

OBDII plug is behind cover above clutch pedal. Plug connection to ECM is through D-bus. This allows communication with scan tools (DISplus, GT1, MoDiC or equivalent), as well as emission related powertrain components.

Troubleshooting

Troubleshooting and fault diagnosis on the engine management system (EMS) is best performed using an electronic scan tool. However, it may be necessary to perform basic tests of EMS main components, fuel system or wiring.

Checking fuel delivery volume and fuel pressure is a fundamental part of troubleshooting and diagnosing the engine management system. Fuel pressure directly influences fuel delivery. Procedures for measuring fuel pressure and fuel volume are given in **160 Fuel Tank and Fuel Pump**.

Warnings and cautions

For personal safety, as well as the protection of sensitive electronic components, adhere to the following warnings and cautions.

WARNING —

- *Gasoline is highly flammable and its vapors are explosive. Do not smoke or work on a car near heaters or other fire hazards when diagnosing and repairing fuel system problems. Have a fire extinguisher available in case of an emergency.*
- *Renew fuel system hoses, clamps and O-rings any time they are removed.*
- *When working on an open fuel system, wear suitable hand protection, as prolonged contact with fuel can cause illnesses and skin disorders.*
- *The ignition system produces high voltages that can be fatal. Avoid contact with exposed terminals. Use extreme caution when working on a car with the ignition switched on or the engine running.*

CAUTION—
- *Disconnecting the battery may erase fault code(s) stored in memory. Check for fault codes prior to disconnecting the battery cables. Read the battery disconnection cautions in* **001 General Warnings and Cautions**.
- *Connect and disconnect the EMS system wiring and test equipment leads only when the ignition is switched off.*
- *Wait at least 5 minutes after turning off the ignition before removing the engine control module (ECM) connector. If the connector is removed before this time, residual power in the system relay may damage the control module.*
- *Fuel system cleaners and other chemical additives other than those specifically recommended by BMW may damage the catalytic converter, the oxygen sensor or other fuel supply components.*
- *Connect test equipment only as specified by BMW or the equipment maker.*
- *Only use a digital multimeter for electrical tests. Only use an LED test light for quick tests.*
- *Fuel system tests may set fault codes (DTCs) in the ECM and illuminate the MIL. After all testing tests is completed, access and clear DTC fault memory using a BMW compatible scan tool. See* **OBD On Board Diagnostics**.
- *Relay positions can vary. Be sure to confirm relay position by identifying the wiring in the socket using the wiring diagrams found at the rear of this manual.*

Component Replacement

The fuel pump is located inside the fuel tank. For replacement procedures, see **160 Fuel Tank and Fuel Pump**.

The fuel filter is a "lifetime" unit installed in the fuel tank. There is no routine interval or procedure for replacing the fuel filter.

Fuel rail and injectors, removing and installing

– Read out engine control module (ECM) fault memory. Then switch off ignition.

◄ Remove fuel rail cover:
- Squeeze locking tabs (**A**) to detach rear of cover.
- Tilt and pull cover forward until plastic clips (**arrows**) detach from fuel rail.

CAUTION—
Plastic clips can break easily.

Cooper S: Remove intercooler. See **113 Cylinder Head Removal and Installation** for complete procedure.

- Remove intercooler cover.
- Remove intercooler sealing bellows clamps and intercooler cover mounting brackets.
- Detach sealing bellows and tilt intercooler to remove.

Detach fuel rail from top of engine:

- Disconnect fuel injector electrical harness connectors (**A**).
- Disconnect harness connector (**B**).
- Working at fuel rail, unclip hoses and ducts. Cut wire ties as needed.
- Disconnect vacuum line at base of fuel pressure regulator (**C**).
- Detach fuel supply line from clips. Unlock quick-connect coupling (**D**) on fuel supply line and detach from fuel pressure regulator. Use special tool 13 5 281to plug fuel line.
- Remove fuel rail mounting bolts (**arrows**).

WARNING —
- *When disconnecting fuel line, fuel will be expelled under pressure. Loosen fuel filler cap to release fuel tank pressure.*
- *Do not smoke or work near heaters or other fire hazards.*
- *Keep a fire extinguisher handy.*
- *Before disconnecting fuel hose, wrap a cloth around hose to absorb any leaking fuel.*
- *Plug all open fuel connections.*

CAUTION —
Do not allow fuel to drip on alternator.

NOTE —
In illustration above, Cooper engine is shown. Cooper S fuel rail layout and connections are similar.

Cooper S: Unhook fuel injector electrical harness from fuel rail mounting brackets.

– Remove fuel rail together with fuel injectors from intake manifold.

CAUTION —
- *Use compressed air to blow away accumulated debris at the base of each fuel injector before pulling out injectors.*
- *Plug fuel injector bores in intake manifold.*

Remove retaining clip (**arrow**) of fuel injector and then remove fuel injector from fuel rail.

– Installation is reverse of removal.
  - Coat fuel injector sealing O-rings with antiseize compound.
  - Clear ECM fault memory.

Tightening torques	
Fuel rail to intake manifold (M8)	25 Nm (18 ft-lb)
Intercooler cover to intercooler	9 Nm (7 ft-lb)
Intercooler cover bracket to intercooler	9 Nm (7 ft-lb)
Upper sealing bellows clamp to lower clamp	9 Nm (7 ft-lb)

– Run engine and check for fuel leaks.

Fuel pressure regulator, removing and installing

– Switch off ignition.

– Remove fuel rail as described in this repair group.

Remove pressure regulator retaining clip in direction of **arrow**. Twist pressure regulator and pull out of fuel rail housing to remove.

> ***CAUTION—***
> *After removing pressure regulator, make sure filter of regulator is not left behind in housing.*

– When reinstalling regulator, replace both sealing O-rings.

– Reinstall fuel rail. Coat fuel injector sealing O-rings with antiseize agent.

– Run engine and check for fuel leaks.

Air filter housing, removing and installing (Cooper)

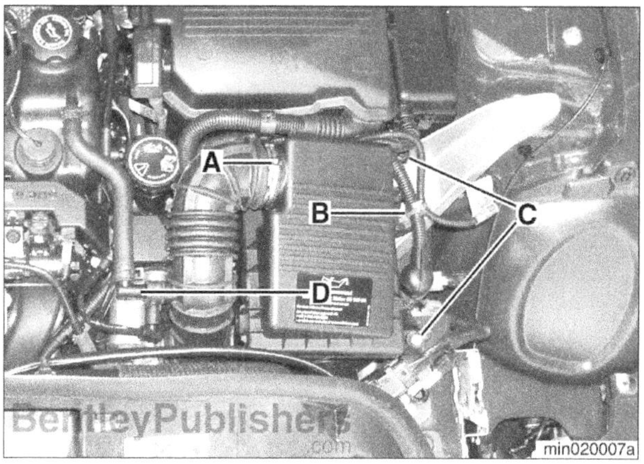

Working at top of engine, loosen or remove air duct clamp (**A**). Detach air duct from air filter housing.
- Detach wire harness (**B**).
- Remove air filter housing mounting screws (**C**).
- Detach crankcase breather hose (**D**) from air duct.

– Remove complete air filter housing.

– When reinstalling, reattach air duct using new clamp.

Tightening torque	
Air filter housing to body (M6)	7 - 10 Nm (5 - 7 ft-lb)

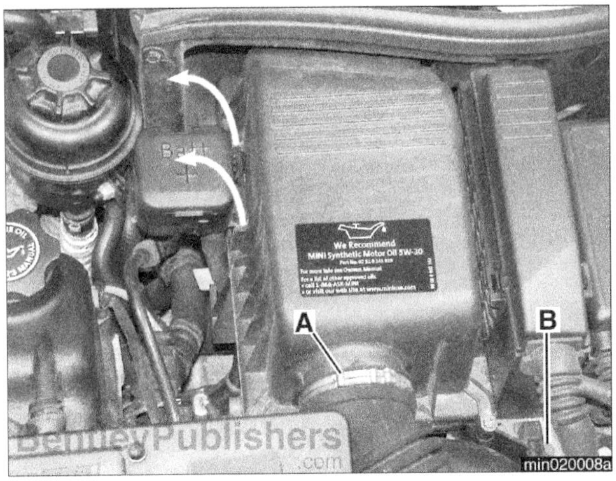

Air filter housing, removing and installing (Cooper S)

Detach air filter housing from throttle housing:

- Pull up on battery positive connection point (Batt+) (**arrows**) to release from right side of air filter housing.
- Loosen or remove hose clamp (**A**) and detach outlet duct from air filter housing.
- Remove air filter housing retaining bolt (**B**).

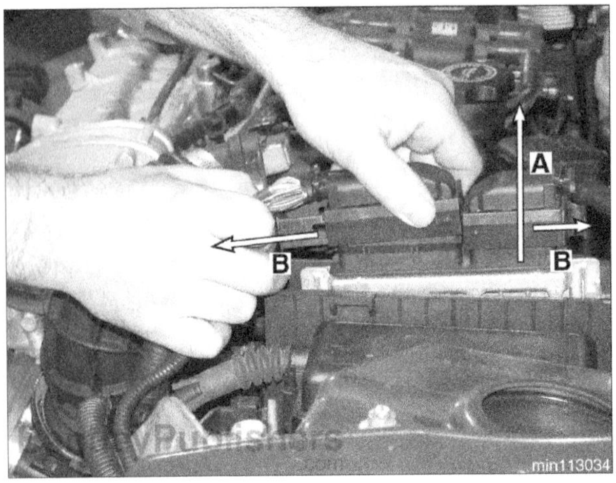

Remove DME control module:

- Remove control module container cover and lift control unit upward (**A**).
- Pull connector locking sliders outward (**B**).
- Disconnect control module connectors.
- Place control module in safe storage location.

NOTE—
There are 2 control module harness connectors.

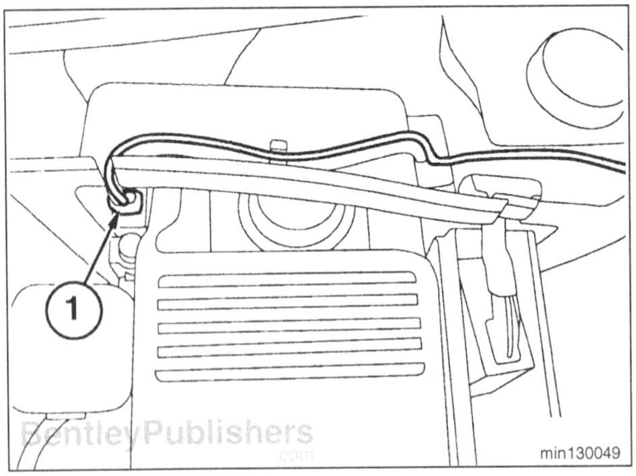

For the John Cooper Works (JCW) air filter assembly, detach harness connector (**1**) and vacuum line near rear of filter assembly. For more JCW upgrade kit information, see **Air filter housing and upgrade kit (JCW)** in this repair group.

- Unclip air filter inlet duct from modular front end. Wriggle duct while pulling it toward front of car to detach from air filter housing.

- Remove air filter housing from engine compartment.

- When reinstalling, reattach air duct using new clamp.

Tightening torque	
Air filter housing to body (M6)	7 - 10 Nm (5 - 7 ft-lb)

- Reinstall ECM and attach connectors.

Air filter housing and upgrade kit (JCW)

A special air filter assembly is available as part of the John Cooper Works (JCW) engine upgrade kit. This air filter assembly has an additional intake flap which opens at 4500 rpm to compliment the additional power demands of the modified JCW cylinder head, supercharger and engine management software.

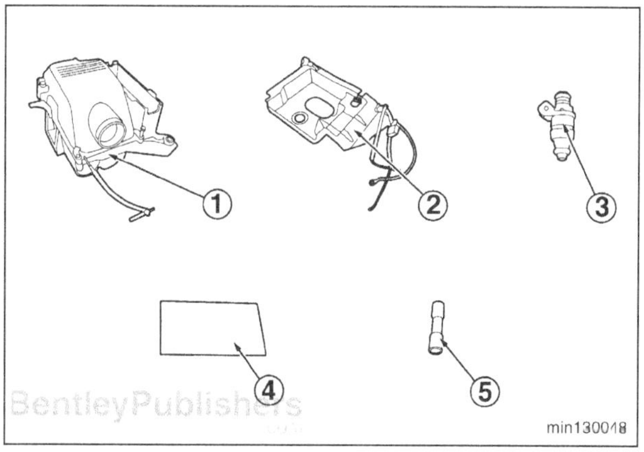

The JCW upgrade kit includes the following components:

1. Air filter housing
2. Rear plenum partition with hole for additional air filter intake
3. Fuel injectors
4. Heat shield
5. Crimp connector

Installation of the JCW upgrade kit requires an update to the engine management software and should be done by an authorized MINI dealer.

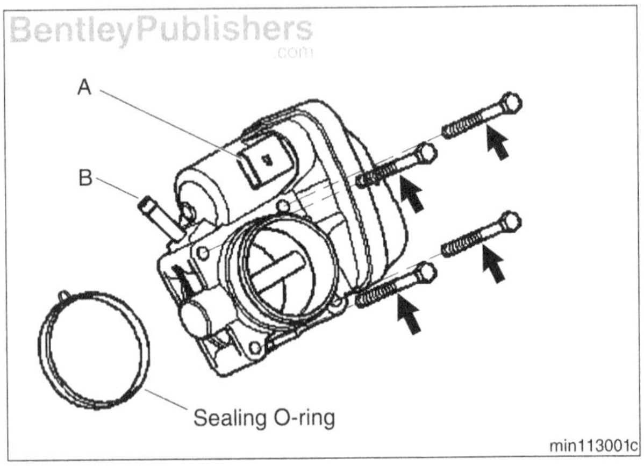

Throttle housing (EDR), removing and installing (Cooper)

- Read out engine control module (ECM) fault memory. Then switch off ignition.
- Remove air filter housing. See **Air filter housing, removing and installing (Cooper)** in this repair group.

< Remove throttle housing:

- Loosen or remove air duct clamp at throttle assembly. Detach air duct.
- Disconnect electrical harness connector (**A**).
- Detach fuel tank vent line (**B**).
- Loosen and remove throttle assembly mounting screws (**arrows**).
- Lift off throttle assembly.

- When reinstalling, replace sealing O-ring.

Tightening torque	
Throttle housing to intake manifold (M6)	9 Nm (7 ft-lb)

- Reinstall air filter housing.

Tightening torque	
Air filter housing to body (M6)	7 - 10 Nm (5 - 7 ft-lb)

NOTE—
If throttle housing is replaced, be sure to reset DME control unit adaptation values using diagnostic scan tool.

Throttle housing (EDR), removing and installing (Cooper S)

- Read out engine control module (ECM) fault memory. Then switch off ignition.

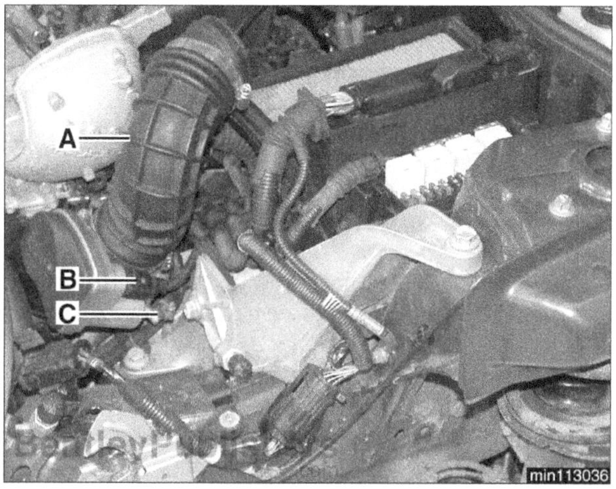

- Remove ECM and air filter housing. See **Air filter housing, removing and installing (Cooper S)** in this repair group.

< Remove throttle housing:

- Loosen or remove air duct clamp at throttle assembly. Detach air duct (**A**).
- Disconnect electrical harness connector (**B**).
- Detach fuel tank vent line (**C**).
- Loosen and remove throttle assembly mounting screws.
- Lift throttle assembly off supercharger intake duct.

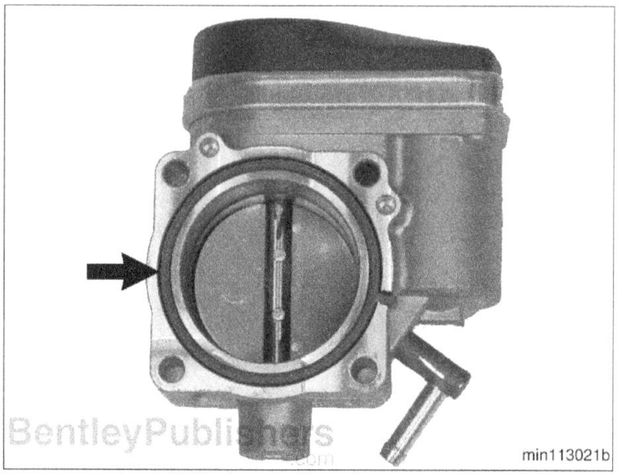

< When reinstalling throttle housing, replace sealing O-ring (**arrow**).

< Make sure locating pins (**arrows**) on supercharger intake duct align with corresponding bores on throttle housing. Check locating pins for damage. Replace if necessary.

Tightening torque	
Throttle housing to supercharger intake duct (M6)	9 Nm (7 ft-lb)

- Reinstall air filter housing.

Tightening torque	
Air filter housing to body (M6)	7 - 10 Nm (5 - 7 ft-lb)

- Reattach air duct using new clamps.

NOTE—
If throttle housing is replaced, be sure to reset DME control unit adaptation values using diagnostic scan tool.

Supercharger oil service

MKi5625a

Owing to the extreme heat and harsh operating conditions, it is a good idea to check the supercharger oil level on a regular basis. MINI does not specifically recommend a supercharger oil service and no oil is available through MINI parts.

The supercharger on the MINI is manufactured by Eaton. General Motors uses a similar Eaton supercharger and the GM supercharger oil (GM Part number **12345982**) can be substituted.

The supercharger has two oil reservoirs; one on the front of the supercharger below the drive pulley and one on the rear behind the water pump. Oil service for each is covered separately.

The reservoir fill plugs are difficult to access, and some disassembly is required to access the rear plug. A small diameter hose may help to facilitate topping off.

Supercharger front oil reservoir, adding oil (W11 engine)

– Remove engine oil dipstick tube.

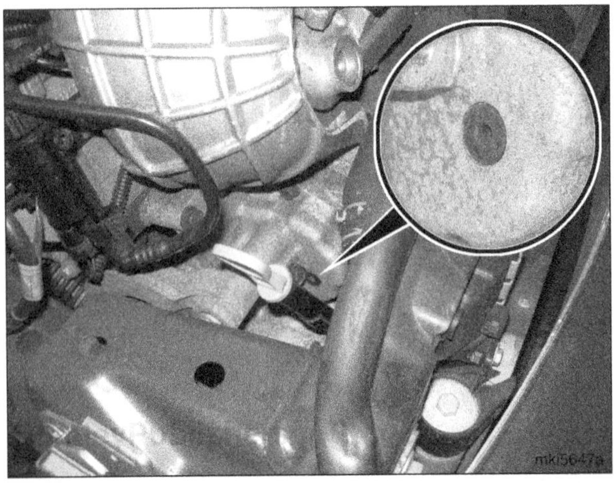
mki5647a

< Using a 3/16 Allen wrench remove oil fill plug (**inset**).

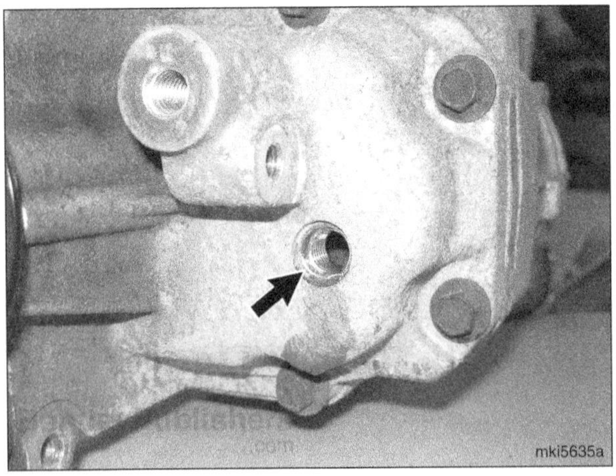
mki5635a

< Fill reservoir with oil until it begins to run out (**arrow**).

– Reinstall plug and tighten.

Supercharge rear oil reservoir, adding oil (W11 engine)

– Remove water pump. See **170 Radiator and Cooling System**.

mki5632a

< Using a 3/16 Allen wrench, remove oil fill plug (**arrow**).

NOTE—
- *Supercharger removed from vehicle for clarity*

mk15939a

< Fill reservoir with oil until it begins to run out (**arrow**).

– Reinstall plug and tighten.

Supercharger, removing and installing

– Working in cargo compartment, lift floor trim cover. Disconnect negative (–) cable from battery. See **121 Battery, Starter, Alternator** for more information.

> ***CAUTION—***
> *Disconnecting the battery may erase fault code(s) stored in memory. Check for fault codes prior to disconnecting the battery cables.*

– Remove intercooler. See **113 Cylinder Head Removal and Installation**.

– Remove throttle housing. See **Throttle housing, removing and installing (Cooper S)** in this repair group.

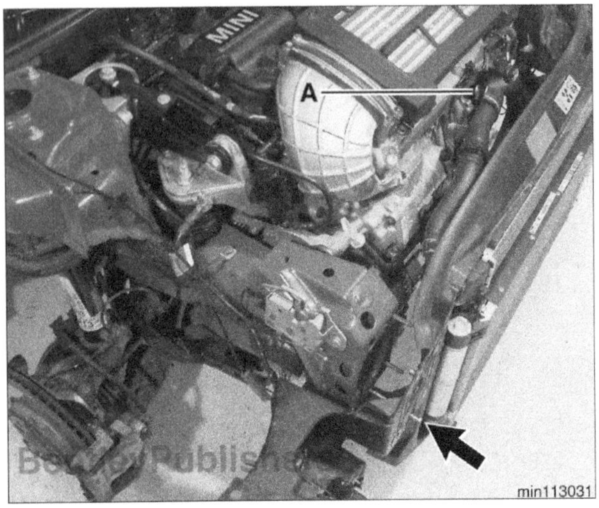

Loosen modular front end (MFE):

- Remove right front wheel housing liner.
- Detach left front wheel housing liner from front bumper cover trim.
- Remove front bumper cover trim.
- Remove bumper.
- Detach radiator upper hose support clamp (**A**) from intake manifold.
- Remove MFE mounting bolts. Install two 100 mm (4 in) M8 bolts (**arrow**) in left and right bumper support members. Slide MFE forward, supported on long bolts.

> ***CAUTION—***
> *For ease of component alignment when reassembling front bumper assembly, do not loosen or remove bumper alignment bosses.*

NOTE—
- *If available, use BMW special tools 11 8 401 and 11 8 402 instead of long bolts to support MFE.*
- *Removal of MFE is a complicated job. It is covered in* **510 Bumpers, External Trim**.

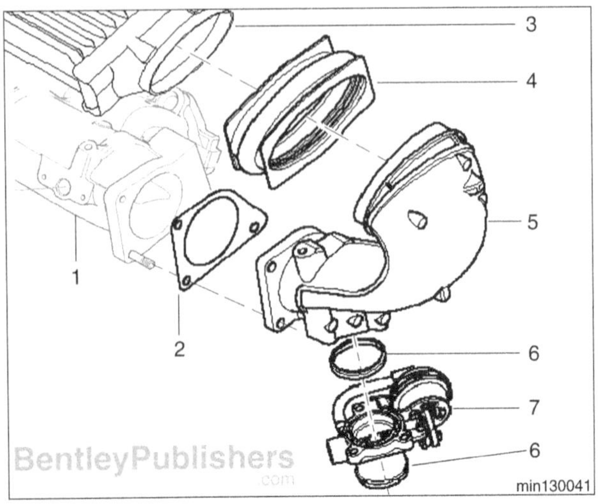

Detach air intake bypass valve from intercooler output air duct. Remove air duct fasteners from intake manifold and remove duct.

1. Intake manifold
2. Gasket
3. Intercooler
4. Intercooler bellows
5. Intercooler output air duct
6. O-ring seal
7. Air intake bypass valve

– Remove intake manifold. See **113 Cylinder Head Removal and Installation**.

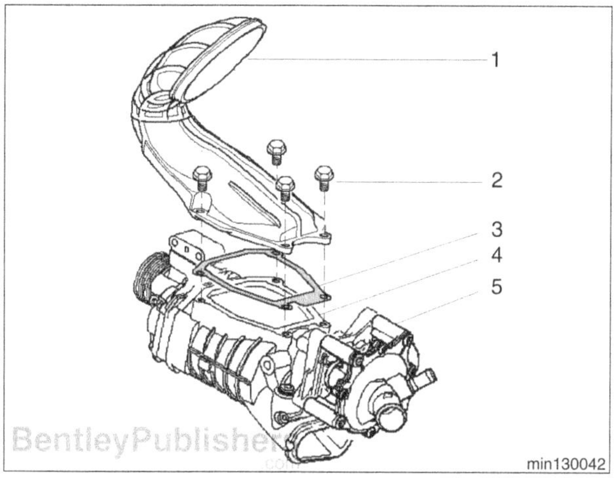

Remove supercharger output air duct.

1. Supercharger output air duct
2. Bolt M8
   - Tighten to 25 Nm (18 ft-lb)
3. Gasket (always replace)
4. Supercharger
5. Coolant pump

– Remove engine accessory belt. See **020 Maintenance**.

NOTE —
Mark direction of belt travel. Reinstall drive belt in same direction of rotation.

Working below left side of engine compartment, loosen lower radiator hose clamp (**arrow**). Detach hose from radiator, allowing coolant to drain into suitable catch pan.

WARNING —
Due to risk of personal injury, be sure the engine is cold before opening any part of the cooling system or removing the coolant reservoir cap.

NOTE —
Save coolant and reuse, or dispose of properly.

WARNING —
Use caution when draining and disposing of engine coolant. Coolant is poisonous and lethal to humans and pets. Pets are attracted to coolant because of its sweet smell and taste. Seek medical attention immediately if coolant is ingested.

– Detach alternator electrical harness. Remove alternator mounting bolts. Remove alternator. See **121 Battery, Starter, Alternator**.

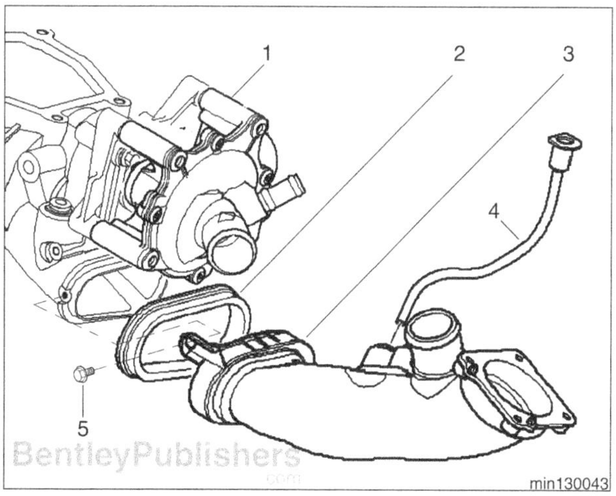

Unclip vacuum lines from supercharger intake air duct. Remove duct from supercharger.

1. Supercharger-coolant pump assembly
2. Profile gasket
3. Supercharger intake air duct
4. Vacuum line
5. Bolt M6
   - Tighten to 9 Nm (7 ft-lb)

– Loosen clamps and detach coolant hoses from rear of coolant pump. Also, for better access, detach coolant hose from thermostat housing at left end of cylinder head.

– Remove dipstick mounting bolt and move dipstick aside.

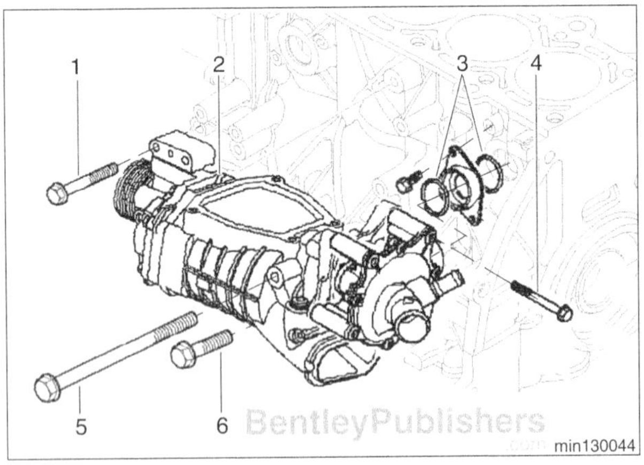

- Remove supercharger mounting bolts at engine block. Remove supercharger.
  1. Bolt M8 x 60 mm
     - Tighten to 25 Nm (18 ft-lb)
  2. Supercharger-coolant pump assembly
  3. Coolant pump O-ring seals
  4. Bolt M8 x 80 mm
     - Tighten to 30 Nm (22 ft-lb)
  5. Bolt M10 x 150 mm
     - Tighten to 45 Nm (33 ft-lb)
  6. Bolt M10 x 40 mm
     - Tighten to 45 Nm (33 ft-lb)

– With supercharger on workbench, remove coolant pump mounting bolts and separate coolant pump from supercharger.

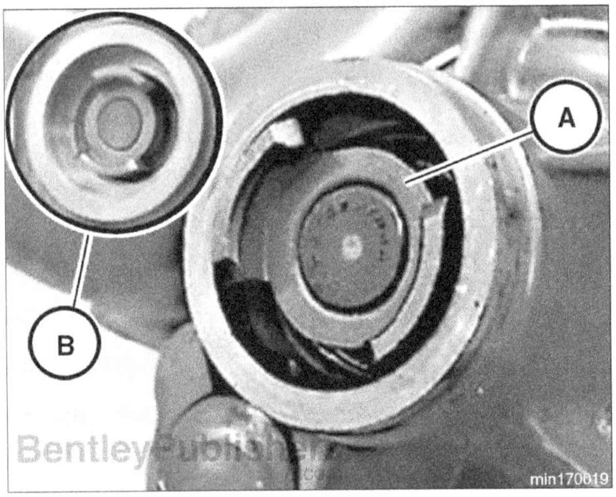

- When reinstalling coolant pump, make sure pump shaft (**A**) is keyed correctly to supercharger shaft (**B**).
  - Item **B** is shown in illustration at a smaller scale than item **A**.

Tightening torque	
Coolant pump to supercharger (M8)	30 Nm (22 ft-lb)

– Supercharger installation is opposite of removal. Note the following:

- Install new coolant pump sealing O-ring prior to placing supercharger-coolant pump assembly up against engine block.
- Fill and bleed cooling system. See **170 Radiator and Cooling System**. Run engine and check for leaks.

Tightening torques	
Alternator harness connector to alternator (M6)	10 Nm (7 ft-lb)
Alternator to engine block (M8)	25 Nm (18 ft-lb)
Intercooler outlet duct to intake manifold (M7)	16 Nm (12 ft-lb)
Supercharger outlet duct to supercharger (M8)	25 Nm (18 ft-lb)
Supercharger to engine block (M8)	25 Nm (18 ft-lb)
Supercharger to engine block (M10)	45 Nm (33 ft-lb)

Supercharger pulley, removing and installing

A common upgrade for the Cooper S is to replace the stock supercharger pulley with a smaller diameter aftermarket pulley. This allows the supercharger to spin faster and therefore provide more boost. This upgrade may void the manufacturer's vehicle warranty.

The procedure given here shows how to remove the stock pulley only. Installation for the various aftermarket pulleys will vary. Always follow manufacturer's installation instructions supplied with pulley.

To remove the stock pulley using a conventional gear puller, the engine must be lifted slightly and the accessory belt tensioner removed.

– Working in cargo compartment, lift floor trim cover. Disconnect negative (–) cable from battery. See **121 Battery, Starter, Alternator** for more information.

> ***CAUTION—***
> *Disconnecting the battery may erase fault code(s) stored in memory. Check for fault codes prior to disconnecting the battery cables.*

– Remove lower sound absorber panel under engine.

– Support weight of engine with hoist or floor jack.

NOTE—
- *Use wooden block or rubber pad under engine to prevent damage from floor jack.*
- *Use of two floor jacks on left and right sides of engine may make removal and installation of engine mounting bolts easier.*

◄ Working underneath car, remove lower engine vibration damper mounting bolt (**arrow**).

Vehicles through mid-2004

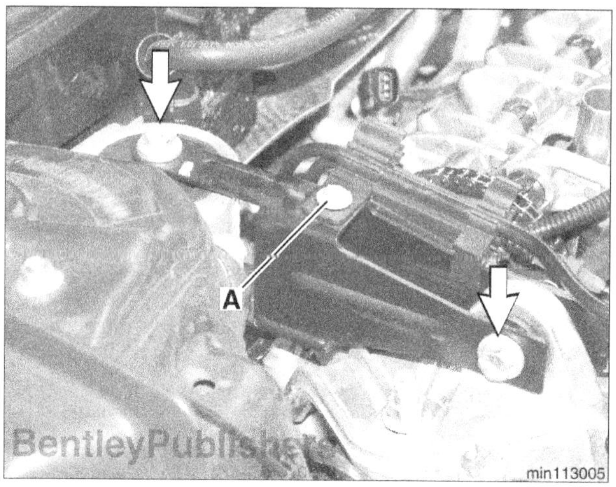

◄ Working on right side of engine, remove bolts from top engine vibration damper bracket (**arrows**), if applicable.

NOTE—
Later model vehicles may not have vibration damper bracket.

– Unclip hoses and remove plastic bracket (**A**).

Continued for all vehicles

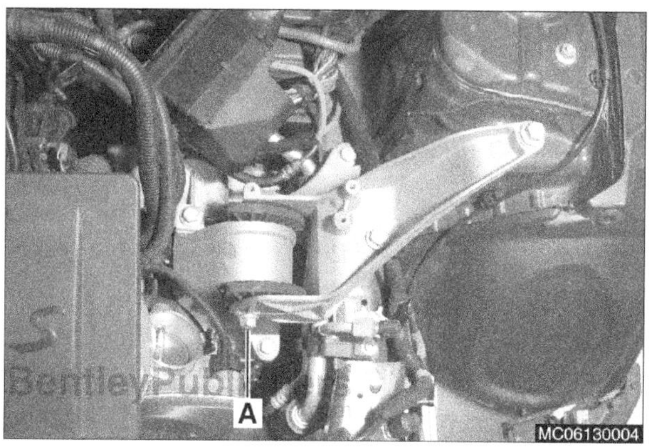

◄ Remove left side engine mount bolt (**A**).

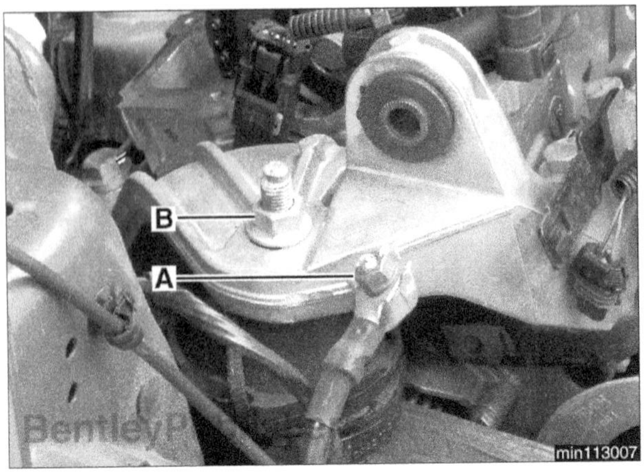

- Remove ground wire (**A**) from engine mount and mounting nut (**B**) from hydraulic mount.
- Remove clamp bolt at top of radiator that holds wiring harness and radiator hose to intake manifold.
- Use hoist or floor jack to slowly lift engine enough to access accessory belt tensioner.
  - Make sure that coil pack plug, throttle body and power steering fan are not damaged while lifting engine.
  - Make sure hoses and wiring harnesses are not damaged while lifting engine.

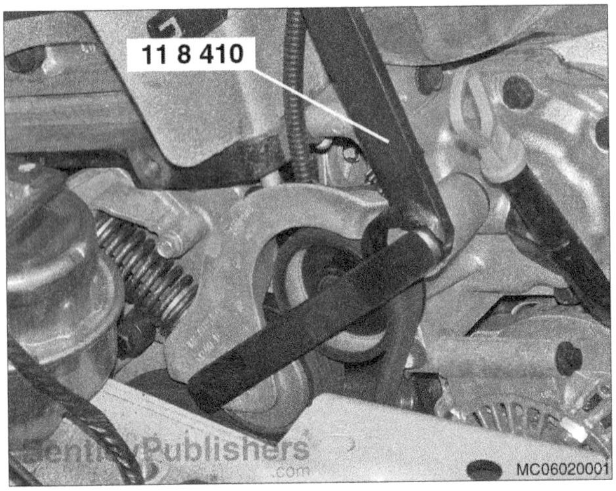

- Use special tool 11 8 410 to release accessory belt tension.

- Use lock pin (**arrow**) (special tool 11 8 470, or equivalent) to lock belt tensioner.

> ***CAUTION—***
> *Belt tensioner is under high tension. Check that lock pin is secure.*

◄ Remove mounting bolts (**arrows**) for accessory belt tensioner.

- Note location of mounting bolts for correct installation later.

◄ Pry out top lip of center cap of supercharger pulley. Remove cap by unthreading it.

– Install a 10 mm (13mm head) bolt into the center of the pulley shaft to protect the internal threads from damage when using gear puller.

◄ Install collar around pulley and gear puller to collar. Remove pulley using gear puller.

- If necessary, block puller (open end wrench shown at **arrow**) from turning using suitable flat stock. Use a rag to protect painted surfaces.

MC06130006

Clean pulley shaft (**arrow**) and follow manufacturer's instructions for installing new pulley.

– If re-installing old pulley, install in reverse order of removal. Tighten fasteners to specification.

Tightening torques	
Belt tensioner to crankcase	45 Nm (33 ft-lb)
Engine mount bracket to engine mount	68 Nm (50 ft-lb)
Engine mount bracket to transmission	38 Nm (28 ft-lb)
Engine mount to body	68 Nm (50 ft-lb)
Ground wire to engine support bracket	13.5 Nm (10 ft-lb)
Lower engine vibration damper to holder	100 Nm (74 ft-lb)
Top engine vibration damper bracket to body	100 Nm (74 ft-lb)
Top engine vibration damper bracket to engine	100 Nm (74 ft-lb)
Transmission mount bolts to transmission	66 Nm (49 ft-lb)
Transmission mount retaining bolt to upper bracket	66 Nm (49 ft-lb)

Sensor Replacement

The manifold pressure sensors, air temperature sensor and coolant temperature sensor are among the most important components of the engine management system.

Testing of these components is generally best accomplished with a BMW or BMW-compatible diagnostic scan tool.

T-MAP sensor, replacing (Cooper)

– Read out engine control module (ECM) fault memory. Then switch off ignition.

– Remove throttle housing. See **Throttle housing, removing and installing (Cooper)** in this repair group.

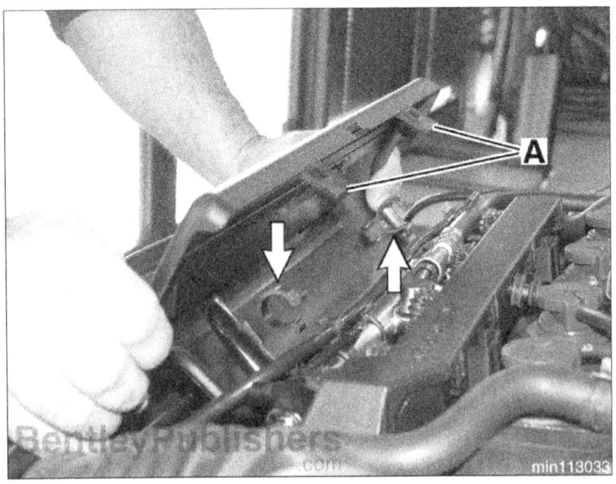

min113033

Remove fuel rail cover:

- Squeeze locking tabs (**A**) to detach rear of cover.
- Tilt and pull cover forward until plastic clips (**arrows**) detach from fuel rail.

> *CAUTION—*
> *Plastic clips can break easily.*

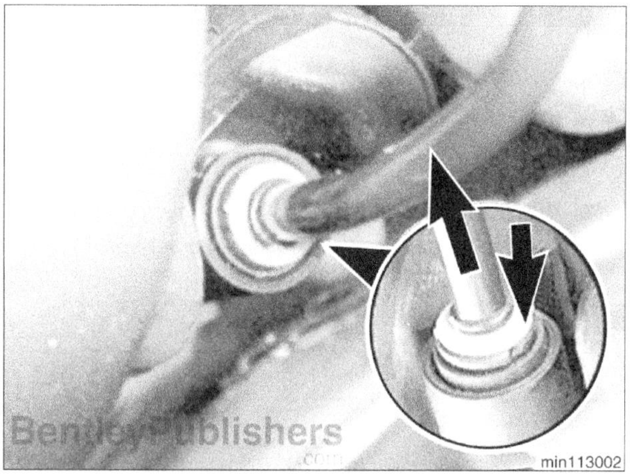

- Working at intake manifold, press brake booster vacuum line locking ring downward to detach line from manifold.
- Unlock electrical harness connector from sensor and detach.

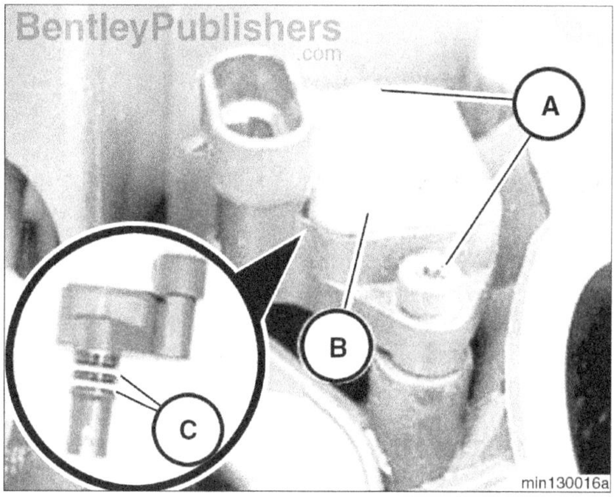

- Release mounting screws (**A**) and remove T-MAP sensor (**B**).
- When reinstalling, replace sealing O-rings (**C**).

Tightening torque	
T-MAP sensor to intake manifold	4 Nm (35 in-lb)

- Clear ECM fault memory.

MAP sensor, replacing (Cooper S)

- Read out engine control module (ECM) fault memory. Then switch off ignition.

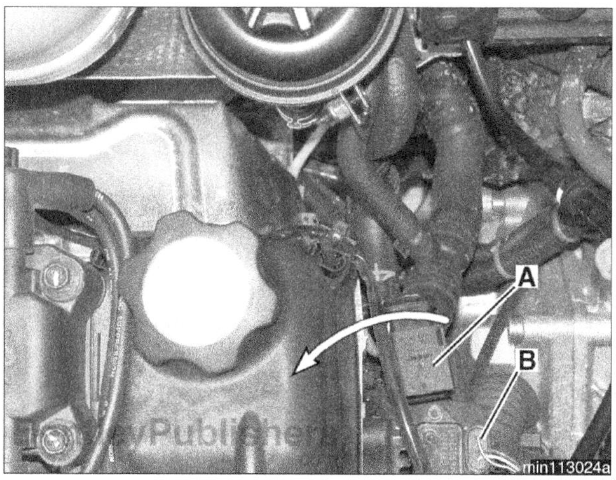

- Working at left end of cylinder head:
  - Disconnect oxygen sensor connector (**A**). Twist in direction of **arrow** to detach from support bracket.
  - Disconnect MAP sensor connector (**B**).
- Remove MAP sensor mounting screws and remove sensor.
- When reinstalling, replace sealing O-rings.

Tightening torque	
MAP sensor to bracket	3 Nm (28 in-lb)

- Clear ECM fault memory.

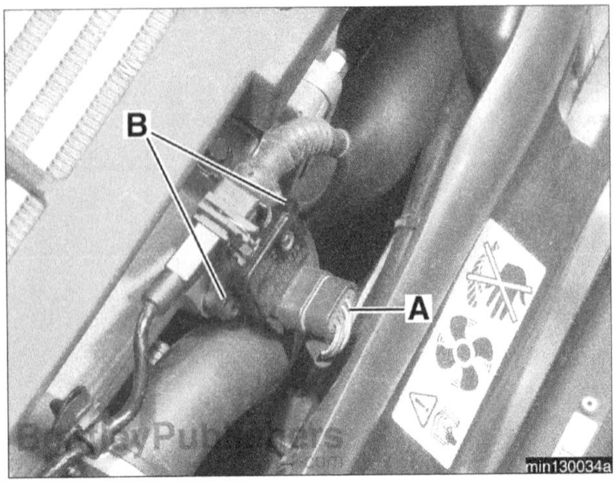

T-MAP sensor, replacing (Cooper S)

- Read out engine control module (ECM) fault memory. Then switch off ignition.

◀ Working at top left of intake manifold:

- Disconnect electrical harness connector (**A**) from T-MAP sensor.
- Remove sensor mounting screws (**B**). Remove sensor.

- When reinstalling, replace sealing O-rings.

Tightening torque	
T-MAP sensor to intake manifold	4 Nm (35 in-lb)

- Clear ECM fault memory.

Engine coolant temperature (ETC) sensor, replacing (Cooper)

- Read out engine control module (ECM) fault memory. Then switch off ignition.

- Remove air filter housing. See **Air filter housing, removing and installing (Cooper)** in this repair group.

- Remove engine control module (ECM), battery and battery box. See **121 Battery, Starter, Alternator**.

◀ Working below left side of engine compartment, loosen lower radiator hose clamp (**arrow**). Detach hose from radiator and drain coolant down to below height of thermostat housing.

> ***WARNING —***
> *Due to risk of personal injury, be sure the engine is cold before opening any part of the cooling system or removing the coolant reservoir cap.*

- Working at left end of cylinder head, detach temperature sensor electrical harness connector.

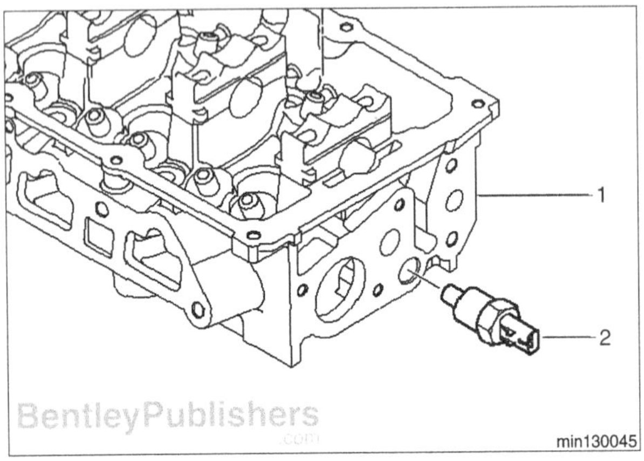

◀ Remove coolant temperature sensor.

1. Cylinder head
2. Engine coolant temperature (ECT) sensor

- Install and torque new ECT sensor.

Tightening torque	
ECT sensor to cylinder head	17 Nm (13 ft-lb)

- Fill and bleed cooling system. See **170 Radiator and Cooling System**. Run engine and check for leaks. Clear ECM fault memory.

Engine coolant temperature (ETC) sensor, replacing (Cooper S)

– Read out engine control module (ECM) fault memory. Then switch off ignition.

– Remove air filter housing. See **Air filter housing, removing and installing (Cooper S)** in this repair group.

◄ Working below left side of engine compartment, loosen lower radiator hose clamp (**arrow**). Detach hose from radiator and drain coolant down to below height of thermostat housing.

> ***WARNING —***
> *Due to risk of personal injury, be sure the engine is cold before opening any part of the cooling system or removing the coolant reservoir cap.*

– Working at left end of cylinder head, detach temperature sensor electrical harness connector.

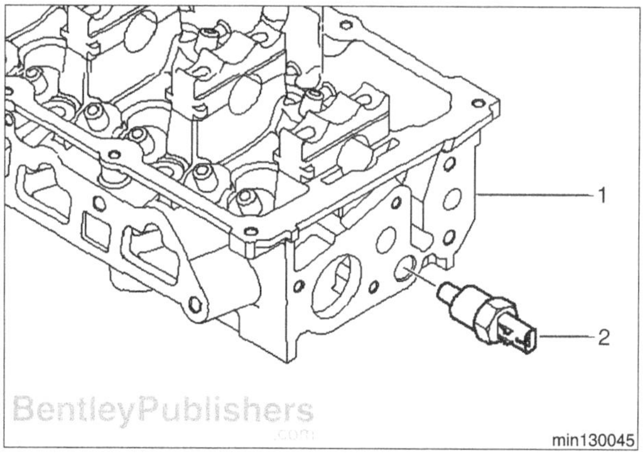

◄ Remove coolant temperature sensor.

1. Cylinder head
2. Engine coolant temperature (ECT) sensor

– Install and torque new ECT sensor.

Tightening torque	
ECT sensor to cylinder head	17 Nm (13 ft-lb)

– Fill and bleed cooling system. See **170 Radiator and Cooling System**. Run engine and check for leaks.

– Clear ECM fault memory.

Electrical Components

Main relay, accessing

Main relay (K6300) is located in engine compartment fuse and relay panel.

> ***CAUTION—***
> *Wait at least 5 minutes after turning off the ignition before removing the main relay. If the relay is removed before this time, a residual power surge may damage the engine control module (ECM).*

Fuel pump relay, accessing

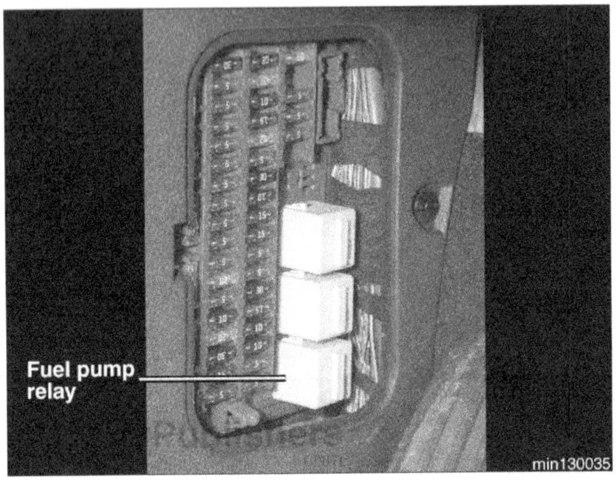

Fuel pump relay is located in passenger compartment fuse and relay panel in left kick panel.

> ***CAUTION—***
> *Wait at least 5 minutes after turning off the ignition before removing the fuel pump relay. If the relay is removed before this time, residual power surge in the main relay may damage the engine control module (ECM).*

Engine control module (ECM), accessing

Engine control module (ECM) is located in engine compartment and shares a plastic housing with battery (Cooper) or air filter assembly (Cooper S).

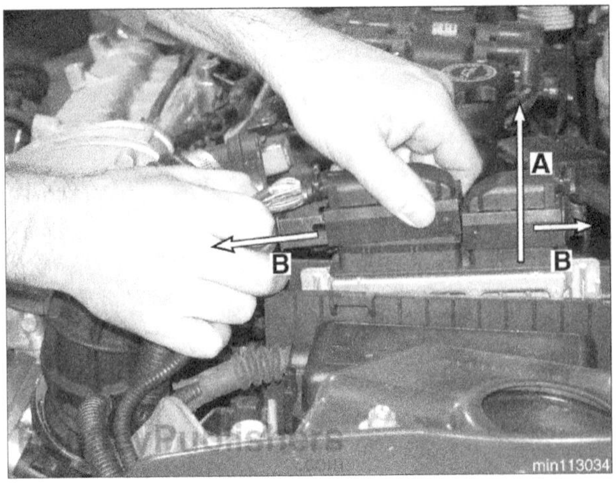

Remove DME control module:

- Remove control module container cover and lift control unit upward (**A**).
- Pull connector locking sliders outward (**B**).
- Disconnect control module connectors.

NOTE—
There are 2 control module harness connectors.

ECM pin assignments

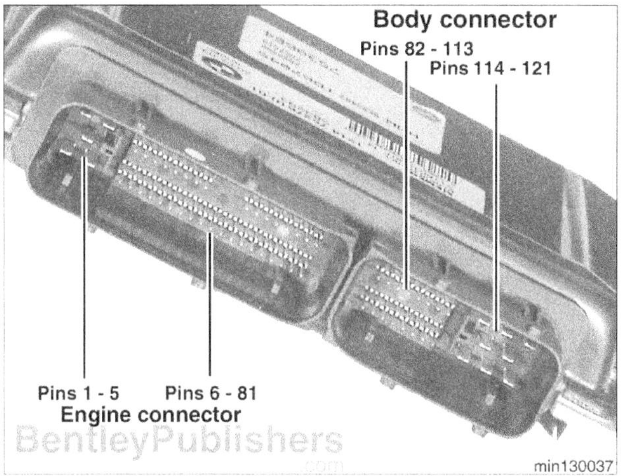

ECM electrical connector is a single molding split into two connectors.

- Engine connector (X6000): 81 pins
  Pins 1 - 5: load terminals
  Pins 6 - 81: signal terminals
- Body connector(X60004): 40 pins
  Pins 82 - 113: signal terminals
  Pins 114 - 121: load terminals

NOTE—
- *In wiring diagrams, the connector pins for X6000 and X60004 are each numbered starting at #1.*
- *In the molded housing of the ECM, the pins are numbered continuously from 1 to 121. Both pin numbering systems are shown in the pin assignment table for X60004. The wiring diagram pin numbers for X60004 are in the* **W.D. pin** *column of* **Table c.** *later in this repair group.*

ECM pin assignments are given in **Table b** and **Table c**. This information may be helpful when diagnosing faults to or from the ECM. If all inputs and wiring are OK but operational problems still exist, ECM itself may be faulty.

Generally, absence of voltage or continuity means there is a wiring or connector problem. Test results with incorrect values do not necessarily mean that a component is faulty.

- Check for loose, broken or corroded connections and wiring before replacing components.
- If results are still incorrect, check component itself.

For engine management system electrical schematics, see **ELE Electrical Wiring Diagrams**.

When making checks at ECM itself, use a breakout box to allow tests to be made with connectors attached to ECM. This also prevents damage to small terminals in connectors. As an alternative, separate harness connector housing so that electrical checks can be made from back of connector.

CAUTION—
- *Wait at least 5 minutes after turning off the ignition before removing the ECM connectors. If the ECM is disconnected before this time, residual power surge in the main relay may damage it.*
- *Always connect or disconnect the ECM connectors and meter probes with the ignition off.*

Table b. EMS2000 ECM pin assignments, connector X6000 (81-pin)			
Pin	**Signal**	**Function**	**Component**
1	not used		
2	not used		
3	Output	Signal, ignition coil 2-3	Plug connector, ignition coils X6150
5	Input	Oxygen sensor heater (precatalyst)	
6	Ground	Oxygen sensor (precatalyst)	
7	Ground	Knock sensor	
8	Input	Knock sensor	
9	Input	Oxygen sensor (precatalyst)	
10	Input	Throttle potentiometer 1	Throttle housing (EDR)
11	Input	Throttle potentiometer 2	Throttle housing (EDR)
12	Ground	Throttle position sensor	Throttle housing (EDR)
13	Output	Voltage supply, throttle position sensor	Throttle housing (EDR)
14	not used		
15	not used		
16	not used		
17	not used		
18	not used		
19	Output	Cyl. 4 fuel injector	
20	Output	Cyl. 2 fuel injector	
21	Output	Cyl. 1 fuel injector	
22	Output	Cyl. 3 fuel injector	
23	not used		
24	Output	Solenoid valve	Fuel tank ventilation valve
25	Input	Transmission fluid temperature sensor	Transmission control module (GIU)
26	Input	Engine coolant temperature sensor	
27	Input	Intake air temperature	T-MAP sensor
28	not used		
29	not used		
30	not used		
31	Input	T-MAP sensor	
32	not used		
33	not used		
34	not used		
35	not used		
36	not used		
37	not used		

Table b. EMS2000 ECM pin assignments, connector X6000 (81-pin)			
Pin	**Signal**	**Function**	**Component**
38	not used		
39	not used		
40	not used		
41	not used		
42	Input	Leakage diagnosis pump (LDP)	Engine plug connector I (X6011)
43	Input	Oxygen sensor heater (post-catalyst)	
44	not used		
45	Ground	Transmission fluid temperature sensor	Transmission control module (GIU)
46	Ground	Electronic ground, coolant temperature sensor	
47	Ground	T-MAP sensor	
48	Ground	Shield, knock sensor	
49	not used		
50	not used		
51	Output	Voltage supply, T-MAP sensor	
52	Input	Terminal 61	Alternator (generator)
53	Input	Engine cooling fan, camshaft sensor	
54	Input	Engine speed sensor signal	Transmission rpm sensor
55	Input	Crankshaft position sensor	
56	not used		
57	not used		
58	not used		
59	not used		
60	not used		
61	Ground	X6454	
62	Ground	X6454	
63	Output	Throttle valve	Throttle housing (EDR)
64	Output	Throttle valve	Throttle housing (EDR)
65	Output	Throttle valve	Throttle housing (EDR)
66	Output	Throttle valve	Throttle housing (EDR)
67	not used		
68	Input	Coolant outlet temperature	Coolant temperature sensor, radiator outlet
69	Ground	Coolant outlet temperature	Coolant temperature sensor, radiator outlet
70	not used		
71	Ground	Catalytic converter	
72	Input	Ground signal, engine cooling fan, camshaft sensor	
73	Input	Engine speed sensor signal ground	Transmission rpm sensor
74	Ground	Crankshaft position sensor	

Table b. EMS2000 ECM pin assignments, connector X6000 (81-pin)			
Pin	Signal	Function	Component
75	not used		
76	not used		
77	not used		
78	not used		
79	not used		
80	Ground	X6454	
81	Ground	X6454	

Table c. EMS2000 ECM pin assignments, connector X60004 (40-pin)				
W.D. pin	ECM pin	Signal	Function	Component
1	82	Output	Voltage supply, driver's wish sensor 1	Pedal position sensor
2	83	Output	Voltage supply, driver's wish sensor 2	Pedal position sensor
3	84	Not used		
4	85	Input	Signal, clutch pedal position switch	Clutch switch module
5	86	Output	Electronic ground, pedal position sensor 1	
6	87	Output	Electronic ground, pedal position sensor2	
7	88	Output	Activation, engine cooling fan	Relay, engine cooling fan
8	89	Not used		
9	90	Input	Pressure sensor signal, air-conditioning system	
10	91	Input	Pressure sensor signal, air-conditioning system	
11	92	Not used		
12	93	Input	Leakage diagnosis pump (LDP)	
13	94	Output	Compressor enable from main relay	A/C compressor
14	95	Output	Activation, engine cooling fan stage 2	Switch unit, 2-speed blower relay, fan, steering control module
15	96	Output	Gearshift lock	Shiftlock relay
16	97	Output	Activation, main relay (K6300)	
17	98	Input	Brake light	Brake light switch
18	99	Input	Brake light test signal	Brake light switch
19	100	Input	Electronic ground, A/C pressure sensor	Pressure sensor, A/C system
20	101	Output	Volute spring (I01002) speed signal	
21	102	Not used		
22	103	Not used		
23	104	Not used		
24	105	Input	Activation, fuel pump relay (K96)	
25	106	Input	Signal, immobilizer control module (EWS) (A836)	

Table c. EMS2000 ECM pin assignments, connector X60004 (40-pin)				
W.D. pin	**ECM pin**	**Signal**	**Function**	**Component**
26	107	Input	Pedal position sensor	
27	108	Input	Pedal position sensor	
28	109	Input	Oxygen sensor (post-catalyst)	
29	110	Input / output	CAN-bus high (X8090)	
30	111	Input / output	CAN-bus low (X8091)	
31	112	Ground	Oxygen sensor (post-catalyst)	
32	113	Input / output	OBDII diagnostic link connector (X19527)	
33	114	Ground	Ground	
34	115	Ground	Ground	
35	116	Not used		
36	117	Input / output	Diagnostic link connector (X9397)	
37	118	Output	Terminal 15 Fuse F34	
38	119	Not used		
39	120	Input	Terminal 87 Fuse F02	
40	121	Input	Terminal 30 Fuse F01	

160 Fuel Tank and Fuel Pump

General

This repair group covers service information for the fuel supply system. Information on the fuel injection system, EMS2000, is covered in **130 Fuel Injection**.

The fuel filter is a "lifetime" unit installed in the fuel tank. There is no routine interval or procedure for replacing the fuel filter.

Special tools

Some of the procedures in this group require the use of special tools.

11 8 480

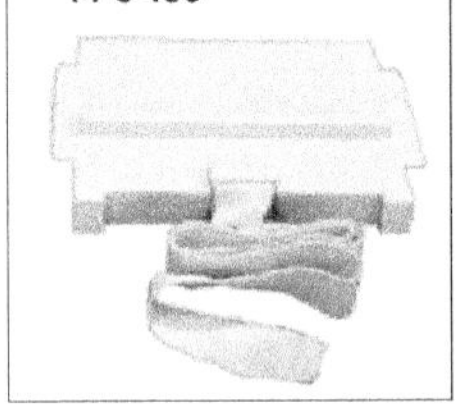

Intercooler protective cover
(BMW special tool 11 8 480)

13 1 130

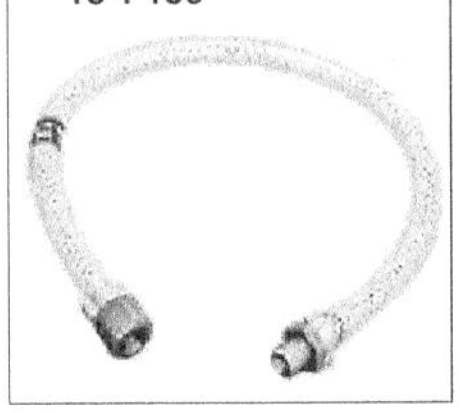

Fuel line adapter
(BMW special tool 13 1 130)

1318

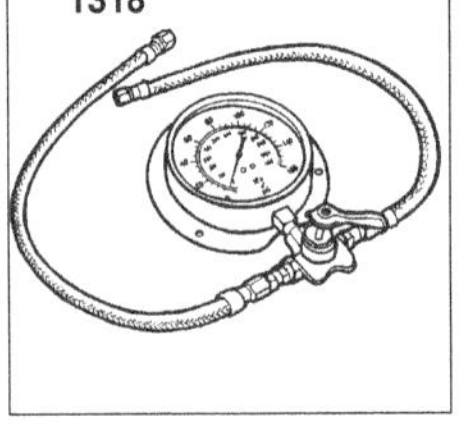

Fuel pressure gauge
(Baum Tools 1318)

13 3 010

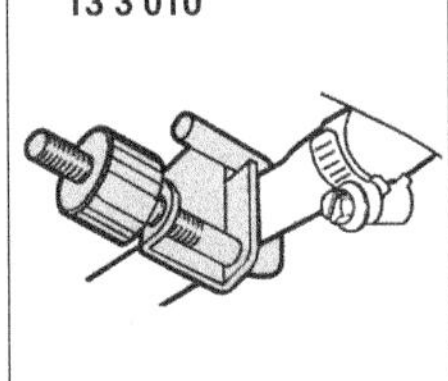

Fuel line clamp
(BMW special tool 13 3 010)

13 5 220

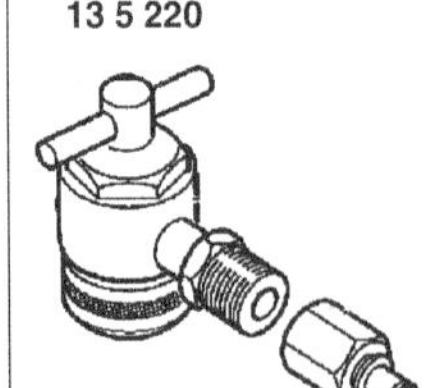

Fuel rail to pressure gauge adapter
(BMW special tool 13 5 220)

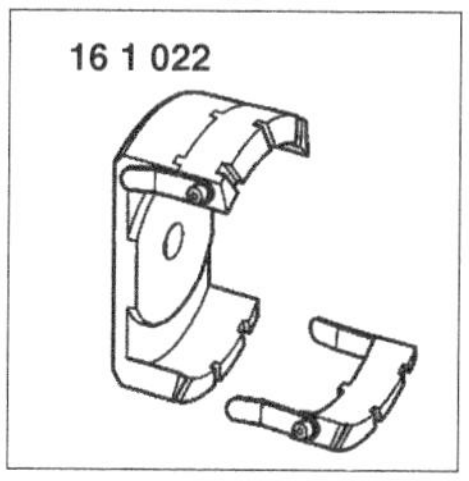

Fuel tank threaded collar spanner (BMW special tool 16 1 022)

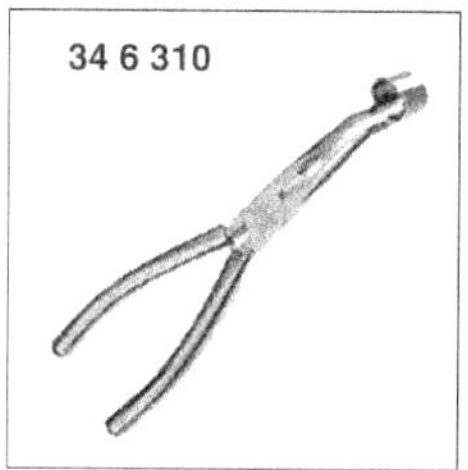

Parking brake cable removal pliers (BMW special tool 34 6 310)

Fuel pump

The electric fuel pump is mounted in the fuel tank in tandem with the left side fuel level sender. The fuel pump delivers fuel at high pressure to the fuel injection system. A pressure regulator maintains system pressure.

NOTE—
Fuel pump removal procedures are given under **Fuel level sender (left side) and fuel pump, removing and installing** *in this repair group.*

Fuel tank evaporative control system

Evaporative control is designed to prevent fuel system evaporative losses from venting into the atmosphere. The components of this system allow control and monitoring of evaporative losses by the on-board diagnostic (OBD II) software incorporated into the engine control module (ECM).

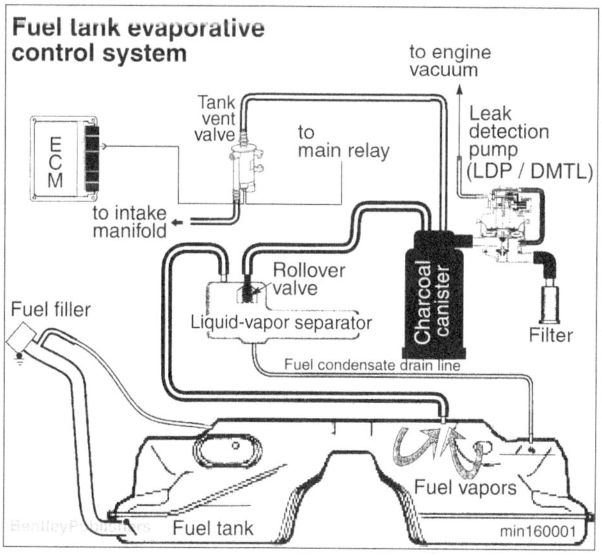

The main components of the evaporative control system and their functions are as follows:

- **Liquid/vapor separator** acts as fuel overflow tank.
- **Activated charcoal canister** stores evaporated fuel.
- **Plumbing** ducts vapors from fuel tank to canister and from canister to intake manifold.
- **Fuel tank vent valve** is controlled by engine control module (ECM).
- **Leak detection pump (LDP or DMTL)** pressurizes fuel tank to monitor system leaks.

Evaporative system troubleshooting

Start troubleshooting and testing of evaporative system by accessing DTCs using a diagnostic scan tool (DISplus, GT1, MoDiC or equivalent).

The ECM sets a diagnostic trouble code (DTC) when it detects a leak in the evaporative system. The malfunction indicator light (MIL) is illuminated upon a second recurrence of the fault. See **OBD On-Board Diagnostics** for further details.

BMW recommends that an independent pressure test of the fuel system be conducted with a chemical leak detector to confirm the presence of the fuel system leak.

Warnings and cautions

Observe the following warnings and cautions when servicing the fuel system.

WARNING —

- *The fuel system is designed to retain pressure even when the ignition is off. When working with the fuel system, loosen the fuel lines slowly to allow residual fuel pressure to dissipate. Avoid spraying fuel. Use shop rags to capture leaking fuel.*
- *Before beginning any work on the fuel system, place a fire extinguisher in the vicinity of the work area.*
- *Fuel is highly flammable. When working around fuel, do not disconnect any wires that could cause electrical sparks. Do not smoke or work near heaters or other fire hazards.*
- *Always unscrew the fuel tank cap to release pressure in the tank before working on the tank or lines.*
- *Do not use a work light with an incandescent bulb near any fuel. Fuel may spray on the hot bulb, causing a fire.*
- *Make sure the work area is properly ventilated.*

> ***CAUTION—***
> - *Prior to disconnecting the battery, read the battery disconnection cautions given in* **001 General Warnings and Cautions**.
> - *Before making any electrical tests with the ignition turned on, disable the ignition system as described in* **120 Ignition System**. *Be sure the battery is disconnected when replacing components.*
> - *To prevent damage to the ignition system or other engine management components, including the ECM, always connect and disconnect wires and test equipment with the ignition off.*
> - *Cleanliness is essential when working with the fuel system. Thoroughly clean the fuel line unions before disconnecting any of the lines.*
> - *Use only clean tools. Keep removed parts clean and sealed or covered with a clean, lint-free cloth, especially if completion of the repair is delayed.*
> - *Do not move the car while the fuel system is open.*
> - *Avoid using high pressure compressed air to blow out lines and components. High pressure can rupture internal seals and gaskets.*
> - *Always replace seals, O-rings and hose clamps.*

FUEL SYSTEM TROUBLESHOOTING

Fuel pressure, testing (Cooper)

– Switch off ignition.

– If using DISplus: Connect BMW special tools 13 1 130 and 13 5 222 to scan tool pressure sensor adapter.

– If using pressure gauge: Attach BMW special tools 13 3 063 and 13 1 130 to gauge. (Both are flexible fuel hoses.)

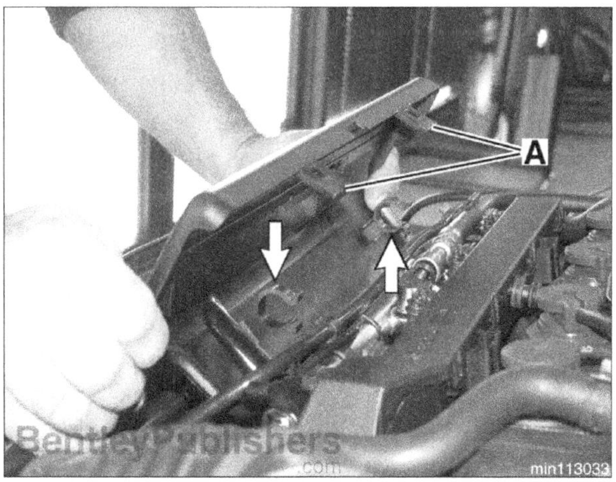

◄ Remove fuel rail cover:

- Squeeze locking tabs (**A**) to detach rear of cover.
- Tilt and pull cover forward until plastic clips (**arrows**) detach from fuel rail.

> ***CAUTION—***
> *Plastic clips can break easily.*

– Remove dust cap from end of fuel rail.

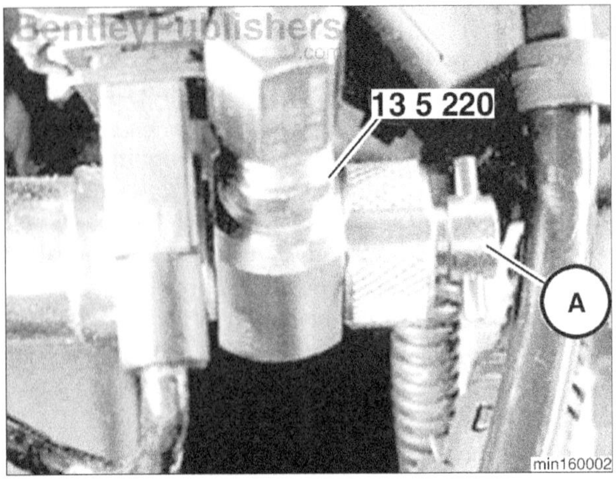

Connect BMW special tool kit 13 5 220 (consisting of 13 5 221 and 13 5 222) to end of vehicle fuel rail and connect to pressure gauge or scan tool fuel pressure fitting.

- Make sure special tool sealing O-ring is correctly placed.
- Make sure non-return valve (**A**) on special tool is backed out so that fuel rail pressure release valve remains closed.

WARNING —
- *When working with fuel line, fuel may be expelled under pressure. Loosen fuel filler cap to release fuel tank pressure.*
- *Do not smoke or work near heaters or other fire hazards.*
- *Keep a fire extinguisher handy.*
- *Before disconnecting fuel hose, wrap a cloth around hose to absorb any leaking fuel.*
- *Plug all open fuel connections.*

– Start engine.

– Screw in non-return valve until pressure gauge indicates pressure.

CAUTION —
Do not screw in non-return valve up to mechanical stop. This could damage measuring valve in fuel rail.

Fuel pressure specification	
Cooper	3.0 ± 0.2 bar (43.5 ± 3 psi)

– Turn off engine.

Fuel pressure, testing (Cooper S)

– Switch off ignition.

– If using DISplus: Connect BMW special tools 13 1 130 and 13 5 222 to scan tool pressure sensor adapter.

– If using pressure gauge: Attach special tools BMW 13 3 063 and 13 1 130 to gauge. (Both are flexible fuel hoses.)

Remove intercooler. See **113 Cylinder Head** for complete procedure.

- Remove intercooler cover.
- Remove intercooler sealing bellows clamps and intercooler cover mounting brackets.
- Detach sealing bellows and tilt intercooler to right to remove.

◄ Remove dust cap (**insert**) from end of fuel rail.

– Connect fuel pressure gauge or scan tool with adapters to end of fuel rail.

> ***WARNING —***
> - *When working with fuel line, fuel may be expelled under pressure. Loosen fuel filler cap to release fuel tank pressure.*
> - *Do not smoke or work near heaters or other fire hazards.*
> - *Keep a fire extinguisher handy.*
> - *Before disconnecting fuel hose, wrap a cloth around hose to absorb any leaking fuel.*
> - *Plug all open fuel connections.*

– Reinstall intercooler and cover with protective shield (BMW special tool 11 8 480).

– Start engine and read off fuel pressure.

Fuel pressure specification	
Cooper S	3.5 ± 0.2 bar (50.8 ± 3 psi)

– Turn off engine.

Fuel Pump and Fuel Level Senders

MINI models use a two-lobed fuel tank. Each lobe of the tank has its own fuel level sending unit.

The left side fuel level sender is integrated with the fuel pump. The right side fuel sender is integrated with the fuel filter. Each sender has a float connected to a variable resistance wiper contact for fuel level. When replacing the tank unit assembly always replace hose clamps, gaskets and O-rings.

To equalize fuel level between the two tank lobes, a siphon pump is installed in the right lobe.

Fuel level sender (left side) and fuel pump, removing and installing

– Drain fuel tank. See **Fuel tank, draining** in this repair group.

– Remove rear seat cushion. See **520 Seats**.

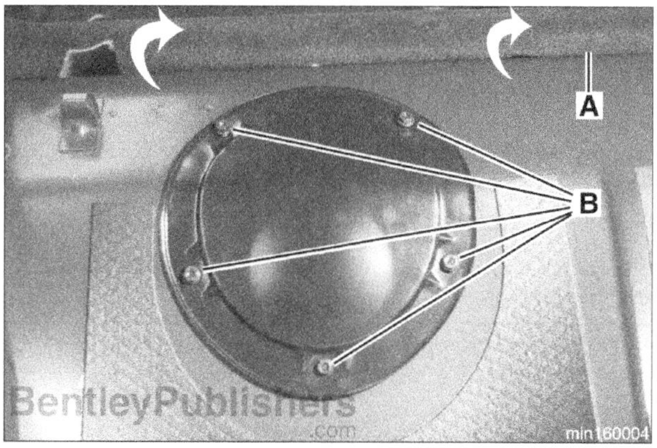

◄ Working under left rear seat:

- Roll trim (**A**) forward.
- Remove left fuel tank unit cover fasteners (**B**). Remove cover.

Pull off fuel tank unit electrical harness connector:

- Slide connector lock in direction **A**.
- Pull off connector in direction **B**.

– Use fuel tank cap spanner, BMW special tool 16 1 020, to unscrew and remove fuel tank locking collar.

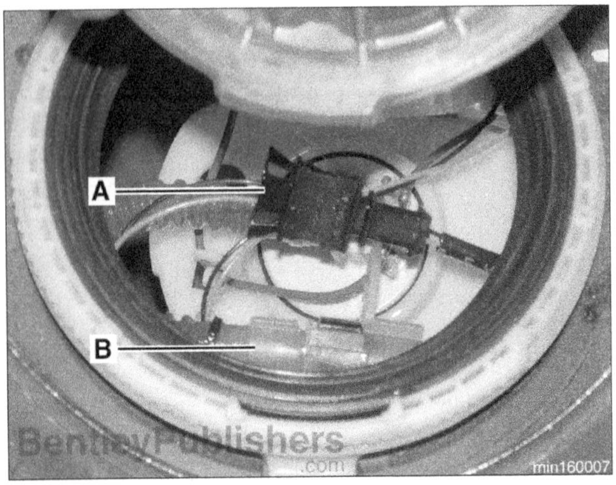

Lift fuel pump / fuel level sender unit slightly and detach electrical harness connector (**A**) and fuel line (**B**).

– Carefully lift pump / sender unit out of tank.

> ***WARNING —***
> *Fuel may be spilled. Do not smoke or work near heaters or other fire hazards.*

To reinstall:

- Use new sealing O-ring at tank collar.
- Reconnect harness connector and hose inside tank.
- Make sure fuel pump assembly locating lug engages tank recess. Then install locking collar.

Tightening torque	
Locking collar to fuel tank	35 Nm (26 ft-lb)

– Reconnect tank unit electrical harness.

– Fill fuel tank with at least 5 liters (1.5 gallons) of fuel. Then start engine and check for leaks.

> ***CAUTION —***
> *The fuel pump will be damaged if you run it without fuel.*

– Install access cover and tighten fasteners.

Tightening torque	
Fuel tank access cover to body (M6)	6 Nm (53 in-lb)

Fuel level sender (right side), removing and installing

– Drain fuel tank. See **Fuel tank, draining** in this repair group.

– Remove rear seat cushion. See **520 Seats**.

– Remove left side fuel tank unit. See **Fuel level sender (left side) and fuel pump, removing and installing** in this repair group.

◀ Working under left rear seat:
- Roll trim (**A**) forward.
- Remove left fuel tank unit cover fasteners (**B**). Remove cover.

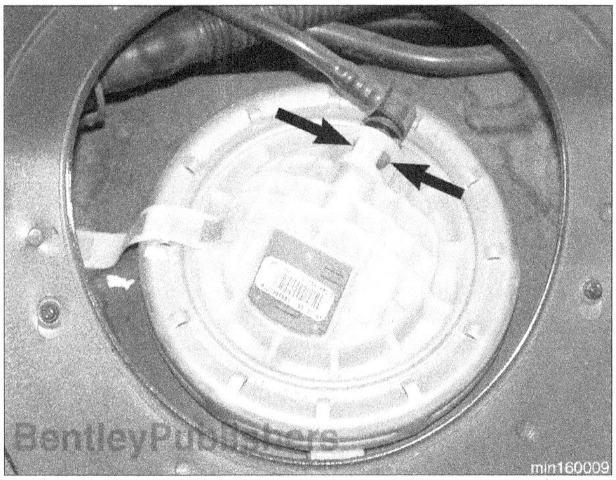

◀ Squeeze plastic tabs on fuel line connector (**arrows**) to detach line from tank.

– Use fuel tank cap spanner, BMW special tool 16 1 020, to unscrew and remove fuel tank locking collar.

– Lift up sender unit slightly and disconnect line fastener. Lift sender unit and lines out of fuel tank.

> ***WARNING —***
> *Fuel may be spilled. Do not smoke or work near heaters or other fire hazards.*

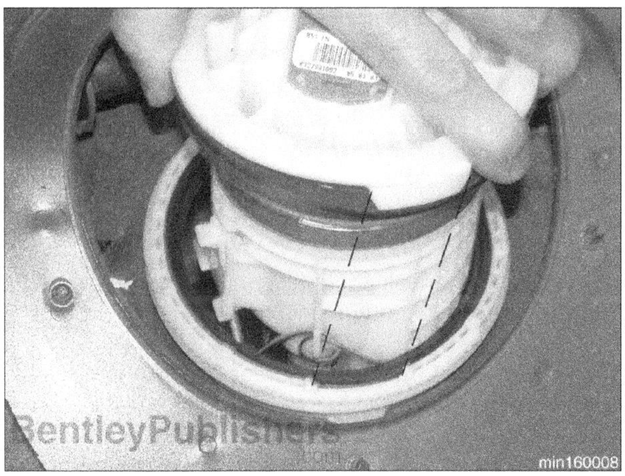

◀ To reinstall:
- Use new sealing O-ring at tank collar.
- Reconnect fuel line inside tank.
- Make sure fuel level sender assembly locating lug engages tank recess. Then install locking collar.

Tightening torque	
Locking collar to fuel tank	35 Nm (26 ft-lb)

– Reconnect fuel hose.

– Fill fuel tank with at least 5 liters (1.5 gallons) of fuel. Then start engine and check for leaks.

> ***CAUTION —***
> *The fuel pump will be damaged if you run it without fuel.*

– Install access cover and tighten fasteners.

Tightening torque	
Fuel tank access cover to body (M6)	6 Nm (53 in-lb)

Fuel Tank and Fuel Lines

The plastic fuel tank is mounted beneath the center of the car (underneath the rear seat). Mounted in the fuel tank are the fuel pump and fuel level sending units. Connecting lines for the evaporative emission control system and expansion tank are also attached to the tank.

Fuel tank capacity	
Cooper, Cooper S Fuel tank Reserve capacity	 50 liters (13.2 US gal) 8 liters (2.1 US gal)

Fuel tank, draining

Except for a small remainder, most of the fuel can be drawn off from the left and right tank halves via the fuel filler pipe. The residual amount is drawn off after removal of the left and right fuel tank units.

Drain fuel tank into a safe storage unit using an approved fuel pumping device.

> ***WARNING —***
> - *Before draining tank, be sure that all hot components, such as the exhaust system, are completely cooled down.*
> - *Fuel may be spilled. Do not smoke or work near heaters or other fire hazards.*

– Start engine and allow to run 10 - 15 seconds to fill fuel compensating siphon assembly. This will allow both lobes of fuel tank to be drawn off through fuel filler pipe.

– Disconnect negative (–) cable from battery.

> ***CAUTION —***
> *Prior to disconnecting the battery, read the battery disconnection cautions given in* **001 General Warnings and Cautions**.

– Remove fuel tank filler cap.

– Slide suction hose into filler neck about 90 cm (35 in), twisting as necessary. Withdraw fuel into storage unit.

– Remove left fuel sender. See **Fuel level sender (left side) and fuel pump, removing and installing** in this repair group.

– Remove right fuel sender. See **Fuel level sender (right side), removing and installing** in this repair group.

– Draw off residual fuel quantity through openings.

– If siphoning mechanism is faulty, drain left lobe via filler neck. Drain right tank lobe separately by removing sender cover and pumping fuel directly out of right lobe.

– Remove suction hose from tank filler neck carefully to avoid damaging filler neck baffle plate.

– After finishing repairs but before starting engine, fill fuel tank with at least 5 liters (1.5 gallons) of fuel.

> ***CAUTION—***
> *The fuel pump will be damaged if you run it without fuel.*

– After reconnecting battery, be sure to reinitialize power windows: Raise window and keep power window switch in raise position for about 5 seconds.

Fuel tank, removing and installing

– Drain fuel tank. See **Fuel tank, draining** in this repair group.

> ***WARNING—***
> - *Before draining tank, be sure that all hot components, such as the exhaust system, are completely cooled down.*
> - *Fuel may be spilled. Do not smoke or work near heaters or other fire hazards.*

– Disconnect negative (–) cable from battery.

> ***CAUTION—***
> *Prior to disconnecting the battery, read the battery disconnection cautions given in* **001 General Warnings and Cautions**.

– Cooper S: Detach rear half of battery positive lead.

– Remove rear center console. See **513 Interior Trim**.

– Remove mounting bracket for rear center console.

– Disconnect parking brake cables at rear brake calipers. See **340 Brakes**.

– Remove rear seat. See **520 Seats**.

– Remove trim panels above fuel tank.

– Remove fuel tank unit covers. Disconnect wiring harness connector from left tank unit. Detach fuel line from right tank unit.

– Detach parking brake cables from parking brake lever. Use special tool 34 6 310 to squeeze cable tips and push out through bottom of car.

– Raise car and support in a safe manner.

> ***CAUTION—***
> *Make sure the car is stable and well supported at all times. Use a professional automotive lift or jack stands designed for the purpose. A floor jack is not adequate support.*

Fuel Tank and Fuel Lines

– Remove rear exhaust from catalyst back. See **180 Exhaust System**.
  - Cooper S: Remove heat shield under battery box in rear.
  - Remove center body brace under exhaust pipe.
  - Remove exhaust.

– Remove rear half of exhaust system heat shield.

– Remove left rear wheel and left rear wheel housing liner.

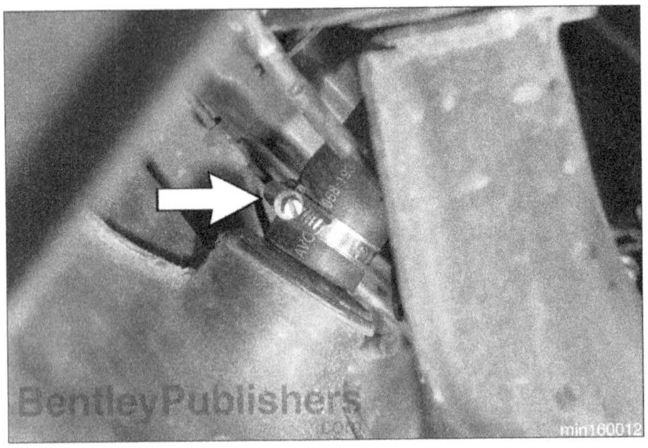

◄ Loosen hose clamp (**arrow**) and detach filler hose from fuel tank.

– Working above left rear wheel housing liner, disconnect vent hose from fuel filler pipe. See **Fuel filler pipe, removing and installing** in this repair group.

– Support fuel tank over a large area with a suitable lift.

◄ On convertible models, remove mounting bolts (**arrows**) and reinforcement supports.

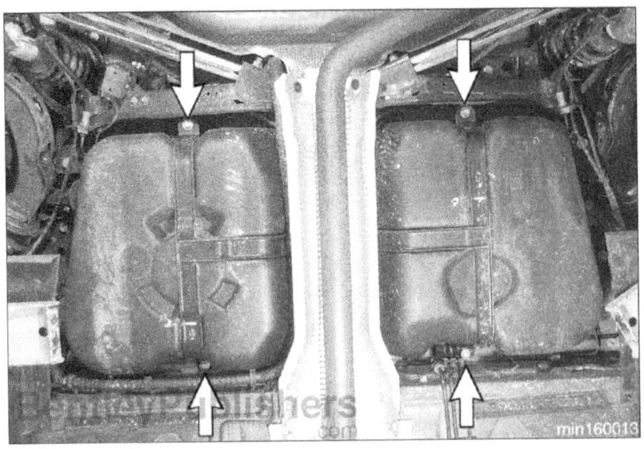

◄ Remove fuel tank mounting bolts (**arrows**).

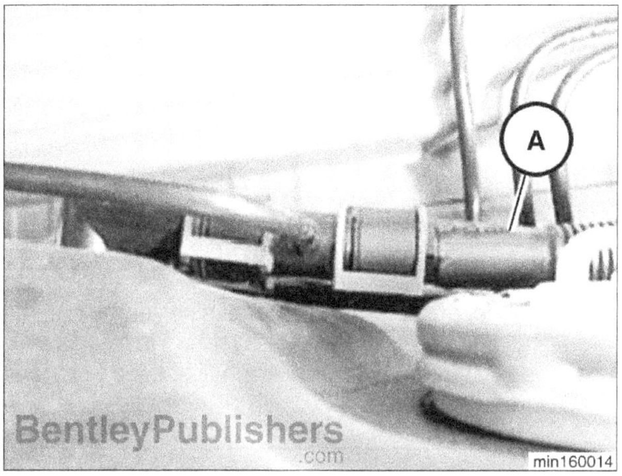

◄ Lower fuel tank and disconnect right vent line (**A**).

– Pull vent line through body.

– Carefully lower fuel tank, checking for any wiring, fuel lines or mechanical parts that might become snagged.

– Installation is reverse of removal:
  - Make sure lines and wiring harness are not trapped or crushed during installation.
  - Always use new seals, gaskets, O-rings, and hose clamps.
  - Inspect hoses and replace any that are chafed, dried out or cracked.
  - Inspect heat shield and replace if corroded.
  - Inspect rubber buffers and liners on fuel tank, support straps and on underside of body. Replace rubber parts that are hardened or damaged.

Tightening torques	
Fuel filler pipe to body (M6)	5 Nm (44 in-lb)
Fuel tank unit cover to body (M6)	6 Nm (53 in-lb)
Fuel tank to subframe and body (M8)	22 - 28 Nm (16 - 21 ft-lb)
Hose clamp 32 - 38 mm dia.	2.5 - 3.5 Nm (22 - 31 in-lb)

– After installation of fuel tank, prior to first engine start-up, check electrical resistance between filler pipe and wheel hub. To insure that filler pipe is correctly grounded, resistance should be no higher than approx. 0.65 Ω.

– After finishing repairs but before starting engine, fill fuel tank with at least 5 liters (1.5 gallons) of fuel.

> ***CAUTION—***
> *The fuel pump will be damaged if you run it without fuel.*

– After reconnecting battery, be sure to reinitialize power windows: Raise window and keep power window switch in raise position for about 5 seconds.

Fuel filler pipe, removing and installing

- Raise car and support in a safe manner.

> ***CAUTION —***
> *Make sure the car is stable and well supported at all times. Use a professional automotive lift or jack stands designed for the purpose. A floor jack is not adequate support.*

- Remove left rear wheel and left rear wheel housing liner.

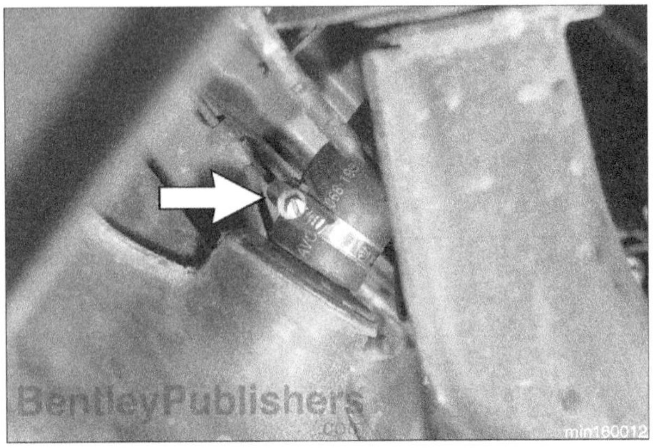

- Loosen or cut hose clamp (**arrow**) and detach filler hose from fuel tank.

> ***WARNING —***
> *Fuel tank must be no more than half full; if necessary, draw off some fuel from fuel tank.*

- Unscrew fuel filler cap from fuel filler pipe.

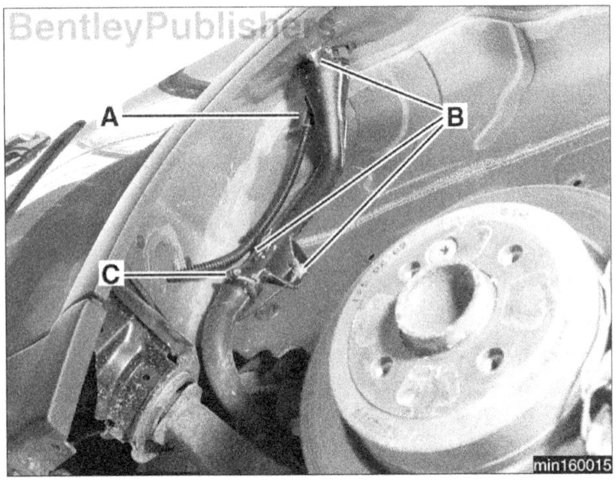

- Detach filler neck assembly:
  - Release quick-release fastener (**A**) and disconnect vent hose from filler pipe.
  - Remove filler neck mounting nuts (**B**).
  - Loosen or cut hose clamp (**C**) and separate filler neck from filler hose.

- Installation is reverse of removal:
  - Make sure vent hose to fuel filler pipe and fuel filler pipe to body cutout are correctly sealed.
  - Replace filler neck mounting nuts.

Tightening torques	
Fuel filler neck to body (M6)	5 Nm (44 in-lb)
Hose clamp 32 - 38 mm dia.	2.5 - 3.5 Nm (22 - 31 in-lb)
Wheel housing liner to fender	3 Nm (27 in-lb)
Wheel to wheel hub	120 ± 10 Nm (89 ± 7 ft-lb)

- After installation of fuel tank, prior to first engine start-up, check electrical resistance between filler pipe and wheel hub. To insure that filler pipe is correctly grounded, resistance should be no higher than approx. 0.65 Ω.

- After finishing repairs but before starting engine, fill fuel tank with at least 5 liters (1.5 gallons) of fuel.

> ***CAUTION —***
> *The fuel pump will be damaged if you run it without fuel.*

LDP / DMTL unit, removing and installing (2002-2004)

The fuel tank leakage diagnosis pump (LDP or DMTL) and the charcoal canister assembly are located behind the right rear wheel housing liner on 2002-2004 models.

- Raise car and support in a safe manner.

> ***CAUTION—***
> *Make sure the car is stable and well supported at all times. Use a professional automotive lift or jack stands designed for the purpose. A floor jack is not adequate support.*

- Remove right rear wheel and right rear wheel housing liner.

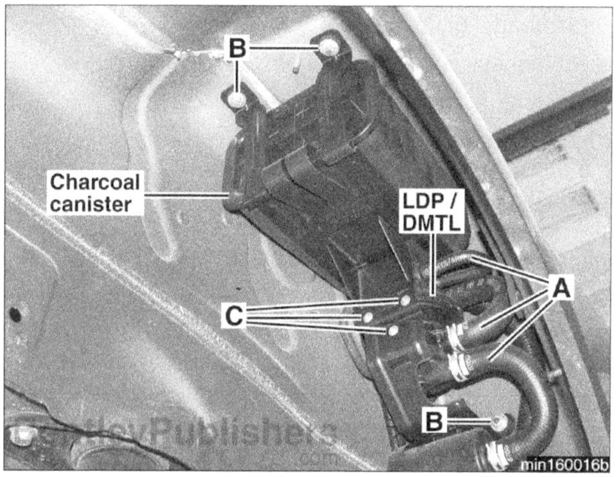

< Working underneath right rear of car:

- Detach hoses (**A**) from charcoal canister and LDP / DMTL pump.
- Remove mounting nuts (**B**).
- Lower LDP / DTML pump and canister assembly.

> ***CAUTION—***
> *Lower assembly carefully to prevent damage to hoses.*

- Working at rear of canister:
  - Detach electrical harness connector.
  - Disconnect main breather hose from rear of canister.
  - Detach vacuum line at quick disconnect.
- Remove canister and LDP / DMTL assembly complete with mounting bracket and place on work bench.
- Separate canister by prying away from mounting bracket.
- Remove LDP / DMTL mounting screws (**C**) and slide pump out of bracket.
- Installation is reverse of removal.

Tightening torques	
Wheel housing liner to fender	3 Nm (27 in-lb)
Wheel to wheel hub	120 ± 10 Nm (89 ± 7 ft-lb)

LDP / DMTL unit, removing and installing (2005-2006)

The fuel tank leakage diagnosis pump (LDP or DMTL) and the charcoal canister assembly are located under the right rear side of vehicle on 2005-2006 models.

– Raise car and support in a safe manner.

> ***CAUTION—***
> *Make sure the car is stable and well supported at all times. Use a professional automotive lift or jack stands designed for the purpose. A floor jack is not adequate support.*

– Remove plastic underbody wind deflector at right rear corner.

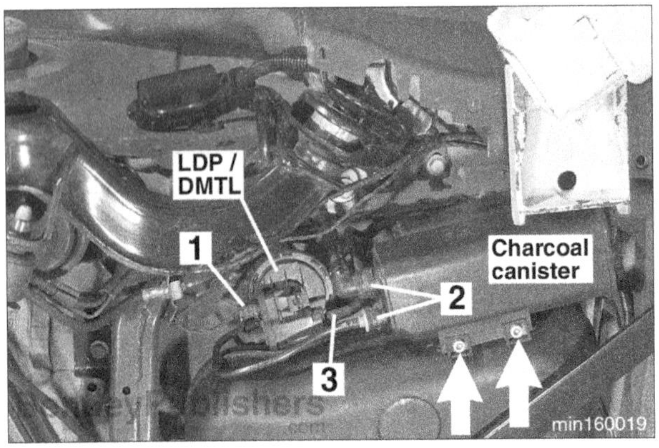

◄ Working underneath right rear of car:

- Disconnect harness connector (**1**) from LDP / DTML pump.
- Detach hoses (**2**) from charcoal canister.
- Detach hose (**3**) from LDP / DTML pump.

– Remove mounting bolts (**arrows**).

– Lower LDP / DTML unit from upper fixtures and remove.

> ***CAUTION—***
> *Lower assembly carefully to prevent damage to hoses.*

– Place assembly on work bench.

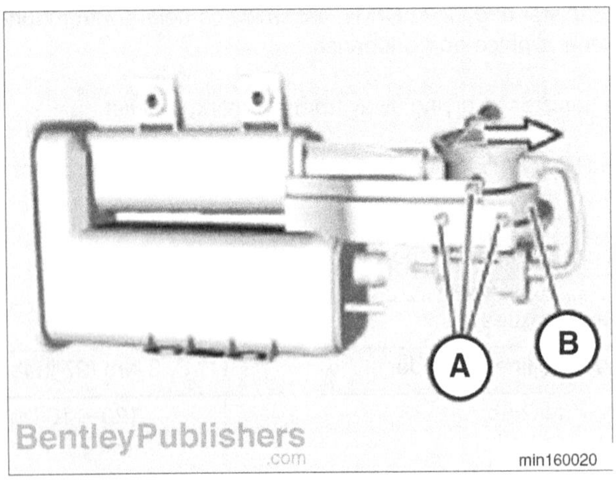

◄ Remove LDP / DMTL (**B**) mounting screws (**A**) and slide pump out of bracket in direction of **arrow**.

– Installation is reverse of removal.

170 Radiator and Cooling System

GENERAL

This section covers component repair information for the engine cooling system.

Heater core replacement is covered in **640 Heating and Air Conditioning**.

Special tools

Special tools may be needed for some radiator and cooling system procedures.

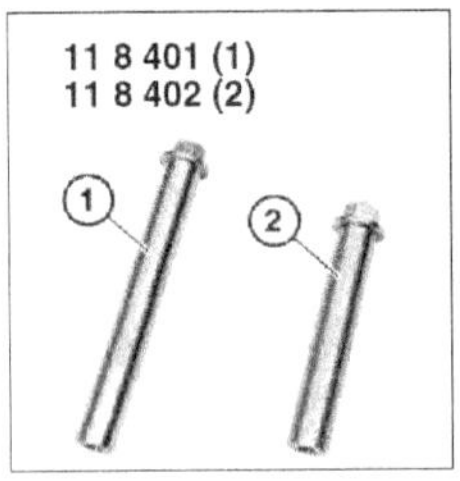

Modular front end extensions (BMW tool no. 11 8 400)

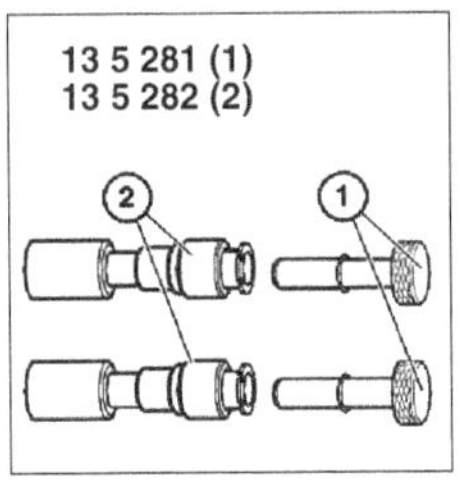

Plugs for fuel lines and fittings (BMW tool no. 13 5 280)

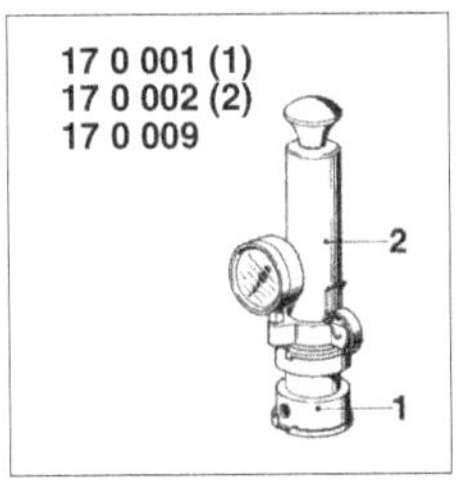

Cooling system pressure tester and adapters (BMW tool no. 17 0 000)

Cooling system overview

Coolant pump

Cooper models use a coolant pump mounted to the front of the engine. Cooper S models use a pump mounted on the rear of the supercharger. These belt-driven pumps circulate coolant through the system whenever the engine is running.

Thermostat

All models use a conventional mechanical thermostat that relies on coolant temperature to operate. While coolant is cold, the thermostat remains closed, and circulating coolant bypasses the radiator for rapid engine warm up. At higher coolant temperature, the thermostat progressively opens to allow coolant flow through the radiator, thus controlling engine temperature.

Radiator and expansion tank

The radiator is a crossflow design. An expansion tank provides for coolant expansion at higher temperatures and easy monitoring of the coolant level.

Transmission fluid cooler

Models with Continuously Variable Transmission (CVT) have a transmission fluid cooler mounted to the front of the radiator. Lines from the CVT to the transmission cooler circulate transmission fluid for additional cooling. Airflow across the cooler is used to cool the transmission fluid.

Models with automatic transmission (Agitronic) have a small transmission fluid cooler mounted on the top of the transmission. Engine coolant lines run to the cooler to keep the transmission fluid cool. Engine coolant and transmission fluid are kept separate in the cooler.

Electric cooling fan

An electric cooling fan mounted behind the radiator operates at two speeds (low and high) depending on conditions. The fan is controlled by the ECM through two relays, a low speed relay and a high speed relay. The low speed relay is mounted in the engine compartment fuse box and is energized any time the fan is operating. The high speed relay is mounted on the fan housing in the relay pack. The relay pack contains the High speed relay and a voltage reducing resistor.

The cooling fan operates on low speed when the A/C is switched on and the system pressure reaches 8 bar.The fan switches to high speed should the AC system pressure rise above 18 bar. For the engine cooling system, the fan operates on low speed at 105°C (221°F). When the temperature drops to 101°C (214°F) the fan will switch off. The fan switches to high speed at 112°C (234°F) and remains on high until the system coolant temperature drops by 4°C (7°F) at which point the system reverts to low speed.

Engine coolant temperature sensor

The engine coolant temperature sensor is located in the cylinder head next to the thermostat housing. The sensor is a two wire type whose signal is sent to the ECM and used as input for the radiator fan, fuel control, ignition timing, and coolant temperature gauge.

Coolant temperature gauge

A dash mounted display incorporating a warning lamp illuminates at an engine temperature equal to or greater than 120°C (240°F). Engine temperature data is provided to the instrument cluster (IKE) by the engine control module (ECM) via the CAN-bus.

Warnings and cautions

Observe the following warnings and cautions when working on the cooling system.

WARNING —
- *At normal operating temperature the cooling system is pressurized. Allow the system to cool as long as possible before opening (a minimum of one hour), then release the cap slowly to allow safe release of pressure.*
- *Releasing cooling system pressure lowers the coolant boiling point and the coolant may boil suddenly. Use heavy gloves and wear eye and face protection to guard against scalding.*
- *Use extreme care when draining and disposing of engine coolant. Coolant is poisonous and lethal to humans and pets. Pets are attracted to coolant because of its sweet smell and taste. Seek medical attention immediately if coolant is ingested.*

CAUTION —
- *Avoid adding cold water to the coolant while the engine is hot or overheated. If it is necessary to add coolant to a hot system, do so only with the engine running and coolant pump turning.*
- *To avoid excess silicate gel precipitation in the cooling system and loss of cooling capacity, use BMW coolant or equivalent low silicate antifreeze.*
- *If oil enters the cooling system, the radiator, expansion tank and heating circuit must be flushed with cleaning agent.*
- *When working on the cooling system, cover the alternator to protect it against coolant drips.*
- *Prior to disconnecting the battery, read the battery disconnection cautions given in* **001 General Warnings and Cautions**.

TROUBLESHOOTING

Begin diagnosing cooling system problems with a thorough visual inspection. If no visual faults are found, check the engine control module (ECM) for stored diagnostic trouble codes (DTCs).

Common cooling system faults can be grouped into one of four categories:

- Cooling system leaks
- Poor coolant circulation
- Radiator cooling fan faults
- Electrical/electronic faults

Cooling system inspection

– Check condition of accessory drive belt.

– Check that accessory drive belt tensioner is functioning properly and that belt tension is correct.

– Check coolant hoses for cracks or softness. Check clamps for looseness. Check coolant level and check for evidence of coolant leaks.

– Check that radiator fins are not blocked with dirt or debris. Clean out radiator using low-pressure water or compressed air. Blow outward, from engine side out.

– To check coolant pump (Cooper):
  - Remove drive belt from coolant pump pulley. See **020 Maintenance**.
  - Firmly grasp opposite sides of pulley and check for play in all directions.
  - Spin pulley and check that shaft runs smoothly without play.

NOTE—
- *Cooper S: Coolant pump shaft is concealed within supercharger. Physical inspection is not possible.*
- *Cooper: Coolant provides lubrication for the pump shaft, so an occasional drop of coolant from the pump is acceptable. If coolant drips steadily from the vent hole, replace the pump.*

– At normal engine operating temperature, cooling system is pressurized. This raises the boiling point of coolant. Leaks may prevent system from becoming pressurized. If visual evidence is inconclusive, pressure test cooling system. See **Cooling system pressure test** in this repair group.

– If cooling system is full of coolant and holds pressure:
  - Use an appropriate scan tool to interrogate the ECM for DTCs related to radiator fan or other cooling system components.
  - Check for loose or worn accessory drive belt.
  - Check for failed thermostat or coolant pump.
  - Check for clogged/plugged radiator or coolant passages.

Cooling system pressure test

A cooling system pressure tester is used to test for coolant leaks, including internal ones. Common sources of internal coolant leaks are a faulty cylinder head gasket, a cracked cylinder head, or a cracked engine block.

> ***WARNING —***
> *At normal operating temperature the cooling system is pressurized. Allow the system to cool before opening. Release the cap slowly to allow safe release of pressure.*

– With the engine cold, install pressure tester (BMW special tools 17 0 002/17 0 009 or equivalent) to radiator fill neck (Cooper) or expansion tank (Cooper S). Pressurize system to specification listed below.
  - Pressure should not drop more than 0.1 bar (1.45 psi) for at least two minutes.
  - If pressure drops rapidly and there is no sign of external leakage, cylinder head gasket may be faulty. Perform compression and leak-down tests. See **113 Cylinder Head Removal and Installation**.
  - Test radiator cap using pressure tester with correct adapter (BMW special tool 17 0 001 or equivalent). Replace faulty cap or cap gasket.

Cooling system test pressures	
Component	**Test pressure**
Radiator	1.5 bar (21.75 psi)
Radiator cap	0.95 - 1.24 bar (14 - 18 psi)

> ***CAUTION —***
> *Exceeding the specified test pressure could damage the radiator or other system components.*

Combustion chamber leak test

If you suspect that combustion chamber pressure is leaking into the cooling system past the cylinder head gasket, use an exhaust gas analyzer to test the vapors rising from the coolant at the expansion tank.

> ***CAUTION —***
> - *Use an extension tube above the reservoir neck to maintain distance between the top of the coolant and the gas analyzer nozzle. The gas analyzer is easily damaged if it is allowed to inhale liquid coolant.*
> - *While running engine to check for causes of overheating, observe coolant temperature carefully in order to avoid engine damage.*

Thermostat test

If the engine overheats or runs too cool and no other cooling system tests indicate trouble, the thermostat may be faulty.

COOLING SYSTEM SERVICE

Coolant, draining and filling

> ***WARNING—***
> *Allow the cooling system to cool before opening or draining the cooling system.*

Coolant, draining

- Unscrew pressure relief cap (**arrow**) from filler neck (Cooper), or expansion tank (Cooper S).

- Working underneath car, remove splash shield and disconnect lower radiator hose (**arrow**).

> ***WARNING—***
> *Use extreme care when draining and disposing of engine coolant. Coolant is poisonous and lethal to humans and pets.*

- Reattach lower radiator hose and tighten hose clamp.
- Reattach splash shield.

Coolant, filling

- Cooper: Working at left of engine, loosen vent screw in heating distribution pipe located in rear of thermostat housing.

- Loosen vent screw (**arrow**) in upper radiator hose.

NOTE—
Cooper engine is illustrated. Cooper S is similar.

- Add coolant at filler neck (Cooper), or expansion tank (Cooper S).
- Tighten vent screw(s) as soon as coolant begins to emerge.
- Cooper: Add coolant until level reaches inner ring of filler neck.
  - Replace pressure cap.
  - Fill overflow tank to MAX mark.
- Cooper S: Add coolant until level reaches MAX mark on expansion tank.

- Start engine and run at idle. Top off coolant to correct level until level stops dropping.
- Replace overflow cap or pressure cap and check system carefully for leaks.

NOTE—
- *Before filling system, turn on ignition and set heating controls to maximum temperature with fan on low.*
- *Add coolant slowly.*
- *Use only BMW approved coolant.*

Electric cooling fan, removing

> ***WARNING—***
> *Allow the cooling system to cool before opening or draining the cooling system.*

Cooling fan and cooling fan cowl are removed as a unit.

- Remove radiator. See **Radiator, removing and installing** in this repair group.
- Unclip fan and fan cowl and slide off radiator.

Thermostat, removing and installing

> ***WARNING—***
> *Allow the cooling system to cool before opening or draining the cooling system.*

The thermostat housing in the Cooper engine is part of the coolant filler neck at the rear of the engine. The Cooper S thermostat housing is in a similar location on the engine, but there is no filler neck at that location.

The illustrations used in this procedure are from a Cooper.

- Cooper: Remove battery and battery box. See **121 Battery, Starter, Alternator** for more information.

> ***CAUTION—***
> *Prior to disconnecting the battery, read the battery disconnection cautions in* **001 General Warnings and Cautions***.*

- Remove engine air intake duct and air filter housing. See **130 Fuel Injection** for more information.
- Drain coolant. See **Coolant, draining and filling** in this repair group.

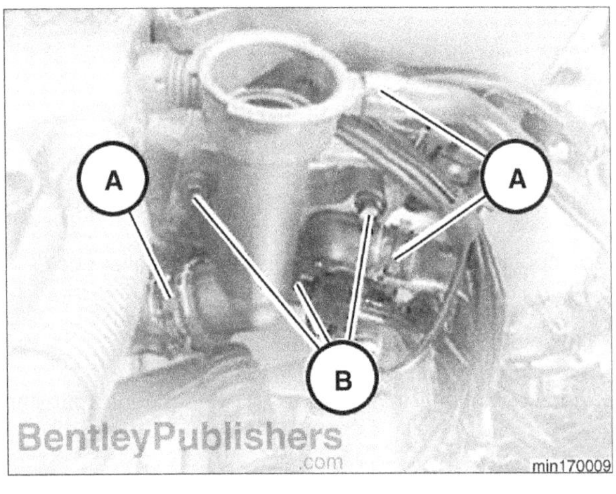

- Working at left side of cylinder head, detach coolant hoses (**A**) from thermostat housing.
- Remove thermostat housing mounting bolts (**B**). Remove housing.

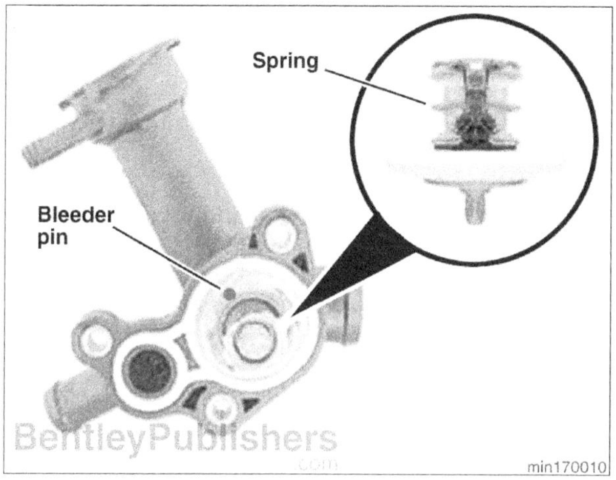

Remove thermostat (**inset**) from housing and make sure housing and cylinder head mating surfaces are clean.

- When replacing thermostat make sure bleeder pin is positioned correctly and that spring side of thermostat is facing cylinder head.
- Replace thermostat and seal. Make sure aligning guides mate.
- Install housing to cylinder head.

Tightening torque	
Thermostat housing to cylinder head	12 Nm (9 ft-lb)

- Remainder of installation is reverse of removal:
  - Fill and bleed cooling system. See **Coolant, draining and filling** in this repair group.
  - Check system for leaks.

Coolant pump, replacing (Cooper)

> ***WARNING —***
> *Allow the cooling system to cool before opening or draining the cooling system.*

- Disconnect battery. See **121 Battery, Starter, Alternator** for more information.
- Drain coolant. See **Coolant, draining and filling** in this repair group.
- Remove alternator. See **121 Battery, Starter, Alternator** for more information.
- Slacken modular front end and pull forward to allow access to front of engine. See **510 Exterior Trim, Bumpers** for more information.

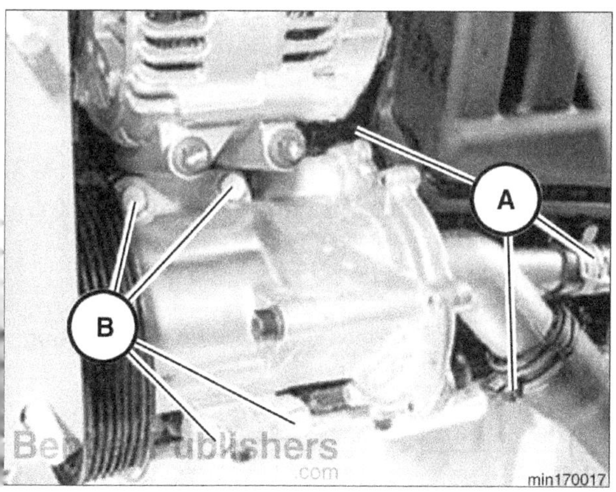

Working at coolant pump:

- Loosen or remove hose clamps (**A**) and remove hoses from coolant pump.
- Remove mounting bolts (**B**) and remove coolant pump.

- Installation is reverse of removal:
  - Fill and bleed cooling system. See **Coolant, draining and filling** in this repair group.
  - Check system for leaks.

Tightening torque	
Coolant pump to crankcase (Cooper)	30 Nm (22 ft-lb)

Coolant pump, replacing (Cooper S)

> ***WARNING —***
> *Allow the cooling system to cool before opening or draining the cooling system.*

– Drain coolant. See **Coolant, draining and filling** in this repair group.

– Remove supercharger. See **130 Fuel Injection** for more information.

Remove coolant pump retaining bolts (**A**) from rear of supercharger and remove coolant pump.

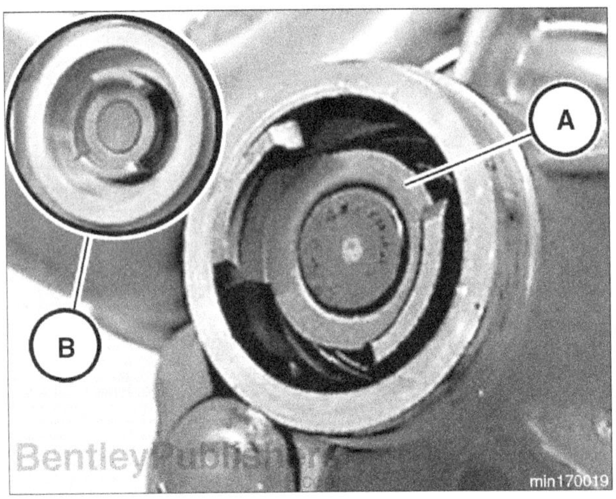

To install, engage coolant pump drive (**A**) with supercharger drive (**B**). Install coolant pump to supercharger.

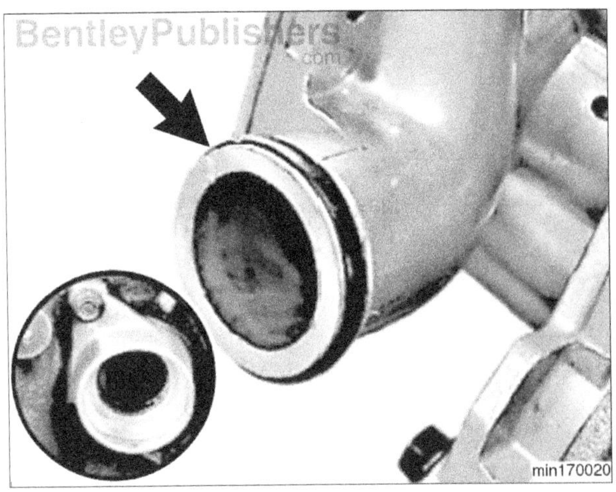

Install new O-ring (**arrow**) to coolant pump outlet pipe. Lubricate O-ring with Syntheso Glep 1® or equivalent.

– Remainder of installation is reverse of removal:

- Fill and bleed cooling system. See **Coolant, draining and filling** in this repair group.
- Check system for leaks.

Tightening torque	
Coolant pump to supercharger (Cooper S)	25 Nm (18 ft-lb)

Engine oil cooler, removing and installing (Cooper S)

> ***WARNING —***
> *Allow the cooling system to cool before opening or draining the cooling system.*

The engine oil cooler is located on the right side of the engine compartment on the oil filter housing. Vehicles with automatic transmission (Agitronic) do not have an engine oil cooler.

– Drain coolant. See **Coolant, draining and filling** in this repair group.

– Remove right front wheel.

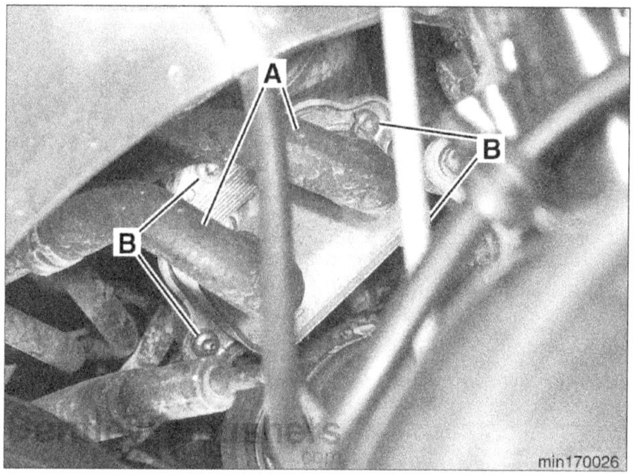

Working at rear of engine:

- Disconnect coolant hoses (**A**) at engine oil cooler. Be prepared to catch any coolant that escapes.
- Remove oil cooler mounting bolts (**B**) and remove oil cooler. Be prepared to catch any dripping oil.

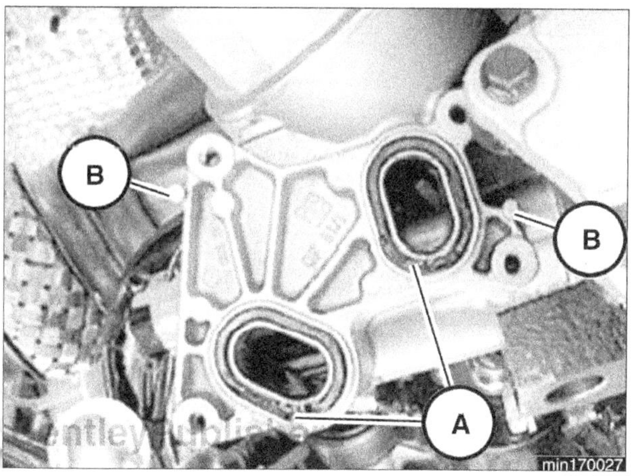

Before reinstalling:

- Remove sealing rings (**A**) and clean mating surfaces on cooler and oil filter housing.
- Install new sealing rings and refit oil cooler. Note cooler location lugs (**B**).

– Check engine oil level.

– Bleed and top-up cooling system.

Tightening torque	
Engine oil cooler to oil filter housing	12 Nm (9 ft-lb)

Transmission fluid cooler, removing and installing (CVT models)

> ***WARNING —***
> *Allow the cooling system to cool before opening or draining the cooling system.*

- Working in front of the radiator, disconnect pipes (**arrows**) from transmission cooler and use special tools (13 5 281 and 13 5 282) to seal cooler and lines. Be prepared to catch dripping transmission fluid.

- Remove transmission cooler mounting screw (**A**). Slide cooler (**arrow**) off mounting bracket to remove.

– Installation is reverse of removal. Check automatic transmission fluid level. See **240 Automatic Transmission**.

Tightening torque	
Transmission fluid cooler to radiator carrier	4 Nm (3 ft-lb)

Transmission fluid cooler, removing and installing (Agitronic models)

– Drain engine coolant as described in **Coolant, draining and filling** in this repair group.

> ***WARNING —***
> *Allow the cooling system to cool before opening or draining the cooling system.*

– Remove air filter assembly and inlet/outlet ducts, as necessary, to access transmission cooler. See **130 Fuel Injection**.

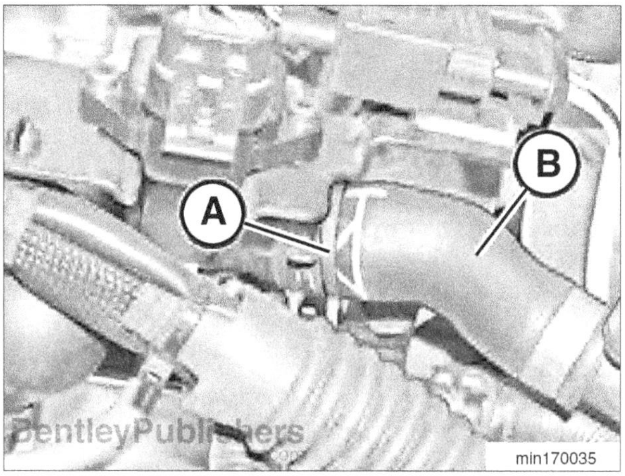

Working on top of transmission, release hose clamp (**A**) and remove coolant hose (**B**).

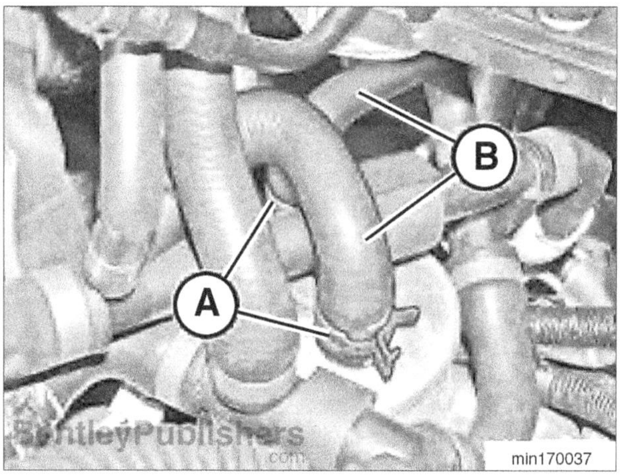

Release hose clamps (**A**) and detach coolant hosed (**B**) from transmission fluid cooler.

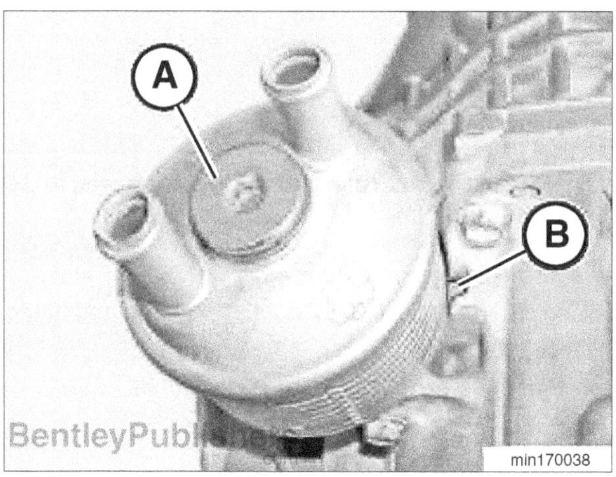

Remove mounting bolt (**A**) and remove transmission fluid cooler.

- When installing, note position of locating lug (**B**).

– If re-using cooler, coat upper and lower O-ring seals with transmission fluid.

– Clean sealing contact surfaces on transmission before installing cooler.

– Installation is reverse of removal. Tighten cooler mounting bolt to specification.

Tightening torque	
Transmission fluid cooler to transmission	35 Nm (26 ft-lb)

– Check automatic transmission fluid level. See **240 Automatic Transmission**.

Radiator, removing and installing

> ***WARNING —***
> *Allow the cooling system to cool before opening or draining the cooling system.*

– Drain coolant. See **Coolant, draining and filling** in this repair group.

– Slacken modular front end and pull forward to allow access to radiator. See **510 Exterior Trim, Bumpers** for more information.

– CVT models: remove oil cooler. See **Transmission oil cooler, removing and installing (CVT models)** in this repair group.

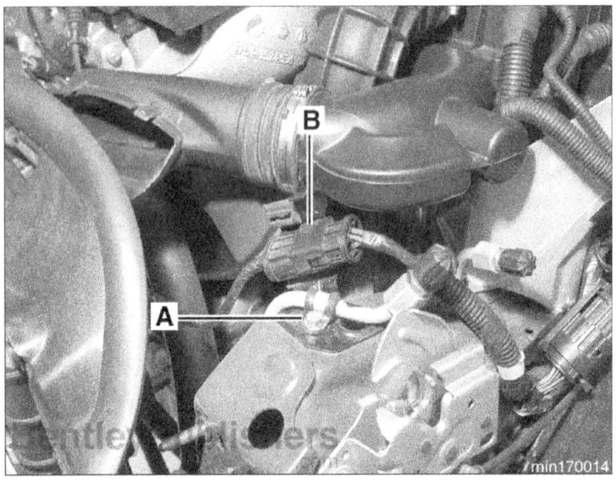

< Working at left frame extension, next to left engine hood latch:

- Remove refrigerant service fitting mounting bolt (**A**).
- Disconnect cooling fan electrical harness connector (**B**).

– Detach upper radiator hose.

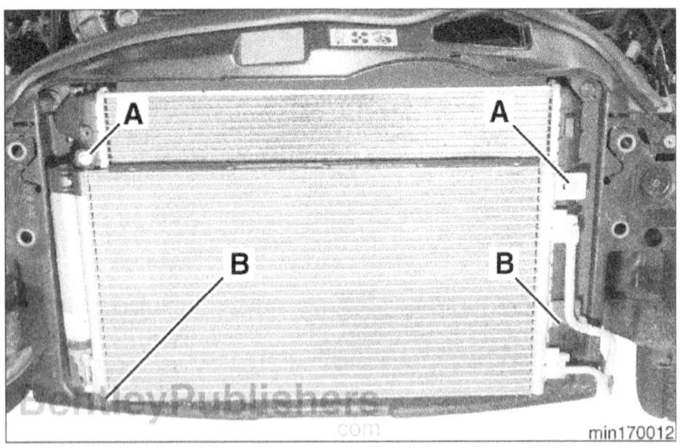

< Remove air-conditioning condenser:

- Remove condenser mounting bolts (**A**).
- Carefully lift condenser out of retaining lugs (**B**). Support condenser and continue with radiator removal.

< Remove radiator retaining pins (**inset**). Lift radiator (**arrows**) to disengage from support lugs.

– Installation is reverse of removal. Remember to:

- Fill and bleed cooling system. See **Coolant, draining and filling** in this repair group.
- Check system for leaks.

Expansion tank, removing and installing

> ***WARNING —***
> *Allow the cooling system to cool before opening or draining the cooling system.*

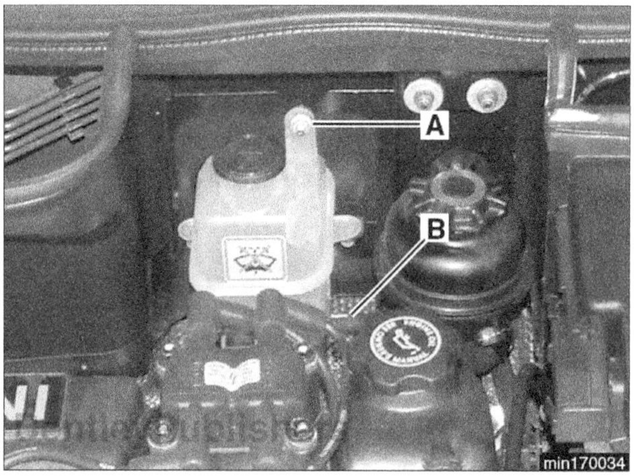

- Cooper: Remove expansion tank mounting nut (**A**).
- Detach hose (**B**) and lift expansion tank from mounting slots. Be prepared to catch any coolant that escapes.
- When replacing:
  - Replace hose clamp(s).
  - Fill and bleed cooling system. See **Coolant, draining and filling** in this repair group.
  - Check system for leaks.

Tightening torque	
Expansion tank to bulkhead	5 Nm (4 ft-lb)

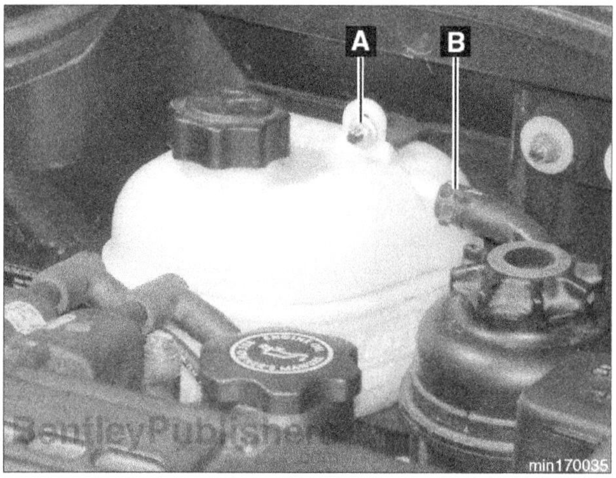

- Cooper S: Remove expansion tank mounting nut (**A**).
- Detach hose (**B**) from top of expansion tank.
- Lift tank out of mounting slots and remove hose from bottom of tank.
- Remove expansion tank. Be prepared to catch any coolant that escapes.
- When replacing:
  - Replace hose clamp(s).
  - Fill and bleed cooling system. See **Coolant, draining and filling** in this repair group.
  - Check system for leaks.

Tightening torque	
Expansion tank to bulkhead	5 Nm (4 ft-lb)

180 Exhaust System

General

The exhaust system is designed to be maintenance free, although regular inspection is warranted due to the harsh operating conditions. Under normal conditions, the catalytic converter does not require replacement unless it is damaged.

Exhaust System

New fasteners, clamps, rubber mounts and gaskets should be used when replacing exhaust components. A liberal application of penetrating oil to the exhaust system nuts and bolts in advance may make removal easier.

> ***WARNING—***
> - *The exhaust system and catalytic converter operate at very high temperatures. Allow components to cool before servicing. Wear protective gloves to prevent burns. Do not use flammable chemicals near a hot catalytic converter.*
> - *Corroded exhaust system components crumble easily and often have exposed sharp edges. To avoid injury, wear eye protection and heavy gloves when working with exhaust parts.*

Exhaust system, removing and installing

Removal and installation procedures given here are similar for all models. The removal of the exhaust system as a complete unit is recommended. Once the complete system is removed from the car it can be worked on more easily.

> ***WARNING—***
> *Exhaust gases are colorless, odorless, and very toxic. Run the engine only in a well-ventilated area. Immediately repair any leaks in the exhaust system or structural damage to the car body that might allow exhaust gases to enter the passenger compartment.*

> ***CAUTION—***
> *Use care not to drag or bang the oxygen sensors. Oxygen sensors can be easily ruined.*

– With the exhaust system fully cold, raise and support car for access to exhaust system.

> ***WARNING—***
> *Do not work under a lifted car unless it is solidly supported on jack stands designed for that purpose. Never work under a car that is supported solely by a jack.*

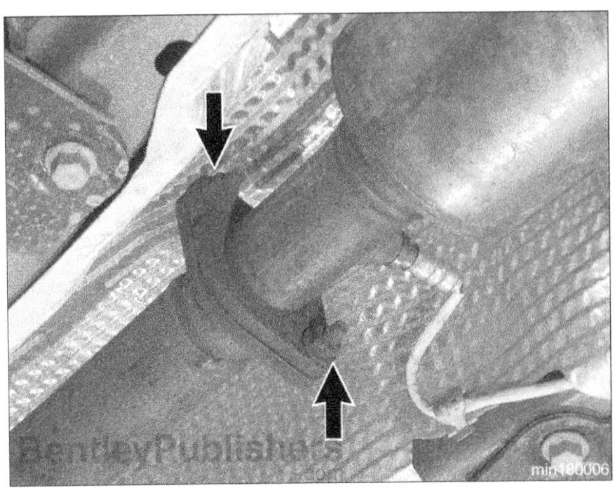

◄ Remove fasteners (**arrows**) that join exhaust system to catalytic converter.

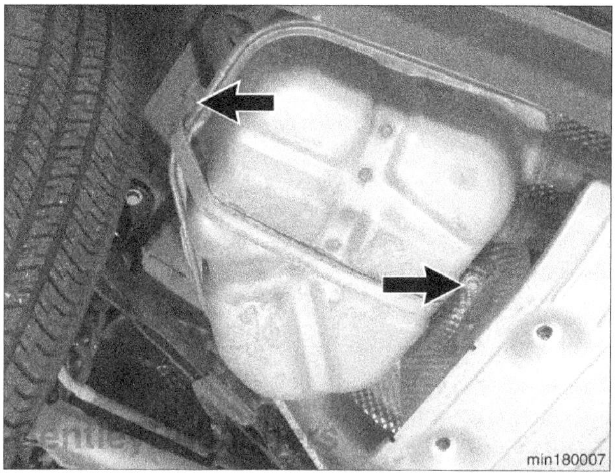

Working at the muffler(s), remove bolts (**arrows**) from lower clamp.

NOTE —
Cooper S models: Remove shield attached to bottom of battery box.

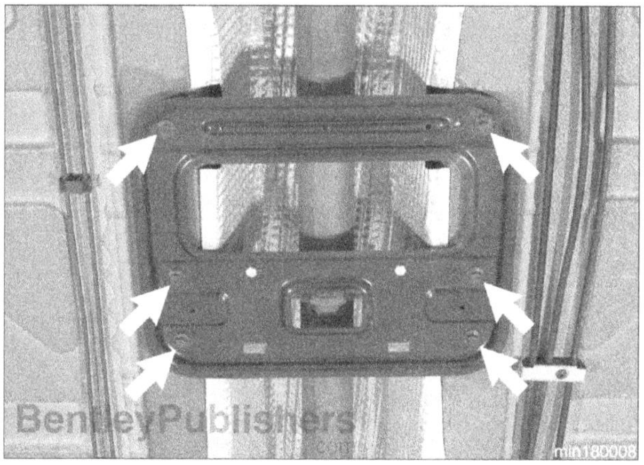

With an assistant supporting exhaust system, Remove bolts (**arrows**) from reinforcement plate.

- Carefully lower exhaust system out from under car.
- Installation is reverse of removal noting the following:
  - Replace all nuts and bolts. Coat threads with copper paste.
  - Replace gasket between exhaust and converter.
  - Replace rubber mount rings.

Tightening torque	
Exhaust system to exhaust manifold (M10)	60 Nm (44 ft-lb)
Lower muffler clamp to upper clamp (M8)	15 Nm (11 ft-lb)
Reinforcement plate to body	14 Nm (10 ft-lb)

EXHAUST MANIFOLD

Exhaust manifold removal procedure is similar for all models. Always use new retaining nuts and gaskets when removing and installing the exhaust manifold.

> *WARNING —*
> *Exhaust gases are colorless, odorless, and very toxic. Run the engine only in a well-ventilated area. Immediately repair any leaks in the exhaust system or structural damage to the car body that might allow exhaust gases to enter the passenger compartment.*

Exhaust manifold, removing and installing

- With the exhaust system fully cold, raise and support car for access to exhaust system.

> *WARNING —*
> *Do not work under a lifted car unless it is solidly supported on jack stands designed for that purpose. Never work under a car that is supported solely by a jack.*

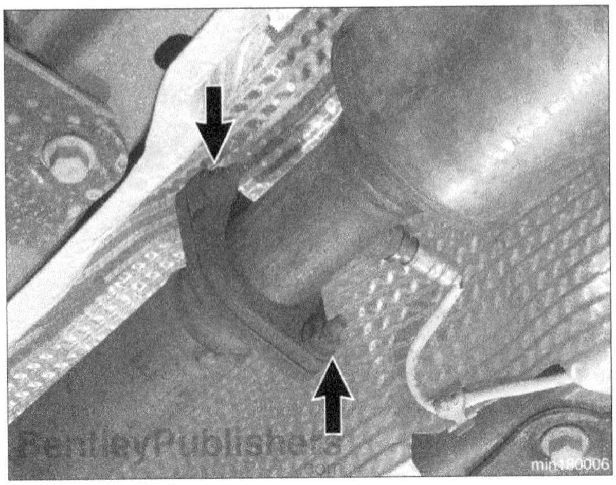

Remove fasteners (**arrows**) that join exhaust system to catalytic converter.

NOTE—
Leave exhaust system suspended from body.

– Disconnect front and rear oxygen sensor electrical connectors.

NOTE—
Do not reverse oxygen sensor connectors. Label connectors to insure proper installation.

> ***CAUTION—***
> - *Use care not to drag or bang the oxygen sensors. Oxygen sensors can be easily ruined.*

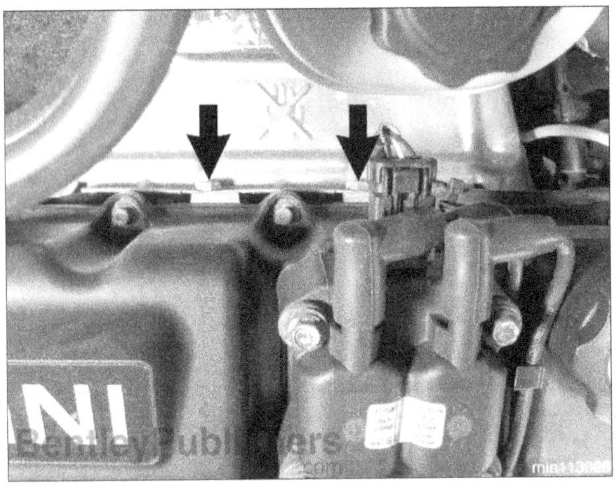

Remove heat shield bolts (**arrows**) where shield attaches to cylinder head.

Remove remaining bolts (**arrows**) holding heat shield to body and remove heat shield.

NOTE—
Note routing of oxygen sensor wiring.

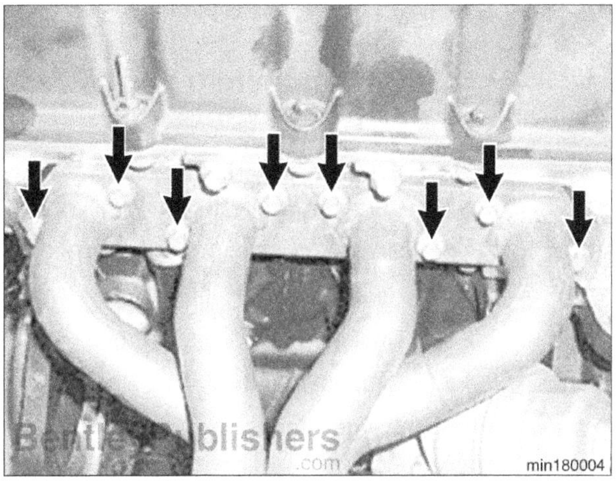

Remove exhaust manifold mounting bolts (**arrows**) from cylinder head. Remove exhaust manifold.

– Installation is reverse of removal noting the following:
  - Replace all nuts and bolts.
  - Clean mating surfaces of cylinder head and manifold.
  - Coat exhaust system to exhaust manifold bolt threads with copper paste.
  - Replace exhaust manifold gasket.

Tightening torque	
Exhaust manifold to cylinder head (M8)	24 Nm (18 ft-lb)
Exhaust system to exhaust manifold (M10)	60 Nm (44 ft-lb)
Heat shield to cylinder head (M8)	13 Nm (10 ft-lb)

Oxygen Sensors

Oxygen sensors are located in front of (pre-catalyst) and behind (post-catalyst) the catalytic converter. Oxygen sensor replacement is similar for all models.

Oxygen sensors, removing and installing

– Disconnect front and rear oxygen sensor electrical connectors.

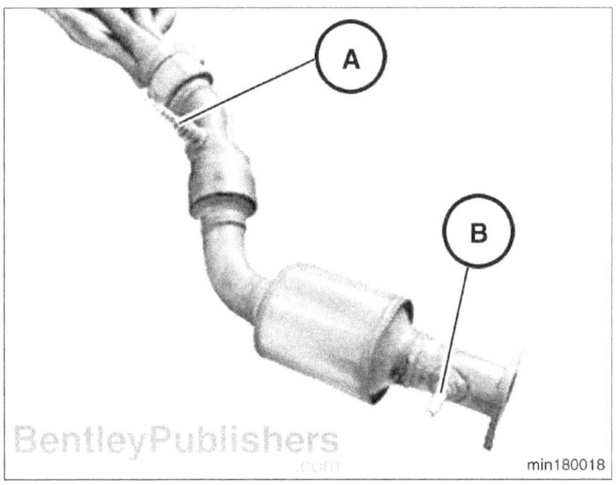

Remove front oxygen sensor (**A**) and rear oxygen sensor (**B**).

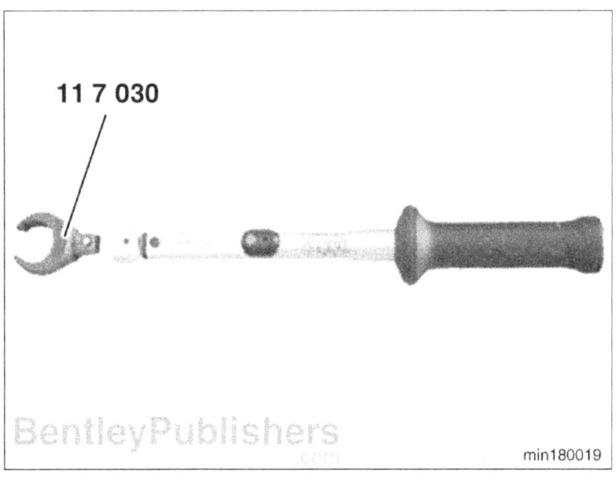

To replace, use special tool (BMW tool no. 11 7 030) or equivalent to tighten oxygen sensor to proper torque. Installation is reverse of removal, noting the following:
  - Oxygen sensors come pre-coated with anti-seize paste.
  - Do not contaminate tip of sensor.

Tightening torque	
Oxygen sensors to exhaust	39 Nm (29 ft-lb)

Exhaust System Diagrams

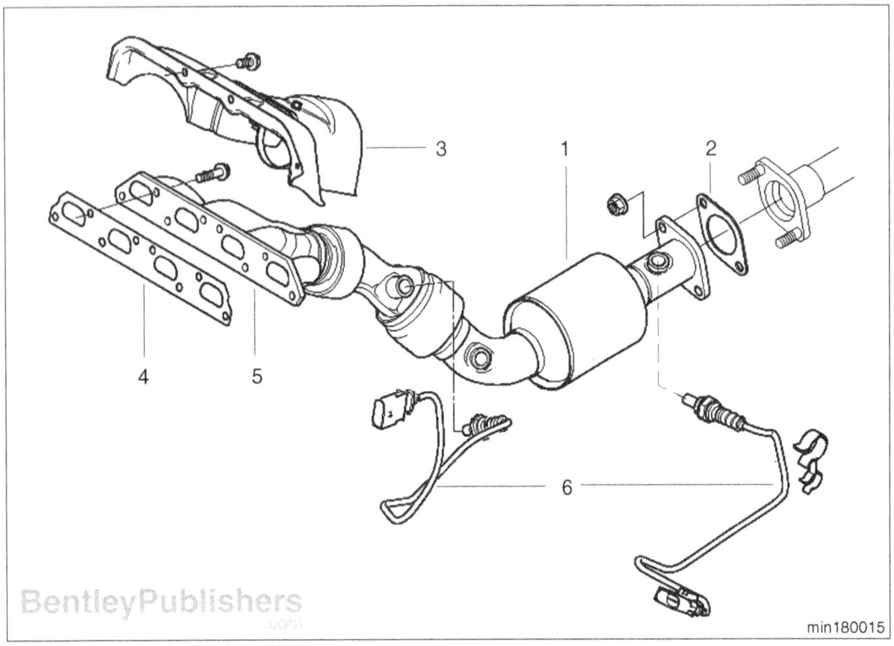

Exhaust manifold and catalytic converter (all models)

1. Catalytic converter
2. Gasket
3. Heat shield
4. Gasket
5. Exhaust manifold
6. Oxygen sensors

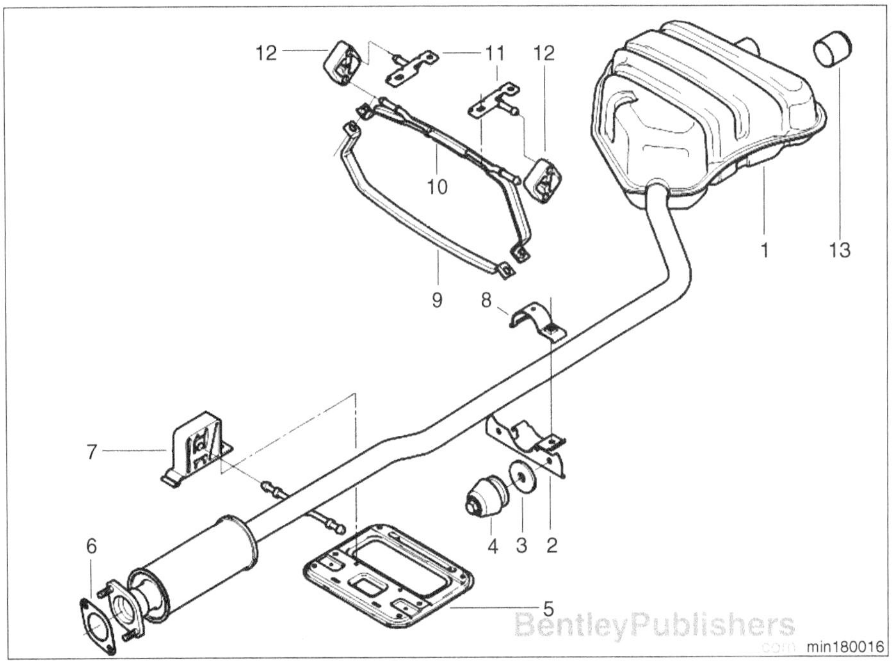

Cooper exhaust system

1. Muffler (rear)
2. Bracket
3. Washer
4. Vibration absorber
5. Support
6. Gasket
7. Rubber mounting
8. Bracket
9. Muffler clamp (lower)
10. Muffler clamp (upper)
11. Rear hanger
12. Rubber ring
13. Tailpipe extension

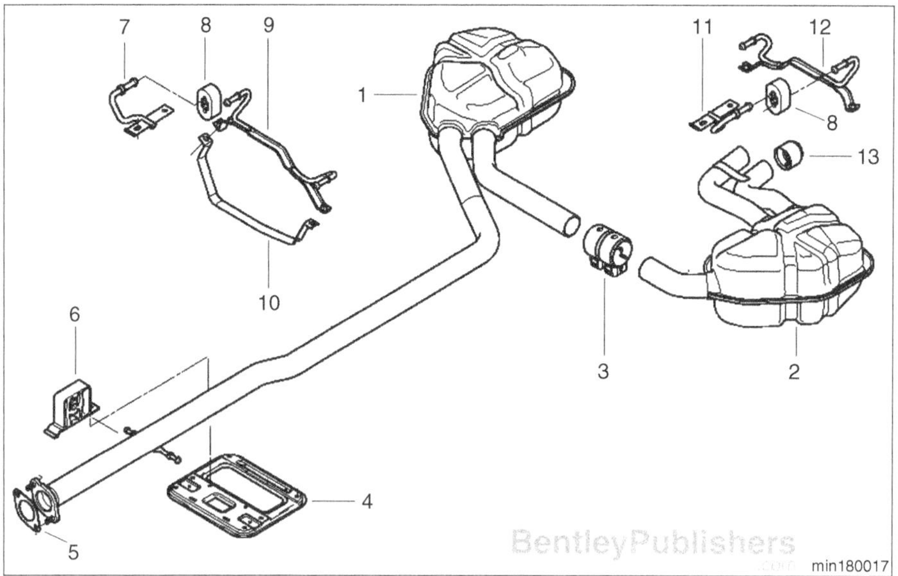

Cooper S exhaust system

1. Right muffler
2. Left muffler
3. Connecting clamp
4. Support
5. Gasket
6. Rubber mounting
7. Hanger bracket
8. Rubber ring
9. Muffler clamp (upper right)
10. Muffler clamp (lower right)
11. Hanger bracket
12. Muffler clamp (upper left)
13. Tailpipe extension

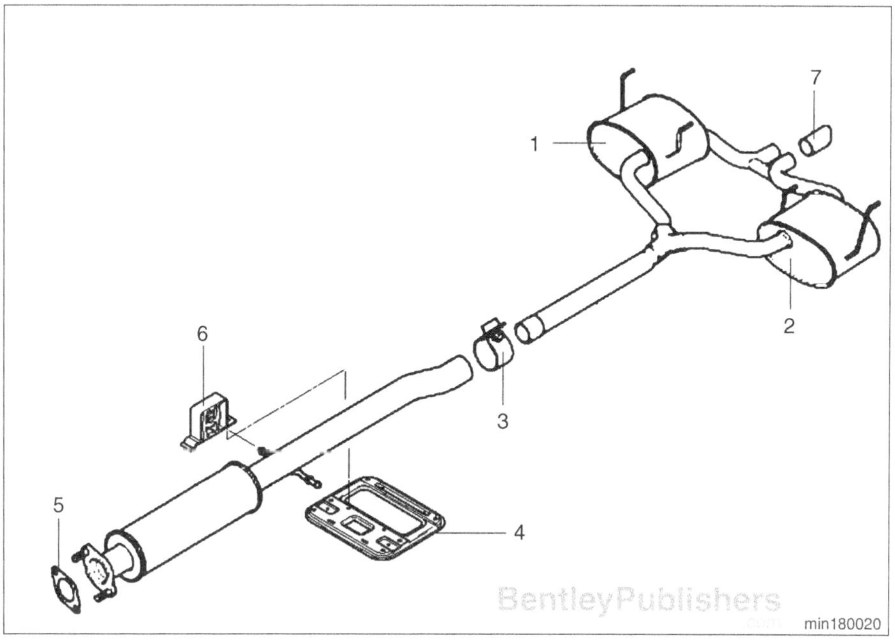

John Cooper Works exhaust system

1. Right muffler
2. Left muffler
3. Connecting clamp
4. Support
5. Gasket
6. Rubber mounting
7. Tailpipe extension

200 Transmission–General

General

This repair group covers general application information for the clutch, manual and automatic transmissions. For repairs related to gearshift linkage, see **250 Gearshift Linkage**. For repairs related to the drive axles, see **310 Front Suspension, Drive Axles**.

Drivetrain

The Cooper is available with either a 5-speed manual or automatic transmission. The Cooper S is available with a 6-speed manual or an automatic (from 2005-on). All transmissions are mounted on left side of engine bay in line with the engine. The final drive/differential assembly is integral with the transmission housing and provides drive to the front wheels.

Manual Transmission

Manual transmission (Cooper 2002-2004)

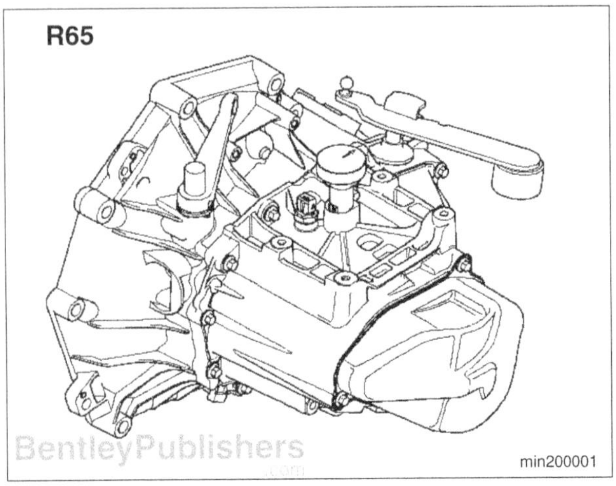

The manual transmission in the 2002-2004 Cooper is known as the R65 and is manufactured by Midlands. It is a 5-speed transmission with a maximum torque input of 160 Nm (118 lb-ft). An identification label is mounted on the side of the transmission. For manual transmission specifications or repair procedures, refer to **230 Manual Transmission**.

A conventional clutch and flywheel are used. For more information, see **210 Clutch**.

Manual transmission (Cooper 2005-2006)

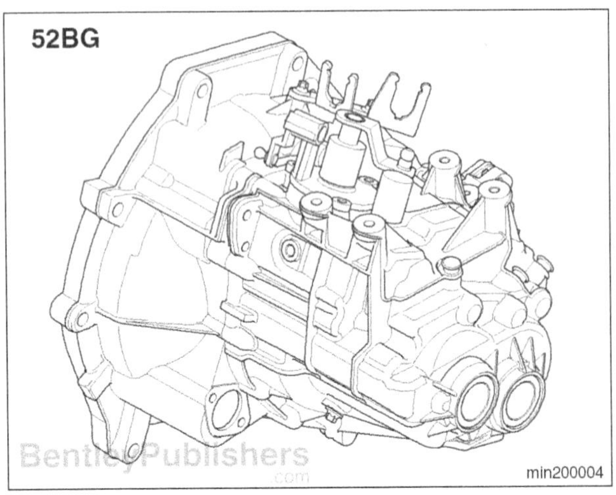

The manual transmission in the 2005-2006 Cooper is known as the 52BG. It is a 5-speed transmission manufactured by Getrag. For manual transmission specifications or repair procedures, refer to **230 Manual Transmission**.

A conventional clutch and flywheel are used. For more information, see **210 Clutch**.

Manual transmission (Cooper S)

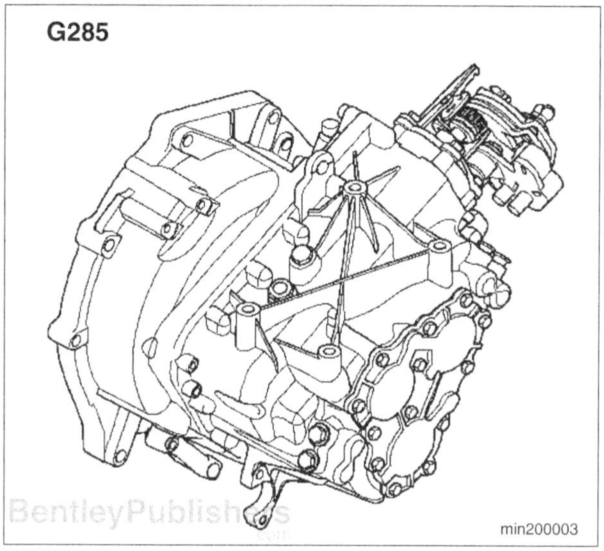

The manual transmission in the Cooper S is known as the G285 and is manufactured by Getrag. It is a 6-speed transmission with a maximum torque input of 210 Nm (155 lb-ft). An identification label is mounted to the side of the transmission. For manual transmission specifications or repair procedures, refer to **230 Manual Transmission**.

The 6-speed manual transmission features an upgraded clutch and pressure plate and a dual mass flywheel. For more information, see **210 Clutch**.

AUTOMATIC TRANSMISSION

Continuously Variable Transmission (CVT) (Cooper)

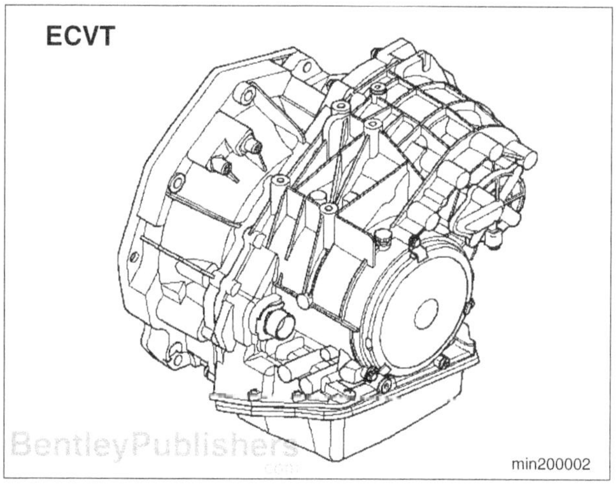

The automatic transmission available in the Cooper is an Electronic Continuously Variable Transmission (ECVT). Commonly known as the CVT, it provides stepless shifting with an infinitely variable gear ratio. This allows full engine performance to be available at any time. An identification label is mounted to the side of the automatic transmission. For more information, see **240 Automatic Transmission**.

Automatic transmission with Agitronic (Cooper S)

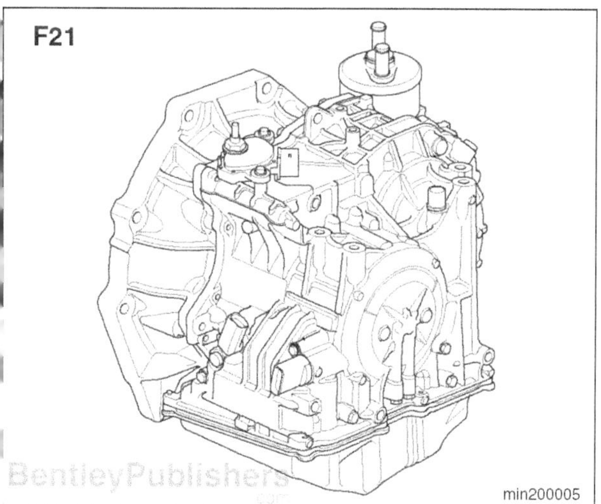

The automatic transmission available in the Cooper S from 2005-on is known as the F21 and is manufactured by Aisin. It is a 6-speed automatic with a maximum torque input of 220 Nm (162 ft-lb) and an Agitronic feature. The Agitronic feature enables individual gear selection in Drive (D) position by means of shift paddles on the steering wheel. The steering wheel shift paddles may also be used with the transmission in Sport (S) position. For more information, see **240 Automatic Transmission**.

TRANSMISSION LUBRICANTS

The lubricant used varies between transmissions and model years. Fluid application information is clearly marked on a label on the transmission.

Manual transmission models use BMW MTF 94 (or equivalent) lubricant. See **230 Manual Transmission**.

CVT equipped models require a special lubricant. MINI specifies ECVT oil type Esso CVT EZL 799. The Agitronic automatic transmission requires Esso JWS-3309 ATF. For more information, see **240 Automatic Transmission**.

> ***CAUTION—***
> *If in doubt as to the type of fluid installed in a particular transmission, consult an authorized MINI dealer service department for the latest in operating fluids.*

210 Clutch

GENERAL

This repair group covers replacement of clutch hydraulic and mechanical components.

MINI models with manual transmissions use a single disc hydraulically actuated clutch. Two versions of the clutch are used:

- Cooper models use a traditional clutch with integral damper springs.
- Cooper S models use an upgraded clutch without damper springs. Torsion forces are absorbed by a special dual mass flywheel.

Special tools

Some special tools and techniques are required for clutch repairs. Be sure to read and understand the procedure before beginning a job.

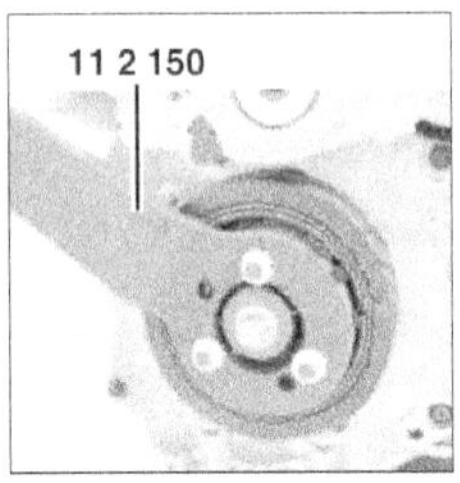

◀ Crankshaft pulley holding tool (Cooper)
(BMW tool no. 11 2 150)

◀ Crankshaft pulley holding tool bolts
(BMW tool no. 11 8 320)

◀ Crankshaft pulley holding tool (Cooper S)
(BMW tool no. 11 8 460)

◀ Clutch alignment tool (Cooper S)
(BMW tool no. 21 2 210)

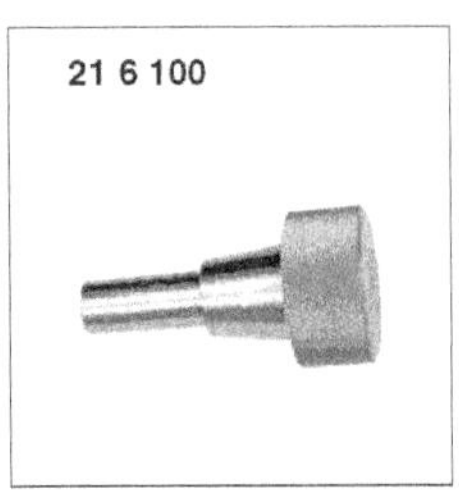

◀ Clutch alignment tool (Cooper)
(BMW tool no. 21 6 100)

21 5 030

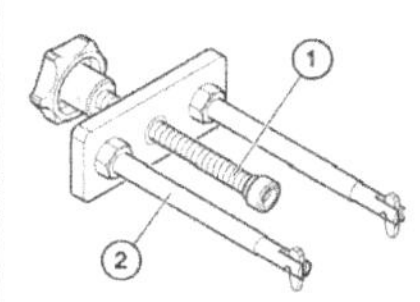

Fixture, used to depress clutch cylinder piston (BMW tool no. 21 5 030)

Clutch Hydraulics

The clutch is hydraulically actuated by master and slave cylinders. A soft or spongy feel to the clutch pedal, long pedal free-play, or grinding noises from the gears while shifting can all indicate problems with the clutch hydraulics. In these circumstances it is best to start with a clutch fluid flush, followed, if necessary, by replacement of hydraulic parts.

NOTE—
The clutch hydraulic system shares the fluid reservoir and fluid with the brake hydraulic system.

Clutch hydraulic system, bleeding and flushing

If the clutch/brake fluid is murky or muddy, or has not been changed within the last two years, flush the system and change the fluid. flushing the old fluid from the clutch lines is done using a brake system pressure bleeder.

- Remove brake fluid reservoir cap. Using a clean syringe, remove brake fluid from reservoir. Refill reservoir with clean DOT 4 brake fluid.

- Cooper: remove battery box to gain access to slave cylinder on top of transmission. See **121 Battery, Starter, Alternator** for more information.

- Cooper S: Raise car and remove splash shield to gain access to slave cylinder on front side of transmission.

> ***WARNING—***
> *Make sure the car is stable and well supported at all times. Use a professional automotive lift or jack stands designed for the purpose. A floor jack is not adequate support.*

- Remove bolts holding slave cylinder to transmission. See **Clutch slave cylinder, replacing** later in this section.

21 5 030

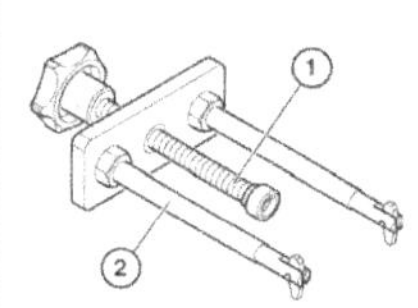

Fit special tool 21 5 030 and tighten center shaft (**1**) fully to stop to compress cylinder piston.

– Attach pressure brake bleeder to fluid reservoir and pump brake bleeder a few times to pressurize hydraulic system.

CAUTION —
- *Do not exceed 1 bar (14 psi) pressure at the fluid reservoir when bleeding or flushing the hydraulic system.*
- *Brake fluid is poisonous, highly corrosive and dangerous to the environment. Wear safety glasses and rubber gloves when working with brake fluid. Do not siphon brake fluid with your mouth. Immediately clean away any fluid spilled on painted surfaces and wash with water, as brake fluid will remove paint.*
- *Always use new brake fluid from a fresh, unopened container. Brake fluid will absorb moisture from the air. This can lead to corrosion problems in the clutch and brake hydraulic systems, and will also lower the brake fluid boiling point. Dispose of brake fluid properly.*

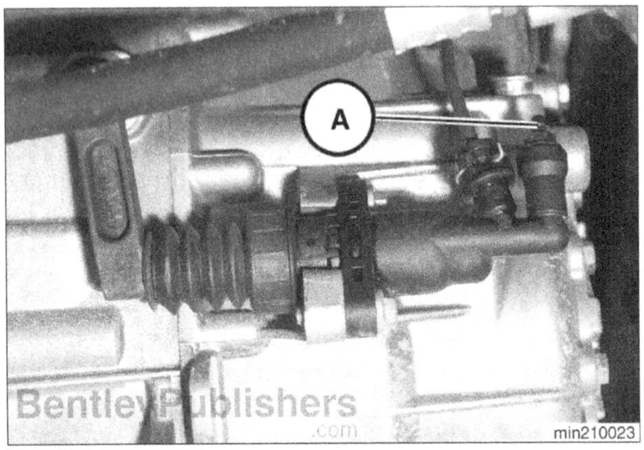

Connect a hose from clutch slave cylinder bleeder valve (**A**) to a container.

- Open bleeder valve and allow brake fluid to expel until clean fluid comes out, free of air bubbles.
- Close bleeder valve and disconnect pressure bleeding equipment from fluid reservoir.
- Remove special tool 21 5 030 and install slave cylinder back on transmission.
- Slowly operate clutch pedal about 10 times. Fill reservoir with clean fluid as necessary.

NOTE —
Cooper S cylinder shown. Cooper cylinder is located on top of transmission.

Tightening torque	
Clutch slave cylinder to transmission	24 Nm (18 lb-ft)

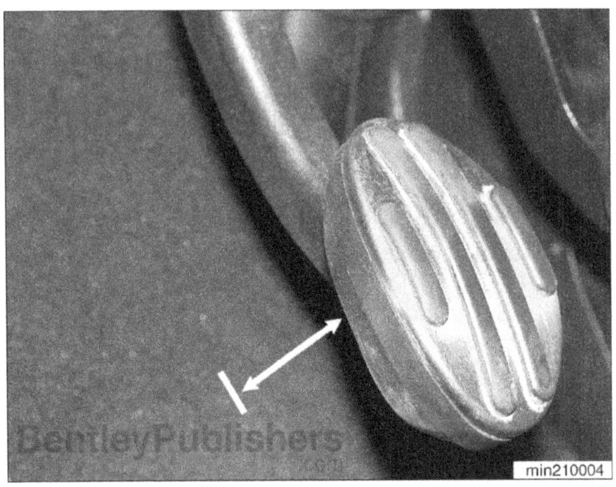

Check clutch pedal free play (**arrow**). Repeat procedure if free play exceeds specification.

Maximum clutch pedal free play	7-10 mm (0.28 - 0.39 in)

Clutch master cylinder, replacing

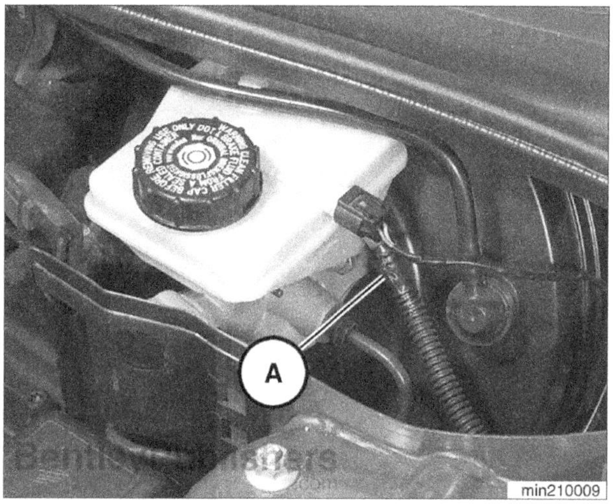

◄ Remove brake fluid reservoir cap. Using a clean syringe, remove brake fluid from reservoir until level falls below clutch master cylinder feed pipe (**A**).

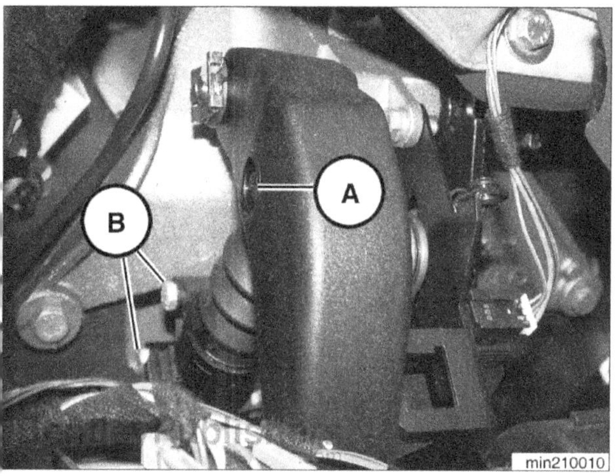

◄ Working at pedal cluster, remove pin (**A**) from clutch pedal. Remove clutch master cylinder retaining bolts (**B**) and remove fluid outlet pipe.

NOTE—
Place pan under hydraulic lines to catch dripping fluid.

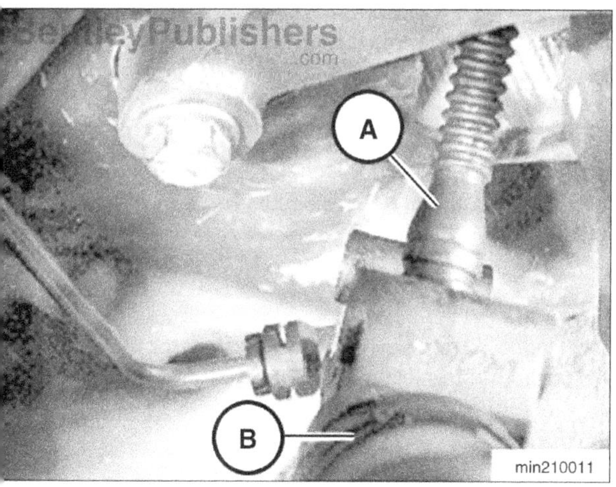

◄ Lower clutch master cylinder to remove fluid inlet pipe (**A**) and remove clutch master cylinder (**B**).

– Installation is reverse of removal. Remember to:

- Replace O-ring on inlet pipe.
- Refill reservoir with clean DOT 4 brake fluid.
- Bleed clutch hydraulic system and check system for leaks. See **Clutch hydraulic system, bleeding and flushing**.

Clutch slave cylinder, replacing

– Cooper: Remove battery box to gain access to slave cylinder on top of transmission. See **121 Battery, Starter, Alternator** for more information.

– Cooper S: Raise car and remove splash shield to gain access to slave cylinder on front side of transmission.

> ***WARNING —***
> *Make sure the car is stable and well supported at all times. Use a professional automotive lift or jack stands designed for the purpose. A floor jack is not adequate support.*

NOTE —
Cooper and Cooper S models use the same slave cylinder mounted in a different position.

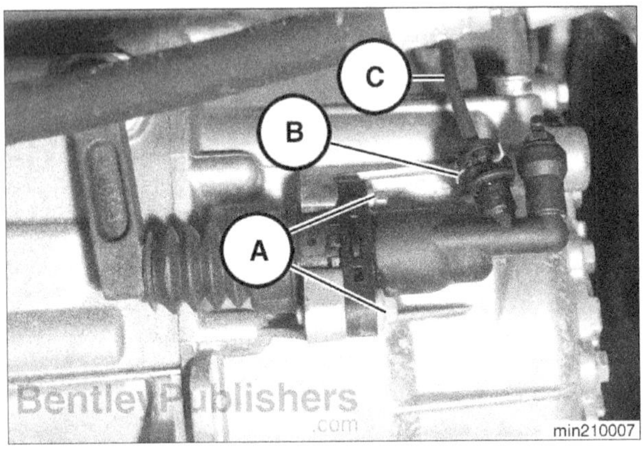

min210007

◄ Remove clutch slave cylinder bolts (**A**) and remove clutch slave cylinder.

– Remove retaining clip (**B**) and remove hydraulic line (**C**).

NOTE —
Have a container ready to catch dripping fluid.

> ***CAUTION —***
> - *Brake fluid is poisonous, highly corrosive and dangerous to the environment. Wear safety glasses and rubber gloves when working with brake fluid. Do not siphon brake fluid with your mouth. Immediately clean away any fluid spilled on painted surfaces and wash with water, as brake fluid will remove paint.*
> - *Always use new brake fluid from a fresh, unopened container. Brake fluid will absorb moisture from the air. This can lead to corrosion problems in the clutch and brake hydraulic systems, and will also lower the brake fluid boiling point. Dispose of brake fluid properly.*

– When replacing clutch slave cylinder, check that sealing ring is properly seated on hydraulic line. A click must be heard when fitting is reinserted. Remember to:

- Replace retaining clip.
- Mount clutch slave cylinder.
- Bleed clutch hydraulic system and check system for leaks. See **Clutch hydraulic system, bleeding and flushing**.

Tightening torque	
Clutch slave cylinder to transmission	24 Nm (18 lb-ft)

CLUTCH MECHANICAL

The transmission must be removed from the engine to access clutch mechanical components. Replace the clutch disc, pressure plate and release bearing during a clutch overhaul.

Be sure to check the bottom of the bellhousing for oil. If engine oil is found, check for a faulty rear crankshaft main oil seal. See **119 Lubrication System** for rear main seal replacement.

Clutch, removing and installing (Cooper)

– Remove transmission. See **230 Manual Transmission** for more information.

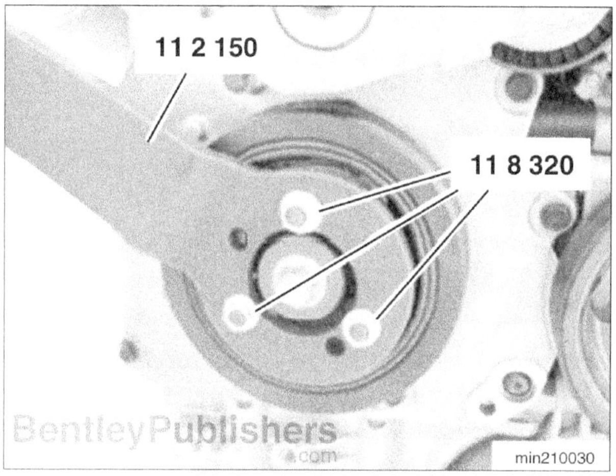

⊲ Fit special tool 11 2 150 and 11 8 320 on front of crankshaft to counterhold flywheel during pressure plate removal.

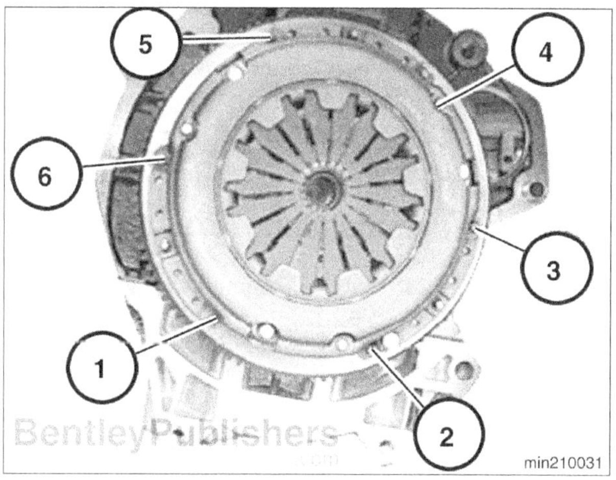

⊲ Gradually loosen pressure plate bolts in a cross pattern (**1 - 6**) and remove bolts.

– Remove pressure plate and clutch disk.

– Clean flywheel and check for wear and damage. If necessary, replace flywheel. See **Flywheel, removing and installing** later in this repair group for more information.

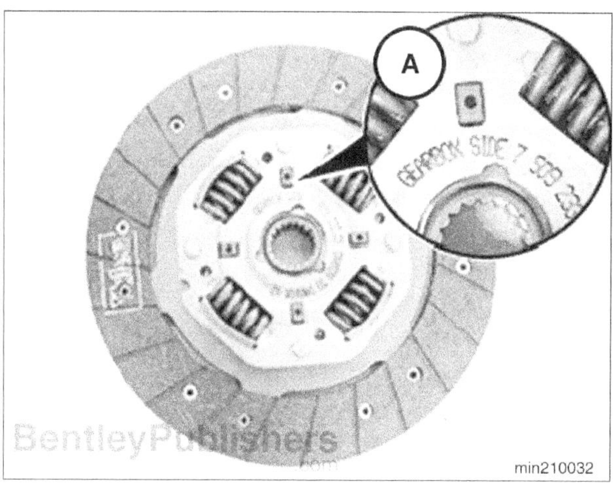

⊲ Note clutch disk direction mark (**A**). Face shown is gearbox side.

– Check clutch disk for damage, rust and wear. Replace clutch disk if clutch material measures 1 mm (0.04 in) or less above rivet head on gearbox side.

– Slide clutch disk onto transmission input shaft and check that clutch moves freely along shaft before continuing with installation.

Install clutch disk and pressure plate back onto flywheel. Use special tool 21 6 100 to align clutch disk.

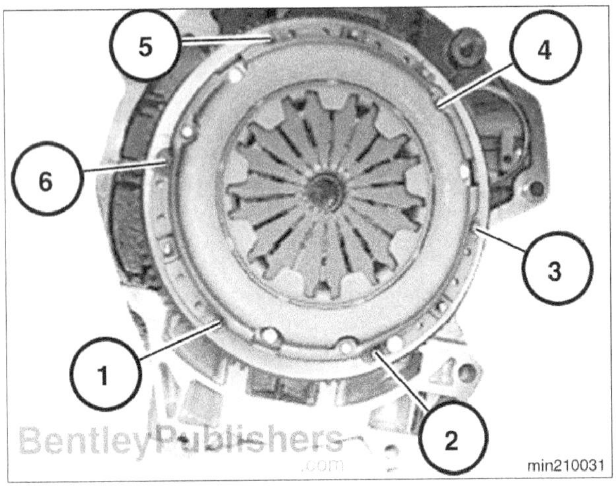

Replace pressure plate retaining bolts and gradually tighten in a cross pattern (**1 - 6**).

NOTE—
During the tightening process, periodically rotate special tool 21 6 100 to help center clutch disk.

Tightening torque	
Clutch pressure plate to flywheel (M7)	20 Nm (15 ft-lb)

– Install transmission. See **230 Manual Transmission** for more information.

Clutch, removing and installing (Cooper S)

– Remove transmission. See **230 Manual Transmission** for more information.

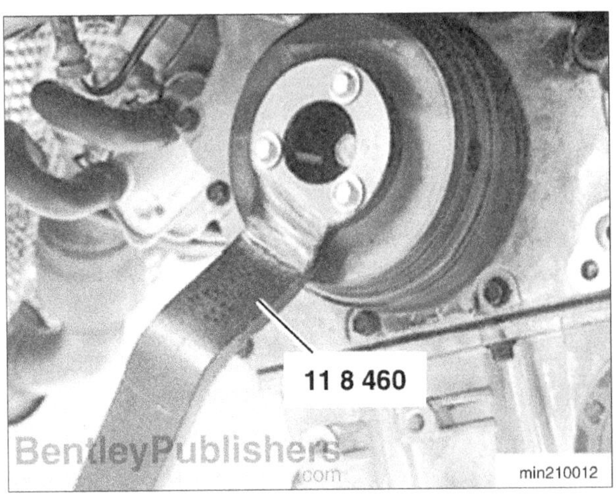

Fit special tool 11 8 460 on front of crankshaft to counterhold flywheel during pressure plate removal.

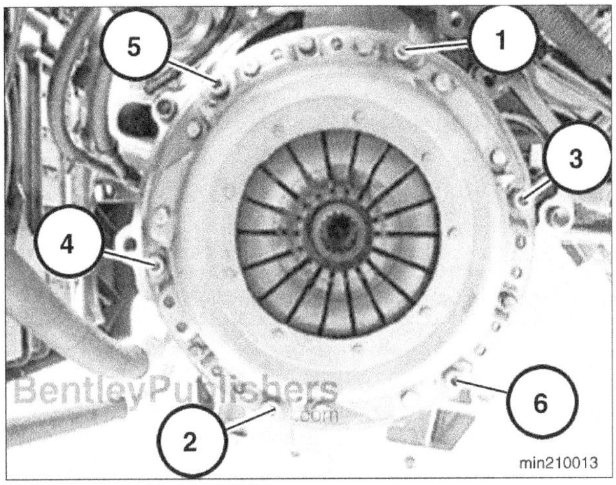

min210013

◄ Gradually loosen pressure plate bolts in a cross pattern (**1 - 6**) and remove bolts.

– Remove pressure plate and clutch disk.

– Clean flywheel and check for wear and damage. If necessary, replace flywheel. See **Flywheel, removing and installing** later in this repair group for more information.

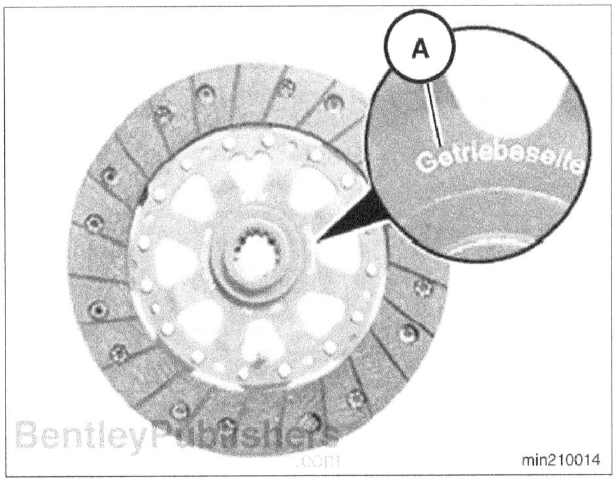

min210014

◄ Note clutch disk direction mark (**A**). Face shown is gearbox side.

– Check clutch disk for damage, rust and wear. Replace clutch disk if clutch material measures 0.2 mm (0.008 in) or less above rivet head on transmission side.

– Slide clutch disk onto transmission input shaft and check that clutch moves freely along shaft before continuing with installation.

min210017

◄ Install clutch disk and pressure plate back onto flywheel. Use special tool 21 2 210 to align clutch disk.

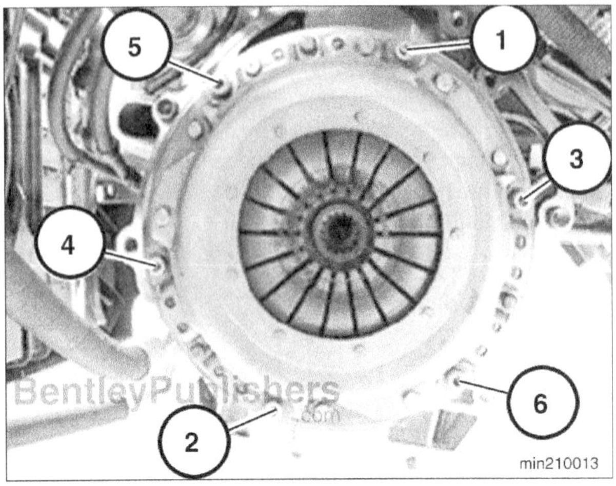

Replace pressure plate retaining bolts and gradually tighten in a cross pattern (**1-6**).

NOTE —
During the tightening process, periodically rotate special tool 21 2 210 to help center clutch disk.

> *CAUTION —*
> *Cooper S pressure plate is fastened using one of two diameter metric bolts. Use the proper torque specification.*

Tightening torque	
Clutch pressure plate to flywheel (M8)	23 Nm (17 ft-lb)
Clutch pressure plate to flywheel (M9)	28 Nm (21 ft-lb)

Flywheel, removing and installing

– Remove transmission. See **230 Manual Transmission** for more information.

– Remove clutch. See **Clutch, removing and installing (Cooper)** or **Clutch, removing and installing (Cooper S)** in this repair group.

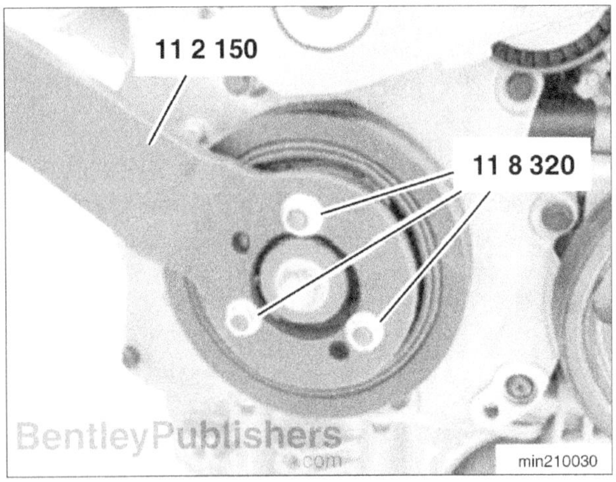

Fit special tool 11 2 150 and 11 8 320 (or special tool 11 8 460 for Cooper S) on front of crankshaft to counterhold flywheel during removal.

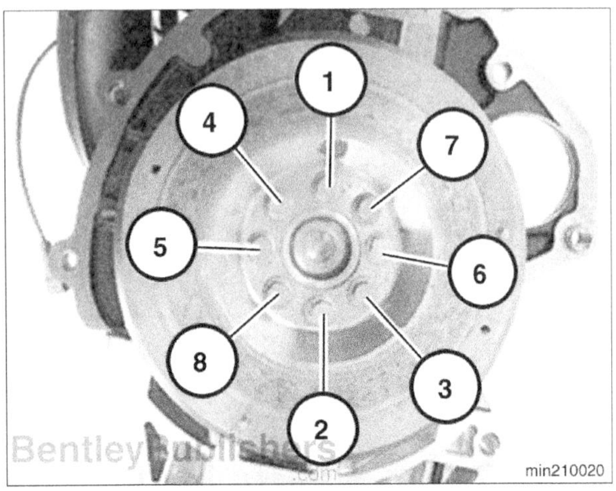

Gradually loosen flywheel bolts in a cross pattern (**1 - 8**) and remove flywheel.

> *CAUTION —*
> *Support flywheel against crankshaft when last bolt is removed. Flywheel is very heavy.*

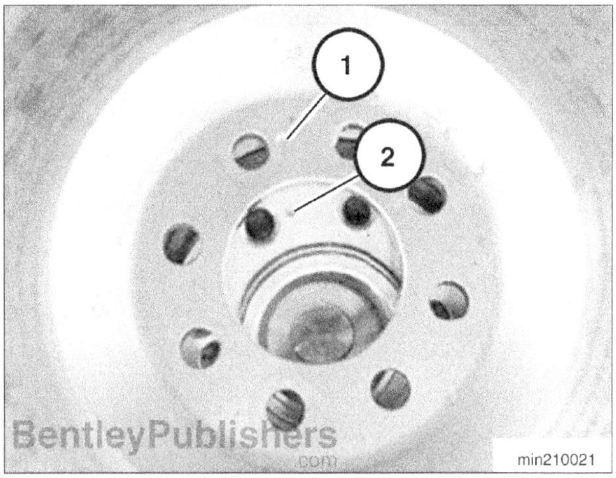

To reinstall, align mark on flywheel (**1**) with mark on crankshaft (**2**).

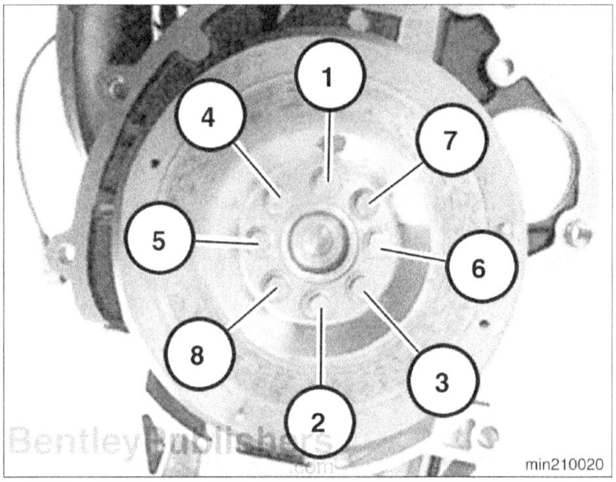

Gradually tighten flywheel bolts in a cross pattern (**1 - 8**) to final torque specification.

NOTE —
Replace flywheel bolts with new.

Tightening torque	
Flywheel to crankshaft (Cooper)	80 Nm (59 ft-lb)
Flywheel to crankshaft (Cooper S)	90 Nm (66 ft-lb)

– Install clutch and pressure plate. See **Clutch, removing and installing** in this repair group.

– Install transmission. See **230 Manual Transmission** for more information.

Clutch release bearing, replacing

– Remove transmission. See **230 Manual Transmission** for more information.

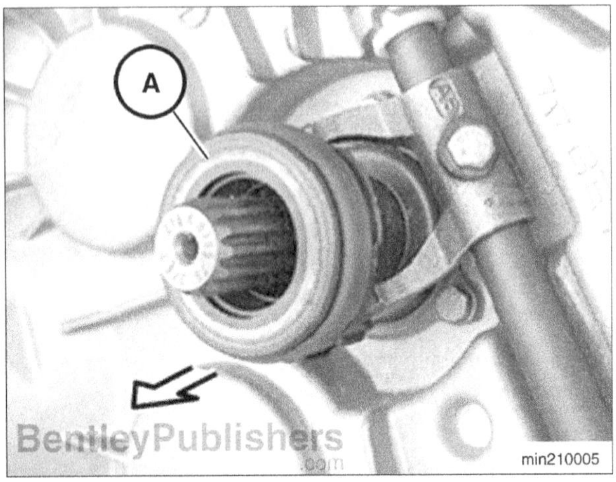

Working inside the transmission bellhousing, remove release bearing (**A**) from release lever. Clean all sliding surfaces.

NOTE —
Cooper S transmission shown. Cooper is similar.

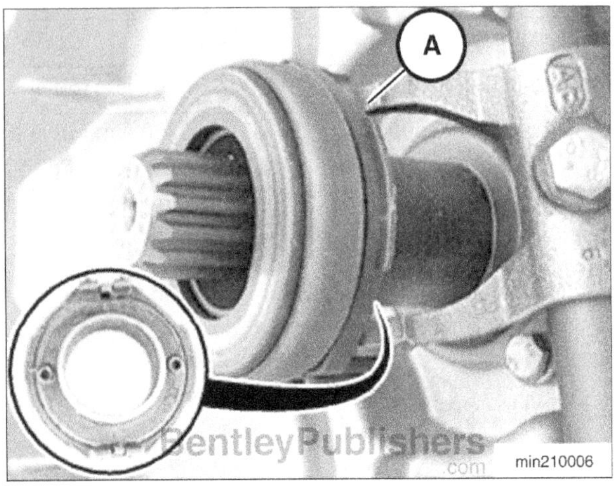

Apply a thin film of Unirex® S2 or equivalent lubricant to inner sleeve of bearing. Attach new bearing to release lever (**A**).

– Install transmission. See **230 Manual Transmission** for more information.

230 Manual Transmission

General

This section covers external transmission service, including removal and installation of the transmission unit. Internal transmission repair is not covered. Special press tools and procedures are required to disassemble and service the internal components of the transmission.

Special tools

BMW special tools are recommended for removal and installation of the manual transmission, as well as the removal and installation of seals at the transmission input, output, and selector shafts. If these tools are unavailable, equivalent tools may be substituted.

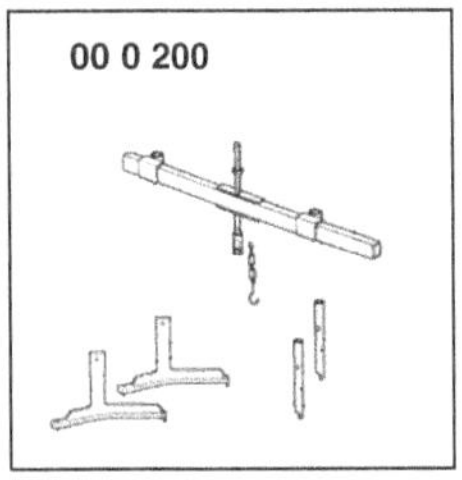

◄ Engine support cross member and attachments (BMW tool 00 0 200)

◄ Seal installation drift (6-speed) (BMW tool 00 5 000)

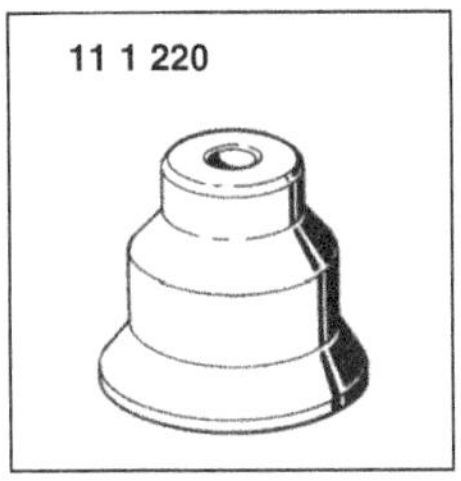

◄ Seal installation bushing (5-speed) (BMW tool 11 1 220)

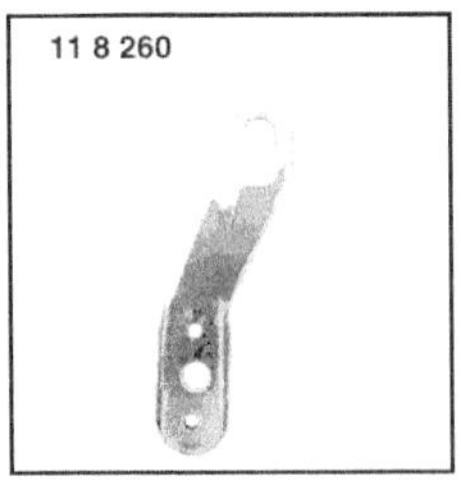

◄ Engine support bracket (BMW tool 11 8 260)

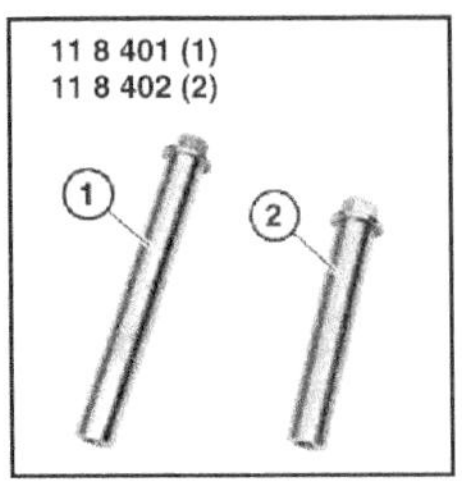

◄ Modular front end extension bolts (BMW tool 11 8 400)

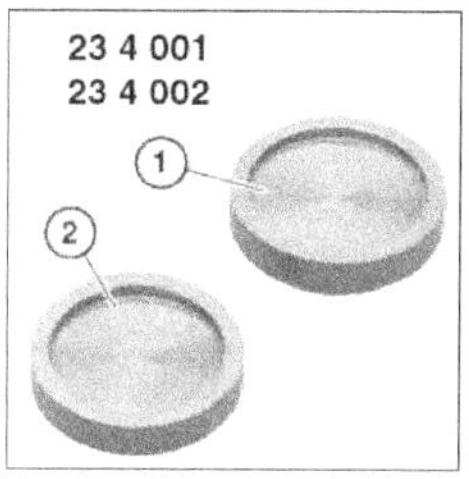

Seal installation bushings (6-speed)
(BMW tool 23 4 000)

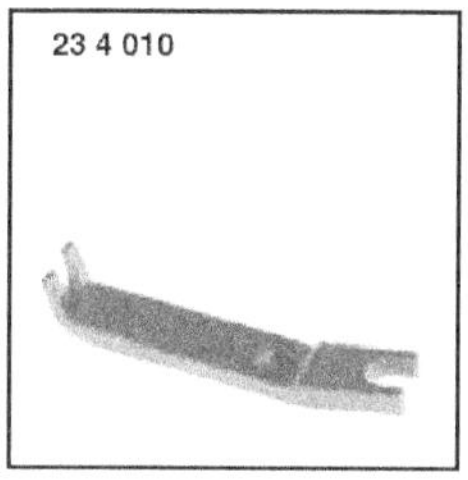

Manual transmission linkage tool
(BMW tool 23 4 010)

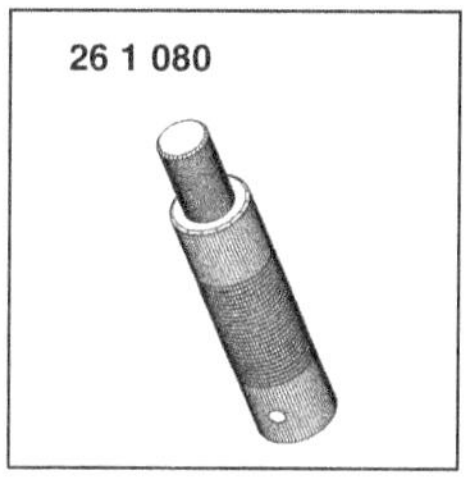

Seal installation drift (5-speed)
(BMW tool 26 1 080)

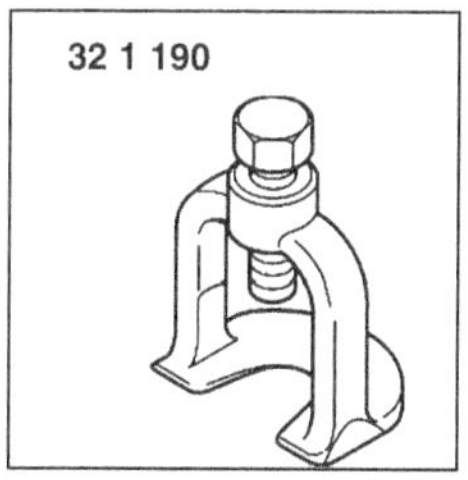

Ball joint removal tool
(BMW tool 32 1 190)

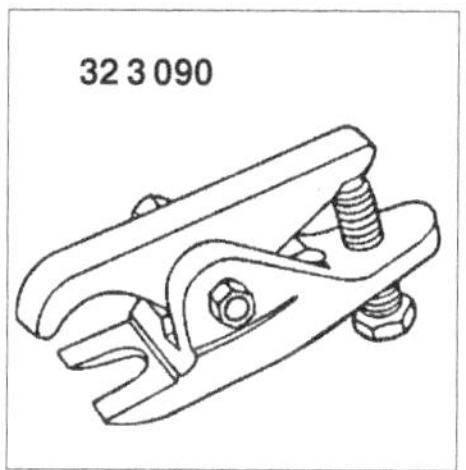

Tie rod removal tool
(BMW tool 32 3 090)

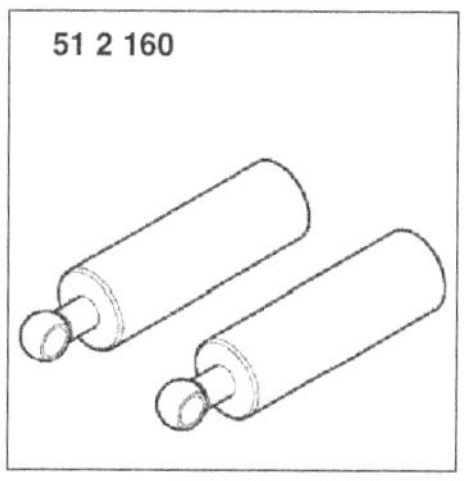

Hood prop extensions
(BMW tool 51 2 160)

Transmission Fluid Service

Manual transmissions installed in MINI models are normally filled with "Lifetime" manual transmission fluid and do not require periodic fluid changes.

Manual transmission fluid	
Lifetime fluid	MTF 94

Transmission fluid level, checking

– Drive vehicle for a few miles to warm transmission.

– Raise and safely support vehicle to access transmission fill plug.

> ***CAUTION—***
> *Make sure the car is stable and well supported at all times. Use a professional automotive lift or jack stands designed for the purpose. A floor jack is not adequate support.*

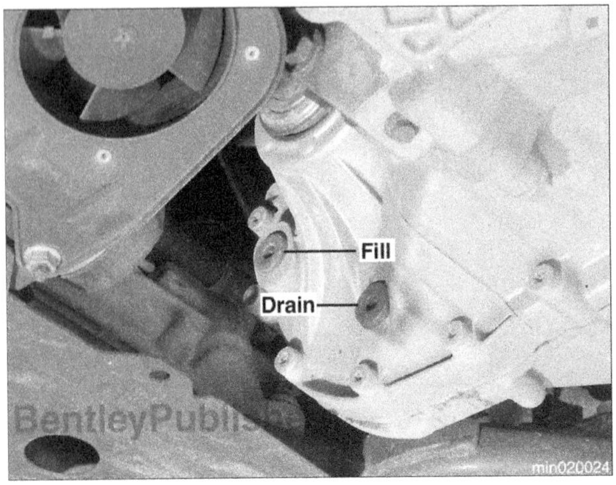

◄ Unscrew transmission fill plug. Make sure fluid is even with bottom of hole or drips slightly.

– Clean plug and reinstall.

> ***NOTE—***
> *Cooper S transmission shown, Cooper is similar with the fill plug located to the right of the drive axle.*

Tightening torque	
Transmission fill plug to transmission: Cooper Cooper S	 20 - 30 Nm (15 - 22 ft-lb) 43 Nm (32 ft-lb)

Transmission fluid, replacing

– Drive vehicle for a few miles to warm transmission.

– Raise and safely support vehicle to access transmission fill plug.

> ***CAUTION—***
> *Make sure the car is stable and well supported at all times. Use a professional automotive lift or jack stands designed for the purpose. A floor jack is not adequate support.*

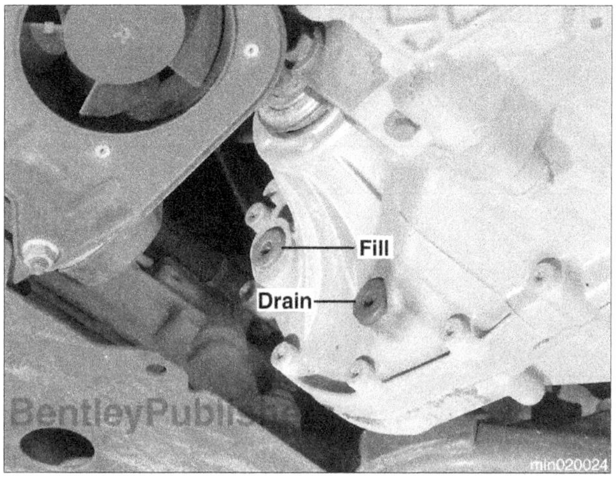

Unscrew transmission drain plug and allow fluid to completely drain into a suitable container.

– Clean drain plug and reinstall.

– Unscrew transmission fill plug and fill transmission until fluid overflows slightly.

– Clean fill plug and reinstall.

NOTE —
Cooper S transmission shown, Cooper is similar with the fill plug located to the right of the drive axle.

Tightening torque	
Transmission drain plug to transmission: Cooper Cooper S	 32 - 40 Nm (24 - 30 ft-lb) 43 Nm (32 ft-lb)
Transmission fill plug to transmission: Cooper Cooper S	 20 - 30 Nm (15 - 22 ft-lb) 43 Nm (32 ft-lb)

TRANSMISSION SERVICE

Back-up light switch, replacing

Cooper back-up light switch is located on top of the transmission. Cooper S back-up light switch is on the bulkhead side of the transmission below the gear selector mechanism. Replacement procedure is similar.

– Cooper: Remove air filter box.

Remove electrical harness connector (**A**) and unscrew back-up light switch.

NOTE —
Cooper location shown. Cooper S switch is on bulkhead side of transmission.

– Replacement is reverse of removal.

Tightening torque	
Back-up light switch to transmission	25 Nm (18.5 ft-lb)

Drive axle seal, replacing

The following BMW (or equivalent) special tools are recommended for seal replacement.

Transmission	BMW tool
5-speed	26 1 080 11 1 220
6-speed	00 5 500 23 4 001 23 4 002

NOTE —
The procedures for left and right drive axle seal replacement show Cooper transmission. Cooper S is similar.

Left drive axle seal

– Remove left drive axle. See **310 Front Suspension, Drive Axles**.

– Be prepared to catch dripping fluid.

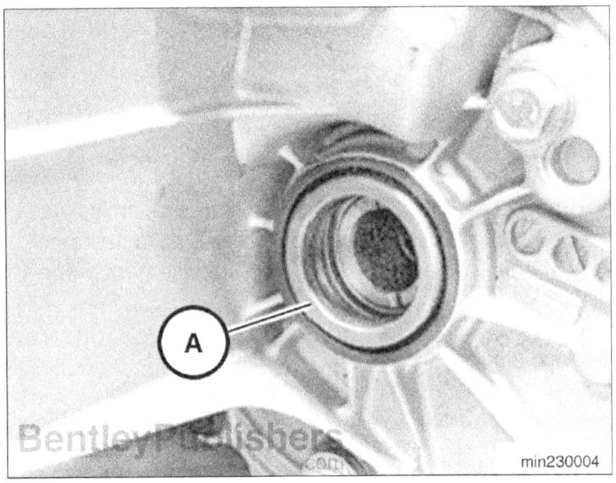

◄ Use a screwdriver to lever out drive axle seal (**A**) from transmission housing. Take care not to damage seal seating surface.

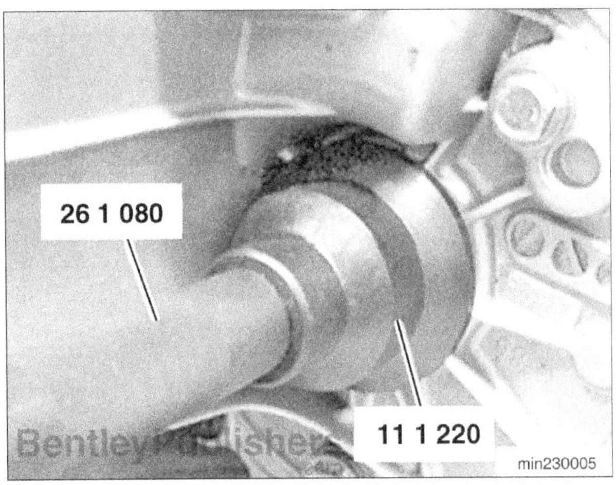

◄ Using special tools 26 1 080 and 11 1 220 tap new seal into case until outer seal ring is flush with transmission casing.

NOTE —
Cooper S seal replacement requires special tools 00 5 500 and 23 4 001 (or equivalent).

– Remainder of installation is reverse of removal. Remember to:
- Push drive axle past resistance of snap ring until an audible snap is heard.
- Fill transmission with approved fluid and check for leaks.

Right drive axle seal

- Remove right drive axle. See **310 Front Suspension, Drive Axles** for more information.
- Be prepared to catch dripping fluid.

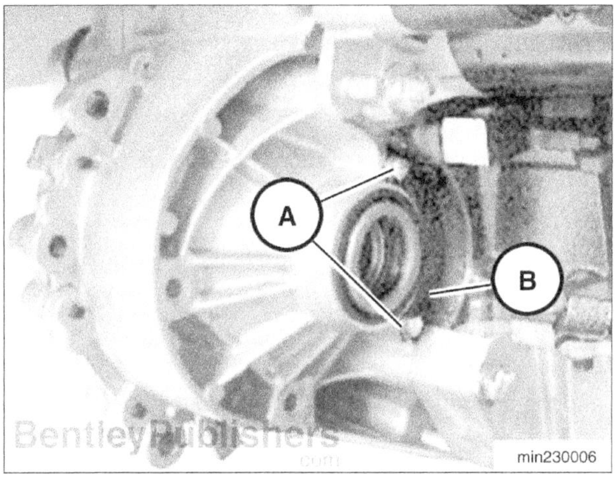

< Remove cover plate bolts (**A**) and remove cover plate (**B**).

- Remove lower engine damper.

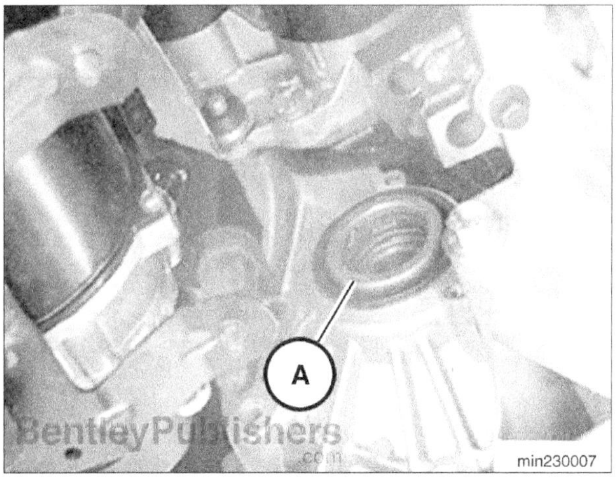

< Use a screwdriver to lever out drive axle seal (**A**) from transmission housing. Take care not to damage seal seating surface.

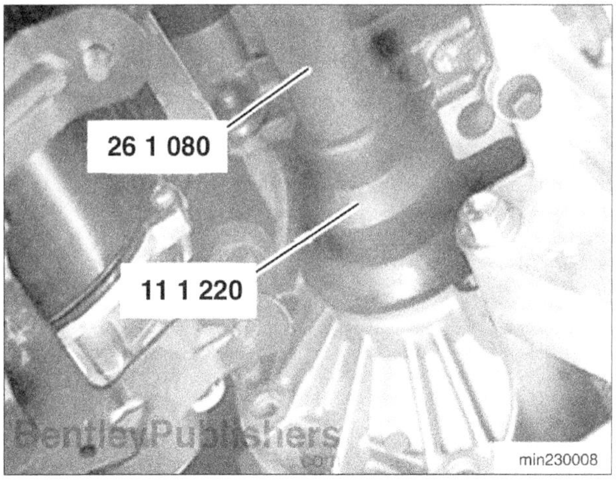

< Using special tools 26 1 080 and 11 1 220 tap new seal into case until outer seal ring is flush with transmission casing.

NOTE —
Cooper S seal replacement requires special tools 00 5 500 and 23 4 002 (or equivalent).

- Remainder of installation is reverse of removal. Remember to fill transmission with approved fluid and check for leaks.

Input shaft seal, replacing

– Remove transmission. See **Transmission removal and installation** in this section.

– Remove clutch release bearing. See **210 Clutch** for more information.

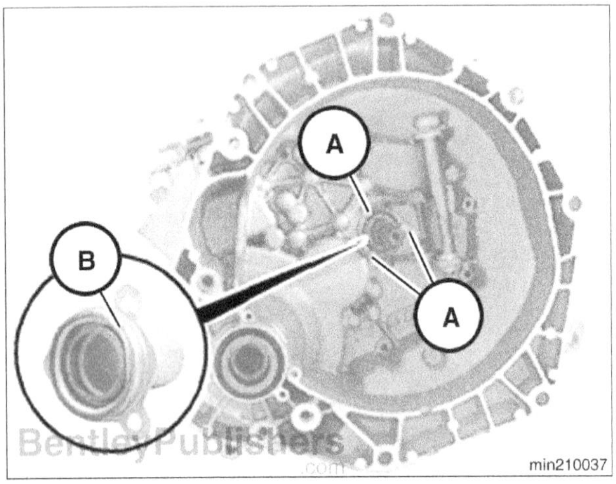

< Remove input shaft seal and guide tube bolts (**A**) and remove guide tube (**B**).

NOTE—
Input shaft seal and guide tube are supplied as an assembly.

– Installation is reverse of removal. Remember to:
- Apply tape to input shaft splines during installation to protect new seal.
- Lubricate seal to ease installation.
- Replace guide tube bolts.
- Check and top off transmission with approved fluid.

Tightening torque	
Input seal assembly to transmission (replace bolts)	6 Nm (4.5 ft-lb)

TRANSMISSION REMOVAL AND INSTALLATION

Removal and installation of the transmission is best accomplished on a lift using a transmission jack. Support engine using appropriate equipment. This allows the engine to pivot on its mounts to aid access to the upper bolts at the bellhousing.

Transmission removal is a complicated job. This section describes transmission removal and installation without engine removal. If additional engine repair is needed, the engine and transmission can be removed as a complete unit with little additional work. See **110 Engine Removal and Installation**.

WARNING—
- *Make sure the car is stable and well supported at all times. Use a professional automotive lift or jack stands designed for the purpose. A floor jack is not adequate support.*
- *The removal of the transmission may upset the balance of the vehicle on a lift.*

Front subframe, removing

Front subframe removal is a necessary prerequisite to transmission removal without engine removal. Upon completion of the front subframe procedure a professional wheel alignment check is required. To remove the engine and transmission as a complete unit, see **110 Engine Removal and Installation**.

– Disconnect negative (-) cable from battery.

> ***CAUTION—***
> *Prior to disconnecting the battery, read the battery disconnection cautions given in* **001 General Warnings and Cautions**.

– Raise car and support in a safe manner.

> ***WARNING—***
> *Make sure the car is stable and well supported at all times. Use a professional automotive lift or jack stands designed for the purpose. A floor jack is not adequate support.*

– Remove front wheels.

– Remove engine splash shield.

– Remove front bumper cover and bumper carrier. See **510 Exterior Trim, Bumpers** for more information.

– Remove crush tubes. See **Modular front end (MFE)** in **510 Exterior Trim, Bumpers** for more information.

– Remove stabilizer links from front stabilizer bar. See **310 Front Suspension** for more information.

– Detach tie rod ends. See **310 Front Suspension** for more information.

– Remove control arms from steering knuckles on front subframe. See **310 Front Suspension** for more information.

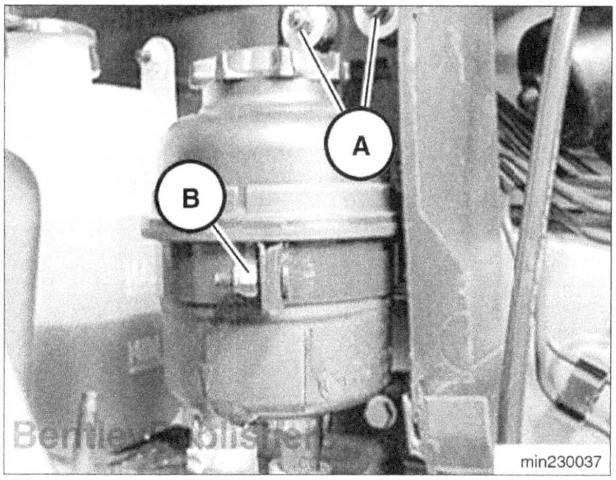

< Working in engine compartment, unscrew power steering reservoir bracket bolts (**A** and **B**) and remove reservoir from bracket.

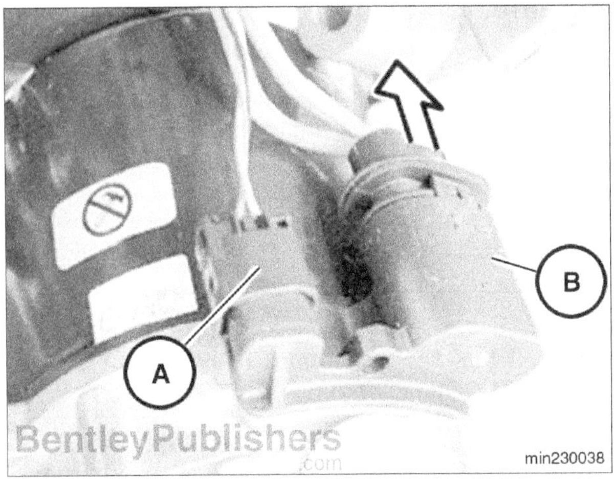

Working underneath car, remove connectors (**A** and **B**) from power steering pump. If necessary, disconnect power steering pump fan and make sure no other wires interfere with subframe removal.

Release nut and bolt from steering spindle and fold back joint in direction of arrow.

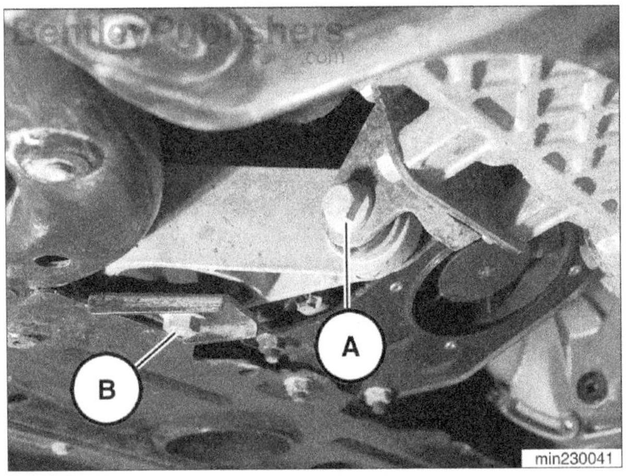

Remove lower engine vibration damper by first removing bolt (**A**).

– Remove bolt (**B**) and remove vibration damper.

> ***WARNING —***
> *Front subframe is heavy. support subframe with a suitable tool (transmission jack) and secure frame to jack to prevent slipping.*

Unclip ABS sensor wire and remove bolt (**A**) holding top of subframe to chassis.

– Repeat for other side of subframe.

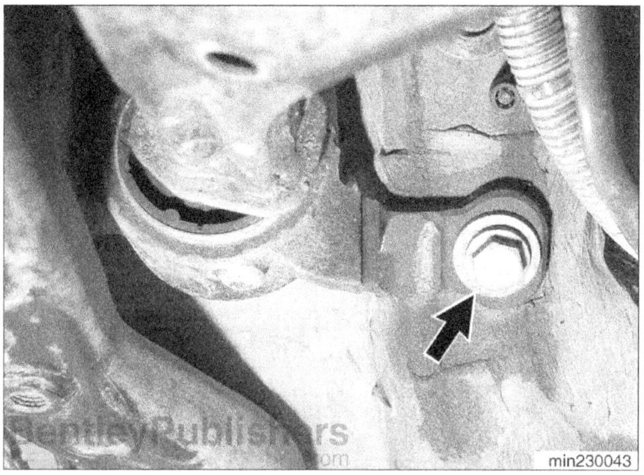

Remove control arm bushing bracket bolt (**arrow**) from control arm bushing bracket.

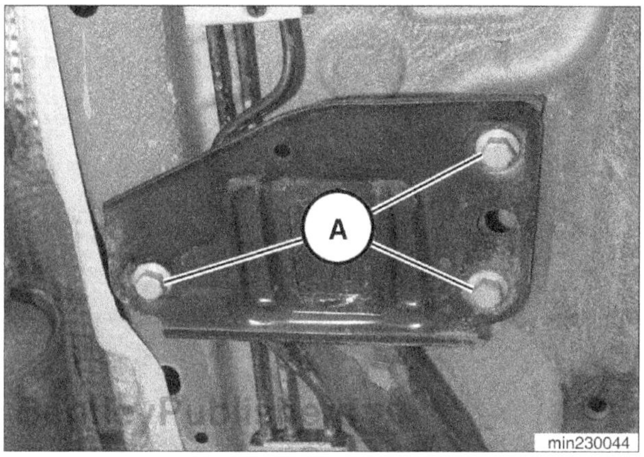

Remove bolts (**A**) at rear of subframe.

> ***WARNING —***
> - *Front subframe is heavy. Support with a suitable tool (transmission jack) and secure frame to jack to prevent slipping.*

- Lower front subframe slowly and guide power steering reservoir down through engine compartment.
- To avoid spilling power steering fluid, use a suitable tool to hold power steering reservoir in a vertical position.
- Installation is reverse of removal. Keep in mind the following:
  - Tighten fastener at control arm bushing bracket only after all other fasteners on front subframe have been tightened down. This will prevent unwanted torsional stress on the control arm bushing.
  - Replace all self-locking fasteners.
  - Perform an electronic wheel alignment check.

Tightening torques	
Control arm bracket to chassis (replace bolts) (tighten last)	59 Nm (44 ft-lb) angle: 90° ± 15°
Front subframe front mount to chassis	100 Nm (74 ft-lb)
Front subframe rear mount to chassis	100 Nm (74 ft-lb)
Lower engine vibration damper to engine bracket	100 Nm (74 ft-lb)
Lower engine vibration damper to front subframe bracket	100 Nm (74 ft-lb)
Lower steering shaft to steering rack	21 Nm (15 ft-lb)
Power steering reservoir bracket to bulkhead	8 Nm (6 ft-lb)
Power steering reservoir to bracket	5.5 Nm (4 ft-lb)

Transmission, removing and installing

– Disconnect negative (-) cable from battery.

> ***CAUTION—***
> *Prior to disconnecting the battery, read the battery disconnection cautions given in* **001 Warnings and Cautions**.

– Cooper: Remove battery housing.

– Cooper S: Remove air filter housing.

– Raise car and support in a safe manner.

> ***WARNING—***
> *Make sure the car is stable and well supported at all times. Use a professional automotive lift or jack stands designed for the purpose. A floor jack is not adequate support.*

– Drain transmission fluid. See **Transmission fluid, replacing** in this section.

– Remove left front wheel arch trim.

– Remove both drive axles and steering knuckles. See **310 Front Suspension, Drive Axles** for more information.

– Remove front subframe. See **Front subframe, removing** in this section.

– Secure engine in installation position with special tool 00 0 200 and accessories.

– Remove top engine mount bracket. See **110 Engine Removal and Installation** for more information.

– Support engine from below with suitable jack.

– Remove bottom engine vibration damper bracket. See **110 Engine Removal and Installation** for more information.

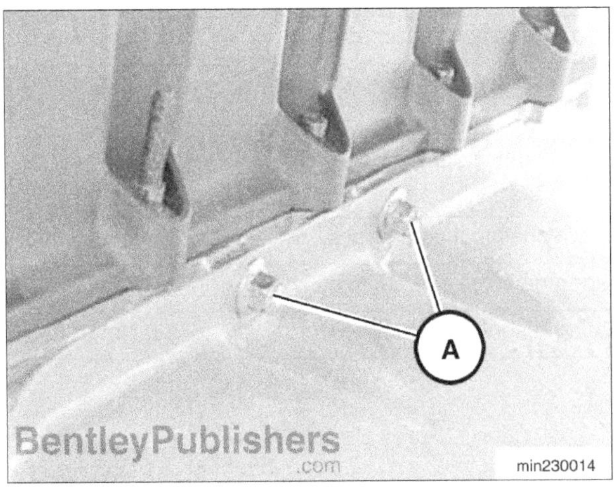

◄ Remove bolts (**A**) and remove exhaust manifold heat shield.

◄ Remove clips from gearshift cables and remove gearshift cables from ball joints with special tool 23 4 010, or by carefully prying off with a pair of screwdrivers.

– Remove gearshift cable bracket bolts and remove gearshift bracket.

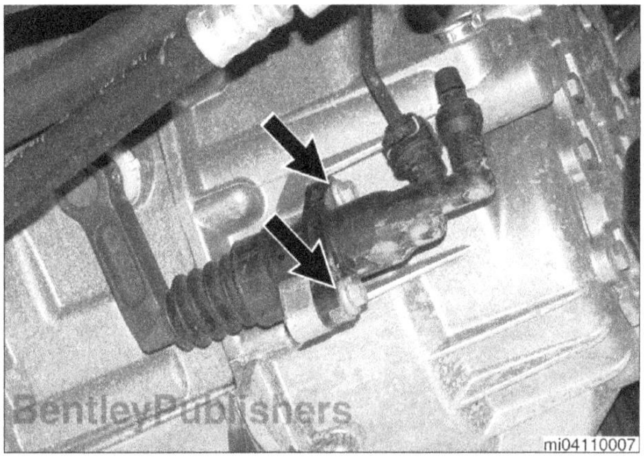

◄ Remove mounting bolts (**arrows**) and remove clutch slave cylinder from top of transmission.

NOTE—
Cooper S transmission shown. Cooper slave cylinder is located on top of transmission.

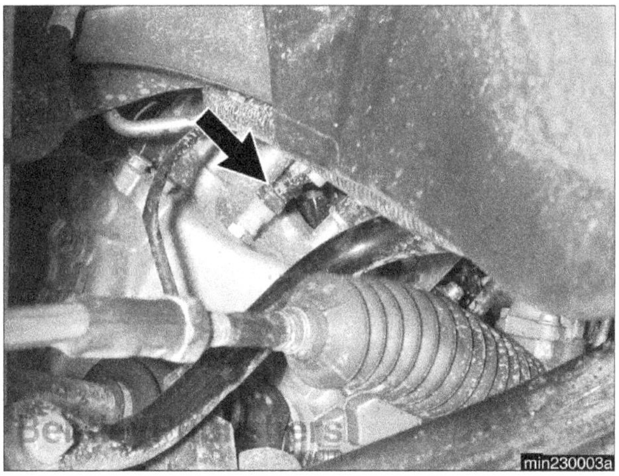

◄ Remove electrical harness connector (**arrow**) from back-up light switch.

NOTE—
Cooper S back-up light switch shown. Cooper switch is on top of transmission.

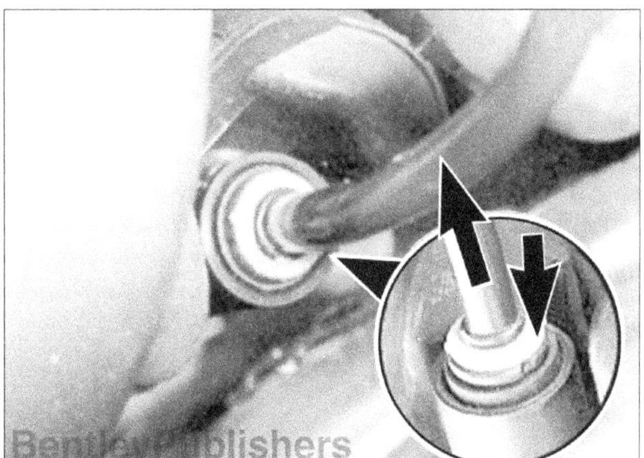

◄ Detach brake booster vacuum line from intake manifold by pressing ring in while pulling hose out (see inset).

– Unclip brake booster vacuum line from any clips securing it to transmission.

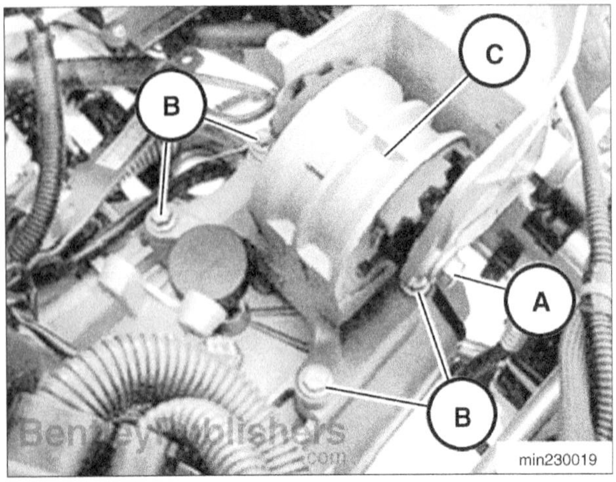

- Remove bolt (**A**) from transmission mount and remove bolts (**B**) securing mount to transmission.
- Remove transmission mount (**C**).
- Carefully lower engine approximately 40 mm (1.5 in).

> ***CAUTION—***
> *Lowering the engine too far may damage the exhaust system or refrigerant lines to the A/C compressor. Make sure no stress is placed on these or other components.*

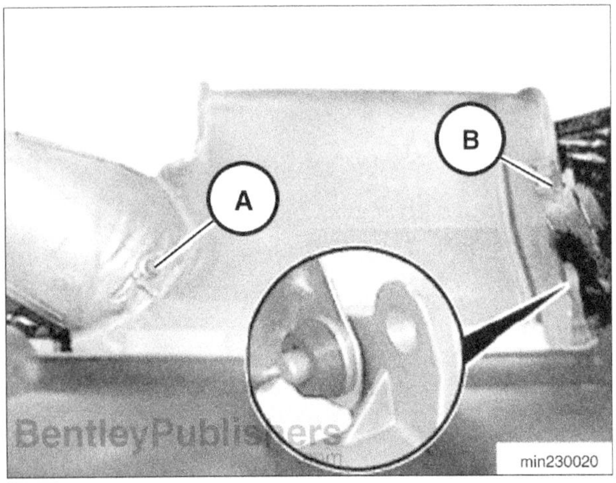

- Remove bolt (**A**) from starter motor heat shield and release wiring harness for oxygen sensor from clip (**B**). Remove heat shield.

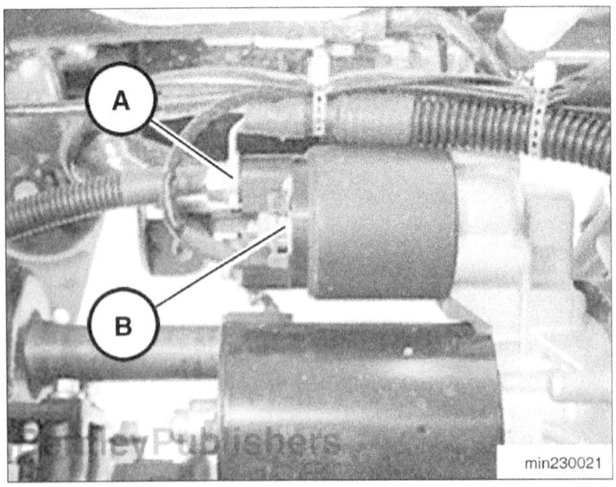

- Remove starter motor electrical connections (**A** and **B**) and make sure wiring harness is clear of transmission area.

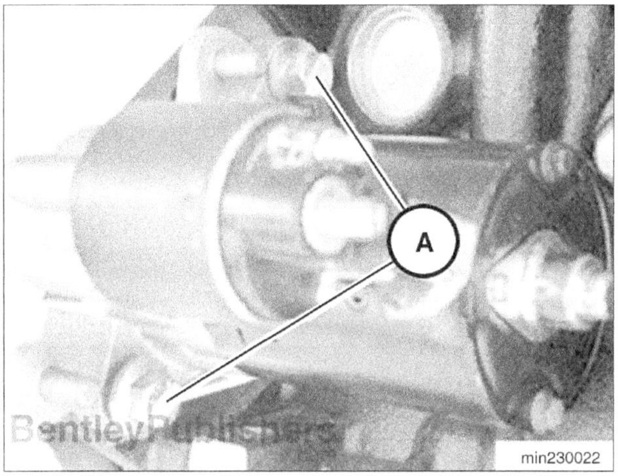

◄ Remove bolts (**A**) and remove starter motor.

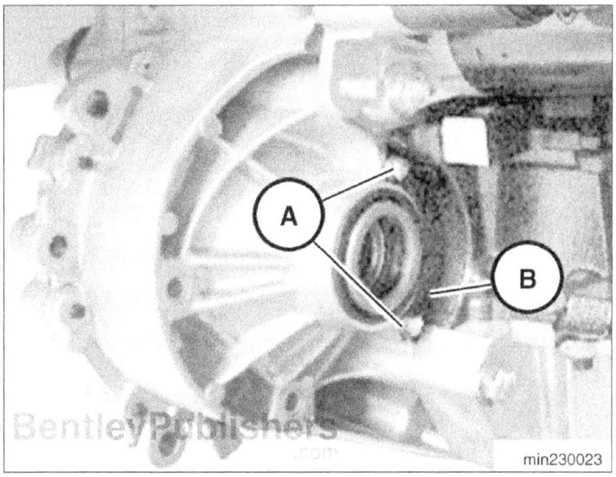

◄ Remove bolts (**A**) and remove cover plate (**B**).

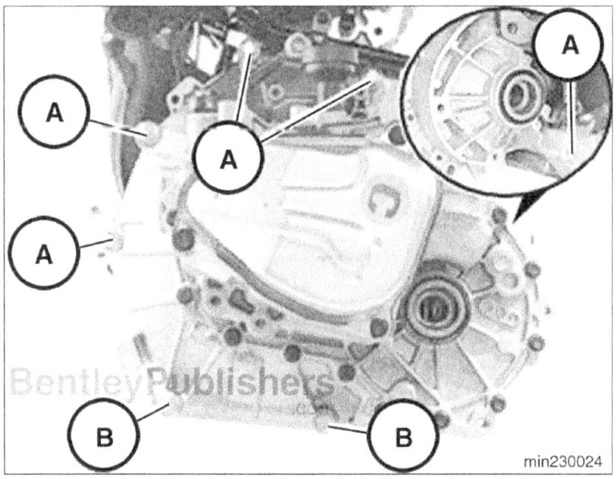

◄ Remove bolts (**A** and **B**). Note bolt on opposite side of transmission (see inset).

– Carefully remove transmission.

> ***CAUTION—***
> *The transmission is heavy. A second person is needed to help remove and install the transmission.*

NOTE—
*Lower bolts (**2**) in transmission sump are shorter than the other bolts.*

– Installation is reverse of removal. Remember to:
  - Fill transmission with approved fluid. See **Transmission Service** in this section.
  - Check transmission for leaks.

Tightening torques	
Alternator connection to starter motor	14 Nm (10 ft-lb)
Cover plate to transmission	9 Nm (6.6 ft-lb)
Clutch slave cylinder to transmission	24 Nm (18 ft-lb)
Engine harness to starter solenoid	8 Nm (6 ft-lb)
Gearshift cable bracket to transmission	22 Nm (16 ft-lb)
Manifold heat shield to cylinder head	8 Nm (6 ft-lb)
Starter motor heat shield to starter motor	9 Nm (6.6 ft-lb)
Starter motor to transmission	85 Nm (63 ft-lb)
Transmission mount to transmission: Cooper Cooper S	 38 Nm (28 ft-lb) 66 Nm (49 ft-lb)
Transmission support bracket to body	68 Nm (50 ft-lb)
Transmission support bracket to transmission mount	68 Nm (50 ft-lb)
Transmission to engine	85 Nm (63 ft-lb)

240 Automatic Transmission

General

This section covers external automatic transmission service, including removal and installation of the transmission unit. Internal transmission repair is not covered.

The automatic transmission available in the Cooper is a Continuously Variable Transmission (CVT). The CVT provides stepless shifting with an infinitely variable gear ratio. This allows full engine performance to be available at any time.

The automatic transmission available in the Cooper S is a 6-speed automatic with Agitronic. The Agitronic feature enables individual gear selection in Drive or Sport mode using shift paddles on the steering wheel.

Automatic transmission fluid cooler removal and installation is covered in **170 Radiator and Cooling System**.

Special tools

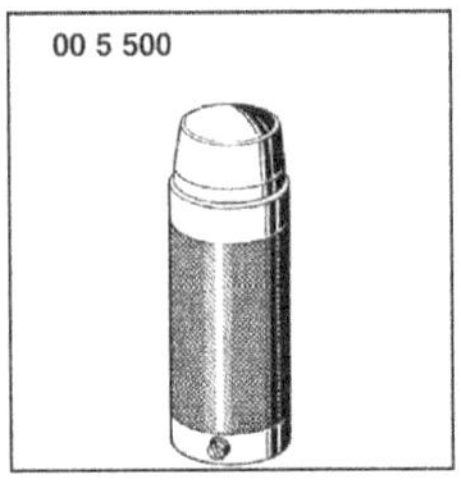

◄ Seal installation drift
(BMW tool 00 5 000)

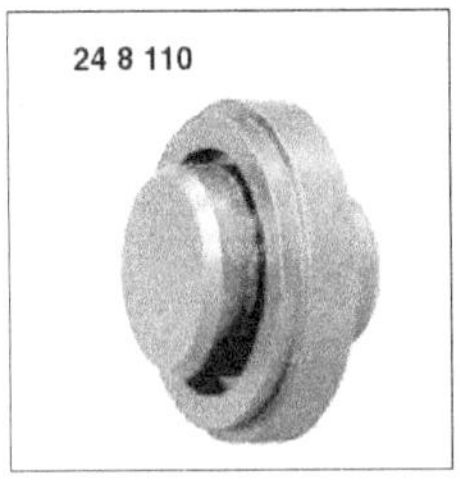

◄ Seal installation bushing
(BMW tool 24 8 110)

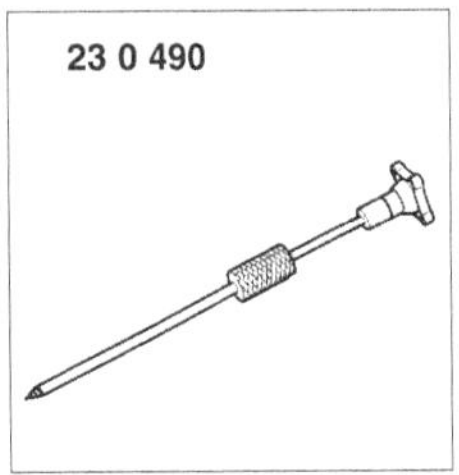

◄ Seal puller
(BMW tool 24 0 490)

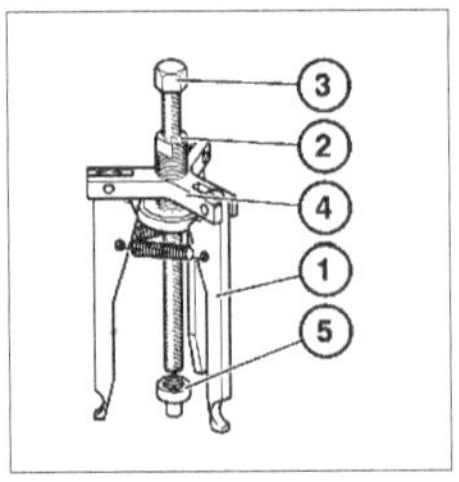

◄ Universal puller for removing input shaft seal
(BMW tool 00 5 010)

- ATF fill plug removal tool
  (BMW 24 4 240)
- Seal installation tool
  (BMW 24 4 300)
- Radial sleeve seal installer
  (BMW tool 24 4 250)
- Torque converter removal tool
  (BMW tool 24 4 260)
- Gear position switch adjustment tool
  (BMW tool 24 4 270)
- Depth gauge
  (BMW special tool 00 2 550)

TRANSMISSION FLUID SERVICE

Transmission fluid level, checking (CVT)

> ***CAUTION—***
> *The CVT is filled with special transmission fluid. DO NOT top up with ATF. Extensive damage to the CVT may result.*

CVT fluid change is scheduled for every **INSPECTION I** service. See **020 Maintenance** for more information.

Continuously Variable Transmission (CVT)			
Model, transmission	**Approximate capacity liter (US qt)**	**Refill capacity liter (US qt)**	**Oil specification**
MINI Cooper GACVT16Z	4.5 (4.8)	4.5 (4.8)	ESSO CVT E2L 799

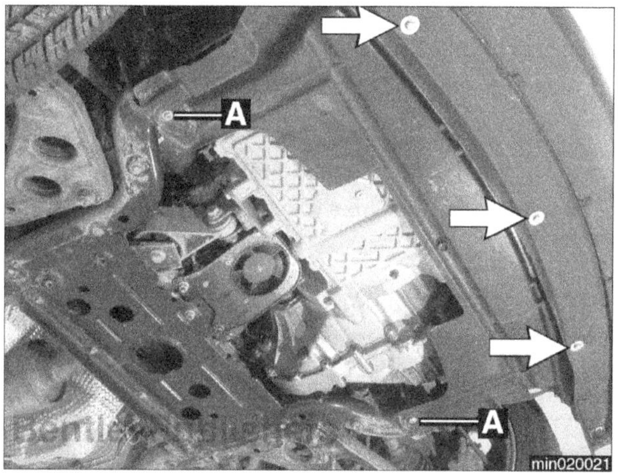

To gain access to transmission fill/level check plug, remove splash shield under engine:

- Remove 3 bolts (**arrows**) at bottom of front bumper cover.
- Unlock 2 splash shield mounting screws (**A**).
- Remove splash shield.

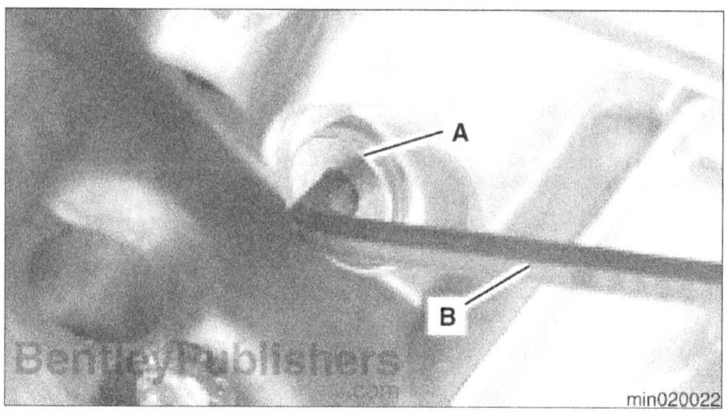

Use 6 mm Allen key (**B**) to remove fill/level check plug (**A**). If transmission is full, some fluid will drip out.

Tightening torque	
Fill/level check plug to transmission (M14)	21 Nm (15 ft-lb)

Transmission fluid, replacing (CVT)

CVT transmission fluid replacement is required at every **Inspection I** service. See **020 Maintenance** for recommended service intervals.

> ***WARNING —***
> *Hot transmission fluid can cause burns. Allow CVT to cool before replacing fluid.*

> ***CAUTION —***
> - *The CVT is filled with special transmission fluid. DO NOT use ordinary ATF. Extensive damage to the CVT may result.*
> - *Comply with all federal, state and local waste disposal regulations.*

– Raise and safely support vehicle.

> ***CAUTION —***
> *Make sure the car is stable and well supported at all times. Use a professional automotive lift or jack stands designed for the purpose. A floor jack is not adequate support.*

– Remove engine splash shield.

– Make sure transmission is in P (Park).

– Remove front wheels and attach front rotors to axle hubs using four bolts and washers.

– Apply foot brake and tighten rotors to hubs using proper torque specification.

Tightening torque	
Front rotors to hubs (M12 x 1.5 x 25 mm)	60 Nm (44 ft-lb)

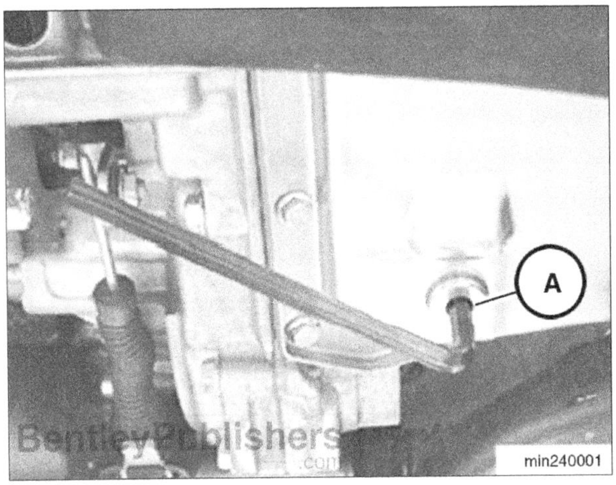

◀ Remove drain plug (**A**) from transmission fluid pan and allow fluid to drain into a suitable catch pan.

– Replace sealing ring and reinstall drain plug.

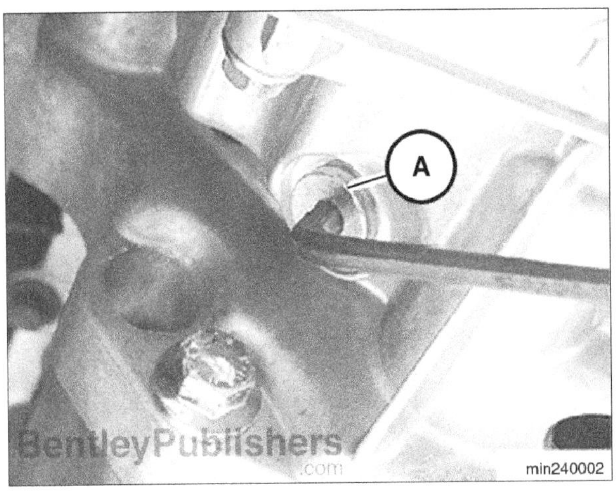

◀ Remove fill/level check plug (**A**) and allow excess fluid to drain into catch pan.

NOTE —
CVT capacity is 4.5 liters (4.8 quarts). Using a suitable pump, note the number of pump strokes required to transfer 4.5 liters.

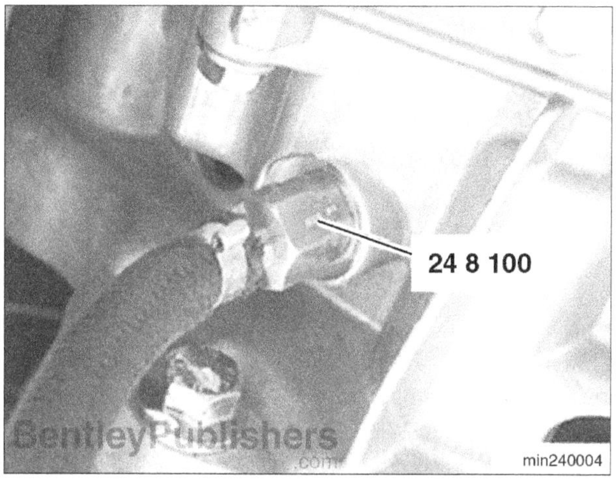

< Use special tool 24 8 100 to connect approved BMW CVT fluid (ESSO CVT E2L 799) to transmission fill hole.

- Pump 4.5 liters of CVT fluid into transmission.
- Close return valve of CVT fluid pump so that oil cannot flow in either direction.
- Lower vehicle maintaining at least 15 cm (6 in) between drive wheels (brake rotors) and ground.

> ***CAUTION—***
> *It is necessary to run engine and shift transmission to complete this procedure. Take special care to:*
> - *Make sure there is sufficient distance between drive wheels and floor to prohibit any possible contact. Drive wheels must rotate at high speed.*
> - *Switch off traction control (if applicable).*
> - *Firmly apply parking brake.*
> - *Attach an exhaust gas extractor.*

- Apply foot brake.
- Start engine.
- Run engine for 10 seconds.
- Apply and maintain foot brake.
- Shift transmission out of P (Park).

NOTE—
A hissing noise may be heard as air circulates through the transmission. This is normal.

- Shift through all CVT modes. Pause 5 seconds in each mode before making next selection.
- Shift transmission to D (Drive).
- Release foot brake.
- Lightly press accelerator pedal (maintaining engine speed of less than 2500 rpm) and allow CVT to progress through all gear ratios.
- Allow transmission to remain in 6th gear for 3 seconds.
- Repeat light acceleration, allowing engine speed to increase to less than 2500 rpm and return to idle twice.
- Gradually apply foot brake to bring drive axles to a complete stop.
- Shift transmission to position R (Reverse).
- Release foot brake.
- Lightly press accelerator pedal (maintaining engine speed of less than 2500 rpm).

- Repeat light acceleration, allowing engine speed to increase to less than 2500 rpm and return to idle twice.
- Gradually apply foot brake to bring drive axles to a complete stop.
- Shift transmission to position P (Park).
- Switch off engine.
- Connect diagnostic equipment to measure CVT fluid temperature.
- Start engine.
- Check transmission fluid temperature.
- Leave engine running and raise vehicle.
- When fluid temperature is between 30°C and 50°C (86°F and 122°F), detach CVT fluid pump and special tool.

> ***CAUTION—***
> *Transmission fluid will be hot. Danger of scalding exists.*

- Allow excess fluid to drip from fill tube. Should fill tube run dry very quickly, CVT is probably under-filled. Reattach special tool 24 8 100 and add more fluid.
- Remove special tool 24 8 100 and wait for fluid drip to slow to ensure CVT is properly filled.
- Use new sealing washer and replace fill/level check plug.
- Lower car and turn off engine.
- Remove bolts holding rotors to hubs and replace road wheels.

> ***CAUTION—***
> *Attach drain and fill plugs properly and use new washers. Otherwise, as transmission temperature increases, CVT fluid will expand causing fluid leaks and transmission damage.*

Tightening torques	
CVT fluid drain plug to transmission	40 Nm (30 ft-lb)
CVT fluid fill plug to transmission	21 Nm (16 ft-lb)
Road wheel to hub	120 ± 10 Nm (89 ± 7 ft-lb)

Transmission fluid level, checking (automatic with Agitronic)

The 6-speed automatic transmission with Agitronic is filled with lifetime fluid from the factory. As a result there is no scheduled automatic transmission fluid change interval. The automatic transmission should be checked for fluid leaks.

Automatic transmission fluid			
Model, transmission	**Approximate capacity liter (US qt.)**	**When replacing oil sump liter (US qt.)**	**Oil specification**
Cooper S Aisin GA6F21	6.0 (4.8)	4.5 (4.8)	ESSO JWS-3309 ATF

The automatic transmission fluid (ATF) is checked by the overflow method which is temperature dependent. ATF level will vary according to temperature.

- Remove sound absorber panel from under engine and transmission.
- Remove heat shield from power steering gear.
- Make gear selector lever is in Park (P).
- Connect diagnostic scan tool to vehicle. See **OBD On Board Diagnostics** for more information.
- Raise car and support in a safe manner.
  - Make sure vehicle is sitting level

> ***WARNING —***
> *Make sure the car is stable and well supported at all times. Use a professional automotive lift or jack stands designed for the purpose. A floor jack is not adequate support.*

min240026

- Use BMW special tool 24 4 240 to remove ATF filler plug on transmission.
- Place a clean drip tray beneath transmission to catch fluid.

min240027

- Remove ATF inspection plug (**1**) from transmission sump.
- Top off ATF through fill hole until it overflows at inspection hole.
- Start engine and run at idle.
- Check and see if any more ATF runs out of inspection hole.
  - If not, continue to add ATF
- Press brake pedal and slowly shift from "P" to "D" twice. Put gear selector back in "P" and release brake pedal.
- Check and see if any ATF runs out of inspection hole.

– Read temperature of ATF with diagnostic scan tool.

Specification	
Operating temperature of ATF	35° - 45°C (95° - 113°F)

– With ATF at operating temperature, top off ATF until it runs out inspection hole.
- A gravity fed ATF filler tool is best to use for filling transmission.

– Install inspection and filler plugs to transmission.
- Replace sealing rings of plugs.

Tightening torque	
ATF fill plug to transmission (M14)	23.5 Nm + 10 Nm (17 ft-lb + 7 ft-lb)
ATF inspection plug to transmission (M10)	27 Nm (20 ft-lb)

Transmission fluid level, replacing (automatic with Agitronic)

The 6-speed automatic transmission with Agitronic is filled with lifetime fluid from the factory. As a result there is no scheduled automatic transmission fluid change interval. If it is necessary to drain the ATF, use a clean container to store and re-use the ATF.

Automatic transmission fluid			
Model, transmission	**Approximate capacity liter (US qt.)**	**When replacing oil sump liter (US qt.)**	**Oil specification**
Cooper S Aisin GA6F21	6.0 (4.8)	4.5 (4.8)	ESSO JWS-3309 ATF

– Remove sound absorber panel from under engine and transmission.

– Remove heat shield from power steering gear.

– Raise car and support in a safe manner.
- Make sure vehicle is sitting level

> ***WARNING —***
> *Make sure the car is stable and well supported at all times. Use a professional automotive lift or jack stands designed for the purpose. A floor jack is not adequate support.*

◄ Remove ATF inspection plug (**1**) from transmission sump.

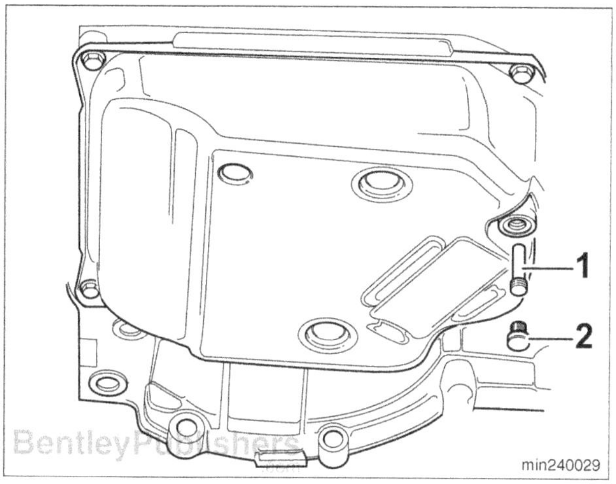

With inspection plug (**2**) removed, unscrew overflow tube (**1**) from inside inspection hole.

– Drain ATF into suitable container.

– When ATF is completely drained, reinstall overflow tube and tighten.

– Hand tighten inspection plug using new seal.

– Fill transmission with ATF through filler hole.
  - A gravity fed ATF filler tool is best to use for filling transmission.

– Check ATF level. See **Transmission fluid level, checking (automatic with Agitronic)** in this repair group.

– Install inspection and filler plugs to transmission.
  - Replace sealing rings of plugs.

Tightening torque	
ATF fill plug to transmission (M14)	23.5 Nm + 10 Nm (17 ft-lb + 7 ft-lb)
ATF inspection plug to transmission (M10)	27 Nm (20 ft-lb)

Transmission Service (CVT)

Back-up light switch, locating

The back-up light switch function for CVT models is controlled by the CVT switch. See **CVT switch**, replacing in this repair group.

CVT switch, replacing

The CVT switch is located on the bulkhead side of the CVT. The switch functions as a back-up light switch as well as the park/neutral switch. The park/neutral function prevents the vehicle from being started when the gear selector is not in "P" or "N".

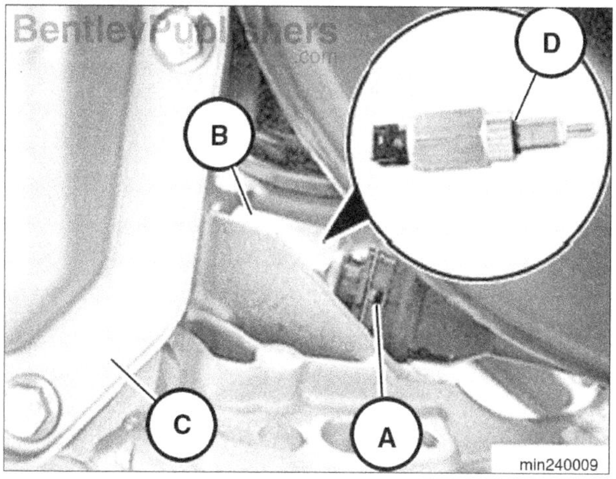

Working above the fluid pan (**C**), remove electrical connector (**A**) from the neutral safety switch (**B**).

– Remove switch. Be prepared to catch dripping fluid.

– Installation is reverse of removal. Remember to replace and lubricate sealing ring (**D**) before installing switch.

Tightening torque	
Neutral safety switch to transmission	12 Nm (9 ft-lb)

Drive axle seals, replacing

Left drive axle seal

– Remove left drive axle. See **310 Front Suspension, Drive Axles**.

NOTE —
Be prepared to catch dripping fluid.

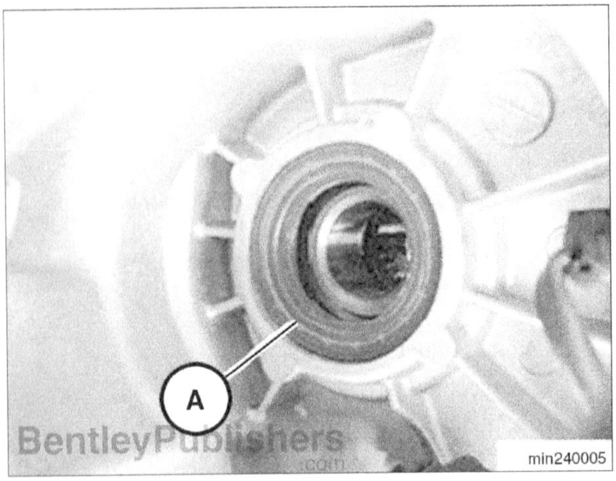

◀ Use a screwdriver to pry out drive axle seal (**A**) from transmission housing. Take care not to damage seal seating surface.

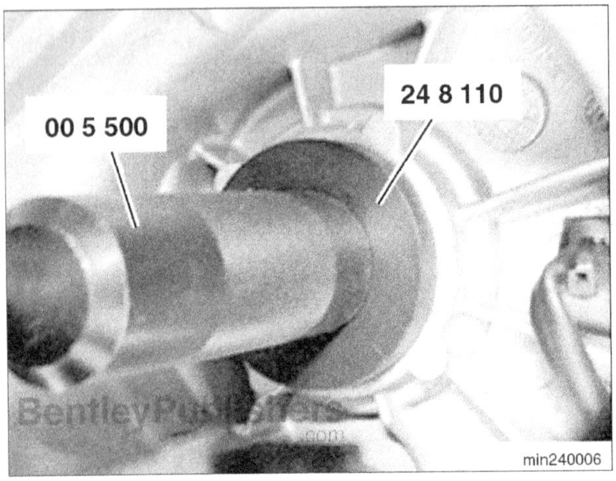

◀ Using special tools 00 5 500 and 24 8 110 tap new seal into case until outer seal ring is flush with transmission casing.

– Remainder of installation is reverse of removal. Remember to:
- Push drive axle past resistance of snap ring until an audible snap is heard.
- Fill transmission with approved fluid and check for leaks.

Right drive axle seal

– Remove right drive axle. See **310 Front Suspension, Drive Axles**.

NOTE —
Be prepared to catch dripping fluid.

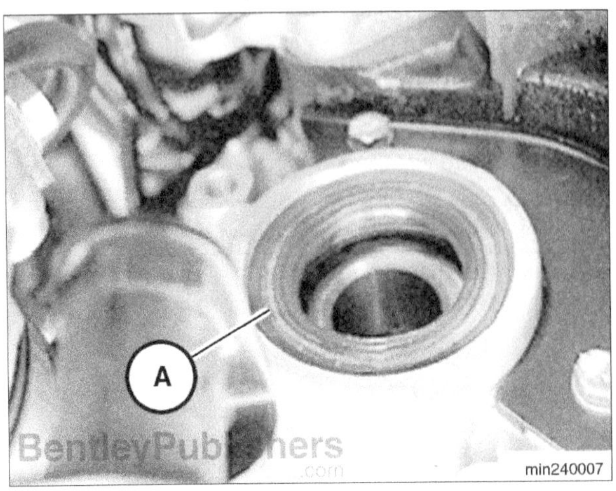

◀ Use a screwdriver to pry out drive axle seal (**A**) from transmission housing. Take care not to damage seal seating surface.

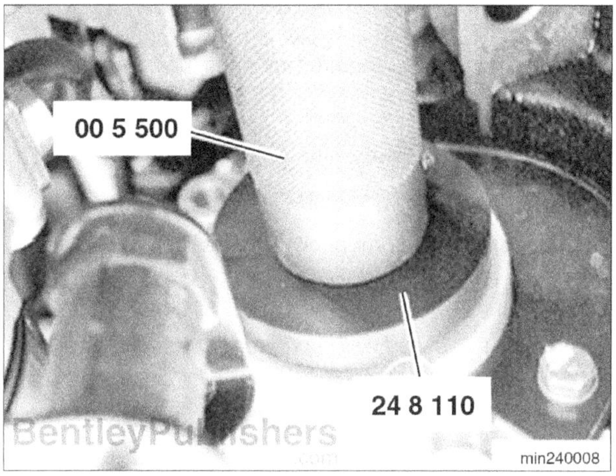

- Using special tools 00 5 500 and 24 8 110 tap new seal into case until outer seal ring is flush with transmission casing.

- Remainder of installation is reverse of removal. Remember to fill transmission with approved fluid and check for leaks.

Transmission control module, replacing

- Disconnect negative (-) cable from battery.

> ***CAUTION—***
> *Prior to disconnecting the battery, read the battery disconnection cautions given in* **001 Warnings and Cautions**.

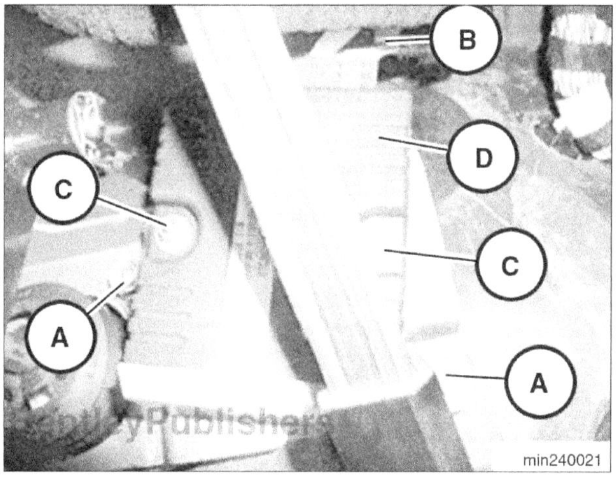

- Working under left side of dashboard at pedal cluster, remove bolts (**A**) from control module mounting bracket and remove electrical connector (**B**) from control unit (**D**).

- Installation is reverse of removal. If replacing control module remove nuts (**C**) holding unit to mounting bracket and reuse bracket.

- Check control module for stored faults. Reset adaptation values. Fix and clear faults if found. See **OBD On Board Diagnostics** for more information.

Tightening torque	
Control module to module bracket	2 Nm (18 in-lb)
Control module bracket to bulkhead	10 Nm (7 ft-lb)

Transmission fluid pan gasket, replacing

Removing the transmission pan requires lowering the front subframe approximately 10 mm (0.4 in).

- Drain transmission fluid. See **Transmission fluid, replacing (CVT)** in this repair group.
- Raise car and support in a safe manner.

> ***WARNING —***
> *Make sure the car is stable and well supported at all times. Use a professional automotive lift or jack stands designed for the purpose. A floor jack is not adequate support.*

- Remove left front wheel and left front wheel arch liner.
- Remove engine splash shield.
- Remove front bumper cover and bumper carrier. See **510 Exterior Trim, Bumpers** for more information.
- Remove crush tubes. See **510 Exterior Trim, Bumpers** for more information.

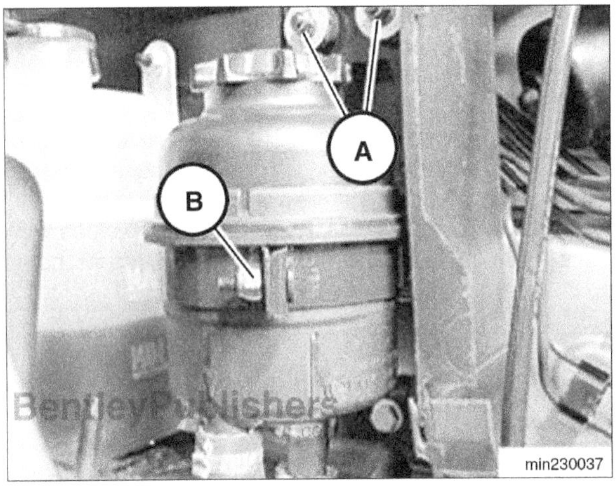

Working in engine compartment, unscrew power steering reservoir bracket bolts (**A** and **B**) and remove reservoir from bracket.

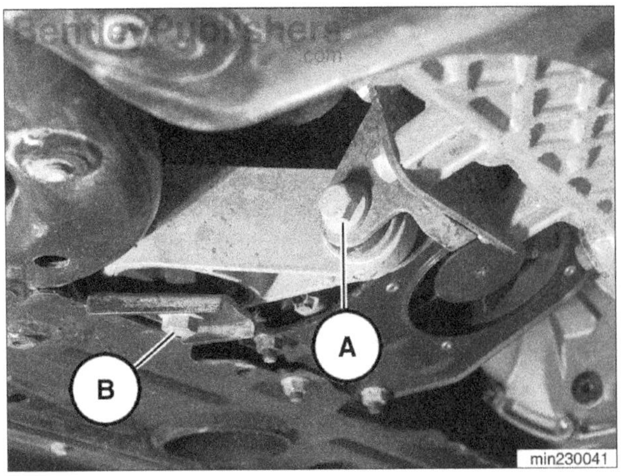

Remove lower engine vibration damper by removing bolts (**A** and **B**).

- Support subframe with a suitable tool (transmission jack) and secure frame to jack to prevent slipping.

∢ Unclip ABS wheel speed sensor wire and remove bolt (**A**) holding top front of subframe to chassis. Repeat for other side of subframe.

– For vehicles with Xenon headlights, disconnect vehicle height sensor at left control arm (as applicable).

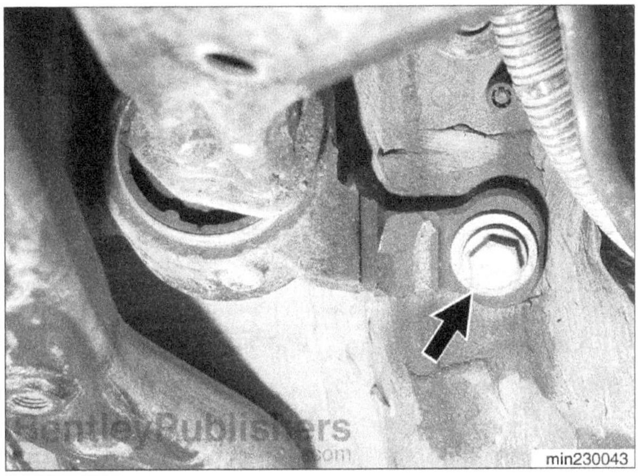

∢ Remove control arm bushing bracket bolt (**arrow**) from control arm bushing bracket. Repeat for other side of subframe.

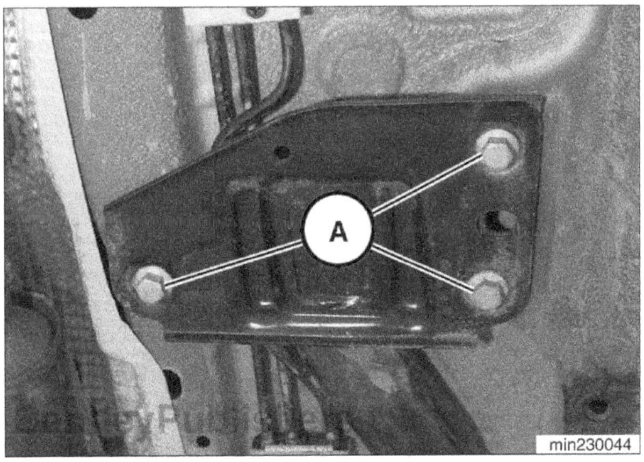

∢ Loosen bolts (**A**) at rear of subframe and lower subframe approximately 10 mm (0.4 in) using hydraulic jack.

> ***WARNING—***
> - *Do not remove rear subframe bolts.*
> - *Front subframe is heavy. Support with a suitable tool (transmission jack) and secure frame to jack to prevent slipping.*

> ***CAUTION—***
> *Lower front subframe slowly making sure no strain is placed on power steering reservoir or other components.*

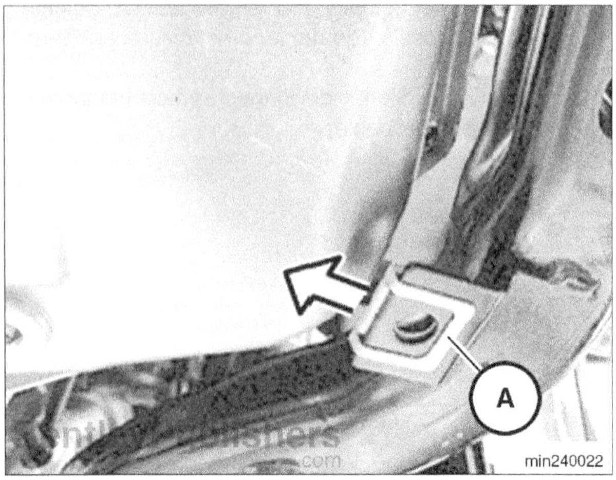

◄ Remove engine splash shield retaining clip (**A**) from front subframe.

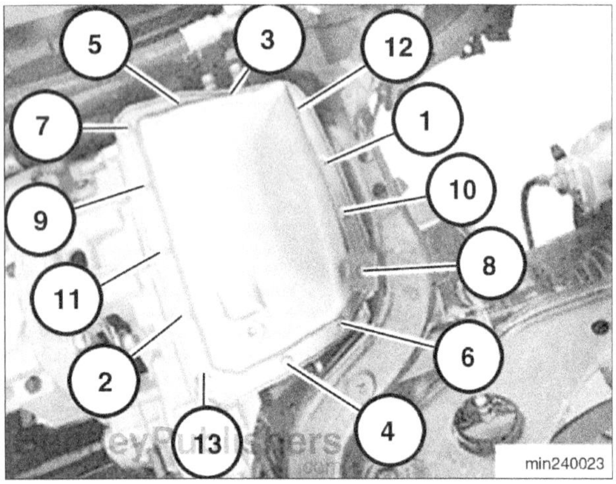

◄ Remove transmission pan retaining bolts in a cross pattern (13 - 1) and remove pan.

NOTE—
Be prepared to catch dripping fluid.

– Remove fluid pan gasket and clean sealing surfaces.

– Installation is reverse of removal. Keep in mind the following:
  - Use new transmission pan gasket.
  - Tighten pan retaining bolts in a cross pattern (1 - 13).
  - Tighten fastener at control arm bushing bracket only after all other fasteners on front subframe have been tightened down. This will prevent unwanted torsional stress on the control arm bushing.
  - Replace all self-locking fasteners.
  - Check and top off transmission with approved fluid.
  - Perform an electronic wheel alignment check.

Tightening torques	
Control arm bracket to chassis (replace bolts) Tighten final 90° with vehicle on ground	59 Nm (44 ft-lb) plus 90° ± 15°
Front subframe front mount to chassis	100 Nm (74 ft-lb)
Front subframe rear mount to chassis	100 Nm (74 ft-lb)
Lower engine vibration damper to engine bracket	100 Nm (74 ft-lb)
Lower engine vibration damper to front subframe bracket	100 Nm (74 ft-lb)
Power steering reservoir bracket to bulkhead	8 Nm (6 ft-lb)
Power steering reservoir to bracket	5.5 Nm (4 ft-lb)
Transmission fluid pan to transmission	9 Nm (6 ft-lb)

Transmission fluid screen, replacing

- The transmission fluid screen is located inside the transmission fluid pan.
- Remove transmission fluid pan. See **Transmission fluid pan gasket, replacing** in this section.
- Pull down on transmission fluid screen to remove. Note position of screen on valve block.
- Clean magnet on fluid screen and transfer magnet to replacement screen.
- Lubricate sealing ring with CVT fluid and replace screen by pressing up into valve block.
- Install transmission fluid pan.
- Remainder of installation is reverse of removal. Fill transmission with approved fluid and check for leaks. See **Transmission fluid, replacing (CVT)** in this repair group.

Vehicle speed sensor, replacing

- Disconnect negative (-) cable from battery.

> ***CAUTION—***
> *Prior to disconnecting the battery, read the battery disconnection cautions given in* **001 Warnings and Cautions**.

- Remove battery box.

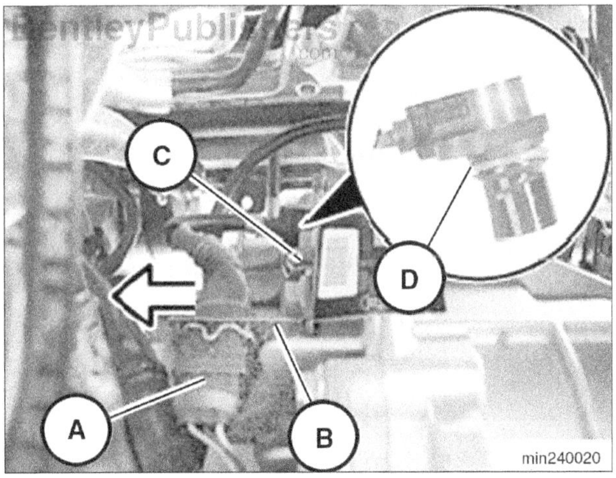

min240020

- ◄ Remove electrical connector (**A**) from speed sensor and remove connector from bracket (**B**).
- Remove speed sensor retaining bolt (**C**) and remove sensor.
- Installation is reverse of removal. Use new sealing washer (**D**) and lubricate with CVT fluid.
- Check ECM for stored faults. Fix and clear faults if found.

Tightening torque	
Speed sensor to transmission	9 Nm (7 ft-lb)

Transmission Service (Automatic with Agitronic)

Replacing the automatic transmission fluid (ATF) strainer requires the engine to be supported and raised from below. Due to the complexity of this job it is not covered in this manual.

Back-up light switch, locating

The back-up lights are directly supplied with voltage (12 V) by the gear position switch. See **Gear position switch, replacing** later in this repair group.

Drive axle seals, replacing

– Drain automatic transmission fluid. See **Transmission fluid, replacing (automatic with Agitronic)** earlier in this repair group.

– Remove drive axle. See **310 Front Suspension, Drive Axles**.

NOTE—
- *Be prepared to catch dripping fluid.*
- *Left side drive axle shown, right side is similar.*

– Use a center punch to drive a small hole into drive axle seal.

> ***CAUTION—***
> *Do not use a drill to create hole in drive axle seal. Metal shavings could result in transmission failure.*

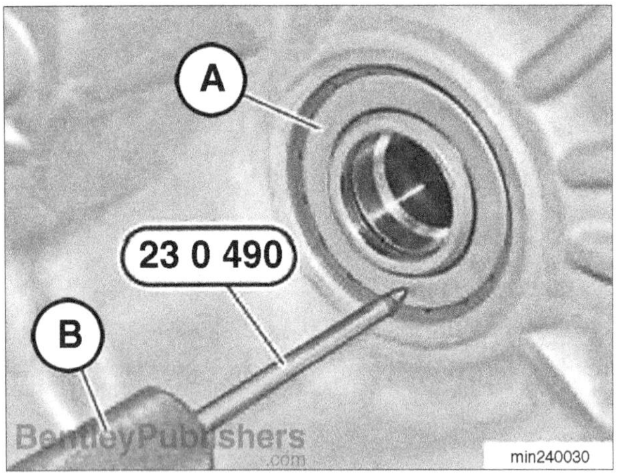

◄ Screw the tip of BMW special tool 23 0 490 into seal (**1**) and pull out using impact weight (**2**).

– Before installing new seal, coat sealing surfaces with clean ATF.

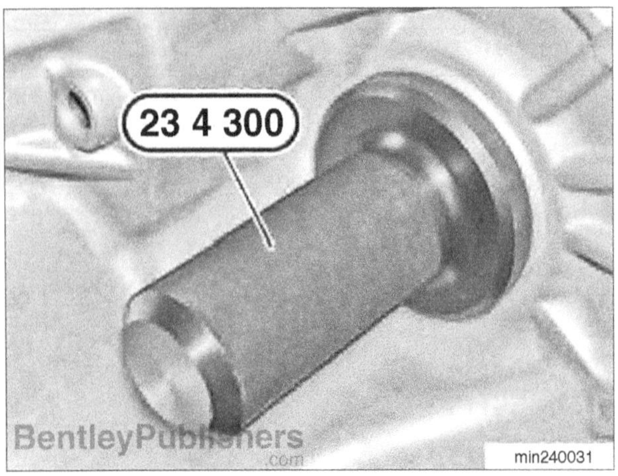

◄ Install seal with BMW special tool 24 4 300.
- Seat seal firmly

– Remaining installation is reverse of removal.

Gear position switch, replacing

NOTE—
The back-up (reverse) lights are supplied with 12V (vehicle voltage) directly by the gear position switch.

– Remove air filter assembly and intake hoses. See **130 Fuel Injection**.

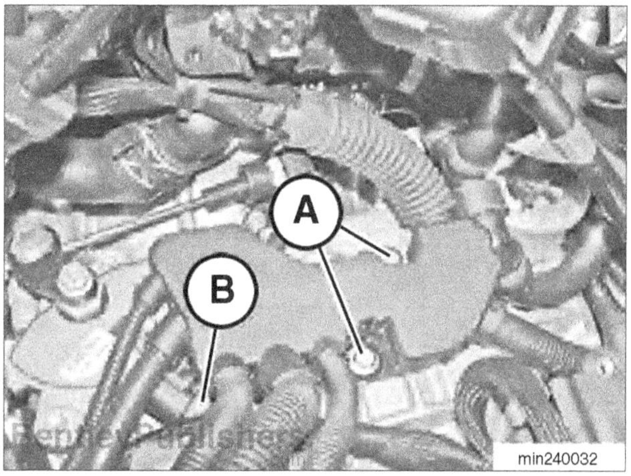

< Remove mounting screws (**A**) and cover.

– Disconnect gear position switch harness connector (**B**).

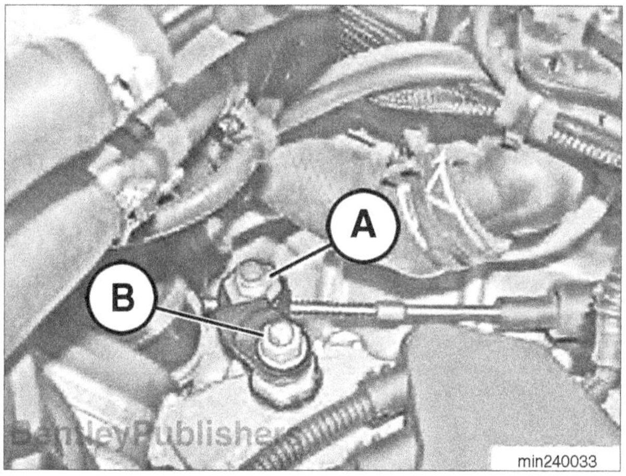

< Release cable lock nut (**A**).

– Counterhold cable bracket and remove mounting nut (**B**). Remove bracket.

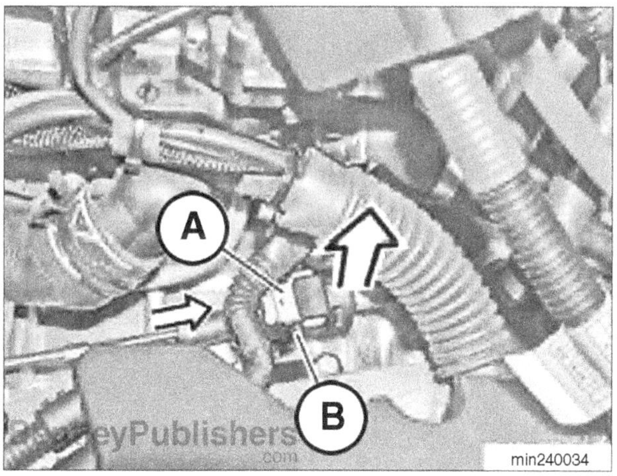

< Squeeze retaining clip (**A**) and remove cable from mounting fixture (**B**).

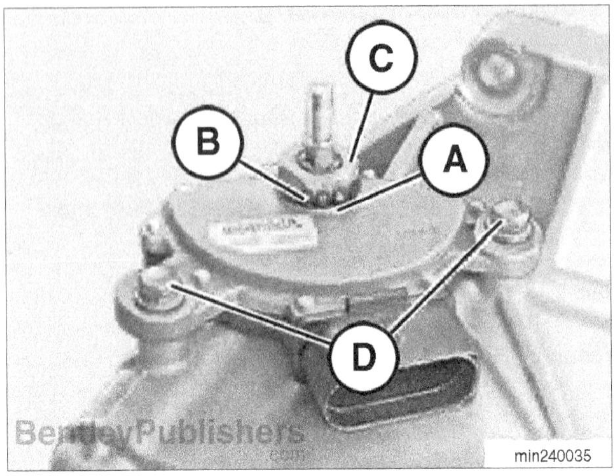

◀ Remove nut (**C**), retaining ring (**B**) and shim (**A**).

– Remove mounting bolts (**D**) and gear position switch.

Installing and adjusting

– Install gear position switch in reverse order or removal. Do not tighten mounting bolts (**D**) completely until switch has been adjusted.

Tightening torques	
Shift cable bracket (M8)	12 Nm (9 ft-lb)
Gear position switch mounting nut (M14)	7 Nm (62 in-lb)
Gear position switch to transmission (M6)	5.5 Nm (48 in-lb)

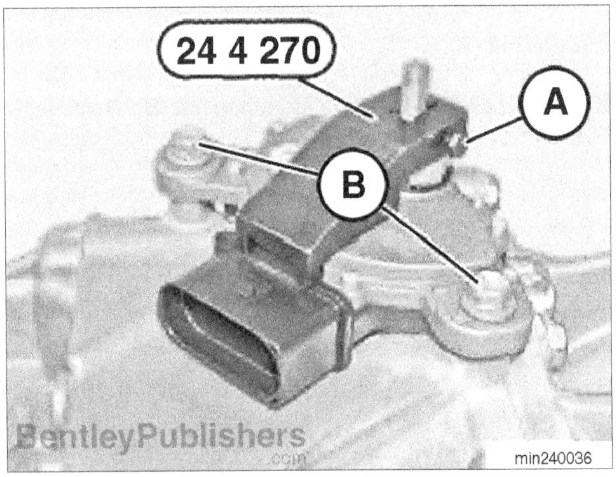

◀ Position BMW special tool 24 4 270 over gear position switch as shown. Tighten screw (**A**).

– Align gear position switch to "N" (neutral). Tighten mounting bolts (**B**).

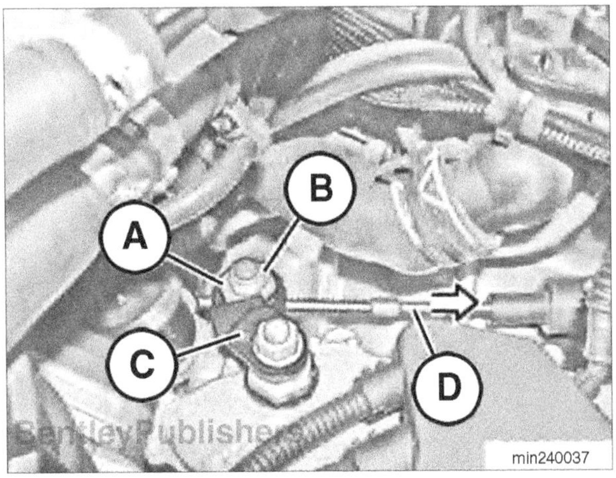

– Make sure selector lever is in "P" (park) position.

◀ Grip clamping sleeve (**A**) and loosen nut (**B**).

– Move shift cable bracket (**C**) park position, then release again.

– Press cable (**D**) in direction of arrow, then release again.

– Grip clamping sleeve (**A**) and tighten cable mounting nut (**B**).

– Remaining installation is the reverse of removal.

– When finished, road test vehicle and note any shifting problems.

Input shaft seal, replacing

– Remove transmission. See **Transmission Removal and Installation** in this repair group.

– Remove torque converter. See **Torque converter, removing and installing** in this repair group.

◄ Remove input shaft seal using universal puller (BMW tool 00 5 010 shown).

– Clean sealing surfaces before installing new seal.

– Lubricate sealing lip of new seal with clean ATF.

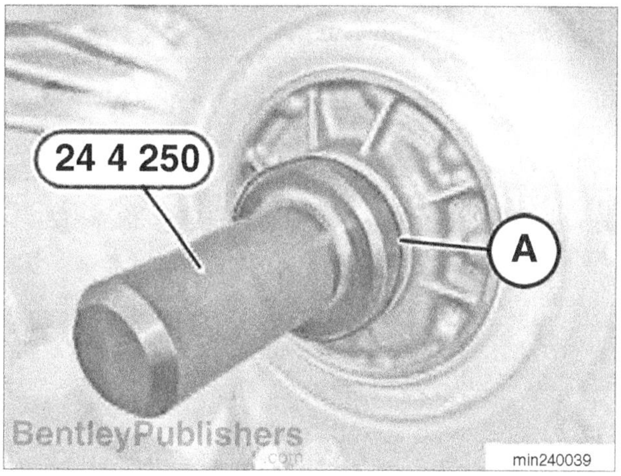

◄ Install new seal using installation tool (BMW tool 24 4 250).

- Place seal (**A**) on installation tool and press seal fully into position.

Transmission control module (EGS), replacing

– Disconnect negative (-) cable from battery.

> ***CAUTION—***
> *Prior to disconnecting the battery, read the battery disconnection cautions given in* **001 Warnings and Cautions**.

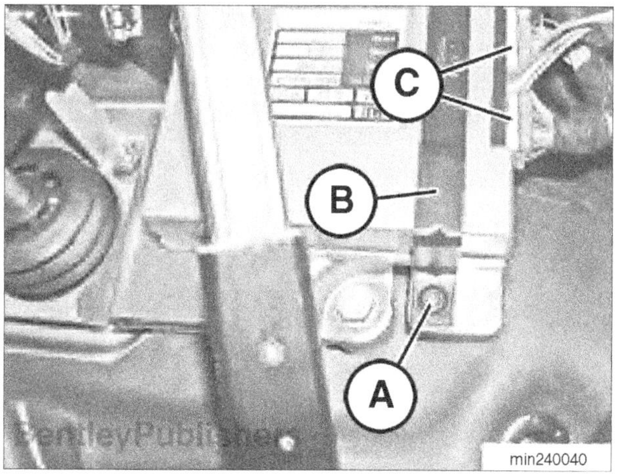

◄ Working under left side of dashboard at pedal cluster, remove mounting bolt (**A**) and bracket (**B**).

– Disconnect harness connector (**C**) and remove control unit.

– Installation is reverse of removal.

– Check control module for stored faults. Reset adaptation values. Fix and clear faults if found. See **OBD On Board Diagnostics** for more information.

Tightening torque	
Control module to module bracket	2 Nm (18 in-lb)
Control module bracket to bulkhead	10 Nm (7 ft-lb)

Vehicle speed sensors

Accessing the input and output vehicle speed sensors requires removal of the transmission hydraulic unit and sump. These operations are beyond the scope of this manual.

TRANSMISSION REMOVAL AND INSTALLATION (CVT)

CVT removal and installation requires engine removal. See **110 Engine Removal and Installation**.

Transmission, removing and installing

– Disconnect negative (-) cable from battery.

> ***CAUTION*** *—*
> *Prior to disconnecting the battery, read the battery disconnection cautions given in* **001 Warnings and Cautions**.

– Remove Engine. See **110 Engine Removal and Installation**.

– Remove throttle valve assembly.

– Remove evaporative purge pipe from clips on injector rail and remove pipe from inspection cover.

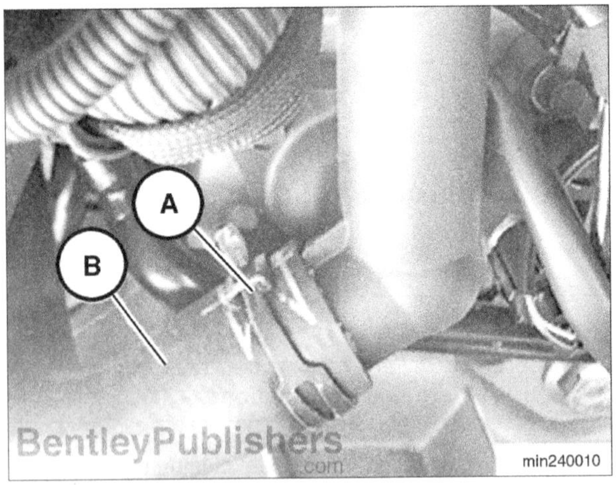

min240010

◄ Remove clip (**A**) from lower radiator hose (**B**) and remove hose from thermostat housing.

min240011

◄ Remove electrical connector from speed sensor (**A**) and neutral safety switch (**B**). Remove wiring harnesses and note wire routing.

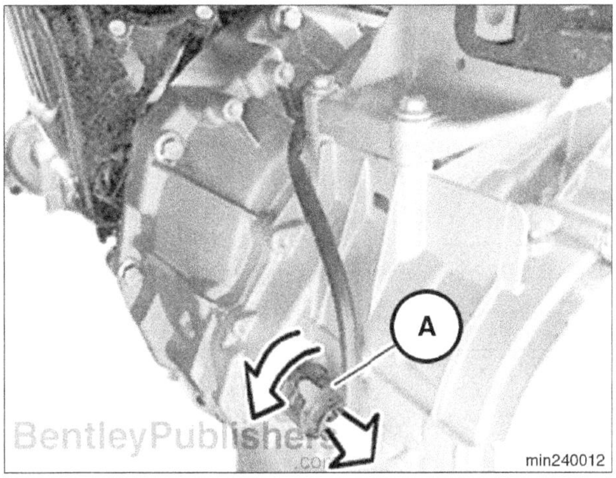

Remove electrical connector (**A**) from transmission switch and remove wiring harness. Note wire routing.

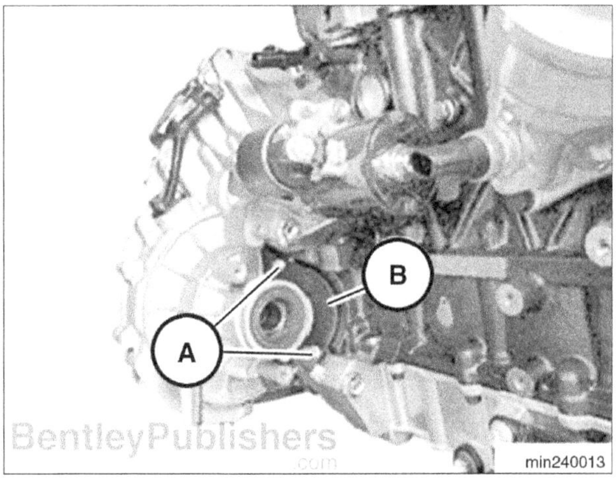

Remove cover plate retaining bolts (**A**) and remove cover plate (**B**).

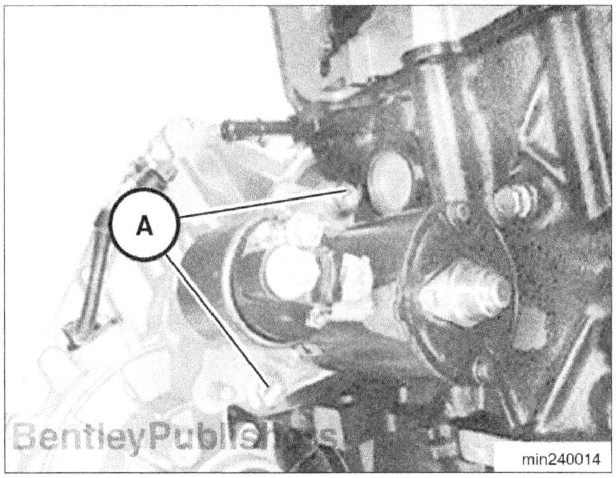

Remove starter motor bolts (**A**) and remove starter motor.

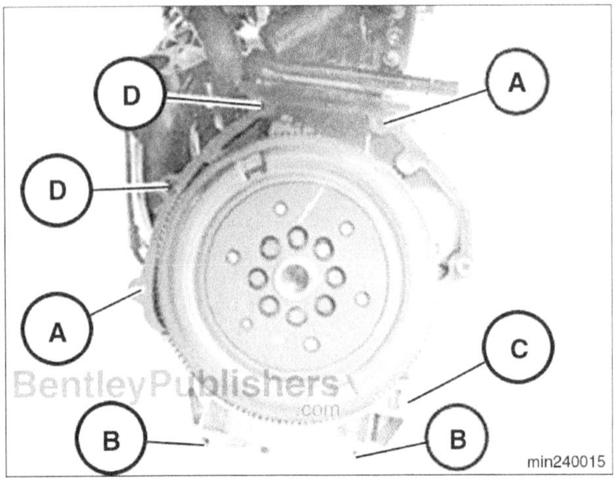

Remove transmission bolts and remove transmission. Note size and location of bolts:

- Bolt **A** = 50 mm (2 in)
- Bolt **B** = 60 mm (2 ¼ in)
- Bolt **C** = 65 mm (2 ½ in)
- Bolt **D** = 75 mm (3 in)

NOTE —
Flywheel side of transmission is shown for clarity. Bolt heads are on opposite side.

> ***CAUTION —***
> *Install bolts in original position. Serious damage may result from incorrect installation.*

– Installation is reverse of removal. Remember to fill transmission with proper fluid and check for leaks.

Tightening torques	
Cover plate to transmission	9 Nm (7 ft-lb)
Starter motor to transmission	85 Nm (63 ft-lb)
Transmission to engine	85 Nm (63 ft-lb)

– Check control module for stored faults. Reset adaptation values. Fix and clear faults if found. See **OBD On Board Diagnostics** for more information.

Transmission Removal and Installation (Automatic with Agitronic)

Automatic transmission removal and installation requires engine removal. See **110 Engine Removal and Installation**.

Transmission, removing and installing

– Disconnect negative (-) cable from battery.

> ***CAUTION —***
> *Prior to disconnecting the battery, read the battery disconnection cautions given in* **001 Warnings and Cautions**.

– Remove Engine. See **110 Engine Removal and Installation**.

– Remove throttle valve assembly.

– Remove evaporative purge pipe from clips on injector rail and remove pipe from inspection cover.

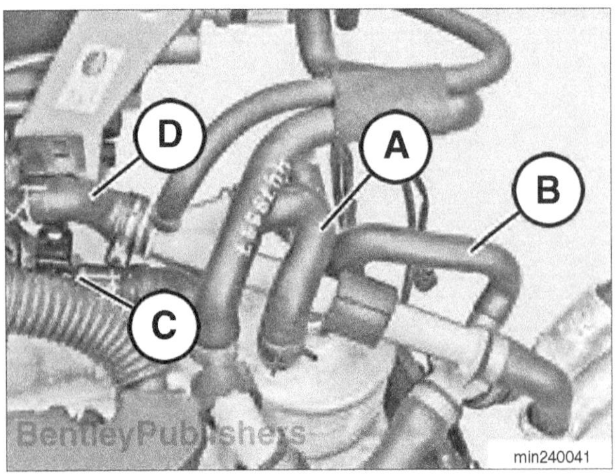

Disconnect coolant hoses (**A** and **B**) from oil cooler. Disconnect coolant hoses (**C** and **D**) from coolant pipe.

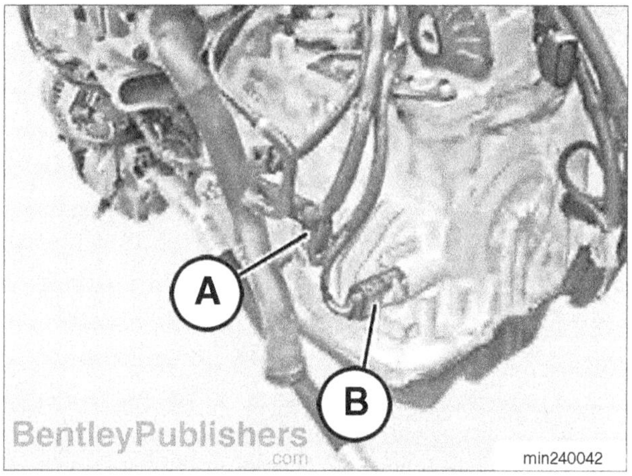

◄ Disconnect harness connectors (**A** and **B**) for transmission control unit.

– Position wiring harness out of way.
  - Pay attention to routing of harness.

◄ Disconnect oil pressure switch plug (**A**).

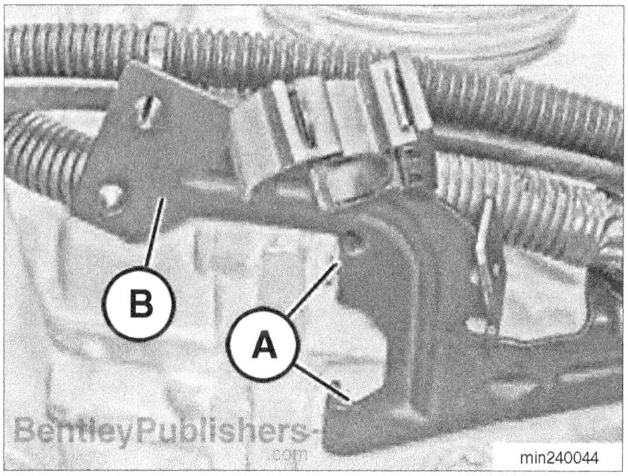

◄ Remove mounting nuts (**A**) and bracket (**B**) with wiring harness.

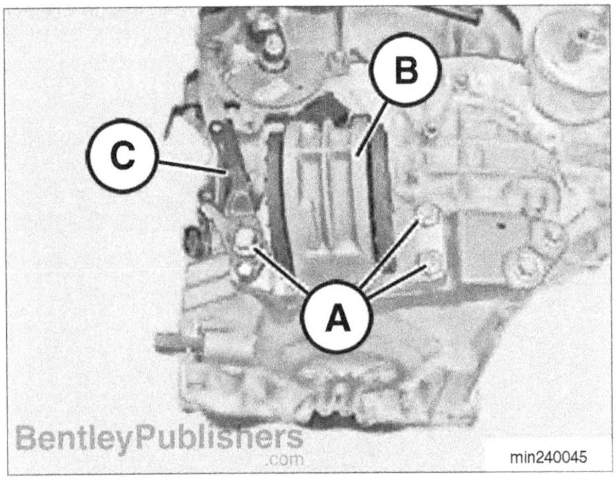

◀ Remove mounting bolts (**A**) and transmission support block (**B**) and bracket (**C**).

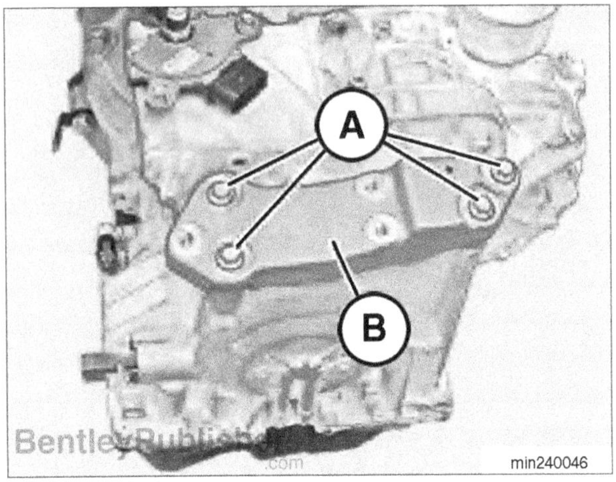

◀ Remove mounting bolts (**A**) and transmission carrier (**B**).

– Remove torque converter-to-driveplate mounting nuts (quantity 6).
  - Mounting nuts are accessed through hole in transmission housing.
  - Turn crankshaft to access each torque converter mounting nut.

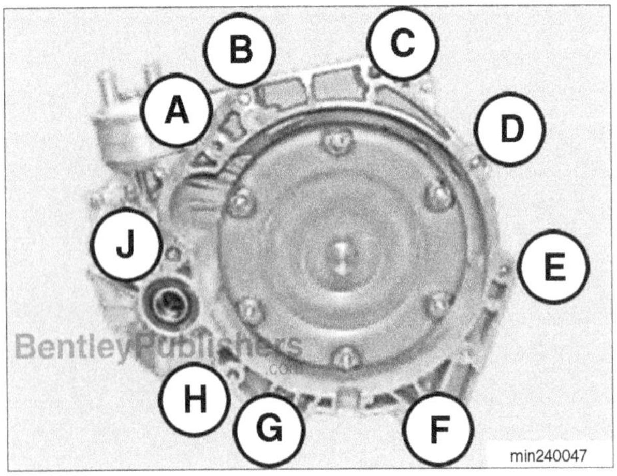

◀ Remove transmission-to-engine bolts (**A** through **J**).

– Installation is reverse of removal. Check that dowel sleeves are correctly seated. Replace any damaged dowel sleeves.

Tightening torques	
Support block to transmission (M10) Cooper S Convertible (M12) Cooper S	 38 Nm (28 ft-lb) 66 Nm (49 ft-lb)
Transmission carrier to transmission (M12)	68 Nm (50 ft-lb)
Transmission support bracket to body (M12)	68 Nm (50 ft-lb)
Starter motor to transmission	85 Nm (63 ft-lb)
Transmission to engine	82 Nm (60 ft-lb)

NOTE—
- *Transmission-to-engine mounting bolts differ in length.*
- *Note installation position.*
- *Installing the wrong bolts may cause serious damage.*

Torque converter, removing and installing

– Remove transmission. See **Transmission, removing and installing** in this repair group.

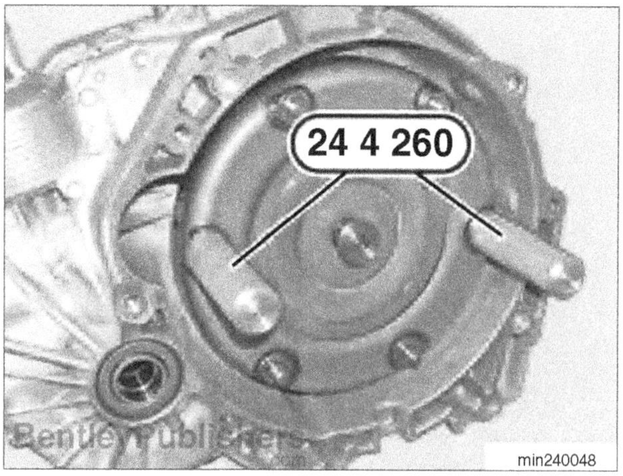

min240048

◄ Install special tools 24 4 260 onto two of the torque converter mounting studs.

– Remove torque converter by pulling on handles of removal tools.

- Place torque converter on clean surface so that fluid does not run out.

Installing

> ***CAUTION—***
> - *Do not damage shaft seal and bearing when installing torque converter.*
> - *If torque converter is not correctly installed, the driver of the pump impeller may be damaged when the transmission is flanged to the engine.*

– To install, push torque converter onto input shaft as far as it will go. While pushing, rotate the torque converter by hand.

- Torque converter hub opening must snap into place in driver of pump impeller.
- Torque converter must be felt to slip inward.

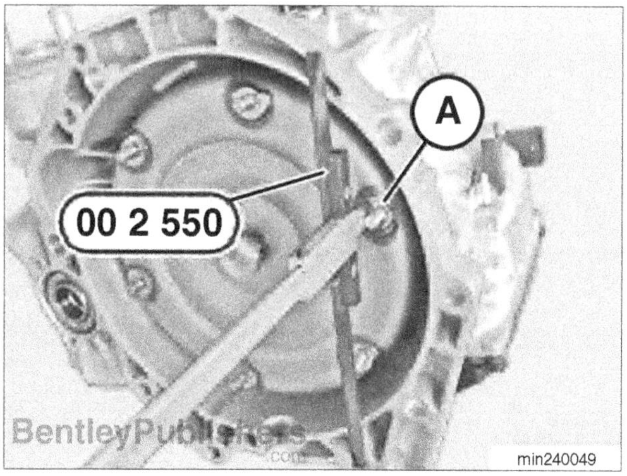

min240049

◄ Use depth gauge (BMW special tool 00 2 550) to measure distance between transmission-to-engine contact surface and torque converter mounting stud shoulder (**A**).

- Re-position torque converter if distance exceeds specification.

Specification	
Torque converter clearance (maximum permissible)	17.5 mm (0.69 in)

250 Gearshift Linkage

GENERAL

This repair group covers transmission gearshift and linkage service for both manual and automatic transmission (CVT or automatic with Agitronic) equipped vehicles.

To gain access to the complete gearshift mechanism it is necessary to remove part of the exhaust system as described in **180 Exhaust System**.

Special tools

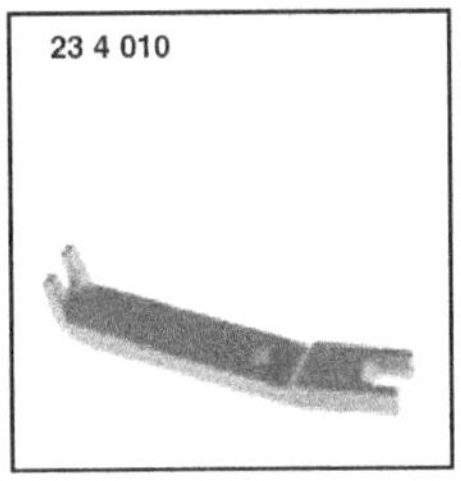

Manual transmission linkage tool (BMW tool 23 4 010)

Automatic transmission shift lock

When the automatic transmission selector lever is in PARK or NEUTRAL and the ignition is switched on, the engine control module (ECM) locks the shifter in position by energizing a shift lock solenoid. The selector lever cannot be moved into any other position until the brake is applied.

TRANSMISSION GEARSHIFT

Removing and installing the gearshift knob and dust boot is similar for all models.

Gearshift knob, removing and installing

- Move manual gearshift to neutral. Position automatic gear selector in Park.

- Remove shift knob by pulling knob straight off shift lever.
  - Manual transmission shown. Shift knob for automatic transmission is removed the same way.

NOTE—
Do not twist knob or locating key can be damaged.

- Installation is reverse of removal. Remember to align gearshift knob with locating key on gearshift lever and push knob straight down on lever without twisting.

Gearshift dust boot, removing and installing

- Remove gearshift knob. See **Gearshift knob, removing and installing** in this repair group.
- Pull dust boot trim up (**arrows**) to release from console.
- Cut wire tie holding boot to shaft and remove dust boot.

Automatic transmission (CVT or Agitronic)

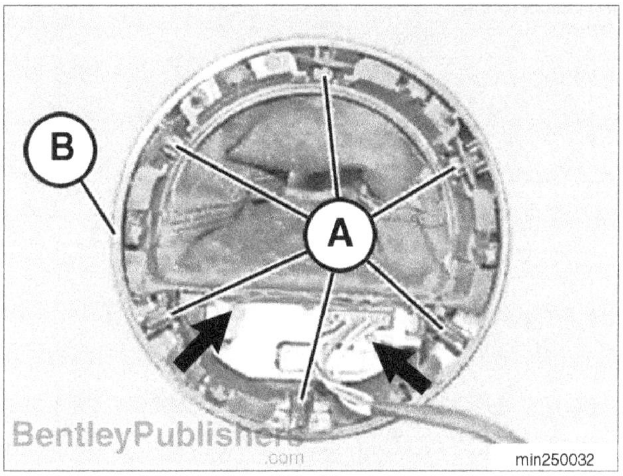

- Disconnect harness connector for gear selector display.
- Unlock detent lugs (**A**) of mounting ring. Remove mounting ring (**B**).
- If replacing dust boot, remove mounting screws (**arrows**) for gear selector display. Transfer gear selector display to new dust boot assembly.
- Install mounting ring.

Continued for all vehicles

- Installation is reverse of removal. Remember to align locating pins on dust boot trim with slots in center console.

Front center console, removing and installing

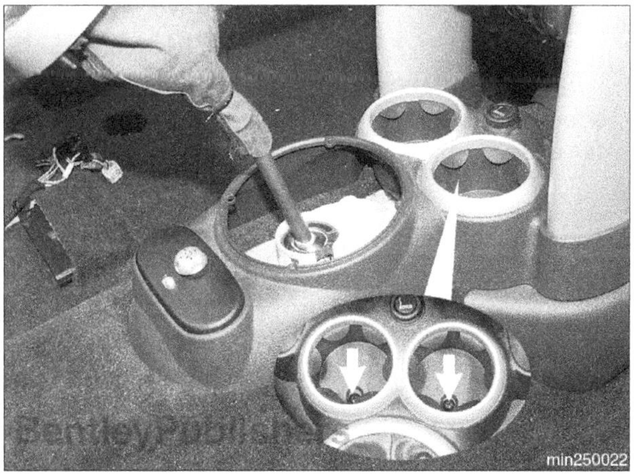

- Remove gearshift knob. See **Gearshift knob, removing and installing** in this repair group.
- Remove gearshift dust boot. See **Gearshift dust boot, removing and installing** in this repair group.
- Remove ashtray, if equipped, by pulling up from cup holder. Remove screws (**inset arrows**) in bottom of cup holders.

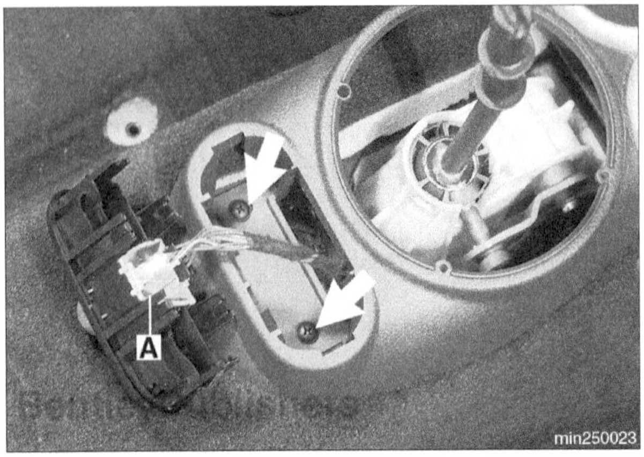

< Gently pry switch plate up and remove electrical connections (**A**) for power mirrors and heated seats (if applicable).

– Remove screws (**arrows**) in base of console.

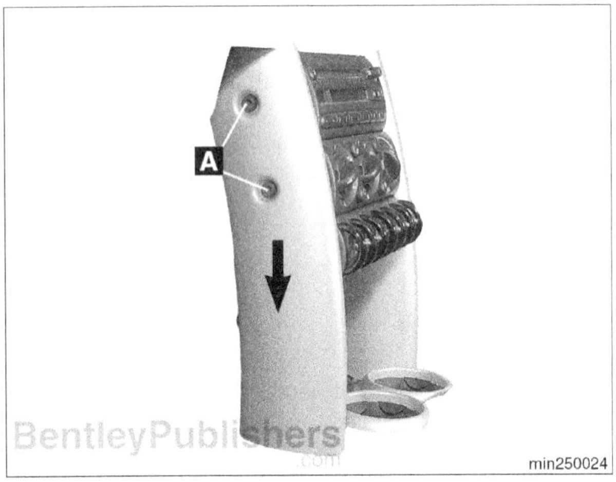

< Remove screws (**A**) on side fascia and pull fascia down slightly. Repeat for other side fascia.

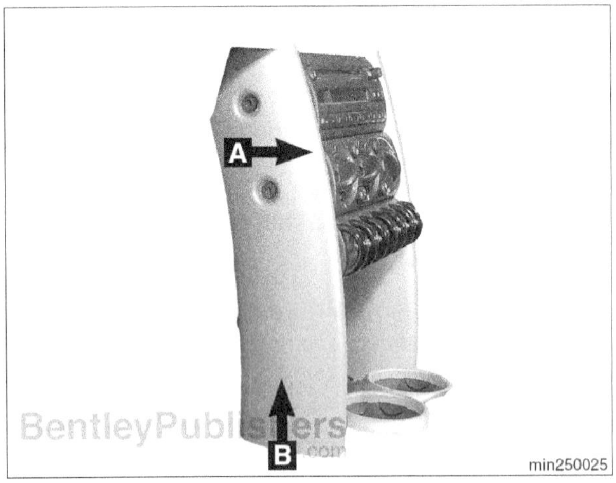

< Slide center console and side fascia (**A**) back as a unit until both side fascia can be pulled out of console (**B**).

– Remove electrical connectors from power/lighter socket and remove center console.

– Installation is reverse of removal. Make sure foam blocks at base of fascia are correctly positioned, and that no electrical wires are pinched or damaged when replacing console.

Gearshift Assembly

Gearshift assembly, removing and installing (manual transmission)

- Remove gearshift knob. See **Gearshift knob, removing and installing** in this repair group.
- Remove gearshift dust boot. See **Gearshift dust boot, removing and installing** in this repair group.
- Raise vehicle to gain access to underside of vehicle.

> ***WARNING —***
> *Make sure car is stable and well supported at all times. Use a professional automotive lift or jack stands designed for the purpose. A floor jack is not adequate support.*

- Remove rear section of exhaust system. See **180 Exhaust System**.

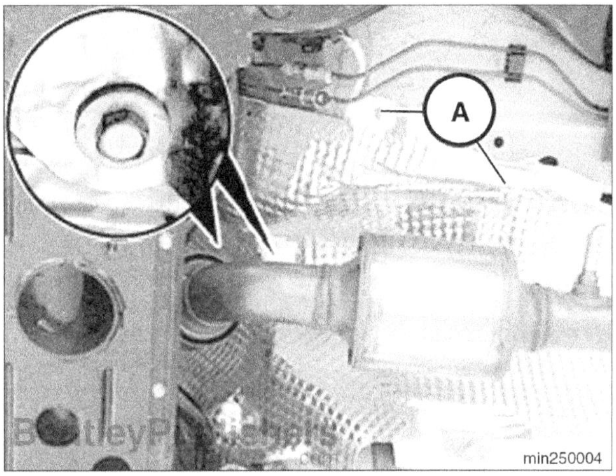

min250004

◄ Remove 4 upper heat shield retaining bolts (**A** and **inset**) from front heat shield.

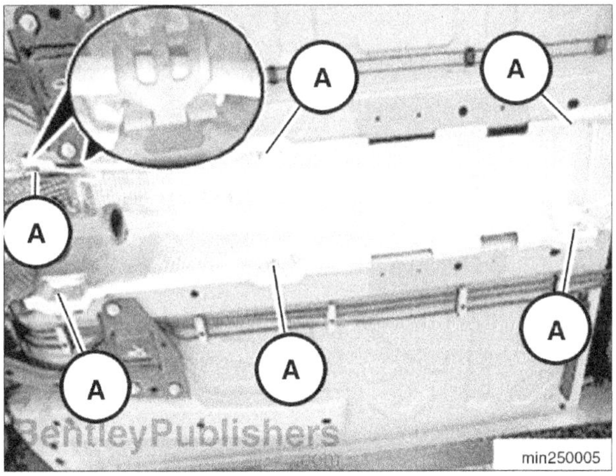

min250005

◄ Remove center heat shield retaining bolts (**A**) and remove center heat shield.

- Release shift cables from clips underneath heat shield.

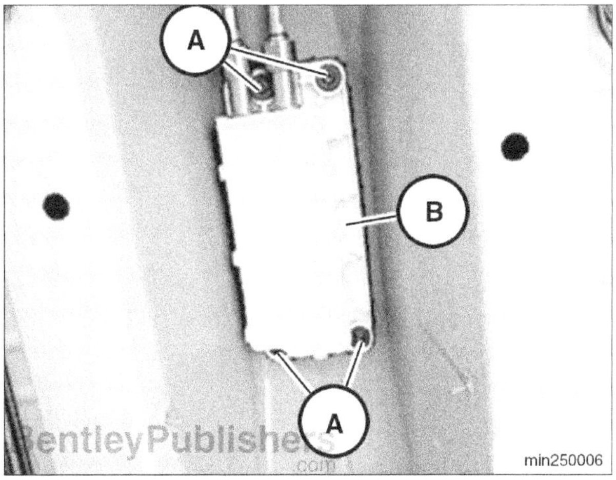

min250006

◄ Remove bolts (**A**) securing gearshift assembly (**B**) to underside of vehicle.

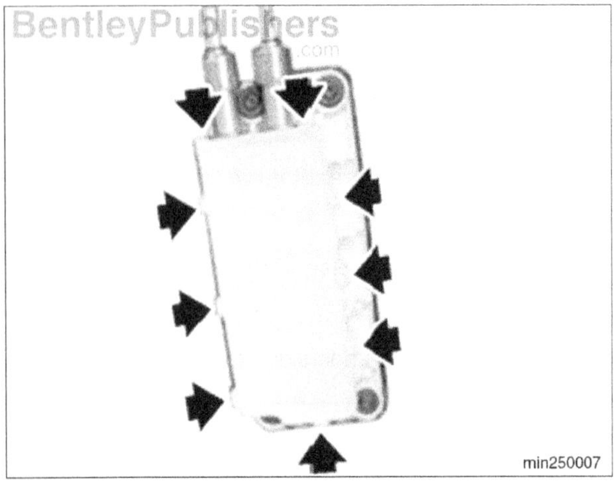

min250007

◀ Carefully release locking tabs (**arrows**) and remove cover from gearshift assembly.

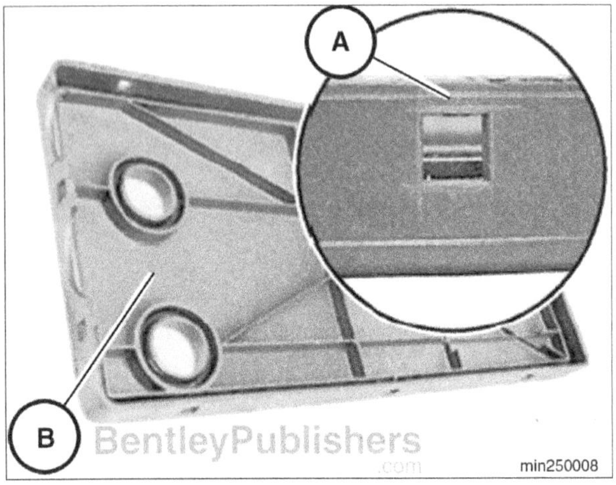

min250008

◀ Make sure that locking tabs (**A**) and seals on cover (**B**) are not damaged. Replace seals or cover if damaged.

> ***WARNING —***
> *Gearshift cover locking tabs and seals must be in good condition to prevent road debris from damaging the gearshift mechanism.*

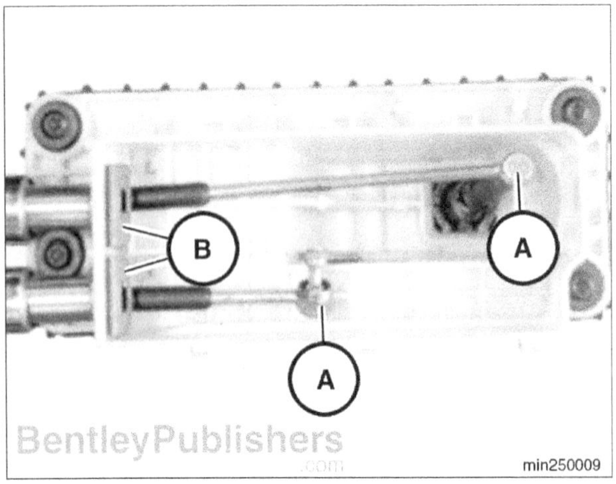

min250009

◀ Release shift cables (**A**) with suitable tool.

– Remove clips (**B**) and remove cables from gearshift assembly.

– Installation of new assembly is reverse of removal.

Shift cables, removing and installing (manual transmission)

– Remove gearshift lever assembly. See **Gearshift assembly, removing and installing (manual transmission)**.

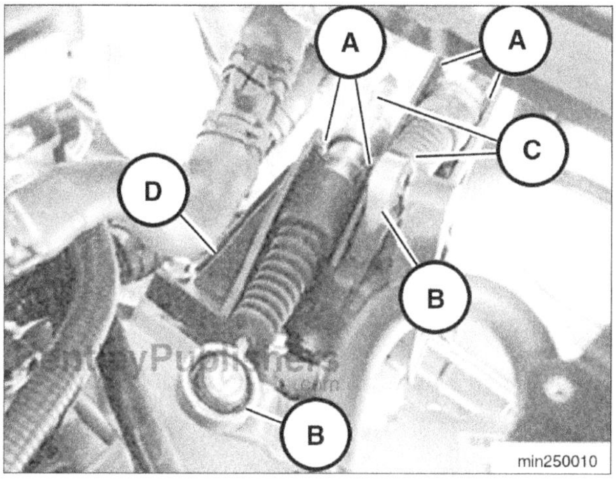

Cooper models:

– Remove battery box. See **121 Battery, Starter and Alternator** for more information.

◄ Working at transmission, remove cable clips (**A**) at shift cable bracket (**D**). Release shift cable ball joints (**B**) using a suitable tool and pull cables (**C**) out of bracket.

– Installation is reverse of removal.

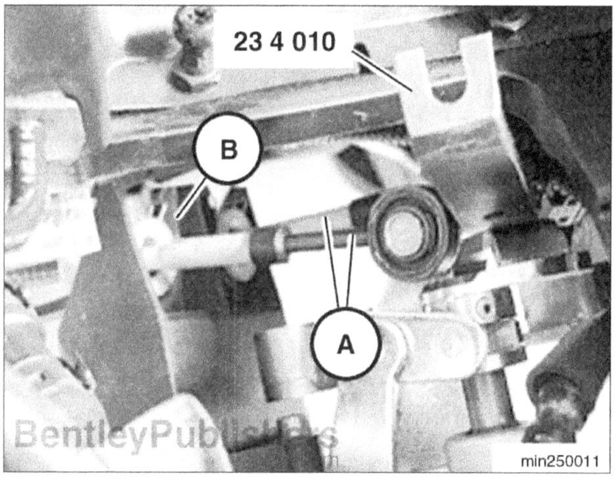

Cooper S models:

– Remove air intake housing. See **130 Fuel Injection** for more information.

◄ Working at transmission, release shift cable ball joints (**A**) using special tool 23 4 010. Release cables from bracket (**B**) and pull cables out of bracket.

– Installation is reverse of removal.

Gearshift assembly, removing and installing (CVT and automatic transmission)

– Remove gearshift knob. See **Gearshift knob, removing and installing** in this repair group.

– Remove gearshift dust boot. See **Gearshift dust boot, removing and installing** in this repair group.

– Remove transmission interlock cable. See **320 Steering and Wheel Alignment**.

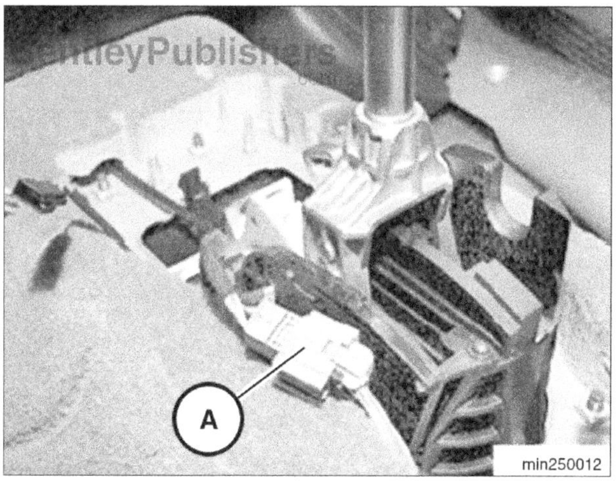

◄ Disconnect harness connector (**A**) from gearshift position switch.

– Raise vehicle to gain access to underside of vehicle.

> ***WARNING—***
> *Make sure car is stable and well supported at all times. Use a professional automotive lift or jack stands designed for the purpose. A floor jack is not adequate support.*

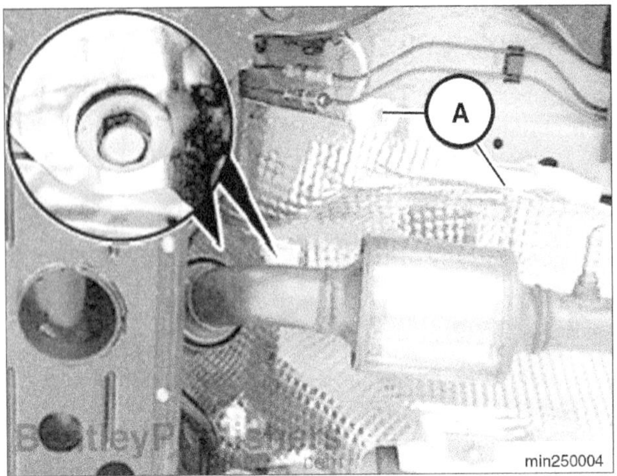

- Remove rear section of exhaust system. See **180 Exhaust System**.
- Remove 4 upper heat shield retaining bolts (**A** and **inset**) from front heat shield.

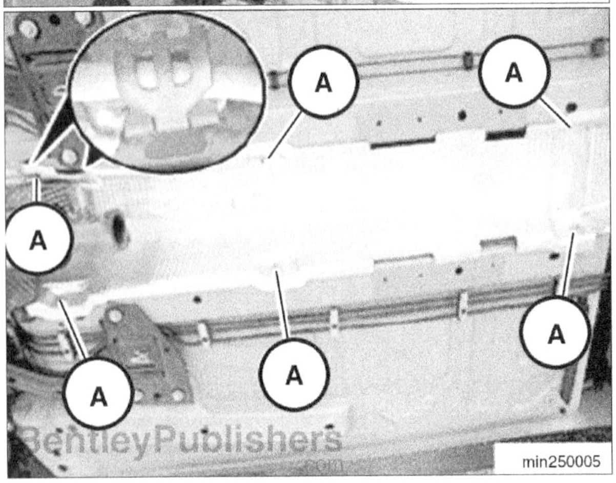

- Remove center heat shield retaining bolts (**A**) and remove center heat shield.
- Release shift cable from clips underneath heat shield.

Automatic transmission with Agitronic (Cooper S)

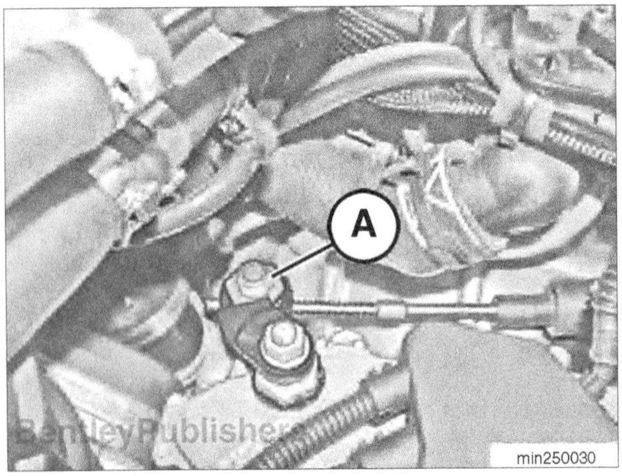

- Loosen shift cable lock nut (**A**).

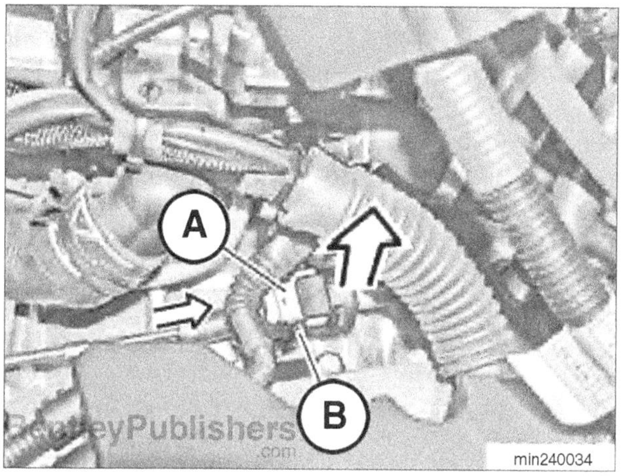

min240034

Squeeze retaining clip (**A**) together and remove shift cable from fixture (**B**).

Continued for all vehicles

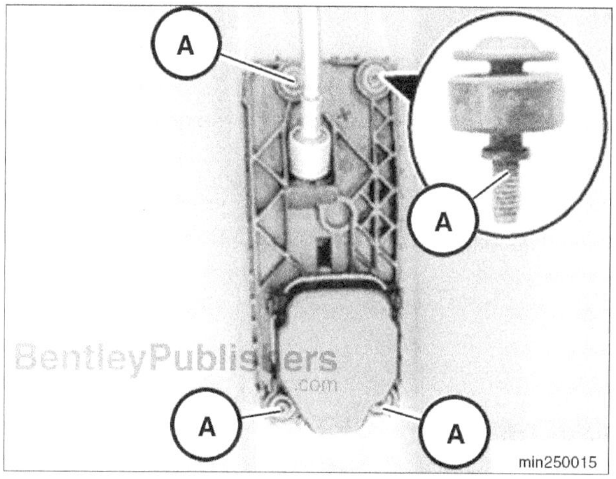

min250015

Remove retaining screws (**A**) from underside of gearshift assembly.

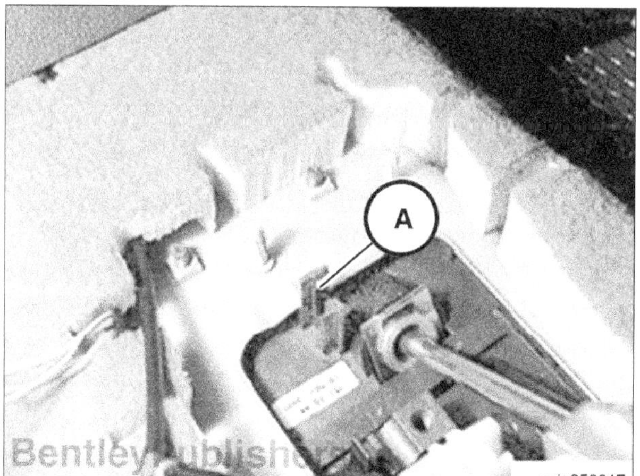

min250017

Working inside car, remove retaining clip (**A**) holding gearshift assembly to floor and remove gearshift assembly.

Installation

- Installation is reverse of removal. Remember to:
  - Make sure gearshift assembly gasket is correctly seated when attaching assembly from underside of vehicle.
  - Replace shift cable (if removed). See **Shift cable, removing (CVT transmission)** in this repair group.
  - Install and adjust transmission interlock cable. See **320 Steering and Wheel Alignment**.
  - Install gearshift trim, dust boot and knob.

Automatic transmission with Agitronic (Cooper S)

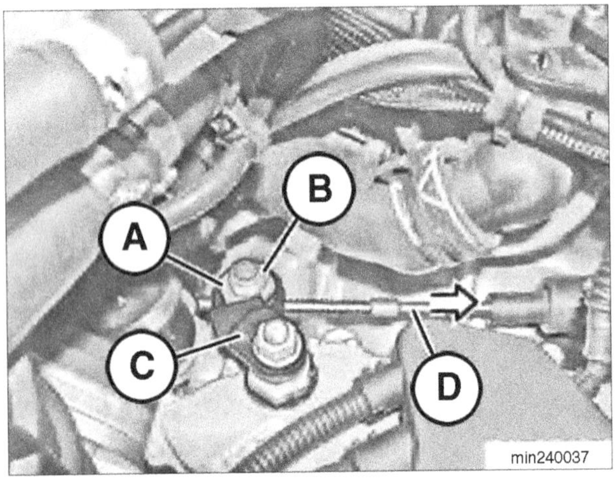

- Make sure selector lever is in "P" (park) position.
- Grip clamping sleeve (**A**) and loosen nut (**B**).
- Move shift cable bracket (**C**) park position, then release again.
- Press cable (**D**) in direction of arrow, then release again.
- Grip clamping sleeve (**A**) and tighten cable mounting nut (**B**).
- Remaining installation is the reverse of removal.
- When finished, road test vehicle and note any shifting problems.

Shift cable, removing (CVT transmission)

- Remove gearshift assembly. See **Gearshift assembly, removing and installing (CVT and automatic transmission)** in this repair group.

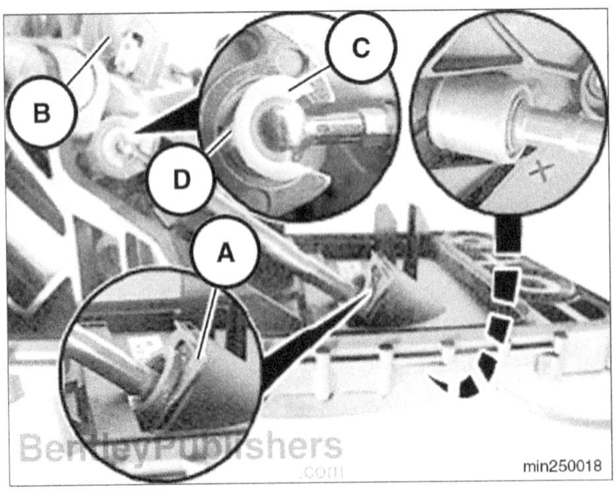

- Remove retaining clip (**A**) holding gearshift cable to assembly.
- Use retaining lug (**D**) to release ball joint locking sleeve (**C**) on gearshift lever (**B**) and pull cable from housing.

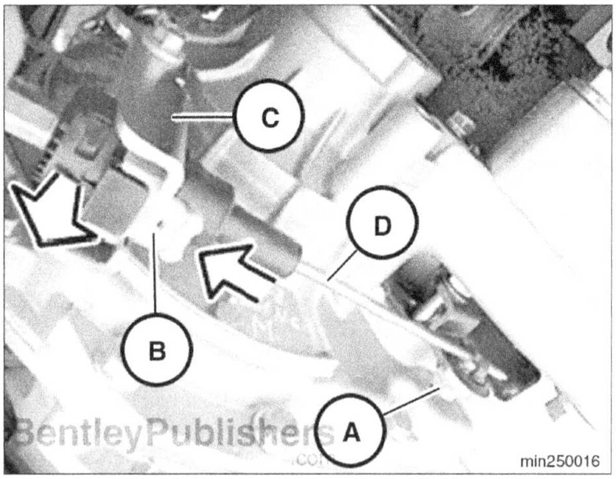

- Loosen cable fastener (**A**) at transmission. Remove retaining clip (**B**) from bracket (**C**) and pull cable (**D**) out of bracket.

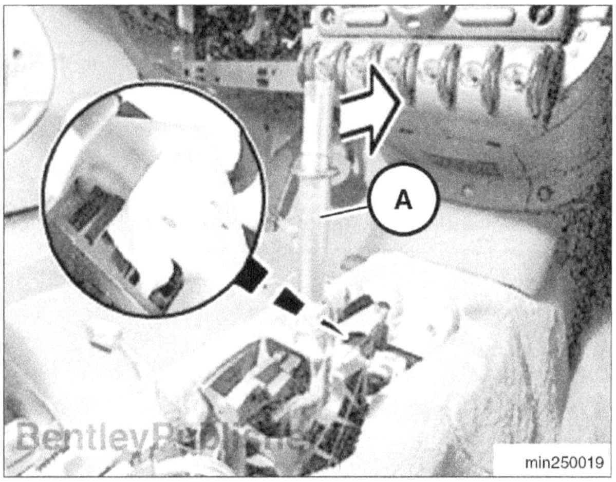

- Installation is reverse of removal. Remember to:
  - Make sure gearshift bracket gasket is correctly seated when attaching from underside of vehicle.
  - Attach shift cable to gearshift cable housing.
  - Move gearshift lever (**A**) to P (Park).

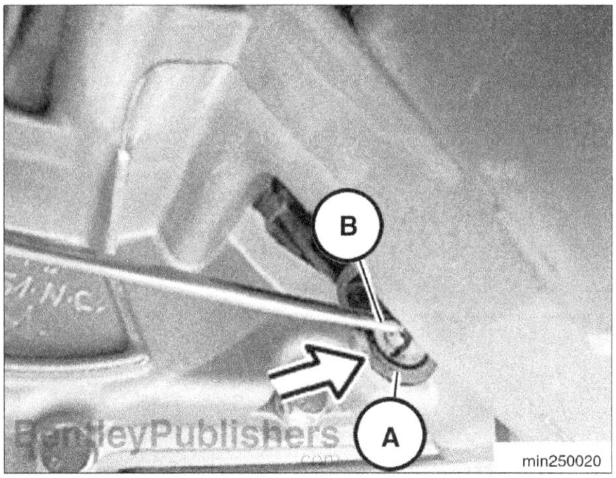

- Attach other end of shift cable to bracket on transmission.
- Insert cable (**A**) in fastener (**B**) and tighten.
- Install and adjust transmission interlock cable. See **320 Steering and Wheel Alignment**.
- Install gearshift trim, dust boot and knob.

Shiftlock cable, adjusting (CVT or automatic with Agitronic)

- Remove front center console. See Front center console, removing in this repair group.
- Move selector lever to “P” Park position.
- Turn ignition key off and remove.

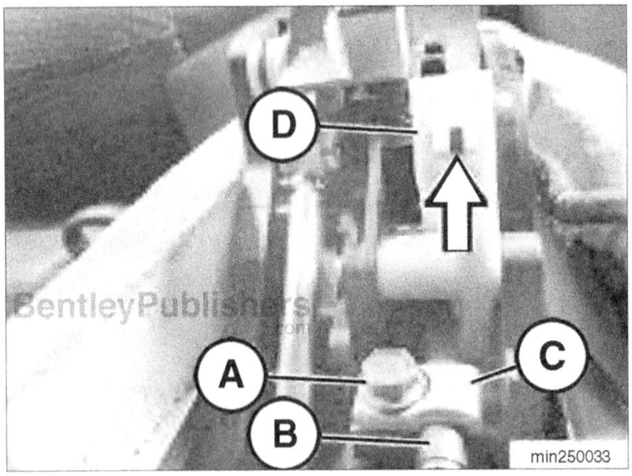

- Unscrew bolt (**A**). The cable (**B**) must be able to move slightly in the mount (**C**).
- Press parking lock lever (**D**) downward as far as it will go.
- Tighten bolt (**A**).

Tightening torque	
Shiftlock cable to shift console	4 - 6 Nm (35 - 53 in-lb)

Check shiftlock cable adjustment:

- With ignition key removed, it should not be possible for the selector lever to moved from the “P” Park position.
- Press brake pedal.
- Turn ignition key to “on” position.
  - It must be possible to shift through all gear positions.
- Move selector lever to every position except “P”.
  - It must not be possible to turn ignition key back to “off” position and remove key.
- Move selector lever to “P” position.
  - It must be possible to turn ignition key to “off” position and remove key.
- If necessary, re-adjust shiftlock cable.

300 Suspension, Steering and Brakes–General

TABLE

300

General

This section covers general information for front and rear suspension, braking and steering systems. Also covered here are the integrated braking and suspension systems including antilock braking (ABS) and traction control (ASC/DSC).

The section also contains wheel and tire specifications.

Vehicle ride height information is at the end of this repair group. Alignment information is in **320 Steering and Wheel Alignment**.

Special tool

To measure front ride height, a special adapter is necessary.

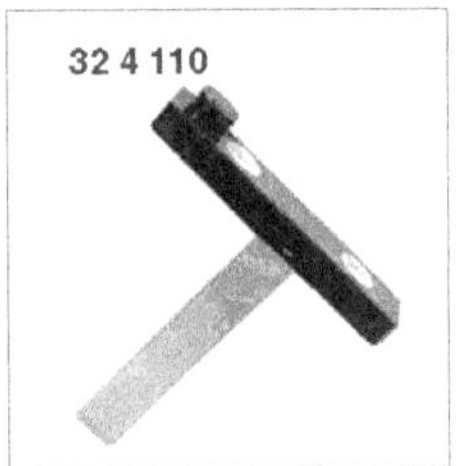

Front ride height measurement adapter (BMW special tool 32 4 110)

Suspension

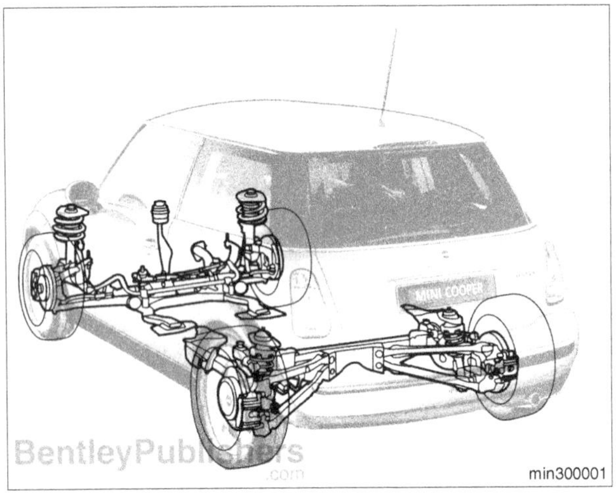

The suspension system provides a firm and responsive ride with good levels of comfort and good acoustic properties. The weight distribution between the front and rear axles is biased to the front due the front wheel drive configuration.

Front suspension

The front suspension features antidive and antisquat geometry due to the 1: 1 ratio of stabilizer bar movement to strut movement. The layout of the front suspension minimizes camber loss due to side forces, thus improving handling and steering response.

The front suspension consists of the following main components:

- MacPherson strut assemblies, including coil springs and swiveling strut top mounts
- Lower control arms
- Front subframe
- Stabilizer bar and links
- Front wheel hubs and bearings

Front drive axles

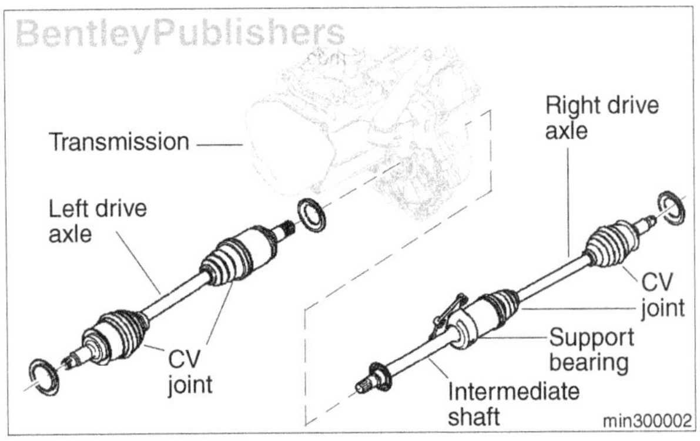

Power to the front (steering) wheels is transmitted through two drive axles (half-shafts) with two constant velocity (CV) joints each. Due to the requirement to steer the wheels, the outer joints are designed with a large degree of articulation.

Two different diameters for inner and outer CV joint are used. The larger size is for the Cooper S and the smaller for Cooper.

Rear suspension

The multilink rear suspension is used for all models of MINI, making it the only car in its class to use such a sophisticated system. Benefits of this suspension are the exceptional handling characteristics and ride comfort, with good acoustic properties.

Main components of the rear suspension include:
- Rear springs
- Rear shock absorbers
- Trailing arms
- Lateral links
- Rear subframe
- Stabilizer bar
- Rear wheel hubs and bearings

JCW suspension kit

John Cooper Works (JCW) produces a full line of factory engineered and approved high performance parts and accessories to improve the performance, handling, functionality and appearance of the MINI. A high performance JCW sports suspension kit is available that can be dealer or owner installed.

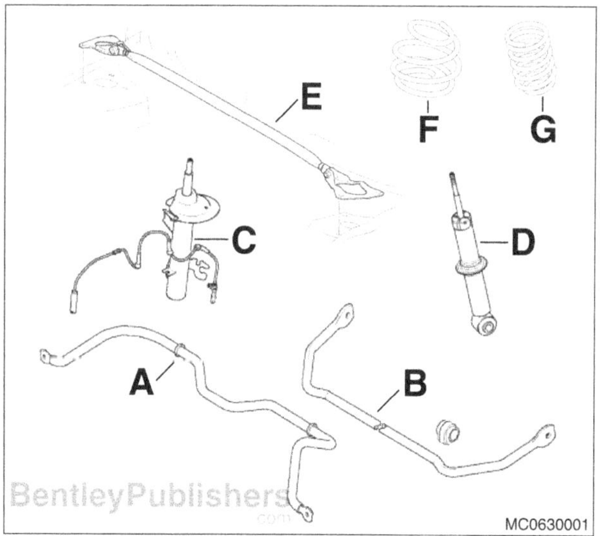

Some components are already part of the Cooper S package and kit components will therefore vary depending on vehicle type. The sport suspension kit can include the following components:

A. Front stabilizer bar, 24 mm
B. Rear stabilizer bar, 17 mm
C. Front strut assembly
D. Rear strut assembly
E. Strut brace
F. Front coil spring
G. Rear coil spring

A sport brake kit with heavy duty pads and larger diameter brake discs along with larger 18 inch wheels with performance tires are suggested to complement the handling improvements of the JCW sport suspension kit. Each JCW package includes specific detailed installation instructions.

Steering

The MINI features a power assisted steering system on all models. The steering gear is of conventional rack-and-pinion design.

The main components of the steering system are:
- Electrohydraulic power steering pump
- Rack and pinion
- Steering column
- Steering wheel

Steering wheel and steering column

The steering column has two collapsible mechanisms and tilt adjustment. A two spoke steering wheel is common to all models although there are small differences depending on the model and equipment. A three spoke steering wheel is available as a John Cooper Works Tuning option.

The steering column has approximately 54 mm of tilt adjustment, manually controlled by a lever underneath the column.

Electrohydraulic power steering (EHPS)

The steering system is powered by an electrohydraulic power steering motor rather than the conventional engine-driven belt. This conserves valuable engine power and provides a 3% savings in fuel economy.

BRAKE SYSTEM

The brake system features discs and calipers at all four wheels. The hydraulic system has a dual circuit that is split diagonally. A brake booster is fitted to all models.

Brake system components include:

- Brake booster
- Master cylinder with reservoir
- Front wheel brake assemblies
- Rear wheel brake assemblies
- Cable operated parking brake

MINI braking systems include Antilock Braking System (ABS) as standard equipment with subsystems Electronic Brake Force Distribution (EBV), Cornering Brake Control (CBC) and Engine Drag Torque Control (MSR).

WHEELS AND TIRES

Tire size is critical to the proper operation of the antilock brake system and traction control system. Numerous styles of wheels in 15, 16, 17 and 18 inch diameters are available from an authorized MINI dealer.

Tire pressures

Factory recommended cold tire inflation pressures vary considerably based on numerous factors including tire size, type, load index, and speed rating; vehicle speed, body style, type, model year, and load. Because of the wide range of possible cold inflation pressures, a sticker is affixed to the vehicle's driver side B-pillar with information applicable to each particular vehicle's characteristics. Similar information is contained in the vehicle owner's manual.

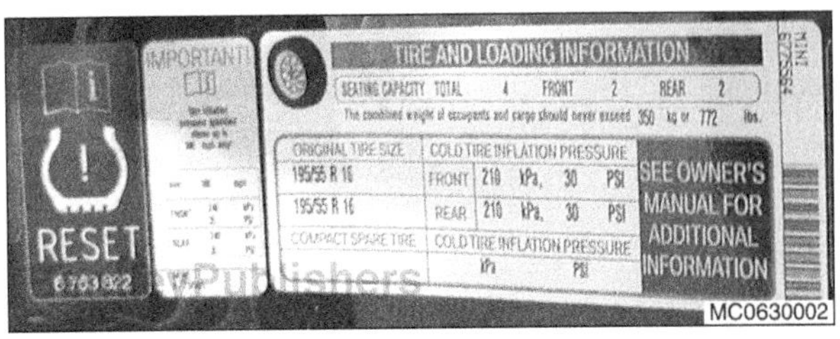

MC0630002

A large sticker located on the driver's side B-pillar provides information on tire pressures for original equipment tires. Smaller additional stickers remind the driver to consult the vehicle owner's manual for information on resetting the flat tire monitor and to limit vehicle speed when pressures are lower.

Tire and wheel information

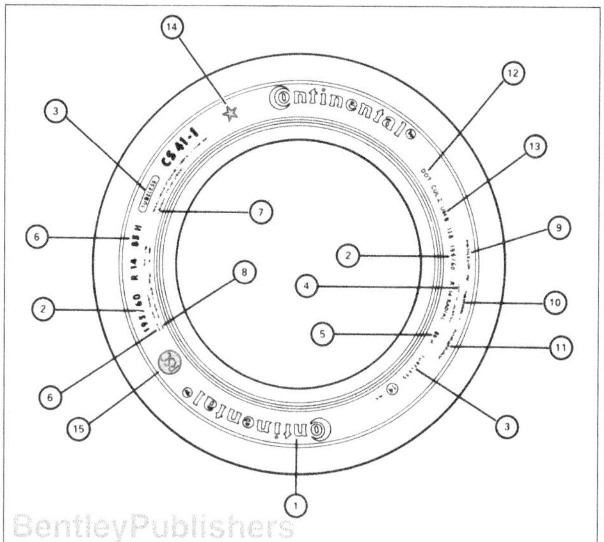

MC0630003

Overview of typical information found on tire sidewall that complies with United States Federal Motor Vehicle Safety Standard (FMVSS) 109 and MINI specifications. Information on USA Department of Transportation (DOT) required uniform tire quality grading is noted.

1. Manufacturers brand name
2. Tire dimensions (size)
3. Type (tubeless)
4. Country of origin
5. Load index and speed rating
6. Load range (A, B, C)
7. Tread and sidewall ply and construction
8. Maximum cold inflation pressure
9. Treadwear index number (DOT)
10. Traction index number (AA, A, B, C) (DOT)
11. Temperature index number (A, B, C) (DOT)
12. USA DOT compliant
13. Manufacturers production and date coding
14. MINI recommendation designation (approval star)
15. Run flat designation (RSC) where applicable

When installing new tires or wheels, the following points should be observed:

- MINI specifies that only approved tires be used. Only radial tires are approved, and retreaded are specifically not recommended.
- MINI specifies that only approved wheels be used. Consult an authorized MINI dealer or the vehicle owner's manual for additional information.
- For safety reasons, general industry guidelines suggest replacing tires in pairs and on the same axle. However, replacement of all tires at the same time is preferred. Only tires of the same type and tread pattern should be installed.
- New tires should be balanced and their rims (wheels) fitted with new valve stems.
- New tires tend to be slippery. General industry guidelines suggest a break-in period of at least 350 miles (560 Km).
- New tires should be the same size as originally installed or of an approved alternative size. If you change sizes, be aware that gear ratios and speedometer readings may change.
- Do not use tires of a lesser speed rating and/or load range than originally installed.
- Wheel and tires may require re-balancing as they accumulate mileage due to settling and wear.

If in doubt about tire application or safety information, consult the vehicle owner's manual or your MINI dealer.

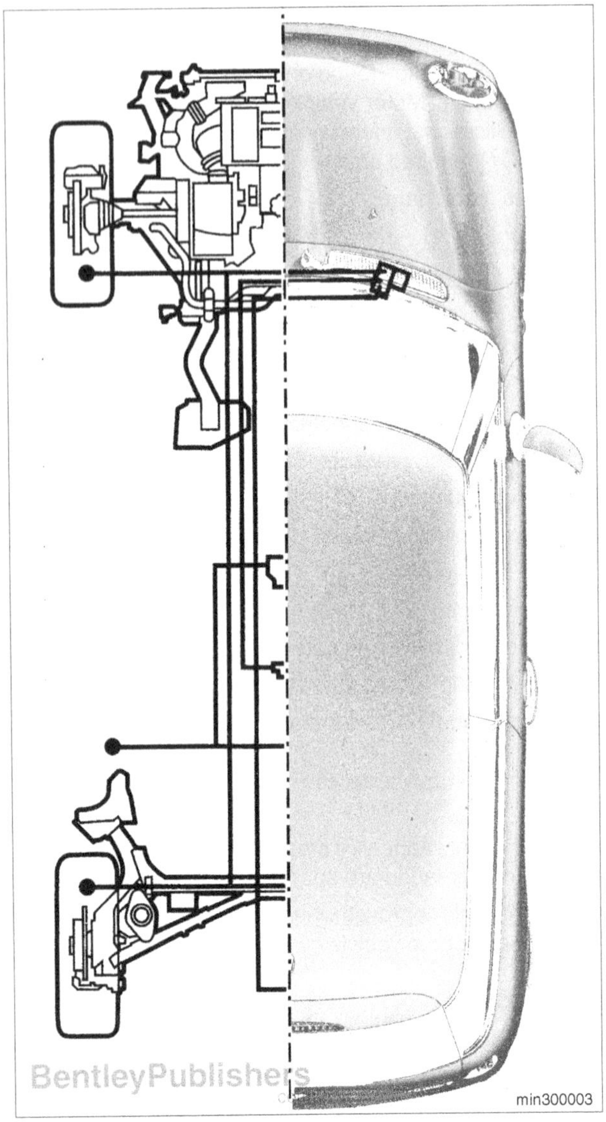

Electronic Brake and Stability Control Systems

Safe vehicle handling and braking is primarily achieved by highly developed, state of the art chassis, suspension and brake design. However, even on a perfectly balanced vehicle, critical situations may occur during braking, accelerating and cornering which result in loss of directional stability. MINI has developed the following systems to assist drivers in these extreme situations:

Antilock Braking System (ABS) prevents lock-up of road wheels during brake application, thus maintaining vehicle stability and steering control under emergency conditions. The system is fitted to all versions of the MINI and cannot be disabled by the driver. ABS always includes the following:

- Electronic Brake Force Distribution (EBV)
- Cornering Brake Control (CBC)
- Engine Drag Torque Control (MSR)

Automatic Stability Control (ASC) prevents the driven front wheels from spinning when engine torque is applied and maintains an equal torque distribution to both wheels. This ensures that optimum traction and directional stability are maintained. ASC is always fitted together with ABS and its subsystems, EBV, CBC and MSR.

Dynamic Stability Control (DSC) interfaces with ABS, EBV, CBC, MSR and ASC but also monitors:

- Steering wheel movements via the steering angle sensor.
- Lateral (sideways) forces on the vehicle via the acceleration sensor in the DSC sensor cluster.
- Vehicle rotational speed via the yaw rate sensor in the DSC sensor cluster.

DSC is not standard on any version of the MINI but is available as an option. For additional details about electronic brake and suspension systems, see **340 Brakes**.

Ride Height

Ride height, measuring

– Preconditions for measuring ride height:
- Correct wheel bearing play. See **310 Front Suspension, Drive Axles** and **330 Rear Suspension.**
- Car in normal loaded position on the ground.

Normal loaded position	
Each front seat	68 kg (150 lb)
Center of rear seat	68 kg (150 lb)
Cargo compartment	14 kg (30 lb)
Fuel tank	full

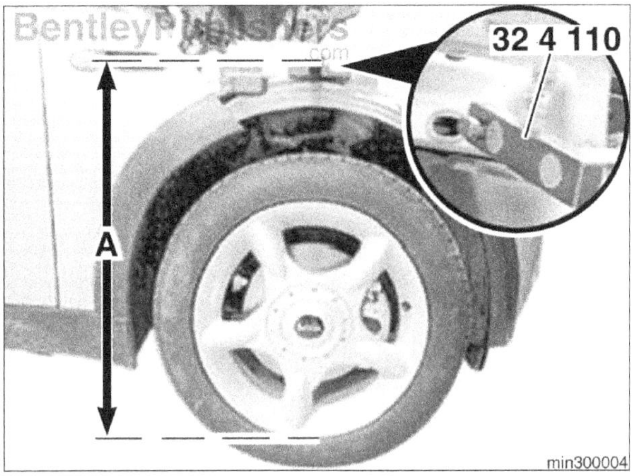

- Measure front ride height (**A**) from lower edge of special tool 32 4 110 to bottom of front wheel rim flange.

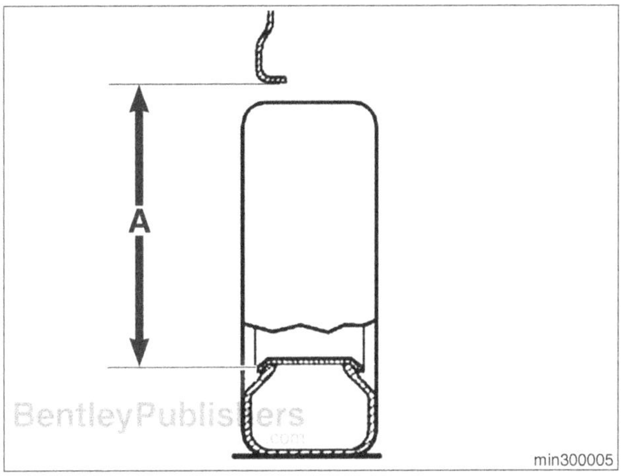

- Measure rear ride height (**A**) from lower edge of wheel arch to rim flange at wheel center.

– Determine a mean value for each wheel after rocking vehicle body up and down and derive a mean value for each axle.

– If ride height is outside specification listed in **Table a**, install new springs. Suspension spring removal and installation is covered in **310 Front Suspension, Drive Axles** and **330 Rear Suspension**.

Table a. Ride height		
Tire size	**Front in mm (in)**	**Rear in mm (in)**
15 in (Cooper only)	660 (25.98)	543 (21.38)
16 in	673 (26.50)	556 (21.89)
17 in	686 (27.01)	571 (22.48)
18 in	699 (27.52)	586 (23.07)
Maximum variation between sides: 10 mm (0.04 in)		
Maximum deviation from specifications: 10 mm (0.04 in)		

310 Front Suspension, Drive Axles

GENERAL

This repair group covers the repair and replacement of components that make up the front suspension and drive axles of MINI cars.

See the following repair groups for additional related information:

- **300 Suspension, Steering and Brakes-General** for a description of the front suspension and components, as well as specifications for setting ride height.
- **320 Steering and Wheel Alignment** for wheel alignment specifications.

Special tools

Special service tools are required for most of the work described in this repair group. Read the procedures through before beginning any job.

Transmission output flange seal protector
(BMW special tool 24 8 120)

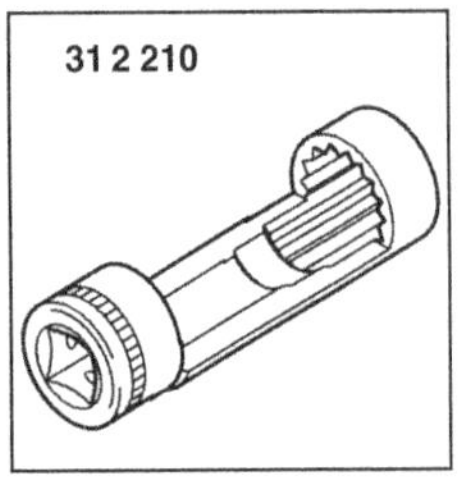

Strut nut socket
(BMW special tool 31 2 210)

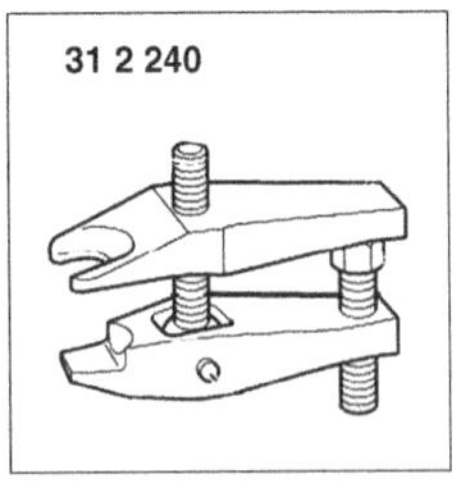

Inner ball joint removal tool
(BMW special tool 31 2 240)

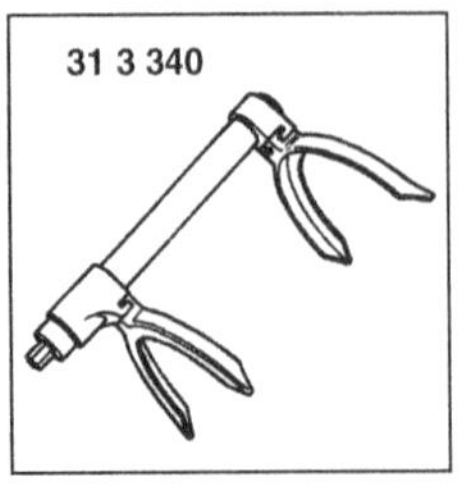

Spring compressor
(BMW special tool 31 3 340)

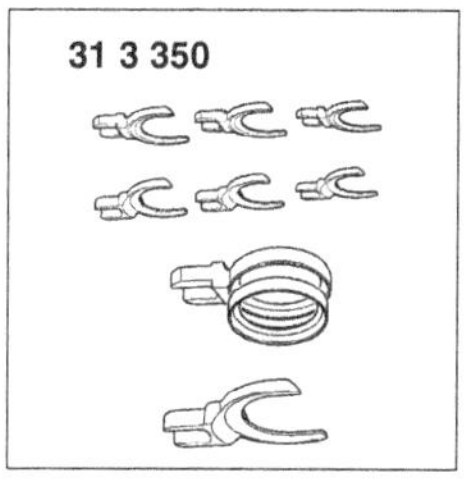

◄ Spring compressor adapters
(BMW special tool 31 3 350)

◄ Control arm removal and installation tool
(BMW special tool 31 5 140)

◄ Control arm bushing removal and installation press tool
(BMW special tool 31 5 150)

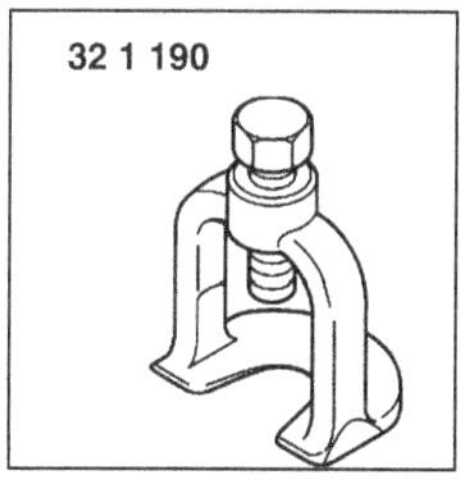

◄ Outer ball joint removal tool
(BMW special tool 32 1 190)

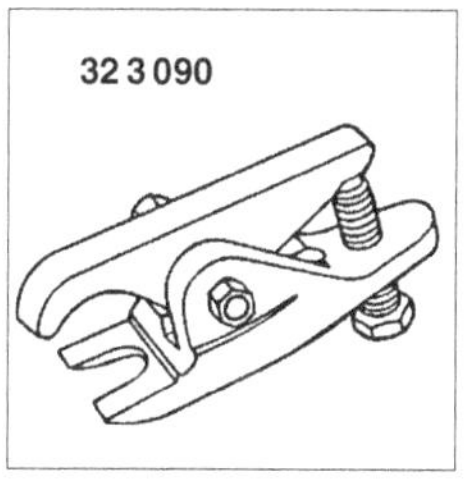

◄ Tie rod ball joint removal tool
(BMW special tool 32 3 090)

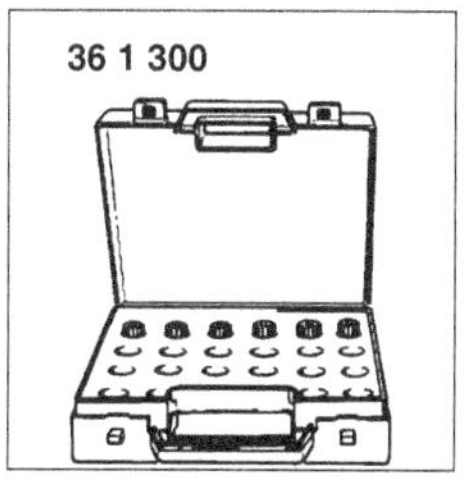

◄ Anti-theft wheel nut removal kit
(BMW special tool 36 1 300)

Front suspension design

Front suspension components are mounted on a rigid subframe which is bolted to the body.

Springs and shock absorbers are of MacPherson variety. Lower control arms are supported by ball joints and bushings. They are attached to the steering knuckles by outer ball joints.

A stabilizer bar is anchored to the subframe and attached at each end to the strut assembly tube.

Wheel bearing are incorporated in the wheel hubs and a part of the steering knuckle assembly.

The transmission drives the left and right drive axles which, in turn, drive the front wheels.

Front suspension components

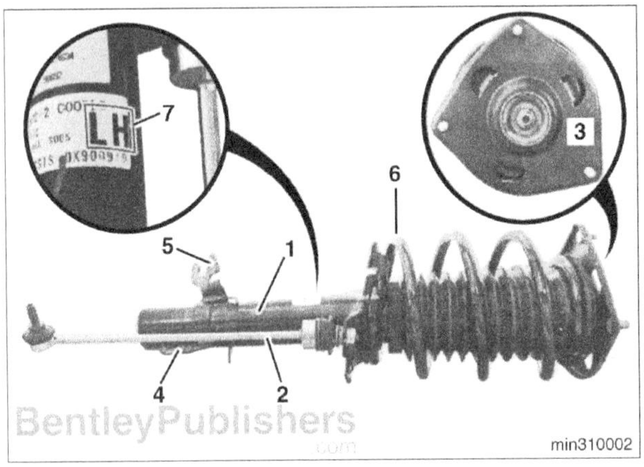

MacPherson strut assemblies control the damping of the front suspension. Each strut assembly has the following features:

1. Strut tube
2. Stabilizer bar link attachment tab
3. Upper strut mount and bearing
4. Strut locating tab
5. Brake hose and ABS sensor bracket
6. Coil spring
7. Left / right identification label

The front dampers (inside the strut tubes) are twin-tube gas pressure shock absorbers. A tab at the bottom of the strut tube determines the correct orientation of the strut assembly; the tab slides into a machined slot in the wheel hub cylindrical sleeve and is secured with a pinch bolt. A label is adhered to the damper for left or right side identification.

Each front coil spring mounts between the lower spring seat on the strut housing and the upper spring seat on the top mount. Isolators at each end of the spring reduce noise transmission from the suspension to the cabin. The coil spring axis is offset to the axis of the damper. This arrangement reduces friction between the damper and piston during cornering.

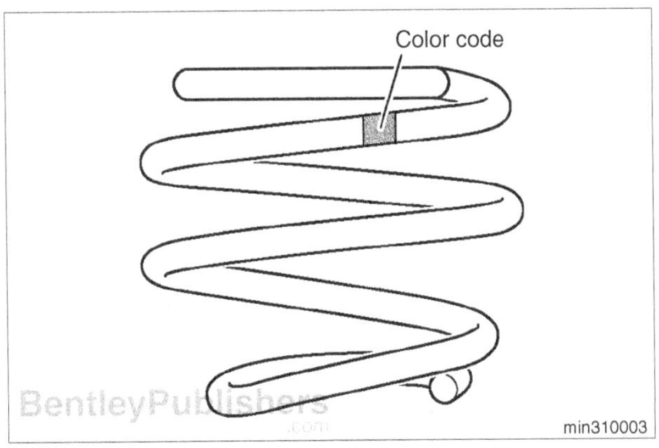

Springs are color coded to suit the suspension type and equipment level of the car.

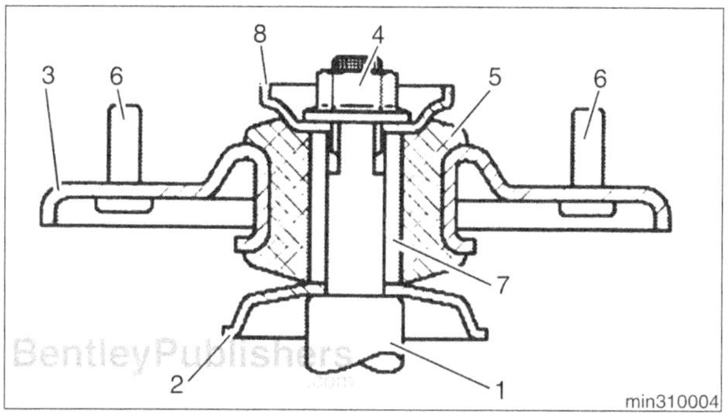

The upper strut mount assembly includes:

1. Threaded damper rod
2. Bump plate
3. Top mounting plate with a bonded rubber bushing and integral metal sleeve
4. Damper rod retaining nut
5. Elastomer rubber
6. Three studs pressed into plate to mount assembly to body
7. Bearing fitted to top spring mount, allowing spring to rotate as steering is operated
8. Rebound plate

The bearing is available as a service part separate from the top mount plate.

There is no provision for camber adjustment at the top front strut mounts.

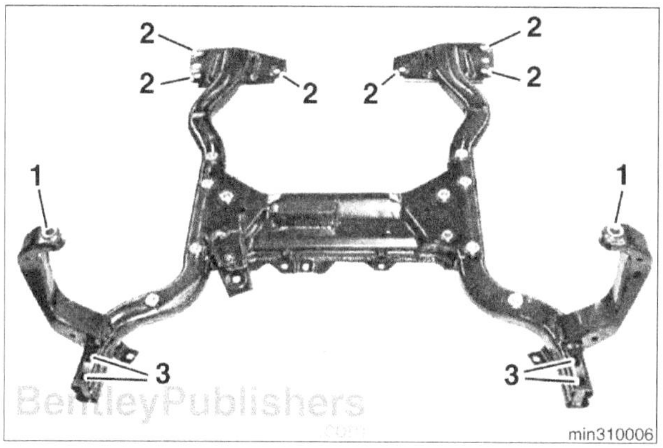

Front subframe is made of hydroformed steel tubing and bolts directly to the body. The subframe provides mounting points for all suspension and steering components with the exception of the upper strut mounts.

1. Front mounting points
2. Rear mounting points
3. Attachment points for crush tubes

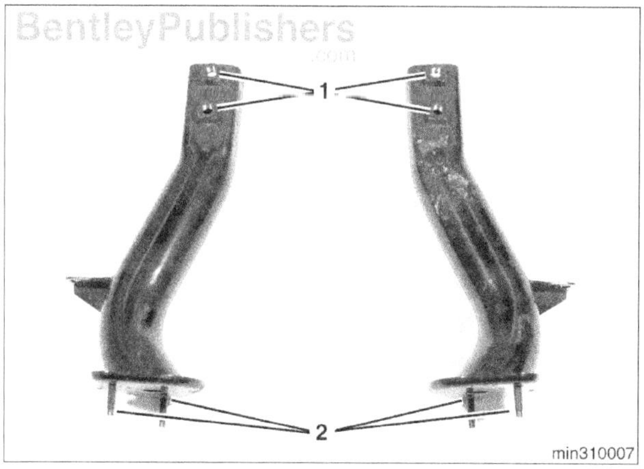

Attached between the front of the subframe and the front bumper are the crush tubes. These tubes are attached with two bolts to the subframe and are designed to deform in an accident.

1. Attachment points to subtrame
2. Attachment points to front bumper

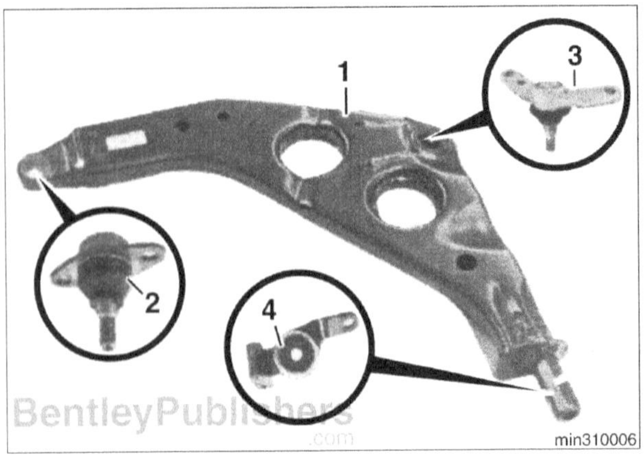

Control arms are pressed steel and link the subframe to the steering knuckle and wheel hub assemblies via two ball joints on each side. Both ball joints are available as separate service parts.

1. Control arm
2. Outer ball joint, attached to steering knuckle
3. Inner ball joint, attached to subframe
4. Control arm bushing, press fit on hexagonal shaft at rear of control arm. Bushing is attached to body with a single mounting and to subframe with two bolts which also secure stabilizer bar.

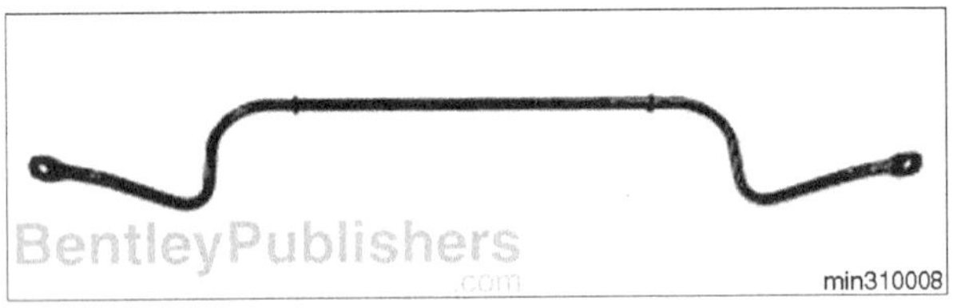

Stabilizer bar and links. Two sizes of front stabilizer bar are used:

- 19 mm (3⁄4 in) for Cooper
- 24 mm (15⁄16 in) for Cooper S
  (also used with JCW tuning kit on Cooper)

Stabilizer bar bushings clamped to the top of the control arm bushing housings are manufactured from low friction PTFE. This material requires no additional lubrication. This allows the stabilizer bar to rotate freely and quietly. It also allows the stabilizer bar to respond quickly to roll inputs.

Washers attached to the stabilizer bar located on the inside of each bushing prevent sideways movement.

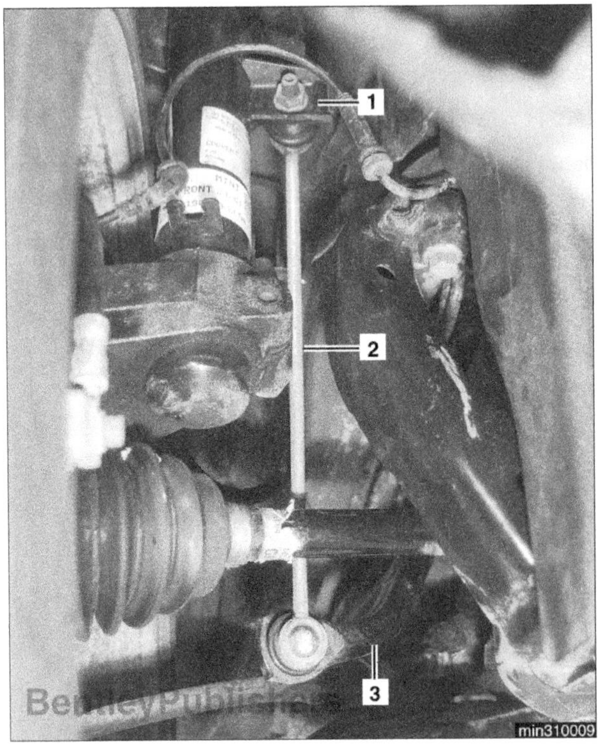

Stabilizer bar links have ball joints fitted at each end. The illustration shows the link design used from 03/2002.

1. Strut connection
2. Stabilizer bar link
3. Stabilizer bar

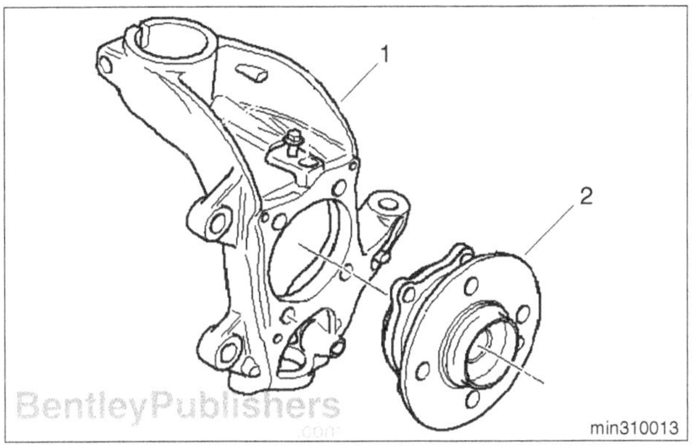

Wheel bearing and hub are an integrated unit bolted to the steering knuckle.

1. Steering knuckle
2. Wheel bearing hub

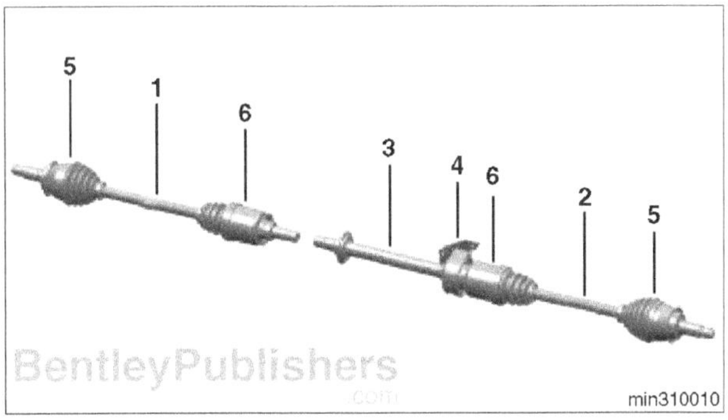

Drive axles articulate at a constant velocity (CV) joint at each end.

1. Left drive axle
2. Right drive axle
3. Intermediate shaft
4. Support bearing
5. Outer CV joint
6. Inner CV joint

To achieve greater articulation, the outer CV joint has six balls in a cage running on convex grooves on the inner race and longitudinal elliptical grooves in the outer race. The outer joint construction allows the joint to turn at the same speed as the shaft when in line and when the joint is turned through any position up to 45°.

1. Outer race
2. Inner race and cage with balls
3. Spring circlip
4. Drive axle shaft

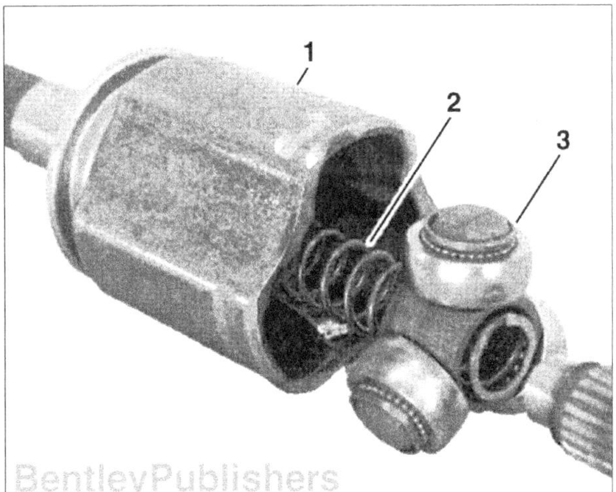

The inner joint is of the tripod type with spherical bearings to reduce sliding resistance. The joint has three bearings supported on needle roller bearings and allows the shaft to slide horizontally inside the outer race. This allows the overall length of the shaft to increase or decrease with suspension travel. A maximum angle of 25° is possible with this type of joint, but the working angle is normally less than 10°.

1. Outer race
2. Retaining spring
3. Inner bearings

Shock Absorbers and Springs

The front suspension shock absorbers are MacPherson struts. The strut is a major component of the front suspension and supports the spring. Most strut assembly components are available as replacement parts.

Front strut, upper strut mount or spring replacement is a two-step procedure:

- Removal of strut assembly from vehicle
- Disassembly and replacement of component on work bench

NOTE —
- *If the shock absorbers have not run for more than 50,000 km (30,000 miles), the manufacturer recommends replacing only the shock absorber that is worn or defective.*
- *Shock absorbers made by different manufacturers may be fitted on the same axle. However, always make sure that the new strut bears the same code as the old. Note identification label on strut tube.*
- *Determine whether a shock absorber needs replacing by checking it on a shock absorber test rig.*
- *Spring version is determined by vehicle type and equipment.*

Front strut assembly, removing and installing

– Raise front of car and support in a safe manner.

> ***WARNING —***
> *Make sure the car is stable and well supported at all times. Use a professional automotive lift or jack stands designed for the purpose. A floor jack is not adequate support.*

– Remove front wheel.

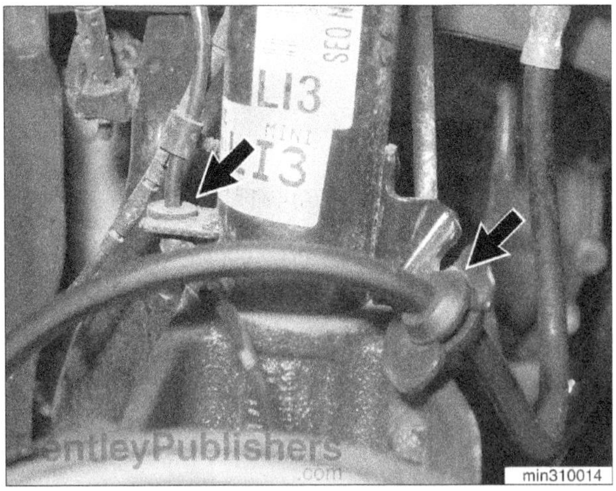

min310014

◄ Release ABS cable and brake hose from their retaining fixtures (**arrows**).

– Unbolt brake caliper and tie back to body.

Models with headlight vertical aim control: Mark position of vehicle level sensor bracket at left control arm before unbolting.

Detach stabilizer bar link from stabilizer bar. Use Allen wrench to counterhold shaft of stabilizer bar link while removing mounting nut (**arrow**).

Detach upper end of stabilizer bar link:

- Models to 03/2002: Use thin open-end wrench to counterhold shaft of stabilizer bar link while removing mounting nut.
- Models from 03/2002: Use Allen wrench to counter hold shaft of stabilizer bar link while removing mounting nut (**arrow**).

Remove steering tie rod mounting nut at steering knuckle. Use BMW special tool 32 3 090 or equivalent to detach tie rod.

Remove outer control arm ball joint mounting nut. Use BMW special tool 32 1 190 or equivalent to detach control arm from steering knuckle.

Support steering knuckle from below. Remove pinch bolt (**arrow**) and slide steering knuckle down and off strut tube.

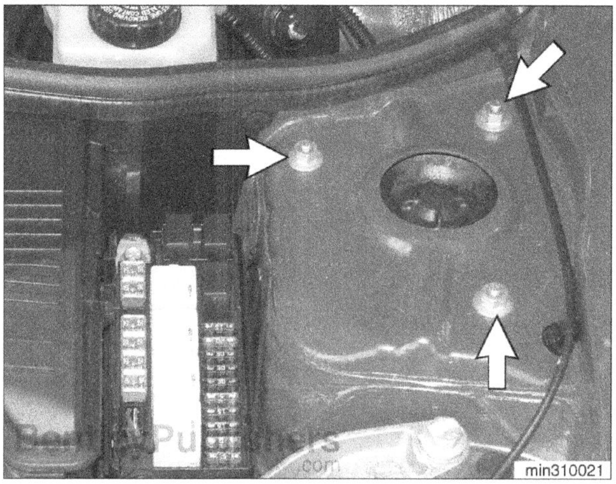

◄ Support strut assembly from below. Working in engine compartment at top of strut tower, remove strut top mounting nuts (**arrows**).

> ***CAUTION—***
> *Do not remove center strut retaining nut.*

– Lower MacPherson strut assembly from car.

– Place assembly on bench to disassemble. See **Front strut, disassembling and assembling** in this repair group.

– When reinstalling, replace self-locking nuts at top mount.

Tightening torque	
Front strut upper mount to strut tower (M8 x 1.25) (replace self-locking nuts)	34 Nm (25 ft-lb)

– When reinstalling steering knuckle to strut, fit locating tab on strut tube to gap in steering knuckle. Slide steering knuckle up until pinch bolt hole in knuckle lines up with dimple in strut tube and pinch bolt can slide in.

Tightening torque	
Front strut to steering knuckle pinch bolt (M12 x 1.5 x 60 mm)	81 Nm (60 ft-lb)

– Remainder of installation is reverse of removal. Keep in mind the following:

- When reinstalling vehicle level sensor bracket to control arm, use previously made position marks to align correctly.
- Replace self-locking nuts.

Tightening torques	
Brake caliper to brake pad carrier (guide bolts, 7 mm Allen)	25 - 30 Nm (18 - 22 ft-lb)
Control arm to outer ball joint at steering knuckle (M12 x 1.5) (replace self-locking nut)	56 Nm (41 ft-lb)
Stabilizer bar link to stabilizer bar (M10 x 1.5) (replace self-locking nut)	56 Nm (41 ft-lb)
Stabilizer bar link to lower spring seat or to strut tube bracket (M10 x 1.5) (replace self-locking nut)	56 Nm (41 ft-lb)
Road wheel to hub	120 ± 10 Nm (89 ± 7 ft-lb)
Tie rod to steering knuckle (M10 x 1.25) (replace self-locking nut)	52 Nm (38 ft-lb)

– Have car professionally aligned when job is complete.

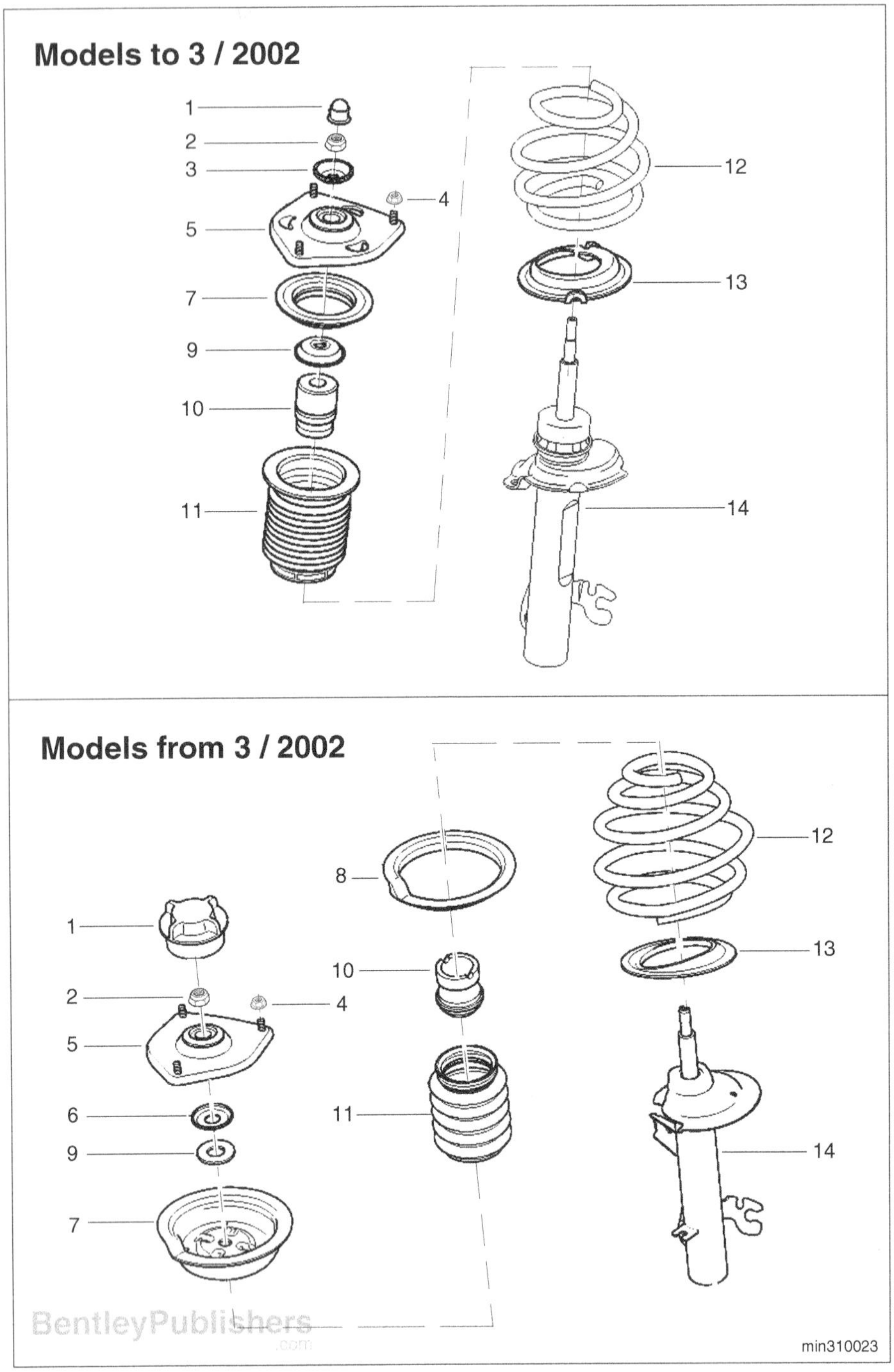

Front strut assembly components

1. **Protective cap**
2. **Center strut retaining nut**
   - M12 x 1.5 (to 03/2002) tighten to 64 Nm (47 ft-lb)
   - M14 x 1.5 (from 03/2002) tighten to 64 Nm (47 ft-lb)
3. **Articulated disk**
4. **Strut top mounting nut**
   - M8 x 1.25 tighten to 34 Nm (25 ft-lb)
5. **Strut top mount**
6. **Dust protection collar**
7. **Upper spring seat**
8. **Upper spring seat pad**
9. **Washer**
10. **Rubber bump stop**
11. **Dust boot**
12. **Spring**
13. **Lower spring seat pad**
14. **Strut**

Front strut, disassembling and assembling

Replacing the strut, upper strut mount or spring requires that the strut assembly first be removed from the car and disassembled. This is described in **Front strut assembly, removing and installing** in this repair group.

For a guide to parts used during component replacement, see **Front strut assembly components** in this repair group.

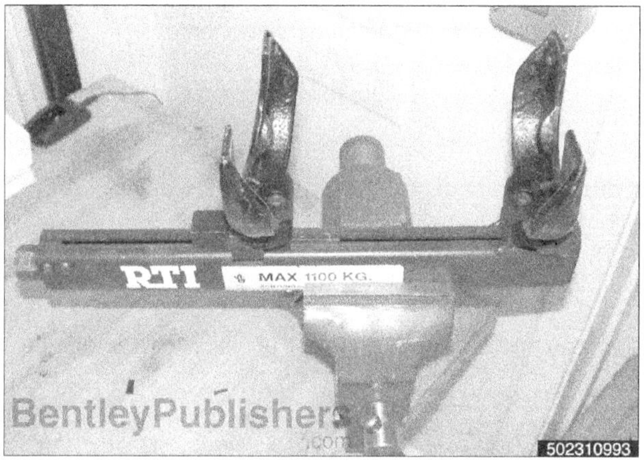

Clamp spring compressor (BMW special tool 31 3 340 and 31 3 350 or equivalent) in shop vise.

WARNING—
- *Do not attempt to disassemble the strut assembly without a spring compressor designed specifically for this job.*
- *Prior to each use, check spring compressor for functionality. Do not use a damaged tool. Do not make any modifications to tool.*
- *Use the correct size of spring retainers when compressing coil spring.*
- *When assembling BMW spring compressor, make sure spring retainer plates are felt and heard snapping into place. Check seating of spring retainers carefully.*

Position coil spring between spring holders so that coils lie in spring holder grooves (**arrows**). Compress spring.

WARNING—
- *When tensioned, the spring coils must rest completely in the spring holder recess.*
- *Never tighten or loosen spring compressor with an impact tool.*
- *Only tighten down the coil springs until stress on the thrust bearing is relieved.*
- *Only loosen strut center nut if spring coils are completely inserted in the spring holder grooves. If necessary, loosen compressor, reposition and recompress.*

NOTE—
- *Models up to 03/2002: Make sure three coils lie between the spring retainers, with the end of the spring under the end of spring retainer.*
- *Models from 03/2002: Make sure two coils lie between the spring retainers, with the end of the spring under the end of spring retainer.*

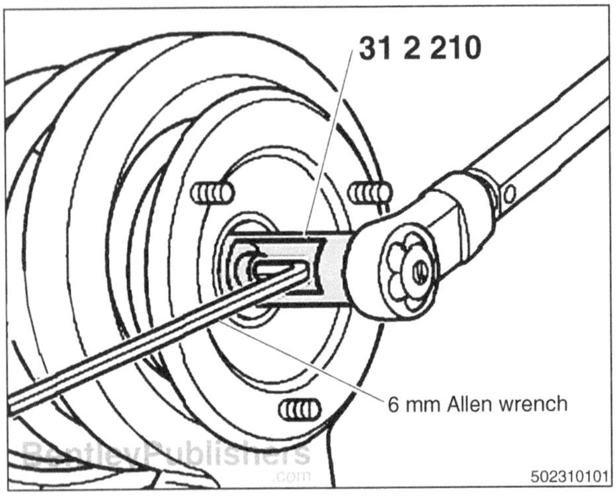

– Models up to 03/2002: Use box wrench to remove strut center nut. Counterhold strut shaft using 6 mm Allen wrench.

◄ Models from 03/2002: Use BMW special tool 31 2 210 or equivalent to remove strut nut. Counterhold strut shaft using 6 mm Allen wrench.

> ***CAUTION—***
> *Do not remove strut nut with impact tool.*

– Remove upper strut bearing and related components.

– If a new coil spring is being installed, relieve tension on spring compressor and remove coil spring.

– Check strut dust boot, rubber stop and spring pads. Replace as necessary.

– Replace strut, upper strut mount or spring, as needed.

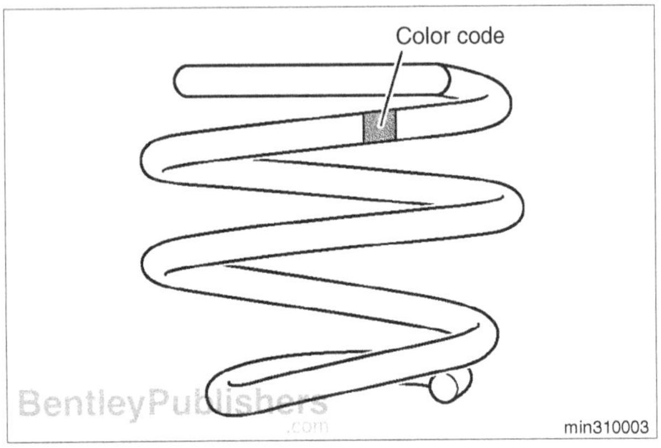

◄ Replace springs in matched pairs only. Use color code to determine match.

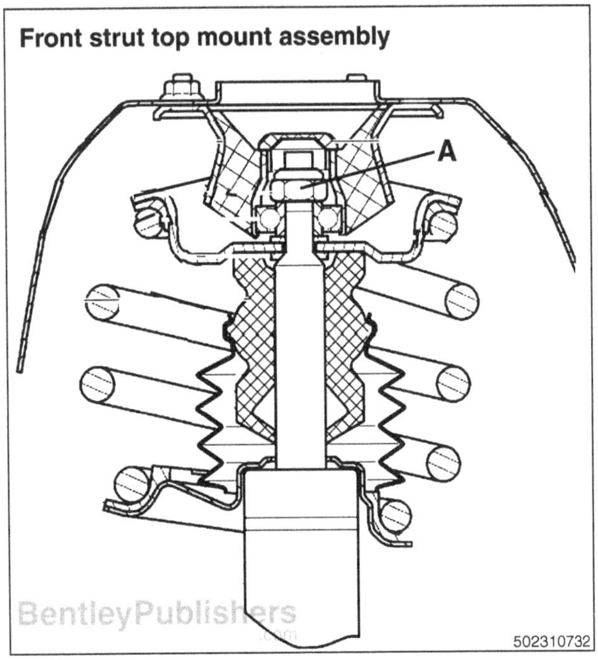

◄ Assembly is reverse of disassembly, noting the following:

- Use a new upper strut self-locking nut (**A**). Tighten nut fully before releasing spring compressor.

> ***CAUTION—***
> *Do not tighten strut nut with impact tool.*

Tightening torque	
Upper strut mount to strut shaft Models to 03/2002: M12 x 1.5 Models from 03/2002: M14 x 1.5 (replace self-locking nut)	64 Nm (47 ft-lb)

Be sure upper and lower spring pads (**arrows**) are correctly installed to spring seats before releasing spring compressor.

- Release spring compressor carefully and evenly, allowing spring to expand slowly.

– Reinstall strut assembly as described in **Strut assembly, removing and installing** in this repair group.

Strut brace

Vehicles with the JCW Tuning package may be equipped with an upper strut tie bar or strut brace to further improve chassis stiffness.

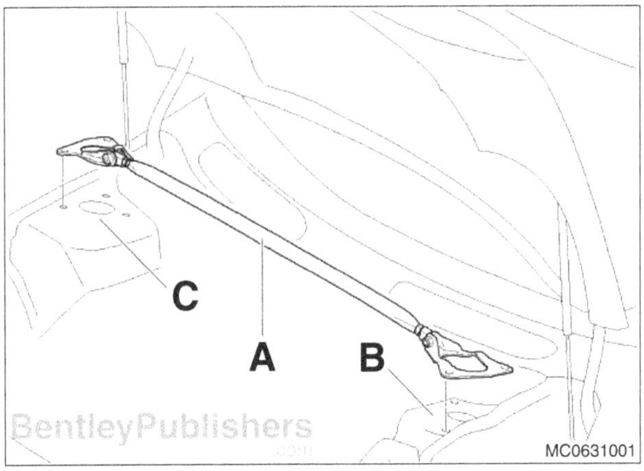

Strut brace (**A**) links left (**B**) and right (**C**) strut towers and is secured by the nuts holding the strut top mount to the tower.

FRONT SUBFRAME

The front subframe provides rigid mounting points for suspension and steering components. The engine and transmission are suspended from the chassis above but are located and stabilized by the front subframe.

For major component attachment points and torque values, see **Front subframe attachment points** in this repair group.

The subframe is not normally subject to wear and should only be replaced if structurally damaged.

> ***CAUTION—***
> *Removal or replacement of the subframe may affect suspension and steering geometry, including front wheel alignment. Make appropriate matching marks during removal and have the front end aligned once repairs are complete.*

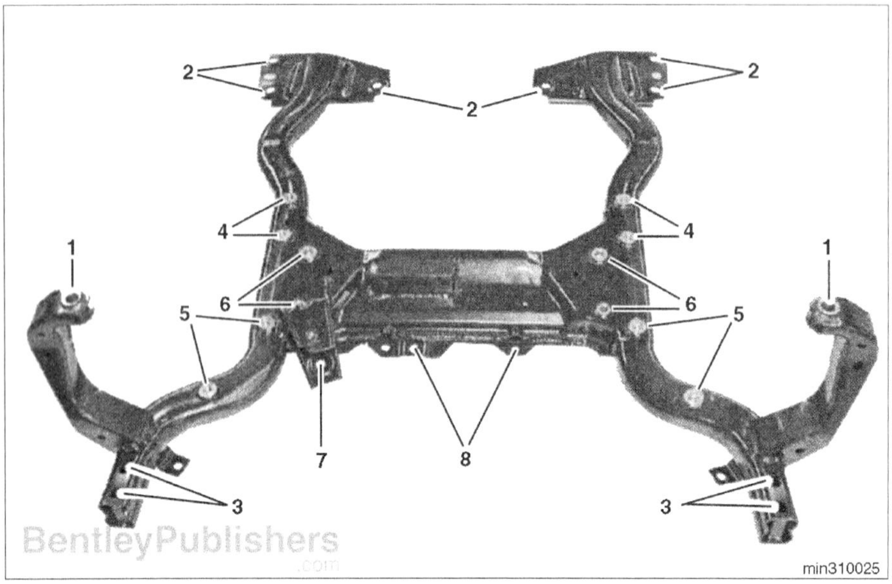

Front subframe attachment points

1. **Front of chassis**
   - M12 x 1.5 x 85
     tighten to 100 Nm (74 ft-lb)
2. **Mid-chassis**
   - M12 x 1.5 x 61.7
     tighten to 100 Nm (74 ft-lb)
3. **Crush tube**
   - M12 x 1.5 x 85
     tighten to 100 Nm (74 ft-lb)
4. **Stabilizer bar anchor / control arm bushing housing**
   - M14 x 1.5 x 80
     tighten to 165 Nm (122 ft-lb)
5. **Control arm ball joint**
   - M12 x 1.5 x 60
     tighten to 100 Nm (74 ft-lb)
6. **Steering rack**
   - Torx M10 x 50
     tighten to 56 Nm (41 ft-lb)
7. **Lower engine vibration damper**
   - M12 x 1.5 x 90
     tighten to 100 Nm (74 ft-lb)
8. **Steering rack cooling fan / steering pump mounting bracket**
   - M8
     tighten to 19 Nm (14 ft-lb)

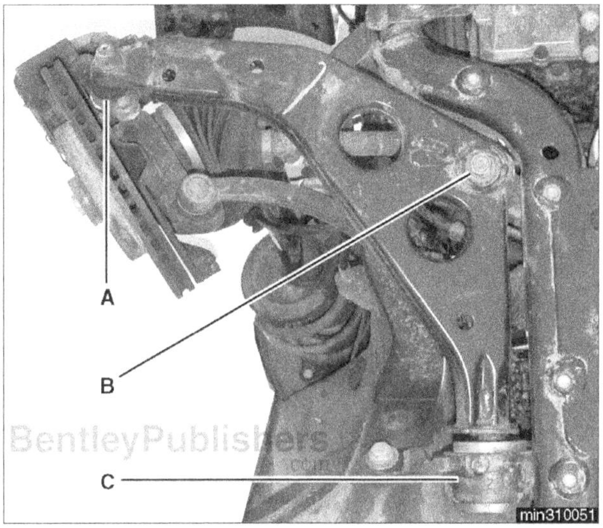

Control Arms

Each front control arm has three attachment points:

- **A** Outer ball joint attached to steering knuckle
- **B** Inner ball joint attached to front subframe
- **C** Bushing and bushing carrier attached to front subframe and body

The ball joints and rear bushing are available as replacement parts.

Control arm, removing and installing

Special tools are necessary for this procedure. See **Special tools** in this repair group.

– Raise front of car and support in a safe manner.

> ***WARNING —***
> *Make sure the car is stable and well supported at all times. Use a professional automotive lift or jack stands designed for the purpose. A floor jack is not adequate support.*

– Remove front wheel.

Models with headlight vertical aim control: Mark position of vehicle level sensor bracket at left control arm before unbolting.

Remove outer ball joint mounting nut. Use BMW special tool 32 1 190 or equivalent to detach control arm from steering knuckle.

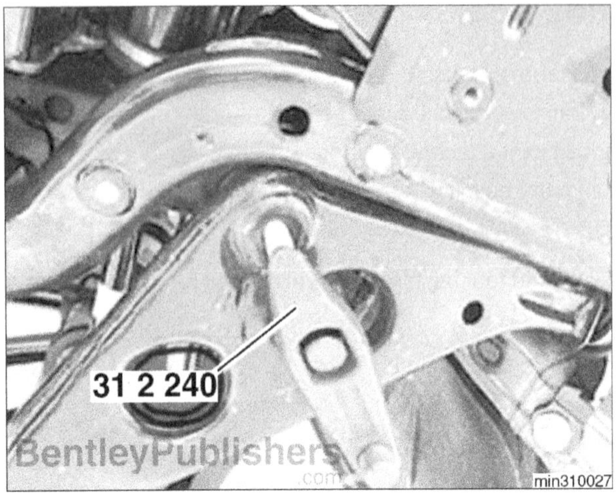

Remove inner ball joint mounting nut. Use BMW special tool 31 2 240 or equivalent to loosen ball joint, which is press-fit into control arm bore.

– Use BMW special tool 31 5 140 to press control arm out of bushing while tilting to release from inner ball joint.

– To install:
  - Grease rubber mount and hexagon of control arm with antiseize agent.
  - Fit hexagonal control arm shaft in correct position on bushing and use BMW special tool 31 5 140 to press in.
  - Tilt control arm to clear inner ball joint shaft.

– Continue pressing in until control arm flange is positioned at edge of bushing carrier.

– Install inner ball joint mounting nut. Replace self-locking nut.

Tightening torque	
Inner ball joint to control arm (M14 x 1.5) (replace self-locking nut)	80 Nm (59 ft-lb)

– Remainder of installation is reverse of removal. Replace self-locking nuts.

Tightening torque	
Outer ball joint to control arm (M12 x 1.5) (replace self-locking nut)	56 Nm (41 ft-lb)
Road wheel to hub	120 ± 10 Nm (89 ± 7 ft-lb)

Outer ball joint, replacing

– Raise front of car and support in a safe manner.

> ***WARNING —***
> *Make sure the car is stable and well supported at all times. Use a professional automotive lift or jack stands designed for the purpose. A floor jack is not adequate support.*

– Remove front wheel.

Models with headlight vertical aim control: Mark position of vehicle level sensor bracket at left control arm before unbolting.

Remove outer ball joint mounting nut. Use BMW special tool 32 1 190 or equivalent to detach control arm from steering knuckle.

Remove ball joint mounting bolts (**arrows**). Remove ball joint.

- Installation is reverse of removal.
  - Replace self-locking fasteners.

Tightening torque	
Outer ball joint to control arm (M12 x 1.5) (replace self-locking nut)	56 Nm (41 ft-lb)
Outer ball joint to steering knuckle (M10 x 20, 10.9 hardness) (replace self-locking bolt)	56 Nm (41 ft-lb)
Road wheel to hub	120 ± 10 Nm (89 ± 7 ft-lb)

Control arm bushings, replacing

Removal of control arm bushing carrier can only be accomplished if the front subframe is lowered. See **Front subframe attachment points** in this repair group.

Using a special press tool, replacement of the control arm rubber bushings is possible on the car, once the control arms have been removed. See **Special tools** in this repair group.

> ***CAUTION—***
> *Replace control arm bushings in pairs.*

- Remove front control arms. See **Control arm, removing and installing** in this repair group.

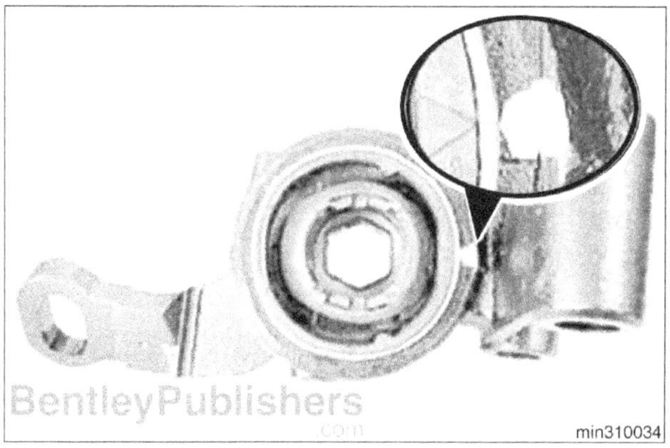

Before pressing out rubber mount, mark alignment of rubber with respect to carrier.

- Use special tools 31 5 150 to press mount out of bushing carrier.
- Installation is reverse of removal. Use special tool to press in mount.

NOTE—
Be sure arrow on the rubber mount points toward previously made mark on bushing carrier.

STABILIZER BAR

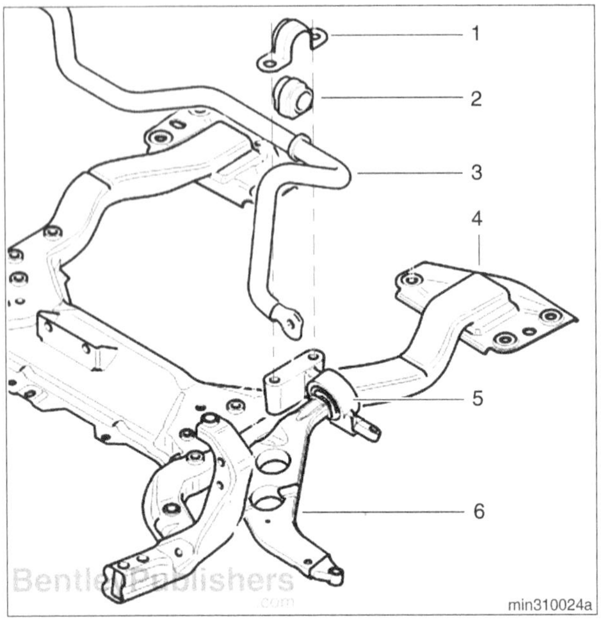

Front stabilizer bar removal requires that the front subframe is lowered. See **Front subframe attachment points** in this repair group.

1. Mounting bracket
2. PTFE bushing
3. Stabilizer bar
4. Front subframe
5. Control arm bushing carrier
6. Control arm

Stabilizer bar links, removing and installing

Detach stabilizer bar link from stabilizer bar. Use Allen wrench to counterhold shaft of stabilizer bar link while removing mounting nut (**arrow**).

Detach upper end of stabilizer bar link:

- Models to 03/2002: Use thin open-end wrench to counterhold shaft of stabilizer bar link while removing mounting nut.
- Models from 03/2002: Use Allen wrench to counter hold shaft of stabilizer bar link while removing mounting nut (**arrow**).

– Installation is reverse of removal.

- Replace self-locking nuts.

Tightening torques	
Stabilizer bar link to stabilizer bar (M10 x 1.5) (replace self-locking nut)	56 Nm (41 ft-lb)
Stabilizer bar link to lower spring seat or to strut tube bracket (M10 x 1.5) (replace self-locking nut)	56 Nm (41 ft-lb)

Front Wheel Bearings

The front wheel bearings are permanently sealed and require no maintenance. The bearing is integral with the wheel hub. The wheel bearing and hub assembly is bolted to the steering knuckle.

NOTE —
Front wheel bearing and wheel hub are not available separately.

Front wheel bearing, removing and installing

– Raise front of car and support in a safe manner.

> ***WARNING —***
> *Make sure the car is stable and well supported at all times. Use a professional automotive lift or jack stands designed for the purpose. A floor jack is not adequate support.*

– Remove front wheel.

min310037

◄ Have an assistant press down on brake pedal to keep wheels from turning. Loosen flanged nut (**arrow**) on wheel hub. Remove nut.

– Remove brake caliper and disc. See **340 Brakes**.

> ***CAUTION —***
> *Hang brake caliper aside using stiff wire to prevent the brake fluid hose from stretching.*

min310038

◄ Undo ABS sensor mounting bolt (**arrow**). Remove sensor to avoid damaging it while removing wheel hub.

Remove four bolts (**arrows**) securing bearing hub to steering knuckle.

– Detach wheel hub from steering knuckle and drive shaft.

– Installation is reverse of removal.
  - Replace self-locking flange nut.
  - Stake nut to lock in place.

Tightening torques	
Bearing hub to steering knuckle (M10 x 30)	56 Nm (41 ft-lb)
Brake caliper to steering knuckle (M12 x 1.5)	110 Nm (81 ft-lb)
Brake disc to bearing hub (Torx M10 x 13)	27 Nm (20 ft-lb)
Drive axle outer CV joint to bearing hub (M22 x 1.5)(replace flange nut)	182 Nm (134 ft-lb)
Road wheel to hub	120 ± 10 Nm (89 ± 7 ft-lb)
Wheel speed sensor to steering knuckle	8 Nm (6 ft-lb)

FRONT DRIVE AXLES

Drive axles use constant-velocity (CV) joints on both ends. The outer joints are of the traditional CV joint variety, whereas the inners are referred to as tripod joints.

An intermediate shaft on the right insures equal length swinging drive axles, reducing the intensity of torque steer. The shaft is one piece with the inner tripod joint cup. It is supported at the right end by a bearing bolted to the engine oil pan.

For replacement parts, the outer CV joint, both inner and outer CV joints boots or complete axles are offered by MINI.

NOTE—
The inner tripod joint, intermediate shaft and intermediate shaft support bearing are not available separately.

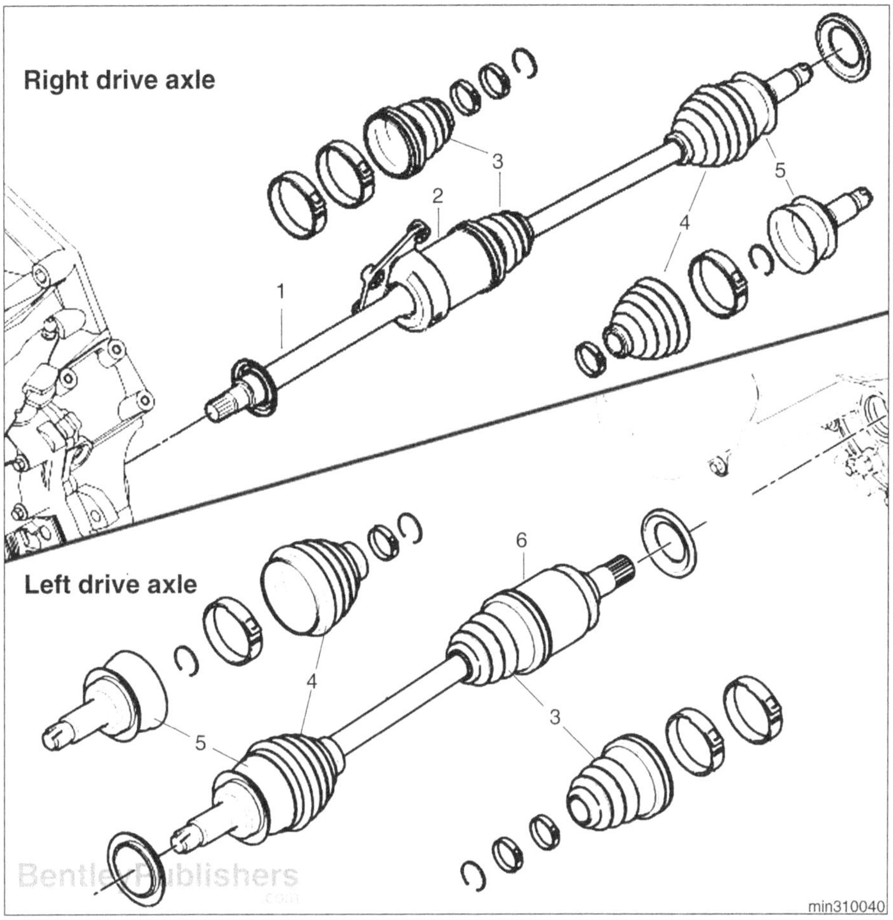

Front drive axle components

1. **Intermediate shaft**
2. **Right inner tripod joint**
3. **Inner CV joint boot**
4. **Outer CV joint boot**
5. **Outer CV joint**
6. **Left inner tripod joint**

Drive axle, removing and installing

– Raise front of car and support in a safe manner.

> ***WARNING —***
> *Make sure the car is stable and well supported at all times. Use a professional automotive lift or jack stands designed for the purpose. A floor jack is not adequate support.*

– Drain transmission oil.

– Remove front wheel.

◄ Have an assistant press down on brake pedal to keep wheels from turning. Loosen flanged nut (**arrow**) on wheel hub. Remove nut.

– Remove brake caliper. See **340 Brakes**.

> ***CAUTION —***
> *Hang brake caliper aside using stiff wire to prevent the brake fluid hose from stretching.*

Remove steering tie rod mounting nut at steering knuckle. Use BMW special tool 32 3 090 or equivalent to detach tie rod.

Undo ABS sensor mounting bolt (**arrow**). Remove sensor and set aside.

Remove outer control arm ball joint mounting nut. Use BMW special tool 32 1 190 or equivalent to detach control arm from steering knuckle.

∢ Right drive axle: Remove intermediate shaft bearing housing mounting bolts (**arrows**).

NOTE —

- *Remove electrohydraulic power steering (EHPS) pump cooling fan for access to bolts. See* **320 Steering and Wheel Alignment**.
- *Lower engine vibration damper removed for illustration purposes.*

– Pry inner drive axle out of transmission using suitable lever.

NOTE —

Be prepared to catch oil drips from transmission.

∢ Support steering knuckle from below. Remove pinch bolt (**arrow**) and slide steering knuckle down and off strut tube.

– Lift out drive axle together with steering knuckle.

– Separate drive axle from steering knuckle. Repair or replace axle components as necessary.

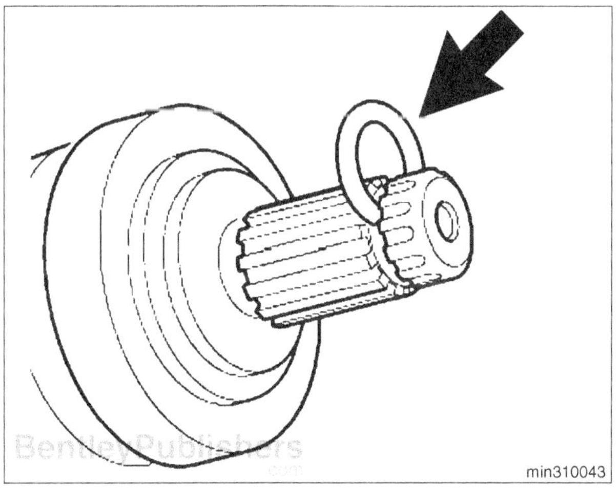

∢ Left drive axle: If reinstalling drive axle with original inner joint, replace spring circlip (**arrow**) on joint shaft.

– Use BMW special tool 24 8 120 to protect transmission output flange seal. Insert drive axle, then remove tool.

– Left drive axle: Push in inner CV joint shaft until spring circlip snaps audibly into place.

– Remainder of assembly is reverse of disassembly. Keep in mind the following:
  - Replace self-locking fasteners.
  - Stake outer drive axle flange nut to lock in place.
  - Fill transmission with oil and check for leaks. See **230 Manual Transmission** or **240 Automatic Transmission**.

Tightening torques	
Brake caliper to steering knuckle (M12 x 1.5)	110 Nm (81 ft-lb)
Drive axle outer CV joint to bearing hub (M22 x 1.5)(replace nut)	182 Nm (134 ft-lb)
Front strut to steering knuckle pinch bolt (M12 x 1.5 x 60 mm)	81 Nm (60 ft-lb)
Intermediate shaft bearing housing to engine oil pan (M8 x 1.25 x 30 mm)	25 Nm (18 ft-lb)
Outer ball joint to control arm (M12 x 1.5) (replace self-locking nut)	56 Nm (41 ft-lb)
Road wheel to hub	120 ± 10 Nm (89 ± 7 ft-lb)
Tie rod end to steering knuckle (M10 x 1.25) (replace self-locking nut)	52 Nm (38 ft-lb)
Wheel speed sensor to steering knuckle	8 Nm (6 ft-lb)

CV joint, outer, removing and installing

Only the outer CV joint is available as a replacement part through MINI.

– To replace outer CV joint, remove drive axle as described in **Drive axle, removing and installing** in this repair group. Clamp axle in workshop vise.

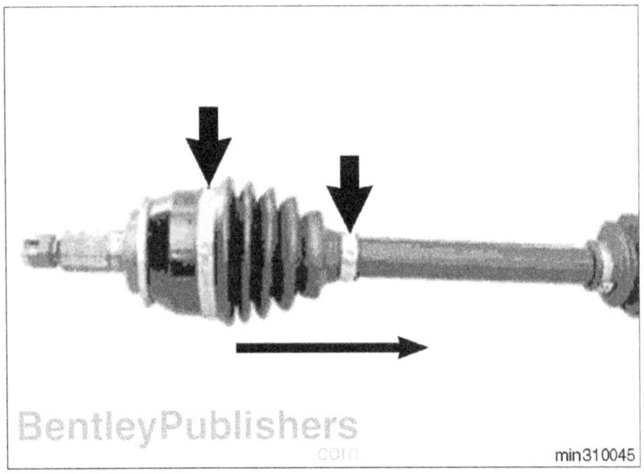

◄ Remove boot from joint.
  - Cut off or loosen boot retaining clamps (**arrows**).
  - Slide boot off joint.

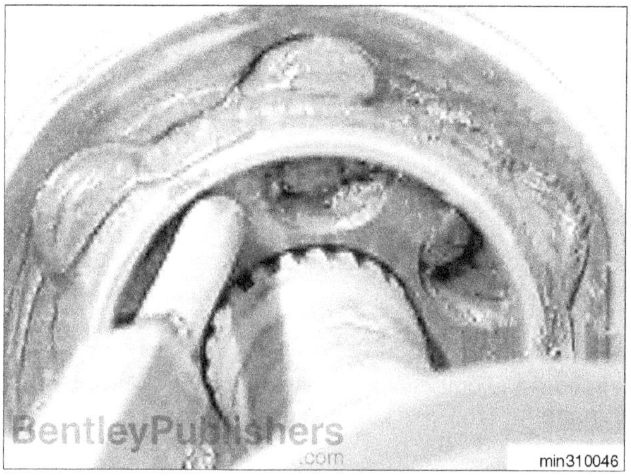

< Use punch and hammer to drive inner CV joint race off axle shaft.

- Remove boot from shaft.

– Slide new boot on shaft.

– Install new spring circlip on end of shaft.

– Generously pack new CV joint with grease and fit on shaft.

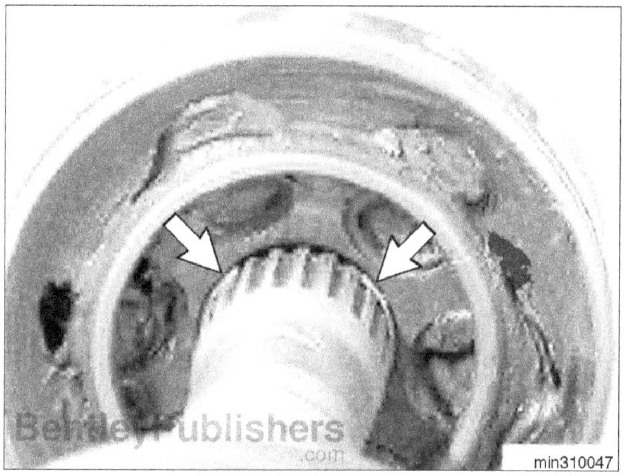

< Make sure that joint rests against new spring circlip (**arrows**). Push circlip into groove and drive joint onto shaft.

< Slide boot on joint and shaft. Make sure seal beading of boot fits in grooves (**arrows**).

– Secure boot clamps.

NOTE —
Before installing the small clamp, be sure to "burp" the boot by flexing CV joint as far over as it will go. A small screw driver inserted between boot and axle shaft will help the process.

CV joint boot, removing and installing

Drive axle CV joint boots are available in case of a split or damaged boot.

Two different types of lubricant are used for drive axle joints. The appropriate type of grease is supplied with each boot kit.

– To replace CV joint boot(s), remove drive axle as described in **Drive axle, removing and installing** in this repair group. Clamp axle in workshop vise.

– Remove outer boot and outer CV joint as described in **CV joint, outer, removing and installing** in the repair group.

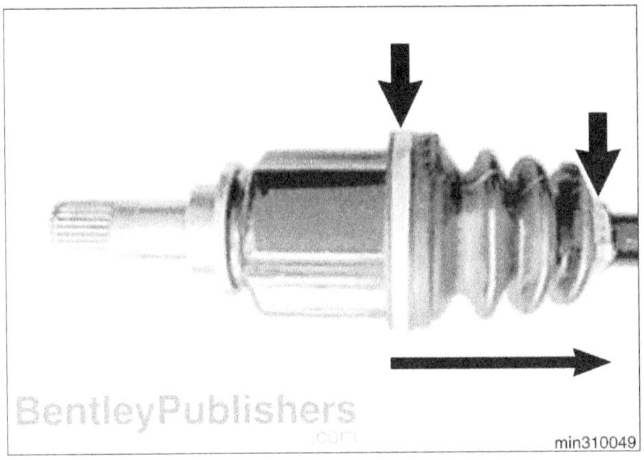

min310049

◄ Inner CV joint boot: Remove boot from joint.

- Cut off or loosen boot retaining clamps (**arrows**).
- Slide boot off joint and off shaft.

– Install new boots as needed.

NOTE —
The CV boot repair kit contains two types of lubricating grease. Make sure you use the correct product for the application:
Outer joint: Graphite-based grease
Inner joint: High-temperature-melting-point grease

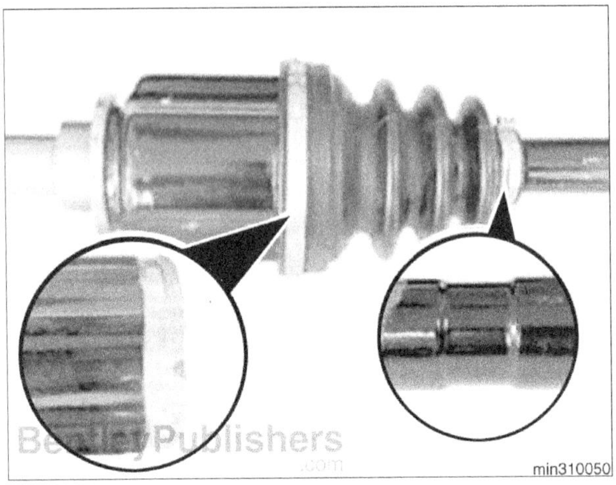

min310050

◄ Slide boot on inner joint. Make sure seal beading of boot fits in grooves (**insets**).

– Secure boot clamps.

– Install outer boot and joint as described in **CV joint, outer, removing and installing** in this repair group.

320 Steering and Wheel Alignment

TABLES

GENERAL

This repair group covers steering wheel and column removal and steering system service, including wheel alignment information.

NOTE—
- *Ignition lock cylinder replacement is covered in this section.*
- *Removal and installation of steering column mounted switches, including ignition switch electrical, is covered in* **612 Switches**.
- *Driver airbag removal and installation is covered in* **721 Airbag System (SRS)**.

Special tools

Special tools are required for some procedures described in this repair group. Read the procedures through before beginning any job.

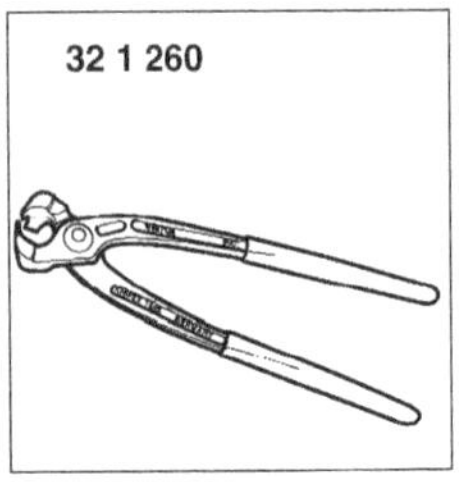

Pliers for cutting and installing steering rack boot clamp (BMW special tool 32 1 260)

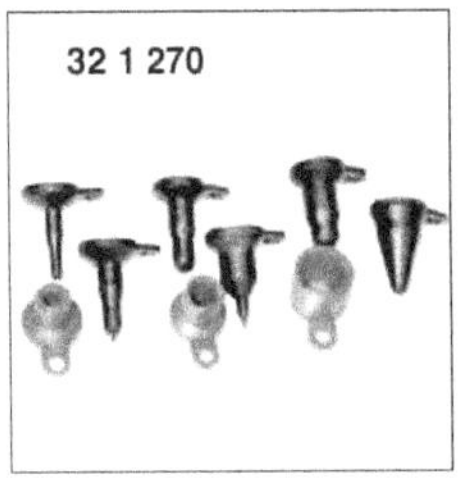

Plug set for fluid lines (BMW special tool 32 1 270)

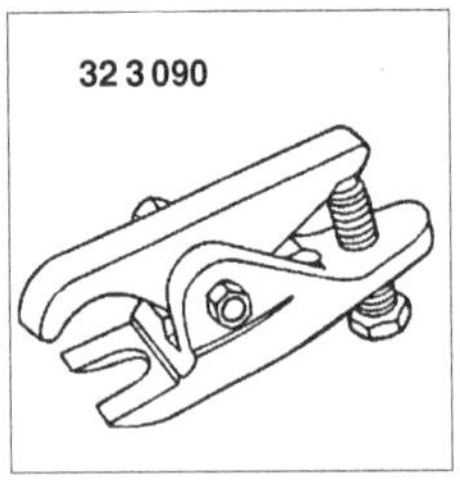

Tie rod ball joint removal tool (BMW special tool 32 3 090)

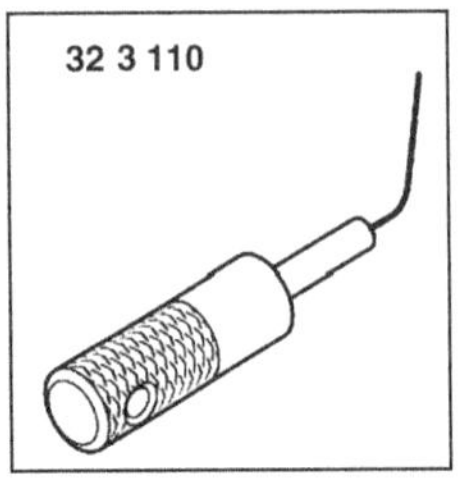

Ignition lock cylinder releasing tool (BMW special tool 32 3 110)

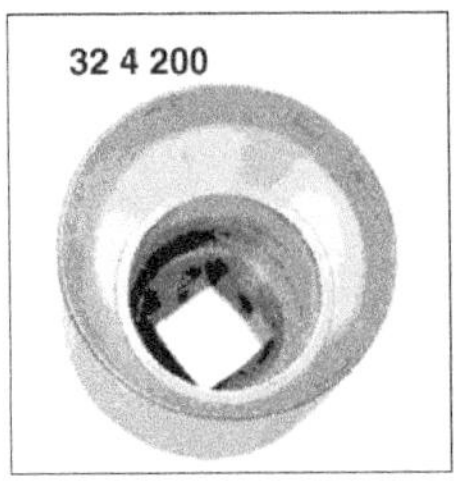

Rear toe adjusting tool (BMW special tool 32 4 200)

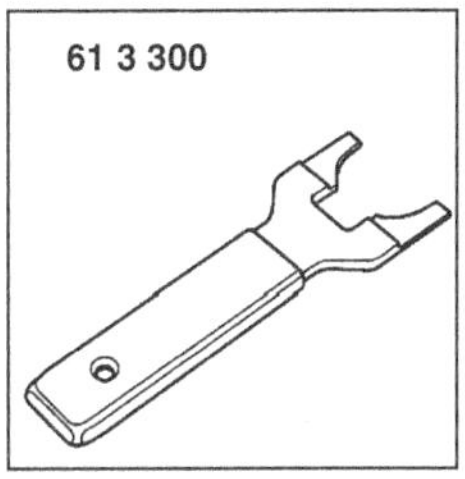

EWS ring antenna removal tool (BMW special tool 61 3 300)

Steering system

The power steering system consists of an electrohydraulic pump, a rack-and-pinion steering rack assembly, and connecting linkage to the road wheels. The steering wheel connects to an adjustable upper steering column. A shaft with two universal joints connects the upper column to the steering rack.

Power steering fluid is supplied from the fluid reservoir to the electrohydraulic (EHPS) pump and to the steering rack via rubber/metal hoses.

The steering rack requires no maintenance other than alignment and periodic inspection for worn. Periodically inspect rubber rack boots and tie-rod end boots for tears or damage. Replace if necessary.

STEERING WHEEL

MINI models are equipped with either a two-spoke or three-spoke steering wheel. The Cooper steering wheel rim is vinyl, while the Cooper S rim is leather trimmed. Leather is an option in Cooper models. The diameter of the steering wheel is 370 mm (14.6 in). Additionally, several leather rimmed three-spoke steering wheels are factory available from JCW Tuning.

Several functions can be performed without the driver having to remove hands from the steering wheel.

- Horn push buttons are located on the outside of the two spokes facing the driver.
- Cruise control buttons (if equipped) are on the left (cruise ON and RESUME) and right (speed increase and decrease plus SET).
- Audio system remote control rocker switches (if equipped) are located on the back of the steering wheel. On the left is station search; on the right is volume control. In the middle of the right side rocker switch is a mode button to change to bass, treble, balance and fade.

Cooper S models equipped with a 6-speed automatic transmission, have a set of paddles (**arrows**) on the steering wheel that can be used to manually shift the transmission when in agitronic mode. These supplement the console-mounted agitronic shift gate on the selector lever.

On all versions, an airbag is incorporated into the center of the steering wheel. The two-stage airbag has a capacity of 57 liters (15 US gal).

The airbag system is referred to as Supplemental Restraint System (SRS). Before working on or around any component or system that involves any part of the airbag system, be sure to read **Airbag warnings** in this repair group and also in **721 Airbag System (SRS)**.

Airbag warnings

> ***WARNING —***
> - *The MINI airbag system is complex. Special precautions must be observed when servicing the system. Serious injury may result if system service is attempted by persons unfamiliar with the MINI airbag system and its approved service procedures. MINI specifies that all inspection and service be performed by an authorized MINI dealer.*
> - *The MINI is equipped with an airbag mounted in the steering wheel. The airbag is an explosive device and should be treated with extreme caution. Always follow the airbag removal procedure as outlined in* **721 Airbag System (SRS)**.
> - *Wait 5 seconds after the battery cable has been disconnected to work on airbag electrical connections. This will allow the system capacitor to discharge.*
> - *Store a removed airbag with horn pad facing up. If stored facing down, accidental deployment of airbag could propel it violently into the air, causing injury.*

Steering wheel, removing and installing

– Center steering wheel. Make sure front wheels are pointed straight ahead.

– Disconnect negative (-) cable from battery.

> ***CAUTION —***
> *Prior to disconnecting the battery, read the battery disconnection cautions given in* **001 General Cautions and Warnings**.

< Loosen 2 airbag mounting fasteners (Torx T30) located at rear of steering wheel. Pull airbag module away from steering wheel.

◄ Gently pry up electrical harness connector locking tabs at rear of airbag. Detach connectors by pulling straight off airbag.

– Remove airbag. Store in safe place with horn pad facing up.

◄ Remove steering wheel center bolt (**arrow**).

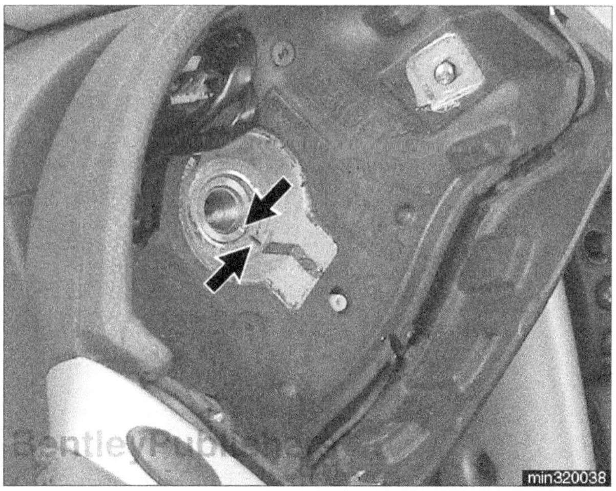

◄ Check for proper alignment of steering wheel hub to steering shaft match marks (**arrows**) before removing steering wheel.

NOTE —
The steering column and steering wheel are match marked at the factory.

– Remove steering wheel.

– When reinstalling steering wheel:
  - Align steering wheel and column match marks.
  - Install steering column center bolt. Do not over-torque.

Tightening torque	
Steering wheel to steering shaft (M14 x 1.5)	45 Nm (33 ft-lb)

– Install airbag, being careful to not pinch electrical wire harnesses and connectors. See **721 Airbag System (SRS)**.

Tightening torque	
Airbag to steering wheel (M6)	10 Nm (7 ft-lb)

STEERING COLUMN

> ***WARNING —***
> *Do not reuse self-locking fasteners. They are designed to be used only once and may fail if reused. Always replace them with new locking fasteners.*

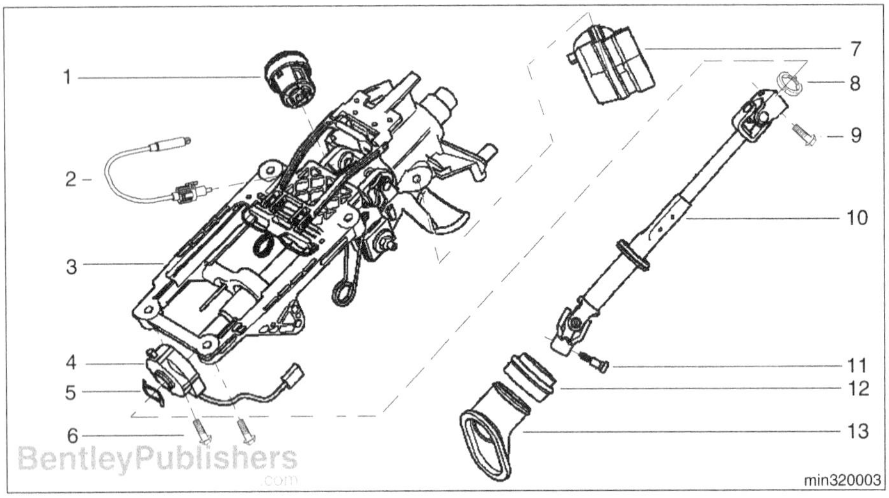

Steering column components

1. **Ignition lock cylinder**
2. **Shift interlock cable (CVT)**
3. **Upper steering column**
4. **Steering angle sensor**
5. **Wavy washer**
6. **Steering column mounting bolt**
   - M8 x 39 mm, replace bolts tighten to 22 Nm (16 ft-lb)
7. **Ignition switch**
8. **Adapter ring**
9. **Pinch bolt and nut**
   - M8 x 35 mm, replace bolt and nut tighten to 30 Nm (22 ft-lb)
10. **Lower steering shaft**
11. **Eccentric pinch bolt and nut**
    - M8, replace bolt and nut tighten to 22 Nm (16 ft-lb)
12. **Collar**
13. **Protective boot**

Ignition lock cylinder, removing and installing

– Remove rubber boot surrounding ignition switch.

◄ Using BMW special tool 61 3 300 with flattened end (**A**) facing ring antenna, pry out ring antenna.

– Disconnect ring antenna electrical harness connector.

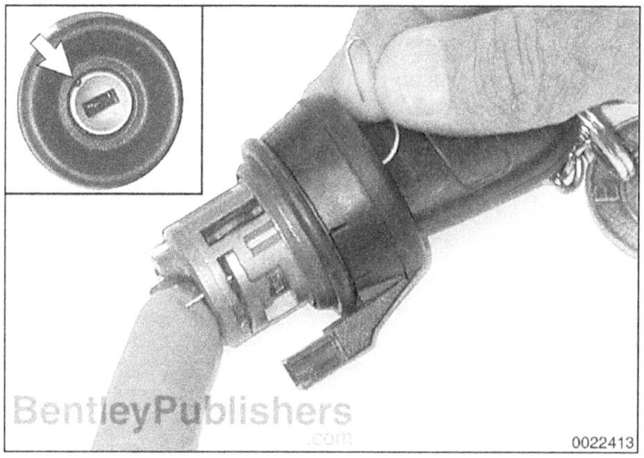

With ignition key in ON position (60° from LOCKED):

- Insert BMW special tool 32 3 110 or a thin piece of stiff wire into opening (**arrow**) in lock cylinder.
- Remove lock cylinder. If necessary, turn tool from side to side.
- Reinstallation is reverse of removal.

NOTE—
Using the valet key during this procedure will provide better access to lock opening.

Steering column upper cover, removing

- Tilt steering column to lowest position.

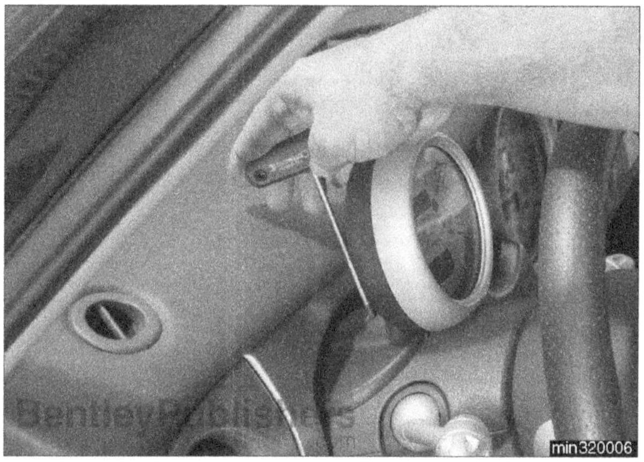

Remove mounting screws at rear of instrument cluster.

- Lift off trim bezel at base of instrument cluster.

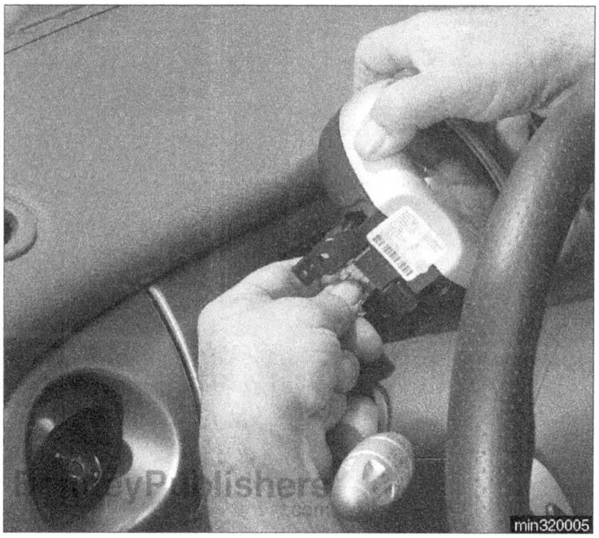

Disconnect electrical harness connector at base of instrument cluster.

- Remove remote instrument cluster.
- Unclip upper cover. Lift upward to remove.
- When reinstalling, make sure upper cover is aligned correctly with lower cover.

Steering column lower cover, removing

– Adjust steering to highest position.

◄ Pull on top edge of left lower dashboard trim to unclip and swing down.

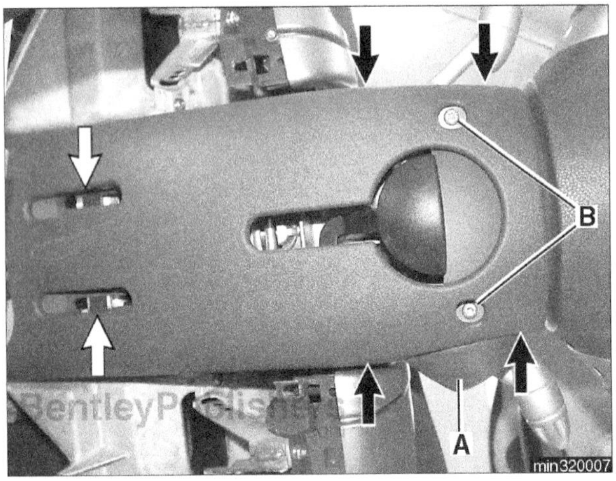

◄ Working underneath steering column:
- Remove rubber boot (**A**) surrounding ignition switch.
- Remove lower steering column cover mounting screws (**B**).
- Release retaining clips (**arrows**) from upper cover and steering column.
- Remove cover.

> ***CAUTION—***
> *To avoid breaking or marring plastic trim, work carefully and use plastic prying tool.*

– When reinstalling, make sure lower cover is aligned correctly with upper cover.

Tightening torque	
Lower steering column cover to steering column	2 Nm (18 in-lb)

Shift interlock cable, removing and installing (CVT only)

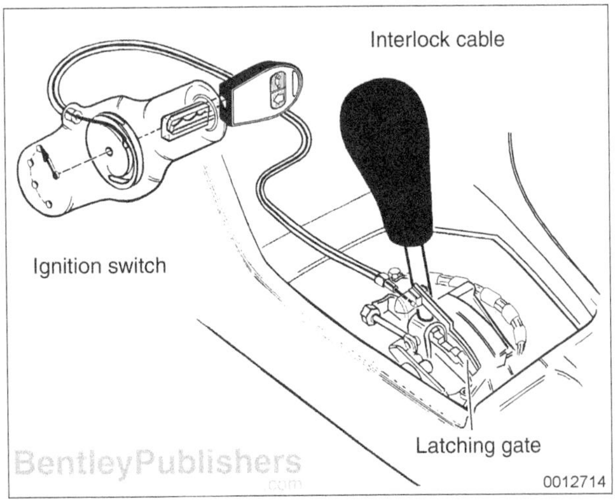

◄ Vehicles with automatic transmission (CVT) use a shift interlock cable between the ignition switch and the gear selector lever. This feature prevents the key from being removed from the ignition lock until the selector lever is in PARK, and locks the lever in PARK when the key is OFF or removed.

– Remove steering column lower cover. See **Steering column lower cover, removing** in this repair group.

– Remove selector knob and disassemble center console to access gear selector mechanism. See **250 Gearshift Linkage**.

– Remove gear selector assembly trim.

Working at gear selector assembly, loosen interlock cable bracket mounting bolt (**inset**). Slide cable out of bracket.

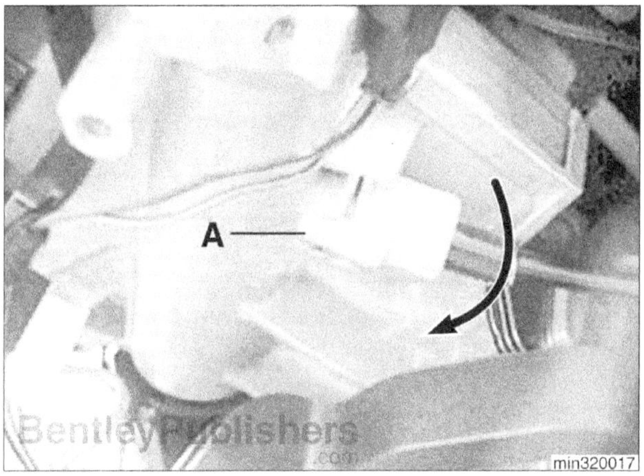

Working below steering column, pry down on interlock cable retaining tab (**A**).

- Twist cable end (**arrow**) to remove from ignition lock.

– Unhook cable from trim lug and remove.

– Installation is reverse of removal.

- Before retightening cable bracket mounting bolt, adjust cable.

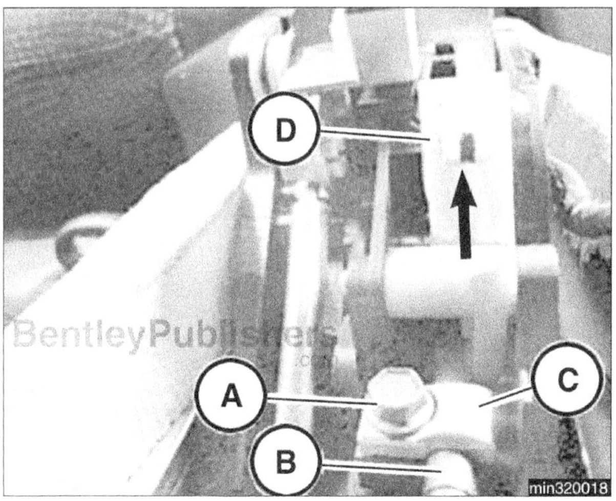

To adjust cable:

- Move selector lever to PARK.
- Turn ignition key OFF and remove.
- Loosen cable bracket bolt (**A**). It must be possible to move cable (**B**) slightly in bracket (**C**).
- Press (**arrow**) interlock lever (**D**) down into limit position.
- Tighten bolt (**A**).

Tightening torque	
Shift interlock cable bracket to selector assembly (M6)	4 - 6 Nm (35 - 53 in-lb)

– Check shift interlock adjustment:

- Ignition key removed or OFF: Selector lever must not move from PARK.
- Turn ignition key ON: It must be possible to select all selector lever positions.
- Select any lever position except PARK: It should not be possible to turn ignition key to OFF and remove it.
- Move selector lever to PARK: It should be possible to turn ignition key back to OFF and remove it.

– If necessary readjust interlock cable.

Lower steering shaft, removing and installing

– Raise front of car and support in a safe manner.

> ***WARNING —***
> *Make sure the car is stable and well supported at all times. Use a professional automotive lift or jack stands designed for the purpose. A floor jack is not adequate support.*

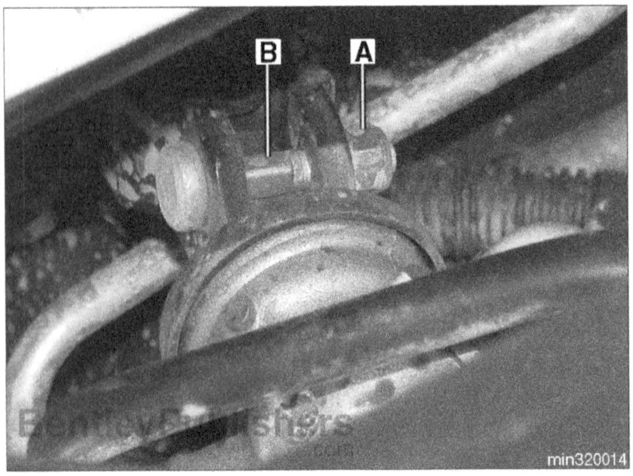

Working underneath front of car, remove lower steering shaft pinch nut (**A**) and remove eccentric bolt (**B**). Pry universal joint off steering rack shaft.

> ***CAUTION —***
> *Once the steering shaft is separated from the steering rack there is no longer a steering end stop and the steering wheel can be turned without restriction. This can damage the airbag contact ring. To prevent this damage:*
> *- Move steering wheel to straight-ahead position.*
> *- Remove ignition key to lock steering.*

– Fold lower universal joint upwards. Push lower half of joint up through protective boot.

– Remove steering column lower cover. See **Steering column lower cover, removing** in this repair group.

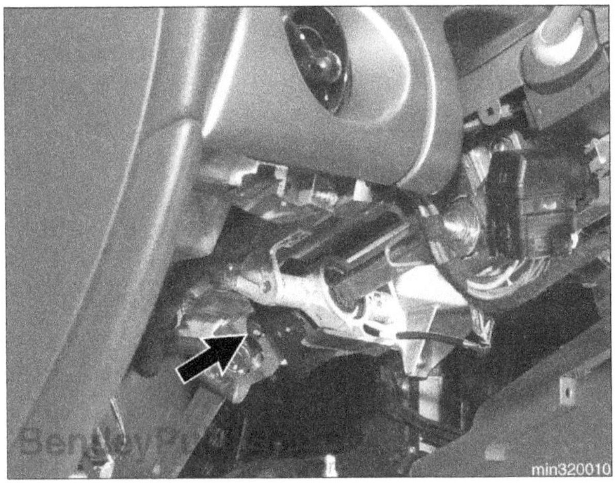

Working inside vehicle under steering column, remove steering shaft upper universal pinch bolt (**arrow**). Pry universal joint off upper steering column.

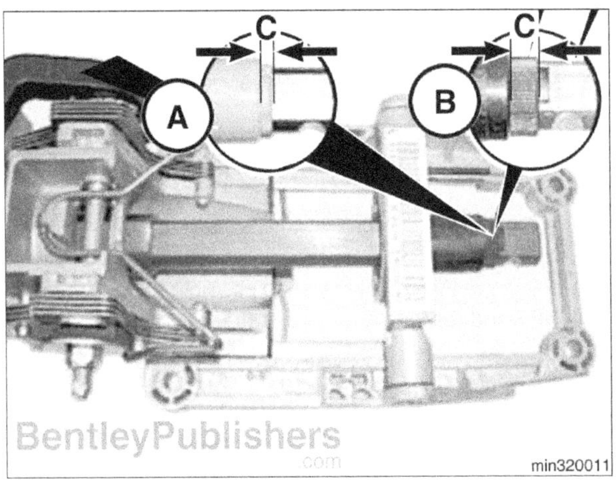

Upper steering columns versions differ:

- **A** (early version): **C** is approx. 3 mm (0.12 in). Color is natural or blue.
- **B** (new version): **C** is approx. 8 mm (0.31 in). Color is black.

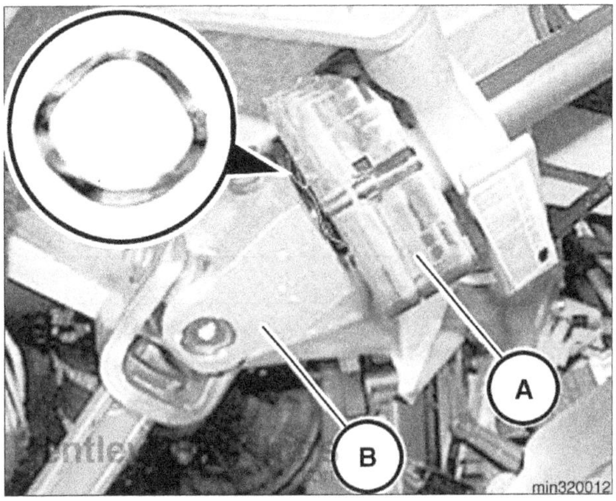

Vehicle with DSC and early steering column:

- Wavy washer (**inset**) is fitted between steering angle sensor (**A**) and lower steering shaft (**B**).
- The wavy washer is not required in vehicles without DSC or with new steering column.

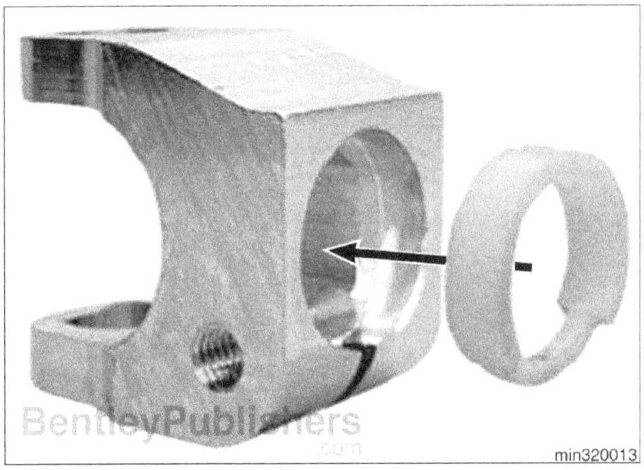

Vehicle with early upper steering column and new lower steering shaft:

- Fit adapter ring inside bore of universal joint.
- Adapter ring and wavy washer are not required for new upper steering column and new lower steering shaft.

When installing, make sure pinch bolt hole in lower steering shaft and locating dimple in upper steering shaft are aligned. When assembled, make sure end of upper steering shaft lines up with **dashed line**.

Tightening torque	
Lower steering shaft universal joint pinch bolt to upper steering shaft (M8 x 35 mm, replace bolt and nut)	30 Nm (22 ft-lb)

– Push lower universal joint down through protective boot.

Working underneath front of car, fit lower universal joint over steering rack shaft.

- Slide in new eccentric bolt and place new nut on it fingertight. Do not tighten.
- Push steering shaft up (**arrow**) as far as it will go (approx. 2 mm or 0.08 in).
- Tighten nut.

Tightening torque	
Lower steering shaft universal joint eccentric pinch bolt to steering rack shaft (M8, replace bolt and nut)	22 Nm (16 ft-lb)

Power Steering System

Steering system warnings and cautions

WARNING—
- *Do not reuse self-locking nuts. They are designed to be used only once and may fail if reused. Always replace them with new locking nuts.*
- *Do not install bolts and nuts coated with undercoating wax, as correct tightening torque cannot be assured. Always clean the threads with solvent before installation, or install new parts.*
- *Do not attempt to weld or straighten any steering components. Always replace damaged parts.*
- *Replace steering rack if any of the following apply:*
  -Visible or otherwise detectable damage to rack
  -Increase in steering torque or steering rack seizure when steering wheel is turned from stop to stop without power assistance
  -Acceptable tolerances for wheel alignment exceeded
  -Fire damage
- *Replace steering rack in case of damage, lasting deformation or fractures to:*
  -Wheels
  -Suspension struts
  -Wheel bearing carriers
  -Control arms
  -Stabilizer bar or links
  -Suspension bolting points on body
  -Front subframe
  -Steering knuckles
  -Steering rack mountings
  -Steering column

CAUTION—
- *The MINI power steering system is filled with Pentosin CHF 11S hydraulic fluid. DO NOT top up with ATF or other types of steering fluid. Mixing hydraulic fluid types will damage the seals in the steering system and cause them to leak.*
- *Maintain clean conditions when working with open power steering fluid lines.*
- *Plug off power steering lines and connections to keep out contamination.*

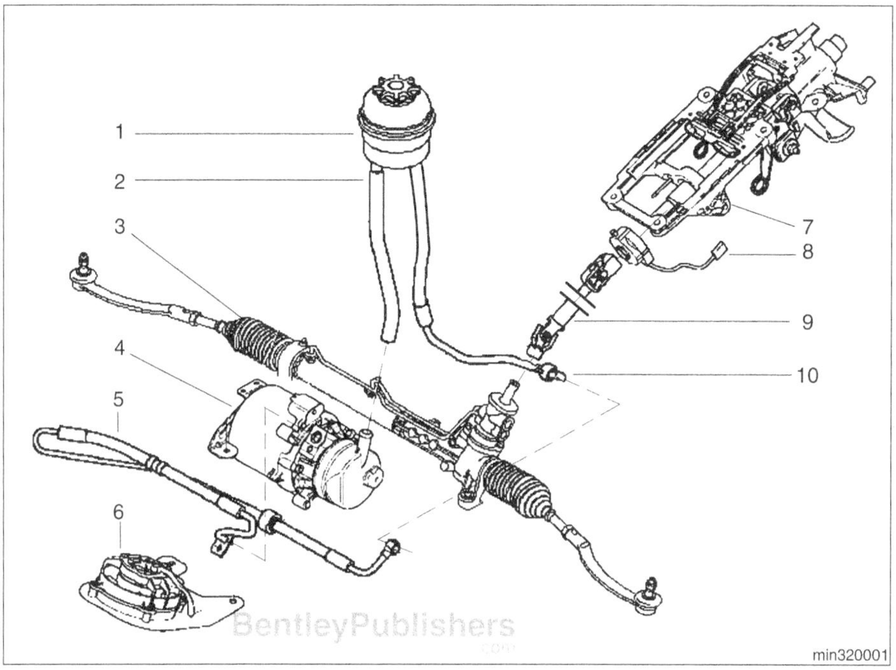

Steering system components

1. Steering fluid reservoir
2. Fluid supply line
3. Steering rack assembly
4. Electrohydraulic steering pump (EHPS)
5. Pressure hose assembly
6. Steering pump cooling fan
7. Upper (adjustable) steering column
8. Steering angle sensor
9. Lower steering shaft assembly
10. Fluid return line

Steering system, bleeding and filling

◀ The power steering fluid reservoir (**arrow**) is located in the rear of the engine compartment at the bulkhead.

> **CAUTION—**
> *Maintain clean conditions when working with power steering fluid.*

Power steering fluid
Recommended fluid: Pentosin CHF 11S

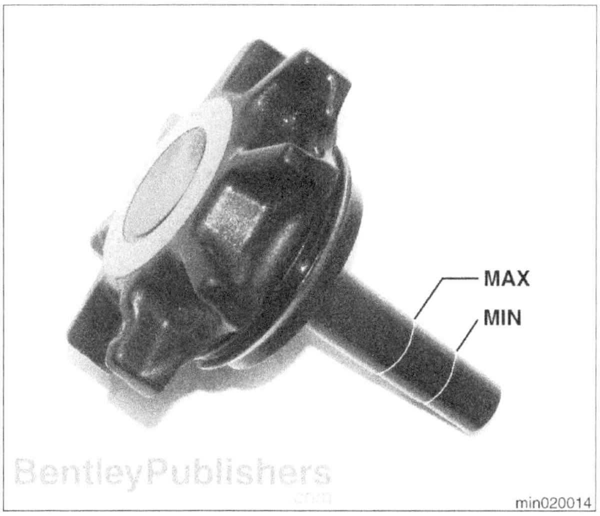

◀ With engine off, fill power steering fluid reservoir with clean fluid. Fill to **MAX** mark on dipstick.

– Start engine. Slowly turn steering wheel from lock to lock a minimum of two times.

– Turn engine off and check fluid level, adding more if necessary.

Electrohydraulic steering pump (EHPS), removing and installing

> ***CAUTION—***
> - *Maintain clean conditions when working with open power steering fluid lines.*
> - *Plug off power steering lines and connections to keep out contamination.*

– Siphon hydraulic fluid out of steering fluid reservoir and discard.

> ***CAUTION—***
> *Hydraulic fluid is a hazardous substance. Be sure to follow local, regional and federal guidelines for disposing of hazardous materials.*

– Raise front of car and support in a safe manner.

> ***WARNING—***
> *Make sure the car is stable and well supported at all times. Use a professional automotive lift or jack stands designed for the purpose. A floor jack is not adequate support.*

– Working underneath engine, remove power steering pump cooling fan. See **Power steering pump fan, removing and installing** in this repair group.

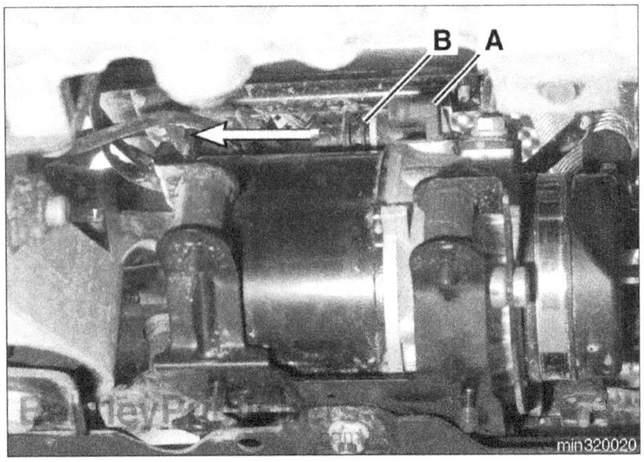

◄ Release pump electrical harness connectors (**A** and **B**).

- Connector **A**: Squeeze lock tab and remove plug.
- Connector **B**: Gently pry off in direction of **arrow**.

◄ Remove hydraulic pressure line mounting bolt (**A**). Detach line from pump.

– Models with manual transmission: Loosen or cut supply line hose clamp (**B**). Detach line from pump.

– Models with automatic transmission (CVT): Detach fluid supply line just below fluid reservoir.

> ***NOTE—***
> *Be prepared to catch dripping fluid.*

- Working from rear and above front subframe, remove pump bracket mounting bolts (**arrows**).

- Remove pump bracket front mounting nut (**arrow**). Remove steering pump with bracket.

- When replacing steering pump:
  - Remove pump mounting bolts (**A**) to detach pump from bracket.
  - Loosen nuts (**B**) on rubber mounts.

– Installation is reverse of removal.

- Replace any damaged rubber mounts.
- Replace fluid supply line hose clamp.
- Replace self-locking fasteners.

Tightening torques	
High pressure line to power steering pump (M8 x 1.25 x 20 mm)	10 Nm (89 in-lb)
Power steering pump bracket to rubber mount (M6)	8 ± 1 Nm (71 ± 9 in-lb)
Power steering pump bracket to subframe (M8) (replace self-locking fasteners)	19 Nm (14 ft-lb)
Power steering pump to mounting bracket (Torx M8)	10 ± 1 Nm (89 ± 9 in-lb)

– Fill system with correct fluid and bleed. See **Power steering system, bleeding and filling** in this repair group.

Steering pump fan, removing and installing

Working underneath engine, remove fan mounting nuts (**arrows**).

Pull fan away from subframe. Pry gently at electrical harness connector locking tab and detach connector.

– Installation is reverse of removal.
  - Reconnect electrical harness connector before installing fan.
  - Replace self-locking nuts.

Tightening torque	
Power steering pump fan to pump bracket (M8) (replace self-locking nuts)	19 Nm (14 ft-lb)

Steering rack, removing and installing

NOTE —
Models to 05/2003: Replace both tie rods if replacing steering rack.

CAUTION —
Maintain clean conditions when working with open power steering fluid lines.

– Siphon hydraulic fluid out of steering fluid reservoir and discard.

CAUTION —
Hydraulic fluid is a hazardous substance. Be sure to follow local, regional and federal guidelines for disposing of hazardous materials.

– Raise front of car and support in a safe manner.

WARNING —
Make sure the car is stable and well supported at all times. Use a professional automotive lift or jack stands designed for the purpose. A floor jack is not adequate support.

– Remove front wheels.

– Remove left stabilizer bar link. See **310 Front Suspension, Drive Axles**.

Paint mark steering tie rod adjusting collar (**arrow**) to simplify resetting of front toe after steering rack replacement.

Remove tie rod mounting nut at right steering knuckle. Use BMW special tool 32 3 090 or equivalent to detach tie rod.

- Repeat for left side.

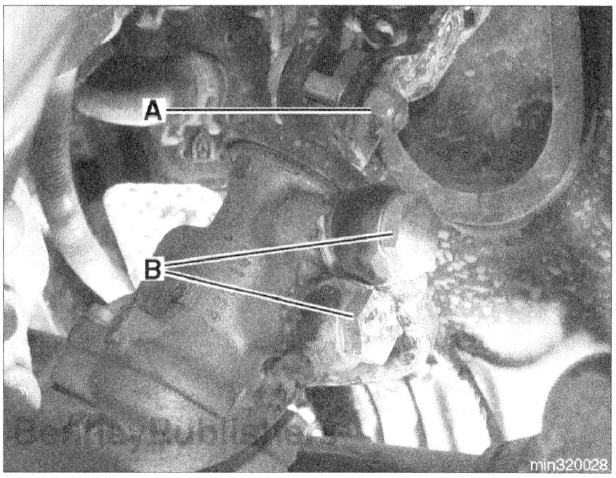

Remove lower steering shaft universal joint pinch bolt and nut (**A**). Pry universal joint off steering rack shaft.

- Remove fluid line banjo bolts (**B**) and detach fluid lines from steering rack.

> ***CAUTION—***
> *Plug off power steering lines and connections to keep out contamination.*

NOTE—
Be prepared to catch dripping fluid.

- Remove steering rack heat shield and heat shield support bracket.

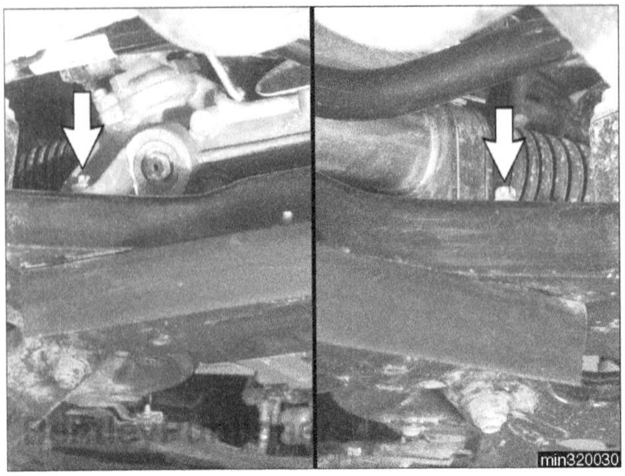

Remove steering rack mounting bolts (**arrows**).

NOTE —
- *There are two bolts on each side. Illustration shows one of each.*
- *Models to 05/2003: Note positions of different length rack mounting bolts.*

– Slide rack to left to remove.

– Models to 05/2003: If replacing rack, replace both tie rods.

– Models from 05/2003: If replacing rack, reuse tie rods if they are not defective. See **Tie rod or rack boot, removing and installing** in this repair group.

– Installation is reverse of removal. Bear in mind the following:
- Ensure that steering rack rubber mount and bracket are fitted and aligned correctly.
- Replace sealing rings at fluid line connections.
- Replace self-locking fasteners.
- Temporarily reset front toe using previously made paint marks on tie rod adjustment collars.

Tightening torques	
Fluid lines to steering rack (replace seals): High pressure (banjo bolt M14) Low pressure (banjo bolt M16)	 34 Nm (25 ft-lb) 40 Nm (30 ft-lb)
Heat shield to steering rack (M6)	8 Nm (6 ft-lb)
Lower steering shaft universal joint eccentric pinch bolt to steering rack shaft (M8, replace bolt and nut)	22 Nm (16 ft-lb)
Road wheel to hub	120 ± 10 Nm (89 ± 7 ft-lb)
Steering rack to subframe (Torx M10)	56 Nm (41 ft-lb)
Tie rod adjusting nut (M14 x 1.5): Lock nut (to 05/2003) Clamping ring (from 05/2003)	 63 Nm (46 ft-lb) 45 Nm (33 ft-lb)
Tie rod end to steering knuckle (M10 x 1.25) (replace self-locking nut)	52 Nm (38 ft-lb)
Tie rod to steering rack	80 ± 8 Nm (59 ± 6 ft-lb)

– Fill system with correct fluid and bleed. See **Power steering system, bleeding and filling** in this repair group.

– Have car professionally aligned.

Tie rod or rack boot, removing and installing

– Raise front of car and support in a safe manner.

> ***WARNING —***
> *Make sure the car is stable and well supported at all times. Use a professional automotive lift or jack stands designed for the purpose. A floor jack is not adequate support.*

– Remove front wheel.

< Remove tie rod mounting nut at right steering knuckle. Use BMW special tool 32 3 090 or equivalent to detach tie rod.

< Paint mark steering tie rod adjusting collar (**arrow**) to simplify resetting of front toe after tie rod replacement.

– Working at inner end of tie rod, loosen or cut rack boot retaining clamps. Slide boot back on tie rod.

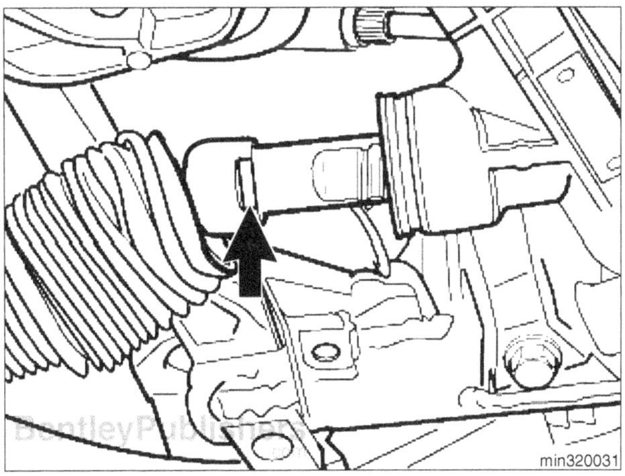

< Detach inner tie rod shaft (**arrow**) from steering rack.

> ***CAUTION —***
> *To avoid damage to steering rack while removing tie rod, turn steering until end of rack is as far as possible inside rack housing.*

– Clean and inspect internal rack shaft including toothed segment.

- Replace rack if polished surface is damaged or rusty.
- Lubricate rack toothed segment with cold-resistant grease.

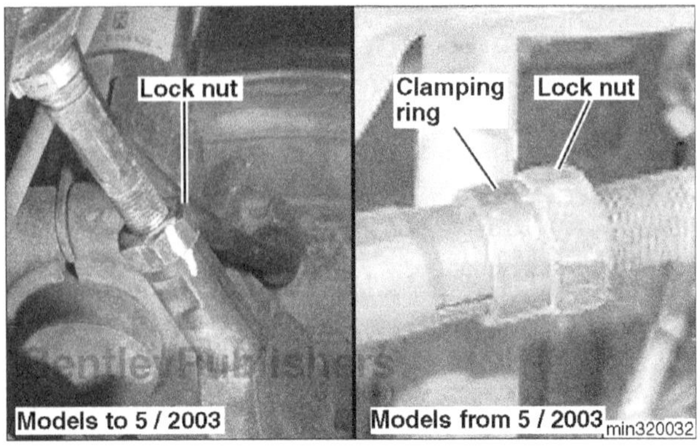

Loosen tie rod lock nut to separate outer tie rod from inner.

- Models to 05/2003: Tie rod secured with lock nut.
- Models from 05/2003: Tie rod secured with clamping ring and lock nut.

When installing new rack boot, grease tie rod at taper (**arrow**) so that small end of boot slides on tie rod, preventing rack book from twisting during toe-in adjustment.

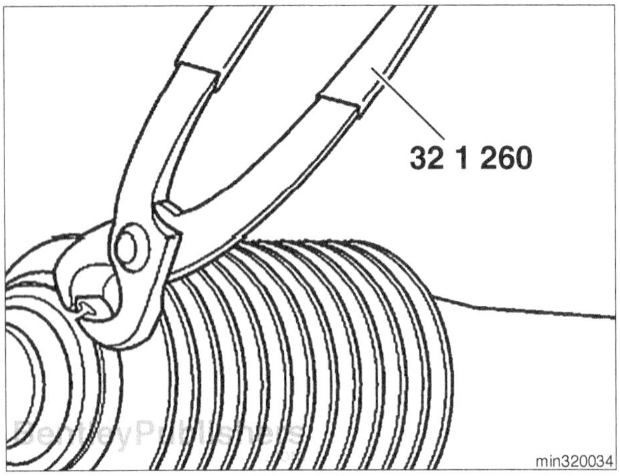

Use BMW special tool 32 1 260 or equivalent to tighten rack boot clamps.

– Remainder of installation is reverse of removal. Bear in mind the following:
  - Replace self-locking fasteners.
  - Temporarily reset front toe using previously made paint marks on tie rod adjustment collar.

Tightening torques	
Road wheel to hub	120 ± 10 Nm (89 ± 7 ft-lb)
Tie rod adjusting nut (M14 x 1.5): Lock nut (to 05/2003) Clamping ring (from 05/2003)	 63 Nm (46 ft-lb) 45 Nm (33 ft-lb)
Tie rod end to steering knuckle (M10 x 1.25) (replace self-locking nut)	52 Nm (38 ft-lb)
Tie rod to steering rack	80 ± 8 Nm (59 ± 6 ft-lb)

– Have car professionally aligned.

Wheel Alignment

Proper handling, stability, tire wear and driving ease depend upon the correct alignment of all four wheels. The front axle is aligned in relation to the rear axle, then the front wheels are aligned in relation to one another. This is known as a four-wheel or thrust-axis alignment. Proper alignment requires computerized alignment equipment.

Alignment specifications only apply under the following conditions:

- Correct wheels and tires are installed, in good condition, and are at the correct inflation pressures.
- All steering and suspension parts and bushings are undamaged and show no signs of abnormal wear.
- Wheel bearings are in good condition.
- Ride height is in accordance with specifications. See **300 Suspension, Steering and Brakes-General.**
- Car is in normal loaded position. See table **Normal loaded position** in this repair group.

Before aligning vehicle, bounce suspension a few times.

See **Table a** for front wheel alignment specifications, **Table b** for rear wheel alignment specifications.

> ***WARNING—***
> *While performing alignment procedures, make sure the car is stable and well supported at all times. Use a professional automotive lift or jack stands designed for the purpose. A floor jack alone is not adequate support.*

Normal loaded position	
Each front seat	68 Kg (150 lb)
Cargo compartment	14 kg (31 lb)
Fuel tank	Full

Table a. Front wheel alignment specifications

Parameter	**Specification**
Toe angle (total)	0° 18' ± 5'
Camber (difference between left/right max. 30')	–30' ± 25'
Track offset angle (20° steer angle on inside wheel)	–1° 16' ± 30'
Caster (difference between left/right max. 30') with ± 10° steer angle with ± 20° steer angle	 4° 47' ± 30' 4° 55' ± 30'
Front wheel offset (max steer angle) inside wheel outside wheel	 approx 36° 10' approx 30° 36'

Table b. Rear wheel alignment specifications	
Parameter	**Specification**
Toe angle (total)	0° 24' ± 8'
Camber (difference between left/right max. 30')	–1° 32' ± 30'
Geometrical axis difference	0° ± 10'

330 Rear Suspension

GENERAL

This repair group covers repair and replacement of components that make up the rear suspension of MINI cars.

See the following repair groups for additional related information:

- **300 Suspension, Steering and Brakes–General** for a description of the rear suspension and components, as well as specifications for setting ride height.
- **320 Steering and Wheel Alignment** for wheel alignment specifications.

Special tools

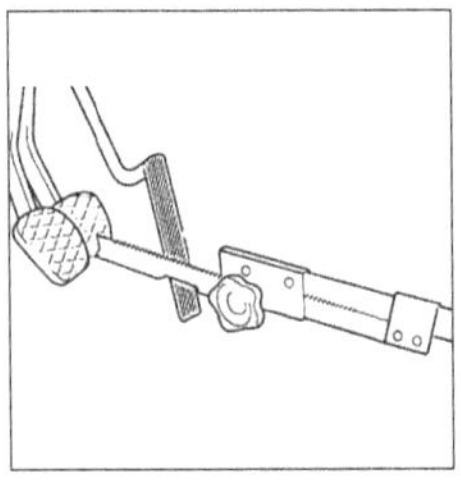

Pedal prop

Spring compressor
(BMW special tool 31 3 340)

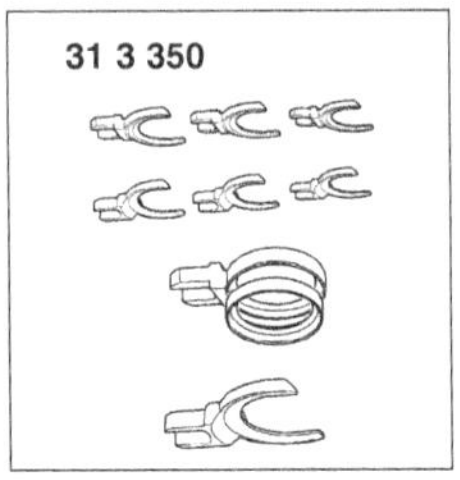

Spring compressor adapters
(BMW special tool 31 3 350)

Rear suspension design

Rear suspension components are mounted on a rigid subframe which is bolted to the body.

Springs and shock absorbers are of MacPherson variety.

Trailing arms are bolted to the chassis ahead of the subframe. Two control arms on each side attach to the trailing arms.

A stabilizer bar is anchored to the subframe and attached at each end to the trailing arms.

Wheel bearing are incorporated in the wheel hubs. The wheel bearing housing and brake caliper carrier are bolted to the trailing arm.

Rear suspension components

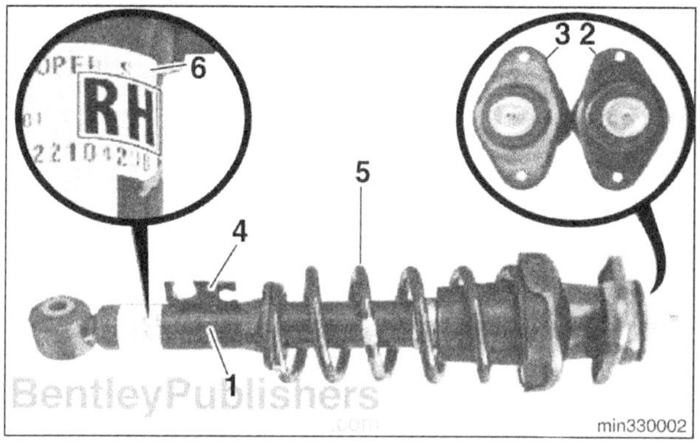

MacPherson strut assemblies control the damping of the rear suspension. Each strut assembly has the following features:

1. Damper (shock absorber)
2. Upper left strut mount
3. Upper right strut mount
4. Brake hose and ABS sensor bracket
5. Coil spring
6. Left / right identification label

Rear shock absorbers are of the twin-tube gas pressure design and are side-specific.

Coil springs, manufactured from silicon steel, are fitted to the shock assembly. The lower end of each spring is seated on a plate welded to the shock absorber tube. Between the spring and shock absorber is an isolator pad. The top of the spring is seated on an isolator pad and a cup clamped between the shock absorber rod and rebound plate by a nut.

The upper strut mounts are side-specific, identified by the letter R or L and attached to the body by two bolts each.

The bottom of the strut is attached to the trailing arm.

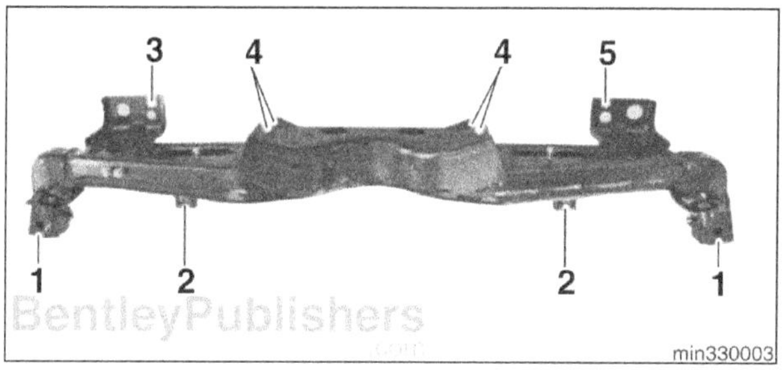

Rear subframe is attached to the body by four bolts. The left rear bolt (**3**) locates the subframe is in the correct lateral position.

1. Front mounting points
2. Fuel tank mounting points
3. Locator bolt, left rear mounting point
4. Control arm mounting points
5. Right rear mounting point

The subframe provides attachment points for the fuel tank straps and rear suspension control arms. Stabilizer bar anchor points are at the top of the subframe.

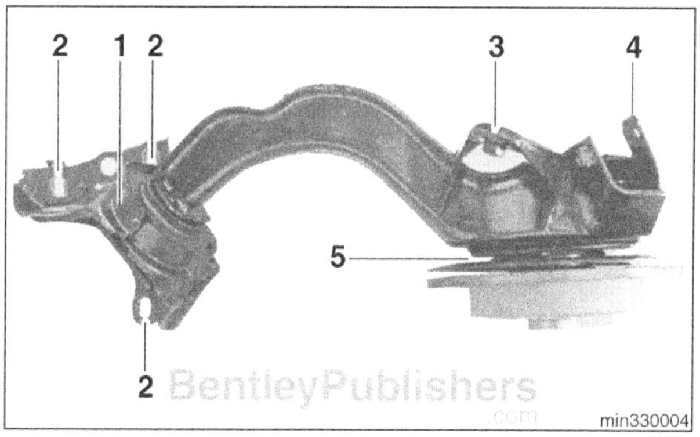

Trailing arms are fabricated from 4 mm (0.16 in) steel.

1. Trailing arm bushing
2. Body mounting points
3. Shock absorber lower mounting point
4. Control arm mounting points
5. Wheel bearing hub

The trailing arm provides rear toe control, counters brake lift and provides a low roll center. The arm also provides attachment for the wheel bearing hub, brake caliper, shock absorber and stabilizer bar link.

The trailing arm has attachment points for the rear upper and lower control arms.

Attached to the front of the trailing arm is a bushing which controls side-force steering performance and enhances ride and noise levels. The bushing housing is attached to the trailing arm by a single bolt. The housing bracket is mounted to the body by three bolts in slotted holes, allowing for rear toe adjustment.

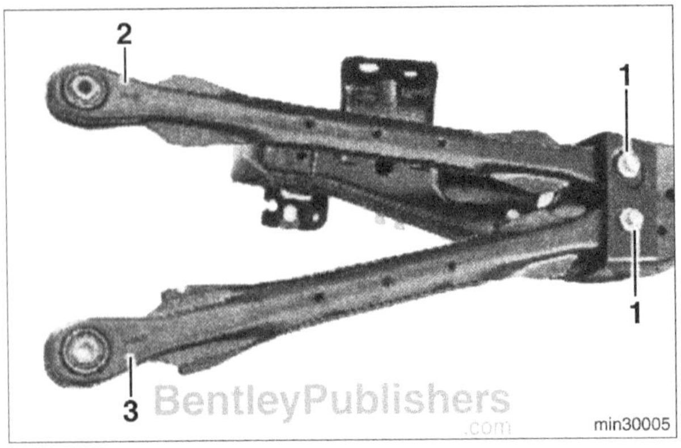

Rear control arms (sometimes referred to as lateral arms or wishbones) are steel with elastokinematic bushings at both ends. They connect the subframe to the trailing arms.

1. Inner mounting points to subframe
2. Upper control arm
3. Lower control arm

The control arms set rear wheel track.

The control arms are not side specific.

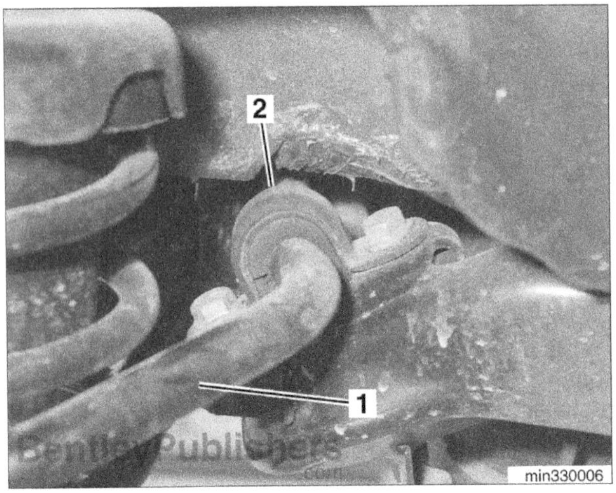

Rear stabilizer bar is attached to the top of the rear subframe by two PTFE bushings secured with clamp plates. The low friction PTFE bushings allow the stabilizer bar to rotate freely and quietly and require no additional lubrication.

1. Stabilizer bar
2. Bushing and clamp mounting to subframe

The stabilizer bar is made of solid spring steel.

The links from the stabilizer bar to trailing arms have ball joints at both ends.

Wheel bearing hub with integrated wheel bearing is bolted to the trailing arm.

Shock Absorbers and Springs

The rear suspension shock absorbers are MacPherson struts. The strut is a major component of the suspension and supports the spring. Most strut assembly components are available as replacement parts.

Rear strut, upper strut mount or spring replacement is a two-step procedure:

- Removal of strut assembly from vehicle
- Disassembly and replacement of component on work bench

NOTE—
- *If the shock absorbers have less than 50,000 km (30,000 miles), the manufacturer recommends replacing only the shock absorber that is worn or defective.*
- *Shock absorbers made by different manufacturers may be fitted on the same axle. However, always make sure that the new strut bears the same code as the old. Note identification label on strut tube.*
- *Determine whether a shock absorber needs replacing by checking it on a shock absorber test rig.*

Rear strut assembly, removing and installing

– Raise rear of car and support in a safe manner.

> ***WARNING —***
> *Make sure the car is stable and well supported at all times. Use a professional automotive lift or jack stands designed for the purpose. A floor jack alone is not adequate support.*

– Working underneath rear of car, remove stone guard at strut lower mounting.

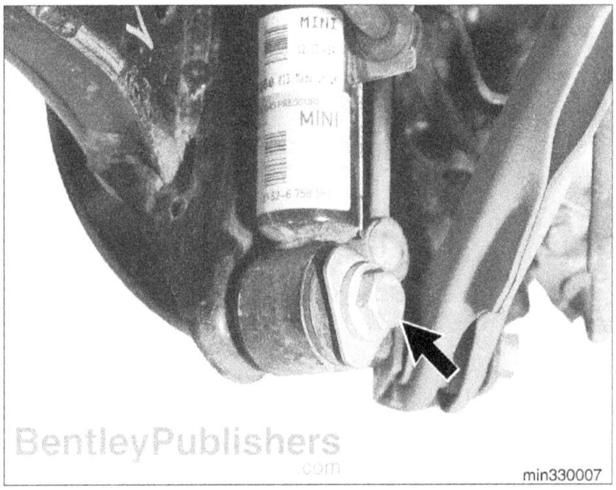

min330007

◄ Support trailing arm from below using an adjustable jackstand. Remove strut lower mounting bolt (**arrow**).

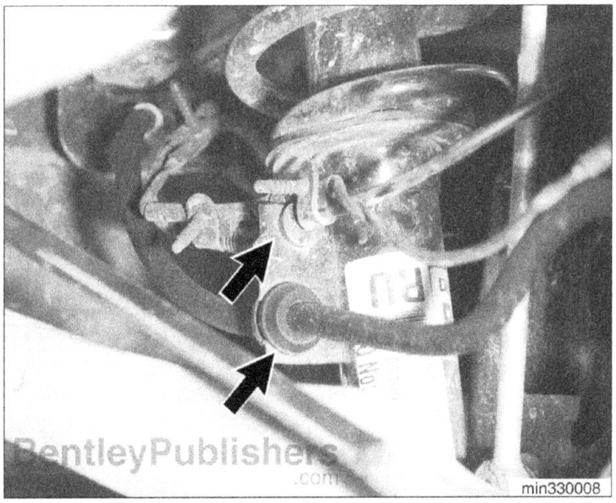

min330008

◄ Pry wheel speed sensor and brake hose protective grommets off strut bracket (**arrows**).

min330009

◄ While supporting strut assembly, remove strut top mounting bolts (**arrows**).

– Lower strut assembly from car and place on bench to disassemble. See **Strut assembly, disassembling and reassembling** in this repair group.

– Installation is reverse of removal. Keep in mind the following:
  - Make sure rubber grommets are correctly seated at strut bracket.
  - Replace self-locking bolts at upper mount.

Tightening torques	
Rear strut lower mount to trailing arm (M14 x 1.5)	140 Nm (103 ft-lb)
Rear strut upper mount to strut tower (M10 x 1.5) (replace self-locking bolts)	56 Nm (41 ft-lb)

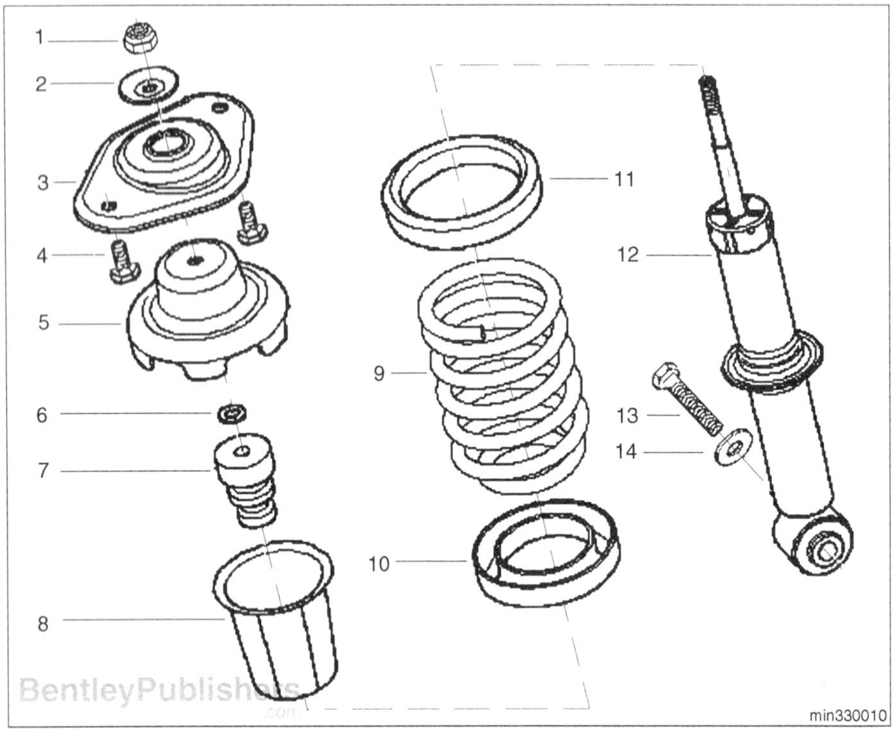

Rear strut assembly components

1. **Center strut retaining nut**
   - M10 x 1 self-locking tighten to 30 Nm (22 ft-lb)
2. **Cupped washer**
3. **Upper mount plate**
4. **Upper mount bolt**
   - M10 x 1.5 x 40 mm self-locking tighten to 56 Nm (41 ft-lb)
5. **Upper spring seat**
6. **Spacer ring**
7. **Rubber bump stop**
8. **Dust boot**
9. **Coil spring**
10. **Lower spring seat pad**
11. **Upper spring seat pad**
12. **Strut**
13. **Lower strut mounting bolt**
    - M14 x 1.5 x 80 mm tighten to 140 Nm (103 ft-lb)
14. **Washer**

Rear strut, disassembling and assembling

Replacing the strut, upper strut mount or spring requires that the strut assembly first be removed from the car and disassembled. See **Rear strut assembly, removing and installing** in this repair group.

For a guide to parts used during component replacement, see **Rear strut assembly components** in this repair group.

◀ Clamp spring compressor (BMW special tool 31 3 340 and 31 3 350 or equivalent) in shop vise.

> ***WARNING —***
> - *Do not attempt to disassemble the strut assembly without a spring compressor designed specifically for this job.*
> - *Prior to each use, check spring compressor for functionality. Do not use a damaged tool. Do not make any modifications to tool.*
> - *Use the correct size of spring retainers when compressing coil spring.*
> - *When assembling BMW spring compressor, make sure spring retainer plates are felt and heard snapping into place. Check seating of spring retainers carefully.*

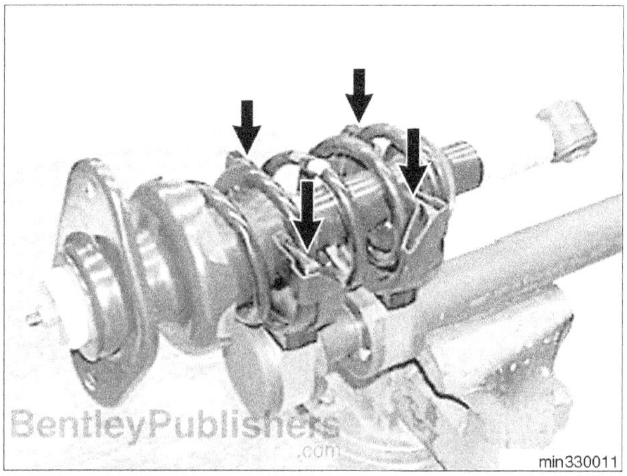

min330011

Position coil spring between spring holders so that coils lie in spring holder grooves (**arrows**). Compress spring.

> ***WARNING —***
> - *When tensioned, the spring coils must rest completely in the spring holder recess.*
> - *Never tighten or loosen spring compressor with an impact tool.*
> - *Only tighten down the coil springs until stress on the thrust bearing is relieved.*
> - *Only loosen strut center nut if spring coils are completely inserted in the spring holder grooves. If necessary, loosen compressor, reposition and recompress.*

> ***CAUTION —***
> *Do not damage dust boot.*

min330012

Remove strut nut. Counterhold strut shaft using 6 mm Allen wrench.

– Remove upper strut mount and related components.

– If new spring are being installed, relieve tension on spring compressor and remove coil spring.

NOTE —
- *Replace springs in pairs.*
- *When replacing springs, match color code at end of spring. These color codes are keyed to vehicle version and options such as air conditioning, automatic transmission etc.*

min330013

When placing spring in spring compressor:
- Smaller end of spring points to bottom of strut.
- Spring end touches stop on spring pad (**A**).
- Spring pad locator tab (**B**) rests in recess of spring seat.

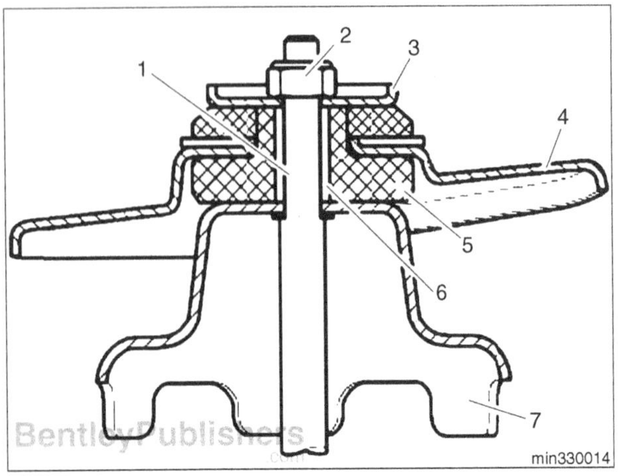

Reinstall upper strut mounting plate. Refer to illustration for order of components.

NOTE—
Make sure that the correct upper mounting plate is installed on each side of the vehicle. Match left and right markings on strut and upper mounting plate.

1. Strut piston rod
2. Center strut nut (M10 x 1.0) tighten to 30 Nm (22 ft-lb)
3. Cupped washer
4. Upper mount plate
5. Elastomer bushing
6. Inner sleeve (steel insert)
7. Upper spring seat

Tightening torque	
Upper strut mount to strut shaft (M10 x 1) (replace self-locking nut)	30 Nm (22 ft-lb)

Rear Subframe

The rear of the fuel tank mounting straps are bolted to the rear subframe, as are the rear suspension control arms and rear stabilizer bar. Subframe removal is a complex job. Read the procedure through before starting the job.

Rear subframe, removing and installing

– Cooper S: Detach battery positive lead and poke end of cable through rear floor grommet.

CAUTION—
Disconnecting the battery cable may erase fault codes stored in engine control module memory, as well as radio presets.

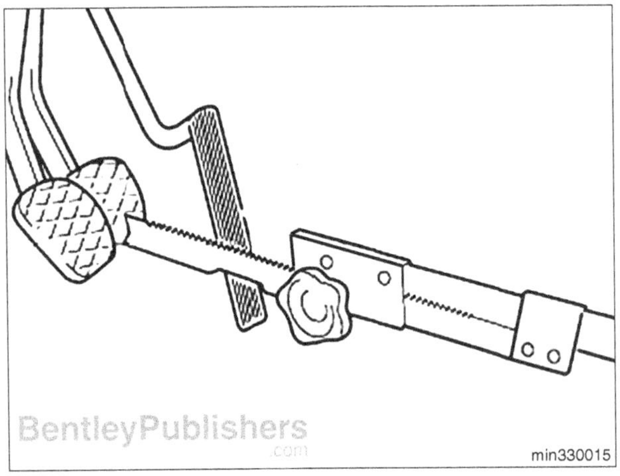

Insert pedal prop and depress brake pedal slightly. This prevents brake fluid from escaping when brake lines are detached.

– Raise car and support in a safe manner.

WARNING—
Make sure the car is stable and well supported at all times. Use a professional automotive lift or jack stands designed for the purpose. A floor jack is not adequate support.

– Remove exhaust system from catalytic converter back. See **180 Exhaust System**.

– Remove exhaust system heat shield.

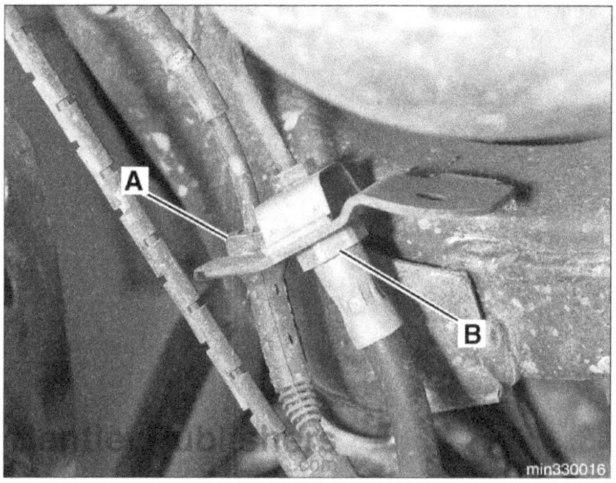

Working underneath rear of car at ends of subframe:

- Pry wheel speed sensor harnesses (**A**) off brackets.
- Separate brake lines (**B**).

NOTE—
Be prepared to catch dripping brake fluid.

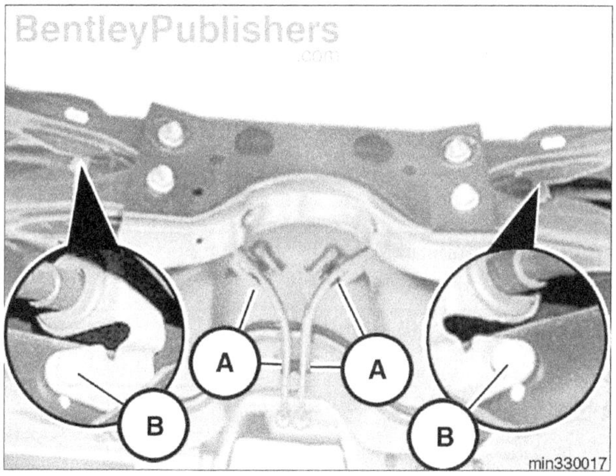

Detach parking brake cables;

- Pry cables off plastic retainers (**A**) on fuel tank.
- Remove cable bracket mounting bolts (**B**) at subframe.
- Detach cable ends from rear brake calipers.
- Push cables through subframe and allow to hang under fuel tank.

Remove stabilizer bar anchor mounting bolts (**arrows**).

– Models with headlight vertical aim control: Working at left end of subframe, remove vehicle level sensor mounting bolts.

– Remove upper and lower control arms. See **Rear Suspension Arms** in this repair group.

While supporting fuel tank from below, remove rear tank strap mounting bolts (**A**).

– While supporting subframe from below, remove subframe mounting bolts (**arrows**). Lower subframe carefully, checking for any wiring, hose or mechanical parts that might get snagged.

NOTE—
Right side is illustrated. Left side is similar.

– Installation is reverse of removal. Bear in mind the following:
- If installing new subframe, clip on new plastic brake line retainers.
- Replace self-locking fasteners.
- Raise subframe into position carefully, checking for any wiring, hose or mechanical parts that might get pinched.
- Tighten control arm mounting bolts only after vehicle is placed on ground and loaded normally.

Tightening torques	
Battery positive clamp to battery (M6)	5 Nm (44 in-lb)
Brake line to brake flex hose (M10 x 1)	14 Nm (10 ft-lb)
Control arm to rear subframe and to trailing arm (M12 x 1.5)	100 Nm (74 ft-lb)
Fuel tank strap to rear subframe (M8)	22 - 28 Nm (16 - 21 ft-lb)
Parking brake cable bracket to subframe (M6)	9 ± 1 Nm (168 ± 9 in-lb)
Rear subframe to body (M12 x 1.5)	100 Nm (74 ft-lb)
Stabilizer bar to rear subframe (M8)	19 Nm (14 ft-lb)
Vehicle level sensor to rear subframe (M6 x 1.0)	8 Nm (71 in-lb)

– Adjust parking brake and bleed brakes. See **340 Brakes**.

REAR SUSPENSION ARMS

This section covers removal, installation and repairs to the rear suspension arms and bushings.

Each side is equipped with:
- A single trailing arm with separately available bushing.
- Two control arms, each with integral rubber bushings.

Trailing arm, removing and installing

– Raise rear of car and support in a safe manner.

WARNING—
Make sure the car is stable and well supported at all times. Use a professional automotive lift or jack stands designed for the purpose. A floor jack alone is not adequate support.

– Remove road wheel.

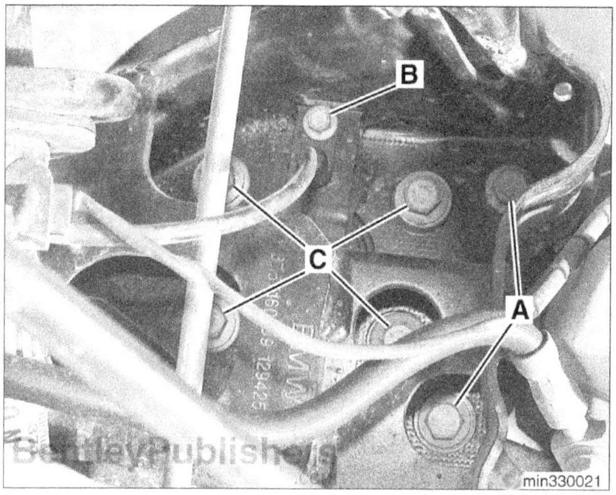

- Working underneath rear of car at wheel hub:
  - Remove brake caliper mounting bolts (**A**). Use stiff wire to hang caliper from body to relieve strain on brake hose.
  - Remove brake disc.
  - Remove wheel speed sensor mounting bolt (**B**) from trailing arm. Detach sensor and set aside.
  - Remove wheel hub mounting bolts (**C**). Remove wheel hub.

– Detach stabilizer bar link from trailing arm.

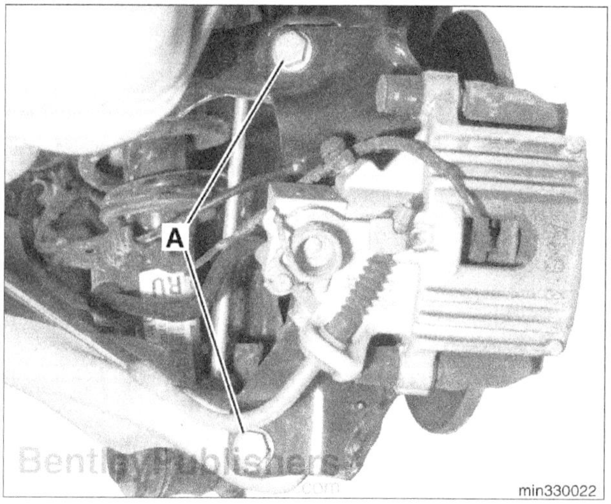

- While supporting trailing arm from below, remove upper and lower control arm mounting bolts (**A**).

- Remove strut lower mounting bolt (**arrow**).

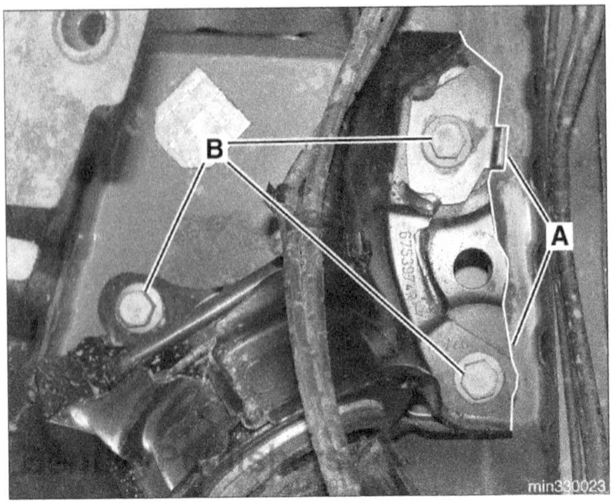

Working at front of trailing arm:

- Mark position of trailing arm bushing with respect to body (**A**) to facilitate subsequent rear toe adjustment.
- Remove lower trailing arm bushing mounting bolts (**B**).

– Installation is reverse of removal. Bear in mind the following:

- Replace self-locking fasteners.
- Align trailing arm bushing bracket with previously made marks on body before tightening mounting bolts.
- Tighten control arm mounting bolts only after vehicle is placed on ground and loaded normally.

Tightening torques	
Brake caliper to trailing arm (M10)	65 Nm (48 ft-lb)
Brake disc to wheel hub (Torx M10 x 13)	27 Nm (20 ft-lb)
Control arm to trailing arm (M12 x 1.5 x 64 mm)	100 Nm (74 ft-lb)
Rear strut lower mount to trailing arm (M14 x 1.5)	140 Nm (103 ft-lb)
Road wheel to hub	120 ± 10 Nm (89 ± 7 ft-lb)
Stabilizer bar link to trailing arm (M10 x 1.5) (replace self-locking nut)	56 Nm (41 ft-lb)
Trailing arm bushing to body (M12 x 1.5 x 66 mm)	100 Nm (74 ft-lb)
Wheel hub to trailing arm (M10 x 35 mm)	56 Nm (41 ft-lb)
Wheel speed sensor to trailing arm (M6)	8 Nm (71 in-lb)

– Have car professionally aligned.

Trailing arm bushing, removing and installing

– Remove trailing arm. See **Trailing arm, removing and installing** in this repair group.

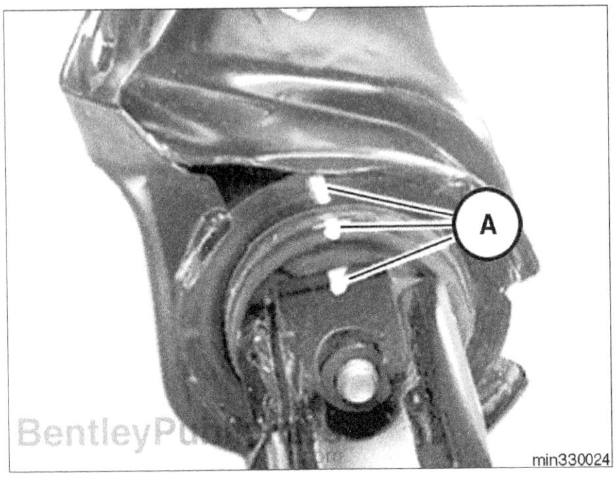

Mark position of trailing arm bushing with respect to trailing arm (**A**) before removal.

– If installing new trailing arm or bushing, tighten bushing mounting bolt only after vehicle suspension is loaded.

NOTE—
In vehicles from 05/2003, a self-tapping bolt is used to secure the bushing to the trailing arm. If the trailing arm is replaced, be sure to tap out the thread before installing the bushing.

Tightening torque	
Trailing arm bushing to trailing arm (M14 x 1.5)	165 Nm (122 ft-lb)

Upper control arm, removing and installing

– Raise rear of car and support in a safe manner.

> *WARNING —*
> *Make sure the car is stable and well supported at all times. Use a professional automotive lift or jack stands designed for the purpose. A floor jack is not adequate support.*

– Remove rear wheel.

– Cooper left side: Remove spare wheel.

– Cooper right side: Remove exhaust system from catalytic converter back. See **180 Exhaust System**.

– Models with headlight vertical aim control: Detach vehicle level sensor from left upper control arm.

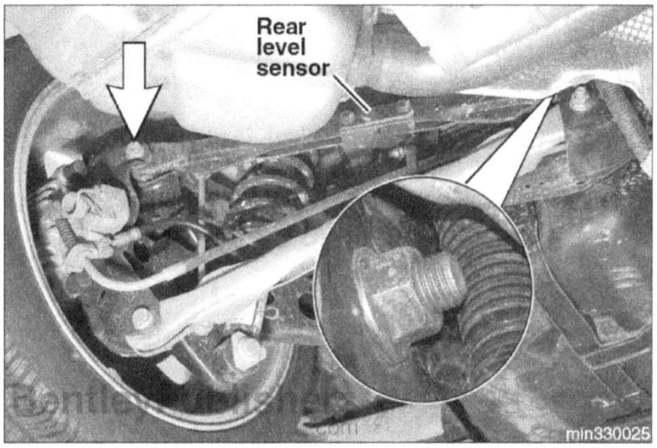

◄ Detach both ends of control arm:
- Remove outer mounting bolt (**arrow**).
- Remove inner mounting nut (**insert**).
- Remove control arm.

– Installation is reverse of removal.
- Replace self-locking fasteners.
- Tighten control arm mounting bolts only after vehicle is placed on ground and loaded normally.

Tightening torque	
Control arm to rear subframe and to trailing arm (M12 x 1.5)	100 Nm (74 ft-lb)
Vehicle level sensor to control arm (Torx self-tapping M8 x 1.25)	19 - 25 Nm (14 - 18 ft-lb)

– Have car professionally aligned.

STABILIZER BAR

Rear stabilizer bar removal requires that you loosen and lower the rear subframe in order to access the stabilizer bar anchors. Read the procedure through before starting the job.

Stabilizer bar, removing and installing

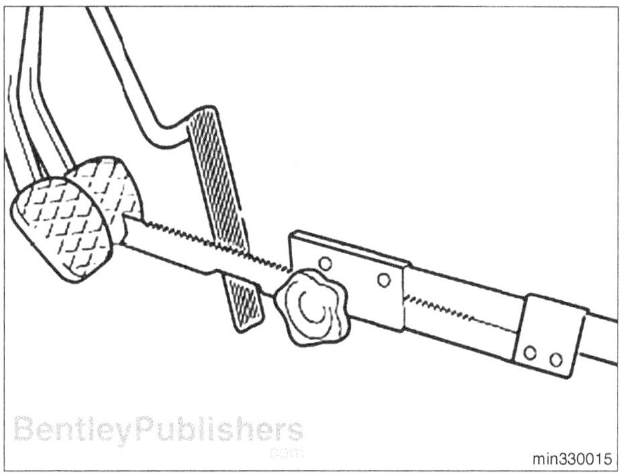

◄ Insert pedal prop and depress brake pedal slightly. This prevents brake fluid from escaping when brake lines are detached.

– Raise rear of car and support in a safe manner.

> *WARNING —*
> *Make sure the car is stable and well supported at all times. Use a professional automotive lift or jack stands designed for the purpose. A floor jack alone is not adequate support.*

– Remove rear wheels.

– Remove exhaust system from catalytic converter back. See **180 Exhaust System**.

– Remove exhaust system heat shield.

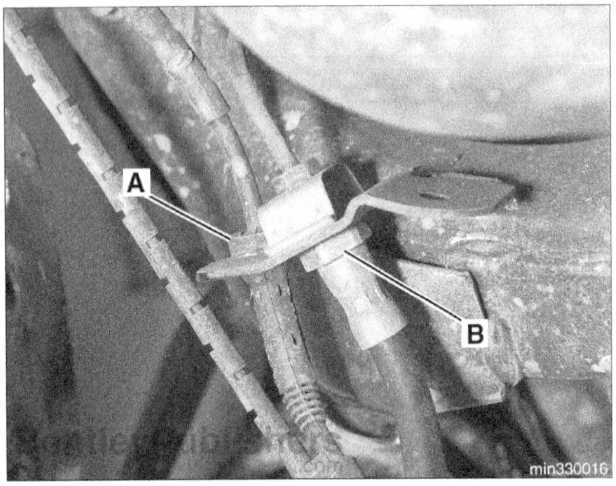

Working underneath rear of car at ends of subframe:

- Pry wheel speed sensor harnesses (**A**) off brackets.
- Separate brake lines (**B**).

NOTE —
Be prepared to catch dripping brake fluid.

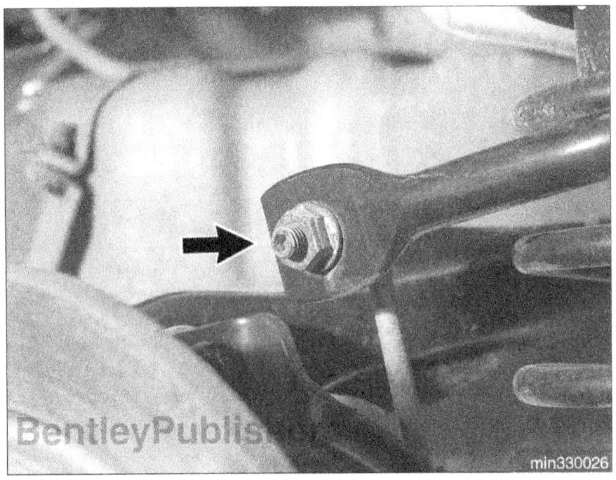

Remove stabilizer bar link mounting nuts (**arrow**). Detach stabilizer bar links from stabilizer bar.

WARNING —
For the next step, support rear subframe and fuel tank from underneath. Use a professional transmission jack or equivalent support designed for the purpose.

Remove rear subframe mounting bolts (**arrows**).

- Carefully lower subframe and fuel tank until stabilizer bar anchor bolts are accessible.

NOTE —
- *Right side is illustrated. Left side is similar.*
- *Cooper S: If necessary, detach battery positive lead and poke end of cable through rear floor grommet.*

CAUTION —
Disconnecting the battery cable may erase fault codes stored in engine control module memory, as well as radio presets.

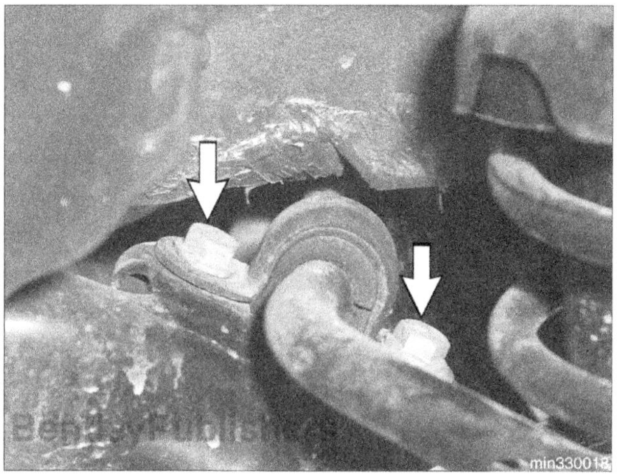

Remove stabilizer bar anchor mounting bolts (**A**). Lift stabilizer bar over rear subframe mounting brackets.

– Replace stabilizer bar rubber bushings:
  - Remove metal brackets and bushings. Discard old bushings.
  - Do not lubricate new bushings.

NOTE —
Low friction PTFE bushings allow the stabilizer bar to rotate freely and quietly and require no additional lubrication.

– Installation is reverse of removal. Bear in mind the following:
  - Raise subframe and fuel tank into position carefully, checking for any wiring, hose or mechanical parts that might get pinched.
  - Stabilizer bar links can be installed only one way, with the ball joints on the inside.
  - Replace self-locking fasteners.

Tightening torques	
Battery positive clamp to battery (M6)	5 Nm (44 in-lb)
Brake line to brake flex hose (M10 x 1)	14 Nm (10 ft-lb)
Rear subframe to body (M12 x 1.5)	100 Nm (74 ft-lb)
Road wheel to hub	120 ± 10 Nm (89 ± 7 ft-lb)
Stabilizer bar to rear subframe (M8)	19 Nm (14 ft-lb)
Stabilizer bar link to stabilizer bar (M10 x 1.5) (replace self-locking nuts)	56 Nm (41 ft-lb)

– Bleed brakes. See **340 Brakes**.

Stabilizer bar link, removing and installing

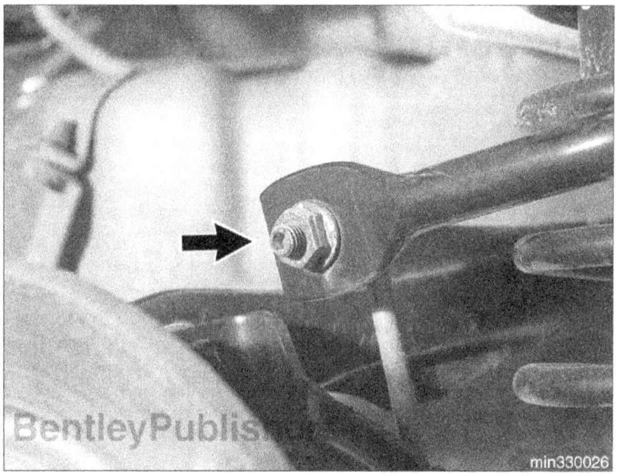

Working underneath rear of car at stabilizer bar end, remove stabilizer bar link mounting nut (**arrow**). Detach stabilizer bar link from stabilizor bar.

Working at trailing arm, remove stabilizer bar link mounting nut (**A**). Detach stabilizer bar link from trailing arm.

– Installation is reverse of removal. Bear in mind the following:
  - Stabilizer bar links can be installed only one way, with the ball joints on the inside.
  - Replace self-locking fasteners.

Tightening torques	
Road wheel to hub	120 ± 10 Nm (89 ± 7 ft-lb)
Stabilizer bar link to stabilizer bar and to trailing arm (M10 x 1.5) (replace nuts)	56 Nm (41 ft-lb)

Rear Wheel Bearings

The rear wheel bearings are permanently sealed and require no maintenance. The bearing is integral with the wheel hub. The wheel bearing and hub assembly is bolted to the trailing arm.

NOTE —
Rear wheel bearing and wheel hub are not available separately.

Rear wheel bearing, removing and installing

– Raise rear of car and support in a safe manner.

> ***WARNING —***
> *Make sure the car is stable and well supported at all times. Use a professional automotive lift or jack stands designed for the purpose. A floor jack alone is not adequate support.*

– Remove rear wheel.

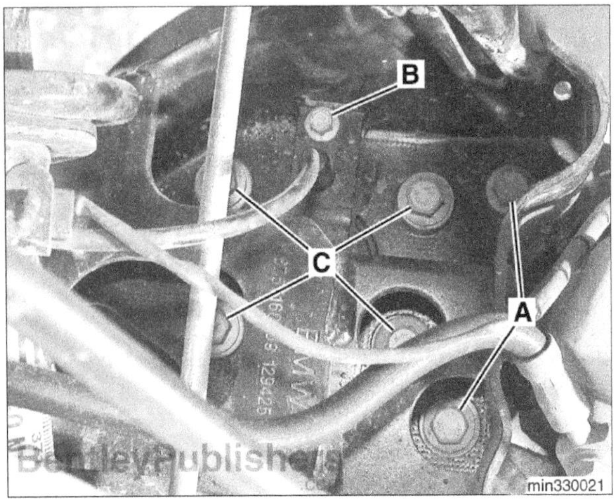

◄ Working underneath rear of car at wheel hub:
  - Remove brake caliper mounting bolts (**A**). Use stiff wire to hang caliper from body to relieve strain on brake hose.
  - Remove brake disc.
  - Remove wheel speed sensor mounting bolt (**B**) from trailing arm. Detach sensor and set aside.
  - Remove wheel hub mounting bolts (**C**). Remove wheel hub.

– Installation is reverse of removal.

Tightening torques	
Brake caliper to trailing arm (M10)	65 Nm (48 ft-lb)
Brake disc to wheel hub (Torx M10 x 13)	27 Nm (20 ft-lb)
Road wheel to hub	120 ± 10 Nm (89 ± 7 ft-lb)
Wheel hub to trailing arm (M10 x 35 mm)	56 Nm (41 ft-lb)
Wheel speed sensor to trailing arm (M6)	8 Nm (71 in-lb)

340 Brakes

TABLES

GENERAL

This repair group covers brake service:

- Brake pads, calipers, and rotors (disks) repairs
- Master cylinder, brake booster, and parking brake repairs
- ABS/ASC and ABS/DSC component replacement

Special tools

Pressure brake bleeder

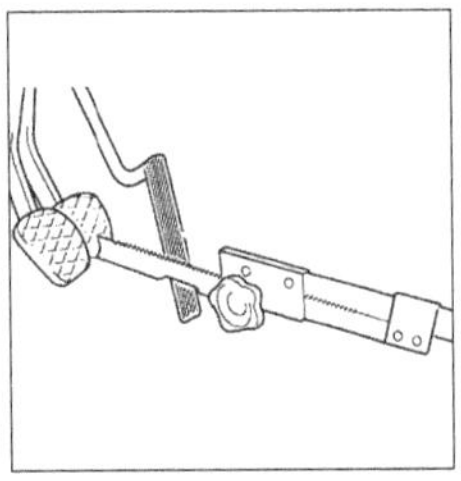

Pedal prop

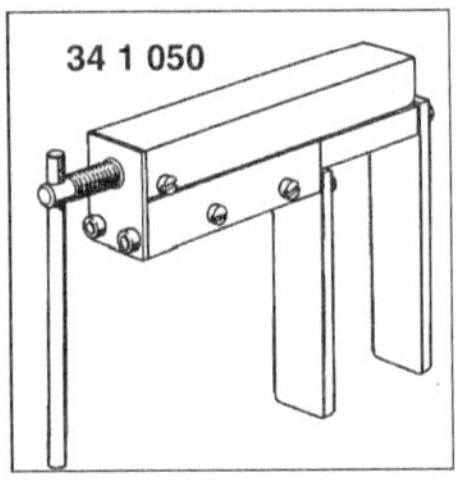

Caliper piston retraction tool (BMW special tool 34 1 050)

Steel wire brush (BMW special tool 34 1 090)

Brake pad lining gauge (BMW special tool 34 1 260)

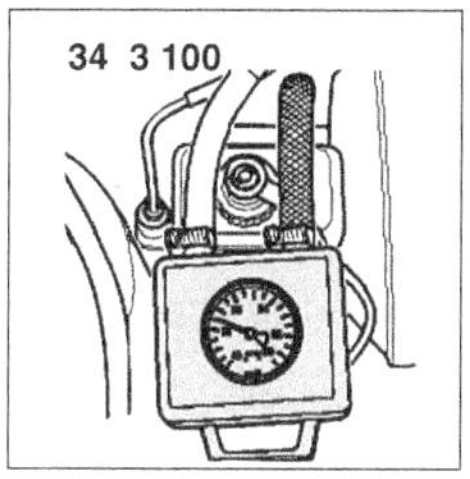

◀ Vacuum tester
(BMW special tool 34 3 100)

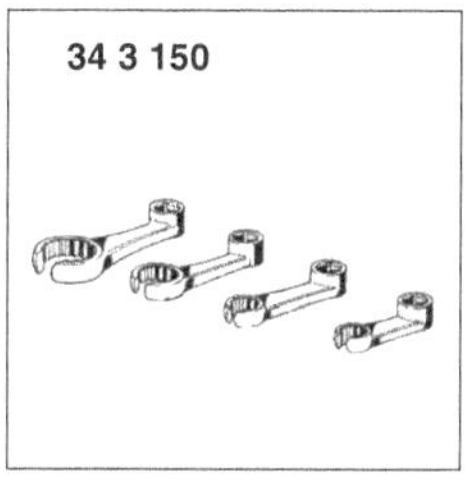

◀ Brake line flare wrench set (12 mm, 14 mm, 17 mm, 19 mm)
(BMW special tool 34 3 150)

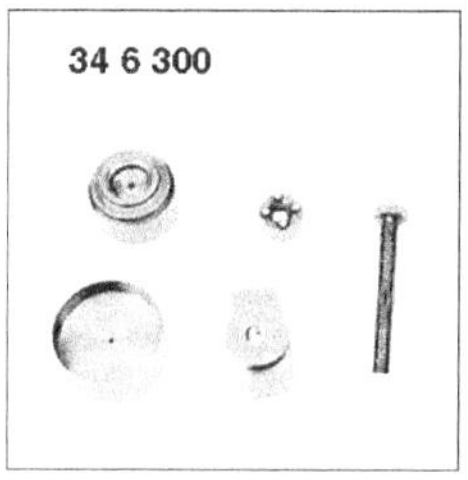

◀ Caliper piston retraction tool
(BMW special tool 34 6 300)

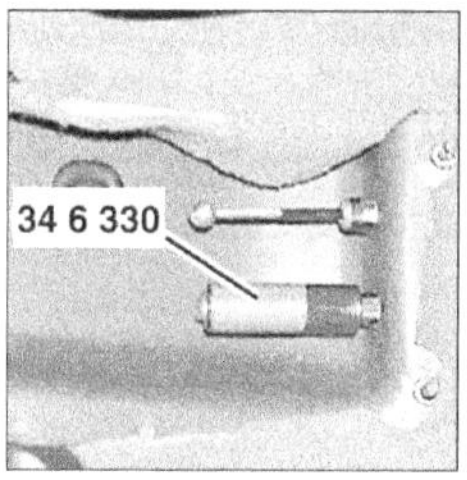

◀ Parking brake cable removal tool
(BMW special tool 34 6 330)

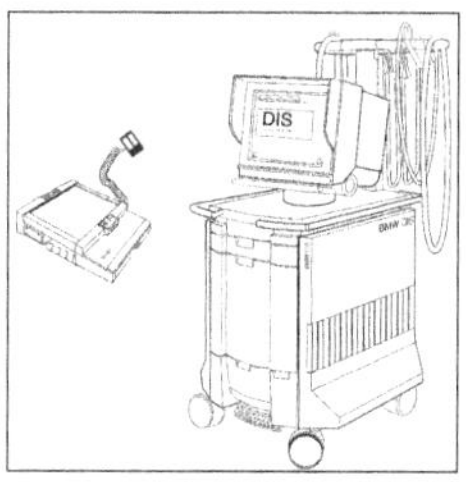

◀ DISplus, GT1 or MoDiC
(BMW scan tools)

MINI brake system

MINI models are equipped with vacuum-assisted four-wheel disc brakes. Single-piston calipers act on vented front and solid rear rotors (discs). A brake pad wear sensor for each axle indicates when brake pads need replacement.

The parking brake system is integrated with the rear brake calipers and rotors.

Electronic braking, stability and traction control features on MINI models are as follows:

- Antilock Braking System (ABS) is standard in all models.
- Automatic Stability Control (ASC) is optional in Cooper models, standard in Cooper S models.
- Dynamic Stability Control (DSC), otherwise known as traction control, is an option on all MINI models.

Antilock Braking System (ABS)

ABS prevents road wheel lock-up during brake application, thus maintaining vehicle stability and steering control under emergency conditions. Fitted to all MINI models, ABS is always active when the vehicle is being driven and cannot be disabled by the driver.

ABS components

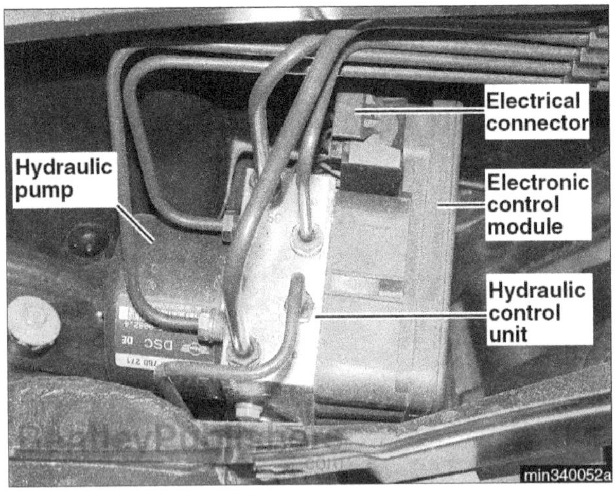

Hydraulic control unit/electronic control module is mounted in the right rear of the engine compartment in an insulated chamber.

- The aluminum hydraulic body contains the valves that control pressure to the individual brake lines.
- The cylindrical part is the electrically driven hydraulic pump.
- The plastic molded part is the electronic control module.

Wheel speed sensors are located at each of the four wheel hubs. The sensors are of the magnetoresistive type.

Brake light switch, mounted to the pedal cluster support bracket, is a Hall effect sensor supplying brake application information to the ABS control module.

Brake fluid level sensor contacts remain closed if hydraulic brake fluid level is correct. If fluid level fall below the minimum required, the switch opens.

ABS warning LED (amber), located in the instrument cluster, is ON during predrive self-check or when the system detects a fault.

ABS operation

The four ABS wheel sensors monitor the rotational speed of each wheel and signal the information to the ABS control module. If, during braking, one or more wheels decelerate at a rate that exceeds a programmed value, the ABS control module activates the ABS hydraulic unit and reduces the hydraulic brake pressure supplied to the affected wheel(s).

During antilock brake operation the driver will experience feed-back in the form of a pulsating brake pedal that is accompanied by solenoid/pump noise.

All versions of MINI are equipped with Teves Mk 60 ABS, which includes the following subfunctions:

Electronic Brake Force Distribution (EBV) controls the distribution of hydraulic brake pressure between the front and rear wheels. This is necessary to optimize the efficiency of the brakes under varying vehicle load conditions. It also allows larger rear brakes to be specified and ensures that front and rear brake linings wear at a similar rate.

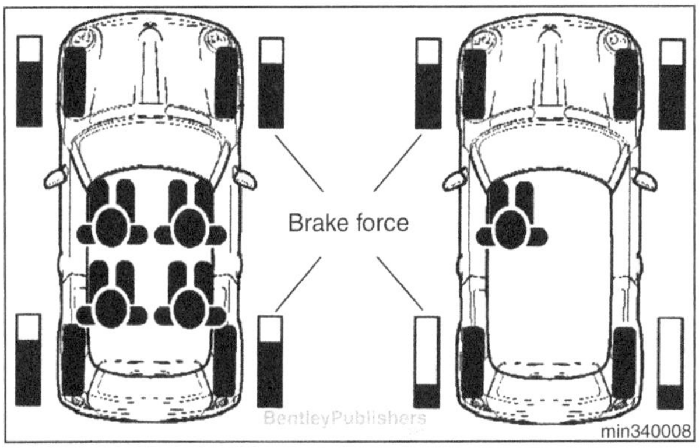

◄ Brake force is applied by the ABS hydraulic pressure pump to individual wheels in unloaded and fully loaded conditions.

EBV is in all MINI models and cannot be disabled by the driver.

Cornering Brake Control (CBC), together with EBV, helps to keep the vehicle under control under all braking, vehicle loading and steering conditions.

When the brakes are applied, weight is transferred to the front axle of the vehicle. If this occurs during cornering or directional change, the self-steering properties of the vehicle have a tendency to turn the vehicle into the curve (slight oversteer). This tendency is normally compensated for by the design characteristics of the suspension and by intuitive steering adjustment by the driver.

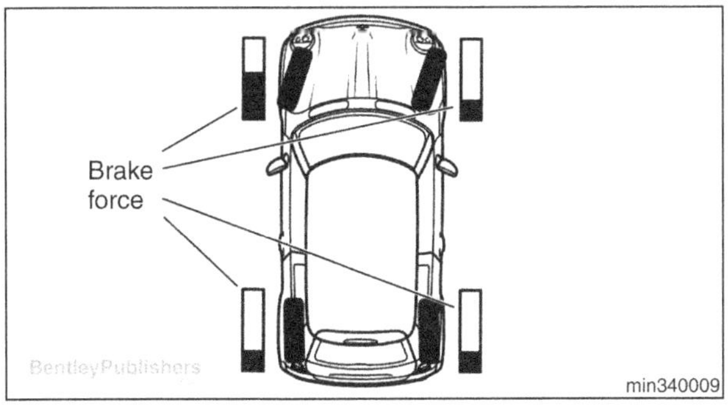

◄ CBC adjusts hydraulic pressure to each individual wheel under these circumstances to optimize directional stability.

NOTE—
CBC operates only when the brakes are applied by the driver, as opposed to DSC which operates even when the brakes are not applied.

CBC is in all MINI models and cannot be disabled by the driver.

Engine Drag Torque Control (MSR) controls vehicle slip during deceleration. If the vehicle is coasting with a low gear selected or during a downshift, excessive engine braking can result in wheel slip and vehicle instability. MSR intervenes by sending a command to the engine management system to cancel deceleration fuel cut-off and open the throttle valve.

Automatic Stability Control (ASC)

ASC prevents the driven front wheels from spinning when engine torque is applied and maintains an equal torque distribution to both wheels. This ensures that optimum traction and directional stability are maintained.

ASC components

The following ASC components differ from ABS components:

Hydraulic control unit/electronic control module, different in function but similar in appearance and in the same locations as the ABS control module.

Interface to engine control module (ECM).

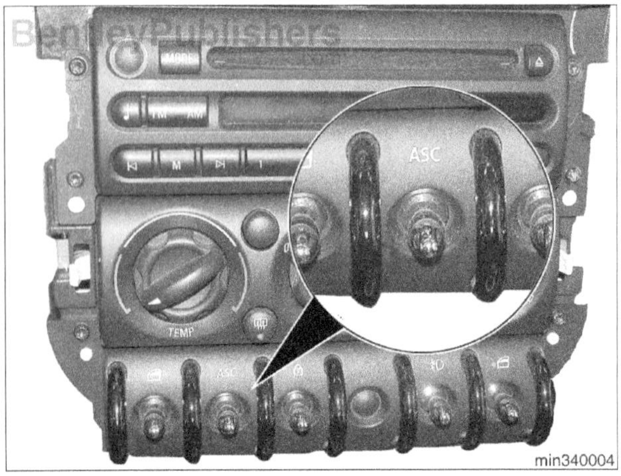

ASC switch in the center console, enabling driver to turn off ASC functions.

ASC warning LED (amber), located in the instrument cluster, illuminates in the noted mode for the following conditions:

- ON during predrive self-check
- ON when system is switched off by driver
- ON when a system fault is detected
- FLASHES when the system is active.

ASC operation

ASC is always installed with ABS, EBV, CBC and MSR.

The main ASC tasks are:

- Preventing loss of traction and possible resultant steering veer when engine torque is applied to driving wheels.
- Minimizing power-induced understeer (front end running wide) when driving through a corner and the front (driven) wheels start to slip due to excessive torque application.

ASC is always active when the vehicle is being driven unless disabled by the driver *via* the ASC switch on the dash.

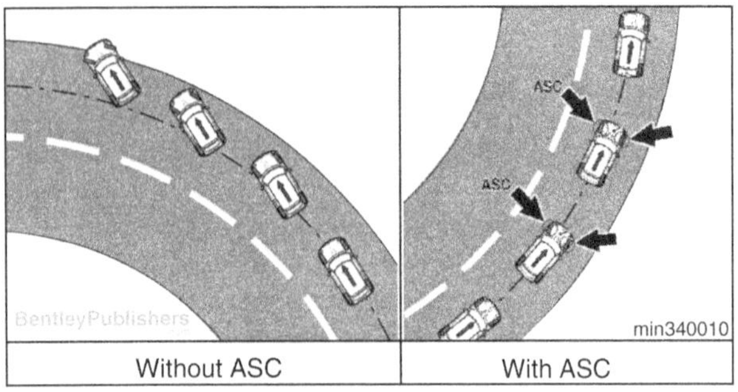

Prevention of wheel spin is achieved by engine torque reduction or brake application.

- To prevent one wheel spinning, braking is applied to that wheel.
- To prevent both wheels spinning, engine torque is reduced.

The ASC control module checks for wheel spin by comparing the speed of each driven (front) wheel with the speed of the rear wheel on the same side. The ASC control module then uses this information to determine whether brake application or engine torque reduction is appropriate.

Engine torque reduction is applied by the ASC control module communicating with the engine control module (ECM).

These processes occur within milliseconds of wheel speed sensors detecting a difference in wheel speeds.

A torque change request from the ASC control module to the ECM takes priority over any other torque change request received by, or generated internally within, the ECM. Torque reduction is limited to the minimum necessary to prevent wheel spin.

The ASC warning light in the instrument cluster flashes during system intervention. The light goes ON if the driver switches ASC OFF, or if the ASC control module detects a system fault.

NOTE —
- *Switch off ASC only under unusual circumstances such as driving away in snow or muddy conditions.*
- *Switch off ASC when snow chains are in use.*

Dynamic Stability Control (DSC)

ABS and ASC detect and control longitudinal (fore and aft) wheel slip but cannot detect lateral (side to side) forces acting on the vehicle. These systems therefore have limited influence on stability when cornering. DSC, supplementing either the ABS or ABS/ASC system, monitors the following:

- Steering wheel movement (*via* steering angle sensor)
- Lateral (side to side) forces (*via* acceleration sensor in DSC motion sensor cluster)
- Vehicle rotational speed (*via* yaw rate sensor in DSC sensor cluster)

Based on sensor inputs, the DSC control module analyzes the desired motion of the vehicle and compares it with the actual motion. If instability is detected, it takes corrective action.

DSC components

The following DSC components are in addition to ABS and ASC components:

Hydraulic control unit/electronic control module, different in function but similar in appearance and in the same locations as the ABS control module.

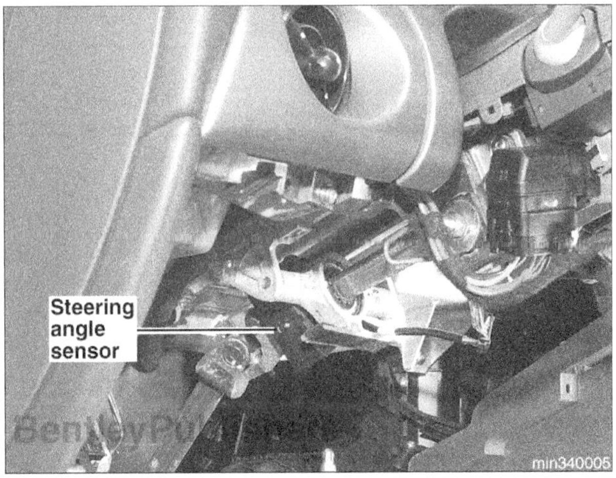

Steering wheel angle sensor, located on the upper steering shaft immediately above the upper steering universal joint, is a potentiometer with two brushes positioned at 90° to each other.

When first fitted on the vehicle, it is necessary to calibrate the sensor with the wheels in straight ahead position. This allows the DSC control module to identify and analyze steering wheel angle.

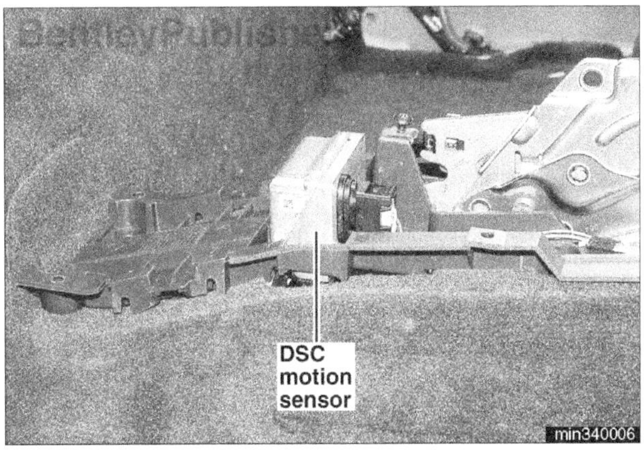

DSC motion sensor cluster, attached by 2 screws to a floor-mounted bracket behind the parking brake handle, contains the transverse acceleration sensor and a yaw (rotation speed) sensor.

The acceleration sensor measures the lateral force acting on the vehicle when cornering.

The yaw sensor measures the speed at which the vehicle is rotating (spinning) around its vertical axis.

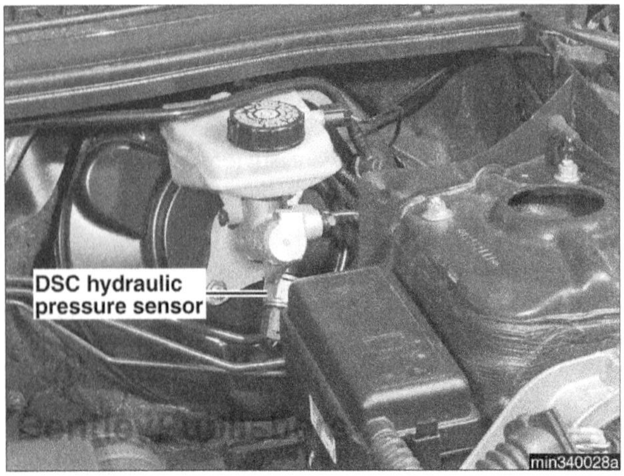

Hydraulic pressure sensor, screwed to the bottom of the brake master cylinder, provides fluid pressure information to the DSC control module in case the driver applies the brakes during DSC intervention.

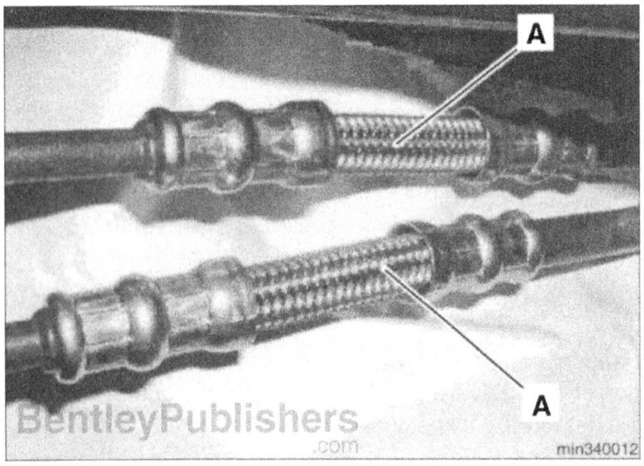

Flexible brake hoses (A) in the hydraulic brake lines from the master cylinder to the hydraulic unit reduce the rigidity of the brake lines and minimize noise transmission.

DSC switch, in the same location as ASC switch in the center console, enables driver to turn off DSC functions.

DSC warning LED (amber), located in the instrument cluster, is ON in the following circumstances:

- During predrive self-check
- When system is switched off by driver
- When a system fault is detected

The LED flashes when the system is active.

DSC operation

DSC is installed as a package with ABS, EBV, CBC, MSR and ASC.

DSC operates whether or not the driver applies the brakes and, if necessary, during ABS intervention.

Understeer or oversteer occur when lateral cornering forces acting on the vehicle overcome the grip of the tires on the road. Understeer is also commonly known as "push" and oversteer as "loose".

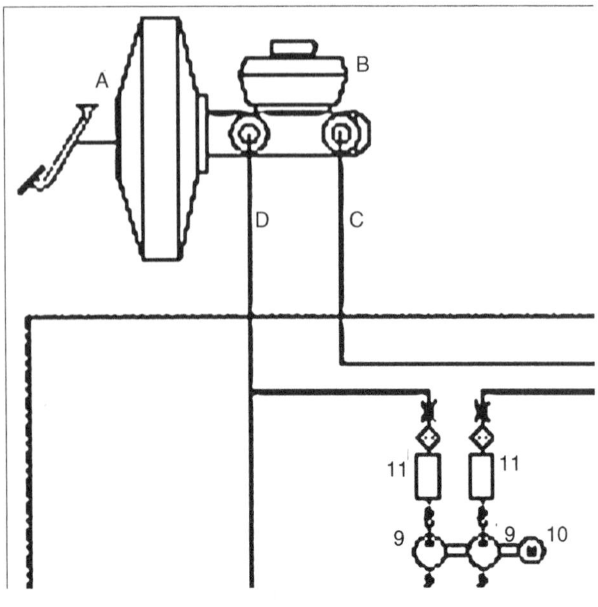

If DSC detects understeer during a turn, it applies brake force to the inside rear wheel, thus applying a counter force to bring the vehicle back onto its original course. A much lower braking force may also be applied to the inside front wheel.

If DSC detects oversteer, it applies brake pressure to the outside front wheel. A much lower braking force may also be applied to the outside rear wheel.

As with ASC, engine torque is also reduced.

The DSC warning light in the center mounted speedometer flashes when the system is operating. This is the same light and symbol as that of a vehicle equipped with the ASC system.

The warning light goes ON if the DSC control unit detects a fault with the system, or if the driver, *via* the switch panel, disables DSC.

Troubleshooting

Brake performance is mainly affected by three things:

- Level and condition of brake fluid
- Ability of brake system to create and maintain pressure
- Condition of friction components

Fluid contaminated by moisture or dirt can corrode the system. Inspect the brake fluid inside the reservoir. If it is dirty or murky, or is more then two years old, replace fluid.

To check the function of the master cylinder hold the brake pedal down hard with the engine running. If the pedal slowly falls to the floor, either the master cylinder is leaking internally, or fluid is leaking externally.

Inspect the rotors for glazing, discoloration and scoring. Steering wheel vibration while braking at speed is often caused by warped rotors, but can also be caused by worn suspension components.

When troubleshooting, keep in mind that tire inflation, wear and temperature can all have an affect on braking. See **300 Suspension, Steering and Brakes–General** for tire size and inflation information.

Table a lists symptoms of brake problems, probable causes, and suggested corrective actions. Unless noted otherwise, relevant repairs are described later in this repair group.

Table a. Brake system troubleshooting

Symptom	Probable cause	Repairs
Brake squeal	Brake pad carriers dirty or corroded or pads loose (poor fit) in pad carrier	Remove brake pads and clean calipers. Use original equipment pads for proper fit and use anti-squeal compound during installation.
	Brake pads heat-glazed or oil-soaked	Replace brake pads. Clean rotors. Replace leaking calipers as required.
	Wheel bearings worn (noise most pronounced when turning)	Replace worn bearings. See **310 Front Suspension, Drive Axles** or **330 Rear Suspension**.
	Incorrectly installed brake pads or other brake parts	Check component installation. Check/replace anti-rattle springs.
Pedal goes to floor when braking	Brake fluid loss due to system leaks	Check fluid level and inspect for signs of leakage.
	Master cylinder or traction control system faulty	Replace master cylinder. Diagnose traction control system using factory or compatible diagnostic tool.
Low pedal after system bleeding	Master cylinder faulty	Replace master cylinder.
Pedal spongy or brakes work only when pedal is pumped	Air in brake fluid	Bleed system using factory or compatible diagnostic tool.
	Master cylinder or traction control system faulty	Replace master cylinder. Diagnose traction control system using factory or compatible diagnostic tool.
Excessive braking effort	Brake pads wet	Use light pedal pressure to dry pads while driving.
	Brake pads heat-glazed or fluid-soaked	Replace brake pads and rotors. Replace leaking calipers.
	Vacuum booster or vacuum hose connections to booster faulty	Inspect vacuum lines. Test vacuum booster and replace as required. Test vacuum check valve for one-way flow.

Table a. Brake system troubleshooting		
Symptom	**Probable cause**	**Repairs**
Brakes pulsate, chatter or grab	Warped brake rotors	Resurface or replace rotors.
	Brake pads worn	Replace brake pads.
	Brake pads heat-glazed or oil-soaked	Clean rotors. Replace leaking calipers.
Uneven braking, car pulls to one side, rear brakes lock	Incorrect tire pressures or worn tires	Inspect tire condition. Check and correct tire pressures.
	Brake pads on one side of car heat-glazed or fluid-soaked	Replace brake pads. Clean rotors. Replace leaking calipers.
	Caliper or brake pads binding	Clean and recondition brakes.
	Worn suspension components	Inspect for worn or damaged suspension components. See **310 Front Suspension, Drive Axles** or **330 Rear Suspension**.
Brakes drag, bind or overheat	Brake caliper or brake pads binding	Clean or replace caliper.
	Master cylinder or traction control system faulty	Replace master cylinder. Diagnose traction control system using factory or compatible diagnostic tool.

Warnings and cautions

WARNING —

- *MINI vehicles require special BMW service equipment to properly bleed the brake hydraulic system. See* **Bleeding brakes** *in this repair group for additional information.*
- *Semi-metallic and metallic brake friction materials in brake pads produce dangerous dust. Wash or vacuum to remove dust. Do not blow away with compressed air.*
- *Brake fluid is poisonous, corrosive and dangerous to the environment. Wear safety glasses and rubber gloves when working with brake fluid. Do not siphon brake fluid with your mouth. Dispose of brake fluid properly.*
- *Do not reuse self-locking nuts or bolts. They are designed to be used only once and may fail if reused. Always replace with new self-locking fasteners.*

> ***CAUTION—***
> - *Immediately clean brake fluid spilled on painted surfaces and wash with water, as brake fluid will remove paint.*
> - *Always use new brake fluid from a fresh, unopened container. Brake fluid absorbs moisture from the air. This leads to corrosion problems in the braking system, and lowers the fluid boiling point.*
> - *Do not allow petroleum products or brake cleaner fluid to enter the brake system. These substances will cause brake failure.*
> - *Route brake lines and hoses so that they do not chafe on the body or other components.*
> - *To prevent damage to brake line couplings, loosen and tighten with special flare wrenches.*
> - *Plug open lines and brake fluid ports to prevent contamination.*
> - *Only tighten brake hoses on front wheels with wheels in straight ahead position.*
> - *If carrying out electric welding work, be sure to disconnect electrical harness connector from electronic control modules.*
> - *Do not expose electronic control modules to high sustained heat, such as in a paint drying booth. Maximum heat exposure:*
>   *-95°C (203°F) for short periods of time*
>   *-85°C (185°F) for long periods of time (approx. 2 hours)*

BLEEDING BRAKES

Brake bleeding is usually done for one of two reasons: either to replace old brake fluid as part of routine maintenance or to expel trapped air in the system that resulted from opening the brake hydraulic system during repairs.

Use factory scan tool (DISplus, GT1 or MoDiC) to bleed the brake system.

> ***WARNING—***
> *The ASC and DSC systems use electronic controls and a sophisticated hydraulic unit. When flushing the brake fluid from the system, use extreme care to not let the brake fluid reservoir run dry. Once air enters the hydraulic unit, it is very difficult to remove using traditional methods. For this reason, bleed brakes using a pressure bleeder and a large supply of fluid.*

> ***CAUTION—***
> *Use factory scan tool (DISplus, GT1 or MoDiC) to recode The ABS/ASC or ABS/DSC control module if one of the following components has been removed or replaced:*
> *- ABS/ASC or ABS/DSC control module*
> *- Throttle-valve assembly*
> *- Throttle potentiometer*

Always use new brake fluid from an unopened container. Bleed the entire system when any part of the hydraulic system has been opened. Be sure to have the special equipment needed before beginning the job.

When bleeding the brakes, start at the wheel farthest from the master cylinder and progress in the following order:

- right rear brake
- left rear brake
- right front brake
- left front brake

Brakes, bleeding

Brake fluid application	
All models	DOT 4

– Connect factory scan tool (DISplus, GT1 or MoDiC) to OBD II diagnostic plug (under left side of dashboard). Call up service function Bleeding ABS/ASC or service function Bleeding ABS/DSC, depending on installed system.

◄ Top off brake fluid in reservoir. Connect pressure bleeder to brake fluid reservoir. Turn on pressure.

> ***CAUTION—***
> *Do not exceed a pressure of 2 bar (29 psi) when pressure bleeding the brake system. Higher pressure may damage the fluid reservoir.*

– Raise vehicle and support in a safe manner.

> ***WARNING—***
> *Make sure the car is stable and well supported at all times. Use a professional automotive lift or jack stands designed for the purpose. A floor jack is not adequate support.*

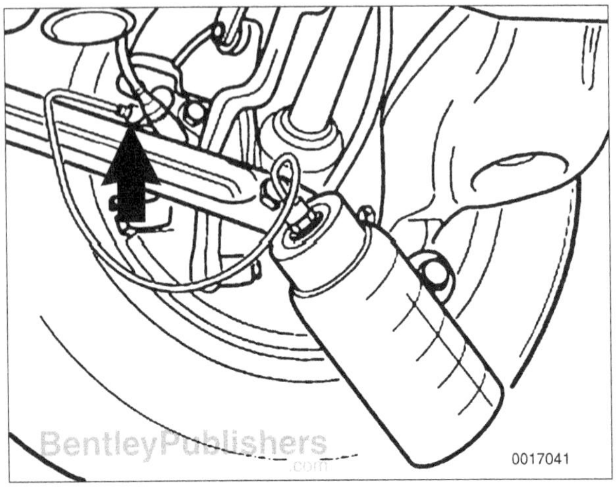

◄ Starting at right rear wheel, connect bleed hose and fluid receptacle to brake caliper bleeder screw (**arrow**).

– Open bleeder valve and flush until clear brake fluid emerges without bubbles. Close bleeder valve.

> ***CAUTION—***
> *Bleeder hose must always remain submersed in clean brake fluid whenever the bleeder valve is open.*

– Follow same procedure to bleed left rear, right front and left front brakes.

– Return to the right rear caliper and follow instructions in DIS to bleed. Repeat on left rear.

– Connect bleed hose and fluid receptacle to brake caliper bleeder screw at right front.

– Open bleeder valve.

– Fully depress brake pedal at least twelve times until clear brake fluid emerges without bubbles.

– Hold brake pedal in fully depressed position and close bleeder valve. Release brake pedal.

– Follow same procedure to bleed left front brakes.

– Remove brake bleeding apparatus. Top up brake fluid. Check to make sure brake fluid reservoir cap seal is intact.

Brake Pads, Calipers and Rotors

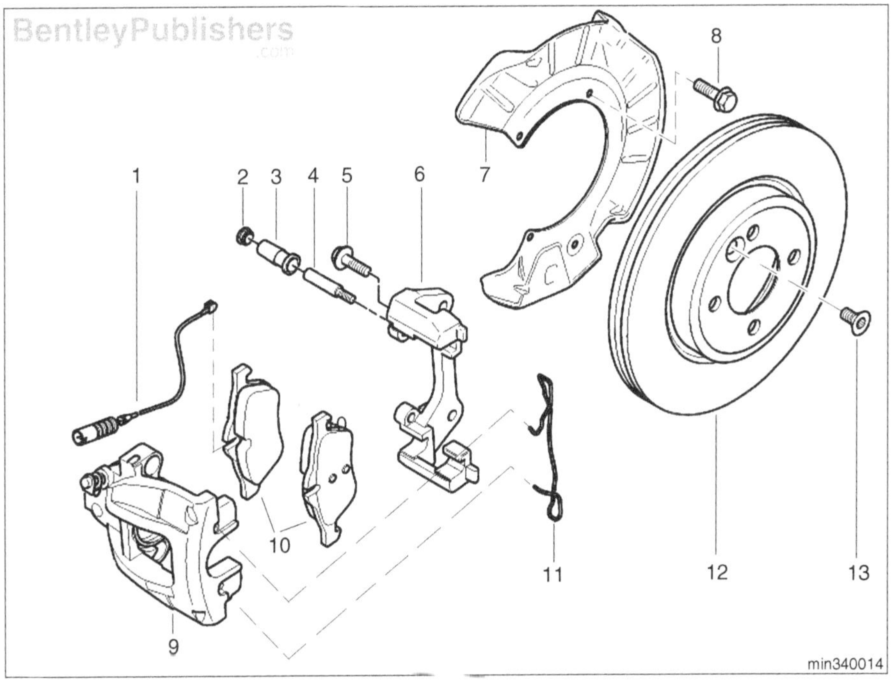

Front brake assembly components

1. **Brake pad wear sensor**
   - Left front only
   - See **Brake pad wear sensor**
2. **Protective cap**
3. **Guide bushing**
4. **Guide bolt**
   - 7 mm Allen
     tighten to 25 - 30 Nm (18 - 22 ft-lb)
5. **Collar bolt**
   - M12 x 1.5
     tighten to 110 Nm (81 ft-lb)
   - See **Brake pad carrier removal**
6. **Brake pad carrier**
7. **Dust shield**
8. **Bolt**
   - Torx M6 x 10
     tighten to 8 Nm (6 ft-lb)
9. **Brake caliper**
   - See **Front caliper removal**
10. **Brake pads**
    - See **Brake pad removal and installation**
11. **Spring clip**
    - See **Front caliper removal**
12. **Brake rotor (disc)**
    - 276 mm diameter
    - See **Brake rotor removal**
    - New: 22 mm (0.866 in)
      Wear limit: 19 mm (0.748 in)
13. **Bolt**
    - Torx M10 x 13
      tighten to 27 Nm (20 ft-lb)

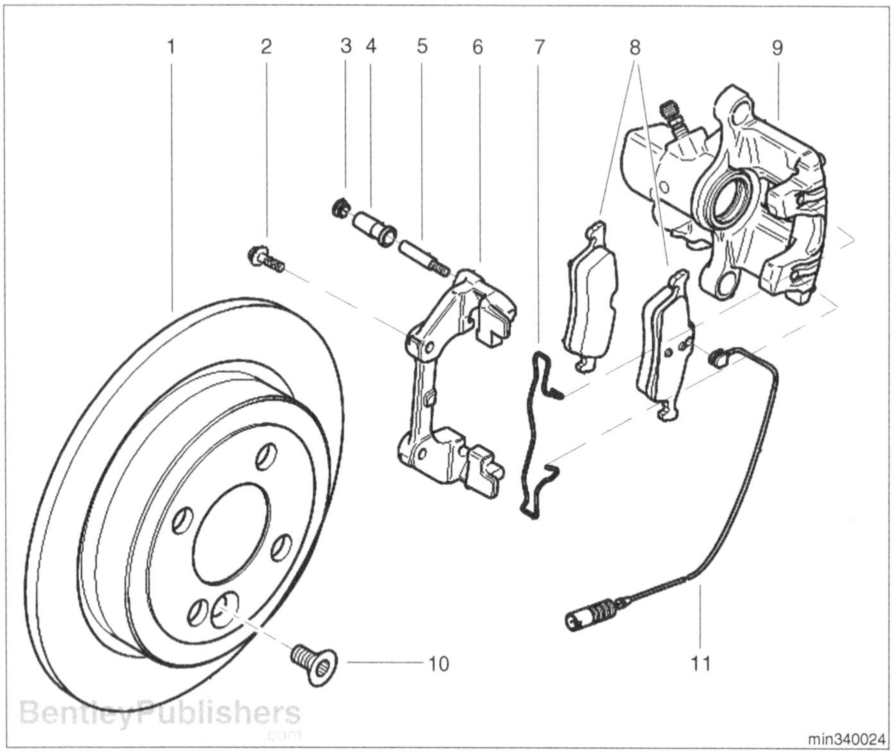

Rear brake assembly components

1. **Brake rotor (disc)**
   - 259 mm diameter
   - See **Brake rotor removal**
   - New: 10 mm (0.394 in)
     Wear limit: 7 mm (0.276 in)
2. **Collar bolt**
   - M10
     tighten to 65 Nm (48 ft-lb)
   - See **Brake pad carrier removal**
3. **Protective cap**
4. **Guide bushing**
5. **Guide bolt**
   - 7 mm Allen
     tighten to 25 - 30 Nm (18 - 22 ft-lb)
6. **Brake pad carrier**
7. **Spring clip**
8. **Brake pads**
   - See **Brake pad removal and installation**
9. **Brake caliper**
   - See **Rear caliper removal**
10. **Bolt**
    - Torx M10 x 13
      tighten to 27 Nm (20 ft-lb)
11. **Brake pad wear sensor**
    - Right rear only
    - See **Brake pad wear sensor**

JCW Tuning brake assembly components

A high performance brake component kit is available from John Cooper Works Tuning (JCW) and MINI to upgrade the standard brake system. Most of the hardware upgrade is for the front brakes, but special duty pads are required for the rear to retain brake balance and are included in the kit.

NOTE—
- *Due to the special heavy duty nature of the friction material components of the JCW Tuning brake kit, a minimum break-in period of 200 miles (322 kilometers) is required. Consult the installation instructions included with the kit for specific details.*
- *Because of the larger diameter of the brake rotors, this kit cannot be installed with certain wheel/tire combinations. Consult an authorized MINI dealer for applications.*
- *Part replacement information and some specifications were not available at time of publication.*

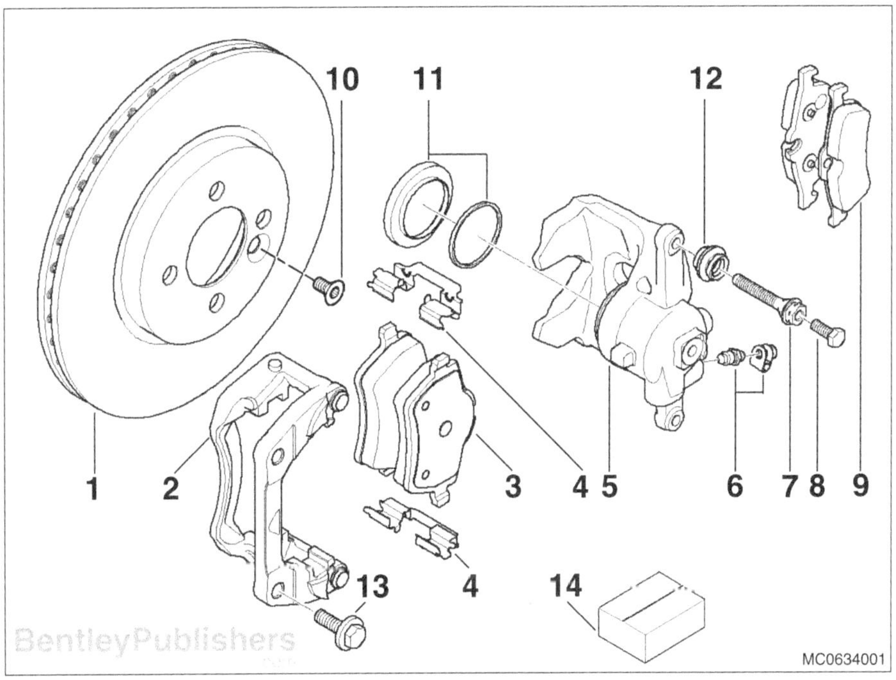

1. **Brake rotor (disc)**
   - 294 mm diameter
   - Internally vented
   - New: 22 mm (0.866 in)
     Wear limit: 19 mm (0.748 in)
2. **Caliper frame**
3. **Front brake pads**
   - JCW front brake pad replacement information not available at time of publication.
4. **Anti-rattle springs**
5. **Brake caliper**
   - Special red powder coated finish
6. **Bleeder valve and cap**
7. **Guide bolt**
8. **Retaining bolt**
   - M8
9. **Rear brake pads**
   - Special material in standard brake pad configuration
   - See **Brake pad removal and installation**
10. **Bolt**
    - Torx M6 x 10
      tighten to 8 Nm (6 ft-lb)
11. **Caliper piston dust boot and seal**
12. **Protective boot**
13. **Mounting bolt**
    - M12 x 1.5
14. **Additional kit components**
    - JCW identification sticker
    - Installation instructions

Brake pads, calipers, rotors, removing and installing

Refer to the illustrations in **Front brake assembly components** or **Rear brake assembly components** and detail notes (references in **bold**) during replacement or repair procedures.

> ***WARNING —***
> - *Although semi-metallic and metallic brake friction materials in brake pads no longer contain asbestos, they produce dangerous dust.*
> - *Treat all brake dust as a hazardous material.*
> - *Do not create dust by grinding, sanding, or cleaning brake friction surfaces with compressed air.*

NOTE —
JCW Tuning front brake pad replacement information not available at time of publication.

– Raise vehicle and support in a safe manner.

> ***WARNING —***
> *Make sure the car is stable and well supported at all times. Use a professional automotive lift or jack stands designed for the purpose. A floor jack is not adequate support.*

Brake pad lining thickness

To determine thickness of outer brake pad without removing road wheel:

- Rotate wheel until inspection bore (**arrow**) in brake pad backing plate can be seen through rim.
- Insert BMW special tool 34 1 260 through rim into bore and read off pad thickness.

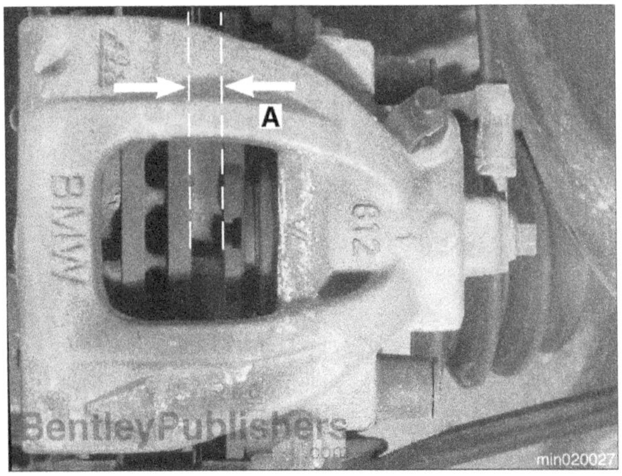

To determine brake pad thickness after removing wheel but without removing brake caliper:

- Measure pad thickness through caliper opening.

Brake pad thickness (A)	
Minimum thickness	3 mm (1/8 in)

Caliper removal

- Remove road wheel.

min340015

- Pry spring clip (**arrows**) off caliper and set aside.

min340016

- Remove protective caps (**arrows**) from caliper guide bolts. Remove caliper guide bolts and remove caliper from pad carrier.

- Brake caliper hydraulic line only needs to be removed if replacing caliper with a new unit. If removing brake caliper to service brake pads or rotors, leave hydraulic line connected.

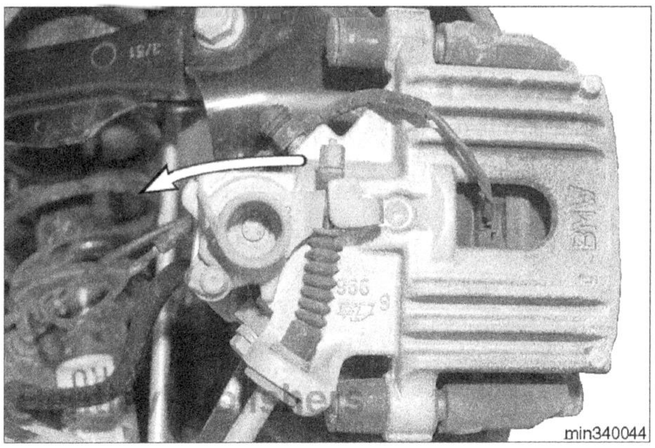

min340044

- Rear caliper: If replacing caliper with a new unit, remove parking brake cable:
  - Detach cable at parking brake handle inside vehicle. See **Parking brake cable, replacing** in this repair group.
  - Detach rear end of cable from rear caliper actuating lever (**arrow**).
  - Pull cable housing down to free from caliper bracket.

- If there is a ridge on rotor edge, press caliper piston slightly into caliper before removing caliper.

> ***CAUTION—***
> - *Pressing caliper piston in may cause brake fluid reservoir to overflow and damage paint work. To prevent this, use a clean syringe to first remove some fluid from reservoir.*
> - *Open caliper bleeder screw only when applying force to the piston. Do not allow air to be drawn in through bleeder screw. Catch expelled fluid in appropriate container.*
> - *Do not let brake caliper hang from brake hose. Suspend it from chassis using stiff wire.*

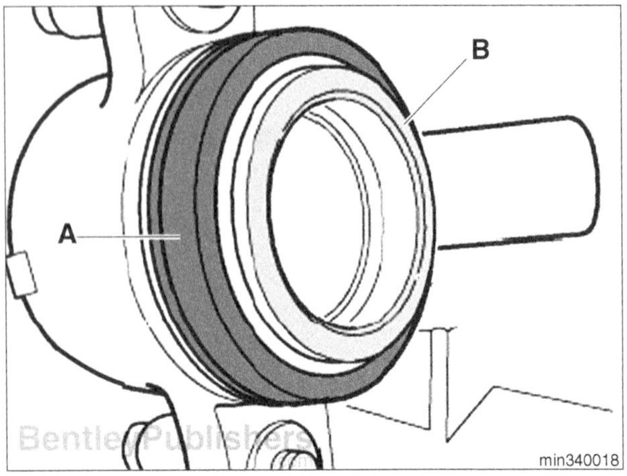

Inspect brake caliper for damage or signs of leakage.

- Check that caliper piston slides smoothly into caliper.
- Check dust seal (**A**) for damage. Replace if necessary.
- Replace caliper if leaks are found.
- Apply thin coat of Plastilube® or equivalent antisqueak compound to flat surface (**B**) of caliper piston.

Front caliper: If reusing caliper and installing new pads, press in caliper piston fully with BMW special tool 34 1 050 or equivalent.

– Rear caliper: If reusing caliper and installing new pads, press in caliper piston fully with BMW special tool kit 34 6 300 or equivalent.

NOTE —
- *The special tool turns the piston while pressing in.*
- *Make sure that the caliper piston dust seal does not twist and get damaged during this process.*

– Thoroughly clean all contact points on caliper and brake pad carrier. Clean guide bolts and make sure they slide freely. Do not lubricate guide bolts.

– When reinstalling caliper, attach spring clip at bottom first, then at top.

Tightening torque	
Caliper to brake pad carrier (7 mm Allen)	25 - 30 Nm (18 - 22 ft-lb)

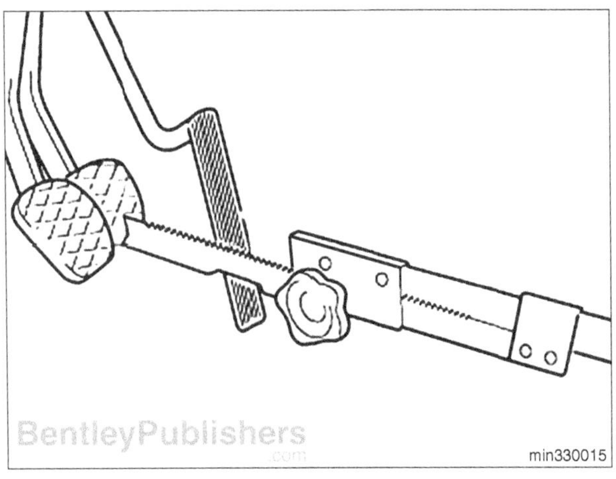

If removing caliper hose, first insert pedal prop and depress brake pedal slightly. This prevents brake fluid from escaping when brake lines are detached.

NOTE —
Be prepared to catch some brake fluid dripping. Dispose of brake fluid properly.

– If detaching caliper hose from metal brake line, be sure to use appropriate sized flare wrenches.
  - Reinstall caliper hose banjo bolt using new copper seals.
  - Tighten banjo bolts with front wheels pointing forward.

Tightening torques	
Caliper hose to caliper: front (M10 banjo bolt) rear (M12 banjo bolt)	 40 Nm (30 ft-lb) 49 Nm (36 ft-lb)
Caliper hose to brake line (M10, M12)	14 Nm (10 ft-lb)

– After assembly, if caliper brake hose was removed or replaced, bleed brake system. See **Bleeding brakes**.

Brake pad wear sensor

◄ Working at left front or right rear brake pad:
  - Carefully pry pad wear sensor from pad.
  - Insert sensor into cutout in new pad where applicable.
  - If brake lining indicator light illuminated prior to brake pad replacement, replace wear sensor.
  - Route pad wear sensor wiring properly to avoid snagging.

Brake pad removal and installation

– Replace brake pads by removing calipers from brake pad carriers but without disconnecting brake fluid hose from caliper or bleeding brakes.

NOTE—
Always replace pads in axle sets.

– Front brake pads: Outer pad seats loosely in caliper. Inner pad is held to caliper piston with a spring clip.

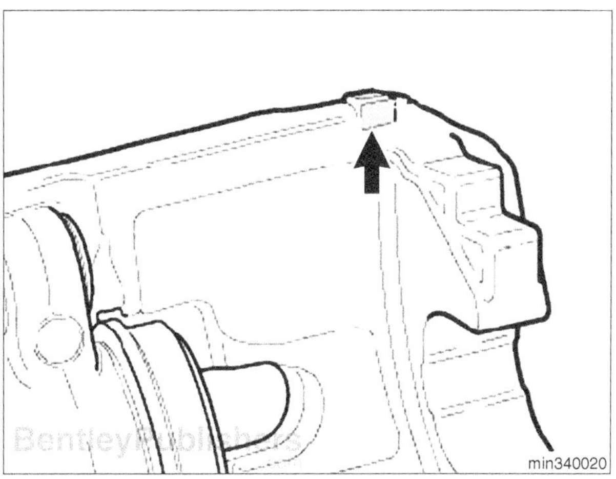

◄ Clean brake pad seat (**arrow**) on caliper. Apply thin coat of Plastilube® or equivalent antisqueak compound.

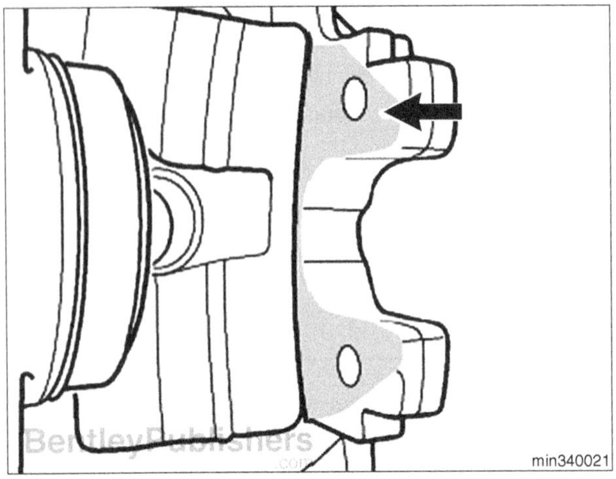

Clean contact face (**arrow**) of brake caliper. Apply thin coating of Plastilube® or equivalent antisqueak compound.

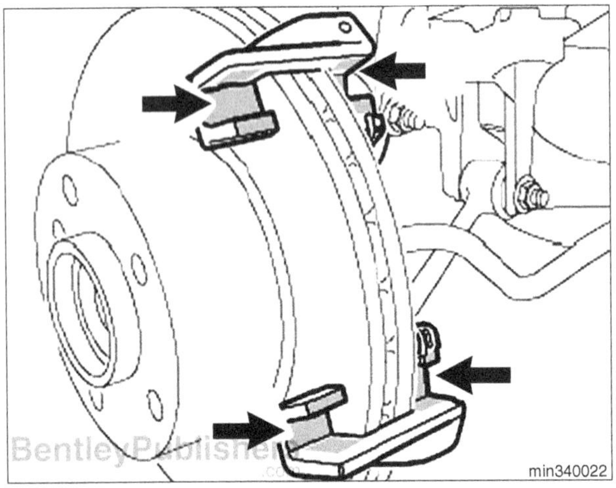

Clean brake pad seats (**arrows**) on pad carrier. Apply thin coating of Plastilube® or equivalent antisqueak compound.

– Do not grease back of brake pads.

– Rear brake pads: Make sure inner pad is correctly seated in caliper piston groove.

Brake pad carrier removal

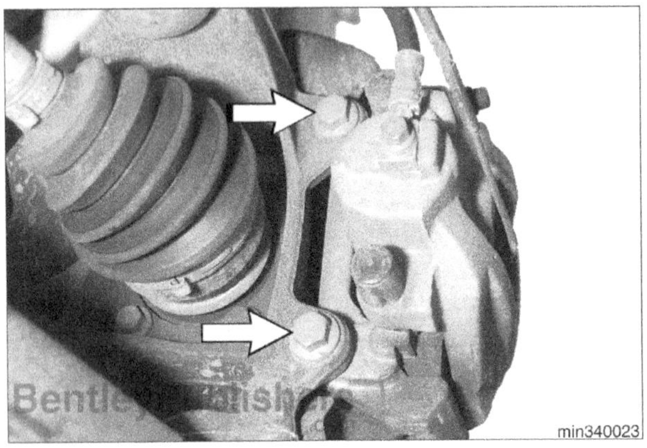

After removing caliper and pads, remove collar bolts (**arrows**) to detach brake pad carrier.

NOTE—
It is possible to remove the pad carrier with brake pads and caliper attached. In that case, be sure to hang caliper and pad carrier assembly securely using stiff wire in order to protect the brake hose from stretching.

Tightening torques	
Brake pad carrier to wheel hub front (M12 x 1.5) rear (M10)	 110 Nm (81 ft-lb) 65 Nm (48 ft-lb)

Brake rotor removal

– Replace brake rotor without disassembling wheel hub or bearing.

NOTE—
- *Machine or replace brake rotors in pairs.*
- *Fit new rotors together with new pads.*

After removal of brake pad carrier, remove brake rotor mounting bolt (**arrow**). Remove rotor.

> ***CAUTION—***
> *Do not strike the rotor friction surface with a hammer or other hard object. If rotor is stuck on wheel hub, carefully tap with a rubber hammer.*

- Inspect rotor for cracks, signs of overheating and scoring.

- On original equipment rotor, the minimum allowable thickness is stamped on rotor hub or edge. Measure friction surface thickness with a micrometer at eight to ten different points and use the smallest measurement recorded. See **Table b**.

Table b. Brake rotor specifications	
Front brake rotor thickness: new wear limit	 22 mm (0.866 in) 19 mm (0.748 in)
Rear brake rotor thickness: new wear limit	 10 mm (0.394 in) 7 mm (0.276 in)
Max. machining limit per friction side	0.8 mm (0.031 in)

- If rotor does not pass minimum thickness requirements or is damaged, replace rotor.

- When reinstalling rotor:
  - Make sure rotor contact surface with wheel hub is clean and smooth. Remove rust traces, if necessary.
  - Replace brake rotor retaining bolt.

Tightening torque	
Brake rotor to wheel hub (Torx M10 x 13)	27 Nm (20 ft-lb)

- After reassembly, pump brake pedal several times to make sure that brake pads are firmly pressed against brake rotors.

BRAKE MASTER CYLINDER

The brake master cylinder is mounted to the front of the vacuum booster on the driver's side bulkhead.

> ***CAUTION—***
> *MINI vehicles are equipped with ABS/ASC or ABS/DSC. In order for the braking and stability controls to work properly, if the brake master cylinder is replaced, make sure that the brake system is bled by an authorized MINI dealer or other qualified repair facility using the BMW scan tool (DISplus, GT1 or MoDiC). See* ***Brake bleeding*** *in this repair group.*

Brake master cylinder, removing and installing

– Using a clean syringe, empty brake fluid reservoir.

> ***CAUTION—***
> *Brake fluid is highly corrosive and dangerous to the environment. Dispose of it properly.*

< Detach brake fluid reservoir connections (**arrows**):

- Electrical connection to fluid level sensor
- Fluid supply hose to clutch master cylinder

NOTE—
Secure clutch fluid hose in vertical position to prevent dripping fluid.

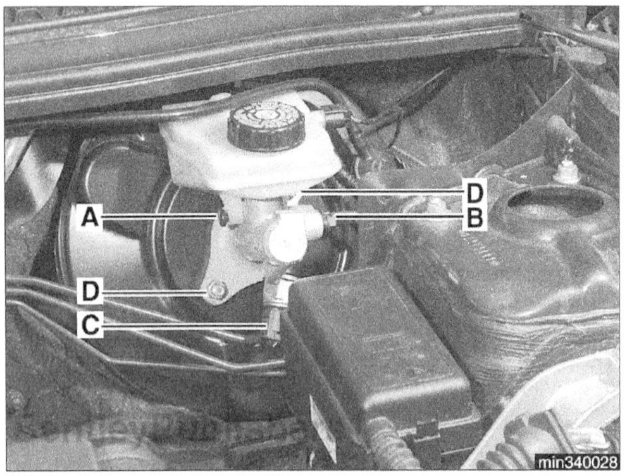

< Disconnect and remove from master cylinder:

- Brake fluid reservoir hold down screw (**A**). Pull fluid reservoir straight up off master cylinder.
- Both brake lines (**B**) on left side of master cylinder.
- Electrical harness connector (**C**) from DSC pressure switch (where applicable).
- Master cylinder mounting nuts (**D**). Discard nuts.

NOTE—
Be prepared to catch dripping brake fluid. Dispose of brake fluid properly.

– Pull master cylinder forward out of brake booster.

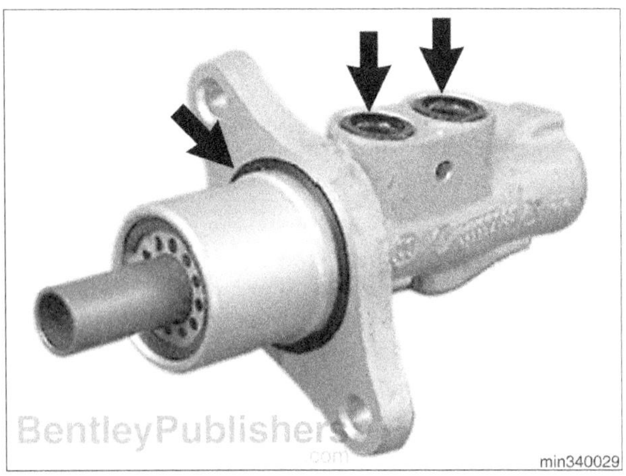

< Prior to reinstallation, install new O-ring seals (**arrows**) on master cylinder.

– During installation, make sure master cylinder pressure rod lines up with brake booster pressure rod.

Tightening torques	
Brake fluid reservoir to master cylinder	3.5 Nm (31 in-lb)
Brake lines to master cylinder	14 Nm (10 ft-lb)
Brake master cylinder to brake booster (M8)(replace self-locking nuts)	21 ± 3 Nm (15 ± 2 ft-lb)

– Top off fluid reservoir and bleed brakes. See **Bleeding brakes** in this repair group.

Brake Booster

The brake booster is mounted to the bulkhead on the driver's side of the engine compartment, directly behind the brake master cylinder. Intake manifold vacuum acts on a large diaphragm in the brake booster to reduce brake pedal effort.

A check valve in the brake booster vacuum line prevents contamination from engine backfires and other sources from entering the brake booster.

The check valve and the molded vacuum hose are permanently connected and are replaced together as a single unit.

Brake booster check valve, removing and installing (Cooper)

– Remove battery and battery box. See **121 Battery, Starter, Alternator**.

> ***CAUTION—***
> *Disconnecting the battery may erase fault code(s) stored in memory. Check for fault codes prior to disconnecting the battery cables.*

– Pump brake pedal several times to reduce vacuum in brake booster.

◄ Gently pry brake booster check valve (**arrow**) out of the rubber grommet in brake booster.

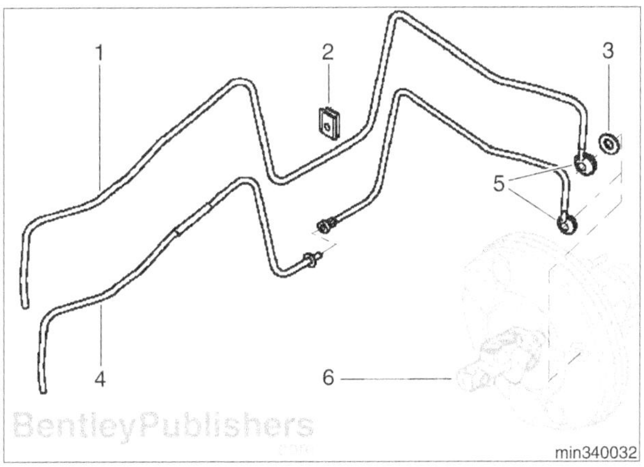

◄ Brake booster vacuum line.

1. Vacuum line (to 09/2002; one piece)
2. Rubber grommet (in plastic bulkhead)
3. Rubber grommet (seal)
4. Vacuum line (from 09/2002; two piece)
5. Check valve
6. Brake booster and master cylinder assembly

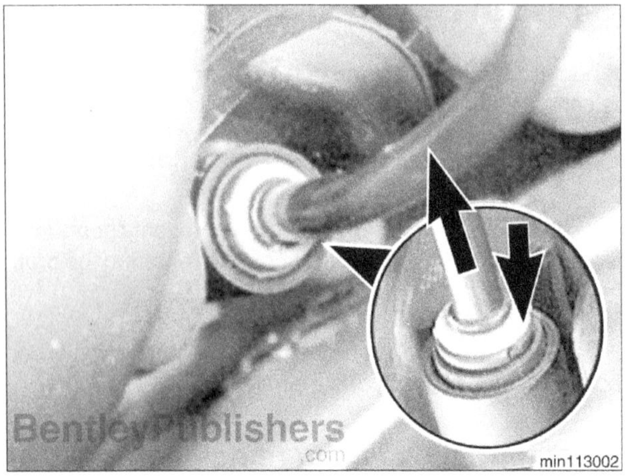

Models to 09/2002:

- Detach brake booster line from manifold by pushing on locking ring (**arrows**).

Models from 09/2002: Working to left of engine, squeeze white tab on vacuum hose to separate two parts of vacuum supply hose.

– During installation, inspect check valve rubber grommet in brake booster. Replace if necessary.

– All models, installation is reverse of removal.

Brake booster check valve, removing and installing (Cooper S)

– Remove air filter housing from engine compartment. See **130 Fuel Injection**.

– Pump brake pedal several times to reduce vacuum in brake booster.

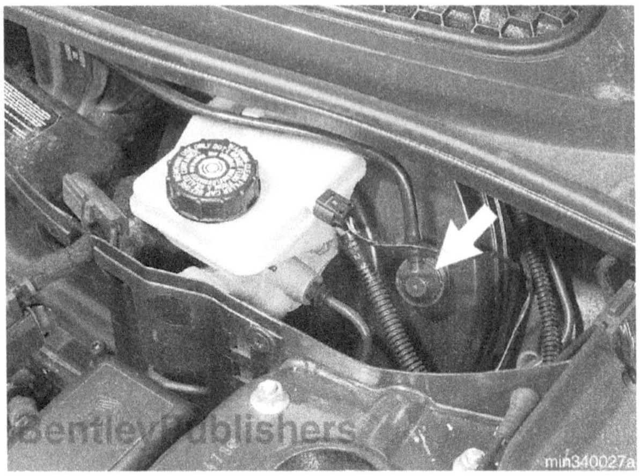

Gently pry brake booster check valve (**arrow**) out of the rubber grommet in brake booster.

Peel off bulkhead rubber seal in front of brake master cylinder enclosure. Remove 5 bulkhead fasteners (**arrows**) and remove plastic bulkhead.

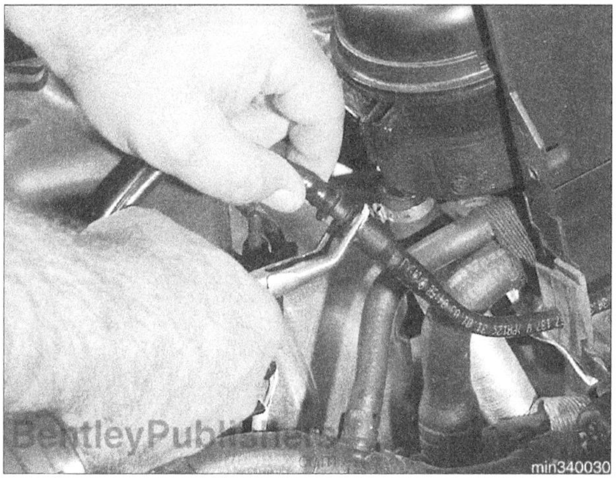

Working to left of engine, squeeze white tab on vacuum hose to separate two parts of vacuum supply hose.

- During installation, inspect check valve rubber grommet in brake booster. Replace if necessary.
- Use new clamp at air cleaner outlet duct.

Brake booster, removing and installing

- Remove both windshield wiper arms.
- Cooper: Remove battery and battery box. See **121 Battery, Starter, Alternator**.

> ***CAUTION—***
> *Disconnecting the battery may erase fault code(s) stored in memory. Check for fault codes prior to disconnecting the battery cables.*

- Cooper S: Remove air filter housing from engine compartment. See **130 Fuel Injection**.

Cooper S: Peel off bulkhead rubber seal in front of brake master cylinder enclosure. Remove 5 bulkhead fasteners (**arrows**) and remove plastic bulkhead.

- Remove brake master cylinder. See **Master cylinder, removing and installing** in this repair group.

> ***CAUTION—***
> - *MINI vehicles are equipped with ABS/ASC or ABS/DSC. In order for the braking and stability controls to work properly if the brake master cylinder is replaced, make sure that the brake system is bled by an authorized MINI dealer or other qualified repair facility using the BMW scan tool (DISplus, GT1 or MoDiC). See* ***Brake bleeding*** *in this repair group.*
> - *Brake fluid is highly corrosive and dangerous to the environment. Dispose of it properly.*

Brake Booster

Working inside vehicle, pull on top edge of left lower dashboard trim to unclip and swing down.

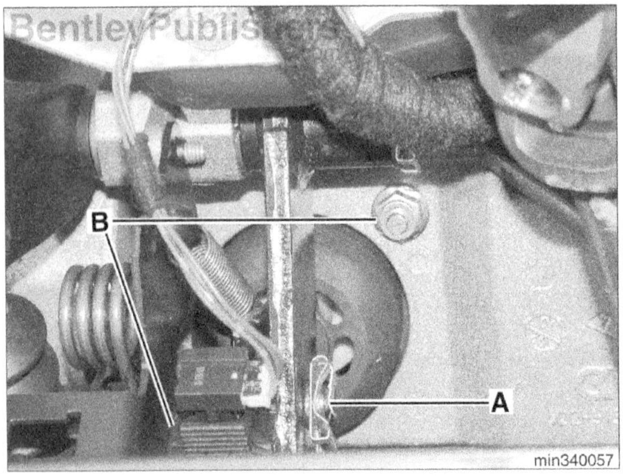

Working under dashboard:

- Remove brake pedal clevis pin clip (**A**) and pull out pin.
- Remove brake booster mounting nuts (**B**).

Working in engine compartment at bulkhead, remove coolant expansion tank and power steering fluid reservoir mounting nuts (**arrows**). Pull both reservoirs forward and secure out of the way.

- Do not detach fluid lines from either reservoir.

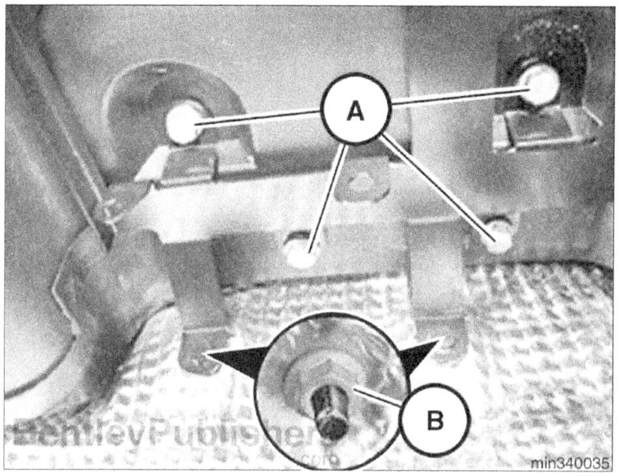

Remove plastic bulkhead mounting screws (**A** and **B**).

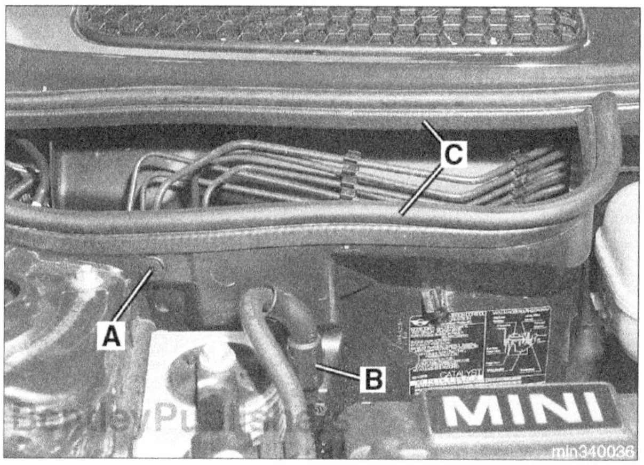

Working at right rear of engine compartment:

- Remove plastic bulkhead retaining screw (**A**).
- Release fuel pipe (**B**) from retaining clip.
- Peel off right side bulkhead rubber seal (**C**).

– Lift out plastic bulkhead.

Detach brake lines (**arrows**) from ABS hydraulic unit to brake master cylinder.

NOTE —
Be prepared to catch dripping brake fluid.

Loosen wiper assembly:

- Remove wiper arms. Working at cowl, below windshield, remove wiper assembly upper mounting nuts at base of wiper pivots.
- Remove wiper assembly lower mounting fastener (**arrow**).
- Slide wiper assembly to right.
- Cooper: Detach ground cable at bulkhead.

– Carefully remove brake booster from bulkhead and tilt away from engine. Remove by pulling upward.

– Installation is reverse of removal.

- Place wiper assembly bracket so that it is flush against rubber grommet.
- Cooper: Reattach ground cable.

Tightening torques	
Brake booster to body (M8)(replace self-locking nuts)	21 ± 3 Nm (15 ± 2 ft-lb)
Brake fluid reservoir to master cylinder	3.5 Nm (31 in-lb)
Brake lines to master cylinder or to ABS hydraulic unit	14 Nm (10 ft-lb)
Brake master cylinder to brake booster (M8)(replace self-locking nuts)	21 ± 3 Nm (15 ± 2 ft-lb)
Coolant expansion tank to plastic bulkhead (M6)	5 Nm (4 ft-lb)
Engine ground cable to body (M8)	19 Nm (14 ft-lb)
Heat shield to plastic bulkhead	7 - 10 Nm (5 - 7 ft-lb)
Steering fluid reservoir to bracket (M6)	8 Nm (6 ft-lb)
Steering fluid reservoir bracket to bulkhead (M8 x 1.25 x 20mm)	19 Nm (14 ft-lb)
Wiper arm to wiper pivot (M8)	18 Nm (13 ft-lb)
Wiper assembly to body	10 Nm (7 ft-lb)

Parking Brake

The parking brake is operated from the parking brake lever by two Bowden cables, with a left-right compensator fitted behind the brake lever and an adjustment nut under the lever in the center console.

The parking brake stops the car via the rear caliper pistons acting directly on the rear brake pads which, in turn, press on the brake rotors (discs).

The parking brake is self-adjusting, with the adjustment being made at the caliper piston. A special tool is required to wind back the piston when fitting new pads.

Parking brake, checking and adjusting

– Check parking brake as follows:
  - Release parking brake lever. Check to see that there is no drag at rear wheels.
  - Pull up brake lever 1 notch. Again check to see that there is no drag. Parking brake warning light, however, may be ON.
  - Pull up lever 2 notches. Check to see that there is a slight drag at each wheel and that parking brake warning light is on.
  - Drive vehicle and pull up on parking brake handle. Check that rear wheels lock up.

> ***WARNING —***
> *Perform this test where there is no other traffic.*

– Adjust parking brake under following circumstances:
  - Brake lever pulls up more than 7 notches.
  - Rear brake pads or rotors are replaced.
  - Parking brake cables or handle are replaced.

< Working inside vehicle at rear console, unclip parking brake handle boot (**A**) and lift off to access end of cable.

– Back off self-locking nut (**B**) until cable is completely slack.

– Raise rear of vehicle and support in a safe manner.

> ***WARNING —***
> *Make sure the car is stable and well supported at all times. Use a professional automotive lift or jack stands designed for the purpose. A floor jack is not adequate support.*

– Remove rear wheels.

– Remove rear brake pads. See **Brake pads, calipers, rotors, removing and installing** in this repair group.

– Remove rear calipers. Press in each rear caliper piston fully with BMW special tool kit 34 6 300 or equivalent.

NOTE —
- *The special tool turns the piston while pressing in.*
- *Make sure that the caliper piston dust seal does not twist and get damaged during this process.*

– Reinstall brake calipers. Reinstall brake pads.

Tightening torque	
Caliper to brake pad carrier (7 mm Allen)	30 - 5 Nm (22 - 4 ft-lb)

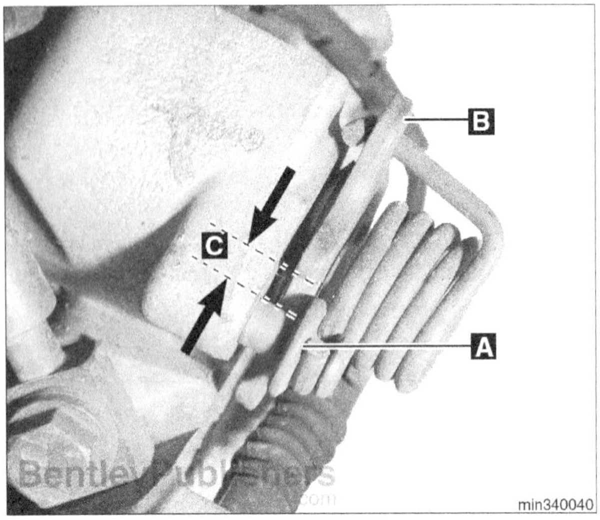

Working inside vehicle, tighten parking brake cable adjusting nut until gap (**C**) at rear caliper is set between cable-actuated lever (**B**) and pin (**A**).

Parking actuator lever	
Gap **C** between actuator lever and notched pin	2 - 3 mm (0.08 - 0.12 in)

- Apply parking brake lever three times.
- Press brake pedal to floor at least three times.
- Check gap at actuator lever. Reset if necessary.
- Check function of parking brake.
- Check function of parking brake warning light.

Parking brake cables, removing and replacing

- Raise rear of vehicle and support in a safe manner.

> ***WARNING —***
> *Make sure the car is stable and well supported at all times. Use a professional automotive lift or jack stands designed for the purpose. A floor jack alone is not adequate support.*

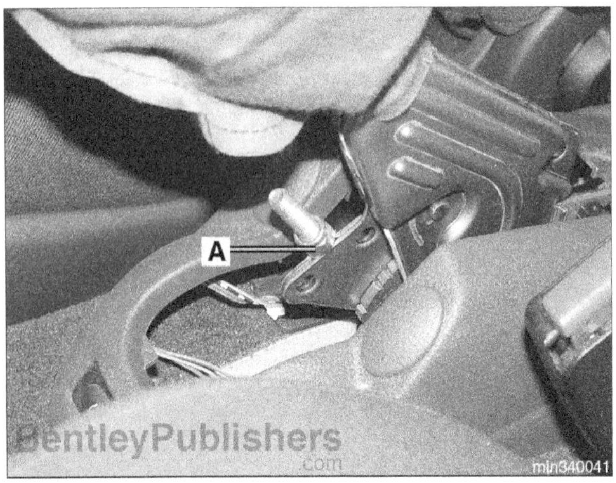

- Remove exhaust system from the catalyst back. See **180 Exhaust System**.
- Remove rear section of exhaust system heat shield.
- Working inside car, remove rear center console and console support bracket. See **513 Interior Trim**.

Working at base of parking brake handle, remove brake cable adjusting nut (**A**).

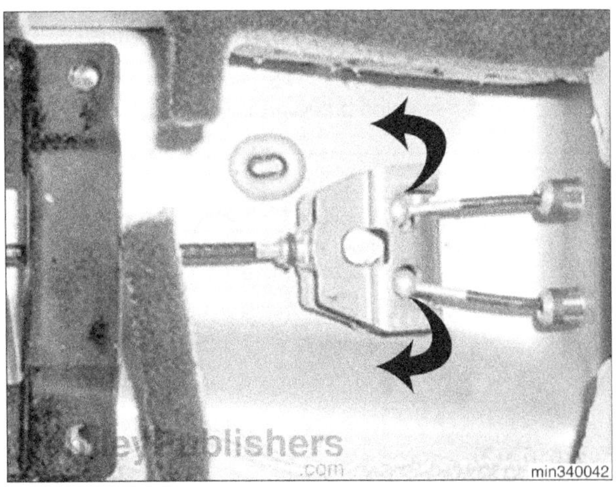

Working behind parking brake handle, detach both brake cables from compensator arm behind parking brake handle.

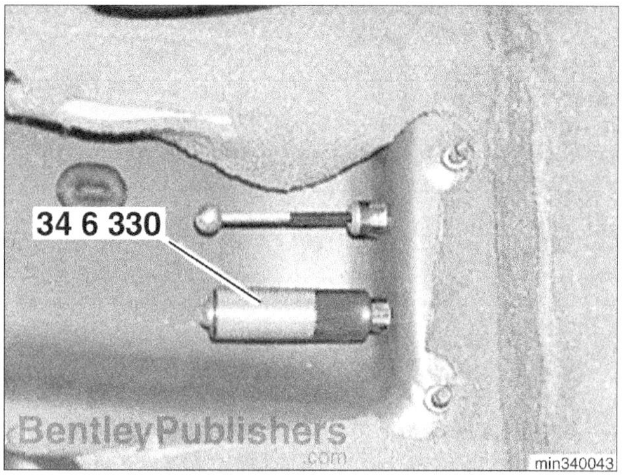

Use BMW special tool 34 6 330 to force parking brake cable barbs free of body. Push cable through.

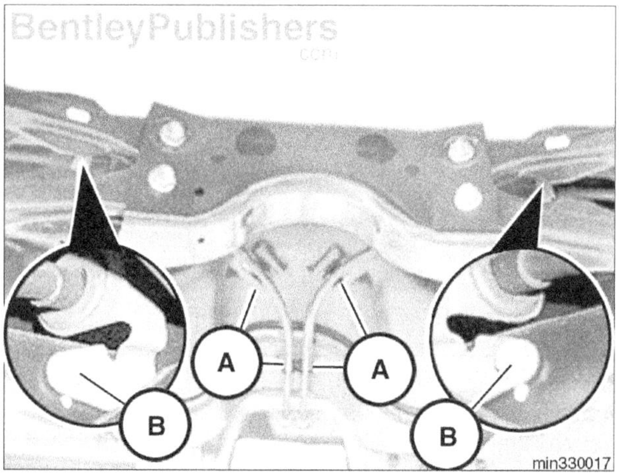

Detach parking brake cables:

- Pry cables off plastic retainers (**A**) on fuel tank.
- Remove cable bracket mounting bolts (**B**) at rear subframe.

Detach rear end of each cable from rear caliper actuating lever (**arrow**).

- Pull cable housing down to free from caliper bracket.

– Installation is reverse of removal. Adjust parking brake.

Tightening torque	
Parking brake cable bracket to subframe (M6)	9 ± 1 Nm (168 ± 9 in-lb)

ABS COMPONENT REPLACEMENT

> ***WARNING —***
> - *The laws of physics cannot be repealed, even with DSC. Practice a responsible driving style. Avoid using the additional margin of safety afforded by traction control as an excuse for taking risks.*
> - *Do not make any modifications to the DSC system. Allow only authorized technicians to service DSC.*

> ***CAUTION —***
> - *If tires are different makes, the ABS/traction control system may over-react. Only fit tires of the same make and tread pattern.*
> - *In adverse conditions, such as rocking the car out of deep snow or other soft surface, or when snow chains are fitted, it is advisable to switch off traction control and allow the car's drive line to operate conventionally.*

MINI vehicles are equipped with Antilock Braking System (ABS) combined with Automatic Stability Control (ABS/ASC) or Dynamic Stability Control (ABS/DSC). This manual will refer to these systems as ABS. ASC or DSC will be specified when necessary.

ABS wheel speed sensor, front, removing and installing

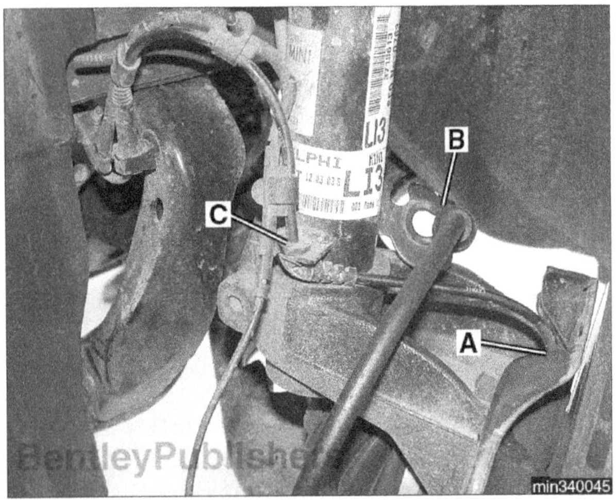

– Raise front of vehicle and support in a safe manner.

> ***WARNING —***
> *Make sure the car is stable and well supported at all times. Use a professional automotive lift or jack stands designed for the purpose. A floor jack alone is not adequate support.*

– Remove road wheel.

◄ Working at top of wheel bearing housing, remove wheel speed sensor mounting screw (**A**). Pull sensor out of housing.
- Pry sensor harness from retaining clip (**B**) on MacPherson strut.
- Detach sensor harness from retaining clip (**C**) on front subframe.

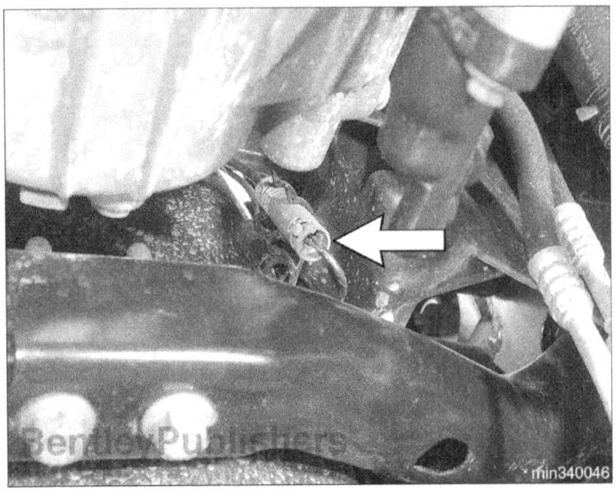

◄ Detach sensor harness from retaining clips on subframe.
- Separate sensor electrical harness connector (**arrow**).

– Installation is reverse of removal. Bear in mind the following:
- Clean wheel speed sensor mounting hole and grease with Staburags® NBU 12/K or equivalent anti-seize compound.
- Make sure plug connections, plastic harness retainers and rubber grommets are fitted correctly.
- Make sure sensor harness is routed correctly.

Tightening torque	
Wheel speed sensor to wheel bearing housing	8 Nm (6 ft-lb)

ABS wheel speed sensor, rear, removing and installing

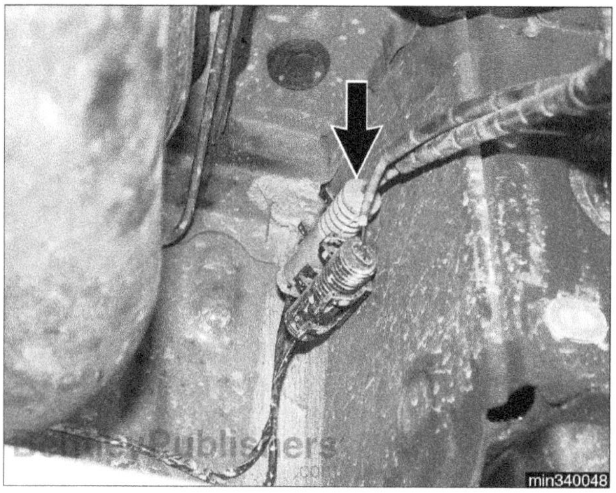

– Raise rear of vehicle and support in a safe manner.

> ***WARNING—***
> *Make sure the car is stable and well supported at all times. Use a professional automotive lift or jack stands designed for the purpose. A floor jack alone is not adequate support.*

◄ Working underneath rear of car, detach sensor harness (**arrow**) from body.

- Detach harness from clip at trailing arm bushing.
- Cut wire ties as necessary.
- Separate sensor electrical harness connector.
- Unclip harness from clip at chassis subframe.
- Pry harness insulating grommet from rear strut bracket.

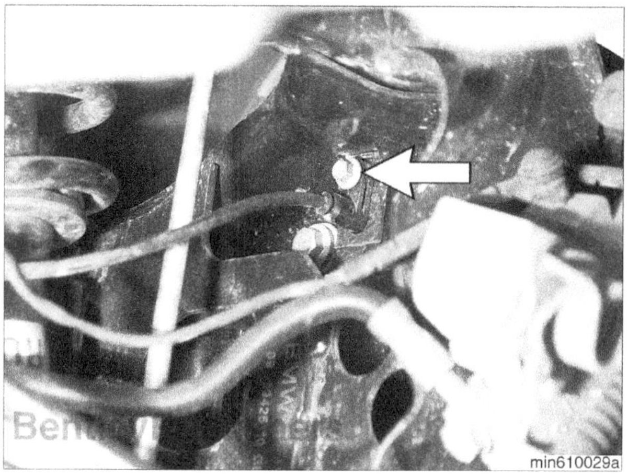

◄ Remove sensor mounting screw (**arrow**) and pull wheel speed sensor out of wheel bearing housing.

– Installation is reverse of removal. Bear in mind the following:

- Clean wheel speed sensor mounting hole and grease with Staburags® NBU 12/K or equivalent anti-seize compound.
- Make sure plug connections, plastic harness retainers and rubber grommets are fitted correctly.
- Make sure sensor harness is routed correctly.
- Use new wire ties as necessary.

Tightening torque	
Wheel speed sensor to wheel bearing housing	8 Nm (6 ft-lb)

DSC motion sensor, removing and installing

The Dynamic Stability Control (DSC) motion sensor combines the functions of the transverse acceleration sensor and the yaw (rotation rate) sensor.

– Remove rear center console. See **513 Interior Trim**.

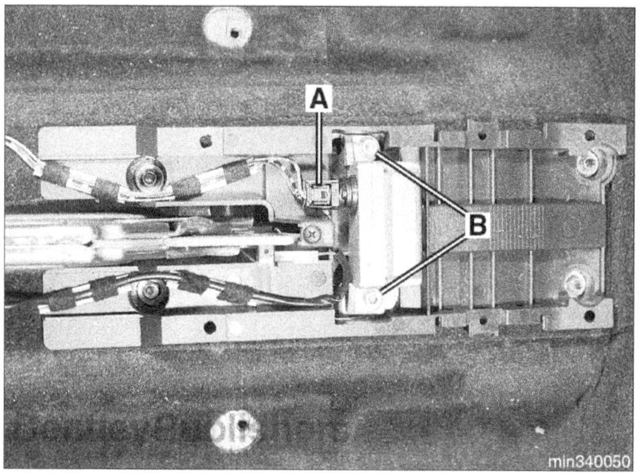

◄ Detach sensor electrical harness connector (**A**).

- Remove sensor mounting screws (**B**).
- Remove motion sensor.

– Installation is reverse of removal.

Tightening torque	
DSC motion sensor to mounting bracket (M6 x 16)	8 Nm (6 ft-lb)

> ***CAUTION—***
> *The motion sensor is vibration sensitive. Make sure mounting screws are torqued correctly and the harness connector is attached firmly. Otherwise there may be DSC malfunctions.*

DSC steering angle sensor, removing and installing

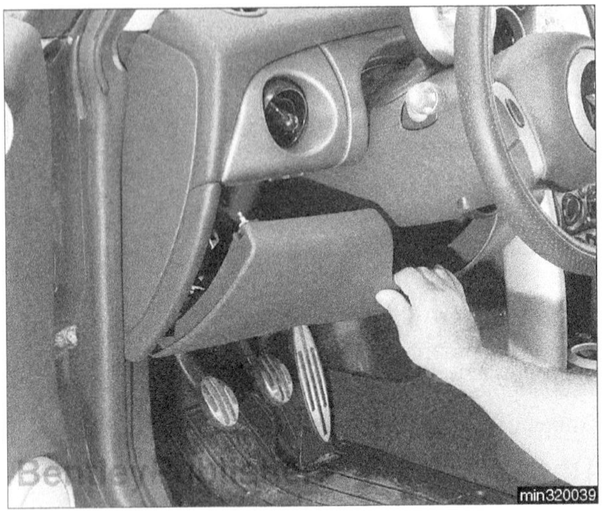

- Working inside vehicle, pull on top edge of left lower dashboard trim to unclip and swing down.

- Remove lower steering column cover. Remove lower steering shaft. See **320 Steering and Wheel Alignment**.

> ***CAUTION—***
> *Once the steering shaft is separated from the steering rack there is no longer a steering end stop and the steering wheel can be turned without restriction. This can damage the airbag contact ring. To prevent this damage, before disconnecting steering shaft:*
> *- Move steering wheel to straight-ahead position.*
> *- Remove ignition key to lock steering.*

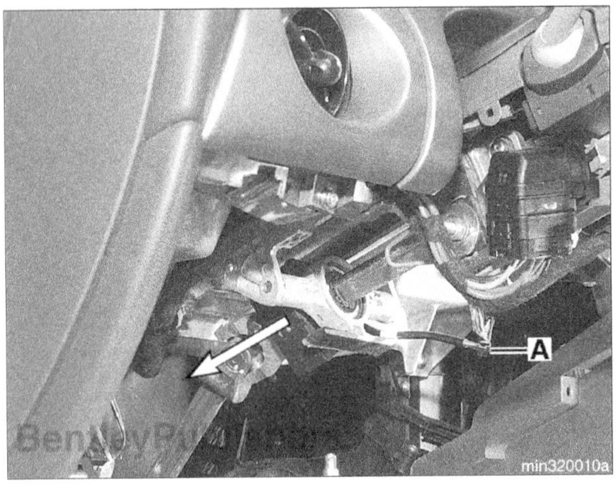

- Working inside vehicle underneath steering column, disconnect steering angle sensor electrical harness connector (**A**).
  - Slide sensor off upper steering shaft in direction of **arrow**.

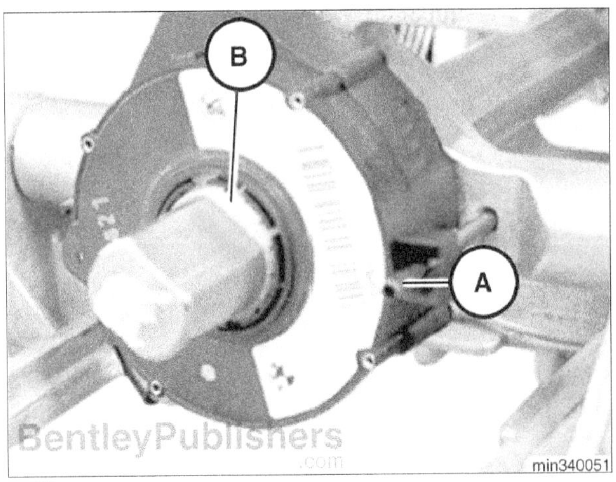

- When installing, make sure steering angle sensor is fitted on locator pin (**A**) and is centered (**B**) on shaft.

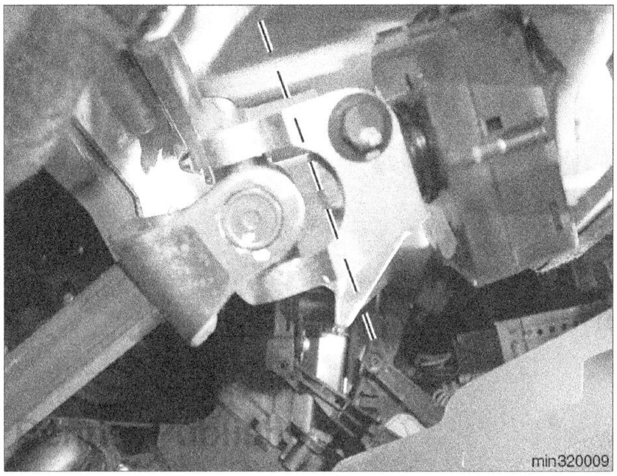
min320009

When installing lower steering shaft, make sure pinch bolt hole in lower shaft and locating dimple in steering shaft are aligned. When assembled, make sure end of upper steering shaft lines up with **dashed line**.

Tightening torque	
Lower steering shaft universal joint pinch bolt to upper steering shaft (M8 x 35 mm, replace bolt and nut)	30 Nm (22 ft-lb)

- Remainder of installation is reverse of removal.
- After completing work, calibrate steering angle sensor using BMW scan tool DISplus, GT1 or MoDiC.

ABS hydraulic unit/control module, removing and installing

The ABS/ASC or ABS/DSC electronic control module is attached to the hydraulic control unit. Remove combined unit and place on workshop bench to separate.

- Before starting work, use BMW scan tool DISplus, GT1 or MoDiC to read out ABS control module fault memory. Print out diagnostic report.

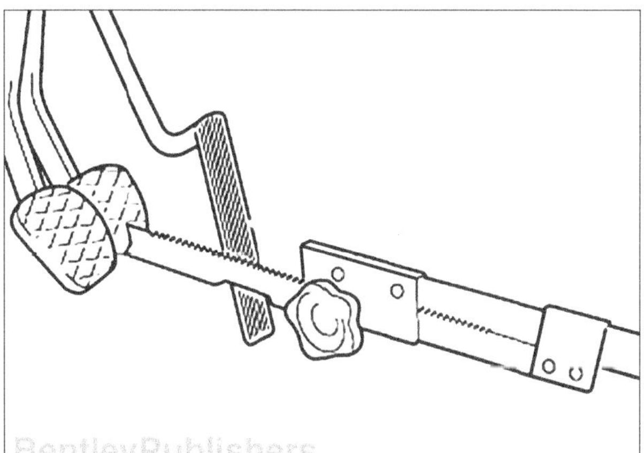
min330015

Insert pedal prop and depress brake pedal slightly. This prevents brake fluid from escaping when brake lines are detached.

min340034

Working in engine compartment at bulkhead, remove coolant expansion tank and power steering fluid reservoir mounting nuts (**arrows**). Pull both reservoirs forward and secure out of the way.

- Do not detach fluid lines from either reservoir.

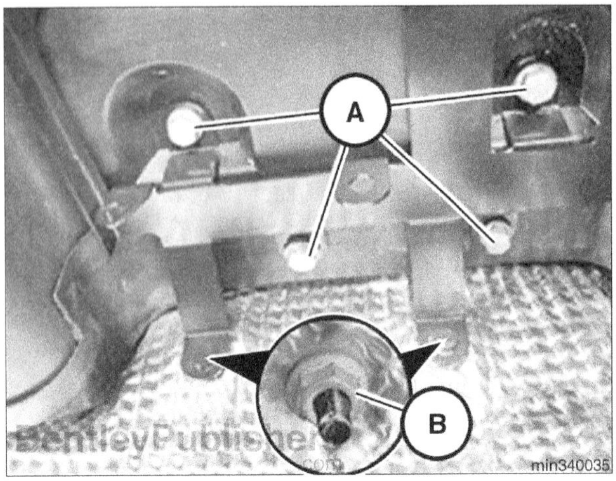

◄ Remove plastic bulkhead mounting screws (**A** and **B**).

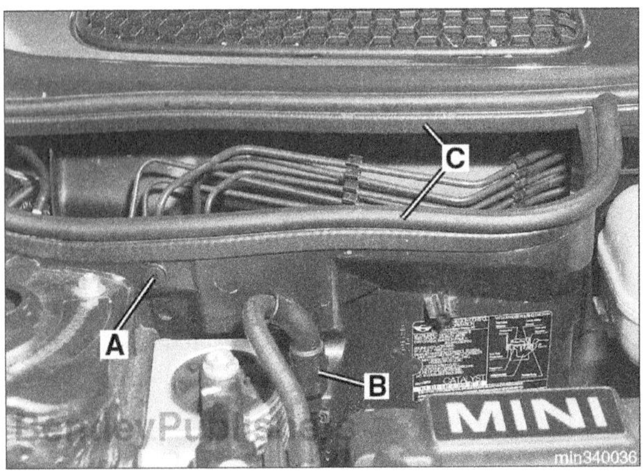

◄ Working at right rear of engine compartment:

- Remove plastic bulkhead retaining screw (**A**).
- Release fuel pipe (**B**) from retaining clip.
- Peel off right side bulkhead rubber seal (**C**).
- Detach plastic bulkhead and pull forward to clear ABS module.

– Detach all brake lines from hydraulic unit.

NOTE—

- *Loosen hydraulic lines with special flare wrenches.*
- *Be prepared to catch dripping brake fluid. Dispose of brake fluid properly.*

◄ Detach electrical harness connector (**A**) from electronic control module.

- Remove hydraulic unit/control module bracket mounting bolts (**arrows**).
- Lift out hydraulic unit/control module and place on work bench.

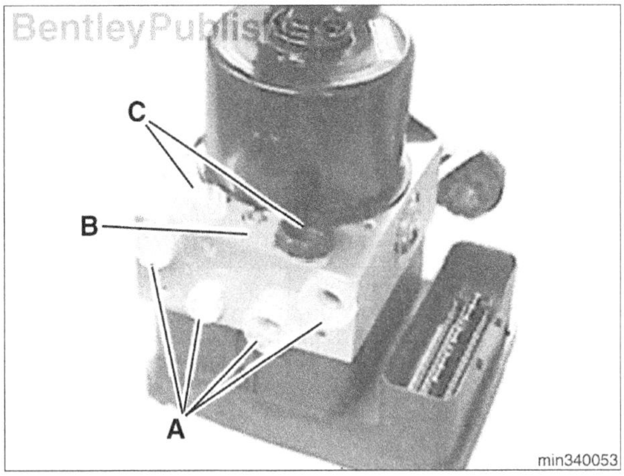

Before separating control module from hydraulic unit, plug off brake line connections (**A** and **C**). This prevents brake fluid from being discharged from hydraulic unit and makes it easier to bleed brakes.

- Secure pump motor with retaining tab (**B**).

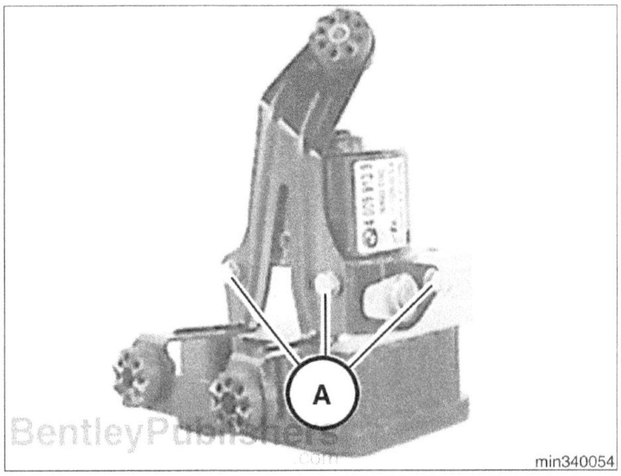

Remove hydraulic unit/control module bracket mounting bolts (**A**).

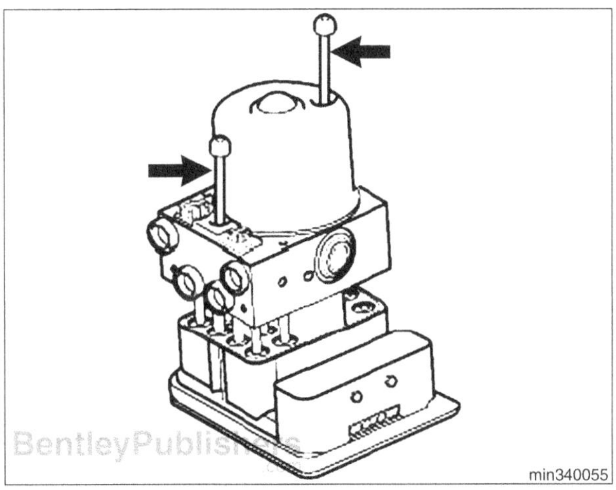

Remove screws (**arrows**) securing hydraulic unit to control module. Separate components from each other.

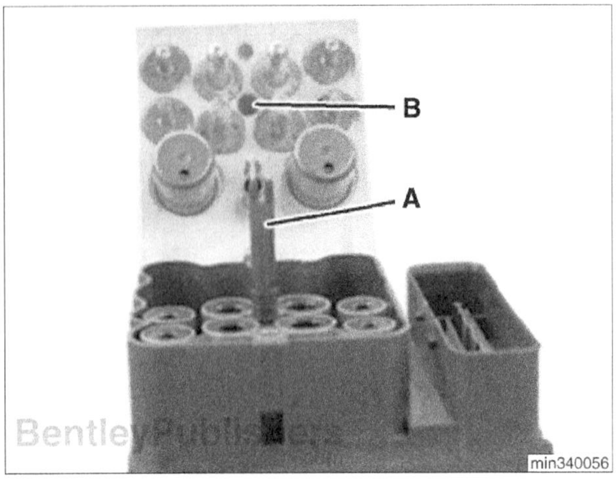

When reassembling components:

- Make sure connector (**A**) is engaged correctly in control module.
- Make sure connector fits in hole (**B**).

Tightening torques	
ABS hydraulic unit/control module to bracket	8 Nm (6 ft-lb)
Electronic control module to hydraulic unit	5 Nm (4 ft-lb)

– When reinstalling in vehicle, make sure that mounting bracket rubber grommets (sound insulation) are fitted correctly. Replace damaged rubber mounts.

Tightening torques	
ABS hydraulic unit/control module bracket to body	8 Nm (6 ft-lb)
Brake lines to ABS hydraulic unit	14 Nm (10 ft-lb)
Coolant expansion tank to plastic bulkhead (M6)	5 Nm (4 ft-lb)
Heat shield to plastic bulkhead	7 - 10 Nm (5 - 7 ft-lb)
Steering fluid reservoir to bracket (M6)	8 Nm (6 ft-lb)
Steering fluid reservoir bracket to bulkhead (M8 x 1.25 x 20 mm)	19 Nm (14 ft-lb)

– After competing work:

- Top off brake fluid reservoir.
- Use BMW scan tool DISplus, GT1 or MoDiC to bleed brakes and perform ABS/ASC or ABS/DSC function tests. See **Bleeding brakes** in this repair group.

400 Body–General

General

This section covers system descriptions and general information for the repair groups found in **4 Body** and **5 Body Equipment**.

Body Assembly

Body dimensions

Body dimensions vary slightly among models. Dimensions are given in inches and (mm).

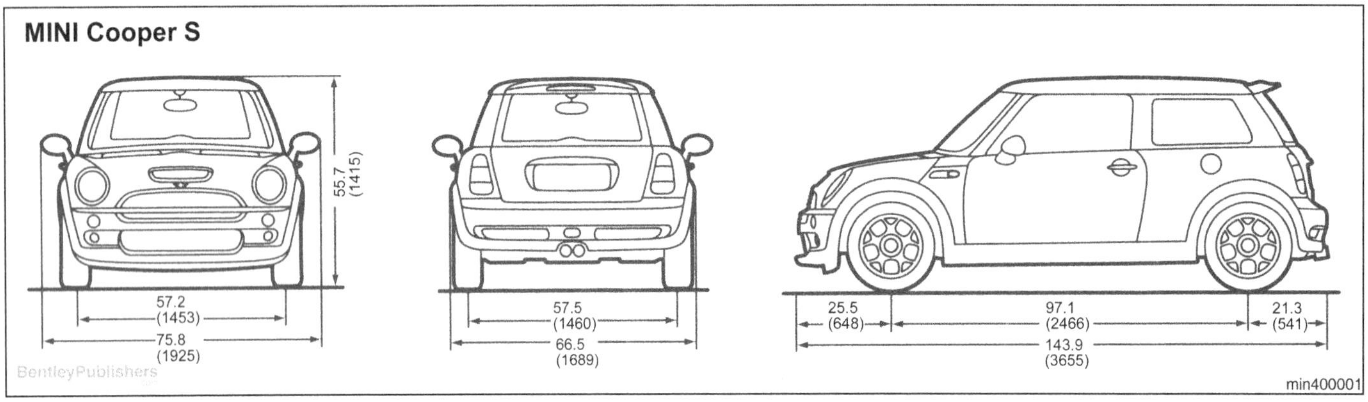

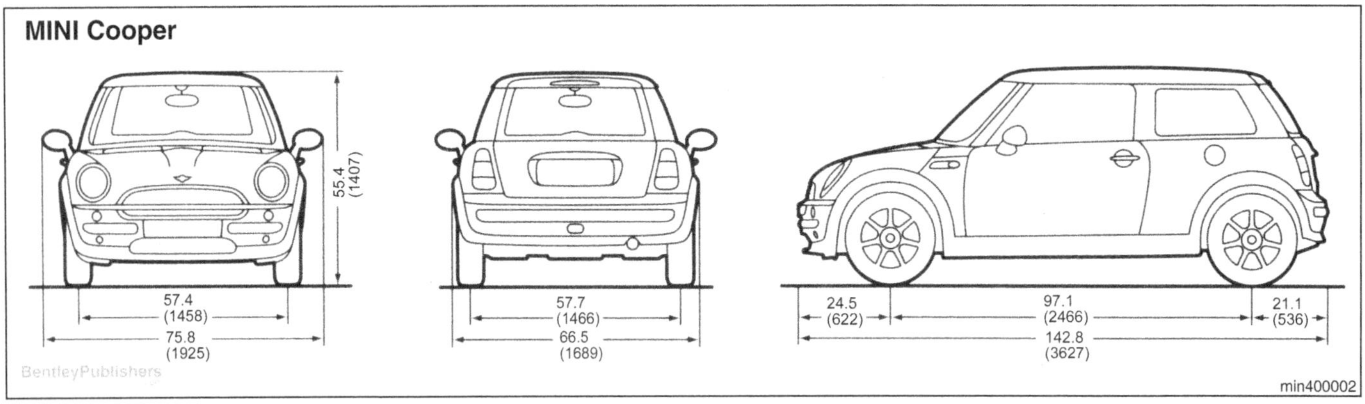

Body shell

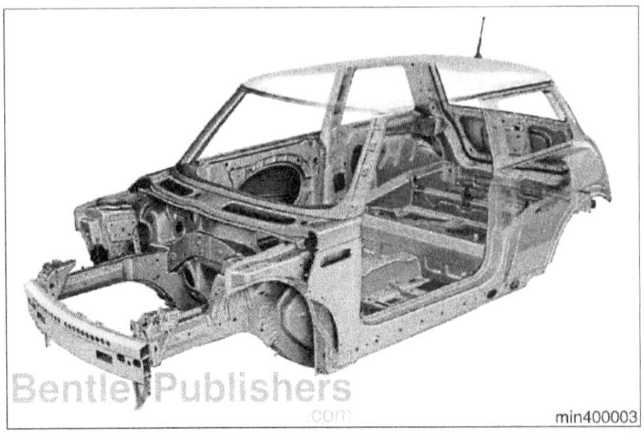

The rigid MINI body shell is constructed of sheet steel partly reinforced by box section longitudinal members and cross beams. Three major sub assemblies come together to produce the one piece floor frame, the front end, main floor and rear end.

The following components add to structural rigidity:

- Cross members
- Front wheelhouse
- Side frame and pillars
- Roof

SAFETY AND SECURITY

Body cage

The following features ensure the safety of the occupants:

- Strong occupant safety cell with roof bow welded above the B-pillar (door pillar).
- Crumple zones at both the front and rear.
- Reinforced at front, rear and side to protect against impacts.
- High torsional rigidity.

Door locks and handles

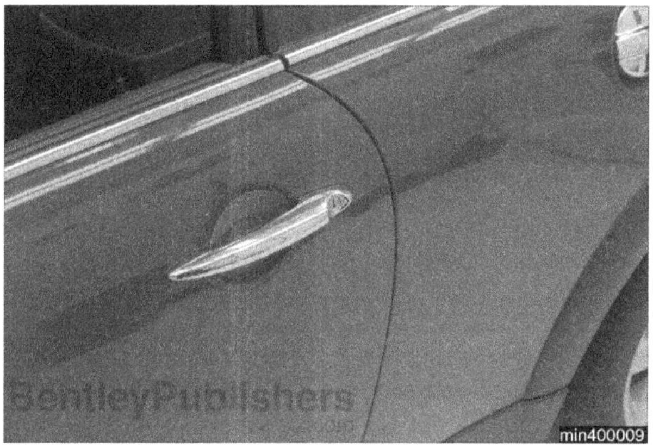

The bow type door handles allow easy door opening, but are secure in accidents.

Electrical components in the locks are fully encapsulated and cannot be picked easily. The door handles, latches and lock assemblies offer reliability and security against theft.

Door position and lock condition are detected by hall sensors.

Door lock repairs are covered in **515 Central Locking, and Anti-theft**.

Seat belts

Each front seat belt assembly has a height-adjustable anchor at the B-pillar.

An automatic pyrotechnic (explosive charge) tensioner tightens the front belt at the buckle upon impact, snugging up lap and shoulder segments of the belt. The tensioners are designed to automatically tension the belts by about 2 inches (55 mm) in case of a collision.

Seat belt service is covered in **720 Seat Belts**.

Airbags

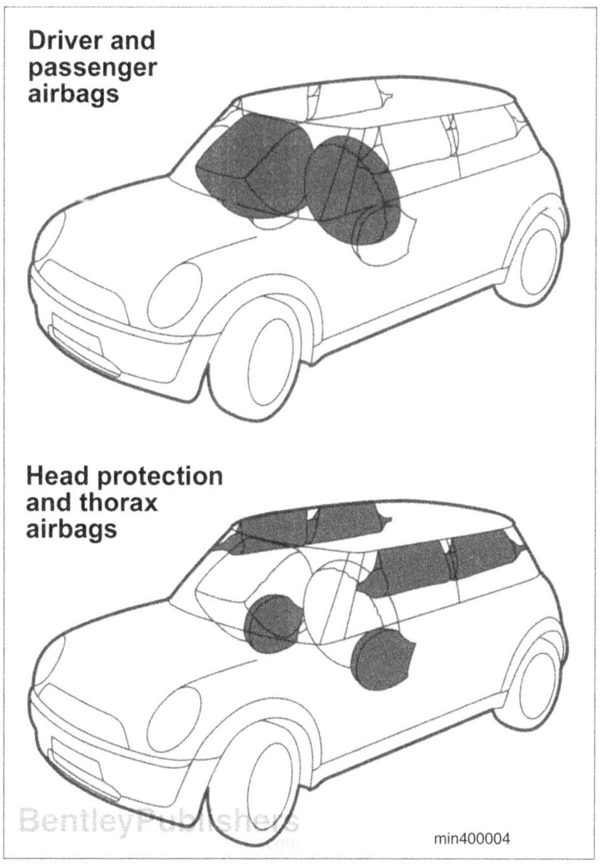

Six airbags are installed in MINI models:

- One airbag each in front of the driver and the passenger
- A thorax airbag in the outboard bolster of each front seat.
- A head protection curtain style airbag on each side, anchored to the sides of the roof.

The airbags are controlled by the multiple restraint system (MRS). The front airbag design includes a passenger seat occupancy sensor to help prevent the unnecessary deployment of the passenger airbag if the seat is unoccupied.

Head protection (AHPS2) airbags stretch across the tops of the front doors and rear windows.

Airbag deployment automatically triggers fuel shut-off, turns on the hazard and interior lights, and unlocks the doors.

> *WARNING —*
> - *An airbags is inflated by an explosive device. Handled improperly or without adequate safeguards, an airbag unit can be very dangerous. Special precautions must be observed prior to any work at or near any of the airbags. See* **721 Airbag System (SRS)**.
> - *Always disconnect the battery and cover the negative battery terminal with an insulator before starting diagnostic, troubleshooting or service work on cars fitted with SRS, and before doing any welding on the car.*

Battery Safety Terminal (BST)

In Cooper S models, due to the location of the battery in the cargo compartment, a pyrotechnic (explosive charge) terminal is installed at the positive battery pole. See **121 Battery, Starter, Alternator** for more details.

Security

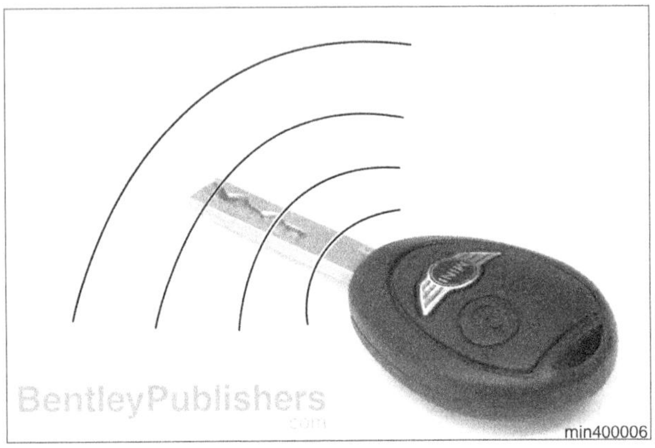

Electronic immobilization (EWS). MINI vehicles incorporate an electronic immobilization system known as EWS. This system uses a wireless communication link between a transponder chip in the ignition key and the ring antenna surrounding the ignition switch. The EWS control module blocks the starting of the vehicle unless the correctly coded ignition key is used.

NOTE —
EWS is sometimes referred to as drive away protection.

Anti-theft alarm (DWA). This is a dealer-installed option for which MINI vehicles are factory-wired. When armed, the system monitors door lock and rear hatch contacts and engine hood locks. It sounds a siren if it detects tampering.

Tilt sensor. If equipped, the tilt sensor monitors the vehicle parked angle when DWA is armed. The siren is activated if the vehicle angle is changed. This helps prevent theft of the car using a ramp truck.

Ultrasonic interior protection. An interior ultrasonic emitter/detector is installed in the rear of the headliner. On convertible models the ultrasonic emitter/detectors are installed behind various interior trim panels. The alarm is triggered if motion is detected inside the locked car.

See **515 Central Locking and Anti-theft** for more details on the anti-theft systems.

INTERIOR FEATURES

Seats

The manually operated front seat is designed with a mechanical memory, so that after being folded and slid forward to allow rear seat access, the front seat returns to its original position.

MINI models are equipped with a thorax protection airbag in the outer bolster of each front seat, and a 50 / 50 split rear seat backrest which can be folded down.

Heated seats are an option in all models.

Seat repairs are covered in **520 Seats**.

Instruments and controls

Tilt steering wheel is manually operated.

Multi-function steering wheel contains two key pads containing controls for the sound system and cruise control.

Instrument cluster and **speedometer** use large easy-to-read analog gauges. Each is removable as a unit without removing the dashboard.

On-Board Computer functions are integrated into the instrument cluster display.

Integrated on-board navigation system, based on Global Positioning System (GPS) technology, is optionally available. When installed, a multi-function monitor is included in the center of the dashboard above the radio.

Park Distance Control (PDC) is an optional system which uses ultra-sonic sensors in the rear bumper trim to warn the driver of approaching too close to obstacles when parking.

Tire pressure warning (RDW) system uses ABS wheel speed sensors to monitor run-flat tire inflation and warn of under inflation.

Body control electronics

MINI cars are equipped with a sophisticated centralized design for body electronics which is self-diagnostic and incorporates many functions into a single General Module (BC1). The consolidation of several systems into a single control module minimizes power requirements and the incorporation of the diagnostic link results in more efficient and accurate troubleshooting.

BC1 controls the following functions:

- Windshield wiper/washer system, with optional Rain Sensor Interface (AIC). See **611 Wipers and Washers**.
- Central locking with power rear hatch release. See **515 Central Locking and Anti-Theft**.
- Keyless entry (FZV).
- Power window control. See **512 Door Windows**.
- Interior lighting.
- Alarm system (DWA).
- Sunroof operation. See **540 Sunroof**.
- Convertible top operation. See **541 Convertible Top**.
- Seat memory. See **520 Seats**.
- Outside rear-view mirror control and heating.

NOTE—
In wiring diagrams, the General Module is referred to as A1.

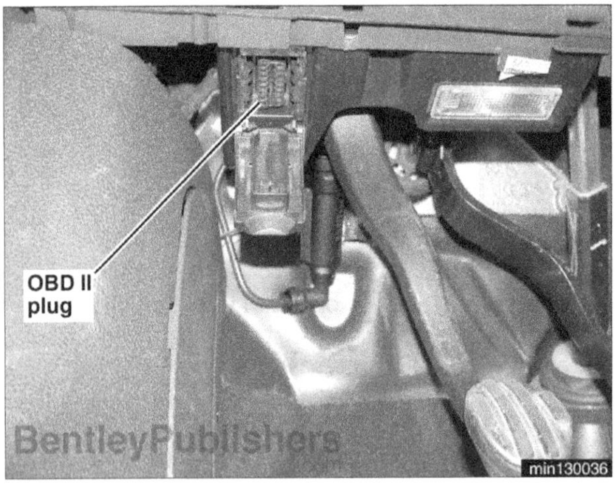

BC1 Diagnostic Trouble Codes (DTCs) are accessible electronically through the OBD II connector under left side of dashboard.

Heating and air-conditioning

MINI models are equipped with either a manual or automatic heating and air-conditioning system.

IHKS is the manually operated heating and air-conditioning system.

IHKA is the integrated automatic climate control system. It uses an 8-speed radial blower motor for good distribution of air. Fresh air enters through the grill at the base of the windshield and into the passenger compartment via the dashboard and footwell vents.

Repair information for the heating and air-conditioning system is in **640 Heating and Air-conditioning**.

410 Fenders, Engine Hood

General

This repair group covers replacement of the front fenders and removal and installation of the engine hood.

NOTE —
The body is painted at the factory after assembly. Realignment of body panels may expose unpainted metal. Paint all exposed metal once work is complete.

Special tools

Most body repairs can be performed using regular automotive service tools. Some BMW special tools are required to set body pieces into service position.

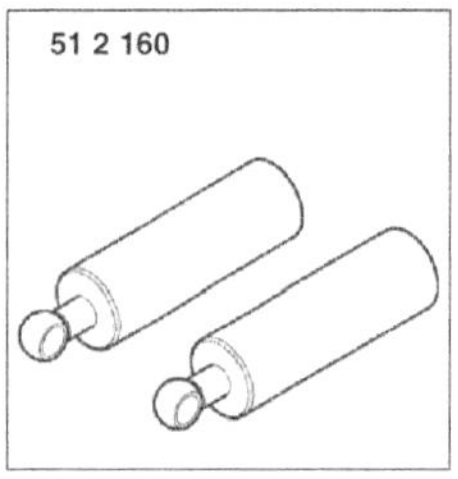

Service position hood prop extensions (BMW tool no. 51 2 160)

Use of plastic feeler gauges is recommended to prevent damaging painted surfaces when measuring panel gaps.

Front Fenders

Front fender, removing and installing

– Raise front of vehicle and remove wheel.

> ***WARNING —***
> *Make sure car is stable and well supported at all times. Use a professional automotive lift or jack stands designed for the purpose. A floor jack is not adequate support.*

– Remove side marker lights by prying on front edge with a plastic prying tool and sliding bezel forward. Remove harness connector for light.

– Remove fender lining.

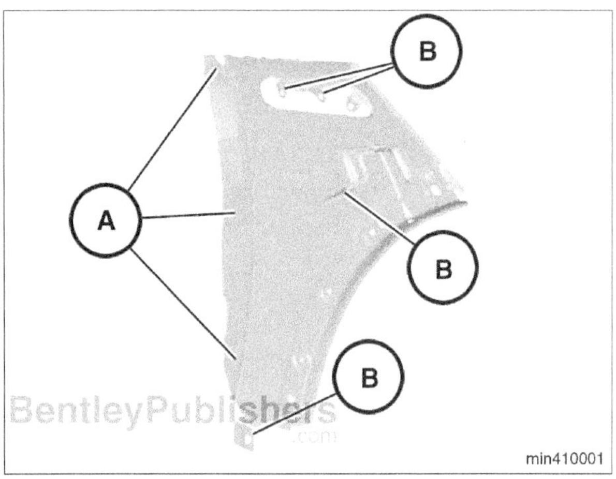

Remove nuts (**A**) and bolts (**B**).

NOTE —
Inside view of left front fender shown to show location of fasteners.

Engine Hood

Hood, raising to service position

> ***CAUTION—***
> - *Do not switch on the windshield wipers with the engine hood raised. As a precaution, remove the wiper motor fuse. See* **ECL Electrical Component Locations**.
> - *The hood is heavy. Before removing the hood supports, be sure to have an assistant help support the hood.*

< Release top of gas strut from hood by prying spring clip (**arrow**) and removing strut from ball joint.

– Repeat for other side while an assistant supports the hood.

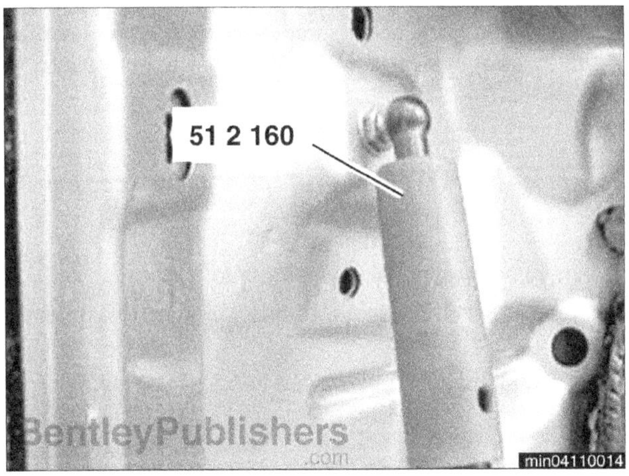

< Slip open end of special tool 51 2 160 over gas strut and press other end onto ball joint.

– Repeat for other side.

NOTE—
If replacing strut, bottom of strut is attached to body in a similar fashion.

Hood, removing and installing

> ***CAUTION—***
> *The hood is heavy. Before removing hood supports and hinges, have an assistant help support the hood.*

- Disconnect negative (-) cable from battery.

> ***CAUTION—***
> *Prior to disconnecting the battery, read the battery disconnection cautions given in* **001 General Warnings and Cautions**.

- Remove top of gas struts from hood. See **Hood, raising to service position** in this repair group.
- Remove sound insulation from underside of hood.
- Remove electrical connectors from headlights and unclip ties holding wires to hood. Make note of wire routing.
- Unbolt chassis ground connection from lower left of hood.
- Remove washer fluid hose from nozzles and unclip ties holding hose to hood. Make note of hose routing.
- Remove clamps holding washer fluid hose to head light washer (if equipped) and unclip ties holding hose to hood. Make note of hose routing.

With an assistant supporting hood, remove hood hinge mounting bolts (**arrows**) and remove hood. Make sure no wires or hoses interfere with removal.

NOTE—
The left lower hinge bolt is the attachment point for the hood ground strap.

- Installation is reverse of removal. Remember to:
  - Tighten hinges in previously installed position to maintain hood alignment.
  - Re-route wires and hoses, and replace ties and clamps.

NOTE—
The body is painted at the factory after assembly. Realignment of body panels may expose unpainted metal. Paint all exposed metal once work is complete.

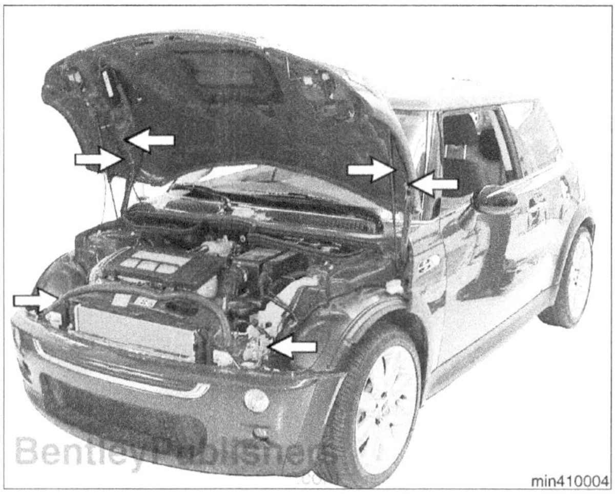
min410004

Hood, aligning

Hood adjustment points are at the top and bottom bolt of each hinge and at each latch (**arrows**). Hood gap specifications are shown in the table below.

- Use plastic feeler gauges to check body gaps.

Hood gap specifications	
Back of hood to cowl	4 ± 0.75 mm
Front of hood to bumper	6 ± 1 mm

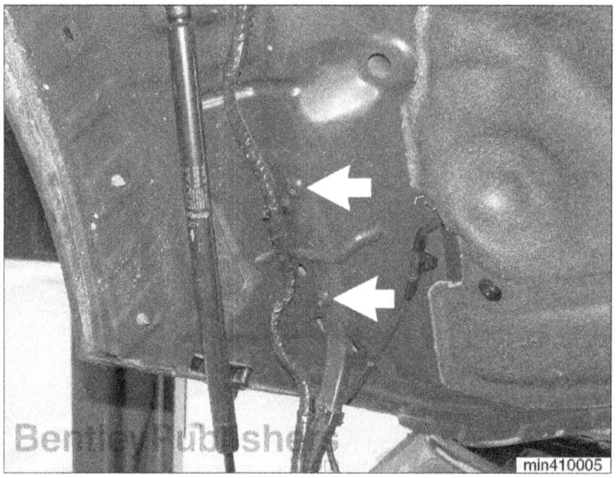
min410005

Loosen hood hinge bolts (**arrows**) enough to allow slight movement and adjust hood according to hood gap specifications.

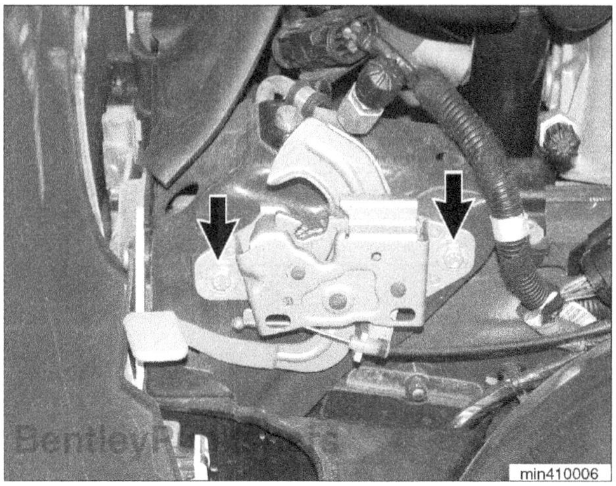
min410006

If alignment is not possible using only hinge bolts, loosen hood latch to achieve final hood gap specification.

– Make sure catch on hood does not scrape against body or latch. See **Hood latch, adjusting** in the this repair group for more information.

NOTE —
The body is painted at the factory after assembly. Realignment of body panels may expose unpainted metal. Paint all exposed metal once work is complete.

Hood latch, adjusting

Loosen bolts (**arrows**) on left side hood latch to adjust position.

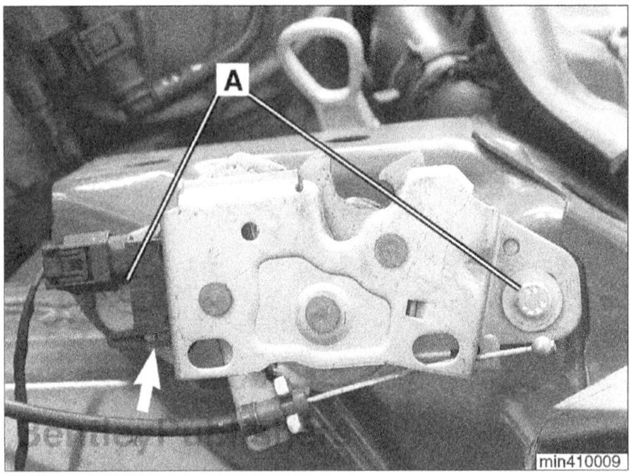

Release electrical connector on right side hood latch by pressing retaining tab (**arrow**) and rotating upward from latch. Loosen bolts (**A**) to adjust latch position.

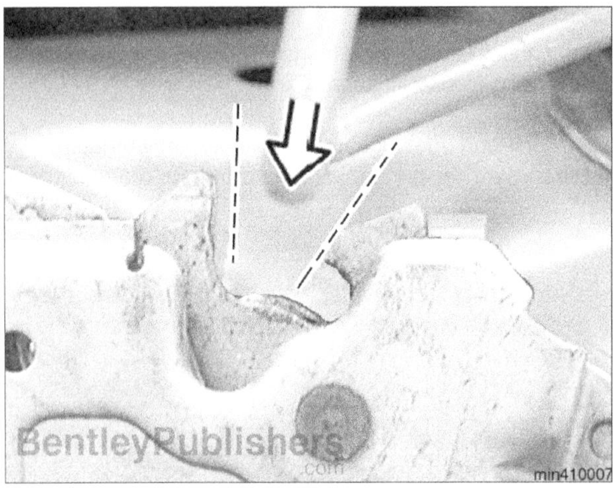

Hood latch is correctly adjusted when each striker (**arrow**) is approximately centered on each release catch, and hood to bumper gap meets the specification below.

- Use plastic feeler gauges to check body gaps.

Hood gap specifications	
Back of hood to cowl	4 ± 0.75 mm
Front of hood to bumper	6 ± 1 mm

Hood latch, removing and installing

Right hood latch

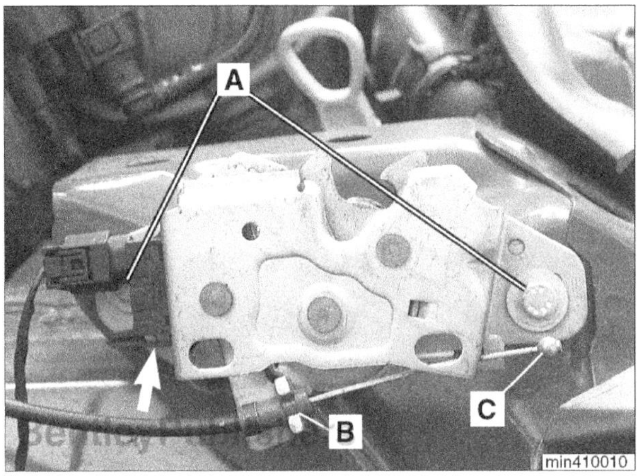

- Release electrical connector on right side hood latch by pressing retaining tab (**arrow**) and rotating upward from latch.
- Remove latch cable (**B**) from bracket and release ball end of cable (**C**) from latch.
- Remove bolts (**A**) and remove latch.
- Installation is reverse of removal. Remember to adjust latch and hood to bumper gap. See **Hood latch, adjusting** in this repair group.

Left hood latch

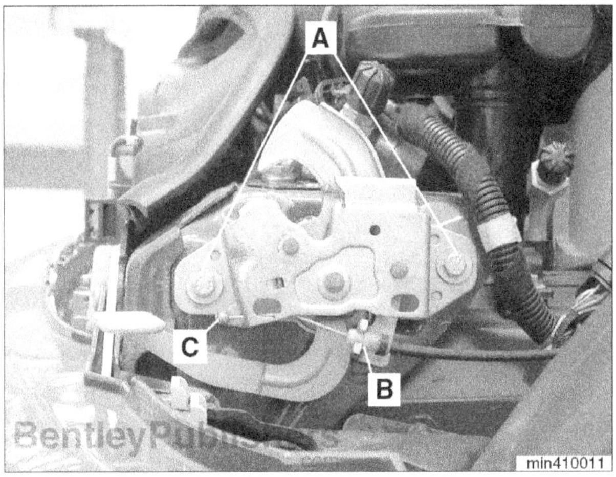

- Remove latch cable (**B**) from bracket and release ball end of cable (**C**) from latch.
- Remove bolts (**A**) and remove latch.
- Installation is reverse of removal. Remember to adjust latch and hood to bumper gap. See **Hood latch, adjusting** in this repair group.

Hood grill, replacing

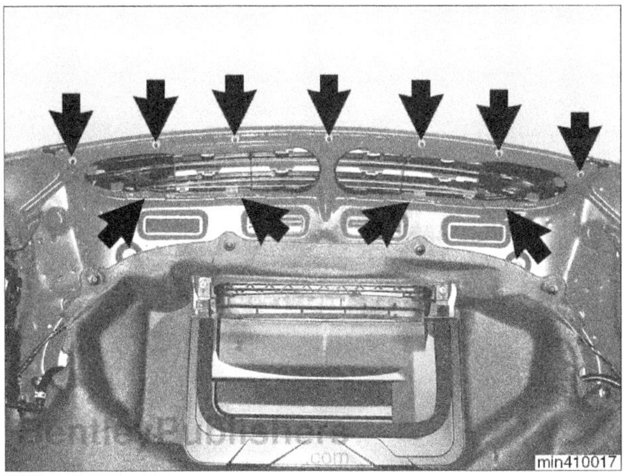

- Working with hood open, remove nuts and washers (**arrows**) from back of grill.

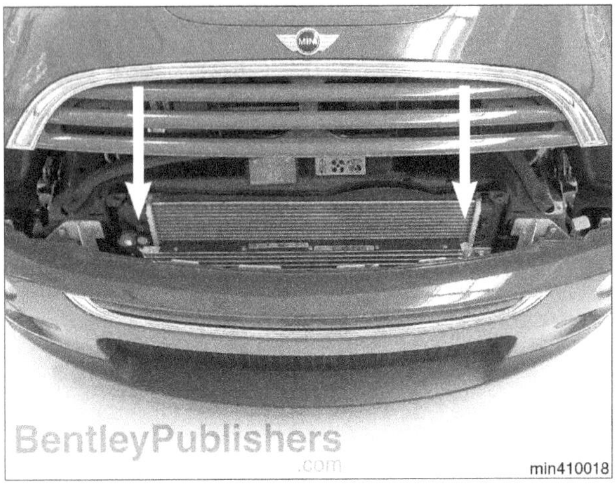

◄ Pull grill down and away from hood to remove.

– Installation is reverse of removal.

Hood scoop trim, replacing (Cooper S)

◄ Working with hood open, remove nuts (**arrows**) from hood scoop trim.

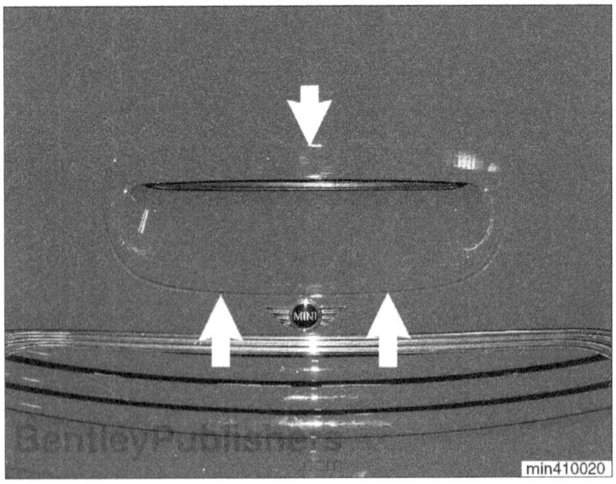

◄ Detach trim ring by carefully prying up lower edge (**lower arrows**) and pulling trim ring forward.

– Installation is reverse of removal. Make sure center clip (**upper arrow**) snaps securely in place and that retaining screws are properly located underneath hood.

411 Doors

General

This repair group covers removal and installation of the doors and door trim panels.

For information on power door windows and door glass replacement, see **512 Door Windows**.

For information on door locks, see **515 Central Locking and Anti-Theft**.

Special tools

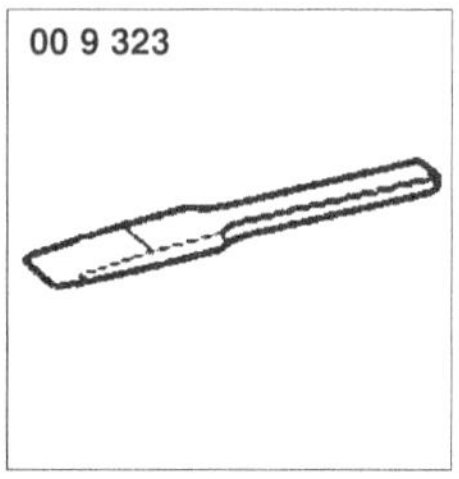

Body trim wedge

- BMW trim panel wedge 00 9 323 from special tool kit 00 9 310

Use of plastic feeler gauges is recommended to prevent damaging painted surfaces when measuring panel gaps.

Doors

Door, removing and installing

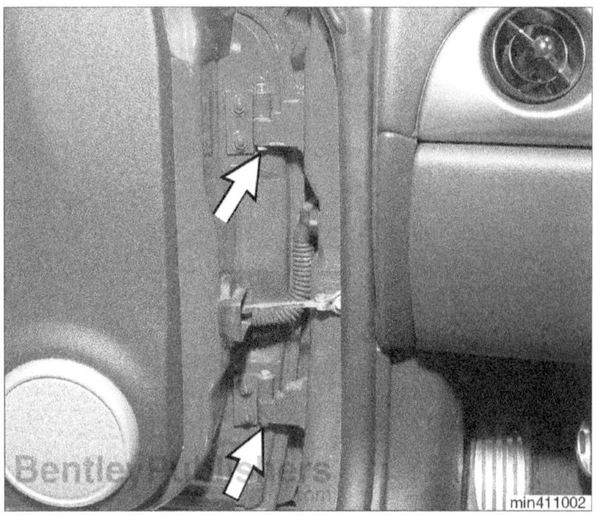

With door fully open, unscrew pin bolts (**arrow**) from top and bottom hinges.

◄ Working at the door post, remove mounting bolt from door check (**arrow**).

> ***CAUTION—***
> *With door check disconnected, door can hyper-extend, damaging door and body panels.*

– Remove door by lifting up and off lower hinge halves.

– Support door on a suitable surface to gain access to wiring harness connector.

> ***CAUTION—***
> *Be careful not to damage door or other painted surfaces. Make sure no load is placed on wiring harness.*

◄ Remove bolt from harness connector (**left arrow**) and slide connector rearward to remove from door jam.

– Separate harness connector (**right arrow**) by pulling release bar rearward.

– Remove door.

– Installation is reverse of removal. Remember to:
  - Align door so that panel gap is equal on both sides. If necessary, check door alignment. See **Door, adjusting** in this repair group.
  - Repair any paint damage and paint any exposed metal.

Door, adjusting

NOTE—
The body is painted at the factory after assembly. Realignment of body panels may expose unpainted metal. Paint all exposed metal once work is complete.

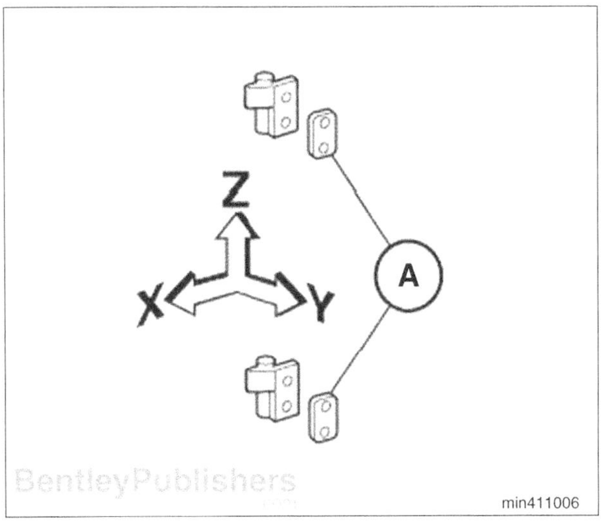

◄ Door adjustments along the X and Z-axis are accomplished by repositioning the door hinges. Door adjustments along the Y-axis are accomplished by installing shims (**A**) behind the hinges.

◄ Loosen door hinge attachment bolts (**arrows**) and adjust door according to door gap specifications.

- Use plastic feeler gauges to check door gap.

Door gap specifications	
Front fender to front door	4 ± 0.75 mm
Front door to rear fender	4 ± 0.75 mm

NOTE—
Check that door lock striker does not scrape against door or interfere with door operation.

◄ If necessary, adjust door lock striker by removing plastic cover and loosening bolts (**arrows**).

– When satisfied with door and striker alignment, tighten bolts securely and touch up any exposed metal surfaces. See **515 Central locking and Anti-Theft** for more information.

Door Trim Panels

Door trim panel, removing and installing

NOTE—
Trim panel plastic clips are easily damaged. Have extra clips on hand before beginning procedure.

– Lower door window.

From 09/2002

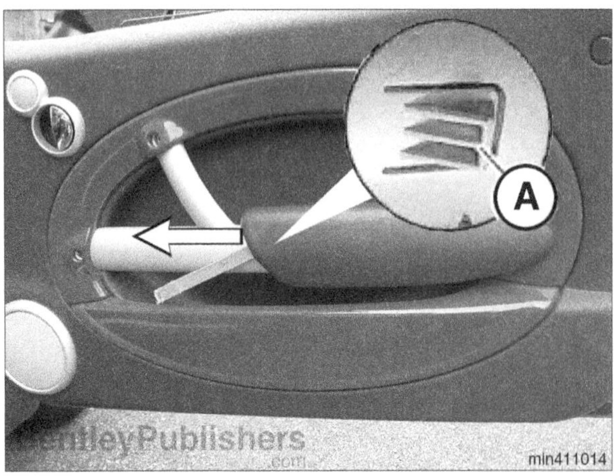

◄ Remove armrest using special tool 00 9 323 (trim panel wedge).

– Use trim panel wedge to pry retaining clip (**A**) outward. Slide armrest forward (**arrow**) to remove. Do not damage retaining clips at edges of armrest when removing.

Continued for all vehicles

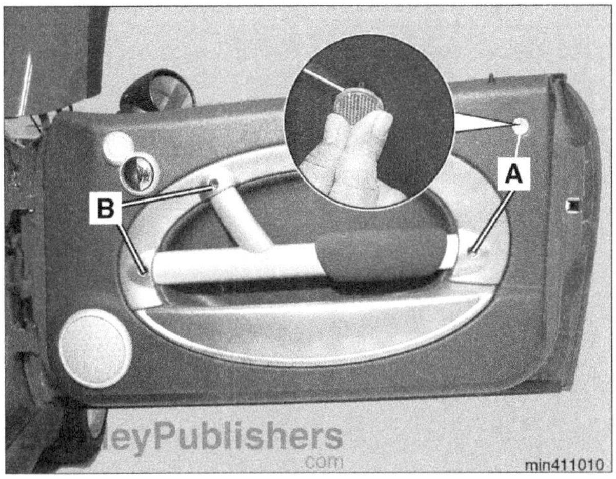

- Remove reflector (**inset**) with plastic prying tool.

- Remove short screws (**A**) and long screws (**B**) from panel.

- Release door trim panel clips (**inset, arrows**) using a plastic prying tool.

  NOTE—
  Clips are color coded and not all interchangeable. Replace clips in previously installed positions.

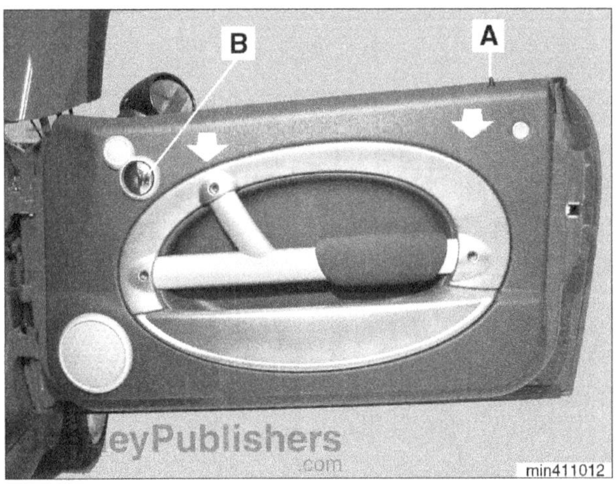

- Raise door trim panel clear of manual door lock (**A**). Release door trim panel from top clips by pulling horizontally (**arrows**).

- Guide door handle (**B**) out of handle trim and remove door trim panel.

- Installation is reverse of removal. Replace any damaged trim clips.

Tightening torque	
Door trim panel to door	3 Nm (2.2 ft-lb)

412 Rear Hatch

GENERAL

This repair group covers removal and installation of the rear hatch and hatch trim panels.

For information on rear hatch lock components and the rear hatch emergency release cable, see **515 Central Locking and Anti-Theft**.

Use of plastic feeler gauges is recommended to prevent damaging painted surfaces when measuring panel gaps.

REAR HATCH (COUPE)

> ***WARNING —***
> *MINI cars are equipped with multiple airbags around the rear hatch area. When servicing the hatch, always disconnect the negative (-) battery terminal. See* **721 Airbag System (SRS)** *for cautions and procedures relating to the airbag system.*

Rear hatch strut, removing and installing

> ***WARNING —***
> *Support hatch with suitable prop rod before removing struts.*

- Release top of gas strut from hatch by prying spring clip (**upper arrow**). Remove strut end from ball joint.
- Remove bottom (**lower arrow**) of gas strut from body in a similar manner.
- Install struts to hatch and body by firmly pressing strut sockets onto ball joints and locking spring clips into position.

Rear hatch, removing and installing

- Disconnect negative (-) battery cable.

> ***CAUTION—***
> *Prior to disconnecting the battery, read the battery disconnection cautions given in* **001 General Cautions and Warnings**.

- Remove top, side, and bottom trim pieces. See **Rear hatch trim** in this repair group.
- Remove electrical connectors and washer fluid hose from wiper motor and remove electrical connectors from license plate lights.
- Remove wiring harness and washer fluid hose from rear hatch.
- Support hatch with a suitable prop and remove hatch struts. See **Rear hatch struts, removing and replacing** in this repair group.

min412002

◄ With an assistant supporting the hatch, remove screws (**arrows**) from hinges and remove hatch. Make sure no wires or hoses interfere with removal.

- Installation is reverse of removal. Remember to:
  - Tighten hinges in previously installed position to maintain hatch alignment.
  - Reroute wires and hoses, and replace wire ties and clamps.

NOTE—
The body is painted at the factory after assembly. Realignment of body panels may expose unpainted metal. Paint all exposed metal once work is complete.

Rear hatch, aligning

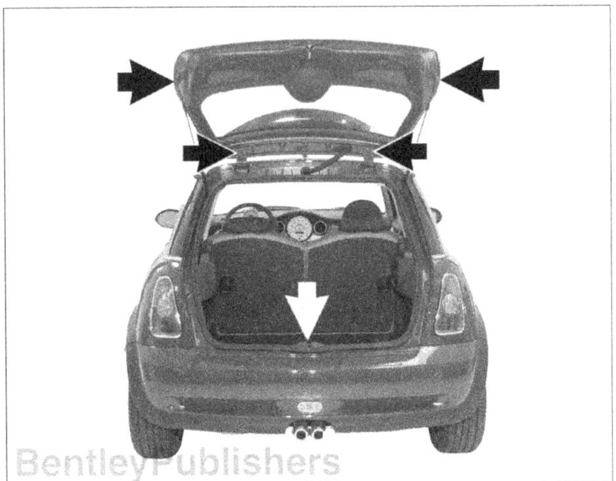
min412003

◄ Hatch adjustment points (**arrows**) are at each hinge, the buffer stops and the latch.

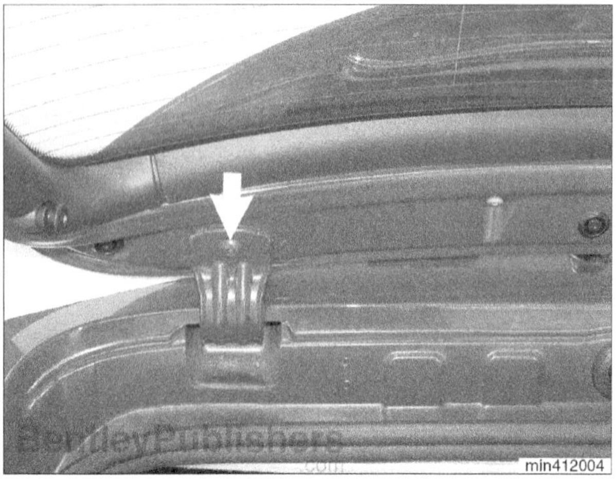

Loosen hatch hinge bolts (**arrow**) enough to allow slight movement and adjust hatch according to gap specifications.

- Use plastic feeler gauges to check body gaps.

Rear hatch gap specifications	
Rear hatch to roof	4 ± 0.75 mm
Rear hatch to bumper	4 ± 0.75 mm
Rear hatch to C-pillar	4 ± 1 mm
Rear hatch to fender	4 ± 0.75 mm

NOTE—
If adjustment range is insufficient, lower hinge bolts can be accessed by lowering the headliner. See **513 Interior Trim** *for more information.*

– Check that rear latch remains properly aligned. See **515 Central Locking and Anti-Theft** for more information.

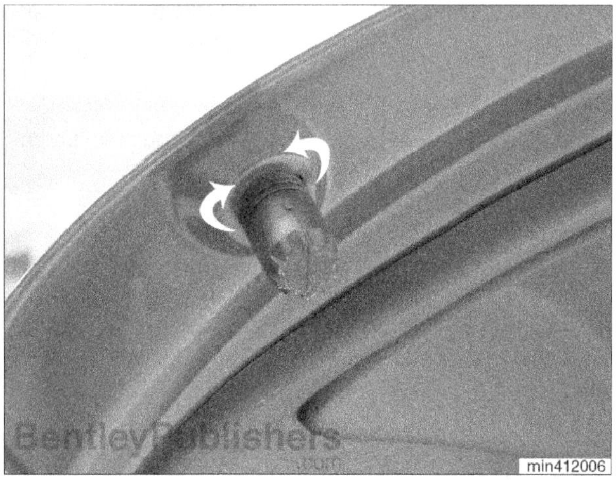

Turn buffer stops in or out (**arrows**) until they make slight contact with the hatch opening to complete alignment.

Rear hatch lower trim, removing

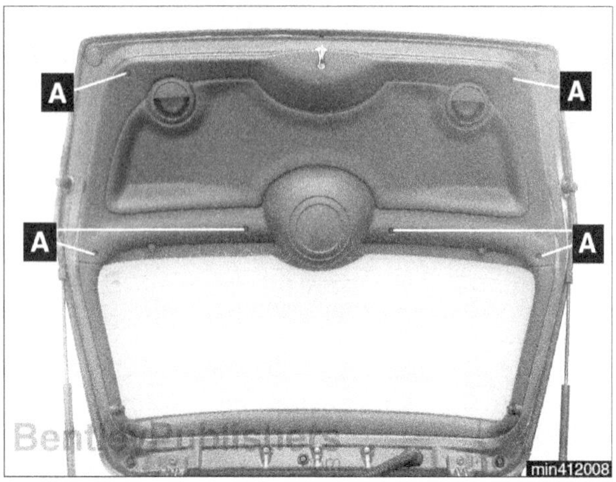

With hatch open, remove trim mounting screws (**A**).

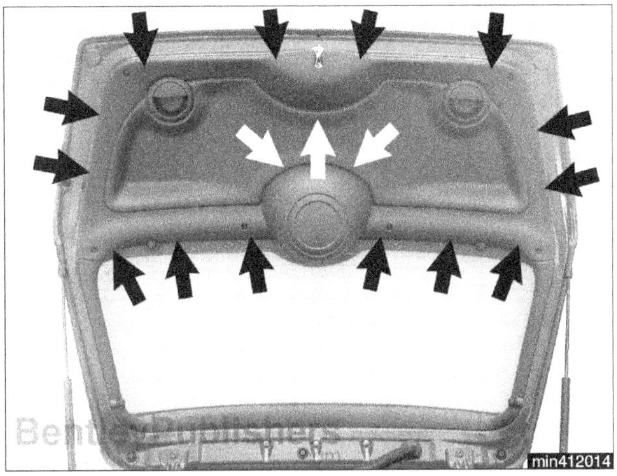

Use a plastic prying tool to release clips (**arrows**) and remove lower trim from hatch.

- Installation is reverse of removal. Replace any damaged clips.

NOTE —
If necessary, apply body tape over clips to reduce trim movement.

Rear hatch side trim, removing

- Partially remove hatch lower trim at bottom of side trim. See **Rear hatch lower trim, removing** in this repair group.

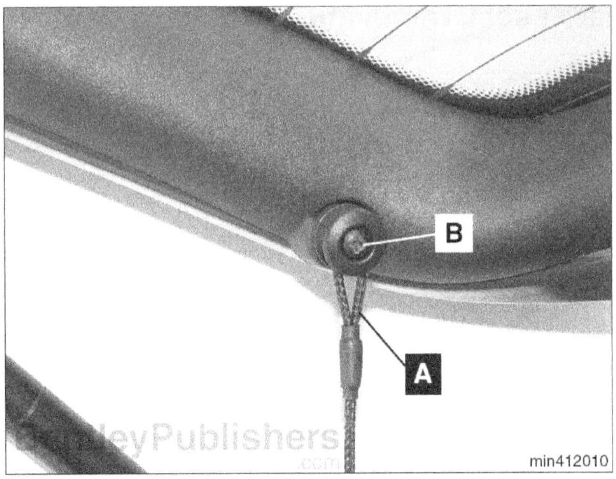

Remove parcel shelf retaining strap (**A**) and remove screw (**B**) from strap button.

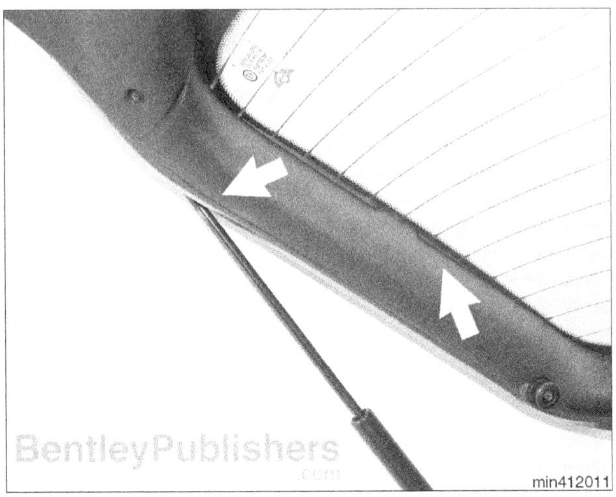

Release trim clips (**arrows**) using a plastic prying tool and remove side trim.

- Installation is reverse of removal. Replace any damaged clips.

Rear hatch upper trim, removing

– Remove left and right hatch side trim. See **Rear hatch side trim, removing** in this repair group.

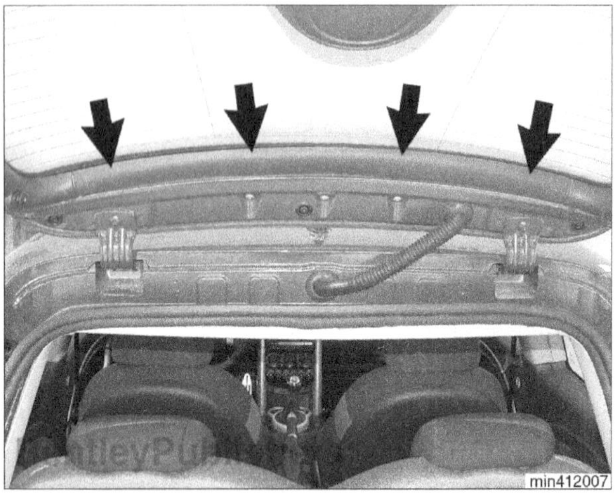

◄ Release trim clips (**arrows**) using a plastic prying tool and remove upper trim.

Rear hatch seal, removing

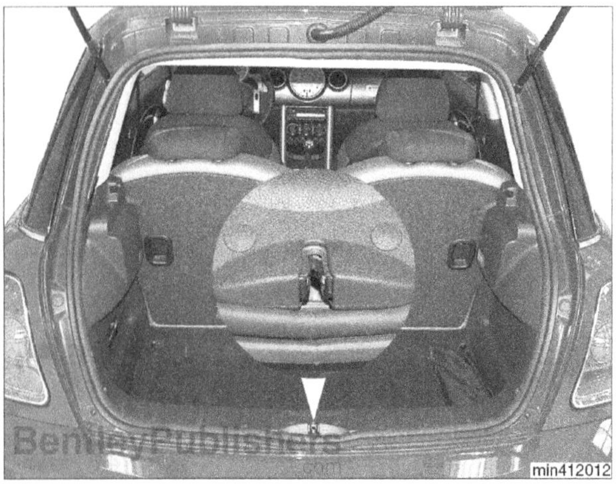

◄ Remove hatch seal by gently pulling away from hatch opening.

– When reinstalling, make sure seal is correctly oriented at latch mechanism (**inset**).

Rear hatch handle, removing

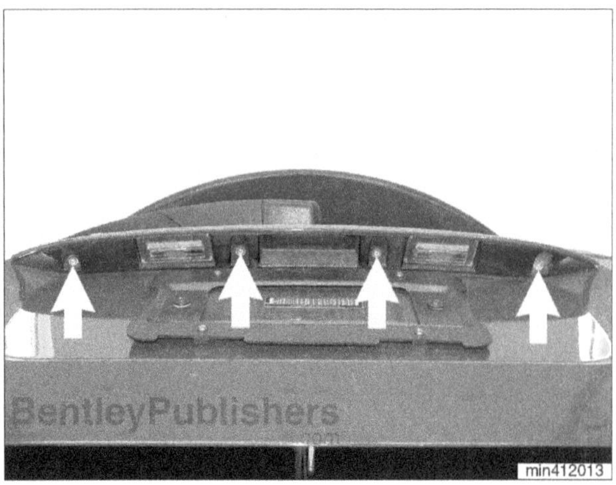

◄ Working outside the car, remove screws (**arrows**) holding handle to rear hatch.

– Detach electrical connectors for license plate lights and remove handle.

Rear spoiler, removing (Cooper S)

Rear spoiler is attached to rear hatch using bolts and adhesive tape.

NOTE—
Spoiler mounting foot must be heated with a hot air blower to aid removal.

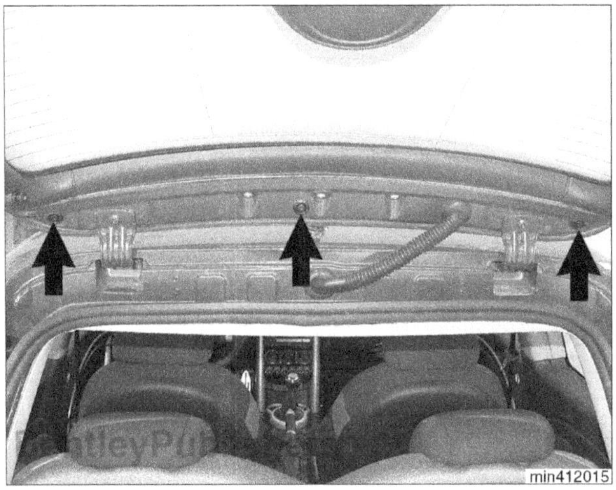

◄ Remove bolt covers (**arrows**) and unbolt spoiler from hatch.

– Heat spoiler mounting foot with a hot air blower to at least 18°C (65°F) to soften adhesive.

– Loosen spoiler from hatch with a plastic prying tool and remove spoiler.

– To install, use mineral oil and a lint-free cloth to remove adhesive from surface of hatch and spoiler.

– Apply new adhesive tape to mounting feet of spoiler and press firmly into place on hatch.

– Tighten nuts securing spoiler to hatch and replace stoppers.

Tightening torque	
Spoiler to rear hatch	8.5 Nm (6.3 ft-lb)

Rear Lid (Convertible)

Rear lid trim panel, removing and installing

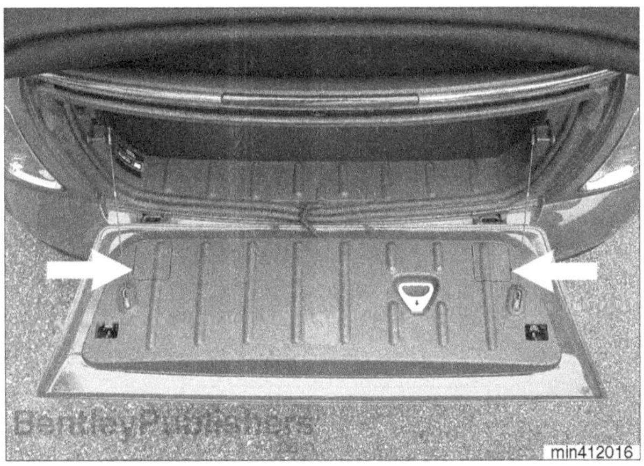

◄ Open rear lid and remove access panels in rear lid trim.
- Pry at **arrows** to remove access panels.

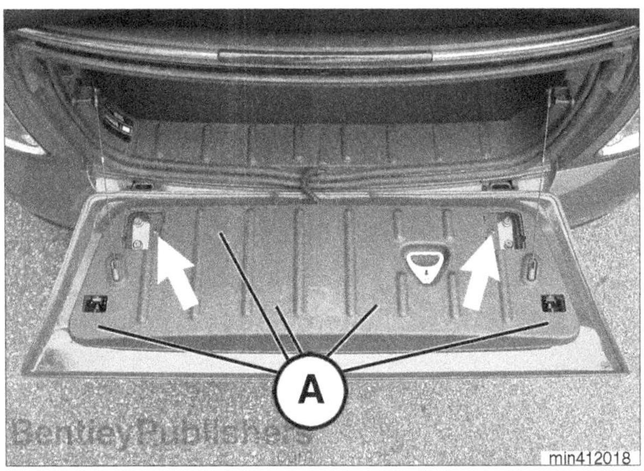

◄ Remove mounting screws (**arrows**) for rear lid trim.

– Carefully pry trim panel from lid and release retaining clips (**A**). Do not remove lid trim panel yet.

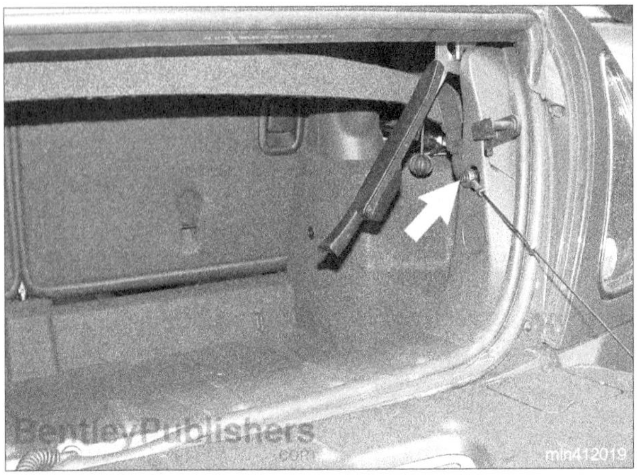

◄ Remove mounting screw (**arrow**) on left and right sides for rear hatch support cables.

> ***WARNING —***
> *Rear hatch retaining cables are under tension. Be careful when removing mounting screw.*

– Let retaining cables carefully retract into rear lid.

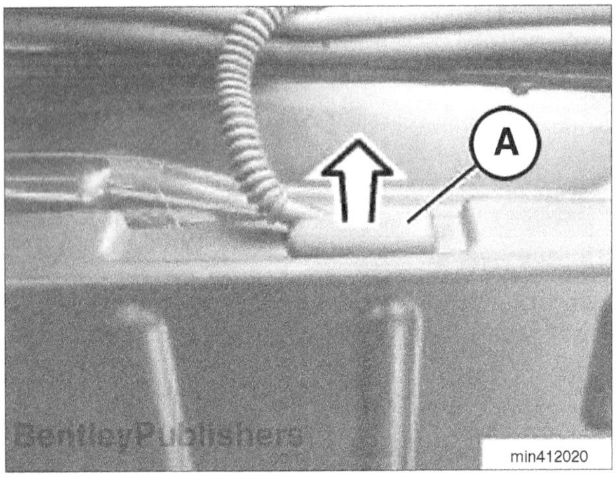

◄ Remove harness connector grommet (**A**) in direction of **arrow**.

– Disconnect harness connector from rear lid.

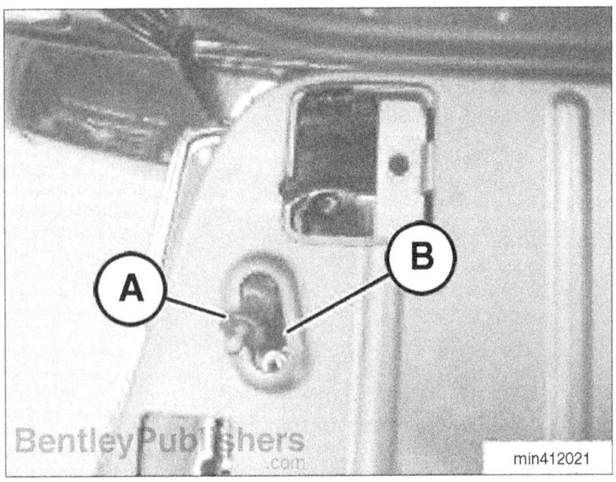

◄ Feed rear lid retaining cable (**A**) through opening (**B**) in trim panel.

- Do left side cable first and then right side.
- Work carefully as rear lid could be damaged.

– Reach under trim panel and disconnect bowden cable from back of emergency release handle.

– Remove rear lid trim panel.

– Installation is the reverse of removal.

Tightening torque	
Rear lid retaining cable to body	18.5 Nm (13 ft-lb)

Rear lid, removing and installing

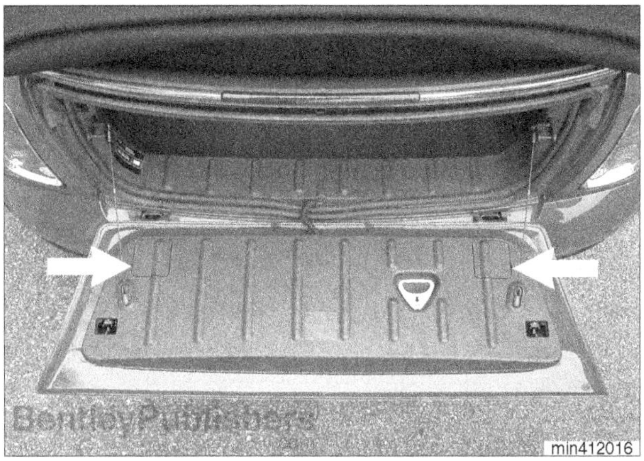

Open rear lid and remove access panels for rear lid hinge mounting bolts.

- Pry at **arrows** to remove access panels.

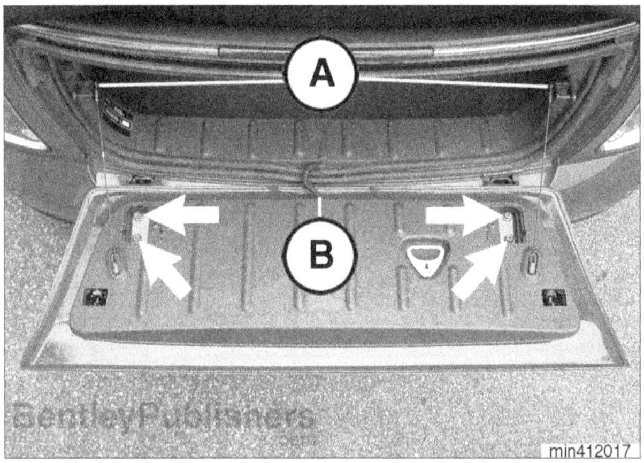

Disconnect rear lid harness connector (**B**) and unscrew rear lid retaining cables (**A**). See **Rear lid trim panel, removing and installing**.

– Remove mounting bolts (**arrows**) for rear lid.

– Remove rear lid.

– Install in reverse order of removal.

Tightening torque	
Rear lid to hinge	18.5 Nm (13 ft-lb)

Rear lid, adjusting

Loosen mounting bolts (**arrows**) for rear lid.

– Adjust rear lid as necessary.

- Check for even gaps around rear lid using plastic feeler gauges.

– Tighten rear lid mounting bolts.

Tightening torque	
Rear lid to hinge	18.5 Nm (13 ft-lb)

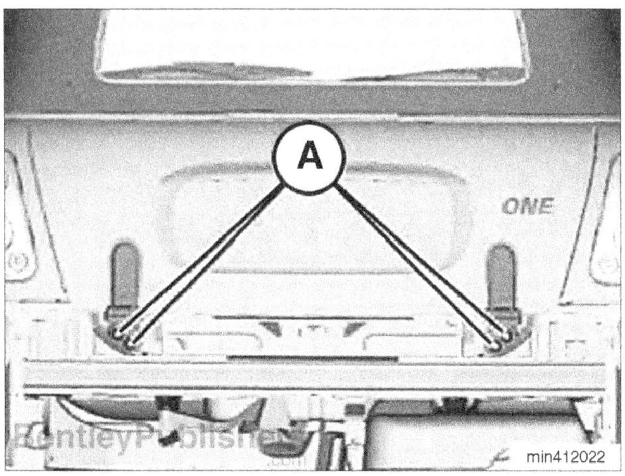

If more adjustment range for the rear lid is needed, do the following:

- Remove rear bumper cover. See **510 Exterior Trim, Bumpers**.
- Loosen lower mounting bolts (**A**) on outer rear lid hinges.
- Adjust rear lid as necessary.
- Tighten mounting bolts.

Tightening torque	
Rear lid hinge to body	18.5 Nm (13 ft-lb)

- If necessary, adjust rear lid lock.

510 Exterior Trim, Bumpers

GENERAL

This repair group includes repair information for the outside rear view mirrors, front and rear bumpers, and the easily removable exterior trim parts.

For additional body repair information, see the following groups:

- Hood center grille and air intake vent (Cooper S) is covered in **410 Fenders, Engine Hood**.
- Door locks are covered in **515 Central Locking and Anti-theft**.
- Rear spoiler (Cooper S) is covered in **412 Rear Hatch**.

Special tools

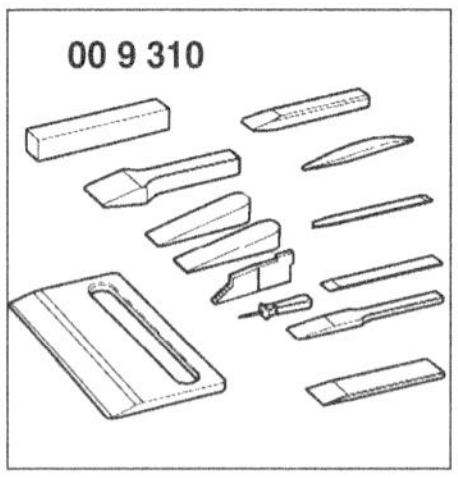

Body trim wedges and prying tools (BMW special tool kit 00 9 310)

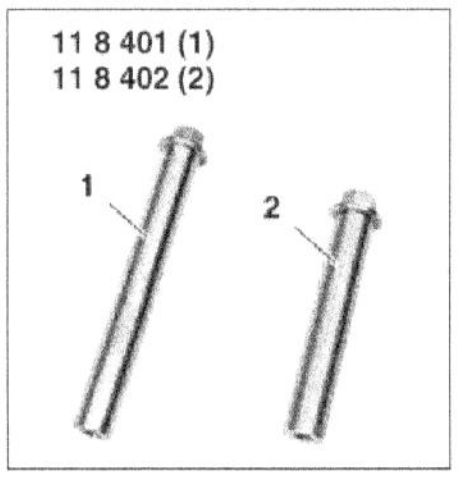

Modular front end supports (BMW special tool 11 8 400)

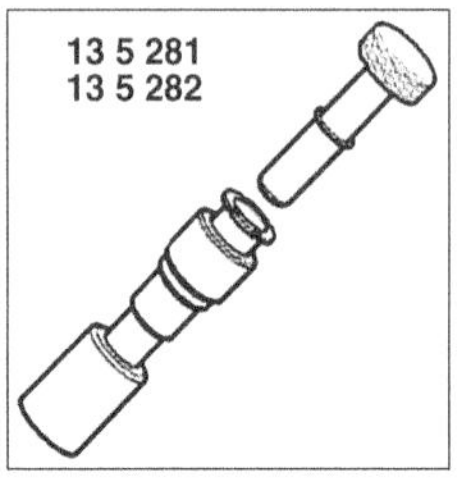

Fluid line plugs (BMW special tools 13 5 281, 13 5 282)

OUTSIDE REAR VIEW MIRRORS

The mirror glass can be separated and replaced. Replacements parts are available from an authorized MINI dealer.

Outside mirror glass, replacing

- Turn door mirror glass (**arrow**) to maximum inward position.
- Pry mirror glass to release plastic lock tabs from mirror actuator assembly.

> ***CAUTION—***
> *If the mirror is not already broken, this process may shatter it. Protect eyes and hands from broken glass.*

- Disconnect glass heating connectors from behind mirror, if applicable.
- When reassembling:
  - Ensure correct engagement of plastic locking tabs on glass to mirror actuator assembly.
  - Check mirror function.

Outside rear view mirror, removing and installing

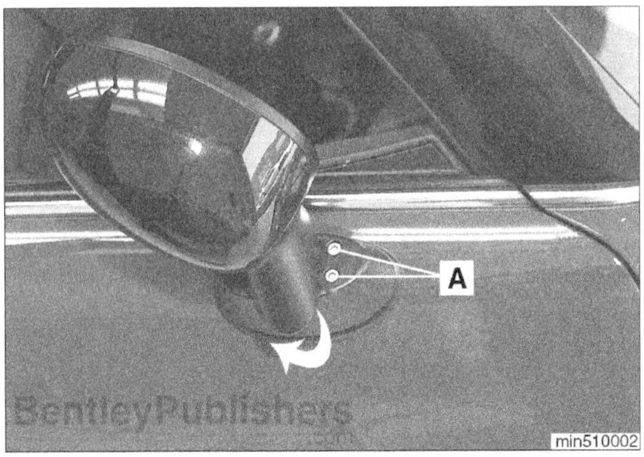

- Rotate mirror (**arrow**) to access front mounting screws (**A**). Remove screws.

- Rotate mirror (**arrow**) to access rear mounting screw (**A**). Remove screw.

Pull mirror away from door and disconnect electrical harness connector (**arrow**) at base of mirror. Remove mirror.

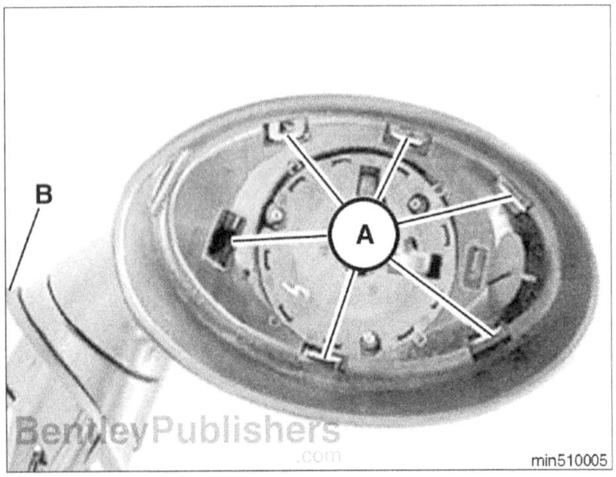

To replace mirror actuator:

- Remove mirror glass. See **Outside mirror class, replacing**.
- Release door mirror cover clips (**A**) and remove cover.
- Remove door mirror gasket (**B**).
- Remove 3 screws securing mirror actuator.
- Disconnect mirror actuator electrical connector.

– Reassembly is reverse of removal.

Tightening torque	
Outside rear view mirror to door (M6 x 65 mm)	4 Nm (35 in-lb)

Front Bumper

The front bumper assembly consists of the following:

- Aluminum bumper carrier bolted to front of chassis and front subframe crush tubes.
- Plastic Modular Front End (MFE), housing for radiator, air-conditioning condenser and (where applicable) CVT (automatic transmission) fluid cooler.
- Plastic bumper cover, which includes upper and lower grille, spoiler and molding trim, and houses foglights and turn signals.

Front bumper cover position adjustment is covered in **Front bumper cover, removing and installing** in this repair group.

Front bumper cover, removing and installing

– Raise front of vehicle and support in a safe manner.

> ***WARNING —***
> *Make sure the car is stable and well supported at all times. Use a professional automotive lift or jack stands designed for the purpose. A floor jack is not adequate support.*

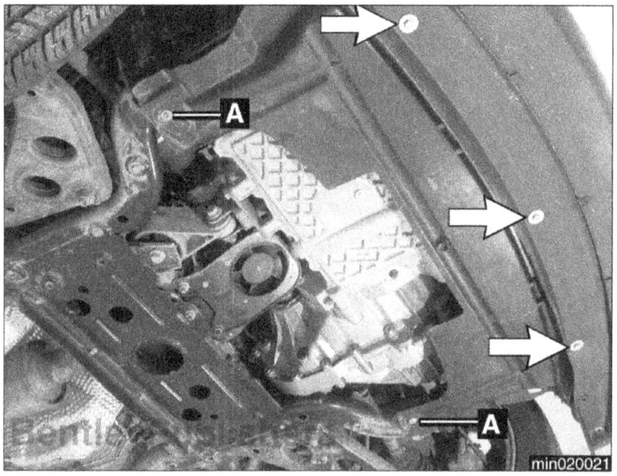

◄ Remove splash shield under engine:

- Remove 3 bolts (**arrows**) at bottom of front bumper cover.
- Unlock 2 splash shield mounting screws (**A**).
- Remove splash shield.

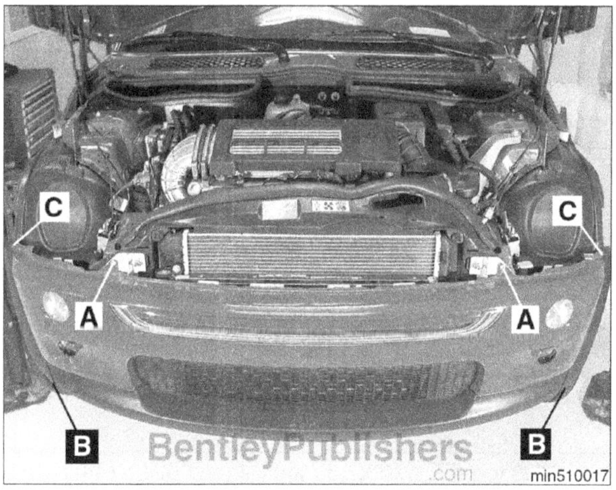

◄ Remove screws (**A**) securing bumper cover to MFE.

- Remove plastic rivets (**B**) securing bumper to lower spoiler from underneath bumper.
- Remove bolts (**C**) securing bumper cover to position adjustment bracket.

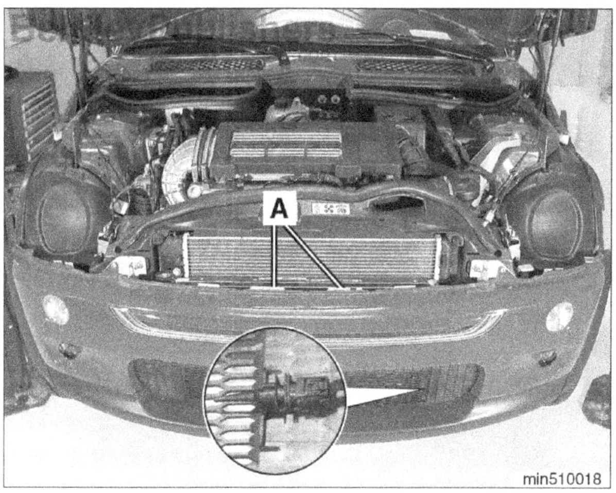

◄ Detach bumper cover clips (**A**) from bumper carrier.

- Disconnect foglight electrical harness connectors.
- Disconnect turn signal electrical harness connectors.
- Disconnect outside temperature sensor connector (**inset**).
- Lift off bumper cover.

– With bumper cover off, repair or replace trim items:
  - Spoiler is secured to bumper by means of locating clips.
  - Upper grill is secured to bumper cover by means of adhesive tape.
  - Lower grill is secured to both bumper cover and front spoiler by ten clips.
  - Front towing eye cover is push fit into lower grille.
  - Bumper side trim is secured with clips.

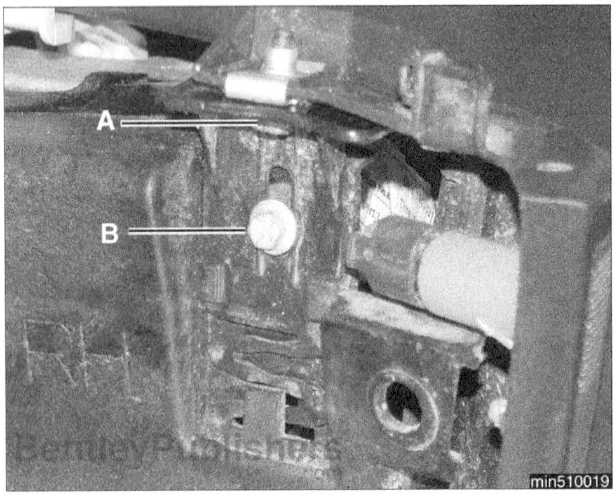

When reinstalling, adjust bumper cover position at adjustment brackets at corners of bumper:
  - Loosen bolt **A** for horizontal adjustment.
  - Loosen bolt **B** for vertical adjustment.

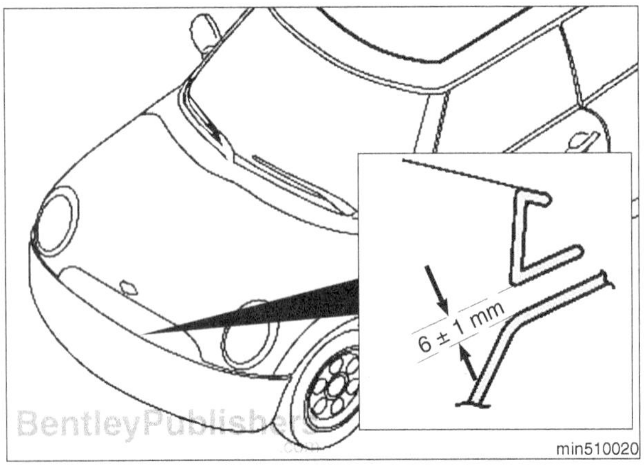

Tighten bumper cover mounting bolts when gap to front hood is set as illustrated.

Front bumper cover position	
Front bumper cover to engine hood gap	6 ± 1 mm (0.24 ± 0.04 in)

Tightening torques	
Engine splash shield to front bumper cover and MFE (M6 x 16 mm)	6 Nm (53 in-lb)
Front bumper adjuster to bumper (M6 x 20 mm)	6 Nm (53 in-lb)
Front bumper cover to MFE (M6 x 16 mm)	5 Nm (44 in-lb)

Front bumper carrier, removing and installing

– Remove front bumper cover. See **Front bumper cover, removing and installing** in this repair group.

min510015

Pull out electrical harness (**A**) for outside temperature sensor through opening in bumper carrier.

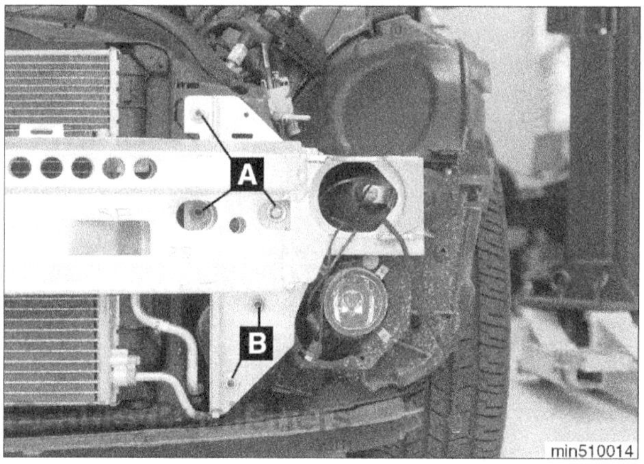

min510014

Remove mounting fasteners (**A**) from vehicle frame bumper support members.

- Remove nuts (**B**) from crush tube studs.
- Remove bumper carrier carefully. Watch for electrical harness connectors or hoses that might become snagged.

min510016

During reinstallation, adjust bumper carrier position if necessary:

- To adjust bumper height or side-to-side position, loosen adjustment bolt (**arrow**). Adjust carrier. Lock guide sleeve in position by tightening adjustment bolt.
- To adjust fore-aft position, fit shims between carrier and frame mounting flange.

Tightening torque	
Front bumper carrier to frame or to crush tube (M8)	22 Nm (16 ft-lb)

Modular Front End (MFE)

A number of procedures, including engine, alternator and radiator repairs, require that the front bumper and modular front end (MFE) be loosened and slid forward on special tools (or long bolts) for clearance purposes. The procedure given here is for sliding the MFE forward. Additional notes at the end of this procedure are useful if the MFE is to be removed from the vehicle altogether.

MFE, loosening

– Disconnect battery negative (-) cable. See **121 Battery, Starter, Alternator**.

> ***CAUTION—***
> *Disconnecting the battery may erase fault code(s) stored in memory. Check for fault codes prior to disconnecting the battery cables.*

– Raise front of vehicle and support in a safe manner.

> ***WARNING—***
> *Make sure the car is stable and well supported at all times. Use a professional automotive lift or jack stands designed for the purpose. A floor jack is not adequate support.*

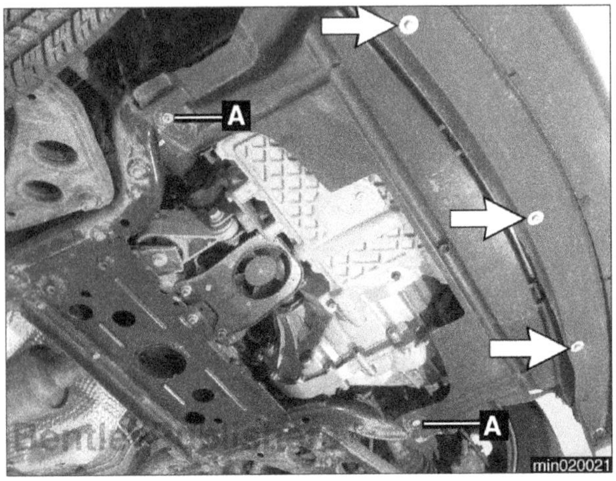

◄ Remove splash shield under engine:

- Remove 3 bolts (**arrows**) at bottom of front bumper cover.
- Unlock 2 splash shield mounting screws (**A**).
- Remove splash shield.

– Cooper: Drain cooling system. See **170 Cooling System**.

> ***WARNING—***
> *Due to risk of personal injury, be sure the engine is cold before opening any part of the cooling system.*

◄ Cooper: Loosen or remove upper radiator hose clamp (**arrow**). Detach hose.

– Remove front bumper cover. See **Front bumper cover, removing and installing** in this repair group.

– Detach both front wheel housing liners from MFE.

◄ CVT (automatic transmission) model: Working in front of radiator, remove transmission fluid cooler mounting screw (**A**). Slide cooler (**arrow**) to release from retaining bracket.

Disconnect CVT (automatic transmission) fluid cooler lines (**arrows**) from cooler. Use BMW fluid line plugs 13 5 281 / 13 5 282 or equivalent to plug fluid lines and cooler connections.

NOTE—
Be prepared to catch dripping fluid.

– Remove front bumper carrier. See **Front bumper carrier, removing and installing** in this repair group.

NOTE—
Bumper carrier removal is not necessary for loosening of MFE, but is useful or necessary for removal of other components attached to the MFE.

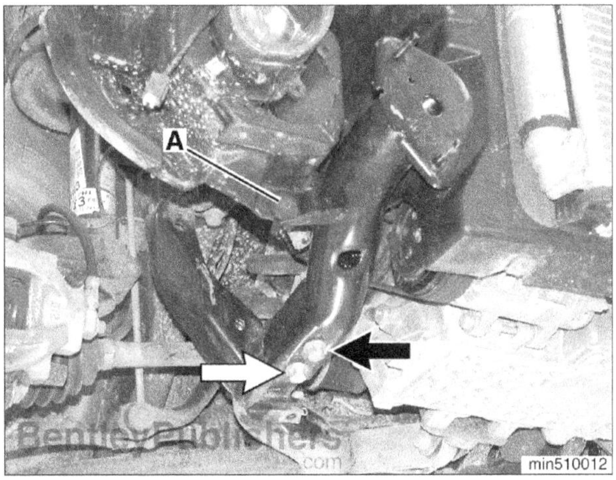

Working underneath car, remove crush tube mounting bolts (**arrows**) at subframe.

- Remove bolt (**A**) mounting each crush tube to MFE.
- Slide crush tubes out.

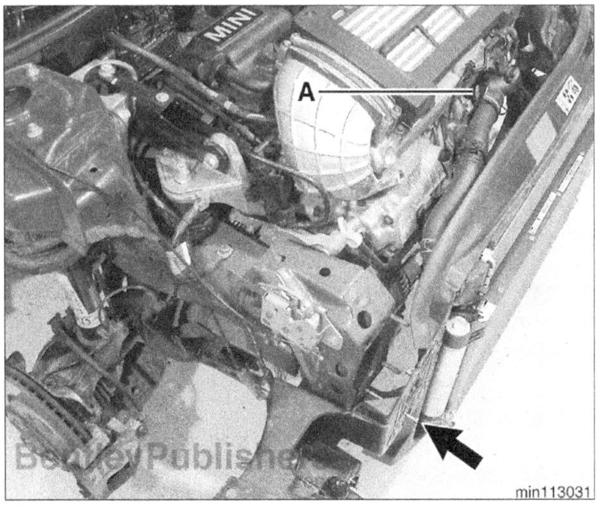

Loosen MFE:

- Cooper S: Detach radiator upper hose support clamp (**A**) from intake manifold.
- Install BMW special tools 11 8 401 and 11 8 402 (**arrow**) in left and right bumper support members. Slide MFE forward, supported on special tools.

> ***CAUTION—***
> *For ease of component alignment when reassembling front bumper assembly, do not loosen or remove bumper alignment bosses.*

NOTE—
Instead of special tools, two 100 mm (4 in) M8 bolts can be used to support the MFE.

– Installation is reverse of removal. Keep in mind the following:
  - Cooper: Replace radiator hose clamp.

Tightening torques	
Crush tube to front subframe (M12 x 1.5 X 85 mm)	100 Nm (74 ft-lb)
Crush tube to MFE (M6)	5 Nm (44 in-lb)
Engine splash shield to front bumper cover and MFE (M6 x 16 mm)	6 Nm (53 in-lb)
Front bumper carrier to frame or to crush tube (M8)	22 Nm (16 ft-lb)
Transmission fluid cooler to radiator	4 Nm (35 in-lb)

– After assembly:
  - CVT (automatic transmission) model: Top off transmission fluid. See **240 Automatic Transmission (CVT)**.
  - Cooper: Top off and bleed cooling system. See **170 Cooling System**.

MFE, removing

To remove MFE from vehicle, additional steps beyond those included in **MFE, loosening** in this repair group are as follows:

– Discharge air-conditioning system. See **640 Heating and Air-conditioning**.

– Cooper S: Drain cooling system. See **170 Cooling System**.

– Disconnect engine cooling fan electrical harness connector(s) at left side of MFE. Detach connector(s) from mounting clip. See **170 Cooling System**.

– Disconnect foglight, horn, outside temperature sensor and ABS wheel speed sensor harnesses and remove or detach from MFE.

– Once MFE, radiator and air-conditioning condenser are removed from vehicle, place assembly on work bench and disassemble as necessary.

– After reassembly, recharge air-conditioning system. See **640 Heating and Air-conditioning**.

Rear Bumper

Rear bumper cover, removing and installing

The rear bumper cover incorporates the back-up light socket and tow hook cover. It may also incorporate park distance control (PDC) sensors, side marker lights and spare tire storage compartment cover.

– Raise rear of vehicle and support in a safe manner.

> ***WARNING —***
> *Make sure the car is stable and well supported at all times. Use a professional automotive lift or jack stands designed for the purpose. A floor jack is not adequate support.*

– Remove both rear wheels.

– Remove threaded plastic rivets holding rear wheel housing liner on both sides. Remove wheel housing liners.

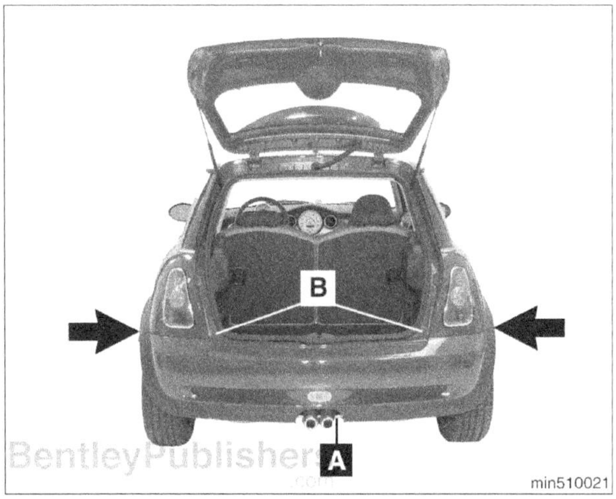

◄ Partially remove fender trim (**arrows**) to allow access to bumper cover outer mounting screws. Remove screws.

– Cooper S: Remove bumper cover lower mounting bolt to the right of the dual exhaust pipes (**A**).

– Cooper: Remove back-up light by carefully prying out with plastic pry tool or flat screwdriver with protective tape on the tip. Remove bumper cover mounting screw from behind back-up light.

– Cooper: Remove spare tire cover mounting screws from underneath bumper.

– All models: Open rear hatch and remove bumper cover mounting screws (**B**).

– Pull back bumper cover and disconnect park distance control (PDC) sensor electrical harness connector, if applicable.

  • Disconnect side marker light electrical harness connectors, if applicable.

– Remove bumper cover.

– Installation is reverse of removal.

Rear bumper carrier, removing and installing

– Remove rear bumper cover. See **Rear bumper cover, removing and installing** in this repair group.

◄ Remove bumper carrier bolts (**arrows**).

– Remove carrier.

– Installation is reverse of removal.

Tightening torque	
Rear bumper carrier to body (M8 x 25 mm)	13 Nm (10 ft-lb)

Exterior Trim

Exterior trim is attached to the body with adhesive tape, plastic clips or fasteners. Be sure to have necessary 2-sided tape and fasteners on hand when reinstalling exterior trim pieces.

MINI emblem, removing and installing

The procedure below applies to both front and rear emblems.

MINI emblems are attached with adhesive tape and located by two location studs each. Once removed, an emblem cannot be reused.

– Use a heat gun to gently apply heat to emblem.

– Use special tool 00 9 322 (part of BMW special body tool kit 00 9 310) to assist removal of emblem. Alternatively, wrap tip of small screwdriver with masking tape and use that to pry at emblem.

– Clean off adhesive tape and residue from body panel.

– Install new emblem at room temperature.

Side skirt, removing and installing

< Peel off door sill cover insert (**arrow**) and discard.

– Remove and discard sill cover clips. Lift off sill cover.

< Remove screw (**A**).

- Remove plastic rivets (**arrows**).

– When reinstalling, replace damaged rivets and clips.

512 Door Windows

GENERAL

This repair group covers door glass and power window motor repair information.

Special tools

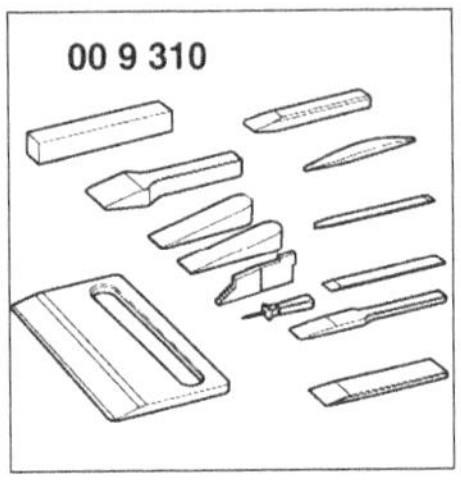

Body trim wedges and prying tools (BMW special tool kit 00 9 310)

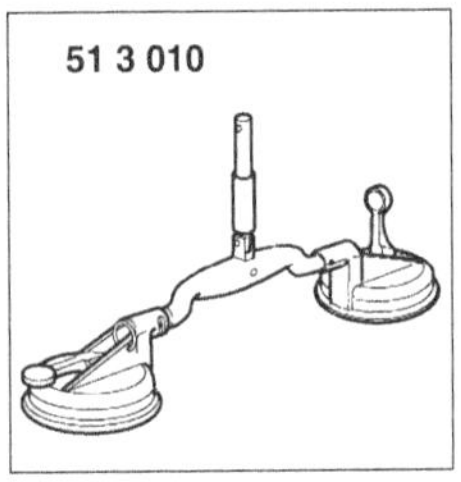

Window glass suction cups (BMW special tool 51 3 010)

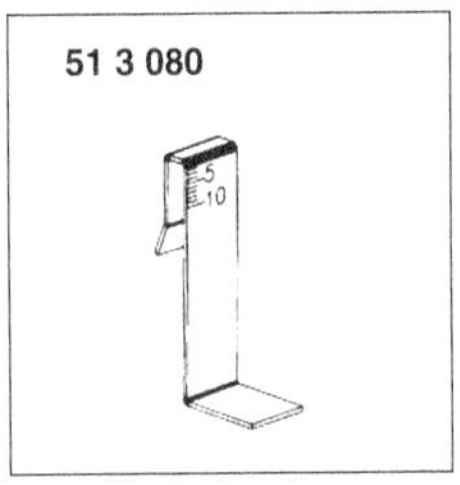

Window seal insertion height gauge (BMW special tool 51 3 080)

Spanner for window mounting toothed washer (BMW special tool 51 3 240)

Electric window control

The General Module (BC1) controls the power windows using internal relays. Window operation from inside the car is possible from the window switches or the interior central locking switch and from outside using the driver door lock cylinder.

NOTE —
The General Module is also referred to as the Body Control Module. In wiring diagrams, the General Module is referred to as A1.

The power window system is represented schematically in the accompanying diagram.

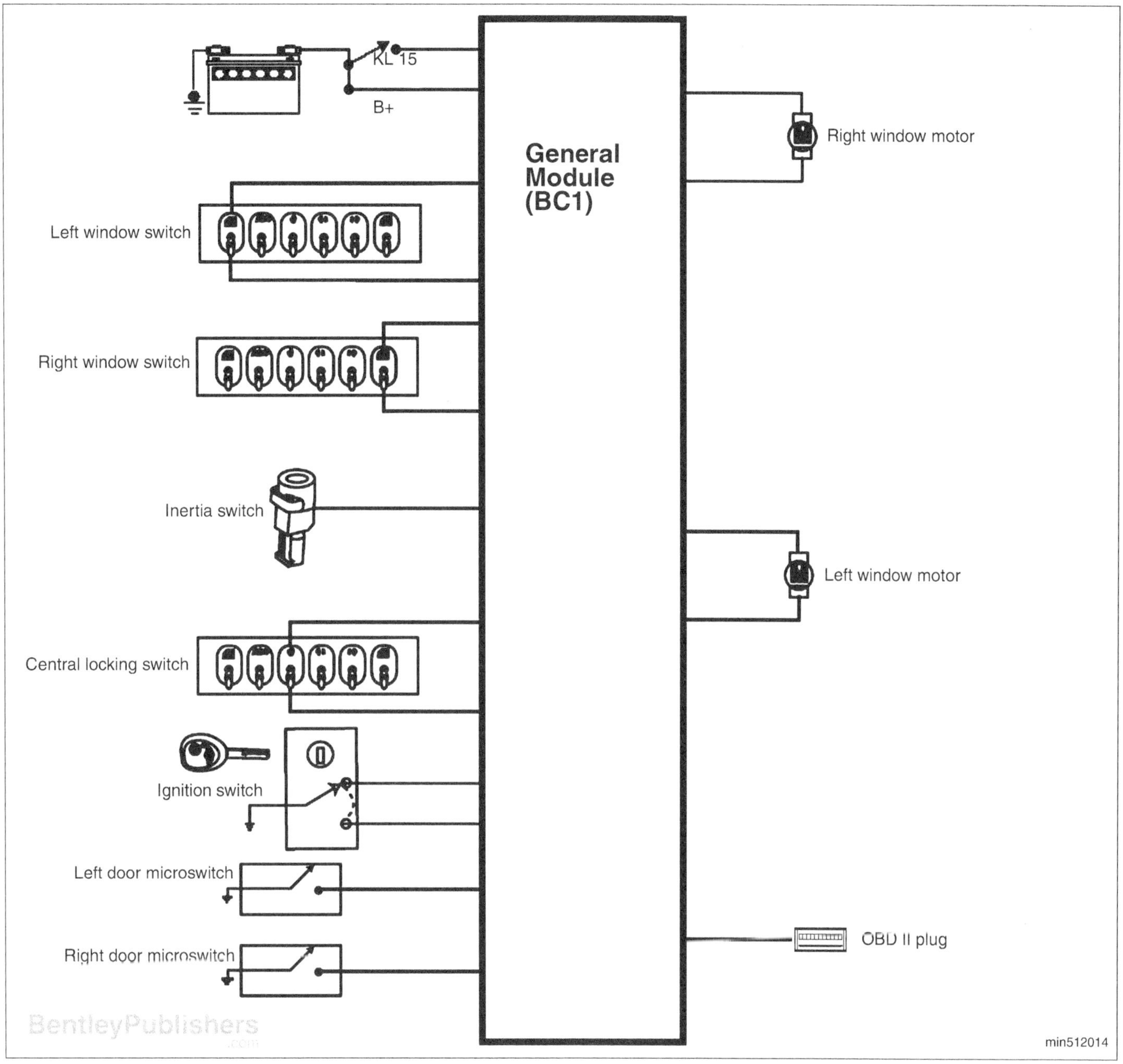

Window drop down feature

When the door is opened, the window is lowered slightly to clear the door seal. When the door is closed, the window is raised fully up into the seal. Convertible top opening and closing also activates this function via data transmission on the K-Bus.

Any time the window or window regulator is loosened or removed, adjust window pretension and position carefully to:

- Prevent glass breakage.
- Avoid high wind noise.
- Avoid water leaks.

WINDOW REPAIRS

Window motor, initializing

If the battery is disconnected, the power window regulator loses its reference point for closed position. Therefore, initialize door windows whenever power to the window motors has been interrupted.

– To initialize window motor, raise window and keep power window switch in raise position for about 5 seconds.

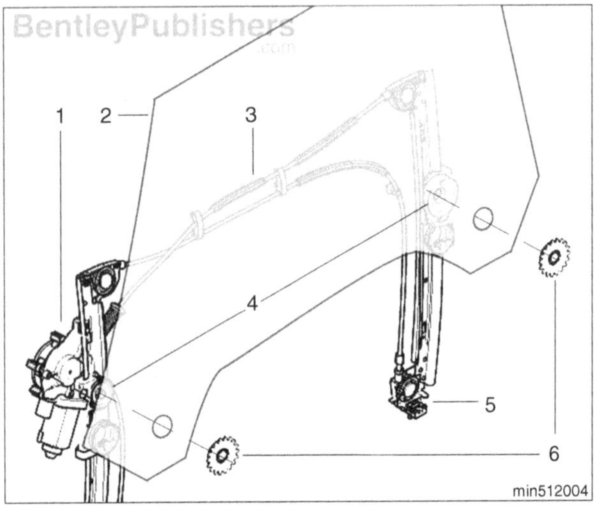

Window glass, removing and installing

The frameless door window is secured to the electric window regulator by 2 large toothed metal washers facing the outside of the door.

1. Electric window motor
2. Glass
3. Window regulator mechanism
4. Plastic window seat on regulator
5. Pretension adjuster
6. Toothed metal washer
   - torque using special tool 51 3 240
   - tighten to 11 Nm (8 ft-lb)

– Lower door window.

Working carefully to avoid scratching paint, pry off window trim strip (**arrows**).

– Remove interior door trim panel. See **411 Doors**.

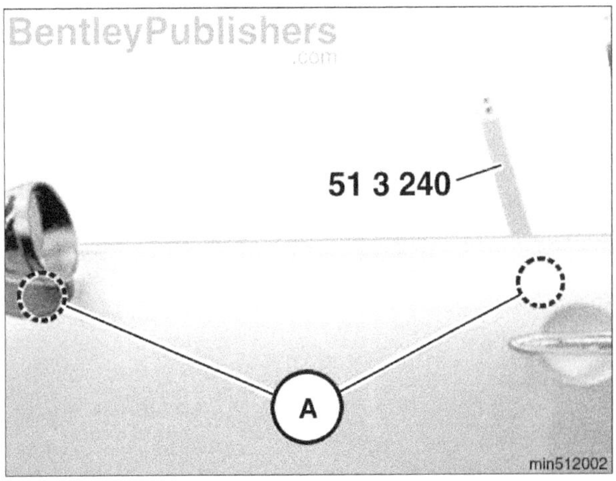

Loosen door glass from window regulator.

- Raise glass for better access to toothed metal washers (**A**) in regulator.
- Use BMW special tool 51 3 240 to loosen washers.

Lower window to height **A** (approximately 22 cm or 8.7 in).

– Disconnect battery negative (-) cable. See **121 Battery, Starter, Alternator**.

> ***CAUTION—***
> *Disconnecting the battery may erase fault code(s) stored in memory. Check for fault codes prior to disconnecting battery cables.*

– Working inside door cavity, reach between glass and outer door skin to remove toothed washers (**B**) from plastic window seats (**C**).

> ***CAUTION—***
> *Support window to prevent falling into door cavity and breaking.*

– Release glass from window regulator and lift out.

– Installation is reverse of removal. Keep in mind the following:
  - Support window securely while installing toothed washers.
  - Tighten toothed washers temporarily.
  - Adjust window glass. See **Window, adjusting** in this repair group.
  - Tighten toothed washers to final torque.

Tightening torque	
Toothed washer to window regulator (tighten with special tool 51 3 240)	11 Nm (8 ft-lb)

– Initialize window motor after reconnecting battery. See **Window motor, initializing** in this repair group.

Window, adjusting

Proper window adjustment reduces wind noise and prevents water leakage.

Window adjustment is in two stages:

- **Pretension adjustment**. With the door closed, the window jams against door seal.
- **Position adjustment**. Closed window lifts slightly and inserts a predetermined distance into door seal when door is closed.

Pretension adjustment

> ***NOTE—***
> *Window pretension can be thought of as the inward rake of the window.*

– Raise door window.

Close front door until door latch clicks to first lock position. There should be a sizeable gap (**A**).
  - Pretension is correct if top edge of glass touches door seal lightly.

> ***NOTE—***
> *Insert sheet of paper between door seal and glass. If paper can be pulled out with light resistance, pretension is correct.*

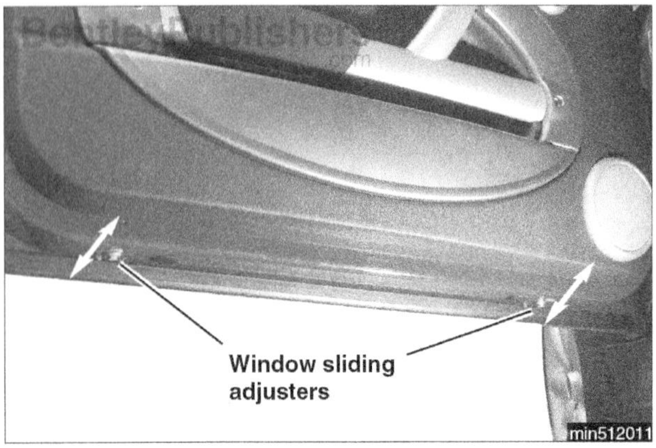

If pretension adjustment is needed:

- Loosen sliding adjusters at bottom of door. Move sliders in or out to decrease or increase pretension.
- Recheck pretension and repeat procedure until correct adjustment is achieved.
- Tighten pretension adjustment bolts.

Tightening torque	
Pretension bolt to bottom of door (M6)	7 Nm (5 ft-lb)

Position adjustment

– Raise door window.

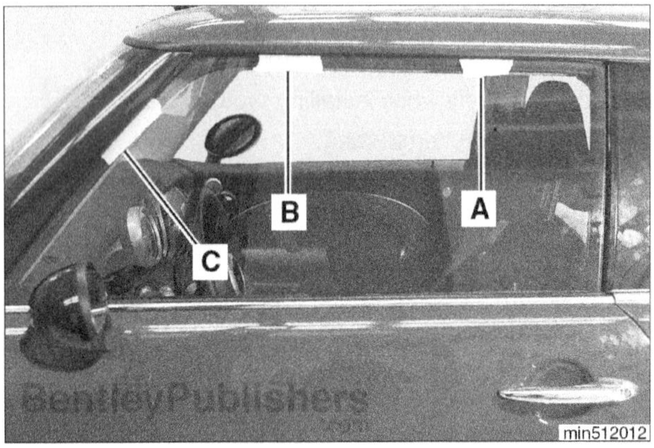

Open door, attach BMW special tool 51 3 080 to window glass at position **A** and close door firmly.

- Record depth of window insertion into door seal.
- Repeat for positions **B** and **C**.

Door window to door seal insertion	
Top of glass (**A**, **B**)	min. 5 mm (0.2 in)
Front of glass at A-pillar (**C**)	min 3.5 mm (0.14 in)

– If window position adjustment is needed, use BMW special tool 51 3 080 or equivalent to measure and mark correct insertion depth at top and front of glass.

NOTE —
Use masking tape to mark glass.

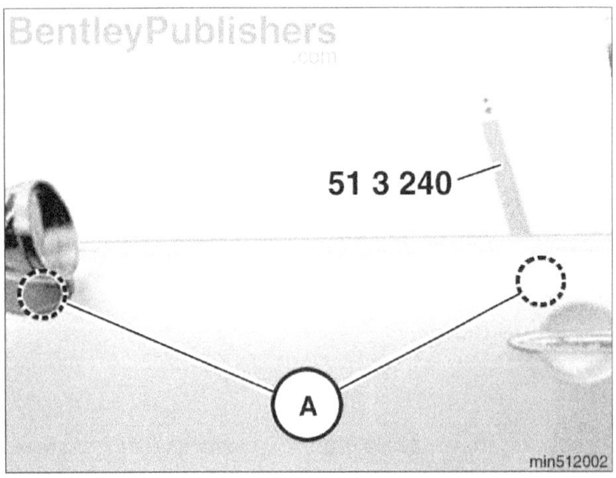

Working carefully to avoid scratching paint, pry off window trim strip. Begin at front of door and proceed to rear.

- Use BMW special tool 51 3 240 to loosen (but do not fully release) toothed metal washers (**A**) in window regulator.

NOTE —
Leave toothed washers with some tension so that window can be slid around but is not loose.

– Attach suction cup tool (BMW special tool 51 3 010 or equivalent) to glass.

- With window at top of travel and door closed fully, adjust glass position for correct insertion into door seal.

– Tighten toothed washers to final torque.

Tightening torque	
Toothed washer to window regulator (tighten with special tool 51 3 240)	11 Nm (8 ft-lb)

– Initialize window motor. See **Window motor, initializing** in this repair group.

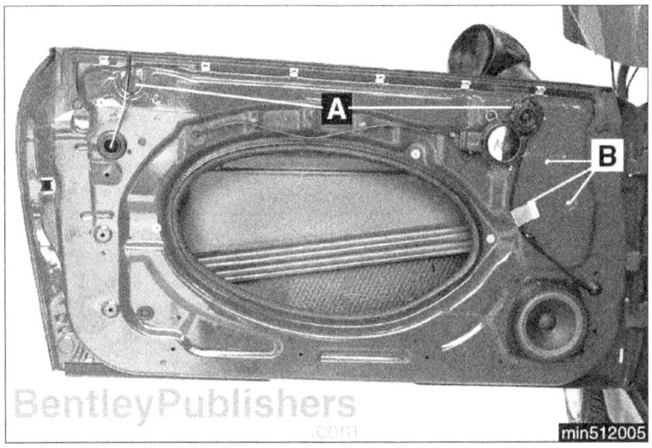

Window regulator, removing and installing

– Remove door window glass. See **Window glass, removing and installing** in this repair group.

– Remove treble speaker in front door. See **650 Radio**.

◁ Remove plastic plugs (**A**).

- Remove window regulator motor mounting screws (**B**).

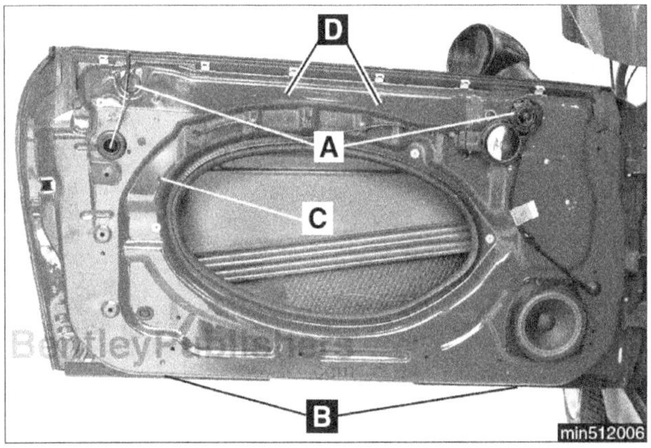

◁ Remove window regulator rails mounting screws (**A**).

- Loosen window pretension slider bolts (**B**).
- Unhook door latch release cables from clips (**C**, **D**).

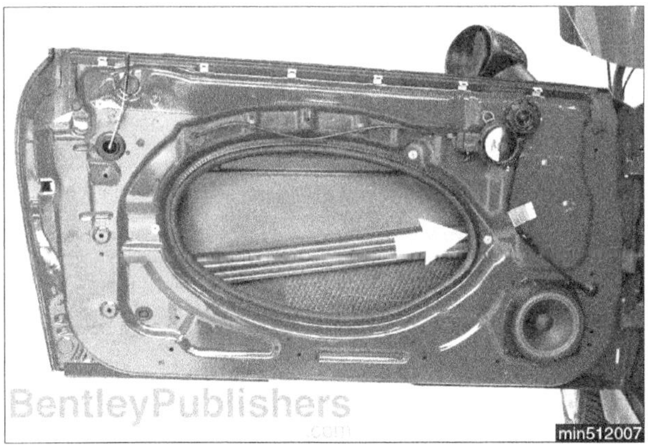

◁ Move front window regulator rail to expose window motor located at **arrow**. Detach electrical harness connector.

– Remove window regulator assembly.

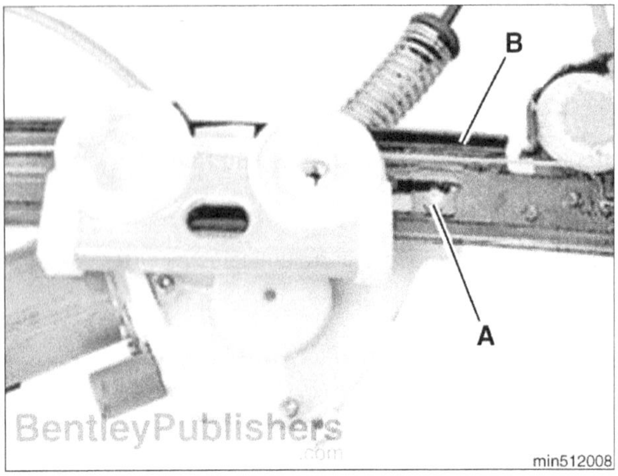

min512008

◀ If replacing window motor:

- Remove motor mounting screw (**A**).
- Detach motor from window regulator rail (**B**).

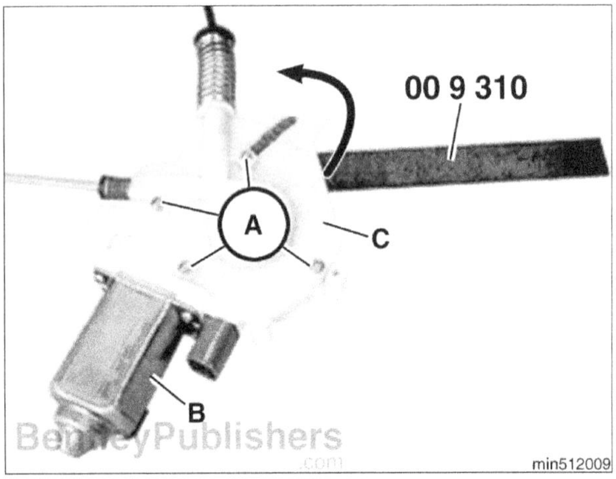

min512009

◀ Loosen motor assembly screws (**A**) (do not release fully).

- Slide in BMW special tool 00 9 310 or equivalent wedge to lift motor (**B**) off mounting plate (**C**). This keeps window regulator cable from flying apart.

– Remove screws (**A**) and separate motor from cable mechanism.

– Installation is reverse of removal. Adjust window. See **Window, adjusting** in this repair group.

Tightening torques	
Pretension bolt to bottom of door (M6)	7 Nm (5 ft-lb)
Toothed washer to window regulator (tighten with special tool 51 3 240)	11 Nm (8 ft-lb)
Window motor to inside door skin	3 Nm (27 in-lb)
Window regulator rail to top of door (M6 x 10 mm)	7 Nm (5 ft-lb)

– Initialize window. See **Window motor, initializing** in this repair group.

513 Interior Trim

GENERAL

This repair group covers interior trim removal and installation procedures.

Refer to the following repair groups for additional information:

- **250 Gearshift Linkage** (shifter boot and bezel)
- **320 Steering and Wheel Alignment** (steering column trim)
- **411 Doors** (interior door panels)
- **612 Switches**
- **620 Instruments**
- **640 Heating and Air-conditioning** (IHKA control panel)
- **650 Radio**
- **721 Airbag System (SRS)**

Special tools

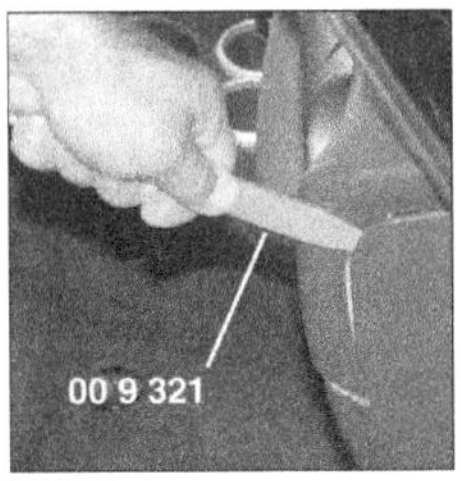

Plastic prying tool
(BMW special tool kit 00 9 321)

Interior trim repairs

Interior trim and finish panels are clipped or screwed into place. Many of the trim retaining clips are designed to be used only once. When removing trim that is held in place with clips, be sure to have spares on hand before beginning the job.

NOTE —
MINI models are equipped with SRS airbags mounted in the steering wheel, dashboard, headliner and seat bolsters. See **Warnings and cautions** *in this repair group and in* **721 Airbag System (SRS)**.

Warnings and cautions

WARNING —
- *Observe special precautions when servicing the Supplemental Restraint System (SRS). Serious injury may result if system service is attempted by persons unfamiliar with SRS and its approved service procedures.*
- *Before performing any work involving airbags, disconnect the negative (–) battery cable. See* **721 Airbag System (SRS)**.
- *Airbags contain a back-up power supply within the SRS control module. Allow a 5 second discharge period after the battery cable has been disconnected before working on airbag components.*

CAUTION —
- *When working on electrical switches or lights, always disconnect the negative (–) cable from the battery and insulate the cable end to prevent accidental reconnection.*
- *Prior to disconnecting the battery, read the battery disconnection cautions given in* **001 General Warnings and Cautions**.
- *To prevent marring the trim when working on interior components, work with plastic prying tools or wrap the tips of screwdrivers and pliers with tape before prying out switches or electrical accessories.*

Center Consoles

The front console houses the shifter mechanism, power mirror controls, seat heater controls (if equipped), cigarette lighter and cupholders.

The rear console houses the parking brake handle and tire pressure warning (RDW) switch (if equipped). The console is placed over the DSC motion sensor (if equipped).

Front console, removing and installing

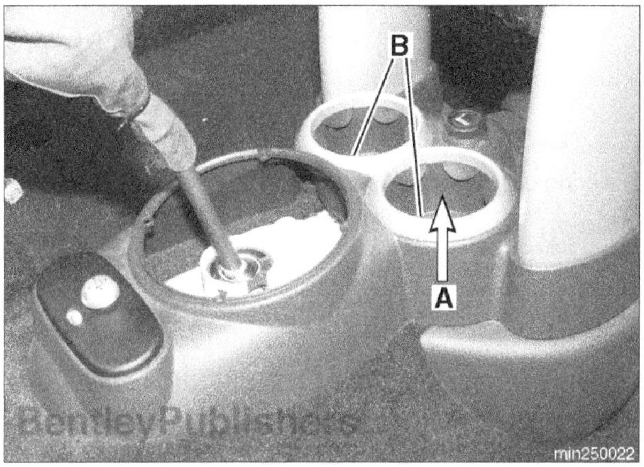

- Remove gearshift knob by pulling straight up. See **250 Gearshift Linkage**.
- Remove gearshift dust boot by unclipping from front console and lifting up. See **250 Gearshift Linkage**.
- ◄ Remove ashtray (**A**) (if installed) by pulling up. Remove screws (**B**) in bottom of cup holders.

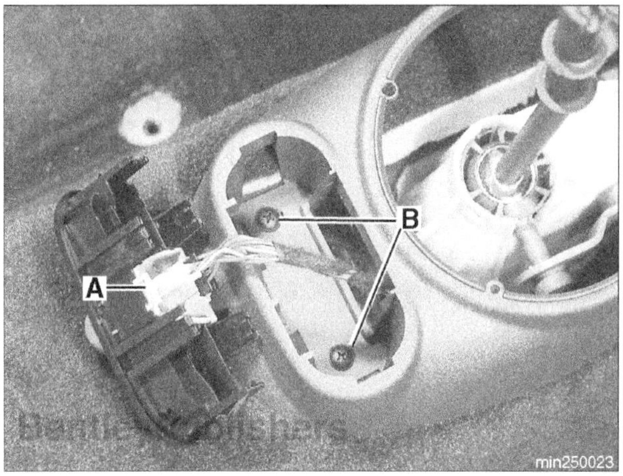

- ◄ Gently pry switch plate up and remove electrical connections (**A**) for power mirrors and seat heaters (if equipped).
- Remove screws (**B**) in base of console.

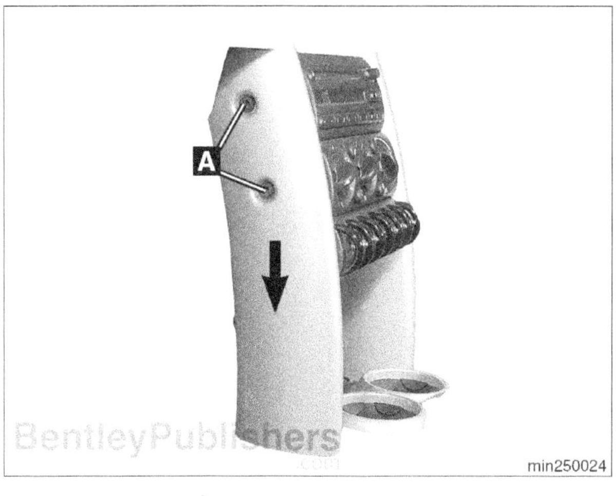

- ◄ Remove screws (**A**) on left side fascia and pull fascia down slightly. Repeat for right side fascia.
- Slide front console and side fascia toward rear of car as a unit until both side fascia can be pulled out of console.
- Remove electrical connectors from lighter socket and remove console.
- Installation is reverse of removal. Make sure foam blocks at base of fascia are correctly positioned, and that no electrical wires are pinched or damaged when replacing console.

Rear console, removing and installing

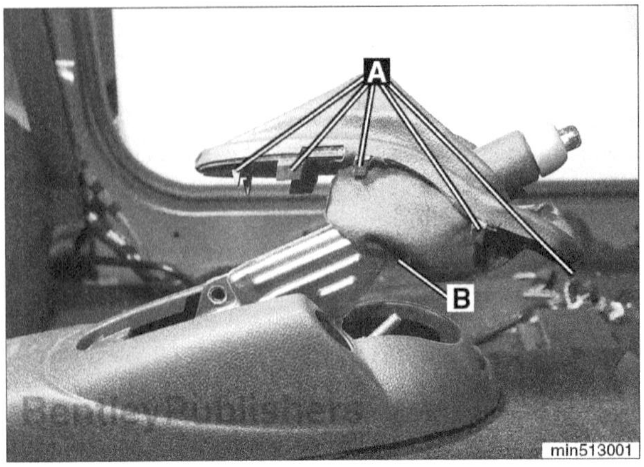

Release parking brake handle boot clips (**A**).

- Peel parking brake boot forward to expose cable tie (**B**) securing boot to parking brake handle.
- Cut off cable tie and discard.

NOTE —
- *Do not pull on parking brake boot to release from console without releasing clips.*
- *Left side clips are illustrated. Right side is symmetrical.*

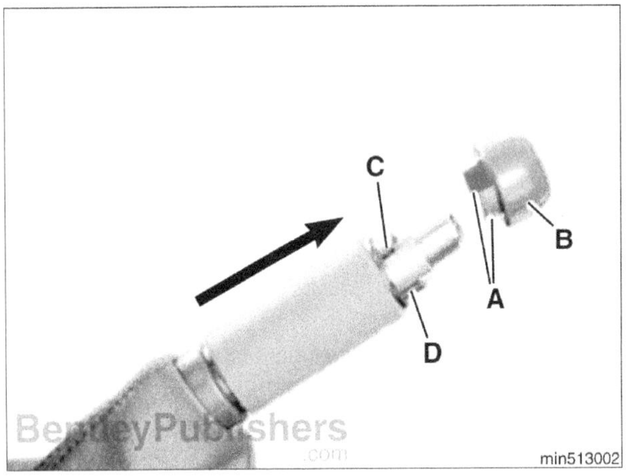

Working at parking brake handle, squeeze brake handle grip end trim clips (**A**).

- Pull off end trim (**B**).
- Release parking brake grip locking tab (**C**). Remove grip.

NOTE —
*Brake handle grip is located by a keyway (**D**) and will only fit one way.*

– Pull off brake handle boot.

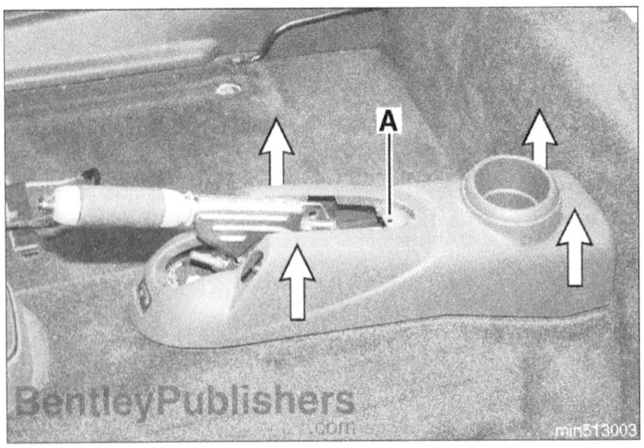

Remove console securing screw (**A**).

- Lift console to release four securing clips (**arrows**).
- Disconnect tire pressure warning (RDW) switch (if installed).
- Lift out console.

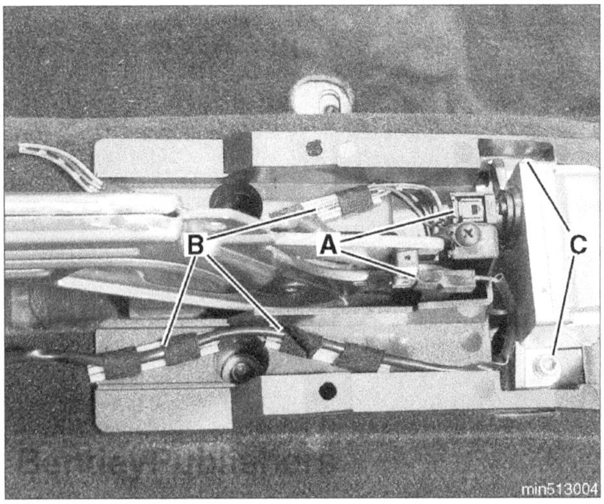

If removing rear console mounting bracket:

- Detach electrical harness connectors (**A**).
- Detach electrical harness retainers (**B**) from bracket.
- Remove DSC motion sensor mounting bolts (**C**) (if equipped) and remove motion sensor.

Remove console bracket mounting nuts (**arrows**).

– Lift out console bracket.

– If replacing console, swap console securing clips and electrical switches over from old unit to new as needed.

– Installation is reverse of removal. Keep in mind the following:
  - Replace wire ties as necessary.
  - Make sure electrical harnesses are not trapped or pinched under console.
  - Make sure parking brake handle grip is positioned correctly on locating keyway.

Tightening torque	
Rear console bracket to center tunnel (M6)	2.6 Nm (23 in-lb)

Center console armrest, removing and installing

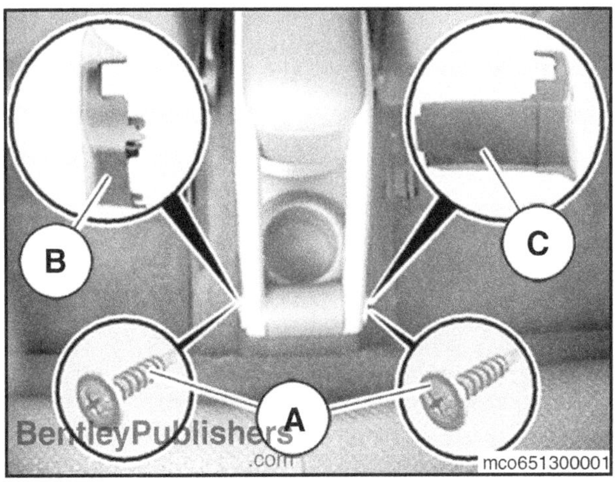

Remove trim cover fasteners (**A**). Remove driver side trim cover (**B**) then passenger side trim cover (**C**).

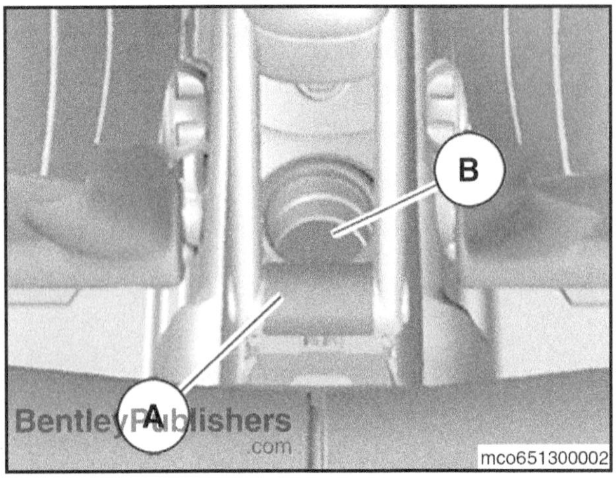

Snap out and remove trim cover (**A**) and fastener (**B**) beneath it.

– Remove rear cup holder.

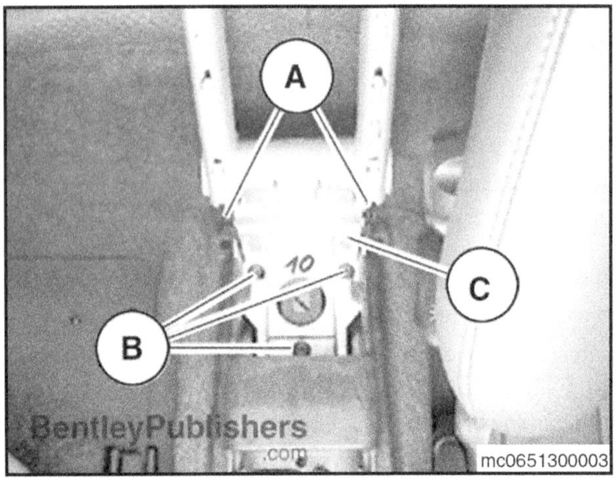

Remove fasteners at side of armrest (**A**) and fasteners (**B**) at front. Remove center armrest (**C**) from vehicle.

– Installation is reverse of removal.

Dashboard

Glove compartment, removing and installing

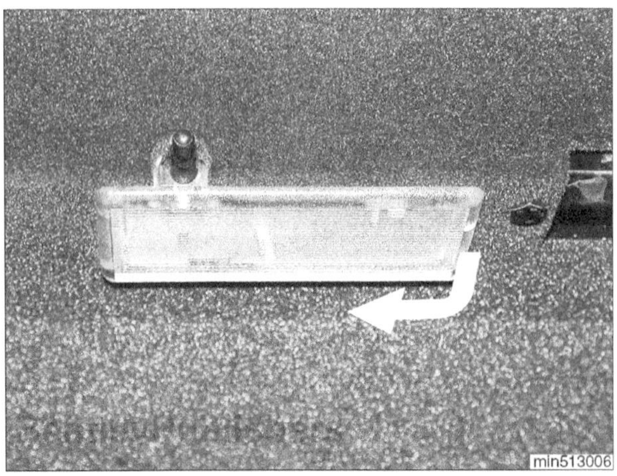

Open glove compartment. Pry out glove compartment light in direction of **arrow**.

– Disconnect electrical harness connection and remove light.

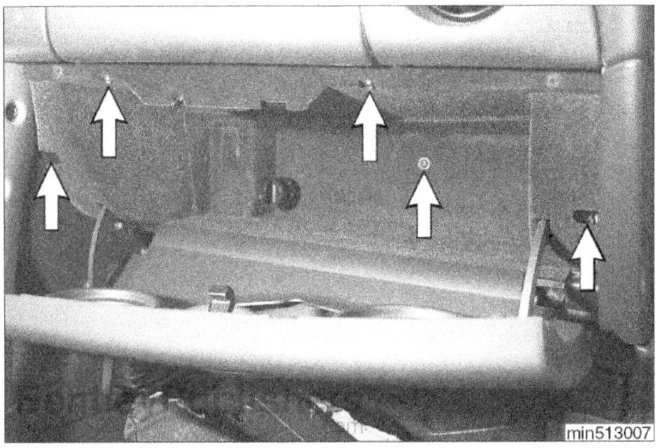

Remove glove compartment mounting screws (**arrows**).

- Pull glove compartment out from under dashboard.
- Detach glove compartment cooling hose.

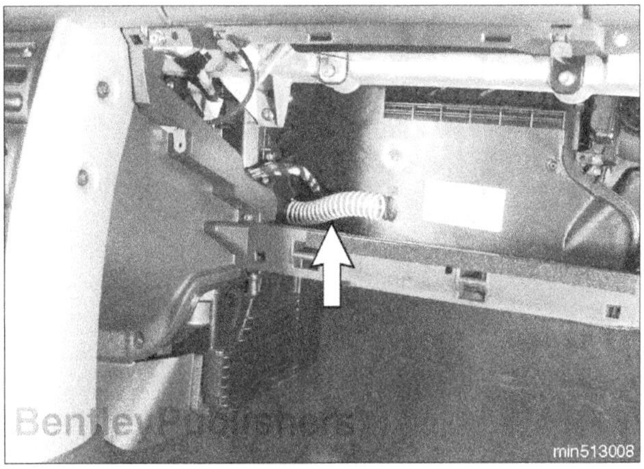

When reinstalling, be sure to reconnect cooling hose (**arrow**) correctly to glove compartment air valve and to air-conditioning vent under dash.

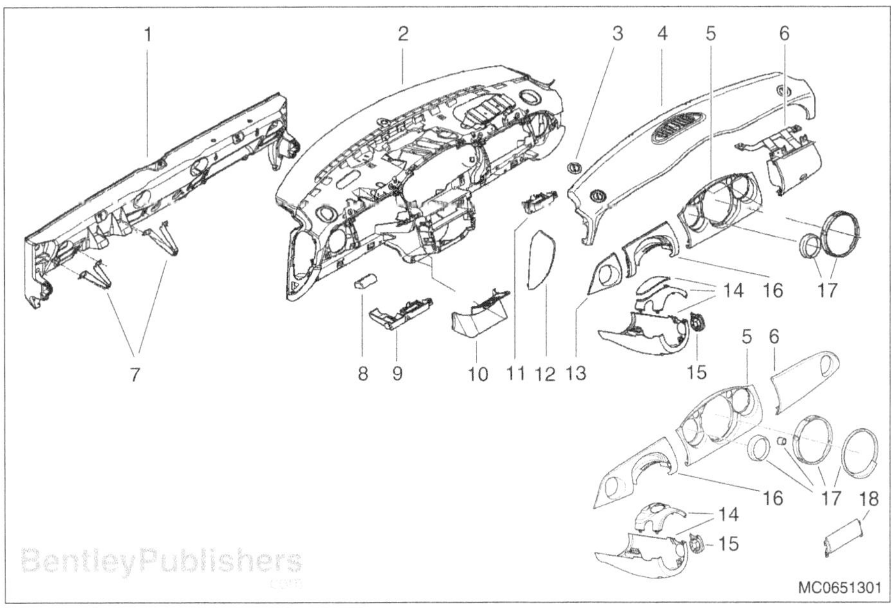

Dashboard components

1. **Dashboard carrier brace**
2. **Dashboard**
3. **Side window defroster vent**
4. **Dashboard cover**
5. **Center dashboard instrument bezel**
6. **Passenger airbag**
7. **Support bracket**
8. **Courtesy light cover**
9. **Left footwell trim**
10. **Left lower dashboard trim**
11. **Right footwell trim**
12. **Left dashboard side cover**
13. **Dashboard vent**
14. **Steering column covers**
15. **Ignition lock trim**
16. **Left upper dashboard trim**
17. **Instrument cluster trim**
18. **Radio delete plate**

Dashboard, removing and installing

Dashboard removal is a complicated job, involving dismantling of air-conditioning and airbag components, among many other systems. Be sure to read the warning and cautions in **001 General Warning and Cautions**, as well as in **640 Heating and Air-conditioning** and **721 Airbag System (SRS)**.

Read the procedure through before starting work.

– Disconnect negative (-) cable from battery.

> ***CAUTION —***
> *Prior to disconnecting the battery, read the battery disconnection cautions given in* **001 General Warnings and Cautions**.

– Drain coolant. See **170 Cooling System**.

– Cooper: Remove battery box. See **121 Battery, Starter, Alternator**.

– Cooper S: Remove air filter housing. See **130 Fuel Injection**.

– Following manufacturer's instructions, connect an approved refrigerant recovery / recycling / recharging unit to A/C system and discharge system.

> ***WARNING —***
> *Do not attempt to discharge or charge the A/C system without proper equipment and training. Damage to the vehicle and personal injury may result.*

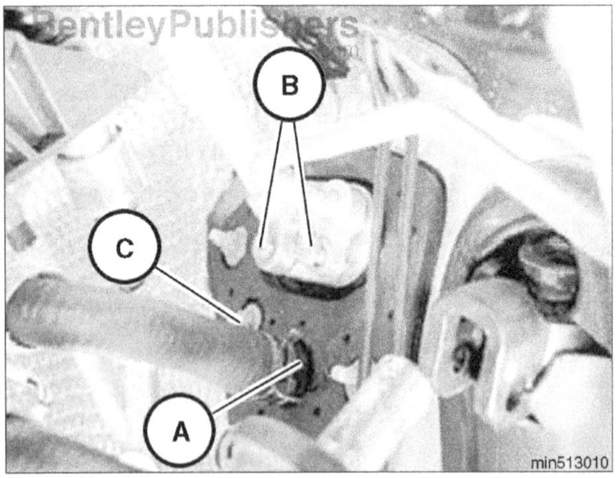

◀ Working at engine compartment rear bulkhead:

- Loosen or remove heater hose clamps (**A**) and detach hoses from heater core connections. Carefully blow compressed air through dual aluminum heater core pipes to remove remaining coolant from heater core.
- Remove bolts (**B**) mounting air-conditioning lines to expansion valve clamping adapter. Detach air-conditioning lines.
- Remove nut (**C**) from heater core locating stud.

– Raise front of car and support in a safe manner.

> ***WARNING —***
> *Make sure the car is stable and well supported at all times. Use a professional automotive lift or jack stands designed for the purpose. A floor jack is not adequate support.*

◀ Working underneath front of car, remove lower steering shaft pinch nut (**A**) and remove eccentric bolt (**B**). Pry universal joint off steering rack shaft.

– Fold lower universal joint upwards. Push lower half of joint up through protective boot.

– Remove left and right interior A-pillar trim. See **A-pillar (windshield pillar) trim, removing** in this repair group.

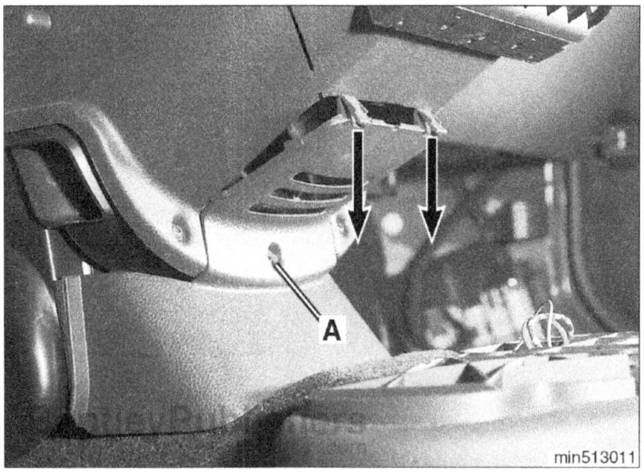

> **CAUTION—**
> *Once the steering shaft is separated from the steering rack there is no longer a steering end stop and the steering wheel can be turned without restriction. This can damage the airbag contact ring. To prevent this damage:*
> *- Move steering wheel to straight-ahead position.*
> *- Remove ignition key to lock steering.*

– Loosen front console and remove side fascia. See **Front console, removing and installing** in this repair group.

– Remove radio. See **650 Radio**.

< Working underneath center of dashboard, remove panel mounting screw (**A**), release clips and remove panel.

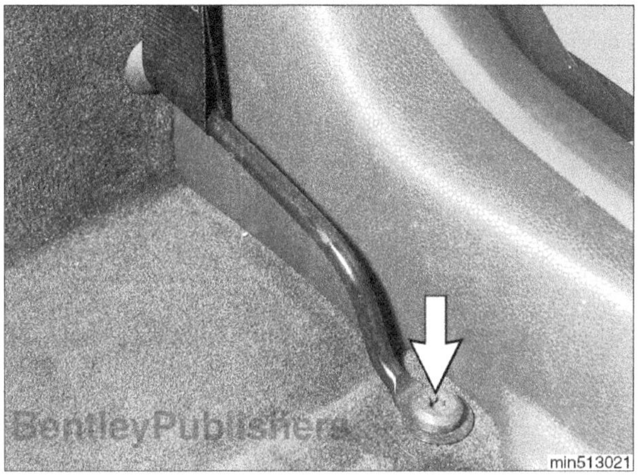

< Remove left and right seat belt slider bar mounting bolt (**arrow**) and remove slider.

– Remove left and right door sill cover strip.

NOTE—
- *Sill covers are retained by plastic clips. Use plastic prying tool to remove.*
- *Left sill cover is retained by plastic rivet in front of left front fuse and relay panel.*

< Working below left front fuse and relay panel in left kick panel area:
- Twist electrical harness connector 45° (**arrow**) to detach assembly from mounting bracket.
- Separate connector.

Working at right kick panel:

- Remove harness connector cover (**A**) from General Module (BC1).
- Detach connector.

NOTE —
The General Module is also referred to as the Body Control Module.

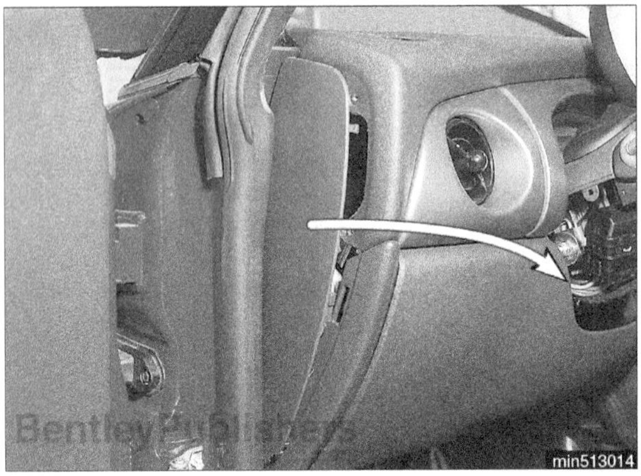

Pry gently with plastic tool at dashboard side covers at left and right. Remove by pulling toward rear of vehicle.

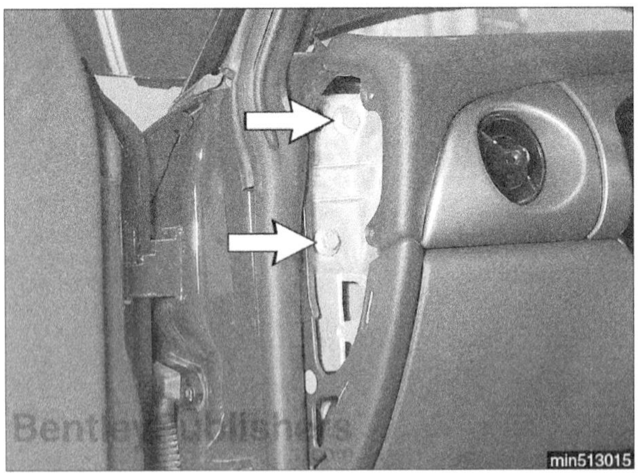

Remove dashboard carrier brace mounting bolts (**arrows**) at left and right.

– Remove steering column lower cover. See **320 Steering and Wheel Alignment**.

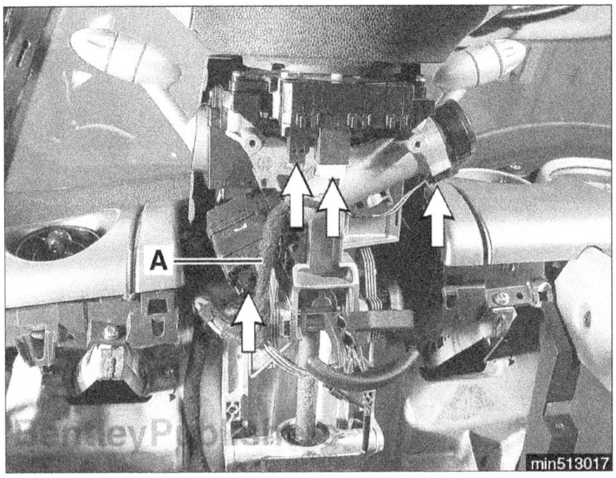

Working underneath steering column:

- Unclip and release harness (**A**) from dashboard.
- Detach electrical harness connectors (**arrows**) at steering column.
- CVT (automatic transmission): Disconnect shift interlock cable from steering column.

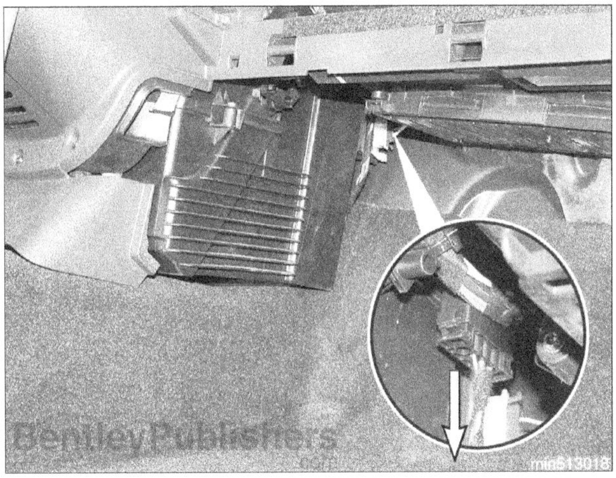

Working at right center underneath dashboard, detach electrical harness connector (**inset**) from blower fan resistor.

Support dashboard. Working underneath steering column, remove lower dashboard mounting bolts (**arrows**).

– Remove dashboard together with steering column, heating and air-conditioning components and dashboard instruments. Check carefully for any wiring, hoses or mechanical parts that might become snagged as dashboard assembly is removed.

> **CAUTION—**
> *Dashboard is heavy. It will require two people to remove from car.*

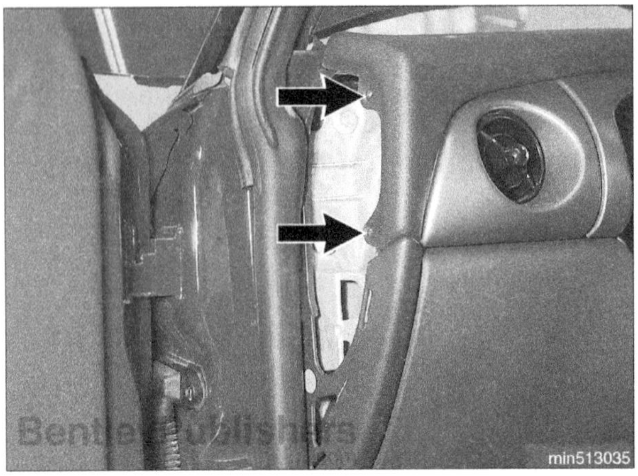

Remove screws (**arrows**) on left and right to separate dashboard cover from dashboard.

NOTE —
The cover can be separated from the dashboard without removing the dashboard from the car.

– Detach instruments and switches from dashboard by unclipping carefully to avoid breakage.

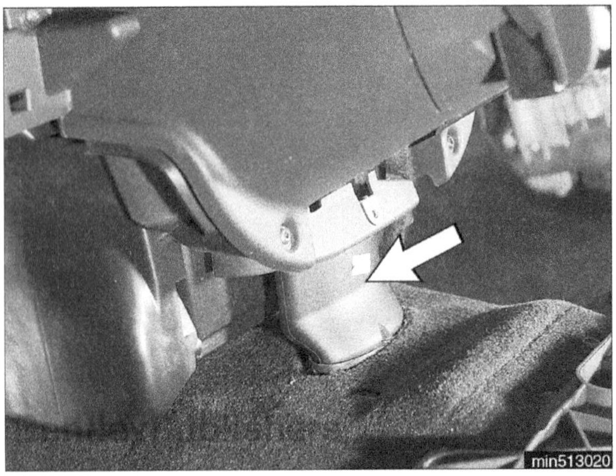

Installation is reverse of removal. Bear in mind the following:

- Ensure rear ventilation duct (**arrow**) is fitted to heating and air-conditioning housing prior to installation and is correctly located.
- Replace air-conditioning line sealing O-rings.
- Replace self-locking fasteners.
- Use Loctite®270 or equivalent thread locking compound when installing seat belt slider bolt.

Tightening torques	
Air-conditioning lines to expansion valve flange (M6 x 20 mm) (replace O-rings)	8 Nm (6 ft-lb)
Dashboard carrier brace to body M10 M6	 48 Nm (35 ft-lb) 8 nm (6 ft-lb)
Heater core to engine compartment bulkhead (M6)	8 Nm (6 ft-lb)
Lower steering column cover to steering column	2 Nm (18 in-lb)
Lower steering shaft universal joint eccentric pinch bolt to steering rack shaft (M8, replace bolt and nut)	22 Nm (16 ft-lb)
Seat belt slider to floor (use Loctite)	40 Nm (30 ft-lb)

– Use new hose clamps at coolant hoses. Fill and bleed cooling system. See **170 Cooling System**.

– Recharge air-conditioning system. See **640 Heating and Air-conditioning**.

– Initialize window motors. See **513 Door Windows**.

Center dashboard face trim, removing and installing

– Working at center of dashboard:
  - Protect instrument panel with tape along upper curved area **D**.
  - Remove trim mounting screws (**A** and **B**). (There is one screw in lower right, none in lower left.)
  - Pull off trim (**C**) around switches.
  - Gently pry face trim off dashboard (**arrows**).

– When installing:
  - Replace any damaged mounting clips on back of trim as necessary.
  - Inspect plastic components for damage. replace as necessary.
  - Fit plastic cap on end of screw (**C**). This prevents damage to passenger airbag.

Left dashboard face trim, removing and installing

Through 06/2004 production

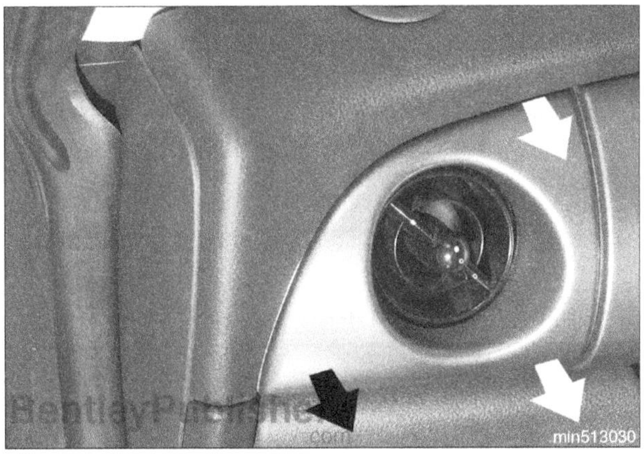

– Pry gently to partially unclip adjacent face trim (above steering column).

– Gently pry face trim off left end of dashboard (**arrows**).

– When installing, replace any damaged mounting clips on back of trim as necessary.

From 07/2004 production

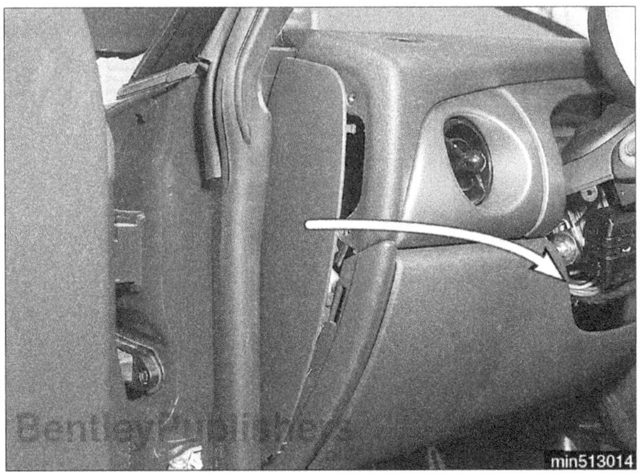

– Remove instruments from steering column, see **620 Instruments**.

– Using a plastic tool, gently pry off left side dashboard side cover and remove by pulling toward rear of vehicle (**arrow**).

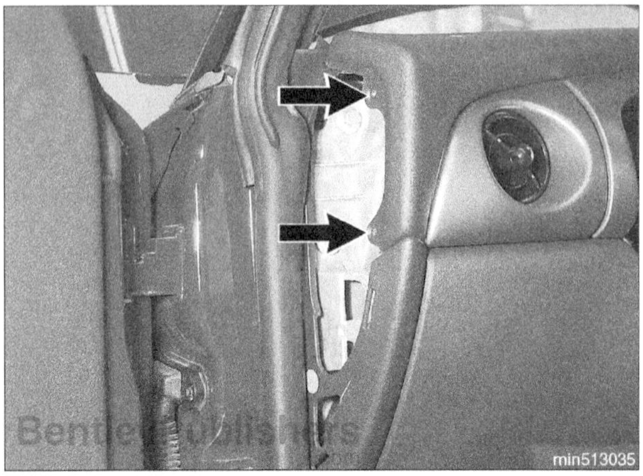

< Remove both screws (**arrows**).

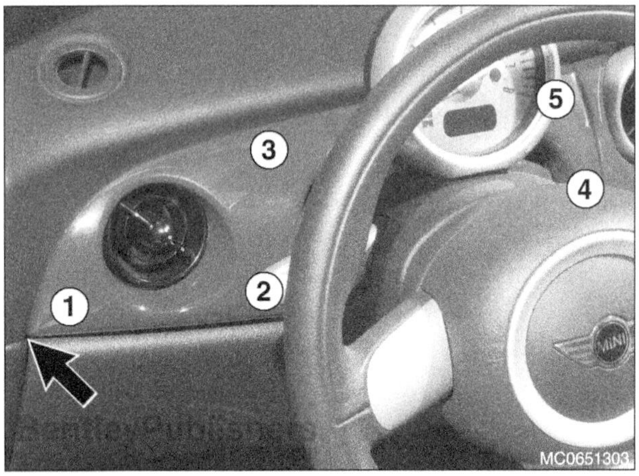

< Pull instrument panel out slightly (**arrow**) and unsnap first three face trim securing clips (**1**, **2**, **3**) one at a time until loose.

– Push instrument panel slightly up, unsnap last two securing clips (**4**, **5**), and remove face trim with a gentle twisting motion.

NOTE—
- *Instrument must be removed from steering column prior to removing face trim.*
- *Adjust steering column to the farthest down and extended position.*
- *Use care, face trim can be easily damaged.*
- *Securing clips can be damaged during removal; replace as required to ensure vibration-free installation.*

– When reinstalling, make sure securing clips are correctly seated.

Right dashboard face trim, removing and installing

Through 06/2004 production

– Open glove compartment door.

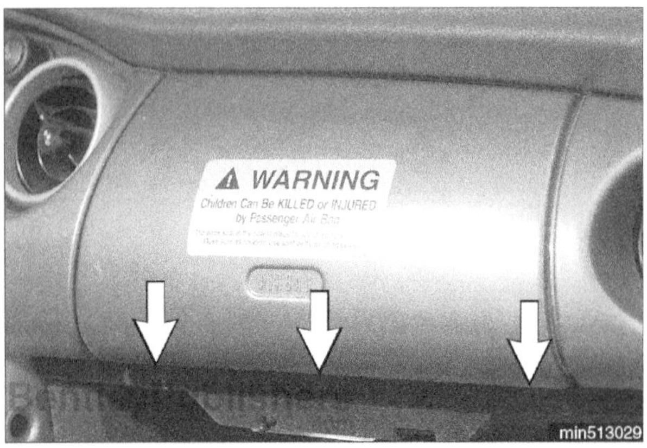

< Remove passenger airbag trim cover bottom mounting screws (**arrows**).

Gently pry face trim off right end of dashboard (**arrows**).

– When installing, replace any damaged mounting clips on back of trim as necessary.

From 07/2004 production

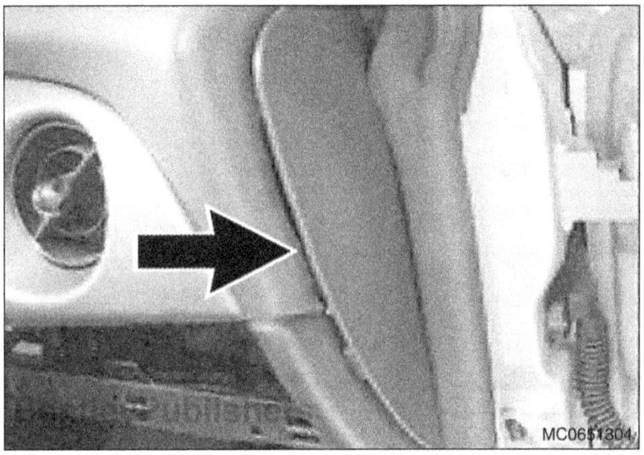

Using a plastic tool, gently pry off right side dashboard side cover and remove by pulling toward rear of vehicle (**arrow**).

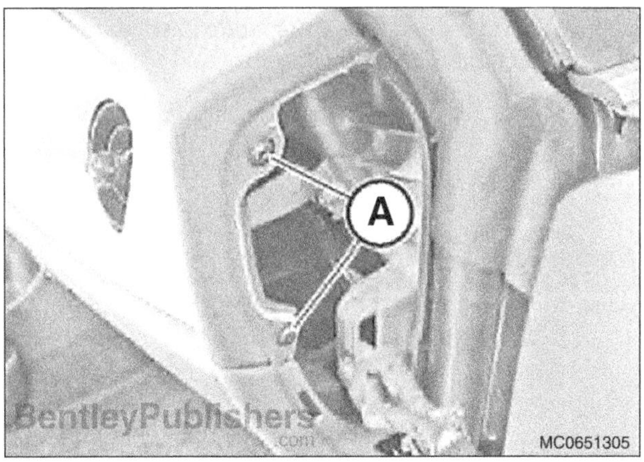

Remove both side panel screws (**A**).

– Open glove box door.

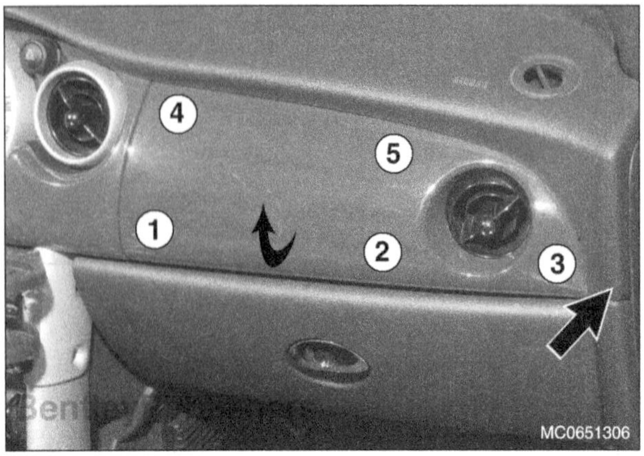

- Pull right side instrument panel out slightly (**right arrow**) and hold in that position.
- Unsnap lower face trim securing clips (**1**, **2**, **3**).
- Unsnap upper face trim securing clips (**4**, **5**) and pull out (**curved arrow**).

NOTE—
- *Use care, face trim can be easily damaged.*
- *Securing clips can be damaged during removal; replace as required to ensure vibration-free installation.*

- When reinstalling, make sure securing clips are correctly seated.

Passenger airbag trim cover, removing and installing

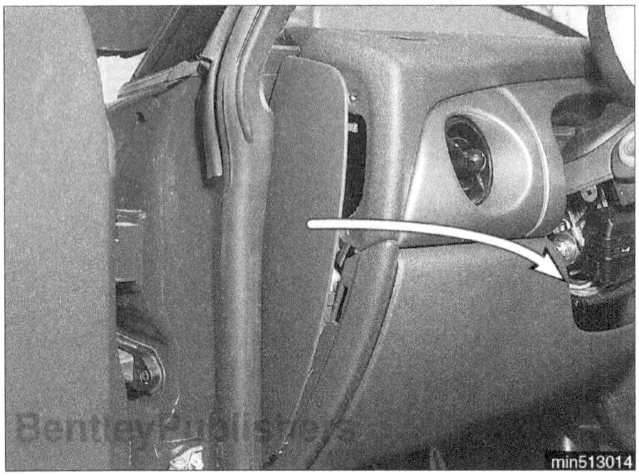

- Pry gently with plastic tool at dashboard side covers at left and right. Remove by pulling toward rear of vehicle.

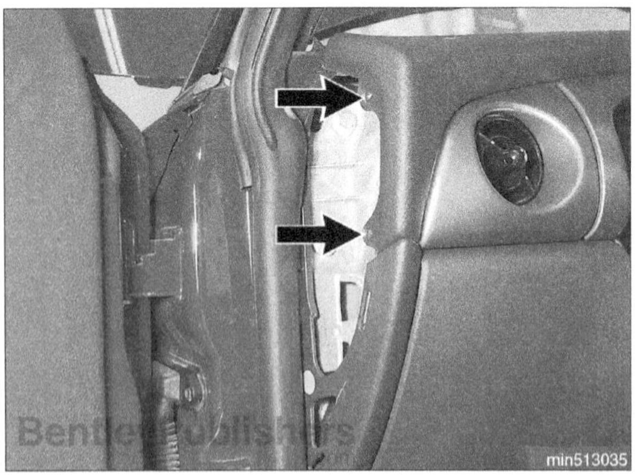

- Remove screws (**A**) on left and right to separate dashboard cover from dashboard. Lift top cover off dashboard.
- Open glove compartment door.

Working at top right of dashboard:

- Remove airbag cover safety strap rod mounting bolts (**A**).
- Slip rod (**B**) out of safety straps.

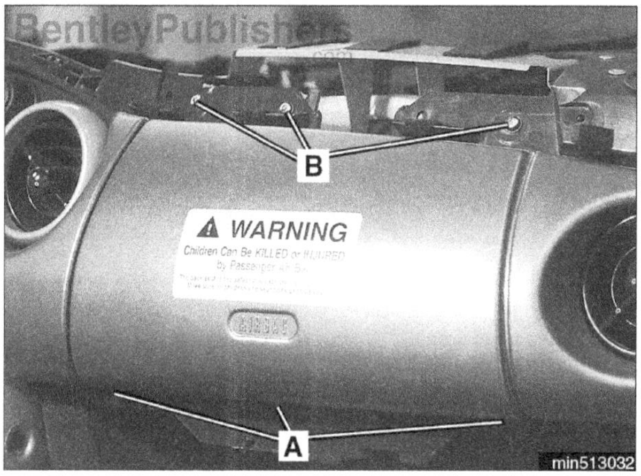

Remove cover mounting screws (**A** and **B**). Pull off cover.

– When installing:
  - Replace faulty mounting clips and plastic nuts as necessary.
  - Make sure safety straps are undamaged.

Interior Trim Panels

Interior door trim panel removal and installation is covered in **411 Doors**.

Rear hatch trim panel removal and installation is covered in **412 Rear Hatch**

Rear inside trim panel, removing

– Remove rear seat and rear seat backrest. See **520 Seats**.

– Right panel: Remove CD changer.

– Remove rear hatch sill cover.

– Remove cargo compartment floor trim.

– Remove cargo compartment light. See **630 Lights**.

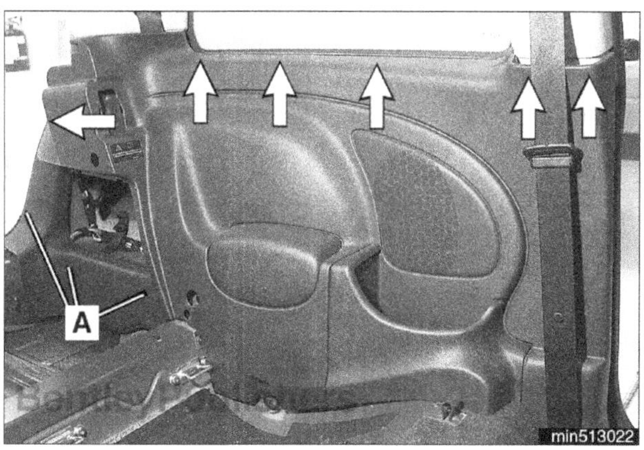

Remove screws (**A**) and pry off trim clips (**arrows**) to remove panel.

Rear inside trim panel storage compartment, removing

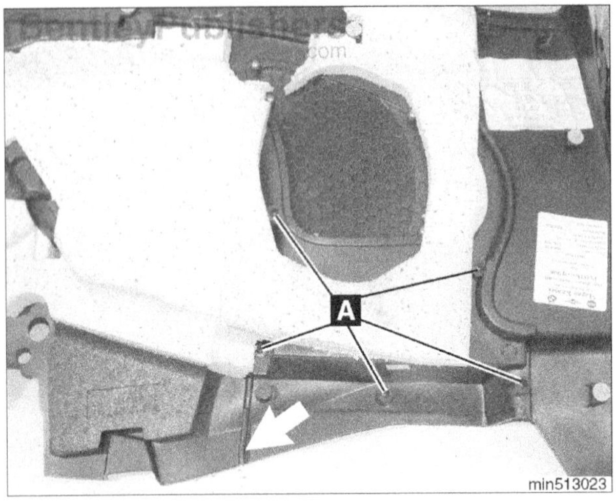

- Remove rear side trim. See **Rear inside trim panel**, **removing** in this repair group.
- Remove screws (**A**) and clip (**arrow**) on back side of trim panel and detach compartment.

A-pillar (windshield pillar) trim, removing

WARNING —
- *Read the warnings and cautions in* **721 Airbag System (SRS)**. *Serious injury may result if system service is attempted by persons unfamiliar with SRS and its approved service procedures.*
- *Do not use sharp-edged tools to remove A-pillar trim. Damage to head protection airbag could result.*

NOTE —
A-pillar trim is destroyed when removed. Be sure to have new trim on hand before attempting to remove.

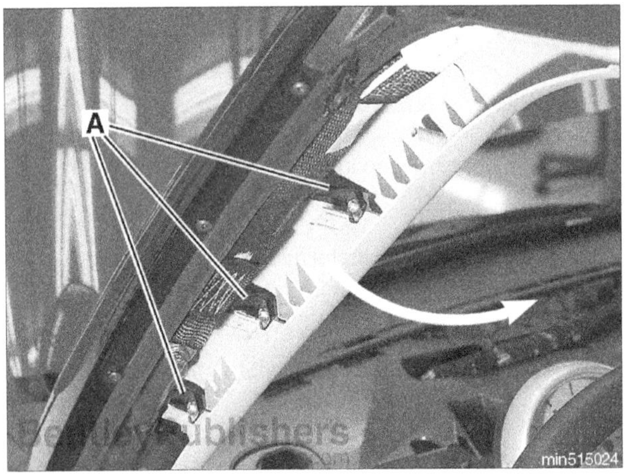

- Peel off door opening seal.
- Release trim by pulling rear of trim toward center of car. Some or all mounting clips will break.
- Cut retaining straps (**A**) and remove trim.
- When reinstalling, make sure door opening seal seats correctly.

B-pillar (door post) trim, removing

WARNING —
- *Read the warnings and cautions in* **721 Airbag System (SRS)**. *Serious injury may result if system service is attempted by persons unfamiliar with SRS and its approved service procedures.*
- *Do not use sharp-edged tools to remove B-pillar trim. Damage to head protection airbag could result.*

- Remove front seatbelt sliding rail and detach seatbelt. See **720 Seat Belts**.

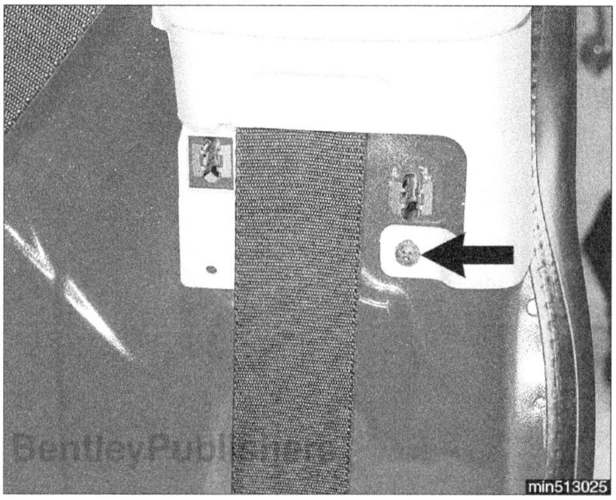

- Partially unclip rear inside trim. See **Rear inside trim panel, removing** in this repair group.
- ◄ Remove B-pillar trim mounting screw (**arrow**).
- Release trim by pulling down.
- Pull seat belt strap out of slot and remove trim.

> ***CAUTION—***
> *Incorrect installation of B-pillar trim may cause breakage of seat belt adjuster, blockage of seat belt movement, or rattling noises.*

- When reinstalling, make sure door opening seal seats correctly.

C-pillar (rear roof pillar) trim, removing

> ***WARNING—***
> - *Read the warnings and cautions in* **721 Airbag System (SRS)**. *Serious injury may result if system service is attempted by persons unfamiliar with SRS and its approved service procedures.*
> - *Do not use sharp-edged tools to remove C-pillar trim. Damage to head protection airbag could result.*

NOTE—
C-pillar trim is destroyed when removed. Be sure to have new trim on hand before attempting to remove.

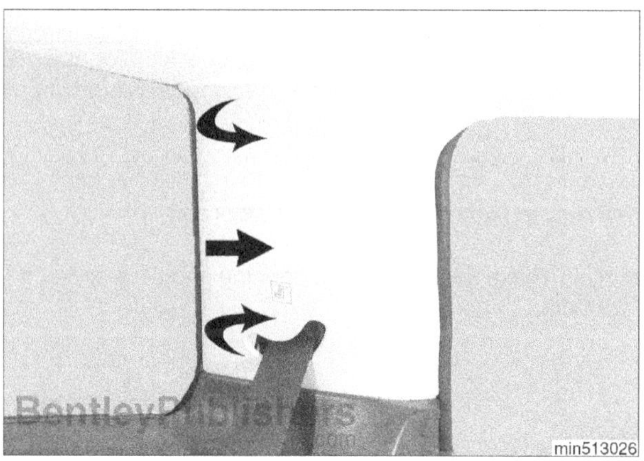

- Remove rear inside trim. See **Rear inside trim panel, removing** in this repair group.
- Working underneath rear seat, remove rear seat belt anchor mounting bolt. See **720 Seat Belts**.
- Peel off rear hatch opening seal.
- ◄ Release trim by pulling (**arrows**) rear of trim toward center of car. Some or all mounting clips will break.
- Cut top retaining strap and pull trim back.
- Pull seat belt strap out of slot and remove trim.
- When reinstalling, make sure rear hatch opening seal seats correctly. Replace any damaged mounting clips.

Headliner, removing

> ***WARNING —***
> *Read the warnings and cautions in* **721 Airbag System (SRS)**.

– Remove front seats. **See 520 Seats**.

– Remove A-pillar trim. See **A-pillar (windshield pillar) trim, removing** in this repair group.

– Remove B-pillar trim. See **B-pillar (door post) trim, removing** in this repair group.

– Remove C-pillar trim. See **C-pillar (rear roof pillar) trim, removing** in this repair group.

– Remove interior rear view mirror.

– Remove sun visors.

– Remove clock from headliner, if applicable. See **620 Instruments**.

– Remove front and rear sunroof trim. See **540 Sunroof**.

– Remove interior light. See **630 Lights**.

– Remove interior grab handles.

– Remove ultrasound module housing. See **515 Central Locking and Anti-theft**.

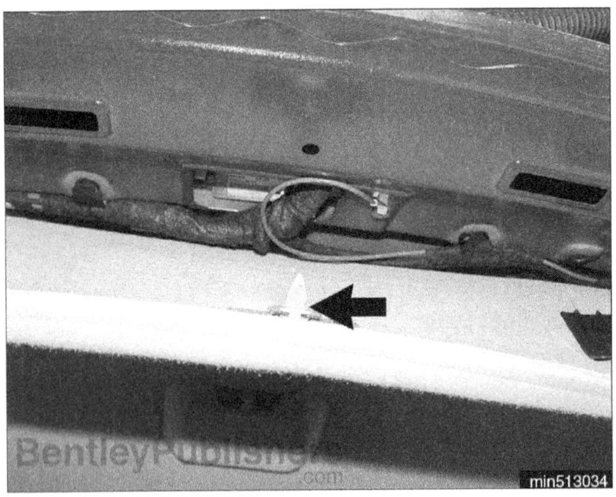

◄ Starting at rear hatch opening, gently pry headliner retaining clips (**arrow**) off roof metal braces.

– Remove headliner through rear hatch opening, checking for any wiring or mechanical parts that might become snagged as headliner is removed.

NOTE —
Headliner will require two people to remove from car.

– When reinstalling, make sure rear hatch opening and door opening seals seat correctly.

515 Central Locking and Anti-theft

515

GENERAL

This repair group covers repair information for door and rear hatch locks. Also covered are electronic immobilizer (EWS) and anti-theft alarm (DWA).

Keep in mind that MINI cars are equipped with sophisticated and self-diagnostic electrical systems. When experiencing malfunctions relating to central locking or anti-theft systems, diagnose the system using BMW service tester DISplus, GT1, MoDiC or equivalent. An advanced diagnostic scan tool can usually pinpoint electrical faults quickly and safely. Consult an authorized MINI dealer for assistance.

Additional information can be found in the following repair groups:

- **600 Electrical System–General**
- **ECL Electrical Component Locations**

Special tools

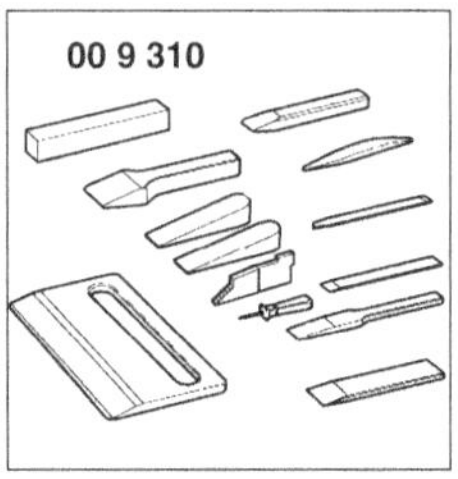

Body trim wedges and prying tools (BMW special tool kit 00 9 310)

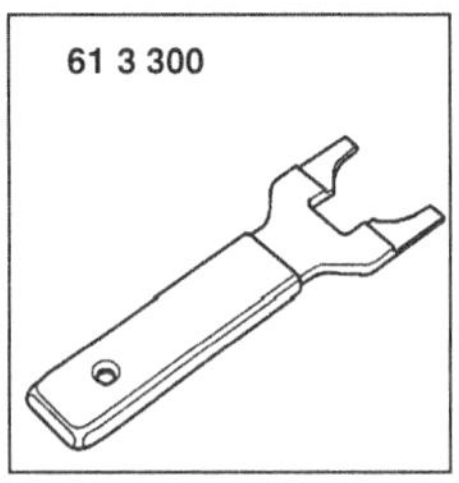

EWS ring antenna removal tool (BMW special tool 61 3 300)

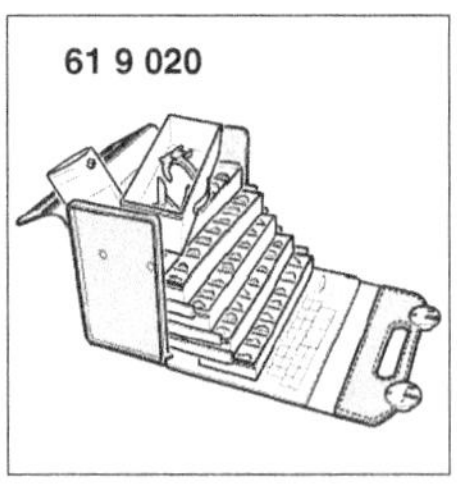

Wiring harness repair kit (BMW special tool 61 9 020)

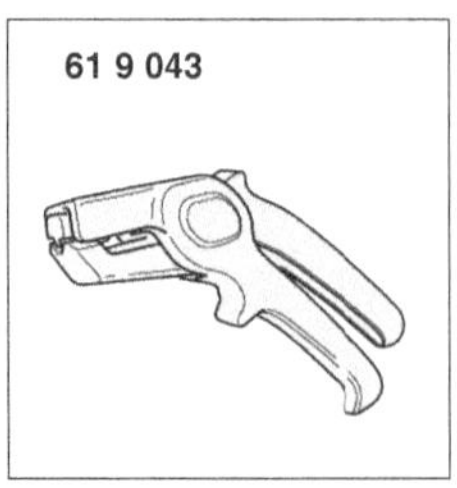

Wire cutter and insulation stripper (BMW special tool 61 9 043)

Central locking

Central locking with keyless remote operation is standard equipment on most MINIs. The system incorporates components from several different systems linked together within the General Module.

Because of the remote operation function, only the driver's door is equipped with a lock cylinder. Central locking performs the following functions depending on vehicle body type:

- Lock / unlock vehicle doors
- Release tailgate
- Lock / unlock fuel filler door
- Convenience opening and closing of windows and sunroof
- Convenience opening and closing of the convertible top (where applicable)

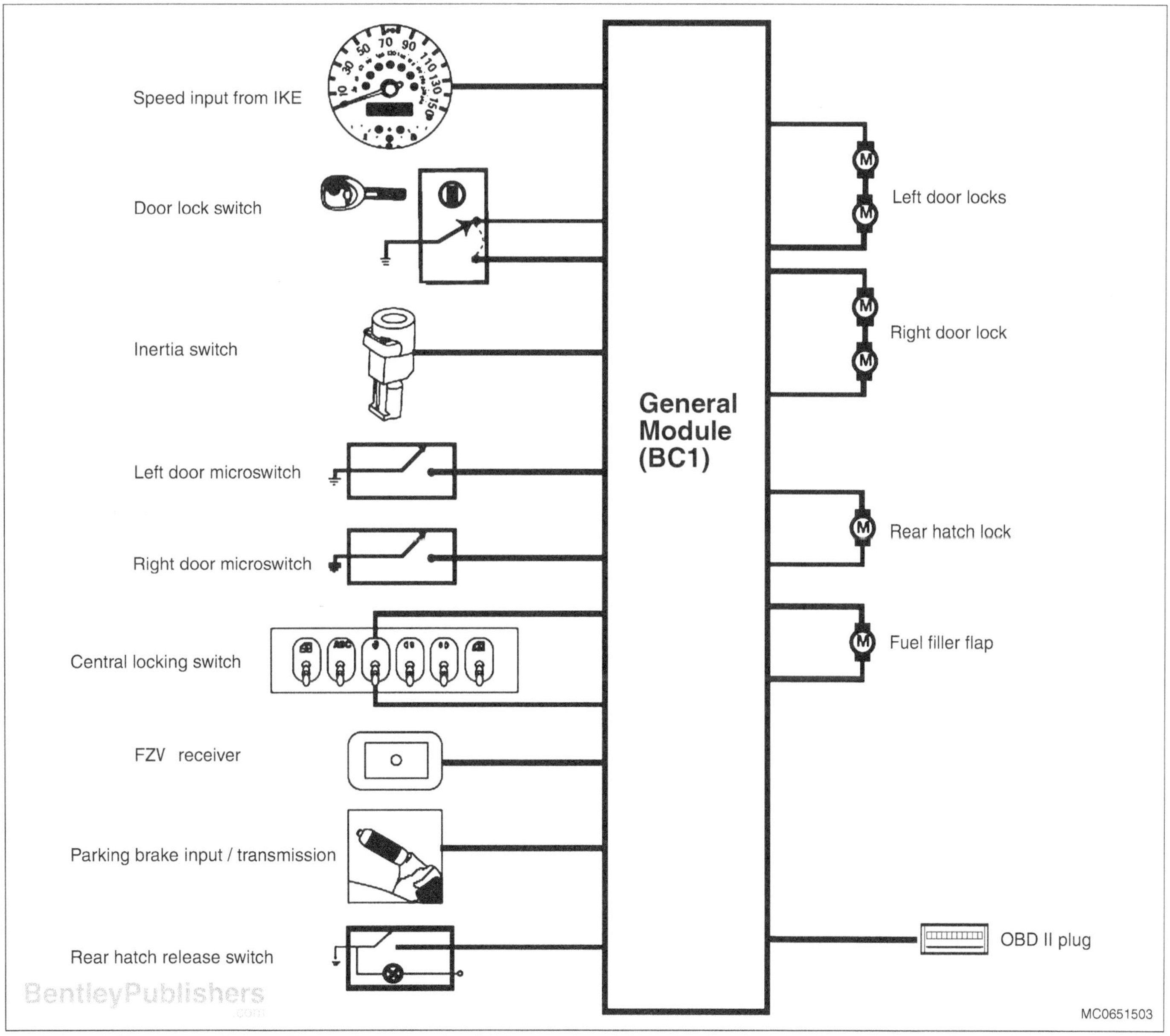

Electronic immobilizer (EWS)

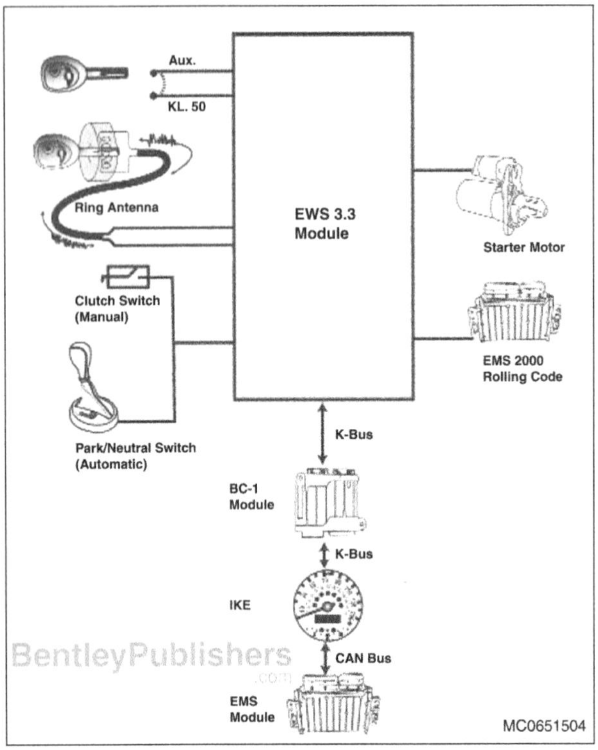

The electronic immobilizer (EWS) makes it impossible to start the engine using any means other than the special keys furnished with the vehicle.

The EWS ignition key is embedded with a computer chip and permanently coded. The key communicates with the vehicle using a transponder in the key and a ring antenna surrounding the steering lock cylinder.

A primary code is programmed into the key and into the vehicle itself. A secondary code is changed every time the vehicle is started. If the key code and EWS control module code do not match, the engine management control module and the starter are disabled.

EWS ignition keys cannot be duplicated. The system is designed to have up to ten keys. Only an authorized MINI dealer can provide replacement keys.

NOTE —
- *EWS is sometimes referred to as driveaway protection.*
- *If a vehicle key is lost or stolen, the electronic authorization for that key can be cancelled using the BMW scan and diagnostic tools DISplus, GT1 or MoDiC.*
- *Force applied to a key can damage the electronic circuitry. A damaged key will not start the engine.*
- *If electronic circuitry in the key is damaged, rendering it unusable, purchase a new key at an authorized MINI dealer and have it recoded.*

Anti-theft alarm (DWA)

The anti-theft system (DWA) is a dealer installed accessory option. DWA monitors door, rear hatch and engine hood contact switches, as well as the status of the ignition switch terminals, tilt sensor and the interior protection system. DWA is controlled by the General Module (BC1). The module and the auxiliary powered siren are installed in a place that is difficult to access.

Two different versions are installed depending on body type. Operation is similar for both types.

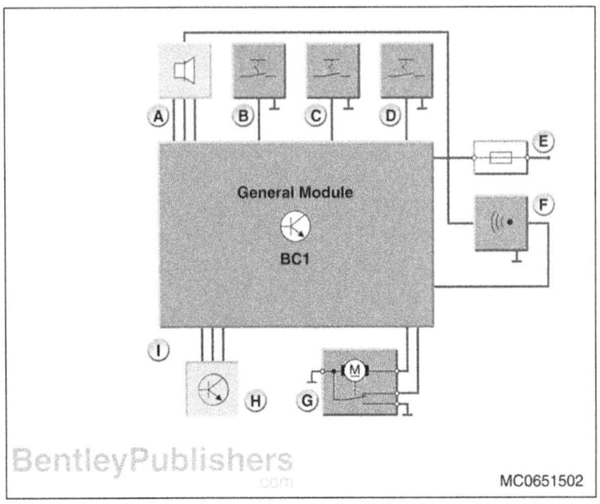

DWA anti-theft as installed in Coupe vehicles.

A. Emergency siren
- Location: Right side rear of engine compartment

B. Steering column stalk with indicator

C. Door microswitch for central locking

D. Hood contact

E. Terminal 30 connection, battery positive (Kl.30)

F. Microwave (ultrasonic) sensor
- Location: Center rear of vehicle on headliner

G. Tailgate central locking

H. Tilt sensor
- Location: Left rear of luggage compartment behind trim panel

I. General Module (BC1)

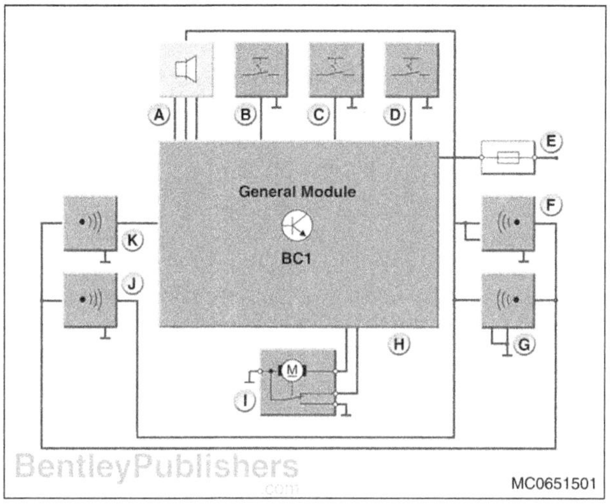

◄ DWA anti-theft as installed in Convertible vehicles.

A. Emergency siren with integral tilt sensor
- Location: Right side rear of engine compartment

B. Steering column stalk with indicator

C. Door microswitch for central locking

D. Hood contact

E. Terminal 30 connection, battery positive (Kl.30)

F. Microwave (ultrasonic) sensor (server 1)
- Location: R/F door behind speaker cover

G. Microwave (ultrasonic) sensor (server 2)
- Location: R/R trim panel behind speaker cover

H. General Module

I. Tailgate central locking

J. Microwave (ultrasonic) sensor (server 3)
- Location: L/R trim panel behind speaker cover

K. Microwave (ultrasonic) sensor with controller
- Location: L/F door behind speaker cover

General module

The General Module (BC1) controls several functions including central locking and anti-theft alarm. It is also linked to other systems including electronic immobilizer and convertible top.

Central locking, anti-theft, and convertible top operation are activated by the use of the key in the drivers door, remote operation (FZV) or from inside the vehicle using the appropriate switch. The General Module is also linked to the Convertible top Module (CVM).

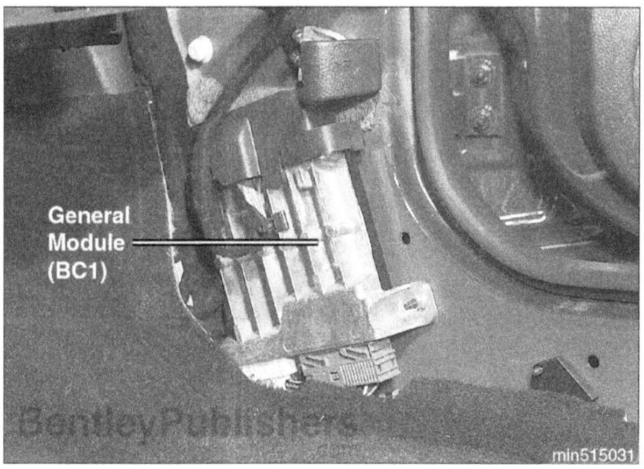

◄ The General Module is located at the base of the A-pillar behind the kick panel.

NOTE —
The General Module is also referred to as the Body Control Module or BC1. In wiring diagrams, the General Module is referred to as A1.

Door Lock Component Replacement

For door and door trim panel removal and installation information, see **411 Doors**.

Door latch, removing

- Remove door trim panel.
- Partially remove rear rail of window regulator. See **512 Door Windows**.
- Remove outside door handle. See 5**15 Central Locking and Anti-theft**.

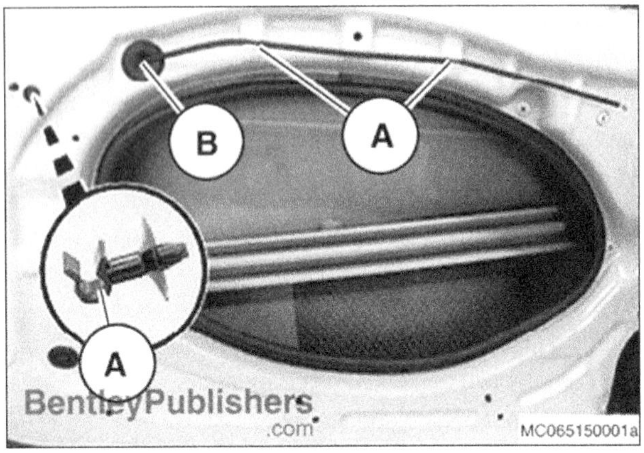

◄ Remove Bowden cable from mounts (**A**) and feed through door through opening (**B**).

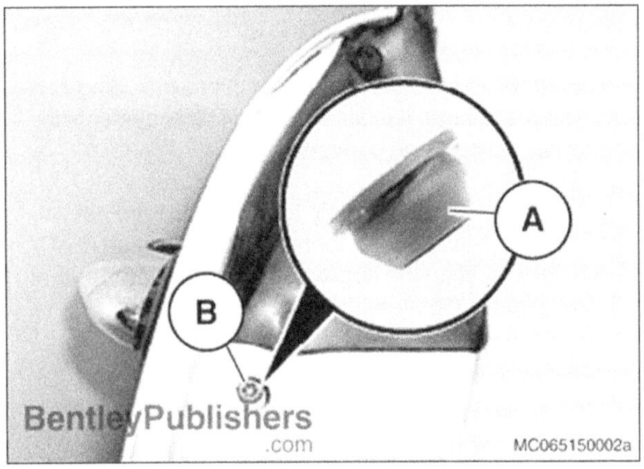

◄ Remove rubber plug (**A**) and fastener (**B**) beneath it.

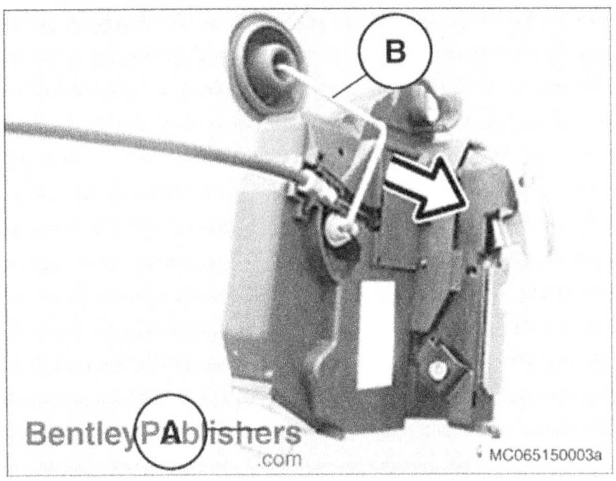

◄ Working inside door and door latch, disconnect electrical connector (**A**) and remove latch linkage (**B**) in direction of arrow.

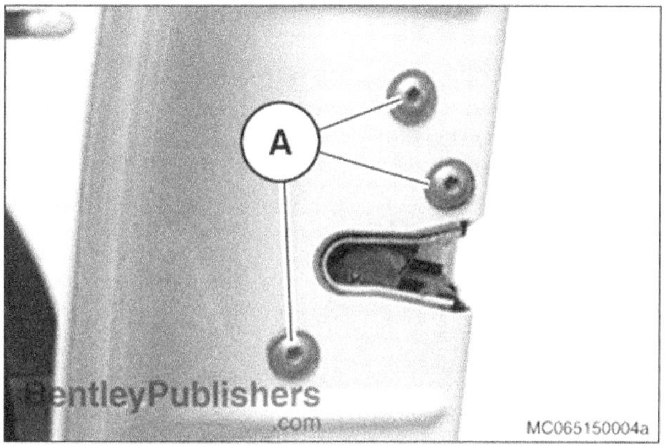

Remove door latch fasteners (**A**).

While supporting door latch from inside door, slide lock cylinder (**A**) in direction of arrow and remove door latch from door.

– Installation is reverse of removal. If replacing door latch use following steps:

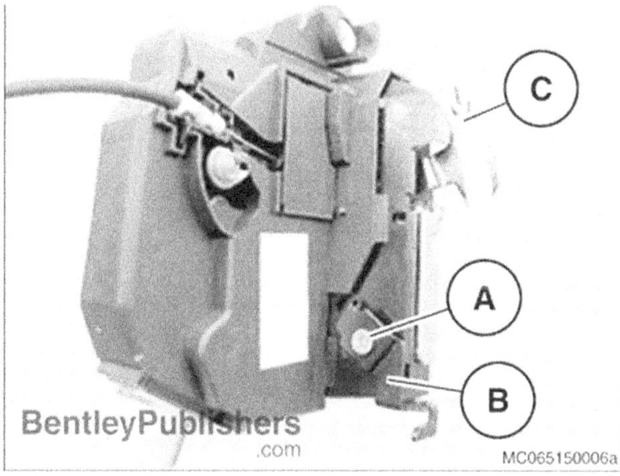

Remove fastener (**A**) and door lock lever (**B**). Remove cable guide (**C**). Install removed parts on new door latch.

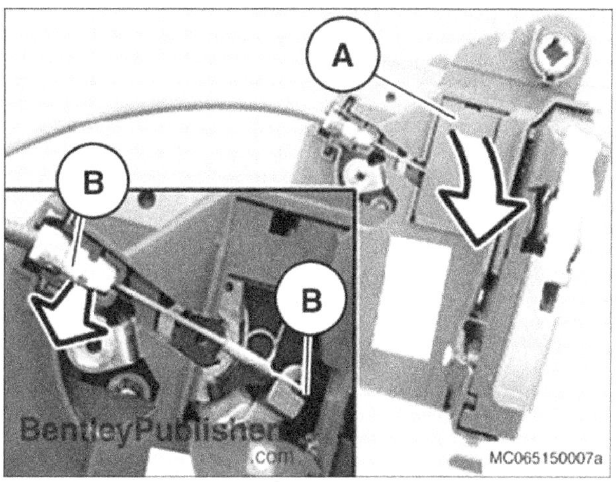

Door striker, replacing

- Remove door striker mounting Torx bolts (**arrows**).
- Remove striker and retrieve gasket.

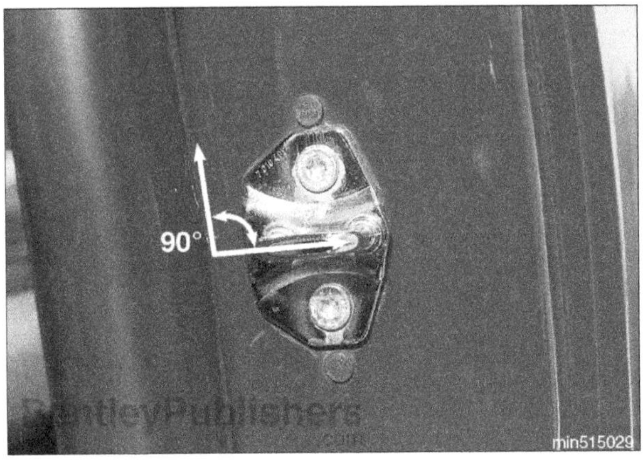

When installing, adjust striker as follows:

- Position striker approx. 90° to door frame inner edge.
- Tighten bolts.
- Close door and make sure striker does not touch or brush against door edge or lock interior. Look for scratches on paint.
- Adjust striker position or door position if necessary.

NOTE—
Door position adjustment is covered in **411 Doors**.

Tightening torque	
Door striker to door jamb	10 Nm (7 ft-lb)

Door handle, removing and installing

- Raise door window.
- Remove inner door trim panel. See **411 Doors**.

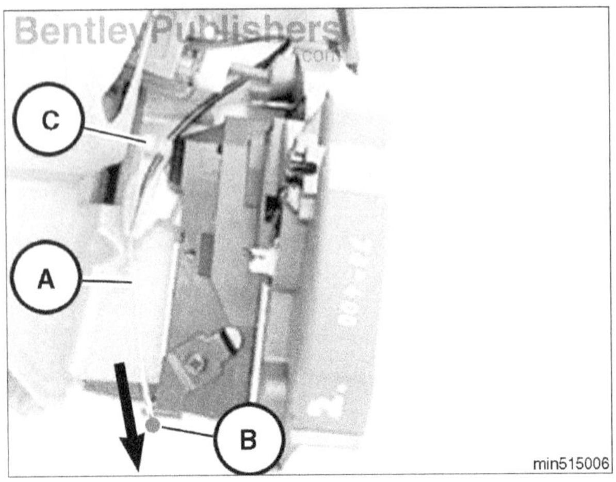

Working inside door cavity between outer door skin and door lock mechanism:

- Pull door handle actuating cable (**A**) down to release from door lock lever (**B**).
- Remove cable from cable guide (**C**).

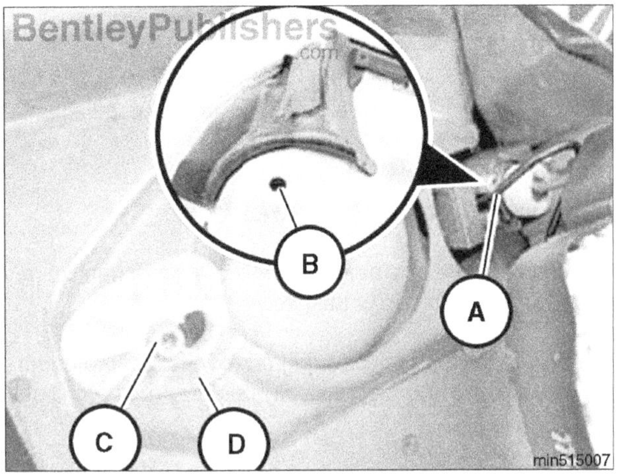

◄ Detach handle from door:

- Remove handle mounting bolt (**A**). Access to bolt is through inner door bore **(B)**, located under inner door cavity strip.
- Remove handle mounting bolt (**C**) and spacer block (**D**).

– Pull door handle off door.

– When installing new handle, fit self-threading bolts to new handle prior to installation.

– Make sure handle actuating cable inside door cavity is correctly routed.

Inside door handle, removing and installing

– Remove inner door trim panel. See **411 Doors**.

– Remove combined treble door speaker and inner door handle frame. See **650 Radio**.

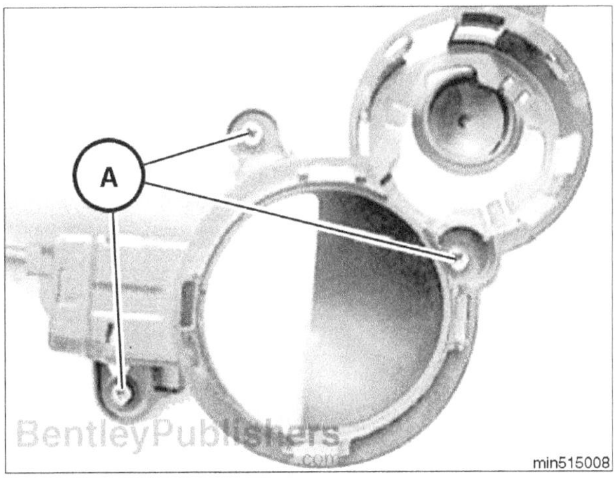

◄ Turn over frame and remove door handle mounting screws (**A**).

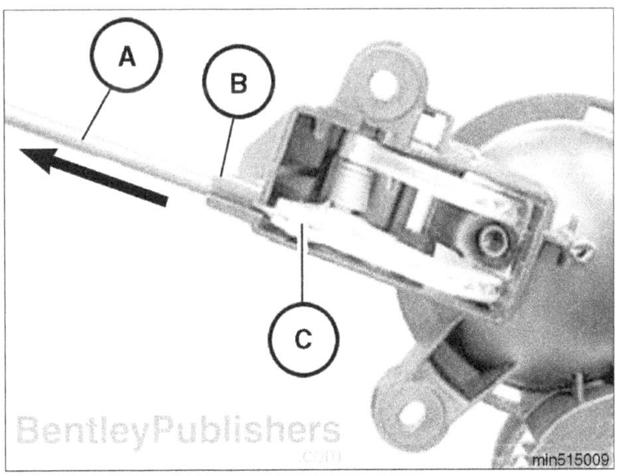

◄ Detach lock release cable sheath (**A**) from inside door handle trim (**B**).

– Detach cable end (**C**) from inside handle.

– Installation is reverse of removal.

Rear Hatch Lock Component Replacement

For rear hatch and rear hatch trim panel removal and installation information, see **412 Rear Hatch**.

Rear hatch lock emergency release

In case of electrical failure or emergency, there is a mechanical release that opens the rear hatch or tailgate.

◄ **Coupe**: The rear hatch lock release ring (**arrow**) is located under the rear seat. Pulling the ring mechanically releases the hatch latches.

◄ **Convertible**: The tailgate release handle (**arrow**) is recessed into the trim panel. Pulling this glow-in-the-dark handle mechanically releases the tailgate latches.

Emergency rear hatch lock release cable, replacing (Coupe)

– Open rear hatch and remove cargo compartment floor trim.

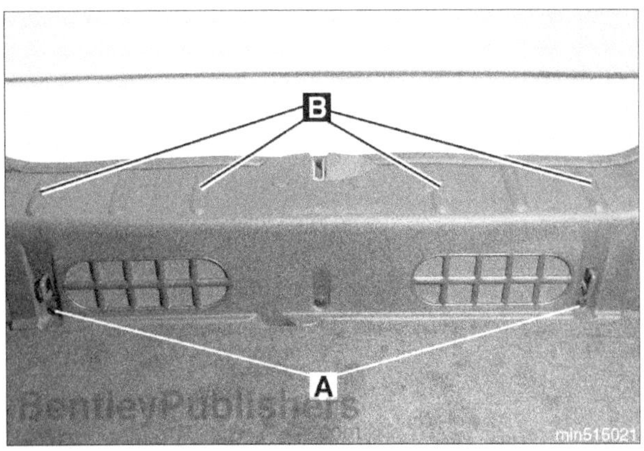

◄ Remove rear hatch sill cover:

- Release expansion rivets (**A**).
- Pry up gently to unsnap clips (**B**) and remove trim.

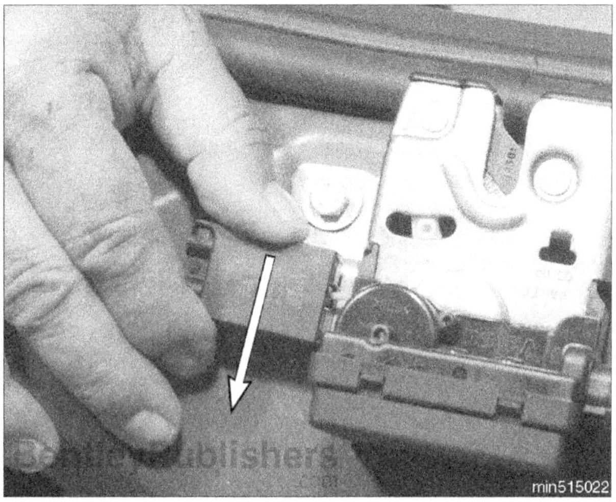

Working at rear hatch lock, push forward (**arrow**) on manual release cable end housing to disengage from bracket.

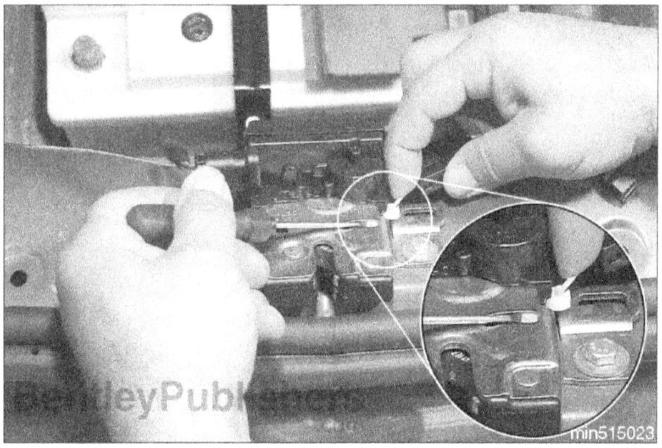

Use small screwdriver to push plastic tab in lock assembly to right. Unhook cable end from lock assembly.

– Remove rear seat and rear seat backrest. See **520 Seats**.

– Remove right rear inside trim panel. See **515 Interior Trim**.

– Unhook emergency release cable from retaining clips.

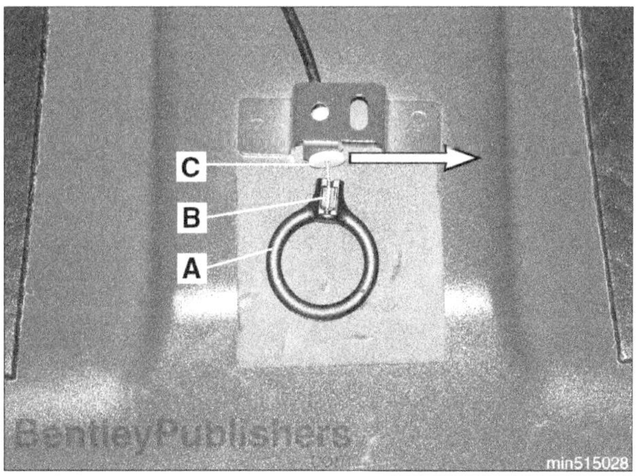

Detach emergency release cable end (**B**) from pull handle (**A**).

- Push plastic cable retainer (**C**) out of metal bracket and remove cable.

– Installation is reverse of removal.

– When replacing interior trim, replace defective or damaged expansion rivets and trim clips.

– Make sure rear hatch rubber seal overlaps trim correctly.

Rear hatch release switch, replacing

Remove handle from rear hatch:

- Hold hatch partially open.
- Remove T20 Torx mounting screws (**arrows**).
- Pull handle away from hatch and disconnect electrical harness connector.

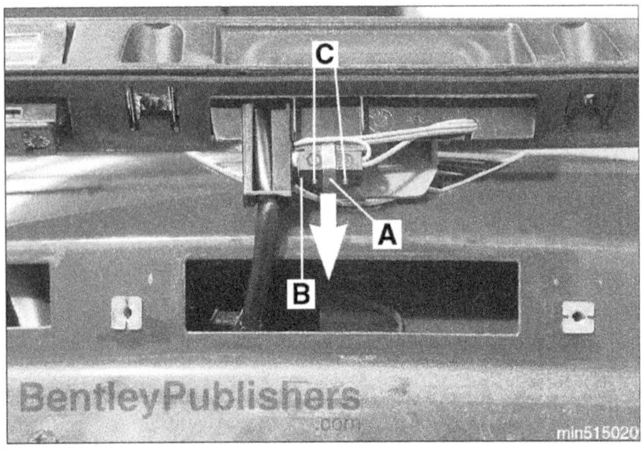

Disassemble release switch assembly:

- Remove clip (**A**) in direction of **arrow**.
- Remove switch (**B**) from connectors (**C**) in direction of **arrow**.

– Cut off switch wires. Cut wires on new switch to length and strip insulation. Install new microswitch using new electrical connectors.

– Installation is reverse of removal. Make sure clip (**A**) is seated correctly.

Rear hatch lock, removing and installing

– Open rear hatch and remove cargo compartment floor trim.

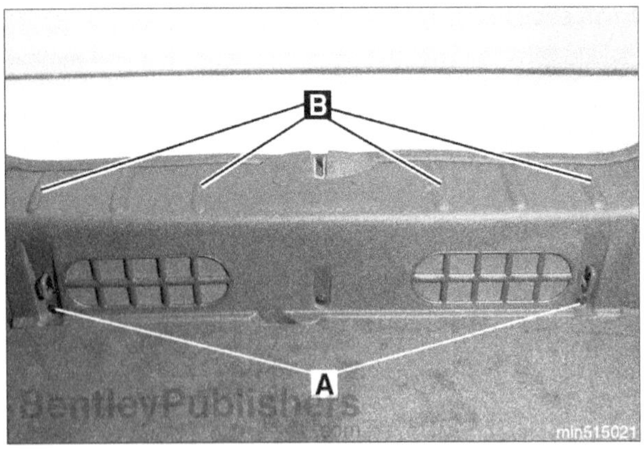

Remove rear hatch sill cover:

- Release expansion rivets (**A**).
- Pry up gently to unsnap clips (**B**) and remove trim.

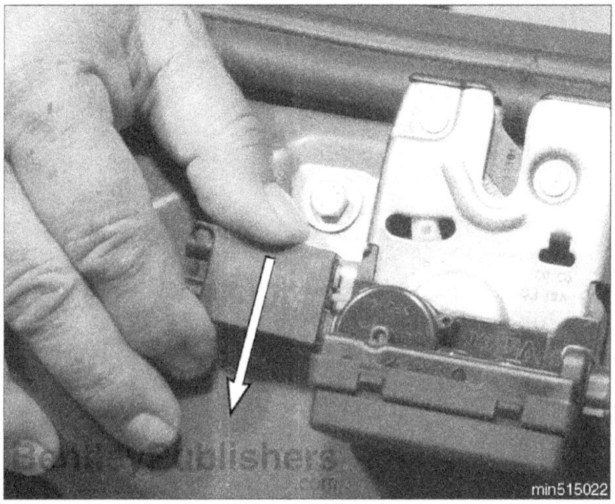

Working at rear hatch lock, push forward (**arrow**) on manual release cable end housing to disengage from bracket.

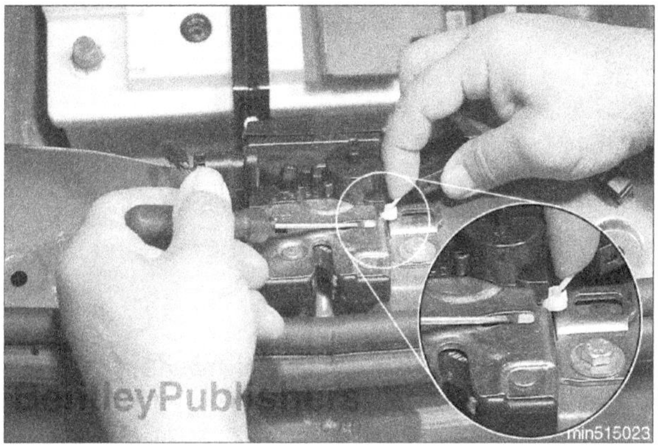

Use small screwdriver to push plastic tab in lock assembly to right. Unhook cable end from lock assembly.

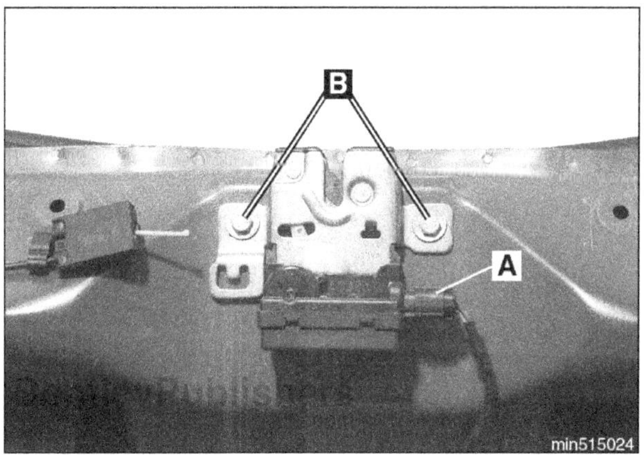

Detach electrical harness connector (**A**) from lock assembly.

- Remove lock mounting screws (**B**) and remove lock.

– Installation is reverse of removal. Adjust lock position. See **Rear hatch lock, adjusting** in this repair group.

Tightening torque	
Hatch lock to body	10 Nm (7 ft-lb)

– Replace defective or damaged expansion rivets and trim clips.

– Make sure rear hatch rubber seal overlaps trim correctly.

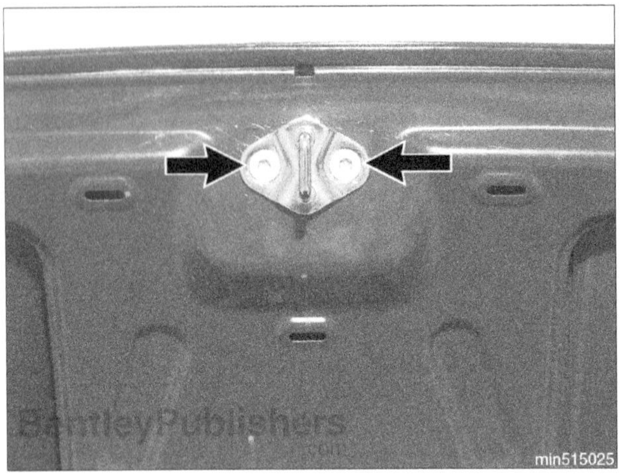

Rear hatch lock striker, removing and installing

- Open rear hatch and remove hatch lower trim panel. See **412 Rear Hatch**.
- Working at bottom of hatch, remove striker mounting Torx screws (**arrows**). Remove striker.
- Installation is reverse of removal.

Tightening torque	
Hatch lock striker to hatch	6 Nm (4 ft-lb)

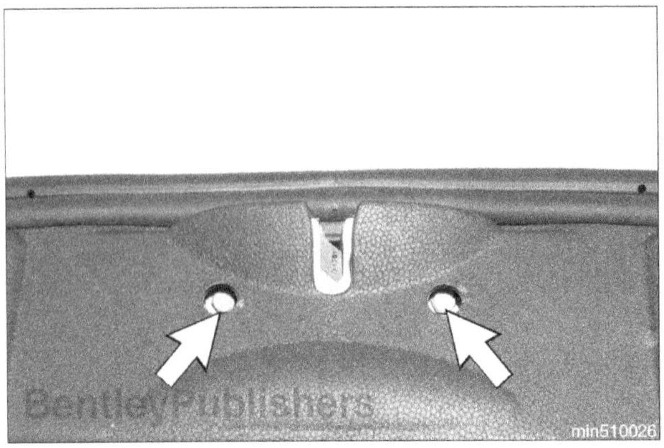

Rear hatch lock, adjusting

- Prior to adjusting rear hatch lock, make sure hatch is aligned correctly. See **412 Rear Hatch**.
- Twist hatch rubber buffers until they bottom out.
- Working at rear hatch sill trim, pry out hatch lock mounting bolt trim plugs.
- Loosen lock mounting bolts (**arrows**) so that lock can just slide.
- Close hatch while holding hatch release switch (on handle strip) depressed to prevent lock from engaging.
  - Allow hatch lock to adjust itself correctly.
  - Open hatch and tighten lock mounting bolts.

Tightening torque	
Hatch lock to body	10 Nm (7 ft-lb)

- Recheck hatch and hatch lock, repeating procedure if necessary.
- Adjust rear hatch rubber buffers for correct body aperture. See **412 Rear Hatch**.

NOTE —
Do not overextend rubber buffers as this can cause rear hatch to close with difficulty or not at all.

ELECTRONIC IMMOBILIZER (EWS) COMPONENT REPLACEMENT

EWS control module, replacing

– Disconnect negative (-) cable from battery.

> ***CAUTION—***
> *Prior to disconnecting the battery, read the battery disconnection cautions given in* **001 General Warnings and Cautions**.

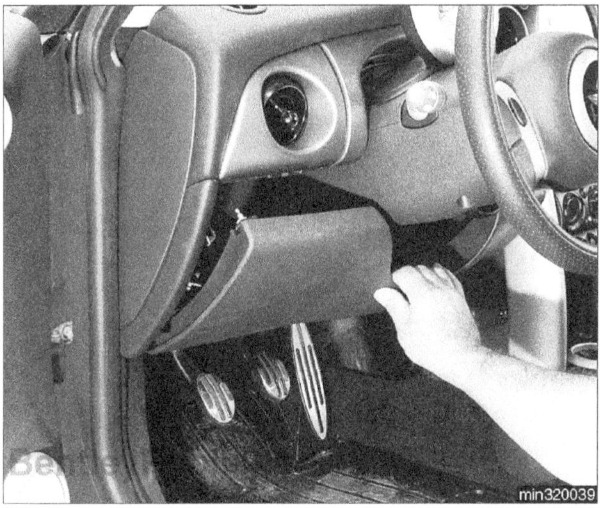

◄ Pull on top edge of left lower dashboard trim to unclip and swing down.

◄ Working under left side of dashboard to left of steering column:

- Disconnect EWS control module electrical harness connector (**A**).
- Remove module mounting nuts (**B**).

NOTE—
Only one of the mounting nuts is visible in photo.

– Installation is reverse of removal.

Tightening torque	
EWS control module to body	3 Nm (26 in-lb)

– Recode EWS control module and ignition keys using DISplus, GT1 or MoDiC.

Ring antenna, removing

– Remove rubber boot surrounding ignition switch.

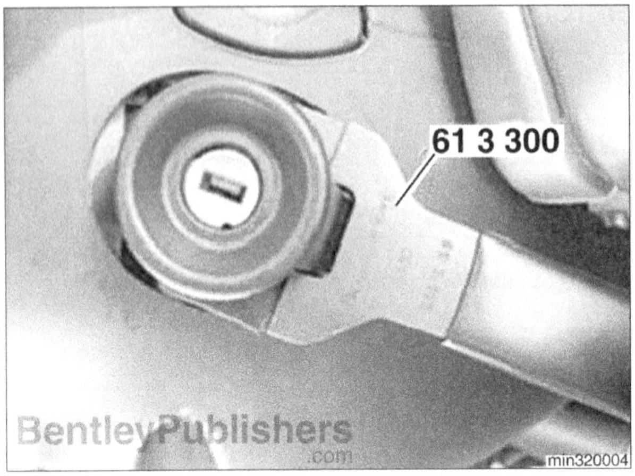

◄ Using BMW special tool 61 3 300 with flattened end (**A**) facing ring antenna, lever out ring antenna.

– Disconnect ring antenna electrical harness connector.

ANTI-THEFT ALARM (DWA) COMPONENT REPLACEMENT

Tilt sensor, removing and installing (Coupe)

– Open rear hatch.

◄ Working at left side cargo compartment trim, release clips (**arrows**) and remove access panel.

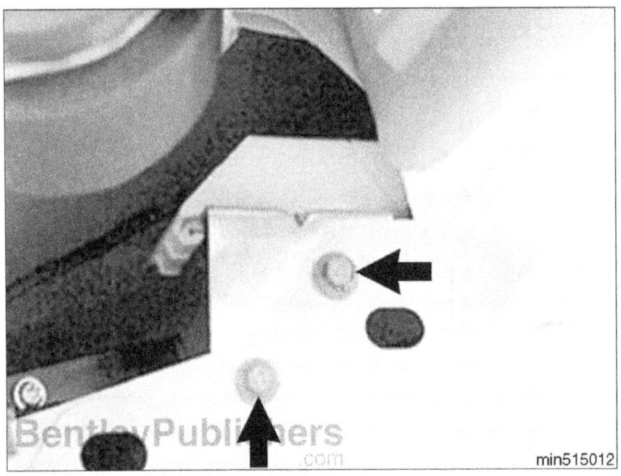

◄ Remove tilt sensor mounting bolts (**arrows**).

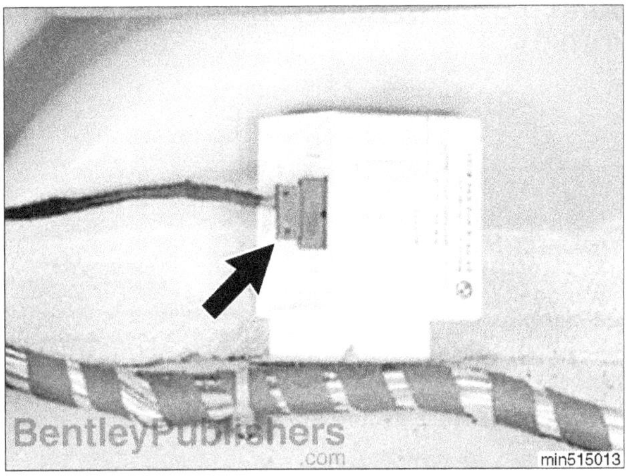

- Disconnect electrical harness connector (**arrow**) and remove tilt sensor.

– Installation is reverse of removal.

Ultrasound module, removing and installing (Coupe)

- Working at headliner, gently pry off ultrasound module housing in direction of **arrow** and disconnect electrical harness connector.

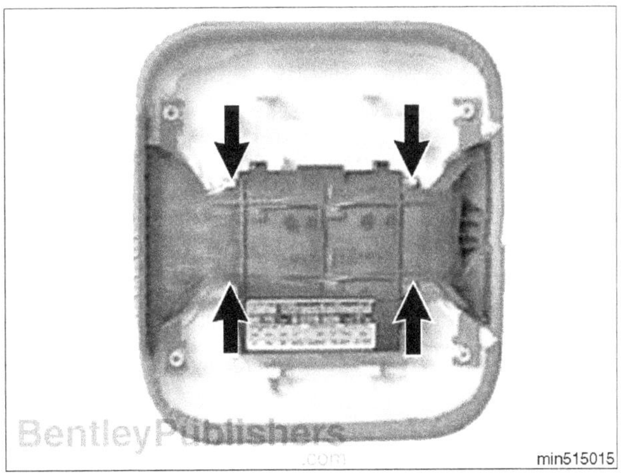

- Pry retaining clips (**arrows**) to remove ultrasound module from housing.

– Installation is reverse of removal.

Ultrasound modules, removing and installing (Convertible)

Front

– Remove interior door trim panel. See **411 Doors**.

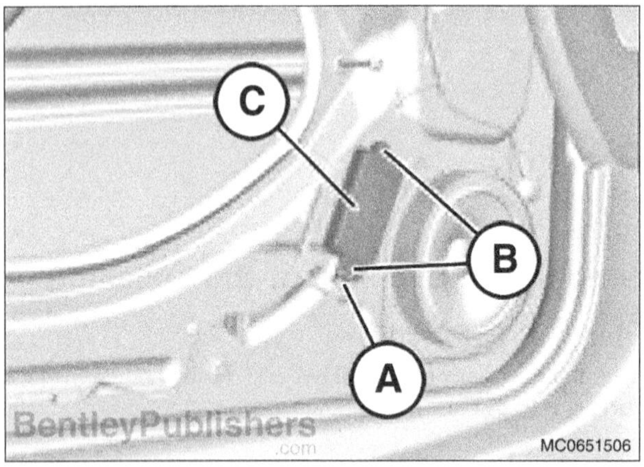

◄ Disconnect electrical harness connector (**A**).

– Remove securing screws (**B**).

– Remove ultrasound module (**C**) from door frame.

– When reinstalling, be sure to secure module (**C**) in same position.

– Balance of installation is reverse of removal.

Rear

– Remove rear inside trim panel. See **513 Interior Trim**.

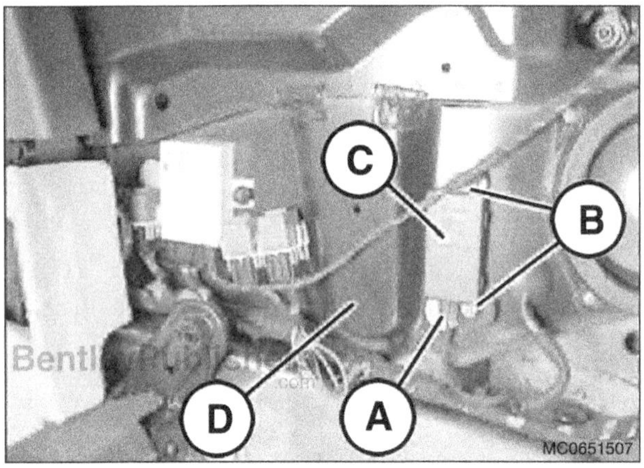

◄ Disconnect electrical harness connector (**A**).

– Remove securing screws (**B**).

– Remove ultrasound module (**C**) from side panel (**D**).

– When reinstalling, be sure to secure module (**C**) in same position.

– Balance of installation is reverse of removal.

Alarm siren, removing and installing

◄ Working at right rear of engine compartment, remove alarm siren bracket mounting nuts (**arrows**).

min515017

- Disconnect siren electrical harness connector (**arrow**).
- Lift siren bracket out of engine compartment.

min515018

- Remove mounting nut (**arrow**) and detach siren from mounting bracket.
- Installation is reverse of removal.

Tightening torques	
Siren mounting bracket to body	9 Nm (7 ft--lb)
Siren to mounting bracket	4 Nm (35 in-lb)

520 Seats

General

This repair group covers removal and installation procedures for front and rear seats. Seat repair procedures are beyond the scope of this manual.

Front seats in MINI models incorporate side ("thorax") airbags, controlled by the MRS control module.

The front seat belt buckle and pretensioner is mounted to the inboard edge of each front seat base. The pretensioner is a pyrotechnic (explosive charge) device controlled by the MRS control module.

Be sure to read the warnings and cautions in **721 Airbag System (SRS)** when working on front seats.

NOTE—
Airbags are also known as the Supplemental Restraint System (SRS). In MINI cars the occupant safety system is known as the Multiple Restraint System (MRS), and includes seat belts, airbags and battery safety terminal (BST).

Front seats

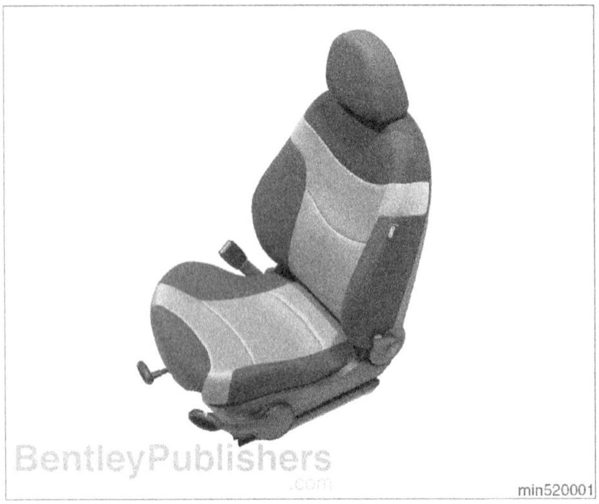

min520001

Front seats on MINI models are manually operated. They are offered in two variants:

- **Basic seat** in MINI Cooper
- **Sports seat** in MINI Cooper S

Front seats are secured to the body by four Torx bolts.

Front seat adjustment

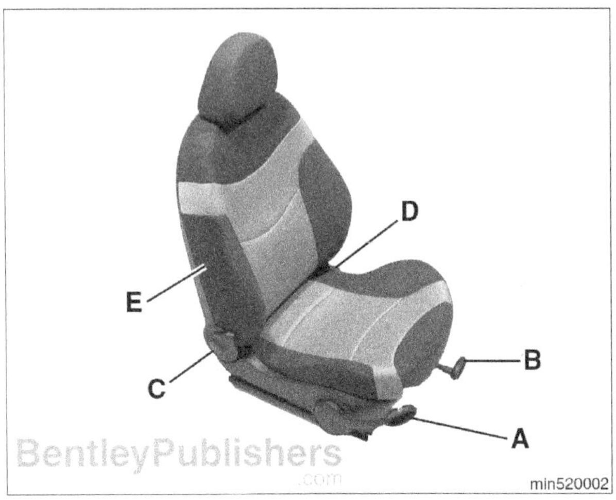

min520002

Front seats, regardless of type (basic or Sports), are manually adjusted as follows (left seat illustrated; right is similar):

- Fore and aft adjustment: Lever **A**
- Height adjustment of the seat: Lever **B** raises or lowers seat cushion height.
- Backrest angle adjustment: Lever **C** pulled up allows position of back rest to be altered.
- Backrest forward tilt: Lever **D** pushed forward allows easy rear seat entry.
- Lumbar support adjustment (optional): Knob **E**

The front seats have position memory with no electronics involved. Once the seat has been set to the desired position, when lever **D** is operated (pushed forward and down), the seat backrest and cushion will move fully forward for easy rear seat entry. When the seat backrest is returned (manually), the entire seat returns to the previously set position and locks into place.

Front seat options

Standard seating covering is cloth or leatherette. Leather is an option in all models.

Lumbar adjustment is offered as an option on Sports seats.

Front seat heating, if equipped, is activated by the seat heating switches, housed in the center console. When seat heating is activated the LED in the switch illuminates.

Rear seat

The rear seat cushion is one piece. The folding backrest is 50 / 50 split.

Rear seat backrest operation

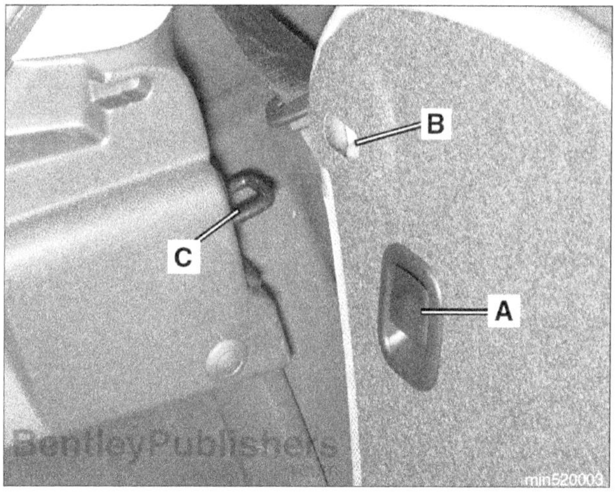

Rear seat backrest may be folded horizontally by releasing lock handle (**A**) in the rear face of the backrest, accessed through the cargo compartment.

The rear seat lock (**B**) is in the seat backrest and the striker (**C**) is secured to the body.

Child safety seat anchors

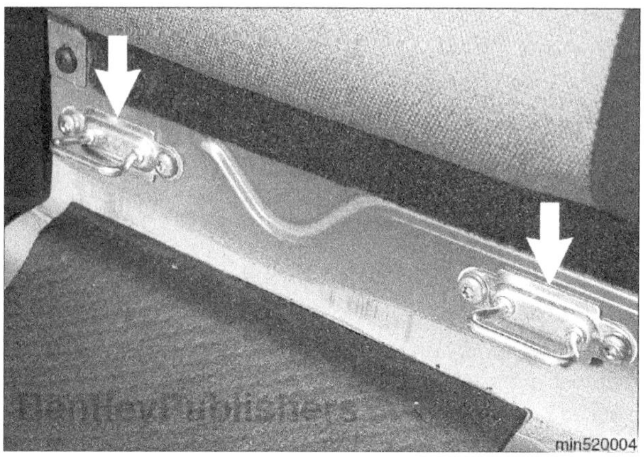

The MINI is equipped with anchors (**arrows**) under the rear seat cushion for child seat retention. The anchor brackets are accessed through trim plugs in the rear of the seat cushion.

Front Seat Components

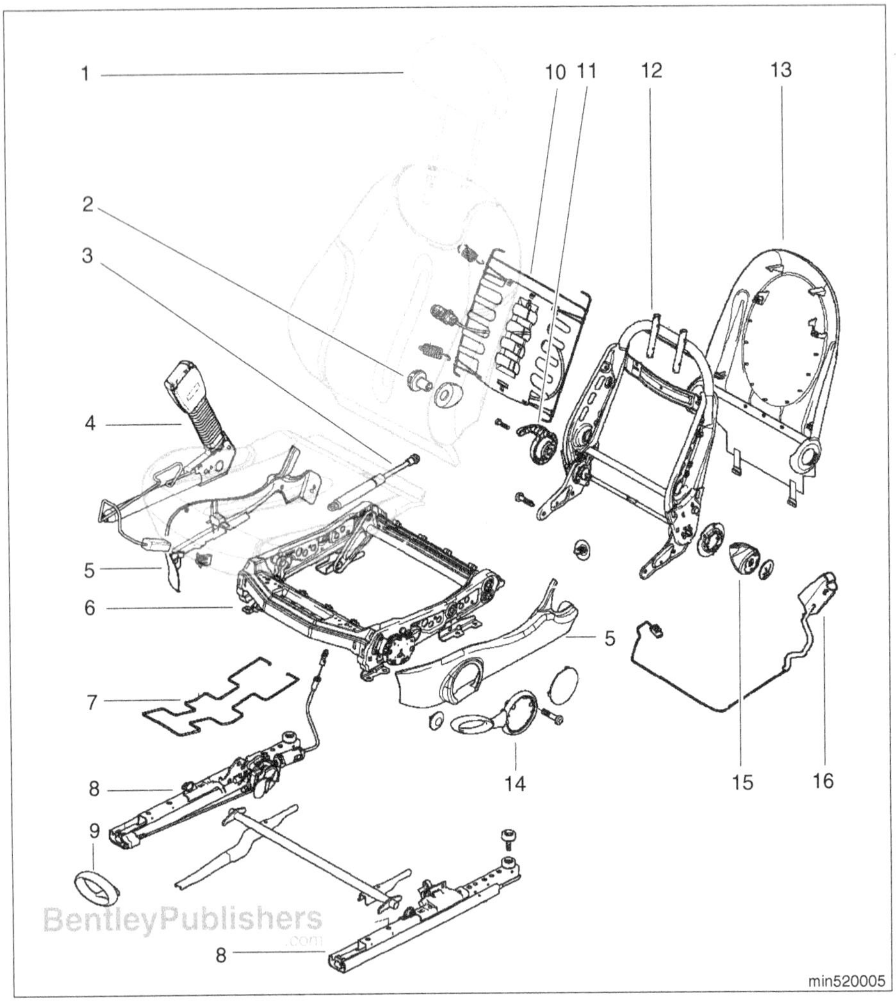

1. Headrest
2. Lumbar support adjuster knob
3. Seat height control air spring (left seat)
4. Pyrotechnic seat belt pretensioner (buckle)
5. Plastic trim
6. Seat cushion frame
7. Seat cushion spring
8. Seat rail
9. Fore-aft control handle
10. Backrest spring
11. Backrest angle control handle
12. Backrest frame
13. Rear panel
14. Seat cushion height handle (left seat)
15. Easy rear seat entry handle
16. Airbag

Front seat, removing and installing

> ***WARNING —***
> - *The front seat is equipped with both an airbag and a pyrotechnic seat belt pretensioner.*
> - *The airbag is inflated by an explosive device. Handled improperly or without adequate safeguards, it is very dangerous. Observe special precautions prior to any work at or near the airbags. Read the warnings and cautions in* **721 Airbag System (SRS)**.
> - *The pyrotechnic seat belt tensioner is a powerful device. Handle with extreme care. Incorrect handling can trigger the tensioner and cause injury.*

– Switch off ignition and remove ignition key.

– Tilt and move front seat forward as far as possible.

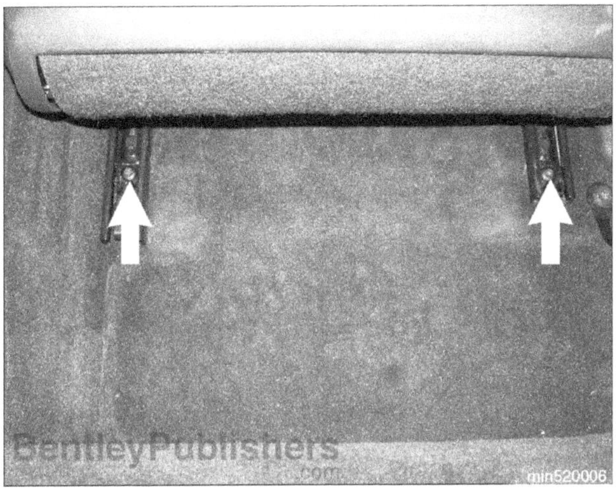

◀ Remove bolts (**arrows**) at rear of seat rails. Discard bolts.

– Move front seat back as far as possible.

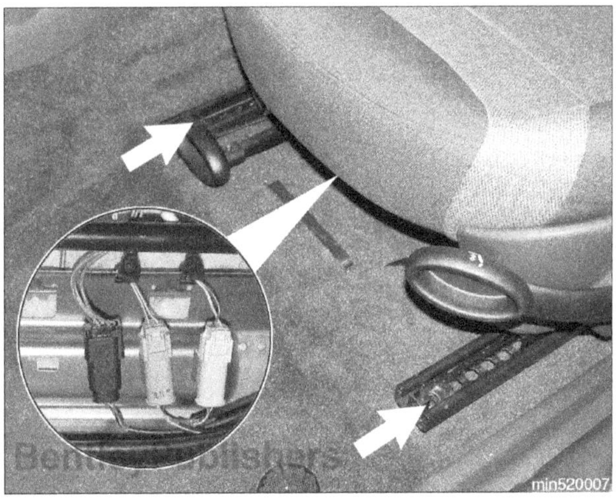

◀ Remove bolts (**arrows**) at front of seat rails. Discard bolts.

– Tilt seat backward and disconnect electrical harness connectors (**inset**) under front of seat.

– Lift out front seat.

> ***CAUTION—***
> *Cover door sill with protective cover to prevent damage from seat rails.*

– Reinstallation is reverse of removal. Bear in mind the following:
  - Make sure carpeting does not become trapped between seat rails and floor near seat mounting bolts.
  - Replace seat mounting bolts. Use Loctite® 270 or equivalent thread locking compound.

Tightening torque	
Front seat to floor (M8 x 40) (replace bolts, use Loctite® 270)	35 Nm (26 ft-lb)

REAR SEAT COMPONENTS

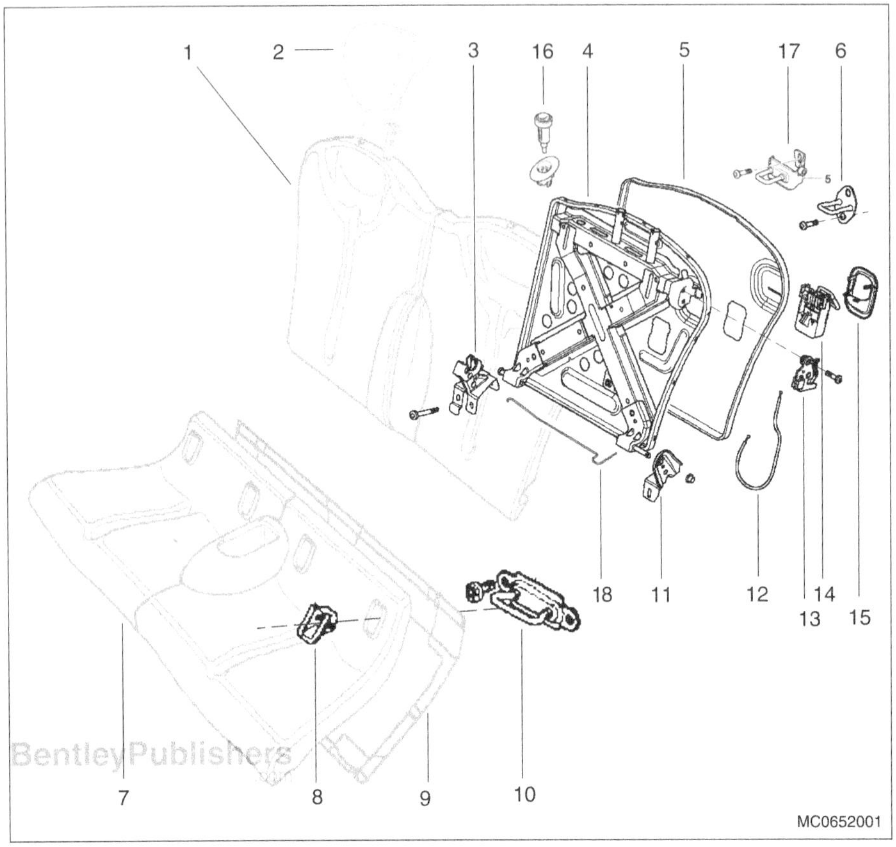

1. **Backrest**
2. **Headrest**
3. **Backrest center mount**
4. **Backrest frame**
5. **Backrest trim**
6. **Backrest striker**
   - Coupe
7. **Seat cushion**
8. **Child seat anchor trim plug**
9. **Seat cushion frame**
10. **Child seat anchor**
11. **Backrest mount**
12. **Bowden cable**
13. **Backrest lock**
14. **Backrest lock handle**
15. **Trim**
16. **Lock cylinder**
    - Convertible
    - On each backrest
17. **Backrest striker**
    - Convertible
18. **Manual release hook tool**
    - Convertible
    - Under seat cushion or stowed with vehicle tool kit

Rear seat cushion, manual folding (Convertible)

In case of an electrical failure, the rear seat backrest can be folded to gain access to the luggage compartment and the release handle on the tailgate.

– Using a vehicle master or ignition key, unlock both lock cylinders on the upper part of the backrest cushion.

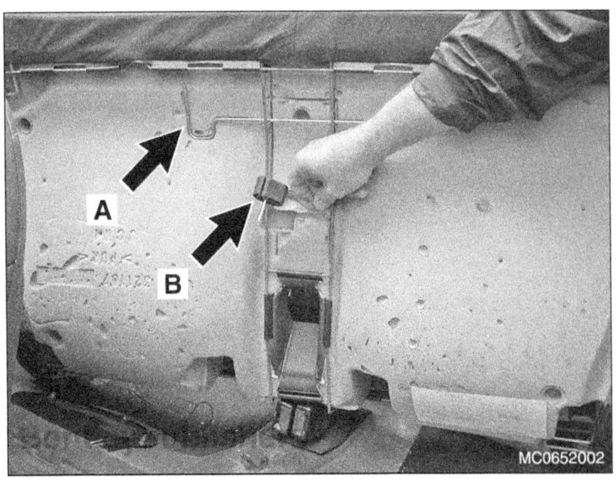

◄ Locate manual release hook tool:

- Remove manual release tool from vehicle tool kit.
- Lift forward edge of rear seat cushion slightly upward to release cushion from retainer clips. Pull cushion forward and up far enough to allow access to the underside. Remove tool (**A**) from underside of seat cushion.

NOTE—
*Second tool (**B**) under seat cushion is used for emergency closing of convertible top and/or folding sunroof and is not applicable in this operation.*

– Return seat cushion to installed position or remove from vehicle.

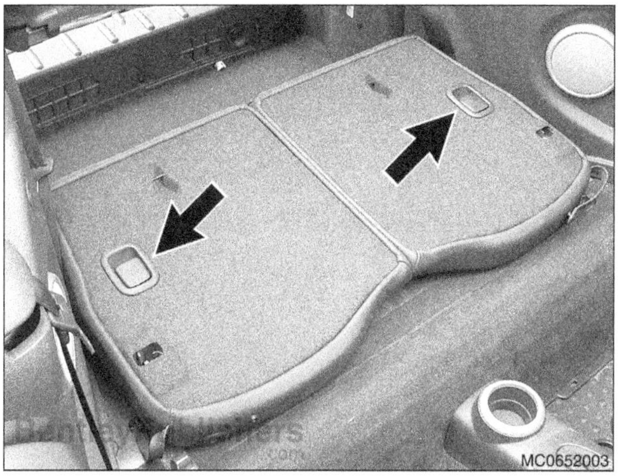

Using the hook tool, reach back down between the convertible top and rear seat backrest and pull handle (**arrows**) releasing backrest for that side. Fold backrest down.

– Repeat for other side as needed.

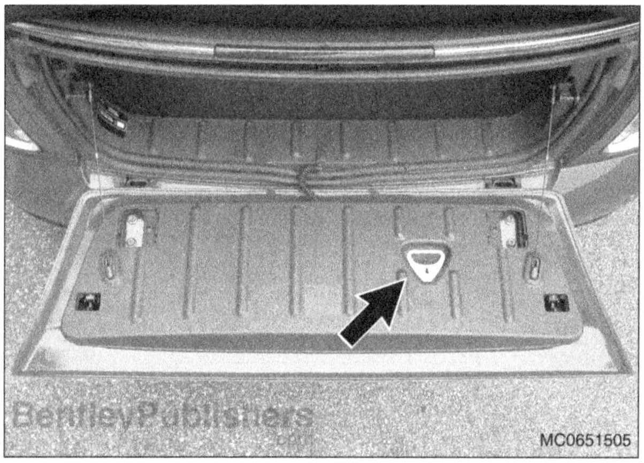

Reposition items in luggage compartment as needed to allow access to tailgate release handle (**arrow**).

– When completed, return hook to vehicle tool kit or seat cushion stowage.

– Install seat cushion or stow tool kit.

Rear seat cushion, removing and installing

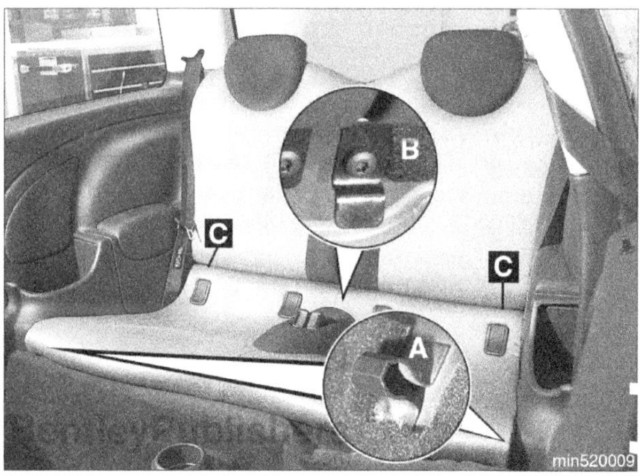

Lift front of rear seat cushion (**A**) to unclip.

– Push cushion towards rear and release from center retaining mount (**B**).

– Lift cushion off side retaining brackets (**C**) and remove from car.

– Installation is reverse of removal.

Rear seat backrest, removing and installing

– Remove rear parcel shelf.

– Fold rear seat backrest forward far enough to allow headrest to pull out.

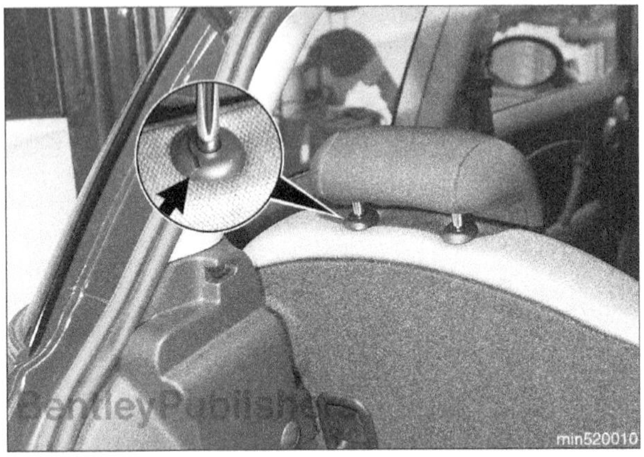

◄ Press headrest release button (**arrow**) and pull out headrest.

– Fold backrest fully down.

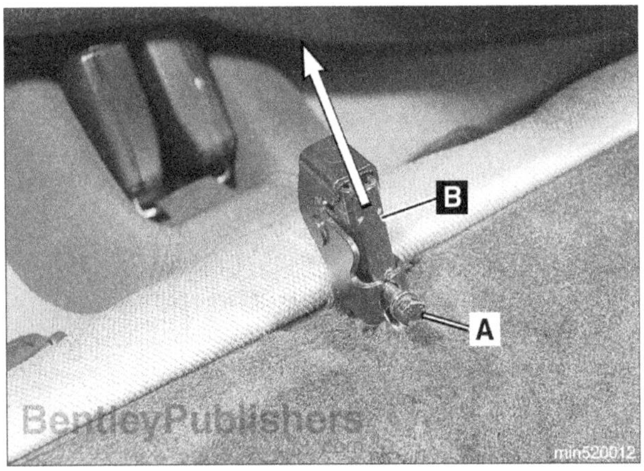

◄ Working at middle of backrest base:
- Remove screw (**A**).
- Open center mount lock (**B**) in direction of **arrow**.
- Detach backrest from center mount.

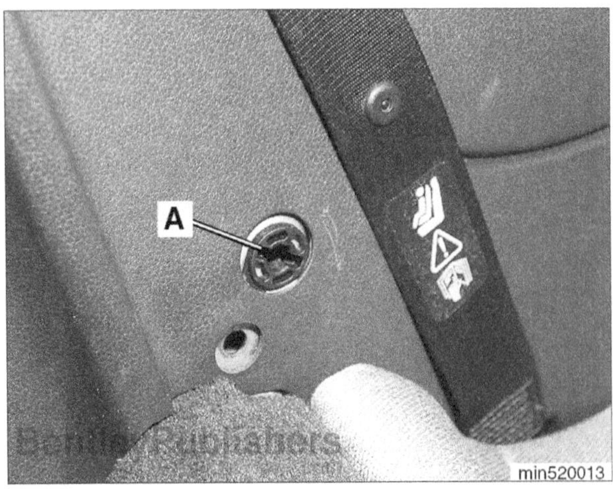

◄ Tilt backrest approx. 45° to remove from outer mount (**A**).

– Installation is reverse of removal. Do not trap seat belt when folding up backrest.

540 Sunroof

GENERAL

This repair group covers the slide-tilt sunroof assembly. For information on the folding sunroof on Convertible MINIs, see **541 Convertible Top**.

Component locations, initialization, sunroof panel replacement, and sunroof height adjustment are covered.

Sunroof assembly removal and sunroof frame gasket replacement are beyond the scope of this manual.

NOTE —
Be sure to check drains in the corners of sunroof assembly if water is entering car through headliner.

Special tools

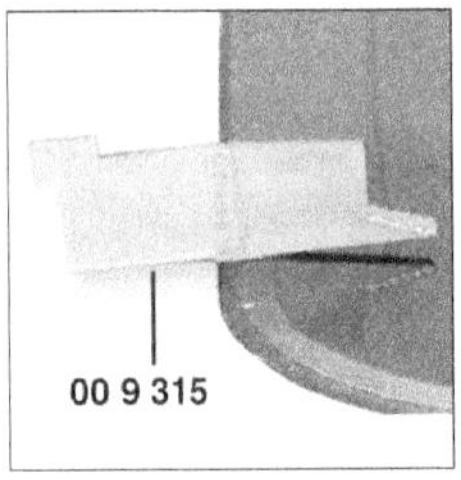

◄ Gap measuring wedge (BMW special tool 00 9 315)

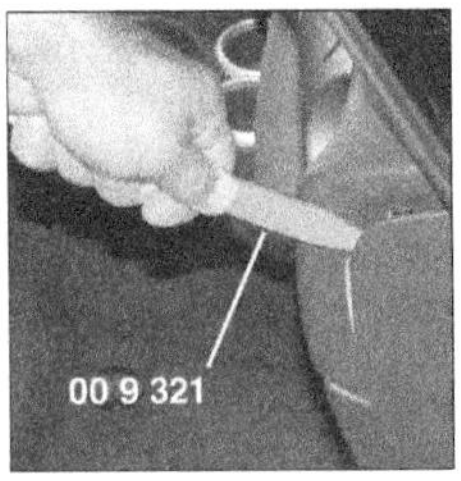

◄ Plastic prying tool (BMW special tool 00 9 321)

Sunroof motor, initializing

Initialize sunroof motor in the following situations:

- Malfunctions such as lack of one-touch function or lack of comfort opening/closing.
- If sunroof motor is removed and reinstalled.
- If power supply to sunroof is interrupted.
- If repairs are performed on slide/tilt mechanism.

– Initialize sunroof motor as follows:

- Close sunroof with switch.
- Hold switch in TILT (raised) position for approx. 15 seconds.
- When roof moves from TILT to CLOSED position, initialization is complete.

Sunroof, emergency closing

In case of electrical or mechanical failure, an open sunroof can be closed using the emergency Allen key from the vehicle tool kit.

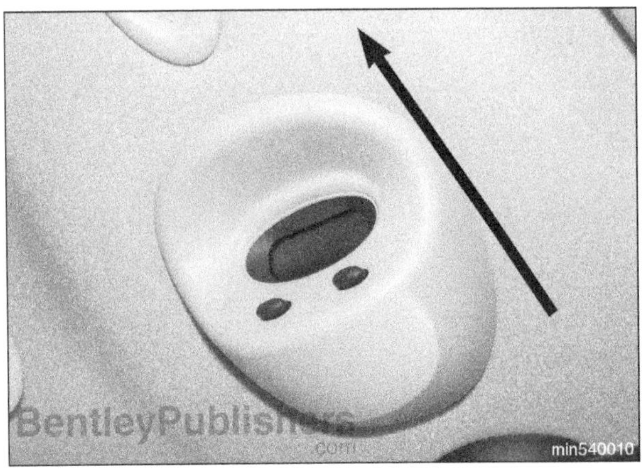

◄ Push clock or trim cover in headliner backward to remove.

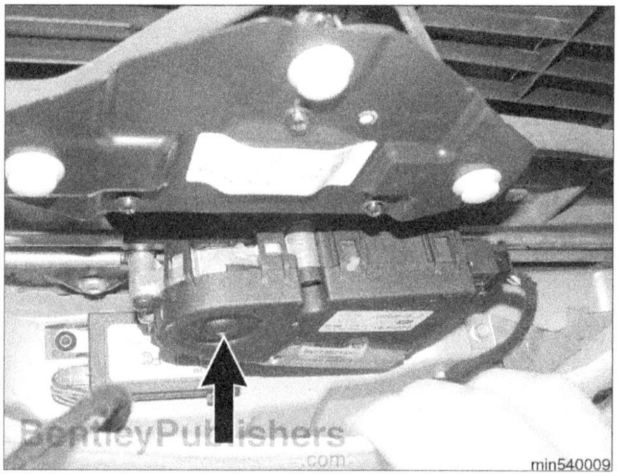

Insert Allen key into drive in sunroof motor and turn to manually close sunroof.

NOTE—
*The headliner has been pulled down to demonstrate location of Allen key hole (**arrow**) in sunroof motor. Allen key can be inserted through opening provided when clock or trim cover is removed.*

Sunroof panel, adjusting

The sunroof is controlled by a set of cables that move the glass panel along guide rails when the motor is operated.

Adjust sunroof panel under the following circumstances:

- Sunroof misaligned with roof
- Sunroof does not close squarely
- Wind noise at high speeds (sunroof closed)
- Sunroof has been removed and reinstalled

Horizontal alignment, checking

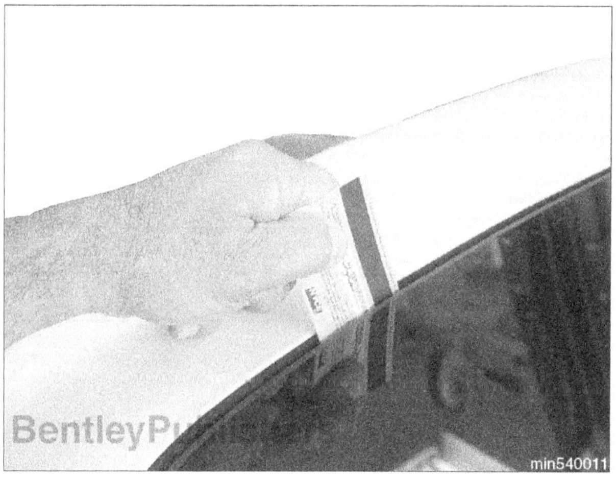

With sunroof fully closed, use a credit card to measure gap. Make sure there is equal resistance to card as it is slid all around perimeter of sunroof.

Vertical adjustment, checking

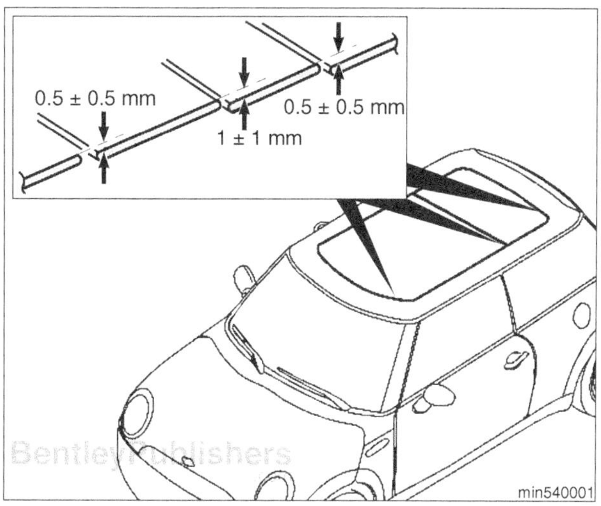

With sunroof fully closed, check height of front and rear of moveable glass panel against roof and stationary panel.

Sunroof height	
Rear	flush / 1 mm (0.04 in) higher than roof
Front	flush / 0.5 mm (0.02 in) lower than roof

NOTE—
BMW special gap wedge tool 00 9 315 may be used to determine height of glass panel.

Adjustment procedure

- Working inside car, release sunroof blind from guide channels by twisting front rail as shown (**arrows**).

- Allow blind to retract as far as possible and allow to hang down.

> ***CAUTION—***
> *Do not allow grease on rail ends to make contact with interior trim.*

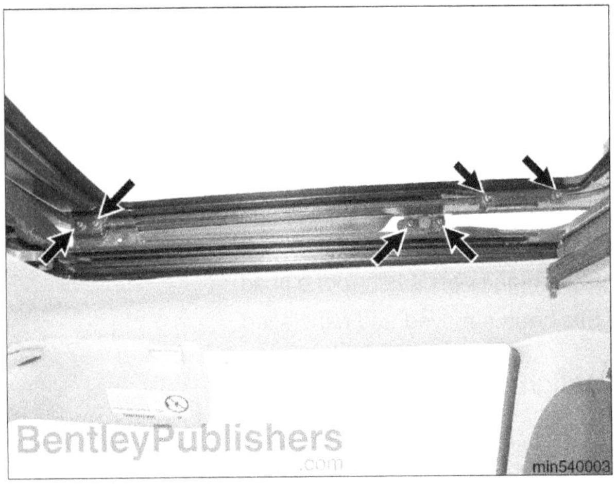

- Open sunroof so that Torx T20 mounting screws (**arrows**) are accessible.

- Adjust one side at a time:
  - Loosen right side screws so that sunroof panel is just able to move.
  - Adjust panel height: Press glass up or down at front or rear.
  - Adjust horizontal alignment: Slide glass forward or backward.
  - Tighten right side screws.

- Repeat on left side.

- Recheck gap and vertical alignment. Readjust if necessary.

Tightening torque	
Glass carrier to rails (front)	4 Nm (35 ft-lb)
Glass carrier to rails (rear and center) (T20 M6)	5 Nm (44 in-lb)

- Check sunroof function.

Sunroof Components

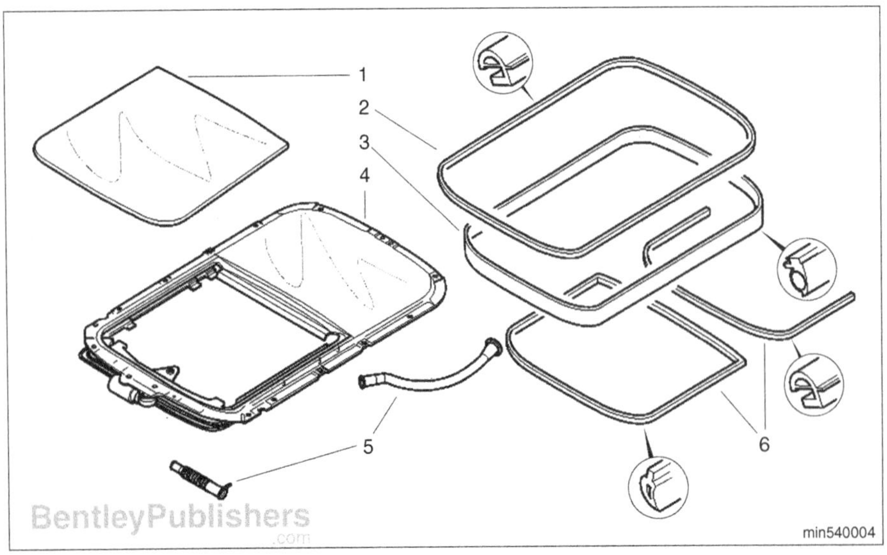

1. **Glass panel**
2. **Glass panel gasket**
3. **Sunroof frame gasket**
4. **Sunroof frame**
5. **Drain hose**
6. **Interior trim**

Sunroof glass panel, removing and installing

Working inside car, release sunroof blind from guide channels by twisting front rail as shown (**arrows**).

- Allow blind to retract as far as possible and allow to hang down.

> ***CAUTION—***
> *Do not allow grease on rail ends to make contact with interior trim.*

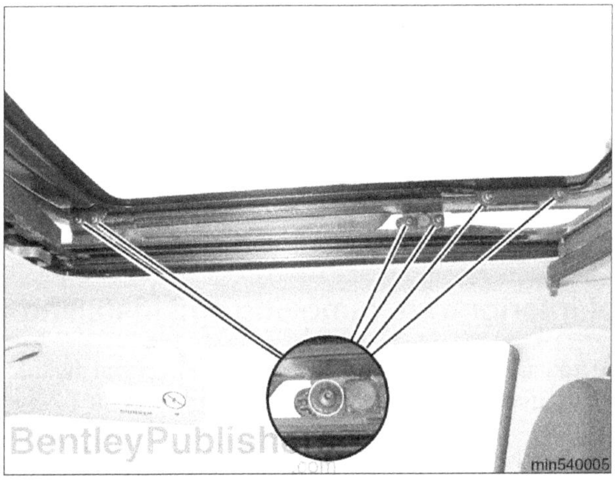

Open sunroof to TILT position so that Torx T20 mounting screws are accessible.

- Mark position of washers (**insert)** to facilitate correct sunroof alignment when installing.
- Remove and discard Torx screws securing glass panel to rear of mechanism.
- Remove and discard Torx screws securing glass lid to center of mechanism.
- Open sunroof sufficiently to gain access to front mounting screws. Remove and discard screws.
- Lift off glass panel.

- Installation is reverse of removal. Bear in mind the following:
  - Use new Torx mounting screws.
  - Align washers to marks made previously.

- Adjust sunroof. See **Sunroof panel, adjusting** in this repair group.

Tightening torque	
Glass carrier to rails (front)	4 Nm (35 ft-lb)
Glass carrier to rails (rear and center) (T20 M6)	5 Nm (44 in-lb)

- Initialize sunroof motor and check sunroof function.

Sunroof switch, removing and installing

Working inside vehicle at headliner, gently pry switch out of sunroof trim.

- Detach electrical harness connector.

- Installation is reverse of removal.
  - Initialize sunroof motor. See **Sunroof motor, initializing** in this repair group.
  - Check sunroof function.

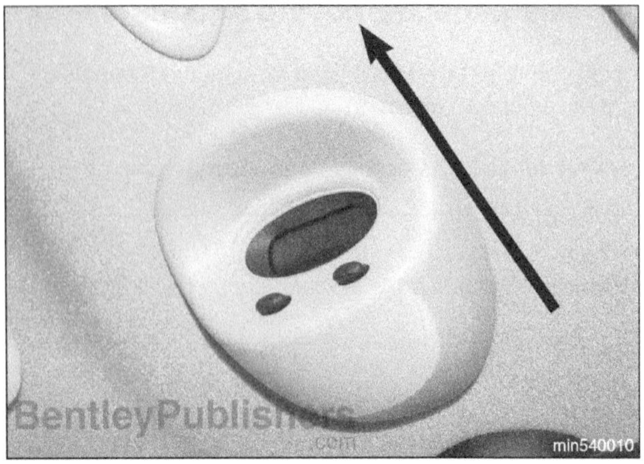

Sunroof motor, removing and installing

- Working inside vehicle, remove sunroof switch. See **Sunroof switch, removing and installing** in this repair group.
- Remove front sunroof interior trim. See **Sunroof interior trim, removing and installing** in this repair group.
- Push clock in headliner backward to remove, if applicable.
- Remove interior rear view mirror.
- Remove left and right sun visor center supports. Fold sun visors out to sides.

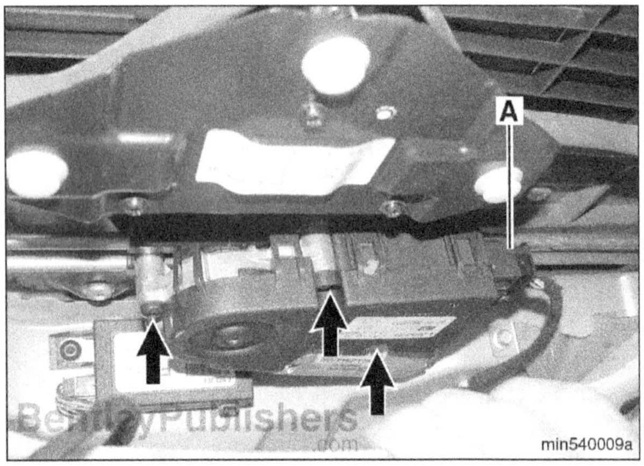

- Working above headliner opening:
  - Remove Torx T25 screws (**arrows**) mounting sunroof motor to sunroof assembly.
  - Detach motor and disconnect electrical harness connector (**A**).
- When installing, make sure motor drive gear and location pegs engage correctly before installing Torx screws.
- Recode motor using BMW scan tools DISplus, GT1 or MoDiC.

Sunroof interior trim, removing and installing

- If removing front trim, remove sunroof switch. See **Sunroof switch, removing and installing** in this repair group.
- Working inside vehicle, release sunroof blind and pull back.

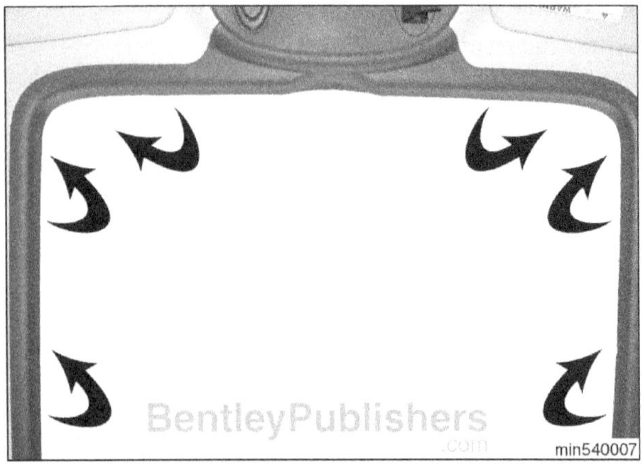

- Peel off sunroof interior trim and remove.
- When installing, make sure trim is correctly seated around headliner and sunroof frame.

Glass panel gasket, removing and installing

- Remove sunroof glass panel. See **Sunroof glass panel, removing and installing** in this repair group.

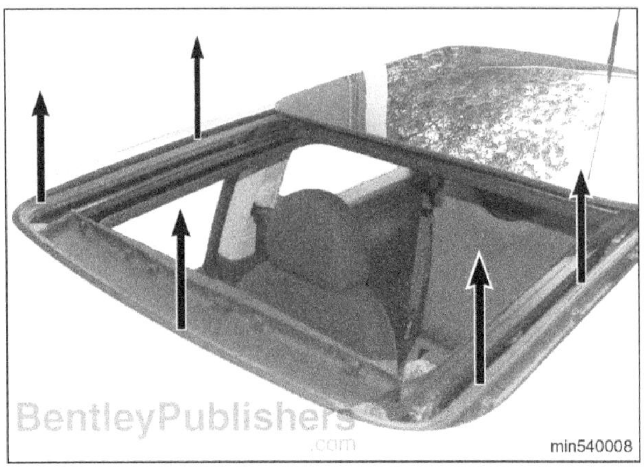

- Working above sunroof, peel gasket off sunroof assembly (**arrows**).
- When installing, begin at rear center.

> ***CAUTION—***
> *Do not pull seal too tightly during installation.*

541 Convertible Top

GENERAL

This repair group covers the automatic electro-hydraulic convertible top. Information on the rollover bar is also included. Two main versions of the convertible top have been installed with the primary difference being the inclusion of stowage or down-locks.

NOTE —
- *The convertible top control module (CVM) is located behind the left rear side trim panel. For the location of other convertible top electric components see* **610 Electrical Component Locations**.
- *The glass rear window in the Convertible is not replaceable separately.*

Convertible top overview

MC0654102

The MINI Convertible top system automatically opens and closes the top using hydraulic cylinders and electric motors. Integrated into the convertible top is a sliding sunroof that folds open a section of the top over the front seat passengers. A rollover bar behind the rear seat headrests complements the heavily reinforced windshield frame providing occupant protection in situations when the vehicle may not be sitting on the wheels.

The convertible top system features:
- Electro-hydraulic operation.
- Opening time of approximately 15 seconds.
- One-touch operation without manual unlocking.
- Integrated folding sunroof with variable opening to any position up to approximately 40 centimeters (15¾ inches).
- Sunroof operation up to a maximum road speed of approximately 75 mph (120 kph).
- Integral glass rear window with defrosting elements.
- Maintenance-free, self-bleeding hydraulic system.
- Emergency manual closing mode.
- Convenience opening mode for the convertible top (but not closing) via the remote control.
- Convenience opening and closing mode for the convertible top via the outside driver's door lock.
- Convertible Top Module (CVM) with fault memory storage and communications via K-bus.
- Top monitoring during operation using Hall sensors and micro-switches.

CONVERTIBLE TOP OPERATION

The convertible top is equipped with system interlocks that allow operation only when the vehicle is stationary. Due to the high electrical load, start engine and allow to idle where possible. The top can also be operated from the remote control or the door handle lock.

Opening

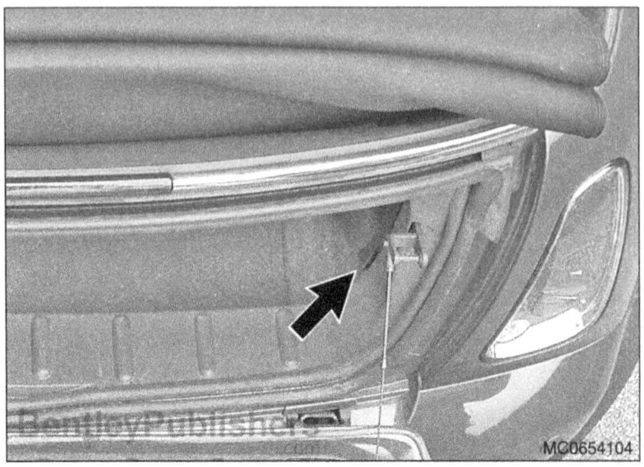

MC0654104

Lock the loading aid handles (**arrow**) and place the luggage compartment shelf behind the rear seat in the lower position. Right side handle shown, left side handle is similar.

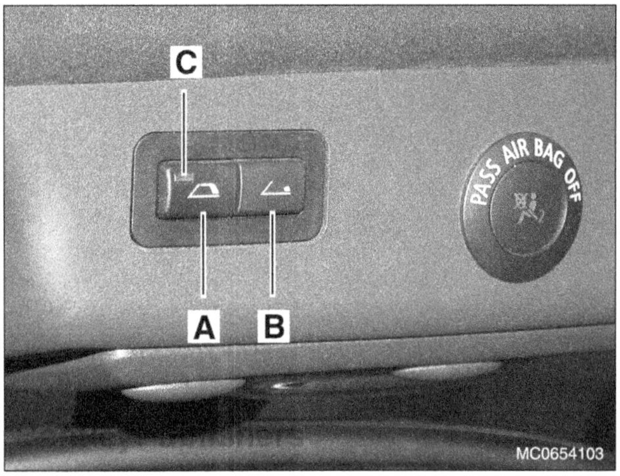

Lift clear protective cover (where equipped) of switch on upper windshield frame and push and hold sliding sunroof/convertible top button (**B**). Pressing and holding button a second time opens the convertible top.

- LED (**C**) will illuminate during operation and stay on until operation is completed.
- Front and rear side windows will automatically move down slightly when button (**B**) is pressed. Side windows will move up if button (**B**) is pressed upon completion of convertible top opening cycle and LED has gone out.
- If button (**B**) is released during cycle, top movement will stop, but will resume when button is again pressed.
- A convertible top cover is not required due to the self-storing nature of the top.

Closing

– Press and hold button (**A**) until convertible top and sliding sunroof are closed.

- If convertible top has been left folded (open) for an extended period of time, it may be necessary to manually assist the closing cycle as it nears completion.
- To accomplish this, unlock the loading aid and finish the cycle. Secure the loading aid when top is fully closed.

Loading aid

A convenient feature of the convertible top is the ability to manually unlock and raise the rear edge of the top approximately 6 inches (153 mm). When the loading aid is used, the rear shelf behind the rear seat is also lifted out of the way.

Moving the rear edge of the convertible top up and out of the way allows for easier access to the luggage compartment and can improve the serviceability of components located in rear areas. While this feature is primarily used when the convertible top is up, a limited amount of movement is also available when the top is down.

– Open tailgate.

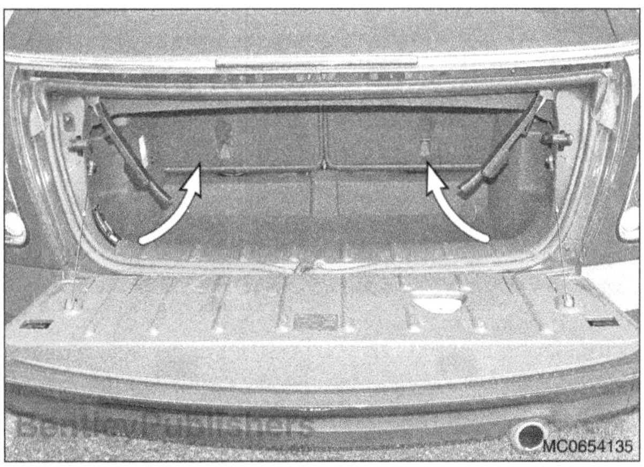

Pull handles on each side (**arrows**) out and up 90° (¼ turn) until they snap into position on the bottom of the convertible top.

– Rear edge of convertible top can now be lifted as needed.

– If necessary for loading, the loading aid position can be maintained.

While gently holding the convertible top up, pull the loading aid handles back down through 90° (¼ turn) and position ends on the tailgate latches as shown (**arrows**) to prop up the rear of the top.

– To return convertible top to normal position and operation, lift up on rear edge of top and pull handles on each side out and up through 90° (¼ turn) until they snap into position on the bottom of the convertible top.

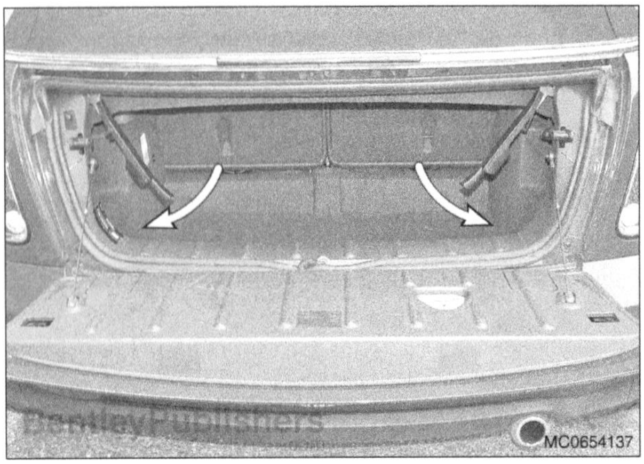

◄ Gently push down on rear edge of convertible top while pulling handles (**arrows**) down 90° (¼ turn) until they snap into position on the sides of the luggage compartment opening.

– Close tailgate.

Convertible top emergency operation

If a failure in the electro-hydraulic system prevents automatic closing of the convertible top, it can be closed manually in an emergency.

NOTE—
- *Only close the convertible top manually if it absolutely necessary. Do not open manually. Do not use manual emergency operation as a substitute for automatic operation.*
- *Emergency closing of the convertible top is best accomplished with two people.*
- *Improper handling can result in damage.*

– Lift forward edge of rear seat cushion slightly upward to release cushion from retainer clips. Pull cushion forward and up far enough to allow access to the underside.

◄ Locate and remove special wrench (**arrow**) attached to bottom of seat cushion.

NOTE—
Special wrench is a 4 mm hex key (Allen) wrench.

– Open tailgate and release loading aid handles.

– Lower side windows slightly.

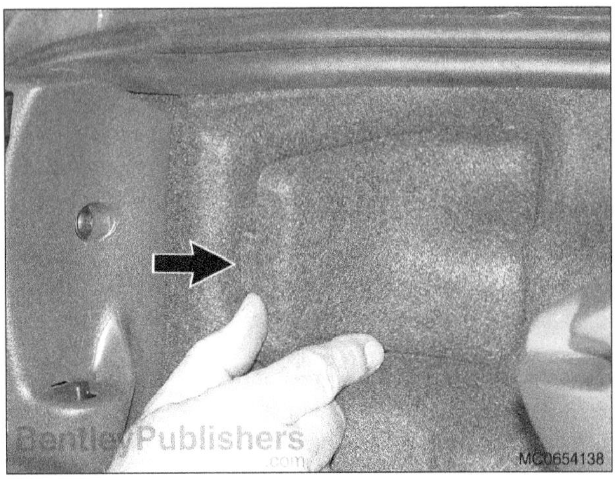

◄ Remove soft molded trim panel (**arrow**) from upper left side of luggage compartment by pulling off by hand or by gently prying. Removal of this panel allows access to hydraulic pump.

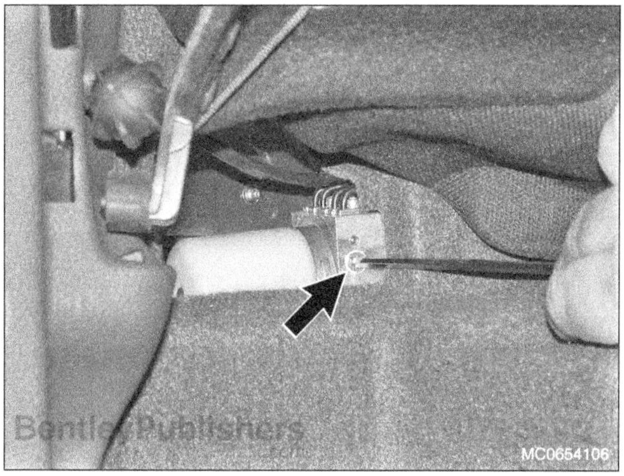

Using a screwdriver or hex key (Allen wrench) from the vehicle tool kit, turn valve screw (**arrow**) on the hydraulic pump 1½ turns counter-clockwise. Remove tool.

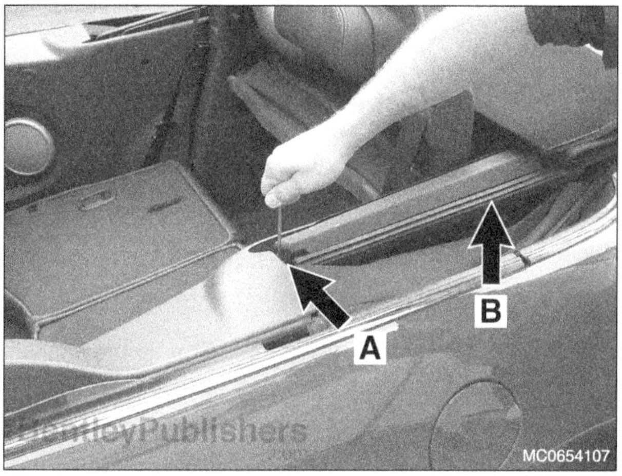

Depending on vehicle model year, it may be necessary to release convertible top from the down-locks.

- Press lever (**arrow A**) with hex key (Allen wrench) or screwdriver to release top and lift side frame up (**arrow B**).
- Hold frame slightly upward or insert a tool into frame to prevent lock from re-engaging.
- Repeat for other side.

With the help of a second person, use both hands on side frames and pull convertible top up and forward out of stowed position on each side (**arrows**). Pull forward to approximately the open sunroof position.

– Working inside the vehicle and with the convertible top at the open sunroof position, carefully pry off trim cover from middle of large center bow of convertible top frame. Unclip forward section of cover and pivot to the rear to remove.

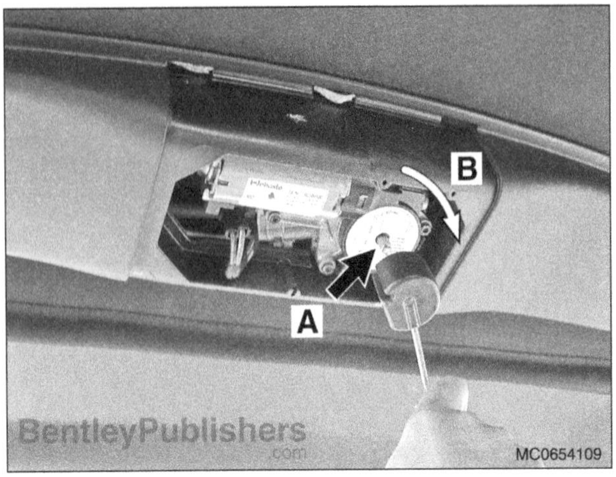

◄ Insert special wrench or 4 mm hex key (Allen wrench) into position in socket (**arrow A**) in middle of mechanism as shown. Be sure hex key is fully inserted into opening in mechanism to insure proper uncoupling of sliding sunroof motor from gearbox. Hex key will move upwards.

– Turn hex key clockwise (**arrow B**) until convertible top is closed and locked in place.

– from mechanism and stow in seat cushion. Slide rear seat cushion back into position.

– Snap removed trim cover back into middle of large center bow of convertible top frame.

– Lock loading aid and close luggage compartment lid. Be sure that soft trim panel is in vehicle.

– Close valve on hydraulic pump.

Hydraulic fluid level, checking or filling

The convertible top hydraulic system is filled with special life-time hydraulic fluid and does not require changing.

– Position vehicle on a level horizontal surface.

– Lower convertible top fully.
  - Both hydraulic cylinders must be fully extended for accurate fluid level measurement.

– Open tailgate and release loading aid.

– Working inside luggage compartment, remove left side trim panel to access hydraulic pump.

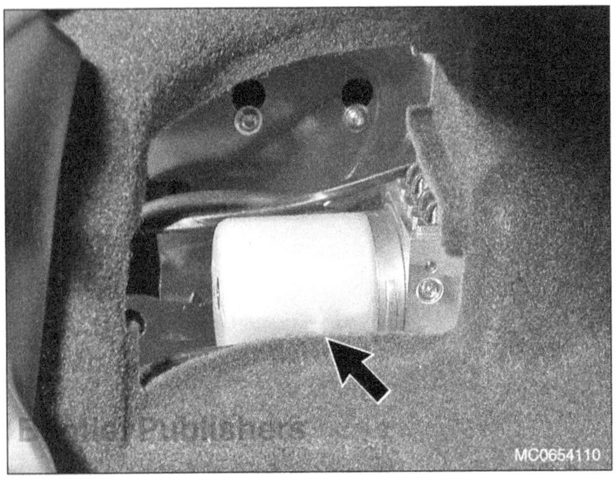

◄ Hydraulic fluid level is correct when fluid level is visible within the + markings on the transparent pump reservoir (**arrow**). Vehicle must be level.
  - If level is not correct, remove hydraulic pump and add or extract fluid as needed.

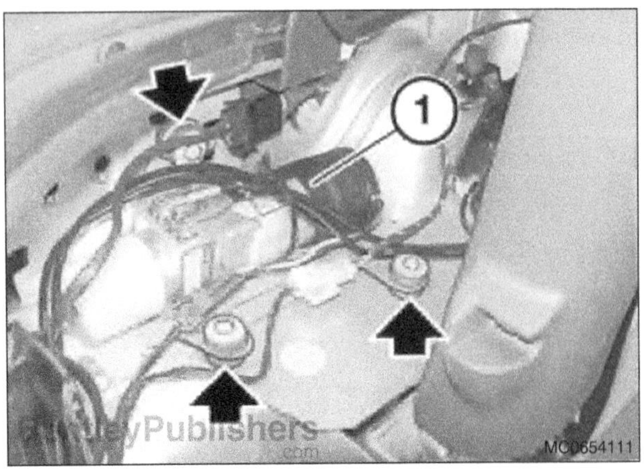

◄ Without disconnecting hydraulic lines, remove mounting bolts (**arrows**).

– Carefully position hydraulic pump (**1**) away from wiring harnesses and position vertically so that reservoir end is up.

– Protect surrounding area and components and take care to avoid hydraulic fluid spillage.

MC0654113

Remove filler plug (**arrow**) with a 4 mm hex key (Allen wrench).

- To fill: pour in appropriate quantity of hydraulic fluid.
- To extract: tilt hydraulic pump and pour out appropriate quantity of fluid.
- Position hydraulic pump horizontally as often as needed to check proper fluid level.

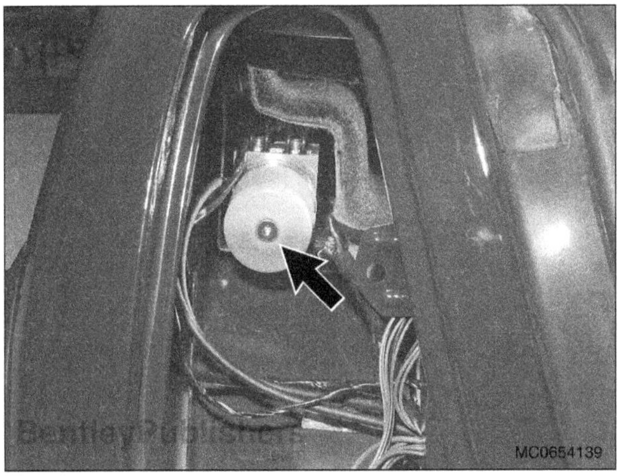

MC0654139

Limited access to the pump reservoir filler plug (**arrow**) can also be obtained by removing the left tail light. Adjust/correct fluid level as appropriate.

NOTE—
If servicing hydraulic fluid with pump/reservoir in the horizontal position, use extra care to avoid spilling hydraulic oil on painted surfaces and soft trim. Place clean shop towels under reservoir to catch overflow and/or spillage.

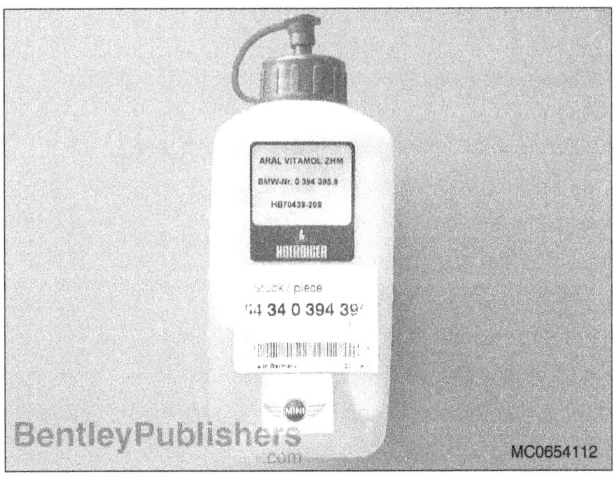

MC0654112

If fluid level is low and needs to be topped up or refilled, use only specified MINI hydraulic oil. Typical package is 250 ml container.

Convertible top hydraulic fluid	
MINI hydraulic fluid	54 34 0 394 395
ARAL hydraulic fluid	VITAMOL ZHM

- When hydraulic fluid is at desired level, install filler plug.
- Work hydraulic pump back into position and tighten bolts.

Tightening torque	
Hydraulic pump mounting bolts	7.6 Nm (67 in-lb)

- If hydraulic system had been opened or if air ingress is suspected, operate convertible top through ten complete cycles to ensure complete self-bleeding of system.
- Recheck hydraulic fluid level and adjust level as required.
- Install left side rear trim panel.

Function checking

Convertible top diagnosis can only be accomplished when normal function conditions have be eliminated as possible sources of any malfunction.

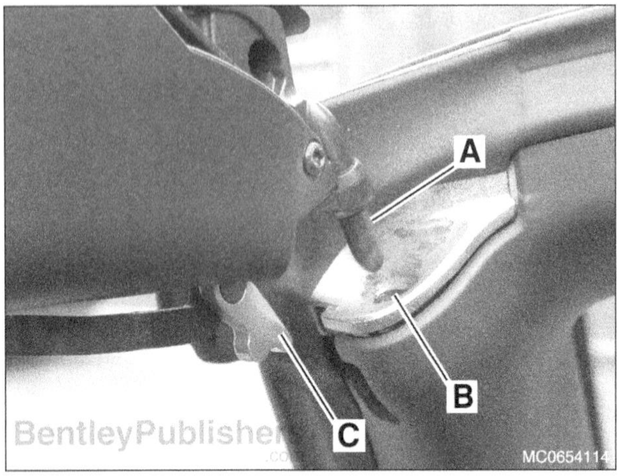

◀ Convertible top is centered by locating pin (**A**) during insertion into striker guide (**B**). Locking is by rotary latch (**C**) on frame.

– Carefully close convertible top until rotary latch (**C**) has moved approximately 1/3 of the way into striker.
  - Rotary latch (**C**) and pin (**A**) must align, close correctly and be virtually silent.
  - Adjust convertible top as needed.

– Open and close convertible top at least three times.
  - Check all hydraulic lines and connections for leakage and correct as required.
  - Check and/or fill hydraulic fluid as needed.

Function conditions, folding roof (sunroof)

◀ Opening (**arrow**) and closing of folding roof is only possible when:
- Battery voltage is sufficient; run engine if necessary.
- Tension brackets are lowered and locked.
- Speed is less than 75 mph (120 kph).
- Outside temperature is between 14° F and 176°F (-10°C and 80°C).
- Ignition switch on (engine off or running).
- Control switch in windshield frame depressed.

Function conditions, convertible top

◀ Opening (**arrow**) and closing of convertible top is only possible when:
- Battery voltage is sufficient; run engine if necessary.
- Folding roof is in proper working condition.
- Front side windows are lowered slightly.
- Rear side windows are lowered completely.
- Tension brackets are lowered and locked.
- Stowage shelf is engaged in lower position.
- Road speed is less than 2.5 mph (4 kph).
- Outside temperature is between 14° F and 176°F (-10°C and 80°C).
- Ignition switch on (engine off or running).
- Control switch in windshield frame depressed.

Convertible Top Components

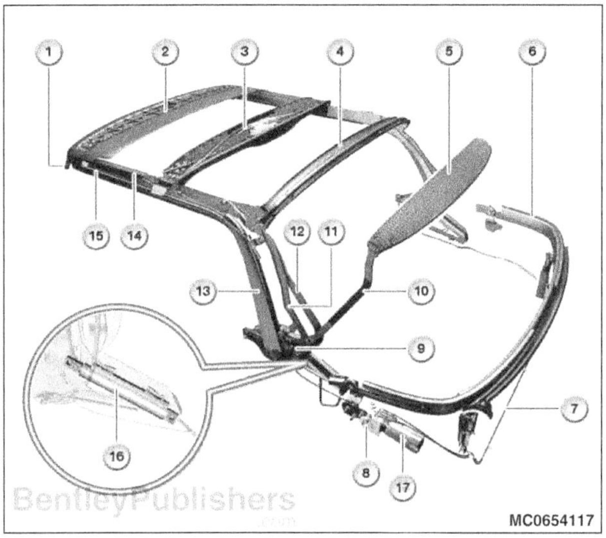

MC0654117

Convertible top frame components

1. Catch hook pin
2. Front bow strut
3. Folding sunroof frame with sunroof and locking mechanism drive
4. Middle bow strut
5. Rear bow strut
6. Tensioning bar
7. Hydraulic lines
8. Hydraulic pump unit with emergency closure valve
9. Main bearing
10. Rear bow strut arm
11. Main pillar
12. Main pillar arm
13. Side seal brace
14. Folding sunroof guide rail
15. Tension cable
16. Hydraulic cylinder (double acting)
17. Hydraulic pump fluid reservoir

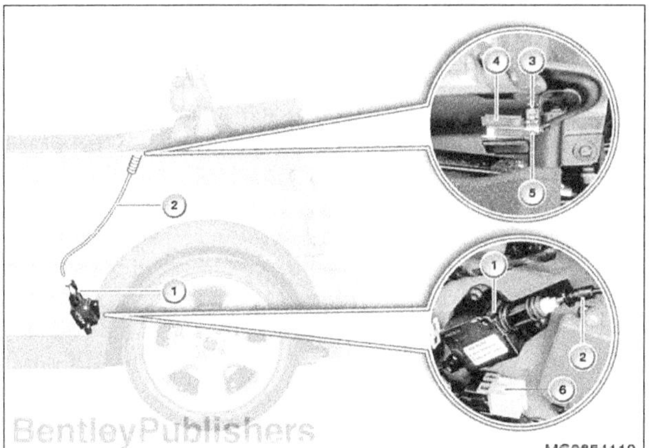

MC0654118

Stowage lock drive (from 09/2005)

1. Stowage lock drive motor
2. Bowden cable
3. Latch spring
4. Latch (down-lock)
5. Microswitch
6. Stowage lock drive motor relay

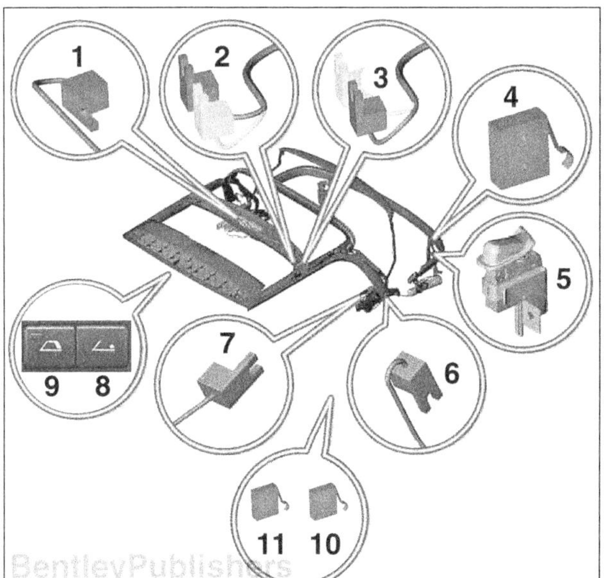

MC0654119

Sensors and switches

1. Hall sensor for folding sunroof - closed position
2. Hall sensor for folding sunroof - open position
3. Hall sensor for catch hook pin - catch hook pin unlocked
4. Microswitch for tensioning bar
5. Microswitch for rear shelf
6. Hall sensor for convertible top - closed position
7. Hall sensor for convertible top - open position
8. Switch for convertible top and folding sunroof - open
9. Switch for convertible top and folding sunroof - close with LED indicator
10. Microswitch for stowage lock - left (from 09/2005)
11. Microswitch for stowage lock - right (from 09/2005)

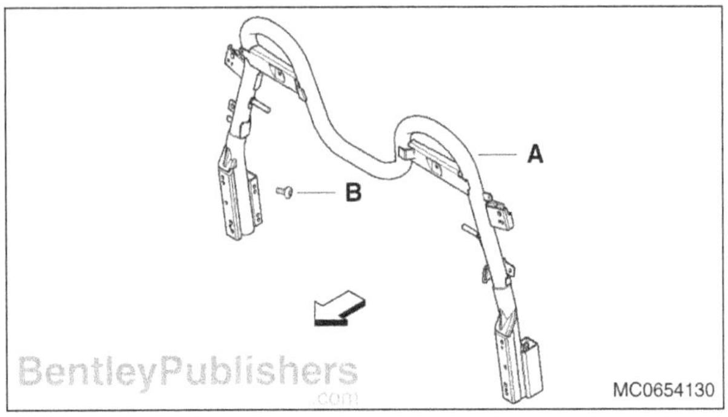

Rollover bar

A rollover bar is attached to the body at the sides and provides additional chassis rigidity and protection. **Arrow** points to front of vehicle.

A. Rollover bar

B. Torx screw(s)

- Always replace
- Tighten to 18 Nm (13.3 ft-lb)

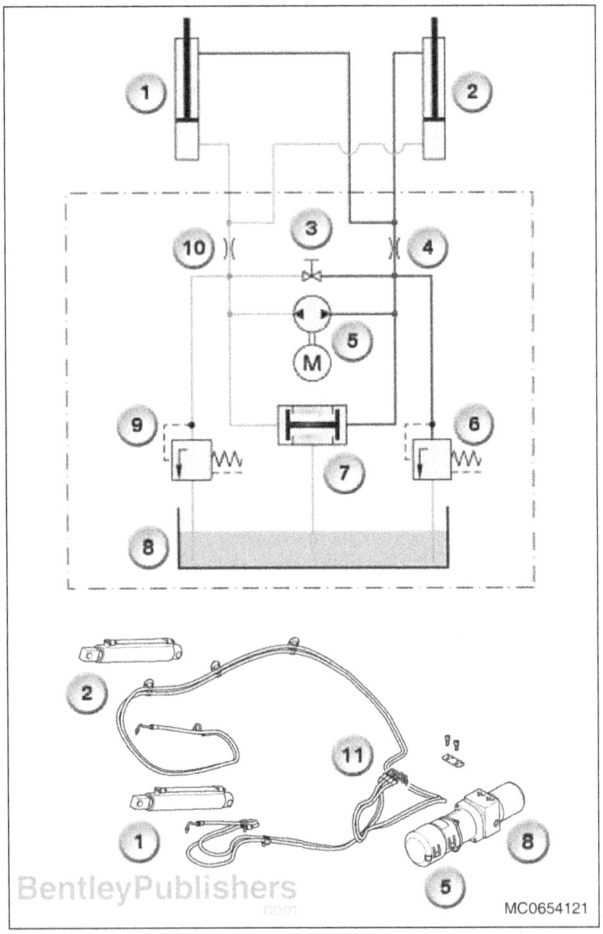

Hydraulic system overview

The convertible top hydraulic system is maintenance free and periodic replacement of fluid is not required. The system will self-bleed during normal use.

An emergency closure valve is built into the pump valve mechanism to disconnect the motor and pump from the system in case of a malfunction. Opening the valve allows fluid flow that is independent of the pump which allows manual movement of the two hydraulic cylinders.

The two hydraulic cylinders are activated together. One set of lines is under pressure during the opening cycle with the other line set used as a return. When the top is closed, the internal switchover valve directs fluid in the opposite direction and the lines used previously for the return become the pressure lines and the previous pressure lines become the return. Switchover is a mechanical function of the pump, but the pump is electronically controlled by the convertible top module (CVM) acting in response to signals from the sensors and switches.

The hydraulic pump is protected against overloading and overheating. A counter in the CVM counts up at the rate of 1 count per second when the pump is switched on and pump valving switches down at the rate of 1 count per 20 seconds when switched off. If the counter exceeds 100, the system counts down to 100 and remains in a sleep mode for approximately 20 minutes. If the system has overheated, the convertible top may only operate to complete a closing cycle and in one direction only.

Counter memory and timer readings are retained with the ignition switched off. They are also retained if the battery is disconnected and cannot be erased by this method.

Components within the dashed line area are components of the convertible top drive and are integral to it.

1. Hydraulic cylinder, left
2. Hydraulic cylinder, right
3. Emergency closure valve
4. Line restrictor left
5. Convertible top drive motor (**M**) with hydraulic pump
6. Pressure limiter valve, right
7. Switchover valve
8. Hydraulic pump fluid reservoir
9. Pressure limiter valve, left
10. Line restrictor, left
11. Hydraulic line set

Electronic Controls

Two convertible top versions are installed in MINIs. Both versions have similar electronic components and functions. Applicable differences are noted with an asterisk (*).

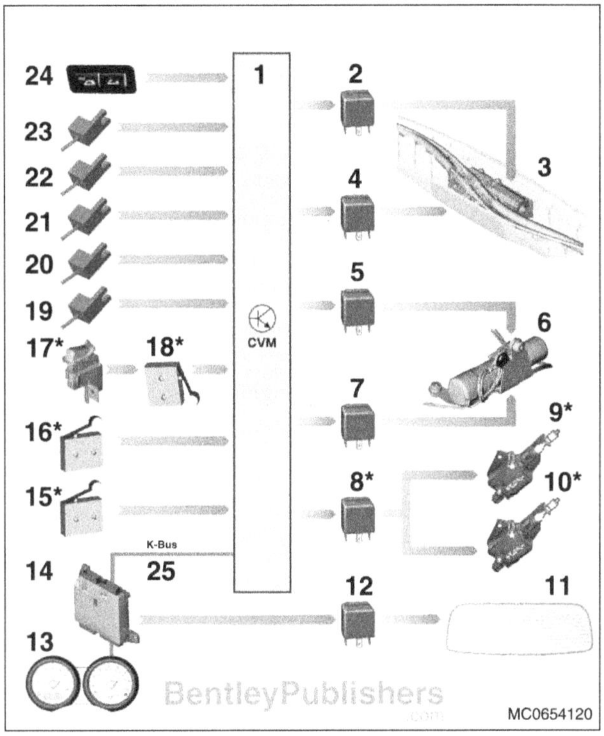

Electronic component overview

1. Convertible top module (CVM)
   - Location: Left rear inner wheel housing, behind trim panels
2. Relay 1 for locking mechanism and folding sunroof
3. Drive motor for locking mechanism and folding sunroof
4. Relay 2 for locking mechanism and folding sunroof
5. Relay 1 for convertible top drive
6. Convertible top drive with motor, hydraulic pump, reservoir, mechanical valve unit, and emergency closure valve.
7. Relay 2 for convertible top drive
8. Stowage lock relay (from 09/2005 only)
9. Stowage lock drive motor, left (from 09/2005 only)
10. Stowage lock drive motor, right (from 09/2005 only)
11. Heated rear window glass
12. Heated rear window glass relay
13. Instrument cluster
14. General module (GM)
15. Microswitch for stowage lock, left (from 09/2005 only)
16. Microswitch for stowage lock, right (from 09/2005 only)
17. Microswitch for rear shelf (through 08/2005 wired with separate inputs, not in series as shown)
18. Microswitch for tensioning bar (through 08/2005 wired with separate inputs, not in series as shown)
19. Hall sensor for convertible top - open position
20. Hall sensor for convertible top - closed position
21. Hall sensor for folding sunroof - closed position
22. Hall sensor for folding sunroof - open position
23. Hall sensor for catch hook pin
24. Control switches for convertible top and folding sunroof with LED indicator
25. K-Bus data communication line

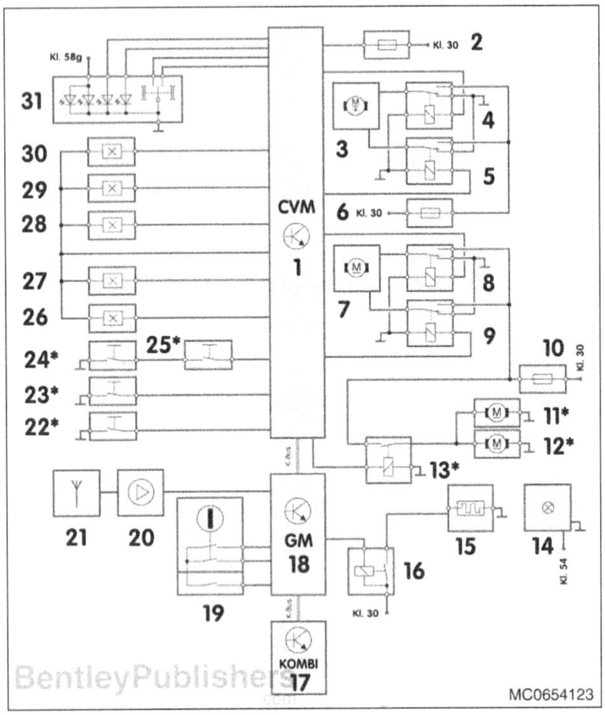

System circuit diagram

Two convertible top versions are installed in MINIs. Both versions are wired in a similar manner as shown in this functional wiring diagram. Applicable differences are noted with an asterisk (*).

1. Convertible top module (CVM)
2. Terminal 30 connection (fused)
3. Drive motor for locking mechanism and folding sunroof
4. Relay 1 for locking mechanism and folding sunroof
5. Relay 2 for locking mechanism and folding sunroof
6. Terminal 30 connection (fused) for relays and drive motor for locking mechanism and folding sunroof
7. Convertible top hydraulic pump drive motor
8. Relay 1 for convertible top drive
9. Relay 2 for convertible top drive
10. Terminal 30 connection (fused) for relays and convertible top hydraulic pump drive motor
11. Stowage lock drive motor, left (from 09/2005 only)
12. Stowage lock drive motor, right (from 09/2005 only)
13. Stowage lock relay (from 09/2005 only)
14. High-mounted brake light in convertible top
15. Heated rear window glass
16. Heated rear window glass relay
17. Instrument cluster (KOMBI)
18. General module (GM)
19. Door lock cylinder, driver's side
20. Antenna amplifier
21. Antenna
22. Microswitch for stowage lock, left (from 09/2005 only)
23. Microswitch for stowage lock, right (from 09/2005 only)
24. Microswitch for rear shelf (through 08/2005 wired with separate inputs, not in series as shown)
25. Microswitch for tensioning bar (through 08/2005 wired with separate inputs, not in series as shown)
26. Hall sensor for convertible top - open position
27. Hall sensor for convertible top - closed position
28. Hall sensor for folding sunroof - closed position
29. Hall sensor for folding sunroof - open position
30. Hall sensor for catch hook pin
31. Control switches for convertible top and folding sunroof with LED indicator

K-Bus. K-Bus data communication line

Kl. 30. Terminal 30 connection, direct battery positive

Kl. 54. Terminal 54 connection, from brake light switch

Kl. 58g. Terminal 58g connection, from instrument panel lights

Body Dimensions

The body dimensions identified in the following illustrations should be checked and adjusted as required before making adjustments to the convertible top.

◄ A plastic feeler gauge prevents damage to painted surfaces and is suitable for measuring panel gaps.

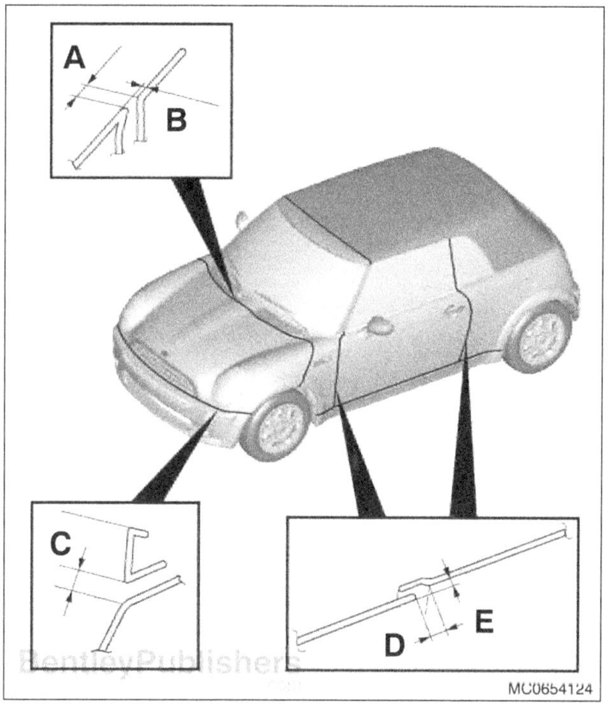

Body gaps, front and side

- **A.** 4.0 mm ± 0.75 mm
- **B.** 0.75 mm ± 0.75 mm
- **C.** 6.0 mm ± 1.0 mm
- **D.** 4.0 mm ± 0.8 mm
- **E.** 0.5 mm ± 0.5 mm

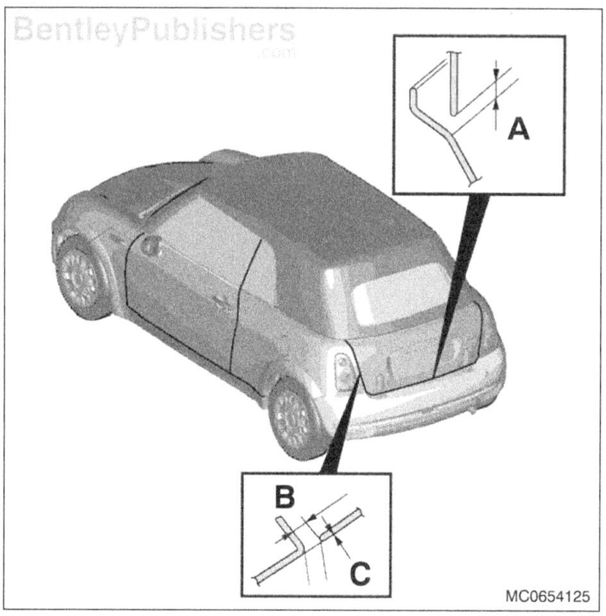

Body gaps, rear

- **A.** 5.0 mm ± 1.3 mm
- **B.** 4.0 mm ± 0.8 mm
- **C.** 0.0 mm ± 0.5 mm

Rear power window

Rear power window regulator, removing and installing

Removing

- Remove rear seat bottom and backrest.
- Open convertible top.

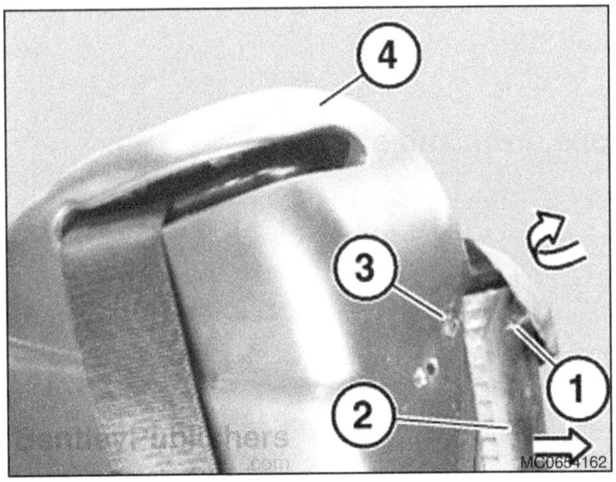

MC0654162

- Unclip rear side trim clip (**1**).
- Detach molded trim (**2**) in direction of arrow from side trim (**4**).
- Remove fastener (**3**).

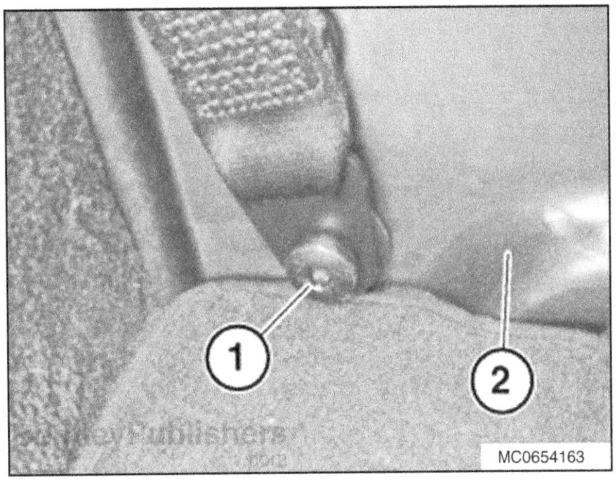

MC0654163

- Remove seat belt fastener (**1**) on entrance cover strip (**2**).

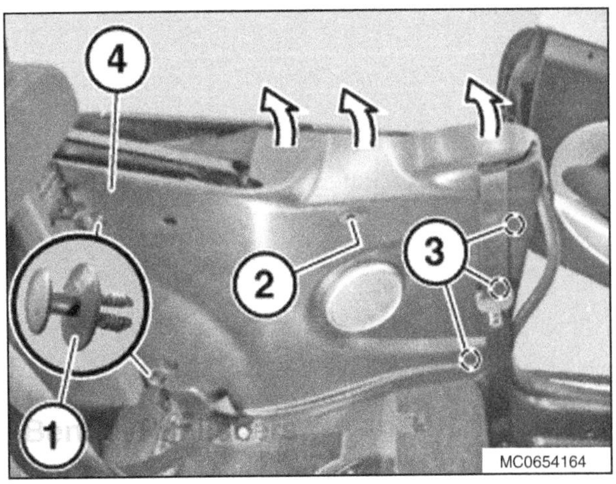

MC0654164

- Lever out expansion rivets (**1**). Release screw (**2**).
- Unclip clips (**3**) and remove side trim (**4**) inwards and upwards.

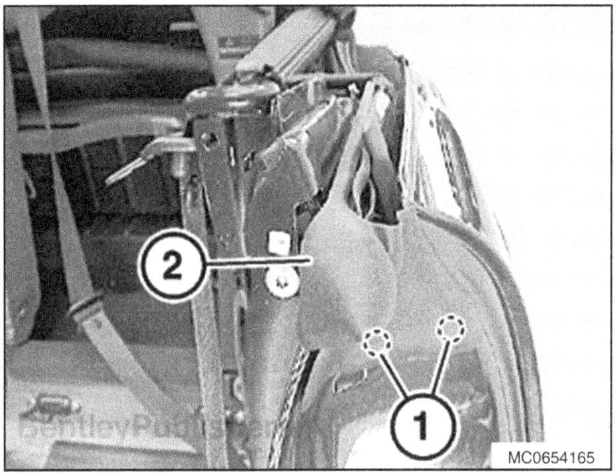

Lever out catches (**1**) on inner window cavity cover strip (**2**).

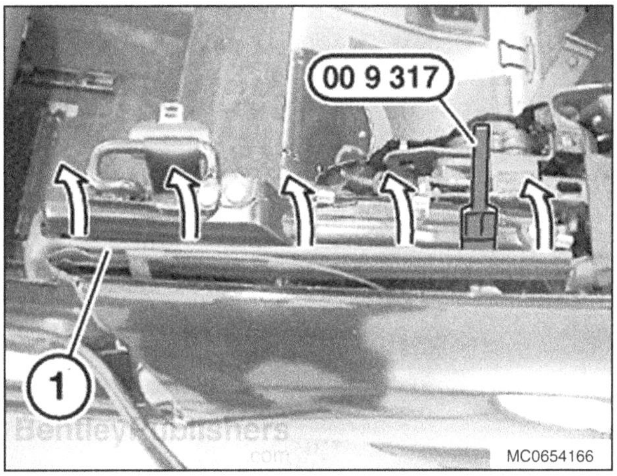

Carefully lever out window cavity cover strip (**1**) with MINI special tool **00 9 317** towards top.

– Remove outer window cavity cover strip.

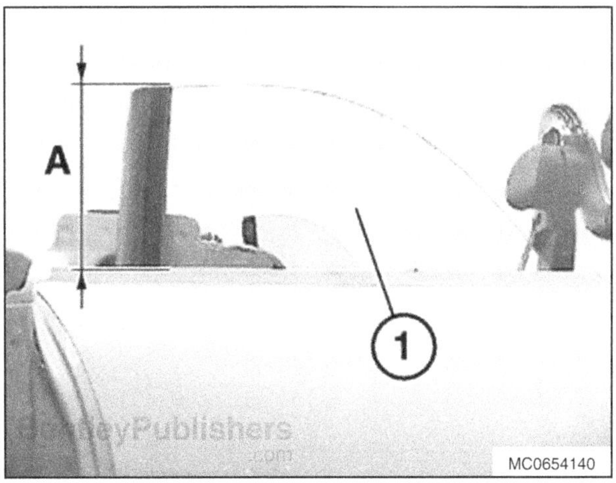

Open side window (**1**) to approximately 150 mm (**A**).

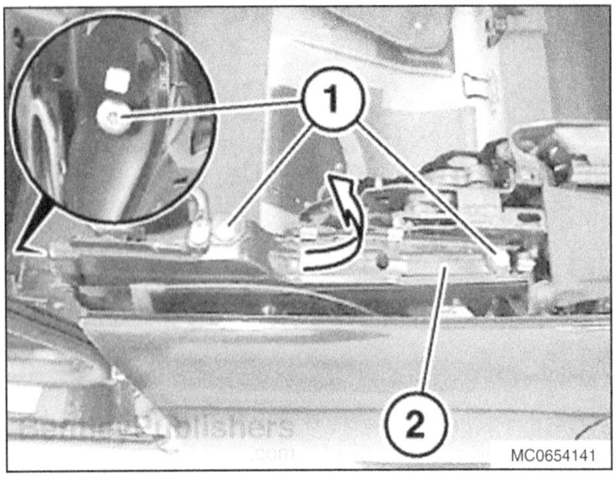

Remove fasteners (**1**) and remove connecting bracket (**2**).

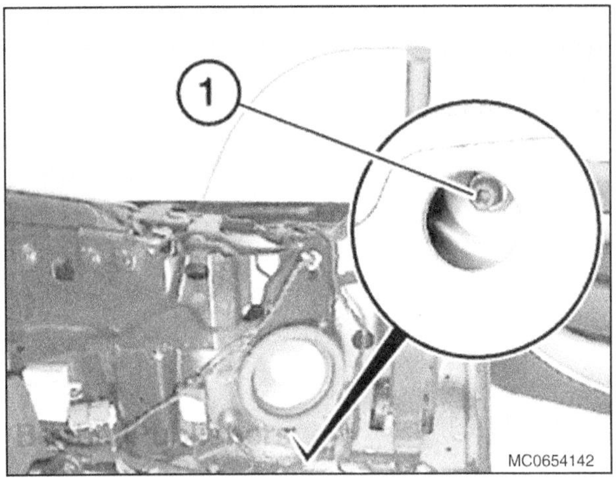

Loosen fastener (**1**), do not remove.

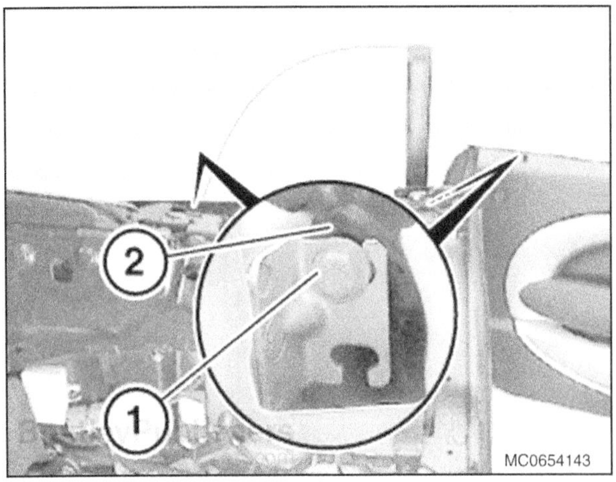

Loosen fastener (**1**) on power window regulator (**2**), do not remove.

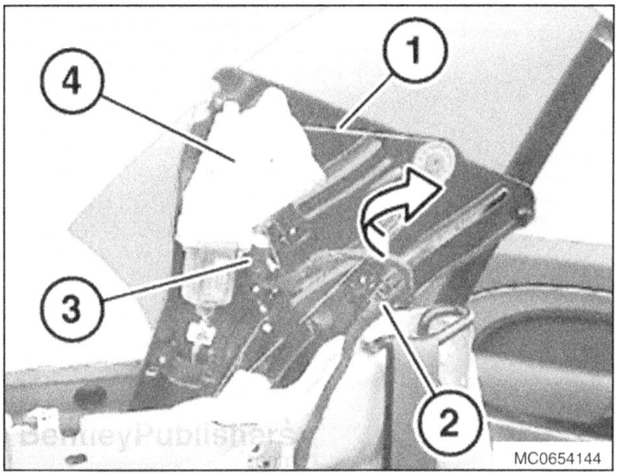

MC0654144

- Pull out power window regulator (**1**) towards top slightly.
- Lift out cable holder (**2**) from power window regulator (**1**).
- Disconnect plug connection (**3**) on motor (**4**).
- Remove power window regulator (**1**).

Installing

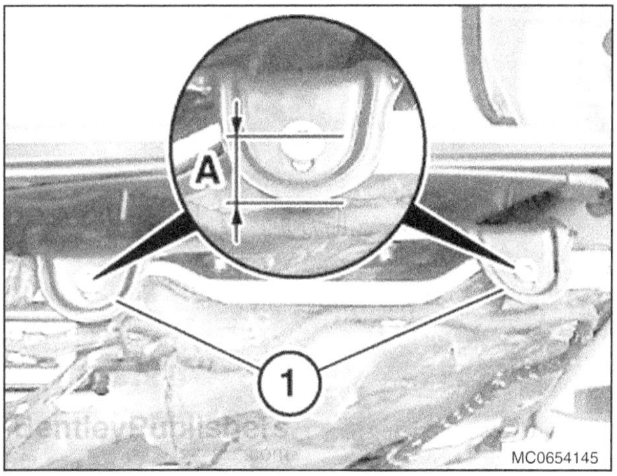

MC0654145

- Preset dimension (**A**) to 18.5 mm at power window regulator (**1**).

> ***CAUTION—***
> Do not operate window regulator before setting dimension **A**

- Failure to preset dimension (**A**) to 18.5 mm at power window regulator (**1**) could result in glass breakage.

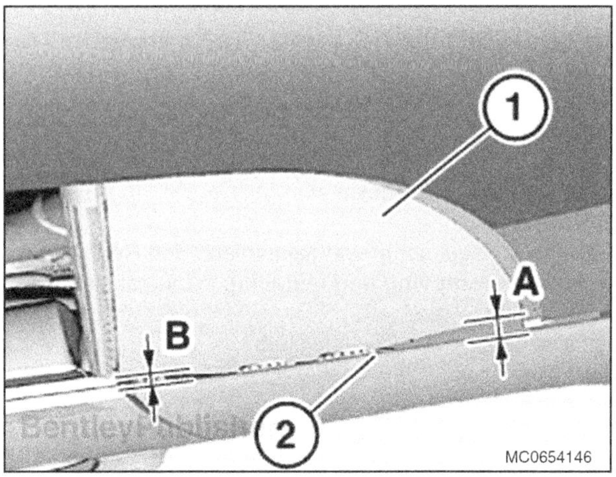

MC0654146

- Close window completely. Check gap dimension (**A** and **B**) between side window glass (**1**) and flange on side panel (**2**) and adjust if necessary.
  - Distance (**A**) = 16 mm
  - Dimension (**B**) = 10 mm

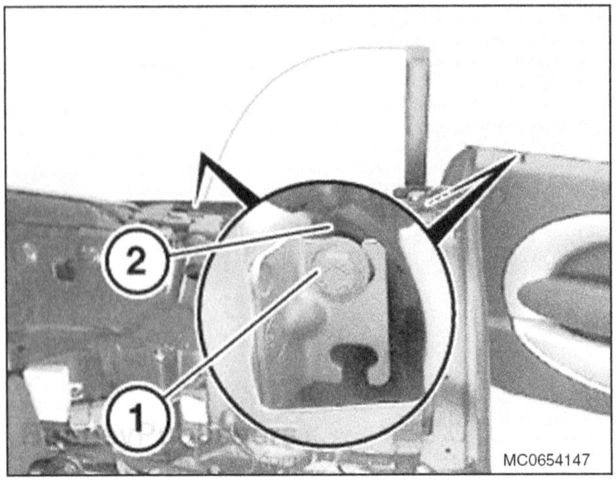

To adjust, loosen (**1**) on power window regulator (**2**).

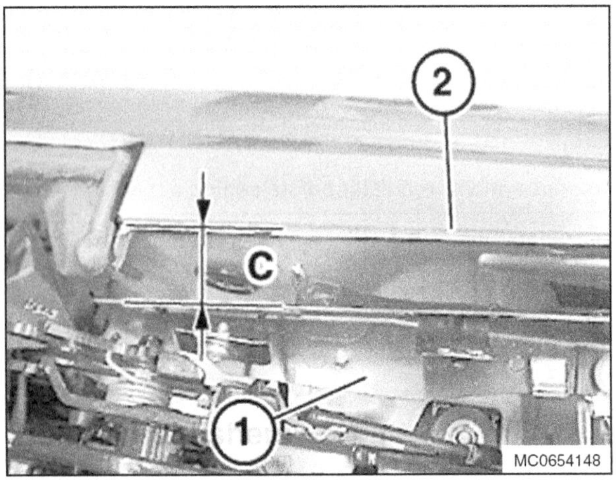

Open soft top and soft top compartment lid. Fit window cavity cover (**1**). Preset dimension (**C**) between window cavity cover (**1**) and flange on side panel (**2**).

- Dimension (C) = 24 mm

– Remainder of installation is reverse of removal, remember to adjust rear window glass before operating motor or glass breakage could occur. See **Rear power window, adjusting** in this repair group.

Tightening torque	
Motor to window regulator	3 Nm (26.5 in-lb)
Window glass to regulator	12 Nm (8.8 ft-lb)
WIndow regulator to body	8 Nm (5.9 ft-lb)
Connecting bracket to body	8 Nm (5.9 ft-lb)
Seat belt to body	36 Nm (26.5 ft-lb)

Rear power window regulator, adjusting

This procedure should be performed with vehicle on flat level surface. Do not raise of jack or hoist.

– Remove outer window cavity cover strip

– Check rear window regulator preset adjustment. See **Rear power window regulator, removing and installing** in this repair group.

– Check that front door is correctly adjusted.

– Check that convertible top is correctly adjusted.

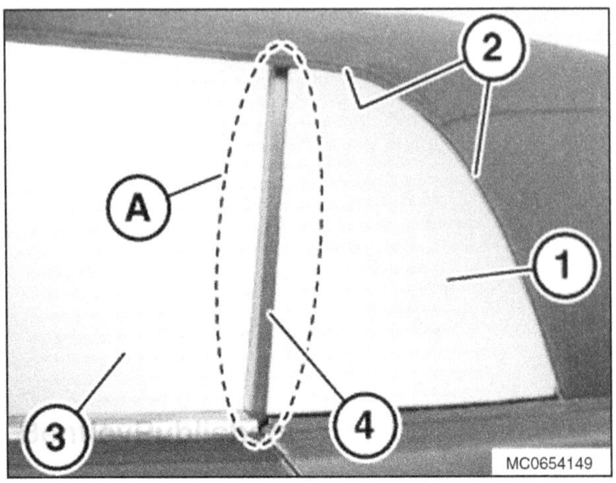

Close convertible top.

- Side window glass (**1**) should rest in seal (**2**) over entire width.
- Door window glass (**3**) should rest in area (**A**) on side window seal (**4**).

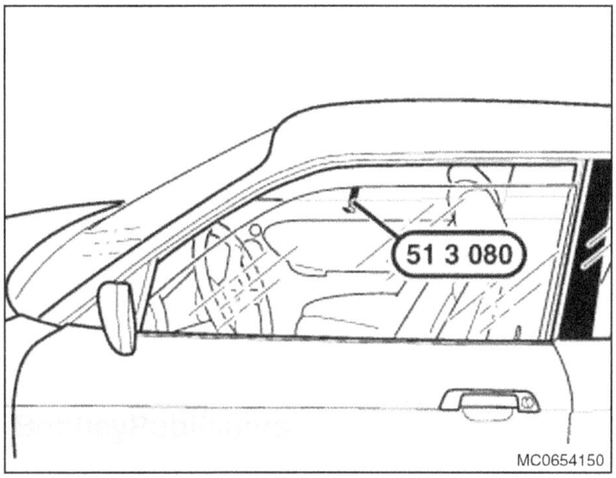

Vertical adjustment (retraction depth)

- Close convertible top.
- Open door window.
- Attach MINI special tool **51 3 080** to side window glass.
- Close side window completely.

NOTE —
- *3 items of MINI special tool **51 3 080** are required for rapid adjustment of retraction depths.*

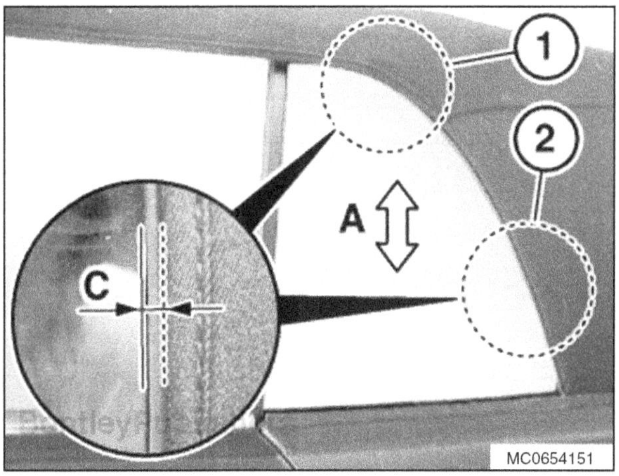

- Measure retraction depth (**A**) at areas (**C**), 6 ± 1.5 mm.
- Measure retraction depth in areas (**1** and **2**), 6 ± 1.5 mm.

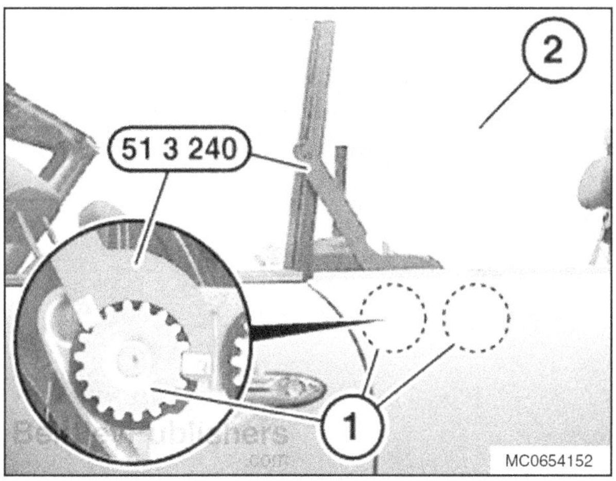

- Slacken washers (**1**) of side window (**2**) with MINI special tool **51 3 240** (do not release fully).
- Adjust height (**A**) of side window (**2**).
- Tighten down washers (**1**) with MINI special tool **51 3 240**.

Tightening torque	
Washers to regulator	12 Nm (8.8 in-lb)

Parallelism (B) (window tilt)

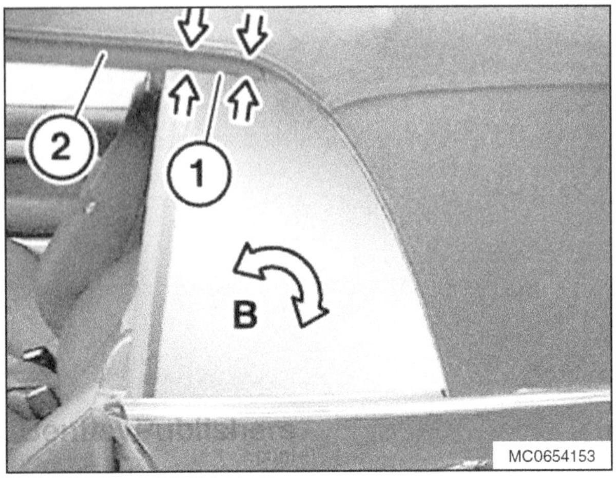

- Side window top edge (**1**) must be parallel to contour of seal (**2**).
- Close side window glass and mark along seal (**2**) at bottom and top with a water-soluble marker pen.
- Clean side window.

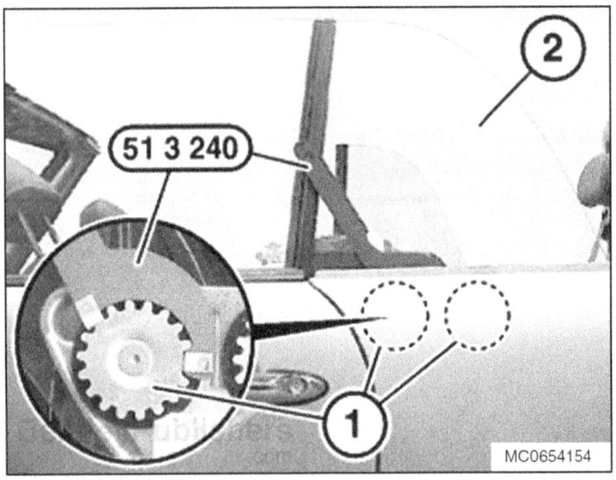

- Slacken washers (**1**) of side window (**2**) with MINI special tool **51 3 240** (do not release fully).
- Adjust side window (**2**) by turning (**B**).
- Tighten down washers (**1**) with special tool **51 3 240** .

Tightening torque	
Washers to regulator	12 Nm (8.8 in-lb)

Longitudinal adjustment (C)

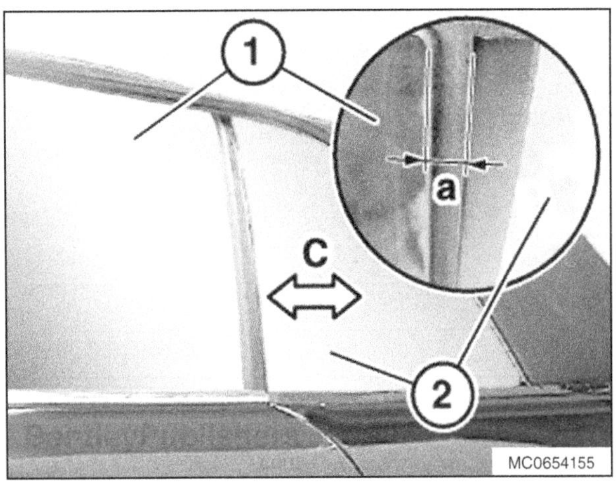

Ideal adjustment:

- Distance (**a**) of door window glass (**1**) to strip (**2**) of side window:
- Dimension (**a**) = 8.8 ± 1.5 mm

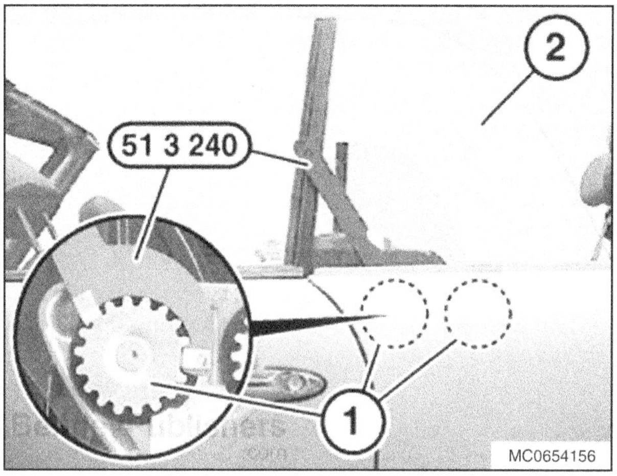

Slacken washers (**1**) of side window (**2**) with MINI special tool **51 3 240** (do not release fully).

– Adjust side window (**2**) by turning (**C**).

– Tighten down washers (**1**) with MINI special tool **51 3 240**.

Tightening torque	
Washers to regulator	12 Nm (8.8 in-lb)

Gap between side window and side panel

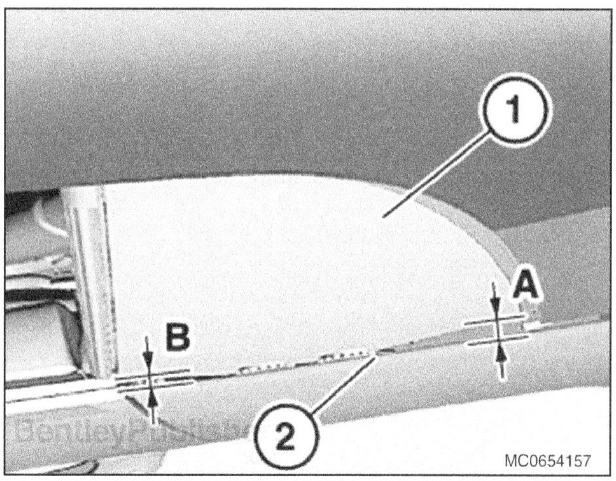

Gap between side window and side panel:

– Gaps (**A** and **B**) between side window (**1**) and side panel (**2**):
- Measurement (**A**) = 10 mm
- Measurement (**B**) = 16 mm

NOTE —
- *Side window glass must run into seal when soft top is closed.*

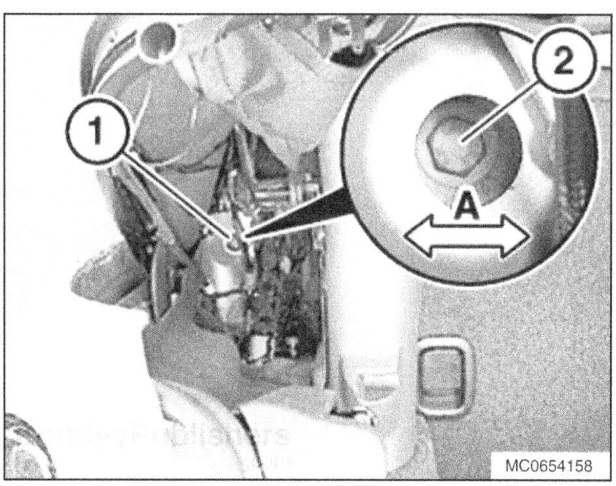

Open EasyLoad, luggage compartment. Remove plug (**1**).

– Slacken screw (**2**) underneath.

– Adjust measurement (**A**) by moving side window glass.

Tightening torque	
WIndow regulator to body	8 Nm (5.9 ft-lb)

NOTE —
- *If setting is not achieved: adjust power window regulator if necessary. See* **Rear power window regulator, removing and installing** *in this repair group.*

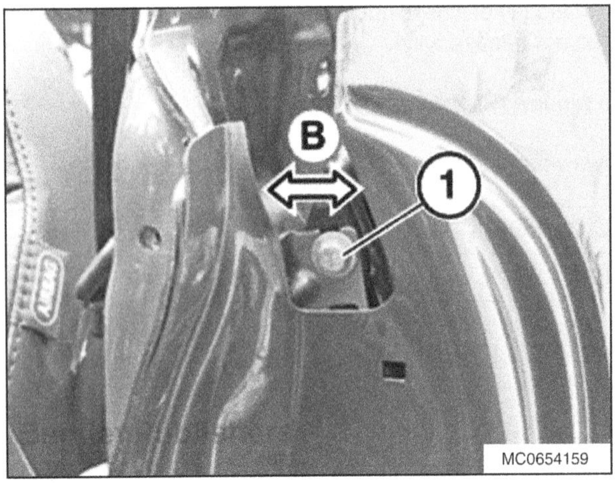

Raise door pillar sealing cloth from inside window cavity cover strip interior. Slacken screw (**1**) underneath. Adjust measurement (**B**) by moving side window glass.

Tightening torque	
WIndow regulator to body	8 Nm (5.9 ft-lb)

NOTE—
- *If setting is not achieved: adjust power window regulator if necessary. See* **Rear power window regulator, removing and installing** *in this repair group.*

Pre-tension (D) of side window inclination

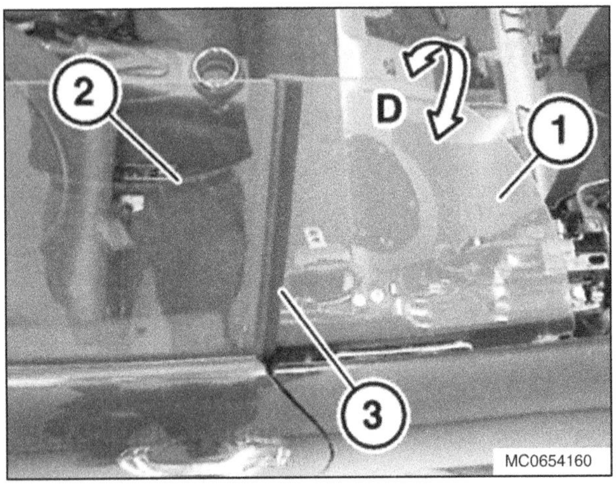

Side window glass (**1**) must run into soft top seal when soft top is closed.

- Door window glass (**2**) must rest firmly on side window seal (**3**) when soft top is open.
- It must be possible to pull out the paper strip with moderate friction resistance.

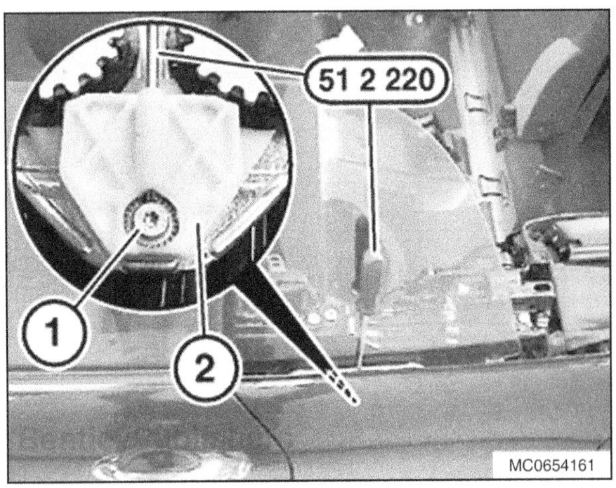

Adjust pre-tension (**D**) by turning toothed knurled nut (**1**) on mounting (**2**) with MINI special tool **51 2 220** .

CONVERTIBLE TOP ADJUSTMENT

Main bearing adjustment

The convertible top main bearing can be adjusted longitudinally (fore/aft) by means of elongated holes in the body retaining plate.

– Position convertible top to the closed position, but do not engage catch pin hook.

– Remove side trim panel to allow access to main bearing.

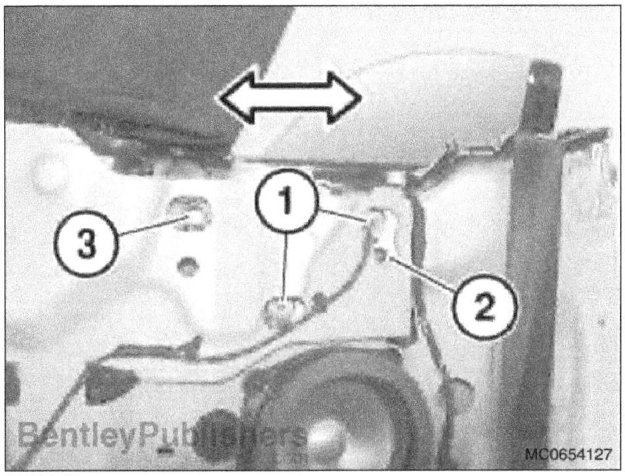

– Loosen securing screw (**2**) slightly; do not completely loosen.

– Insert nuts (**1**) and screw (**3**). Ensure convertible top remains moveable (**arrow**).

– Fully close and lock convertible top.

– With convertible top closed and secured, check and adjust gap dimensions as needed.

– When convertible top is properly adjusted, tighten securing screw (**2**) and nuts (**1**).

Tightening torque	
Convertible top main bearing securing screw (**2**)	10 Nm (7.4 ft-lb)
Convertible top nuts (**1**)	18.5 Nm (13.6 ft-lb)

– Operate convertible top through a full opening and closing cycle and check for proper operation and adjustment.

– Install removed trim panels as needed.

Tension bar bearing adjustment

The convertible top tension bar bearing can be adjusted longitudinally (fore/aft) and vertically (up/down) by means of elongated holes on the tension bar bearing.

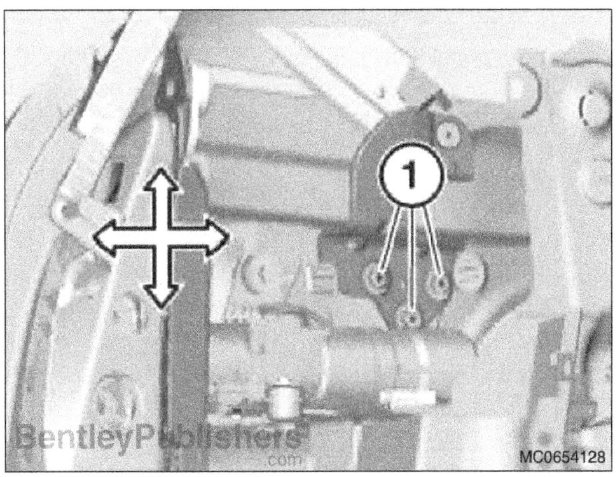

– Screw in nuts (**1**) until they contact screw heads. Ensure bearing remains movable (**arrows**).

– Fully close and lock loading aid.

– With loading aid fully closed and locked, check and adjust gap dimensions as needed.

– When tension bar bearing is properly adjusted, tighten nuts (**1**).

Tightening torque	
Tension bar nuts (**1**)	18.5 Nm (13.6 ft-lb)

Front bow base plate adjustment

The convertible top front bow can be adjusted vertically (up/down) and laterally (side/side) by means of elongated holes in the base plate.

– Remove roof antenna.

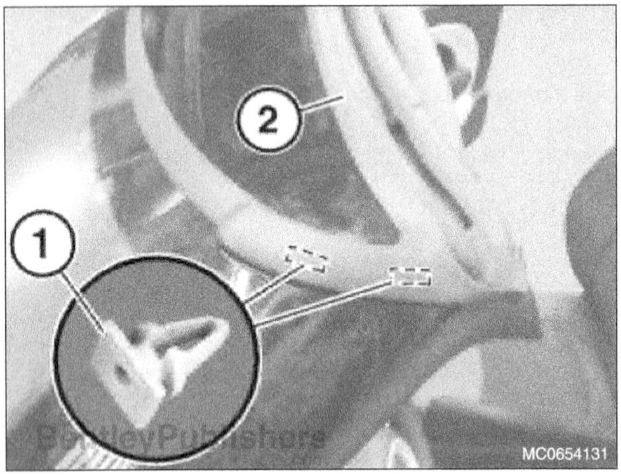

◄ Working from left side, unclip cowl panel seal (**2**) retaining clips (**1**) as shown. Repeat for right side.

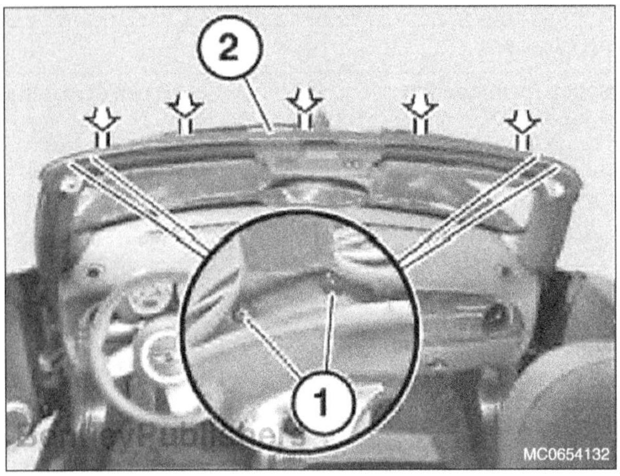

◄ Loosen screws (**1**) at both ends of cowl panel seal (**2**).

– Pull off seal (**2**) at front door openings and feed seal out at upper cowl panel to remove.

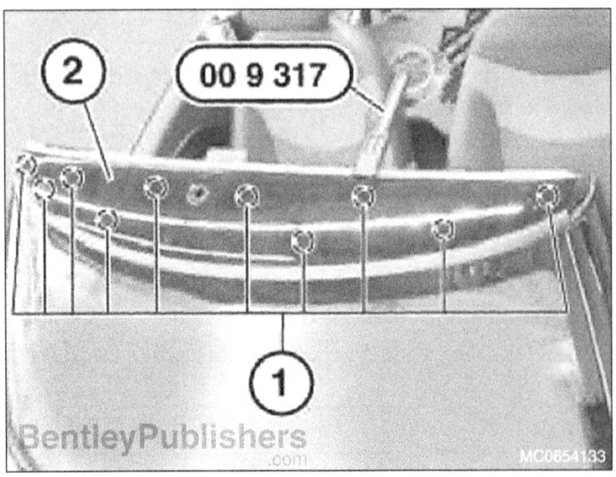

◄ Unclip securing clips (**1**) on upper cowl panel trim (**2**) using tool 00 9 317 or equivalent.

– Remove upper cowl panel trim.

– Fully close and lock convertible top.

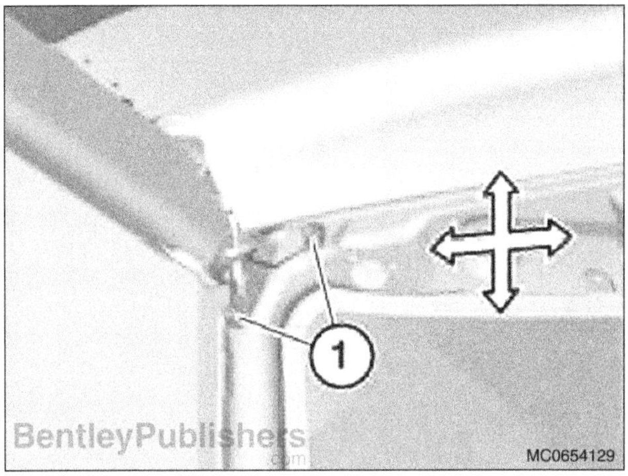

Loosen securing screws (**1**) slightly; do not completely loosen.

- With convertible top closed and secured, check and adjust gap dimensions by moving base plate as needed (**arrows**).
- When front bow base plate is properly adjusted, tighten securing screws (**1**).

Tightening torque	
Base plate securing screws (**1**)	7.6 Nm (5.6 ft-lb)

- Open convertible top and install removed components. Installation is the reverse of removal noting the following points.

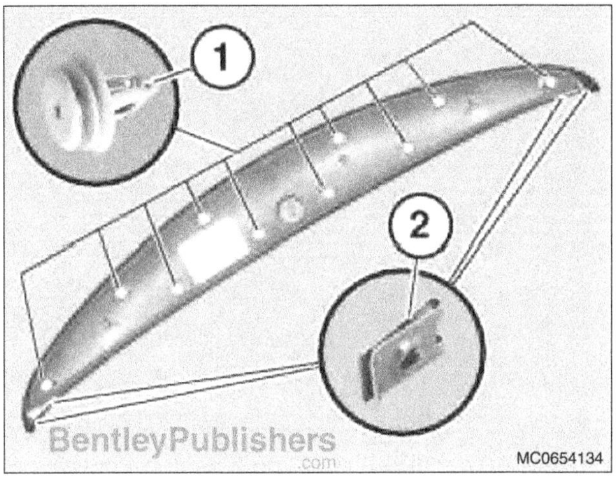

Before installing upper cowl trim panel inspect and replace securing clips (**1**) and retainers (**2**) as needed.

- Before installing cowl panel seal, inspect and replace clips as needed.

600 Electrical System–General

TABLE

600

GENERAL

This repair group presents a brief description of the principal parts of the electrical system. Also covered here are basic electrical system troubleshooting tips.

For additional information on electrical systems and components, refer to the following repair groups:

- **121 Battery, Starter, Alternator**
- **400 Body–General**
- **512 Door Windows**
- **515 Central Locking and Anti-theft**
- **ECL Electrical Component Locations**
- **ELE Electrical Wiring Diagrams**

Electrical test equipment

For measuring voltage, current, or resistance use a digital multimeter. An analog meter (swing-needle) may draw enough current to damage sensitive electronic components.

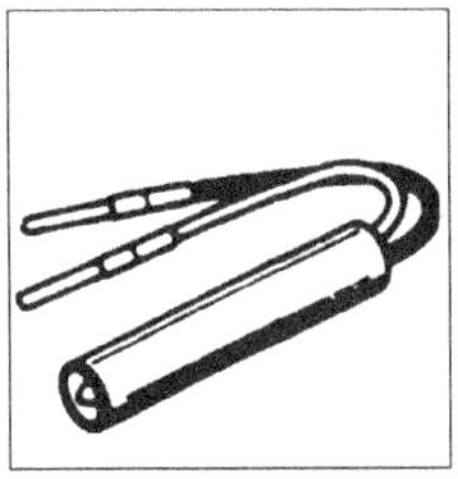

An LED test light is a safe, inexpensive tool that can be used to perform many simple electrical tests that would otherwise require a digital multimeter. The LED indicates when voltage is present between any two test-points in a circuit.

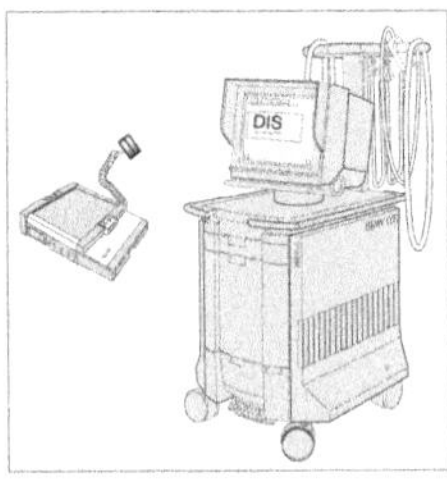

The integrated safety, comfort, security and handling systems on MINI models are designed with self-diagnostic capabilities. The quickest way to diagnose many problems is to start with a read out of Diagnostic Trouble Codes (DTCs) using one of the BMW scan tools (DISplus, GT1, MoDiC) or equivalent. See **OBD On-Board Diagnostics**.

Electrical wiring diagrams

Electrical wiring diagrams are located at the back of this manual. Wiring schematics and electrical information are also available on the official MINI subscription site maintained by BMW. For more information, log on to:

http://www.minitechinfo.com

MINI acronyms

Acronyms used in the wiring diagrams and in this manual are summarized in the table **Common MINI acronyms**.

Common MINI acronyms

Acronym	Component or system	Acronym	Component or system
A/C	air-conditioning	IHKS	manual heating and air-conditioning
AIC	rain sensor	IKE	instrument cluster control module
ABS	antilock brakes	KL 15	battery positive, ignition switch on "run"
AHPS2	head protection system airbag	KL 30	battery positive "hot" all the time
AIC	rain sensor	KL 31	battery/chassis ground
ASC	Automatic Stability Control	KL 50	ignition switch start position
BC1	General Module (Body Control Module)	LDP	fuel tank leak diagnosis pump
BST	Battery Safety Terminal	LWR	headlight vertical aim control relay
CAN	Controller Area Network (bus)	LWS	steering angle sensor
CBC	cornering brake control	MAP	manifold absolute pressure
CVT	continuously variable transmission	MFL	multifunction steering wheel
DLC	diagnostic link connector	MRS	multiple restraint system
DME	digital motor electronics	NTC	negative temperature coefficient resistor
DSC	Dynamic Stability Control	OBC	On-board computer
DTC	diagnostic trouble code	OBD II	second generation on-board diagnostics
DWA	anti-theft alarm system	PDC	park distance control
ECM	engine control module	PWG	accelerator pedal position sensor
EHPS	electrohydraulic power steering	RAM	random access memory
ECT	engine coolant temperature	RDW	tire pressure control
EMS	engine management system	SHD	sunroof module
EWS	electronic immobilizer	SII	service interval indicator
FZV	keyless entry	SRS	supplemental restraint system (airbag)
GIU	transmission control module	TDC	top dead center
IHKA	automatic heating and air-conditioning	T-MAP	temperature / absolute manifold pressure

Electrical system safety precautions

Please read the warnings and cautions in this section before doing any work on your electrical system.

WARNING —
- *Airbags, front seat tensioners and the Battery Safety Terminal (BST) are pyrotechnic (explosive charge) devices. Making repairs without the proper knowledge and special test equipment may cause serious personal injury. See* **721 Airbag System (SRS)**.
- *The ignition system of the car operates at lethal voltages. People with pacemakers or weak hearts should not expose themselves to the ignition system electric currents. Take extra precautions when working on the ignition system or when servicing the engine while it is running or the key is ON. See* **120 Ignition System** *for additional ignition system warnings and cautions.*
- *Keep hands, clothing and other objects clear of the electric radiator cooling fan when working on a warm engine. The fan may start at any time, even when the ignition is switched OFF.*

CAUTION —
- *Always turn off the engine and disconnect the negative (–) cable from the battery before removing any electrical components. Disconnecting the battery may erase fault code(s) stored in control module memory. Check for fault codes using special BMW diagnostic equipment.*
- *Prior to disconnecting the battery, read the battery disconnection cautions given in* **001 General Warnings and Cautions**.
- *Connect and disconnect ignition system wires, multiple connectors and ignition test equipment leads only while the ignition is OFF.*
- *Do not disconnect the battery with the engine running.*
- *Do not quick-charge the battery (for boost starting) for longer than one minute, and do not exceed 16.5 vdc at the battery with the boosting cables attached. Wait at least one minute before boosting the battery a second time.*
- *Many solid-state control modules are static sensitive. Static discharge will permanently damage them. Always handle the modules using proper static prevention equipment and techniques.*
- *Always switch a digital multimeter to the appropriate function and range before making test connections.*

CAUTION—
- *To avoid damaging harness connectors or relay panel sockets, use jumper wires with flat-blade connectors that are the same size as the connector or relay terminals.*
- *Do not try to start the engine of a car which has been heated above 176°F (80°C), (for example, in a paint drying booth). Allow it to cool to normal temperature.*
- *Disconnect the battery before doing any electric welding on the car.*
- *Do not wash the engine while it is running, or any time the ignition is switched on.*
- *Choose test equipment carefully. Use a digital multimeter with at least 10 mΩ input impedance, or an LED test light. An analog meter (swing-needle) or a test light with a normal incandescent bulb may draw enough current to damage sensitive electronic components.*
- *Do not use an ohmmeter to measure resistance on solid state components such as control modules or time delay relays.*
- *Always disconnect the battery before making resistance (ohm) measurements on a circuit.*

MINI ELECTRICAL SYSTEM

The main MINI wiring harness is vehicle specific depending on optional equipment. In addition there are three other modular harnesses which connect to the main body harness:

- Two door harnesses
- Dashboard harness

Integral to the main body harness are the multiple restraint system (MRS) harness and the various buses. Repairs may be made on most harnesses depending on extent of damage, although no part of the MRS harness is repairable.

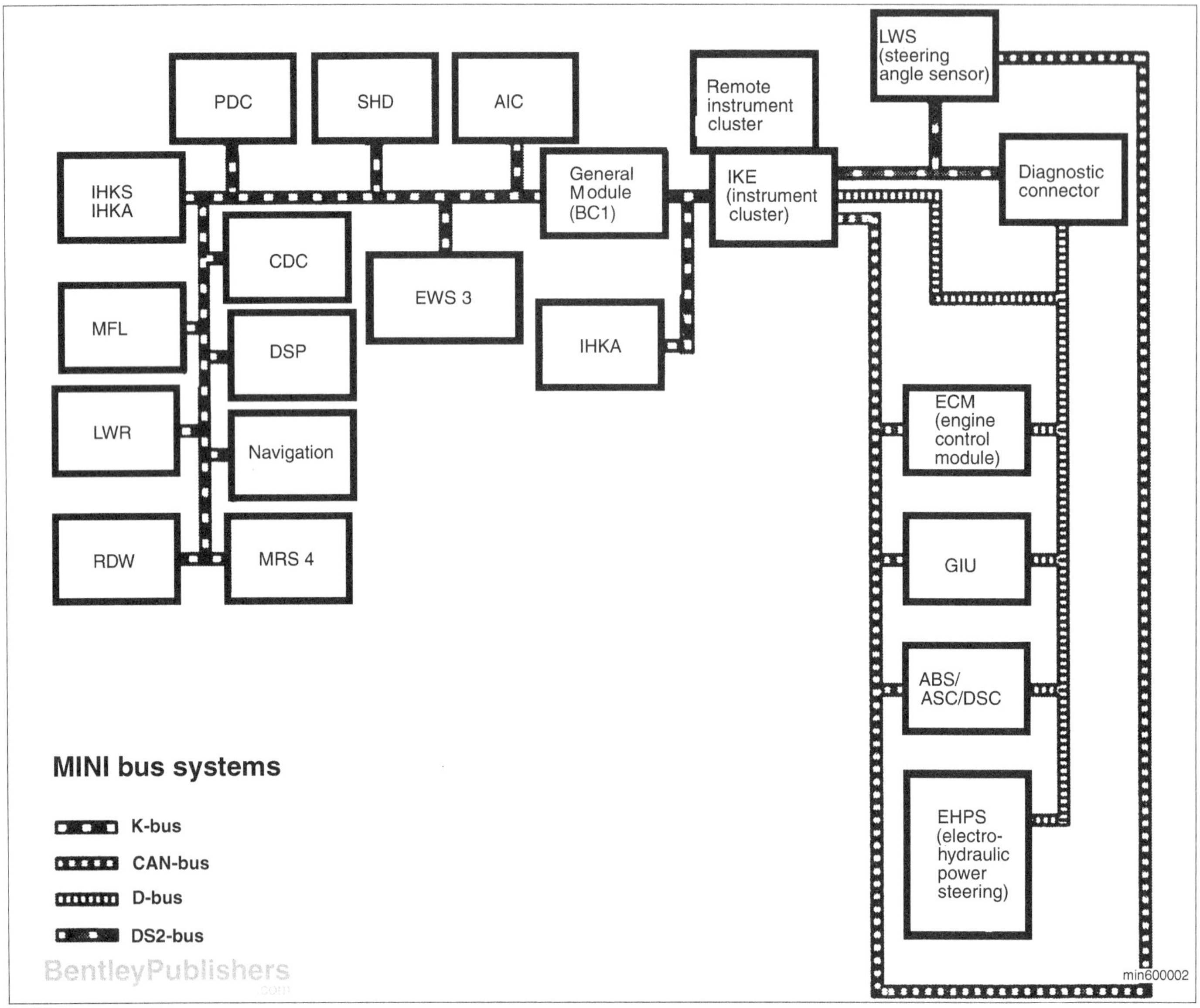

Bus systems

MINI wiring is a network which uses a number of interconnected bus systems:

- CAN-bus
- K-bus
- Diagnostic buses D-bus and DS2-bus

Most control modules transmit and receive messages on this network.

The IKE (instrument cluster) acts as the communication gateway, enabling transfer of data from one bus system to another. The IKE contains a microprocessor which converts and processes all signals into the format required for transmission on to another bus system.

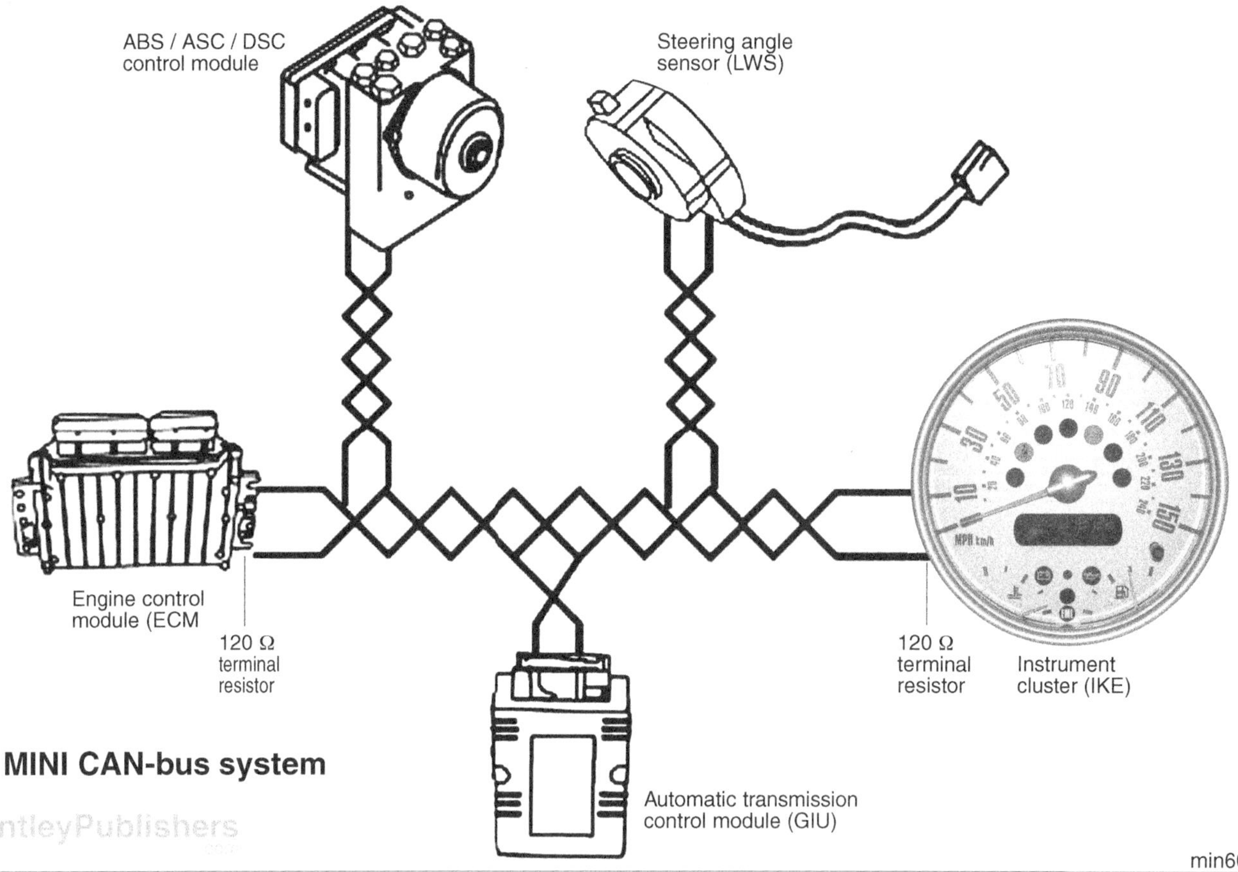

MINI CAN-bus system

CAN-bus

The high speed serial data CAN-bus system is linked by an unshielded twisted pair of wires: yellow / black and yellow / brown. The wires are twisted to minimize electromagnetic interference. Both wires carry information. For CAN-bus to operate, both signals must be present.

CAN-bus, the fastest of the bus systems, is used for vehicle systems where the speed of exchange of information is vital for performance:

- Engine management
- Automatic transmission
- Stability control (ASC or DSC)

The CAN system consists of the main bus length and shorter stubs. The main bus length terminates at the engine control module (ECM) and the instrument cluster (IKE). The shorter stubs must be as short as possible and no longer than one meter. Any untwisted portion of the bus may not be longer than four centimeters.

For correct operation, CAN-bus must terminate at both ends with a control module resistance of 120 Ω. These terminations ensure that bit errors due to signal reflections are avoided.

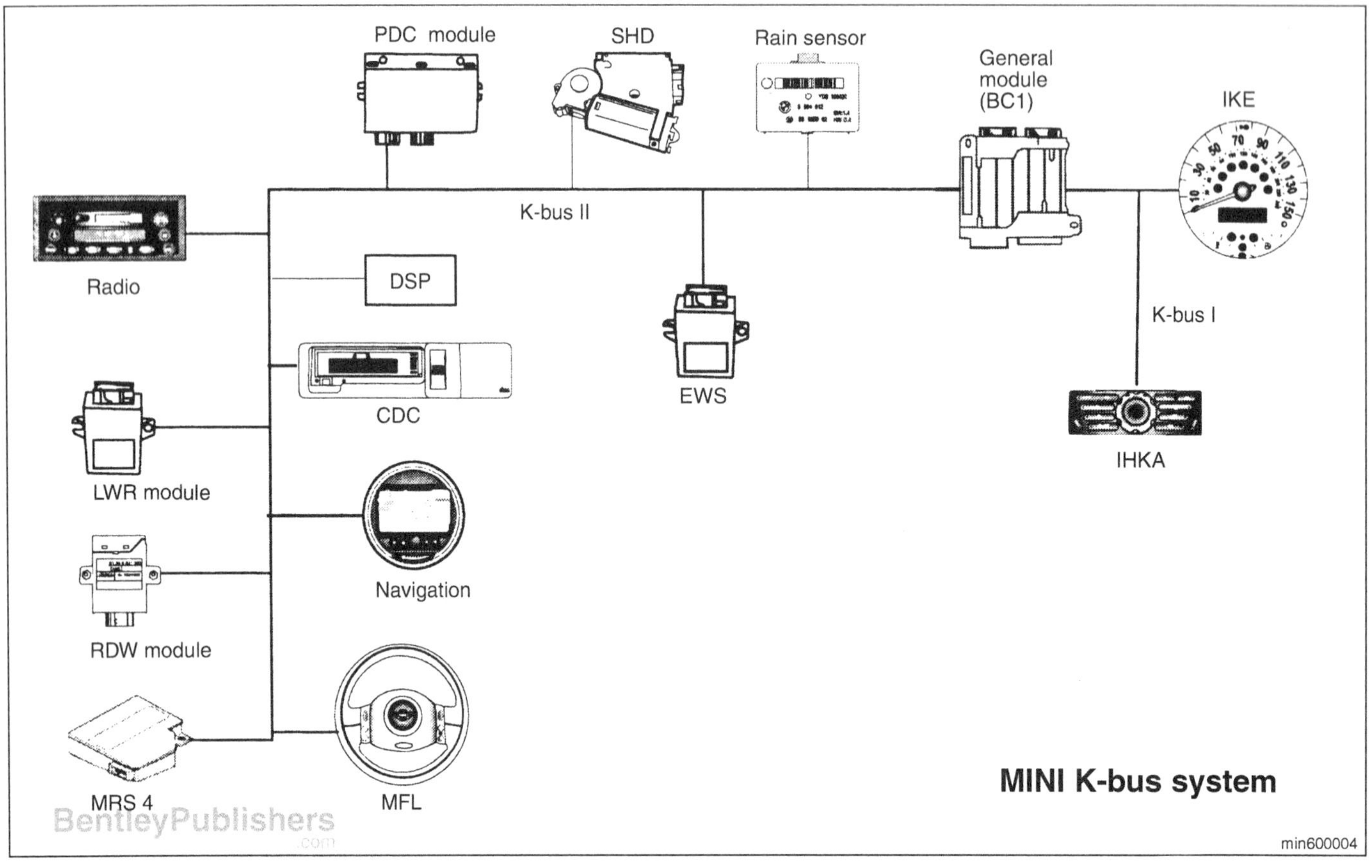

K-bus

The K-bus system, using a single white / red / yellow wire, is a relatively slow bus system. It is used by body-related components to connect with other body electronic systems and the instrument cluster (IKE).

K-bus is primarily an event-driven system. Messages are sent after a request has been made for component operation or status change.

The General Module (Body Control Module or BC1) has the highest priority on K-bus. Messages from the BC1 override any other K-bus message transmitted simultaneously.

All control modules on K-bus receive all messages, but only the module addressed in the message will accept and react to the data. K-bus remains operational in case of a disconnected or failed control module. This is referred to as a "tree" structure, with each control module occupying a branch.

K-bus provides the diagnostic connection to the control modules located on the bus (except for IKE).

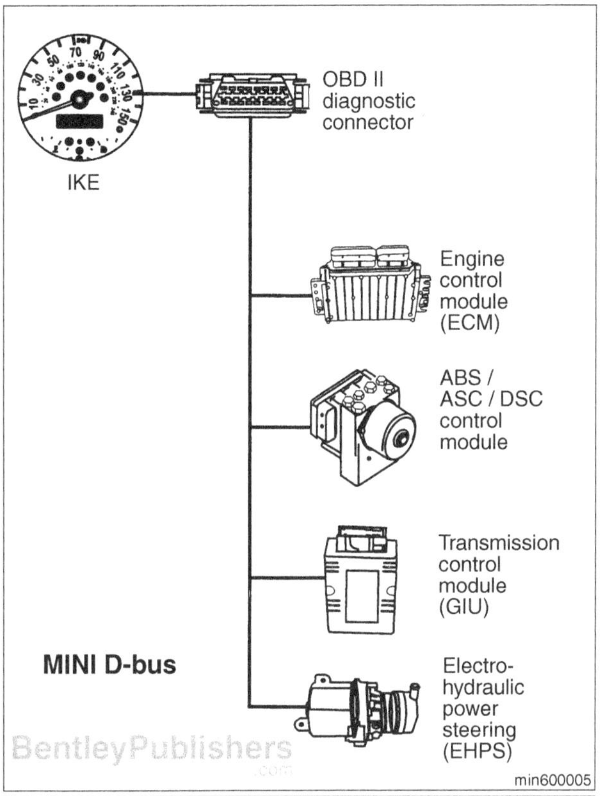

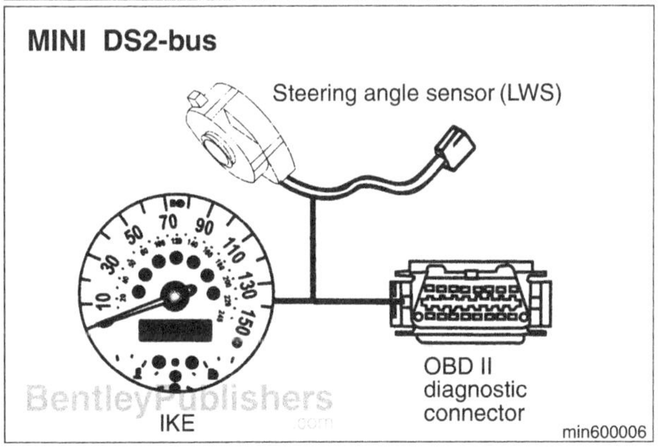

Diagnostic buses

The diagnostic bus system is made up of two separate bus systems. These allow BMW scan tools DISplus, GT1 or MoDiC to communicate with the vehicle control modules via the OBD II diagnostic socket above the driver footwell:

- **D-bus** allows communication with power-train (emission) related components. (This is also known as the ISO 9141-2 OBD II network.)
- **DS2-bus** connects to control modules that do not affect emissions and to the instrument cluster (IKE), which enables communication with all control modules on K-bus.

D-bus uses serial communication to transmits data between DISplus (or equivalent scan tool) and control modules on the network. The control module subject to diagnosis is selected by sending an interrogation to its network address. The module responds by transmitting information such as the contents of fault memory.

Electrical Troubleshooting

Four things are required for current to flow in any electrical circuit: a voltage source, wires or connections to transport the voltage, a load or device that uses the electricity, and a connection to ground.

Most problems can be found using a digital multimeter (volt/ohm/amp meter) to check the following:

- Voltage supply
- Breaks in the wiring (infinite resistance/no continuity)
- A path to ground that completes the circuit

Electric current is logical in its flow, always moving from the voltage source toward ground. Electrical faults can usually be located through a process of elimination. When troubleshooting a complex circuit, separate the circuit into smaller parts. General tests outlined below may be helpful in finding electrical problems. The information is most helpful when used with wiring diagrams.

Be sure to analyze the problem. Use wiring diagrams to determine the most likely cause. Get an understanding of how the circuit works by following the circuit from ground back to the power source.

When making test connections at connectors and components, use care to avoid spreading or damaging the connectors or terminals. Some tests may require jumper wires to bypass components or connections in the wiring harness. When connecting jumper wires, use blade connectors at the wire ends that match the size of the terminal being tested. The small internal contacts are easily spread apart, and this can cause intermittent or faulty connections that can lead to more problems.

Voltage and voltage drops

Wires, connectors, and switches that carry current are designed with very low resistance so that current flows with a minimum loss of voltage. A voltage drop is caused by higher than normal resistance in a circuit. This additional resistance actually decreases or stops the flow of current. A voltage drop can be noticed by problems ranging from dim headlights to sluggish wipers. Some common sources of voltage drops are corroded or dirty switches, dirty or corroded connections or contacts, and loose or corroded ground wires and ground connections.

A voltage drop test is a good test to make if current is flowing through the circuit but the circuit is not operating correctly. A voltage drop test will help to pinpoint a corroded ground strap or a faulty switch. Normally, there should be less than 1 vdc drop across most wires or closed switches. A voltage drop across a connector or short cable should not exceed 0.5 vdc.

NOTE —
- *A voltage drop test is generally more accurate than a simple resistance check because the resistances involved are often too small to measure with most ohmmeters. For example, a resistance as small as 0.02 Ω would results in a 3 vdc drop in a typical 150 amp starter circuit (150 amps x 0.02 Ω =3 vdc).*
- *Keep in mind that voltage with the key on and voltage with the engine running are not the same. With the ignition on and the engine off (battery voltage), voltage should be approximately 12.6 vdc. With the engine running (charging voltage), voltage should be approximately 14.0 vdc. Measure voltage at the battery with the ignition on and then with the engine running to get exact measurements.*

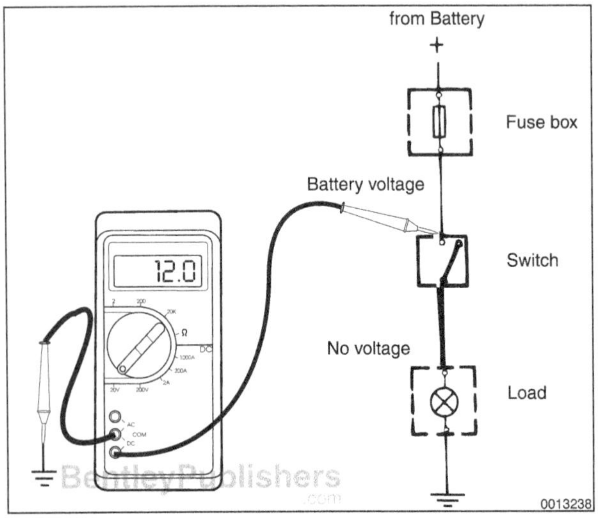

Voltage, measuring

◄ Connect digital multimeter negative lead to a reliable ground point on car.

NOTE —
The negative (-) battery terminal is always a good ground point.

– Connect digital multimeter positive lead to point in circuit you wish to measure.

NOTE —
The voltage reading should not deviate more than 1 vdc from voltage at the battery. If voltage drop is more than this, check for a corroded connector or loose ground wire.

Voltage drop, testing

Voltage drop can only be checked when there is a load on the circuit, such as when operating the starter motor or turning on the headlights. A digital multimeter should be used to ensure accurate readings.

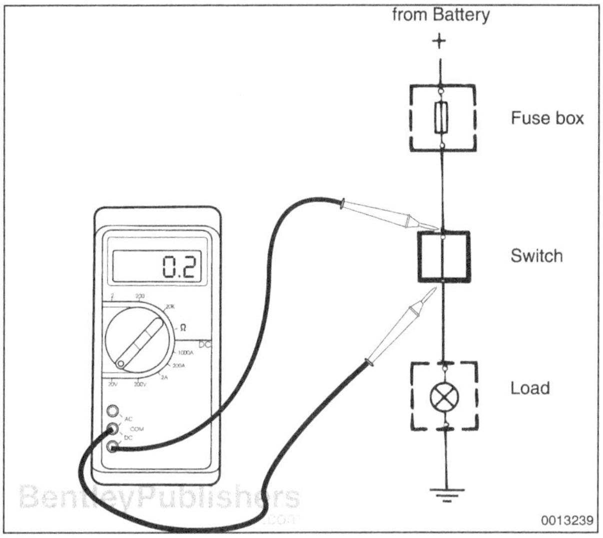

- Connect digital multimeter positive lead to positive (+) battery terminal or a positive power supply close to battery source.
- Connect digital multimeter negative lead to other end of cable or switch being tested.
- With power on and circuit working, meter shows voltage drop (difference between two points). This value should not exceed 1 volt.

NOTE—
- *The maximum voltage drop in an automotive circuit, as recommended by the Society of Automotive Engineers (SAE), is as follows:*
- *0 vdc for small wire connections*
- *0.1 vdc for high current connections*
- *0.2 vdc for high current cables*
- *0.3 vdc for switch or solenoid contacts*
- *On longer wires or cables, the drop may be slightly higher. In any case, a voltage drop of more than 1.0 vdc usually indicates a problem.*

Continuity, checking

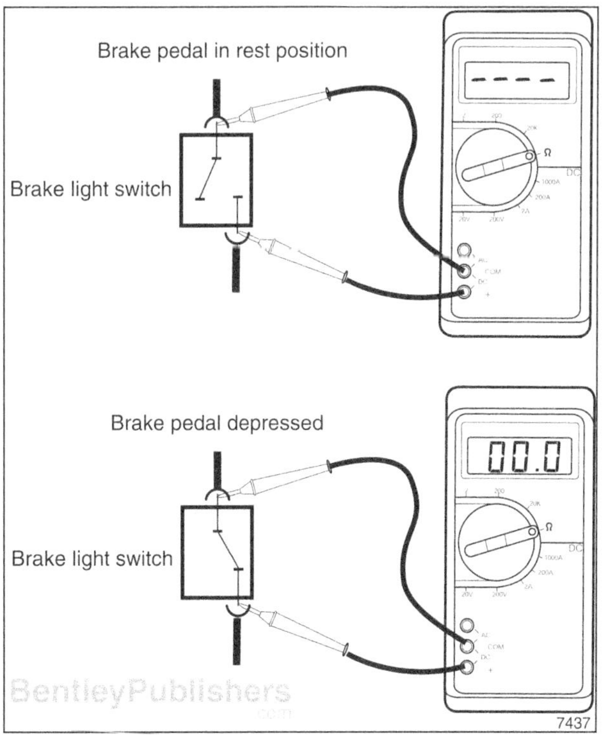

Continuity tests can be used to check a circuit or switch. Because most automotive circuits are designed to have little or no resistance, a circuit or part of a circuit can be easily checked for faults using an ohmmeter. An open circuit or a circuit with high resistance will not allow current to flow. A circuit with little or no resistance allows current to flow easily.

When checking continuity, the ignition should be off. On circuits that are powered at all times, the battery should be disconnected. Using the appropriate wiring diagram, a circuit can be easily tested for faulty connections, wires, switches, relays and engine sensors by checking for continuity.

- For example, to test brake light switch for continuity:
  - With brake pedal in rest position (switch open) there is no continuity (infinite Ω).
  - With pedal depressed (switch closed) there is continuity (0Ω).

Short circuits

Short circuits are exactly what the name implies. The circuit takes a shorter path than it was designed to take. The most common short that causes problems is a short to ground where the insulation on a positive (+) wire wears away and the metal wire is exposed. When the wire rubs against a metal part of the car or other ground source, the circuit is shorted to ground. If the exposed wire is live (positive battery voltage), a fuse will blow and the circuit may be damaged.

Shorts to ground can be located with a digital multimeter. Short circuits are often difficult to locate and vary in nature. Short circuits can be found using a logical approach based on knowledge of the current path.

NOTE —
- *On circuits protected with high rating fuses (25 amp and greater), wires or circuit components may be damaged before the fuse blows. Always check for damage before replacing fuses of this rating.*
- *When replacing blown fuses, use only fuses having the correct rating. Always confirm by checking for the correct fuse rating printed on the fuse panel cover.*

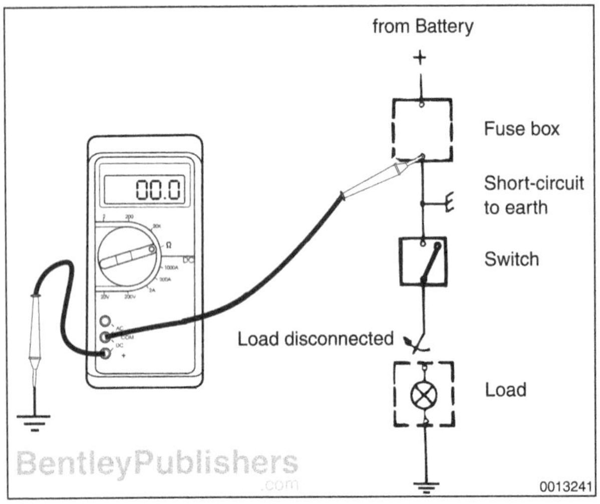

Short circuit, testing with ohmmeter

– Remove blown fuse from circuit and disconnect cables from battery. Disconnect harness connector from circuit load or consumer.

◄ Using an ohmmeter, connect one test lead to load side of fuse terminal (terminal leading to circuit) and the other test lead to ground.

– If there is continuity to ground, there is a short to ground.

– If there is no continuity, work from wire harness nearest to fuse/relay panel and move or wiggle wires while observing meter. Continue to move down harness until meter displays a reading. This is the location of short to ground.

– Visually inspect wire harness at this point for any faults. If no faults are visible, carefully slice open harness cover or wire insulation for further inspection. Repair any faults found.

Short circuit, testing with voltmeter

– Remove blown fuse from circuit. Disconnect harness connector from circuit load or consumer.

NOTE —
Most fuses power more than one consumer. Be sure all consumers are disconnected when checking for a short circuit.

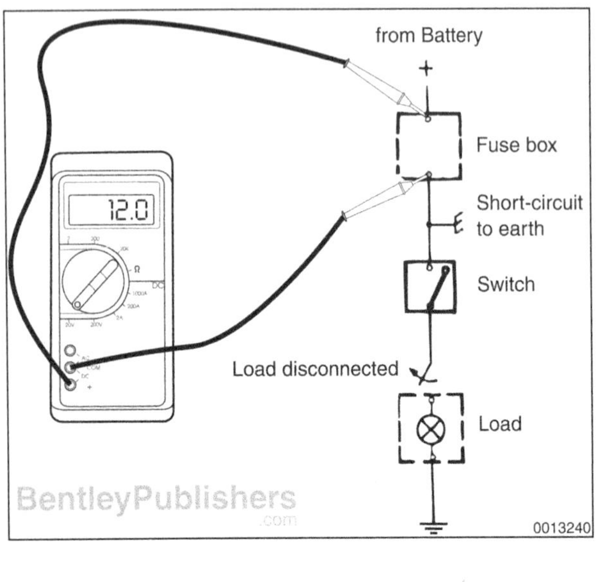

◄ Using a digital multimeter, connect test leads across fuse terminals. Make sure power is present in circuit. If necessary turn key on.

– If voltage is present at voltmeter, there is a short to ground.

– If voltage is not present, work from wire harness nearest to fuse/relay panel and move or wiggle wires while observing meter. Continue to move down harness until meter displays a reading. This is the location of short to ground.

– Visually inspect wire harness at this point for any faults. If no faults are visible, carefully slice open harness cover or wire insulation for further inspection. Repair any faults found.

611 Wipers and Washers

611

General

This repair group covers removal and installation of the wiper blades, the front and rear wiper assemblies and the window and headlight washer systems.

Special tools

Some BMW special tools are recommended for the procedures in this repair group.

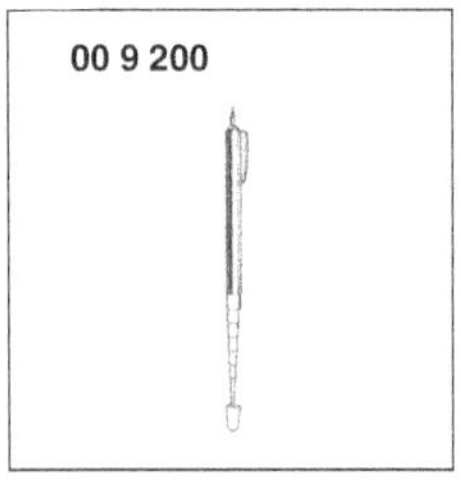

◄ Washer nozzle adjusting tool (BMW tool no. 00 9 200)

◄ Wiper arm angle adjusting tool (BMW tool no. 00 9 220)

◄ Wiper blade angle gauge (BMW tool no. 00 9 230)

◄ Wiper arm puller (BMW tool no. 61 2 100)

Wiper Controls

The General Module (BC1), with external relays, controls all wiper and washer functions including the headlight washer.

NOTE —
The General Module is also referred to as the Body Control Module.

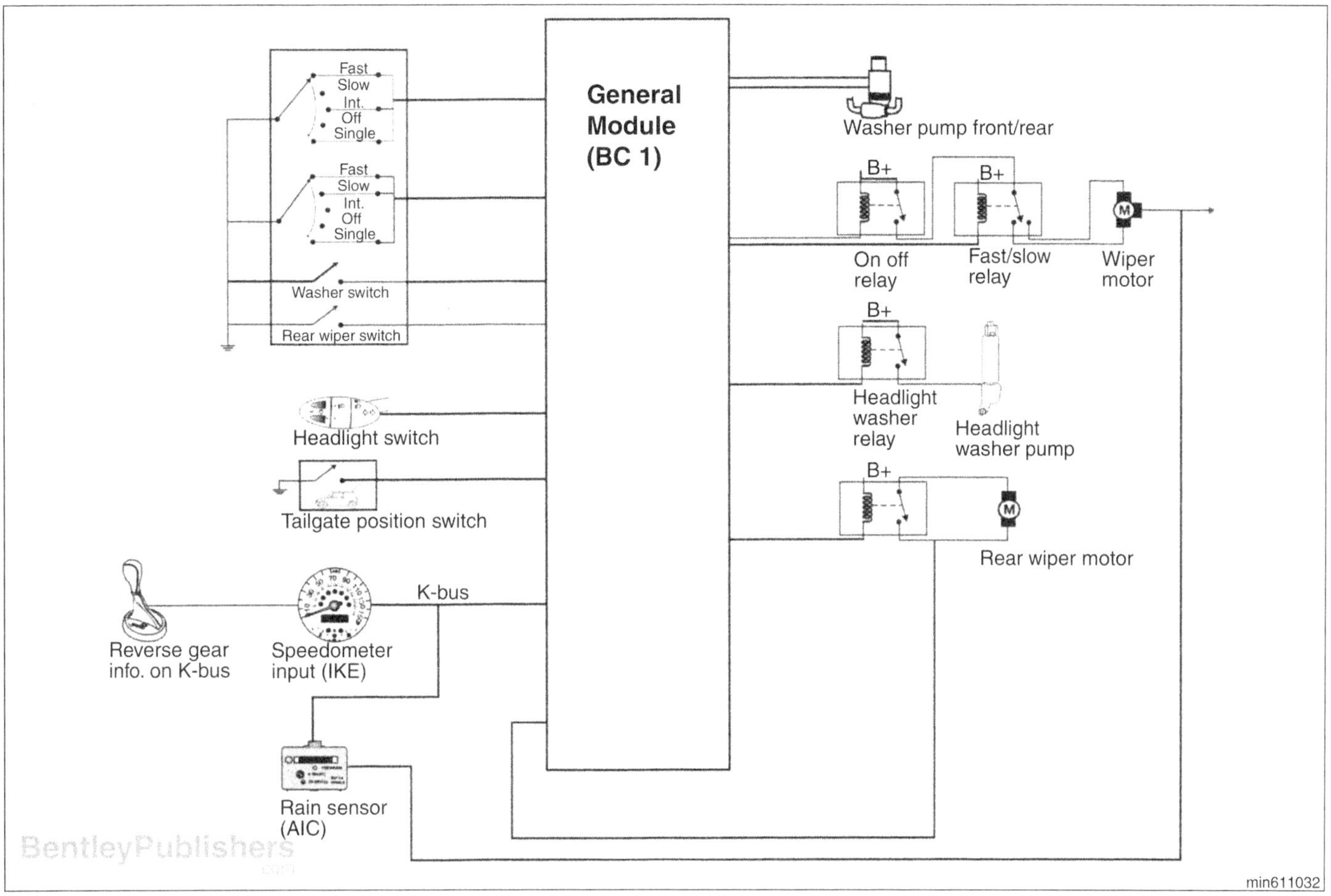

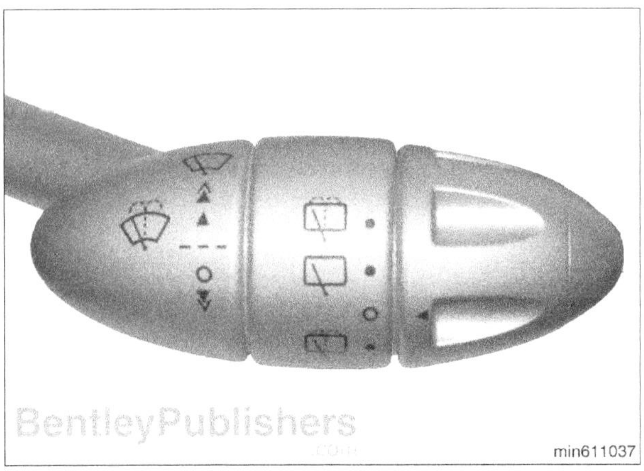

Wiper switch

All wiper and washer function requests begin at the wiper stalk switch, located on the right of the steering column. The stalk switch sends front and rear wipe, wash and speed requests to the General Module (BC1). Headlight wash requests are sent to the BC1 with every fifth windshield wash request, providing the headlights are on.

NOTE —
The General Module is also referred to as the Body Control Module. In wiring diagrams, the General Module is referred to as A1.

Wiper relays

Front wipers are controlled by two replaceable relays in the engine compartment fuse and relay panel (fuse and relay panel 3).

- **R/P relay** runs the wipers when energized and places them in PARK when de-energized.
- **F/S relay** selects fast or slow wiper operation.

Rear wipers are controlled by a non replaceable relay on the back of the A-pillar fuse and relay panel (fuse and relay panel 2 behind left kick panel). See **ECL Electrical Component Locations**.

Headlight washers are controlled by a replaceable relay in the A-pillar fuse and relay panel (fuse and relay panel 2 behind left kick panel). See **ECL Electrical Component Locations**.

WIPERS

Wiper blade cleaning problems

Common cleaning problems with wipers include streaking or sheeting, water drops after wiping, and blade chatter.

Streaking is usually caused when wiper blades are coated with road film or car wash wax. Clean the blades using soapy water. If cleaning does not cure the problem, replace the blades. MINI recommends replacing the wiper blades twice a year, before and after the cold season.

Water drops that remain behind after wiping are usually caused by oil, road film, or diesel exhaust residue on the glass. Use an alcohol or ammonia solution or other nonabrasive cleaner to clean the windshield.

Chatter may be caused by dirty or worn blades, or by wiper arms that are out of alignment. Clean the blades and windshield as described above. Adjust the wiper arm so that there is even pressure along the blade, and so that the blade at rest is perpendicular to the windshield. If problems persist, replace the blades and wiper arms.

NOTE—
MINI has available a special tool/gauge (BMW special tool 00 9 230) to align the angle of the wiper arm to the windshield glass.

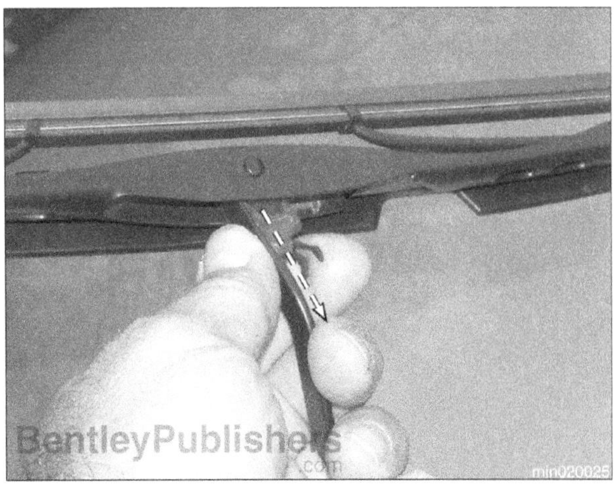

Wiper blade, replacing (front)

To replace wiper blade:

- Pivot wiper arm off windshield.
- Position wiper blade approximately perpendicular to wiper arm.
- Remove wiper blade from wiper arm by depressing tab (**arrow**) and sliding blade out of arm.
- Slide new blade onto arm and lock fastener into place.

Wiper blade, replacing (rear)

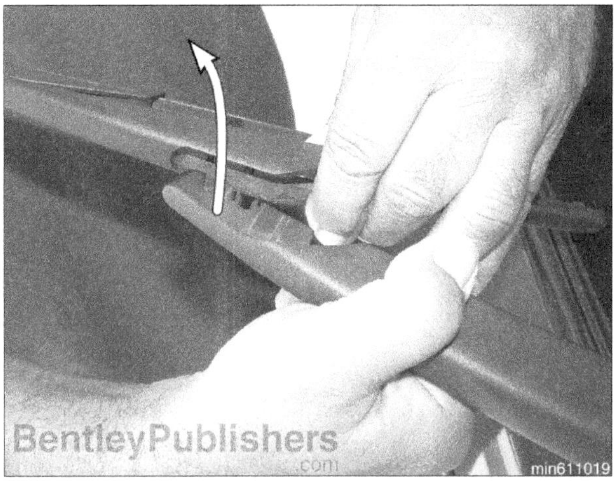

- Raise wiper arm and push blade in direction of **arrow** to unclip.
- To install, snap blade firmly into place.

Wiper arm, replacing (front)

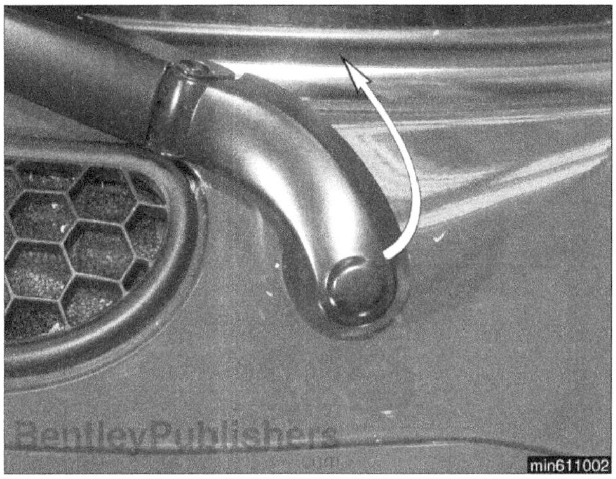

- Lift off protective cap from base of wiper arm (**arrow**).
- Make sure clips on underside of cap are not damaged. Replace if necessary.

- Remove nut (**arrow**) securing wiper arm to wiper pivot shaft.
- Remove wiper arm from pivot shaft using BMW special tool 61 2 100 or other suitable puller.

NOTE —
Do not bend wiper arm while removing.

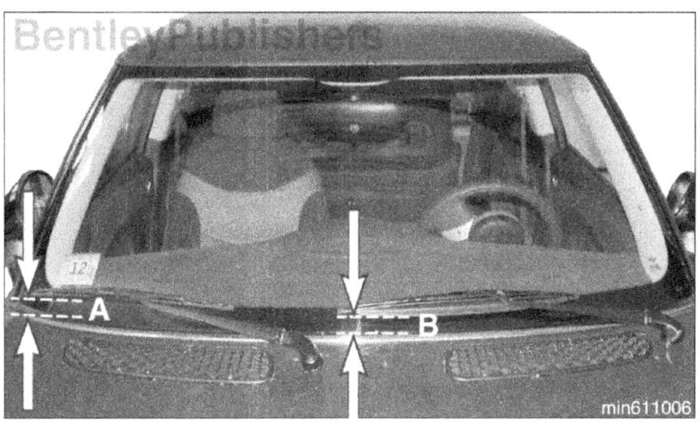

- Installation is reverse of removal. Make sure wiper motor is in PARK and:
  - Set right wiper arm to windshield frame distance (**A**) at 59 mm (2.3 in).
  - Set left wiper arm to windshield distance (**B**) at 60 mm (2.4 in).
  - Tighten nut and replace wiper arm nut protective cap.

Tightening torque	
Wiper arm to wiper shaft	18 Nm (13 ft-lb)

Wiper arm, replacing (rear)

- Flip open protective cap/washer jet at base of rear wiper arm.
- Remove mounting nut (**arrow**). Remove wiper arm.

NOTE—
If necessary, use special tool 61 2 100 or another puller to remove arm from shaft.

- Check that protective cap and washer jet are not damaged. Replace if necessary.

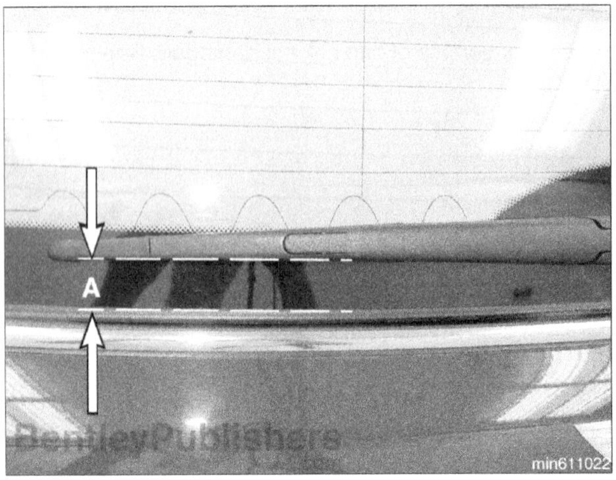

- Installation is reverse of removal. Remember to:
  - Make sure wiper motor is in PARK.
  - Position wiper arm so that distance from wiper tip to lower edge of rear window (**A**) is 44 mm (1.75 in).

Tightening torque	
Rear wiper arm to wiper motor	13 Nm (10 ft-lb)

WIPER MOTOR AND LINKAGE

Wiper motor and linkage, removing and installing (front)

- Working at base of windshield, remove cowl vents. Pry front edge with a plastic pry tool and slide vent forward.
- Remove wiper arms. See **Wiper arm, replacing (front)**.
- Remove fasteners securing wiper arm shafts to cowl.
- Remove side marker lights by prying on front edge of bezel with a plastic tool. Slide bezel forward to release. Remove electrical connector to light.
- Remove A-pillar outside trim by gently lifting front edge with a plastic prying tool to release trim clips. Carefully pry back edge of trim from adhesive tape and remove trim.

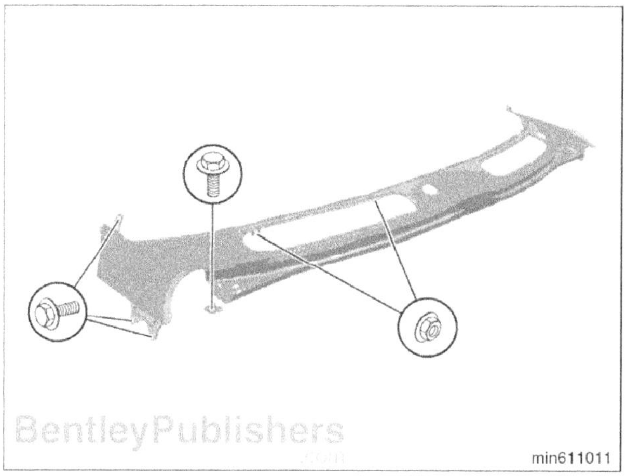

min611011

◄ Remove cowl panel fasteners (**inserts**). Remove cowl panel.

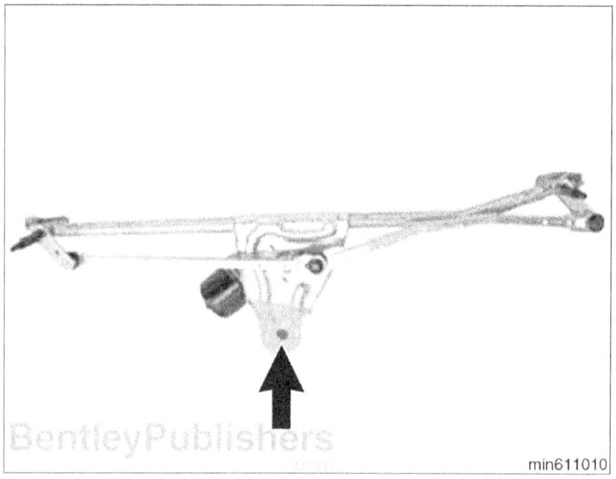

min611010

◄ Remove wiper assembly lower mounting nut (**arrow**) by reaching down to base of assembly at rear engine compartment bulkhead. Remove wiper assembly.

– Installation is reverse of removal. When installing wiper arms, make sure wiper motor is in PARK. See **Wiper arm, replacing (front)** in this repair group.

Tightening torques	
Front wiper arm to wiper motor	18 Nm (13 ft-lb)
Front wiper assembly to cowl or bulkhead	10 Nm (7 ft-lb)

Wiper motor, removing and installing (rear)

– Remove rear wiper arm. See **Wiper arm, replacing (rear)** in this repair group.

– Remove lower hatch trim. See **412 Rear Hatch**.

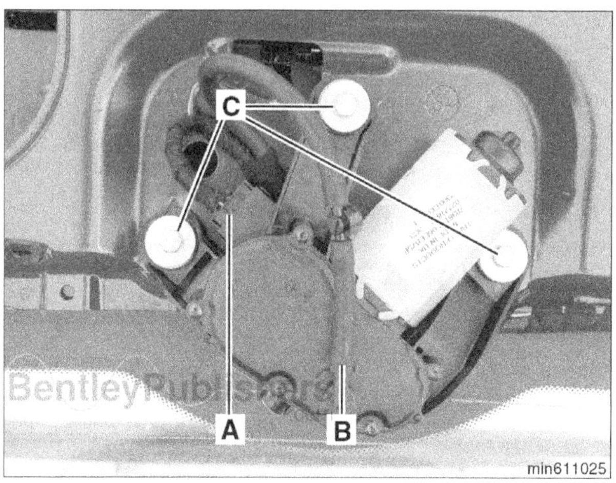

min611025

◄ Working at rear wiper motor assembly:

- Detach electrical connector (**A**).
- Detach washer fluid hose (**B**).
- Remove mounting bolts (**C**)
- Remove wiper motor.

– Installation is reverse of removal.

- Make sure wiper motor shaft and grommet are not damaged.
- When installing wiper arm, make sure wiper motor is in PARK. See **Wiper arm, replacing (rear)** in this repair group.

Tightening torque	
Rear wiper motor to rear hatch	10 Nm (7 ft-lb)

Windshield and Headlight Washers

Windshield washer nozzle, replacing (front)

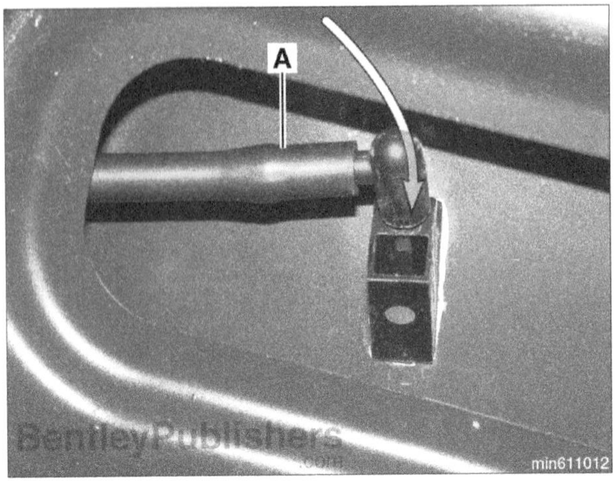

– Open engine hood and remove sound insulation from under hood.

Working at washer nozzle:

- Disconnect washer hose (**A**).
- Remove electrical connector for heated nozzle (if equipped).
- Squeeze tab (**arrow**) to release nozzle from hood.

Windshield washer nozzle, adjusting (front)

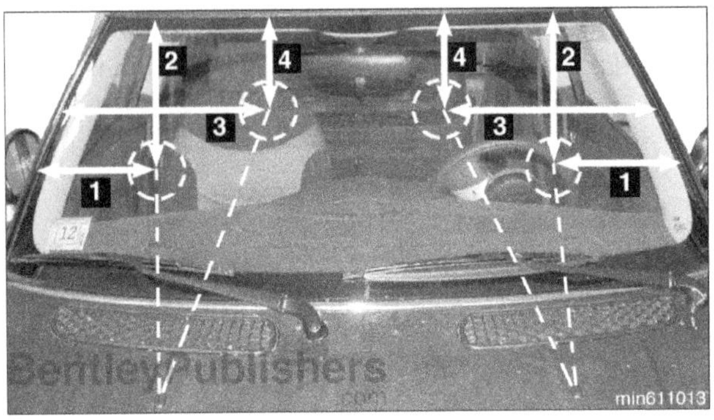

Use BMW special tool 00 9 200 (or equivalent) to adjust washer nozzles so that spray contacts the windshield in accordance with the following values.

Measurement **1**	240 mm (9.5 in)
Measurement **2**	340 mm (13.5 in)
Measurement **3**	510 mm (20.0 in)
Measurement **4**	230 mm (9.0 in)

Washer nozzle, adjusting (rear)

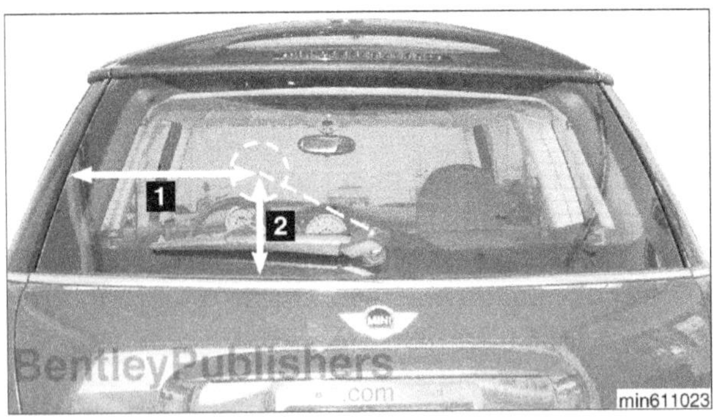

Adjust rear washer nozzle so that spray hits contact point.

Measurement **1**	370 mm (14.5 in)
Measurement **2**	90 mm (3.5 in)

Adjust nozzle (**arrow**) using screwdriver in slot.

> ***CAUTION—***
> *Do not insert pointed objects in washer nozzle.*

Washer pump, replacing

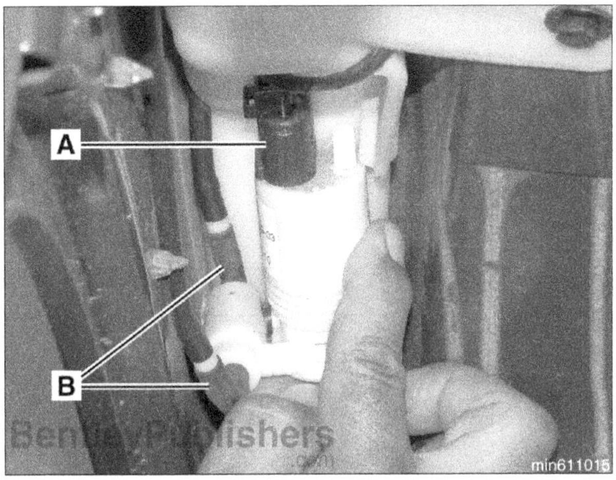

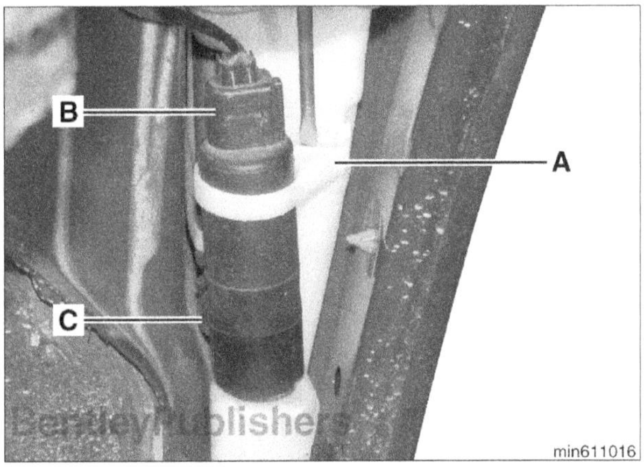

NOTE—
Windshield washer pump is located behind right front fender. Headlight washer pump is located behind left front fender. Replacement is similar.

– Turn front wheels to gain access to rear of front fender lining. Release fender lining fasteners to gain access to washer pump.

– Drain washer fluid reservoir and have a container ready to catch dripping fluid.

< Right side:
- Release pump from clip and lift out of reservoir.
- Detach electrical harness connector (**A**).
- Detach washer fluid hoses (**B**).

< Left side:
- Gently pry plastic pump mount (**A**) off reservoir. Lift pump out of reservoir.
- Release electrical harness connector (**B**).
- Detach washer fluid hose (**C**).

– Installation is reverse of removal. Remember to:
- Replace washer pump sealing ring/strainer.
- Coat sealing ring with antiseize paste.
- Make sure hoses are not kinked.
- Refill washer fluid reservoir.

Washer reservoir, replacing

NOTE—
Windshield washer reservoir is located behind right front fender. Headlight washer reservoir is located behind left front fender. Replacement is similar.

– Remove front fender lining.

– Drain washer fluid reservoir and have a container ready to catch dripping fluid.

– Remove washer fluid pump. See **Washer pump, replacing** in this repair group.

– Remove side marker lights by prying on front edge of bezel with a plastic tool. Slide bezel forward to release. Remove electrical connector to light.

< Working in side marker light cavity, remove washer reservoir top mounting bolt (**arrow**).

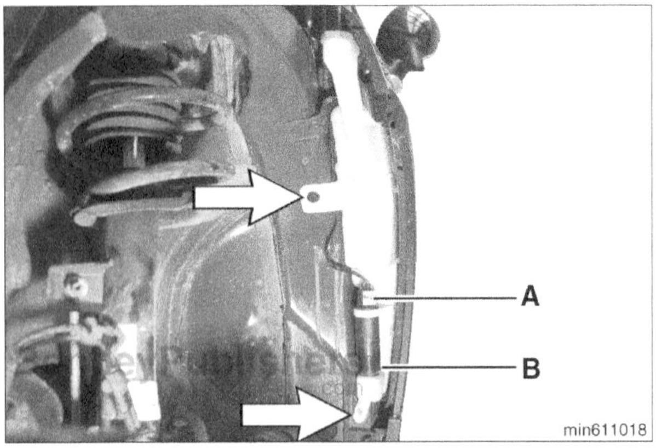

Remove washer reservoir lower mounting fasteners (**arrows**).

– Detach wiring harness connector (**A**) and washer hose(s) (**B**) and remove reservoir.

– Installation is reverse of removal. Remember to:
  - Replace washer pump sealing ring/strainer.
  - Coat sealing ring with antiseize paste.
  - Make sure hoses are not kinked.
  - Refill washer reservoir.

Tightening torque	
Washer reservoir to fender	4 Nm (3 ft-lb)

Headlight washer assembly, replacing

Working underneath engine hood, extend headlight washer nozzle plunger and hold in extended position with BMW special tool 00 9 341 or other suitable tool.

NOTE —
In illustration, a pair of locking pliers are shown holding plunger extended.

Working above engine hood, next to headlight lens, carefully pry chrome cover off plunger with a small screwdriver.

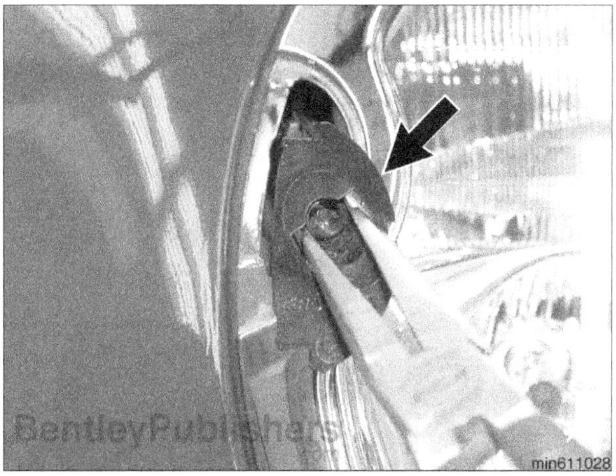

Use needle-nosed pliers to squeeze retaining lugs together in nozzle. This allows removal of external plastic flange (**arrow**).

– Remove BMW special tool 00 9 341 (or locking pliers) and allow washer nozzle plunger to retract.

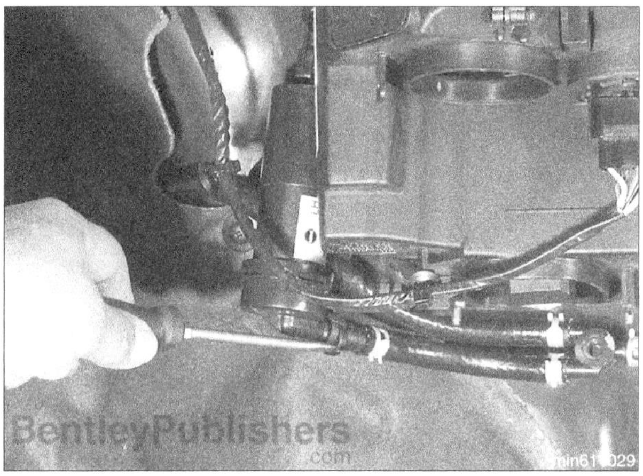

Release washer fluid hose connector and detach hose from nozzle assembly.

NOTE—
Release clip by squeezing or by prying gently with small screwdriver.

– Slide washer nozzle toward rear of engine hood to release from headlight assembly mount.

NOTE—
To allow nozzle to slide freely off mount, use fingernail or small screwdriver to release nozzle mount locking clip. See next illustration.

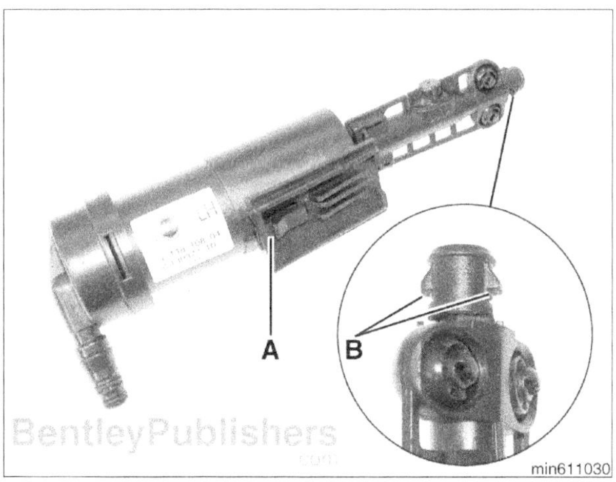

If reinstalling old nozzle, make sure the locking clips are not damaged:

- Nozzle mount locking clip (**A**)
- External flange retaining lugs (**B**)

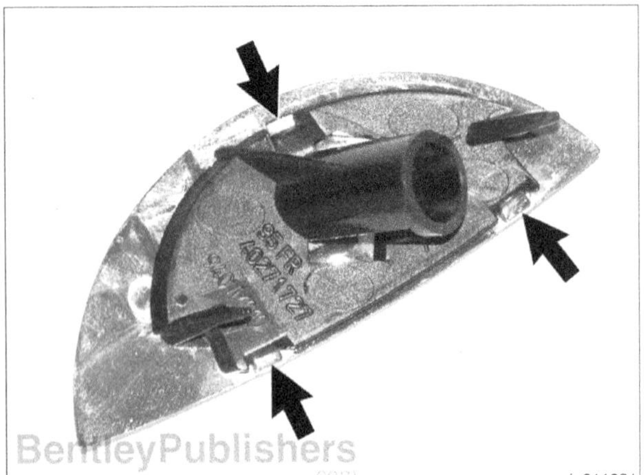

min611031

- Inspect chrome cover to make sure lugs (**arrows**) are not damaged.

- Installation is reverse of removal. Use special tool 00 9 341 or other suitable tool to keep nozzle plunger extended when installing external flange.

612 Switches

General

This repair group covers replacement of electrical switches at the steering wheel, steering column, dashboard, pedal cluster, front console and rear console.

See also the following sections:
- **230 Manual Transmission**
- **240 Automatic Transmission**
- **250 Gearshift Linkage**
- **515 Central Locking and Anti-Theft**
- **520 Seats**
- **540 Sunroof**
- **630 Lights**
- **ECL Electrical Component Locations**

NOTE—
Instrument lighting and hazard warning switch are integral with the instrument carrier and not available separately.

CAUTION—
- *When working on electrical switches, always disconnect the negative (-) cable from the battery. Prior to disconnecting the battery, read the battery disconnection cautions given in* **001 General Warnings and Cautions**.
- *To prevent marring the trim when working on interior components, work with plastic prying tools or wrap the tips of screwdrivers and pliers with tape before prying out switches or electrical accessories.*

Special tools

Some BMW special tools are recommended for the procedures in this repair group.

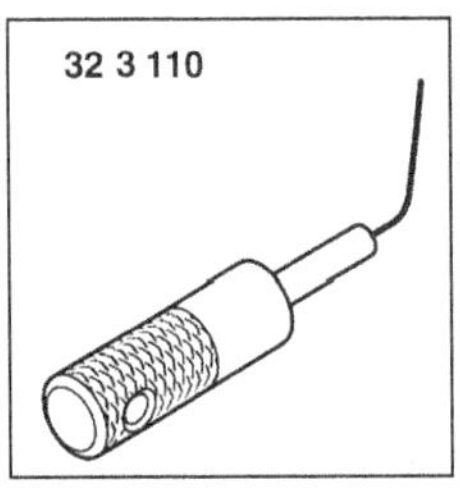

Ignition lock cylinder releasing tool (BMW tool no. 32 3 110)

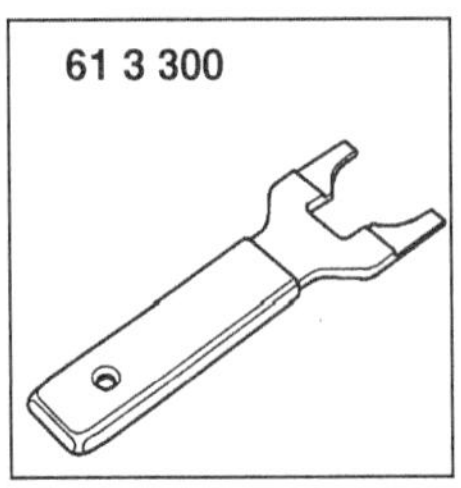

EWS ring antenna removal tool (BMW tool no. 61 3 300)

Steering Wheel Switches

Steering wheel switches vary depending on MINI model and optional equipment installed. They include horn buttons, and may include cruise control, sound system and Steptronic shift switches.

Steering wheel switches, removing and installing

Procedure for removal of steering wheel switches is similar for various MINI steering wheels.

> ***WARNING —***
> *Read the warnings and cautions in* **721 Airbag System (SRS)**. *Serious injury may result if system service is attempted by persons unfamiliar with SRS and its approved service procedures.*

– Remove driver airbag. See **320 Steering and Wheel Alignment**.

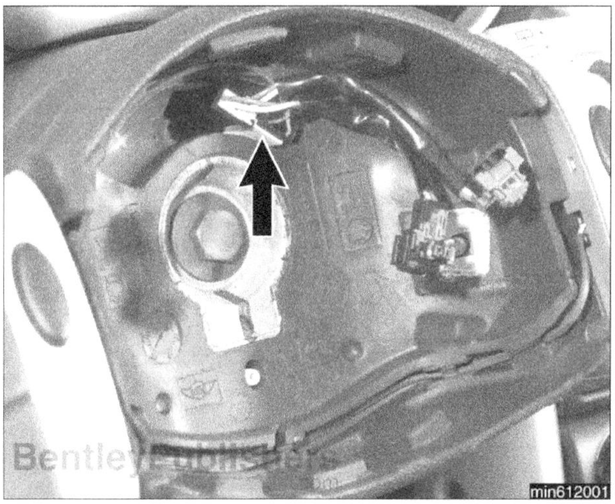

Detach electrical connector(s) (**arrow**) from steering wheel contact ring.

NOTE —
Standard steering wheel is illustrated. If equipped with multifunction (MFL) steering wheel, there is another electrical connector to detach.

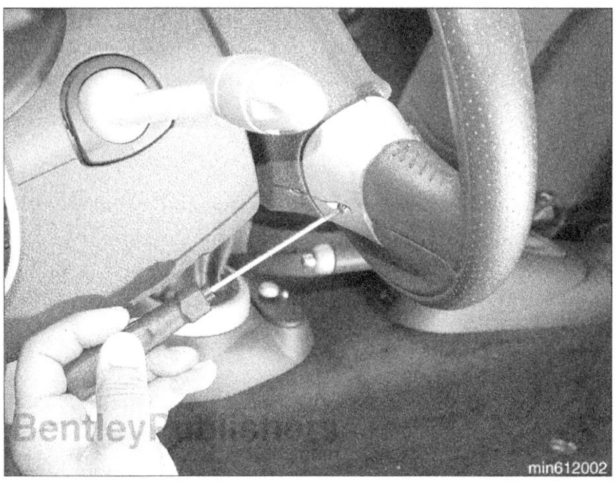

Remove switch mounting screws from back of steering wheel.

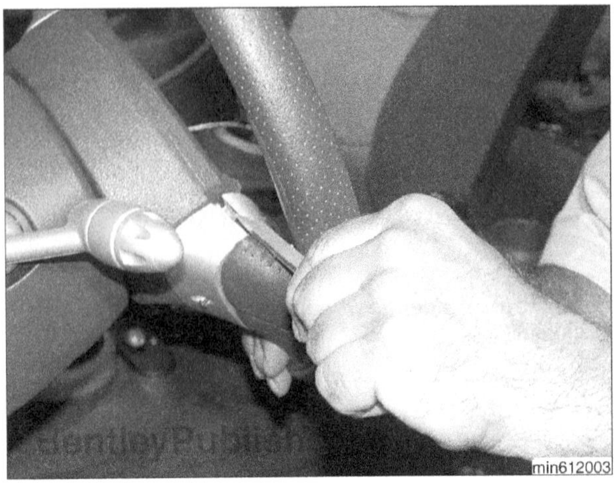

Use plastic prying tool to separate front and rear switch housing on each side.

– Unhook switch assembly wire harness from steering wheel. Remove switches.

NOTE —
The switch assembly on models with MFL steering wheel differs from those without MFL. For further information, see an authorized MINI dealer parts department.

– Installation is reverse of removal. Make sure all wires are correctly routed to prevent electrical shorts when replacing switches and driver airbag.

Steering Column Switches

EWS ring antenna, removing and installing

– Remove rubber boot surrounding ignition switch.

Use BMW special tool 61 3 300 to gently pry out ring antenna.

– Detach electrical connector and remove ring antenna.

Ignition lock cylinder, removing and installing

– Remove EWS ring antenna. See **EWS ring antenna, removing and installing**.

With ignition key in ON position (60° from LOCKED):

- Insert BMW special tool 32 3 110 or a thin piece of stiff wire into opening (**arrow**) in lock cylinder.
- Remove lock cylinder. If necessary, turn tool from side to side.
- Installation is reverse of removal.

NOTE —
Using the valet key during this procedure will provide better access to lock opening.

Ignition switch, removing and installing

- Disconnect negative (-) battery terminal.

> ***CAUTION—***
> *Prior to disconnecting the battery, read the battery disconnection cautions given in* **001 General Warnings and Cautions**.

- Pull on top edge of left lower dashboard trim to unclip and allow trim to swing down.

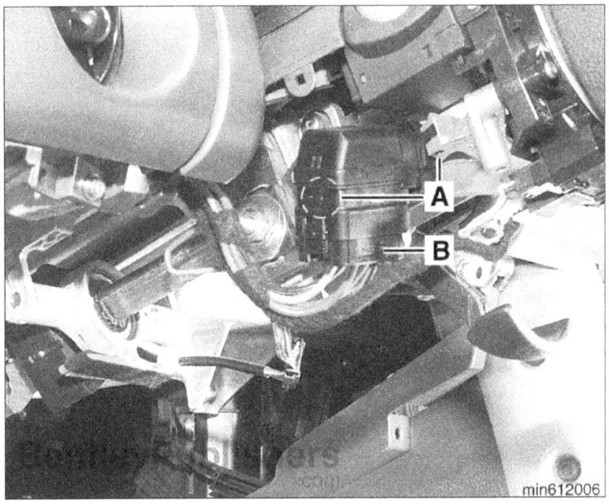

- Working at left side of steering column:
  - Clear sealing material from switch mounting screw heads (**A**) and remove screws.
  - Detach ignition switch electrical harness connector (**B**) and remove switch.
- Installation is reverse of removal. Use varnish to secure screw heads.

NOTE—
Correctly align ignition switch with ignition lock.

> ***CAUTION—***
> *After replacing ignition switch, make sure lock cylinder and electrical systems function properly.*

Windshield wiper switch, removing and installing

Procedure is similar to turn signal/headlight switch replacement. See **Turn signal/headlight switch, removing and installing** in this repair group.

Turn signal / headlight switch, removing and installing

- Remove upper and lower steering column trim. See **320 Steering and Wheel Alignment**.

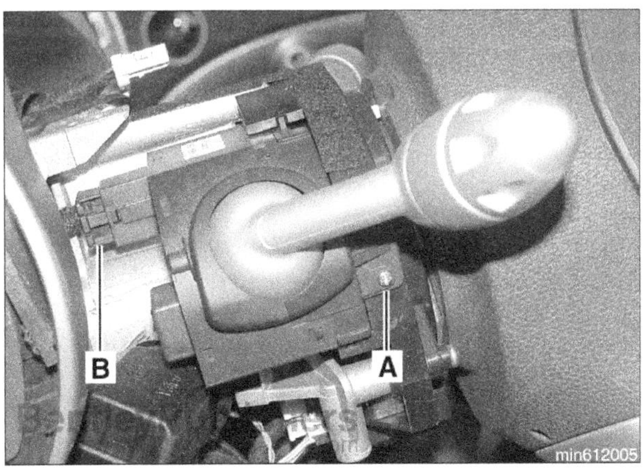

- Remove retaining screw (**A**) from front of stalk.
- Slide switch out and detach electrical harness connector (**B**).
- Installation is reverse of removal.

Pedal Cluster Switches

Brake light switch, removing and installing

- Pull on top edge of left lower dashboard trim to unclip and allow trim to swing down.

min612010

◄ Working above pedal cluster, detach brake light switch electrical harness connector (**A**).

- Pull switch from housing (**B**).
- Installation is reverse of removal. Insert switch as far as possible into housing to allow for self-adjustment.

Clutch switch

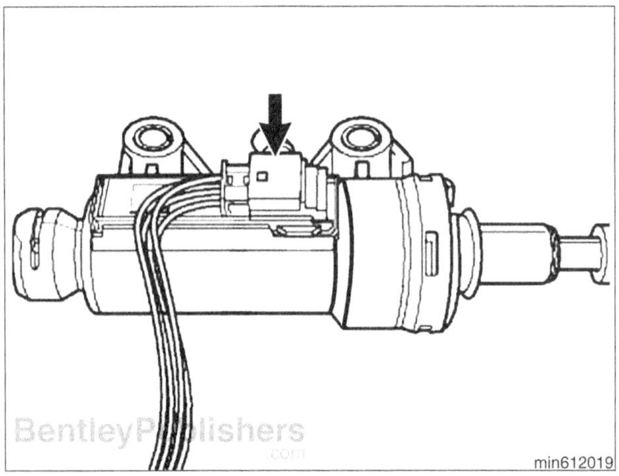
min612019

◄ The clutch pedal position switch (**arrow**) is located on the clutch master cylinder, under the left side dashboard, ahead of the clutch pedal pushrod.

Other Interior Switches

Fascia switch module, removing and installing

Removal of the fascia switch module is required to gain access to several switches for standard and optional MINI equipment. Included among these are switches for windows, door locks, fog lights and stability control (ASC / DSC).

- Remove front console trim. See **513 Interior Trim**.

min612011

◄ Remove screws (**arrows**) and pull switch module forward to release clips.

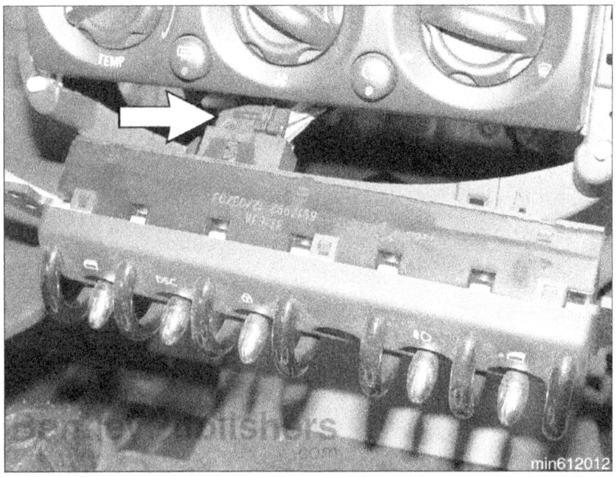

Detach electrical connector (**arrow**) using connector release lever and remove switch module.

NOTE—
Individual toggle switches are not replaceable. Switch module must be replaced as a complete unit.

– Installation is reverse of removal.

NOTE—
If model is equipped with optional telephone kit, the connector is identical to the switch module connector. Be sure of connector placement when installing.

Seat heater switches

If equipped, the seat heater switches are located in the front center console bezel. Replacement procedure is similar to outside mirror switch replacement. See **Outside mirror switch, removing and installing** in this repair group.

Outside mirror switch, removing and installing

Working at front center console, use plastic prying tool to release switch block from console.

– Detach electrical connectors and remove switch block.

NOTE—
Illustration shows version with mirror control switch only. Seat heater switches, if equipped, are part of same switch block.

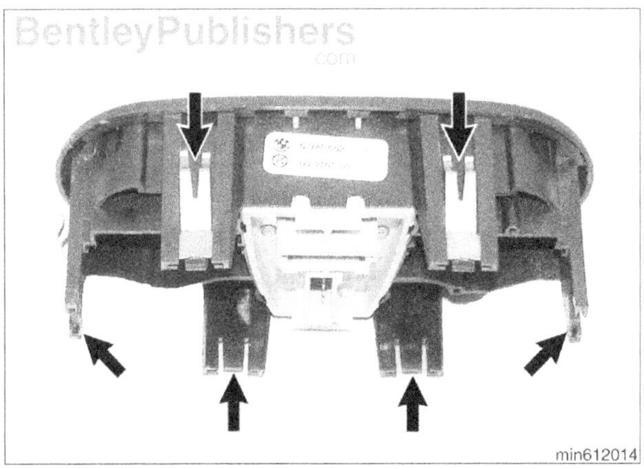

Installation is reverse of removal. Make sure metal clips (**arrows**) are in place.

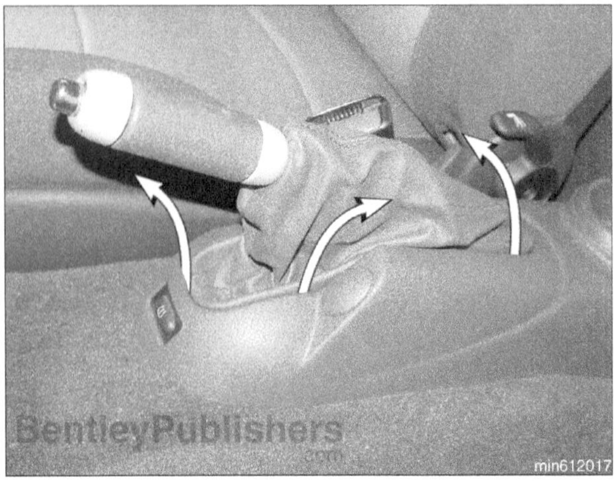

Tire pressure warning switch (RDW), removing and installing

- Release clips securing parking brake boot to rear console. Fold boot upwards (**arrows**) to gain access to switch.
- Push tire pressure warning switch forward and out of rear console.

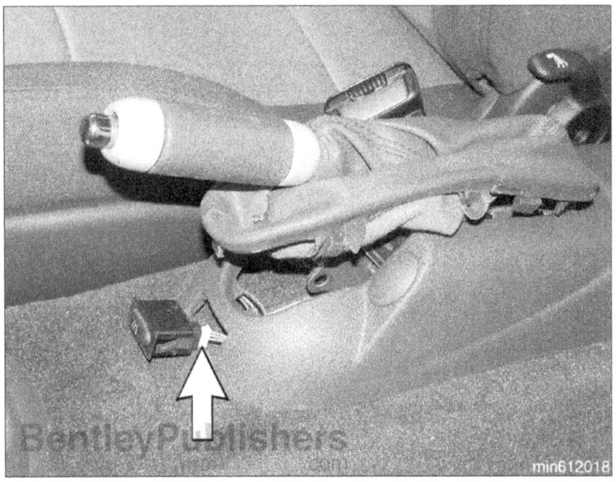

- Detach electrical harness connector (**arrow**) and remove switch.
- Installation is reverse of removal.

CONVERTIBLE SWITCHES

Overhead console switches, replacing

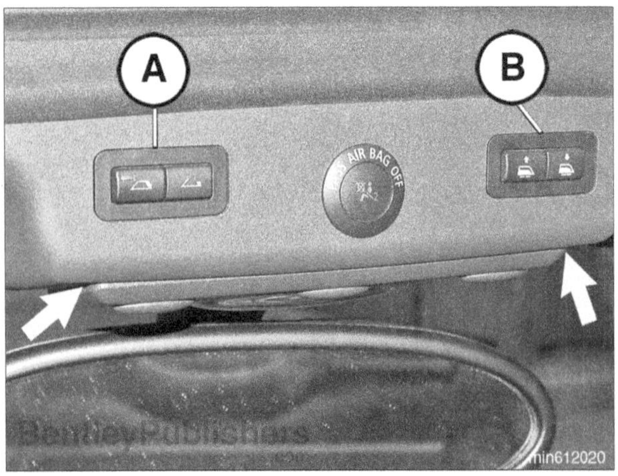

- Carefully pry (**arrows**) out and remove courtesy light console above interior rear view mirror.
- Reach behind switch and press out from overhead console. Disconnect harness connector and remove switch.
    - **A** = Convertible top up/down switches
    - **B** = Front window up/down switches
- Install in reverse order of removal.

Luggage compartment light switch, replacing

- Use plastic pry tool to pry out luggage compartment light switch (**arrow**).
  - Pry at top portion of switch.
- Disconnect harness connector and remove switch.
- Install in reverse order of removal.

Rear lid unlock switch, replacing

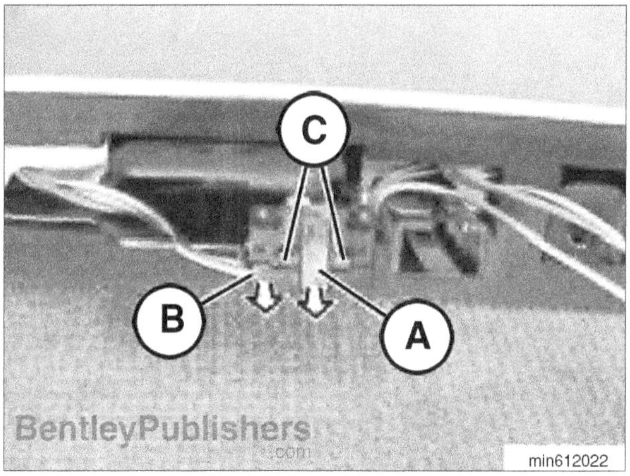

- Remove rear hatch outer handle.
- Detach clip (**A**) in direction of **arrow**.
- Remove switch (**B**) in direction of **arrow** from retaining clips (**C**).
- When installing, make sure clip (**A**) is correctly seated.

Convertible top tension bar switch, replacing

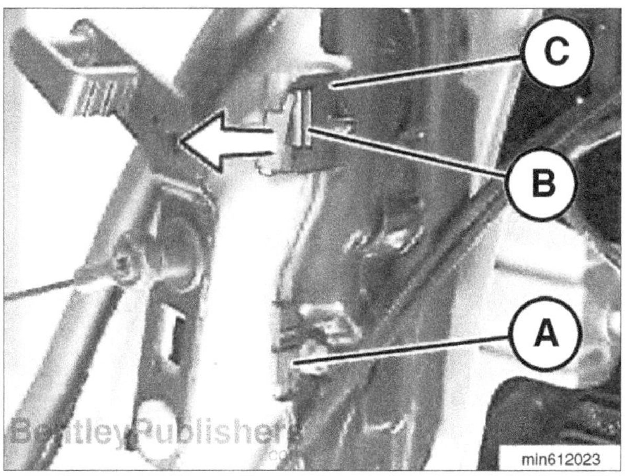

- Remove rear trim panel from left side of rear lid opening.
- Disconnect harness connector (**A**).
- Remove switch (**B**) in direction of **arrow** from holder (**C**).
- Install in reverse order of removal.

Convertible top rear window shelf switch, replacing

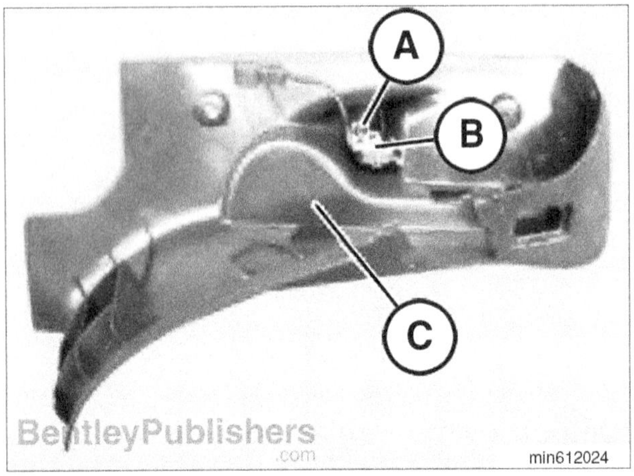

- Remove rear trim panel from left side of rear lid opening.
- Release mounting screw (**A**).
- Remove switch (**B**) from panel for rear lid opening (**C**).
- Install in reverse order of removal.

620 Instruments

GENERAL

This repair group covers removal and installation of the instrument cluster, instrument carrier and clock.

INSTRUMENT CLUSTER AND CARRIER

MINI models have a tachometer cluster with indicator and warning lights mounted on the steering column and a speedometer with indicator and warning lights mounted in a carrier in the center of the dashboard.

MINI models equipped with optional navigation system have a dual pod cluster containing a tachometer and speedometer mounted on the steering column and a navigation display mounted in the center of the dashboard.

MINI models equipped with optional cockpit chrono pack have a dual pod cluster containing a tachometer and speedometer mounted on the steering column and a multifunction gauge cluster in the center of the dashboard.

NOTE —
Instrument lighting and hazard warning switch are integral with the instrument carrier and not available separately.

Instrument cluster, removing and installing

Removal procedure is similar for standard and navigation equipped MINI models.

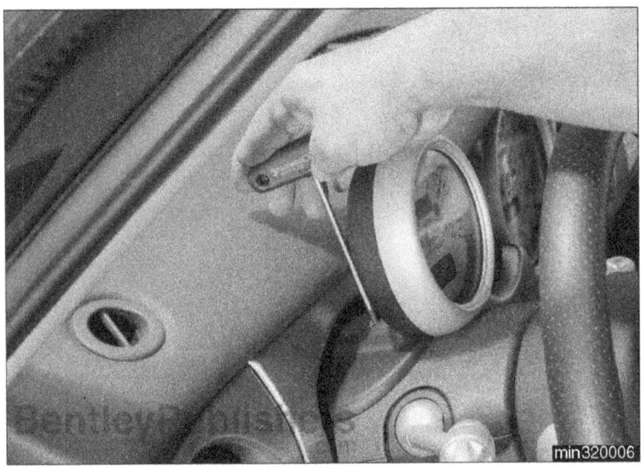

◄ Working at the steering column, remove screws (**1**) at back of instrument cluster.

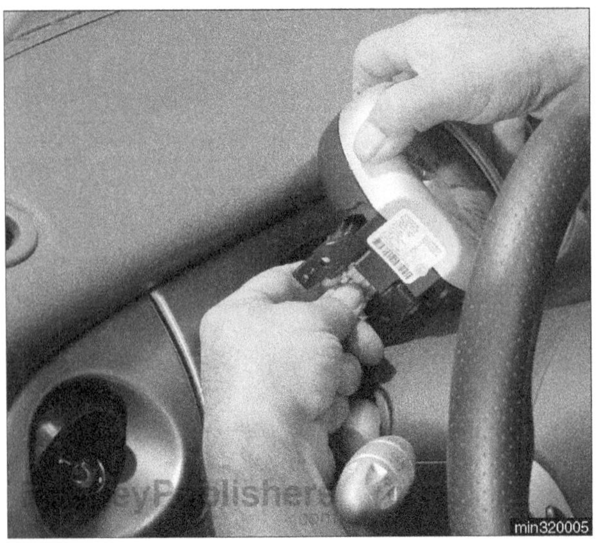

◄ Release electrical connector (**1**) and remove cluster.

– Installation is reverse of removal.

NOTE —
Instrument cluster IKE memory functions (such as odometer reading) are retained in the center dashboard display.

Instrument carrier, removing and installing

– Remove center dashboard face trim. See **513 Interior Trim**.

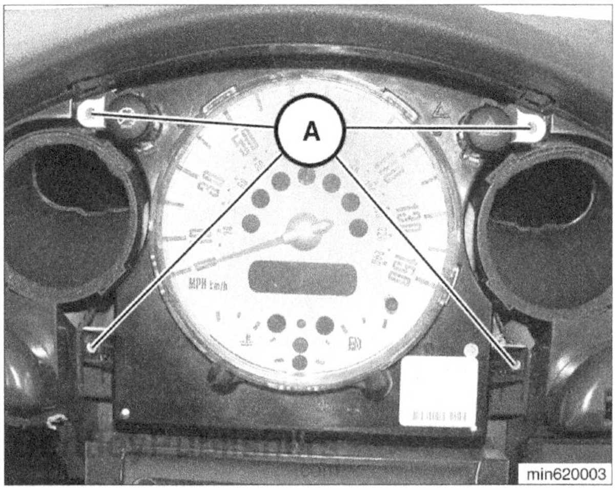

◄ Remove screws (**A**) holding instrument carrier to dashboard.

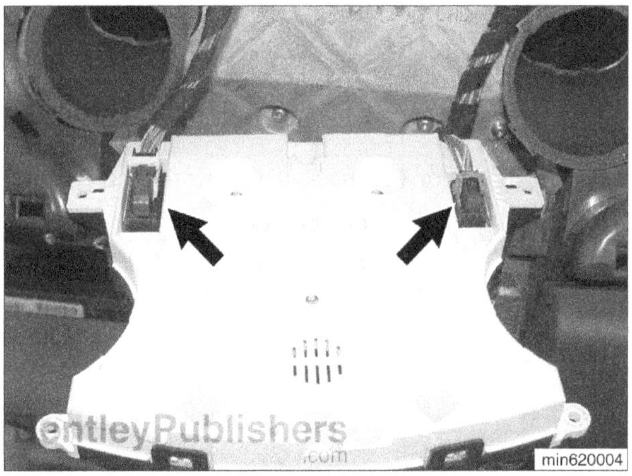

◄ Unlock electrical connectors (**arrows**) and detach. Remove instrument carrier.

– Remove screws holding speedometer and remove speedometer from carrier.

– Installation is reverse of removal. If necessary, use BMW service tester (DIS, GT1 or MoDiC) to synchronize speedometer with vehicle.

Navigation system carrier, removing and installing

– Remove center dashboard face trim. See **513 Interior Trim**.

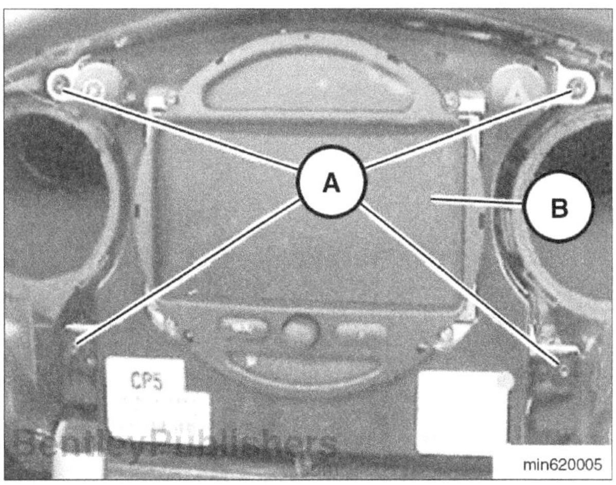

◄ Remove screws (**A**) holding navigation carrier and display (**B**) to dashboard.

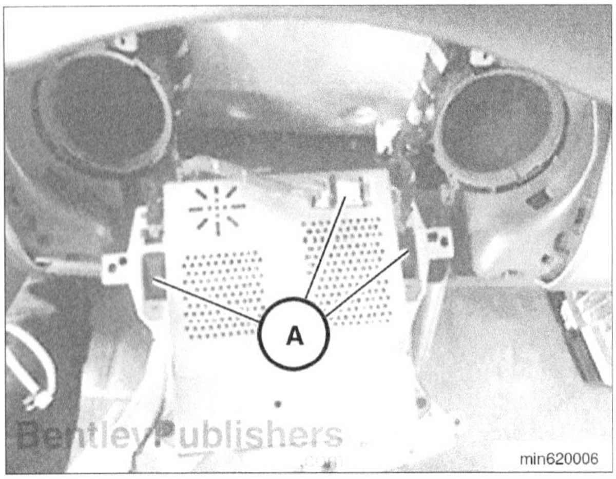

min620006

- Unlock electrical connectors (**A**) and detach. Remove navigation carrier.
- Remove screws holding navigation display and remove display from carrier.
- Installation is reverse of removal. If necessary, use BMW service tester (DIS, GT1 or MoDiC) to synchronize display with vehicle.

CLOCK

Roof mounted clock, removing and installing

Convertible models and some later model coupes do not have the roof mounted clock. In these vehicles the clock has been relocated to the instrument cluster.

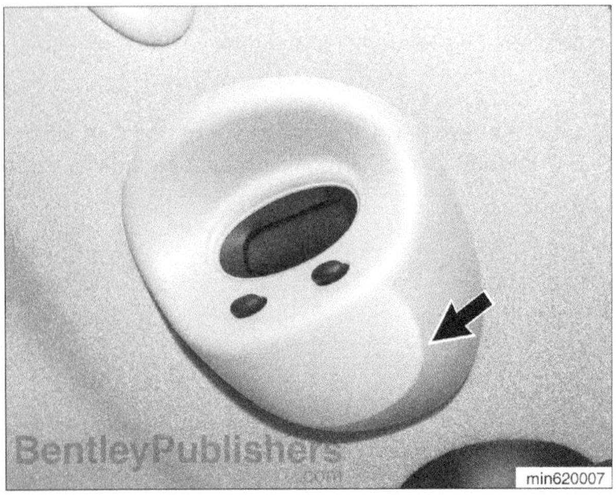
min620007

- Working at front of headliner, use a plastic prying tool to release clock housing in direction of **arrow**.
- Release electrical connector from back of clock and remove clock housing.

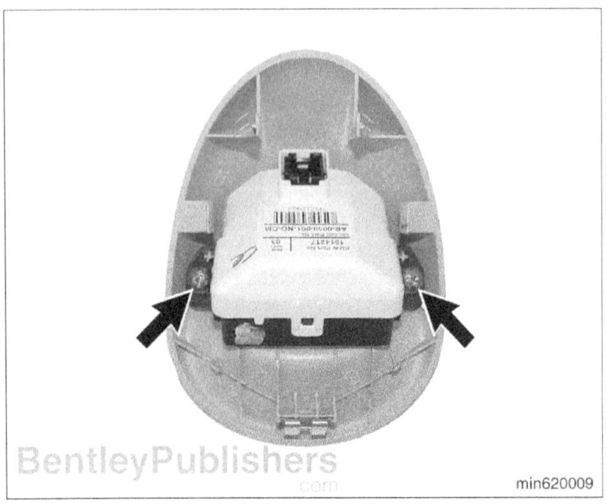
min620009

- Remove screws (**arrows**) and remove clock from housing.
- Installation is reverse of removal.

630 Lights

General

This repair group covers exterior and interior lighting. For information on the headlight switch, see **612 Switches**.

MINI exterior lighting is controlled by the General Module (BC1) except for back-up lights and optional xenon headlights. Xenon headlight operation is signaled by the BC1, but controlled by individual modules at each headlight.

MINI interior lighting is also controlled by the BC1. Interior lighting is activated by a signal from the door contacts, tailgate switch, key remote, inertia switch or manually from the overhead light switch.

Special tools

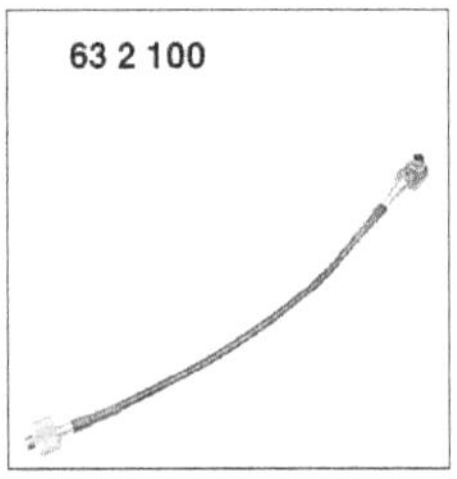

Headlight adjustment tool
(BMW special tool 63 2 100)

Bulb applications

Bulb applications for MINI models are shown below.

Bulb location	Type and rating
Headlights	
• Low beam (halogen)	H7 55W
• Low beam (xenon)	D2-R
• High beam	H7 55W
Foglights	
• Front	H7 55W
• Rear	21W
Turn signals and taillights	
• Back-up	21W
• Brake/Taillight	21W/5W
• Licence plate	5W
• Side marker	5W
• Turn signal (front and rear)	21W
Interior lights	
• Footwell	5W
• Glove compartment	5W
• Cargo compartment	5W
• Interior light	5W
• Reading light	5W

HEADLIGHTS

The headlight assembly was redesigned for 2005 and later MINI models (from 07/2004). The information in this section covers both early and late headlight assemblies.

Headlight aim, adjusting

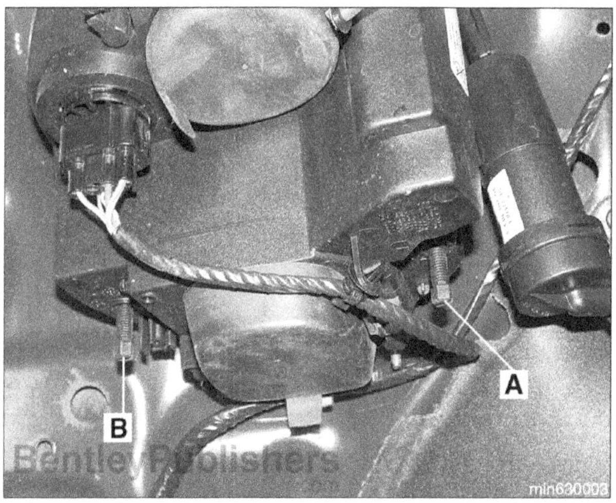

NOTE —
Right headlight shown.

Headlight adjustment is accomplished by turning vertical adjustment screw (**A**) and horizontal adjustment screw (**B**).

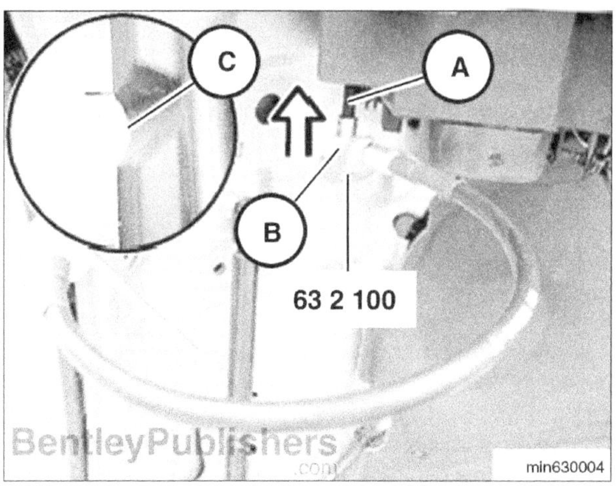

Use special tool 63 2 100 to adjust headlight with hood closed:

- Attach tool to adjustment screw (**A**) and lock collar (**B**).
- Route tool through notch (**C**) between hood and fender liner.
- Close hood.

– Adjust headlight by turning special tool.

– If necessary, repeat procedure for remaining adjustment screw.

– If necessary, repeat procedure for opposite headlight.

Headlight leveling module (LWR), removing and installing

MINI models equipped with xenon headlights include a headlight leveling module (LWR). This system automatically adjusts the vertical position of the headlights to compensate for vehicle load angle changes.

LWR monitors vehicle load via two Hall effect sensors mounted to the front and rear suspension members. When an adjustment to the angle of the headlight beams is necessary, LWR simultaneously activates two stepper motors (one in each headlight assembly) to change the vertical aim of the headlights.

– Disconnect negative (-) battery terminal.

> ***CAUTION—***
> *Prior to disconnecting the battery, read the battery disconnection cautions given in* **001 General Warnings and Cautions**

– Remove glove compartment. See **513 Interior Trim**.

< Remove nuts (**arrows**) securing module to inside of dash.

– Release electrical connector and remove module.

– Installation is reverse of removal. After replacement, LWR module must be coded using a diagnostic scan tool (DISplus, GT1, MoDiC, etc.).

Tightening torque	
LWR module to dashboard	4 Nm (3 ft-lb)

Headlight bulb, replacing (halogen)

Halogen bulb replacement is similar for high and low beam.

– Working with hood open, remove protective rubber cap from rear of headlight unit.

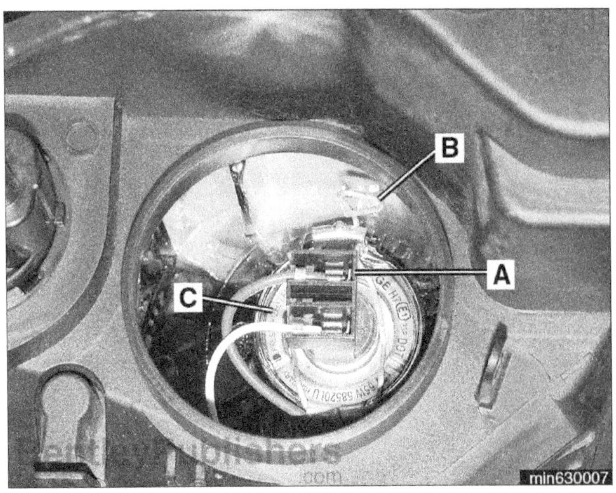

< Remove electrical connector (**A**) from bulb.

– For vehicles produced up to 07/2004, release retaining clip (**B**) holding bulb to housing and remove bulb (**C**).

– For vehicles produced from 07/2004, unclip bulb (**C**) toward rear from retaining catch.

– Replace bulb, making sure it is correctly seated before fastening retaining clip and protective cap.

NOTE—
- *Do not touch bulb glass with bare fingers.*

Headlight bulb, replacing (xenon to 07/2004)

Xenon equipped models use xenon low beams and halogen high beams. For high beam replacement, see **Headlight bulb, replacing (halogen)**.

– Disconnect negative (-) cable from battery.

> ***CAUTION—***
> *Prior to disconnecting the battery, read the battery disconnection cautions given in* **001 General Warnings and Cautions**

– Working with hood open, remove protective rubber cap from rear of headlight unit.

◄ Turn xenon ignition unit in direction of **arrow** and release from bulb.

◄ Release retaining clips (**arrows**) and remove bulb assembly.

– Replace bulb, making sure it is correctly seated before fastening retaining clip.

– Install ignition unit and protective cap.

NOTE—
Do not touch bulb glass with bare fingers.

Headlight bulb, replacing (xenon from 07/2004)

Xenon equipped models use xenon low beams and halogen high beams. For high beam replacement, see **Headlight bulb, replacing (halogen)**.

– Disconnect negative (-) cable from battery.

> ***CAUTION—***
> *Prior to disconnecting the battery, read the battery disconnection cautions given in* **001 General Warnings and Cautions**

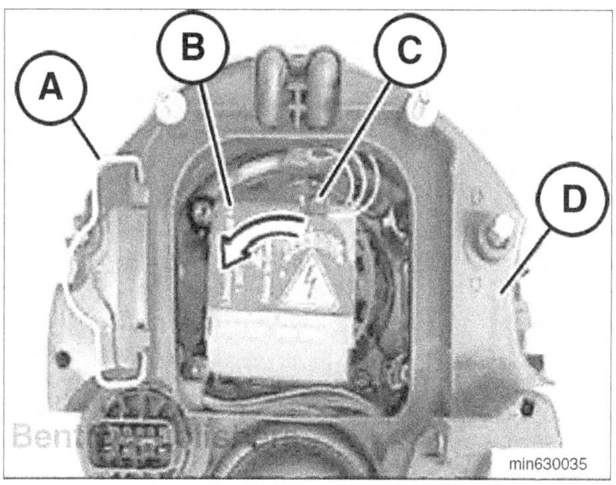

◄ Working with hood open, release lock (**A**) and remove protective cap from rear of headlight unit.

– Turn xenon ignition unit (**B**) in direction of **arrow** and release from bulb.

- Harness connector (**C**) should automatically release when rotating ignition unit (**B**).

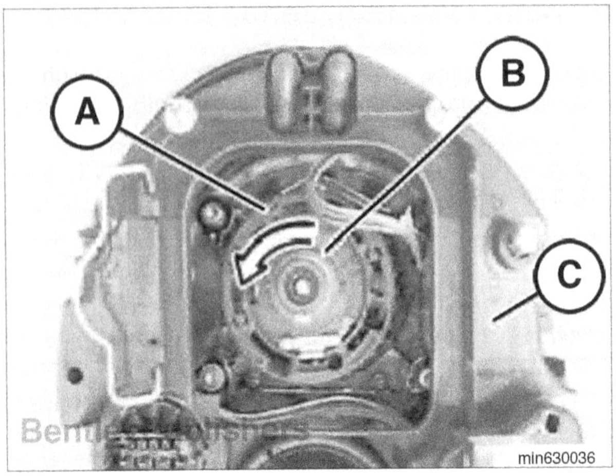

Unlock retaining ring (**A**) by rotating in direction of **arrow**.

– Remove bulb (**B**) from housing (**C**).

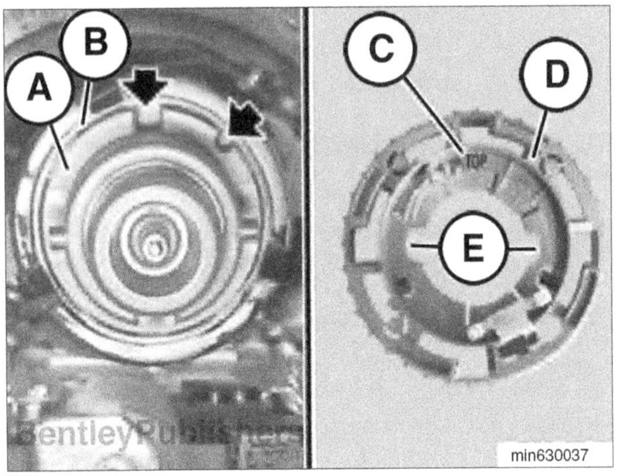

Install new bulb (**A**), making sure it is correctly seated (**arrows**) in headlight housing (**B**).

NOTE—
Do not touch bulb glass with bare fingers.

– Install retaining ring (**D**) with "Top" marking (**C**) at top. Apertures (**E**) should slide over bulb fixtures for ignition unit.

– Install ignition unit and attach harness connector.

– Install protective cover on rear of headlight assembly and lock in place.

Headlight assembly, removing and installing (all with halogen and xenon to 07/2004)

– Remove headlight washer assembly (if equipped). See **611 Wipers and Washers**.

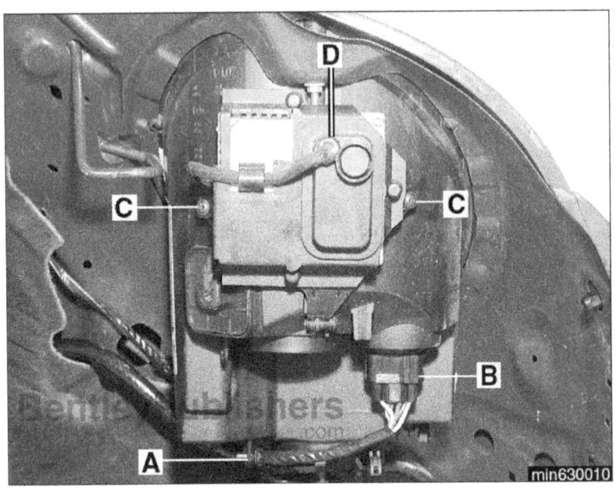

All models: Release electrical connector (**A**) from headlight and release wire harness from clip (**B**).

– Models with xenon headlights: Release screws (**C**) on xenon control unit (**D**) and pivot control unit down.

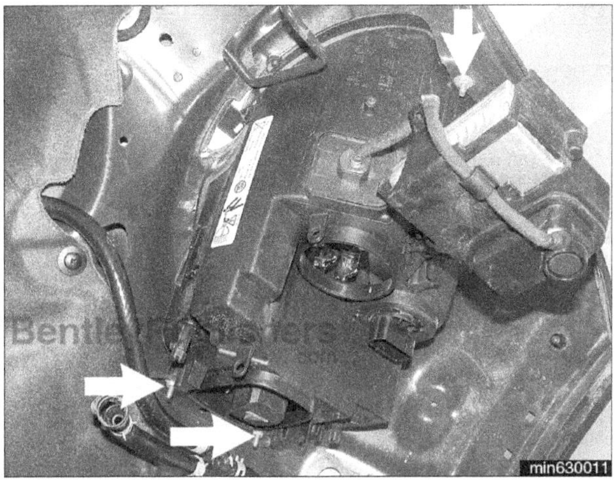

Remove nuts (**arrows**) securing headlight assembly to hood and remove assembly.

– Installation is reverse of removal. If replacing headlight assembly, transfer headlight bulbs to new unit.

– If necessary, adjust headlights. See **Headlight aim, adjusting** in this repair group.

Headlight assembly, removing and installing (xenon from 07/2004)

– Remove headlight washer assembly (if equipped). See **611 Wipers and Washers**.

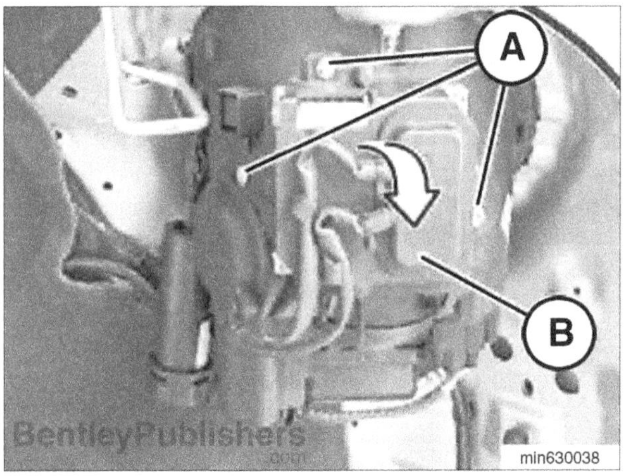

Working at back of headlight assembly, remove mounting screws (**A**) and fold down (**arrow**) control unit holder (**B**).

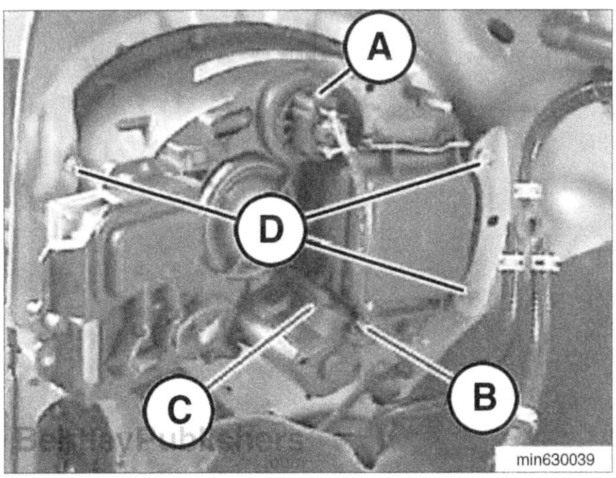

Disconnect harness connector (**A**) and release wiring harness retaining clip (**B**).

– Remove mounting nuts (**D**) and headlight assembly (**C**).

– Installation is reverse of removal. If replacing headlight assembly, transfer headlight bulbs to new unit.

– If necessary, adjust headlights. See **Headlight aim, adjusting** in this repair group.

Xenon control unit, removing and installing

– Disconnect negative (-) cable from battery.

> ***CAUTION —***
> *Prior to disconnecting the battery, read the battery disconnection cautions given in* **001 General Warnings and Cautions**.

Up to 07/2004

◄ Working with the hood open, remove screws (**arrows**) securing control unit cover to headlight assembly.

– Swing cover down to expose control unit.

From 07/2004

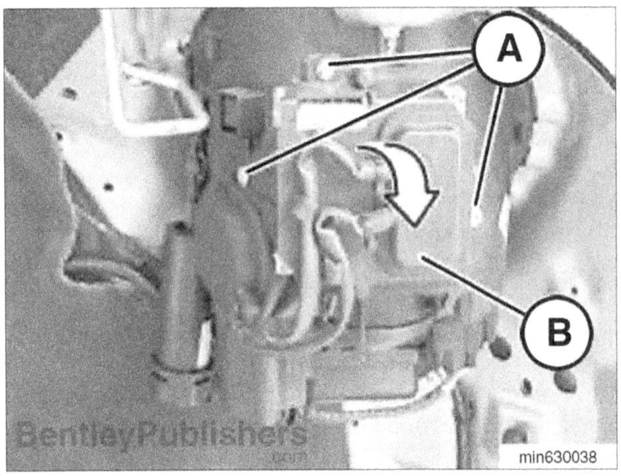

◄ Working at back of headlight assembly, remove mounting screws (**A**) and fold down (**arrow**) control unit holder (**B**).

Continued for all vehicles

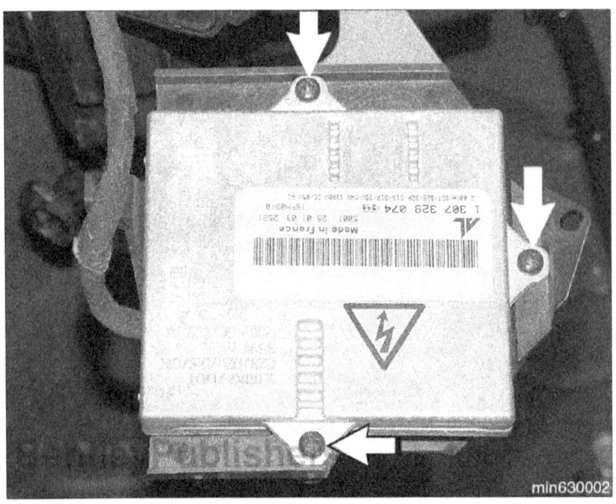

◄ Remove screws (**arrows**) and remove control unit from cover.

– Installation is reverse of removal.

Exterior Lights

Back-up light (to 07/2004) and rear fog light (from 07/2004), replacing

For the 2005 model year (as of 07/2004), the back-up lights were integrated into the taillight assembly. The optional rear fog light is now located in the center of the rear bumper cover, where the back-up light had previously been. The procedure below applies to both lights, as applicable.

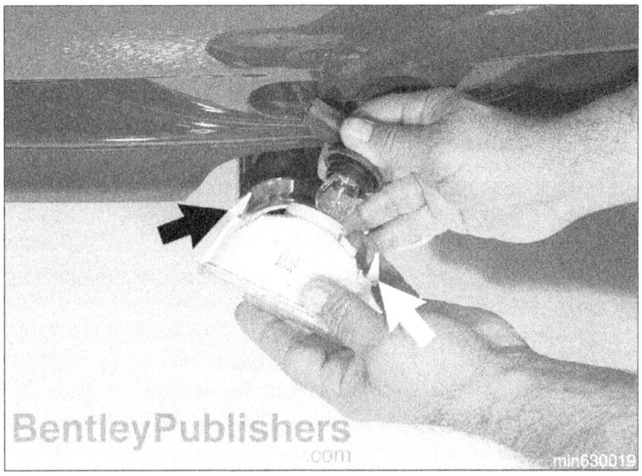

- Working behind rear bumper, release retaining clips (**arrows**) and push bulb holder out of bumper.
- Turn bulb holder 90° to remove from housing.
- Replace bulb.

Center brake light, removing

CAUTION—
Before removing screws, secure center brake light to body with masking tape to avoid damaging light or paint finish.

Coupe models

- Working with rear hatch open, remove screws (**arrows**) holding center brake light to rear hatch.

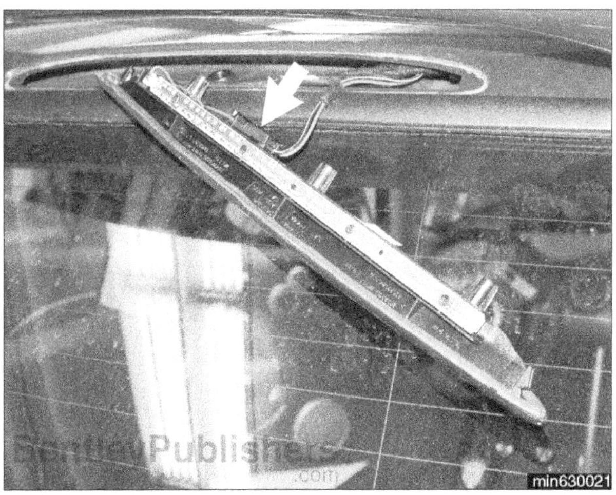

- Close hatch and remove brake light assembly.
- Release clip (**arrow**) to detach electrical connector.
- Remove light assembly.

NOTE—
The center brake light is an LED assembly that must be replaced as a complete unit.

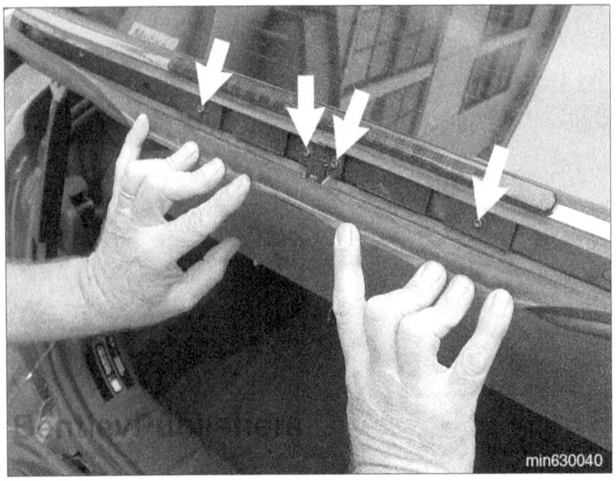

min630040

Convertible models

- Open rear lid and lift rear of convertible top up. Use loading aids to support rear of convertible top. See **541 Convertible Top**.
- ◄ Pull back rubber gasket and remove mounting screws (**arrows**) holding center brake light to convertible top.
- Disconnect harness connector and remove brake light assembly.

NOTE —
The center brake light is an LED assembly that must be replaced as a complete unit.

min630012

Foglight aim, adjusting

- ◄ Working with hood open, turn adjusting screw (**arrow**) to aim foglight.

Foglight assembly, removing and installing

- Remove front wheel and remove front fender lining.
- ◄ Release electrical connector (**A**) from foglight bulb holder.
- Turn bulb holder in direction of **arrow** and remove.

min630013

Remove screws (**arrows**) and remove foglight assembly.

– Installation is reverse of removal. If necessary, adjust foglight. See **Foglight aim, adjusting** in this repair group.

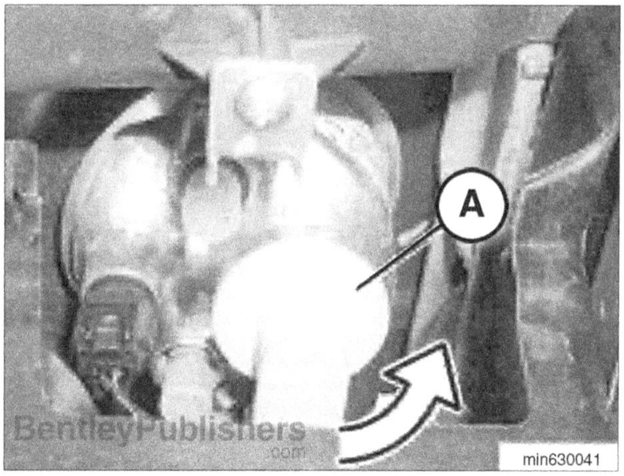

Front turn signal bulb, replacing

– Open front hood.

Reach into the opening for the turn signal indicators from above and rotate bulb holder (**A**) in direction of **arrow** to remove.

– Remove and replace bulb.

Front turn signal assembly, removing

– Remove front bumper cover. See **510 Exterior Trim, Bumpers**.

Working on inside of bumper cover, remove screws (**arrows**) and remove signal light assembly.

License plate light, removing

- Holding the rear hatch partially open, use a plastic prying tool at **arrow** to pry out lens.
- Remove bulb.

Side marker lights, removing

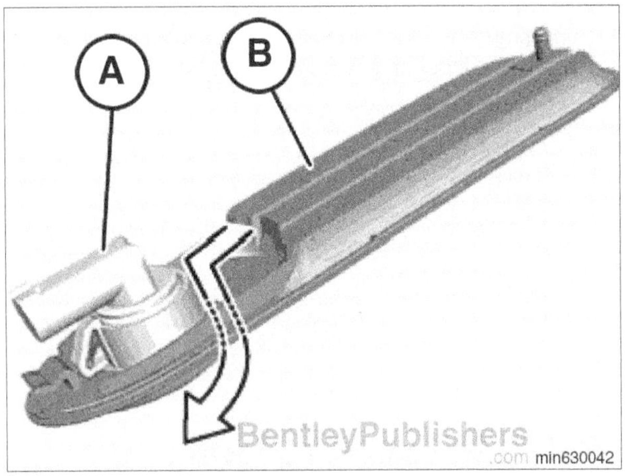

- Remove side marker lights by prying on front edge with a plastic prying tool and sliding bezel toward rear. Remove light bezel (**A**) in direction of **arrow** from plastic trim (**B**).
- Remove harness connector and replace light bulb.
- Install in reverse order of removal.

Taillight assembly, removing and installing

- The taillight assembly was redesigned for 2005 and later MINI models (from 07/2004). The new taillight assembly now has a reverse light in the bottom inner corner. The information in this section covers both early and late taillight assemblies.

Working with rear hatch/lid open, release clips (**arrows**) and remove side access panel.

- Coupe shown. Convertible has similar small access panels in the sides of the cargo compartment liner.

Remove nut (**arrow**) securing taillight assembly to body.

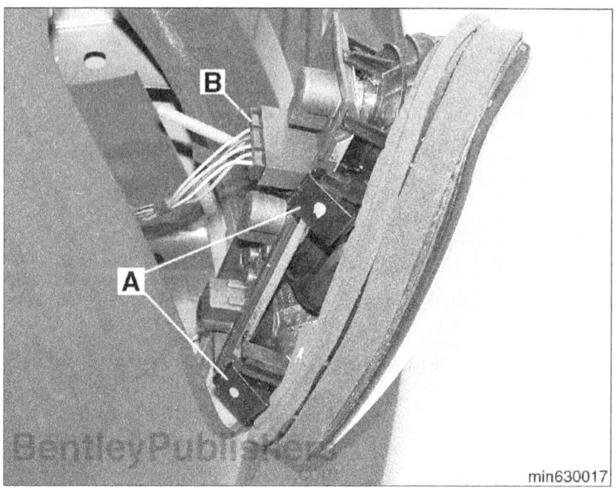

Release retaining clips (**A**) and tilt assembly away from body.

– Disconnect harness connector (**B**) and remove taillight assembly.

– Installation is reverse of removal. Make sure taillight gasket is properly located and that body mating surface is clean.

Taillight bulbs, replacing

Removal of the taillight assembly is not necessary to replace the bulbs on the coupe models. Access to the taillight bulb holders is limited on convertible models, therefore removal of the taillight assembly is recommended. Taillight assemblies are shown removed here for clarity.

Up to 07/2004

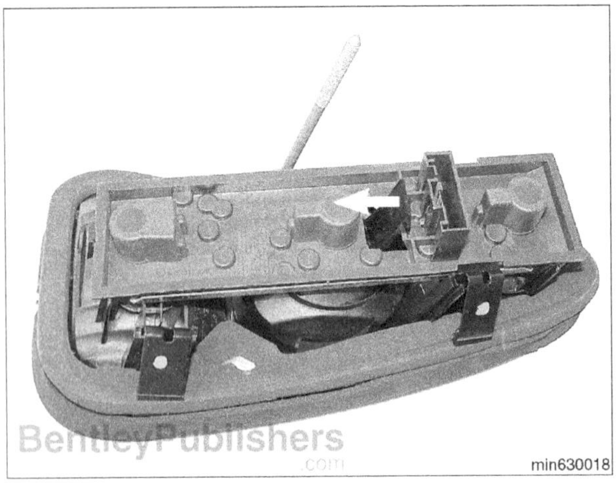

- Release clip (**arrow**) and remove bulb holder.

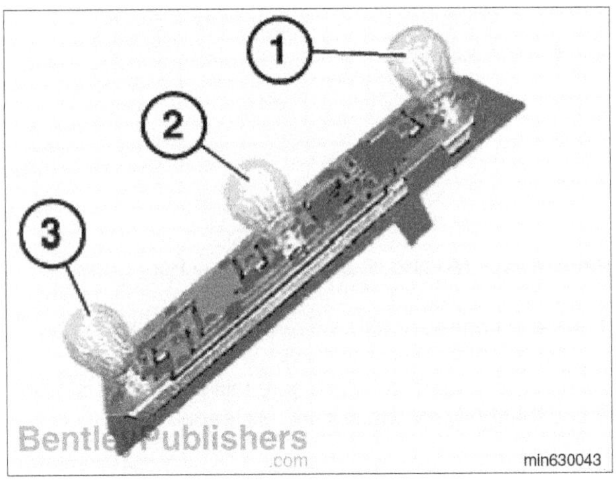

- Replace bulbs as needed.
  1. Fog light bulb (optional)
  2. Turn signal bulb
  3. Brake/taillight bulb

From 07/2004

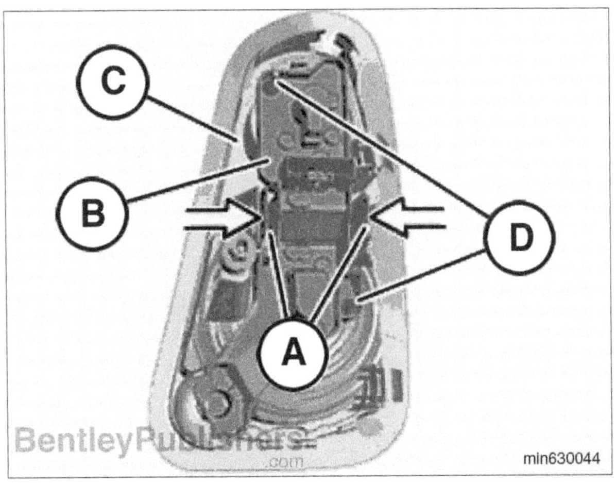

- Squeeze retaining clips (**A**) and remove bulb holder (**B**).
- Be sure to properly locate bulb holder on guide pins (**D**) when installing.

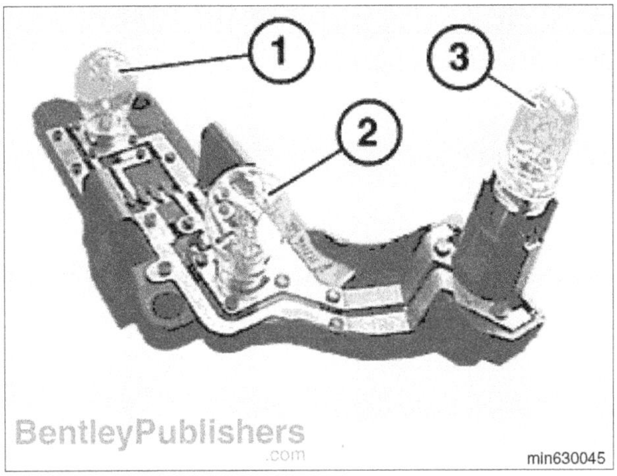

Replace bulbs as needed.

1. Turn signal bulb
2. Brake/taillight bulb
3. Back-up light bulb

Interior Lights

Cargo compartment light, removing

Use a plastic prying tool to release luggage compartment light from side of cargo compartment.

– Release electrical connectors and remove light.

Ceiling light, removing

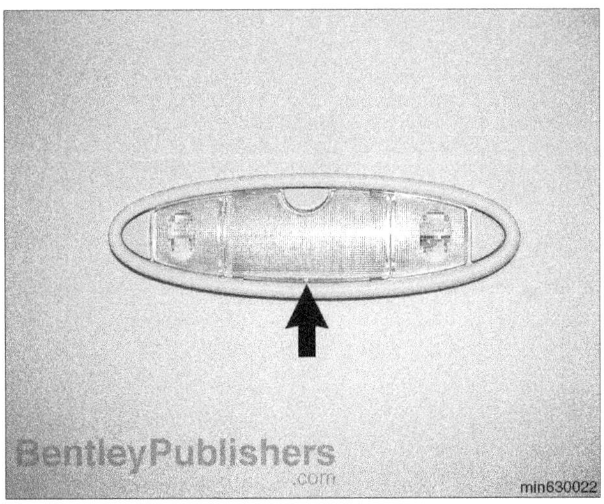

Use a plastic prying tool at **arrow** to release cover from ceiling light module.

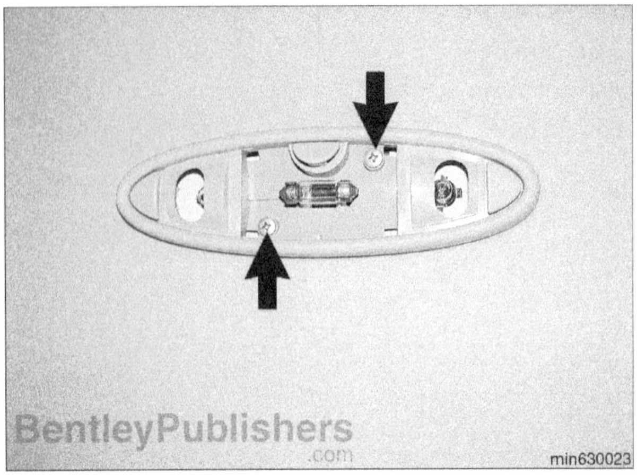

Remove screws (**arrows**) holding light module to headliner.

– Release electrical connector from back of module and remove ceiling light.

Footwell light, removing

Use a plastic prying tool to release footwell light from footwell trim.

– Release electrical connector and remove footwell light.

Glove compartment light, removing

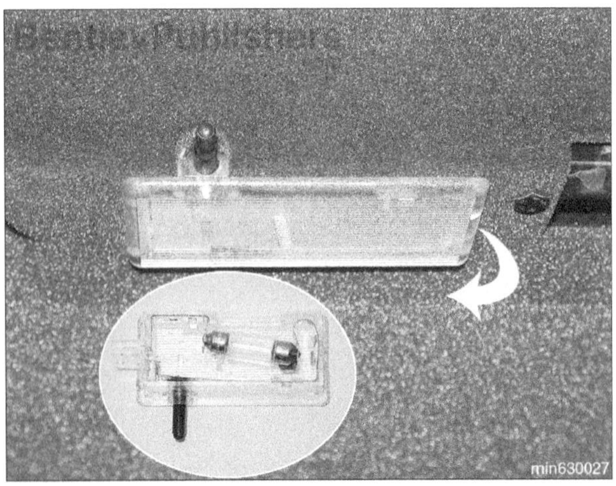

Use a plastic prying tool to release glove compartment light from glove compartment.

– Release electrical connector and remove light.

640 Heating and Air-conditioning

640

General

This repair group covers the heating and air-conditioning (A/C) system. Many of the procedures require that the A/C refrigerant charge be evacuated. A/C system recharging procedures are beyond the scope of this manual. See **A/C system warnings and cautions** in this repair group.

For information on the engine cooling system, see **170 Radiator and Cooling System**.

Ventilation microfilter replacement is covered in **020 Maintenance**.

Special tools

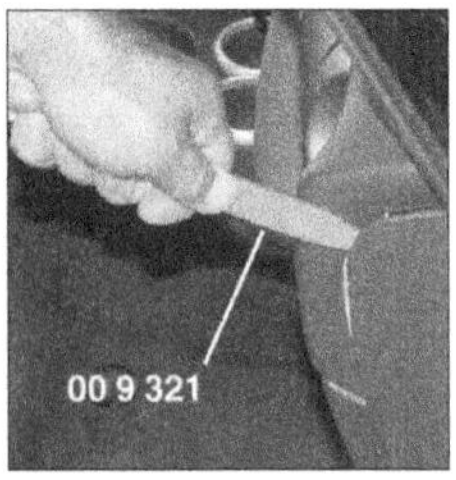

◄ Plastic prying tool
(BMW special tool 00 9 321)

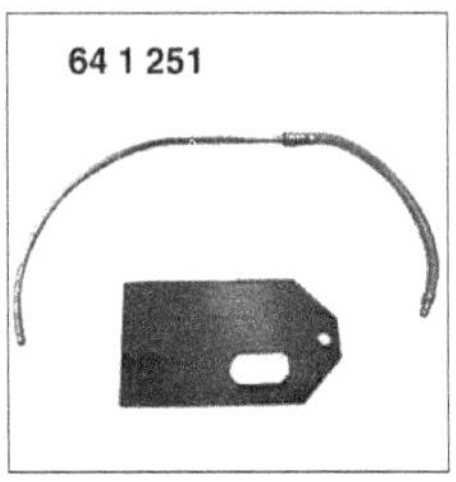

◄ A/C vent spray wand
(BMW special tool 64 1 251)

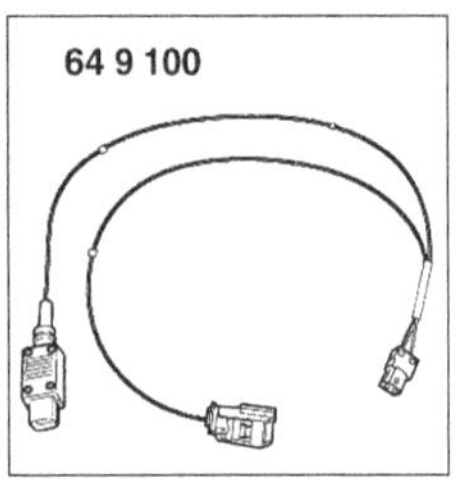

◄ Test lead
(BMW special tool 64 9 100)

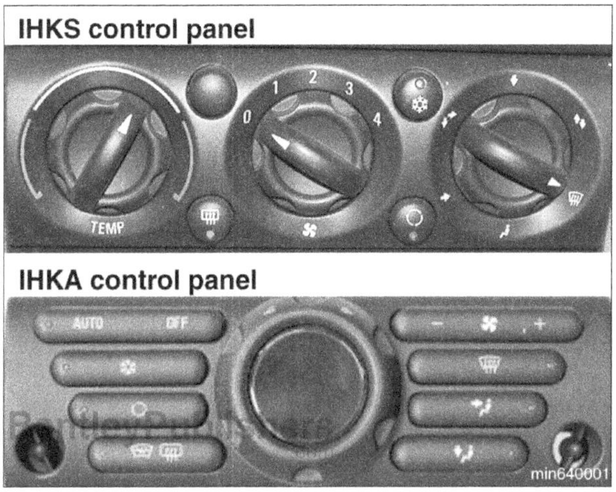

Heating and air-conditioning systems

Two heating and A/C systems are available on MINI models:

- **IHKS**: Manually controlled system. Driver input is required to regulate temperature, air direction and air speed.
- **IHKA**: Optional automatic climate control. System controls air outlet temperature, air direction and air speed according to program set by driver. Automatic functions may be overridden by driver.

Both systems consist of a refrigerant system, a heater system and a control system. The refrigerant system is the same on both. Heater assembly and control systems are different.

Refrigerant system (IHKS and IHKA)

The refrigerant system provides dehumidified cool air to the vehicle interior. The system is a sealed, closed loop, filled with R134a refrigerant as the heat transfer medium. Oil is added to the refrigerant to lubricate the internal components of the compressor.

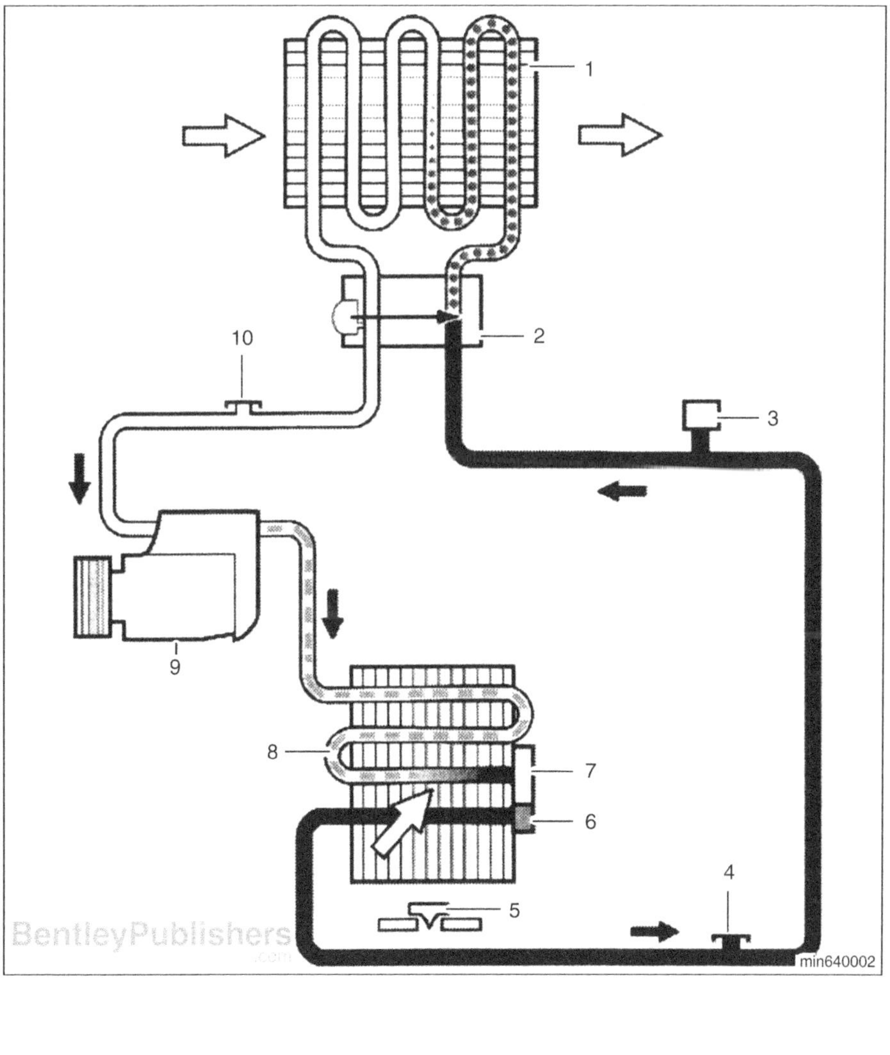

Refrigerant system components

1. **Evaporator**
2. **Expansion valve**
3. **Pressure sensor**
4. **High pressure service port**
5. **Cooling fan**
6. **Filter**
7. **Desiccant**
8. **Condenser**
9. **Compressor**
10. **Low pressure service port**

Heating system

The heating and ventilation system controls the temperature and distribution of air supplied to the vehicle interior. The system consists of microfilter housing, heater assembly, distribution ducts, refrigerant system and control panel.

Fresh air enters through vents beneath the windshield and flows into the heater assembly. A blend flap mixes the warm air passing the heat exchanger with the cool air and distributes it into the vehicle interior. Flow-through vents incorporated in the cargo compartment enable the air to exit the vehicle interior.

Fresh or recirculated air passes through the filter into the heater assembly where an electrical variable speed blower forces the air through the system. Depending on the settings on the control panel, the air is then heated or cooled and supplied through the distribution ducts to face, defrost and floor level outlets.

IHKS heating and air distribution

The IHKS system allows manual selection of inlet air source, outlet air temperature, air distribution and blower speed. Components of the IHKS heating and air distribution system include:

- Microfilter
- Heater assembly (including air-conditioning system)
- Blower motor and resistor pack
- Heater core
- Air flow control flaps
- Distribution ducts
- Outlet vents

IHKA heating and air distribution

The IHKA system differs from the IHKS system as follows:

- Two servo motors controlled by the IHKA module operate the temperature blend flap and air distribution mechanism. Flap position is monitored by the IHKA using potentiometers integrated in the servomotors.
- A heater core temperature sensor, located on the right front of the heater core, measures the temperature of the heated air.
- A rocker switch and resistor control the blower motor at eight speeds. The blower speed is controlled manually via the rocker switch or automatically in AUTO mode.

NOTE—
Troubleshoot IHKA operation using BMW scan tools DISplus, GT1 or MoDiC.

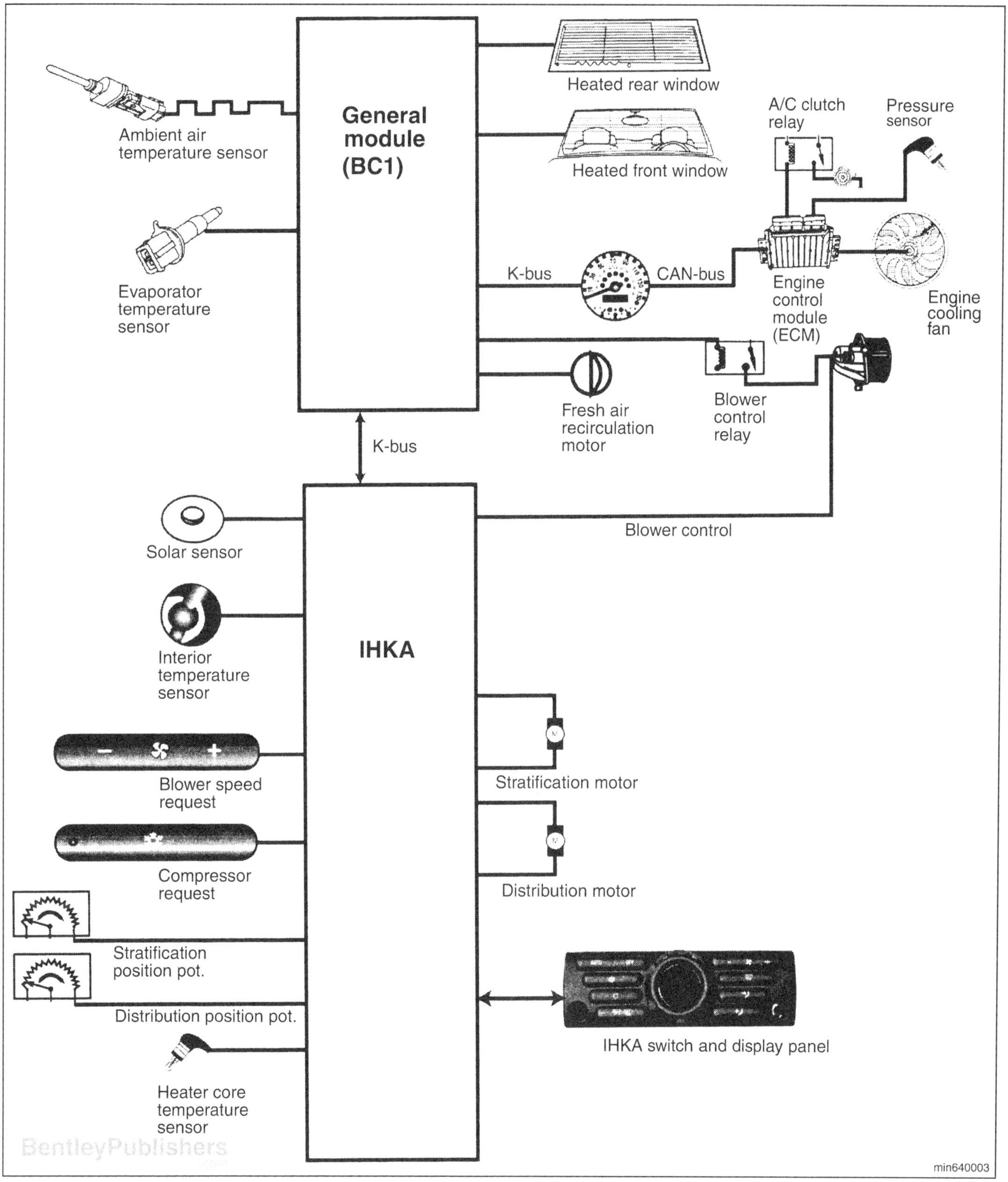
General module (BC1)
Heated rear window
Heated front window
A/C clutch relay
Pressure sensor
Ambient air temperature sensor
Evaporator temperature sensor
K-bus
CAN-bus
Engine control module (ECM)
Engine cooling fan
Fresh air recirculation motor
Blower control relay
K-bus
Solar sensor
Blower control
Interior temperature sensor
IHKA
Blower speed request
Stratification motor
Compressor request
Distribution motor
Stratification position pot.
Distribution position pot.
IHKA switch and display panel
Heater core temperature sensor
min640003

A/C system warnings and cautions

WARNING —

- *Wear hand and eye protection (gloves and goggles) when working around the A/C system. If refrigerant does come in contact with your skin or eyes:*
  -Do not rub skin or eyes.
  -Immediately flush skin or eyes with cool water for 15 minutes.
  -Rush to a doctor or hospital. Do not attempt to treat yourself.
- *Work in a well ventilated area. Switch on building exhaust/ventilation system when working on the refrigerant system.*
- *Do not expose any component of the A/C system to high temperatures (above 80°C/176°F) or open flames. Excessive heat causes a pressure increase which could burst the system.*
- *Keep refrigerant away from open flames. Poisonous gas is produced if it burns. Do not smoke near refrigerant gases for the same reason.*
- *The A/C system is filled with refrigerant gas which is under pressure. Pressurized refrigerant in the presence of oxygen may form a combustible mixture. Do not introduce compressed air into any refrigerant container (full or empty).*
- *Electric welding near refrigerant hoses causes R-134a to decompose. Discharge system before welding.*

CAUTION —

- *US law requires that any person who services a motor vehicle air-conditioner is properly trained and certified, and uses approved refrigerant recycling equipment. Technicians must complete an EPA-approved recycling course to be certified.*
- *It is recommended that all A/C service be left to an authorized MINI dealer or other qualified A/C service facility.*
- *State and local governments may have additional requirements regarding A/C servicing. Always comply with state and local laws.*
- *Do not top off a partially charged refrigerant system. Discharge system, evacuate and then recharge system.*
- *Do not use R-12 refrigerant, refrigerant oils or system components in R-134a system. Component damage and system contamination results.*
- *The mixture of refrigerant oil (PAG oil) and refrigerant (R-134a) attacks some metals and alloys (for example, copper) and breaks down certain hose materials. Use only hoses and pipes that are identified with a green mark (stripe) or the lettering "R-134a".*
- *Immediately plug open connections on A/C components and pipes to prevent dirt and moisture contamination.*
- *Do not steam clean A/C condensers or evaporators. Use only cold water or compressed air.*

Heating and A/C Controls

Heating and air-conditioning control panel, removing and installing

– Remove radio. See **650 Radio**.

– Remove front console fascia trim. See **513 Interior Trim**.

◄ Remove switch module mounting screws (**arrows**). Lower module.

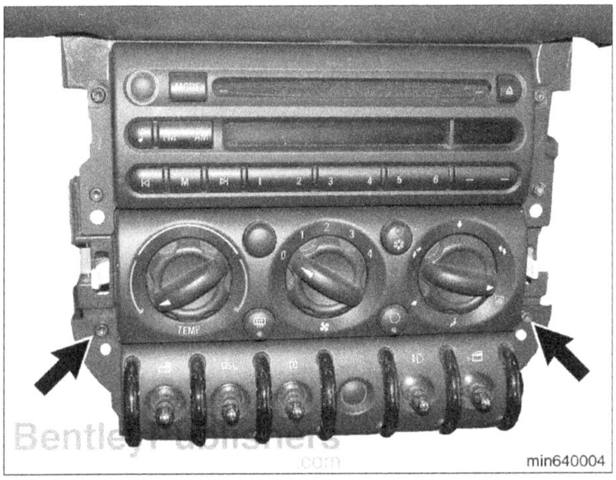

min640004

◄ Disconnect switch module electrical harness connector (**arrow**).

min640005

◄ Working at control panel:

- Remove securing clips (**A**).
- Slide control panel inward to clear locating pins (**B**).

NOTE—
IHKS control panel illustrated. IHKA panel is similar.

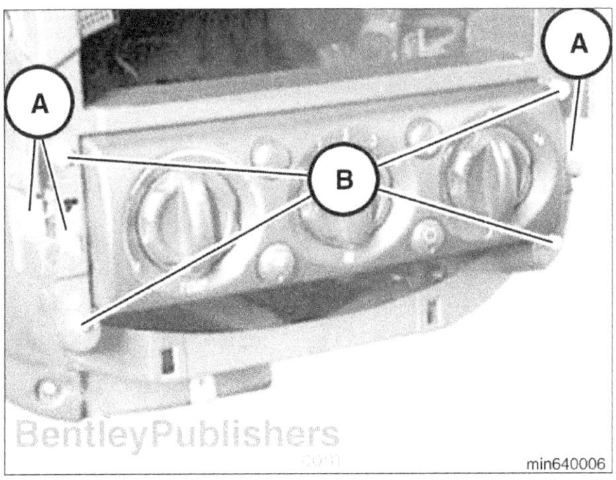

min640006

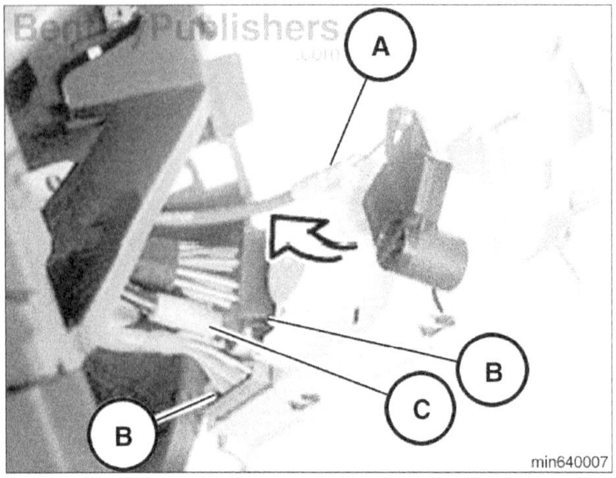

IHKS: Pull back left side of control panel to gain access to rear of panel.

- Disconnect temperature control cable (**A**).
- Disconnect electrical harness connectors (**B**).
- Disconnect air distribution control cable (**C**).

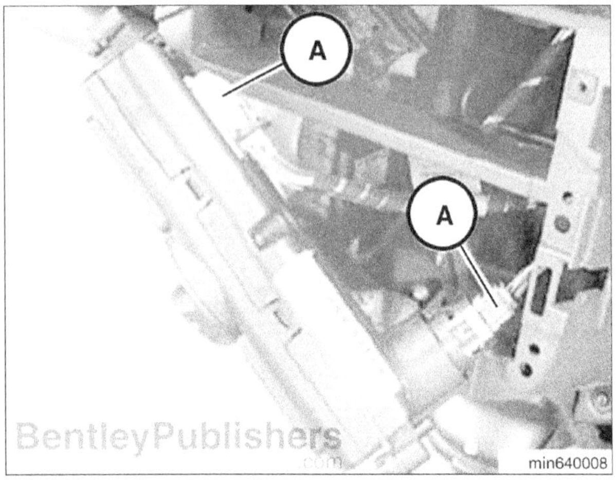

IHKA: Pull out control panel slightly to gain access to rear of panel.

- Disconnect electrical harness connectors (**A**).

– Installation is reverse of removal.

Interior temperature sensor, removing (IHKA)

– Remove IHKA control panel. See **Heating and air-conditioning control panel, removing and installing** in this repair group.

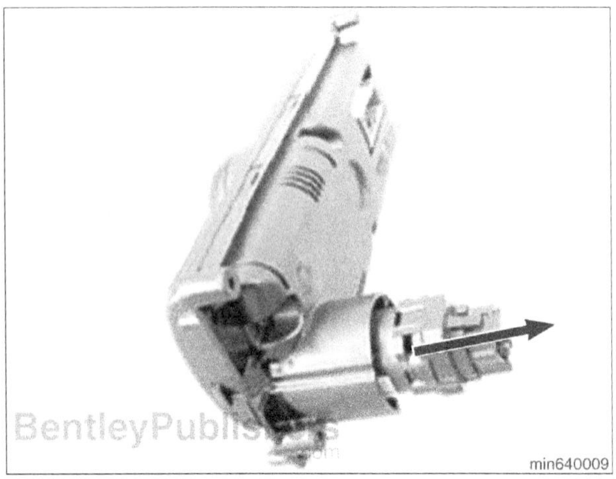

Working at back of control panel, pull out interior temperature sensor and fan assembly (**arrow**).

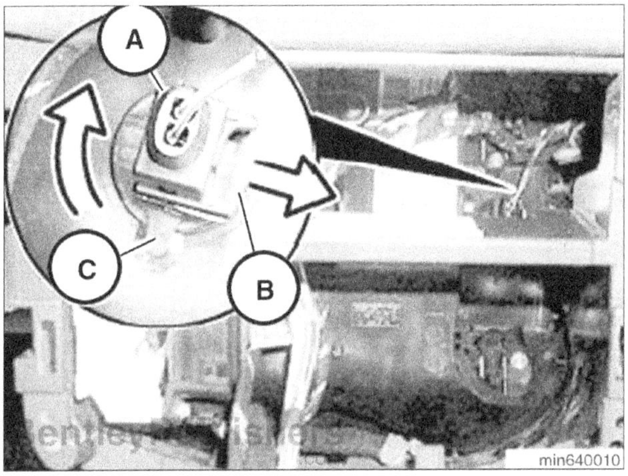

Heater core temperature sensor, removing (IHKA)

– Remove IHKA control panel. See **Heating and air-conditioning control panel, removing and installing** in this repair group.

◄ Working inside radio opening in dashboard:

- Rotate temperature sensor (**A**) clockwise (approximately 45°) to release.
- Squeeze electrical harness connector retaining clip (**B**) and disconnect harness.
- When installing, make sure sensor indent (**C**) is correctly located.

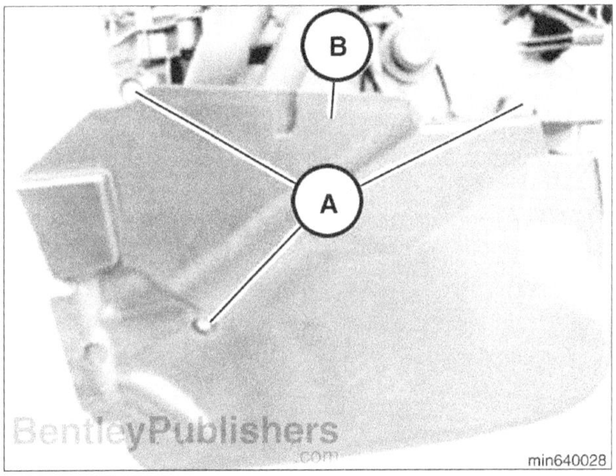

A/C evaporator temperature sensor, removing

◄ Working underneath left side of dashboard, remove cover on left side of heating and A/C housing:

- Remove three mounting screws (**A**).
- Carefully wriggle cover (**B**) backward past throttle pedal assembly.

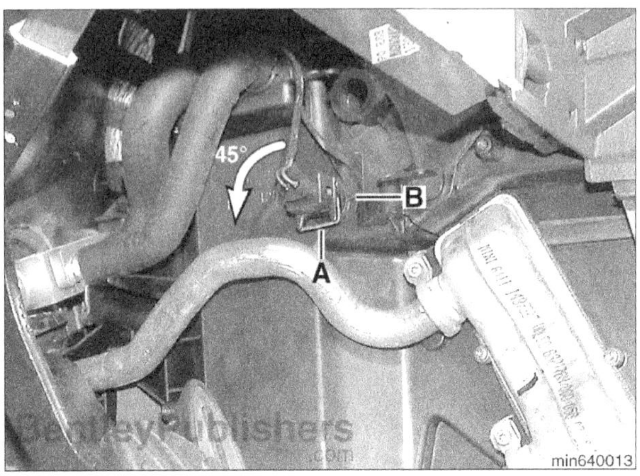

◄ Working at heater and A/C housing:

- Rotate evaporator sensor counterclockwise (approximately 45°) to release.
- Squeeze electrical harness connector retaining clip (**A**) and disconnect harness.
- When installing, make sure sensor indent (**B**) is correctly located.

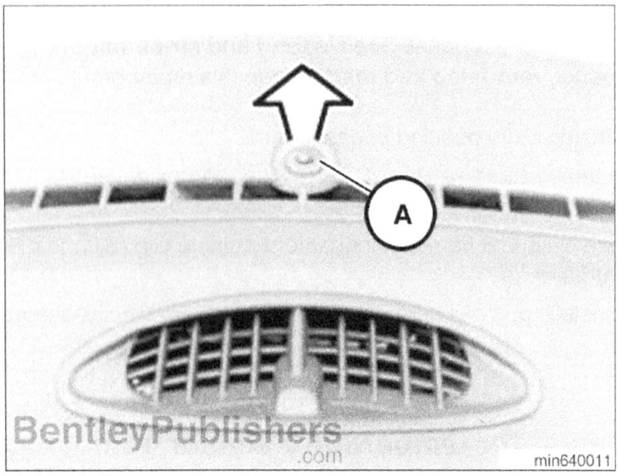

Solar sensor, removing and installing (IHKA)

Working at top center of dashboard, gently pry out solar sensor (**A**).

- Disconnect electrical harness connector and remove solar sensor.

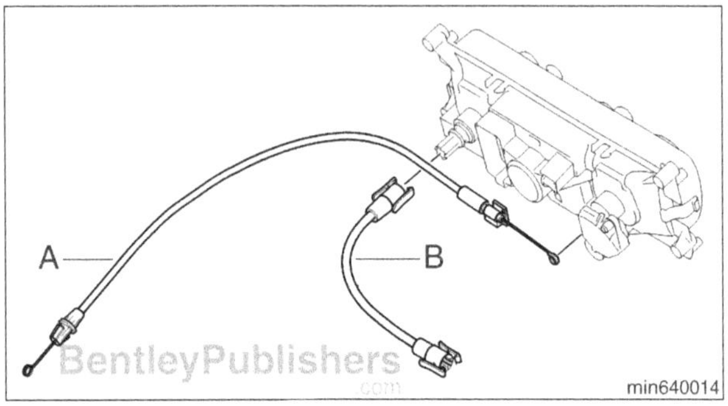

VENTS AND AIR VALVES

IHKS uses mechanically operated valves and flaps to control and direct heating, ventilation and air-conditioning air flow.

A bowden cable (**A**) and a flexible shaft (**B**) couple the rotary knobs on the dashboard mounted control panel to valves in the heating and A/C housing.

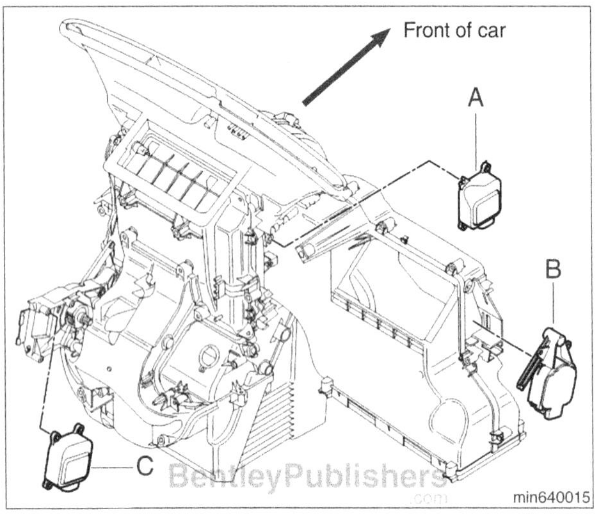

IHKA uses three electrically operated air distribution servo motors in heating and A/C housing:

- **A**: Temperature control
- **B**: Recirculation control
- **C**: Air distribution control

Temperature control servo motor, removing and installing

– Remove glove compartment. See **513 Interior Trim**.

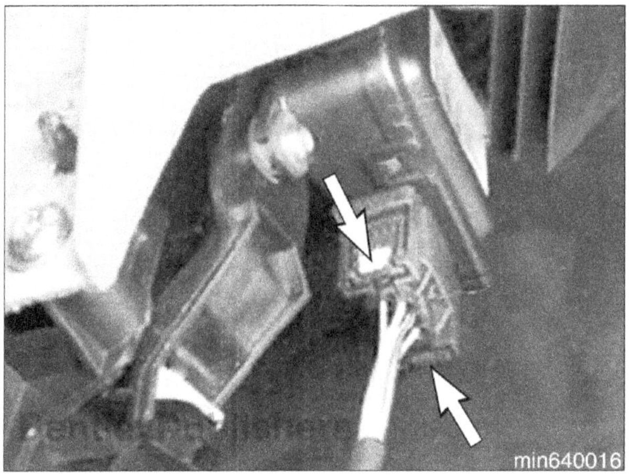

◄ Reach above glove compartment to squeeze servo motor electrical harness connector retaining tabs (**arrows**). Detach connector.

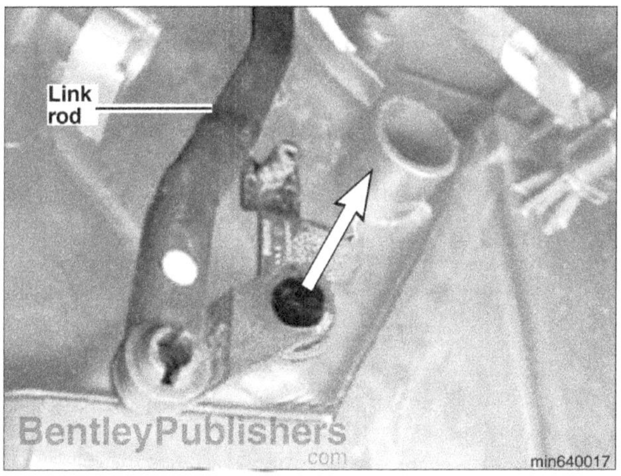

◄ Pull off control lever (**arrow**).

– Loosen cable and disengage link rod.

– Release servo motor and bracket and lower.

– Separate link rod from servo motor arm.

– Remove servo motor screws to detach from bracket.

– When installing, be sure link rod and servo motor engage and seat correctly.

Recirculation control servo motor, removing and installing

– Remove glove compartment. See **513 Interior Trim**.

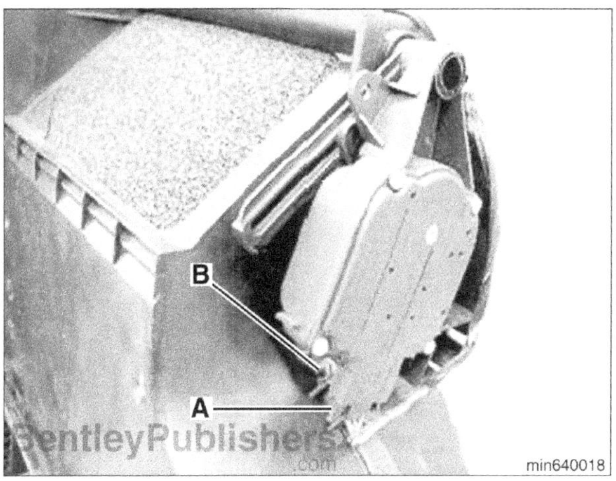

◄ Working behind glove compartment:

- Detach electrical harness connector (**A**).
- Remove mounting screw (**B**) and pull servo away from flap mechanism.

– When installing, be sure servo is seated in keyway correctly.

Air distribution control servo motor, removing and installing

- Remove IHKA control panel. See **Heating and air-conditioning control panel, removing and installing** in this repair group.

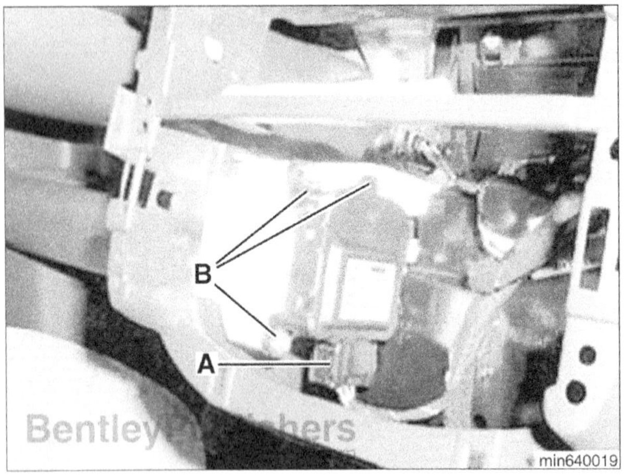

- Working inside IHKA control panel opening:
  - Detach electrical harness connector (**A**).
  - Remove mounting screws (**B**) and pull servo away from flap mechanism.

- When installing, be sure servo is seated correctly.

Dashboard center vent, removing and installing

- Remove center dashboard face trim. See **513 Interior Trim**.

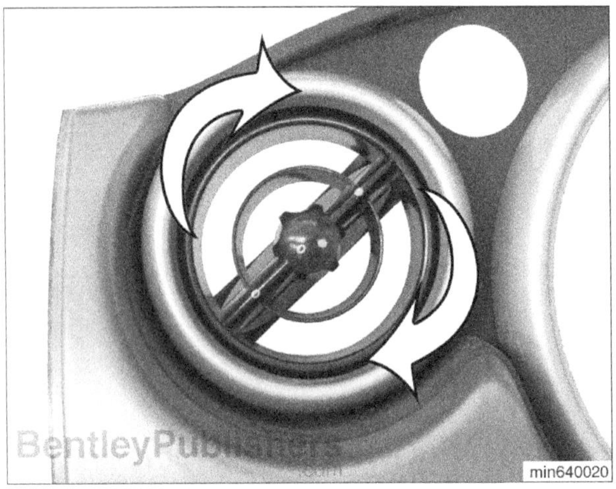

- Turn trim ring of air vent (**arrows**) and pull off.

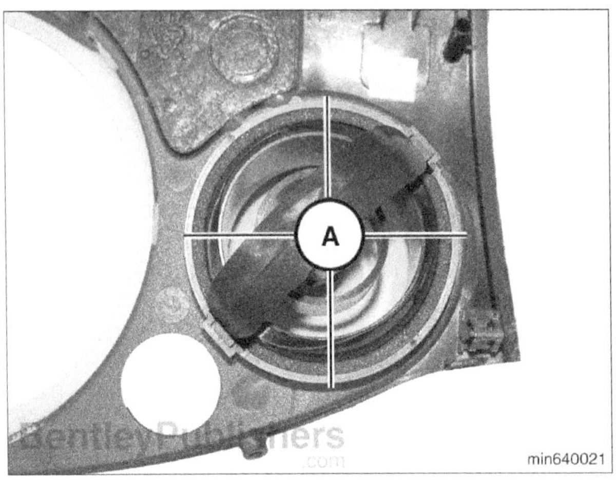

- Release air vent retaining clips (**A**) and remove air vent.

- When installing, make sure vent snaps into place correctly.

Dashboard outer fresh air vent, removing and installing

– Remove outer dashboard face trim (left or right). See **513 Interior Trim**.

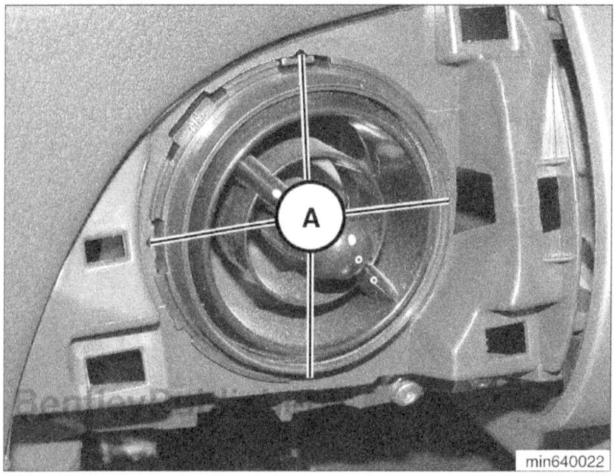

◄ Release air vent retaining clips (**A**) and remove air vent.

– When installing, make sure vent snaps into place correctly.

BLOWER COMPONENTS

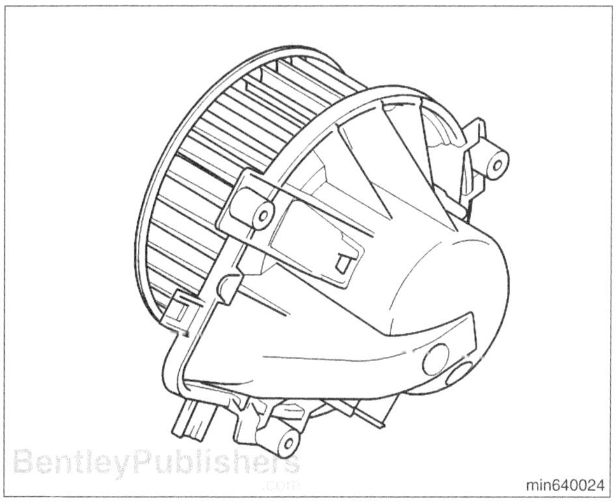

◄ The blower is installed in the driver's side of the heater and A/C housing and consists of an open-hub centrifugal fan powered by an electric motor.

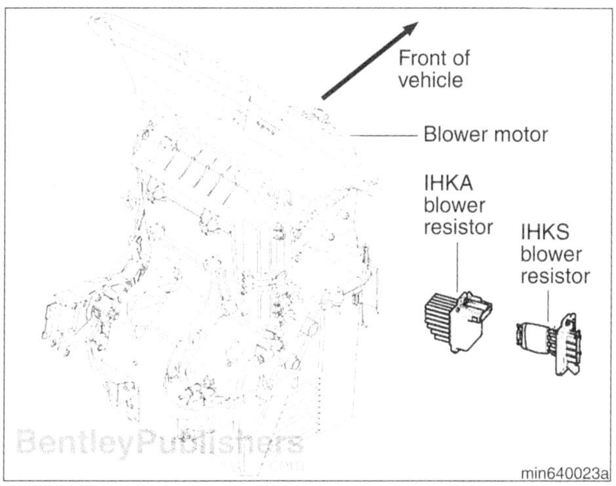

◄ The blower resistor pack is installed in the air outlet from the blower, so that any heat generated within the resistor pack is dissipated by the airflow. Power is supplied to the resistor pack via an external relay controlled by the General Module (BC1).

Blower motor operation

IHKS

A rotary switch on the control panel and the resistor pack control the four blower speeds. At position 4 the resistor pack is bypassed allowing maximum blower voltage and speed.

IHKA

If the climate control system is in AUTO mode, blower speed is regulated automatically. In MANUAL mode, 8 blower speeds are available. These correspond to approximately 20% to 100% of battery voltage powering the blower.

The blower relay is energized by the General Module (BC1) whenever the ignition switch is in ACCESSORY or RUN position. The IHKA control module modulates the final output stage to regulate voltage across the blower motor and control blower speed.

In automatic blower and fresh air mode, blower speed is corrected to compensate for increase in inlet air ram effect as vehicle speed increases.

In AUTOMATIC mode, if temperature is set to LO or HI, the blower runs at maximum speed with correction only for vehicle speed. However, if a specific temperature is set, blower speed is corrected to compensate for coolant temperature, external air temperature, and solar heating.

Blower speed adjustments are as follows:

- During warm-up, to avoid blowing excessive amounts of cold air into the passenger compartment, blower speed is limited if the heater core air temperature is below approximately 60°C (140°F) or exterior temperature is below -10°C (14°F).
- During cool down, to purge air ducts of hot air, blower speed is limited for approximately 5 seconds after the system is switched on. Blower speed is then progressively increased over approximately 10 seconds.
- As passenger compartment temperature approaches the selected temperature, blower speed is progressively reduced until it stabilizes at the selected temperature.
- If air-conditioning is on, blower speed increases with increasing solar heating.

Blower motor, removing and installing

Blower motor removal and installation is a complicated job, involving removal of airbag components and the complete dashboard. Be sure to read airbag warnings and cautions in **721 Airbag System (SRS)**.

– Following manufacturer's instructions, connect an approved refrigerant recovery / recycling / recharging unit to A/C system and discharge system.

> ***WARNING —***
> *Do not attempt to discharge or charge the A/C system without proper equipment and training. Damage to the vehicle and personal injury may result.*

– Remove complete dashboard. See **513 Interior Trim**.

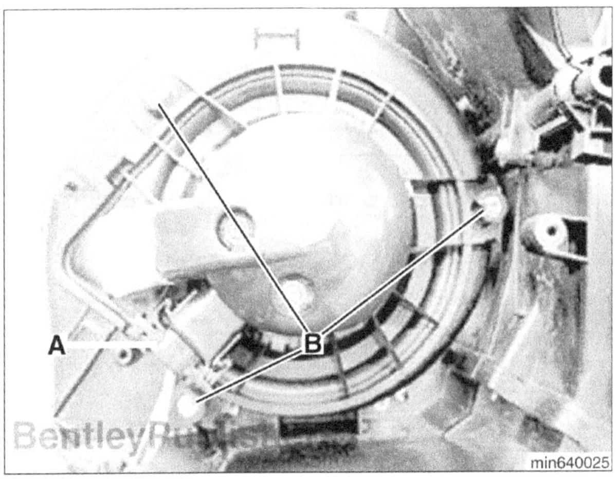

< Working at left side of heating and A/C housing:

- Detach blower electrical harness connector (**A**).
- Remove blower mounting screws (**B**) and remove blower from housing.

– Installation is reverse of removal.

Blower motor resistor, removing and installing

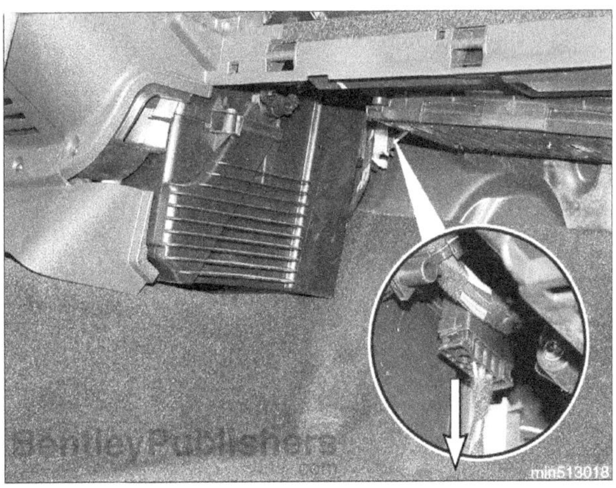

– Remove glove compartment. See **513 Interior Trim**.

< Reach in next to right side of heating and A/C housing:

- Detach electrical harness connector (**inset**).
- Remove resistor mounting screw and pull unit out of housing.

– Installation is reverse of removal.

NOTE —

- *In the IHKS system, the blower resistor may also be referred to as "power stage", "final stage" or "output stage."*
- *Models up to 3/2002 with IHKA: Blower resistor may be on left side of IHKA housing. Removal procedure is similar.*

HEATER COMPONENTS

Engine coolant flows through the heater core constantly. There is no heater valve. If there is heated air in the vehicle at all times, then the the temperature control flap may have failed, preventing it from blocking off air flow through the heater core.

Heater core, replacing

> ***WARNING —***
> *Due to risk of personal injury, be sure the engine is cold before opening any part of the cooling system or removing the coolant reservoir cap.*

– Drain coolant. See **170 Radiator and Cooling System**.

– Cooper: Remove battery box. See **121 Battery, Starter, Alternator**.

> ***CAUTION —***
> *Prior to disconnecting the battery, read the battery disconnection cautions given in* **001 General Warnings and Cautions**.

– Cooper S: Remove air filter housing. See **130 Fuel Injection**.

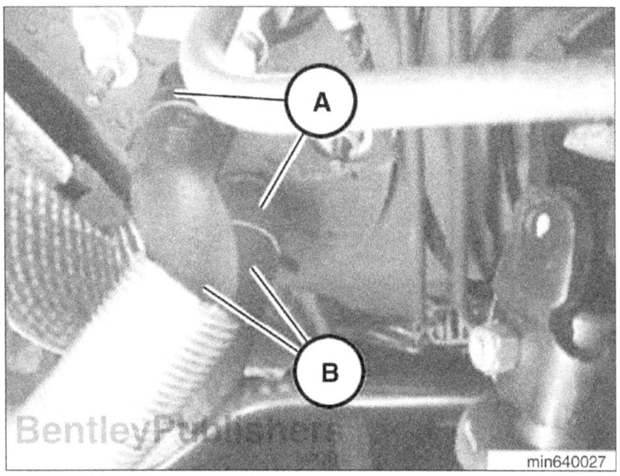

Working inside engine compartment at bulkhead:

- Loosen or remove heater hose clamps (**A**).
- Detach coolant hoses (**B**). Carefully blow compressed air through dual aluminum heater core pipes to remove remaining coolant from heater core.

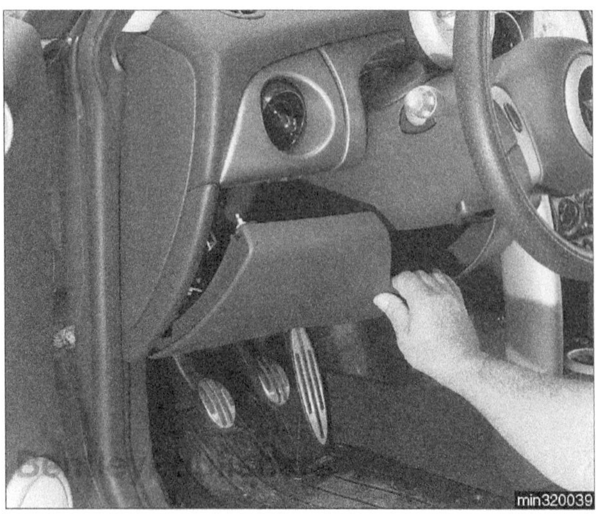

Working under left side of dashboard, pull on top edge of left lower dashboard trim to unclip and swing down.

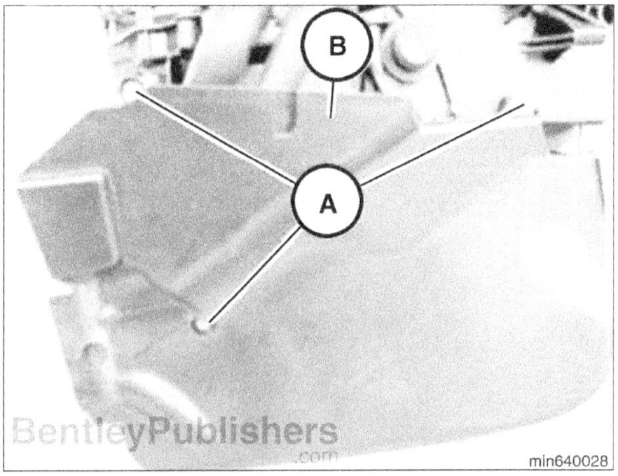

Remove cover on left side of heating and A/C housing:

- Remove three mounting screws (**A**).
- Carefully wriggle cover (**B**) backward past throttle pedal assembly.

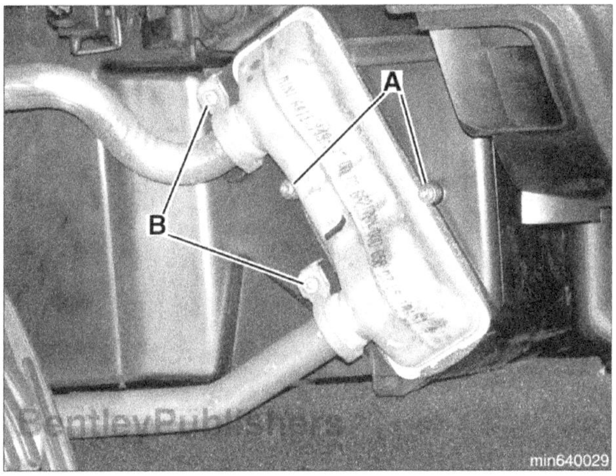

Remove heater core:

- Remove heater core mounting screws (**A**).
- Loosen coolant pipe clamps (**B**) and disconnect pipes.
- Slide heater out past pipes.

– Installation is reverse of removal. Be sure to:
  - Seat heater core correctly.
  - Replace coolant pipe clamp sealing O-rings. Coat new sealing rings with antiseize paste.
  - Seat sealing rings and clamps correctly.
  - Fill and bleed cooling system and check for leaks, See **170 Radiator and Cooling System**.

Heating and A/C housing, removing and installing

Heater housing removal and installation is a complicated job. The procedure involves the following:

- Draining of cooling system
- Discharging of air-conditioning refrigerant
- Removal of airbag components
- Removal of complete dashboard

Read through complete procedure before starting the job. Be sure to read airbag warnings and cautions in **721 Airbag System (SRS)**.

– Following manufacturer's instructions, connect an approved refrigerant recovery / recycling / recharging unit to A/C system and discharge system.

> ***WARNING —***
> *Do not attempt to discharge or charge the A/C system without proper equipment and training. Damage to the vehicle and personal injury may result.*

– Remove complete dashboard. See **513 Interior Trim**. Place dashboard assembly on work bench.

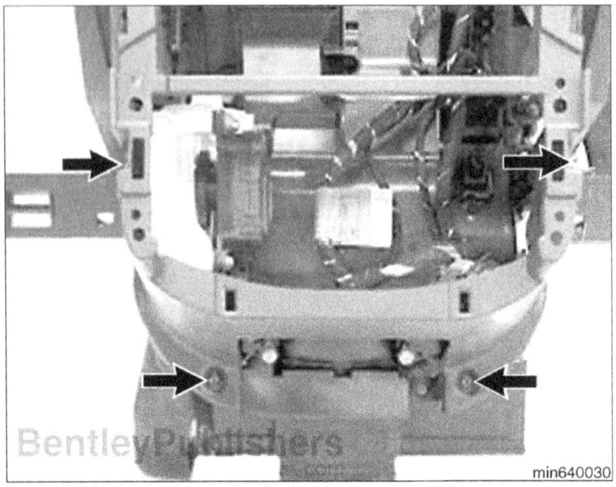

Working at rear of center console, remove mounting screws (**arrows**).

Working inside radio opening, detach electrical harness connector (**arrow**).

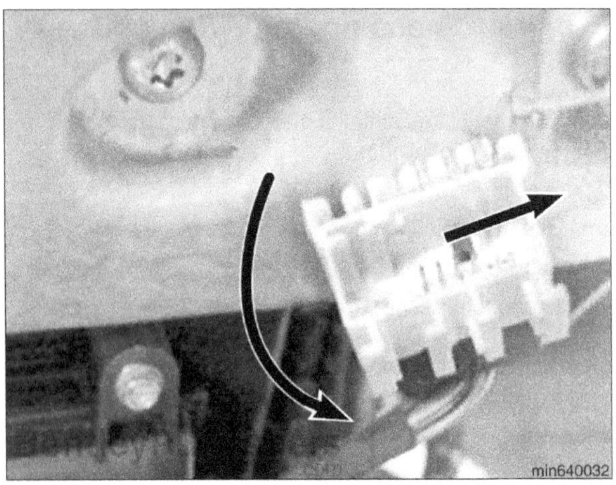

Twist and pull off (**arrows**) harness connector to detach from dashboard carrier brace.

Remove heater housing screws (**arrows**) at center.

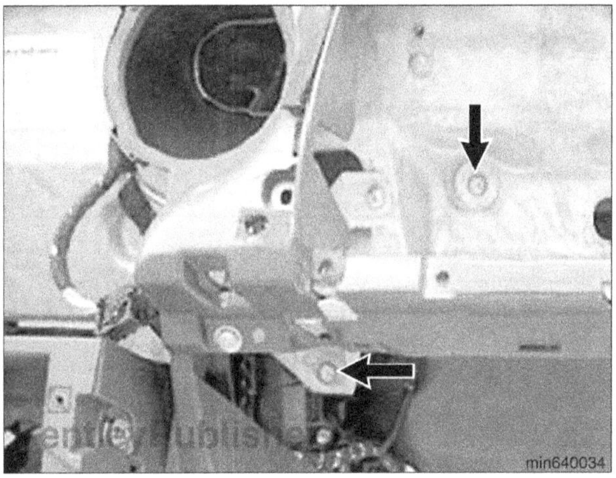

Remove heater housing screws (**arrows**) at right.

Models with IHKA: Disconnect solar sensor electrical harness connector (**arrow**).

- Separate heater and A/C housing from dashboard assembly.

> ***CAUTION—***
> *Dashboard and heater housing are bulky and will require two people to separate.*

- Installation is reverse of removal.

Tightening torques	
Heater housing to dashboard carrier brace	4 Nm (35 in-lb)
Heater housing to lower dashboard	3 Nm (27 in-lb)

- Use new hose clamps at coolant hoses. Fill and bleed cooling system. See **170 Radiator and Cooling System**.
- Recharge A/C system following equipment manufacturer's instructions.
- Initialize window motors. See **512 Door Windows**.

AIR-CONDITIONING COMPONENTS

This section covers removal and installation of air-conditioning refrigerant components. A/C testing and diagnosis, refrigerant evacuation and recharge are not covered here.

The A/C condenser, compressor, receiver drier, pressure sensor, expansion valve and refrigerant pipes are accessible from the engine compartment.

The A/C evaporator is in the heater and A/C housing. Access to the evaporator is only possible by removing the dashboard assembly.

Be sure to read **A/C system warning and cautions** in this repair group.

NOTE—
The air conditioning system is filled with R-134a refrigerant.

A/C compressor, removing and installing

- Disconnect battery.

> ***CAUTION—***
> *Prior to disconnecting the battery, read the battery disconnection cautions in* **001 General Warnings and Cautions**.

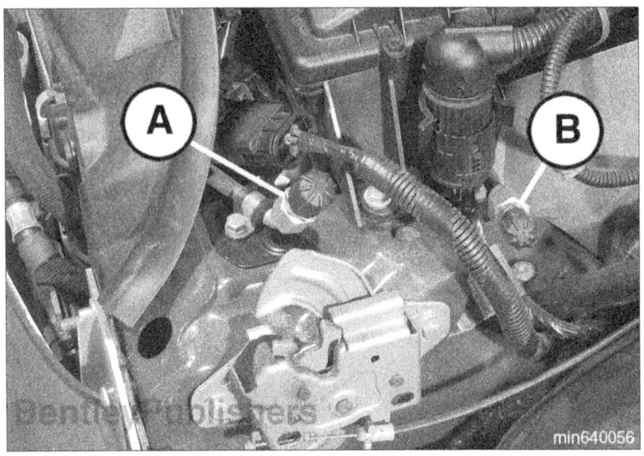

- Following manufacturer's instructions, connect an approved refrigerant recovery / recycling / recharging unit to A/C system and discharge system. High pressure (**A**) and low pressure (**B**) A/C lines are shown in MINI Cooper engine compartment.

> ***WARNING—***
> *Do not attempt to discharge or charge the A/C system without proper equipment and training. Damage to the vehicle and personal injury may result.*

- Raise front of car and support in a safe manner.

> ***WARNING—***
> *Make sure the car is stable and well supported at all times. Use a professional automotive lift or jack stands designed for the purpose. A floor jack is not adequate support.*

- Remove splash shield from under engine.

- Mark engine drive belt with direction of rotation. Then remove belt. See **020 Maintenance**.

Models with CVT

Remove transmission fluid cooler mounting screw (**A**).

- Slide cooler right (**arrow**) to disengage from bracket.
- Tie cooler out of the way.

Continued for all models

Loosen modular front end:

- Detach front wheel housing liners from front bumper cover trim.
- Remove front bumper cover trim.
- Remove front bumper carrier.
- Working underneath car, loosen and remove crush tubes that support modular front end to front subframe.
- Cooper S: Detach radiator upper hose support clamp (**A**) from intake manifold.
- Remove modular front end mounting bolts. Install two 100 mm (4 in) M8 bolts (**arrow**) in left and right bumper support members. Slide modular front end forward, supported on long bolts.

> ***CAUTION—***
> *For ease of component alignment when reassembling front bumper assembly, do not loosen or remove bumper alignment bosses.*

NOTE—

- *If available, use BMW special tools 11 8 401 and 11 8 402 instead of long bolts to support modular front end.*
- *Removal of modular front end is a complicated job. It is covered in* **510 Bumpers, External Trim**.

– Remove A/C condenser. See **A/C condenser, removing and installing** in this repair group.

Pry out dowel pins securing radiator to modular front end.

– Release radiator from lower mounting brackets to gain access to A/C compressor upper mounting bolts and refrigerant pipes.

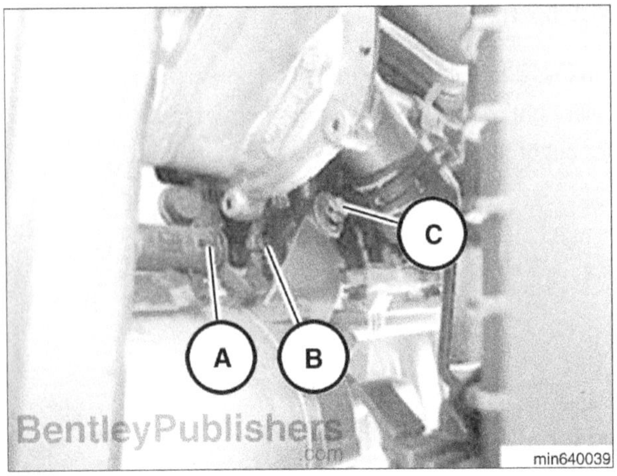

Working above A/C compressor, at right front of car:

- Disconnect compressor clutch electrical harness connector (**A**).
- Detach harness clip (**B**) from bracket.
- Remove compressor upper right mounting bolt (**C**).

NOTE—
Cooper engine, with coolant pump located above A/C compressor, is illustrated. In Cooper S, layout is similar but the alternator is located above the A/C compressor.

Working above A/C compressor at front of car: Remove compressor upper left mounting bolt (**arrow**).

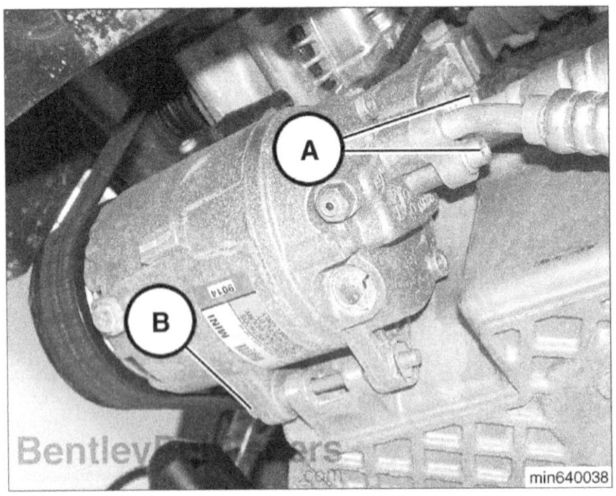

Working underneath vehicle at A/C compressor:

- Remove refrigerant pipe mounting screws (**A**) at compressor.
- Detach pipes from compressor.
- Remove compressor lower mounting bolt (**B**).
- Remove compressor.

– Installation is reverse of removal.
  - Always replace O-rings when reconnecting refrigerant pipes.
  - If installing a new compressor, or if A/C system has been open for more than 24 hours, replace receiver/drier unit. See **A/C receiver drier, removing and installing** in this repair group.
  - Install and tension engine drive belt, noting previously made direction mark. See **020 Maintenance**.

Tightening torques	
A/C compressor to engine (M8)	25 Nm (18 ft-lb)
A/C pipes to compressor (M8 x 25) (replace sealing O-rings)	20 Nm (15 ft-lb)
Transmission fluid cooler to condenser	4 Nm (35 in-lb)

– Recharge A/C system following equipment manufacturer's instructions.

A/C system refrigerant	
R-134a capacity	415 ± 10 grams (14.6 ± 0.4 oz)

NOTE—
Refer to rating plate on compressor to determine quantity of refrigerant oil required.

– When starting a new compressor for the first time, carry out the following break-in procedure:
  - Switch on A/C system.
  - Set all dashboard air vents to OPEN.
  - Start engine and allow to idle.
  - Set blower to at least 75% output.
  - Run A/C for at least 2 minutes at idle speed.

> ***CAUTION—***
> *The new compressor may be damaged if run at higher than idle speed at start-up.*

A/C condenser, removing and installing

The A/C condenser is located in front of the radiator.

– Following manufacturer's instructions, connect an approved refrigerant recovery / recycling / recharging unit to A/C system and discharge system.

> ***WARNING—***
> *Do not attempt to discharge or charge the A/C system without proper equipment and training. Damage to the vehicle and personal injury may result.*

– Remove front bumper cover and carrier. See **510 Exterior Trim, Bumpers**.

◄ Models with CVT:
  - Remove transmission fluid cooler mounting screw (**A**).
  - Slide cooler right (**arrow**) to disengage from bracket.
  - Tie cooler out of the way.

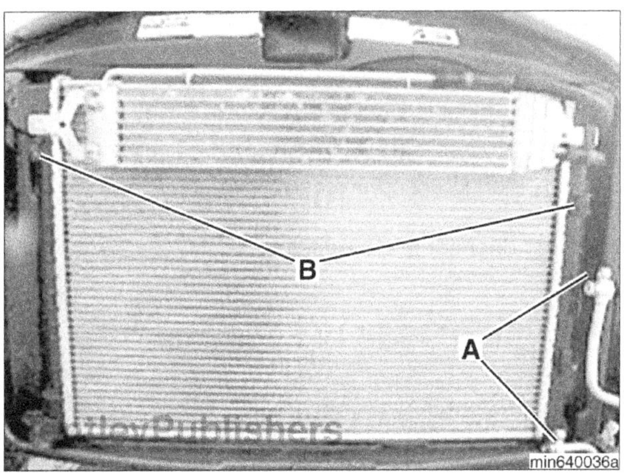

◄ Working at front of condenser:
  - Remove screws (**A**) securing refrigerant pipes to condenser. Detach pipes.
  - Remove condenser mounting screws (**B**).
  - Lift out condenser.

– Installation is reverse of removal.
  - Always replace O-rings when reconnecting refrigerant pipes.
  - If A/C system has been open for more than 24 hours, replace receiver/drier unit. See **A/C receiver drier, removing and installing** in this repair group.

Tightening torques	
A/C condenser to radiator (M6 x 12 mm)	6 Nm (53 in-lb)
A/C refrigerant pipes to condenser (M6) (replace sealing O-rings)	8 Nm (6 ft-lb)
Transmission fluid cooler to radiator	4 Nm (35 in-lb)

– Recharge A/C system following equipment manufacturer's instructions.

A/C system refrigerant	
R-134a capacity	415 ± 10 grams (14.6 ± 0.4 oz)

A/C receiver drier, removing and installing

The receiver / drier is mounted in the right front corner of the car, next to the A/C condenser.

CAUTION—
Replace the receiver / drier if:
- *There is dirt in the A/C system.*
- *The compressor has seized or has been replaced.*
- *The condenser or evaporator are replaced.*
- *The A/C system is leaking and there is no more refrigerant.*
- *The A/C system was open for 24 hours or more.*

– Following manufacturer's instructions, connect an approved refrigerant recovery / recycling / recharging unit to A/C system and discharge system.

WARNING—
Do not attempt to discharge or charge the A/C system without proper equipment and training. Damage to the vehicle and personal injury may result.

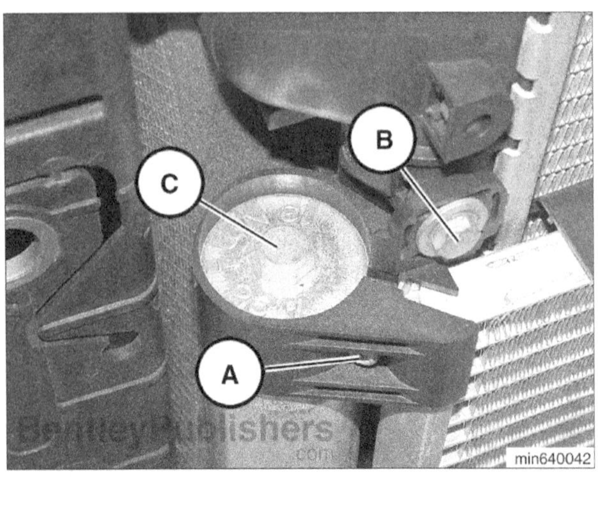

◄ Open engine hood. Working at right front of vehicle:
- Remove receiver drier plastic housing screw (**A**).
- Remove housing mounting bolt (**B**).
- Use hex boss at top to unscrew receiver / drier (**C**).

– Remove receiver / drier complete with housing.

– Installation is reverse of removal. Be sure to replace sealing O-rings.

Tightening torque	
A/C receiver drier to condenser (replace sealing O-rings)	13 Nm (10 ft-lb)

– Recharge A/C system following equipment manufacturer's instructions.

A/C system refrigerant	
R-134a capacity	415 ± 10 grams (14.6 ± 0.4 oz)

A/C expansion valve, removing and installing

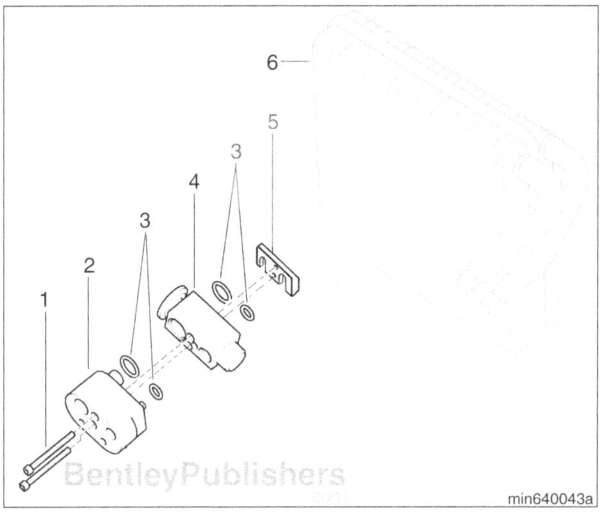

The expansion valve is fitted to the A/C evaporator refrigerant pipes at the engine compartment bulkhead.

1. Clamping bolt
   - Tighten to 8 Nm (6 ft-lb)
2. Clamping adapter
3. Sealing O-rings
4. Expansion valve
5. Threaded clamping plate
6. A/C evaporator

– Following manufacturer's instructions, connect an approved refrigerant recovery / recycling / recharging unit to A/C system and discharge system.

> ***WARNING —***
> *Do not attempt to discharge or charge the A/C system without proper equipment and training. Damage to the vehicle and personal injury may result.*

– Cooper: Remove battery box. See **121 Battery, Starter, Alternator**.

> ***CAUTION —***
> *Prior to disconnecting the battery, read the battery disconnection cautions given in* **001 General Warnings and Cautions**.

– Cooper S: Remove air filter housing. See **130 Fuel Injection**.

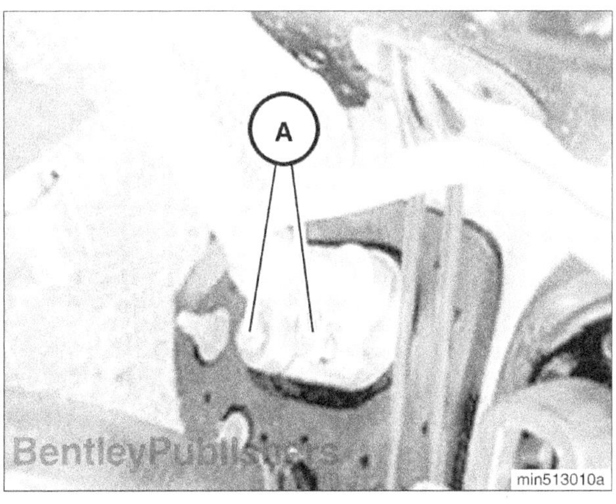

Working at engine compartment rear bulkhead: Remove bolts (**A**) mounting air-conditioning lines to expansion valve clamping adapter. Detach air-conditioning lines.

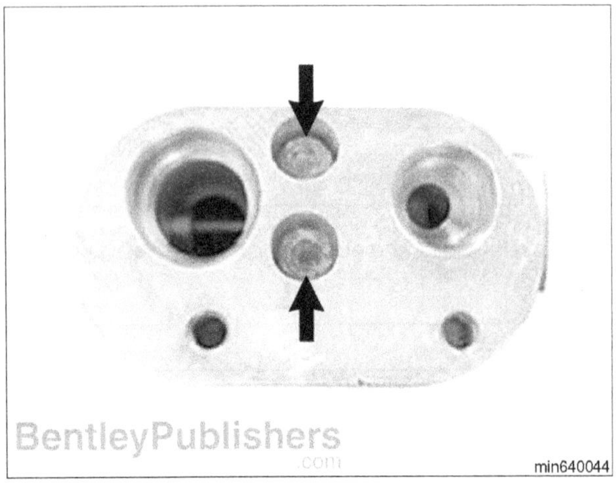

Remove clamping bolts (**arrows**) and remove adapter block.

Working under left side of dashboard, pull on top edge of left lower dashboard trim to unclip and swing down.

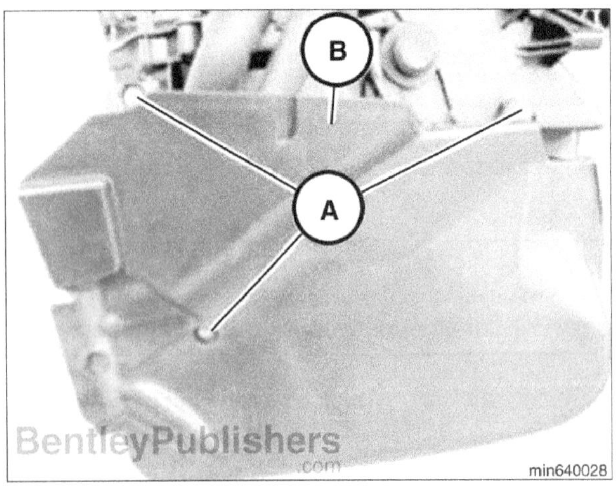

Remove cover on left side of heating and A/C housing:

- Remove three mounting screws (**A**).
- Carefully wriggle cover (**B**) backward past throttle pedal assembly.

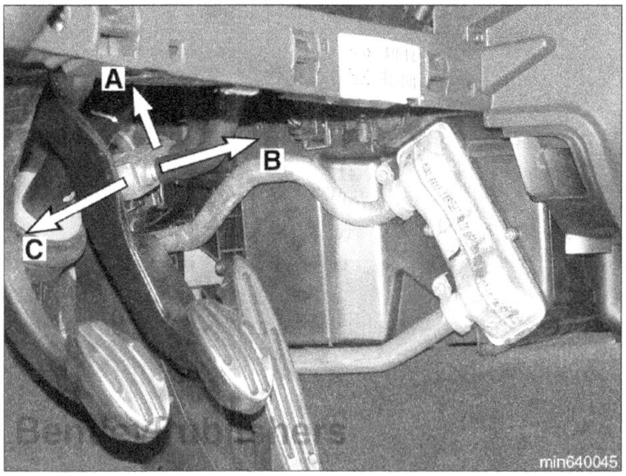

Working inside vehicle at engine compartment bulkhead:

- Pull out threaded clamping plate (**A**).
- Pull backward on expansion valve refrigerant pipes (**B**).
- Remove expansion valve (**C**).

– Installation is reverse of removal. Be sure to replace sealing O-rings.

Tightening torques	
A/C refrigerant pipes to expansion valve adapter (M6 x 20 mm) (replace sealing O-rings)	8 Nm (6 ft-lb)
Expansion valve clamping bolt	6 Nm (53 in-lb)

– Recharge A/C system following equipment manufacturer's instructions.

A/C system refrigerant	
R-134a capacity	415 ± 10 grams (14.6 ± 0.4 oz)

A/C pressure sensor, accessing, removing and installing

The A/C pressure sensor is located in the refrigerant pipe near the rear engine compartment bulkhead.

– Cooper: Remove battery box. See **121 Battery, Starter, Alternator**.

> ***CAUTION—***
> *Prior to disconnecting the battery, read the battery disconnection cautions given in* **001 General Warnings and Cautions**.

– Cooper S: Remove air filter housing. See **130 Fuel Injection**.

– If it becomes to necessary to remove the sensor, be sure to discharge A/C system by using an approved refrigerant recovery / recycling / recharging unit.

> ***WARNING—***
> *Do not attempt to discharge or charge the A/C system without proper equipment and training. Damage to the vehicle and personal injury may result.*

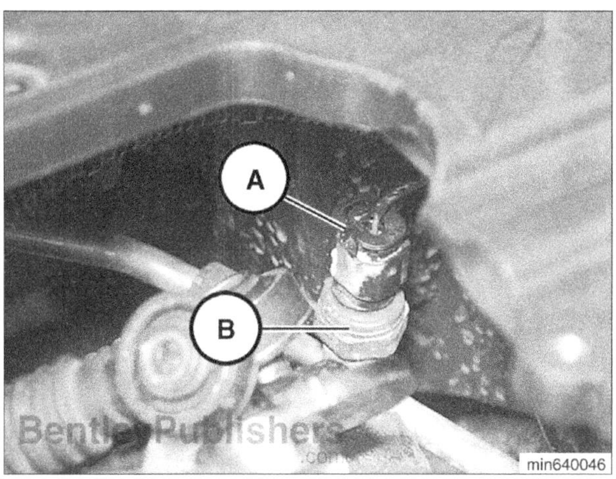

To remove sensor:

- Detach electrical harness connector (**A**).
- Unscrew switch (**B**) from refrigerant pipe.

– Installation is reverse of removal.

Tightening torque	
A/C pressure sensor to refrigerant pipe	10 Nm (7 ft-lb)

– Recharge A/C system following equipment manufacturer's instructions.

A/C system refrigerant	
R-134a capacity	415 ± 10 grams (14.6 ± 0.4 oz)

A/C evaporator, removing and installing

A/C evaporator removal and installation is a complicated job. The procedure involves the following:

- Draining cooling system
- Discharging air-conditioning refrigerant
- Removing airbag components
- Removing dashboard

Read through complete procedure before starting the job. Be sure to read airbag warnings and cautions in **721 Airbag System (SRS)**.

– Following manufacturer's instructions, connect an approved refrigerant recovery / recycling / recharging unit to A/C system and discharge system.

> ***WARNING —***
> *Do not attempt to discharge or charge the A/C system without proper equipment and training. Damage to the vehicle and personal injury may result.*

– Remove heater and A/C housing. See **Heating and A/C housing, removing and installing** in this repair group. Place housing on work bench.

– Remove evaporator temperature sensor. See **A/C evaporator temperature sensor, removing** in this repair group.

◄ Working at right end of heater housing:

- Disconnect recirculation servo motor electrical harness connector (**A**).
- Detach wiring harness (**B**) from air intake duct.

– IHKA: Remove temperature control servo motor at right center of housing.

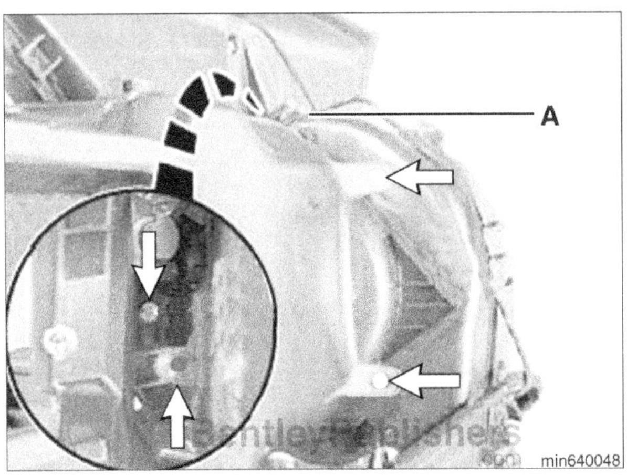

◄ Working in opening (**inset**) of heater housing:

- Remove air intake duct mounting screws (**arrows**).
- Release retaining clip (**A**).
- Remove air intake duct.

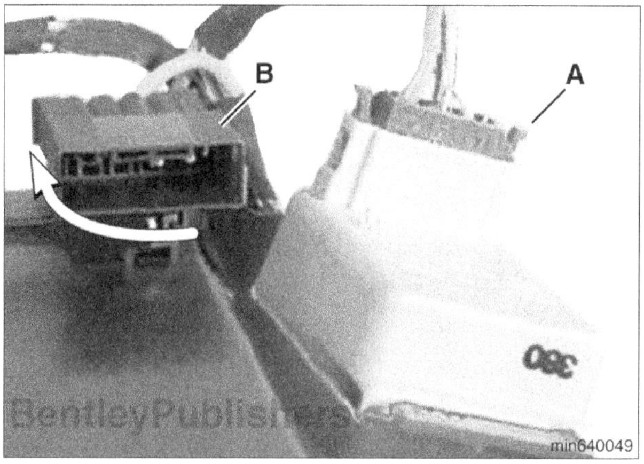

Working at right side of housing:

- Detach blower resistor pack electrical harness connector (**A**).
- Twist (**arrow**) heater housing harness connector (**B**) to remove.

Detach air distribution flap link rods (**arrows**).

Working at center of housing:

- Release electrical harness from clip (**A**).
- IHKA: Remove electrical harness connector (**B**) from air distribution servo motor.
- Remove housing screws (**arrows**).

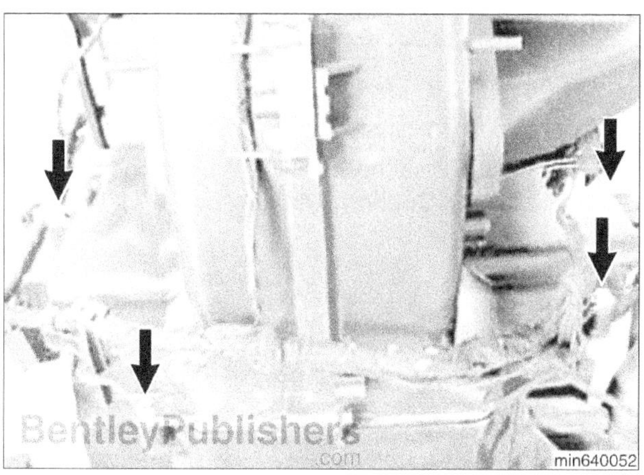

Remove housing screws (**arrows**).

Remove housing screws (**arrows**).

Tilt top half of heater housing to expose air flap link rod (**A**).

- Release link rod from air flap (**B**).
- Lay top of housing aside.

NOTE—
Heater core removed for clarity in illustration.

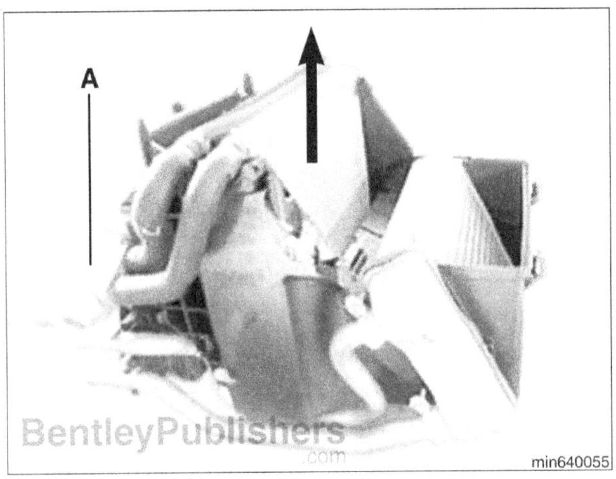

Working at evaporator and heater core (lower) housing:

- Remove expansion valve (**A**) from refrigerant pipes.
- Slide (**arrow**) evaporator out of housing.

– Installation is reverse of removal. Be sure to replace sealing O-rings.

Tightening torque	
Expansion valve clamping bolt	6 Nm (53 in-lb)

– Recharge A/C system following equipment manufacturer's instructions.

A/C system refrigerant	
R-134a capacity	415 ± 10 grams (14.6 ± 0.4 oz)

– Use new hose clamps at coolant hoses. Fill and bleed cooling system. See **170 Radiator and Cooling System**.

650 Radio

650

GENERAL

This repair group covers MINI factory installed radio, CD changer, and speaker removal. Radio antenna and navigation computer removal is also included.

Radio installations can vary by year and model. For information on aftermarket sound equipment, contact the original installer.

Sound system

The standard MINI radio features a stereo receiver and CD player. An optional upgraded sound system features a Digital Sound Processing (DSP) amplifier and Harmon Kardon speakers. The standard and upgraded radio receivers are the same. Radio and speaker removal procedures are also the same. The radio receiver is prewired for an optional cargo compartment mounted CD changer. A satellite radio is also an option on later model MINIs.

Sound system components are interconnected via K-bus. Theft proofing of the radio by code is not required as the radio will not function without the K-bus connection and a valid signal from the instrument cluster.

MINI models are equipped with a diversity antenna that uses the rear window defroster element along with the roof mounted aerial to improve radio reception.

Navigation system

Models equipped with MINI navigation system feature a dashboard mounted display and a navigation computer mounted under the right front seat. For information on the dashboard display, see **620 Instruments**.

Radio and CD Changer

Radio, removing and installing

- Remove front console and console side fascia. See **513 Interior Trim**.

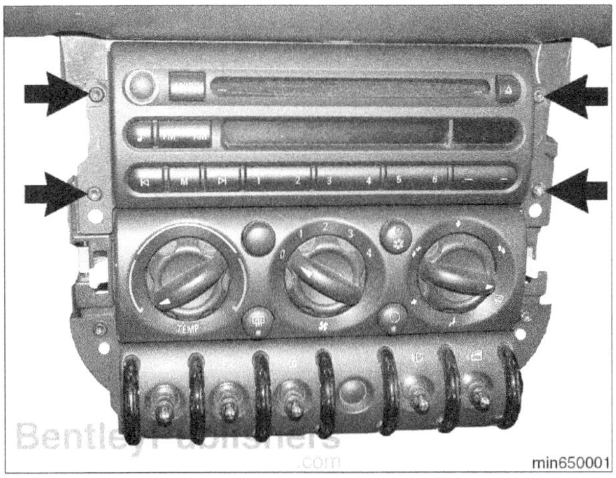

- Remove screws (**arrows**) and pull radio forward out of console.

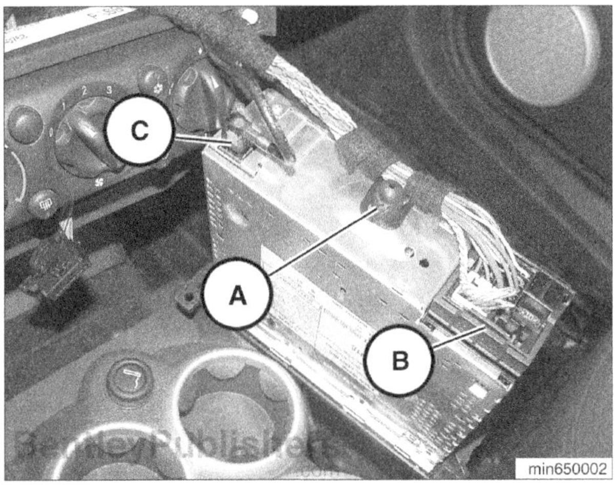

- Remove radio harness retaining plug (**A**) and electrical connector (**B**).
- Remove antenna connector (**C**) and remove radio.
- Installation is reverse of removal.

CD changer, removing and installing

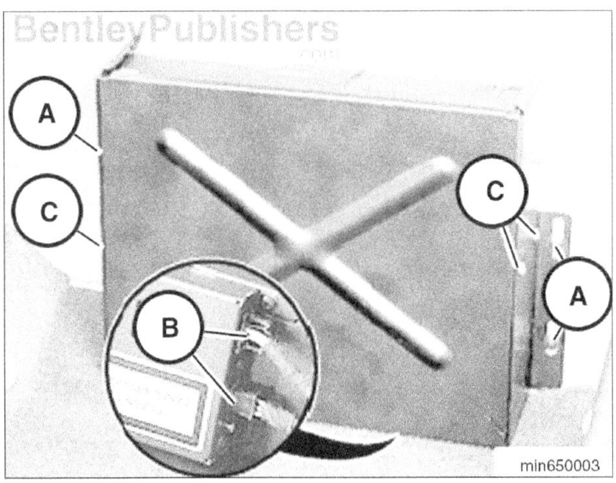

- Working in right side of cargo compartment, remove mounting bracket screws (**A**) and release electrical connectors (**B**) on changer.
- Remove screws (**C**) securing CD changer to mounting bracket and remove changer.
- Installation is reverse of removal.

DSP amplifier, locating

The Digital Sound Processing (DSP) amplifier is located on the right side of the cargo compartment behind the rear lights.

SPEAKERS

Tweeter, removing

– Remove door trim panel. See **411 Doors**.

◄ Release clips (**arrows**) securing tweeter to housing.

– Release electrical connector and remove tweeter.

NOTE—
Models equipped with upgraded Harmon Kardon speakers include a combined tweeter/midrange speaker in this location.

Front speaker, removing

– Remove door trim panel. See **411 Doors**.

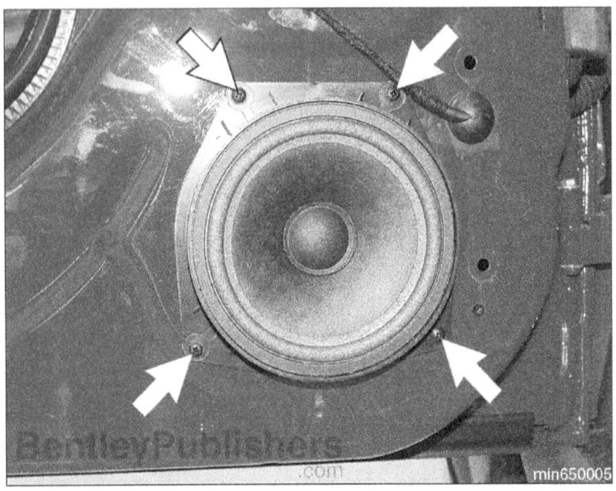

◄ Remove screws (**arrows**) securing front speaker to door.

– Release electrical connector and remove speaker.

Rear speaker, removing

– Remove rear inside trim panel. See **515 Interior trim**.

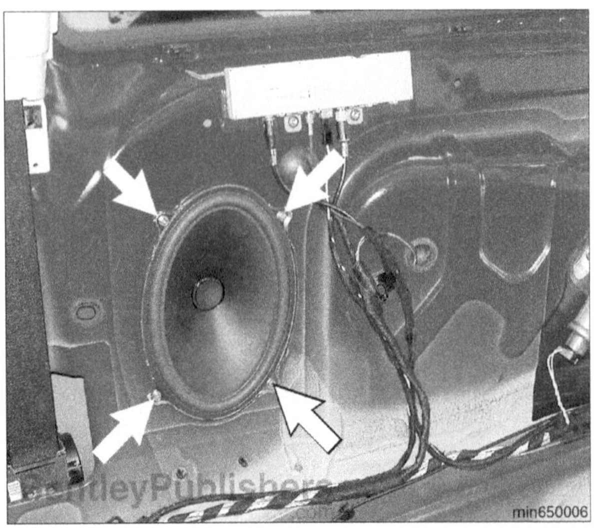

◄ Remove screws (**arrows**) securing rear speaker to body.

– Release electrical connector and remove speaker.

ANTENNA

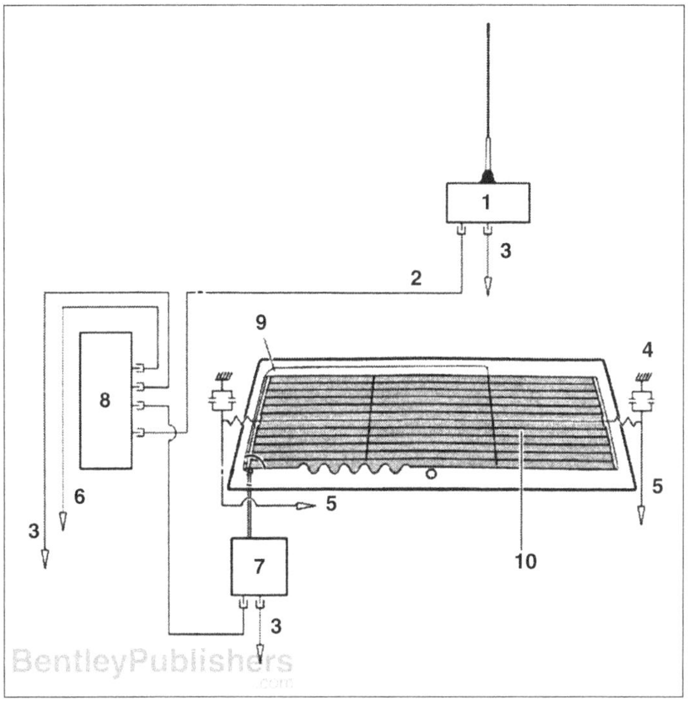

Diversity antenna components

1. **Aerial assembly**
2. **To diversity amplifier**
3. **To radio ON**
4. **Filter**
5. **To rear defrost wiring**
6. **To radio RF**
7. **Antenna amplifier**
8. **Diversity amplifier**
9. **FM2 aerial elements**
10. **FM3 aerial elements**

Roof antenna, removing and installing

– Remove door post trim, rear roof pillar trim, and rear grab handles. See **513 Interior Trim** for more information.

– Remove rear sunroof trim (if equipped). See **540 Sunroof, Rear Spoiler**.

◄ Unscrew aerial from base and remove.

Rotate cover (**arrows**) and remove from base.

Remove nut (**A**) and seal holding aerial mount to roof.

- Working inside rear hatch, carefully lower rear edge of headliner to gain access to aerial base. See **513 Interior trim** from more information.

> ***CAUTION—***
> *Headliner creases easily. Use special care when lowering headliner from roof.*

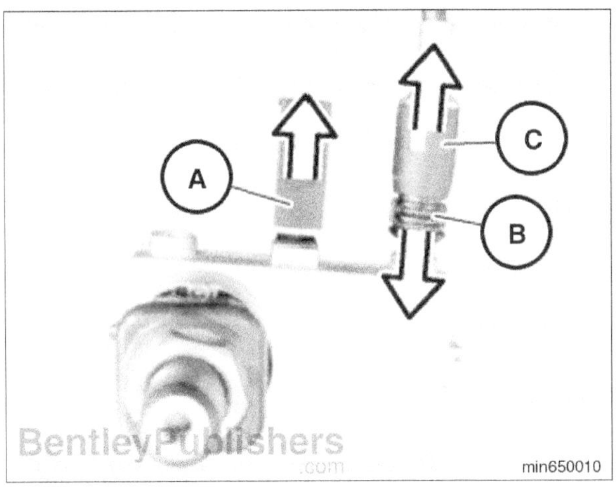

Remove aerial base from roof and release electrical connector (**A**).

- Pull clip (**B**) in direction shown and release connector (**C**). Remove aerial.
- Installation is reverse of removal. Remember to:
  - Position long side of aerial base rearward.
  - Replace any broken trim clips.

Tightening torque	
Aerial mount to roof	3.5 Nm (2.6 ft-lb)

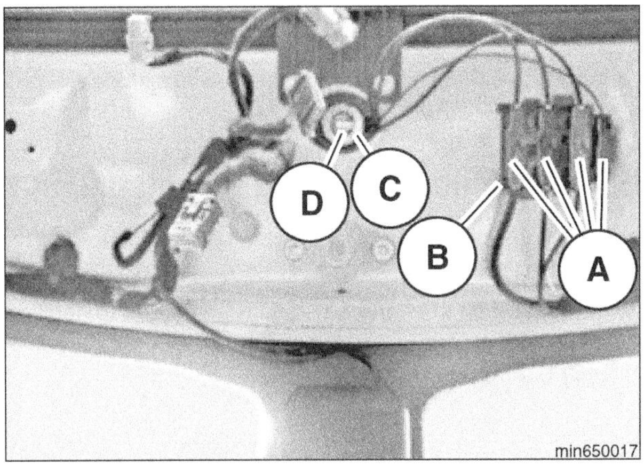

Antenna, removing and installing (Convertible)

- Remove front roof liner trim.
- Remove roof antenna rod.

Working at top center of windshield:
- Detach harness connectors (**A**) from bracket (**B**).
- Separate harness connectors (**A**).
- Remove antenna mounting nut (**C**) from base of antenna (**D**).
- Pull antenna out from above.

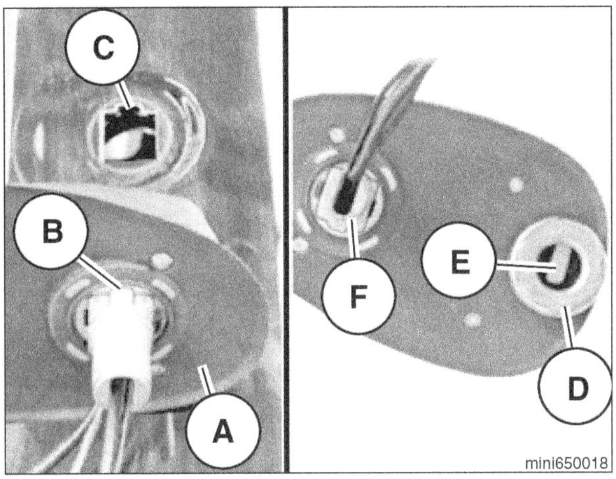

When installing:
- Position antenna base (**A**) in such a way that tab (**B**) engages in groove (**C**).
- Engage nut (D) so that tab (E) engages in associated recess (F) of antenna base.
- Before reattaching harness connectors, make sure harness is routed as before.

Tightening torque	
Antenna to roof	7 Nm (5 ft-lb)

Antenna amplifier, removing

Antenna amplifier is located inside and to right of rear hatch.

- Remove rear hatch trim. See **412 Rear Hatch**.
- Release electrical connectors and remove screws holding amplifier to rear hatch.

Diversity amplifier, removing and installing

- Remove right rear inside trim panel. See **515 Interior trim**.

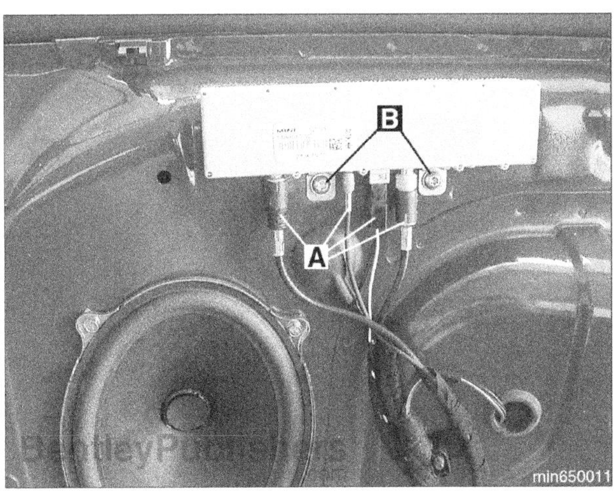

Release electrical connectors (**A**) from antenna amplifier.

- Remove screws (**B**) and remove amplifier.

Tightening torque	
Diversity antenna amplifier to body	4.6 Nm (3.4 ft-lb)

NAVIGATION COMPUTER

Models equipped with the optional navigation system include a navigation computer with CD drive located under the right front seat.

Navigation computer, removing and installing

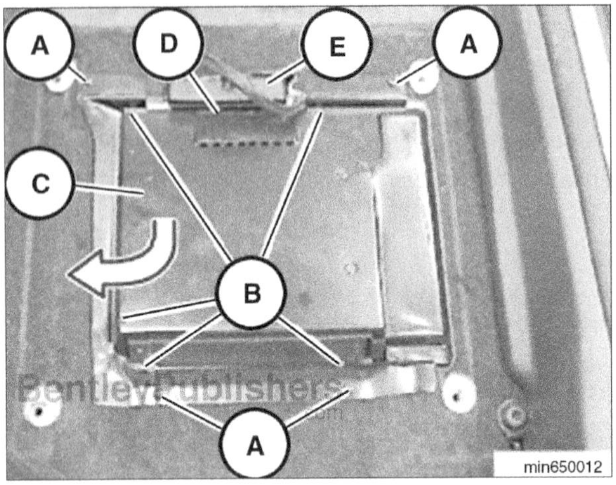

- Remove right front seat. See **520 Seats**.
- Remove screws (**A** and **B**) securing navigation enclosure to floor.
- Lift enclosure (**C**) out from underneath air duct (**E**).

NOTE —
*When installing, make sure tab (**D**) fits below air duct.*

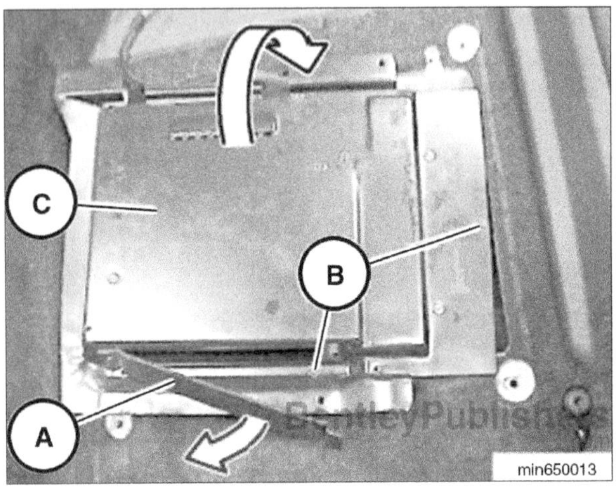

- Open flap (**A**) and remove screws (**B**). Move enclosure cover (**C**) forward.

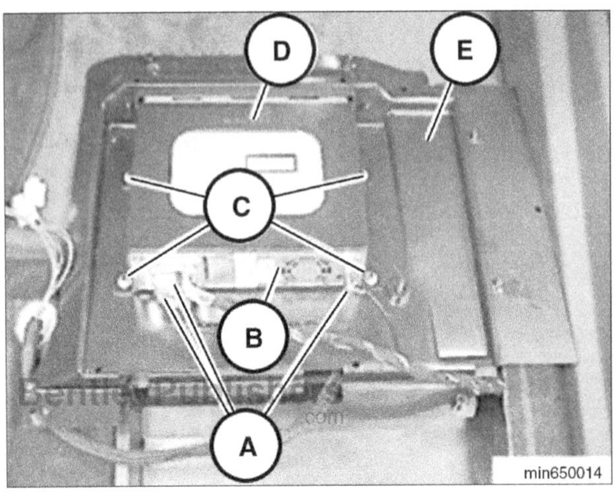

- Release electrical connectors (**A**) from back of navigation computer (**B**).
- Remove navigation computer (**D**) complete with enclosure cover (**E**).
- If replacing, remove fasteners (**C**) and remove computer and mounting bracket from enclosure cover.

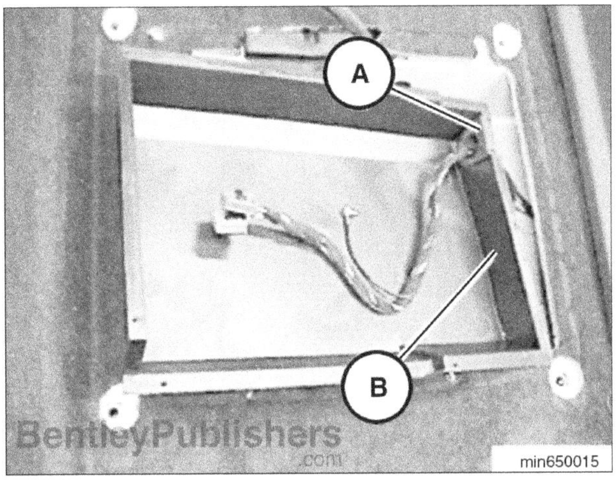

Make sure harness grommet (**A**) remains in position on enclosure cover (**B**).

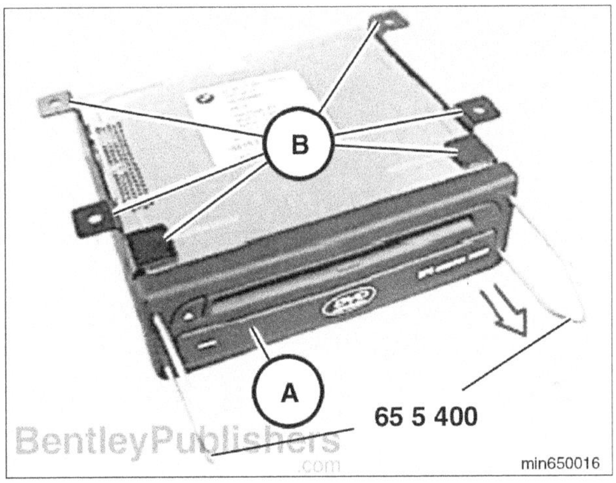

Remove navigation computer (**A**) from mounting bracket (**B**) using special tool 65 5 400. Remove navigation computer.

– Installation is reverse of removal. If necessary, use BMW service tester (DIS, GT1 or MoDiC) to code computer.

720 Seat Belts

General

This repair group covers repairs to the seat belts systems.

Multiple restraint system (MRS) electronics coordinate the operation of front seat belt pyrotechnic tensioners with the deployment of air-bags. For more details about MRS, see the following:

- **121 Battery Starter, Alternator**
- **600 Electrical System–General**
- **721 Airbag System (SRS)**

For diagnostic work on MRS, use BMW service tester DISplus, GT1, MoDiC or equivalent tool.

Special tools

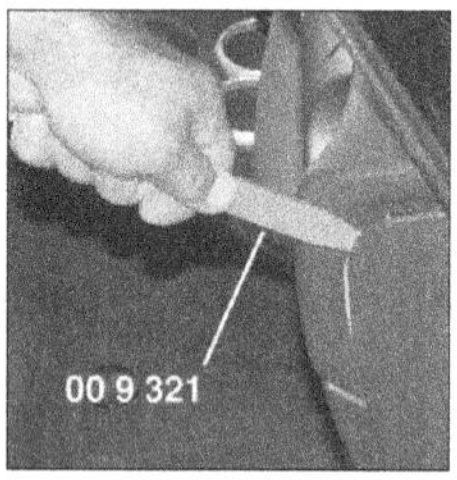

Plastic prying tool
(BMW special tool 00 9 321)

Troubleshooting

Make sure that the seat belt reel locks when pulled quickly and sharply. The automatic reel does not require any servicing and should never be opened.

In case of trouble, see the table **Seat belt troubleshooting**.

Seat belt troubleshooting

Problem	Probable cause	Repairs
Damage to seat belt.	Accident in which bumpers were permanently deformed.	Replace complete automatic seat belt. Also check and replace, if necessary, seat belt mounting on car body and seat belt mounting on seat runner.
Belt creased, unraveled, pinched, cut or melted. Belt buckle or belt lock plastic casing worn, damaged or missing.	Seat belts aged or worn.	Replace complete automatic seat belt.
Seat belt buckle is not ejected by spring pressure when red button on seat belt lock is pressed.	Seat belt buckle mechanism worn or damaged.	Replace seat belt buckle.
Seat belt automatic reel does not lock when pulled out suddenly.	Automatic reel defective.	Replace reel assembly.
Seat belt automatic reel jams when pulled out.	Automatic reel loose. Return spring broken inside reel.	Tighten reel mounting bolt. Replace reel assembly.
Seat belt does not retract automatically.	Automatic reel loose. Return spring broken inside reel.	Tighten reel mounting bolt. Replace reel assembly.
Automatic belt squeaks when fastened or unfastened.	Excessive friction in belt guides.	Replace reel assembly.
	Automatic reel loose.	Tighten reel mounting bolt.
	Return spring broken inside reel.	Replace reel assembly.
Seat belt pyrotechnic tensioner triggered.	Accident triggered pyrotechnic deployment.	Replace complete automatic seat belt. Also check retaining bracket of belt tensioner for twist.

Warnings and cautions

WARNING —
- *Pyrotechnic seat belt tensioners are powerful devices and should be handled with extreme care. Incorrect handling can trigger the tensioner and cause injury.*
- *MINI recommends that all repair or replacement work on pyrotechnic devices must be carried out by a qualified MINI technician.*
- *Be sure to disconnect the battery and wait 5 seconds before attempting to work on pyrotechnic devices.*
- *Pyrotechnic devices cannot be repaired. Always replace them.*
- *Never treat pyrotechnic components with cleaning agents or grease.*
- *Do not expose pyrotechnic components to temperatures above 75°C (167°F).*
- *Do not fire a pyrotechnic gas generator prior to disposal. It must be fired by a special disposal company or shipped back to the manufacturer in the packaging of the new component.*
- *During body straightening and welding with an electric arc welder, always disconnect the battery and the connection to the pyrotechnic gas generators.*
- *For maximum protection from injury, replace seat belts as a set (including all hardware), if they are subject to occupant loading in a collision.*
- *Do not modify or repair seat belts. Do not modify seat belt mounting points.*
- *Do not bleach or dye seat belt webbing. Webbing that is severely faded or redyed will not meet the strength requirements of a collision and must be replaced.*
- *Clean belts with a luke-warm soap solution only.*
- *Periodically inspect seat belts for webbing defects such as cuts or pulled threads.*
- *Immediately after replacing a damaged or worn seat belt, destroy the old belt to prevent it from being used again.*

CAUTION —
- *When working on electrical switches or lights, always disconnect the negative (–) cable from the battery and insulate the cable end to prevent accidental reconnection.*
- *Prior to disconnecting the battery, read the battery disconnection cautions given in* **001 General Warnings and Cautions**.
- *To prevent marring trim when working on interior components, work with plastic prying tools or wrap the tips of screwdrivers and pliers with tape before prying out panels.*

Front Seat Belts

Front seat belts differ slightly between Coupe and Convertible in the area of belt and pivot/loop mounting. Coupe removing and installing information shown, Convertible is similar.

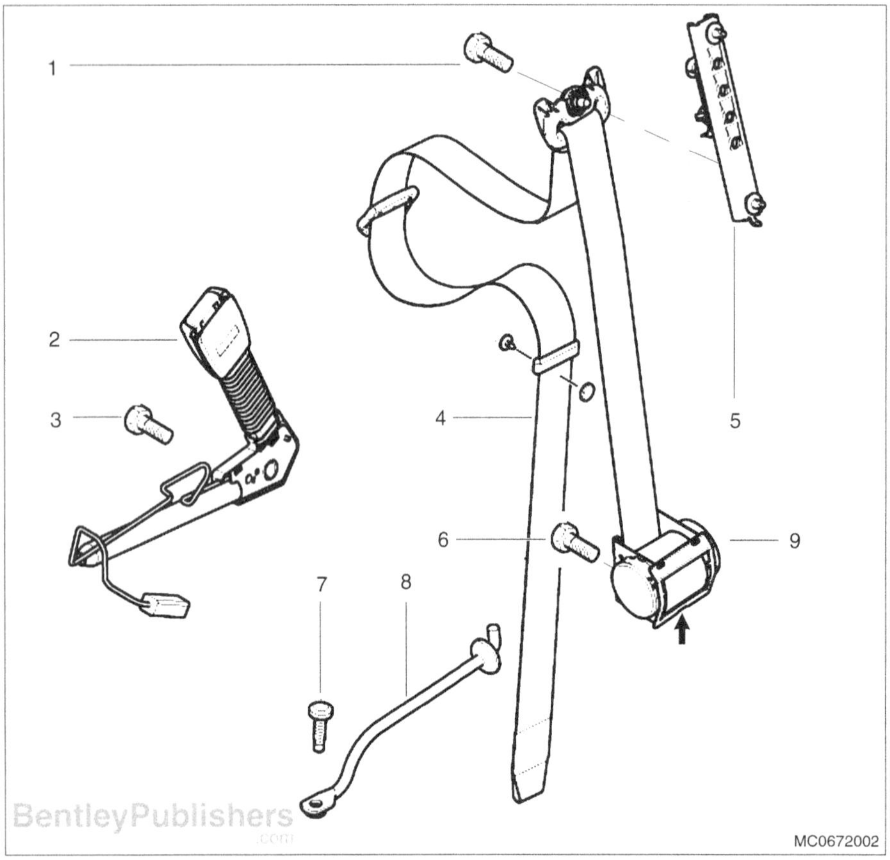

Front seat belt assembly (Coupe)

1. **Seat belt loop mounting bolt**
   - M10
     tighten to 32 Nm (24 ft-lb)
     use Loctite®270
2. **Pyrotechnic tensioner and seat belt buckle assembly**
3. **Seat belt buckle mounting bolt**
   - M10
     tighten to 45 Nm (33 ft-lb)
     use Loctite®270
4. **Seat belt**
5. **Seat belt height adjuster**
6. **Seat belt reel mounting bolt**
   - 7/16"
     tighten to 29 Nm (21 ft-lb)
     use Loctite®270
7. **Seat belt slider bar mounting bolt**
   - 7/16"
     tighten to 40 Nm (30 ft-lb)
     use Loctite®270
8. **Seat belt slider bar**
9. **Automatic seat belt reel**
   - Note locating/alignment tab slot on bottom inside (**arrow**)

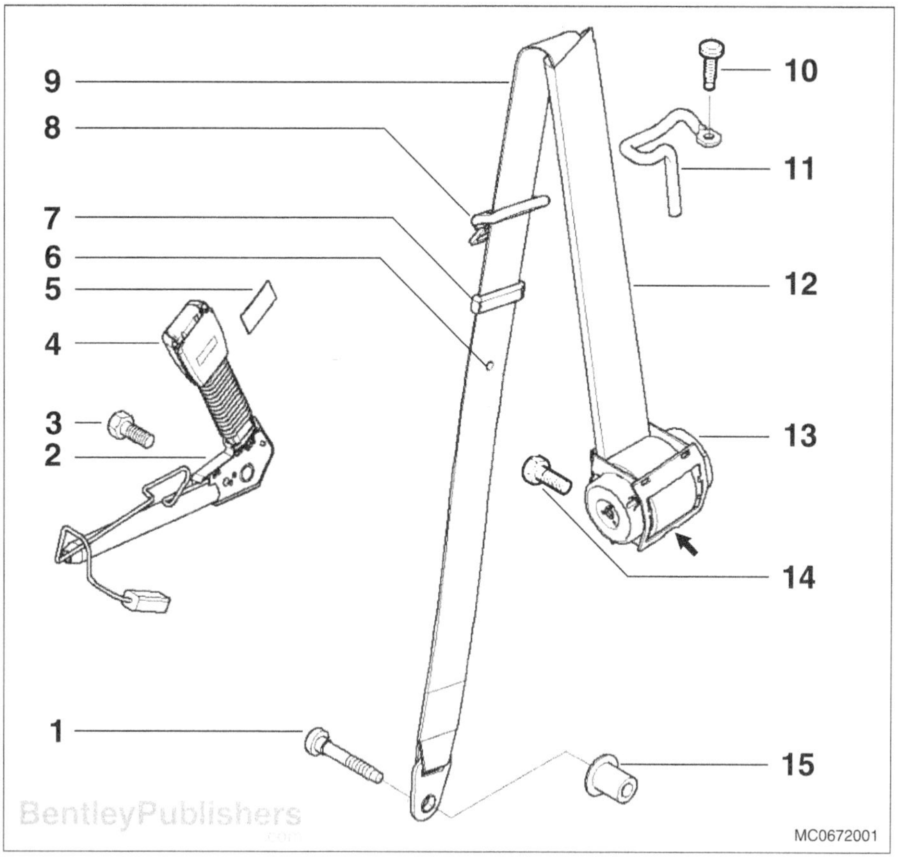

Front seat belt assembly (Convertible)

1. **Lower seat belt mounting bolt**
   - 7/16"
     tighten to 40 Nm (30 ft-lb)
     use Loctite®270
2. **Mounting bracket**
3. **Seat belt buckle mounting bolt**
   - M10
     tighten to 45 Nm (33 ft-lb)
     use Loctite®270
4. **Pyrotechnic tensioner and seat belt buckle assembly**
5. **Anti-scuffing pad**
6. **Limit/stop button**
7. **Seat belt latch limiting clip**
8. **Seat belt latch**
9. **Seat belt**
10. **Upper seat belt pivot bar bolt**
    - M10
      tighten to 45 Nm (18 ft-lb)
      use Loctite®270
11. **Upper seat belt pivot bar**
12. **Seat belt guide**
13. **Automatic seat belt reel**
    - Note locating/alignment tab slot on bottom inside (**arrow**)
14. **Automatic seat belt reel mounting bolt**
    - 7/16"
      tighten to 31 Nm (23 ft-lb)
      use Loctite®270
15. **Spacer bushing**

Front seat belt, removing and installing (Coupe)

– Remove rear side trim panel. See **513 Interior Trim**.

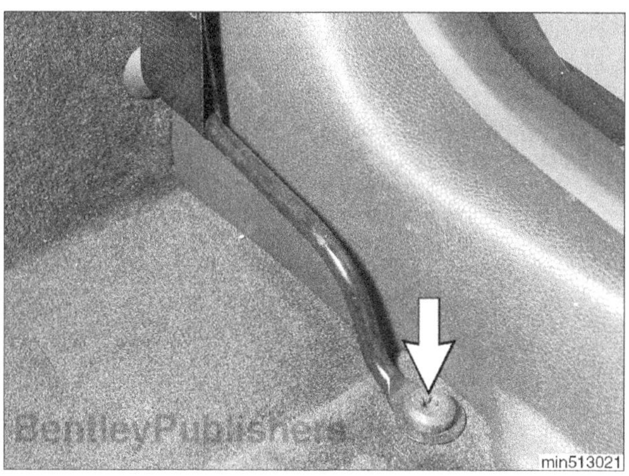

◄ Remove seat belt slider bar mounting bolt (**arrow**) and remove slider.

– Remove B-pillar (door post) trim. See **513 Interior Trim**.

> ***WARNING —***
> - *Read warnings and cautions in* **721 Airbag System (SRS)**. *Serious injury may result if system service is attempted by persons unfamiliar with SRS and its approved service procedures.*
> - *Do not use sharp-edged tools to remove B-pillar trim. Damage to head protection airbag could result.*

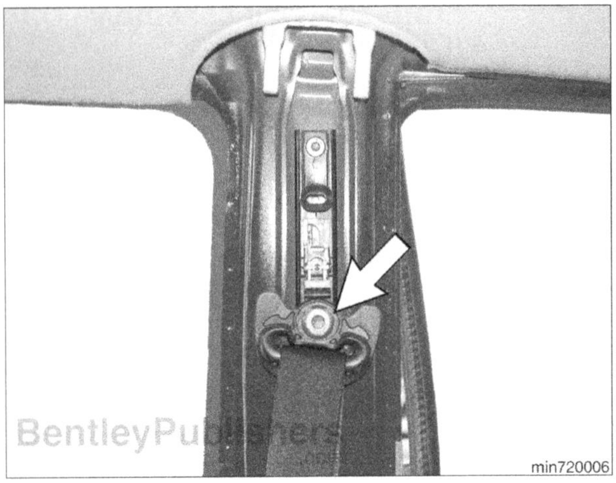

Working at top of B-pillar, remove mounting bolt (**arrow**) from seat belt height adjuster.

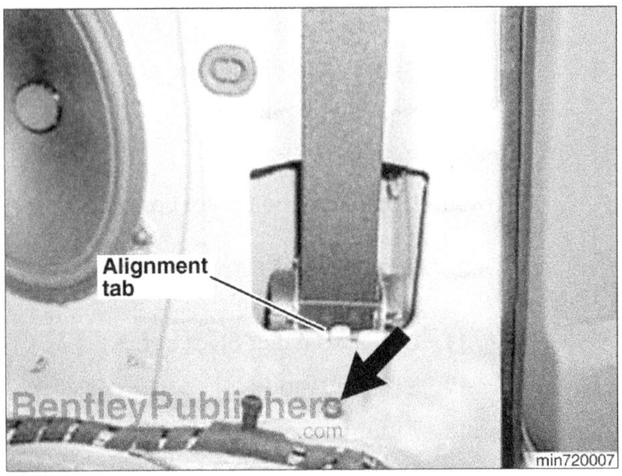

Working at bottom of B-pillar, remove seat belt reel mounting bolt (**arrow**). Pull reel and belt out of side cavity.

– Installation is reverse of removal.
  - Make sure reel alignment tab is in correct position.
  - Use Loctite®270 or equivalent thread-locking compound on bolt threads.

Tightening torque	
Seat belt slider bar to floor (7/16", use Loctite)	40 Nm (30 ft-lb)
Seat belt to adjuster (M 10, use Loctite)	32 Nm (24 ft-lb)
Seat belt reel to B-pillar (7/16", use Loctite)	29 Nm (21 ft-lb)

Front seat belt buckle, removing and installing (Coupe)

The front seat belt buckles are equipped with pyrotechnic (explosive charge) automatic tensioners. These tensioners are designed to automatically retract and tension the seat belt by 55 mm (approximately 2 inches) in case of accident.

– Disconnect negative (–) cable from battery and cover negative terminal with insulating material.

> ***WARNING —***
> *After disconnecting the battery, wait 5 seconds before beginning work on airbag components.*

> ***CAUTION —***
> *Prior to disconnecting the battery, read the battery disconnection cautions given in* **001 General Warnings and Cautions**.

– Remove front seat as described in **520 Seats**.

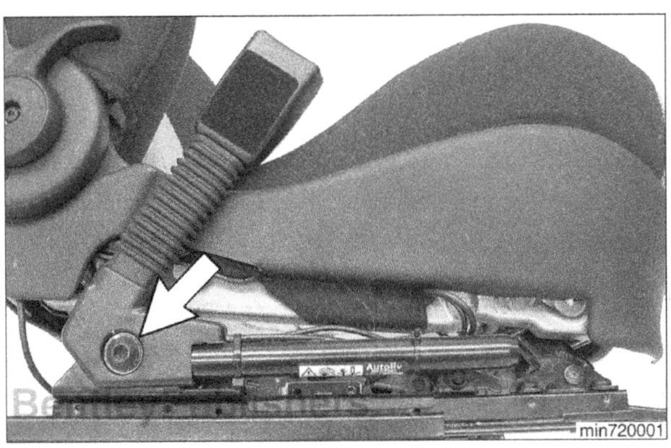

Working at seat rail, remove tensioner mounting bolt (**arrow**).

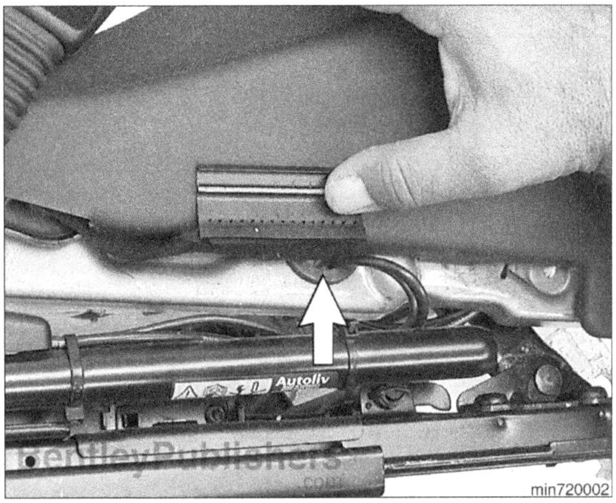

◄ Unclip plastic trim and peel up fabric clip from lip of seat. This allows tensioner electrical harness to pass through grommet (**arrow**).

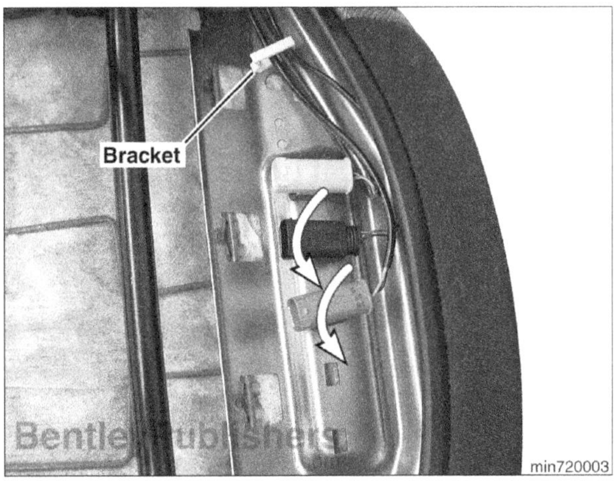

◄ Working underneath seat cushion at front, rotate electrical harness connectors 45° counterclockwise (**arrows**) to detach.

NOTE —
Two connectors are attached to tensioner harness.

– Detach electrical harness from brackets and pass through grommet in side of seat.

– Remove seat belt buckle and tensioner assembly.

– Installation is reverse of removal. Use Loctite®270 or equivalent thread-locking compound on bolt threads.

Tightening torque	
Seat belt buckle and tensioner assembly to seat rail (M10)	45 Nm (33 ft-lb)

Front seat belt height adjuster, removing and installing (Coupe)

– Remove B-pillar (door post) trim. See **513 Interior Trim**.

◄ Working at top of B-pillar, remove mounting bolt (**arrow**) from seat belt height adjuster.

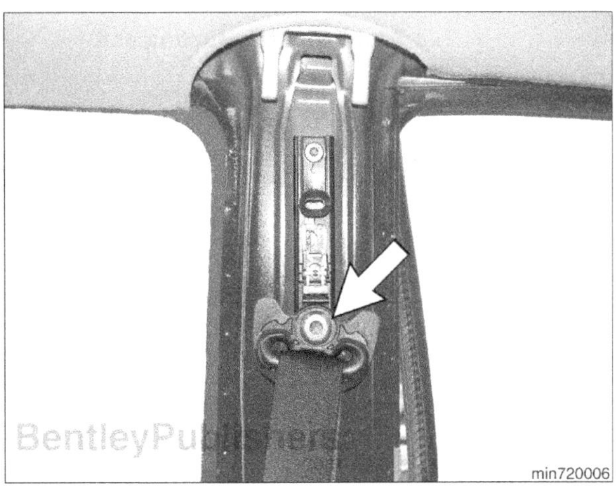

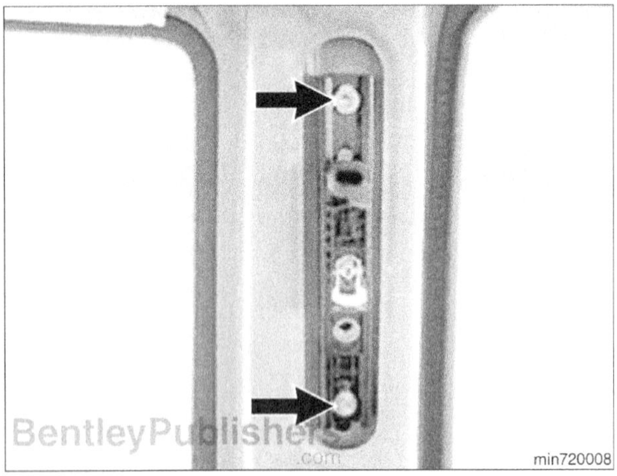

◀ Remove seat belt adjuster mounting bolts (**arrows**). Remove adjuster.

– Installation is reverse of removal. Use Loctite®270 or equivalent thread-locking compound on bolt threads.

Tightening torque	
Front seat belt adjuster to body (M8, use Loctite)	25 Nm (18 ft-lb)
Front seat belt to adjuster (M 10, use Loctite)	32 Nm (24 ft-lb)
Front seat belt to B-pillar (7/16", use Loctite)	29 Nm (21 ft-lb)

Rear Seat Belts

Rear seat belts differ slightly between Coupe and Convertible in the area of reel mounting. Coupe removing and installing information shown, Convertible is similar.

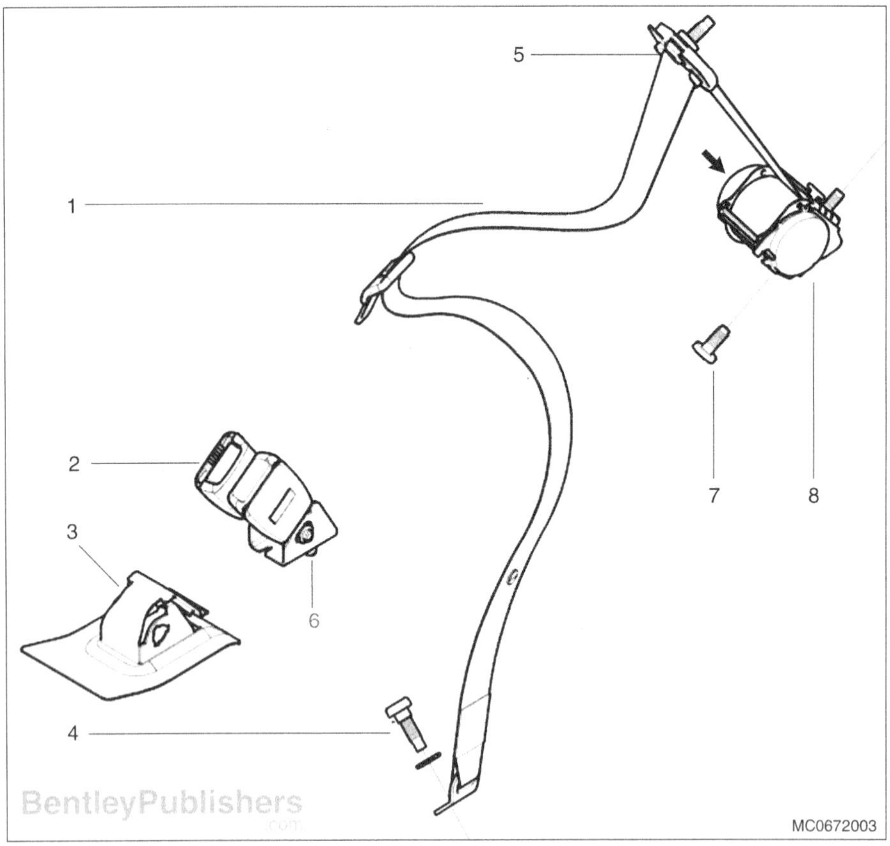

Rear seat belt assembly

1. **Seat belt**
2. **Seat belt buckle**
3. **Seat belt buckle cover**
4. **Seat belt lower mounting bolt**
   - 7/16"
     tighten to 40 Nm (30 ft-lb)
     use Loctite®270
5. **Seat belt loop mounting bolt**
   - Coupe - mounted on C-pillar
   - Convertible - mounted to roll bar
   - 7/16"
     tighten to 40 Nm (30 ft-lb)
     use Loctite®270
6. **Seat belt buckle mounting bolt**
   - M10
     tighten to 40 Nm (30 ft-lb)
     use Loctite®270
7. **Seat belt reel mounting bolt**
   - 7/16"
     tighten to 40 Nm (30 ft-lb)
     use Loctite®270
8. **Automatic seat belt reel**
   - Coupe - mounted at base of C-pillar
   - Convertible - mounted at base of roll bar
   - Note locating/alignment tab slot on bottom inside (**arrow**)

Rear seat belt, removing and installing (Coupe)

– Remove C-pillar (rear roof pillar) trim. See **513 Interior Trim**.

> ***WARNING —***
> - *Read the warnings and cautions in* **721 Airbag System (SRS)**. *Serious injury may result if system service is attempted by persons unfamiliar with SRS and its approved service procedures.*
> - *Do not use sharp-edged tools to remove C-pillar trim. Damage to head protection airbag could result.*

NOTE —
C-pillar trim is destroyed when removed. Be sure to have new trim on hand before attempting to remove.

◄ Working above rear seat backrest at C-pillar, remove seat belt loop mounting bolt (**arrow**).

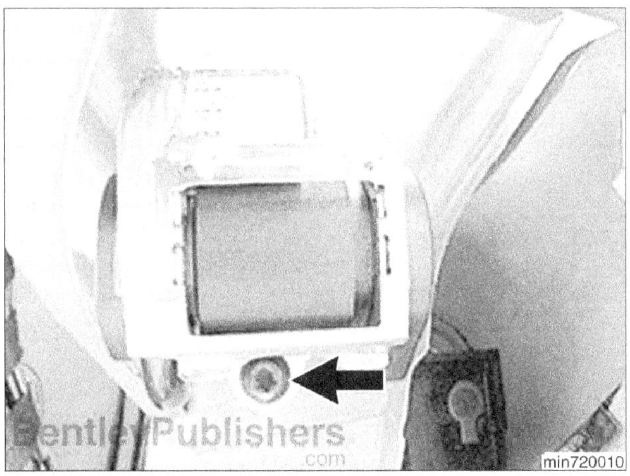

◄ Remove seat belt reel mounting bolt (**arrow**).

– Lift up rear seat cushion and remove seat belt lower mounting bolt. Remove belt from car.

– Installation is reverse of removal. Use Loctite®270 or equivalent thread-locking compound on bolt threads.

Tightening torques	
Rear seat belt loop to C-pillar (7/16", use Loctite)	40 Nm (30 ft-lb)
Rear seat belt lower attachment to body (7/16", use Loctite)	40 Nm (30 ft-lb)
Rear seat belt reel to C-pillar (7/16", use Loctite)	40 Nm (30 ft-lb)

Rear set belt buckle, removing and installing

– Remove back seat. See **520 Seats**.

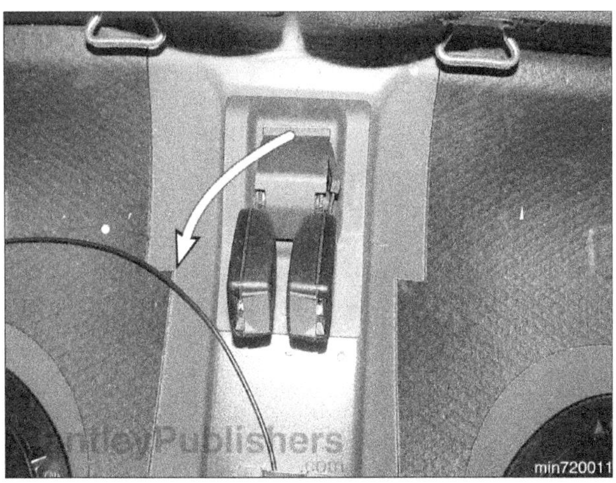

◄ Unclip rear belt buckle cover (**arrow**).

Rear Seat Belts

Remove belt buckle mounting bolt (**arrow**). Remove buckle.

– Installation is reverse of removal. Use Loctite®270 or equivalent thread-locking compound on bolt threads.

Tightening torque	
Rear seat belt buckle to body (M10, use Loctite)	40 Nm (30 ft-lb)

721 Airbag System (SRS)

GENERAL

This repair group covers airbag components and replacement. Airbag system repair and fault diagnosis is not covered. Be sure that diagnostics, component testing, and airbag system repair are carried out by a trained MINI service technician.

Airbag systems are also known generically as Supplemental Restraint Systems (SRS) and are part of the MINI Multiple Restraint System (MRS). For more details about MRS, see the following:

- **121 Battery Starter, Alternator**
- **600 Electrical System–General**
- **720 Seat Belts**
- **ELE Electrical Wiring Diagrams**

Individual airbag system components can only be tested electronically when installed in the car. Use BMW service tester DISplus, GT1 or MoDiC or equivalent tool for diagnostic work.

Special tools

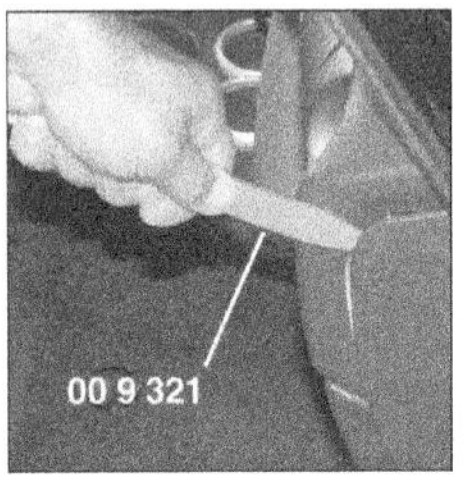

◀ Plastic prying tool
(BMW special tool kit 00 9 321)

Multiple Restraint Systems

Two different Multiple Restraint Systems (MRS) versions are installed in MINIs depending on model year and body type.

MRS 4 - this system provides enhanced passive protection for vehicle occupants in case of serious collision and is considered passive because it operates automatically and provides protection without occupant preconditional interaction. However, seatbelt use is integral to full protection. Components of the MRS 4 system include the following:

- MRS control unit
- Driver's airbag module
- Front passenger's airbag module
- Driver's side (thorax) airbag module
- Front passenger's side (thorax) airbag module
- Advanced Head Protection System (AHPS) airbag, left side
- Advanced Head Protection System (AHPS) airbag, right side
- Driver's seat belt pre-tensioner
- Front passenger's seat belt pre-tensioner
- MRS indicator/warning light in instrument cluster
- Crash sensor, left side B-pillar
- Crash sensor, right side B-pillar
- Seat belt buckle switches
- Slip ring (rotary coupler)
- Battery safety terminal
- Inertia sensor (through 08/2002 production only)

MRS 5 - this system provides protection in much the same way as the MRS 4, but differs due to decreased reaction time and increased accuracy provided by the inclusion of door pressure sensors and up-front longitudinal acceleration sensors. In addition to the components of the MRS 4 system, the MRS 5 system adds the following:

- Crash sensor upfront, left longitudinal acceleration
- Crash sensor upfront, right longitudinal acceleration
- Door pressure sensor, inside left door
- Door pressure sensor, inside right door
- Passenger's seat occupancy detector (OC-3 Mat)
- Passenger's seat airbag warning light (status LED)

Airbag system components

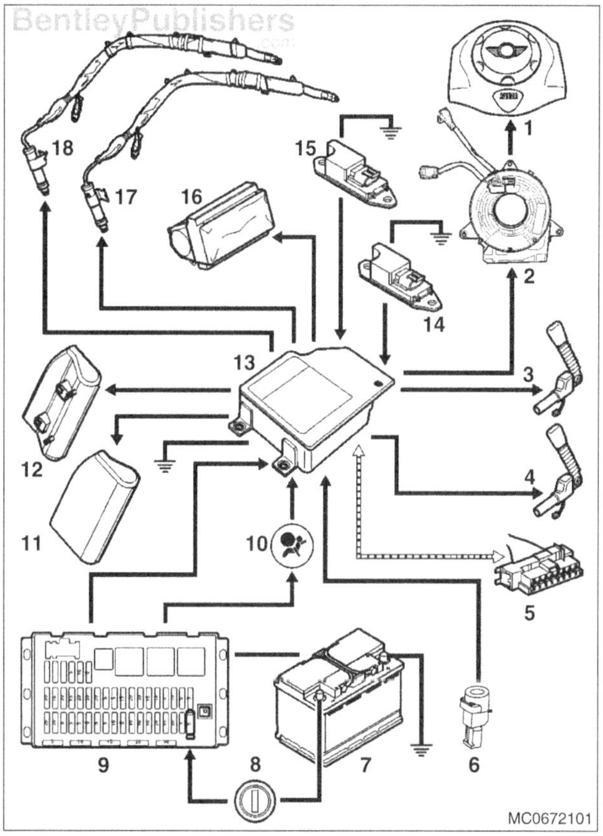

Primary components of MRS 4 systems with locations are:

1. Driver airbag module
   - In steering wheel
2. Slip ring (rotary coupler)
   - Between steering wheel and steering column
3. Driver's seat belt pre-tensioner
   - On driver's seat frame, integral with buckle
4. Passenger's seat belt pre-tensioner
   - On passenger's seat frame, integral with buckle
5. Data link connector
   - Under left side padded dashboard, above pedal cluster
6. Inertia sensor (through 08/2002 production only)
   - Under hood, near left side bulkhead/fender junction
7. Battery (trunk-mounted batteries only)
   - With Battery Safety Terminal (BST) on positive terminal
8. Ignition switch
9. Passenger compartment fuse box
10. MRS indicator/warning light in instrument cluster
11. Front passenger's side (thorax) airbag module
    - On front passenger's seat backrest frame
12. Driver's side (thorax) airbag module
    - On driver's seat backrest frame
13. MRS control unit
    - On body tunnel between front seats under carpet
14. Crash sensor
    - On body at base of left B-pillar
15. Crash sensor
    - On body at base of right B-pillar
16. Passenger airbag module
    - In padded dashboard above glove box
17. Advanced Head Protection System (AHPS) airbag, left side
    - On body above left door and rear window, igniter/gas generator behind left inside panel under rear side windows
18. Advanced Head Protection System (AHPS) airbag, right side
    - On body above right door and rear window, igniter/gas generator behind right inside panel under rear side windows

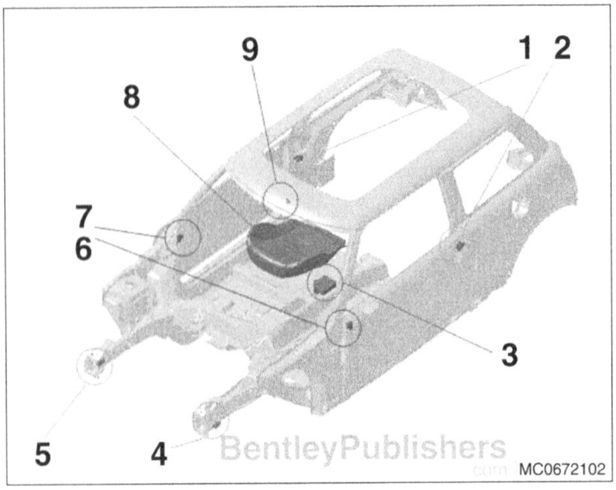

Most of the MRS 4 system components are carried over to the MRS 5 system. Components of MRS 5 systems include:

1. Crash sensor (carried over from MRS 4 system)
   - On body at base of right B-pillar
2. Crash sensor (carried over from MRS 4 system)
   - On body at base of left B-pillar
3. MRS control unit
   - On body tunnel between front seats under carpet
4. Crash sensor upfront, left longitudinal acceleration
   - On left front longitudinal member at bumper mounting
5. Crash sensor upfront, right longitudinal acceleration
   - On right front longitudinal member at bumper mounting
6. Door pressure sensor
   - Inside left door, near forward edge
7. Door pressure sensor
   - Inside right door, near forward edge
8. Passenger's seat occupancy detector (OC-3 Mat)
   - In passenger's seat cushion
9. Passenger's seat airbag warning light (status LED)
   - In upper windshield frame

The airbag indicator warning light (**arrow**) of MRS 5 systems is located in the center of the upper windshield frame on both Coupe and Convertible. The airbag indicator light is activated and illuminates in yellow when the front passenger airbag and the front passenger's side (thorax) airbags are deactivated.

Principles of operation

The main airbag system impact sensor is a deceleration detection device contained in the MRS control module. A second sensor is used to confirm a crash condition. In case of a crash, the MRS module initiates airbag module and/or seat belt tensioner deployment.

For side impacts, the side impact crash sensors under the rear seat and in the doors (where applicable) provide additional inputs to the control module. For front angled impacts, the control module acts in cooperation with the side sensors and applicable upfront sensors to determine which airbags and seat belt tensioners to deploy. For rear impacts, the control module uses a rear impact trigger threshold to determine that a severe rear collision has occurred and deploys all seat belt tensioners to restrain vehicle occupants.

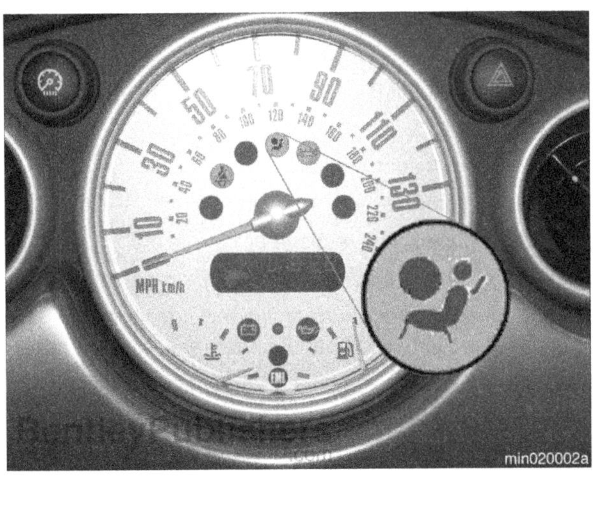

The red MRS warning LED, located in the center dashboard instrument cluster, is illuminated for a short period after switching the ignition ON. If a fault is present within the MRS, the LED illuminates and remain ON until the ignition is switched OFF. In this case, MRS may not operate in the event of a collision.

Airbag deactivation

On MRS 4 systems, it is possible to deactivate the passenger airbag. See an authorized MINI dealer service department for more information. On MRS 5 systems, consult the vehicle's glovebox owner's manual.

> ***WARNING —***
> *Disabling an airbag may lead to serious injury in case of an accident.*

Triggered airbags

The following is recommended after airbag deployment:

MRS control module. Fault memory can be cleared and reset twice. Replace module after 3rd incident.

Cables and plugs. Check for damage, replace if necessary.

Airbag modules. Replace if deployed.

Steering wheel. Replace if damaged.

Steering column. Replace if damaged.

Passenger airbag cover. Replace if airbag deployed.

Passenger airbag support rod. Replace if damaged.

A-pillar, B-pillar, C-pillar trim, headliner. Replace if head protection airbags deployed.

Warnings and cautions

When working on airbags, airbag system components or other pyrotechnic safety devices, observe the following precautions to prevent personal injury.

> ***WARNING —***
> - *If the airbag warning light is on, there is a risk that the airbags will not be triggered in case of an accident. Be sure to have the system inspected and repaired immediately.*
> - *Airbag(s) are inflated by an explosive device. Handled improperly or without adequate safeguards, the system can be very dangerous. Special precautions must be observed prior to any work at or near the airbags.*
> - *The airbag is a vehicle safety system. Serious injury may result if system service is attempted by persons unfamiliar with the MINI MRS and its approved service procedures. MINI advises that all inspection and service be performed by an authorized MINI dealer.*
> - *Disconnect the battery and cover the negative (–) battery terminal with an insulator before starting electrical work, and before doing any welding on the car.*
> - *After disconnecting the battery, wait 5 seconds before beginning work on airbag components.*

> ***WARNING —***
> - *If an airbag has been activated due to an accident, MINI specifies that airbag components be replaced. For more information on post-collision airbag service, see* **Triggered airbags** *in this repair group or consult an authorized MINI dealer.*
> - *Do not fire an airbag unit prior to disposal. It must be fired by a special disposal company or shipped back to MINI in the packaging of the new components.*
> - *When removing a fired airbag unit, avoid contact with skin; wear gloves. In case of skin contact, wash with water.*
> - *Do not allow airbag system components to come in contact with cleaning solutions or grease. Never subject airbag components to temperatures above 167°F (75°C). When reconnecting the battery, no person should be inside the vehicle.*
> - *Always place an airbag unit that has been removed from its packaging with the padded side facing upward. Do not leave an airbag unit unattended.*
> - *If the airbag unit or airbag control module has been dropped from a height of ½ meter (1½ ft) or more, do not install the component in a vehicle.*

> ***CAUTION —***
> - *When working on electrical switches or lights, always disconnect the negative (–) cable from the battery and insulate the cable end to prevent accidental reconnection.*
> - *Prior to disconnecting the battery, read the battery disconnection cautions given in* **001 General Warnings and Cautions**.
> - *To prevent marring trim when working on interior components, work with plastic prying tools or wrap the tips of screwdrivers and pliers with tape before prying out panels.*

AIRBAG ELECTRONICS

MRS control module

The MRS control module (**arrow**) is located between the front seat under the carpet. To access:

- Remove front seats. See **520 seats**.
- Remove rear console. See **513 Interior Trim**
- Lift up carpet.

Side impact crash sensor

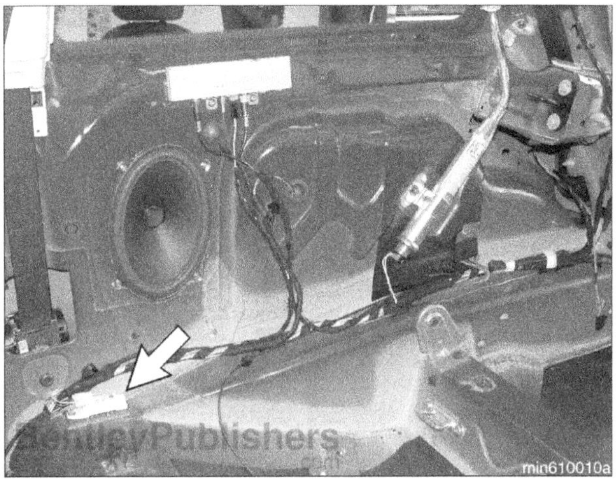

The side impact crash sensors (**arrow**) are installed under the rear seat. To access, lift up rear seat cushion.

AIRBAGS

Replacement of the thorax airbag (in the front seat) is beyond the scope of this manual. Consult an authorized MINI dealer.

Driver airbag, removing and installing

- Disconnect negative (–) cable from battery.

> ***WARNING—***
> *After disconnecting the battery, wait 5 seconds before beginning work on airbag components.*

> ***CAUTION—***
> *Prior to disconnecting the battery, read the battery disconnection cautions given in* **001 General Warnings and Cautions**.

Loosen two airbag mounting fasteners (Torx T30) located at rear of steering wheel. Pull airbag module away from steering wheel.

Gently pry up electrical harness connector locking tabs at rear of airbag. Detach connectors by pulling straight off airbag.

- Remove airbag. Store in safe place with horn pad facing up.
- Installation is reverse of removal. Be careful to not pinch electrical wire harnesses and connectors. See **721 Airbag System (SRS)**.

Tightening torque	
Airbag to steering wheel (M6)	10 Nm (7 ft-lb)

Passenger airbag, removing and installing

– Disconnect negative (–) cable from battery and cover negative terminal with insulating material.

> ***WARNING —***
> *After disconnecting the battery, wait 5 seconds before beginning work on airbag components.*

> ***CAUTION —***
> *Prior to disconnecting the battery, read the battery disconnection cautions given in* **001 General Warnings and Cautions**.

– Working at right side of dashboard, remove glove compartment and passenger airbag trim cover. See **513 Interior Trim**.

NOTE —
If the airbag has been deployed, replace the trim cover.

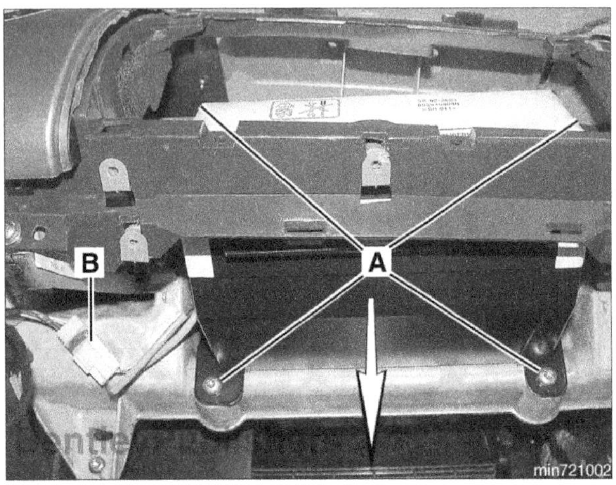

Working at passenger airbag module:

- Remove module mounting screws (**A**).
- Disconnect airbag electrical harness connector (**B**).
- Slide airbag module downward (**arrow**) to remove.

Tightening torque	
Passenger airbag to dashboard carrier	10 Nm (7 ft-lb)

Advanced head protection system airbag

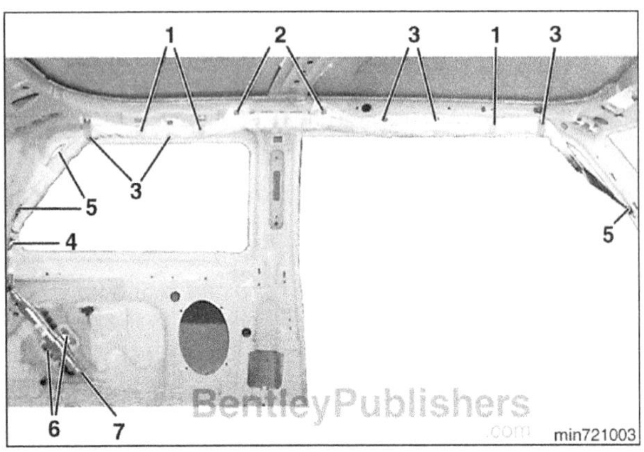

The Advanced Head Protection System airbag assembly (AHPS2) mounting hardware is as follows:

1. Securing clips
2. B-pillar retaining strap mounting bolts
3. Airbag mounting screws
4. Bracket mounting screw
5. A-pillar and C-pillar tether mounting screws
6. Gas generator self-locking nuts
7. Electrical harness connector

– Disconnect negative (–) cable from battery and cover negative terminal with insulating material.

> ***WARNING —***
> *After disconnecting the battery, wait 5 seconds before beginning work on airbag components.*

> ***CAUTION—***
> *Prior to disconnecting the battery, read the battery disconnection cautions in* **001 General Warnings and Cautions**.

– Replacement of a head protection airbag is an extensive operation, including removal of the following (see **513 Interior Trim** for details):

- Complete headliner
- A-pillar (windshield pillar) trim
- B-pillar (door pillar) trim
- C-pillar (rear hatch pillar) trim
- Rear side panels

NOTE—
A-pillar trim and C-pillar trim are one time use components. They must be replaced if removed.

Tightening torques	
Airbag to base of roof	6 Nm (4 ft-lb)
Airbag to B-pillar (use loctite®270)	6 Nm (4 ft-lb)
Bracket to C-pillar	3.5 Nm (30 in-lb)
Gas generator tube to body (replace self-locking nuts)	6 Nm (4 ft-lb)
Tether to A-pillar and C-pillar	11 Nm (8 ft-lb)

ECL Electrical Component Locations

TABLES

ECL

GENERAL

This repair group covers fuse, relay and control module location information. Ground points and other component locations are also covered.

Keep in mind that electrical equipment and accessories installed vary depending on model and model year. Always confirm that the proper electrical component has been identified.

For additional MINI electrical system information, see **600 Electrical System-General**.

Special tools

Some special tools are needed to test and repair the electrical systems in the car.

◀ Automotive digital multimeter

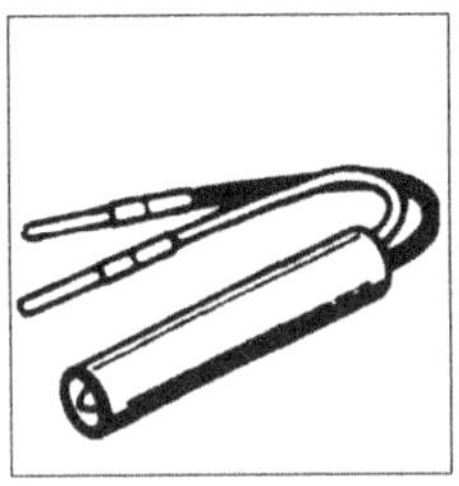

◀ LED test light
(Baum tool 1115)

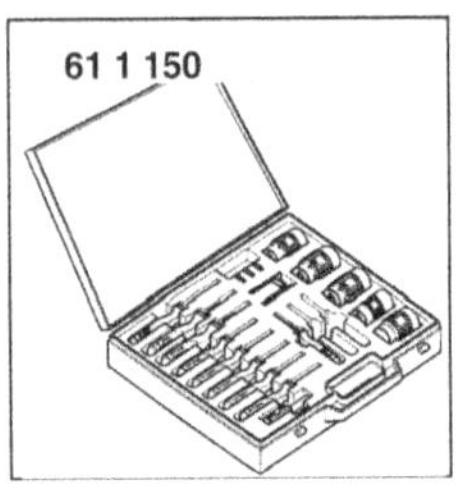

◀ Wiring harness repair kit
(BMW special tool 61 1 150)

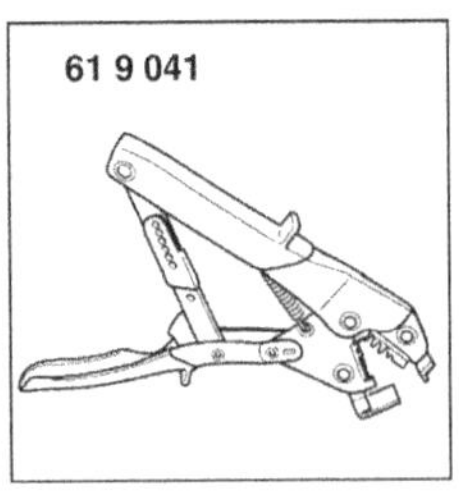

◀ Wire end crimp tool
(BMW special tool 61 9 041)

Electrical system safety precautions

Please read the following warnings and cautions before doing any work on the electrical system.

WARNING—
The Battery Safety Terminal (BST), pyrotechnic seat belt tensioners and airbags utilize explosive devices and must be handled with extreme care. Refer to the warnings and cautions in **121 Battery, Alternator, Starter**; **720 Seat Belts**; *and* **721 Airbag System (SRS)**.

CAUTION—
- *Prior to disconnecting the battery, read the battery disconnection cautions in* **001 General Warnings and Cautions**.
- *Relay and fuse positions are subject to change and may vary from car to car. If questions arise, an authorized MINI dealer is the best source for the most accurate and up-to-date information.*
- *Always switch the ignition off and remove the negative (-) battery cable before removing any electrical components. Connect and disconnect ignition system wires, multiple connectors, and ignition test equipment leads only while the ignition is switched off.*
- *Only use a digital multimeter for electrical tests.*

ELECTRICAL COMPONENTS

The **Electrical component location table** is a cross-referenced listing of electrical components in MINI vehicles. Where available, photos of components in this repair group are referred to in the fourth column.

NOTE—
- *Every attempt has been made in this repair manual to standardize component names.*
- *The* **Component identifier** *column in the table identifies the component by the alphanumeric code used in wiring diagrams.*
- *No MINI vehicle is equipped with all components listed or illustrated.*

Electrical component location table

Component	Comp. identifier	Location	Refer to
A/C		*see also* Heating and A/C entries	
A/C compressor		Right lower front of engine	
A/C compressor clutch	Y8099	On A/C compressor, right lower front of engine	
A/C compressor relay	K19	Left engine compartment, fuse and relay panel 3	Fig. 6
A/C evaporator temperature sensor	B14	Under left side of dashboard, in heating and A/C housing	Fig. 33
A/C pressure sensor	B8	Left engine compartment, underneath battery (Cooper) or air filter housing (Cooper S)	Fig. 20
ABS / ASC or ABS / DSC control module	A7, A65	Right rear engine compartment	Fig. 17 Fig. 18

Electrical Components

Electrical component location table			
Component	**Comp. identifier**	**Location**	**Refer to**
ABS / ASC or ABS / DSC control switch		Fascia (center dashboard) switch console	
ABS / DSC motion sensor	B9801	Rear center console, under parking brake handle boot	Fig. 42
ABS / DSC charge pressure sensor	B76	On brake master cylinder, bottom left	Fig. 22
ABS / DSC steering angle sensor	A33	On steering column above universal joint	Fig. 37
ABS wheel speed sensor	B1, B2, B3, B4	Wheel hub (right front, left front) (right rear, left rear)	Fig. 28 Fig. 29
Accelerator pedal position sensor (PWG)	R10	Underneath accelerator pedal	
Additional instrument cluster	A9001	Top of steering column	
AIC		*see* Rain sensor (AIC)	
Air distribution servo motor		*see* Heating and A/C servo motors (IHKA)	
Airbag		*see also* Multiple restraint system (MRS) control module	
Airbag contact spring (volute spring)	I01002	Behind steering wheel	
Airbag crash sensors, B-pillar	A173, A174	Base of B-pillar (left, right)	
Airbag crash sensors, front	B10508, B10509	Front longitudinal frame member, near bumper mount (left, right)	
Airbag crash sensor, head protection airbags (left, right)	S71, S72	Underneath rear seat (left, right)	
Airbag, driver (MRS 4) Airbag, driver (MRS 5)	G5 G5a	In steering wheel In steering wheel	Fig. 43
Airbag, head protection	G17, G18	Seat back outer side bolster (left, right)	
Airbag, front passenger deactivation light	H23a	Upper windshield frame, center	
Airbag, side (MRS 4) Airbag, side (MRS 5)	G14, G15 G14a, G15a	Seat back outer side bolster (left, right) Seat back outer side bolster (left, right)	
Airbag gas generator, head protection	G17, G18 G17a, G18a	Luggage compartment behind side trim panel (left, right) Luggage compartment behind side trim panel (left, right)	Fig. 43
Airbag, passenger (MRS 4) Airbag, passenger (MRS 5)	G6 G6a	In dashboard	
Alternator (generator)	G6524	Right front of engine	
Amplifier		*see* Radio	
Antenna, roof, combined	I01200	Coupe - roof, Convertible - rear fender	
Antenna, diversity	N8	Rear window	Fig. 45
Antenna, diversity connector block	A421	Behind side panel, cargo compartment	Fig. 43
Antenna, FZV (keyless entry)		Above headliner	Fig. 41
Antenna, telephone and navigation	I01183	Center of roof, rear	
Antenna amplifier, main	I01190	Under rear spoiler	
Antenna amplifier, wave trap 1 (lockout circuit 1)	I01010	Left side of rear window frame, behind trim panel	

Electrical component location table			
Component	**Comp. identifier**	**Location**	**Refer to**
Antenna amplifier, wave trap 2 (lockout circuit 2)	I01010	Right side of rear window frame, behind trim panel	Fig. 45
Anti-theft (DWA), engine hood lock sensor	S19	In right engine hood lock	Fig. 9 Fig. 17 Fig. 18
Anti-theft (DWA), microwave sensors (convertible)	B14143 B14144 B14145 B14146	In driver's door In passenger door Left rear behind trim panel Right rear behind trim panel	Fig. 48
Anti-theft (DWA), ultrasound sensor	A121	In headliner above back seat	
Anti-theft (DWA) siren	H1	Right rear engine compartment	
Anti-theft (DWA) tilt sensor	B28	Behind left trim panel, cargo compartment	
ASC		*see* ABS / ASC entries	
Automatic transmission central shift unit	I01154	Front console near selector lever	
Automatic transmission control module (GIU)	A7000	Above brake pedal, bolted to brake master cylinder flange	
Automatic transmission (CVT) switch	S6575	Transmission, lower left	Fig. 26
Automatic transmission fluid temperature sensor		In valve block	
Automatic transmission gear position switch	Y8505a	Under shift console	
Automatic transmission shift indicator light	E82	Front console near selector lever	
Automatic transmission shift lock relay	I01148	Left footwell, fuse and relay panel 2	
Automatic transmission shift lock solenoid	S235	In front console, underneath shifter boot	
Automatic transmission shiftlock selector lever lock	Y19	Under shift console	
Automatic transmission speed (RPM) sensor	Y8516	In differential housing	
B+ jumper connector		Center engine compartment, next to battery (Cooper) or air filter housing (Cooper S)	Fig. 17 Fig. 18
Back-up light switch, Cooper	S8511	On transmission underneath air filter housing	
Back-up light switch, Cooper S	S8511	Transmission rear	Fig. 21
Back-up light switch, CVT	S8511	Transmission left bottom	
Battery, Cooper	G1	Engine compartment, in battery box	Fig. 17
Battery, Cooper S	G1	Underneath floor trim, cargo compartment	
Battery Safety Terminal (BST) generator (Cooper S)	B19, B19a	At battery positive (+) terminal, underneath floor trim, cargo compartment	
BC1		*see* General Module (BC1)	
Blower motor		*see* Heating and A/C blower	

Electrical Components

Electrical component location table			
Component	**Comp. identifier**	**Location**	**Refer to**
Body control module		*see* General Module (BC1)	
Brake fluid level switch	B18	On brake master cylinder, left rear engine compartment	Fig. 22
Brake light switch	S29	Above brake pedal	Fig. 38
Brake pad wear sensor, left front	B16	Inboard left front brake pad	
Brake pad wear sensor, right rear	B17	Inboard right rear brake pad	
Camshaft position sensor	B6204	Right side cylinder head	Fig. 17 Fig. 18
Cargo compartment light	E32	Left rear of cargo compartment	Fig. 52
CD changer	N22	Behind right side panel, cargo compartment	
Central locking		*see* FZV entries	
Cigarette lighter relay	I01151	Behind left footwell trim	
Clock	A110	Headliner front	
Clutch position switch module	S805	On clutch master cylinder	
Compressor		*see* A/C	
Convertible cross brace microswitch	I11385	Convertible top frame, left rear, near cargo compartment	Fig. 52
Convertible rear power window switches	S40	Roof console, above interior mirror	Fig. 49
Convertible rear window microswitch	I11386	Convertible top frame, left rear, near cargo compartment	Fig. 52
Convertible top catch hall sensor	I01199	Folding sunroof frame, front left	Fig. 50
Convertible top “closed” microswitch	I01188	Convertible top frame, left rear, near cargo compartment	Fig. 53
Convertible top control button	I13045	Roof console, above interior mirror	Fig. 49
Convertible top control module	A96	Left rear inner wheel housing, behind trim panel	Fig. 47
Convertible top “folded down” microswitch	I01187	Convertible top frame, behind driver	Fig. 51
Convertible top relay 1	K18363	Left rear inner wheel housing, behind trim panel and rear seat belt reel	Fig. 47
Convertible top relay 2	K18364	Left rear inner wheel housing, behind trim panel and rear seat belt reel	Fig. 47
Convertible top storage compartment, drive motor	I13148 I13149	Right rear inner wheel housing, behind trim panel Left rear inner wheel housing, behind trim panel	Fig. 48
Convertible top storage compartment lock microswitch	I13073 I13076	Storage compartment, under convertible frame, left Storage compartment, under convertible frame, right	Fig. 54
Convertible top storage compartment lock relay	K405	Left rear inner wheel housing, behind trim panel	Fig. 48
Cooling fan		*see* Engine cooling fan	
Crankshaft position sensor	B6203	Front of engine near bellhousing	Fig. 23
Crash sensor		*see* Airbag entries	
CVT		*see* Automatic transmission (CVT) entries	
Diversity antenna		*see* Antenna	
DLC		*see* OBD II diagnostic connector (16 pin DLC)	

Electrical component location table			
Component	**Comp. identifier**	**Location**	**Refer to**
DME control module (ECM)	A6000	Engine compartment next to battery (Cooper) or air filter housing (Cooper S)	Fig. 17 Fig. 18
DME main relay	K6300	Left engine compartment, fuse and relay panel 3	Fig. 6
Door entrance lights	E88, E89	Interior door panel (left, right)	
Door lock		*see* Lock entries	
Door window motor		*see* Window motor	
DSC		*see* ABS / DSC	
EHPS (electrohydraulic power steering) fan relay	K69923	Left footwell, next to fuse panel 2	Fig. 5
EHPS (electrohydraulic power steering) cooling fan and control module	I01180	Behind oil pan	Fig. 25
EHPS (electrohydraulic power steering) pump	Y6975	Behind oil pan	Fig. 25
Electrochromic rear view mirror	A22	In headliner	
Electronic immobilizer		*see* EWS entries	
Emergency release, fuel filler flap		Above left rear access panel, behind rear seat	Fig. 46
Emergency release, rear hatch		Under center of rear seat cushion	Fig. 44
Engine compartment fusebox		*see* Fuse and relay panel 3 (engine compartment)	
Engine control module (ECM)		*see* DME control module (ECM)	
Engine coolant temperature (ECT) sensor	B6048	Left end of cylinder head	Fig. 17 Fig. 18
Engine cooling fan	M135	Behind radiator	
Engine cooling fan relay (stage 1)	I01135	Left engine compartment, fuse and relay panel 3 (from March, 2003)	Fig. 6
Engine cooling fan (stage 2)	I01152	Left engine compartment, fuse and relay panel 3	Fig. 6
Electric cooling fan switching unit (2 stage blower)	I01153	Front of engine, behind radiator	
Evaporator temperature sensor		*see* A/C entries	
EWS control module	A836	Under left side of dashboard	
EWS diode, starter motor control	V554	Left footwell behind trim panel	
EWS ring antenna (toroidal coil)	L1	Around ignition switch	
Fascia switch module		Center dashboard, below heating and A/C control panel	
Fresh / recirculated air flap servo motor		*see* Heating and A/C servo motors (IHKA)	
Front foglight relay	K47	Left engine compartment, fuse and relay panel 3	Fig. 6
Front washer pump relay		*see* Washer pump relay	
Fuel filler door lock	M16	Behind left trim panel, cargo compartment	
Fuel injectors	Y6101 - Y6104	At front of cylinder head (cylinder 1 on right)	
Fuel pump	M2	In fuel tank, under left rear seat	Fig. 44

Electrical Components

Electrical component location table			
Component	**Comp. identifier**	**Location**	**Refer to**
Fuel pump relay	K96	Left footwell, fuse and relay panel 2	Fig. 4
Fuel tank vent valve	Y6550	Right side of engine	Fig. 17 Fig. 18
Fuse and relay panel 2 (car interior)	A47	Left footwell behind cover	Fig. 2
Fuse and relay panel 3 (engine compartment)	A48	Left engine compartment, next to left strut tower	Fig. 3
Fuse panel 1 (Cooper S)		Fusible link, next to battery, underneath floor trim, cargo compartment	Fig. 1
Fuses		*see also* **Fuse locations** in this repair group	
Fusible link		*see* Fuse panel 1 (Cooper S)	
FZV antenna		*see* Antenna entries	
FZV receiver and control module		Above front headliner	Fig. 41
FZV relay		Left footwell behind cover	Fig. 4
Gas generator		*see* Airbag entries *see also* Battery Safety Terminal (BST) (Cooper S)	
General Module (BC1)	A1	Behind right footwell trim	Fig. 40
Generator		*see* Alternator	
Grounds		*see* **Ground locations** in this repair group	Fig. 8
Hazard warning switch	S241	Dashboard, right of center instrument	
Headlight vertical aim control module (LWR)	A53	Behind glove compartment	
Headlight vertical aim control front sensor	B42	Left front control arm	Fig. 30
Headlight vertical aim control rear sensor	B64	Rear suspension subframe and upper control arm, left side	Fig. 31
Headlight washer pump		*see* Washer entries	
Headlight washer pump relay	K6	Left footwell, near fuse and relay panel 2	Fig. 4
Heated windshield washer nozzle		*see* Washer entries	
Heater blower relay		*see* Heating and A/C blower entries	
Heater core temperature sensor	B11	Behind radio in center of dash	Fig. 35
Heating and A/C		*see also* A/C entries	
Heating and A/C blower motor	M30	Underneath dashboard, left of center	
Heating and A/C blower relay	K4	Left engine compartment, fuse and relay panel 3 (up March, 2003) Left footwell, fuse and relay panel 2 (from March 2004)	Fig. 6 Fig. 4
Heating and A/C blower output stage (IHKA to 3/2002)		IHKA housing, left of center	
Heating and A/C blower output stage (IHKA from 3/2002)		Behind glove compartment, right of center	Fig. 35
Heating and A/C blower output stage	N2	Behind center console, right	Fig. 36
Heating and A/C blower resistor (IHKS)		Behind glove compartment, right of center	Fig. 35

Electrical component location table			
Component	**Comp. identifier**	**Location**	**Refer to**
Heating and A/C control module (IHKA)	A11	Integrated with control panel	
Heating and A/C control panel	I01175	Center dashboard, above switch panel	Fig. 34
Heating and A/C interior temperature sensor, fan (IHKA)	I01157	In heating and A/C control panel	Fig. 34
Heating and A/C (IHKA) servo motor, air distribution	M150	Inside center left IHKA housing	
Heating and A/C (IHKA) servo motor, recirculation control	M111	Behind glove compartment	
Heating and A/C (IHKA) servo motor, temperature control	I01155	Behind glove compartment	
Height sensor		*see* Headlight vertical aim control	
Hood contact switch		*see* Anti-theft (DWA), engine hood lock sensor	
Horn (left, right)	H2, H3	Under front fender (left, right)	
Horn relay	K2	Left footwell, next to fuse panel 2	Fig. 5
Horn switch	S4	On steering wheel	
Ignition and starting switch	S2	On steering column, right	Fig. 37
Ignition coil	T6150	Rear of cylinder head cover, top	Fig. 17 Fig. 18
Impact switch	I01149	Left side engine compartment	
Instrument cluster	A2	Center of dashboard	
Intake air pressure sensor	B60632	Air duct before supercharger	
Intake air temperature sensor	B60631	Air filter duct	
Intake pressure sensor, MAP sensor (Cooper S)	B6032	Left end of cylinder head	Fig. 18
Intake pressure sensor, T-MAP sensor (Cooper)	B6031	Next to throttle housing	
Intake pressure sensor, T-MAP sensor (Cooper S)	B6063	Front of cylinder head, just below intercooler	Fig. 18
Interior motion sensor		*see* Anti-theft (DWA) ultrasound sensor	
Interior temperature sensor and fan (IHKA)		*see* Heating and A/C entries	
Keyless entry		*see* FZV entries	
Knock sensor	B6240	Front of engine, under intake manifold	Fig. 23
LDP (leakage diagnosis pump)	M119	Behind right rear wheel housing liner	
Lock contact switch, rear lid (convertible)	S213	Left rear cargo compartment	Fig. 52
Lock, driver door	S47	Driver door	
Lock, fuel filler door	M16	Left rear panel	
Lock, passenger door	S49	Passenger door	

Electrical Components

Electrical component location table			
Component	**Comp. identifier**	**Location**	**Refer to**
Lock, rear hatch	M17	Rear hatch sill	
Lockout circuit		*see* Antenna amplifier, wave trap	
Magnetic clutch		*see* A/C compressor clutch	
MAP sensor		*see* Intake pressure sensor entries	
Midrange speaker		*see* Speakers	
Multiple restraint system (MRS) control module	A12	Center tunnel underneath carpet, between parking brake and shift lever	Fig. 42
Navigation computer	A112	Mounted to bracket, underneath right front seat	
OBD II diagnostic connector (16 pin DLC)		Left footwell top trim	Fig. 39
Oil pressure switch	B6231	Right rear of engine on oil filter housing	Fig. 24
On-board monitor and control unit	A196	Dashboard center	
Outside mirror adjuster	S17	Front console, behind shifter	
Outside mirror control module	A208	Left footwell, behind trim	
Outside temperature sensor	B13	Left lower front bumper cover	Fig. 27
Oxygen sensor, precatalyst	B62001	Front exhaust pipe, in front of catalyst	
Oxygen sensor, post-catalyst	B62002	Exhaust pipe, in back of catalyst	
Parking brake warning switch	S31	Under rear console, at rear of parking brake handle	Fig. 42
PDC (park distance control) module and warning speaker	A81	Right rear cargo compartment, behind trim panel	
PDC (park distance control) sensors	B34, B35, B36, B37	Rear bumper cover	
Power steering		*see* EHPS (electrohydraulic power steering) entries	
Pressure sensor (A/C)		*see* A/C entries	
Pyrotechnic tensioner		*see* Seat belt, front	
Radio amplifier	A18	Behind left trim panel, cargo compartment	
Radio control module	N9	Front console	
Rain sensor (AIC)	B57	Windshield top	
Reading light, front (convertible)	E34	Above interior mirror	Fig. 49
Rear foglight		In left taillight assembly (up to July 2004) Center of lower rear bumper cover (from July 2004)	
Rear hatch lock	M17	Rear hatch sill	
Rear lid lock contact switch	S213	Rear lid lock	
Rear speaker		*see* Speakers	
Rear washer pump		*see* Washer entries	
Rear window defroster relay	K13	Left footwell, fuse and relay panel 2	
Rear wiper motor		*see* Wiper motor	
Remote locking		*see* FZV entries	
Reversing light switch		*see* Back-up light switch entries	

Electrical component location table			
Component	**Comp. identifier**	**Location**	**Refer to**
Ring antenna		*see* EWS entries	
Seat belt buckle contact (MRS 5)	S58a, S59a	Front seat belt buckle (left, right)	
Seat belt hall sensor, front (MRS 4)	S58, S59	On seat belt buckle (left, right)	
Seat belt tensioner (pyrotechnic) (MRS 4)	G12, G13	In front seat belt buckle (left, right)	
Seat belt tensioner generator (pyrotechnic) (MRS 5)	G12a, G13a	Base of front seat, attached to buckle (left, right)	
Seat heating coil, front	E57	Front seat cushion and backrest	
Seat heating switch, front	S53	Front console, behind shifter	
Seat occupancy recognition	A113a	Passenger seat cushion	
Servo motor		*see* Heating and A/C servo motor entries	
SHD		*see* Sunroof (SHD) entries	
Shift indicator light		*see* Automatic transmission entries	
Shift lock		*see* Automatic transmission entries	
Signal horn, burglar alarm system		*see* Anti-theft siren	
Siren		*see* Anti-theft entries	
Slide / tilt sunroof		*see* Sunroof (SHD)	
Solar sensor	B66	Top of dashboard, center	
Speaker, midrange	H54, H55	Bottom front of door	
Speaker, rear	H45, H46	Behind side rear trim	
Speaker, tweeter	H50, H59	Top front of door	
Starter motor	M6510	Rear of engine	
Steering angle sensor		*see* ABS / DSC entries	
Steering		*see* EHPS (electrohydraulic power steering) entries	
Steering wheel switches	I01000, I01001	In steering wheel	
Sunroof (SHD) control module	A33	Above headliner, front	
Sunroof (SHD) motor		Above headliner, front	
Sunroof (SHD) switch	S38	In headliner, front	
Sunroof, sliding canvas "closed" hall sensor	I01197	Folding sunroof frame, near motor	Fig. 50
Sunroof, sliding canvas drive motor	I11301	Folding sunroof frame, center, under trim panel	Fig. 50
Sunroof, sliding canvas "open" hall sensor	I01198	Folding sunroof frame, left front	Fig. 50
Sunroof, sliding canvas relay 1	K16761	Left rear inner wheel housing, behind trim panel	Fig. 47
Sunroof, sliding canvas relay 2	K16762	Left rear inner wheel housing, behind trim panel	Fig. 47
Tailgate lock		*see* Rear hatch	
Telephone and navigation antenna		*see* Antenna entries	

Electrical Components

Electrical component location table			
Component	**Comp. identifier**	**Location**	**Refer to**
Telephone eject box	A117	Right of center dashboard	
Telephone, hands free microphone	B400	Above interior mirror, driver's side	Fig. 49
Tilt sensor		*see* Anti-theft entries	
Tire pressure warning switch (RDW)		In rear console below parking brake handle	
T-MAP sensor		*see* Intake pressure sensor entries	
Toroidal coil		*see* EWS ring antenna	
Transmission fluid temperature sensor (CVT)		*see* Automatic transmission (CVT) entries	
Turn signal / headlight dimmer switch	S7	Steering column left	
Tweeter		*see* Speaker entries	
Ultrasound sensor		*see* Anti-theft entries	
Volute spring		*see* Airbag contact spring	
Washer fluid level switch	S136	Washer fluid reservoir	
Washer, headlight nozzle		Underside of engine hood at headlight assembly	Fig. 19
Washer, headlights pump	M7	Behind left front wheel housing liner, in washer tank	
Washer, heated windshield nozzle	E51, E52	Underside of engine hood	
Washer pump relay		Left footwell, behind trim panel	
Washer, windshield or rear window pump		Behind right front wheel housing liner, in washer tank	
Wavetrap		*see* Antenna amplifier	
Wheel speed sensor		*see* ABS wheel speed sensor	
Window motor, front	M21, M23	In door cavity (left, right)	
Window motor, rear (convertible)	M98, M99	Behind rear seat side trim panel (left, right)	Fig. 51
Window switches, rear power (convertible)	S40	Above interior mirror	Fig. 49
Windshield heating field (left, right)	I01143, I01144	Behind windshield trim panel (left, right)	
Windshield heating field relay		Behind footwell trim panel (left, right)	
Windshield washer pump	M4	*see* Washer entries	
Wiper motor, rear window	M94	In rear hatch behind trim panel	
Wiper motor relay, fast / slow	I01136	Left engine compartment, fuse and relay panel 3	Fig. 6
Wiper motor relay, on / off	I01137	Left engine compartment, fuse and relay panel 3	Fig. 6
Wiper motor relay, rear	I01142	Behind left footwell trim	Fig. 5
Wiper motor, windshield	M3	Underneath left cowl vent grill	Fig. 17, Fig. 18
Wiper / washer switch	S5	On steering column, right	
Xenon headlight beam adjuster		Next to headlight assembly, underneath engine hood (left, right)	Fig. 19
Xenon headlight ignition trigger	I01179	In headlight assembly, underneath engine hood (left, right)	Fig. 19

Fuse locations

There are 2 locations for fuses in Cooper models and 3 locations in Cooper S models:

- Cooper S: Fuse F100 (fusible link) in cargo compartment near battery
- Cooper and Cooper S: Fuse and relay panel 2 fuses F1 – F42 behind left kick panel
- Cooper and Cooper S: Fuse and relay panel 3 fuses FL1 – FL12 fuses F01 – F010 in engine compartment, next to left strut tower

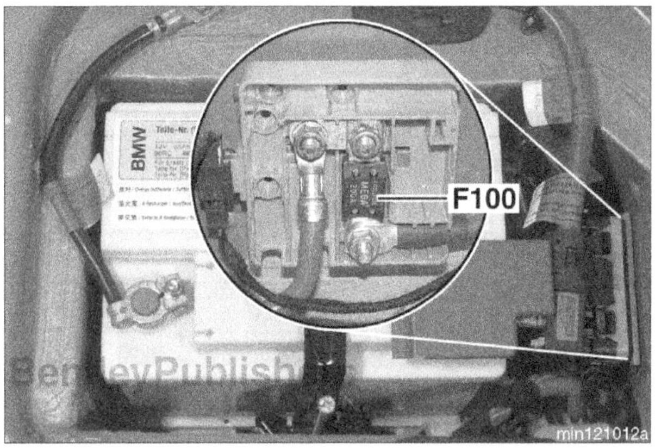

Fig. 1 F100 fuse

Cooper S: In cargo compartment, next to battery.

- 250 amp fusible link supplies power to fuse panel 3.

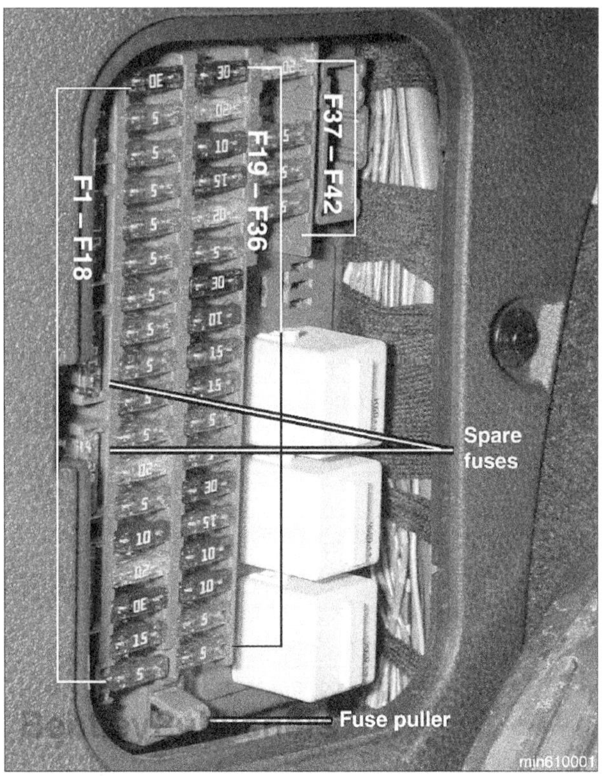

Fig. 2 Fuse and relay panel 2 (fuses F1 – F42)

Behind cover in left kick panel.

- Horizontal fuses are active.
- Vertical fuses are spares.
- Fuse puller at bottom left.
- Fuse location guide in cover.

Fuse panel 2, behind left kick panel		
Fuse	**Rating in amps**	**Protected circuit**
F1	30	Right window motor
F2	5	Steering angle sensor
F3	5	Clock
F4	5	General Module (BC1)
F5	5	Instruments
F6	5	Brake lights; clutch switch; EWS
F7	5	Rain sensor; General Module (BC1)
F8	5	Turn signal; General Module (BC1)
F9	5	Instruments
F10	5	Navigation; sound system; on-board monitor
F11	5	Front and rear window washers
F12	20	Seat heaters
F13	5	Automatic transmission; back-up lights
F14	10	General Module (BC1); interior lights

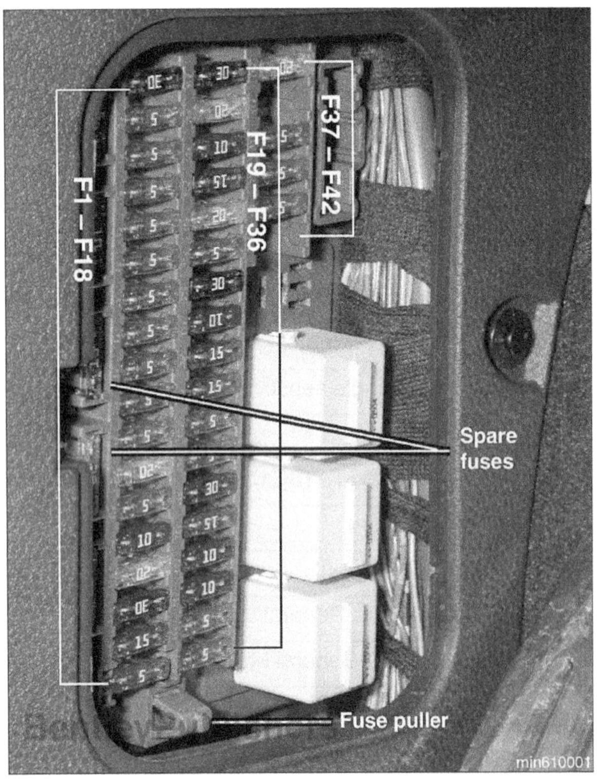

Fuse panel 2, behind left kick panel		
Fuse	**Rating in amps**	**Protected circuit**
F15	20	Sunroof (glass or canvas)
F16	30	Diversity antenna
F17	15	Rear window wiper
F18	5	Airbag system (MRS)
F19	30	Left window motor
F20	20	Fuel pump
F21	10	Instruments
F22	15	Navigation; on-board monitor; telephone eject box
F23	20	Front and rear window washers
F24	5	EWS; alarm siren; ultrasonic sensor
F25	30	Headlight washers
F26	10	Gearshift lock (CVT)
F27	15	Sound system
F28	15	Horns
F29	5	Air-conditioning control panel
F30	5	Instruments
F31	30	Heating and air-conditioning blower
F32	15	Cigarette lighter
F33	10	ABS / ASC / DSC
F34	10	DME main relay
F35	5	Electric outside rear-view mirrors; heated windshield washer nozzles, electrochromic rear view mirror
F36	5	Heated windshield
F37	20	Central locking
F38		
F39	5	Alternator; electric power steering pump (EHPS)
F40	5	Instruments; front console switch panel; steering angle sensor
F41	5	Park distance control (PDC); electric power steering pump (EHPS)
F42		

Fig. 3 Fuse and relay panel 3 (fuses FL1 – FL12 and F01 – F010)

In engine compartment, next to left strut tower.

- High current fuses FL1 - FL5 underneath panel
- High current fuses FL6 - FL12 on right
- Fuses F1 - F10 on left; indicated in wiring diagrams as F01 - F010.

To access fuses FL1 - FL5:

- Disconnect battery negative (-) terminal.
- Cooper: Remove engine control module (ECM) and battery box. See **121 Battery, Starter, Alternator**.
- Cooper S: Remove engine control module and air filter housing. See **130 Fuel Injection**.
- Detach current supply cable lug at rear of fuse panel 3
- Remove fuse panel mounting screws.
- Lift panel and turn over. Remove bottom cover.

High current fuses (with FL numbers) supply power to components listed below through regular fuses.

Fuse panel 3, next to left strut tower		
Fuse	**Rating in amps**	**Protected circuit**
FL1	50	General Module (BC1); heating and air-conditioning blower; cigarette lighter
FL2	50	General module (BC1); steering angle sensor; clock; instruments; navigation; on-board monitor; window washers; EWS; anti-theft; washers; sound system; horns
FL3	40	Convertible top control
FL4	100	Electric power steering pump (EHPS)
FL5	50	Sunroof; diversity antenna; rear window wiper
FL6	40	ABS / ASC / DSC
FL7	50	Ignition switch
FL8	50	General Module (BC1); lights
FL9	50	Engine cooling fan
FL10	50	Rear window defogger
FL11	50	Heating and air-conditioning blower
FL12	50	General Module (BC1); lights

Fuse panel 3, next to left strut tower		
Fuse	**Rating in amps**	**Protected circuit**
F01	5	Engine control module (ECM), convertible top control module
F02	20	Engine control module; ignition coil; fuel injectors; crankshaft position sensor
F03	15	Engine control
F04	15	Engine control
F05	5	Engine cooling fan
F06	30	ABS / ASC / DSC
F07	30	Air-conditioning compressor
F08	30	Heating and A/C blower; electric power steering pump (EHPS) cooling fan
F09	20	Windshield wipers
F010	15	Parking lights

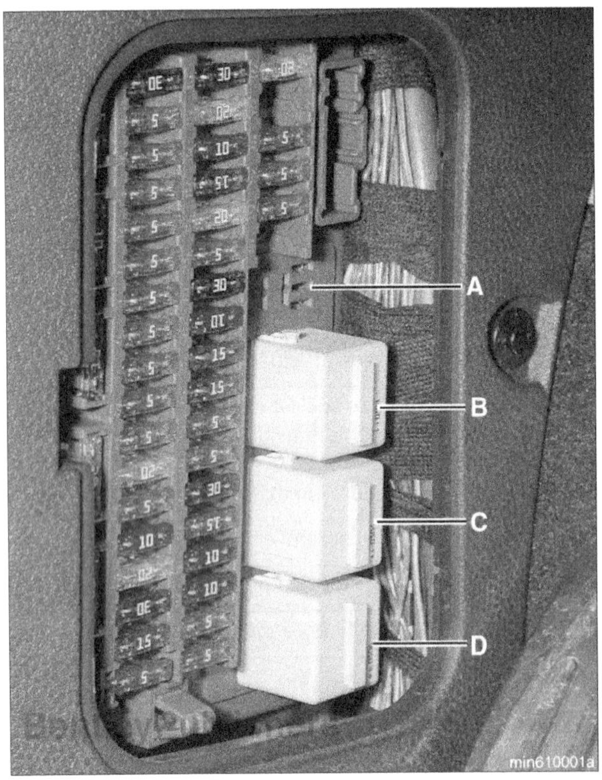

Relay panels

Fig. 4 Fuse and relay panel 2 (front)

◄ Behind cover in left kick panel.

A Shiftlock relay (I01148)

B Headlight washer relay (K6)

C Heating and A/C blower relay (K4) (from March 2003)

D Fuel pump relay (K96)

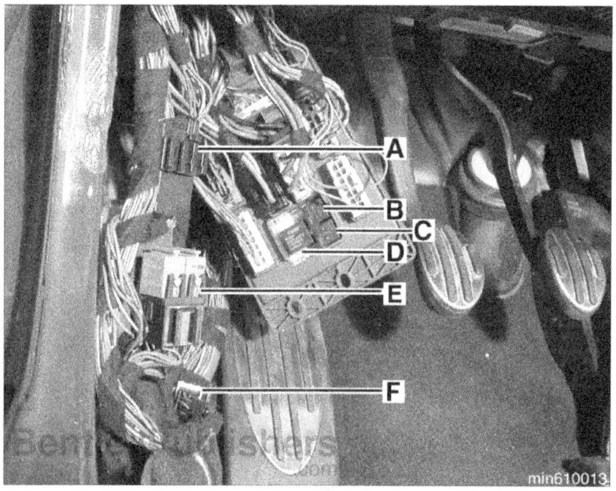

Fig. 5 Fuse and relay panel 2 (rear)

Behind cover in left kick panel, remove fuse panel and turn over.

- **A** Instrument cluster connector
- **B** Front cigarette lighter relay (I01151)
- **C** Rear window defogger relay (K13)
- **D** Rear window wiper relay (I01142)
  Horn relay (K42)
- **E** Windshield washer pump relay, front (K5)
- **F** EHPS (electrohydraulic power steering) fan relay (K69923)
  Windshield defroster relay (K416)

NOTE —
Connector **A** *and relay* **E** *are accessed once relay panel is moved aside.*

Fig. 6 Fuse and relay panel 3

Left side engine compartment, next to left strut tower.

- **A** Front wiper fast / slow relay (I01136)
- **B** Front wiper ON / OFF relay (I01137)
- **C** Front foglight relay (K47)
- **D** Heating and A/C blower relay (K4) (up to March 2003)
  Engine cooling fan relay (stage II) (I01135) (from March 2003)
- **E** Engine cooling fan relay (stage 1) (I01152)
- **F** A/C compressor relay (K19)
- **G** DME main relay (K6300)

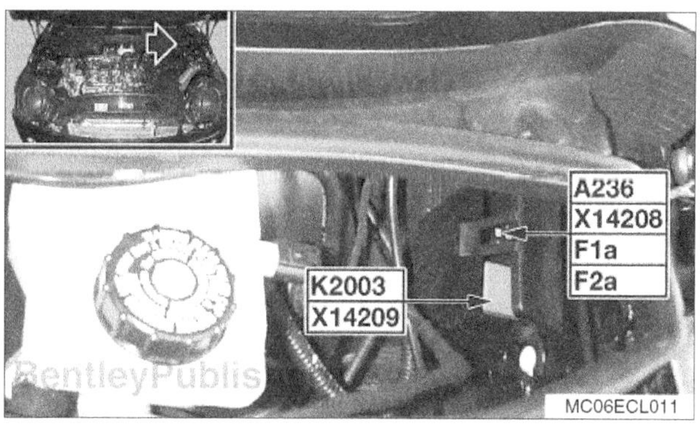

Fig. 7 Fuse and relay panel 4

Left rear engine compartment, under plenum cover.

- **F1a** Fuse 1a, 15A, Automatic transmission control (Aisin)
- **F2a** Fuse 2a, 30A, Diesel only
- **K2003** DDE main relay, Diesel only

Ground locations

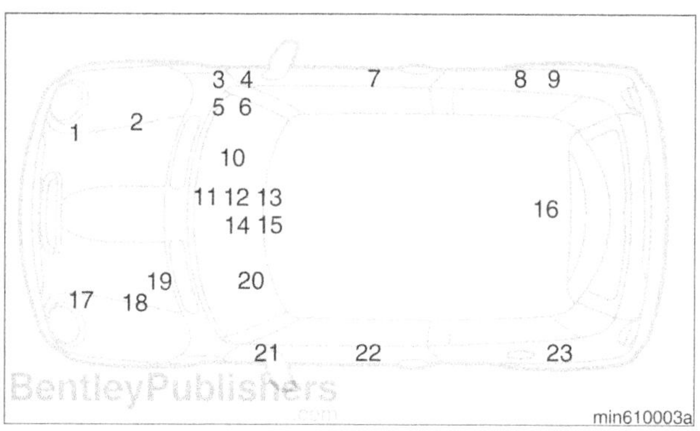

Fig. 8 Grounds overview

Grounds are distributed throughout the vehicle body. Many are found under the interior carpets or behind trim panels. Several components grounds are often ganged into a ground bus.

Lugs and connectors attached to ground are susceptible to damage and corrosion. Clean or renew as necessary.

Electrical Components

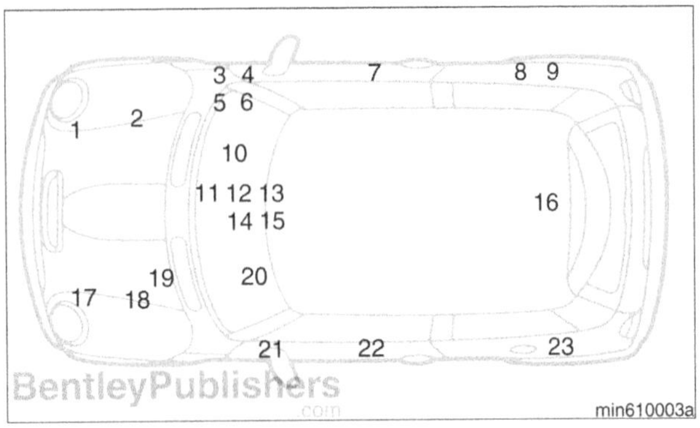

In the table **Ground locations**, the No. column refers to numbers in the accompanying illustration.

Ground locations			
No.	**Designation**	**Location**	**Function / notes**
1	X165	At frame extension, right	Ground bus
2		At frame extension, right	Body to engine ground cable
3	X3486	Lower A-pillar at door sill, right	Ground bus
4	X151	Lower A-pillar at door sill, right	Ground bus
5	X923	Behind glove compartment	
6	X11337	Behind glove compartment	Ground bus
7	X179	At door sill right, right	Ground bus
8	X18723	Under right rear window	Antenna diversity connector
9	X494	Behind trim panel, right rear	Ground bus
10	X01116	Behind glove compartment	Heating and A/C system
11	X11339	Behind dashboard	
12	X11353	Behind dashboard	
13	X01108	Ahead of front console	IHKA servo motors
14	X3454	Ahead of front console	Ground bus
15	X46 X18170	Ahead of front console	Airbags
16	X13016	Above rear headliner	Stereo / diversity antenna
17	X175	At frame extension, left	Ground bus
18	X 167 X6454	At frame extension, left	DME ground lugs
19	X173	Underneath engine compartment fuse and relay panel 3	Electrohydraulic power steering (EHPS) pump
20	X6465	Behind instrument cluster	Ground bus
21	X1108	Lower A-pillar at door sill, left	Ground bus
22	X13230	At door sill, left	Ground bus
23	X490	Behind trim panel, left rear	Ground bus

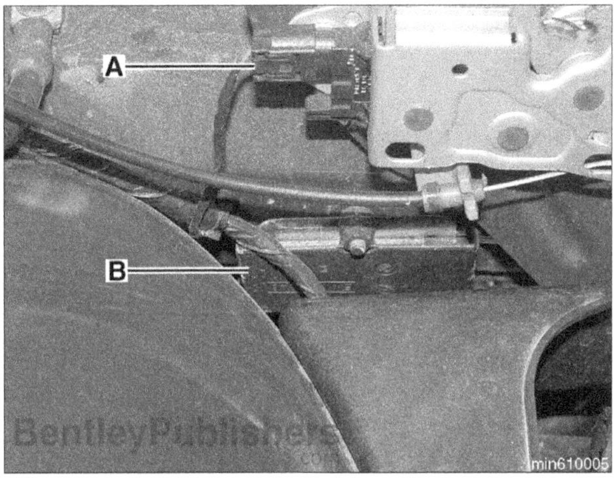

Fig. 9 Ground bus, right front

At right frame extension, engine compartment.

A Engine hood lock electrical connector

B Ground bus X165

Fig. 10 Ground bus, left front

At left frame extension, engine compartment.

A Ground bus X175

Fig. 11 Engine to body ground, right front

At right frame extension, engine compartment.

A Ground lug

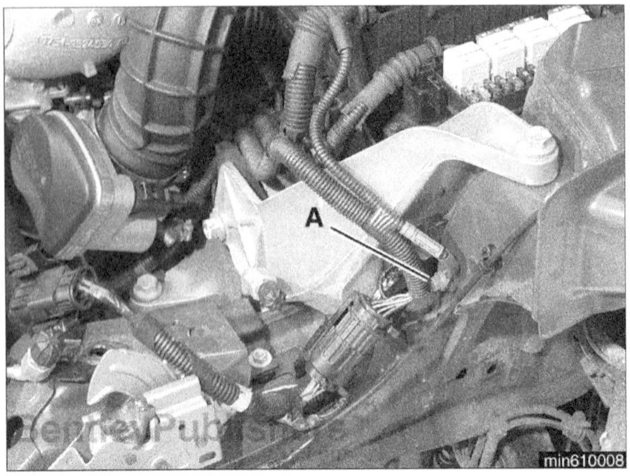

Fig. 12 DME grounds, left front

At left frame extension, engine compartment.

A Ground lug
X167
X6454

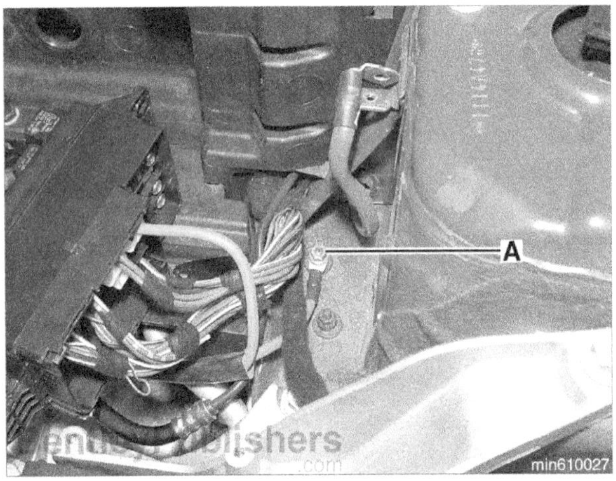

Fig. 13 Electrohydraulic power steering (EHPS) pump ground

Underneath engine compartment fuse and relay panel 3.

A EHPS ground X173

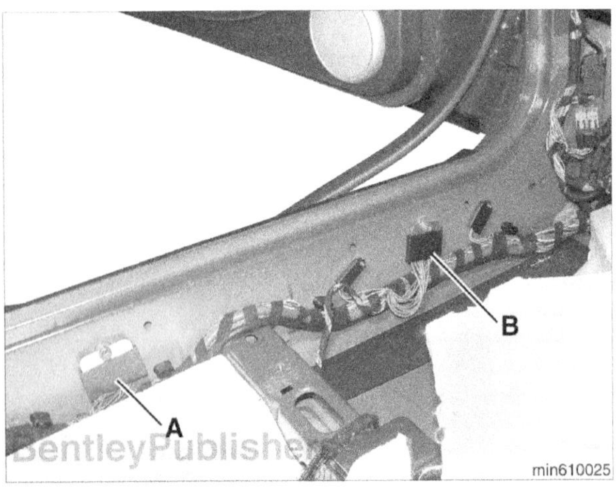

Fig. 14 Ground buses, left door sill

Just inside left door sill. Right is similar.

A Ground bus X13230

B Ground bus X1108

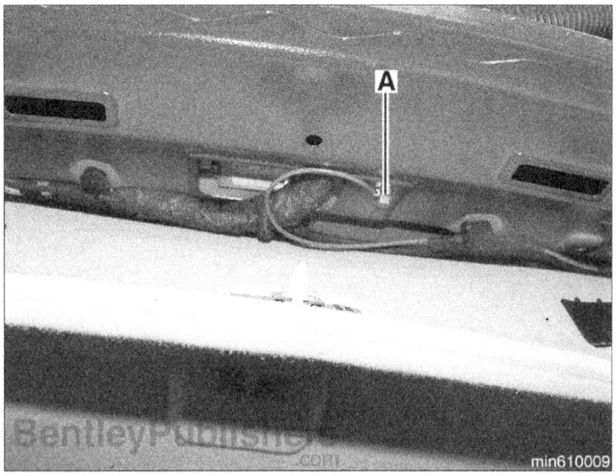

Fig. 15 Sound system and diversity antenna ground

Above headliner at rear hatch.

A Ground X13016

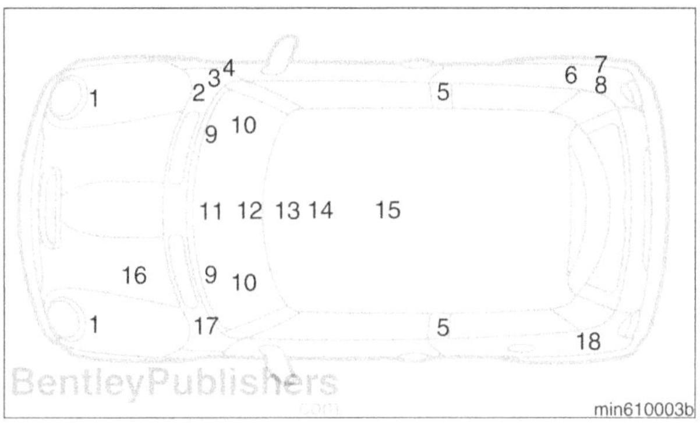

Component location illustrations

Fig. 16 Components overview

In addition to the relays identified in Fig. 2, Fig. 3, Fig. 4, Fig. 5, or Fig. 6, control modules, relays and sensors are found in a variety of locations.

1. Xenon light motor, in engine hood
2. Alarm (DWA) siren, in right side cowl
3. Headlight vertical aim control (LWR) relay
4. General Module (BC1) (Body Control Module)
5. Head protection airbag crash sensor
6. Park Distance Control (PDC) gong
7. PDC module
8. Tire Pressure Warning (RDW)
9. Automatic transmission control (GIU)
10. Remote cluster
11. Instrument cluster (IKE)
12. Rain sensor (AIC)
13. FZV receiver
14. Multiple Restraint System (MRS 4) control module
15. DSC motion sensor
16. Engine control module (ECM)
17. Electronic immobilizer (EWS) control module
18. Tilt sensor

Engine compartment components

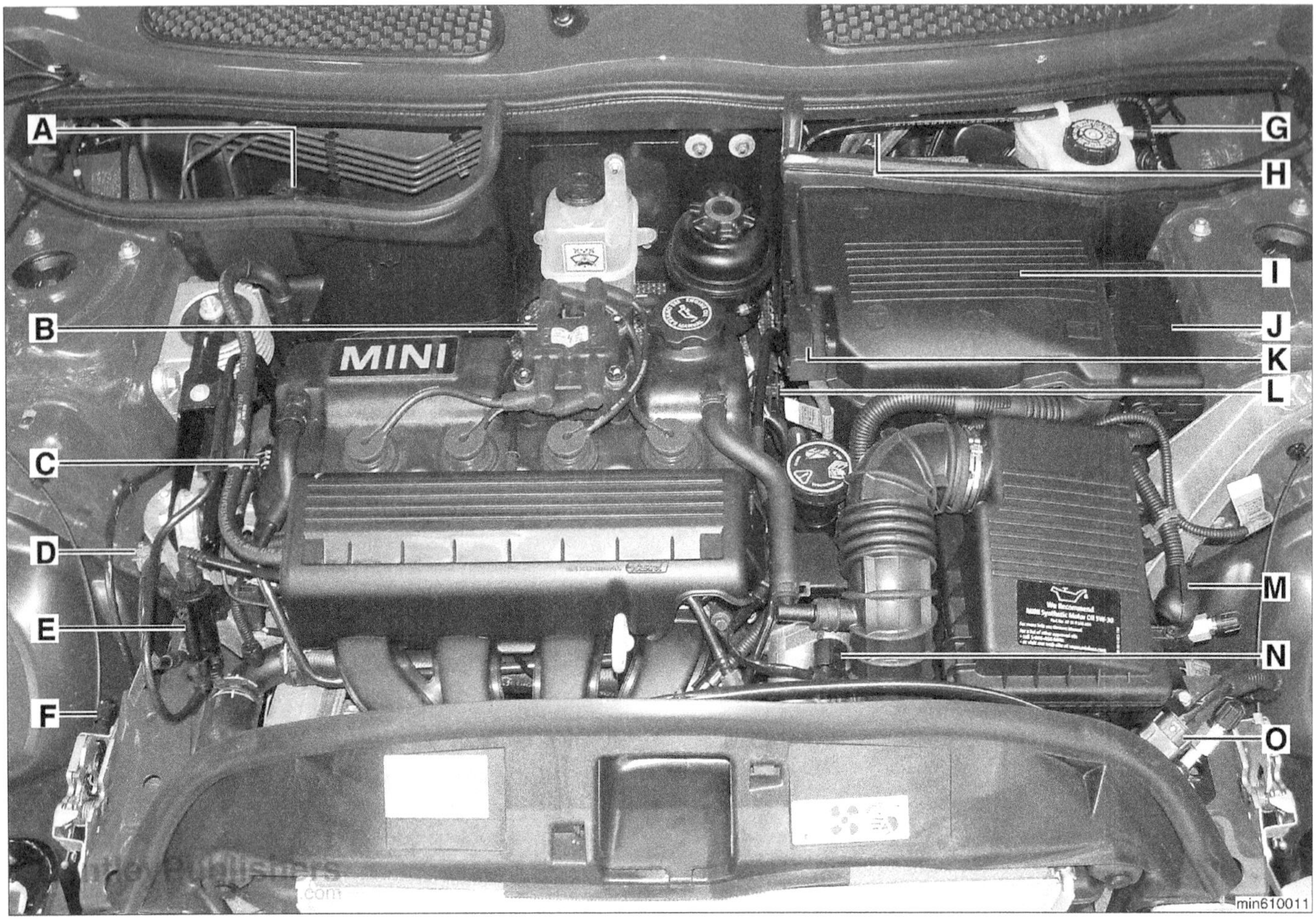

Fig. 17 Cooper engine compartment

A ABS control module
B Ignition coil
C Camshaft position sensor
D Engine ground cable lug
E Fuel tank vent valve
F Right engine hood lock electrical connector
G Brake fluid level sensor
H Windshield wiper assembly
I Battery box and engine control module (ECM) housing
J Fuse and relay panel 3
K B+ connector
L Engine coolant temperature (ECT) sensor
M Main engine harness connector
N Throttle housing
O Engine cooling fan connector

Fig. 18 Cooper S engine compartment

- **A** ABS control module
- **B** Ignition coil
- **C** Camshaft position sensor
- **D** Engine ground cable lug
- **E** Fuel tank vent valve
- **F** Right engine hood lock electrical connector
- **G** Brake fluid level sensor
- **H** Windshield wiper assembly
- **I** B+ connector
- **J** Engine control module (ECM) housing
- **K** Fuse and relay panel 3
- **L** Engine coolant temperature (ECT) sensor
- **M** Oxygen sensor harness connector
- **N** MAP sensor
- **O** Main engine harness connector
- **P** Engine cooling fan connector
- **Q** T-MAP sensor

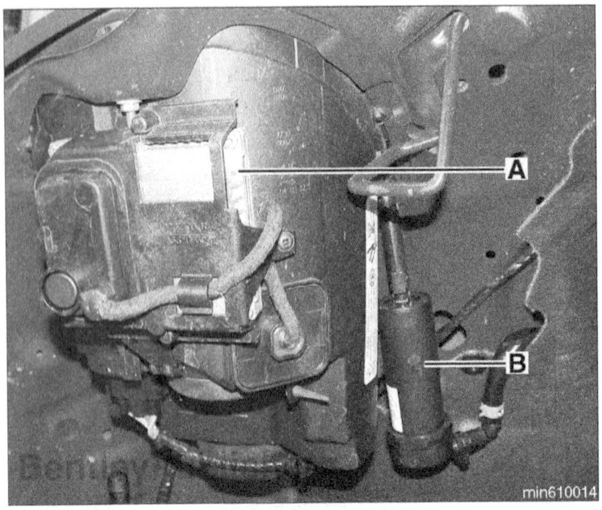

Fig. 19 Headlight assembly

Underneath engine hood.

A Xenon headlight ignition module

B Headlight washer nozzle

Fig. 20 A/C pressure sensor

Underneath battery box (Cooper) or air filter housing (Cooper S).

A A/C pressure sensor

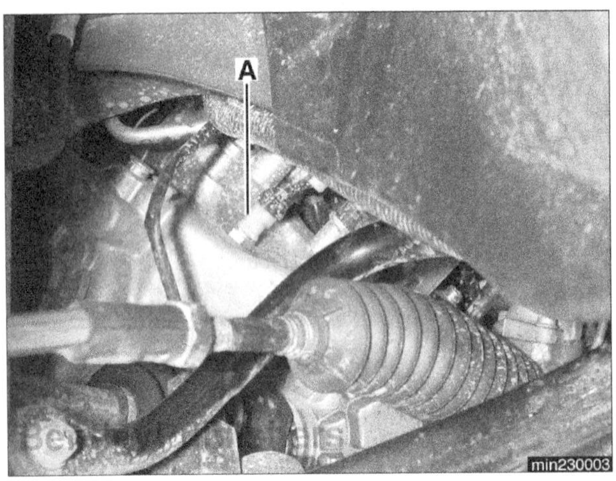

Fig. 21 Back-up light switch (Cooper S)

On rear of transmission.

A Back-up light switch

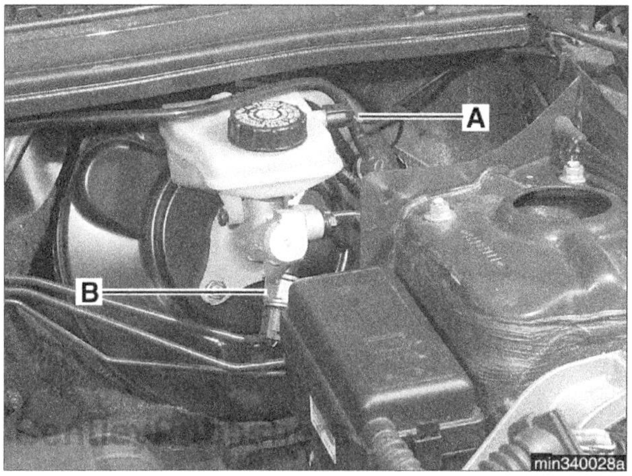

Fig. 22 Brake switches

At brake master cylinder and fluid reservoir,

- **A** Brake fluid level switch
- **B** DSC charge pressure sensor

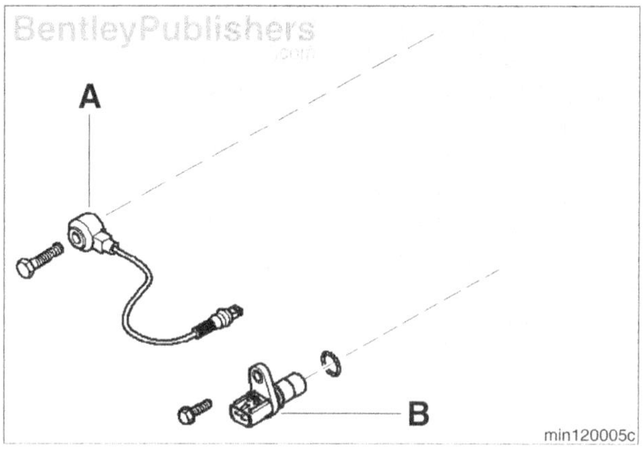

Fig. 23 DME sensors

At front of engine block.

- **A** Knock sensor
- **B** Crankshaft position sensor

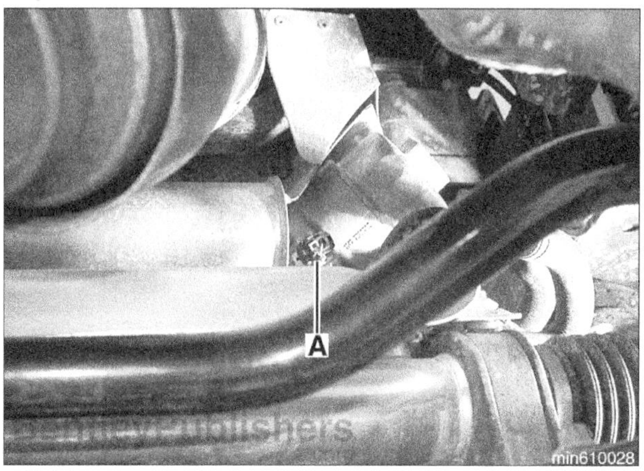

Fig. 24 Oil pressure switch

At oil filter housing, rear of engine.

- **A** Oil pressure switch

Components underneath vehicle

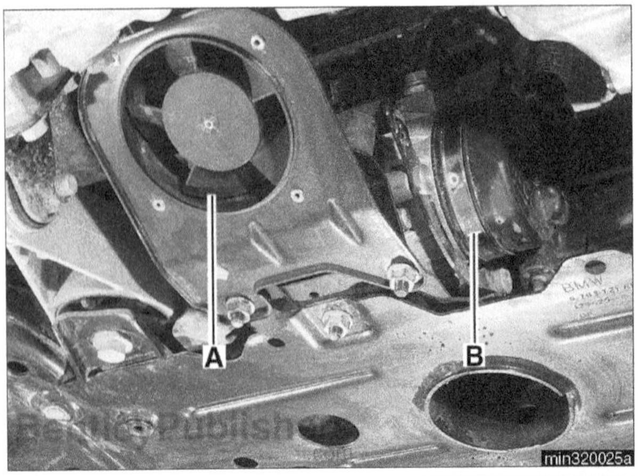

Fig. 25 Electrohydraulic power steering (EHPS) pump

◀ Underneath engine, behind oil pan.

A Power steering pump cooling fan

B Electrohydraulic power steering (EHPS) pump

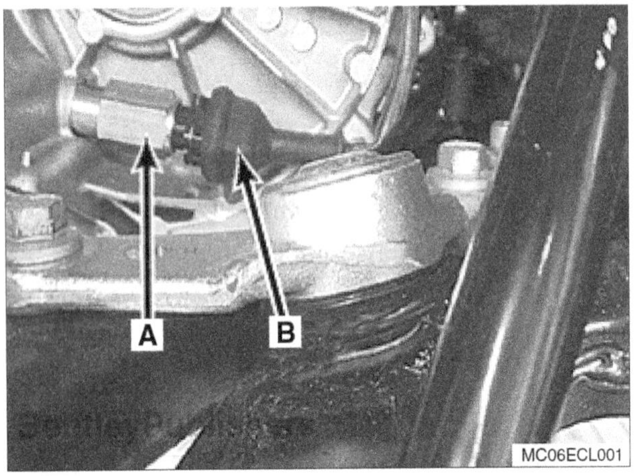

Fig. 26 Automatic transmission (CVT) switch

◀ Transmission lower left.

A CVT switch

B CVT switch harness connector

Fig. 27 Outside temperature sensor

◀ Underneath front bumper cover, left side.

A Outside temperature sensor

Fig. 28 Front wheel speed sensor

◀ Top of front wheel hub.

A ABS front wheel speed sensor

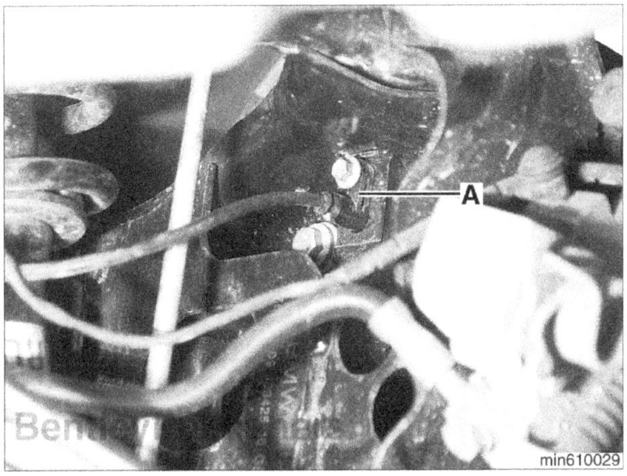

Fig. 29 Rear wheel speed sensor

Rear of rear wheel hub.

A ABS rear wheel speed sensor

Fig. 30 Headlight vertical aim sensor, front

Left front control arm.

A Front height sensor

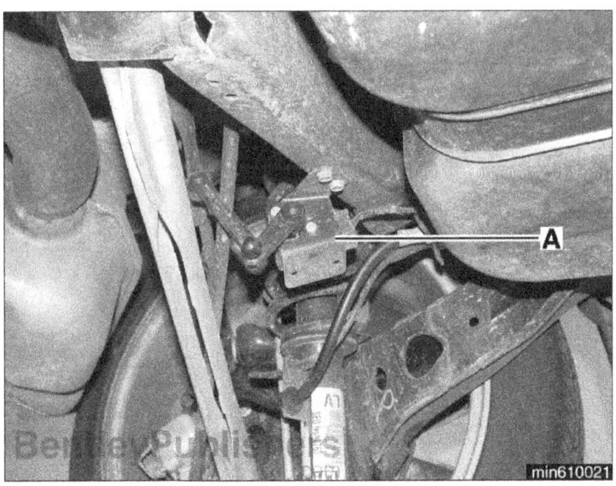

Fig. 31 Headlight vertical aim sensor, rear

Rear suspension subframe.

A Front height sensor

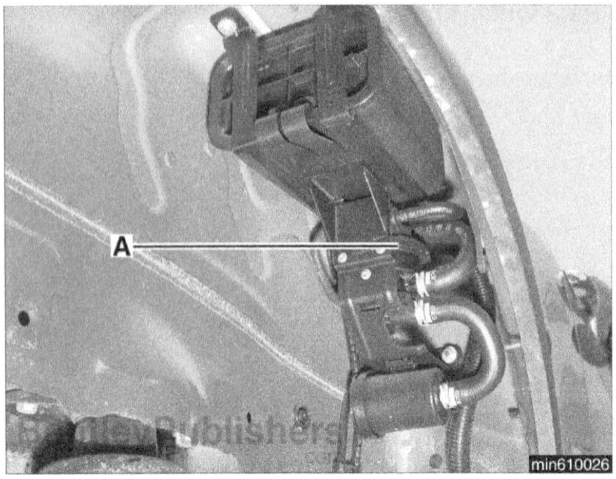

Fig. 32 Leak Diagnosis Pump (LDP)

Under right rear wheel housing trim.

A LDP assembly

Heating and A/C components (vehicle interior)

Fig. 33 A/C evaporator temperature sensor

Left side of heating and A/C housing, under dashboard.

A A/C evaporator temperature sensor

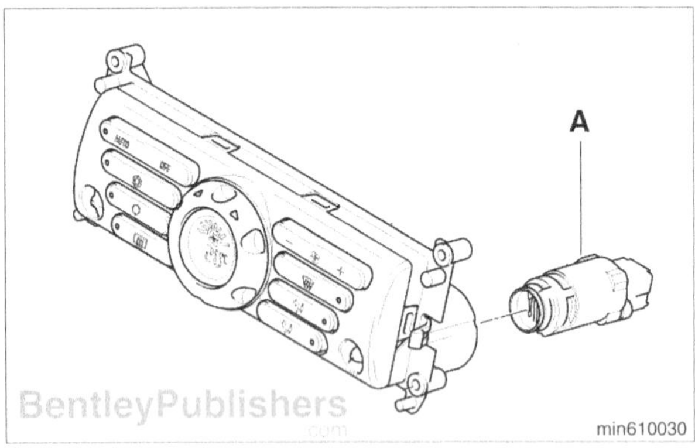

Fig. 34 IHKA interior temperature sensor

Back of IHKA control panel, center of dashboard.

A Temperature sensor and fan

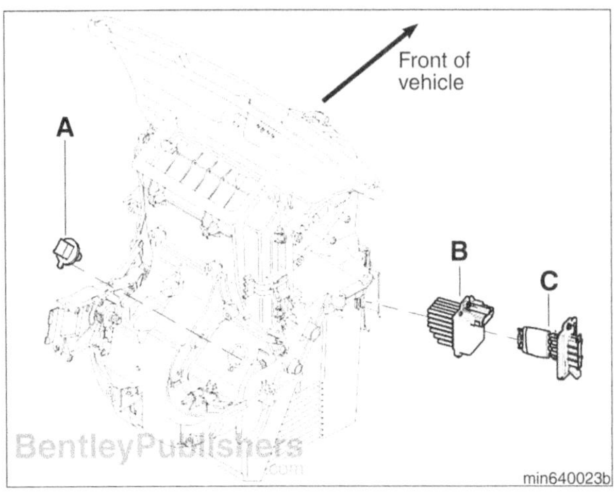

Fig. 35 Heating and A/C housing components

Behind glove compartment, right side of heater and A/C housing.

A Heater core temperature sensor

B IHKA power stage (from 3/2002)

C IHKS resistor pack

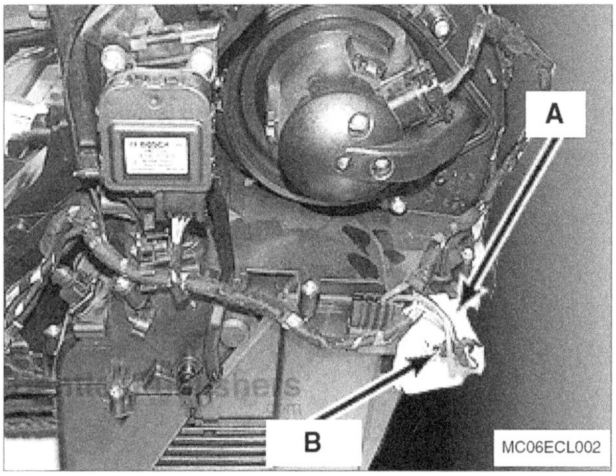

Fig. 36 Heating and A/C blower output stage

◄ Behind center console, right side.

- **A** Blower output stage
- **B** Harness connector

Components under dashboard

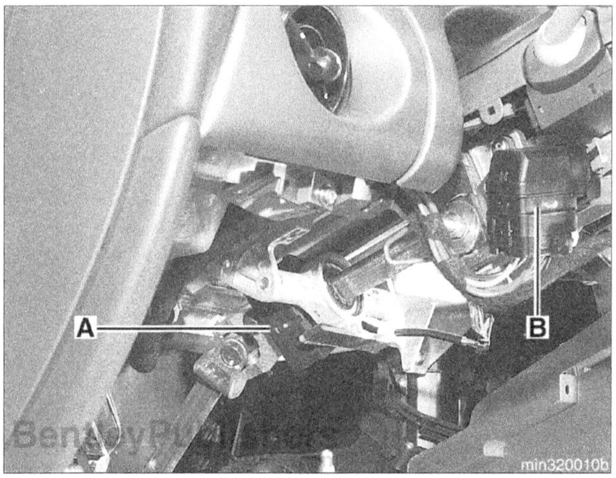

Fig. 37 Steering column components

◄ Under steering column.

- **A** DSC steering angle sensor
- **B** Ignition switch assembly

Fig. 38 Brake light switch

◄ At brake pedal clevis pin, under left dashboard.

- **A** Brake light switch

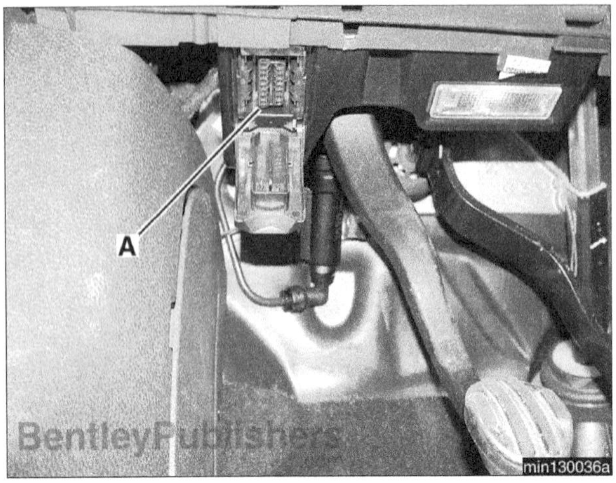

Fig. 39 OBD II diagnostic connector

Above pedal cluster, under left dashboard.

A OBD II diagnostic connector (16 pin DLC)

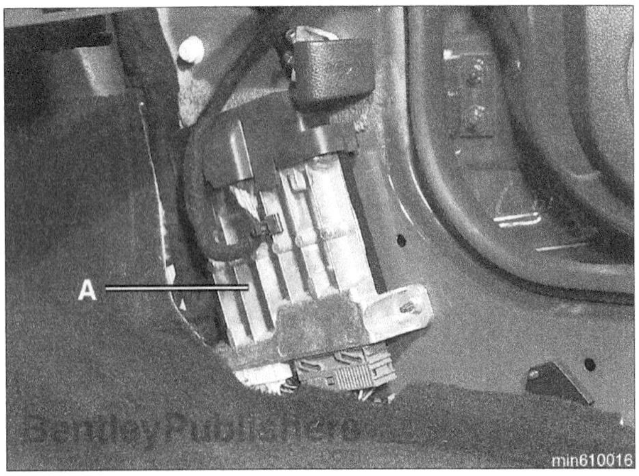

Fig. 40 General Module (BC1)

Behind right footwell trim.

A General Module (BC1) (Body Control module)

NOTE —
In wiring diagrams, the General Module is referred to as A1.

Vehicle interior components

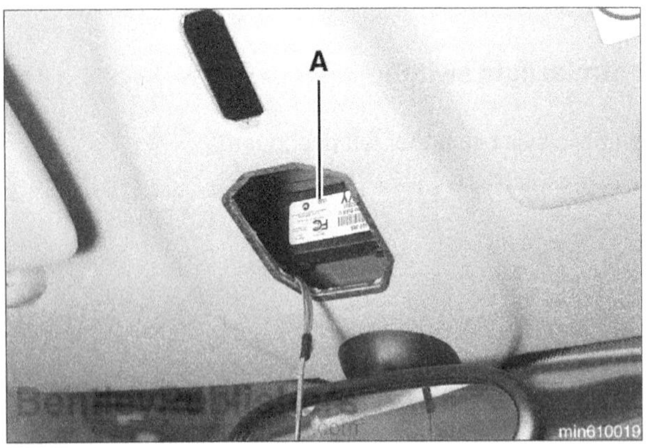

Fig. 41 Above headliner

Behind clock (as applicable).

A FZV receiver and control module

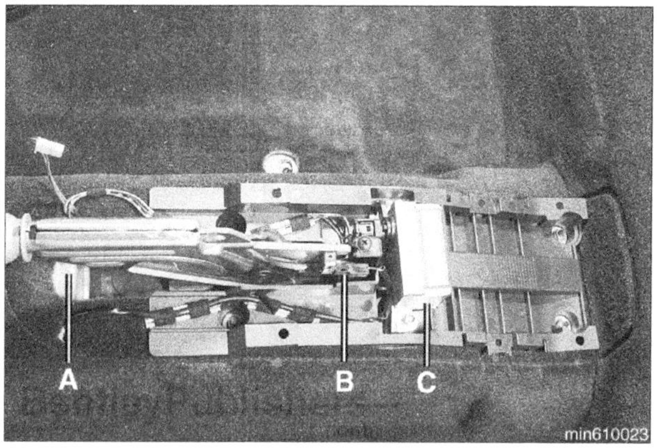

Fig. 42 At parking brake handle

Underneath rear console.

- **A** MRS control module (under carpet)
- **B** Parking brake warning switch
- **C** DSC motion sensor

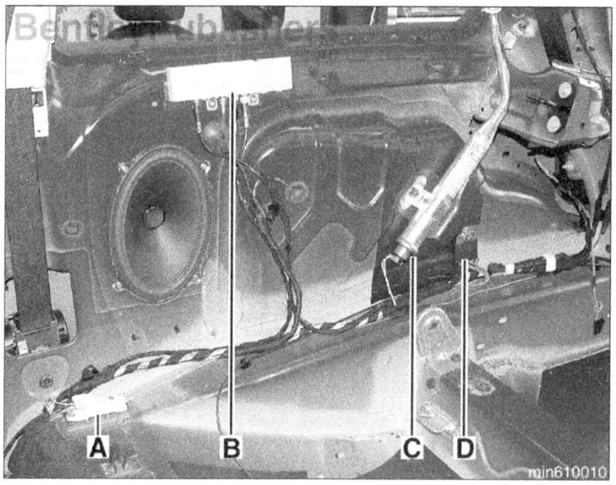

Fig. 43 Right rear components

Behind right side panel.

- **A** Head protection airbag crash sensor
- **B** Diversity antenna junction box
- **C** Gas generator, head protection airbag
- **D** Ground bus X494

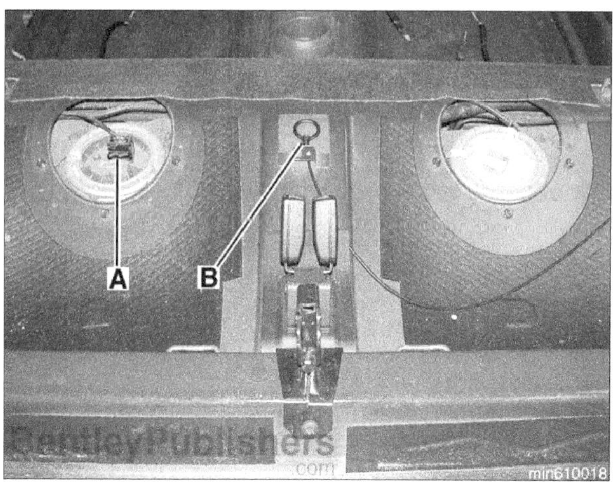

Fig. 44 Rear components

Under rear seat cushion.

- **A** Fuel pump and left fuel level sender
- **B** Rear hatch emergency inside release

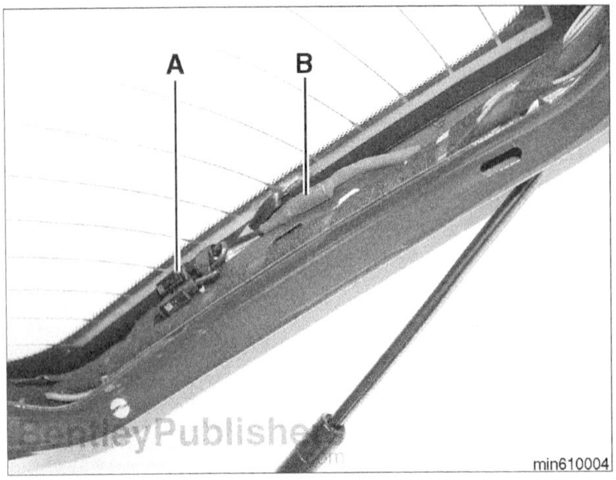

Fig. 45 Diversity antenna, rear hatch window

Right side, rear hatch window.

- **A** Rear window heater and diversity antenna connector
- **B** Wavetrap 2, diversity antenna

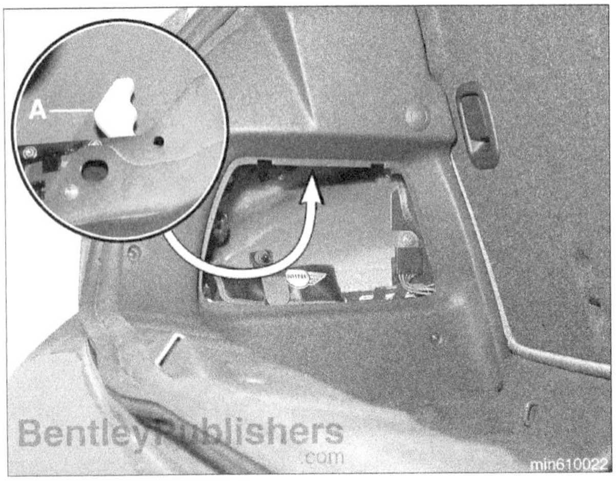

Fig. 46 Fuel flap emergency release

Above left rear access panel, behind rear seat.

- **A** Fuel flap release handle

NOTE—
*Reach up into cavity (**inset**) and pull backward on handle.*

Convertible top and additional interior components

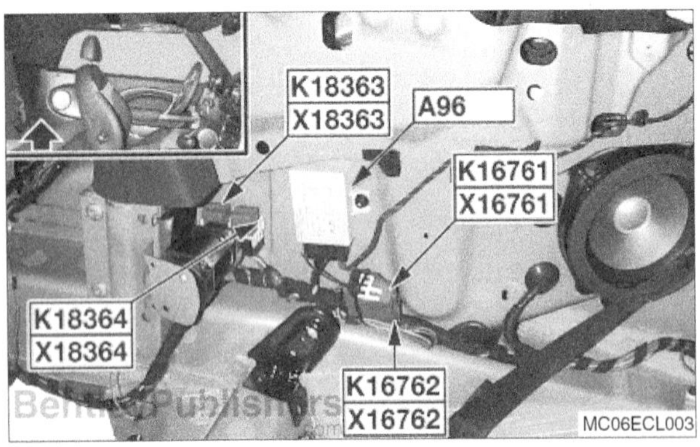

Fig. 47 Convertible top control module and relays

Behind left rear trim panel.

- **A96** Convertible top control module
- **K16761** Sliding canvas sunroof relay 1
- **K16762** Sliding canvas sunroof relay 2
- **K18363** Convertible top relay 1
- **K18364** Convertible top relay 2

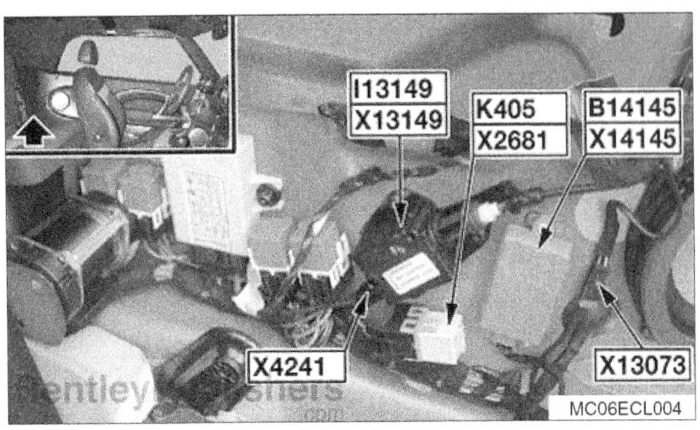

Fig. 48 Convertible top storage compartment lock and microwave sensor

Behind left rear trim panel.

- **B14145** Microwave sensor
- **I13149** Drive motor for convertible top storage compartment lock, left
- **K405** Convertible top storage compartment lock relay

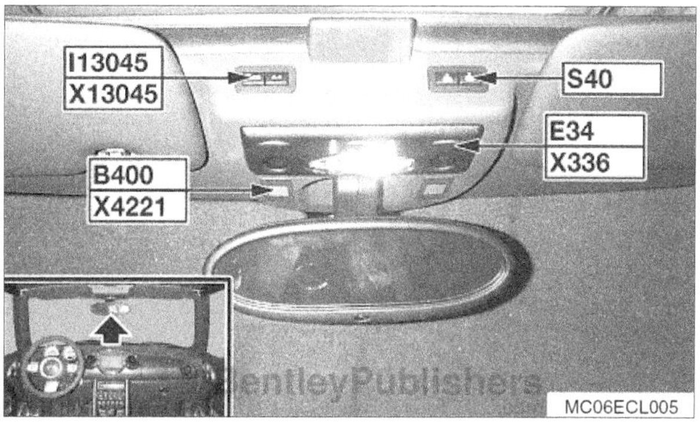

Fig. 49 Components above interior rear view mirror

Above interior rear view mirror.

- **B400** Hands-free microphone
- **E34** Reading light, front
- **I13045** Convertible top control button
- **S40** Convertible rear power window switch

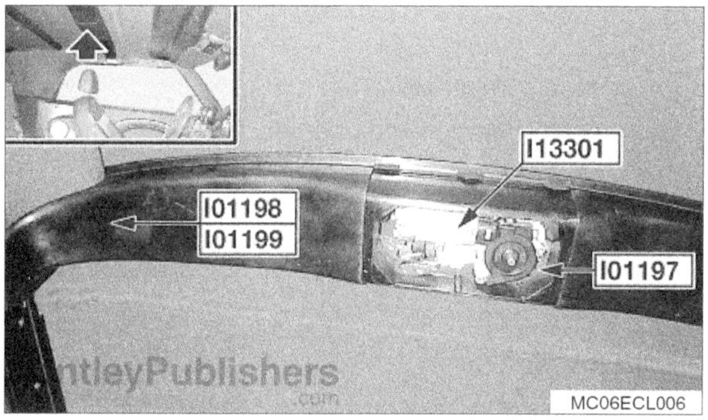

Fig. 50 Sliding canvas sunroof components

Folding sunroof frame, front.

- **I01197** Sliding canvas sunroof “closed” hall sensor
- **I01198** Sliding canvas sunroof “open” hall sensor
- **I01199** Convertible top catch hall sensor
- **I13301** Drive unit for sliding canvas sunroof

Fig. 51 Convertible top and rear window regulator motor

Convertible top storage compartment, left rear.

- **I01187** Convertible top “folded down” microswitch
- **M98** Window regulator motor, left rear

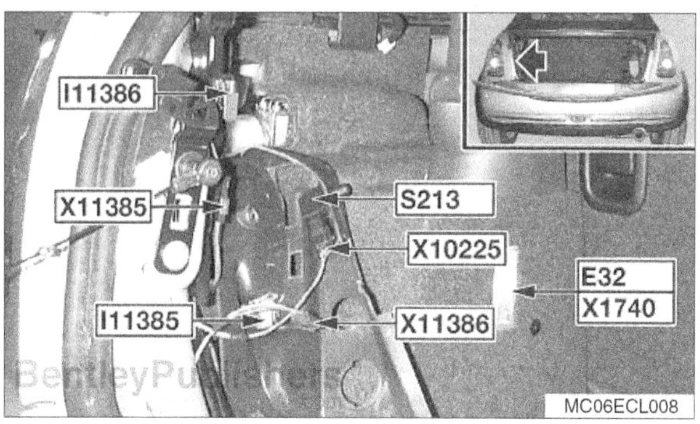

Fig. 52 Convertible top and rear cargo compartment

Cargo compartment, left rear.

- **E32** Cargo compartment light
- **I11385** Convertible top cross brace microswitch
- **I11386** Convertible top rear window shelf microswitch
- **S213** Rear lid lock contact switch

Fig. 53 Convertible top closed microswitch

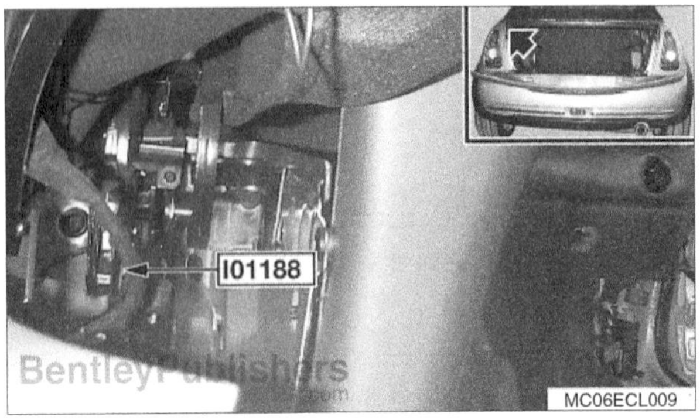

Cargo compartment, left rear.

- **I01188** Convertible top “closed” microswitch

Fig. 54 Convertible top storage compartment lock microswitch

Convertible top storage compartment, left rear.

- **I13073** Convertible top storage compartment lock microswitch, left

ELE Electrical Wiring Diagrams

NOTE—

- *Common MINI acronyms are explained in* **600 Electrical System—General**.
- *Airbags systems, formerly known by the acronym SRS (supplemental restraint system), are currently referred to by the BMW acronym MRS (multiple restraint system).*
- *Every attempt has been made in this repair manual to standardize component names. However, in some cases, the same component may appear with different names. Regardless of the component name, the alphanumeric component designation used on the wiring diagrams is consistent throughout.*

GENERAL

The wiring schematics given in this section represent detailed electrical circuit information for MINI Cooper and Cooper S models. Each diagram shows the wiring, connectors, terminals, and electrical or electronic components of the circuit. It also identifies the wires by color or terminal coding.

The wiring diagrams are identified by primary system, model (as applicable) and date of manufacture (as applicable). The date of manufacture can be found on the vehicle data sticker. This sticker is located on the rear edge of the driver's door. See **020 Maintenance** for more information.

Wiring diagrams are necessary to troubleshoot and repair electrical or electronic circuits. Remember that electrical and electronic circuits often have more than one source of power and/or ground. In many cases there is a switched ground. Take time to study the schematics of the entire system to understand the circuit logic prior to circuit troubleshooting.

When working on electrical or electronic circuits, observe the following warnings and cautions:

> ***WARNING —***
> *On cars equipped with airbags and pyrotechnic seatbelt retractors, special precautions apply to any electrical testing or repair. These components are explosive devices and must be handled with extreme care. Before starting any work, refer to the warnings and cautions in* **720 Seat Belts** *and* **721 Airbag System (SRS)**.

> ***CAUTION —***
> - *Prior to disconnecting the battery, read the battery disconnection cautions in* **001 General Warnings and Cautions.**
> - *Connect and disconnect ignition system wires, multiple connectors and ignition test equipment leads only while ignition is switched off. Keep clothing, hands, and feet dry if possible.*
> - *Always switch a test meter to the appropriate function and range before making test connections.*
> - *Always switch the ignition off and disconnect the negative (–) battery cable before removing any electrical components.*

Special tools

< Automotive digital multimeter

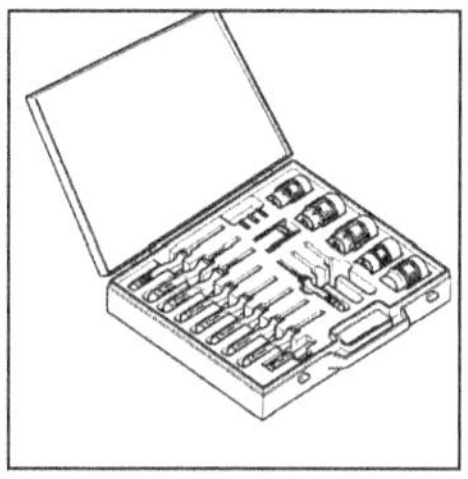

< Wiring harness end repair tools
(BMW tool no. BMW 61 1 150)

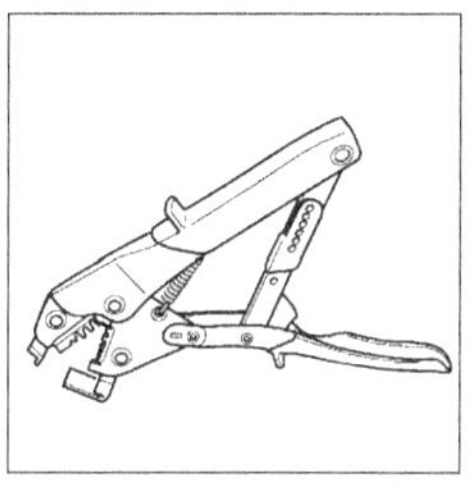

< Wire end crimp tool
(BMW tool no. BMW 61 9 041)

Circuit and terminal descriptions

MINI identifies electrical circuits, including junctions and grounds, with unique designations, most of which follow the German DIN standard. See **Table a**. For example, if a relay terminal is labeled '30', it tells you that positive (+) voltage is supplied to that terminal at all times directly from the battery.

Electrical components are identified in the schematics using a letter followed by a number. For example, A6000 is the DME control module. The letter A identifies the component as an electronic control module. **Table b** lists the component alpha-codes.

The table **Electrical wiring diagram component identifier** is a detailed listing of electrical components from the wiring diagrams, arranged by alphanumeric code.

General

Table a. Electrical terminal designations (according to DIN 72 552 standard)

Number	Circuit Description
1	Low voltage switched terminal of coil
4	High voltage center terminal of coil
+X	Originates at ignition switch. Supplies power when ignition switch is in PARK, RUN or START position.
15	Originates at ignition switch. Supplies power when ignition switch is in RUN or START position.
30	Battery positive (+) voltage. Supplies power whenever battery is connected. (Not dependent on ignition switch position, unfused)
31	Ground, battery negative (-) terminal
50	Supplies power from battery to starter solenoid when ignition switch is in START position only.
+54	Originates at ignition switch. Supplies power when ignition switch is in the RUN position only.
85	Ground side (-) of relay coil
86	Power-in side (+) of relay coil
87	Power out when the relay coil is energized (usually terminal 87 connects to terminal 30 power when relay is energized)
D	Alternator warning light and field energizing circuit

Table b. BMW electrical component designation

Letter	Function	Example
A	Electronic control modules, electronic assemblies	A6000 (DME engine control module)
B	Sensors, transducers	B48 (right rear acceleration sensor)
E	Lights, electric heaters	E46 (left trunk-lid light)
F	Fuses	F3 (fuse 3 in fuse panel)
G	Power supply	G6524 (alternator), G1 (battery)
H	Warning lights, signal indicators	H10 (chime module)
I	Components of outside manufacturers	I01032 (aerial amplifier)
K	Relays	K6300 (DME relay)
L	Coils	L1 (EWS toroidal coil)
M	Electric motors, actuators	M6510 (starter)
N	Amplifiers, controllers, control units	N22 (CD changer)
R	Resistors	R11 (blower ballast resistor)
S	Switches, coding plugs	S805 (clutch switch module)
T	Ignition coils	T6151 (cylinder 1 ignition coil)
U	Radio / interference suppression	U4 (telephone)
V	Diodes, potentiometers	V554 (starter motor control diode)
W	Aerials, shields	W18 (Bluetooth aerial)
X	Connectors; splice connectors; grounds	X165 (right front frame extension ground)
Y	Electromechanical components	Y6100 (fuel injectors, solenoids)
Z	Interference suppression coils	Z400 (noise suppression coil)

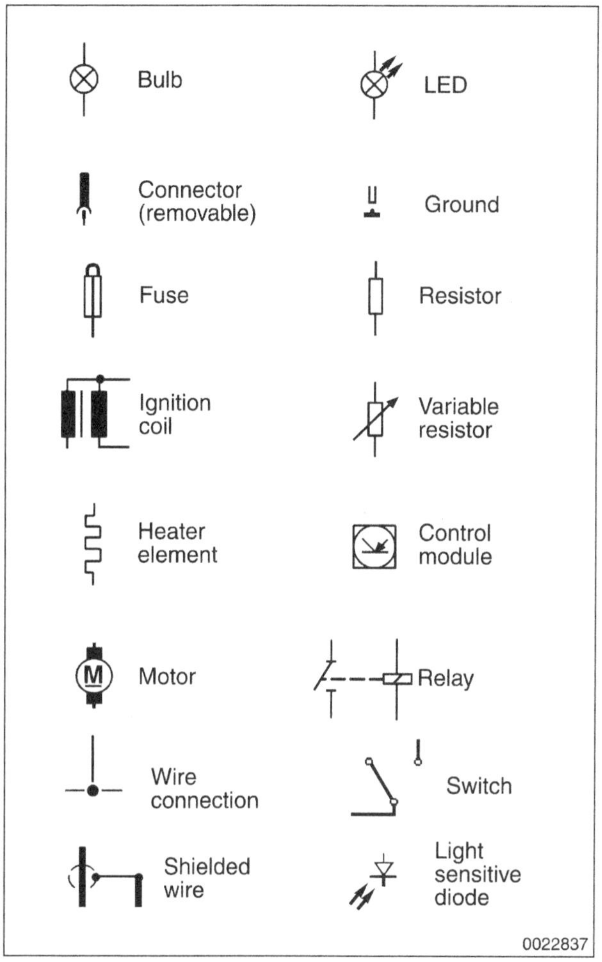

Electrical schematic symbols

The schematics utilize simplified electrical symbols.

Wire insulation colors in this section are given with German color abbreviations. Wire sizes follow the DIN (European) convention.

NOTE —
For example, an 0.5 wire is ½ mm^2 in cross-section area. This corresponds to approx. SAE 16 gauge wire.

Sample wiring schematic

A sample wiring diagram is shown below. It identifies many of the elements found in the diagrams within this section. Note that a dashed line indicates that only a partial view is given and more information on that component or circuit can be found elsewhere in the schematics.

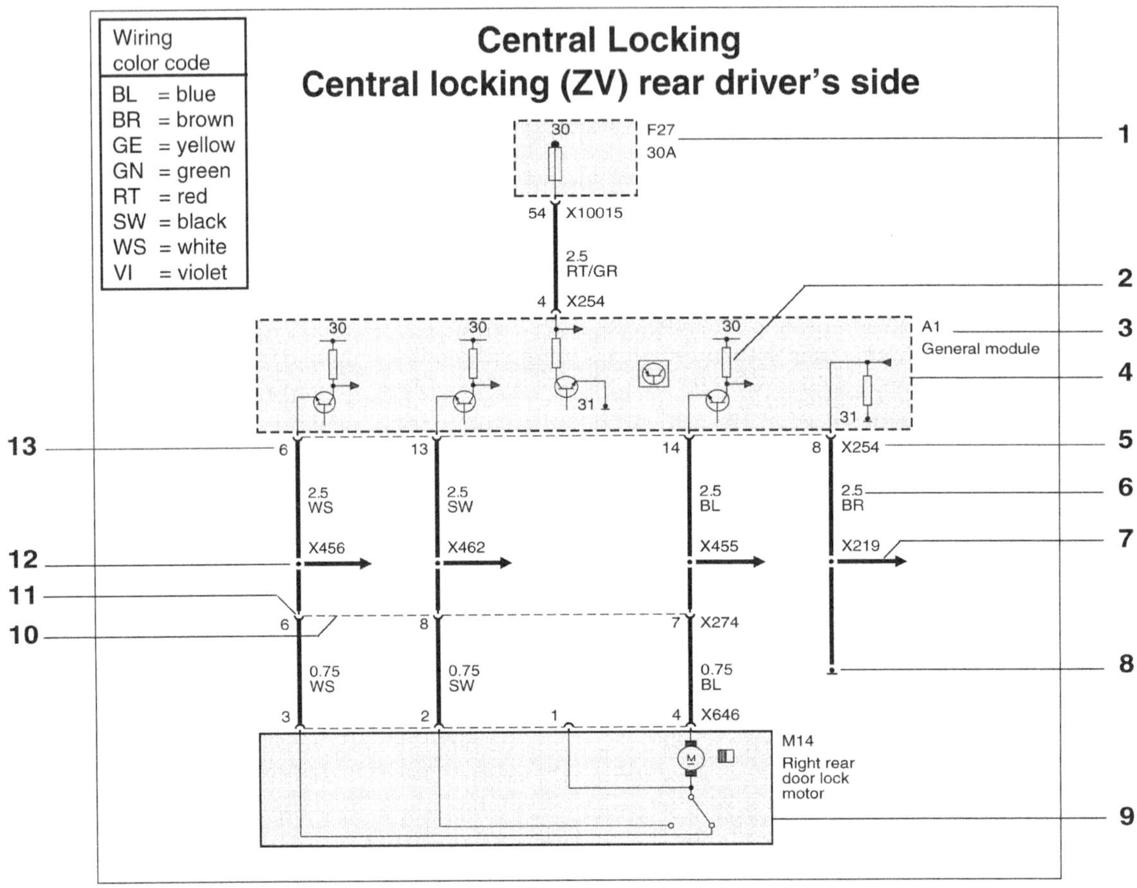

1 - Fuse location. F27 is a fuse in position 27, and is rated 30 amp.

2 - Indicates internal component circuitry in the electronic control module (A1, General Module).

3 - Alpha-numeric designation and name of component. A1 is the General Module.

4 - Dashed border: Indicates that only a portion of component is shown. Complete component is not shown.

5 - Connector alpha-numeric designation.

6 - Wire size (in mm^2) and color.

7 - Arrow indicates that wiring continues to additional diagram.

8 - Ground.

9 - Solid border: Complete component illustrated.

10 - Dashed line represents a multipin connector.

11 - Connector is removable.

12 - Indicates a splice in wiring harness.

13 - Indicates connector pin number. Example: pin 6 in connector X254

mineleOC

Electrical wiring diagram component identifier

Comp. identifier	Name
A1	General Module (BC1)
A2	Instrument cluster control module
A7	ABS / ASC control module
A11	Heating and A/C control module
A12	Multiple restraint system (MRS 4) control module
A18	Amplifier
A22	Electrochromic rear view mirror
A33	Sunroof control module
A47	Fuse and relay panel 2 (left kick panel)
A48	Fuse and relay panel 3 (engine compartment)
A53	Headlight vertical aim control module
A57	Rain sensor
A65	ABS / DSC control module
A81	PDC (park distance control)
A96	Convertible soft top control module
A110	Digital clock
A112	Navigation computer
A113a	Seat occupancy recognition
A117	Telephone eject box
A121	Interior protection (ultrasound) module
A173	Airbag sensor, left B-pillar
A174	Airbag sensor, right B-pillar
A194	Keyless entry antenna / control module
A196	On-Board monitor
A197	Video control module
A208	Mirror folding control module
A421	Diversity antenna
A502	Tire pressure warning (RDW) control module
A836	Electronic immobilizer (EWS) control module
A6000	DME control module (ECM)
A7000	Transmission control module (GIU)
A9001	Additional instrument cluster
A14147	Siren with tilt sensor
B1, B2, B3, B4	ABS wheel speed sensor

Comp. identifier	Name
B8	A/C pressure sensor
B11	Heater core temperature sensor
B13	Outside temperature sensor
B14	Evaporator temperature sensor
B16	Left front brake pad sensor
B17	Right rear brake pad sensor
B18	Brake fluid level switch
B28	Tilt monitoring module
B34	Left outer park distance control sensor
B35	Left center park distance control sensor
B36	Right center park distance control sensor
B37	Right outer park distance control sensor
B42	Front load sensor, headlight vertical aim
B57	Rain sensor (AIC)
B64	Rear load sensor, headlight vertical aim
B66	Solar sensor
B76	Hydraulic brake pressure sensor (DSC)
B400	Microphone, hands-free
B6031	Temperature / manifold absolute pressure (T-MAP) sensor
B6032	Manifold absolute pressure (MAP) sensor
B6048	Engine coolant temperature (ECT) sensor
B6063	Intake vacuum sensor
B6203	Crankshaft position sensor
B6204	Camshaft position sensor
B6231	Oil pressure switch
B6240	Knock sensor
B9801	DSC motion sensor
B10508	Airbag front sensor, left
B10509	Airbag front sensor, right
B14143, B14144, B14145, B14146	Antitheft (DWA) microwave sensors (convertible)
B60631	Intake air temperature sensor
B60632	Intake air pressure sensor (before supercharger)
B62001	Oxygen sensor, precatalyst
B62002	Oxygen sensor, post-catalyst

Comp. identifier	Name
E7	Headlight, left
E8	Headlight, right
E15	Parking light, left
E19	Parking light, right
E21	Side marker light, left
E22	Side marker light, right
E26	Foglight, left
E27	Foglight, right
E28	Cigar lighter, front
E32	Cargo compartment light
E34	Reading light, front
E35	Make-up mirror light, left
E36	Make-up mirror light, right
E42	Glove compartment light
E43	License plate light
E46	Tail light, left
E47	Tail light, right
E51, E52	Heated windshield washer nozzles
E56	Seat back heater, left front
E57	Seat heating, left front
E58	Seat heating, right front
E59	Seat back heating, right front
E61	Footwell light, left front
E62	Footwell light, right front
E70	Side marker light, left rear
E71	Side marker light, right rear
E82	Shift indicator light
E88	Door entrance light, left
E89	Door entrance light, right
E108	Interior light, rear
E117	Door open light, left
E118	Door open light, right
G1	Battery
G5	Airbag, driver
G6	Airbag, passenger
G12	Seat belt tensioner, driver side
G12a	Seat belt tensioner generator, driver's side

Comp. identifier	Name
G13	Seat belt tensioner, passenger side
G13a	Seat belt tensioner generator, passenger side
G14	Side airbag, driver (MRS 4)
G14a	Side airbag, driver (MRS 5)
G15	Side airbag, passenger (MRS 4)
G15a	Side airbag, passenger (MRS 5)
G17	Head protection airbag, left (MRS 4)
G17a	Head protection airbag, left (MRS 5)
G18	Head protection airbag, right (MRS 4)
G18a	Head protection airbag, right (MRS 5)
G19, G19a	Generator, battery safety terminal
G27	Seat belt tensioner, left rear
G28	Seat belt tensioner, right rear
G6524	Alternator (generator)
H1	Horn, alarm system
H2	Horn, left
H3	Horn, right
H4	Turn signal light, right front
H7	Turn signal light, left front
H8	Supplementary turn signal light, right front
H9	Supplementary turn signal light, left front
H10	Chime module
H23a	Passenger airbag deactivation light
H28	Back-up lights
H34	Brake light, center
H40	Speaker, park distance control
H45	Speaker, right rear
H46	Speaker, left rear
H50	Speaker, right front tweeter
H54	Speaker, right front mid-range
H55	Speaker, left front mid-range
H59	Speaker, left front tweeter
H16026	Rear foglight
I01000	Switch block, MFL, left
I01001	Switch block, MFL, right
I01002	Contact spring (volute spring)

Comp. identifier	Name
I01009	Rear window defogger coil
I01010	Wave trap 1
I01020	Wave trap 2
I01032	Antenna amplifier
I01058	AM / FM antenna, right side window
I01066	Steering wheel button
I01135	Engine cooling fan relay, stage 1 (from March 2003)
I01136	Wiper fast / slow relay
I01137	Wiper motor relay
I01142	Wiper motor relay, rear
I01143	Windshield heating, left
I01144	Windshield heating, right
I01148	Shift lock relay
I01149	Impact switch (crash sensor)
I01150	Charging socket relay, rear
I01151	Cigar lighter relay
I01152	Engine cooling fan relay, stage 2
I01153	Electric cooling fan switching unit, 2 stage blower
I01154	Central shift unit
I01155	Mixing flap servo
I01157	Interior temperature sensor
I01158	AM / FM antenna
I01175	Heating and A/C control panel
I01178	Xenon headlight ignition module, left
I01179	Xenon headlight ignition module, right
I01180	Fan control module, power steering
I01183	Telephone and navigation module
I01187	Convertible top folded down microswitch
I01188	Convertible top closed microswitch
I01190	AM / FM antenna amplifier
I01192	Universal electronics charging and hands-free module (ULF)
I01197	Sliding canvas sunroof closed hall sensor
I01198	Sliding canvas sunroof open hall sensor
I01199	Convertible top catch hall sensor
I01200	Combined roof antenna

Comp. identifier	Name
I11385	Convertible cross brace microswitch
I11386	Convertible rear window shelf microswitch
I13045	Convertible top control button
I13073	Convertible top storage compartment lock microswitch, left
I13076	Convertible top storage compartment lock microswitch, right
I13148	Drive unit for storage compartment lock, right
I13149	Drive motor for storage compartment lock, left
I13301	Drive motor, lock and sliding canvas sunroof
K2	Horn relay
K4	Heater and A/C blower relay
K5	Front washer pump front relay
K6	Headlight washer pump relay
K13	Rear window defroster relay
K19	A / C compressor relay
K47	Front foglight relay
K96	Fuel pump relay
K119	Rear washer pump relay
K405	Convertible top storage compartment lock relay
K416	Heated windshield relay
K6300	DME main relay
K16761	Relay 1 for lock and sliding canvas sunroof
K16762	Relay 2 for lock and sliding canvas sunroof
K18363	Convertible top relay 1
K18364	Convertible top relay 2
K69923	Fan relay, electrohydraulic power steering
L1	Ring antenna (toroidal coil) (EWS)
M2	Fuel pump
M3	Wiper motor
M4	Window washer pump
M7	Headlights washer pump
M16	Fuel filler door lock
M17	Rear hatch/tailgate lock
M21	Window motor, driver door
M23	Window motor, passenger door
M30	Blower motor

Comp. identifier	Name
M94	Wiper motor, rear
M98	Window regulator motor, left rear
M99	Window regulator motor, right rear
M101	Convertible top drive
M111	Fresh air recirculated air flap servo
M119	LDP (leakage diagnosis pump)
M135	Engine cooling fan
M150	Air distribution servo motor
M6510	Starter motor
N2	Blower output stage
N8	Antenna amplifier AM / FM
N9	Radio control module
N22	CD changer
R10	Accelerator pedal position sensor
R11	Blower resistor
R33	Steering angle sensor
S2	Ignition and starting switch
S4	Horn switch
S5	Wiper / washer switch
S7	Turn signal / headlight dimmer switch
S17	Mirror adjustment switch
S19	Hood contact switch
S29	Brake light switch
S31	Parking brake warning switch
S38	Sunroof switch
S40	Convertible rear power window switch
S47	Driver door lock
S49	Passenger door lock
S53	Seat heater switch, driver
S54	Seat heater switch, passenger
S58	Seat belt hall sensor
S58a	Seat belt buckle contact, driver
S59	Seat belt hall sensor

Comp. identifier	Name
S59a	Seat belt buckle contact, passenger
S71	Crash sensor, head protection airbags, left
S72	Crash sensor, head protection airbags, right
S73	Glove compartment light switch
S75	Blower switch
S136	Washer fluid level switch
S213	Rear lid lock contact switch
S222	Headlight high / low beam switch
S235	Shift lock switch
S241	Switch panel
S273	Tire pressure warning (RDW) switch
S559	Airbag deactivation switch
S805	Clutch switch module
S6575	Automatic transmission (CVT) switch
S8511	Back-up light switch
T6150	Ignition coil
V551	Diode
V554	Diode, starter motor control
Y5	Outside mirror, driver
Y6	Outside mirror, passenger
Y19	Shiftlock selector lever lock
Y103	Door window motor
Y6101	Fuel injector, cylinder 1
Y6102	Fuel injector, cylinder 2
Y6103	Fuel injector, cylinder 3
Y6104	Fuel injector, cylinder 4
Y6390	Electronic throttle control housing
Y6550	Solenoid valve, fuel tank vent valve
Y6975	Steering control module
Y8099	A/C compressor clutch
Y8505	Transmission valve unit
Y8505a	Gear position switch
Y8516	Transmission RPM sensor

ABS/ASC (A7)
(up to Sept. 2003)
(page 1 of 3)

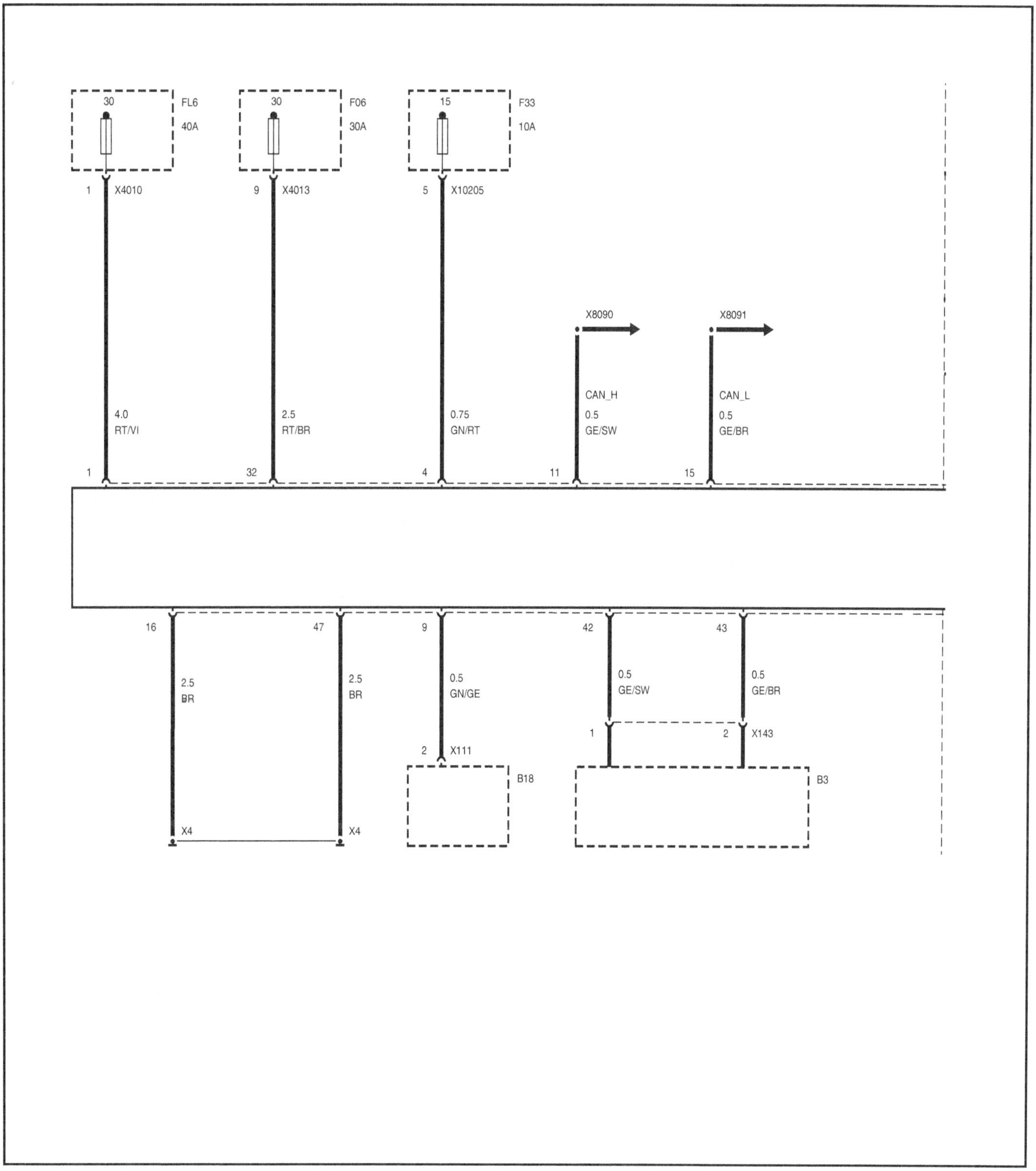

ABS/ASC (A7)
(up to Sept. 2003)
(page 2 of 3)

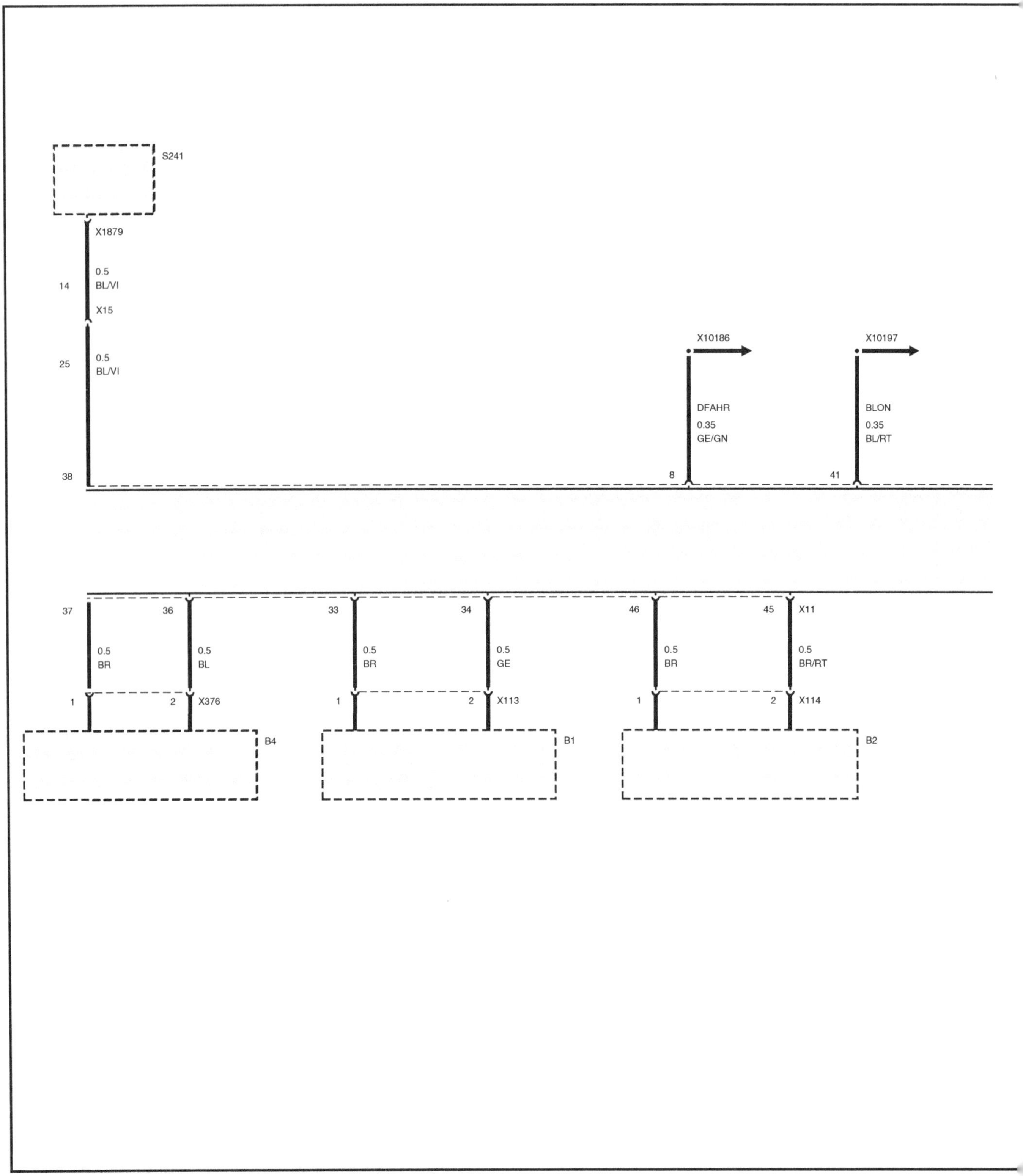

ABS/ASC (A7)
(up to Sept. 2003)
(page 3 of 3)

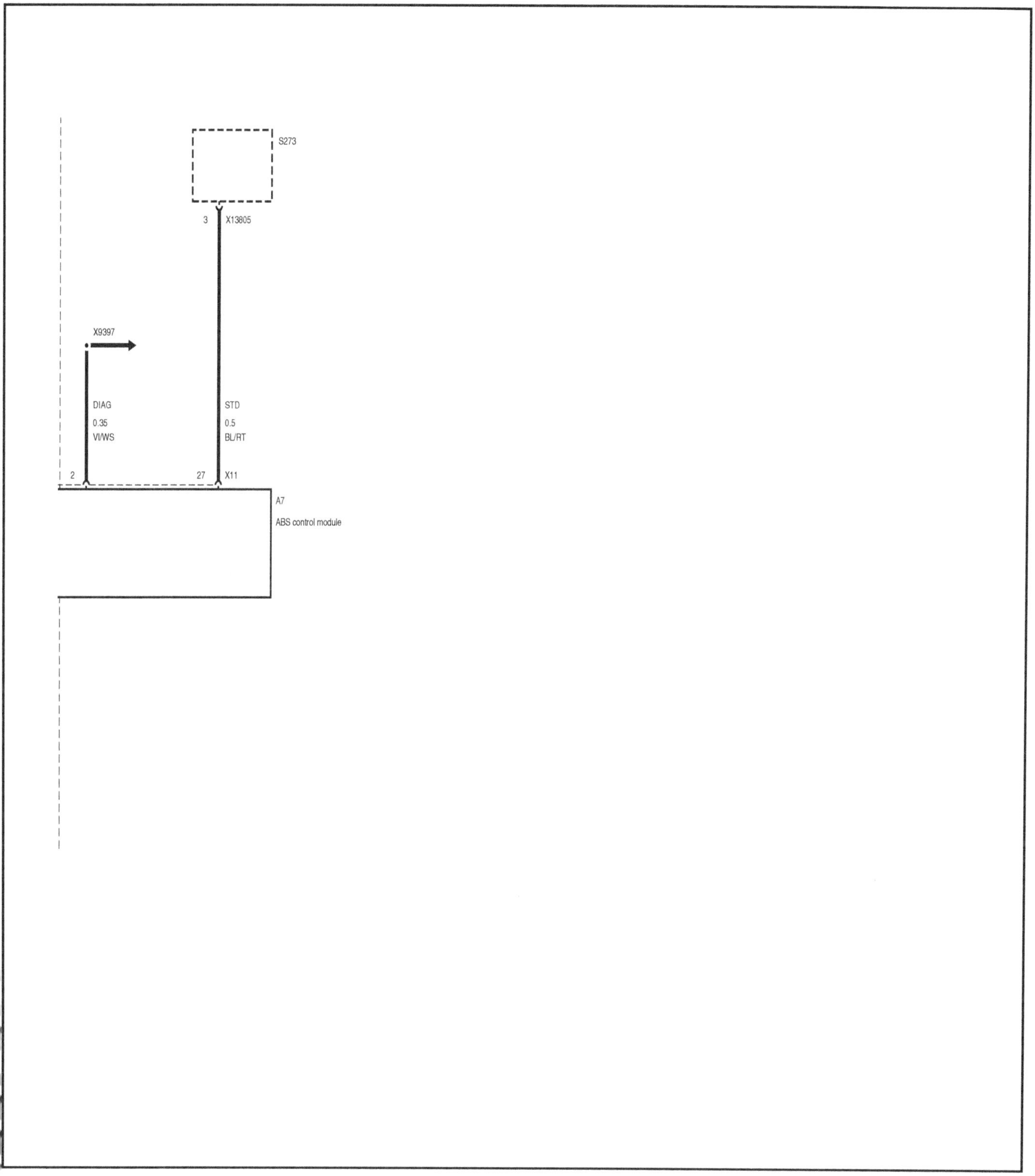

ABS/ASC (A7)
(from Sept. 2003)
(page 1 of 2)

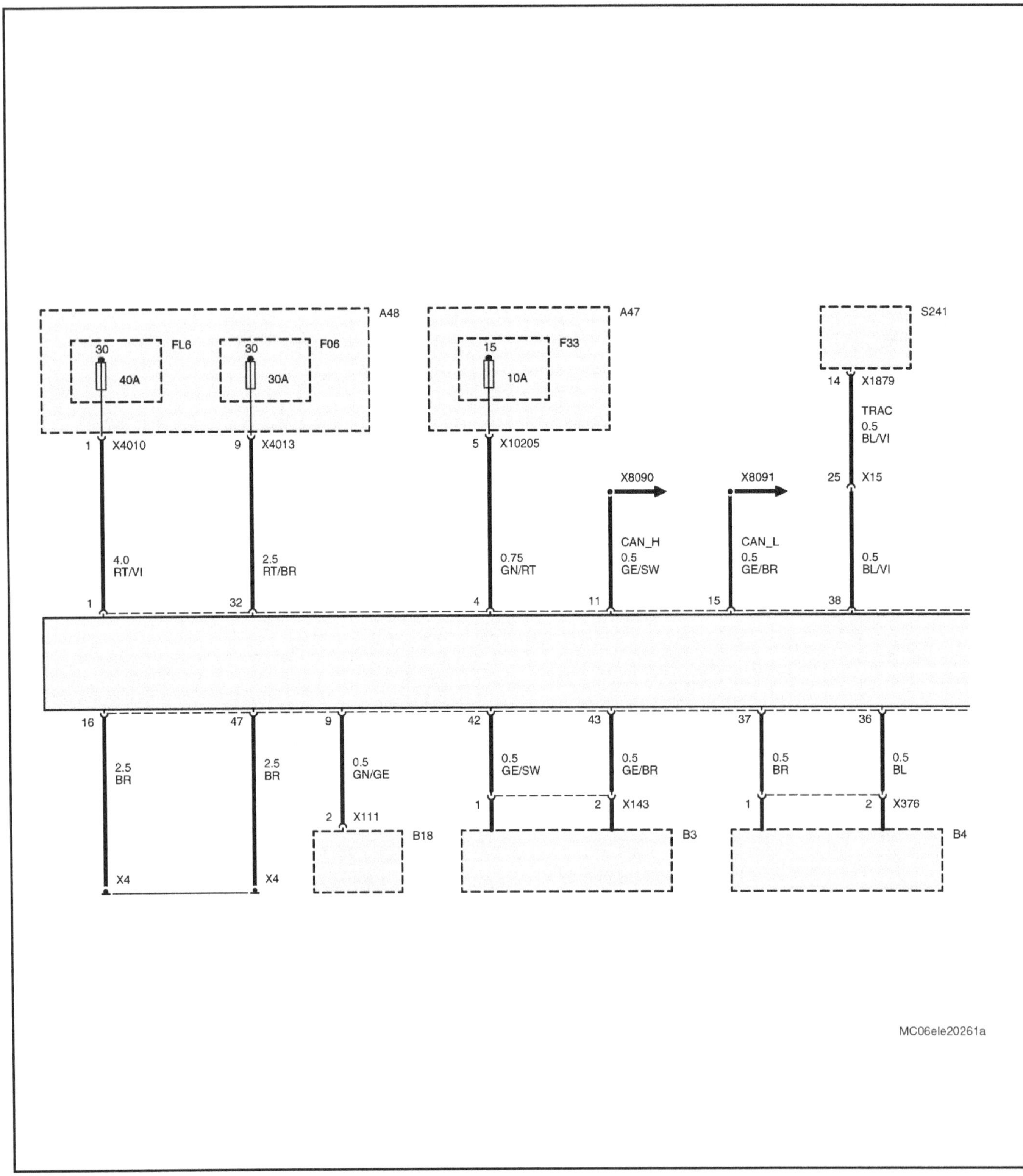

ABS/ASC (A7)
(from Sept. 2003)
(page 2 of 2)

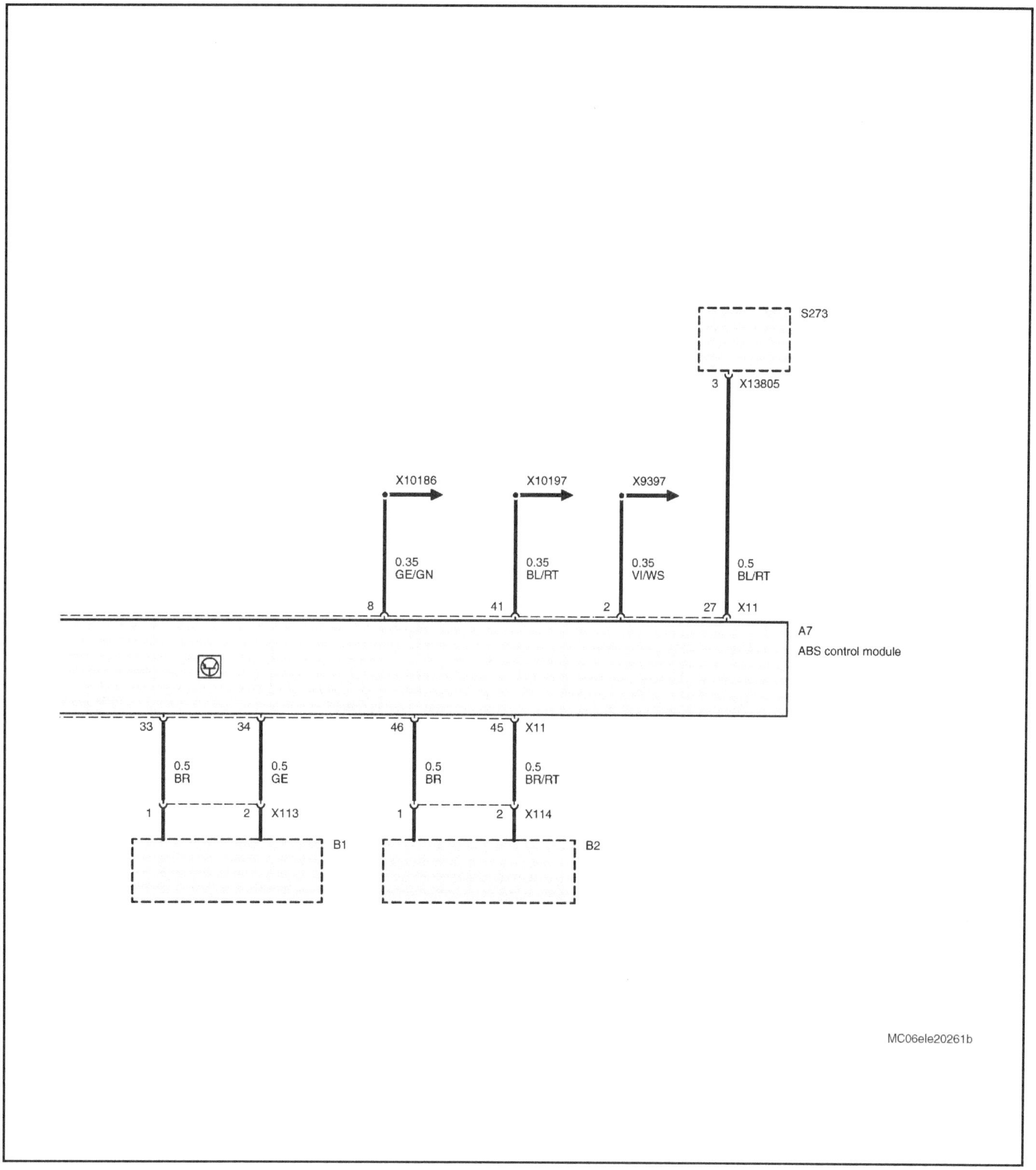

ABS/DSC (A65)
(from March 2003)
(page 1 of 3)

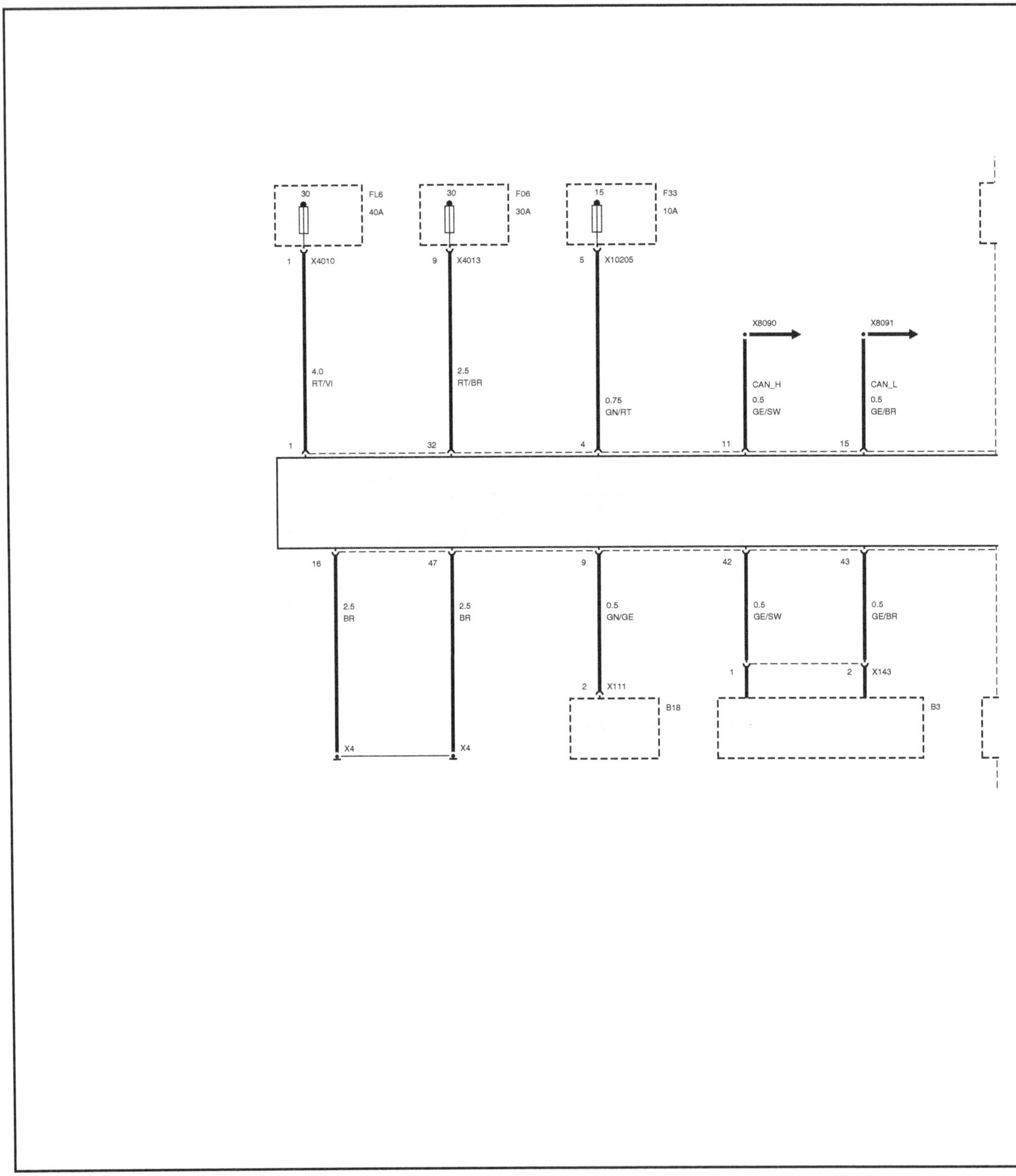

ABS/DSC (A65)
(from March 2003)
(page 2 of 3)

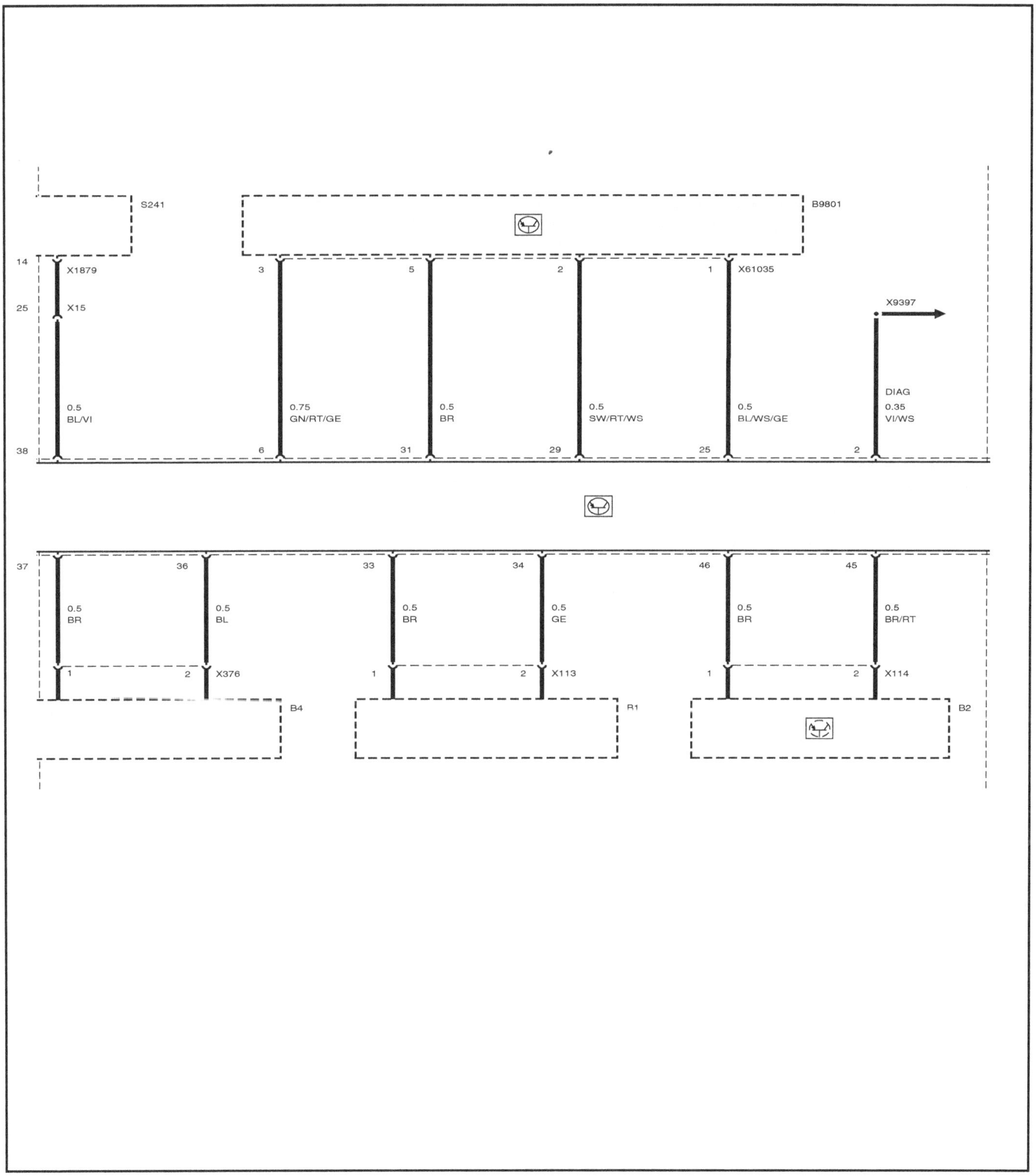

ABS/DSC (A65)
(from March 2003)
(page 3 of 3)

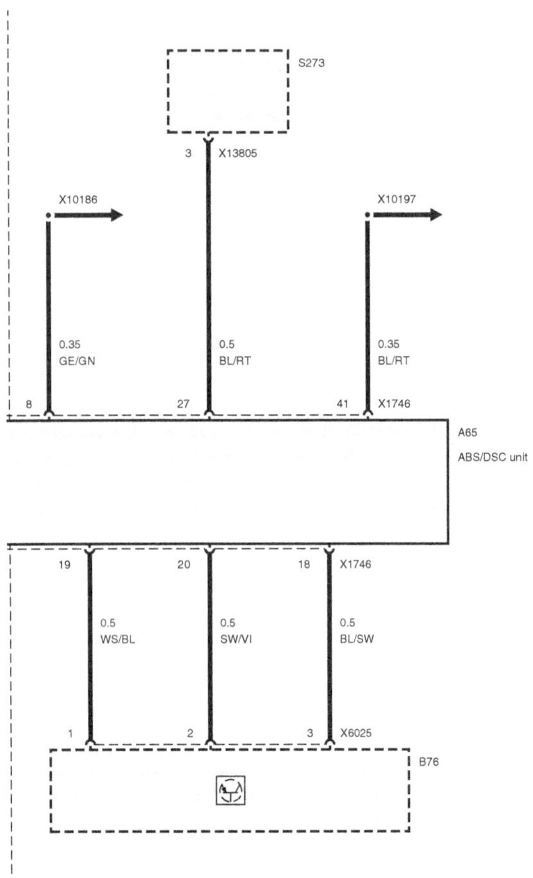

Airbag system
MRS 4 (up to July 2004)
(page 1 of 3)

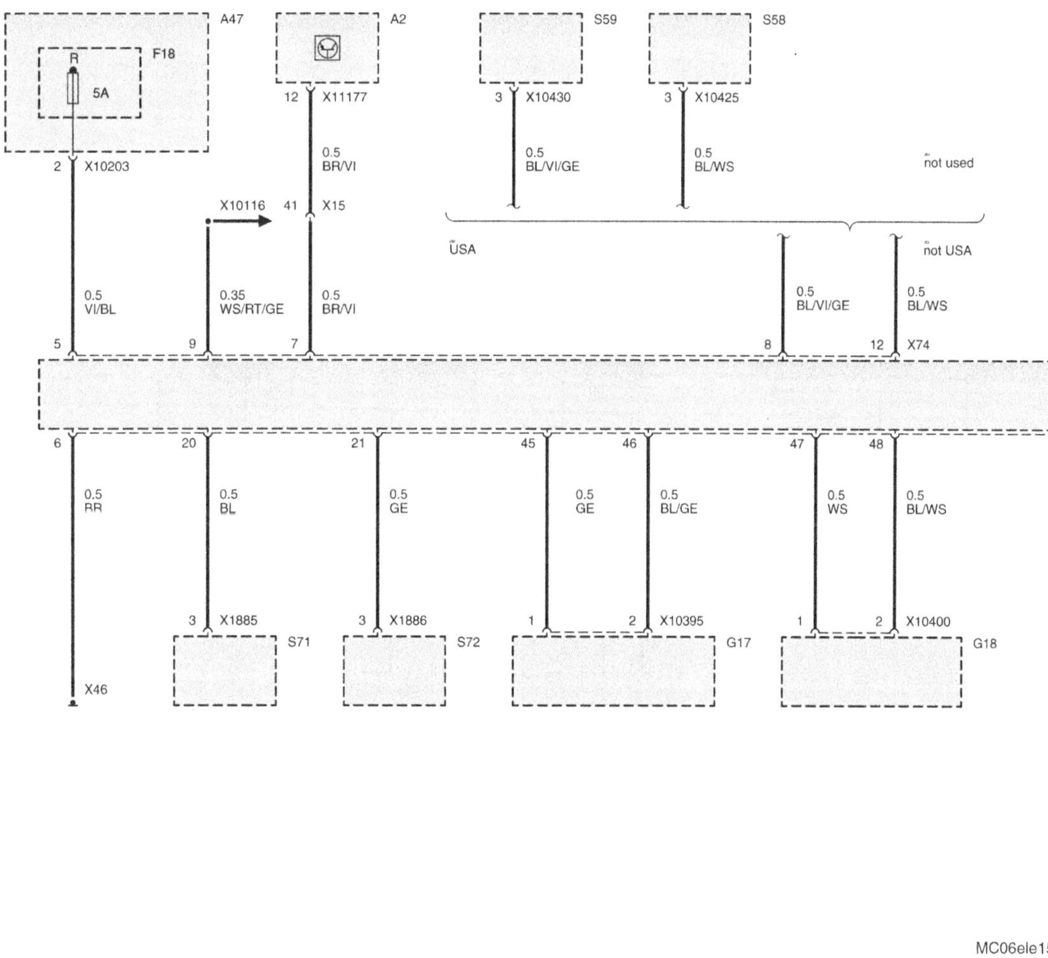

Airbag system
MRS 4 (up to July 2004)
(page 2 of 3)

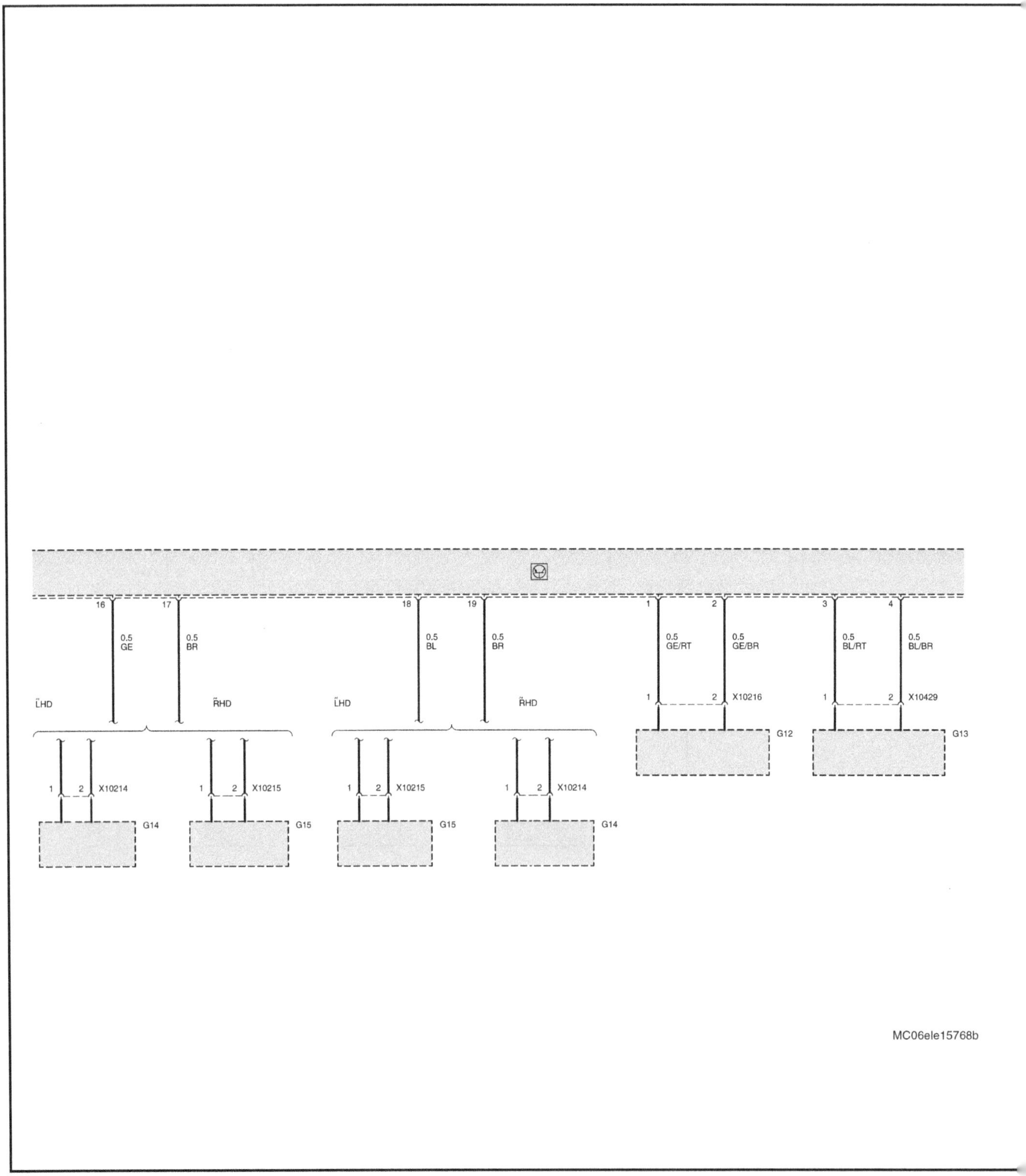

Airbag system

MRS 4 (up to July 2004)

(page 3 of 3)

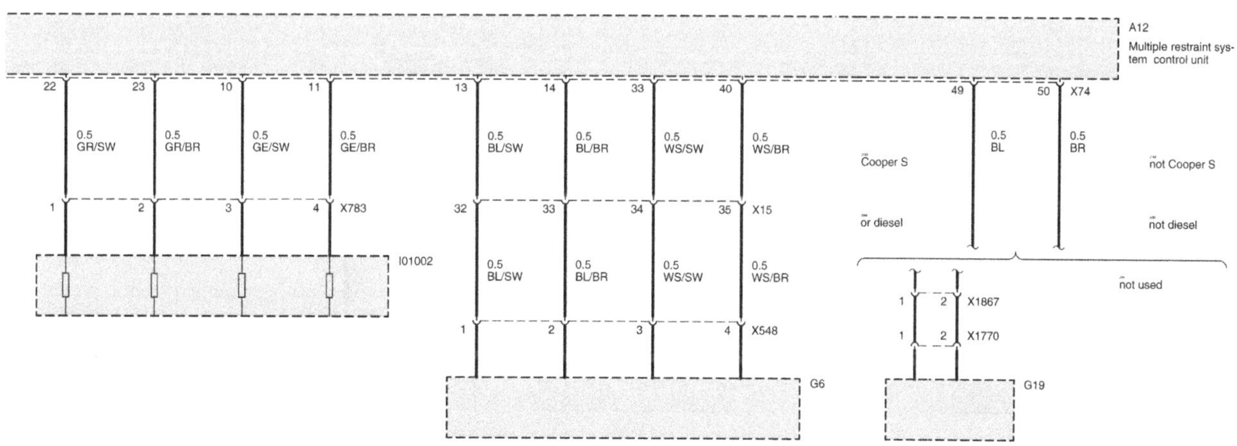

MC06ele15768c

Airbag system
MRS 4 (from July 2004)
(page 1 of 3)

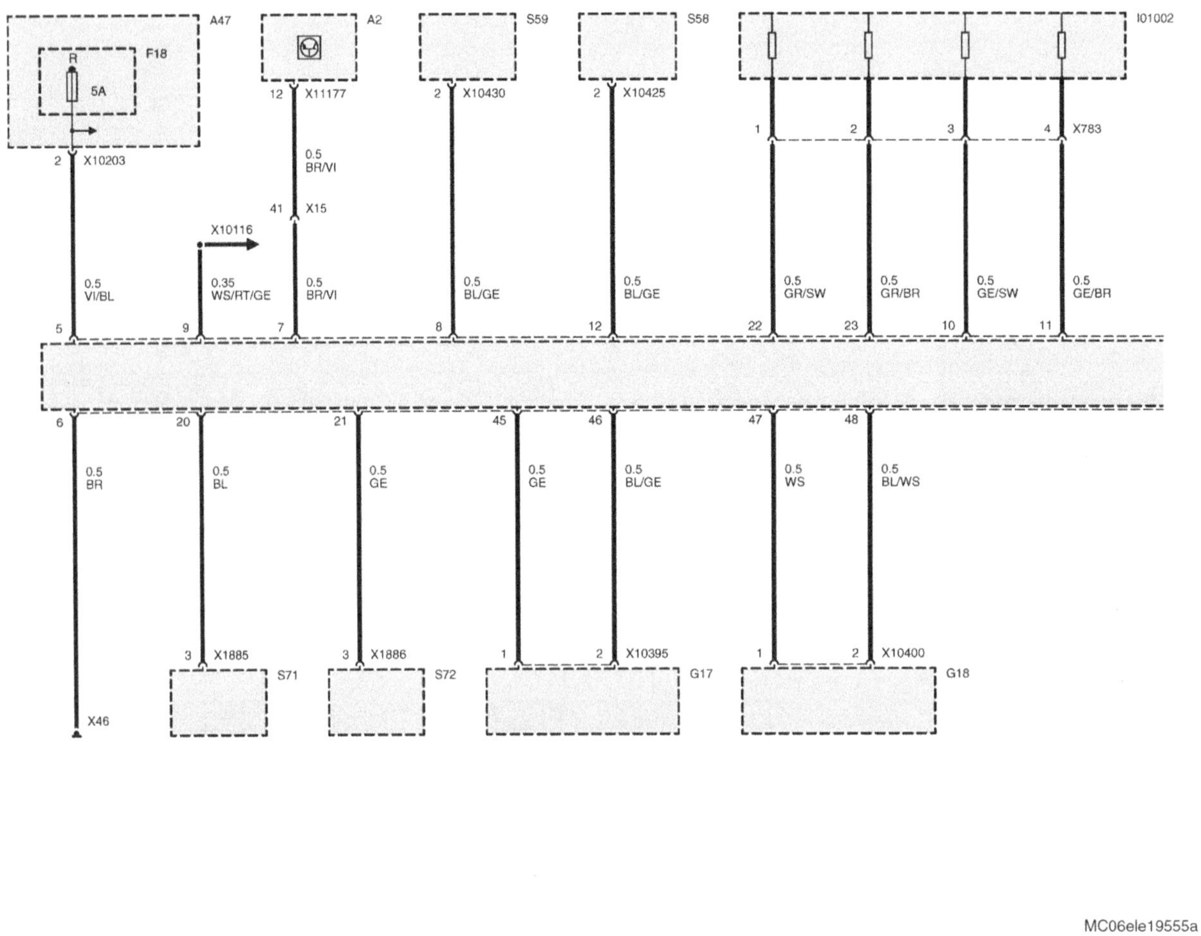

Airbag system
MRS 4 (from July 2004)
(page 2 of 3)

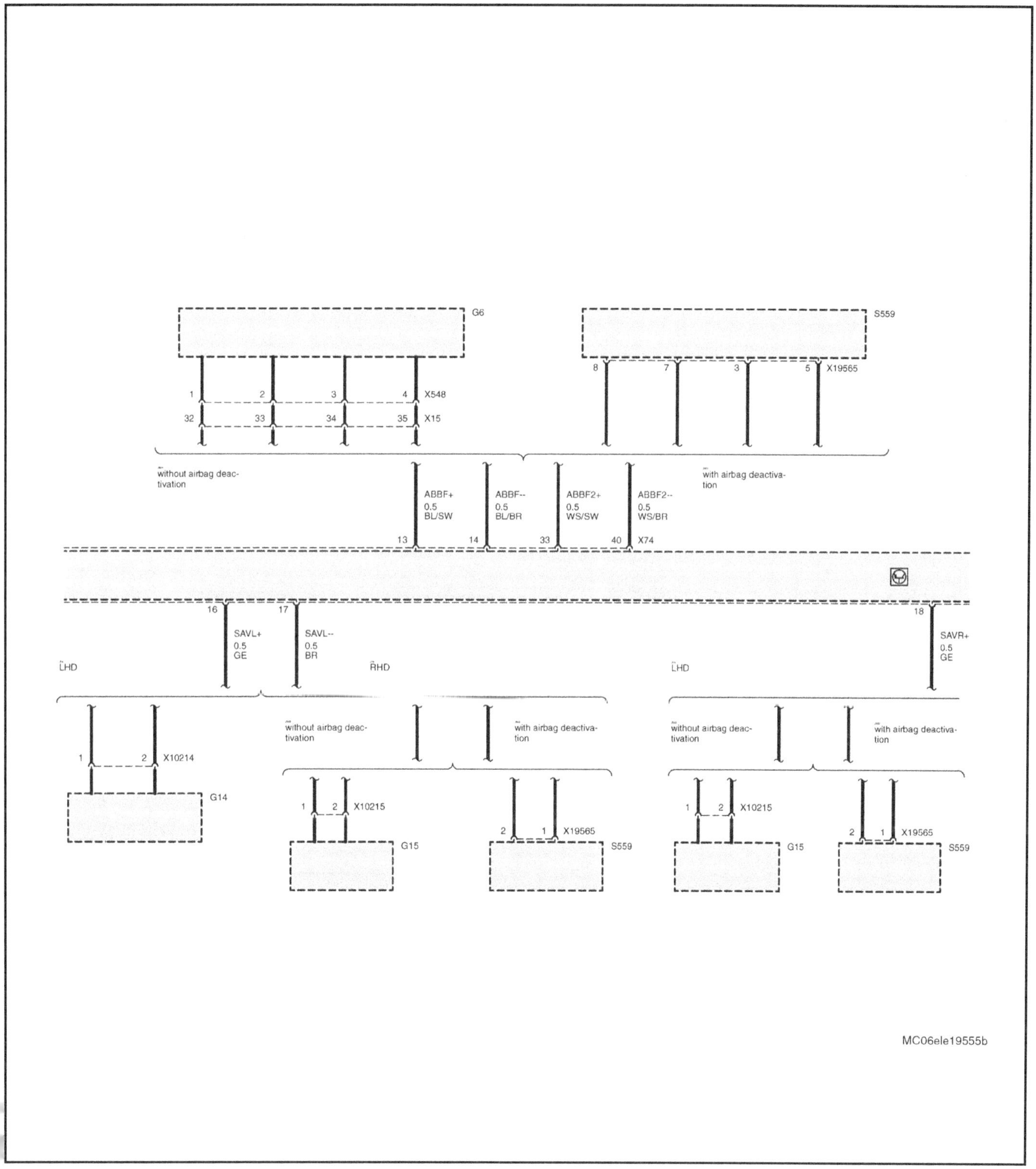

Airbag system
MRS 4 (from July 2004)
(page 3 of 3)

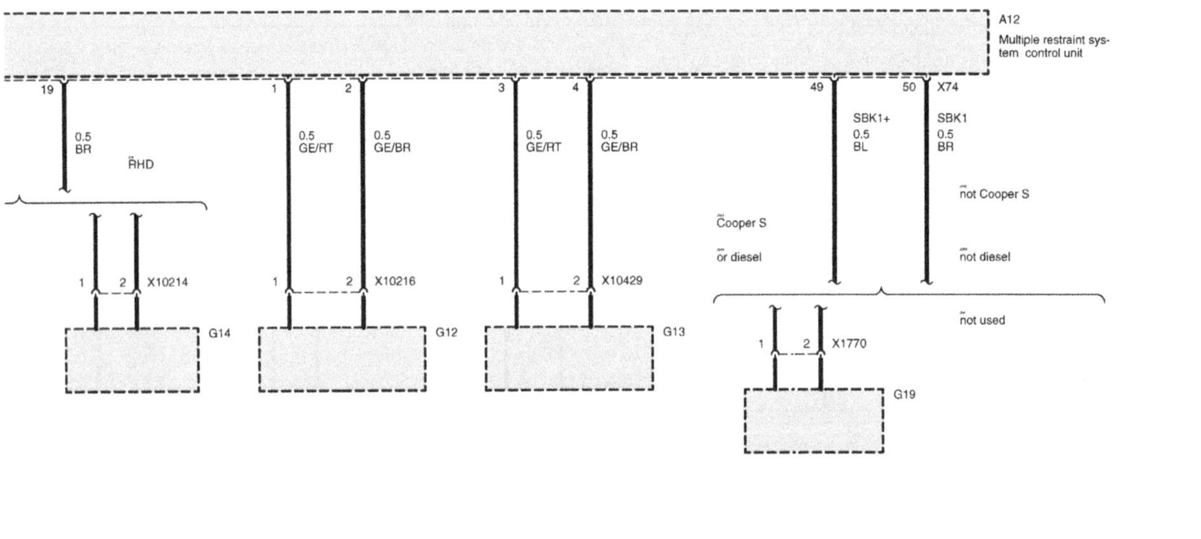

Airbag system
MRS 5
(page 1 of 3)

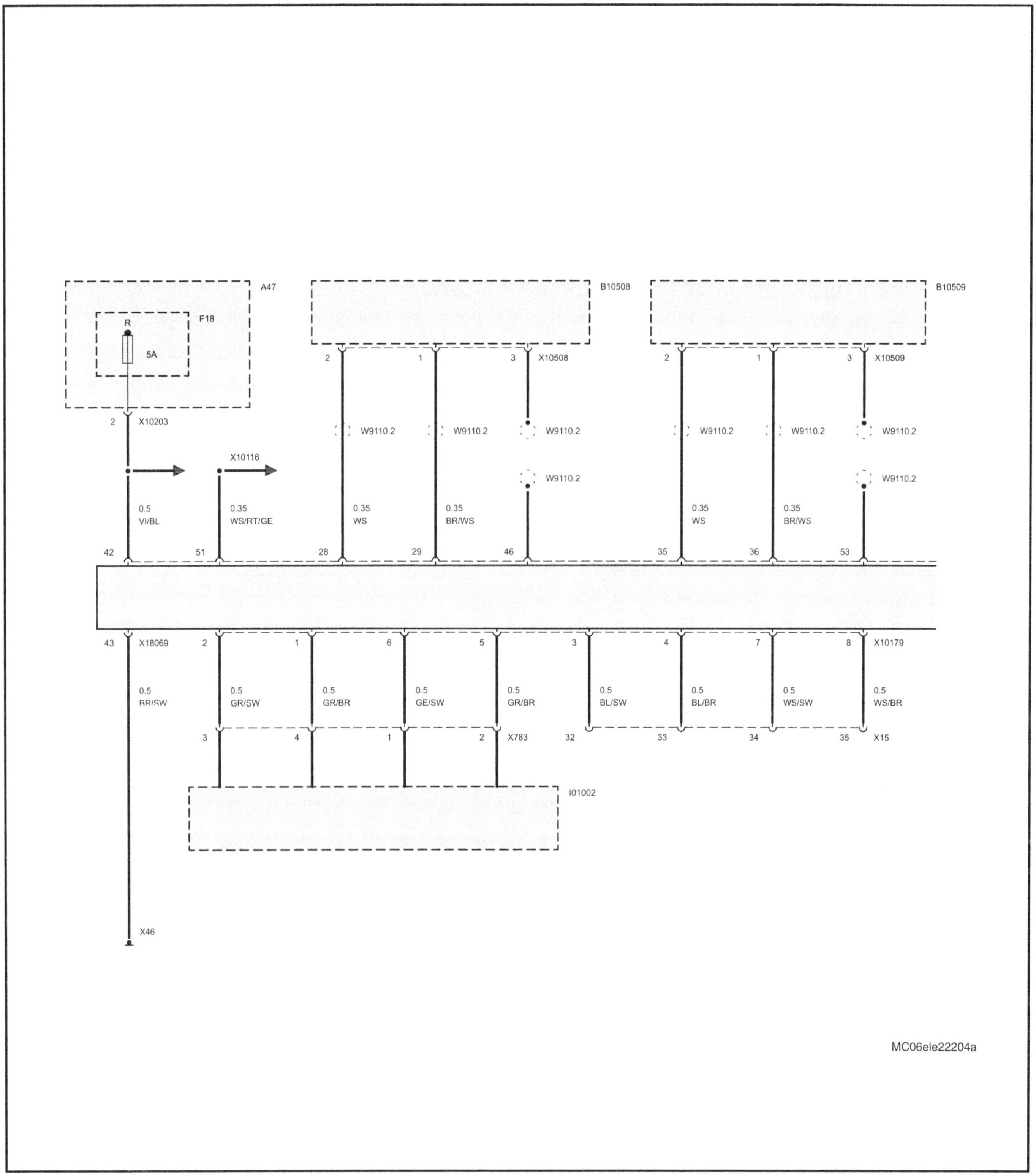

Airbag system
MRS 5
(page 2 of 3)

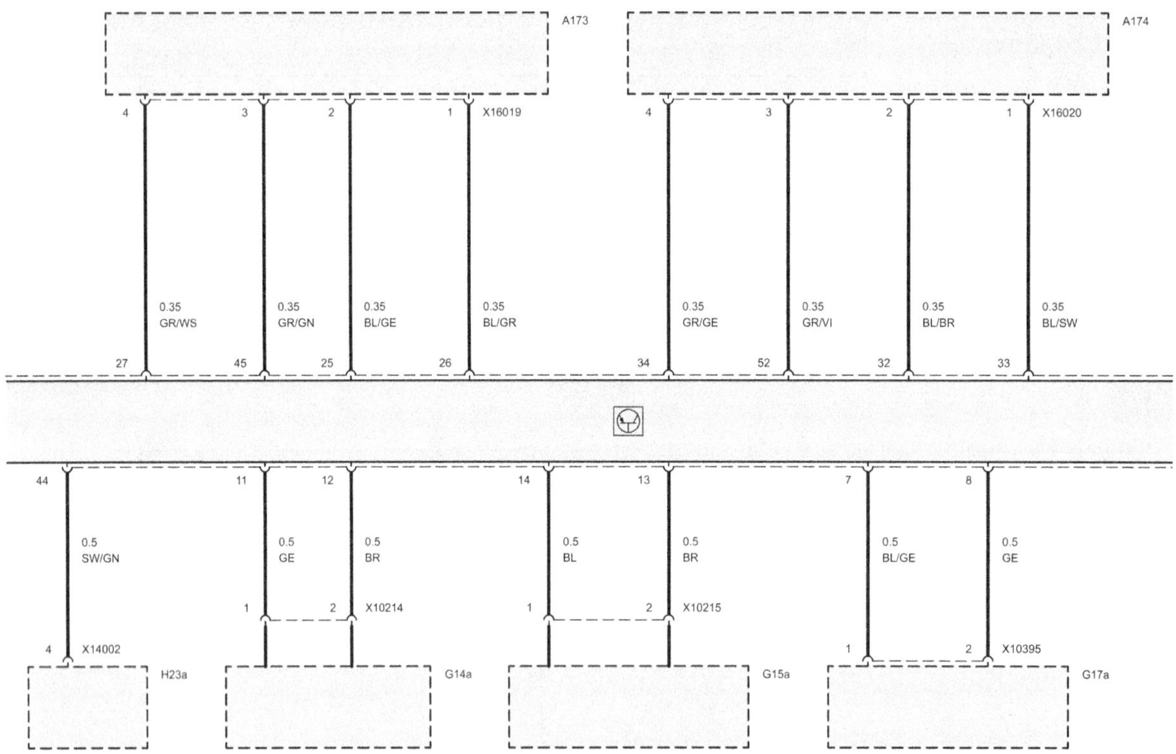

Airbag system
MRS 5
(page 3 of 3)

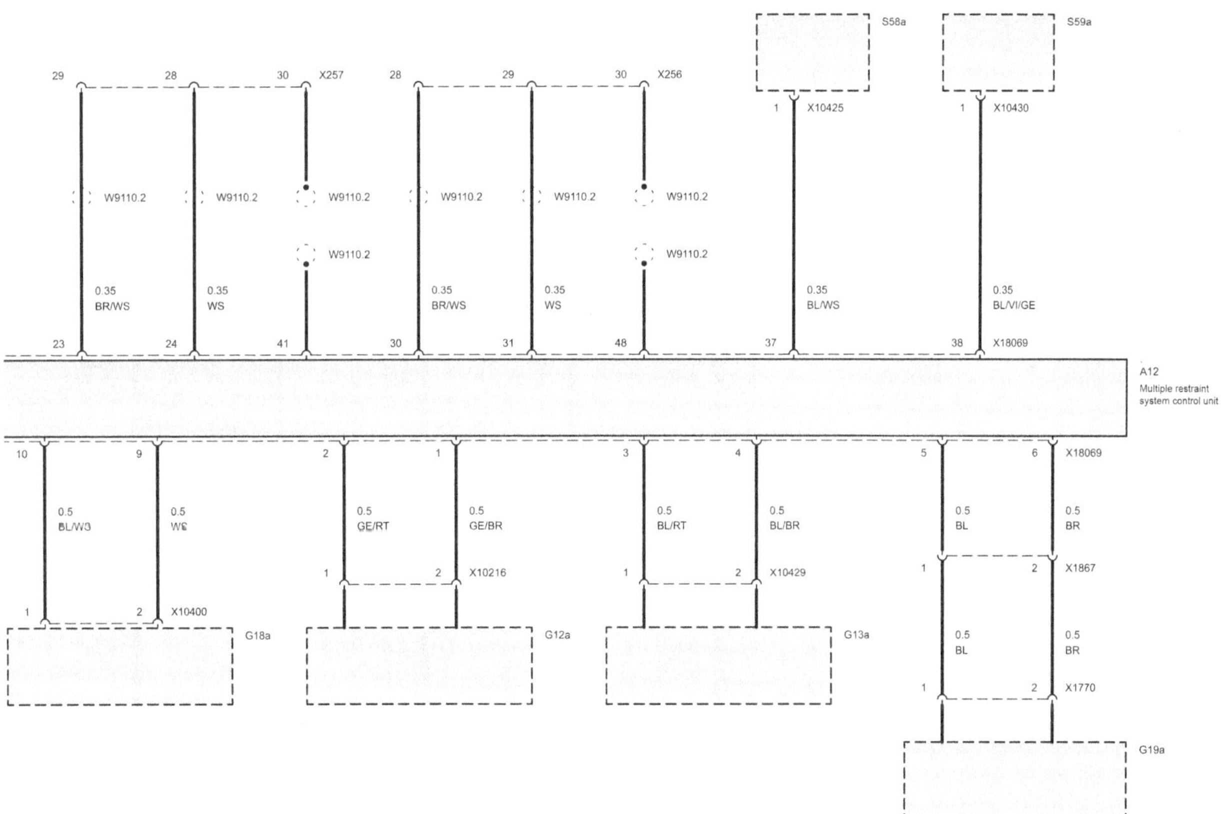

MC06ele22204c

Automatic transmission (Aisin)
Electronics (A7000) (GIU), Cooper S (up to Sept. 2005)
(page 1 of 3)

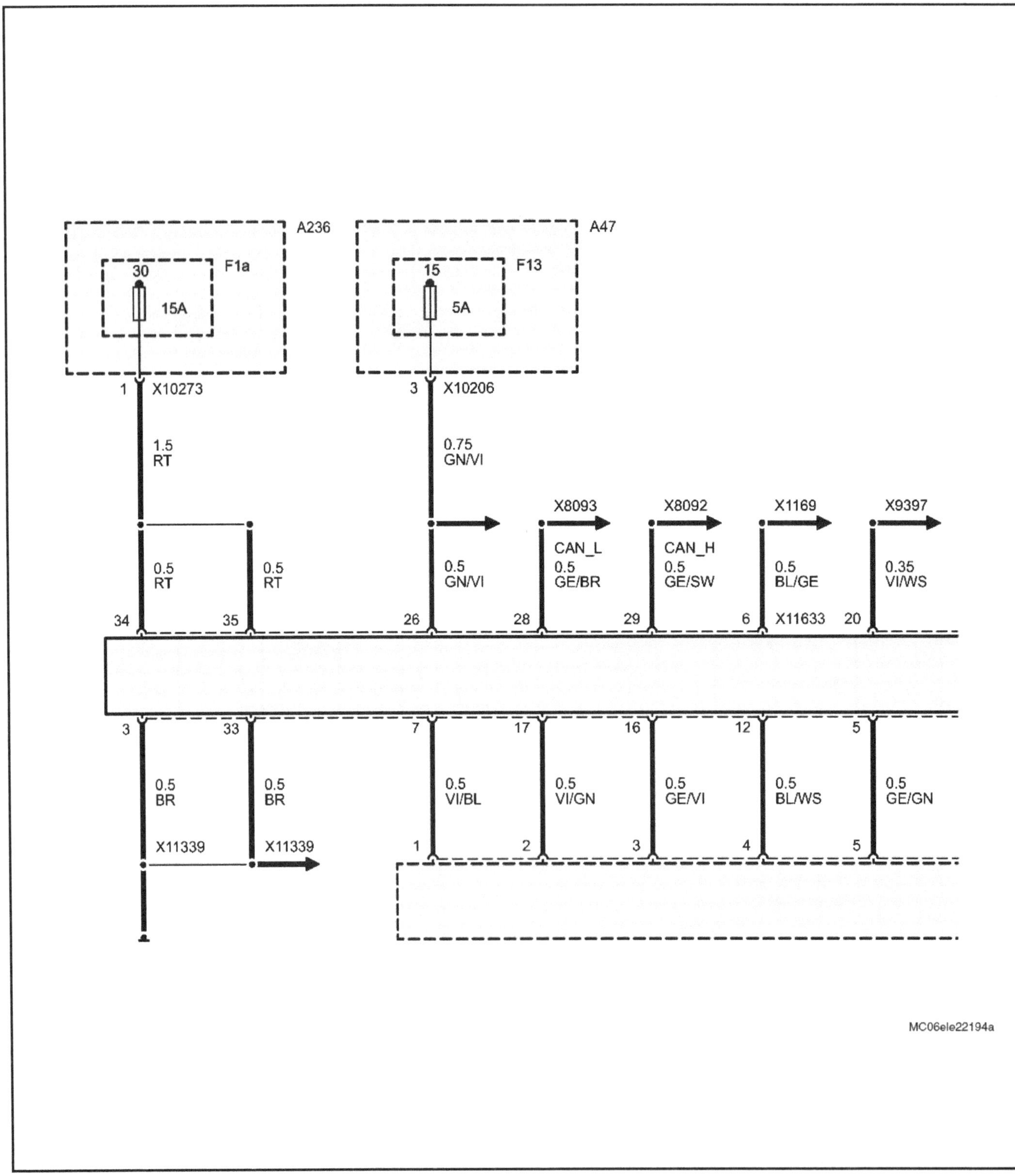

Automatic transmission (Aisin)
Electronics (A7000) (GIU), Cooper S (up to Sept. 2005)
(page 2 of 3)

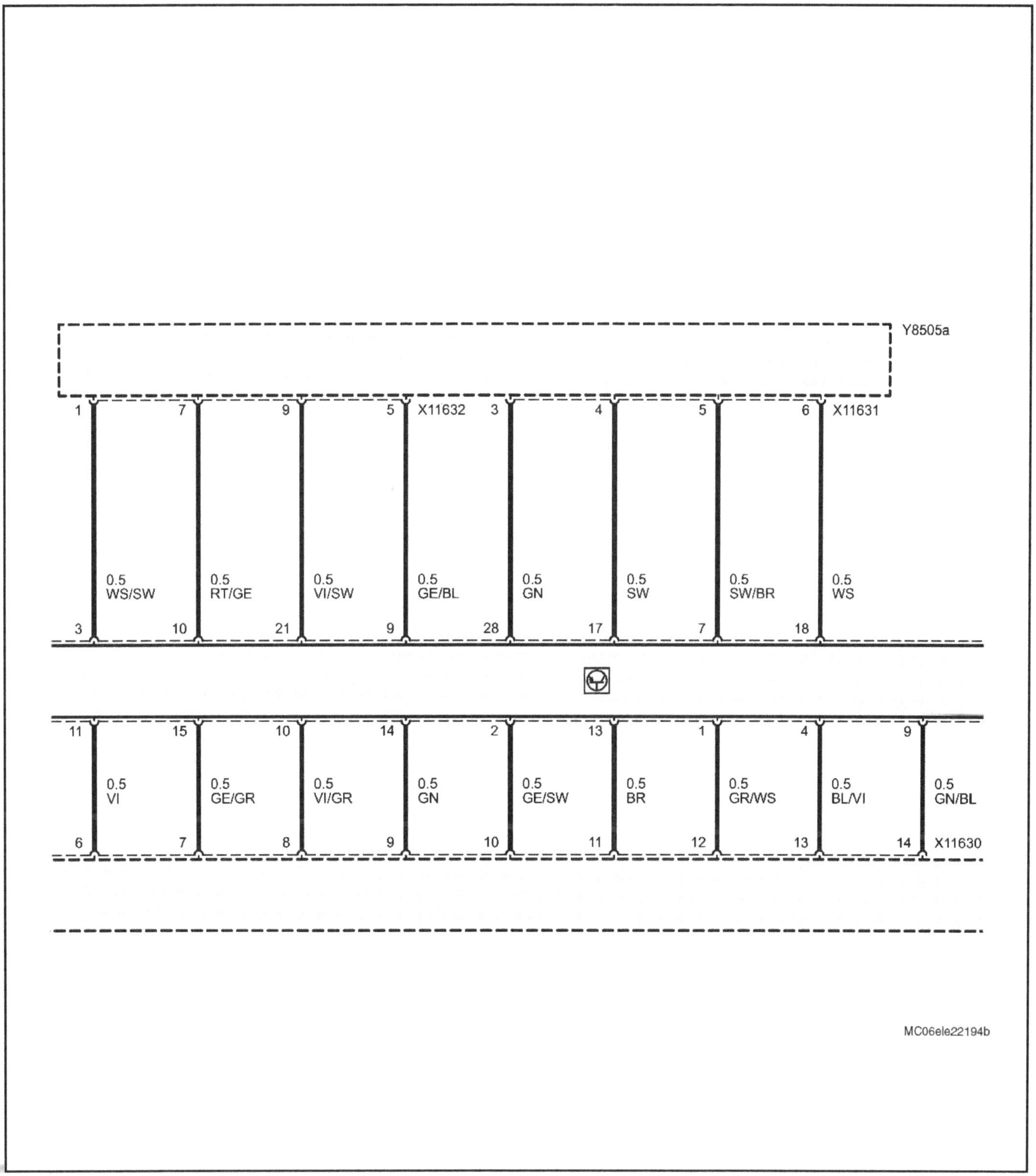

Automatic transmission (Aisin)
Electronics (A7000) (GIU), Cooper S (up to Sept. 2005)
(page 3 of 3)

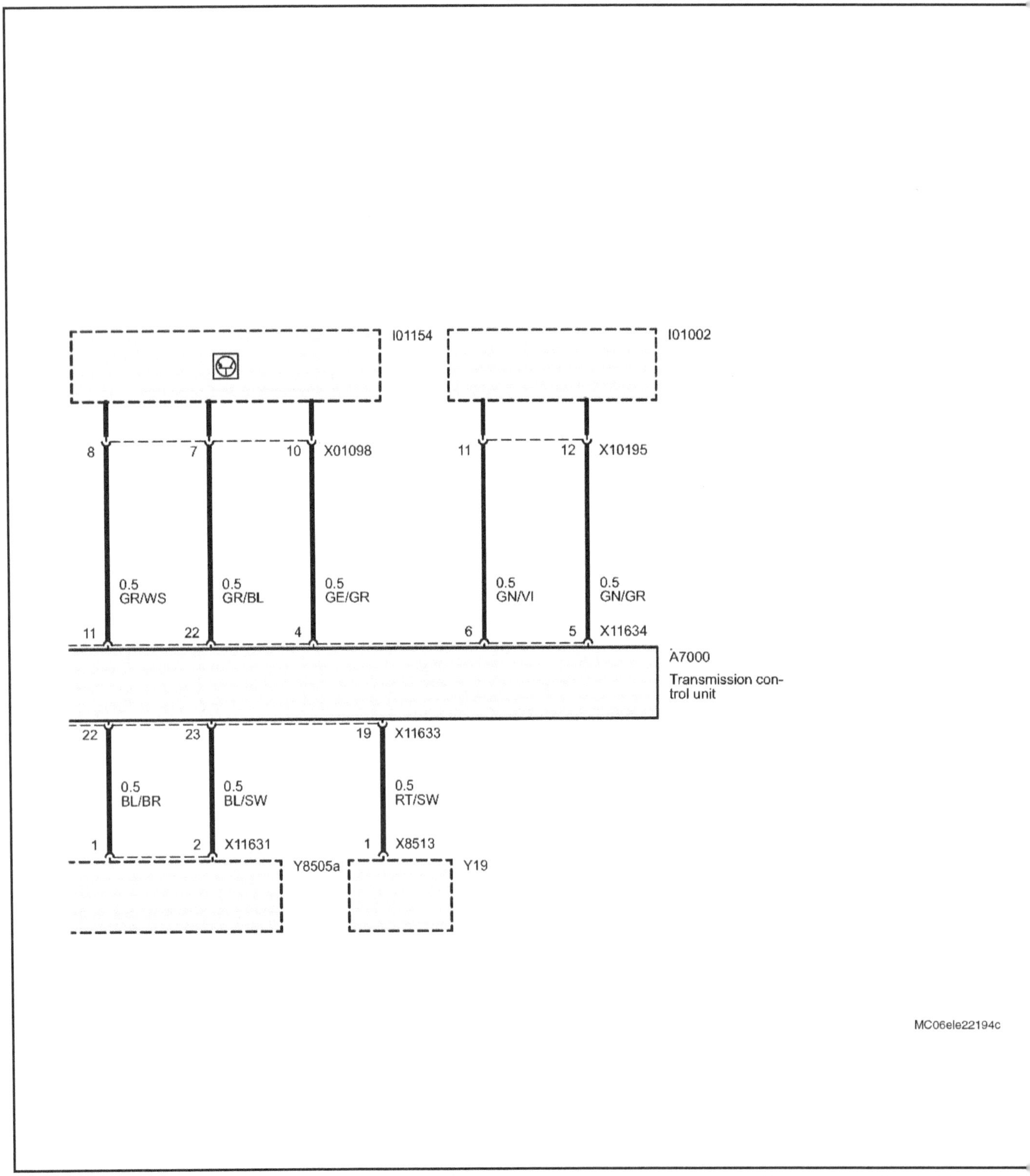

Automatic transmission (Aisin)
Electronics (A7000) (GIU), Cooper S (from Sept. 2005)
(page 1 of 3)

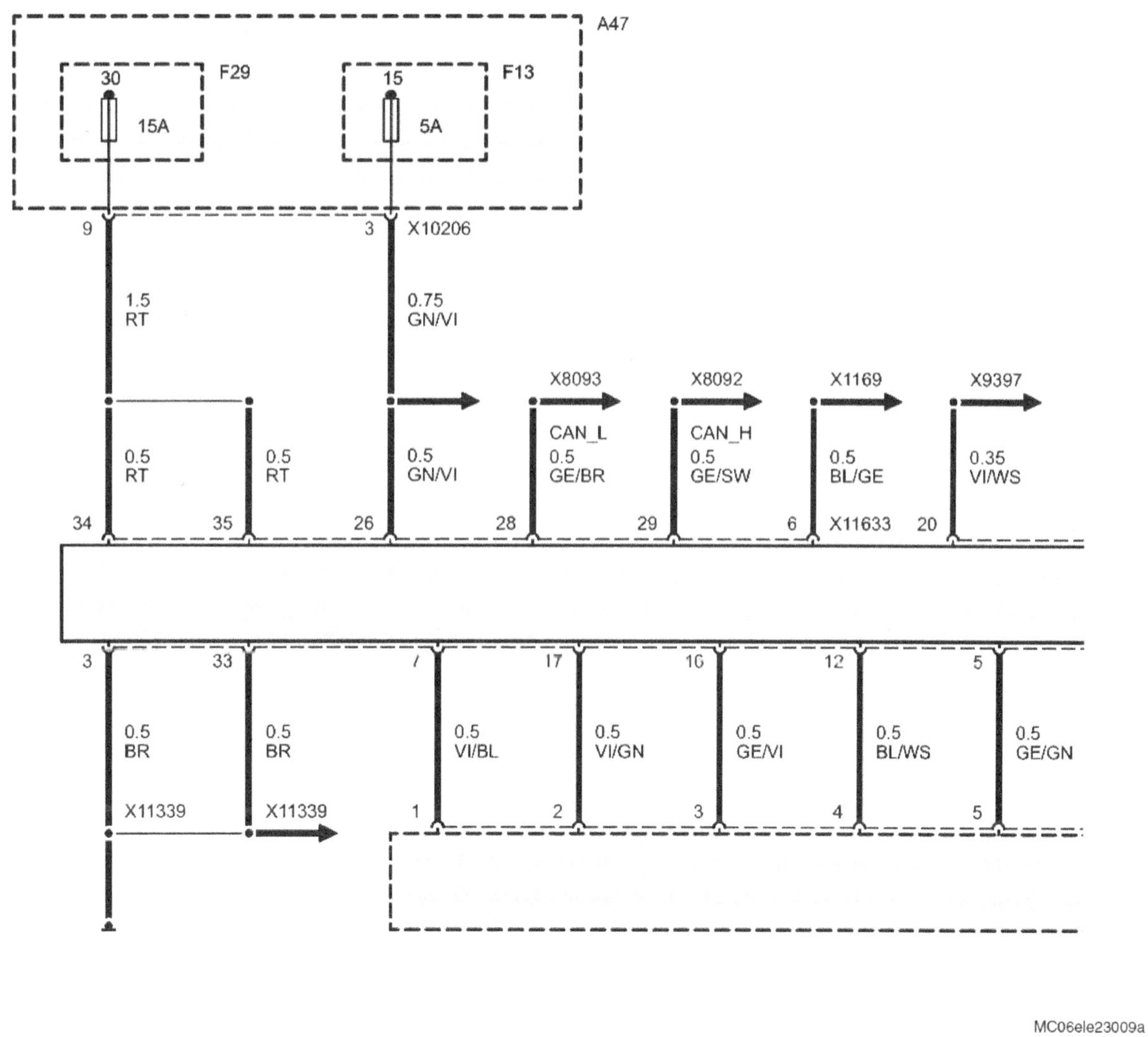

Automatic transmission (Aisin)
Electronics (A7000) (GIU), Cooper S (from Sept. 2005)
(page 2 of 3)

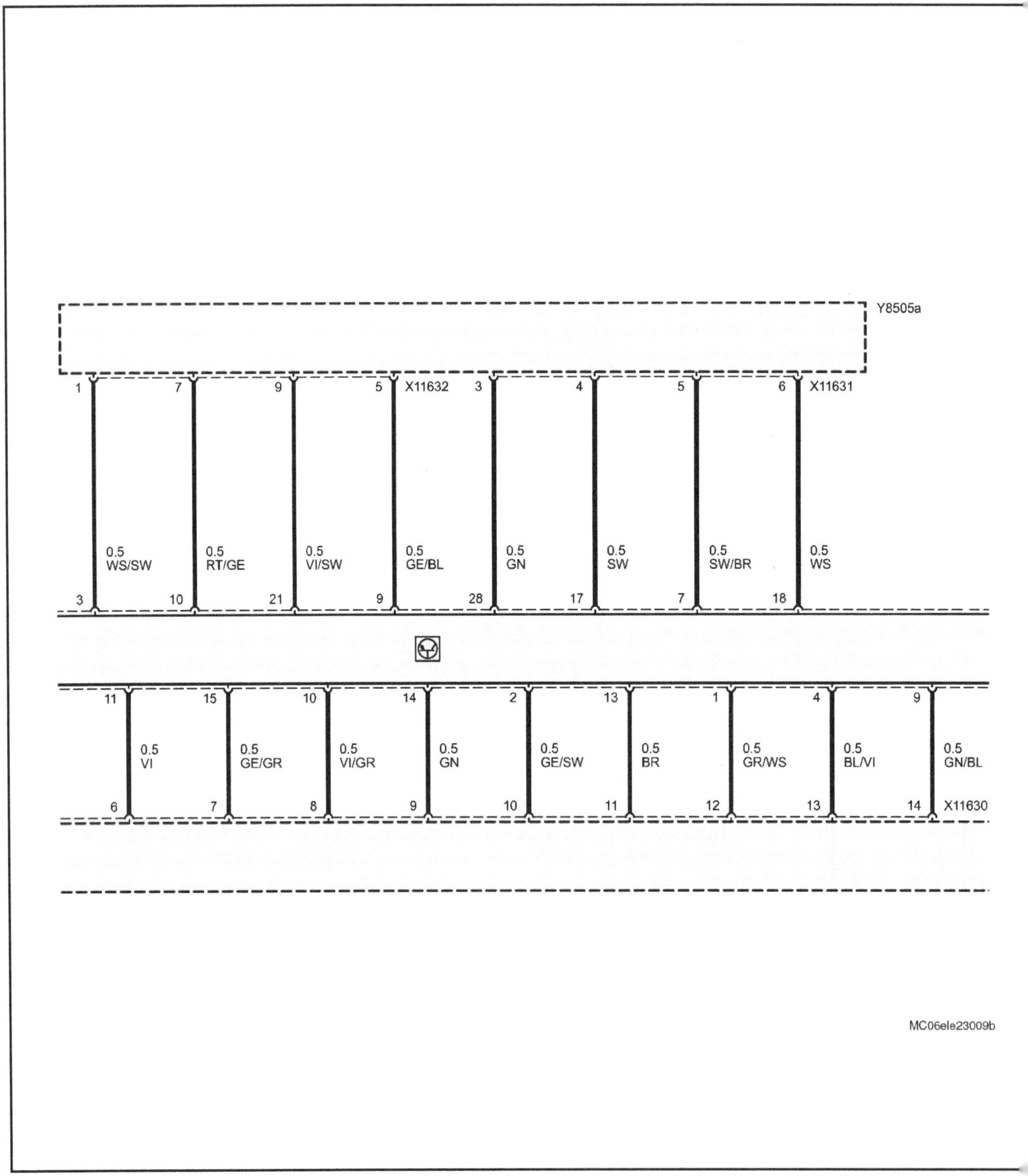

Automatic transmission (Aisin)
Electronics (A7000) (GIU), Cooper S (from Sept. 2005)
(page 3 of 3)

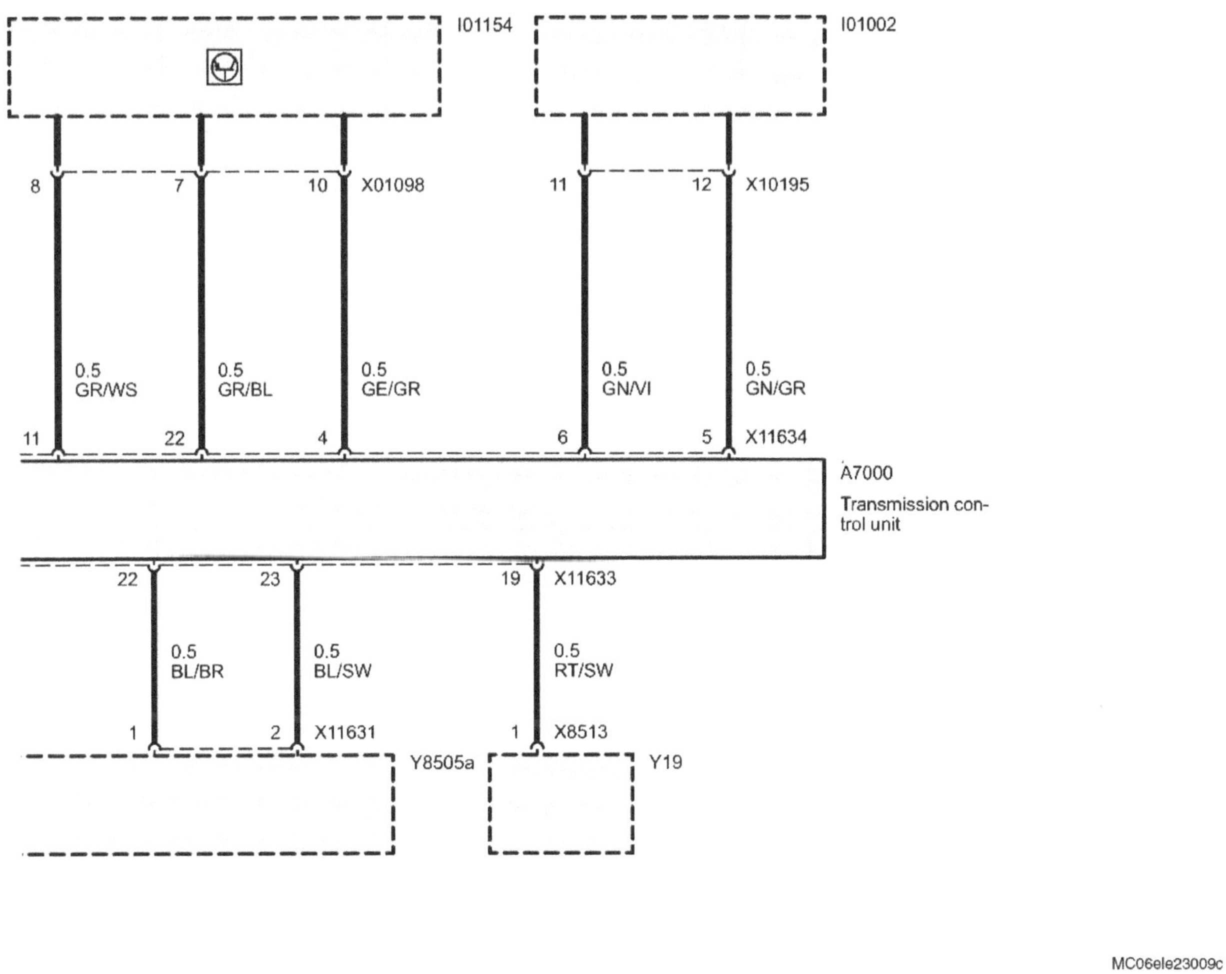

Automatic transmission (Aisin)
Gear position switch
(page 1 of 3)

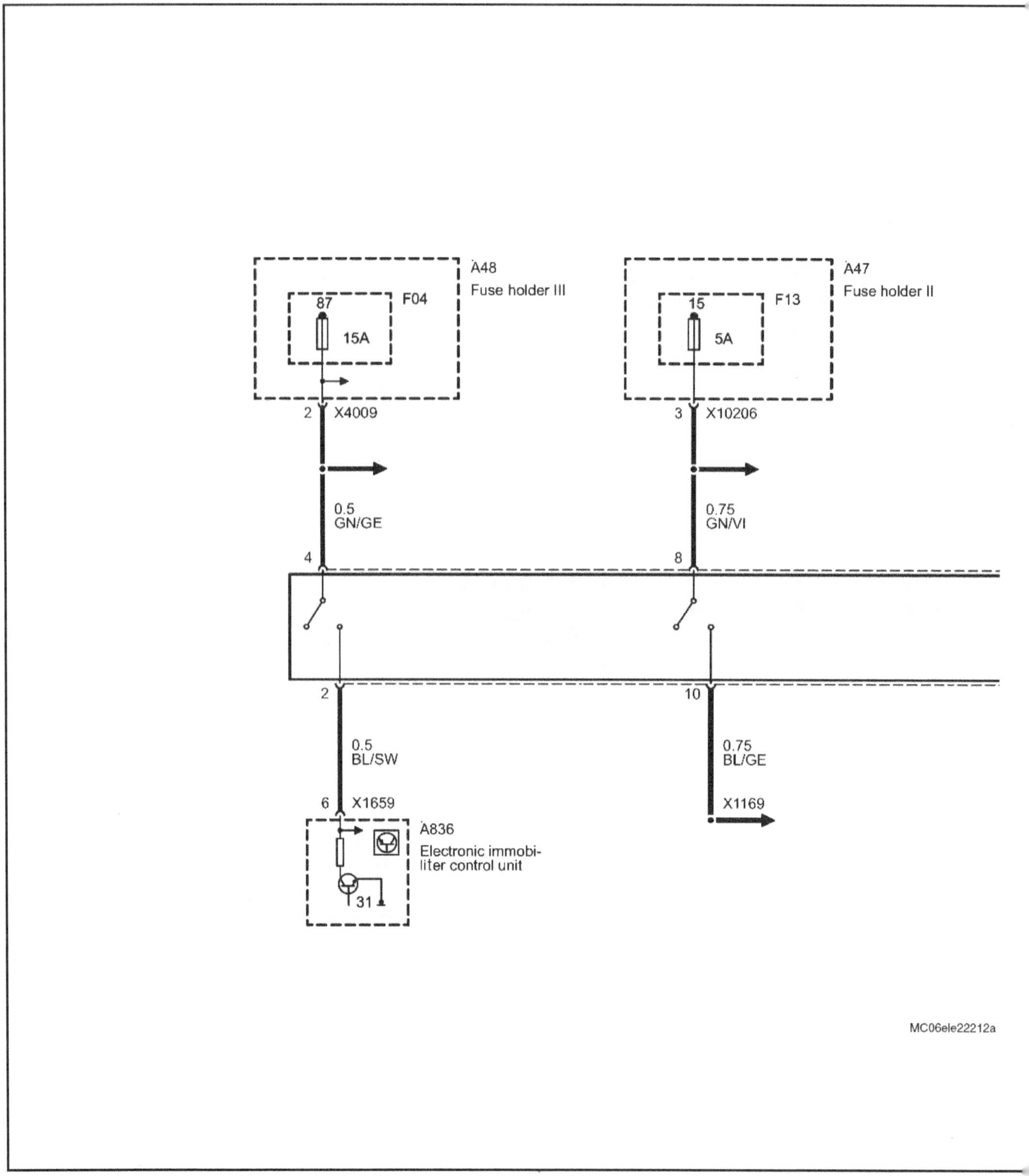

Automatic transmission (Aisin)
Gear position switch
(page 2 of 3)

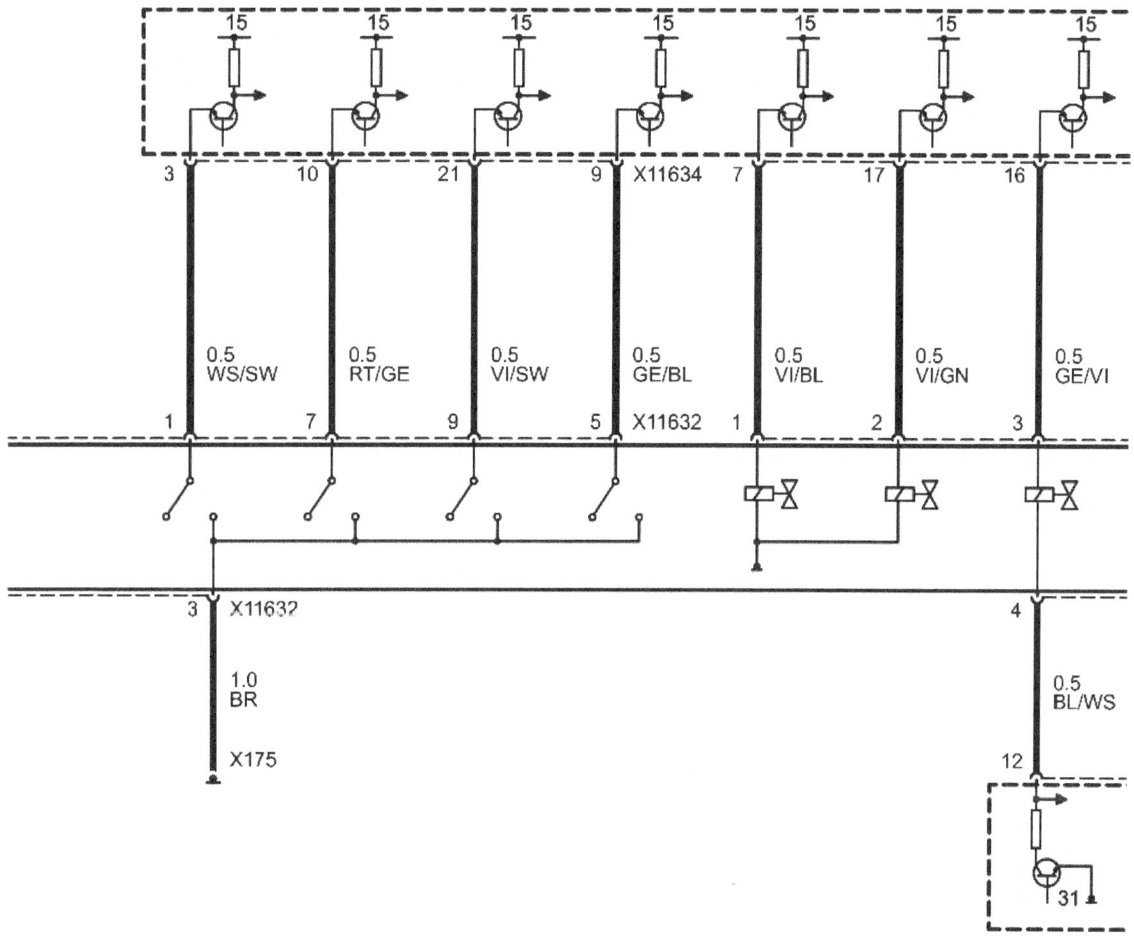

Automatic transmission (Aisin)
Gear position switch
(page 3 of 3)

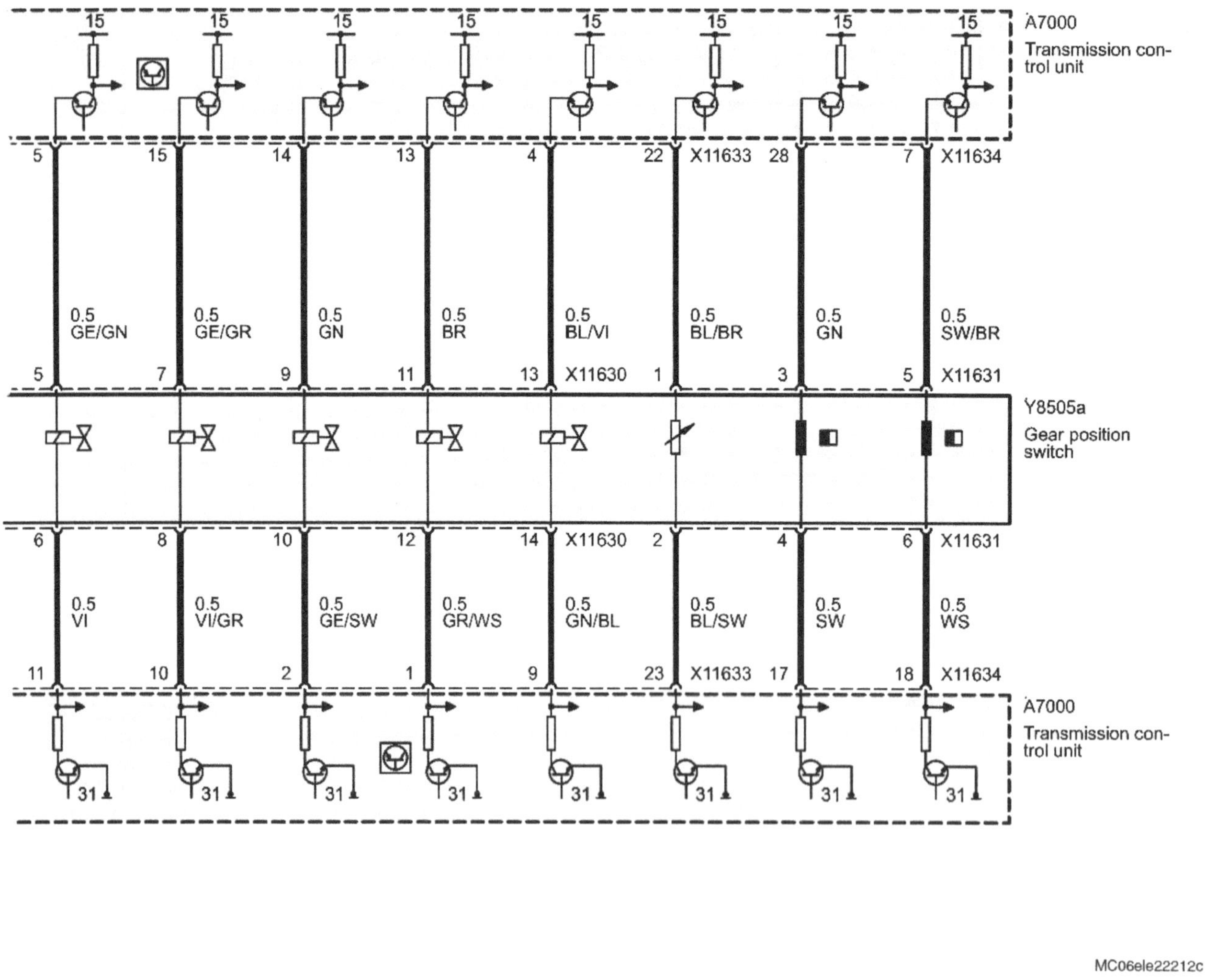

Automatic transmission (CVT)
Shift lock (up to Sept. 2005)
(page 1 of 1)

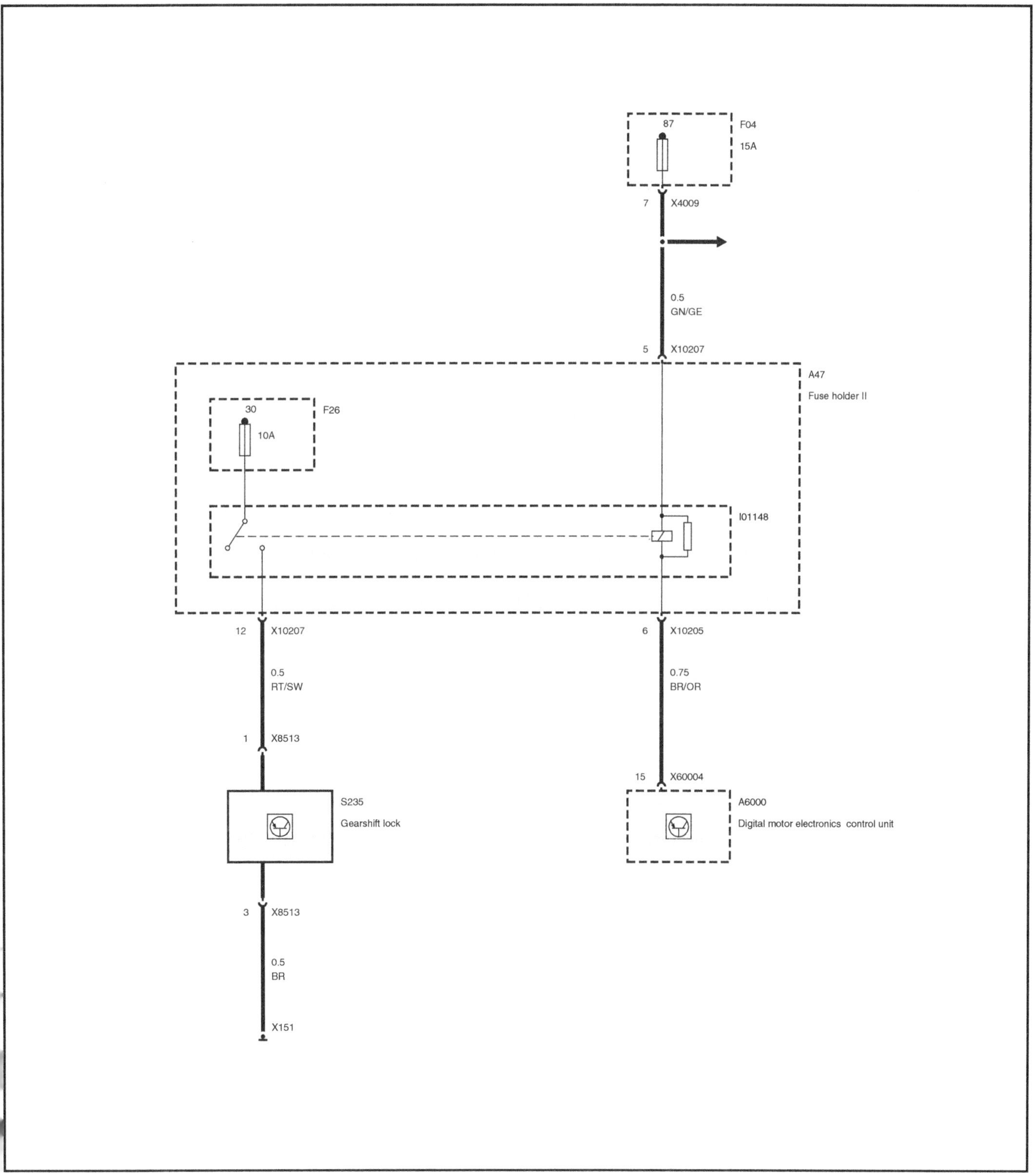

Automatic transmission (CVT)
Shift lock (from Sept. 2005)
(page 1 of 1)

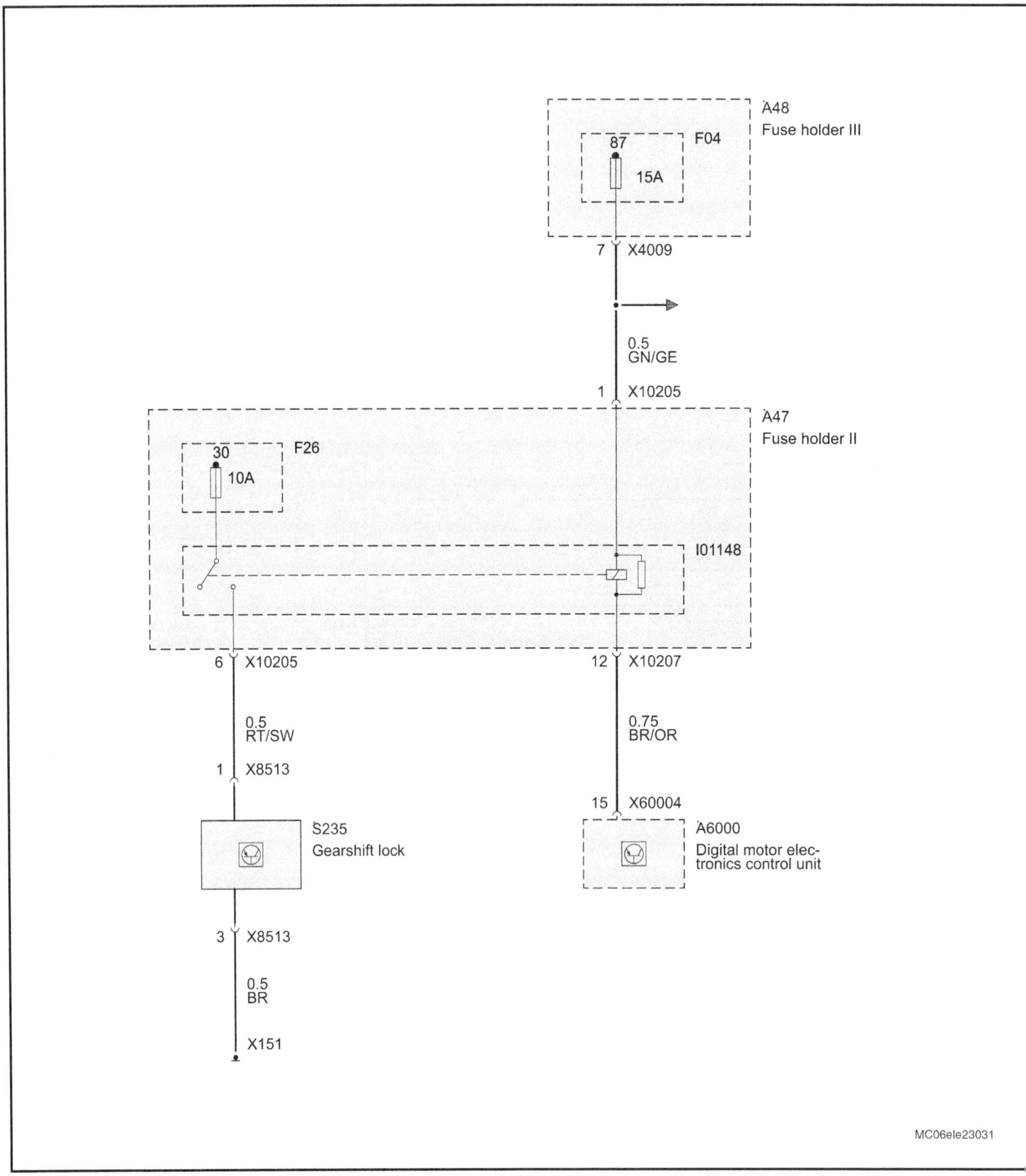

Automatic transmission (CVT)
Solenoid valve unit
(page 1 of 2)

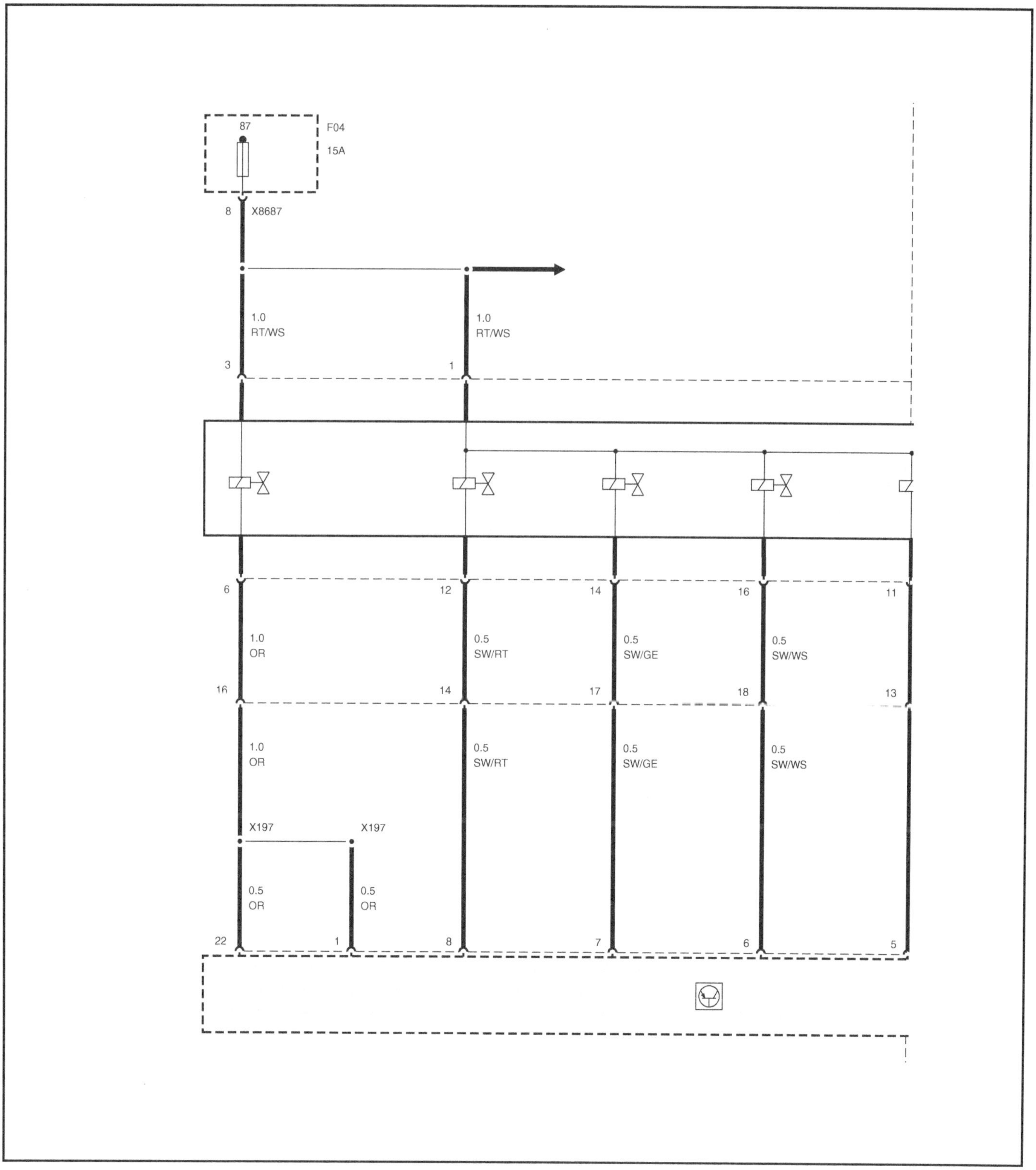

Automatic transmission (CVT)
Solenoid valve unit
(page 2 of 2)

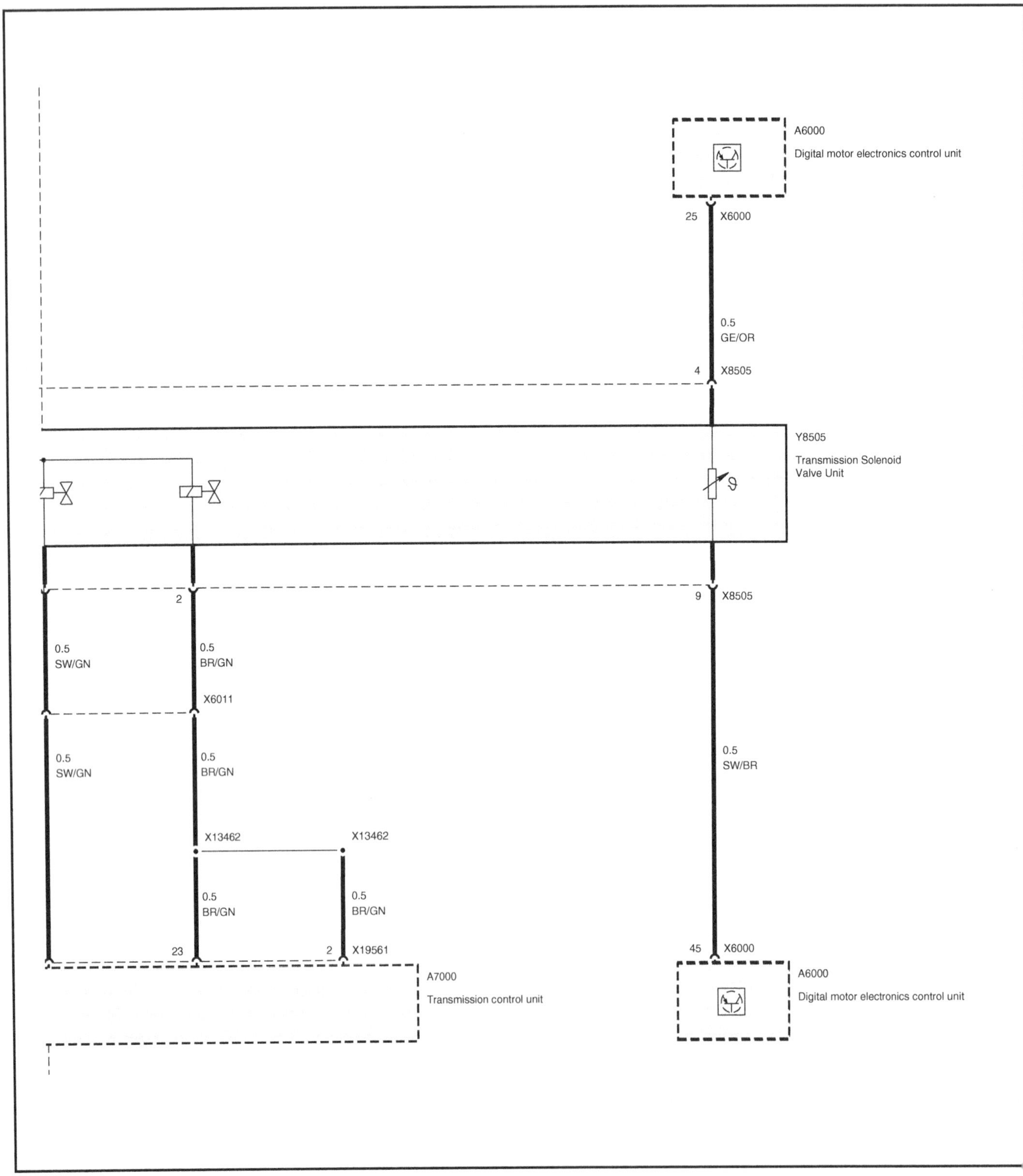

Automatic transmission (CVT)
Transmission speed sensor
(page 1 of 1)

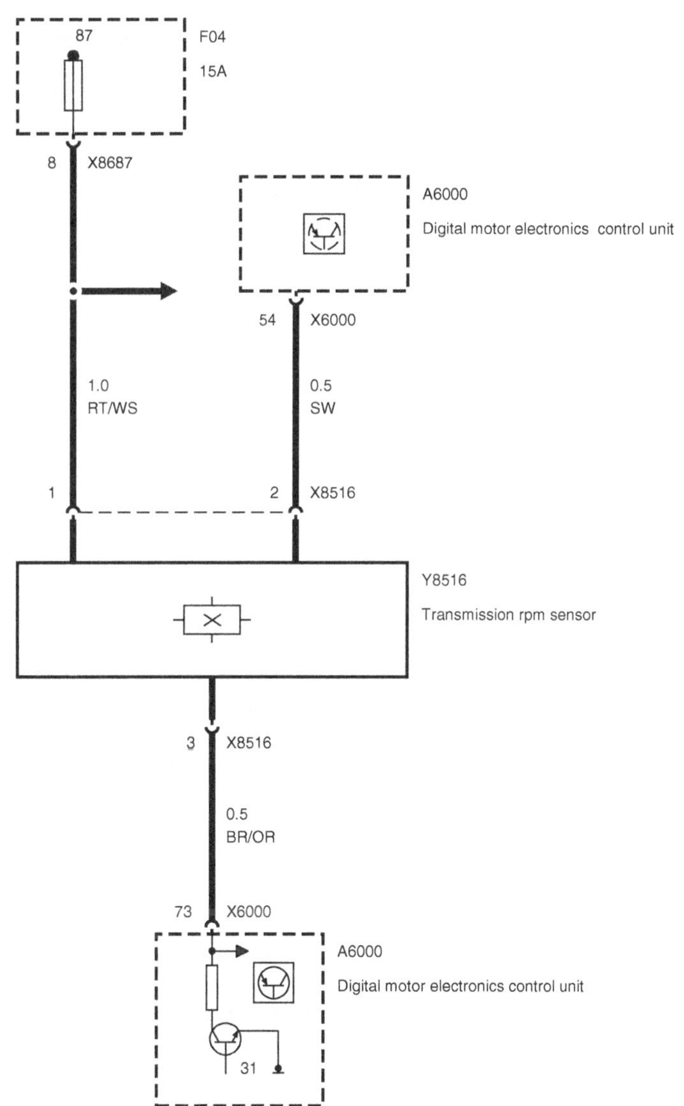

Automatic transmission (CVT)
Electronics (A7000) (GIU)
(page 1 of 3)

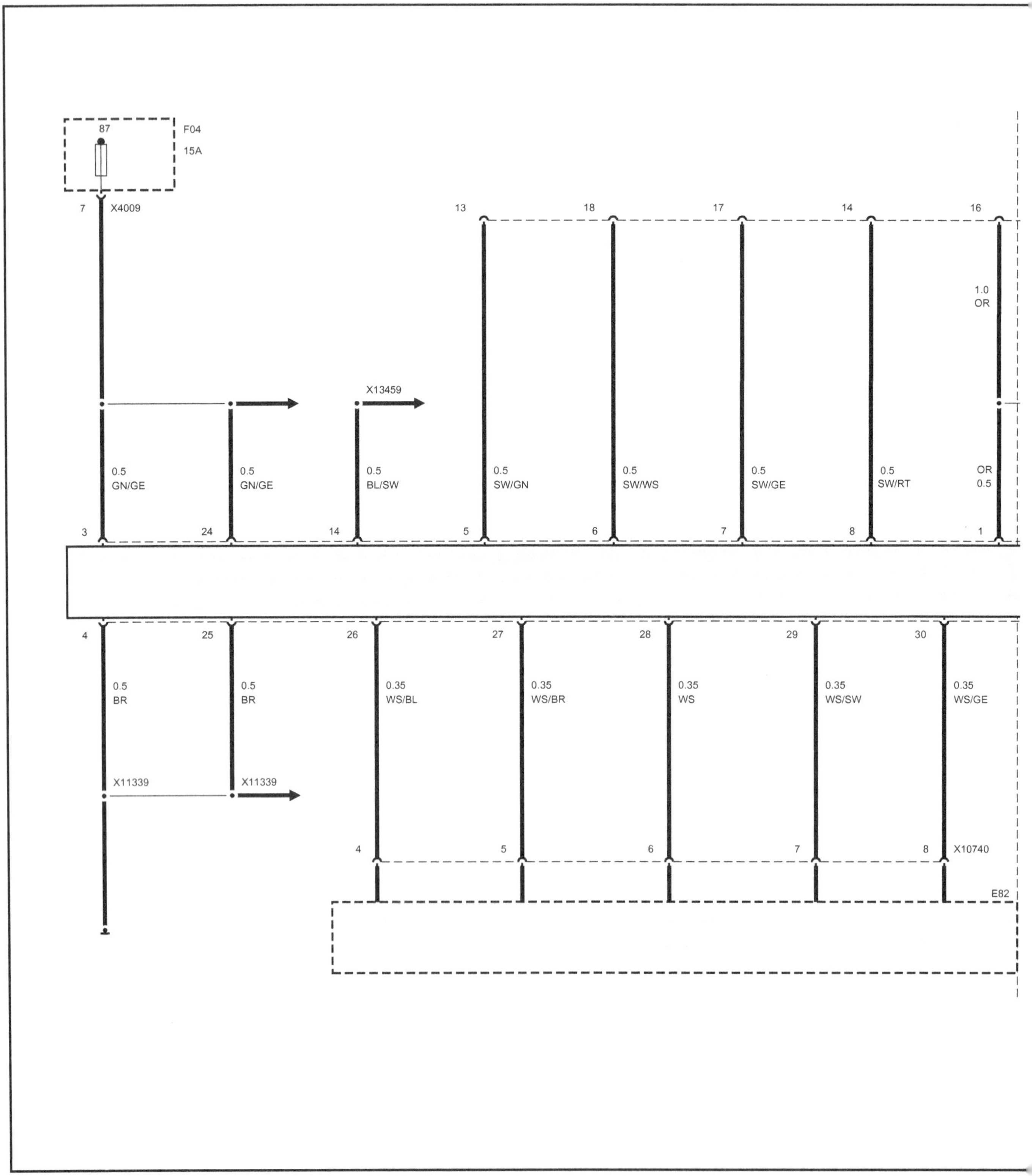

Automatic transmission (CVT)

Electronics (A7000) (GIU)

(page 2 of 3)

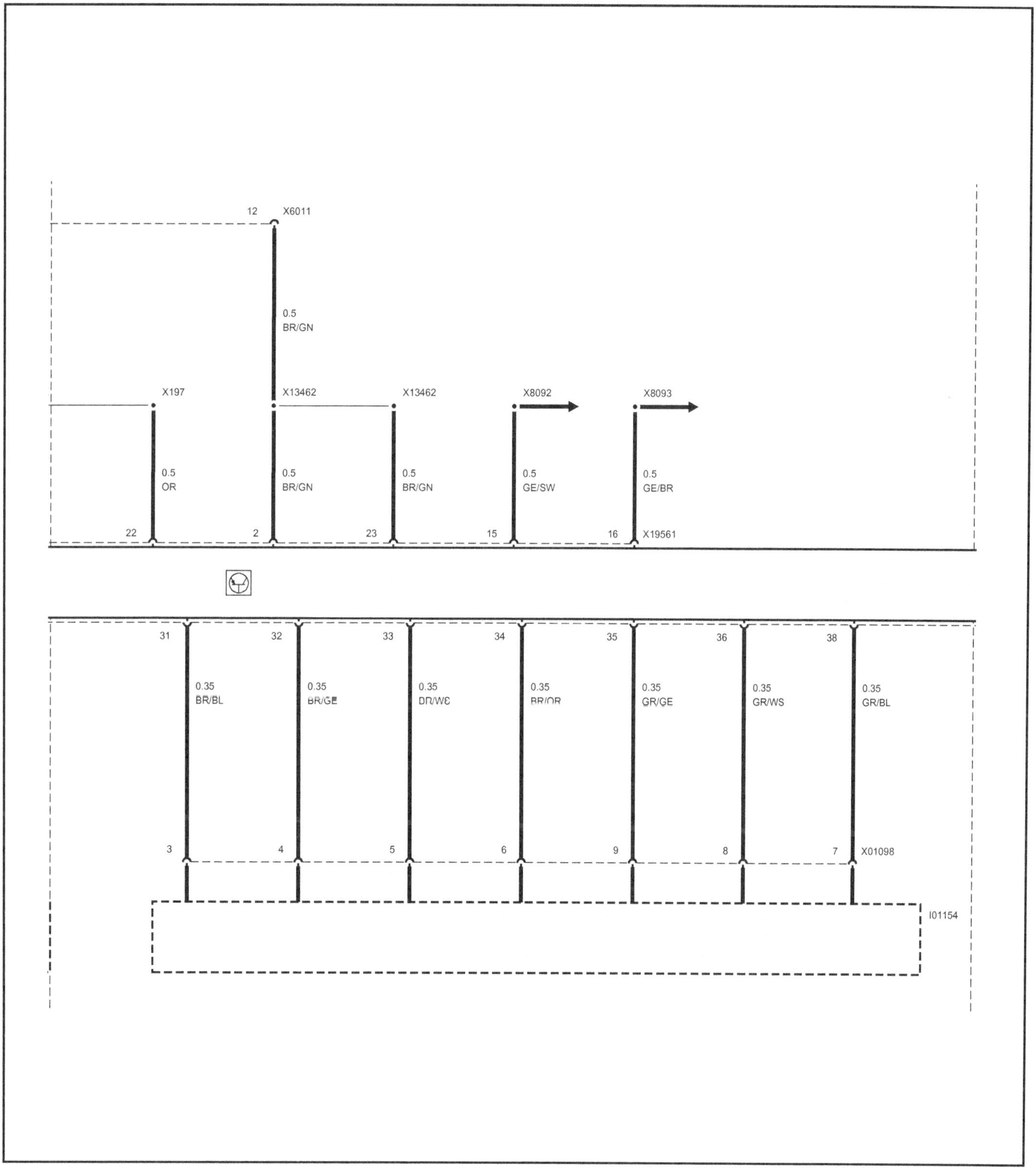

Automatic transmission (CVT)
Electronics (A7000) (GIU)
(page 3 of 3)

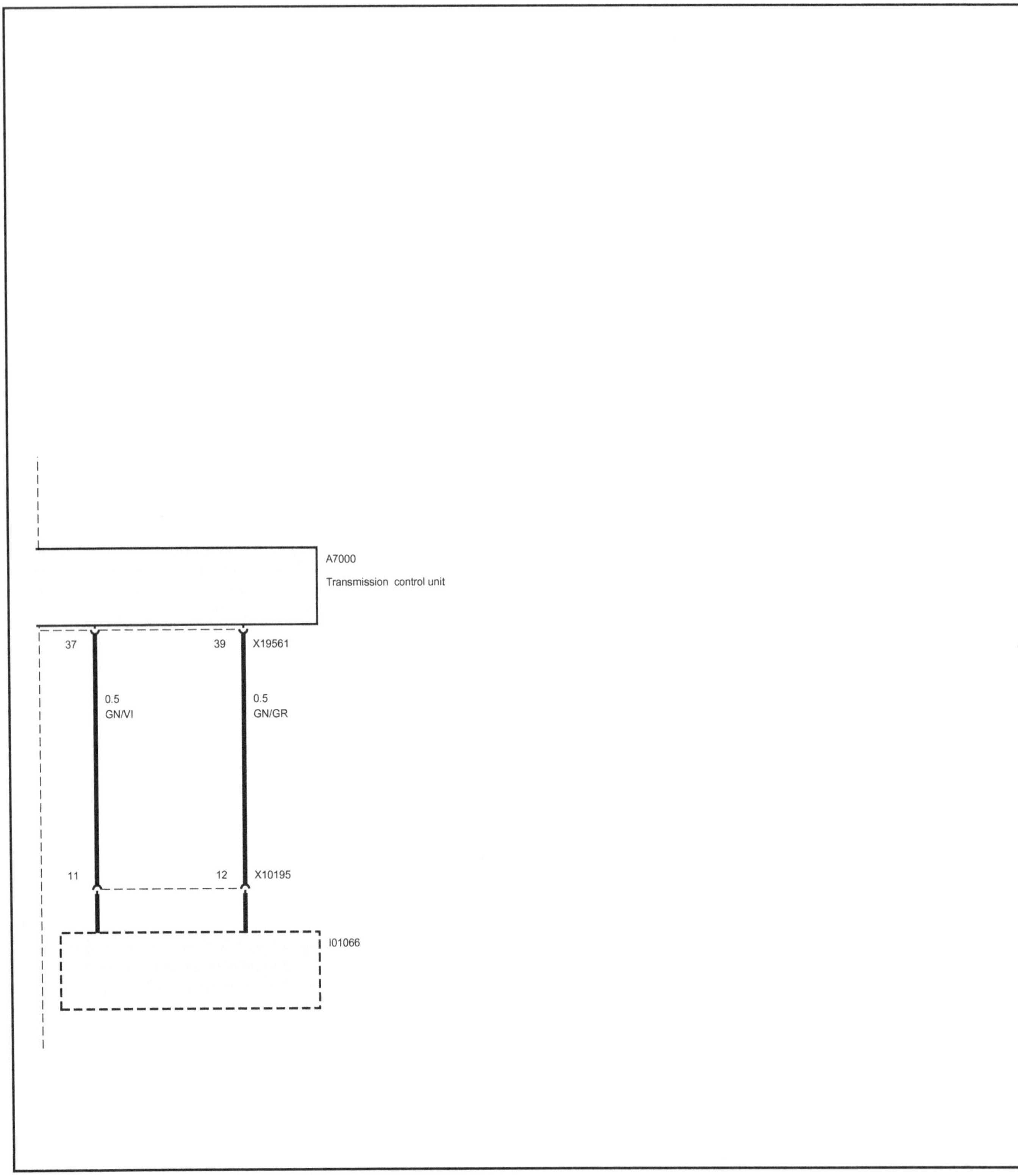

Automatic transmission (CVT)
Gearshift and electronics (A7000) (GIU)
(page 1 of 3)

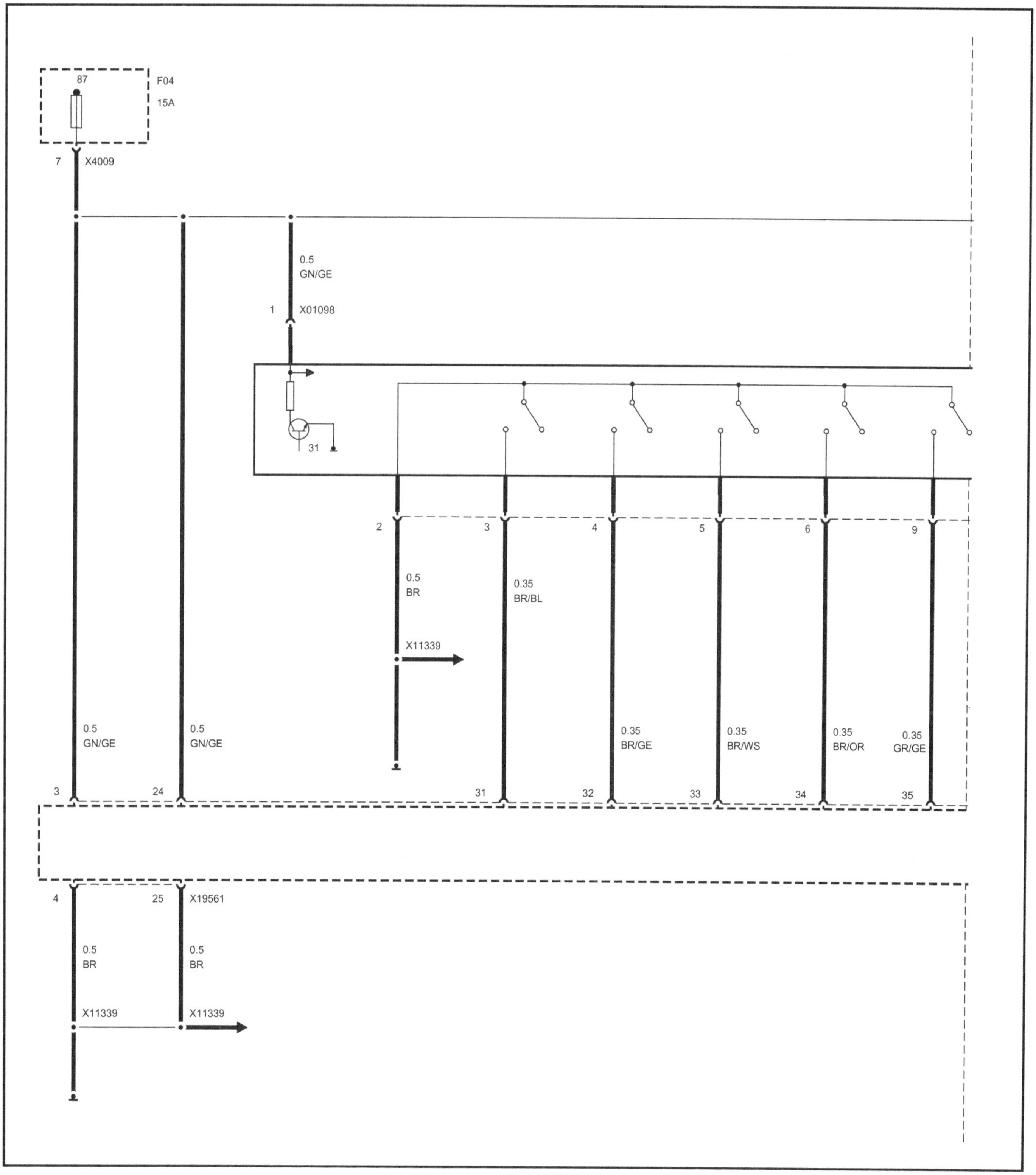

Automatic transmission (CVT)
Gearshift and electronics (A7000) (GIU)
(page 2 of 3)

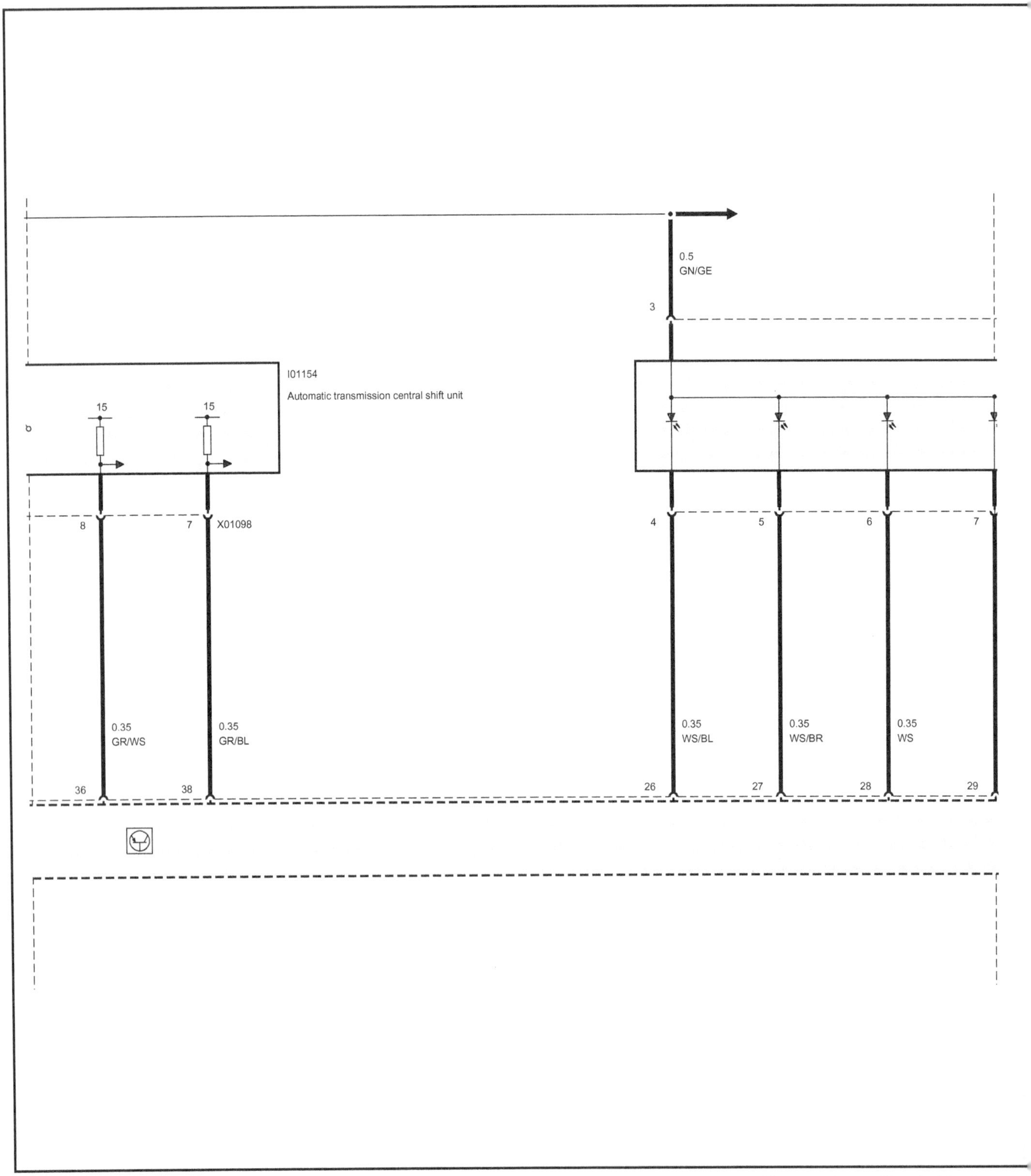

Automatic transmission (CVT)
Gearshift and electronics (A7000) (GIU)
(page 3 of 3)

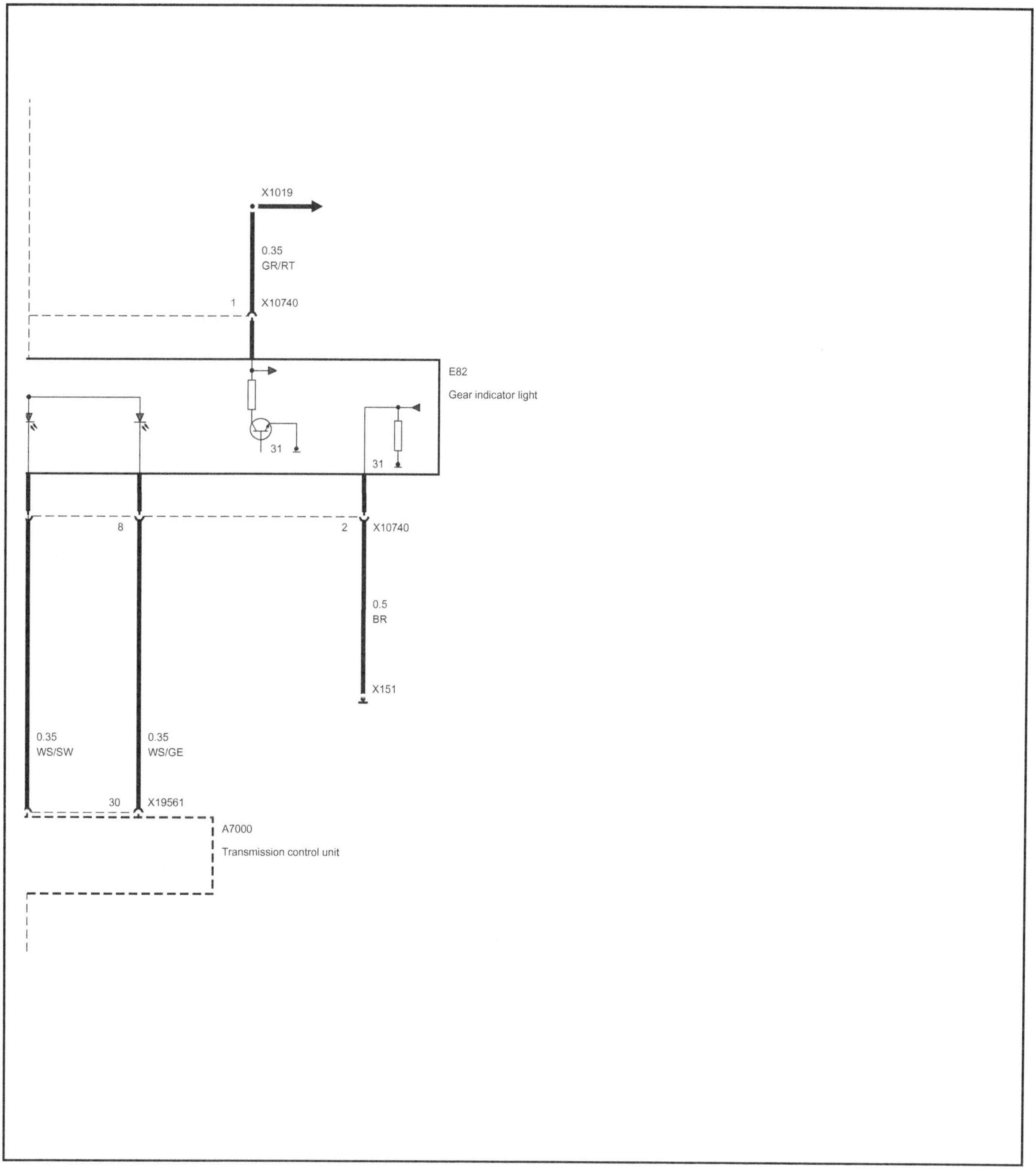

Battery (G1)
Cooper
(page 1 of 2)

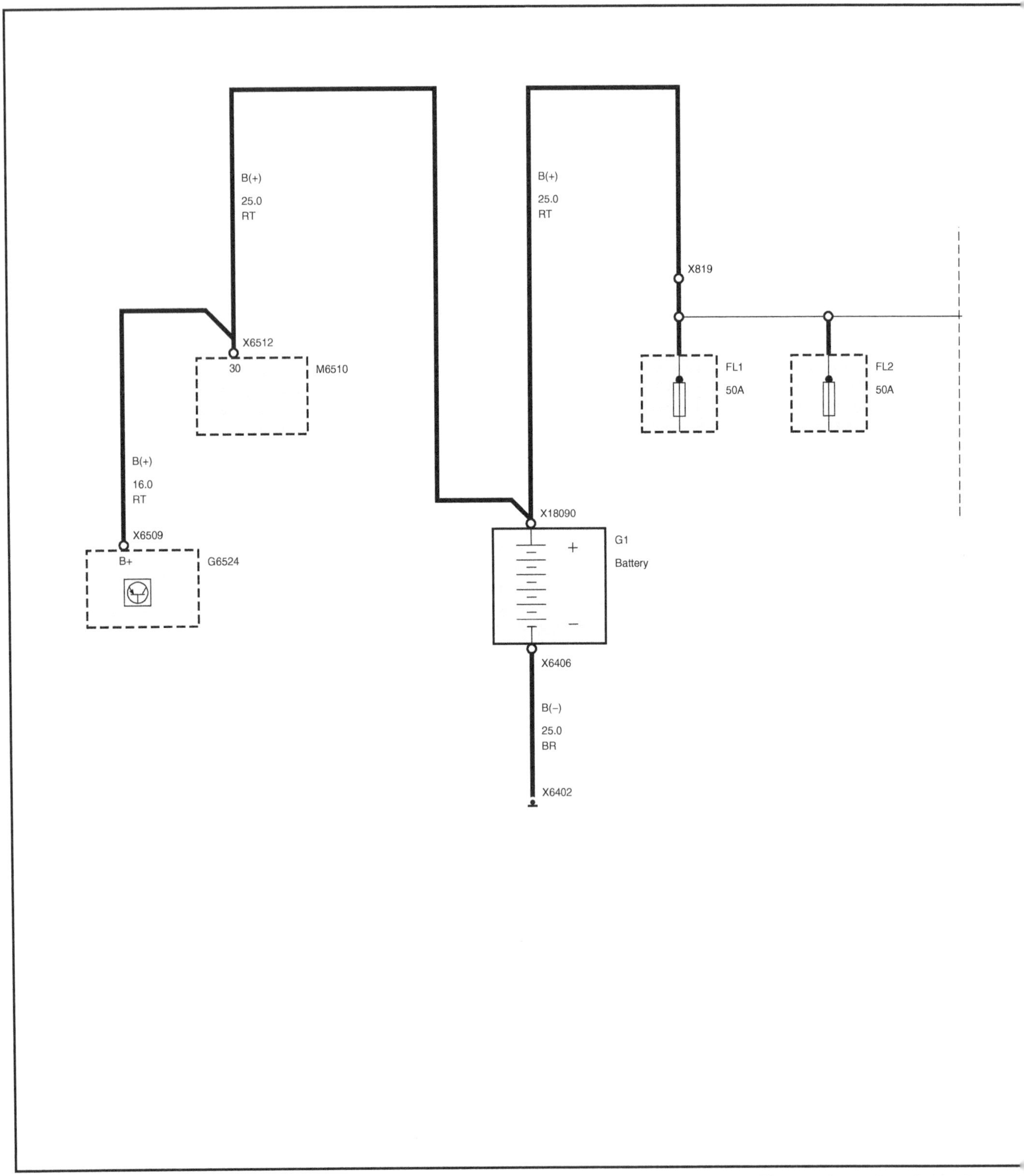

Battery (G1)
Cooper
(page 2 of 2)

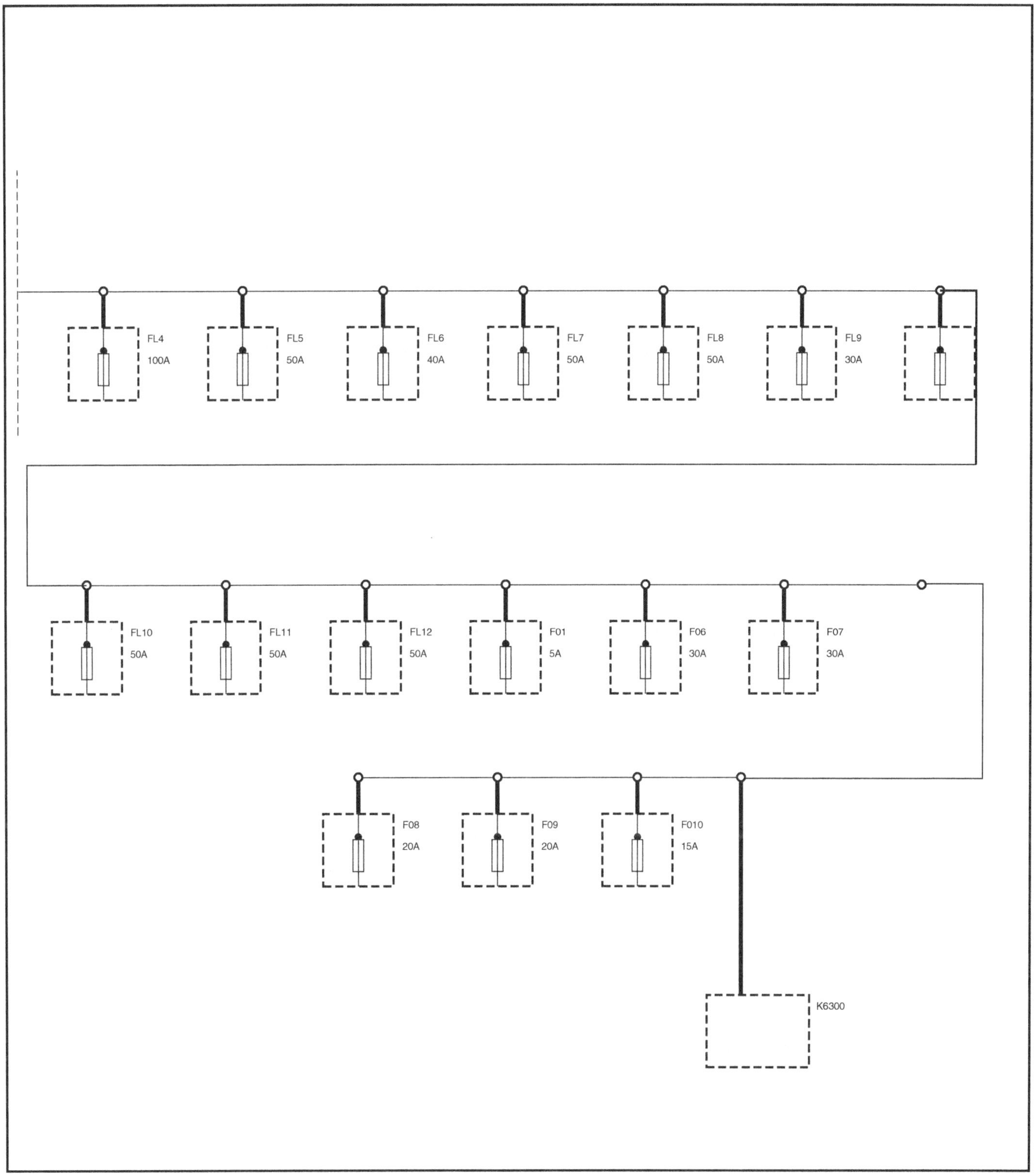

Battery (G1)
Cooper S
(page 1 of 3)

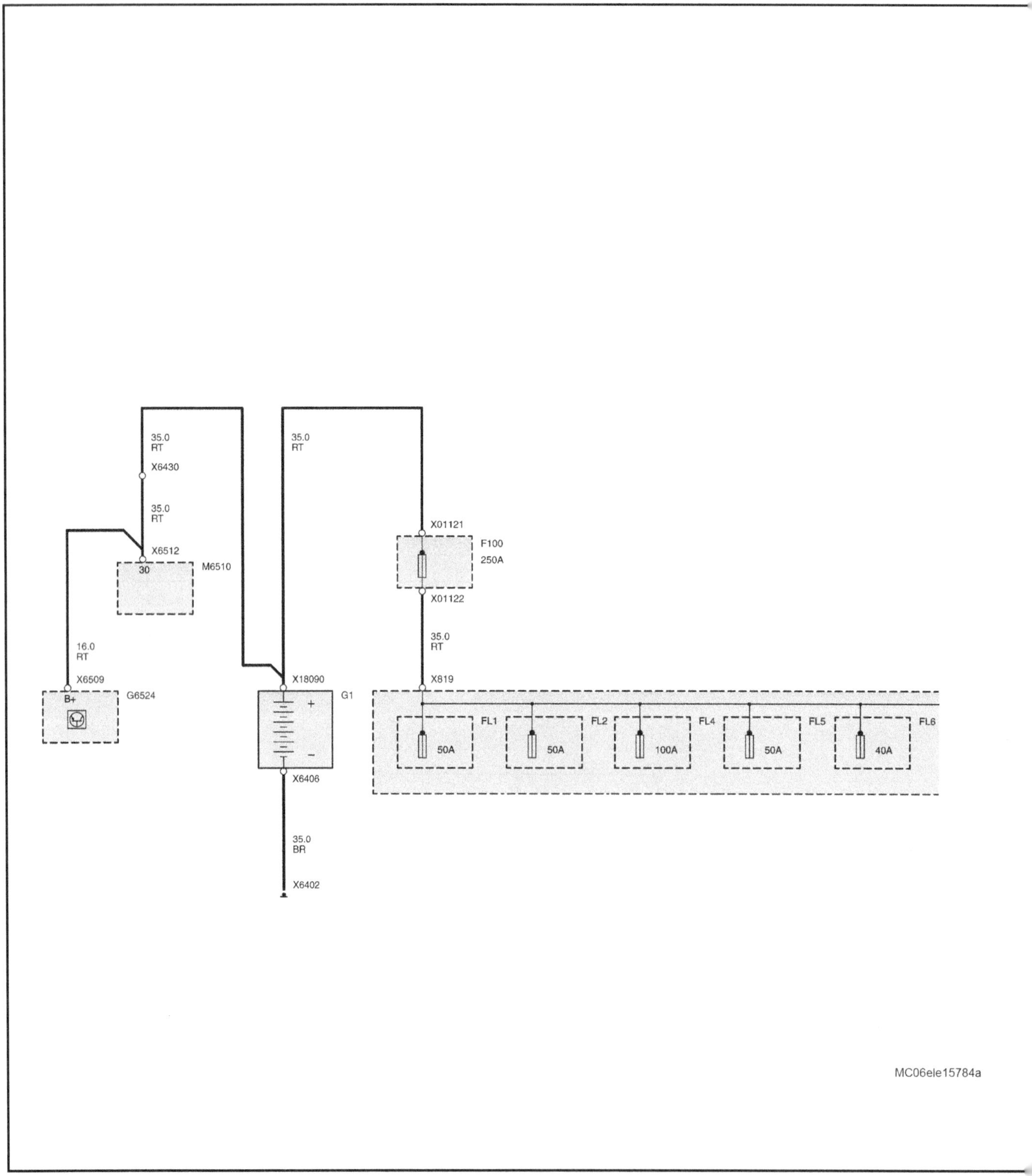

Battery (G1)
Cooper S
(page 2 of 3)

Battery (G1)
Cooper S
(page 3 of 3)

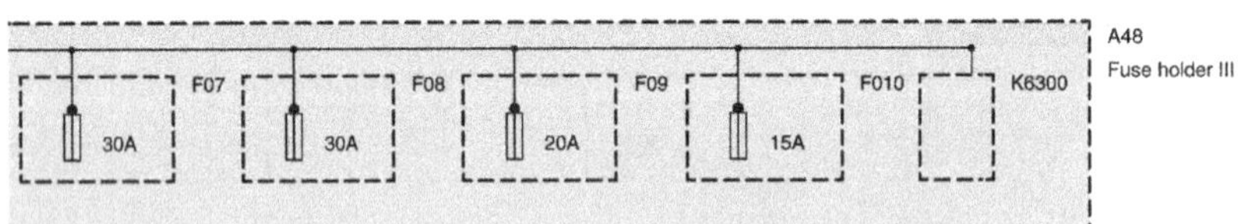

MC06ele15784c

Brake light / clutch switches
(page 1 of 3)

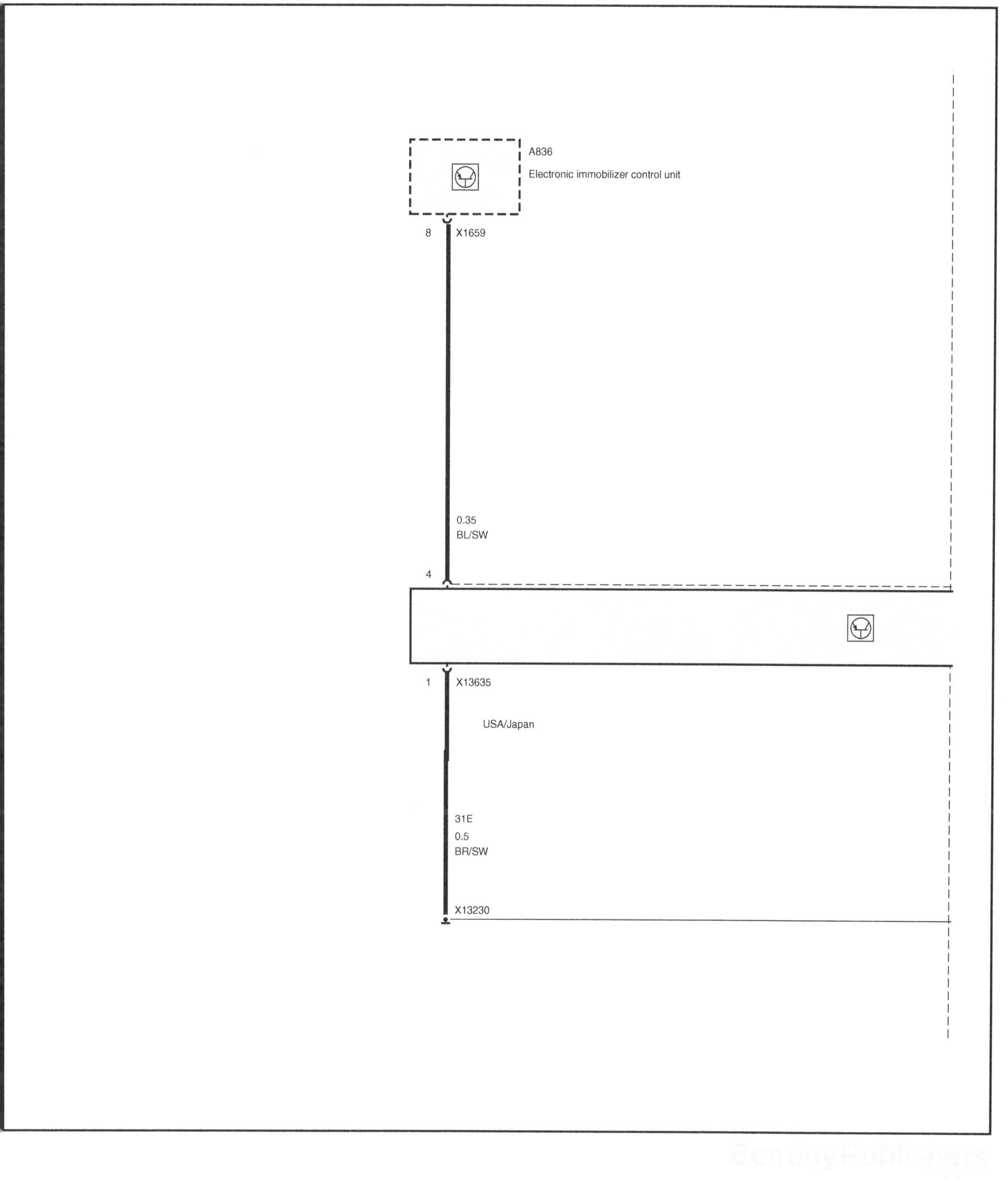

Brake light / clutch switches

(page 2 of 3)

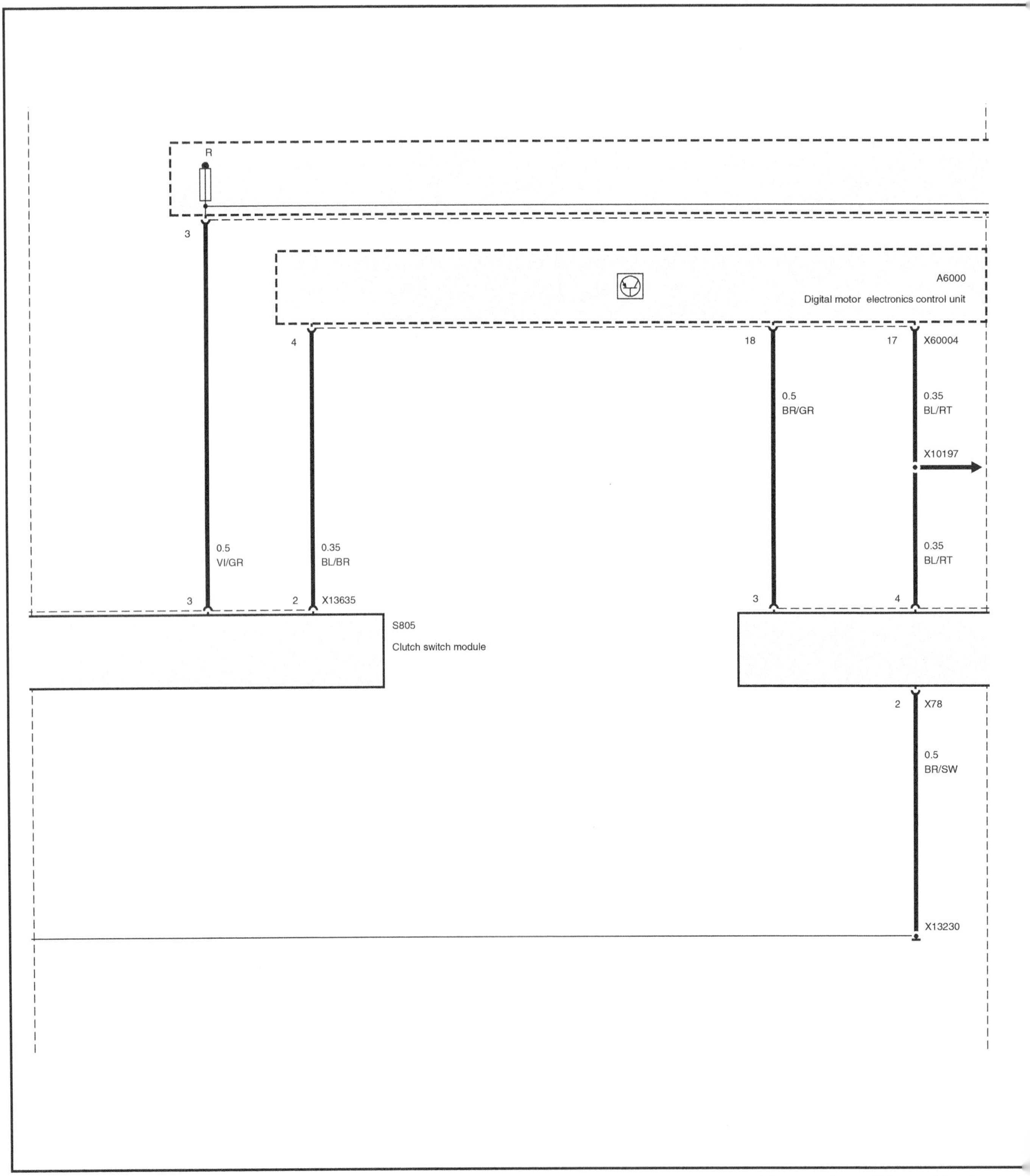

Brake light / clutch switches

(page 3 of 3)

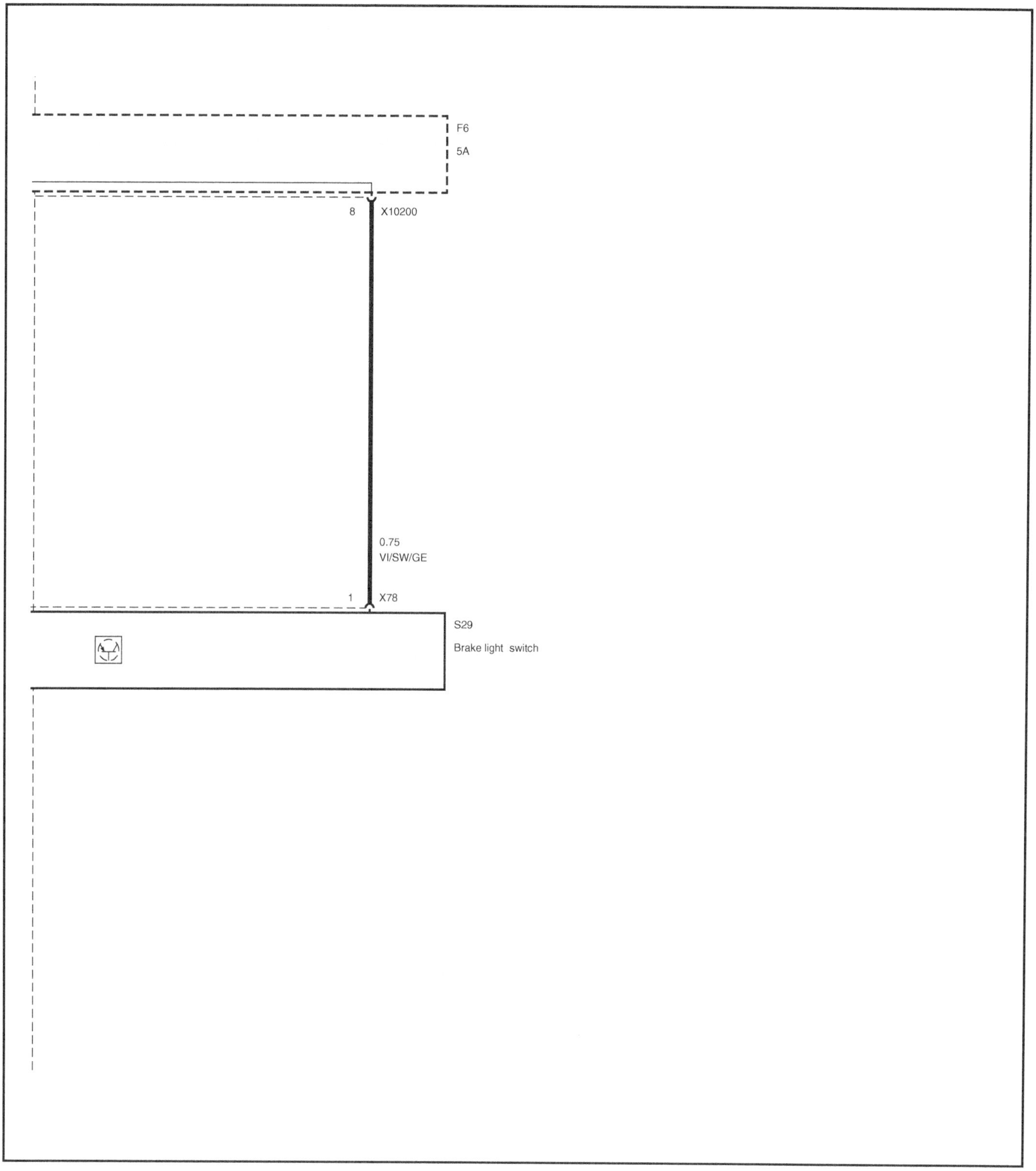

Bus system
K-bus (from March 2003 to Sept. 2003)
(page 1 of 4)

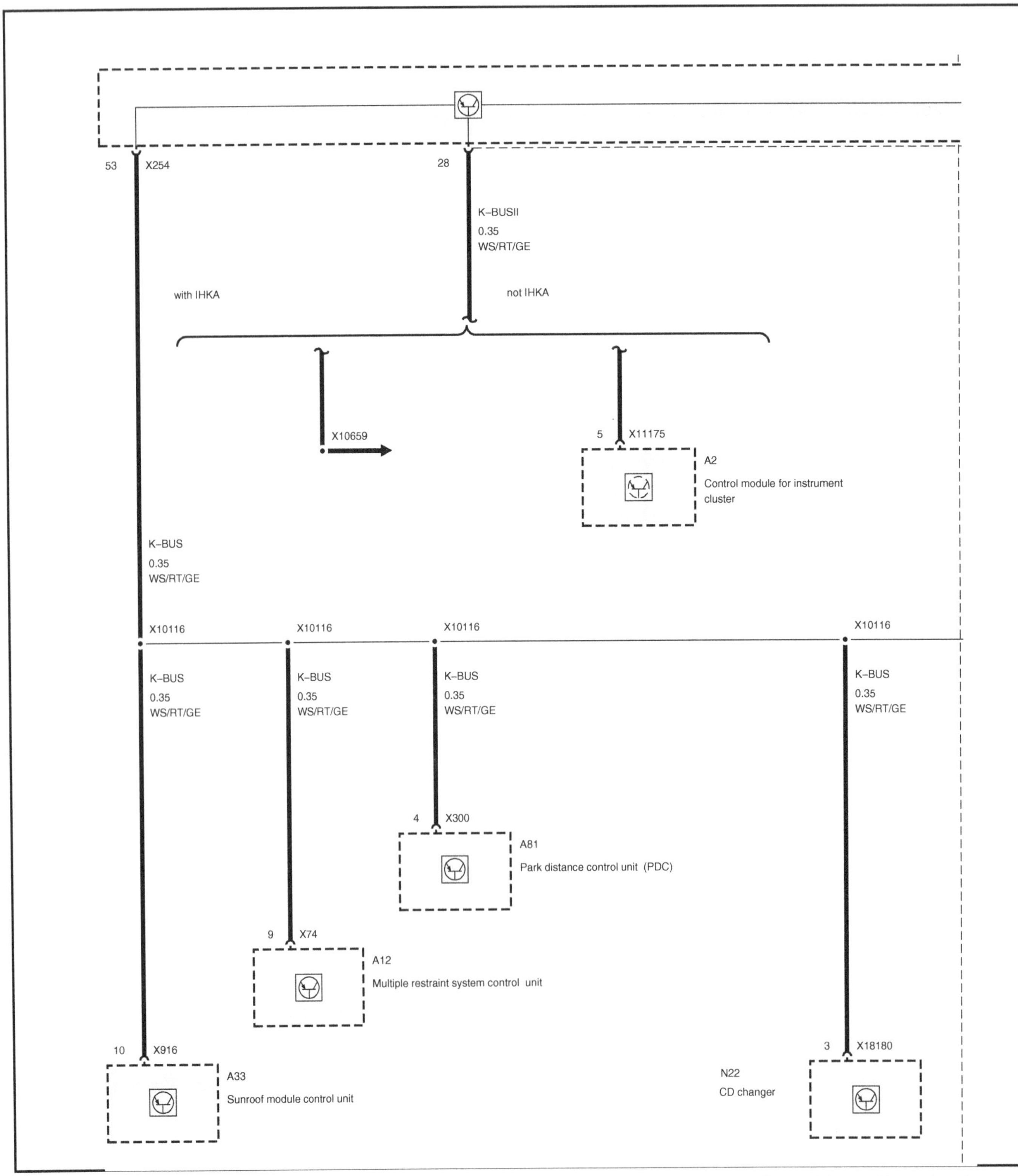

Bus system
(from March 2003 to Sept. 2003)
(page 2 of 4)

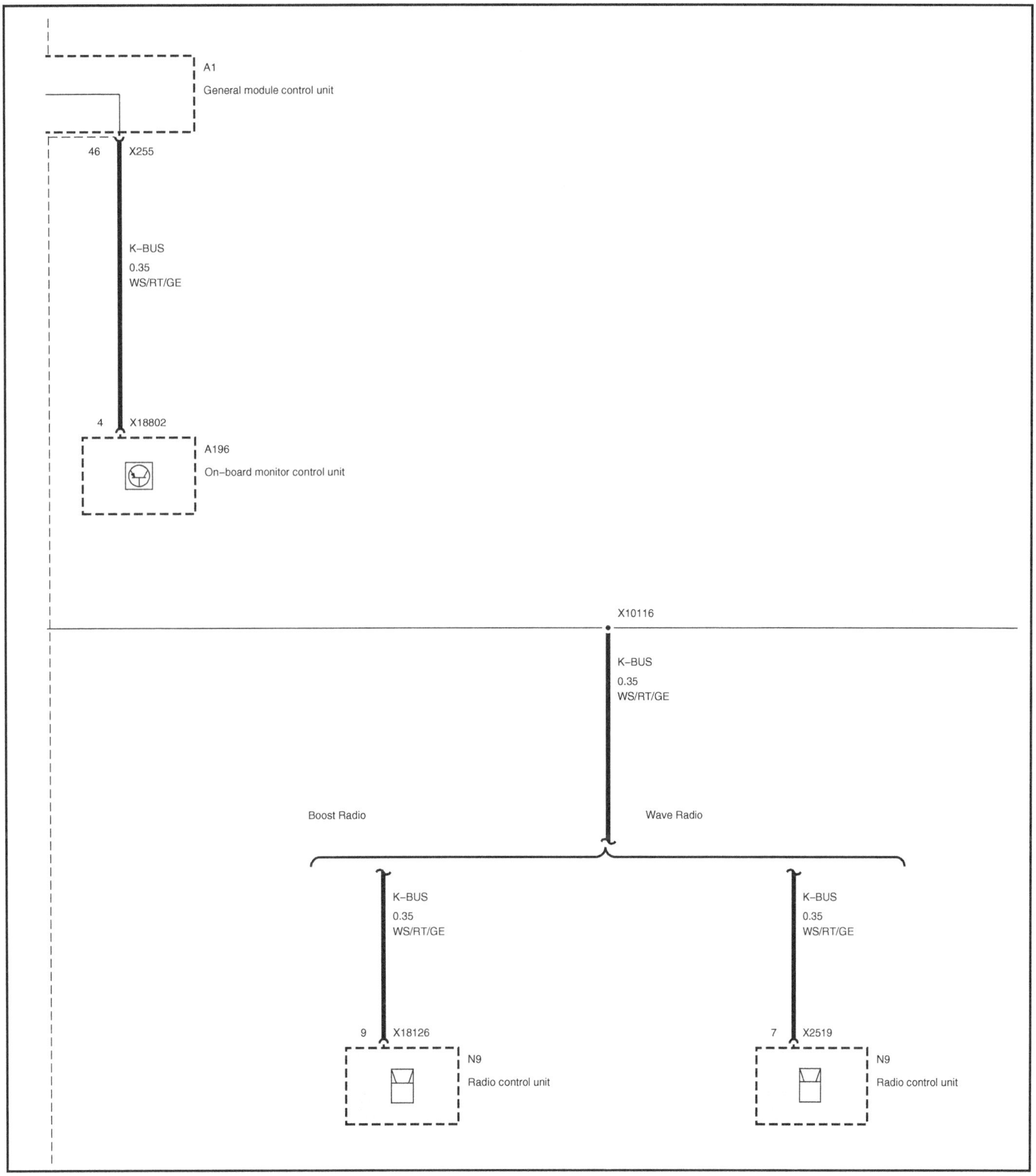

Bus system
(from March 2003 to Sept. 2003)
(page 3 of 4)

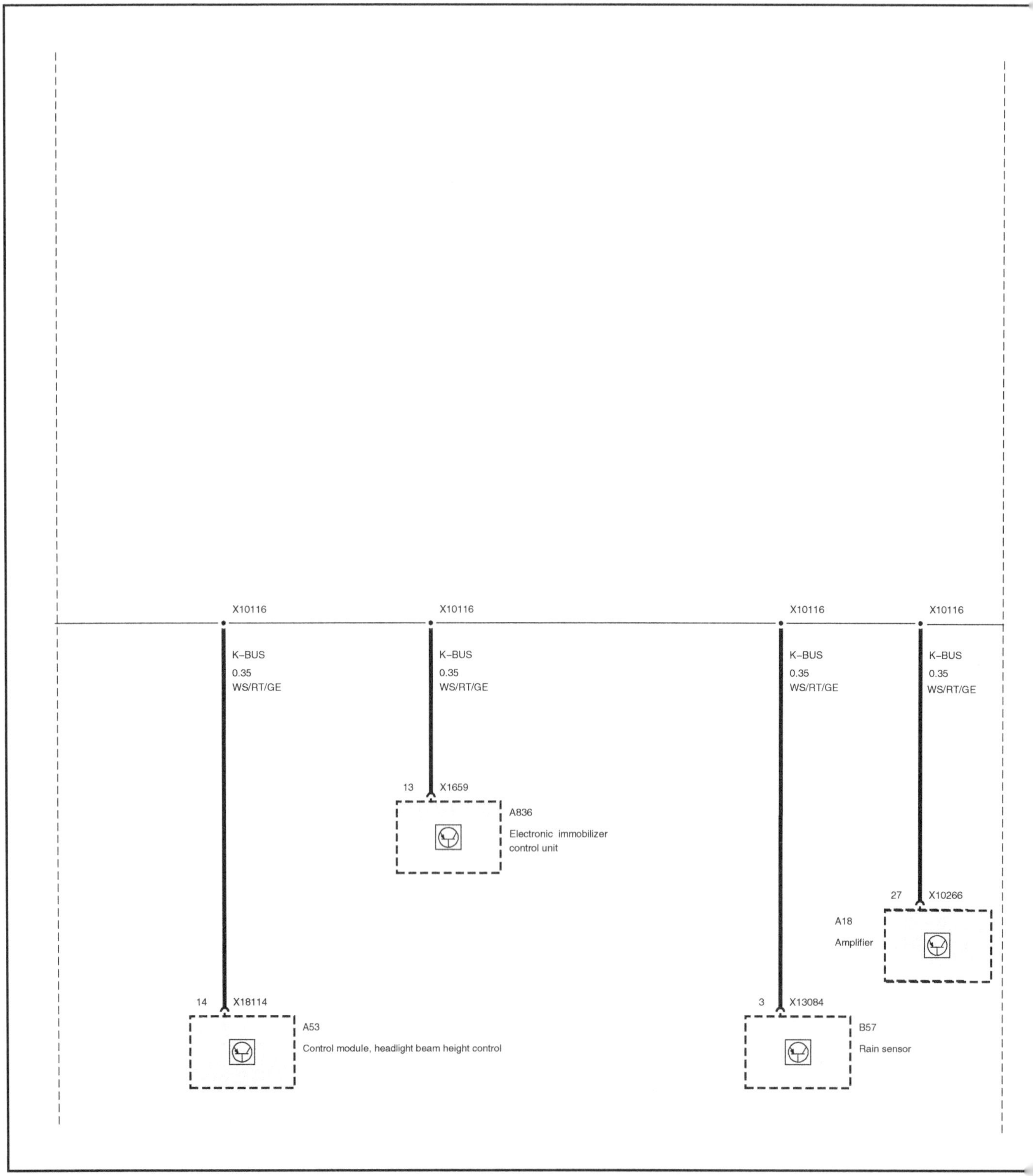

Bus system
(from March 2003 to Sept. 2003)
(page 4 of 4)

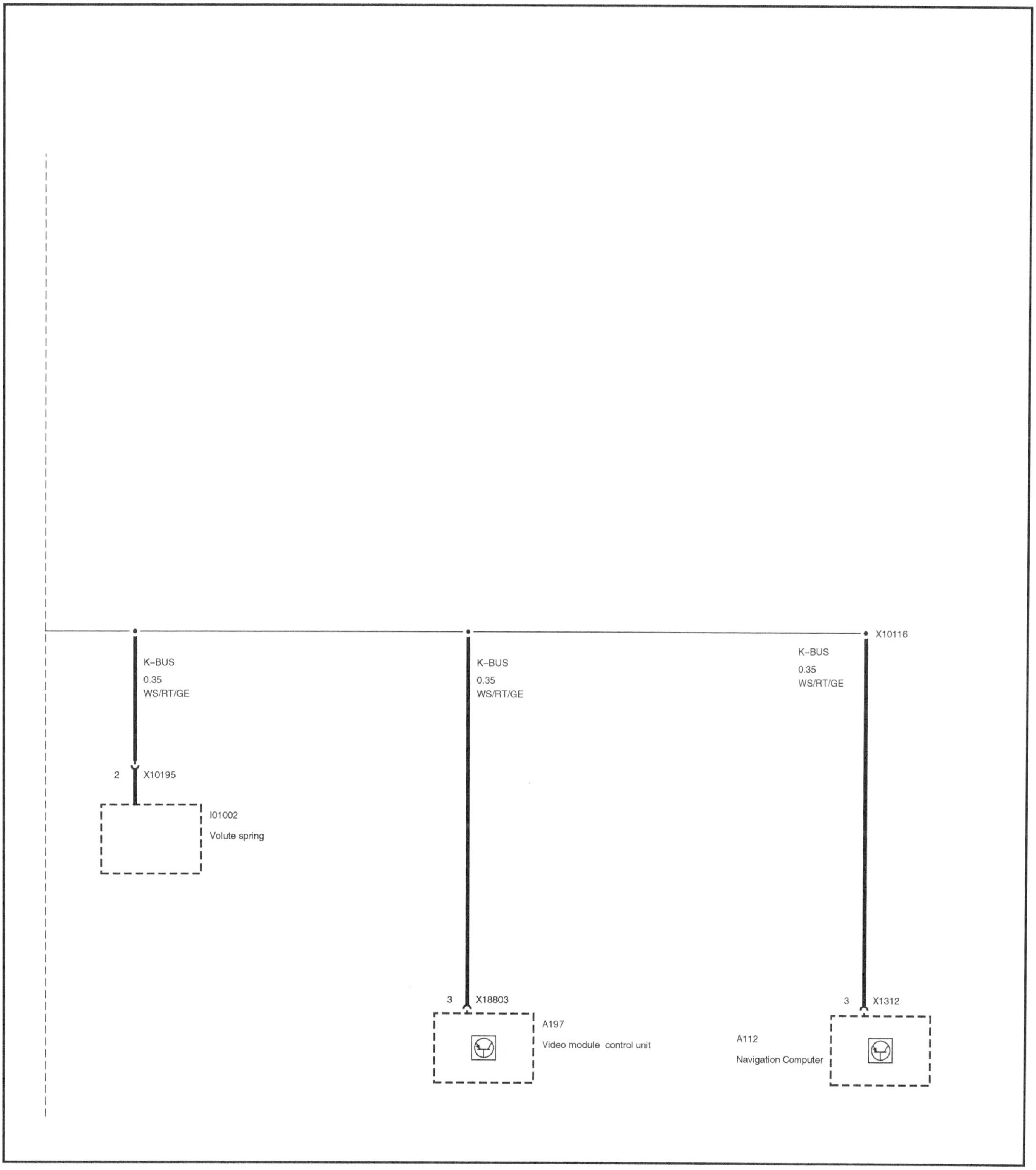

Bus system
K-bus (from Sept. 2003 to Jan. 2005)
(page 1 of 2)

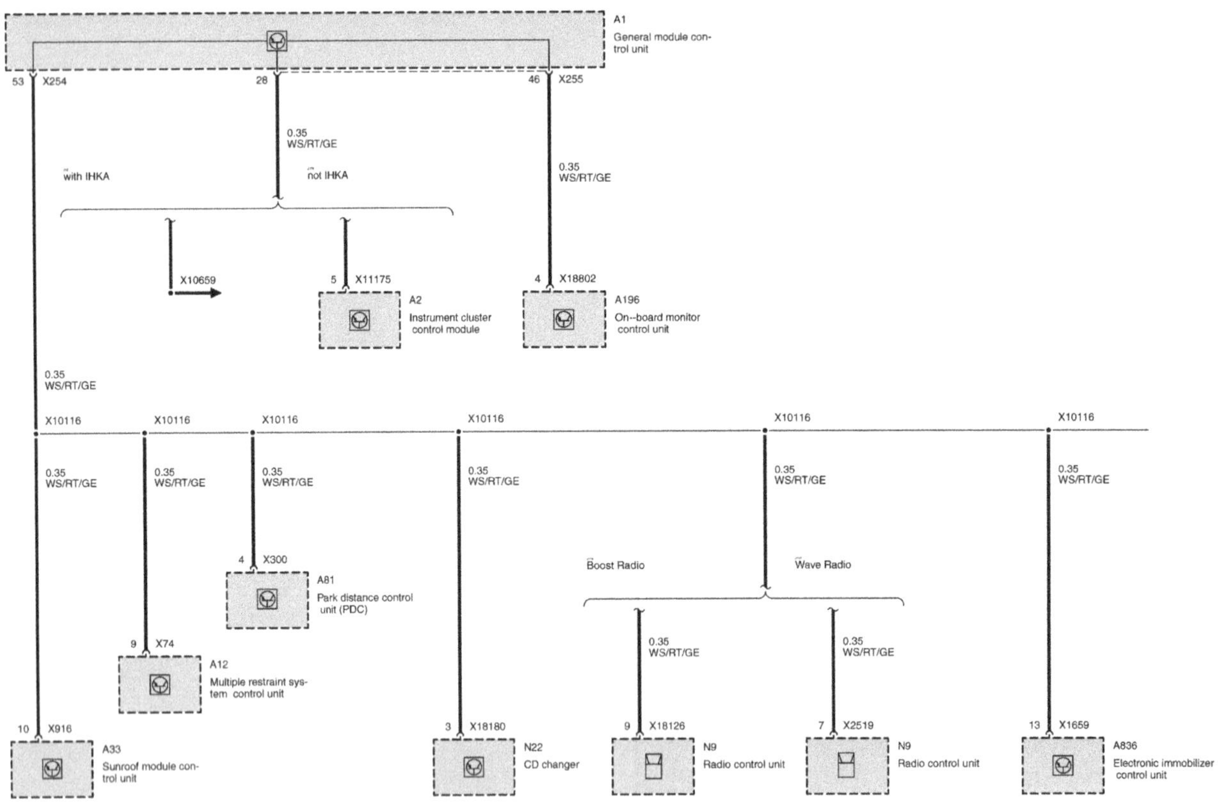

MC06ele20253a

Bus system

K-bus (from Sept. 2003 to Jan. 2005)

(page 2 of 2)

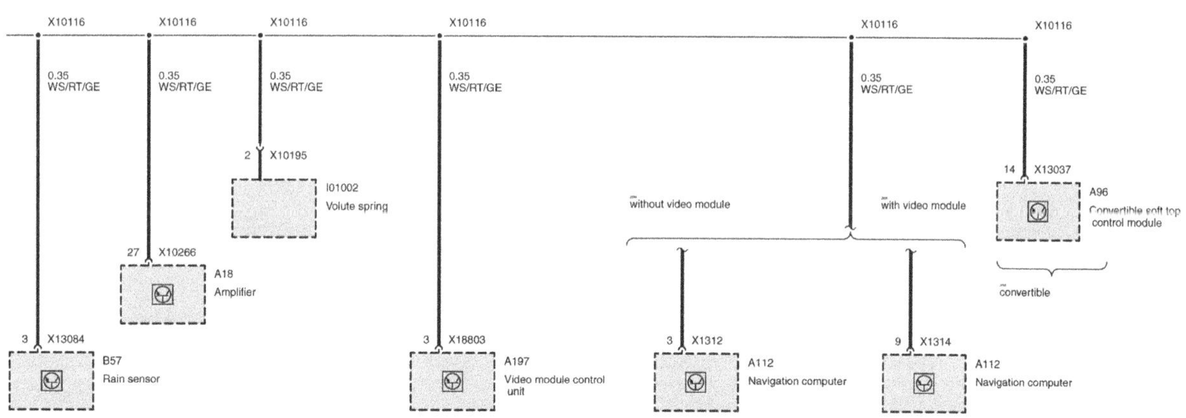

MC06ele20253b

Bus system

K-bus (from Jan. 2005)

(page 1 of 3)

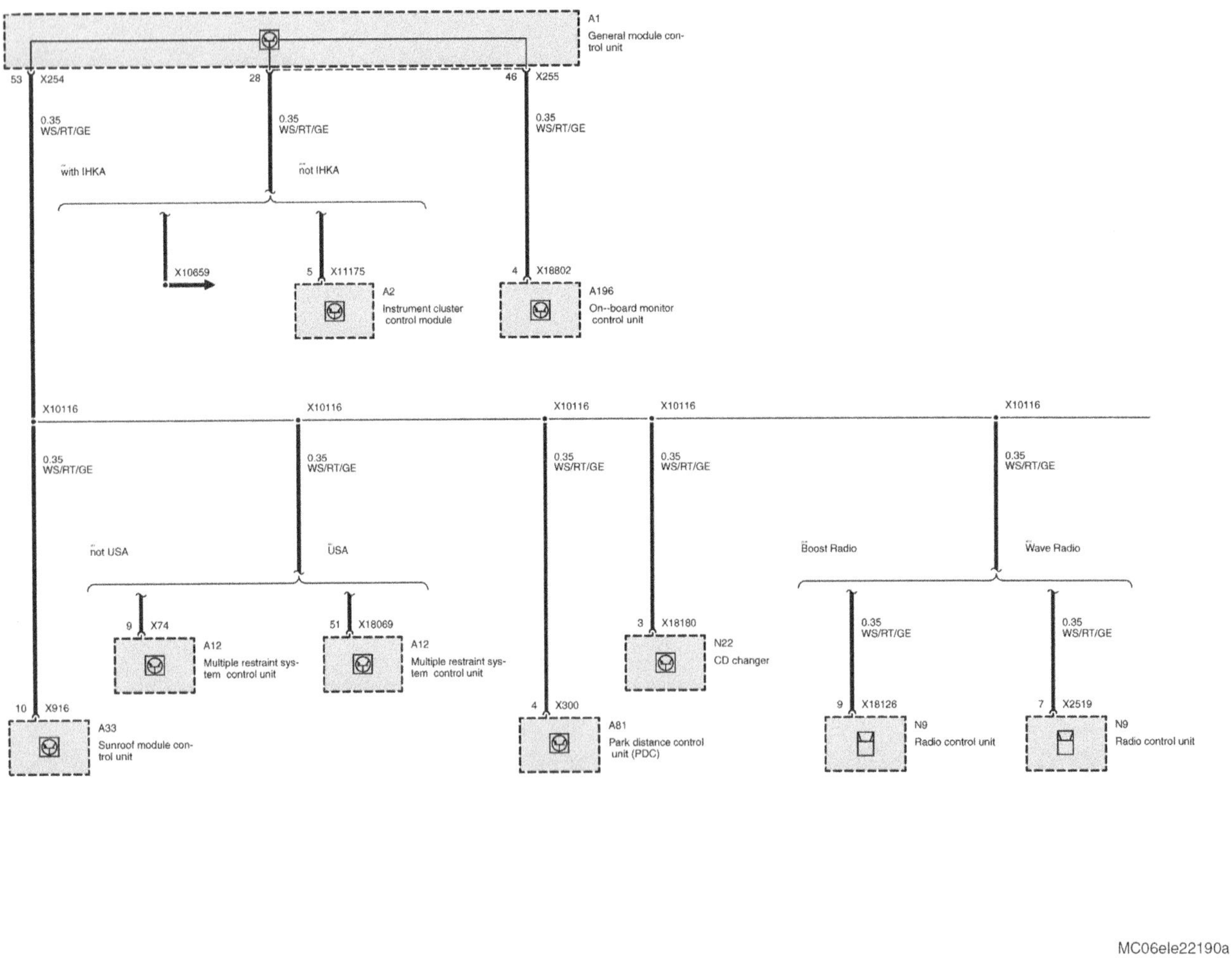

Bus system

K-bus (from Jan. 2005)

(page 2 of 3)

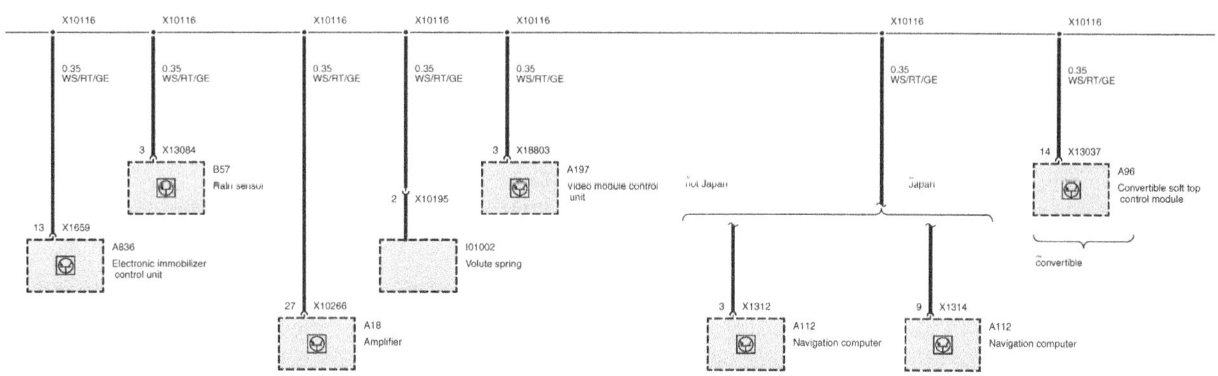

MC06ele22190b

Bus system
K-bus (from Jan. 2005)
(page 3 of 3)

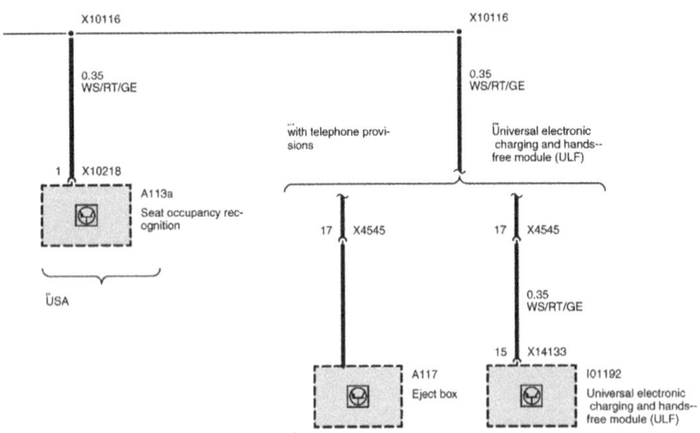

MC06ele22190c

Bus system

CAN-bus (up to Sept. 2003)

(page 1 of 1)

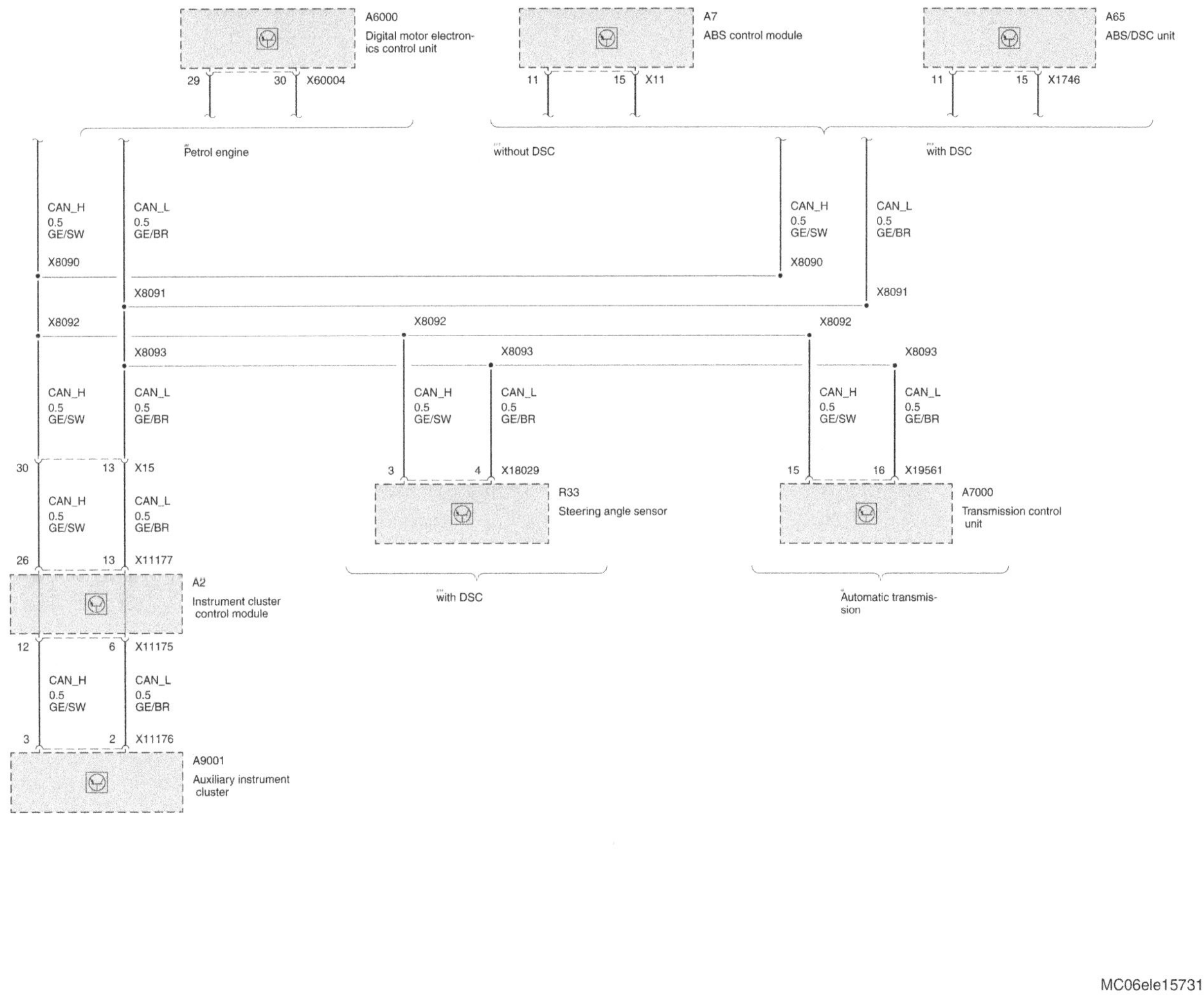

Bus system

CAN-bus (from Sept. 2003 to Sept. 2005)

(page 1 of 2)

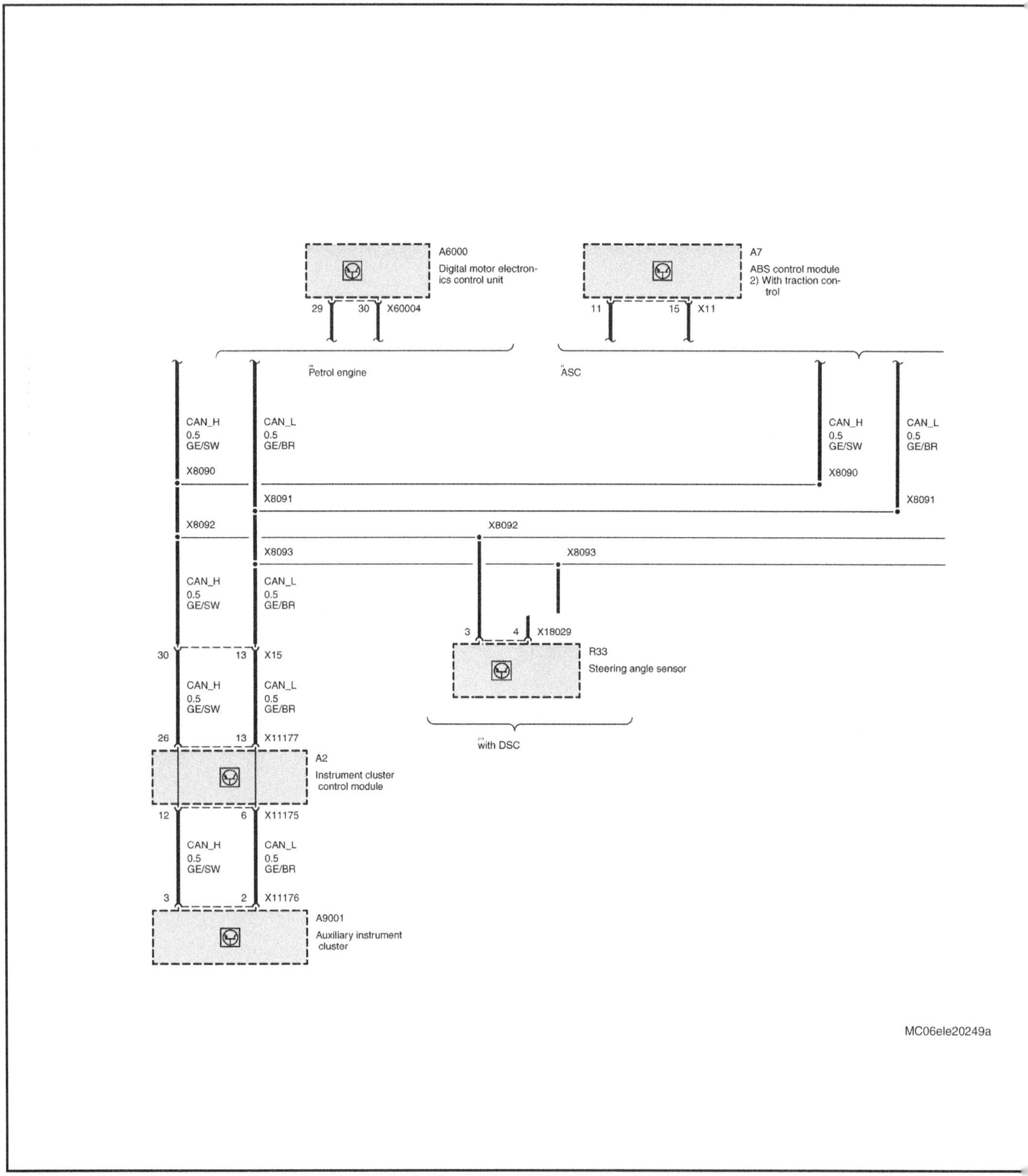

Bus system

CAN-bus (from Sept. 2003 to Sept. 2005)

(page 2 of 2)

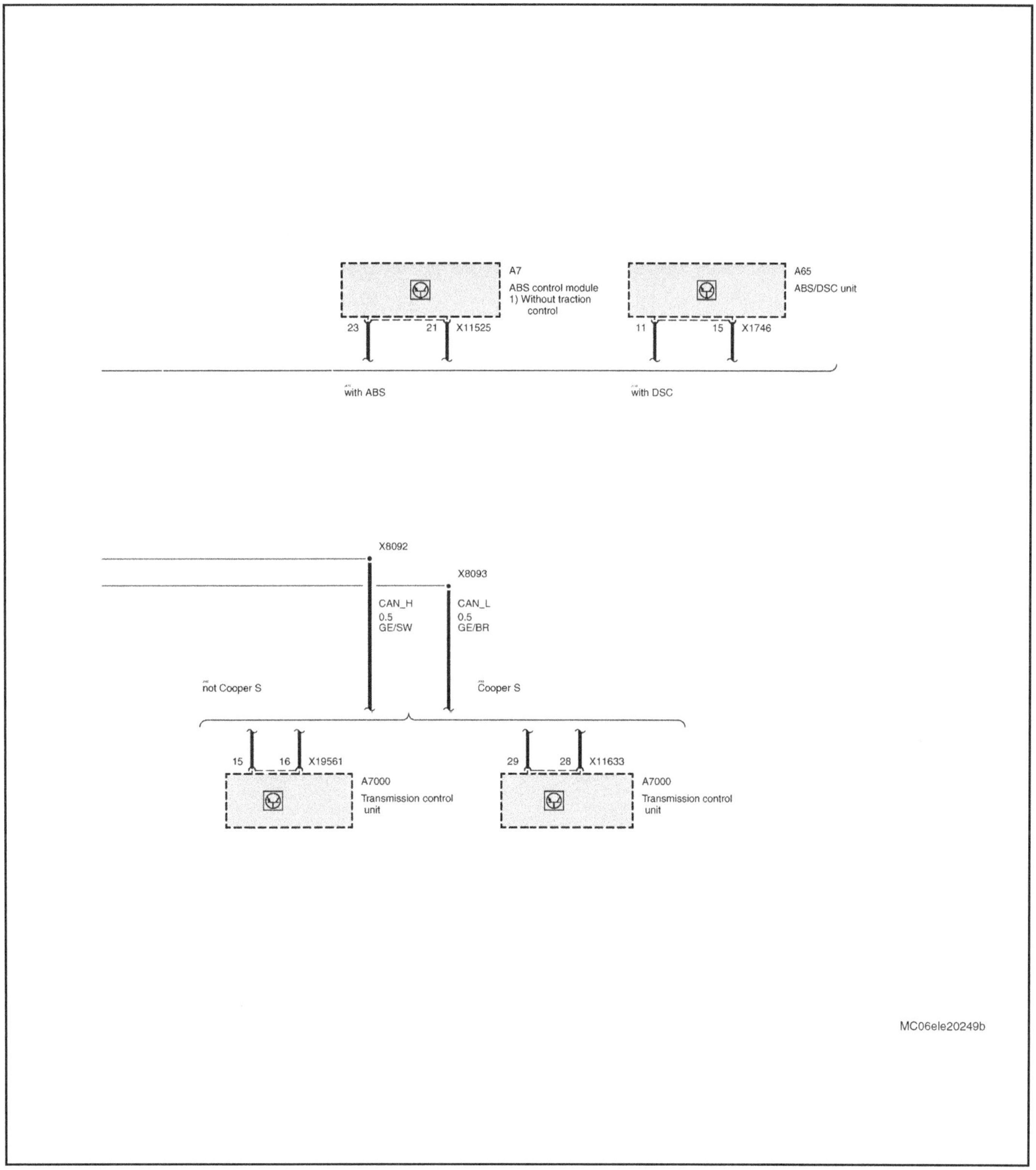

Bus system

CAN-bus (from Sept. 2005)

(page 1 of 2)

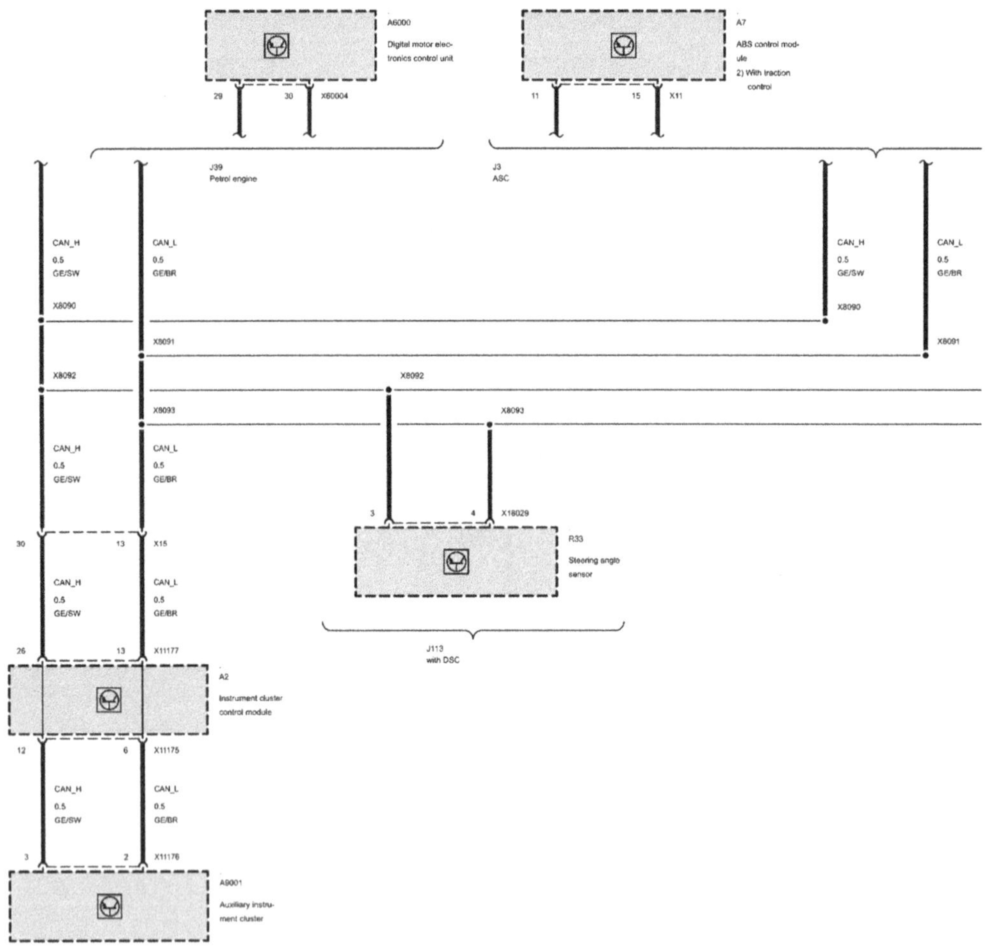

MC06ele15734a

Bus system

CAN-bus (from Sept. 2005)

(page 2 of 2)

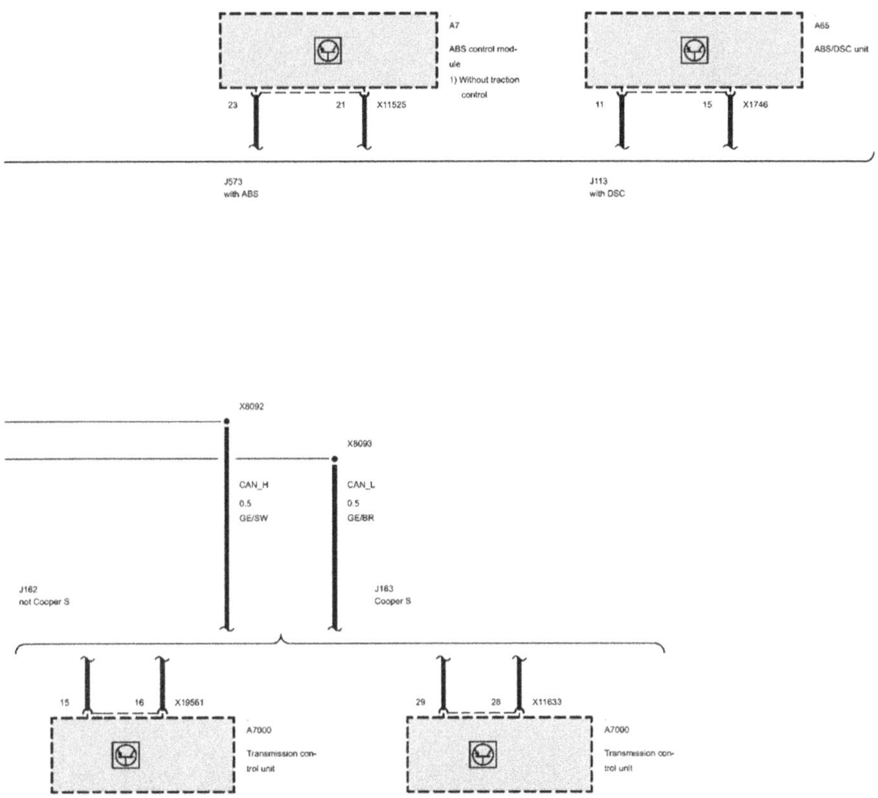

MC06ele15734b

Central locking
Driver's door (up to March 2002)
(page 1 of 1)

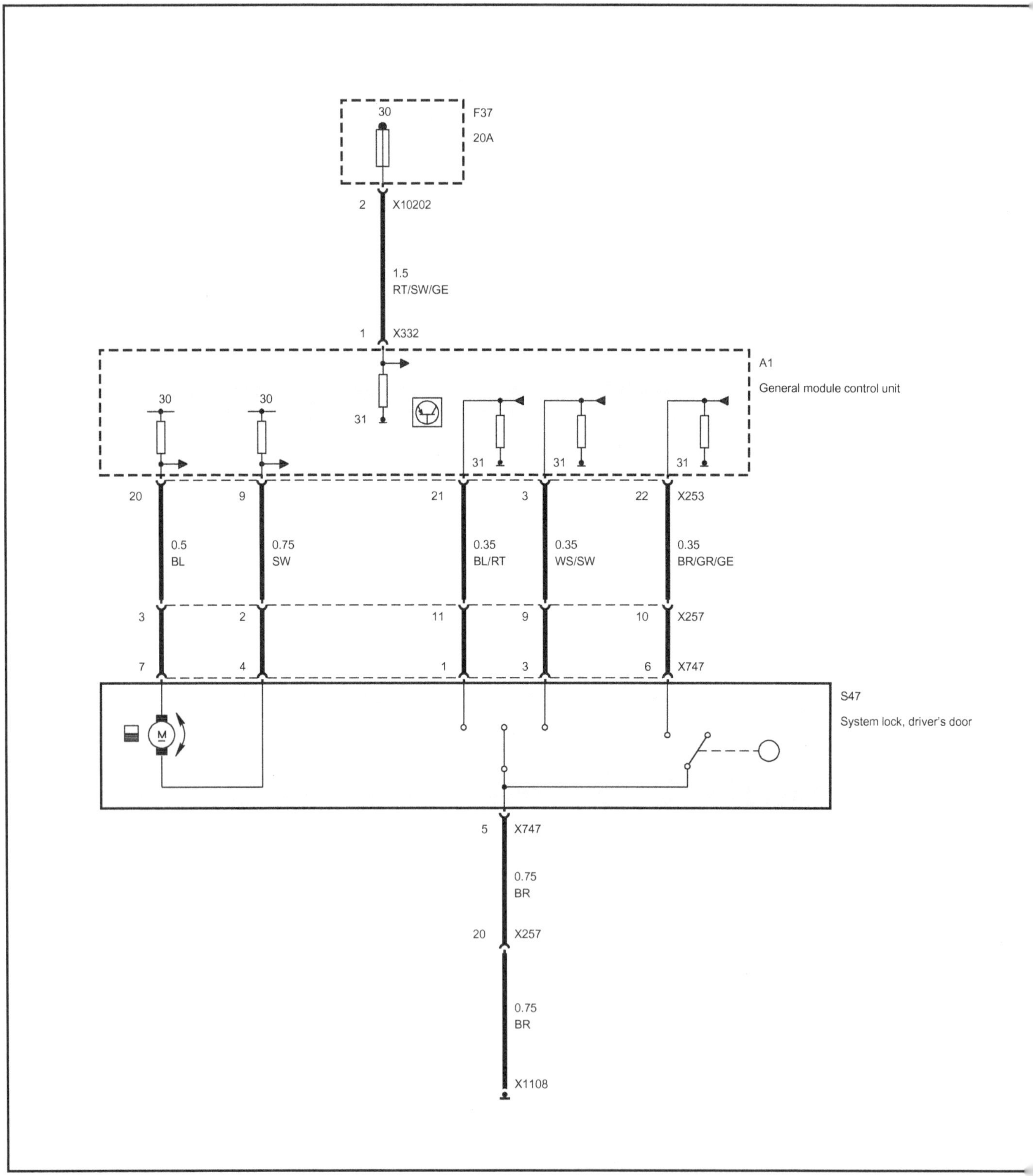

Central locking
Driver's door (from July 2004 to Sept. 2005)
(page 1 of 1)

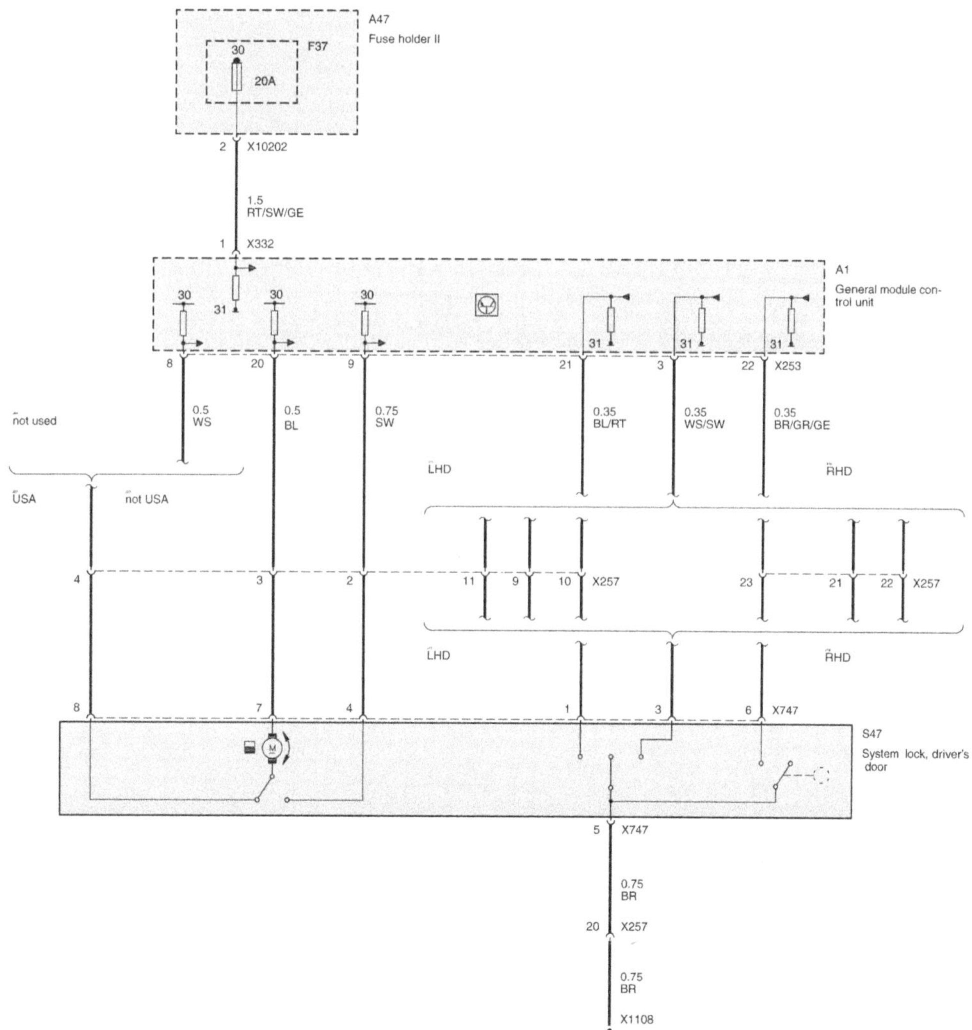

MC06ele21201

Central locking
Passenger door (up to March 2002)
(page 1 of 1)

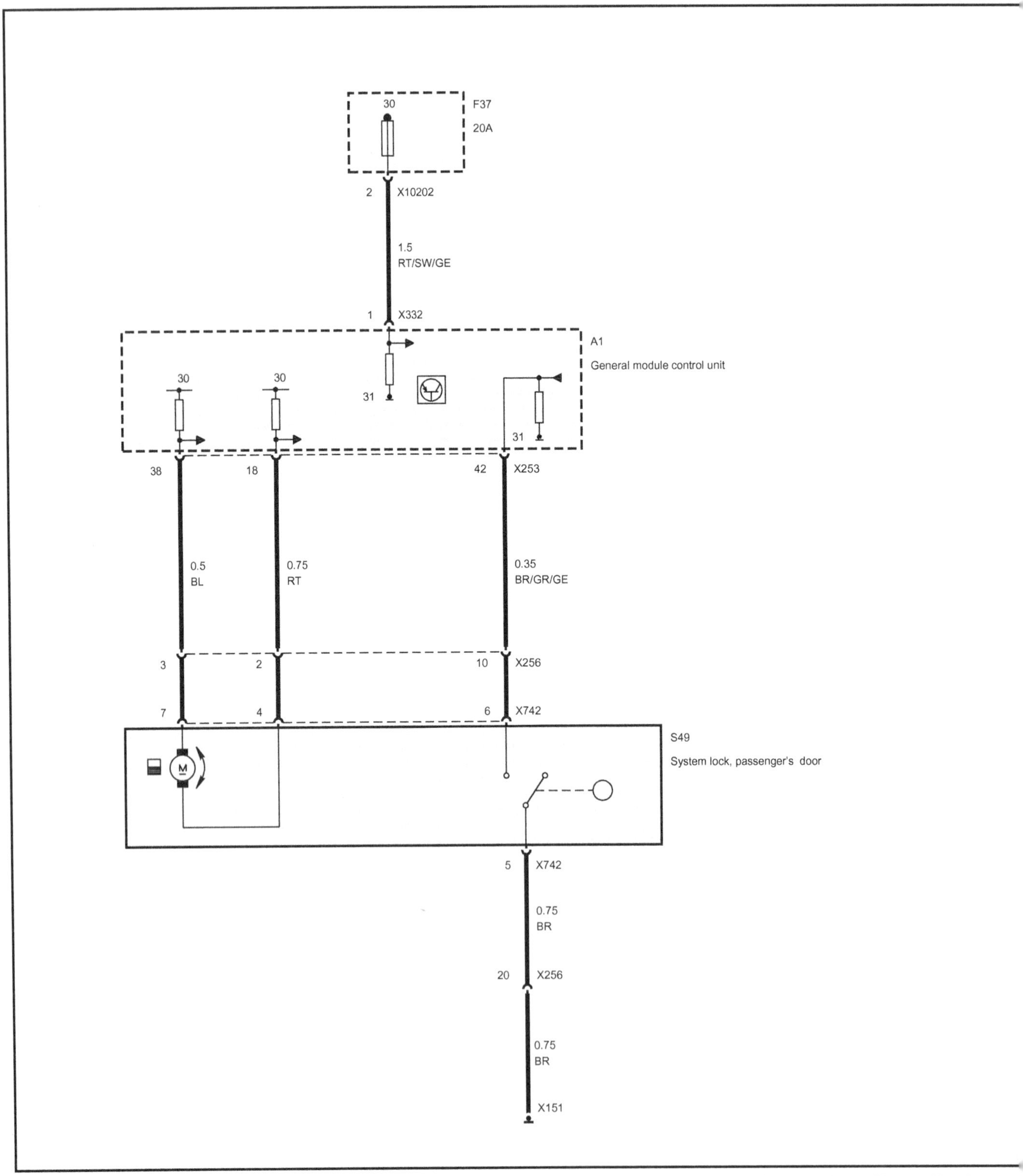

Central locking

Passenger door (from July 2004 to Sept. 2005)

(page 1 of 1)

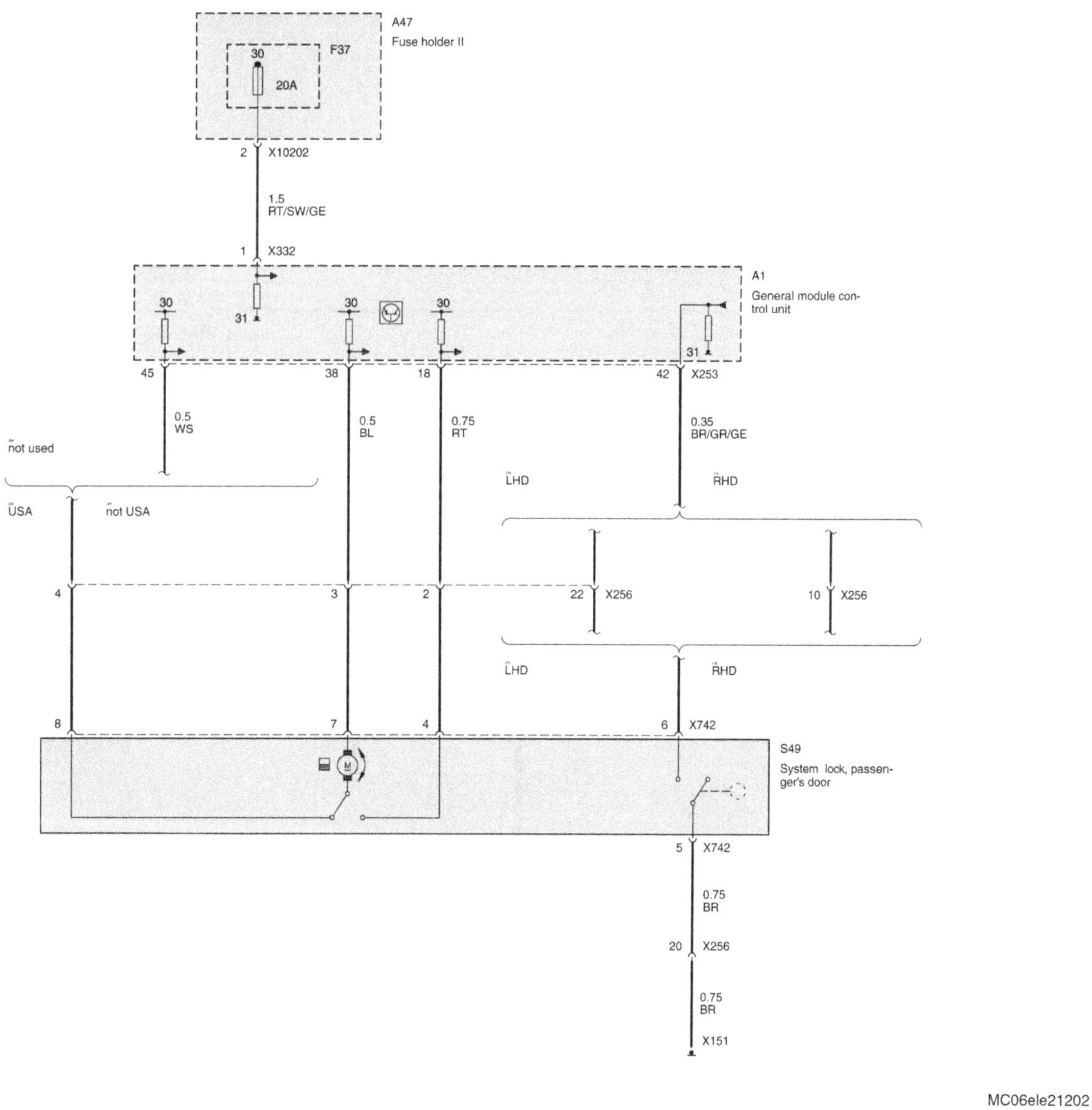

Central locking
Fuel filler flap
(page 1 of 1)

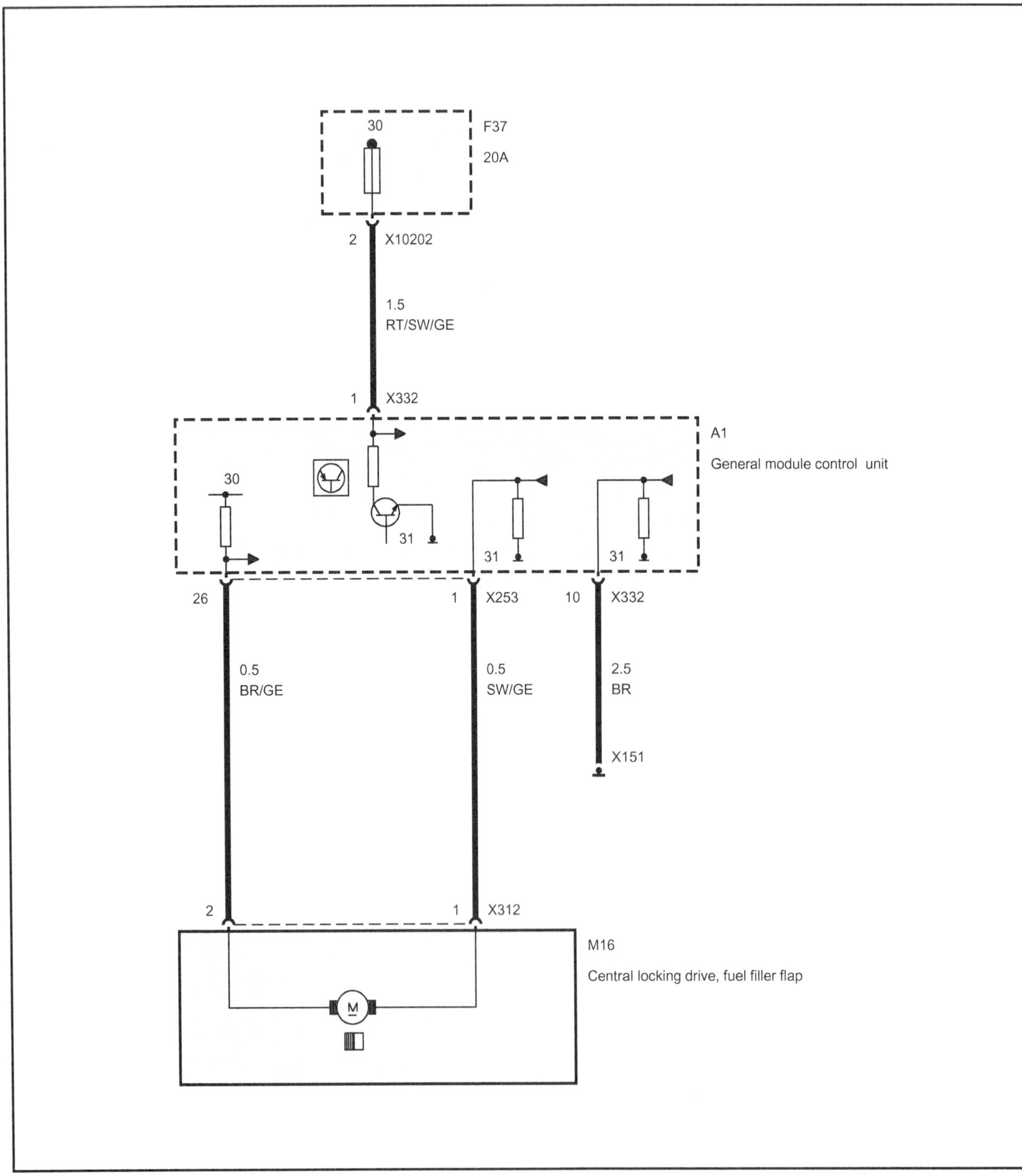

Central locking

Rear hatch (coupe)

(page 1 of 1)

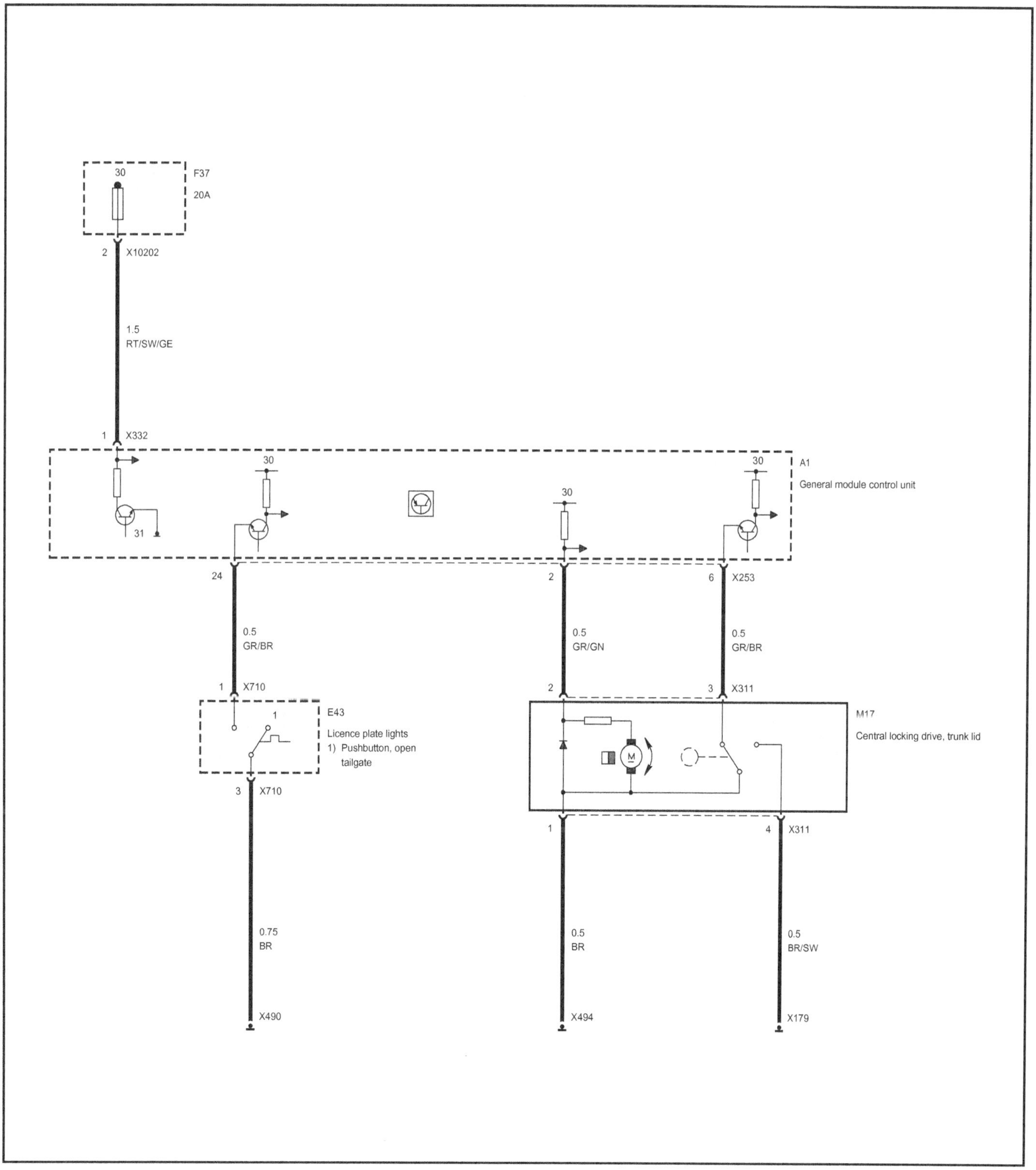

Central locking
Rear hatch (convertible)
(page 1 of 1)

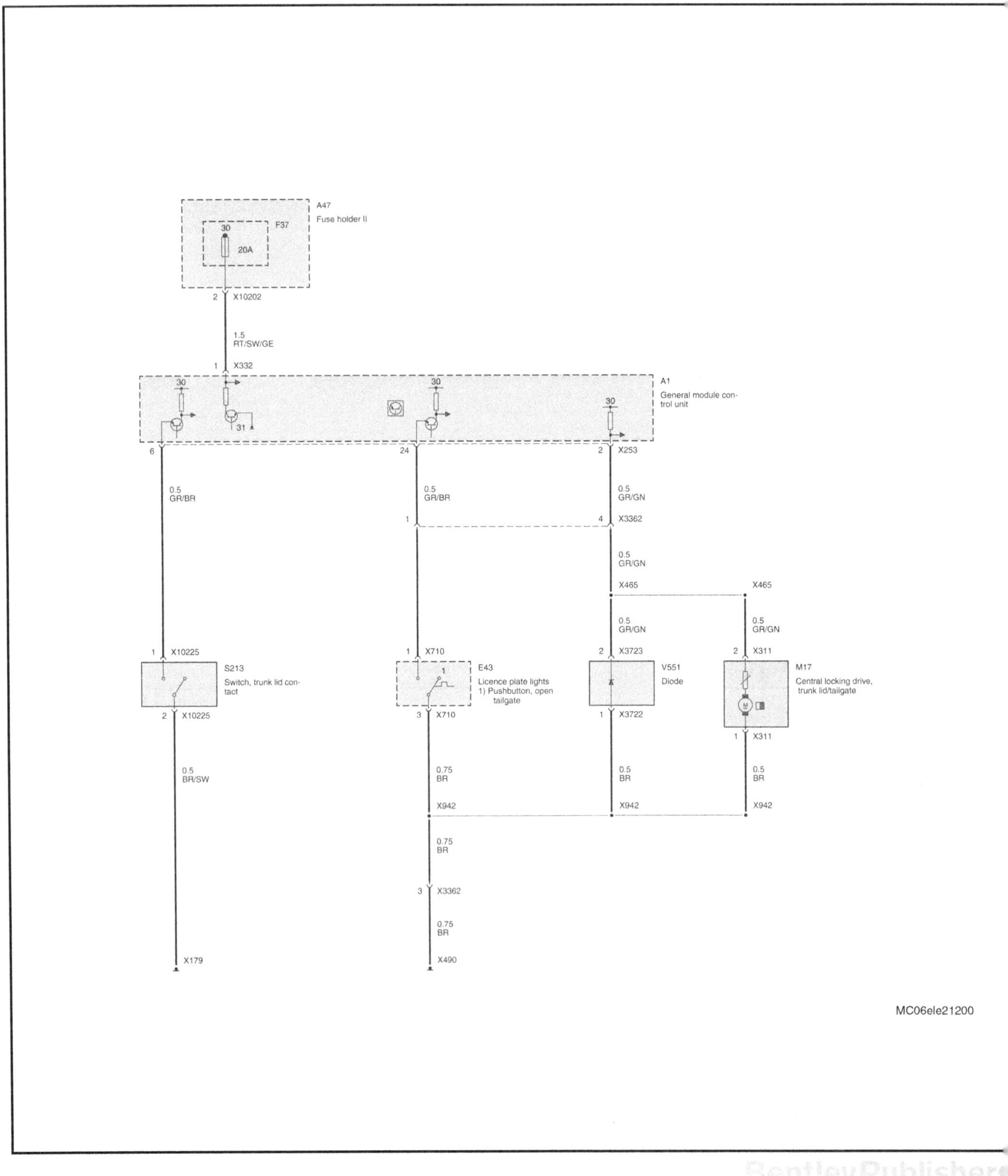

Central locking
Switch panel
(page 1 of 1)

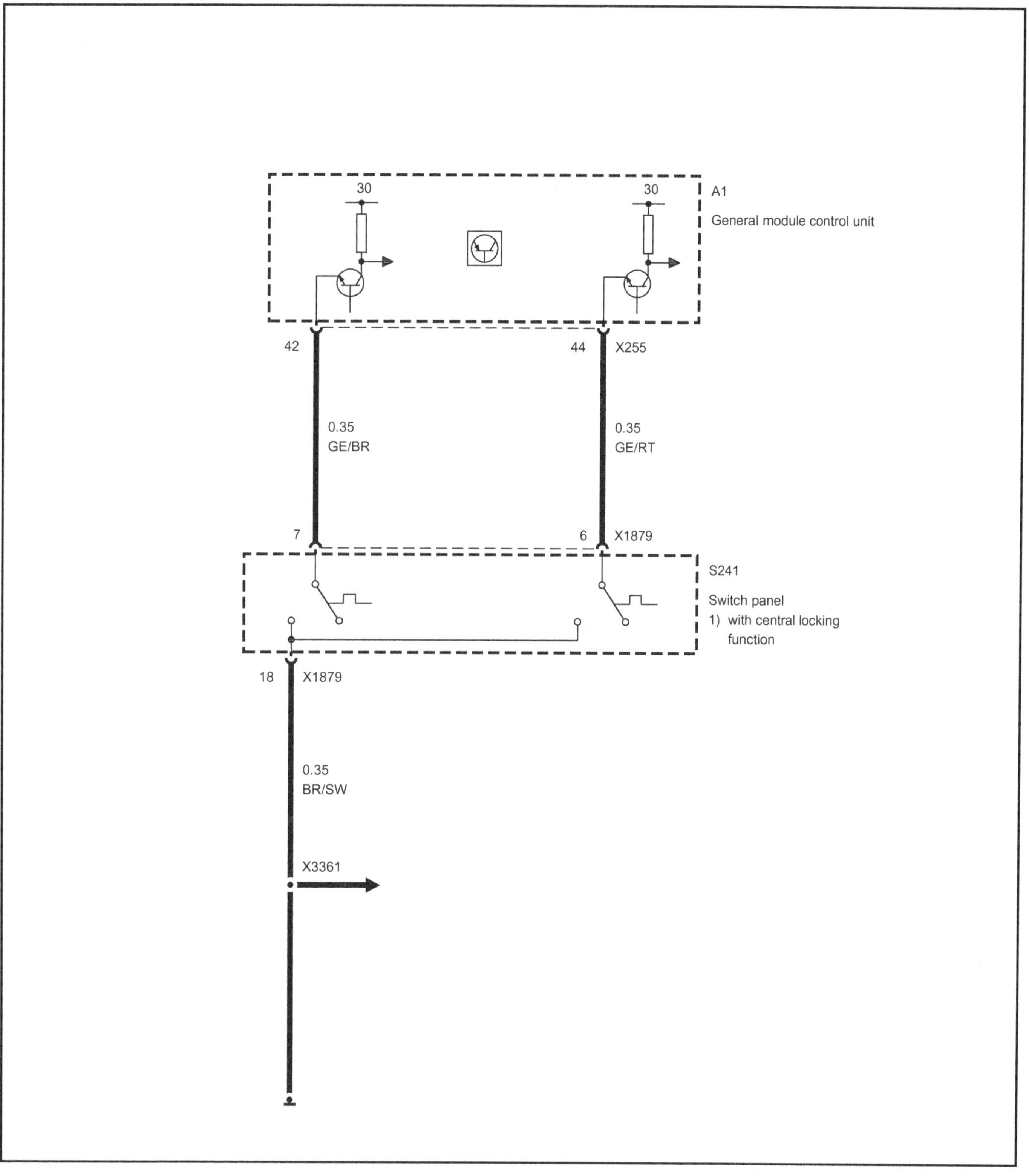

Charging system
Cooper alternator (Denso)
(page 1 of 1)

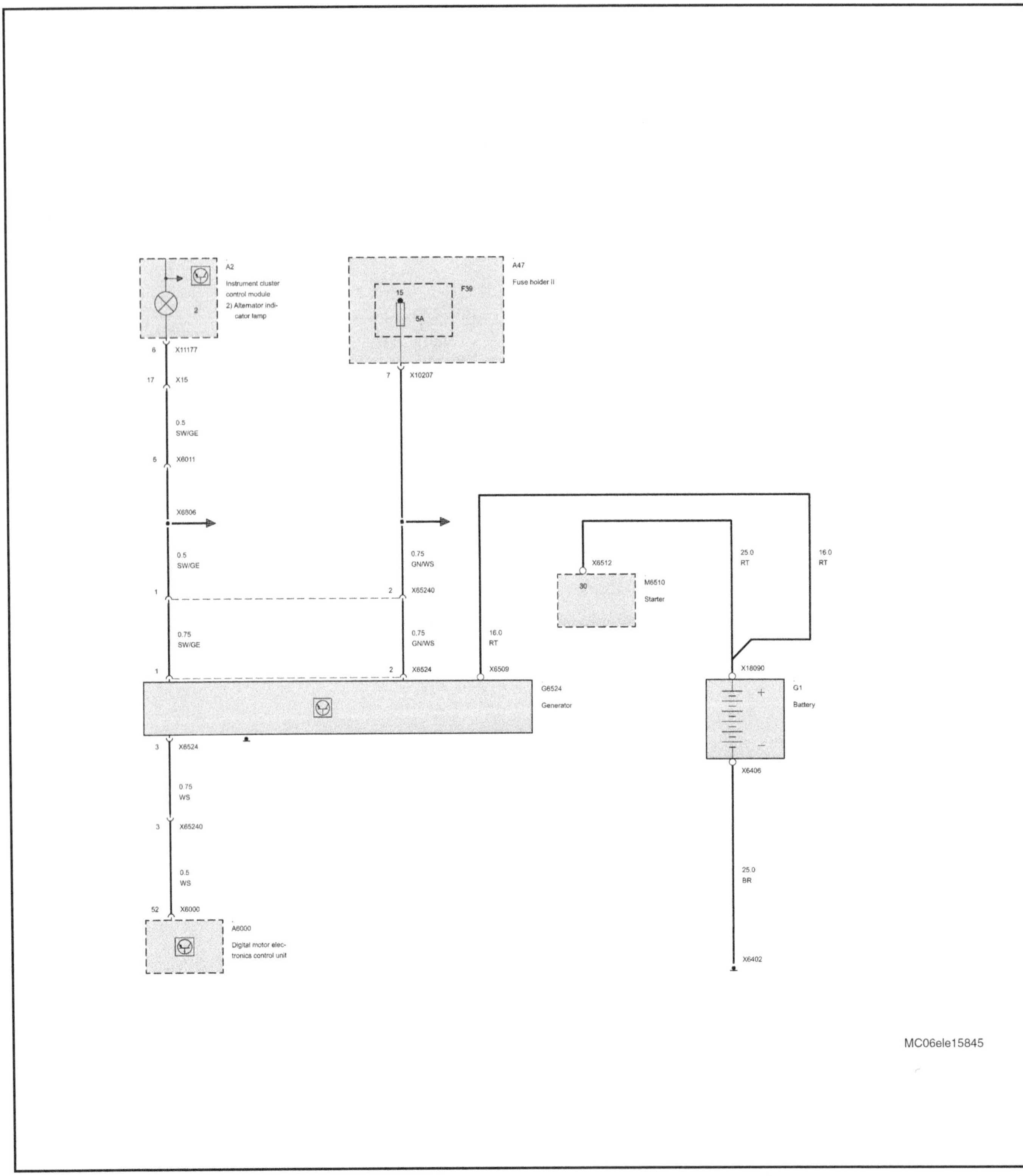

MC06ele15845

Charging system
Cooper alternator (Valeo)
(page 1 of 1)

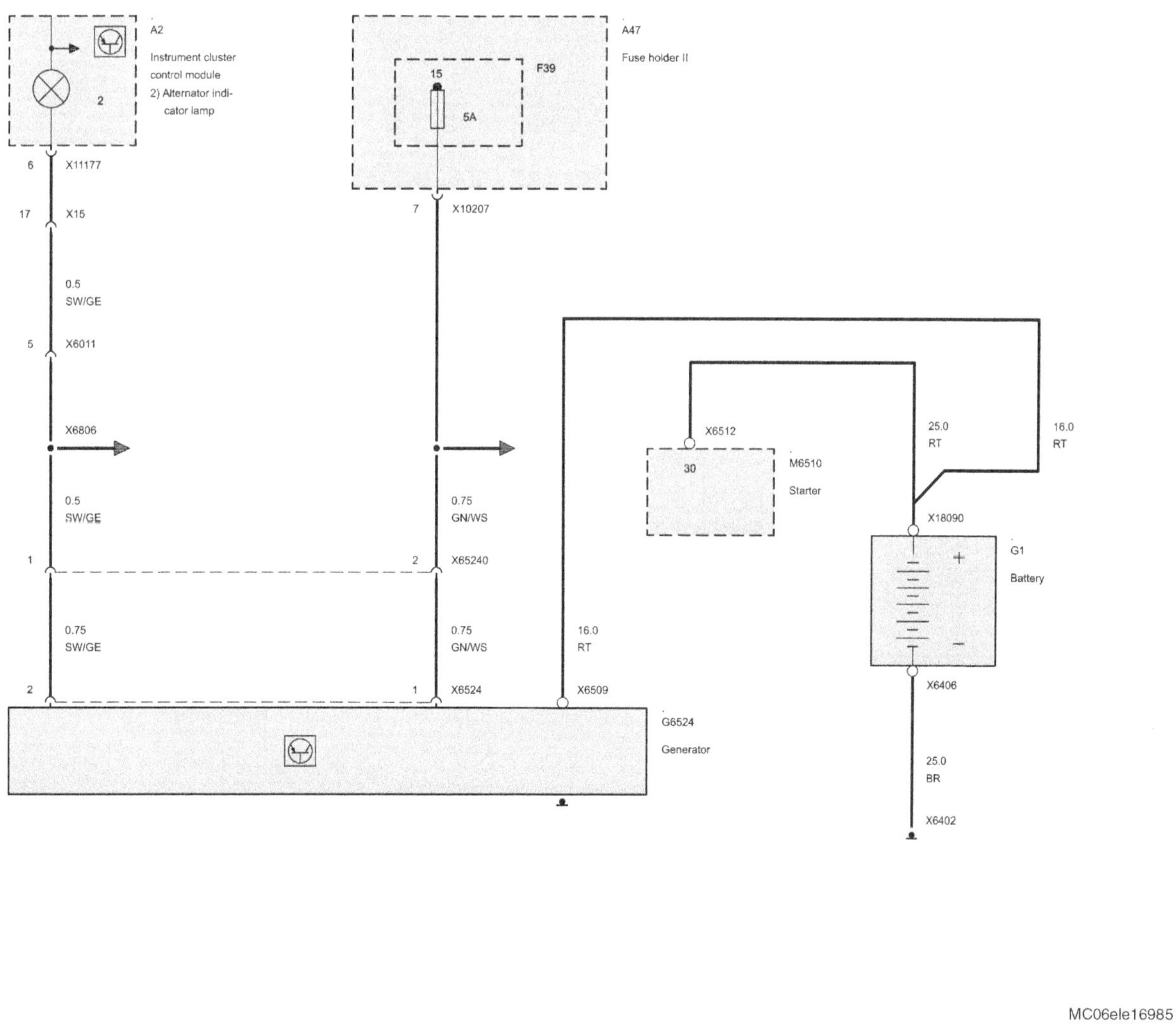

Charging system
Cooper S alternator
(page 1 of 1)

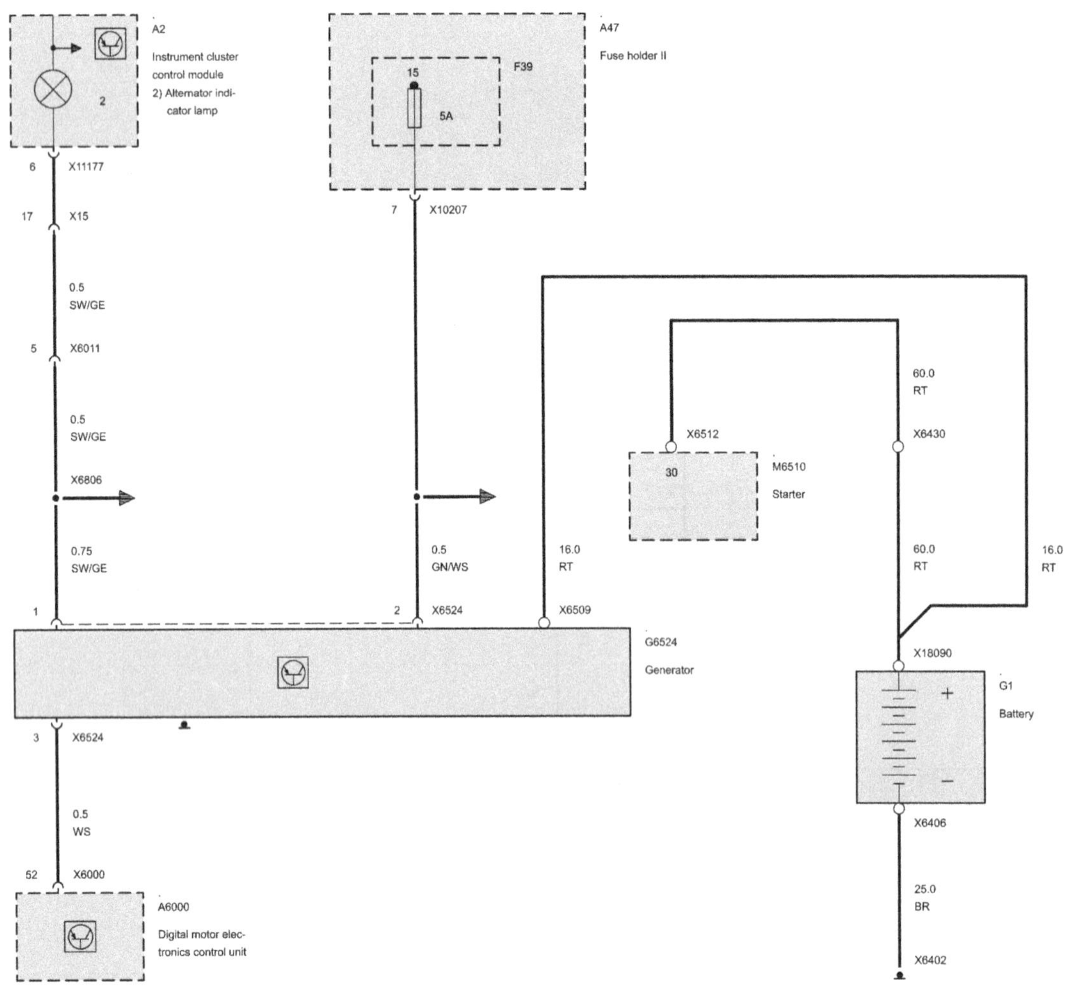

MC06ele16986

Convertible top

Electric drive (up to Sept. 2005)

(page 1 of 2)

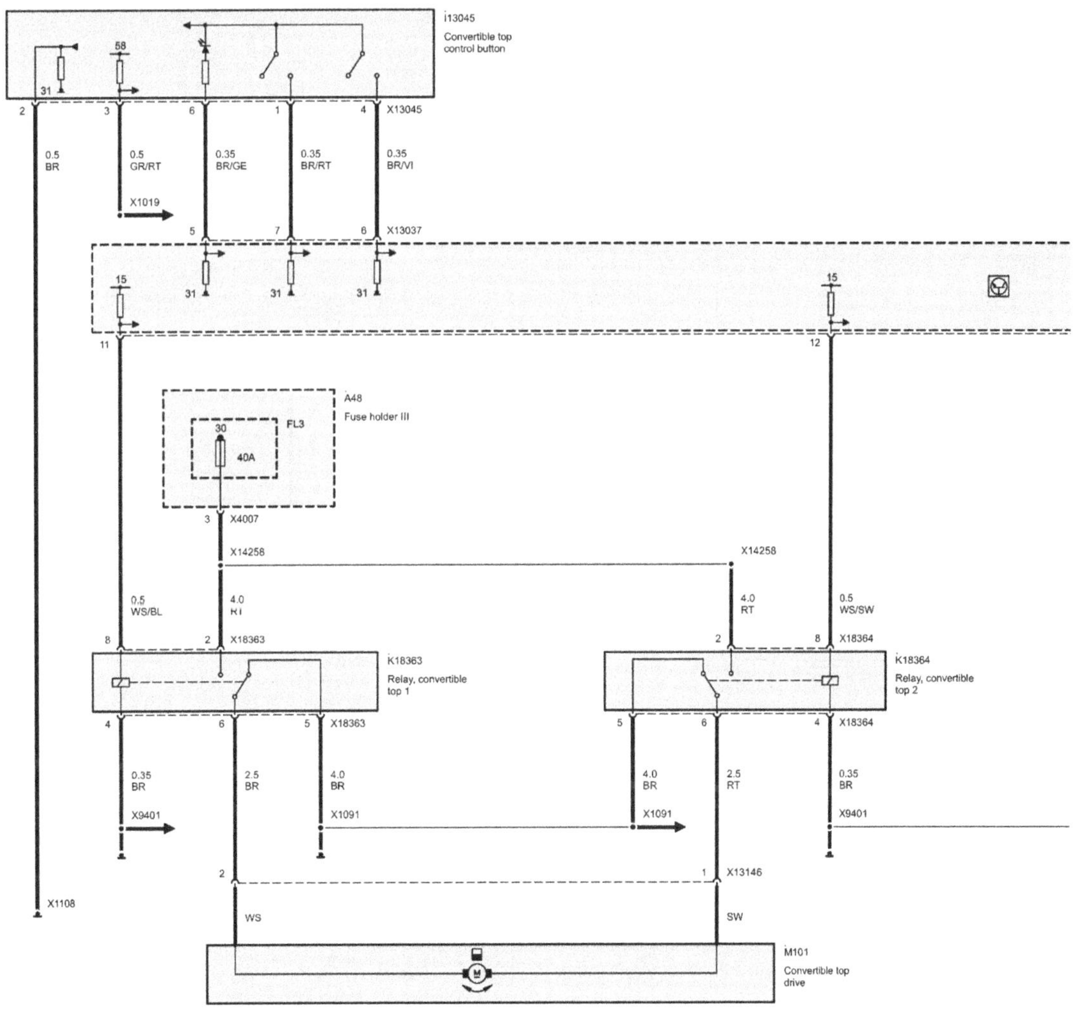

MC06ele21305a

Convertible top

Electric drive (up to Sept. 2005)

(page 2 of 2)

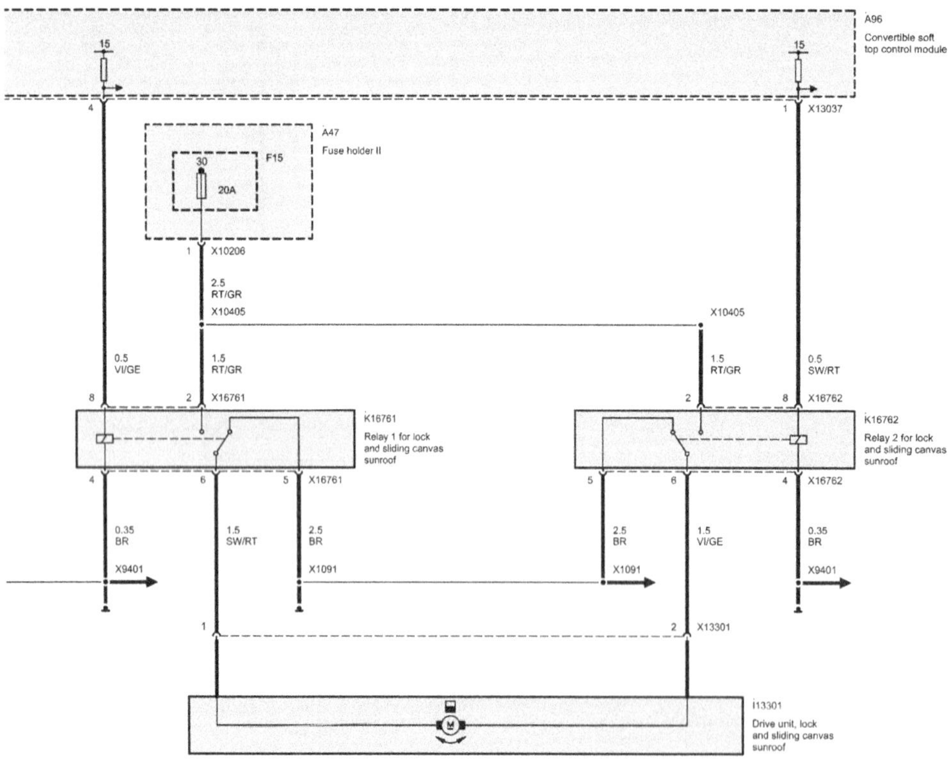

MC06ele21305b

Convertible top
Electric drive (from Sept. 2005)
(page 1 of 2)

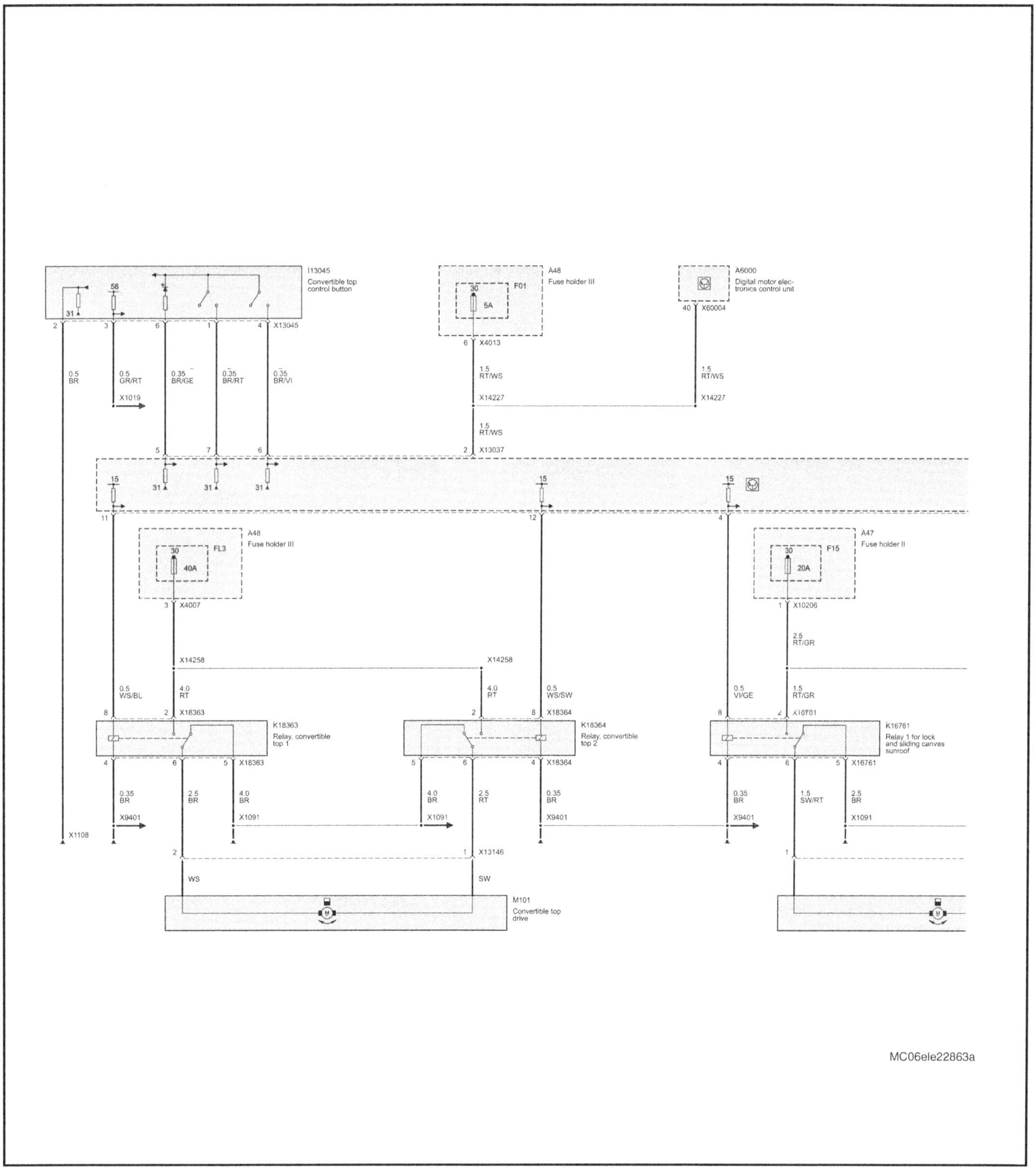

Convertible top
Electric drive (from Sept. 2005)
(page 2 of 2)

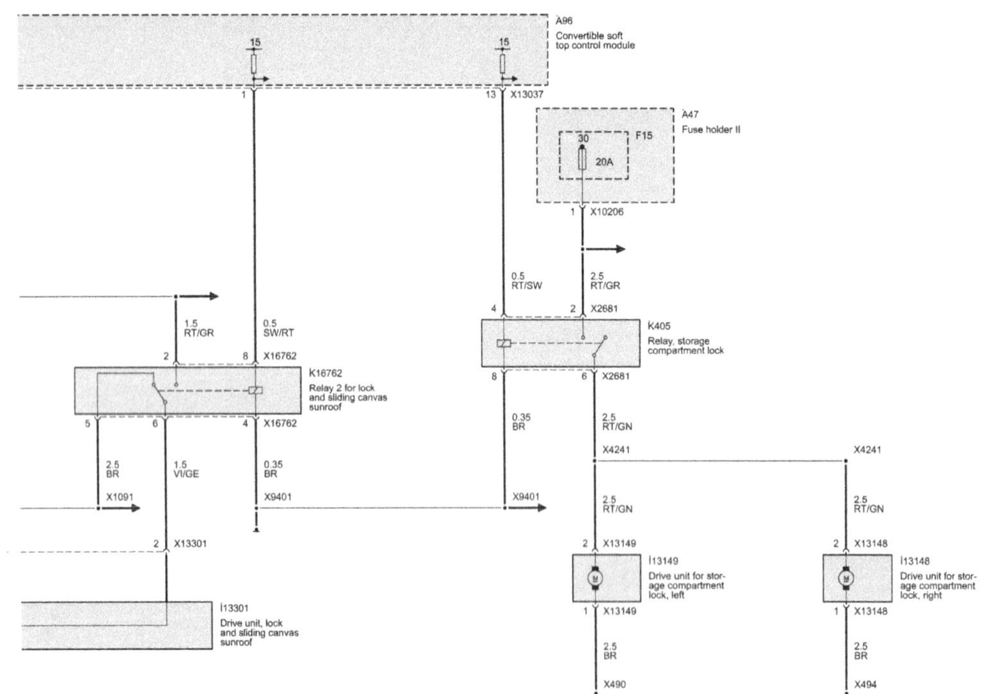

MC06ele22863b

Convertible top
Power supply
(page 1 of 1)

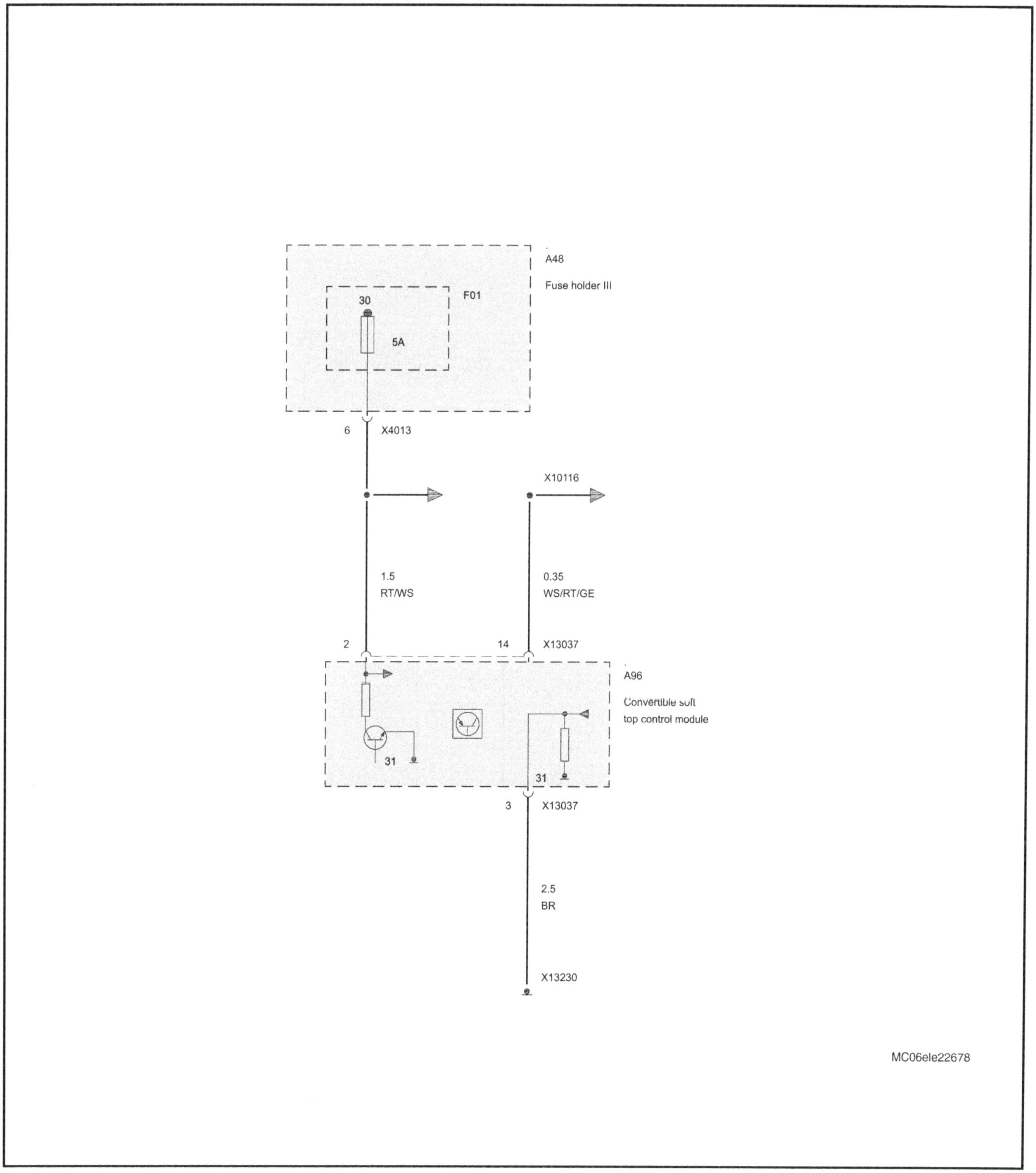

Convertible top
Sensors (up to Sept. 2005)
(page 1 of 2)

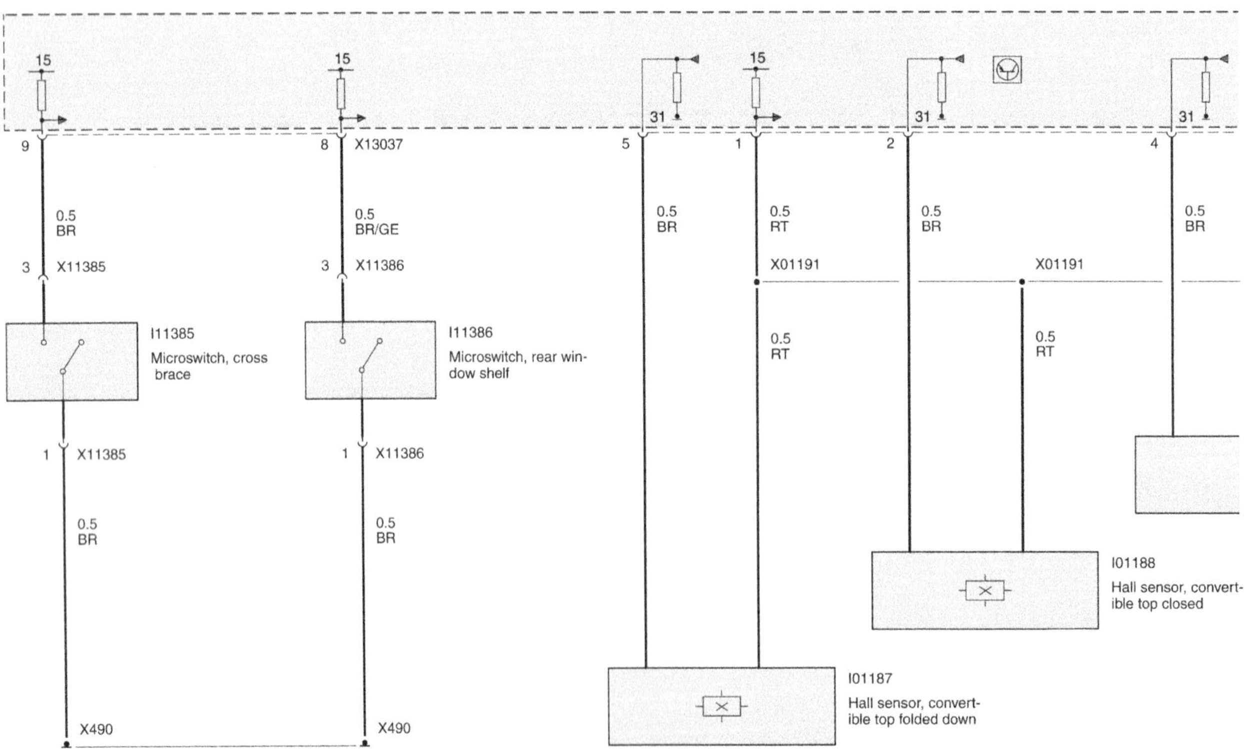

MC06ele21304a

Convertible top
Sensors (up to Sept. 2005)
(page 2 of 2)

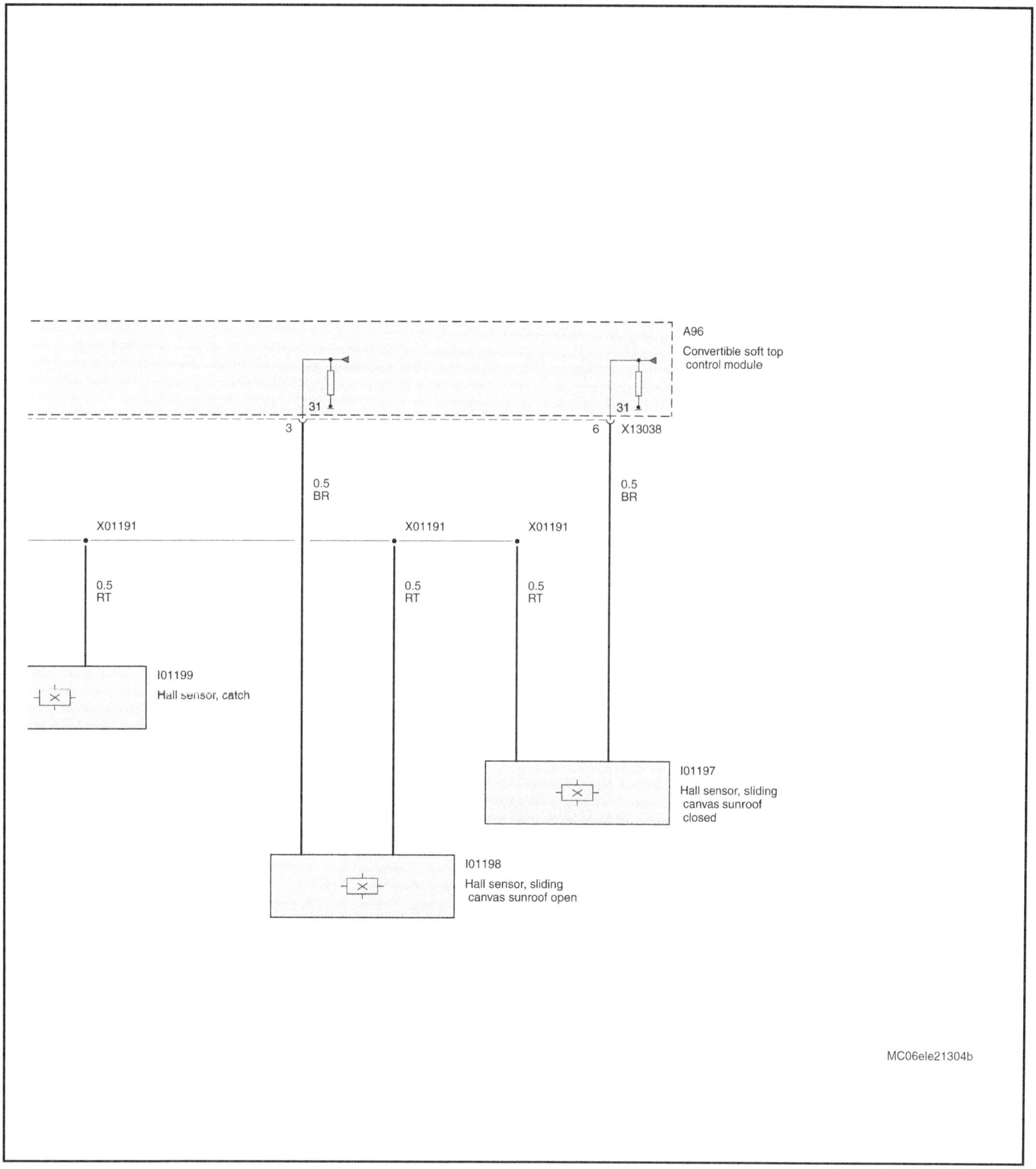

Convertible top
Sensors (from Sept. 2005)
(page 1 of 3)

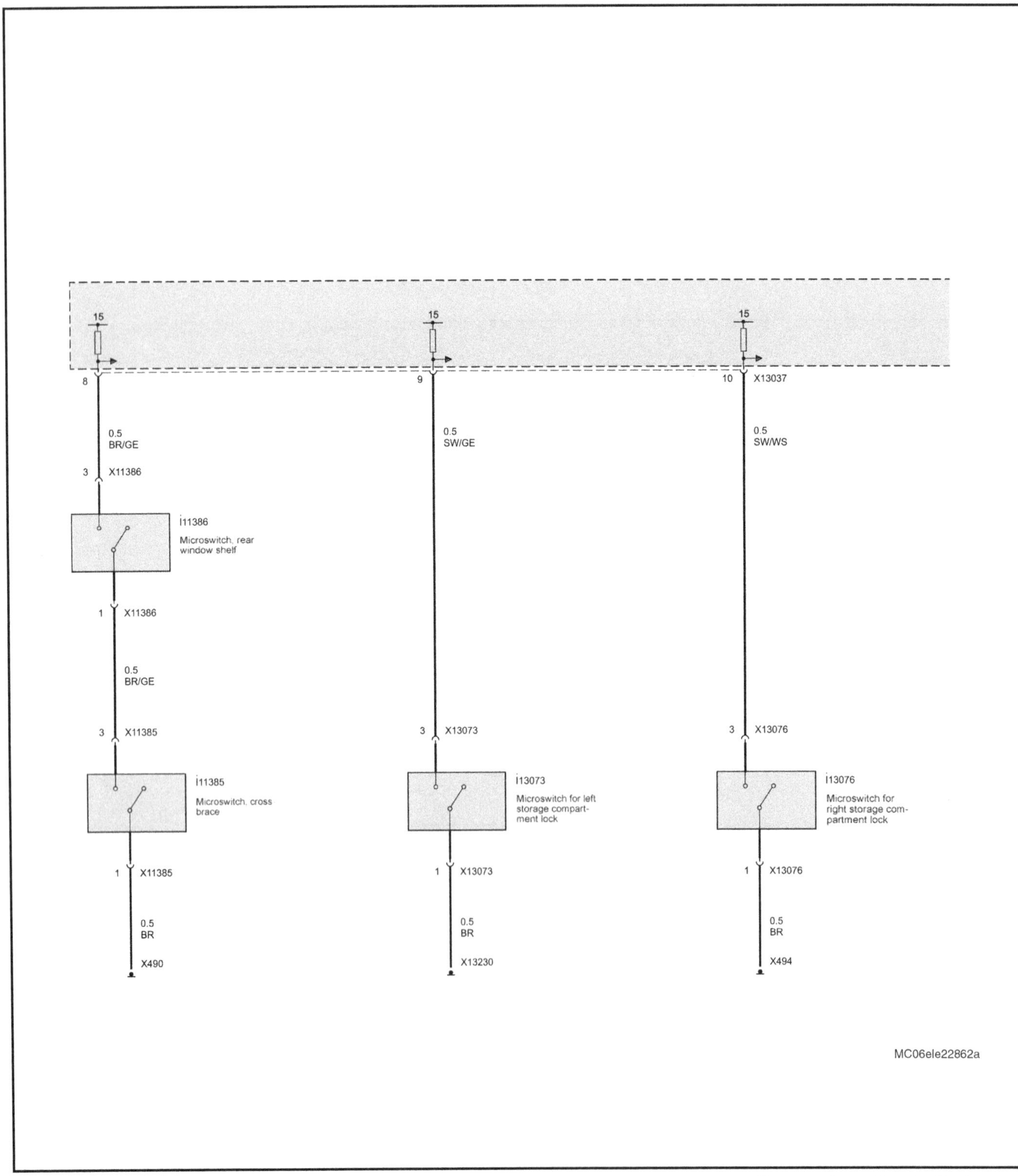

Convertible top

Sensors (from Sept. 2005)

(page 2 of 3)

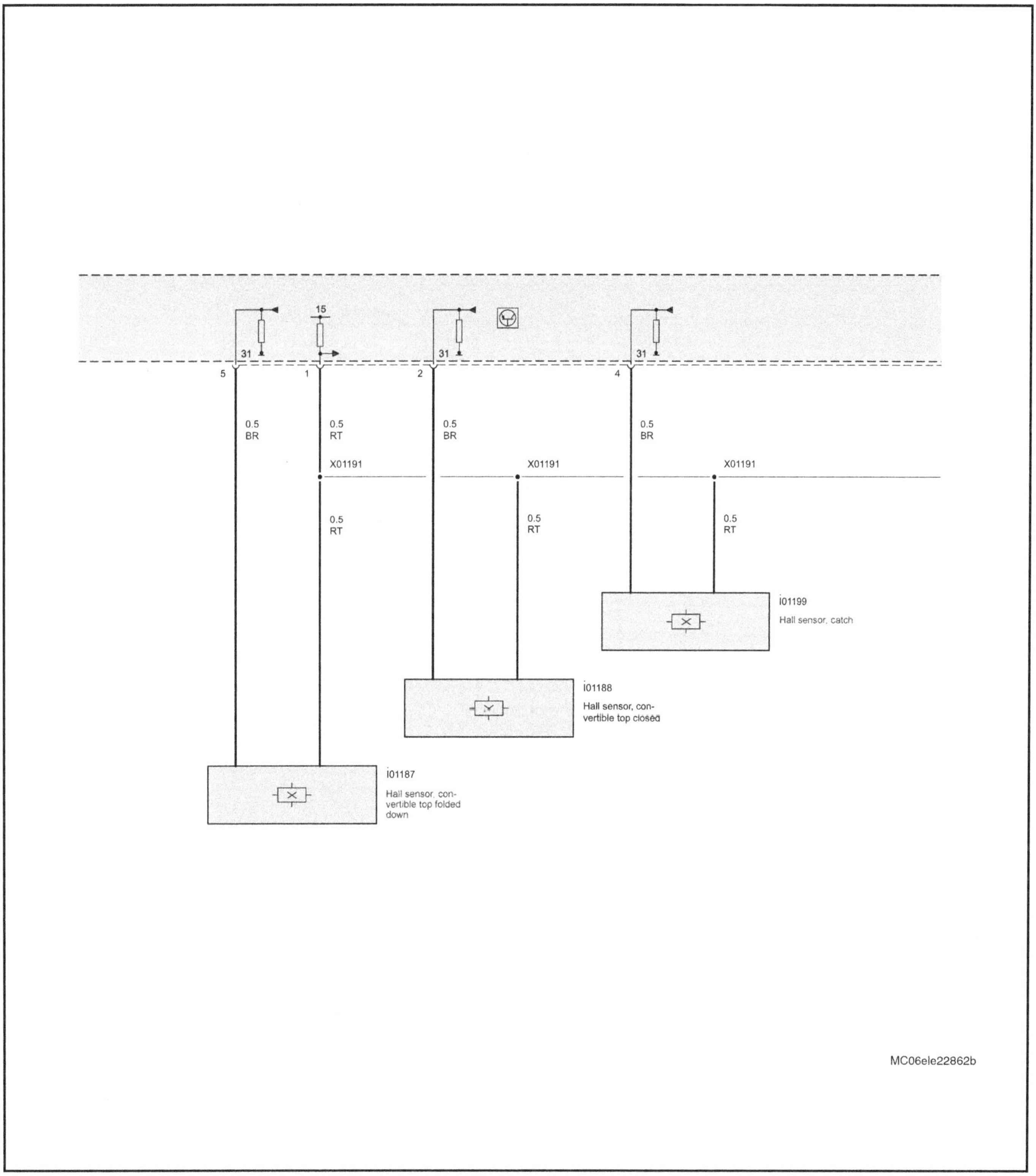

Convertible top
Sensors (from Sept. 2005)
(page 3 of 3)

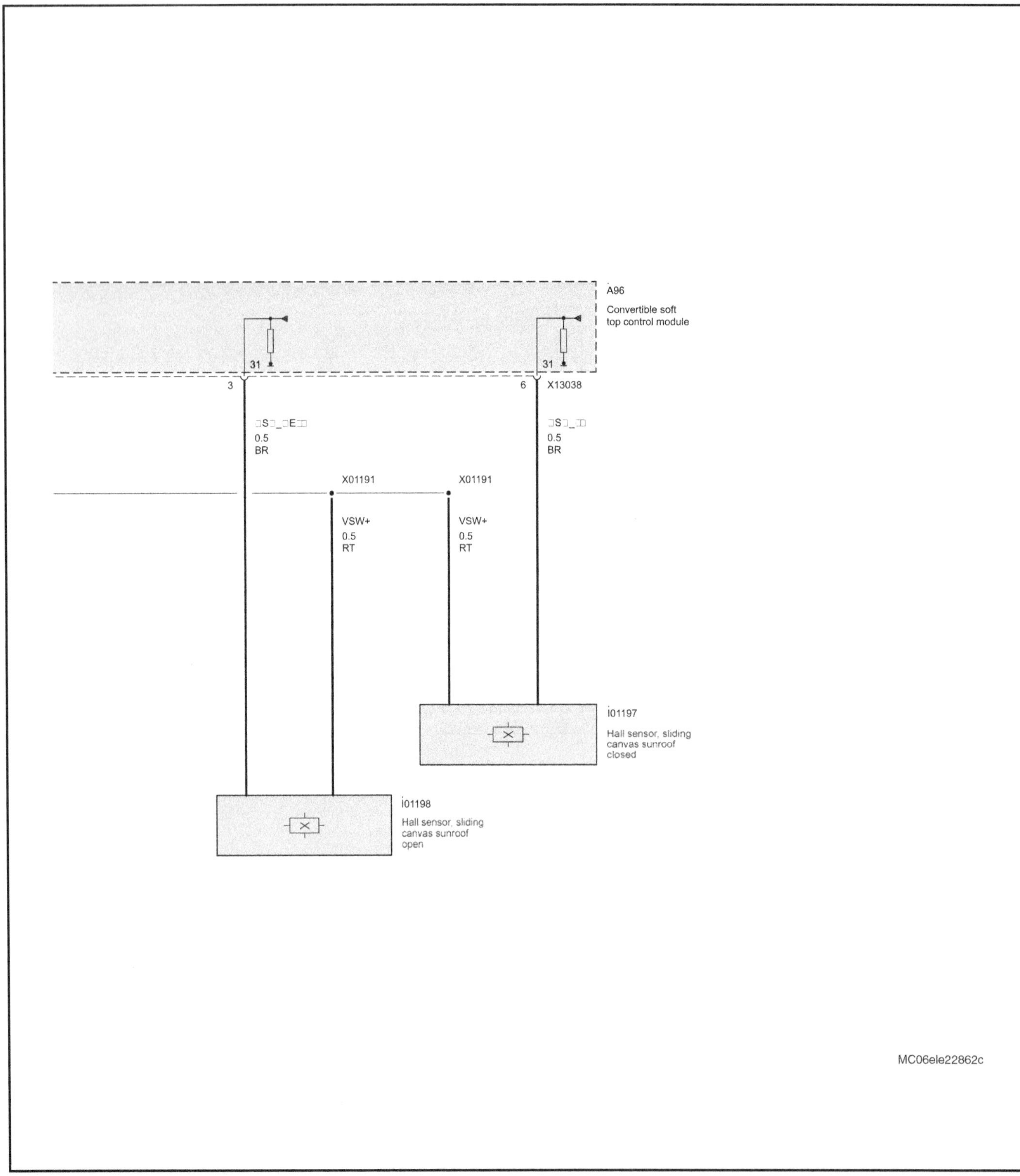

Electrochromic rear view mirror
(up to July 2004)
(page 1 of 1)

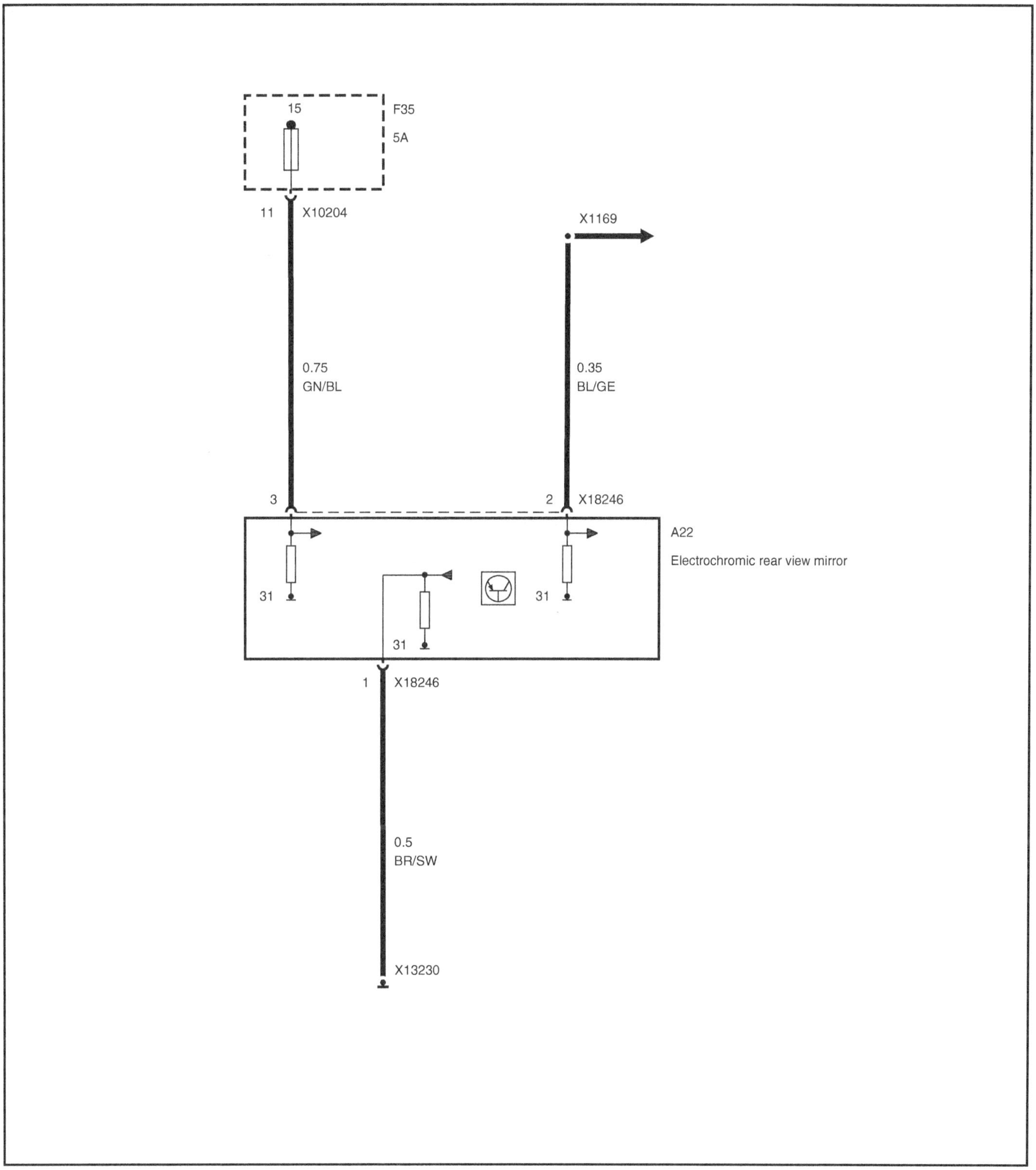

Electrochromic rear view mirror
(from July 2004)
(page 1 of 1)

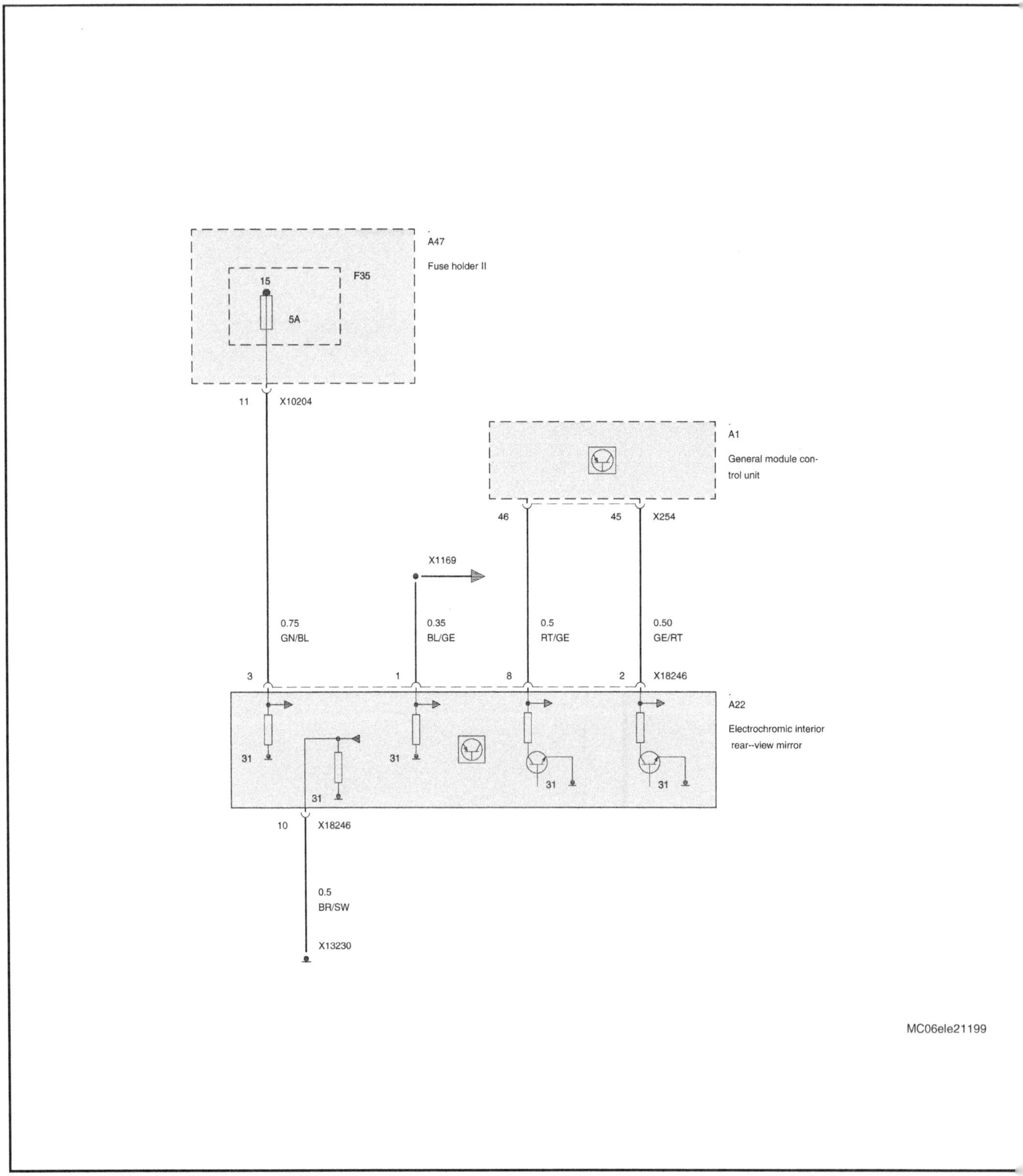

Electronic immobilizer (EWS) (A836)

Cooper (from March 2002)

(page 1 of 2)

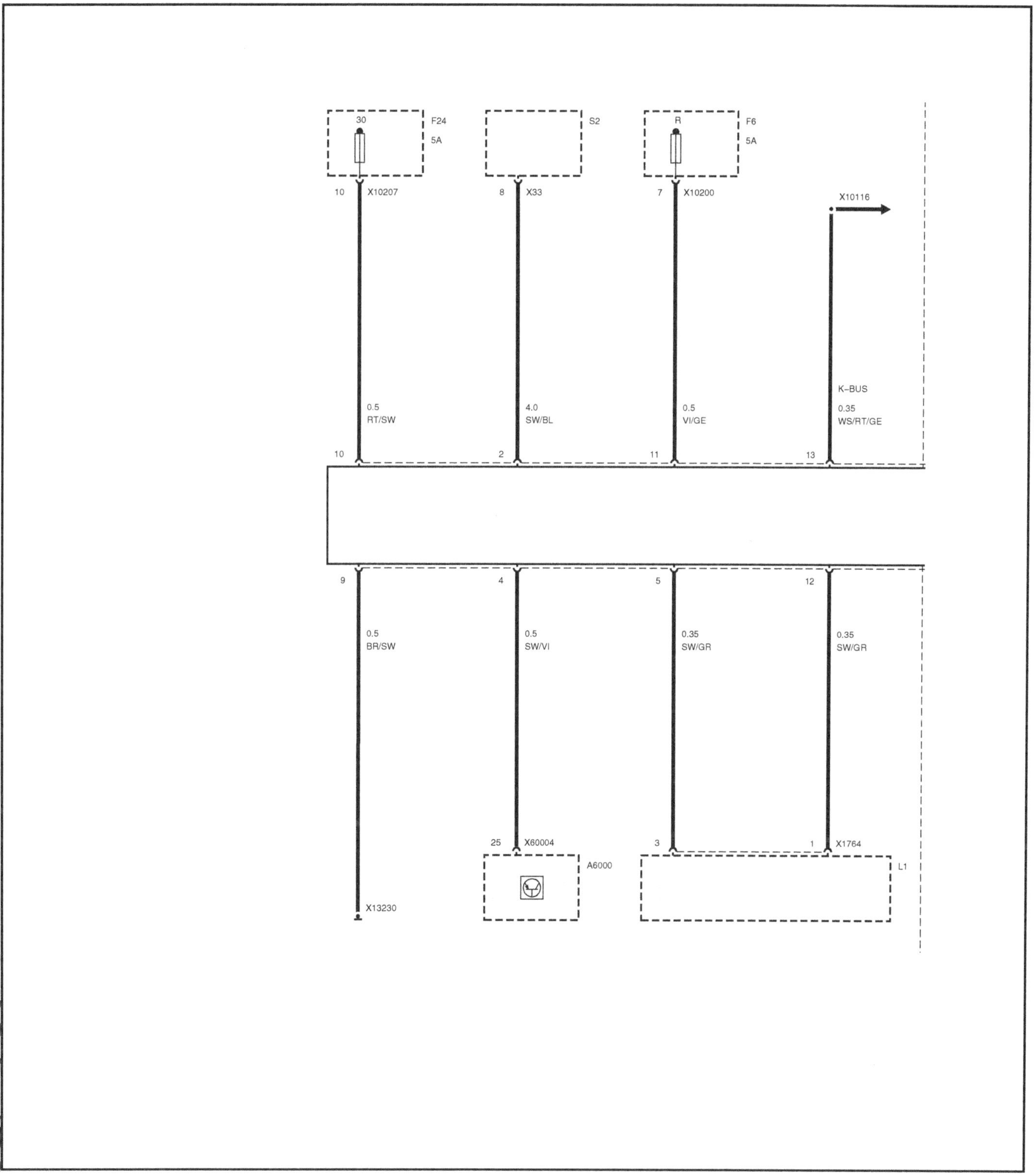

Electronic immobilizer (EWS) (A836)
Cooper (from March 2002)
(page 2 of 2)

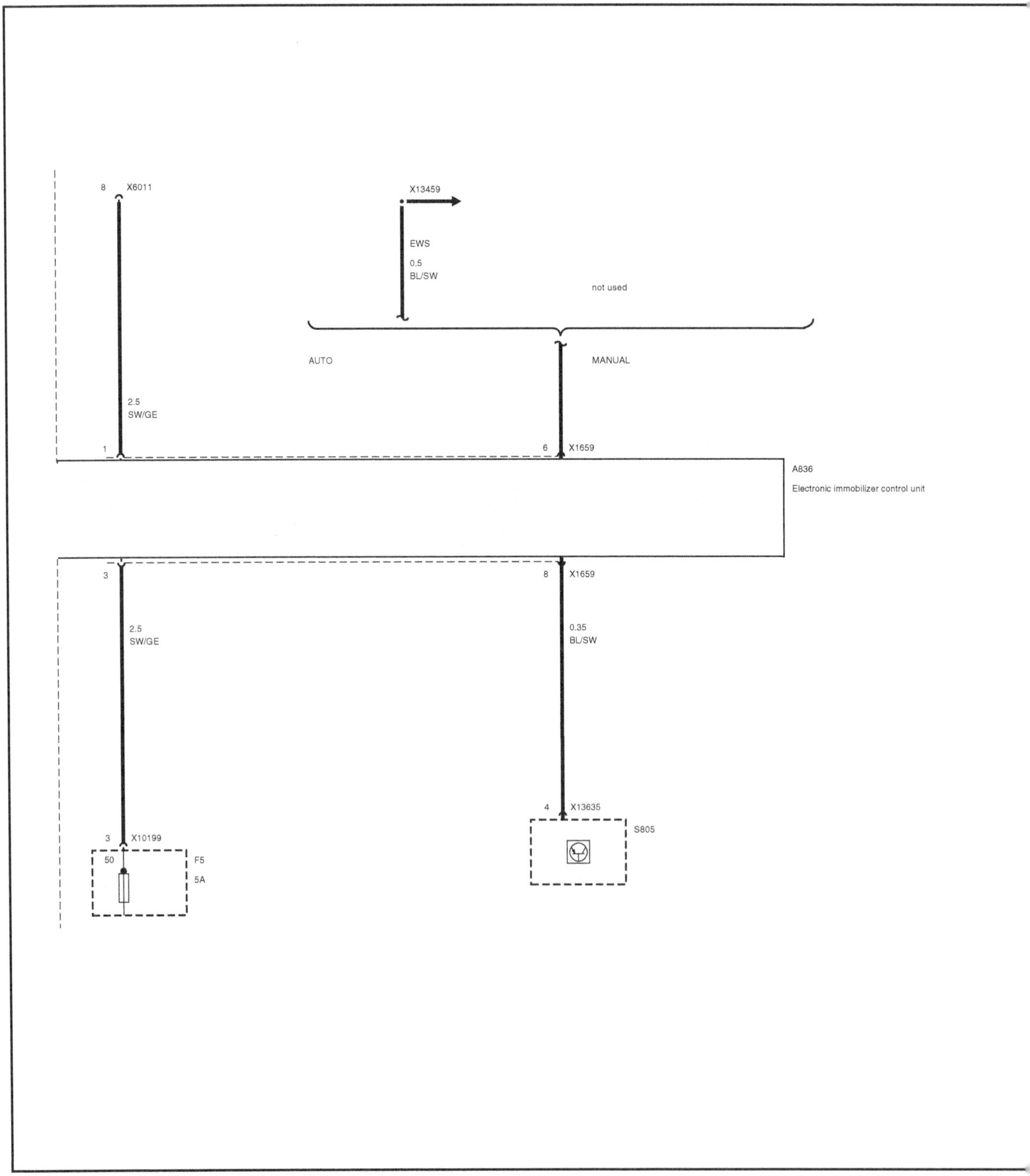

Electronic immobilizer (EWS) (A836)
Cooper S (from Sept. 2002)
(page 1 of 2)

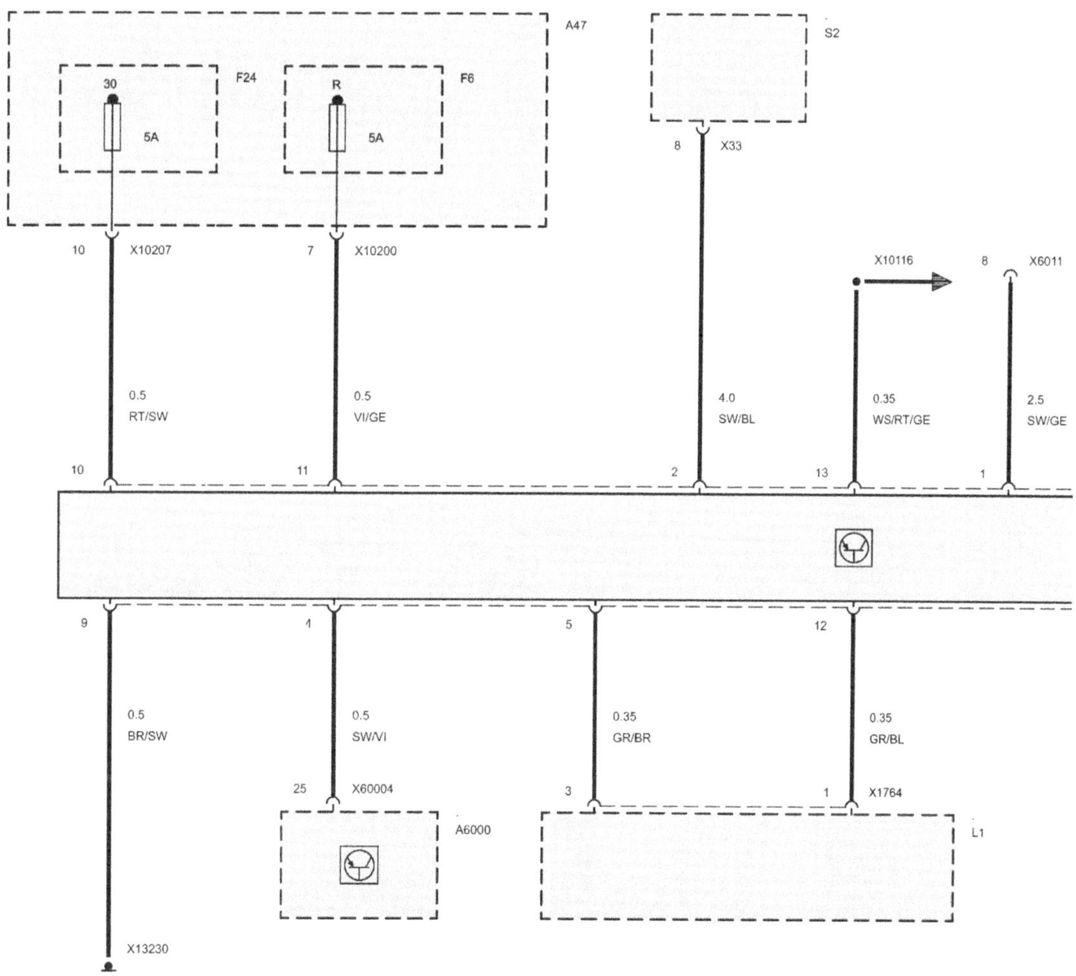

MC06ele22197

Electronic immobilizer (EWS) (A836)
Cooper S (from Sept. 2002)
(page 2 of 2)

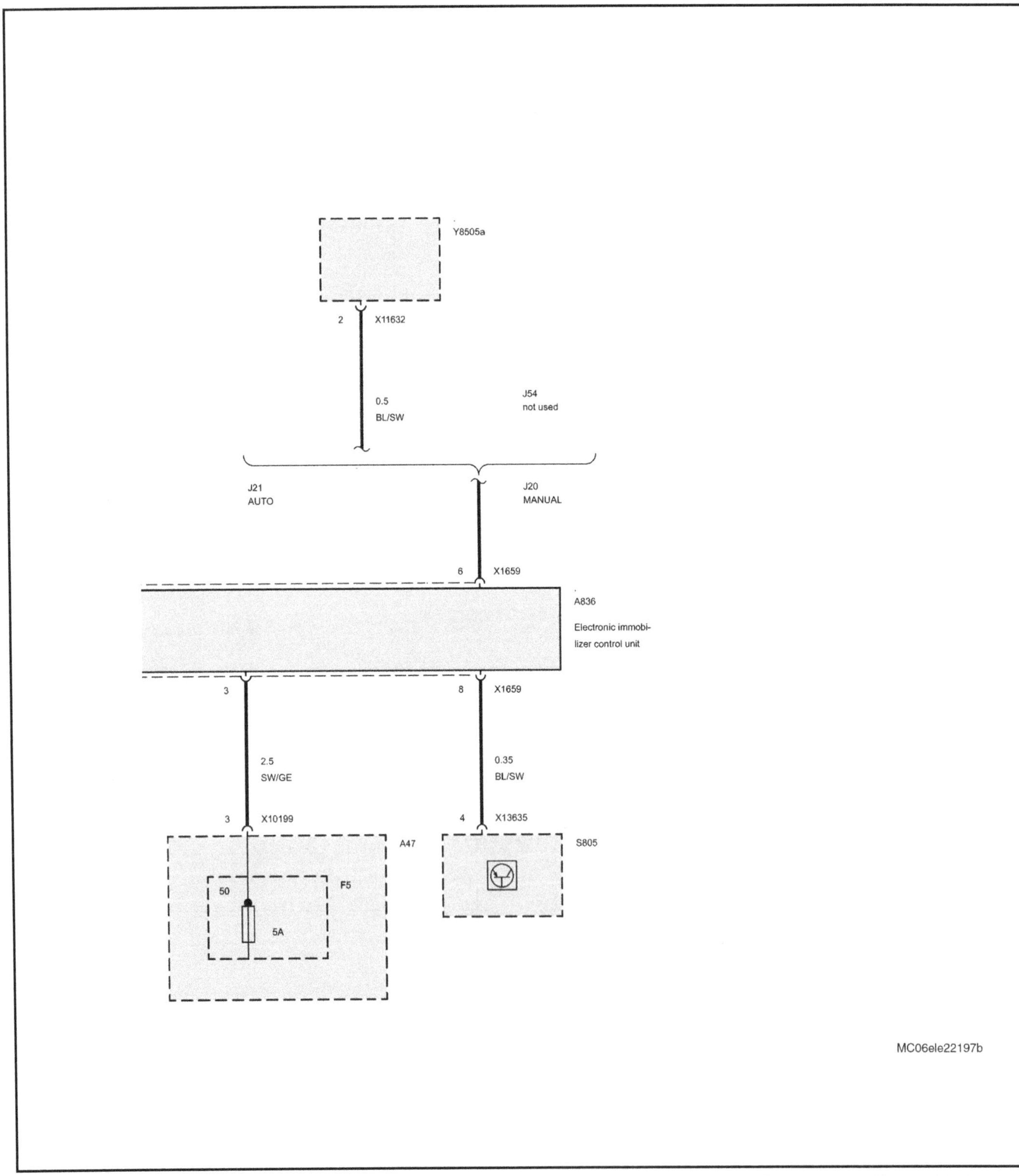

Engine cooling

Cooper (up to March 2003)

(page 1 of 1)

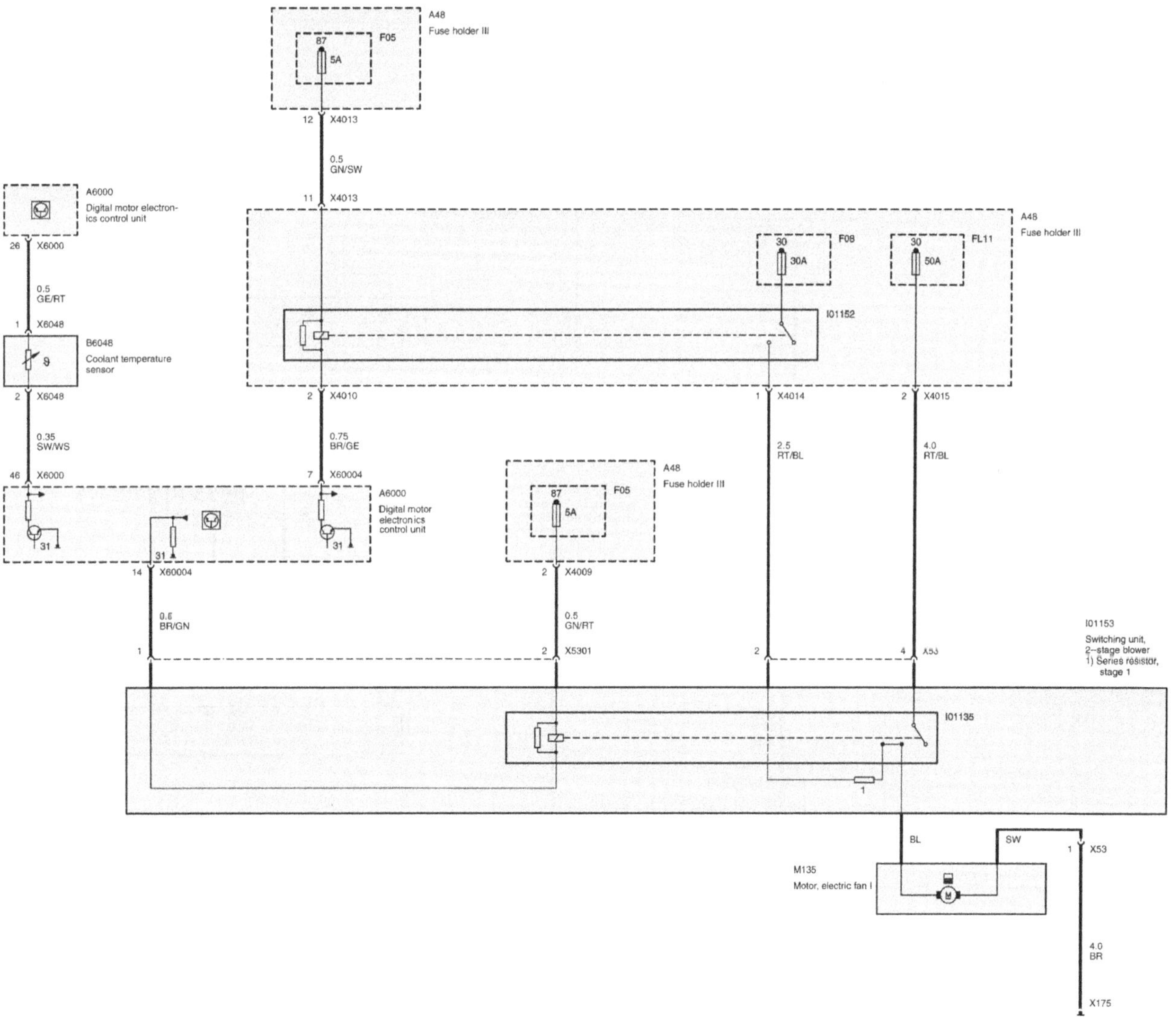

MC06ele15989

Engine cooling
Cooper and Cooper S (from March 2003)
(page 1 of 2)

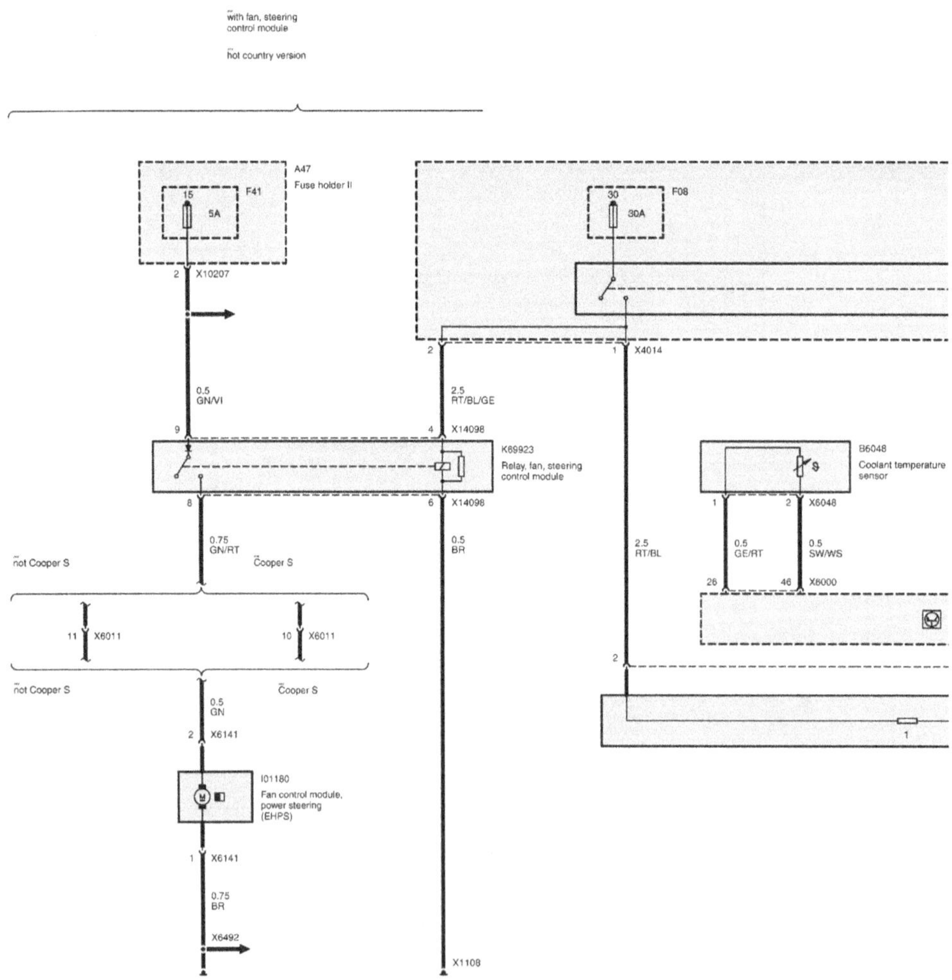

MC06ele19592a

Engine cooling

Cooper and Cooper S (from March 2003)

(page 2 of 2)

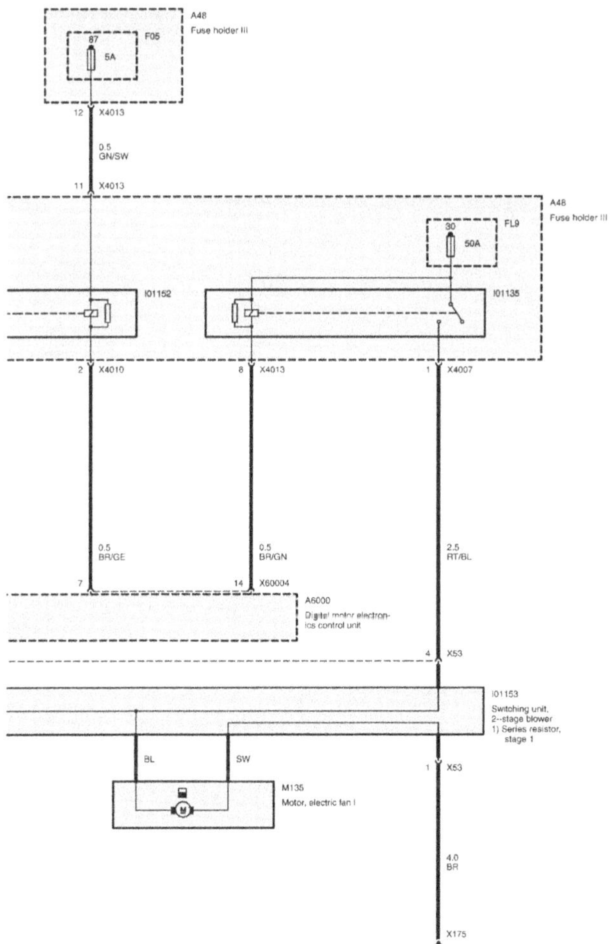

MC06ele19592b

Engine management (DME)

Cooper (from March 2003 to Sept. 2005)

(page 1 of 6)

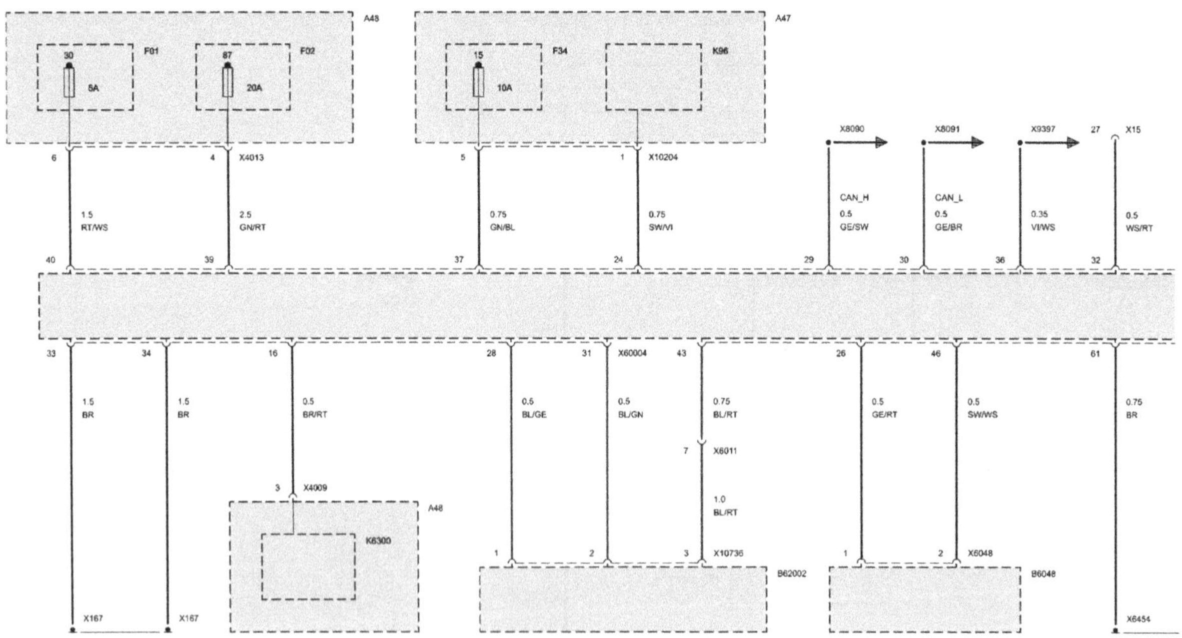

MC06ele19552a

Engine management (DME)

Cooper (from March 2003 to Sept. 2005)

(page 2 of 6)

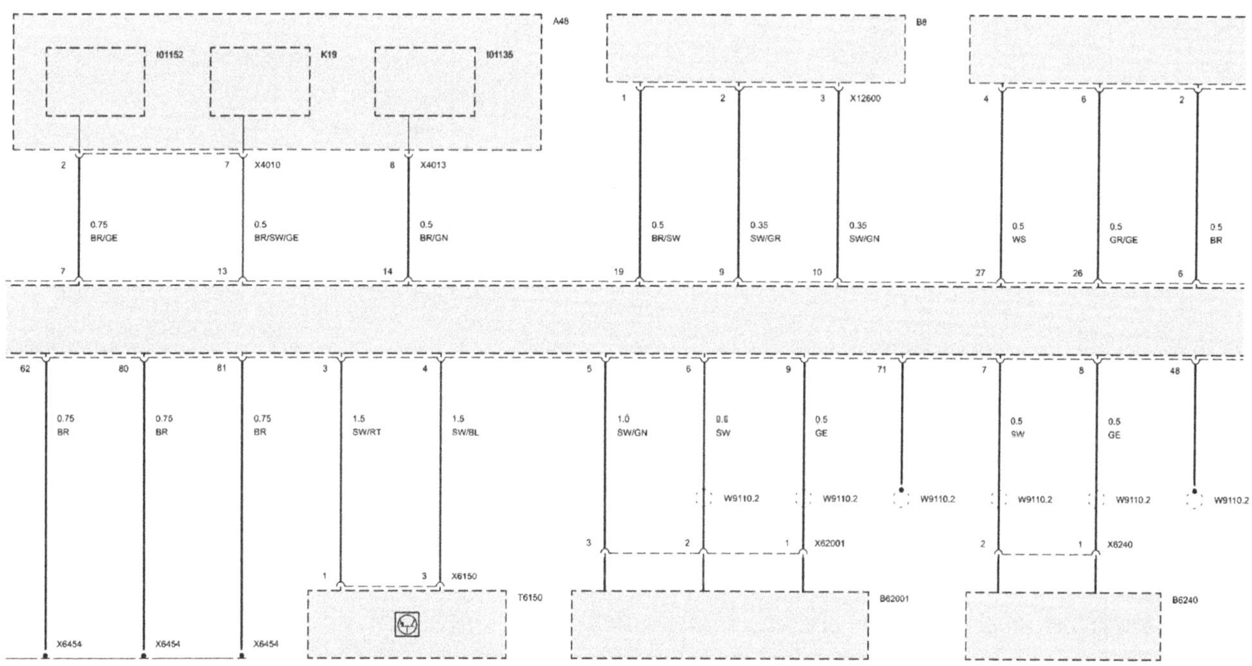

MC06ele19552b

Engine management (DME)

Cooper (from March 2003 to Sept. 2005)

(page 3 of 6)

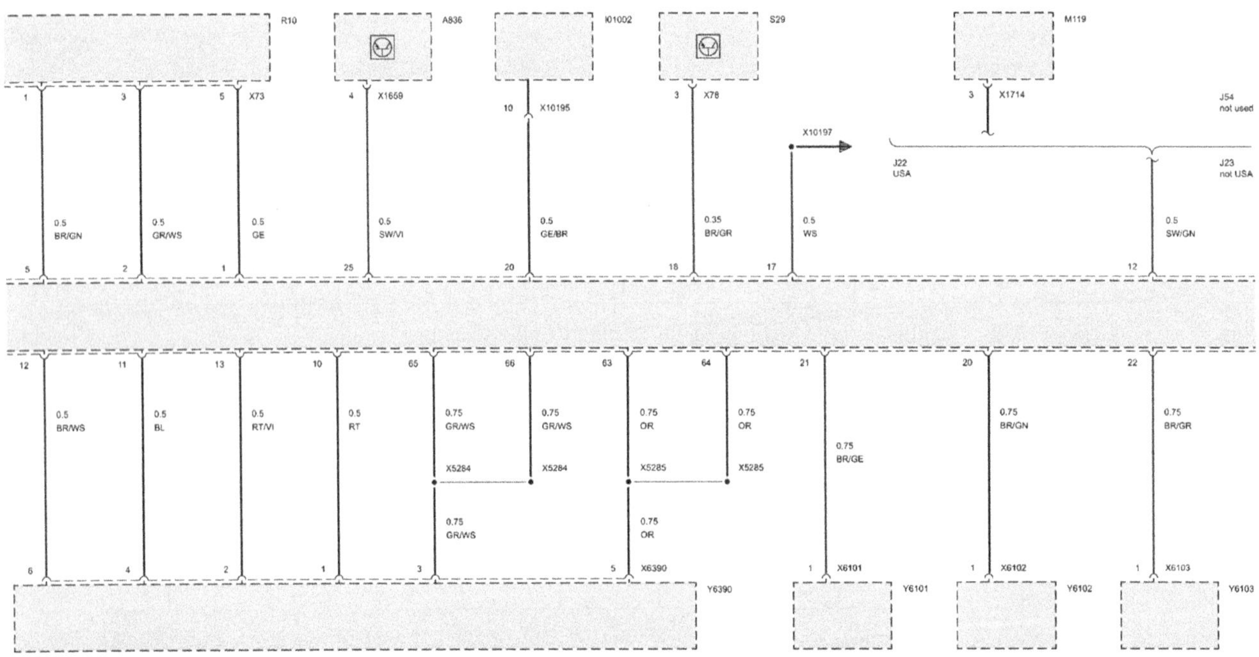

MC06ele19552c

Engine management (DME)
Cooper (from March 2003 to Sept. 2005)
(page 4 of 6)

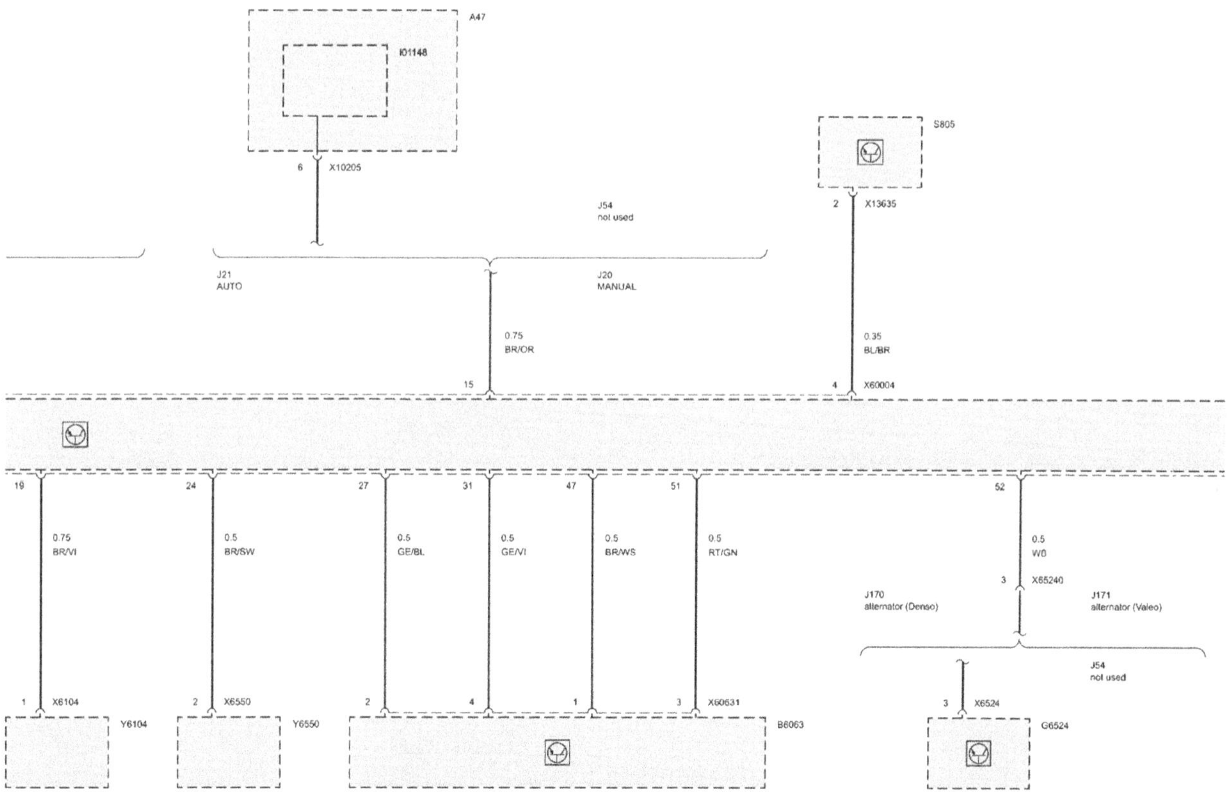

MC06ele19552d

Engine management (DME)

Cooper (from March 2003 to Sept. 2005)

(page 5 of 6)

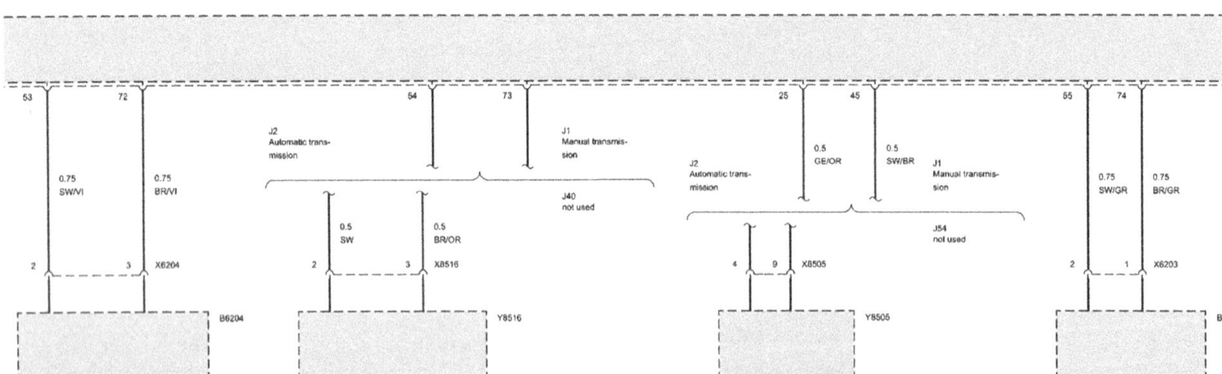

MC06ele19552e

Engine management (DME)

Cooper (from March 2003 to Sept. 2005)

(page 6 of 6)

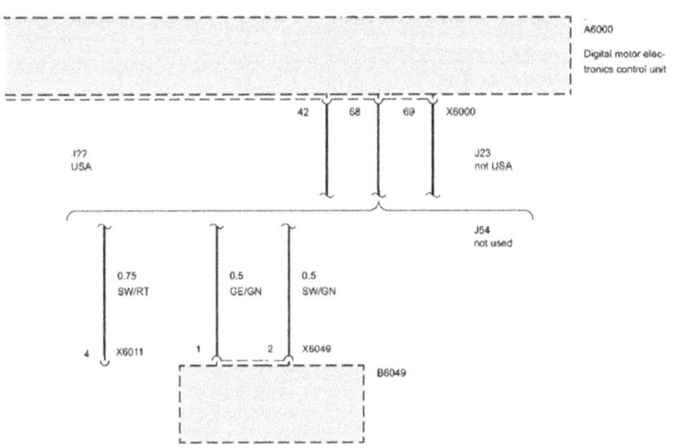

MC06ele19552f

Engine management (DME)
Cooper S (from March 2003 to Sept. 2005)
(page 1 of 6)

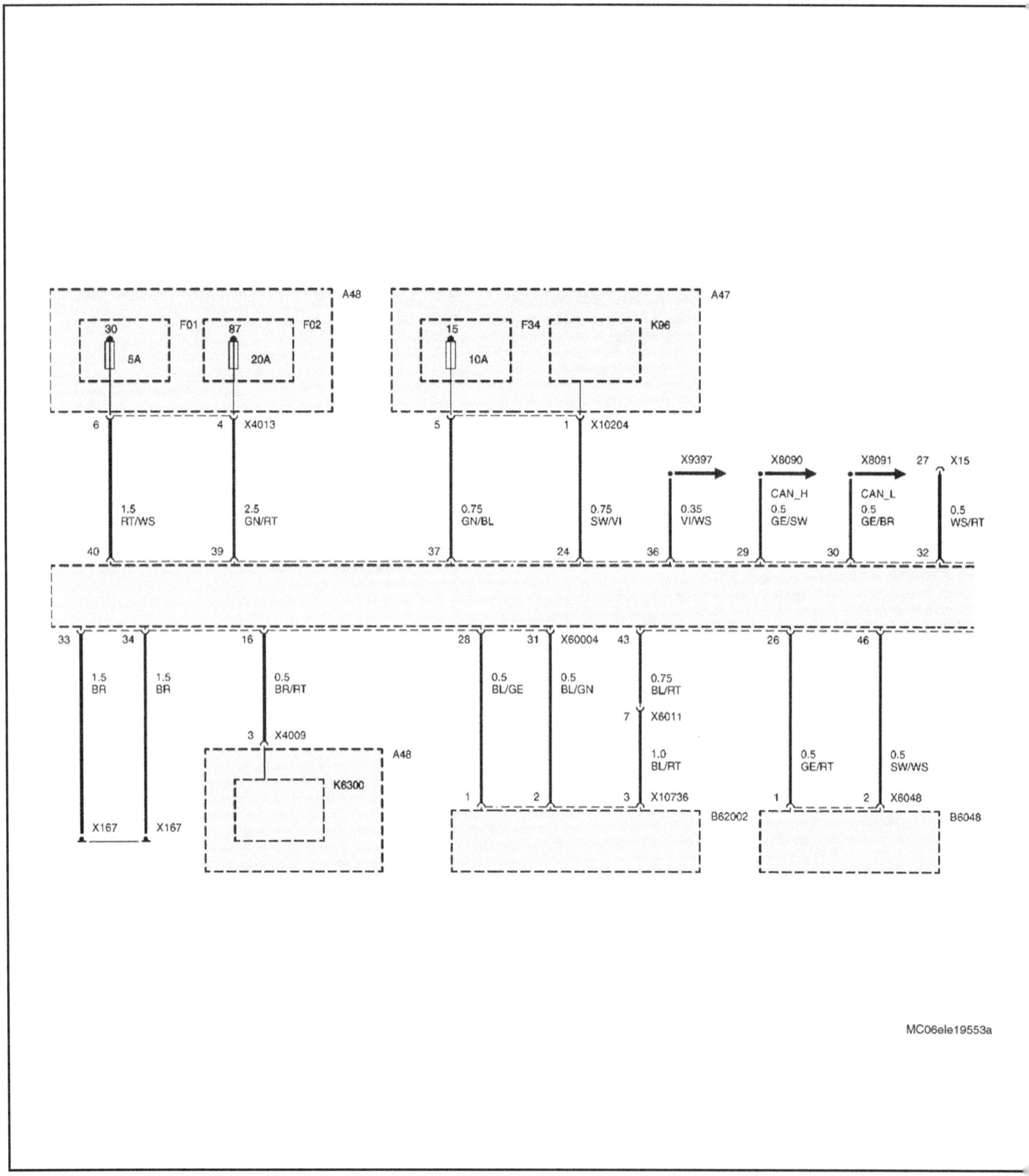

Engine management (DME)
Cooper S (from March 2003 to Sept. 2005)
(page 2 of 6)

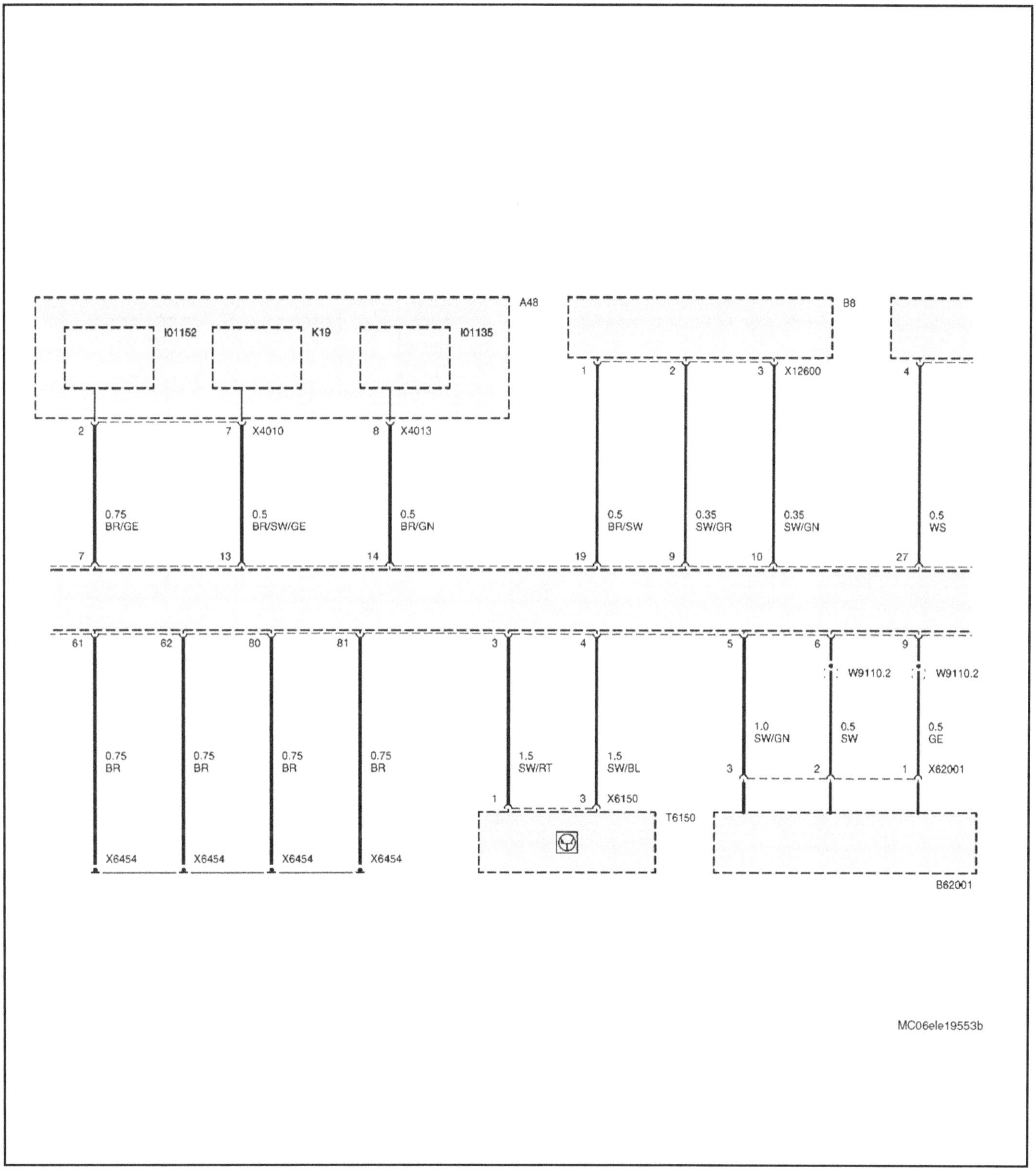

Engine management (DME)
Cooper S (from March 2003 to Sept. 2005)
(page 3 of 6)

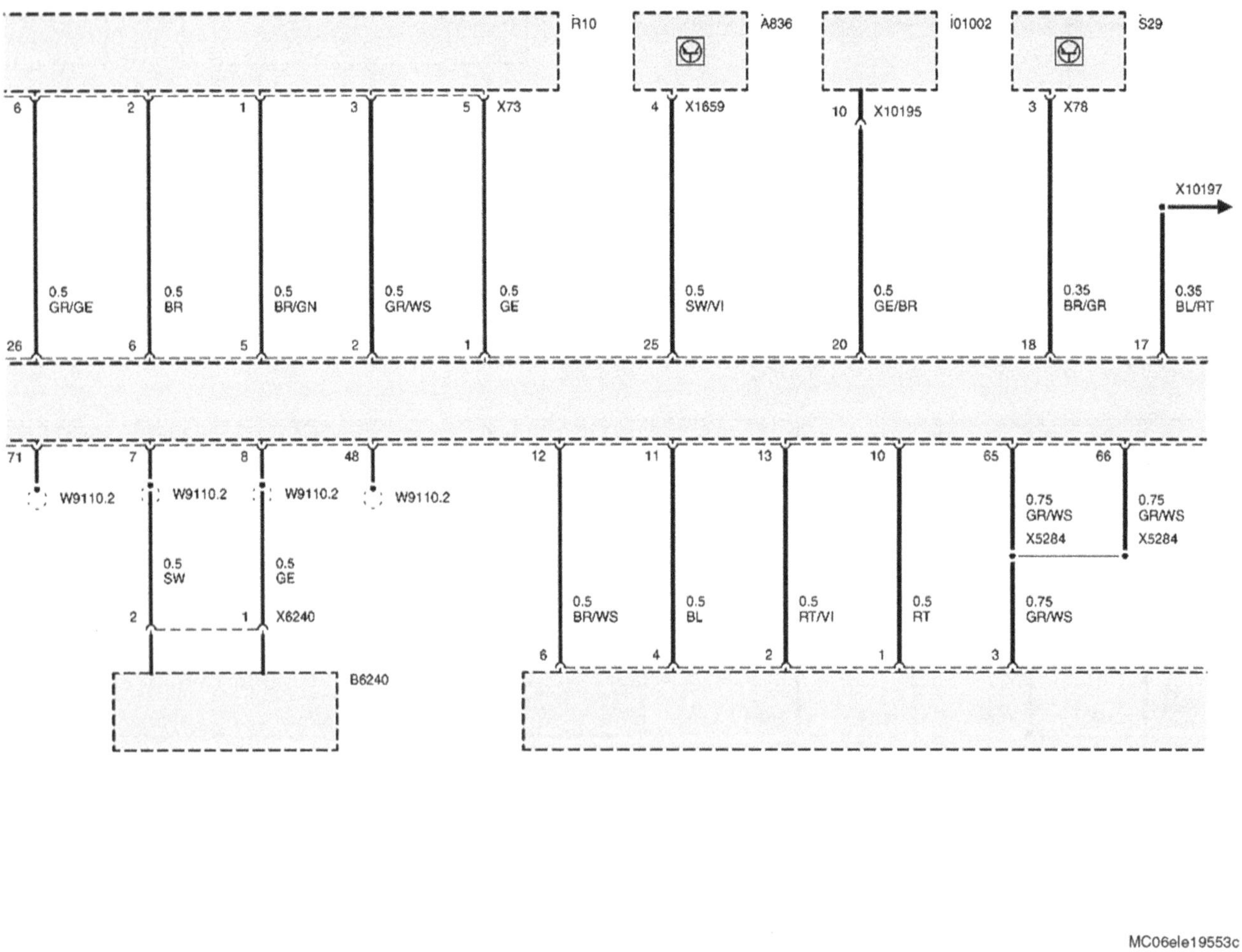

Engine management (DME)
Cooper S (from March 2003 to Sept. 2005)
(page 4 of 6)

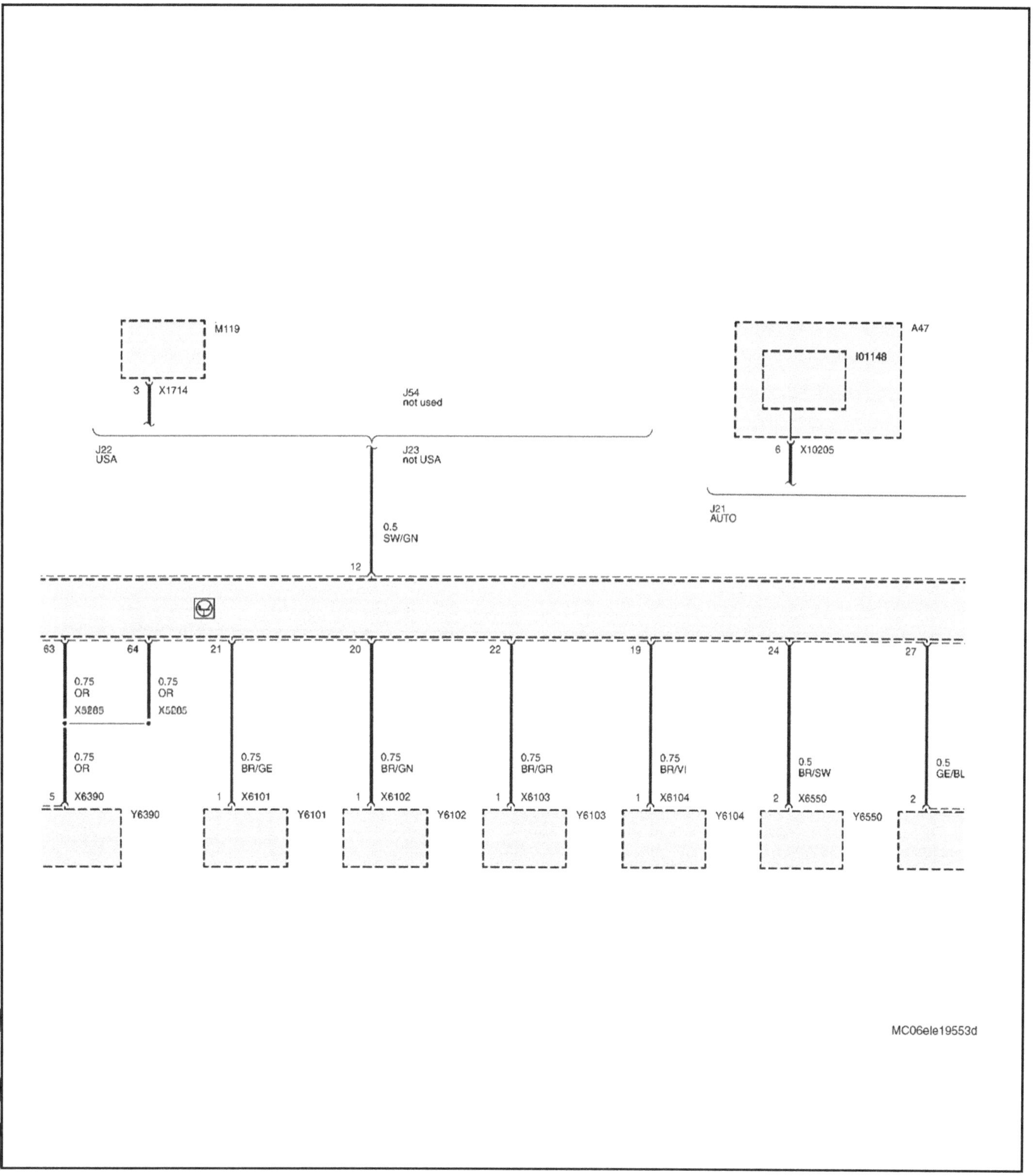

Engine management (DME)
Cooper S (from March 2003 to Sept. 2005)
(page 5 of 6)

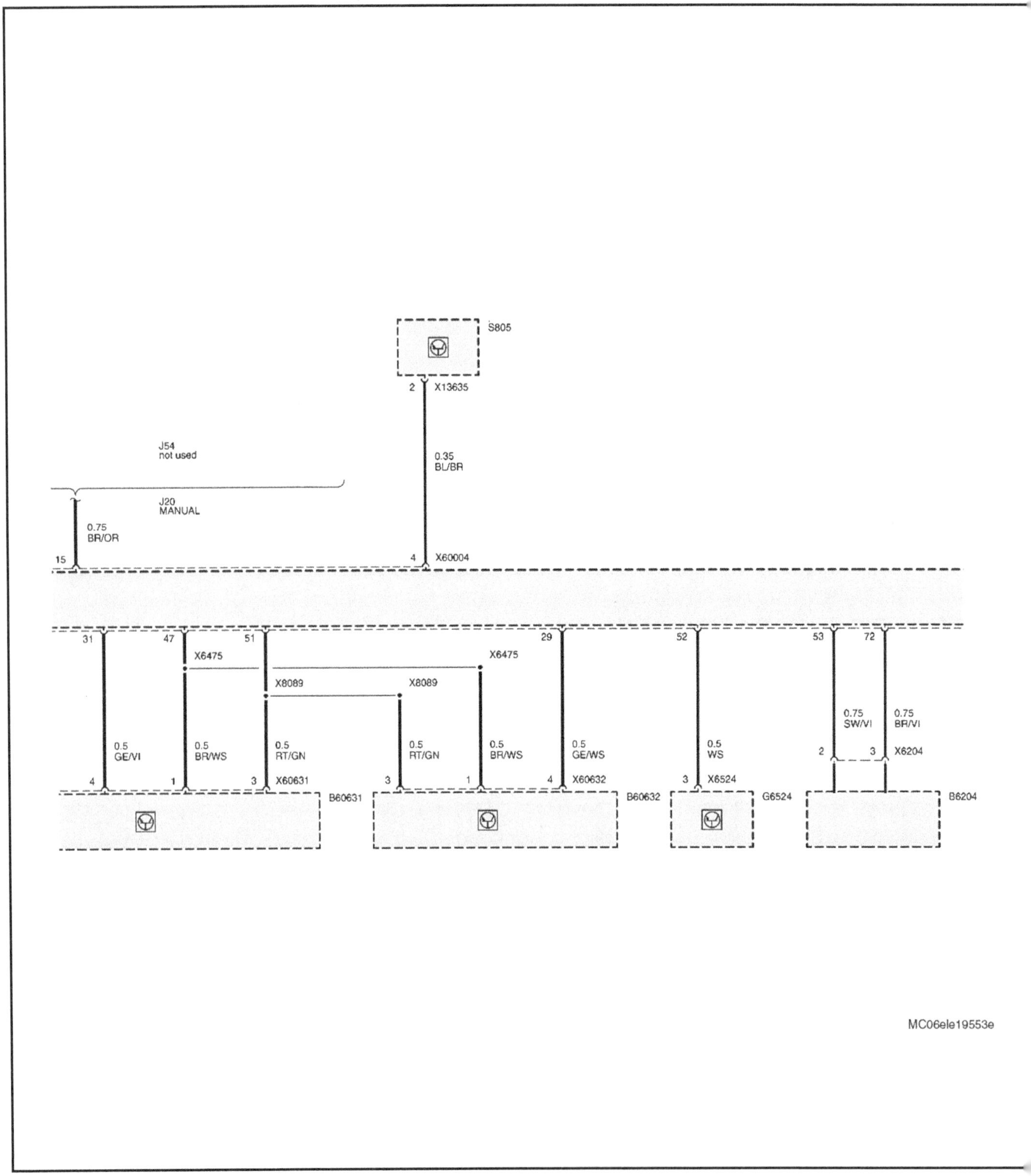

Engine management (DME)
Cooper S (from March 2003 to Sept. 2005)
(page 6 of 6)

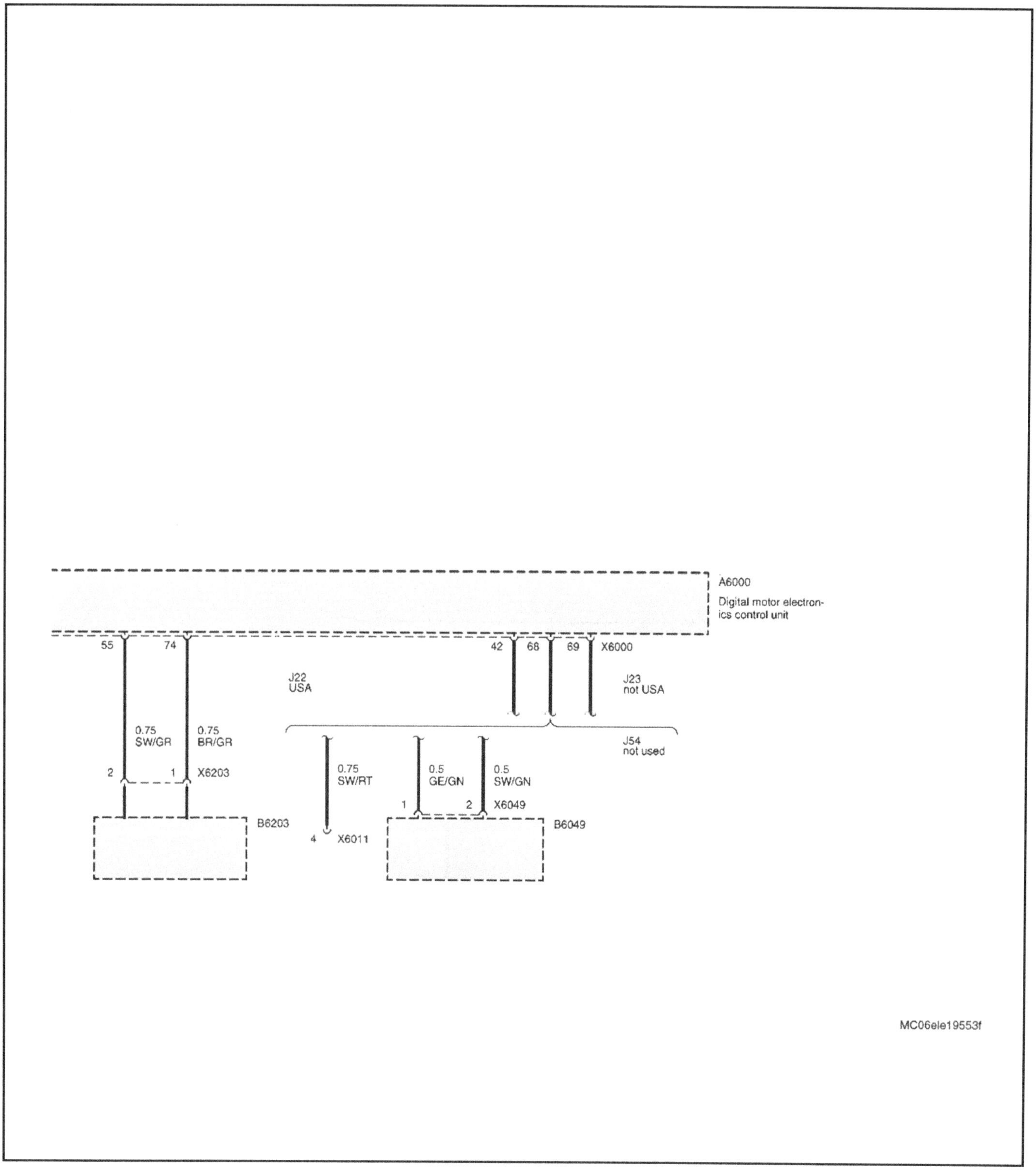

Engine management (DME)
Fuel pump relay (K96) (up to Sept. 2002)
(page 1 of 1)

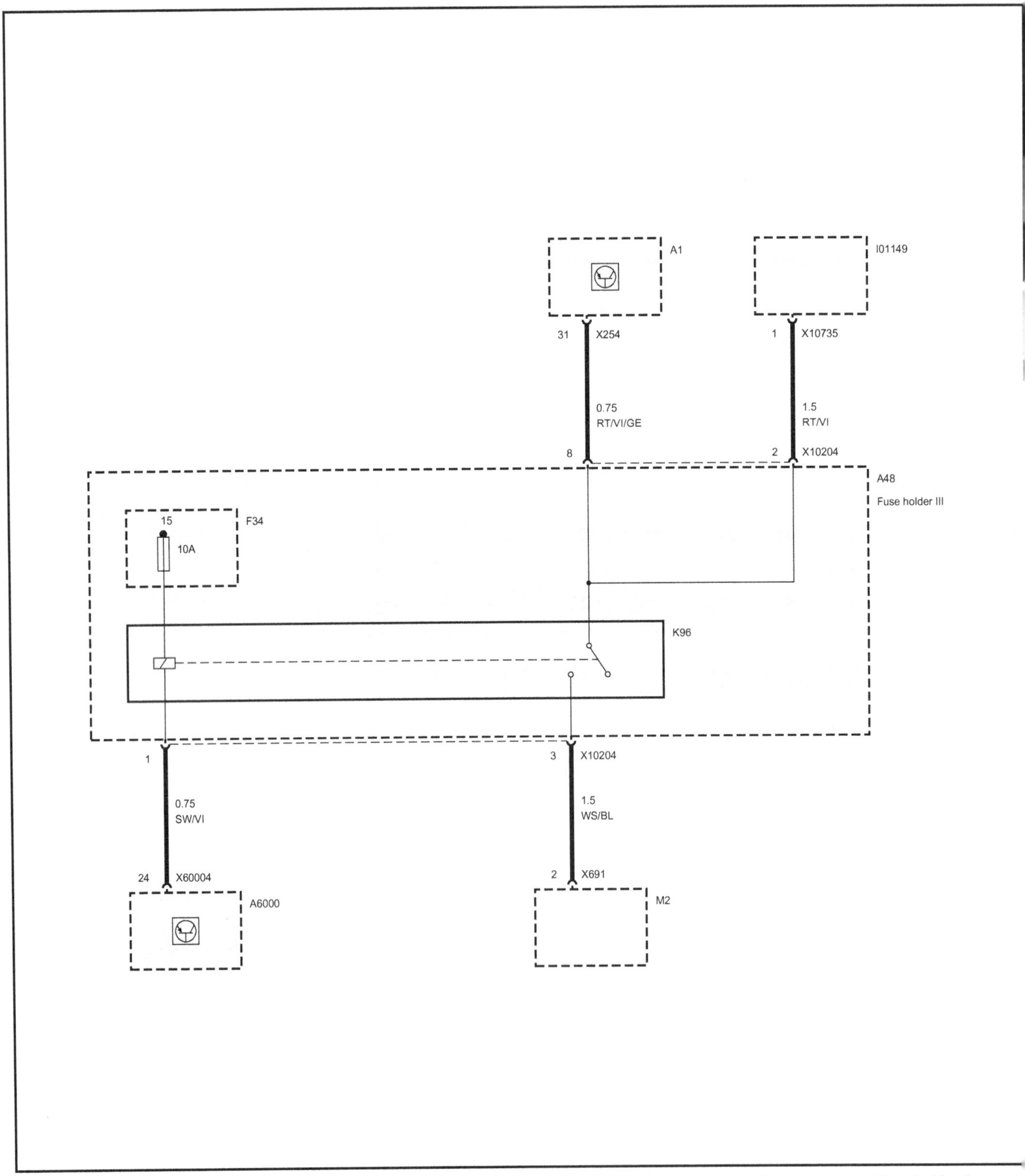

Engine management (DME)
Fuel pump relay (K96) (from Sept. 2002 to July 2004)
(page 1 of 1)

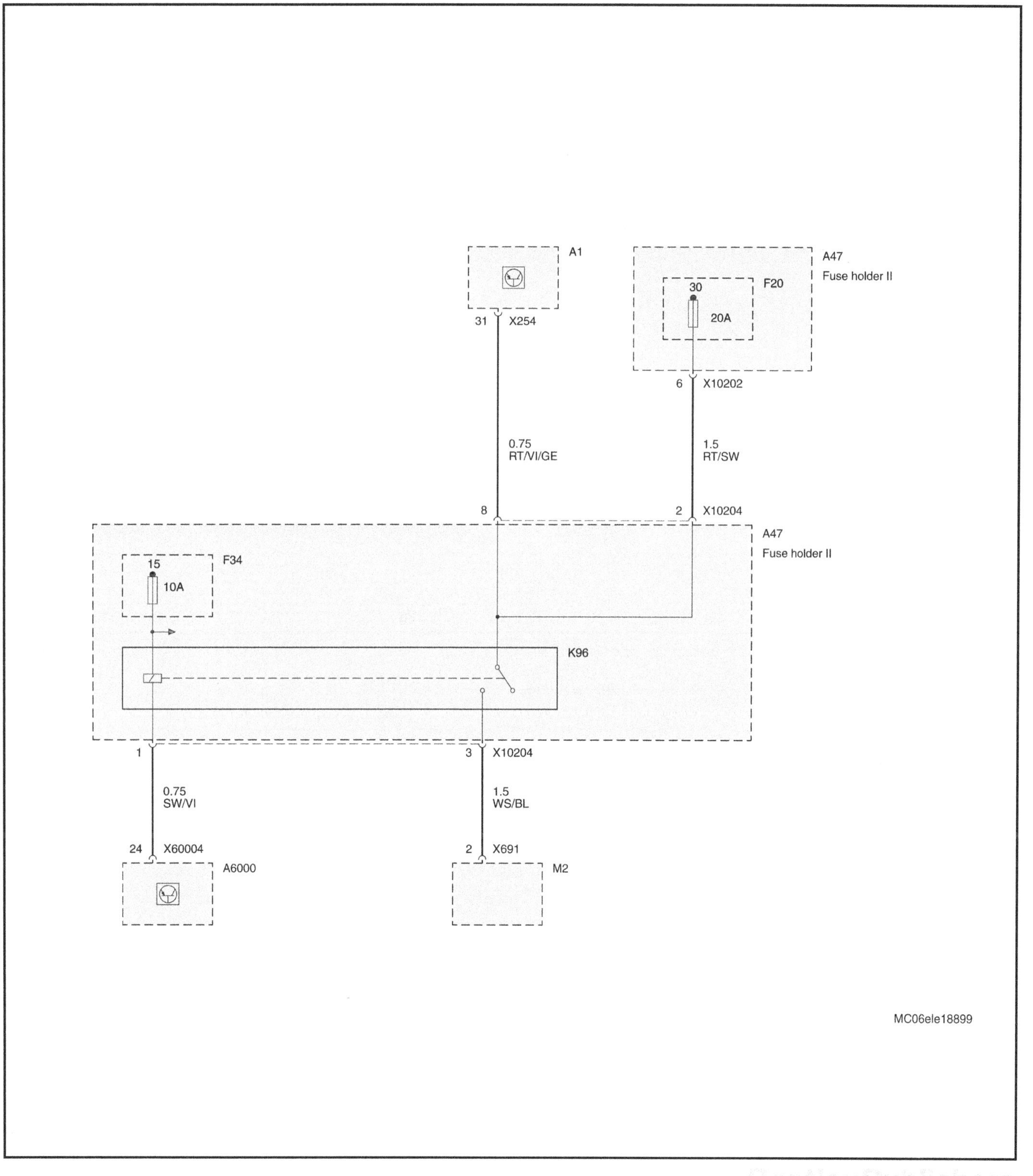

Engine management (DME)

Fuel pump relay (K96) (from July 2004 to Sept. 2005)

(page 1 of 1)

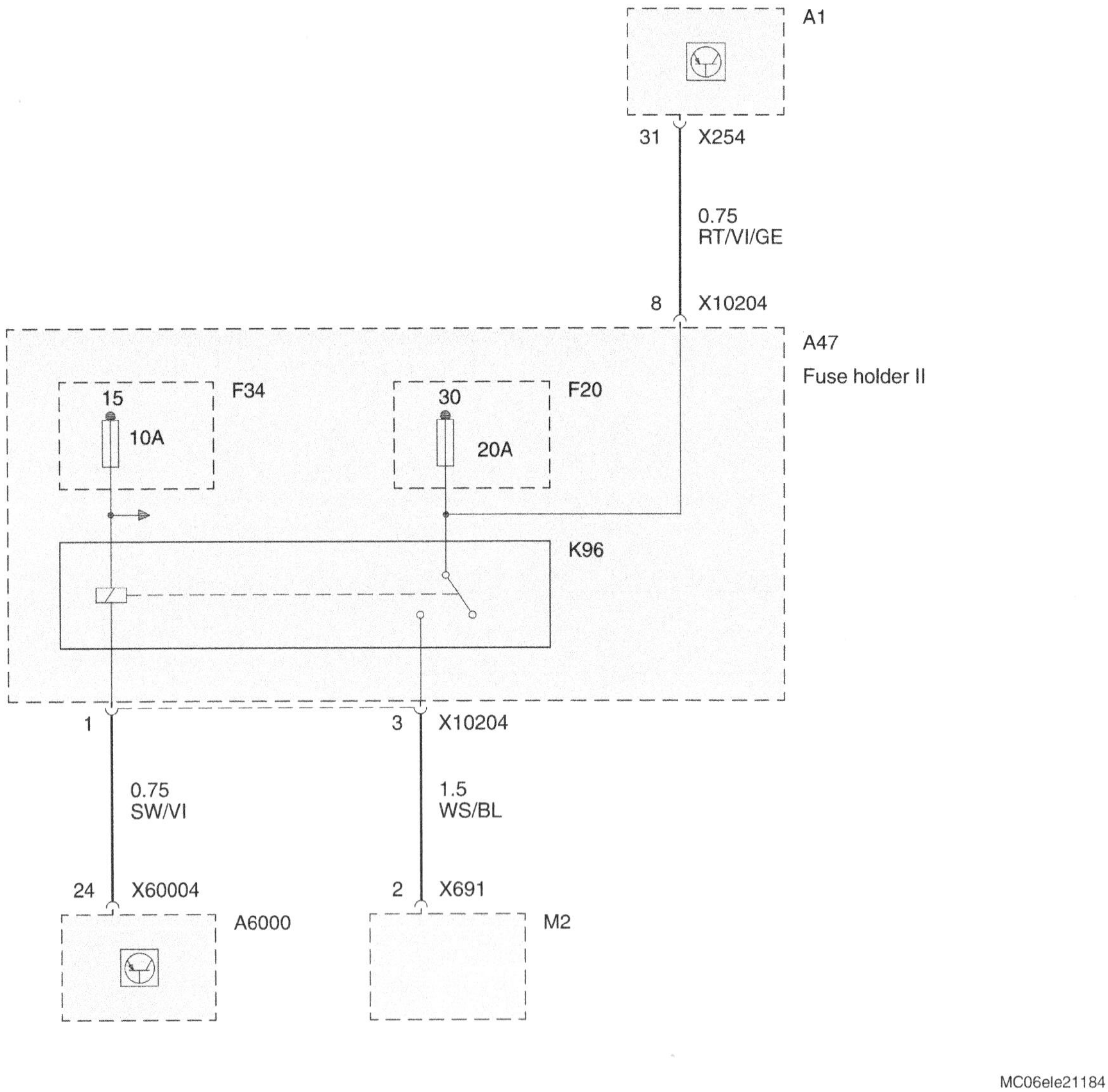

Engine management (DME)
Fuel pump relay (K96) (from Sept. 2005)
(page 1 of 1)

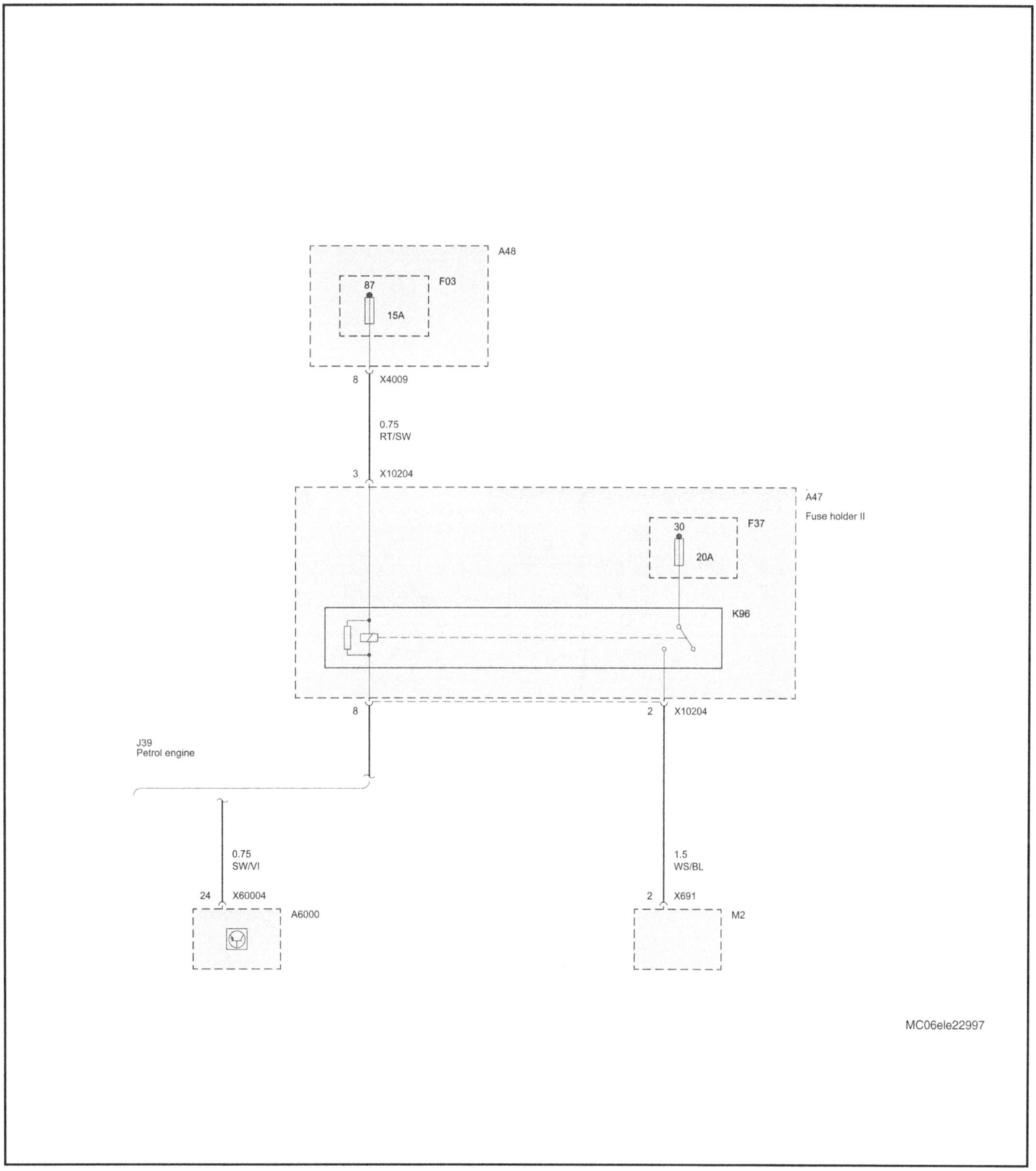

Engine management (DME)
Engine coolant temperature (ECT) sensor (B6048)
(page 1 of 1)

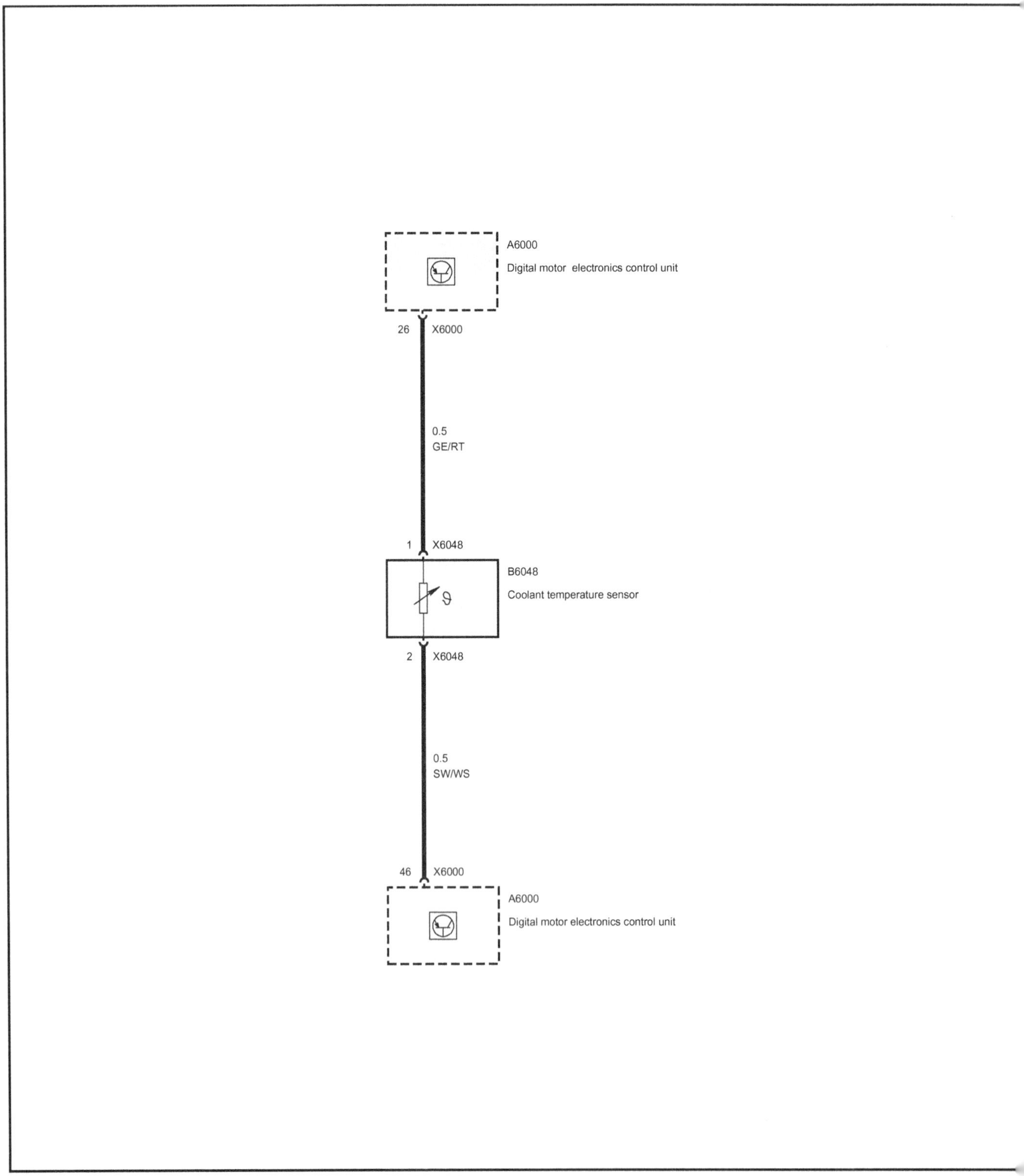

Engine management (DME)
DME relay (K6300)
(page 1 of 2)

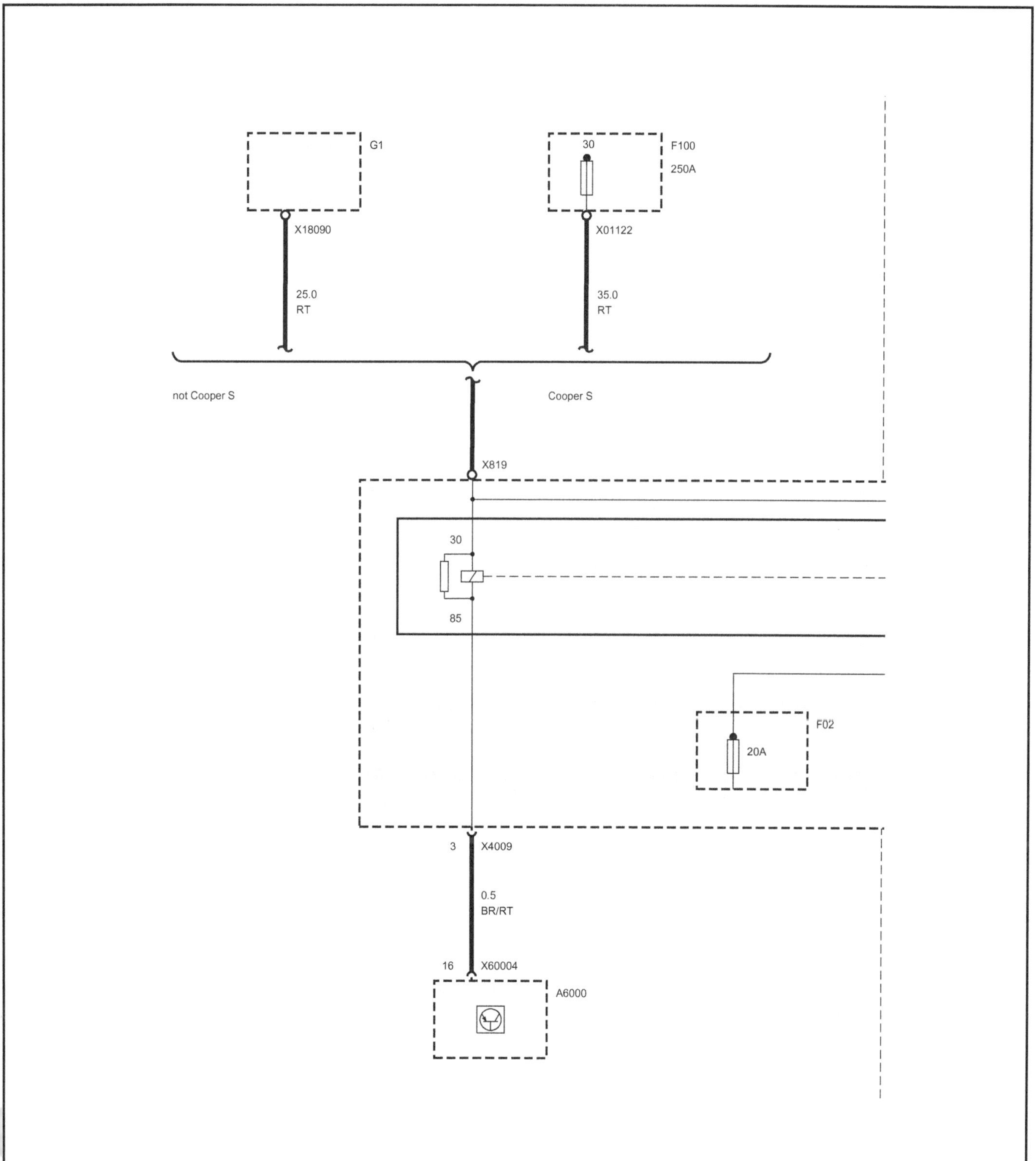

Engine management (DME)
DME relay (K6300)
(page 2 of 2)

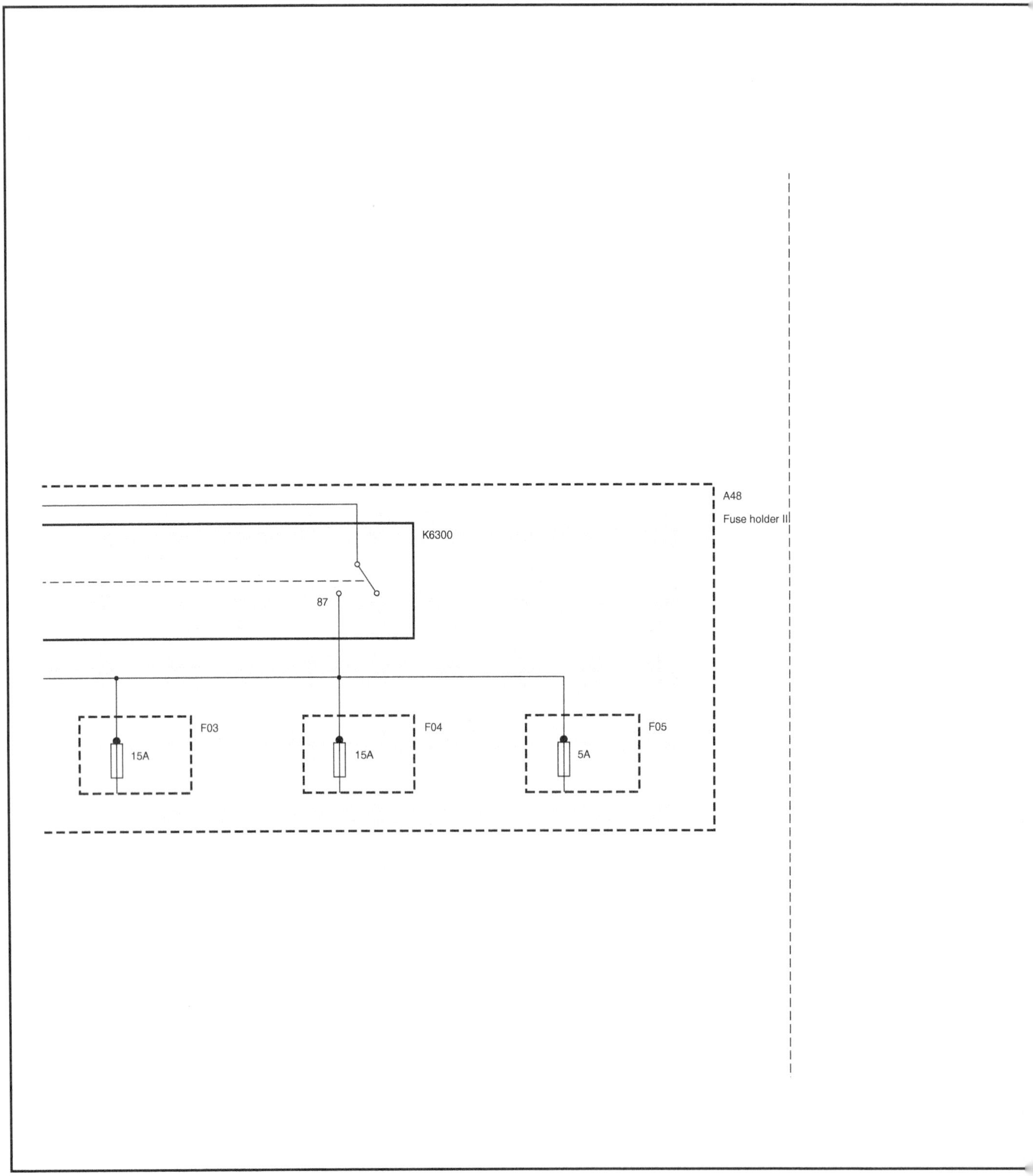

Engine management (DME)
Knock sensor (B6240)
(page 1 of 1)

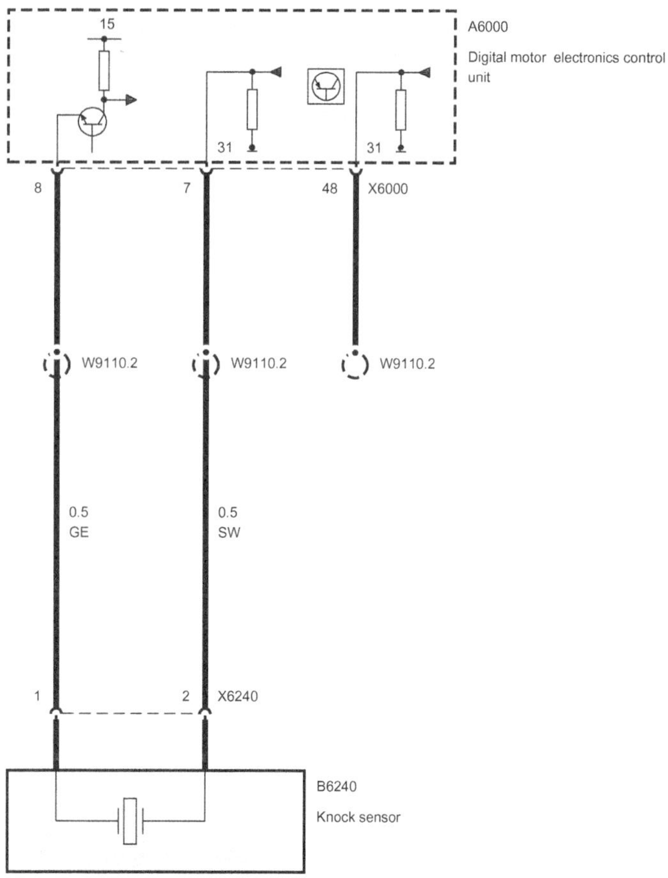

Engine management (DME)
Oxygen sensor (precatalyst)
(page 1 of 1)

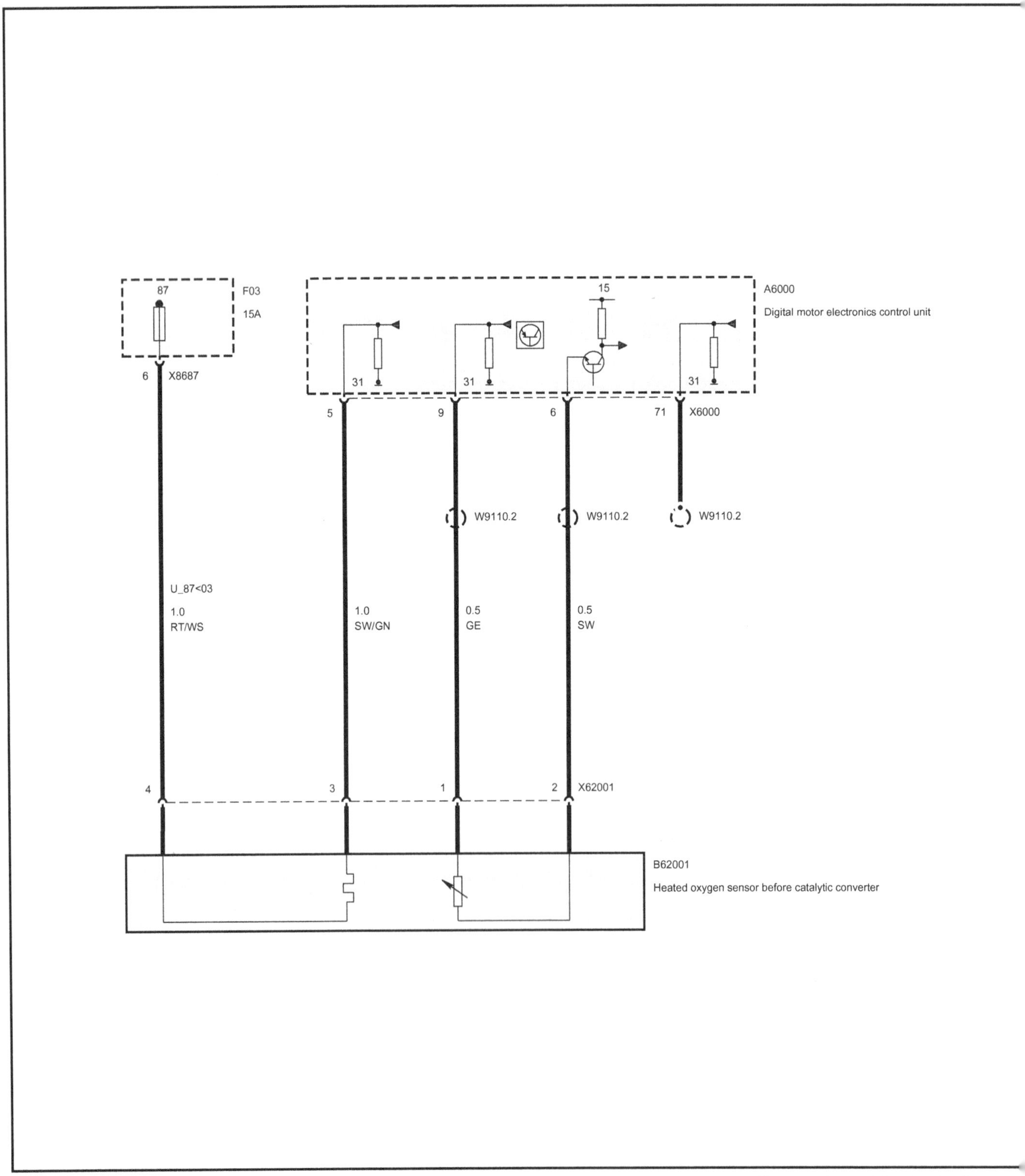

Engine management (DME)
Oxygen sensor (post-catalyst)
(page 1 of 1)

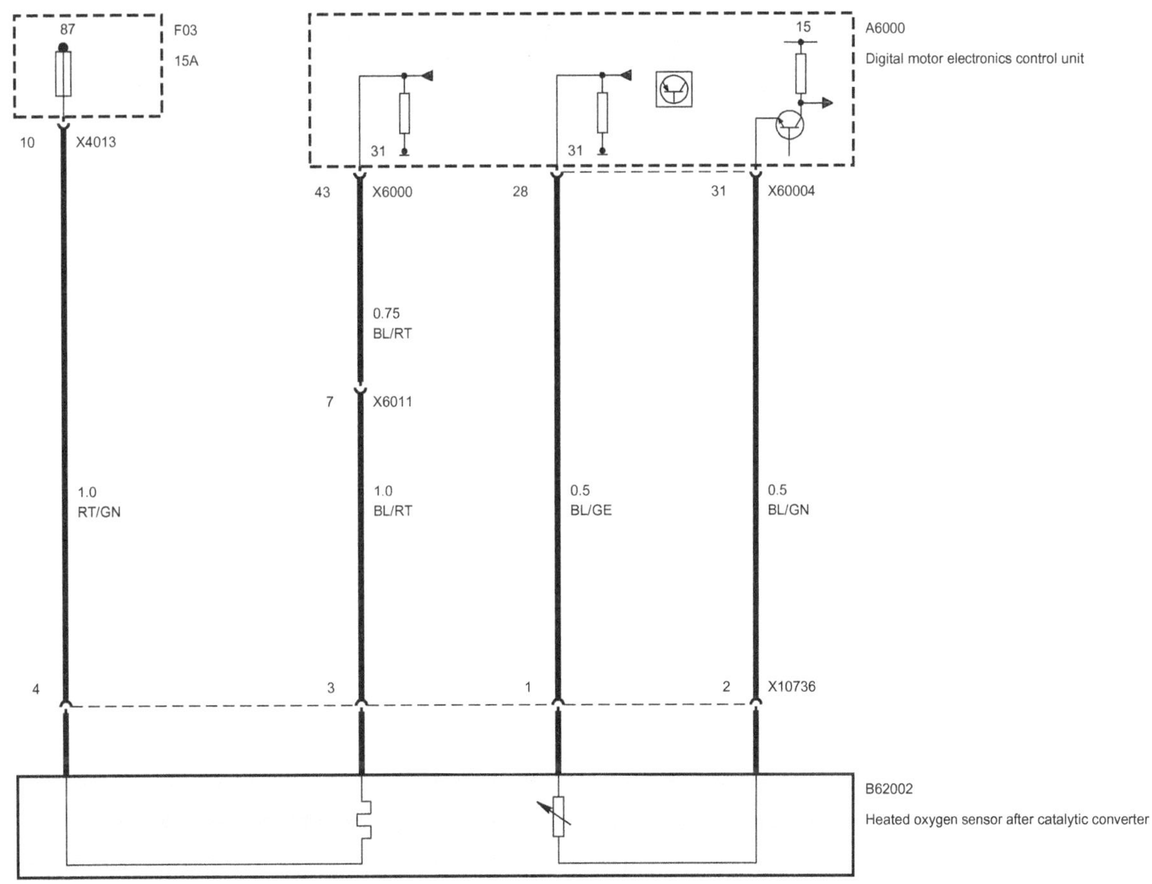

Engine management (DME)
Fuel injectors
(page 1 of 2)

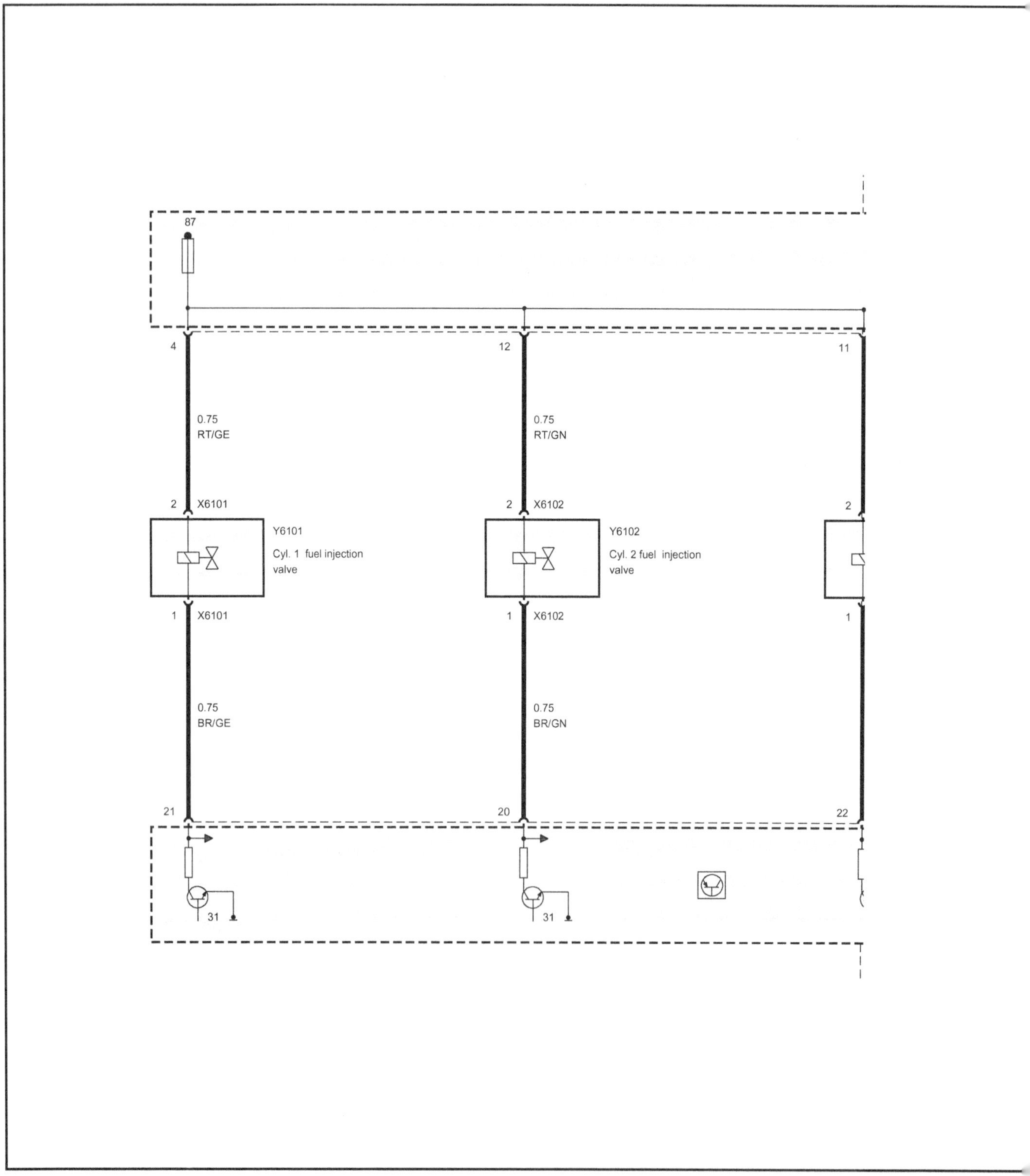

Engine management (DME)
Fuel injectors
(page 2 of 2)

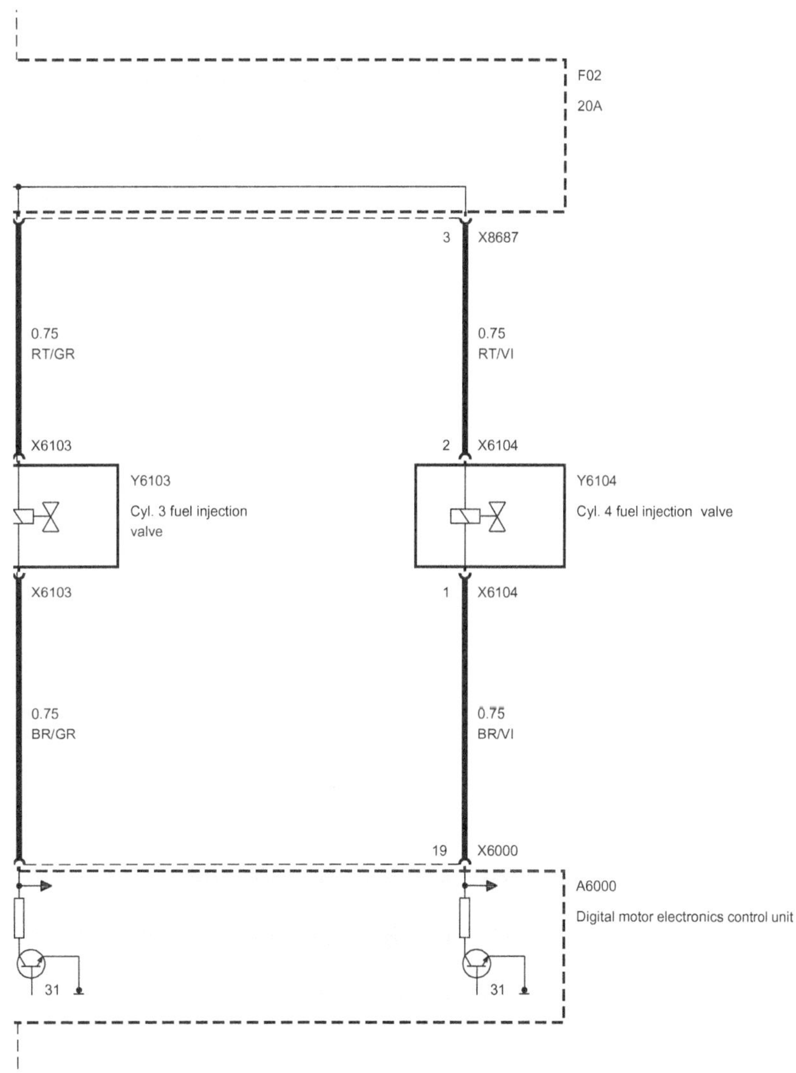

Engine management (DME)
Camshaft position sensor (B6204)
(page 1 of 1)

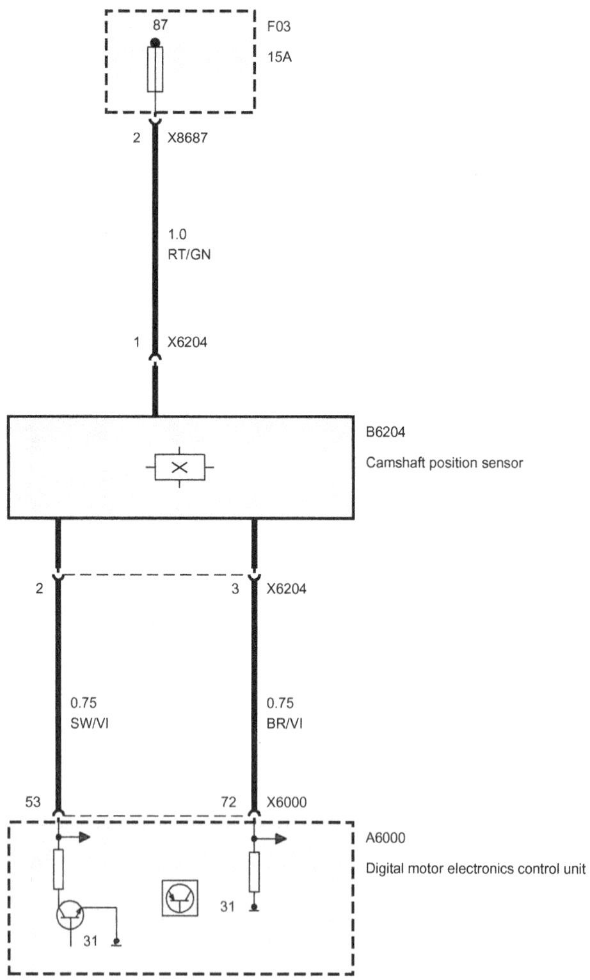

Engine management (DME)
Crankshaft position sensor (B6203)
(page 1 of 1)

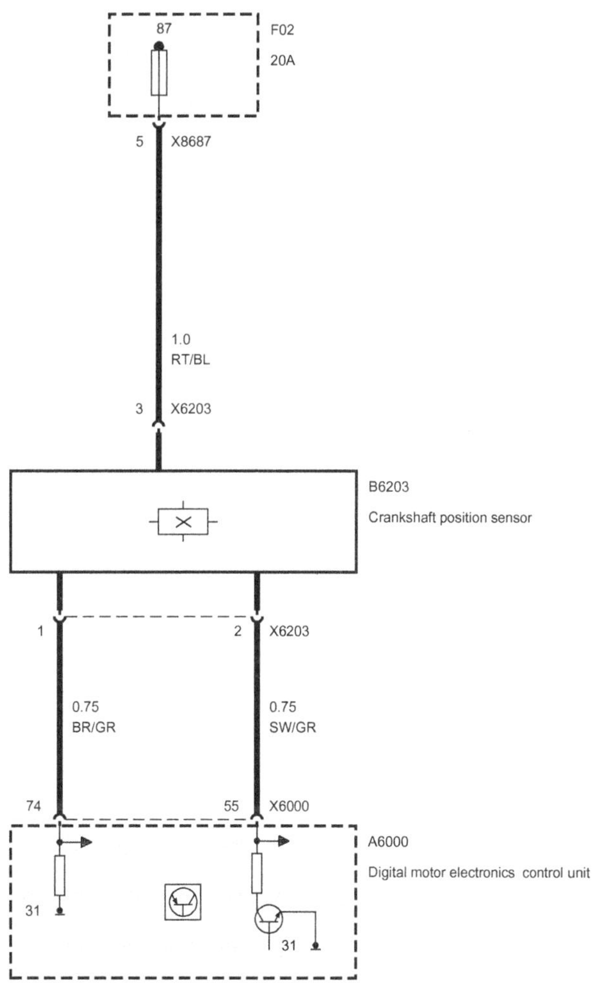

Engine management (DME)
Ignition coil (T6150)
(page 1 of 1)

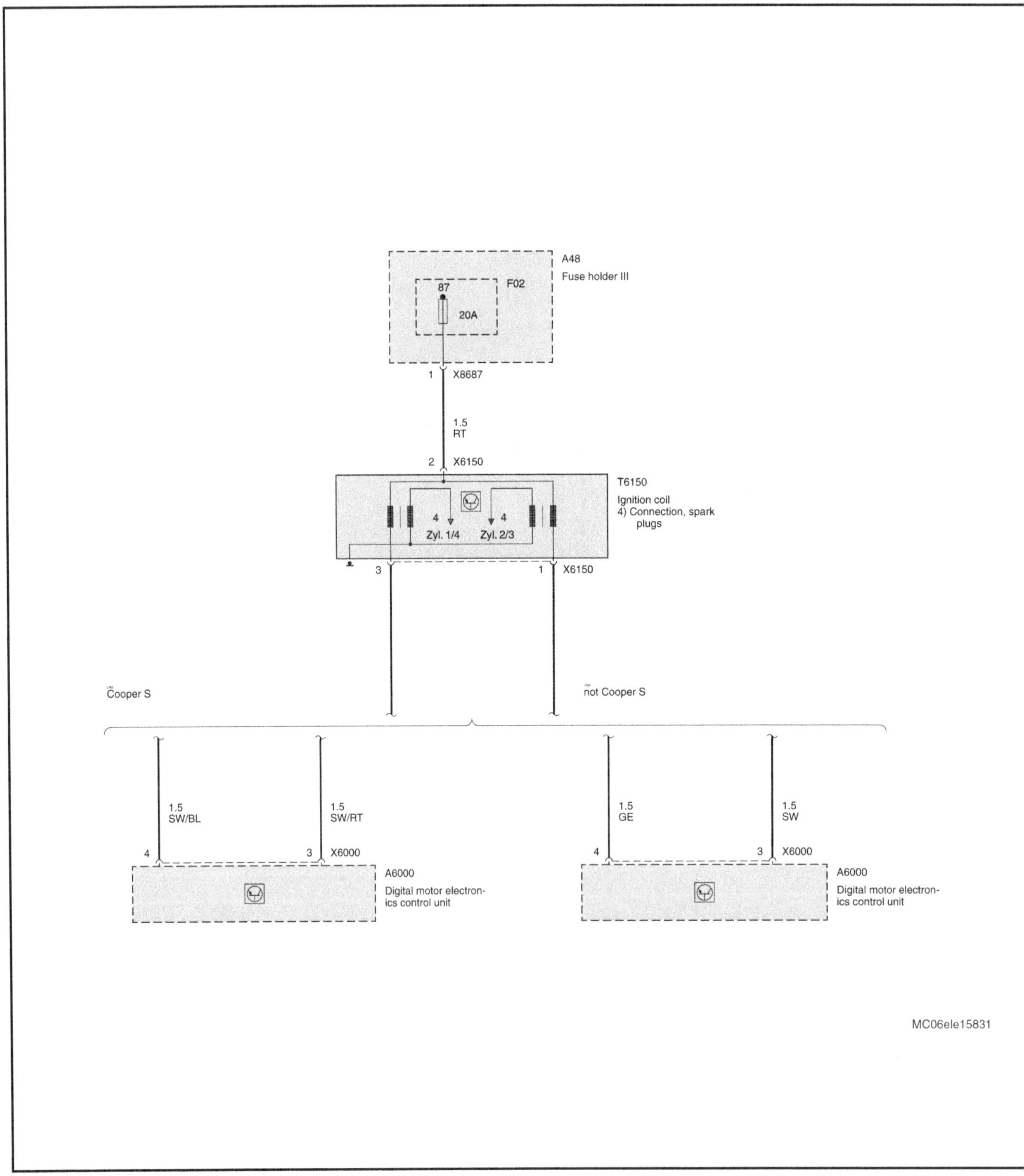

Engine management (DME)
Accelerator pedal position sensor (R10)
(page 1 of 1)

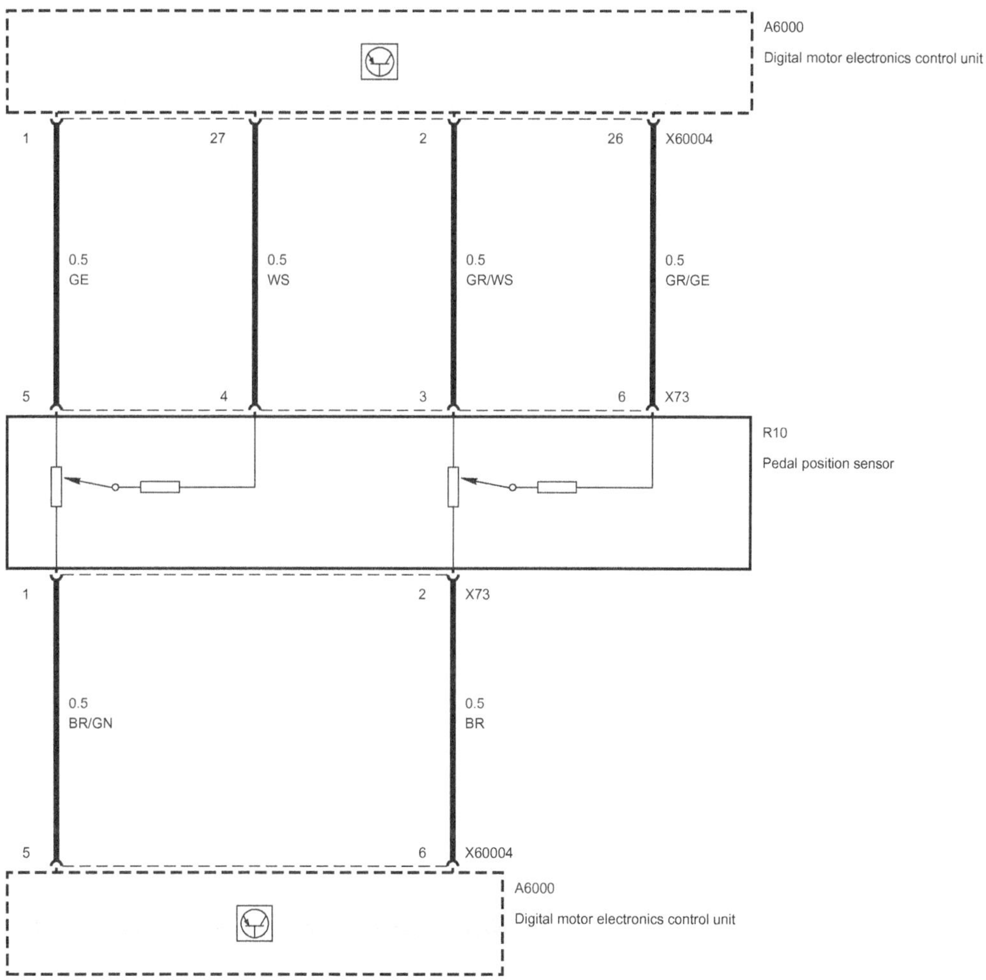

Engine management (DME)
Tank leakage diagnosis pump (LDP/DMTL) (M119)
(page 1 of 1)

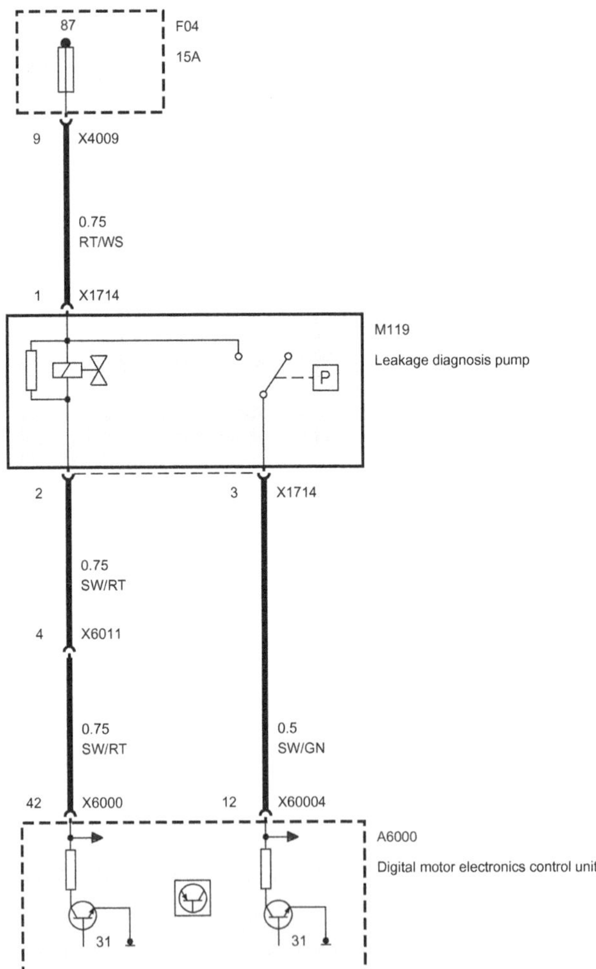

Engine management (DME)

Fuel changeover, tank vent (Y6550)

(page 1 of 1)

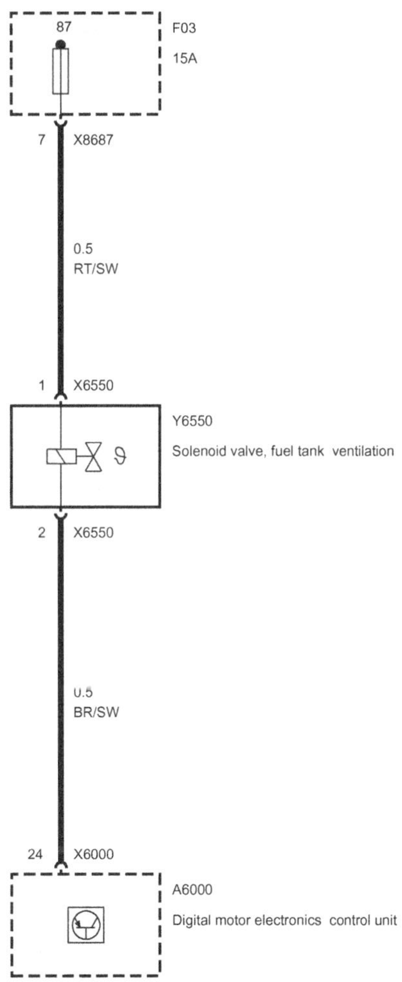

Exterior lights
Back-up lights (up to July 2004)
(page 1 of 1)

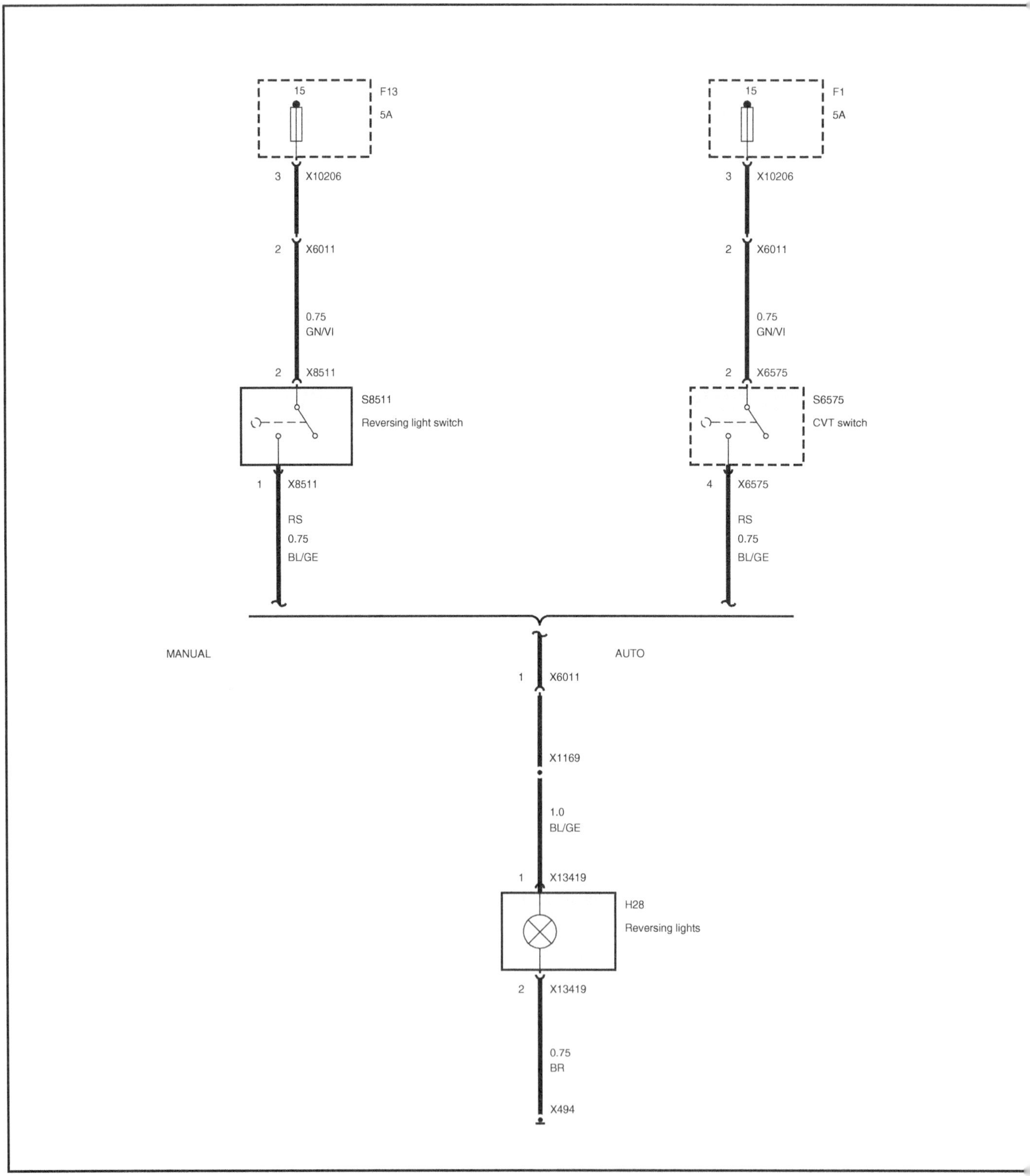

Exterior lights

Back-up lights (from July 2004)

(page 1 of 1)

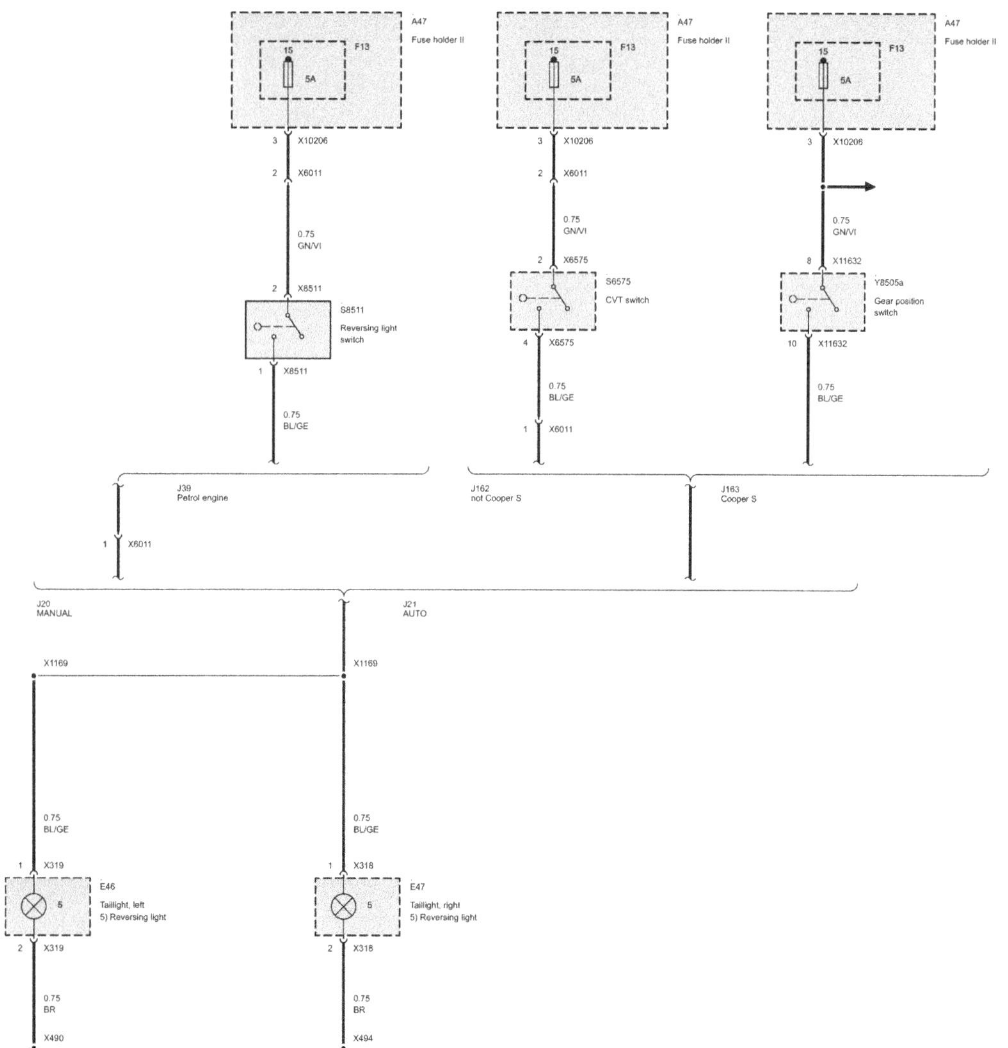

MC06ele21205

Exterior lights

Courtesy and door handle lights (from July 2004)

(page 1 of 1)

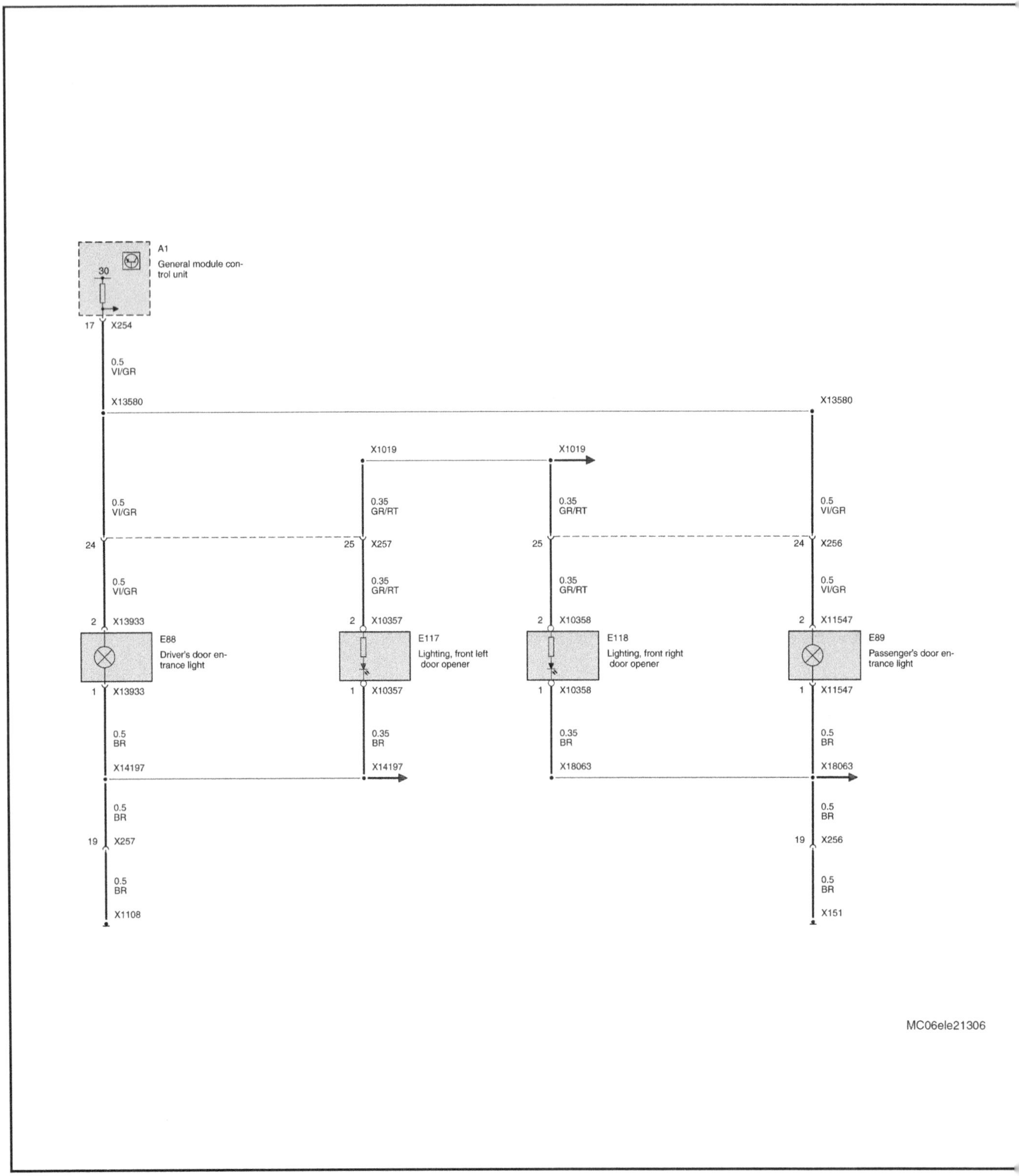

Exterior lights
Front foglights
(page 1 of 2)

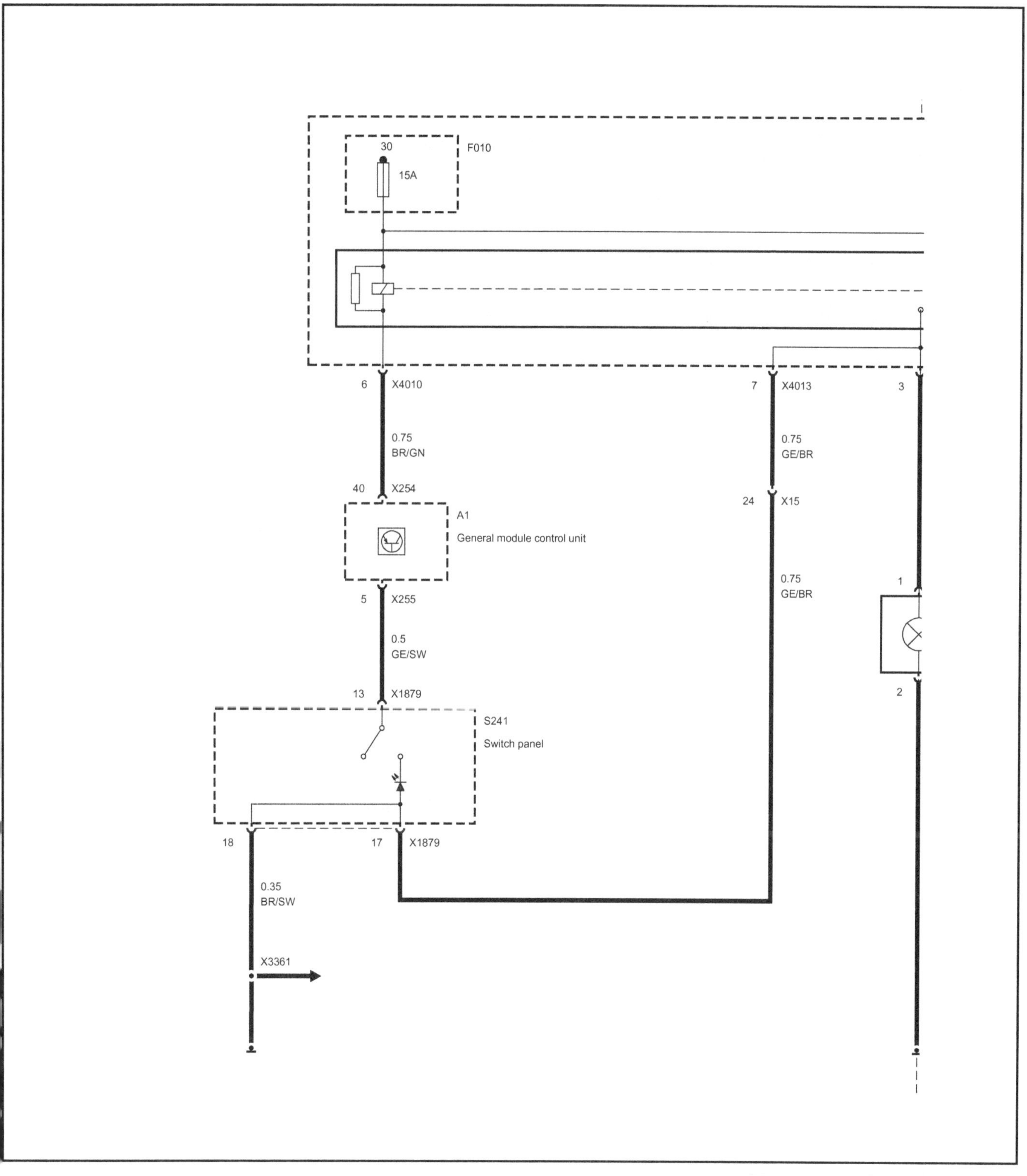

Exterior lights
Front foglights
(page 2 of 2)

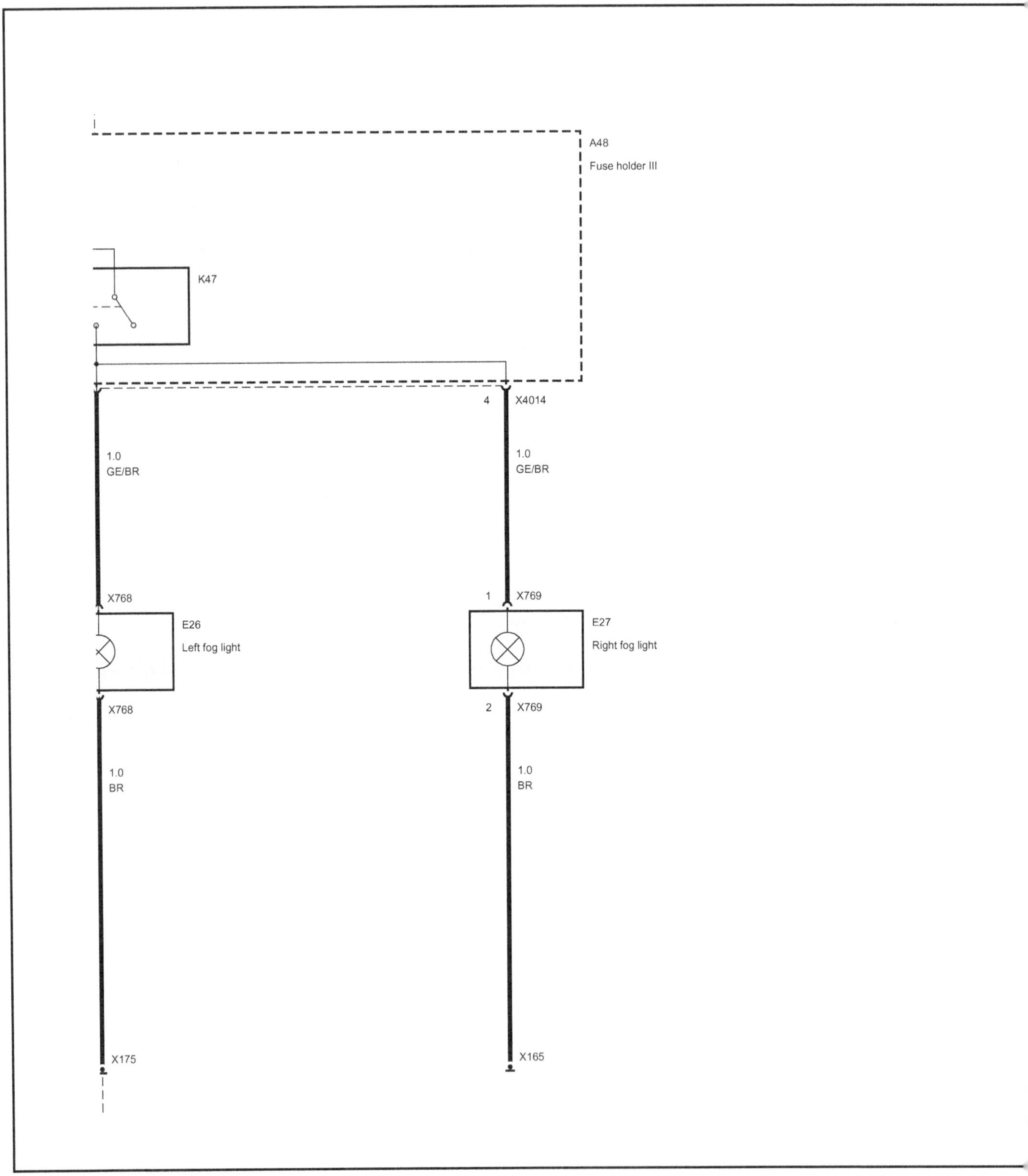

Exterior lights
Headlight vertical aim control (A53)
(page 1 of 3)

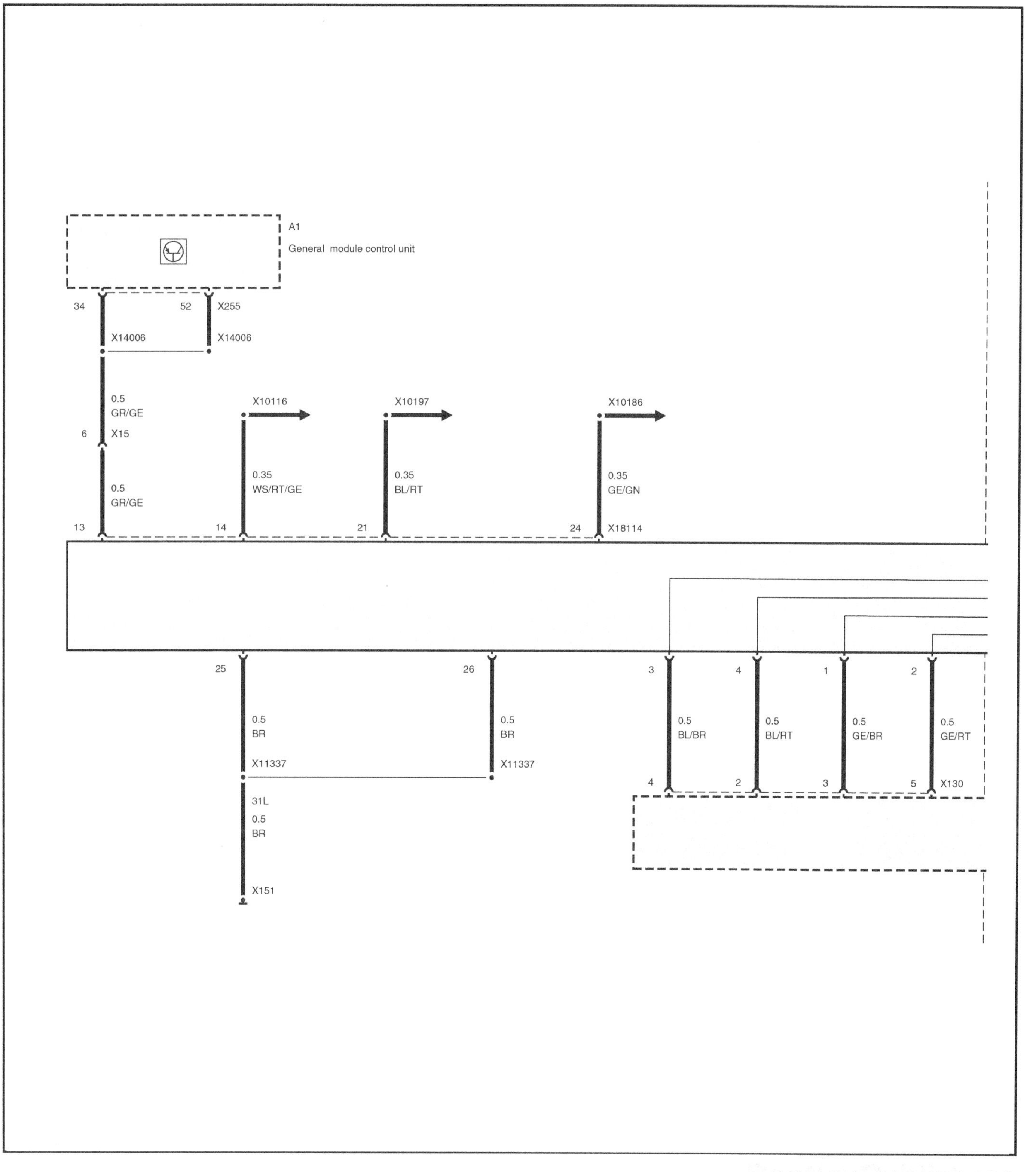

Exterior lights
Headlight vertical aim control (A53)
(page 2 of 3)

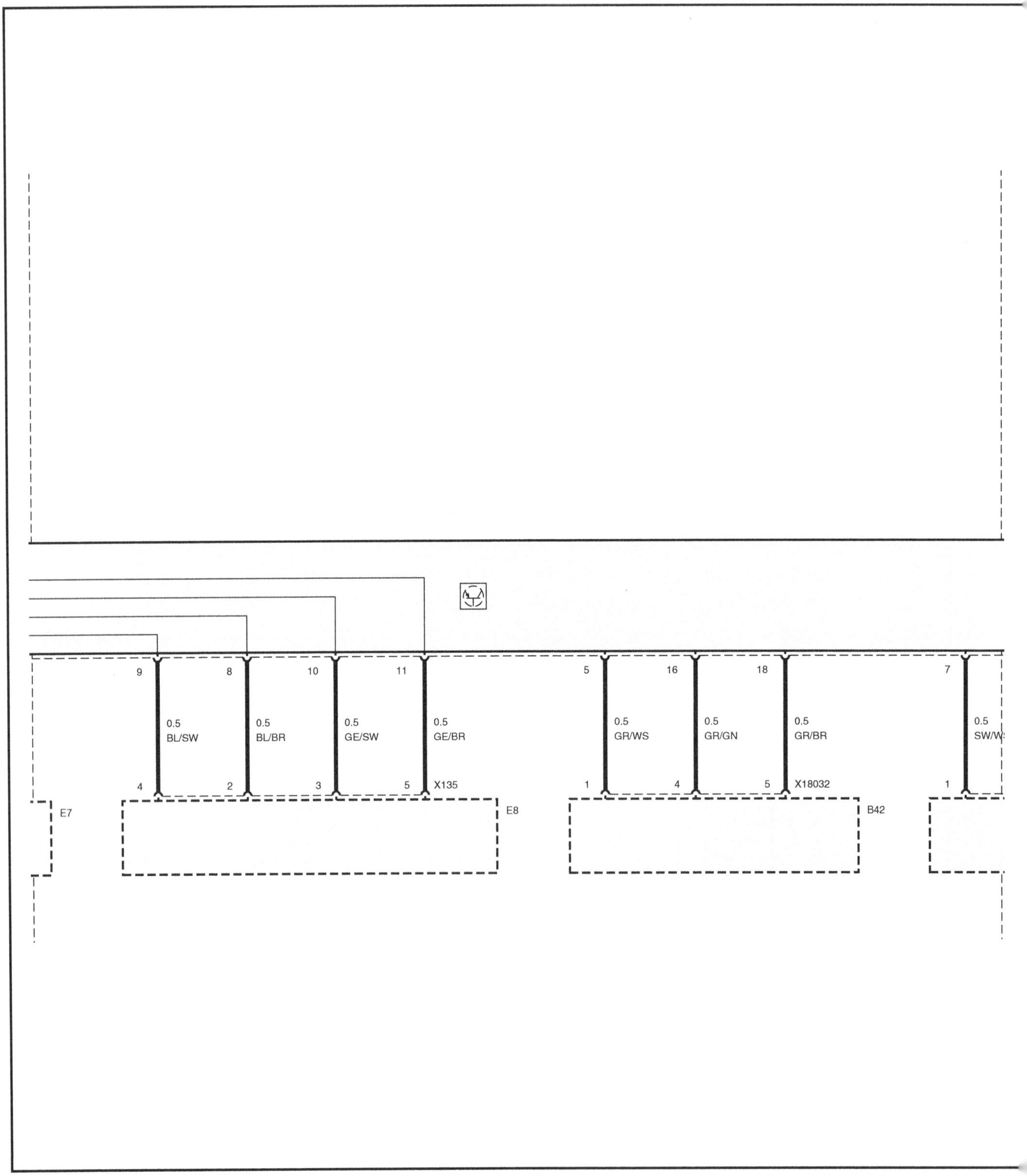

Exterior lights

Headlight vertical aim control (A53)

(page 3 of 3)

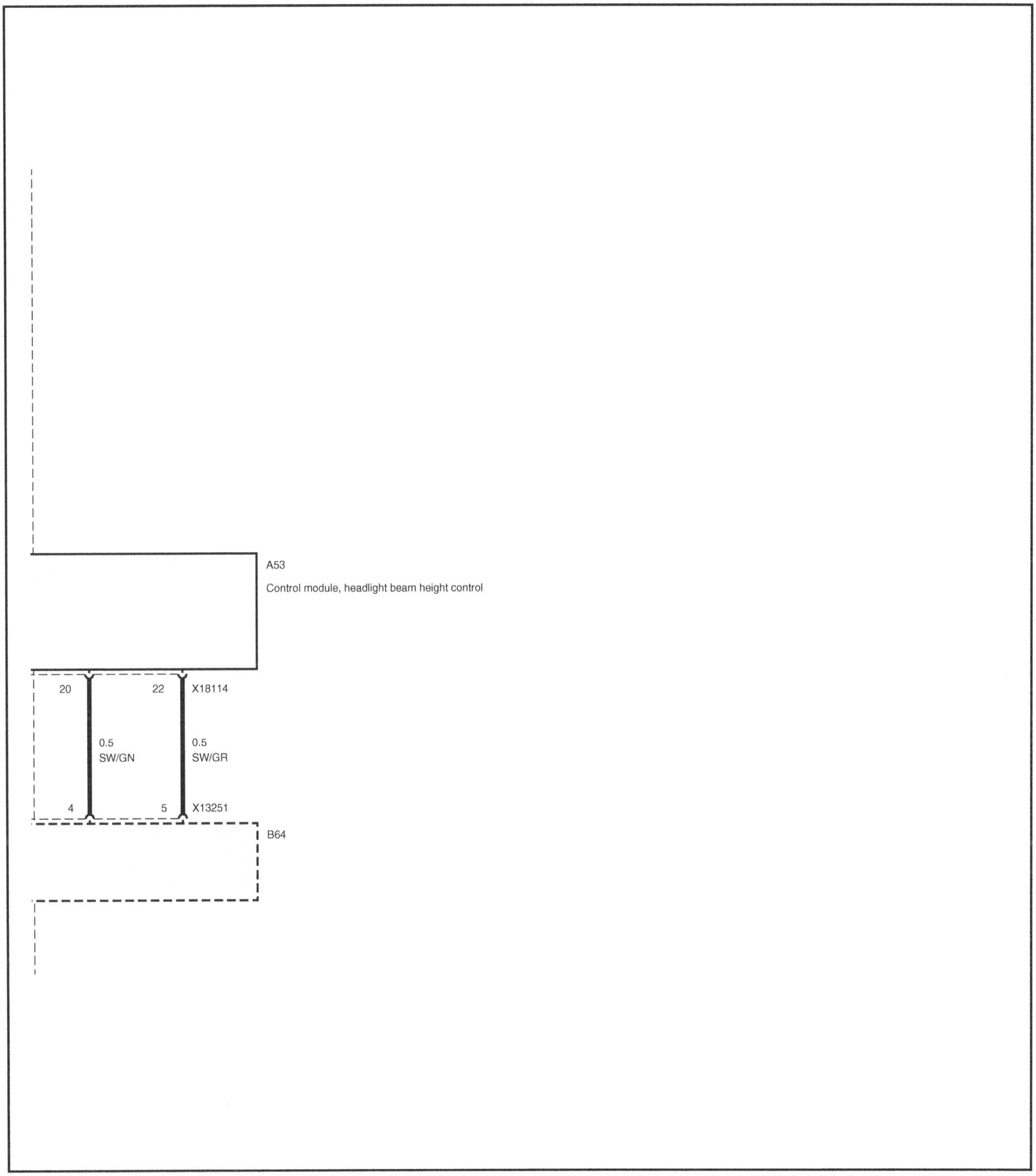

Exterior lights
Headlights, halogen
(page 1 of 1)

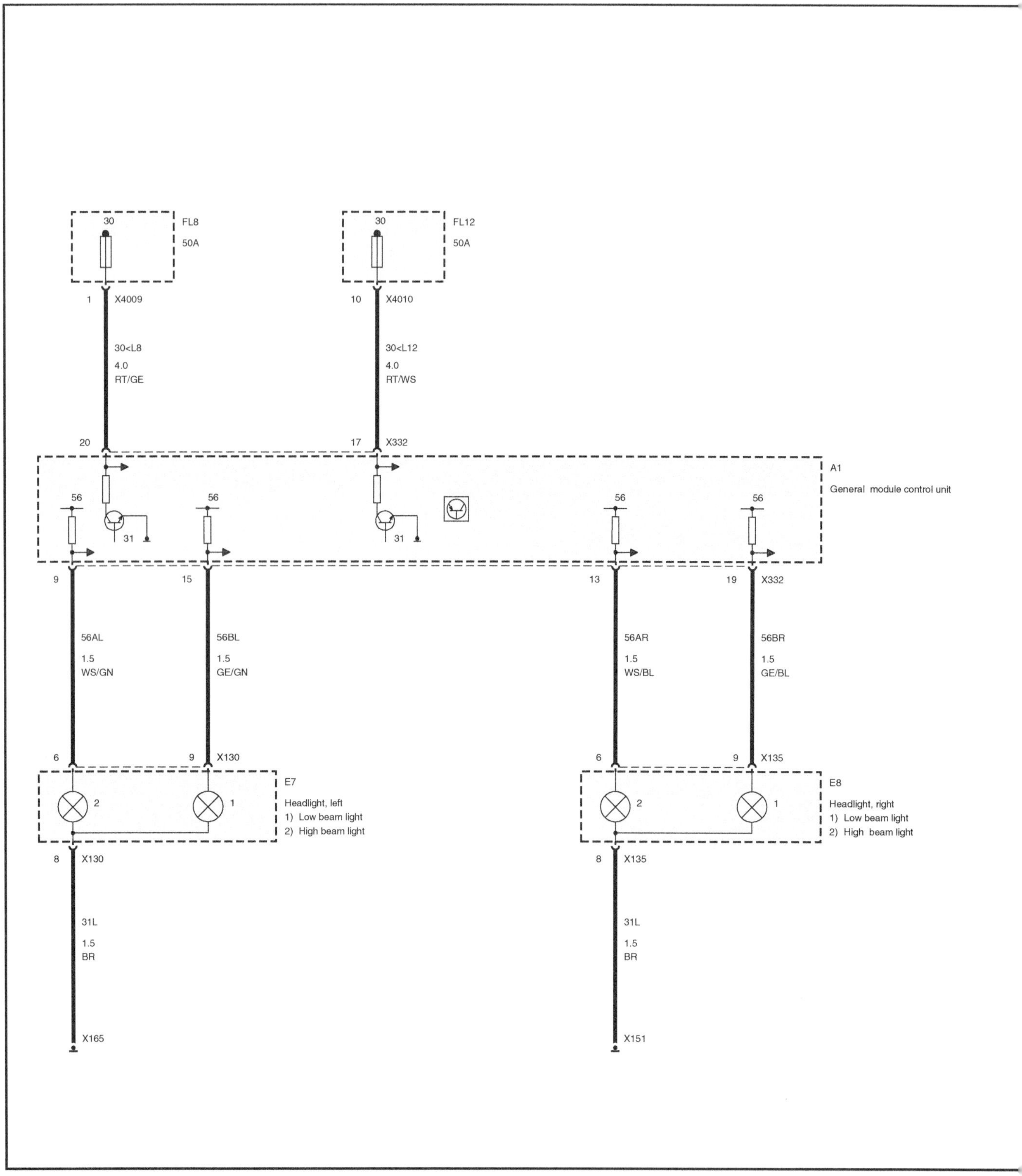

Exterior lights
Headlights, xenon
(page 1 of 4)

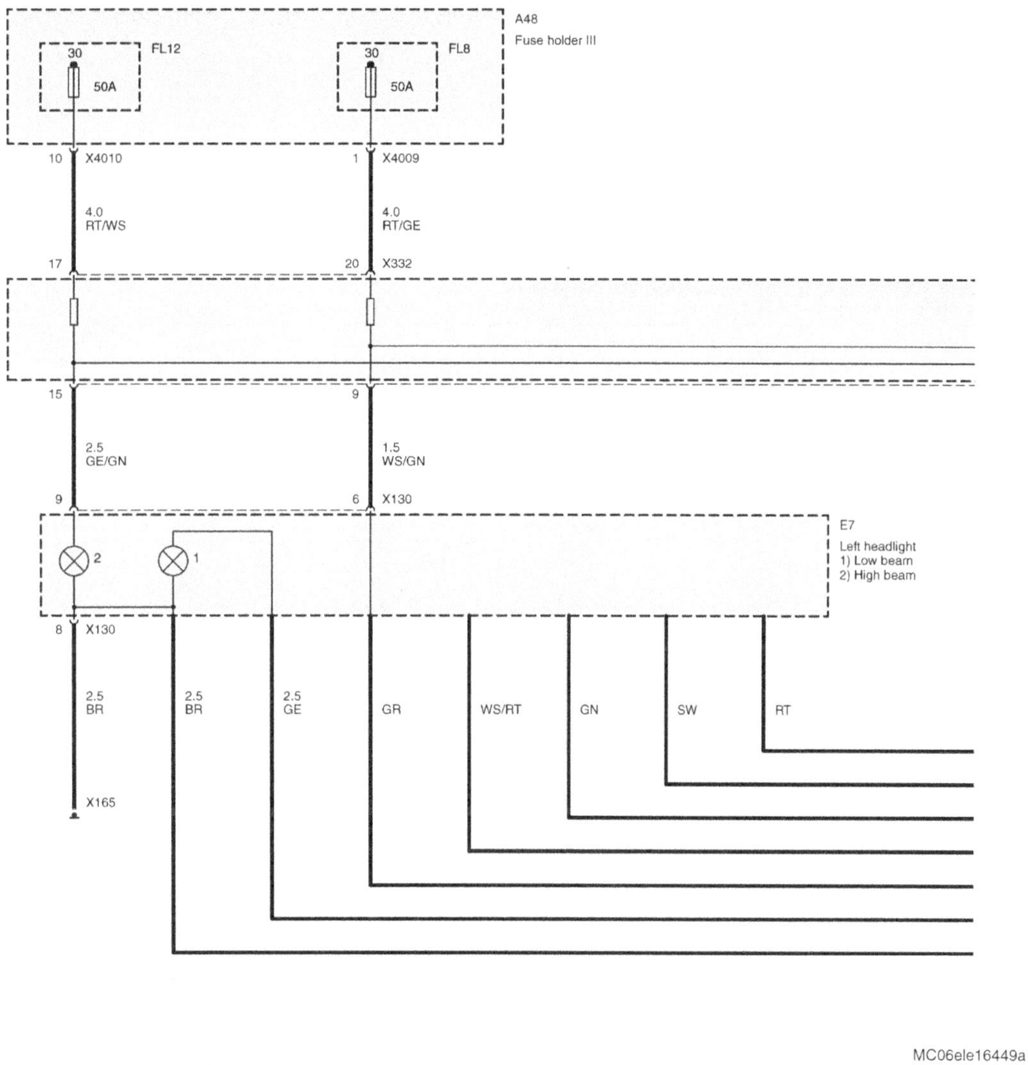

Exterior lights
Headlights, xenon
(page 2 of 4)

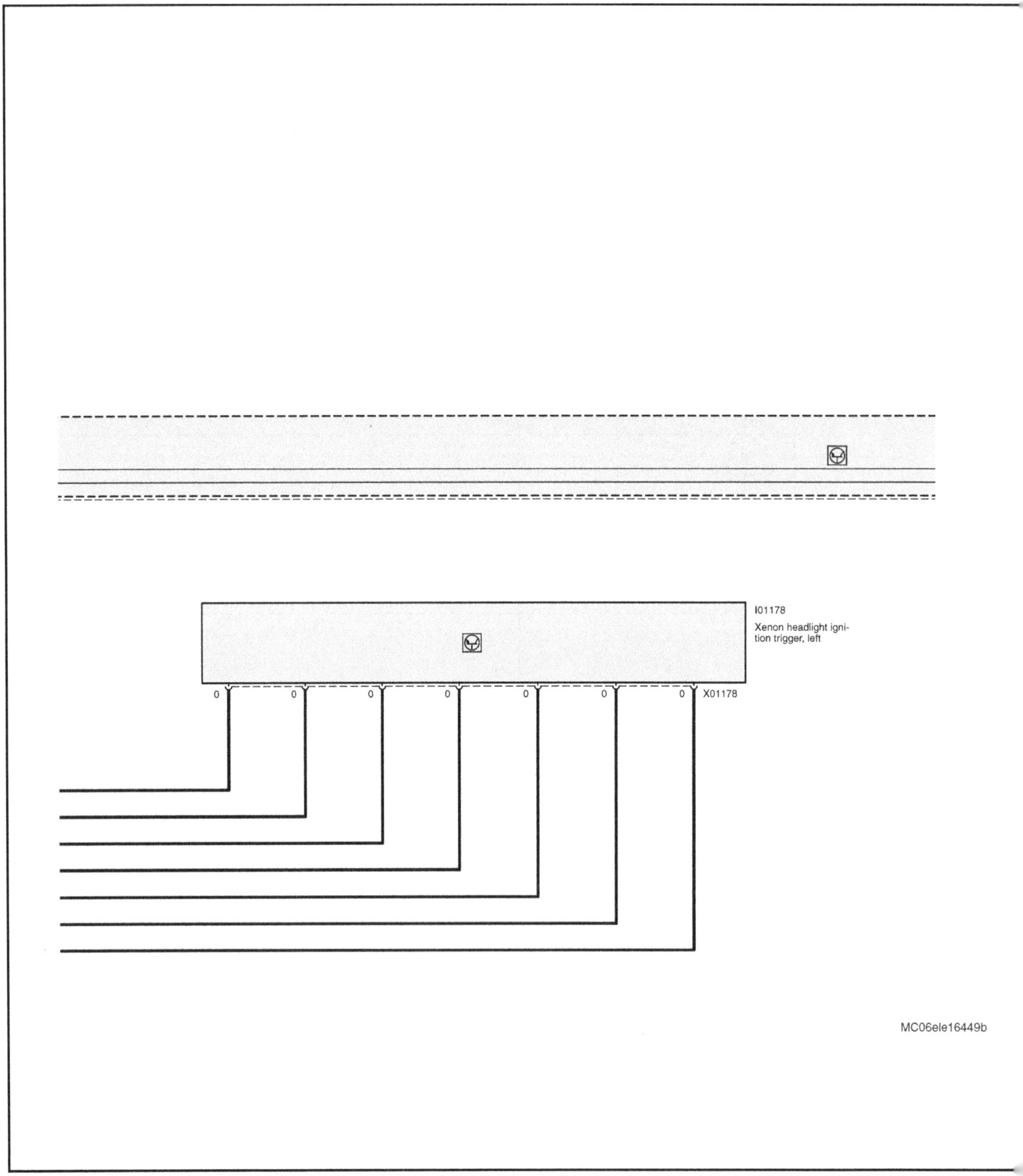

Exterior lights
Headlights, xenon
(page 3 of 4)

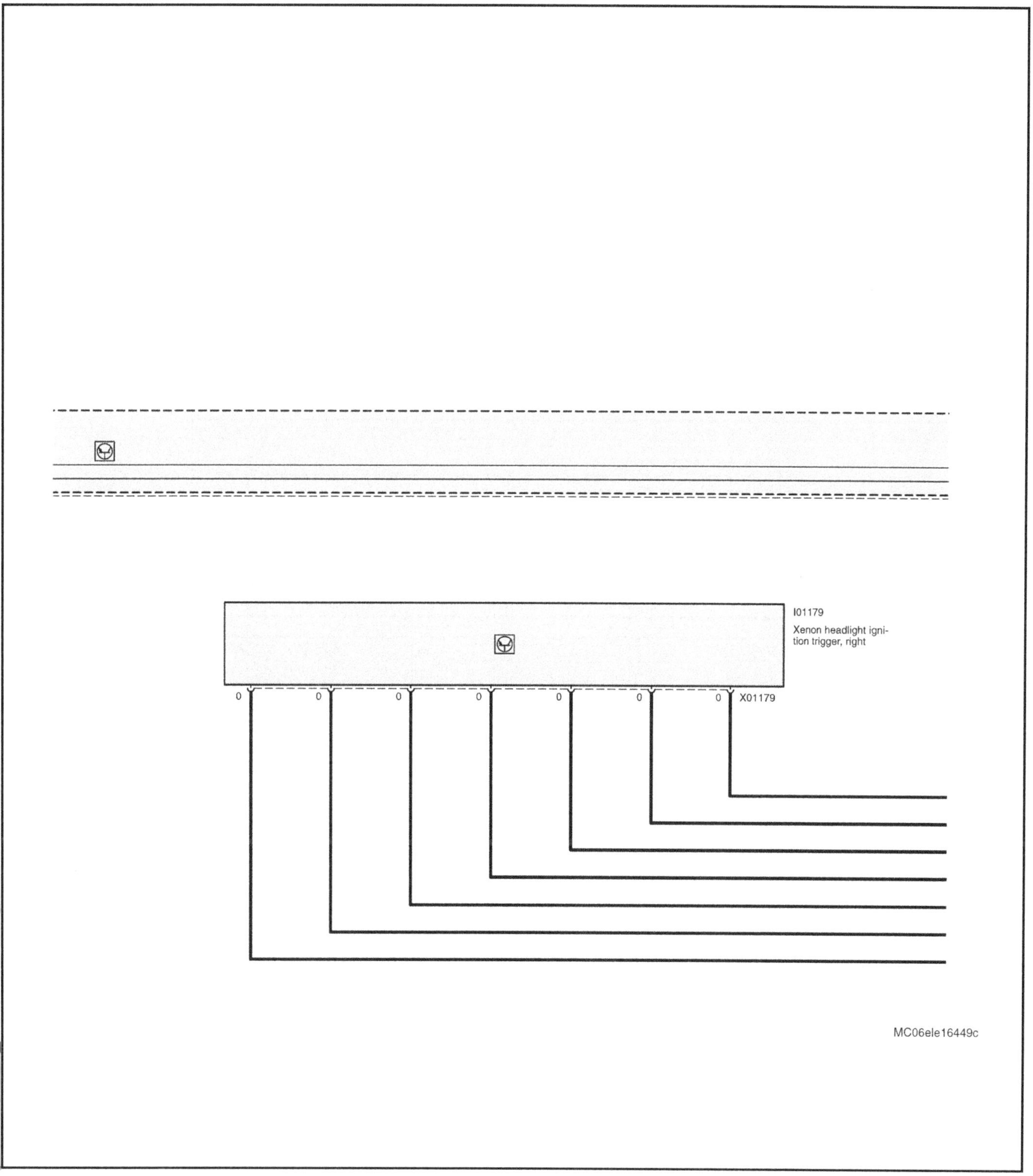

Exterior lights
Headlights, xenon
(page 4 of 4)

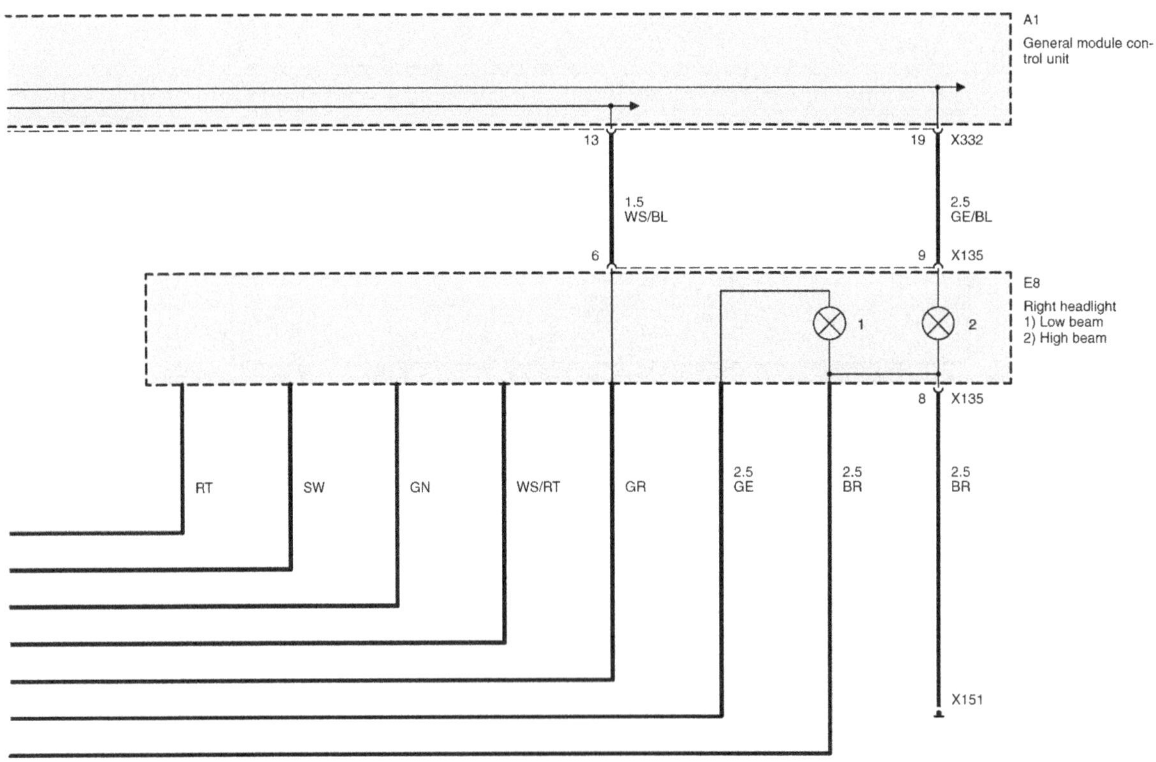

MC06ele16449d

Exterior lights
License plate
(page 1 of 1)

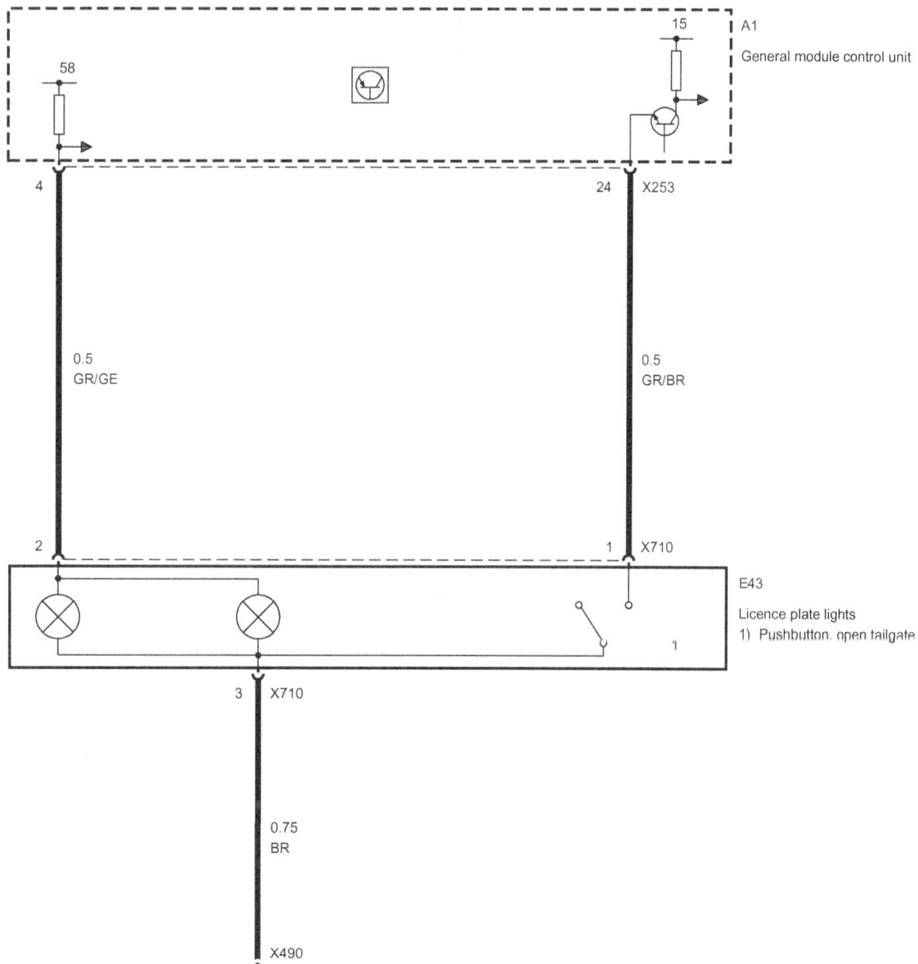

Exterior lights
Parking lights
(page 1 of 3)

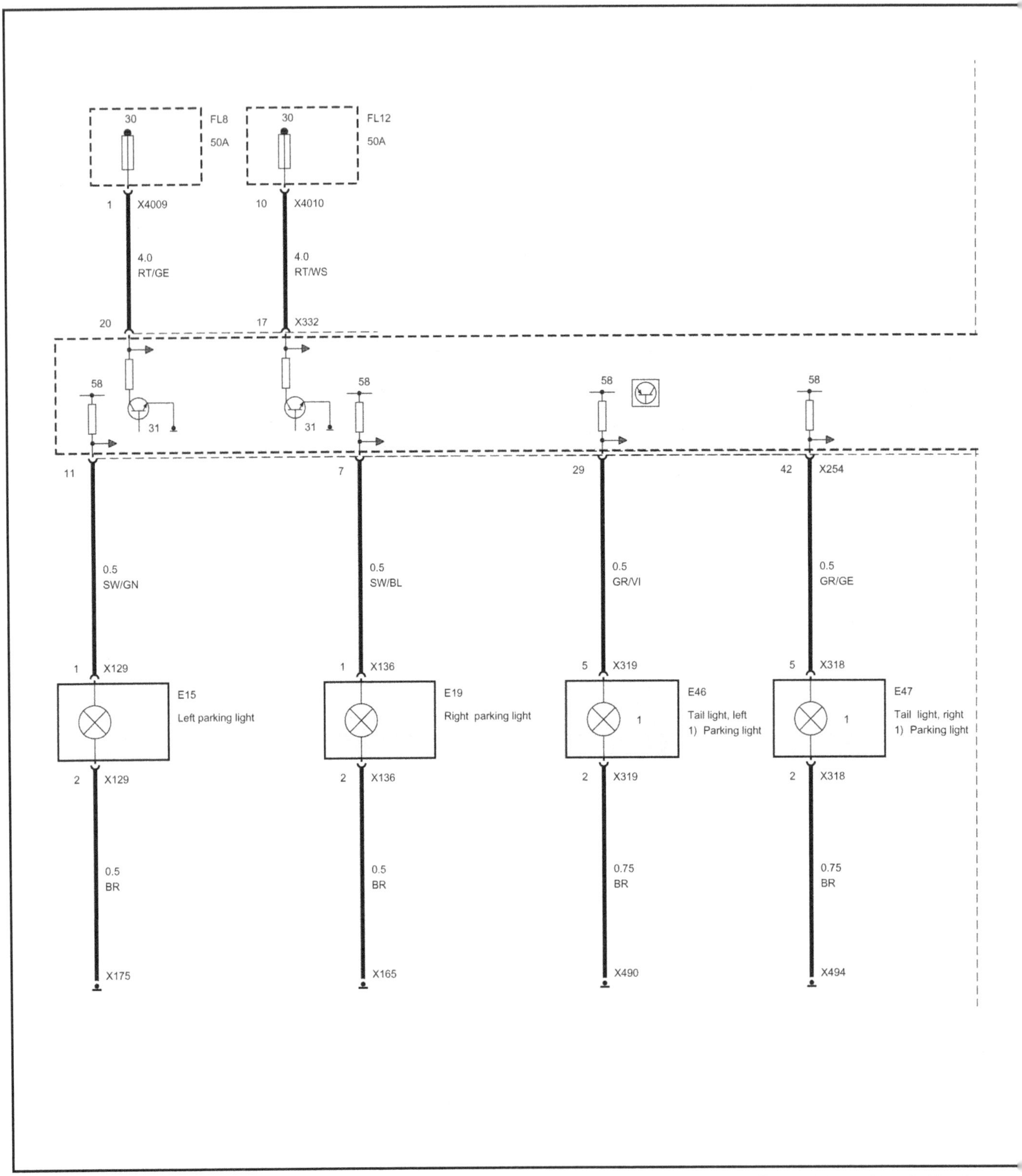

Exterior lights
Parking lights
(page 2 of 3)

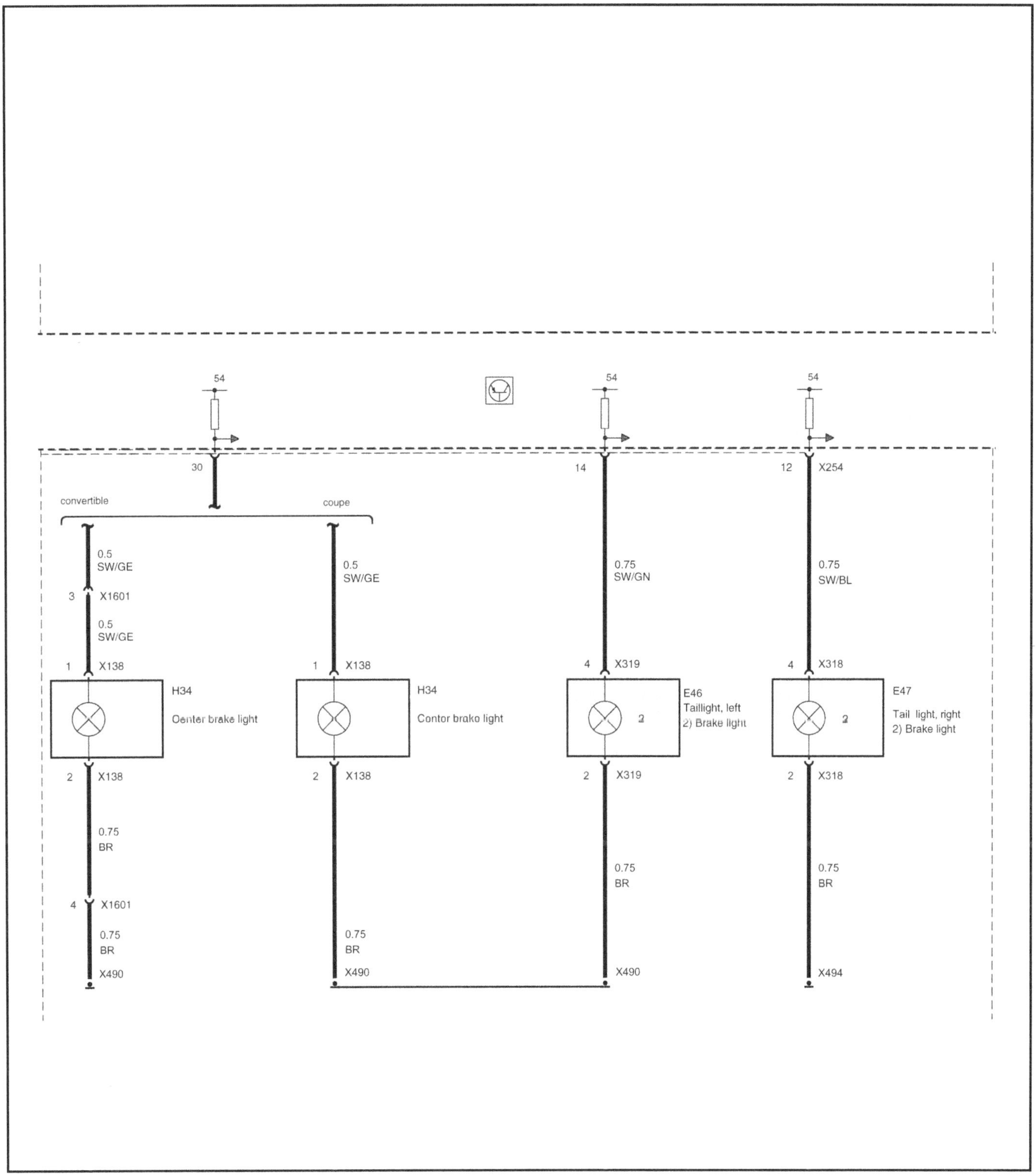

Exterior lights
Parking lights
(page 2 of 3)

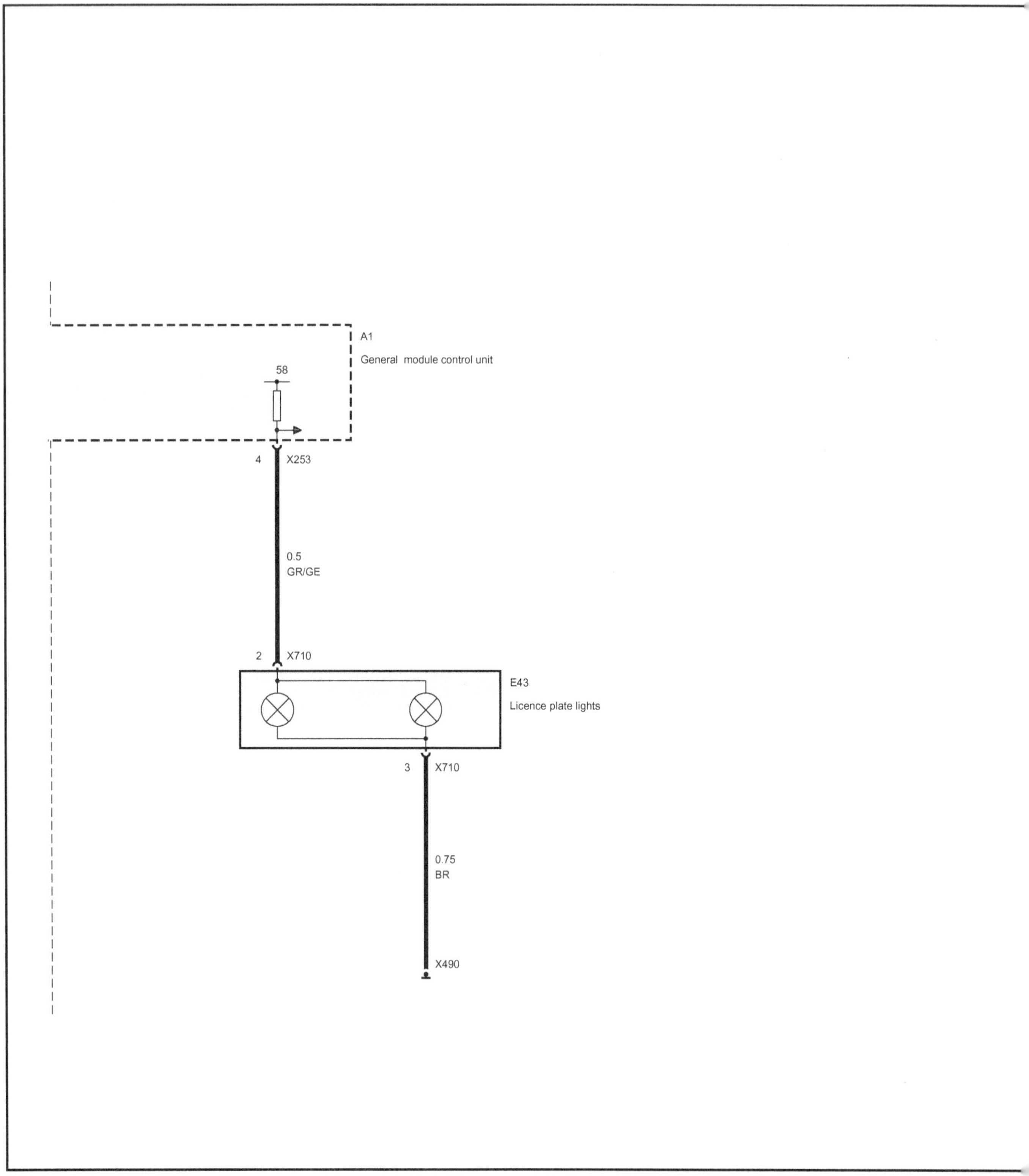

Exterior lights
Rear foglight (from July 2004)
(page 1 of 1)

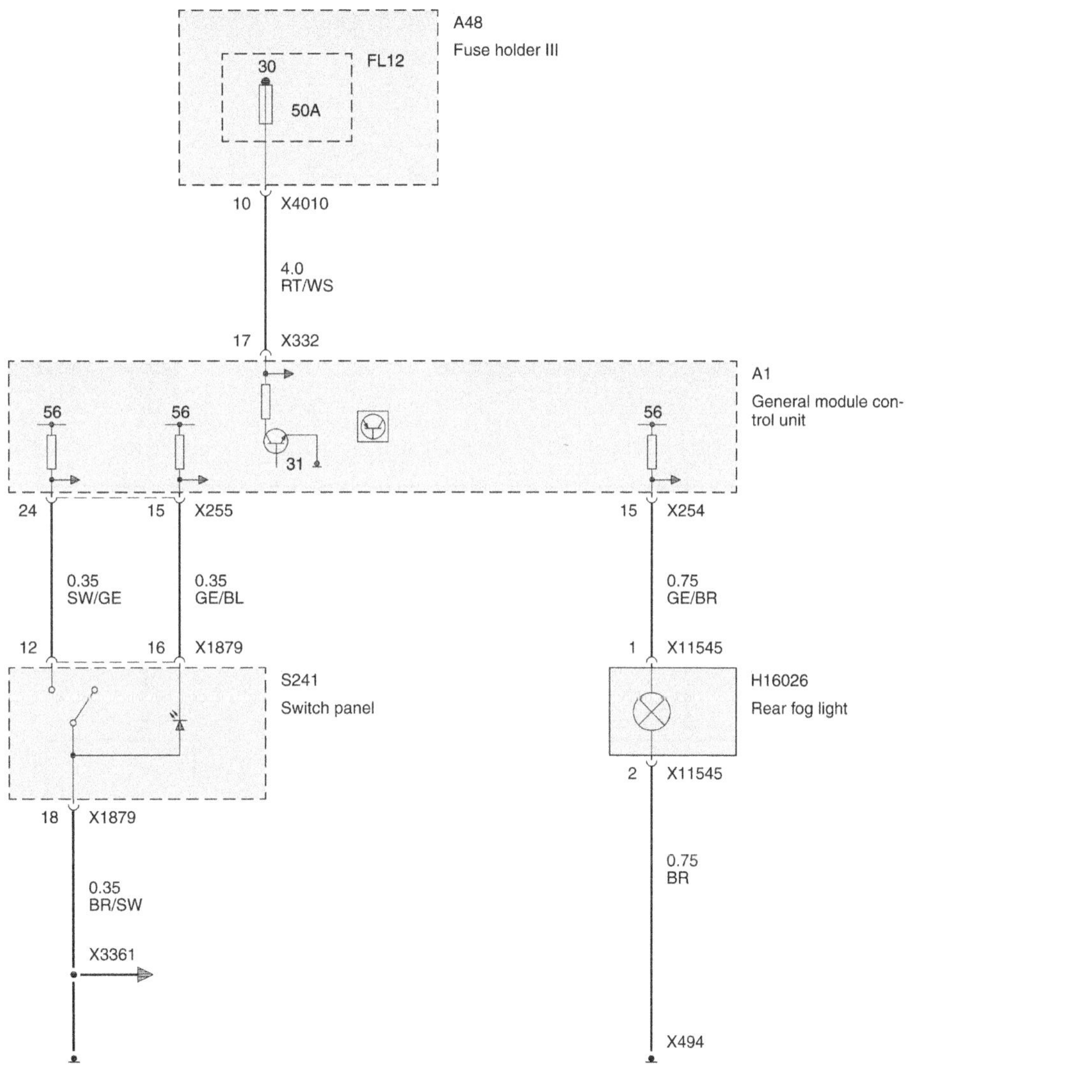

Exterior lights
Side marker lights
(page 1 of 2)

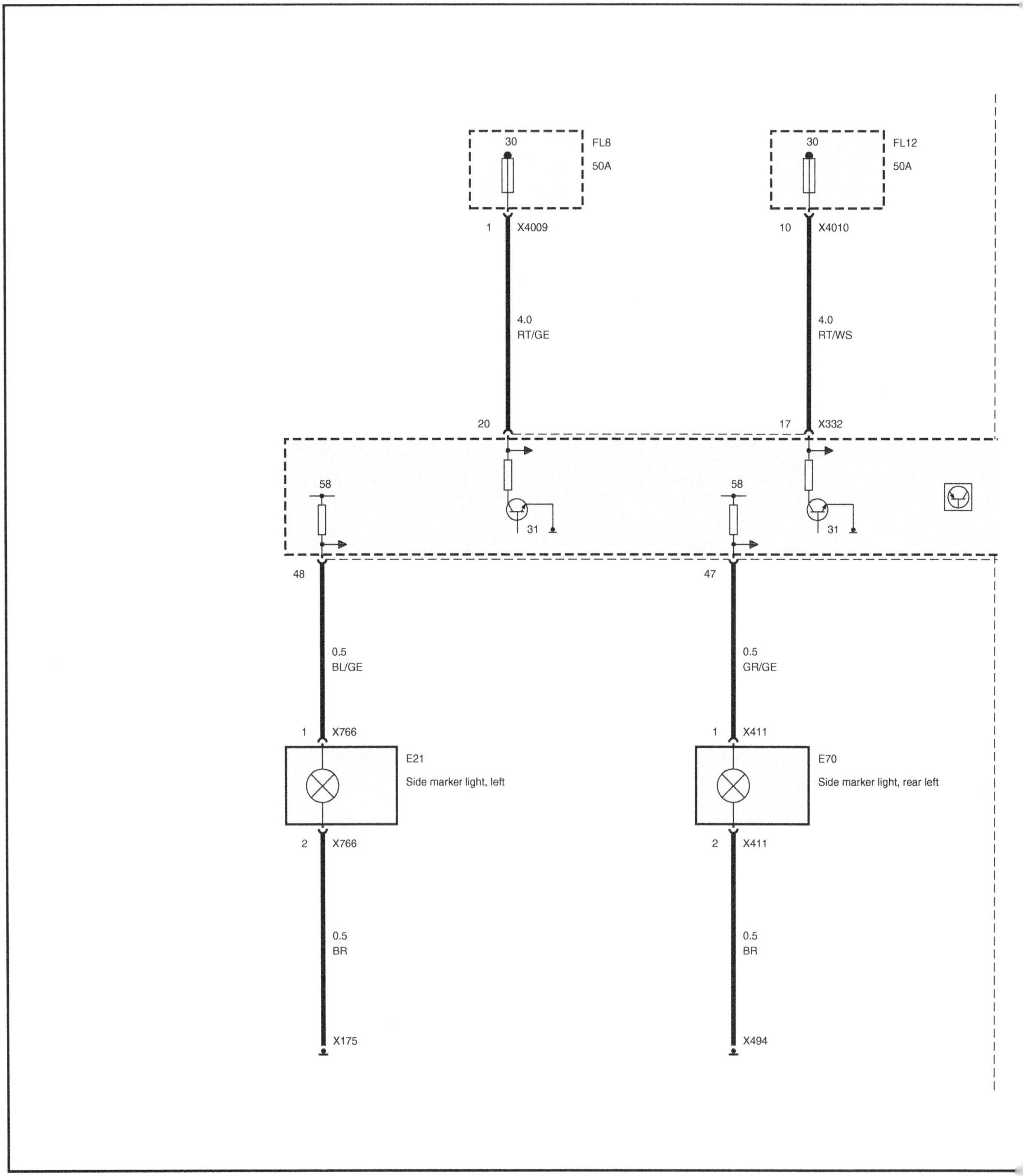

Exterior lights
Side marker lights
(page 2 of 2)

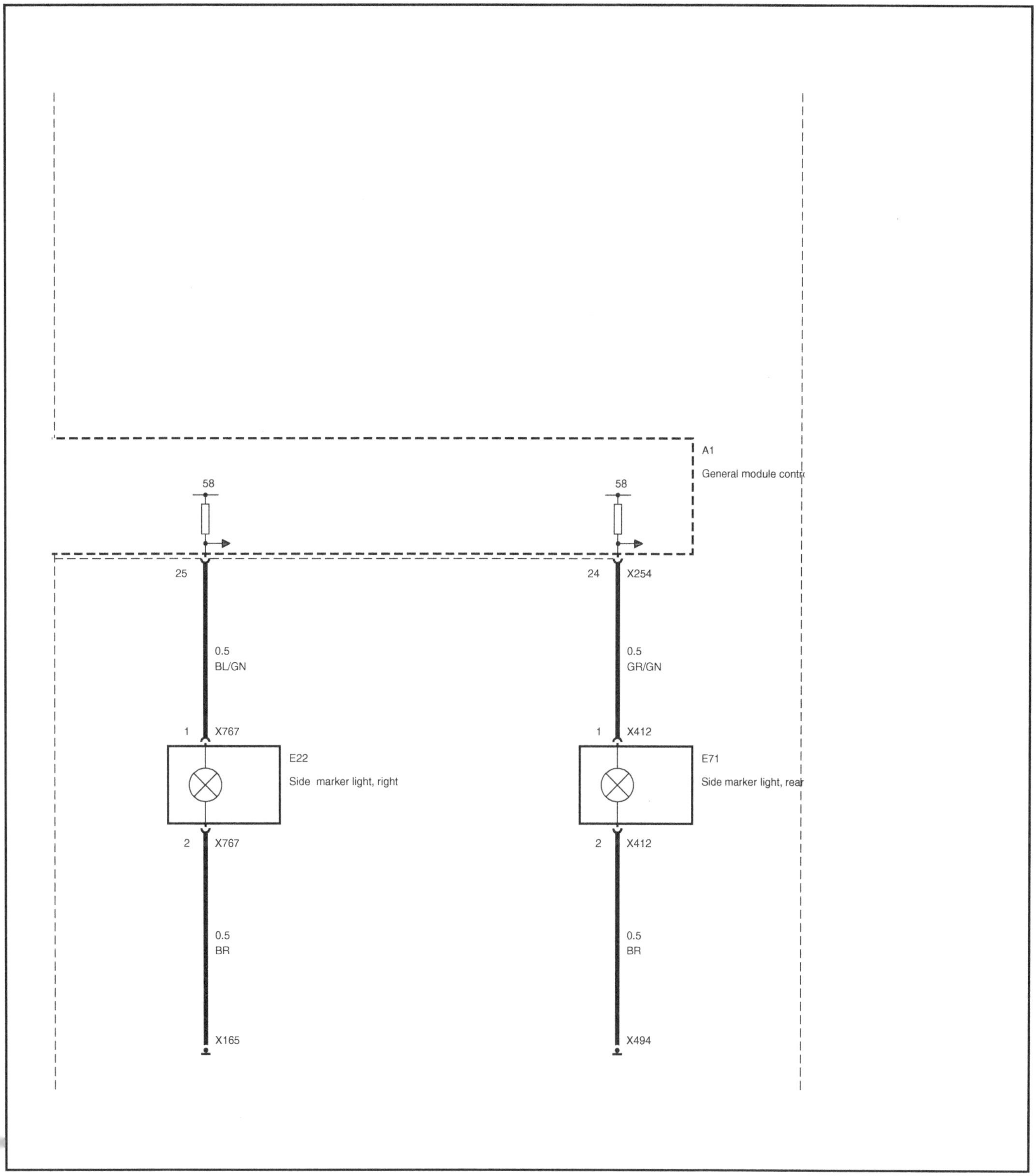

Exterior lights
Stalk switch, turn indicator (up to Jan. 2005)
(page 1 of 1)

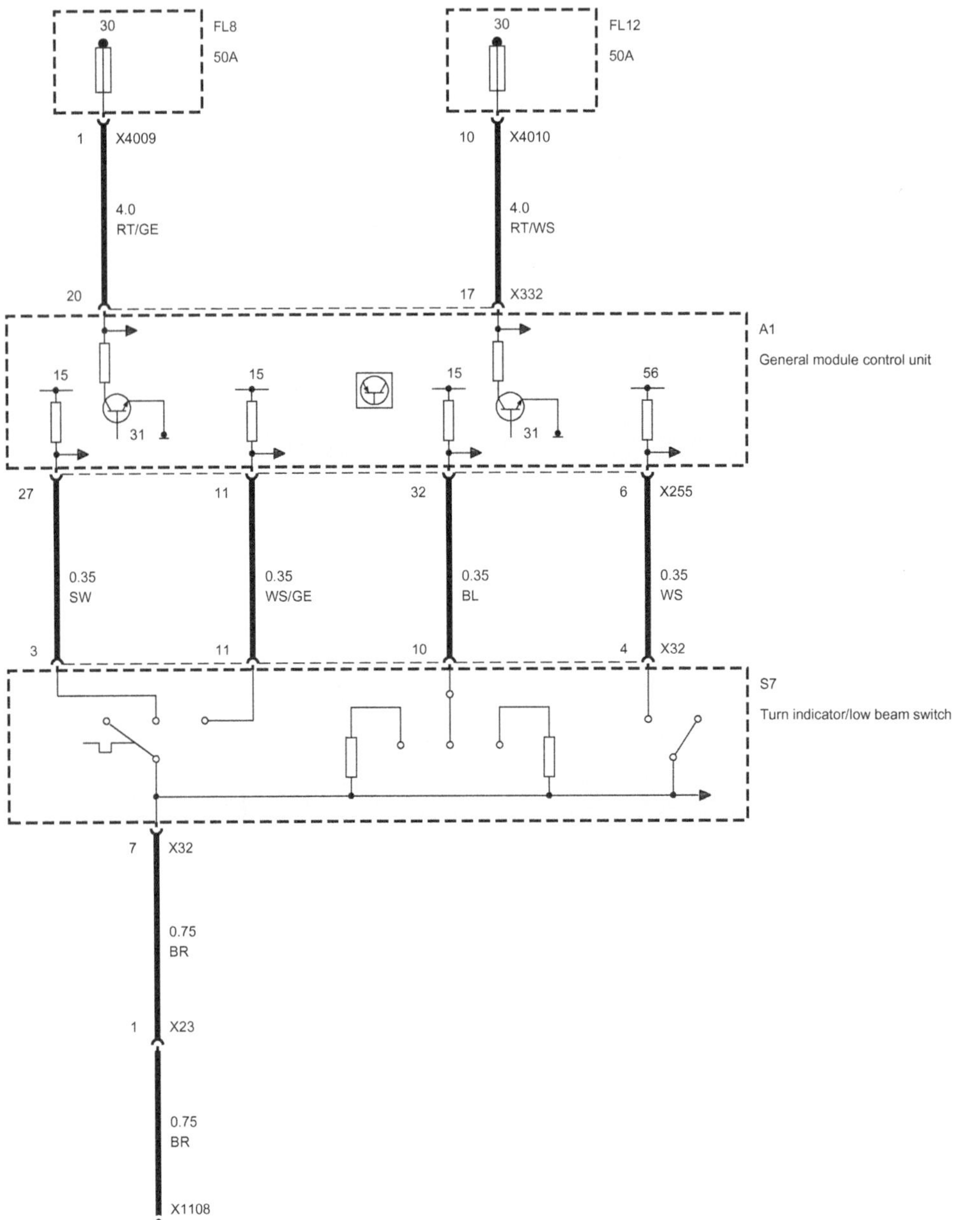

Exterior lights

Stalk switch, turn indicator (from Jan. 2005)

(page 1 of 1)

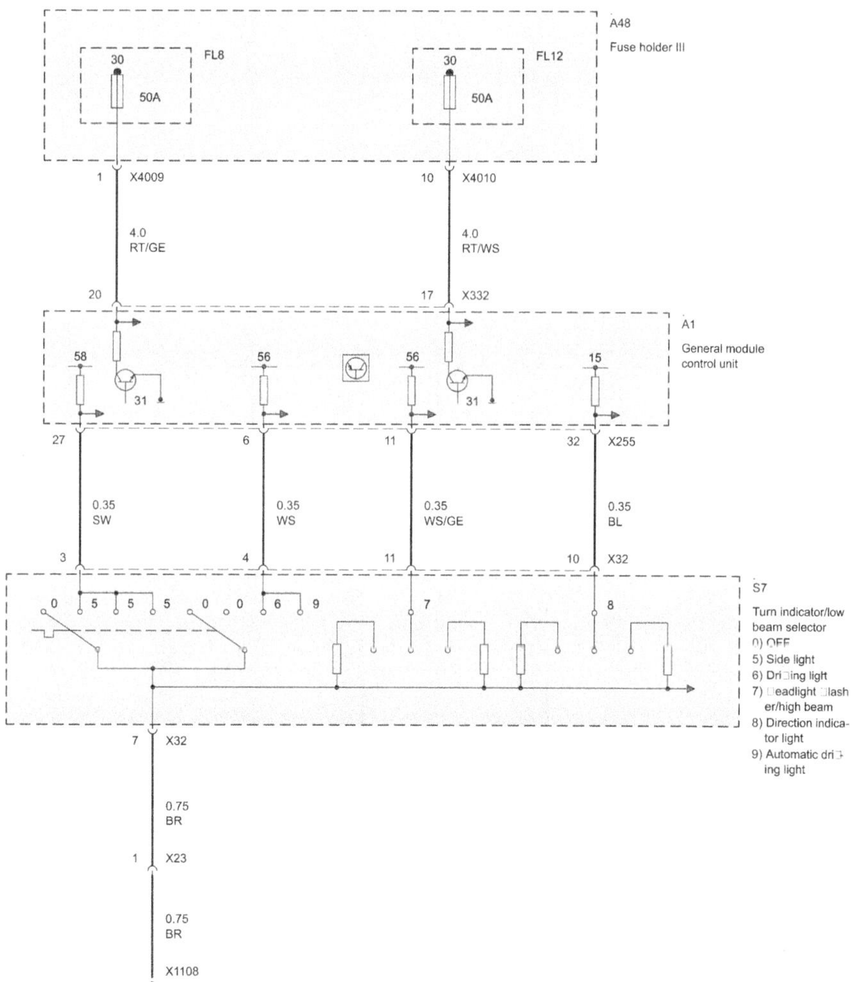

MC06ele22401

General Module (A1)
Control module A (up to Sept. 2005)
(page 1 of 2)

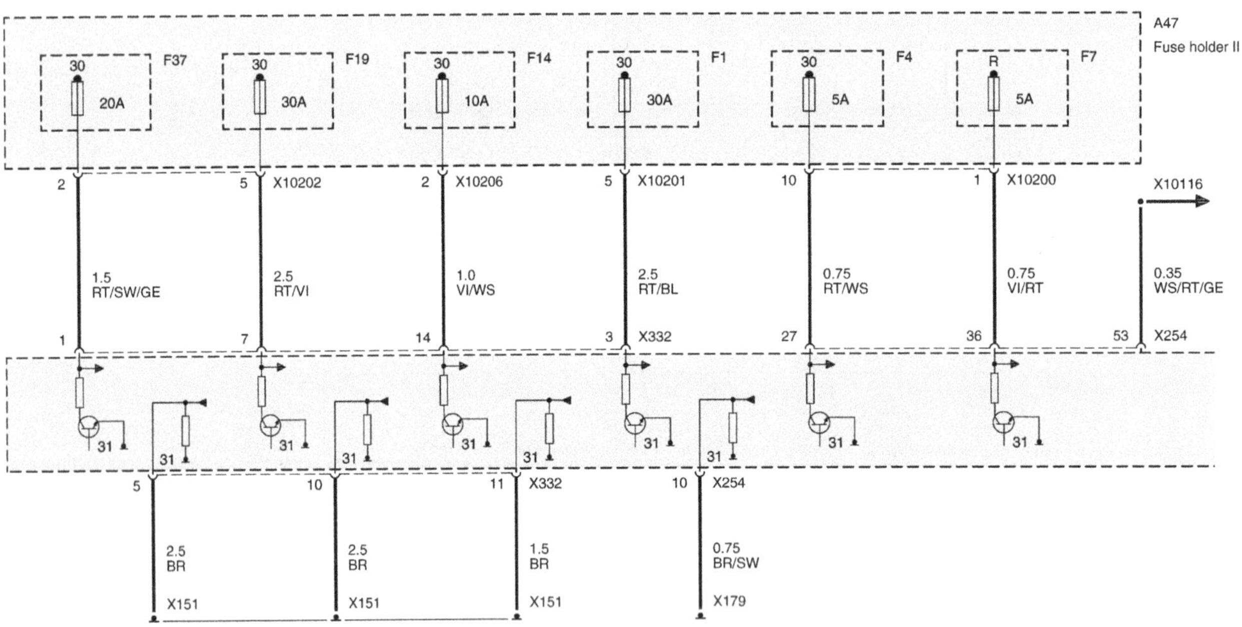

General Module (A1)
Control module A (up to Sept. 2005)
(page 2 of 2)

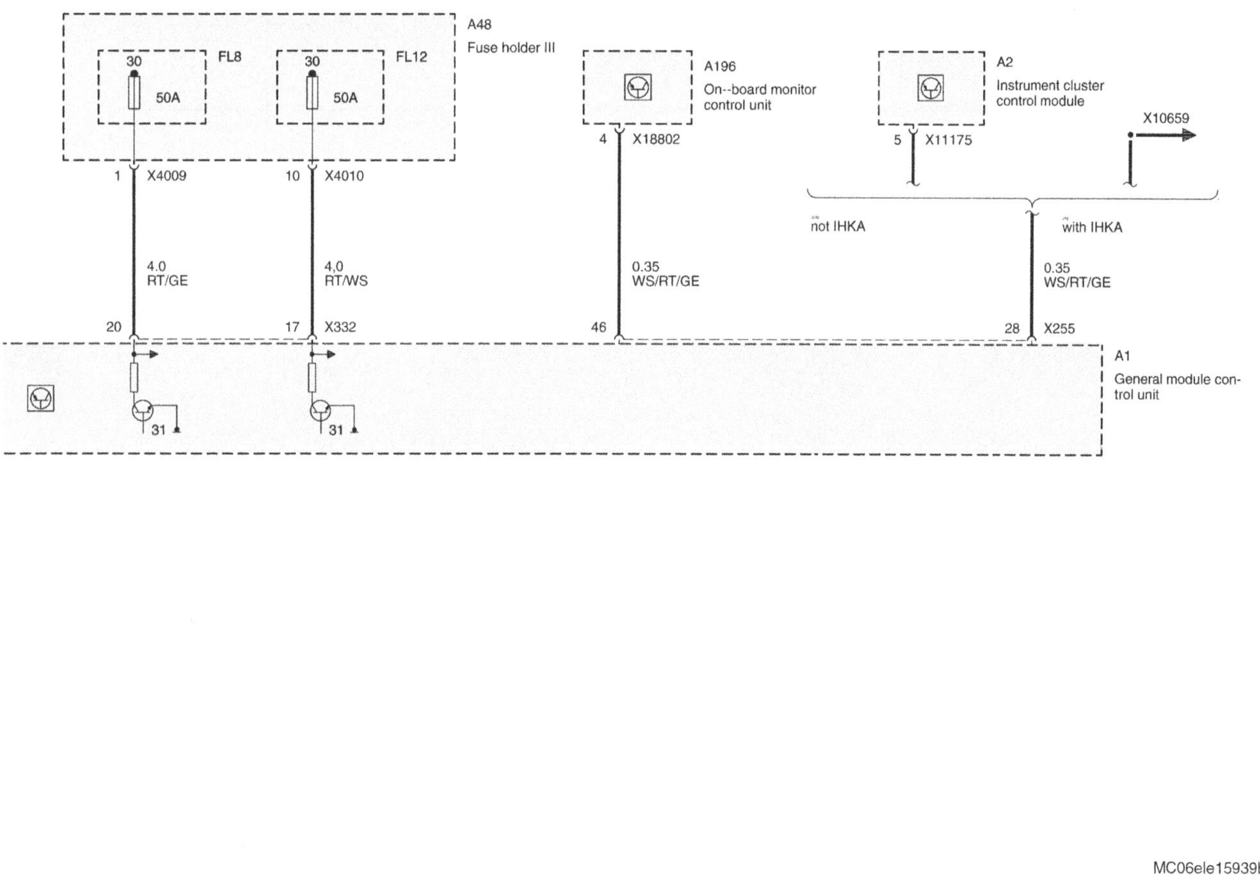

General Module (A1)

Control module A (from Sept. 2005)

(page 1 of 2)

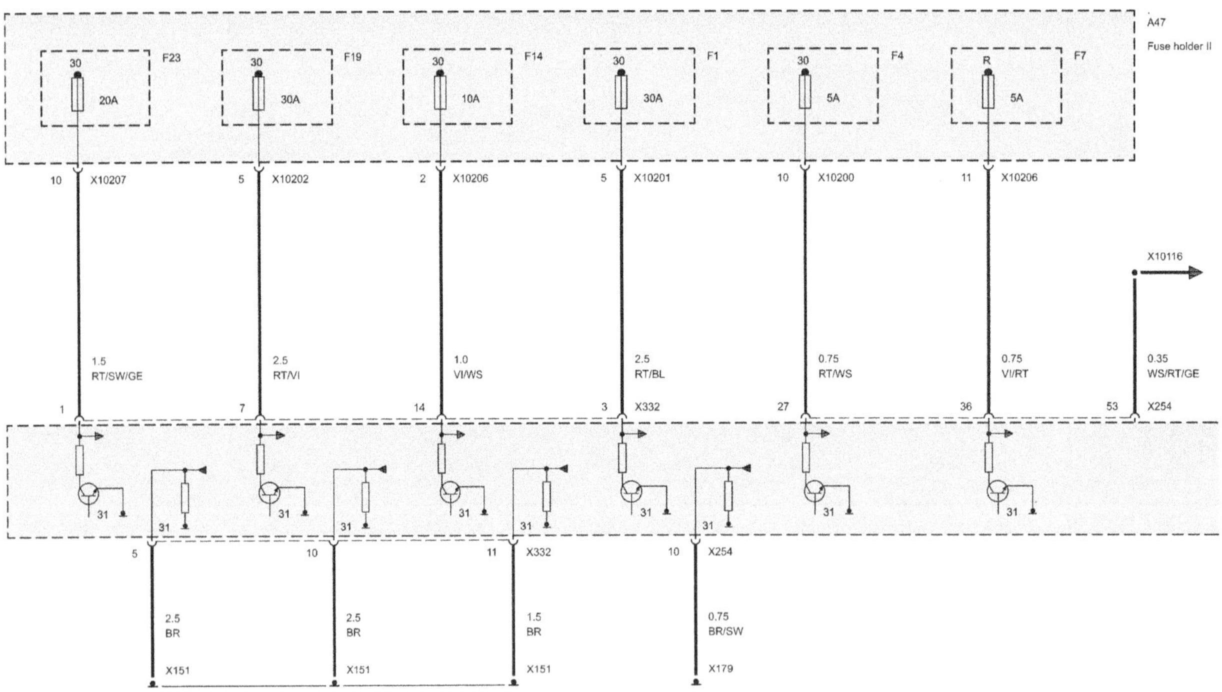

MC06ele23058a

General Module (A1)
Control module A (from Sept. 2005)
(page 2 of 2)

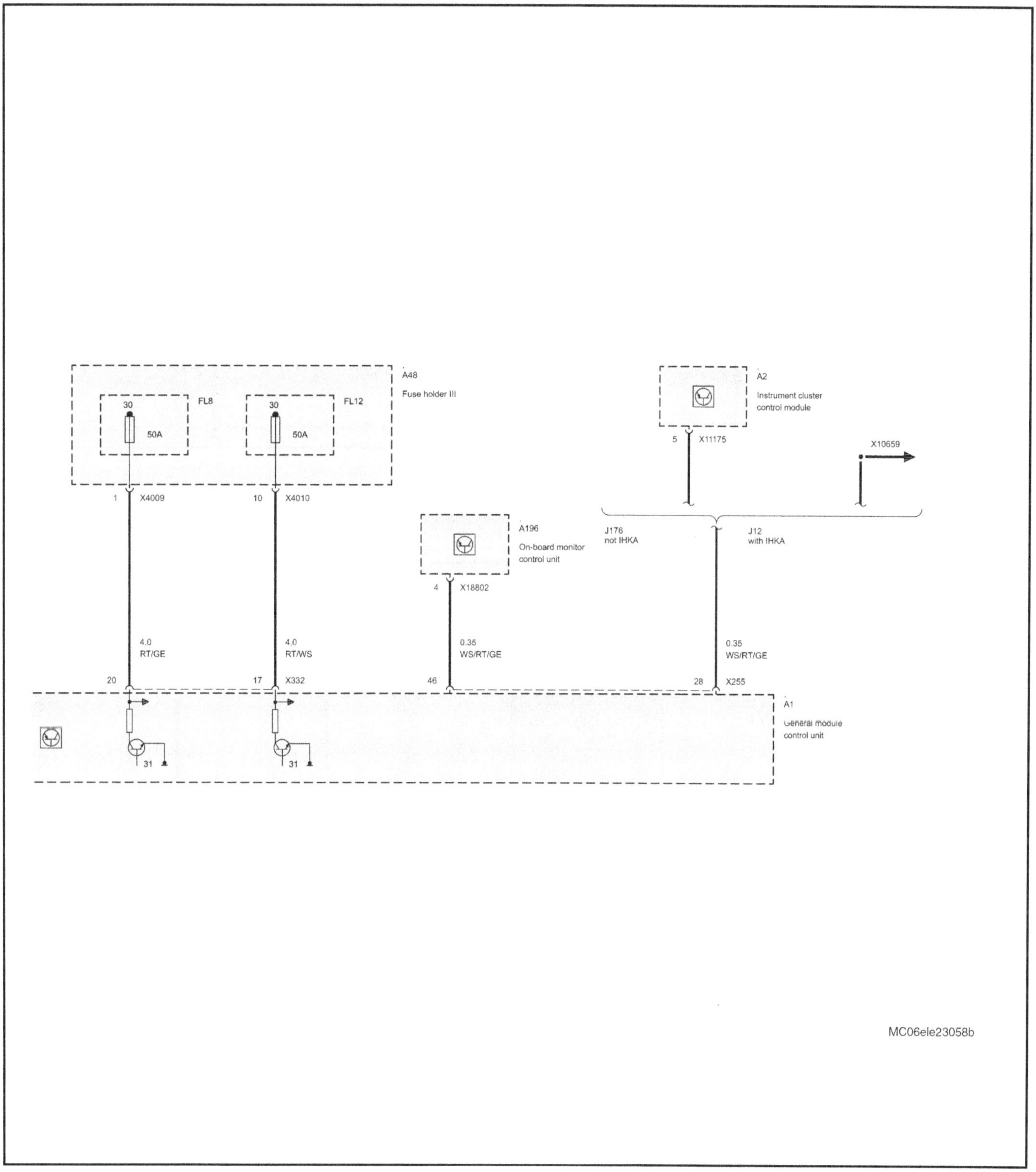

General Module (A1)
Control module B (up to March 2002)
(page 1 of 4)

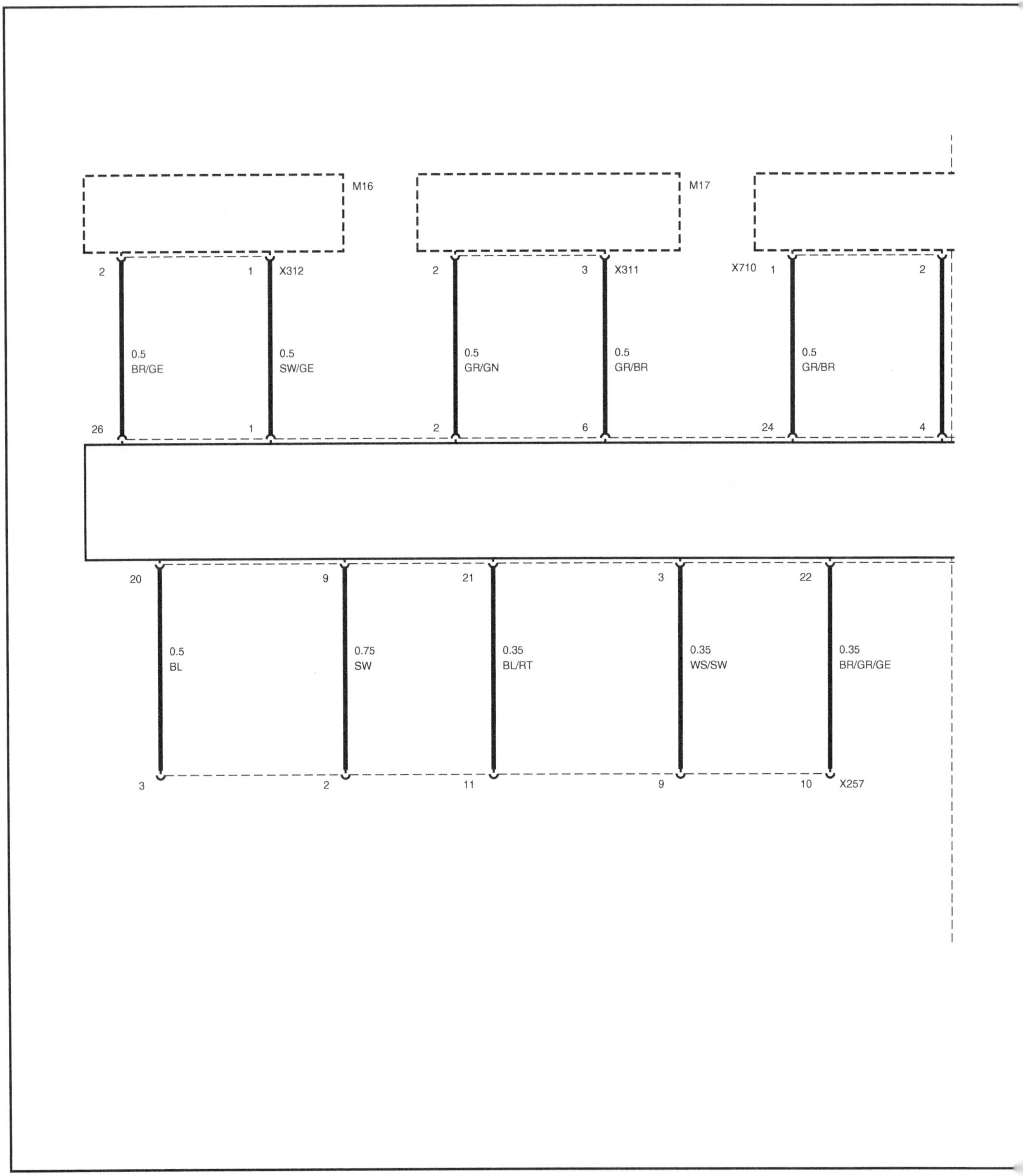

General Module (A1)
Control module B (up to March 2002)
(page 2 of 4)

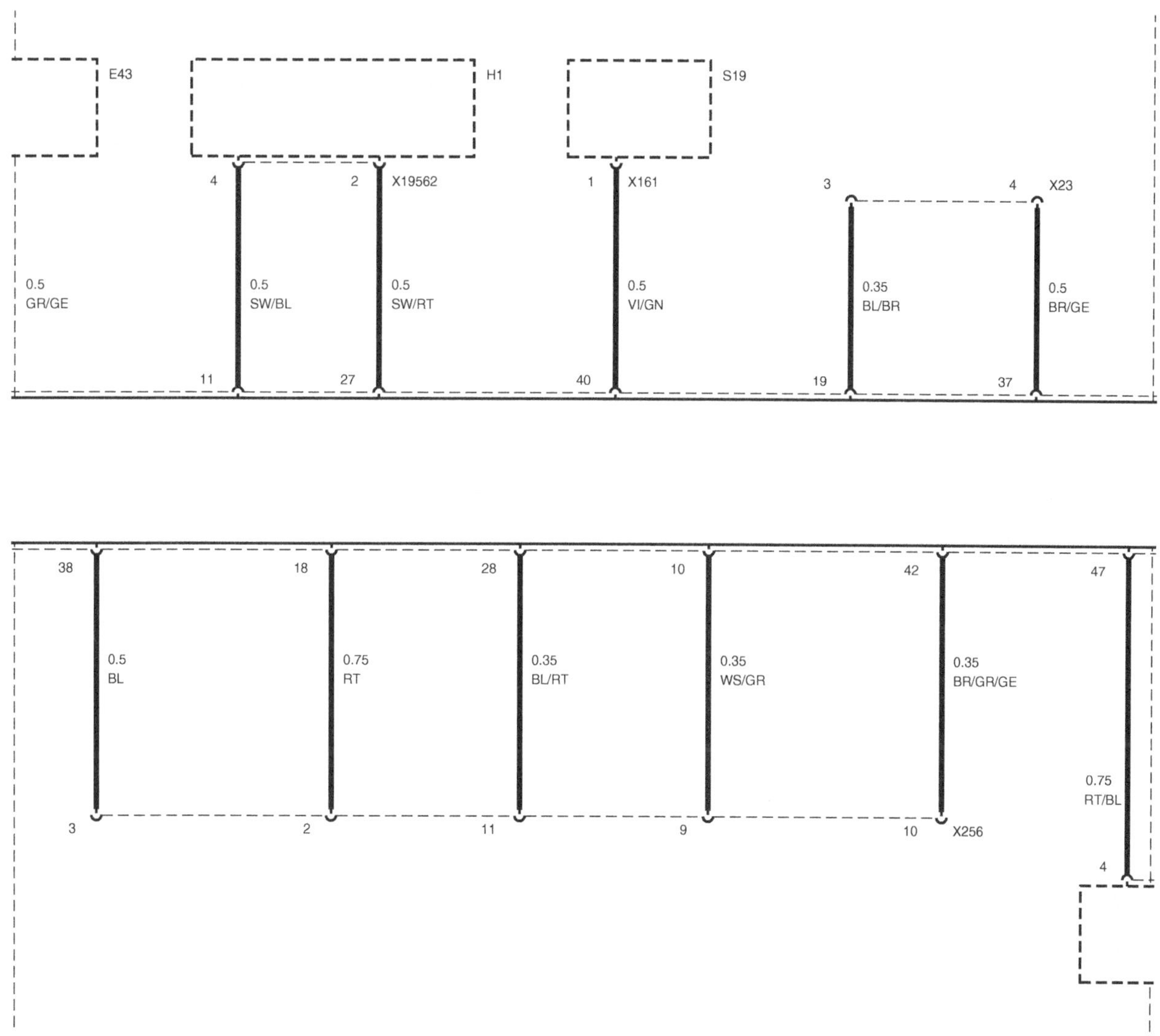

General Module (A1)
Control module B (up to March 2002)
(page 3 of 4)

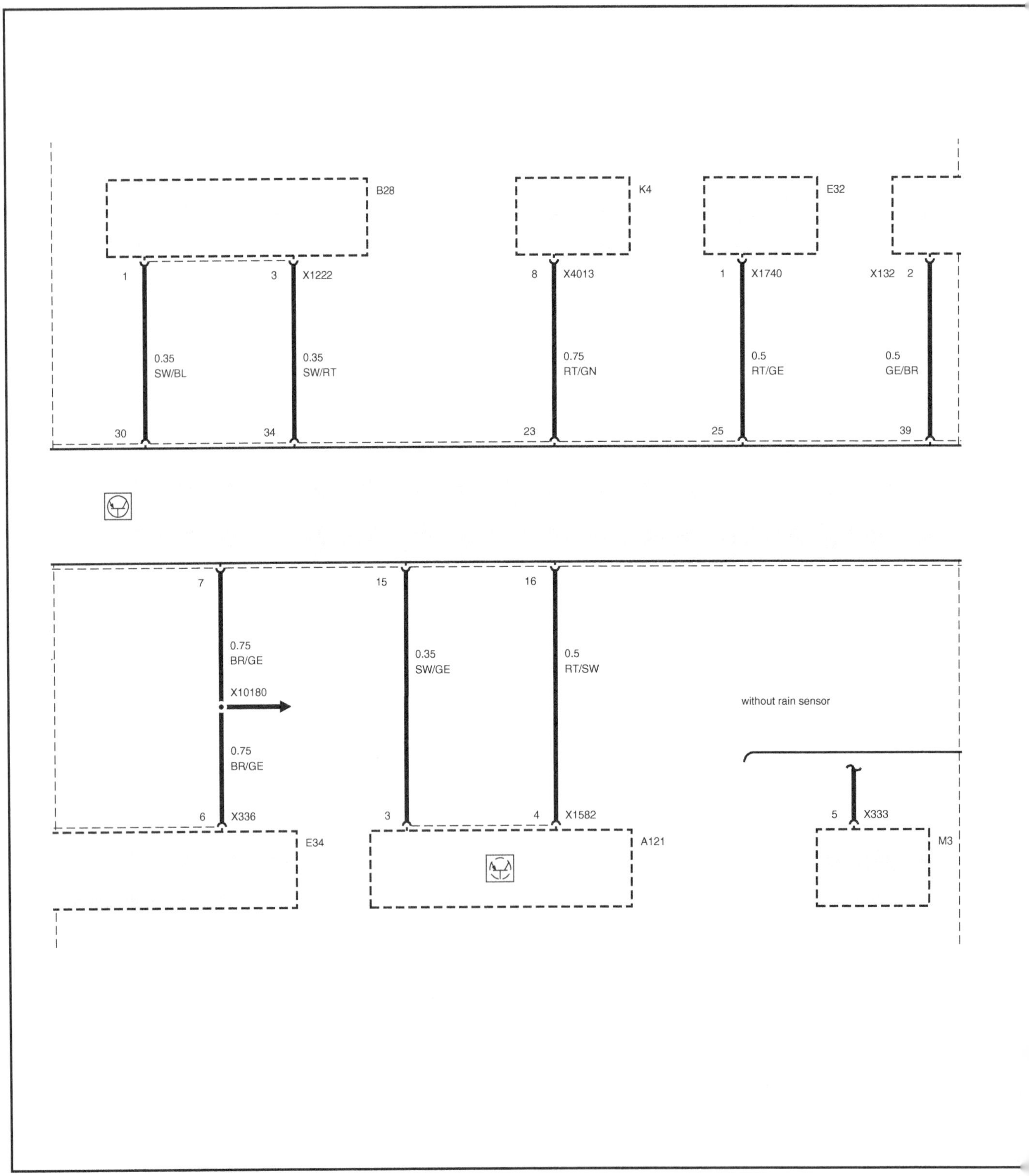

General Module (A1)

Control module B (up to March 2002)

(page 4 of 4)

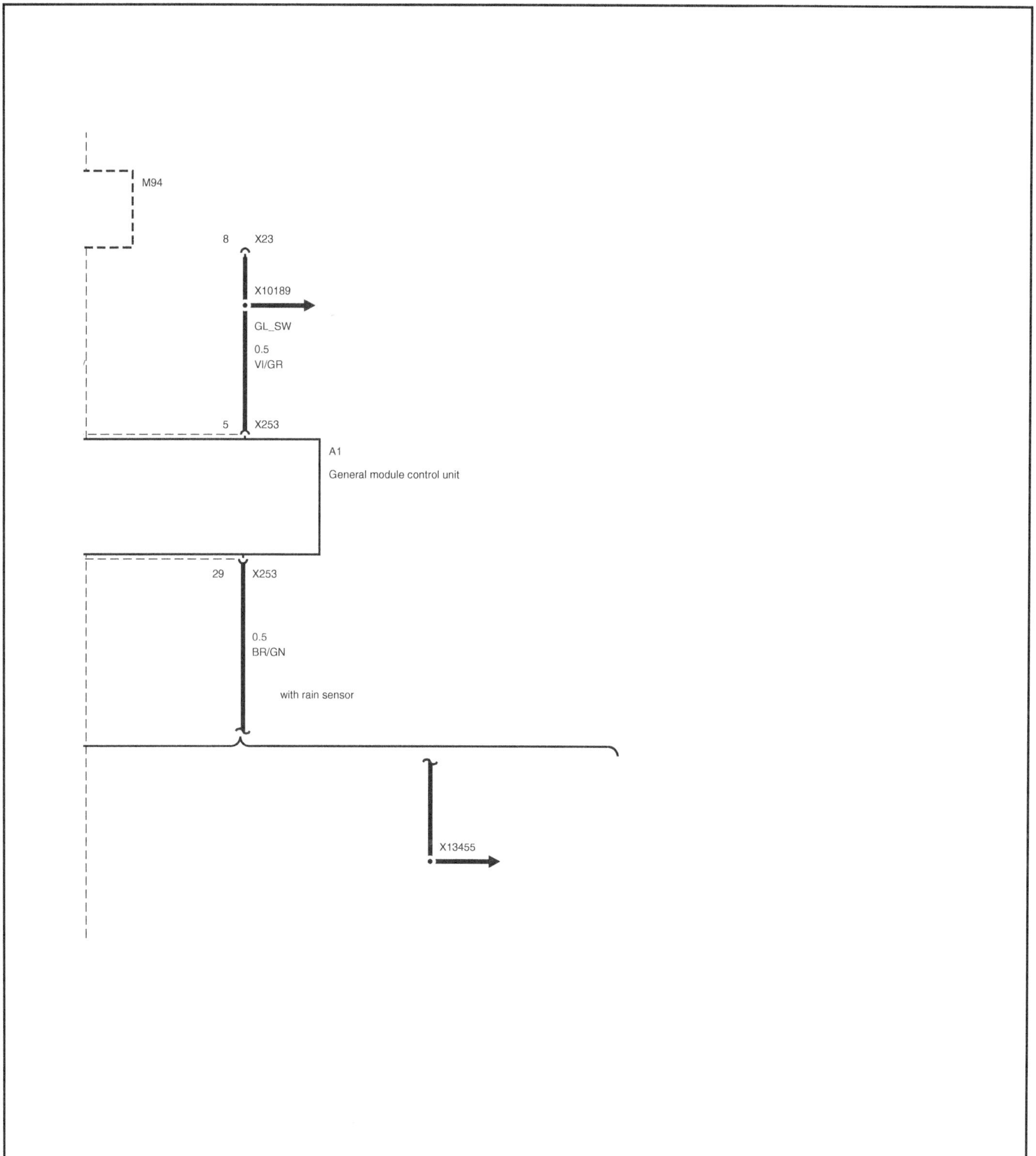

General Module (A1)

Control module B (from July 2004)

(page 1 of 3)

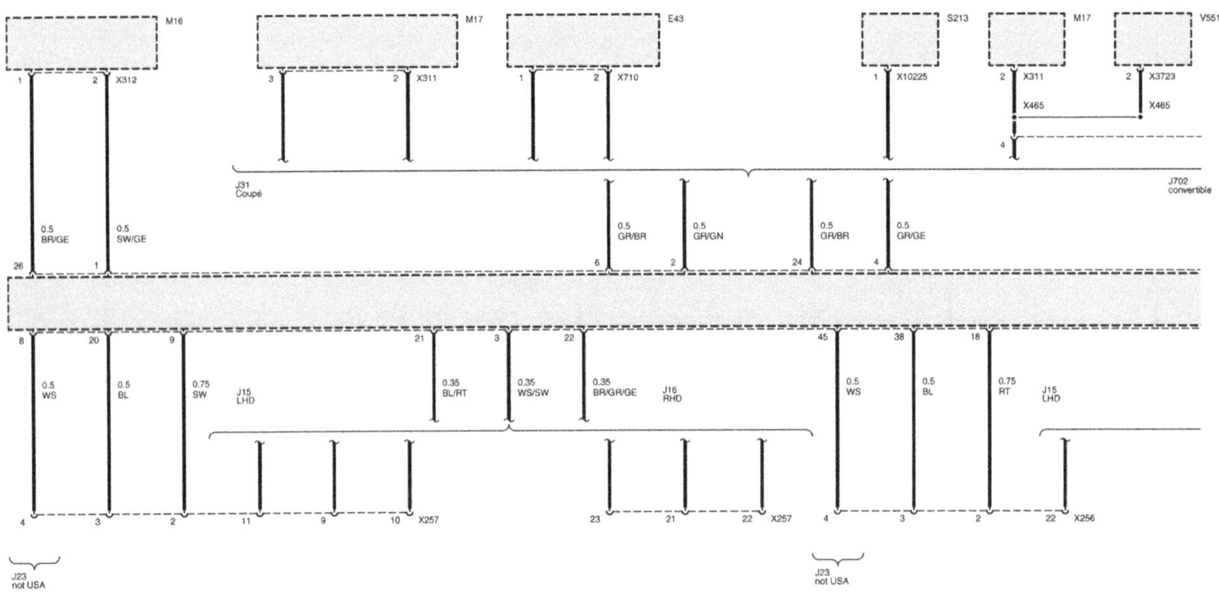

MC06ele21180a

General Module (A1)
Control module B (from July 2004)
(page 2 of 3)

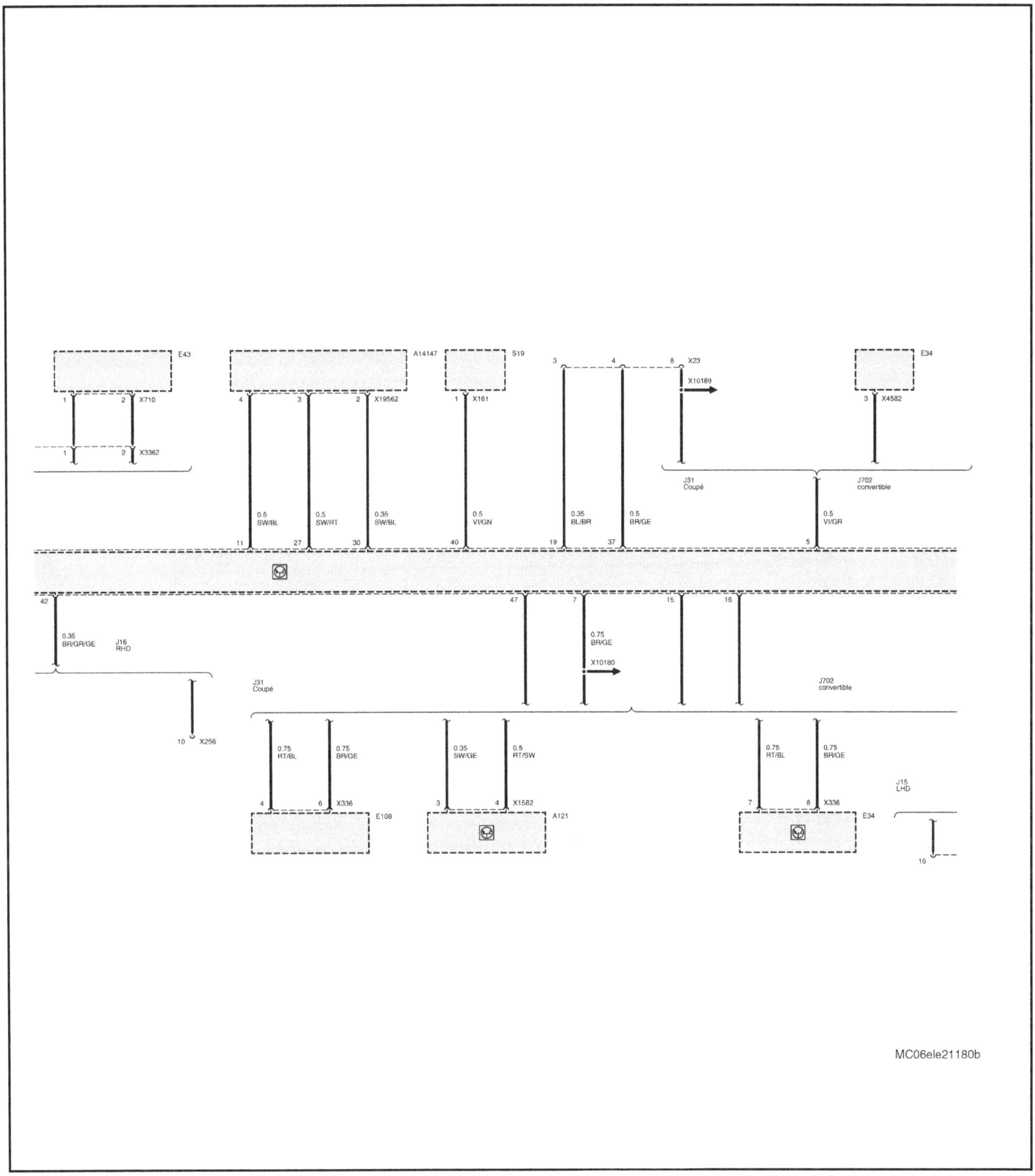

General Module (A1)
Control module B (from July 2004)
(page 3 of 3)

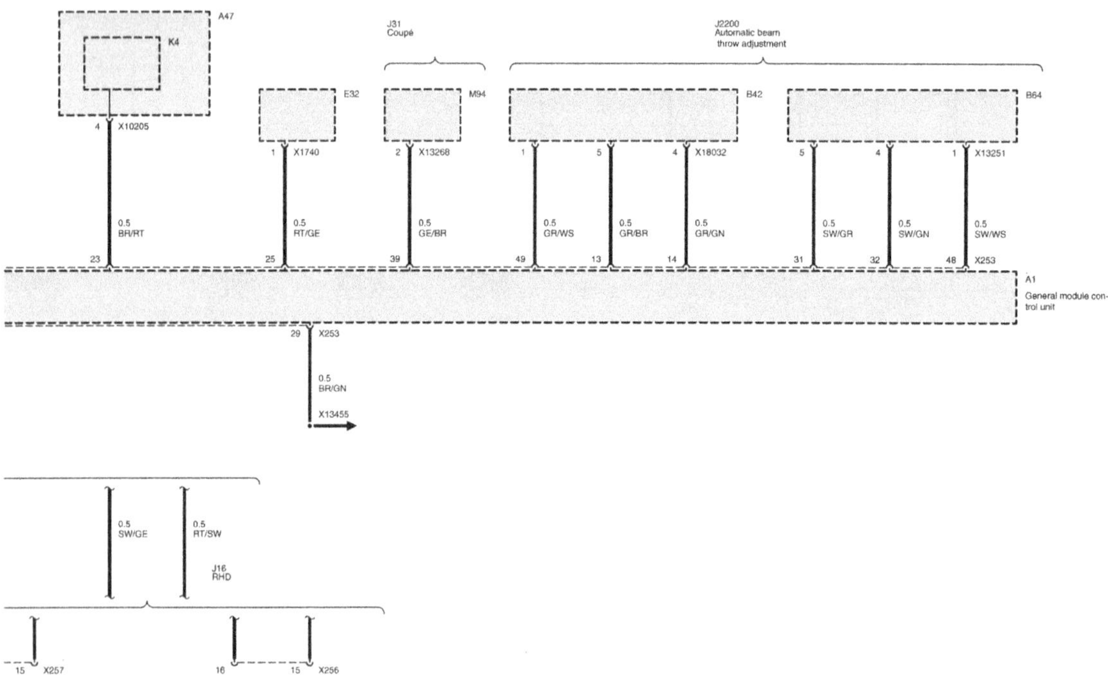

MC06ele21180c

General Module (A1) – control module B

Control module B (from Jan. 2005)

(page 1 of 3)

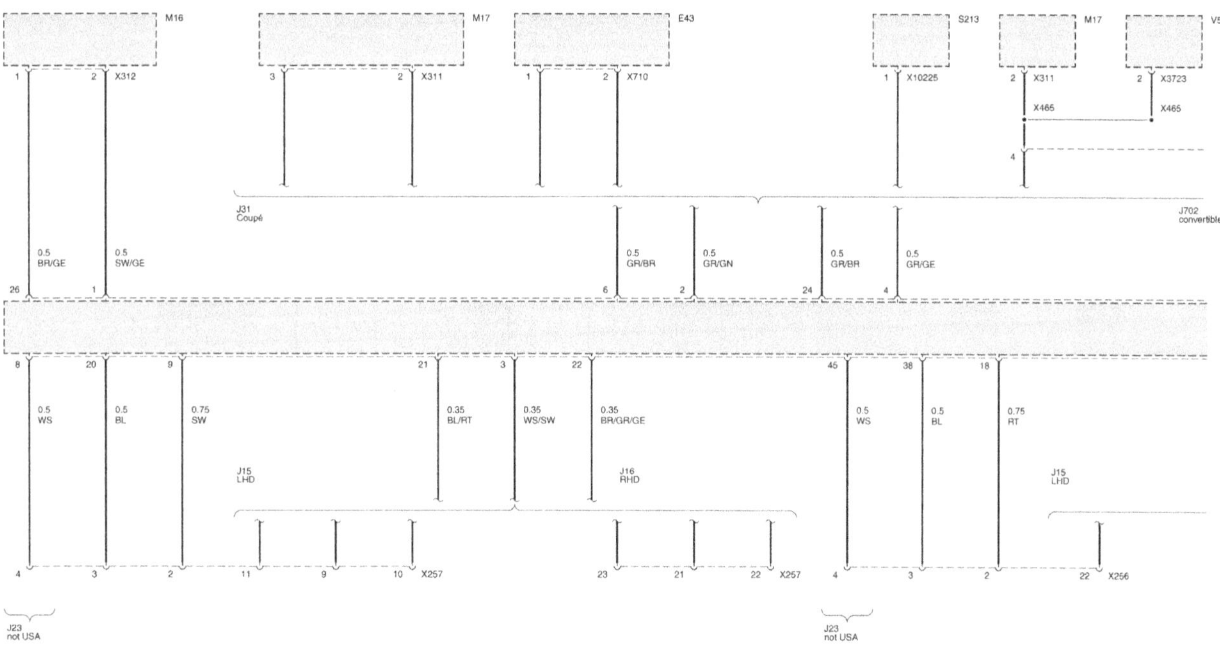

MC06ele22203a

General Module (A1) – control module B
Control module B (from Jan. 2005)
(page 2 of 3)

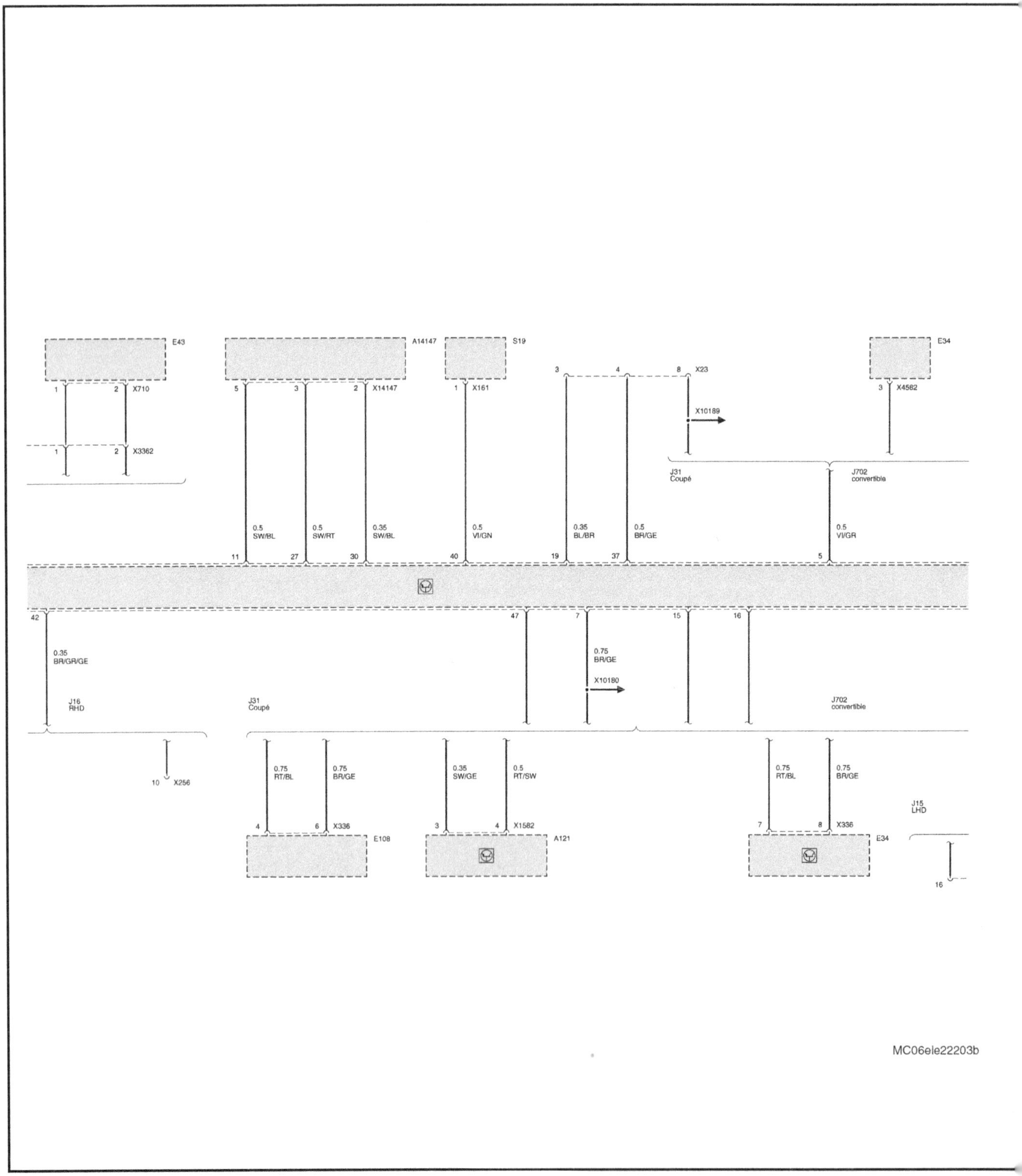

General Module (A1) – control module B

Control module B (from Jan. 2005)

(page 3 of 3)

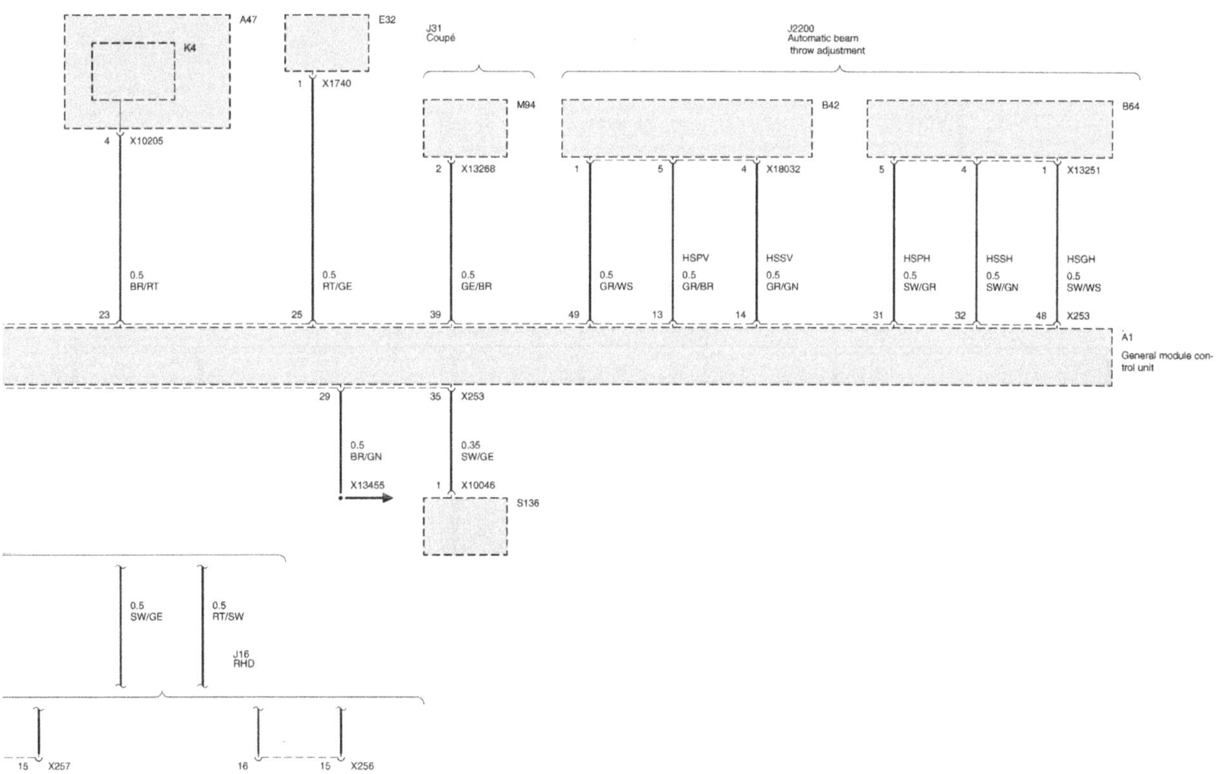

MC06ele22203c

General Module (A1)

Control module C, without Xenon headlights (up to Sept. 2003)

(page 1 of 5)

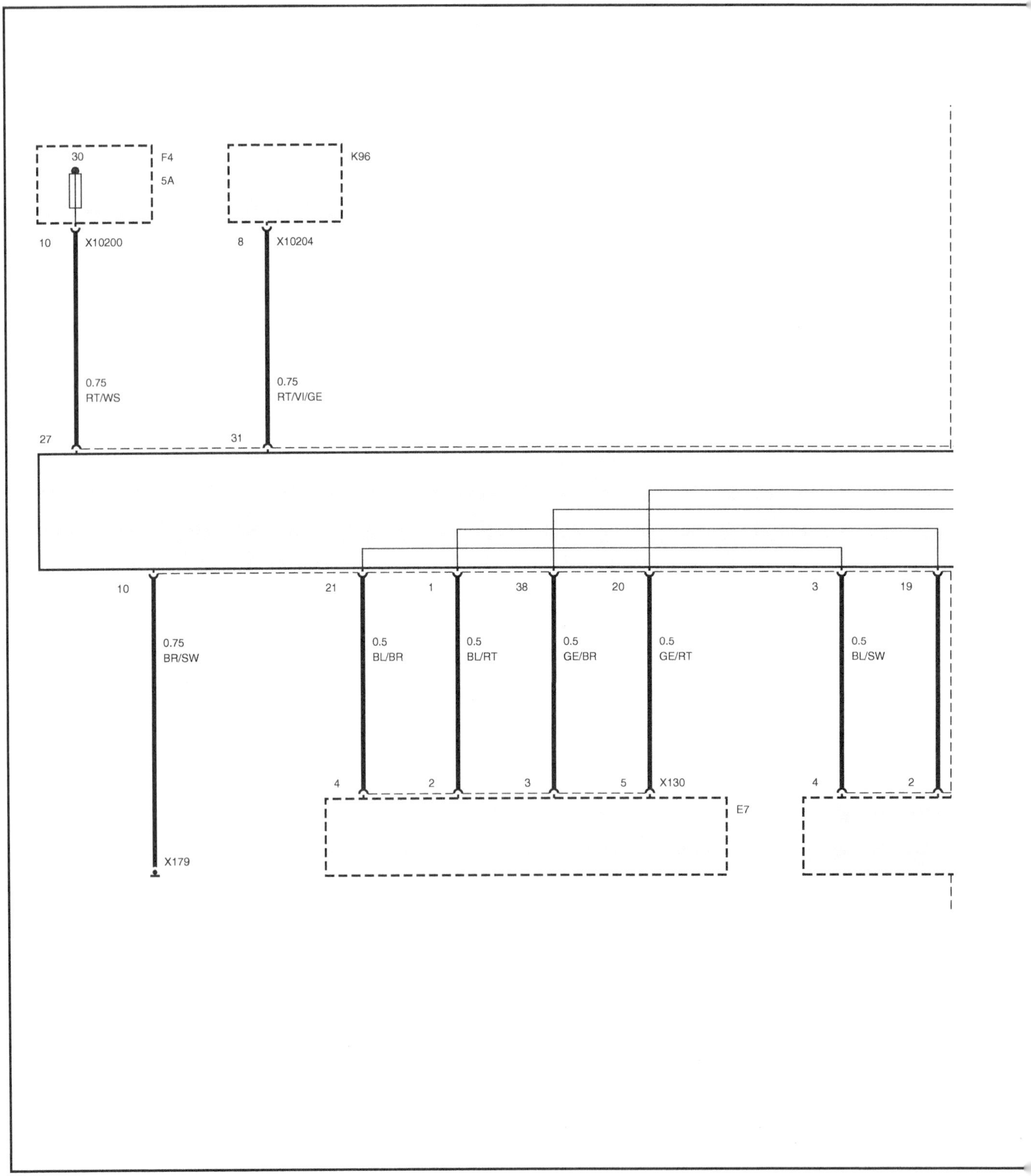

General Module (A1)

Control module C, without Xenon headlights (up to Sept. 2003)

(page 2 of 5)

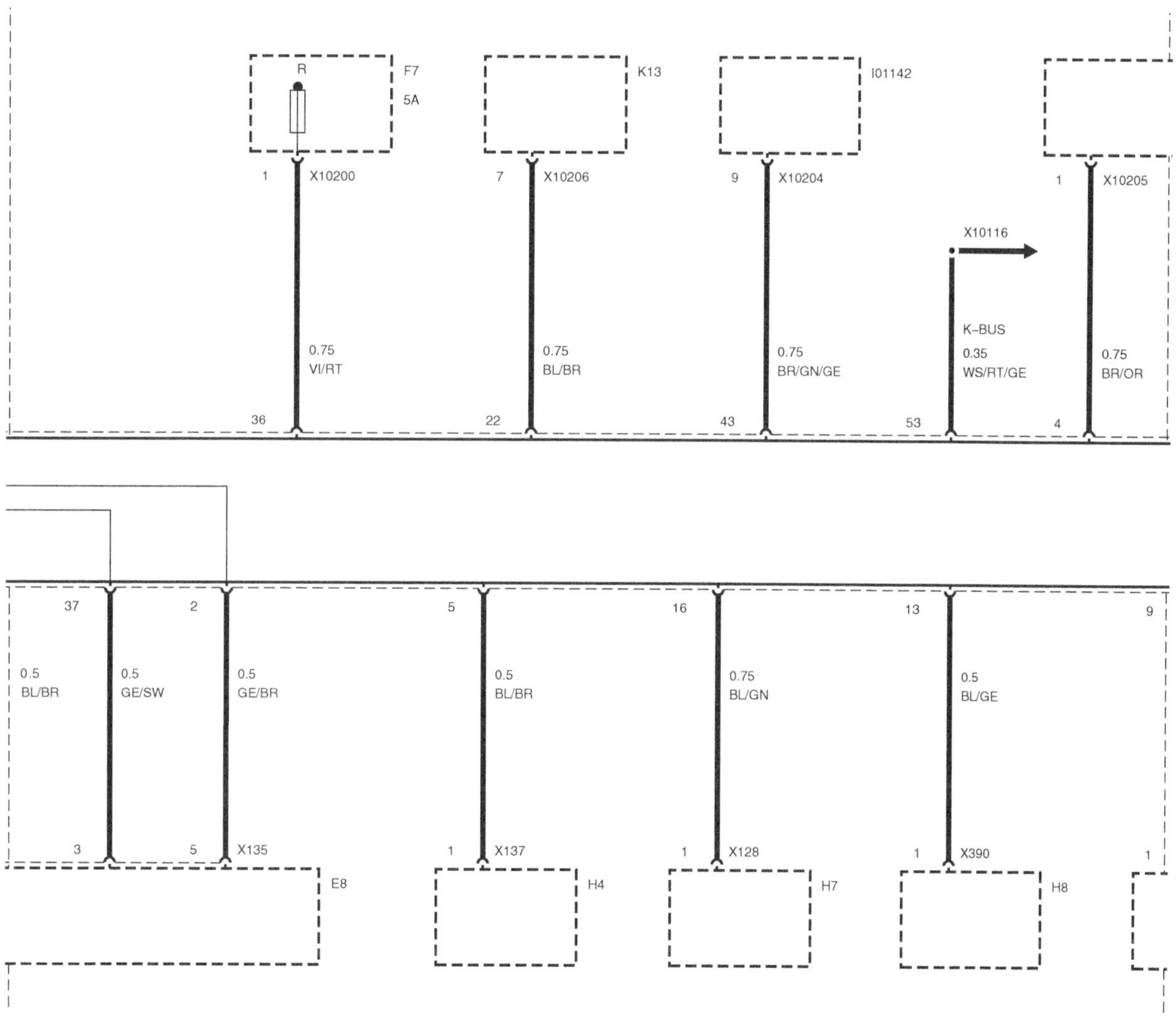

General Module (A1)

Control module C, without Xenon headlights (up to Sept. 2003)

(page 3 of 5)

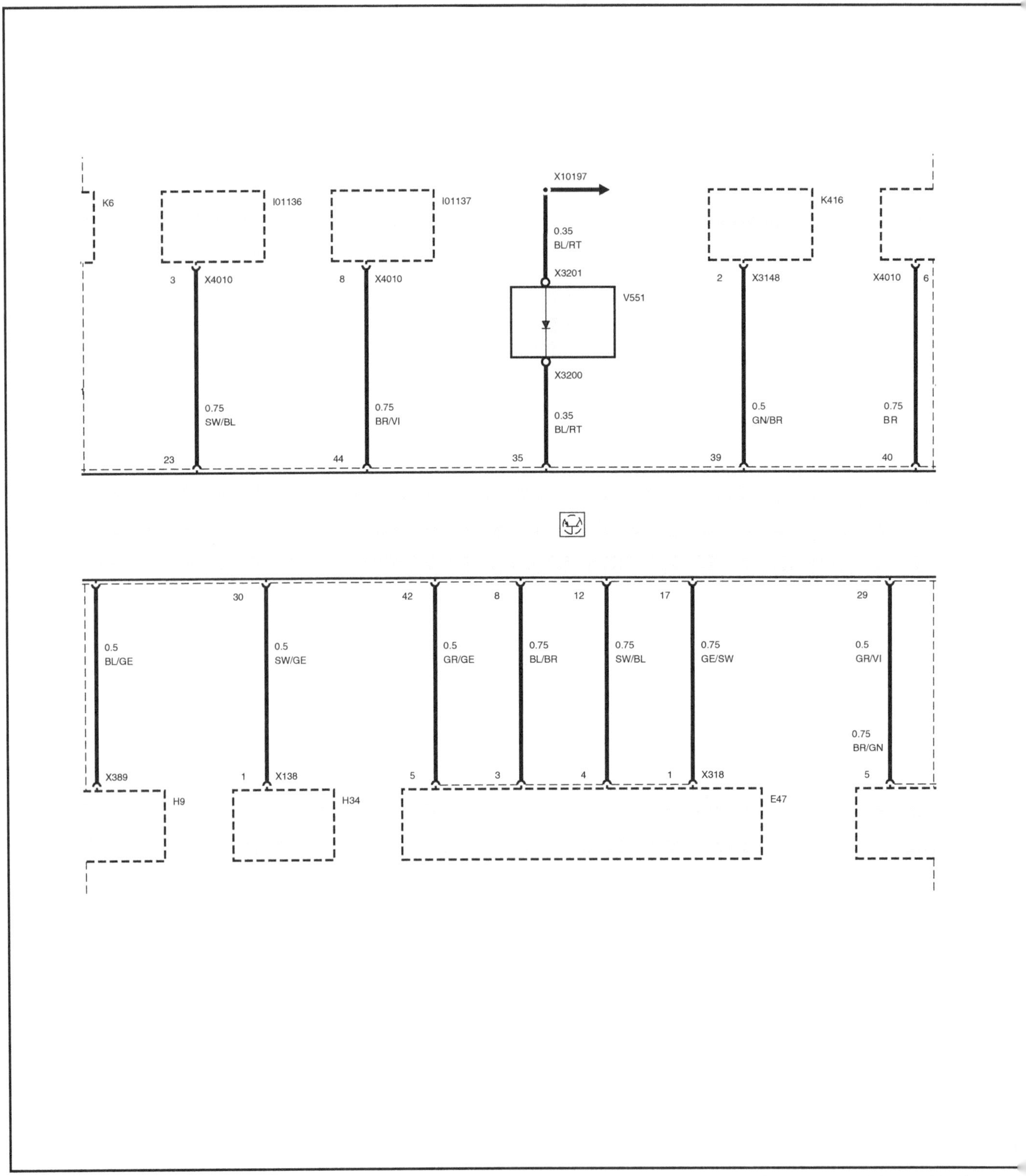

General Module (A1)

Control module C, without Xenon headlights (up to Sept. 2003)

(page 4 of 5)

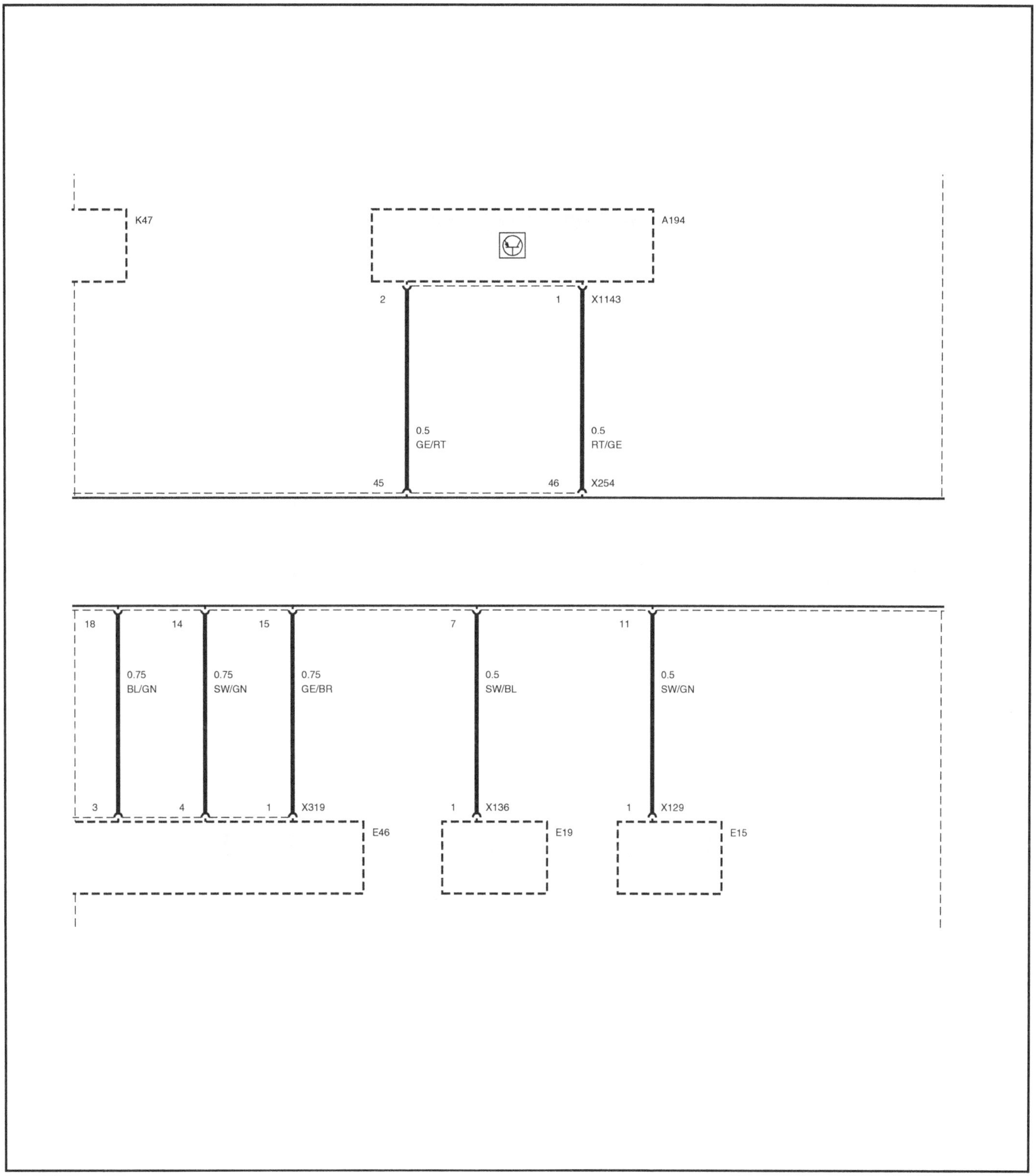

General Module (A1)

Control module C, without Xenon headlights (up to Sept. 2003)

(page 5 of 5)

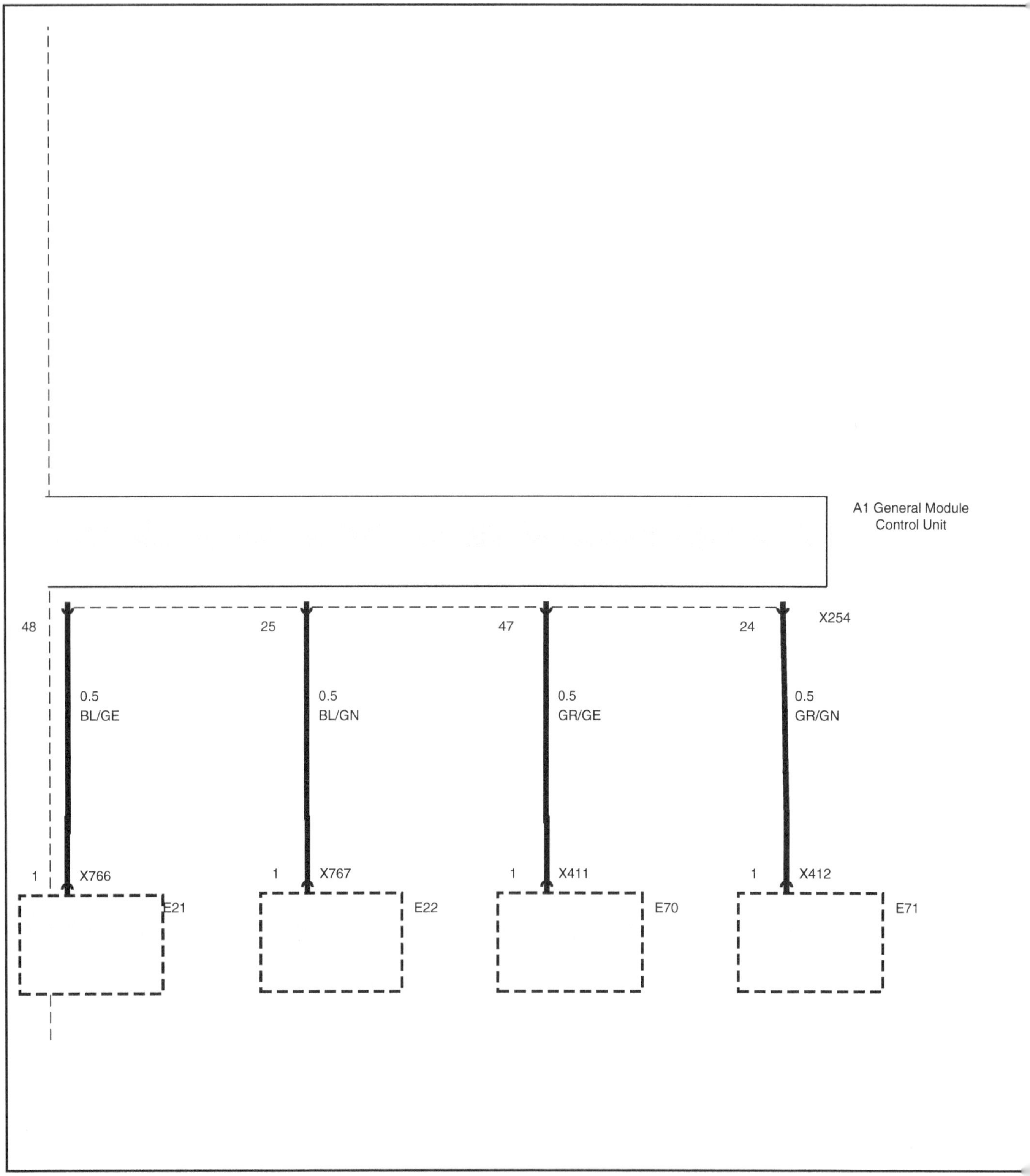

General Module (A1)

Control module C, with Xenon headlights (from July 2004 to Sept. 2005)

(page 1 of 3)

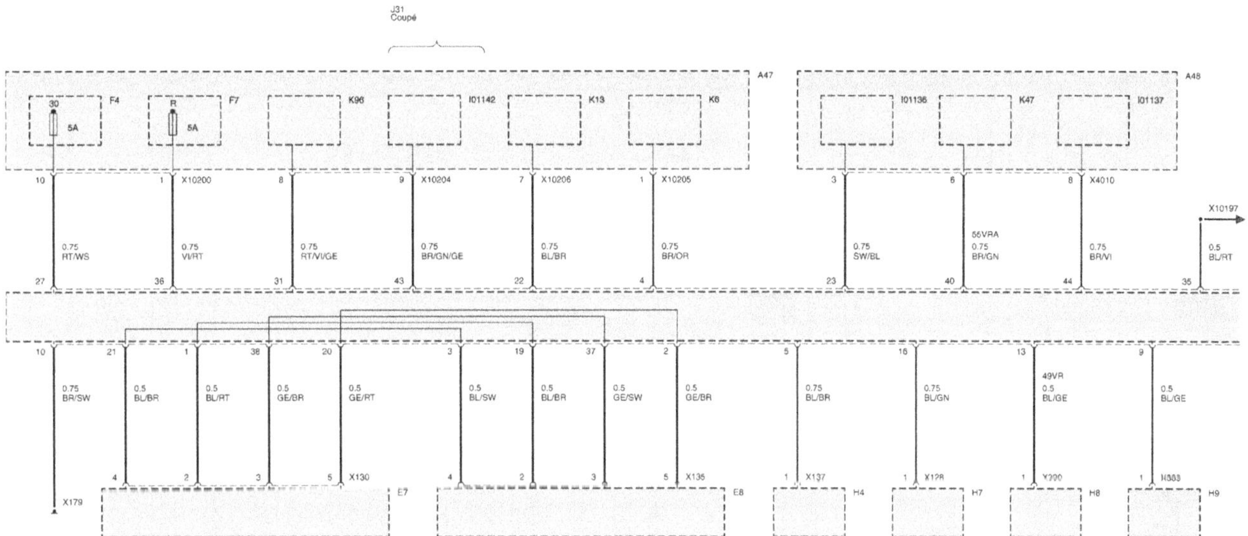

MC06ele21181a

General Module (A1)

Control module C, with Xenon headlights (from July 2004 to Sept. 2005)

(page 2 of 3)

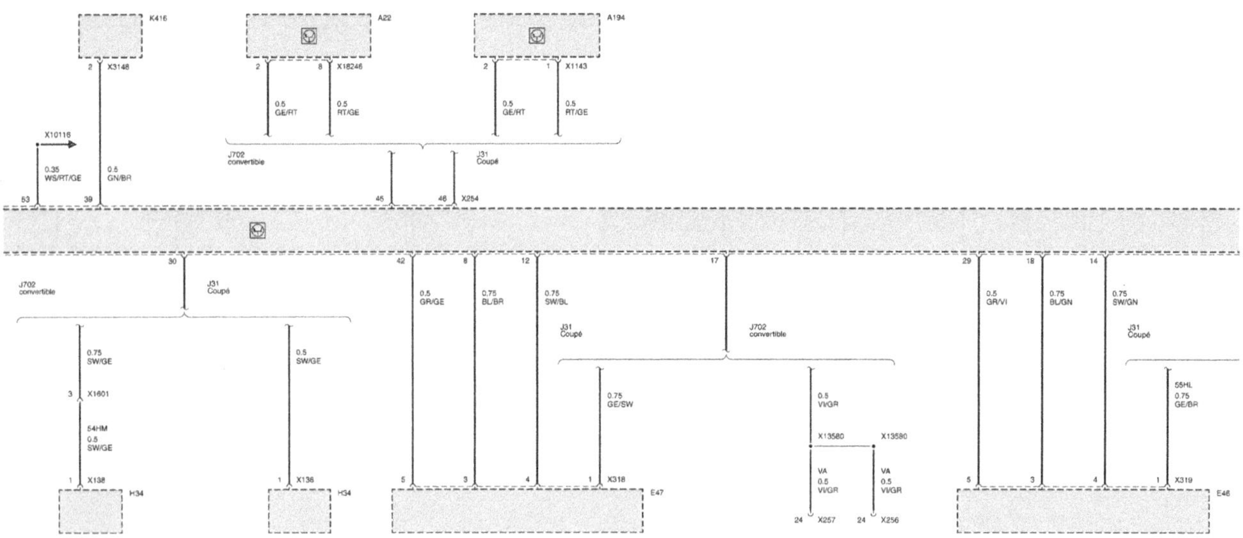

MC06ele21181b

General Module (A1)

Control module C, with Xenon headlights (from July 2004 to Sept. 2005)

(page 3 of 3)

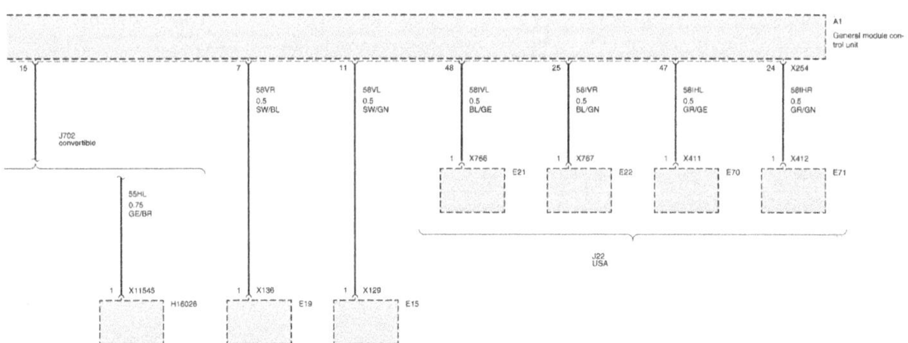

MC06ele21181c

General Module (A1)

Control module D, with IHKA (up to Sept. 2003)

(page 1 of 3)

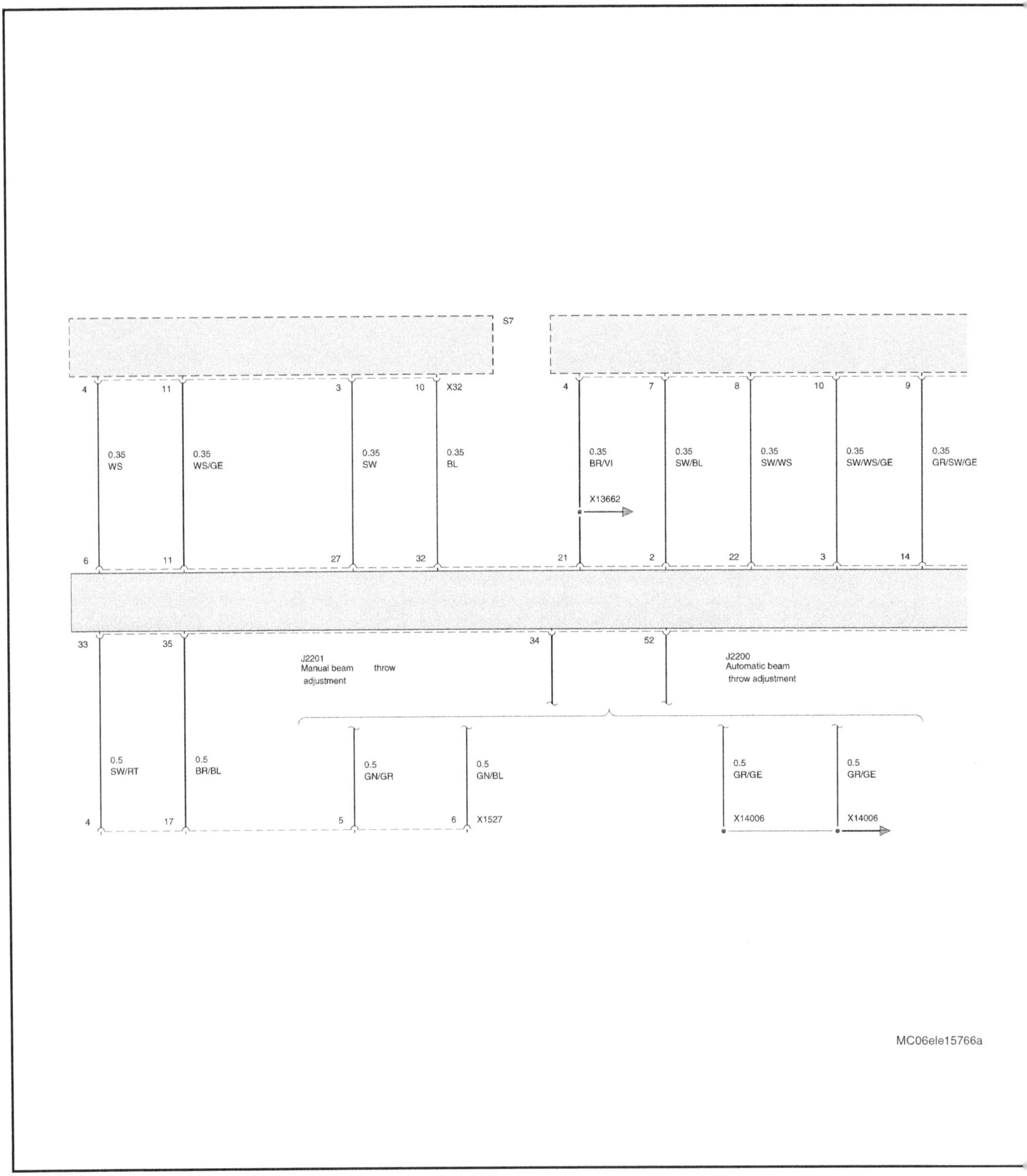

General Module (A1)

Control module D, with IHKA (up to Sept. 2003)

(page 2 of 3)

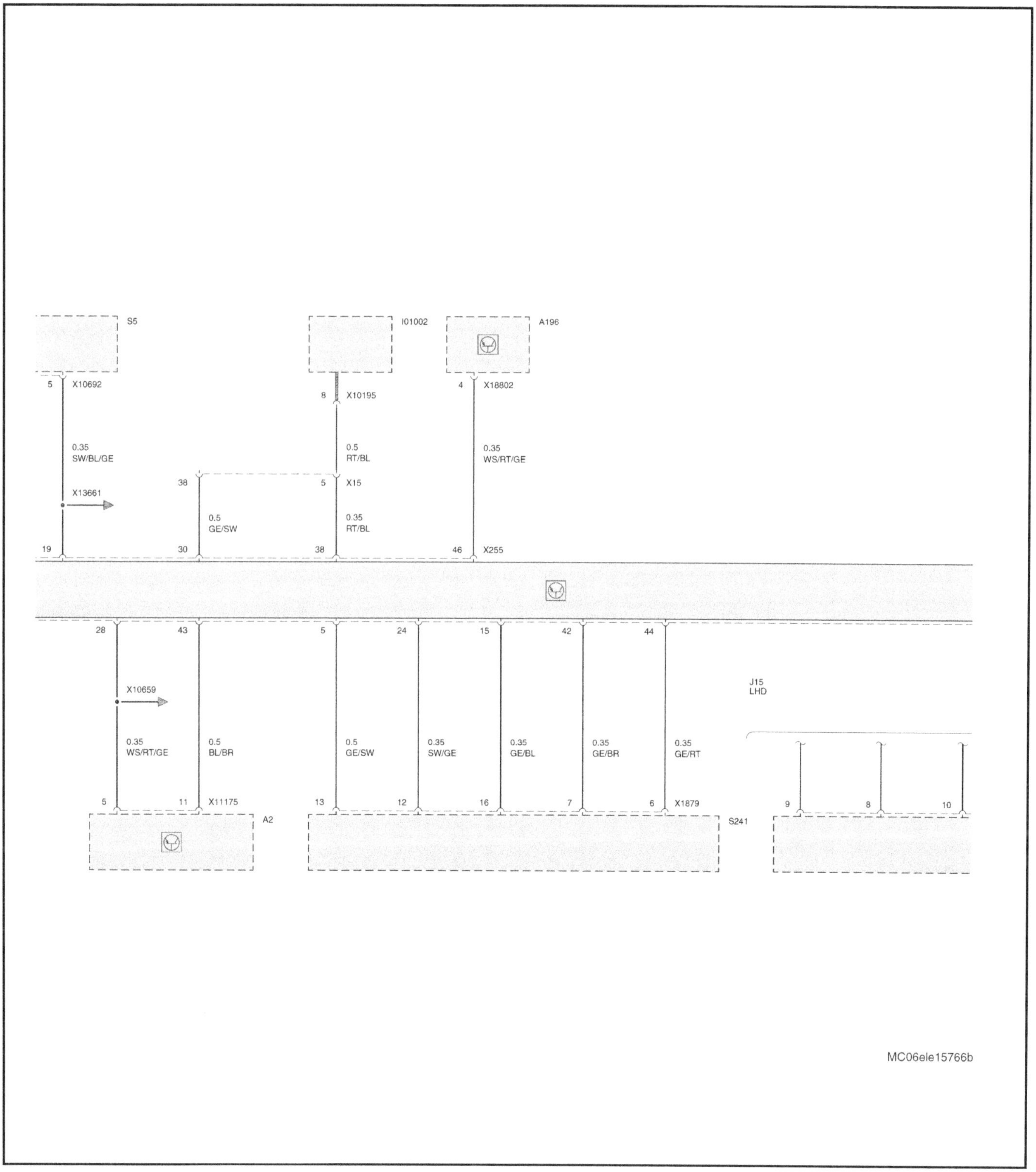

General Module (A1)
Control module D, with IHKA (up to Sept. 2003)
(page 3 of 3)

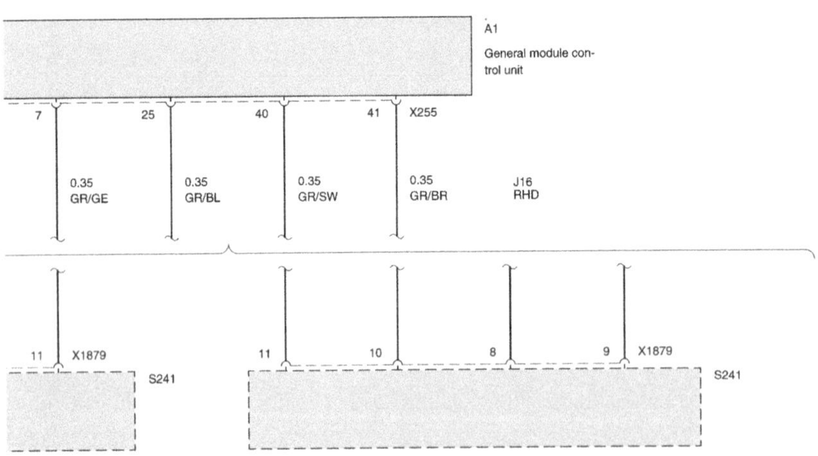

MC06ele15766c

General Module (A1)
Control module D (from Jan. 2005)
(page 1 of 2)

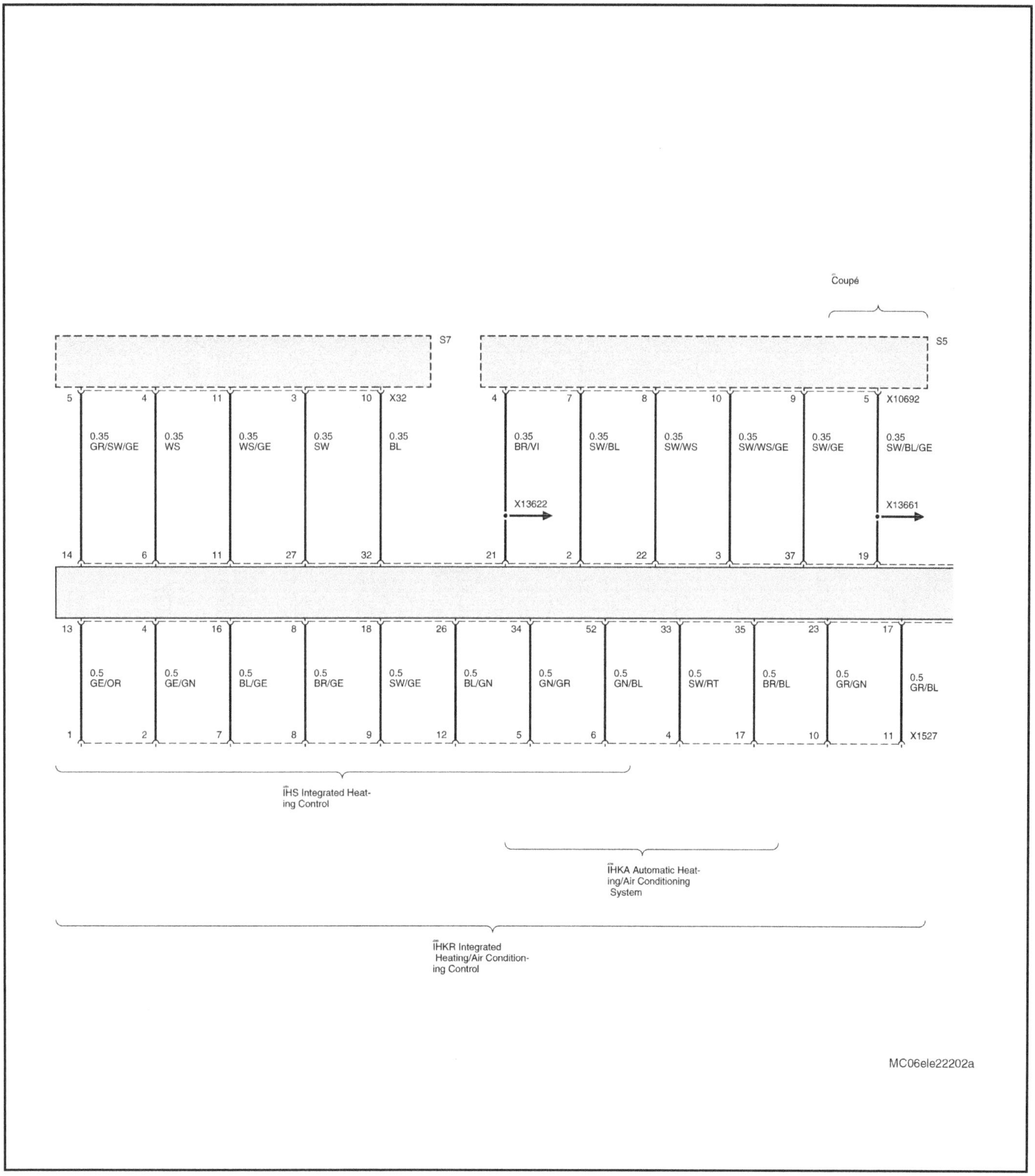

General Module (A1)
Control module D (from Jan. 2005)
(page 2 of 2)

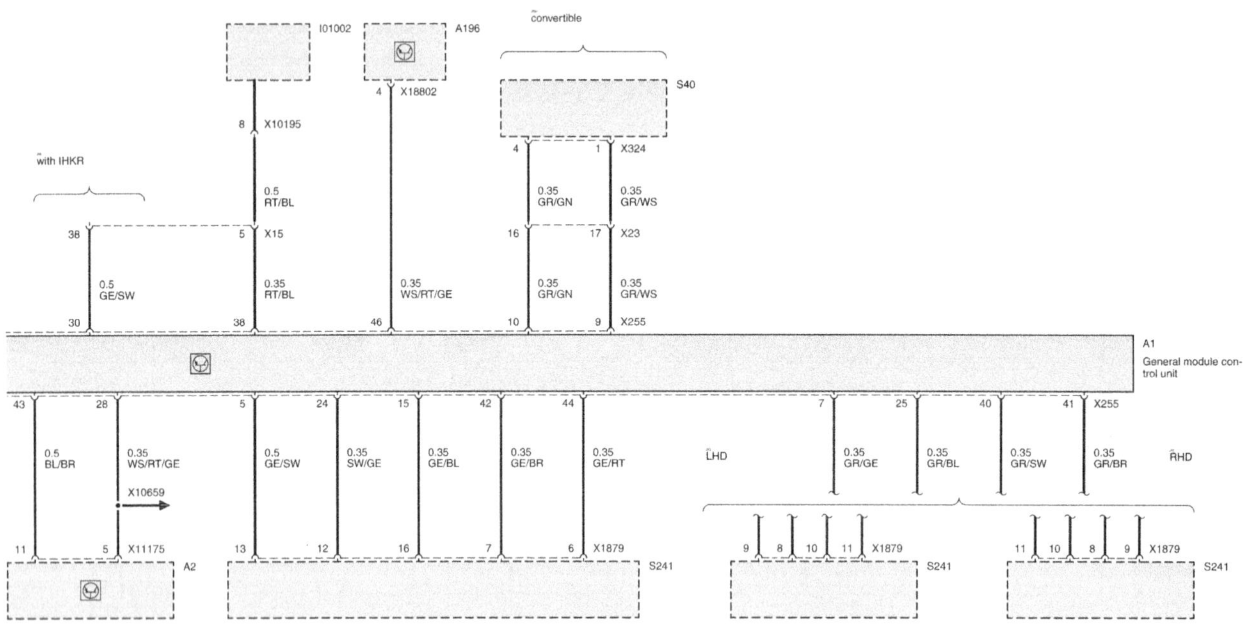

MC06ele22202b

Heated seats
Driver's seat
(page 1 of 1)

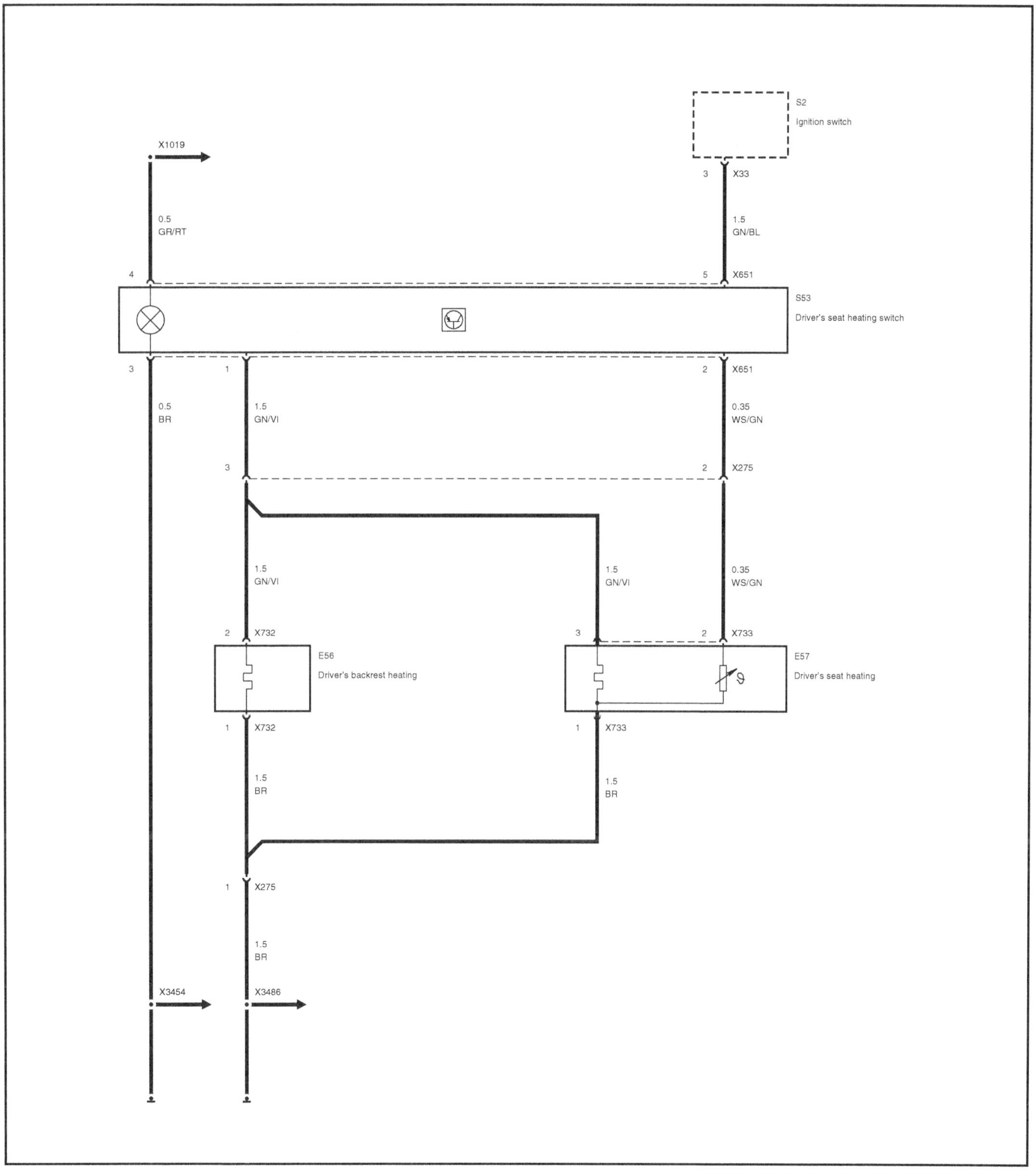

Heated seats
Passenger seat
(page 1 of 1)

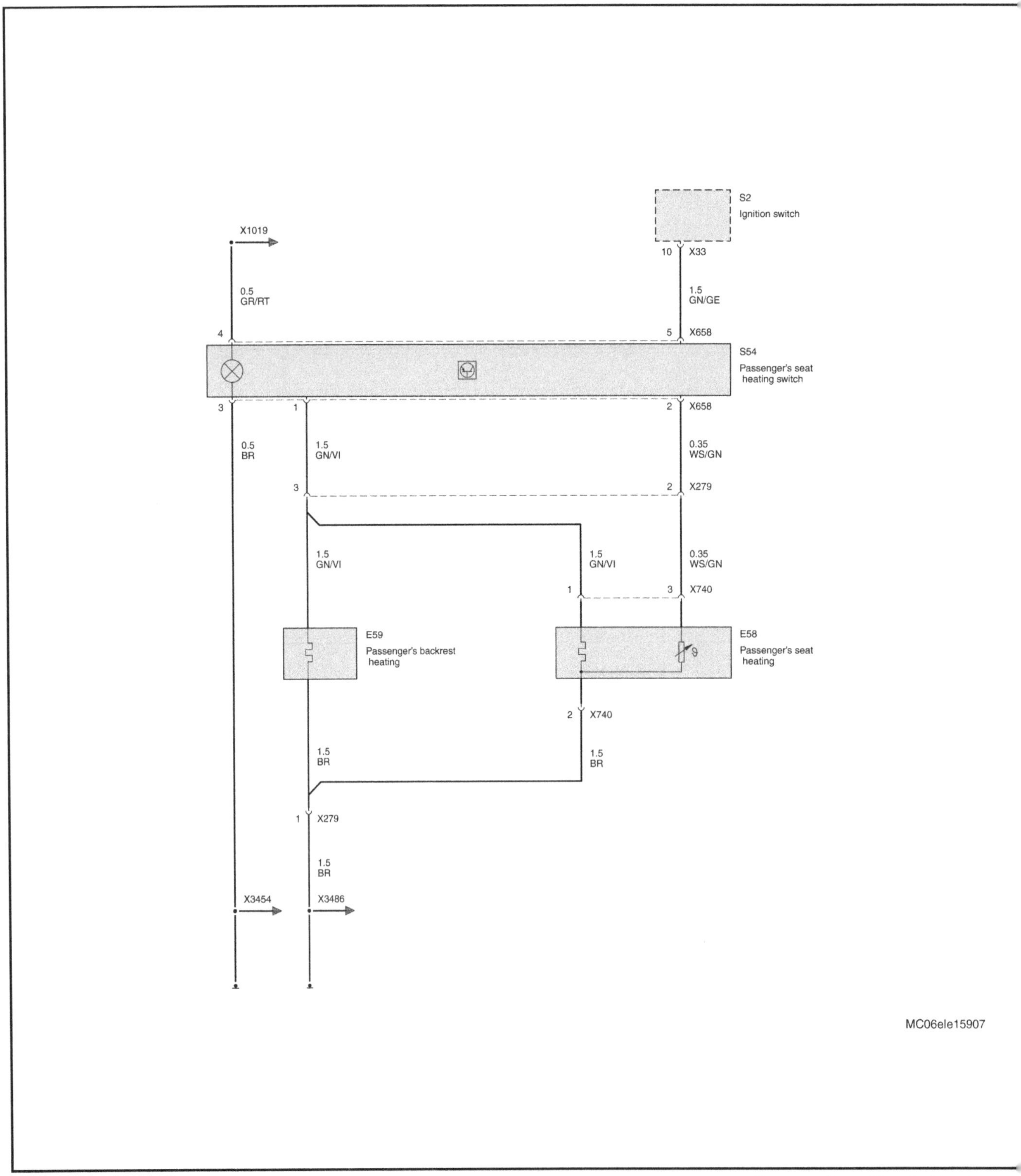

Heating and A/C

Power supply (W10 engine to March 2002)

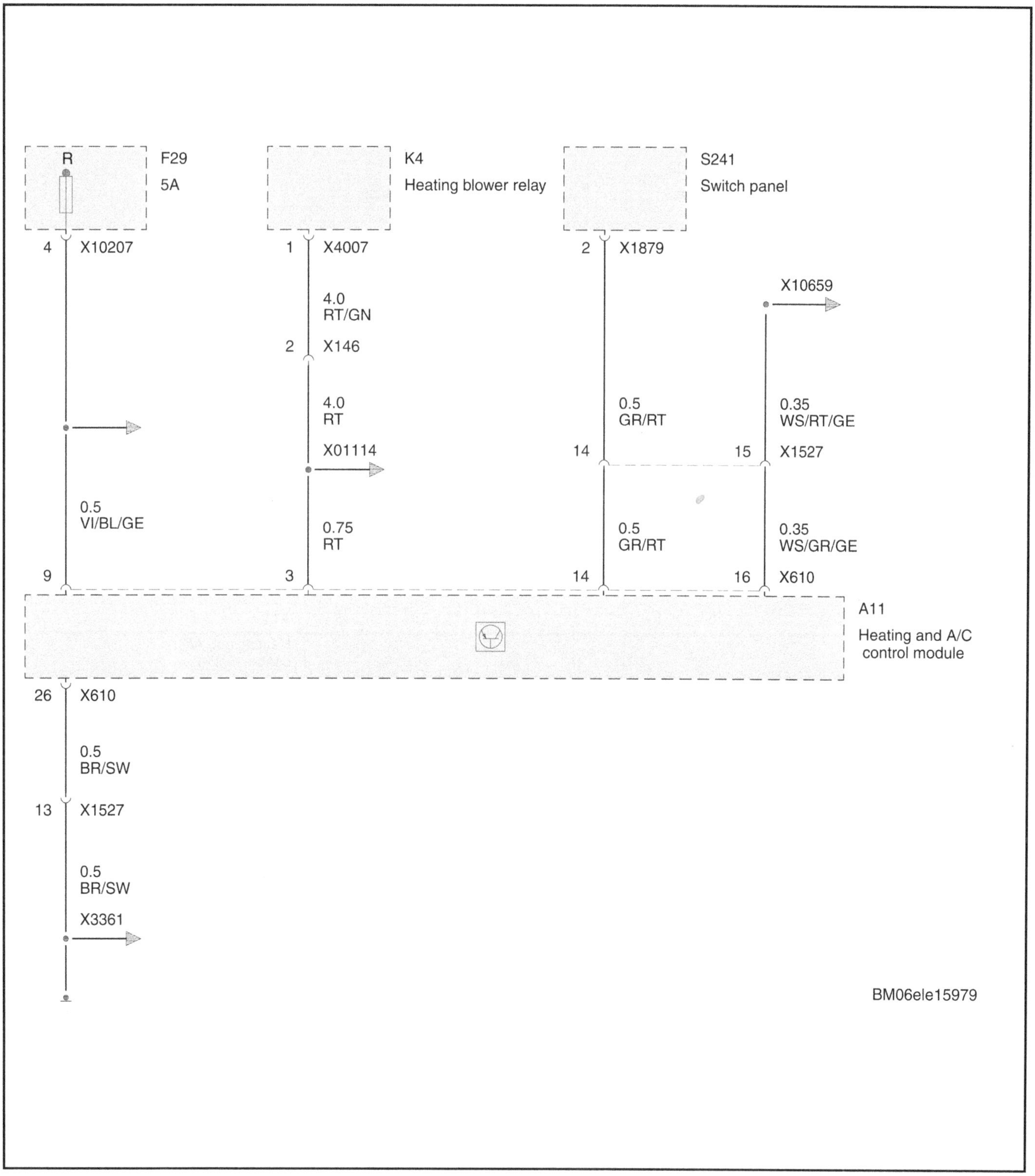

Heating and A/C
Power supply (from March 2002 to March 2003)

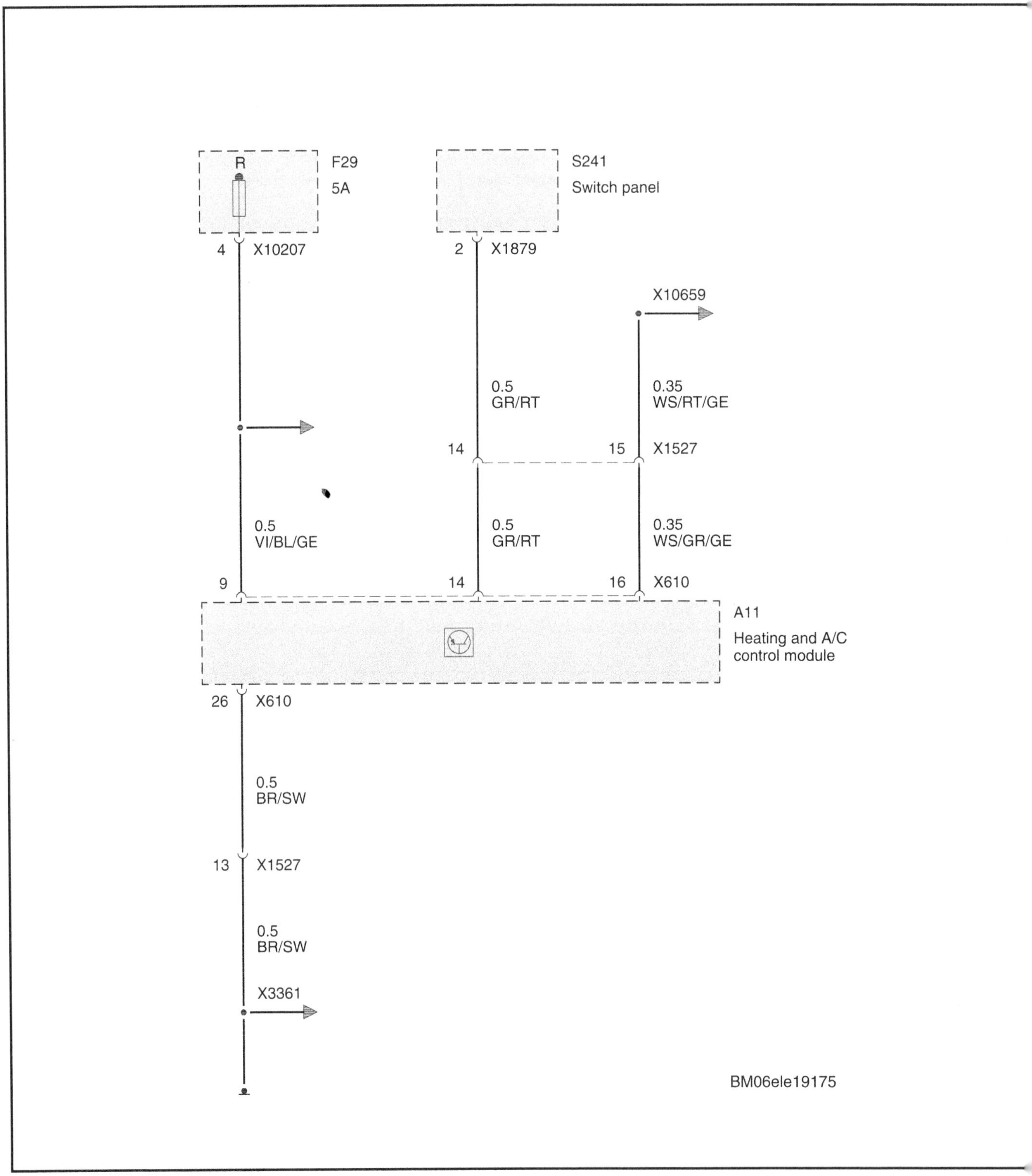

Heating and A/C

Power supply (from March 2003 to Sept. 2005)

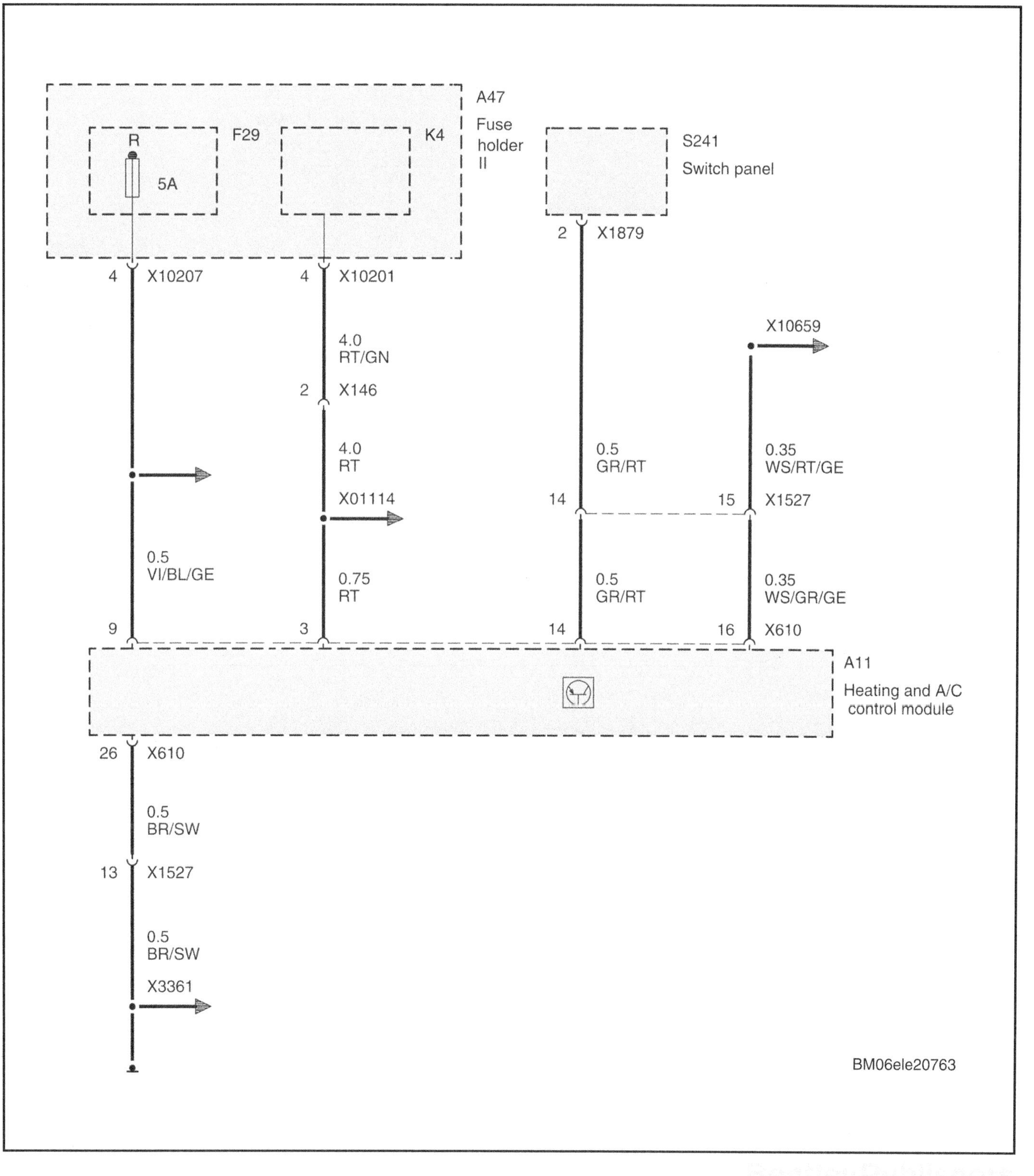

Heating and A/C

Power supply (from Sept. 2005)

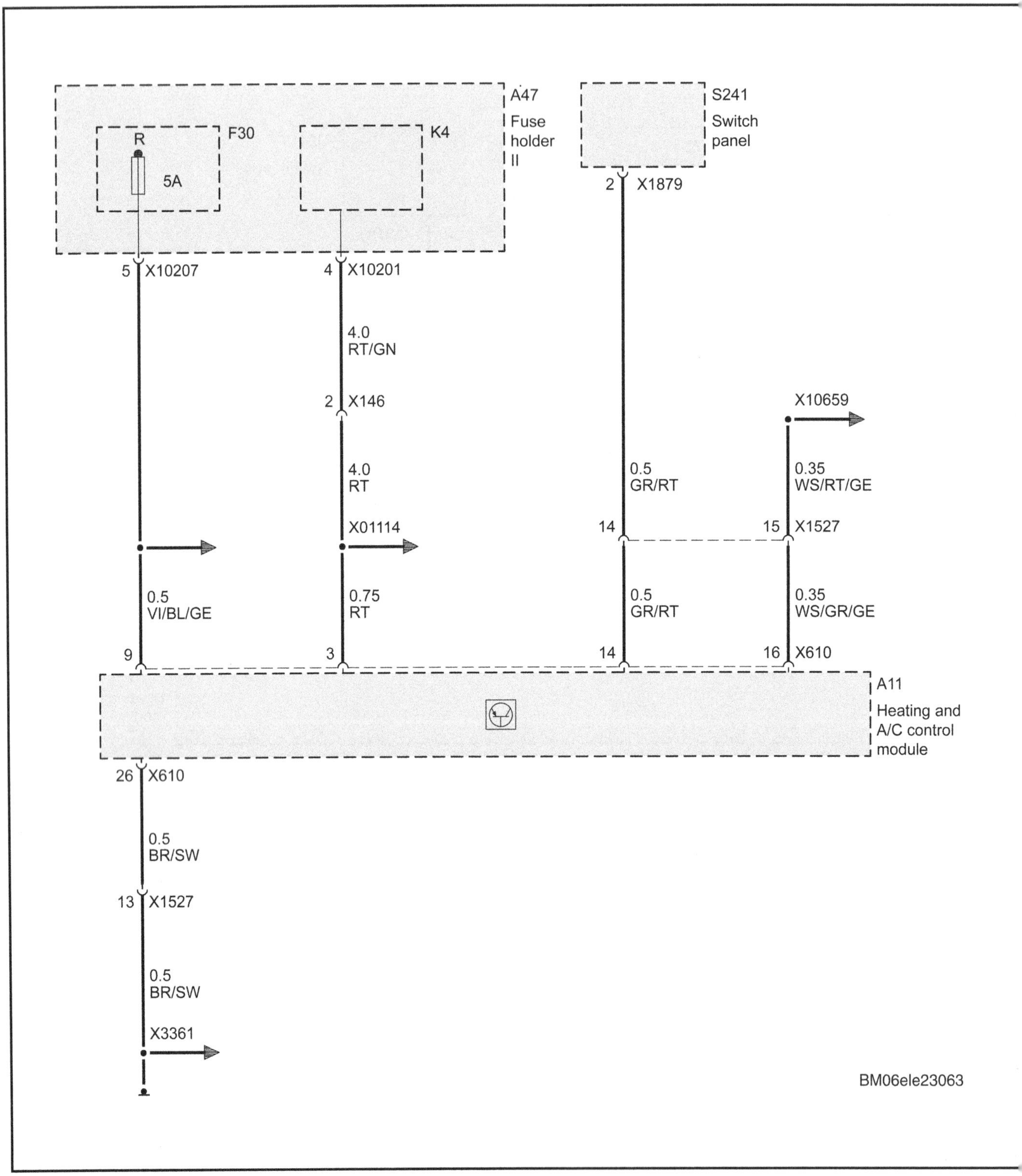

Heating and A/C
IHKA blower control (W10 engine to March 2002)

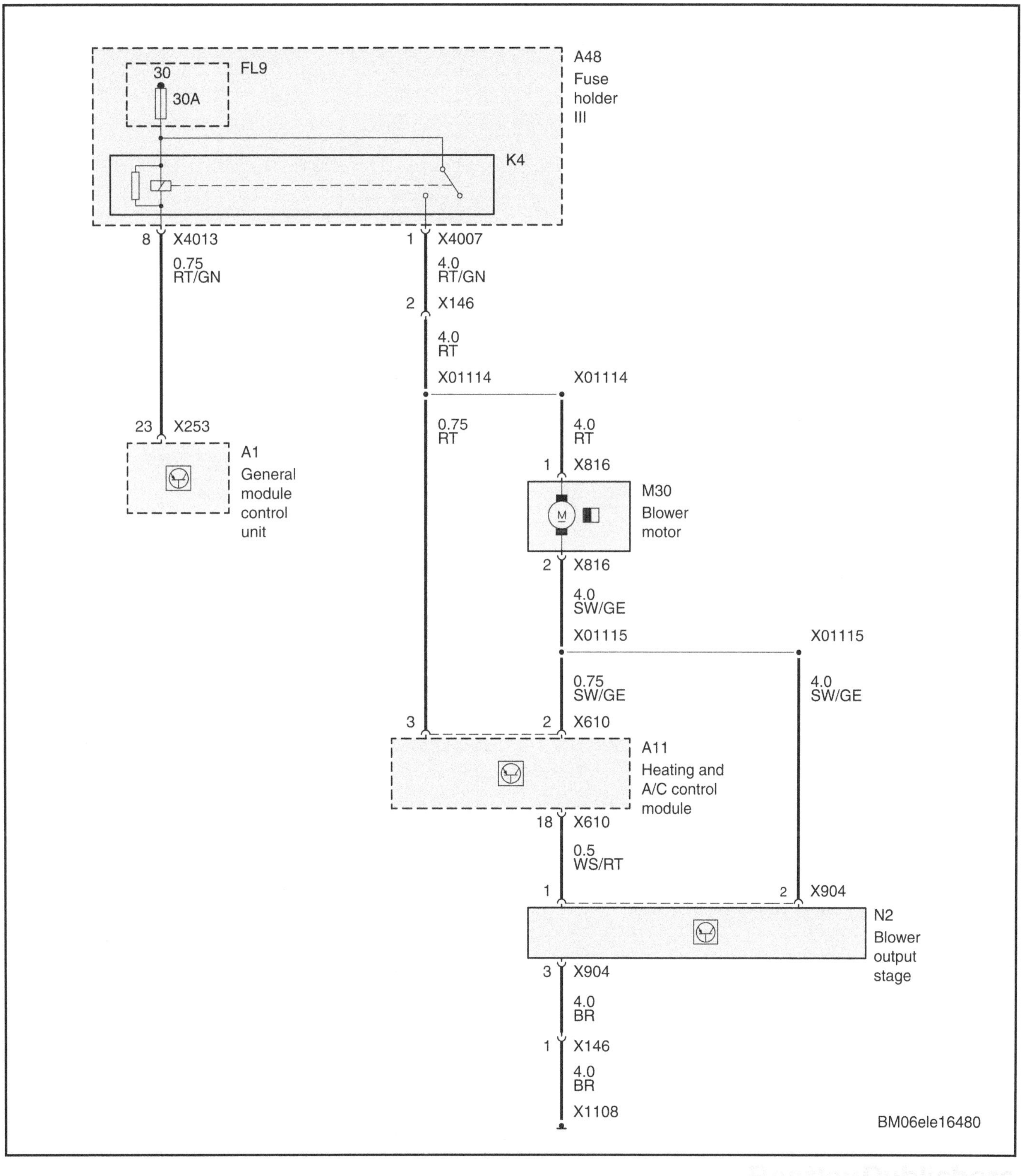

Heating and A/C

IHKA blower control (W11 engine to March 2003) and (W10 engine from March 2002 to March 2003)

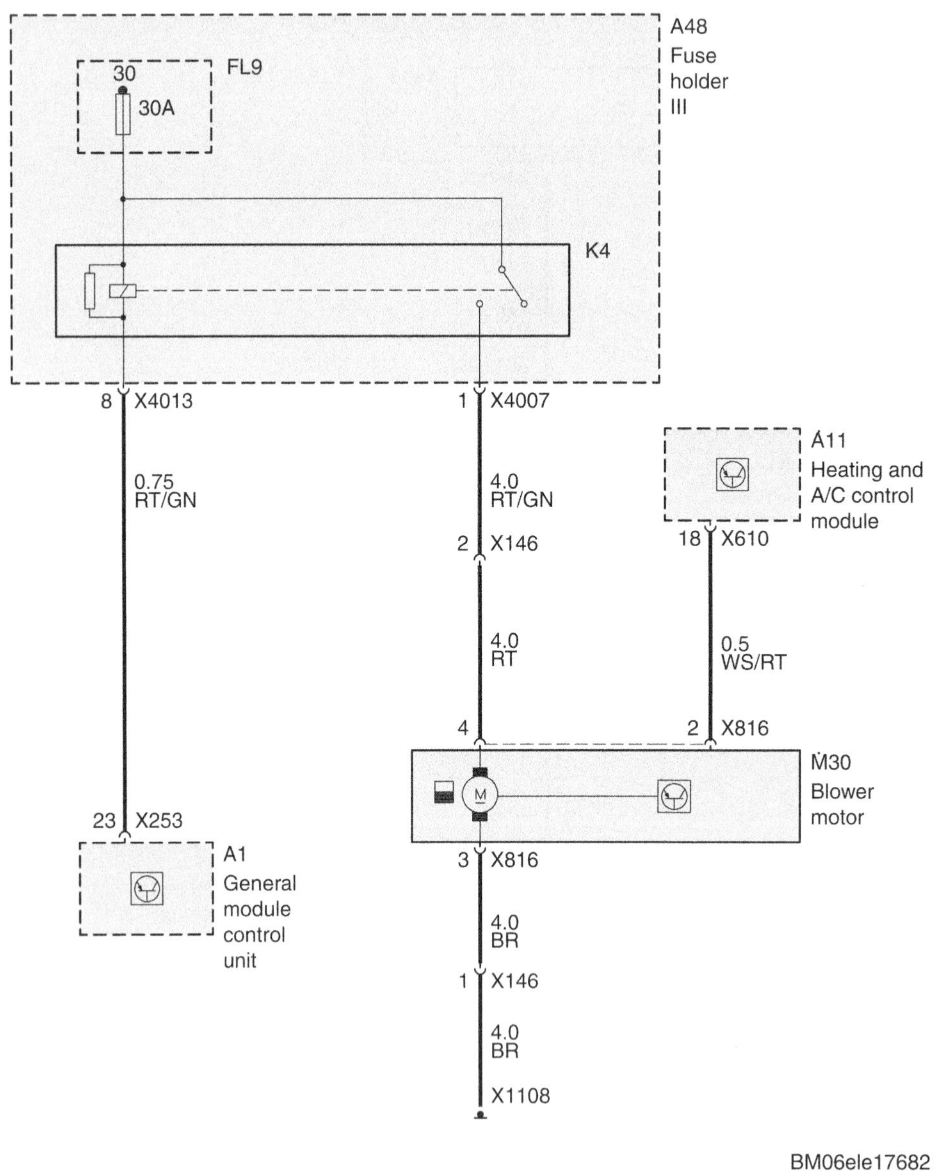

Heating and A/C

IHKA blower control (from March 2003)

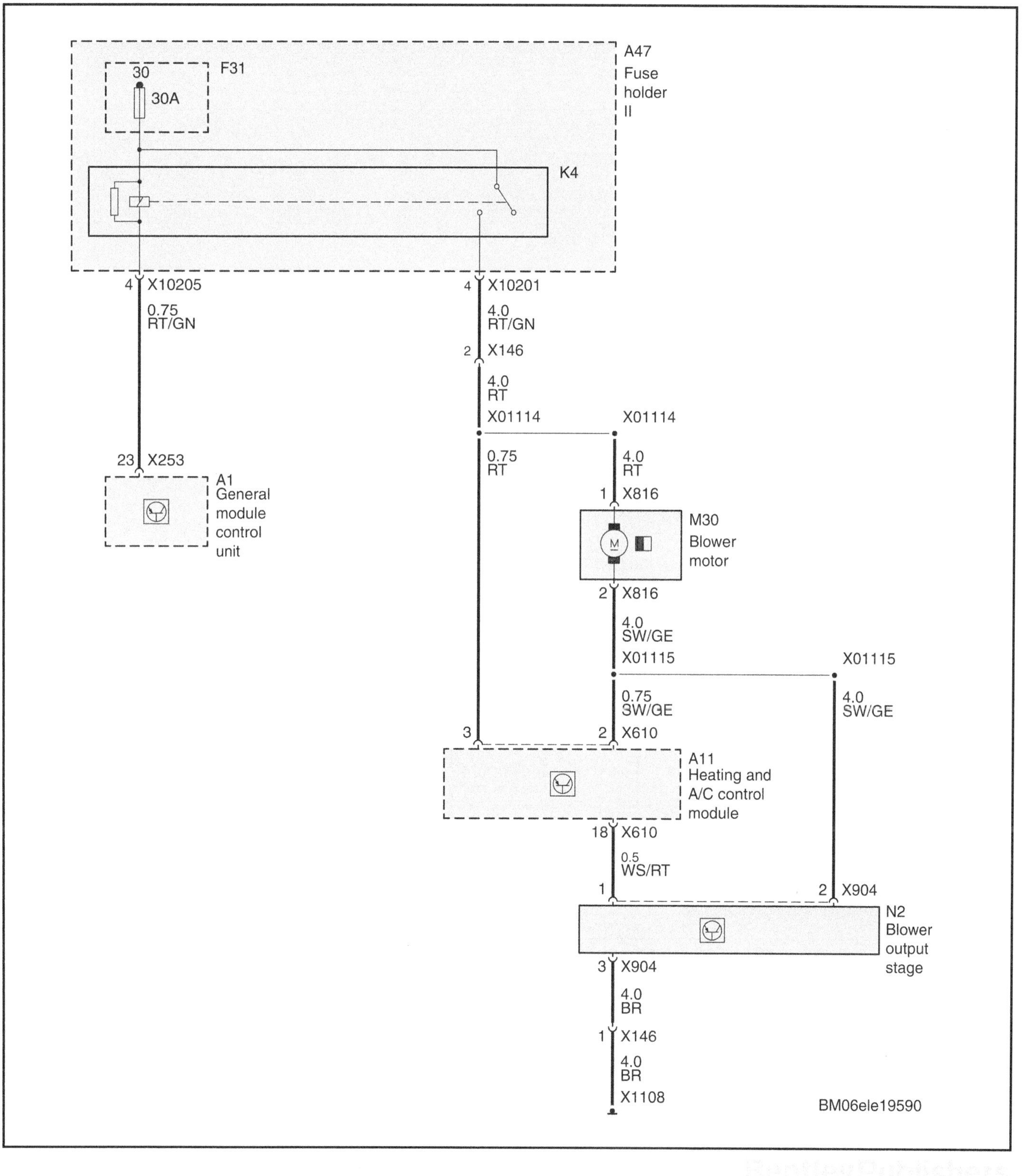

Heating and A/C
IHKR blower control (to March 2003)

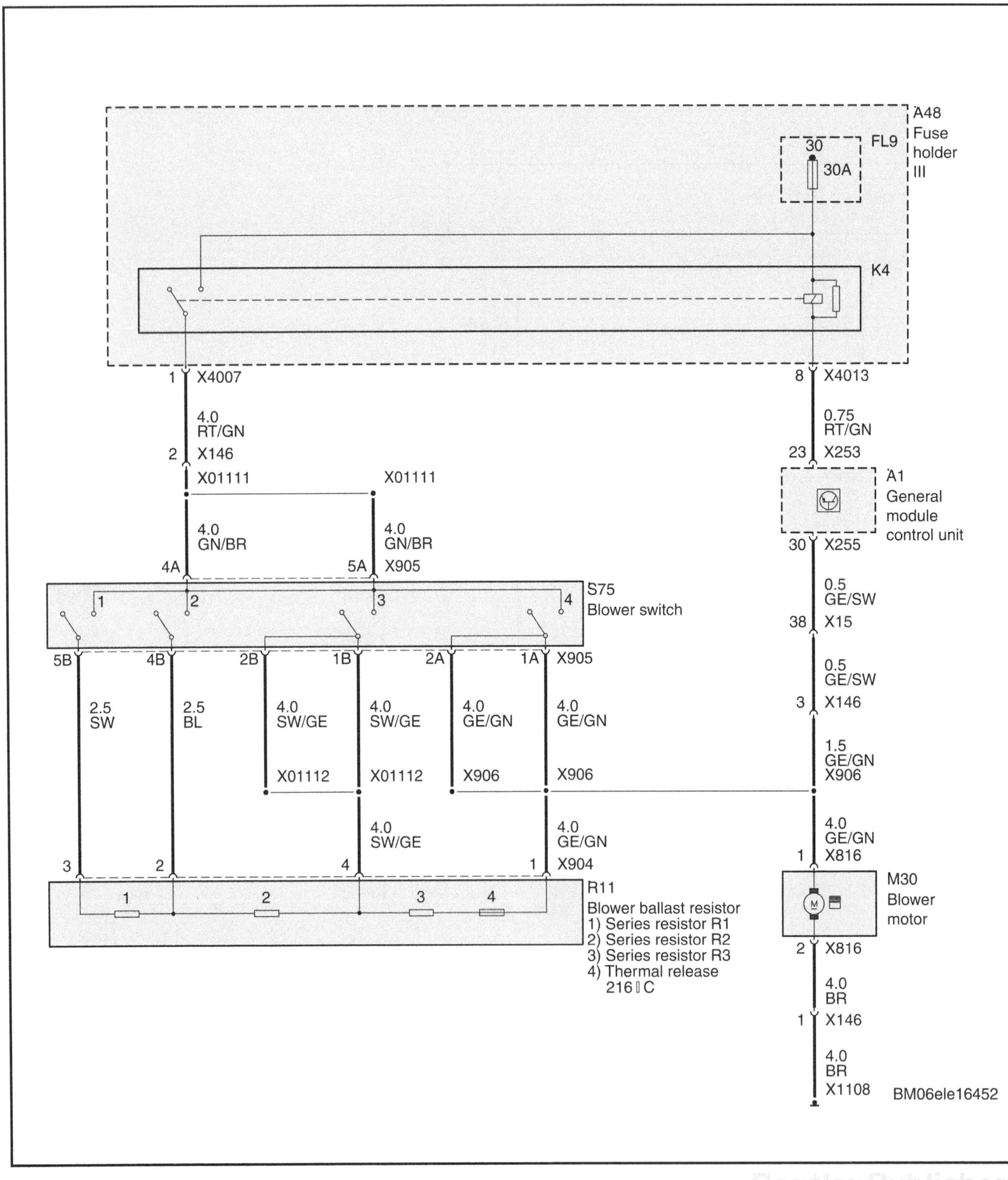

Heating and A/C

IHKR blower control (from March 2003)

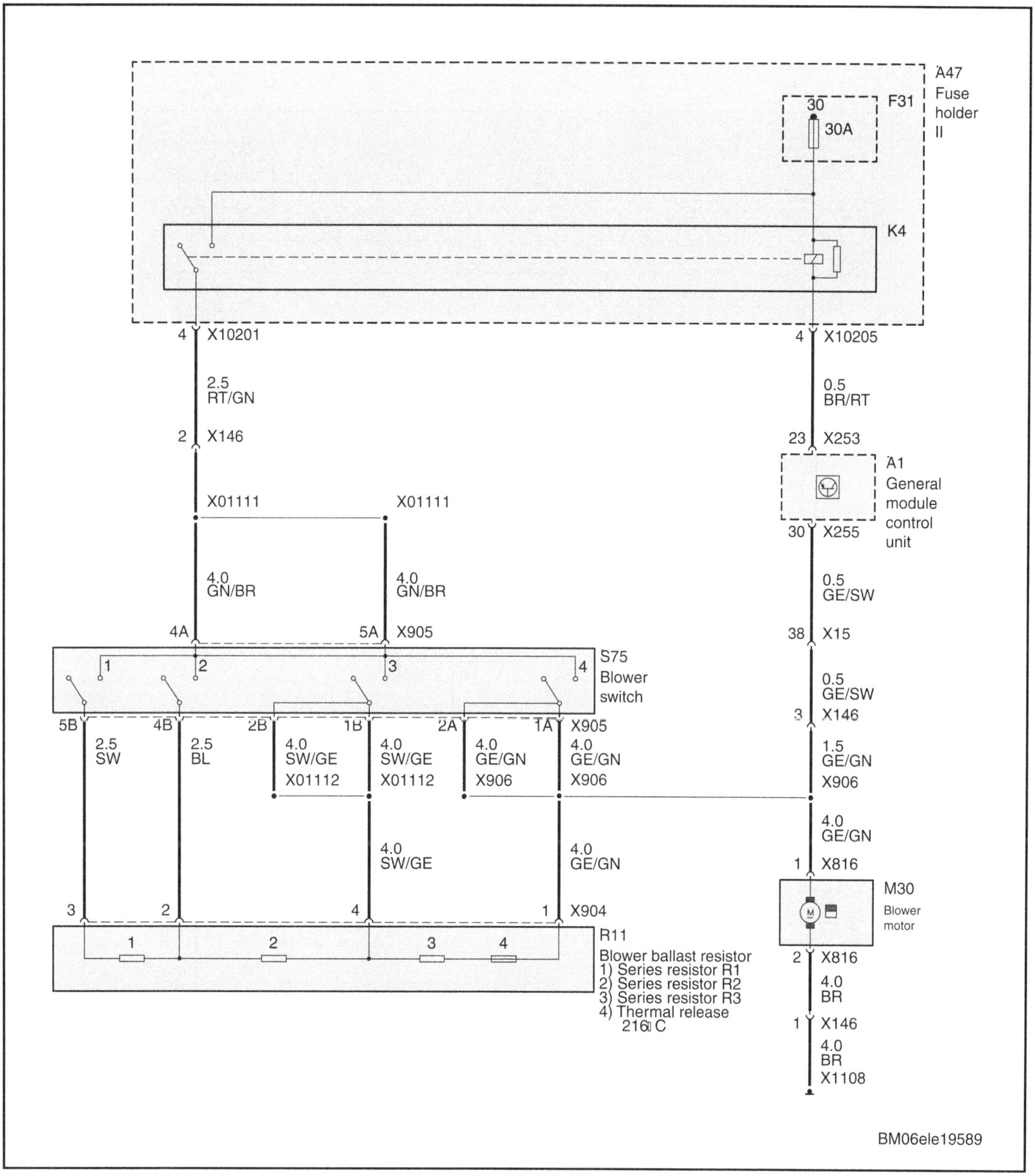

Heating and A/C

IHKS blower control (to March 2003)

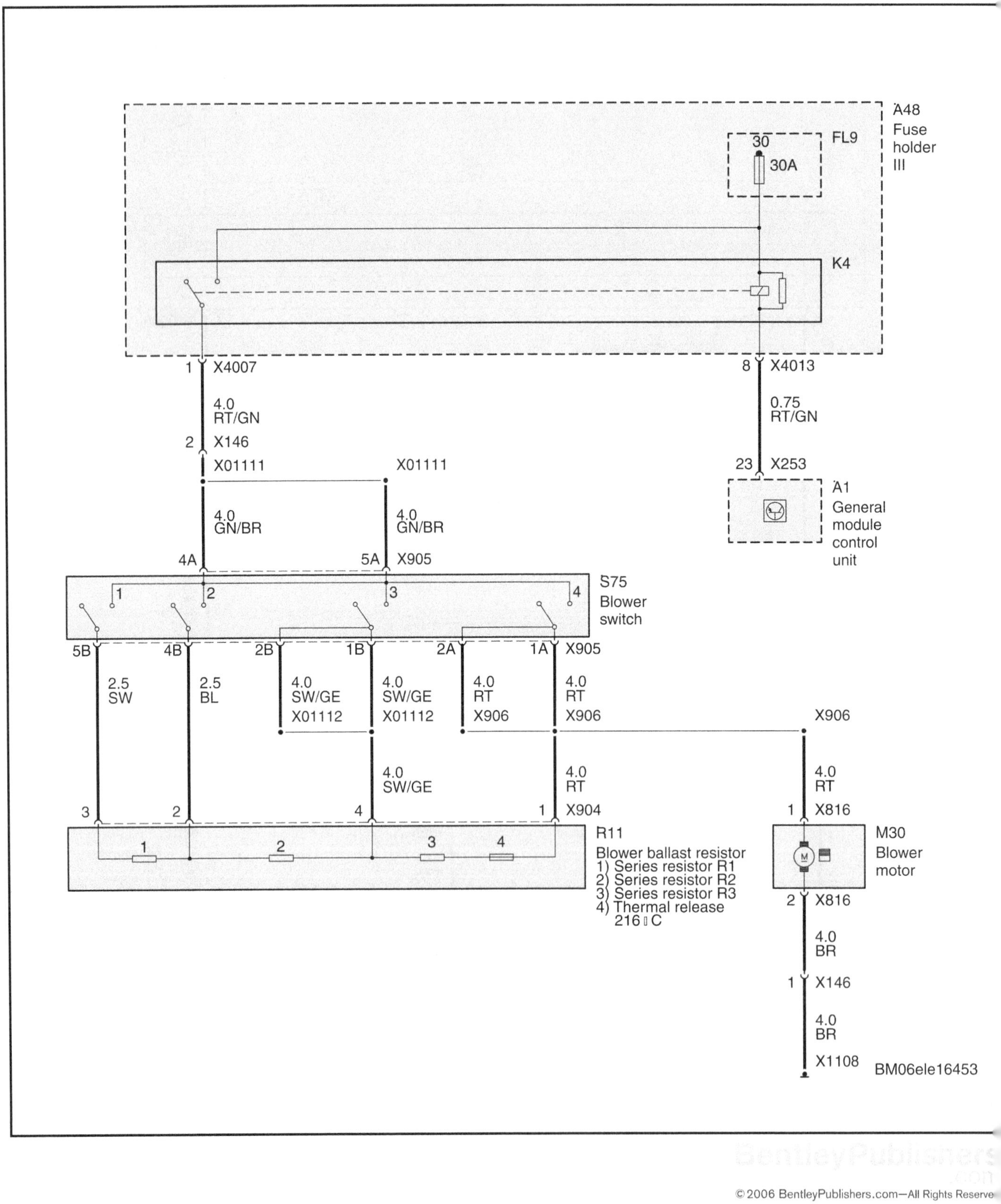

Heating and A/C

IHKS blower control (from March 2003)

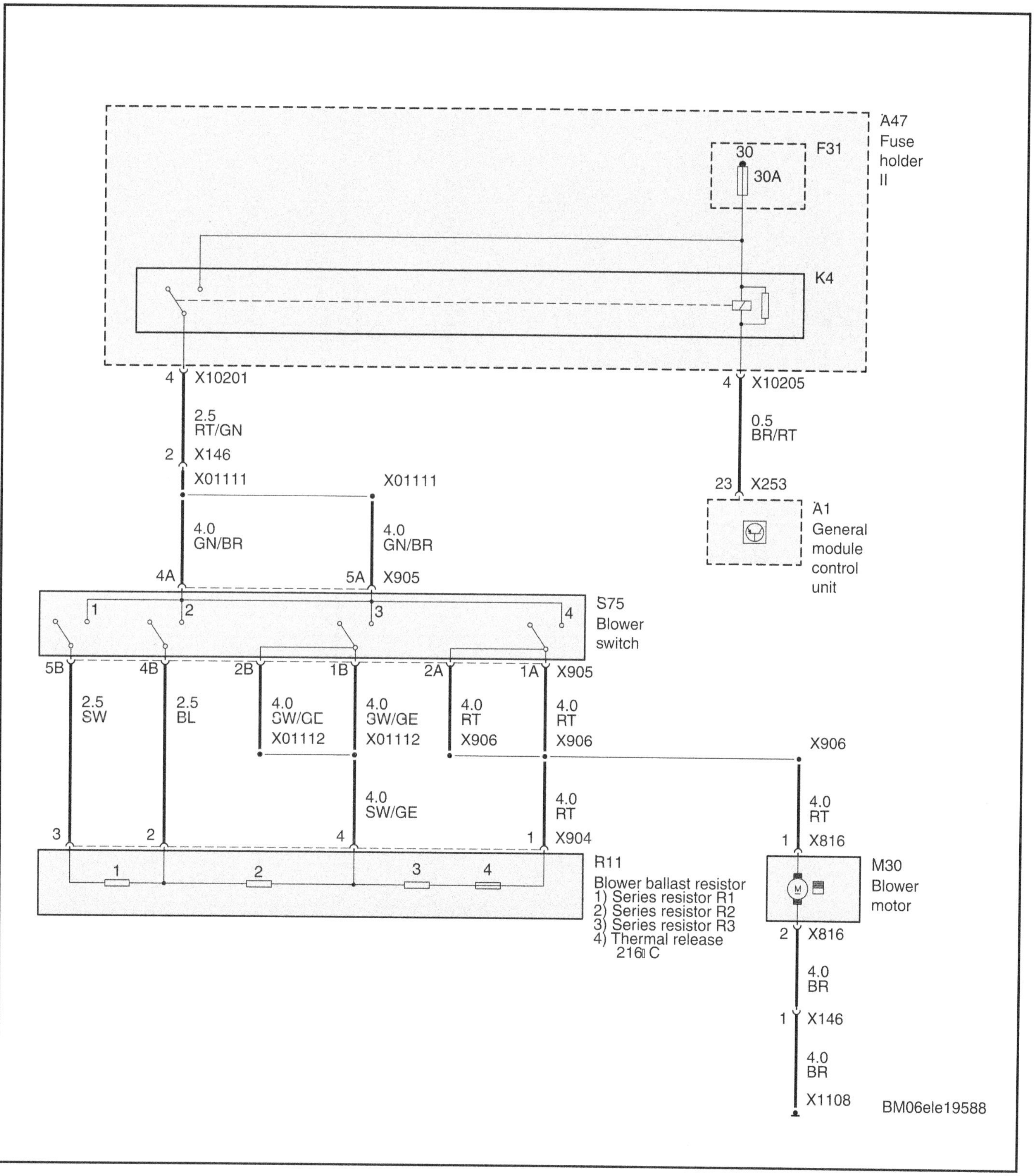

Heating and A/C
Compressor control
(page 1 of 1)

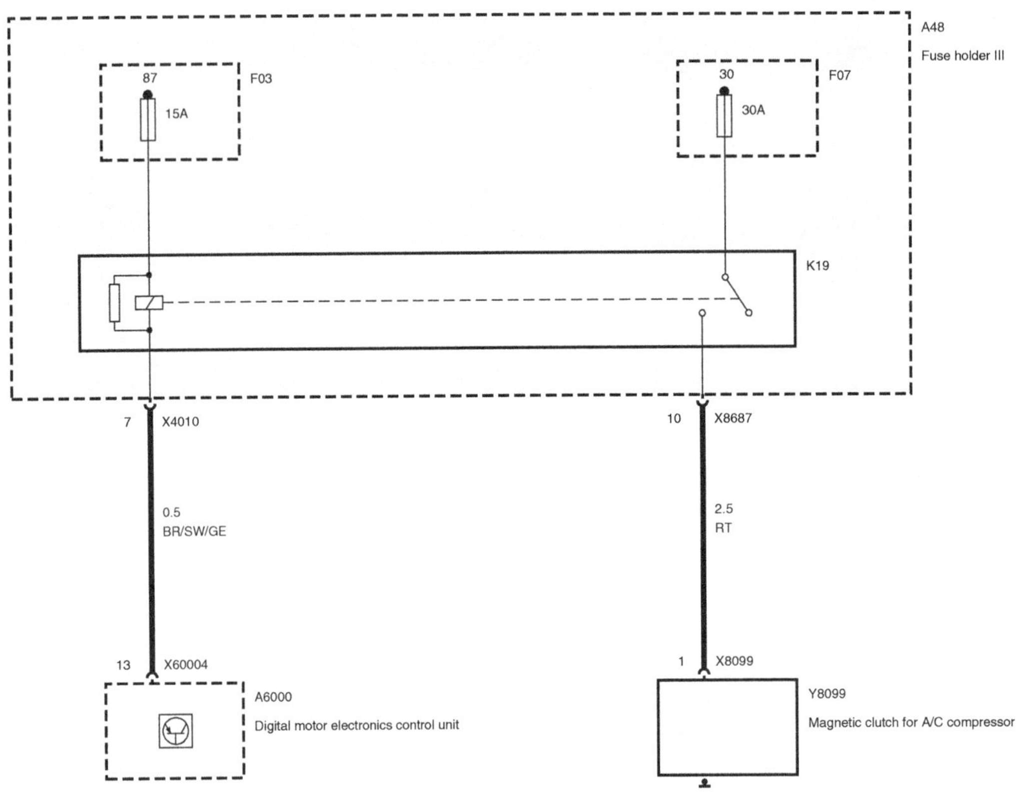

Heating and A/C
Pressure sensor
(page 1 of 1)

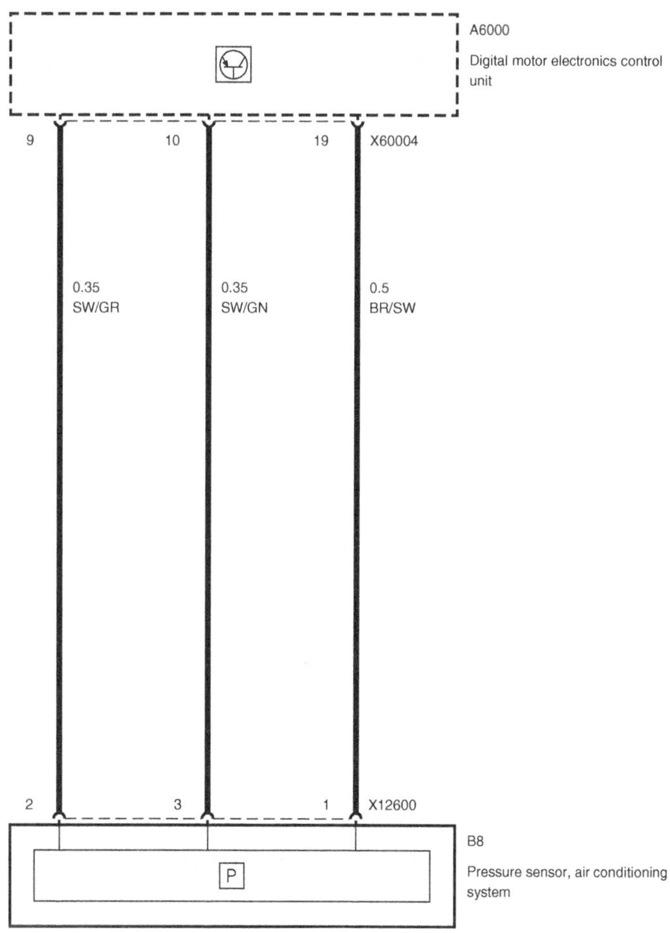

Horns
(up to Sept. 2005)
(1 of 1)

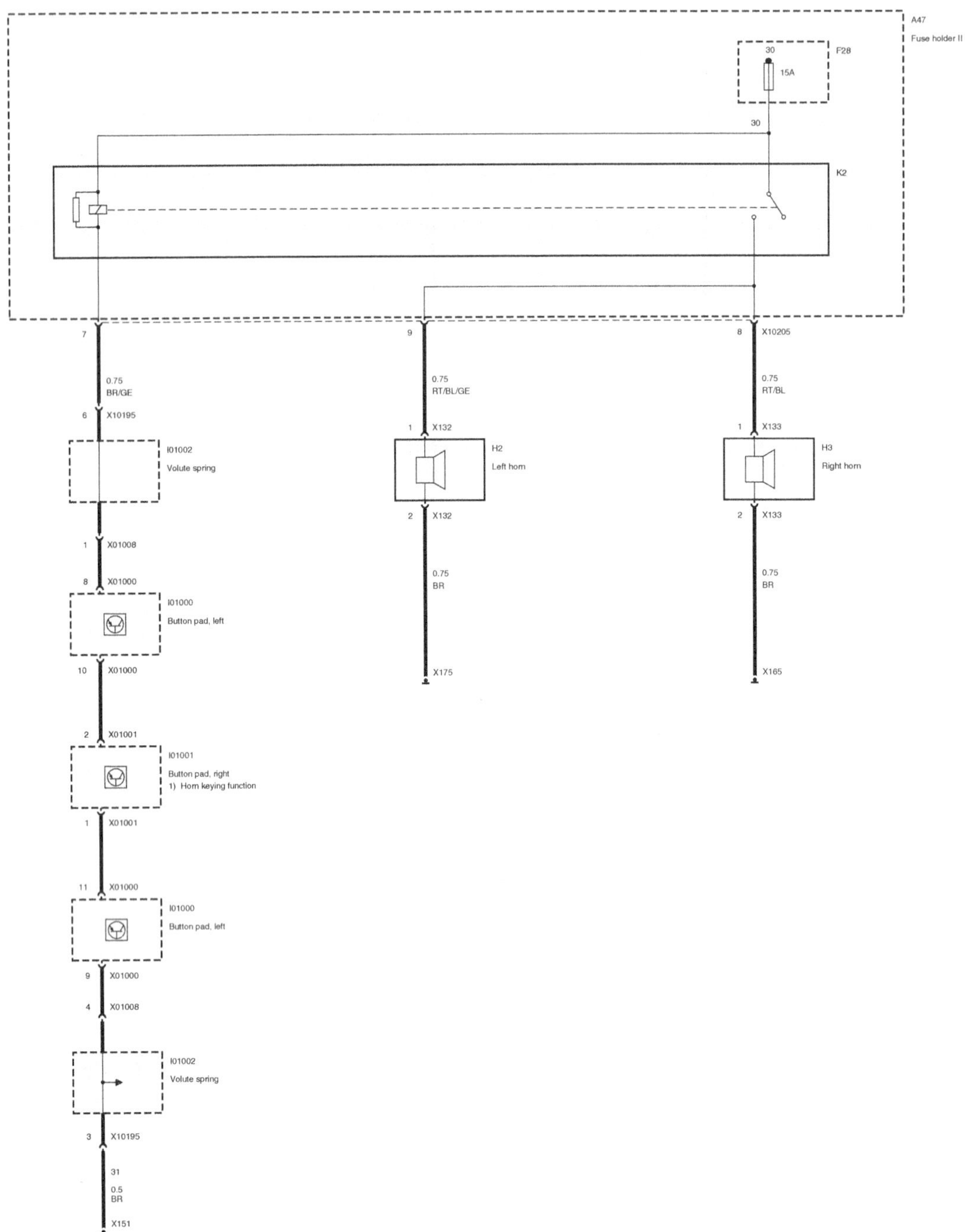

Horns
(from Sept. 2005)
(1 of 1)

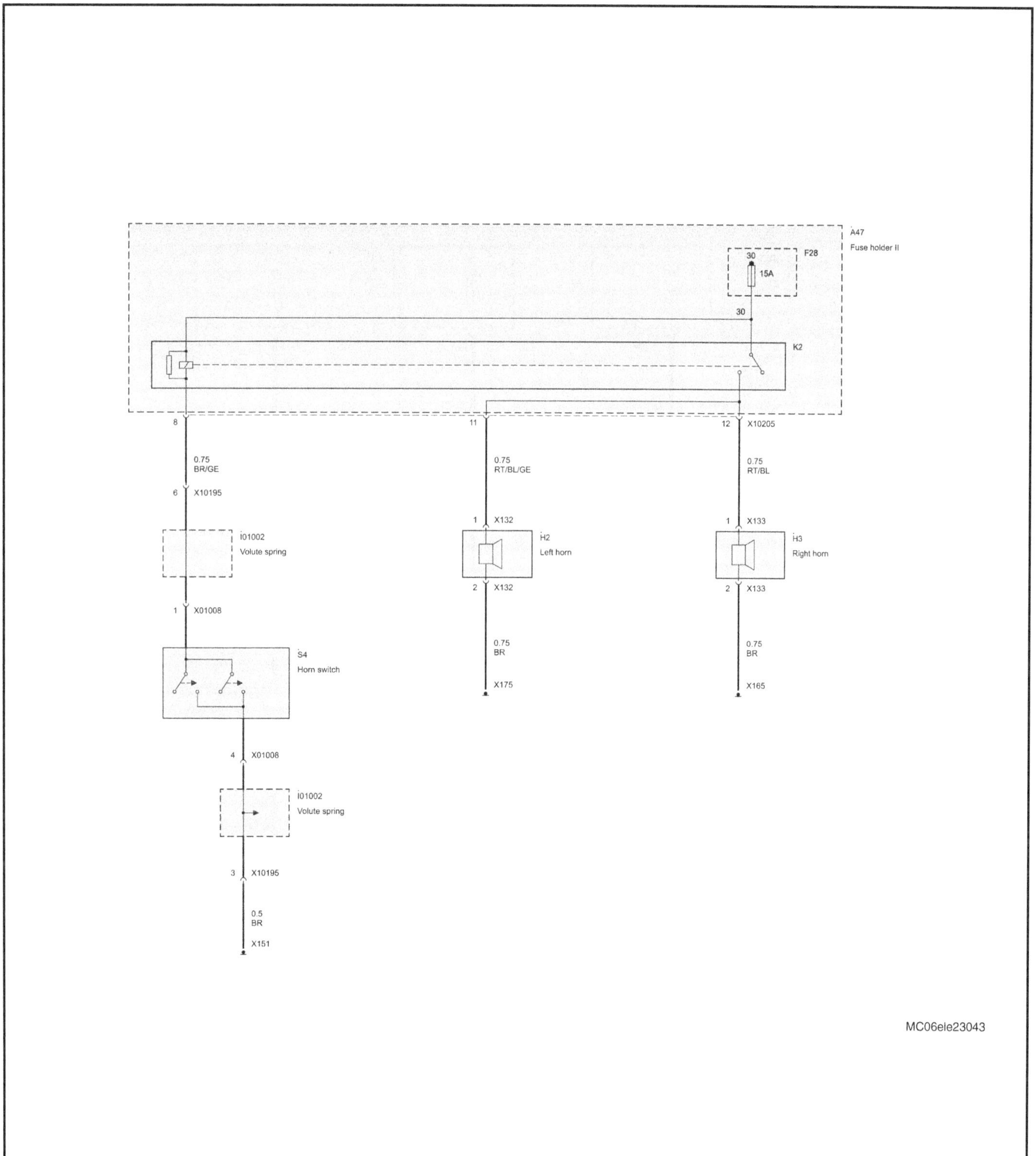

Instruments

Instrument cluster (A2) (up to March 2003)

(page 1 of 4)

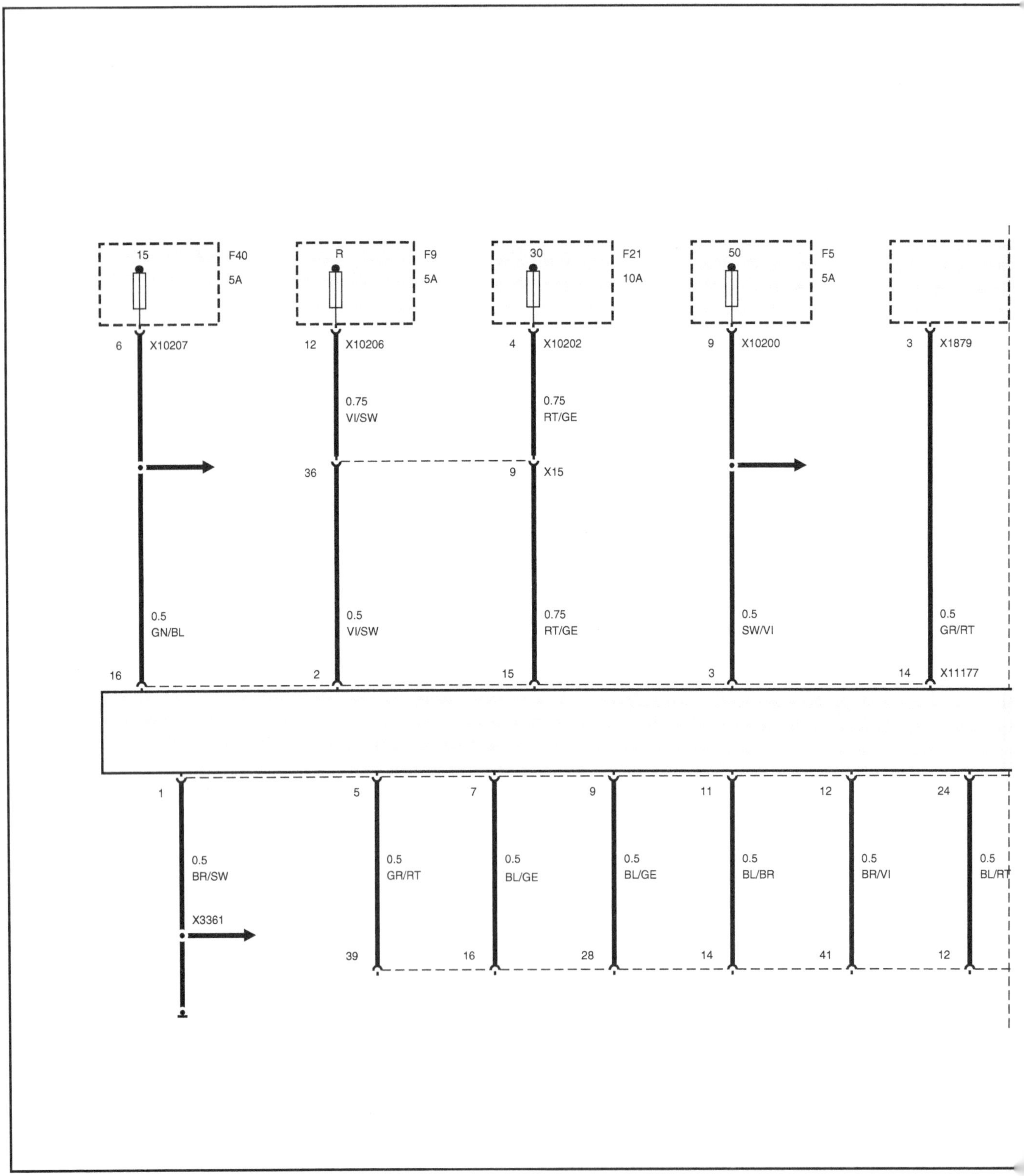

Instruments

Instrument cluster (A2) (up to March 2003)

(page 2 of 4)

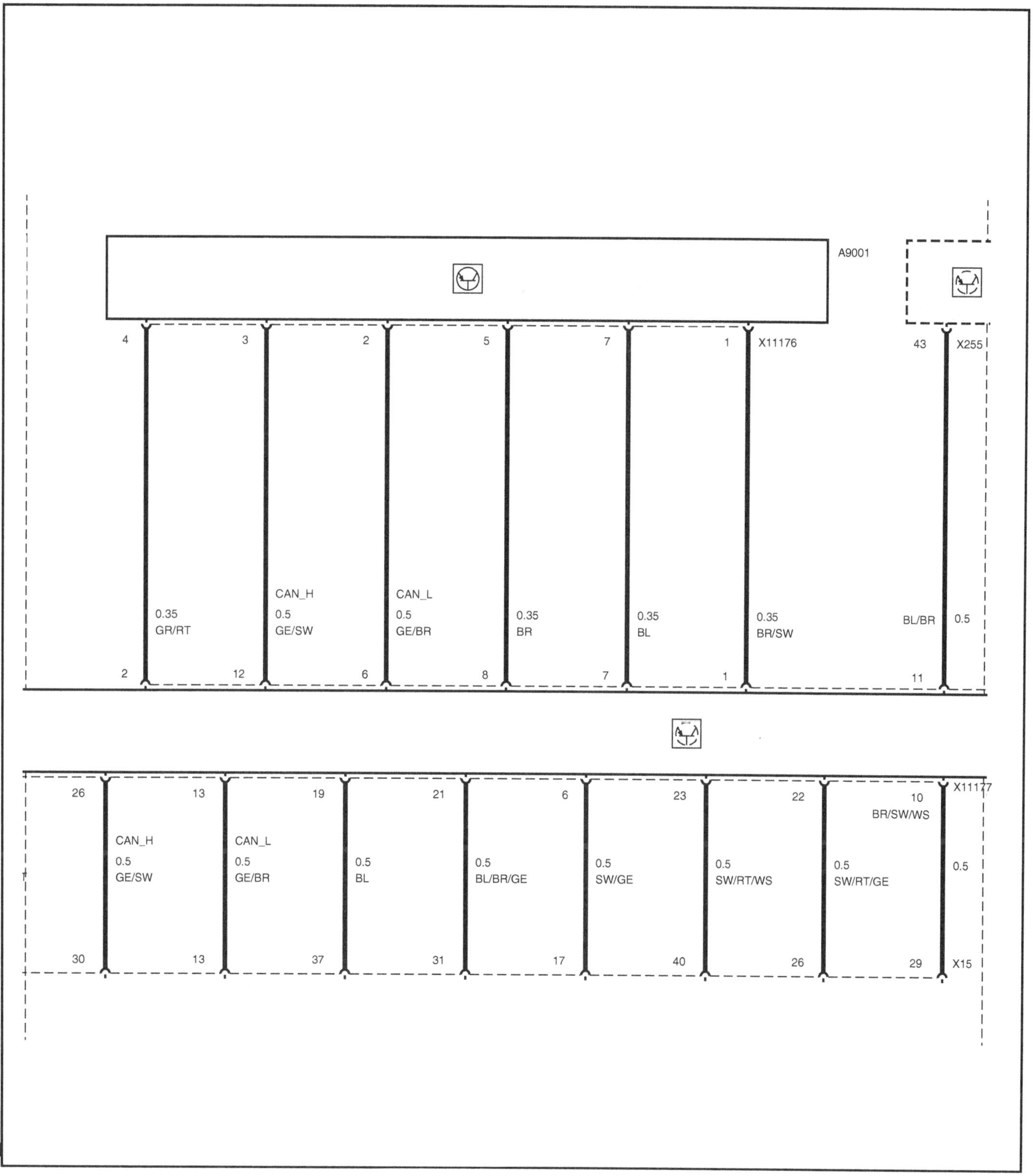

Instruments

Instrument cluster (A2) (up to March 2003)

(page 3 of 4)

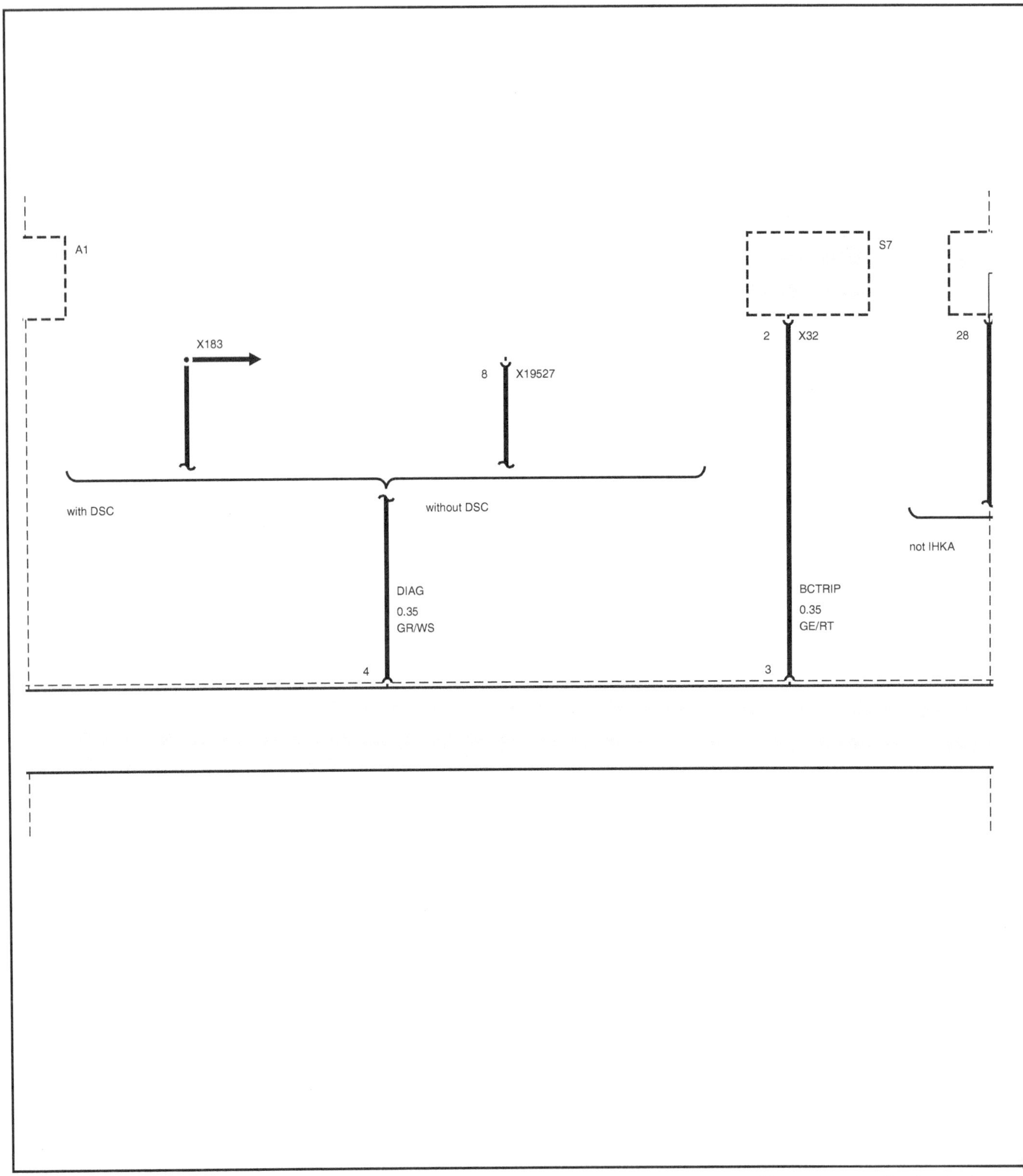

Instruments

Instrument cluster (A2) (up to March 2003)

(page 4 of 4)

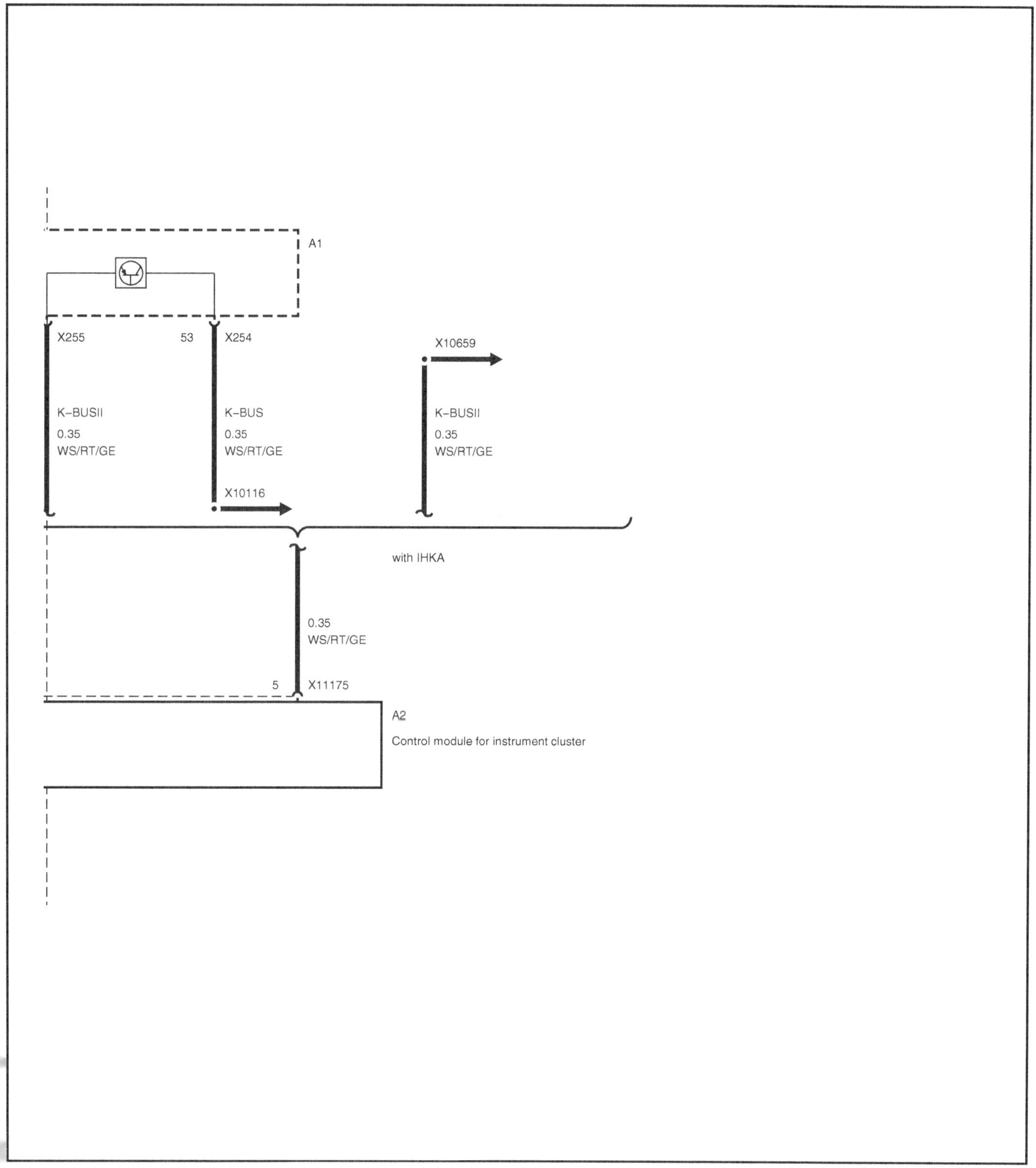

Instruments

Instrument cluster (A2) (from March 2003 to Jan. 2005)

(page 1 of 3)

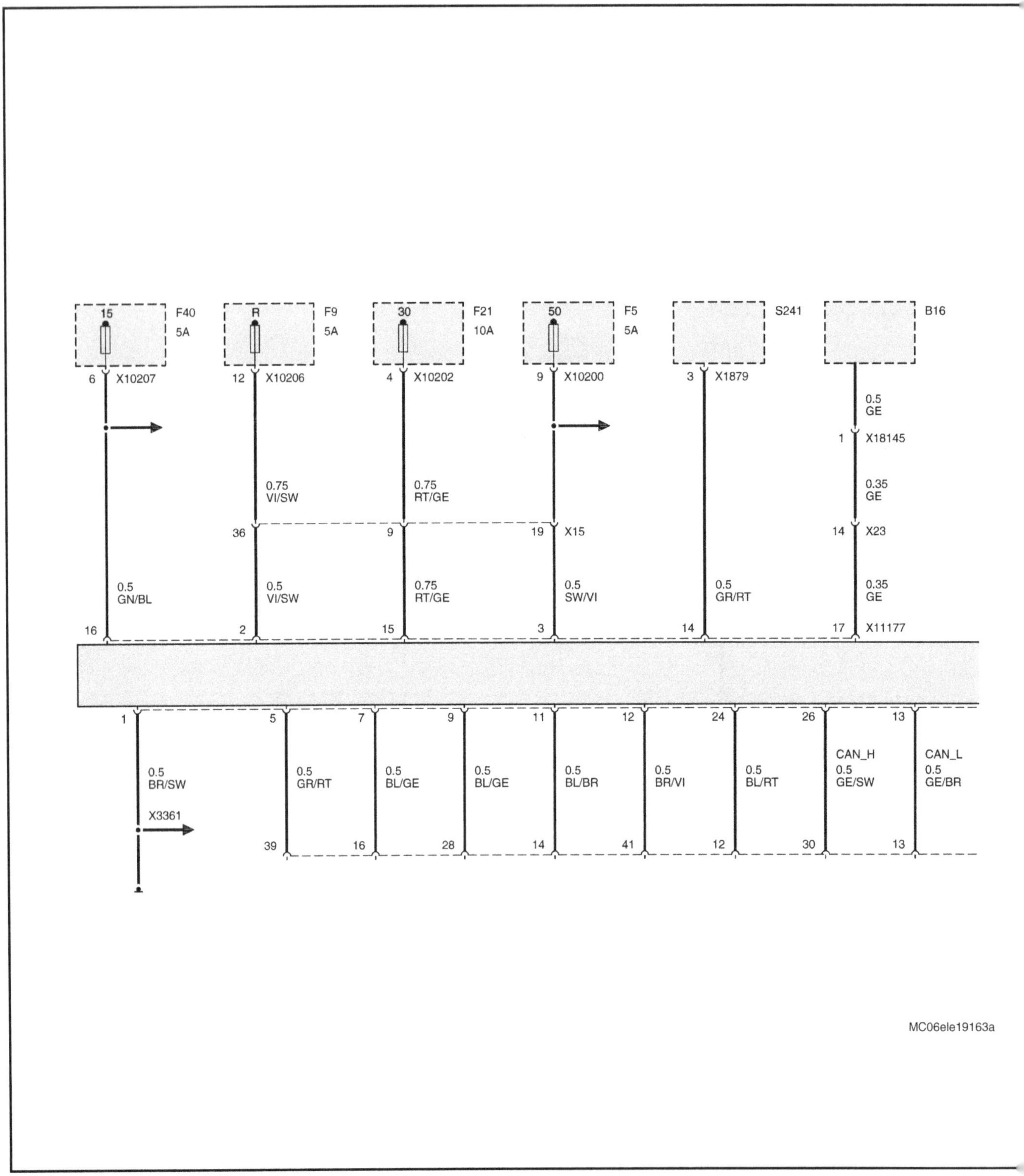

Instruments

Instrument cluster (A2) (from March 2003 to Jan. 2005)

(page 2 of 3)

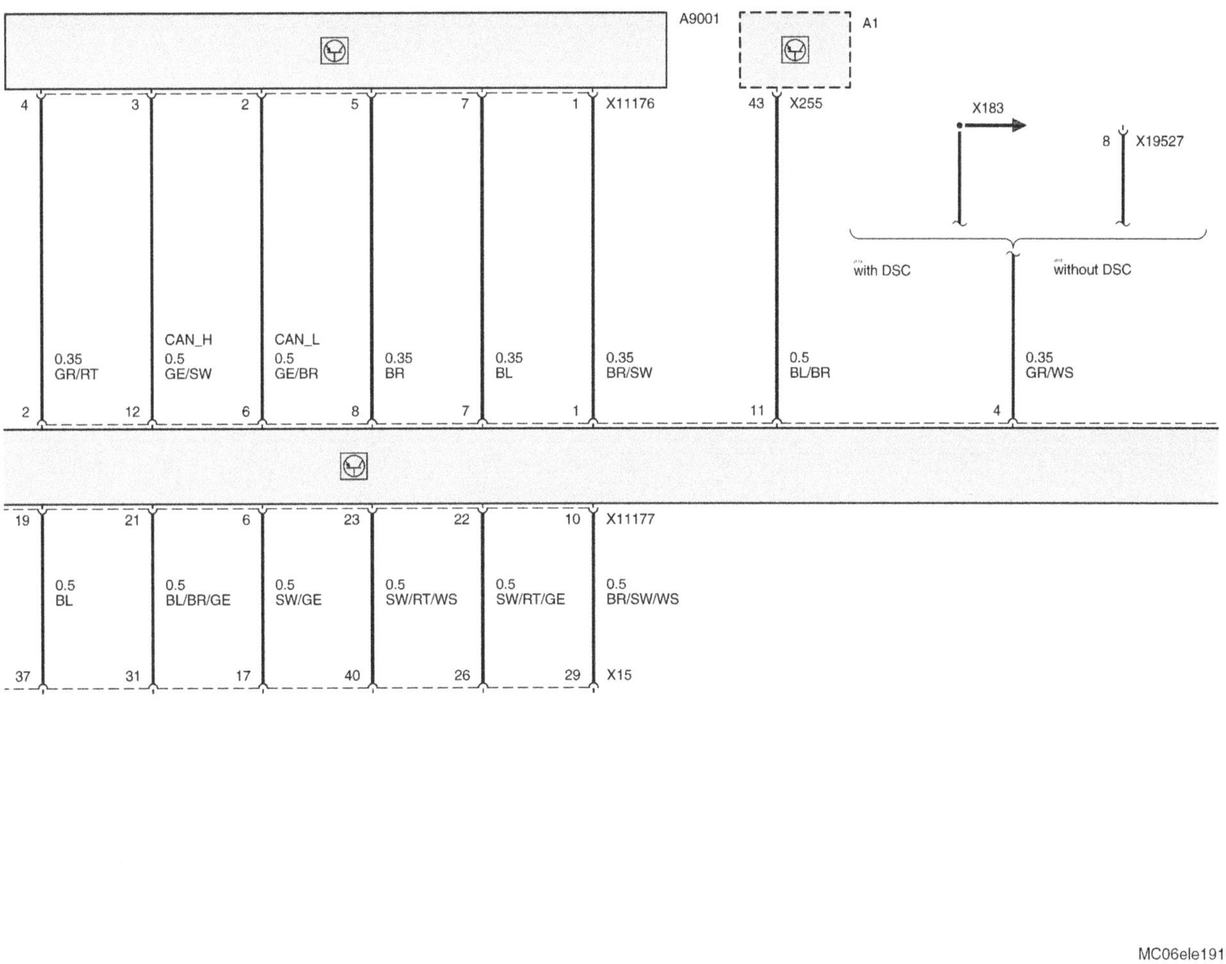

Instruments

Instrument cluster (A2) (from March 2003 to Jan. 2005)

(page 3 of 3)

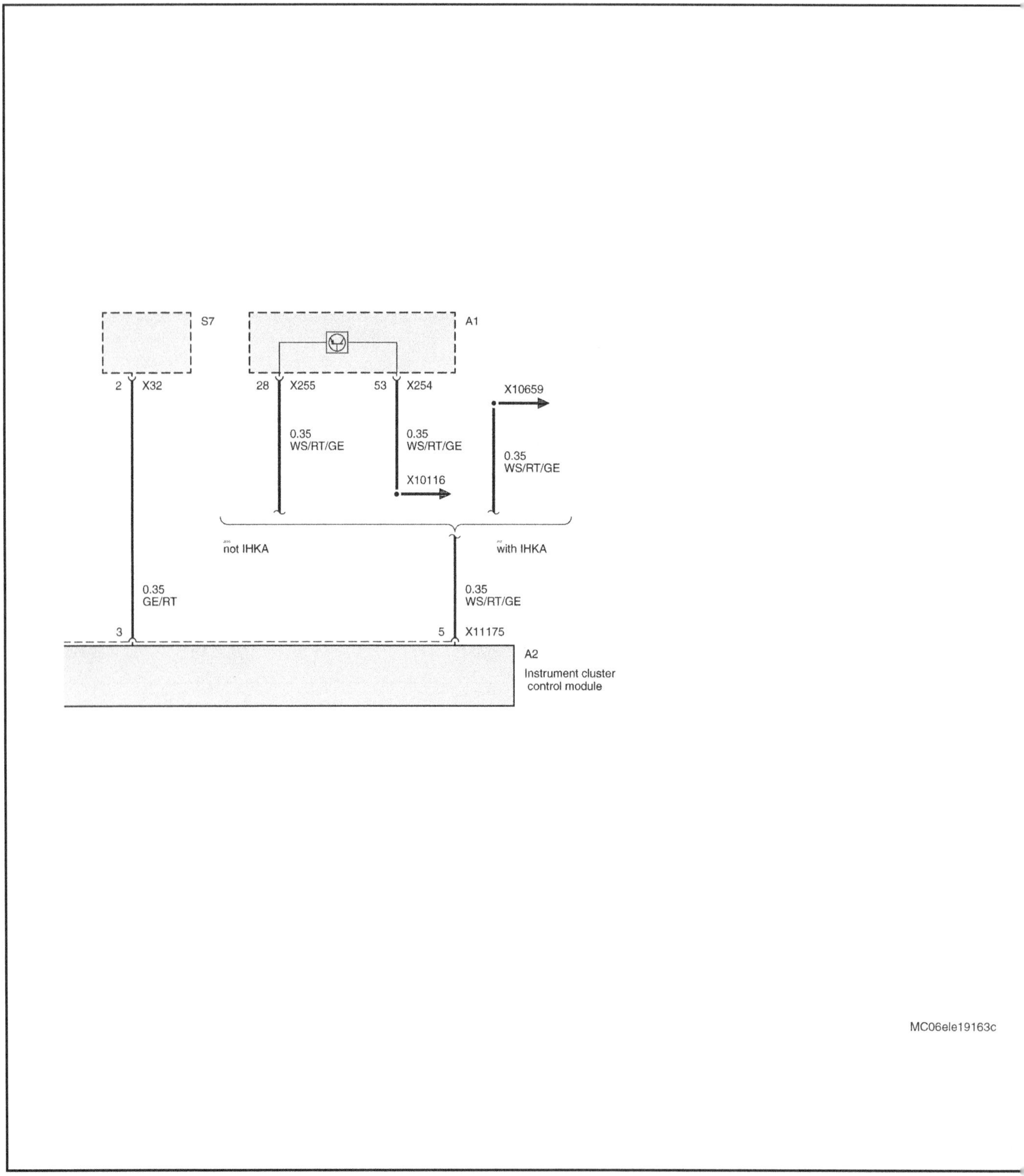

Instruments

Instrument cluster (A2) (from Jan. 2005)

(page 1 of 3)

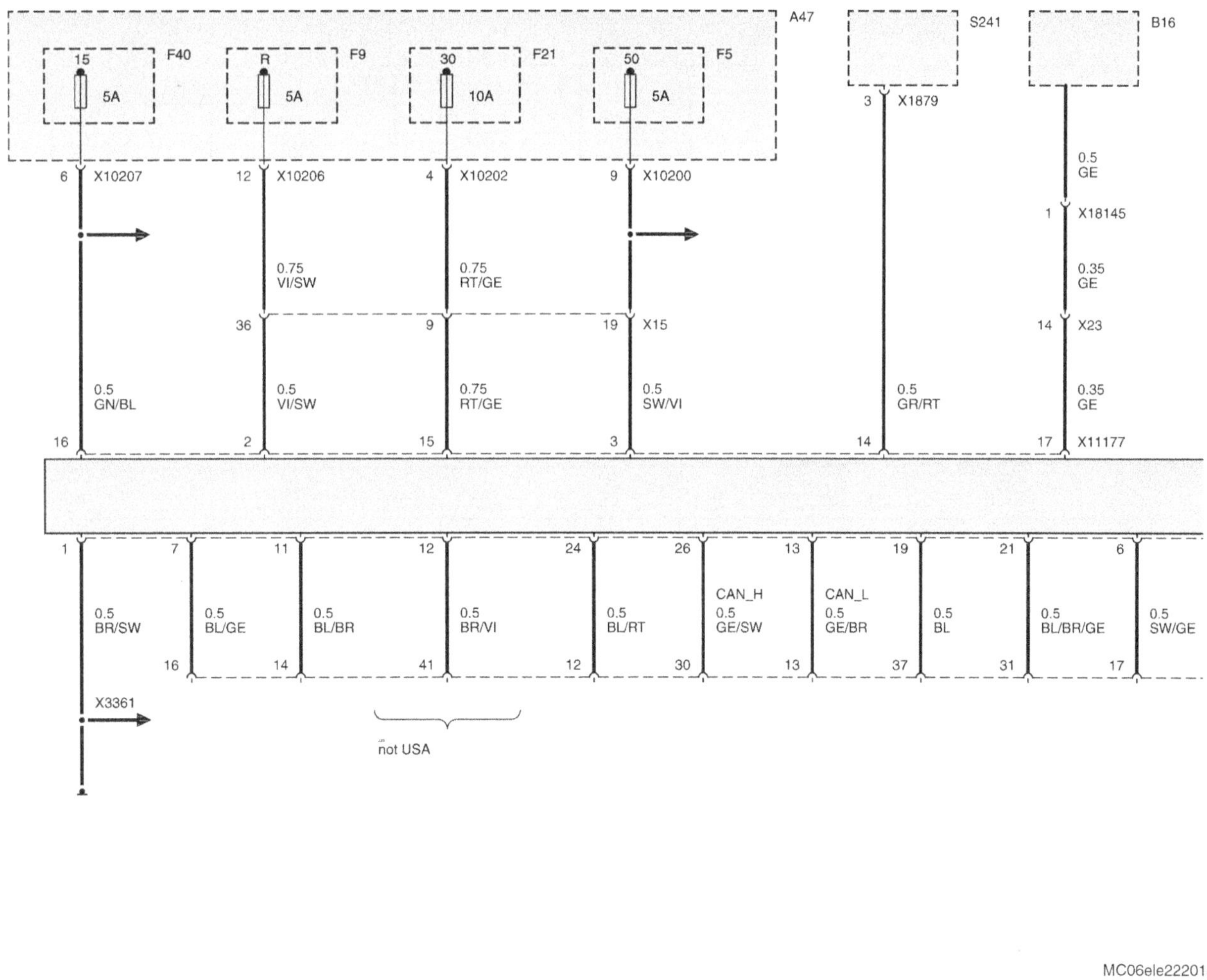

Instruments

Instrument cluster (A2) (from Jan. 2005)

(page 2 of 3)

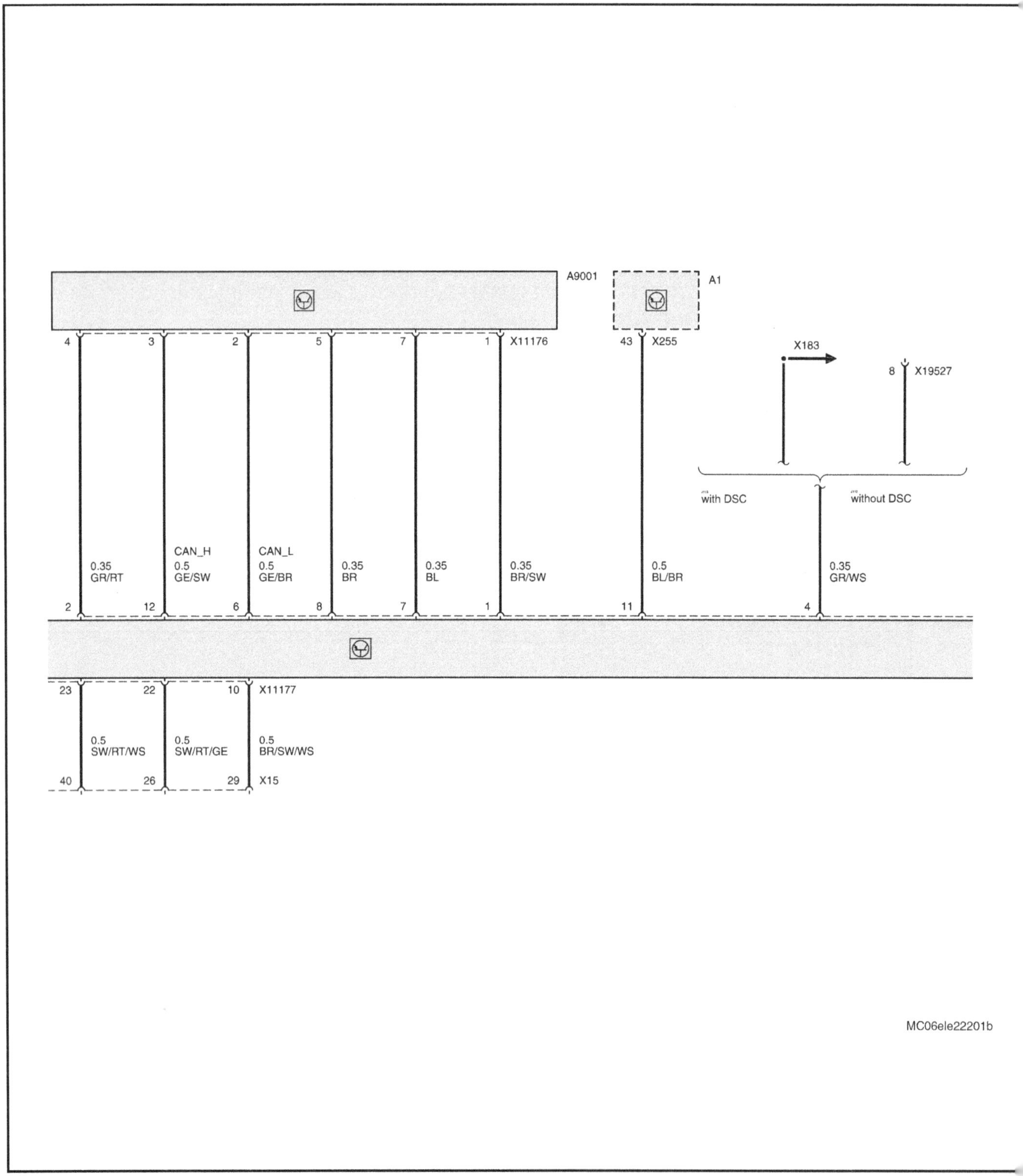

Instruments

Instrument cluster (A2) (from Jan. 2005)

(page 3 of 3)

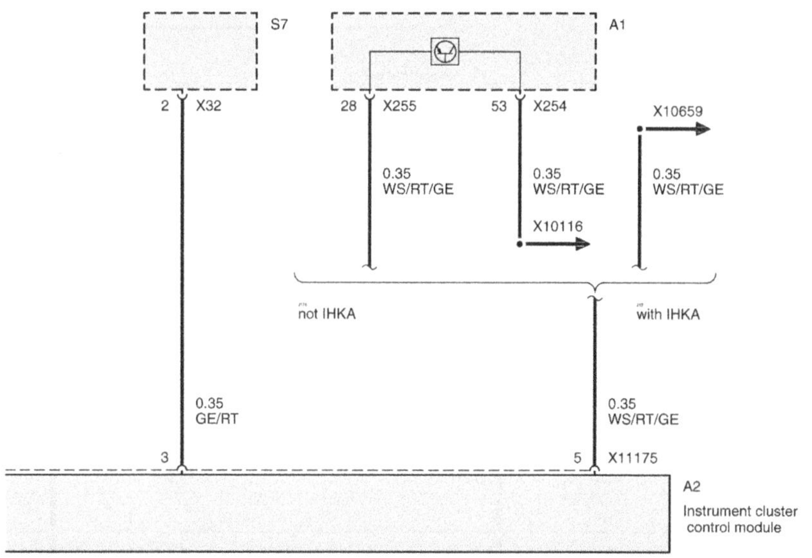

Instruments

Intermediate connector (X15) (up to Sept. 2003)

(page 1 of 9)

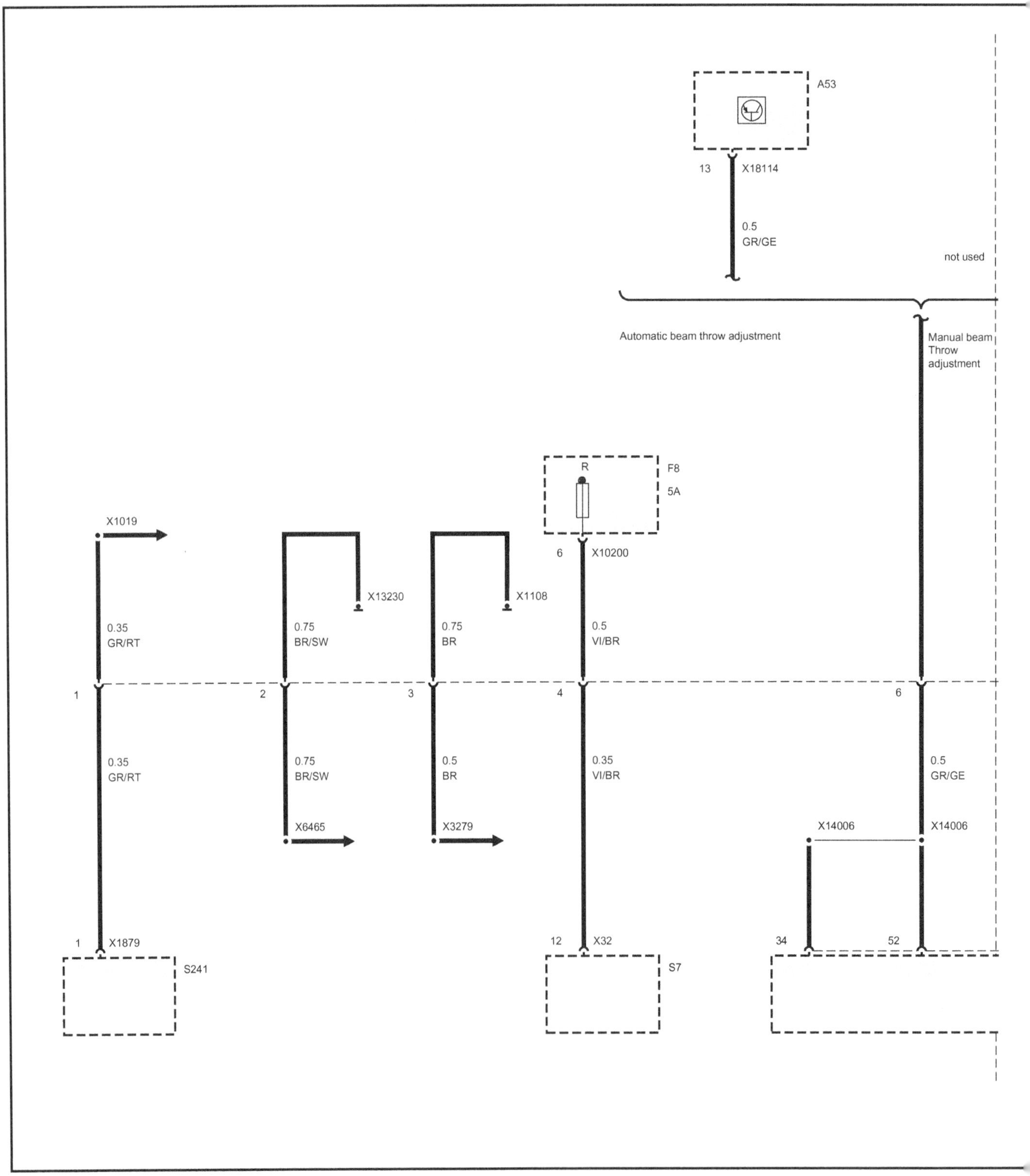

Instruments

Intermediate connector (X15) (up to Sept. 2003)

(page 2 of 9)

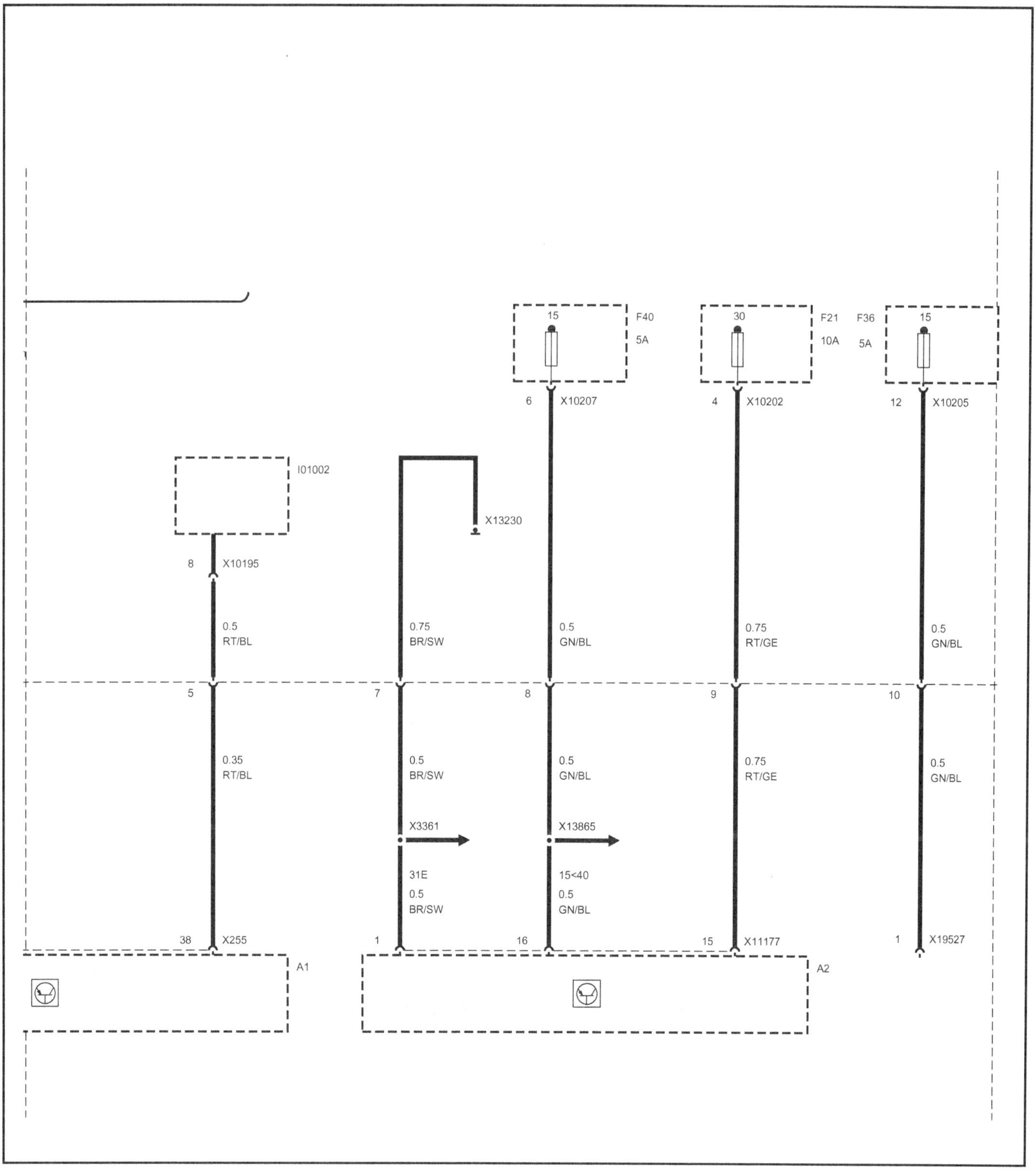

Instruments

Intermediate connector (X15) (up to Sept. 2003)

(page 3 of 9)

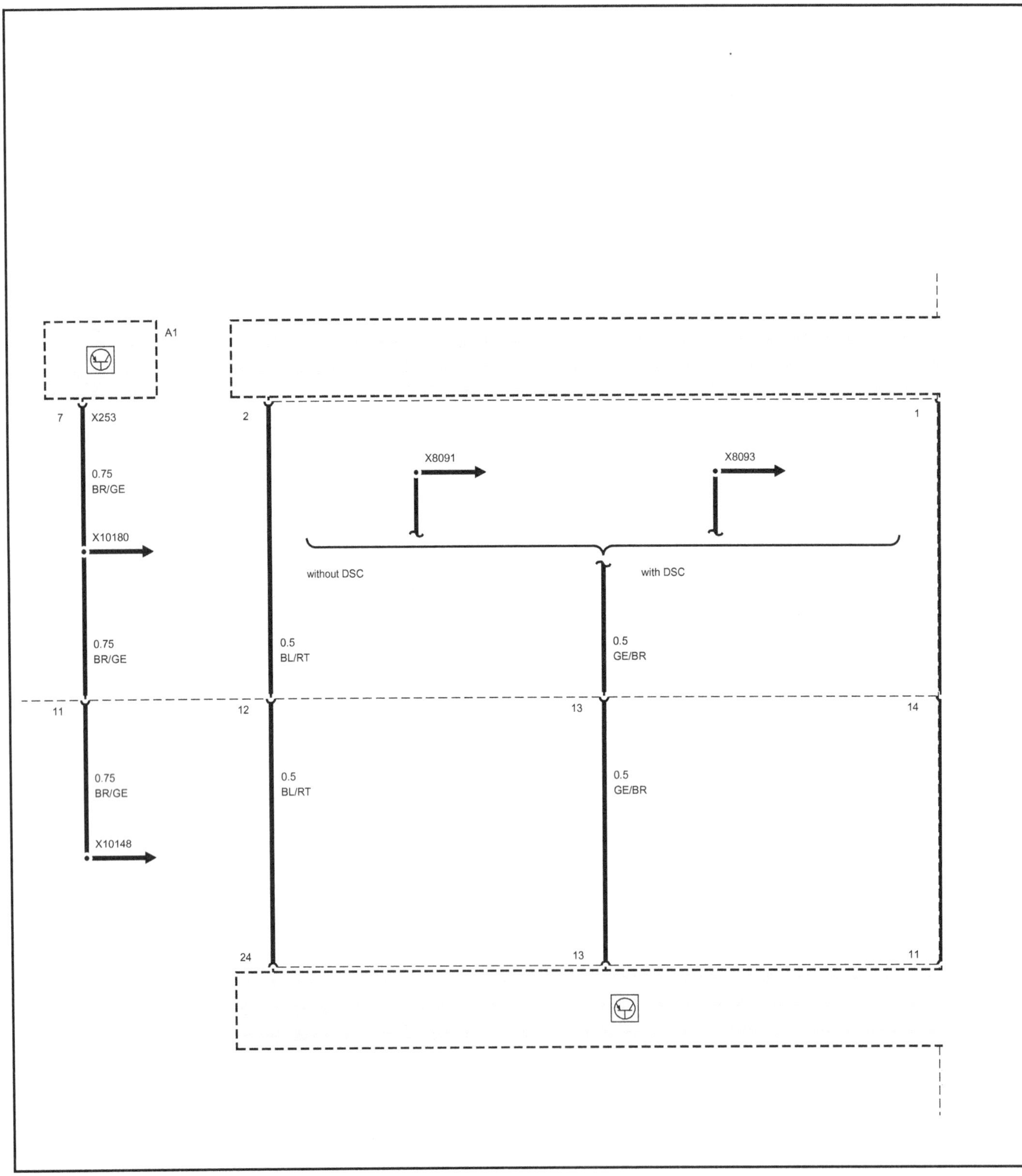

Instruments
Intermediate connector (X15) (up to Sept. 2003)
(page 4 of 9)

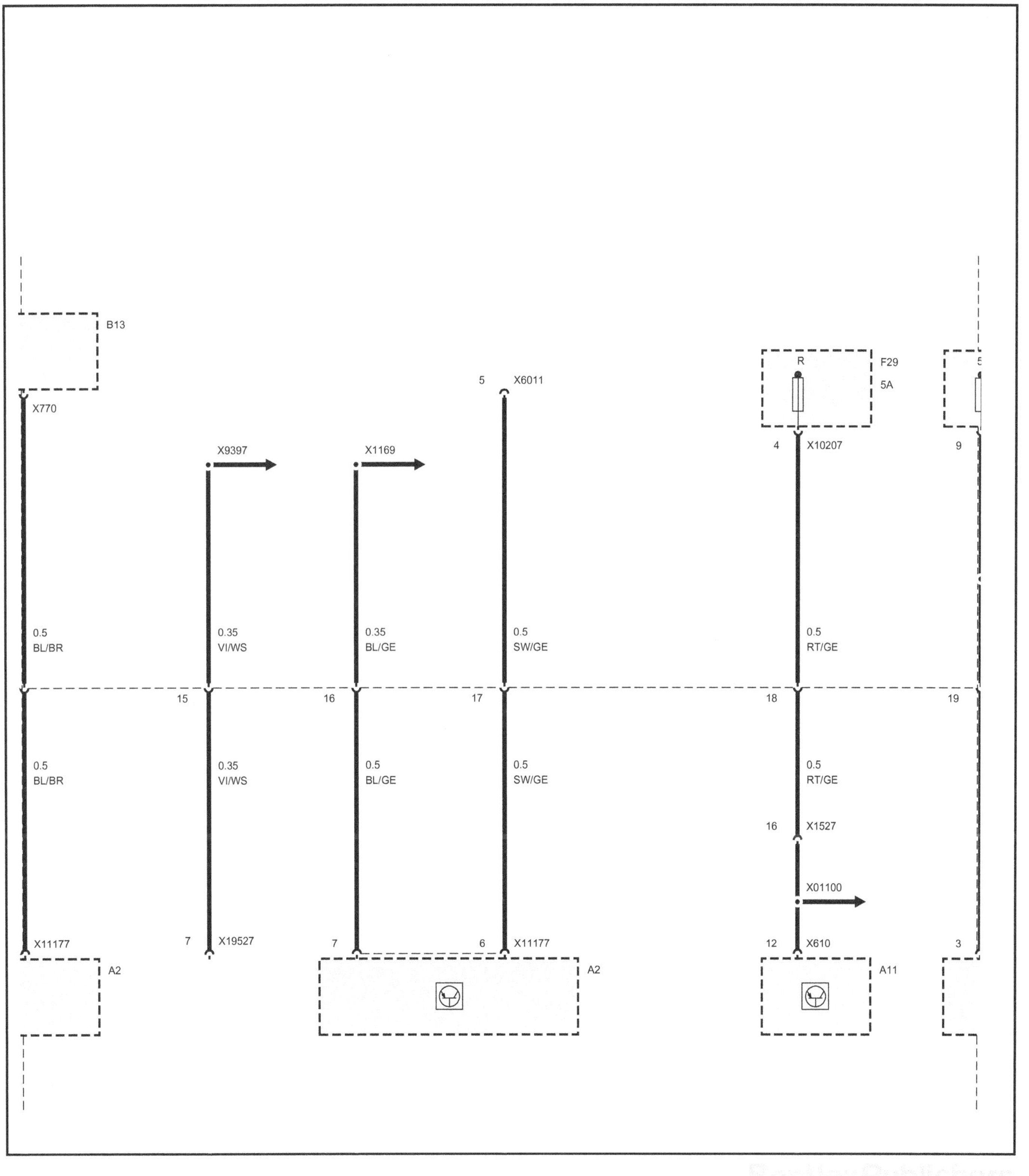

Instruments
Intermediate connector (X15) (up to Sept. 2003)
(page 5 of 9)

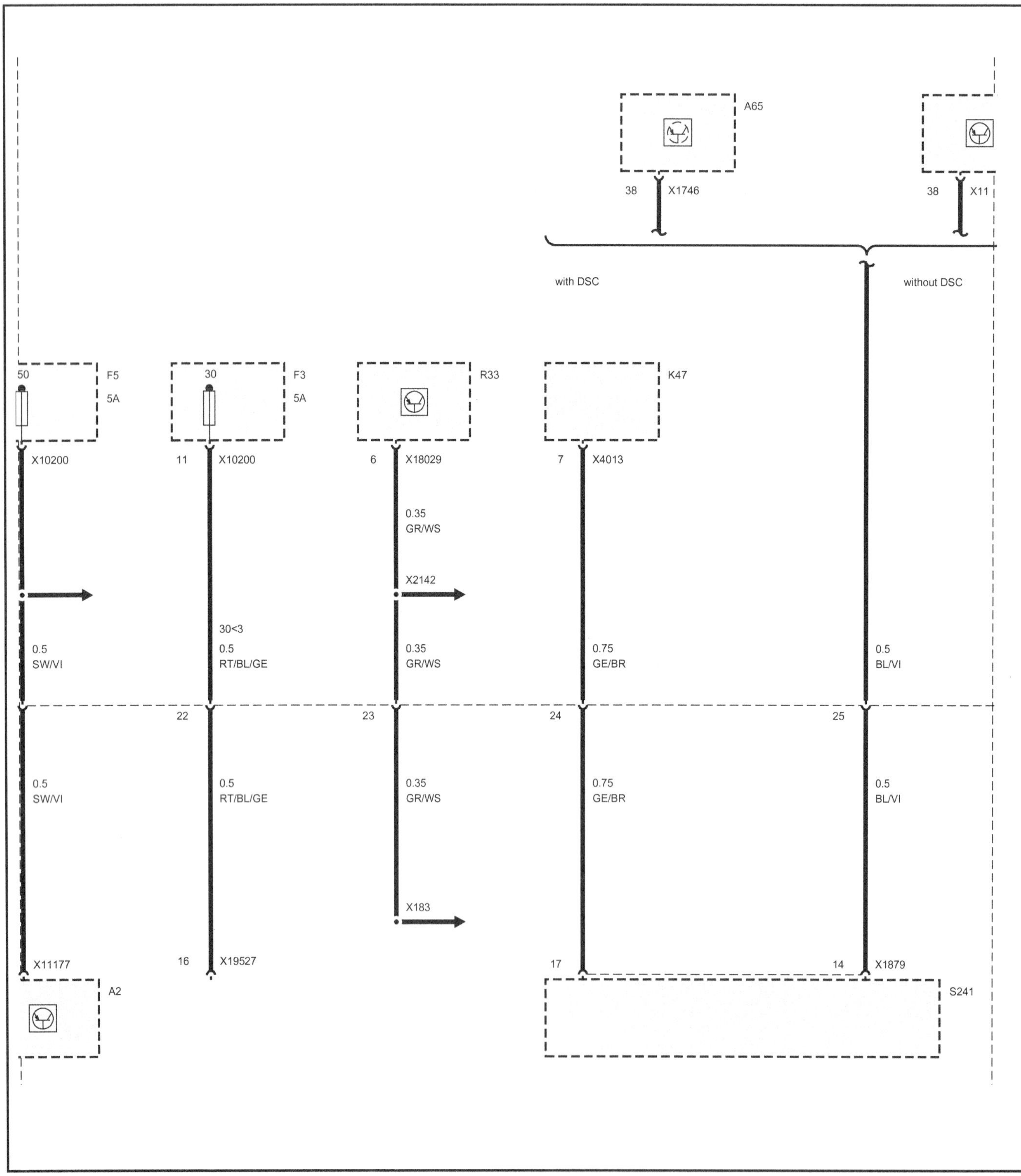

Instruments

Intermediate connector (X15) (up to Sept. 2003)

(page 6 of 9)

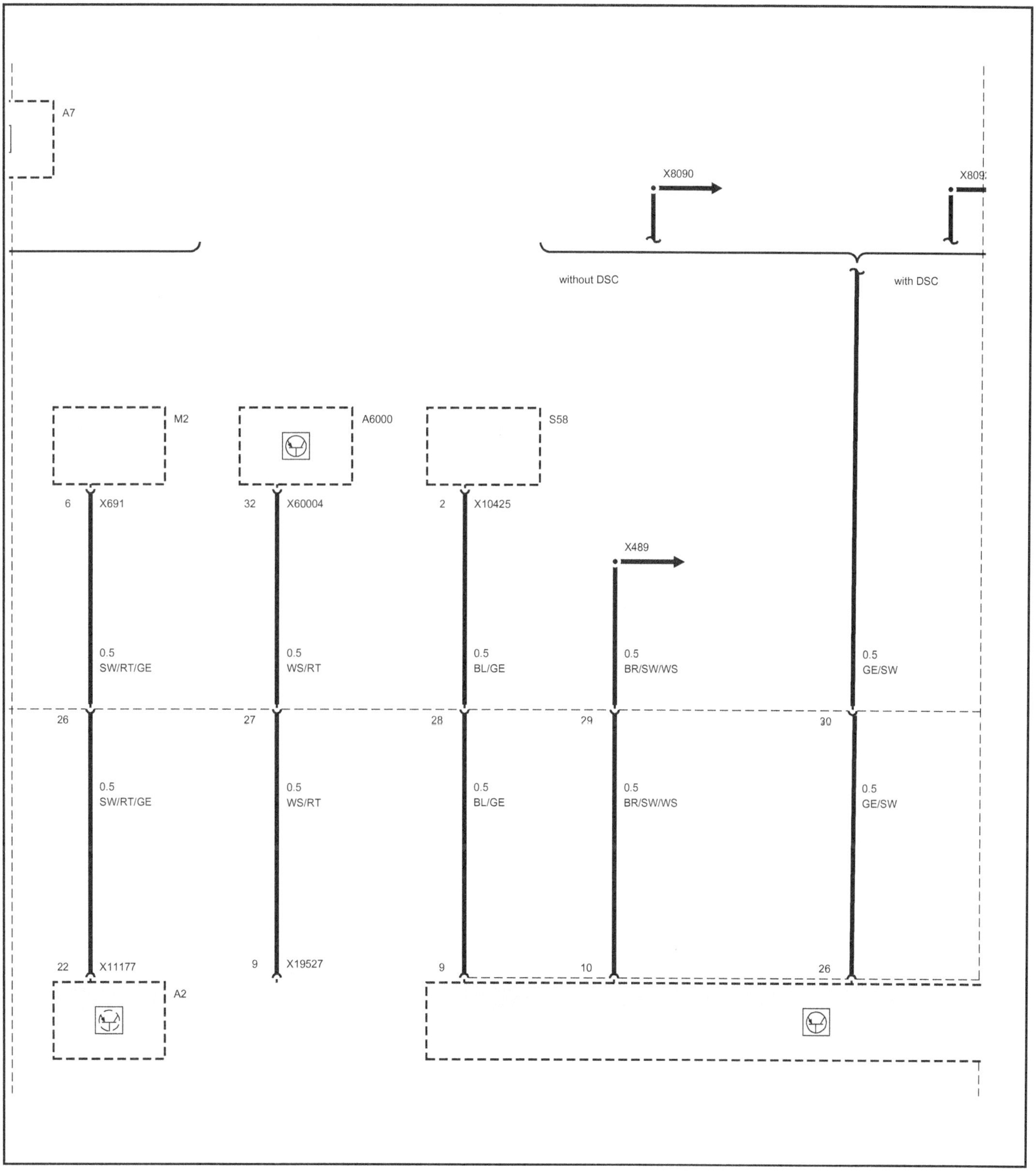

Instruments

Intermediate connector (X15) (up to Sept. 2003)

(page 7 of 9)

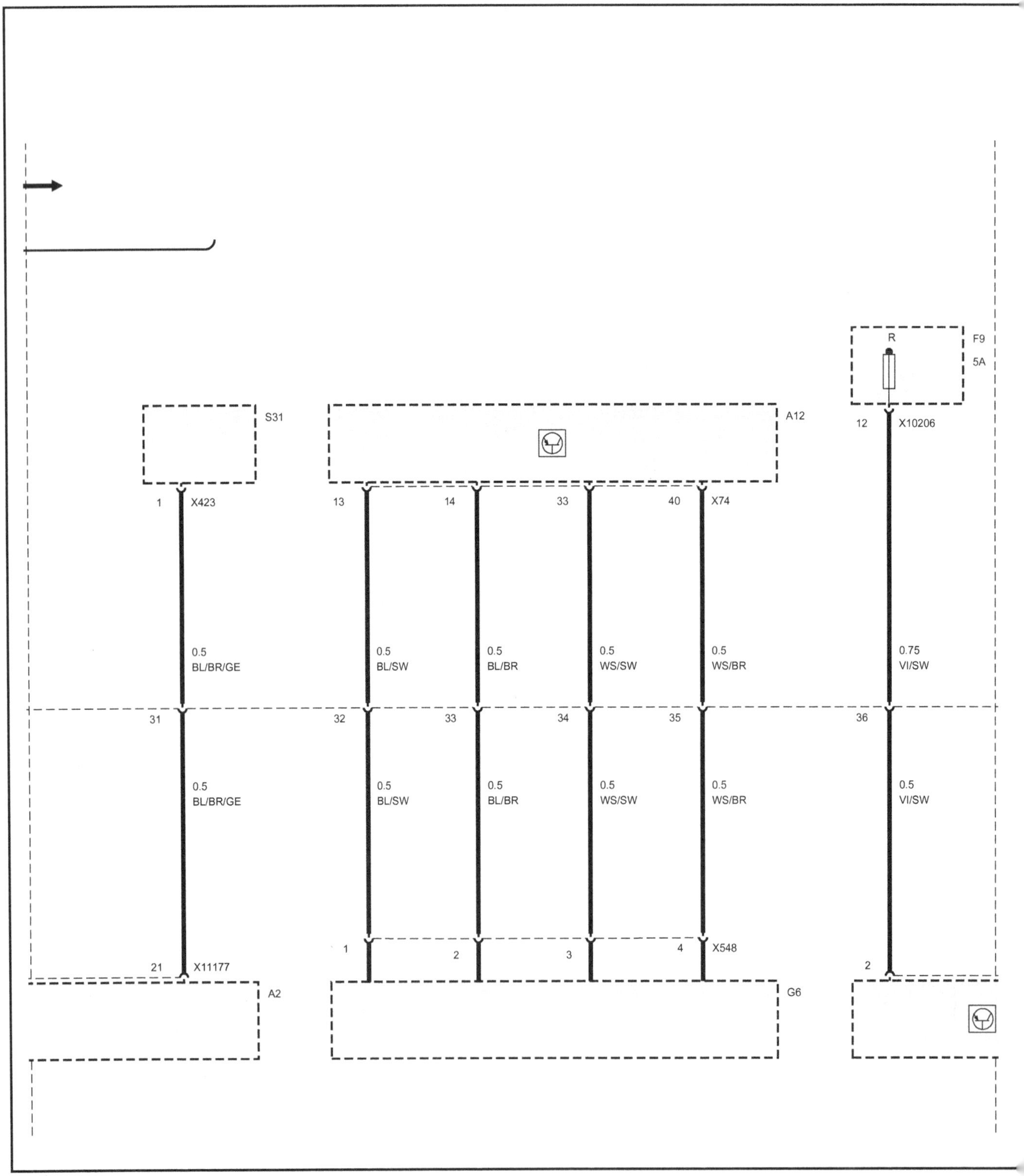

Instruments

Intermediate connector (X15) (up to Sept. 2003)

(page 8 of 9)

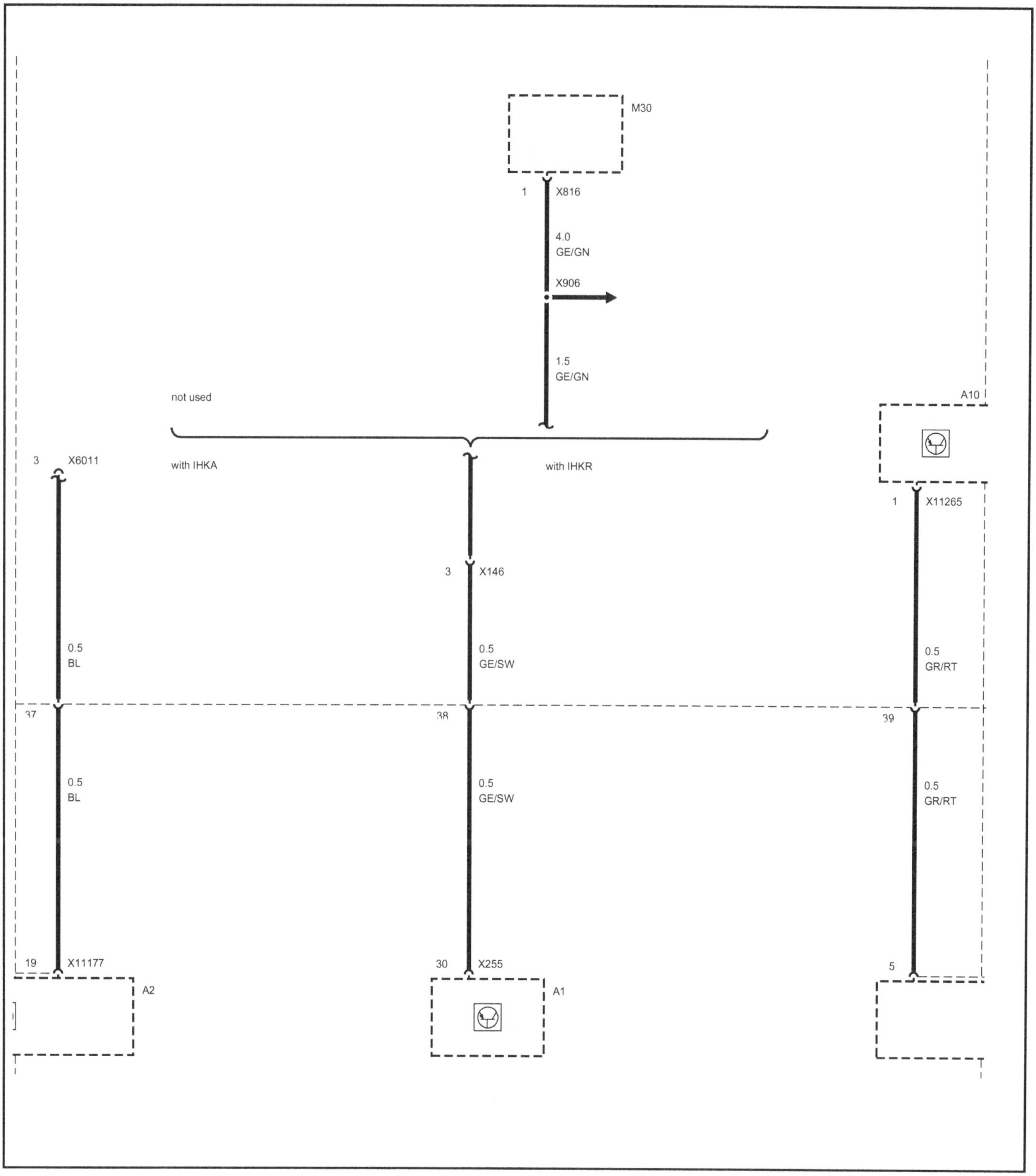

Instruments

Intermediate connector (X15) (up to Sept. 2003)

(page 9 of 9)

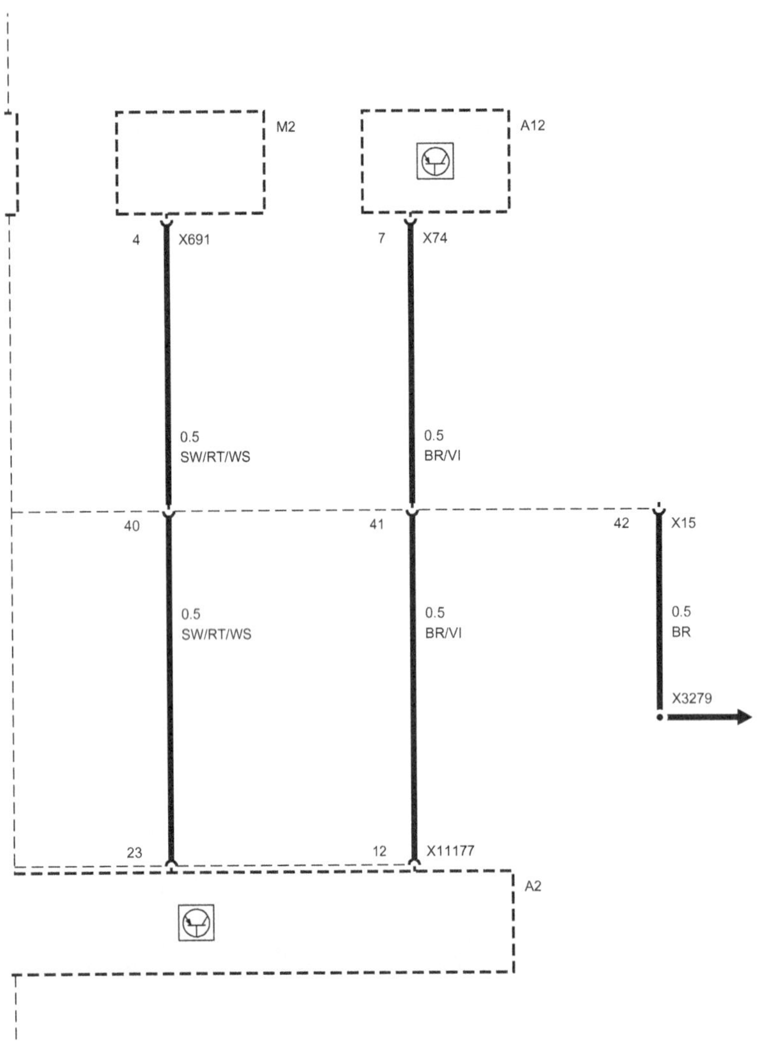

Instruments
Seat belt warning light
(page 1 of 1)

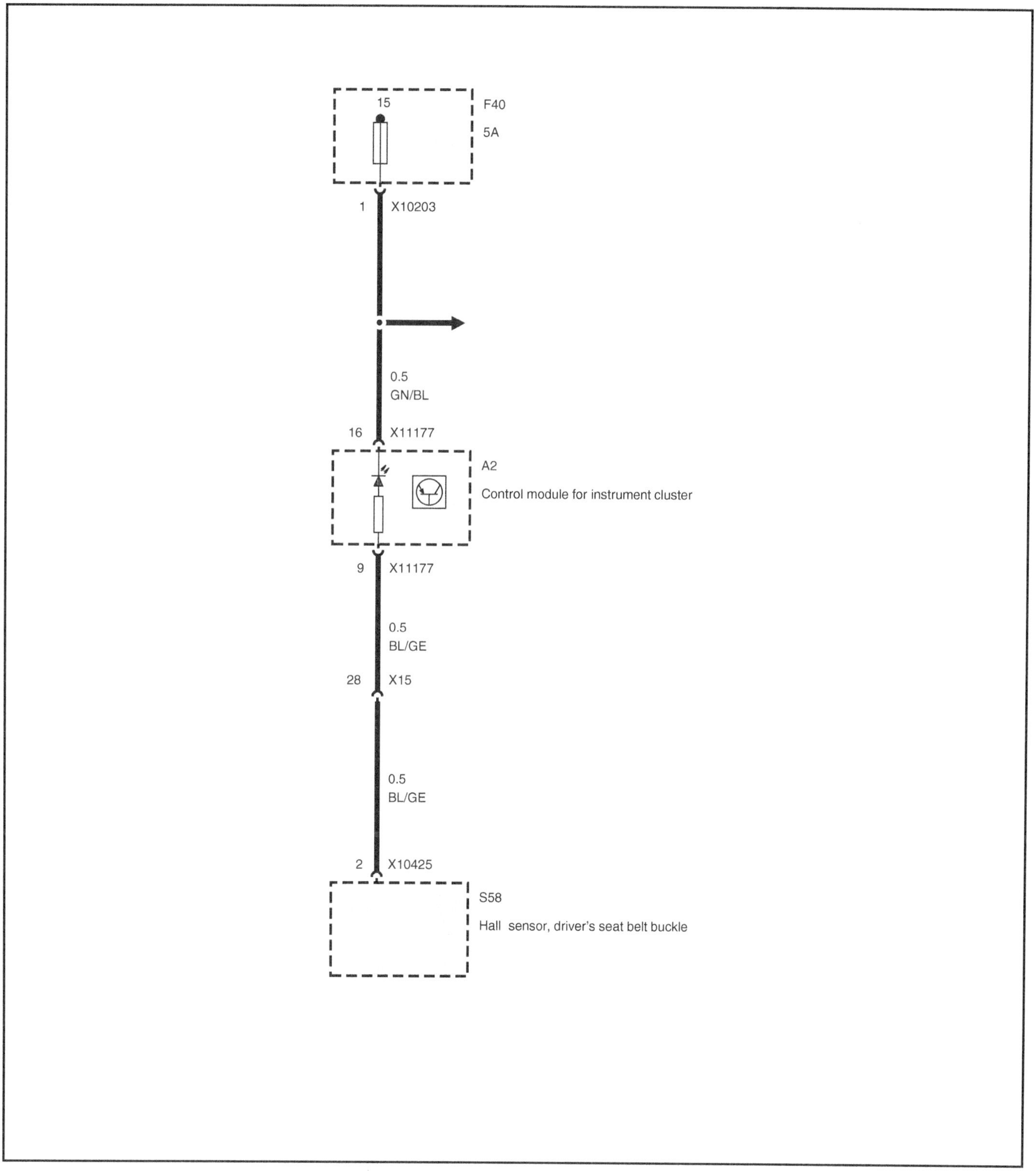

Instruments
Hazard warning
(page 1 of 2)

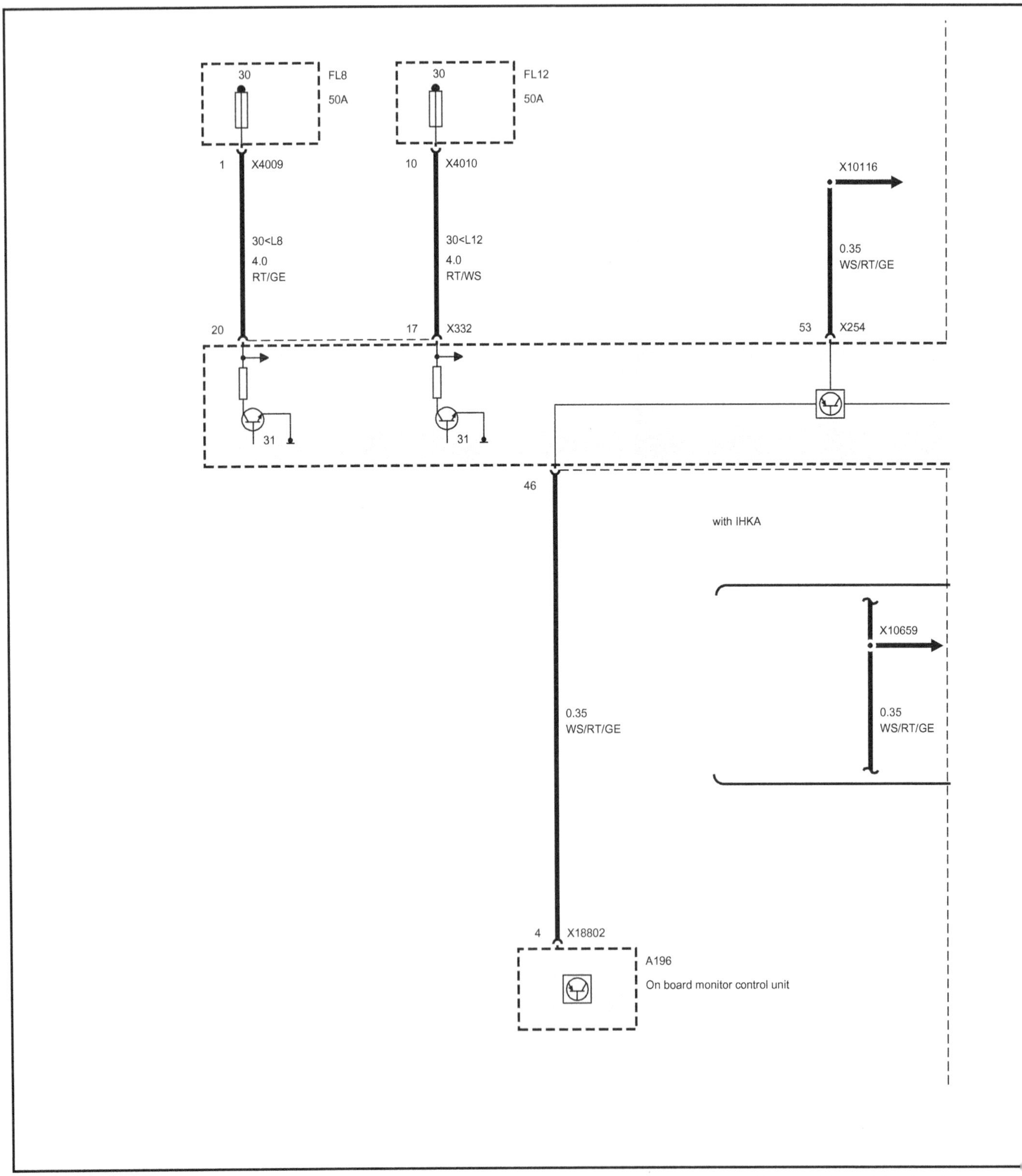

Instruments
Hazard warning
(page 2 of 2)

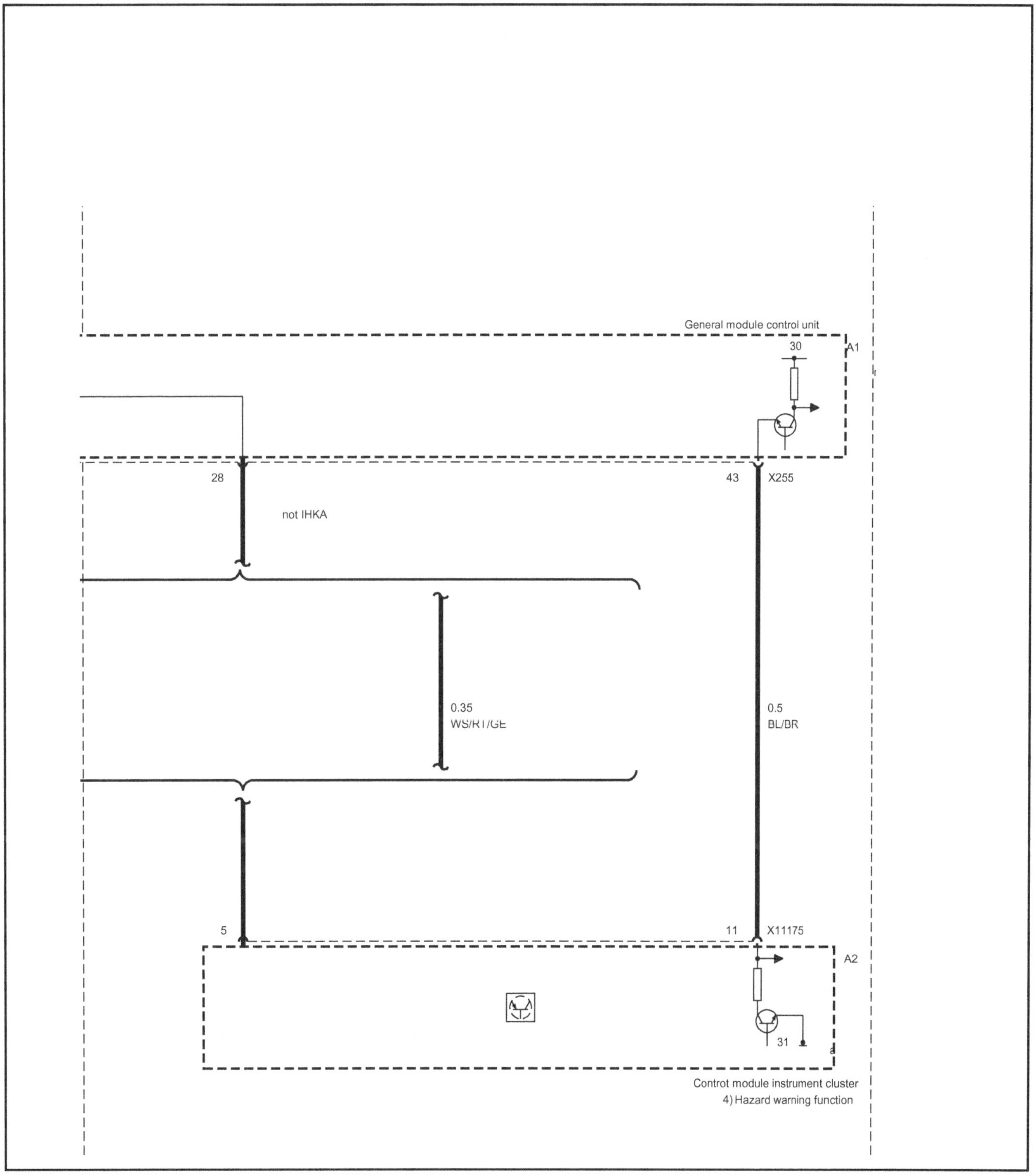

Instruments
Fuel level sensor
(page 1 of 1)

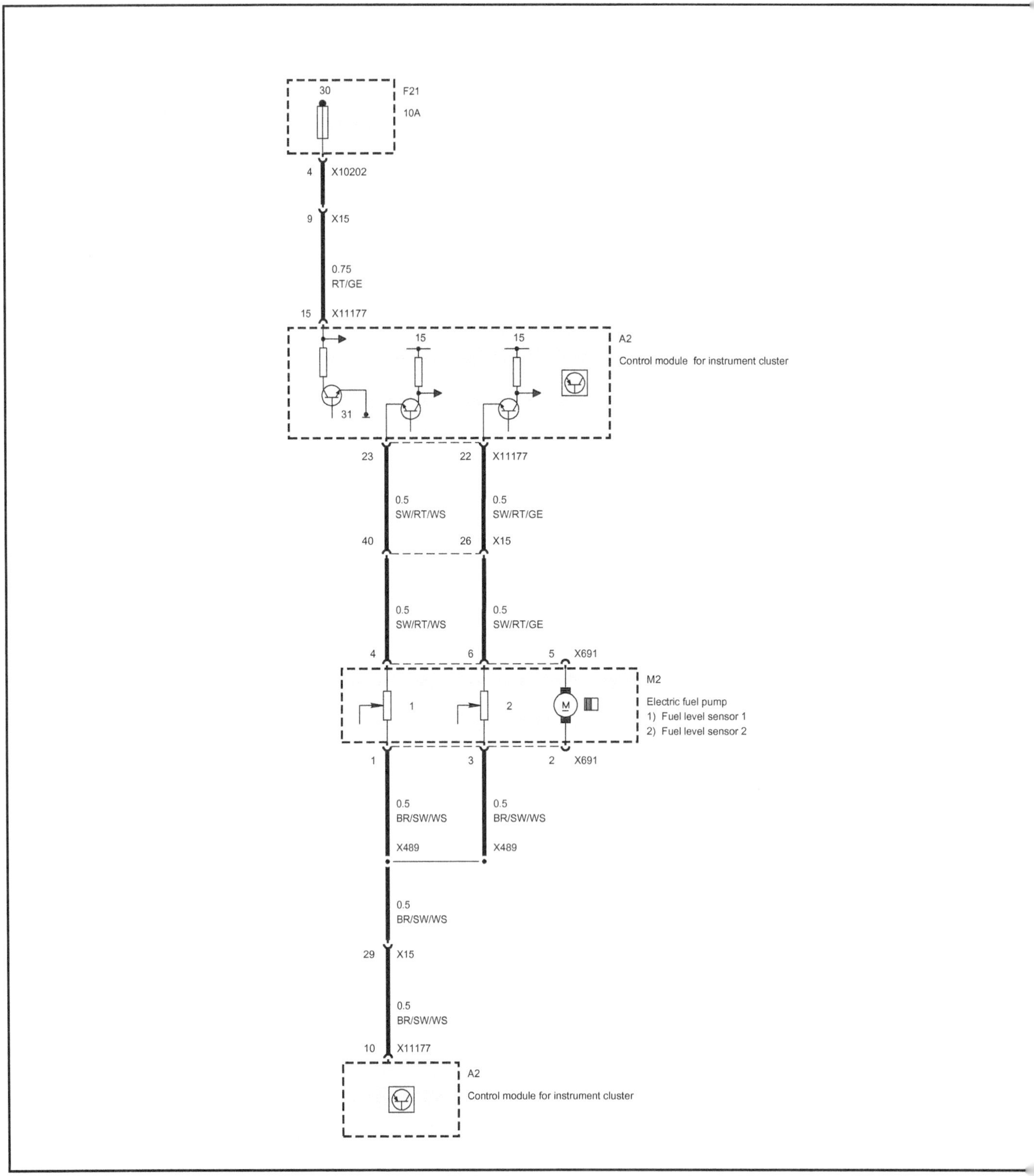

Instruments
Charge indicator light
(page 1 of 1)

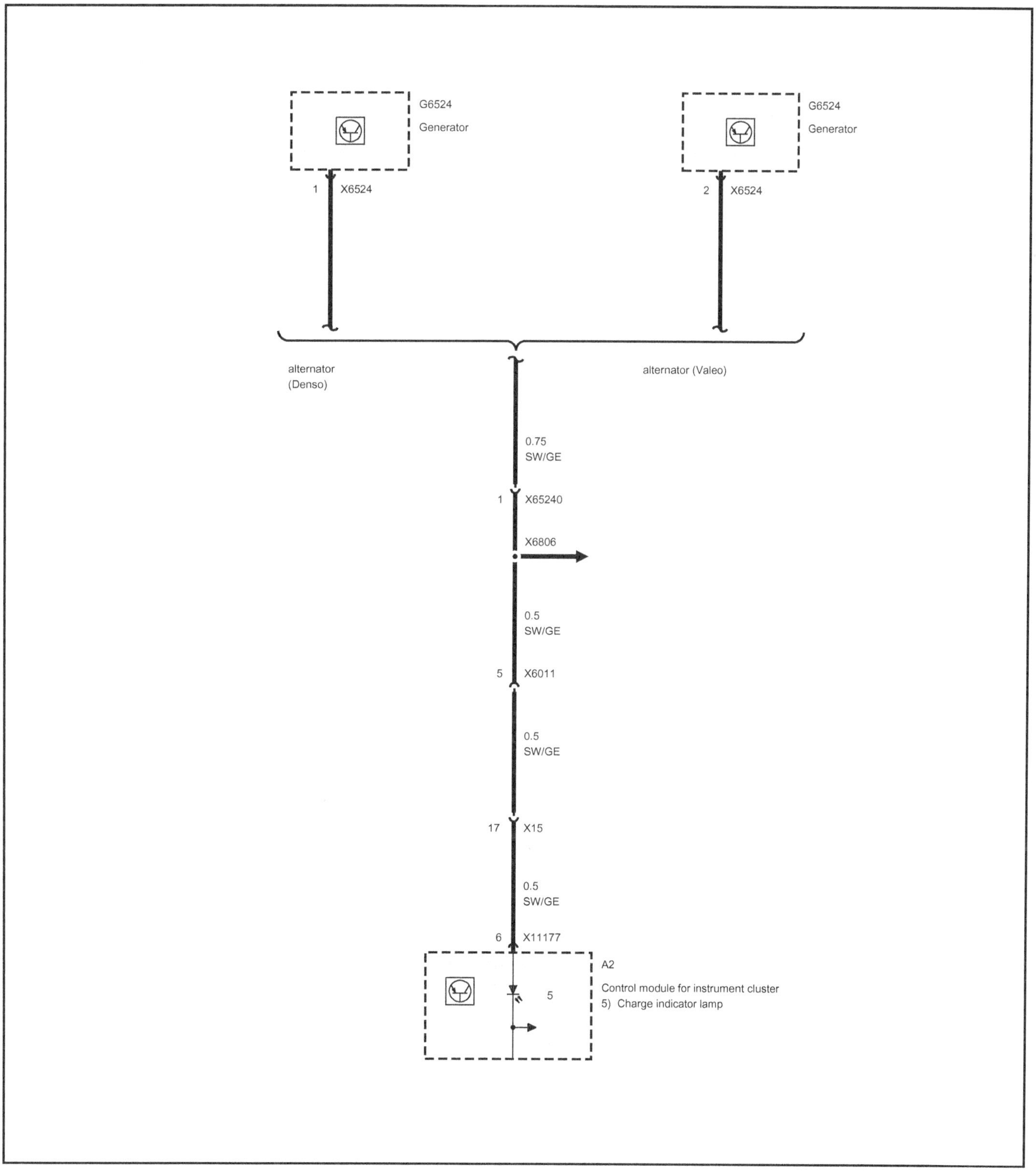

Instruments
Oil pressure indicator
(page 1 of 1)

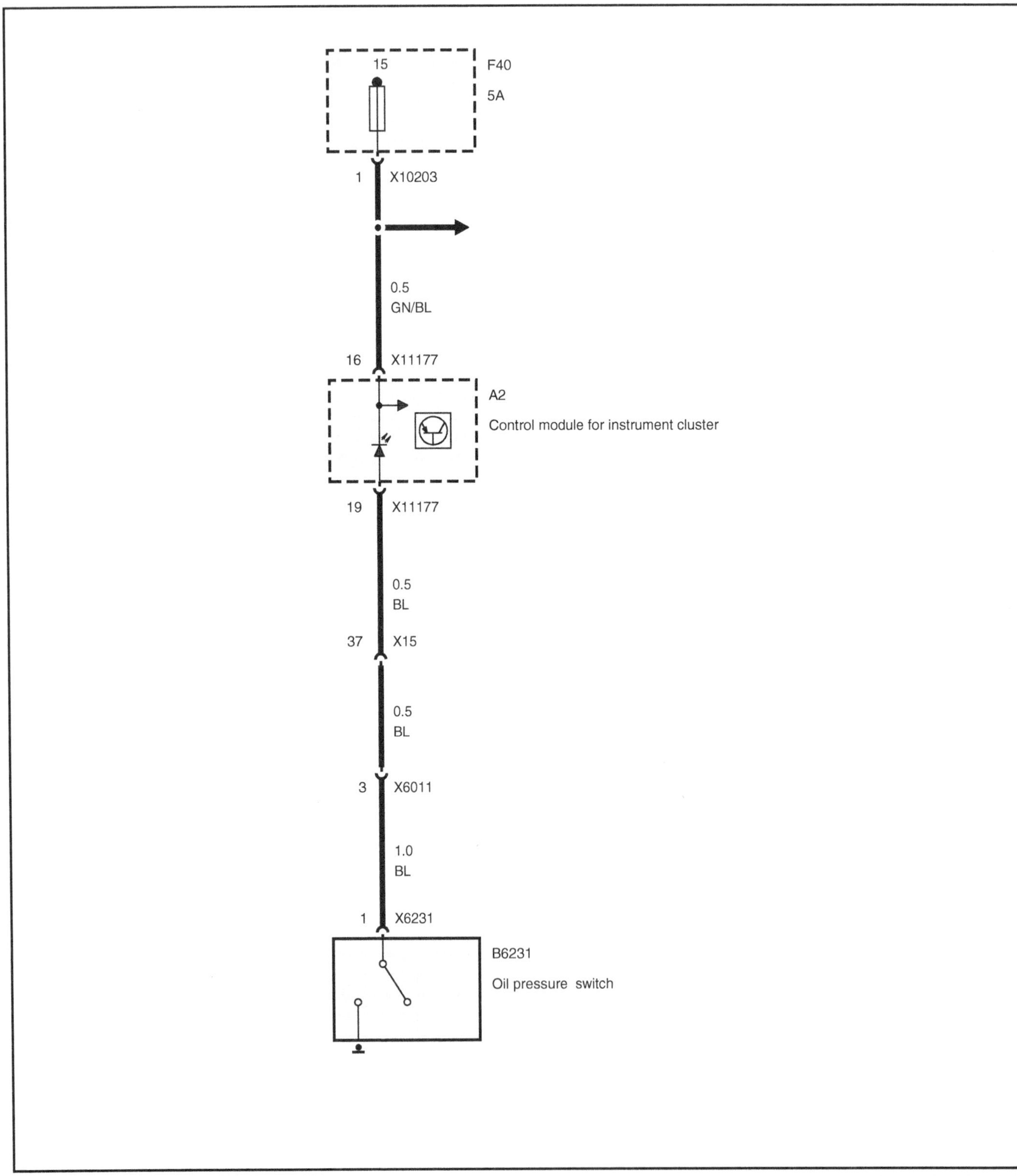

Instruments
Illumination (terminal 58G)
(page 1 of 2)

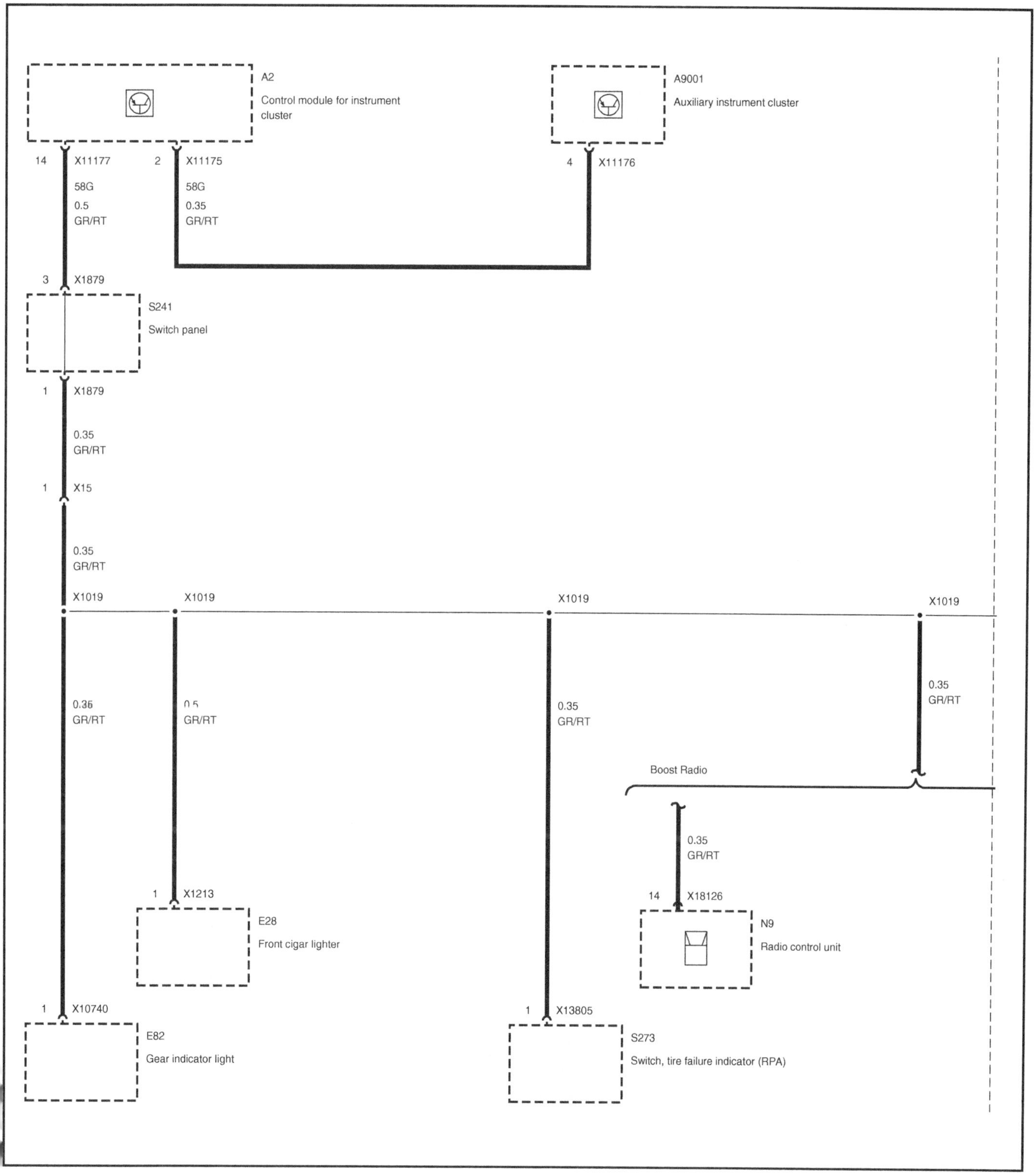

Instruments
Illumination (terminal 58G)
(page 2 of 2)

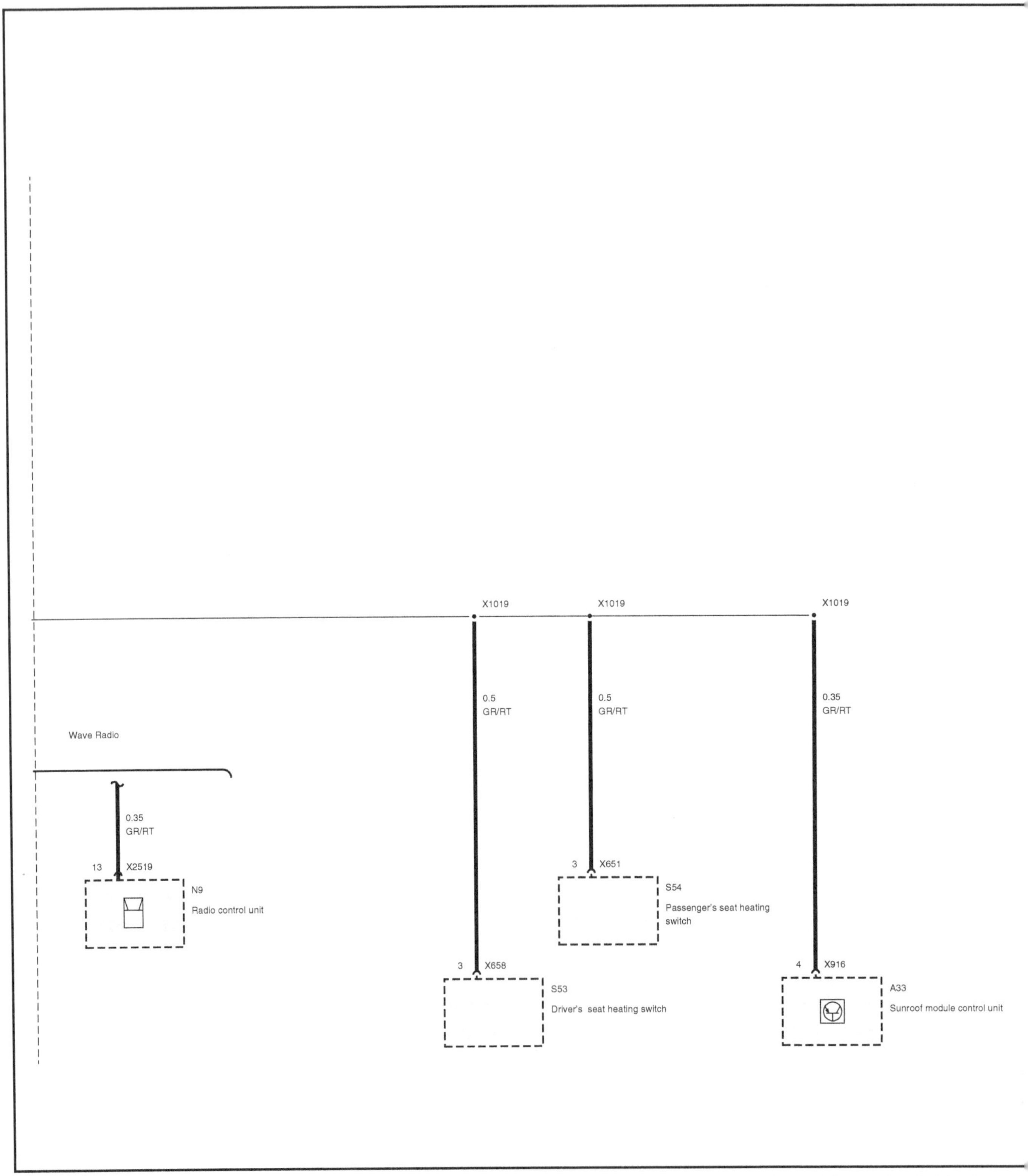

Interior lights
Interior reading light (coupe)
(page 1 of 1)

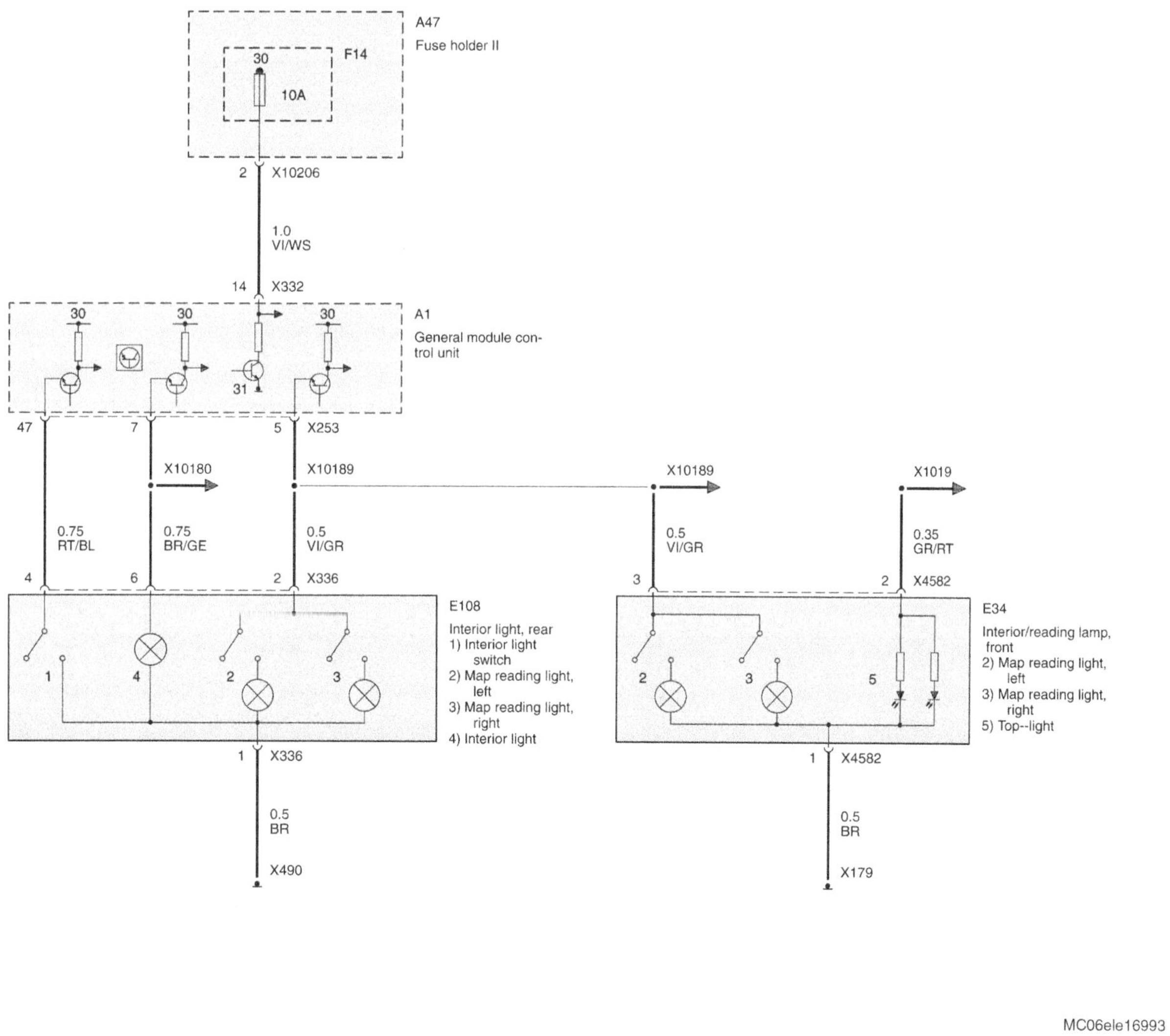

Interior lights
Interior reading light (convertible)
(page 1 of 1)

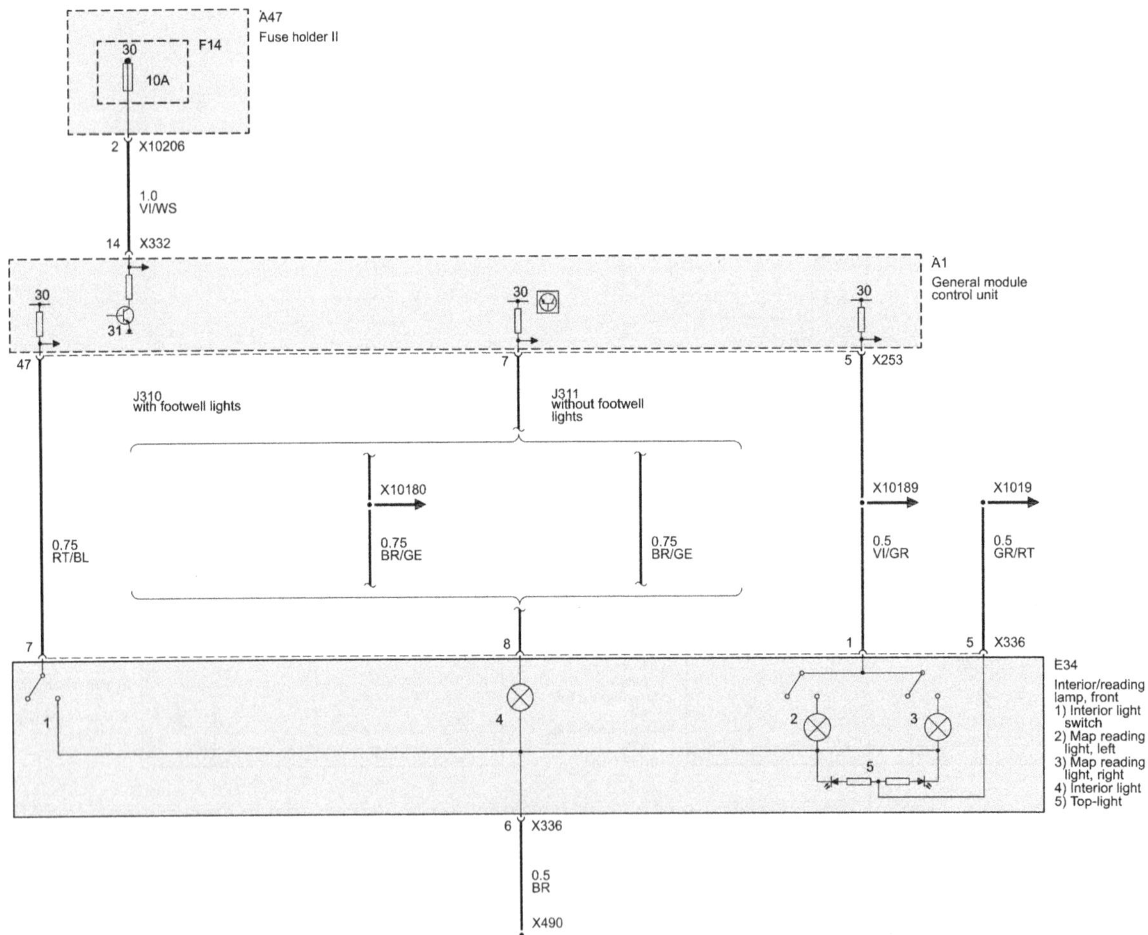

MC06ele15966

Interior lights
Glove compartment light
(page 1 of 1)

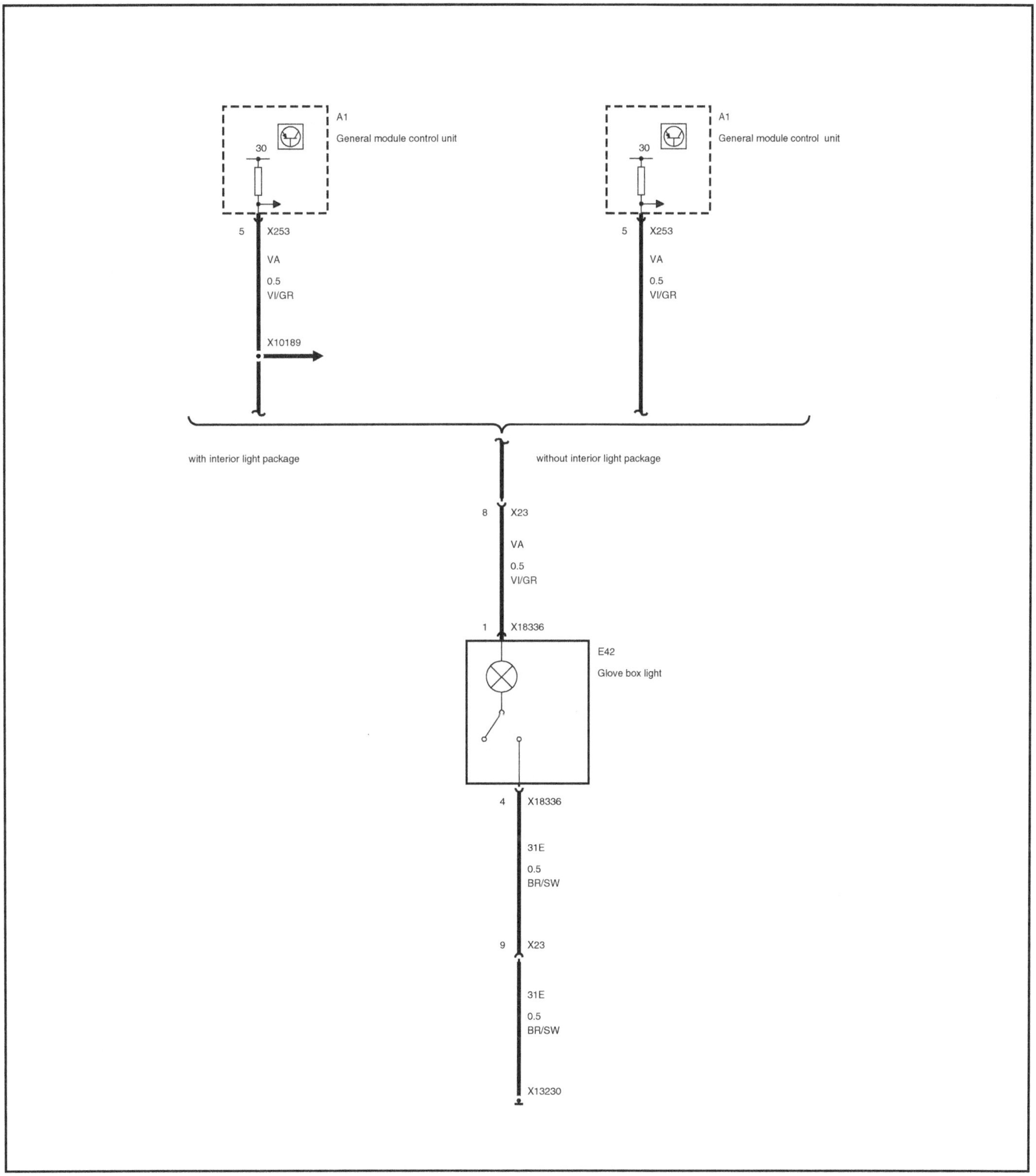

Interior lights
Footwell lights
(page 1 of 1)

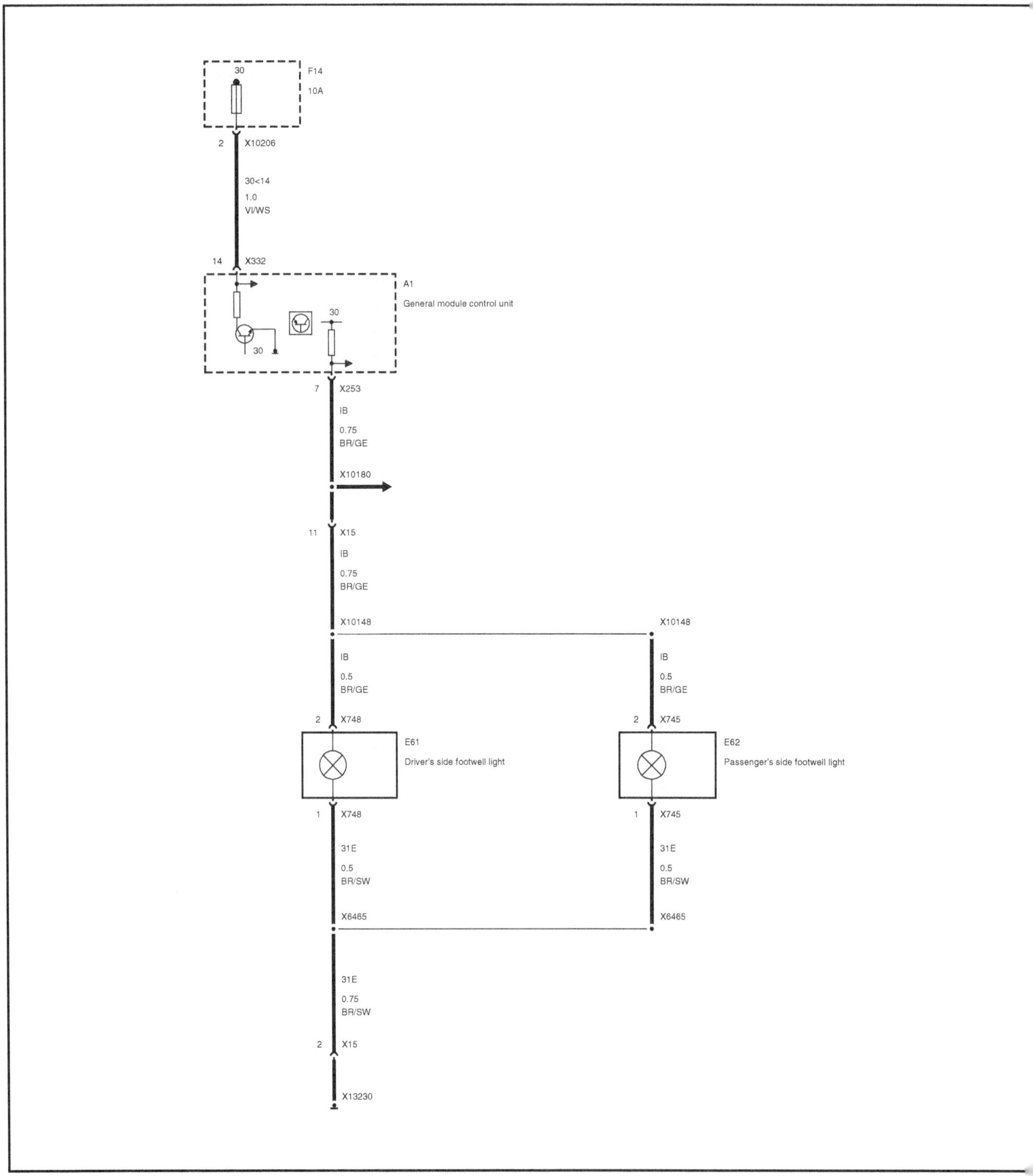

Interior lights
Ashtray light
(page 1 of 1)

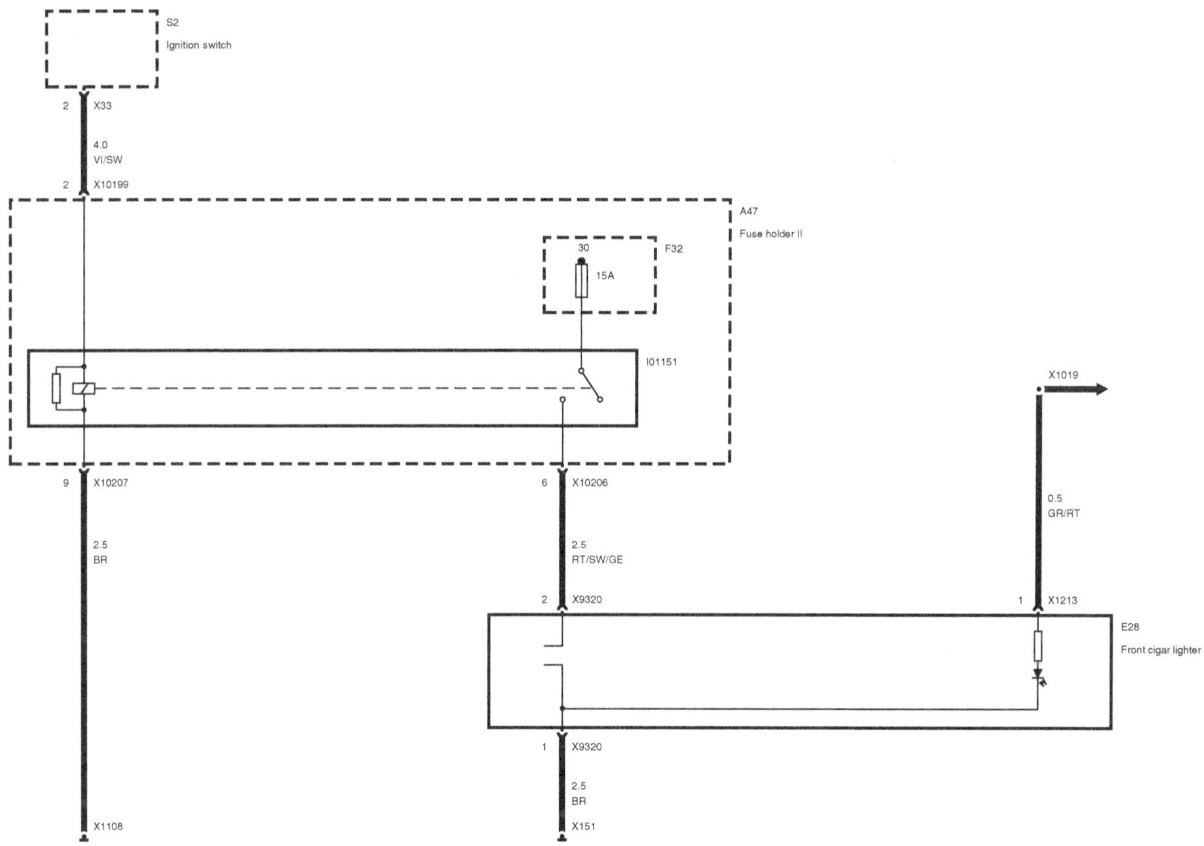

Interior lights
Mirror lights
(page 1 of 1)

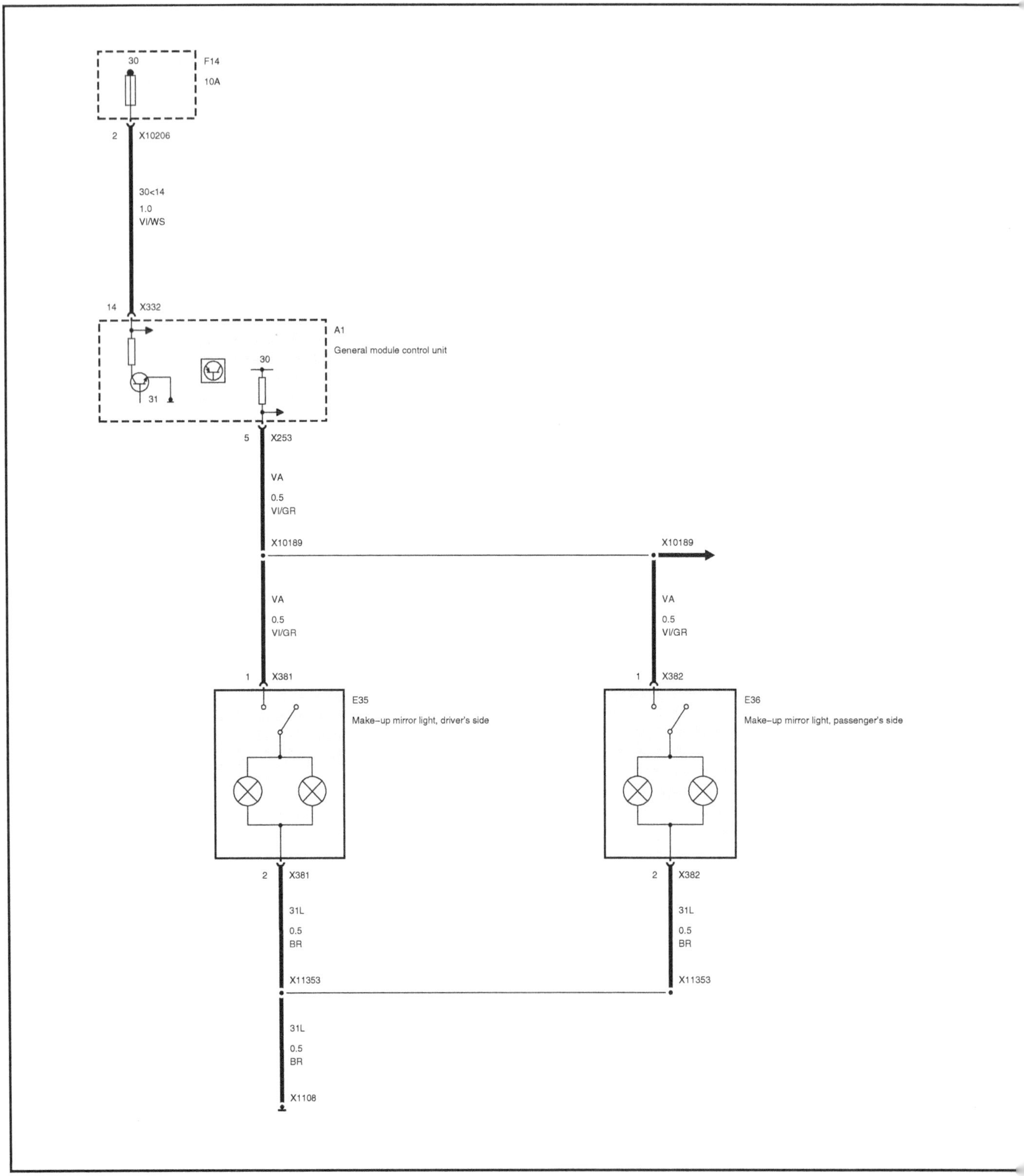

Interior lights
Cargo compartment lights (coupe)
(page 1 of 1)

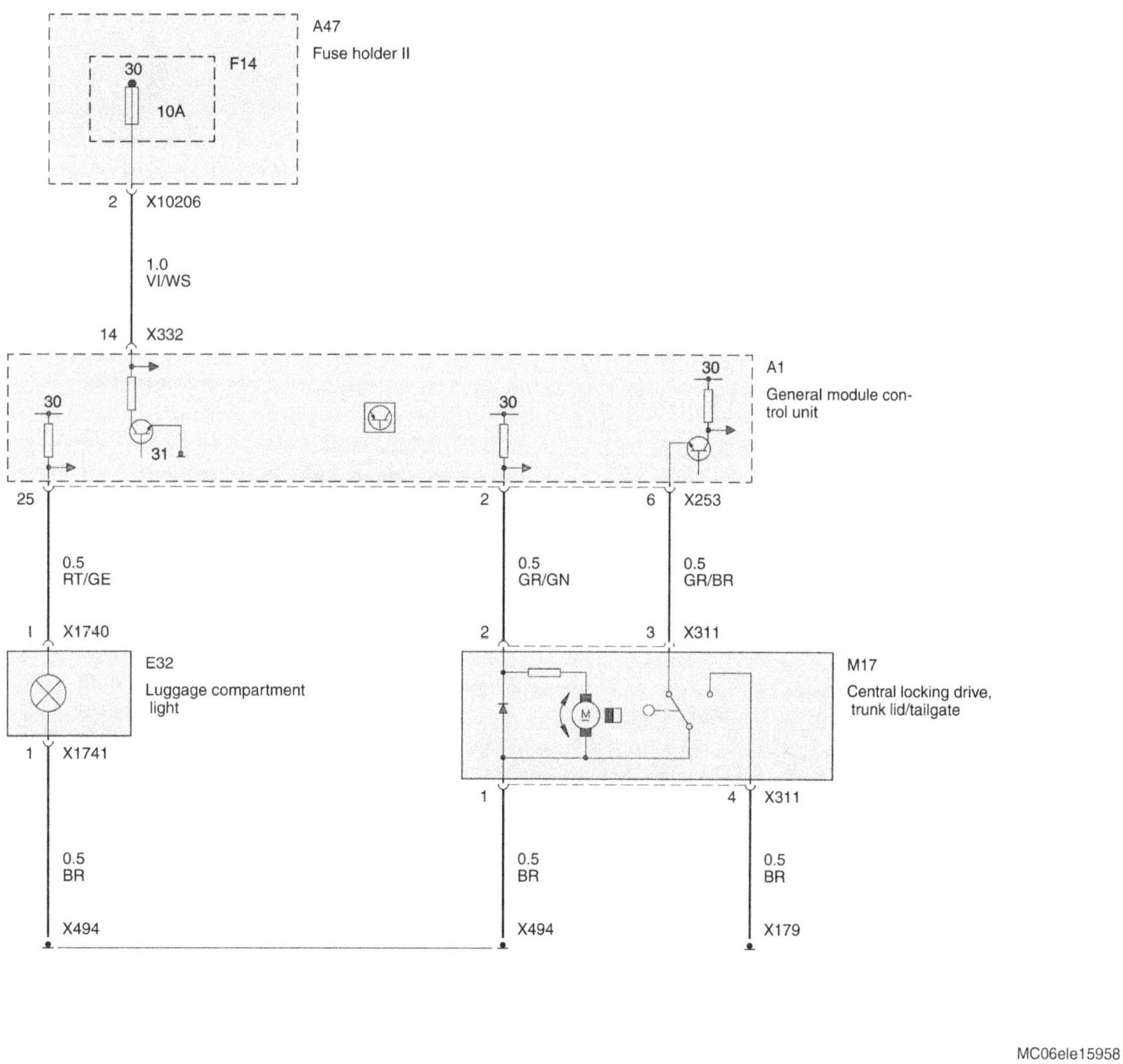

Interior lights
Cargo compartment lights (convertible)
(page 1 of 1)

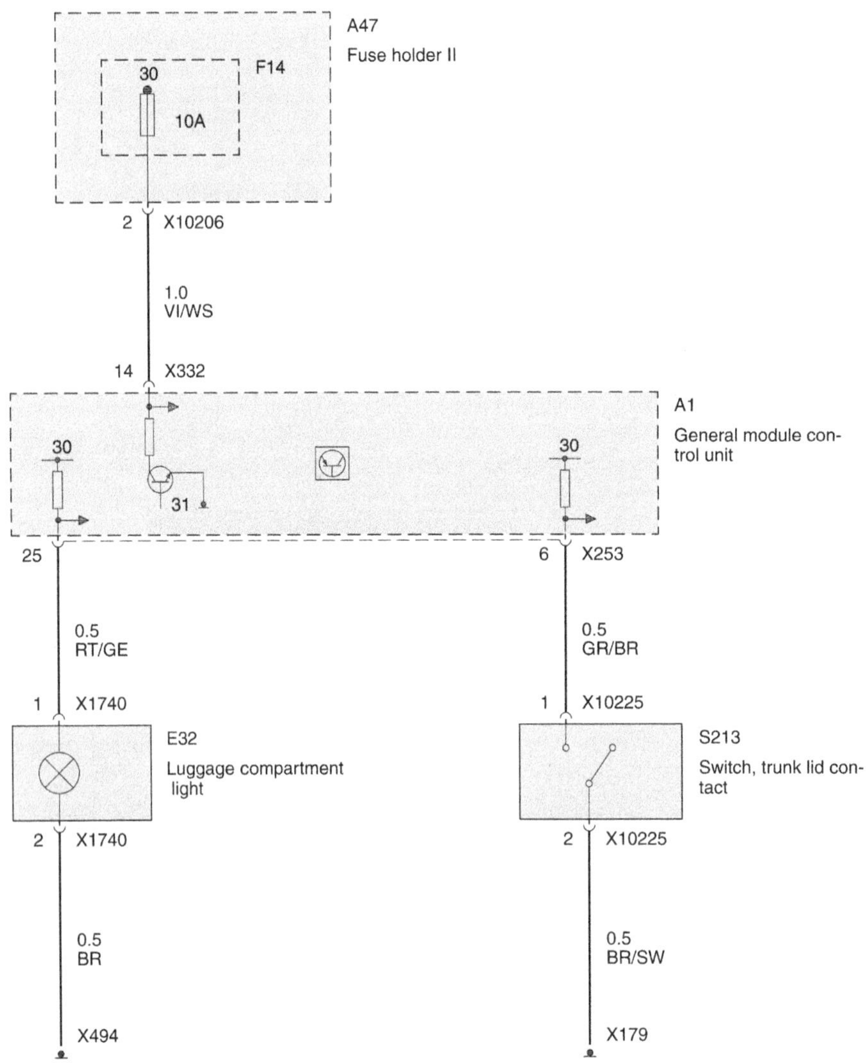

MC06ele21206

Mirror control
(up to July 2004)
(page 1 of 1)

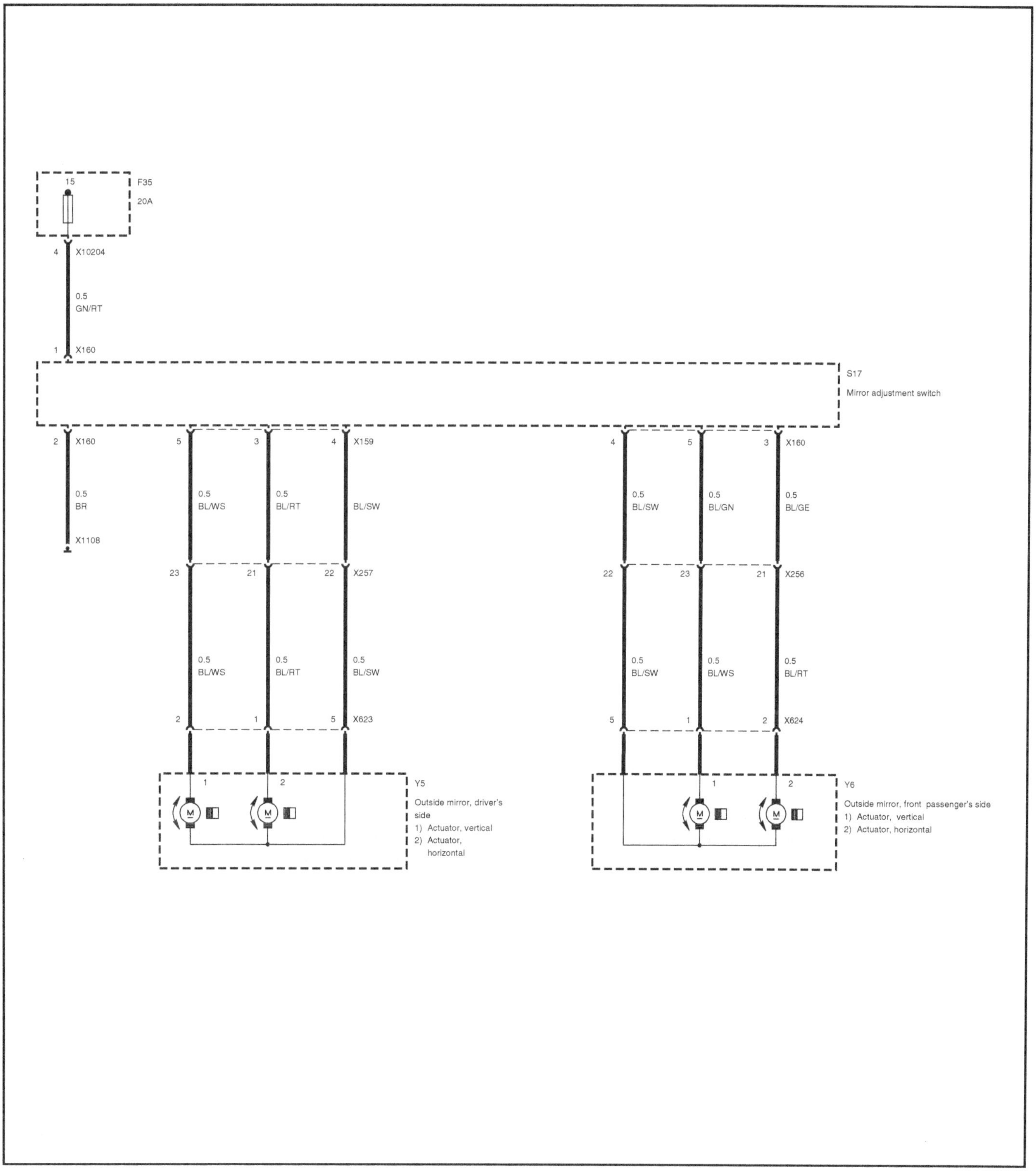

Mirror control
(from July 2004)
(page 1 of 1)

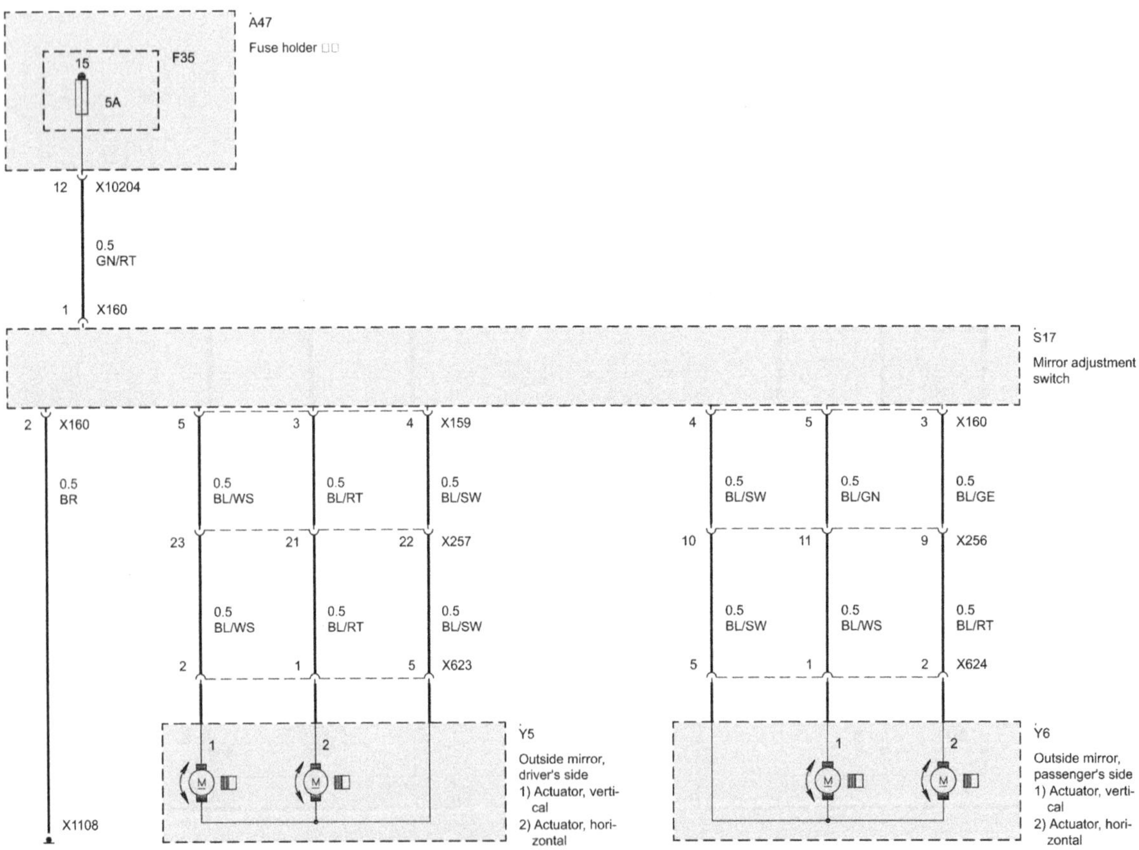

MC06ele23034

Mirror heater
(page 1 of 1)

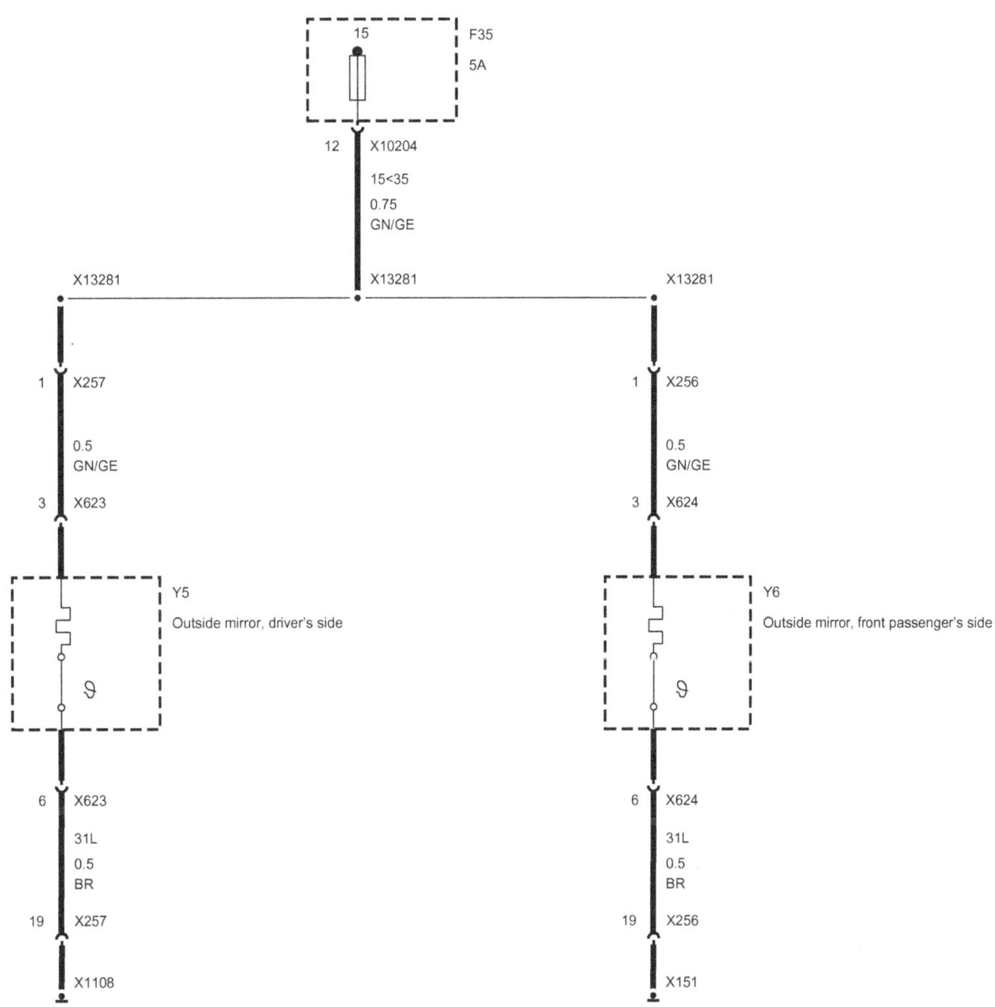

Multifunction (MFL) steering wheel
MFL I
(page 1 of 3)

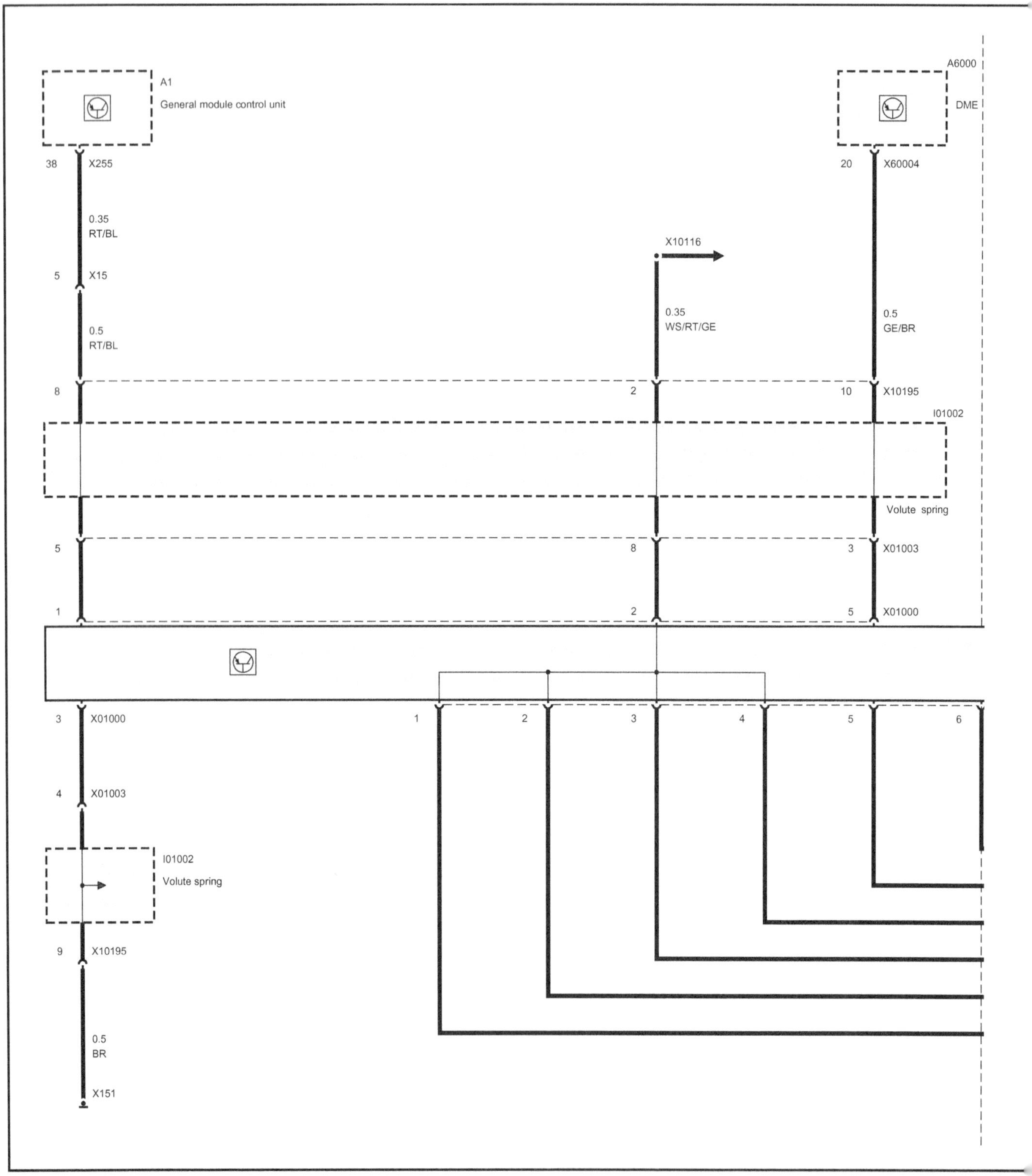

Multifunction (MFL) steering wheel

MFL I

(page 2 of 3))

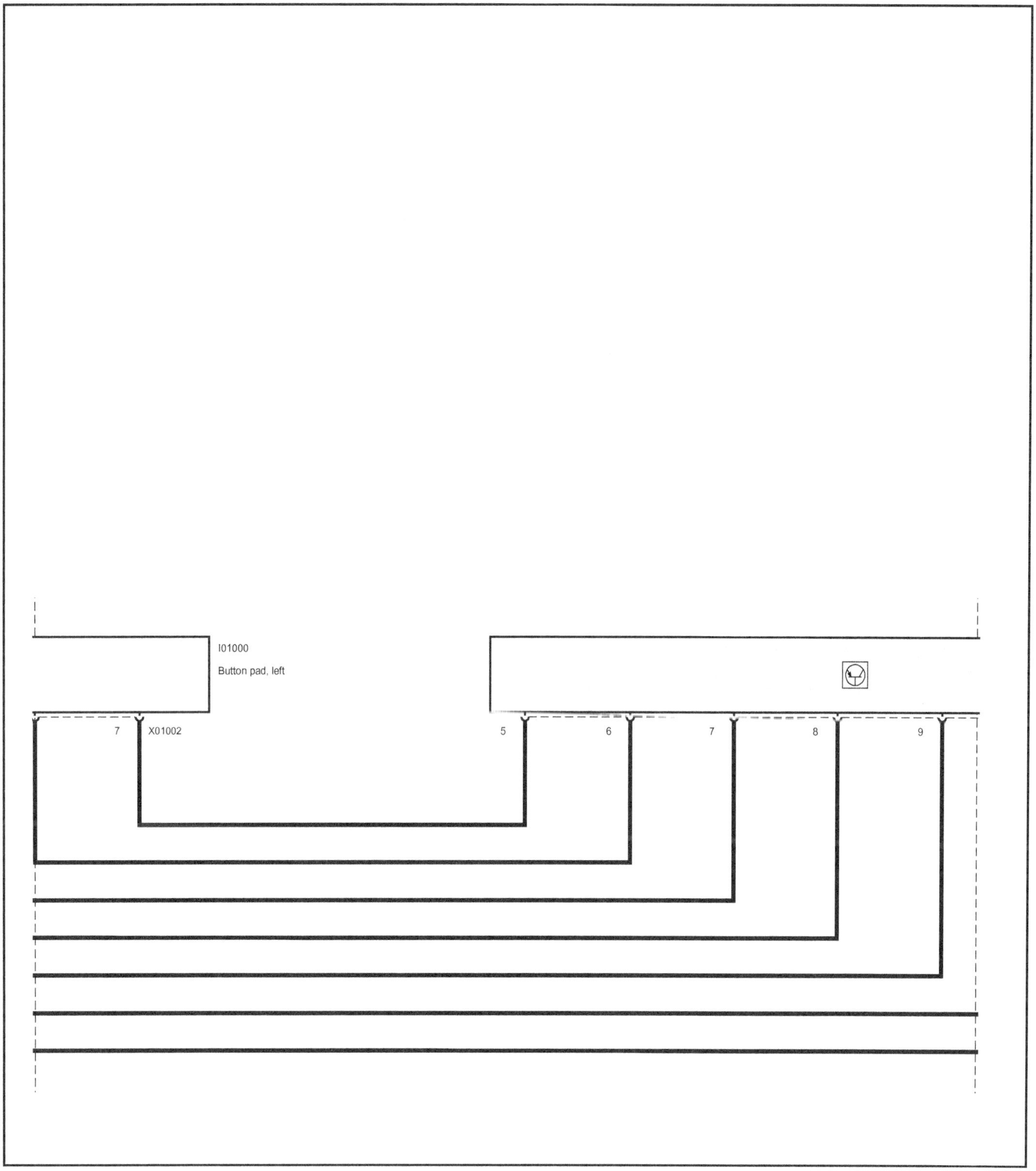

Multifunction (MFL) steering wheel
MFL I
(page 3 of 3)

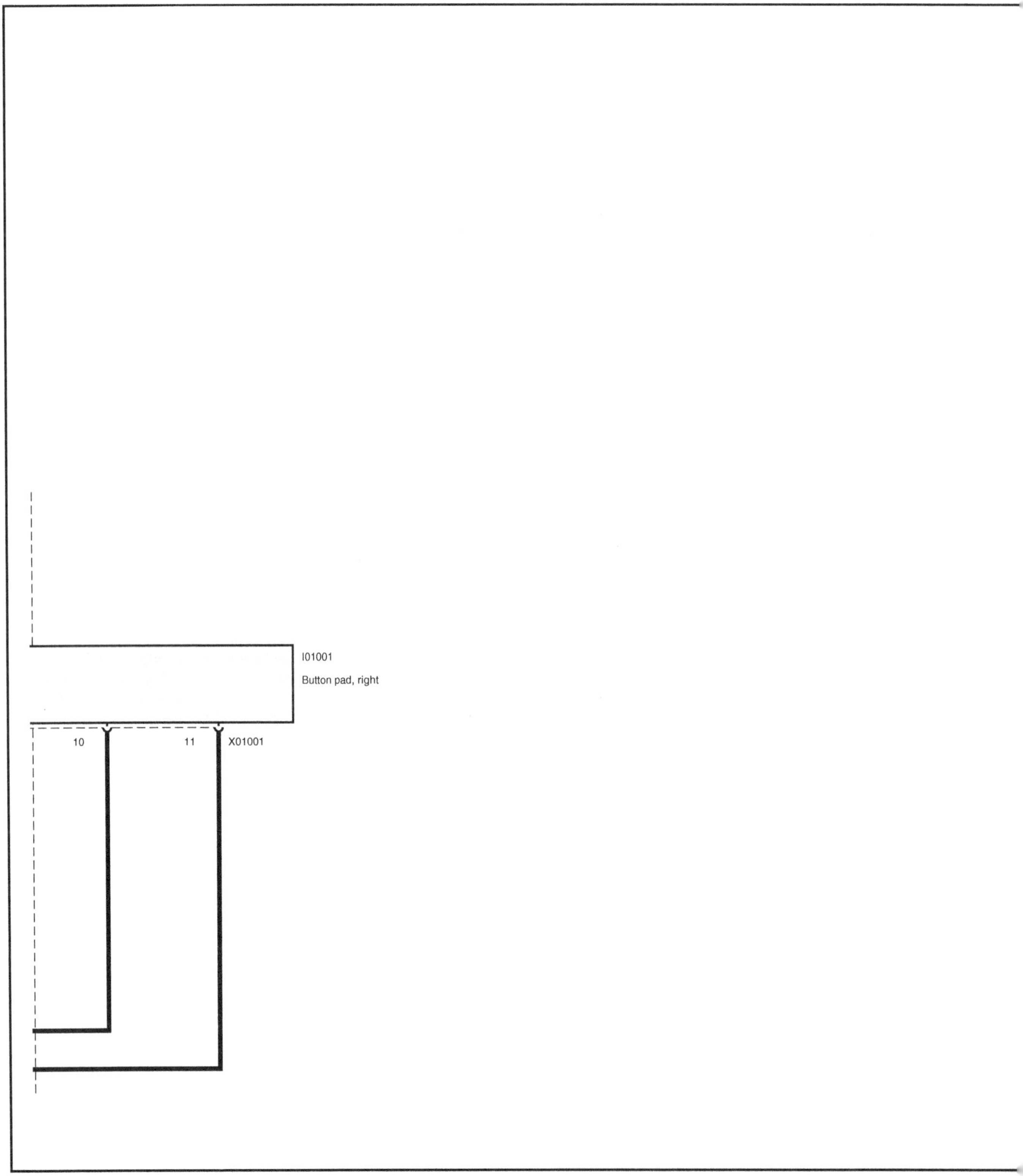

Multifunction (MFL) steering wheel
MFL II
(page 1 of 2)

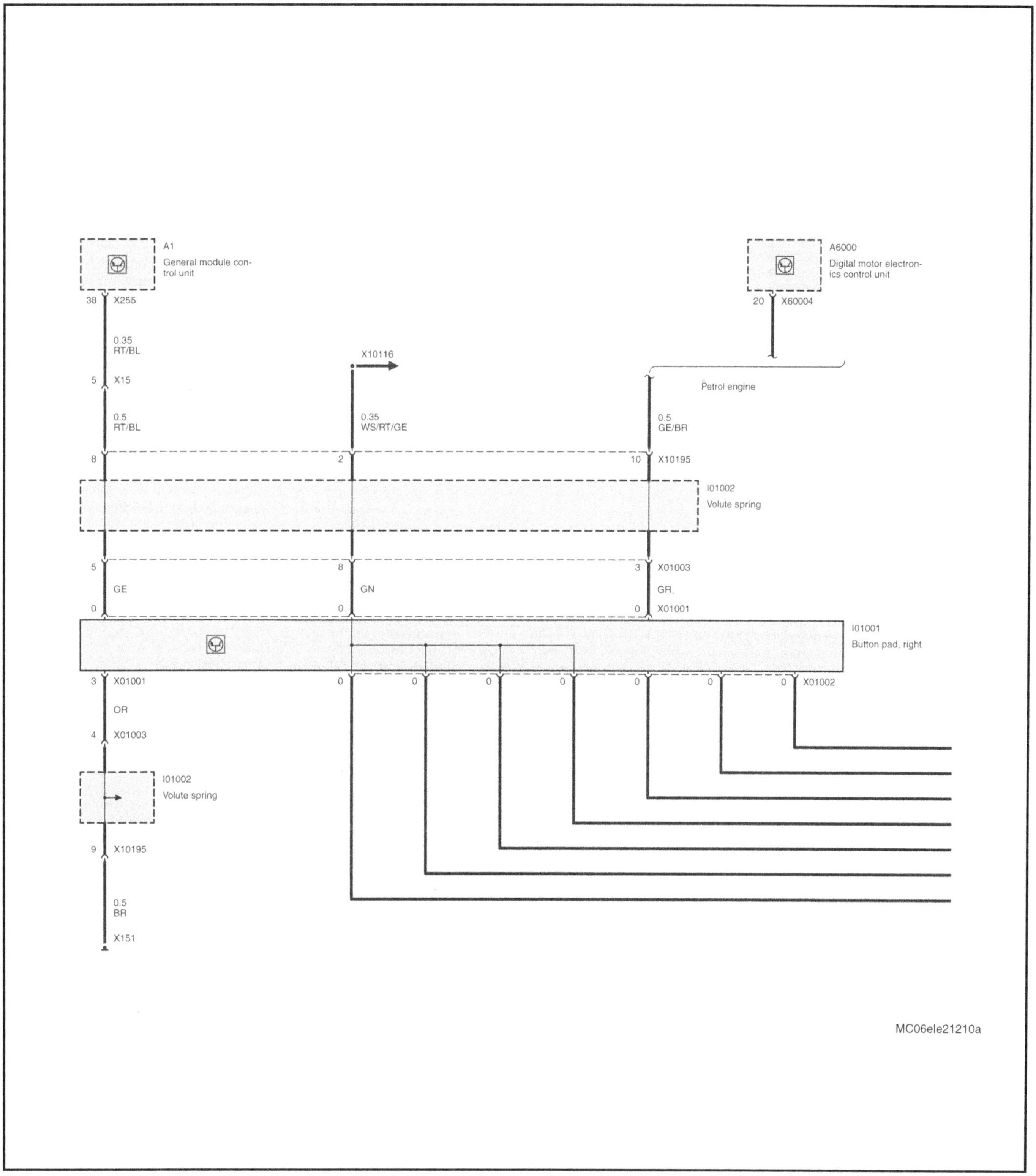

Multifunction (MFL) steering wheel
MFL II
(page 2 of 2)

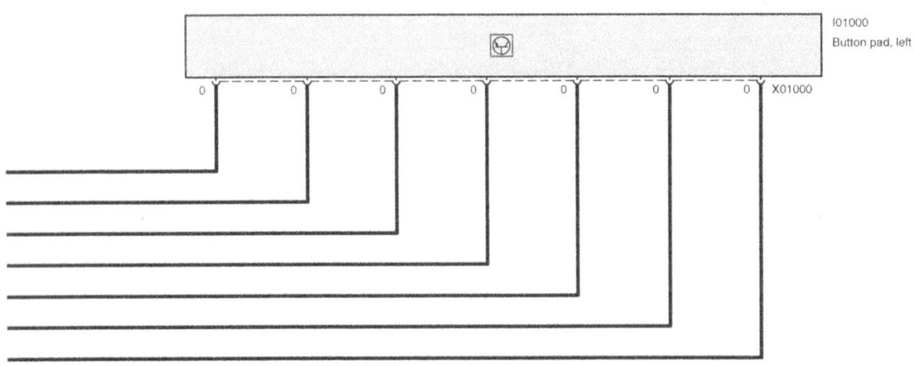

MC06ele21210b

Multifunction (MFL) steering wheel
Automatic transmission (Aisin) upshift/downshift
(page 1 of 1)

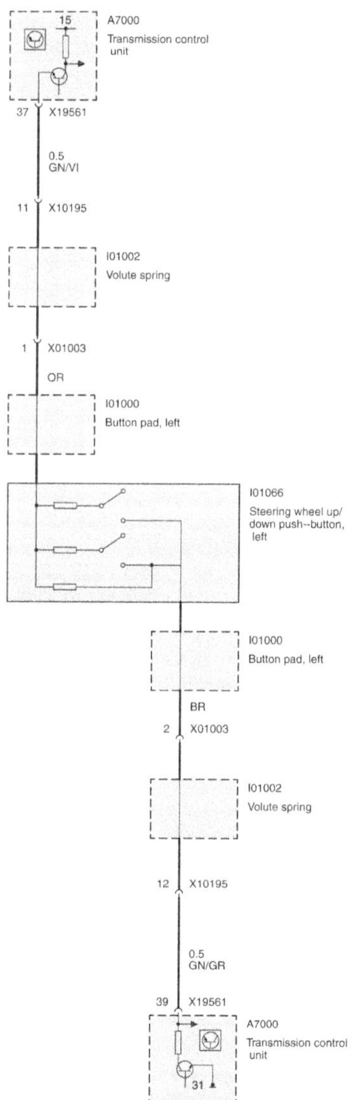

MC06ele15852

Navigation system
(from March 2002)
(page 1 of 2)

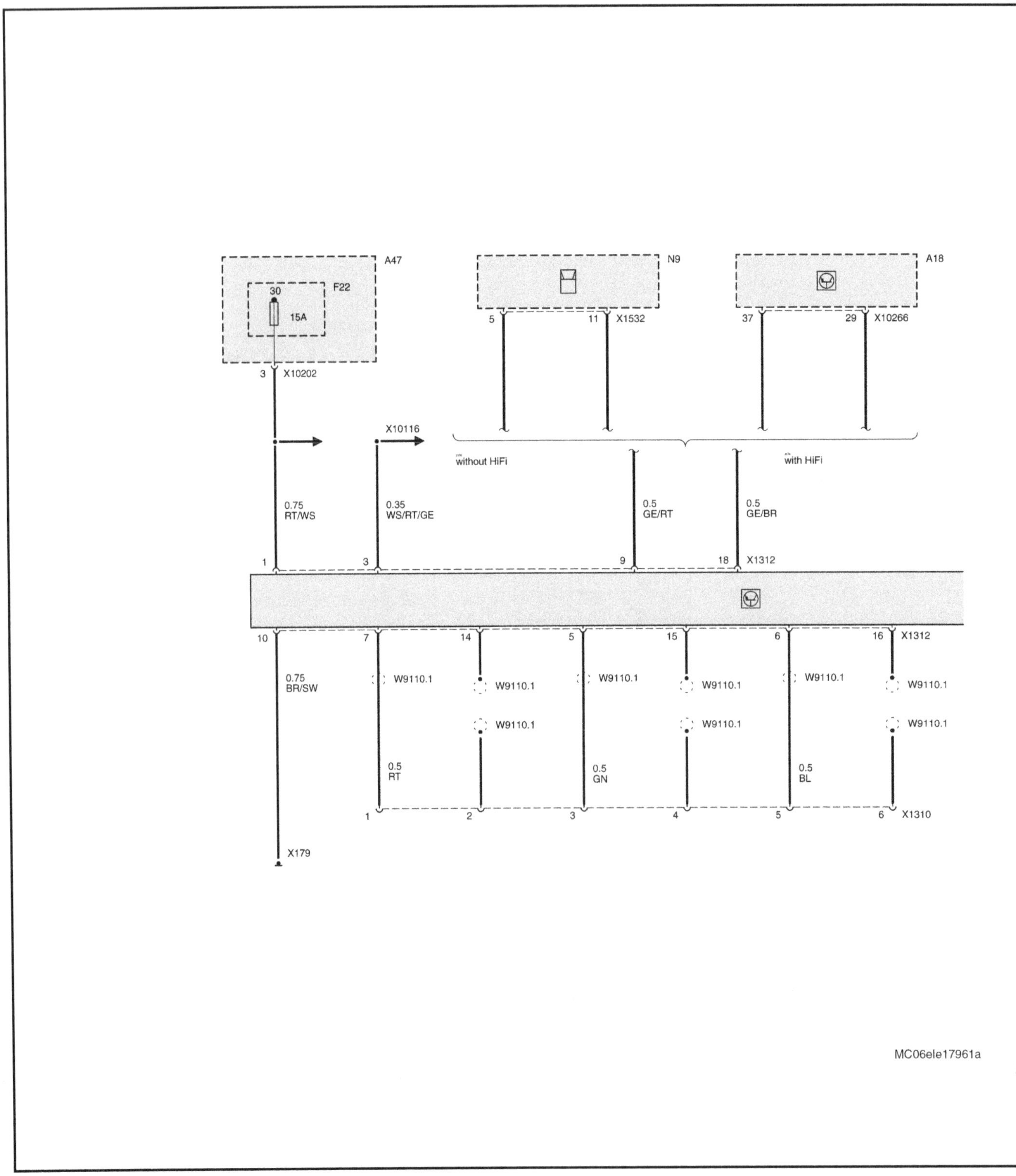

Navigation system
(from March 2002)
(page 2 of 2)

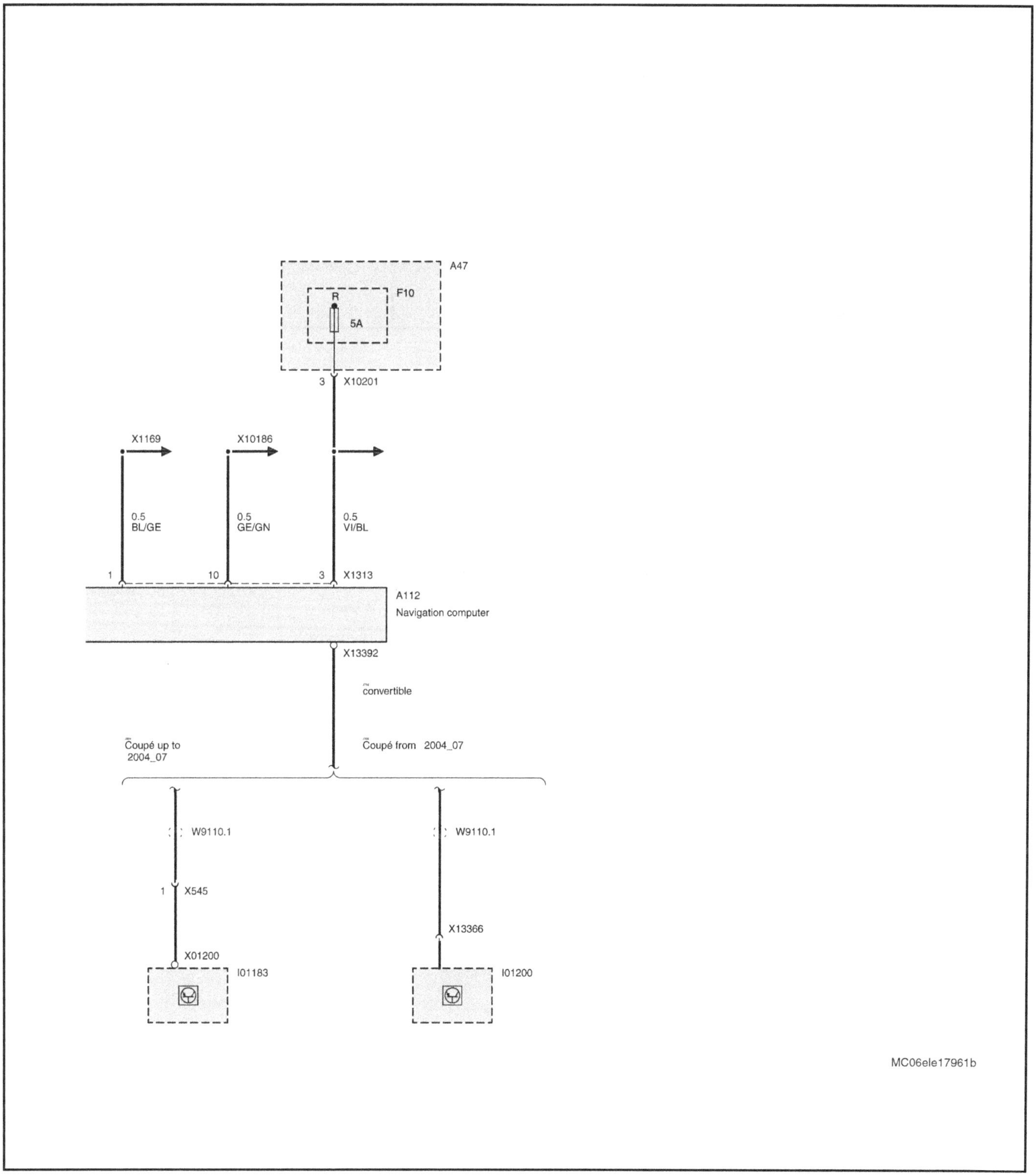

On-board computer
(page 1 of 1)

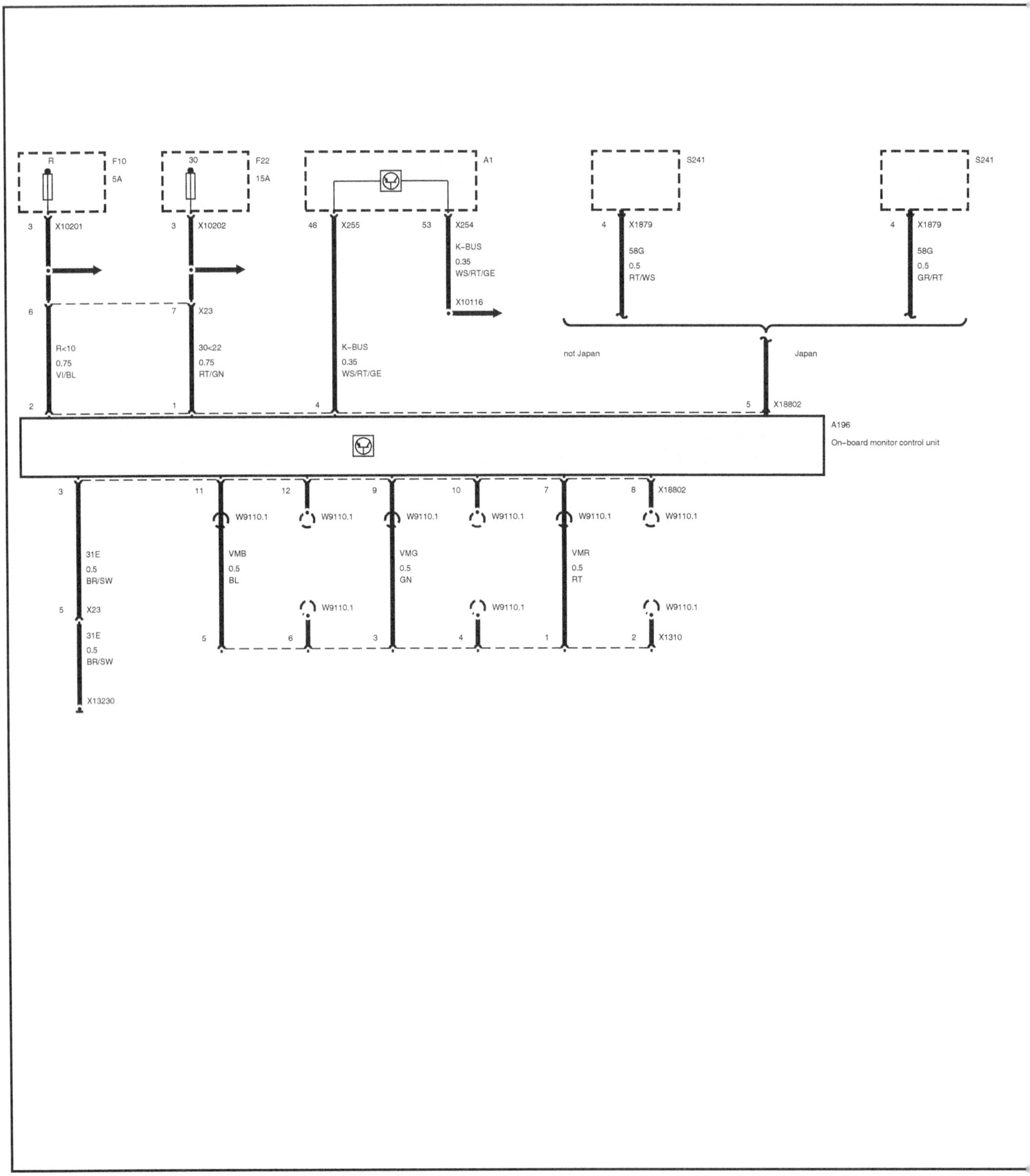

Park distance control (A81)

(page 1 of 4)

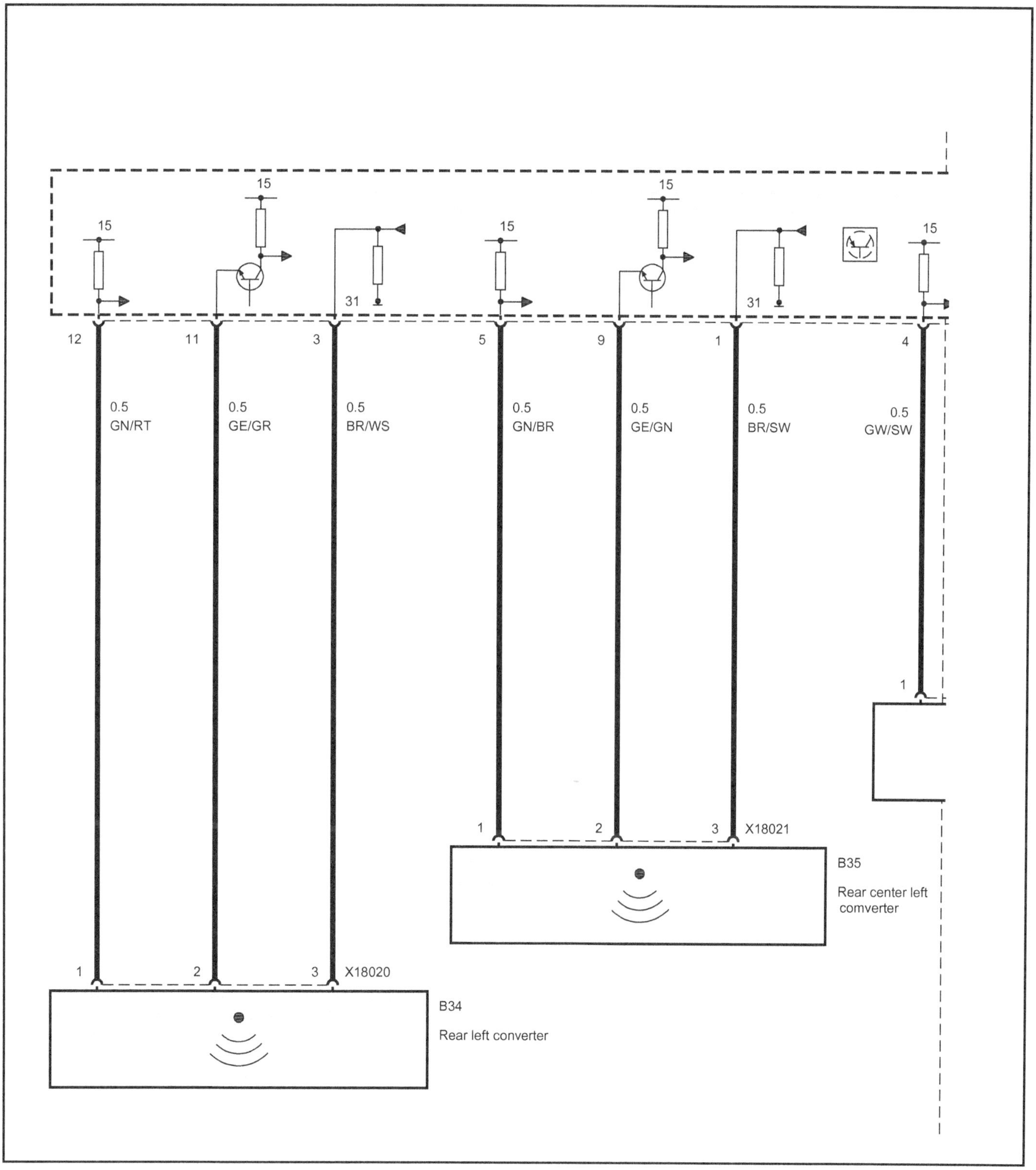

Park distance control (A81)
(page 2 of 4)

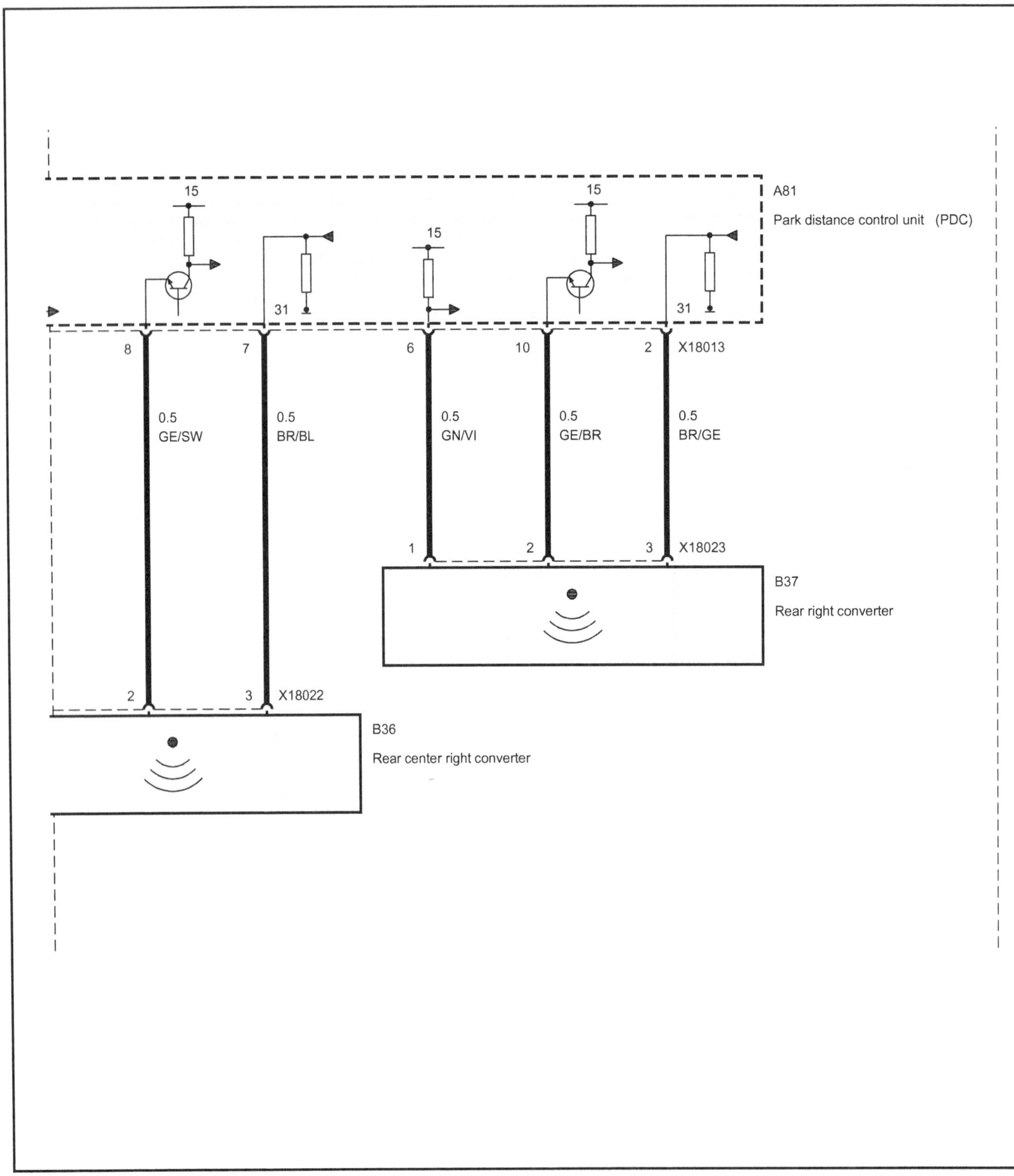

Park distance control (A81)

(page 3 of 4)

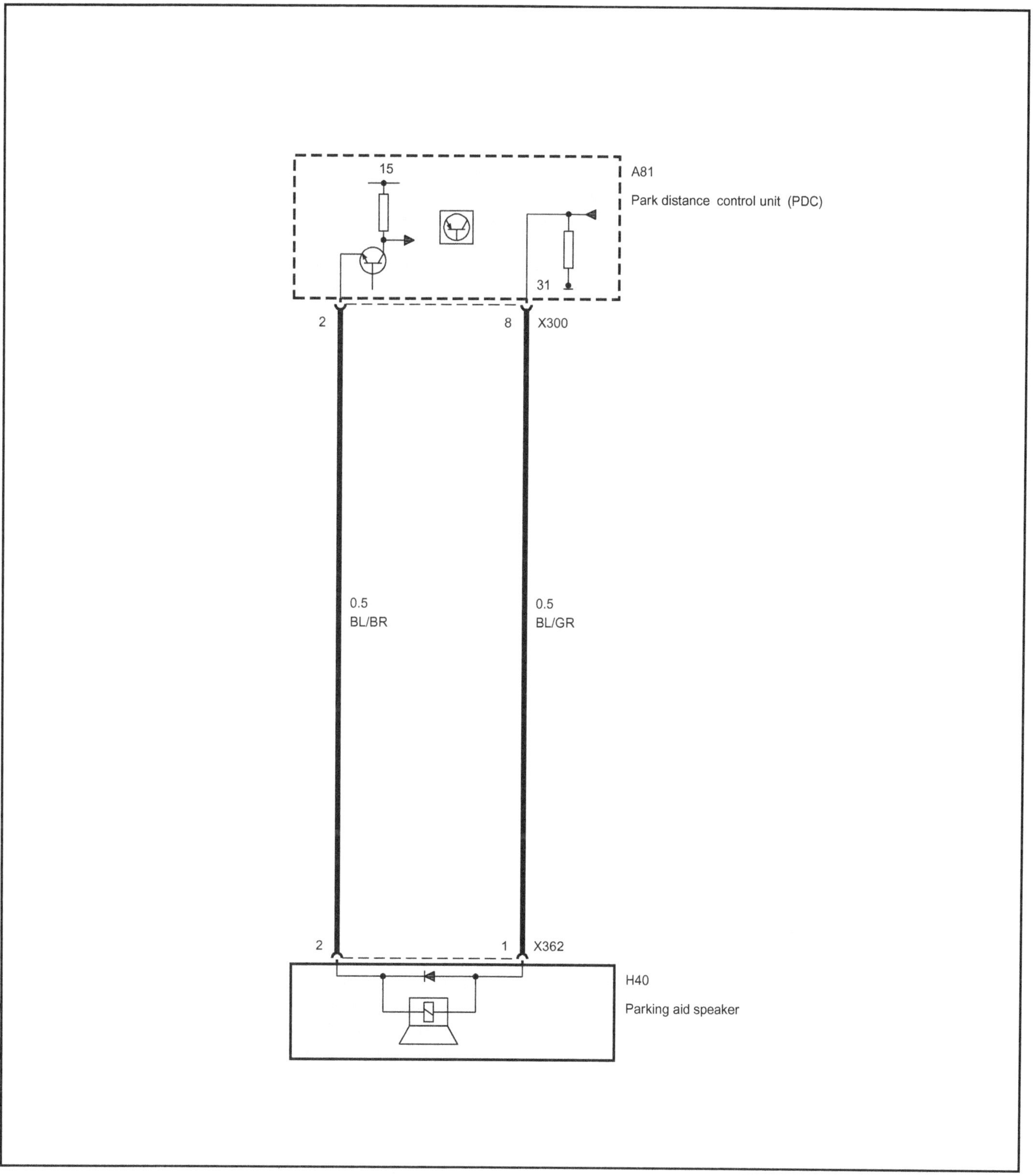

Park distance control (A81)
(page 4 of 4)

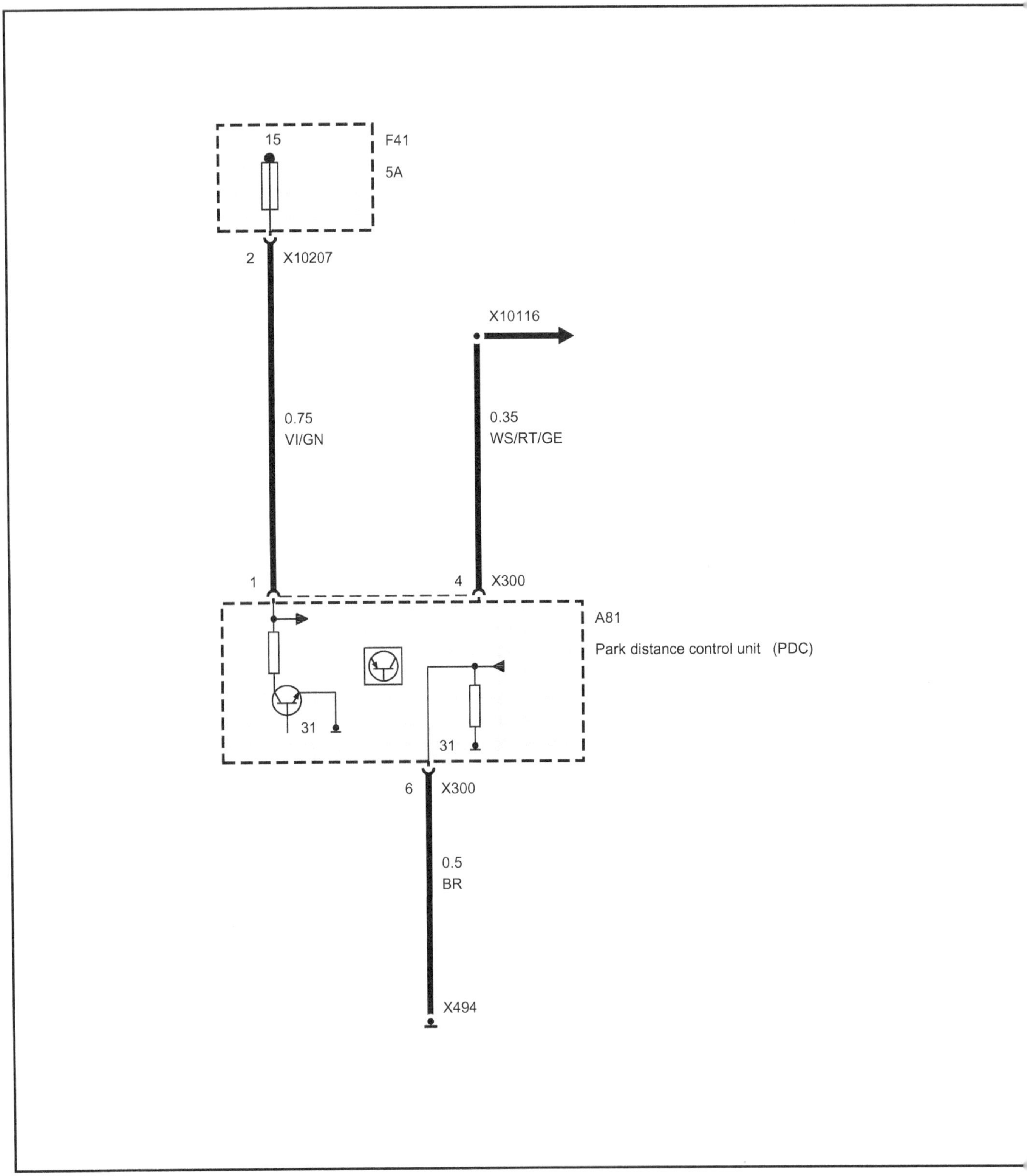

Power steering
Electrohydraulic control (EHPS)
(page 1 of 1)

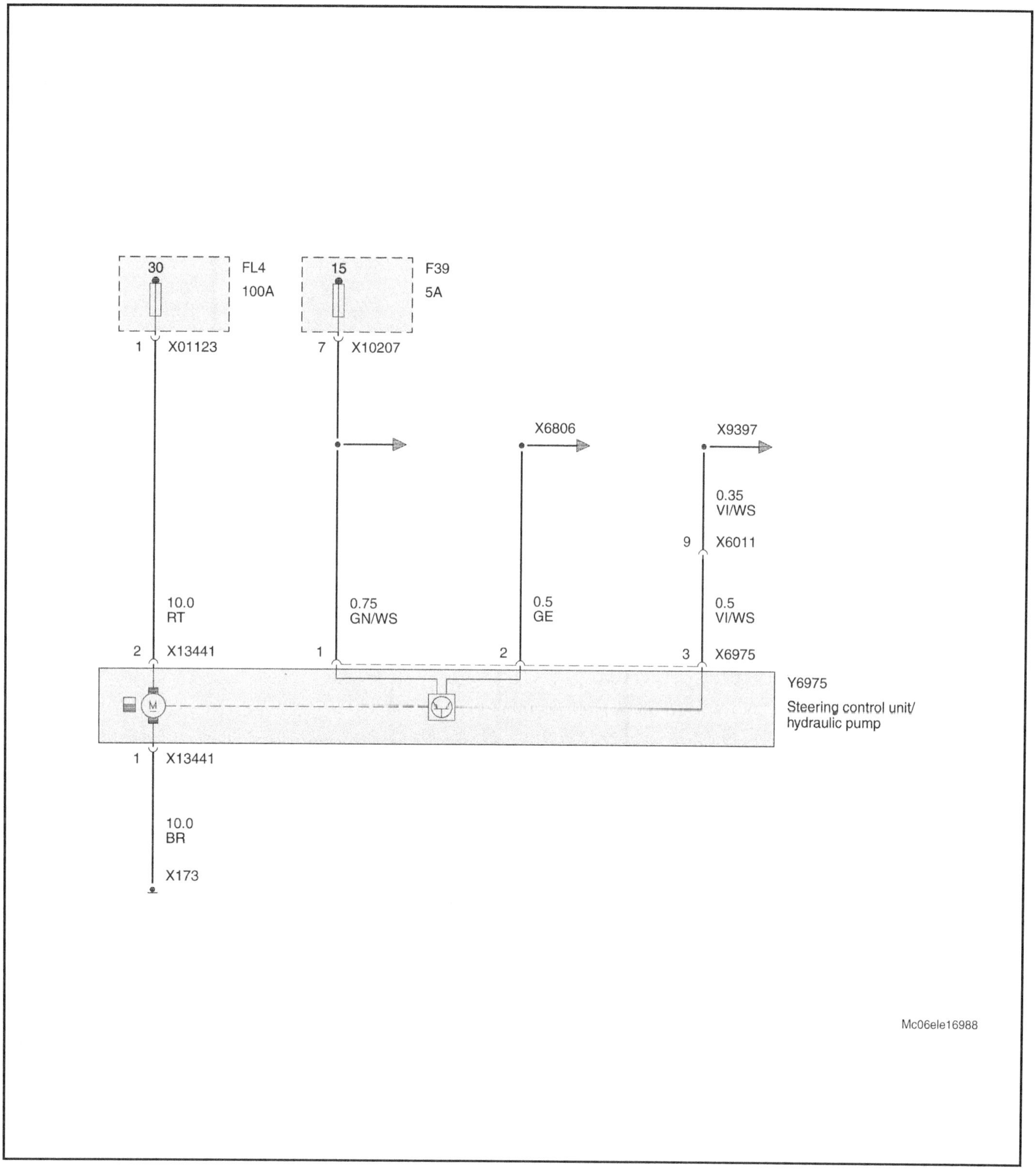

Power windows
Coupe
(page 1 of 2)

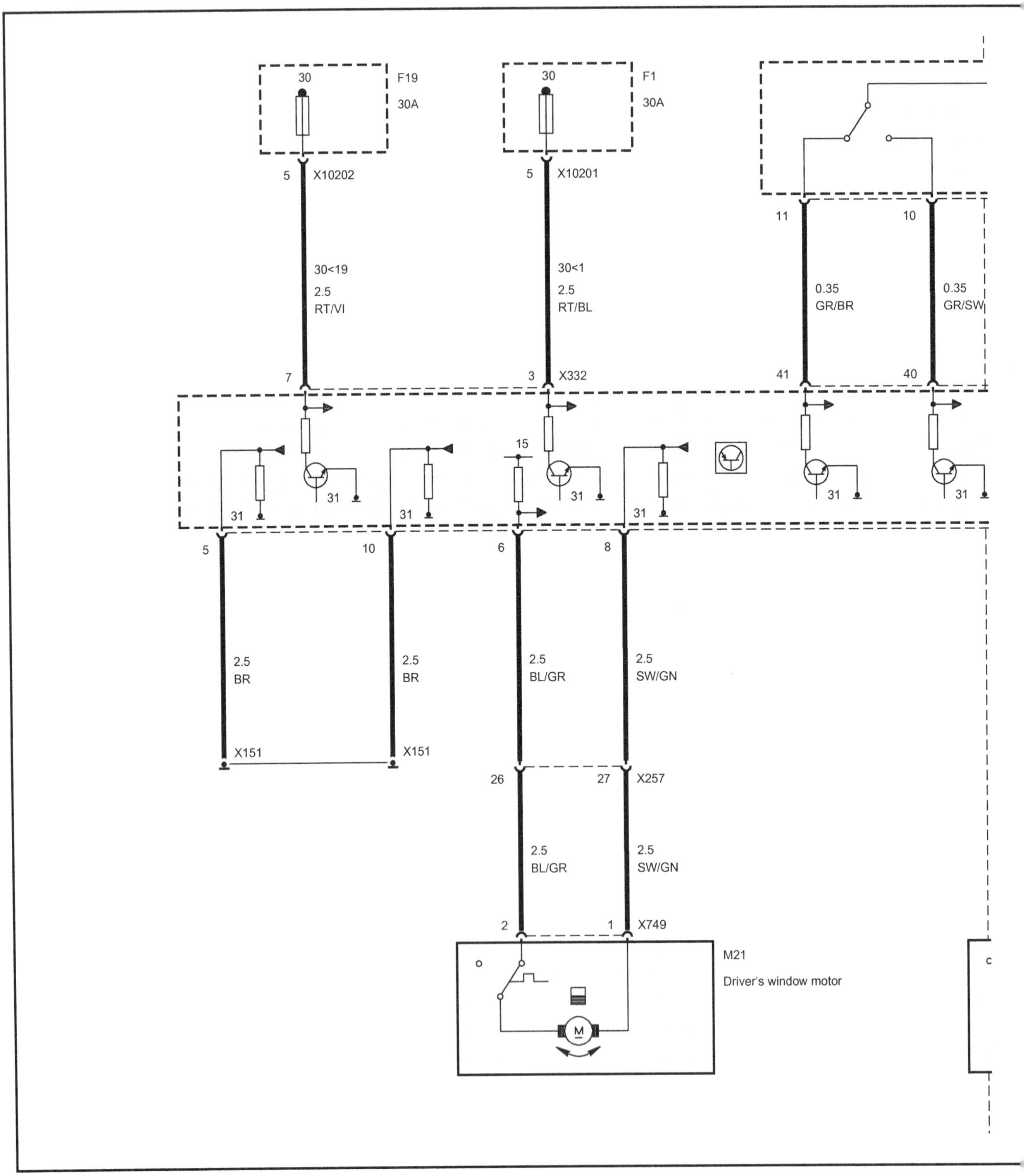

Power windows
Coupe
(page 2 of 2)

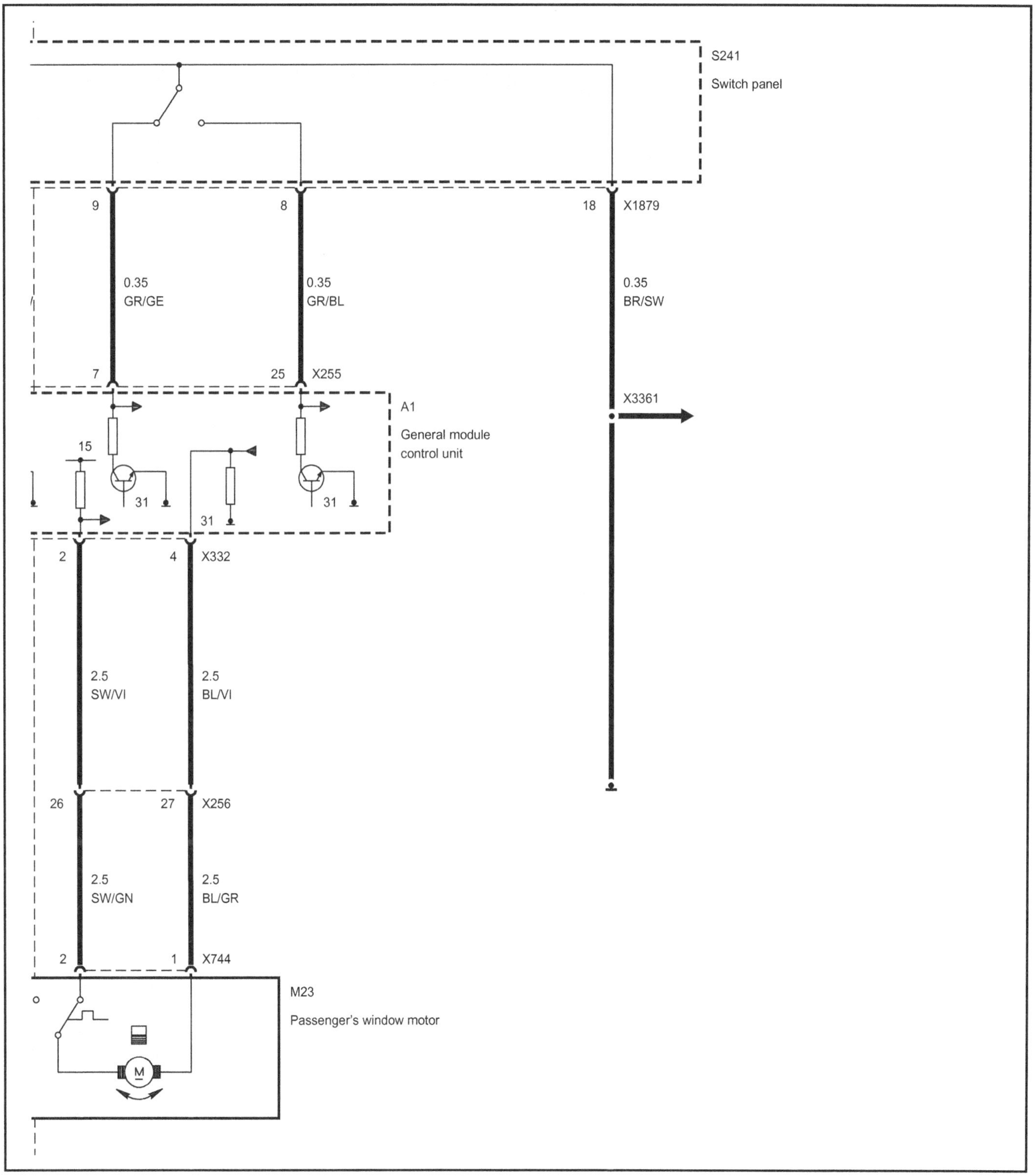

Power windows
Convertible
(page 1 of 2)

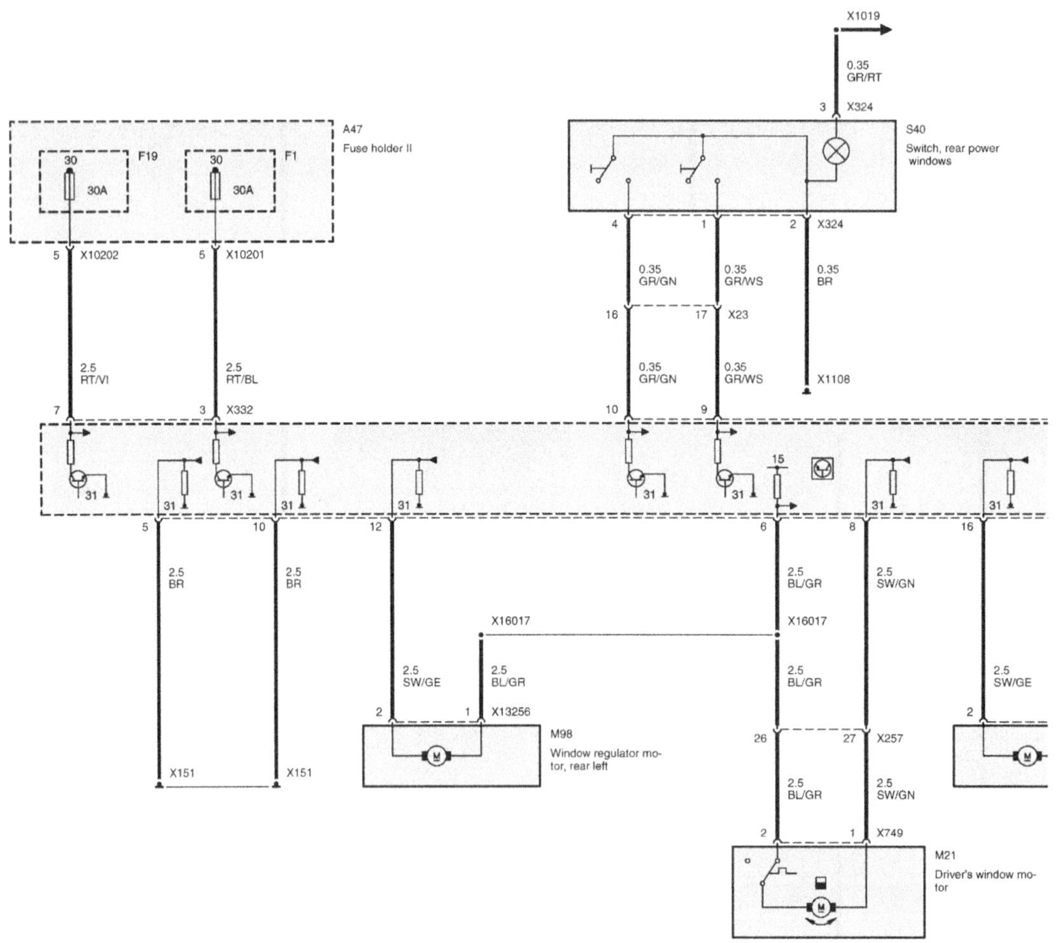

MC06ele21301a

Power windows
Convertible
(page 2 of 2)

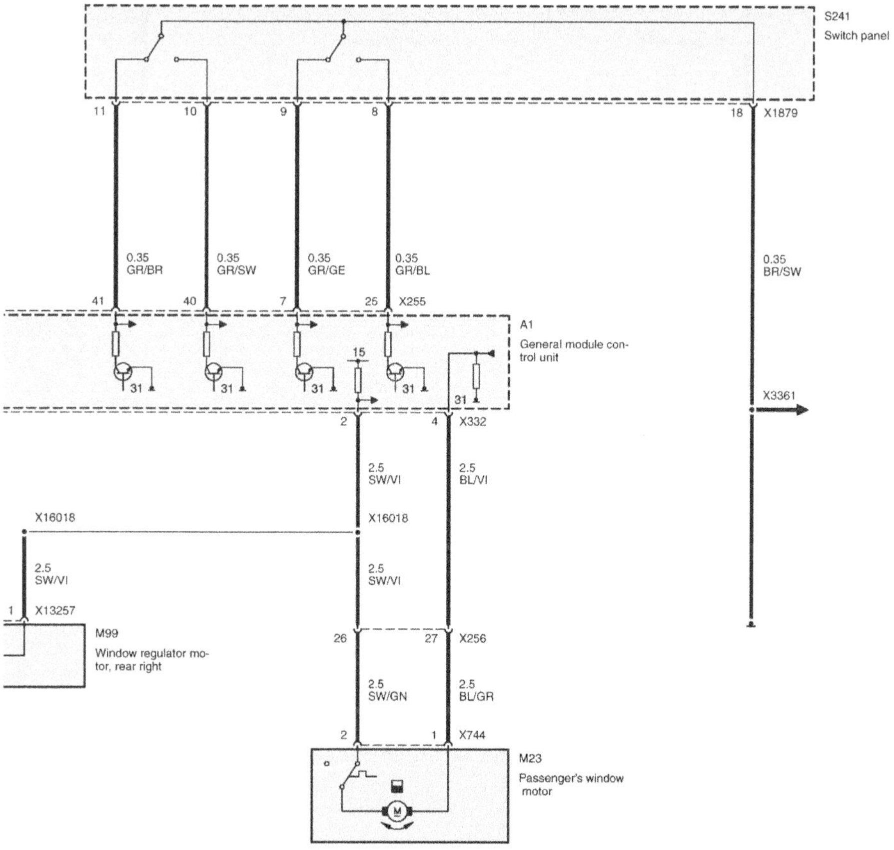

MC06ele21301b

Radio / CD changer
Stereo (up to Jan. 2005)
(page 1 of 3)

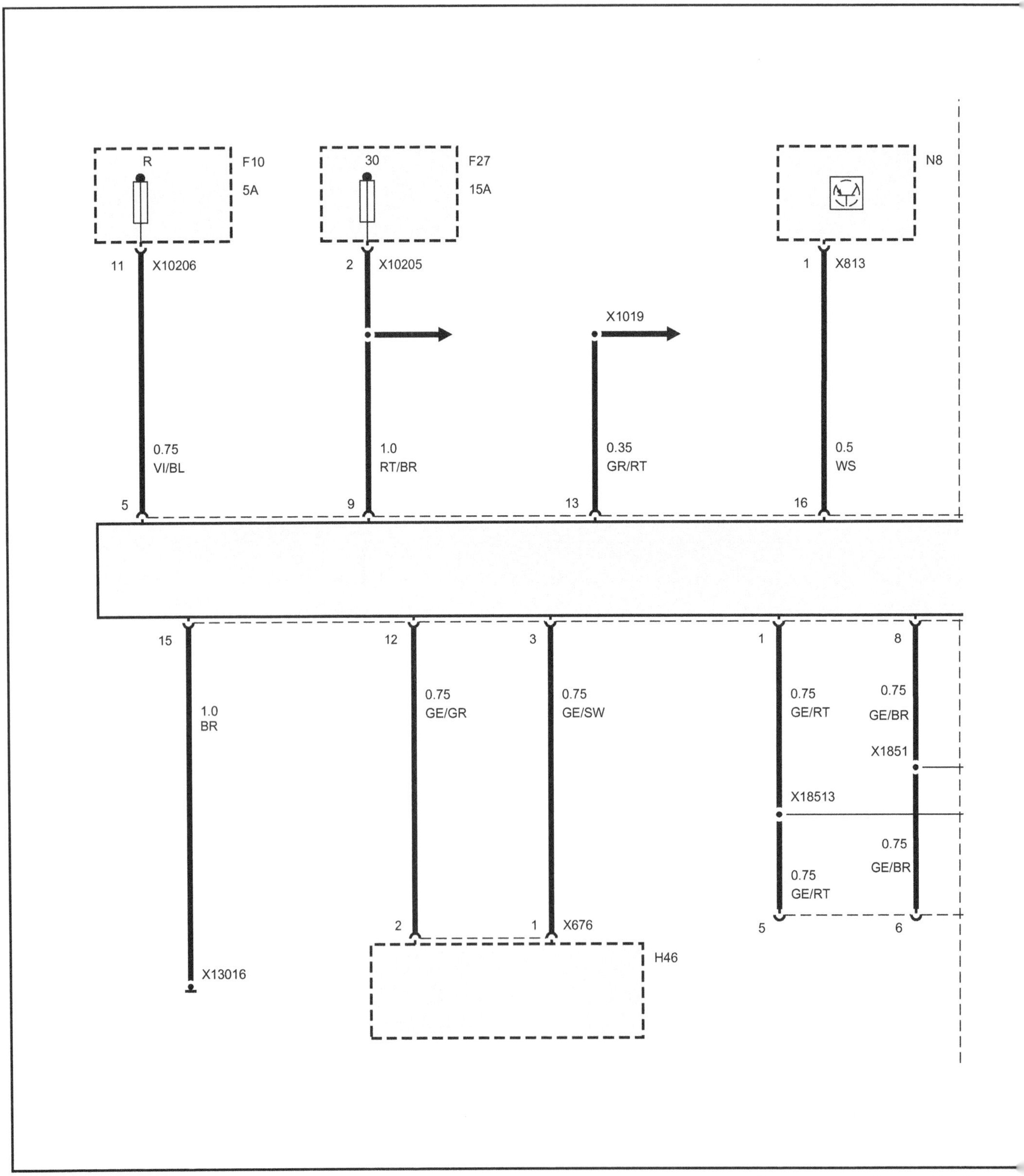

Radio / CD changer
Stereo (up to Jan. 2005)
(page 2 of 3)

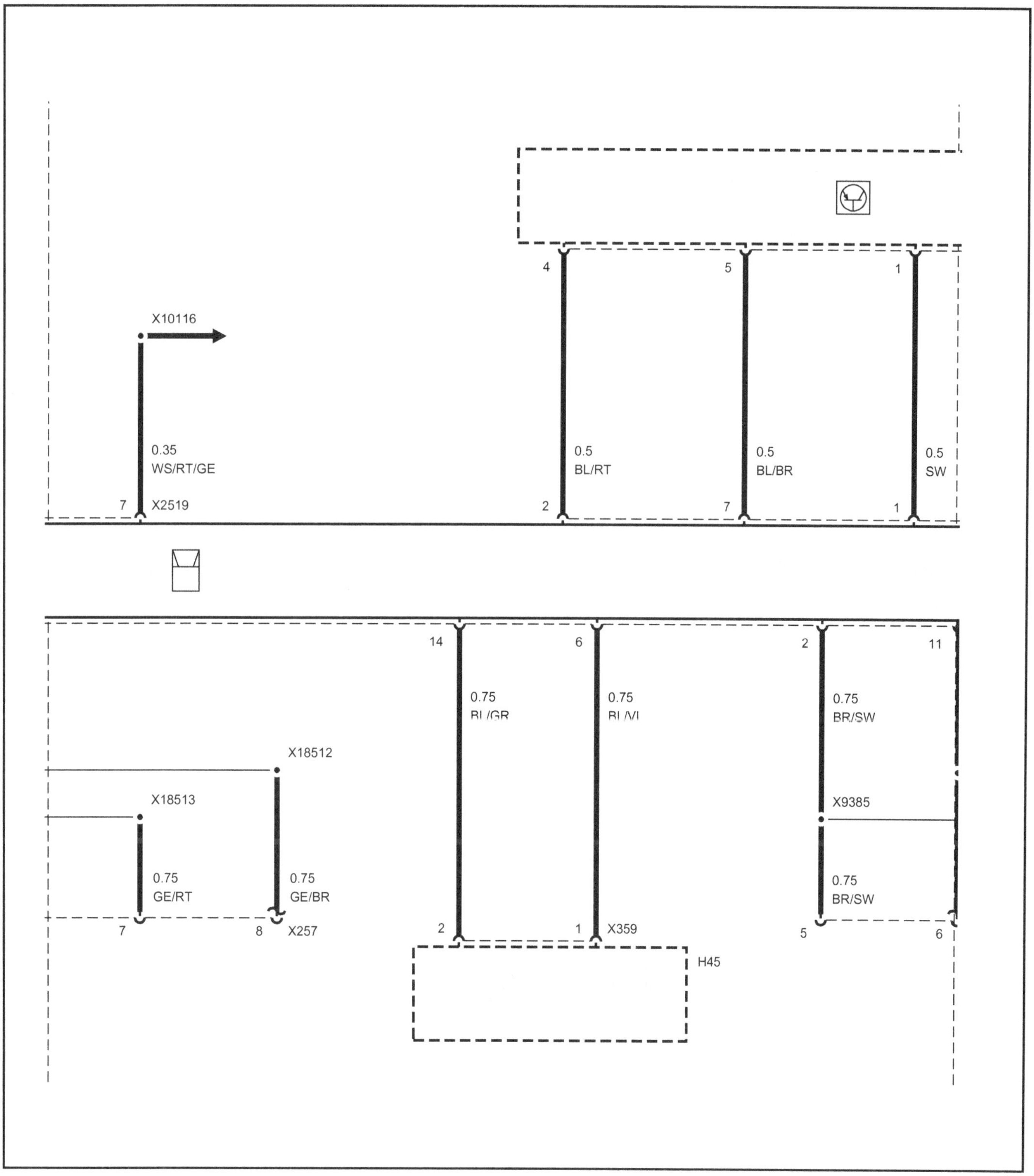

Radio / CD changer
Stereo (up to Jan. 2005)
(page 1 of 3)

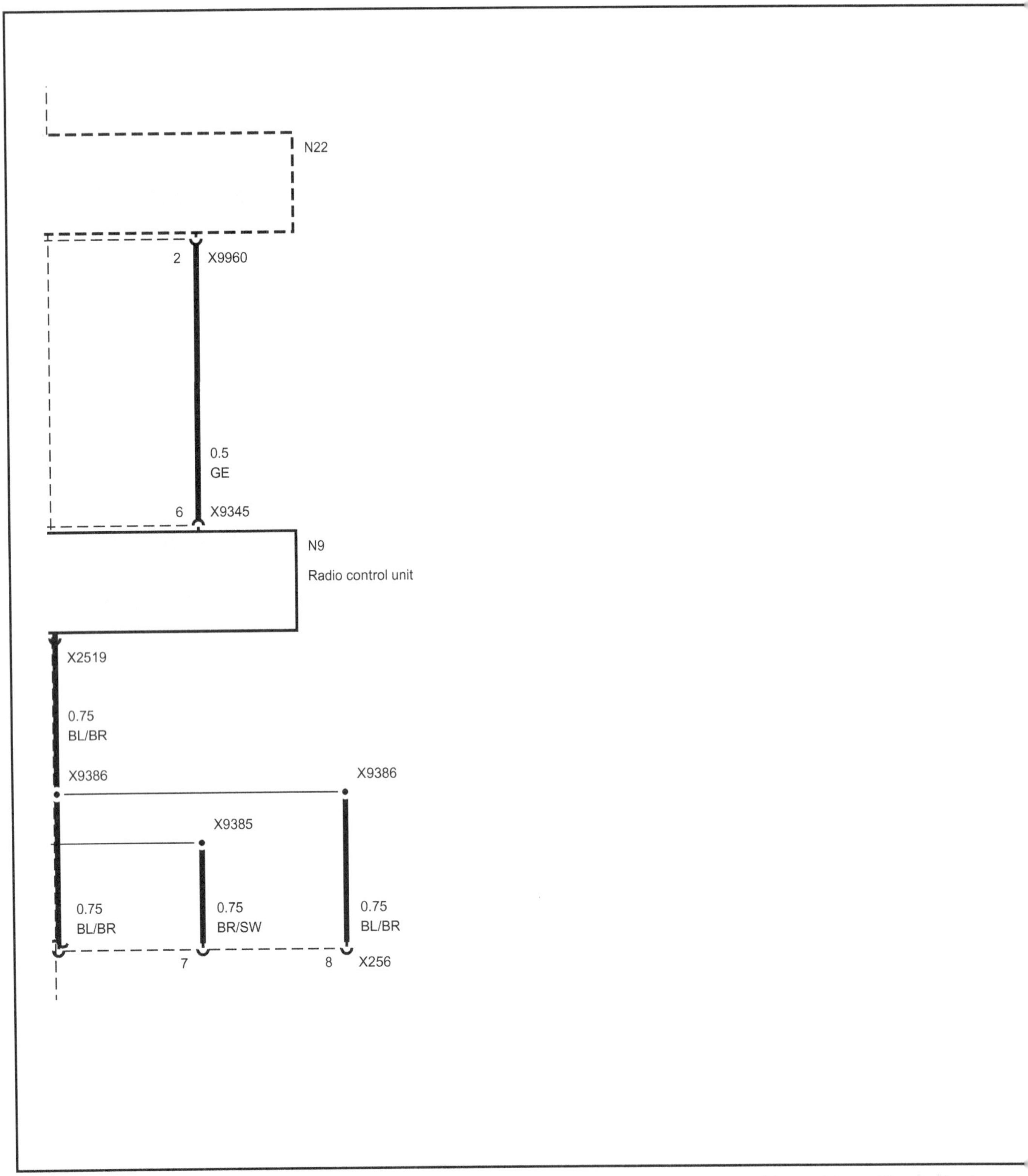

Radio / CD changer
CD changer
(page 1 of 1)

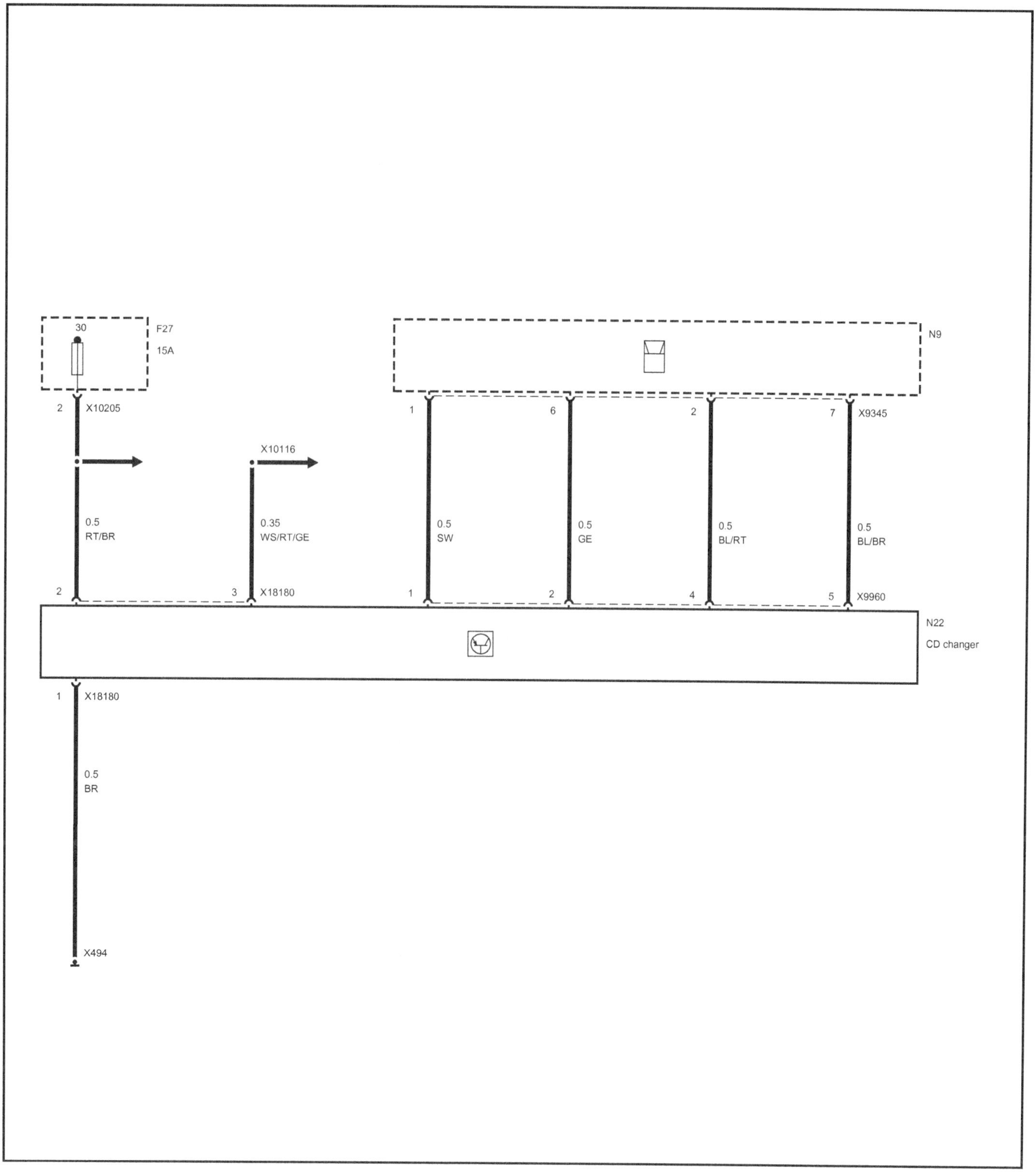

Radio / CD changer

Stereo (from Jan. 2005)

(page 1 of 3)

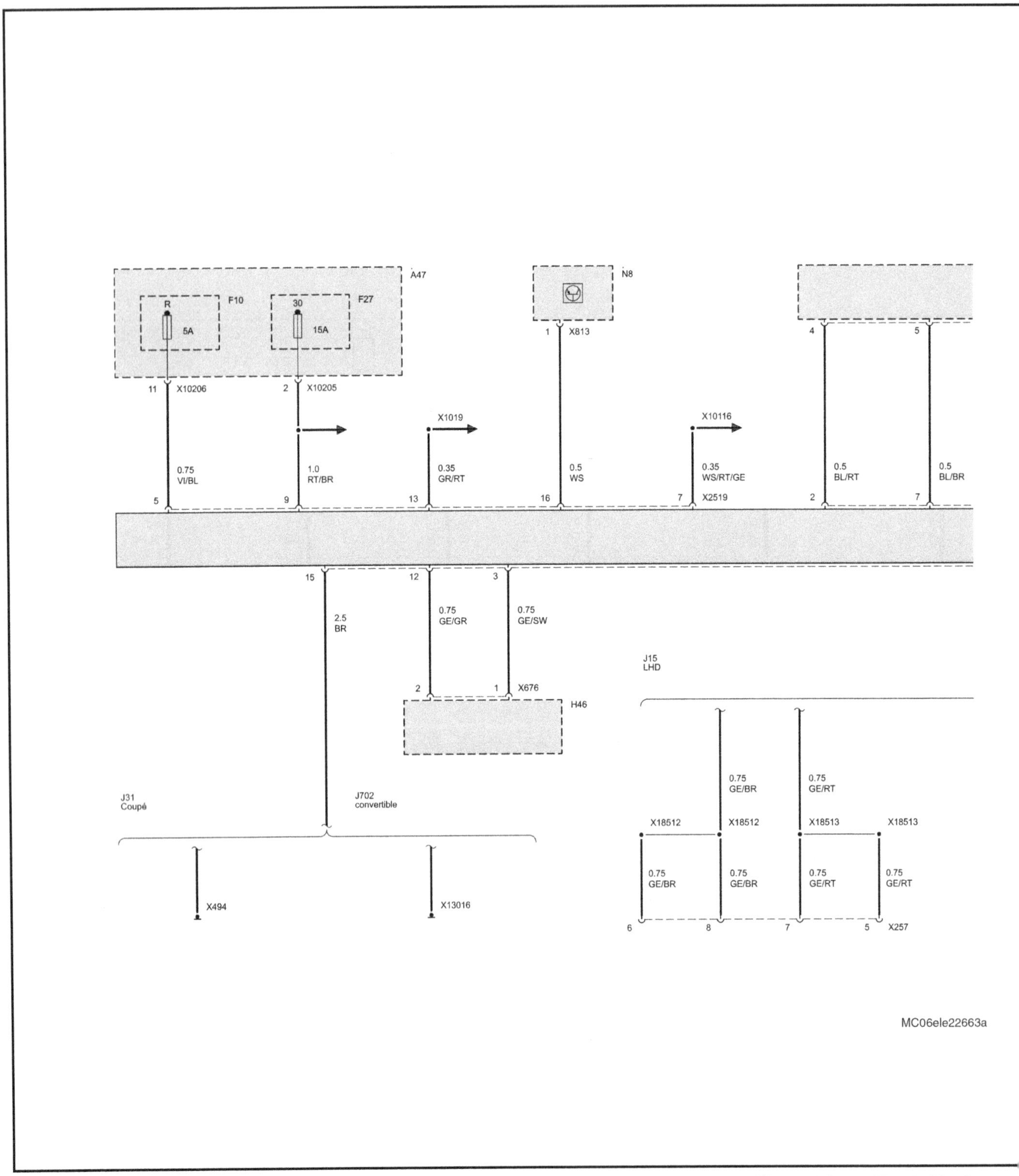

Radio / CD changer

Stereo (from Jan. 2005)

(page 2 of 3)

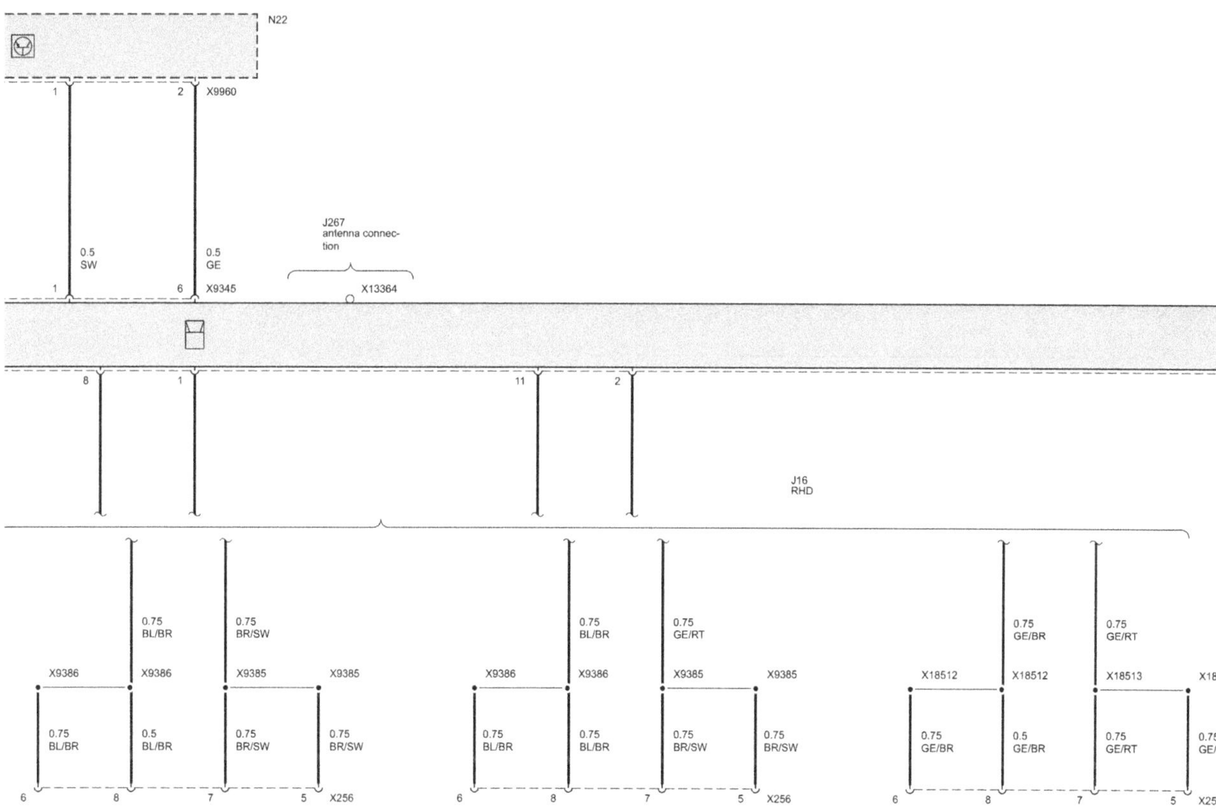

MC06ele22663b

Radio / CD changer
Stereo (from Jan. 2005)
(page 3 of 3)

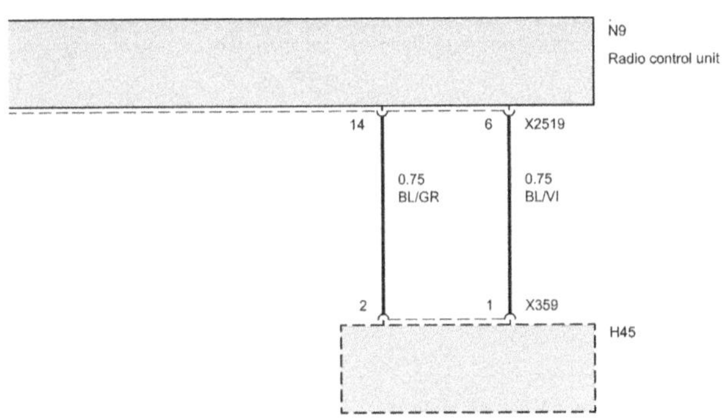

MC06ele22663c

Radio / CD changer

Stereo with digital sound processor (DSP) (from Jan. 2005)

(page 1 of 2)

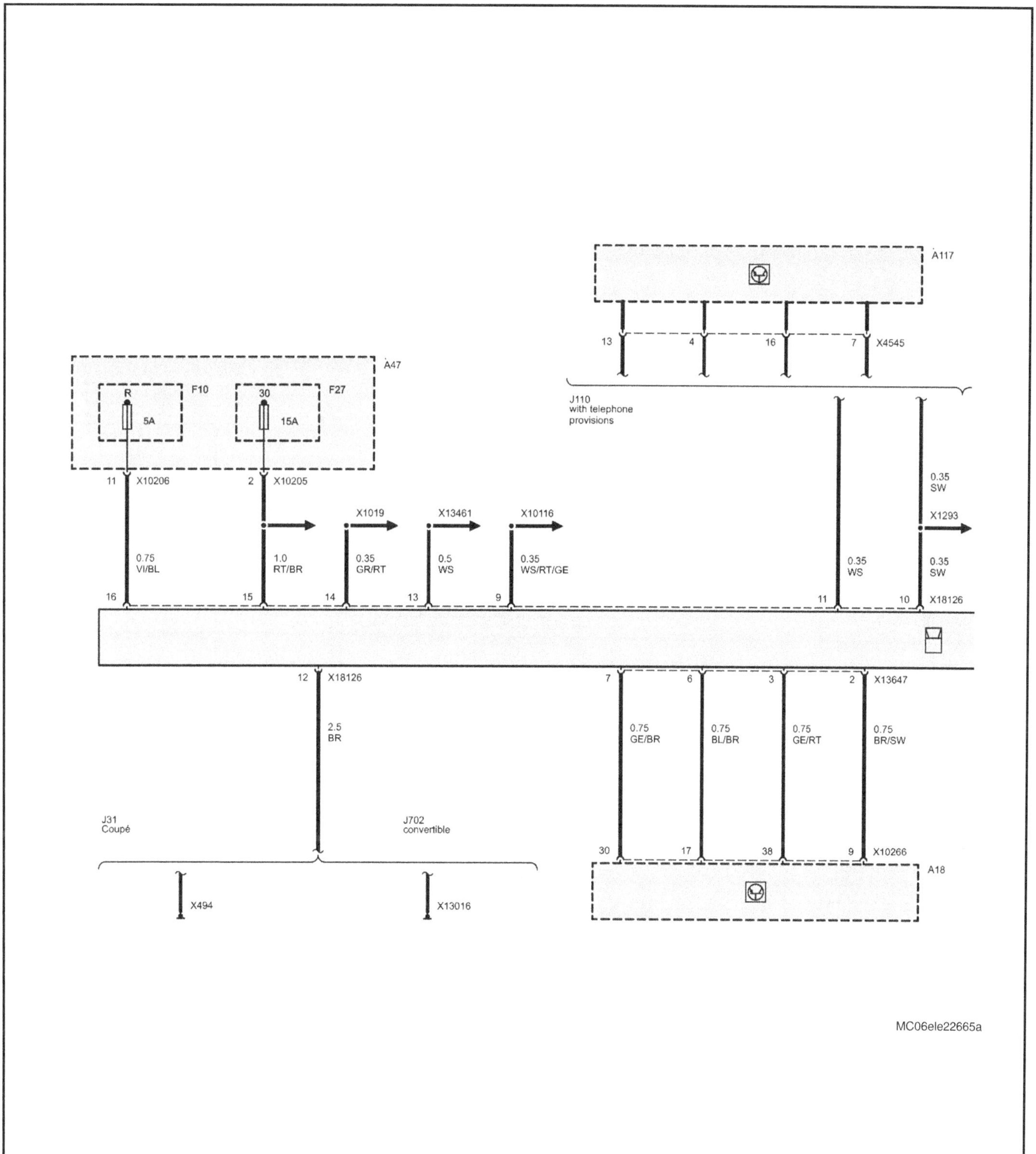

Radio / CD changer

Stereo with digital sound processor (DSP) (from Jan. 2005)

(page 2 of 2)

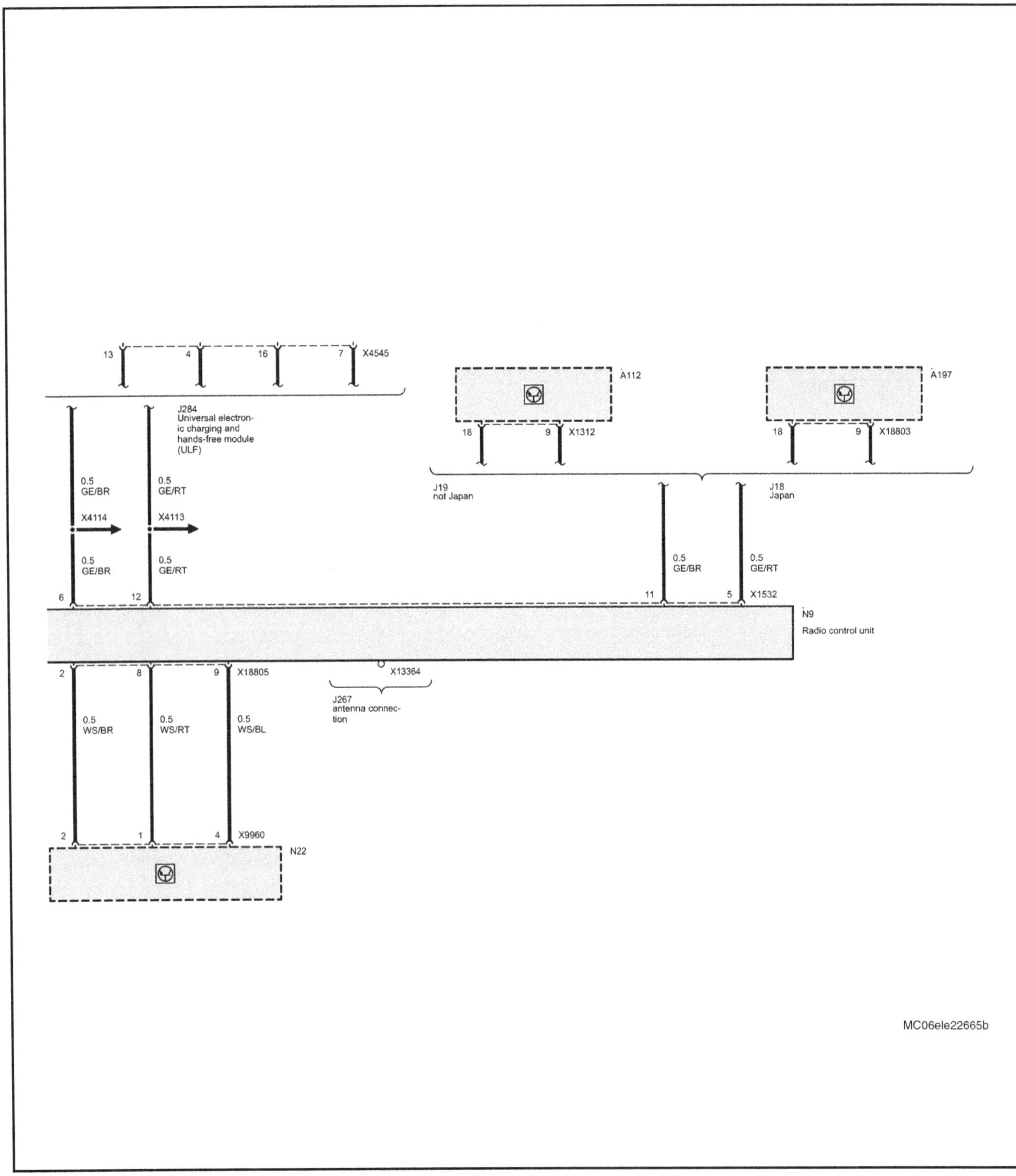

Rear window defogger
With IHKA (up to Sept. 2005)
(page 1 of 2)

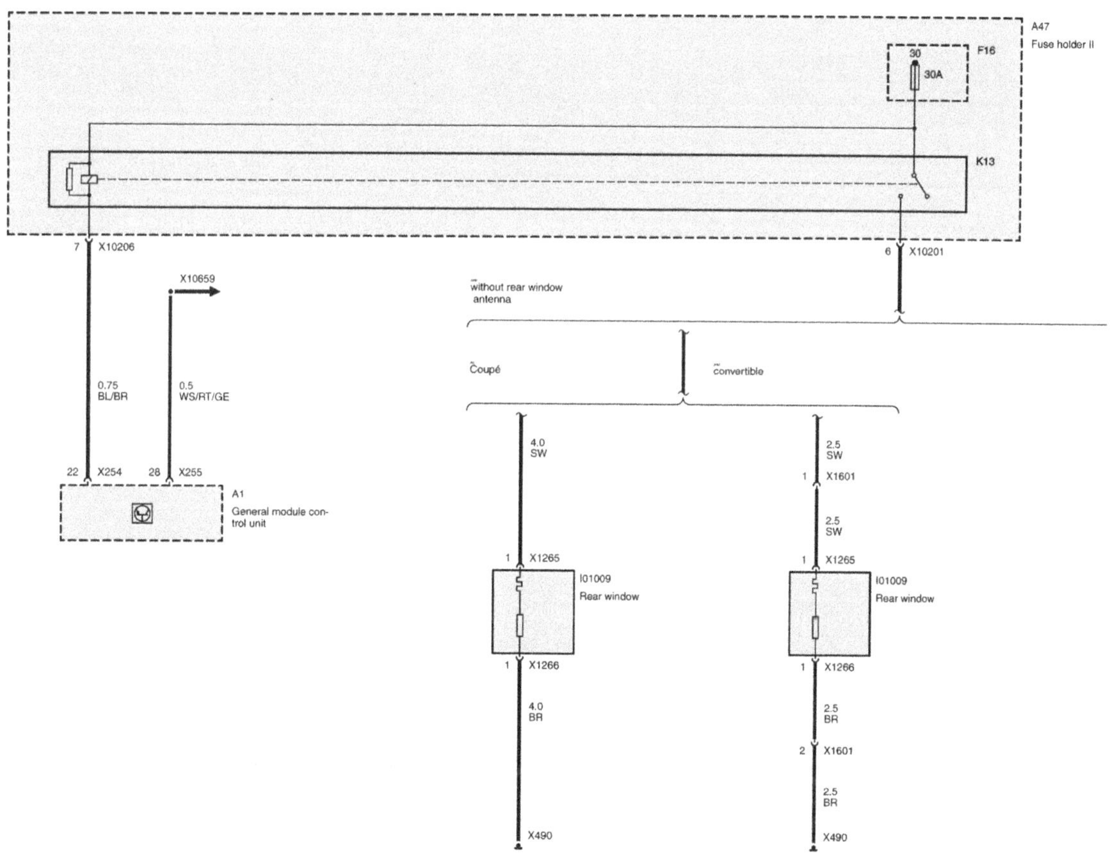

MC06ele15972a

Rear window defogger
With IHKA (up to Sept. 2005)
(page 2 of 2)

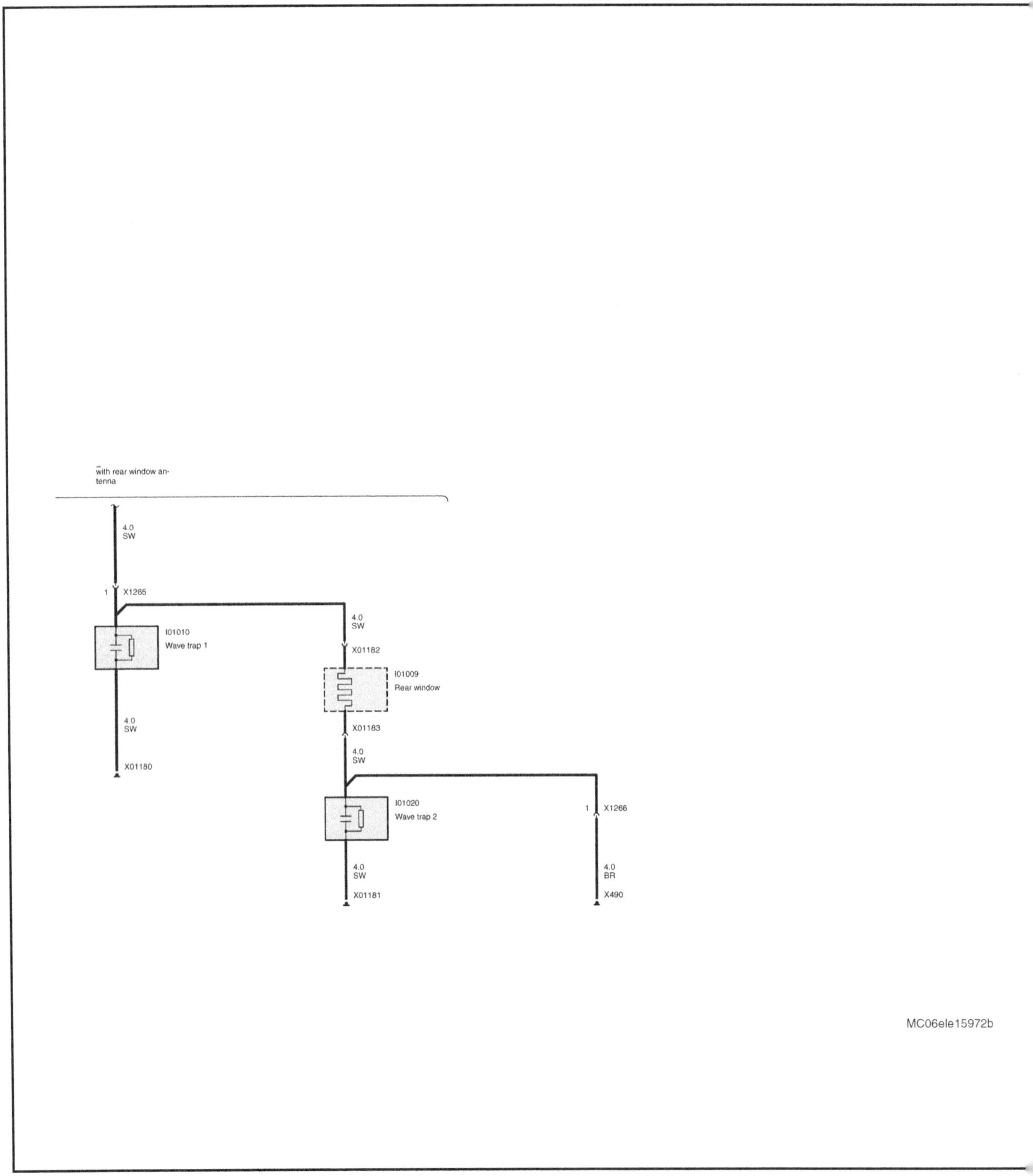

Rear window defogger
Without IHKA (up to Sept. 2005)
(page 1 of 2)

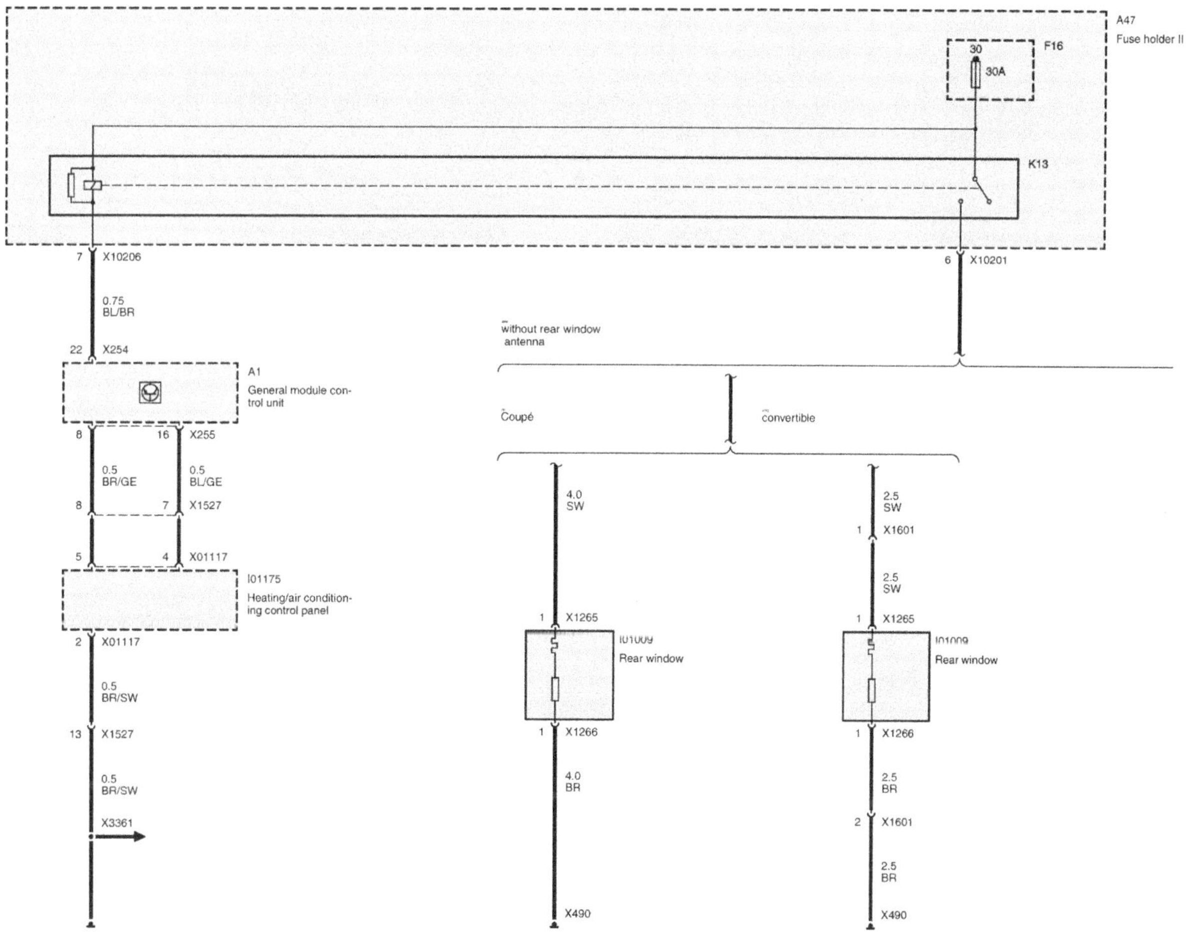

Rear window defogger
Without IHKA (up to Sept. 2005)
(page 2 of 2)

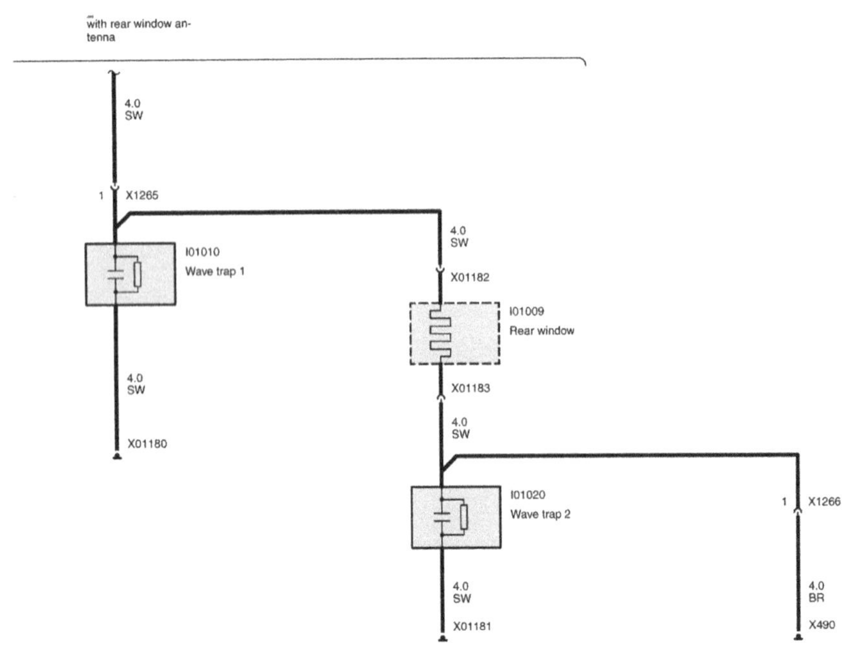

MC06ele15971b

Starter (M6510)
(page 1 of 1)

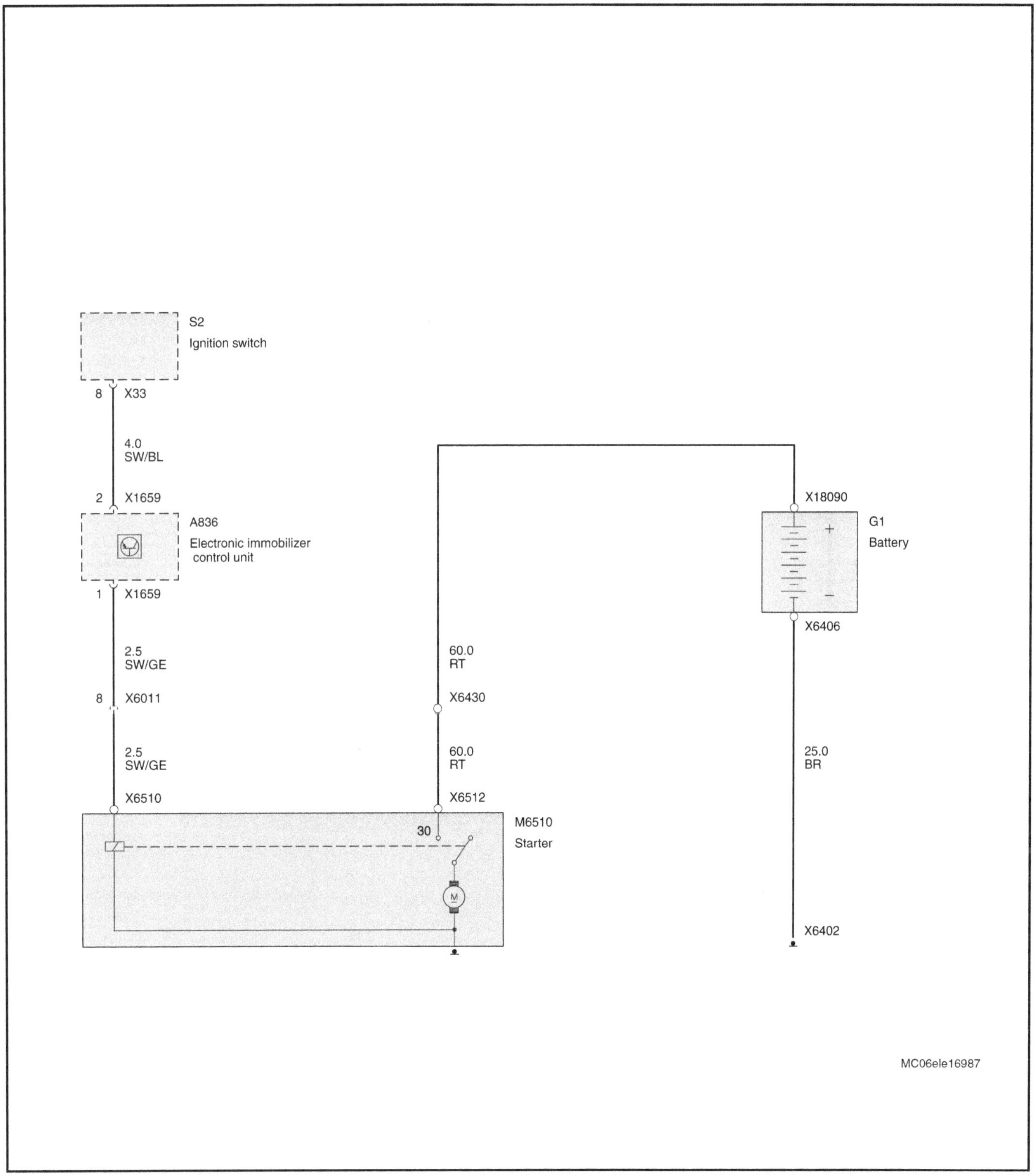

Sunroof
(page 1 of 1)

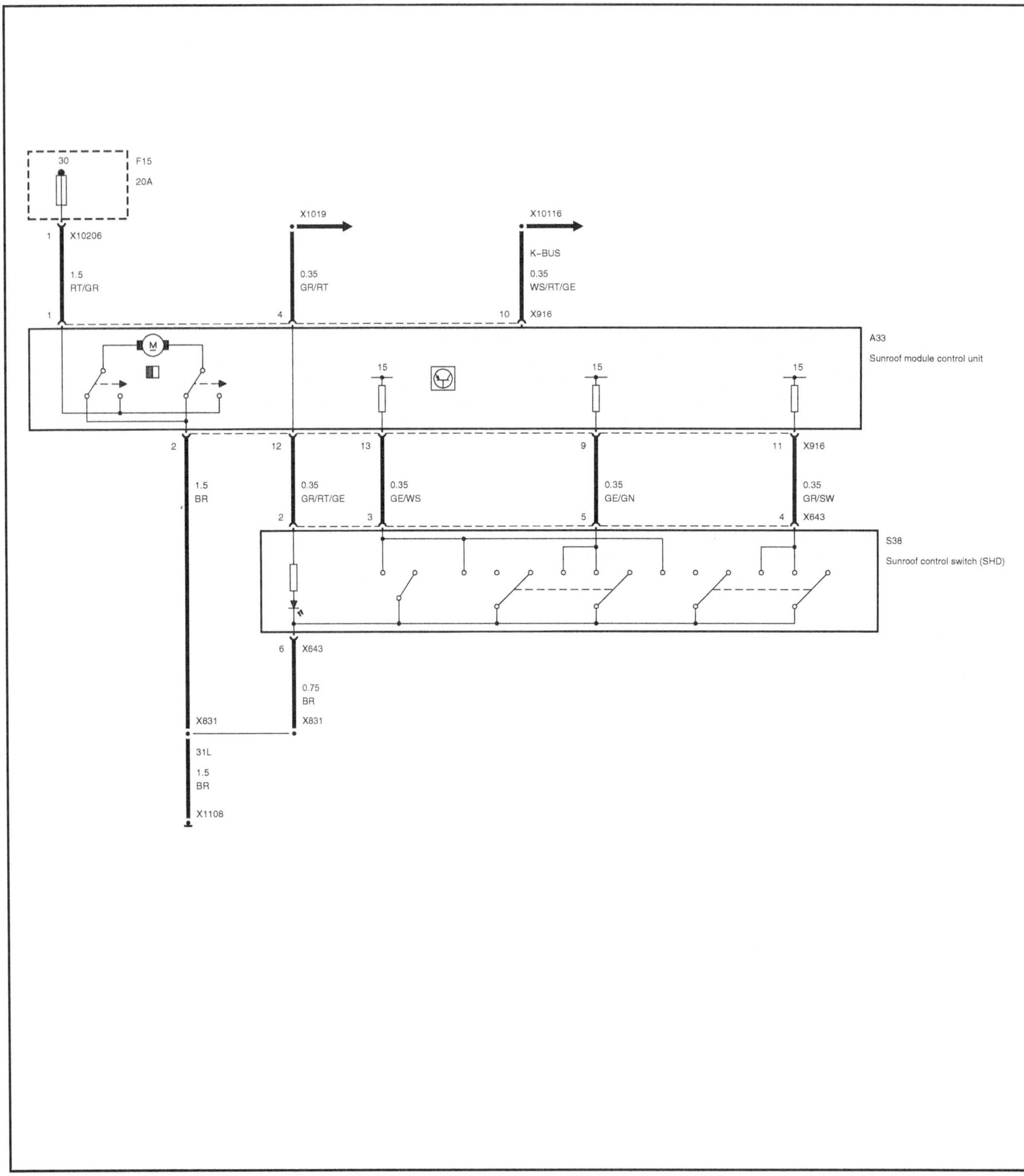

Wipers / washers

Windshield wipers, coupe (March 2002 to Sept. 2005)

(page 1 of 4)

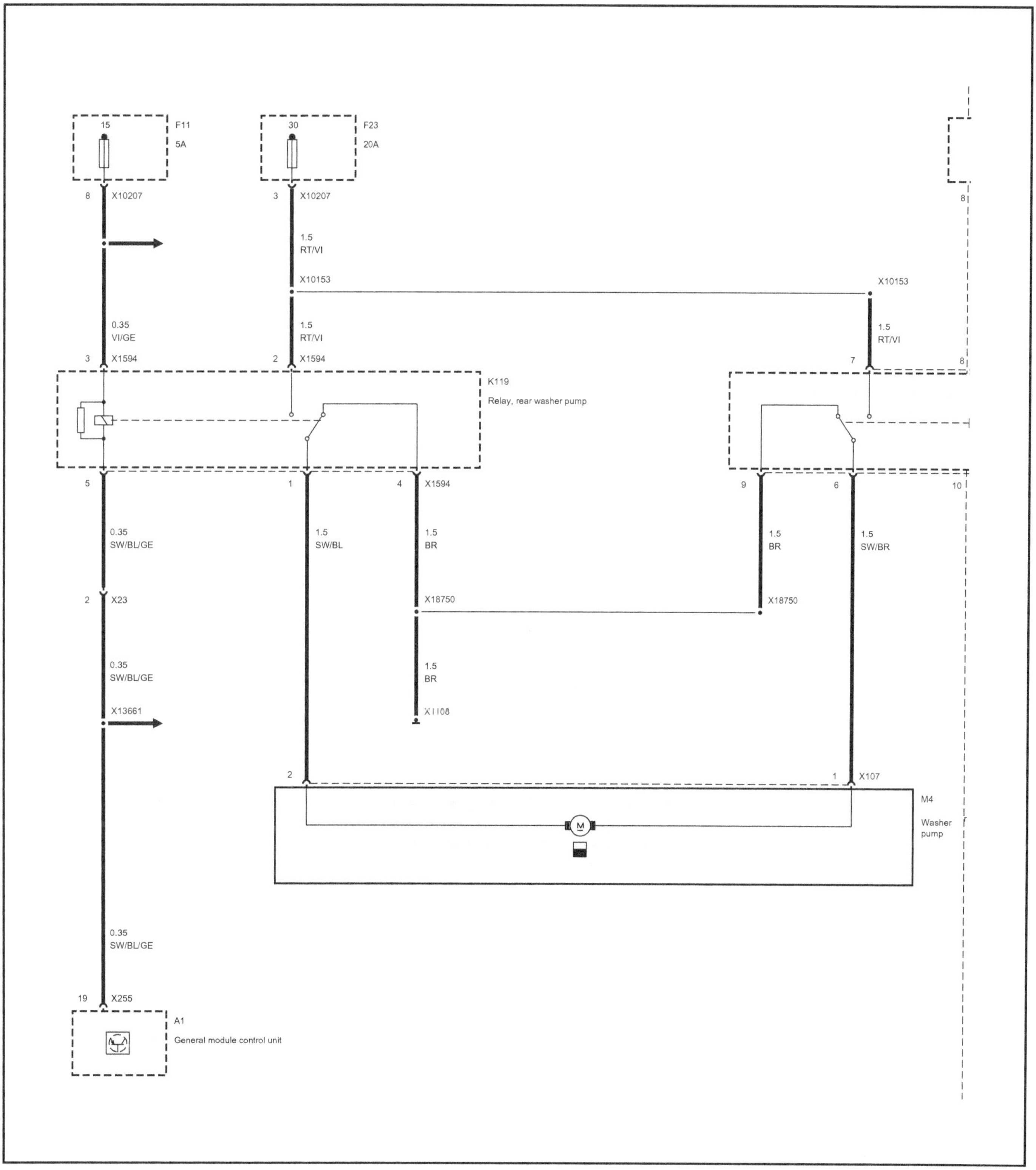

Wipers / washers

Windshield wipers, coupe (March 2002 to Sept. 2005)

(page 2 of 4)

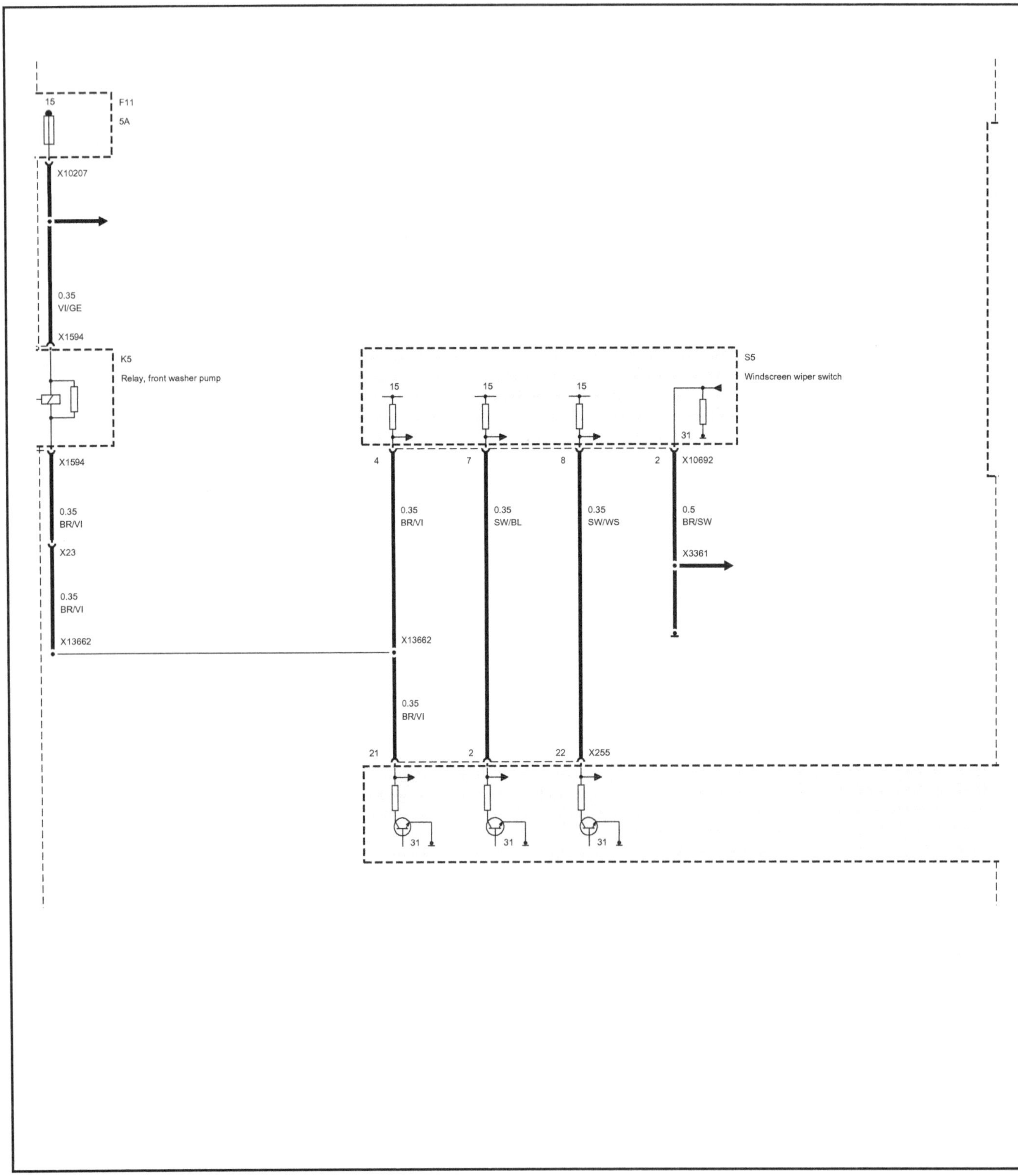

Wipers / washers

Windshield wipers, coupe (March 2002 to Sept. 2005)

(page 3 of 4)

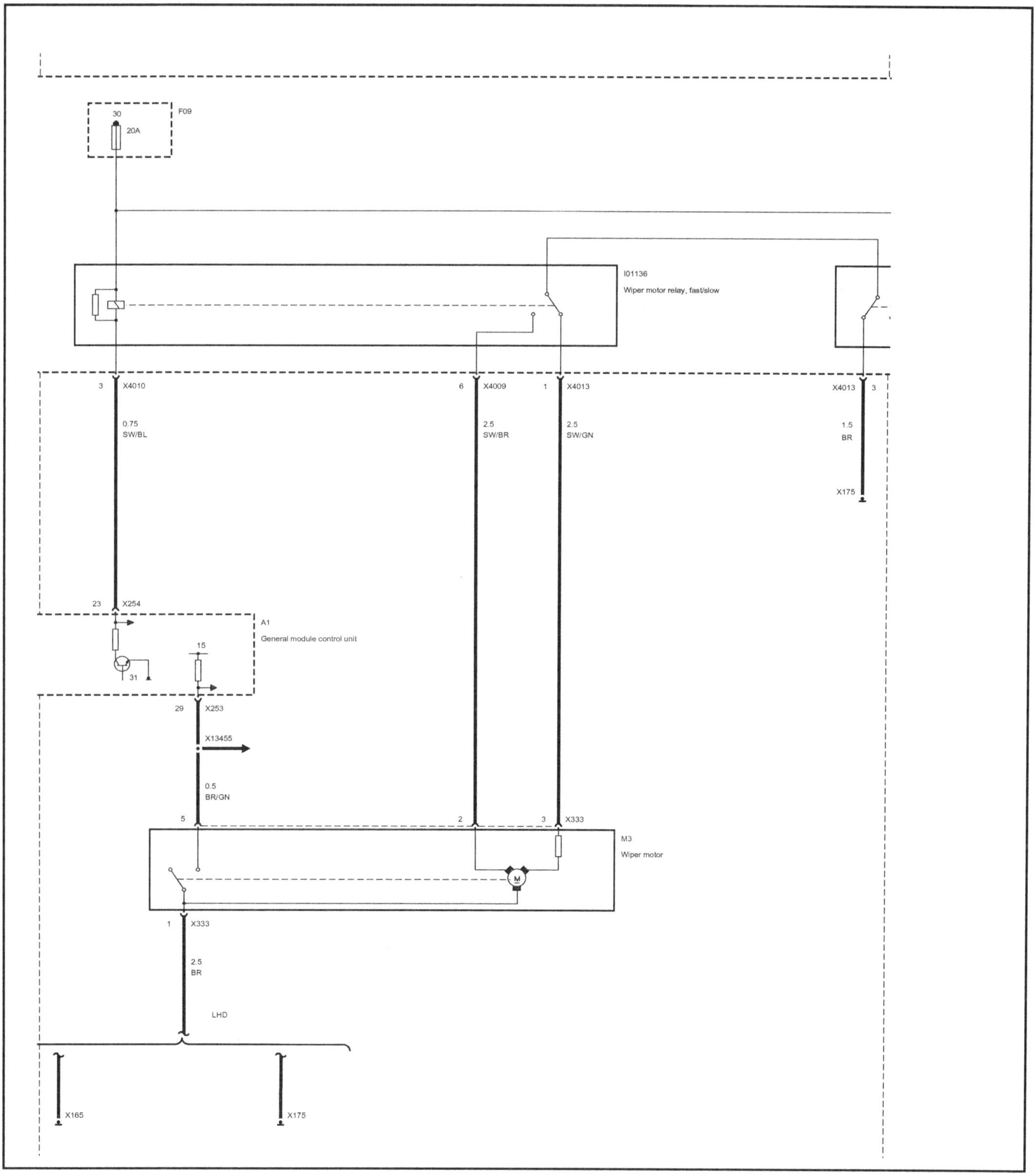

Wipers / washers

Windshield wipers, coupe (March 2002 to Sept. 2005)

(page 4 of 4)

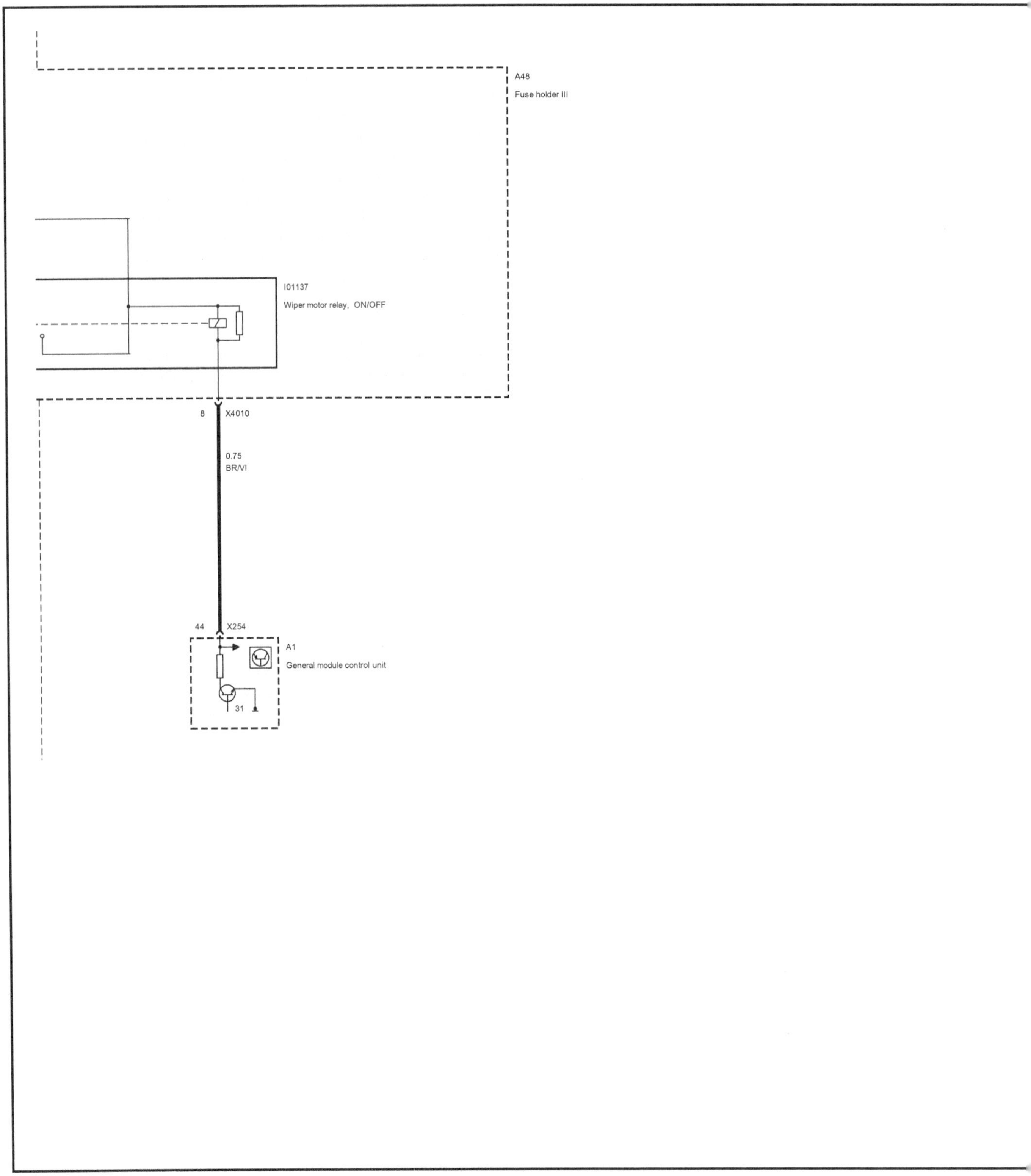

Wipers / washers

Windshield wipers, convertible (from Sept. 2005)

(page 1 of 2)

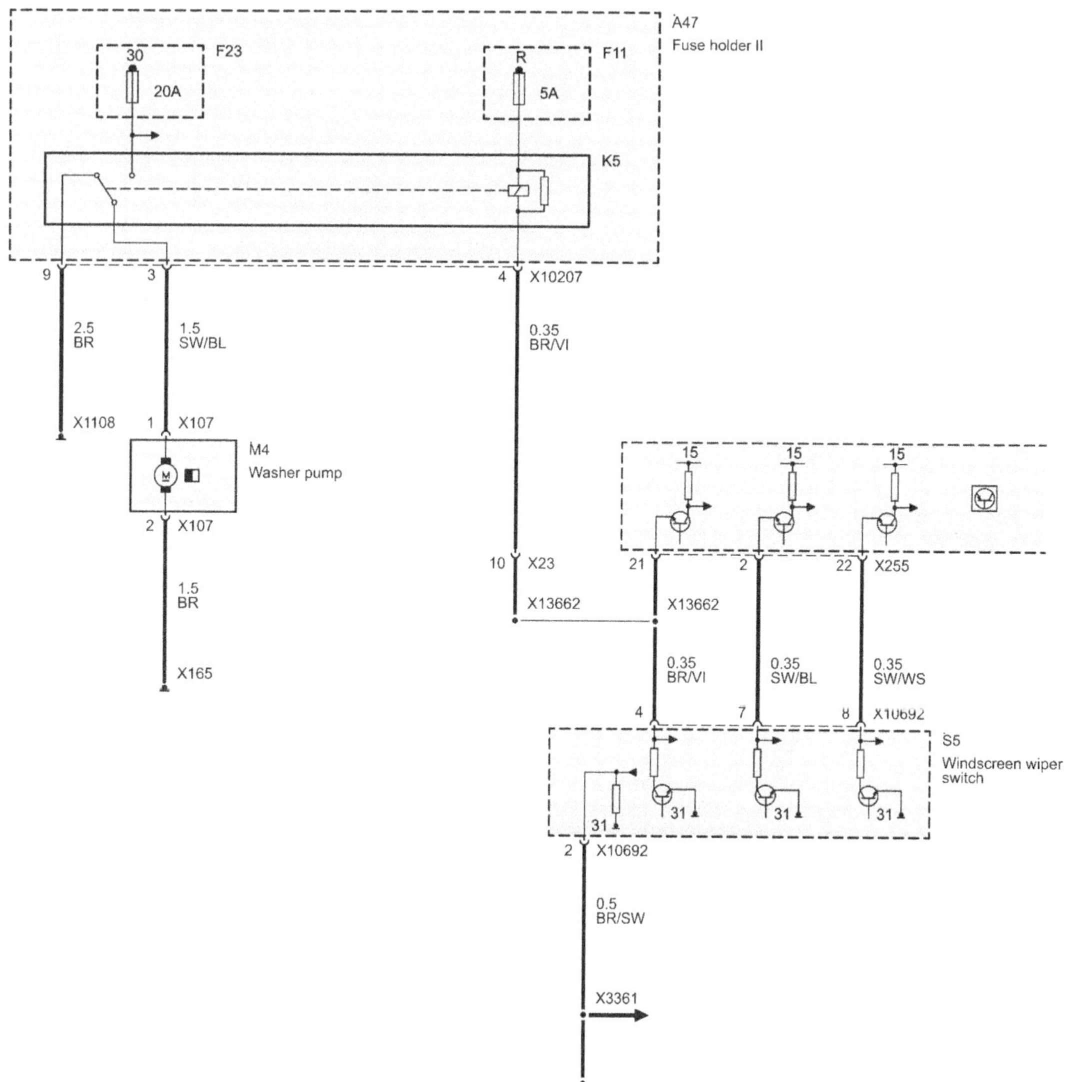

Wipers / washers

Windshield wipers, convertible (from Sept. 2005)

(page 2 of 2)

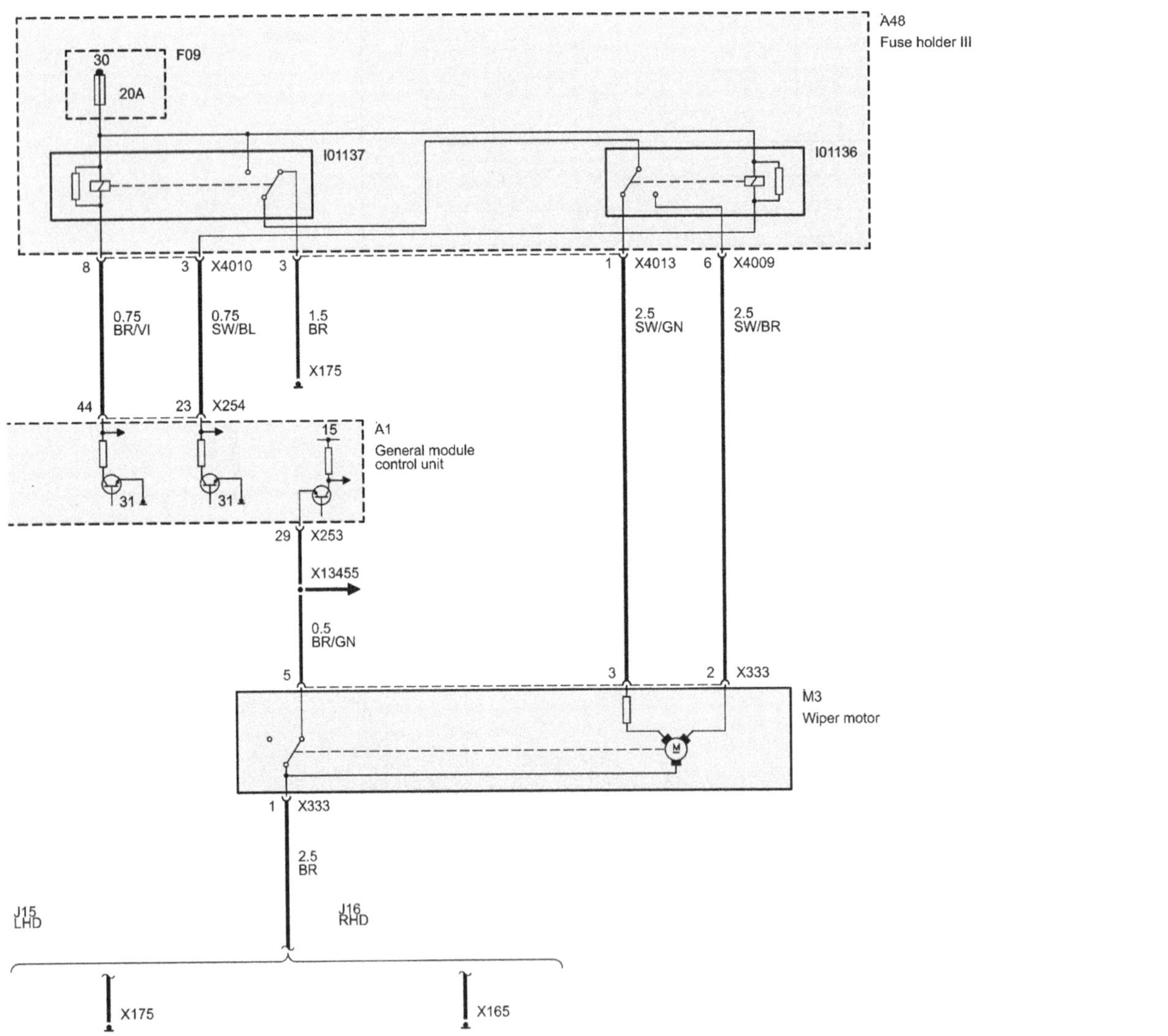

Wipers / washers
Rear wiper, coupe
(page 1 of 4)

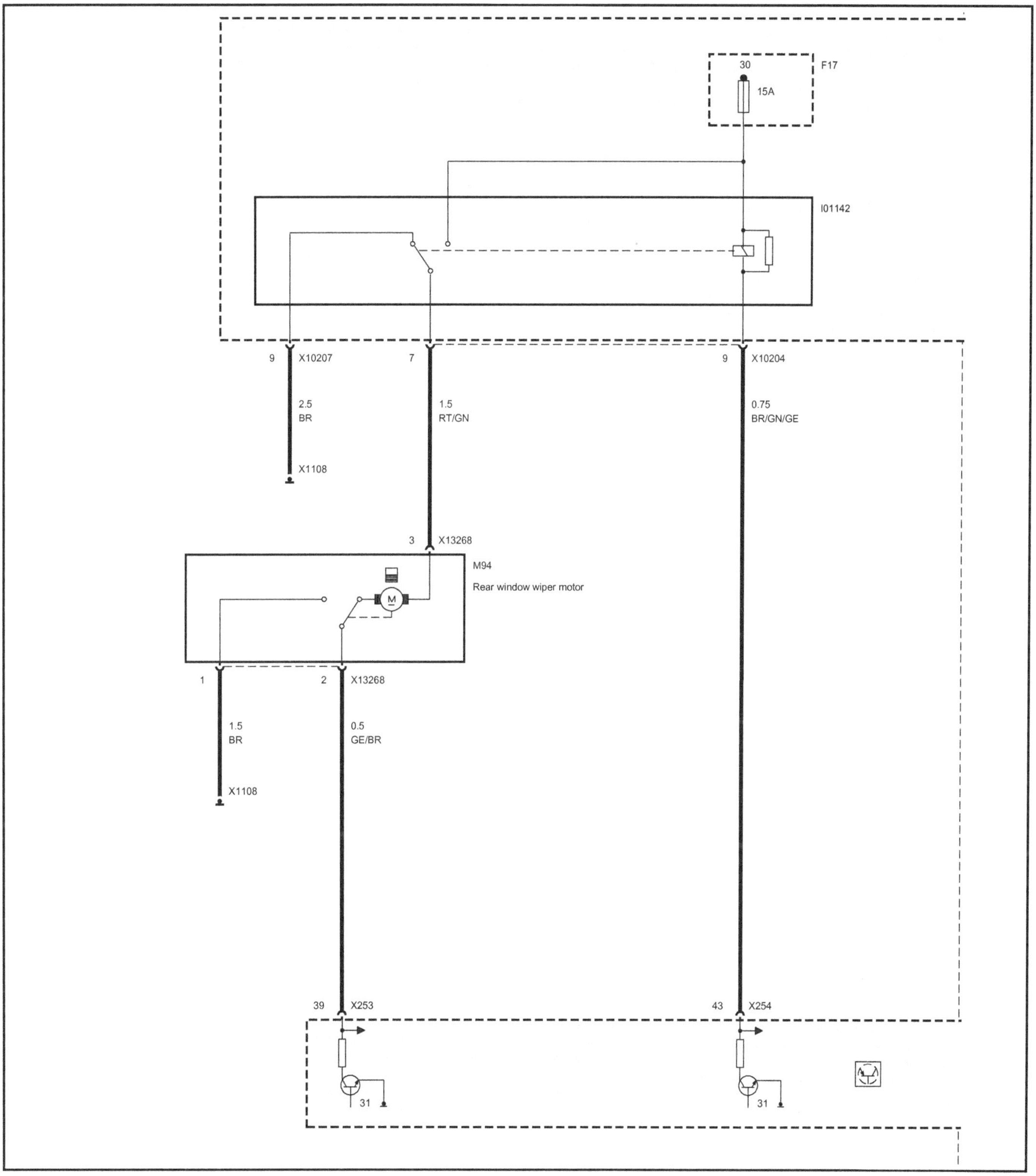

Wipers / washers

Rear wiper, coupe

(page 2 of 4)

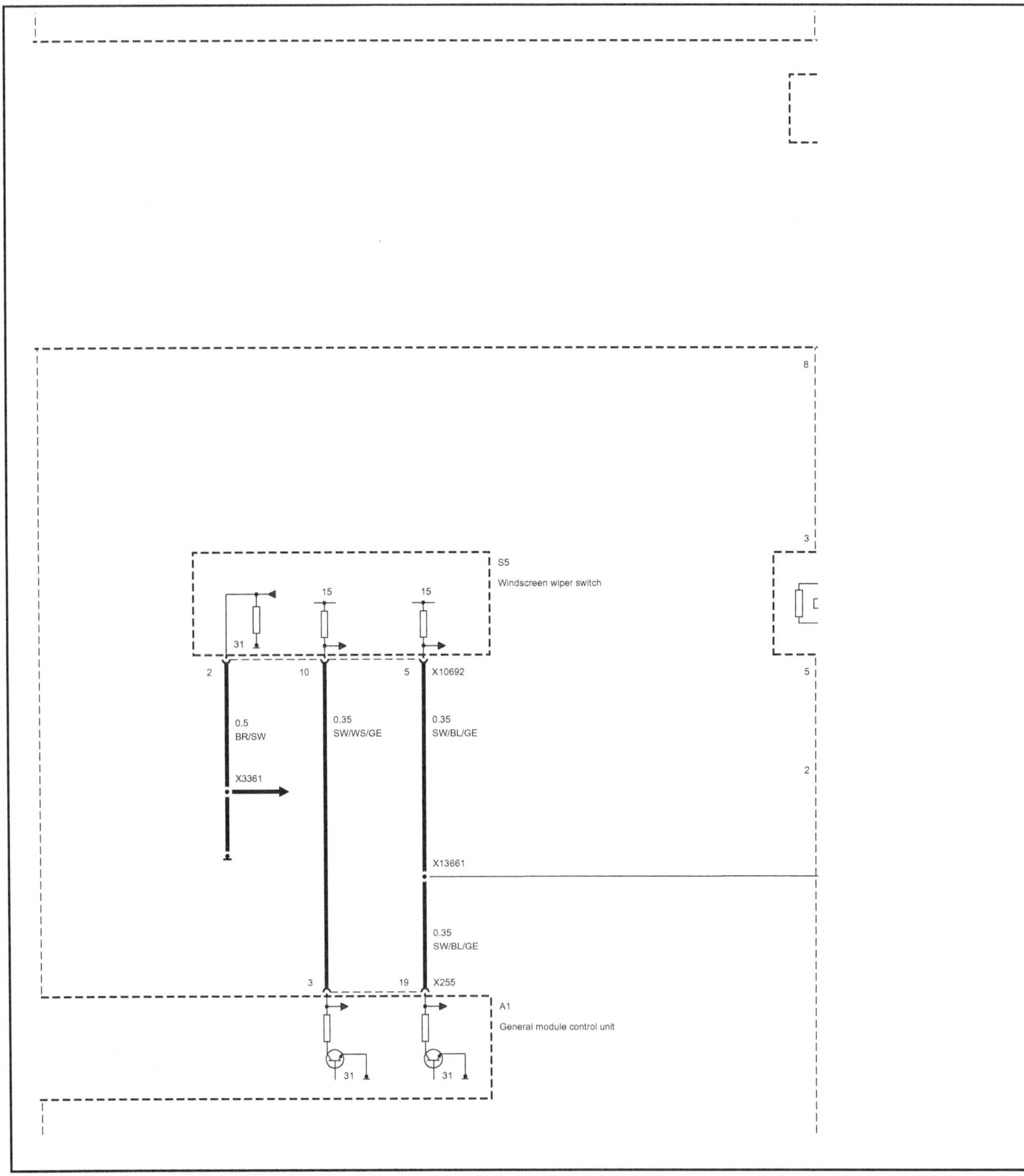

Wipers / washers
Rear wiper, coupe
(page 3 of 4)

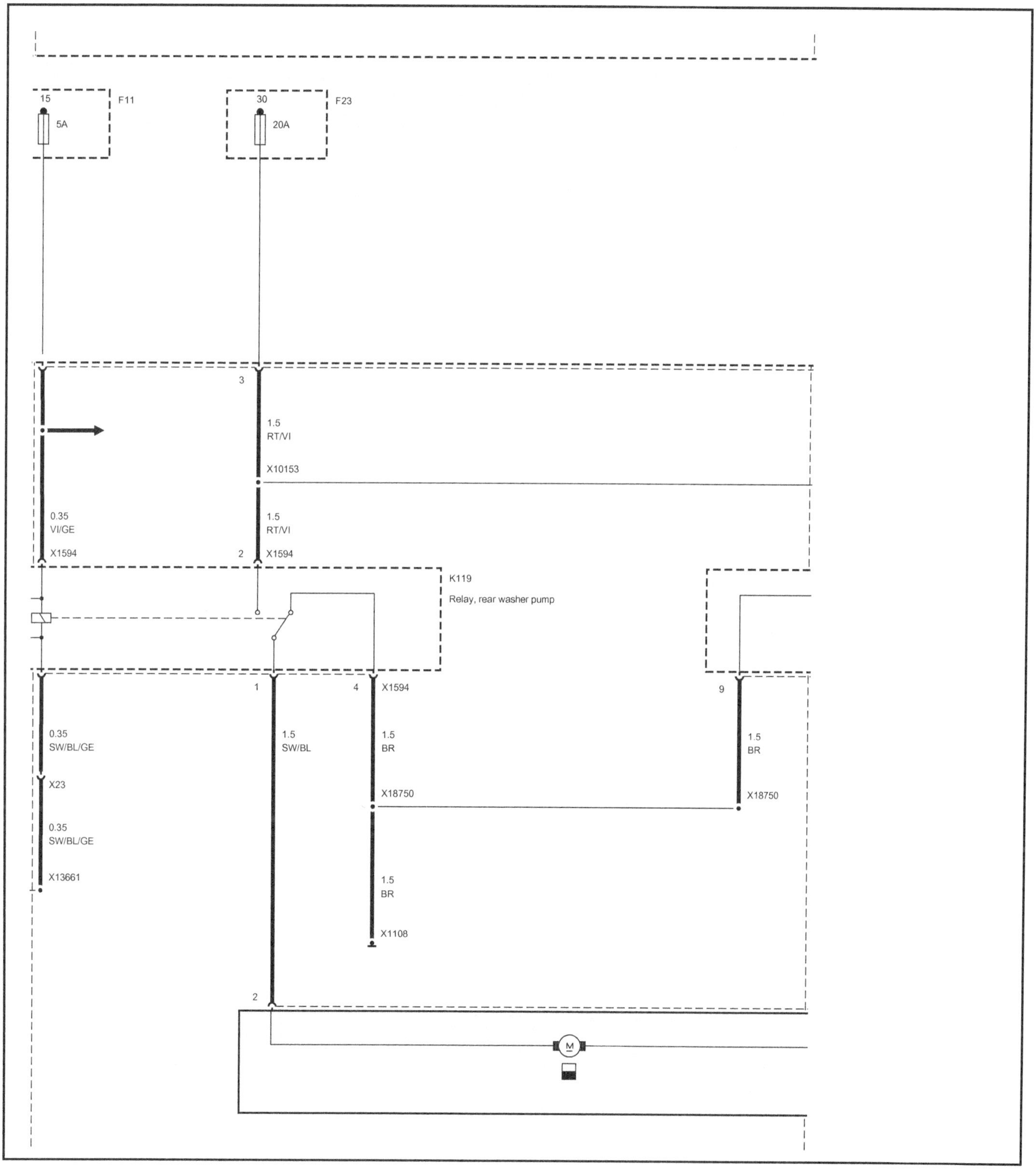

Wipers / washers
Rear wiper, coupe
(page 4 of 4)

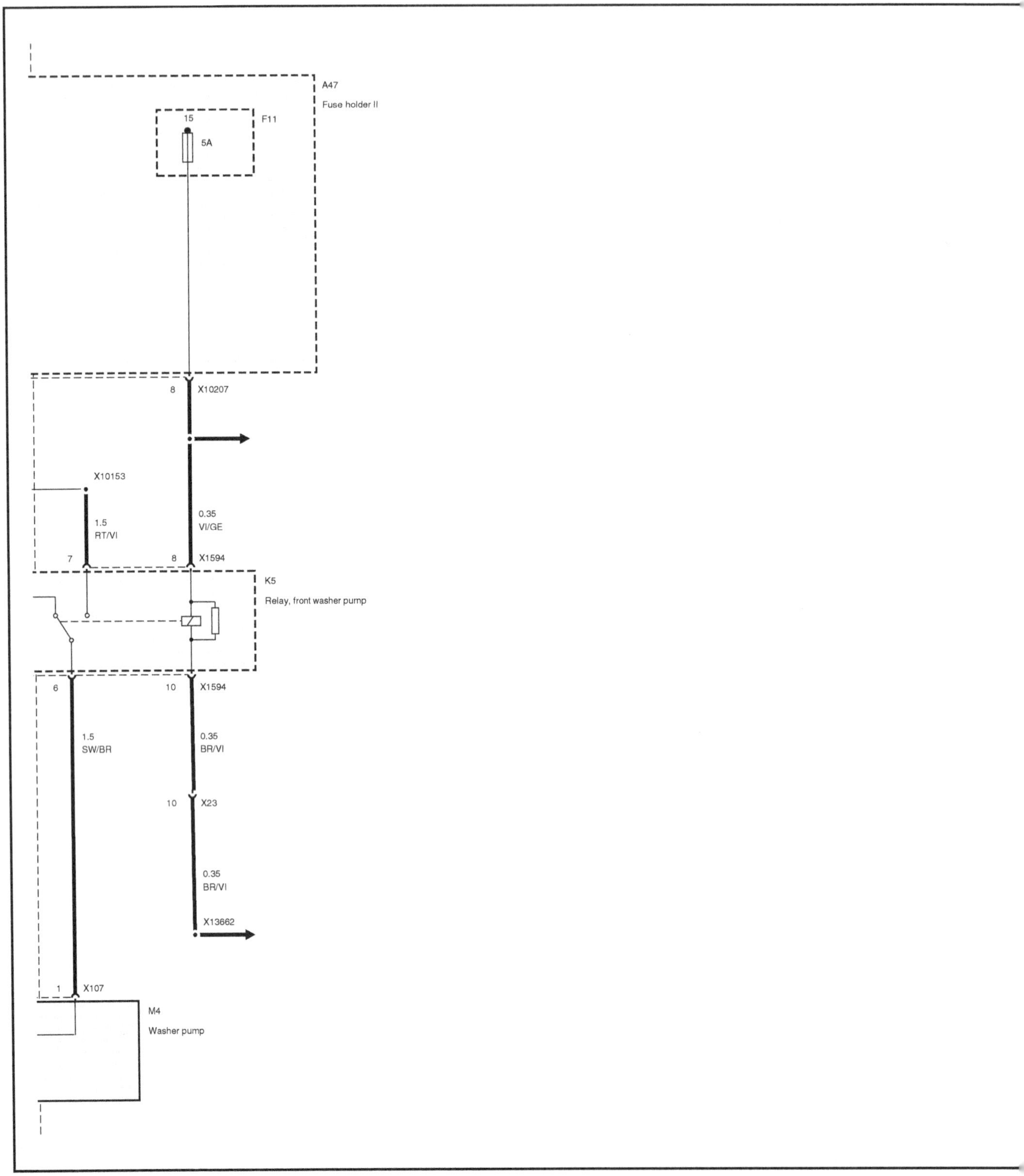

Wipers / washers
Windshield washer nozzle heaters
(page 1 of 1)

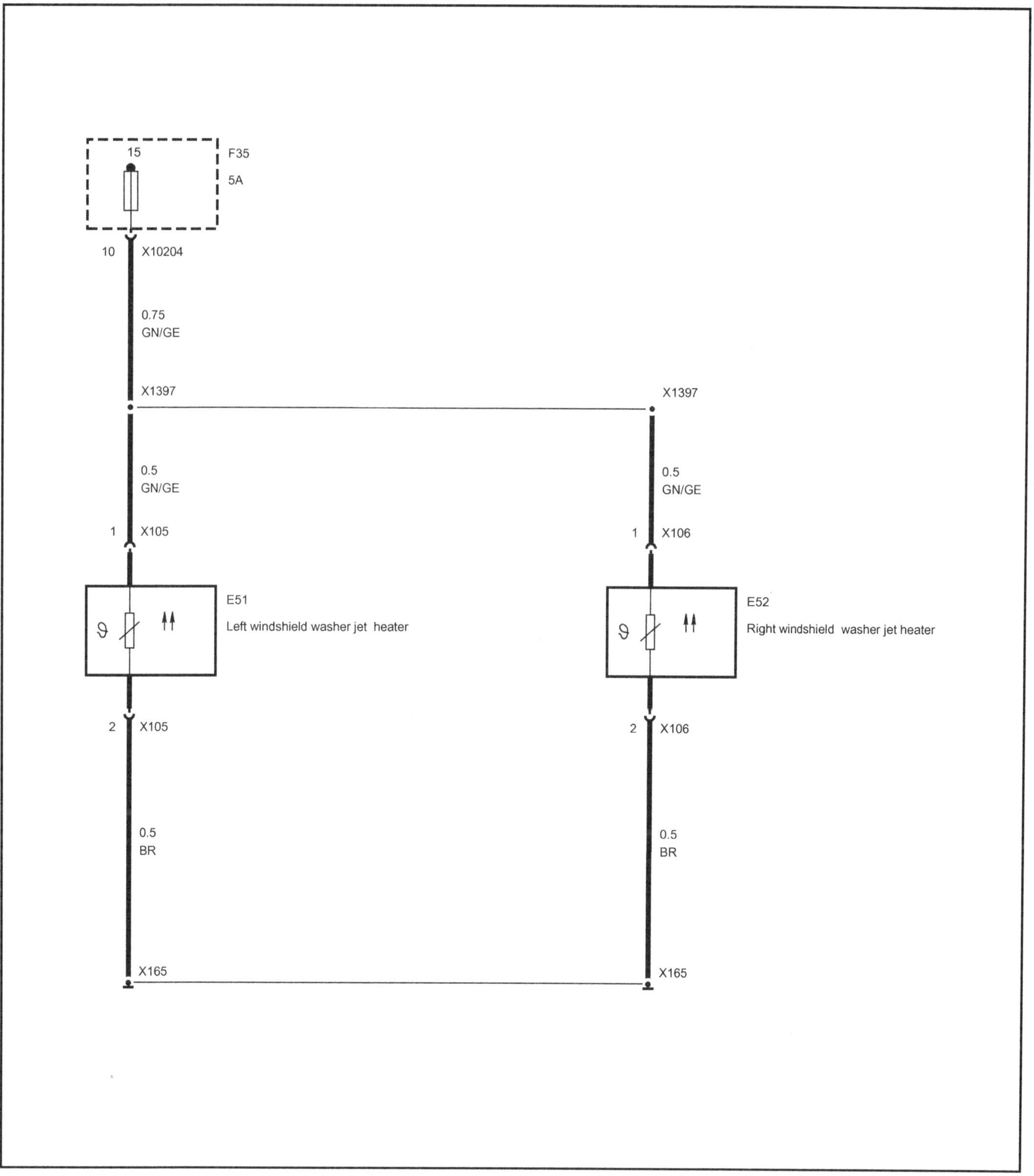

Wipers / washers
Wiper switch
(page 1 of 1)

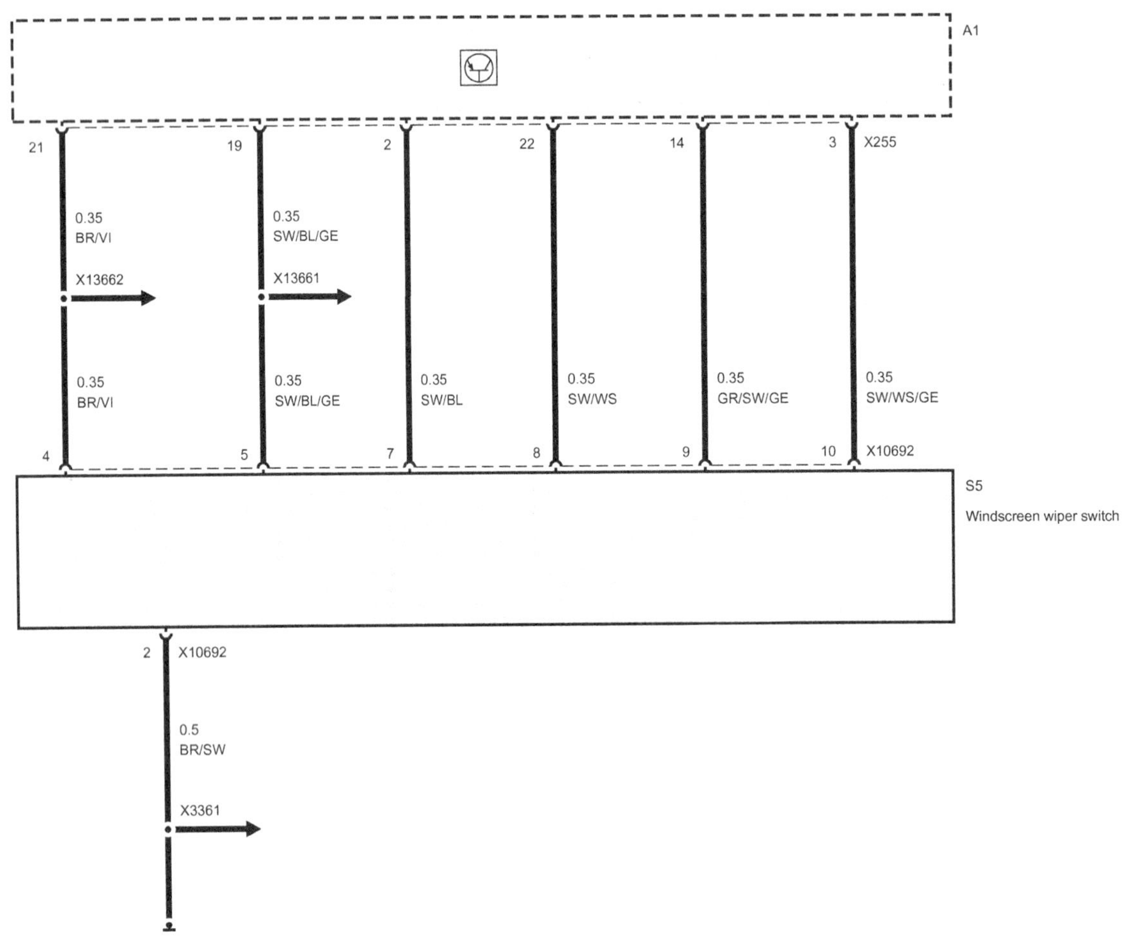

OBD On-Board Diagnostics

TABLES

General

This chapter outlines the fundamentals and equipment requirements of the SAE (Society of Automotive Engineers) and CARB (California Air Resource Board) On-Board Diagnostics II (OBD II) standard as it applies to MINI vehicles. Also covered here is a comprehensive listing of MINI and OBD II diagnostic trouble codes (DTCs).

On-Board Diagnostics II (OBD II)

OBD II standards have been applied to all passenger vehicles sold in the United States from model year 1996.

OBD capabilities are incorporated into engine software to monitor virtually every component that can affect vehicle emissions. Each emission-influencing component is checked by a diagnostic routine to verify that it is functioning properly. If a problem or malfunction is detected, the OBD II system illuminates the malfunction indicator light (MIL) on the instrument panel.

The OBD II system also stores important information about the detected malfunction in the engine control module (ECM) so that a repair technician can find the problem and repair it accurately.

NOTE —
- *Specialized OBD II scan tool equipment is needed to access the fault memory and OBD II data.*
- *The OBD II fault memory (including the MIL) can only be reset using a special scan tool. Removing the connector from the engine control module (ECM) or disconnecting the battery will not erase the fault memory.*

Malfunction indicator light (MIL)

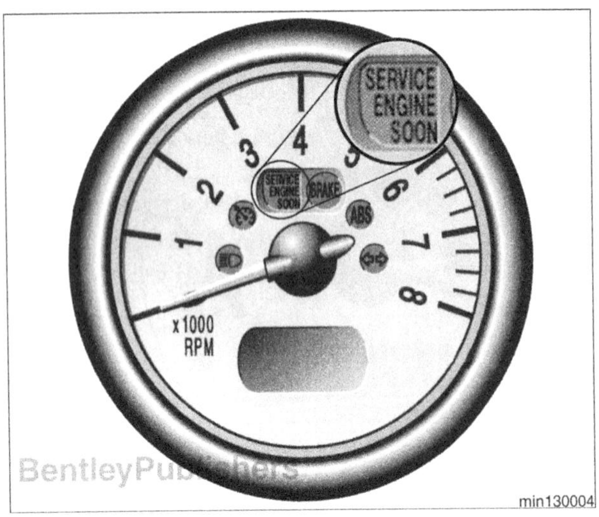

min130004

A bulb check function illuminates the malfunction indication light (MIL) when the ignition key is in the ON position before cranking the engine.

The OBD II system illuminates the MIL when emission levels exceed 1.5 times the Federal standards or when an emission-related system malfunction occurs.

Conditions that illuminate the MIL:
- Engine management system fault detected for **two** consecutive OBD II drive cycles. (See **Drive Cycle** in this repair group.)
- Catalyst damaging fault.
- Malfunction in component(s) monitoring emissions.
- Manufacturer-defined specifications exceeded.
- Implausible sensor input signal.
- Misfire.
- Leak detected in fuel tank evaporative system.
- No purge flow detected from purge valve / evaporative system.
- Failure of DME to enter closed-loop operation within specified time.
- Engine management system enters "limp home" mode.

Additional information on MIL:

- A fault code is stored within the engine control module (ECM) upon the first occurrence of a fault being identified.
- Two **complete** consecutive drive cycles with the fault present illuminate the MIL. The exception to the two-fault requirement is a catalyst-damaging fault which turns the light on immediately.
- If the second drive cycle is not completed and the fault not checked, the engine control module (ECM) counts the third drive cycle as the next consecutive drive cycle. The MIL is illuminated if the fault is still present.
- Once the MIL is illuminated, it remains illuminated unless the system is tested and found to be without faults through three complete consecutive drive cycles.
- In all cases except catalyst-damaging fault(s), the fault code is cleared from memory automatically if the system is checked through **40** consecutive drive cycles without the fault being detected.
- In case of a catalyst-damaging fault, the fault code is cleared from memory automatically if the system is checked through **80** consecutive drive cycles without the fault being detected.

Scan tool and scan tool display

With a generic OBD II scan tool connected to the 16-pin OBD II plug, diagnostic trouble codes (DTCs) can be obtained, along with the conditions associated with the illumination of the MIL. A more advanced or MINI-dedicated scan tool accesses additional proprietary information.

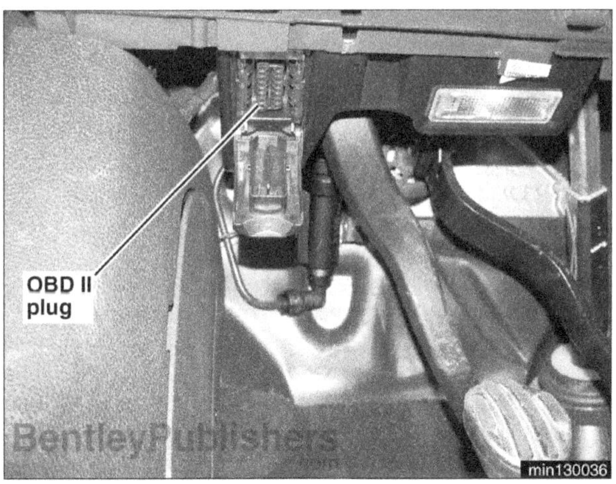

OBD II standards mandate that the 16-pin OBD II plug must be located within three (3) feet of the driver and must not require any tools to be exposed. The MINI OBD II plug is located under the left side of dash board.

NOTE—
- *In addition to the MINI / BMW factory scan tools, Baum, Snap-on, and Assenmacher are specialized tool companies that may offer MINI-specific scan tools. Check with the tool manufacturer for availability.*
- *Generic OBD II scan tool software programs and handheld units are readily available. Though limited in capabilities compared to the factory tools, they are nonetheless powerful diagnostic devices which read live data streams, freeze-frame data and other valuable diagnostic information.*

A simple aftermarket DTC reader is available from Peake Research. This tool is capable of checking for DTCs as well as turning off the illuminated MIL.

Diagnostic monitors

Diagnostic monitors run tests and checks on specific systems, components, or functions.

A complete drive cycle (see **Drive cycle** in this repair group) is required for the tests to be valid. The diagnostic monitor signals the engine control module (ECM) of the loss or impairment of the signal or component and determines if a signal or sensor is faulty based on 3 conditions:

- Signal or component shorted to ground
- Signal or component shorted to B+
- Signal or component missing (open circuit)

The OBD II system monitors all emission control systems in the vehicle. Not all vehicles have a full complement of emission control systems. For example, a vehicle may not be equipped with secondary air injection, so naturally no secondary air readiness code would be present.

OBD II requires monitoring of the following:

- Oxygen sensor(s)
- Catalytic converter(s)
- Engine misfire
- Fuel tank evaporative control system (EVAP)
- Secondary air
- Fuel system

Oxygen sensor monitoring: When drive conditions allow, response rate and switching time of each oxygen sensor is monitored. The oxygen sensor heater function is also monitored. The OBD II system differentiates between precatalyst and post-catalyst oxygen sensors. In order for the oxygen sensors to be effectively monitored, the system must be in closed loop operation.

Catalyst monitoring: Comparing precatalyst and post-catalyst oxygen sensor output yields information about oxygen storage in the catalyst. This, in turn, indicates catalyst efficiency.

Misfire detection: This strategy monitors crankshaft speed fluctuations to determine if an engine misfire occurs. The system identifies a misfire, as well as other pertinent misfire information such as:

- Specific cylinder(s) misfiring
- Severity of misfire event
- Relevance to vehicle emissions
- Potential of damage to catalyst.

Misfire detection is an on-going process only disabled under certain limited conditions.

Fuel system monitoring: This monitor checks fuel delivery for proper engine operation based on programmed data (long and short term fuel trim). If too much or not enough fuel is delivered over a predetermined time, a DTC is set and the MIL is turned on.

NOTE—
Fuel trim refers to adjustments to programmed fuel delivery. Long-term fuel trim refers to gradual adjustments to fuel delivery calibration. Long term fuel trim adjustments compensate for gradual changes that occur over time.

Fuel system monitoring monitors the calculated injection time (ti) in relation to engine speed, load, and precatalyst oxygen sensor signals.

Evaporative system (EVAP) monitoring: This monitor checks the fuel storage system and related fuel lines for leaks. It can detect very small leaks anywhere in the system. A pressure test is performed on the EVAP system on a continuous basis if the drive cycle allows.

Drive cycle

The OBD II drive cycle is a specific driving routine used by the EPA to test vehicle emissions. The purpose of the drive cycle is to run all of the emission-related on-board diagnostics over a broad range of driving conditions.

A drive cycle is considered complete when all of the diagnostic monitors have run their tests without interruption. For a drive cycle to be initiated, the vehicle must be started cold and brought up to 160°F and at least 40°F above its original starting temperature.

Once a drive cycle is completed, the system status or inspection/maintenance (I/M) readiness codes are set to “Yes.” When all codes are set to “Yes” the system is described as having established “readiness”.

System status codes will be set to “No” in the following cases:

- Battery or ECM is disconnected.
- DTCs have been erased after completion of repairs but drive cycle has not be completed.

A scan tool can be used to determine I/M readiness.

I/M readiness codes

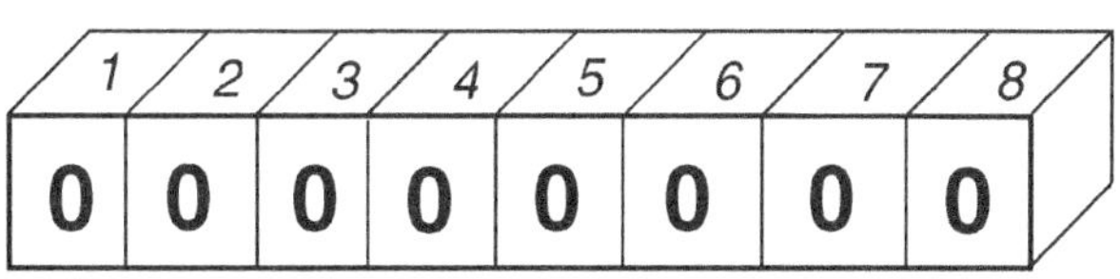

1. EGR monitoring
2. Oxygen sensor heater monitoring
3. Oxygen sensor monitoring
4. Air conditioning
5. Secondary air monitoring
6. Evaporative system monitoring
7. Catalyst heating
8. Catalyst efficiency monitoring

Readiness codes

The inspection/maintenance (I/M) readiness codes are mandated as part of OBD II. The readiness code is stored after complete diagnostic monitoring of all components and systems has been carried out. The readiness code function was put into place to prevent manipulation of I/M emission test procedures by clearing faults codes or disconnecting the ECM or battery.

The readiness code can be displayed using an aftermarket scan tool. The code is binary:

- 0 means “yes”, the function is ready
- 1 means “no”, the function is not ready.

When all zeros are displayed, the system has established its readiness.

Diagnostic trouble codes (DTCs)

The SAE mandates a 5-digit diagnostic trouble code (DTC) standard. Emission related DTCs start with the letter "P" for power train and are commonly referred to as "P-codes". When the malfunction indicator light (MIL) is illuminated it
indicates that a DTC has been stored:

- DTCs are stored as soon as they occur, whether or not the MIL illuminates.
- DTCs store and display a time stamp.
- DTCs record the current fault status.

DTC digit interpretation	
1st digit P B C	 powertrain body chassis
2nd digit 0 1	 SAE MINI
3rd digit (in P-codes) 0 1 2 3 4 5 6 7	 total system air/fuel induction fuel injection ignition system or missfire auxiliary emission control vehicle speed & idle control ECM inputs/outputs transmission
4th - 5th digits	individual circuits or components

DTC example: P 0 3 0 6

- **P:** A powertrain problem
- **0**: SAE sanctioned or generic
- **3:** Related to an ignition system/misfire
- **06** Misfire has been detected at cylinder #6

Freeze frame data: DTCs provide a "freeze frame" or snap-shot of vehicle performance or emissions fault at the moment that the fault first occurs. This information is accessible through generic scan tools.

Freeze frame data contains, but is not limited to, the following:

- Engine load (calculated)
- Engine RPM
- Short and long-term fuel trim
- Vehicle speed
- Coolant temperature
- Intake manifold pressure
- Open/closed loop operation
- Fuel pressure (if available)
- DTC

P-CODES (DTCS) AND MINI FAULT CODES

MINI fault codes (FC) expand on the SAE sanctioned DTCs and are accessible primarily through BMW (or BMW specific) diagnostic scan tools. Below is a comprehensive listing of SAE P-codes, the available corresponding MINI FCs and an explanation of their meanings.

Cooper DTCs

Table a. Cooper DTCs from 1/1/2002 to 8/31/2002

P-code	MINI FC	P-code text
P0030	48	HO2S Heater Control Circuit (Bank 1 Sensor 1)
P0031	49	HO2S Heater Control Circuit Low (Bank 1 Sensor 1)
P0032	50	HO2S Heater Control Circuit High (Bank 1 Sensor 1)
P0036	54	HO2S Heater Control Circuit (Bank 1 Sensor 2)
P0037	55	HO2S Heater Control Circuit Low (Bank 1 Sensor 2)
P0038	56	HO2S Heater Control Circuit High (Bank 1 Sensor 2)
P0107	263	Manifold Absolute Pressure/Barometric Pressure Circuit Low Input
P0108	264	Manifold Absolute Pressure/Barometric Pressure Circuit High Input
P0112	274	Intake Air Temperature Sensor 1 Circuit Low
P0113	275	Intake Air Temperature Sensor 1 Circuit High
P0114	276	Intake Air Temperature Sensor 1 Circuit Intermittent
P0116	278	Engine Coolant Temperature Circuit Range/Performance
P0117	279	Engine Coolant Temperature Circuit Low
P0118	280	Engine Coolant Temperature Circuit High
P0119	281	Engine Coolant Temperature Circuit Intermittent
P0122	290	Throttle/Pedal Position Sensor/Switch 'A' Circuit Low
P0123	291	Throttle/Pedal Position Sensor/Switch 'A' Circuit High
P0125	293	Insufficient Coolant Temperature for Closed Loop Fuel Control
P0128	296	Coolant Thermostat (Coolant Temperature Below Thermostat Regulating Temperature)
P0130	304	O2 Sensor Circuit (Bank 1 Sensor 1)
P0131	305	O2 Sensor Circuit Low Voltage (Bank 1 Sensor 1)
P0132	306	O2 Sensor Circuit High Voltage (Bank 1 Sensor 1)
P0133	307	O2 Sensor Circuit Slow Response (Bank 1 Sensor 1)
P0135	309	O2 Sensor Heater Circuit (Bank 1 Sensor 1)
P0136	310	O2 Sensor Circuit (Bank 1 Sensor 2)
P0137	311	O2 Sensor Circuit Low Voltage (Bank 1 Sensor 2)
P0138	312	O2 Sensor Circuit High Voltage (Bank 1 Sensor 2)
P0141	321	O2 Sensor Heater Circuit (Bank 1 Sensor 2)
P0171	369	System Too Lean (Bank 1)
P0172	370	System Too Rich (Bank 1)
P0201	513	Injector Circuit/Open - Cylinder 1
P0202	514	Injector Circuit/Open - Cylinder 2
P0203	515	Injector Circuit/Open - Cylinder 3
P0204	516	Injector Circuit/Open - Cylinder 4

P-codes (DTCs) and MINI fault codes

Table a. Cooper DTCs from 1/1/2002 to 8/31/2002

P-code	MINI FC	P-code text
P0222	546	Throttle/Pedal Position Sensor/Switch 'B' Circuit Low
P0223	547	Throttle/Pedal Position Sensor/Switch 'B' Circuit High
P0261	609	Cylinder 1 Injector Circuit Low
P0262	610	Cylinder 1 Injector Circuit High
P0264	612	Cylinder 2 Injector Circuit Low
P0265	613	Cylinder 2 Injector Circuit High
P0267	615	Cylinder 3 Injector Circuit Low
P0268	616	Cylinder 3 Injector Circuit High
P0270	624	Cylinder 4 Injector Circuit Low
P0271	625	Cylinder 4 Injector Circuit High
P0300	768	Random/Multiple Cylinder Misfire Detected
P0301	769	Cylinder 1 Misfire Detected
P0302	770	Cylinder 2 Misfire Detected
P0303	771	Cylinder 3 Misfire Detected
P0304	772	Cylinder 4 Misfire Detected
P0313	787	Misfire Detected with Low Fuel
P0324	804	Knock Control System Error
P0326	806	Knock Sensor 1 Circuit Range/Performance (Bank 1 or Single Sensor)
P0335	821	Crankshaft Position Sensor 'A' Circuit
P0336	822	Crankshaft Position Sensor 'A' Circuit Range/Performance
P0340	832	Camshaft Position Sensor 'A' Circuit (Bank 1 or Single Sensor)
P0341	833	Camshaft Position Sensor 'A' Circuit Range/Performance (Bank 1 or Single Sensor)
P0351	849	Ignition Coil 'A' Primary/Secondary Circuit
P0352	850	Ignition Coil 'B' Primary/Secondary Circuit
P0420	1056	Catalyst System Efficiency Below Threshold (Bank 1)
P0441	1089	Evaporative Emission System Incorrect Purge Flow
P0442	1090	Evaporative Emission System Leak Detected (small leak)
P0443	1091	Evaporative Emission System Purge Control Valve Circuit
P0444	1092	Evaporative Emission System Purge Control Valve Circuit Open
P0445	1093	Evaporative Emission System Purge Control Valve Circuit Shorted
P0455	1109	Evaporative Emission System Leak Detected (large leak)
P0456	1110	Evaporative Emission System Leak Detected (very small leak)
P0500	1280	Vehicle Speed Sensor 'A'
P0506	1286	Idle Air Control System RPM Lower Than Expected
P0507	1287	Idle Air Control System RPM Higher Than Expected
P0601	1537	Internal Control Module Memory Check Sum Error
P0603	1539	Internal Control Module Keep Alive Memory (KAM) Error
P0604	1540	Internal Control Module Random Access Memory (RAM) Error
P0705	1797	Transmission Range Sensor 'A' Circuit Malfunction (PRNDL Input)
P0815	2069	Upshift Switch Circuit
P0816	2070	Downshift Switch Circuit

Table a. Cooper DTCs from 1/1/2002 to 8/31/2002		
P-code	**MINI FC**	**P-code text**
P1106	4358	Manifold Air Pressure Sensor Too Low at Engine Stop
P1107	4359	Manifold Air Pressure Sensor Too Low at Idle Engine Running
P1108	4360	Manifold Air Pressure Sensor Too Low at Full Load for Low Engine Speed
P1109	4361	Manifold Air Pressure Too High in Deceleration
P1122	4386	Pedal Position Sensor 1 Low Input
P1123	4387	Pedal Position Sensor 1 High Input
P1125	4389	Throttle Position Sensor A and B Range/Performance Small Error
P1126	4390	Throttle Position Sensor A and B Range/Performance Large Error
P1143	4419	O2 Sensor Activity Check Signal Too High (Bank 1 Sensor 2)
P1144	4420	O2 Sensor Activity Check Signal Too Low (Bank 1 Sensor 2)
P1222	4642	Pedal Position Sensor 2 Low Input
P1223	4643	Pedal Position Sensor 2 High Input
P1224	4644	Pedal Position Sensor 1 and 2 Range/Performance Error
P1226	4646	Throttle Malfunction (Flap Malfunction)
P1229	4649	Throttle Sensor Adaptation Failure
P1320	4896	Flywheel Adaptation for Misfire Detection Range
P1321	4897	Flywheel Adaptation for Misfire Detection Performance
P1366	4966	Ignition Coil 'A' Primary/Secondary Circuit Low
P1367	4967	Ignition Coil 'B' Primary/Secondary Circuit Low
P1436	5174	Leakage Diagnostic Pump Open Circuit
P1437	5175	Leakage Diagnostic Pump Range/Performance
P1442	5186	Leakage Diagnostic Pump Control Circuit Signal Low
P1443	5187	Leakage Diagnostic Pump Control Circuit Signal High
P1475	5237	Leakage Diagnostic Pump Reed Switch Did not Close
P1476	5238	Leakage Diagnostic Pump Clamped Tube
P1477	5239	Leakage Diagnostic Pump Reed Switch Did Not Open
P1570	5488	Electronic Control Module Sensor Supply A Low Output
P1571	5489	Electronic Control Module Sensor Supply A High Output
P1572	5490	Electronic Control Module Sensor Supply A Noisy Signal
P1573	5491	Electronic Control Module Sensor Supply B Low Output
P1574	5492	Electronic Control Module Sensor Supply B High Output
P1575	5493	Electronic Control Module Sensor Supply B Noisy Signal
P1600	5632	External Control Module Random Access Memory (RAM) Error
P1607	5639	CAN-Version
P1611	5649	Serial Communication Link Transmission Control Module
P1612	5650	Serial Communication Link Instrument Pack
P1613	5651	Serial Communication Link ASC (Automatic Stability Control)
P1615	5653	Electronic Control Module Processor SPI-Bus Failure
P1617	5655	Electronic Control Module H Bridge Controller
P1679	5753	Electronic Throttle Control Monitor Level 2/3 Torque Loss Calculation
P1680	5760	Electronic Throttle Control Monitor Level 2/3 ADC Processor Fault
P1681	5761	Electronic Throttle Control Monitor Level 2/3 Engine Speed Calculation Error

P-codes (DTCs) and MINI fault codes

Table a. Cooper DTCs from 1/1/2002 to 8/31/2002		
P-code	**MINI FC**	**P-code text**
P1682	5762	Electronic Throttle Control Monitor Level 2/3 Idle Speed 'A' Calculation Fault
P1683	5763	Electronic Throttle Control Monitor Level 2/3 Idle Speed 'B' Calculation Fault
P1684	5764	Electronic Throttle Control Monitor Level 2/3 Clutch Torque Min Error
P1685	5765	Electronic Throttle Control Monitor Level 2/3 Clutch Torque Max Error
P1686	5766	Electronic Throttle Control Monitor Level 2/3 Pedal Position Sensor Diagnostic Error
P1687	5767	Electronic Throttle Control Monitor Level 2/3 Throttle Position Sensor Diagnostic Error
P1688	5768	Electronic Throttle Control Monitor Level 2/3 Mass Air Flow Calculation
P1689	5769	Electronic Throttle Control Monitor Level 2/3 Torque Calculation Error
P1691	5777	Electronic Throttle Control Monitor Level 2/3 Motorized Throttle Control Engine Speed Limitation Error
P1692	5778	Electronic Throttle Control Monitor Level 2/3 Motorized Throttle Control and Fuel Injection Switch Off 'A'
P1693	5779	Electronic Throttle Control Monitor Level 2/3 Motorized Throttle Control and Fuel Injection Switch Off 'B'
P1698	5784	Transmission Control Module Control Error
P1699	5785	Transmission Control Module Checksum Error
P1705	5893	Transmission Control Module LED Output Open Circuit
P1706	5894	Transmission Control Module LED Output Short Circuit
P1739	5945	Clutch Solenoid Communication Error
P1741	5953	Clutch Solenoid Open Circuit
P1742	5954	Clutch Solenoid Short Circuit
P1749	5961	Secondary Pressure Solenoid Communication Error
P1751	5969	Secondary Pressure Solenoid Open Circuit
P1752	5970	Secondary Pressure Solenoid Short Circuit
P1785	6021	Transmission Ratio Control Actuator Circuit Malfunction
P1786	6022	Transmission Ratio Control Actuator Circuit Range/Performance
P1787	6023	Transmission Ratio Control Actuator Open Circuit
P1788	6024	Transmission Ratio Control Actuator Short Circuit
P1789	6025	Transmission Ratio Control Actuator Communication Error
P1815	6165	Wheel Plus Switch Error Low Input
P1816	6166	Wheel Minus Switch Error Low Input
P2096	8342	Post Catalyst Fuel Trim System Too Lean (Bank 1)
P2097	8343	Post Catalyst Fuel Trim System Too Rich (Bank 1)

Table b. Cooper DTCs from 9/1/2002 to 1/10/2003		
P-code	**MINI FC**	**P-code text**
P0030	48	HO2S Heater Control Circuit (Bank 1 Sensor 1)
P0031	49	HO2S Heater Control Circuit Low (Bank 1 Sensor 1)
P0032	50	HO2S Heater Control Circuit High (Bank 1 Sensor 1)
P0036	54	HO2S Heater Control Circuit (Bank 1 Sensor 2)
P0037	55	HO2S Heater Control Circuit Low (Bank 1 Sensor 2)
P0038	56	HO2S Heater Control Circuit High (Bank 1 Sensor 2)
P0107	263	Manifold Absolute Pressure/Barometric Pressure Circuit Low Input
P0108	264	Manifold Absolute Pressure/Barometric Pressure Circuit High Input
P0112	274	Intake Air Temperature Sensor 1 Circuit Low
P0113	275	Intake Air Temperature Sensor 1 Circuit High
P0114	276	Intake Air Temperature Sensor 1 Circuit Intermittent
P0116	278	Engine Coolant Temperature Circuit Range/Performance
P0117	279	Engine Coolant Temperature Circuit Low
P0118	280	Engine Coolant Temperature Circuit High
P0119	281	Engine Coolant Temperature Circuit Intermittent
P0122	290	Throttle/Pedal Position Sensor/Switch 'A' Circuit Low
P0123	291	Throttle/Pedal Position Sensor/Switch 'A' Circuit High
P0125	293	Insufficient Coolant Temperature for Closed Loop Fuel Control
P0128	296	Coolant Thermostat (Coolant Temperature Below Thermostat Regulating Temperature)
P0130	304	O2 Sensor Circuit (Bank 1 Sensor 1)
P0131	305	O2 Sensor Circuit Low Voltage (Bank 1 Sensor 1)
P0132	306	O2 Sensor Circuit High Voltage (Bank 1 Sensor 1)
P0133	307	O2 Sensor Circuit Slow Response (Bank 1 Sensor 1)
P0135	309	O2 Sensor Heater Circuit (Bank 1 Sensor 1)
P0136	310	O2 Sensor Circuit (Bank 1 Sensor 2)
P0137	311	O2 Sensor Circuit Low Voltage (Bank 1 Sensor 2)
P0138	312	O2 Sensor Circuit High Voltage (Bank 1 Sensor 2)
P0141	321	O2 Sensor Heater Circuit (Bank 1 Sensor 2)
P0171	369	System Too Lean (Bank 1)
P0172	370	System Too Rich (Bank 1)
P0201	513	Injector Circuit/Open - Cylinder 1
P0202	514	Injector Circuit/Open - Cylinder 2
P0203	515	Injector Circuit/Open - Cylinder 3
P0204	516	Injector Circuit/Open - Cylinder 4
P0222	546	Throttle/Pedal Position Sensor/Switch 'B' Circuit Low
P0223	547	Throttle/Pedal Position Sensor/Switch 'B' Circuit High
P0261	609	Cylinder 1 Injector Circuit Low
P0262	610	Cylinder 1 Injector Circuit High
P0264	612	Cylinder 2 Injector Circuit Low
P0265	613	Cylinder 2 Injector Circuit High
P0267	615	Cylinder 3 Injector Circuit Low
P0268	616	Cylinder 3 Injector Circuit High

Table b. Cooper DTCs from 9/1/2002 to 1/10/2003		
P-code	**MINI FC**	**P-code text**
P0270	624	Cylinder 4 Injector Circuit Low
P0271	625	Cylinder 4 Injector Circuit High
P0300	768	Random/Multiple Cylinder Misfire Detected
P0301	769	Cylinder 1 Misfire Detected
P0302	770	Cylinder 2 Misfire Detected
P0303	771	Cylinder 3 Misfire Detected
P0304	772	Cylinder 4 Misfire Detected
P0313	787	Misfire Detected with Low Fuel
P0324	804	Knock Control System Error
P0326	806	Knock Sensor 1 Circuit Range/Performance (Bank 1 or Single Sensor)
P0335	821	Crankshaft Position Sensor 'A' Circuit
P0336	822	Crankshaft Position Sensor 'A' Circuit Range/Performance
P0340	832	Camshaft Position Sensor 'A' Circuit (Bank 1 or Single Sensor)
P0341	833	Camshaft Position Sensor 'A' Circuit Range/Performance (Bank 1 or Single Sensor)
P0351	849	Ignition Coil 'A' Primary/Secondary Circuit
P0352	850	Ignition Coil 'B' Primary/Secondary Circuit
P0420	1056	Catalyst System Efficiency Below Threshold (Bank 1)
P0441	1089	Evaporative Emission System Incorrect Purge Flow
P0442	1090	Evaporative Emission System Leak Detected (small leak)
P0443	1091	Evaporative Emission System Purge Control Valve Circuit
P0444	1092	Evaporative Emission System Purge Control Valve Circuit Open
P0445	1093	Evaporative Emission System Purge Control Valve Circuit Shorted
P0455	1109	Evaporative Emission System Leak Detected (large leak)
P0456	1110	Evaporative Emission System Leak Detected (very small leak)
P0500	1280	Vehicle Speed Sensor 'A'
P0506	1286	Idle Air Control System RPM Lower Than Expected
P0507	1287	Idle Air Control System RPM Higher Than Expected
P0601	1537	Internal Control Module Memory Check Sum Error
P0603	1539	Internal Control Module Keep Alive Memory (KAM) Error
P0604	1540	Internal Control Module Random Access Memory (RAM) Error
P0705	1797	Transmission Range Sensor 'A' Circuit Malfunction (PRNDL Input)
P0815	2069	Upshift Switch Circuit
P0816	2070	Downshift Switch Circuit
P1106	4358	Manifold Air Pressure Sensor Too Low at Engine Stop
P1107	4359	Manifold Air Pressure Sensor Too Low at Idle Engine Running
P1108	4360	Manifold Air Pressure Sensor Too Low at Full Load for Low Engine Speed
P1109	4361	Manifold Air Pressure Too High in Deceleration
P1122	4386	Pedal Position Sensor 1 Low Input
P1123	4387	Pedal Position Sensor 1 High Input
P1125	4389	Throttle Position Sensor A and B Range/Performance Small Error
P1126	4390	Throttle Position Sensor A and B Range/Performance Large Error

Table b. Cooper DTCs from 9/1/2002 to 1/10/2003		
P-code	**MINI FC**	**P-code text**
P1143	4419	O2 Sensor Activity Check Signal Too High (Bank 1 Sensor 2)
P1144	4420	O2 Sensor Activity Check Signal Too Low (Bank 1 Sensor 2)
P1222	4642	Pedal Position Sensor 2 Low Input
P1223	4643	Pedal Position Sensor 2 High Input
P1224	4644	Pedal Position Sensor 1 and 2 Range/Performance Error
P1226	4646	Throttle Malfunction (Flap Malfunction)
P1229	4649	Throttle Sensor Adaptation Failure
P1320	4896	Flywheel Adaptation for Misfire Detection Range
P1321	4897	Flywheel Adaptation for Misfire Detection Performance
P1366	4966	Ignition Coil 'A' Primary/Secondary Circuit Low
P1367	4967	Ignition Coil 'B' Primary/Secondary Circuit Low
P1436	5174	Leakage Diagnostic Pump Open Circuit
P1437	5175	Leakage Diagnostic Pump Range/Performance
P1442	5186	Leakage Diagnostic Pump Control Circuit Signal Low
P1443	5187	Leakage Diagnostic Pump Control Circuit Signal High
P1475	5237	Leakage Diagnostic Pump Reed Switch Did not Close
P1476	5238	Leakage Diagnostic Pump Clamped Tube
P1477	5239	Leakage Diagnostic Pump Reed Switch Did Not Open
P1570	5488	Electronic Control Module Sensor Supply A Low Output
P1571	5489	Electronic Control Module Sensor Supply A High Output
P1572	5490	Electronic Control Module Sensor Supply A Noisy Signal
P1573	5491	Electronic Control Module Sensor Supply B Low Output
P1574	5492	Electronic Control Module Sensor Supply B High Output
P1575	5493	Electronic Control Module Sensor Supply B Noisy Signal
P1600	5632	External Control Module Random Access Memory (RAM) Error
P1607	5639	CAN-Version
P1611	5649	Serial Communication Link Transmission Control Module
P1612	5650	Serial Communication Link Instrument Pack
P1613	5651	Serial Communication Link ASC (Automatic Stability Control)
P1615	5653	Electronic Control Module Processor SPI-Bus Failure
P1617	5655	Electronic Control Module H Bridge Controller
P1679	5753	Electronic Throttle Control Monitor Level 2/3 Torque Loss Calculation
P1680	5760	Electronic Throttle Control Monitor Level 2/3 ADC Processor Fault
P1681	5761	Electronic Throttle Control Monitor Level 2/3 Engine Speed Calculation Error
P1682	5762	Electronic Throttle Control Monitor Level 2/3 Idle Speed 'A' Calculation Fault
P1683	5763	Electronic Throttle Control Monitor Level 2/3 Idle Speed 'B' Calculation Fault
P1684	5764	Electronic Throttle Control Monitor Level 2/3 Clutch Torque Min Error
P1685	5765	Electronic Throttle Control Monitor Level 2/3 Clutch Torque Max Error
P1686	5766	Electronic Throttle Control Monitor Level 2/3 Pedal Position Sensor Diagnostic Error
P1687	5767	Electronic Throttle Control Monitor Level 2/3 Throttle Position Sensor Diagnostic Error
P1688	5768	Electronic Throttle Control Monitor Level 2/3 Mass Air Flow Calculation
P1689	5769	Electronic Throttle Control Monitor Level 2/3 Torque Calculation Error

Table b. Cooper DTCs from 9/1/2002 to 1/10/2003		
P-code	**MINI FC**	**P-code text**
P1691	5777	Electronic Throttle Control Monitor Level 2/3 Motorized Throttle Control Engine Speed Limitation Error
P1692	5778	Electronic Throttle Control Monitor Level 2/3 Motorized Throttle Control and Fuel Injection Switch Off 'A'
P1693	5779	Electronic Throttle Control Monitor Level 2/3 Motorized Throttle Control and Fuel Injection Switch Off 'B'
P1698	5784	Transmission Control Module Control Error
P1699	5785	Transmission Control Module Checksum Error
P1705	5893	Transmission Control Module LED Output Open Circuit
P1706	5894	Transmission Control Module LED Output Short Circuit
P1739	5945	Clutch Solenoid Communication Error
P1741	5953	Clutch Solenoid Open Circuit
P1742	5954	Clutch Solenoid Short Circuit
P1749	5961	Secondary Pressure Solenoid Communication Error
P1751	5969	Secondary Pressure Solenoid Open Circuit (M52: System Voltage Input High)
P1752	5970	Secondary Pressure Solenoid Short Circuit
P1785	6021	Transmission Ratio Control Actuator Circuit Malfunction
P1786	6022	Transmission Ratio Control Actuator Circuit Range/Performance
P1787	6023	Transmission Ratio Control Actuator Open Circuit
P1788	6024	Transmission Ratio Control Actuator Short Circuit
P1789	6025	Transmission Ratio Control Actuator Communication Error
P1815	6165	Wheel Plus Switch Error Low Input
P1816	6166	Wheel Minus Switch Error Low Input
P2096	8342	Post Catalyst Fuel Trim System Too Lean (Bank 1)
P2097	8343	Post Catalyst Fuel Trim System Too Rich (Bank 1)

Table c. Cooper DTCs from 1/11/2003		
PCode	**MINI-FC**	**PCode text**
P0030	48	HO2S Heater Control Circuit (Bank 1 Sensor 1)
P0031	49	HO2S Heater Control Circuit Low (Bank 1 Sensor 1)
P0032	50	HO2S Heater Control Circuit High (Bank 1 Sensor 1)
P0036	54	HO2S Heater Control Circuit (Bank 1 Sensor 2)
P0037	55	HO2S Heater Control Circuit Low (Bank 1 Sensor 2)
P0038	56	HO2S Heater Control Circuit High (Bank 1 Sensor 2)
P0053	83	HO2S Heater Resistance (Bank 1 Sensor 1)
P0054	84	HO2S Heater Resistance (Bank 1 Sensor 2)
P0101		Mass or Volume Air Flow Circuit Range/Performance
P0102		Mass or Volume Air Flow Circuit Low Input
P0103		Mass or Volume Air Flow Circuit High Input
P0107	263	Manifold Absolute Pressure/Barometric Pressure Circuit Low Input
P0108	264	Manifold Absolute Pressure/Barometric Pressure Circuit High Input
P0111		Intake AIr Temperature Range/Performance
P0112	274	Intake Air Temperature Sensor 1 Circuit Low
P0113	275	Intake Air Temperature Sensor 1 Circuit High
P0114	276	Intake Air Temperature Sensor 1 Circuit Intermittent
P0116	278	Engine Coolant Temperature Circuit Range/Performance
P0117	279	Engine Coolant Temperature Circuit Low
P0118	280	Engine Coolant Temperature Circuit High
P0119	281	Engine Coolant Temperature Circuit Intermittent
P0121		Throttle/Pedal Position Sensor/Switch 'A' Circuit Range/Performance
P0122	290	Throttle/Pedal Position Sensor/Switch 'A' Circuit Low
P0123	291	Throttle/Pedal Position Sensor/Switch 'A' Circuit High
P0125	293	Insufficient Coolant Temperature for Closed Loop Fuel Control
P0128	296	Coolant Thermostat (Coolant Temperature Below Thermostat Regulating Temperature)
P0130	304	O2 Sensor Circuit (Bank 1 Sensor 1)
P0131	305	O2 Sensor Circuit Low Voltage (Bank 1 Sensor 1)
P0132	306	O2 Sensor Circuit High Voltage (Bank 1 Sensor 1)
P0133	307	O2 Sensor Circuit Slow Response (Bank 1 Sensor 1)
P0134		O2 Sensor Circuit No Activity Detected (Bank 1 Sensor 1)
P0135		O2 Sensor Heater Circuit Malfunction (Bank 1 Sensor 1)
P0136	310	O2 Sensor Circuit (Bank 1 Sensor 2)
P0137	311	O2 Sensor Circuit Low Voltage (Bank 1 Sensor 2)
P0138	312	O2 Sensor Circuit High Voltage (Bank 1 Sensor 2)
P0139		O2 Sensor Circuit Slow Response (Bank 1 Sensor 2)
P0140		O2 Sensor Circuit Slow Response (Bank 1 Sensor 2)
P0141		O2 Sensor Heater Circuit Malfunction (Bank 1 Sensor 2)
P0170		Fuel Trim Malfunction (Bank 1)
P0171	369	System Too Lean (Bank 1)
P0172	370	System Too Rich (Bank 1)

P-codes (DTCs) and MINI fault codes

Table c. Cooper DTCs from 1/11/2003		
PCode	**MINI-FC**	**PCode text**
P0201	513	Injector Circuit/Open - Cylinder 1
P0202	514	Injector Circuit/Open - Cylinder 2
P0203	515	Injector Circuit/Open - Cylinder 3
P0204	516	Injector Circuit/Open - Cylinder 4
P0221		Throttle/Pedal Position Sensor/Switch ‘B’ Circuit Range/Performance
P0222	546	Throttle/Pedal Position Sensor/Switch ‘B’ Circuit Low
P0223	547	Throttle/Pedal Position Sensor/Switch ‘B’ Circuit High
P0261	609	Cylinder 1 Injector Circuit Low
P0262	610	Cylinder 1 Injector Circuit High
P0264	612	Cylinder 2 Injector Circuit Low
P0265	613	Cylinder 2 Injector Circuit High
P0267	615	Cylinder 3 Injector Circuit Low
P0268	616	Cylinder 3 Injector Circuit High
P0270	624	Cylinder 4 Injector Circuit Low
P0271	625	Cylinder 4 Injector Circuit High
P0300	768	Random/Multiple Cylinder Misfire Detected
P0301	769	Cylinder 1 Misfire Detected
P0302	770	Cylinder 2 Misfire Detected
P0303	771	Cylinder 3 Misfire Detected
P0304	772	Cylinder 4 Misfire Detected
P0313	787	Misfire Detected with Low Fuel
P0324	804	Knock Control System Error
P0325		Knock Sensor 1 Circuit, Electrical Fault in Circuit
P0326	806	Knock Sensor 1 Circuit Range/Performance
P0335	821	Crankshaft Position Sensor ‘A’ Circuit
P0336	822	Crankshaft Position Sensor ‘A’ Circuit Range/Performance
P0339		Crankshaft Position Sensor ‘A’ Circuit Intermittent
P0340	832	Camshaft Position Sensor ‘A’ Circuit (Bank 1 or Single Sensor)
P0341	833	Camshaft Position Sensor ‘A’ Circuit Range/Performance (Bank 1 or Single Sensor)
P0344		Camshaft Position Sensor ‘A’ Circuit Intermittent (Bank 1 or Single Sensor)
P0365		Camshaft Position Sensor ‘B’ Circuit Malfunction (Bank 1)
P0369		Camshaft Position Sensor Exhaust Cam Malfunction
P0412		Secondary Air Injection System Switching Valve Malfunction
P0413		Secondary Air Injection System Switching Valve ‘A’ Circuit Open
P0414		Secondary Air Injection System Switching Valve ‘A’ Circuit Shorted
P0420	1056	Catalyst System Efficiency Below Threshold (Bank 1)
P0441	1089	Evaporative Emission System Incorrect Purge Flow
P0442	1090	Evaporative Emission System Leak Detected (small leak)
P0443	1091	Evaporative Emission System Purge Control Valve Circuit
P0444	1092	Evaporative Emission System Purge Control Valve Circuit Open
P0445	1093	Evaporative Emission System Purge Control Valve Circuit Shorted
P0455	1109	Evaporative Emission System Leak Detected (large leak)

Table c. Cooper DTCs from 1/11/2003

PCode	MINI-FC	PCode text
P0456	1110	Evaporative Emission System Leak Detected (very small leak)
P0491		Secondary Air Injection System Insufficient Flow (Bank 1)
P0500	1280	Vehicle Speed Sensor
P0505		Idle Control System Malfunction - Idle Speed not Plausible
P0506	1286	Idle Air Control System RPM Lower Than Expected
P0507	1287	Idle Air Control System RPM Higher Than Expected
P0600		Serial Communication Link (Data Bus) Message Missing
P0601	1537	Internal Control Module Memory Check Sum Error
P0603	1539	Internal Control Module Keep Alive Memory (KAM) Error
P0604	1540	Internal Control Module Random Access Memory (RAM) Error
P0638	1592	Throttle Actuator Control Range/Performance (Bank 1)
P0642	1602	Sensor Reference Voltage 'A' Circuit Low
P0643	1603	Sensor Reference Voltage 'A' Circuit High
P0652	1618	Sensor Reference Voltage 'B' Circuit Low
P0653	1619	Sensor Reference Voltage 'B' Circuit High
P0705	1797	Transmission Range Sensor 'A' Circuit Malfunction (PRNDL Input)
P0815	2069	Upshift Switch Circuit
P0816	2070	Downshift Switch Circuit
P1083		Fuel Control Limit Mixture Too Lean (Bank 1 Sensor 1)
P1084		Fuel Control Limit Mixture Too Rich (Bank 1 Sensor 1)
P1087		O2 Sensor Circuit Slow Response in Lean Control Range (Bank 1 Sensor 1)
P1088		O2 Sensor Circuit Slow Response in Rich Control Range (Bank 1 Sensor 1)
P1089		O2 Sensor Circuit Slow Response in Lean Control Range (Bank 1 Sensor 2)
P1090		Pre Catalyst Fuel Trim Control Too Lean (Bank 1)
P1092		Pre Catalyst Fuel Trim Control Too Rich (Bank 1)
P1106	4358	Manifold Air Pressure Sensor Too Low at Engine Stop
P1107	4359	Manifold Air Pressure Sensor Too Low at Idle Engine Running
P1108	4360	Manifold Air Pressure Sensor Too Low at Full Load for Low Engine Speed
P1109	4361	Manifold Air Pressure Too High in Deceleration
P1111		Engine Coolant Temperature Sensor Radiator Outlet Low Input
P1112		Engine Coolant Temperature Sensor Radiator Outlet High Input
P1120		Pedal Position Sensor Circuit
P1121		Pedal Position Sensor 1 Range/Performance Problem
P1122		Pedal Position Sensor 1 Low Input
P1123		Pedal Position Sensor 1 High Input
P1125	4389	Throttle Position Sensor A and B Range/Performance Small Error
P1126	4390	Throttle Position Sensor A and B Range/Performance Large Error
P1134		O2 Sensor Heater Circuit Signal Intermittent (Bank 1 Sensor 1)
P1135		O2 Sensor Heater Circuit Low Voltage (Bank 1 Sensor 1)
P1136		O2 Sensor Heater Circuit High Voltage (Bank 1 Sensor 1)
P1137		O2 Sensor Heater Circuit Signal Intermittent (Bank 1 Sensor 2)
P1138		O2 Sensor Heater Circuit Low Voltage (Bank 1 Sensor 2)

Table c. Cooper DTCs from 1/11/2003		
PCode	**MINI-FC**	**PCode text**
P1139		O2 Sensor Heater Circuit High Voltage (Bank 1 Sensor 2)
P1143		O2 Sensor Activity Check Signal Too High (Bank 1 Sensor 2)
P1144		O2 Sensor Activity Check Signal Too Low (Bank 1 Sensor 2)
P1171		Ambient Pressure Sensor Variant Recognition Value in Boost Range Implausible
P1172		Ambient Pressure Sensor Variant Recognition Error Value Stored in Boot Range
P1173		Ambient Pressure Sensor Variant Recognition Learning Failed
P1178		O2 Sensor Pre Cat (Bank 1) - Switching Time Too Slow
P1188		Fuel Trim (Bank 1), O2 Control Adaptation Limit - Multiplicative
P1190		Pre Catalyst Fuel Trim Control (Bank 1)
P1192		Post Catalyst Fuel Trim System (Bank 1)
P1222		Pedal Position Sensor 2 Low Input
P1223		Pedal Position Sensor 2 High Input
P1229	4649	Throttle Sensor Adaptation Failure
P1314		Fuel Mixture Deviation Detected with Low Fuel
P1320	4896	Flywheel Adaptation for Misfire Detection Range
P1321	4897	Flywheel Adaptation for Misfire Detection Performance
P1342		Misfire During Start Cylinder 1
P1343		Misfire Cylinder 1 with Fuel Cut-Off
P1344		Misfire During Start Cylinder 2
P1345		Misfire Cylinder 2 with Fuel Cut-Off
P1346		Misfire During Start Cylinder 3
P1347		Misfire Cylinder 3 with Fuel Cut-Off
P1348		Misfire During Start Cylinder 4
P1349		Misfire Cylinder 4 with Fuel Cut-Off
P1403		ECU internal test
P1413		Secondary Air Injection Pump Relay Control Circuit Signal Low
P1414		Secondary Air Injection Pump Relay Control Circuit Signal High
P1423		Secondary Air Injection (Bank 1) - Flow too Low
P1444		Diagnostic Module Tank Leakage (DM-TL) Pump Control Open Circuit
P1445		Diagnostic Module Tank Leakage (DM-TL) Pump Control Circuit Signal Low
P1446		Diagnostic Module Tank Leakage (DM-TL) Pump Control Circuit Signal High
P1447		Diagnostic Module Tank Leakage (DM-TL) Pump Control Current Too High during Switching Solenoid Test
P1448		Diagnostic Module Tank Leakage (DM-TL) Pump Current Too Low
P1449		Diagnostic Module Tank Leakage (DM-TL) Pump Current Too High
P1451		Diagnostic Module Tank Leakage (DM-TL) Switching Solenoid Control Circuit Signal Low
P1452		Diagnostic Module Tank Leakage (DM-TL) Switching Solenoid Control Circuit Signal High
P1453		Secondary Air Injection System - Relay or Pump
P1475	5237	Leakage Diagnostic Pump Reed Switch Did not Close
P1476	5238	Leakage Diagnostic Pump Clamped Tube
P1477	5239	Leakage Diagnostic Pump Reed Switch Did Not Open
P1478		Range Check Voltage Supply Pedal Sensor 1 and Throttle Position Sensors

Table c. Cooper DTCs from 1/11/2003		
PCode	**MINI-FC**	**PCode text**
P1500		Idle Speed Control Valve Stuck Open
P1501		Idle Speed Control Valve Stuck Closed
P1502		Idle Speed Control Valve Closing Solenoid Control Circuit Signal High
P1503		Idle Speed Control Valve Closing Solenoid Control Circuit Signal Low
P1504		Idle Speed Control Valve Closing Solenoid Control Open Circuit
P1506		Idle Speed Control Valve Opening Solenoid Control Circuit Signal High
P1507		Idle Speed Control Valve Opening Solenoid Control Circuit Signal Low
P1508		Idle Speed Control Valve Opening Solenoid Control Open Circuit
P1509		Idle Control Valve Opening Coil - Malfunction
P1512		DISA (Differentiated Intake Manifold) Control Circuit Signal Low
P1513		DISA (Differentiated Intake Manifold) Control Circuit Signal High
P1523		'A' Camshaft Position Actuator Signal Low (Bank 1)
P1524		'A' Camshaft Position Actuator Control Circuit Signal High (Bank 1)
P1529		'B' Camshaft Position Actuator Control Circuit Signal Low (Bank 1)
P1530		'B' Camshaft Position Actuator Control Circuit Signal High (Bank 1)
P1550		Idle Control Valve Closing Coil - Malfunction
P1572	5490	Electronic Control Module Sensor Supply A Noisy Signal
P1575	5493	Electronic Control Module Sensor Supply B Noisy Signal
P1600	5632	External Control Module Random Access Memory (RAM) Error
P1602		Control Module Self Test - Control Module Defective
P1603		Control Module Self Test - Torque Monitoring
P1604		Control Module Self Test - Speed Monitoring
P1607	5639	CAN-Version
P1611	5649	Serial Communication Link Transmission Control Module
P1612	5650	Serial Communication Link Instrument Pack
P1613	5651	Serial Communication Link ASC (Automatic Stability Control)
P1615	5653	Electronic Control Module Processor SPI-Bus Failure
P1617	5655	Electronic Control Module H Bridge Controller
P1619		Map Cooling Thermostat Control Circuit Signal Low
P1620		Map Cooling Thermostat Control Circuit Signal High
P1624		Pedal Position Sensor Potentiometer Supply Channel 1 Electrical
P1625		Pedal Position Sensor Potentiometer Supply Channel 2 Electrical
P1632		Throttle Valve Adaptation Conditions Not Met
P1633		Throttle Valve Adaptation Limp-Home Positions Unknown
P1634		Throttle Valve Adaptation Spring Test Failed
P1635		Throttle Valve Adaptation Lower Mechanical Stop not Adapted
P1636		Throttle Valve Control Circuit Mechanically; pulse width not plausible
P1637		Throttle Valve Position Control, Control Deviation
P1638		Throttle Valve Position Control Throttle Stuck Temporarily
P1639		Throttle Valve Position Control Throttle Stuck Permanently
P1675		Throttle Valve Actuator Start Test Re-Adaptation Required
P1679	5753	Electronic Throttle Control Monitor Level 2/3 Torque Loss Calculation

Table c. Cooper DTCs from 1/11/2003		
PCode	**MINI-FC**	**PCode text**
P1680	5760	Electronic Throttle Control Monitor Level 2/3 ADC Processor Fault
P1681	5761	Electronic Throttle Control Monitor Level 2/3 Engine Speed Calculation Error
P1682	5762	Electronic Throttle Control Monitor Level 2/3 Idle Speed 'A' Calculation Fault
P1683	5763	Electronic Throttle Control Monitor Level 2/3 Idle Speed 'B' Calculation Fault
P1684	5764	Electronic Throttle Control Monitor Level 2/3 Clutch Torque Min Error
P1685	5765	Electronic Throttle Control Monitor Level 2/3 Clutch Torque Max Error
P1686	5766	Electronic Throttle Control Monitor Level 2/3 Pedal Position Sensor Diagnostic Error
P1687	5767	Electronic Throttle Control Monitor Level 2/3 Throttle Position Sensor Diagnostic Error
P1688	5768	Electronic Throttle Control Monitor Level 2/3 Mass Air Flow Calculation
P1689	5769	Electronic Throttle Control Monitor Level 2/3 Torque Calculation Error
P1691	5777	Electronic Throttle Control Monitor Level 2/3 Motorized Throttle Control Engine Speed Limitation Error
P1692	5778	Electronic Throttle Control Monitor Level 2/3 Motorized Throttle Control and Fuel Injection Switch Off 'A'
P1693	5779	Electronic Throttle Control Monitor Level 2/3 Motorized Throttle Control and Fuel Injection Switch Off 'B'
P1694		Throttle Valve Actuator Start Test Spring Test and Limp-Home Position Failed
P1698	5784	Transmission Control Module Control Error
P1699	5785	Transmission Control Module Checksum Error
P1705	5893	Transmission Control Module LED Output Open Circuit
P1706	5894	Transmission Control Module LED Output Short Circuit
P1739	5945	Clutch Solenoid Communication Error
P1741	5953	Clutch Solenoid Open Circuit
P1742	5954	Clutch Solenoid Short Circuit
P1749	5961	Secondary Pressure Solenoid Communication Error
P1751	5969	Secondary Pressure Solenoid Open Circuit
P1752	5970	Secondary Pressure Solenoid Short Circuit
P1785	6021	Transmission Ratio Control Actuator Circuit Malfunction
P1786	6022	Transmission Ratio Control Actuator Circuit Range/Performance
P1787	6023	Transmission Ratio Control Actuator Open Circuit
P1788	6024	Transmission Ratio Control Actuator Short Circuit
P1789	6025	Transmission Ratio Control Actuator Communication Error
P1815	6165	Wheel Plus Switch Error Low Input
P1816	6166	Wheel Minus Switch Error Low Input
P2088		'A' Camshaft Position Actuator Control Circuit Low (Bank 1)
P2089		'A' Camshaft Position Actuator Control Circuit High (Bank 1)
P2090		'B' Camshaft Position Actuator Control Circuit Low (Bank 1)
P2091		'B' Camshaft Position Actuator Control Circuit High (Bank 1)
P2096	8342	Post Catalyst Fuel Trim System Too Lean (Bank 1)
P2097	8343	Post Catalyst Fuel Trim System Too Rich (Bank 1)
P2122	8482	Throttle/Pedal Position Sensor/Switch 'D' Circuit Low Input
P2123	8483	Throttle/Pedal Position Sensor/Switch 'D' Circuit High Input
P2127	8487	Throttle/Pedal Position Sensor/Switch 'E' Circuit Low Input
P2128	8488	Throttle/Pedal Position Sensor/Switch 'E' Circuit High Input
P2138	8504	Throttle/Pedal Position Sensor/Switch 'D' / 'E' Voltage Correlation

Table c. Cooper DTCs from 1/11/2003

PCode	MINI-FC	PCode text
P2228		Barometric Pressure Circuit Low
P2229		Barometric Pressure Circuit Low
P2270	8816	O2 Sensor Signal Stuck Lean (Bank 1 Sensor 2)
P2271	8817	O2 Sensor Signal Stuck Rich (Bank 1 Sensor 2)
P2300	8960	Ignition Coil 'A' Primary Control Circuit Low
P2301	8961	Ignition Coil 'A' Primary Control Circuit High
P2303	8963	Ignition Coil 'B' Primary Control Circuit Low
P2304	8964	Ignition Coil 'B' Primary Control Circuit High
P2400	9216	Evaporative Emission System Leak Detection Pump Control Circuit/Open
P2401	9217	Evaporative Emission System Leak Detection Pump Control Circuit Low
P2402	9218	Evaporative Emission System Leak Detection Pump Control Circuit High
P2404	9220	Evaporative Emission System Leak Detection Pump Sense Circuit Range/Performance

Cooper S DTCs

Table d. Cooper S DTCs from 1/1/2002 to 8/31/2002		
P-code	**MINI FC**	**P-code text**
P0030	48	HO2S Heater Control Circuit (Bank 1 Sensor 1)
P0031	49	HO2S Heater Control Circuit Low (Bank 1 Sensor 1)
P0032	50	HO2S Heater Control Circuit High (Bank 1 Sensor 1)
P0036	54	HO2S Heater Control Circuit (Bank 1 Sensor 2)
P0037	55	HO2S Heater Control Circuit Low (Bank 1 Sensor 2)
P0038	56	HO2S Heater Control Circuit High (Bank 1 Sensor 2)
P0107	263	Manifold Absolute Pressure/Barometric Pressure Circuit Low Input
P0108	264	Manifold Absolute Pressure/Barometric Pressure Circuit High Input
P0112	274	Intake Air Temperature Sensor 1 Circuit Low
P0113	275	Intake Air Temperature Sensor 1 Circuit High
P0114	276	Intake Air Temperature Sensor 1 Circuit Intermittent
P0116	278	Engine Coolant Temperature Circuit Range/Performance
P0117	279	Engine Coolant Temperature Circuit Low
P0118	280	Engine Coolant Temperature Circuit High
P0119	281	Engine Coolant Temperature Circuit Intermittent
P0122	290	Throttle/Pedal Position Sensor/Switch 'A' Circuit Low
P0123	291	Throttle/Pedal Position Sensor/Switch 'A' Circuit High
P0125	293	Insufficient Coolant Temperature for Closed Loop Fuel Control
P0128	296	Coolant Thermostat (Coolant Temperature Below Thermostat Regulating Temperature)
P0130	304	O2 Sensor Circuit (Bank 1 Sensor 1)
P0131	305	O2 Sensor Circuit Low Voltage (Bank 1 Sensor 1)
P0132	306	O2 Sensor Circuit High Voltage (Bank 1 Sensor 1)
P0133	307	O2 Sensor Circuit Slow Response (Bank 1 Sensor 1)
P0135	309	O2 Sensor Heater Circuit (Bank 1 Sensor 1)
P0136	310	O2 Sensor Circuit (Bank 1 Sensor 2)
P0137	311	O2 Sensor Circuit Low Voltage (Bank 1 Sensor 2)
P0138	312	O2 Sensor Circuit High Voltage (Bank 1 Sensor 2)
P0141	321	O2 Sensor Heater Circuit (Bank 1 Sensor 2)
P0171	369	System Too Lean (Bank 1)
P0172	370	System Too Rich (Bank 1)
P0201	513	Injector Circuit/Open - Cylinder 1
P0202	514	Injector Circuit/Open - Cylinder 2
P0203	515	Injector Circuit/Open - Cylinder 3
P0204	516	Injector Circuit/Open - Cylinder 4
P0222	546	Throttle/Pedal Position Sensor/Switch 'B' Circuit Low
P0223	547	Throttle/Pedal Position Sensor/Switch 'B' Circuit High
P0261	609	Cylinder 1 Injector Circuit Low
P0262	610	Cylinder 1 Injector Circuit High
P0264	612	Cylinder 2 Injector Circuit Low
P0265	613	Cylinder 2 Injector Circuit High

Table d. Cooper S DTCs from 1/1/2002 to 8/31/2002		
P-code	**MINI FC**	**P-code text**
P0267	615	Cylinder 3 Injector Circuit Low
P0268	616	Cylinder 3 Injector Circuit High
P0270	624	Cylinder 4 Injector Circuit Low
P0271	625	Cylinder 4 Injector Circuit High
P0300	768	Random/Multiple Cylinder Misfire Detected
P0301	769	Cylinder 1 Misfire Detected
P0302	770	Cylinder 2 Misfire Detected
P0303	771	Cylinder 3 Misfire Detected
P0304	772	Cylinder 4 Misfire Detected
P0313	787	Misfire Detected with Low Fuel
P0324	804	Knock Control System Error
P0326	806	Knock Sensor 1 Circuit Range/Performance (Bank 1 or Single Sensor)
P0335	821	Crankshaft Position Sensor 'A' Circuit
P0336	822	Crankshaft Position Sensor 'A' Circuit Range/Performance
P0340	832	Camshaft Position Sensor 'A' Circuit (Bank 1 or Single Sensor)
P0341	833	Camshaft Position Sensor 'A' Circuit Range/Performance (Bank 1 or Single Sensor)
P0351	849	Ignition Coil 'A' Primary/Secondary Circuit
P0352	850	Ignition Coil 'B' Primary/Secondary Circuit
P0420	1056	Catalyst System Efficiency Below Threshold (Bank 1)
P0441	1089	Evaporative Emission System Incorrect Purge Flow
P0442	1090	Evaporative Emission System Leak Detected (small leak)
P0443	1091	Evaporative Emission System Purge Control Valve Circuit
P0444	1092	Evaporative Emission System Purge Control Valve Circuit Open
P0445	1093	Evaporative Emission System Purge Control Valve Circuit Shorted
P0455	1109	Evaporative Emission System Leak Detected (large leak)
P0456	1110	Evaporative Emission System Leak Detected (very small leak)
P0500	1280	Vehicle Speed Sensor 'A'
P0506	1286	Idle Air Control System RPM Lower Than Expected
P0507	1287	Idle Air Control System RPM Higher Than Expected
P0601	1537	Internal Control Module Memory Check Sum Error
P0603	1539	Internal Control Module Keep Alive Memory (KAM) Error
P0604	1540	Internal Control Module Random Access Memory (RAM) Error
P0705	1797	Transmission Range Sensor 'A' Circuit Malfunction (PRNDL Input)
P0815	2069	Upshift Switch Circuit
P0816	2070	Downshift Switch Circuit
P1106	4358	Manifold Air Pressure Sensor Too Low at Engine Stop
P1107	4359	Manifold Air Pressure Sensor Too Low at Idle Engine Running
P1108	4360	Manifold Air Pressure Sensor Too Low at Full Load for Low Engine Speed
P1109	4361	Manifold Air Pressure Too High in Deceleration
P1122	4386	Pedal Position Sensor 1 Low Input
P1123	4387	Pedal Position Sensor 1 High Input

Table d. Cooper S DTCs from 1/1/2002 to 8/31/2002		
P-code	**MINI FC**	**P-code text**
P1125	4389	Throttle Position Sensor A and B Range/Performance Small Error
P1126	4390	Throttle Position Sensor A and B Range/Performance Large Error
P1143	4419	O2 Sensor Activity Check Signal Too High (Bank 1 Sensor 2)
P1144	4420	O2 Sensor Activity Check Signal Too Low (Bank 1 Sensor 2)
P1222	4642	Pedal Position Sensor 2 Low Input
P1223	4643	Pedal Position Sensor 2 High Input
P1224	4644	Pedal Position Sensor 1 and 2 Range/Performance Error
P1226	4646	Throttle Malfunction (Flap Malfunction)
P1229	4649	Throttle Sensor Adaptation Failure
P1237	4663	Secondary Upstream Manifold Air Pressure Sensor Low Input
P1238	4664	Secondary Upstream Manifold Air Pressure Sensor High Input
P1239	4665	Secondary Upstream Manifold Air Pressure Sensor Too Low at Engine stop
P1240	4672	Secondary Upstream Manifold Air Pressure Sensor Too Low at Idle Engine Running
P1241	4673	Secondary Upstream Manifold Air Pressure Sensor Too Low at Full Load for Low Engine Speed
P1242	4674	Secondary Upstream Manifold Air Pressure Sensor Too High in Deceleration
P1320	4896	Flywheel Adaptation for Misfire Detection Range
P1321	4897	Flywheel Adaptation for Misfire Detection Performance
P1366	4966	Ignition Coil 'A' Primary/Secondary Circuit Low
P1367	4967	Ignition Coil 'B' Primary/Secondary Circuit Low
P1436	5174	Leakage Diagnostic Pump Open Circuit
P1437	5175	Leakage Diagnostic Pump Range/Performance
P1442	5186	Leakage Diagnostic Pump Control Circuit Signal Low
P1443	5187	Leakage Diagnostic Pump Control Circuit Signal High
P1475	5237	Leakage Diagnostic Pump Reed Switch Did not Close
P1476	5238	Leakage Diagnostic Pump Clamped Tube
P1477	5239	Leakage Diagnostic Pump Reed Switch Did Not Open
P1570	5488	Electronic Control Module Sensor Supply A Low Output
P1571	5489	Electronic Control Module Sensor Supply A High Output
P1572	5490	Electronic Control Module Sensor Supply A Noisy Signal
P1573	5491	Electronic Control Module Sensor Supply B Low Output
P1574	5492	Electronic Control Module Sensor Supply B High Output
P1575	5493	Electronic Control Module Sensor Supply B Noisy Signal
P1600	5632	External Control Module Random Access Memory (RAM) Error
P1607	5639	CAN-Version
P1611	5649	Serial Communication Link Transmission Control Module
P1612	5650	Serial Communication Link Instrument Pack
P1613	5651	Serial Communication Link ASC (Automatic Stability Control)
P1615	5653	Electronic Control Module Processor SPI-Bus Failure
P1617	5655	Electronic Control Module H Bridge Controller
P1679	5753	Electronic Throttle Control Monitor Level 2/3 Torque Loss Calculation
P1680	5760	Electronic Throttle Control Monitor Level 2/3 ADC Processor Fault
P1681	5761	Electronic Throttle Control Monitor Level 2/3 Engine Speed Calculation Error

Table d. Cooper S DTCs from 1/1/2002 to 8/31/2002

P-code	MINI FC	P-code text
P1682	5762	Electronic Throttle Control Monitor Level 2/3 Idle Speed 'A' Calculation Fault
P1683	5763	Electronic Throttle Control Monitor Level 2/3 Idle Speed 'B' Calculation Fault
P1684	5764	Electronic Throttle Control Monitor Level 2/3 Clutch Torque Min Error
P1685	5765	Electronic Throttle Control Monitor Level 2/3 Clutch Torque Max Error
P1686	5766	Electronic Throttle Control Monitor Level 2/3 Pedal Position Sensor Diagnostic Error
P1687	5767	Electronic Throttle Control Monitor Level 2/3 Throttle Position Sensor Diagnostic Error
P1688	5768	Electronic Throttle Control Monitor Level 2/3 Mass Air Flow Calculation
P1689	5769	Electronic Throttle Control Monitor Level 2/3 Torque Calculation Error
P1691	5777	Electronic Throttle Control Monitor Level 2/3 Motorized Throttle Control Engine Speed Limitation Error
P1692	5778	Electronic Throttle Control Monitor Level 2/3 Motorized Throttle Control and Fuel Injection Switch Off 'A'
P1693	5779	Electronic Throttle Control Monitor Level 2/3 Motorized Throttle Control and Fuel Injection Switch Off 'B'
P1698	5784	Transmission Control Module Control Error
P1699	5785	Transmission Control Module Checksum Error
P1705	5893	Transmission Control Module LED Output Open Circuit
P1706	5894	Transmission Control Module LED Output Short Circuit
P1739	5945	Clutch Solenoid Communication Error
P1741	5953	Clutch Solenoid Open Circuit
P1742	5954	Clutch Solenoid Short Circuit
P1749	5961	Secondary Pressure Solenoid Communication Error
P1751	5969	Secondary Pressure Solenoid Open Circuit
P1752	5970	Secondary Pressure Solenoid Short Circuit
P1785	6021	Transmission Ratio Control Actuator Circuit Malfunction
P1786	6022	Transmission Ratio Control Actuator Circuit Range/Performance
P1787	6023	Transmission Ratio Control Actuator Open Circuit
P1788	6024	Transmission Ratio Control Actuator Short Circuit
P1789	6025	Transmission Ratio Control Actuator Communication Error
P1815	6165	Wheel Plus Switch Error Low Input
P1816	6166	Wheel Minus Switch Error Low Input
P2096	8342	Post Catalyst Fuel Trim System Too Lean (Bank 1)
P2097	8343	Post Catalyst Fuel Trim System Too Rich (Bank 1)

Table e. Cooper S DTCs from 9/1/2002 to 1/10/2003

PCode	MINI-FC	PCode text
P0030	48	HO2S Heater Control Circuit (Bank 1 Sensor 1)
P0031	49	HO2S Heater Control Circuit Low (Bank 1 Sensor 1)
P0032	50	HO2S Heater Control Circuit High (Bank 1 Sensor 1)
P0036	54	HO2S Heater Control Circuit (Bank 1 Sensor 2)
P0037	55	HO2S Heater Control Circuit Low (Bank 1 Sensor 2)
P0038	56	HO2S Heater Control Circuit High (Bank 1 Sensor 2)
P0107	263	Manifold Absolute Pressure/Barometric Pressure Circuit Low Input
P0108	264	Manifold Absolute Pressure/Barometric Pressure Circuit High Input
P0112	274	Intake Air Temperature Sensor 1 Circuit Low
P0113	275	Intake Air Temperature Sensor 1 Circuit High
P0114	276	Intake Air Temperature Sensor 1 Circuit Intermittent
P0116	278	Engine Coolant Temperature Circuit Range/Performance
P0117	279	Engine Coolant Temperature Circuit Low
P0118	280	Engine Coolant Temperature Circuit High
P0119	281	Engine Coolant Temperature Circuit Intermittent
P0122	290	Throttle/Pedal Position Sensor/Switch 'A' Circuit Low
P0123	291	Throttle/Pedal Position Sensor/Switch 'A' Circuit High
P0125	293	Insufficient Coolant Temperature for Closed Loop Fuel Control
P0128	296	Coolant Thermostat (Coolant Temperature Below Thermostat Regulating Temperature)
P0130	304	O2 Sensor Circuit (Bank 1 Sensor 1)
P0131	305	O2 Sensor Circuit Low Voltage (Bank 1 Sensor 1)
P0132	306	O2 Sensor Circuit High Voltage (Bank 1 Sensor 1)
P0133	307	O2 Sensor Circuit Slow Response (Bank 1 Sensor 1)
P0135	309	O2 Sensor Heater Circuit (Bank 1 Sensor 1)
P0136	310	O2 Sensor Circuit (Bank 1 Sensor 2)
P0137	311	O2 Sensor Circuit Low Voltage (Bank 1 Sensor 2)
P0138	312	O2 Sensor Circuit High Voltage (Bank 1 Sensor 2)
P0141	321	O2 Sensor Heater Circuit (Bank 1 Sensor 2)
P0171	369	System Too Lean (Bank 1)
P0172	370	System Too Rich (Bank 1)
P0201	513	Injector Circuit/Open - Cylinder 1
P0202	514	Injector Circuit/Open - Cylinder 2
P0203	515	Injector Circuit/Open - Cylinder 3
P0204	516	Injector Circuit/Open - Cylinder 4
P0222	546	Throttle/Pedal Position Sensor/Switch 'B' Circuit Low
P0223	547	Throttle/Pedal Position Sensor/Switch 'B' Circuit High
P0261	609	Cylinder 1 Injector Circuit Low
P0262	610	Cylinder 1 Injector Circuit High
P0264	612	Cylinder 2 Injector Circuit Low
P0265	613	Cylinder 2 Injector Circuit High
P0267	615	Cylinder 3 Injector Circuit Low
P0268	616	Cylinder 3 Injector Circuit High

Table e. Cooper S DTCs from 9/1/2002 to 1/10/2003		
PCode	**MINI-FC**	**PCode text**
P0270	624	Cylinder 4 Injector Circuit Low
P0271	625	Cylinder 4 Injector Circuit High
P0300	768	Random/Multiple Cylinder Misfire Detected
P0301	769	Cylinder 1 Misfire Detected
P0302	770	Cylinder 2 Misfire Detected
P0303	771	Cylinder 3 Misfire Detected
P0304	772	Cylinder 4 Misfire Detected
P0313	787	Misfire Detected with Low Fuel
P0324	804	Knock Control System Error
P0326	806	Knock Sensor 1 Circuit Range/Performance (Bank 1 or Single Sensor)
P0335	821	Crankshaft Position Sensor 'A' Circuit
P0336	822	Crankshaft Position Sensor 'A' Circuit Range/Performance
P0340	832	Camshaft Position Sensor 'A' Circuit (Bank 1 or Single Sensor)
P0341	833	Camshaft Position Sensor 'A' Circuit Range/Performance (Bank 1 or Single Sensor)
P0351	849	Ignition Coil 'A' Primary/Secondary Circuit
P0352	850	Ignition Coil 'B' Primary/Secondary Circuit
P0420	1056	Catalyst System Efficiency Below Threshold (Bank 1)
P0441	1089	Evaporative Emission System Incorrect Purge Flow
P0442	1090	Evaporative Emission System Leak Detected (small leak)
P0443	1091	Evaporative Emission System Purge Control Valve Circuit
P0444	1092	Evaporative Emission System Purge Control Valve Circuit Open
P0445	1093	Evaporative Emission System Purge Control Valve Circuit Shorted
P0455	1109	Evaporative Emission System Leak Detected (large leak)
P0456	1110	Evaporative Emission System Leak Detected (very small leak)
P0500	1280	Vehicle Speed Sensor 'A'
P0506	1286	Idle Air Control System RPM Lower Than Expected
P0507	1287	Idle Air Control System RPM Higher Than Expected
P0601	1537	Internal Control Module Memory Check Sum Error
P0603	1539	Internal Control Module Keep Alive Memory (KAM) Error
P0604	1540	Internal Control Module Random Access Memory (RAM) Error
P0705	1797	Transmission Range Sensor 'A' Circuit Malfunction (PRNDL Input)
P0815	2069	Upshift Switch Circuit
P0816	2070	Downshift Switch Circuit
P1106	4358	Manifold Air Pressure Sensor Too Low at Engine Stop
P1107	4359	Manifold Air Pressure Sensor Too Low at Idle Engine Running
P1108	4360	Manifold Air Pressure Sensor Too Low at Full Load for Low Engine Speed
P1109	4361	Manifold Air Pressure Too High in Deceleration
P1122	4386	Pedal Position Sensor 1 Low Input
P1123	4387	Pedal Position Sensor 1 High Input
P1125	4389	Throttle Position Sensor A and B Range/Performance Small Error
P1126	4390	Throttle Position Sensor A and B Range/Performance Large Error

Table e. Cooper S DTCs from 9/1/2002 to 1/10/2003		
PCode	**MINI-FC**	**PCode text**
P1143	4419	O2 Sensor Activity Check Signal Too High (Bank 1 Sensor 2)
P1144	4420	O2 Sensor Activity Check Signal Too Low (Bank 1 Sensor 2)
P1222	4642	Pedal Position Sensor 2 Low Input
P1223	4643	Pedal Position Sensor 2 High Input
P1224	4644	Pedal Position Sensor 1 and 2 Range/Performance Error
P1226	4646	Throttle Malfunction (Flap Malfunction)
P1229	4649	Throttle Sensor Adaptation Failure
P1237	4663	Secondary Upstream Manifold Air Pressure Sensor Low Input
P1238	4664	Secondary Upstream Manifold Air Pressure Sensor High Input
P1239	4665	Secondary Upstream Manifold Air Pressure Sensor Too Low at Engine stop
P1240	4672	Secondary Upstream Manifold Air Pressure Sensor Too Low at Idle Engine Running
P1241	4673	Secondary Upstream Manifold Air Pressure Sensor Too Low at Full Load for Low Engine Speed
P1242	4674	Secondary Upstream Manifold Air Pressure Sensor Too High in Deceleration
P1320	4896	Flywheel Adaptation for Misfire Detection Range
P1321	4897	Flywheel Adaptation for Misfire Detection Performance
P1366	4966	Ignition Coil 'A' Primary/Secondary Circuit Low
P1367	4967	Ignition Coil 'B' Primary/Secondary Circuit Low
P1436	5174	Leakage Diagnostic Pump Open Circuit
P1437	5175	Leakage Diagnostic Pump Range/Performance
P1442	5186	Leakage Diagnostic Pump Control Circuit Signal Low
P1443	5187	Leakage Diagnostic Pump Control Circuit Signal High
P1475	5237	Leakage Diagnostic Pump Reed Switch Did not Close
P1476	5238	Leakage Diagnostic Pump Clamped Tube
P1477	5239	Leakage Diagnostic Pump Reed Switch Did Not Open
P1570	5488	Electronic Control Module Sensor Supply A Low Output
P1571	5489	Electronic Control Module Sensor Supply A High Output
P1572	5490	Electronic Control Module Sensor Supply A Noisy Signal
P1573	5491	Electronic Control Module Sensor Supply B Low Output
P1574	5492	Electronic Control Module Sensor Supply B High Output
P1575	5493	Electronic Control Module Sensor Supply B Noisy Signal
P1600	5632	External Control Module Random Access Memory (RAM) Error
P1607	5639	CAN-Version
P1611	5649	Serial Communication Link Transmission Control Module
P1612	5650	Serial Communication Link Instrument Pack
P1613	5651	Serial Communication Link ASC (Automatic Stability Control)
P1615	5653	Electronic Control Module Processor SPI-Bus Failure
P1617	5655	Electronic Control Module H Bridge Controller
P1679	5753	Electronic Throttle Control Monitor Level 2/3 Torque Loss Calculation
P1680	5760	Electronic Throttle Control Monitor Level 2/3 ADC Processor Fault
P1681	5761	Electronic Throttle Control Monitor Level 2/3 Engine Speed Calculation Error
P1682	5762	Electronic Throttle Control Monitor Level 2/3 Idle Speed 'A' Calculation Fault
P1683	5763	Electronic Throttle Control Monitor Level 2/3 Idle Speed 'B' Calculation Fault

Table e. Cooper S DTCs from 9/1/2002 to 1/10/2003		
PCode	**MINI-FC**	**PCode text**
P1684	5764	Electronic Throttle Control Monitor Level 2/3 Clutch Torque Min Error
P1685	5765	Electronic Throttle Control Monitor Level 2/3 Clutch Torque Max Error
P1686	5766	Electronic Throttle Control Monitor Level 2/3 Pedal Position Sensor Diagnostic Error
P1687	5767	Electronic Throttle Control Monitor Level 2/3 Throttle Position Sensor Diagnostic Error
P1688	5768	Electronic Throttle Control Monitor Level 2/3 Mass Air Flow Calculation
P1689	5769	Electronic Throttle Control Monitor Level 2/3 Torque Calculation Error
P1691	5777	Electronic Throttle Control Monitor Level 2/3 Motorized Throttle Control Engine Speed Limitation Error
P1692	5778	Electronic Throttle Control Monitor Level 2/3 Motorized Throttle Control and Fuel Injection Switch Off 'A'
P1693	5779	Electronic Throttle Control Monitor Level 2/3 Motorized Throttle Control and Fuel Injection Switch Off 'B'
P1698	5784	Transmission Control Module Control Error
P1699	5785	Transmission Control Module Checksum Error
P1705	5893	Transmission Control Module LED Output Open Circuit
P1706	5894	Transmission Control Module LED Output Short Circuit
P1739	5945	Clutch Solenoid Communication Error
P1741	5953	Clutch Solenoid Open Circuit
P1742	5954	Clutch Solenoid Short Circuit
P1749	5961	Secondary Pressure Solenoid Communication Error
P1751	5969	Secondary Pressure Solenoid Open Circuit
P1752	5970	Secondary Pressure Solenoid Short Circuit
P1785	6021	Transmission Ratio Control Actuator Circuit Malfunction
P1786	6022	Transmission Ratio Control Actuator Circuit Range/Performance
P1787	6023	Transmission Ratio Control Actuator Open Circuit
P1788	6024	Transmission Ratio Control Actuator Short Circuit
P1789	6025	Transmission Ratio Control Actuator Communication Error
P1815	6165	Wheel Plus Switch Error Low Input
P1816	6166	Wheel Minus Switch Error Low Input
P2096	8342	Post Catalyst Fuel Trim System Too Lean (Bank 1)
P2097	8343	Post Catalyst Fuel Trim System Too Rich (Bank 1)

P-codes (DTCs) and MINI fault codes

Table f. Cooper S DTCs from 1/11/2003		
PCode	**MINI-FC**	**PCode text**
P0030	48	HO2S Heater Control Circuit (Bank 1 Sensor 1)
P0031	49	HO2S Heater Control Circuit Low (Bank 1 Sensor 1)
P0032	50	HO2S Heater Control Circuit High (Bank 1 Sensor 1)
P0036	54	HO2S Heater Control Circuit (Bank 1 Sensor 2)
P0037	55	HO2S Heater Control Circuit Low (Bank 1 Sensor 2)
P0038	56	HO2S Heater Control Circuit High (Bank 1 Sensor 2)
P0053	83	HO2S Heater Resistance (Bank 1 Sensor 1)
P0054	84	HO2S Heater Resistance (Bank 1 Sensor 2)
P0101		Mass or Volume Air Flow Circuit Range/Performance
P0102		Mass or Volume Air Flow Circuit Low Input
P0103		Mass or Volume Air Flow Circuit High Input
P0107	263	Manifold Absolute Pressure/Barometric Pressure Circuit Low Input
P0108	264	Manifold Absolute Pressure/Barometric Pressure Circuit High Input
P0111		Intake AIr Temperature Range/Performance
P0112	274	Intake Air Temperature Sensor 1 Circuit Low
P0113	275	Intake Air Temperature Sensor 1 Circuit High
P0114	276	Intake Air Temperature Sensor 1 Circuit Intermittent
P0116	278	Engine Coolant Temperature Circuit Range/Performance
P0117	279	Engine Coolant Temperature Circuit Low
P0118	280	Engine Coolant Temperature Circuit High
P0119	281	Engine Coolant Temperature Circuit Intermittent
P0121		Throttle/Pedal Position Sensor/Switch 'A' Circuit Range/Performance
P0122	290	Throttle/Pedal Position Sensor/Switch 'A' Circuit Low
P0123	291	Throttle/Pedal Position Sensor/Switch 'A' Circuit High
P0125	293	Insufficient Coolant Temperature for Closed Loop Fuel Control
P0128	296	Coolant Thermostat (Coolant Temperature Below Thermostat Regulating Temperature)
P0130	304	O2 Sensor Circuit (Bank 1 Sensor 1)
P0131	305	O2 Sensor Circuit Low Voltage (Bank 1 Sensor 1)
P0132	306	O2 Sensor Circuit High Voltage (Bank 1 Sensor 1)
P0133	307	O2 Sensor Circuit Slow Response (Bank 1 Sensor 1)
P0134		O2 Sensor Circuit No Activity Detected (Bank 1 Sensor 1)
P0135		O2 Sensor Heater Circuit Malfunction (Bank 1 Sensor 1)
P0136	310	O2 Sensor Circuit (Bank 1 Sensor 2)
P0137	311	O2 Sensor Circuit Low Voltage (Bank 1 Sensor 2)
P0138	312	O2 Sensor Circuit High Voltage (Bank 1 Sensor 2)
P0139		O2 Sensor Circuit Slow Response (Bank 1 Sensor 2)
P0140		O2 Sensor Circuit Slow Response (Bank 1 Sensor 2)
P0141		O2 Sensor Heater Circuit Malfunction (Bank 1 Sensor 2)
P0170		Fuel Trim Malfunction (Bank 1)
P0171	369	System Too Lean (Bank 1)
P0172	370	System Too Rich (Bank 1)

Table f. Cooper S DTCs from 1/11/2003		
PCode	**MINI-FC**	**PCode text**
P0201	513	Injector Circuit/Open - Cylinder 1
P0202	514	Injector Circuit/Open - Cylinder 2
P0203	515	Injector Circuit/Open - Cylinder 3
P0204	516	Injector Circuit/Open - Cylinder 4
P0221		Throttle/Pedal Position Sensor/Switch 'B' Circuit Range/Performance
P0222	546	Throttle/Pedal Position Sensor/Switch 'B' Circuit Low
P0223	547	Throttle/Pedal Position Sensor/Switch 'B' Circuit High
P0261	609	Cylinder 1 Injector Circuit Low
P0262	610	Cylinder 1 Injector Circuit High
P0264	612	Cylinder 2 Injector Circuit Low
P0265	613	Cylinder 2 Injector Circuit High
P0267	615	Cylinder 3 Injector Circuit Low
P0268	616	Cylinder 3 Injector Circuit High
P0270	624	Cylinder 4 Injector Circuit Low
P0271	625	Cylinder 4 Injector Circuit High
P0300	768	Random/Multiple Cylinder Misfire Detected
P0301	769	Cylinder 1 Misfire Detected
P0302	770	Cylinder 2 Misfire Detected
P0303	771	Cylinder 3 Misfire Detected
P0304	772	Cylinder 4 Misfire Detected
P0313	787	Misfire Detected with Low Fuel
P0324	804	Knock Control System Error
P0325		Knock Sensor 1 Circuit, Electrical Fault in Circuit
P0326	806	Knock Sensor 1 Circuit Range/Performance
P0335	821	Crankshaft Position Sensor 'A' Circuit
P0336	822	Crankshaft Position Sensor 'A' Circuit Range/Performance
P0339		Crankshaft Position Sensor 'A' Circuit Intermittent
P0340	832	Camshaft Position Sensor 'A' Circuit (Bank 1 or Single Sensor)
P0341	833	Camshaft Position Sensor 'A' Circuit Range/Performance (Bank 1 or Single Sensor)
P0344		Camshaft Position Sensor 'A' Circuit Intermittent (Bank 1 or Single Sensor)
P0365		Camshaft Position Sensor 'B' Circuit Malfunction (Bank 1)
P0369		Camshaft Position Sensor Exhaust Cam Malfunction
P0412		Secondary Air Injection System Switching Valve Malfunction
P0413		Secondary Air Injection System Switching Valve 'A' Circuit Open
P0414		Secondary Air Injection System Switching Valve 'A' Circuit Shorted
P0420	1056	Catalyst System Efficiency Below Threshold (Bank 1)
P0441	1089	Evaporative Emission System Incorrect Purge Flow
P0442	1090	Evaporative Emission System Leak Detected (small leak)
P0443	1091	Evaporative Emission System Purge Control Valve Circuit
P0444	1092	Evaporative Emission System Purge Control Valve Circuit Open
P0445	1093	Evaporative Emission System Purge Control Valve Circuit Shorted
P0455	1109	Evaporative Emission System Leak Detected (large leak)

P-codes (DTCs) and MINI fault codes

Table f. Cooper S DTCs from 1/11/2003		
PCode	**MINI-FC**	**PCode text**
P0456	1110	Evaporative Emission System Leak Detected (very small leak)
P0491		Secondary Air Injection System Insufficient Flow (Bank 1)
P0500	1280	Vehicle Speed Sensor
P0505		Idle Control System Malfunction - Idle Speed not Plausible
P0506	1286	Idle Air Control System RPM Lower Than Expected
P0507	1287	Idle Air Control System RPM Higher Than Expected
P0600		Serial Communication Link (Data Bus) Message Missing
P0601	1537	Internal Control Module Memory Check Sum Error
P0603	1539	Internal Control Module Keep Alive Memory (KAM) Error
P0604	1540	Internal Control Module Random Access Memory (RAM) Error
P0638	1592	Throttle Actuator Control Range/Performance (Bank 1)
P0642	1602	Sensor Reference Voltage 'A' Circuit Low
P0643	1603	Sensor Reference Voltage 'A' Circuit High
P0652	1618	Sensor Reference Voltage 'B' Circuit Low
P0653	1619	Sensor Reference Voltage 'B' Circuit High
P0705	1797	Transmission Range Sensor 'A' Circuit Malfunction (PRNDL Input)
P0815	2069	Upshift Switch Circuit
P0816	2070	Downshift Switch Circuit
P1083		Fuel Control Limit Mixture Too Lean (Bank 1 Sensor 1)
P1084		Fuel Control Limit Mixture Too Rich (Bank 1 Sensor 1)
P1087		O2 Sensor Circuit Slow Response in Lean Control Range (Bank 1 Sensor 1)
P1088		O2 Sensor Circuit Slow Response in Rich Control Range (Bank 1 Sensor 1)
P1089		O2 Sensor Circuit Slow Response in Lean Control Range (Bank 1 Sensor 2)
P1090		Pre Catalyst Fuel Trim Control Too Lean (Bank 1)
P1092		Pre Catalyst Fuel Trim Control Too Rich (Bank 1)
P1106	4358	Manifold Air Pressure Sensor Too Low at Engine Stop
P1107	4359	Manifold Air Pressure Sensor Too Low at Idle Engine Running
P1108	4360	Manifold Air Pressure Sensor Too Low at Full Load for Low Engine Speed
P1109	4361	Manifold Air Pressure Too High in Deceleration
P1111		Engine Coolant Temperature Sensor Radiator Outlet Low Input
P1112		Engine Coolant Temperature Sensor Radiator Outlet High Input
P1120		Pedal Position Sensor Circuit
P1121		Pedal Position Sensor 1 Range/Performance Problem
P1122		Pedal Position Sensor 1 Low Input
P1123		Pedal Position Sensor 1 High Input
P1125	4389	Throttle Position Sensor A and B Range/Performance Small Error
P1126	4390	Throttle Position Sensor A and B Range/Performance Large Error
P1134		O2 Sensor Heater Circuit Signal Intermittent (Bank 1 Sensor 1)
P1135		O2 Sensor Heater Circuit Low Voltage (Bank 1 Sensor 1)
P1136		O2 Sensor Heater Circuit High Voltage (Bank 1 Sensor 1)
P1137		O2 Sensor Heater Circuit Signal Intermittent (Bank 1 Sensor 2)
P1138		O2 Sensor Heater Circuit Low Voltage (Bank 1 Sensor 2)

Table f. Cooper S DTCs from 1/11/2003		
PCode	**MINI-FC**	**PCode text**
P1139		O2 Sensor Heater Circuit High Voltage (Bank 1 Sensor 2)
P1143		O2 Sensor Activity Check Signal Too High (Bank 1 Sensor 2)
P1144		O2 Sensor Activity Check Signal Too Low (Bank 1 Sensor 2)
P1171		Ambient Pressure Sensor Variant Recognition Value in Boost Range Implausible
P1172		Ambient Pressure Sensor Variant Recognition Error Value Stored in Boot Range
P1173		Ambient Pressure Sensor Variant Recognition Learning Failed
P1178		O2 Sensor Pre Cat (Bank 1) - Switching Time Too Slow
P1188		Fuel Trim (Bank 1), O2 Control Adaptation Limit - Multiplicative
P1190		Pre Catalyst Fuel Trim Control (Bank 1)
P1192		Post Catalyst Fuel Trim System (Bank 1)
P1222		Pedal Position Sensor 2 Low Input
P1223		Pedal Position Sensor 2 High Input
P1229	4649	Throttle Sensor Adaptation Failure
P1237	4663	Secondary Upstream Manifold Air Pressure Sensor Low Input
P1238	4664	Secondary Upstream Manifold Air Pressure Sensor High Input
P1239	4665	Secondary Upstream Manifold Air Pressure Sensor Too Low at Engine stop
P1240	4672	Secondary Upstream Manifold Air Pressure Sensor Too Low at Idle Engine Running
P1241	4673	Secondary Upstream Manifold Air Pressure Sensor Too Low at Full Load for Low Engine Speed
P1242	4674	Secondary Upstream Manifold Air Pressure Sensor Too High in Deceleration
P1314		Fuel Mixture Deviation Detected with Low Fuel
P1320	4896	Flywheel Adaptation for Misfire Detection Range
P1321	4897	Flywheel Adaptation for Misfire Detection Performance
P1342		Misfire During Start Cylinder 1
P1343		Misfire Cylinder 1 with Fuel Cut-Off
P1344		Misfire During Start Cylinder 2
P1345		Misfire Cylinder 2 with Fuel Cut-Off
P1346		Misfire During Start Cylinder 3
P1347		Misfire Cylinder 3 with Fuel Cut-Off
P1348		Misfire During Start Cylinder 4
P1349		Misfire Cylinder 4 with Fuel Cut-Off
P1403		ECU internal test
P1413		Secondary Air Injection Pump Relay Control Circuit Signal Low
P1414		Secondary Air Injection Pump Relay Control Circuit Signal High
P1423		Secondary Air Injection (Bank 1) - Flow too Low
P1444		Diagnostic Module Tank Leakage (DM-TL) Pump Control Open Circuit
P1445		Diagnostic Module Tank Leakage (DM-TL) Pump Control Circuit Signal Low
P1446		Diagnostic Module Tank Leakage (DM-TL) Pump Control Circuit Signal High
P1447		Diagnostic Module Tank Leakage (DM-TL) Pump Control Current Too High during Switching Solenoid Test
P1448		Diagnostic Module Tank Leakage (DM-TL) Pump Current Too Low
P1449		Diagnostic Module Tank Leakage (DM-TL) Pump Current Too High
P1451		Diagnostic Module Tank Leakage (DM-TL) Switching Solenoid Control Circuit Signal Low
P1452		Diagnostic Module Tank Leakage (DM-TL) Switching Solenoid Control Circuit Signal High

P-codes (DTCs) and MINI fault codes

Table f. Cooper S DTCs from 1/11/2003

PCode	MINI-FC	PCode text
P1453		Secondary Air Injection System - Relay or Pump
P1475	5237	Leakage Diagnostic Pump Reed Switch Did not Close
P1476	5238	Leakage Diagnostic Pump Clamped Tube
P1477	5239	Leakage Diagnostic Pump Reed Switch Did Not Open
P1478		Range Check Voltage Supply Pedal Sensor 1 and Throttle Position Sensors
P1500		Idle Speed Control Valve Stuck Open
P1501		Idle Speed Control Valve Stuck Closed
P1502		Idle Speed Control Valve Closing Solenoid Control Circuit Signal High
P1503		Idle Speed Control Valve Closing Solenoid Control Circuit Signal Low
P1504		Idle Speed Control Valve Closing Solenoid Control Open Circuit
P1506		Idle Speed Control Valve Opening Solenoid Control Circuit Signal High
P1507		Idle Speed Control Valve Opening Solenoid Control Circuit Signal Low
P1508		Idle Speed Control Valve Opening Solenoid Control Open Circuit
P1509		Idle Control Valve Opening Coil - Malfunction
P1512		DISA (Differentiated Intake Manifold) Control Circuit Signal Low
P1513		DISA (Differentiated Intake Manifold) Control Circuit Signal High
P1523		'A' Camshaft Position Actuator Signal Low (Bank 1)
P1524		'A' Camshaft Position Actuator Control Circuit Signal High (Bank 1)
P1529		'B' Camshaft Position Actuator Control Circuit Signal Low (Bank 1)
P1530		'B' Camshaft Position Actuator Control Circuit Signal High (Bank 1)
P1550		Idle Control Valve Closing Coil - Malfunction
P1572	5490	Electronic Control Module Sensor Supply A Noisy Signal
P1575	5493	Electronic Control Module Sensor Supply B Noisy Signal
P1600	5632	External Control Module Random Access Memory (RAM) Error
P1602		Control Module Self Test - Control Module Defective
P1603		Control Module Self Test - Torque Monitoring
P1604		Control Module Self Test - Speed Monitoring
P1607	5639	CAN-Version
P1611	5649	Serial Communication Link Transmission Control Module
P1612	5650	Serial Communication Link Instrument Pack
P1613	5651	Serial Communication Link ASC (Automatic Stability Control)
P1615	5653	Electronic Control Module Processor SPI-Bus Failure
P1617	5655	Electronic Control Module H Bridge Controller
P1619		Map Cooling Thermostat Control Circuit Signal Low
P1620		Map Cooling Thermostat Control Circuit Signal High
P1624		Pedal Position Sensor Potentiometer Supply Channel 1 Electrical
P1625		Pedal Position Sensor Potentiometer Supply Channel 2 Electrical
P1632		Throttle Valve Adaptation Conditions Not Met
P1633		Throttle Valve Adaptation Limp-Home Positions Unknown
P1634		Throttle Valve Adaptation Spring Test Failed
P1635		Throttle Valve Adaptation Lower Mechanical Stop not Adapted
P1636		Throttle Valve Control Circuit Mechanically; pulse width not plausible

Table f. Cooper S DTCs from 1/11/2003		
PCode	**MINI-FC**	**PCode text**
P1637		Throttle Valve Position Control, Control Deviation
P1638		Throttle Valve Position Control Throttle Stuck Temporarily
P1639		Throttle Valve Position Control Throttle Stuck Permanently
P1675		Throttle Valve Actuator Start Test Re-Adaptation Required
P1679	5753	Electronic Throttle Control Monitor Level 2/3 Torque Loss Calculation
P1680	5760	Electronic Throttle Control Monitor Level 2/3 ADC Processor Fault
P1681	5761	Electronic Throttle Control Monitor Level 2/3 Engine Speed Calculation Error
P1682	5762	Electronic Throttle Control Monitor Level 2/3 Idle Speed 'A' Calculation Fault
P1683	5763	Electronic Throttle Control Monitor Level 2/3 Idle Speed 'B' Calculation Fault
P1684	5764	Electronic Throttle Control Monitor Level 2/3 Clutch Torque Min Error
P1685	5765	Electronic Throttle Control Monitor Level 2/3 Clutch Torque Max Error
P1686	5766	Electronic Throttle Control Monitor Level 2/3 Pedal Position Sensor Diagnostic Error
P1687	5767	Electronic Throttle Control Monitor Level 2/3 Throttle Position Sensor Diagnostic Error
P1688	5768	Electronic Throttle Control Monitor Level 2/3 Mass Air Flow Calculation
P1689	5769	Electronic Throttle Control Monitor Level 2/3 Torque Calculation Error
P1691	5777	Electronic Throttle Control Monitor Level 2/3 Motorized Throttle Control Engine Speed Limitation Error
P1692	5778	Electronic Throttle Control Monitor Level 2/3 Motorized Throttle Control and Fuel Injection Switch Off 'A'
P1693	5779	Electronic Throttle Control Monitor Level 2/3 Motorized Throttle Control and Fuel Injection Switch Off 'B'
P1694		Throttle Valve Actuator Start Test Spring Test and Limp-Home Position Failed
P1698	5784	Transmission Control Module Control Error
P1699	5785	Transmission Control Module Checksum Error
P1705	5893	Transmission Control Module LED Output Open Circuit
P1706	5894	Transmission Control Module LED Output Short Circuit
P1739	5945	Clutch Solenoid Communication Error
P1741	5953	Clutch Solenoid Open Circuit
P1742	5954	Clutch Solenoid Short Circuit
P1749	5961	Secondary Pressure Solenoid Communication Error
P1751	5969	Secondary Pressure Solenoid Open Circuit
P1752	5970	Secondary Pressure Solenoid Short Circuit
P1785	6021	Transmission Ratio Control Actuator Circuit Malfunction
P1786	6022	Transmission Ratio Control Actuator Circuit Range/Performance
P1787	6023	Transmission Ratio Control Actuator Open Circuit
P1788	6024	Transmission Ratio Control Actuator Short Circuit
P1789	6025	Transmission Ratio Control Actuator Communication Error
P1815	6165	Wheel Plus Switch Error Low Input
P1816	6166	Wheel Minus Switch Error Low Input
P2088		'A' Camshaft Position Actuator Control Circuit Low (Bank 1)
P2089		'A' Camshaft Position Actuator Control Circuit High (Bank 1)
P2090		'B' Camshaft Position Actuator Control Circuit Low (Bank 1)
P2091		'B' Camshaft Position Actuator Control Circuit High (Bank 1)
P2096	8342	Post Catalyst Fuel Trim System Too Lean (Bank 1)
P2097	8343	Post Catalyst Fuel Trim System Too Rich (Bank 1)

Table f. Cooper S DTCs from 1/11/2003		
PCode	**MINI-FC**	**PCode text**
P2122	8482	Throttle/Pedal Position Sensor/Switch 'D' Circuit Low Input
P2123	8483	Throttle/Pedal Position Sensor/Switch 'D' Circuit High Input
P2127	8487	Throttle/Pedal Position Sensor/Switch 'E' Circuit Low Input
P2128	8488	Throttle/Pedal Position Sensor/Switch 'E' Circuit High Input
P2138	8504	Throttle/Pedal Position Sensor/Switch 'D' / 'E' Voltage Correlation
P2228		Barometric Pressure Circuit Low
P2229		Barometric Pressure Circuit Low
P2270	8816	O2 Sensor Signal Stuck Lean (Bank 1 Sensor 2)
P2271	8817	O2 Sensor Signal Stuck Rich (Bank 1 Sensor 2)
P2300	8960	Ignition Coil 'A' Primary Control Circuit Low
P2301	8961	Ignition Coil 'A' Primary Control Circuit High
P2303	8963	Ignition Coil 'B' Primary Control Circuit Low
P2304	8964	Ignition Coil 'B' Primary Control Circuit High
P2400	9216	Evaporative Emission System Leak Detection Pump Control Circuit/Open
P2401	9217	Evaporative Emission System Leak Detection Pump Control Circuit Low
P2402	9218	Evaporative Emission System Leak Detection Pump Control Circuit High
P2404	9220	Evaporative Emission System Leak Detection Pump Sense Circuit Range/Performance

MINI Cooper

2002 ▸ 2006

Cooper, Cooper S

Diagnosis Without Guesswork

- Diagnostic tools and techniques
- Component descriptions and operation
- Circuit wiring schematics
- Pinout charts, waveforms, and nominal values
- Bus system operation and testing
- OBD II P-code (DTC) lookup

[B] BentleyPublishers™
.com

DIAGNOSIS WITHOUT GUESSWORK™

MINI Cooper

2002 ▸ 2006
Cooper, Cooper S

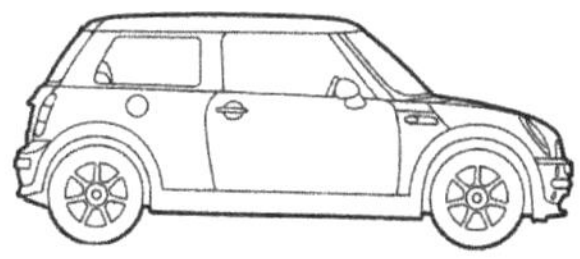

Diagnosis Without Guesswork

[B] BentleyPublishers™
.com

DIAGNOSIS WITHOUT GUESSWORK™

Introduction to Diagnosis Without Guesswork

Diagnosis Without Guesswork covers 2002 through 2006 MINI Cooper and Cooper S models and is specifically designed to cover models built for sale in the United States. It is not intended to be a service manual and should be used solely as a reference in conjunction with MINI repair and service information.

It has been written for the professional technician as well as the MINI enthusiast. It is a DTC-based source of engine management and on-board system diagnosis. Equipped with only a basic code reader, the content is designed to guide the user from DTC to diagnosis in an easy to use format.

For the MINI technician with advanced automotive test equipment, the circuit descriptions, pinout charts, wiring schematics, scope traces, and real-world nominal values will prove essential in diagnosing problem cars.

Even if all you own is a generic OBD-II scan tool, a DVOM and a set of jumper wires, this diagnostic information will help provide fast answers and guidance as it steps you through the diagnostic process.

As a result of publishing our MINI repair manuals, we have been asked for DTC-based diagnostic information. The information found in this appendix is a response to that feedback.

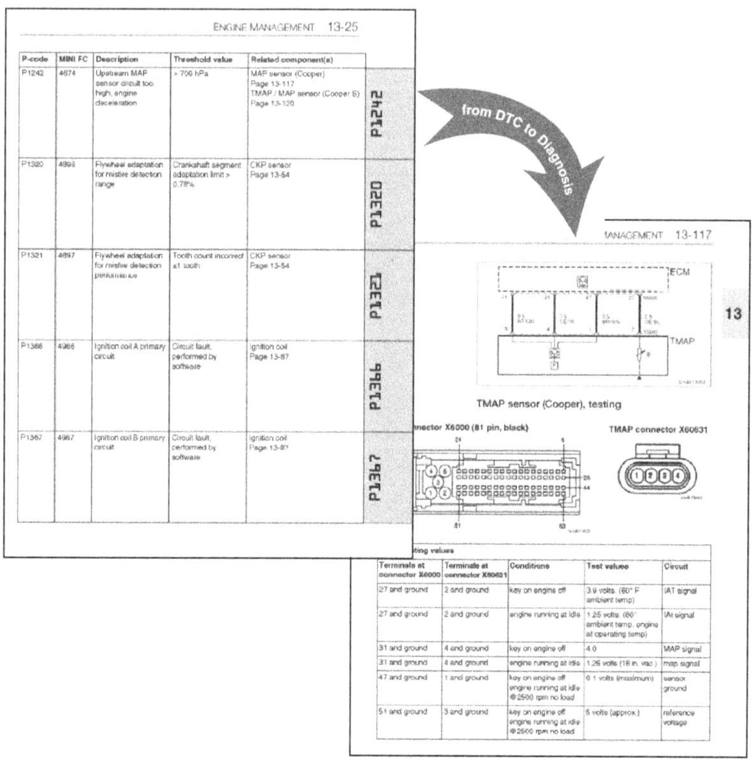
ENGINE MANAGEMENT 13-25

P-code	MINI FC	Description	Threshold value	Related component(s)	
P1242	4674	Upstream MAP sensor circuit too high, engine deceleration	> 700 hPa	MAP sensor (Cooper) Page 13-117 TMAP / MAP sensor (Cooper S) Page 13-120	P1242
P1320	4896	Flywheel adaptation for misfire detection range	Crankshaft segment adaptation limit > 0.78%	CKP sensor Page 13-54	P1320
P1321	4897	Flywheel adaptation for misfire detection performance	Tooth count incorrect ±1 tooth	CKP sensor Page 13-54	P1321
P1366	4966	Ignition coil A primary circuit	Circuit fault, performed by software	Ignition coil Page 13-87	P1366
P1367	4967	Ignition coil B primary circuit	Circuit fault, performed by software	Ignition coil Page 13-87	P1367

MANAGEMENT 13-117

TMAP sensor (Cooper), testing

nector X6000 (81 pin, black)

TMAP connector X60631

ting values

Terminals at connector X6000	Terminals at connector X60631	Conditions	Test values	Circuit
27 and ground	2 and ground	key on engine off	3.9 volts. (60° F ambient temp)	IAT signal
27 and ground	2 and ground	engine running at idle	1.25 volts. (60° ambient temp, engine at operating temp)	IAt signal
31 and ground	4 and ground	key on engine off	4.0	MAP signal
31 and ground	4 and ground	engine running at idle	1.26 volts (18 in. vac.)	map signal
47 and ground	1 and ground	key on engine off engine running at idle @2500 rpm no load	0.1 volts (maximum)	sensor ground
51 and ground	3 and ground	key on engine off engine running at idle @2500 rpm no load	5 volts (approx.)	reference voltage

An OBD-II scan tool is invaluable for on-board diagnostics, such as clearing fault codes, and resetting the Check Engine light.
See **OBD On-Board Diagnostics**

Factory scan tool being used to check power supply (Terminal 30)
See **65 Electronic Immobilization System (EWS)**

Backprobing ECM connector to check for voltage to LDP solenoid.
See **13 Engine Management**

01 Diagnostic Tools and Techniques

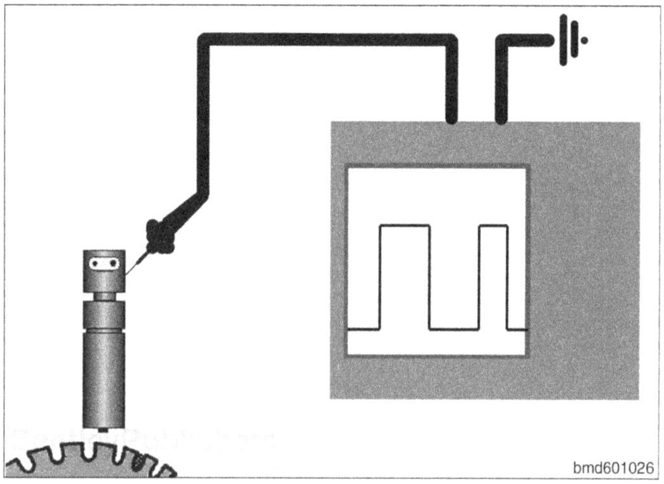

Table of Contents

GENERAL

This chapter discusses fault diagnosis and troubleshooting techniques from a technician's perspective. Also covered here are the essential tools needed to diagnosis MINI vehicles. Tools listed are used as examples and an equivalent tool may be substituted.

Use of advanced diagnostic equipment, including the factory scan tool, a lab scope, and a current probe are a few of the tools that are necessary to take full advantage of the information in this book. A generic OBD II scan tool, a good DVOM and assorted jumper wires is the minimum required tool list. Note that an ODB II scan tool is limited to powertrain systems only.

MINI-dedicated (Autologic) scan tool

Proper upkeep of your diagnostic tools and equipment is as important as the equipment itself. Cables, leads and clamps are some examples of the parts that wear quickly. A faulty piece of test equipment causes false readings and test results, leading the technician down the wrong path. Check your equipment regularly and replace worn or faulty parts.

Warnings and Cautions

Please read these warnings and cautions before proceeding with diagnostic and repair work.

WARNINGS—

- Read **WARNINGS— important safety notice** on the copyright page. Read each procedure and the **WARNINGS** and **CAUTIONS** that accompany the procedure. Also review posted corrections at **www.BentleyPublishers.com/errata/** before beginning work.
- If any procedure, tightening torque, wear limit, specification or data presented in this manual does not appear to be appropriate for a specific application, contact the publisher or the vehicle manufacturer for clarification before using the information in question.
- Do not reuse worn or deformed fasteners. Many fasteners are designed to be used only once and become unreliable and may fail when used a second time. This includes, but is not limited to, nuts, bolts, washers, self-locking nuts or bolts, snap rings and cotter pins. Replace these fasteners with new parts.
- If you lack the skills, tools and equipment, or a suitable workshop for any procedure described in this book, leave such repairs to an authorized MINI dealer or other qualified shop.
- Before starting a job, be sure that you have the necessary tools and parts on hand. Read all the instructions thoroughly, and do not attempt shortcuts. Use tools appropriate to the work and use only replacement parts meeting MINI specifications.
- Observe good workshop practices. Wear goggles when operating machine tools or working with battery acid. Wear gloves or other protective clothing whenever working with harmful substances. Wear a dust mask or respirator to lower your exposure to harmful dusts, fumes, vapors, and/or gases. Protect ears whenever exposed to continuous loud noise.
- Do not run the engine unless the work area is well ventilated. Carbon monoxide kills.
- Disconnect the battery negative (-) terminal whenever you work on the fuel system or the electrical system.
- Vehicles covered by this manual are equipped with a supplemental restraint system (SRS).These explosive devices, handled improperly or without adequate safeguards, can be accidentally activated and can cause serious injury.
- The ignition system produces high voltages that can be fatal. Avoid contact with exposed terminals. Use extreme care when working on a vehicle with the engine running or the ignition on.

Shop safety

Safety is very important in the automotive repair shop. Knowing and following safety rules can help prevent injury and damage.

An automotive repair shop poses many dangers that can result in serious injury or even death. Running engines, moving cars, cars on jack stands or on a lift, and open fuel lines are just a few of the daily working situations that all have the potential to cause a catastrophic event.

Basic shop safety rules:

- Always follow safe operating instructions from equipment and tool manufacturers and safety instructions from the vehicle manufacturers for specific maintenance and repair procedures.
- Follow shop rules and OSHA guidelines for personal safety.
- Follow shop rules and EPA guidelines for disposal or recycling of used oil, antifreeze, refrigerants and wastes.
- Keep your work area clean and orderly. Do not allow spare parts or other material to accumulate on the shop floor or in work areas.
- Rely on your judgment and knowledge of safety to guide your work. If you are unsure about the safe execution of the job at hand, request assistance from your supervisor or someone with specific experience.
- Be familiar with the locations and operation of fire extinguishers. In case of a fire, sound an alarm and, if possible, get help to extinguish the fire.
- Do not use an air hose for cleaning yourself. Do not point an air hose at anyone. Use a special cleaning gun when cleaning with air; an approved safety gun must not exceed 30 pounds per square inch (psi).
- Do not use an air hose for dusting off brake shoes and parts. Use an approved HEPA vacuum device or wet-wash method.
- Wear seat belts when test driving a customer's vehicle.

THE FLOW OF DIAGNOSIS

A technician's approach

It is important to approach each diagnostic problem you encounter with the same strategy. An engine misfire and an electrical fault may need different test procedures, but the same flow of diagnosis can be used to find the root cause quickly. With the correct information, techniques and training, quick work can be made of locating most faults.

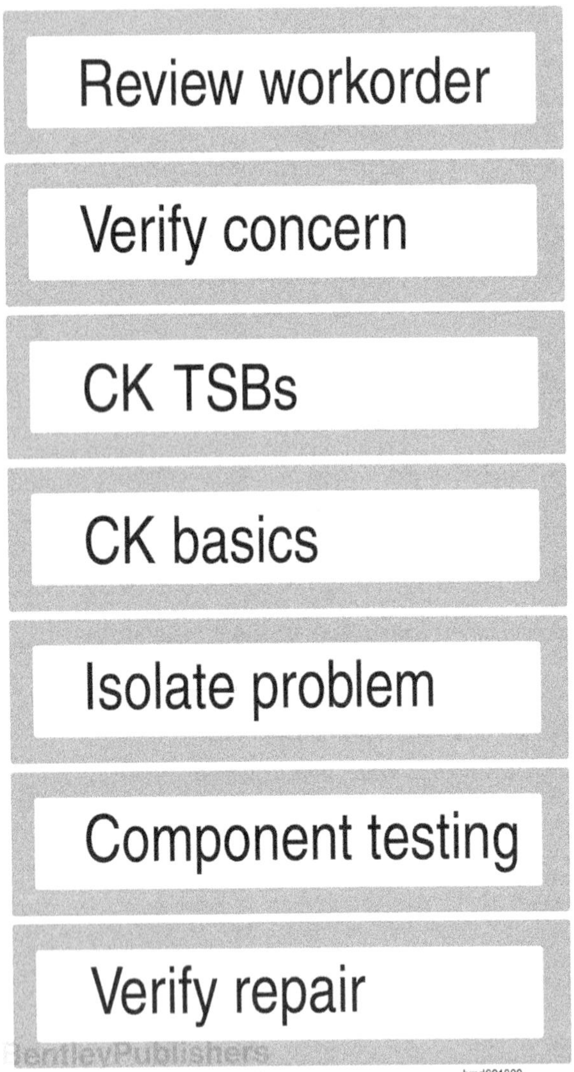

bmd601200

Develop an organized approach to problem solving. This is also known as the flow of diagnosis.

If the vehicle is not repaired efficiently, the productivity of the technician suffers. Taking time at the beginning to plan can help give structure to what may appear to be an overwhelming situation.

Review vehicle work order

A good work order should clearly state the customer's description of the symptom or problem. When filling out the work order, it is important to ask detailed questions about the concern. Here are some examples of questions you may need to ask.

- When does it occur?
- What engine temperature?
- At idle or under load?
- What are the weather conditions?

The answers to these questions will give you a solid starting point for your flow of diagnosis.

Verify customer's concern

Whenever possible, personally verify the complaint. Having experienced the malfunction, you are less likely to attempt repairing non-existent faults.

To verify the customer complaint, run the engine or road test the vehicle to verify that the complaint is a deviation from normal operation. If the concern can not be duplicated, check for any stored faults or applicable technical service bulletins.

bmd601014

There may be multiple symptoms that are created with one problem. An example is a defective thermostat. If the thermostat is stuck open, the heater output can be insufficient. Also, if the engine never reaches operating temperature, fuel mileage and performance suffer. Which one of these concerns (symptoms) would lead you to the problem faster?

Remember that the customer may only complain about one symptom. It is the job of the technician to be a detective and carefully observe. There may be another symptom that directly points to the root cause of the problem.

Check technical service bulletins (TSBs)

Check all available service information. This includes engine control module (ECM) software or manual updates, warranty, and recall information. For the most up-to- date information, the manufacture's website is a reliable source.

http://www.minitechinfo.com

bmd6010015b

Difficult-to-diagnose problems are often described in periodical publications by experienced technicians who have encountered like symptoms in similar vehicles.

Sample TSB

SI M 12 01 07 Engine Electrical Systems	June 2007 Technical Service

SUBJECT

Unstable Idle when Stopping, usually at Higher Ambient Temperatures

MODEL

R52, R53 with W11 and Automatic Transmission produced from 01/2005 to 08/2006

SITUATION

When driving in stop and go traffic, or when idling at a stop, the idle speed may fluctuate or become unstable. This may be more prominent at higher ambient temperatures and under a load, such as with the A/C on.

CAUSE

DME software calibration

CORRECTION

On a customer complaint basis only, please reprogram the DME (MS5150) using Progman V26.0 or higher.

PROCEDURE

1. Connect an approved battery charger and DK diagnostic head.
2. Start new Progman session and select "MINI" series.
3. Select "Load Software", and then answer "NO" to the query as to whether any of the control modules have been replaced.
4. Select DME from the list of the control modules.
5. Select "Programming" and then "Manual Entry" from the submenu.
6. Enter the Programmed number **7581561** first and the basic number **7557395** second using the virtual keyboard in Progman.
7. The measures plan will be determined and displayed.
8. Select "Accept" and "Start" to start programming.

Important: After the DME programming ends, continue to follow the Progman screen prompts to "Turn Off Ignition", wait 25 seconds, and then "Turn On" ignition to complete the DME coding and measures plan successfully. If necessary, determine and clear any fault codes using the GT1 diagnostic path.

WARRANTY INFORMATION

Covered under the terms of the MINI New Vehicle Limited Warranty.

Please refer to the latest KSD for all applicable labor operations and allowances.

If the appropriate labor operation is not contained in KSD, then a work time labor operation should be used.

Defect Code	**10 32 01 47 00**

Check the basics

Begin diagnostics by checking the basics. For example, is battery voltage correct, are chassis ground connections good. Inspect for disconnected or damaged hoses and wiring.

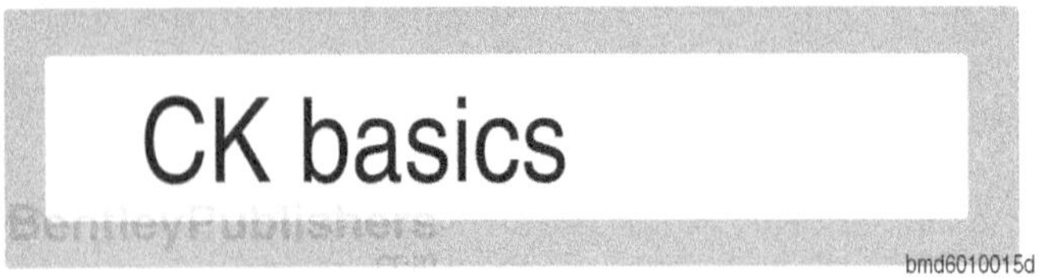

bmd6010015d

Diagnosing problems in modern automobiles can often be frustrating and confusing, especially when it involves computer controlled systems. The fact is, automotive electronics have proven quite reliable, and the greatest number of problems with new cars are the same kinds of problems that occurred on older non-computer controlled cars.

bmd601002

Isolate problem

Before suspecting a computer problem, perform a careful visual inspection. Check under the hood for the same kinds of problems you would look for on a non-computer controlled engine. These include fluid leaks, vacuum leaks, dirty filters, overheating, oil burning, poor connections or loose wires, bad ignition coil, spark plug boots, worn out spark plugs, restricted exhaust system, worn mechanical parts, exhaust leaks, and other familiar kinds of problems. Be thorough! You may save a lot of time.

bmd6010015e

Certain basic faults may be undetectable by the ECM self-diagnostic system and can actually interfere with self-checking and fault memory operation. Low battery voltage, for example, can cause all kinds of faults to set.

Batteries in a state of partial discharge can have a dramatic effect on the ECM and related electrical components. Make sure B+ connections are in perfect condition for trouble-free electronic system operation.

bmd601016

Sophisticated automotive systems are easily misunderstood. This can lead to misinformed repairs and misleading diagnostic strategies. When troubleshooting, compare actual system operation to nominal system operation as described in the appropriate chapter. If possible compare problem vehicle system operation with a known good vehicle.

Vehicle repair history can provide explanations of unusual complaints which seem to elude normal troubleshooting attempts. Incorrect components or unapproved repairs can have subtle influences on seemingly unrelated systems.

BMD601024

Fuses prevent possible damage to electrical components and wiring harnesses. Make sure fuses and their connections are without dynamic resistance. Dynamic resistance can only be checked using the voltage drop method of testing. See **Test Techniques**, given later in this chapter.

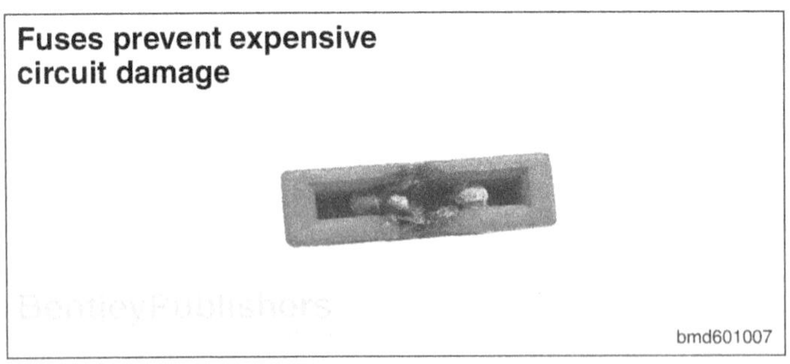

bmd601007

Check all chassis ground connections (B-, KL31) and verify that they are in perfect condition. A poor common ground can cause seemingly unrelated systems to influence one another.

High current systems that encounter a poor ground can backfeed through other electrical systems, causing unusual operation and inexplicable component failure. Check grounds dynamically using the voltage drop technique.

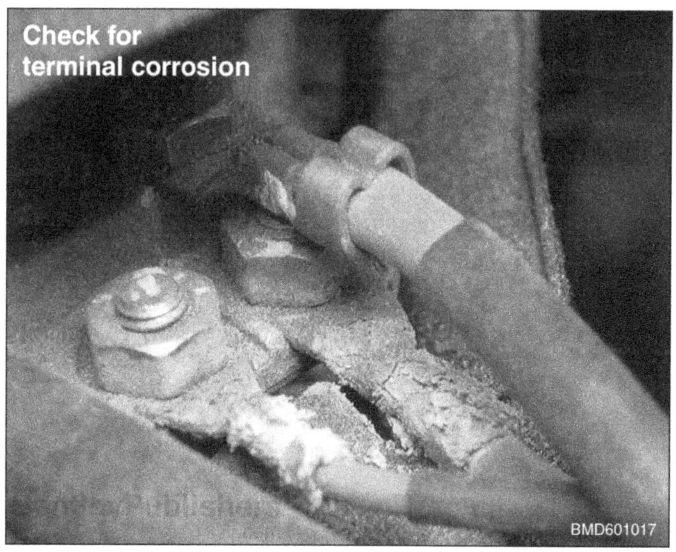

Fuel pump and main relays, as well as their plug connections, can be a source of intermittent operation which do not set a code in ECM fault memory.

Check that fuel delivery begins at once when cranking and pressure in the fuel rail is within specifications.

Adaptive engine management systems efficiently compensate for conditions which may affect the combustion process. However, when troubleshooting idle quality or driveability complaints, it is still necessary to consider the following:

- Fuel injector spray pattern quality
- Leaks in engine air intake
- Fuel tank evaporative purge system faults
- Excessive engine oil dilution
- Substandard fuel or unapproved additives
- Piston or valve carbon build-up

Component testing

Once the fault has been isolated to a component group or specific part of the engine management system (i.e. fuel delivery or ignition), testing at the component level will be necessary.

bmd6010015f

The first step is to review scan tool data (fault memory and live data) and identify a PID (Parameter ID) value or output control to direct you to a component group. Individual component testing will require specific wiring diagrams and component information. Reviewing the theory and operation of a component can aid in diagnosing it efficiently.

Make circuit tests to check the diagnosis you made during isolation of the symptom. Keep in mind that a logical, simple procedure is the key to efficient troubleshooting. Test for the most likely cause of failure first. Try to make tests at points that are easily accessible. It may be necessary to perform your tests during the occurrence of the symptom. A second technician should operate the vehicle while the tests are performed.

WARNING —
- *Avoid contact with hot metal parts.*
- *Do not remove the radiator cap when the engine is hot.*
- *Do not operate the engine indoors without proper ventilation or exhaust extraction.*
- *Do not smoke while working on the vehicle.*

A few test conventions / conditions are used through out this book:

KOEO (key on engine off) is with key in run position. A power supply connected to the vehicle battery maintains a steady voltage of 14.2 volts. Use a high quality vehicle power supply.

NOTE —
- Having constant, consistent voltage during testing ensures accurate results during circuit testing. A drop or fluctuation in battery voltage could appear as a failed sensor or component. This could lead to unnecessary replacement of good parts.

CAUTION —
- *A vehicle power supply or battery charger must not produce more than 120 mv A/C. Test your equipment for A/C ripple.*

KOER (key on engine running) is with the engine running at operating temperature (**arrow**). If engine does not reach operating temperature in a reasonable amount of time, address that issue first.

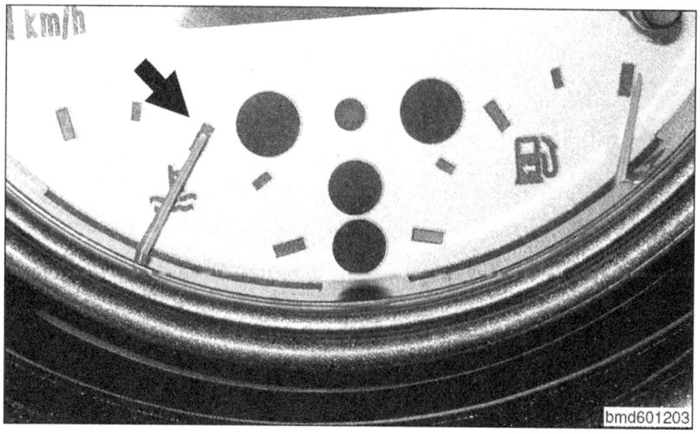

bmd601203

An engine that does not reach operating temperature may not go into closed loop fuel control, fuel mileage may suffer and a lack of power may be noticeable (applied colder temp engine fuel / timing map). DTCs may also be present.

> ***WARNING —***
> - *Avoid contact with hot metal parts.*
> - *Do not remove the radiator cap when the engine is hot.*
> - *Do not operate the engine indoors without proper ventilation.*
> - *Do not smoke while working on the vehicle.*

Electrical testing performed at a set engine speed above idle is performed using an adjustable throttle pedal depressor. Vehicle in Neutral or Park, no load. Parking brake applied, with wheels blocked. Be sure engine cooling fan and cooling system are operating normally before holding engine above idle for testing.

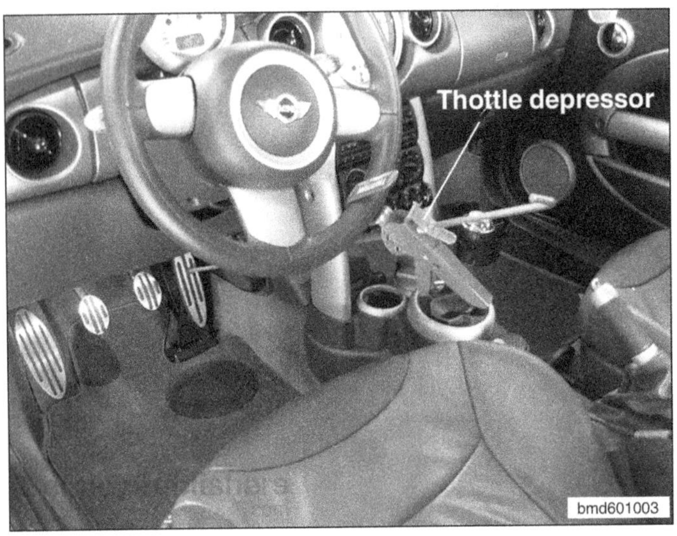

> ***WARNING —***
> - *Do not run engine at a high rpm for an extended period of time.*
> - *Do not leave vehicle unattended during this process.*
> - *Keep clear of moving parts in engine compartment.*
> - *Cooling fan could start at any time.*

Verify repair

The process of verifying a repair is similar to verifying the symptom. Follow the same test procedure you did when isolating the fault. A test drive, operation of repaired component or a review of the scan data may be needed. Make sure no new problems turn up and the original problem does not recur.

bmd6010015g

If the component you repaired is functioning normally and the symptom is no longer present, perform a final check for trouble codes (stored and pending) to verify there are no other faults present.

One item may fail and mask another that is faulty. For example a failed oxygen sensor may not allow a catalyst monitor to run. Once the oxygen sensor is repaired the catalyst may fail the monitor and set the MIL. The customer sees the MIL illuminated and may feel the vehicle was not repaired correctly.

The verification step can save you time and face with your customer. Also check with repair information to see if the component you repaired requires and adaptation value to be reset. Failure to do so may set the MIL with the same or a similar fault.

SPECIAL TOOLS

An extensive list of tools is required to test and diagnose all the systems and components on the MINI.

Diagnostic scan tools

The factory tool at the time this manual was printed was the Siemens Group Tester One (GT1). This tool functions as a diagnostic scan tool, but also has DVOM functions and an oscilloscope. This is a powerful diagnostic tool, capable of assembling a diagnostic test plan (guided fault finding) based on the faults stored in the vehicle memory.

Scan tool (Siemens GT1)

bmd601001

There are also a limited number of MINI-enhanced aftermarket scan tools. The Diagnos Autologic and the Snap-On Modis are two such examples.

Lastly, an SAE-generic OBD II scanner is also a capable diagnostic tool. Many offers can be found on the internet for reasonably priced hand-held and laptop variants.

Diagnos Autologic

Snap-on MODIS

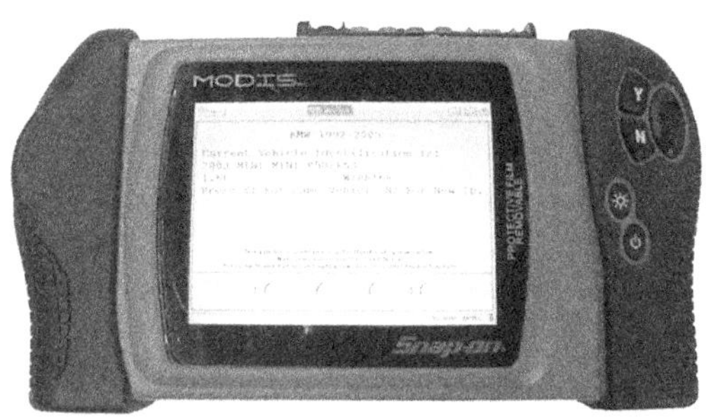

bmd601004

Power supply

A power supply is a stand-alone, constant voltage, high-amperage power station or an intelligent battery charger with an electronically controlled power supply function. It is a different tool than the traditional battery charger.

Consistent, clean voltage is absolutely necessary when coding or programming the MINI. Transformer-based battery chargers are now considered inadequate for this task.

A power supply insures clean DC voltage and eliminates AC. It also provides faster, more efficient and safer battery charging. Once proper battery voltage is obtained a power supply floats the charge to maintain that voltage.

70-amp ACCTIVA Professional Power supply

bmd6010022

> ***CAUTION —***
> - *Non-regulated power supplies, such as transformer-based battery chargers, have been known to cause problems during coding or programming. A/C voltage from the wall outlet can be passed on and damage the electronic control units. Unstable voltage levels during programming will cause errors during the download or aborting of the process.*

MINI specifications for power supply units:

- Microprocessor-controlled measuring and monitoring.
- Power supply mode (dual function power supply/chargers). In power supply mode the charger monitors amperage draw of the vehicle and adjusts itself accordingly. For example, charger outputs 32 amps in response to a 32-amp draw and does not allow the battery draw down.
- 250 Ah rating, with 10 Ah or higher resolution.
- Power supply output voltage: min 12.4 V, max 14.4 V, with 0.02 V or higher resolution.
- Power supply output current: 45 A, with 0.05 A or higher resolution.
- Non-sparking due to delayed switch-on (battery chargers).

Deutronic Power Supply

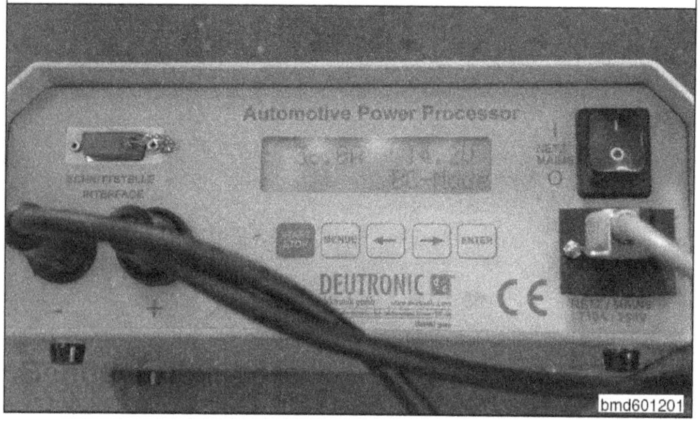

Digital volt / ohm Meter (DVOM)

The ability to measure voltage, current flow, and resistance is important in the diagnosis of electrical problems. Without the results of these measurements, troubleshooting in an electrical system is a futile process.

The instrument most commonly used to make electrical measurements is called a digital volt / ohm meter (DVOM).

Basic DVOMs are capable of measuring:

- AC voltage
- Millivolts
- Conductance
- Continuity
- Amps/milliamps
- DC voltage
- Resistance
- Capacitance
- Diode test
- Microamps

Advanced DVOM's add:

- Frequency
- RPM
- Duty cycle

- Pulse width

DVOM - Fluke 87

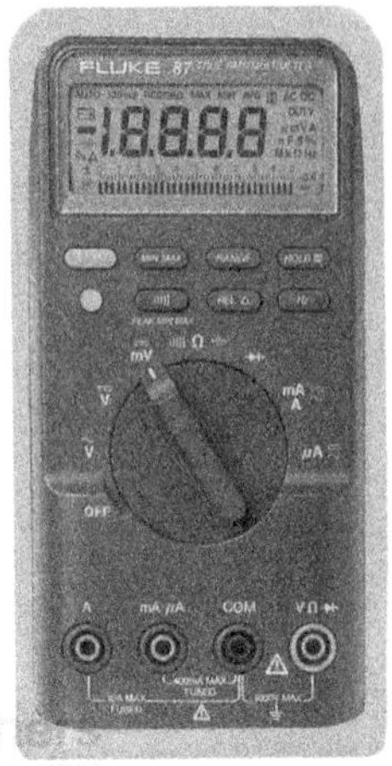

BMD601010

The DVOM is a tool for accurate electrical measurements. Even though accurate measurements are the key to electrical diagnosis, the following four factors determine the effectiveness of the measurements

- Accuracy of the measuring instrument.
- Correct installation in the circuit of the measuring instrument.
- Ability of the technician to read the instrument.
- Skill of the technician in interpreting the results.

Only one of the factors depends on the DVOM (e.g. accuracy). The rest depend on the ability of the technician to read and interpret the results.

Choosing a DVOM

A good choice of a DVOM is the factory scan tool, as the measuring system contains a highly accurate DVOM. If the factory tester is not available, a Fluke 87 DVOM or equivalent can be used.

Choosing a handheld DVOM from a reputable manufacturer, however, leaves the factory scan tool free to perform other tasks that a DVOM cannot do (e.g. retrieval of fault codes).

Several factors need to be considered when choosing a DVOM, one of which is input impedance. Impedance is the combined resistance to current created by the resistance, capacitance and inductance of the meter. Impedance is measured in ohms / volt.

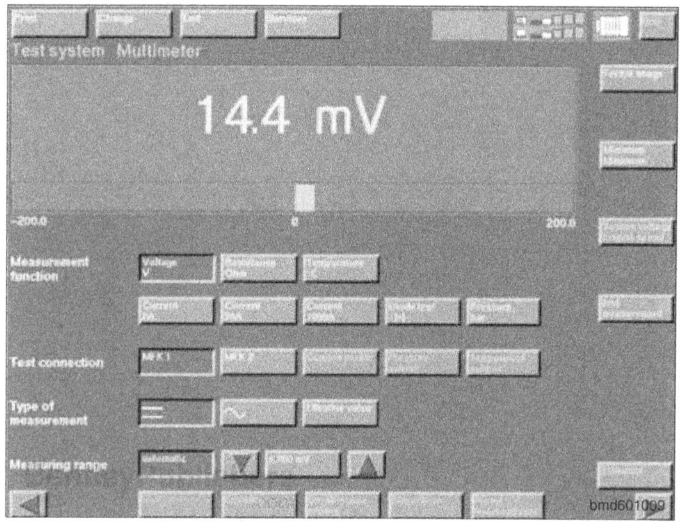

DVOMs with the highest ohms / volt impedance are the most accurate. More importantly, using a meter with high impedance does not cause damage to sensitive electronic circuitry.

> ***CAUTION —***
> - *Using an older type meter with low impedance values (20,000 to 30,000 ohms / volt) can damage modern electronic circuits and components or give inaccurate readings.*
> - *Avoid test lights for the same reason. It lowers the total resistance of the circuit and causes increased current flow.*

When connecting a meter across a circuit to measure voltage, connect it in parallel. This adds parallel resistance. The total resistance in a parallel circuit is less than the lowest resistance in that circuit (Ohm's Law). Using a meter with low impedance reduces the total resistance of the circuit and allow more current to flow.

A meter with low impedance can draw enough current to cause inaccurate measurement, voltage drops or damage sensitive electronic circuit boards. A high impedance meter draws little current and provides accurate readings.

Other factors in choosing the proper DVOM are:

- Cost
- Features

Fluke 87: 10 mega-ohm impedance

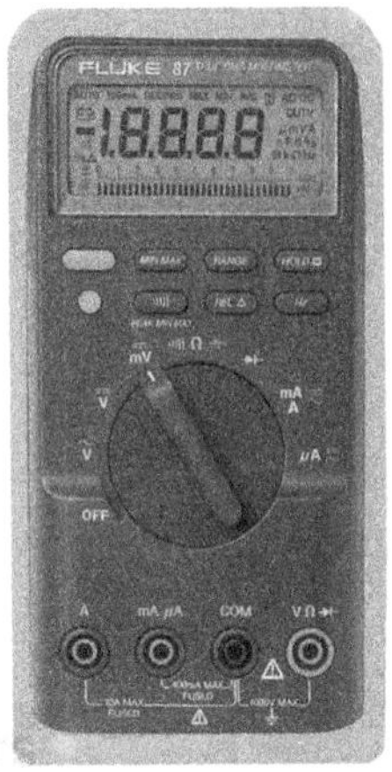

bmd601011

Basic DVOMs are reasonably priced. These basic models may be more than sufficient for use in MINI repair centers, given the availability of the factory tool for advanced measurement and scope functions.

Advanced features and price go hand in hand. The more features, the higher the cost. Some of those features may be worth the increase in cost (e.g. frequency, duty cycle, oscilloscope and graphing). Other features may not (e.g. pulse width).

Choose a DVOM wisely based on personal preference and cost. Like many other tools it is valuable in the diagnosis and repair of MINIs.

Experience has shown if the technician is not comfortable with the DVOM or confident in the results of the measurements, the DVOM will not be used. Considering the technology in MINI automobiles, diagnosis with a quality DVOM makes repairing the problem correctly and expediently a more manageable task.

Diagnostic lab scope / graphing multimeter

A lab scope or a graphing multimeter is needed when diagnosing automotive signals. It is important to have the ability to see the voltage pattern from a sensor or an actuator over time. Using a lab scope or a graphing multimeter enables the technician to do this.

A lab scope is comparable to a camera that captures electrical signals and displays them for us to observe and interpret. A graphing multimeter is a digital multimeter with the ability to show any signal it can display over time. For example; voltage, amperage, frequency, and resistance. This can be very useful when looking for an intermittent failure, or viewing a sensor input.

ATS laptop-based lab scope

bmd6010023

A standard digital multimeter can only display voltage or other signals in the format of digits. The disadvantage of the DVOM is that it averages the signal it is monitoring. If viewing a crankshaft position sensor that has an amplitude of 5 volts, you will read around 2.5 volts on a DVOM. A labscope displays the entire signal clearly for the technician to interpret. This is what makes a lab scope or a graphing multimeter a powerful tool.

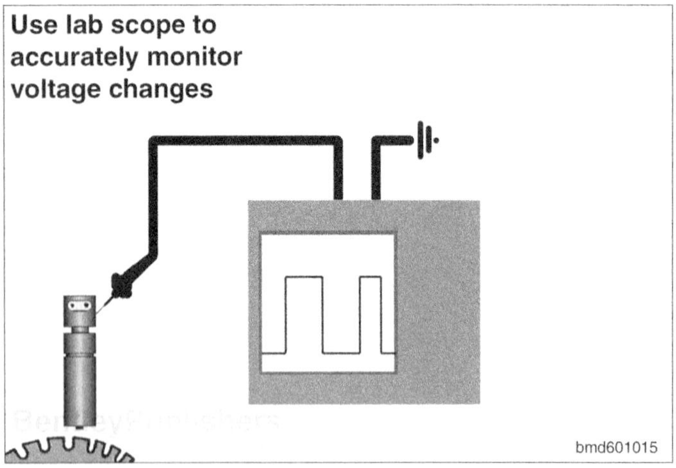

Back probes and test leads

During testing it is important not to pierce wires while monitoring a signal. A break in the insulation leads to future problems. Use back probes and test leads in conjunction with your meter or scope. If absolutely necessary to pierce a wire, be sure to seal the hole using liquid electrical tape when testing is complete.

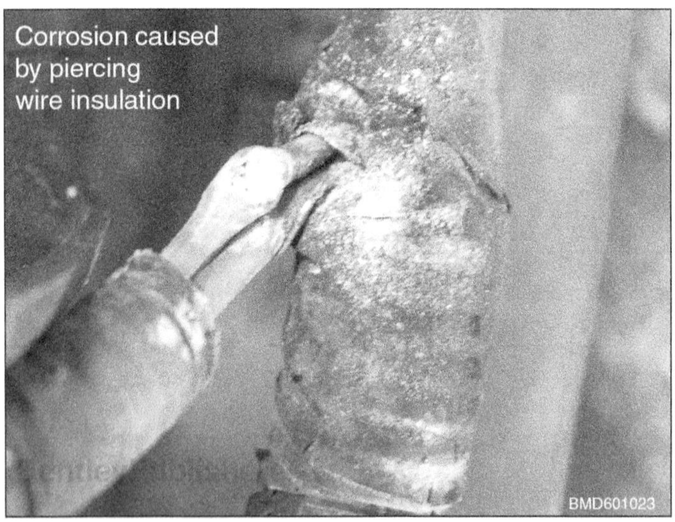

Many different types of extensions / adapters are available for your DVOM or scope. When back probing a connector use the proper back probe adapter. Do not use a paper clip, cotter pin or other object that may be too large and cause damage to the terminals.

DVOM test lead adapters and backprobes

bmd601007

Using a high quality test lead for your DVOM or scope provides a more accurate reading. Use a shielded lead where there may be high EMI (electro magnetic interference). If you need to extend the DVOM or scopes leads, avoid using a jumper wire. Use a test lead extension. The secure fit of the extension leaves less chance for an incorrect reading or accidental shorts. There are times when a jumper wire may be needed. Avoid using a coiled jumper wire; these can have a higher resistance than a straight wire test lead and can lead to incorrect readings.

Jumper wire **Test lead extensions**

BMD601209

Current probes

A good quality low and high current probe is useful for testing multiple components, such as fuel pump, fuel injectors, solenoids, etc. Current probes allow you to measure the amount of current in a circuit without breaking into it.

Current probes

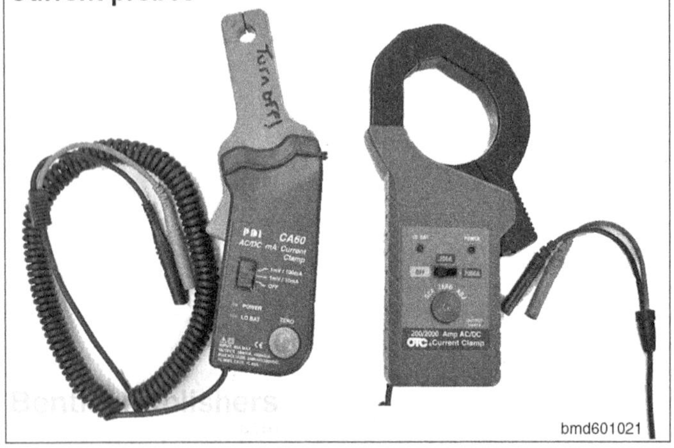

bmd601021

Fuel pressure gauge

A good quality fuel pressure gauge is needed to test the fuel delivery system.

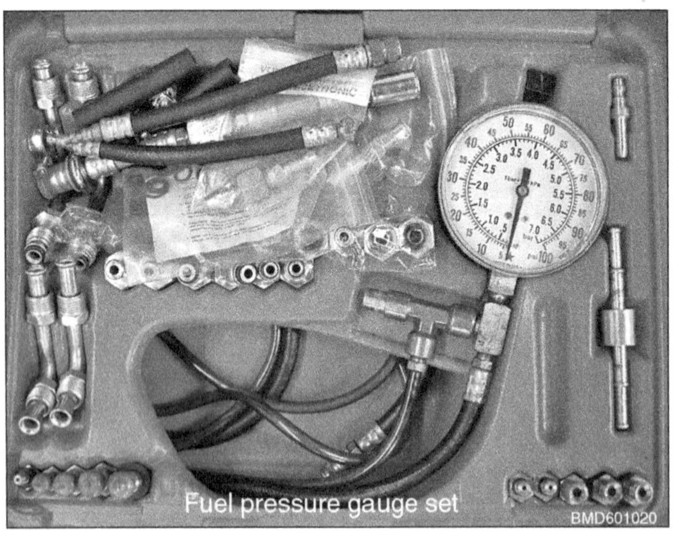

Fuel pressure gauge set

BMD601020

Break-out box (BOB) and test adapters

When servicing the engine control system, use a factory break-out box with the correct adapters to ensure the small terminal connectors and the harnesses are not be damaged. A substitute is high quality back-probes.

Break-out Box
(p/n 61 4 380)

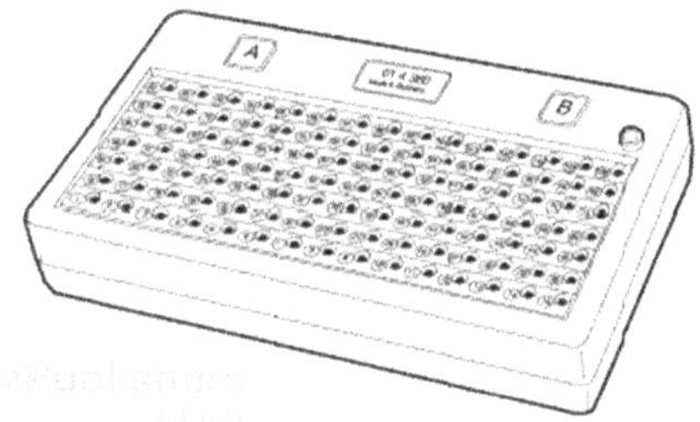

V-adapter cables for the ECM are:

- V-adapter cable P/N 12 7 240 (40 -pin)
- V-adapter cable P/N 12 7 250 (81 - pin)
- Break-out box (BOB) P/N 61 4 380 (105 - pin)
- Break-out box (BOB) P/N 61 4 390 (60 - pin)

Adapter cable
(p/n 12 7 240)

Adapter cable
(p/n 12 7 250)

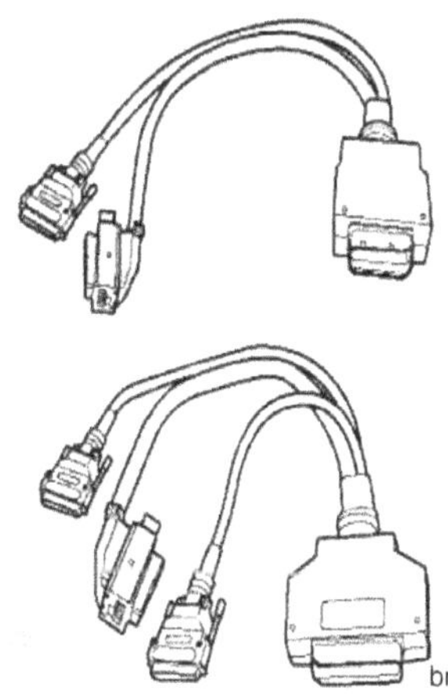

bmd601210

> ***CAUTION —***
> - *When installing the BOB and V-adapter cables to the ECM, make sure the ignition is switched off for at least 5 minutes before making connections.*

TEST TECHNIQUES

Voltage drop testing

Voltage drop is the amount of voltage that is used as electrical current passes through a load. All source voltage is used up in the circuit. The sum of the voltage drops throughout the circuit equals the source voltage. Measure voltage drop by measuring the voltage before the load and the voltage as it leaves the load. The difference between these two voltage readings is the voltage drop.

When more than one load is present in the circuit, the voltage is divided or shared between loads.The total voltage drops though all loads equals source voltage. The higher the resistance of a load, the higher the voltage drop. This means each load can have a different amount of voltage drop.

As current flows through a circuit, the presence of a load in that circuit causes the voltage to drop as it passes through the load. The difference in voltage on each side of the load is called voltage drop. When current flows through the circuit below, voltage drops across loads 1 and 2 (the example is a series circuit).

Voltage drop in series circuit

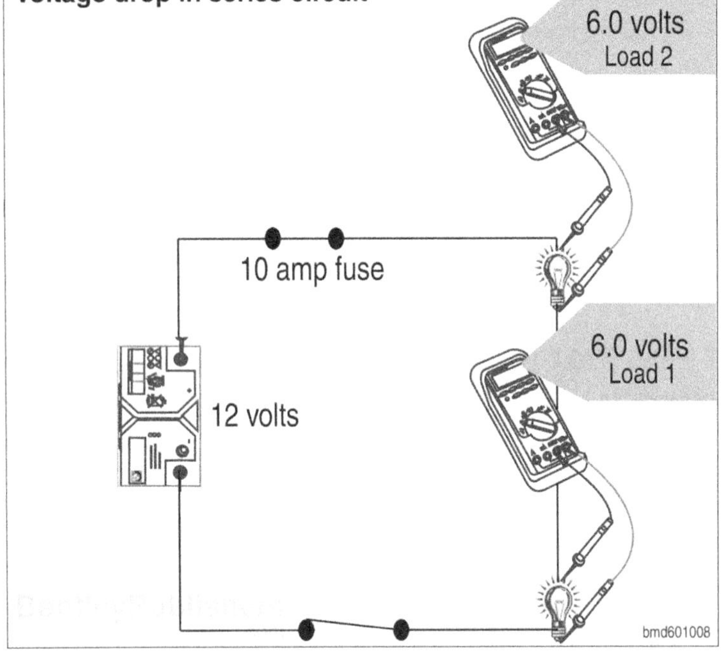

Parasitic draw test and specifications

Parasitic draw may occur permanently or intermittently, and cause the battery to discharge prematurely. Some common causes of increased closed circuit current are faulty control units, a stuck relay, the installation of a non-approved accessory, or a faulty electrical component such as a radiator fan final stage resistor.

- Check and test battery using an approved MINI battery tester. If necessary, recharge or replace battery.
- Remove negative cable and install battery isolation switch, OTC battery isolation switch - #7645, or equivalent.
- Close switch. Connect DVOM in series across switch.

Battery isolation switch

BMD601006

- Battery installed in the trunk: Open trunk and turn lock to locked position using screwdriver or similar (simulates trunk lid being closed). Keep hood closed.
- Battery installed in engine compartment: Open hood and pull front lid contact switch fully up and lock in this position (workshop position, simulates closed front lid). Keep trunk closed.
- With exception of trunk / hood, keep all other doors and lids closed.

– To simulate normal parasitic draw conditions: Switch ignition ON and activate all electrical consumers, including any accessories. Switch ignition OFF. Open and close the driver's door (simulates somebody getting out). Lock car, arming alarm if installed.

NOTE —

- In some cases, a short test drive may be needed to duplicate a closed circuit current problem.

– Ignition off, all loads off. With DMM in series with battery isolation switch, open switch. Monitor current draw.

– In general, investigate any parasitic draw consistently over 50 mA.

Closed circuit current	
R50, R52, R53	40 milliamps after 16 minutes

– If the nominal milliamp reading is not achieved after the appropriate time, begin isolation of the faulty circuit. With use of a power distribution wiring diagram, remove circuit fuses until current draw drops below specification.

Exhaust back pressure testing

Checking exhaust back pressure requires a special gauge designed for connecting to the exhaust system.

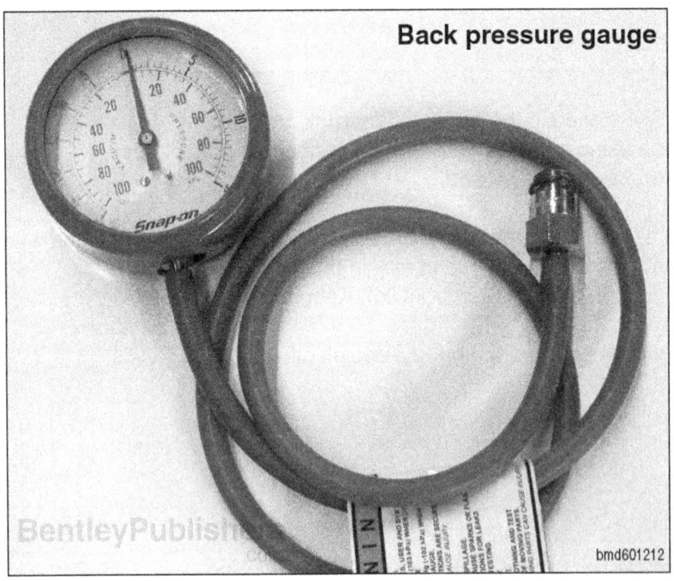

Back pressure gauge

Increased exhaust back pressure can cause multiple driveability complaints. Some examples are low power, fuel trim faults (indicating rich mixture), and pinging or knock. Increased back pressure is normally caused by a restricted exhaust system.

Make sure the engine is at operating temperature before measurements are taken.

– Remove pre catalytic converter oxygen sensor as outlined in appropriate repair manual.

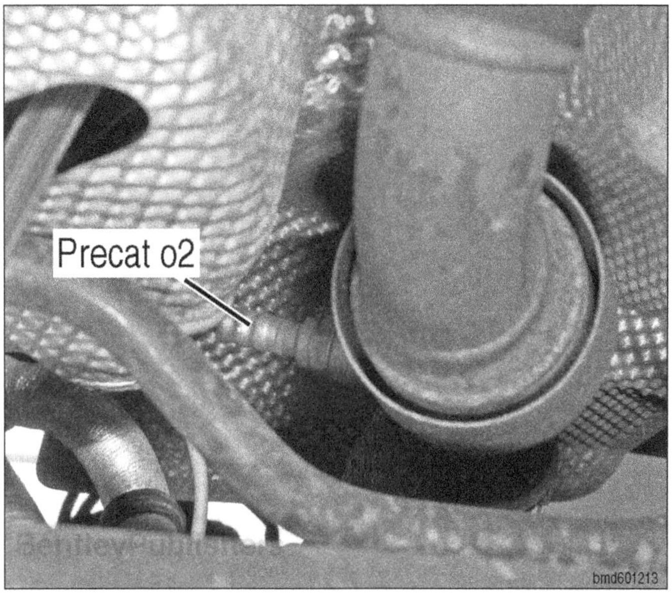

– Install exhaust back pressure gauge hose adapter in place of precatalytic converter oxygen sensor.

NOTE —
• Lightly tighten (snug) the tool fitting. Do not over tighten.

– Mount gauge in upright position.

NOTE—
- Incorrect test readings will result if gauge is not in an upright position, for example: hung up side down, laid flat, etc.

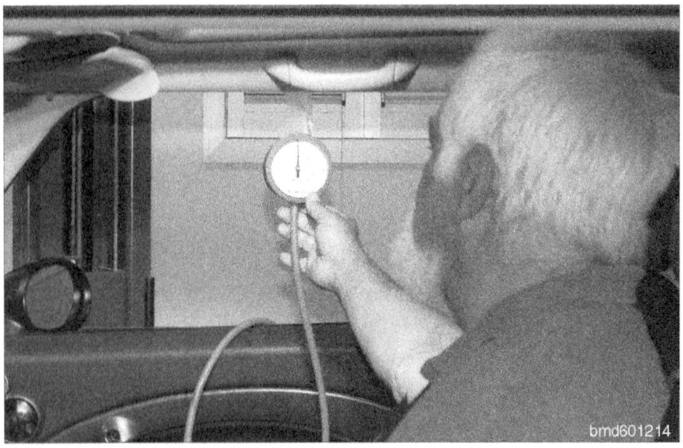

– With parking brake on and vehicle in PARK or NEUTRAL, start engine.

– Raise and briefly hold engine speed at 2500 rpm. With engine speed stabilized at 2500 rpm gauge should remain below 1.5 psi.

– Turn engine OFF and remove gauge adapter. Reinstall oxygen sensor and tighten.

Tightening torque	
Oxygen sensor to exhaust system	55 Nm (41 ft-lb)

NOTE—
- A restricted exhaust system typically indicates readings greater than 2 psi.
- A non-restricted exhaust may momentarily indicate a reading up to approx. 2 psi when revving the engine rapidly from idle speed to 3000-4000 rpm or under load.
- A gauge reading greater than 1.5 psi is normal under these conditions due to momentary pressure build up and does not indicate a restricted exhaust.

Evaluating smooth running values

The smooth running values of individual cylinders are used for troubleshooting an engine misfire. This test is done on the factory scan tool.

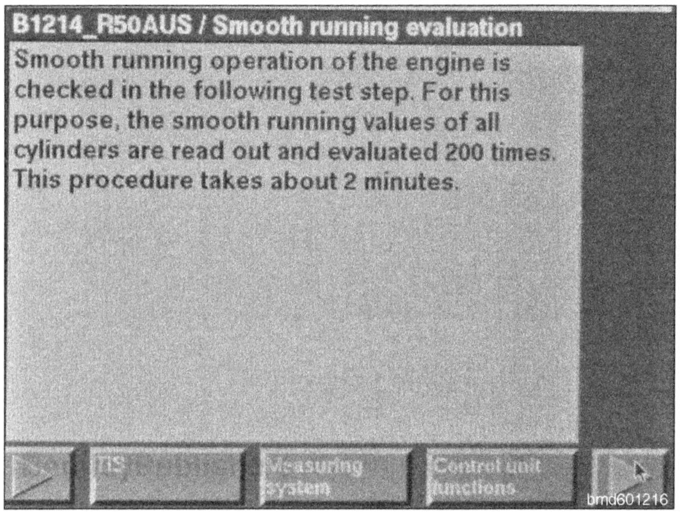

Make sure engine runs at idle speed for at least 3 minutes to ensure that correct values are set. Smooth running can only be evaluated with the engine running at idle speed (cold or hot). An indication of the combustion quality of individual cylinders can be obtained by evaluating crankshaft acceleration, measured at the crankshaft position/rpm sensor. A weak individual cylinder with poor combustion can be detected in this way.

An increase in the smooth-running values may be caused by various factors (e.g. misfiring, secondary air, mixture deviations, faults in fuel supply, low compression).

If smooth running values are exceeded within a certain number of combustion cycles, a fault is set and stored.

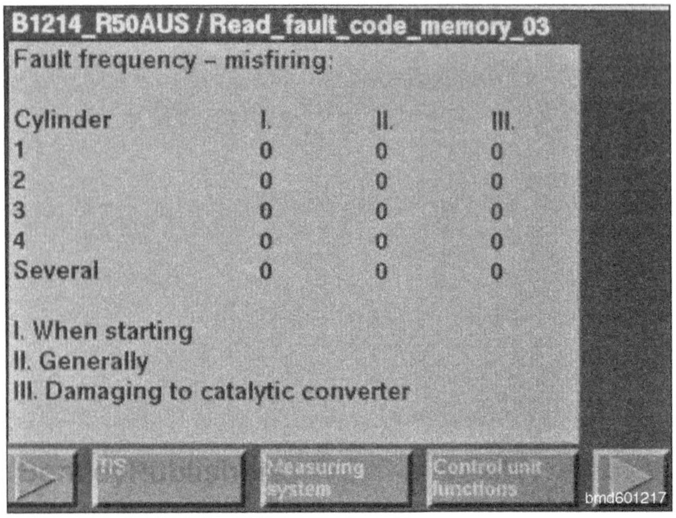

Evaluating smooth running values

A smooth running value less than 2.5 is an acceptable reading. When the values are in a range of 2.5 to 7.0 this is considered a critical range. Once over 2.5 there is a noticeable decrease in smoothness of the engine, i.e. rough idle. A value greater than 7.5 is considered not ok. This usually indicates a cylinder misfire that is catalyst damaging. A fault code should be stored.

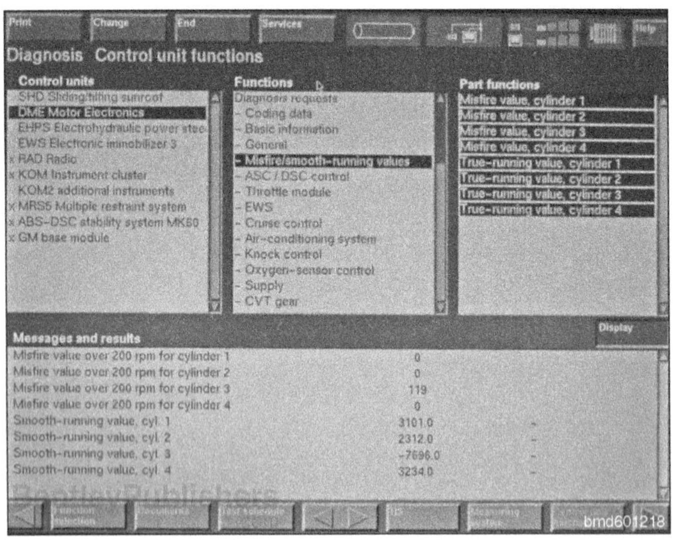

Compression testing

Prior to doing a compression check, read out ECM fault memory, check all stored fault data, rectify faults, clear fault memory. Switch ignition OFF.

NOTE —
- Making the compression check outlined here (disconnecting coil pack and cranking engine) induces a fault code. Clear stored faults when testing is complete.

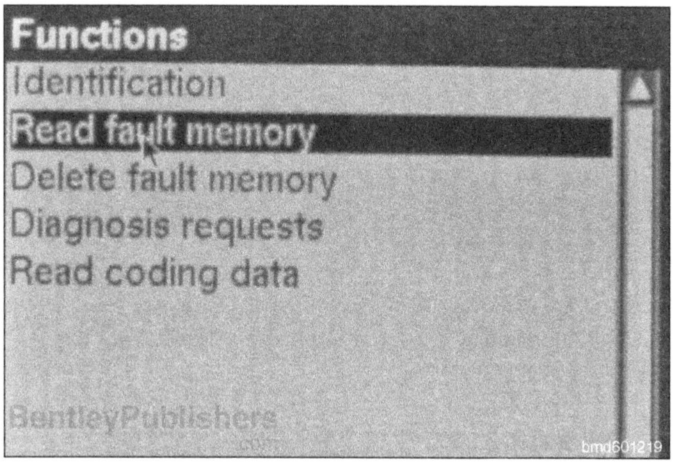

> ***CAUTION —***
> - *Interrupt power supply to ignition coils to prevent shock from high voltage ignition system.*

– Disable fuel system

– Start and run engine until it stalls. This step eliminates residual fuel from injectors and fuel lines.

- Switch ignition OFF, working at coil pack harness connector, slide connector lock (**arrow**) and disconnect.

bmd601019

- Remove spark plugs and keep in order for reinstallation. Screw compression gauge into cylinder. Check compression adaptor hose seal is in perfect condition.

bmd601018

- Depress accelerator pedal fully and actuate starter until compression stops rising. Record reading. Repeat above steps for each cylinder.

Adaptations

Adaptation makes it possible for the ECM to learn certain values from components and equipment variants, thus making it possible to compensate for component tolerances. Adaptation is also normally necessary when installing replacement ECM components.

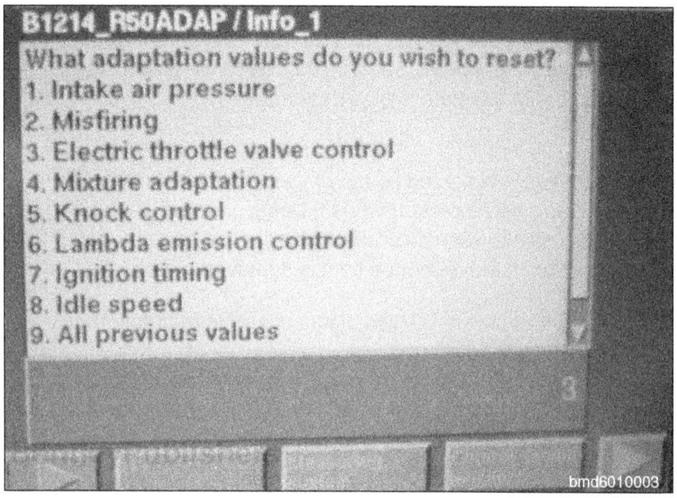

bmd6010003

These adaptation values can be reset to their original base status with the **Reset adaptation values** service function in the factory scan tool.

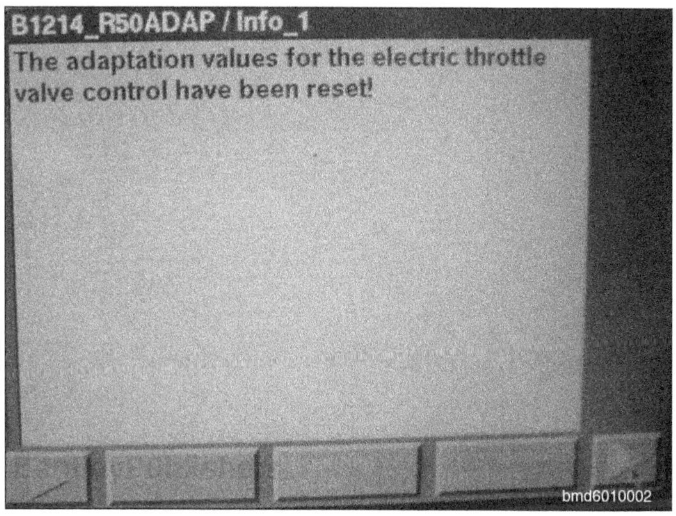

bmd6010002

Failure to reset adaptations after a repair can result in symptoms similar to the original fault. In some cases a fault code does not clear until the adaptations are reset. There have also been cases where the adaptations must be reset multiple times. In order to confirm the fault has been repaired, reset adaptations and attempt to duplicate fault.

Example: Mixture adaptation

The fuel-air mixture requires a certain period of time until it reaches the oxygen sensor in the form of exhaust gas. This time decreases as load and engine speed increase. For this reason, the response time of the emission (Lambda) control system is also dependent on load and engine speed.

Fuel-air mixture deviations detected by the oxygen sensor result in adaptation values (learned correction values) being stored. By way of the adaptations, fuel injection can be brought close to the nominal value in advance. A reduction in the response time is achieved in this way.

If the basic injection values were too low at idle speed to maintain the ideal fuel-air mixture, Lambda control would have to constantly increase the injection time. In this case, an adaptation value is stored, which corrects the basic value. The emission (Lambda) control then only needs to do fine adjustments.

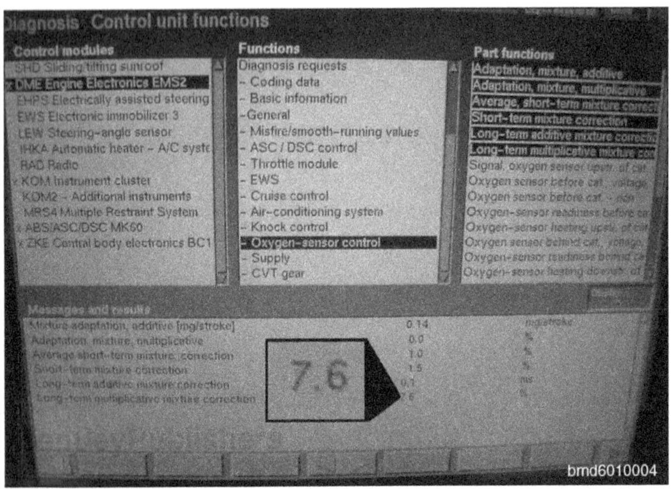

Instrument cluster (IKE) self-test

In addition to fault memory and diagnostic link, the base instrument cluster contains a series of test functions that can be accessed to check for various functions and values. The test functions are displayed in the mileage LCD block. There are a total of 21 test functions.

- Test 1 and 2 are always unlocked.
- Tests 3-21 are only accessible after unlocking the test function.
- Test 19 is the unlock function for accessing the displays.

NOTE —
- Read through the procedure at least once beforehand.

– Switch ignition ON: ignition switch in position 2.

– Press trip recorder reset button for 5 seconds.

– Press trip recorder reset button again immediately until tESt 19. appears in display.

– Wait briefly until - _L_off and _L_on appear alternately in display.

– With _L_off, press reset button once to unlock cluster test functions.

- Press trip recorder reset button again immediately until tESt 21. appears. Wait until 21.0 appears in display.
- Press trip recorder reset button once -> Reset/self-test is initiated.

- To start test functions, press trip distance reset button for 5 seconds. tESt 1 appears in display. Pressing reset button again takes you to different test numbers.
- To access locked tests (3 to 21), scroll to tESt 19. Display alternates between _L_oFF and _L_oN. Pressing reset button opens the lock. The display then jumps to test 0 and the corresponding test number group can be selected immediately by pressing reset button again.
- The following graphics illustrate additional self test functions.

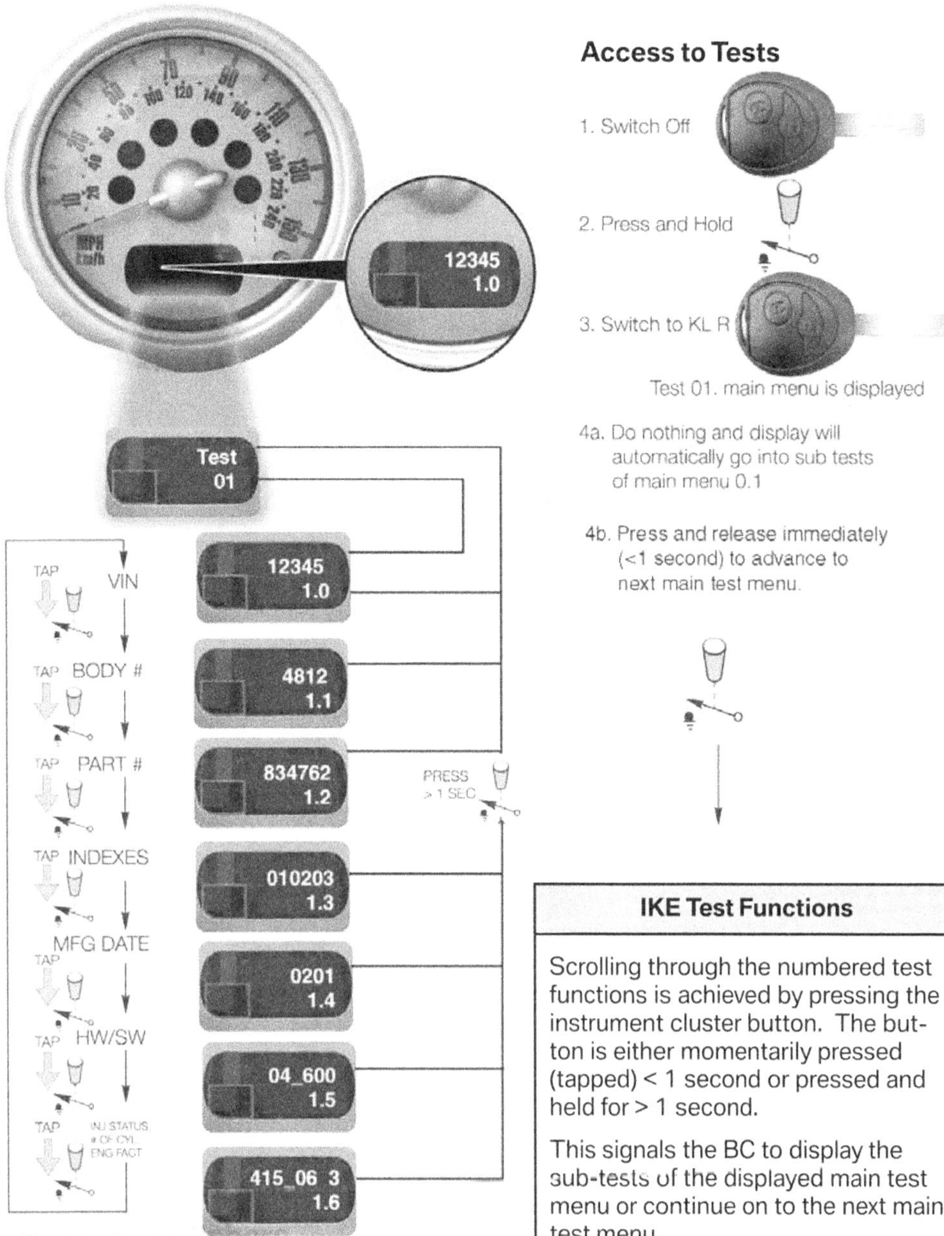

IKE Test Functions

Scrolling through the numbered test functions is achieved by pressing the instrument cluster button. The button is either momentarily pressed (tapped) < 1 second or pressed and held for > 1 second.

This signals the BC to display the sub-tests of the displayed main test menu or continue on to the next main test menu.

bmd6010024

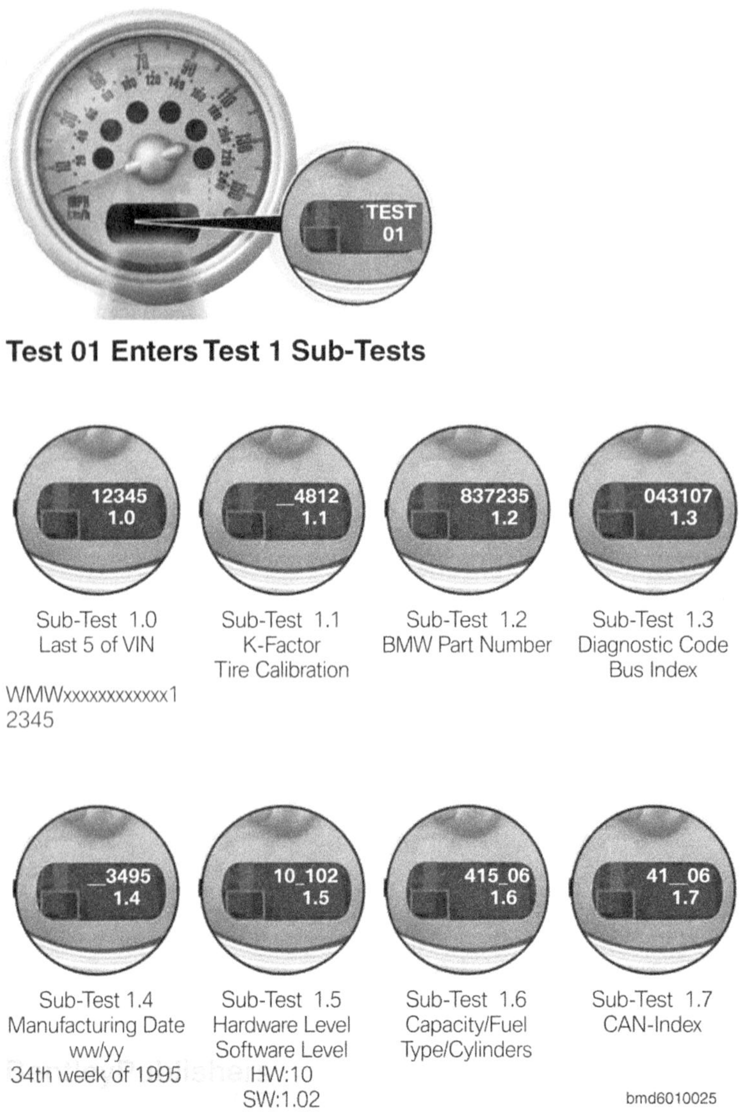
TEST
01
Test 01 Enters Test 1 Sub-Tests
12345
1.0
_4812
1.1
837235
1.2
043107
1.3
Sub-Test 1.0
Last 5 of VIN
WMWxxxxxxxxxxxx1
2345
Sub-Test 1.1
K-Factor
Tire Calibration
Sub-Test 1.2
BMW Part Number
Sub-Test 1.3
Diagnostic Code
Bus Index
_3495
1.4
10_102
1.5
415_06
1.6
41_06
1.7
Sub-Test 1.4
Manufacturing Date
ww/yy
34th week of 1995
Sub-Test 1.5
Hardware Level
Software Level
HW:10
SW:1.02
Sub-Test 1.6
Capacity/Fuel
Type/Cylinders
Sub-Test 1.7
CAN-Index
bmd6010025

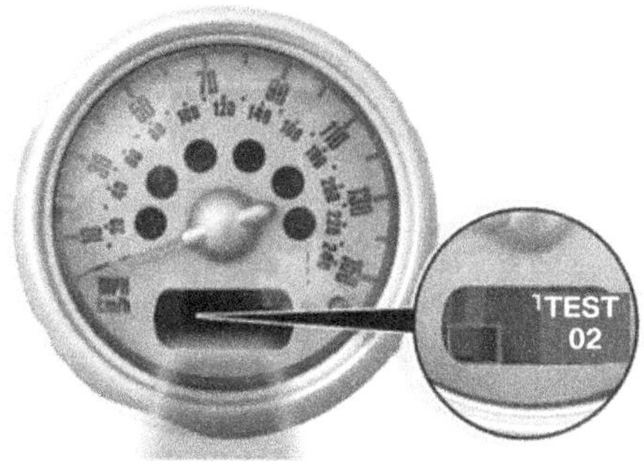

Test 02 Enters Test 2 Sub-Tests

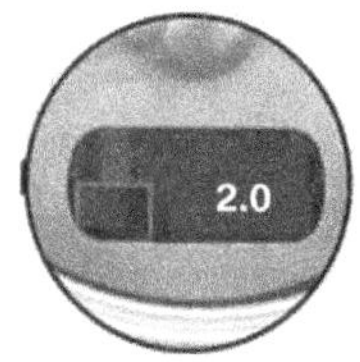

Sub-Test 2.0
Activates all gauges, LED's and indicators

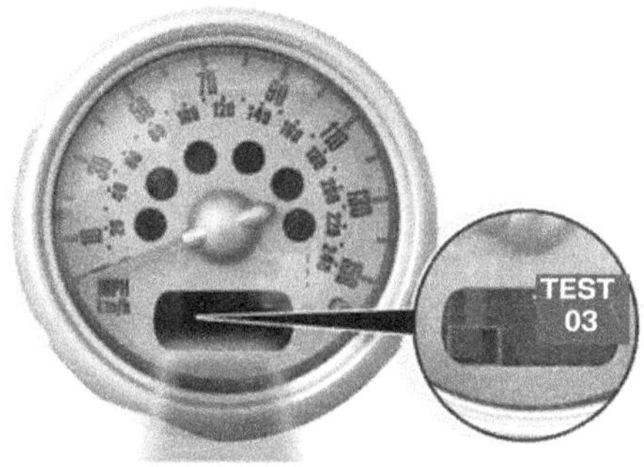

Test 03 Enters Test 3 Sub-Tests

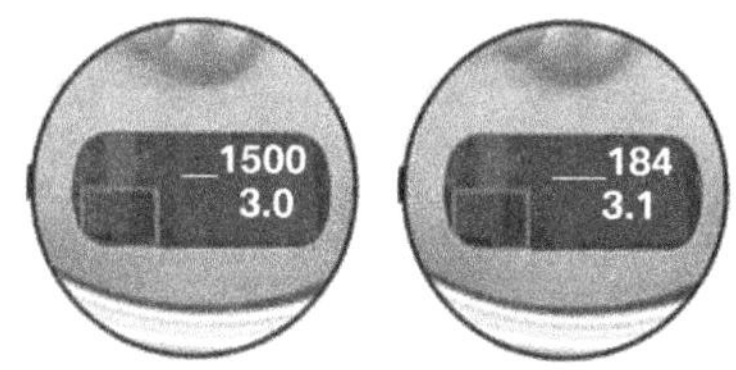

Sub-Test 3.0
Fuel used Since Last SIA Reset
1500 Liters of Fuel Used

Sub-Test 3.1
Days Elapsed Since Last SIA Reset
Not Used in USA

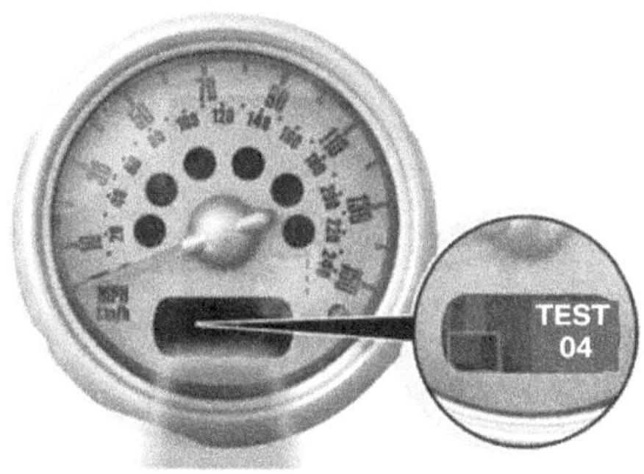

Test 04 Enters Test 4 Sub-Tests

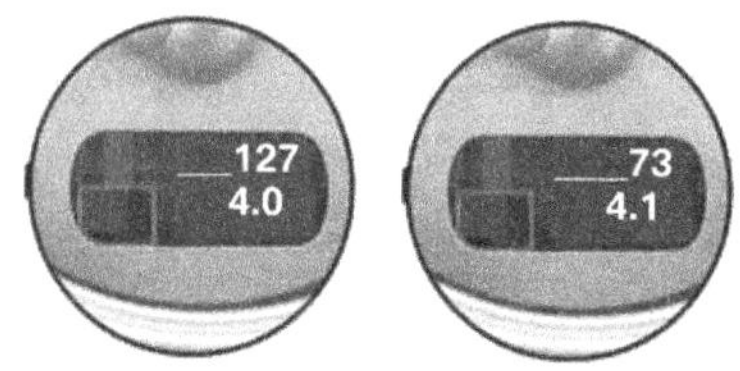

Sub-Test 4.0
Instantaneous Fuel Consumption
L/100km
12.7

Sub-Test 4.1
Instantaneous Fuel Consumption
L/Hour
7.3

bmd6010026

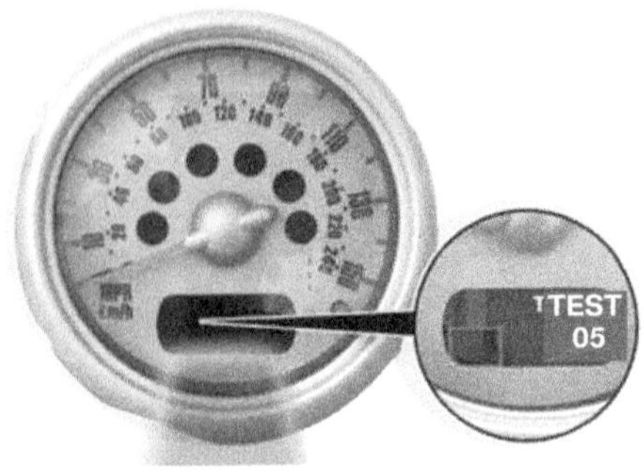

Test 05 Enters Test 5 Sub-Tests

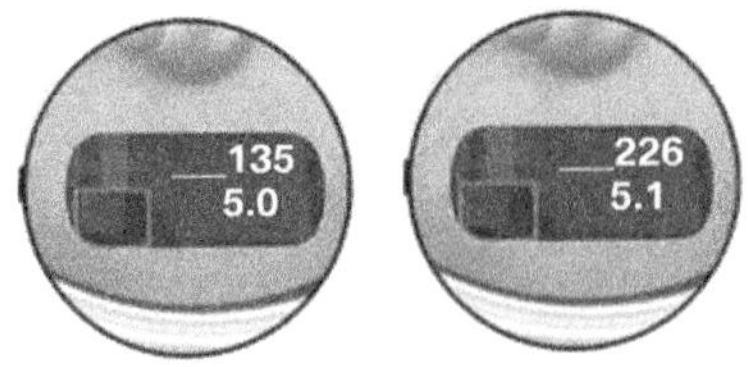

Sub-Test 5.0
Average Fuel
Consumption
L/100
13.5

Sub-Test 5.1
Distance Remaining
on Present Fuel
km
226km

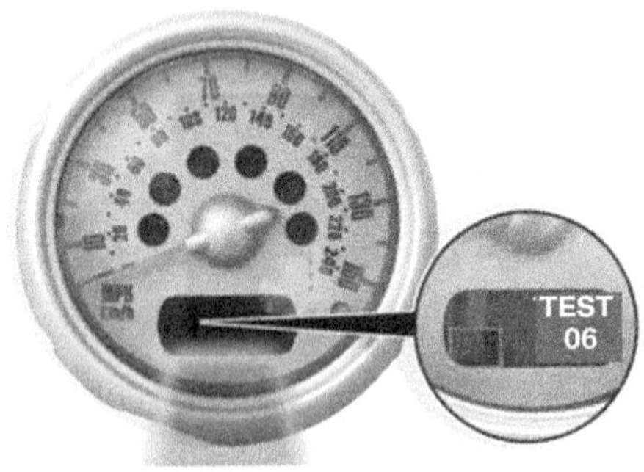

Test 06 Enters Test 6 Sub-Tests

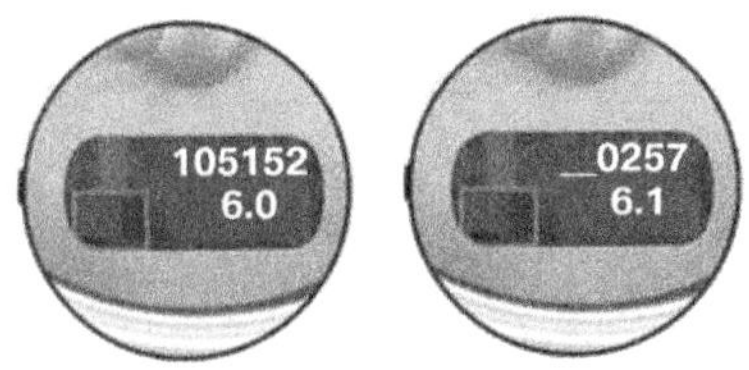

Sub-Test 6.0
Fuel Level Sensor
Readings
LH/RH
Left 10.5 Liters
Right 15.2 Liters

Sub-Test 6.1
Total Tank
Volume in Liters
25.7 Liters

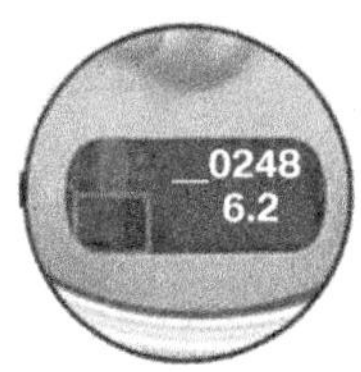

Sub-test 6.2
Tank Sender Status

bmd6010027

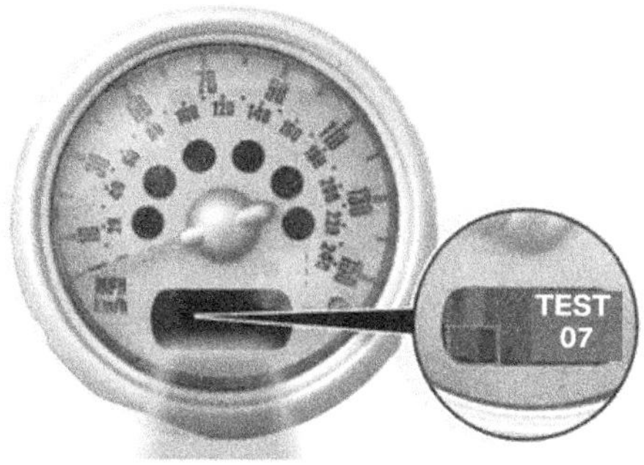

Test 07 Enters Test 7 Sub-Tests

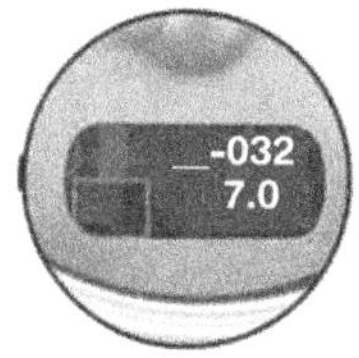

Sub-Test 7.0
Engine Coolant
Temperature °C
-32°C

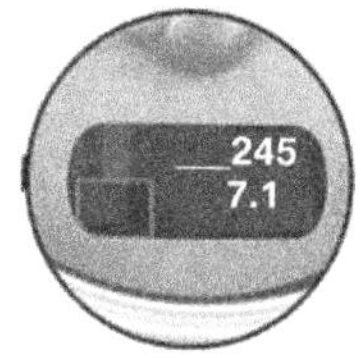

Sub-Test 7.1
Ambient Air
Temperature °C
24.5°C
In 5° increments

Sub-Test 7.2
Engine Speed
RPM
3525 RPM

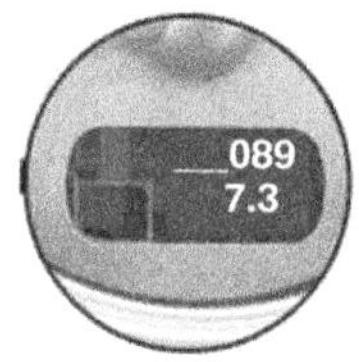

Sub-Test 7.3
Instantaneous Road
Speed
km/h
89 km/h

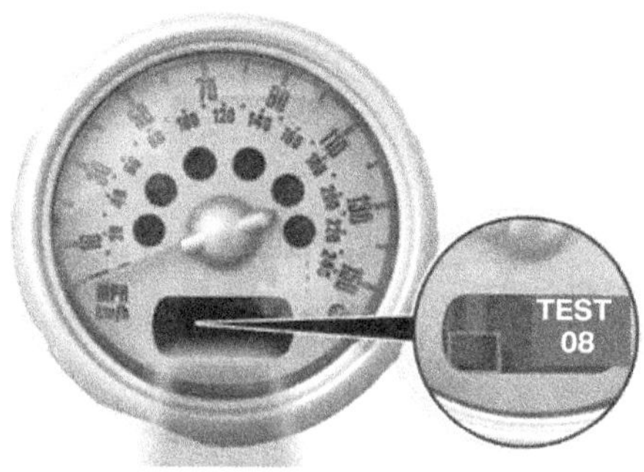

Test 08 Enters Test 8 Sub-Tests

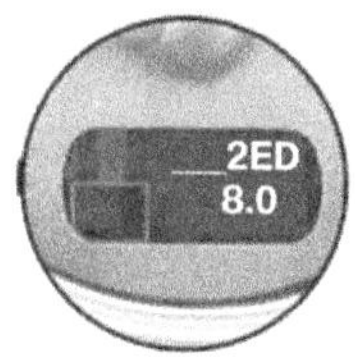

Sub-Test 8.0
HEXIDECIMAL
Information
System Voltage

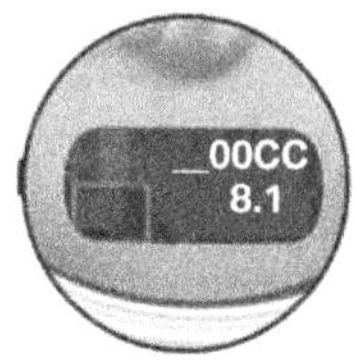

Sub-Test 8.1
HEXIDECIMAL
Information
Fuel Level Sensors

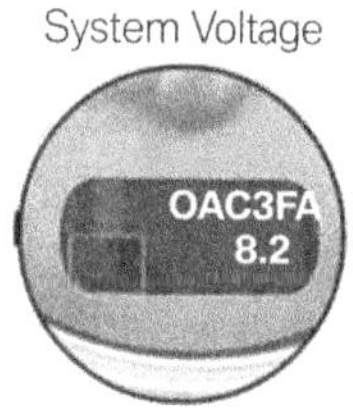

Sub-Test 8.2
HEXIDECIMAL
Information
Phototransistor

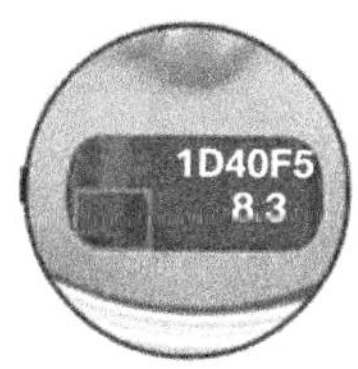

Sub-Test 8.3
HEXIDECIMAL
Information
Coolant
Temp/Ambient

bmd6010028

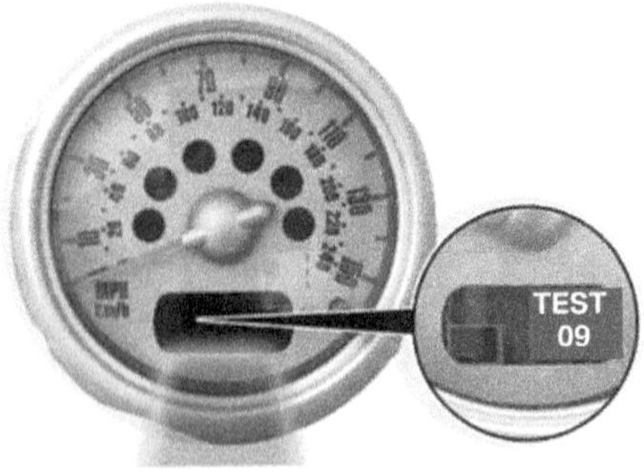

Test 09 Enters Test 9 Sub-Tests

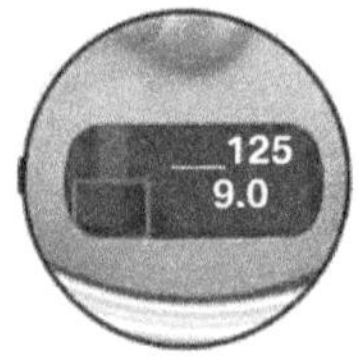

Sub-Test 9.0
System Voltage
12.5 volts

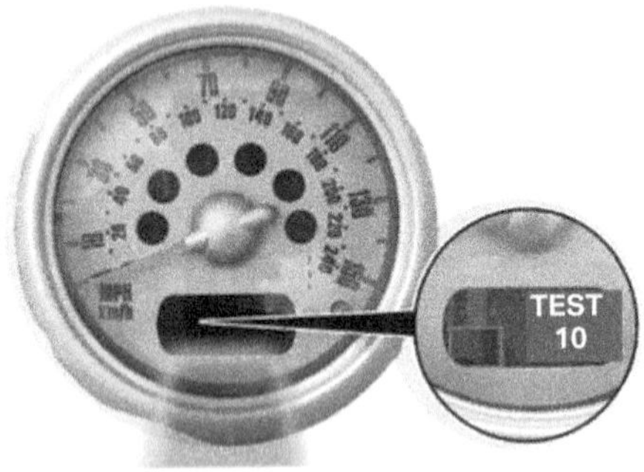

Test 10 Enters Test 10 Sub-Tests

(Test 10 information concerns the Remote Cluster.)

Sub-Test 10.0
Not Used

Sub-Test 10.0
Not Used

Sub-Test 10.2
BMW Part Number
Remote Cluster

Sub-Test 10.3
Diagnostics Code
Bus Index

Sub-Test 10.4
Manufacturing Date
wwyy
Week 34
Year 1995

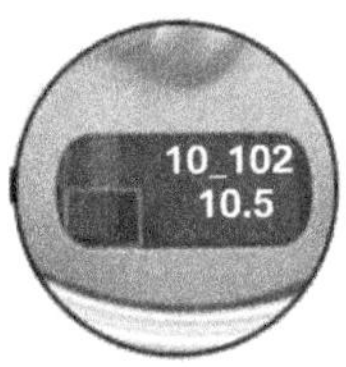

Sub-Test 10.5
Hardware and
Software Level
HW: 10
SW:1.02

Sub-Test 10.6
Not Used

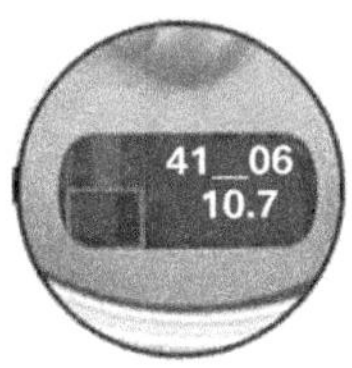

Sub-Test 10.7
CAN-Index

bmd6010029

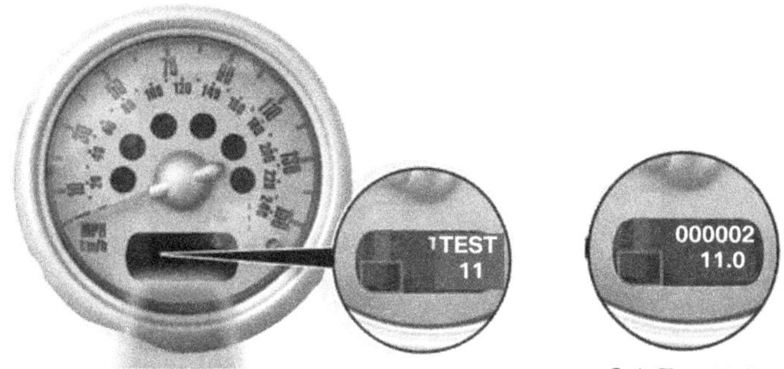

Test 11 Enters Test 11 Sub-Tests

Sub-Test 11.0
Market Code
US = 02

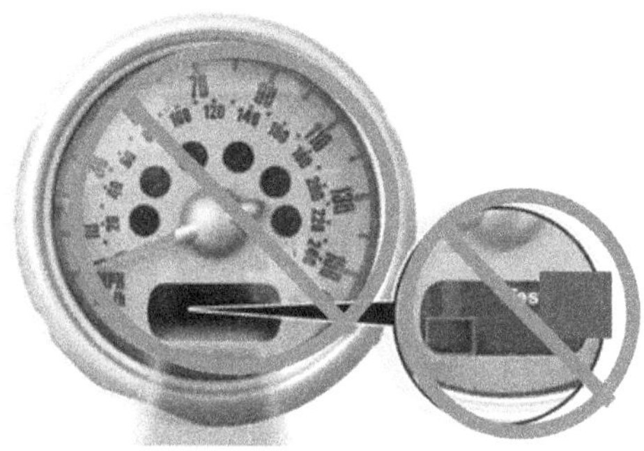

Test 12 Not Used

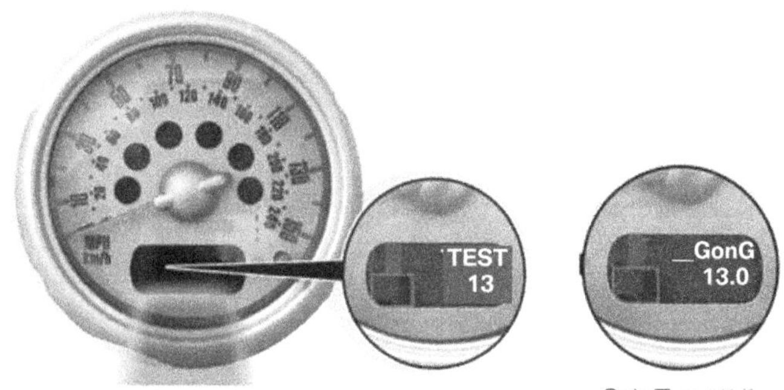

Test 13 Enters Test 13 Sub-Tests

bmd6010030

Sub-Test 13.0
Test the Audible
Warning Speaker

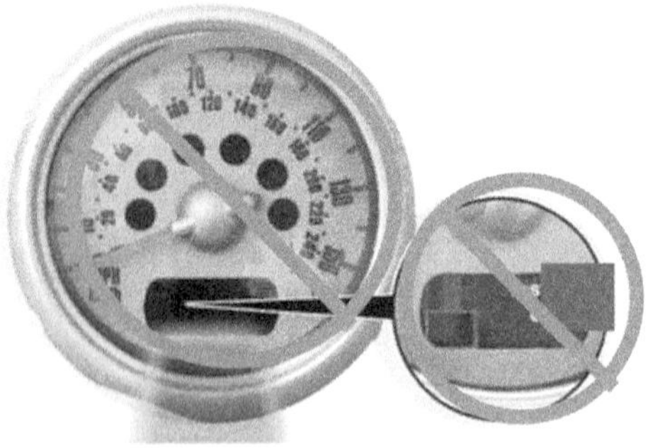

Test 14 Error Data Not Used for Diagnostics

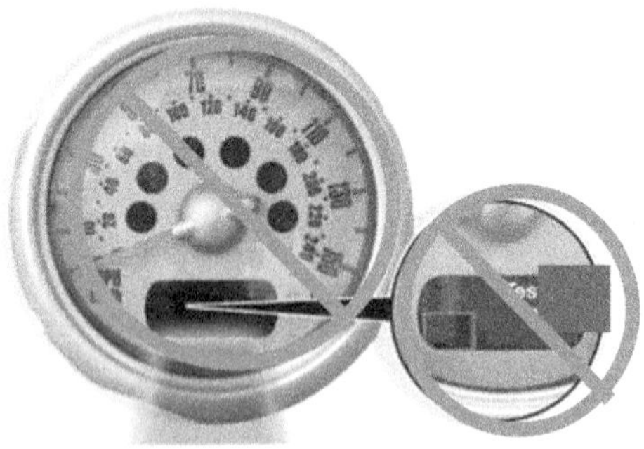

Tests 15, 16,17,18 Not Used for Diagnostics

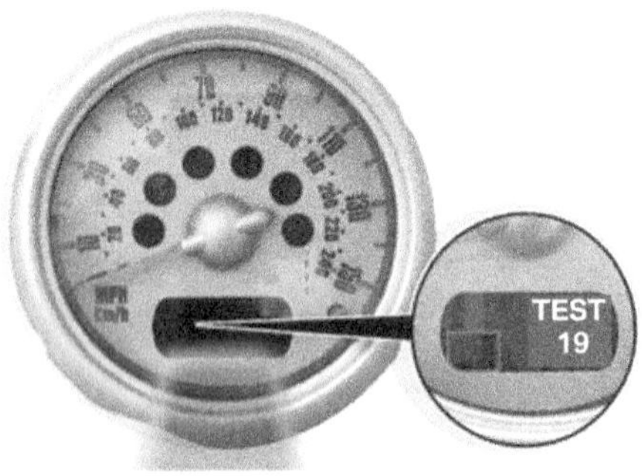

Test 19 Enters Test 19 Sub-Tests

Sub-Test 19.0
Lock On
Self Test
Capabilities Not
Available

Sub-Test 19.0
Lock Off
All Available Tests
Accessible

bmd6010031

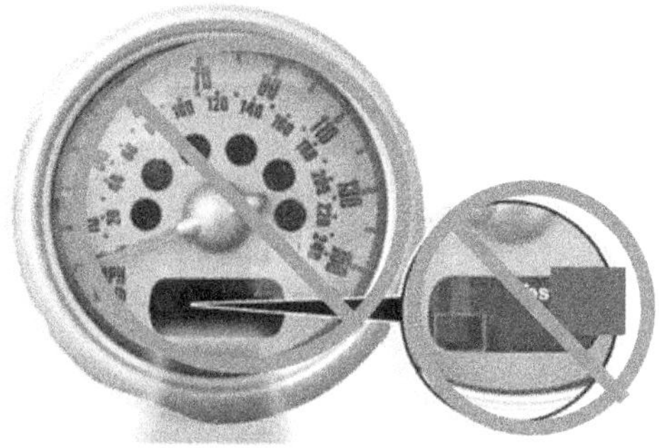

Test 20 Not Used

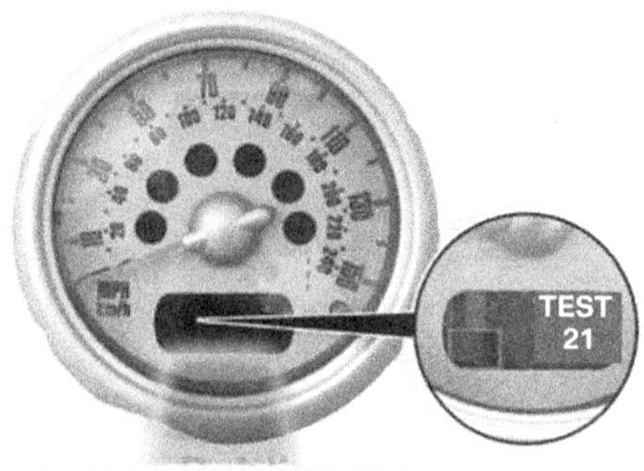

Test 21 Enters Test 21 Sub-Tests

bmd6010032

Sub-Test 21
Software Reset

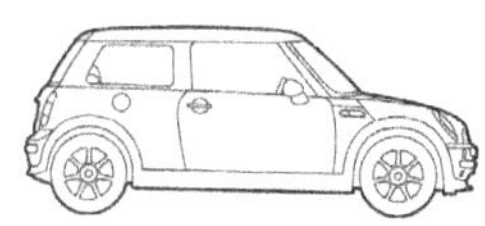

Notes:

13 Engine Management

Table of Contents

Table of Contents

P-code	MINI FC	Description	Threshold value	Related component(s)
P0030	48	O_2 sensor heater circuit, B1S1	Incorrect voltage	Pre-catalyst O_2 sensor bank 1, (heater component of heated oxygen sensor) Page 13-105
P0031	49	O_2 sensor heater circuit, (B1S1)	Circuit low	Pre-catalyst O_2 sensor bank 1, (heater component of heated oxygen sensor) Page 13-105
P0032	50	O_2 sensor heater circuit, (B1S1)	Circuit high	Pre-catalyst O_2 sensor bank 1, (heater component of heated oxygen sensor) Page 13-105
P0036	54	O_2 sensor heater circuit, (B1S2)	Incorrect voltage	Post-catalyst O_2 sensor bank 1, (heater component of heated oxygen sensor) Page 13-106
P0037	55	O_2 sensor heater circuit, (B1S2)	Circuit low	Post-catalyst O_2 sensor bank 1, (heater component of heated oxygen sensor) Page 13-106

P0030
P0031
P0032
P0036
P0037

P-code	MINI FC	Description	Threshold value	Related component(s)
P0038	56	O_2 sensor heater circuit, (B1S2)	Circuit high	Post-catalyst O_2 sensor bank 1, (heater component of heated oxygen sensor) Page 13-106
P0107	263	MAP / BARO pressure circuit low Cooper Cooper S, upstream of supercharger	< 0.07 volts	MAP sensor (Cooper) Page 13-117 TMAP / MAP sensor (Cooper S) Page 13-120
P0108	264	MAP / BARO pressure circuit high Cooper Cooper S, upstream of supercharger	> 4.89 volts	MAP sensor (Cooper) Page 13-117 TMAP / MAP sensor (Cooper S) Page 13-120
P0112	274	IAT sensor circuit low	< 0.07 volts	MAP sensor (Cooper) Page 13-117 TMAP / MAP sensor (Cooper S) Page 13-120
P0113	275	IAT sensor circuit high	> 4.89 volts	MAP sensor (Cooper) Page 13-117 TMAP / MAP sensor (Cooper S) Page 13-120

P0038
P0107
P0108
P0112
P0113

P-code	MINI FC	Description	Threshold value	Related component(s)
P0114	276	IAT sensor circuit	Intermittent	MAP sensor (Cooper) Page 13-117 TMAP / MAP sensor (Cooper S) Page 13-120
P0116	278	ECT sensor circuit	Range / performance	ECT sensor Page 13-73 ECT sensor resistance Page 13-74
P0117	279	ECT sensor circuit low	< 0.07 volts	ECT sensor Page 13-73 ECT sensor resistance Page 13-74
P0118	280	ECT sensor circuit high	> 4.89 volts	ECT sensor Page 13-73 ECT sensor resistance Page 13-74
P0119	281	ECT sensor circuit	Intermittent	ECT sensor Page 13-73 ECT sensor resistance Page 13-74 Thermostat monitor *See also* **OBD On-board Diagnostics (OBD II)**

P-code	MINI FC	Description	Threshold value	Related component(s)
P0122	290	Accelerator pedal position (PWG) sensor A circuit low	< 0.02 volts	Accelerator pedal position (PWG) sensor Page 13-44
P0123	291	Accelerator pedal position (PWG) sensor A circuit high	> 4.9 volts	Accelerator pedal position (PWG) sensor Page 13-44
P0125	293	Insufficient temperature for closed loop	Modeled temperature difference. 150 seconds for -7°C (19°F), 75 seconds 10°C (50°F)	Engine does not reach operating temperature in allotted time. Thermostat monitor *See* **OBD On-board Diagnostics (OBD II)**
P0128	296	Thermostat	Engine temperature below regulating threshold, 91°C (195°)	Thermostat monitor *See* **OBD On-board Diagnostics (OBD II)**
P0130	304	O_2 sensor B1S1 circuit	0.371-0.469 volts for 10 seconds	Pre-catalyst O_2 sensor Page 13-102 Thermostat monitor *See also* **OBD On-board Diagnostics (OBD II)**

P-code	MINI FC	Description	Threshold value	Related component(s)
P0131	305	O_2 sensor B1S1 circuit low	0.03 volts for 25 seconds	Pre-catalyst O_2 sensor Page 13-102 Oxygen sensor monitor *See also* **OBD On-board Diagnostics (OBD II)**
P0132	306	O_2 sensor B1S1 circuit high	1.39 volts for 10 seconds	Pre-catalyst O_2 sensor Page 13-102 Oxygen sensor monitor *See also* **OBD On-board Diagnostics (OBD II)**
P0133	307	O_2 sensor B1S1 slow response	Sum of switch times = > 3 seconds (manual trans) Sum of switch times = > 3 seconds (auto trans)	Pre-catalyst O_2 sensor Page 13-102 Oxygen sensor monitor *See also* **OBD On-board Diagnostics (OBD II)**
P0135	309	O_2 sensor heater circuit, (B1S1) calculated resistance	< 1.5 ohms > 25 ohms for 8 seconds	Pre-catalyst O_2 sensor bank 1, (heater component of HO2S) Page 13-105
P0136	310	O_2 sensor B1S2 circuit	0.371-0.469 volts for 7 seconds	Post-catalyst O_2 sensor bank 1 Page 13-103 Oxygen sensor monitor *See also* **OBD On-board Diagnostics (OBD II)**

P0131
P0132
P0133
P0135
P0136

P-code	MINI FC	Description	Threshold value	Related component(s)
P0137	311	O_2 sensor B1S2 circuit low	< 0.03 for 15 seconds	Post-catalyst O_2 sensor bank 1 Page 13-103 Oxygen sensor monitor *See also* **OBD On-board Diagnostics (OBD II)**
P0138	312	O_2 sensor B1S2 circuit high	> 1.39 volts for 10 seconds	Post-catalyst O_2 sensor bank 1 Page 13-103 Oxygen sensor monitor *See also* **OBD On-board Diagnostics (OBD II)**
P0141	321	O_2 sensor heater circuit, (B1S2) calculated resistance	< 1.5 ohms > 25 ohms for 8 seconds	Post-catalyst O_2 sensor bank 1, (heater component of HO2S) Page 13-106
P0171	369	System too lean	> 27% for > 40 seconds inside a 320 second block	Fuel delivery system *See* **16 Fuel Supply** Fuel system monitoring *See* **OBD On-board Diagnostics (OBD II)**
P0172	370	System too rich	> -27% for > 40 seconds inside a 320 second block	Fuel delivery system *See* **16 Fuel Supply** Fuel system monitoring *See* **OBD On-board Diagnostics (OBD II)**

P0137
P0138
P0141
P0171
P0172

P-code	MINI FC	Description	Threshold value	Related component(s)
P0201	513	Injector 1 circuit open	< 0.03 for 15 seconds	Fuel injectors Page 13-81
P0202	514	Injector 2 circuit open	< 0.03 for 15 seconds	Fuel injectors Page 13-81
P0203	515	Injector 3 circuit open	< 0.03 for 15 seconds	Fuel injectors Page 13-81
P0204	516	Injector 4 circuit open	< 0.03 for 15 seconds	Fuel injectors Page 13-81
P0222	546	TPS B circuit low	< 0.09 volts	Electronic throttle position actuator (EDR) feedback Page 13-64

P0201
P0202
P0203
P0204
P0222

P0223 P0261 P0262 P0264 P0265

P-code	MINI FC	Description	Threshold value	Related component(s)
P0223	547	TPS B circuit high	> 4.9 volts	Electronic throttle position actuator (EDR) feedback Page 13-64
P0261	609	Injector 1 circuit low	< 0.03 for 15 seconds	Fuel injectors Page 13-81
P0262	610	Injector 1 circuit high	> 12 volts for 10 seconds	Fuel injectors Page 13-81
P0264	612	Injector 2 circuit low	< 0.03 for 15 seconds	Fuel injectors Page 13-81
P0265	613	Injector 2 circuit high	> 12 volts for 10 seconds	Fuel injectors Page 13-81

P-code	MINI FC	Description	Threshold value	Related component(s)
P0267	615	Injector 3 circuit low	< 0.03 for 15 seconds	Fuel injectors Page 13-81
P0268	616	Injector 3 circuit high	> 12 volts for 10 seconds	Fuel injectors Page 13-81
P0270	624	Injector 4 circuit low	< 0.03 for 15 seconds	Fuel injectors Page 13-81
P0271	625	Injector 4 circuit high	> 12 volts for 10 seconds	Fuel injectors Page 13-81
P0300	768	Random misfire detected	Sum of misfire 3% of crankshaft revolutions (first 2000 engine revolutions) 7-20% of crankshaft revolutions (catalyst damaging)	Fuel injectors Page 13-81 Fuel delivery system *See* **16 Fuel Supply** Ignition coil Page 13-87 Engine compression *See* **01 Diagnostic Tools and Techniques**

P-code	MINI FC	Description	Threshold value	Related component(s)
P0301	769	Cylinder 1 misfire detected	Sum of misfire 3% of crankshaft revolutions (first 2000 engine revolutions) 7-20% of crankshaft revolutions (catalyst damaging)	Fuel injectors Page 13-81 Fuel delivery system *See* **16 Fuel Supply** Ignition coil Page 13-87 Engine compression *See* **01 Diagnostic Tools and Techniques**
P0302	770	Cylinder 2 misfire detected	Sum of misfire 3% of crankshaft revolutions (first 2000 engine revolutions) 7-20% of crankshaft revolutions (catalyst damaging)	Fuel injectors Page 13-81 Fuel delivery system *See* **16 Fuel Supply** Ignition coil Page 13-87 Engine compression *See* **01 Diagnostic Tools and Techniques**
P0303	771	Cylinder 3 misfire detected	Sum of misfire 3% of crankshaft revolutions (first 2000 engine revolutions) 7-20% of crankshaft revolutions (catalyst damaging)	Fuel injectors Page 13-81 Fuel delivery system *See* **16 Fuel Supply** Ignition coil Page 13-87 Engine compression *See* **01 Diagnostic Tools and Techniques1**
P0304	772	Cylinder 4 misfire detected	Sum of misfire 3% of crankshaft revolutions (first 2000 engine revolutions) 7-20% of crankshaft revolutions (catalyst damaging)	Fuel injectors Page 13-81 Fuel delivery system *See* **16 Fuel Supply** Ignition coil Page 13-87 Engine compression *See* **01 Diagnostic Tools and Techniques**
P0313	787	Misfire detected with low fuel level	Fuel level < 10 liters	Low fuel level

P0301
P0302
P0303
P0304
P0313

P-code	MINI FC	Description	Threshold value	Related component(s)
P0324	804	Knock control system error	n/a	Knock sensor Page 13-93 ECM Page 13-66
P0326	806	Knock sensor circuit range / performance	< 0.04 volts for 5 seconds	Knock sensor Page 13-93
P0335	821	Crankshaft position sensor no signal	No crankshaft signal seen after 10 camshaft edges	CKP sensor Page 13-54
P0336	822	Crankshaft position sensor circuit range / performance	Reluctor tooth error > 2	CKP sensor Page 13-54
P0340	832	Camshaft position sensor circuit	No signal	CKP sensor Page 13-54

P0324
P0326
P0335
P0336
P0340

P-code	MINI FC	Description	Threshold value	Related component(s)
P0341	833	Camshaft position sensor alignment to crankshaft position sensor	Outside allowable window	CMP sensor Page 13-50 CKP sensor testing Page 13-54
P0351	849	Ignition coil A primary circuit	Circuit fault, performed by software	Ignition coil Page 13-87
P0352	850	Ignition coil B primary circuit	Circuit fault, performed by software	Ignition coil Page 13-87
P0420	1056	Catalyst efficiency low	A switch ratio > 23 after 8 half periods of controlled stimuli diagnosis	Catalyst monitor *See* **OBD On-Board Diagnostics (OBD II)**
P0441	1089	Evaporative purge, incorrect flow	Blocked flow for > 1.1 seconds	Purge solenoid Page 13-77 Purge solenoid leak test Page 13-75

P0341
P0351
P0352
P0420
P0441

P-code	MINI FC	Description	Threshold value	Related component(s)
P0442	1090	Evaporative small leak detected	Leakage over 1mm > 1.1 seconds	Evap system testing Page 13-75
P0443	1091	Evaporative purge control solenoid circuit	Circuit fault	Purge solenoid Page 13-77 Purge solenoid leak test Page 13-75
P0444	1092	Evaporative purge control solenoid circuit open	< 0.03 for 15 seconds	Purge solenoid Page 13-77 Purge solenoid leak test Page 13-75
P0445	1093	Evaporative purge control solenoid circuit shorted	> 12 volts for 10 seconds	Purge solenoid Page 13-77 Purge solenoid leak test Page 13-75
P0455	1109	Evaporative large leak detected	< 52 seconds for 92 fast LDP pulses	Evap system testing Page 13-75 LDP reed switch Page 13-109

P-code	MINI FC	Description	Threshold value	Related component(s)
P0456	1110	Evaporative very small leak detected	Leakage over 0.5mm < 4.7 seconds	Evap system testing Page 13-75
P0500	1280	Vehicle speed sensor A	No signal from left and right front wheel speed sensors.	Data stream ABS / DSC module
P0506	1286	RPM lower than expected	-100 RPm for idle setpoint	Idle speed control Page13-61
P0507	1287	RPM higher than expected	+100 RPm for idle setpoint	Idle speed control Page13-61
P0601	1537	Internal ECM fault	Checksum error	ECM Page 13-66 ECM power and grounds Page 13-68

P0456
P0500
P0506
P0507
P0601

P-code	MINI FC	Description	Threshold value	Related component(s)
P0603	1539	Internal ECM fault	KAM error	ECM Page 13-66 ECM power and grounds Page 13-68
P0604	1540	Internal ECM fault	RAM error	ECM Page 13-66 ECM power and grounds Page 13-68
P0705	1797	Transmission range sensor A circuit	Signal from gearbox interface unit	ECM pinout Page 13-125
P0712		Oil temp sensor circuit to battery or open	> 4.96 volts	Transmission oil temperature sensor Page 13-122
P0713		Oil temp sensor circuit short to ground	< 0.025 volts	Transmission oil temperature sensor Page 13-122

P-code	MINI FC	Description	Threshold value	Related component(s)
P0815	2069	Upshift switch circuit	Switch closed too long, > 100 seconds	ECM pinout Page 13-125
P0816	2070	Downshift switch circuit	Switch closed too long, > 100 seconds	ECM pinout Page 13-125
P1106	4358	MAP signal too low, engine stopped	< 600 hPa	MAP sensor (Cooper) Page 13-117 TMAP / MAP sensor (Cooper S) Page 13-120
P1107	4359	MAP signal too low, engine running at idle	< 120 hPa	MAP sensor (Cooper) Page 13-117 TMAP / MAP sensor (Cooper S) Page 13-120
P1108	4360	MAP signal too low at full load for lower engine speed	< 600 hPa	MAP sensor (Cooper) Page 13-117 TMAP / MAP sensor (Cooper S) Page 13-120

P-code	MINI FC	Description	Threshold value	Related component(s)
P1109	4361	MAP signal too high during decel	> 600 hPa	MAP sensor (Cooper) Page 13-117 TMAP / MAP sensor (Cooper S) Page 13-120
P1122	4386	Accelerator pedal position (PWG) sensor 1 low input	< 0.4 volts for 150 seconds	Accelerator pedal position (PWG) sensor Page 13-44
P1122	4386	Accelerator pedal position (PWG) sensor 2 low input	< 0.3 volts for 150 seconds	Accelerator pedal position (PWG) sensor Page 13-44
P1123	4387	Accelerator pedal position (PWG) sensor 1 high input	> 4.9 volts for 150 seconds	Accelerator pedal position (PWG) sensor Page 13-44
P1123	4387	Accelerator pedal position (PWG) sensor 2 high input	> 2.6 volts for 150 seconds	Accelerator pedal position (PWG) sensor Pagc 13-44

P-code	MINI FC	Description	Threshold value	Related component(s)
P1124		Accelerator pedal position (PWG) sensor plausibility error	> 12 volts at 25% Accelerator pedal position (PWG) sensor 1 > 11 volts at 78% Accelerator pedal position (PWG) sensor 2	Accelerator pedal position (PWG) sensor Page 13-44
P1125	4389	Plausibility error	Delta TPS 1 & 2 > 5°	Electronic throttle position (EDR) feedback Page 13-64
P1126	4390	Plausibility error	Delta TPS 1 & 2 > 18°	Electronic throttle position (EDR) feedback Page 13-64
P1143	4419	O_2 sensor signal too high (B1S1)	Check signal too high	Post-catalyst O_2 sensor bank 1 Page 13-103
P1144	4420	O_2 sensor signal too low (B1S1)	Check signal too low	Post-catalyst O_2 sensor bank 1 Page 13-103

P1124
P1125
P1126
P1143
P1144

P-code	MINI FC	Description	Threshold value	Related component(s)
P1222	4642	Accelerator pedal position (PWG) sensor 2 circuit low	< 0.3 volts for 150 seconds	Accelerator pedal position (PWG) sensor Page 13-44
P1223	4643	Accelerator pedal position (PWG) sensor 2 circuit high	> 2.6 volts for 150 seconds	Accelerator pedal position (PWG) sensor Page 13-44
P1224	4644	Accelerator pedal position (PWG) sensor plausibility error	Delta TPS 1 & 2 > 5°	Accelerator pedal position (PWG) sensor Page 13-44
P1226	4646	Electronic throttle valve malfunction	Mechanical fault, (jamming)	Electronic throttle actuator (EDR) Page 13-58
P1229	4649	Throttle sensor adaptation failure	Plausibility error	Adaptations *See* **01 Diagnostic Tools and Techniques**

P1237 P1238 P1239 P1240 P1241

P-code	MINI FC	Description	Threshold value	Related component(s)
P1237	4663	Upstream MAP sensor circuit low	< 0.03 for 150 seconds	MAP sensor (Cooper) Page 13-117 TMAP / MAP sensor (Cooper S) Page 13-120
P1238	4664	Upstream MAP sensor circuit high	> 4.9 volts for 150 seconds	MAP sensor (Cooper) Page 13-117 TMAP / MAP sensor (Cooper S) Page 13-120
P1239	4665	Upstream MAP sensor circuit low, engine stopped	< 600 hPa	MAP sensor (Cooper) Page 13-117 TMAP / MAP sensor (Cooper S) Page 13-120
P1240	4672	Upstream MAP sensor circuit low, engine running at idle	< 120 hPa	MAP sensor (Cooper) Page 13-117 TMAP / MAP sensor (Cooper S) Page 13-120
P1241	4273	Upstream MAP sensor circuit low, engine at full load or low engine speed	< 600 hPa	MAP sensor (Cooper) Page 13-117 TMAP / MAP sensor (Cooper S) Page 13-120

P-code	MINI FC	Description	Threshold value	Related component(s)
P1242	4674	Upstream MAP sensor circuit too high, engine deceleration	> 700 hPa	MAP sensor (Cooper) Page 13-117 TMAP / MAP sensor (Cooper S) Page 13-120
P1320	4896	Flywheel adaptation for misfire detection range	Crankshaft segment adaptation limit > 0.78%	CKP sensor Page 13-54
P1321	4897	Flywheel adaptation for misfire detection performance	Tooth count incorrect ±1 tooth	CKP sensor Page 13-54
P1366	4966	Ignition coil A primary circuit	Circuit fault, performed by software	Ignition coil Page 13-87
P1367	4967	Ignition coil B primary circuit	Circuit fault, performed by software	Ignition coil Page 13-87

P-code	MINI FC	Description	Threshold value	Related component(s)
P1436	5174	LDP circuit open	Circuit fault, performed by software	LDP Page 13-109
P1437	5175	LDP range performance	Circuit fault, performed by software	LDP Page 13-109
P1442	5186	LDP circuit low	< 0.03 for 15 seconds	LDP Page 13-109
P1443	5187	LDP circuit high	> 12 volts for 10 seconds	LDP Page 13-109
P1475	5237	LDP reed switch did not close	Reed switch level edge from low to high after deactivation of solenoid	LDP Page 13-109

P-code	MINI FC	Description	Threshold value	Related component(s)
P1476	5238	LDP line clamped	Time period of any 5 first pump cycles > 5 seconds	LDP Page 13-109 Restriction in EVAP line (s)
P1477	5239	LDP reed switch did not open	Reed switch level continuously low	LDP Page 13-109
P1570	5488	ECM 5 volts output low	< 4.2 volts	ECM Page 13-66 ECM power and grounds Page 13-68 ECM pinout Page 13-125
P1571	5489	ECM 5 volt output high	> 5.8 volts	ECM Page 13-66 ECM power and grounds Page 13-68 ECM pinout Page 13-125
P1572	5490	ECM 5 volt output noisy	> 0.7 volts	ECM Page 13-66 ECM power and grounds Page 13-68 ECM pinout Page 13-125

	P-code	MINI FC	Description	Threshold value	Related component(s)
P1273	P1573	5491	ECM 5 volt output low	< 4.2 volts	ECM Page 13-66 ECM power and grounds Page 13-68 ECM pinout Page 13-125
P1574	P1574	5492	ECM 5 volt output high	> 5.8 volts	ECM Page 13-66 ECM power and grounds Page 13-68 ECM pinout Page 13-125
P1575	P1575	5493	ECM 5 volt output noisy	> 0.7 volts	ECM Page 13-66 ECM power and grounds Page 13-68 ECM pinout Page 13-125
P1600	P1600	5632	Internal ECM fault	RAM error	ECM Page 13-66 ECM power and grounds Page 13-68 ECM pinout Page 13-125
P1607	P1607	5639	CAN bus fault	CAN version	Bus systems *See* **61 Bus Systems**

P-code	MINI FC	Description	Threshold value	Related component(s)
P1611	5649	Communication fault with trans control module	No message from TCM	Bus systems *See* **61Bus Systems**
P1612	5650	Communication fault with instrument cluster	No message from instrument cluster	Bus systems *See* **61Bus Systems**
P1613	5651	Communication fault with ABS / DSC module	No message from ABS / DSC module	Bus systems *See* **61Bus Systems**
P1615	5653	ECM processor bus fault	SPI-bus system fault	Bus systems *See* **61Bus Systems**
P1617	5655	ECM H-bridge driver	Circuit fault, performed by software. EDR high current draw - jammed throttle body.	Performed by software See **13 Engine Management**

P-code	MINI FC	Description	Threshold value	Related component(s)
P1679	5753	Monitoring of torque loss	Limit exceeded in threshold map during 360 milliseconds	Electronic throttle actuator (EDR) Page 13-56 Supercharger operation Page 13-113
P1680	5760	Monitoring of A / D convertor. Error between main and monitored values	PVS ratio hysteresis exceeds 0.273 volts during 360 milliseconds	Electronic throttle actuator (EDR) Page 13-56
P1681	5761	Monitoring of engine speed. Calculation error	Engine speed difference exceeds 448 RPM during 480 milliseconds	Electronic throttle actuator (EDR) Page 13-56 Supercharger operation Page 13-113
P1682	5762	Monitoring of proportional derivative part of idle speed controller	Maximum limit exceeded during 480 milliseconds	Electronic throttle actuator (EDR) Page 13-56 Idle speed Page 13-61
P1683	5763	Monitoring of integral part of idle speed	Maximum limit exceeded during 480 milliseconds	Electronic throttle actuator (EDR) Page 13-56 Idle speed Page 13-61

P1679
P1680
P1681
P1682
P1683

P-code	MINI FC	Description	Threshold value	Related component(s)
P1684	5764	Monitoring of minimum torque at clutch	Minimum torque clutch limit exceeded during 240 milliseconds	Electronic throttle actuator (EDR) Page 13-56 Supercharger operation Page 13-113
P1685	5764	Monitoring of maximum torque at clutch	Maximum torque clutch limit exceeded during 240 milliseconds	Electronic throttle actuator (EDR) Page 13-56 Supercharger operation Page 13-113
P1686	5766	Monitoring of Accelerator pedal position (PWG) sensor	Fault in Accelerator pedal position (PWG) sensor value checks	Accelerator pedal position (PWG) sensor Page 13-41 ECM Page 13-66
P1687	5767	Monitoring of throttle position	Fault in TPS calculation	Electronic throttle actuator (EDR) Page 13-62 ECM Page 13-66
P1688	5768	Monitoring of mass air flow	Low MAF limit exceeded n threshold map	Electronic throttle actuator (EDR) Page 13-56 Supercharger operation Page 13-113

P1689 P1691 P1692 P1693 P1698

P-code	MINI FC	Description	Threshold value	Related component(s)
P1689	5769	Monitoring of actual indicated engine torque	Fault in torque calculation	Electronic throttle actuator (EDR) Page 13-56 Supercharger operation Page 13-113
P1691	5777	Monitoring of engine speed limit	Engine speed limitation error	Electronic throttle actuator (EDR) Page 13-56 Supercharger operation Page 13-113
P1692	5778	Monitoring of engine speed limit	Engine in limp home with fuel injector cut-off A	Electronic throttle actuator (EDR) Page 13-56 Supercharger operation Page 13-113
P1693	5779	Monitoring of engine speed limit	Engine in limp home with fuel injector cut-off B	Electronic throttle actuator (EDR) Page 13-56 Supercharger operation Page 13-113
P1698	5784	Transmission control module (TCM) error	TCM internal failure	TCM *See* **24 Automatic Transmission**

P-code	MINI FC	Description	Threshold value	Related component(s)
P1699	5785	Transmission control module (TCM) error	TCM checksum error	TCM *See* **24 Automatic Transmission**
P1705	5893	TCM LED open circuit	Plausibility	TCM *See* **24 Automatic Transmission**
P1706	5894	TCM circuit short		TCM *See* **24 Automatic Transmission**
P1739	5945	Communication error	TCM / TCU solenoid	TCM *See* **24 Automatic Transmission**
P1741	5953	Clutch solenoid circuit open		TCM *See* **24 Automatic Transmission**

P1699
P1705
P1706
P1739
P1741

P-code	MINI FC	Description	Threshold value	Related component(s)
P1742	5954	Clutch solenoid circuit short		TCM *See* **24 Automatic Transmission**
P1749	5961	Communication error	Secondary pressure solenoid communication fault	TCM *See* **24 Automatic Transmission**
P1751	5969	Secondary pressure solenoid circuit open		TCM *See* **24 Automatic Transmission**
P1752	5970	Secondary pressure solenoid circuit short		TCM *See* **24 Automatic Transmission**
P1785	6021	Trans ratio control actuator circuit fault		TCM *See* **24 Automatic Transmission**

P-code	MINI FC	Description	Threshold value	Related component(s)
P1786	6022	Trans ratio control actuator circuit	range performance	TCM *See* **24 Automatic Transmission**
P1787	6023	Trans ratio control actuator circuit open		TCM *See* **24 Automatic Transmission**
P1788	6024	Trans ratio control actuator circuit short		TCM *See* **24 Automatic Transmission**
P1789	6025	Trans ratio control actuator	Communication error	TCM *See* **24 Automatic Transmission**
P1815	6165	Steering wheel plus switch circuit low		TCM *See* **24 Automatic Transmission**

P1786
P1787
P1788
P1789
P1815

P1816
P2096
P2097

P-code	MINI FC	Description	Threshold value	Related component(s)
P1816	6166	Steering wheel plus switch circuit high		TCM *See* **24 Automatic Transmission**
P2096	8342	Post catalyst fuel trim, system too lean	< -0.8	Fuel injection monitoring. *See* **OBD On-Board Diagnostics (OBD II)** Fuel delivery. *See* **16 Fuel Supply** Post cat O_2 sensor 13-103 Note: Reset fuel trim adaptations after repair Adaptations *See* **01 Diagnostic Tools and Techniques**
P2097	8343	Post catalyst fuel trim, system too rich	> 1.1 Depending on engine speed and mass air flow	Fuel injection monitoring. *See* **OBD On-Board Diagnostics (OBD II)** Fuel delivery. *See* **16 Fuel Supply** Post cat O_2 sensor 13-103 Note: Reset fuel trim adaptations after repair Adaptations *See* **01 Diagnostic Tools and Techniques**

GENERAL

This chapter covers diagnosis and testing for the engine management system used on 2002 through 2006 MINI Cooper and MINI Cooper S. Be sure to read **01 Diagnostic Tools and Techniques** for general diagnostic information.

WARNINGS

Please read the following warnings and adhere to them when servicing the engine management system.

WARNINGS—

- Read **WARNINGS— important safety notice** on the copyright page. Read each procedure and the **WARNINGS** and **CAUTIONS** that accompany the procedure. Review any additions and corrections at **www.BentleyPublishers.com/errata/** before beginning work.
- Before starting a job, be sure that you have the necessary tools and parts on hand. Read all the instructions thoroughly, and do not attempt shortcuts. Use tools appropriate to the work and use only replacement parts meeting MINI specifications.
- Observe good workshop practices. Wear goggles when operating machine tools or working with battery acid. Wear gloves or other protective clothing whenever working with harmful substances. Wear a dust mask or respirator to lower your exposure to harmful dusts, fumes, vapors, and gases. Protect ears whenever exposed to continuous loud noise.
- Do not run the engine unless the work area is well ventilated. Carbon monoxide kills.
- Disconnect the battery negative (-) terminal whenever you work on the fuel system or the electrical system.
- Vehicles covered by this manual are equipped with a supplemental restraint system (SRS).These explosive devices, handled improperly or without adequate safeguards, can be accidentally activated and can cause serious injury.
- The ignition system produces high voltages that can be fatal. Avoid contact with exposed terminals. Use extreme care when working on a vehicle with the engine running or the ignition ON.

A/C COMPRESSOR CONTROL

When the ignition is turned ON, the body controller 1 (BC1) outputs a compressor ON / OFF request to the ECM every 10 (± 0.1) seconds. When the ignition switch is first turned to position 2, the request is set to compressor OFF. When the engine is running and the A/C switch is ON, BC1 illuminates the indicator LED in the switch and changes request to compressor ON.

When it detects the request has changed to compressor ON, the ECM energizes the compressor clutch relay, located in the engine bay fuse box, to supply battery power to the compressor clutch.

Once the compressor ON request is granted, the A/C switch LED remains illuminated until the request is cancelled by the operator or the engine stops. The ECM can de-energize the compressor clutch relay, to disengage the compressor clutch, when conditions warrant.

After receiving a compressor ON request, the ECM outputs a compressor clutch status message to the BC1 to advise if the request was granted or not.

If the compressor ON request was refused, the BC1 flashes the LED at ½ second intervals and repeats the compressor ON request until it is granted or cancelled by:

- Turning A/C switch off (compressor OFF request)
- Turning blower OFF (compressor OFF request)
- Turning ignition switch to position 0

Hard acceleration can cause the compressor clutch to be disengaged. After three occurrences in a single ignition cycle the ECM disregards further hard acceleration occurrences and leaves the compressor clutch engaged. The air-conditioning is automatically suspended if:

- Engine speed is above 6016 rpm.
- Evaporator temperature falls below 2°C (to prevent freezing).
- Coolant temperature goes above 118°C (to protect the engine).
- A/C system pressure goes above 30 bar (to protect the system).
- A/C system pressure goes below 1.6 bar (to protect the system).
- Accelerator pedal is fully depressed (continuous full pedal demand) for more than 5 seconds.
- Accelerator pedal is depressed rapidly (instantaneous full pedal demand) for more than 2 seconds.
- Engine speed is below 500 rpm (engine stall).

System reverts to normal operation once the reason for suspension is removed.

NOTE —
- During cranking the A/C compressor is disengaged.
- On IHKA systems, compressor requests are handled via CAN-bus from the IHKA.

ABS/ASC/DSC

The ECM receives road speed signals from the ABS / ASC / DSC module for cruise functions (if equipped) and maximum vehicle speed limit. Road speed signals are sent to the ECM over CAN-bus. Requests for torque modification are also received from ABS module over CAN-bus.

If either of the driven wheels lose traction, the ABS control module requests a change in engine torque output to regain traction.

This occurs in either of the following situations:

- Wheels have lost traction due to excess torque being generated by the engine; the ABS module requests a torque decrease under these circumstances.
- Wheels have broken traction due to excess engine braking when on low grip surface; the ABS module requests a torque increase under these circumstances. This is called the engine drag torque control (MSR) system.

Once traction has been regained, the brake or torque intervention is removed allowing the engine to return to its appropriate torque output as determined by other inputs such as driver demand or cruise control.

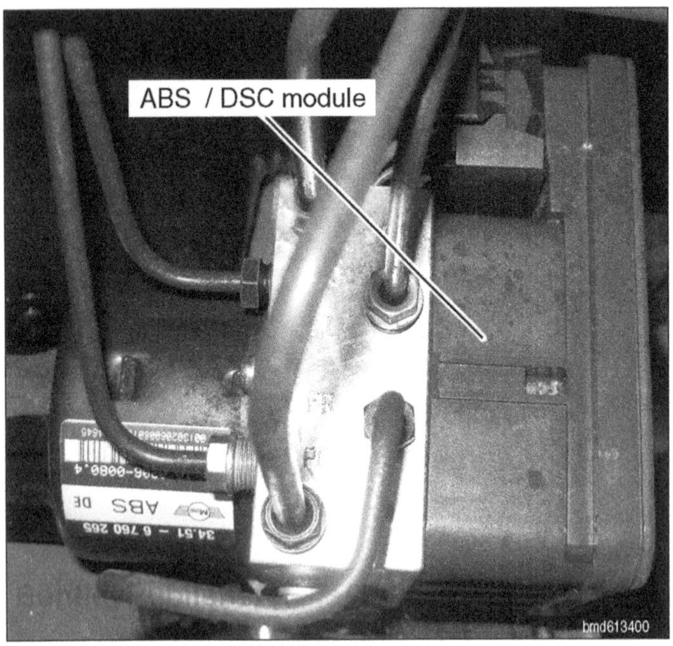

ACCELERATOR PEDAL POSITION (PWG) SENSOR

The accelerator pedal position (PWG) sensor is mounted in the passenger compartment, clipped to a bracket bolted to the floor. The PWG sensor is monitored by the ECM for input or speed requests.

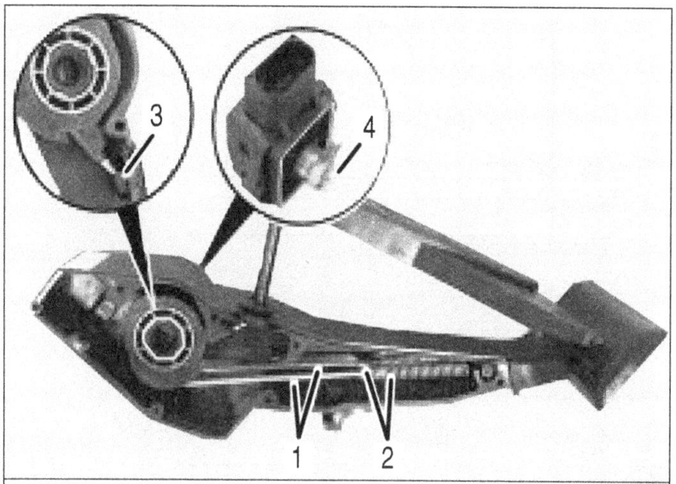

1. Twin Cables
2. Return Springs
3. Ball Joint
4. Throttle lever for angle Hall sensors

bmd613401

The PWG sensor uses two hall sensors with different voltage characteristics and independent power supplies to provide the driver's input request to the ECM. The Hall sensors receive power (5v) and ground from the ECM and produce linear voltage signals as the pedal is pressed from the LL position (idle) to the VL position (full throttle).

- Hall sensor 1 - 0.5 to 4.5 volts driver request
- Hall sensor 2 - 0.5 to 2.0 volts plausibility check

The PWG sensor is monitored by the ECM for pedal angle position and rate of movement. As the accelerator is moved, a rising voltage signal from the hall sensors requests pedal position and acceleration rate. The ECM increases the volume of fuel injected into the engine, advances ignition timing and opens the throttle valve. Full throttle position indicates maximum acceleration to the ECM and effects air conditioning compressor activation.

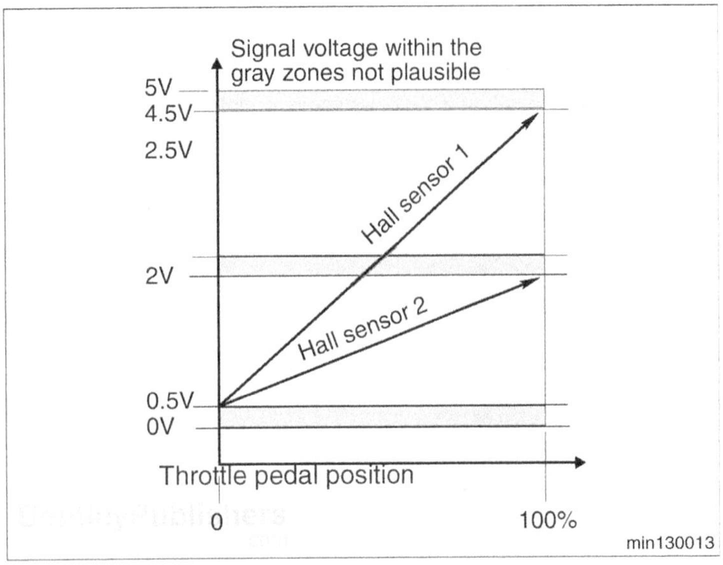

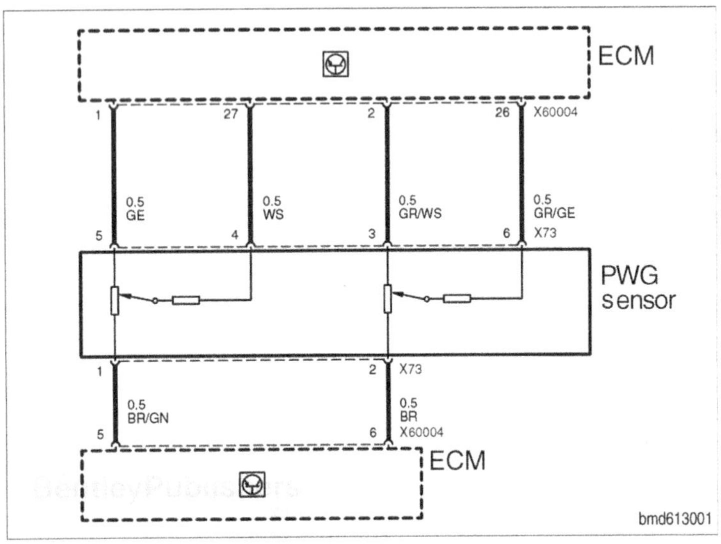

A decrease in voltage signals the ECM to activate fuel shut off if the rpm is above idle speed (coasting). The throttle valve is closed (idle speed is maintained).

The ECM monitors engine idle speed in addition to accelerator pedal position and throttle position voltage. If voltage values change (mechanical wear of throttle plate or linkage), the ECM adjusts the throttle plate to maintain correct idle speed.

The ECM learns throttle angle voltage at idle speed. The Hall sensors are non-adjustable. If the throttle housing / accelerator pedal module is replaced, perform adaptations using a factory or enhanced scan tool. If this is not performed, the vehicle may not start or runs in fail-safe mode. If this input is defective, a fault code is stored and the MIL and / or EML light are illuminated. Only limited engine operation is possible.

NOTE —

- In case of discrepancy between the 2 pedal position sensors, the channel giving the smallest value is then selected (torque request is under-estimated).

PWG sensor failure modes

If defect in one of the Hall effect pedal position sensors is detected, maximum engine speed will be limited.

If a defect in both of the Hall effect pedal position sensors is detected, engine speed is limited to approximately 1300 rpm and reduced to 1000 rpm when the brake pedal is depressed.

PWG sensor, testing

ECM connector X60004 (40-pin, black)

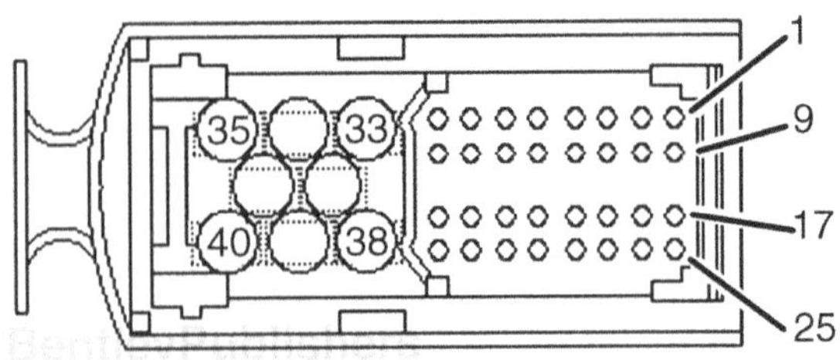

bmd613402

PWG sensor connector (X73)

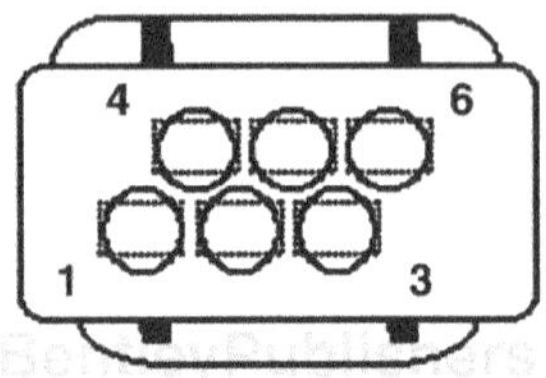

bmd613403

PWG sensor testing values				
Terminals at connector X60004	**Terminals at connector X73**	**Conditions**	**Test values**	**Circuit**
1 and ground	5 and ground	key on engine off engine running at idle @2500 rpm (no load	5 volts (approx.)	reference voltage
2 and ground	3 and ground	key on engine off engine running at idle @2500 rpm (no load	5 volts (approx.)	reference voltage
5 and ground	1 and ground	key on engine off engine running at idle @2500 rpm (no load	0.1 volts (maximum)	sensor ground
6 and ground	2 and ground	key on engine off engine running at idle @2500 rpm (no load	0.1 volts (maximum)	sensor ground
26 and ground	6 and ground	key on engine off	0.4 to 2.0 volts while pushing throttle from idle to full throttle	PWG sensor 2 signal
26 and ground	6 and ground	engine running at idle	0.4 volts	PWG sensor 2 signal
26 and ground	6 and ground	@ 2500 rpm, (no load)	0.5 volts	PWG sensor 2 signal
27 and ground	4 and ground	key on engine off	0.7 to 4 volts while pushing throttle from idle to full throttle	PWG sensor 1 signal
27 and ground	4 and ground	engine running at idle	0.7 volts	PWG sensor 1 signal
27 and ground	4 and ground	@ 2500 rpm, (no load)	1.0 volts	PWG sensor 1 signal

PWG sensor - dual trace waveform

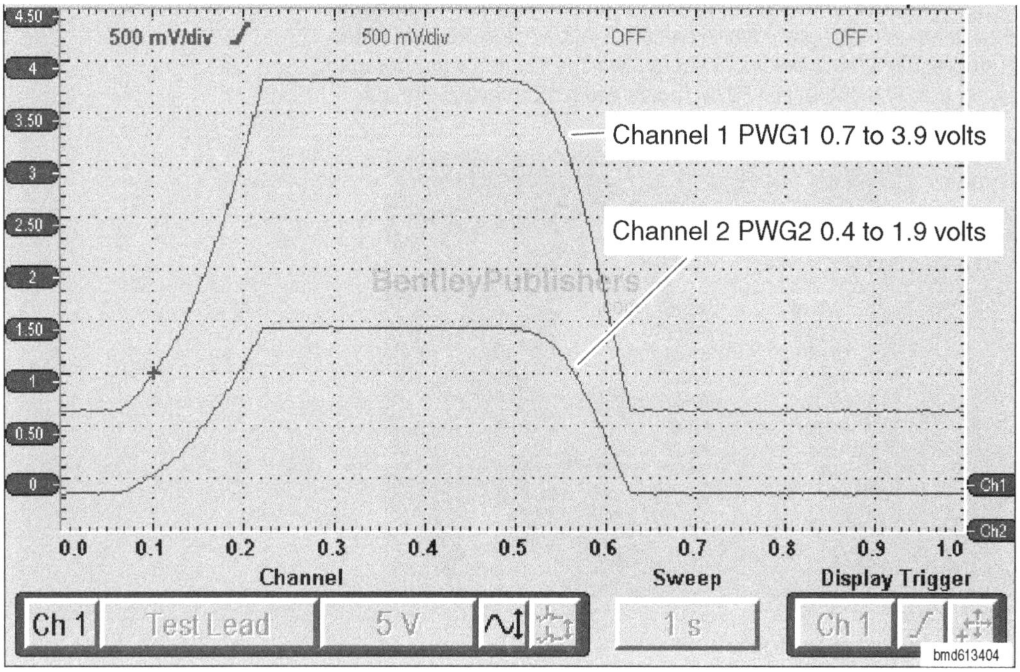

ALTERNATOR

A dedicated output from the alternator is provided to the ECM to determine electrical load on the engine. When electrical loads are switched ON the generator requires more electrical energy, which i creates a greater load on the engine. The output is a PWM (pulse-width modulated) signal.

Alternator failure symptoms

If the PWM signal fails, the engine may exhibit idle speed instability as the electrical loads are increased.

Alternator, testing

ECM connector X6000 (81 pin, black)

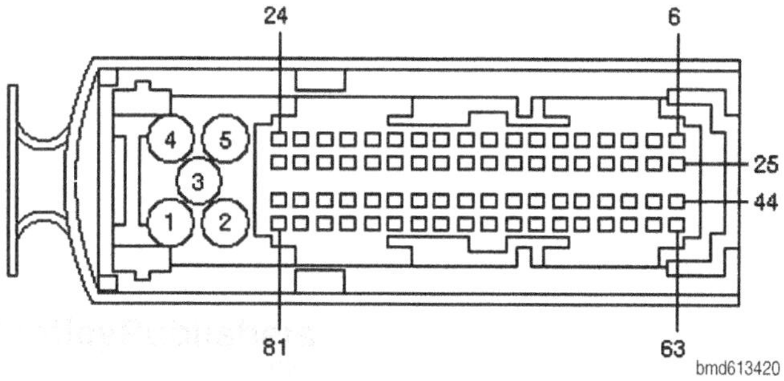

Alternator testing values (at ECM connector X6000)			
Terminals	**Conditions**	**Test values**	**Circuit**
52 and ground	key on engine off	10 volts	alternator pwm signal
52 and ground	engine running at idle	6.0 volts	alternator pwm signal

Alternator PWM signal at idle

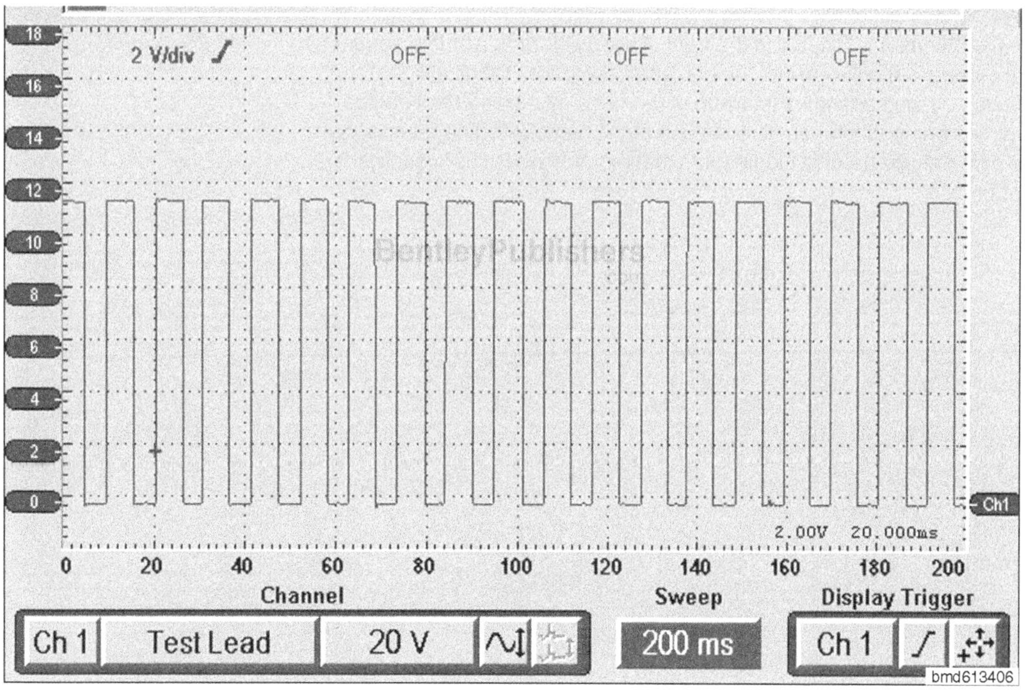

CAMSHAFT POSITION (CMP) SENSOR

The signal from the camshaft position (CMP) sensor enables the ECM to detect the position of the camshaft in relation to the position of the crankshaft. This allows the ECM to synchronize fuel injection and ignition spark.

The CMP sensor is located in the front of the cylinder head. The sensor reluctor ring is machined into the camshaft sprocket. A digital signal is provided by the sensor in the range of 0-5 volts. The Hall-effect device produces one pulse for each revolution of the camshaft. The reluctor is half-moon shaped with a single tooth that extends over 180° of camshaft rotation.

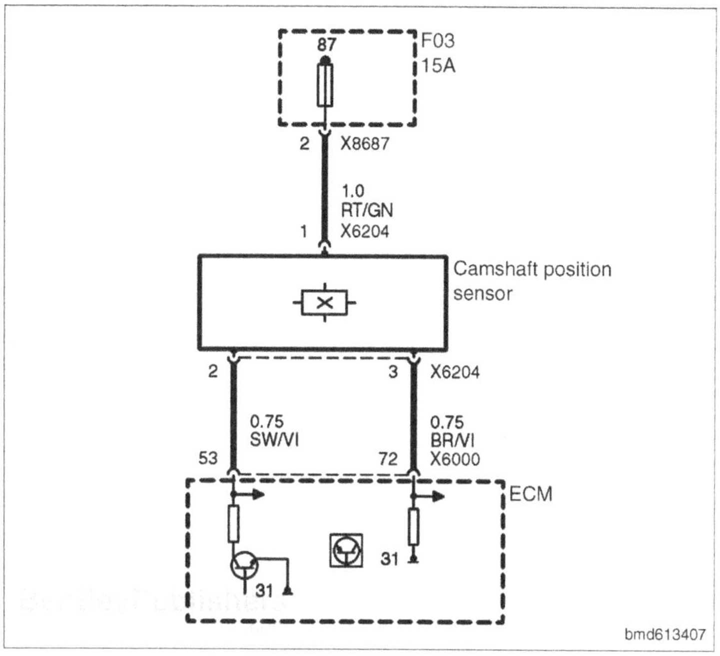

CMP sensor failure symptoms

The engine control module switches to emergency operation function. Camshaft sensor failure causes the injection system to default into a semi-sequential mode. The possible deterioration of emission values causes MIL to illuminate.

The engine continues to run if the cam shaft sensor fails while driving. The engine will restart and run and a fault will be stored.

The CMP signal displays 1 pulse (rising and falling) per engine revolution. A plausibility diagnosis is performed to compare CMP and CKP signals. The CMP pulse must fall in a gap of the camshaft reluctor in order to declare the CMP signal as valid. If an error is detected the ECM defaults to sequential fuel injection. It runs with a constant injection phase of 180° of crankshaft position and the engine runs in open loop.

CMP sensor, testing

ECM connector X6000 (81 pin, black)

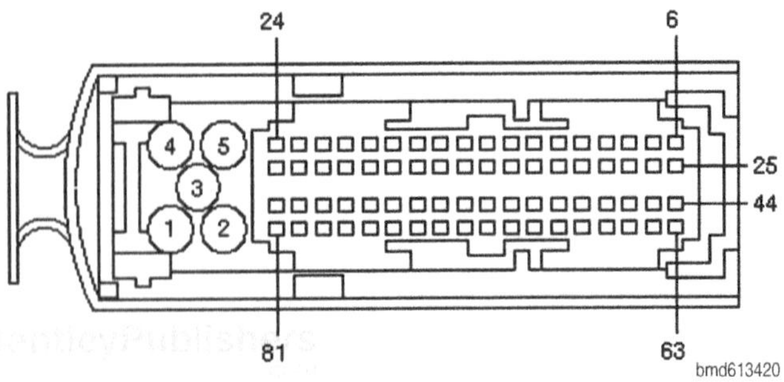

CMP connector X6204

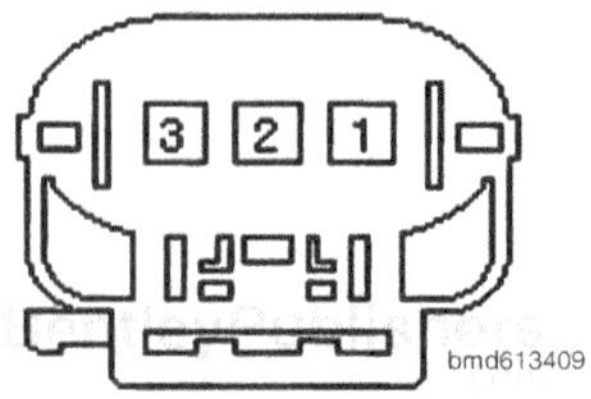

CMP testing values (at ECM connector X6000)			
Terminals	**Conditions**	**Test values**	**Circuit**
53 and ground	key on engine off	0.2 volts	CMP sensor signal
53 and ground	engine running at idle	2.5 volts	CMP sensor signal
72 and ground	key on engine off, engine running at idle, @2500 rpm, (no load)	0.1 volts	CMP sensor ground

CMP testing values (at CMP connector X6204)			
Terminals	**Conditions**	**Test values**	**Circuit**
1 and ground	key on engine off, engine running at idle @2500 rpm, (no load)	battery volts	voltage supply
2 and ground	key on engine off	0.2 volts	CMP sensor signal
2 and ground	engine running at idle	2.5 volts	CMP sensor signal
3 and ground	key on engine off, engine running at idle @2500 rpm, (no load)	0.1 volts (maximum)	CMP sensor ground

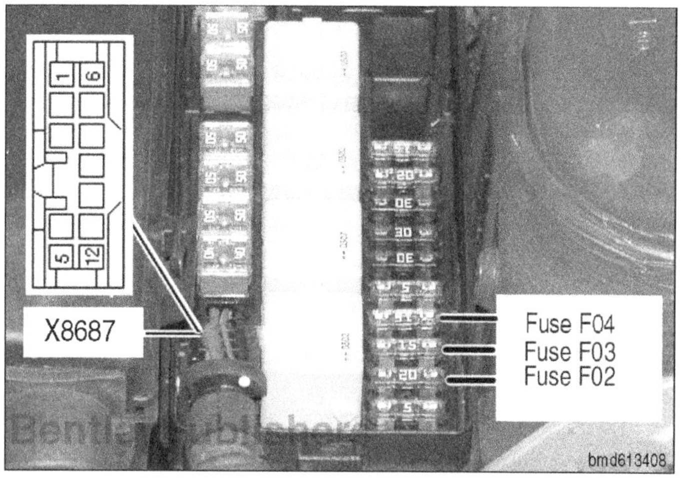

CMP testing values (at fuse holder 3 connector X8687)			
Terminals	**Conditions**	**Test values**	**Circuit**
Fuse F03 and ground	key on engine off, engine running at idle, @2500 rpm, (no load)	battery volts	voltage supply
2 and ground	key on engine off, engine running at idle, @2500 rpm, (no load)	battery volts	voltage supply

Camshaft position sensor waveform at idle

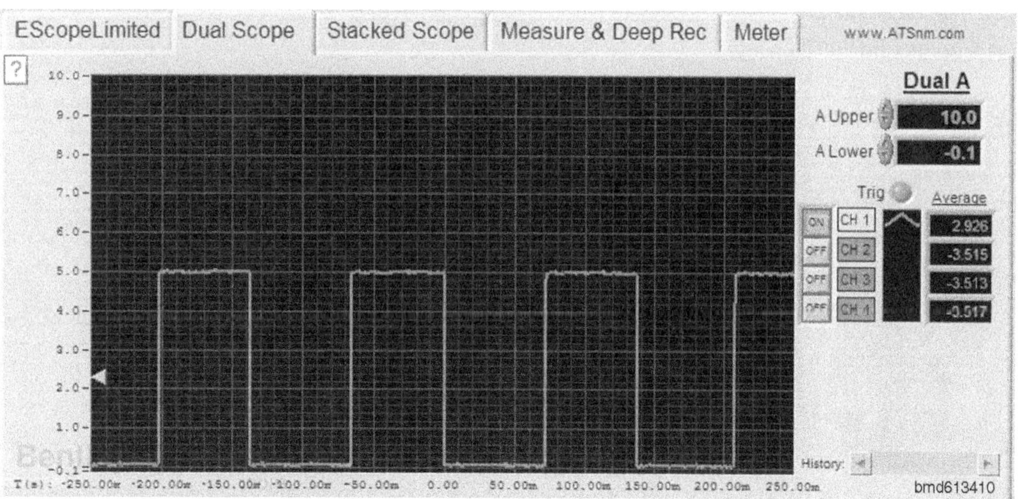

CRANKSHAFT POSITION (CKP) SENSOR

The crankshaft position (CKP) sensor is a primary input to the ECM. It is used to determine the speed and position of the engine crankshaft. The sensor is a Hall effect device. A digital electrical signal is created as the reluctor ring teeth pass the sensor. The reluctor ring tooth pattern consists of 58 targets and a space of two missing teeth grouped together (6-tooth intervals).

The CKP sensor is in the right front of the cylinder block.

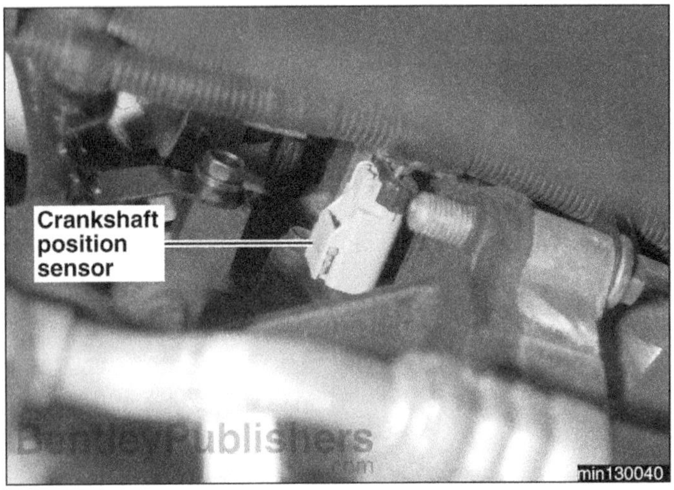

The missing teeth are essential for correct engine operation. They are used by the ECM as an angular reference point. This information is used together with information from the camshaft sensor to determine correct timing for spark and fuel delivery.

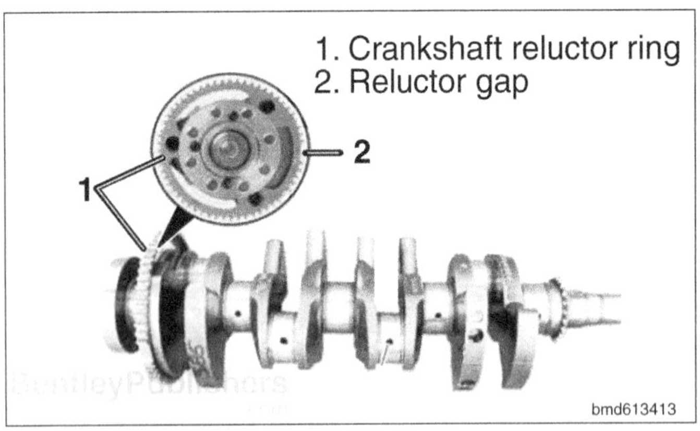

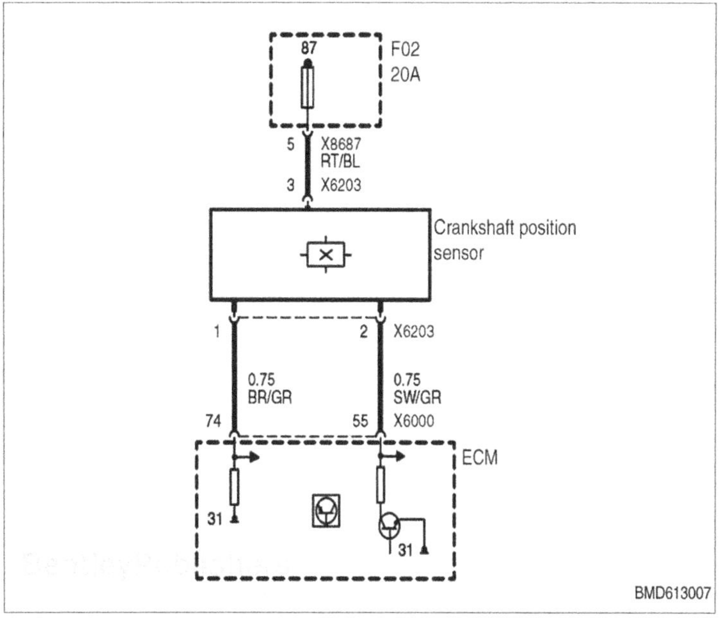

During cranking the digital signal produced by the CKP is sent to the ECM. The ECM expects to receive a signal of 58 targets and two missing targets. If this signal is received the ECM synchronizes itself to the engine. (The ECM knows crankshaft position.) If a faulty or implausible signal is received, the ECM does not allow injection or ignition and the car will not start. There is no crankshaft back-up facility on the ECM because the camshaft sensor target only produces one pulse per revolution.

CKP sensor failure symptoms

If the synchronization is not successfully concluded, injector pulse is interrupted.

An open or short in the circuit wiring causes a no-start condition.

CKP sensor, testing

ECM connector X6000 (81 pin, black)

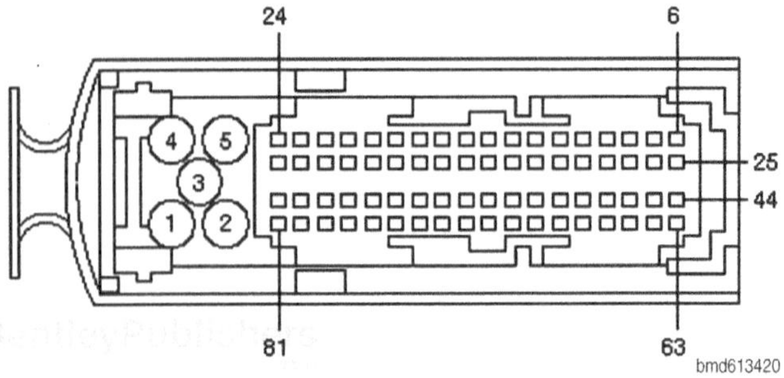

CKP connector X6203

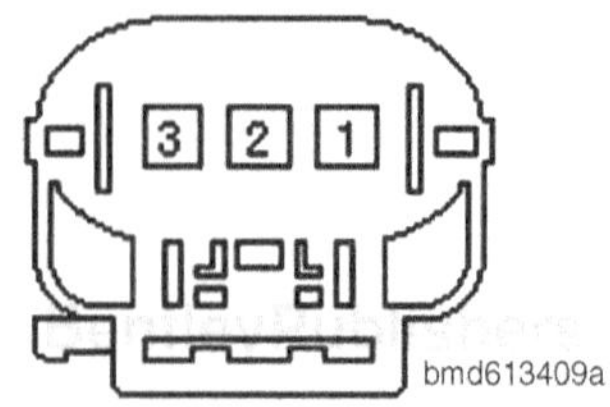

CKP sensor testing values				
Terminals at connector X6000	**Terminals at connector X6203**	**Conditions**	**Test values**	**Circuit**
55 and ground	2 and ground	key on engine off	0.2 volts	CKP signal
55 and ground	2 and ground	engine running at idle	2.5 volts	CKP signal
74 and ground	1 and ground	key on engine off, engine running at idle, @2500 rpm, (no load)	0.1 volts	sensor ground
N-A	3 and ground	key on engine off, engine running at idle @2500 rpm, (no load)	battery volts	voltage supply

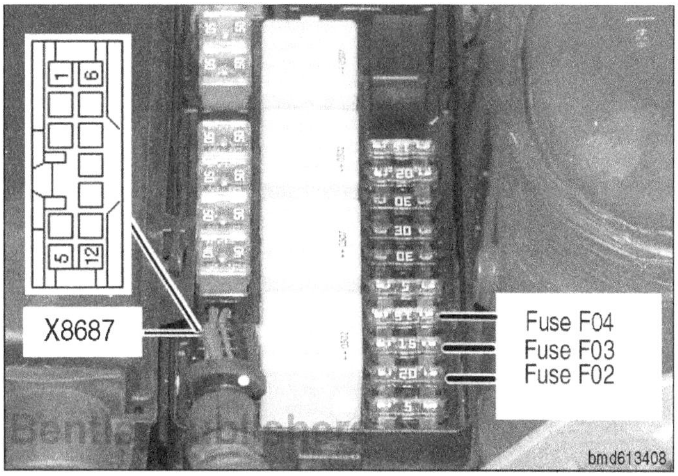

CKP sensor testing values (at fuse holder 3 connector X8687)			
Terminals	**Conditions**	**Test values**	**Circuit**
Fuse F02 and ground	key on engine off, engine running at idle, @2500 rpm, (no load)	battery volts	voltage supply
5 and ground	key on engine off, engine running at idle, @2500 rpm, (no load)	battery volts	voltage supply

Camshaft / crankshaft sync waveform at idle

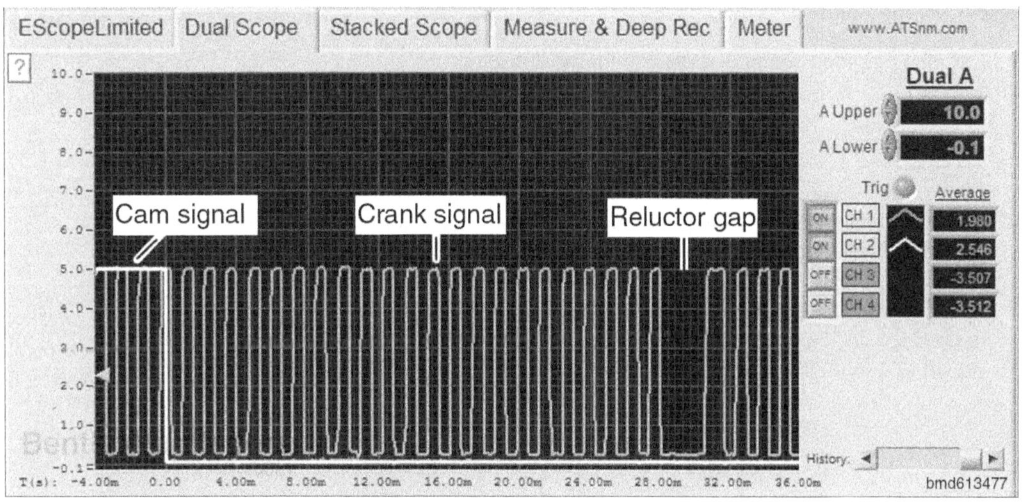

ELECTRONIC THROTTLE (EDR) ACTUATOR

The electronic throttle actuator (EDR) is used to adjust engine load based on throttle position requests received by the ECM from the accelerator pedal position (PWG) sensor.

A D/C-motor electrically positions the throttle plate from idle to full load. An H-bridge circuit in the ECM allows the D/C motor to be driven open and closed. The feedback of the position of the throttle plate and the position of the D/C motor is achieved via 2 potentiometers which are integrated in the throttle body.

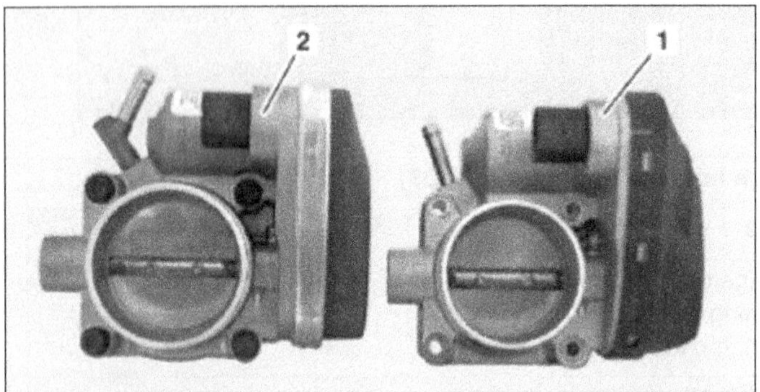

1. Cooper - 52 mm
2. Cooper S - 57 mm

bmd613415

A variable duty cycle fixed frequency signal is sent to the EDR by the ECM. Position and movement of the throttle plates is confirmed through dual feedback of potentiometer control functions.

The ECM also provides power and ground for the feedback potentiometers.

NOTE —
- MINI uses a brake switch with two Hall switches. If the two brake signals are seen to be inconsistent at any time, a fault condition is assumed. The accelerator pedal activation signal is set to zero (no throttle demand) and the engine remains at idle. If this occurs while in a cruise control active mode, cruise is disabled for the remainder of that journey. The system registers a fault disallowing further operation.

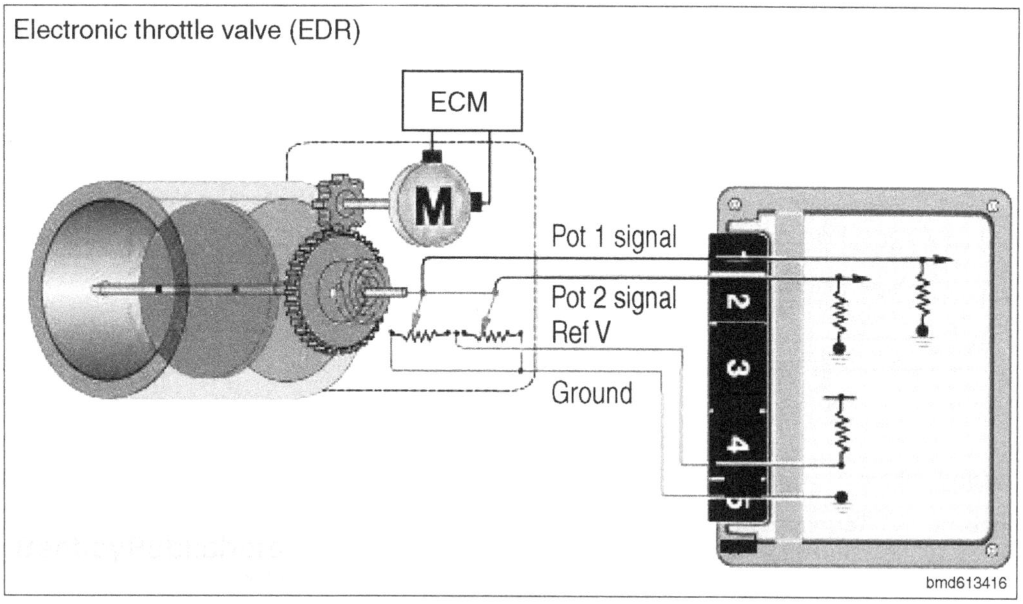
Electronic throttle valve (EDR)
ECM
M
Pot 1 signal
Pot 2 signal
Ref V
Ground
1
2
3
4
5
bmd613416

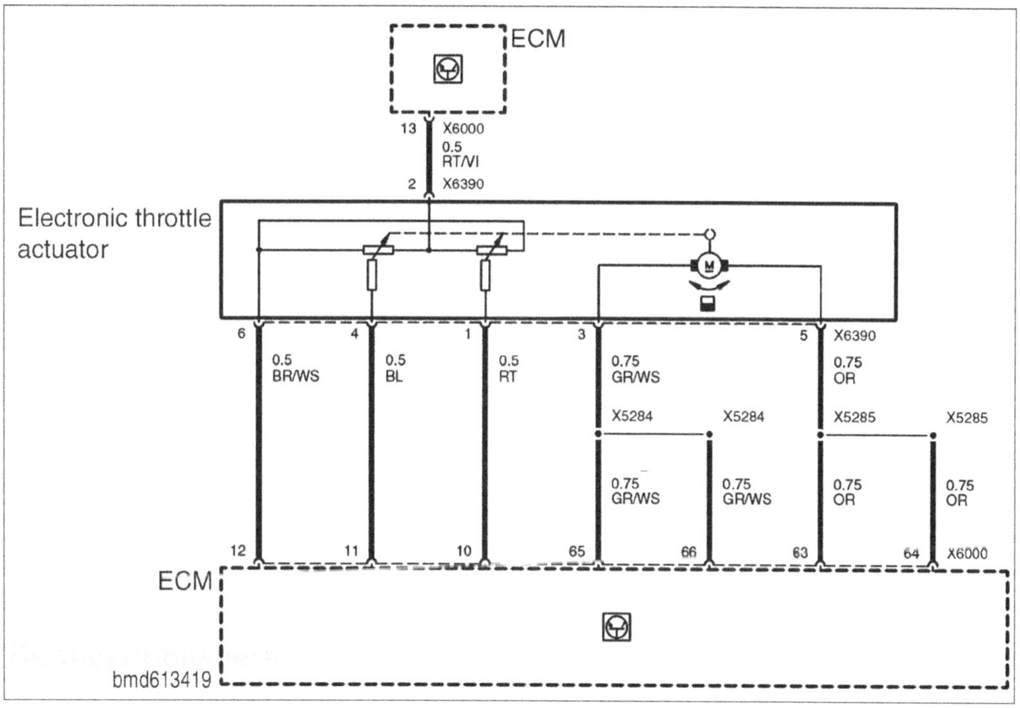
ECM
13 X6000
0.5
RT/VI
2 X6390
Electronic throttle actuator
6
4
1
3
5 X6390
0.5
BR/WS
0.5
BL
0.5
RT
0.75
GR/WS
0.75
OR
X5284
X5284
X5285
X5285
0.75
GR/WS
0.75
GR/WS
0.75
OR
0.75
OR
12
11
10
65
66
63
64 X6000
ECM
bmd613419

Electronic throttle actuator, testing

ECM connector X6000 (81 pin, black)

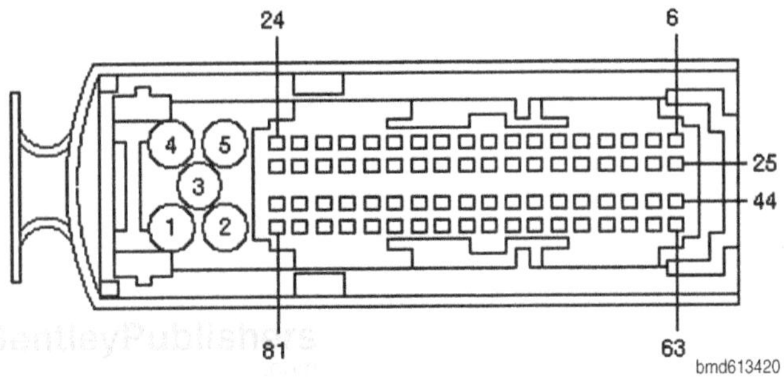

EDR connector X6390

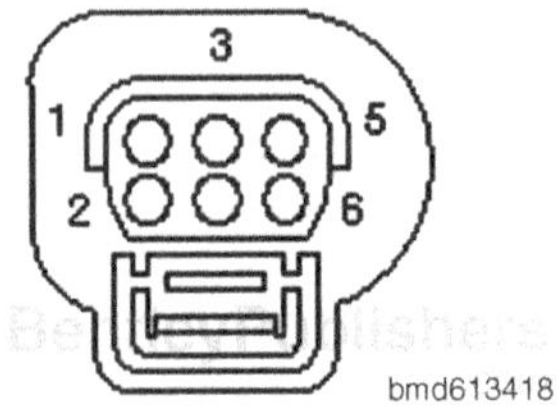

EDR testing values				
Terminals at connector X6000	**Terminals at connector X6390**	**Conditions**	**Test values**	**Circuit**
63 and ground	5 and ground	key on engine off	0 volts	motor control
63 and ground	5 and ground	engine running at idle	2.6 volts	motor control
64 and ground		key on engine off	0 volts	motor control
64 and ground		engine running at idle	2.6 volts	motor control
65 and ground	3 and ground	key on engine off	0 volts	motor control
65 and ground	3 and ground	engine running at idle	0.2 volts	motor control
66 and ground		key on engine off	0 volts	motor ground
66 and ground		engine running at idle	0.2 volts	motor ground

EDR waveform at idle

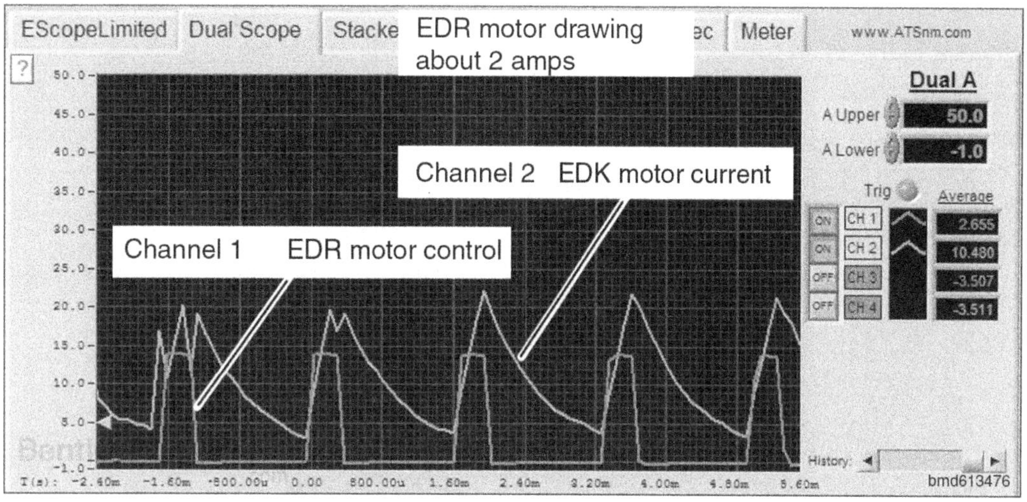

EDR waveform at idle

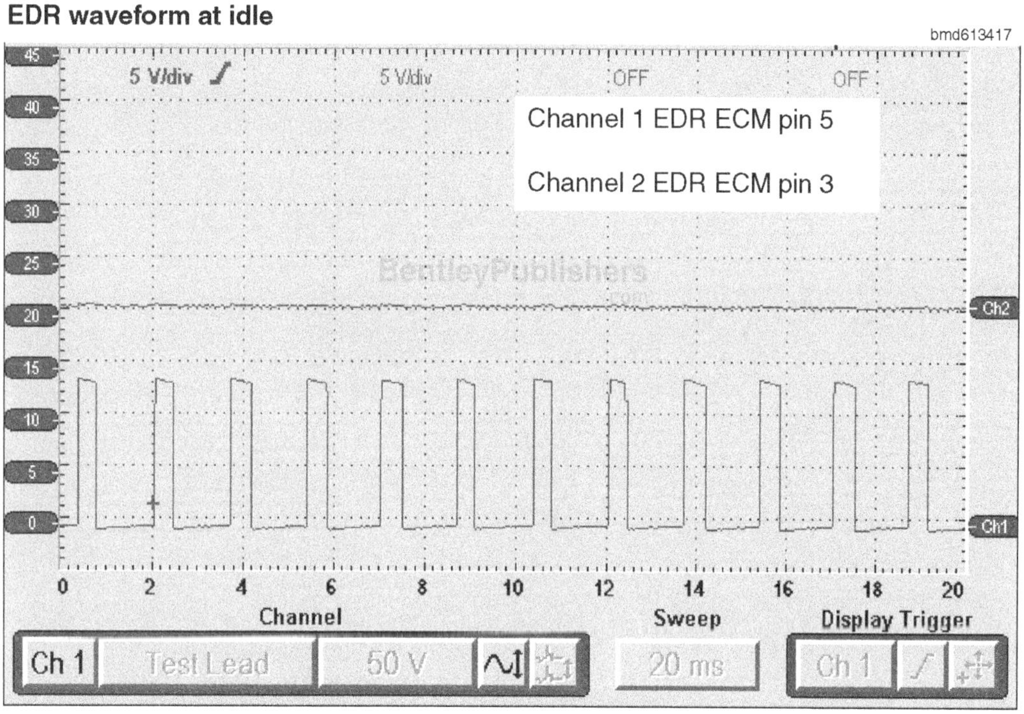

EDR waveform during snap throttle

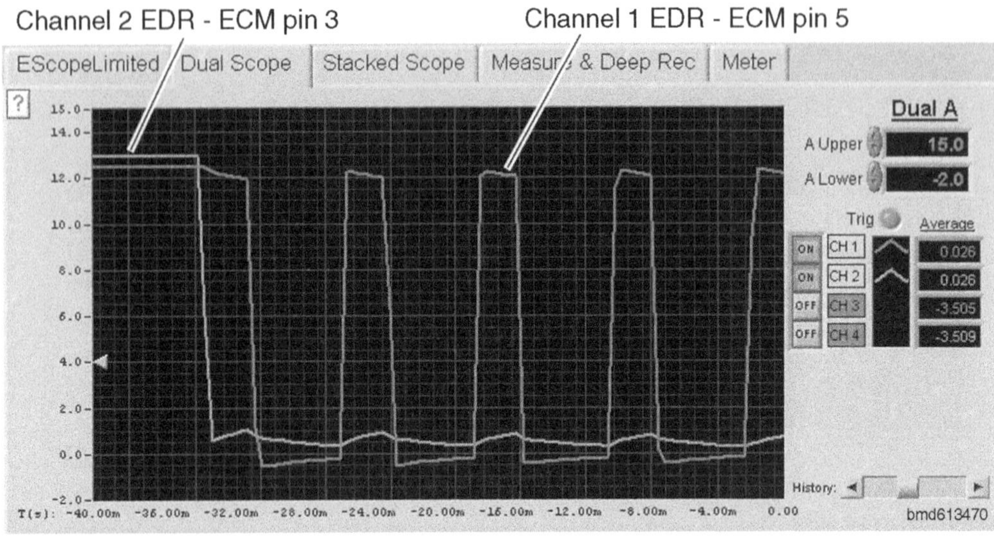

Idle speed control

Idle speed is controlled by the ECM and EDR without the aid of an idle speed motor. For smooth driveability of the vehicle, engine speed should remain constant when at idle no matter the loads on the engine. The driver should not detect a fluctuation in engine idle speed under the following conditions:

- Engine cold start (increase in idle speed setting).
- Switching electrical loads on/off (e.g. headlights, blower motor).
- Turning steering wheel to full lock (increased load from the electro-hydraulic power steering pump).
- A/C compressor engagement.

When there is a sudden change in electrical demand on the alternator (i.e. front and rear heated windshield and air-conditioning on), there is a corresponding rapid increase in the mechanical loading that the alternator exerts upon the engine.

This has a significant effect on engine idle speed stability. There is a delay between the electrical demand being made on the alternator and the mechanical demand being made on the engine. During this period the alternator transmits a signal to the ECM to indicate a change in electrical load. The signal is PWM and allows the ECM to control the throttle demand at idle to prevent flares and dips in the engine speed due to alternator loading or unloading.

> ***NOTE —***
> - Target idle speed setting for all MINI models is 750 rpm.

Idle speed jack feature

Engine idle speed increases to a predetermined set speed if the ECM system voltage falls below a certain threshold. Engine speed remains at this elevated point until system voltage has recovered.

EML warning LED

If the ECM detects an engine safety-related fault (not emissions related) in the ECM itself, the EML warning light illuminates in the IKE to show there is a fault with the drive-by-wire system. (This could be caused by a faulty throttle motor, gearbox or a sticking throttle flap.) The warning LED is amber in color and activated through a CAN-bus message to the IKE.

ELECTRONIC THROTTLE (EDR) FEEDBACK

The EDR throttle plate position is monitored by two integrated potentiometers. The potentiometers provide D/C voltage feedback signals as input to the ECM for throttle and idle control functions.

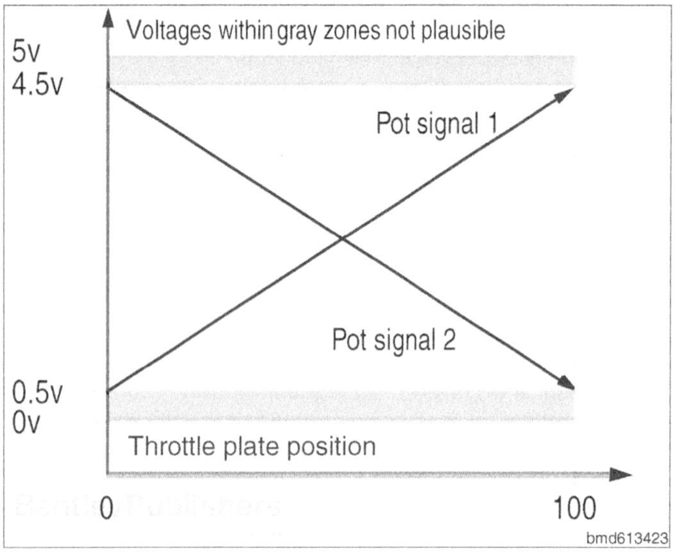

Potentiometer 1 provides a signal ranging from 0.5 v (LL) to 4.5 v (VL). Potentiometer 2 simultaneously provides a signal ranging from 4.5 v (LL) to 0.5 v (VL). Potentiometer signal 1 is the primary signal while signal 2 is used as a plausibility cross-check through the total range of throttle plate movement.

View of throttle feedback potentiometer inside EDR

EDR failure symptoms

If there is an open or short in signal 1, signal 2 is used as a temporary substitute providing fail-safe operation (faults stored).

If plausibility errors are detected between Pot 1 and Pot 2, one of two emergency modes is activated depending on severity of fault.

Emergency mode 1: This limits the dynamic operation if one of the potentiometers fail. The engine can slowly reach maximum speed with limited power.

Emergency mode 2: If another fault is encountered in addition to emergency mode 1 or if the plausibility is affected, emergency mode 2 is activated by the ECM. Emergency mode 2 can be initiated by simultaneously pressing both the accelerator pedal and the brake pedal while in emergency mode 1 or if a fault is encountered in brake light switch diagnosis. When in emergency 2 operation mode, the engine runs very roughly. Throttle response is limited. This mode enables limp home functions.

Additional monitoring

The EDR monitoring system can detect a jammed or binding throttle valve as well as a broken link spring. The fault is detected by ECM monitoring of feedback potentiometers from the EDR in relation to the pulse width modulation signal activating the EDR.

Electronic throttle feedback, testing

ECM connector X6000 (81 pin, black)

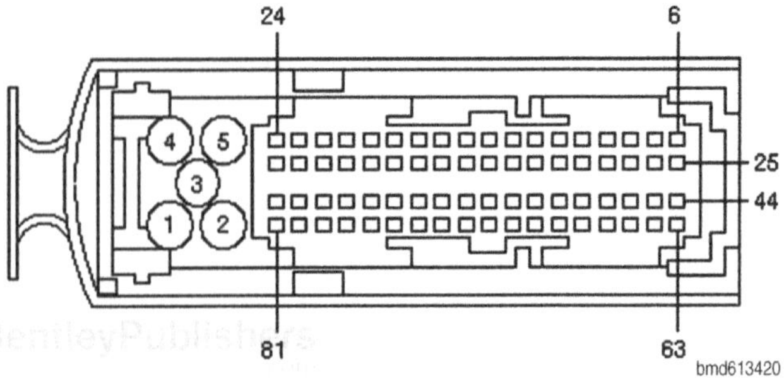

EDR connector X6390

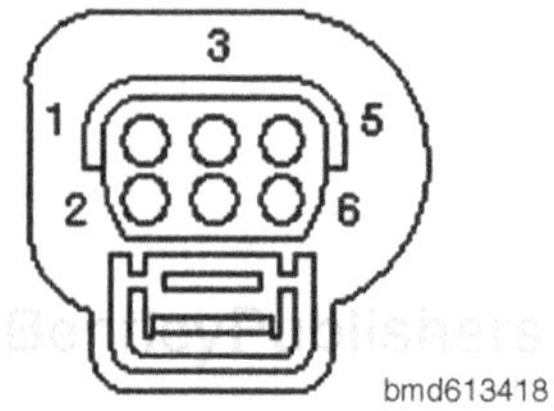

EDR testing values				
Terminals at connector X6000	**Terminals at EDR connector X6390**	**Conditions**	**Test values**	**Circuit**
10 and ground	1 and ground	key on engine off	0.8 to 4.5 volts while pushing throttle from idle to full throttle	TPS 1
10 and ground	1 and ground	engine running at idle	0.6 volts	TPS 1
10 and ground	1 and ground	@2500 rpm (no load)	0.7 volts	TPS 1
11 and ground	4 and ground	key on engine off	4.0 to 0.3 volts while pushing throttle from idle to full throttle	TPS 2
11 and ground	4 and ground	engine running at idle	4.4 volts	TPS 2
11 and ground	4 and ground	@2500 rpm (no load)	4.3 volts	TPS 2
12 and ground	6 and ground	key on engine off engine running at idle @2500 rpm, (no load	0.1 volts (maximum)	sensor ground
13 and ground	2 and ground	key on engine off engine running at idle @2500 rpm, (no load	5 volts (approx.)	reference voltage

TPS 1 and 2 waveform during closed, wide open and closed throttle operation

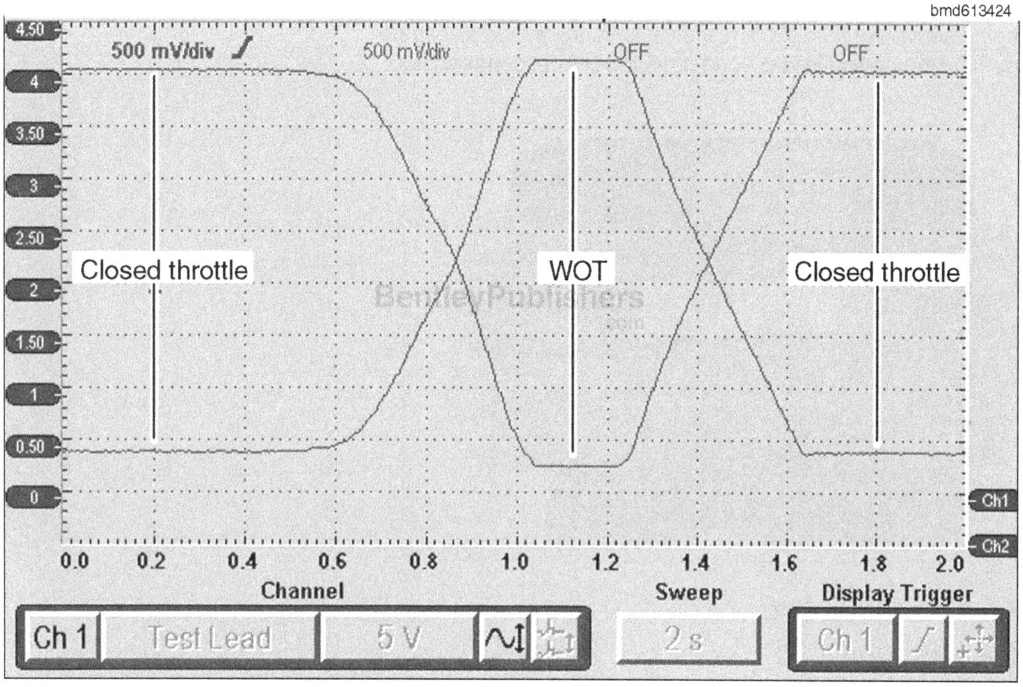

ENGINE CONTROL MODULE (ECM)

The engine control module (ECM) receives and processes all engine management inputs and issues corresponding control commands.

bmd613426

The ECM has two connectors:

- X60004 40-pin, black
- X6000 81-pin, black

ECM location	
MINI Cooper	Left compartment of battery box next to engine compartment fuse and relay panel
MINI Cooper S	Left of air filter housing

In addition to engine controls, the ECM has direct control over the following systems:

- Drive-by-wire throttle control
- Automatic transmission
- Cruise control
- Air-conditioning clutch relay
- Engine cooling fan relays

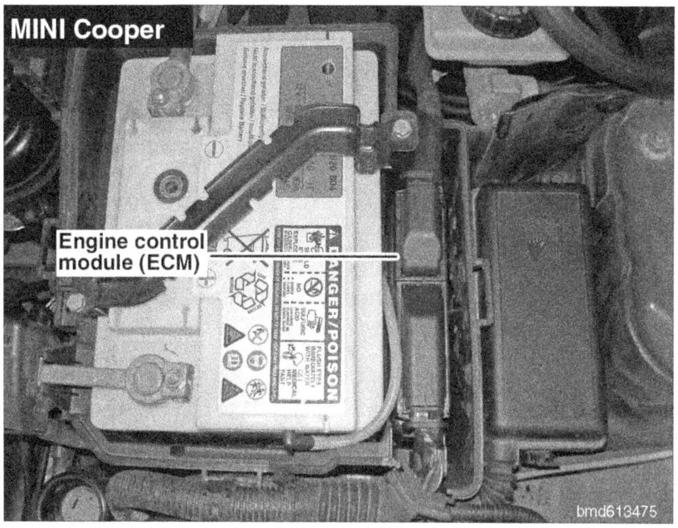
MINI Cooper
Engine control module (ECM)
bmd613475

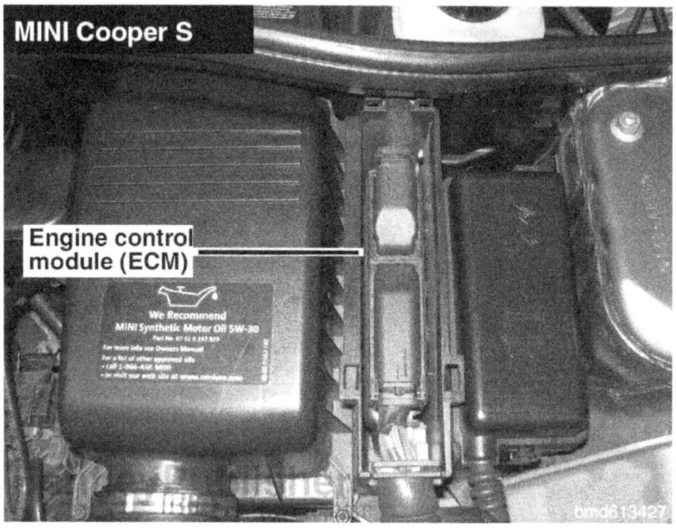
MINI Cooper S
Engine control module (ECM)
We Recommend
MINI Synthetic Motor Oil 5W-30
bmd613427

ECM power supply and grounds

The ECM monitors battery voltage for fluctuations. The ECM adjusts output functions to compensate for lower and higher voltages.

Based on battery voltage, the ECM has the ability to adjust:

- Fuel injection pulse width.
- Ignition system dwell.

When the ignition switch is ON (terminal KL15) or in the start position (terminal KL50), fuse F34 is provided with power. F34 supplies the wake up or ON signal to the ECM.

Battery voltage (KL30) is supplied through fuse F01 to provide constant memory.

F02, as provided by the main relay, supplies operating power to the ECM.

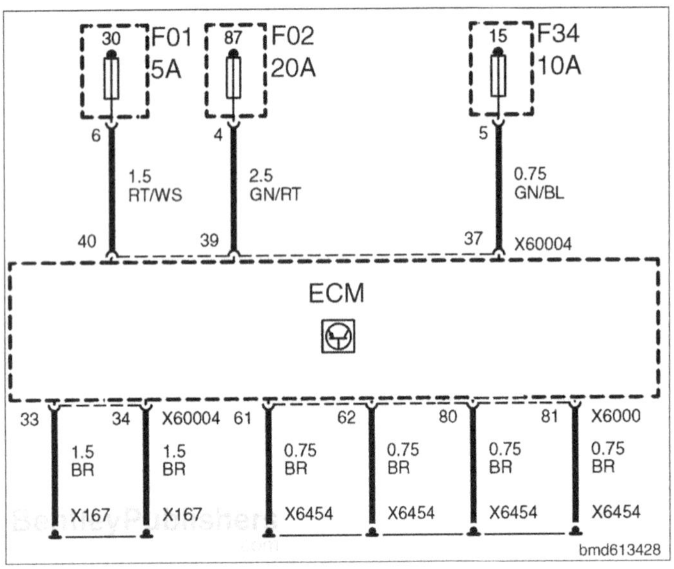

ECM grounds are present at both ECM connectors.

Connector X 6000: Pins 61, 62, 80, and 81 supply ground to the ECM.

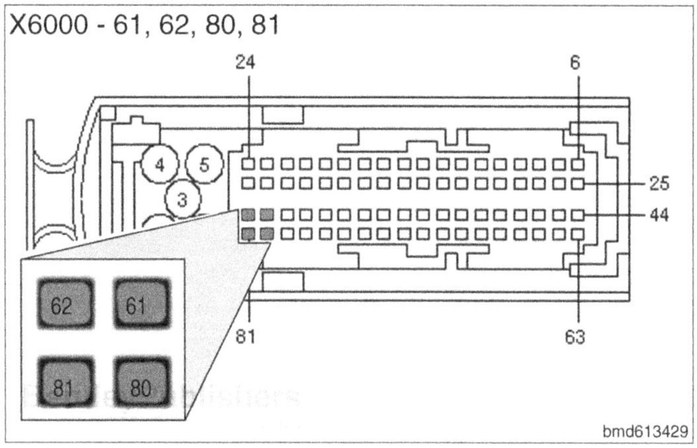

Ground pins 61, 62, 80, and 81 meet at a common grounding point, X6454, (left side of engine compartment). These grounds also share the grounding point with X167 (left side of engine compartment).

Connector X6004: Pins 33 and 34 supply ground to the ECM.

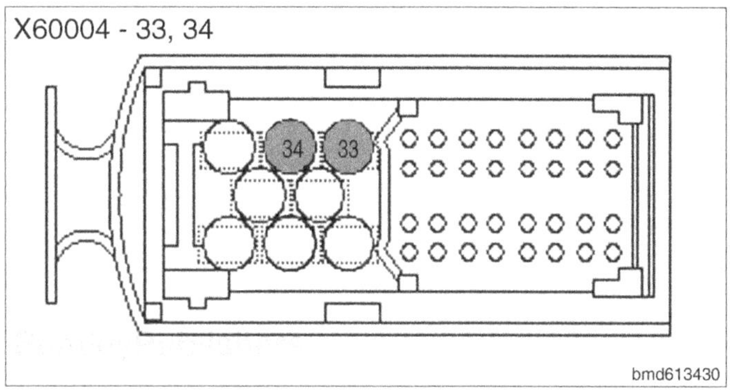

Grounds from pins 33 and 34 meet at a common chassis ground X167, (left side of engine compartment). These grounds also share the grounding point with X6454.

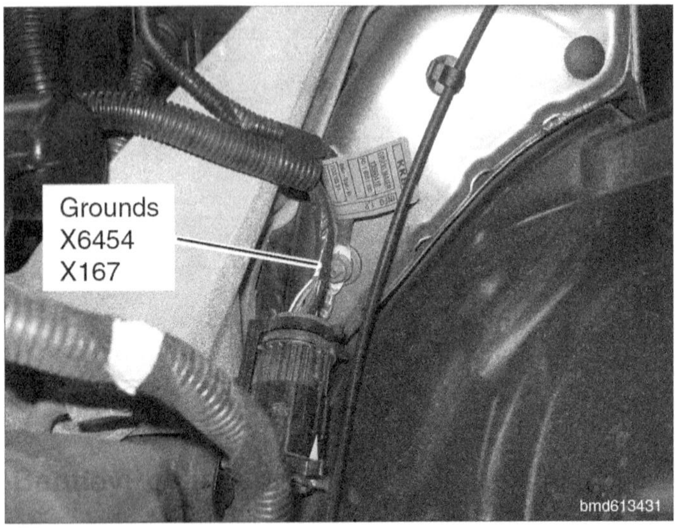

NOTE —

- Test grounds using the voltage drop method. A maximum voltage drop of 0.1 volts for all ECM chassis grounds is acceptable. See **01 Diagnostic Tools and Techniques**.

ENGINE COOLANT TEMPERATURE (ECT) SENSOR

The ECT sensor monitors engine coolant temperature. The signal is used for fuel control and ignition timing.

The signal is also used to control the radiator fan speed as well as supplying a signal on the CAN-bus for the coolant temperature gauge in the instrument cluster.

The ECT sensor is a negative temperature coefficient (NTC) thermistor sensor located in the coolant system. As temperature increases resistance across the sensor decreases. This causes sensor voltage to go from a high voltage reading to a low voltage reading.

A 5 volt reference signal is supplied to the ECT sensor by the ECM. High voltage readings at the ECM indicate high sensor resistance, or low temperature. Low voltage readings indicate low sensor resistance or high temperature.

The sensor is located in the cylinder head next to the thermostat housing. It is accessed easier with the air box removed.

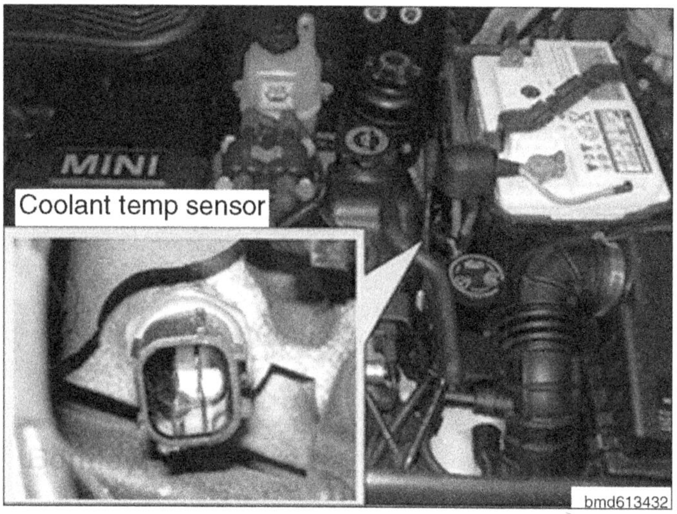

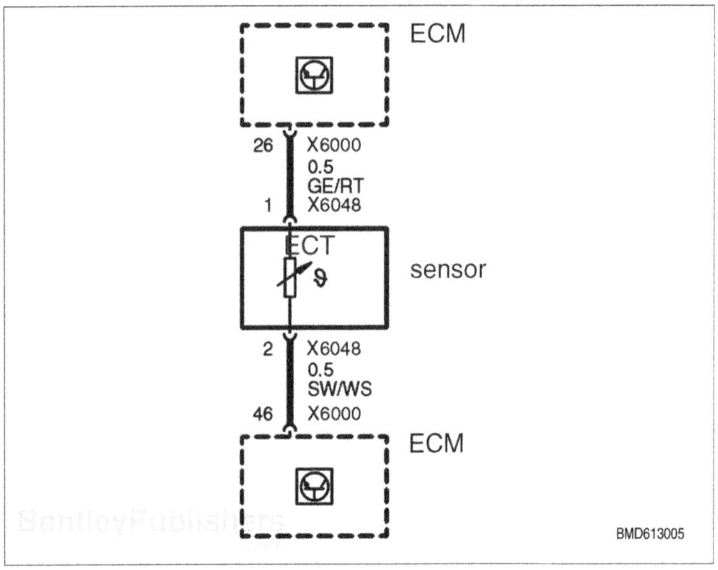

ECT sensor failure symptoms

In the event an engine coolant temperature sensor failure, a fault code is set in the ECM and a substitute value is used.

If the ECT sensor output signal fails open, shorted to ground, shorted to 12 v or shorted to 5 v, the ECM will replace the temperature value with a default value.

If the sensor ground is shorted to 12 v or shorted to 5 v, the temperature gauge indicates full hot and the engine cooling fan goes to high.

When ECT sensor faults are present, the vehicle may be difficult to start, may run lean when the engine is cold or slightly rich when the engine is hot (default to emergency program mode). Driveability of the vehicle is likely affected.

ECT sensor, testing

ECM connector X6000 (81 pin, black)

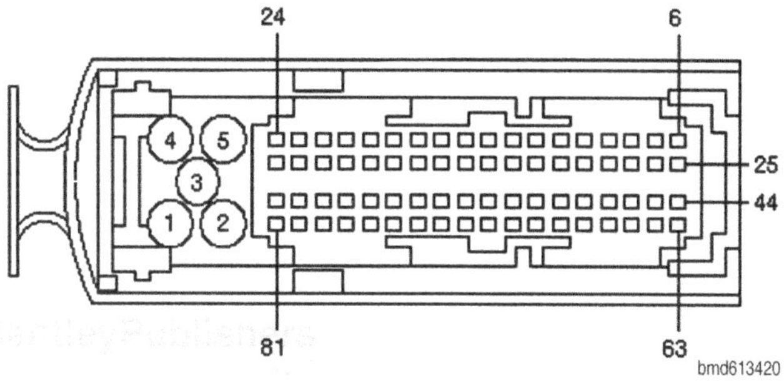

ECT sensor connector X6048

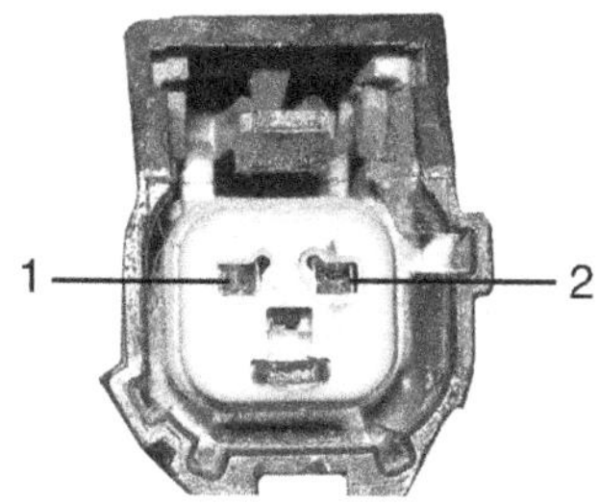

CTS testing values				
Terminals at connector X6000	**Terminals at connector X6048**	**Conditions**	**Test values**	**Circuit**
26 and ground	2 and ground	key on engine off	3.9 volts, (60* F ambient temp)	ECT sensor signal
26 and ground	2 and ground	engine running at idle	0.58 volts (operating temp.)	ECT sensor signal
46 and ground	1 and ground	key on engine off engine running at idle @2500 rpm (no load)	0.1 volts (maximum)	sensor ground

NOTE —

- A calculated coolant temperature is used based on coolant temperature at start, engine speed and load while running, time spent in idle and fuel shut-off. When actual engine temperature (TCO) reaches a calculated threshold closed loop is activated.

ECT sensor resistance

Engine coolant temperature sensor resistance				
Temperature		**Minimum**	**Typical**	**Maximum**
° C	**° F**	**Resistance (K Ω)**	**Resistance (K Ω)**	**Resistance (K Ω)**
-40	-40	291.49	336.6	381.71
0	32	29.33	32.66	35.99
30	86	7.37	8.06	8.75
50	122	3.33	3.60	3.88
60	140	2.31	2.49	2.67
70	158	1.63	1.75	1.87
80	176	1.17	1.255	1.34
90	194	0.86	0.915	.097
100	212	0.64	0.68	0.72
110	230	0.48	0.51	0.54
120	248	0.37	0.39	0.32
130	266	0.28	0.30	0.32
140	284	0.21	0.23	0.25
150	302	0.017	0.185	0.20

EVAPORATIVE EMISSION LEAK TEST

If an evaporative emission leak is suspected, confirm the leak using the EVAP system test function of the factory or equivalent scan tool. Then use the following tests to pinpoint the leak.

Leak tests are performed with engine switched OFF.

> ***CAUTION —***
> - *Due to combustible fumes in the evaporative emission system, DO NOT not induce oxygen into the system.*

NOTE —
- Perform leak tests in calm air. Smoke from leak tester can be blown around, hindering your diagnosis.

Purge solenoid, leak testing

– Disconnect electrical connector to purge solenoid (**A**).

– Remove vacuum line from activated charcoal canister to solenoid (**B**).

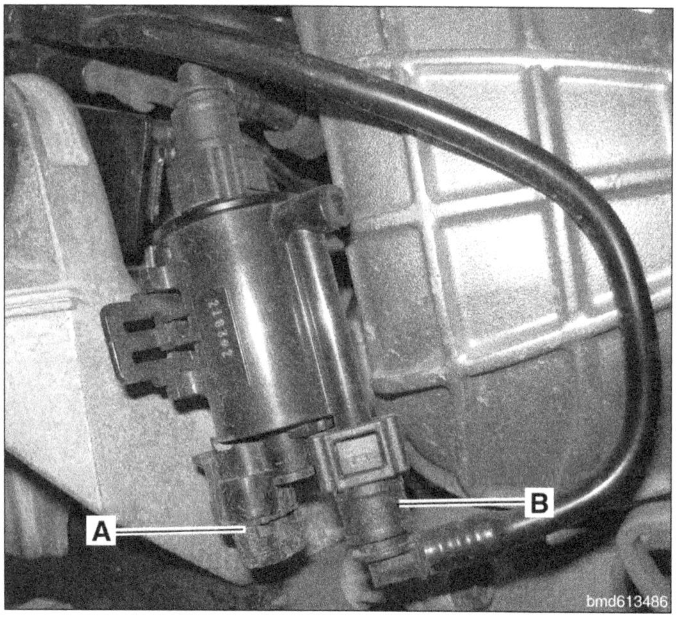

- Install vacuum gauge to disconnected vacuum line.
- Run engine at idle. There should be 0 inches of vacuum present. Solenoid is normally closed.

System smoke test

- Remove vacuum line at purge solenoid (activated charcoal canister side).
- Connect smoke tester to disconnected vacuum line.
- Close or seal EVAP vent solenoid (integrated in LDP).

NOTE —
- The vent solenoid is integrated in the LDP. To close vent solenoid with a scan tool, vacuum must be supplied to LDP. To manually seal vent, cap off EVAP vent line.

- Remove fuel filler cap.
- Fill EVAP system with smoke until it exits the fuel filler neck.
- Install fuel filler cap.
- Continue to induce smoke until system reaches 1 psi.

> ***CAUTION —***
> - *Do not exceed 1 psi.*

- Inspect EVAP components to locate leak, indicated by emerging smoke.

NOTE —
- Smoke exiting a very small leak is easier to find with lower pressure. If you see smoke in an area but cannot pinpoint it, lower the pressure in the system.

- Check for fault codes and clear and reset ECM control module memory.

NOTE —
- An incorrectly secured fuel filler cap sets a fault code for the EVAP system. To avoid a fault code, a switch is installed in the fuel filler neck. This switch checks whether the fuel tank is closed correctly.

EVAPORATIVE EMISSION PURGING

Evaporative emission purging is regulated by the ECM via the evaporative emission purge valve, a solenoid that regulates purge flow from the active carbon canister into the intake manifold. The ECM main relay provides operating voltage, and the ECM controls the valve by switching the ground circuit.

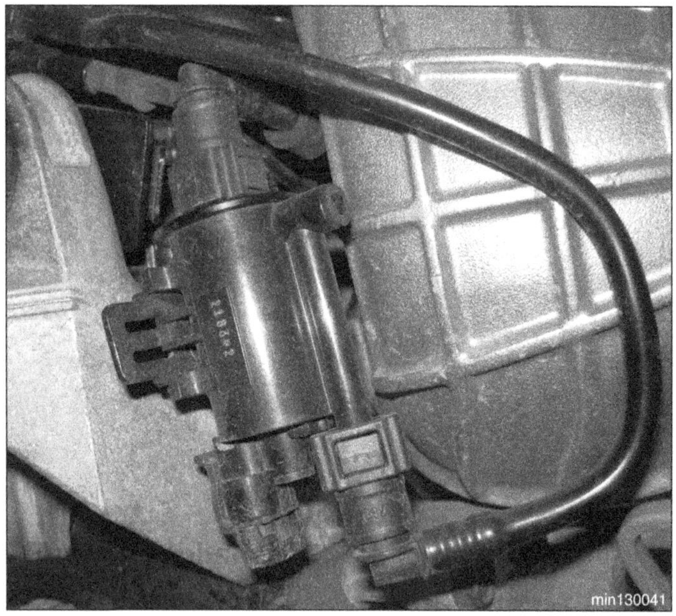

The valve is powered open and closed by an internal spring. The purging process takes place when:

- Oxygen sensor control is active.
- Engine coolant temperature is > 67°C.
- Engine load is present.

The evaporative emission purge valve is opened in stages to moderate purging.

During stage 1, the purge valve opens for 10 ms (milliseconds) and then closes for 150 ms. Stages continue with increased opening times (up to 16 stages) until the valve is completely open.

The valve now starts to close in 16 stages in reverse order. This staged process takes 6 minutes to complete. The function is inactive for 1 minute, then starts the process all over again.

During the purging process, the valve is completely opened during full throttle operation and is completely closed during deceleration fuel cutoff.

An evaporative purge system flow check is performed by the ECM when oxygen sensor control and purging is active. When the evaporative emission valve is open, the ECM detects a rich /lean shift as monitored by the oxygen sensors, indicating the valve is functioning properly.

If the ECM does not detect a rich / lean shift, a second step is performed when the vehicle is stationary and the engine is at idle speed. The ECM opens and closes the valve (abruptly) several times and monitors the engine rpm for changes. If there are no changes, a fault code is set.

Evaporative emission purge valve waveform at idle

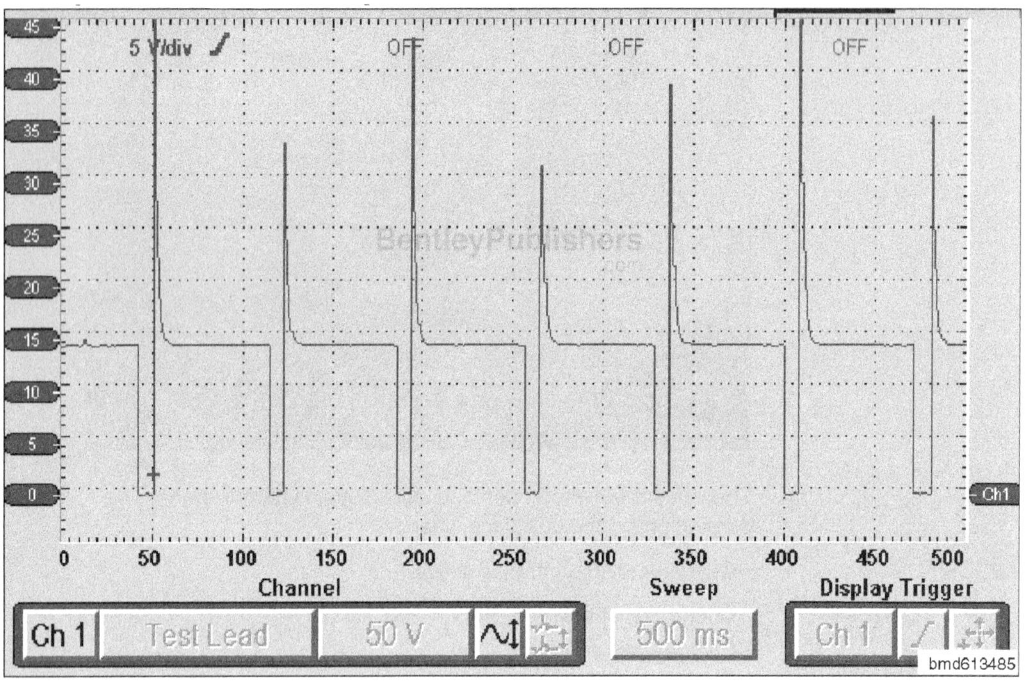

ELECTRONIC IMMOBILIZATION SYSTEM (EWS)

The ECM communicates with the EWS system (also know as drive-away protection) to provide immobilization of the vehicle. To start the engine, the correct key must be used which transmits the correct code magnetically to the coil in the steering column. This code is then fed to the immobilization unit, which transmits various codes to the ECM to enable the engine to start.

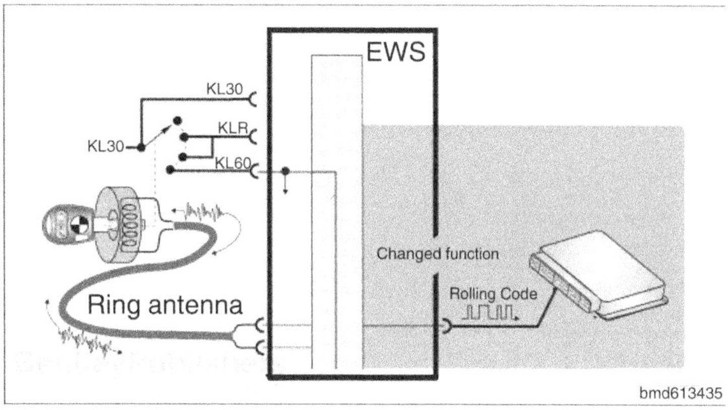

bmd613435

EWS signal, testing

ECM connector X60004 (40-pin, black)

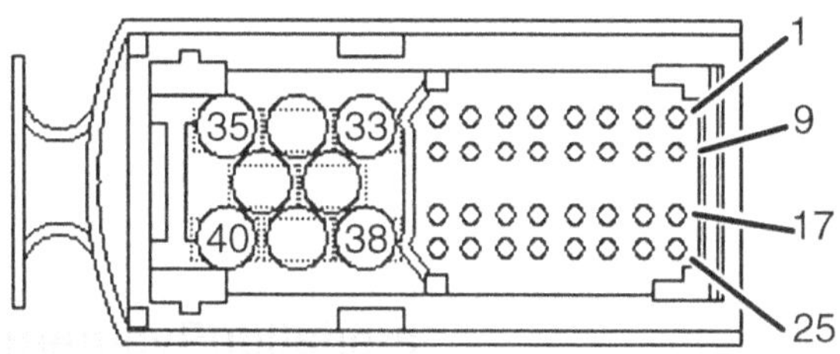

bmd613402

EWS testing values (at ECM connector x60004)			
Terminals	**Conditions**	**Test values**	**Circuit**
25 and ground	key on engine off	7.5 to 8.0 volts	EWS signal
25 and ground	engine running at idle	7.5 volts	EWS signal

EWS signal waveform key on engine off

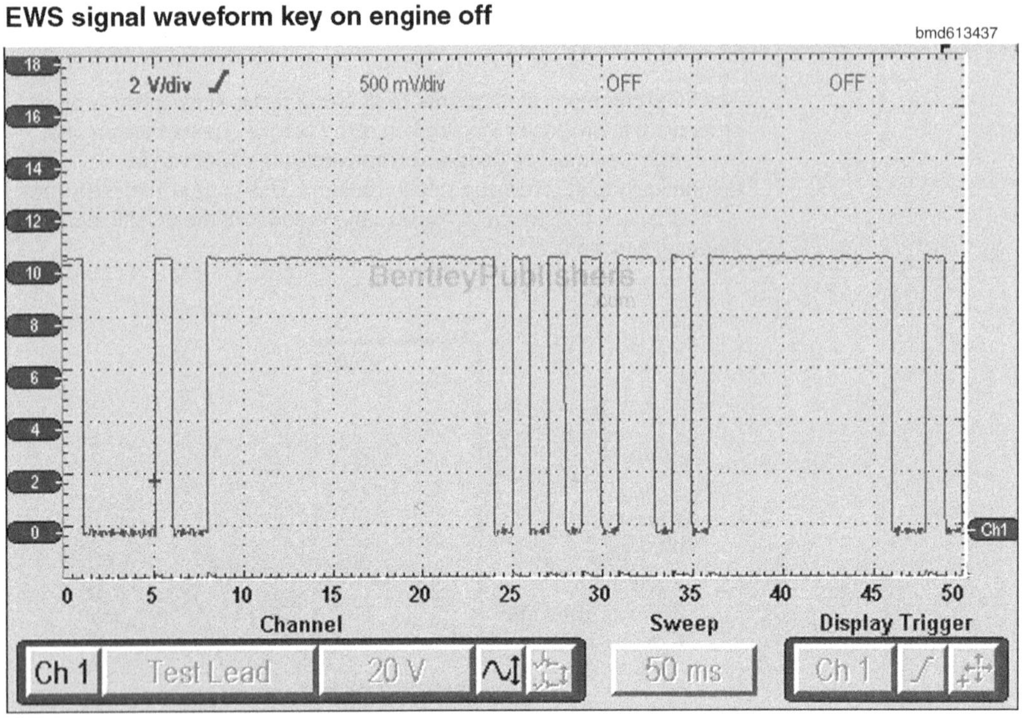

EWS signal waveform engine running at idle

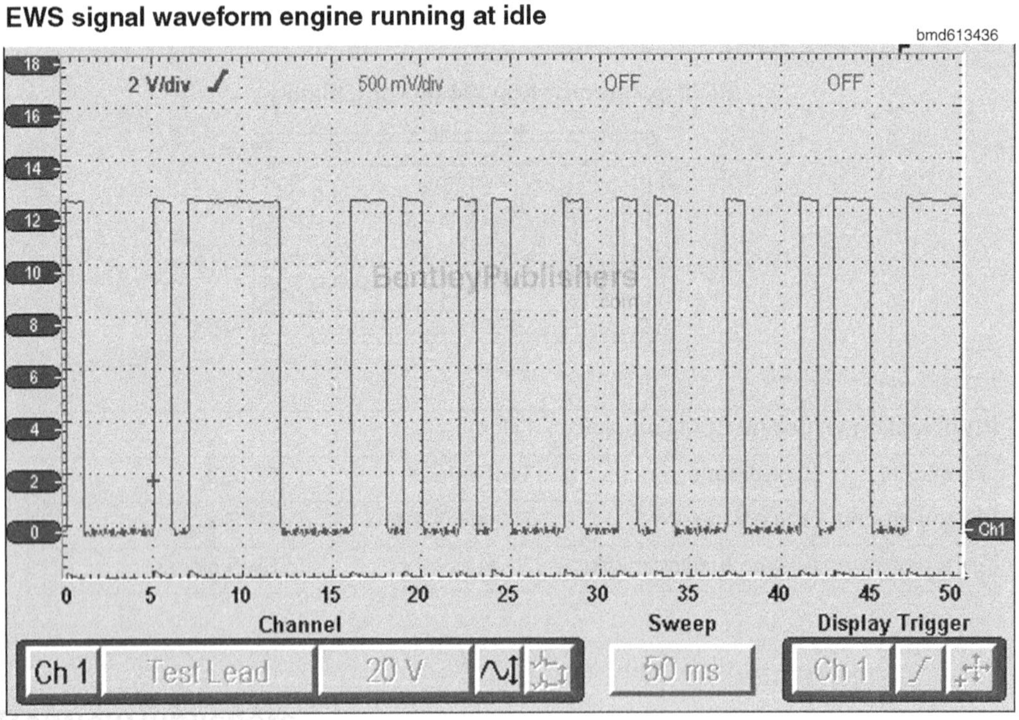

FUEL INJECTORS

Fuel Injectors are electronically controlled solenoid valves that provide precisely metered and atomized fuel into the engine intake ports.

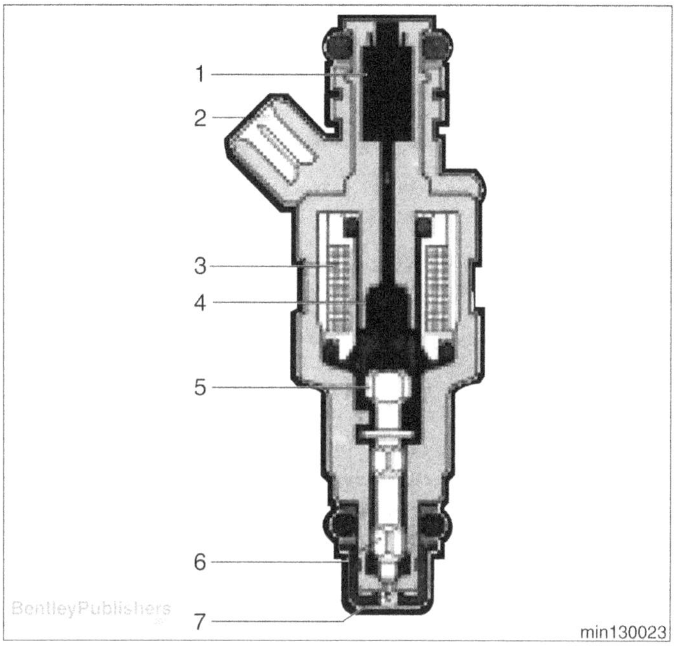

1. Fuel strainer
2. Electrical connector
3. Solenoid winding
4. Closing spring
5. Solenoid armature
6. Needle valve
7. Pintle

The fuel Injectors receive voltage from the main relay. The ECM activates ground, via a transistor.There are two types of injectors used on the MINI.

- MINI COOPER uses an injector with 4 holes (Bosch).
- MINI COOPER S uses an injector with 2 holes and a higher flow rate (Siemens).

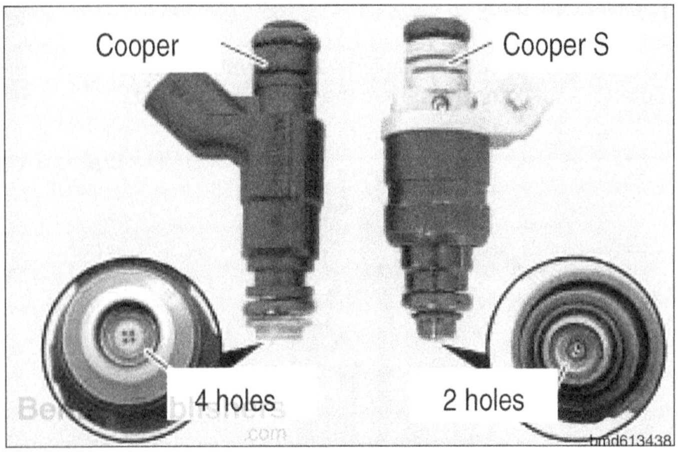

Catalytic converter protection

The ECM protects the catalytic converter from damage by deactivating injectors. If the ECM detects a misfire (ignition, injection or combustion) it can selectively deactivate the injector output transistor for that cylinder. The injector does not open, preventing unburned fuel from entering the exhaust system. There are four individual injector circuits in the ECM that can deactivate one or multiple injectors. This limits engine power, but protects the catalytic converter.

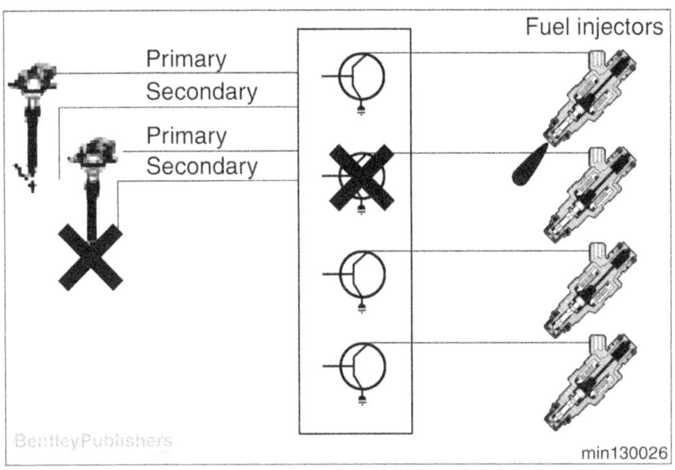

Fuel injector waveform engine running at idle

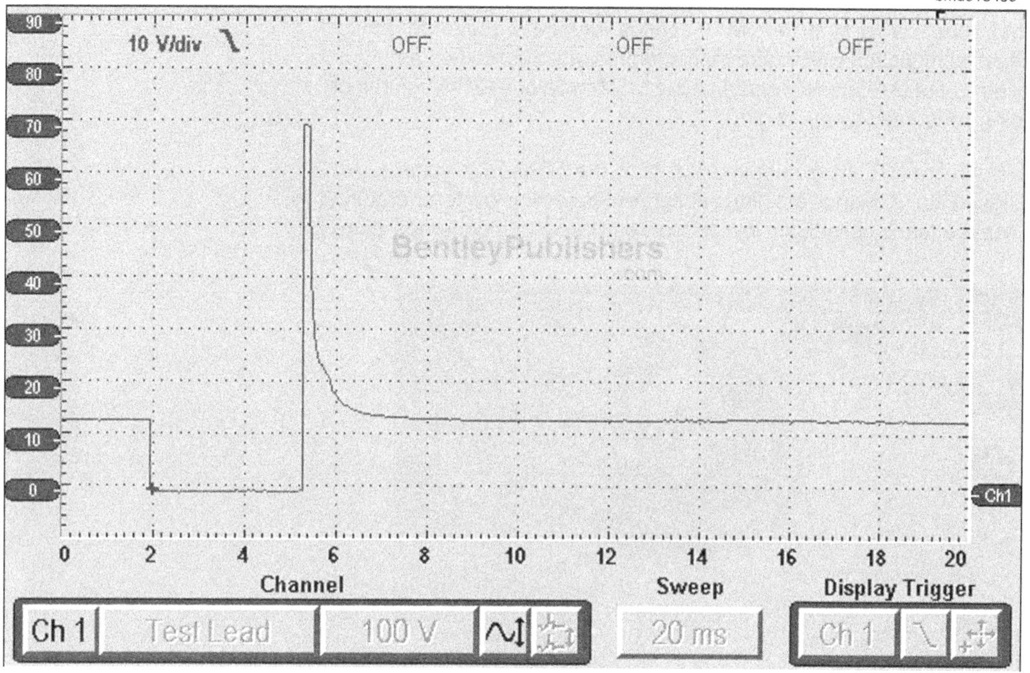

Fuel injector current waveform engine running at idle

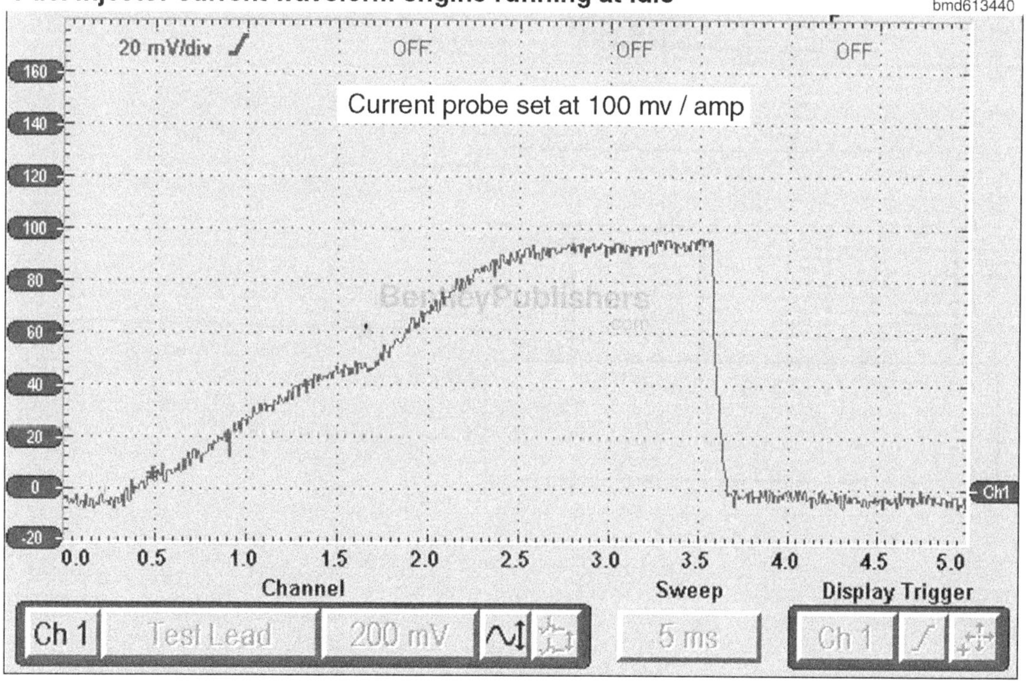

INTAKE AIR TEMPERATURE (IAT) SENSOR

The IAT sensor is built into the TMAP sensor. Air temperature information is used in conjunction with intake manifold pressure information as supplied by the MAP sensor enabling the ECM to calculate the volume of air being consumed by the engine.

The sensor is a NTC type, supplied 5 volts by the ECM. High voltage readings at the ECM indicate high sensor resistance, or low temperature. Low voltage readings indicate low sensor resistance or high temperature.

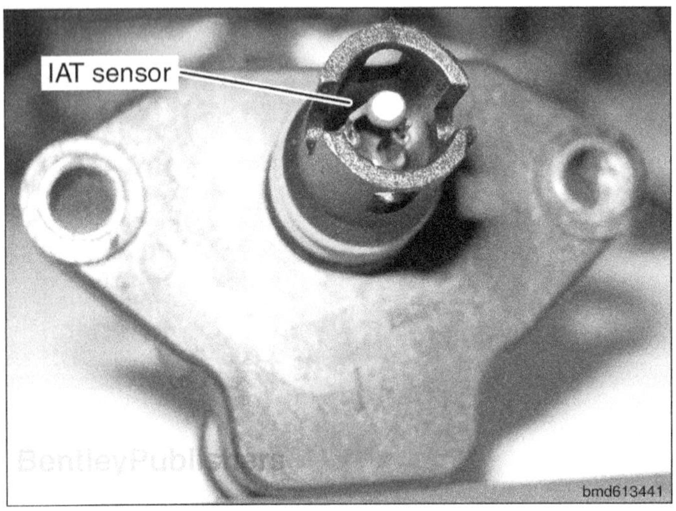

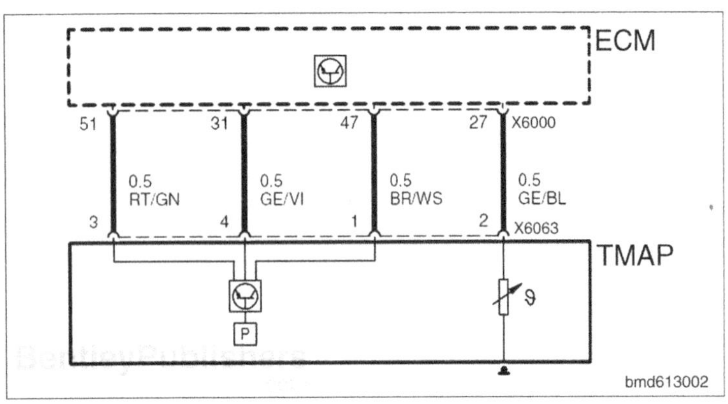

Failure symptoms:

- Problems with cold starting, poor driveability and an increase in emissions.
- Possible lack of power, particularly noticeable in hot ambient temperatures.

IAT sensor, testing

ECM connector X6000 (81 pin, black)

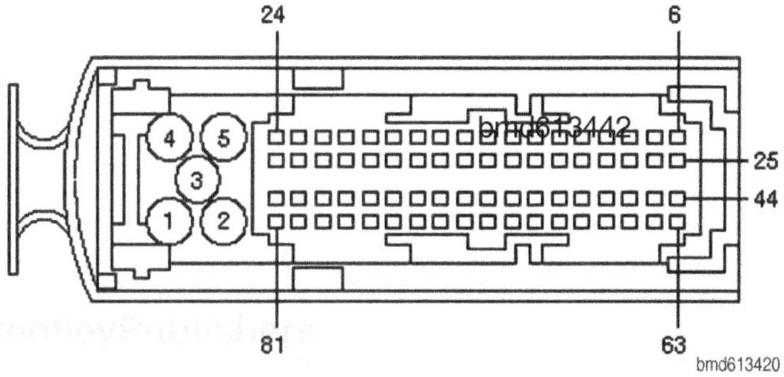

TMAP connector X60631

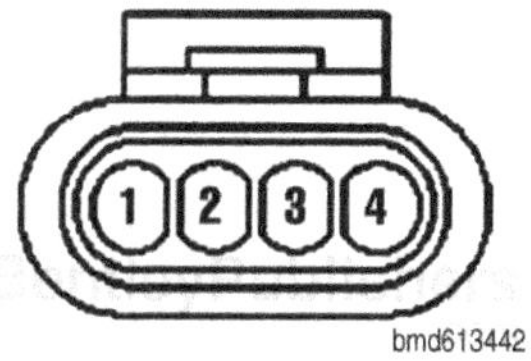

IAT testing values				
Terminals at connector X6000	**Terminals at connector X60631**	**Conditions**	**Test values**	**Circuit**
27 and ground	7 and ground	key on engine off	3.9 volts, (60* F ambient temp)	IAT signal
27 and ground	2 and ground		1.25 volts, (60* F ambient temp, engine at operating temp)	IAT signal,
47 and ground	1 and ground	key on engine off engine running at idle	0.1 volts (maximum)	sensor ground

IGNITION COIL

MINI Cooper and Cooper S models use an ignition coil pack containing two coils. Each ignition coil has two secondary voltage outputs to supply ignition voltage for two cylinders. The coils use a wastespark, which means that ignition takes place in two cylinders at the same time. One spark plug ignites during the compression stroke and the other during the exhaust stroke.

The high voltage required to ignite the mixture in the combustion chambers is supplied by the stored energy in the ignition coils. The stored energy contributes to the ignition duration, ignition current and rate of high voltage increase. The coil circuit, including primary and secondary components, consists of:

- Coil assembly
- Primary winding
- Secondary winding
- Resistor
- Spark plug
- ECM final stage transistor

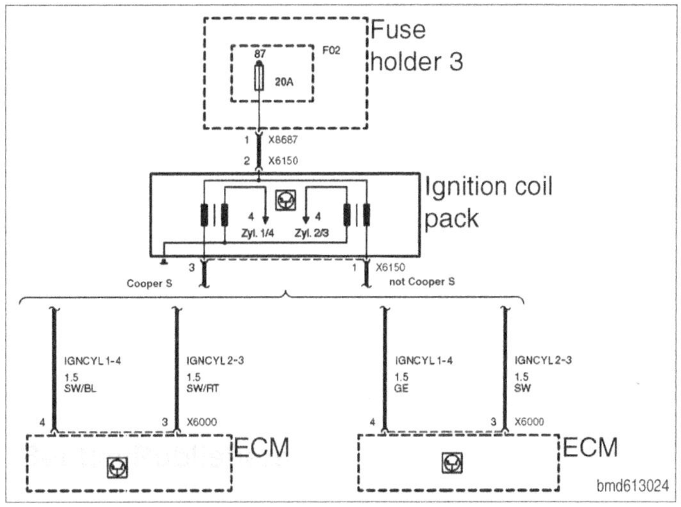

Spark delivered on the firing stroke provides ignition. When the cylinder is on the firing stroke, fuel is in the cylinder and cylinder pressures are high. The spark has a higher resistance and requires higher voltage to jump the plug gap.

Ignition coil, testing

Check secondary ignition system visually and with a suitable engine analyzer or scope. Check components for tight connections and freedom from carbon tracking, moisture and corrosion.

When measuring voltage on cylinders 1 and 2, remember that voltage is flowing from the outer electrode to the center and back to the coil. This will be indicated by negative voltage readings on the oscilloscope. Voltage on cylinders 1 and 2 is approximately 20% higher than on cylinders 3 and 4 as it takes more energy for the spark to travel from the outer to the center electrode.

Secondary output in the 7 - 9 kv range are normal during firing of the cylinder on the compression stroke and 3 - 5 kv for the cylinder on the wasted spark stroke. A defective coil influences two cylinders, whereas a defective plug wire only influences one cylinder.

Secondary ignition waveform

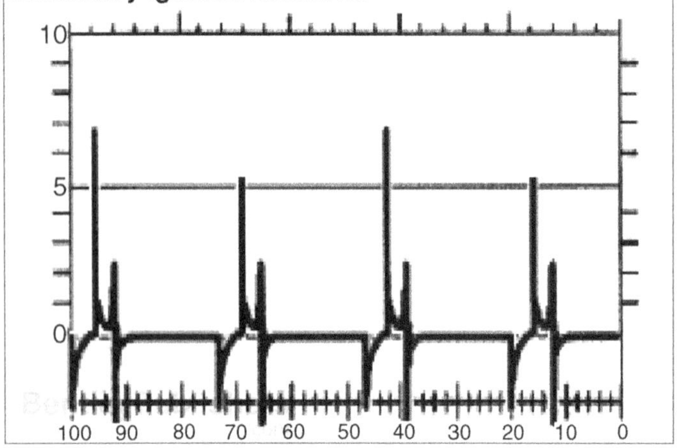

ECM connector X6000 (81 pin, black)

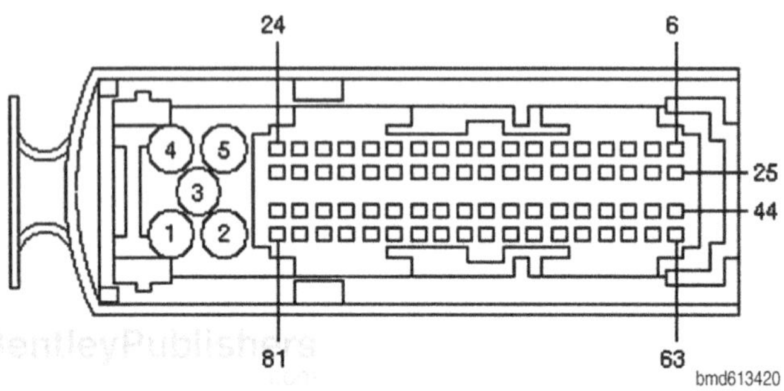

Coil connector X6150

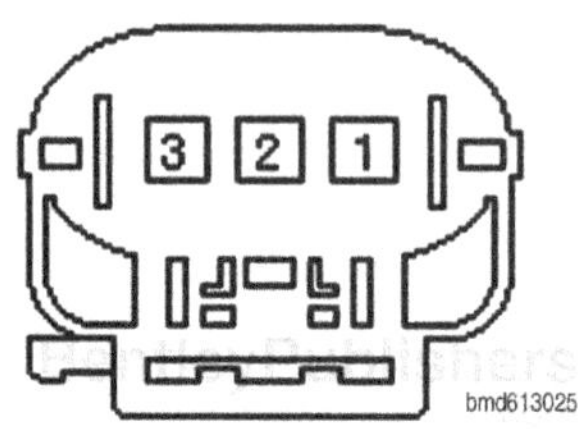

Coil testing values (at ECM connector X6000)				
Terminals at connector X6000	**Terminals at connector X6150**	**Conditions**	**Test values**	**Circuit**
3 and ground	1 and ground	key on engine off	battery volts	coil trigger
3 and ground	1 and ground	engine running at idle @2500 rpm, (no load)	ignition primary signal	coil trigger
N-A	2 and ground	key on engine off engine running at idle @2500 rpm, (no load)	battery volts	voltage supply
4 and ground	3 and ground	key on engine off	battery volts	coil trigger
4 and ground	3 and ground	engine running at idle @2500 rpm, (no load)	ignition primary signal	coil trigger

Fuse panel connector X8687

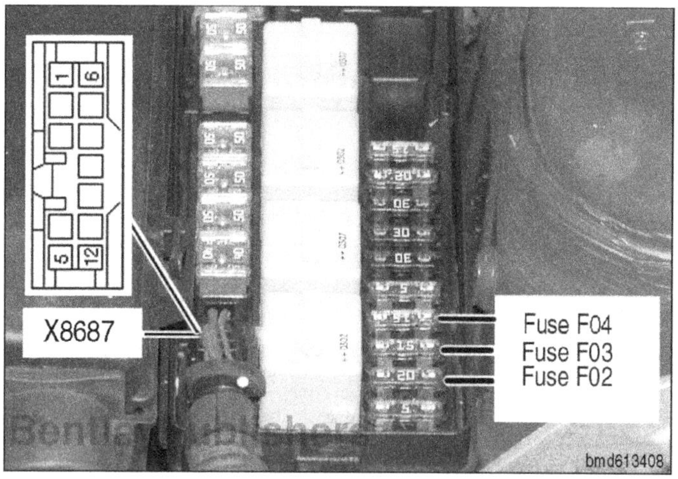

Coil testing values (at fuse panel connector X8687)			
Terminals	**Conditions**	**Test values**	**Circuit**
1 and ground	key on engine off engine running at idle @2500 rpm, (no load)	battery volts	voltage supply
Fuse F02 and ground	key on engine off engine running at idle @2500 rpm, (no load)	battery volts	voltage supply

Ignition coil trigger waveform at idle

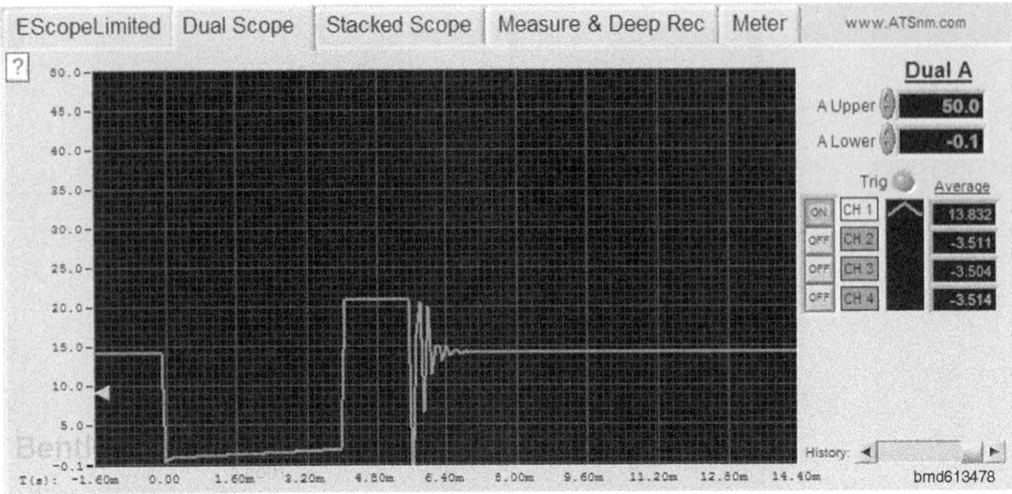

Ignition coil current waveform engine running at idle

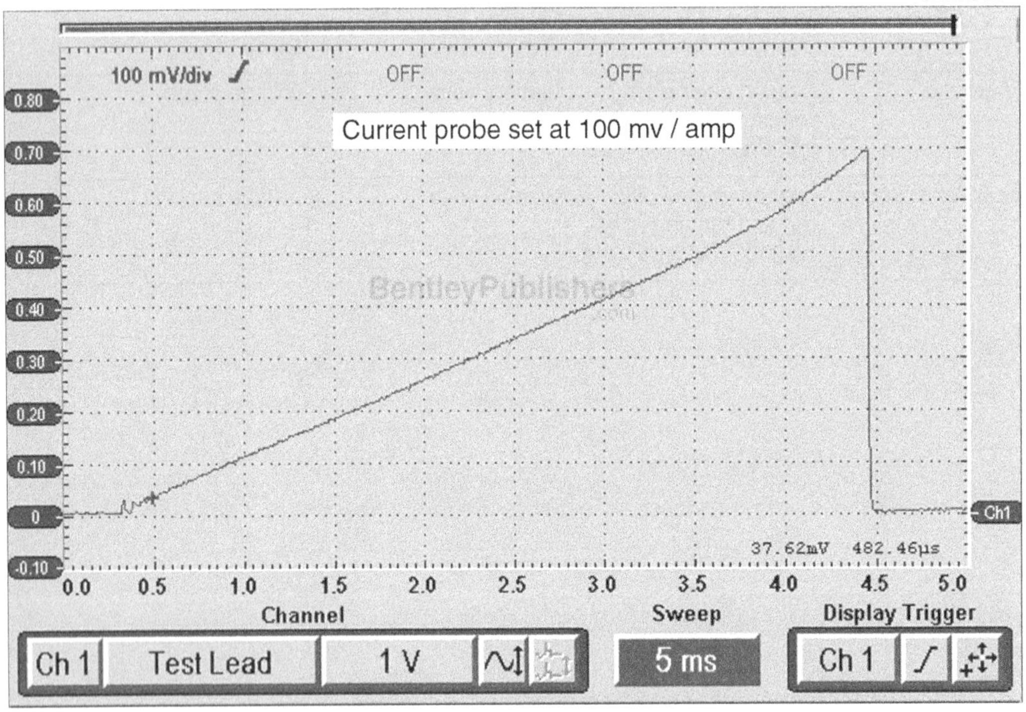

KNOCK SENSOR

A single knock sensor is bolted to the block to prevent detonation (pinging) from damaging the engine. The ECM retards the ignition timing (cylinder selective) based on the input from this sensor.

Detonation occurs due to:

- High compression ratio
- Poor quality fuel
- High level of cylinder filling
- Maximum timing advance curve
- High intake and engine temperatures
- Combustion chamber carbon build-up

The knock sensor is a piezo-electric accelerometer producing an output voltage proportional to mechanical vibration (knock) produced by the engine. Knock is caused by high-pressure waves of uncontrolled spontaneous combustion of gasses in the cylinder.

The ECM receives a signal from the knock sensor, filters out any noise and calculates if the engine is knocking. During knock the crystal oscillation increases. The ECM compares the signal received to stored signal.

From the camshaft and crankshaft signals supplying information regarding the position of the engine in its cycle, the ECM is able to determine which cylinder is knocking and retards the ignition, on that particular cylinder until the knock is eliminated. It then gradually advances the ignition again towards the original setting until knock resumes.

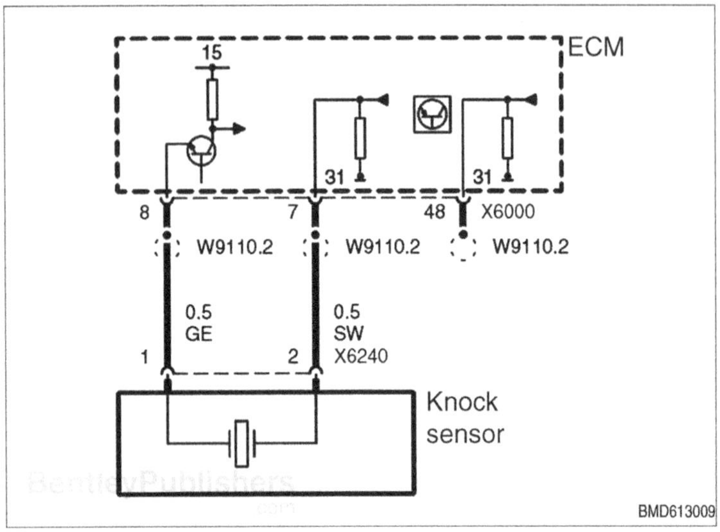

Knock sensor fault symptoms

If a knock sensor fault is set, the ECM can no longer detect and correct for engine knock. A default setting is used to ensure safe engine operation (safe ignition value). The engine lacks power and fuel consumption is affected.

Knock sensor, testing

ECM connector X6000 (81 pin, black)

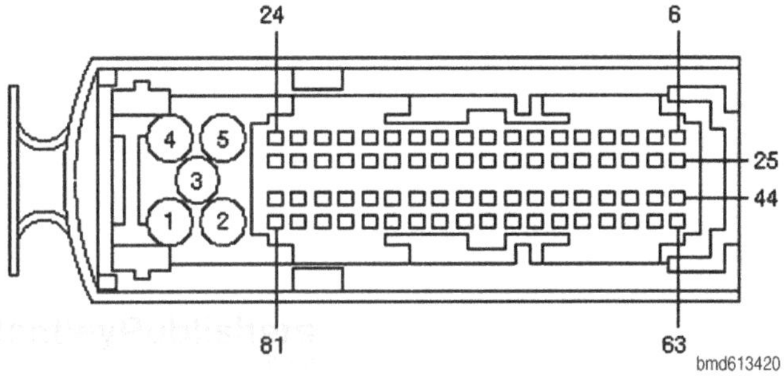

Knock sensor connector X6240

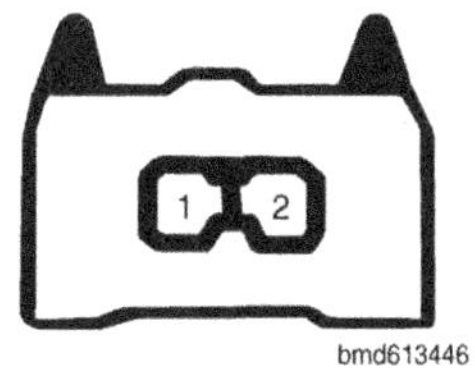

Knock sensor testing values				
Terminals at connector X6000	**Terminals at connector X6240**	**Conditions**	**Test values**	**Circuit**
7 and ground	2 and ground	key on engine off engine running at idle @2500 rpm, (no load)	0.1 volts (maximum)	sensor ground
8 and ground	1 and ground	key on engine off engine running at idle @2500 rpm, (no load)	2.5 volts	knock sensor signal
48 and ground	N-A	key on engine off engine running at idle @2500 rpm, (no load)	0 volts	knock sensor signal shield

Acceptable range shown on top waveform. Unacceptable range shown on bottom.

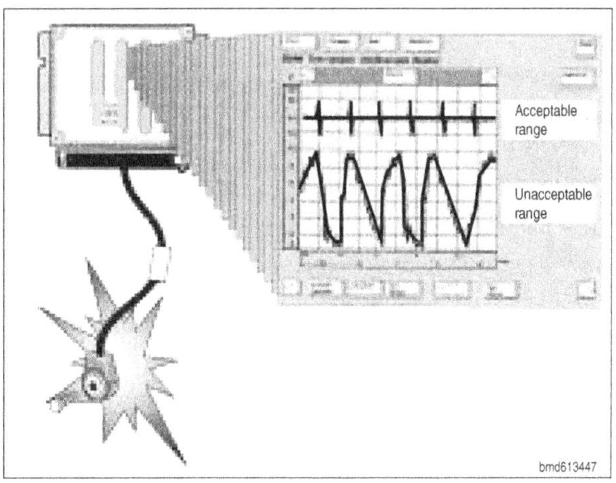

MAIN RELAY (K6300)

The main relay is energized whenever the ignition switch is ON (via terminal KL15). The main relay supplies power to the fuel injector valves, the camshaft sensor and the crankshaft sensor as well as other system components.

The main relay (K6300) is located in the fuse box in the engine compartment.

The relay has an additional protective function whereby it is not activated in the event of incorrect battery polarity.

Upon receiving the wake up signal from the ignition switch (KL15, fuse 34), the ECM provides a switched ground signal via terminal 97 to the main relay. This energizes the main relay, providing operating power to the ECM.

Output of the main relay is monitored so that if the relay is energized and power is not received by the ECM, a fault code is set.

Engine management fuses:

- F02 - ECM, fuel injectors, crankshaft sensor, ignition coils
- F03 - camshaft sensor, 02 heaters, engine cooling fan, A/C compressor relay, purge system
- F04 - automatic transmission controls
- F05 - engine coolant fan

The main relay supplies power to the following components:

- ECM
- Evaporative control leak detection pump
- Oxygen sensor heaters
- Fuel injectors
- Auxiliary cooling fan
- Crankshaft sensor
- Camshaft sensor
- Ignition coils
- ECVT / TCM

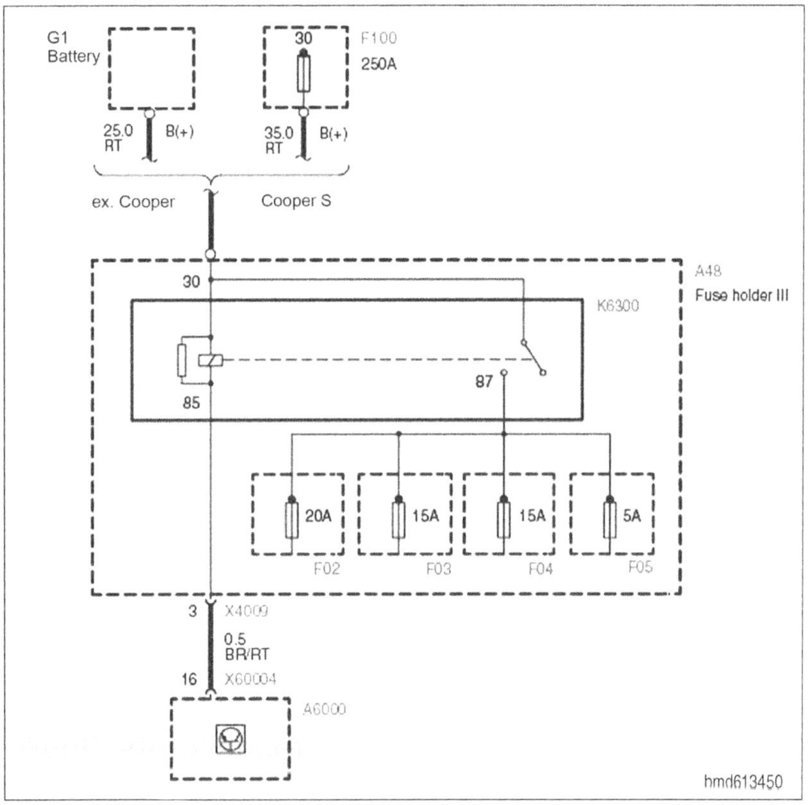

NOTE —

- The main relay remains energized for 5 minutes after the key is shut off. To prevent damaging electronic components, wait a minimum of 5 minutes before disconnecting connections to ECM. Alternately remove the main relay.

Main relay, testing

– Check for sufficient voltage supply at battery.

– Test for voltage at fuse F34 (with key ON).

– Check for voltage at KL30 terminals of main relay (large red wire, see wiring diagram).

– Remove relay and connect relay adapter in line. Check that relay operates by testing at rear of adapter.

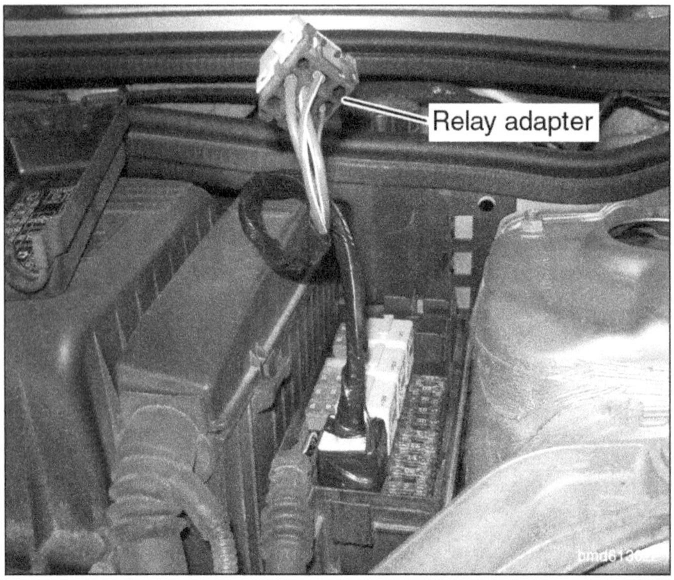

CAUTION —

- *Always check wiring diagrams for complete description of power supplies.*

ECM connector X60004 (40-pin, black)

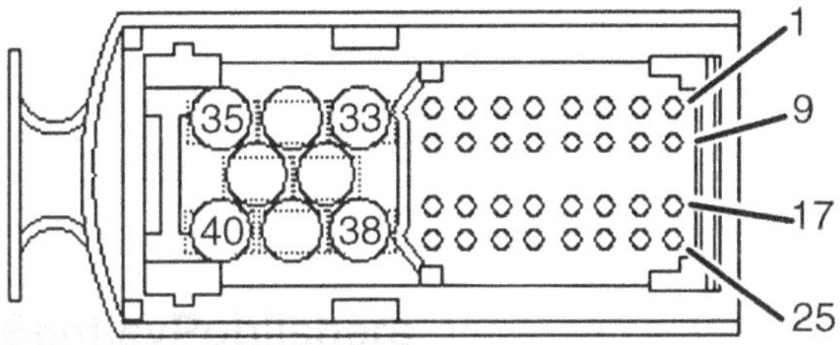

bmd613402

Main relay testing values (at ECM connector X60004)			
Terminals	**Conditions**	**Test values**	**Circuit**
16 and ground	key on engine off engine running at idle @2500 rpm, (no load)	0.9 volts, (relay activated by ECM)	relay control, ground

Main relay failure symptoms

Main relay activation is monitored by the ECM. Faults are set if the main relay is energized and power is not received by the ECM.

An open circuit in the main relay results in no voltage to the ECM and the other components supplied by the main relay. A permanently engaged main relay results in an excessive closed current draw and eventually a dead battery.

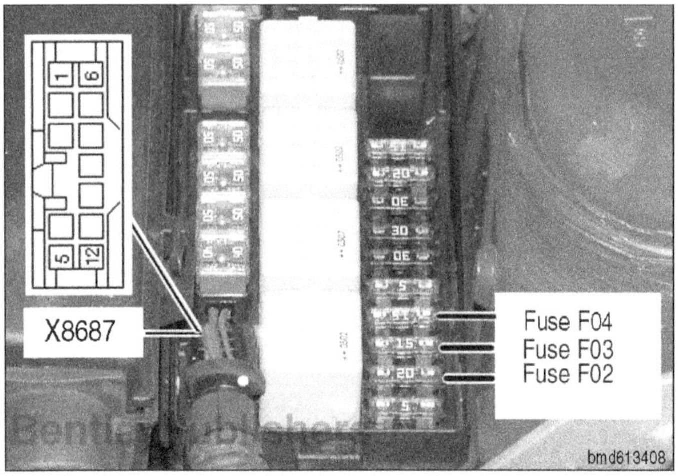

Main relay testing values (at fuse holder 3)			
Terminals	**Conditions**	**Test values**	**Circuit**
F02 and ground F03 and ground F04 and ground	key on engine off engine running at idle @2500 rpm, (no load)	battery volts, (relay activated by ECM)	relay output

OBD II DIAGNOSTIC CONNECTOR

The OBD II diagnostic connection to the ECM is through the D-bus. This allows communication with scan tools and emission related powertrain components. The D-shaped connector is located in the driver's side footwell, to the left of the pedals.

OBD II diagnostic connector pinout

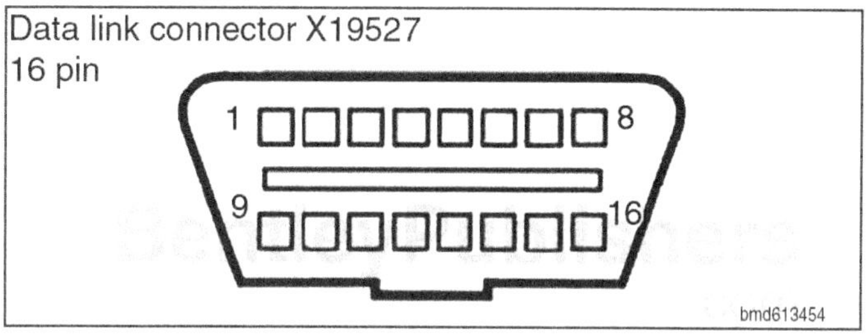

OBD II connector testing	
Terminals	**Circuit**
1 and ground	battery volts (with key on)
2 and ground	not used
3 and ground	not used
4 and ground	0.1 volts (maximum) ground
5 and ground	0.1 volts (maximum) ground
6 and ground	not used
7 and ground	signal diagnostic link
8 and ground	diagnosis bus TXD
9 and ground	engine speed
pins 10 through 15	not used
16 and ground	battery volts, (hot at all times)

OXYGEN SENSOR

The MINI COOPER and MINI COOPER S are equipped with two oxygen sensors. One is positioned upstream and the other downstream of the catalyst.

The precatalyst oxygen sensor measures the residual oxygen content of the exhaust gas. The post-catalyst sensor is used to monitor catalyst performance. The sensor produces a low voltage (0-1000 mv) proportional to the oxygen content that allows the ECM to monitor the air/fuel ratio.

The precatalytic converter oxygen sensor causes an increase in exhaust emissions if the response time is reduced or the fuel-air ratio in the closed regulating circuit is too far away from the efficiency window of the catalytic converter. A malfunction is indicated if the adaptive values of the mixture control function lie outside the specified limits.

Oxygen sensor operation

The tip of the sensor contains microporous platinum coatings (electrodes) which conducts current. The platinum electrodes are separated by a solid electrolyte which conducts oxygen ions. The platinum conductors are covered with a highly porous ceramic coating and the entire tip is encased in a ventilated metal cage.

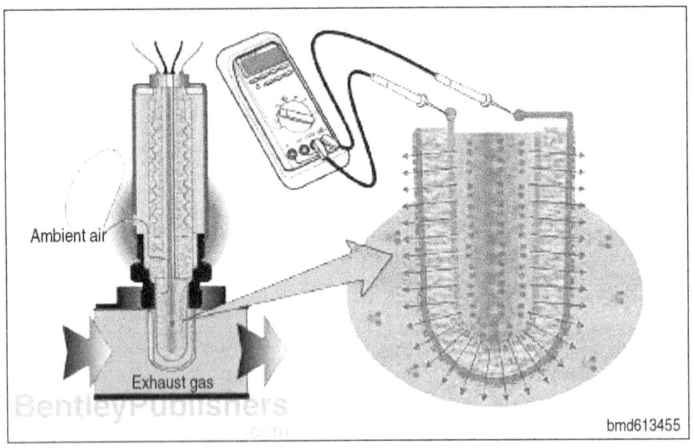

The sensor body (external) has a small vent opening in the housing that allows ambient air to enter the inside of the tip. Ambient air contains a constant oxygen content (21%) and the exhaust stream has a much lower oxygen content. The oxygen ions (which contain small electrical charges) are "purged" through the solid electrolyte by the hot exhaust gas flow. The electrical charges (low voltage) are conducted by the platinum electrodes to the sensor signal wire that is monitored by the ECM.

If the exhaust has a lower oxygen content (rich mixture), there is a large ion migration through the sensor generating a higher voltage (950 mv). If the exhaust has a higher oxygen content (lean mixture), there is a small ion migration through the sensor generating a lower voltage (80 mv).

This voltage signal is constantly changing due to combustion variations and normal exhaust pulsations.

The ECM monitors the length of time the sensors are operating in lean, rich (including the time of rise and fall) and rest conditions. The evaluation period of the sensors is over a predefined number of oscillation cycles. This conductivity is efficient when the oxygen sensor is hot (250° - 300° C). For this reason, the sensor contains a heating element. The heated sensor reduces warm up time, and retains heat during low engine speed when the exhaust temperature is cooler.

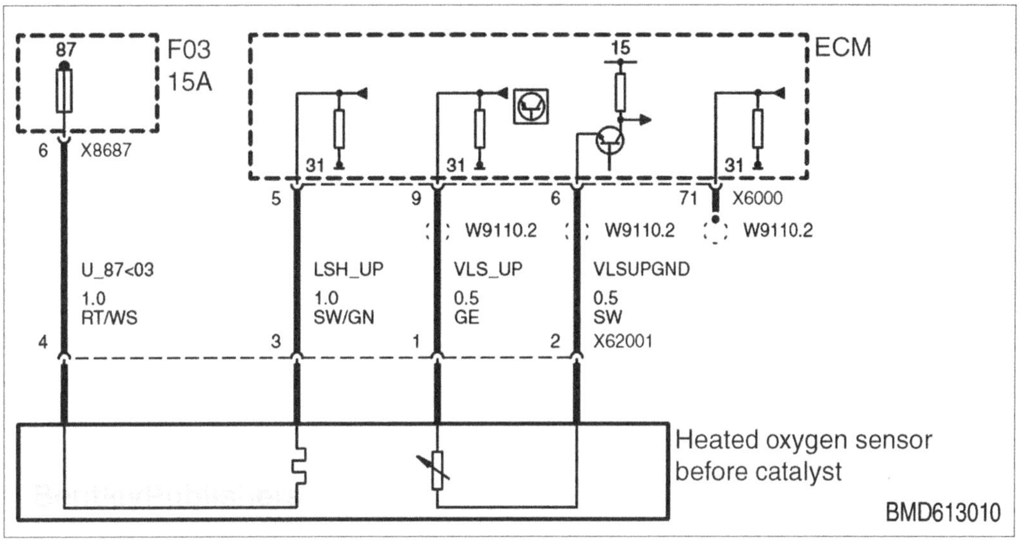

Oxygen sensor failure symptoms

In case of a precatalyst oxygen sensor failure, a fault code is stored in the ECM. The fuel injection switches to preactivation. The car can display poor running characteristics and rough idling. High exhaust emissions activates the MIL.

A post-catalyst oxygen sensor failure illuminates the MIL.

Precatalyst oxygen sensor, testing

ECM connector X6000 (81 pin, black)

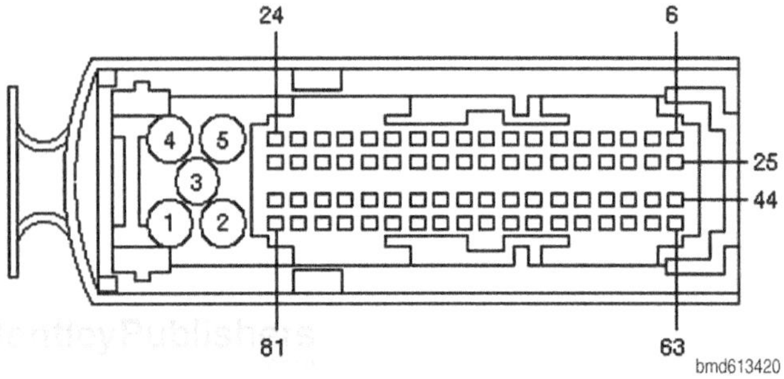

Pre-oxygen sensor connector X62001

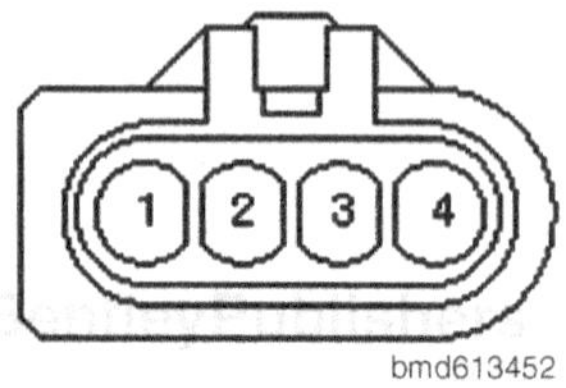

Precatalyst oxygen sensor testing values				
Terminals at connector X6000	**Terminals at connector X62001**	**Conditions**	**Test values**	**Circuit**
6 and ground	2 and ground	key on engine off engine running at idle @2500 rpm (no load)	0.1 volts (maximum)	sensor ground
9 and ground	1 and ground	key on engine off	0.4 volts (sensor circuit reference voltage)	sensor signal
9 and ground	1 and ground	engine running at idle @2500 rpm no load	0.1 to 1.0 volts	sensor signal

Post-catalyst oxygen sensor, testing

ECM connector X60004 (40-pin, black)

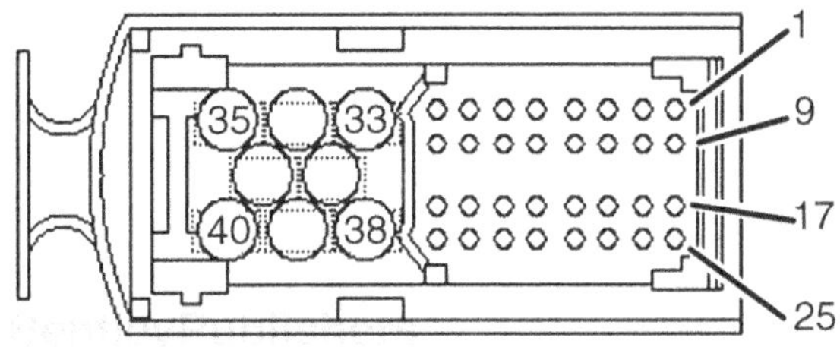

Post-oxygen sensor connector X10736

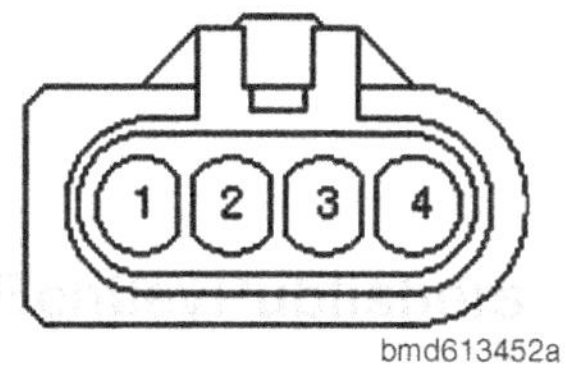

Post - catalyst oxygen sensor testing values				
Terminals at connector X60004	**Terminals at connector X10736**	**Conditions**	**Test values**	**Circuit**
28 and ground	1 and ground	key on engine off	0.4 volts (sensor circuit reference voltage)	sensor signal
28 and ground	1 and ground	engine running at idle @2500 rpm no load	0.1 to 1.0 volts	sensor signal
31 and ground	2 and ground	key on engine off engine running at idle @2500 rpm no load	0.1 volts (maximum)	sensor ground

Oxygen sensor waveform engine running at idle

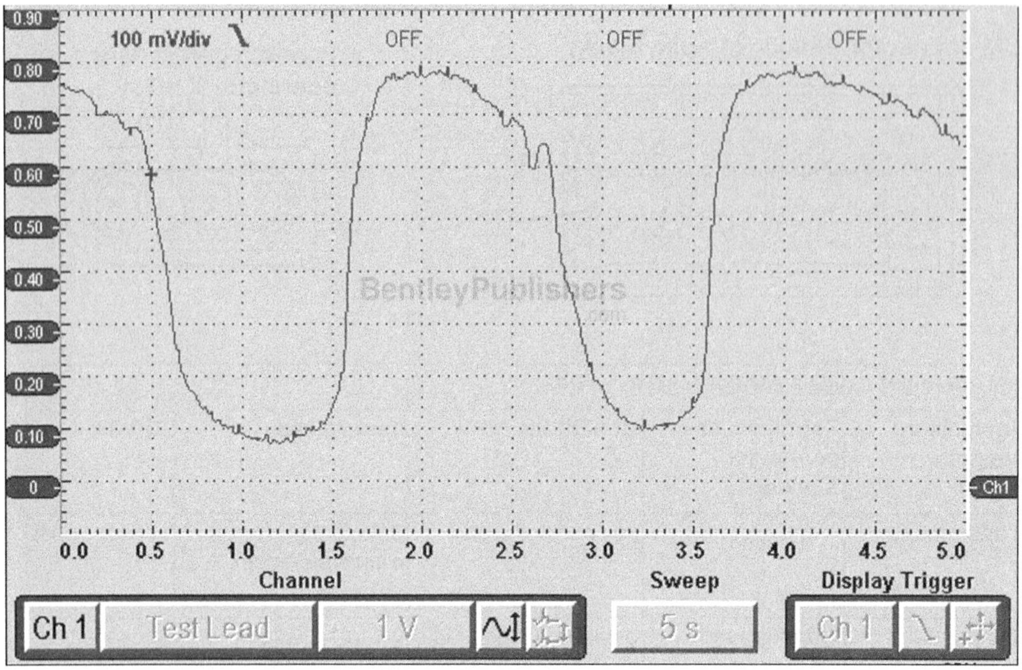

Oxygen sensor heating

The oxygen sensor conductivity is efficient when it is hot (250°- 300° C). For this reason, the sensors contain heating elements. Heated sensors reduce warm-up time, and retain heat during low engine speed when the exhaust temperature is cooler. OBD II requires monitoring of the oxygen sensor heating function and heating elements.

The two oxygen sensor heating circuits receive operating voltage from the main relay when KL15 is switched ON. Each of the sensor heaters is controlled through separate final stage transistors.

The sensor heaters are controlled with a pulse width modulated (PWM) ground during a cold start. This allows the sensors to be brought up to operating temperature without the possibility of thermal shock. The duty cycle is then varied to maintain the heating of the sensors.

When the engine is decelerating (closed throttle), the ECM increases the duty cycle of the heating elements to compensate for decreased exhaust temperature.

Precatalyst oxygen sensor heater circuit, testing

ECM connector X6000 (81 pin, black)

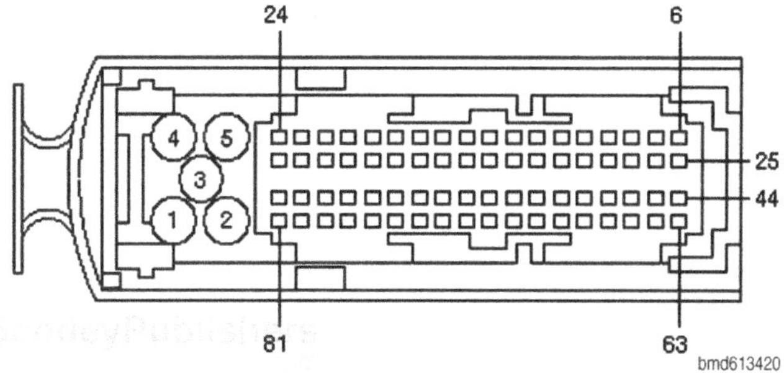

Pre-oxygen sensor connector X62001

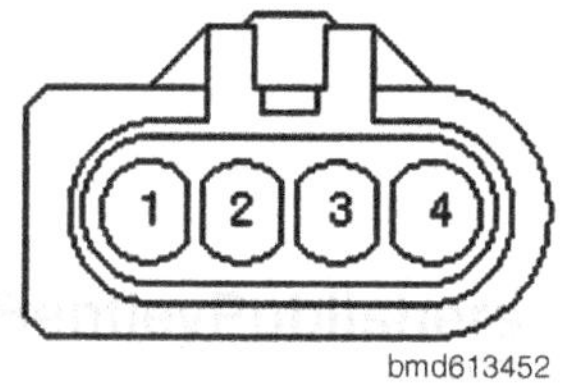

Precatalyst oxygen sensor heater testing values				
Terminals at connector X6000	**Terminals at connector X62001**	**Conditions**	**Test values**	**Circuit**
5 and ground	3 and ground	key on engine off	battery volts	heater control
5 and ground	3 and ground	engine running at idle @2500 rpm no load	6.0 volts, (dependant on battery volts and heater duty cycle)	heater control
N-A	4 and ground	key on engine off engine running at idle @2500 rpm, (no load)	battery volts	voltage supply

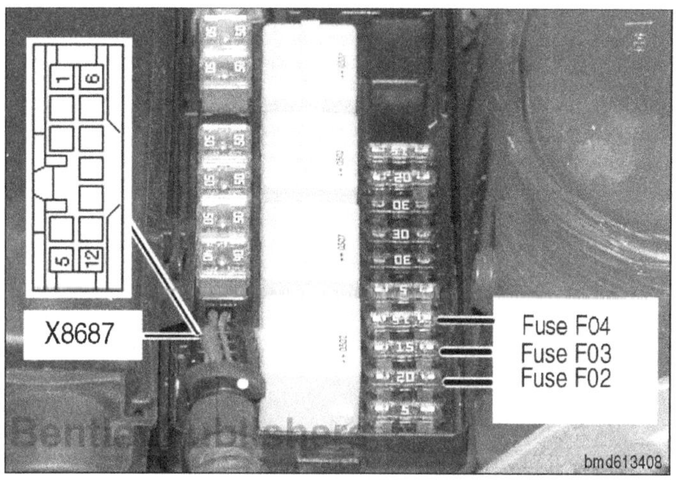

Precatalyst oxygen sensor heater testing values (at fuse holder 3 connector X8687)			
Terminals	**Conditions**	**Test values**	**Circuit**
6 and ground	key on engine off engine running at idle @2500 rpm (no load)	battery volts	voltage supply
fuse F03 and ground	key on engine off engine running at idle @2500 rpm (no load)	battery volts	voltage supply

Post-catalyst oxygen sensor heater circuit, testing

ECM connector X6000 (81 pin, black)

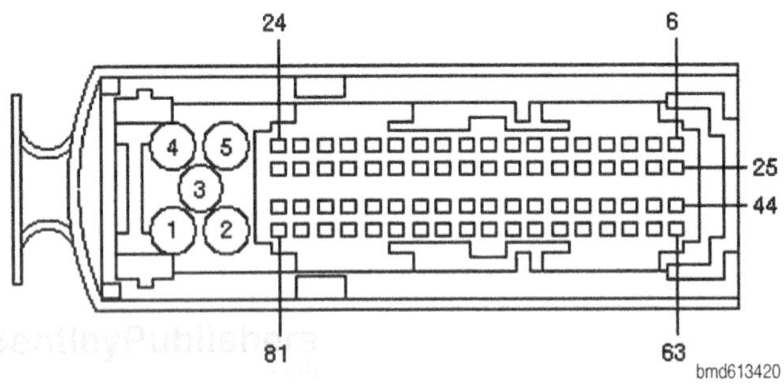

Post-oxygen sensor connector X10736

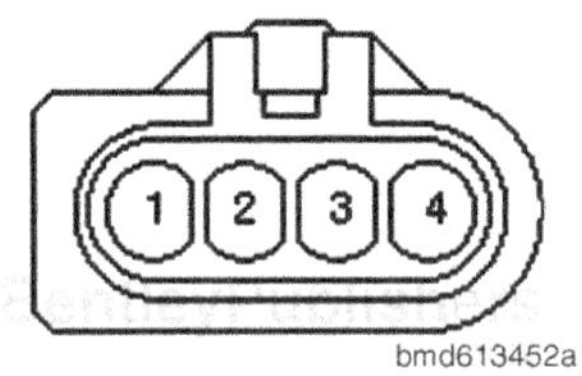

Post-catalyst oxygen sensor heater testing values				
Terminals at connector X6000	**Terminals at connector X10738**	**Conditions**	**Test values**	**Circuit**
43 and ground	3 and ground	key on engine off	battery volts	heater control
43 and ground	3 and ground	engine running at idle @2500 rpm (no load)	6.0 volts, (dependant on battery volts and heater duty cycle)	heater control
N-A	4 and ground	key on engine off engine running at idle @2500 rpm, (no load)	battery volts	voltage supply

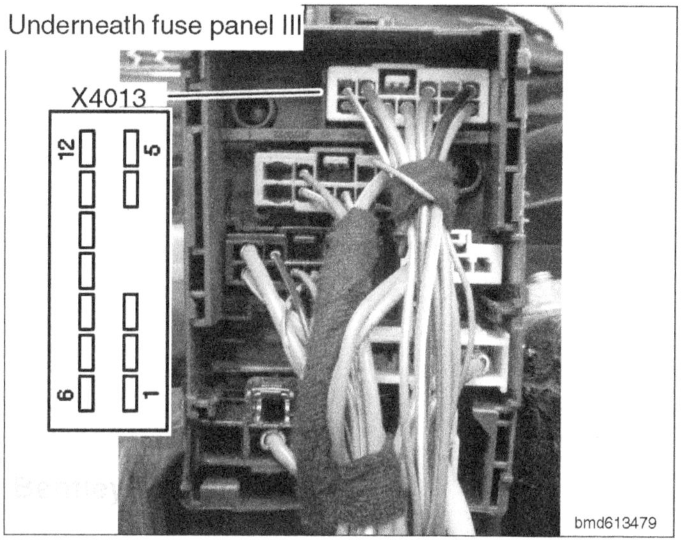

Post-catalyst oxygen sensor heater testing values (at fuse panel III connector X4013)			
Terminals	**Conditions**	**Test values**	**Circuit**
10 and ground	key on engine off engine running at idle @2500 rpm (no load)	battery volts	voltage supply
fuse F03 and ground	key on engine off engine running at idle @2500 rpm (no load)	battery volts	voltage supply

Oxygen sensor heater waveform at idle

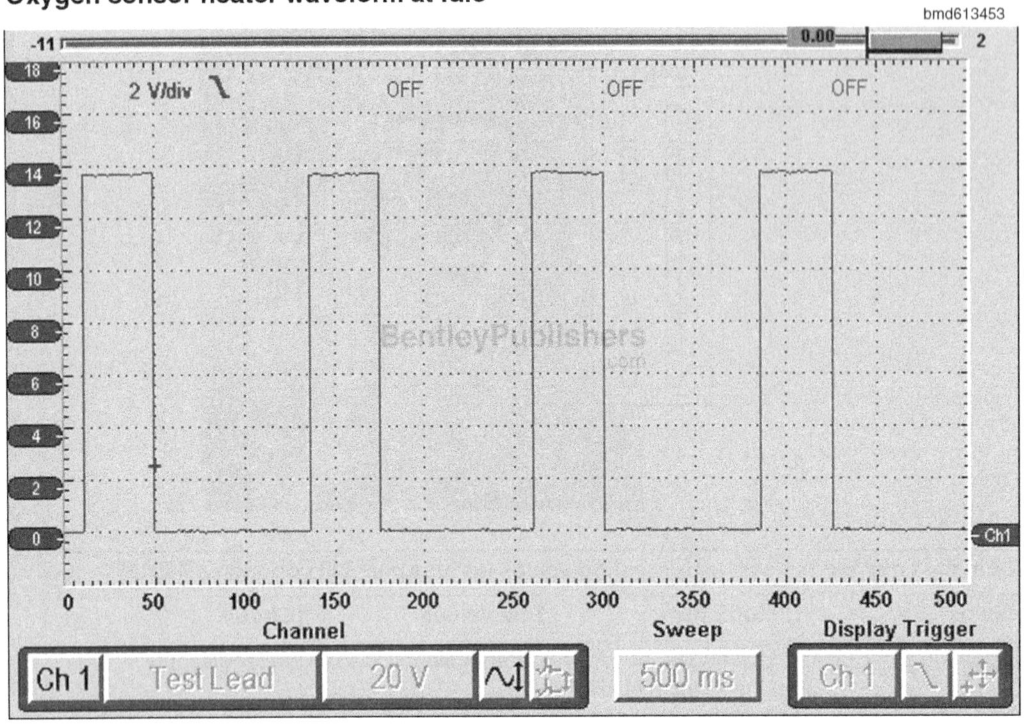

LEAK DETECTION PUMP (LDP) REED SWITCH

Vapors containing hydrocarbons form in the vehicles fuel tank. To prevent hydrocarbons from venting to atmosphere, legislation in the USA requires on-board monitoring of the fuel system sealing on all vehicles powered by an internal combustion engine. Beginning with model year 2000 these regulations were tightened and call for detection of a 0.5 mm (0.02 inch) leak.

The EVAP system leak detection Pump (LDP) is located above the right rear inner fender liner. The LDP is an electrically/vacuum-actuated device that pressurizes the evaporative emission system for the purpose of detecting leaks and verifying charcoal canister purge valve integrity. It has an integrated canister vent valve (CVV) that controls the atmospheric venting of the fuel vapor storage canister.

The LDP assembly is only replaceable as a complete unitized component. It is separate from the charcoal canister.

The upper chamber of the LDP contains an integrated reed switch that produces a switched high/low voltage signal monitored by the ECM. The switch is opened by the magnetic interruption of the metal rod connected to the diaphragm when in top dead center position.

The repetitive up/down stroke confirms to the ECM that the valve is functioning and the basis for determining if a leak is present in the system. The ECM monitors the length of time it takes for the reed switch to open, which is opposed by pressure under the diaphragm in the lower chamber. If this component or its circuits are defective, a fault code sets and the MIL illuminates.

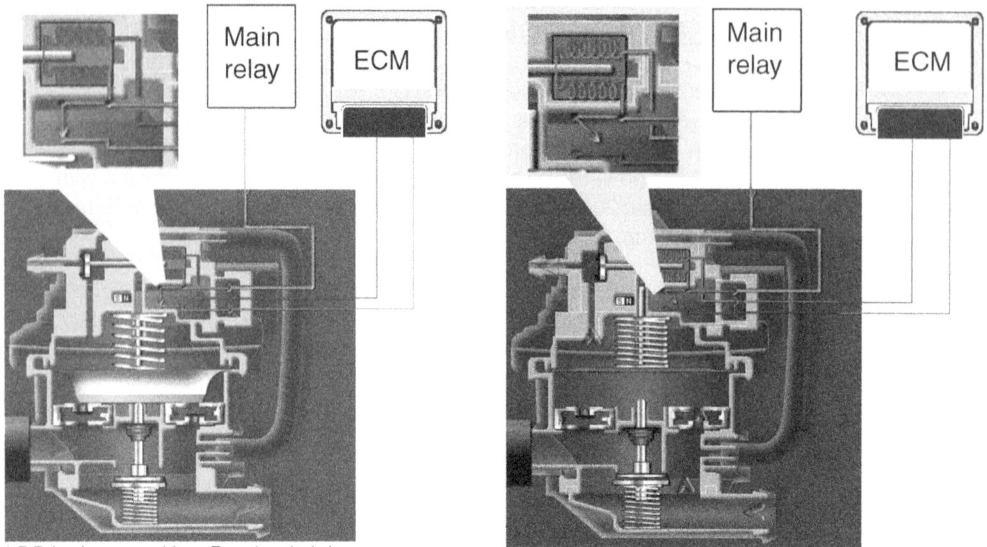

LDP in down position. Reed switch is closed.

LDP in up position. Reed switch is open.

bmd613458

LDP reed switch, testing

The LDP solenoid is toggled to ground by the ECM. Confirm solenoid activation by monitoring voltage at connector X6000 pin 42. Battery volts is present when solenoid is OFF.

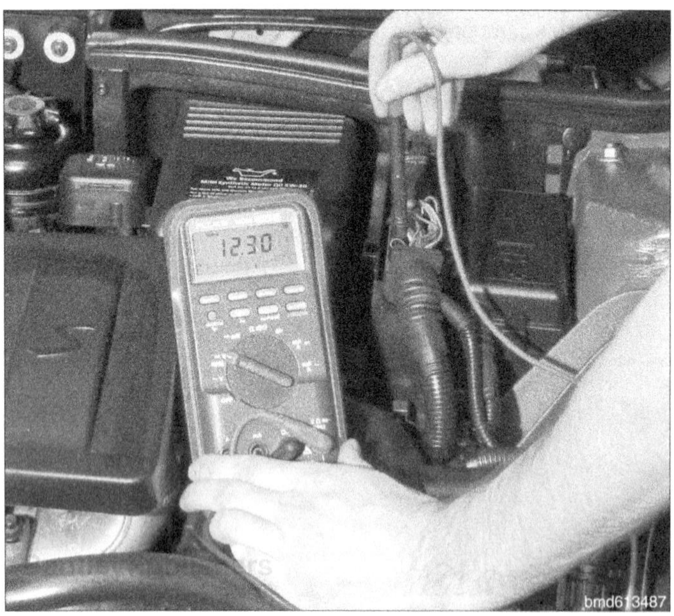

bmd613487

The reed switch toggles a 12 volt signal. When the LDP diaphram is in the up position, the reed switch is open and the switch voltage is 0.

When the LDP diaphram is in the down position, the reed switch is closed and switch voltage is battery volts.

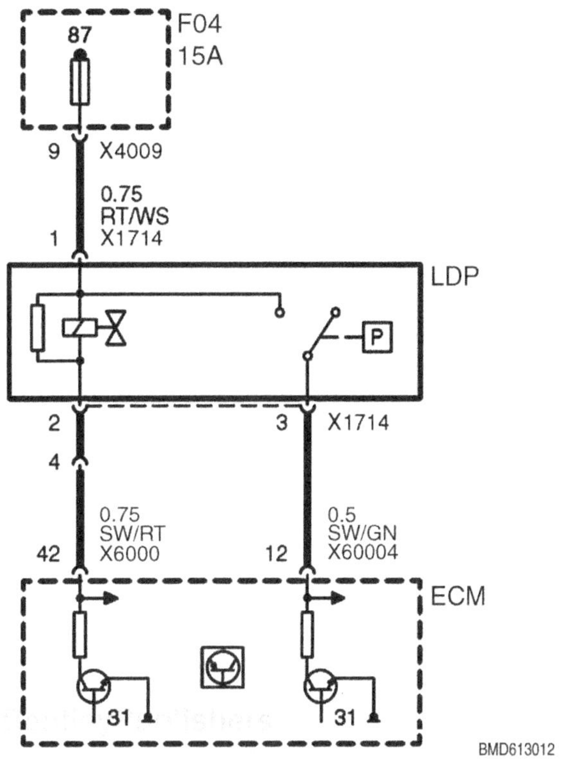

ECM connector X60004 (40-pin, black)

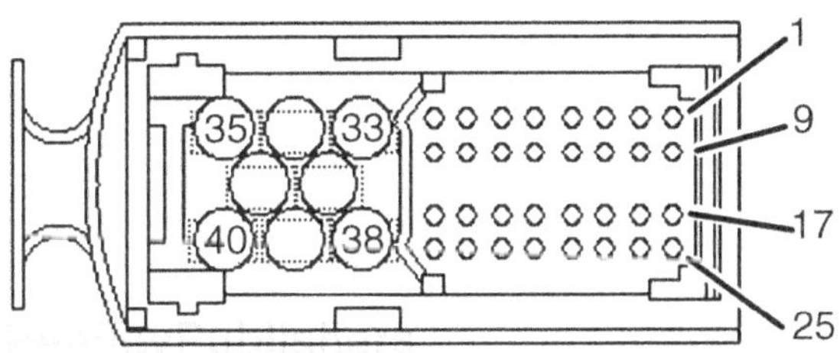

LDP connector X1714

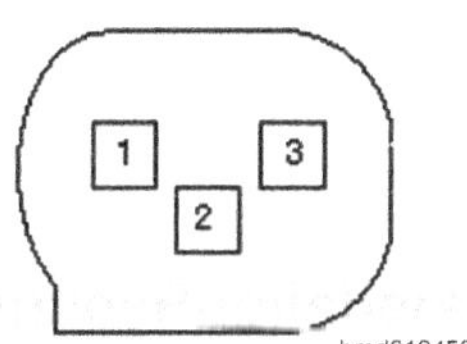

LDP reed switch testing values				
Terminals at connector X60004	**Terminals at connector X1714**	**Conditions**	**Test values**	**Circuit**
N-A	1 and ground	key on engine off engine running at idle @2500 rpm (no load)	battery volts	voltage supply
12 and ground	3 and ground	switch open	0.1 volts	switch signal
12 and ground	3 and ground	switch closed	battery volts	switch signal

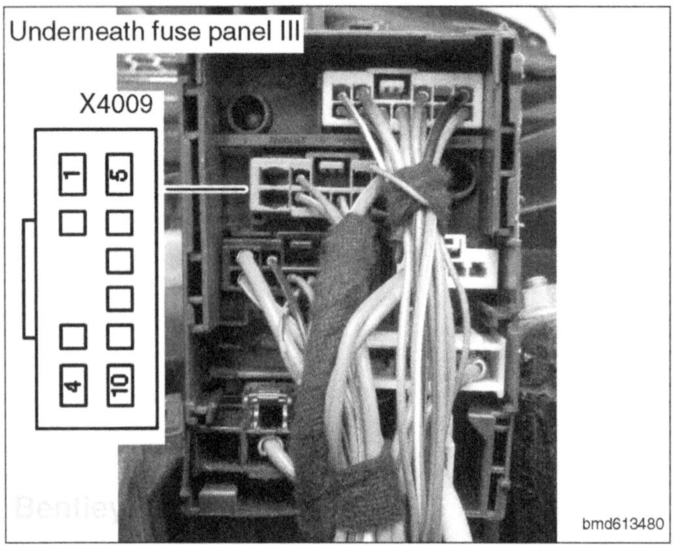

LDP reed switch testing values (at fuse panel III connector X4009)			
Terminals	**Conditions**	**Test values**	**Circuit**
9 and ground	key on engine off engine running at idle @2500 rpm (no load)	battery volts	voltage supply
fuse F04 and ground	key on engine off engine running at idle @2500 rpm (no load)	battery volts	voltage supply

SUPERCHARGER OPERATION

The supercharger is a positive displacement pump. Its purpose is to increase air pressure and density in the intake manifold. The supercharger is matched to the engine by its displacement and belt ratio (driven from the crankshaft).

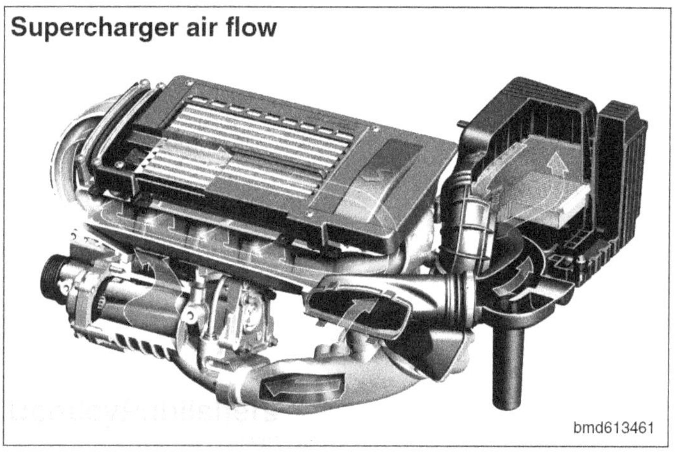

If a fault is suspected with the supercharger, inspect crankshaft pulley. The rubber dampening material between crankshaft connection and drive belt can fail, causing the drive belt to rotate slower than the crankshaft. A failed pulley can cause low boost and multiple driveability complaints.

To inspect, mark pulley with chalk line from crankshaft bolt to drive belt. Run engine and inspect marks. Chalk marks misalign if pulley fails.

The supercharger incorporates a bypass valve. This is actuated by a vacuum tube near the throttle body and recirculates the supercharger air when boost is not required. During typical driving conditions the intake manifold is under pressure for only 5% of the time. For the remaining time, the intake manifold is under vacuum (negative pressure), allowing for better fuel economy and a quieter ride.

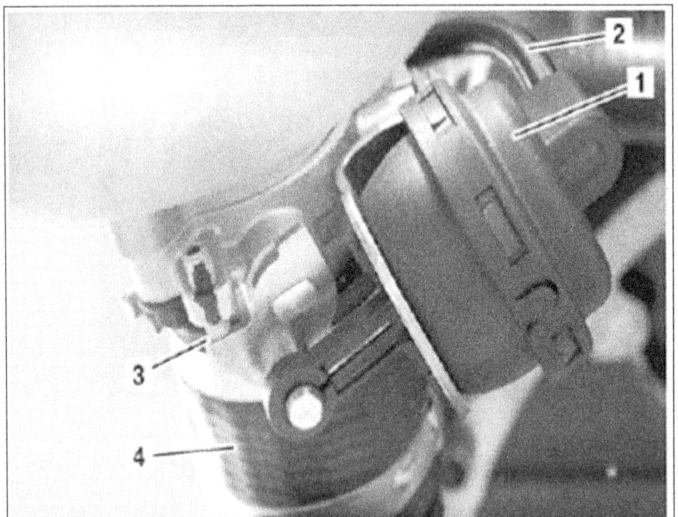

1. Vacuum unit
2. Vacuum supply
3. Butterfly stop
4. Outlet duct to intake

bmd613462

Supercharger bypass valve, testing

The bypass valve is a normally closed. When vacuum is applied, the valve opens and boost pressure is recirculated. Valve operation can be checked with a hand operated vacuum pump.

If the valve fails to return fully to the closed position, a low boost / low power situation is present. When closed the butterfly lever should rest on the stop screw.

The image below shows a faulty valve that is not returning to the correct closed position (failed internal return spring). This internal spring works against the vacuum diaphram to hold the butterfly closed.

bmd613463

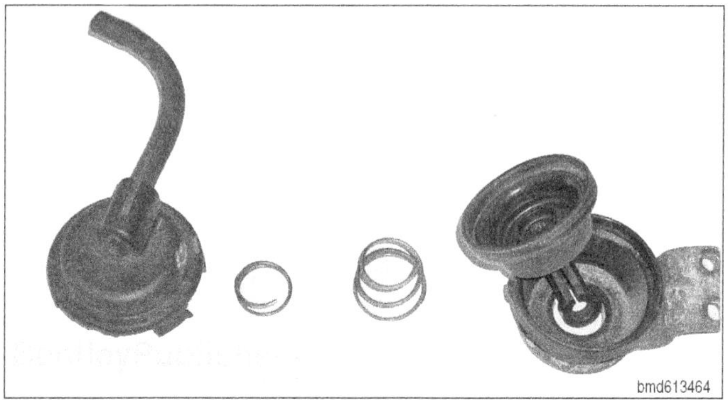

bmd613464

TMAP SENSOR (COOPER)

TMAP (temperature and manifold absolute pressure) sensor contains a manifold absolute pressure (MAP) sensor and an intake air temperature sensor.

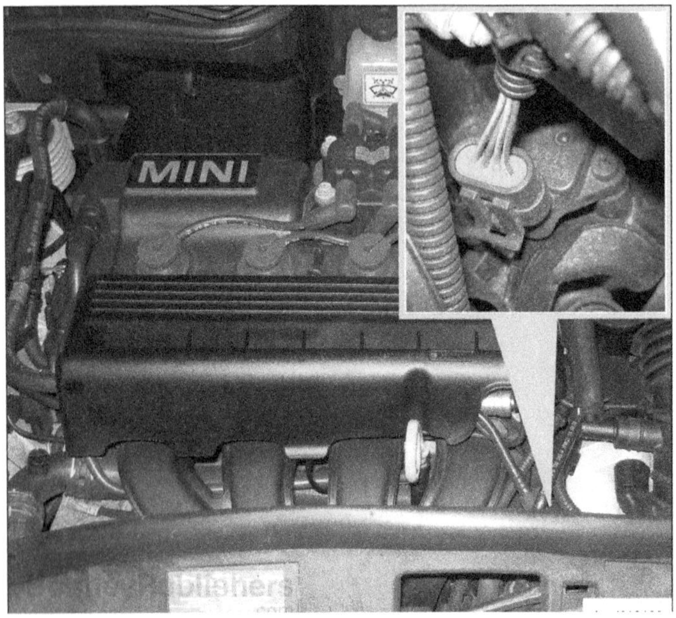

MAP sensor

The MAP sensor is a piezo resistive pressure gauge that supplies an analog input to the ECM corresponding to the absolute manifold pressure. The signal is used to determine engine load, used for internal engine control via the ECM and for ASC/DSC (traction control). This sensor has a maximum pressure range of 120 kpa (17.4 psi).

The MAP sensor measures the changes in the intake manifold pressure which result from engine load (intake manifold vacuum) and rpm changes and converts these into a voltage output.

By monitoring the sensor output voltage, the ECM knows the manifold pressure. At idle manifold pressure is lower (high vacuum) and output voltage is about 1 to 2 volts, at higher pressure or at wide open throttle (lower vacuum) output voltage is about 4 volts. The ECM uses the MAP sensor to control fuel delivery and ignition timing.

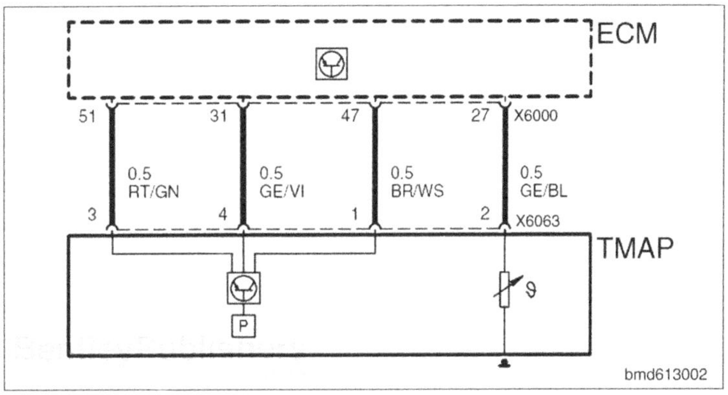

TMAP sensor (Cooper), testing

ECM connector X6000 (81 pin, black)

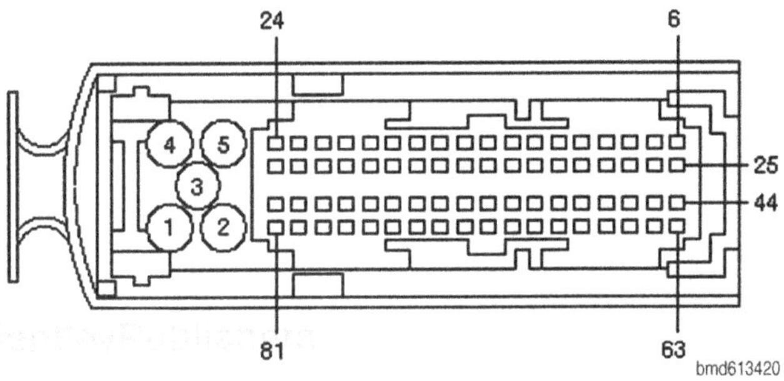

TMAP connector X60631

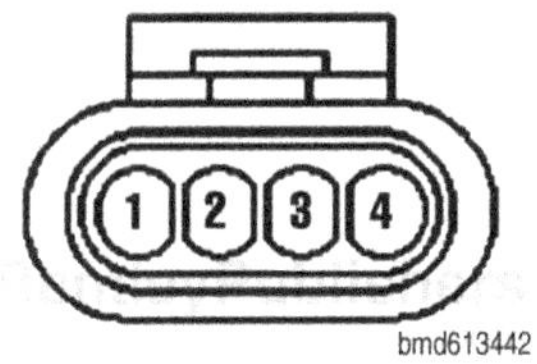

TMAP testing values				
Terminals at connector X6000	**Terminals at connector X60631**	**Conditions**	**Test values**	**Circuit**
27 and ground	2 and ground	key on engine off	3.9 volts. (60° F ambient temp)	IAT signal
27 and ground	2 and ground	engine running at idle	1.25 volts. (60° ambient temp, engine at operating temp)	IAt signal
31 and ground	4 and ground	key on engine off	4.0	MAP signal
31 and ground	4 and ground	engine running at idle	1.25 volts (18 in. vac.)	map signal
47 and ground	1 and ground	key on engine off engine running at idle @2500 rpm no load	0.1 volts (maximum)	sensor ground
51 and ground	3 and ground	key on engine off engine running at idle @2500 rpm no load	5 volts (approx.)	reference voltage

TMAP AND MAP SENSORS (COOPER S)

In the MINI Cooper S, the pressure differential is measured across the supercharger to determine the manifold air density. To measure this differential two sensors are fitted, one on either side of the supercharger.

The TMAP sensor is fitted on the manifold pressure side and has the same hardware specification as the MINI Cooper sensor but the maximum pressure range is 250 kpa.

The MAP sensor is fitted upstream of the TMAP sensor between the supercharger and the throttle body and measures atmospheric pressure. This sensor is the same as the TMAP sensor but does not sense temperature.

The TMAP sensor is exposed to higher than atmospheric pressures as produced by the supercharger.

The job of the MAP sensor is the measurement of barometric pressure. It also measures the absolute pressure in the intake manifold. Absolute pressure is barometric pressure minus the vacuum created by the pistons. So, if the barometer is reading 1.5 Bar at sea level and the manifold vacuum (gauge) is 1.0 Bar at idle, the manifold absolute pressure would be 0.5 Bar.

The ECM compares the voltage outputs from the MAP sensor (ahead of the supercharger) to the outputs received from the TMAP sensor (subject to increased pressures from the supercharger) and calculates air volume drawn into the engine.

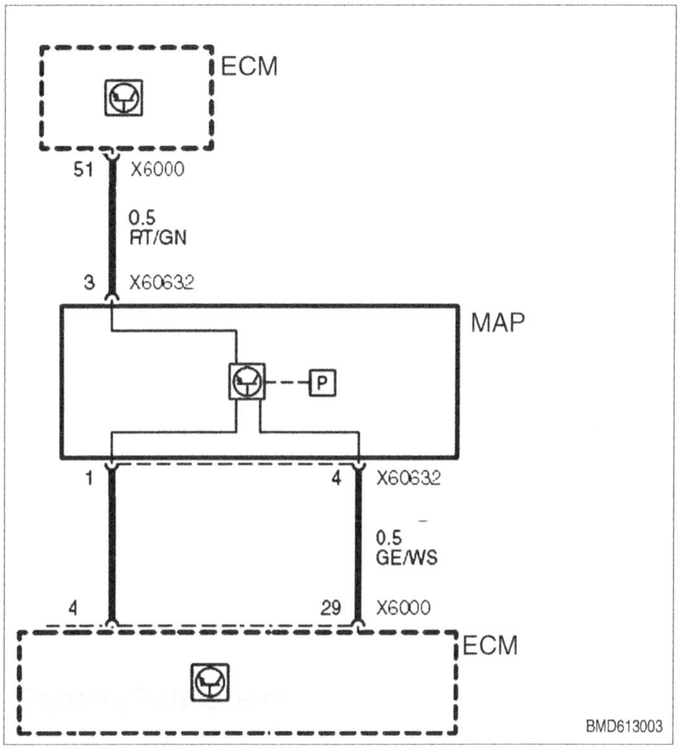
ECM
51
X6000
0.5
RT/GN
3
X60632
MAP
P
1
4
X60632
0.5
GE/WS
4
29
X6000
ECM
BMD613003

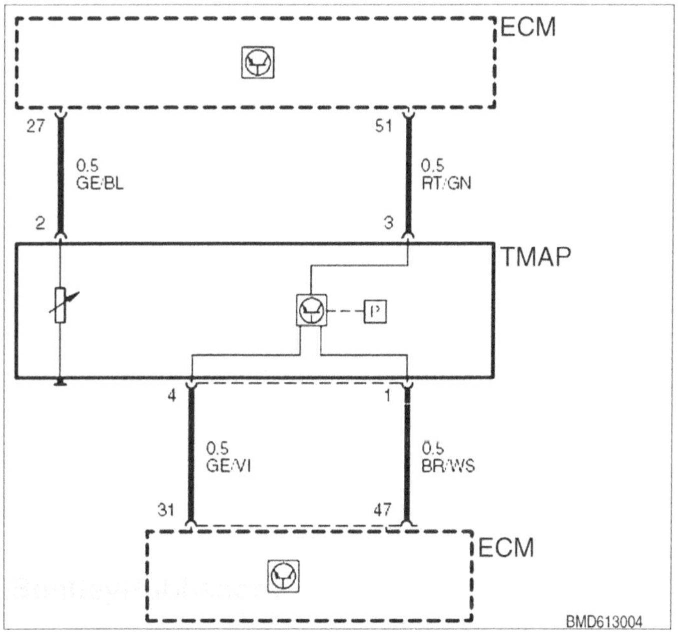
ECM
27
51
0.5
GE/BL
0.5
RT/GN
2
3
TMAP
P
4
1
0.5
GE/VI
0.5
BR/WS
31
47
ECM
BMD613004

TMAP and MAP sensors failure symptoms

Engine difficult to start, stalls at idle and misfires.

TMAP and MAP sensor (Cooper S), testing

Check status on scan tool (0 - 5 volts). Ensure that voltage changes and does not remain constant at 0 or 5 volts. Voltage reading of 0.6-1.5 volts indicates high vacuum condition (Idle or no load).

Voltage reading of 4 volts indicates low vacuum condition (full throttle).

ECM connector X6000 (81 pin, black)

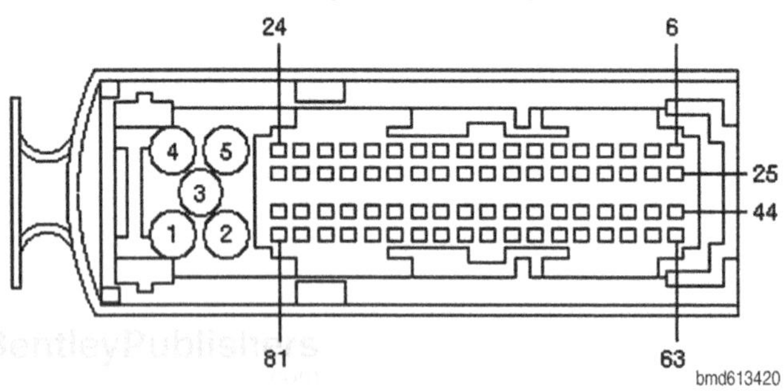

TMAP connector X60631

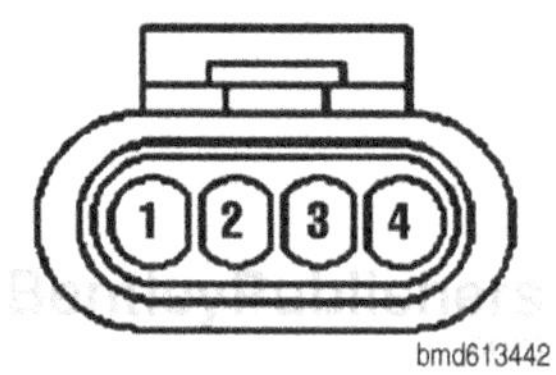

TMAP (after compressor) testing values				
Terminals at connector X6000	**Terminals at connector X60631**	**Conditions**	**Test values**	**Circuit**
27 and ground	2 and ground	key on engine off	3.9 volts. (60° F ambient temp)	IAT signal
27 and ground	2 and ground	engine running at idle	1.25 volts. (60° ambient temp, engine at operating temp)	IAt signal
31 and ground	4 and ground	key on engine off	1.90 volts	tmap signal
31 and ground	4 and ground	engine running at idle	0.9 volts, (18 in. vac.)	tmap signal
31 and ground	4 and ground	@2500 rpm, (no load)	0.6 volts (22 in. vac.)	tmap signal
31 and ground	4 and ground	bench testing 5 psi	2.5 volts	tmap signal
47 and ground	1 and ground	key on engine off engine running at idle @2500 rpm no load	0.1 volts (maximum)	sensor ground
51 and ground	3 and ground	key on engine off engine running at idle @2500 rpm no load	5 volts (approx.)	reference voltage

ECM connector X6000 (81 pin, black)

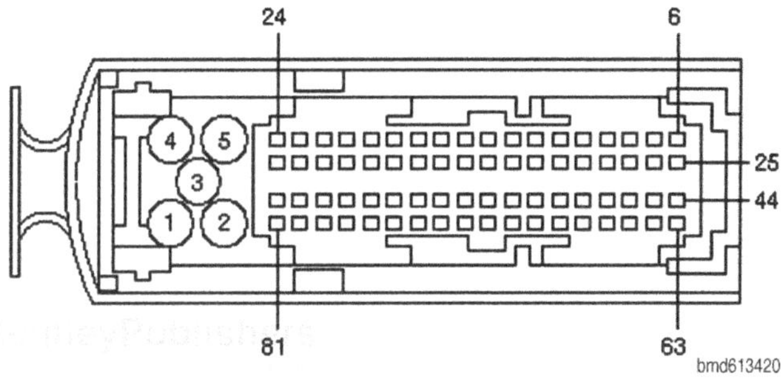

TMAP connector X60632

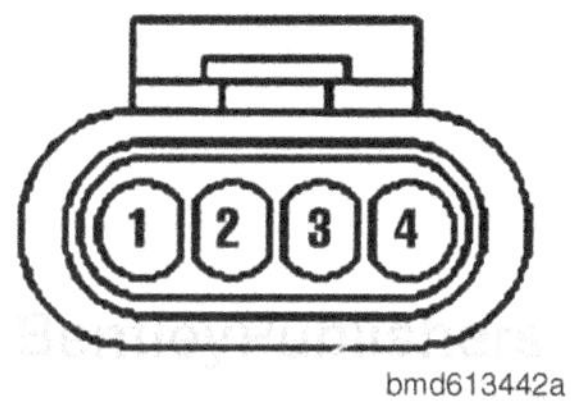

MAP sensor (before compressor) testing values				
Terminals at connector X6000	**Terminals at connector X60632**	**Conditions**	**Test values**	**Circuit**
29 and ground	4 and ground	key on engine off	4.0 volts	map signal
29 and ground	4 and ground	engine running at idle	1.8 volts (18" vacuum)	map signal
29 and ground	4 and ground	@2500 rpm, (no load)	1.1 volts (22" vacuum)	tmap signal
47 and ground	1 and ground	key on engine off engine running at idle @2500 rpm no load	0.1 volts maximum	sensor ground, (spliced with tmap)
51 and ground	3 and ground	key on engine off engine running at idle @2500 rpm no load	5 volts (approx.)	reference voltage

In certain conditions, the MAP (manifold pressure) sensor is checked for plausibility (coherent value vs. engine speed and throttle opening). These conditions are:

- MAP too low, engine stopped (in these conditions, MAP cannot be lower than the minimum ambient pressure value).
- MAP too low at idle, engine running (in these conditions, the engine cannot be run with too low manifold pressure).
- MAP too low at full load, low engine speed (in these conditions, MAP cannot be lower than the minimum ambient pressure value).
- MAP too high in deceleration (ECM calibration is tuned so that MAP target value is 200 hPa in deceleration). In case of error on MAP acquisition, the MAP information is built up by using engine speed and throttle position information.

TRANSMISSION OIL TEMPERATURE SENSOR

The transmission oil temperature sensor, a two wire sensor located in the valve block body, continuously monitors the temperature of the oil. Should oil temperature rise above preset parameters, the ECM reduces the amount of slip within the clutch to reduce oil temperature.

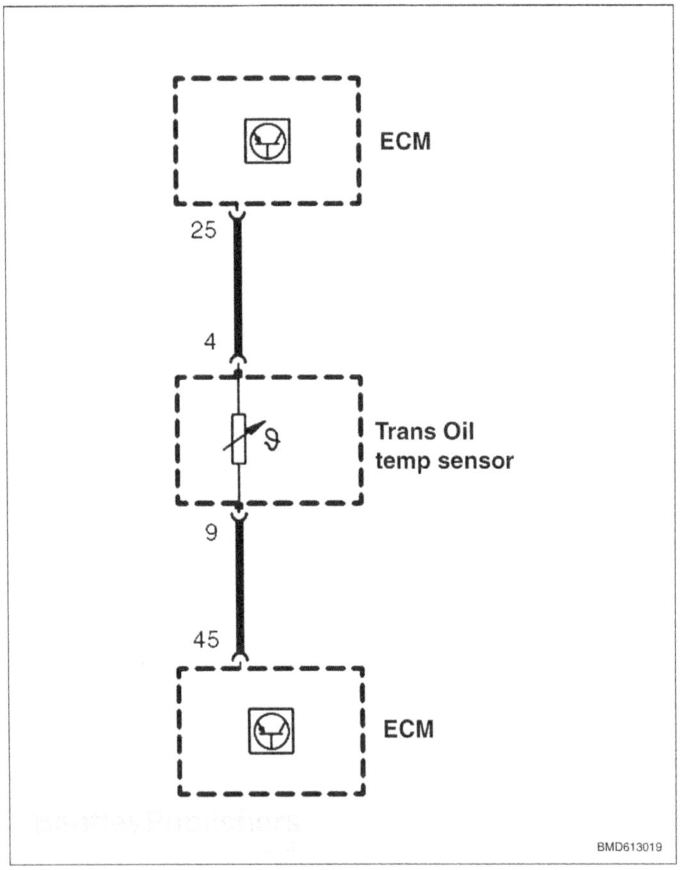

FUEL INJECTION MONITORING

Fuel injection control monitoring is performed by the ECM for OBD II requirements. Faults with the fuel injectors and / or control circuits are stored in memory. This monitoring includes:

- Closed loop operation
- Oxygen sensor feedback

These additional corrections are factored into the calculated injection time. If the correction factor exceeds set limits, a fault is stored in memory. When the criteria for OBD II monitoring is achieved, the MIL is illuminated.

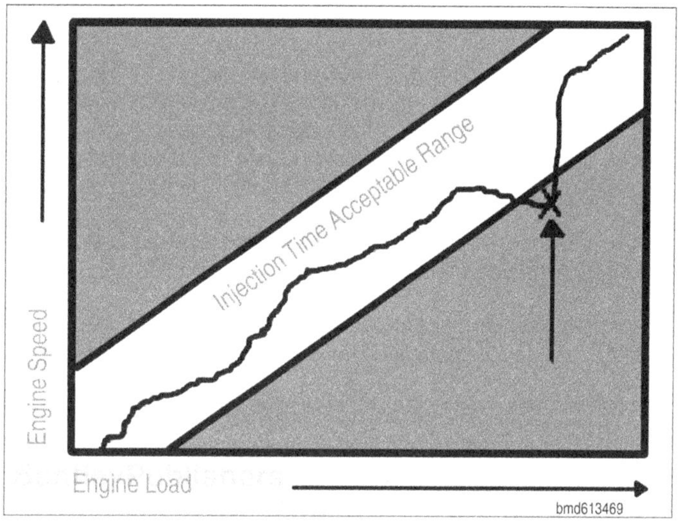

Adaptation values are stored by the ECM in order to maintain an ideal air / fuel ratio. Within the range of adaptation, the ECM modifies the injection rate during two engine operation conditions:

- Idle and low load mid range speeds (additive adaptation).
- Operation under normal load to higher load at higher engine speeds (multiplicative adaptation).

These values indicate how the ECM is compensating for a less than ideal initial air/fuel ratio.

If the adaptation value is greater than 0.0 ms additive or 0% multiplicative, the ECM is trying to richen the mixture.

If the adaptation value is less then 0.0 ms additive or 0% multiplicative, the ECM is trying to lean out the mixture.

Adaptation makes it possible for the engine control module to learn certain values from components and equipment variants, thus making it possible to compensate for certain component tolerances.

The fuel-air mixture formed in the intake tract requires a certain period of time until it reaches the oxygen sensor in the form of exhaust gas. This time decreases as load and engine speed increase. For this reason, the response time of the mixture control system is also dependent on load and engine speed.

Fuel-air mixture deviations detected by the oxygen sensor result in adaptation values (learned correction values) being stored. By way of the adaptations, the injection can be brought close to the nominal value in advance. A reduction in the response time is achieved in this way.

For example, mixture control would have to constantly increase the injection time if the basic injection values of the control map for the engine control module are too low at idle speed in order to maintain the ideal fuel-air mixture. In this case, an adaptation value is learned, which corrects the basic injection value. Mixture control then only needs to undertake the fine adjustment.

Longer operation between idle speed and partial load is necessary in order to learn the mixture adaptation values.

These adaptation values can be reset to their original status with the Reset adaptation values service function and they must be relearned afterwards.

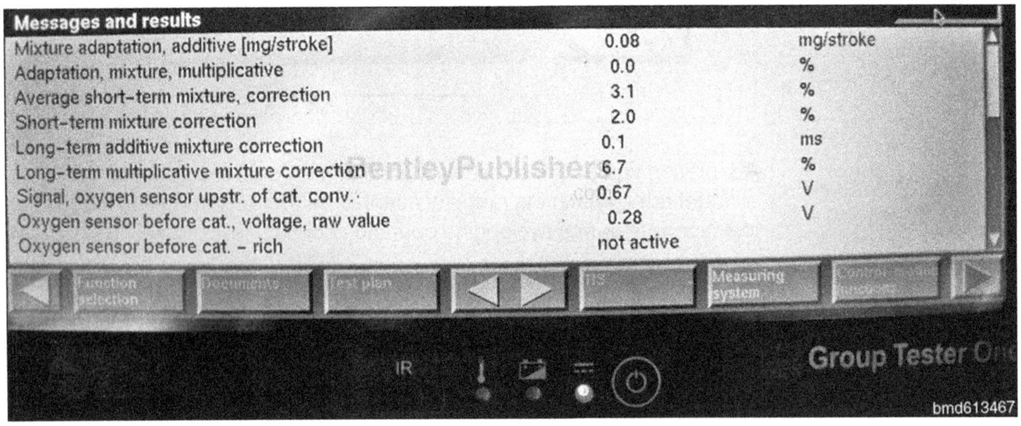

An additive fuel adaptation malfunction is most likely caused by a fault at idle, such as a vacuum leak.

A multiplicative fuel adaptation malfunction is most likely be cause by an item affecting the engine throughout all rpm ranges, such as fuel pressure, dirty injectors, etc.

ECM PINOUT

ECM connector X60004 (40-pin, black)

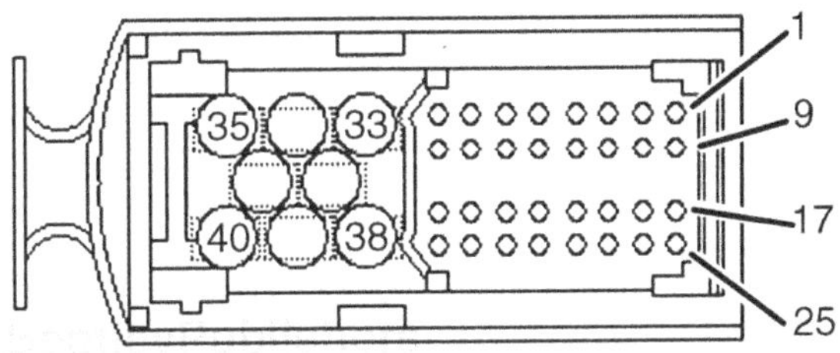

Engine Control Module (ECM) Pinout - connector X60004

PIN	Wire color	Signal description	KOEO	Idle
1	yel	PWG sensor 1 ref. voltage	5 volts	5 volts
2	gry/wht	PWG sensor 2 ref. voltage	5 volts	5 volts
3		not used		
4	blu/brn	Clutch position switch	0.02 volts, battery volts clutch depressed	0.02 volts, battery volts clutch depressed
5	brn/grn	PWG sensor 1 signal ground	0.01 volts	0.01 volts
6	brn	PWG sensor 2 signal ground	0.01 volts	0.01 volts
7	brn/yel	Activation, cooling fan	battery volts	battery volts fan off, 0.2 volts fan activated
8		not used		
9	blk/gry	A/C pressure sensor signal >1.5 volts activates cooling fan	0.8 volts	0.9 volts A/C off, 1.2 volts A/C on
10	blk/grn	A/C pressure sensor ref. volts	5 volts	5 volts
11		not used		
12	blk/grn	Signal LDP reed switch, toggles to ground during leak test	0.0 volts	12 volts
13	brn/blk/yel	A/C compressor activation, 0.2 volts compressor on	battery volts	battery volts
14	brn/grn	Activation, cooling fan stage 2, 0.2 volts fan on	battery volts	battery volts
15		Manul tran not used		
15	brn/org	Gearshift lock signal		

Engine Control Module (ECM) Pinout - connector X60004 (continued)

PIN	Wire color	Signal description	KOEO	Idle
16	brn/red	Activation, main relay	0.9 volts, (relay on)	0.9 volts, (relay on)
17	blu/red	Brake switch signal, Vbr brake depressed	0.1 volts	0.1 volts
18	brn/gry	Brake switch test signal, Vbr brake depressed	0.1 volts	0.1 volts
19	brn/blk	A/C pressure switch sensor ground	0.1 volts	0.1 volts
20	yel/brn	Speed signal, (volute spring)	9.4 volts	9.2 volts
21		not used		
22		not used		
23		not used		
24	blk/vio	Activation, fuel pump relay	battery volts	0.9 volts
25	blk/vio	EWS signal	8.0 volts	7.5 volts
26	gry/yel	PWG sensor 2 signal	0.4 volts C/T, 1.95 volts WOT	0.4 volts
27	wht	APWG sensor 1 signal	0.7 volts C/T, 3.9 volts WOT	0.7 volts
28	blu/yel	Post cat O_2 sensor signal	0.4 volts, (reading ref volts)	0.1 to 1.0 volts
29	yel/blk	CAN high	2.7 volts	2.7 volts
30	yel/brn	CAN low	2.3 volts	2.3 volts
31	blu/grn	Post cat O_2 sensor ground	0.1 volts	0.1 volts
32	wht/red	Diagnosis line, (obd II connector)	battery volts	
33	brn	ECM ground	0.1 volts	0.1 volts
34	brn	ECM ground	0.1 volts	0.1 volts
35		not used		
36	vio/wht	Data bus	battery volts	
37	grn/blu	Terminal 15	battery volts	battery volts
38		not used		
39	grn/red	Terminal 87	battery volts	battery volts
40	red/wht	Terminal 30	battery volts	battery volts

ECM connector X6000 (81 pin, black)

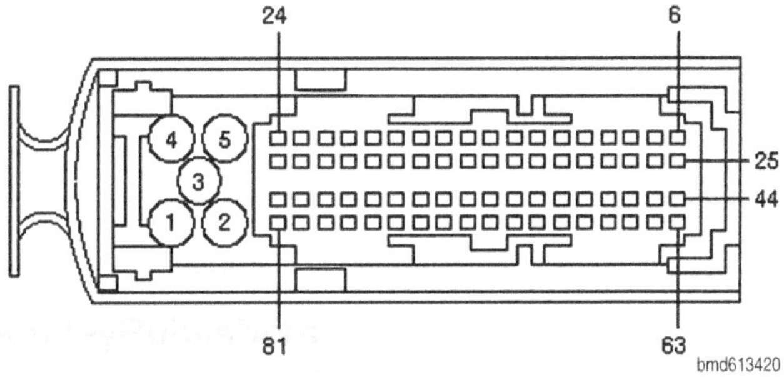

Engine Control Module (ECM) Pinout - connector X6000

PIN	Wire color	Signal description	KOEO	Idle
1		not used		
2		not used		
3	blk/red	ignition coil 2-3 trigger	battery volts	primary trigger
4	blk/blu	ignition coil 1-4 trigger	battery volts	primary trigger
5	blk/grn	precatalyst O_2 sensor heater	battery volts	battery volts square wave
6	blk	precatalyst O_2 sensor ground	0.1 volts (maximum)	0.1 volts (maximum)
7	blk	knock sensor ground	0.1 volts (maximum)	0.1 volts (maximum)
8	yel	knock sensor signal	2.5 volts	2.5 volts
9	yel	precatalyst O_2 sensor signal	0.4 volts	0 - 1.0 volts
10	red	edk tps 1 signal	0.9 volts	0.6 volts
11	blu	edk tps 2 signal	4 volts	4.3 volts
12	brn/wht	edk tps ground	0.1 volts (maximum)	0.1 volts (maximum)
13	red/vio	edk tps reference voltage	5 volts	5 volts
14		not used		
15		not used		
16		not used		
17		not used		
18		not used		
19	brn/vio	injector #4 trigger	battery volts	injector waveform
20	brn/grn	injector #2 trigger	battery volts	injector waveform
21	brn/yel	injector #1 trigger	battery volts	injector waveform

Engine Control Module (ECM) Pinout - connector X6000 (continued)

PIN	Wire color	Signal description	KOEO	Idle
22	brn/gry	injector #3 trigger	battery volts	injector waveform
23		not used		
24	brn/blk	purge solenoid trigger	battery volts	battery volts square wave
25		not used		
26	yel/red	coolant temp sensor signal	4 volts	0.6 volts
27	yel/blu	iat signal	2.5 volts	1.2 volts
28		not used		
29		not used		
30		not used		
31	yel/vio	tmap signal after compressor	2 volts	0.9 volt
32		not used		
33		not used		
34		not used		
35		not used		
36		not used		
37		not used		
38		not used		
39		not used		
40		not used		
41		not used		
42	blk/red	LDP solenoid trigger	battery volts	battery volts square wave when activated
43	blu/red	post-catalyst O_2 heater control	battery volts	battery volts square wave
44		not used		
45		not used		
46	blk/wht	coolant temp sensor ground	0.1 volts (maximum)	0.1 volts (maximum)
47	brn/wht	IAT sensor ground	0.1 volts (maximum)	0.1 volts (maximum)
48	NCA	knock sensor shield	0.1 volts (maximum)	0.1 volts (maximum)
49		not used		
50		not used		

Engine Control Module (ECM) Pinout - connector X6000 (continued)

PIN	Wire color	Signal description	KOEO	Idle
51	red/grn	TMAP reference voltage	5 volts (approx.)	5 volts (aprox.)
52	wht	alternator load signal	10 volts	6.5 volts
53	blk/vio	camshaft sensor signal	0.25 volts (approx.)	2.5 volts (approx.)
54		not used		
55	blk/gry	crankshaft sensor signal	0.25 (approx.)	2.5 volts (approx)
56		not used		
57		not used		
58		not used		
59		not used		
60		not used		
61	brn	ground		
62	brn	ground		
63	org	EDK throttle motor control	0.01 volts	2.6 volts
64	org	EDK throttle motor control	0.01 volts	2.6 volts
65	yel	EDK throttle motor control	0.01 volts	0.2 volts
66	yel	EDK throttle motor control	0.01 volts	0.2 volts
67		not used		
68		not used		
69		not used		
70		not used		
71	NCA	O_2 sensor shield	0.1 volts (maximum)	0.1 volts (maximum)
72	brn/vio	camshaft sensor ground	0.1 volts (maximum)	0.1 volts (maximum)
73		not used		
74	brn/gry	crankshaft sensor ground	0.1 volts (maximum)	0.1 volts (maximum)
75		not used		
76		not used		
77		not used		
78		not used		
79		not used		
80	brn	ground	0.1 volts (maximum)	0.1 volts (maximum)
81	brn	ground	0.1 volts (maximum)	0.1 volts (maximum)

SCAN TOOL DATA

The following tables are based on live data acquired from the factory scan (Seimen's GT1). Other scan tool output may be similar.

A/C data			
PID	**KOEO value**	**Idle value**	**Type / description**
Air conditioning system	not active	not active	status changes as per request
Compressor relay activated	not active	not active	status changes as per request
Refrigerant pressure	6.9	6.9	bar
A/C compressor status	OFF	OFF	status changes as per request
Signal, refrigerant pressure	1.12	1.12	volts
Request, A/C readiness	No	No	status changes as per request
Evaporator temperature	42.8	42.8	°C

ASC / DSC data			
PID	**KOEO value**	**Idle value**	**Type / description**
ASC or DSC - torque increase	not active	not active	
ASC or DSC - torque reduction	not active	not active	
ASC or DSC - request, torque reduction	not active	not active	
ASC or DSC - requests authorized	not active	not active	
ASC or DSC - deactivated, engine fault	active	active	
MSR - request, torque increase	not active	not active	

EWS data			
PID	**KOEO value**	**Idle value**	**Type / description**
Communication fault	No	No	
Incorrect answer received	No	No	
Incorrect message received	No	No	
Random code initialized	No	No	
No answer from immobilizer	No	No	
No successful communication	No	No	
Status, immobilizer	blocked	blocked	
Immobilizer -- counter engine disabled	24	24	
Counter, unsuccessful comm.	3	3	

General data			
PID	**KOEO value**	**Idle value**	**Type / description**
Terminal 15	active	active	
Engine OFF	active	not active	
Battery volts	12	13.9	volts / vehicle battery volts
main relay	active	active	
Relay, fuel pump	not active	active	
Engine speed	0	745	rpm
Intake air temp sensor	51.8	49.5	°C
Coolant temperature	100	95.2	°C
Intake - air pressure	990.69	463.23	hPa
Intake - air pressure before compressor	990.69	466.38	hPa
Idle speed	not active	active	
Engine part load	not active	not active	
Engine full load	not active	not active	
Engine running	not active	active	
Engine synchronized	not active	active	
Pedal position sensor 1	0.0	0.0	%
Pedal position sensor 2	0.0	0.0	%
Pedal position sensor	0.0	0.0	%
Pedal position sensor voltage1	0.74	0.74	volts
Pedal position sensor voltage 2	0.37	0.37	volts
Manual transmission gear reduction	0.00	0.00	
Engine load signal	0.0	33.7	%
Throttle position 1	7.6	3.0	°
Throttle position 2	7.6	3.1	°
Throttle valve position	7.5	2.8	°
Throttle voltage 1	0.86	0.65	volts
Throttle voltage 2	4.17	4.39	volts
Signal, oxygen sensor upstream	0.41	0.20	volts / should switch between 0.0 - 1.0 volts, engine running
Signal, oxygen sensor downstream	0.41	0.72	volts / should switch between 0.0 - 1.0 volts, engine running
Oxygen sensor control	not active	active	
Duty factor, tank venting valve	0.0	52.9	%

General data (continued)			
PID	**KOEO value**	**Idle value**	**Type / description**
Tank leak diagnosis pressure switch	active	active	
Electric fan - stage 1	not active	not active	
Electric fan - stage 2	not active	not active	
Electric fan - stage 3	not active	not active	
Clutch blocked	not active	not active	
Brake light switch 1	not active	not active	
Brake light switch 2	not active	not active	
Alternator load	25.50	90.68	%
Specified speed, idle controller	750	750	rpm
Vehicle speed	0	0	km/h
Engine MIL status	ON	OFF‘	
EML lamp	ON	OFF	
Fuel level	0 - 49	0 - 49	liters

Knock control data			
PID	**KOEO value**	**Idle value**	**Type / description**
Knock control adaptation	0.00	0.00	°
Knock sensor voltage	1.1	0.8	volts
Knock value	0.0	37.0	
Engine load signal	0.0	33.7	%
Spark advance angle, specified value	72.00	3.8	°
Spark advance angle, cylinder 1	19.9	3.8	°
Spark advance angle, cylinder 2	19.9	3.8	°
Spark advance angle, cylinder 3	19.9	3.8	°
Spark advance angle, cylinder 4	19.9	3.8	°

Oxygen sensor control data			
PID	**KOEO value**	**Idle value**	**Type / description**
Mixture adaptation, additive mg / stroke	0.02	0.02	mg stroke
Mixture adaptation, multiplicative	0.0	0.0	%
Short term mixture correction	0.0	3.1	%

Oxygen sensor control data (continued)			
PID	**KOEO value**	**Idle value**	**Type / description**
Average short term mixture correction	0.0	2.1	%
Long term additive correction	0.1	0.1	ms / acceptable range ± 0.3
Long term multiplicative correction	3.0	4.1	% / acceptable range ± 10%
Signal, oxygen sensor upstream	0.34	0.20	volts / should switch between 0.0 - 1.0 volts, engine running
Signal, oxygen sensor downstream	0.34	0.72	volts / should switch between 0.0 - 1.0 volts, engine running
Oxygen sensor upstream - rich	not active	not active	
Oxygen sensor readiness, upstream	not active	active	
Oxygen sensor heating, upstream	0	62	%
Oxygen sensor downstream, raw voltage value	0.39	0.20	volts
Oxygen sensor readiness, downstream	not active	active	
Oxygen sensor heating, downstream	0	49	%
Oxygen sensor controller	0.00	2.96	%
Load signal	0.0	31.0	%
Duration of injection	0.00	1.92	ms
Air mass	2.8	11.0	kg/h
Throttle valve position	7	2	%
Spark advance angle	72	0	°cr
Engine speed	0	760	rpm
Solenoid tank ventilation	0	49	%
Signal oxygen sensor downstream	0.39	0.17	volts

Throttle module data			
PID	**KOEO value**	**Idle value**	**Type / description**
Intake air pressure specified value	178	457.3	hPa
Intake air pressure specified value before compressor	211	493.4	hPa
Idle specified value	750	750	rpm
Limp home operation, throttle position	7.7	7.7	°
Throttle specified value	8.5	2.8	°
Lower mechanical stop adaptation 1	0.5	0.5	volts
Lower mechanical stop adaptation 2	4.5	4.5	volts
Lower electrical stop adaptation potentiometer 1	0.5	0.5	volts
Lower electrical stop adaptation potentiometer 2	4.5	4.5	volts
Emergency operation adaptation potentiometer 1	0.9	0.9	volts
Emergency operation adaptation potentiometer 2	4.2	4.2	volts
Throttle adaptation	not active	not active	
Pedal position sensor angle 1	0.0	0.0	°
Pedal position sensor angle 2	0.0	0.0	°
Pedal travel sensor angle (°PT)	0	0	°
Throttle angle 1	7.6	2.8	°
Throttle angle 2	7.6	3.0	°
Load signal	0.0	32.2	%

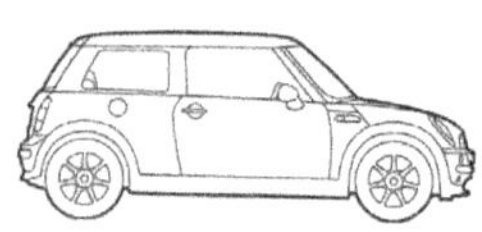

16 Fuel Supply

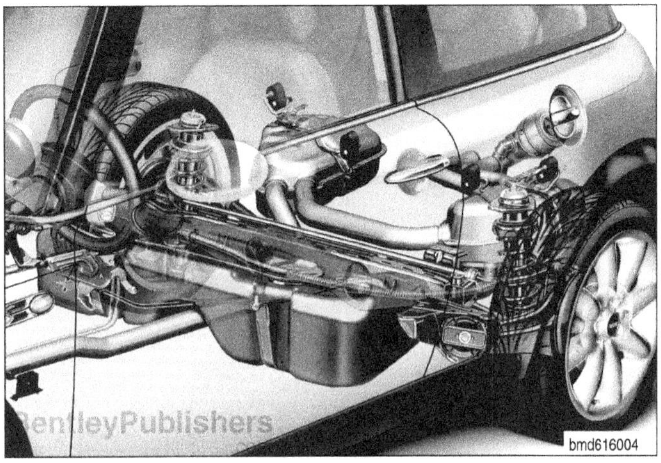

Table of Contents

GENERAL

MINI Cooper models use a saddle shaped fuel tank with one fuel pump and a syphon jet pump to transfer fuel between sides. A lifetime fuel filter is installed in the right side of the tank with the level sender. The fuel pump is relay powered and activated by the engine control module (ECM).

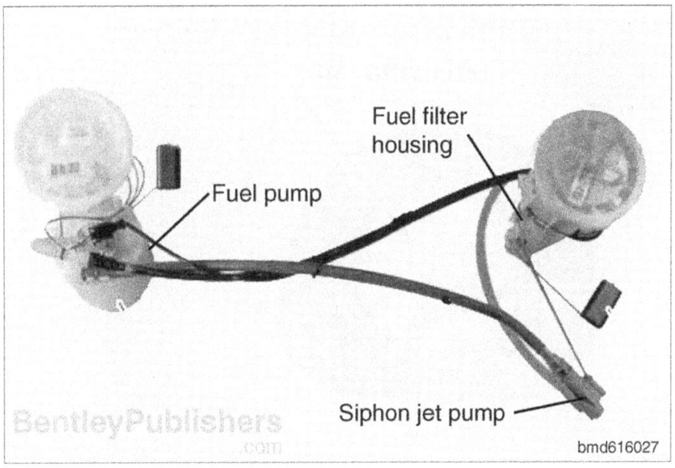

System operation

The pump is initially energized when ignition is switched to position 2. It remains energized for a set time to pressurize the fuel system. The fuel pump relay remains energized if a running engine is detected. Otherwise it switches OFF after a time delay has elapsed.

Once engine revolutions are detected by the ECM, a ground signal energizes to the fuel pump relay to run the fuel pump.

The relay is switched OFF immediately if the engine stops. The relay gets its B+ feed from the ignition switch through fuse 20 and is energized when the ECM provides a ground on pin 105.

NOTE —

- On 2002 models (up to 09-2008), an inertia switch (I01149) is installed in the B+ line to the relay between the fuse and the fuel pump relay. See **Inertia switch** for testing information.

Fuel pump

The fuel pump, mounted in the left side of fuel tank, picks up fuel through the fuel filter and passes it to the right side of the tank. In the right tank the fuel is passed through a pressure regulator where a fuel pressure of 3.5 bar (50 psi) is maintained. Excess pumped fuel is returned from the right side tank to the left side tank through a syphon jet.

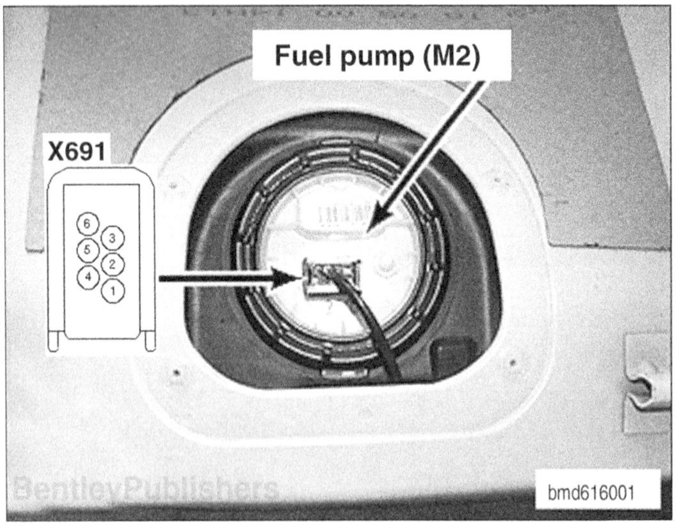

Fuel at 3.5 bar (50 psi) is sent to the fuel rail in the engine compartment. The fuel rail contains a pressure damper / fuel pressure regulator (**arrow**) to smooth out fluctuations in fuel pressure during high load situations.

FUEL SUPPLY TESTS

Fuel pressure testing

– Switch ignition OFF.

– Cooper S: remove intercooler.
  - Remove intercooler cover retaining bolts.
  - Remove intercooler cover.
  - Remove sealing ring bracket bolts and remove upper brackets.
  - Remove intercooler cover mounting bracket bolts and remove brackets.
  - Remove intercooler from right hand side and store in a safe place.
  - Access schrader valve.

bmd616014

> ***CAUTION —***
> - *The fuel in the fuel lines is under pressure (approx. 3 bar / 43 psi) Catch and dispose of escaping fuel.*

- Cooper:
  - Remove fuel injector rail cover.
  - Access schrader valve.

- All models: Remove dust cap from schrader valve on fuel rail and connect fuel pressure gauge to valve.

- Start engine.

- Measure fuel pressure.

Fuel pressure specifications
Cooper 3.0 ± 0.2 bar (43 ± 3 psi)
Cooper S 3.5 ± 0.2 bar (50 ± 3 psi)

- Switch engine off. Disconnect fuel pressure gauge.

- On Cooper S models, install intercooler O-rings dry; do not lubricate.

Pressure retaining feature

Fuel pressure is retained in the system for a period of time after the engine is switched OFF. This enhances engine restart and prevents vapor lock issues. Upon engine shutdown, the spring-loaded fuel pressure damper / regulator closes in conjunction with a one-way valve in the fuel pump to maintain pressure in the system.

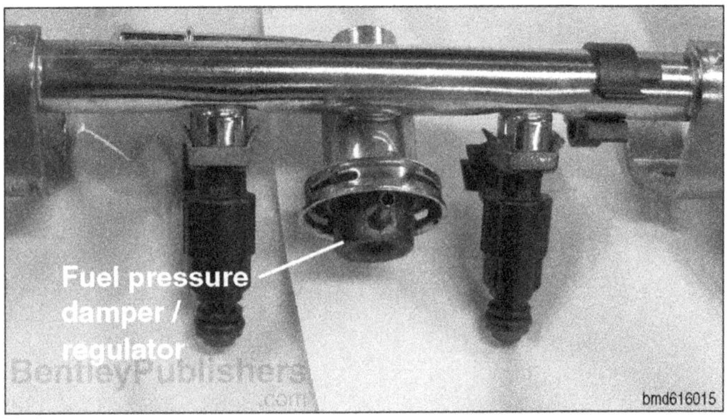

If fuel pressure reading is incorrect, confirm that the proper fuel pressure regulator is installed. The control function of the fuel pressure regulator should be checked under all operating conditions, including the in-tank fuel pressure regulator. Make sure the fuel pump is able to generate a higher fuel pressure than the pressure regulated by the pressure regulator.

Fuel pressure dampner / regulators	
Cooper S W11	part # 13311521393
Cooper W10	part # 13531487238

NOTE —
- Check with an authorized MINI dealer for latest part numbers and supersessions.

Fuel volume and quality testing

When testing fuel volume, activate the fuel pump with engine OFF. Test flow against normal system pressure. Pump fuel into suitable container and measure amount of fuel over a given period of time. This is the fuel pump volume. Check with the latest repair information for volume specification.

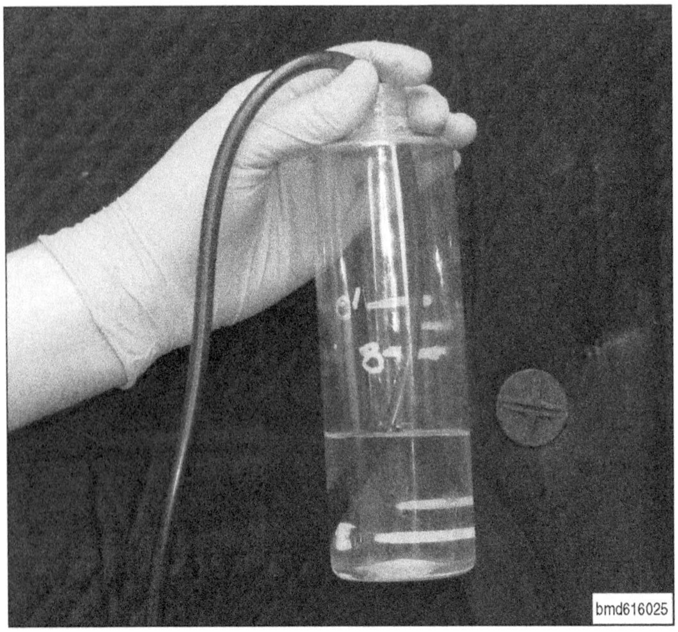

Visually check quality of fuel. Contaminated fuel may have a unique color or odor, contain sediments or floating matter, appear cloudy or separate, (after settling) or contain water.

Check ethanol content of the fuel. Most filling stations use a gasoline-ethanol blend with a maximum of 10% ethanol. Anything over 10% can affect the driveability of the vehicle.

To check ethanol content; drain 9 oz. of fuel into a suitable container, add 1 oz. of water. Shake container. This mixes the ethanol and water.

Once fuel settles, check to see that you have no more than 2 oz. of water.

Top Tier gasoline

Deposit-control additives have been required by the EPA in all gasoline since 1995. However, many gasoline manufacturers have actually reduced the concentration level of detergent additives by up to 50% in recent years. Low content of cleaning additives results in an excessive accumulation of deposits on fuel injectors, intake valves, exhaust manifold or inside the combustion chamber.

Due to deposit build-up, the drives may experience various driveability problems such as cold start problems or rough idle, increased emissions with malfunction indicator light (MIL) illumination, reduced engine performance and poor fuel economy.

In response to this issue, a special detergent gasoline called Top Tier detergent gasoline was developed, and approved by four automotive companies (BMW, GM, Honda and Toyota).

Usage of Top Tier gasoline helps keep engines cleaner, and reduces deposits-related concerns.

MINI recommends using Top Tier gasoline (min. octane rating AKI 91 with alcohol content of less than 10%). Only the exclusive usage of Top Tier gasoline provides the full benefit of reducing deposits build-up. For more information on Top Tier gasoline, visit the official website: **www.toptiergas.com**.

For more information on alcohol fuels, refer to MINI Service Information Bulletin no. SI M13 01 06 (Alcohol Fuel Blends in MINI Vehicles).

FUEL SYSTEM ELECTRICAL TESTS

Multiple fuel pump circuit configurations are implemented on the cars covered by this book. Be sure to identify the build date on the subject vehicle prior to testing. Simplified circuit diagrams are given at the rear of this chapter under **Fuel Pump Wiring Diagrams**.

Most all tests given here are done with the fuel pump relay installed and all harness connectors connected. Make checks by back probing the wire terminals. When testing at the ECM connector, remove the plastic connector housing to gain access to the wiring.

The fuel pump relay is located in Fuse relay panel 2. This panel located in the driver's footwell, left hand side.

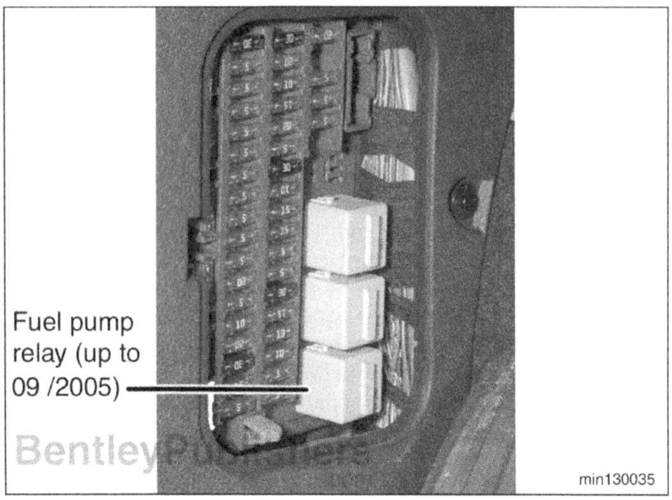

Fuel pump relay testing (Coupe up to 07/2004)

The following checks are made at the ECM connector X60004.

ECM connector X60004 (40-pin, black)

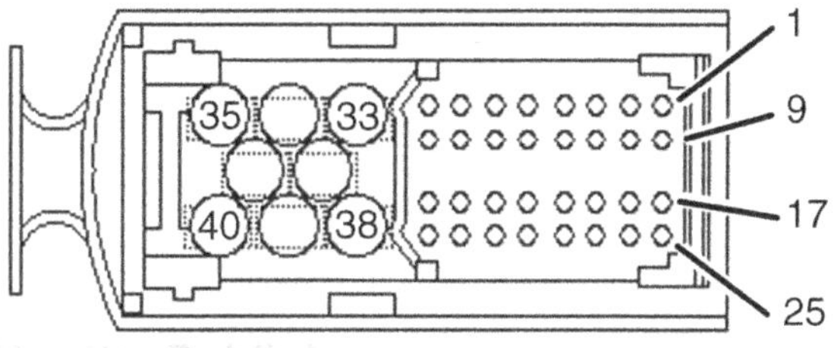

Fuel pump relay testing values (at EMS2000 connector X60004)			
Terminals	**Conditions**	**Test values**	**Circuit**
24 and ground	key ON engine OFF	0.9 volts for approx. 5 sec., then battery volts	relay control circuit (relay activated briefly by ECM)
24 and ground	engine running at idle @2500 rpm (no load)	0.9 volts	relay control circuit (relay activated by ECM)

The following checks are made at the back side of Fuse Relay Panel II at connector X10204.

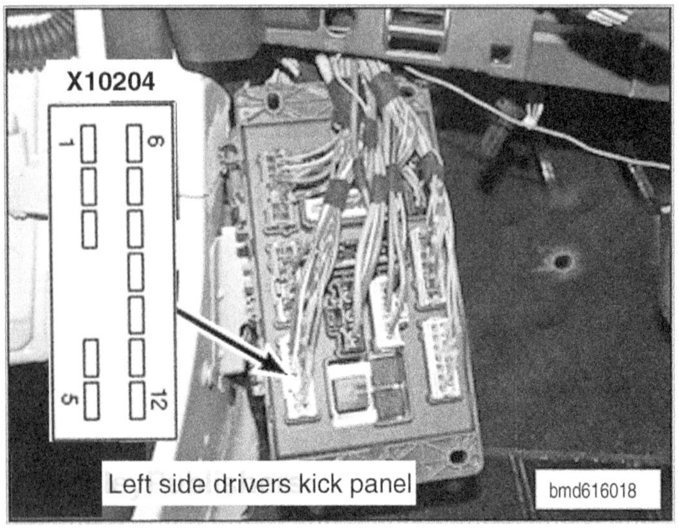

Fuel pump relay testing values (at Fuse Relay Panel 2- connector X10204)			
Terminals	**Conditions**	**Test values**	**Circuit**
1 and ground	key ON engine OFF	0.9 volts for approx. 5 sec., then battery volts	relay control circuit (relay activated briefly by ECM)
1 and ground	engine running at idle @2500 rpm (no load)	0.9 volts	relay control circuit (relay activated by ECM)
2 and ground	key on engine off engine running at idle @2500 rpm (no load)	battery volts	voltage supply (via relay) to fuel pump
3 and ground	key ON engine OFF	battery volts for approx. 5 sec., then 0 - 0.1 volts	voltage supply to pump
3 and ground	engine running at idle @2500 rpm (no load)	battery volts	voltage supply to pump
5 and ground	key ON engine OFF engine running at idle @2500 rpm (no load)	battery volts	ECM voltage output
8 and ground	key ON engine OFF engine running at idle @2500 rpm (no load)	battery volts	general module voltage input

Fuel pump relay testing (Coupe from 07/2004 > 09/2005, Convertible up to 09/2005)

The following checks are made at the ECM connector X60004.

ECM connector X60004 (40-pin, black)

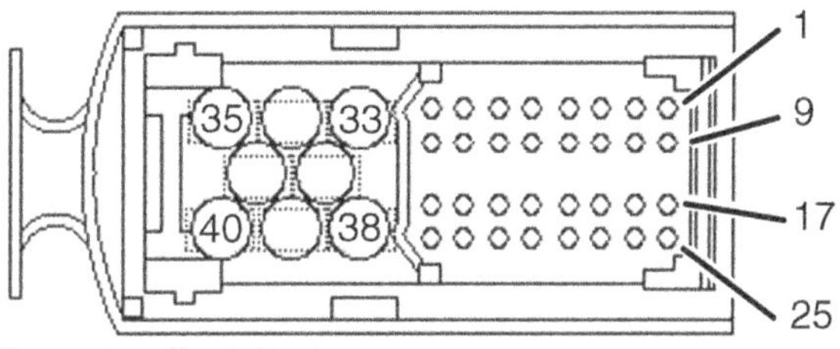

Fuel pump relay testing values (at EMS2000 connector X60004)			
Terminals	**Conditions**	**Test values**	**Circuit**
24 and ground	key ON engine OFF	0.9 volts for approx. 5 sec., then battery volts	relay control circuit (relay activated briefly by ECM)
24 and ground	engine running at idle @2500 rpm (no load)	0.9 volts	relay control circuit (relay activated by ECM)

The following checks are made at the back side of Fuse Relay Panel II at connector X10204.

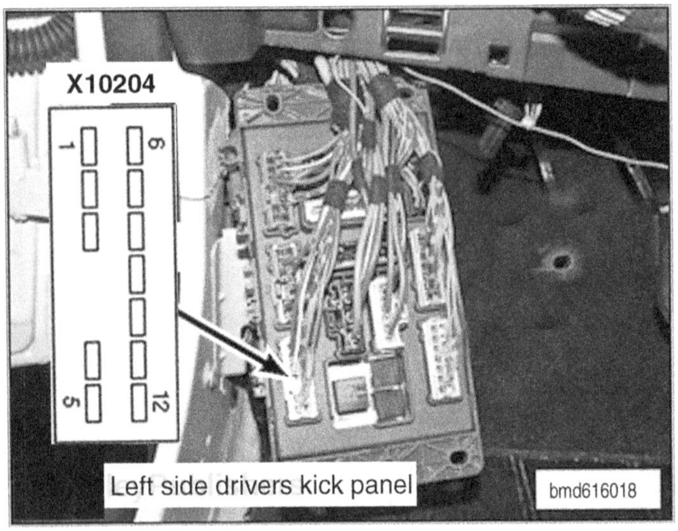

Fuel pump relay testing values (at Fuse Relay Panel 2 - connector X10204)			
Terminals	**Conditions**	**Test values**	**Circuit**
1 and ground	key ON engine OFF	0.9 volts for approx. 5 sec., then battery volts	relay control circuit (relay activated briefly by ECM)
1 and ground	engine running at idle @2500 rpm (no load)	0.9 volts	relay control circuit (relay activated by ECM)
3 and ground	initial key ON engine OFF	battery volts for approx. 5 sec., then 0 - 0.1 volts	voltage supply to pump
3 and ground	engine running at idle @2500 rpm (no load)	battery volts	voltage supply to pump
5 and ground	key ON engine OFF engine running at idle @2500 rpm (no load)	battery volts	ECM voltage output
8 and ground	key ON engine OFF engine running at idle @2500 rpm (no load)	battery volts	general module voltage input

Fuel pump relay testing (Coupe and Convertible from 09/2005)

The following checks are made at the ECM connector X60004.

ECM connector X60004 (40-pin, black)

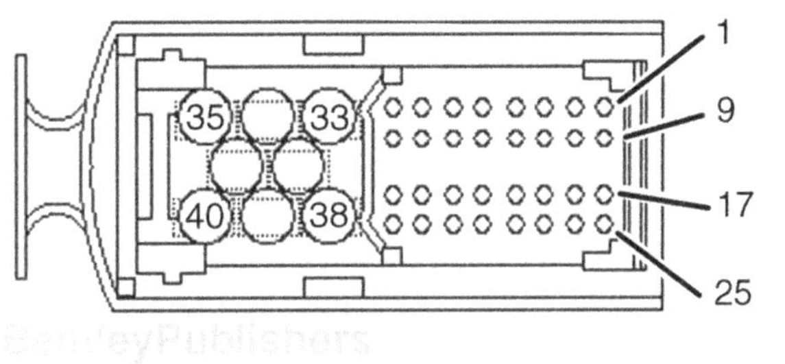

Fuel pump relay testing values (at ECM connector X60004)			
Terminals	**Conditions**	**Test values**	**Circuit**
24 and ground	key ON engine OFF	0.9 volts for approx. 5 sec., then battery volts	relay control circuit (relay activated briefly by ECM)
24 and ground	engine running at idle @2500 rpm (no load)	0.9 volts	relay control circuit (relay activated by ECM)

The following checks are made at the back side of Fuse Relay Panel II at connector X10204.

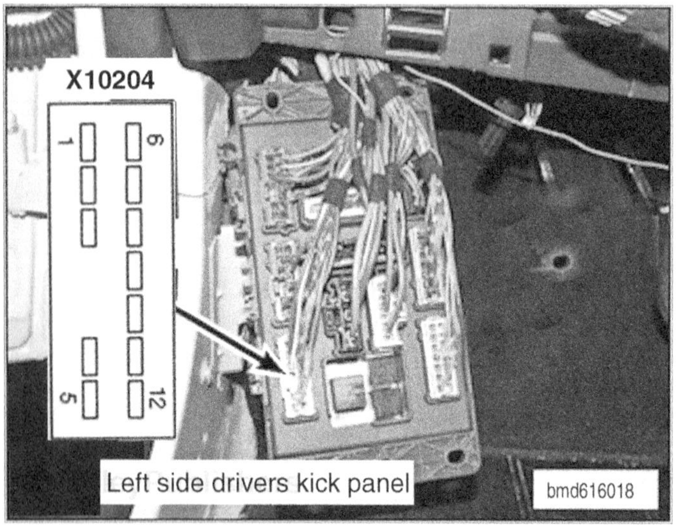

Fuel pump relay testing values (Fuse Relay Panel 2 - connector X10204)			
Terminals	**Conditions**	**Test values**	**Circuit**
2 and ground	key ON engine OFF	battery volts for approx. 5 sec., then 0 - 0.1 volts	voltage supply to pump (relay activated by ECM)
2 and ground	engine running at idle @2500 rpm (no load)	battery volts	voltage supply to pump (relay activated by ECM)
3 and ground	key ON engine OFF engine running at idle @2500 rpm (no load)	battery volts	voltage supply to relay
8 and ground	key ON engine OFF	0.9 volts for approx. 5 sec., then battery volts	relay control circuit (relay activated briefly by ECM)
8 and ground	engine running at idle @2500 rpm (no load)	0.9 volts	relay control circuit (relay activated by ECM)

Inertia switch testing

The inertia switch is installed on early MINIs (up to 09/2002).

The switch is located in the left rear of the engine compartment, in front of the strut tower. In the event of an impact greater than 14 g, the switch opens and interrupts power to the relay. The inertia switch has to be reset manually once it is triggered.I

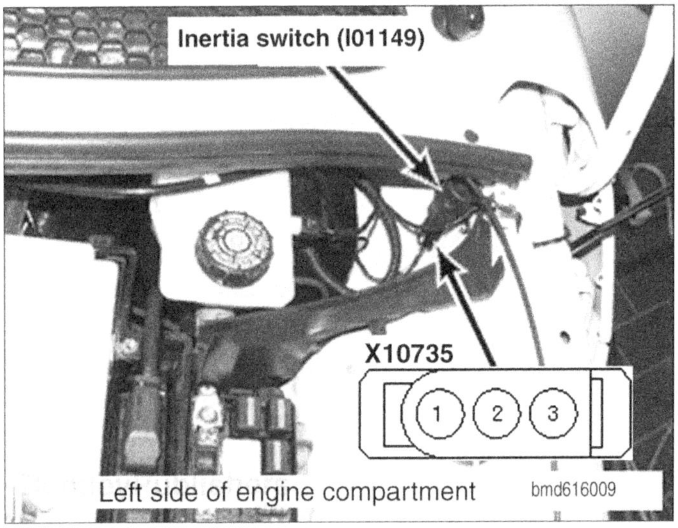

Left side of engine compartment

Inertia switch testing values (at connector x10735)			
Terminals	**Conditions**	**Test values**	**Circuit**
1 and ground	key ON engine OFF engine running at idle @2500 rpm (no load)	battery volts	inertia switch output voltage
3 and ground	key ON engine OFF engine running at idle @2500 rpm (no load)	battery volts	voltage input to inertia switch

Fuel pump circuit testing

Access to fuel pump is beneath rear seat, left hand side. Remove access cover to reveal pump and pump connector.

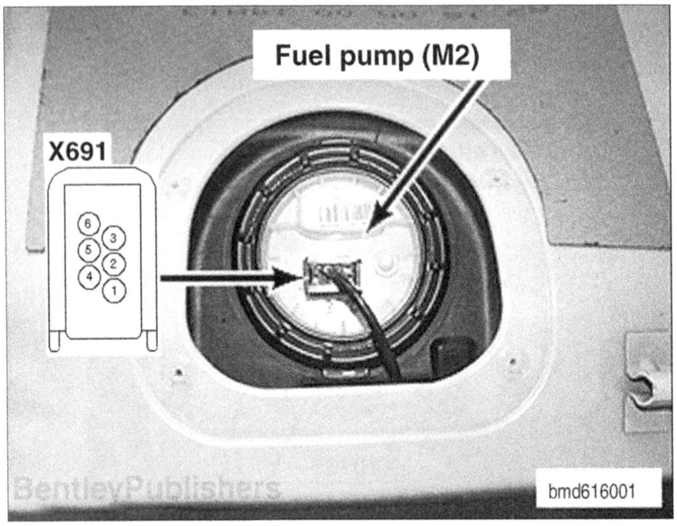

Fuel pump testing values (at connector x691)			
Terminals	**Conditions**	**Test values**	**Circuit**
2 and ground	key ON engine OFF	battery volts for approx. 5 sec., then 0 - 0.1 volts	voltage supply
2 and ground	engine running at idle @2500 rpm (no load)	battery volts	voltage supply
5 and ground	key ON engine OFF engine running at idle @2500 rpm (no load)	0 - 0.1 volts	fuel pump ground

Fuel pump current testing

Fuel pump current is a good indicator of fuel pump performance. Current can indicate worn brushes, higher than normal circuit resistance, even restrictions.

Required equipment to test fuel pump current draw.

- Labscope
- Low current probe
- Fused jumper wires

When testing fuel pump current draw, connect to the high current side of circuit. You can test on either the battery positive or ground side of circuit. The easiest place to do this is at the fuel pump fuse.

It is also necessary to run the fuel pump by bypassing the fuel pump relay.

– Remove fuel pump fuse.

- 2002 - 2005 models: Fuse 20 (20 amp)
- 2006 models: Fuse 37 (20 amp)

– Using fused jumper wire, carefully connect to fuse terminals in fuse panel. Connect low amp probe to jumper wire as shown.

> ***WARNING —***
> - *Terminals are live at all times. Use care when connecting wires.*

– Remove fuel pump relay and jump relay terminals using fused jumper wise to run fuel pump. See **Fuel Pump Wiring Diagrams** at rear of this chapter.

NOTE —
- The current probe was set at 100 mv / amp.

Fuel pump current taken during pump intialization

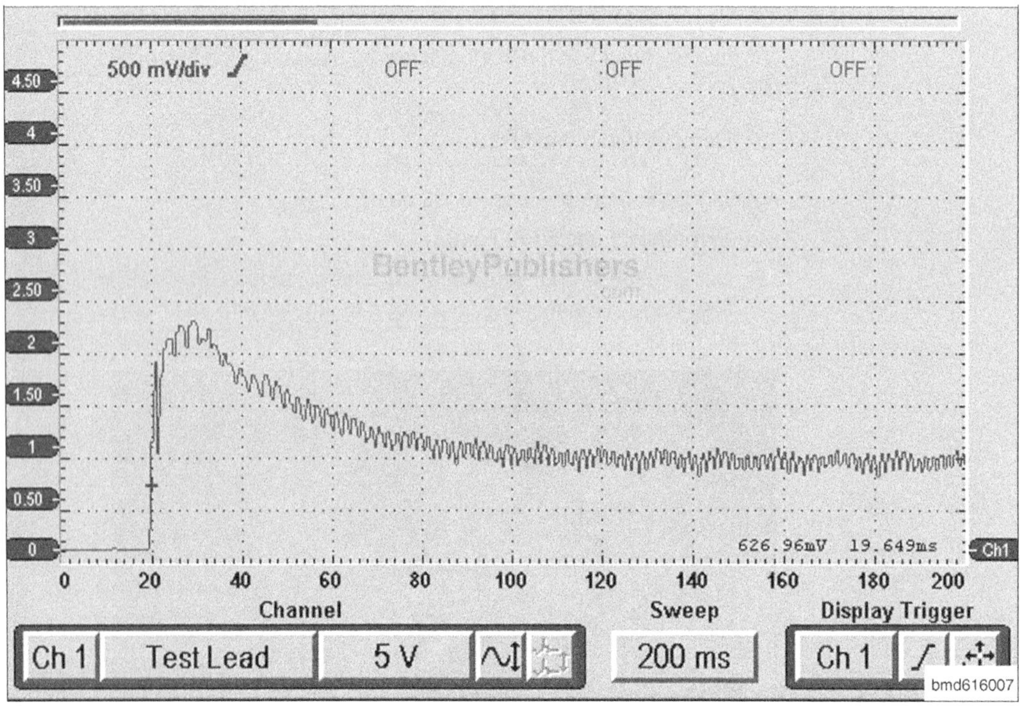

Fuel pump current waveform captured at idle

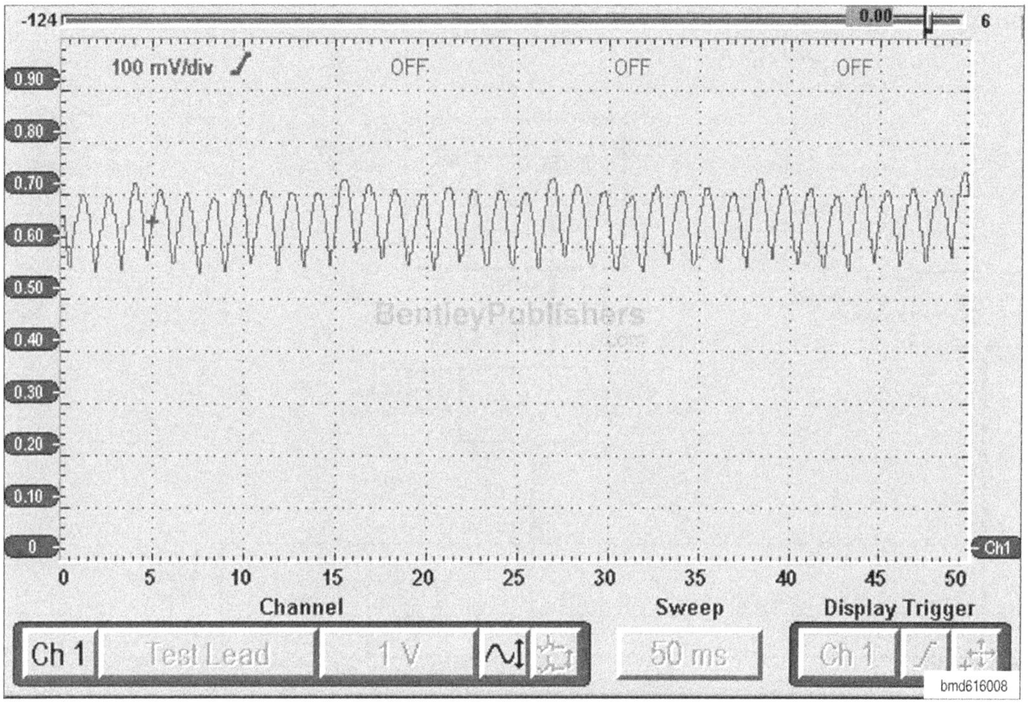

FUEL PUMP WIRING DIAGRAMS

Fuel pump wiring schematic (Coupe to 09/2002)

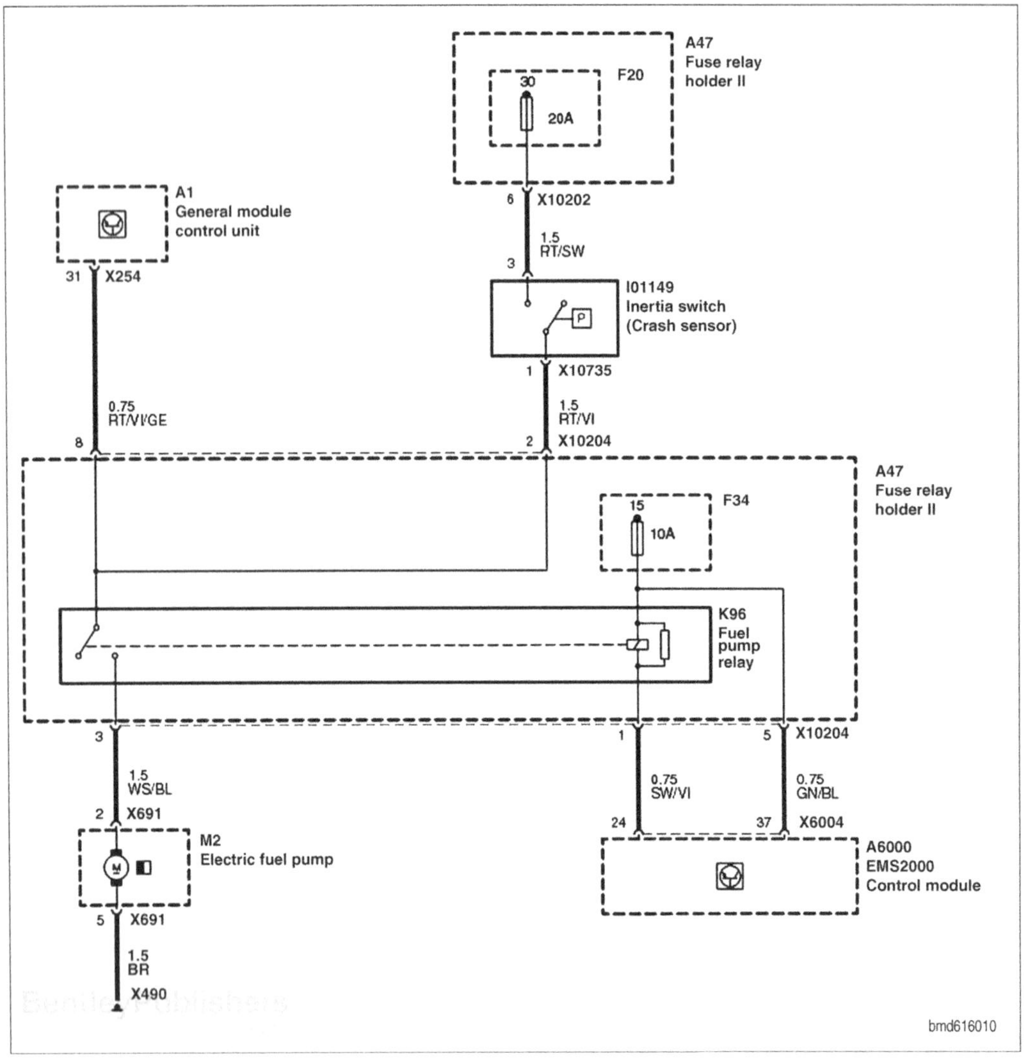

Fuel pump schematic (Coupe 09/2002 - 07/2004)

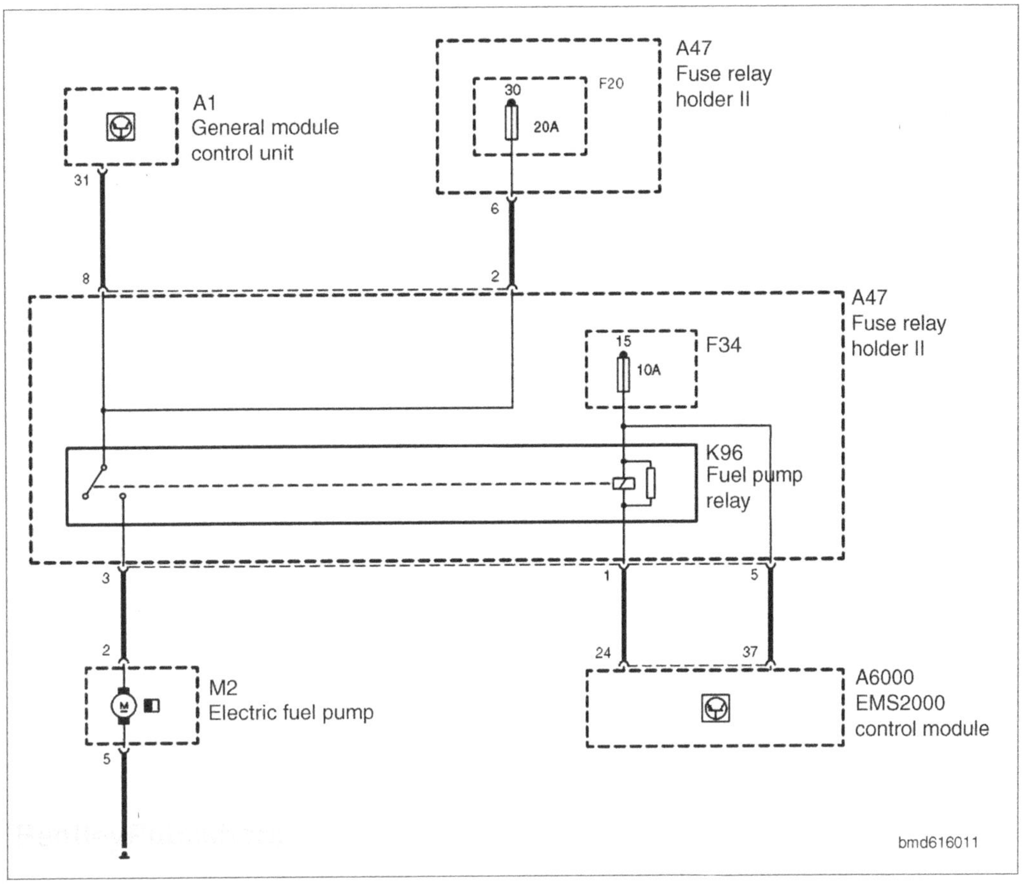

Fuel pump schematic (Convertible 7/2004 - 9/2005)

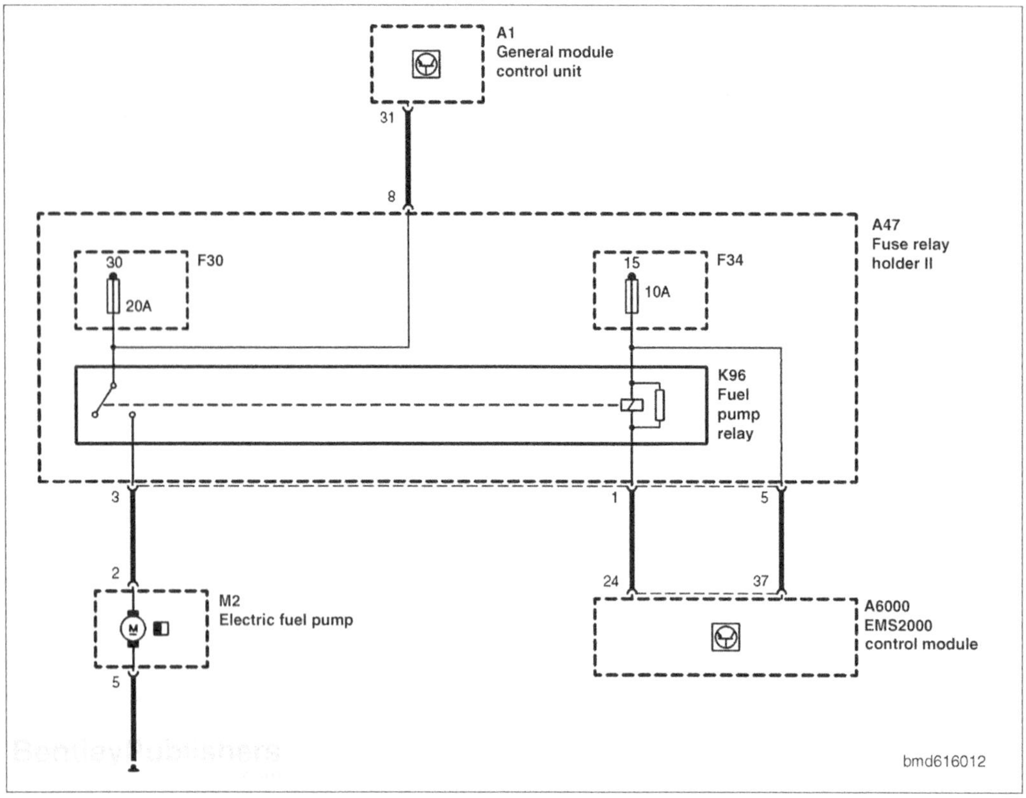

Fuel pump schematic (07/2004 - 09/2005)

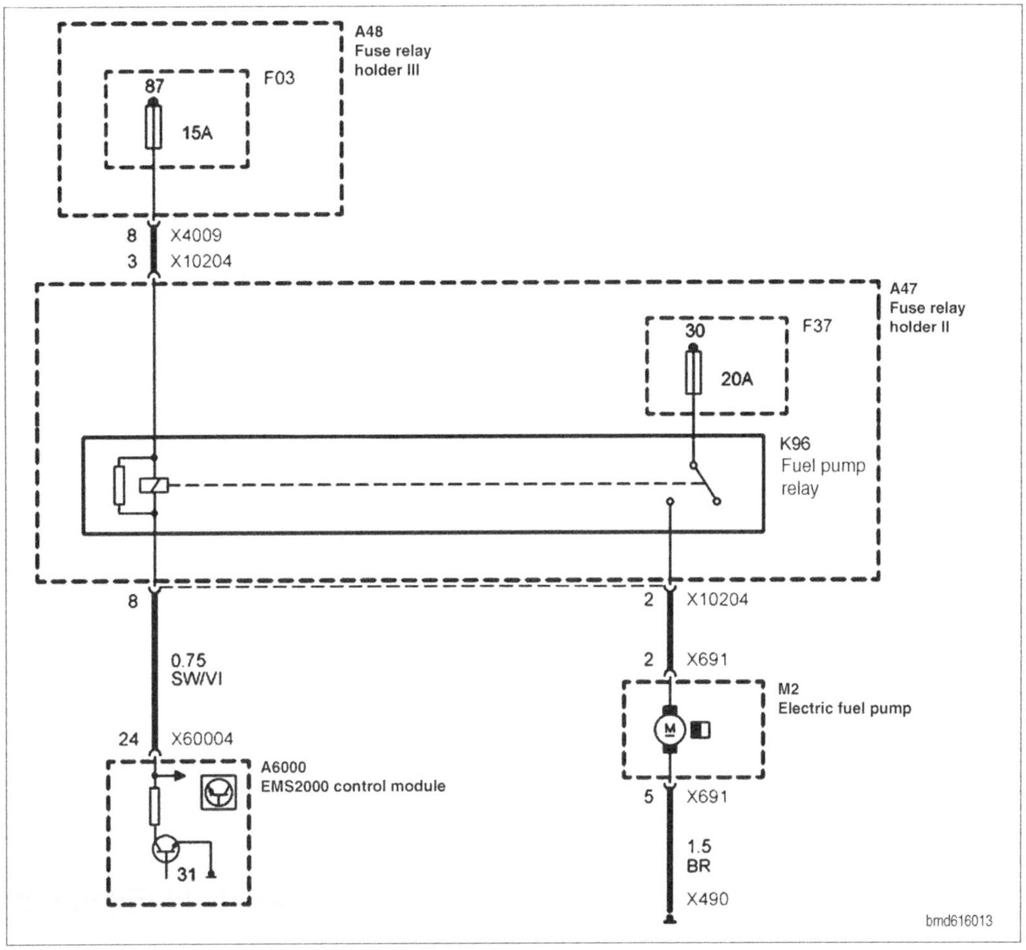

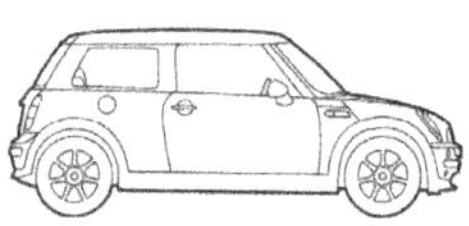

Notes:

24 Automatic Transmission

bmd624003

Table of Contents

GENERAL

This chapter describes the various automatic transmissions used in MINI automobiles.

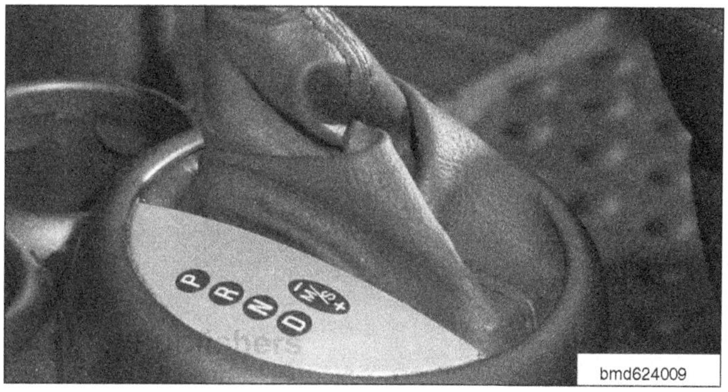

bmd624009

Automatic transmission applications	
Cooper	ECVT
Cooper S (from 01 / 2005)	6-speed AISIN

ECVT DESCRIPTION

Cooper models are equipped with the ECVT (electronic continuously variable transmission).

The origins of the continuously variable transmission (CVT) manufactured by ZF dates back to 1974 with a revolutionary rubber drive belt. The MK 1 Cooper (2002-2006) uses a new generation of ECVT incorporating a steel drive belt.

ECVT cutaway

The advantages of a CVT automatic transmission are:

- Low engine revolutions at constant speeds
- Improved emission control and fuel consumption
- Low noise, vibration and harshness levels
- Smooth acceleration
- Flexible driving on mountain roads

ECVT principle of operation

Unlike conventional planetary automatic transmissions that provide a limited number of gear ratios, the ECVT, as its name suggests, continuously varies the gear ratio.

The ECVT uses a primary pulley and a secondary pulley. Both pulleys have one fixed half and one mobile half, controlled by hydraulic pressure. The position of the drive belt on the pulleys determines the ratio. If the mobile half of the pulley is close to its opposite half then the drive belt is forced to travel around the outer circumference. When the pulley is open wide then this circumference is reduced. The primary and secondary pulley mobile halves are opposed, so when the drive belt diameter is reduced on the primary pulley, it increases on the secondary pulley.

To pull away, a low ratio is required. To provide this, the primary pulley is open, allowing the drive belt to sit down into the pulley and forcing it to run around the outer part of the closed secondary pulley.

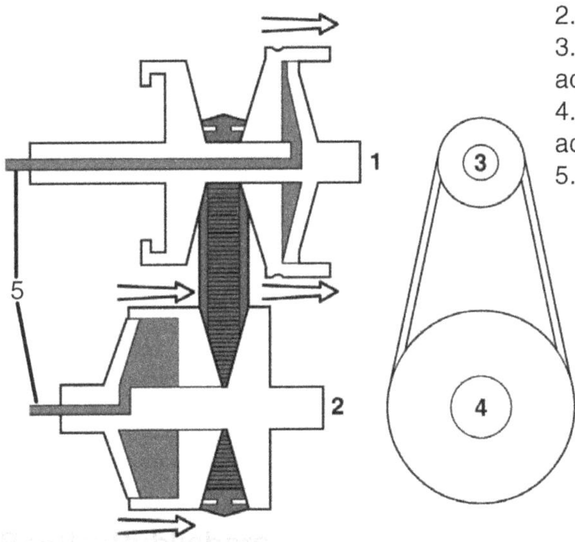

1. Input from the engine
2. Output to the wheels
3. Drive pulley at start of acceleratation (pulling away from rest)
4. Driven pulley at start of acceleration (pulling away from rest)
5. Hydraulic pressure

ECVT Pulleys in Low position

bmd624201

As vehicle speed increases, a higher gear ratio is required. To do this, the primary pulley gradually moves towards its fixed partner, increasing the pulley circumference. At the same time the secondary pulley is forced apart reducing pulley diameter and therefore creating a higher gear ratio.

If acceleration continues to take place, further up-shifts may be made until the drive pulley diameter is as large as possible and the driven pulley diameter is as small as possible. Therefore, for every revolution of the drive pulley the driven pulley revolves several times.

An overdrive ratio is obtained when the primary pulley is fully closed and the secondary pulley is fully open. The secondary pulley is now forced to rotate approximately two and a half times for every turn of the primary pulley.

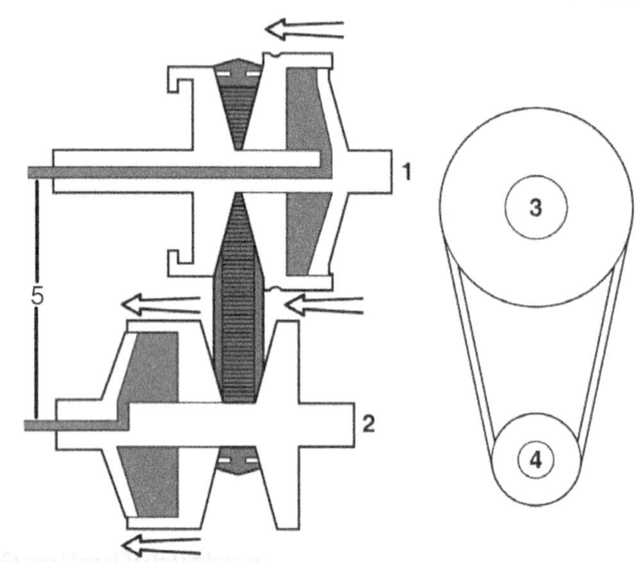

ECVT pulleys in high position

Drive plate

The engine is connected to the transmission via a torsional damper instead of the torque converter used by a conventional automatic transmission.

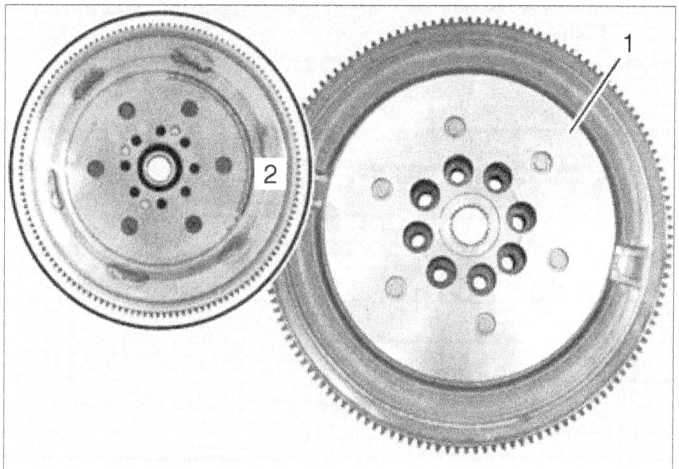

1. Drive plate, transmission side
2. Drive plate, engine side

bmd624203

Electronic controls

Transmission control is incorporated into the engine management ECM. The ECM does not control the transmission ratio directly but does provide all of the programing relating to the required position of the ratio control motor. It also provides the programing for how fast it should be operated.

The ECM receives inputs from the main sensors of this system, communicates with the gearbox interface unit (TCM) to control the transmission, accepts driver inputs and provides information to the driver via the instrument cluster. The TCM acts as a slave and interpreter for the ECM.

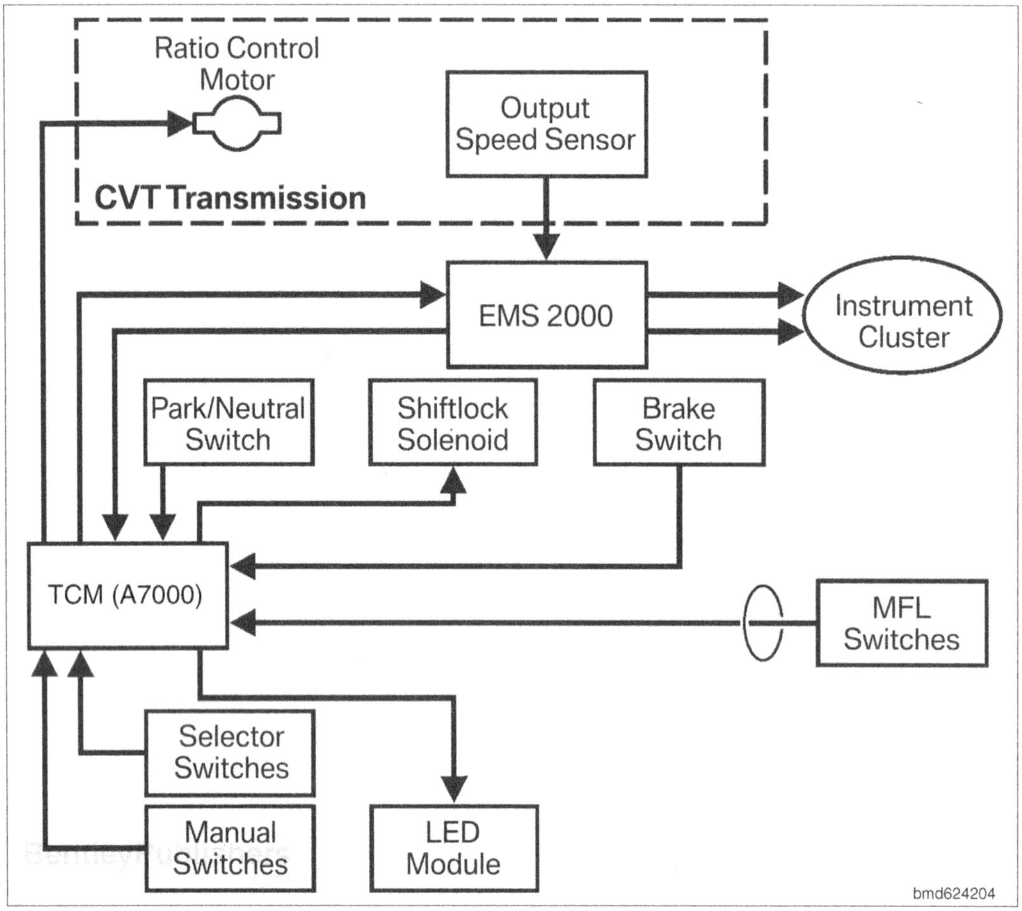

The ECM can control the transmission so that the input shaft speed, relative to the output shaft speed, is fixed in one of six ratios. This gives the effect that the vehicle has a six speed manual transmission with a sequential gear change.

The system protects the transmission, while in manual mode, against shifts that could be potentially dangerous or could damage the engine; for example, shifting to first gear at 90 mph, or shifting to top gear at 10 mph.

In addition, if the driver does not shift up, the next gear is automatically selected when engine speed reach approximately 6000 rpm. Equally, if the driver does not shift down when reducing vehicle speed, the system performs the down-shift automatically thus ensuring the transmission is in the appropriate gear when throttle is applied. This prevents a feeling of excessive clutch slip should the throttle pedal be reapplied.

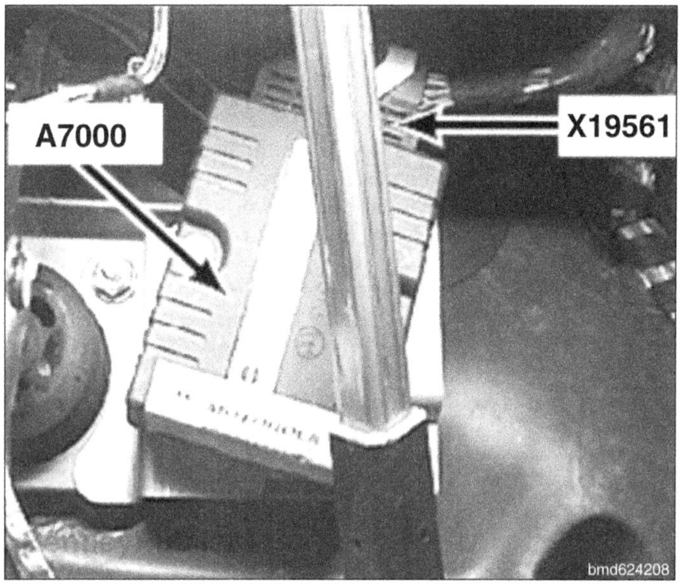

Driving and ambient conditions can influence the pulley positions. Ambient conditions include:

- Oil temperature
- Clutch slip
- Hydraulic balancing of the controlling cylinders
- Hydraulic pressure within the control lines

The primary and secondary pulleys alter their position to maintain the requested transmission input/output ratio.

All inputs and outputs of the ECVT control system pass through the ECM and the TCM. The ECM monitors the speed of the transmission output shaft and communicates with the TCM to select the correct gear ratio to suit the current driving conditions.

The TCM illuminates park, reverse, neutral, or sport LEDs display next to the gear selector lever. The ECM provides the output for the instrument cluster speed display.

NOTE —

- The ECM is programmed with the maximum torque that the belt can transfer across all possible ratio ranges. It is extremely important that the belt is not allowed to slip on the pulleys, thus causing excessive wear.

Fault mode

When the ECM or gear box interface (GIU) detects a fault, the ECM tries to position the ratio control motor so that engine speed in most driving conditions is around 2800 - 3200 rpm. In this position the vehicle still has reasonable driving characteristics. For certain failure modes where the ECM cannot command the position of the motor, the GIU sets the motor position to 130 steps (full range of travel is 0 - 214 steps). In this case, engine speed in most driving conditions is 3750 - 4000 rpm.

The ECM instructs the instrument cluster to display EP or the engine malfunction indicator light, (MIL) depending on legislative requirements.

There are certain faults that when stored do not cause the ECM to default the transmission into limp home position.

These are:

- Gear lever + (upshift) switch failure
- Gear lever - (downshift) switch failure
- Steering wheel + (upshift) switch failure (if fitted)
- Steering wheel - (downshift) switch failure (if fitted)
- Shift interlock system fault
- Center console LED fault

The limp home mode is not necessary for these failures because the gearbox control is not compromised; it is only necessary to warn the driver. The ECM will not operate the sequential gear changes in manual mode if these switches are faulty.

ECVT adaptation

The transmission hydraulic/mechanical characteristics are mapped inside the ECM. The ECM monitors actual engine speed relative to mapped engine speed and learns a new profile of actual transmission characteristics. The ECM also monitors the amount of deviation from the mapped program. As long as this difference is within tolerance, the ECM accepts the value and learns from it. If the actual value goes beyond the adaptive tolerance, the ECM performs a reset. If the value still exceeds the adaptive tolerance band, the ECM stores a fault code and places the transmission into its default position.

For adaptation of the ECVT transmission, carry out the following steps.

- If available: Using BMW scan tool clear adaptations. X will be displayed.

Clutch adaptation

Press brake pedal in the following steps.

- Start cold engine. At increased idle speed, shift to N for 10 seconds, then shift to D for 10 seconds. XP is indicated in P position in display.

- Repeat following ten times: Shift to N for 3 seconds, then shift to D for 3 seconds. XP is indicated in P position in display.

- Repeat following ten times: Shift to N for 3 seconds, then shift to R for 3 seconds. XP is indicated in P position in display.

Step-up ratio adaptation

- Accelerate vehicle to above 60 kph, (37 mph). Release accelerator pedal and allow vehicle to coast. Transmission begins to adapt itself for 4500, 4000, ..., 1500 and 1400 rpm. P is indicated in P position in display.

- With engine at operating temperature and idle speed at 800 rpm, repeat first two steps of clutch adaptation.

- Turn ignition lock to position 0.

If it is necessary to disconnect the battery or the ECM, wait at least 10 minutes to allow software to learn adaptation.

NOTE—

- If the instrument cluster display continues to display the X (adaptation required) character and the transmission does not operate normally, repeat the above process to give the software another chance to learn the remaining points.
- If either the ECM or transmission is changed during the service life of the vehicle, reset ECVT adaptation so the ECM begins learning from the new base point.

6-SPEED AUTOMATIC TRANSMISSION

From January 2005, a 6-speed automatic transmission (manufacturer: AISIN AW CO. LTD, Japan) was available in MINI Cooper S and the MINI Cooper S Convertible.

bmd624010

The 6-speed automatic transmission is characterized by its short gearshift times (¼ second) in conjunction with Agitronic.

Agitronic enables individual gear selection in drive position D by means of shift paddles on the steering wheel. The driver can shift manually in drive position D without having to use the selector lever to switch to Steptronic mode beforehand. There are four innovative shift modes associated with Agitronic.

Drive Position "D"	Normal	Automatic gear shifting. Perfect combination of comfort, sportiness and fuel consumption.
	Agitronic	Individual gear selection by means of shift paddles on the steering wheel. If there is no acceleration or manual gearshift, the automatic transmission shifts back to the normal drive position D after a short time.
Drive Position "S"	Sport	Automatically provides extremely sporty configuration for maximum agility and ultra-quick response by the car.
	Steptronic	Manual gear shifting by means of shift paddles on the steering wheel or the selector lever without exiting from automatic.

bmd624011

Technical data

6-speed (AISIN) transmission	
Gearbox	6-speed planetary gearbox (6-speed automatic transmission) with hydrodynamic torque converter for front-wheel drive
Transmission Output	max. torque 220 Nm
Transmission Ratios	1st gear: 4.040 4th gear: 1.160 2nd gear: 2.370 5th gear: 0.850 3rd gear: 1.550 6th gear: 0.670 Reverse gear: 3.193
Lever Position	P-R-N-D-M/S and Steptronic
Control	electrohydraulic
Weight	with transmission oil and torque converter: 82.6 kg (63 lbs.)
Towability	50 km at up to 50 km/h
Maintenance	Maintenance-free with life-time oil filling

bmd624012

Emergency operation program

If a fault occurs the transmission solenoids and pressure control valves are not controlled by the transmission control unit. In this case, an emergency operation program is activated and 3rd gear selected to make it possible to continue driving slowly. The display on the instrument cluster shows the letters EP (emergency program).

The following changes were made in order to use the 6-speed transmission:

- Engine oil cooler omitted due to reduced rpm range.
- Adaptation of coolant lines with a connection for transmission oil cooler.
- Crankshaft flange adapted to 6-speed flywheel.
- Engine mounting location. The engine was moved 7 mm up and 10 mm to the right based on size of 6-speed transmission.
- Hydro transmission mounts.
- Use of a suction jet pump to boost vacuum pressure at brake servo.
- Air intake duct with connection for suction jet pump.
- Exhaust manifold adaptation owing to revised engine position.
- Extended body wiring harness for transmission control.
- A/C refrigerant lines redesigned due to engine mounting change.
- Drive axles revised.
- Underbody paneling of engine compartment has an air duct for power steering cooling.

6-speed transmission components

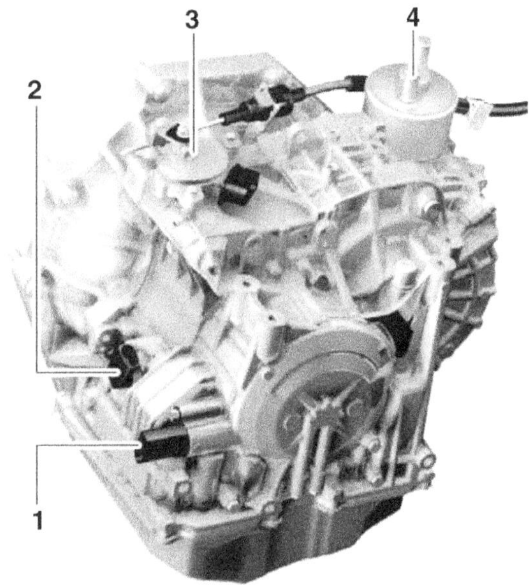

1 Multiple pin connector 1 to transmission control module
2 Multiple pin connector 2 to transmission control module
3 Selector lever position switch
4.Transmission oil cooler

bmd624014

Inhibitor switch

The inhibitor switch on the transaxle case activates the neutral start function (which permits the engine to start only in the P or N ranges), lights up the backup lamps in the R range, and transmits the A/T shift position as a signal to the TCM to perform gear shift control.

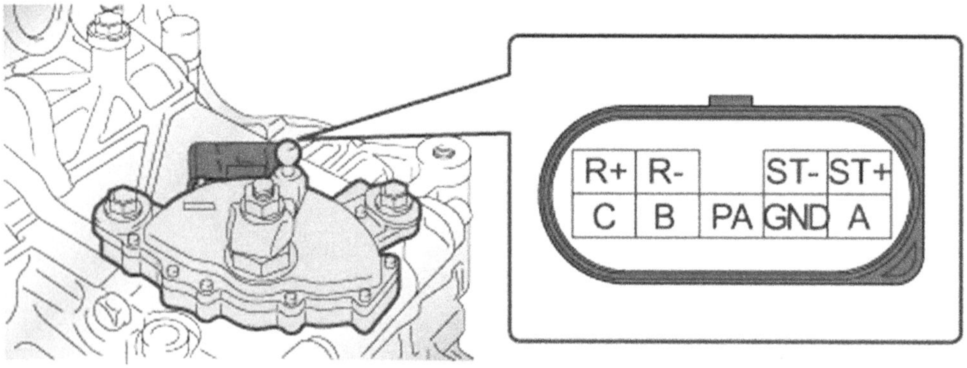

	Starter Circuit		Reverse Circuit		Position Circuit				
Range	ST+	ST-	R+	R-	GND	A	B	C	PA
P	○	○			○	○	—	—	○
R			○	○	○	○	○		
N	○	○			○	—	○	—	○
D					○	—	○	○	
S					○	○	○	○	○
Polarity	+	-	+	-	-	+	+	+	+

min240041

TRANSMISSION CONTROL MODULE (TCM) PINOUTS

ECVT

Connector X19561 (42-pin, black)

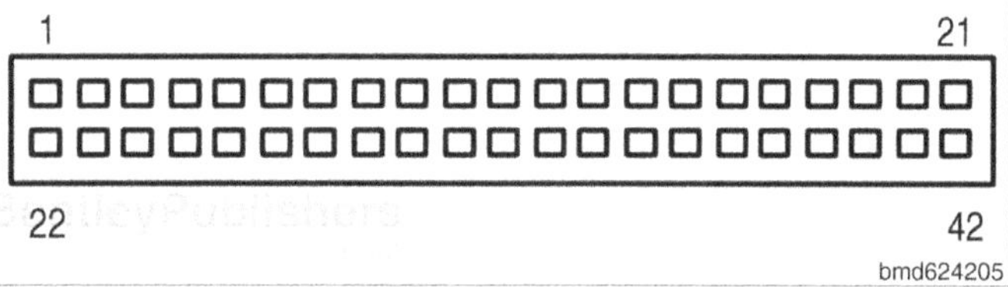

Pin	Wire color	Circuit description	KOEO	Notes
1	or	signal solenoid valve, clutch	battery volts	solenoid not activated
2	brn/grn	signal solenoid valve, clutch	battery volts	solenoid not activated
3	grn/yel	terminal 87	battery volts	
4	brn	ground	0.1 volts (maximum)	
5	blk/grn	signal, drive stage solenoid	7 volts	
6	blk/wht	signal, drive stage solenoid valve	11.5 volts	
7	blk/yel	signal, drive stage solenoid	11.5 volts	
8	blk/red	signal, drive stage solenoid	7 volts	
9		not used		
10		not used		
11		not used		
12		not used		
13		not used		
14	blu/blk	signal, EWS	battery volts	
15	yel/blk	CAN high	2.7 volts	
16	yel/brn	CAN low	2.3 volts	

Pin	Wire color	Circuit description	KOEO	Notes
17		not used		
18-21		not used		
22	or	signal solenoid valve, clutch	battery volts	solenoid not activated
23	brn/grn	signa,l drive stage solenoid	battery volts	solenoid not activated
24	grn/yel	terminal 87	battery volts	
25	brn	ground	0.1 volts (maximum)	
26	wht/blu	signal LED P	1.05 volts (in park)	2.5 volts (not in park)
27	wht/brn	signal LED R	1.05 volts (in reverse)	2.5 volts (not in reverse)
28	wht	signal LED N	1.05 volts (in neutral)	2.5 volts (not in neutral)
29	wht/blk	signal LED D	1.05 volts (in drive)	2.5 volts (not in drive)
30	wht/yel	signal LED MAN/SP	1.05 (in man / sp)	9.3 volts (not in man / sp)
31	brn/blu	signal selector lever position P	0.03 volts (in park)	0.3 volts (not in park)
32	brn/yel	signal selector lever position R	0.03 volts (in reverse)	0.3 volts (not in reverse)
33	brn/wht	signal selector lever position N	0.03 volts (in neutral)	0.3 volts (not in neutral)
34	brn/or	signal selector lever position D	0.03 volts (in drive)	0.3 volts (not in drive)
35	gry/yel	signal selector lever position Sport	0.03 volts (in sport)	0.3 volts (not in sport)
36	gry/wht	signal tiptronic UP	0.03 volts (shift up)	0.3 volts (no shift button depressed)
37	grn/vio	signal tiptronic UP	0.03 volts (shift up)	0.3 volts (no shift button depressed)
38	gry/blu	signal tiptronic Down	0.03 volts (shift down)	0.3 volts (no shift button depressed)
39	grn/gry	signal tiptronic Down	0.03 volts (shift down)	0.3 volts (no shift button depressed)
40-42		not used		

6-speed (AISIN) transmission

Connector X11633 (35-pin, black)

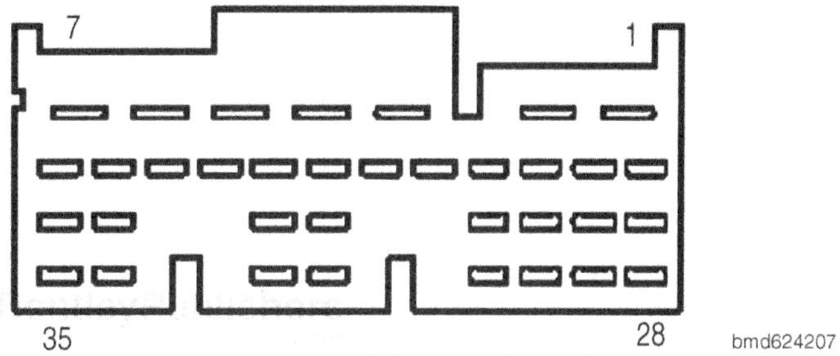

Pin	Wire color	Circuit description	KOEO	Notes
1	gry/wht	Lock up control solenoid, SLU-	battery volts	TCM ground controlled solenoid
2	yel/blk	Shift control solenoid assembly, SLB1-	battery volts	TCM ground controlled solenoid
3	brn	Ground	0.1 volts (maximum)	
4	blu/vio	Line pressure control solenoid, SLT+	battery volts	solenoid voltage supply
5	yel/grn	Shift control solenoid assembly, LSLC2+	battery volts	solenoid voltage supply
6	blu/yel	reverse gear signal	0 volts	battery volts supplied in reverse, (rev. switch)
7	vio/blu	3 way solenoid assembly, S1	0 volts	battery volts supplied by TCM to activate solenoid
8		not used		
9	grn/blu	Line pressure control solenoid, SLT-	battery volts	TCM ground-controlled solenoid
10	vio/gry	Shift control solenoid assembly, SLC3-	battery volts	TCM ground-controlled solenoid
11	vio	Shift control solenoid assembly, SLC2-	battery volts	TCM ground-controlled solenoid
12	blu/wht	Shift control solenoid assembly, SLC1-	battery volts	TCM ground-controlled solenoid

Pin	Wire color	Circuit description	KOEO	Notes
13	brn	Lock up control solenoid, SLU+	battery volts	solenoid voltage supply
14	grn	Shift control solenoid assembly, SLB1+	battery volts	solenoid voltage supply
15	yel/gry	Shift control solenoid assembly, SLC3+	battery volts	solenoid voltage supply
16	yel/vio	Shift control solenoid assembly, SLC1+	battery volts	solenoid voltage supply
17	vio/grn	3 way solenoid assembly, S2	0 volts	battery volts supplied by TCM to activate solenoid
18		not used		
19	red/blk	Signal, shift lever lock selector, SFL	0 volts	battery volts supplied by TCM to activate solenoid
20		not used		
21		not used		
22	blu/brn	Oil temperature sensor, OT	battery volts, (varies dependant on oil temp)	trans oil temp sensor positive
23	blu/blk	Oil temperature sensor, OTG	0.1 volts	trans oil temp sensor ground
24		not used		
25		not used		
26	grn/vio	terminal 15	battery volts	Ignition voltage
27		not used		
28	yel/brn	CAN low	2.7 volts	
29	yel/blk	CAN high	2.3 volts	
30		not used		
31		not used		
32		not used		
33	brn	ground	0.1 volts (maximum)	
34	red	terminal 30	battery volts	battery volts
35	red	terminal 30	battery volts	battery volts

Connector X11634 (35-pin, natural)

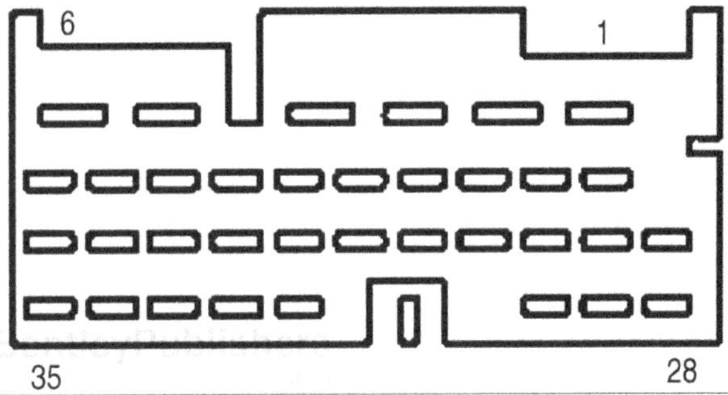

bmd624206

Pin	Wire color	Circuit description	KOEO	Notes
1		not used		
2		not used		
3	wht/blk	Position switch signal A	battery volts	toggled to ground by trans range switch
4	yel/gry	Position switch signal M	0.1 volts (maximum)	TCM ground
5	grn/gry	Paddle switch, TP-	Toggled to ground through resistors	paddle shifter input
6	grn/vio	Paddle switch, TP+	Toggled to ground through resistors	paddle shifter input
7	blk/brn	Output speed sesnor, NOUT+	battery volts	speed sensor signal, hall signal
8		not used		
9	yel/blu	Position switch PA signal	battery volts	toggled to ground by trans range switch
10	red/yel	Position switch B signal	battery volts	toggled to ground by trans range switch
11	gry/wht	Manual shift switch +	battery volts	manual upshift toggles to ground
12-16		not used		
17	blk	Input speed sensor ground, NIN-	0 volts	ground
18	wht	Output speed sensor ground, NOUT-	0 volts	ground
19		not used		

Pin	Wire color	Circuit description	KOEO	Notes
20	vio/wht	K-bus	battery volts	Data waveform
21	vio/blk	Position switch C signal	battery volts	toggled to ground by trans range switch
22	gry/blu	Manual shift switch -	battery volts	manual downshift switched to ground
23-27		not used		
28	grn	Input speed sensor +, NIN+	battery volts	speed sensor signal, hall signal
29-35		not used		

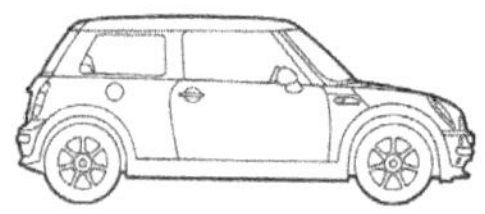

32 Electro-Hydraulic Power Steering (EHPS)

Table of Contents

GENERAL

MINI Cooper MK 1 models are equipped with electro-hydraulic power steering (EHPS) which uses an electric motor to drive an hydraulic pump. EHPS replaces the more traditional engine driven pump design. All EHPS electronic controls are incorporated in the pump unit.

The EHPS pump unit weighs 5.5 kg (12.1 lbs). It is mounted by four rubber insulators and positioned in front of the steering rack. The hydraulic pump displaces 1.25 cm^3 / revolution, developing a pressure of 105 bar (1523 psi) at a max. pump motor speed of 4,200 rpm.

The hydraulic pump incorporates pressure relief valve. The smaller metal pipe from the pump is the high pressure side. There is a small reservoir fitted to the end of the pump with a remote reservoir fitted to the bulkhead. The system is filled with Pentosin CHF 11 S

CAUTION—
- *Always use Pentosin 11 S in the EHPS system.*

EHPS System

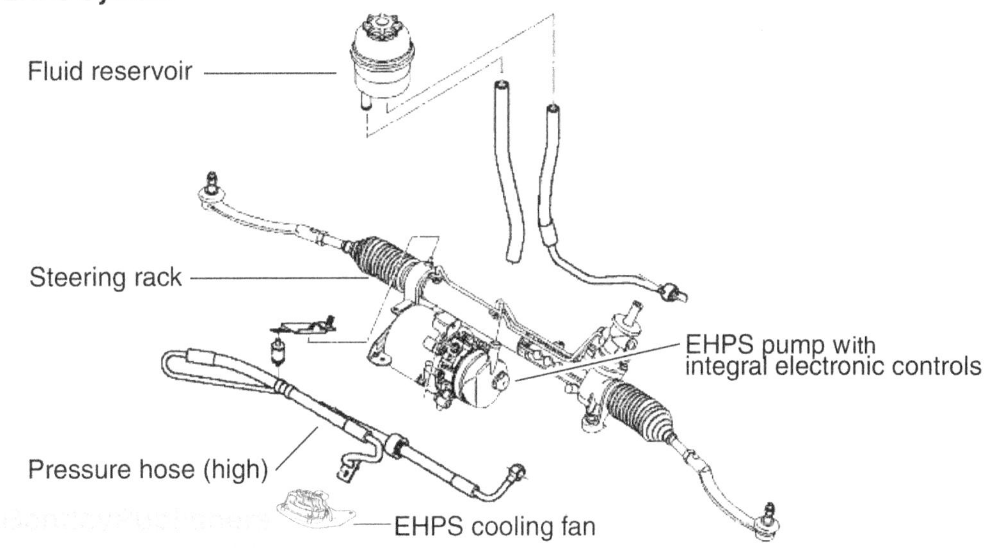

System operation

The pump motor operates only with the engine running.The pump receives an engine speed signal to operate the pump. The speed signal is supplied by the alternator on terminal 61.

The pump's electric motor is controlled by the EHPS control module, which is integral to the pump. EHPS steering pump operates between 9 and 16 V. Outside this range, there is no power-assisted steering.

There are two modes of operation for the EHPS: no steering assistance and steering assistance.

With the engine running and no steering assistance required, the pump idles at approximately 80% capacity (3,500 rpm).

Movement of the steering wheel changes the hydraulic pressure within the circuit, which in turn affects the operation of the pump. This is identified at the EHPS control module by an increase in current. The pump speed is then increased to 4,500 rpm which is 100% capacity.

The current draw of the electric pump motor with an idling pump is normally between 7 and 8 amps.

EHPS pump disassembled

1. Reservoir
2. Motor housing
3. Motor armature
4. EHPS control module
5. EHPS oil pump

BMD632104

When the EHPS control module detects the end of the steering maneuver or, for example, detects that the steering stop has been hit, the current is briefly reduced to 30 A.

Power steering pump power consumption	
Standby mode (idle)	7.0 amps
Average consumption	11.5 amps
Maximum assist	120 amps

If the pump approaches an overheat condition, the control module reduces output to 80%. If pump temperature continues to rise, the pump shuts down to protect itself (electronics) from damage. This happens if the power steering cooling fan fails.

The EHPS control module has diagnostic capabilities. The following faults can be stored in fault memory.

- Power supply fault
- Electric motor fault
- Fault caused by excess temperature
- Control module fault

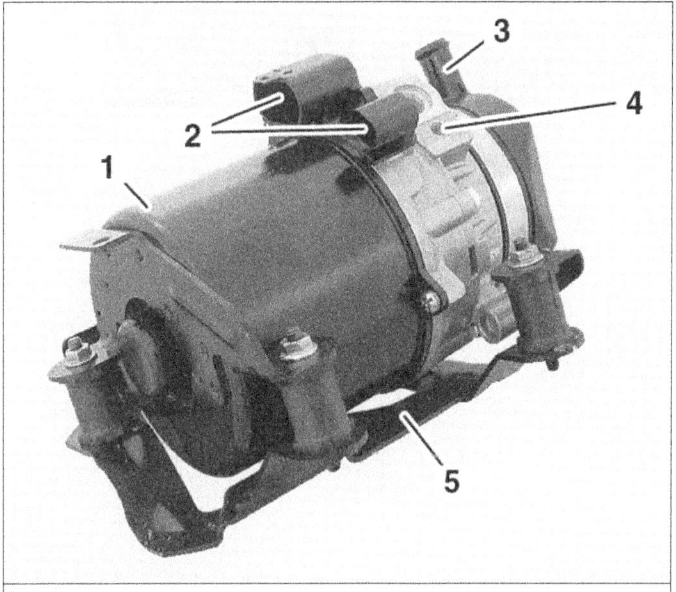

1. Steering pump
2. Electrical connectors
3. Oil inlet
4. Oil outlet
5. Support bracket

bmd632001

A small electric fan used to cool off the EHPS, mounted on rubber bushing to the subframe below the pump unit. There is no connection nor communication between the EHPS pump and the cooling fan. The fan power is supplied from a relay that is activated by the same wire that supplies power to the engine cooling fan. When the ECM activates the engine cooling fan, the EHPS fan is also activated.

The ECM activates the high speed fan relay when the engine temperature reaches 122°C (251°F). Once the engine temperature drops by 4°C (7°F) the engine cooling fan will run in low speed and the power steering fan will deactivate.

POWER STEERING PUMP TESTING

EHPS system fault codes and data stream can be retrieved using the factory or an enhanced scan tool.

When using the factory or an enhanced scan tool with the ability to display power steering pump data stream, the following information is available.

EHPS system scan tool data				
Test condition	**Motor current**	**Voltage PWM**	**Rotational pump speed**	**Motor control PWM**
Key ON engine OFF	0 amps	7 - 7.5 volts	2600 rpm	0%
Engine running, steering inactive	11 - 12 amps	11 - 11.5 volts	3600 - 3800 rpm	83.5%
Engine running, steering active	30 - 40 amps	battery voltage ± 10%	4000 - 4500 rpm	100%

Voltage checks

The EHPS control module has two connections at the pump. The 2-pin connector contains the voltage supply of the electric motor. The 3-pin connector contains the cable for terminal 15, 61 and the D bus for the EHPS control module.

2-pin connector (voltage supply)

- KL 31 (term. 1, black wire)
- KL 30 (term. 2, red wire)

3-pin connector

- KL 15 (term. 1, green/white wire)
- KL 61 used to determine if the engine is running (term. 2, yellow wire)
- D-bus (diagnosis) used for communication with the control electronics integrated in the pump (term. 3, violet/white wire)

The following checks are done with the connectors (X13441 and x6975) connected to the pump unit. The simplified wiring diagram is provided as a guide.

The pump does not run until the engine starts and the voltage on terminal 2 (X13441) exceeds static battery voltage.

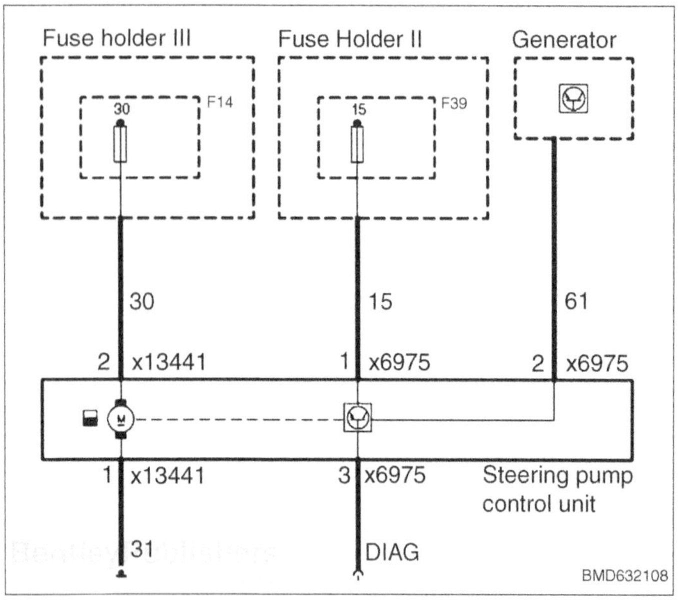

Connector x13441 (2-pin)

BMD632005

Power steering pump test values (at pump connector x13441)			
Terminals	**Conditions**	**Test values**	**Circuit**
1 and ground	key ON engine OFF engine running at idle @2500 rpm, (no load)	0.1 volts (max.)	engine ground
2 and ground	key OFF key ON engine OFF engine running at idle @2500 rpm (no load)	BV (battery voltage)	voltage supply to pump (hot at all times)

Connector x6975 (3-pin)

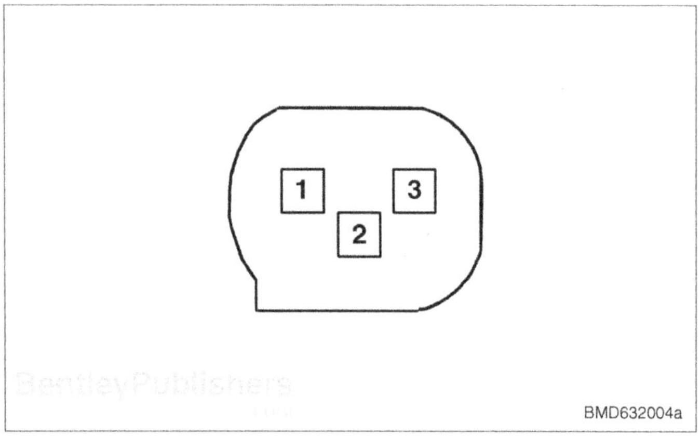

Power steering pump test values (at pump connector x6975)			
Terminals	**Conditions**	**Test values**	**Signal/Circuit**
1 and ground	key ON engine OFF engine running at idle @2500 rpm, (no load)	BV (battery voltage)	Voltage supply to pump, (term. 15)
2 and ground	key ON engine OFF engine running at idle @2500 rpm (no load)	BV (battery voltage)	Signal from alternator (charging system voltage)*
3 and ground	key ON engine OFF engine running at idle @2500 rpm (no load)	BV (battery voltage)	D-bus - no activity on bus line
		6.5 - 7 volts	D-bus - activity on bus line

*An alternate place to access this signal is at the alternator connector. Alternator pin assignment vary depending on vehicle application. Check the appropriate wiring diagram to determine correct pin assignment.

NOTE —

- Variation of pump wiring to alternator:
  All Cooper S models, pin 1 at alternator connector. For Cooper models it is dependant on alternator brand. Valeo is on pin 2. Denso is be on pin 1.

Simplified alternator schematic

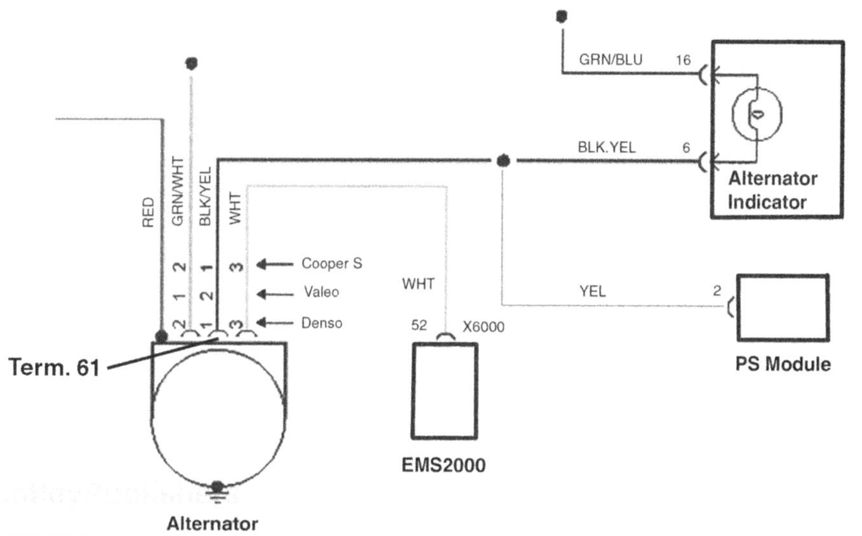

bmd632013

Power steering pump - data bus waveform(s)

Waveform signal from connector x6975 (pin 3) - no activity on bus

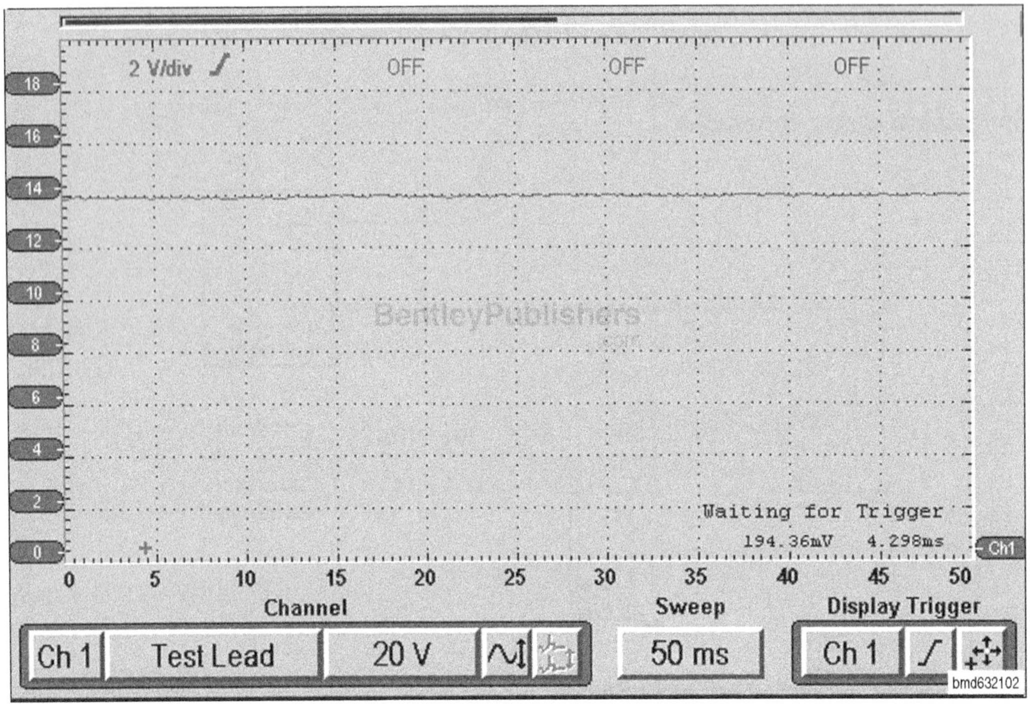

Waveform signal from connector x6975 (pin 3) - with activity on bus

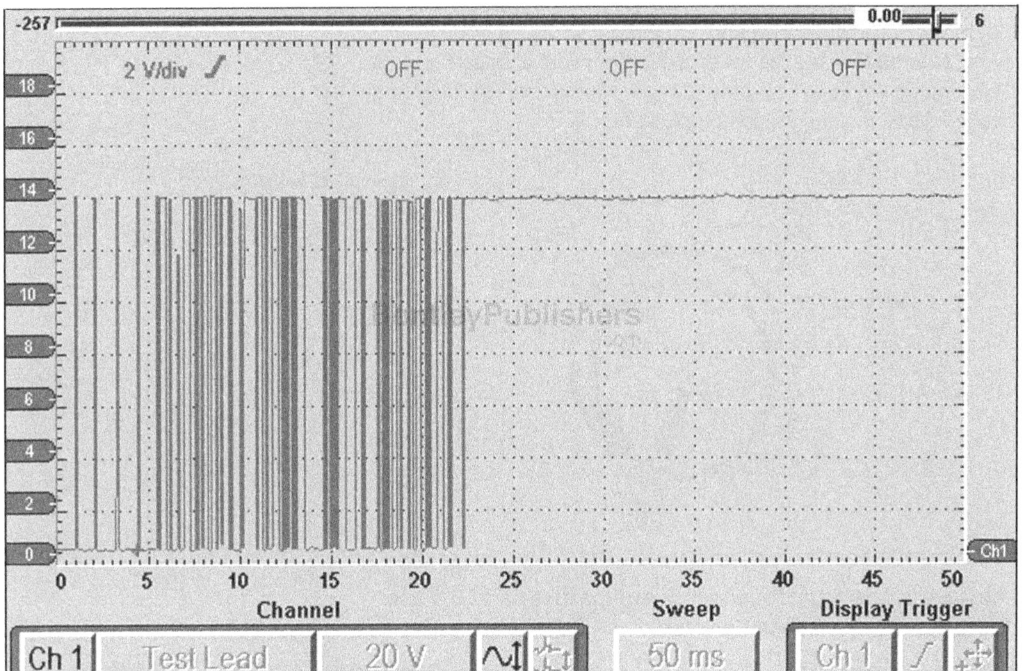

bmd632007

Pump current test

For this test, current is measured at the power steering pump fuse (F4, 100 amp) block. A current probe and two jumper wires (at least 8 gauge) are required to make the test. One jumper wire is bolted in place of the 100 amp bolt-in fuse. The other jumper wire is used to reconnect or extend the disconnected power supply cable.

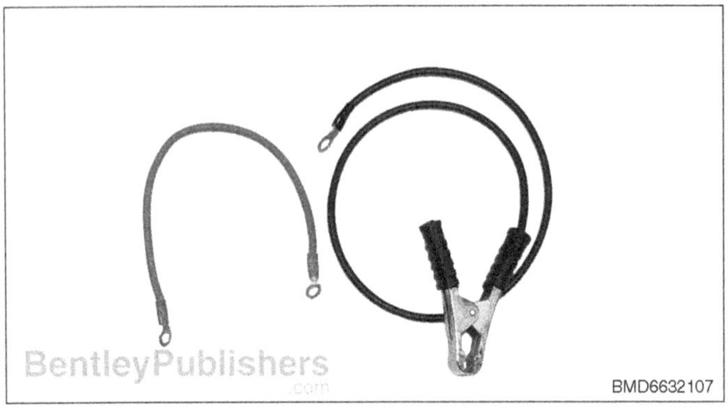

The pump fuse is located on the under side of the fuse panel 3. Fuse panel 3 is in the engine compartment, next to the left strut tower.

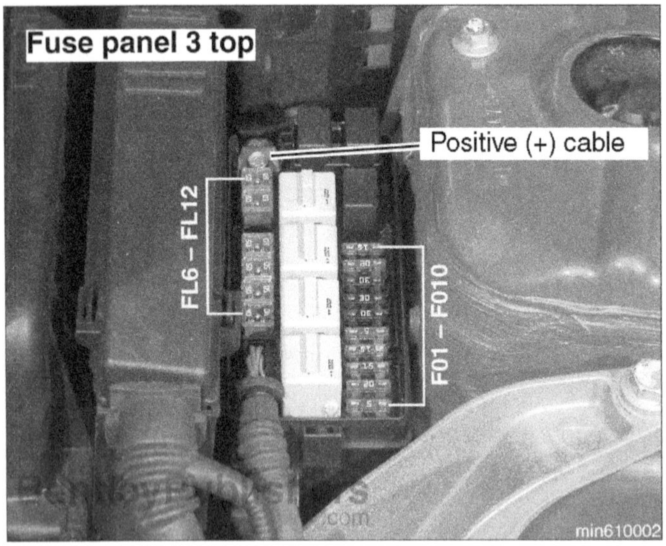

– To access 100 amp fuse (FL4):

- Disconnect battery negative (-) terminal.
- Cooper: Remove engine control module (ECM) and battery box.
- Cooper S: Remove engine control module and air filter housing.
- Disconnect positive (+) supply cable at rear of fuse panel 3.
- Remove fuse panel mounting screws.
- Lift panel and turn over. Remove bottom cover.

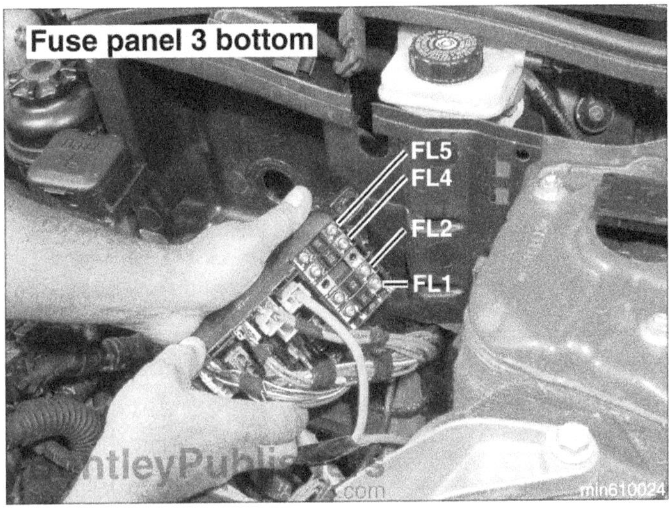

– Remove fuse FL4 and connect fused jumper wire across fuse terminals.

– Bolt one end of second jumper wire to fuse panel where positive (+) cable was earlier removed. Connected other end to disconnected positive (+) cable.

> ***CAUTION —***
> - *Be sure all positive connections are fused and insulated from bare metal before reconnecting battery negative cable.*

– Reconnect negative battery cable.

– Connect amp probe.

– Start engine and measure pump current.

The waveform below depicts good pump current taken at 83.5% output (as displayed on scan tool at idle). We suggest testing pump current only at 100% output, when the steering wheel is turned. This waveform is being included for illustration only. Current probe setting of 100 mv equals 1 amp.

Power steering pump current taken at idle (no steering activity)

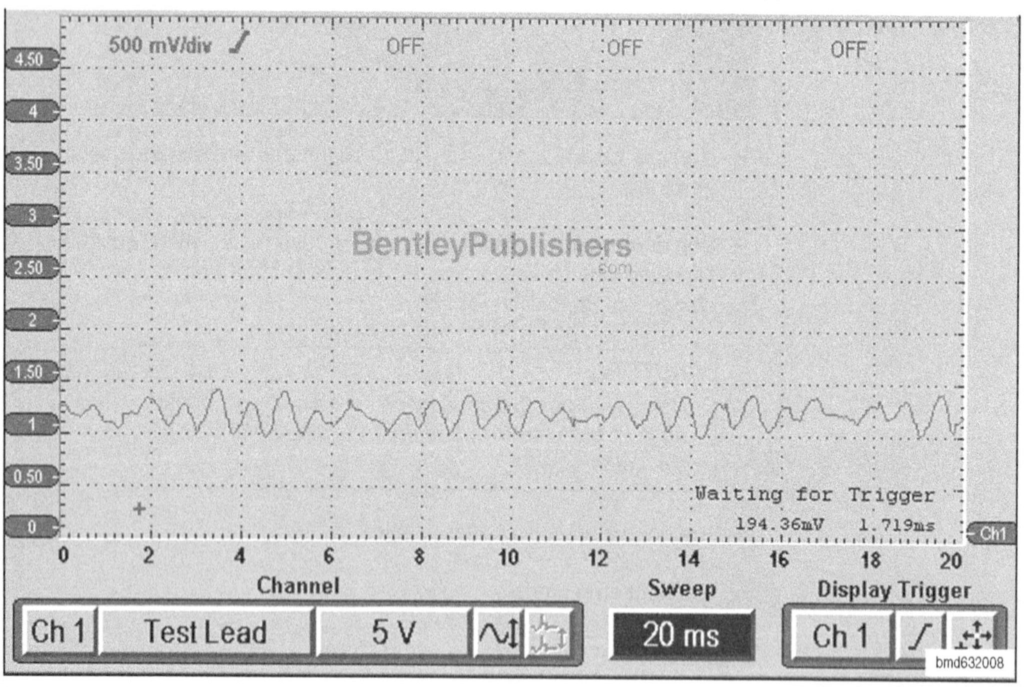

PS pump motor current taken at idle, with steering activity. The recommended current test procedure is done at 100% output. Current probe setting of 100 mv equals 1 amp.

Power steering pump current taken at idle (with steering activity)

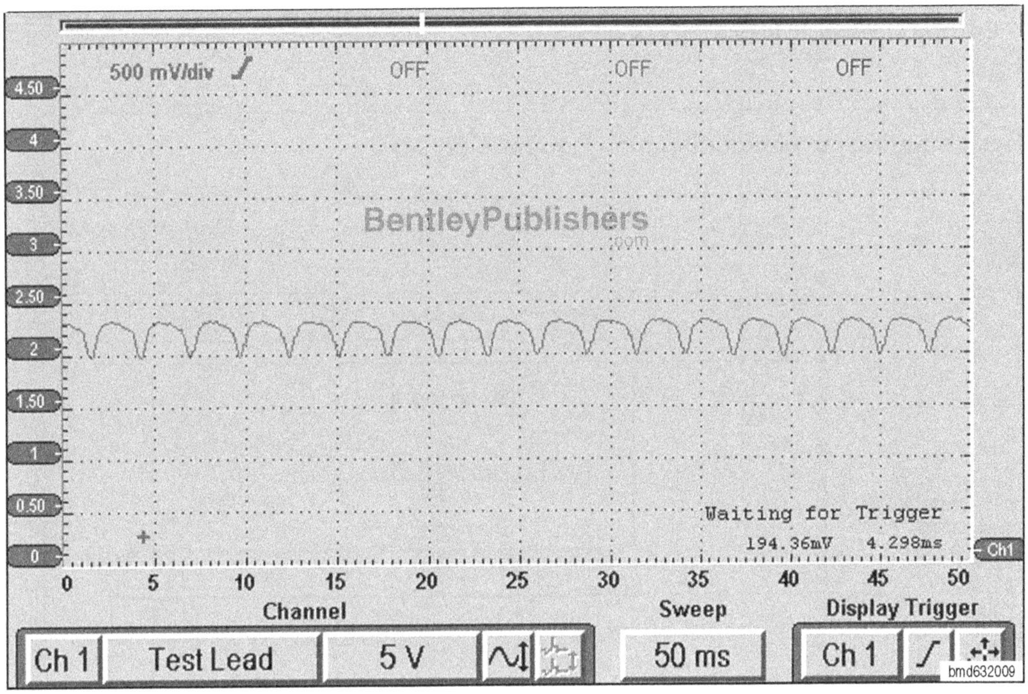

Power steering fan testing

NOTE —

- When testing the fan circuit, review the applicable wiring diagram for the specific test vehicle. Fan wiring and activation circuits vary depending on year and model.
- Below is an example from a 2004 Cooper S.

Power steering fan circuit diagram - Cooper S (2004)

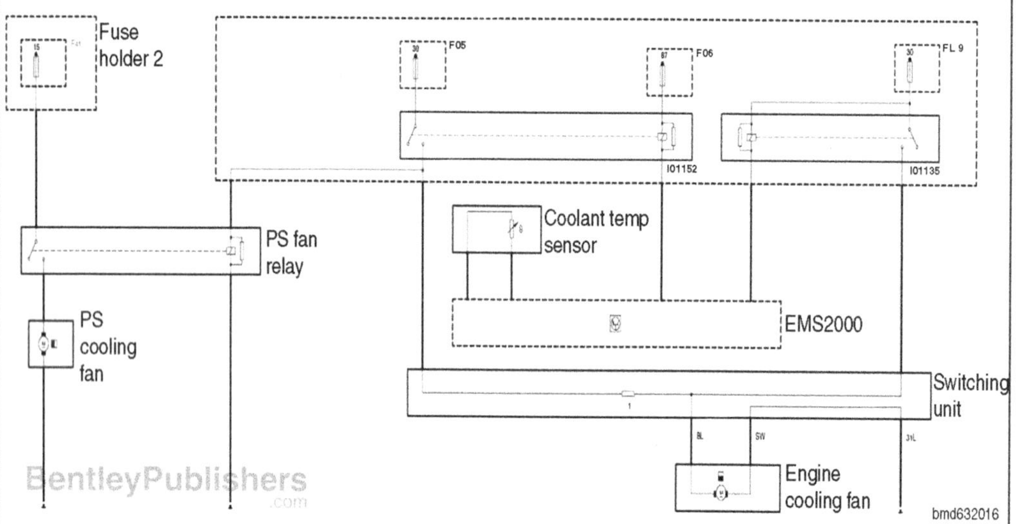

bmd632016

Testing power steering cooling fan at connector x6141

BMD632012

Power steering fan test values (at pump connector x6141)			
Terminals	**Conditions**	**Test values**	**Signal/Circuit**
1 and ground	key ON engine OFF engine running at idle @2500 rpm, (no load)	0.1 volts (max.)	fan ground
2 and ground	fan off	0.1 volts	relay controlled voltage supply
2 and ground	fan activated	BV (battery voltage)	relay controlled voltage supply

Current probe setting of 100 mv equals 1 amp.
When ON, the power steering cooling fan draws 2-3 amps

Power steering fan current - fan activated

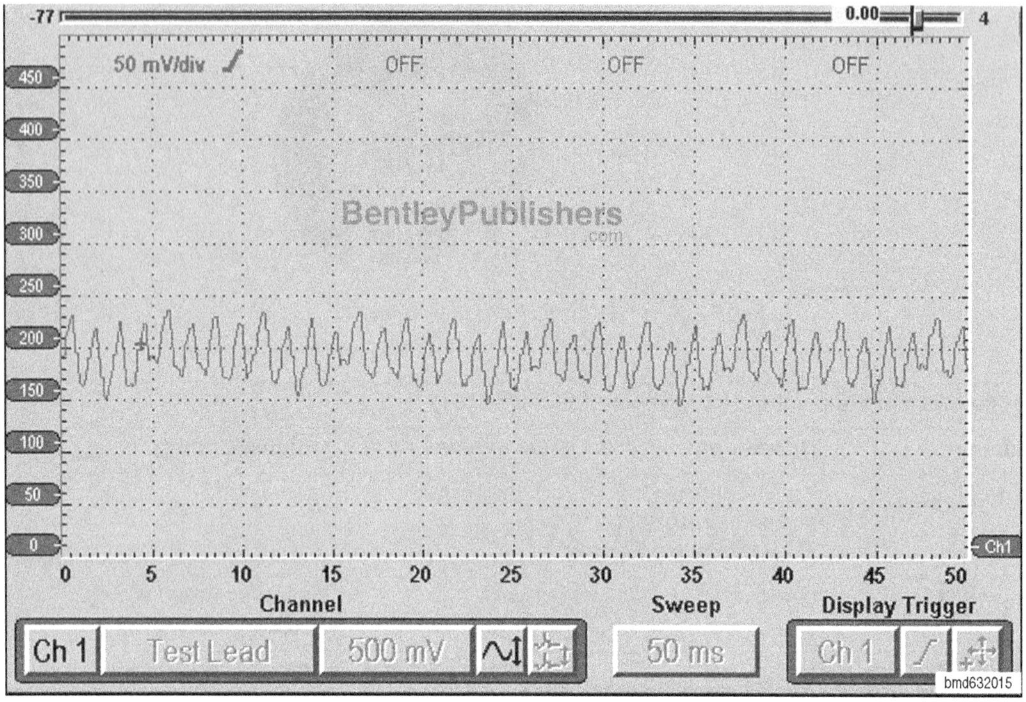

36 Tire Pressure Monitoring (RDW)

bmd636008

Table of Contents

GENERAL

The MINI tire pressure monitoring system (RDW) alerts the driver to changes in tire air pressure by monitoring rotational speed of the tires. RDW interfaces with the ABS/DSC system to receive wheel speed information. Only rotational speed is monitored, not tire pressure.

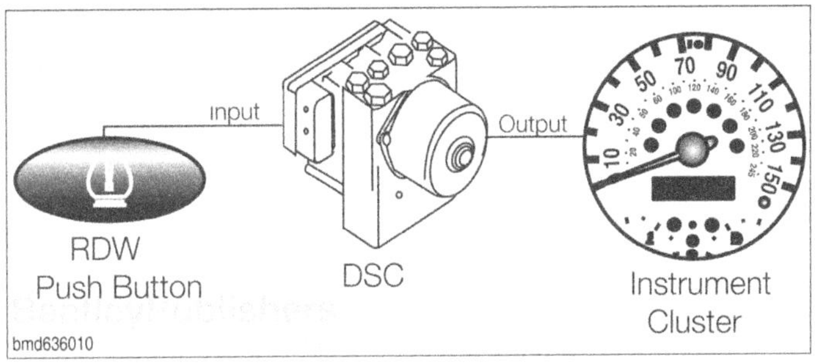

RDW compares the wheel speeds of diagonally-opposing wheels to calculate the average speed and thus detect if there has been a loss of pressure. Extreme driving situations such as heavy acceleration or rapid cornering are detected by the software and corrected for.

The purpose of the RDW system is to warn the driver of potentially hazardous conditions caused by tire air loss. Under-inflation is the number one cause of tire damage. It results in unnecessary tire stress, irregular wear, loss of vehicle control and accidents. A tire can lose up to half of its air pressure and not appear to be flat.

Under inflated tires are major causes for blowouts. An under inflated tire runs hot due to the action of the tire as it rolls under the car. The sidewalls become squashed outward, and It is this movement that causes the tire to generate heat and fail.

The driver is responsible for ensuring that tire pressure is set correctly. The system can only monitor the pressure set at the moment of initialization compared to the warning values stored in the control module. See **Initializing RDW System** in this chapter.

> ***WARNING —***
> - *The RDW system is not a substitute for regular checks of tire air pressure. It is a warning system for loss of tire pressure that occurs while driving.*

Obtain correct cold pressure (normal pressure when the tire is cold) from the owners manual.

> ***CAUTION —***
> - *Cold air pressure means that the vehicle has been driven less than one mile.*

SYSTEM COMPONENTS

RDW system components:

- RDW control module (up to 03/2002)
- ABS, ABS/DSC control module (from 03/2002)

> ***NOTE —***
> - As of 03/2002, the RDW control function is integrated within the traction control module.

- RDW reset switch
- Display in instrument cluster
- Wheel speed sensors (from ABS/DSC)

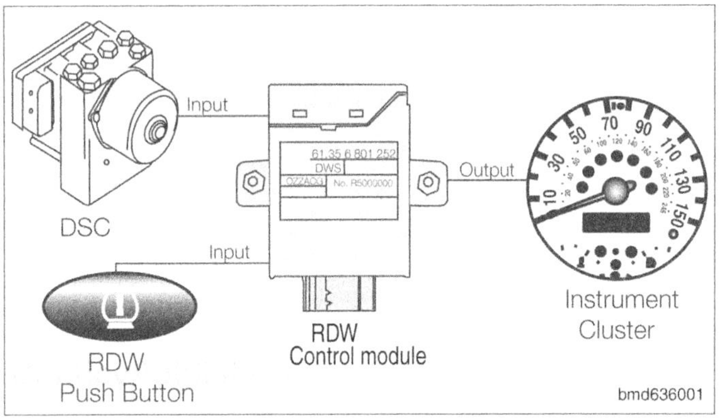

RDW control

On the early cars (up to 03/2002), The RDW control module (A502) is located in the right rear quarter panel.

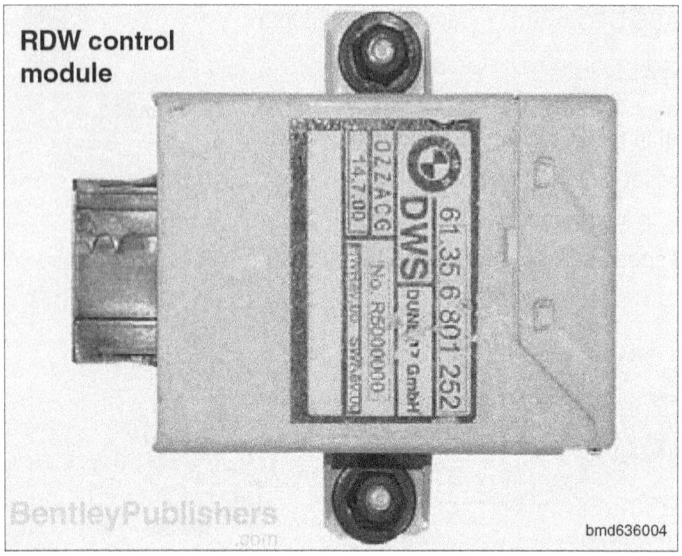

On cars built after 03/2002, the RDW function is integrated in the ABS (A7) or the ABS/DCS (A65) control module, located in the rear engine compartment cowl area.

RDW reset switch

The RDW reset switch in the center console (**arrow**) is required for the initialization procedure after adjusting tire pressures.

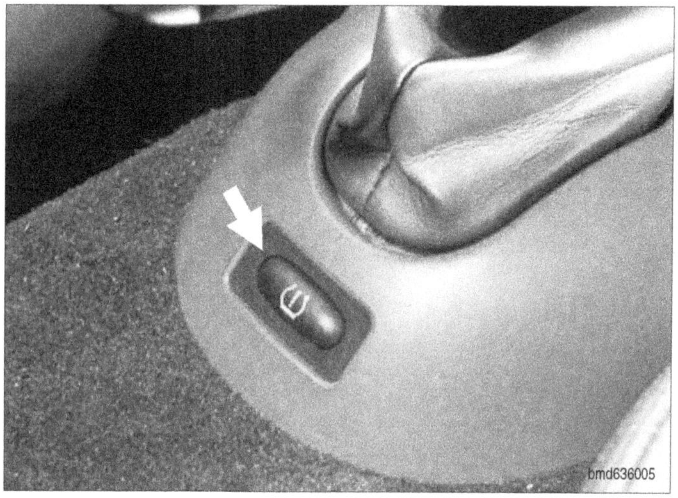

Alternate button location (later models).

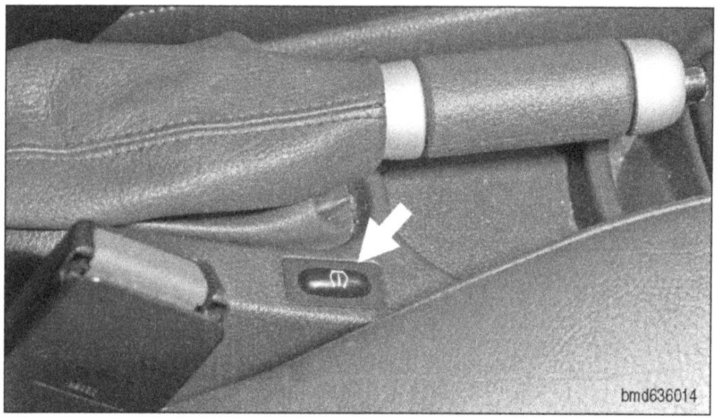

RDW warning light

When the ignition is switched on, the control module performs a self-test and the amber RDW warning LED illuminates. The LED remains lit unless a system OK message is received via K-bus from the control module.

If low tire pressure is detected, the instrument cluster provides an audible warning to the driver and the RDW LED flashes.

Wheel speed sensors

The RDW system makes use of the ABS/DSC wheel speed sensors. Wheel speed information is via the ABS/DSC control module. On cars built up 03/2002, this wheel speed signal signal is sent to the RDW control module.

The RDW system receives wheel speed signals for all four wheels.

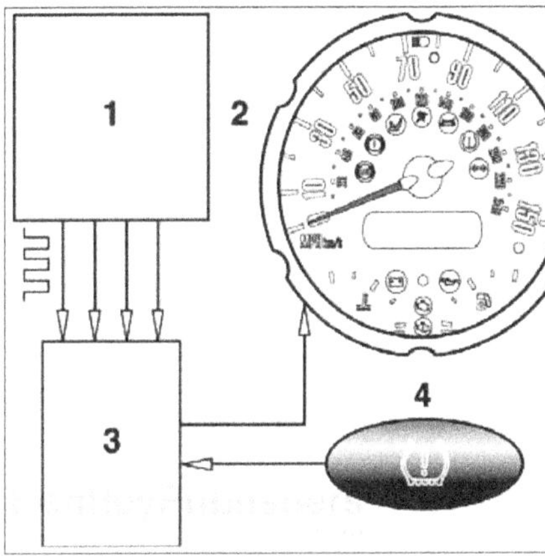

RDW control

1. ABS/DSC Control Unit
2. Instrument Cluster
3. RDW Control Unit (up to 02/2003 only)
4. RDW Push Button

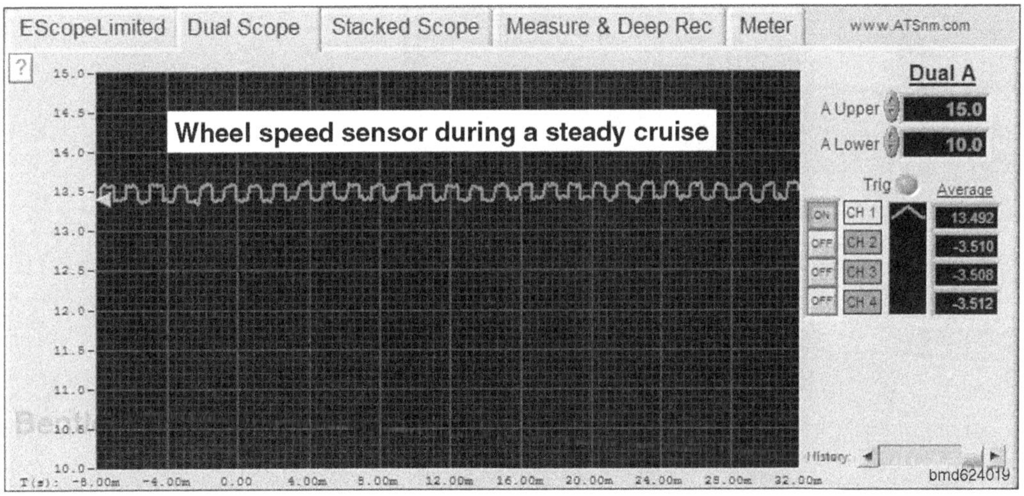

INITIALIZING RDW SYSTEM

Without initialization, the RDW system may issue false warnings or may fail to issue a warning when necessary. False warnings may be given if initialization is not performed after tire pressures are changed or a wheel / wheel position is changed. False warnings may also be given if tires with large differences in the degree of ageing/wear are fitted on the same axle.

> ***NOTE —***
> - If a space saver spare tire is installed, the RDW cannot compensate for the difference in diameters of the wheels. In this case, initialization cannot be completed.

Initialize the system if:

- Tire pressures are adjusted (must be set cold)
- Tire positions are altered (interchanged on the same axle or between axles)
- Tires are changed (new tires for old tires, summer tires for winter tires etc.)

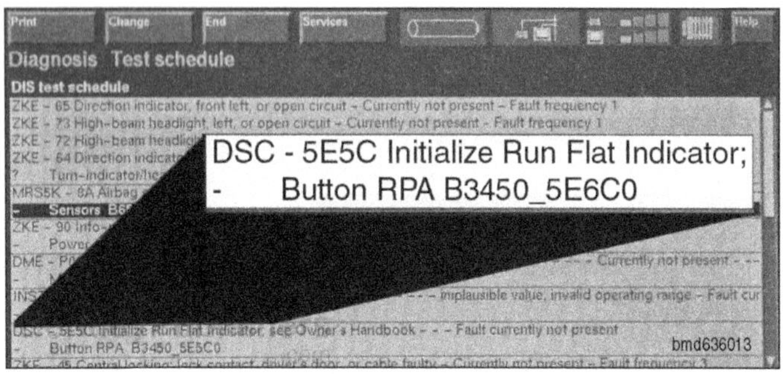

This initialization process can be performed manually using the RDW reset switch (**arrow**).

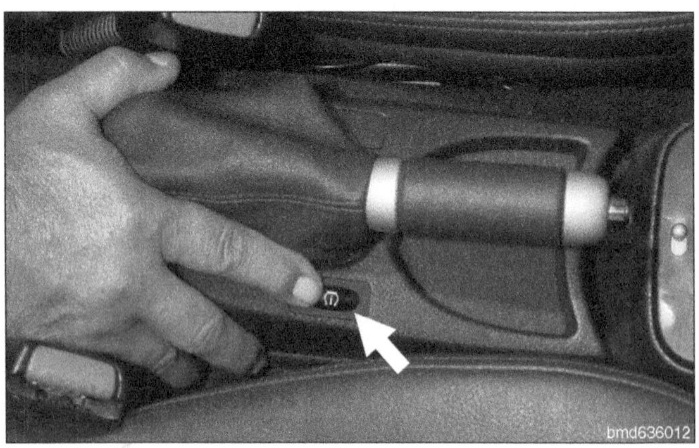

- Switch ON ignition (engine OFF) and then press and hold RDW switch for longer than 4 seconds. Tire warning lamp will turn OFF.

- The control module enters learn mode (initialization). The system then enters a learning phase. Depedning on preset threshold values, this phase can last anywhere from 45 minutes to several hours depending on the driving style and number of data rejections.
- The initialization routine can be interrupted as often as desired; interim results are stored in the control unit. The only way to verify if the system is fully initialized is via diagnosis requests using the factory scan tool.

RDW SYSTEM WIRING DIAGRAMS

Wiring schematic (up to 03/2002)

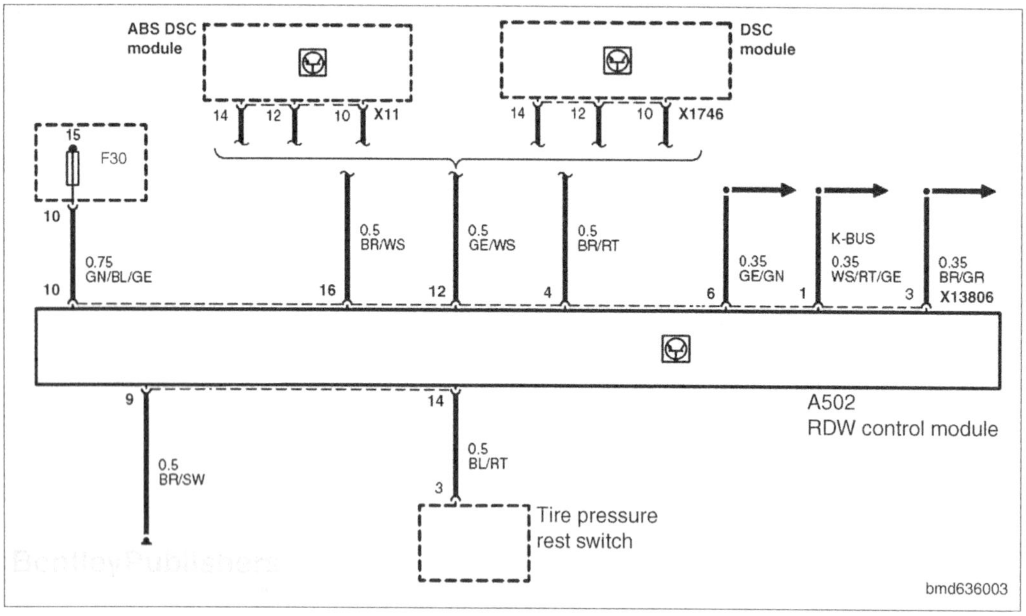

Wiring schematic (from 03/2002)

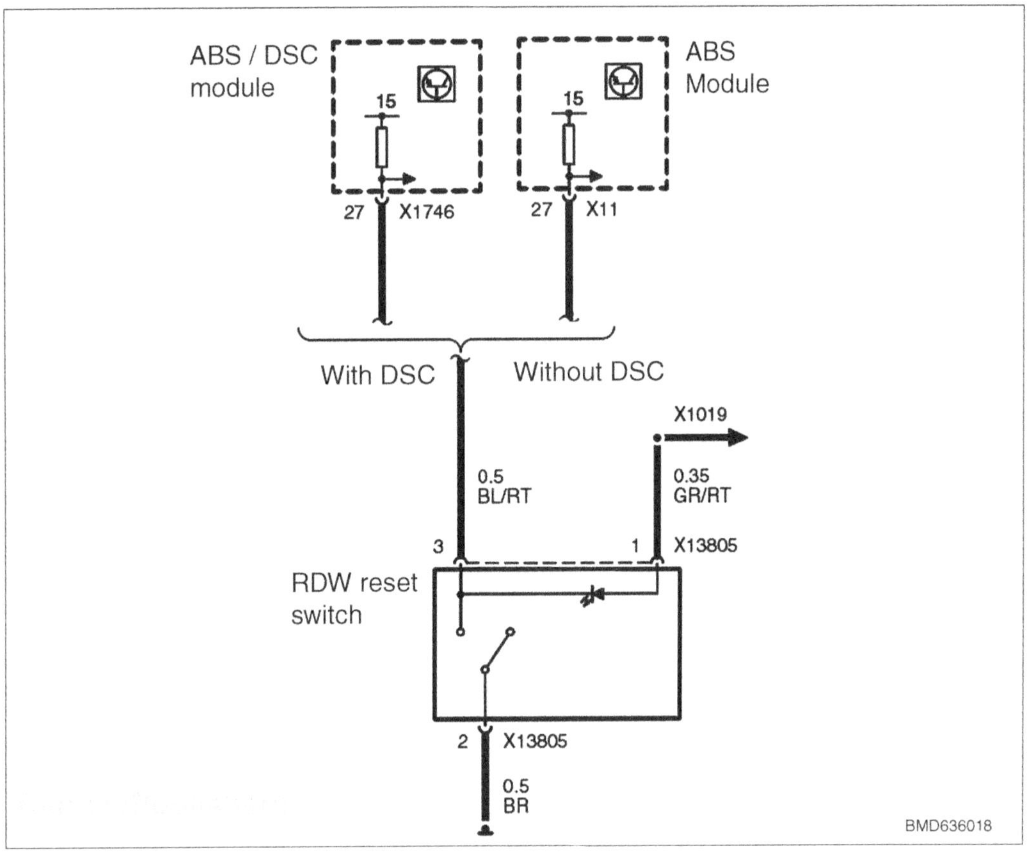

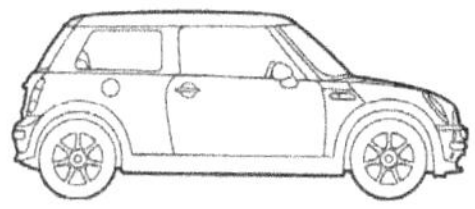

Notes:

61 Bus Systems

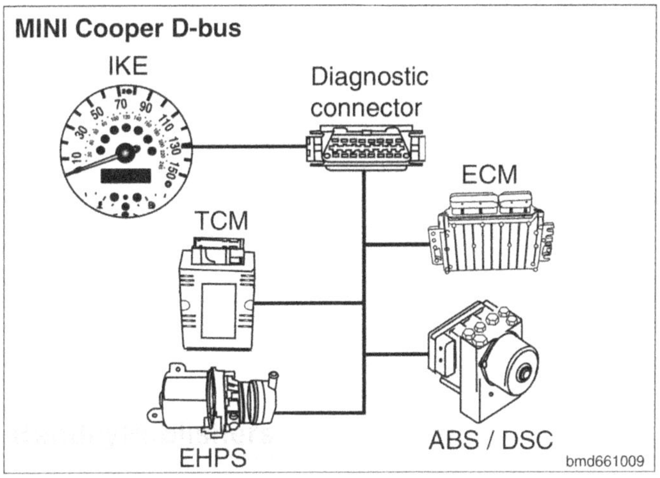

Table of Contents

GENERAL

This chapter describes the various bus systems used in MINI vehicles.

BUS SYSTEMS

In simple terms, the bus system is a group signal line system for exchange of data or messages. A bus makes it possible to connect a central module (master) with a large number of independent control modules. All modules are connected in parallel to the bus line.

The bus systems in use on the MINI Cooper are the CAN-bus, the K-bus and the diagnostic bus (D- and DS2 bus).

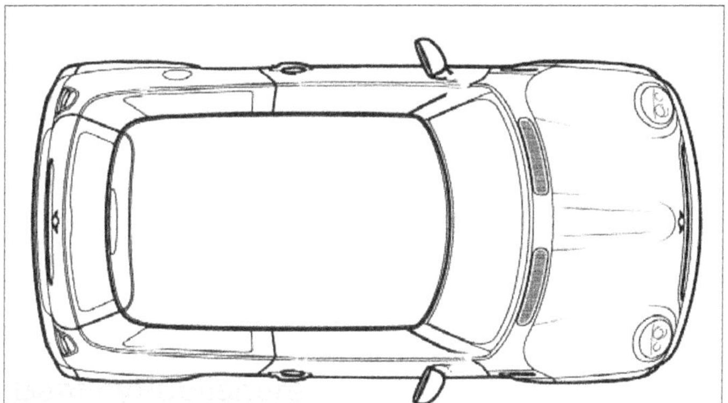

bmd661223

In the bus system, control modules are linked to each other, allowing communication and exchange of information. Through this multiplexing method, harness bulk and complexity is kept to a minimum.

Multiplexing is a technique that uses the same wiring for communication among several systems. The systems are all inter-connected, forming a network of communication lines. The information from sensors and switches are converted into digital signals by the system control module and relayed serially throughout the system network, which is also referred to as a data bus-system.

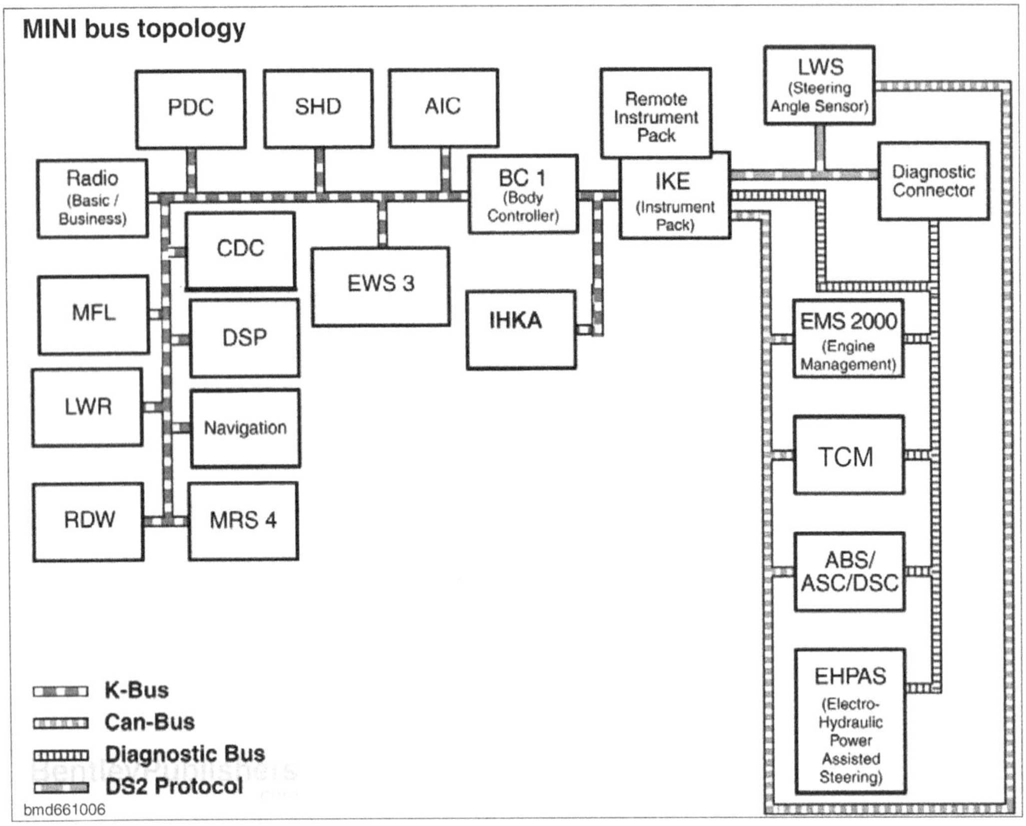

The IKE (instrument cluster electronics) acts as the communication gateway, enabling transfer of data from one bus system to another. The IKE contains a microprocessor which converts and processes all signals into the format required for transmission to another bus system.

Control modules transmit and receive commands and information on the buses as digital messages. Modules connected to the same bus use a common protocol (format) and baud rate (transmission speed in kbit / s) for the messages they transmit. The protocol and baud rate varies from bus to bus.

Messages transmitted on the bus systems are made up of binary digits, referred to as bits. The bit can be either 0 or 1. A combination of bits that make up a signal is transmitted on the bus system. Each control module converts the digital signal back to an analog value for its own use.

Each message contains coded information indicating the message beginning and ending, and an identifier or address to route the message to the correct destination. The signal also contains a check function which a receiving control module uses to check the plausibility of the signal.

The transmitting control module monitors the return of the signal on the bus from the receiving control module. This way the transmitting control module confirms the signal received is the same as the one transmitted and any faults can be dealt with.

A priority system of control module or message importance is established providing a smooth flow of information and ensuring that important messages are processed first.

Information from a single sensor is passed on to many control modules reducing the number of sensors needed. As an example, raw vehicle speed is sent to the DSC module, and a processed road speed is sent to the IKE over the CAN-bus. The IKE converts this signal into a K-bus message and transmits this via K-bus to BC1 (body controller 1) for the speed dependent wiper control and the radio for the speed dependent volume adjustment.

K-bus

The K-bus (body bus) system is a serial communication network and uses the wire color white / red / yellow (white wire / red stripe / yellow tracer).

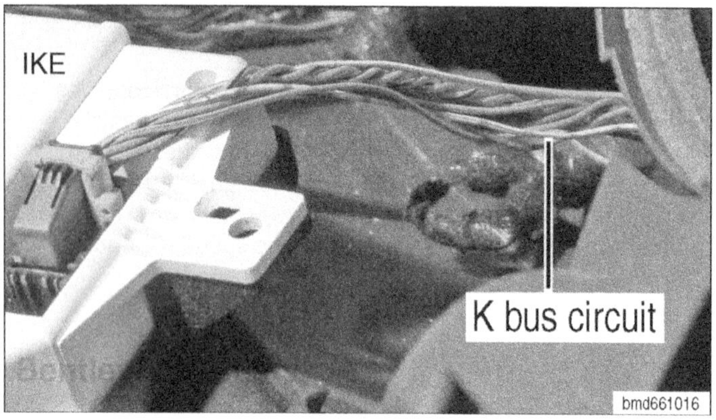

The K-bus system is divided into two segments named K-bus I and K-bus II. These two buses are essentially the same bus spliced inside the BC1.

K-bus communication occurs at 9600 bits per second. The K-bus switches between 0 - 12 volts (high / low) and has a low impedance, making it resistant to electromagnetic interference. Most K-bus users are active on the bus when the ignition is switched to the run position.

K-bus signal during communication

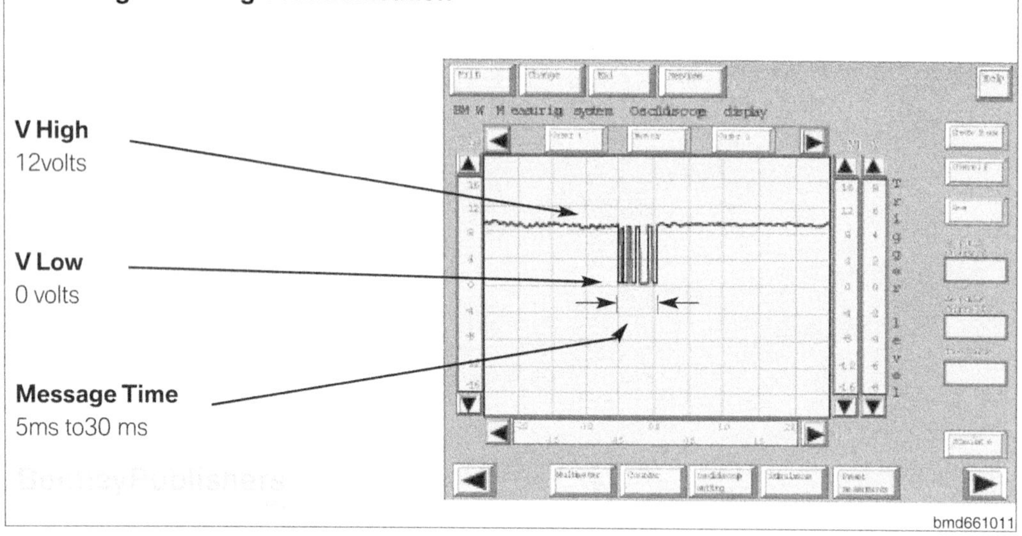

The K-bus network enters sleep mode 60 seconds after the ignition is switched OFF. While in sleep mode the K-bus rests at 12 volts. BC1 is the main body controller, supplying power and establishing message sequencing.

The K-bus system is primarily an event-driven system. Messages are sent after a request has been made for component operation or status change.

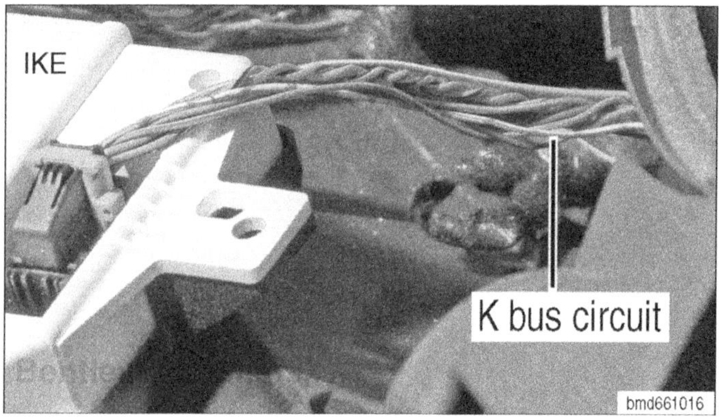

Each of the K-bus users are continually monitoring messages on the system. When a user needs to transmit, it waits until there is no activity (12 volts) and then transmits its message.

The sender then listens to the system for its own message to be returned by the receiver. If it doesn't receive its own message, a user with higher priority has also transmitted and its message was ignored. The lower priority user then has to wait until the bus becomes inactive and retransmits. When it receives its own message back (correctly) successful transmission is confirmed.

All control modules on the K-bus receive all messages, but only the module addressed in the message accepts and reacts to the data. The K-bus remains operational in case of a disconnected or failed control module. This parallel circuit is referred to as a tree structure with each control module occupying a branch.

The K-bus provides the diagnostic connection to the control modules located on the bus (except for the IKE).

Rules for operation:

- Only one module speaks at a time (serial communication)
- Everybody speaks at same speed
- Messages are acknowledged by recipient
- Message is repeated if addressed module fails to respond
- BC1 has priority
- Message aborted after 5 failed attempts

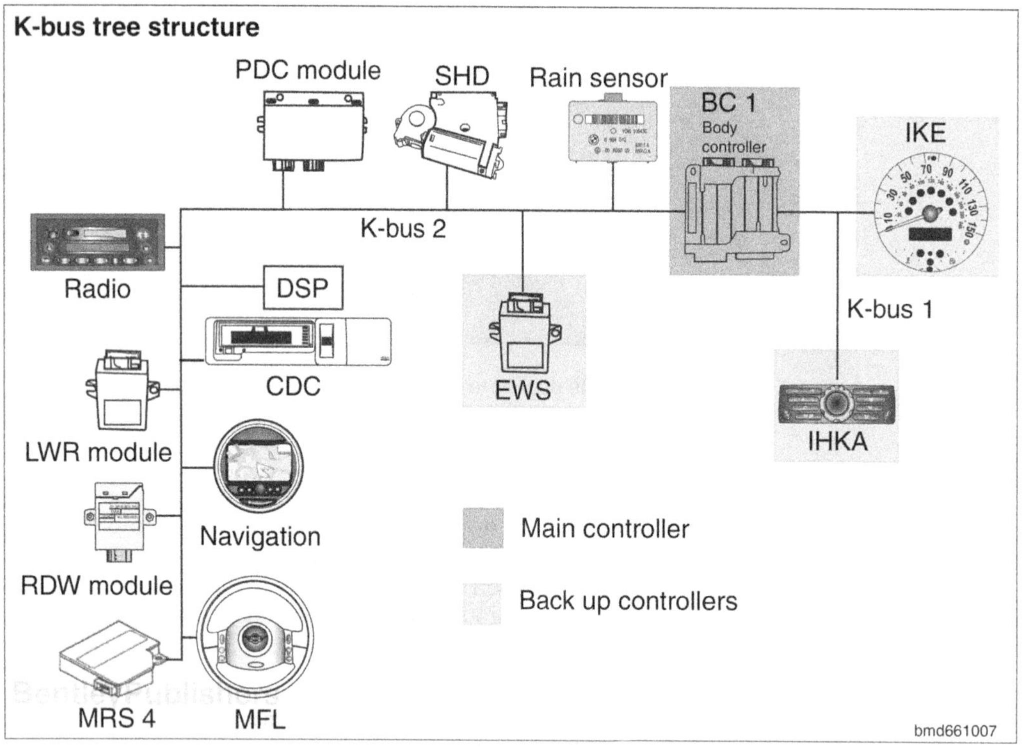

K-bus with no activity. Idles high at battery volts.

bmd632006

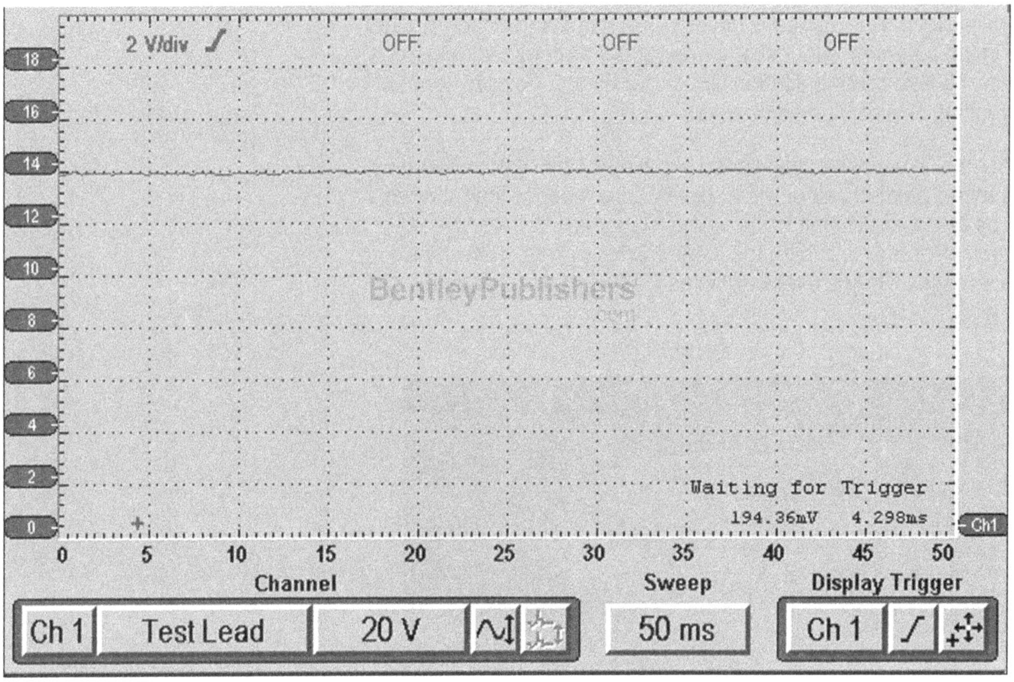

K-bus with normal bus activity. 0 to battery volts - square wave.

bmd632007

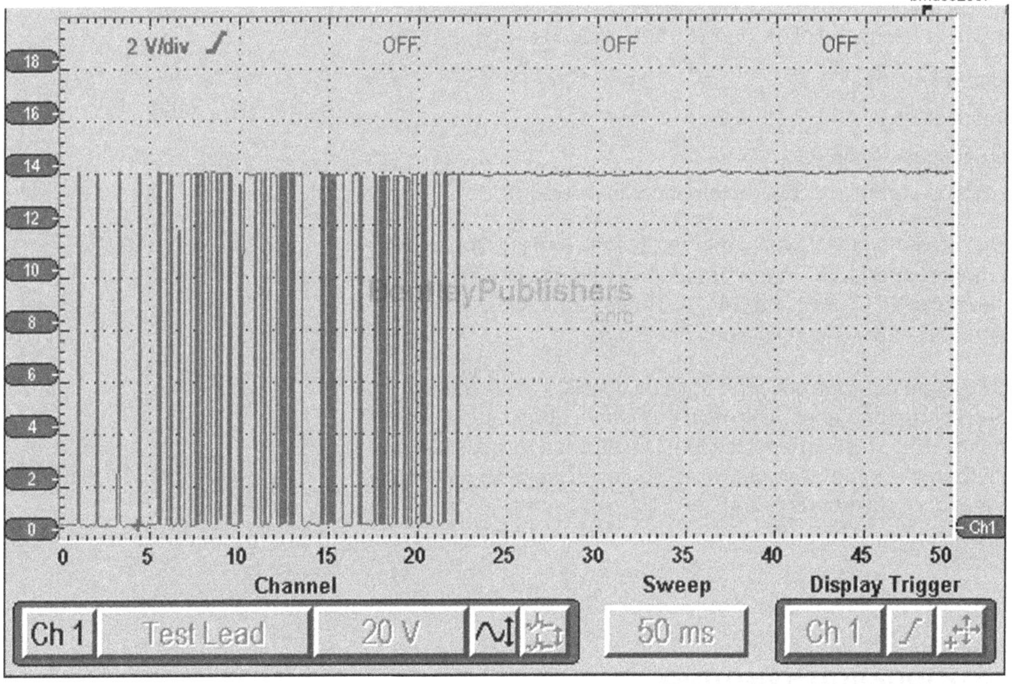

CAN-bus

The CAN protocol was originally developed by Intel and Bosch in 1988 for use in the automotive industry to provide a standardized, reliable and cost-effective communication bus for a car electronics to combat the increasing size of wiring harnesses.

The CAN-bus system is a high speed serial data bus system linked by an unshielded twisted pair of wires, yellow / black and yellow / brown. The wires are twisted to minimize electromagnetic interference. Both wires carry information and for the CAN-bus to operate, both signals must be present. They are referred to as CAN-high (CAN-H) and CAN-low (CAN-L).

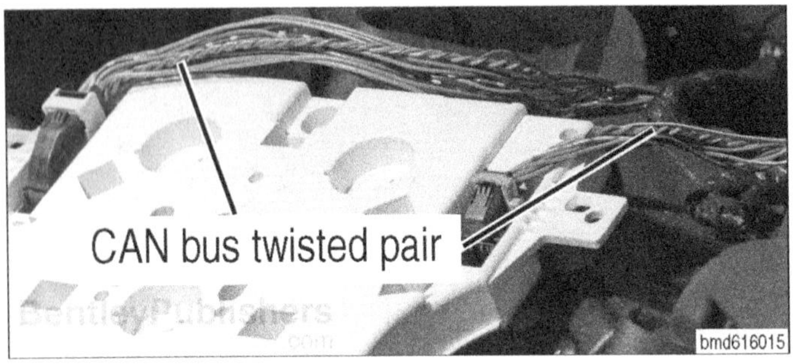

The CAN-bus system is the fastest of the bus systems used in MINI automobiles, transmitting at 500,000 bits per second. It is used for systems where the speed of exchange of information is vital;

- DME control module
- Instrument cluster
- ASC/DSC control module
- Transmission control module (automatic transmission)

For correct operation CAN-bus terminates at both ends (ECM and IKE or instrument cluster) with a control module resistance of 120 ohms, connected between CAN-H and CAN-L. These terminations ensure that bit errors due to signal reflections are avoided.

The CAN system uses a linear topology that consists of the main bus length and shorter branches. The main bus length terminates at the ECM and the IKE. The shorter stubs must be as short as possible and no longer than one meter. Any untwisted portion of the bus must not be longer than four centimeters (1.6 in).

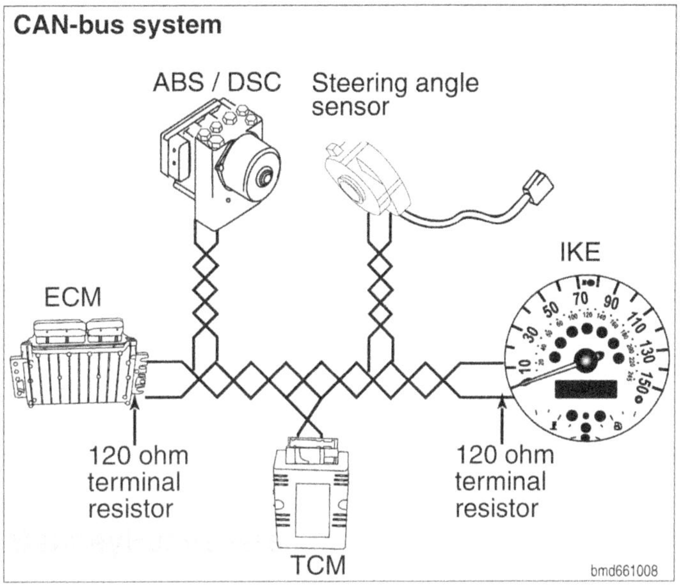

The CAN-bus consists of a twisted pair of wires:

- CAN-high (CAN-H, yellow and black)
- CAN-low (CAN-L, yellow and brown)

Both Can-H and CAN-L have a standing voltage of 2.5 volts. CAN-L is pulled low to approximately 1.5 volts while CAN-H is pulled high to approximately 3.5 volts.

With CAN-H and CAN-L both at 2.5 volts, there is no potential difference (voltage) between them. This is known as the recessive state and is equivalent to logic 1. With CAN-H pulled to 3.5 volts and CAN-L pulled to 1.5 volts there is a potential difference of 2 volts between them. This is known as the dominant state and is equivalent to logic 0. CAN-H and CAN-L always switch together. These two states (0 or 1) are the only two possible.

When a control module transmits a signal, it is made up of a series of dominant and recessive states. The signal is a combination of the two possible states, in effect 0 and 1, hence a digital signal.

Data transmitted from any subscriber on a CAN-bus does not contain addresses. Instead, the content of the message (RPM, TD, Temp, etc.) is labeled with an identifier code unique throughout the CAN. All of the subscribers receive the message and each one checks the message to see if it is relevant to its own operation.

Correctly operating CAN system

CAN-L (Low)

Flat line should average 2.5 volts with signal being pulled low for communication.

CAN-H (High)

Flat line should average 2.5 volts with signal being pulled high for communication.

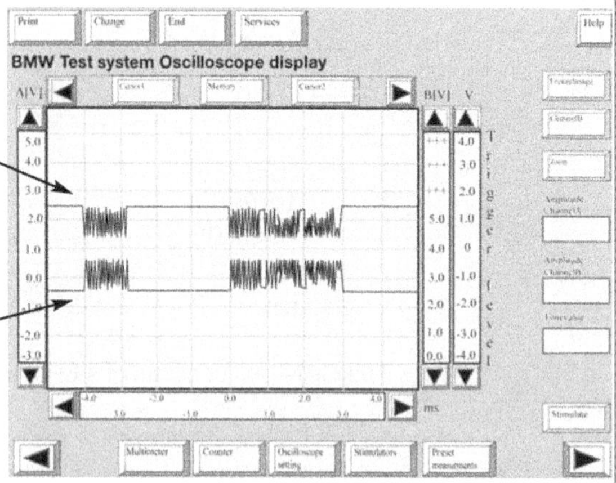

bmd661012

If the message is relevant then it is processed, If not, it is ignored. The identifier code also determines the priority of the message. In a case where two control modules attempt to send a message over a free bus line, the message with the higher priority is transmitted first. The protocol of CAN ensures that no message is lost, but stored by the responsible control module and then retransmitted later when it is possible.

CAN-bus waveform captured during normal operation

bmd613425

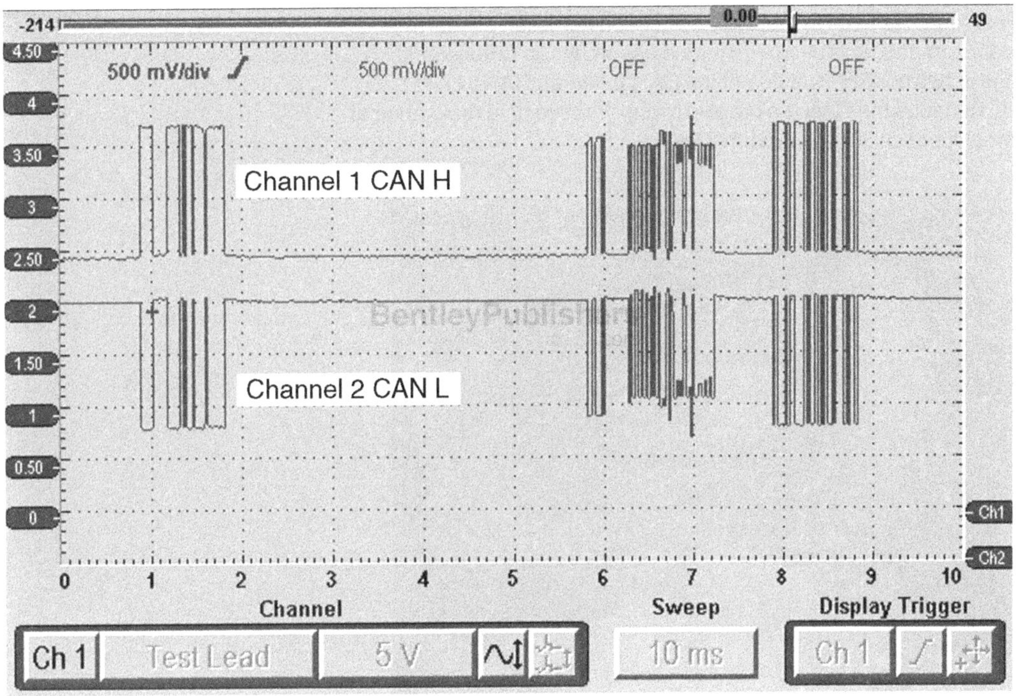

CAN-bus terminating resistors

Terminal resistors used in the CAN-bus circuit establish the correct impedance to ensure fault free communication. A 120 Ω resistor is installed in two control modules of the CAN between CAN-H and CAN-L. Because the CAN is a parallel circuit, the effective resistance of the complete circuit is 60 Ω (Ohms' law).

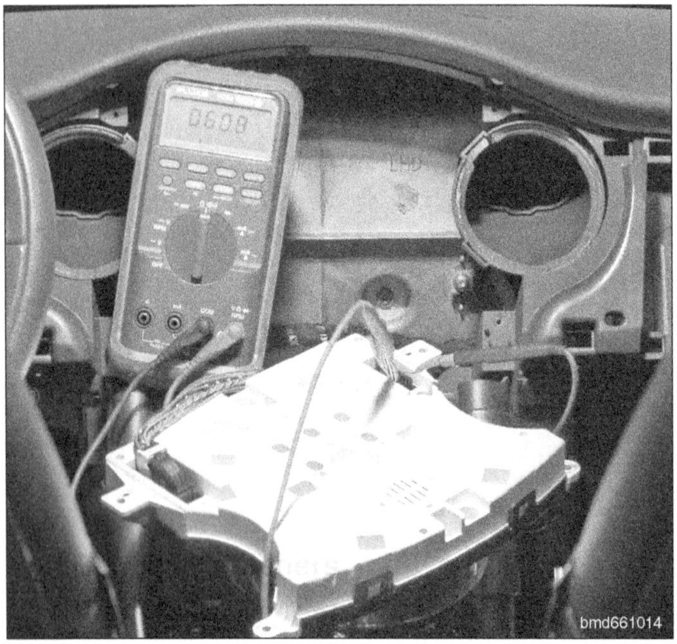

The resistance is measured by connecting the appropriate adapter to any of the modules on the CAN and measuring the resistance between CAN-L and CAN-H. The resistance should be 60 Ω. The CAN-bus is very stable and can continue to communicate if the resistance on the CAN-bus is not completely correct; however, sporadic communication faults occurs.

Modules that do not have the terminal resistor can be checked by disconnecting the module and checking the resistance directly between the pins for CAN-H and CAN-L. The value at these control modules should be between 10 and 50 kΩ.

> ***CAUTION—***
> - *When checking the resistance of the CAN-bus, remember to disconnect the power to the circuit and to leave all modules connected.*

D-bus (diagnostic bus)

The diagnostic bus system is made up of two subsystems (D-bus and DS2-bus) that allow scan tools to communicate with the vehicle control modules via the diagnostic socket.

The D-bus is connected to powertrain (emissions) related control modules via ISO 9141-2 and is only active when a scan tool is connected to the diagnostic socket and communicating.

MINI Cooper D-bus

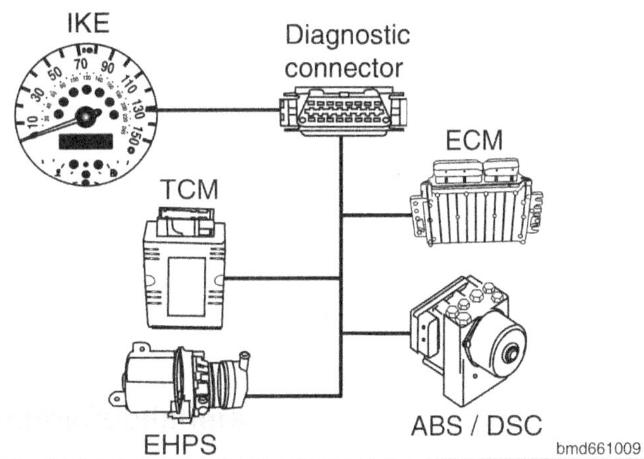

The DS2-bus connects to control modules that do not affect emissions and to the instrument cluster (IKE) to enable communication with all the control modules on the K-bus system.

MINI Cooper DS2-bus

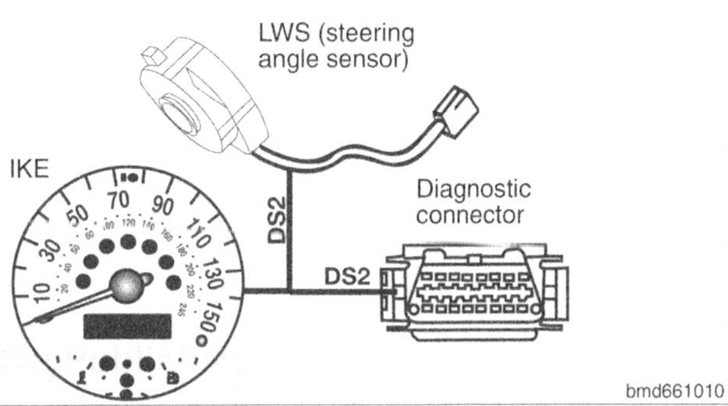

The D-bus system uses a single wire transmitting at 9,600 bits per second. The diagnostic line is used by scan tools to interrogate each control module on the network and helps in the fault finding process. It uses a protocol very similar to that of the K-bus system and is accessed via the 16-pin diagnostic connector located in the driver's footwell (under the dashboard, close to the A-pillar).

The D-bus system is a serial communications bus that transmits data between scan tool and the connected control modules. The control module subject to diagnosis is selected by sending a diagnosis telegram to the control module address.

The normal operating voltage of the D-bus system is 12 volts even though it can support communication with as low as 2 volts. The voltage measurement is taken from each data line connection to ground. Each module on the D-bus provides its own voltage used for communication.

D-bus during normal activity

bmd661229

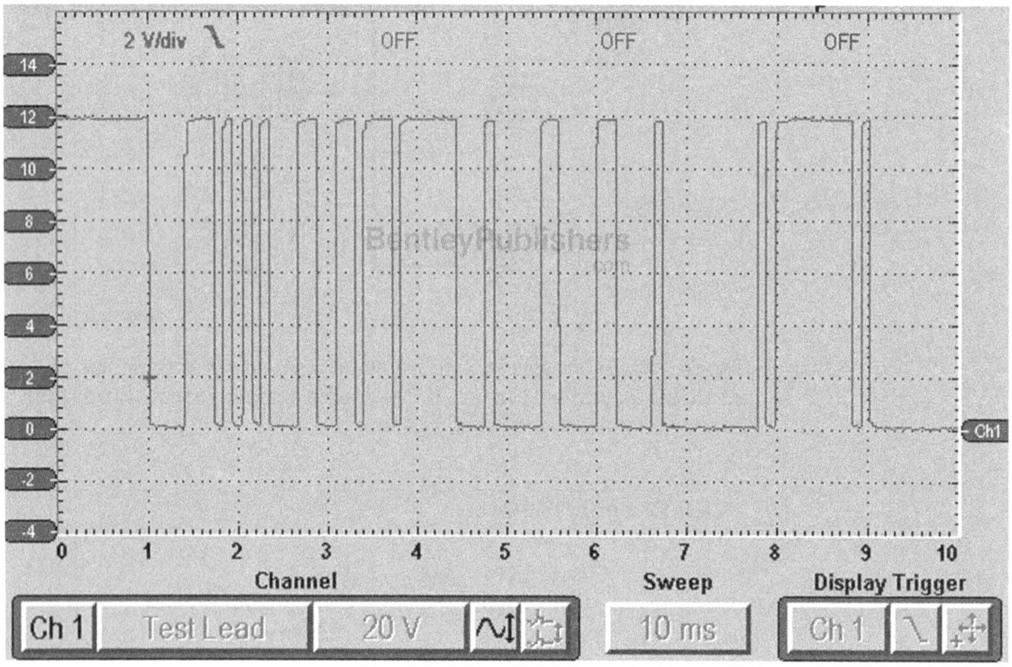

BUS SYSTEM DIAGNOSIS

Bus system communication faults range from a faulty individual module to a failure of the entire bus system.

As an example, if you can't communicate with the ASC / DSC module, determine what modules share the affected bus. Can you communicate with the other modules on the bus? What modules have communication fault codes set?

The answers to these questions can help you determine if the fault affects the module used in the example or the entire bus it resides on.

If you can communicate with all other modules on the bus except the module used in the example, the fault is isolated to that module. Gain access to the module and test the bus system signal to it. If the signal matches the signal at the other modules on the same bus, check the power and ground supply to the problem module.

If the entire bus is faulty, check if the signal present on the bus is good at each module on that bus. Once the affected branch of the bus system is determined, isolate the modules and wiring for that individual circuit.

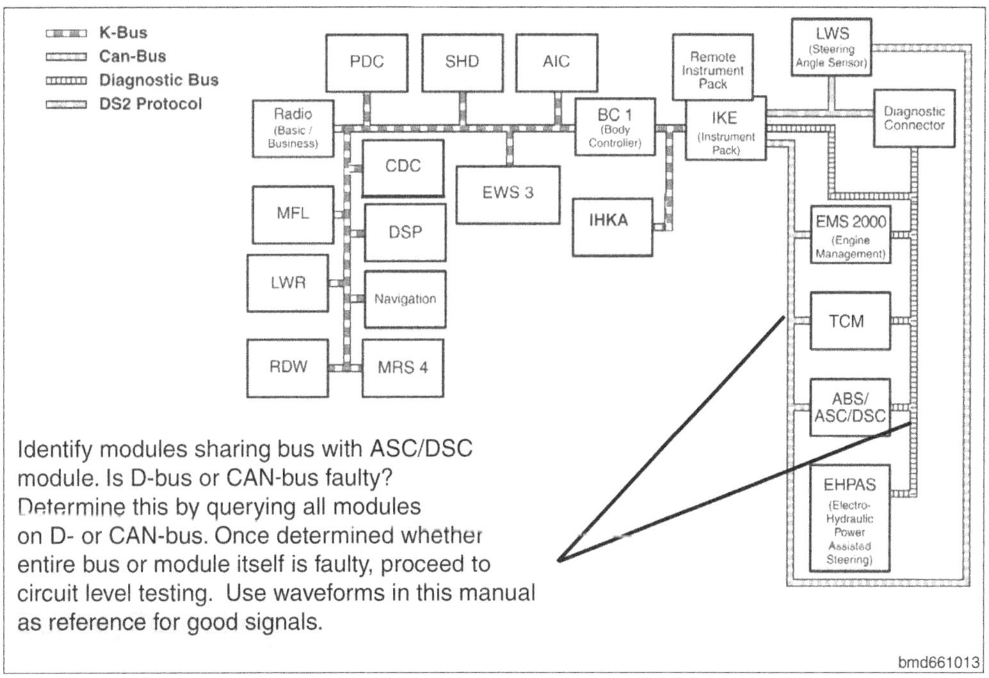

Notes:

65 Electronic Immobilization System (EWS)

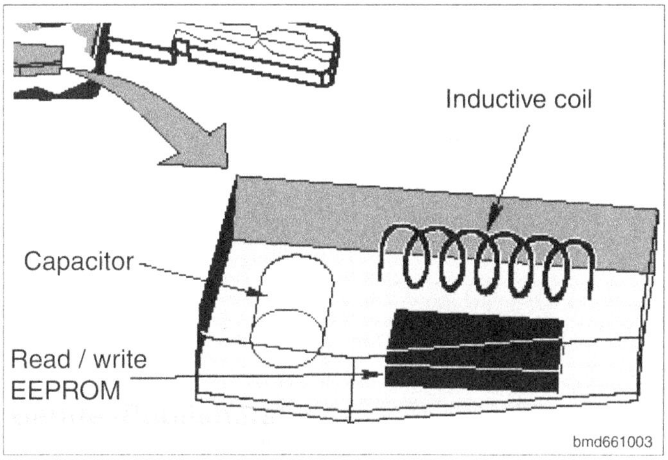

Table of Contents

ELECTRONIC IMMOBILIZATION SYSTEM (EWS 3.3)

MINI vehicles are equipped with and electronic immobilizer (or drive-away protection) known as EWS 3.3. This system provides vehicle theft protection. The EWS control module communicates with the ECM and the transponder in the key to enable/disable vehicle starting.

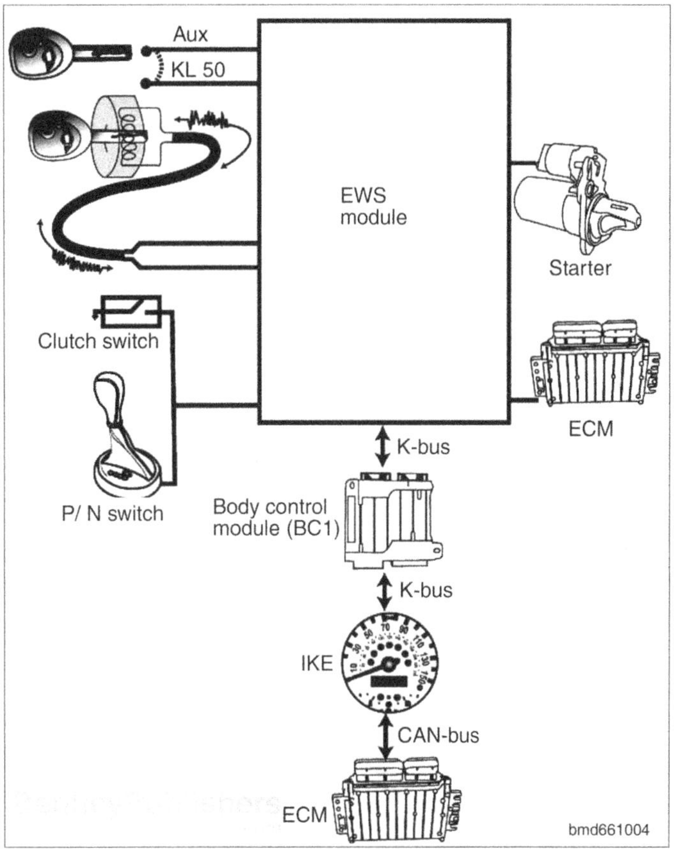

A transponder chip is integrated in each vehicle key. A ring antenna (coil) is mounted around the ignition lock. The transponder chip is powered by the EWS 3.3 control module via this coil. no battery is required for this function.

The key sends data to the EWS control module. If this data is correct, EWS enables the starter by way of a relay in the control module and additionally sends a coded start enable signal via data bus to the ECM.

EWS control module

The EWS control module is located on the left A-pillar under the dashboard.

The EWS control module has a rolling code assigned to it during manufacturing. The rolling code, which is burned in the permanent memory of the module and cannot be overwritten or changed, is transferred to the ECM on a unidirectional signal line during engine start-up.

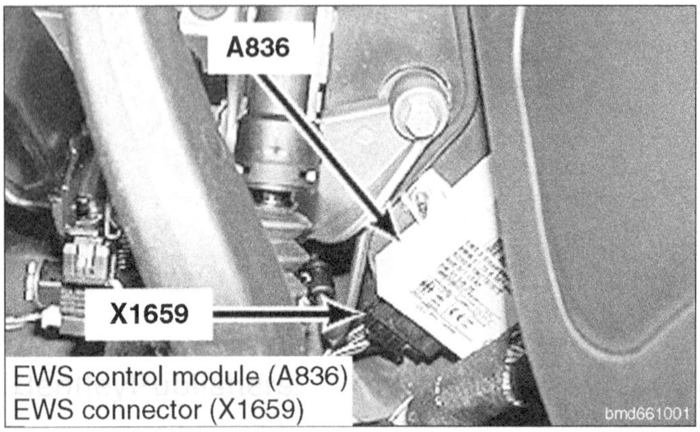

EWS control module (A836)
EWS connector (X1659)

Engine control module (ECM)

The ECM does not allow the engine to start if a valid rolling code is not received. The EWS system disables ignition, fuel injection and the starter motor.

Ring antenna

The ring antenna (coil) is mounted on the ignition lock and serves as an antenna for communication between the transponder chip (in the key) and EWS control module.

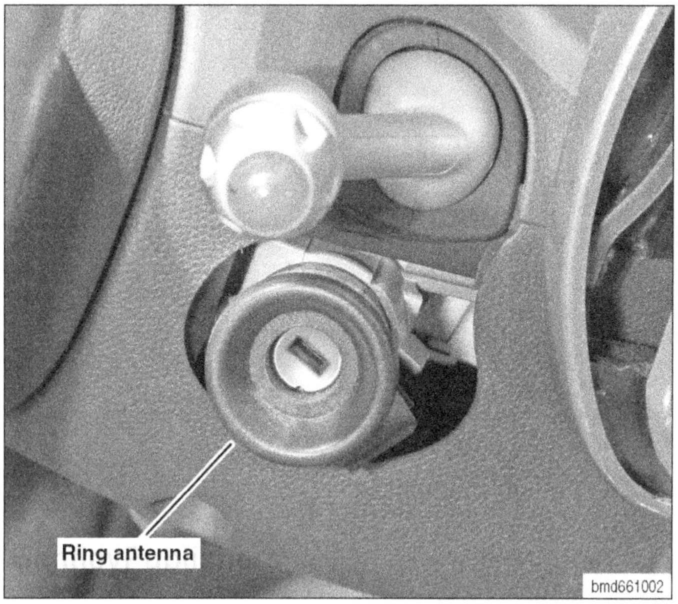

The ring antenna is used to power the key for communication with the EWS control module. The 125 kHz AM signal sent to the antenna ring induces a voltage build up in the key coil and powers up the transponder. Data to and from the key is passed through the ring antenna. The ring antenna is also used to charge the battery in the remote key.

Key with transponder

The key with transponder contains a single 3 volt battery for remote functions. Battery operation is not required for key recognition. The key transponder communicates with the EWS control module, exchanging password, key Identification Information and changing codes through the ring antenna.

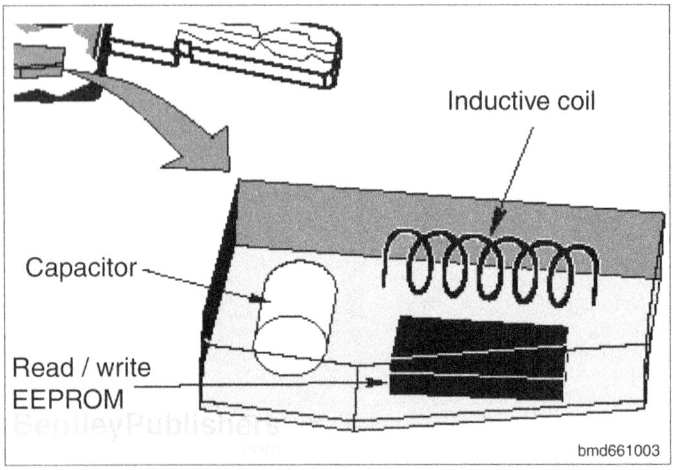

If a fault occurs during communication between the EWS control module and individual keys, these errors are stored in the fault memory (separate fault codes for individual keys).

EWS starting sequence

The starting sequence of the EWS 3.3 is as follows:

- Key is inserted into lock cylinder and switched ON. EWS control module is powered through KL R and sends a 125 kHz AM signal to ring antenna. AM signal induces voltage in key coil and powers up transponder.
- Key transponder sends key identification code to EWS module. EWS module verifies key identification code and checks to see if key is enabled. If key is correct and enabled, a password is sent to transponder (125 kHz AM signal) through ring antenna.
- When transponder accepts password, it releases changing code which it received from EWS control module during last start-up.
- EWS control module compares changing code received from transponder with code stored in memory. If they match, process is allowed to continue.
- EWS control module looks at other inputs for correct status (e.g., code function not active, transmission in P or N or clutch depressed, engine speed below specified rpm) and energizes internal relay to begin starter operation.
- While energizing internal starter relay, EWS control module calculates a stored code from rolling code table and sends calculated results to ECM.
- On receipt of rolling code, ECM calculates its own stored code and compares the results with code it received from EWS control module.
- If codes match, electronic immobilization is released and injection and ignition are enabled and engine starts.
- If codes do NOT match, ECM rolls forward to next code according to rolling code table and makes same calculations. ECM continues this forward roll-up a maximum of 200 times or until a match is found. Failure to find a match results in engine cranking but not starting.
- When ignition is switched OFF and no engine rpm signal is present in both ECM and control module, each module automatically rolls forward to next predetermined code based on rolling code table. This new code is used for next starting sequence.

EWS power supply

For safety reasons, the EWS control module is supplied redundantly with power via terminal 30 and terminal R.

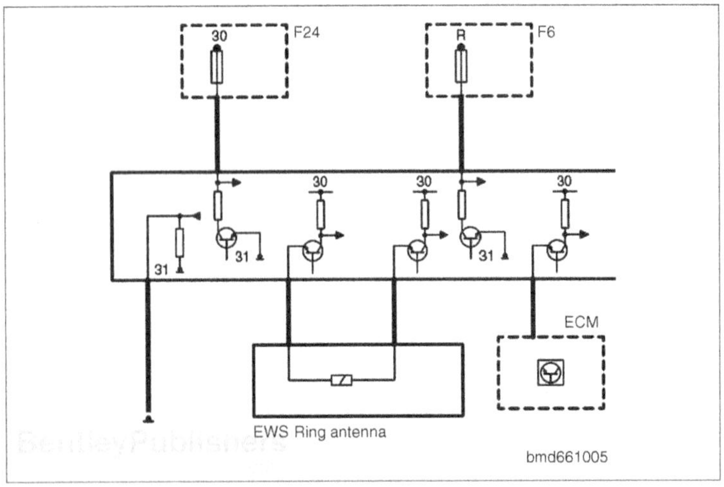

Failure of terminal 30 or terminal R does not cause a noticeable malfunction while starting the engine. Diagnosis of the EWS control module is also still possible.

If terminal 30 fails, diagnosis with the EWS control module is then only possible with terminal R switched on (ignition lock in position 1 or 2).

In addition, failure of terminal 30 results in failure of the keyless entry remote control. The reason for this lies in the following function sequence:

- When terminal 30 is free of defects, the EWS control module detects when the key is removed from the ignition lock after switching off terminal R. This status is sent via the K-bus to the central body electronics control module (BC1). The control module executes the button commands of the key-chain transmitter only when the ignition key is not inserted in the ignition lock.
- When terminal 30 is defective, the EWS control module can no longer detect the key status after switching OFF terminal R. Consequently, no key commands are executed.

Diagnosis of terminal 30

EWS power supply can be checked using the factory scan tool.

To check terminal 30, the data communication between tester and EWS control module is monitored with terminal R switched OFF. If communication with the EWS control module is no longer possible with terminal R switched OFF, this indicates a fault in terminal 30 functions (fuse or cable).

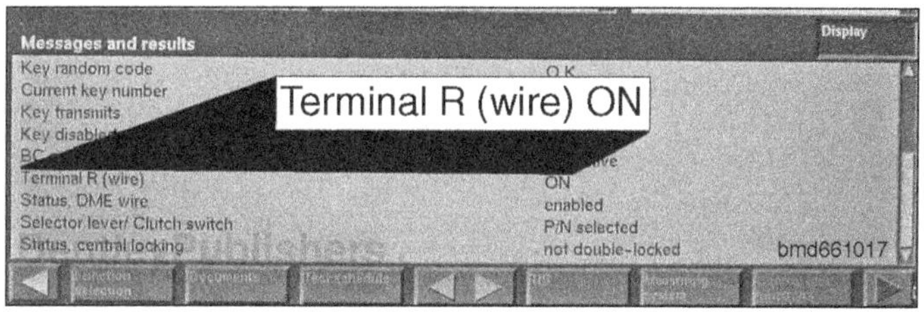

EWS COMPONENT REPLACEMENT

Keys

4 new keys are supplied with the vehicle. A replacement keys can be obtained from an authorized MINI dealer. This key is not a copy of the lost key, but rather a new key. No more than 6 replacement keys can be made for a total of 10.

NOTE —
- Loss of all 10 keys results in the need for a new EWS control module.

There is a noticeable start delay of 1 - 2 seconds when a new key is inserted in the ignition lock for the very first time. From then on, the start procedure should take place without delay.

By providing the necessary data, it is possible to obtain a new EWS control module from an authorized MINI dealer and new transponder keys allocated to it with the previous mechanical key code.

EWS control module

The EWS control module is VIN-specific. The module is supplied with the same rolling code table as the original. Once coded, the scan tool software resets the current rolling code in the ECM back to rolling code 1, providing synchronization of both modules.

EWS module

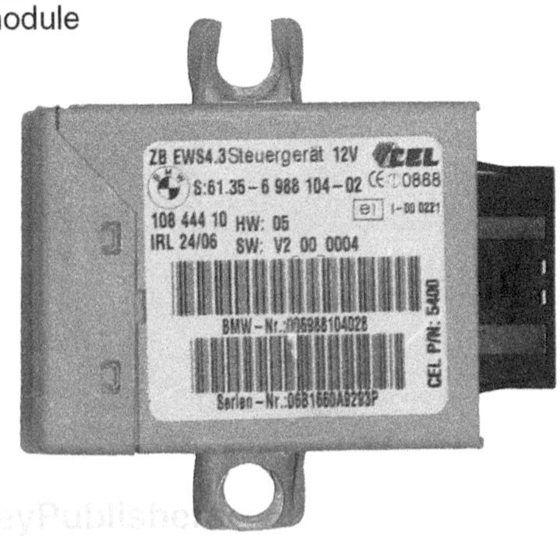

bmd661224

Engine control module

A replacement ECM is available off the shelf but must be programmed for the specific vehicle. After programming, the factory scan tool informs the EWS control module that a new ECM is installed. The next time the ignition is switched ON, the EWS control module sends the entire rolling code table to the ECM and resets it to rolling code 1.

The ECM automatically burns the rolling code table into its memory. Once the table is burned into ECM memory, it can NOT be changed. For this reason once a ECM is married to the vehicle it does not work in any other vehicle. Under certain conditions alignment of the ECM and EWS control modules may still be necessary. The alignment procedure only resets the code table to code 1 but does not change the rolling code table.

EWS ALIGNMENT

In specific cases, it is possible that synchronization between the ECM control module and EWS control module may deviate from each other. In this case perform the matching procedure. The variable codes of both control modules are reset to the common starting value via the factory scan tool.

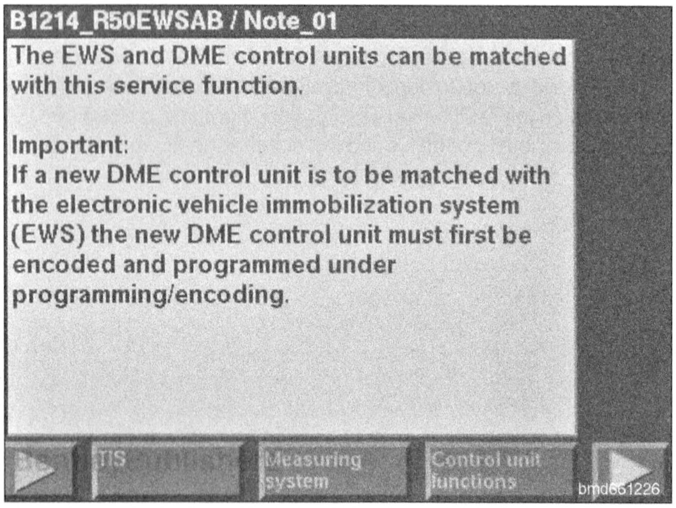

Also perform this alignment procedure when a new engine control module is installed. In this case, the matching procedure adapts the basic code of the EWS control module from the ECM control module. This alignment procedure is only possible once.

With this matching restriction, it is therefore not possible to align an ECM control module, adapted to another vehicle with its own EWS control module, to a new EWS control module.

EWS KEY SERVICE

The key number and the status of the key is determined through the factory scan tool (test module key status).

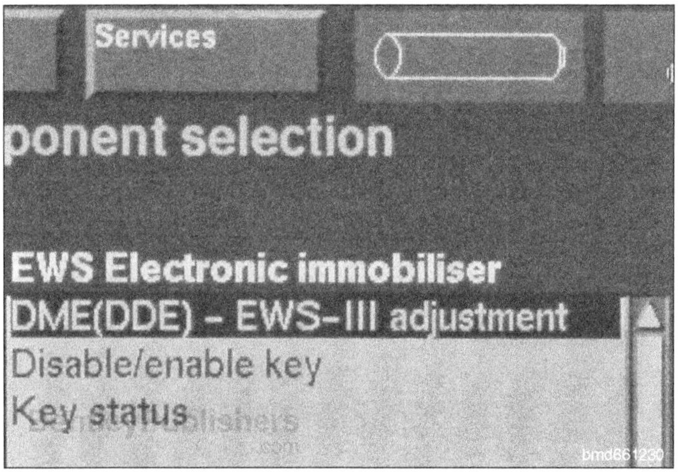

In order to display key status, insert a valid, enabled and intact key in the ignition lock and switch the ignition lock to position 1 or 2.

Key number

The EWS number of the key is permanently programed in the key. This number is entered during initial programming of the key at the factory and cannot be changed. This number is not identical to the key number of the remote key. There is no fixed allocation of the key number to a particular type of key (main key, spare key, workshop key). It is therefore necessary to determine what key number is missing before disabling a key.

Key status

The following key status information is displayed:

- Active for the key that is inserted in the ignition lock.
- Used at least once.
- Not yet used.

Whether a key is enabled or disabled is indicated in the test module Service functions - disable/enable key.

Disabling and enabling keys

Lost or stolen keys may be deactivated or made to not operate the starter functions. The service function menu of the factory scan tool for EWS contains a bar / release code function that activates and deactivates keys.

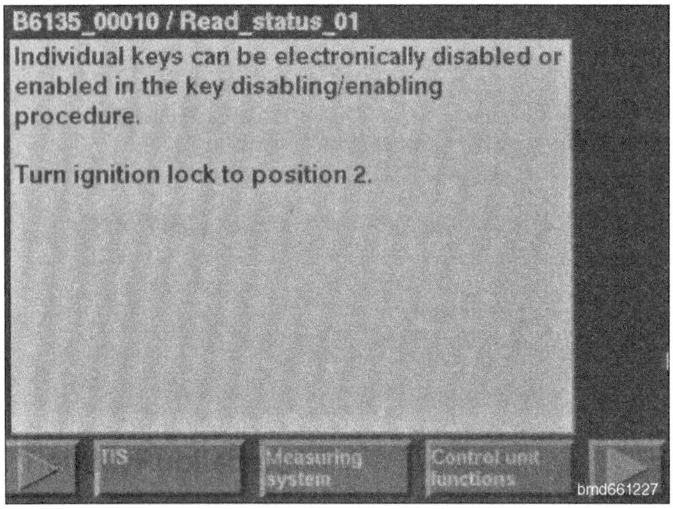

Any key may be deactivated except the key in the ignition at the time of the procedure. The lost or stolen key can be identified by the identification of the remaining keys. There is no limit to the number of times a key can be activated or deactivated.

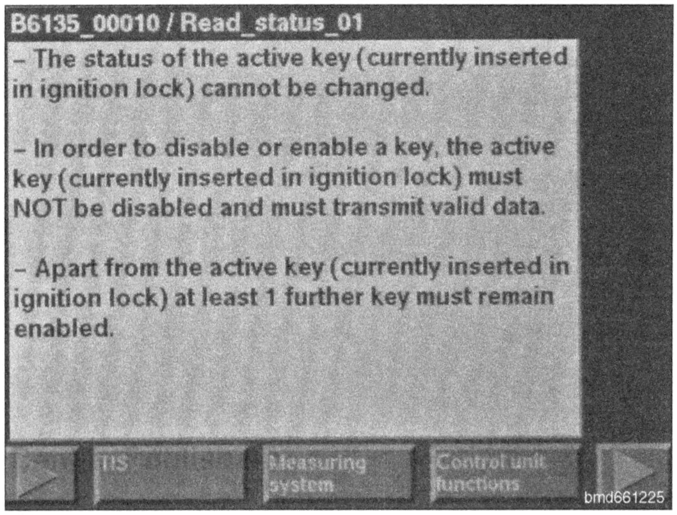

Pay attention to the following points:

- Before disabling keys, determine what keys are missing (key numbers). This is determined in the test module Service functions - key status. For this check, the vehicle owner must present all keys belonging to the vehicle.
- In order to disable or enable a key, the key inserted in the ignition lock must be enabled and transmit valid data.
- The active key (key in ignition lock) cannot be disabled.
- At least two keys should remain enabled.
- Do NOT disable new keys which have not yet been delivered.

NOTE —
- A disabled key is valid (engine start possible) for 10 seconds after removing an enabled key.

Remote transmitter (key) programming

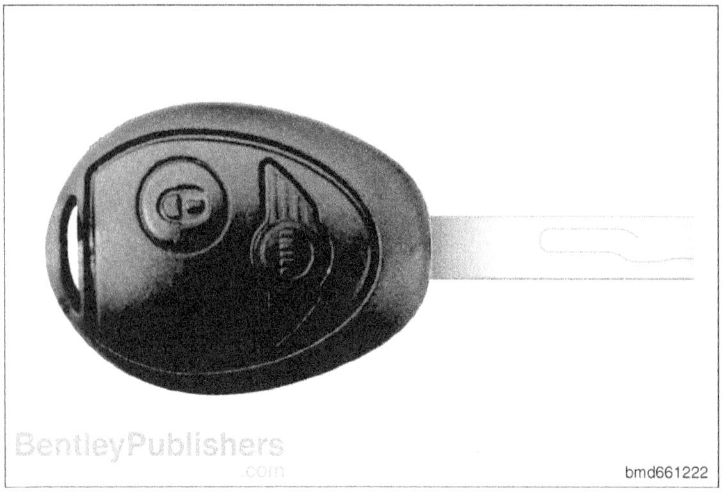

bmd661222

- Close all doors.
- Switch ignition switch to KL-R. Switch OFF within 5 seconds and remove key.
- Within 30 seconds, press and hold unlock button and at same time press and release lock button 3 times. Completed within 10 seconds.
- Release both buttons. Doors lock / unlock to signal a successful initialization.
- If additional keys need to be initialized, repeat steps above within 30 seconds.

Switching ignition to KL-R completes initialization.

EWS ELECTRICAL PLAN

EWS control module pinout

EWS control module - pinout

Pin	Type	Description	Connection / notes
1	Output	Signal, starter terminal 50	Starter
2	Input	Start signal terminal 50	Ignition switch
3	Output	Signal, starter terminal 50	Fuse F5
4	Output	Signal vehicle immobilization	ECM
5	Output	Signal EWS ring antenna	EWS ring antenna
6	Output	Automatic transmission control	Transmission control module
7	not used		
8	Input	Manual transmission clutch position	Clutch switch module (S805)
9	Ground	Ground	Ground point
10	Input	Terminal 30	Fuse F24
11	Input	Terminal R	Fuse F6
12	Output	Signal EWS ring antenna	EWS ring antenna
13	Input / output	K-bus signal	Connector X10116

EWS wiring diagram

NOTE —

- The following diagram is based on 2003 - 2005 models. Wire colors vary slightly on earlier and later models.

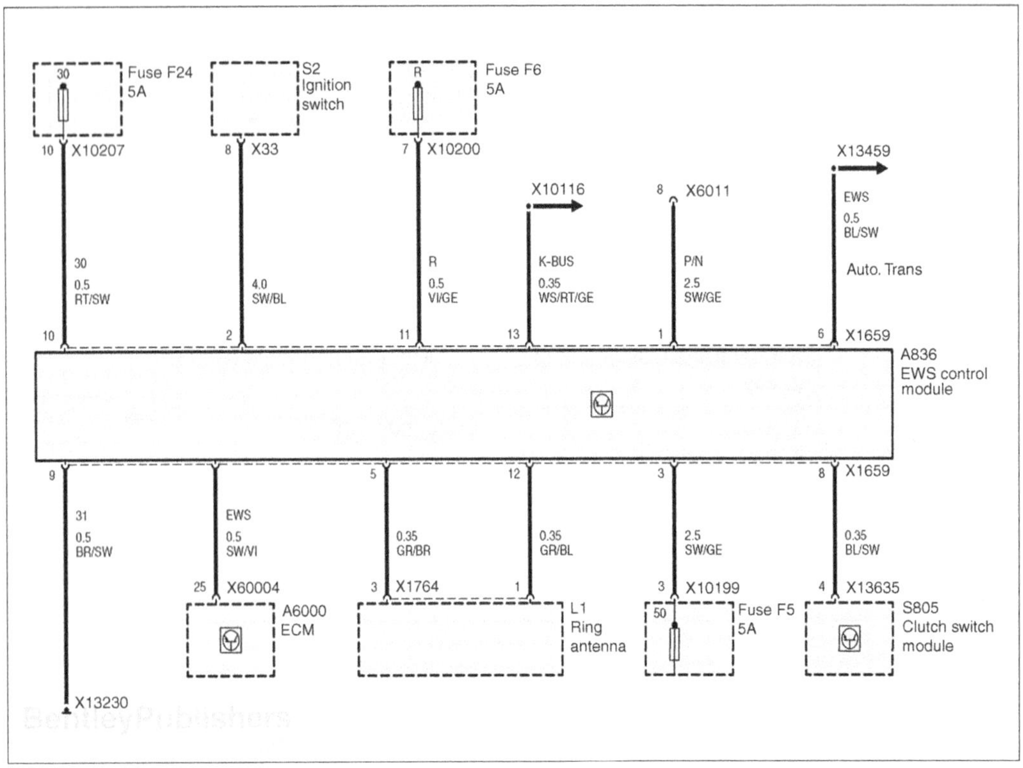

OBD On-Board Diagnostics (OBD II)

Table of Contents

GENERAL

This chapter covers fundamentals and equipment requirements of the SAE (Society of Automotive Engineers) and CARB (California Air Resource Board) ODB II standards as they apply to MINI vehicles.

Also covered here is a comprehensive list of Mode 6 test data.

The engine control module (ECM) software monitors components that can affect vehicle emissions. Each emission-influencing component is checked by a diagnostic routine to verify that it is functioning properly.

If a problem or malfunction is detected, the OBD II system illuminates the malfunction indicator light (MIL) in the instrument cluster. The OBD II system also stores important information about the detected malfunction in ECM memory.

OBD II diagnostic connector

The MINI OBD II diagnostic connector (also called data link connector) is located under the left side of dashboard. Communication between the scan tool and emission related powertrain components is done through this connector.

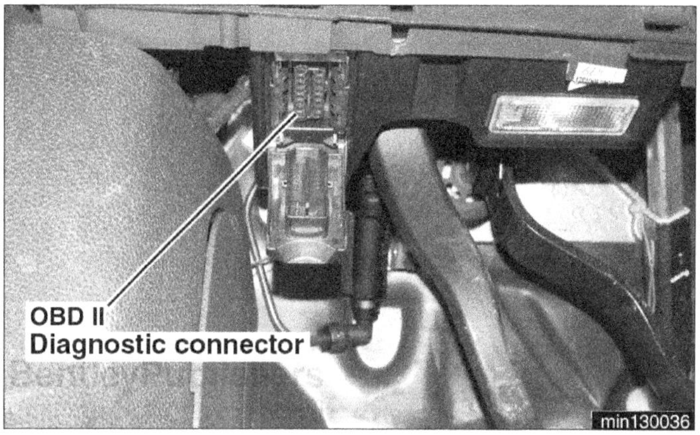

With a generic OBD II scan tool connected, diagnostic trouble codes (DTCs) can be obtained, along with the conditions associated with the illumination of the MIL. A more advanced or MINI-dedicated scan tool accesses additional proprietary information.

The OBD II connection to the engine control module (ECM) is through the D-bus.

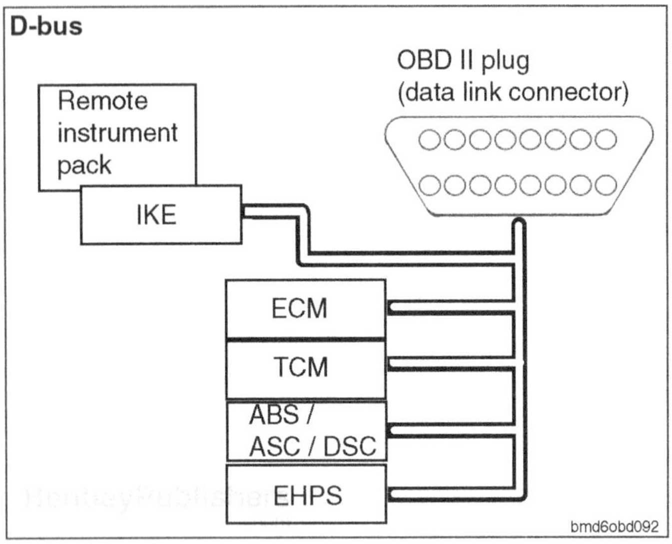

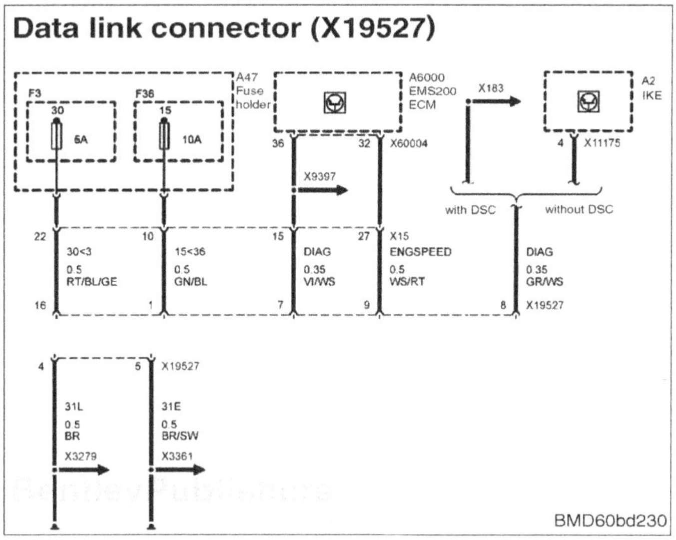

DTC structure

SAE (Society of Automotive Engineers) established the DTCs (diagnostic trouble codes) used for OBD II systems. DTCs are designed to be identified by their alpha/numeric structure. The SAE has designated the emission related DTCs to start with the letter P for powertrain related systems, hence the nickname P-codes.

DTCs are stored whenever the MIL is illuminated.

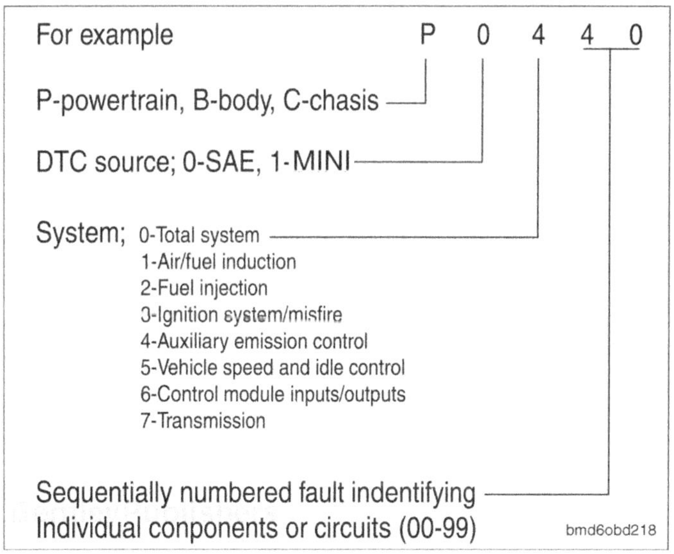

CARB / EPA regulations require universal diagnostic access to DTCs via a standardized DLC (diagnostic link connector) using a standardized scan tool.

DTCs provide one set of environmental operating conditions when a fault is stored. This single freeze frame or snapshot refers to a block of the vehicle's environmental conditions for a specific time when the fault first occurred. The information which is stored is defined by SAE and is limited in scope. This information may not even be specific to the type of fault.

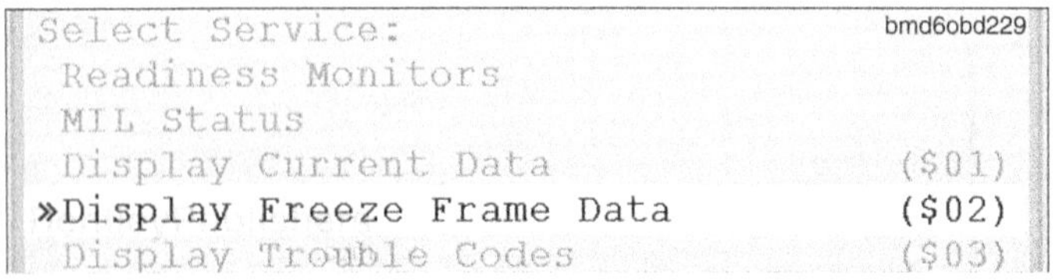

bmd6obd229

Malfunction indicator light (MIL)

The following provides examples of when and how the MIL is illuminated based on the driver's drive cycle.

- A fault code is stored in the ECM upon the first occurrence of a fault in the system being checked.
- The MIL is not illuminated until the completion of the second consecutive driving cycle where the previously faulted system is again monitored and a fault is still present or a catalyst damaging fault has occurred.
- If the second drive cycle was not complete and the specific function was not checked, the ECM counts the third drive cycle as the next consecutive drive cycle. The MIL is illuminated if the function is checked and the fault is still present.
- If there is an intermittent fault, two complete consecutive drive cycles with the fault present are required for the MIL to be illuminated.
- Once the MIL is illuminated it remains illuminated unless the specific function is checked without fault through three complete consecutive drive cycles.
- The fault code will also be cleared from memory automatically if the specific function is checked through 40 consecutive drive cycles

without the fault being detected or with the use of either the factory or generic scan tool.

- In order to clear a catalyst damaging fault from memory, the condition must be evaluated for 80 consecutive cycles without the fault reoccurring.

A fault code and the conditions associated with its setting can often be obtained prior to the illumination of the MIL. This is known as a pending code.

OBD II MONITOR FUNCTIONS

Misfire monitor

According to OBD II regulations, the ECM must have the ability to determine an engine misfire and identify the specific cylinder(s) the severity of the misfire and whether it is emissions relevant or catalyst damaging.

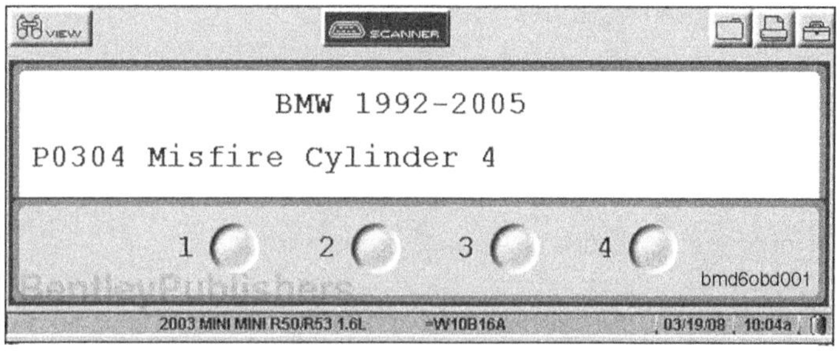

To accomplish this, the ECM monitors the crankshaft for acceleration by the impulse wheel segments of cylinder specific firing order. The misfire / engine roughness calculation is derived from differences in the period duration of individual increment gear segments.

Misfire induces a decrease of the engine speed and therefore a variation in the segment period. Misfire detection is based on the observation of this variation of segment period.

Main causes of misfiring:

- Injector problem (shut-off).
- Fuel pressure problem.
- Fuel combustion problems and ignition cut-off.

Depending on the level of misfire rate measured, the ECM illuminates the MIL, deactivates the specific fuel injector to the particular cylinder, and switches oxygen sensor control to open loop.

The ECM also determines the severity of the misfire and whether it is emissions relevant or catalyst damaging based on monitoring crankshaft acceleration.

- Within an interval of 1000 crankshaft revolutions, the ECM adds detected misfire events for each cylinder. If the sum of all cylinder misfire incidents exceeds the predetermined value, a fault code is stored and the MIL is illuminated.
- If more than one cylinder is misfiring, all misfiring cylinders are specified and the individual fault codes for each misfiring cylinder, or a fault for multiple cylinders is stored. The MIL is illuminated.

Catalyst damaging

Within an interval of 200 crankshaft revolutions, the detected number of misfiring events is calculated for each cylinder. The ECM monitors this based on load / rpm. If the sum of cylinder misfire incidents exceeds a predetermined value, a catalyst damaging fault code is stored and the MIL is illuminated.

If the cylinder misfire count exceeds the predetermined threshold, the ECM takes the following measures:

- Oxygen sensor control is switched to open loop.
- Cylinder selective fault code is stored.
- If more than one cylinder is misfiring, a fault code for all individual cylinders and for multiple cylinders is stored.
- Fuel injector to the respective cylinder(s) is deactivated.

Limits for misfire strategy application:

Variation in engine torque caused by reasons other than misfiring must be recognized to inhibit misfiring monitoring. Some examples:

- Negative torque
- Trailing throttle / acceleration transition
- Ignition retardation
- Air-conditioning compressor activation
- Rough road detection
- Cylinder shut-off (e.g. for engine speed limitation)

Evaporative system monitoring

This monitoring function uses an leakage diagnosis pump (LDP) to generate pressure in the fuel tank system to isolate a jamming tank ventilation valve or system leakage. The LDP is an electrically actuated pressure-generating device.

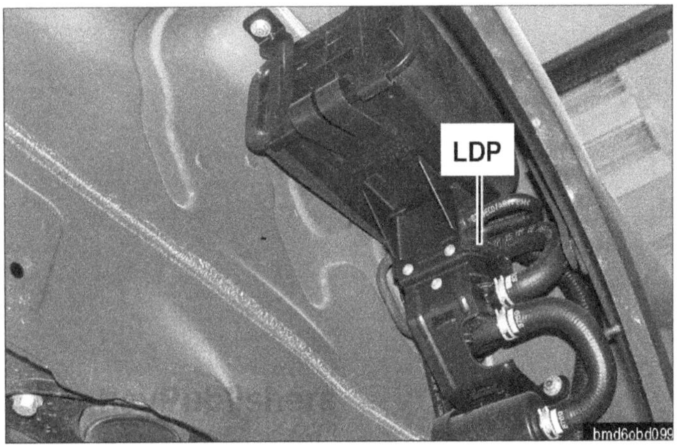

Evaporative system leak monitoring

Evaporative leak detection is performed by two main phases:

- Tank system over pressurizing
- Leak magnitude measurement

During leak detection, the carbon canister purge valve and the canister vent valve (CVV) are closed.

The ECM causes the pump diaphragm to cycle at fixed frequency and for a fixed stroke number. During pumping, air is drawn from outside into the fuel tank system, and the system pressure increases.

Once the tank system overpressure phase is finished, the leak measurement phase starts. The diaphragm stroke is limited by the top of the diaphragm chamber and a position defined by a reed switch level. The overpressure slows diaphragm movement. The pump cycle duration depends on tank pressure. So the length of this time shows the overpressure level.

If there is a leak, the cycling time or pulse interval stabilizes at a rate which compares to the leakage loss. If there is no leak in the system the cycling time or pulse interval becomes longer. The pulse interval is measured by the ECM, which determines whether or not the leak exceeds a defined threshold. Several pulse interval measurements are carried out to secure the test.

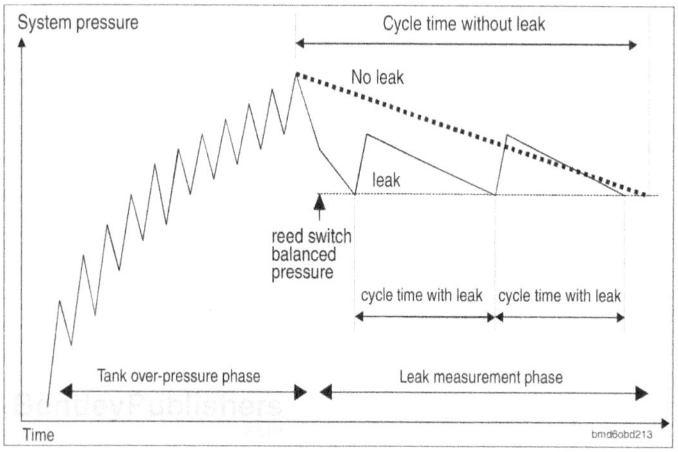

Evap canister purge monitoring

When the tank system is tight (leak measured is smaller than defined threshold), the purge valve is opened. Each time the reed switch level is reached, the LDP performs a pump stroke to maintain pressure in the tank system.

If the purge valve is not blocked, the cycling time or pulse interval becomes shorter. In this case the purge valve operates correctly.

If the canister purge valve is blocked in a closed position or the connection tube canister / valve is pinched, the cycling time or pulse interval remains long.

The pulse interval is measured by the ECM, which determines whether or not the purge flow exceeds a defined threshold. Several pumping cycles are carried out to secure the test.

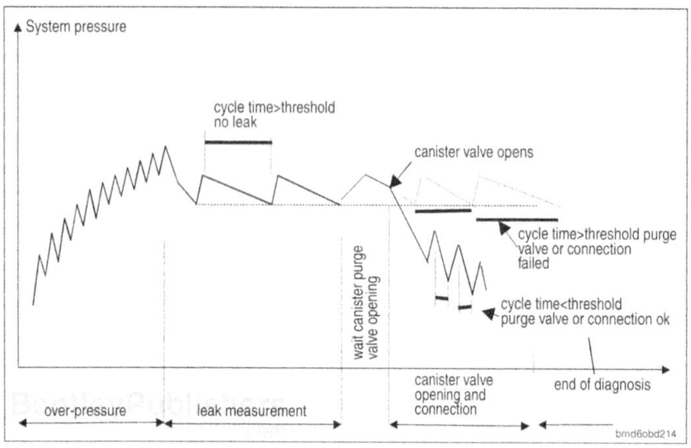

EVAP system monitor flow chart

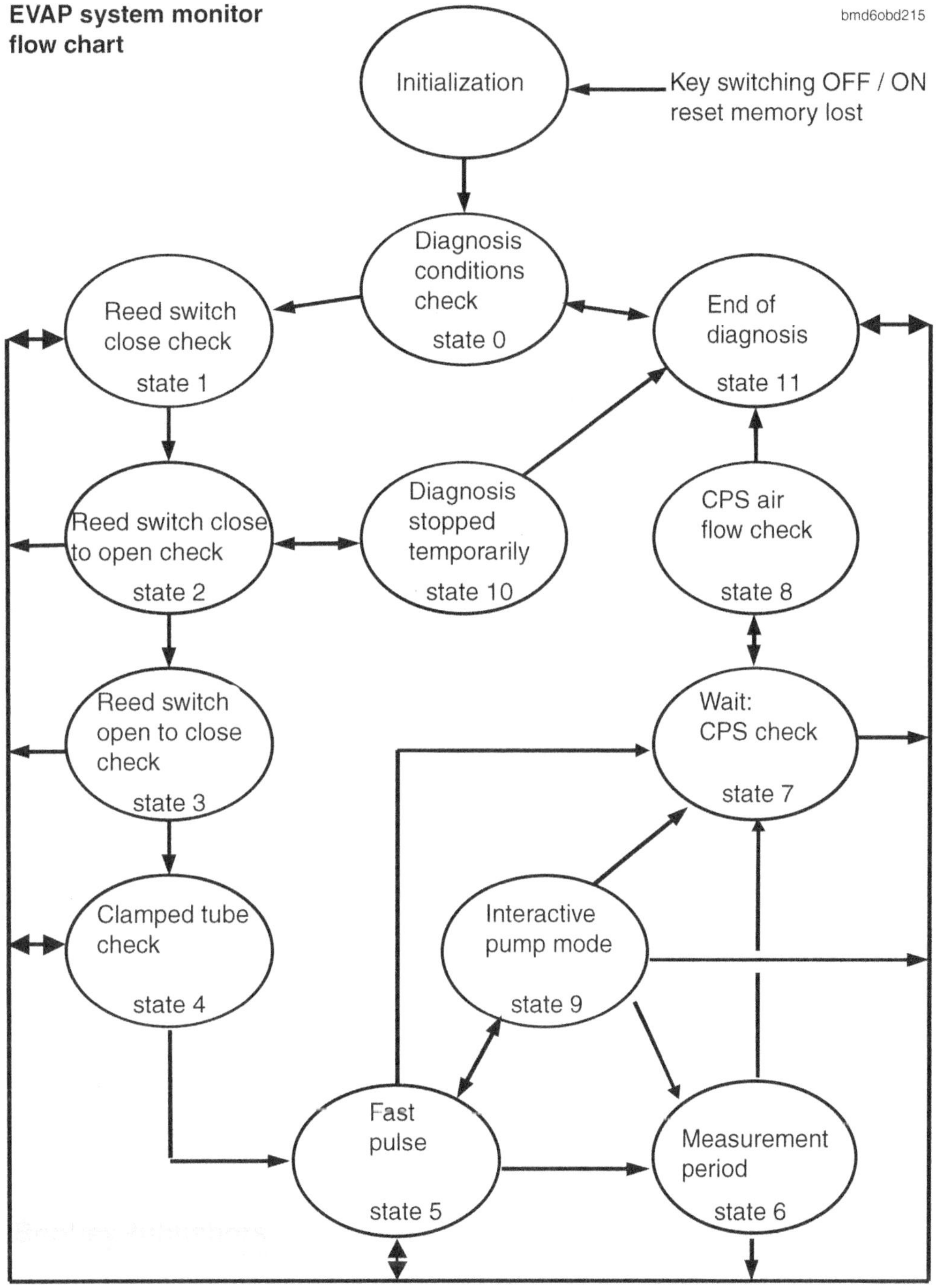

Oxygen sensor monitor

Oxygen sensor monitoring is a sequential diagnosis made during steady state conditions. The monitoring is composed of two main phases:

- Measurement
- Diagnosis

Measurement phase

The measurement algorithm is based on the period measurement (starting from lean to rich sensor transition). To avoid non-representative measurement, the period is valid only if the sensor has been below a low threshold and above a high threshold between 2 consecutive lean / rich transitions. If one of the diagnostic conditions is not met, the test is stopped.

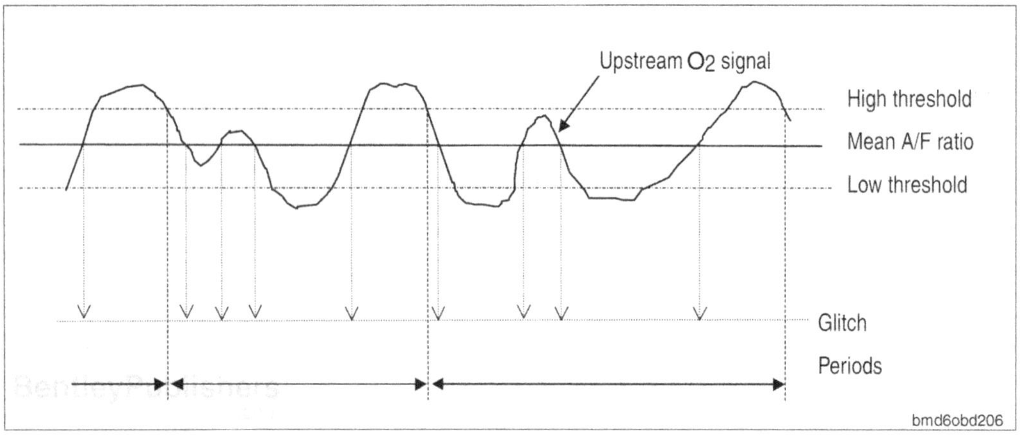

Diagnosis phase

The sum of the periods (rich to lean transition time) is compared to predetermined limit values to determine a failure.

Catalyst monitor

This monitoring function checks the efficiency of the catalytic converter by monitoring post-catalyst oxygen sensor activity. If the efficiency of the catalyst is reduced, the activity of the post-catalyst oxygen sensor increases accordingly.

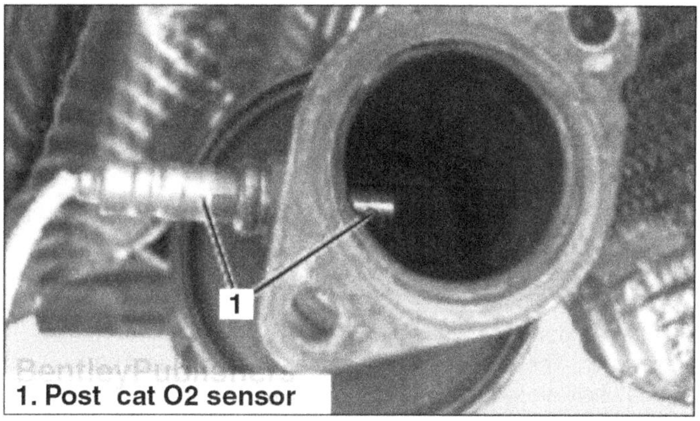

1. Post cat O2 sensor

Efficiency of catalyst operation is determined by evaluating oxygen consumption of the catalytic converters using the pre and post oxygen sensor signals. A properly operating catalyst consumes most of the O_2 (oxygen) that is present in the exhaust gas (input to catalyst). The gases that flow into the catalyst are converted from CO, HC and NO_x to CO_2, H_2O and N_2 respectively.

The signal of the post cat O_2 sensor is evaluated over the course of several pre cat O_2 sensor ECM oscillations. During the evaluation period, the signal of the post-catalyst sensor must remain within a relatively constant voltage range (700-800 mV). The post cat. O_2 voltage remains high with a very slight fluctuation. This indicates a further lack of oxygen when compared to the pre catalyst sensor.

High catalytic converter efficiency

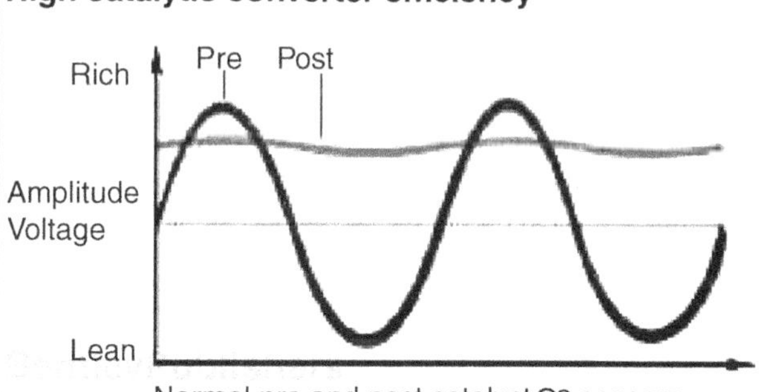

If the post O_2 sensor signal decreases in voltage and / or increases in fluctuation, a fault code is set for catalyst efficiency and the MIL illuminates when the OBD II criteria is achieved.

Low catalytic converter efficiency

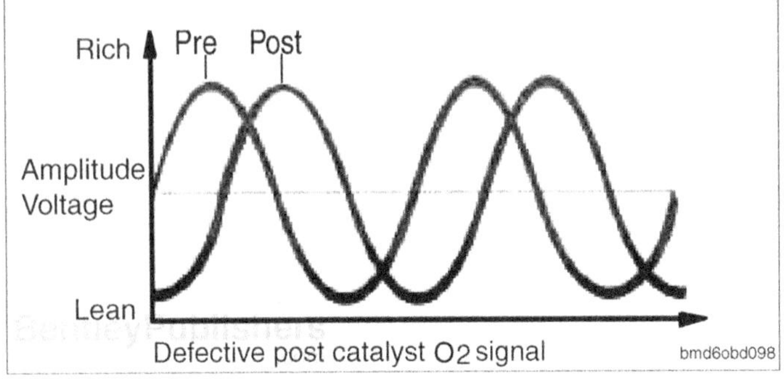

Catalyst monitoring is performed by the ECM under oxygen sensor closed loop operation. The changing air / fuel ratio in the exhaust gas results in lambda oscillations at the pre catalyst sensor. These oscillations are dampened by the oxygen storage activity of the catalyst and are reflected at the post-catalyst sensor as a fairly stable signal (indicating oxygen has been consumed).

Conditions for catalyst monitoring:

- Closed loop operation
- Engine coolant temperature: operating temp.
- Vehicle road speed: 3 - 50 mph (5 to 80 kph)
- Catalyst temperature (calculated)*: 350 to 650 °C (660 to 1200 °F)
- Throttle angle deviation: steady throttle
- Engine speed deviation: steady / stable engine speed
- Average lambda value deviation: steady / stable load

*Catalyst temperature is a theoretical calculation performed by the ECM that is a function of load / air mass and time.

Catalyst efficiency is monitored once per trip while the vehicle is in closed loop operation.

> ***NOTE —***
> - The catalyst monitor is your best tool for catalyst efficiency fault diagnosis. Confirm fuel injection control is in acceptable range, exhaust system is sealed and no other faults are stored. Perform drive cycle to generate readiness code. A stored code P0420 indicates a defective catalyst. Check for TSBs before replacing catalyst.

Fuel system monitor

Fuel system monitor continuously monitors the fuel delivery system. This monitor is continuously performed if conditions are fulfilled.

The fuel system monitor checks if the sum of short term fuel trim (only based on upstream O_2 sensor voltage monitoring) and long term fuel trim (one additive and one multiplicative term) are within a specified range. If out of this range, a failure is detected.

Various fuel system problems can occur and affect the exhaust emissions in different ways. For example:

- Fuel-pressure regulator vacuum hose disconnected. This can lead to exhaust emission problems. However, due to the homogeneous mixture, it has no influence on the catalytic converter.
- Cylinder distribution problems due to injector failure. This affects the catalytic converter as the mixture is not homogeneous. In general, this means that the volume of exhaust gas to be reduced increases, leading to a reduction in efficiency.

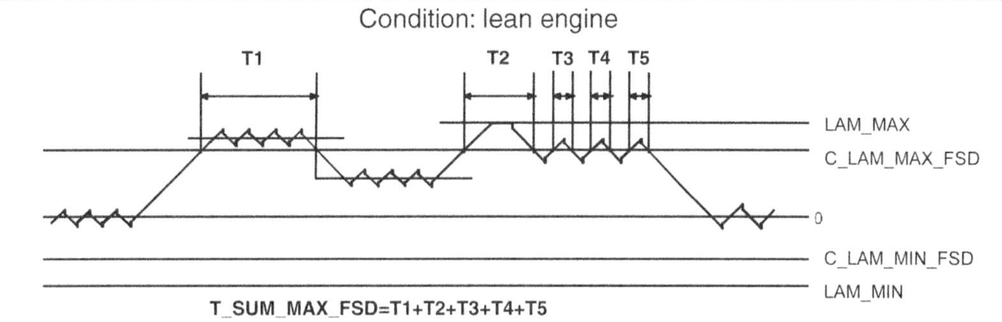

LAM_MAX = restriction for rich limit
LAM_MIN = restriction for lean limit
C_LAM_MAX_FSD = threshold for rich exceeding
C_LAM_MIN_FSD = threshold for lean exceeding

bmd6obd204

Thermostat monitor

This function is used to monitor the coolant thermostat for jamming open. This could cause higher exhaust emissions because the engine does not reach its normal operating temperature.

The coolant thermostat ensures a quick engine warm-up after start. The thermostat is closed after engine start to limit coolant circulation until the thermostat regulating temperature is reached.

If the thermostat sticks open, coolant circulation will not be limited after start. That means an increase of the engine warm-up time and an increase in emsissions as well.

To monitor the thermostat function, a calculated value for coolant temperature is used. This monitoring is only used for diagnosing a thermostat stuck in the open position. The actual coolant temperature is compared to the calculated coolant temperature, When the calculated temperature reaches normal operating temperature it is then compared to the actual coolant temperature. If there is a large discrepancy, the thermostat is determined to be stuck open and a fault is set.

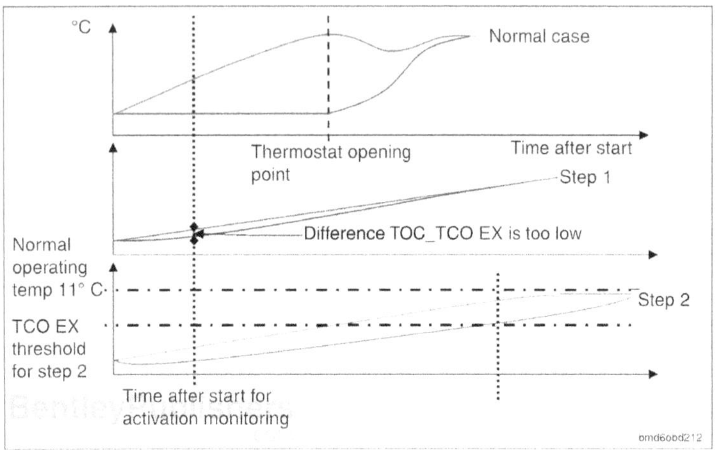

MODE 6

Mode 6 description

Mode 6 is a part of the 9 modes of OBD II.

Mode 01	Data stream
Mode 02	Freeze-frame data (if DTCs are present)
Mode 03	Diagnostic trouble codes
Mode 04	Clear codes and freeze-frame data
Mode 05	Oxygen sensor monitors
Mode 06	Non-continuous monitors
Mode 07	Continuous monitors
Mode 08	Bi-directional communication (on-board tests)
Mode 09	Vehicle VIN, PCM calibration, etc.

bmd6obd226

OBD II monitor test parameters and results can be viewed using the Mode 6 function of your scan tool. Mode 6 data is the results of on-board diagnostic monitoring tests of specific components / systems that are not continuously monitored. These systems include the catalytic converter, EVAP system, oxygen sensor, oxygen sensor heater and thermostat.

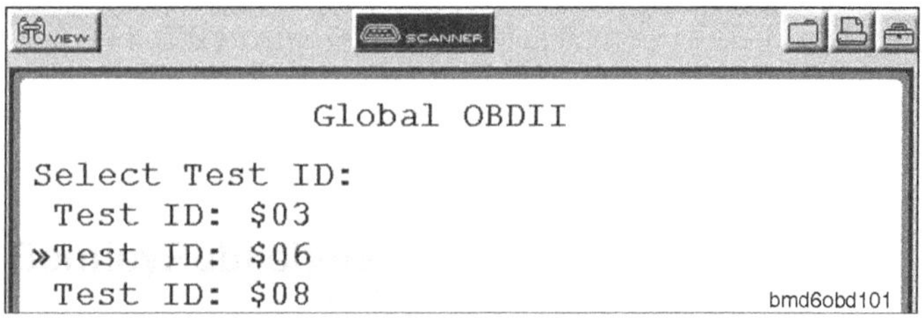

bmd6obd101

Mode 6 data is stored in the ECM until it is overwritten by new test results. The test results will also be erased when codes are cleared.

Mode 6 data is output from the ECM in hexadecimal form. Hexadecimal is a number system in which 16 is the base number. Employing the numerals 0 through 9 and representing digits greater than 9 with the letters A through F. The $ sign indicates a hexadecimal number to follow. Depending on the software in the scan tool, the output could be displayed in hexadecimal numbers (no conversion), raw data (conversion to decimal numbers), or measurable values (applying conversion factor to raw data).

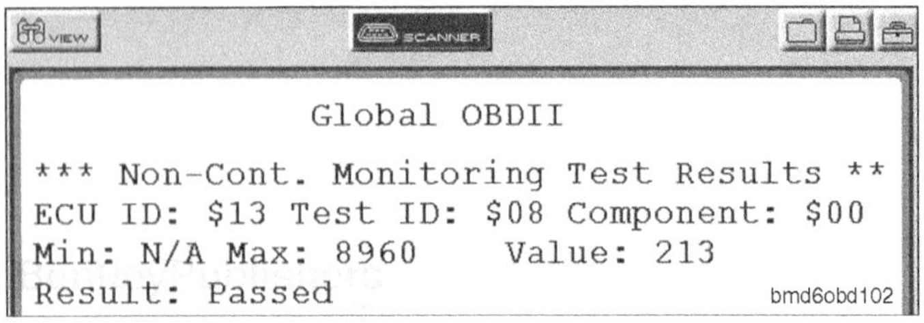

bmd6obd102

Mode 6 test results are broken down into TIDs, (test ids). A TID is a hexidecimal number representing a system level test. Once you gain access to the appropriate TID, you will find CIDs, (component Ids). A component id is a hexadecimal number representing a component level test within a system TID. This is where the indication for a pass or fail is found. CID' enable you to view individual component tests of an emission monitor.

Before requesting Mode 6 information from the ECU, ensure that the monitor for the information you are accessing has run. If the monitor has not run, the Mode 6 data will not be accurate. Individual monitors can be viewed with your scan tool under readiness codes, or readiness monitors.

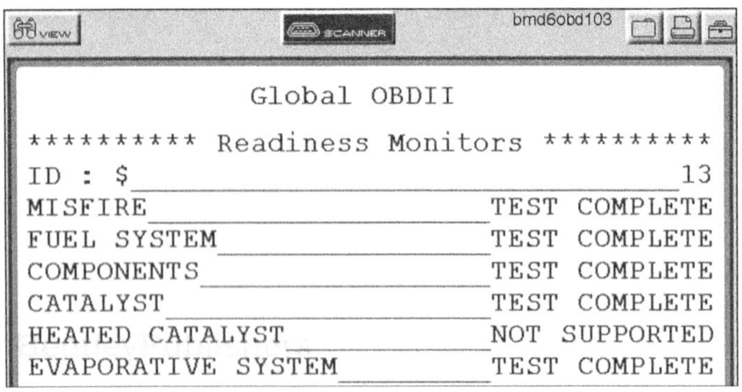

bmd6obd103

Mode 6 test parameters

The following Mode 6 tables can be used in conjuction with a scan tool to identify test results of non-continuous emission monitors.

Examples below depict Mode 6 information as seen with a Snap On MODIS®.

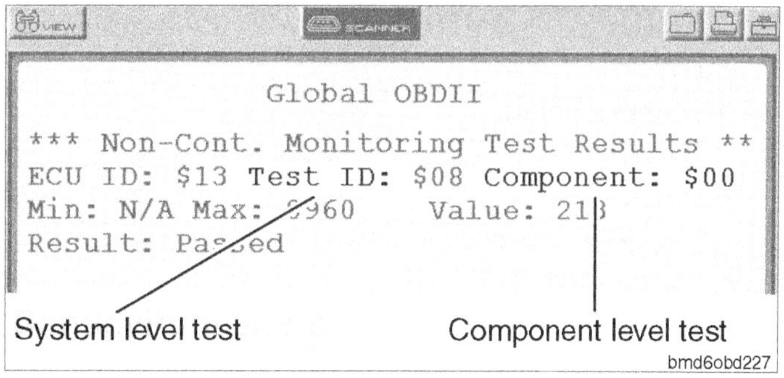

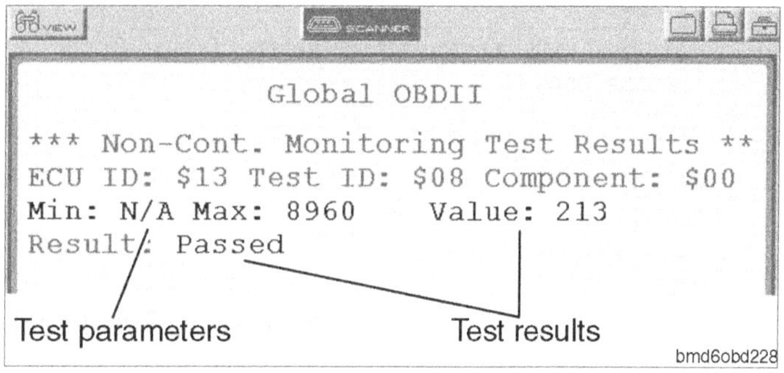

Mode 6 information (2002 model year)

Monitored Component		Possible range	Hex code factor
TID 1 - catalyst monitoring			
CID 1	catalyst monitoring of catalyst 1 sum of the determined ratios	0...65535	0...FFFF H
CID 2	catalyst monitoring of catalyst sum of the determined ratios 2	0...65535	0...FFFF H
TID 2 - o2 sensor monitoring			
CID 1	weighted period time (maximum value) of O_2 sensor 1 exhaust bank 1	0...65535 ms	0...FFFF H
CID 2	weighted period time (maximum value) of O_2 sensor 1 exhaust bank 2	0...65535 ms	0...FFFF H
CID 3	weighted period time (minimum value) of O_2 sensor 1 exhaust bank 1	0...65535 ms	0...FFFF H
CID 4	weighted period time (minimum value) of O_2 sensor 1 exhaust bank 2	0...65535 ms	0...FFFF H
CID 5	weighted switching time for rich to lean of O_2 sensor 1 exhaust bank 1	0... 100 ms	0...3200 H
CID 6	weighted switching time for rich to lean of O_2 sensor 1 exhaust bank 2	0... 100 ms	0...3200 H
CID 7	weighted switching time for lean to rich of O_2 sensor 1 exhaust bank 1	0... 100 ms	0...3200 H
CID 8	weighted switching time for lean to rich of O_2 sensor 1 exhaust bank 2	0... 100 ms	0...3200 H
TID 5 - evaporative system monitoring			
CID 1	functional check: change of trim	0... 2.0	0...FFFF H
CID 2	functional check: change of idle speed + air mass	0... 3200%	0...7FFF H
CID 3	reference current (upper boundary)	0... 590 mA	0... FFFFH
CID 4	reference current (lower boundary)	0... 590 mA	0... FFFFH
CID 5	minimal pump current after switching solenoid	0... 590 mA	0... FFFFH
CID 6	pump current gross leak	0... 590 mA	0... FFFFH
CID 7	pump current fine leak	0... 590 mA	0... FFFFH
CID 8	pump current short cu.	0... 590 mA	0... FFFFH

Mode 6 Information (2003 / 2004 model year)

Monitored Component		Possible range	Hex code factor
TID 1 catalyst monitoring			
CID 1	catalyst monitoring of catalyst 1 sum of the determined ratios	0...255.996	0...FFFF H
CID 2	catalyst monitoring of catalyst sum of the determined ratios 2	0...255.996	0...FFFF H
TID 2 o2 sensor monitoring			
CID 1	weighted lean period time of O_2 sensor 1 exhaust bank 1	0...255.996 s	0...FFFF H
CID 2	weighted rich period time of O_2 sensor 1 exhaust bank 1	0...255.996 s	0...FFFF H
CID 3	weighted switching time for rich to lean of O_2 sensor 1 exhaust bank 1	0...255.996 s	0...FFFF H
CID 4	weighted switching time for lean to rich of O_2 sensor 1 exhaust bank 1	0...255.996 s	0...FFFF H
CID 5	weighted lean period time of O_2 sensor 1 exhaust bank 2	0...255.996 s	0...FFFF H
CID 6	weighted rich period time of O_2 sensor 1 exhaust bank 2	0...255.996 s	0...FFFF H
CID 7	weighted switching time for rich to lean of O_2 sensor 1 exhaust bank 2	0...255.996 s	0...FFFF H
CID 8	weighted switching time for lean to rich of O_2 sensor 1 exhaust bank 2	0...255.996 s	0...FFFF H
TID 5 evaporative system monitoring			
CID 2	pump period time while leakage monitoring	0... 655.35 s	0... FFFFH
CID 7	changing of fuel trim after opening the purge valve	0... 31.25%	0... 5000H
TID 6 O2 sensor heater monitoring			
CID 1	count of ok-tests O_2 sensor 1 heater exhaust bank 1	0... 65535	
CID 2	count of ok-tests O_2 sensor 1 heater exhaust bank 2	0... 65535	0... FFFFH
CID 3	count of ok-tests O_2 sensor 2 heater exhaust bank 1	0... 65535	0... FFFFH
CID 4	count of ok-tests O_2 sensor 2heater exhaust bank 2	0... 65535	0... FFFFH

Mode 6 Information (2005 model year)

	Monitored Component	Possible Range	Hex Code Factor
TID 3 O2 sensor monitoring			
	Sum of period of times during monitoring of upstream O_2 sensor	0... 65535 s	0... FFFFH
	Front O_2 result on scan tool, 100= front O_2 result in seconds		
	Monitored Component	**Possible Range**	**Hex Code Factor**
TID 6 Evaporative system monitoring			
	Pump period time while leakage monitoring	0... 655.35 s	0... FFFFH
	Evap result on scantool,100= evap result in seconds		
TDI 8 Catalyst system monitoring			
	Catalyst diagnostic result	0... 255	0... FFFFH
	Catalyst result on scan tool, 256= catalyst diagnostic result (no units)		
TID 9 Thermostat monitoring			
	TCO value when modeled temperature reaches threshold	-48... 142.5 °c	0... FE H
	Thermostat result on scan tool, *0.75-48=thermostat result (C°)		

Mode 6 information (2006 model year)

Monitored Component		Possible range / scaling	Bit
TID 03 O2 sensor monitoring			
	Upstream O_2 sensor bank 1 sensor 1 response time	0...655.35 seconds	0, 01
81	Downstream O_2 sensor rich to lean switching time	0... 1992	0, 0078125
82	Downstream O_2 sensor pedal up fuel cut	0... 5 volts	0, 00488
TID 06 Evaporative system monitoring			
	Leakage over 0.5 mm	0... 655.35 seconds	0, 01
TID 08 catalyst monitoring			
	Oxygen storage capacity	0... 255	0, 0039

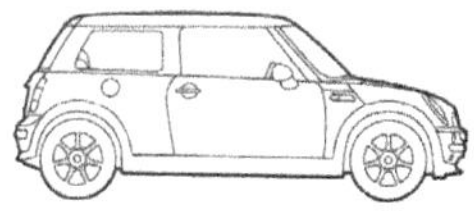

Notes:

> **WARNING**
>
> *Your common sense, good judgement, and general alertness are crucial to safe and successful service work. Before attempting any work on your MINI, be sure to read* **001 General Warnings and Cautions** *and the copyright page at the front of the manual. Review these warnings and cautions each time you prepare to work on your MINI. Please also read any warnings and cautions that accompany the procedures in the manual.*

A

B

> ***WARNING***
>
> *Your common sense, good judgement, and general alertness are crucial to safe and successful service work. Before attempting any work on your MINI, be sure to read* **001 General Warnings and Cautions** *and the copyright page at the front of the manual. Review these warnings and cautions each time you prepare to work on your MINI. Please also read any warnings and cautions that accompany the procedures in the manual.*

> ***WARNING***
>
> *Your common sense, good judgement, and general alertness are crucial to safe and successful service work. Before attempting any work on your MINI, be sure to read* **001 General Warnings and Cautions** *and the copyright page at the front of the manual. Review these warnings and cautions each time you prepare to work on your MINI. Please also read any warnings and cautions that accompany the procedures in the manual.*

G

H

I

WARNING

Your common sense, good judgement, and general alertness are crucial to safe and successful service work. Before attempting any work on your MINI, be sure to read **001 General Warnings and Cautions** *and the copyright page at the front of the manual. Review these warnings and cautions each time you prepare to work on your MINI. Please also read any warnings and cautions that accompany the procedures in the manual.*

WARNING

Your common sense, good judgement, and general alertness are crucial to safe and successful service work. Before attempting any work on your MINI, be sure to read **001 General Warnings and Cautions** *and the copyright page at the front of the manual. Review these warnings and cautions each time you prepare to work on your MINI. Please also read any warnings and cautions that accompany the procedures in the manual.*

U

V

W

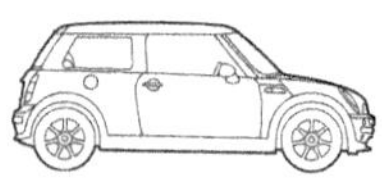

Selected Books and Repair Information From Bentley Publishers

Engineering

The Hack Mechanic Guide to European Automotive Electrical Systems *Rob Siegel*

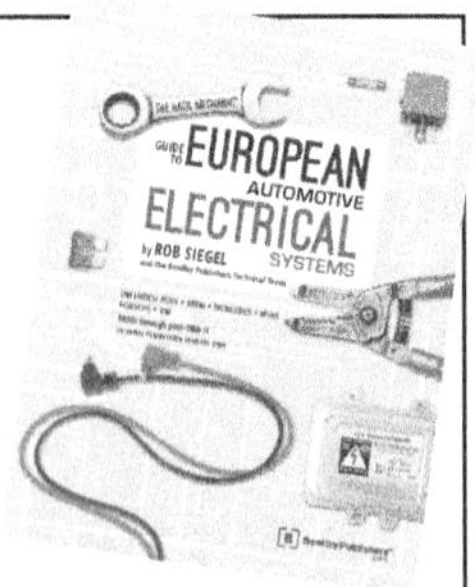

ISBN 978-0-8376-1751-0

Mechanical Ignition Handbook: The Hack Mechanic Guide to Vintage Ignition Systems *Rob Siegel*
ISBN 978-0-8376-1767-1

Physics for Gearheads
Randy Beikmann
ISBN 978-0-8376-1615-5

Bosch Automotive Handbook
Robert Bosch GmbH
ISBN 978-0-8376-1732-9

Bosch Fuel Injection and Engine Management *Charles O. Probst, SAE*
ISBN 978-0-8376-0300-1

Maximum Boost: Designing, Testing, and Installing Turbocharger Systems
Corky Bell ISBN 978-0-8376-0160-1

Motorsports

Alex Zanardi: My Sweetest Victory
Alex Zanardi and Gianluca Gasparini
ISBN 978-0-8376-1249-2

The Unfair Advantage *Mark Donohue and Paul van Valkenburgh*
ISBN 978-0-8376-0069-7

Equations of Motion - Adventure, Risk and Innovation
William F. Milliken
ISBN 978-0-8376-1570-7

Audi

Audi A4 Service Manual: 2002-2008, 1.8L Turbo, 2.0L Turbo, 3.0L, 3.2L
Bentley Publishers
ISBN 978-0-8376-1574-5

Audi TT Service Manual: 2000-2006, 1.8L turbo, 3.2 L *Bentley Publishers*
ISBN 978-0-8376-1625-4

BMW

Memoirs of a Hack Mechanic
Rob Siegel ISBN 978-0-8376-1720-6

BMW 4 Series (F32, F33, F36) Service Manual: 2014-2016 *Bentley Publishers*
ISBN 978-0-8376-1765-7

BMW X3 (E83) Service Manual: 2004-2010 *Bentley Publishers*
ISBN 978-0-8376-1731-2

BMW X5 (53) Service Manual: 2000-2006 *Bentley Publishers*
ISBN 978-0-8376-1643-8

BMW 3 Series (F30, F31, F34) Service Manual: 2012-2015 *Bentley Publishers*
ISBN 978-0-8376-1752-7

BMW 3 Series (E90, E91, E92, E93) Service Manual: 2006-2011
Bentley Publishers
ISBN 978-0-8376-1723-7

BMW 3 Series (E46) Service Manual: 1999-2005 *Bentley Publishers*
ISBN 978-0-8376-1657-5

BMW 3 Series (E36) Service Manual: 1992-1998 *Bentley Publishers*
ISBN 978-0-8376-1709-1

BMW 3 Series (E30) Service Manual: 1984-1990 *Bentley Publishers*
ISBN 978-0-8376-1647-6

BMW Z3 (E36/7) Service Manual: 1996-2002 *Bentley Publishers*
ISBN 978-0-8376-1617-9

BMW 5 Series (E60, E61) Service Manual: 2004-2010 *Bentley Publishers*
ISBN 978-0-8376-1689-6

Corvette

Corvette: America's Star-Spangled Sports Car *Karl Ludvigsen*
ISBN 978-0-8376-1659-9

Corvette by the Numbers: The Essential Corvette Parts Reference 1955-1982 *Alan Covin*
ISBN 978-0-8376-0228-2

Zora Arkus-Duntov: The Legend Behind Corvette *Jerry Burton*
ISBN 978-08376-0858-7

Porsche

Porsche 911 (996) Service Manual: 1999-2005 *Bentley Publishers*
ISBN 978-0-8376-1710-7

Porsche 911 (993) Service Manual: 1995-1998 *Bentley Publishers*
ISBN 978-0-8376-1719-0

Porsche Boxster Service Manual: 1997-2004 *Bentley Publishers*
ISBN 978-0-8376-1645-2

Porsche 911 Carrera Service Manual: 1984-1989 *Bentley Publishers*
ISBN 978-0-8376-1696-4

Porsche 911 SC Service Manual: 1978-1983 *Bentley Publishers*
ISBN 978-0-8376-1705-3

Porsche: Excellence Was Expected
Karl Ludvigsen ISBN 978-0-8376-0235-6

Porsche — Origin of the Species
Karl Ludvigsen ISBN 978-0-8376-1331-4

Volkswagen

Volkswagen Rabbit, GTI Service Manual: 2006-2009 *Bentley Publishers*
ISBN 978-0-8376-1664-3

Volkswagen Jetta Service Manual: 2005-2010 *Bentley Publishers*
ISBN 978-0-8376-1616-2

Volkswagen Jetta, Golf, GTI Service Manual: 1999-2005 *Bentley Publishers*
ISBN 978-0-8376-1678-0

Volkswagen Jetta, Golf, GTI: 1993-1999, Cabrio: 1995-2002 Service Manual *Bentley Publishers*
ISBN 978-0-8376-1660-5

Volkswagen GTI, Golf, Jetta Service Manual: 1985-1992 *Bentley Publishers*
ISBN 978-0-8376-1637-7

Volkswagen Corrado Repair Manual: 1990-1994 *Bentley Publishers*
ISBN 978-0-8376-1699-5

Volkswagen Passat, Passat Wagon Service Manual: 1998-2005
Bentley Publishers
ISBN 978-0-8376-1669-8

MINI Repair Manuals

MINI Cooper Service Manual: 2007-2013 *Bentley Publishers*
ISBN 978-0-8376-1730-5

MINI Cooper Diagnosis Without Guesswork: 2002-2006 *Bentley Publishers* ISBN 978-0-8376-1639-1

BentleyPublishers®
Automotive Reference

Bentley Publishers has published service manuals and automobile books since 1950. For more information, please contact Bentley Publishers at 1734 Massachusetts Avenue, Cambridge, MA 02138 USA, or visit our web site at
BentleyPublishers.com